CONTENTS IN BRIEF

CONTENTS

PREFACE

The eighth edition of ***Fundamentals of Analytical Chemistry,*** like its predecessors, is an introductory textbook designed primarily for a one- or two-semester course for chemistry majors. Since the publication of the seventh edition, the scope of analytical chemistry has continued to evolve, and in this edition we have included many applications to biology, medicine, materials science, ecology, forensic science, and other related fields. The widespread use of computers for instructional purposes has led us to incorporate many spreadsheet applications, examples, and exercises. Our companion book, *Applications of Microsoft® Excel in Analytical Chemistry,* provides students with a tutorial guide for applying spreadsheets to analytical chemistry and introduces many additional spreadsheet operations. We also have added many current topics, such as atomic and molecular mass spectrometry, field-flow fractionation, and chiral chromatography. We have revised many older treatments to incorporate contemporary instrumentation and techniques. We recognize that courses in analytical chemistry vary from institution to institution and depend upon the available facilities, the time allocated to analytical chemistry in the chemistry curriculum, and the desires of individual instructors. We have therefore designed the eighth edition of *Fundamentals of Analytical Chemistry* so that instructors can tailor the text to meet their needs and students can encounter the material on several levels, in descriptions, pictorials, illustrations, and interesting features.

Objectives

Our major objective of this text is to provide a thorough background in those chemical principles that are particularly important to analytical chemistry. Second, we want students to develop an appreciation for the difficult task of judging the accuracy and precision of experimental data and to show how these judgments can be sharpened by the application of statistical methods. Our third aim is to introduce a wide range of techniques that are useful in modern analytical chemistry. Additionally, our hope is that with the help of this book, students will develop the skills necessary to solve analytical problems in a quantitative manner, particularly with the aid of the spreadsheet tools that are so commonly available. Finally, we aim to teach those laboratory skills that will give students confidence in their ability to obtain high-quality analytical data.

Coverage and Organization

The material in this text covers both fundamental and practical aspects of chemical analysis. Users of earlier editions will find that we have organized this edition in a somewhat different manner than its predecessors. In particular, we have organized the chapters into Parts that group together related topics. There are seven major Parts to the text that follow the brief introduction in Chapter 1.

- **Part I** covers the tools of analytical chemistry and comprises seven chapters. Chapter 2 discusses the chemicals and equipment used in analytical laboratories and includes many photographs of analytical operations. A new Chapter 3, "Using Spreadsheets in Analytical Chemistry," is a tutorial introduction to the use of spreadsheets in analytical chemistry. Chapter 4 reviews the basic calculations of analytical chemistry, including expressions of chemical concentration and stoichiometric relationships. Chapters 5, 6, and 7 present topics in statistics and data analysis that are important in analytical chemistry and incorporate extensive use of spreadsheet calculations. Analysis of Variance, ANOVA, is a new topic included in Chapter 7. A new Chapter 8, "Sampling, Standardization, and Calibration," consolidates coverage of sampling, sample handling, external and internal standards, and standard additions, and includes new coverage of calibration and standardization.
- **Part II** covers the principles and application of chemical equilibrium systems in quantitative analysis. Chapter 9 covers the fundamentals of chemical equilibria. Chapter 10 discusses the effect of electrolytes on equilibrium systems. The systematic approach for attacking equilibrium problems in complex systems is the subject of Chapter 11.
- **Part III** brings together several chapters dealing with classical gravimetric and volumetric analytical chemistry. Chapter 12 covers gravimetric analysis. In Chapters 13 through 17, we consider the theory and practice of titrimetric methods of analysis, including acid/base titrations, precipitation titrations, and complexometric titrations. In these chapters, advantage is taken of the systematic approach to equilibria and the use of spreadsheets in calculations.
- **Part IV** is devoted to electrochemical methods. After an introduction to electrochemistry in Chapter 18, Chapter 19 describes the many uses of electrode potentials. Oxidation/reduction titrations are the subject of Chapter 20, while Chapter 21 presents the use of potentiometric methods to obtain concentrations of molecular and ionic species. Chapter 22 considers the bulk electrolytic methods of electrogravimetry and coulometry, while Chapter 23 discusses voltammetric methods including linear sweep and cyclic voltammetry, anodic stripping voltammetry, and polarography.
- **Part V** covers spectroscopic methods of analysis. Basic material on the nature of light and its interaction with matter is presented in Chapter 24. Spectroscopic instruments and their components are described in Chapter 25. The various applications of molecular absorption spectrometric methods are covered in some detail in Chapter 26, while Chapter 27 is concerned with molecular fluorescence spectroscopy. Chapter 28 discusses various atomic spectrometric methods, including atomic mass spectrometry, plasma emission spectrometry, and atomic absorption spectroscopy.
- **Part VI** comprises five chapters dealing with kinetics and analytical separations. Kinetic methods of analysis are covered in Chapter 29. Chapter 30 introduces analytical separations including the various chromatographic methods. Chapter 31 discusses gas chromatography, while high-performance liquid chromatography is presented in Chapter 32. The final chapter in this section, Chapter 33, "Miscellaneous Separation Methods," is new to this edition and includes coverage of supercritical fluid chromatography, capillary electrophoresis, and field-flow fractionation.

- The final **Part VII** consists of four chapters dealing with the practical aspects of analytical chemistry. Real samples are considered and compared with ideal samples in Chapter 34. Methods for preparing samples are discussed in Chapter 35, while techniques for decomposing and dissolving samples are covered in Chapter 36. Chapter 37 provides detailed procedures for 57 laboratory experiments, covering many of the principles and applications discussed in previous chapters. This chapter is only available as an Adobe Acrobat® PDF file on the **Analytical Chemistry CD-ROM** enclosed in this book or on our Web site at **http://chemistry.brookscole.com/skoogfac/.**

Flexibility

Because the text is divided into Parts, there is a good deal of flexibility in the use of material. Many of the Parts can stand alone or be taken in a different order. For example, some instructors may want to cover spectroscopic methods prior to electrochemical methods or separations prior to spectroscopic methods.

Highlights

This edition incorporates many features and methods intended to enhance the learning experience for the student and to provide a versatile teaching tool for the instructor.

Important Equations. Equations that we feel are most important have been highlighted with a color screen for emphasis and ease of review.

Mathematical Level. Generally, the principles of chemical analysis developed here are based on college algebra. Some of the concepts presented require basic differential and integral calculus.

Worked Examples. A large number of worked examples serve as aids in understanding the concepts of analytical chemistry. As in the seventh edition, we follow the practice of including units in chemical calculations and using the factor-label method to check their correctness. The examples also are models for the solution of problems found at the end of most of the chapters. Many of these use spreadsheet calculations, as described next.

New! **Spreadsheet Calculations.** Throughout the book we have introduced spreadsheets for problem solving, graphical analysis, and many other applications. Microsoft® Excel has been adopted as the standard for these calculations, but the instructions could be readily adapted to other programs. Several chapters have tutorial discussions of how to enter values, formulas, and built-in functions. Many other examples worked in detail are presented in our companion book, *Applications of Microsoft® Excel in Analytical Chemistry.* We have attempted to document each stand-alone spreadsheet with working formulas and entries.

New! **Spreadsheet Summaries.** References to our companion book *Applications of Microsoft® Excel in Analytical Chemistry* are given as Spreadsheet Summaries in the text. These are intended to direct the user to examples, tutorials and elaborations of the text topics.

Questions and Problems. An extensive set of questions and problems is included at the end of most chapters. Answers to approximately half of the problems are given at the end of the book. Many of the problems are best solved using spreadsheets. These are identified by a spreadsheet icon placed in the margin next to the problem.

New! **Challenge Problems.** Most of the chapters have a challenge problem at the end of the regular questions and problems. Such problems are intended to be open-ended, research-type problems that are more challenging than normal. These problems may consist of multiple steps, dependent on one another, or may involve library or Web searches to find information. We hope these challenge problems stimulate discussion and extend the topics of the chapter into new areas. We encourage instructors to use them in innovative ways, such as group projects, inquiry-driven learning assignments, and case study discussions.

Features. A series of boxed and highlighted Features are found throughout the text. These essays contain interesting applications of analytical chemistry to the modern world, derivation of equations, explanations of more difficult theoretical points, or historical notes. Examples include Breath Alcohol Analyzers (Chapter 7), Antioxidants (Chapter 20), Fourier Transform Spectroscopy (Chapter 25), LC/MS and LC/MS/MS (Chapter 32), and Capillary Electrophoresis in DNA Sequencing (Chapter 33).

Illustrations and Photos. We feel strongly that photographs, drawings, pictorials, and other visual aids greatly assist the learning process. Hence, we have included new and updated visual materials to aid the student. Most of the drawings are done in two colors to increase the information content and to highlight important aspects of the figures. Photographs and color plates taken exclusively for this book by renowned chemistry photographer Charles Winters are intended to illustrate concepts, equipment, and procedures that are difficult to illustrate with drawings.

Expanded Figure Captions. Where appropriate, we have attempted to make the figure captions quite descriptive so that reading the caption provides a second level of explanation for many of the concepts. In some cases, the figures can stand by themselves much in the manner of a *Scientific American* figure.

New! **Interviews.** Each Part begins with an interview of a noted analytical scientist: Dick Zare (Stanford University), Sylvia Daunert (University of Kentucky), Larry Faulkner (University of Texas), Allen Bard (University of Texas), Gary Hieftje (Indiana University), Isiah Warner (Lousiana State University), and Julie Leary (University of California, Berkeley). The interviews are informal question-and-answer sessions designed to provide information about the scientists and their backgrounds, their reasons for choosing analytical chemistry, their thoughts on the importance of the field, their research areas, and other interesting topics. It is hoped that these interviews will add interest to the subject matter by personalizing some of the topics covered.

New! **Web Works.** At the end of most of the chapters we have included a brief Web Works feature. In this feature, we ask the student to find information on the Web, do online searches, visit the Web sites of equipment manufacturers, or solve analytical problems. These Web Works and the links given are intended to stimulate student interest in exploring the information available on the World Wide Web. These links will be updated regularly on the Brooks/Cole Web site, **http://chemistry.brookscole.com/skoogfac/.**

Glossary. The glossary at the end of the book defines the most important terms, phrases, techniques, and operations used in the text. The glossary is intended to provide students with quick access to meanings, without having to search through the text.

Appendixes and Endpapers. Included in the appendixes are an updated guide to the literature of analytical chemistry, tables of chemical constants, electrode potentials, and recommended compounds for the preparation of standard materials; sections on the use of logarithms and exponential notation, and on normality and equivalents (terms that are not used in the text itself); and a derivation of the propagation of error equations. The inside front and back covers of this book provide a full-color chart of chemical indicators, a table of molar masses of compounds of particular interest in analytical chemistry, a table of international atomic masses, and a periodic table. In addition, the book has a tear-out reference card for Microsoft® Excel.

Changes in the Eighth Edition

Readers of the seventh edition will find that the eighth edition has numerous changes in content as well as style and format.

Content. Several changes in content have been made to strengthen the book.

- New and exciting chapter opening introductions, accompanied by applied photos, present a relevant example of one of the chapter topics. Examples include stalagmites and stalactites as an illustration of an equilibrium process (Chapter 9), the effects of acid rain (Chapter 16), and the oxidation/reduction properties of chlorophyll (Chapter 19).
- Many chapters have been strengthened by adding spreadsheet examples, applications, and problems. The new Chapter 3 gives tutorials on the construction and use of spreadsheets. Many other tutorials are included in our supplement, *Applications of Microsoft® Excel in Analytical Chemistry.*
- The chapters on statistics (Chapters 5–7) have been updated and brought into conformity with the terminology of modern statistics. Analysis of Variance (ANOVA) is included in Chapter 7. ANOVA is easy to perform with modern spreadsheet programs and quite useful in analytical problem solving.
- A new Chapter 8 consolidates material on sampling and integrates material on calibration and standardization. Methods such as external standards, internal standards, and standard additions are presented in this chapter, and their advantages and disadvantages are discussed.
- The chapter on precipitation titrimetry has been eliminated, and some of the material is included in Chapter 13 on titrimetric methods.
- Chapters 18, 19, 20, and 21 on electrochemical cells and cell potentials have been extensively revised to clarify the discussion and introduce the free energy of cell processes. Chapter 23 has been altered to decrease the emphasis on classical polarography. It now includes a discussion of cyclic voltammetry.
- Chapter 28 in this edition covers atomic mass spectrometry, including inductively coupled plasma mass spectrometry. Flame photometry has been deemphasized.
- In Part VI, Chapter 30 is now a general introduction to separations. It includes solvent extraction and precipitation methods, an introduction to chromatography, and a new section on solid-phase extraction. Chapter 31 contains new material on molecular mass spectrometry and gas chromatography/mass spectrometry. Chapter 32 includes new sections on affinity chromatography and chiral chromatography. A section on LC/MS has been added. A new Chapter 33, "Miscellaneous Separation Methods," has been included. It introduces capillary electrophoresis and field-flow fractionation.
- **Style and Format.** To make the text more readable and student-friendly, we have continued to change style and format.
- We have attempted to use shorter sentences, a more active voice, and a more conversational writing style in each chapter.
- More descriptive figure captions are used whenever appropriate to allow a student to understand the figure and its meaning without having to alternate between text and caption.
- Molecular models are liberally used in most chapters to stimulate interest in the beauty of molecular structures and to reinforce structural concepts and descriptive chemistry presented in general chemistry and upper-level courses.

- Photographs, taken specifically for this text, are used whenever appropriate to illustrate important techniques, apparatus, and operations.
- Marginal notes are used throughout to emphasize recently discussed concepts or to reinforce key information.
- A running marginal glossary reinforces key terminology.

A Full Support Package for Students

- **Solutions Manual.** Written by Gary Kinsel, University of Texas, Arlington, the solutions manual contains worked-out solutions for all the starred problems in the text. For added value and convenience, the Student Solutions Manual can be packaged with the text. Contact your Thomson • Brooks/Cole sales representative for more information.
- **Spreadsheet Applications.** *Applications of Microsoft® Excel in Analytical Chemistry,* by Stanley R. Crouch and F. James Holler, treats in detail the spreadsheet approaches summarized in the text. This supplement contains 16 chapters that lead the student from basic concepts and operations to using spreadsheets for simulations, curve fitting, data smoothing, curve resolution, and many other topics. Topics in this companion book are correlated with topics in the text. See pages xvii and xviii for a correlation chart. Summaries in the text point to specific chapters and sections in the companion book. For added value and convenience, this ancillary can be packaged with the text. Contact your Thomson • Brooks/Cole representative for details.
- **Interactive Analytical Chemistry CD-ROM.** Developed by William J. Vining, University of Massachusetts, Amherst, in conjunction with the text authors, this CD-ROM is packaged free with every copy of the book. Prompted by icons with captions in the text, students explore the corresponding Intelligent Tutors, Guided Simulations, and Media-based Exercises. This CD-ROM includes tutorials on statistics, equilibria, spectrophotometry, electroanalytical chemistry, chromatography, atomic absorption spectroscopy, and gravimetric and combustion analysis. Also included on the CD-ROM as an Adobe Acrobat® PDF file is Chapter 37, "Selected Methods of Analysis." Students will be able to print only those experiments that they will perform, and the printed sheets can be easily used in the laboratory.
- **Brooks/Cole Book Companion Web Site** at **http://chemistry.brookscole.com/skoogfac/.** The Web site includes a set of updated links to the Web sites mentioned in the Web Works, problems, and other places in this text. Instructors may download spreadsheets developed in this book as well as those from *Applications of Microsoft® Excel in Analytical Chemistry.* Instructors may download graphics files containing all of the figures from the text to aid in preparing PowerPoint® presentations. The Chapter 37 PDF file is also included on the Web site.
- **InfoTrac® College Edition.** Every new copy of this book is packaged with four months of free access to InfoTrac College Edition. This online resource features a comprehensive database of reliable, full-length articles from thousands of top academic journals and popular sources.

Additional Support for Professors

The following ancillaries are available to qualified adopters. Please consult your Thomson • Brooks/Cole sales representative for details.

- **Instructor's Manual.** Complete solutions to all of the problems in the text are available on the Instructor's password-protected companion site for this text located at **http://chemistry.brookscole.com/skoogfac.**
- **Multimedia Manager for *Fundamentals of Analytical Chemistry,* Eighth Edition: A Microsoft® PowerPoint® Link Tool.** The Multimedia Manager is a digital library and presentation, dual-platform CD-ROM. Included is a library of resources valuable to instructors, such as text art and tables, in a variety of e-formats that are easily exported into other software packages. You can also customize your own presentation by importing your personal lecture slides or other material you choose.
- **Overhead Transparencies.** A set of 100 color overhead transparencies is available to assist instructors in presenting student lectures.
- **MyCourse 2.1.** A new, free, online course builder that offers a simple solution to creating a custom course Web site where professors can assign, track, and report on student progress. Contact your Thomson • Brooks/Cole sales representative for details or visit **http://mycourse.thomsonlearning.com** for a free demo.
- ***Applications of Microsoft® Excel in Analytical Chemistry,*** a clear and concise companion to *Fundamentals of Analytical Chemistry,* eighth edition, and *Analytical Chemistry, An Introduction,* seventh edition, provides students and professors with a valuable resource of the most useful spreadsheet methods.
- **Correlation of Spreadsheet Supplement to Texts.** The following chart lists cross-references to *Fundamentals of Analytical Chemistry,* Eighth Edition, and *Analytical Chemistry: An Introduction,* Seventh Edition.

Applications of Microsoft® Excel in Analytical Chemistry	*Fundamentals of Analytical Chemistry,* Eighth edition	*Analytical Chemistry, An Introduction,* Seventh edition
Chapter 1 Excel Basics	Chapter 3 Using Spreadsheets in Analytical Chemistry	Section 2J Using Spreadsheets in Analytical Chemistry
Chapter 2 Basic Statistical Analysis with Excel	Chapters 5, 6 Basic Statistics	Chapters 5, 6 Basic Statistics
Chapter 3 Statistical Tests with Excel	Chapter 7 Statistical Data Treatment	Chapter 7 Statistical Analysis
Chapter 4 Least Squares and Calibration Methods	Section 8C Standardization and Calibration	Section 7D The Least Squares Method
Chapter 5 Equilibrium Activity and Solving Equations	Section 9B Chemical Equilibrium Chapter 10 Electrolyte Effects	Section 4B Chemical Equilibrium Chapter 9 Electrolyte Effects
Chapter 6 The Systematic Approach to Equilibria: Solving Many Equations	Chapter 11 Complex Equilibrium Calculations	Chapter 10 Complex Equilibrium Calculations
Chapter 7 Titrations and Graphical Representations	Chapter 13 Titrimetric Methods and Precipitation Titrations Chapter 14 Neutralization Titrations	Chapter 11 Titrations Chapter 12 Precipitation Titrations Section 15B-2 Neutralization Titrations
Chapter 8 Polyfunctional Acids and Bases	Chapter 15 Polyfunctional Acids and Bases	Chapter 13 Polyfunctional Acids and Bases

(continues)

Applications of Microsoft® Excel in Analytical Chemistry	*Fundamentals of Analytical Chemistry,* **Eighth edition**	*Analytical Chemistry, An Introduction,* **Seventh edition**
Chapter 9 Complexometric Titrations	Chapter 17 Complexation Reactions and Titrations	Chapter 15 Complexation Titrations
Chapter 10 Potentiometry and Redox Titrations	Chapter 18 Introduction to Electrochemistry Chapter 19 Standard Electrode Potentials Chapter 20 Oxidation/ReductionTitrations Chapter 21 Potentiometry	Chapter 16 Elements of Electrochemistry Chapter 17 Using Electrode Potentials Chapter 18 Oxidation/Reduction Titrations Chapter 19 Potentiometry
Chapter 11 Dynamic Electrochemistry	Chapter 22 Electrogravimetry and Coulometry Chapter 23 Voltammetry	Chapter 20 Other Electro-analytical Methods
Chapter 12 Spectroscopic Methods	Chapter 24 Introduction to Spectrochemical Methods Chapter 25 Optical Instrumentation Chapter 26 Molecular Absorption Spectroscopy Chapter 27 Molecular Fluorescence Spectroscopy	Chapter 21 Spectroscopic Methods of Analysis Chapter 22 Instruments for Measuring Absorption Chapter 23 Spectroscopic Methods
Chapter 13 Kinetic Methods	Chapter 29 Kinetic Methods of Analysis	Section 23A-2 Quantitative UV/Visible Spectrophotometry
Chapter 14 Chromatography	Chapter 30 Introduction to Analytical Separations Section 31C-2 Quantitative GC	Chapter 24 Introduction to Analytical Separations Section 25A-7 GC Applications
Chapter 15 Electrophoresis and Other Separation Methods	Section 30E-7 Column Resolution Section 32H Comparison of HPLC and GC Chapter 33 Miscellaneous Separation Methods	Section 24F-9 Column Resolution Chapters 26 SFC, CE and other Separation Methods
Chapter 16 Data Processing with Excel	Chapter 7 Statistical Data Treatment	Chapter 7 Statistical Analysis

Acknowledgments

We wish to acknowledge with thanks the comments and suggestions of many reviewers who critiqued the seventh edition prior to our writing or who evaluated the current manuscript in various stages.

Joseph Aldstadt,
University of Wisconsin, Milwaukee
Stephen Brown,
University of Delaware
James Burlitch,
Cornell University
Michael DeGrandpre,
University of Montana
Simon Garrett,
Michigan State University
Carol Lasko,
Humboldt State University

Tingyu Li,
Vanderbilt University
Joseph Maloy,
Seton Hall University
Howard Lee McLean,
Rose-Hulman Institute of Technology
Frederick Northrup,
Northwestern University
Peter Palmer,
San Francisco State University
Reginald Penner,
University of California, Irvine
Jeanette Rice,
Georgia Southern University
Alexander Scheeline,
University of Illinois, Urbana-Champaign
James Schenk,
Washington State University
Maria Schroeder,
United States Naval Academy
Manuel Soriaga,
Texas A&M University
Keith Stevenson,
University of Texas
Larry Taylor,
Virginia Technical Institute
Robert Thompson,
Oberlin College
Richard Vachet,
University of Massachusetts
Joseph Wang,
New Mexico State University

We especially acknowledge the assistance of Professor David Zellmer, California State University, Fresno, who reviewed several chapters and served as the accuracy reviewer for the entire manuscript. In addition, we appreciate the comments and suggestions of Professor Gary Kinsel of the University of Texas, Arlington, who prepared the solutions manual, and Professor Scott Van Bramer, of Widener University, who checked all the solutions to the problems. We are grateful to Professor Bill Vining of the University of Massachusetts, who prepared the accompanying CD-ROM, and we are fortunate to have worked with Charles D. Winters, who contributed many of the new photos in the text and in the color plates.

Our writing team enjoys the services of a superb technical reference librarian, Ms. Maggie Johnson. She assisted us in many ways in the production of this book, including checking references, performing literature searches, and providing background information for many of the Features. We appreciate her competence, enthusiasm, and good humor.

We are grateful to the many staff members of Brooks/Cole/Thomson Learning who provided excellent support during the production of this text. Senior Developmental Editor Sandi Kiselica has done a superb job of organizing this project, maintaining continuity, and making many important comments and suggestions. Bonnie Boehme of Nesbitt Graphics is simply the best copy editor that we have ever had. Her keen eye and consummate editorial skills have contributed greatly to the quality of the text. Alyssa White has done an excellent job coordinating the ancillary materials, and Jane Sanders, our photo researcher, has shown style and good humor in handling the various tasks associated with acquiring the many new photos in the book.

Douglas A. Skoog
Donald M. West
F. James Holler
Stanley R. Crouch

Fundamentals of

Analytical Chemistry

Eighth Edition

CHAPTER 1

The Nature of Analytical Chemistry

Analytical chemistry is a measurement science consisting of a set of powerful ideas and methods that are useful in all fields of science and medicine. An exciting illustration of the power and significance of analytical chemistry occurred on July 4, 1997, when the Pathfinder *spacecraft bounced to a halt on Ares Vallis, Mars, and delivered the* Sojourner *rover from its tetrahedral body to the Martian surface. The world was captivated by the* Pathfinder *mission. As a result, the numerous World Wide Web sites tracking the mission were nearly overwhelmed by millions of Internet surfers who closely monitored the progress of tiny* Sojourner *in its quest for information on the nature of the Red Planet. The key experiment aboard* Sojourner *used the APXS, or alpha proton X-ray spectrometer, which combines the three advanced instrumental techniques of Rutherford backscattering spectroscopy, proton emission spectroscopy, and X-ray fluorescence. The APXS data were collected by* Pathfinder *and transmitted to Earth for further analysis to determine the identity and concentration of most of the elements of the periodic table.*[1] *The determination of the elemental composition of Martian rocks permitted*

[1]For detailed information regarding the APXS instrumentation aboard the *Sojourner,* please refer to our Web page at **http://chemistry.brookscole.com/skoogfac/.** From the Chapter Resources menu, choose Web Works. Locate the Chapter 1 section and find the links to a general description of the *Sojourner* instrument package, an article describing the detailed operation of the APXS instrument, and results of the elemental analysis of various Martian rocks.

geologists to rapidly identify them and compare them with terrestrial rocks. The Pathfinder *mission is a spectacular example illustrating the application of analytical chemistry to practical problems. The experiments aboard the spacecraft and the data from the mission also illustrate how analytical chemistry draws on science and technology in such widely diverse disciplines as nuclear physics and chemistry to identify and determine the relative amounts of substances in samples of matter.*

The Pathfinder *example demonstrates that both qualitative information and quantitative information are required in an analysis.* ***Qualitative analysis*** *establishes the chemical identity of the species in the sample.* ***Quantitative analysis*** *determines the relative amounts of these species, or* **analytes,** *in numerical terms. The data from the APXS spectrometer on* Sojourner *contain both types of information. Note that chemical separation of the various elements contained in the rocks was unnecessary in the APXS experiment. More commonly, a separation step is a necessary part of the analytical process. As we shall see, qualitative analysis is often an integral part of the separation step, and determining the identity of the analytes is an essential adjunct to quantitative analysis. In this text, we shall explore quantitative methods of analysis, separation methods, and the principles behind their operation.*

Qualitative analysis reveals the *identity* of the elements and compounds in a sample.

Quantitative analysis indicates the *amount* of each substance in a sample.

Analytes are the components of a sample that are to be determined.

1A THE ROLE OF ANALYTICAL CHEMISTRY

Analytical chemistry is applied throughout industry, medicine, and all the sciences. Consider a few examples. The concentrations of oxygen and of carbon dioxide are determined in millions of blood samples every day and used to diagnose and treat illnesses. Quantities of hydrocarbons, nitrogen oxides, and carbon monoxide present in automobile exhaust gases are measured to assess the effectiveness of smog-control devices. Quantitative measurements of ionized calcium in blood serum help diagnose parathyroid disease in humans. Quantitative determination of nitrogen in foods establishes their protein content and thus their nutritional value. Analysis of steel during its production permits adjustment in the concentrations of such elements as carbon, nickel, and chromium to achieve a desired strength, hardness, corrosion resistance, and ductility. The mercaptan content of household gas supplies is monitored continually to ensure that the gas

Mars landscape. Courtesy of NASA

has a sufficiently noxious odor to warn of dangerous leaks. Farmers tailor fertilization and irrigation schedules to meet changing plant needs during the growing season, gauging these needs from quantitative analyses of the plants and the soil in which they grow.

Quantitative analytical measurements also play a vital role in many research areas in chemistry, biochemistry, biology, geology, physics, and the other sciences. For example, quantitative measurements of potassium, calcium, and sodium ions in the body fluids of animals permit physiologists to study the role of these ions in nerve signal conduction as well as muscle contraction and relaxation. Chemists unravel the mechanisms of chemical reactions through reaction rate studies. The rate of consumption of reactants or formation of products in a chemical reaction can be calculated from quantitative measurements made at equal time intervals. Materials scientists rely heavily on quantitative analyses of crystalline germanium and silicon in their studies of semiconductor devices. Impurities in these devices are in the concentration range of 1×10^{-6} to 1×10^{-9} percent. Archaeologists identify the source of volcanic glasses (obsidian) by measuring concentrations of minor elements in samples taken from various locations. This knowledge in turn makes it possible to trace prehistoric trade routes for tools and weapons fashioned from obsidian.

Many chemists, biochemists, and medicinal chemists devote much time in the laboratory gathering quantitative information about systems that are important and interesting to them. The central role of analytical chemistry in this enterprise and many others is illustrated in Figure 1-1. All branches of chemistry draw on the ideas and techniques of analytical chemistry. Analytical chemistry has a similar function with respect to the many other scientific fields listed in the diagram. Chemistry is often called *the central science;* its top center position and the central position of analytical chemistry in the figure emphasize this importance. The interdisciplinary nature of chemical analysis makes it a vital tool in medical, industrial, government, and academic laboratories throughout the world.

1B QUANTITATIVE ANALYTICAL METHODS

We compute the results of a typical quantitative analysis from two measurements. One is the mass or the volume of sample being analyzed. The second is the measurement of some quantity that is proportional to the amount of analyte in the sample, such as mass, volume, intensity of light, or electrical charge. This second measurement usually completes the analysis, and we classify analytical methods according to the nature of this final measurement. **Gravimetric methods** determine the mass of the analyte or some compound chemically related to it. In a **volumetric method,** the volume of a solution containing sufficient reagent to react completely with the analyte is measured. **Electroanalytical methods** involve the measurement of such electrical properties as potential, current, resistance, and quantity of electrical charge. **Spectroscopic methods** are based on measurement of the interaction between electromagnetic radiation and analyte atoms or molecules or on the production of such radiation by analytes. Finally, a group of miscellaneous methods includes the measurement of such quantities as mass-to-charge ratio of molecules by mass spectrometry, rate of radioactive decay, heat of

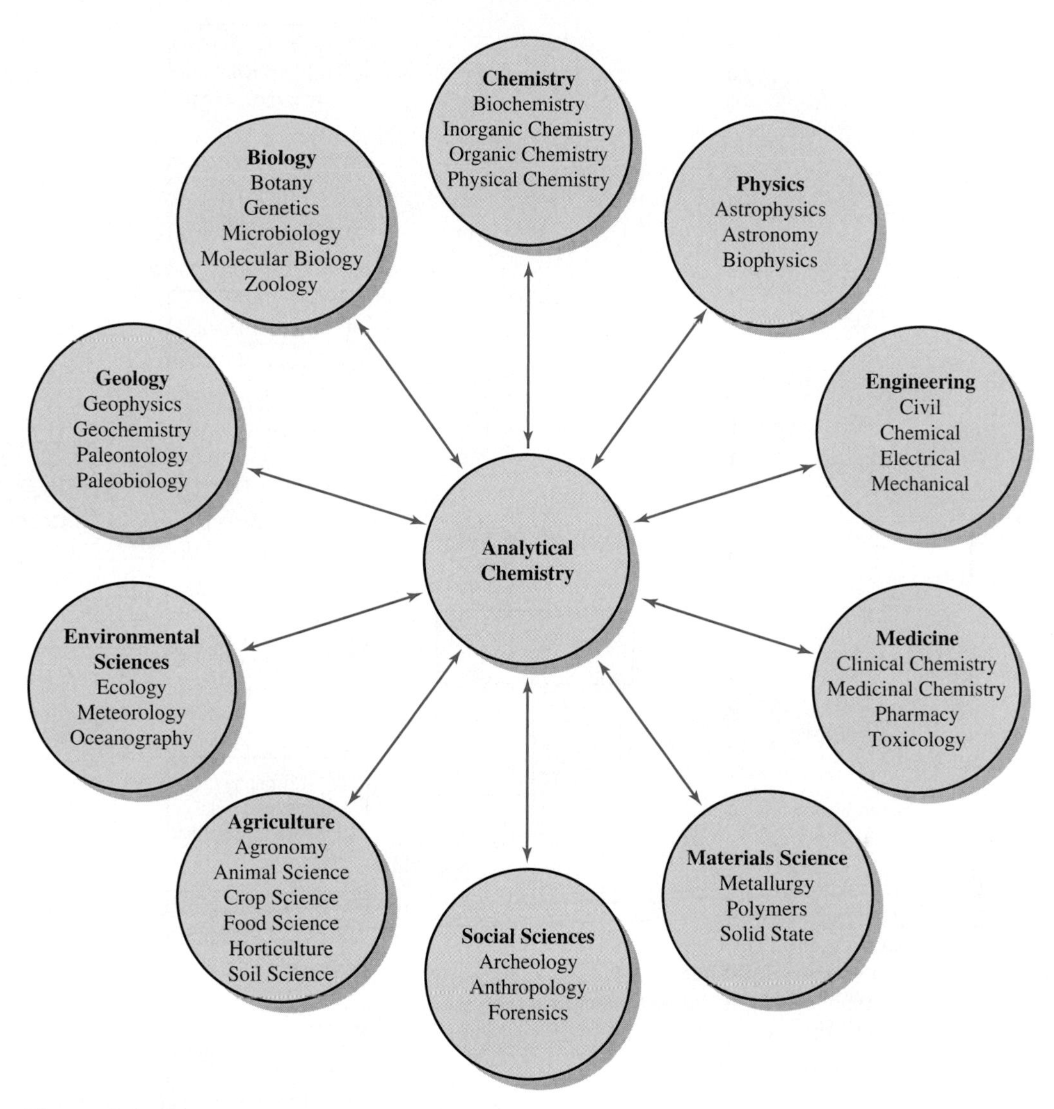

Figure 1-1 The relationship between analytical chemistry, other branches of chemistry, and the other sciences. The central location of analytical chemistry in the diagram signifies its importance and the breadth of its interactions with many other disciplines.

reaction, rate of reaction, sample thermal conductivity, optical activity, and refractive index.

1C A TYPICAL QUANTITATIVE ANALYSIS

A typical quantitative analysis involves the sequence of steps shown in the flow diagram of Figure 1-2. In some instances, one or more of these steps can be omitted. For example, if the sample is already a liquid, we can avoid the dissolution step. The first 29 chapters of this book focus on the last three steps in Figure 1-2. In

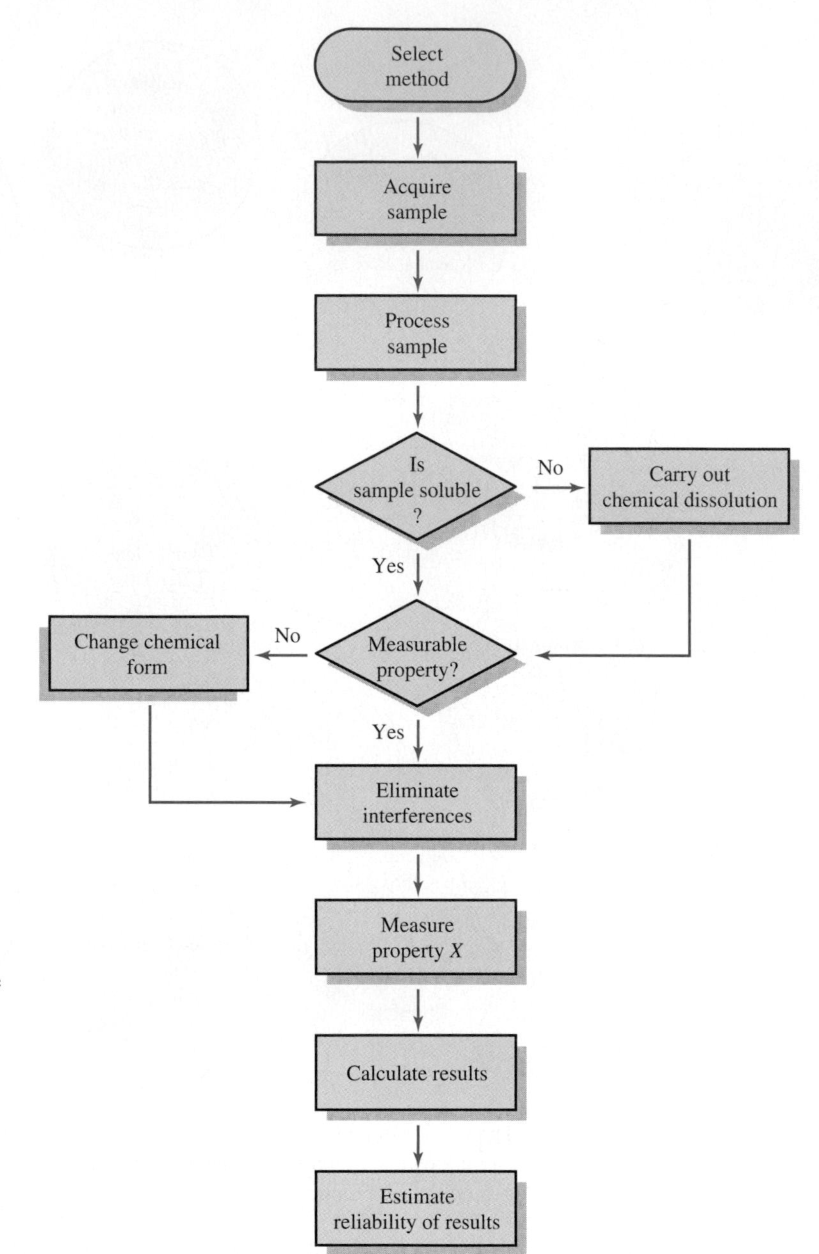

Figure 1-2 Flow diagram showing the steps in a quantitative analysis. There are a number of possible paths through the steps in a quantitative analysis. In the simplest example represented by the central vertical pathway, we select a method, acquire and process the sample, dissolve the sample in a suitable solvent, measure a property of the analyte, calculate the results, and estimate the reliability of the results. Depending on the complexity of the sample and the chosen method, various other pathways may be necessary.

the measurement step, we measure one of the physical properties mentioned in Section 1B. In the calculation step, we find the relative amount of the analyte present in the samples. In the final step, we evaluate the quality of the results and estimate their reliability.

In the paragraphs that follow, you will find a brief overview of each of the nine steps shown in Figure 1-2. We then present a case study to illustrate these steps in solving an important and practical analytical problem. The details of the case study foreshadow many of the methods and ideas you will explore in your study of analytical chemistry.

s. harris
"TODAY EVERYONE HAS TO KNOW 'WHAT'S IN THE FOOD?', 'WHAT'S IN THE WATER?', 'WHAT'S IN THE AIR?' THIS IS TRULY THE 'GOLDEN AGE OF ANALYTICAL CHEMISTRY.'"

1C-1 Choosing a Method

The essential first step in any quantitative analysis is the selection of a method, as depicted in Figure 1-2. The choice is sometimes difficult and requires experience as well as intuition. One of the first questions to be considered in the selection process is the level of accuracy required. Unfortunately, high reliability nearly always requires a large investment of time. The selected method usually represents a compromise between the accuracy required and the time and money available for the analysis.

A second consideration related to economic factors is the number of samples that will be analyzed. If there are many samples, we can afford to spend a significant amount of time in preliminary operations such as assembling and calibrating instruments and equipment and preparing standard solutions. If we have only a single sample, or just a few samples, it may be more appropriate to select a procedure that avoids or minimizes such preliminary steps.

Finally, the complexity of the sample and the number of components in the sample always influence the choice of method to some degree.

1C-2 Acquiring the Sample

As illustrated in Figure 1-2, the next step in a quantitative analysis is to acquire the sample. To produce meaningful information, an analysis must be performed on a sample that has the same composition as the bulk of material from which it was taken. When the bulk is large and **heterogeneous,** great effort is required to get a representative sample. Consider, for example, a railroad car containing 25 tons of silver ore. The buyer and seller of the ore must agree on a price, which will be based primarily on the silver content of the shipment. The ore itself is inherently heterogeneous, consisting of many lumps that vary in size as well as silver content.

A material is **heterogeneous** if its constituent parts can be distinguished visually or with the aid of a microscope. Coal, animal tissue, and soil are heterogeneous materials.

An **assay** is the process of determining how much of a given sample is the material indicated by its name. For example, a zinc alloy is assayed for its zinc content, and its assay is a particular numerical value.

► We *analyze* samples and we *determine* substances. For example, a blood sample is analyzed to determine the concentrations of various substances such as blood gases and glucose. We therefore speak of the determination of blood gases or glucose, *not* the analysis of blood gases or glucose.

The **assay** of this shipment will be performed on a sample that weighs about one gram. For the analysis to have significance, this small sample must have a composition that is representative of the 25 tons (or approximately 22,700,000 g) of ore in the shipment. Isolation of one gram of material that accurately represents the average composition of the nearly 23,000,000 g of bulk sample is a difficult undertaking that requires a careful, systematic manipulation of the entire shipment. **Sampling** is the process of collecting a small mass of a material whose composition accurately represents the bulk of the material being sampled. We explore the details of sampling in Chapter 8.

The collection of specimens from biological sources represents a second type of sampling problem. Sampling of human blood for the determination of blood gases illustrates the difficulty of acquiring a representative sample from a complex biological system. The concentration of oxygen and carbon dioxide in blood depends on a variety of physiological and environmental variables. For example, inappropriate application of a tourniquet or hand flexing by the patient may cause blood oxygen concentration to fluctuate. Because physicians make life-and-death decisions based on results of blood gas determinations, strict procedures have been developed for sampling and transporting specimens to the clinical laboratory. These procedures ensure that the sample is representative of the patient at the time it is collected and that its integrity is preserved until the sample can be analyzed.

Many sampling problems are easier to solve than the two just described. Whether sampling is simple or complex, however, the analyst must be sure that the laboratory sample is representative of the whole before proceeding with an analysis. Sampling is frequently the most difficult step and the source of greatest error. The reliability of the final results of analysis will never be any greater than the reliability of the sampling step.

1C-3 Processing the Sample

The third step in an analysis is to process the sample, as shown in Figure 1-2. Under certain circumstances, no sample processing is required prior to the measurement step. For example, once a water sample is withdrawn from a stream, a lake, or an ocean, its pH can be measured directly. Under most circumstances, we must process the sample in any of a variety of different ways. The first step is often the preparation of a laboratory sample.

Preparing Laboratory Samples

A solid laboratory sample is ground to decrease particle size, mixed to ensure homogeneity, and stored for various lengths of time before analysis begins. Absorption or desorption of water may occur during each step, depending on the humidity of the environment. Because any loss or gain of water changes the chemical composition of solids, it is a good idea to dry samples just before starting an analysis. Alternatively, the moisture content of the sample can be determined at the time of the analysis in a separate analytical procedure.

Liquid samples present a slightly different but related set of problems during the preparation step. If such samples are allowed to stand in open containers, the solvent may evaporate and change the concentration of the analyte. If the analyte is a gas dissolved in a liquid, as in our blood gas example, the sample container must be kept inside a second sealed container, perhaps during the entire analytical procedure, to prevent contamination by atmospheric gases. Extraordinary measures, including sample manipulation and measurement in an inert atmosphere, may be required to preserve the integrity of the sample.

Defining Replicate Samples

Most chemical analyses are performed on replicate samples whose masses or volumes have been determined by careful measurements with an analytical balance or with a precise volumetric device. Replication improves the quality of the results and provides a measure of reliability. Quantitative measurements on **replicates** are usually averaged, and various statistical tests are performed on the results to establish reliability.

Replicate samples, or **replicates,** are portions of a material of approximately the same size that are carried through an analytical procedure at the same time and in the same way.

Preparing Solutions: Physical and Chemical Changes

Most analyses are performed on solutions of the sample made with a suitable solvent. Ideally, the solvent should dissolve the entire sample, including the analyte, rapidly and completely. The conditions of dissolution should be sufficiently mild so that loss of the analyte cannot occur. In our flow diagram of Figure 1-2, we ask whether the sample is soluble in the solvent of choice. Unfortunately, many materials that must be analyzed are insoluble in common solvents. Examples include silicate minerals, high-molecular-weight polymers, and specimens of animal tissue. Under this circumstance, we must follow the flow diagram to the box on the right and carry out some rather harsh chemistry. Conversion of the analyte in such materials into a soluble form is often the most difficult and time-consuming task in the analytical process. The sample may require heating with aqueous solutions of strong acids, strong bases, oxidizing agents, reducing agents, or some combination of such reagents. It may be necessary to ignite the sample in air or oxygen or to perform a high-temperature fusion of the sample in the presence of various fluxes. Once the analyte is made soluble, we then ask whether the sample has a property that is proportional to analyte concentration and that we can measure. If it does not, other chemical steps may be necessary to convert the analyte to a form suitable for the measurement step, as we see in Figure 1-2. For example, in the determination of manganese in steel, manganese must be oxidized to MnO_4^- before the absorbance of the colored solution is measured (see Chapter 26). At this point in the analysis, it may be possible to proceed directly to the measurement step, but more often than not, we must eliminate interferences in the sample before making measurements, as illustrated in the flow diagram.

1C-4 Eliminating Interferences

Once we have the sample in solution and have converted the analyte to an appropriate form for measurement, the next step is to eliminate substances from the sample that may interfere with measurement (see Figure 1-2). Few chemical or physical properties of importance in chemical analysis are unique to a single chemical species. Instead, the reactions used and the properties measured are characteristic of a group of elements of compounds. Species other than the analyte that affect the final measurement are called **interferences,** or **interferents.** A scheme must be devised to isolate the analytes from interferences before the final measurement is made. No hard and fast rules can be given for eliminating interferences;

An **interference** or **interferent** is a species that causes an error in an analysis by enhancing or attenuating (making smaller) the quantity being measured.

Techniques or reactions that work for only one analyte are said to be **specific.** Techniques or reactions that apply for only a few analytes are **selective.**

The **matrix,** or **sample matrix,** is all of the components in the sample containing an analyte.

indeed, resolution of this problem can be the most demanding aspect of an analysis. Chapters 30 through 33 describe separation methods.

1C-5 Calibrating and Measuring Concentration

All analytical results depend on a final measurement X of a physical or chemical property of the analyte, as shown in Figure 1-2. This property must vary in a known and reproducible way with the concentration c_A of the analyte. Ideally, the measurement of the property is directly proportional to the concentration. That is,

$$c_A = kX$$

where k is a proportionality constant. With two exceptions, analytical methods require the empirical determination of k with chemical standards for which c_A is known.[2] The process of determining k is thus an important step in most analyses; this step is called a **calibration.** We examine calibration in some detail in Chapter 8.

The process of determining *k*, an important step in most analyses, is termed a **calibration.**

1C-6 Calculating Results

Computing analyte concentrations from experimental data is usually relatively easy, particularly with modern calculators or computers. This step is depicted in the next-to-last block of Figure 1-2. These computations are based on the raw experimental data collected in the measurement step, the characteristics of the measurement instruments, and the stoichiometry of the analytical reaction. Samples of these calculations appear throughout this book.

1C-7 Evaluating Results by Estimating Their Reliability

As Figure 1-2 implies, analytical results are incomplete without an estimate of their reliability. The experimenter must provide some measure of the uncertainties associated with computed results if the data are to have any value. Chapters 5, 6, and 7 present detailed methods for carrying out this important final step in the analytical process.

► An analytical result without an estimate of reliability is of no value.

1D AN INTEGRAL ROLE FOR CHEMICAL ANALYSIS: FEEDBACK CONTROL SYSTEMS

Analytical chemistry is usually not an end in itself, but is part of a bigger picture in which we may use analytical results to help maintain or improve a patient's health, to control the amount of mercury in fish, to regulate the quality of a product, to determine the status of a synthesis, or to find out whether there is life on Mars. Chemical analysis is the measurement element in all of these examples and in

[2]The two exceptions are gravimetric methods, which are discussed in Chapter 12, and coulometric methods, which are considered in Chapter 22. In both these methods, k can be computed from known physical constants.

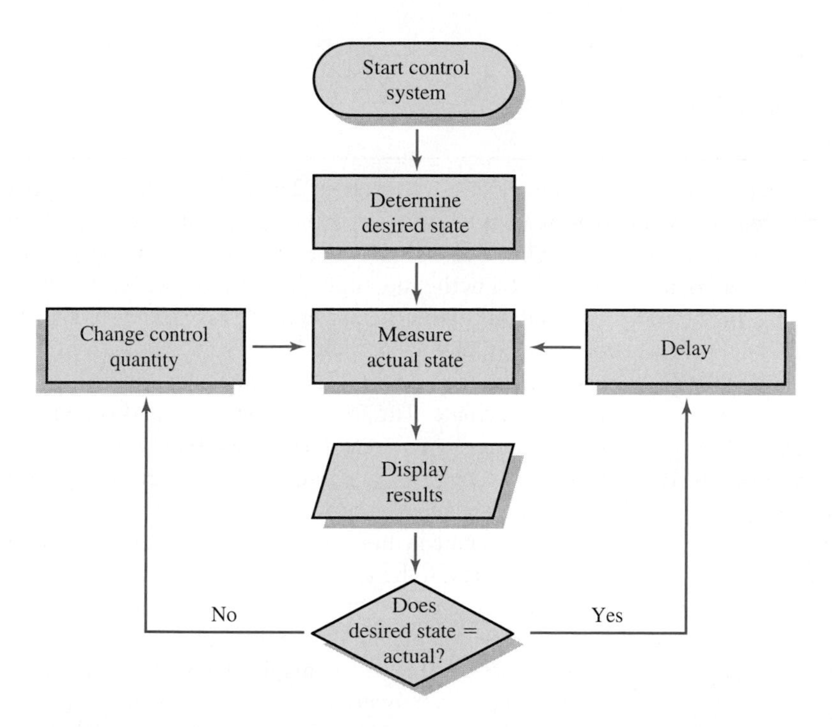

Figure 1-3 Feedback system flow diagram. The desired state is determined, the actual state of the system is measured, and the two states are compared. The difference between the two states is used to change a controllable quantity that results in a change in the state of the system. Quantitative measurements are again performed on the system, and the comparison is repeated. The new difference between the desired state and the actual state is again used to change the state of the system if necessary. The process provides continuous monitoring and feedback to maintain the controllable quantity, and thus the actual state, at the proper level. The text describes the monitoring and control of blood glucose concentration as an example of a feedback control system.

many other cases. Consider the role of quantitative analysis in determining and controlling glucose concentration in blood. The system flow diagram of Figure 1-3 illustrates the process. Patients with insulin-dependent diabetes mellitus develop hyperglycemia, which manifests itself as a blood glucose concentration above the normal value of 60 to 95 mg/dL. We begin our example by determining that the desired state is a blood glucose level below 95 mg/dL. Many patients must monitor their blood glucose levels by periodically submitting samples to a clinical laboratory for analysis or by measuring the levels themselves using a handheld electronic glucose monitor.

The first step in the monitoring process is to determine the actual state by collecting a blood sample from the patient and measuring the blood glucose level. The results are displayed, and then the actual state is compared with the desired (see Figure 1-3). If the measured blood glucose level is above 95 mg/dL, the patient's insulin level, which is a controllable quantity, is increased by injection or oral administration. After a delay to allow the insulin time to take effect, the glucose level is measured again to determine if the desired state has been achieved. If the level is below the threshold, the insulin level has been maintained, so no insulin is required. After a suitable delay time, the blood glucose level is measured again, and the cycle is repeated. In this way, the insulin level in the patient's blood, and thus the blood glucose level, is maintained at or below the critical threshold, which keeps the metabolism of the patient under control.

The process of continuous measurement and control is often referred to as a **feedback system,** and the cycle of measurement, comparison, and control is called a **feedback loop.** These ideas find wide application in biological and biomedical systems, mechanical systems, and electronics. From measuring and controlling the concentration of manganese in steel to maintaining the proper level of chlorine in a swimming pool, chemical analysis plays a central role in a broad range of systems.

FEATURE 1-1

Deer Kill: A Case Study Illustrating the Use of Analytical Chemistry to Solve a Problem in Toxicology

The tools of modern analytical chemistry are widely applied in environmental investigations. In this feature, we describe a case study in which quantitative analysis was used to determine the agent that caused deaths in a population of white-tailed deer inhabiting a wildlife preserve of a national recreational area in Kentucky. We begin with a description of the problem and then show how the steps illustrated in Figure 1-2 were used to solve the analytical problem. This case study also shows how chemical analysis is used in a broad context as an integral part of a feedback control system, as depicted in Figure 1-3.

The Problem

The incident began when a park ranger found a dead white-tailed deer near a pond in the land between the Lakes National Recreation Area in western Kentucky. The park ranger enlisted the help of a chemist from the state veterinary diagnostic laboratory to find the cause of death so that further deer kills might be prevented.

The ranger and the chemist investigated the site where the badly decomposed carcass of the deer had been found. Because of the advanced state of decomposition, no fresh organ tissue samples could be gathered. A few days after the original inquiry, the ranger found two more dead deer near the same location. The chemist was summoned to the site of the kill, where he and the ranger loaded the deer onto a truck for transport to the veterinary diagnostic laboratory. The investigators then conducted a careful examination of the surrounding area to find clues to the cause of death.

White-tailed deer have proliferated in many parts of the country.

The search covered about 2 acres surrounding the pond. The investigators noticed that grass surrounding nearby power line poles was wilted and discolored. They speculated that a herbicide might have been used on the grass. A common ingredient in herbicides is arsenic in any one of a variety of forms, including arsenic trioxide, sodium arsenite, monosodium methanearsenate, and disodium methanearsenate. The last compound is the disodium salt of methanearsenic acid, $CH_3AsO(OH)_2$, which is very soluble in water and thus finds use as the active ingredient in many herbicides. The herbicidal activity of disodium methanearsenate is due to its reactivity with the sulfhydryl (S—H) groups in the amino acid cysteine. When cysteine in plant enzymes reacts with arsenical compounds, the enzyme function is inhibited and the plant eventually dies. Unfortunately, similar chemical effects occur in animals as well. The investigators therefore collected samples of the discolored dead grass for testing along with samples from the organs of the deer. They planned to analyze the samples to confirm the presence of arsenic and, if present, to determine its concentration in the samples.

Selecting a Method

A scheme for the quantitative determination of arsenic in biological samples is found in the published methods of the Association of Official Analytical Chemists (AOAC).[3] This method involves the distillation of arsenic as arsine, which is then determined by colorimetric measurements.

Processing the Sample: Obtaining Representative Samples

Back at the laboratory, the deer were dissected and the kidneys were removed for analysis. The kidneys were chosen because the suspected pathogen (arsenic) is rapidly eliminated from an animal through its urinary tract.

Processing the Sample: Preparing a Laboratory Sample

Each kidney was cut into pieces and homogenized in a high-speed blender. This step reduced the size of the tissue pieces and homogenized the resulting laboratory sample.

[3] *Official Methods of Analysis,* 15th ed., p. 626. Washington, DC: Association of Official Analytical Chemists, 1990.

*Excel Shortcut Keystrokes for the PC**

**Macintosh equivalents, if different, appear in square brackets*

TO ACCOMPLISH THIS TASK	TYPE THESE KEYSTROKES
Alternate between displaying cell values and displaying cell formulas	**Ctrl+`** (Single Left Quotation Mark) **[⌘+`]**
Calculate all sheets in all open workbooks	**F9**
Calculate the active worksheet	**Shift+F9**
Cancel an entry in a cell or formula bar	**Esc**
Complete a cell entry and move down in the selection	**Enter [Return]**
Complete a cell entry and move to the left in the selection	**Shift+Tab**
Complete a cell entry and move to the right in the selection	**Tab**
Complete a cell entry and move up in the selection	**Shift+Enter**
Copy a formula from the cell above the active cell into the cell or the formula bar	**Ctrl+'** (Apostrophe) **[⌘+']**
Copy a selection	**Ctrl+C [⌘+C]**
Copy the value from the cell above the active cell into the cell or the formula bar	**Ctrl+Shift+"** (Quotation Mark) **[⌘+Shift+"]**
Create names from row and column labels	**Ctrl+Shift+F3**
Cut a selection	**Ctrl+X [⌘+X]**
Define a name	**Ctrl+F3 [⌘+F3]**
Delete text to the end of the line	**Ctrl+Delete [Ctrl+Option+Del]**
Delete the character to the left of the insertion point, or delete the selection	**Backspace [Delete]**
Delete the character to the right of the insertion point, or delete the selection	**Delete [Del]**
Display the Formula Palette after you type a valid function name in a formula	**Ctrl+A**
Edit a cell comment	**Shift+F2**
Edit the active cell	**F2 [None]**
Edit the active cell and then clear it, or delete the preceding character in the active cell as you edit the cell contents	**Backspace [Delete]**
Enter a formula as an array formula	**Ctrl+Shift+Enter**
Fill down	**Ctrl+D [⌘+D]**
Fill the selected cell range with the current entry	**Ctrl+Enter [None]**
Fill to the right	**Ctrl+R [⌘+R]**
Insert the argument names and parentheses for a function, after you type a valid function name in a formula	**Ctrl+Shift+A**
Insert the AutoSum formula	**Alt+=** (Equal Sign) **[⌘+Shift+T]**
Move one character up, down, left, or right	**Arrow Keys**
Move to the beginning of the line	**Home**
Paste a name into a formula	**F3 [None]**
Paste a selection	**Ctrl+V [⌘+V]**
Repeat the last action	**F4 Or Ctrl+Y [⌘+Y]**
Start a formula	= (Equal Sign)
Start a new line in the same cell	**Alt+Enter [⌘+Option+Enter]**
Undo	**Ctrl+Z [⌘+Z]**

Microsoft® Excel Toolbars

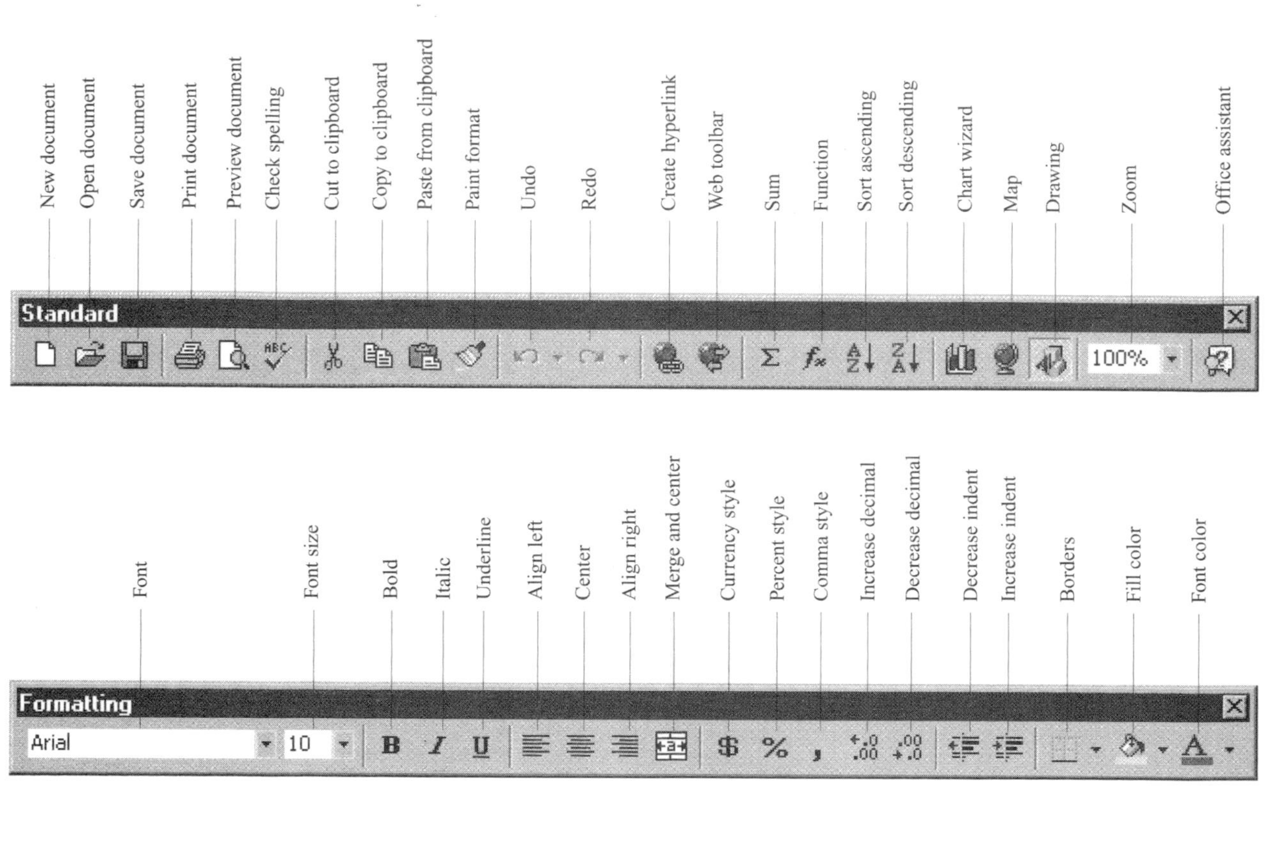
New document
Open document
Save document
Print document
Preview document
Check spelling
Cut to clipboard
Copy to clipboard
Paste from clipboard
Paint format
Undo
Redo
Create hyperlink
Web toolbar
Sum
Function
Sort ascending
Sort descending
Chart wizard
Map
Drawing
Zoom
Office assistant
Standard
100%
Font
Font size
Bold
Italic
Underline
Align left
Center
Align right
Merge and center
Currency style
Percent style
Comma style
Increase decimal
Decrease decimal
Decrease indent
Increase indent
Borders
Fill color
Font color
Formatting
Arial
10

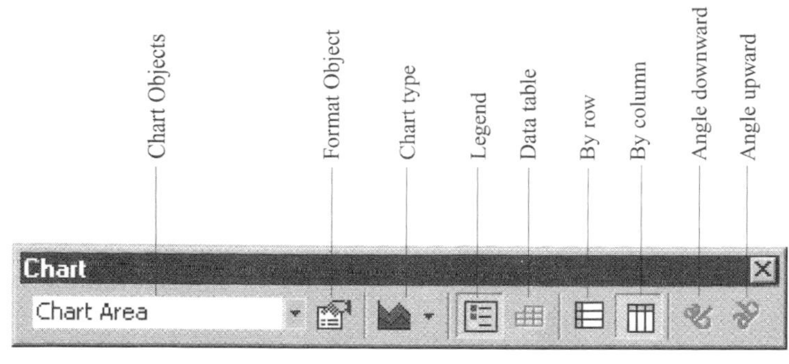
Chart Objects
Format Object
Chart type
Legend
Data table
By row
By column
Angle downward
Angle upward
Chart
Chart Area

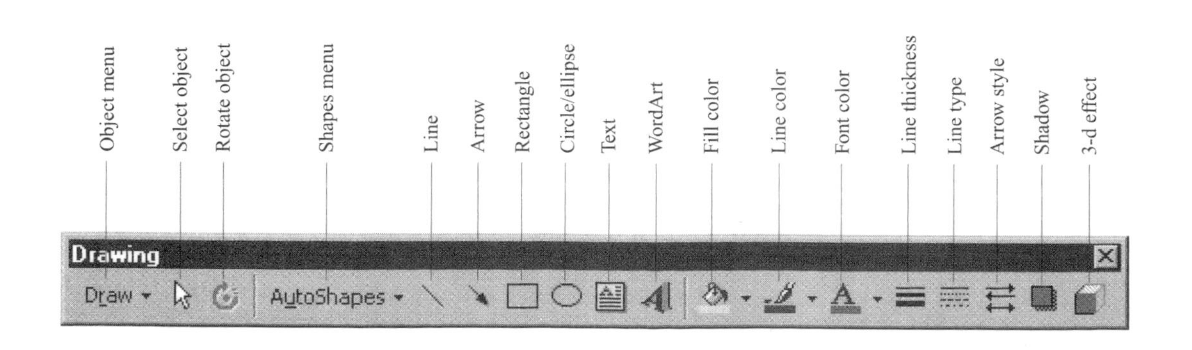
Object menu
Select object
Rotate object
Shapes menu
Line
Arrow
Rectangle
Circle/ellipse
Text
WordArt
Fill color
Line color
Font color
Line thickness
Line type
Arrow style
Shadow
3-d effect
Drawing
Draw
AutoShapes

Processing the Sample: Defining Replicate Samples
Three 10-g samples of the homogenized tissue from each deer were placed in porcelain crucibles.

Throughout this text, we will present models of molecules that are important in analytical chemistry. Here we show arsine, AsH_3. Arsine is an extremely toxic, colorless gas with a noxious garlic odor. Analytical methods involving the generation of arsine must be carried out with caution and proper ventilation.

Doing Chemistry: Dissolving the Samples
To obtain an aqueous solution of the analyte for analysis, it was necessary to dry ash the sample in air to convert its organic matrix to carbon dioxide and water. This process involved heating each crucible and sample cautiously over an open flame until the sample stopped smoking. The crucible was then placed in a furnace and heated at 555°C for 2 hours. Dry ashing served to free the analyte from organic material and convert it to arsenic pentoxide. The dry solid in each sample crucible was then dissolved in dilute HCl, which converted the As_2O_5 to soluble H_3AsO_4.

Eliminating Interferences
Arsenic can be separated from other substances that might interfere in the analysis by converting it to arsine, AsH_3, a toxic, colorless gas that is evolved when a solution of H_3AsO_3 is treated with zinc. The solutions resulting from the deer and grass samples were combined with Sn^{2+}, and a small amount of iodide ion was added to catalyze the reduction of H_3AsO_4 to H_3AsO_3 according to the following reaction:

$$H_3AsO_4 + SnCl_2 + 2HCl \rightarrow H_3AsO_3 + SnCl_4 + H_2O$$

The H_3AsO_3 was then converted to AsH_3 by the addition of zinc metal as follows:

$$H_3AsO_3 + 3Zn + 6HCl \rightarrow AsH_3(g) + 3ZnCl_2 + 3H_2O$$

The entire reaction was carried out in flasks equipped with a stopper and delivery tube so that the arsine could be collected in the absorber solution, as shown in Figure 1F-1. The arrangement ensured that interferences were left in the reaction flask and that only arsine was collected in the absorber in special transparent containers called cuvettes.

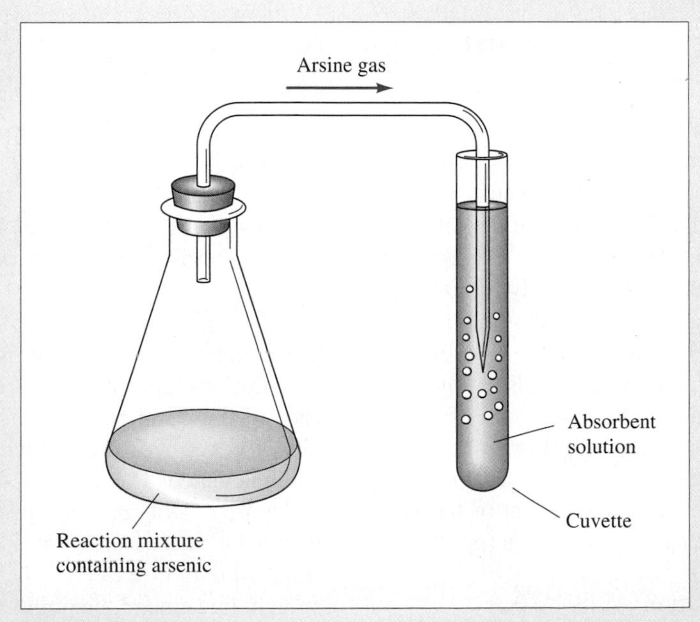

Figure 1F-1 An easily constructed apparatus for generating arsine, AsH_3.

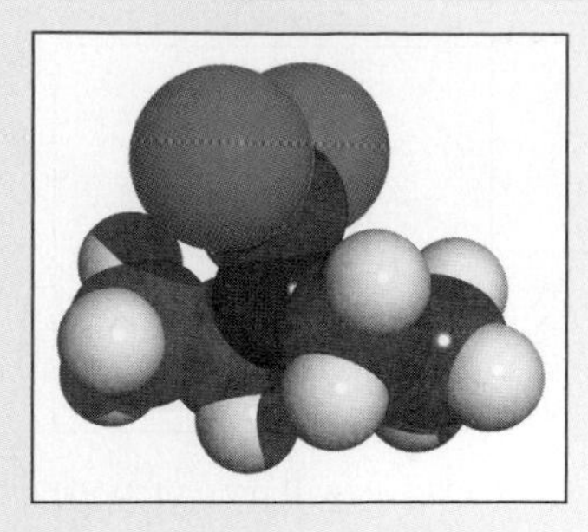

Molecular model of diethyldithiocarbamate. This compound is an analytical reagent used in determining arsenic, as illustrated in this feature.

Arsine bubbled into the solution in the cuvette, reacting with silver diethyldithiocarbamate to form a colored complex compound according to the following equation:

$$AsH_3 + 6Ag^+ + 3\left[\begin{array}{c} C_2H_5 \\ \quad \end{array} \!\!\! > N-C \begin{array}{c} \nearrow S \\ \searrow S \end{array}\right]^- \longrightarrow As\left[\begin{array}{c} C_2H_5 \\ C_2H_5 \end{array} \!\!\! > N-C \begin{array}{c} // S \\ \backslash S \end{array}\right]_3 + 6Ag + 3H^+$$

red

Measuring the Amount of the Analyte

The amount of arsenic in each sample was determined by using an instrument called a spectrophotometer, to measure the intensity of the red color formed in the cuvettes. As discussed in Chapter 26, a spectrophotometer provides a number called **absorbance** that is directly proportional to the color intensity, which is also proportional to the concentration of the species responsible for the color. To use absorbance for analytical purposes, a calibration curve must be generated by measuring the absorbance of several solutions that contain known concentrations of analyte. The upper part of Figure 1F-2 shows that the color becomes more intense as the arsenic content of the standards increases from 0 to 25 parts per million (ppm).

Calculating the Concentration

The absorbances for the standard solutions containing known concentrations of arsenic are plotted to produce a calibration curve, shown in the lower part of Figure 1F-2. Each vertical line between the upper and lower parts of Figure 1F-2 tics a solution to its corresponding point on the plot. The color intensity of each solution is represented by its absorbance, which is plotted on the vertical axis of the calibration curve. Note that the absorbance increases from 0 to about 0.72 as the concentration of arsenic increases from 0 to 25 ppm. The concentration of arsenic in each standard solution corresponds to the vertical grid lines of the calibration curve. This curve is then used to determine the concentration of the two unknown solutions shown on the right. We first find the absorbances of the unknowns on the absorbance axis of the plot and then read the corresponding concentrations on the concentration axis. The lines leading from the cuvettes to the calibration curve show that the concentrations of arsenic in the two deer were 16 ppm and 22 ppm, respectively.

Arsenic in the kidney tissue of an animal is toxic at levels above about 10 ppm, so it was probable that the deer were killed by ingesting an arsenic compound. The tests also showed that the samples of grass contained about 600 ppm arsenic. This very high level of arsenic suggested that the grass had been sprayed with an arsenical herbicide. The investigators concluded that the deer had probably died as a result of eating the poisoned grass.

Estimating the Reliability of the Data

The data from these experiments were analyzed using the statistical methods described in Chapters 5, 6, and 7. For each of the standard arsenic solutions and the deer samples, the average of the three absorbance measurements was calculated. The average absorbance for the replicates is a more reliable measure of the concentration of arsenic than a single measurement. Least-squares analysis of the standard data (see Section 8C) was used to find the best straight line among the points and to calculate the concentrations of the unknown samples along with their statistical uncertainties and confidence limits.

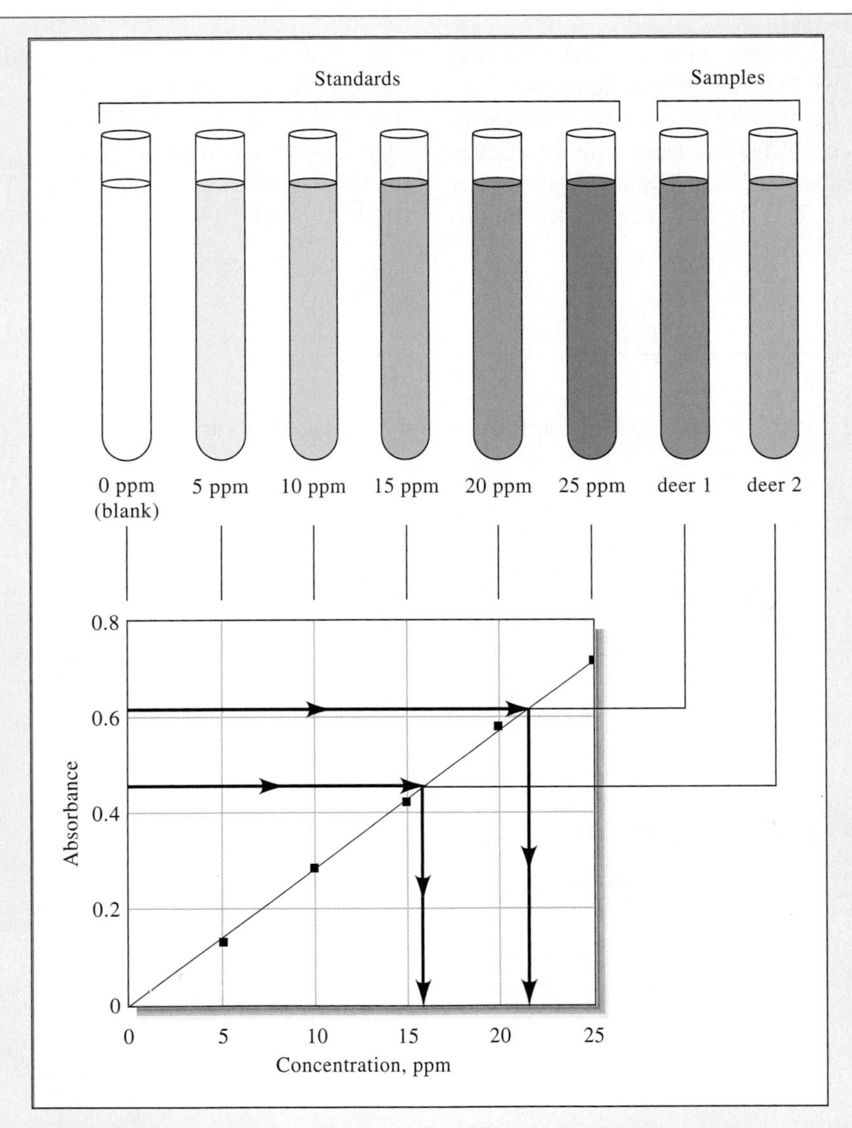

Figure 1F-2 Constructing and using a calibration curve to determine the concentration of arsenic. The absorbances of the solutions in the cuvettes are measured using a spectrophotometer. The absorbance values are then plotted against the concentrations of the solutions in the cuvettes, as illustrated in the graph. Finally, the concentrations of the unknown solutions are read from the plot, as shown by the dark arrows.

In this analysis, the formation of the highly colored product of the reaction served both to confirm the probable presence of arsenic and to provide a reliable estimate of its concentration in the deer and in the grass. On the basis of their results, the investigators recommended that the use of arsenical herbicides be suspended in the wildlife area to protect the deer and other animals that might eat plants there.

This case study illustrates how chemical analysis is used to identify and determine quantities of hazardous chemicals in the environment. Many of the methods and instruments of analytical chemistry are used routinely to provide vital information in environmental and toxicological studies of this type. The system flow diagram of Figure 1-3 may be applied to this case study. The desired state is a concentration of arsenic below the toxic level. Chemical analysis is used to determine the actual state, or the concentration of arsenic in the environment, and this value is compared with the desired concentration. The difference is then used to determine appropriate actions (such as decreased usc of arsenical pesticides) to ensure that deer are not poisoned by excessive amounts of arsenic in the environment, which in this example is the controlled system.

InfoTrac College Edition

For additional readings, go to InfoTrac College Edition, your online research library, at

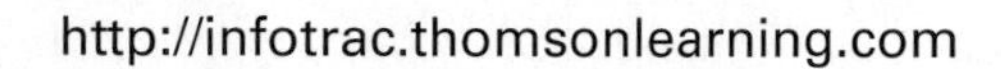

PART I

Tools of Analytical Chemistry

A conversation with *Richard N. Zare*

Courtesy of Liz Edlund

Richard Zare's career choice was made when a chemistry professor inadvertently introduced him to spectroscopy.[1] Once he realized the power of the technique—and what he could do to improve on it—he'd found his life's work. Zare introduced laser techniques to chemical analysis and has used them to study important chemical problems. He has also overseen the development of new separation techniques. Zare has had several academic appointments, including the past 20 years at Stanford University. He has been the recipient of many honorary degrees and prizes, most notably the 1983 National Medal of Science, to honor his work in laser-induced fluorescence, and the 1999 Welsh Award in Chemistry, a lifetime achievement award.

Q: Were you encouraged by your family to become a chemist?

A: My father studied to be a chemist but dropped out of graduate school when he married my mother during the Depression. We had many chemistry books around the house but I was told that they lead only to unhappiness and I shouldn't read them. This only encouraged me to look at them, and I used to read them with a flashlight under the covers of my bed. My parents did not permit me to have a chemistry set, so I formed a close relationship with the local pharmacist, who supplied me with chemicals that today would not be possible. With them I assembled various pyrotechnics, and once set the basement on fire.

Q: How were you introduced to spectroscopy?

A: At Harvard I took a course on quantitative analysis for which we had to do the gravimetric analysis of calcium in limestone. But the instructor told us that we were wasting our time; any sensible person would use atomic spectroscopy. I asked what it was and he told me to read a small book written by Gerhard Herzberg, who would later win a Nobel Prize for spectroscopy. I did, and that summer at home I made my own carbon arc for taking atomic spectra of various compounds.

Q: You have worked a lot with lasers. How has this shaped your career?

A: When I was a graduate student, lasers were just being developed and physicists were calling them a solution in search of a problem. I had a clear idea what they might be good for. First I used them to take the first fluorescence spectra of molecules. Later I pioneered the use of laser-induced fluorescence and resonance-enhanced multiphoton ionization as detection schemes that identify the internal state distribution of reaction products. I was among the first to use lasers to prepare reagents in specific internal states so that their reactivity could be studied as a function of the type and amount of internal motion. I also developed the use of polarized excitation and detection, which provides information on the geometry of the transition state region.

A turning point in my career was when I gave a talk at an American Chemical Society meeting. I had developed a technique for detecting reaction products from molecules formed in crossed molecular beams. Dr. Larry Seitz, from the Department of Agriculture, wandered into this session by mistake. He asked whether I could detect aflatoxin, a poisonous metabolite found in moldy grains. I said that I could if I could put it in the gas phase and it fluoresced. I did not understand that aflatoxin decomposes upon heating. We corresponded, and I became intrigued with the question, "How could you detect aflatoxin when you can't vaporize it?" This led me to think about making chromatographic separations using laser-induced fluorescence as a detector. It was then only a short step to becoming interested in all types of separation techniques and all the types of detectors that could be coupled to them. Thus, a hybrid physical chemist and analytical chemist was born.

I regard myself as a frustrated inventor. I keep arguing with myself, "Isn't there a better way to do something?" and I try things out. I'm very much interested in advances of instrumentation, how they change the ability to analyze chemicals, and how they need to work with increasingly smaller quantities of material.

Q: You believe in the value of spectroscopy. What is it about spectroscopy that makes it so valuable?

A: I see spectroscopy as the use of absorption, emission, or scattering of electromagnetic radiation by matter to study qualitatively

[1]Spectroscopy is the science of the interaction of matter with electromagnetic radiation, as described in Chapters 24–28.

or quantitatively the nature of matter and the processes it undergoes. The matter can be atoms, molecules, atomic or molecular ions, or solids. The interaction of radiation with matter can cause redirection of the radiation or transitions between energy levels of the atoms and molecules, or both. More subtle effects involve not only the color or wavelength of the radiation but its change in intensity and in the polarization of the light. It is by spectroscopy that we are able to see so much of the world, including that which we cannot touch, like analyzing starlight for what it tells us about stars.

I regard myself as a frustrated inventor. I keep arguing with myself, "Isn't there a better way to do something?" and I try things out.

Q: The use of cavity ring-down spectroscopy has especially intrigued you. Can you describe this technique?

A: For a long time people have been looking at absorption by placing a sample between the light source and the detector and noticing the attenuation of the intensity of the light beam as a function of wavelength. Nearly everything has an absorption feature, but absorption is not very sensitive because the light source fluctuates in time. The way around this problem is to put your sample between two mirrors and send a pulse of light into this optical cavity. The light will bounce back and forth between the mirrors, each time traversing the sample. What a detector sees is a train of light pulses that come out of the end mirror, with each pulse having less intensity than the last. The optical cavity is actually an energy storage device, and the rate at which it loses energy, called its rate of ring-down, depends on the quality of the mirrors and the absorption of the sample but not on the intensity of the light pulse. If you put into the cavity a big or a little pulse or even a series of irreproducible pulses, they all ring down at the same rate. Thus, by measuring the ring-down rate, we are able to make absorption measurements more precisely. I use this technique to study ions in plasmas as well as analytes in liquids.

Q: What kind of work have you focused on at the molecular and cellular level?

A: I'm interested in analyzing the chemical constituents of cells: how cells communicate with each other, how they respond when they're chemically stimulated, and how the individual compartments within cells work. Presently, I'm striving to miniaturize separation devices for chemical analysis, using a capillary format or the channels of a microchip. As we move to these tiny devices, a premium is placed on the ability to detect what you have.

We're working on receptors and how they change conformation when a ligand, either an agonist or an antagonist, binds to them. Recently we showed that the molecular recognition event of the receptor triggers a biochemical cascade that amplifies the presence of the analyte. Amplification can also be achieved by opening an ion channel in the cell membrane to let a large number of ions flow across the membrane, which can then be detected with the patch-clamp technique. The sensitivity is so high that the binding of a single ligand to the receptor gives a detectable signal.

Q: Tell us about your use of lasers in spectrometry. What kinds of interesting studies have you done?

A: We also developed laser-desorption laser-ionization mass spectrometry for the analysis of adsorbates on surfaces, such as interplanetary dust particles and meteoritic samples. We use one laser to rapidly heat the sample and evaporate molecules from the surface. A second laser intercepts the rising plume of molecules and ionizes those that absorb that color of light. We then weigh the ions using a mass spectrometer. We have analyzed graphite particles extracted from meteorites and found polycyclic aromatic molecules (PAHs). The PAHs have C^{12} to C^{13} isotope ratios that match closely the graphite grains, which are believed to be the remnants of the "star dust" from which our solar system condensed some 4.5 billion years ago. These are the first interstellar molecules observed directly in the laboratory.

Recently we've been using laser ionization mass spectrometry to look at contaminated dredge sediments to understand the nature of environmental pollutants, like PAHs and polychlorinated biphenyls (PCBs). We've found that like binds with like; most of the contamination goes into coal particles. This raises important questions of the proper remediation for contaminated sites. Now they store the sediment, but it might be better to add coal and keep it sequestered. Until you know what's there, you can't make a rational policy.

Q: Even with all this research, you have found time to teach many students. Can you briefly state your philosophy and goals of teaching?

A: I've taught freshman chemistry so many times that the course is graded on an absolute basis. The advantage is that the students are not competing, so they can work together and teach each other. The lab integrates with the class. We synthesize a compound, purify it, and look at some of its physical properties, both structural and dynamic. I want the students to become active problem solvers and to understand that problems do not come with the names of subdisciplines on them. In university teaching so much knowledge is "dis-integrated" into courses taught in different departments, whereas the solution of real problems requires "re-integration" of that knowledge, often in a new way. ■

CHAPTER 2

Chemicals, Apparatus, and Unit Operations of Analytical Chemistry

LWA-Dann Jardif/Corbis

At the heart of analytical chemistry is a core set of operations and equipment that is necessary for laboratory work in the discipline and that serves as the foundation for its growth and development. In this photo, a student carries out an operation in the process of determining nitrogen in a sample of organic matter by the Kjeldahl method. Although this method was developed over a century ago, it is still used widely in agriculture and soil science.

In this chapter, we shall introduce the tools, techniques, and chemicals that are used by analytical chemists. The development of these tools began over two centuries ago and continues today. As the technology of analytical chemistry has improved with the advent of electronic analytical balances, automated titrators, and computer-controlled instruments, the speed, convenience, accuracy, and precision of analytical methods have generally improved as well. For example, the determination of the mass of a sample that required 5 to 10 minutes 40 years ago is now accomplished in a few seconds. Computations that took 10 to 20 minutes using tables of logarithms may now be carried out almost instantaneously with a computer spreadsheet. Our experience with such magnificent technological innovations often elicits impatience with the sometimes tedious techniques of classical analytical chemistry. It is this impatience that drives the quest to develop better methodologies. Indeed, basic methods have often been modified in the interest of speed or convenience without sacrificing accuracy or precision.

We must emphasize, however, that many of the unit operations encountered in the analytical laboratory are timeless. These tried and true operations have gradually evolved over the past two centuries. From time to time, the directions given in this chapter may seem somewhat didactic. Although we attempt to explain why unit operations are carried out in the way that we describe, you may be tempted to modify a procedure or skip a step here or there to save time and effort. We must caution you against modifying techniques and procedures unless you have discussed your proposed modification with your instructor and have considered its consequences carefully. Such modifications may cause unanticipated results, including unacceptable levels of accuracy or precision; in a worst-case scenario, a serious accident could result. Today, the time required to prepare a carefully standardized solution of sodium hydroxide is about the same as it was 100 years ago.

Mastery of the tools of analytical chemistry will serve you well in chemistry courses and in related scientific fields. In addition, your efforts will be rewarded with the considerable satisfaction of having completed an analysis with high standards of good analytical practice and with levels of accuracy and precision consistent with the limitations of the technique.

2A SELECTING AND HANDLING REAGENTS AND OTHER CHEMICALS

The purity of reagents has an important bearing on the accuracy attained in any analysis. It is therefore essential that the quality of a reagent be consistent with its intended use.

2A-1 Classifying Chemicals

Reagent Grade

Reagent-grade chemicals conform to the minimum standards set forth by the Reagent Chemical Committee of the American Chemical Society (ACS)[1] and are used wherever possible in analytical work. Some suppliers label their products with the maximum limits of impurity allowed by the ACS specifications; others print actual concentrations for the various impurities.

Primary-Standard Grade

The qualities required of a **primary standard,** in addition to extraordinary purity, are discussed in Section 13A-2. Primary-standard reagents have been carefully analyzed by the supplier, and the assay is printed on the container label. The National Institute of Standards and Technology is an excellent source for primary standards. This agency also provides **reference standards,** which are complex substances that have been exhaustively analyzed.[2]

◀ The National Institute of Standards and Technology (NIST) is the current name of what was formerly the National Bureau of Standards.

Special-Purpose Reagent Chemicals

Chemicals that have been prepared for a specific application are also available. Included among these are solvents for spectrophotometry and high-performance liquid chromatography. Information pertinent to the intended use is supplied with these reagents. Data provided with a spectrophotometric solvent, for example, might include its absorbance at selected wavelengths and its ultraviolet cutoff wavelength.

2A-2 Rules for Handling Reagents and Solutions

A high-quality chemical analysis requires reagents and solutions of known purity. A freshly opened bottle of a reagent-grade chemical can ordinarily be used with

[1]Committee on Analytical Reagents, *Reagent Chemicals,* 9th ed. Washington, DC: American Chemical Society, 2000.

[2]The Standard Reference Materials Program (SRMP) of the NIST provides thousands of reference materials for sale. The NIST maintains a catalog and price list of these materials at a Web site that is linked to the main NIST Web site at www.nist.gov. Standard reference materials may be purchased online.

confidence; whether this same confidence is justified when the bottle is half empty depends entirely on the way it has been handled after being opened. The following rules should be observed to prevent the accidental contamination of reagents and solutions.

1. Select the best grade of chemical available for analytical work. Whenever possible, pick the smallest bottle that will supply the desired quantity.
2. Replace the top of every container *immediately* after removal of the reagent; do not rely on someone else to do this.
3. Hold the stoppers of reagent bottles between your fingers; never set a stopper on a desk top.
4. *Unless specifically directed otherwise, never return any excess reagent to a bottle.* The money saved by returning excesses is seldom worth the risk of contaminating the entire bottle.
5. Unless directed otherwise, never insert spatulas, spoons, or knives into a bottle that contains a solid chemical. Instead, shake the capped bottle vigorously or tap it gently against a wooden table to break up an encrustation; then pour out the desired quantity. These measures are occasionally ineffective, and in such cases a clean porcelain spoon should be used.
6. Keep the reagent shelf and the laboratory balance clean and neat. Clean up any spills immediately, even though someone else is waiting to use the same chemical or reagent.
7. Observe local regulations concerning the disposal of surplus reagents and solutions.

2B CLEANING AND MARKING OF LABORATORY WARE

A chemical analysis is ordinarily performed in duplicate or triplicate. Thus, each vessel that holds a sample must be marked so that its contents can be positively identified. Flasks, beakers, and some crucibles have small etched areas on which semipermanent markings can be made with a pencil.

Special marking inks are available for porcelain surfaces. The marking is baked permanently into the glaze by heating at a high temperature. A saturated solution of iron(III) chloride, although not as satisfactory as the commercial preparation, can also be used for marking.

► Unless you are directed otherwise, do not dry the interior surfaces of glassware or porcelain ware.

Every beaker, flask, or crucible that will contain the sample must be thoroughly cleaned before being used. The apparatus should be washed with a hot detergent solution and then rinsed—initially with copious amounts of tap water and finally with several small portions of deionized water.[3] Properly cleaned glassware will be coated with a uniform and unbroken film of water. *It is seldom necessary to dry the interior surface of glassware before use;* drying is ordinarily a waste of time at best and a potential source of contamination at worst.

An organic solvent, such as benzene or acetone, may be effective in removing grease films. Chemical suppliers also market preparations for eliminating such films.

[3]References to deionized water in this chapter and Chapter 37 apply equally to distilled water.

2C EVAPORATING LIQUIDS

It is frequently necessary to decrease the volume of a solution that contains a nonvolatile solute. Figure 2-1 illustrates how this is done. The ribbed cover glass permits vapors to escape and protects the remaining solution from accidental contamination. Using glass hooks to provide space between the rim of the beaker and a conventional cover glass is less satisfactory than using the ribbed cover glass shown.

Evaporation is frequently difficult to control because of the tendency of some solutions to overheat locally. The **bumping** that results can be sufficiently vigorous to cause partial loss of the solution. Careful and gentle heating will minimize the danger of such loss. Where their use is permissible, glass beads will also minimize bumping.

Some unwanted species can be eliminated during evaporation. For example, chloride and nitrate can be removed from a solution by adding sulfuric acid and evaporating until copious white fumes of sulfur trioxide are observed (this operation must be performed in a hood). Urea is effective in removing nitrate ion and nitrogen oxides from acidic solutions. Ammonium chloride is best removed by adding concentrated nitric acid and evaporating the solution to a small volume. Ammonium ion is rapidly oxidized when it is heated; the solution is then evaporated to dryness.

Organic constituents can frequently be eliminated from a solution by adding sulfuric acid and heating to the appearance of sulfur trioxide fumes (in a hood); this process is known as **wet ashing.** Nitric acid can be added toward the end of heating to hasten oxidation of the last traces of organic matter.

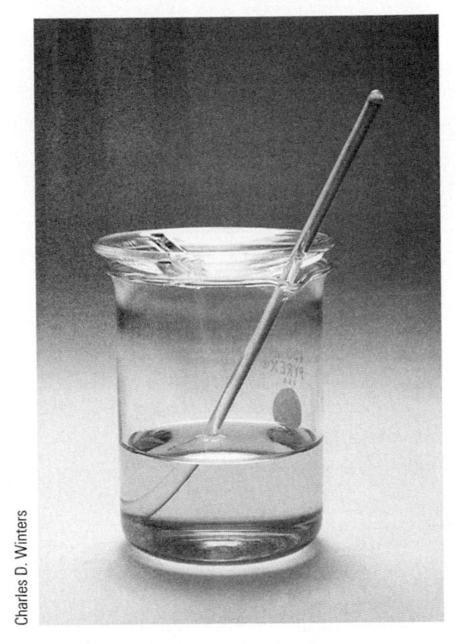

Charles D. Winters

Figure 2-1 Arrangement for the evaporation of a liquid.

Bumping is sudden, often violent boiling that tends to spatter solution out of its container.

Wet ashing is the oxidation of the organic constituents of a sample with oxidizing reagents such as nitric acid, sulfuric acid, hydrogen peroxide, aqueous bromine, or a combination of these reagents.

2D MEASURING MASS

In most analyses, an *analytical balance* must be used to obtain highly accurate masses. Less accurate *laboratory balances* are also used for mass measurements when the demands for reliability are not critical.

2D-1 Types of Analytical Balances

By definition, an **analytical balance** is an instrument for determining mass with a maximum capacity that ranges from 1 g to a few kilograms with a precision of at least 1 part in 10^5 at maximum capacity. The precision and accuracy of many modern analytical balances exceed 1 part in 10^6 at full capacity.

The most commonly encountered analytical balances **(macrobalances)** have a maximum capacity ranging between 160 and 200 g. With these balances, measurements can be made with a standard deviation of ±0.1 mg. **Semimicroanalytical balances** have a maximum loading of 10 to 30 g with a precision of ±0.01 mg. A typical **microanalytical balance** has a capacity of 1 to 3 g and a precision of ±0.001 mg.

The analytical balance has undergone a dramatic evolution over the past several decades. The traditional analytical balance had two pans attached to either end of a lightweight beam that pivoted about a knife edge located in the center of the beam. The object to be weighed was placed on one pan; sufficient standard masses were then added to the other pan to restore the beam to its original position. Weighing with such an **equal-arm balance** was tedious and time consuming.

An **analytical balance** has a maximum capacity that ranges from 1 g to several kilograms and a precision at maximum capacity of at least 1 part in 10^5.

A **macrobalance** is the most common type of analytical balance; it has a maximum load of 160 to 200 g and a precision of 0.1 mg.

A **semimicroanalytical balance** has a maximum load of 10 to 30 g and a precision of 0.01 mg.

A **microanalytical balance** has a maximum load of 1 to 3 g and a precision of 0.001 mg, or 1 μg.

The first **single-pan analytical balance** appeared on the market in 1946. The speed and convenience of weighing with this balance were vastly superior to what could be realized with the traditional equal-arm balance. Consequently, this balance rapidly replaced the latter in most laboratories. The single-pan balance is currently being replaced by the **electronic analytical balance,** which has neither a beam nor a knife edge. This type of balance is discussed in Section 2D-2. The single-pan balance is still used in some laboratories, but the speed, ruggedness, convenience, accuracy, and capability for computer control and data logging of electronic balances ensure that the mechanical single-pan analytical balance will soon disappear from the scene. The design and operation of a single-pan balance are discussed briefly in Section 2D-3.

2D-2 The Electronic Analytical Balance[4]

To **levitate** means to cause an object to float in air.

A **servo system** is a device in which a small electric signal causes a mechanical system to return to a null position.

Figure 2-2 shows a diagram and a photo of an electronic analytical balance. The pan rides above a hollow metal cylinder that is surrounded by a coil that fits over the inner pole of a cylindrical permanent magnet. An electric current in the coil produces a magnetic field that supports or **levitates** the cylinder, the pan and indicator arm, and whatever load is on the pan. The current is adjusted so that the level of the indicator arm is in the null position when the pan is empty. Placing an object on the pan causes the pan and indicator arm to move downward, which increases the amount of light striking the photocell of the null detector. The increased current from the photocell is amplified and fed into the coil, creating a larger magnetic field, which returns the pan to its original null position. A device such as this, in which a small electric current causes a mechanical system to maintain a null position, is called a **servo system.** The current required to keep the pan and object in the null position is directly pro-

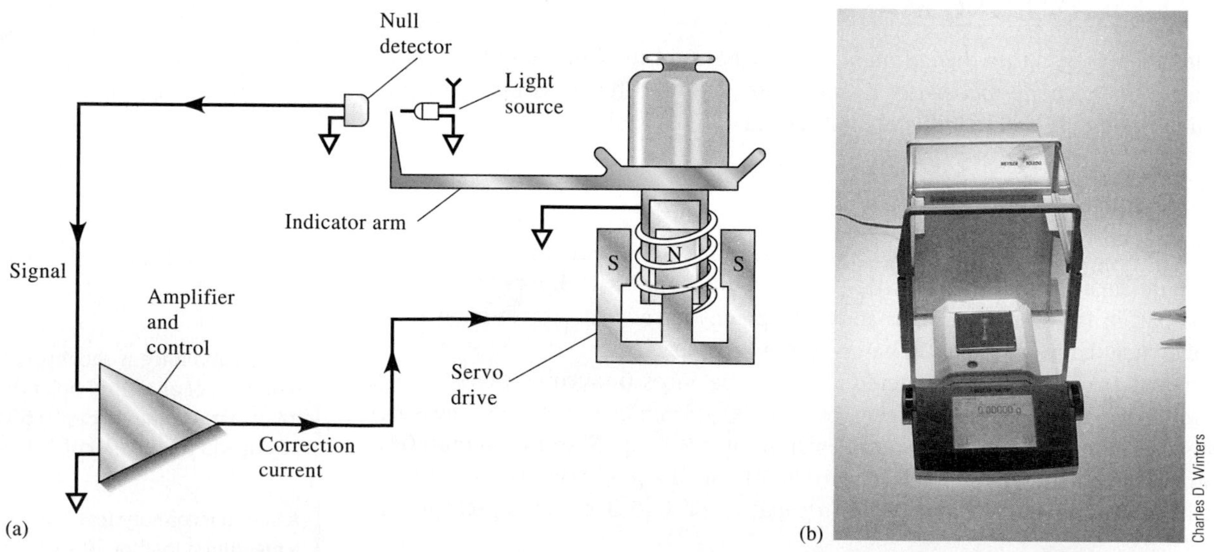

Figure 2-2 Electronic analytical balance. (a) Block diagram. (b) Photo of electronic balance. [*(a) Reprinted from R. M. Schoonover,* Anal. Chem., ***1982,*** 54, *973A. Published 1982 American Chemical Society.*]

[4]For a more detailed discussion, see R. M. Schoonover, *Anal. Chem.,* **1982,** *54,* 973A; K. M. Lang, *Amer. Lab.,* **1983,** *15* (3), 72.

portional to the mass of the object and is readily measured, digitized, and displayed. The calibration of an electronic balance involves using a standard mass and adjusting the current so that the mass of the standard is exhibited on the display.

Figure 2-3 shows the configurations for two electronic analytical balances. In each, the pan is tethered to a system of constraints known collectively as a **cell.** The cell incorporates several **flexures** that permit limited movement of the pan and prevent torsional forces (resulting from off-center loading) from disturbing the alignment of the balance mechanism. At null, the beam is parallel to the gravitational horizon and each flexure pivot is in a relaxed position.

Figure 2-3a shows an electronic balance with the pan located below the cell. Higher precision is achieved with this arrangement than with the top-loading design shown in Figure 2-3b. Even so, top-loading electronic balances have a precision that equals or exceeds that of the best mechanical balances and additionally provides unencumbered access to the pan.

A **tare** is the mass of an empty sample container. Taring is the process of setting a balance to read zero in the presence of the tare.

Electronic balances generally feature an automatic **taring control** that causes the display to read zero with a container (such as a boat or weighing bottle) on the pan. Most balances permit taring up to 100% of the capacity of the balance.

Some electronic balances have dual capacities and dual precisions. These features permit the capacity to be decreased from that of a macrobalance to that of a semimicrobalance (30 g) with a corresponding gain in precision to 0.01 mg. These types of balances are effectively two balances in one.

◀ Photographs of a modern electronic balance are shown in color plates 19 and 20.

A modern electronic analytical balance provides unprecedented speed and ease of use. For example, one instrument is controlled by touching a single bar at various positions along its length. One position on the bar turns the instrument on or off, another automatically calibrates the balance against a standard mass or pair of masses, and a third zeros the display, either with or without an object on the pan. Reliable mass measurements are obtainable with little or no instruction or practice.

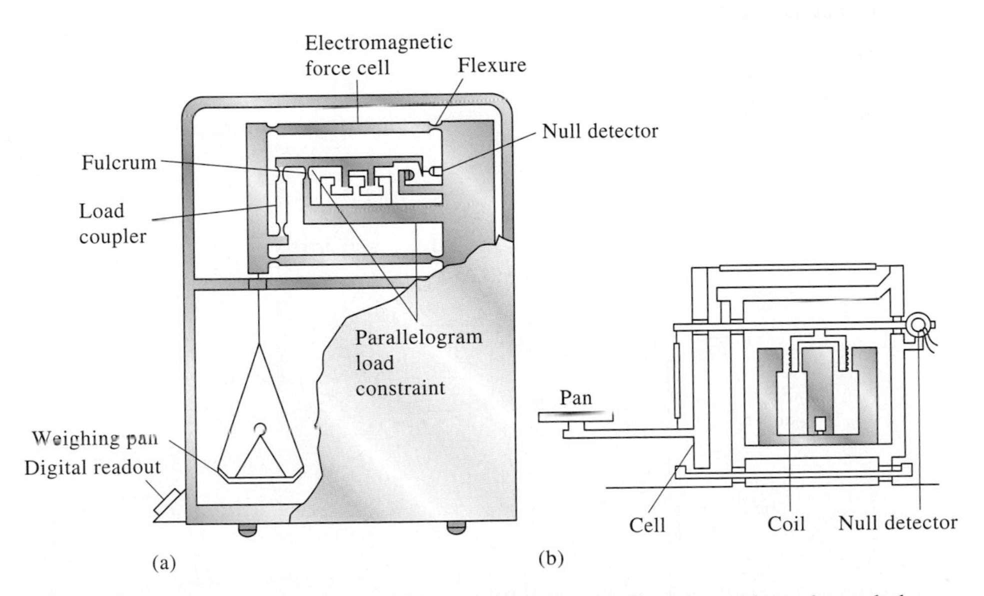

Figure 2-3 Electronic analytical balances. (a) Classical configuration with pan beneath the cell. (b) A top-loading design. Note that the mechanism is enclosed in a windowed case. [*(a) Reprinted from R. M. Schoonover,* Anal. Chem., ***1982,*** 54, *973A. Published 1982 American Chemical Society. (b) Reprinted from K. M. Lang,* Amer. Lab., ***1983,*** 15*(3), 72. Copyright 1983 by International Scientific Communications, Inc.*]

2D-3 The Single-Pan Mechanical Analytical Balance

Components

Although they differ considerably in appearance and performance characteristics, all mechanical balances, equal-arm as well as single-pan, have several common components. Figure 2-4 is a diagram of a typical single-pan mechanical balance. Fundamental to this instrument is a lightweight **beam** that is supported on a planar surface by a prism-shaped **knife edge** *(A)*. Attached to the left end of the beam is a pan for holding the object to be weighed and a full set of masses held in place by hangers. These masses can be lifted from the beam one at a time by a mechanical arrangement that is controlled by a set of knobs on the exterior of the balance case. The right end of the beam holds a counterweight of such size as to just balance the pan and masses on the left end of the beam.

The two **knife edges** in a mechanical balance are prism-shaped agate or sapphire devices that form low-friction bearings with two planar surfaces contained in **stirrups** also of agate or sapphire.

A second knife edge *(B)* is located near the left end of the beam and support as a second planar surface, which is located in the inner side of a **stirrup** that couples the pan to the beam. The two knife edges and their planar surfaces are fabricated from extraordinarily hard materials (agate or synthetic sapphire) and form two bearings that permit motion of the beam and pan with a minimum of friction. The performance of a mechanical balance is critically dependent on the perfection of these two bearings.

► To avoid damage to the knife edges and bearing surfaces, the arrest system for a mechanical balance should be engaged at all times other than during actual weighing.

Single-pan balances are also equipped with a **beam arrest** and a **pan arrest.** The beam arrest is a mechanical device that raises the beam so that the central knife edge no longer touches its bearing surface and simultaneously frees the stirrup from contact with the outer knife edge. The purpose of both arrest mechanisms is to prevent damage to the bearings while objects are being placed on or removed from the pan. When engaged, the pan arrest supports most of the mass of the pan and its contents and thus prevents oscillation. Both arrests are controlled by a lever mounted on the outside of the balance case and should be engaged whenever the balance is not in use.

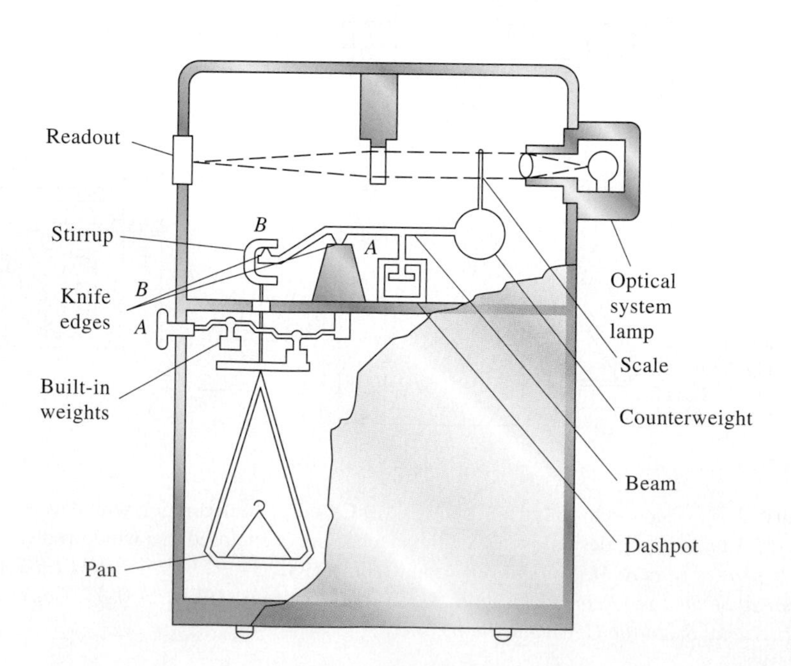

Figure 2-4 Single-pan mechanical analytical balance. *(From R. M. Schoonover,* Anal. Chem., ***1982,*** 54, *973A. Published 1982 American Chemical Society.)*

An **air damper** (also known as a **dashpot**) is mounted near the end of the beam opposite the pan. This device consists of a piston that moves within a concentric cylinder attached to the balance case. Air in the cylinder undergoes expansion and contraction as the beam is set in motion; the beam rapidly comes to rest as a result of this opposition to motion.

Protection from air currents is needed to permit discrimination between small differences in mass (<1 mg). An analytical balance is thus always enclosed in a case equipped with doors to permit the introduction or removal of objects.

Weighing with a Single-Pan Balance

The beam of a properly adjusted balance assumes an essentially horizontal position with no object on the pan and all of the masses in place. When the pan and beam arrests are disengaged, the beam is free to rotate around the knife edge. Placing an object on the pan causes the left end of the beam to move downward. Masses are then removed systematically one by one from the beam until the imbalance is less than 100 mg. The angle of deflection of the beam with respect to its original horizontal position is directly proportional to the additional mass that must be removed to restore the beam to its original horizontal position. The optical system shown in the upper part of Figure 2-4 measures this angle of deflection and converts this angle to milligrams. A **reticle,** which is a small transparent screen mounted on the beam, is scribed with a scale that reads 0 to 100 mg. A beam of light passes through the scale to an enlarging lens, which in turn focuses a small part of the enlarged scale onto a frosted glass plate located on the front of the balance. A vernier makes it possible to read this scale to the nearest 0.1 mg.

Precautions in Using an Analytical Balance

An analytical balance is a delicate instrument that you must handle with care. Consult with your instructor for detailed instructions on weighing with your particular model of balance. Observe the following general rules for working with an analytical balance regardless of make or model:

1. Center the load on the pan as well as possible.
2. Protect the balance from corrosion. Objects to be placed on the pan should be limited to nonreactive metals, nonreactive plastics, and vitreous materials.
3. Observe special precautions (see Section 2E-6) for the weighing of liquids.
4. Consult the instructor if the balance appears to need adjustment.
5. Keep the balance and its case scrupulously clean. A camel's-hair brush is useful for removing spilled material or dust.
6. Always allow an object that has been heated to return to room temperature before weighing it.
7. Use tongs or finger pads to prevent the uptake of moisture by dried objects.

2D-4 Sources of Error in Weighing

Correction for Buoyancy[5]

A **buoyancy error** will affect data if the density of the object being weighed differs significantly from that of the standard masses. This error has its origin in the

A **buoyancy error** is the weighing error that develops when the object being weighed has a significantly different density than the masses.

[5]For further information, see R. Battino and A. G. Williamson, *J. Chem. Educ.*, **1984,** *64,* 51.

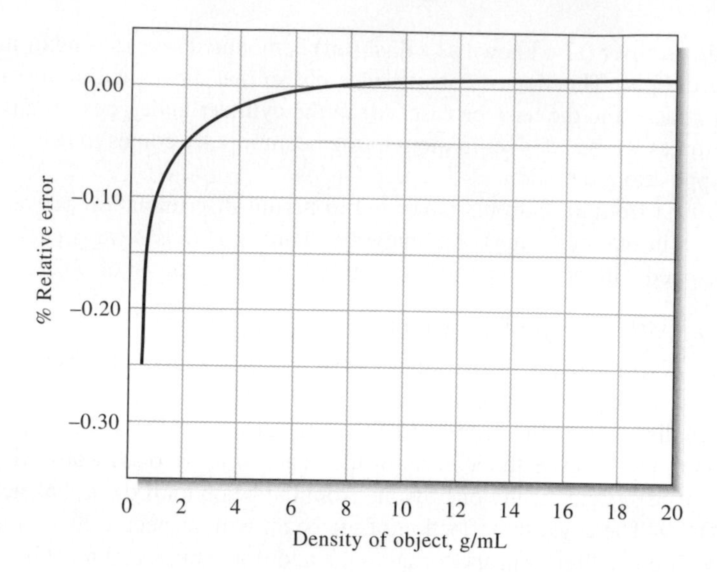

Figure 2-5 Effect of buoyancy on weighing data (density of weights = 8 g/cm^3). Plot of relative error as a function of the density of the object weighed.

difference in buoyant force exerted by the medium (air) on the object and on the masses. Buoyancy corrections for electronic balances[6] may be accomplished with the equation

$$W_1 = W_2 + W_2\left(\frac{d_{air}}{d_{obj}} - \frac{d_{air}}{d_{wts}}\right) \tag{2-1}$$

where W_1 is the corrected mass of the object, W_2 is the mass of the standard masses, d_{obj} is the density of the object, d_{wts} is the density of the masses, and d_{air} is the density of the air displaced by them; d_{air} has a value of 0.0012 g/cm^3.

The consequences of Equation 2-1 are shown in Figure 2-5, in which the relative error due to buoyancy is plotted against the density of objects weighed in air against stainless steel masses. Note that this error is less than 0.1% for objects that have a density of 2 g/cm^3 or greater. It is thus seldom necessary to apply a correction to the mass of most solids. The same cannot be said for low-density solids, liquids, or gases, however, for these, the effects of buoyancy are significant and a correction must be applied.

The density of masses used in single-pan balances (or to calibrate electronic balances) ranges from 7.8 to 8.4 g/cm^3, depending on the manufacturer. Use of 8 g/cm^3 is adequate for most purposes. If greater accuracy is required, the specifications for the balance to be used should be consulted for the necessary density data.

[6]Air buoyancy corrections for single-pan mechanical balances are somewhat different from those for electronic balances. For a thorough discussion of the differences in the corrections, see M. R. Winward et al., *Anal. Chem.*, **1977,** *49,* 2126.

EXAMPLE 2-1

A bottle weighed 7.6500 g empty and 9.9700 g after introduction of an organic liquid with a density of 0.92 g/cm^3. The balance was equipped with stainless steel masses (d = 8.0 g/cm^3). Correct the mass of the sample for the effects of buoyancy.

The apparent mass of the liquid is 9.9700 − 7.6500 = 2.3200 g. The same buoyant force acts on the container during both weighings; thus, we need to consider only the force that acts on the 2.3200 g of liquid. By substituting 0.0012 g/cm^3 for d_{air}, 0.92 g/cm^3 for d_{obj}, and 8.0 g/cm^3 for d_{wts} in Equation 2-1, we find that the corrected mass is

$$W_1 = 2.3200 + 2.3200\left(\frac{0.0012}{0.92} - \frac{0.0012}{8.0}\right) = 2.3227 \text{ g}$$

Temperature Effects

Attempts to weigh an object whose temperature is different from that of its surroundings will result in a significant error. Failure to allow sufficient time for a heated object to return to room temperature is the most common source of this problem. Errors due to a difference in temperature have two sources. First, convection currents within the balance case exert a buoyant effect on the pan and object. Second, warm air trapped in a closed container weighs less than the same volume at a lower temperature. Both effects cause the apparent mass of the object to be low. This error can amount to as much as 10 or 15 mg for a typical porcelain filtering crucible or a weighing bottle (Figure 2-6). Heated objects must always be cooled to room temperature before being weighed.

◀ Always allow heated objects to return to room temperature before you attempt to weigh them.

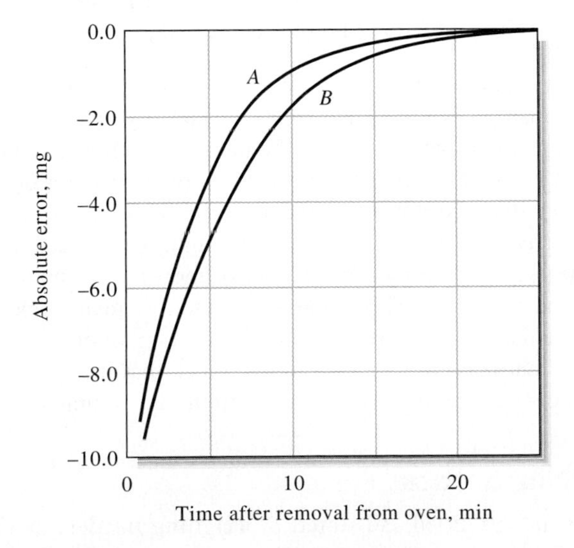

Figure 2-6 Effect of temperature on weighing data. Absolute error as a function of time after the object was removed from a 110°C drying oven. *A:* porcelain filtering crucible. *B:* weighing bottle containing about 7.5 g of KCl.

Other Sources of Error

A porcelain or glass object will occasionally acquire a static charge sufficient to cause a balance to perform erratically; this problem is particularly serious when the relative humidity is low. Spontaneous discharge frequently occurs after a short period. A low-level source of radioactivity (such as a photographer's brush) in the balance case will provide sufficient ions to relieve the charge. Alternatively, the object can be wiped with a faintly damp chamois.

The optical scale of a single-pan mechanical balance should be checked regularly for accuracy, particularly under loading conditions that require the full scale range. A standard 100-mg mass is used for this check.

2D-5 Auxiliary Balances

► Use auxiliary laboratory balances for weighings that do not require great accuracy.

Balances that are less precise than analytical balances find extensive use in the analytical laboratory. These offer the advantages of speed, ruggedness, large capacity, and convenience; they should be used whenever high sensitivity is not required.

Top-loading auxiliary balances are particularly convenient. A sensitive top-loading balance will accommodate 150 to 200 g with a precision of about 1 mg—an order of magnitude less than a macroanalytical balance. Some balances of this type tolerate loads as great as 25,000 g with a precision of ±0.05 g. Most are equipped with a taring device that brings the balance reading to zero with an empty container on the pan. Some are fully automatic, require no manual dialing or mass handling, and provide a digital readout of the mass. Modern top-loading balances are electronic.

A triple-beam balance with a sensitivity less than that of a typical top-loading auxiliary balance is also useful. This is a single-pan balance with three decades of masses that slide along individual calibrated scales. The precision of a triple-beam balance may be one or two orders of magnitude less than that of a top-loading instrument but is adequate for many weighing operations. This type of balance offers the advantages of simplicity, durability, and low cost.

2E EQUIPMENT AND MANIPULATIONS ASSOCIATED WITH WEIGHING

The mass of many solids changes with humidity, owing to their tendency to absorb weighable amounts of moisture. This effect is especially pronounced when a large surface area is exposed, as with a reagent chemical or a sample that has been ground to a fine powder. The first step in a typical analysis, then, involves drying the sample so that the results will not be affected by the humidity of the surrounding atmosphere.

Drying or ignition to constant mass is a process in which a solid is cycled through heating, cooling, and weighing steps until its mass becomes constant to within 0.2 to 0.3 mg.

A sample, a precipitate, or a container is brought to constant mass, by a cycle that involves heating (ordinarily for one hour or more) at an appropriate temperature, cooling, and weighing. This cycle is repeated as many times as needed to obtain successive masses that agree within 0.2 to 0.3 mg of one another. The establishment of constant mass provides some assurance that the chemical or physical processes that occur during the heating (or ignition) are complete.

2E-1 Weighing Bottles

Solids are conveniently dried and stored in **weighing bottles,** two common varieties of which are shown in Figure 2-7. The ground-glass portion of the cap-style bottle shown on the left is on the outside and does not come into contact with the contents; this design eliminates the possibility of some of the sample becoming entrained on and subsequently lost from the ground-glass surface.

Plastic weighing bottles are available; ruggedness is the principal advantage of these bottles over their glass counterparts.

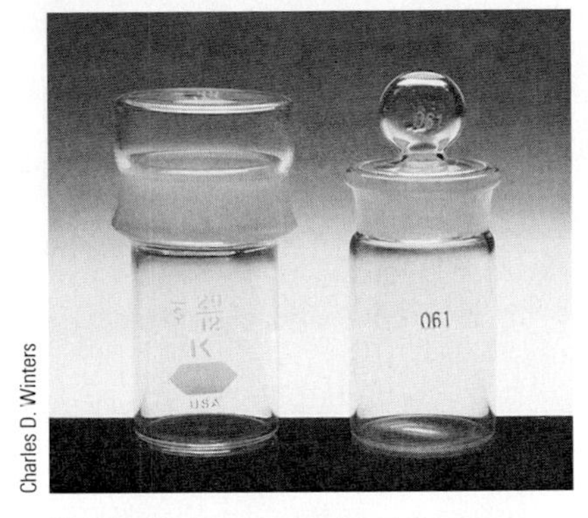

Figure 2-7 Typical weighing bottles.

2E-2 Desiccators and Desiccants

Oven drying is the most common way of removing moisture from solids. This approach is not appropriate for substances that decompose or for those from which water is not removed at the temperature of the oven.

To minimize the uptake of moisture, dried materials are stored in **desiccators** while they cool. Figure 2-8 shows the components of a typical desiccator. The base section contains a chemical drying agent, such as anhydrous calcium chloride, calcium sulfate (Drierite), anhydrous magnesium perchlorate (Anhydrone or Dehydrite), or phosphorus pentoxide. The ground-glass surfaces are lightly coated with grease.

A **desiccator** is a device for drying substances or objects.

When removing or replacing the lid of a desiccator, use a sliding motion to minimize the likelihood of disturbing the sample. An airtight seal is achieved by slight rotation and downward pressure on the positioned lid.

When placing a heated object in a desiccator, the increase in pressure as the enclosed air is warmed may be sufficient to break the seal between lid and base. Conversely, if the seal is not broken, the cooling of heated objects can cause a partial vacuum to develop. Both of these conditions can cause the contents of the desiccator to be physically lost or contaminated. Although it defeats the purpose of the desiccator somewhat, allow some cooling to occur before the lid is seated. It is also

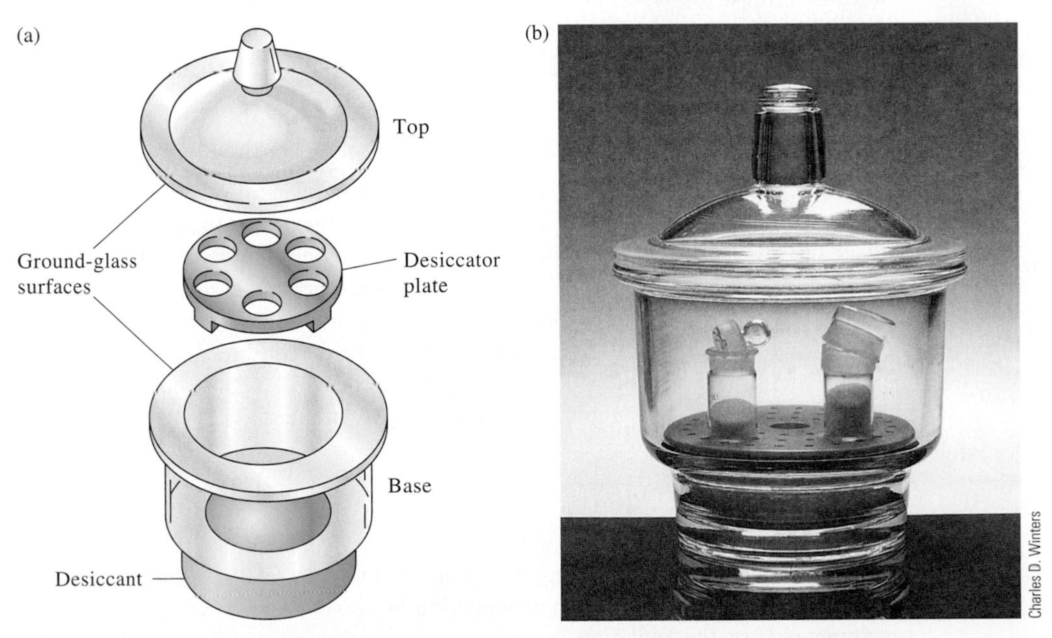

Figure 2-8 (a) Components of a typical desiccator. The base contains a chemical drying agent, which is usually covered with a wire screen and a porcelain plate with holes to accommodate weighing bottles or crucibles. (b) Photo of desiccator containing weighing bottles with dry solids.

Charles D. Winters

Figure 2-9 Arrangement for the drying of samples.

Charles D. Winters

Figure 2-10 Quantitative transfer of solid sample. Note the use of tongs to hold the weighing bottle and a paper strip to hold the cap to avoid contact between glass and skin.

helpful to break the seal once or twice during cooling to relieve any excessive vacuum that develops. Finally, lock the lid in place with your thumbs while moving the desiccator from one place to another.

Very hygroscopic materials should be stored in containers equipped with snug covers, such as weighing bottles; the covers remain in place while in the desiccator. Most other solids can be safely stored uncovered.

2E-3 Manipulating Weighing Bottles

Heating at 105°C to 110°C for 1 hour is sufficient to remove the moisture from the surface of most solids. Figure 2-9 shows the recommended way to dry a sample. The weighing bottle is contained in a labeled beaker with a ribbed cover glass. This arrangement protects the sample from accidental contamination and also allows for free access of air. Crucibles containing a precipitate that can be freed of moisture by simple drying can be treated similarly. The beaker holding the weighing bottle or crucible to be dried must be carefully marked for identification.

Avoid touching dried objects with your fingers because detectable amounts of water or oil from the skin may be transferred to the objects. Instead, use tongs, chamois finger cots, clean cotton gloves, or strips of paper to handle dried objects for weighing. Figure 2-10 shows how a weighing bottle is manipulated with strips of paper.

2E-4 Weighing by Difference

Weighing by difference is a simple method for determining a series of sample masses. First the bottle and its contents are weighed. One sample is then transferred from the bottle to a container; gentle tapping of the bottle with its top and slight rotation of the bottle provide control over the amount of sample removed. Following transfer, the bottle and its residual contents are weighed. The mass of the sample is the difference between the two weighings. It is essential that all the solid removed from the weighing bottle be transferred without loss to the container.

2E-5 Weighing Hygroscopic Solids

Hygroscopic substances rapidly absorb moisture from the atmosphere and therefore require special handling. You need a weighing bottle for each sample to be weighed. Place the approximate amount of sample needed in the individual bottles and heat for an appropriate time. When heating is complete, quickly cap the bottles and cool in a desiccator. Weigh one of the bottles after opening it momentarily to relieve any vacuum. Quickly empty the contents of the bottle into its receiving vessel, cap immediately, and weigh the bottle again along with any solid that did not get transferred. Repeat for each sample and determine the sample masses by difference.

2E-6 Weighing Liquids

The mass of a liquid is always obtained by difference. Liquids that are noncorrosive and relatively nonvolatile can be transferred to previously weighed containers with snugly fitting covers (such as weighing bottles); the mass of the container is subtracted from the total mass.

A volatile or corrosive liquid should be sealed in a weighed glass ampoule. The ampoule is heated, and the neck is then immersed in the sample; as cooling occurs, the liquid is drawn into the bulb. The ampoule is then inverted and the neck sealed off with a small flame. The ampoule and its contents, along with any glass removed during sealing, are cooled to room temperature and weighed. The ampoule is then transferred to an appropriate container and broken. A volume correction for the glass of the ampoule may be needed if the receiving vessel is a volumetric flask.

2F FILTRATION AND IGNITION OF SOLIDS

2F-1 Apparatus

Simple Crucibles

Simple crucibles serve only as containers. Porcelain, aluminum oxide, silica, and platinum crucibles maintain constant mass—within the limits of experimental error—and are used principally to convert a precipitate into a suitable weighing form. The solid is first collected on a filter paper. The filter and contents are then transferred to a weighed crucible, and the paper is ignited.

Simple crucibles of nickel, iron, silver, and gold are used as containers for the high-temperature fusion of samples that are not soluble in aqueous reagents. Attack by both the atmosphere and the contents may cause these crucibles to suffer mass changes. Moreover, such attack will contaminate the sample with species derived from the crucible. The crucible whose products will offer the least interference in subsequent steps of the analysis should be used.

Filtering Crucibles

Filtering crucibles serve not only as containers but also as filters. A vacuum is used to hasten the filtration; a tight seal between crucible and filtering flask is accomplished with any of several types of rubber adaptors (see Figure 2-11; a complete filtration train is shown in Figure 2-16). Collection of a precipitate with a filtering crucible is frequently less time consuming than with paper.

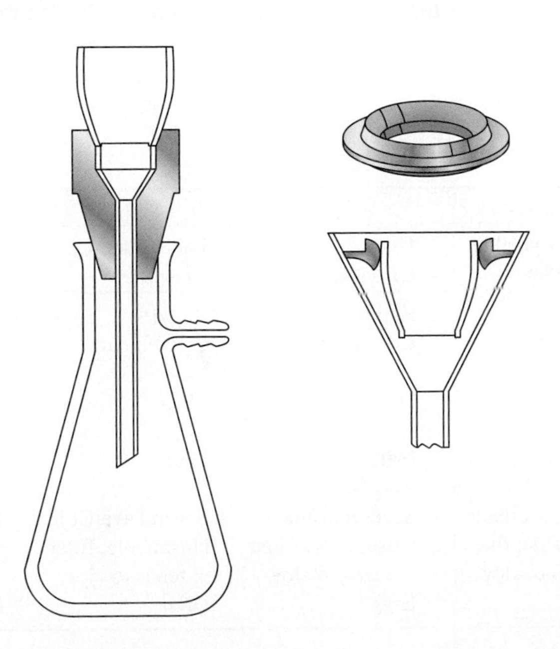

Figure 2-11 Adaptors for filtering crucibles.

Sintered-glass (also called **fritted-glass**) crucibles are manufactured in fine, medium, and coarse porosities (marked *f, m,* and *c*). The upper temperature limit for a sintered-glass crucible is ordinarily about 200°C. Filtering crucibles made entirely of quartz can tolerate substantially higher temperatures without damage. The same is true for crucibles with unglazed porcelain or aluminum oxide frits. The latter are not as costly as quartz.

A **Gooch crucible** has a perforated bottom that supports a fibrous mat. Asbestos was at one time the filtering medium of choice for a Gooch crucible; current regulations concerning this material have virtually eliminated its use. Small circles of glass matting have now replaced asbestos; they are used in pairs to protect against disintegration during the filtration. Glass mats can tolerate temperatures in excess of 500°C and are substantially less hygroscopic than asbestos.

Filter Paper

Paper is an important filtering medium. Ashless paper is manufactured from cellulose fibers that have been treated with hydrochloric and hydrofluoric acids to remove metallic impurities and silica; ammonia is then used to neutralize the acids. The residual ammonium salts in many filter papers may be sufficient to affect the analysis for nitrogen by the Kjeldahl method (see Section 37C-11).

All papers tend to pick up moisture from the atmosphere, and ashless paper is no exception. It is thus necessary to destroy the paper by ignition if the precipitate collected on it is to be weighed. Typically, 9- or 11-cm circles of ashless paper leave a residue that weighs less than 0.1 mg, an amount that is ordinarily negligible. Ashless paper can be obtained in several porosities.

Gelatinous precipitates, such as hydrous iron(III) oxide, clog the pores of any filtering medium. A coarse-porosity ashless paper is most effective for filtering such solids, but even here clogging occurs. This problem can be minimized by mixing a dispersion of ashless filter paper with the precipitate prior to filtration. Filter paper pulp is available in tablet form from chemical suppliers; if necessary, the pulp can be prepared by treating a piece of ashless paper with concentrated hydrochloric acid and washing the disintegrated mass free of acid.

Table 2-1 summarizes the characteristics of common filtering media. None satisfies all requirements.

TABLE 2-1

Comparison of Filtering Media for Gravimetric Analyses

Characteristic	Paper	Gooch Crucible, Glass Mat	Glass Crucible	Porcelain Crucible	Aluminum Oxide Crucible
Speed of filtration	Slow	Rapid	Rapid	Rapid	Rapid
Convenience and ease of preparation	Troublesome, inconvenient	Convenient	Convenient	Convenient	Convenient
Maximum ignition temperature, °C	None	>500	200–500	1100	1450
Chemical reactivity	Carbon has reducing properties	Inert	Inert	Inert	Inert
Porosity	Many available	Several available	Several available	Several available	Several available
Convenience with gelatinous precipitates	Satisfactory	Unsuitable; filter tends to clog	Unsuitable; filter tends to clog	Unsuitable; filter tends to clog	Unsuitable; filter tends to clog
Cost	Low	Low	High	High	High

Heating Equipment

Many precipitates can be weighed directly after being brought to constant mass in a low-temperature drying oven. Such an oven is electrically heated and capable of maintaining a constant temperature to within 1°C (or better). The maximum attainable temperature ranges from 140°C to 260°C, depending on make and model; for many precipitates, 110°C is a satisfactory drying temperature. The efficiency of a drying oven is greatly increased by the forced circulation of air. The passage of predried air through an oven designed to operate under a partial vacuum represents an additional improvement.

Microwave laboratory ovens are currently quite popular. Where applicable, these greatly shorten drying cycles. For example, slurry samples that require 12 to 16 hours for drying in a conventional oven are reported to be dried within 5 to 6 minutes in a microwave oven.[7] The time needed to dry silver chloride, calcium oxalate, and barium sulfate precipitates for gravimetric analysis is also shortened significantly.[8]

An ordinary heat lamp can be used to dry a precipitate that has been collected on ashless paper and to char the paper as well. The process is conveniently completed by ignition at an elevated temperature in a muffle furnace.

Burners are convenient sources of intense heat. The maximum attainable temperature depends on the design of the burner and the combustion properties of the fuel. Of the three common laboratory burners, the Meker burner provides the highest temperatures, followed by the Tirrill and Bunsen types.

A heavy-duty electric furnace **(muffle furnace)** is capable of maintaining controlled temperatures of 1100°C or higher. Long-handled tongs and heat-resistant gloves are needed for protection when transferring objects to or from such a furnace.

2F-2 Filtering and Igniting Precipitates

Preparation of Crucibles

A crucible used to convert a precipitate to a form suitable for weighing must maintain—within the limits of experimental error—a constant mass throughout drying or ignition. The crucible is first cleaned thoroughly (filtering crucibles are conveniently cleaned by backwashing on a filtration train) and then subjected to the same regimen of heating and cooling as that required for the precipitate. This process is repeated until constant mass (page 30) has been achieved, that is, until consecutive weighings differ by 0.3 mg or less.

◀ Backwashing a filtering crucible is done by turning the crucible upside down in the adaptor (Figure 2-11) and sucking water through the inverted crucible.

Filtering and Washing Precipitates

The steps involved in filtering an analytical precipitate are **decantation, washing,** and **transfer.** In decantation, as much supernatant liquid as possible is passed through the filter while the precipitated solid is kept essentially undisturbed in the beaker where it was formed. This procedure speeds the overall filtration rate by delaying the time at which the pores of the filtering medium become clogged with precipitate. A stirring rod is used to direct the flow of decantate (Figure 2-12). When flow ceases, the drop of liquid at the end of the pouring spout is collected with the stirring rod and returned to the beaker. Wash liquid is next added to the

Decantation is the process of pouring a liquid gently so as to not disturb a solid in the bottom of the container.

[7] D. G. Kuehn, R. L. Brandvig, D. C. Lundean, and R. H. Jefferson, *Amer. Lab.*, **1986,** *18*(7), 31. See also *Anal. Chem.*, **1986,** *58,* 1424A; E. S. Beary, *Anal. Chem.*, **1988,** *60,* 742.

[8] R. Q. Thompson and M. Ghadradhi, *J. Chem. Educ.*, **1993,** *70,* 170.

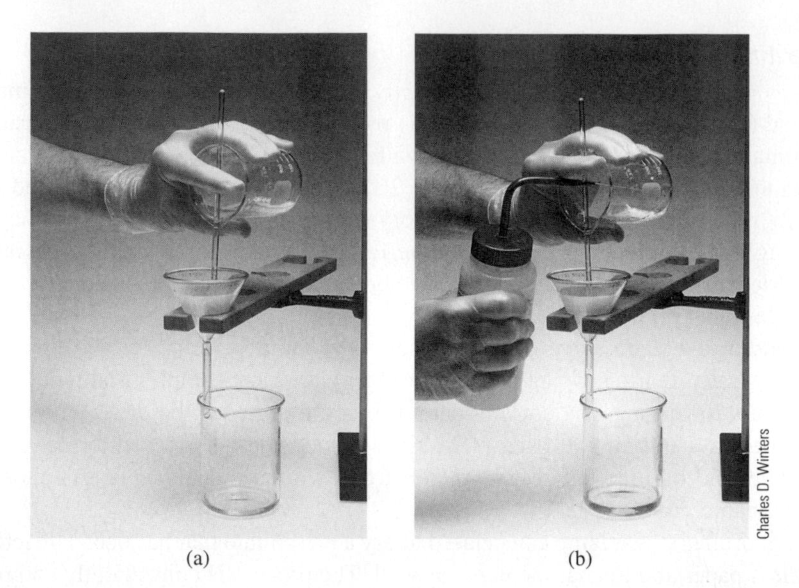

Figure 2-12 (a) Washing by decantation. (b) Transferring the precipitate.

beaker and thoroughly mixed with precipitate. The solid is allowed to settle, and then this liquid is also decanted through the filter. Several such washings may be required, depending on the precipitate. Most washing should be carried out *before* the bulk of the solid is transferred; this technique results in a more thoroughly washed precipitate and a more rapid filtration.

The transfer process is illustrated in Figure 2-12b. The bulk of the precipitate is moved from beaker to filter by directed streams of wash liquid. As in decantation and washing, a stirring rod provides direction for the flow of material to the filtering medium.

The last traces of precipitate that cling to the inside of the beaker are dislodged with a **rubber policeman,** which is a small section of rubber tubing that has been crimped on one end. The open end of the tubing is fitted onto the end of a stirring rod and is wetted with wash liquid before use. Any solid collected with it is combined with the main portion on the filter. Small pieces of ashless paper can be used to wipe the last traces of hydrous oxide precipitates from the wall of the beaker; these papers are ignited along with the paper that holds the bulk of the precipitate.

Creeping is a process in which a solid moves up the side of a wetted container or filter paper.

Many precipitates possess the exasperating property of **creeping,** or spreading over a wetted surface against the force of gravity. Filters are never filled to more than three quarters of capacity, to prevent the possible loss of precipitate through creeping. The addition of a small amount of nonionic detergent, such as Triton X-100, to the supernatant liquid or wash liquid can help minimize creeping.

► Do not permit a gelatinous precipitate to dry until it has been washed completely.

A gelatinous precipitate must be completely washed before it is allowed to dry. These precipitates shrink and develop cracks as they dry. Further additions of wash liquid simply pass through these cracks and accomplish little or no washing.

2F-3 Directions for Filtering and Igniting Precipitates

Preparation of a Filter Paper

Figure 2-13 shows the sequence for folding and seating a filter paper in a 60-deg funnel. The paper is folded exactly in half (a), firmly creased, and folded again (b). A triangular piece from one of the corners is torn off parallel to the second fold (c). The paper is then opened so that the untorn quarter forms a cone (d). The cone is fitted into the funnel, and the second fold is creased (e). Seating is completed by

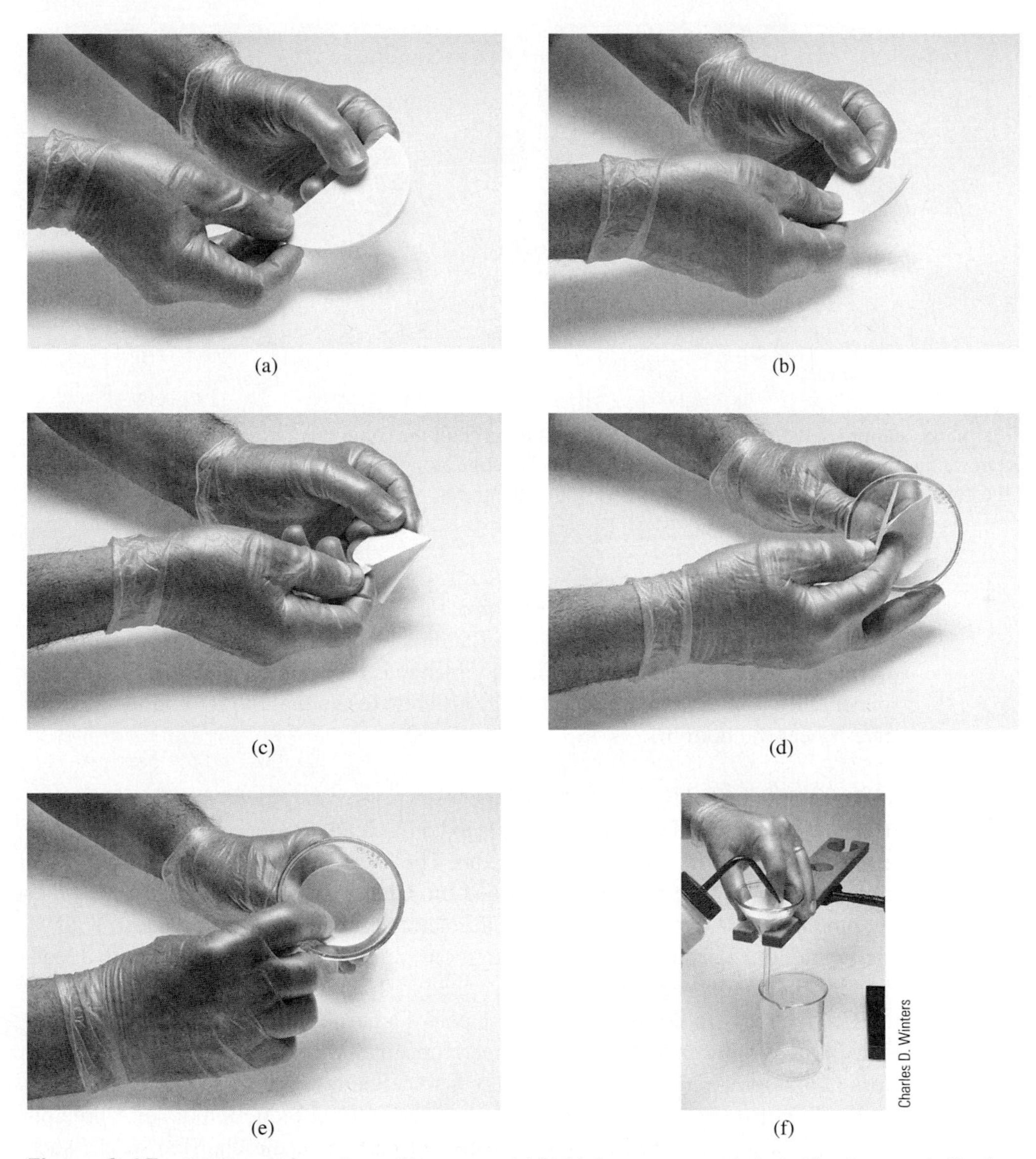

Figure 2-13 Folding and seating a filter paper. (a) Fold the paper exactly in half and crease it firmly. (b) Fold the paper a second time. (c) Tear off one of the corners on a line parallel to the second fold. (d) Open the untorn half of the folded paper to form a cone. (e) Seat the cone firmly into the funnel. Then (f) moisten the paper slightly and gently pat the paper into place.

dampening the cone with water from a wash bottle and *gently* patting it with a finger (f). There will be no leakage of air between the funnel and a properly seated cone; in addition, the stem of the funnel will be filled with an unbroken column of liquid.

Transferring Paper and Precipitate to a Crucible

After filtration and washing have been completed, the filter and its contents must be transferred from the funnel to a crucible that has been brought to constant mass. Ashless paper has very low wet strength and must be handled with care during the transfer. The danger of tearing is lessened considerably if the paper is allowed to dry somewhat before it is removed from the funnel.

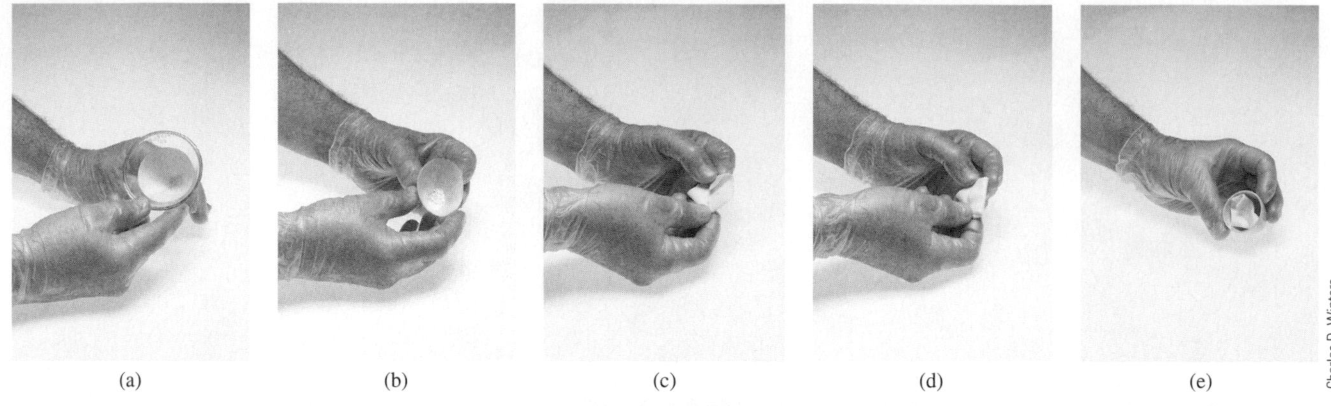

Figure 2-14 Transferring a filter paper and precipitate from a funnel to a crucible. (a) Pull the triple-thick portion of the cone to the opposite side of the funnel. (b) Remove the filter cone from the funnel, and flatten the cone along its upper edge. (c) Fold the corners inward. (d) Fold the top edge of the cone toward the tip to enclose the precipitate in the paper. (e) Gently ease the folded paper and its contents into the crucible.

Figure 2-14 illustrates the transfer process. The triple-thick portion of the filter paper is drawn across the funnel (a) to flatten the cone along its upper edge (b); the corners are next folded inward (c); the top edge is then folded over (d). Finally, the paper and its contents are eased into the crucible (e) so that the bulk of the precipitate is near the bottom.

Ashing Filter Papers

If a heat lamp is used, the crucible is placed on a clean, nonreactive surface, such as a wire screen covered with aluminum foil. The lamp is then positioned about 1 cm above the rim of the crucible and turned on. Charring takes place without further attention. The process is considerably accelerated if the paper is moistened with no more than one drop of concentrated ammonium nitrate solution. The residual carbon is eliminated with a burner, as described in the next paragraph.

► You should have a burner for each crucible. You can tend to the ashing of several filter papers at the same time.

Considerably more attention must be paid if a burner is used to ash a filter paper. The burner produces much higher temperatures than a heat lamp. Thus, mechanical loss of precipitate may occur if moisture is expelled too rapidly in the initial stages of heating or if the paper bursts into flame. Also, partial reduction of some precipitates can occur through reaction with the hot carbon of the charring paper; such reduction is a serious problem if reoxidation following ashing is inconvenient. These difficulties can be minimized by positioning the crucible as illustrated in Figure 2-15. The tilted position allows for the ready access of air; a clean crucible cover should be available to extinguish any flame.

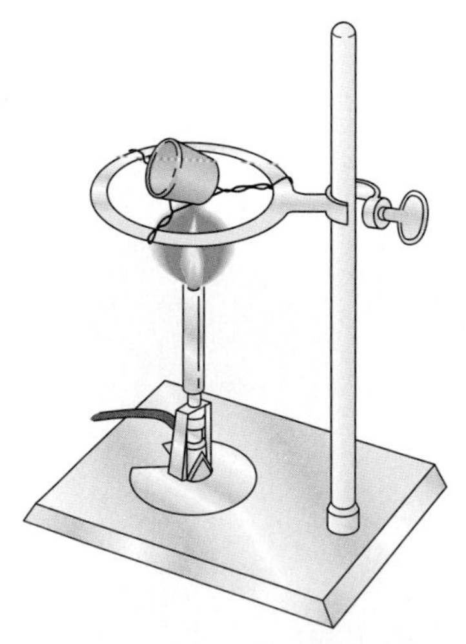

Figure 2-15 Ignition of a precipitate. Proper crucible position for preliminary charring is shown.

Heating should begin with a small flame. The temperature is gradually increased as moisture is evolved and the paper begins to char. The amount of smoke given off indicates the intensity of heating that can be tolerated. Thin wisps are normal. A significant increase in smoke indicates that the paper is about to flash and that heating should be temporarily discontinued. Any flame should be immediately extinguished with a crucible cover. (The cover may become discolored from the condensation of carbonaceous products; these products must ultimately be removed from the cover by ignition to confirm the absence of entrained particles of precipitate.) When no further smoking can be detected, heating is increased to eliminate the residual carbon. Strong heating, as necessary, can then be undertaken.

This sequence ordinarily precedes the final ignition of a precipitate in a muffle furnace, where a reducing atmosphere is equally undesirable.

Using Filtering Crucibles

A vacuum filtration train (Figure 2-16) is used when a filtering crucible can be used instead of paper. The trap isolates the filter flask from the source of vacuum.

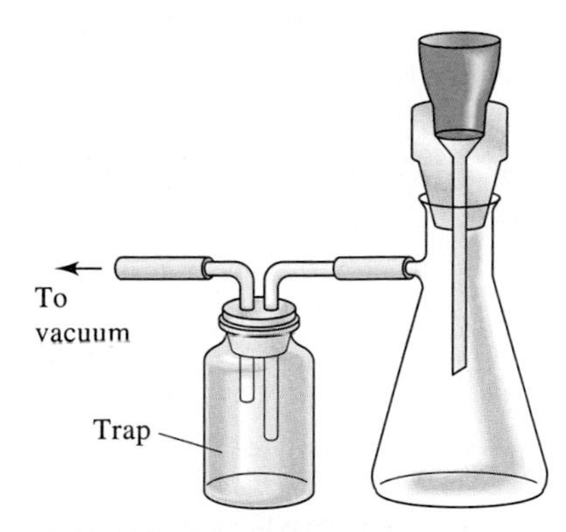

Figure 2-16 Train for vacuum filtration. The trap isolates the filter flask from the source of vacuum.

2F-4 Rules for Manipulating Heated Objects

Careful adherence to the following rules will minimize the possibility of accidental loss of a precipitate:

1. Practice unfamiliar manipulations before putting them to use.
2. *Never* place a heated object on the benchtop; instead, place it on a wire gauze or a heat-resistant ceramic plate.
3. Allow a crucible that has been subjected to the full flame of a burner or to a muffle furnace to cool momentarily (on a wire gauze or ceramic plate) before transferring it to the desiccator.
4. Keep the tongs and forceps used to handle heated objects scrupulously clean. In particular, do not allow the tips to touch the benchtop.

2G MEASURING VOLUME

The precise measurement of volume is as important to many analytical methods as the precise measurement of mass.

2G-1 Units of Volume

The unit of volume is the **liter** (L), defined as one cubic decimeter. The **milliliter** (mL) is one one-thousandth of a liter (0.001 L) and is used when the liter represents an inconveniently large volume unit. The microliter (μL) is 10^{-6} L or 10^{-3} mL.

The **liter** is one cubic decimeter. The **milliliter** is 10^{-3} L.

2G-2 The Effect of Temperature on Volume Measurements

The volume occupied by a given mass of liquid varies with temperature, as does the device that holds the liquid during measurement. Most volumetric measuring devices are made of glass, which fortunately has a small coefficient of expansion. Consequently, variations in the volume of a glass container with temperature need not be considered in ordinary analytical work.

The coefficient of expansion for dilute aqueous solutions (approximately 0.025%/°C) is such that a 5°C change has a measurable effect on the reliability of ordinary volumetric measurements.

EXAMPLE 2-2

A 40.00-mL sample is taken from an aqueous solution at 5°C; what volume does it occupy at 20°C?

$$V_{20^\circ} = V_{5^\circ} + 0.00025(20 - 5)(40.00) = 40.00 + 0.15 = 40.15 \text{ mL}$$

Volumetric measurements must be referred to some standard temperature; this reference point is ordinarily 20°C. The ambient temperature of most laboratories is sufficiently close to 20°C to eliminate the need for temperature corrections in volume measurements for aqueous solutions. In contrast, the coefficient of expansion for organic liquids may require corrections for temperature differences of 1°C or less.

2G-3 Apparatus for Precisely Measuring Volume

► Glassware types include Class A and Class B. Class A glassware is manufactured to the highest tolerances from Pyrex, borosilicate, or Kimax glass (see tables on pages 41 and 42). Class B (economy ware) tolerances are about twice those of Class A.

Volume may be measured reliably with a **pipet,** a **buret,** or a **volumetric flask.**

Volumetric equipment is marked by the manufacturer to indicate not only the manner of calibration (usually TD for "to deliver" or TC for "to contain") but also the temperature at which the calibration strictly applies. Pipets and burets are ordinarily calibrated to deliver specified volumes, whereas volumetric flasks are calibrated on a to-contain basis.

Pipets

Pipets permit the transfer of accurately known volumes from one container to another. Common types are shown in Figure 2-17; information concerning their use is given in Table 2-2. A **volumetric,** or **transfer,** pipet (Figure 2-17a) delivers a sin-

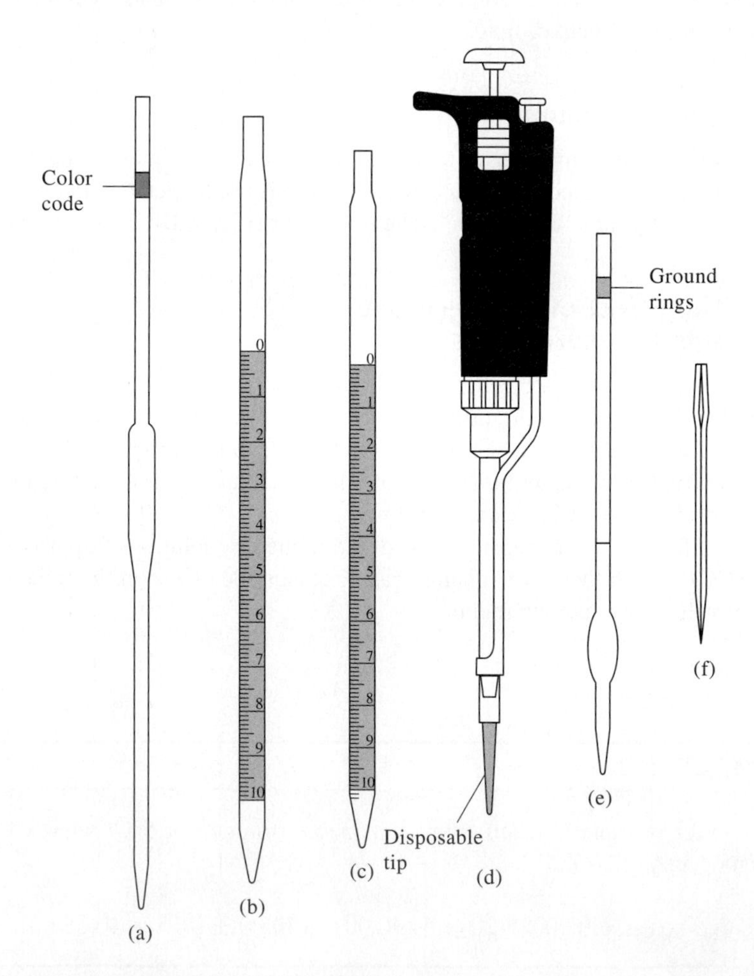

Figure 2-17 Typical pipets: (a) volumetric pipet, (b) Mohr pipet, (c) serological pipet, (d) Eppendorf micropipet, (e) Ostwald–Folin pipet, (f) lambda pipet.

TABLE 2-2

Characteristics of Pipets

Name	Type of Calibration*	Function	Available Capacity, mL	Type of Drainage
Volumetric	TD	Delivery of fixed volume	1–200	Free
Mohr	TD	Delivery of variable volume	1–25	To lower calibration line
Serological	TD	Delivery of variable volume	0.1–10	Blow out last drop†
Serological	TD	Delivery of variable volume	0.1–10	To lower calibration line
Ostwald–Folin	TD	Delivery of fixed volume	0.5–10	Blow out last drop†
Lambda	TC	Containment of fixed volume	0.001–2	Wash out with suitable solvent
Lambda	TD	Delivery of fixed volume	0.001–2	Blow out last drop†
Eppendorf	TD	Delivery of variable or fixed volume	0.001–1	Tip emptied by air displacement

*TD, to deliver; TC, to contain.
†A frosted ring near the top of pipets indicates that the last drop is to be blown out.

gle, fixed volume between 0.5 and 200 mL. Many such pipets are color coded by volume for convenience in identification and sorting. **Measuring pipets** (Figure 2-17b and c) are calibrated in convenient units to permit delivery of any volume up to a maximum capacity ranging from 0.1 to 25 mL.

Volumetric and measuring pipets are filled to a calibration mark at the outset; the manner in which the transfer is completed depends on the particular type. Because an attraction exists between most liquids and glass, a small amount of liquid tends to remain in the tip after the pipet is emptied. This residual liquid is never blown out of a volumetric pipet or from some measuring pipets; it is blown out of other types of pipets (Table 2-2).

Handheld Eppendorf micropipets (Figure 2-17d and Figure 2-18a) deliver adjustable microliter volumes of liquid. With these pipets, a known and adjustable volume of air is displaced from the plastic disposable tip by depressing the pushbutton on the top of the pipet to a first stop. This button operates a spring-loaded piston that forces air out of the pipet. The volume of displaced air can be varied by a locking digital micrometer adjustment located on the front or top of the device. The plastic tip is then inserted into the liquid, and the pressure on the button released, causing liquid to be drawn into the tip. The tip is then placed against the walls of the receiving vessel, and the pushbutton is again depressed to the first stop. After 1 second, the pushbutton is depressed further to a second stop, which completely empties the tip. The range of volumes and precision of typical pipets of this type are shown in the margin. The accuracy and precision of automatic pipets depend somewhat on the skill and experience of the operators and thus should be calibrated for critical work.[9]

Numerous *automatic* pipets are available for situations that call for the repeated delivery of a particular volume. In addition, motorized, computer-controlled microliter pipets are now available (see Figure 2-18b). These devices are programmed to function as pipets, dispensers of multiple volumes, burets, and means for diluting samples. The volume desired is entered on a keypad and is displayed on an LCD panel. A motor-driven piston dispenses the liquid. Maximum volumes range from 10 to 2500 μL.

Tolerances, Class A Transfer Pipets

Capacity, mL	Tolerances, mL
0.5	±0.006
1	±0.006
2	±0.006
5	±0.01
10	±0.02
20	±0.03
25	±0.03
50	±0.05
100	+0.08

Range and Precision of Typical Eppendorf Micropipets

Volume Range, μL	Standard Deviation, μL
1–20	<0.04 @ 2 μL
	<0.06 @ 20 μL
10–100	<0.10 @ 15 μL
	<0.15 @ 100 μL
20–200	<0.15 @ 25 μL
	<0.30 @ 200 μL
100–1000	<0.6 @ 250 μL
	<1.3 @ 1000 μL
500–5000	<3 @ 1.0 mL
	<8 @ 5.0 mL

[9]M. Connors and R. Curtis, *Amer. Lab. News Ed.*, June 1999, pp 21–22.

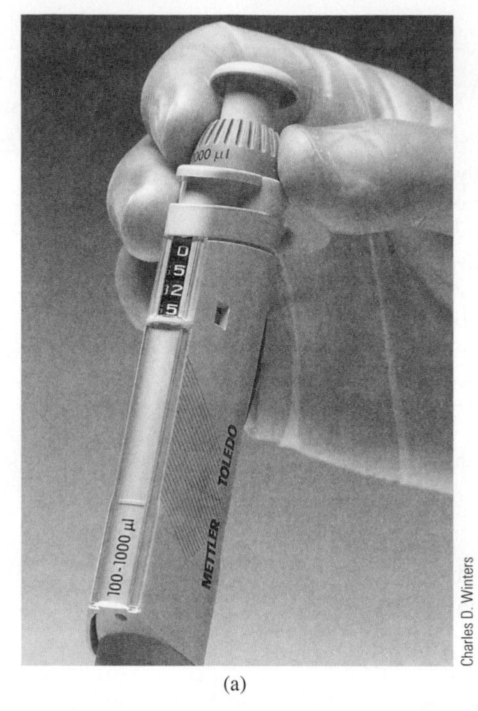

Figure 2-18 (a) Variable-volume automatic pipet, 100–1000 µL. At 100 µL, accuracy is 3.0% and precision is 0.6%. At 1000 µL, accuracy is 0.6% and precision is 0.2%. Volume is adjusted using the thumbwheel as shown. Volume shown is 525 µL. (b) A handheld, battery-operated, computer-controlled, motorized pipet.

Tolerances, Class A Burets

Volume, mL	Tolerances, mL
5	±0.01
10	±0.02
25	±0.03
50	±0.05
100	±0.20

Tolerances, Class A Volumetric Flasks

Capacity, mL	Tolerances, mL
5	±0.02
10	±0.02
25	±0.03
50	±0.05
100	±0.08
250	±0.12
500	±0.20
1000	±0.30
2000	±0.50

Burets

Burets, like measuring pipets, make it possible to deliver any volume up to the maximum capacity of the device. The precision attainable with a buret is substantially greater than the precision with a pipet.

A buret consists of a calibrated tube to hold titrant plus a valve arrangement by which the flow of titrant is controlled. This valve is the principal source of difference among burets. The simplest pinchcock valve consists of a close-fitting glass bead inside a short length of rubber tubing that connects the buret and its tip (Figure 2-19a); only when the tubing is deformed does liquid flow past the bead.

A buret equipped with a glass stopcock for a valve relies on a lubricant between the ground-glass surfaces of stopcock and barrel for a liquid-tight seal. Some solutions, notably bases, cause glass stopcocks to freeze when they are in place for long periods; therefore, thorough cleaning is needed after each use. Valves made of Teflon are commonly encountered; these are unaffected by most common reagents and require no lubricant (Figure 2-19b).

Volumetric Flasks

Volumetric flasks (Figure 2-20) are manufactured with capacities ranging from 5 mL to 5 L and are usually calibrated to contain a specified volume when filled to a line etched on the neck. They are used for the preparation of standard solutions and for the dilution of samples to a fixed volume prior to taking aliquots with a pipet. Some are also calibrated on a to-deliver basis; these are readily distinguished by two reference lines on the neck. If delivery of the stated volume is desired, the flask is filled to the upper line.

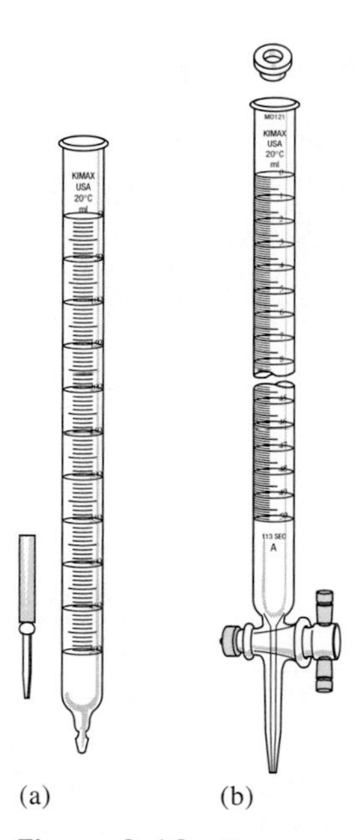

Figure 2-19 Burets: (a) glass-bead valve, (b) Teflon valve.

Figure 2-20 Typical volumetric flasks.

2G-4 Using Volumetric Equipment

Volume markings are blazed on clean volumetric equipment by the manufacturer. An equal degree of cleanliness is needed in the laboratory if these markings are to have their stated meanings. Only clean glass surfaces support a uniform film of liquid. Dirt or oil causes breaks in this film; the presence of breaks is a certain indication of an unclean surface.

Cleaning

A brief soaking in a warm detergent solution is usually sufficient to remove the grease and dirt responsible for water breaks. Prolonged soaking should be avoided because a rough area or ring is likely to develop at a detergent/air interface. This ring cannot be removed and causes a film break that destroys the usefulness of the equipment.

After being cleaned, the apparatus must be thoroughly rinsed with tap water and then with three or four portions of distilled water. It is seldom necessary to dry volumetric ware.

A **meniscus** is the curved surface of a liquid at its interface with the atmosphere.

Avoiding Parallax

The top surface of a liquid confined in a narrow tube exhibits a marked curvature, or **meniscus.** It is common practice to use the bottom of the meniscus as the point of reference in calibrating and using volumetric equipment. This minimum can be established more exactly by holding an opaque card or piece of paper behind the graduations (Figure 2-21).

In reading volumes, the eye must be at the level of the liquid surface to avoid an error due to **parallax,** a condition that causes the volume to appear smaller than its

Parallax is the apparent displacement of a liquid level or of a pointer as an observer changes position. Parallax occurs when an object is viewed from a position that is not at a right angle to the object.

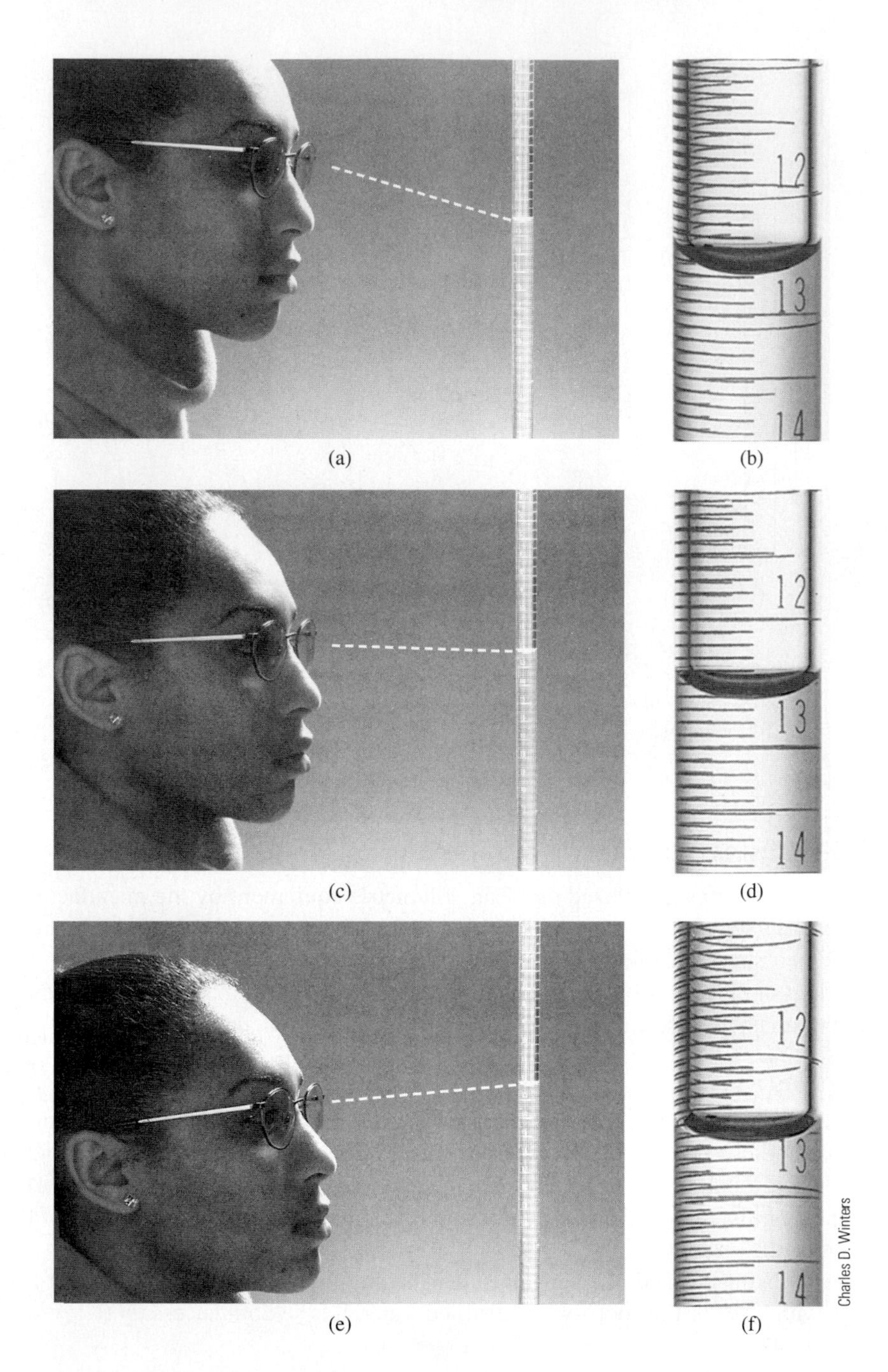

Figure 2-21 Reading a buret. (a) The student reads the buret from a position above a line perpendicular to the buret and makes a reading (b) of 12.58 mL. (c) The student reads the buret from a position along a line perpendicular to the buret and makes a reading (d) of 12.62 mL. (e) The student reads the buret from a position above a line perpendicular to the buret and makes a reading (f) of 12.67 mL. To avoid the problem of parallax, buret readings should be made consistently along a line perpendicular to the buret, as shown in (c) and (d).

actual value if the meniscus is viewed from above and larger if the meniscus is viewed from below (Figure 2-21).

2G-5 Directions for Using a Pipet

The following directions are appropriate specifically for volumetric pipets but can be modified for using other types as well.

Liquid is drawn into a pipet through the application of a slight vacuum. *The mouth should never be used for suction because of the risk of accidentally ingest-*

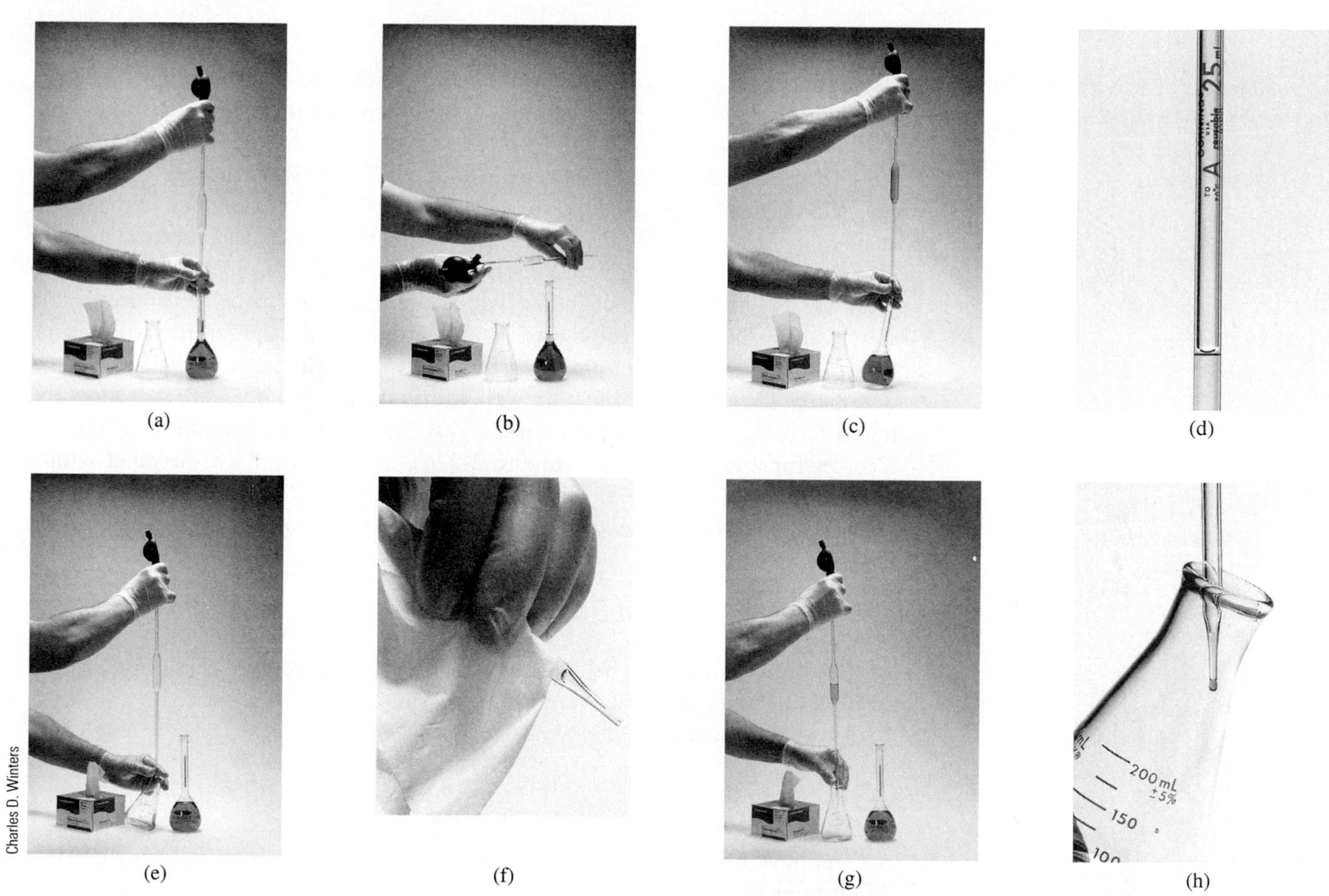

Figure 2-22 Dispensing an aliquot. (a) Draw a small amount of the liquid into the pipet and (b) wet the interior surface of the glass by tilting and rotating the pipet. Repeat this procedure two more times. Then (c) while holding the tip of the pipet against the inside surface of the volumetric flask, allow the liquid level to descend until the bottom of the meniscus is aligned with the line etched on the stem of the pipet (d). Remove the pipet from the volumetric flask and tilt it (e) until liquid is drawn slightly up into the pipet and (f) wipe the tip with a lintless tissue as shown. Then while holding the pipet vertically, (g) allow the liquid to flow into the receiving flask until just a small amount of liquid remains in the inside of the tip and a drop remains on the outside. Finally, tilt the flask slightly as shown in (h), and touch the tip of the pipet to the inside of the flask. When this step is completed, a small amount of liquid will remain in the pipet. *Do not remove this remaining liquid.* The pipet is calibrated to reproducibly deliver its rated volume when this liquid remains in the tip.

ing the liquid being pipetted. Instead, a rubber suction bulb (Figure 2-22a) or a rubber tube connected to a vacuum source should be used.

Cleaning

Use a rubber bulb to draw detergent solution to a level 2 to 3 cm above the calibration mark of the pipet. Drain this solution and then rinse the pipet with several portions of tap water. Inspect for film breaks; repeat this portion of the cleaning cycle if necessary. Finally, fill the pipet with distilled water to perhaps one third of its capacity and carefully rotate it so that the entire interior surface is wetted. Repeat this rinsing step at least twice.

An **aliquot** is a measured fraction of the volume of a liquid sample.

Measuring an Aliquot

Use a rubber bulb to draw a small volume of the liquid to be sampled into the pipet and thoroughly wet the entire interior surface. Repeat with *at least* two additional portions. Then carefully fill the pipet to a level somewhat above the graduation mark (Figure 2-22). Quickly replace the bulb with a *forefinger* to arrest the outflow of liquid (Figure 2-22b). Make certain there are no bubbles in the bulk of the liquid or foam at the surface. Tilt the pipet slightly from the vertical and wipe the exterior free of adhering liquid (Figure 2-22c). Touch the tip of the pipet to the wall of a glass vessel (*not* the container into which the aliquot is to be transferred), and slowly allow the liquid level to drop by partially releasing the forefinger (Note 1). Halt further flow as the bottom of the meniscus coincides exactly with the graduation mark. Then place the pipet tip well within the receiving vessel, and allow the liquid to drain. When free flow ceases, rest the tip against the inner wall of the receiver for a full 10 seconds (Figure 2-22d). Finally, withdraw the pipet with a rotating motion to remove any liquid adhering to the tip. *The small volume remaining inside the tip of a volumetric pipet should not be blown or rinsed into the receiving vessel* (Note 2).

Notes

1. The liquid can best be held at a constant level if the forefinger is *faintly* moist. Too much moisture makes control impossible.
2. Rinse the pipet thoroughly after use.

2G-6 Directions for Using a Buret

A buret must be scrupulously clean before it is used; in addition, its valve must be liquid-tight.

Cleaning

Thoroughly clean the tube of the buret with detergent and a long brush. Rinse thoroughly with tap water and then with distilled water. Inspect for water breaks. Repeat the treatment if necessary.

Lubricating a Glass Stopcock

Carefully remove all old grease from a glass stopcock and its barrel with a paper towel and dry both parts completely. Lightly grease the stopcock, taking care to avoid the area adjacent to the hole. Insert the stopcock into the barrel and rotate it vigorously with slight inward pressure. A proper amount of lubricant has been used when (1) the area of contact between stopcock and barrel appears nearly transparent, (2) the seal is liquid-tight, and (3) no grease has worked its way into the tip.

Notes

1. Grease films that are unaffected by cleaning solution may yield to such organic solvents as acetone or benzene. Thorough washing with detergent should follow such treatment. The use of silicone lubricants is not recommended; contamination by such preparations is difficult—if not impossible—to remove.
2. So long as the flow of liquid is not impeded, fouling of a buret tip with stopcock grease is not a serious matter. Removal is best accomplished with organic solvents. A stoppage during a titration can be freed by *gentle* warming of the tip with a lighted match.

3. Before a buret is returned to service after reassembly, it is advisable to test for leakage. Simply fill the buret with water and establish that the volume reading does not change with time.

◀ Buret readings should be estimated to the nearest 0.01 mL.

Filling

Make certain the stopcock is closed. Add 5 to 10 mL of the titrant, and carefully rotate the buret to wet the interior completely. Allow the liquid to drain through the tip. *Repeat this procedure at least two more times.* Then fill the buret well above the zero mark. Free the tip of air bubbles by rapidly rotating the stopcock and permitting small quantities of the titrant to pass. Finally, lower the level of the liquid just to or somewhat below the zero mark. Allow for drainage (≈1 min), and then record the initial volume reading, estimating to the nearest 0.01 mL.

Titration

Figure 2-23 illustrates the preferred method for manipulating a stopcock; when you position your hand as shown, your grip on the stopcock tends to keep the stopcock firmly seated. Be sure the tip of the buret is well within the titration flask. Introduce the titrant in increments of about 1 mL. Swirl (or stir) constantly to ensure thorough mixing. Decrease the size of the increments as the titration progresses; add titrant dropwise in the immediate vicinity of the end point (Note 2). When it appears that only a few more drops are needed to reach the end point, rinse the walls of the container (Note 3). Allow the titrant to drain from the inner wall of the buret (at least 30 seconds) at the completion of the titration. Then record the final volume, again to the nearest 0.01 mL.

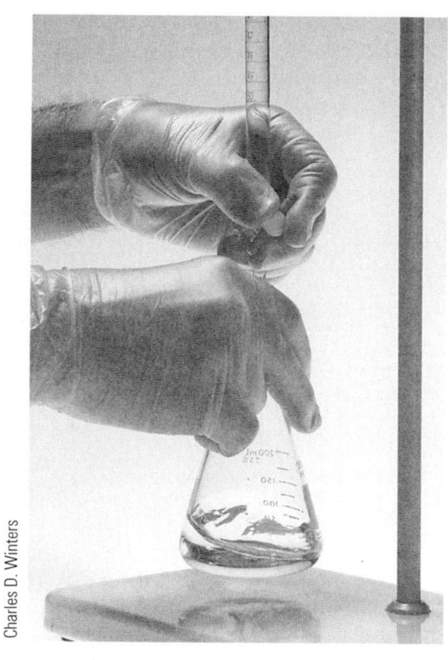
Charles D. Winters

Figure 2-23 Recommended method for manipulating a buret stopcock.

Notes

1. When unfamiliar with a particular titration, many workers prepare an extra sample. No care is taken with its titration, since its functions are to reveal the nature of the end point and to provide a rough estimate of titrant requirements. This deliberate sacrifice of one sample frequently results in an overall saving of time.
2. Increments smaller than one drop can be taken by allowing a small volume of titrant to form on the tip of the buret and then touching the tip to the wall of the flask. This partial drop is then combined with the bulk of the liquid as in Note 3.
3. Instead of being rinsed toward the end of a titration, the flask can be tilted and rotated so that the bulk of the liquid picks up any drops that adhere to the inner surface.

2G-7 Directions for Using a Volumetric Flask

Before being put into use, volumetric flasks should be washed with detergent and thoroughly rinsed. Only rarely do they need to be dried. If required, however, drying is best accomplished by clamping the flask in an inverted position. Insertion of a glass tube connected to a vacuum line hastens the process.

Direct Weighing into a Volumetric Flask

The direct preparation of a standard solution requires the introduction of a known mass of solute to a volumetric flask. Use of a powder funnel minimizes the possibility of losing solid during the transfer. Rinse the funnel thoroughly; collect the washings in the flask.

The foregoing procedure may be inappropriate if heating is needed to dissolve the solute. Instead, weigh the solid into a beaker or flask, add solvent, heat to dissolve the solute, and allow the solution to cool to room temperature. Transfer this solution quantitatively to the volumetric flask, as described in the next section.

Quantitative Transfer of Liquid to a Volumetric Flask

► The solute should be completely dissolved *before* you dilute to the mark.

Insert a funnel into the neck of the volumetric flask; use a stirring rod to direct the flow of liquid from the beaker into the funnel. With the stirring rod, tip off the last drop of liquid on the spout of the beaker. Rinse both the stirring rod and the interior of the beaker with distilled water and transfer the washings to the volumetric flask, as before. Repeat the rinsing process *at least* two more times.

Diluting to the Mark

After the solute has been transferred, fill the flask about half full and swirl the contents to hasten solution. Add more solvent and again mix well. Bring the liquid level almost to the mark, and allow time for drainage (≈1 min); then use a medicine dropper to make any necessary final additions of solvent (see Note below). Firmly stopper the flask, and invert it repeatedly to ensure thorough mixing. Transfer the contents to a storage bottle that either is dry or has been thoroughly rinsed with several small portions of the solution from the flask.

Note

If, as sometimes happens, the liquid level accidentally exceeds the calibration mark, the solution can be saved by correcting for the excess volume. Use a self-stick label to mark the location of the meniscus. After the flask has been emptied, carefully refill to the manufacturer's etched mark with water. Use a buret to determine the additional volume needed to fill the flask so that the meniscus is at the gummed-label mark. This volume must be added to the nominal volume of the flask when calculating the concentration of the solution.

2H CALIBRATING VOLUMETRIC GLASSWARE

Volumetric glassware is calibrated by measuring the mass of a liquid (usually distilled or deionized water) of known density and temperature that is contained in (or delivered by) the volumetric ware. In carrying out a calibration, a buoyancy correction must be made (Section 2D-4), since the density of water is quite different from that of the masses.

The calculations associated with calibration, while not difficult, are somewhat involved. The raw weighing data are first corrected for buoyancy with Equation 2-1. Next, the volume of the apparatus at the temperature of calibration *(T)* is obtained by dividing the density of the liquid at that temperature into the corrected mass. Finally, this volume is corrected to the standard temperature of 20°C, as in Example 2-2.

Table 2-3 is provided to help with buoyancy calculations. Corrections for buoyancy with respect to stainless steel or brass mass (the density difference between the two is small enough to be neglected) and for the volume change of water and of glass containers have been incorporated into these data. Multiplication by the appropriate factor from Table 2-3 converts the mass of water at temperature T to (1) the corresponding volume at that temperature or (2) the volume at 20°C.

TABLE 2-3

Volume Occupied by 1.000 g of Water Weighed in Air Against Stainless Steel Weights*

	Volume, mL	
Temperature, *T*, °C	**At *T***	**Corrected to 20°C**
10	1.0013	1.0016
11	1.0014	1.0016
12	1.0015	1.0017
13	1.0016	1.0018
14	1.0018	1.0019
15	1.0019	1.0020
16	1.0021	1.0022
17	1.0022	1.0023
18	1.0024	1.0025
19	1.0026	1.0026
20	1.0028	1.0028
21	1.0030	1.0030
22	1.0033	1.0032
23	1.0035	1.0034
24	1.0037	1.0036
25	1.0040	1.0037
26	1.0043	1.0041
27	1.0045	1.0043
28	1.0048	1.0046
29	1.0051	1.0048
30	1.0054	1.0052

*Corrections for buoyancy (stainless steel weights) and change in container volume have been applied.

EXAMPLE 2-3

A 25-mL pipet delivers 24.976 g of water weighed against stainless steel mass at 25°C. Use the data in Table 2-3 to calculate the volume delivered by this pipet at 25°C and at 20°C.

At 25°C: $V = 24.976\ \text{g} \times 1.0040\ \text{mL/g} = 25.08\ \text{mL}$

At 20°C: $V = 24.976\ \text{g} \times 1.0037\ \text{mL/g} = 27.07\ \text{mL}$

2H-1 General Directions for Calibration

All volumetric ware should be painstakingly freed of water breaks before being calibrated. Burets and pipets need not be dry; volumetric flasks should be thoroughly drained and dried at room temperature. The water used for calibration should be in thermal equilibrium with its surroundings. This condition is best established by drawing the water well in advance, noting its temperature at frequent intervals, and waiting until no further changes occur.

Although an analytical balance can be used for calibration, weighings to the nearest milligram are perfectly satisfactory for all but the very smallest volumes. Thus a top-loading balance is more convenient to use than an analytical balance. Weighing bottles or small, well-stoppered conical flasks can serve as receivers for the calibration liquid.

Calibrating a Volumetric Pipet

Determine the empty mass of the stoppered receiver to the nearest milligram. Transfer a portion of temperature-equilibrated water to the receiver with the pipet, weigh the receiver and its contents (again, to the nearest milligram), and calculate the mass of water delivered from the difference in these masses. With the aid of Table 2-3, calculate the volume delivered. Repeat the calibration several times; calculate the mean volume delivered and its standard deviation.

Calibrating a Buret

Fill the buret with temperature-equilibrated water and make sure that no air bubbles are trapped in the tip. Allow about 1 minute for drainage; then lower the liquid level to bring the bottom of the meniscus to the 0.00-mL mark. Touch the tip to the wall of a beaker to remove any adhering drop. Wait 10 minutes and recheck the volume; if the stopcock is tight, there should be no perceptible change. During this interval, weigh (to the nearest milligram) a 125-mL conical flask fitted with a rubber stopper.

Once tightness of the stopcock has been established, slowly transfer (at about 10 mL/min) approximately 10 mL of water to the flask. Touch the tip to the wall of the flask. Wait 1 minute, record the volume that was apparently delivered, and refill the buret. Weigh the flask and its contents to the nearest milligram; the difference between this mass and the initial value gives the mass of water delivered. Use Table 2-3 to convert this mass to the true volume. Subtract the apparent volume from the true volume. This difference is the correction that should be applied to the apparent volume to give the true volume. Repeat the calibration until agreement within ± 0.02 mL is achieved.

Starting again from the zero mark, repeat the calibration, this time delivering about 20 mL to the receiver. Test the buret at 10-mL intervals over its entire volume. Prepare a plot of the correction to be applied as a function of volume delivered. The correction associated with any interval can be determined from this plot.

Calibrating a Volumetric Flask

Weigh the clean, dry flask to the nearest milligram. Then fill to the mark with equilibrated water and reweigh. With the aid of Table 2-3, calculate the volume contained.

Calibrating a Volumetric Flask Relative to a Pipet

The calibration of a volumetric flask relative to a pipet provides an excellent method for partitioning a sample into aliquots. These directions pertain to a 50-mL pipet and a 500-mL volumetric flask; other combinations are equally convenient.

Carefully transfer ten 50-mL aliquots from the pipet to a dry 500-mL volumetric flask. Mark the location of the meniscus with a gummed label. Cover with a label varnish to ensure permanence. Dilution to the label permits the same pipet to deliver precisely a one-tenth aliquot of the solution in the flask. Note that recalibration is necessary if another pipet is used.

21 THE LABORATORY NOTEBOOK

A laboratory notebook is needed to record measurements and observations concerning an analysis. The book should be permanently bound with consecutively numbered pages (if necessary, the pages should be hand numbered before any entries are made). Most notebooks have more than ample room; there is no need to crowd entries.

The first few pages should be saved for a table of contents that is updated as entries are made.

21-1 Maintaining a Laboratory Notebook

1. *Record all data and observations directly into the notebook in ink.* Neatness is desirable, but you should not achieve neatness by transcribing data from a sheet of paper to the notebook or from one notebook to another. The risk of misplacing—or incorrectly transcribing—crucial data and thereby ruining an experiment is unacceptable.
2. Supply each entry or series of entries with a heading or label. A series of weighing data for a set of empty crucibles should carry the heading "empty crucible mass" (or something similar), for example, and the mass of each crucible should be identified by the same number or letter used to label the crucible.
3. Date each page of the notebook as it is used.
4. *Never* attempt to erase or obliterate an incorrect entry. Instead, cross it out with a single horizontal line and locate the correct entry as nearby as possible. Do not write over incorrect numbers; with time, it may become impossible to distinguish the correct entry from the incorrect one.
5. *Never* remove a page from the notebook. Draw diagonal lines across any page that is to be disregarded. Provide a brief rationale for disregarding the page.

◀ Remember that you can discard an experimental measurement *only if you have certain knowledge that you made an experimental error.* Thus, you must carefully record experimental observations in your notebook as soon as they occur.

◀ An entry in a laboratory notebook should never be erased but should be crossed out instead.

21-2 Notebook Format

The instructor should be consulted concerning the format to be used in keeping the laboratory notebook.[10] One convention involves using each page consecutively for the recording of data and observations as they occur. The completed analysis is then summarized on the next available page spread (that is, left and right facing pages). As shown in Figure 2-24, the first of these two facing pages should contain the following entries:

1. The title of the experiment ("The Gravimetric Determination of Chloride").
2. A brief statement of the principles on which the analysis is based.
3. A complete summary of the weighing, volumetric, and/or instrument response data needed to calculate the results.
4. A report of the best value for the set and a statement of its precision.

[10]See also Howard M. Kanare, *Writing the Laboratory Notebook.* Washington, DC 20036: The American Chemical Society, 1985.

08

Gravimetric Determination of Chloride

The chloride in a soluble sample was precipitated as AgCl and weighed as such.

Sample masses	1	2	3
Mass bottle plus sample, g	27.6115	27.2185	26.8105
-less bottle, g	27.2185	26.8105	26.4517
mass sample, g	0.3930	0.4080	0.3588
Crucible masses, empty	~~20.7925~~	~~22.8311~~	~~21.2488~~
	20.7926	22.8311	~~21.2482~~
			21.2483
Crucible masses, with AgCl, g	~~21.4294~~	~~23.4920~~	~~21.8324~~
	~~21.4297~~	~~23.4914~~	21.8323
	21.4296	23.4915	
Mass of AgCl, g	0.6370	0.6604	0.5840
Percent Cl-	40.10	40.04	40.27
Average percent Cl-		40.12	
Relative standard deviation	3.0 *parts per thousand*		

Date Started 1-10-03

Date Completed 1-16-03

Figure 2-24 Laboratory notebook data page.

The second page should contain the following items:

1. Equations for the principal reactions in the analysis.
2. An equation showing how the results were calculated.
3. A summary of observations that appear to bear on the validity of a particular result or the analysis as a whole. *Any such entry must have been originally recorded in the notebook at the time the observation was made.*

2J SAFETY IN THE LABORATORY

Work in a chemical laboratory necessarily involves a degree of risk; accidents can and do happen. Strict adherence to the following rules will go far toward preventing (or minimizing the effect of) accidents.

1. At the outset, learn the location of the nearest eye fountain, fire blanket, shower, and fire extinguisher. Learn the proper use of each, and do not hesitate to use this equipment should the need arise.

2. ***Wear eye protection at all times.*** The potential for serious and perhaps permanent eye injury makes it mandatory that adequate eye protection be worn at all times by students, instructors, and visitors. Eye protection should be donned before entering the laboratory and should be used continuously until it is time to leave. Serious eye injuries have occurred to people performing such innocuous tasks as computing or writing in a laboratory notebook; such incidents usually result from someone else's loss of control over an experiment. Regular prescription glasses are not adequate substitutes for eye protection approved by the Office of Safety and Health Administration (OSHA). Contact lenses should *never* be used in the laboratory because laboratory fumes may react with them and have a harmful effect on the eyes.
3. Most of the chemicals in a laboratory are toxic; some are very toxic, and some—such as concentrated solutions of acids and bases—are highly corrosive. Avoid contact between these liquids and the skin. In the event of such contact, *immediately* flood the affected area with copious quantities of water. If a corrosive solution is spilled on clothing, remove the garment immediately. Time is of the essence; do not be concerned about modesty.
4. *NEVER* perform an unauthorized experiment. Unauthorized experiments are grounds for disqualification at many institutions.
5. Never work alone in the laboratory; be certain that someone is always within earshot.
6. Never bring food or beverages into the laboratory. Do not drink from laboratory glassware. Do not smoke in the laboratory.
7. Always use a bulb or other device to draw liquids into a pipet; *NEVER* use your mouth to provide suction.
8. Wear adequate foot covering (no sandals). Confine long hair with a net. A laboratory coat or apron will provide some protection and may be required.
9. Be extremely tentative in touching objects that have been heated; hot glass looks just like cold glass.
10. Always fire-polish the ends of freshly cut glass tubing. *NEVER* attempt to force glass tubing through the hole of a stopper. Instead, make sure that both tubing and hole are wet with soapy water. Protect your hands with several layers of towel while inserting glass into a stopper.
11. Use fume hoods whenever toxic or noxious gases are likely to be evolved. Be cautious in testing for odors; use your hand to waft vapors above containers toward your nose.
12. Notify your instructor immediately in the event of an injury.
13. Dispose of solutions and chemicals as instructed. It is illegal to flush solutions containing heavy metal ions or organic liquids down the drain in many localities; alternative arrangements are required for the disposal of such liquids.

InfoTrac College Edition

For additional readings, go to InfoTrac College Edition, your online research library, at

http://infotrac.thomsonlearning.com

CHAPTER 3

Using Spreadsheets in Analytical Chemistry

© Henry Blackham/Corbis

From the ways that we deal with our finances, using software applications such as Quicken, to our modes of communication with our friends, relatives, and colleagues, using Eudora and Microsoft Outlook, the personal computer has revolutionized nearly every aspect of our lives. Physical chemists use applications such as Hyperchem and Gaussian to carry out quantum calculations. Biological chemists and organic chemists use molecular mechanics programs such as Spartan to build and investigate the properties of molecules. Inorganic chemists exploit ChemDraw to visualize molecules. Certain programs transcend specialization and are used in a broad range of fields. In analytical chemistry and most other chemical and scientific pursuits, spreadsheet programs provide a means for storing, analyzing, and organizing numerical and textual data. Microsoft Excel is an example of this type of program.

The personal computer revolution of the past 20 years has produced many useful tools for students, chemists, and other scientists. One of the best examples of these applications is the spreadsheet, which is versatile, powerful, and easy to use. Spreadsheets are used for record keeping, mathematical calculations, statistical analysis, curve fitting, data plotting, financial analysis, database management, and a variety of other tasks limited only by the imagination of the user. State-of-the-art spreadsheet programs have many built-in functions to assist in accomplishing the computational tasks of analytical chemistry. Throughout this text, we present examples to illustrate some of these tasks and spreadsheets for performing them. You will find many more examples, more elaborate explanations of spreadsheet methodology, and expanded treatments of some of the theory of analytical chemistry in the companion text Applications of Microsoft® Excel in Analytical Chemistry.[1] *There is an increasing need to build and maintain analytical chemistry databases with information that has been collected in the laboratory, downloaded, or imported from the Internet, or e-mailed to us by our colleagues. This need often requires reformatting the resulting data tables to suit our purposes. In this chapter we show how this is done using a few basic string functions, then show how the resulting data tables can be used to make useful calculations. In later chapters we explore how to process and display large amounts of data using Excel's built-in numerical, statistical, and graphing functions.*

[1]For more information on the use of spreadsheets in chemistry, see F. J. Holler and S. R. Crouch, *Applications of Microsoft® Excel in Analytical Chemistry.* Belmont, CA: Brooks/Cole, 2003.

3A KEEPING RECORDS AND MAKING CALCULATIONS

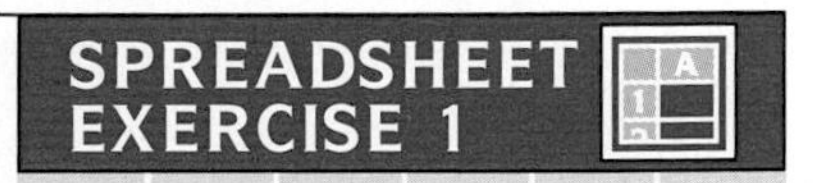

The most popular spreadsheet programs include Microsoft Excel, Lotus 1-2-3, and Quattro Pro. Because of its wide availability and general utility, we have chosen to illustrate our examples using Microsoft Excel on the PC. Although the syntax and commands for other spreadsheet applications are somewhat different from those for Excel, the general concepts and principles of operation are similar. The examples we present may be accomplished using any of the popular spreadsheet applications; the precise instructions must be modified if an application other than Excel is used.[2] In our examples, we will assume that Excel is configured with default options as delivered from the manufacturer unless we specifically note otherwise.

It is our belief that we learn best by doing, not by reading about doing. Although software manufacturers have made great strides in writing manuals for their products, it is still generally true that when we know enough to read a software manual efficiently, we no longer need the manual. With that in mind, we have designed a series of spreadsheet exercises that evolve in the context of analytical chemistry. We introduce commands and syntax only when they are needed for a particular task, so if you need more detailed information, please consult the Excel help screens or your software documentation. Help is available at the click of a mouse button from within Excel by clicking on **Help/Microsoft Excel Help,** or it may be called up by typing the `F1` key. In addition, the latest versions of Microsoft Office, which include Excel, feature a pull-down menu in the upper right-hand corner of the screen that permits you to type questions and obtain context-sensitive help.

3A-1 Getting Started

In this book, we will assume that you are familiar with Windows. If you need assistance with Windows, please consult the Windows guide *Getting Started* or use the on-line help facility available in Windows. In most versions of Windows, for example, you can get help by opening the **Start** menu and clicking on **Help.** To illustrate spreadsheet use, we will use Excel to carry out the functions of the laboratory notebook page that is depicted in Figure 2-25. To begin, we must start Excel by double-clicking on its icon, shown in the margin, on the computer desktop. Alternatively, in recent versions of Windows and Microsoft Office, click on **Start/Programs/Microsoft Excel** on the toolbar. The window shown in Figure 3-1 then appears.

The window contains a **worksheet** consisting of a grid of **cells** arranged in rows and columns. The rows are labeled 1, 2, 3, and so on, and the columns are labeled A, B, C, and so forth, in the margin of the worksheet. Each cell has a unique location specified by its address. For example, the **active cell,** which is surrounded by a dark outline in Figure 3-1, has the address A1. The address of the active cell is always displayed in the box just above the first column of the displayed worksheet in the **formula bar.** You may verify this display of the active cell by clicking on various cells of the worksheet.

[2] D. Diamond and V. C. A. Hanratty, *Spreadsheet Applications in Chemistry Using Microsoft Excel.* New York: John Wiley & Sons, 1997; H. Freiser, *Concepts & Calculations in Analytical Chemistry: A Spreadsheet Approach.* Boca Raton, FL: CRC Press, 1992; R. de Levie, *Principles of Quantitative Chemical Analysis.* New York: McGraw-Hill, 1997; R. de Levie, *A Spreadsheet Workbook for Quantitative Chemical Analysis.* New York: McGraw-Hill, 1992.

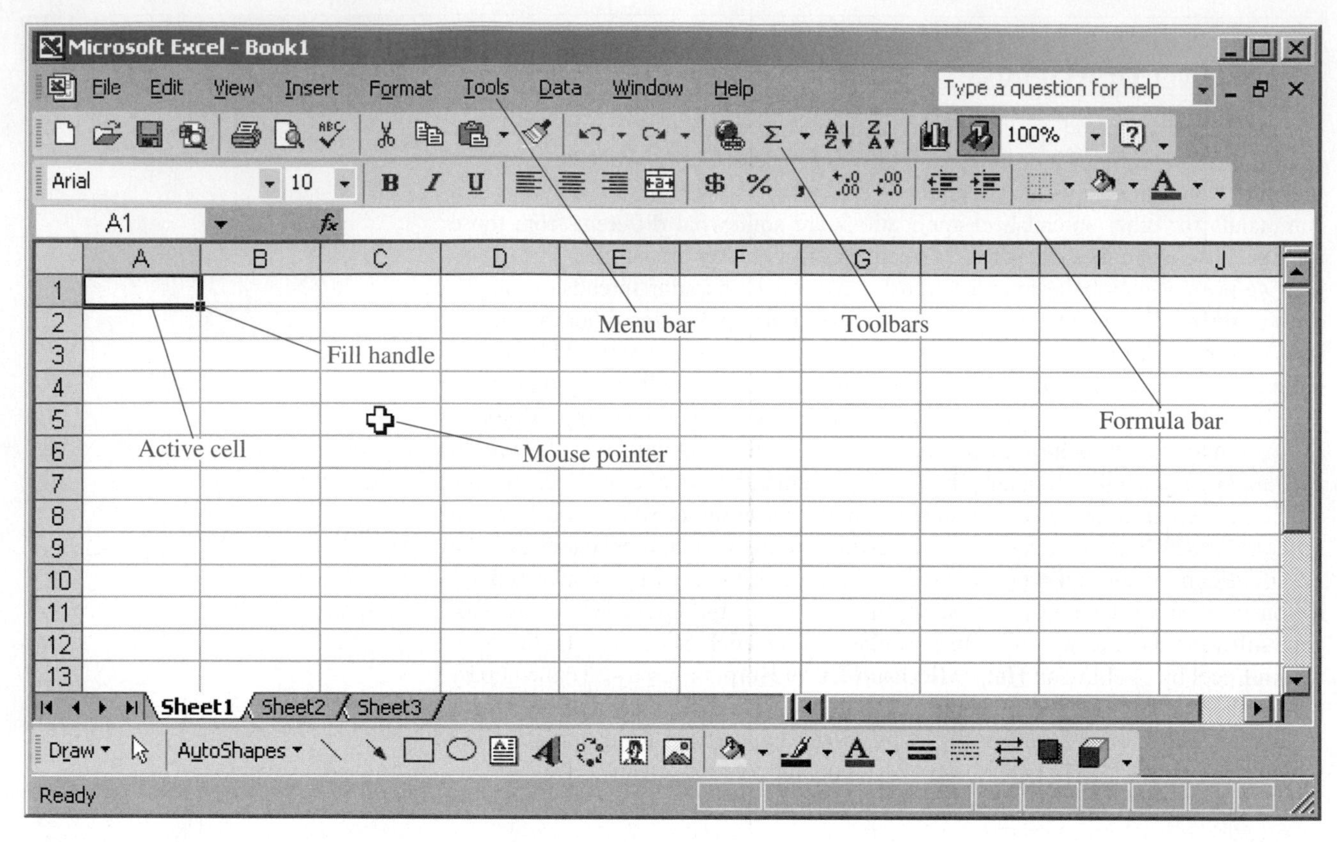

Figure 3-1 The opening window in Microsoft Excel. Note the location of the menu bar, the toolbars, the active cell, and the mouse pointer.

Entering Text in the Spreadsheet

Cells may contain text, numbers, or formulas. We will begin by typing some text into the worksheet. Click on cell A1, and type **`Gravimetric Determination of Chloride`** followed by the Enter key **`[↵]`**. Notice that the active cell is now A2, so you may type **`Samples [↵]`**. As you type, the data that you enter appear in the formula bar. If you make a mistake, just click the mouse in the formula bar, and make necessary corrections using the backspace key or the delete key. Continue to type text into the cells of column A as shown below.

```
Mass of bottle plus sample, g[↵]
Mass of bottle less sample, g[↵]
Mass of sample, g[↵]
[↵]
Crucible masses, with AgCl, g[↵]
Crucible masses, empty, g[↵]
Mass of AgCl, g[↵]
[↵]
% Chloride[↵]
Mean % Chloride[↵]
Standard Deviation, % Chloride[↵]
RSD, parts per thousand[↵]
```

	A	B	C	D
1	Gravimetric Determination of Chloride			
2	Samples			
3	Mass of bottle plus sample, g			
4	Mass of bottle less sample, g			
5	Mass of sample, g			
6				
7	Crucible masses, with AgCl, g			
8	Crucible masses, empty, g			
9	Mass of AgCl, g			
10				
11	% Chloride			
12	Mean % Chloride			
13	Standard deviation, % Chloride			
14	RSD, parts per thousand			
15				

Figure 3-2 The appearance of the worksheet after entering the labels.

When you have finished entering the text, the worksheet should appear as shown in Figure 3-2.

Changing the Width of a Column

Notice that the labels that you typed into column A are wider than the column. You can change the width of the column by placing the mouse pointer on the boundary between column A and column B in the column head, as in Figure 3-3a, and dragging the boundary to the right so that all of the text shows in the column, as in Figure 3-3b.

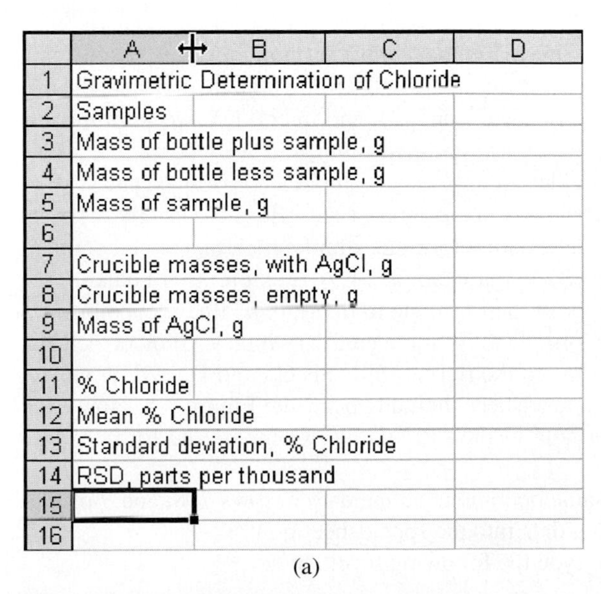

	A	B	C	D
1	Gravimetric Determination of Chloride			
2	Samples			
3	Mass of bottle plus sample, g			
4	Mass of bottle less sample, g			
5	Mass of sample, g			
6				
7	Crucible masses, with AgCl, g			
8	Crucible masses, empty, g			
9	Mass of AgCl, g			
10				
11	% Chloride			
12	Mean % Chloride			
13	Standard deviation, % Chloride			
14	RSD, parts per thousand			
15				
16				

(a)

	A	B	C
1	Gravimetric Determination of Chloride		
2	Samples		
3	Mass of bottle plus sample, g		
4	Mass of bottle less sample, g		
5	Mass of sample, g		
6			
7	Crucible masses, with AgCl, g		
8	Crucible masses, empty, g		
9	Mass of AgCl, g		
10			
11	% Chloride		
12	Mean % Chloride		
13	Standard deviation, % Chloride		
14	RSD, parts per thousand		
15			
16			

(b)

Figure 3-3 Changing the column width. (a) Place the mouse pointer on the boundary between column A and column B, and drag to the right to the position shown in (b).

	A	B	C	D
1	Gravimetric Determination of Chloride			
2	Samples	1	2	3
3	Mass of bottle plus sample, g	27.6115	27.2185	26.8105
4	Mass of bottle less sample, g	27.2185	26.8105	26.4517
5	Mass of sample, g	0.3930		
6				
7	Crucible masses, with AgCl, g			
8	Crucible masses, empty, g			
9	Mass of AgCl, g			

Figure 3-4 Sample data entry.

Entering Numbers into the Spreadsheet

Now let us enter some numerical data into the spreadsheet. Click on cell B2 and type

```
1[↵]
27.6115[↵]
27.2185[↵]
```

In cell B5, we want to calculate the difference between the data in cells B3 and B4, so we type

```
=b3-b4[↵]
```

► Excel formulas always begin with an equals sign [=].

The expression you just typed is called a **formula.** In Excel, formulas begin with an equals sign [=] followed by the desired numerical expression. Notice that the difference between the contents of cell B3 and cell B4 is displayed in cell B5. Now continue entering data until your spreadsheet looks like the one in Figure 3-4.

Filling Cells Using the Fill Handle

The formulas for cells C5 and D5 are identical to the formula in cell B5 except that the cell references for the data are different. In cell C5, we want to compute the difference between the contents of cells C3 and C4, and in cell D5, we want the difference between D3 and D4. We could type the formulas in cells C5 and D5 as we did for cell B5, but Excel provides an easy way to duplicate formulas, and it automatically changes the cell references to the appropriate values for us. To duplicate a formula in cells adjacent to an existing formula, simply click on the cell containing the formula, which is cell B5 in our example, then click on the fill handle (see Figure 3-1) and drag the corner of the rectangle to the right so that it encompasses the cells where you want the formula to be duplicated. Try it now. Click on cell B5, click on the fill handle, and drag to the right to fill cells C5 and D5. When you let up on the mouse button, the spreadsheet should appear like Figure 3-5. Now click on cell B5, and view the formula in the formula bar. Compare the formula with those in cells C5 and D5.

We want to perform the same operations on the data in rows 7, 8, and 9 of Figure 3-6, so enter the remaining data into the spreadsheet now.

Now click on cell B9, and type the following formula:

```
=b7-b8[↵]
```

	A	B	C	D
1	Gravimetric Determination of Chloride			
2	Samples	1	2	3
3	Mass of bottle plus sample, g	27.6115	27.2185	26.8105
4	Mass of bottle less sample, g	27.2185	26.8105	26.4517
5	Mass of sample, g	0.3930	0.4080	0.3588

Figure 3-5 Using the fill handle to copy formulas into adjacent cells of a spreadsheet. In this example, we clicked on cell B5, clicked on the fill handle, and dragged the rectangle to the right to fill cells C5 and D5. The formulas in cells B5, C5, and D5 are identical, but the cell references in the formulas refer to data in columns B, C, and D, respectively.

Again click on cell B9, click on the fill handle, and drag through columns C and D to copy the formula to cells C9 and D9. The mass of silver chloride should now be calculated for all three crucibles.

◄ The fill handle permits you to copy the contents of a cell to other cells either horizontally or vertically, but not both. Just click on the fill handle and drag from the current cell to the last cell where you want the original cell copied.

3A-2 Making Complex Calculations with Excel

As we shall learn in Chapter 12, the equation for finding the % chloride in each of the samples is

$$\%\text{ chloride} = \frac{\dfrac{\text{mass AgCl}}{\text{molar mass AgCl}} \times \text{molar mass Cl}}{\text{mass sample}} \times 100\%$$

$$= \frac{\dfrac{\text{mass AgCl}}{143.321\text{ grams/mol}} \times 35.4527\text{ grams/mol}}{\text{mass sample}} \times 100\%$$

Our task is now to translate this equation into an Excel formula and type it into cell B11 as shown:

```
=B9*35.4527*100/143.321/B5[↵]
```

Once you have typed the formula, click on cell B11 and drag on the fill handle to copy the formula into cells C11 and D11. The % chloride for samples 2 and 3 should now appear in the spreadsheet, as in Figure 3-7.

We will complete and document the spreadsheet in Chapter 6 after we have explored some of the important calculations of statistical analysis. For now, click on **File/Save As ...** in the menu bar, enter a file name such as `grav_chloride,` and save the Excel spreadsheet on a floppy disk or other medium for retrieval and

	A	B	C	D
1	Gravimetric Determination of Chloride			
2	Samples	1	2	3
3	Mass of bottle plus sample, g	27.6115	27.2185	26.8105
4	Mass of bottle less sample, g	27.2185	26.8105	26.4517
5	Mass of sample, g	0.3930	0.4080	0.3588
6				
7	Crucible masses, with AgCl, g	21.4296	23.4915	21.8323
8	Crucible masses, empty, g	20.7926	22.8311	21.2483
9	Mass of AgCl, g			

Figure 3-6 Entering the data into the spreadsheet in preparation for calculating the mass of dry silver chloride in the crucibles.

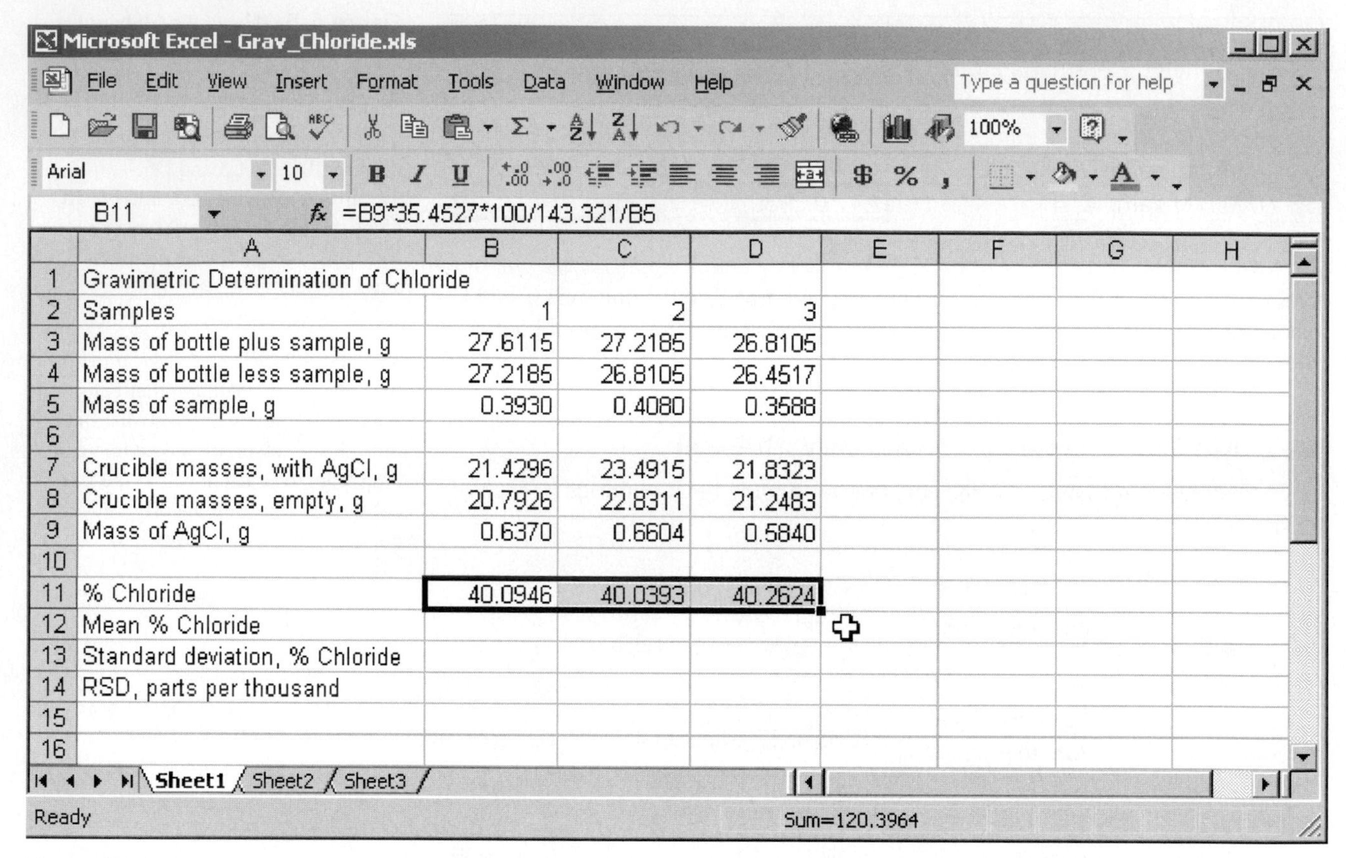

	A	B	C	D
1	Gravimetric Determination of Chloride			
2	Samples	1	2	3
3	Mass of bottle plus sample, g	27.6115	27.2185	26.8105
4	Mass of bottle less sample, g	27.2185	26.8105	26.4517
5	Mass of sample, g	0.3930	0.4080	0.3588
6				
7	Crucible masses, with AgCl, g	21.4296	23.4915	21.8323
8	Crucible masses, empty, g	20.7926	22.8311	21.2483
9	Mass of AgCl, g	0.6370	0.6604	0.5840
10				
11	% Chloride	40.0946	40.0393	40.2624
12	Mean % Chloride			
13	Standard deviation, % Chloride			
14	RSD, parts per thousand			
15				
16				

Figure 3-7 Completing the calculation of percent chloride. Type the formula in cell B11, click on the fill handle, and drag to the right through cell D11.

editing later. Excel will automatically append the file extension **`.xls`** to the file name so that it will appear as **`grav_chloride.xls`** on the disk.

In constructing this spreadsheet, we have learned some of the basics of spreadsheet operation, including typing text into a spreadsheet, changing column widths with the mouse, duplicating cells with the fill handle, and entering formulas into a spreadsheet.

SPREADSHEET EXERCISE 2

3B CALCULATING MOLAR MASSES USING EXCEL

In this exercise, we will learn how to import data from an external data source, to manipulate the data to obtain the desired numerical values for molar masses of the elements, to look up the appropriate values of the molar masses of the elements, and finally, to calculate molar masses of compounds. The Excel functions necessary to accomplish these tasks will serve us well in carrying out a variety of other data manipulations and computations.

3B-1 Importing Data from Web Pages

The development of the Internet and the resulting capability for widespread storage and retrieval of textual and numerical data have made it quite easy to gain access to

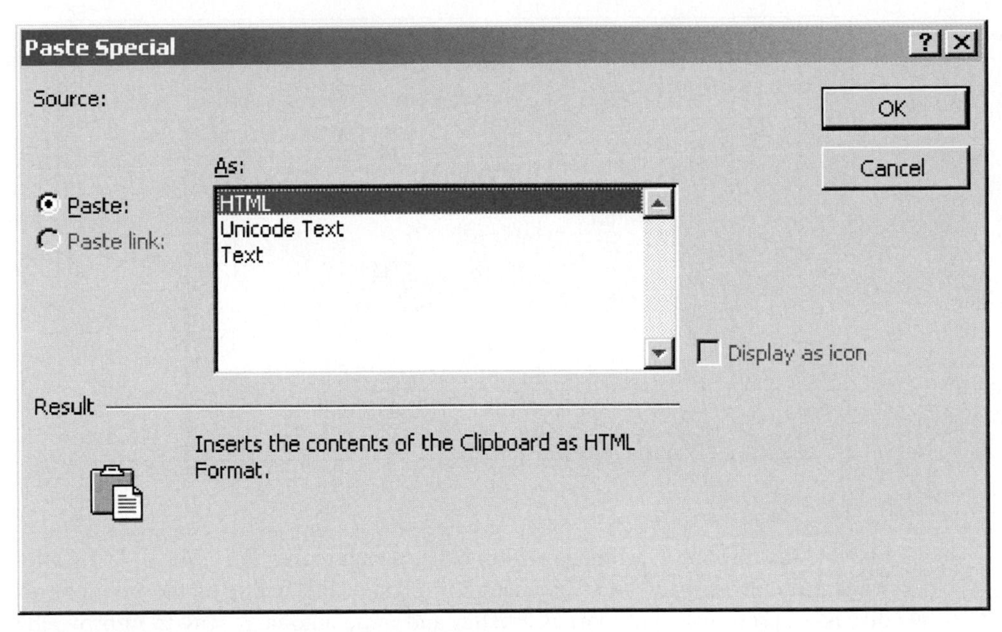

Figure 3-8 Option window for the **Edit/Paste Special...** command in Excel.

the latest values for molar masses, universal constants of nature, and raw data from the scientific literature. The most important potential advantage of direct import of numerical data is the elimination of human transcription errors. In addition, if you require more than a few values, importing them may save considerable time. If only a few values need to be imported, the easiest way to get the data into a spreadsheet is to use the basic editing features of all Windows programs. For example, you may simply highlight the desired number or string of characters, click on **Edit/Copy** (or the keyboard shortcut `[Ctrl-c]`), and then place the cursor in the desired location in your spreadsheet and click on **Edit/Paste** `[Ctrl-v]` or **Edit/Paste Special...**. You can test this function by using your Web browser to surf to **www.google.com** searching for documents containing the keywords "iupac atomic weights," and navigating to the Web site of the International Union of Pure and Applied Chemistry (IUPAC). This site has a Web page containing a table of the latest atomic weights.[3] Use your mouse to highlight the entire table, including the five headings above the columns, and type `[Ctrl-c]`. This action copies the table of data to the clipboard of your computer. Then switch to Excel, click in cell A1 in an empty worksheet, and click on **Edit/Paste Special...**, and a window similar to the one in Figure 3-8 will appear.

◀ Type [Alt-tab] to switch between programs.

Click on **HTML** so that it is highlighted and then **OK,** and your spreadsheet should appear as shown in Figure 3-9. HTML refers to hypertext markup language, which is the language used to encode many Web pages. When you copied the data from the table, hidden HTML instructions were included that allow Excel to arrange the data in your worksheet in much the same way they originally appeared on the Web page. If your version of Excel does not have the **Edit/Paste Special.../HTML** option, then click on cell A1, just paste the table `[Ctrl-v]`, and you should get similar results.

◀ Importing data often minimizes typing errors and may save a lot of time.

[3]T. B. Coplen, *Pure Appl. Chem.*, **2001,** *73,* 667–683.

	A	B	C	D	E	F
1	At No	Symbol	Name	Atomic Wt	Notes	
2	1	H	Hydrogen	1.00794(7)	1, 2, 3	
3	2	He	Helium	4.002602(2)	1, 2	
4	3	Li	Lithium	[6.941(2)]	1, 2, 3, 4	
5	4	Be	Beryllium	9.012182(3)		
6	5	B	Boron	10.811(7)	1, 2, 3	
7	6	C	Carbon	12.0107(8)	1, 2	

Figure 3-9 Table of atomic weight data from the IUPAC Web site pasted into Excel as HTML (hypertext markup language). The table is highlighted and somewhat hard to read because it has not yet been formatted in Excel.

Note that the text is wrapped within cells, which makes the table of data somewhat difficult to read. You can manually change the width of the columns as discussed previously, but there is a better and more automatic way to improve the appearance and readability of the table. The table of data should already be highlighted, so click on **Format/Cells...,** click on the **Alignment** tab, click twice to uncheck the **Wrap** button, and click on **OK.** Finally, click on **Format/Column.../AutoFit Selection,** and your worksheet should be formatted and readable as shown in Figure 3-10. Scroll down the worksheet, and notice that each column is now exactly the correct width to accommodate the largest number of characters in the column and that no cells have wrapped text. Text in the cells is left justified, and numerical data are right justified.

3B-2 Dealing with Character Strings

Generally, Excel is able to recognize the type of data that have been entered or imported into its cells. For example, in cell A2, Excel has recognized the number 1, and so it is right justified in the cell. In fact, all of the atomic numbers in column A are correctly recognized as numerical data. In cell C2, Excel recognizes that the cell contains only alphabetical characters, which are left justified. Note also that cell E2 has a small triangle in the upper left-hand corner, indicating that there is a problem with the cell. If you click on cell E2, a small box appears containing an

	A	B	C	D	E	F
1	At No	Symbol	Name	Atomic Wt	Notes	
2	1	H	Hydrogen	1.00794(7)	1, 2, 3	
3	2	He	Helium	4.002602(2)	1, 2	
4	3	Li	Lithium	[6.941(2)]	1, 2, 3, 4	
5	4	Be	Beryllium	9.012182(3)		
6	5	B	Boron	10.811(7)	1, 2, 3	
7	6	C	Carbon	12.0107(8)	1, 2	

Figure 3-10 Formatted IUPAC table of atomic weights.

exclamation point, and if you position your cursor over the box, another box appears telling you that there is some confusion as to the type of data in the cell. Excel interprets the data (1, 2, 3) as a date. We will not be using the data in column E, so we will ignore errors of this type in the column. Numbers in column E without commas are interpreted as numerical.

Let us now focus on the atomic weights (masses) in column D and note some important characteristics of the data. From now on, and throughout this book, we will refer to **atomic masses** rather than atomic weights. First, Excel has interpreted the data in column D as text rather than numerical data. This happens because there is a digit in parentheses at the end of each line. This digit is the uncertainty in the last place of the atomic mass. For example, we might write the atomic mass of hydrogen as 1.00794 ± 0.00007 rather than 1.00794(7). As we shall learn in Chapter 6, uncertainties in atomic masses can be used to compute the uncertainty in any results that are derived from atomic masses, such as molar masses of compounds. Although it is a relatively simple matter to cut (`[Ctrl-x]` or **Edit/Cut**) and paste the uncertainty into another cell, it could just be deleted from each cell if it is not to be used. To see how Excel interprets the atomic masses without the parenthetical uncertainty, click on D2, copy the data, click on cell G2, and paste the data into the cell. Then click in the formula bar, use the backspace or delete key to remove the characters "(7)", and depress the `[Enter]` key. Notice that now Excel interprets the atomic mass of hydrogen as numerical data, and the number 1.00794 is right justified in the cell. It would be straightforward but rather tedious to perform these operations on all of the 113 atomic masses in the table; furthermore, there would be many opportunities to delete the wrong characters and create errors in the table. Fortunately, Excel has many built-in functions that allow us to treat situations such as that encountered here.

Find and Replace

One way to remove, or strip, the parenthetical uncertainties in the table of atomic masses is to use Excel's **Find/Replace** function. We will illustrate this approach with a few of the entries. Copy the atomic masses from hydrogen through copper, including the uncertainties, to column F, cells F2:F30. Now highlight cells F2:F30. Go to the **Edit** menu and choose **Replace.** This should bring up the **Find and Replace** window shown below. Make sure the **Replace** tab is selected as shown.

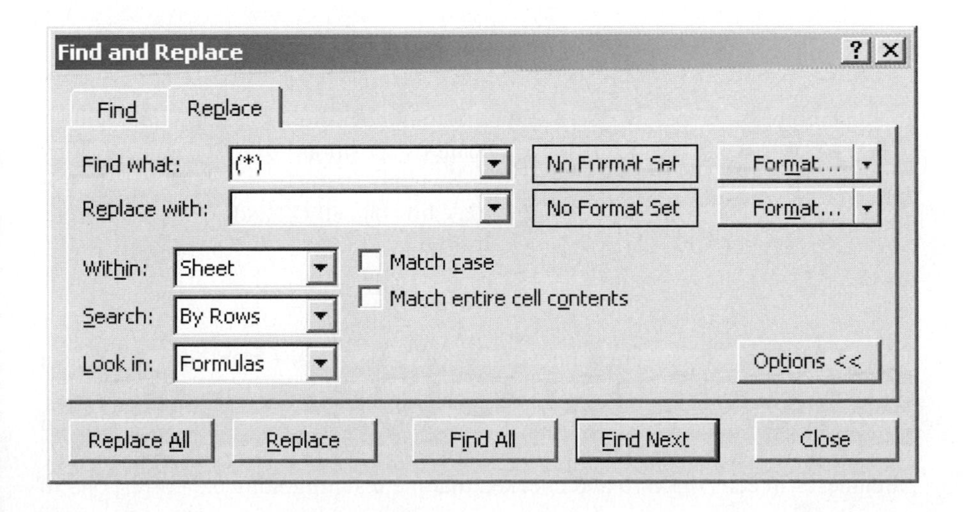

Type (*) in the **Find what:** box and leave the **Replace with:** box blank. Here the asterisk is a wild card. Choosing (*) as your search text means that the parentheses and anything enclosed will be found and replaced with nothing in this case. In the **Search** box, choose **By Columns.** Now click on **Replace All.** Note that the 29 entries have been stripped of the parenthetical characters. Now delete the 29 entries in column F, since we will not explore this approach further.

Although the **Find/Replace** function works well for these data, it is not as generally useful as the built-in functions for manipulating strings of alphabetical characters and numbers. These functions are called **string functions.** We will use string functions to strip the parenthetical uncertainties from the data in column D and produce a column of numerical atomic masses.

A **string** is a group of alphabetical and/or numerical characters.

The FIND Function

The Excel worksheet function **FIND(find_text,within_text,start_num)** permits us to find the position in a text string of any alphanumerical character that we specify. For example, in our table of atomic masses, it would be useful to know where the parentheses are in each of the text strings of column D. Consider once again the atomic mass of hydrogen represented in cell D2 as the text string "1.00794(7)". We surround the string with quotation marks because Excel recognizes characters in quotation marks to be strings. If we count the characters in the string from left to right, we find that the left parenthesis is in the eighth position and the right parenthesis is in the tenth. Excel permits us to automatically find the position of the left parenthesis by using the function FIND("(","1.00794(7)",1) where the string "(" is **find_text,** the string "1.00794(7)" is **within_text,** and the number 1 is **start_num,** which is the character position in the string where we would like Excel to begin counting. If **start_num** is omitted, it is assumed to be the first character of the string. Test the function by clicking on cell G2 and typing

► The FIND function locates a character or string of characters within another string and reveals its location. The formula `=FIND ("&", "Law & Order")` produces the number 5 because "&" is in the fifth position in the string "Law & Order".

```
=FIND("(","1.00794(7)",1)[↵]
```

which produces the number 8 in cell G2, indicating that the left parenthesis occurs in the eighth character position of the atomic mass. Now click on cell G3, and type

```
=FIND(")","1.00794(7)",1)[↵]
```

and the number 10 appears in cell G3, indicating the position of the right parenthesis. We can use the FIND function to locate the position of any character in any string. Now instead of typing the string, we may use its cell reference, which in the case of the atomic mass of hydrogen is D2. Click on cell G2 and type

```
=FIND("(",D2)[↵]
```

and once again the number 8 appears in the cell. Note that **start_num** has been omitted because we wish to begin the search with the first character in the string. Now use the fill handle to copy cell G2 to cells G3:G10, and your worksheet should appear as shown in Figure 3-11. After you have observed the results of finding the parentheses in cells G3:G10 and checked that the resulting numbers correspond to

	A	B	C	D	E	F	G
1	At No	Symbol	Name	Atomic Wt	Notes		
2	1	H	Hydrogen	1.00794(7)	1, 2, 3		8
3	2	He	Helium	4.002602(2)	1, 2		9
4	3	Li	Lithium	[6.941(2)]	1, 2, 3, 4		7
5	4	Be	Beryllium	9.012182(3)			9
6	5	B	Boron	10.811(7)	1, 2, 3		7
7	6	C	Carbon	12.0107(8)	1, 2		8
8	7	N	Nitrogen	14.0067(2)	1, 2		8
9	8	O	Oxygen	15.9994(3)	1, 2		8
10	9	F	Fluorine	18.9984032(5)			11

Figure 3-11 Worksheet showing the results of using the FIND function to locate the position of the left parenthesis in each of the atomic masses of cells D2 through D10.

the positions of the left parentheses in cells D2:D10, depress the `Delete` key to clear column G.

The MID Function

Now that we have learned how to find characters within strings, we can use the Excel function **MID(text,start_num,num_chars)** to extract the numerical atomic mass data from the strings of column D. The variable **text** is the string of interest, **start_num** is the character position where we would like the extraction to begin, and **num_chars** is the number of characters that we want extracted from the string. In our example, the starting position is always 1 because the strings all begin with the first digit of the atomic mass. The number of characters will be determined by the FIND function, which will locate the right parenthesis for us as before. We can try it by clicking on cell F2 and typing

◄ The MID function produces a string from a second string by specifying the position of the first character of the sought-for string and the desired number of characters. For example, `=MID("Oh, Brother, Where Art Thou?",5,7)` gives the string "Brother". Note that spaces count as characters.

```
=MID(D2,1,FIND("(",D2))[↵]
```

You will notice that the atomic mass of hydrogen appears in cell F2, but we do not quite have it right yet because the left parenthesis appears at the end of the string. This difficulty is easily fixed by typing −1 at the end of the FIND function, which subtracts one from the character position of the left parenthesis to give the last character position of the atomic mass. Click on cell F2, then click in the formula bar at the end of the FIND function, and change the cell contents to the following:

```
=MID(D2,1,FIND("(",D2)−1)[↵]
```

Now the atomic mass of hydrogen appears as 1.00794 in the cell. All that remains is to click on the fill handle of cell F2 and drag to the bottom of the table to extract the atomic masses from the strings. Your worksheet should now appear as shown in Figure 3-12. Scan column F, and you will see that some atomic masses are still not displayed correctly. The atomic mass of lithium appears as [6.941, and others (elements 43, 61, 84–89, and 93–114) appear with the notation #VALUE!, which indicates an error because there are no parentheses in these strings. For now, you can

	A	B	C	D	E	F
1	At No	Symbol	Name	Atomic Wt	Notes	
2	1	H	Hydrogen	1.00794(7)	1, 2, 3	1.00794
3	2	He	Helium	4.002602(2)	1, 2	4.002602
4	3	Li	Lithium	[6.941(2)]	1, 2, 3, 4	[6.941
5	4	Be	Beryllium	9.012182(3)		9.012182
6	5	B	Boron	10.811(7)	1, 2, 3	10.811
7	6	C	Carbon	12.0107(8)	1, 2	12.0107
8	7	N	Nitrogen	14.0067(2)	1, 2	14.0067
9	8	O	Oxygen	15.9994(3)	1, 2	15.9994
10	9	F	Fluorine	18.9984032(5)		18.9984032

Figure 3-12 Extracting atomic masses from strings. Most of the atomic masses appear correctly in column F except for lithium and several others that contain square brackets.

just copy the string values from column D, paste them into column F, and individually edit the strings so that no parentheses, square brackets, or uncertainties appear in column D. Problems at the end of the chapter ask you to devise formulas to make these conversions automatically. Excel has functions that permit you to perform checks on results of functions so that when errors occur, you can perform automatic corrections. We will save the discussion of these functions for later.

3B-3 Using VLOOKUP to Locate Data in a Worksheet

The ultimate goal in this exercise is to calculate molar masses of compounds in a relatively straightforward and automatic fashion. Since we have the symbols for all of the elements in column B of our worksheet and the corresponding atomic masses of the elements in column D, it would be quite useful if there were a way to look up a given atomic mass by just specifying its symbol. Excel provides a convenient means to accomplish this task. The function **VLOOKUP(lookup_value, table_array,col_index_num,range_lookup)** finds **lookup_value** in the first column of a section of a worksheet specified by **table_array** and returns the corresponding contents in the column indicated by **col_index_num.** Let us now use this function to look up the atomic mass of fluorine. Begin by clicking in cell G1 and typing

► VLOOKUP("mass", A1:B5,2,FALSE) scans the first column of the array, extending from cell A1 to B5 (see below) for the string "mass", and returns the value in column 2 corresponding to the string. In this example, we obtain 0.357. If an exact match is not found, an error occurs.

	A	B
1	volume	2
2	temperature	300
3	mass	0.357
4	moles	0.5
5	gas constant	0.0821

```
Element[→]
No. Atoms[→]
At. Mass[↵]
=VLOOKUP("F",B2:F114,5,FALSE)[↵]
```

Your worksheet should look like the one shown in Figure 3-13, with the molar mass of fluorine displayed in cell I2. Excel has looked up the atomic mass of fluorine (specified by its symbol "F" as the **lookup_value**) in the rectangular region of the worksheet specified by the variable **table_array,** which in this example is B2:F114. This region, or **array,** contains the atomic symbols in the first column of the array (column B in the worksheet) and the extracted atomic masses in the fifth

	A	B	C	D	E	F	G	H	I
1	At No	Symbol	Name	Atomic Wt	Notes		Element	No. Atoms	At. Mass
2	1	H	Hydrogen	1.00794(7)	1, 2, 3	1.00794			18.9984032
3	2	He	Helium	4.002602(2)	1, 2	4.002602			
4	3	Li	Lithium	[6.941(2)]	1, 2, 3, 4	6.941			
5	4	Be	Beryllium	9.012182(3)		9.012182			
6	5	B	Boron	10.811(7)	1, 2, 3	10.811			
7	6	C	Carbon	12.0107(8)	1, 2	12.0107			
8	7	N	Nitrogen	14.0067(2)	1, 2	14.0067			
9	8	O	Oxygen	15.9994(3)	1, 2	15.9994			
10	9	F	Fluorine	18.9984032(5)		18.9984032			

Figure 3-13 Using VLOOKUP to look up and display the atomic mass of fluorine.

column (column F in the worksheet). Hence, **col_index_num** is set to 5 in the function to indicate that we want the atomic mass in the fifth column of the array. Excel assumes that the lookup value is contained in the first column of the array. The logical variable **range_lookup,** which is set to FALSE here, tells Excel that the match between the atomic symbol being sought and the result must be *exact.* If this variable is set to TRUE, VLOOKUP will find an approximate match. If no match is found, an error results. Try several different element symbols in the VLOOKUP function in cell I2, and note the results.

Now we will generalize the lookup function so that we can look up the atomic mass of any element by simply typing its symbol into a cell. Click on cell I2, click in the formula bar, and edit the contents to read as follows:

```
=VLOOKUP(G2,B2:F114,5,FALSE)[↵]
```

The error condition #N/A then appears in cell I2 because G2 is blank and thus contains no element symbol. Click on cell G2, and type **Fe.** The atomic mass of iron now appears in cell I2. Try typing several other element symbols in cell G2, and note the results. When you are satisfied that the LOOKUP function is working properly, click on cell I2, and copy the contents into cell I3, using the fill handle. Then type various element symbols in cell G3, and your worksheet should look something like the one shown in Figure 3-14.

G	H	I
Element	No. Atoms	At. Mass
Fe		55.845
S		32.066

Figure 3-14 The atomic masses of any two elements may be looked up by typing their symbols in cells G2 and G3.

3B-4 Making the Calculation

The last step in our exercise is to create formulas that will calculate the molar mass of a compound from the atomic masses looked up by the functions in cells G2 and G3. We will confine ourselves to binary compounds for now and leave more complex cases for the problems at the end of the chapter. Let us calculate the molar mass of NaCl. Begin by clicking on cell G2, and type

```
Na[→]
1[↵]
1[←]
Cl[↵]
```

Figure 3-15 Calculation of the molar mass of NaCl. The worksheet is general for binary compounds. Type the symbol for the first element in cell G1 and the number of atoms of the element in H1. Type the symbol and the number of atoms of the second element in G2 and H2. The molar mass of the compound is displayed in cell J4.

G	H	I	J
Element	No. Atoms	At. Mass	Mass
Na	1	22.989770	22.98977
Cl	1	35.453	35.453
			58.44277

Your worksheet should now display the atomic masses of Na and Cl in cells I2 and I3, respectively, and the number 1 in cells H2 and H3 to indicate the number of atoms of each element in the formula of NaCl. Now click on cell J1 and enter the following.

```
Mass[↵]
=H2*I2[↵]
```

Copy the formula from J2 to J3 using the fill handle, and enter the following equation in cell J4.

```
=J2+J3[↵]
```

This formula adds the contents of cells J2 and J3, which contain the total mass of Na and Cl, and displays the molar mass of NaCl in cell J4. Your worksheet should now be similar to the one in Figure 3-15. Test this worksheet with several binary compounds in the table of molar masses on the inside back cover of this textbook. Check the molar masses from the worksheet against those you find in the table. Note that Excel has no automatic way to keep track of significant figures, and so molar masses calculated using your worksheet must be rounded to reflect only those digits that are significant. In Chapter 6, we shall explore methods for dealing with significant figures in computed results. To finish this activity, give your worksheet a descriptive file name and save to a disk for future use.

► Our worksheet works for two elements only. What changes must you make to extend it to more than two elements? Is there a way to implement the worksheet for any number of elements? You may find that as you fill your Excel toolbox, there may be better and more sophisticated ways to compute a molar mass.

In this chapter, we have begun to explore the use of spreadsheets in analytical chemistry. We have examined many of the basic operations of spreadsheet use, including data entry, data import, string handling, and basic calculations. In other spreadsheets in this book and in *Applications of Microsoft® Excel in Analytical Chemistry,* we will build on the techniques that we have acquired and learn much more about Excel that will be useful in our study of analytical chemistry and related fields.

WEB WORKS

Direct your Web browser to a search engine, and perform a search to locate an HTML table of water density as a function of temperature in at least one-degree intervals over the range of 15°C to 30°C. Use key words "water temperature density table g/mL." Copy the table, and paste it into a worksheet as HTML so that the data are displayed in an array of cells. Save the worksheet in a file for later retrieval in Problem 3-10.

Questions and Problems

*3-1. Describe the use of the following Excel functions after reading about them in the Excel help facility.
(a) SQRT
(b) SUM
(c) PI
(d) FACT
(e) EXP
(f) LOG

3-2. Use the Excel help facility to look up the use of the COUNT function. Use the function to determine the number of data in each column of the spreadsheet of Figure 3-7. The count function is quite useful for determining the number of data entered into a given area of a spreadsheet.

3-3. Prepare a spreadsheet similar to the one shown in Figure 3-7 for the gravimetric determination of nickel using dimethylglyoxime. See Section 37B-3 for details. Use the worksheet from Problem 3-9 to calculate the molar mass of $Ni(DMG)_2$ if it is available.

*3-4. Write an Excel formula using the FIND and MID functions to eliminate the square brackets and the uncertainty from the atomic mass of lithium in the IUPAC table and display the numeric characters of the atomic weight.

3-5. Devise an Excel formula for elements 43, 61, 84 to 89, and 93 to 114 that will automatically remove the square brackets from the IUPAC table of atomic masses and eliminate the #VALUE! error described in Section 3B-2.

*3-6. Devise an Excel formula using the FIND and MID functions that will automatically display the uncertainties of the atomic masses in the IUPAC table.

3-7. Use the worksheet of Figure 3-15 to calculate the molar masses of the following compounds.
(a) HCl
(b) NH_3
(c) ZnS
(d) AgCl
(e) $PbCl_2$
(f) Bi_2O_3
(g) Al_2O_3

3-8. Modify the Excel worksheet of Figure 3-15 to compute the molar mass of compounds containing (a) three elements and (b) five elements.

3-9. Modify the worksheet of Figure 3-15 to calculate the molar masses of the following compounds.
(a) Na_2SO_4
(b) $Ba(IO_3)_2$
(c) CaC_2O_4
(d) $KMnO_4$
(e) $K_4Fe(CN)_6$
(f) $Na_2S_2O_3 \cdot 5H_2O$

3-10. **Challenge Problem.** Equation 2-1 permits the calculation of air buoyancy corrections for mass data. Suppose that you are calibrating a 100-mL pipet by weighing aliquots of water on an analytical balance, and you want to prepare a worksheet to correct your masses of water for buoyancy at various laboratory temperatures. The final column of your worksheet should contain the percent error in the weighing as a function of temperature. As a starting point, use the table of water densities that you obtained from this chapter's Web Works. Alternatively, you may look up the data in the *CRC Handbook of Chemistry and Physics* or another reference source, and type them into your worksheet. Use the ideal gas law to calculate the density of air at temperatures from 15°C to 30°C at intervals of one degree. Assume that air is 78% nitrogen and 22% oxygen, the density of the standard masses used to calibrate your balance is 8.0 g/cm^3, and atmospheric pressure is 1 atm.
(a) Do your results indicate that it is necessary to make air buoyancy corrections when calibrating a pipet? Justify your decision.
(b) Is temperature an important variable in making buoyancy corrections during pipet calibration? Explain your answer.
(c) What other role does your table of water density versus temperature play in the calibration of a pipet?
(d) If you made 10 replicate determinations of the mass of water delivered by a 100-mL pipet, and the average *apparent* mass of water delivered at 19°C was 99.736 g, what is the uncorrected volume of the pipet?
(e) What is the volume of the pipet corrected for air buoyancy in your weighings?

*Answers are provided at the end of the book for questions and problems marked with an asterisk.

(f) An additional factor that determines the volume of liquid delivered by a pipet is the expansion and contraction of the glass as the temperature changes. The volume of a glass vessel at a given temperature T is given by $V_T = V_{20}[1 + a(T - 20°C)]$, where V_{20} is the volume of the vessel at 20°C and a is the coefficient of cubic expansion of the glass. The coefficient of cubic expansion varies with the type of glass, but for typical borosilicate glass, a = 0.000010 mL/mL/°C. Add a column to your worksheet that corrects for expansion of the glass with temperature, and comment on this effect relative to other effects you have investigated.

(g) What is the true volume of the 100-mL pipet?

InfoTrac College Edition

For additional readings, go to InfoTrac College Edition, your online research library, at

http://infotrac.thomsonlearning.com

CHAPTER 4

Calculations Used in Analytical Chemistry

Avogadro's number is one of the most important of all physical constants and is central to the study of chemistry. A worldwide effort is under way to determine this important number to 1 part in 100 million. Several spheres like the one shown in the photo have been fabricated specifically for this task, and it is claimed they are the most perfect spheres in the world. The diameter of the 10-cm sphere is uniform to within 40 nm. By measuring the diameter, the mass, the molar mass of silicon, and the spacing between silicon atoms, it is possible to calculate Avogadro's number. Once determined, this number may be used to provide a new standard mass—the silicon kilogram. For more information, see Problem 4-39 and Web Works.

In this chapter, we describe several methods used to compute the results of a quantitative analysis. We begin by presenting the SI system of units and the distinction between mass and weight. We then discuss the mole, a measure of the amount of a chemical substance. Next, we consider the various ways that concentrations of solutions are expressed. Finally, we treat chemical stoichiometry. You have probably encountered much of the material in this chapter in your general chemistry courses.

4A SOME IMPORTANT UNITS OF MEASUREMENT

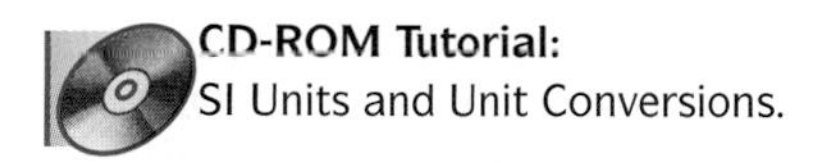

CD-ROM Tutorial: SI Units and Unit Conversions.

4A-1 SI Units

Scientists throughout the world have adopted a standardized system of units known as the **International System of Units (SI).** This system is based on the seven fundamental base units shown in Table 4-1. Numerous other useful units, such as volts, hertz, coulombs, and joules, are derived from these base units.

◄ SI is the acronym for the French "Système International d'Unités."

To express small or large measured quantities in terms of a few simple digits, prefixes are used with these base units and other derived units. As shown in Table 4-2, these prefixes multiply the unit by various powers of 10. For example, the wavelength of yellow radiation used for determining sodium by flame photometry is about 5.9×10^{-7} m, which can be expressed more compactly as 590 nm (nanometers); the volume of a liquid injected onto a chromatographic column is often roughly 50×10^{-6} L, or 50 μL (microliters); or the amount of memory on some computer hard disks is about 20×10^{9} bytes, or 20 Gbytes (gigabytes).

The **ångstrom unit Å** is a non-SI unit of length that is widely used to express the wavelength of very short radiation such as X-rays (1 Å = 0.1 nm = 10^{-10} m). Thus, typical X-radiation lies in the range of 0.1 to 10 Å.

In analytical chemistry, we often determine the amount of chemical species from mass measurements. For such measurements, metric units of kilograms (kg), grams (g), milligrams (mg), or micrograms (μg) are used. Volumes of liquids are measured in units of liters (L), milliliters (mL), and sometimes microliters (μL).

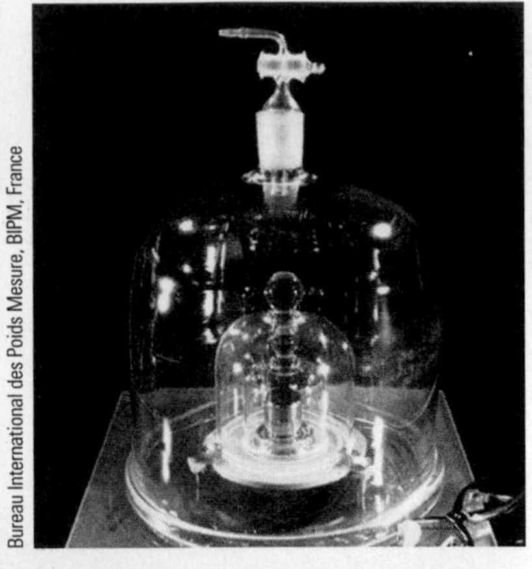

For more than a century, the kilogram has been defined as the mass of a single platinum-iridium standard housed in a laboratory in Sèvres, France. Unfortunately, the standard is quite imprecise relative to other standards such as the meter, which is defined to be the distance that light travels in 1/299792458 of a second. A worldwide consortium of metrologists is working on determining Avogadro's number to 1 part in 100 million, and this number may then be used to define the standard kilogram as 1000/12 of Avogadro's number of carbon atoms. For more on this project, see the chapter opening photo and Problem 4-39.

TABLE 4-1

SI Base Units

Physical Quantity	Name of Unit	Abbreviation
Mass	kilogram	kg
Length	meter	m
Time	second	s
Temperature	kelvin	K
Amount of substance	mole	mol
Electric current	ampere	A
Luminous intensity	candela	cd

The liter, the SI unit of volume, is defined as exactly 10^{-3} m^3. The milliliter is defined as 10^{-6} m^3, or 1 cm^3.

4A-2 The Distinction Between Mass and Weight

It is important to understand the difference between mass and weight. **Mass** is an invariant measure of the amount of matter in an object. **Weight** is the force of attraction between an object and its surroundings, principally the earth. Because gravitational attraction varies with geographical location, the weight of an object depends on where you weigh it. For example, a crucible *weighs* less in Denver than in Atlantic City (both cities are at approximately the same latitude) because the attractive force between the crucible and the earth is smaller at the higher altitude of Denver. Similarly, the crucible *weighs* more in Seattle than in Panama (both cities are at sea level) because the earth is somewhat flattened at the poles, and the force of attraction increases measurably with latitude. The *mass* of the crucible, however, remains constant regardless of where you measure it.

Mass m is an invariant measure of the amount of matter. Weight w is the force of gravitational attraction between that matter and Earth.

TABLE 4-2

Prefixes for Units

Prefix	Abbreviation	Multiplier
yotta-	Y	10^{24}
zetta-	Z	10^{21}
exa-	E	10^{18}
peta-	P	10^{15}
tera-	T	10^{12}
giga-	G	10^{9}
mega-	M	10^{6}
kilo-	k	10^{3}
hecto-	h	10^{2}
deca-	da	10^{1}
deci-	d	10^{-1}
centi-	c	10^{-2}
milli-	m	10^{-3}
micro-	μ	10^{-6}
nano-	n	10^{-9}
pico-	p	10^{-12}
femto-	f	10^{-15}
atto-	a	10^{-18}
zepto-	z	10^{-21}
yocto-	y	10^{-24}

Weight and mass are related by the familiar expression

$$w = mg$$

where w is the weight of an object, m is its mass, and g is the acceleration due to gravity.

A chemical analysis is always based on mass so that the results will not depend on locality. A balance is used to compare the mass of an object with the mass of one or more standard masses. Because g affects both unknown and known equally, the mass of the object is identical to the standard masses with which it is compared.

The distinction between mass and weight is often lost in common usage, and the process of comparing masses is ordinarily called *weighing*. In addition, the objects of known mass as well as the results of weighing are frequently called *weights*. Always bear in mind, however, that analytical data are based on mass rather than weight. Therefore, throughout this text we will use mass rather than weight to describe the amounts of substances or objects. On the other hand, for lack of a better word, we will use "weigh" for the act of determining the mass of an object. Also, we will often say "weights" to mean the standard masses used in weighing.

Courtesy of the National Aeronautics and Space Administration.

Photo of Edwin "Buzz" Aldrin taken by Neil Armstrong in July 1969. Armstrong's reflection may be seen in Aldrin's visor. The suits worn by Armstrong and Aldrin during the Apollo 11 mission to the Moon appear to be massive. But because the mass of the Moon is only 1/81 that of Earth and the acceleration due to gravity is only 1/6 that on Earth, the weight of the suits on the Moon was only 1/6 of their weight on Earth. The mass of the suits, however, was identical in both locations.

4A-3 The Mole

The **mole** (abbreviated mol) is the SI unit for the amount of a chemical species. It is always associated with a chemical formula and represents Avogadro's number (6.022×10^{23}) of particles represented by that formula. The **molar mass** ($\mathcal{M}$) of a substance is the mass in grams of 1 mol of that substance. Molar masses are calculated by summing the atomic masses of all the atoms appearing in a chemical formula. For example, the molar mass of formaldehyde, CH_2O, is

$$\mathcal{M}_{CH_2O} = \frac{1\ \cancel{\text{mol C}}}{\text{mol CH}_2\text{O}} \times \frac{12.0\text{g}}{\cancel{\text{mol C}}} + \frac{2\ \cancel{\text{mol H}}}{\text{mol CH}_2\text{O}} \times \frac{1.0\text{ g}}{\cancel{\text{mol H}}} + \frac{1\ \cancel{\text{mol O}}}{\text{mol CH}_2\text{O}} \times \frac{16.0\text{ g}}{\cancel{\text{mol O}}}$$

$$= 30.0\text{ g/mol CH}_2\text{O}$$

and that of glucose, $C_6H_{12}O_6$, is

$$\mathcal{M}_{C_6H_{12}O_6} = \frac{6\ \cancel{\text{mol C}}}{\text{mol C}_6\text{H}_{12}\text{O}_6} \times \frac{12.0\text{ g}}{\cancel{\text{mol C}}} + \frac{12\ \cancel{\text{mol H}}}{\text{mol C}_6\text{H}_{12}\text{O}_6} \times \frac{1.0\text{ g}}{\cancel{\text{mol H}}} + \frac{6\ \cancel{\text{mol O}}}{\text{mol C}_6\text{H}_{12}\text{O}_6} \times \frac{16.0\text{ g}}{\cancel{\text{mol O}}} = 180.0\text{ g/mol C}_6\text{H}_{12}\text{O}_6$$

Thus, 1 mol of formaldehyde has a mass of 30.0 g and 1 mol of glucose has a mass of 180.0 g.

A **mole** of a chemical species is 6.022×10^{23} atoms, molecules, ions, electrons, ion pairs, or subatomic particles.

FEATURE 4-1

Atomic Mass Units and the Mole

The masses for the elements listed in the table inside the back cover of this text are *relative masses* in terms of *atomic mass units* (amu) or *daltons*. The atomic mass unit is based on a relative scale in which the reference is the ^{12}C carbon isotope, which is *assigned* a mass of exactly 12 amu. Thus, the amu is by definition 1/12 of the mass of one neutral ^{12}C atom. The *molar mass* $\mathcal{M}$ of ^{12}C is then

(continued)

► CHALLENGE: Show that the following interesting and useful relationship is correct: 1 mol of atomic mass units = 6.022×10^{23} amu = 1 g.

defined as the mass in *grams* of 6.022×10^{23} atoms of the carbon-12 isotope, or exactly 12 g. Likewise, the molar mass of any other element is the mass in grams of 6.022×10^{23} atoms of that element and is numerically equal to the atomic mass of the element in amu units. Thus, the atomic mass of naturally occurring oxygen is 15.9994 amu; its molar mass is 15.9994 g.

► The number of moles n_X of a species X of molar mass $\mathcal{M}_X$ is given by

$$n_X = \frac{\text{mass}_X}{\mathcal{M}_X}$$

$$\text{amount X} = n_X = \frac{\text{g X}}{\text{g X/mol X}}$$

$$= \text{g X} \times \frac{\text{mol X}}{\text{g X}}$$

The number of millimoles is given by

$$\text{amount X} = n_X = \frac{\text{g X}}{\text{g X/mmol X}}$$

$$= \text{g X} \times \frac{\text{mmolX}}{\text{g X}}$$

In making calculations of this kind, you should include all units as we do throughout this chapter. This practice often reveals errors in setting up equations.

Charles D. Winters

Approximately one mole of each of several different elements. Clockwise from the upper left we see 64 g of copper beads, 27 g of crumpled aluminum foil, 207 g of lead shot, 24 g of magnesium chips, 52 g of chromium chunks, and 32 g of sulfur powder. The beakers in the photo have a volume of 50 mL.

► 1 mmol = 10^{-3} mol

4A-4 The Millimole

Sometimes it is more convenient to make calculations with millimoles (mmol) rather than moles; the millimole is 1/1000 of a mole. The mass in grams of a millimole, the millimolar mass (m$\mathcal{M}$), is likewise 1/1000 of the molar mass.

4A-5 Calculating the Amount of a Substance in Moles or Millimoles

The two examples that follow illustrate how the number of moles or millimoles of a species can be determined from its mass in grams or from the mass of a chemically related species.

CD-ROM Tutorial: Calculating Amounts in Moles.

EXAMPLE 4-1

How many moles and millimoles of benzoic acid ($\mathcal{M}$ = 122.1 g/mol) are contained in 2.00 g of the pure acid?

If we use HBz to represent benzoic acid, we can write that 1 mol of HBz has a mass of 122.1 g. Thus,

$$\text{amount of HBz} = n_{\text{HBz}} = 2.00\ \text{g}\ \cancel{\text{HBz}} \times \frac{1\ \text{mol HBz}}{122.1\ \text{g}\ \cancel{\text{HBz}}} \qquad (4\text{-}1)$$
$$= 0.0164\ \text{mol HBz}$$

To obtain the number of millimoles, we divide by the millimolar mass (0.1221 g/mmol). That is,

$$\text{amount HBz} = 2.00\ \text{g}\ \cancel{\text{HBz}} \times \frac{1\ \text{mmol HBz}}{0.1221\ \text{g}\ \cancel{\text{HBz}}} = 16.4\ \text{mmol HBz}$$

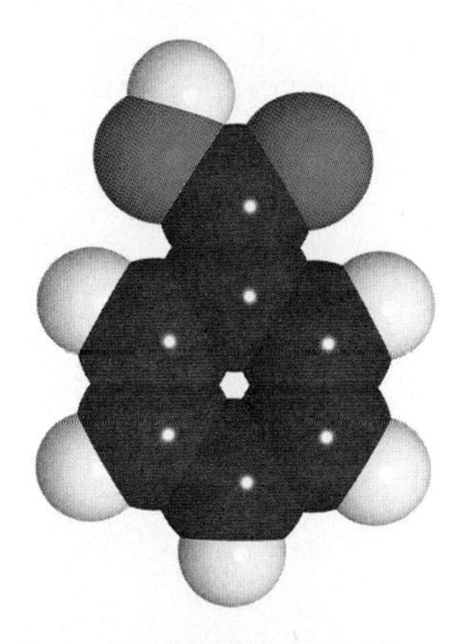

Molecular model of benzoic acid, C_6H_5COOH. Benzoic acid occurs widely in nature, particularly in berries. It finds broad use as a preservative in foods, fats, and fruit juices; as a mordant for dying fabric; and as a standard in calorimetry and in acid/base analysis.

EXAMPLE 4-2

How many grams of Na^+ (22.99 g/mol) are contained in 25.0 g of Na_2SO_4 (142.0 g/mol)?

The chemical formula tells us that 1 mol of Na_2SO_4 contains 2 mol of Na^+. That is,

$$\text{amount Na}^+ = n_{\text{Na}^+} = \text{no. } \cancel{\text{mol Na}_2\text{SO}_4} \times \frac{2\ \text{mol Na}^+}{\cancel{\text{mol Na}_2\text{SO}_4}}$$

To obtain the number of moles of Na_2SO_4 we proceed as in Example 4-1:

$$\text{amount Na}_2\text{SO}_4 = n_{\text{Na}_2\text{SO}_4} = 25.0\ \text{g}\ \cancel{\text{Na}_2\text{SO}_4} \times \frac{1\ \text{mol Na}_2\text{SO}_4}{142.0\ \text{g}\ \cancel{\text{Na}_2\text{SO}_4}}$$

Combining this equation with the first leads to

$$\text{amount Na}^+ = n_{\text{Na}^+} = 25.0\ \text{g}\ \cancel{\text{Na}_2\text{SO}_4} \times \frac{1\ \cancel{\text{mol Na}_2\text{SO}_4}}{142.0\ \text{g}\ \cancel{\text{Na}_2\text{SO}_4}} \times \frac{2\ \text{mol Na}^+}{\cancel{\text{mol Na}_2\text{SO}_4}}$$

To obtain the mass of sodium in 25.0 g of Na_2SO_4, we multiply the number of moles of Na^+ by the molar mass of Na^+, or 22.99 g. That is,

$$\text{mass Na}^+ = \text{no. } \cancel{\text{mol Na}^+} \times \frac{22.99\ \text{g Na}^+}{\cancel{\text{mol Na}^+}}$$

Substituting the previous equation gives the number of grams of Na^+:

$$\text{mass Na}^+ = 25.0\ \text{g}\ \cancel{\text{Na}_2\text{SO}_4} \times \frac{1\ \cancel{\text{mol Na}_2\text{SO}_4}}{142.0\ \text{g}\ \cancel{\text{Na}_2\text{SO}_4}} \times \frac{2\ \cancel{\text{mol Na}^+}}{\cancel{\text{mol Na}_2\text{SO}_4}} \times \frac{22.99\ \text{g Na}^+}{\cancel{\text{mol Na}^+}}$$
$$= 8.10\ \text{g Na}^+$$

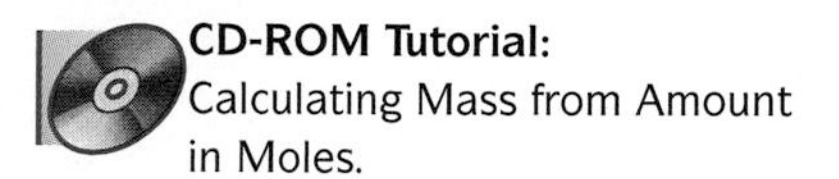
CD-ROM Tutorial:
Calculating Mass from Amount in Moles.

FEATURE 4-2

The Factor-Label Approach to Example 4-2

Some students and instructors find it easier to write out the solution to a problem so that units in the denominator of each succeeding term eliminate the units in the numerator of the preceding one until the units of the answer are obtained. This method has been referred to as the **factor-label method, dimensional analysis,** or the **picket fence method.** For instance, in Example 4-2, the units of the answer are g Na^+ and the units given are g Na_2SO_4. Thus, we can write

$$25.0\ \text{g}\ \cancel{Na_2SO_4} \times \frac{\text{mol}\ Na_2SO_4}{142.0\ \text{g}\ \cancel{Na_2SO_4}}$$

First eliminate mol Na_2SO_4

$$25.0\ \text{g}\ \cancel{Na_2SO_4} \times \frac{\cancel{\text{mol}\ Na_2SO_4}}{142.0\ \text{g}\ \cancel{Na_2SO_4}} \times \frac{2\ \text{mol}\ Na^+}{\cancel{\text{mol}\ Na_2SO_4}}$$

and then eliminate mol Na^+. That is,

$$25.0\ \text{g}\ \cancel{Na_2SO_4} \times \frac{\cancel{\text{mol}\ Na_2SO_4}}{142.0\ \text{g}\ \cancel{Na_2SO_4}} \times \frac{2\ \cancel{\text{mol}\ Na^+}}{\cancel{\text{mol}\ Na_2SO_4}} \times \frac{22.99\ \text{g}\ Na^+}{\cancel{\text{mol}\ Na^+}} = 8.10\ \text{g}\ Na^+$$

4B SOLUTIONS AND THEIR CONCENTRATIONS

4B-1 Concentration of Solutions

Chemists express the concentration of species in solution in several ways. The most important ways are described in this section.

Molar Concentration

The **molar concentration** c_X of a solution of a chemical species X is the number of moles of that species that is contained in 1 L of the solution (*not 1 L of the solvent*). The unit of molar concentration is **molarity,** M, which has the dimensions of mol L^{-1}. Molarity also expresses the number of millimoles of a solute per milliliter of solution.

$$c_X = \frac{\text{no. mol solute}}{\text{no. L solution}} = \frac{\text{no. mmol solute}}{\text{no. mL solution}} \tag{4-2}$$

EXAMPLE 4-3

Calculate the molar concentration of ethanol in an aqueous solution that contains 2.30 g of C_2H_5OH (46.07 g/mol) in 3.50 L of solution.

Because molarity is the number of moles of solute per liter of solution, both of these quantities will be needed. The number of liters is given as 3.50, so all we need to do is convert the number of grams of ethanol to the corresponding number of moles.

$$\text{amount}\ C_2H_5OH = n_{C_2H_5OH} = 2.30\ \text{g}\ \cancel{C_2H_5OH} \times \frac{1\ \text{mol}\ C_2H_5OH}{46.07\ \text{g}\ \cancel{C_2H_5OH}}$$

$$= 0.04992\ \text{mol}\ C_2H_5OH$$

To obtain the molar concentration, $c_{C_2H_5OH}$, we divide by the volume. Thus,

$$c_{C_2H_5OH} = \frac{2.30 \text{ g } \cancel{C_2H_5OH} \times \dfrac{1 \text{ mol } C_2H_5OH}{46.07 \text{ } \cancel{\text{g } C_2H_5OH}}}{3.50 \text{ L}}$$

$$= 0.0143 \text{ mol } C_2H_5OH/L = 0.0143 \text{ M}$$

Analytical molarity is the total number of moles of a solute, regardless of its chemical state, in 1 L of solution. The analytical molarity describes how a solution of a given molarity can be prepared.

Equilibrium molarity is the molar concentration of a particular species in a solution.

Analytical Molarity The **analytical molarity** of a solution gives the *total* number of moles of a solute in 1 L of the solution (or the total number of millimoles in 1 mL). That is, the analytical molarity specifies a recipe by which the solution can be prepared. For example, a sulfuric acid solution that has an analytical concentration of 1.0 M can be prepared by dissolving 1.0 mol, or 98 g, of H_2SO_4 in water and diluting to exactly 1.0 L.

Equilibrium Molarity The **equilibrium molarity** expresses the molar concentration of a particular species in a solution at equilibrium. To state the species molarity, it is necessary to know how the solute behaves when it is dissolved in a solvent. For example, the species molarity of H_2SO_4 in a solution with an analytical concentration of 1.0 M is 0.0 M because the sulfuric acid is entirely dissociated into a mixture of H^+, HSO_4^-, and SO_4^{2-} ions; essentially no H_2SO_4 molecules as such are present in this solution. The equilibrium concentrations and thus the species molarity of these three ions are 1.01, 0.99, and 0.01 M, respectively.

◀ Some chemists prefer to distinguish between species and analytical concentrations in a different way. They use **molar concentration** for species concentration and **formal concentration** (F) for analytical concentration. Applying this convention to our example, we can say that the formal concentration of H_2SO_4 is 1.0 F, whereas its molar concentration is 0.0 M.

Equilibrium molar concentrations are often symbolized by placing square brackets around the chemical formula for the species, so for our solution of H_2SO_4 with an analytical concentration of 1.0 M, we can write

$$[H_2SO_4] = 0.00 \text{ M} \qquad [H^+] = 1.01 \text{ M}$$
$$[HSO_4^-] = 0.99 \text{ M} \qquad [SO_4^{2-}] = 0.01 \text{ M}$$

◀ In this example the *analytical molarity* of H_2SO_4 is given by $c_{H_2SO_4} = [SO_4^{2-}] + [HSO_4^-]$ because these are the only two sulfate-containing species in the solution.

EXAMPLE 4-4

Calculate the analytical and equilibrium molar concentrations of the solute species in an aqueous solution that contains 285 mg of trichloroacetic acid, Cl_3CCOOH (163.4 g/mol), in 10.0 mL (the acid is 73% ionized in water).

As in Example 4-3, we calculate the number of moles of Cl_3CCOOH, which we designate as HA, and divide by the volume of the solution, 10.0 mL, or 0.01000 L. Thus,

$$\text{amount HA} = n_{HA} = 285 \text{ } \cancel{\text{mg HA}} \times \frac{1 \text{ } \cancel{\text{g HA}}}{1000 \text{ } \cancel{\text{mg HA}}} \times \frac{1 \text{ mol HA}}{163.4 \text{ } \cancel{\text{g HA}}}$$
$$= 1.744 \times 10^{-3} \text{ mol HA}$$

The analytical molar concentration, c_{HA}, is then

$$c_{HA} = \frac{1.744 \times 10^{-3} \text{ mol HA}}{10.0 \text{ } \cancel{\text{mL}}} \times \frac{1000 \text{ } \cancel{\text{mL}}}{1 \text{ L}} = 0.174 \frac{\text{mol HA}}{\text{L}} = 0.174 \text{ M}$$

(continued)

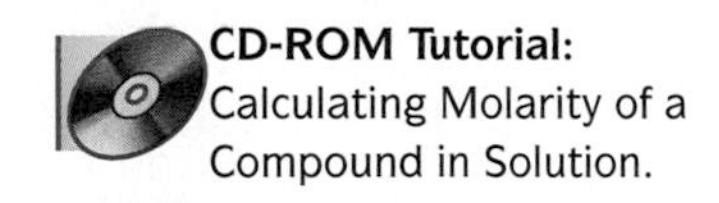
CD-ROM Tutorial: Calculating Molarity of a Compound in Solution.

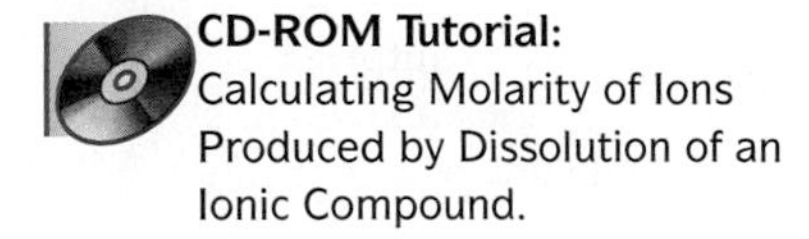
CD-ROM Tutorial: Calculating Molarity of Ions Produced by Dissolution of an Ionic Compound.

Molecular model of trichloroacetic acid, Cl_3CCOOH. The rather strong acidity of trichloroacetic acid is usually ascribed to the inductive effect of the three chlorine atoms attached to the end of the molecule opposite the acidic proton. Electron density is withdrawn away from the carboxylate group so that the trichloroacetate anion that is formed when the acid dissociates is stabilized. The acid is used in protein precipitation and in dermatological preparations for the removal of undesirable skin growths.

In this solution, 73% of the HA dissociates, giving H^+ and A^-:

$$HA \rightleftharpoons H^+ + A^-$$

The species molarity of HA is then 27% of c_{HA}. Thus,

$$[HA] = c_{HA} \times (100 - 73)/100 = 0.174 \times 0.27 = 0.174 \text{ mol/L}$$
$$= 0.047 \text{ M}$$

The species molarity of A^- is equal to 73% of the analytical concentration of HA. That is,

$$[A] = \frac{73 \text{ mol } A^-}{100 \cancel{\text{mol HA}}} \times 0.174 \frac{\cancel{\text{mol HA}}}{\text{L}} = 0.127 \text{ M}$$

Because 1 mole of H^+ is formed for each mole of A^-, we can also write

$$[H^+] = [A^-] = 0.127 \text{ M}$$

EXAMPLE 4-5

Describe the preparation of 2.00 L of 0.108 M $BaCl_2$ from $BaCl_2 \cdot 2H_2O$ (244.3 g/mol).

To determine the number of grams of solute to be dissolved and diluted to 2.00 L, we note that 1 mol of the dihydrate yields 1 mol of $BaCl_2$. Therefore, to produce this solution we will need

$$2.00 \cancel{\text{L}} \times \frac{0.108 \text{ mol } BaCl_2 \cdot 2H_2O}{\cancel{\text{L}}} = 0.216 \text{ mol } BaCl_2 \cdot 2H_2O$$

► The number of moles of the species A in a solution of A is given by

$$\text{no. mol A} = n_A = c_A \times V_A = \frac{\text{mol}}{\text{L}} \times \text{L}$$

where V_A is the volume of the solution in liters.

The mass of $BaCl_2 \cdot 2H_2O$ is then

$$0.216 \cancel{\text{mol } BaCl_2 \cdot 2H_2O} \times \frac{244.3 \text{ g } BaCl_2 \cdot 2H_2O}{\cancel{\text{mol } BaCl_2 \cdot 2H_2O}} = 52.8 \text{ g } BaCl_2 \cdot 2H_2O$$

Dissolve 52.8 g of $BaCl_2 \cdot 2H_2O$ in water and dilute to 2.00 L.

EXAMPLE 4-6

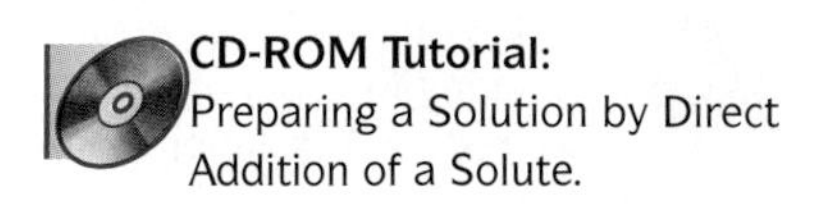

CD-ROM Tutorial:
Preparing a Solution by Direct Addition of a Solute.

Describe the preparation of 500 mL of 0.0740 M Cl^- solution from solid $BaCl_2 \cdot 2H_2O$ (244.3 g/mol).

$$\text{mass } BaCl_2 \cdot 2H_2O = \frac{0.0740 \cancel{\text{mol } Cl^-}}{\cancel{\text{L}}} \times 0.500 \cancel{\text{L}} \times \frac{1 \cancel{\text{mol } BaCl_2 \cdot 2H_2O}}{2 \cancel{\text{mol } Cl^-}}$$
$$\times \frac{244.3 \text{ g } BaCl_2 \cdot 2H_2O}{\cancel{\text{mol } BaCl_2 \cdot 2H_2O}} = 4.52 \text{ g } BaCl_2 \cdot 2H_2O$$

Dissolve 4.52 g of $BaCl_2 \cdot 2H_2O$ in water and dilute to 0.500 L or 500 mL.

Percent Concentration

Chemists frequently express concentrations in terms of percent (parts per hundred). Unfortunately, this practice can be a source of ambiguity because percent composition of a solution can be expressed in several ways. Three common methods are

$$\text{weight percent (w/w)} = \frac{\text{weight solute}}{\text{weight solution}} \times 100\%$$

$$\text{volume percent (v/v)} = \frac{\text{volume solute}}{\text{volume solution}} \times 100\%$$

$$\text{weight/volume percent (w/v)} = \frac{\text{weight solute, g}}{\text{volume solution, mL}} \times 100\%$$

◀ Weight percent would be more properly called mass percent and abbreviated m/m. The term "weight percent" is so widely used in the chemical literature, however, that we will use it throughout this text.

Note that the denominator in each of these expressions refers to the *solution* rather than to the solvent. Note also that the first two expressions do not depend on the units employed (provided, of course, that there is consistency between numerator and denominator). In the third expression, units must be defined because the numerator and denominator have different units that do not cancel. Of the three expressions, only weight percent has the virtue of being temperature independent.

Weight percent is frequently employed to express the concentration of commercial aqueous reagents. For example, nitric acid is sold as a 70% solution, which means that the reagent contains 70 g of HNO_3 per 100 g of solution (see Example 4-10).

Volume percent is commonly used to specify the concentration of a solution prepared by diluting a pure liquid compound with another liquid. For example, a 5% aqueous solution of methanol *usually* describes a solution prepared by diluting 5.0 mL of pure methanol with enough water to give 100 mL.

Weight/volume percent is often employed to indicate the composition of dilute aqueous solutions of solid reagents. For example, 5% aqueous silver nitrate *often* refers to a solution prepared by dissolving 5 g of silver nitrate in sufficient water to give 100 mL of solution.

To avoid uncertainty, always specify explicitly the type of percent composition being discussed. If this information is missing, the user must decide intuitively which of the several types is involved. The potential error resulting from a wrong choice is considerable. For example, commercial 50% (w/w) sodium hydroxide contains 763 g of the reagent per liter, which corresponds to 76.3% (w/v) sodium hydroxide.

◀ You should always specify the type of percent when reporting concentrations in this way.

Parts per Million and Parts per Billion

For very dilute solutions, parts per million (ppm) is a convenient way to express concentration:

$$c_{ppm} = \frac{\text{mass of solute}}{\text{mass of solution}} \times 10^6 \text{ ppm}$$

where c_{ppm} is the concentration in parts per million. Obviously, the units of mass in the numerator and denominator must agree. For even more dilute solutions, 10^9 ppb rather than 10^6 ppm is employed in the foregoing equation to give the results in parts per billion (ppb). The term *parts per thousand* (ppt) is also encountered, especially in oceanography.

◀ A handy rule in calculating parts per million is to remember that for dilute aqueous solutions whose densities are approximately 1.00 g/mL, 1 ppm = 1.00 mg/L. That is,

$$c_{ppm} = \frac{\text{mass solute (mg)}}{\text{volume solution (L)}} \tag{4-3}$$

$$c_{ppb} = \frac{\text{mass solute (g)}}{\text{mass solution (g)}} \times 10^9 \text{ ppb}$$

$$= 1.00\ \mu\text{g/L}$$

EXAMPLE 4-7

What is the molarity of K^+ in a solution that contains 63.3 ppm of $K_3Fe(CN)_6$ (329.3 g/mol)?

Because the solution is so dilute, it is reasonable to assume that its density is 1.00 g/mL. Therefore, according to Equation 4-2,

$$63.3 \text{ ppm } K_3Fe(CN)_6 = 63.3 \text{ mg } K_3Fe(CN)_6/L$$

$$\frac{\text{no. mol } K_3Fe(CN)_6}{L} = \frac{63.3 \text{ mg } K_3Fe(CN)_6}{L} \times \frac{1 \text{ g } K_3Fe(CN)_6}{1000 \text{ mg } K_3Fe(CN)_6} \times \frac{1 \text{ mol } K_3Fe(CN)_6}{329.3 \text{ g } K_3Fe(CN)_6}$$

$$= 1.922 \times 10^{-4} \frac{\text{mol}}{L} = 1.922 \times 10^{-4} \text{ M}$$

$$[K^+] = \frac{1.922 \times 10^{-4} \text{ mol } K_3Fe(CN)_6}{L} \times \frac{3 \text{ mol } K^+}{1 \text{ mol } K_3Fe(CN)_6}$$

$$= 5.77 \times 10^{-4} \frac{\text{mol } K^+}{L} = 5.77 \times 10^{-4} \text{ M}$$

Solution-Diluent Volume Ratios

The composition of a dilute solution is sometimes specified in terms of the volume of a more concentrated solution and the volume of solvent used in diluting it. The volume of the former is separated from that of the latter by a colon. Thus, a 1:4 HCl solution contains four volumes of water for each volume of concentrated hydrochloric acid. This method of notation is frequently ambiguous in that the concentration of the original solution is not always obvious to the reader. Moreover, under some circumstances 1:4 means dilute one volume with three volumes. Because of such uncertainties, you should avoid using solution-diluent ratios.

p-Functions

► The most well known p-function is pH, which is the negative logarithm of $[H^+]$.

Scientists frequently express the concentration of a species in terms of its **p-function,** or **p-value.** The p-value is the negative logarithm (to the base 10) of the molar concentration of that species. Thus, for the species X,

$$pX = -\log [X]$$

As shown by the following examples, p-values offer the advantage of allowing concentrations that vary over ten or more orders of magnitude to be expressed in terms of small positive numbers.

EXAMPLE 4-8

Calculate the p-value for each ion in a solution that is 2.00×10^{-3} M in NaCl and 5.4×10^{-4} M in HCl.

$$pH = -\log [H^+] = -\log (5.4 \times 10^{-4}) = 3.27$$

To obtain pNa, we write

$$pNa = -\log (2.00 \times 10^{-3}) = -\log 2.00 \times 10^{-3} = 2.699$$

The total Cl^- concentration is given by the sum of the concentrations of the two solutes:

$$[Cl^-] = 2.00 \times 10^{-3}\,M + 5.4 \times 10^{-4}\,M$$
$$= 2.00 \times 10^{-3}\,M + 0.54 \times 10^{-3}\,M = 2.54 \times 10^{-3}\,M$$

$$pCl = -\log 2.54 \times 10^{-3} = 2.595$$

Molecular model of HCl. Hydrogen chloride is a gas consisting of heteronuclear diatomic molecules. The gas is extremely soluble in water; when a solution of the gas is prepared, only then do the molecules dissociate to form aqueous hydrochloric acid, which consists of H_3O^+ and Cl^- ions.

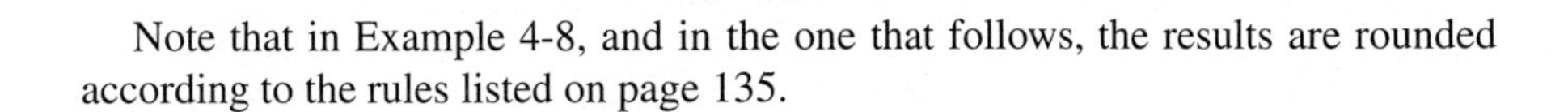

Note that in Example 4-8, and in the one that follows, the results are rounded according to the rules listed on page 135.

EXAMPLE 4-9

Calculate the molar concentration of Ag^+ in a solution that has a pAg of 6.372.

$$pAg = -\log [Ag^+] = 6.372$$
$$\log [Ag^+] = -6.372$$
$$[Ag^+] = 4.246 \times 10^{-7} \approx 4.25 \times 10^{-7}$$

4B-2 Density and Specific Gravity of Solutions

Density and specific gravity are terms often encountered in the analytical literature. The **density** of a substance is its mass per unit volume, whereas its **specific gravity** is the ratio of its mass to the mass of an equal volume of water at 4°C. Density has units of kilograms per liter or grams per milliliter in the metric system. Specific gravity is dimensionless and so is not tied to any particular system of units. For this reason, specific gravity is widely used in describing items of commerce (see Figure 4-1). Since the density of water is approximately 1.00 g/mL and since we employ the metric system throughout this text, density and specific gravity are used interchangeably. The specific gravities of some concentrated acids and bases are given in Table 4-3.

Density expresses the mass of a substance per unit volume. In SI units, density is expressed in units of kg/L or, alternatively, g/mL.

Specific gravity is the ratio of the mass of a substance to the mass of an equal volume of water.

EXAMPLE 4-10

Calculate the molar concentration of HNO_3 (63.0 g/mol) in a solution that has a specific gravity of 1.42 and is 70.5% HNO_3 (w/w).

Let us first calculate the grams of acid per liter of concentrated solution

$$\frac{\text{g } HNO_3}{\text{L reagent}} = \frac{1.42 \cancel{\text{kg reagent}}}{\text{L reagent}} \times \frac{10^3 \cancel{\text{g reagent}}}{\cancel{\text{kg reagent}}} \times \frac{70.5 \text{ g } HNO_3}{100 \cancel{\text{g reagent}}} = \frac{1001 \text{ g } HNO_3}{\text{L reagent}}$$

(continued)

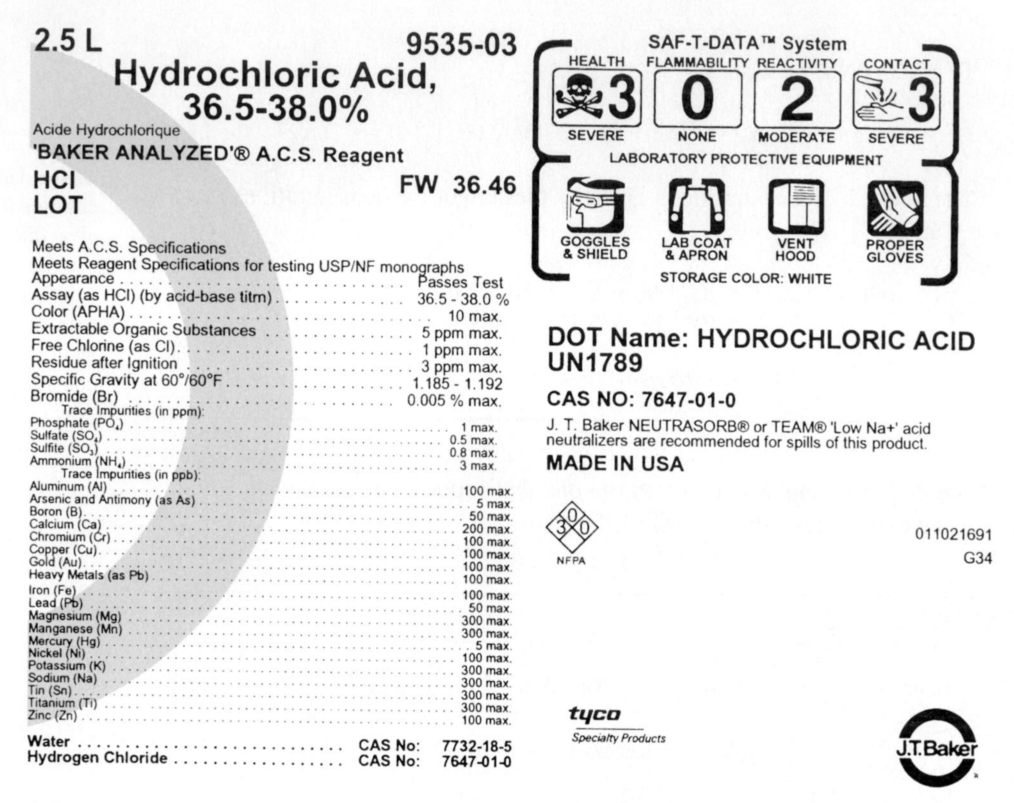

Figure 4-1 Label from a bottle of reagent-grade hydrochloric acid. Note that the specific gravity of the acid over the temperature range of 60° to 80°F is specified on the label. *(Label provided by Mallinckrodt Baker, Inc., Phillipsburg, NJ 08865)*

Then

$$c_{HNO_3} = \frac{1001 \text{ g HNO}_3}{\text{L reagent}} \times \frac{1 \text{ mol HNO}_3}{63.0 \text{ g HNO}_3} = \frac{15.9 \text{ mol HNO}_3}{\text{L reagent}} \approx 16 \text{ M}$$

TABLE 4-3

Specific Gravities of Commercial Concentrated Acids and Bases

Reagent	Concentration, % (w/w)	Specific Gravity
Acetic acid	99.7	1.05
Ammonia	29.0	0.90
Hydrochloric acid	37.2	1.19
Hydrofluoric acid	49.5	1.15
Nitric acid	70.5	1.42
Perchloric acid	71.0	1.67
Phosphoric acid	86.0	1.71
Sulfuric acid	96.5	1.84

EXAMPLE 4-11

Describe the preparation of 100 mL of 6.0 M HCl from a concentrated solution that has a specific gravity of 1.18 and is 37% (w/w) HCl (36.5 g/mol).

Proceeding as in Example 4-10, we first calculate the molarity of the concentrated reagent. We then calculate the number of moles of acid that we need for the diluted solution. Finally, we divide the second figure by the first to obtain the volume of concentrated acid required. Thus, to obtain the molarity of the concentrated reagent, we write

$$c_{HCl} = \frac{1.18 \times 10^3 \cancel{\text{g reagent}}}{\text{L reagent}} \times \frac{37 \cancel{\text{g HCl}}}{100 \cancel{\text{g reagent}}} \times \frac{1 \text{ mol HCl}}{36.5 \cancel{\text{g HCl}}} = 12.0 \text{ M}$$

The number of moles HCl required is given by

$$\text{no. mol HCl} = 100 \cancel{\text{mL}} \times \frac{1 \cancel{\text{L}}}{1000 \cancel{\text{mL}}} \times \frac{6.0 \text{ mol HCl}}{\cancel{\text{L}}} = 0.600 \text{ mol HCl}$$

Finally, to obtain the volume of concentrated reagent, we write

$$\text{vol concd reagent} = 0.600 \cancel{\text{mol HCl}} \times \frac{1 \text{ L reagent}}{12.0 \cancel{\text{mol HCl}}}$$
$$= 0.0500 \text{ L or } 50.0 \text{ mL}$$

Thus dilute 50 mL of the concentrated reagent to 600 mL.

The solution to Example 4-11 is based on the following useful relationship, which we will be using countless times:

$$V_{concd} \times c_{concd} = V_{dil} \times c_{dil} \qquad (4\text{-}4)$$

where the two terms on the left are the volume and molar concentration of a concentrated solution that is being used to prepare a diluted solution having the volume and concentration given by the corresponding terms on the right. This equation is based on the fact that the number of moles of solute in the diluted solution must equal the number of moles in the concentrated reagent. Note that the volumes can be in milliliters or liters as long as the same units are used for both solutions.

► Equation 4-4 can be used with L and mol/L or mL and mmol/mL. Thus,

$$\text{L}_{concd} \times \frac{\text{mol}_{concd}}{\text{L}_{concd}} = \text{L}_{dil} \times \frac{\text{mol}_{dil}}{\text{L}_{dil}}$$

$$\text{mL}_{concd} \times \frac{\text{mmol}_{concd}}{\text{mL}_{concd}} = \text{mL}_{dil} \times \frac{\text{mmol}_{dil}}{\text{mL}_{dil}}$$

CD-ROM Tutorial: Preparing a Solution by Dilution of a Concentrated Stock Solution of Known Concentration.

4C CHEMICAL STOICHIOMETRY

Stoichiometry is defined as the quantitative relationship among reacting chemical species. This section provides a brief review of stoichiometry and its applications to chemical calculations.

The **stoichiometry** of a reaction is the relationship among the number of moles of reactants and products as shown by a balanced equation.

4C-1 Empirical Formulas and Molecular Formulas

An **empirical formula** gives the simplest whole number ratio of atoms in a chemical compound. In contrast, a **molecular formula** specifies the number of atoms in a molecule. Two or more substances may have the same empirical formula but different

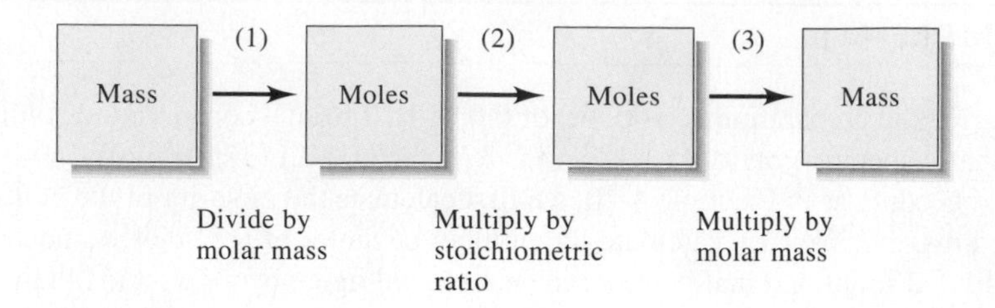

Figure 4-2 Flow diagram for making stoichiometric calculations. (1) When the mass of a reactant or product is given, the mass is first converted to the number of moles, using the molar mass. (2) The stoichiometric ratio given by the chemical equation for the reaction is then used to find the number of moles of another reactant that combine with the original substance or the number of moles of product that form. (3) Finally, the mass of the other reactant or the product is computed from its molar mass.

molecular formulas. For example, CH_2O is both the empirical and the molecular formula for formaldehyde; it is also the empirical formula for such diverse substances as acetic acid, $C_2H_4O_2$, glyceraldehyde, $C_3H_6O_3$, and glucose, $C_6H_{12}O_6$, as well as more than 50 other substances containing 6 or fewer carbon atoms. The empirical formula is obtained from the percent composition of a compound. The molecular formula requires, in addition, a knowledge of the molar mass of the species.

A **structural formula** provides additional information. For example, the chemically different ethanol and dimethyl ether share the same molecular formula C_2H_6O. Their structural formulas, C_2H_5OH and CH_3OCH_3, reveal structural differences between these compounds that are not shown in their common molecular formula.

4C-2 Stoichiometric Calculations

A balanced chemical equation gives the combining ratios, or stoichiometry—in units of moles—of reacting substances and their products. Thus, the equation

► Often the physical state of substances appearing in equations is indicated by the letters *(g)*, *(l)*, *(s)*, and *(aq)*, which refer to gaseous, liquid, solid, and aqueous solution states, respectively.

$$2NaI(aq) + Pb(NO_3)_2(aq) \rightarrow PbI_2(s) + 2NaNO_3(aq)$$

indicates that 2 mol of aqueous sodium iodide combine with 1 mol of aqueous lead nitrate to produce 1 mol of solid lead iodide and 2 mol of aqueous sodium nitrate.[1]

Example 4-12 demonstrates how the weight in grams of reactants and products in a chemical reaction are related. As shown in Figure 4-2, a calculation of this type is a three-step process involving (1) transformation of the known mass of a substance in grams to a corresponding number of moles, (2) multiplication by a factor that accounts for the stoichiometry, and (3) reconversion of the data in moles back to the metric units called for in the answer.

EXAMPLE 4-12

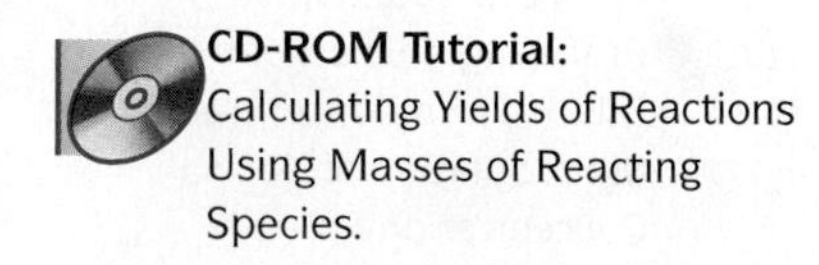

CD-ROM Tutorial: Calculating Yields of Reactions Using Masses of Reacting Species.

(a) What mass of $AgNO_3$ (169.9 g/mol) is needed to convert 2.33 g of Na_2CO_3 (106.0 g/mol) to Ag_2CO_3? (b) What mass of Ag_2CO_3 (275.7 g/mol) will be formed?

(a) $Na_2CO_3(aq) + 2AgNO_3(aq) \rightarrow Ag_2CO_3(s) + 2NaNO_3(aq)$

Step 1. no. mol $Na_2CO_3 = n_{Na_2CO_3} = 2.33 \text{ g } \cancel{Na_2CO_3} \times \dfrac{1 \text{ mol } Na_2CO_3}{106.0 \text{ g } \cancel{Na_2CO_3}}$

$= 0.02198 \text{ mol } Na_2CO_3$

[1]Here it is advantageous to depict the reaction in terms of chemical compounds. If we wish to focus on reacting species, the net ionic equation is preferable:

$$2I^-(aq) + Pb^{2+}(aq) \rightarrow PbI_2(s)$$

Step 2. The balanced equation reveals that

$$\text{no. mol AgNO}_3 = n_{\text{AgNO}_3} = 0.02198 \text{ mol } \cancel{\text{Na}_2\text{CO}_3} \times \frac{2 \text{ mol AgNO}_3}{1 \text{ mol } \cancel{\text{Na}_2\text{CO}_3}}$$
$$= 0.04396 \text{ mol AgNO}_3$$

Here the stoichiometric factor is (2 mol $AgNO_3$)/(1 mol Na_2CO_3).

Step 3. $$\text{mass AgNO}_3 = 0.04396 \cancel{\text{ mol AgNO}_3} \times \frac{169.9 \text{ g AgNO}_3}{\cancel{\text{mol AgNO}_3}}$$
$$= 7.47 \text{ g AgNO}_3$$

(b) no. mol Ag_2CO_3 = no. mol Na_2CO_3 = 0.02198 mol

$$\text{mass Ag}_2\text{CO}_3 = 0.02198 \cancel{\text{ mol Ag}_2\text{CO}_3} \times \frac{275.7 \text{ g Ag}_2\text{CO}_3}{\cancel{\text{mol Ag}_2\text{CO}_3}}$$
$$= 6.06 \text{ g Ag}_2\text{CO}_3$$

EXAMPLE 4-13

What mass of Ag_2CO_3 (275.7 g/mol) is formed when 25.0 mL of 0.200 M $AgNO_3$ are mixed with 50.0 mL of 0.0800 M Na_2CO_3?

Mixing these two solutions will result in one (and only one) of three possible outcomes, specifically:

(a) An excess of $AgNO_3$ will remain after reaction is complete.
(b) An excess of Na_2CO_3 will remain after reaction is complete.
(c) An excess of neither reagent will exist (that is, the number of moles of Na_2CO_3 is exactly equal to twice the number of moles of $AgNO_3$).

As a first step, we must establish which of these situations applies by calculating the amounts of reactants (in chemical units) available at the outset.

Initial amounts are

$$\text{amount AgNO}_3 = n_{\text{AgNO}_3} = 25.0 \cancel{\text{ mL AgNO}_3} \times \frac{1 \cancel{\text{ L AgNO}_3}}{1000 \cancel{\text{ mL AgNO}_3}}$$
$$\times \frac{0.200 \text{ mol AgNO}_3}{\cancel{\text{L AgNO}_3}} = 5.00 \times 10^{-3} \text{ mol AgNO}_3$$

$$\text{no. mol Na}_2\text{CO}_3 = n_{\text{Na}_2\text{CO}_3} = 50.0 \cancel{\text{ mL Na}_2\text{CO}_3} \times \frac{1 \cancel{\text{ L Na}_2\text{CO}_3}}{1000 \cancel{\text{ mL Na}_2\text{CO}_3}}$$
$$\times \frac{0.0800 \text{ mol Na}_2\text{CO}_3}{\cancel{\text{L Na}_2\text{CO}_3}} = 4.00 \times 10^{-3} \text{ mol Na}_2\text{CO}_3$$

Because each CO_3^{2-} ion reacts with two Ag^+ ions, $2 \times 4.00 \times 10^{-3} = 8.00 \times 10^{-3}$ mol $AgNO_3$ is required to react with the Na_2CO_3. Since we have insufficient $AgNO_3$, situation (b) prevails and the amount of Ag_2CO_3 produced will be limited by the amount of $AgNO_3$ available. Thus,

$$\text{mass Ag}_2\text{CO}_3 = 5.00 \times 10^{-3} \cancel{\text{ mol AgNO}_3} \times \frac{1 \cancel{\text{ mol Ag}_2\text{CO}_3}}{2 \cancel{\text{ mol AgNO}_3}} \times \frac{275.7 \text{ g Ag}_2\text{CO}_3}{\cancel{\text{mol Ag}_2\text{CO}_3}}$$
$$= 0.689 \text{ g Ag}_2\text{CO}_3$$

CD-ROM Simulation: The Control of Mass of Product in a Chemical Reaction by the Amounts of Reacting Species Present.

EXAMPLE 4-14

What will be the analytical molar Na_2CO_3 concentration in the solution produced when 25.0 mL of 0.200 M $AgNO_3$ are mixed with 50.0 mL of 0.0800 M Na_2CO_3?

We have seen in the previous example that formation of 5.00×10^{-3} mol of $AgNO_3$ will require 2.50×10^{-3} mol of Na_2CO_3. The number of moles of unreacted Na_2CO_3 is then given by

$$n_{Na_2CO_3} = 4.00 \times 10^{-3} \text{ mol } Na_2CO_3 - 5.00 \times 10^{-3} \cancel{\text{mol } AgNO_3} \times \frac{1 \text{ mol } Na_2CO_3}{2 \cancel{\text{mol } AgNO_3}}$$
$$= 1.50 \times 10^{-3} \text{ mol } Na_2CO_3$$

By definition the molarity is the number of moles of Na_2CO_3/L. Thus,

$$c_{Na_2CO_3} = \frac{1.50 \times 10^{-3} \text{ mol } Na_2CO_3}{(50.0 + 25.0) \cancel{\text{mL}}} \times \frac{1000 \cancel{\text{mL}}}{1 \text{ L}} = 0.0200 \text{ M } Na_2CO_3$$

In this chapter, we have reviewed many of the basic chemical concepts and skills necessary for effective study of analytical chemistry. In the remaining chapters of this book, you will build on this firm foundation as you explore methods of chemical analysis.

WEB WORKS

This chapter opened with a photo of a nearly perfect silicon sphere that is being used to determine Avogadro's number. Use your Web browser to connect to **http://chemistry.brookscole.com/skoogfac/.** From the Chapter Resources Menu, choose Web Works. Locate the Chapter 4 section and click on the link to the Australian National Measurement Laboratory. Read the article on Avogadro's number and the silicon kilogram. What factors limit accuracy in the determination of this number? What are the present and ultimate uncertainties in the measurement of the molar mass of silicon, the number of atoms per unit cell, the mass, the volume, and the lattice parameter of silicon?

QUESTIONS AND PROBLEMS

4-1. Define
*(a) millimole.
(b) molar mass.
*(c) millimolar mass.
(d) parts per million.

4-2. What is the difference between species molarity and analytical molarity?

***4-3.** Give two examples of units derived from the fundamental base SI units.

4-4. Simplify the following quantities using a unit with an appropriate prefix:
*(a) 3.2×10^5 Hz.
(b) 4.56×10^{-8} g.
*(c) 8.43×10^5 μmol.
(d) 6.5×10^6 s.
*(e) 8.96×10^4 nm.
(f) 72,000 g.

*4-5. How many Na^+ ions are contained in 5.43 g of Na_3PO_4?

4-6. How many K^+ ions are contained in 6.76 mol of K_3PO_4?

*4-7. Find the number of moles of the indicated species in
(a) 4.96 g of B_2O_3.
(b) 333 mg of $Na_2B_4O_7 \cdot 10H_2O$.
(c) 8.75 g of Mn_3O_4.
(d) 167.2 mg of CaC_2O_4.

4-8. Find the number of millimoles of the indicated species in
(a) 57 mg of P_2O_5.
(b) 12.92 g of CO_2.
(c) 40.0 g of $NaHCO_3$.
(d) 850 mg of $MgNH_4PO_4$.

*4-9. Find the number of millimoles of solute in
(a) 2.00 L of 3.25×10^{-3} M $KMnO_4$.
(b) 750 mL of 0.0555 M KSCN.
(c) 250 mL of a solution that contains 5.41 ppm of $CuSO_4$.
(d) 3.50 L of 0.333 M KCl.

4-10. Find the number of millimoles of solute in
(a) 175 mL of 0.320 M $HClO_4$.
(b) 15.0 L of 8.05×10^{-3} M K_2CrO_4.
(c) 5.00 L of an aqueous solution that contains 6.75 ppm of $AgNO_3$.
(d) 851 mL of 0.0200 M KOH.

*4-11. What is the mass in milligrams of
(a) 0.777 mol of HNO_3?
(b) 500 mmol of MgO?
(c) 22.5 mol of NH_4NO_3?
(d) 4.32 mol of $(NH_4)_2Ce(NO_3)_6$ (548.23 g/mol)?

4-12. What is the mass in grams of
(a) 7.1 mol of KBr?
(b) 20.1 mmol of PbO?
(c) 3.76 mol of $MgSO_4$?
(d) 9.6 mmol of $Fe(NH_4)_2(SO_4)_2 \cdot 6H_2O$?

4-13. What is the mass in milligrams of solute in
*(a) 26.0 mL of 0.250 M sucrose (342 g/mol)?
*(b) 2.92 L of 4.76×10^{-3} M H_2O_2?
(c) 656 mL of a solution that contains 4.96 ppm of $Pb(NO_3)_2$?
(d) 6.75 mL of 0.0619 M KNO_3?

4-14. What is the mass in grams of solute in
*(a) 450 mL of 0.164 M H_2O_2?
*(b) 27.0 mL of 8.75×10^{-4} M benzoic acid (122 g/mol)?
(c) 3.50 L of a solution that contains 21.7 ppm of $SnCl_2$?
(d) 21.7 mL of 0.0125 M $KBrO_3$?

4-15. Calculate the p-value for each of the indicated ions in the following:
*(a) Na^+, Cl^-, and OH^- in a solution that is 0.0335 M in NaCl and 0.0503 M in NaOH.
(b) Ba^{2+}, Mn^{2+}, and Cl^- in a solution that is 7.65×10^{-3} M in $BaCl_2$ and 1.54 M in $MnCl_2$.
*(c) H^+, Cl^-, and Zn^{2+} in a solution that is 0.600 M in HCl and 0.101 M in $ZnCl_2$.
(d) Cu^{2+}, Zn^{2+}, and NO_3^- in a solution that is 4.78×10^{-2} M in $Cu(NO_3)_2$ and 0.104 M in $Zn(NO_3)_2$.
*(e) K^+, OH^-, and $Fe(CN)_6^{4-}$ in a solution that is 2.62×10^{-7} M in $K_4Fe(CN)_6$ and 4.12×10^{-7} M in KOH.
(f) H^+, Ba^{2+}, and ClO_4^- in a solution that is 3.35×10^{-4} M in $Ba(ClO_4)_2$ and 6.75×10^{-4} M in $HClO_4$.

4-16. Calculate the molar H_3O^+ ion concentration of a solution that has a pH of
*(a) 4.76. *(c) 0.52. *(e) 7.32. *(g) −0.31.
(b) 4.58. (d) 13.62. (f) 5.76. (h) −0.52.

4-17. Calculate the p-functions for each ion in a solution that is
*(a) 0.0200 M in NaBr.
(b) 0.0100 M in $BaBr_2$.
*(c) 3.5×10^{-3} M in $Ba(OH)_2$.
(d) 0.040 M in HCl and 0.020 M in NaCl.
*(e) 6.7×10^{-3} M in $CaCl_2$ and 7.6×10^{-3} M in $BaCl_2$.
(f) 4.8×10^{-8} M in $Zn(NO_3)_2$ and 5.6×10^{-7} M $Cd(NO_3)_2$.

4-18. Convert the following p-functions to molar concentrations:
*(a) pH = 9.67. *(e) pLi = −0.221.
(b) pOH = 0.135. (f) pNO_3 = 7.77.
*(c) pBr = 0.034. *(g) pMn = 0.0025.
(d) pCa = 12.35. (h) pCl = 1.020.

*4-19. Sea water contains an average of 1.08×10^3 ppm of Na^+ and 270 ppm of SO_4^{2-}. Calculate
(a) the molar concentrations of Na^+ and SO_4^{2-} given that the average density of sea water is 1.02 g/mL.
(b) the pNa and pSO_4 for sea water.

4-20. Average human blood serum contains 18 mg of K^+ and 365 mg of Cl^- per 100 mL. Calculate
(a) the molar concentration for each of these species; use 1.00 g/mL for the density of serum.
(b) pK and pCl for human serum.

*4-21. A solution was prepared by dissolving 5.76 g of $KCl \cdot MgCl_2 \cdot 6H_2O$ (277.85 g/mol) in sufficient water to give 2.000 L. Calculate

(a) the molar analytical concentration of $KCl \cdot MgCl_2$ in this solution.
(b) the molar concentration of Mg^{2+}.
(c) the molar concentration of Cl^-.
(d) the weight/volume percentage of $KCl \cdot MgCl_2 \cdot 6H_2O$.
(e) the number of millimoles of Cl^- in 25.0 mL of this solution.
(f) ppm K^+.
(g) pMg for the solution.
(h) pCl for the solution.

4-22. A solution was prepared by dissolving 1210 mg of $K_3Fe(CN)_6$ (329.2 g/mol) in sufficient water to give 775 mL. Calculate
(a) the molar analytical concentration of $K_3Fe(CN)_6$.
(b) the molar concentration of K^+.
(c) the molar concentration of $Fe(CN)_6^{3-}$.
(d) the weight/volume percentage of $K_3Fe(CN)_6$.
(e) the number of millimoles of K^+ in 50.0 mL of this solution.
(f) ppm $Fe(CN)_6^{3-}$.
(g) pK for the solution.
(h) $pFe(CN)_6$ for the solution.

***4-23.** A 6.42% (w/w) $Fe(NO_3)_3$ (241.86 g/mol) solution has a density of 1.059 g/mL. Calculate
(a) the molar analytical concentration of $Fe(NO_3)_3$ in this solution.
(b) the molar NO_3^- concentration in the solution.
(c) the mass in grams of $Fe(NO_3)_3$ contained in each liter of this solution.

4-24. A 12.5% (w/w) $NiCl_2$ (129.61 g/mol) solution has a density of 1.149 g/mL. Calculate
(a) the molar concentration of $NiCl_2$ in this solution.
(b) the molar Cl^- concentration of the solution.
(c) the mass in grams of $NiCl_2$ contained in each liter of this solution.

***4-25.** Describe the preparation of
(a) 500 mL of 4.75% (w/v) aqueous ethanol (C_2H_5OH, 46.1 g/mol).
(b) 500 g of 4.75% (w/w) aqueous ethanol.
(c) 500 mL of 4.75% (v/v) aqueous ethanol.

4-26. Describe the preparation of
(a) 2.50 L of 21.0% (w/v) aqueous glycerol ($C_3H_8O_3$, 92.1 g/mol).
(b) 2.50 kg of 21.0% (w/w) aqueous glycerol.
(c) 2.50 L of 21.0% (v/v) aqueous glycerol.

***4-27.** Describe the preparation of 750 mL of 6.00 M H_3PO_4 from the commercial reagent that is 86% H_3PO_4 (w/w) and has a specific gravity of 1.71.

4-28. Describe the preparation of 900 mL of 3.00 M HNO_3 from the commercial reagent that is 70.5% HNO_3 (w/w) and has a specific gravity of 1.42.

***4-29.** Describe the preparation of
(a) 500 mL of 0.0750 M $AgNO_3$ from the solid reagent.
(b) 1.00 L of 0.285 M HCl, starting with a 6.00 M solution of the reagent.
(c) 400 mL of a solution that is 0.0810 M in K^+, starting with solid $K_4Fe(CN)_6$.
(d) 600 mL of 3.00% (w/v) aqueous $BaCl_2$ from a 0.400 M $BaCl_2$ solution.
(e) 2.00 L of 0.120 M $HClO_4$ from the commercial reagent [71.0% $HClO_4$ (w/w), sp gr 1.67].
(f) 9.00 L of a solution that is 60.0 ppm in Na^+, starting with solid Na_2SO_4.

4-30. Describe the preparation of
(a) 5.00 L of 0.0500 M $KMnO_4$ from the solid reagent.
(b) 4.00 L of 0.250 M $HClO_4$, starting with an 8.00 M solution of the reagent.
(c) 400 mL of a solution that is 0.0250 M in I^-, starting with MgI_2.
(d) 200 mL of 1.00% (w/v) aqueous $CuSO_4$, from a 0.365 M $CuSO_4$ solution.
(e) 1.50 L of 0.215 M NaOH from the concentrated commercial reagent [50% NaOH (w/w), sp gr 1.525].
(f) 1.50 L of a solution that is 12.0 ppm in K^+, starting with solid $K_4Fe(CN)_6$.

***4-31.** What mass of solid $La(IO_3)_3$ (663.6 g/mol) is formed when 50.0 mL of 0.250 M La^{3+} are mixed with 75.0 mL of 0.302 M IO_3^-?

4-32. What mass of solid $PbCl_2$ (278.10 g/mol) is formed when 200 mL of 0.125 M Pb^{2+} are mixed with 400 mL of 0.175 M Cl^-?

***4-33.** Exactly 0.2220 g of pure Na_2CO_3 was dissolved in 100.0 mL of 0.0731 M HCl.
(a) What mass in grams of CO_2 were evolved?
(b) What was the molarity of the excess reactant (HCl or Na_2CO_3)?

4-34. Exactly 25.0 mL of a 0.3757 M solution of Na_3PO_4 were mixed with 100.00 mL of 0.5151 M $HgNO_3$.
(a) What mass of solid Hg_3PO_4 was formed?
(b) What is the molarity of the unreacted species (Na_3PO_4 or $HgNO_3$) after the reaction was complete?

***4-35.** Exactly 75.00 mL of a 0.3132 M solution of Na_2SO_3 were treated with 150.0 mL of 0.4025 M $HClO_4$ and boiled to remove the SO_2 formed.

(a) What was the mass in grams of SO_2 that was evolved?

(b) What was the concentration of the unreacted reagent (Na_2SO_3 or $HClO_4$) after the reaction was complete?

4-36. What mass of $MgNH_4PO_4$ precipitated when 200.0 mL of a 1.000% (w/v) solution of $MgCl_2$ were treated with 40.0 mL of 0.1753 M Na_3PO_4 and an excess of NH_4^+? What was the molarity of the excess reagent (Na_3PO_4 or $MgCl_2$) after the precipitation was complete?

***4-37.** What volume of 0.01000 M $AgNO_3$ would be required to precipitate all of the I^- in 200.0 mL of a solution that contained 24.32 ppt KI?

4-38. Exactly 750.0 mL of a solution that contained 480.4 ppm of $Ba(NO_3)_2$ were mixed with 200.0 mL of a solution that was 0.03090 M in $Al_2(SO_4)_3$.

(a) What mass of solid $BaSO_4$ was formed?

(b) What was the molarity of the unreacted reagent [$Al_2(SO_4)_3$ or $Ba(NO_3)_2$]?

4-39. Challenge Problem. According to Kenny et al.[2], Avogadro's number N_A may be calculated from the following equation using measurements on a sphere fabricated from an ultrapure single crystal of silicon.

$$N_A = \frac{n\mathcal{M}_{Si}(4/3)\pi r^3}{ma^3}$$

where

N_A = Avogadro's number
n = the number of atoms per unit cell in the crystal lattice of silicon
$\mathcal{M}_{Si}$ = the molar mass of silicon
r = the radius of the silicon sphere
m = the mass of the sphere
a = the crystal lattice parameter = $d(220)\sqrt{2^2 + 2^2 + 0^2}$

(a) Derive the equation for Avogadro's number.

(b) From the data assembled by Kenny et al., in the table below, calculate the density of silicon and its uncertainty. You may wish to delay the uncertainty calculations until you have studied Chapter 6.

Variable	Value	Uncertainty
Sphere radius, m	0.046817226	0.0000000015
Sphere mass, kg	1.001132893	0.000000075
Molar mass, kg	0.028085521	0.000000004
Lattice spacing $d(220)$, m	$192015.585 \times 10^{-15}$	0.010×10^{-15}
Atoms/unit cell	7.99999992	0.00000001

(c) Calculate Avogadro's number and its uncertainty.

(d) Which of the variables in the table have the most significant influence on the value that you calculated and why?

(e) What experimental methods were used to make the measurements shown in the table?

(f) Comment on experimental variables that might contribute to the uncertainty in each measurement.

(g) Suggest ways that the determination of Avogadro's number might be improved.

(h) Look up the accepted value and its uncertainty (1998 or later) for Avogadro's number at the NIST Web site on fundamental physical constants, and compare them with your computed values. What is the error in your value for Avogadro's number? Use Google to locate the NIST Web site.

(i) What technological innovations of the past several decades have led to the easy availability of ultrapure silicon?

[2]M. J. Kenny et al., *IEEE Trans. Instrum. Meas.*, **2001,** *50,* 587.

InfoTrac College Edition

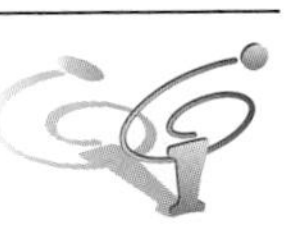

For additional readings, go to InfoTrac College Edition, your online research library, at

http://infotrac.thomsonlearning.com

CHAPTER 5

Errors in Chemical Analyses

© Roger Viollet/Getty Images

Errors can sometimes be calamitous, as this picture of the famous train accident at Montparnasse station in Paris illustrates. A train from Granville, France, on October 22, 1895, crashed through the platform and the station wall because the brakes failed. The engine fell 30 feet into the street below, killing a woman. Fortunately, no one on the train was seriously hurt, although the passengers were badly shaken.

Errors in chemical analyses are seldom this dramatic, but they may have equally serious effects, as described in this chapter. Among other applications, analytical results are often used in the diagnosis of disease, in the assessment of hazardous wastes and pollution, in the solving of major crimes, and in the quality control of industrial products. Errors in these results can have serious personal and societal effects. This chapter considers the various types of errors encountered in chemical analyses and the methods we can use to detect them.

The term **error** has two slightly different meanings. First, error refers to the difference between a measured value and the "true" or "known" value. Second, error often denotes the estimated uncertainty in a measurement or experiment.

Measurements invariably involve errors and uncertainties. Only a few of these are due to mistakes on the part of the experimenter. More commonly, ***errors*** *are caused by faulty calibrations or standardizations or random variations and uncertainties in results. Frequent calibrations, standardizations, and analyses of known samples can sometimes be used to lessen all but the random errors and uncertainties. In the limit, however, measurement errors are an inherent part of the quantized world in which we live. Because of this, it is impossible to perform a chemical analysis that is totally free of errors or uncertainties. We can only hope to minimize errors and estimate their size with acceptable accuracy.*[1] *In this and the next two chapters, we explore the nature of experimental errors and their effects on the results of chemical analyses.*

The effect of errors in analytical data is illustrated in Figure 5-1, which shows results for the quantitative determination of iron. Six equal portions of an aqueous solution with a "known"[2] *concentration of 20.00 ppm of iron(III) were analyzed in*

[1]Unfortunately, many people do not understand these truths. For example, when asked by a defense attorney in a celebrated homicide case what the rate of error in a blood test was, the assistant district attorney replied that their testing laboratories had no percentage of error because "they have not committed any errors" (*San Francisco Chronicle,* June 29, 1994, p. 4).

[2]Although actual concentrations can never be "known" exactly, in many situations we are quite certain of the value, as, for example, when it is derived from a high-quality reference standard.

exactly the same way. Note that the results range from a low of 19.4 ppm to a high of 20.3 ppm of iron. The average, or **mean,** *value, $\bar{x}$, of the data is 19.78 ppm, which rounds to 19.8 ppm (see Section 6D-1 for rounding numbers and the significant figures convention).*

The symbol **ppm** stands for parts per million, that is, 20.00 parts of iron(III) per million parts of solution.

Every measurement is influenced by many uncertainties, which combine to produce a scatter of results like that in Figure 5-1. Because measurement uncertainties can never be completely eliminated, measurement data can give us only an estimate of the "true" value. *However, the probable magnitude of error in a measurement can often be evaluated. It is then possible to define limits within which the true value of a measured quantity lies with a given level of probability.*

◀ Measurement uncertainties cause replicate results to vary.

Although it is not always easy to estimate the reliability of experimental data, it is important to do so whenever we collect laboratory results, because data of unknown quality are worthless. *On the other hand, results that do not seem especially accurate may be of considerable value if the limits of uncertainty are known.*

Unfortunately, there is no simple and widely applicable method for determining the reliability of data with absolute certainty. Often, estimating the quality of experimental results requires as much effort as collecting the data. Reliability can be assessed in several ways. Experiments designed to reveal the presence of errors can be performed. Standards of known composition can be analyzed and the results compared with the known composition. A few minutes in the library to consult the chemical literature can be profitable. Calibrating equipment usually enhances the quality of data. Finally, statistical tests can be applied to the data. Because none of these options is perfect, we must ultimately make judgments *as to the probable accuracy of our results. These judgments tend to become harsher and less optimistic with experience. The quality assurance of analytical methods and the ways to validate and report results are further discussed in Section 8D-3.*

One of the first questions to answer before beginning an analysis is, "What maximum error can I tolerate in the result?" *The answer to this question often determines the method chosen and the time required to complete the analysis. For example, experiments to determine whether the mercury concentration in a river water sample exceeds a certain value can often be done more rapidly than those to determine the specific concentration accurately. To increase the accuracy of a determination by 10-fold may take hours, days, or even weeks of added labor.* No one can afford to waste time generating data that are more reliable than is necessary for the job at hand.

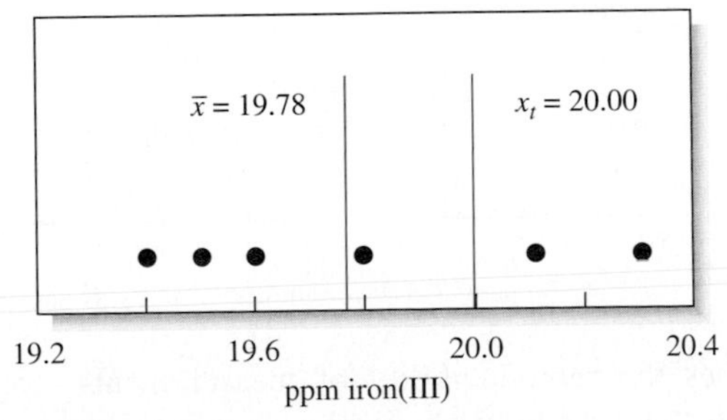

Figure 5-1 Results from six replicate determinations of iron in aqueous samples of a standard solution containing 20.0 ppm iron(III).

5A SOME IMPORTANT TERMS

Replicates are samples of about the same size that are carried through an analysis in *exactly* the same way.

Because one analysis gives no information about the variability of results, chemists usually carry two to five portions **(replicates)** of a sample through an entire analytical procedure. Individual results from a set of measurements are seldom the same (see Figure 5-1), so we usually consider the "best" estimate to be the central value for the set. We justify the extra effort required to analyze several samples in two ways. First, the central value of a set should be more reliable than any of the individual results. Usually, the mean or the median is used as the central value for a set of replicate measurements. Second, an analysis of the variation in data allows us to estimate the uncertainty associated with the central result.

5A-1 The Mean and Median

The **mean** of two or more measurements is their average value.

The most widely used measure of central value is the **mean,** $\bar{x}$. The mean, also called the **arithmetic mean,** or the **average,** is obtained by dividing the sum of replicate measurements by the number of measurements in the set:

► The symbol Σx_i means to add all of the values x_i for the replicates.

$$\bar{x} = \frac{\sum_{i=1}^{N} x_i}{N} \tag{5-1}$$

where x_i represents the individual values of x making up the set of N replicate measurements.

The **median** is the middle value in a set of data that has been arranged in numerical order. The median is used advantageously when a set of data contains an **outlier,** a result that differs significantly from others in the set. An outlier can have a significant effect on the mean of the set but has no effect on the median.

The **median** is the middle result when replicate data are arranged according to increasing or decreasing value. There are equal numbers of results that are larger and smaller than the median. For an odd number of results, the median can be evaluated directly. For an even number, the mean of the middle pair is used (see Example 5-1).

In ideal cases, the mean and median are identical, but when the number of measurements in the set is small, the values often differ, as shown in Example 5-1.

EXAMPLE 5-1

Calculate the mean and median for the data shown in Figure 5-1.

$$\text{mean} = \bar{x} = \frac{19.4 + 19.5 + 19.6 + 19.8 + 20.1 + 20.3}{6}$$
$$= 19.78 \approx 19.8 \text{ ppm Fe}$$

Because the set contains an even number of measurements, the median is the average of the central pair:

$$\text{median} = \frac{19.6 + 19.8}{2} = 19.7 \text{ ppm Fe}$$

5A-2 Precision

Precision is the closeness of results to others obtained in exactly the same way.

Precision describes the reproducibility of measurements—in other words, the closeness of results that have been obtained *in exactly the same way.* Generally, the precision of a measurement is readily determined by simply repeating the measurement on replicate samples.

Three terms are widely used to describe the precision of a set of replicate data: **standard deviation, variance,** and **coefficient of variation.** These three are functions of how much an individual result x_i differs from the mean, which is called the **deviation from the mean** d_i.

$$d_i = |x_i - \bar{x}| \qquad (5\text{-}2)$$

◀ Note that deviations from the mean are calculated without regard to sign.

The relationship between deviation from the mean and the three precision terms is given in Section 6B.

5A-3 Accuracy

Accuracy indicates the closeness of the measurement to the true or accepted value and is expressed by the *error.* Figure 5-2 illustrates the difference between accuracy and precision. Note that accuracy measures agreement between a result and the accepted value. *Precision,* on the other hand, describes the agreement among several results obtained in the same way. We can determine precision just by measuring replicate samples. Accuracy is often more difficult to determine because the true value is usually unknown. An accepted value must be used instead. Accuracy is expressed in terms of either absolute or relative error.

Accuracy is the closeness of a measured value to the true or accepted value.

◀ The term "absolute" has a different meaning here than it does in mathematics. An absolute value in mathematics means the magnitude of a number *ignoring its sign.* As we use it, the absolute error is the difference between an *experimental result and an accepted value including its sign.*

Absolute Error

The **absolute error** E in the measurement of a quantity x is given by the equation

$$E = x_i - x_t \qquad (5\text{-}3)$$

where x_t is the true or accepted value of the quantity. If we return to the data displayed in Figure 5-1, the absolute error of the result immediately to the left of the true value of 20.00 ppm is −0.2 ppm Fe; the result at 20.10 ppm is in error by

The **absolute error** of a measurement is the difference between the measured value and the true value. The sign of the absolute error tells you whether the value in question is high or low. If the measurement result is low, the sign is negative; if the measurement result is high, the sign is positive.

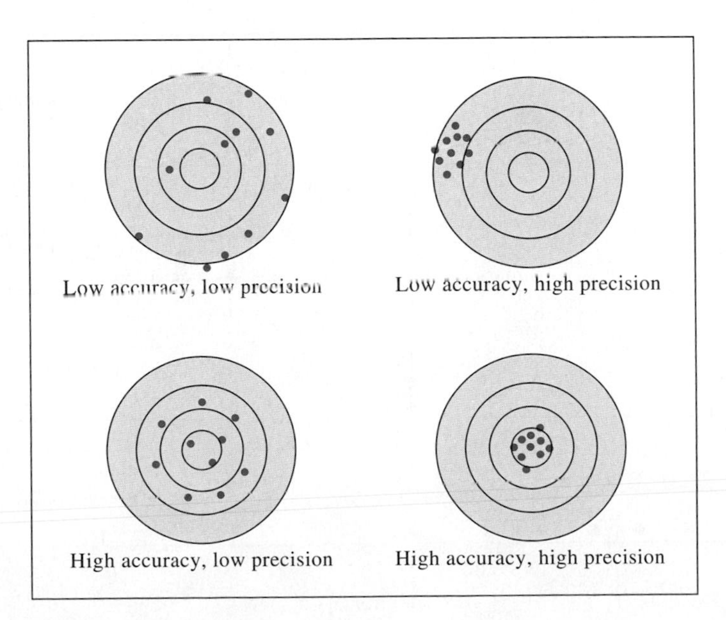

Figure 5-2 Illustration of accuracy and precision using the pattern of darts on a dartboard. Note that we can have very precise results (upper right) with a mean that is not accurate, and an accurate mean (lower left) with data points that are imprecise.

+0.1 ppm Fe. Note that we retain the sign in stating the absolute error. The negative sign in the first case shows that the experimental result is smaller than the accepted value, while the positive sign in the second case shows that the experimental result is larger than the accepted value.

Relative Error

► The **relative error** of a measurement is the absolute error divided by the true value. Relative error may be expressed in percent, parts per thousand, or parts per million, depending on the magnitude of the result.

Often, the **relative error** E_r is a more useful quantity than the absolute error. The percent relative error is given by the expression

$$E_r = \frac{x_i - x_t}{x_i} \times 100\% \tag{5-4}$$

Relative error is also expressed in parts per thousand (ppt). For example, the relative error for the mean of the data in Figure 5-1 is

$$E_r = \frac{19.8 - 20.0}{20.0} \times 100\% = -1\%, \text{ or } -10 \text{ ppt}$$

5A-4 Types of Errors in Experimental Data

The precision of a measurement is readily determined by comparing data from carefully replicated experiments. Unfortunately, an estimate of the accuracy is not as easy to obtain. To determine the accuracy, we have to know the true value, which is usually what we are seeking in the analysis.

Results can be precise without being accurate and accurate without being precise. The danger of assuming that precise results are also accurate is illustrated in Figure 5-3, which summarizes the results for determining nitrogen in two pure compounds. The dots show the absolute errors of replicate results obtained by four analysts. Note that analyst 1 obtained relatively high precision and high accuracy. Analyst 2 had poor precision but good accuracy. The results of analyst 3 are sur-

benzyl isothiourea hydrochloride

nicotinic acid

Small amounts of a vitamin, nicotinic acid, which is often called *niacin,* occur in all living cells, and it is essential in the nutrition of mammals. It is used in the prevention and treatment of pellagra.

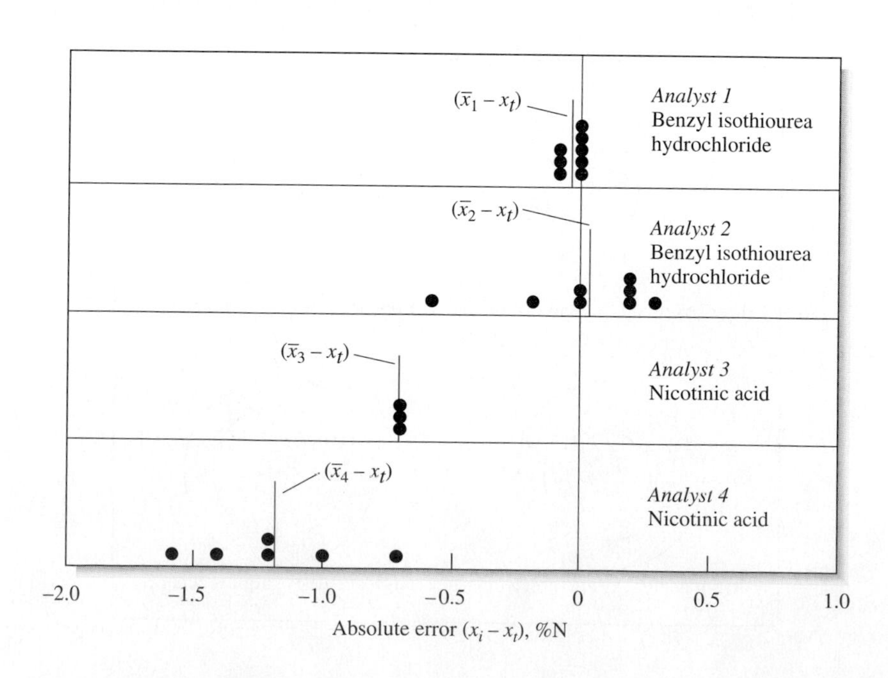

Figure 5-3 Absolute error in the micro-Kjeldahl determination of nitrogen. Each dot represents the error associated with a single determination. Each vertical line labeled $(x_i - x_t)$ is the absolute average deviation of the set from the true value. (Data from C. O. Willits and C. L. Ogg, *J. Assoc. Offic. Anal. Chem.,* **1949,** *32,* 561. With permission.)

prisingly common. The precision is excellent, but there is significant error in the numerical average for the data. Both the precision and the accuracy are poor for the results of analyst 4.

Figures 5-1 and 5-3 suggest that chemical analyses are affected by at least two types of errors. One type, called **random** (or **indeterminate**) **error,** causes data to be scattered more or less symmetrically around a mean value. Refer again to Figure 5-3, and notice that the scatter in the data, and thus the random error, for analysts 1 and 3 is significantly less than that for analysts 2 and 4. In general, then, the random error in a measurement is reflected by its precision. Random errors are discussed in detail in Chapter 6.

Random, or **indeterminate, errors** affect measurement precision.

A second type of error, called **systematic** (or **determinate**) **error,** causes the mean of a data set to differ from the accepted value. For example, the mean of the results in Figure 5-1 has a systematic error of about −0.2 ppm Fe. The results of analysts 1 and 2 in Figure 5-3 have little systematic error, but the data of analysts 3 and 4 show systematic errors of about −0.7% and −1.2% nitrogen. In general, a systematic error in a series of replicate measurements causes all the results to be too high or too low. An example of a systematic error is the unsuspected loss of a volatile analyte while heating a sample.

Systematic, or **determinate, errors** affect the accuracy of results.

A third type of error is **gross error.** Gross errors differ from indeterminate and determinate errors. They usually occur only occasionally, are often large, and may cause a result to be either high or low. They are often the product of human errors. For example, if part of a precipitate is lost before weighing, analytical results will be low. Touching a weighing bottle with your fingers after its empty mass is determined will cause a high mass reading for a solid weighed in the contaminated bottle. Gross errors lead to **outliers,** results that appear to differ markedly from all other data in a set of replicate measurements. There is no evidence of a gross error in Figures 5-1 and 5-3. Had one of the results shown in Figure 5-1 occurred at, say, 21.2 ppm Fe, it might have been an outlier.

An **outlier** is an occasional result in replicate measurements that differs significantly from the rest of the results.

Various statistical tests can be performed to determine if a result is an outlier (see Section 7D).

5B SYSTEMATIC ERRORS

Systematic errors have a definite value and an assignable cause, and are of the same magnitude for replicate measurements made in the same way. Systematic errors lead to **bias** in measurement results. Note that bias affects all of the data in a set in the same way and that it bears a sign.

Bias measures the systematic error associated with an analysis. It has a negative sign if it causes the results to be low and a positive sign otherwise.

5B-1 Sources of Systematic Errors

There are three types of systematic errors: (1) **Instrumental errors** are caused by nonideal instrument behavior, by faulty calibrations, or by use under inappropriate conditions. (2) **Method errors** arise from nonideal chemical or physical behavior of analytical systems. (3) **Personal errors** result from the carelessness, inattention, or personal limitations of the experimenter.

Instrument Errors

All measuring devices are potential sources of systematic errors. For example, pipets, burets, and volumetric flasks may hold or deliver volumes slightly different from those indicated by their graduations. These differences arise from using glassware at a temperature that differs significantly from the calibration temperature, from distortions in container walls due to heating while drying, from errors in the original calibration, or from contaminants on the inner surfaces of the containers. Calibration eliminates most systematic errors of this type.

Electronic instruments are subject to instrumental systematic errors. These can have many sources. For example, errors may emerge as the voltage of a battery-operated power supply decreases with use. Errors can also occur if instruments are not calibrated frequently or calibrated incorrectly. The experimenter may also use an instrument under conditions in which errors are large. For example, a pH meter used in strongly acidic media is prone to an acid error, as discussed in Chapter 20. Temperature changes cause variation in many electronic components, which can lead to drifts and errors. Some instruments are susceptible to noise induced from the alternating current (ac) power lines, and this noise may influence precision and accuracy. In many cases, errors of these types are detectable and correctable.

Method Errors

The nonideal chemical or physical behavior of the reagents and reactions on which an analysis is based often introduces systematic method errors. Such sources of nonideality include the slowness of some reactions, the incompleteness of others, the instability of some species, the nonspecificity of most reagents, and the possible occurrence of side reactions that interfere with the measurement process. For example, a common method error in volumetric analysis results from the small excess of reagent required to cause an indicator to undergo the color change that signals completion of the reaction. The accuracy of such an analysis is thus limited by the very phenomenon that makes the titration possible.

Another example of method error is illustrated by the data in Figure 5-3, in which the results by analysts 3 and 4 show a negative bias that can be traced to the chemical nature of the sample, nicotinic acid. The analytical method used involves the decomposition of the organic samples in hot concentrated sulfuric acid, which converts the nitrogen in the samples to ammonium sulfate. Often a catalyst, such as mercuric oxide or a selenium or copper salt, is added to speed the decomposition. The amount of ammonia in the ammonium sulfate is then determined in the measurement step. Experiments have shown that compounds containing a pyridine ring, such as nicotinic acid (see structure, page 94), are incompletely decomposed by the sulfuric acid. With such compounds, potassium sulfate is used to raise the boiling temperature. Samples containing N—O or N—N linkages must be pretreated or subjected to reducing conditions.[3] Without these precautions, low results are obtained. It is highly likely the negative errors, $(\bar{x}_3 - x_t)$ and $(\bar{x}_4 - x_t)$, in Figure 5-3 are systematic errors that can be blamed on incomplete decomposition of the samples.

► Of the three types of systematic errors encountered in a chemical analysis, method errors are usually the most difficult to identify and correct.

Errors inherent in a method are often difficult to detect and are thus the most serious of the three types of systematic error.

Personal Errors

Many measurements require personal judgments. Examples include estimating the position of a pointer between two scale divisions, the color of a solution at the end point in a titration, or the level of a liquid with respect to a graduation in a pipet or

[3] J. A. Dean, *Analytical Chemistry Handbook,* Section 17, p. 17.4. New York: McGraw-Hill, 1995.

buret (see Figure 6-5, page 134). Judgments of this type are often subject to systematic, unidirectional errors. For example, one person may read a pointer consistently high, another may be slightly slow in activating a timer, and a third may be less sensitive to color changes. An analyst who is insensitive to color changes tends to use excess reagent in a volumetric analysis. Analytical procedures should always be adjusted so that any known physical limitations of the analyst cause negligibly small errors.

◀ Color blindness is a good example of a limitation that could cause a personal error in a volumetric analysis. A famous color-blind analytical chemist enlisted his wife to come to the laboratory to help him detect color changes at end points of titrations.

A universal source of personal error is *prejudice,* or *bias.* Most of us, no matter how honest, have a natural tendency to estimate scale readings in a direction that improves the precision in a set of results. Alternatively, we may have a preconceived notion of the true value for the measurement. We then subconsciously cause the results to fall close to this value. Number bias is another source of personal error that varies considerably from person to person. The most frequent number bias encountered in estimating the position of a needle on a scale involves a preference for the digits 0 and 5. Also common is a prejudice favoring small digits over large and even numbers over odd.

◀ Digital and computer displays on pH meters, laboratory balances, and other electronic instruments eliminate number bias because no judgment is involved in taking a reading. However, many of these produce results with more figures than are significant. The rounding of insignificant figures can also produce bias (see Section 6D-1).

5B-2 The Effect of Systematic Errors on Analytical Results

◀ To preserve the integrity of the collected data, persons who make measurements must constantly guard against personal bias.

Systematic errors may be either **constant** or **proportional.** The magnitude of a constant error stays essentially the same as the size of the quantity measured is varied. With constant errors, the absolute error is constant with sample size, but the relative error varies when sample size is changed. Proportional errors increase or decrease according to the size of the sample taken for analysis. With proportional errors the absolute error varies with sample size, but the relative error stays constant with changing sample size.

Constant Errors

The effect of a constant error becomes more serious as the size of the quantity measured decreases. The effect of solubility losses on the results of a gravimetric analysis, shown in Example 5-2, illustrates this behavior.

EXAMPLE 5-2

Suppose that 0.50 mg of precipitate is lost as a result of being washed with 200 mL of wash liquid. If the precipitate weighs 500 mg, the relative error due to solubility loss is $-(0.50/500) \times 100\% = -0.1\%$. Loss of the same quantity from 50 mg of precipitate results in a relative error of -1.0%.

The excess of reagent required to bring about a color change during a titration is another example of constant error. This volume, usually small, remains the same regardless of the total volume of reagent required for the titration. Again, the relative error from this source becomes more serious as the total volume decreases. One way of reducing the effect of constant error is to increase the sample size until the error is tolerable.

Constant errors are independent of the size of the sample being analyzed. **Proportional errors** decrease or increase in proportion to the size of the sample.

Proportional Errors

A common cause of proportional errors is the presence of interfering contaminants in the sample. For example, a widely used method for determining copper is based on the reaction of copper(II) ion with potassium iodide to give iodine (see Sections 20B-2, 37H-3, and 37H-4). The quantity of iodine is then measured and is proportional to the amount of copper. Iron(III), if present, also liberates iodine from potassium iodide. Unless steps are taken to prevent this interference, high results are observed for the percentage of copper because the iodine produced will be a measure of the copper(II) and iron(III) in the sample. The size of this error is fixed by the *fraction* of iron contamination, which is independent of the size of sample taken. If the sample size is doubled, for example, the amount of iodine liberated by both the copper and the iron contaminant is also doubled. Thus, the magnitude of the reported percentage of copper is independent of sample size.

5B-3 Detection of Systematic Instrument and Personal Errors

Some systematic instrument errors can be found and corrected by calibration. Periodic calibration of equipment is always desirable because the response of most instruments changes with time as a result of wear, corrosion, or mistreatment. Many systematic instrument errors involve interferences in which a species present in the sample affects the response of the analyte. Simple calibration does not compensate for these effects. Instead, the methods described in Section 8C-3 can be used when such interference effects exist.

Most personal errors can be minimized by care and self-discipline. It is a good habit to check instrument readings, notebook entries, and calculations systematically. Errors due to limitations of the experimenter can usually be avoided by carefully choosing the analytical method.

► After entering a reading into the laboratory notebook, many scientists habitually make a second reading and then check this against what has been entered, to ensure the correctness of the entry.

5B-4 Detection of Systematic Method Errors

Bias in an analytical method is particularly difficult to detect. We may take one or more of the following steps to recognize and adjust for a systematic error in an analytical method.

Analysis of Standard Samples

The best way of estimating the bias of an analytical method is by the analysis of **standard reference materials,** materials that contain one or more analytes at known concentration levels. Standard reference materials are obtained in several ways.

Standard reference materials (SRMs) are substances sold by the National Institute of Standards and Technology (NIST) and certified to contain specified concentrations of one or more analytes.

Standard materials can sometimes be prepared by synthesis. Here, carefully measured quantities of the pure components of a material are mixed in such a way as to produce a homogeneous sample whose composition is known from the quantities taken. The overall composition of a synthetic standard material must approximate closely the composition of the samples to be analyzed. Great care must be taken to ensure that the concentration of analyte is known exactly. Unfortunately, a synthetic standard may not reveal unexpected interferences, so the accuracy of determinations may not be known. Hence, this approach is not often practical.

Standard reference material can be purchased from a number of governmental and industrial sources. For example, the National Institute of Standards and Technology (NIST) (formerly the National Bureau of Standards) offers more than 1300 standard reference materials, including rocks and minerals, gas mixtures, glasses, hydrocarbon mixtures, polymers, urban dusts, rainwaters, and river sediments.[4] The concentration of one or more of the components in these materials has been determined in one of three ways: (1) by analysis with a previously validated reference method; (2) by analysis with two or more independent, reliable measurement methods; or (3) by analysis by a network of cooperating laboratories that are technically competent with and thoroughly knowledgeable about the material being tested. Several commercial supply houses also offer analyzed materials for method testing.[5]

Often, analysis of standard reference materials gives results that differ from the accepted value. The question then becomes one of establishing whether such a difference is due to bias or to random error. In Section 7B-1, we demonstrate a statistical test that can help you answer this question.

Photo courtesy of the National Institute of Standards and Technology.

Standard reference materials from NIST.

Independent Analysis

If standard samples are not available, a second independent and reliable analytical method can be used in parallel with the method being evaluated. The independent method should differ as much as possible from the one under study. This minimizes the possibility that some common factor in the sample has the same effect on both methods. Here again, a statistical test must be used to determine whether any difference is a result of random errors in the two methods or due to bias in the method under study (see Section 7B-2).

◀ In using SRMs it is often difficult to separate bias from ordinary random error.

Blank Determinations

A **blank** contains the reagents and solvents used in a determination, but no analyte. Often, many of the sample constituents are added to simulate the analyte environment, often called the **sample matrix.** In a blank determination, all steps of the analysis are performed on the blank material. The results are then applied to correct the sample measurements. Blank determinations reveal errors due to interfering contaminants from the reagents and vessels used in the analysis. Blanks are also used to correct titration data for the volume of reagent needed to cause color change in an indicator.

A **blank** solution contains the solvent and all of the reagents in an analysis. Whenever feasible, blanks may also contain added constituents to simulate the sample matrix.

The term **matrix** refers to the collection of all the constituents in the sample.

Variation in Sample Size

Example 5-2 on page 97 demonstrates that as the size of a measurement increases, the effect of a constant error decreases. Thus, constant errors can often be detected by varying the sample size.

[4]See U.S. Department of Commerce, *NIST Standard Reference Materials Catalog,* 1998–99 ed., NIST Special Publication 260-98-99. Washington, D.C.: U.S. Government Printing Office, 1998. For a description of the reference material programs of NIST, see R. A. Alvarez, S. D. Rasberry, and G. A. Uriano, *Anal. Chem.,* **1982,** *54,* 1226A; see also http://www.nist.gov.

[5]For example, in the clinical and biological sciences area, see Sigma Chemical Co., 3050 Spruce St., St. Louis, MO 63103, or Bio-Rad Laboratories, 1000 Alfred Nobel Dr., Hercules, CA 94547.

A MEAN CALCULATION

In this spreadsheet exercise, we learn to calculate the mean of a data set. First, we define formulas to calculate the mean, and then we use the built-in functions of Excel to accomplish the task.

Entering the Data

We start Excel with a clean spreadsheet. In cell B1, enter the heading **Data[↵]**. Now enter in column B under the heading, the data x_i given in Example 5-1. Click on cell A11, and type

```
Total[↵]
N[↵]
Mean[↵]
```

Your worksheet should now look like the following.

	A	B	C	D
1		Data		
2		19.4		
3		19.5		
4		19.6		
5		19.8		
6		20.1		
7		20.3		
8				
9				
10				
11	Total			
12	N			
13	Mean			
14				
15				

Finding the Mean

Click on cell B11, and type

```
=SUM(B2:B7)[↵]
```

This formula calculates the sum of the values in cells B2 through B7 and displays the result in cell B11. Now, in cell B12, type

```
=COUNT(B2:B7)[↵]
```

The COUNT function counts the number of nonzero cells in the range B2:B7 and displays the result in cell B12. Since we have found the sum of the values and the number N of data points, we can find the mean $\bar{x}$ by typing the following formula in cell B13.

```
=B11/B12[↵]
```

At this point in the exercise, your worksheet should appear as shown below.

	A	B	C	D
1		Data		
2		19.4		
3		19.5		
4		19.6		
5		19.8		
6		20.1		
7		20.3		
8				
9				
10				
11	Total	118.7		
12	N	6		
13	Mean	19.78333		
14				
15				

We will discuss in Section 6D-3 how to round data such as the mean to retain only significant figures.

Using Excel's Built-in Functions

Excel has built-in functions to compute many of the quantities that are of interest to us. Now we shall see how to use them to calculate the mean, or in Excel's syntax, the average. Click on cell C13 and type

=AVERAGE(B2:B7)[↵]

Notice that the mean determined using the built-in AVERAGE function is identical to the value in cell B13 that you determined by typing a formula. Before proceeding or terminating your Excel session, save your file to a disk as **average.xls**.

Finding the Deviations from the Mean

With the definition given in Equation 5-2, we can now use Excel to determine the deviation from the mean of each of the data in our worksheet. Click on cell C2 and type

=ABS(B2−$B$13)[↵]

This formula computes the absolute value **ABS()** of the difference between our first value in B2 and the mean value in B13. The formula is a bit different from those we have used previously. We have typed a dollar sign, **$**, before the B and before the 13 in the second cell reference. This type of cell reference is called an **absolute reference.** It means that no matter where we might copy the contents of the cell C2, the reference will always pertain to cell B13. The other type of cell reference that we consider here is the **relative reference,** exemplified by B2. The reason we use a relative reference for B2 and an absolute reference for B13 is that we want to copy the formula in C2 into cells C3–C7, and we want the mean **$B$13** to be subtracted from each of the successive data in column B. We now copy the formula by clicking on cell C2, clicking on the fill handle, and dragging the rectangle through C7. When you release the mouse button, your worksheet should look like the one below.

	A	B	C	D
1		Data		
2		19.4	0.383333	
3		19.5	0.283333	
4		19.6	0.183333	
5		19.8	0.016667	
6		20.1	0.316667	
7		20.3	0.516667	
8				
9				
10				
11	Total	118.7		
12	N	6		
13	Mean	19.78333	19.78333	
14				
15				

Now click on cell C3, and notice that it contains the formula **`=ABS(B3-$B$13)`**. Compare this formula with the one in cell C2 and in cells C4 through C7. The absolute cell reference **`$B$13`** appears in all cells. As you can see, we have accomplished our task of calculating the deviation from the mean for all of the data. Now we will edit the formula in cell C13 to find the mean deviation of the data.

Editing Formulas

To edit the formula to calculate the mean deviation of the data, click on C13, and then click on the formula in the formula bar. Use the arrow keys, **`[←]`** and **`[→]`**, and either the **`[Backspace]`** or the **`[Delete]`** key to replace both Bs in the formula with Cs so that it reads **`=AVERAGE(C2:C7)`**. Finally, press **`[↵]`**, and the mean deviation will appear in cell C13. Type the label **`Deviation`** in cell C1 so that your worksheet appears as shown below.

	A	B	C	D
1		Data	Deviation	
2		19.4	0.383333	
3		19.5	0.283333	
4		19.6	0.183333	
5		19.8	0.016667	
6		20.1	0.316667	
7		20.3	0.516667	
8				
9				
10				
11	Total	118.7		
12	N	6		
13	Mean	19.78333	0.283333	
14				
15				

Save the file by clicking on the save icon in the toolbar, by clicking File/Save, or by typing `[Ctrl+S]`.

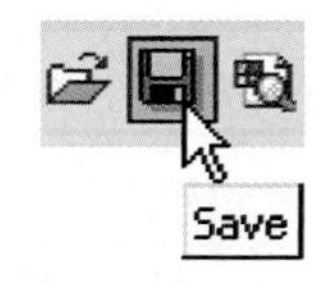

In this exercise, we have learned to calculate a mean, using both the built-in Excel AVERAGE function and a formula of our own design. In Chapter 6, we will use STDEV and other functions to complete our analysis of the data from the gravimetric determination of chloride that we began in Chapter 2. You may now close Excel by typing File/Exit or proceed to Chapter 6 to continue with the spreadsheet exercises.

WEB WORKS

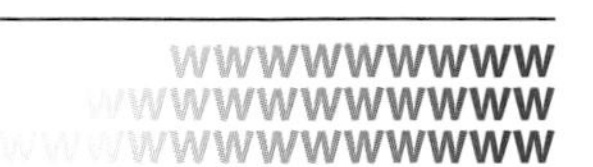

Statistical methods are extremely important, not only in chemistry but in all walks of life. Newspapers, magazines, television, and the World Wide Web bombard us with confusing and often misleading statistics. Browse to **http://chemistry.brookscole.com/skoogfac/.** From the Chapter Resources menu, choose Web Works, and locate the Chapter 5 section. There you will find a link to a Web site that contains an interesting presentation of statistics for writers. Use the links there to look up the definitions of mean and median. You will find some nice examples using salaries that clarify the distinction between the two measures of central tendency, show the utility of comparing the two, and point out the importance of using the appropriate measure for a particular data set. For the nine salaries given, which is larger, the mean or the median? Why are these so different in this case?

QUESTIONS AND PROBLEMS

5-1. Explain the difference between
*(a) constant and proportional error.
(b) random and systematic error.
*(c) mean and median.
(d) absolute and relative error.

***5-2.** Suggest some sources of random error in measuring the width of a 3-m table with a 1-m metal rule.

***5-3.** Name three types of systematic errors.

5-4. Describe at least three systematic errors that might occur while weighing a solid on an analytical balance.

***5-5.** Describe at least 3 ways in which a systematic error might occur while using a pipet to transfer a known volume of liquid.

5-6. How are systematic method errors detected?

***5-7.** What kinds of systematic errors are detected by varying the sample size?

5-8. A method of analysis yields weights for gold that are low by 0.4 mg. Calculate the percent relative error caused by this uncertainty if the weight of gold in the sample is
*(a) 700 mg.
(b) 450 mg.
*(c) 250 mg.
(d) 40 mg.

5-9. The method described in Problem 5-8 is to be used for the analysis of ores that assay about 1.2% gold. What minimum sample weight should be taken if the relative error resulting from a 0.4-mg loss is not to exceed
*(a) −0.2%? (b) −0.5%?
*(c) −0.8%? (d) −1.2%?

5-10. The color change of a chemical indicator requires an overtitration of 0.04 mL. Calculate the percent relative error if the total volume of titrant is
*(a) 50.00 mL. (b) 10.0 mL.
*(c) 25.0 mL. (d) 40.0 mL.

5-11. A loss of 0.4 mg of Zn occurs in the course of an analysis for that element. Calculate the percent relative error due to this loss if the weight of Zn in the sample is
*(a) 40 mg. (b) 175 mg.
*(c) 400 mg. (d) 600 mg.

5-12. Find the mean and median of each of the following sets of data. Determine the deviation from the mean for each data point within the sets and find the mean deviation for each set. Use a spreadsheet if you like.
*(a) 0.0110 0.0104 0.0105
(b) 24.53 24.68 24.77 24.81 24.73
*(c) 188 190 194 187
(d) 4.52×10^{-3} 4.47×10^{-3}
4.63×10^{-3} 4.48×10^{-3}
4.53×10^{-3} 4.58×10^{-3}
*(e) 39.83 39.61 39.25 39.68
(f) 850 862 849 869 865

5-13. Challenge Problem. Richards and Willard[6] determined the atomic mass of lithium and collected the following data.

Experiment	Molar Mass, g/mol
1	6.9391
2	6.9407
3	6.9409
4	6.9399
5	6.9407
6	6.9391
7	6.9406

(a) Find the mean atomic mass determined by these workers.
(b) Find the median atomic mass.
(c) Assuming that the currently accepted value for the atomic mass of lithium is the true value, calculate the absolute error and the percent relative error of the mean value determined by Richards and Willard.
(d) Find in the chemical literature at least three values for the atomic mass of lithium determined since 1910 and arrange them chronologically in a table or spreadsheet along with the values since 1817 given in the table on page 10 of the paper by Richards and Willard. Construct a graph of atomic mass versus year to illustrate how the atomic mass of lithium has changed over the past two centuries. Suggest possible reason(s) why the value changes abruptly in about 1830.
(e) The incredibly detailed experiments described by Richards and Willard suggest that it is unlikely that major changes in the atomic mass of lithium will occur. Discuss this assertion in light of your calculation in part c.
(f) What factors have led to changes in atomic mass since 1910?
(g) How would you determine the accuracy of an atomic mass?

[6]T. W. Richards and H. H. Willard, *J. Am. Chem. Soc.*, **1910,** *32,* 4.

InfoTrac College Edition

For additional readings, go to InfoTrac College Edition, your online research library, at

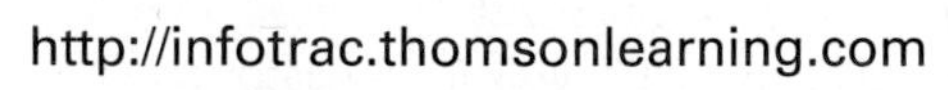

http://infotrac.thomsonlearning.com

CHAPTER 6

Random Errors in Chemical Analysis

The probability distributions discussed in this chapter are fundamental to the use of statistics for judging the reliability of data and for testing various hypotheses. The quincunx shown in the photo is a mechanical device that forms a normal probability distribution. Every 10 minutes 30,000 balls drop from the center top of the machine, which contains a regular pattern of pegs to randomly deflect the balls. Each time a ball hits a peg, it has a 50:50 chance of falling to the right or to the left. After each ball passes through the array of pegs, it drops into one of the vertical "bins" of the transparent case. The height of the column of balls in each bin is proportional to the probability of a ball falling into a given bin. The smooth curve shown traces out the probability distribution.

Boston Museum of Science

All measurements contain random errors. In this chapter, we consider the sources of random errors, the determination of their magnitude, and their effects on computed results of chemical analyses. We also introduce the significant figure convention and illustrate its use in reporting analytical results.

6A THE NATURE OF RANDOM ERRORS

Random, or indeterminate, errors exist in every measurement. They can never be totally eliminated and are often the major source of uncertainty in a determination. Random errors are caused by the many uncontrollable variables that are an inevitable part of every analysis. Most contributors to random error cannot be positively identified. Even if we can identify sources of uncertainty, it is usually impossible to measure them because most are so small that they cannot be detected individually. The accumulated effect of the individual uncertainties, however, causes replicate measurements to fluctuate randomly around the mean of the set. For example, the scatter of data in Figures 5-1 and 5-3 is a direct result of the accumulation of small random uncertainties. We have replotted the Kjeldahl nitrogen data from Figure 5-3 as a three-dimensional plot in Figure 6-1 in order to better see the precision and accuracy of each analyst. Notice that the random error in the results of analysts 2 and 4 is much larger than that seen in the results of analysts 1 and 3. The results of analyst 3 show good precision, but poor accuracy. The results of analyst 1 show excellent precision and good accuracy.

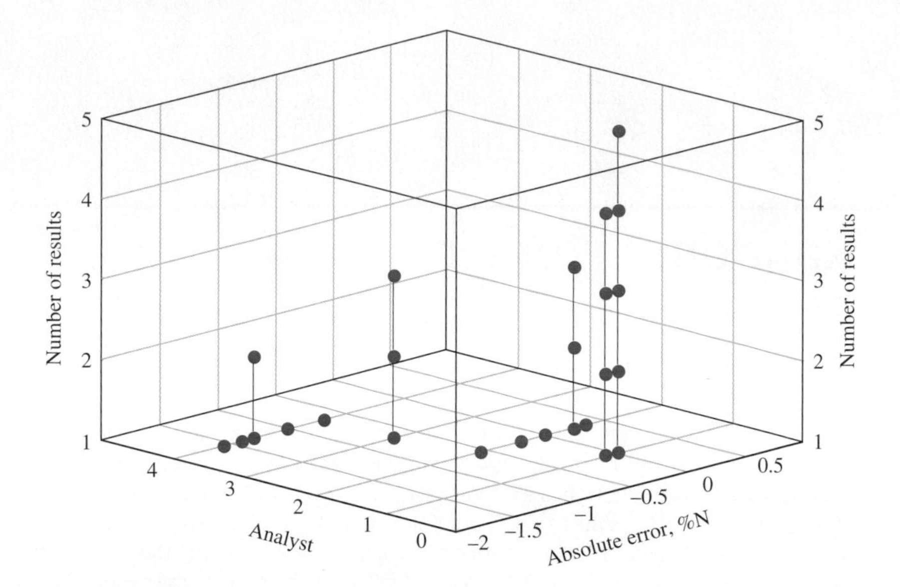

Figure 6-1 Three-dimensional plot showing absolute error in Kjeldahl nitrogen determination for four different analysts. Note that the results of analyst 1 are both precise and accurate. The results of analyst 3 are precise, but the absolute error is large. The results of analysts 2 and 4 are both imprecise and inaccurate.

6A-1 Random Error Sources

We can get a qualitative idea of the way small undetectable uncertainties produce a detectable random error in the following way. Imagine a situation in which just four small random errors combine to give an overall error. We will assume that each error has an equal probability of occurring and that each can cause the final result to be high or low by a fixed amount $\pm U$.

Table 6-1 shows all the possible ways the four errors can combine to give the indicated deviations from the mean value. Note that only one combination leads to

TABLE 6-1

Possible Combinations of Four Equal-Sized Uncertainties

Combinations of Uncertainties	Magnitude of Random Error	Number of Combinations	Relative Frequency
$+U_1+U_2+U_3+U_4$	$+4U$	1	1/16 = 0.0625
$-U_1+U_2+U_3+U_4$ $+U_1-U_2+U_3+U_4$ $+U_1+U_2-U_3+U_4$ $+U_1+U_2+U_3-U_4$	$+2U$	4	4/16 = 0.250
$-U_1-U_2+U_3+U_4$ $+U_1+U_2-U_3-U_4$ $+U_1-U_2+U_3-U_4$ $-U_1+U_2-U_3+U_4$ $-U_1+U_2+U_3-U_4$ $+U_1-U_2-U_3+U_4$	0	6	6/16 = 0.375
$+U_1-U_2-U_3-U_4$ $-U_1+U_2-U_3-U_4$ $-U_1-U_2+U_3-U_4$ $-U_1-U_2-U_3+U_4$	$-2U$	4	4/16 = 0.250
$-U_1-U_2-U_3-U_4$	$-4U$	1	1/16 = 0.0625

a deviation of $+4\ U$, four combinations give a deviation of $+2\ U$, and six give a deviation of $0\ U$. The negative errors have the same relationship. This ratio of 1:4:6:4:1 is a measure of the probability for a deviation of each magnitude. If we make a sufficiently large number of measurements, we can expect a frequency distribution like that shown in Figure 6-2a. Note that the y-axis in the plot is the relative frequency of occurrence of the five possible combinations.

Figure 6-2b shows the theoretical distribution for 10 equal-sized uncertainties. Again we see that the most frequent occurrence is zero deviation from the mean. At the other extreme, a maximum deviation of $10\ U$ occurs only about once in 500 measurements.

When the same procedure is applied to a very large number of individual errors, a bell-shaped curve like that shown in Figure 6-2c results. Such a plot is called a **Gaussian curve,** or **normal error curve.**

◀ In our example, all the uncertainties have the same magnitude. This restriction is not necessary to derive the equation for a Gaussian curve.

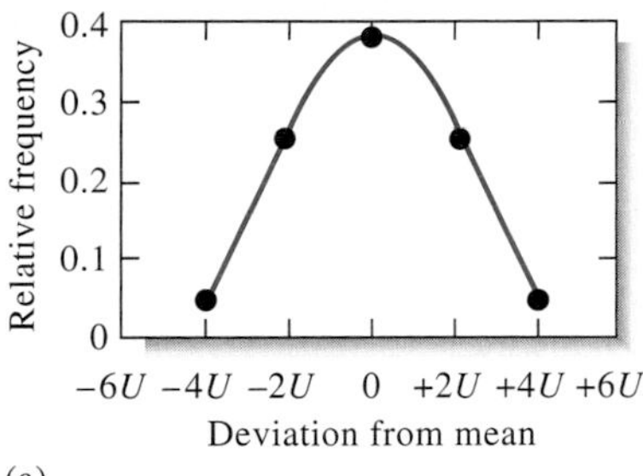

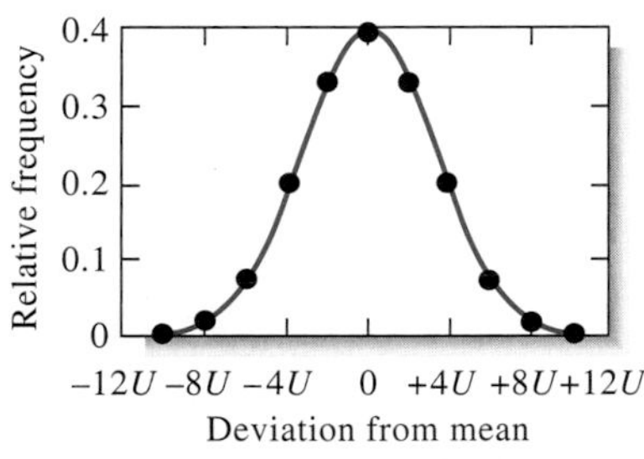

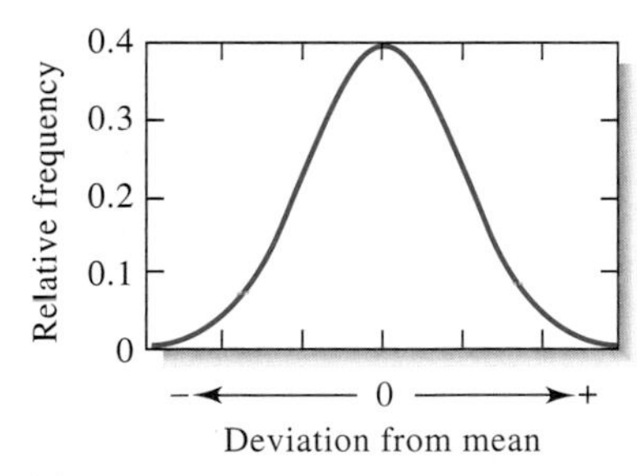

Figure 6-2 Frequency distribution for measurements containing (a) 4 random uncertainties; (b) 10 random uncertainties; (c) a very large number of random uncertainties.

6A-2 Distribution of Experimental Results

From experience with many determinations, we find that the distribution of replicate data from most quantitative analytical experiments approaches that of the Gaussian curve shown in Figure 6-2c. As an example, consider the data in the spreadsheet in Table 6-2 for the calibration of a 10-mL pipet.[1] In this experiment a small flask and stopper were weighed. Ten milliliters of water were transferred to the flask with the pipet, and the flask was stoppered. The flask, the stopper, and the water were weighed again. The temperature of the water was also measured to determine its density. The mass of the water was then calculated by taking the difference between the two masses. The mass of water divided by its density is the volume delivered by the pipet. The experiment was repeated 50 times.

In Table 6-2, the mean is calculated with Excel's function `=AVERAGE()` as described in the Spreadsheet Exercise in Section 5B-4. Note that since the data are in different columns we use `=AVERAGE(B3:B19,E3:E19,H3:H18)` in the calculation. The median is calculated using Excel's function `=MEDIAN()`. The standard deviation function in Excel is described in Section 6B-3. The maximum value is found with `=MAX()` and the minimum value with `=MIN()`. The spread is the maximum value minus the minimum value. The data in Table 6-2 are typical of those obtained by an experienced worker weighing to the nearest milligram (which corresponds to 0.001 mL) on a top-loading balance and being careful to avoid systematic error. Even so, the results vary from a low of 9.969 mL to a high of 9.994 mL. This 0.025-mL **spread** of data results directly from an accumulation of all random uncertainties in the experiment.

The information in Table 6-2 is easier to visualize when the data are rearranged into frequency distribution groups, as in Table 6-3. Here, we tabulate the number of data points falling into a series of adjacent 0.003-mL ranges and calculate the percentage of measurements in each range. Note that 26% of the results occur in the volume range from 9.981 to 9.983 mL. This is the group containing the mean and median value of 9.982 mL. Note also that more than half the results are within ±0.004 mL of this mean.

The **spread** in a set of replicate measurements is the difference between the highest and lowest result.

[1]See Section 37A-4 for an experiment on calibration of a pipet.

TABLE 6-2

	A	B	C	D	E	F	G	H
1	Replicate Data for the Calibration of a 10-mL Pipet*							
2	**Trial**	**Volume, mL**		**Trial**	**Volume, mL**		**Trial**	**Volume, mL**
3	1	9.988		18	9.975		35	9.976
4	2	9.973		19	9.980		36	9.990
5	3	9.986		20	9.994		37	9.988
6	4	9.980		21	9.992		38	9.971
7	5	9.975		22	9.984		39	9.986
8	6	9.982		23	9.981		40	9.978
9	7	9.986		24	9.987		41	9.986
10	8	9.982		25	9.978		42	9.982
11	9	9.981		26	9.983		43	9.977
12	10	9.990		27	9.982		44	9.977
13	11	9.980		28	9.991		45	9.986
14	12	9.989		29	9.981		46	9.978
15	13	9.978		30	9.969		47	9.983
16	14	9.971		31	9.985		48	9.980
17	15	9.982		32	9.977		49	9.984
18	16	9.983		33	9.976		50	9.979
19	17	9.988		34	9.983			
20	*Data listed in the order obtained							
21	Mean	9.982		Maximum	9.994			
22	Median	9.982		Minimum	9.969			
23	Std. Dev.	0.0056		Spread	0.025			

A **histogram** is a bar graph such as that shown by plot *A* in Figure 6-3.

The frequency distribution data in Table 6-3 are plotted as a bar graph, or **histogram** (labeled *A* in Figure 6-3). We can imagine that as the number of measurements increases, the histogram approaches the shape of the continuous curve shown as plot *B* in Figure 6-3. This plot shows a Gaussian curve, or normal error curve, which applies to an infinitely large set of data. The Gaussian curve has the same mean (9.982 mL), the same precision, and the same area under the curve as the histogram.

Variations in replicate measurements, such as those in Table 6-2, result from numerous small and individually undetectable random errors that are attributable to uncontrollable variables in the experiment. Such small errors ordinarily tend to cancel one another and thus have a minimal effect on the mean value. Occasionally, however, they occur in the same direction to produce a large positive or negative net error.

TABLE 6-3

Frequency Distribution of Data from Table 6-2

Volume Range, mL	Number in Range	% in Range
9.969–9.971	3	6
9.972–9.974	1	2
9.975–9.977	7	14
9.978–9.980	9	18
9.981–9.983	13	26
9.984–9.986	7	14
9.987–9.989	5	10
9.990–9.992	4	8
9.993–9.995	1	2
	Total = 50	Total = 100%

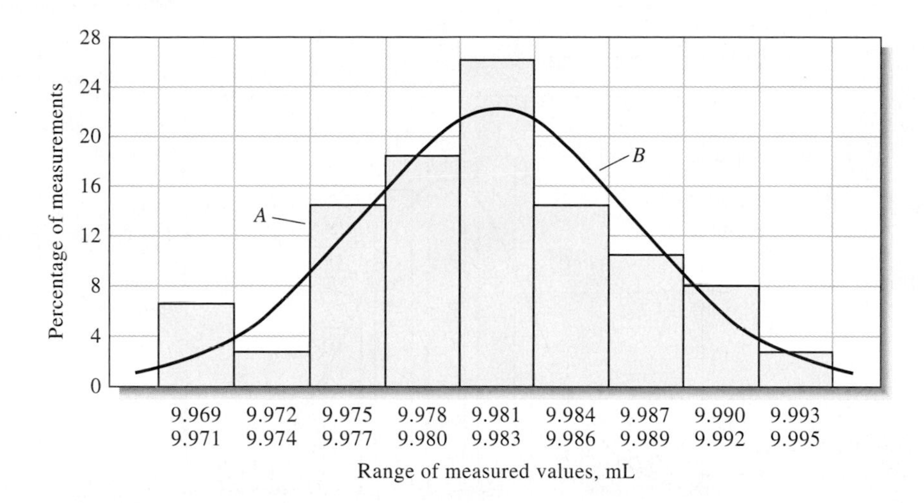

Figure 6-3 A histogram *(A)* showing distribution of the 50 results in Table 6-3 and a Gaussian curve *(B)* for data having the same mean and standard deviation as the data in the histogram.

A **Gaussian,** or **normal error, curve** is a curve that shows the symmetrical distribution of data around the mean of an infinite set of data such as the one in Figure 6-2c.

Sources of random uncertainties in the calibration of a pipet include (1) visual judgments, such as the level of the water with respect to the marking on the pipet and the mercury level in the thermometer; (2) variations in the drainage time and in the angle of the pipet as it drains; (3) temperature fluctuations, which affect the volume of the pipet, the viscosity of the liquid, and the performance of the balance; and (4) vibrations and drafts that cause small variations in the balance readings. Undoubtedly, there are many other sources of random uncertainty in this calibration process that we have not listed. Even the simple process of calibrating a pipet is affected by many small and uncontrollable variables. The cumulative influence of these variables is responsible for the observed scatter of results around the mean.

FEATURE 6-1

Flipping Coins: A Student Activity to Illustrate a Normal Distribution

If you flip a coin 10 times, how many heads will you get? Try it, and record your results. Repeat the experiment. Are your results the same? Ask friends or members of your class to perform the same experiment and tabulate the results. The table below contains the results obtained by several classes of analytical chemistry students over the period from 1980 to 1998.

Number of Heads	0	1	2	3	4	5	6	7	8	9	10
Frequency	1	1	22	42	102	104	92	48	22	7	1

Add your results to those in the table, and plot a histogram similar to the one shown in Figure 6F-1. Find the mean and the standard deviation (see Section 6B-3) for your results and compare them with the values shown in the plot. The smooth curve in the figure is a normal error curve for an infinite number of trials with the same mean and standard deviation as the data set. Note that the mean of 5.06 is very close to the value of 5 that you would predict based on the laws of probability. As the number of trials increases, the histogram approaches the shape of the smooth curve and the mean approaches 5.

(continued)

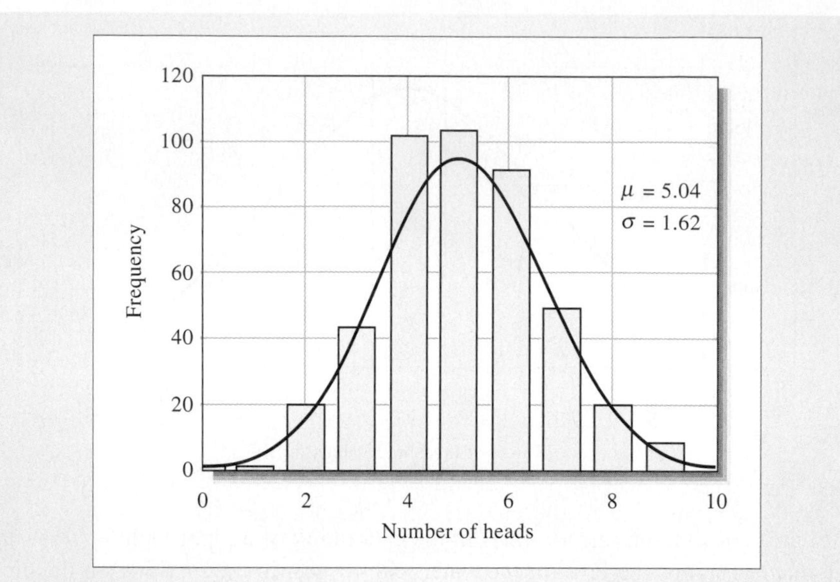

Figure 6F-1 Results of a coin-flipping experiment by 395 students over an 18-year period.

6B STATISTICAL TREATMENT OF RANDOM ERROR

► Statistical analysis reveals only information that is already present in a data set. That is, *no new information is created* by statistical treatments. Statistical methods do allow us, however, to categorize and characterize data in different ways and to make objective and intelligent decisions about data quality and interpretation.

We can use statistical methods to evaluate the random errors discussed in the preceding section. Ordinarily, we base statistical analyses on the assumption that random errors in analytical results follow a Gaussian, or normal, distribution, such as that illustrated in curve *B* in Figure 6-3 or in Figure 6-2c. Analytical data can follow distributions other than the Gaussian distribution. For example, experiments in which there is either a successful outcome or a failure produce data that follow the binomial distribution. Radioactive or photon-counting experiments produce results that follow the Poisson distribution. However, we often use a Gaussian distribution to approximate these distributions. The approximation becomes better in the limit of a large number of experiments. Thus, we base this discussion entirely on normally distributed random errors.

6B-1 Samples and Populations

A **population** is the collection of all measurements of interest to the experimenter, while a **sample** is a subset of measurements selected from the population.

Typically in a scientific study, we infer information about a **population** or **universe** from observations made on a subset or **sample.** The population is the collection of all measurements of interest and must be carefully defined by the experimenter. In some cases the population is finite and real, while in other cases the population is hypothetical or conceptual in nature.

As an example of a real population, consider a production run of multivitamin tablets producing hundreds of thousands of tablets. We usually would not have the time or resources to test all the tablets for quality control purposes. Hence, we select a sample of tablets for analysis according to statistical sampling principles (see Section 8B). We then infer the characteristics of the population from those of the sample.

In many of the cases encountered in analytical chemistry, the population is conceptual. Consider, for example, the determination of calcium in a community water supply to measure water hardness. Here the population is the very large, nearly infinite, number of measurements that could be made if we analyzed the entire water supply. Likewise, in determining glucose in the blood of a diabetic patient, we could hypothetically make an extremely large number of measurements if we used the entire blood supply. The subset of the population selected for analysis in both these cases is the sample. Again, we infer characteristics of the population from those of the selected sample.

Statistical laws have been derived for populations; often they must be modified substantially when applied to a small sample because a few data points may not represent the entire population. In the discussion that follows, we first describe the Gaussian statistics of populations. Then we show how these relationships can be modified and applied to small samples of data.

◀ Do not confuse the *statistical sample* with the *analytical sample.* Four analytical samples analyzed in the laboratory represent a single statistical sample. This is an unfortunate duplication of the term sample.

6B-2 Properties of Gaussian Curves

Figure 6-4a shows two Gaussian curves in which we plot the relative frequency y of various deviations from the mean versus the deviation from the mean. As shown in the margin, curves such as these can be described by an equation that contains just two parameters, the **population mean** μ and the **population standard deviation** σ. The term **parameter** refers to quantities such as μ and σ that define a population or distribution. This is in contrast to quantities such as the data values x that are variables. The term **statistic** refers to an estimate of a parameter that is made from a sample of data, as discussed below. The sample mean and the sample standard deviation are examples of statistics that estimate parameters μ and σ, respectively.

◀ The equation for a Gaussian curve has the form

$$y = \frac{e^{-(x-\mu)^2/2\sigma^2}}{\sigma\sqrt{2\pi}}$$

The Population Mean μ and the Sample Mean $\bar{x}$

Statisticians find it useful to differentiate between the **sample mean** and the **population mean.** The sample mean $\bar{x}$ is the arithmetic average of a limited sample drawn from a population of data. The sample mean is defined as the sum of the measurement values divided by the number of measurements, as given by Equation 5-1, page 92. In that equation, N represents the number of measurements in the sample set. The

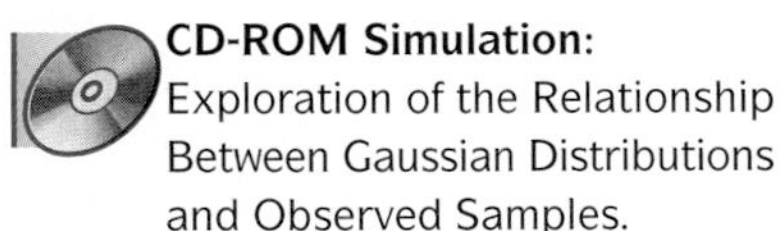

CD-ROM Simulation: Exploration of the Relationship Between Gaussian Distributions and Observed Samples.

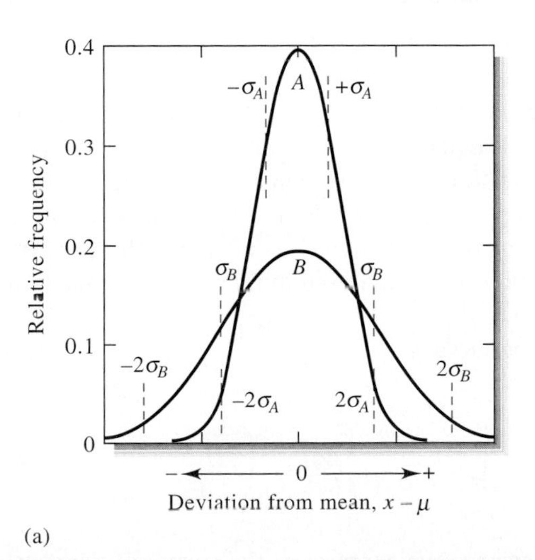

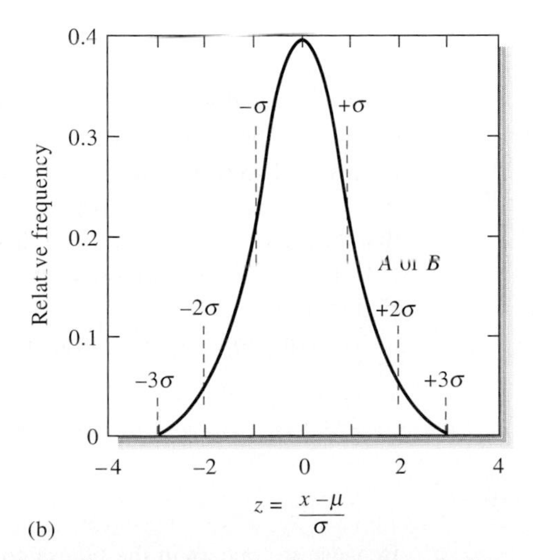

Figure 6-4 Normal error curves. The standard deviation for curve B is twice that for curve A; that is, $\sigma_B = 2\sigma_A$. (a) The abscissa is the deviation from the mean in the units of measurement. (b) The abscissa is the deviation from the mean in units of σ. Thus, the two curves A and B are identical here.

► The sample mean $\bar{x}$ is found from

$$\bar{x} = \frac{\sum_{i=1}^{N} x_i}{N}$$

where N is the number of measurements in the sample set. The same equation is used to calculate the population mean μ

$$\mu = \frac{\sum_{i=1}^{N} x_i}{N}$$

where N is now the total number of measurements in the population.

► When there is no systematic error, the population mean μ is the true value of the measured quantity.

► The quantity $(x_i - \mu)$ in Equation 6-1 is the deviation of data values x_i from the mean μ of a population; compare with Equation 6-4, which is for a sample of data.

► The quantity z represents the deviation of a result from the population mean relative to the standard deviation. It is commonly given as a variable in statistical tables, since it is a dimensionless quantity.

population mean μ, in contrast, is the true mean for the population. It is also defined by Equation 5-1 with the added provision that N represents the total number of measurements in the population. *In the absence of systematic error, the population mean is also the true value for the measured quantity.* To emphasize the difference between the two means, the sample mean is symbolized by $\bar{x}$ and the population mean by μ. More often than not, particularly when N is small, $\bar{x}$ differs from μ because a small sample of data may not exactly represent its population. In most cases we do not know μ and must infer its value from $\bar{x}$. The probable difference between $\bar{x}$ and μ decreases rapidly as the number of measurements making up the sample increases; ordinarily, by the time N reaches 20 to 30, this difference is negligible. Note that the sample mean $\bar{x}$ is a statistic that estimates the population parameter μ.

The Population Standard Deviation (σ)

The **population standard deviation** σ, which is a measure of the *precision* of a population of data, is given by the equation

$$\sigma = \sqrt{\frac{\sum_{i=1}^{N} (x_i - \mu)^2}{N}} \qquad (6\text{-}1)$$

where N is the number of data points making up the population.

The two curves in Figure 6-4a are for two populations of data that differ only in their standard deviations. The standard deviation for the data set yielding the broader but lower curve B is twice that for the measurements yielding curve A. The breadth of these curves is a measure of the precision of the two sets of data. Thus, the precision of the data set leading to curve A is twice as good as that of the data set represented by curve B.

Figure 6-4b shows another type of normal error curve in which the x-axis is now a new variable z, defined as

$$z = \frac{(x - \mu)}{\sigma} \qquad (6\text{-}2)$$

Note that z is the deviation of a data point from the mean relative to one standard deviation. That is, when $x - \mu = \sigma$, z is equal to one; when $x - \mu = 2\sigma$, z is equal to two; and so forth. Since z is the deviation from the mean relative to the standard deviation, a plot of relative frequency versus z yields a single Gaussian curve that describes all populations of data regardless of standard deviation. Thus, Figure 6-4b is the normal error curve for both sets of data used to plot curves A and B in Figure 6-4a.

The equation for the Gaussian error curve is

$$y = \frac{e^{-(x-\mu)^2/2\sigma^2}}{\sigma\sqrt{2\pi}} = \frac{e^{-z^2/2}}{\sigma\sqrt{2\pi}} \qquad (6\text{-}3)$$

Because it appears in the Gaussian error curve expression, the square of the standard deviation σ^2 is also important. This quantity is called the **variance** (see Section 6B-5).

A normal error curve has several general properties: (a) The mean occurs at the central point of maximum frequency, (b) there is a symmetrical distribution of positive and negative deviations about the maximum, and (c) there is an exponential decrease in frequency as the magnitude of the deviations increases. Thus, small uncertainties are observed much more often than very large ones.

Areas under a Gaussian Curve

Feature 6-2 shows that, regardless of its width, 68.3% of the area beneath a Gaussian curve for a population lies within one standard deviation ($\pm 1\sigma$) of the mean μ. Thus, roughly 68.3% of the values making up the population will lie within these bounds. Furthermore, approximately 95.4% of all data values are within $\pm 2\sigma$ of the mean and 99.7% are within $\pm 3\sigma$. The vertical dashed lines in Figure 6-4 show the areas bounded by $\pm 1\sigma$, $\pm 2\sigma$, and $\pm 3\sigma$.

Because of area relationships such as these, the standard deviation of a population of data is a useful predictive tool. For example, we can say that the chances are 68.3 in 100 that the random uncertainty of any single measurement is no more than $\pm 1\sigma$. Similarly, the chances are 95.4 in 100 that the error is less than $\pm 2\sigma$, and so forth. The calculation of areas under the Gaussian curve is described in Feature 6-2.

FEATURE 6-2

Calculating the Areas Under the Gaussian Curve

We often refer to the area under a curve. In the context of statistics, it is important to be able to determine the area under the Gaussian curve between defined limits. The area under the curve between a pair of limits gives the probability of a measured value occurring between those limits. A practical question arises: How do we determine the area under a curve? Equation 6-3 describes the Gaussian curve in terms of the population mean μ and the standard deviation σ, or the variable z. Suppose we want to know the area under the curve between -1σ and $+1\sigma$ of the mean. In other words, we want the area from $\mu - \sigma$ to $\mu + \sigma$.

We can perform this operation using calculus because integration of an equation gives the area under the curve described by the equation. In this case, we wish to find the definite integral from $-\sigma$ to $+\sigma$.

$$\text{area} = \int_{-\sigma}^{\sigma} \frac{e^{-(x-\mu)^2/2\sigma^2}}{\sigma\sqrt{2\pi}}\,dx$$

It is easier to use the form of Equation 6-3 with the variable z, so our equation becomes

$$\text{area} = \int_{-1}^{1} \frac{e^{-z^2/2}}{\sqrt{2\pi}}\,dz$$

Since there is no closed form solution, the integral must be evaluated numerically. The result is

$$\text{area} = \int_{-1}^{1} \frac{e^{-z^2/2}}{\sqrt{2\pi}}\,dz = 0.683$$

(continued)

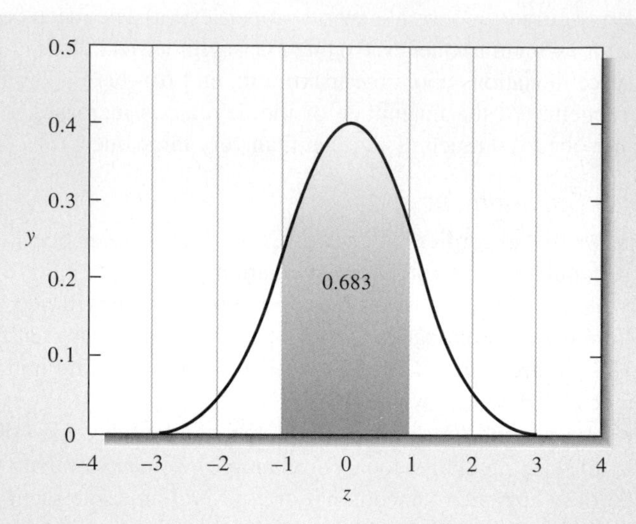

Curve showing area of 0.683.

Likewise, if we want to know the area under the Gaussian curve 2σ on either side of the mean, we evaluate the following integral.

$$\text{area} = \int_{-2}^{2} \frac{e^{-z^2/2}}{\sqrt{2\pi}} dz = 0.954$$

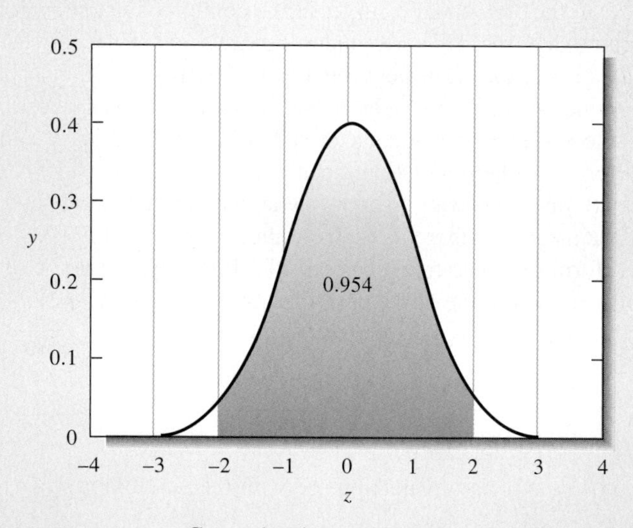

Curve showing area of 0.954.

For $\pm 3\sigma$, we have

$$\text{area} = \int_{-3}^{3} \frac{e^{-z^2/2}}{\sqrt{2\pi}} dz = 0.997$$

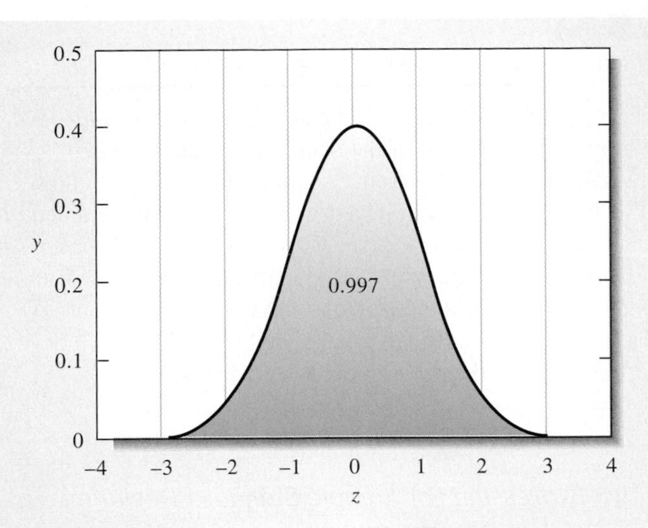

Curve showing area of 0.997.

Finally, it is important to know the area under the entire Gaussian curve, so we find the following integral.

$$\text{area} = \int_{-\infty}^{\infty} \frac{e^{-z^2/2}}{\sqrt{2\pi}}\, dz = 1$$

We can see from the integrals that the areas under the Gaussian curve for one, two, and three standard deviations from the mean are, respectively, 68.3%, 95.4%, and 99.7% of the total area under the curve.

6B-3 The Sample Standard Deviation: A Measure of Precision

Equation 6-1 must be modified when it is applied to a small sample of data. Thus, the **sample standard deviation** s is given by the equation

$$s = \sqrt{\frac{\sum_{i=1}^{N} (x_i - \bar{x})^2}{N - 1}} = \sqrt{\frac{\sum_{i=1}^{N} d_i^2}{N - 1}} \qquad (6\text{-}4)$$

Equation 6-4 applies to small sets of data. It says, "Find the deviations from the mean d_i, square them, sum them, divide the sum by $N - 1$, and take the square root." The quantity $N - 1$ is called **the number of degrees of freedom.** Scientific calculators usually have the standard deviation function built in. Many can find the population standard deviation σ as well as the sample standard deviation, s. For any small data set, you should use the sample standard deviation, s.

where the quantity $(x_i - \bar{x})$ represents the deviation d_i of value x_i from the mean $\bar{x}$. Note that Equation 6-4 differs from Equation 6-1 in two ways. First, the sample mean, $\bar{x}$, appears in the numerator in place of the population mean, μ. Second, N in Equation 6-1 is replaced by the **number of degrees of freedom** $(N - 1)$. When $N - 1$ is used instead of N, s is said to be an unbiased estimator of the population standard deviation σ. If this substitution is not used, the calculated s will be less on average than the true standard deviation σ; that is, s will have a negative bias (see Feature 6-3).

The **sample variance** s^2 is also important in statistical calculations. It is an estimate of the population variance σ^2, as discussed in Section 6B-5.

FEATURE 6-3

The Significance of the Number of Degrees of Freedom

The number of degrees of freedom indicates the number of *independent* results that enter into the computation of the standard deviation. When μ is unknown, two quantities must be extracted from a set of replicate data: $\bar{x}$ and s. One degree of freedom is used to establish $\bar{x}$ because, with their signs retained, the sum of the individual deviations must be zero. Thus, when $N - 1$ deviations have been computed, the final one is known. Consequently, only $N - 1$ deviations provide an *independent* measure of the precision of the set. Failure to use $N - 1$ in calculating the standard deviation for small samples results in values of s that are, on average, smaller than the true standard deviation σ.

An Alternative Expression For Sample Standard Deviation

To calculate s with a calculator that does not have a standard deviation key, the following rearrangement of Equation 6-4 is easier to use than directly applying Equation 6-4:

$$s = \sqrt{\frac{\sum_{i=1}^{N} x_i^2 - \frac{\left(\sum_{i=1}^{N} x_i\right)^2}{N}}{N - 1}} \tag{6-5}$$

Example 6-1 illustrates the use of Equation 6-5 to calculate s.

EXAMPLE 6-1

The following results were obtained in the replicate determination of the lead content of a blood sample: 0.752, 0.756, 0.752, 0.751, and 0.760 ppm Pb. Calculate the mean and the standard deviation of this set of data.

To apply Equation 6-5, we calculate Σx_i^2 and $(\Sigma x_i)^2/N$.

Sample	x_i	x_i^2
1	0.752	0.565504
2	0.756	0.571536
3	0.752	0.565504
4	0.751	0.564001
5	0.760	0.577600
	$\Sigma x_i = 3.771$	$\Sigma x_i^2 = 2.844145$

$$\bar{x} = \frac{\Sigma x_i}{N} = \frac{3.771}{5} = 0.7542 \approx 0.754 \text{ ppm Pb}$$

$$\frac{(\Sigma x_i)^2}{N} = \frac{(3.771)^2}{5} = \frac{14.220441}{5} = 2.8440882$$

Substituting into Equation 6-5 leads to

$$s = \sqrt{\frac{2.844145 - 2.8440882}{5 - 1}} = \sqrt{\frac{0.0000568}{4}} = 0.00377 \approx 0.004 \text{ ppm Pb}$$

Note in Example 6-1 that the difference between Σx_i^2 and $(\Sigma x_i)^2/N$ is very small. If we had rounded these numbers before subtracting them, a serious error would have appeared in the computed value of s. To avoid this source of error, *never round a standard deviation calculation until the very end.* Furthermore, and for the same reason, never use Equation 6-5 to calculate the standard deviation of numbers containing five or more digits. Use Equation 6-4 instead.[2] Many calculators and computers with a standard deviation function use a version of Equation 6-5 internally in the calculation. You should always be alert for roundoff errors when calculating the standard deviation of values that have five or more significant figures.

◀ Any time you subtract two large, approximately equal numbers, the difference will usually have a relatively large uncertainty.

When you make statistical calculations, remember that because of the uncertainty in $\bar{x}$, a sample standard deviation may differ significantly from the population standard deviation. As N becomes larger, $\bar{x}$ and s become better estimators of μ and σ.

◀ As $N \rightarrow \infty$, $\bar{x} \rightarrow \mu$, and $s \rightarrow \sigma$

Standard Error of the Mean

The probability figures for the Gaussian distribution calculated as areas in Feature 6-2 refer to the probable error for a *single* measurement. Thus, it is 95.4% probable that a single result from a population will lie within $\pm 2\sigma$ of the mean μ. If a series of replicate results, each containing N measurements, are taken randomly from a population of results, the mean of each set will show less and less scatter as N increases. The standard deviation of each mean is known as the **standard error of the mean** and is given the symbol s_m. The standard error is inversely proportional to the square root of the number of data points N used to calculate the mean, as given by Equation 6-6.

The **standard error of the mean,** s_m, is the standard deviation of a set of data divided by the square root of the number of data points in the set.

$$s_m = \frac{s}{\sqrt{N}} \qquad (6\text{-}6)$$

Equation 6-6 tells us that the mean of 4 measurements is more precise by $\sqrt{4} = 2$ than individual measurements in the data set. For this reason, averaging results is often used to improve precision. However, the improvement to be gained by averaging is somewhat limited because of the square root dependence seen in Equation 6-6. For example, to increase the precision by a factor of 10 requires 100 times as many measurements. It is better, if possible, to decrease s than to keep averaging more results, since s_m is directly proportional to s, but only inversely proportional to the *square root of N.* The standard deviation can sometimes be decreased by being more precise in individual operations, by changing the procedure, and by using more precise measurement tools.

[2]In most cases, the first two or three digits in a set of data are identical to each other. As an alternative, then, to using Equation 6-4, these identical digits can be dropped and the remaining digits used with Equation 6-5. For example, the standard deviation for the data in Example 6-1 could be based on 0.052, 0.056, 0.052, and so forth (or even 52, 56, 52, etc.).

COMPUTING THE STANDARD DEVIATION

In this exercise, we will calculate the standard deviation, the variance, and the relative standard deviation of two sets of data. We begin with the spreadsheet and data from the Spreadsheet Exercise in Chapter 5. The standard deviation s is given by the equation

$$s = \sqrt{\frac{\sum_{i=1}^{N}(x_i - \bar{x})^2}{N - 1}}$$

and the variance is s^2.

Finding the Variance

If you are continuing the Spreadsheet Exercise from Chapter 5, begin with the data on your computer screen. Otherwise, retrieve the file **`average.xls`** from your disk by clicking on File/Open. Make cell D1 the active cell, and type

```
Deviation^2[↵]
```

Cell D2 should now be the active cell, and your worksheet should appear as follows:

	A	B	C	D	E
1		Data	Deviation	Deviation^2	
2		19.4	0.383333		
3		19.5	0.283333		
4		19.6	0.183333		
5		19.8	0.016667		
6		20.1	0.316667		
7		20.3	0.516667		
8					
9					
10					
11	Total	118.7			
12	N	6			
13	Mean	19.78333	0.283333		
14					

Now type

```
=C2^2[↵]
```

and the square of the deviation in cell C2 appears in D2. Copy this formula into the other cells in column D by once again clicking on cell D2, clicking on the fill handle, and dragging the fill handle through cell D7. You have now calculated the squares of the deviations of each of the data from the mean value in cell B13.

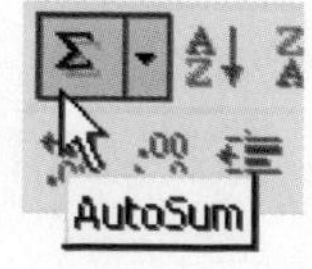

A Shortcut for Performing a Summation

To find the variance, we must find the sum of the squares of the deviations, so now click on cell D11, and then click on the AutoSum icon shown in the margin.

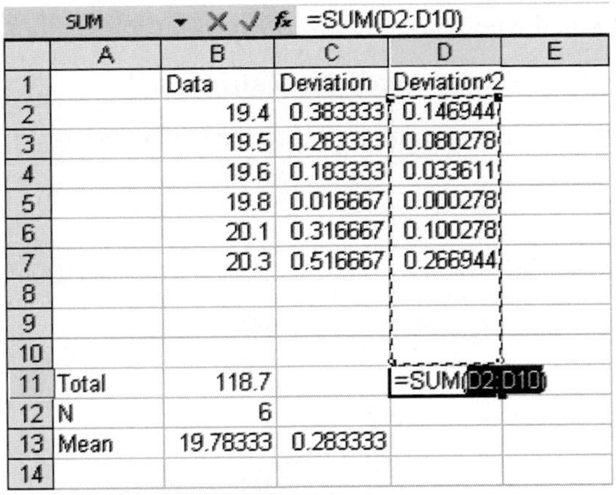

SUM ▾ ✕ ✓ fx =SUM(D2:D10)

	A	B	C	D	E
1		Data	Deviation	Deviation^2	
2		19.4	0.383333	0.146944	
3		19.5	0.283333	0.080278	
4		19.6	0.183333	0.033611	
5		19.8	0.016667	0.000278	
6		20.1	0.316667	0.100278	
7		20.3	0.516667	0.266944	
8					
9					
10					
11	Total	118.7		=SUM(D2:D10)	
12	N	6			
13	Mean	19.78333	0.283333		
14					

The dashed box shown above now surrounds the column of data in cells D2–D10, which appear as arguments of the SUM function in cell D11 and in the formula bar. Note that Excel assumes that you want to add all of the numerical data above the active cell and automatically completes the formula. When you type [↵], the sum of the squares of the deviations appears in cell D11. Since cells D8–D10 are blank, they contribute zero to the sum, and so there is no harm in leaving the references to D8–D10 in the formula. Be aware, however, that references to blank cells could pose difficulty under certain circumstances. You can always resize the box to include only the data of interest.

The final step in calculating the variance is to divide the sum of the squares of the deviations by the number of degrees of freedom, which is $N - 1$. We shall type the formula for carrying out this last calculation in cell D12. Before proceeding, type the label `Variance` in F12. Now click on D12, and type

```
=D11/(B12-1)[↵]
```

The variance is calculated and appears in the cell. Notice that you must enclose the difference B12 − 1 in parentheses so that Excel computes the number of degrees of freedom before the division is carried out. If we had not enclosed the number of degrees of freedom, B12 − 1, in parentheses, Excel would have divided D11 by B12 and then subtracted 1, which is incorrect. To illustrate this point, suppose D11 = 12 and B12 = 3. If we leave off the parentheses, D11/B12 − 1 = 3, but if we put them in, D11/(B12 − 1) = 6. The order of mathematical operations in Excel is extremely important. Remember that just as in algebra, Excel performs exponentiation before multiplication and division, and it performs multiplication and division before addition and subtraction. As in the present example, we can change the order of operations by properly placing parentheses. The order that Excel uses in evaluating various mathematical and logical operations is shown in the margin.

Order of Operations

Order	Operator	Description
1	–	Negation
2	%	Percent
3	^	Exponentiation
4	* and /	Multiplication and division
5	+ and –	Addition and subtraction
6	=, <, >, <=, >=, <>	Comparison

Finding the Standard Deviation

Our next step is to calculate the standard deviation by extracting the square root of the variance. Click on D13, and type

```
=SQRT(D12)[↵]
```

Then click on F13, and type

`Standard Deviation[↵]`

Your worksheet should then appear similar to the following.

	A	B	C	D	E	F	G
1		Data	Deviation	Deviation^2			
2		19.4	0.383333	0.146944			
3		19.5	0.283333	0.080278			
4		19.6	0.183333	0.033611			
5		19.8	0.016667	0.000278			
6		20.1	0.316667	0.100278			
7		20.3	0.516667	0.266944			
8							
9							
10							
11	Total	118.7		0.628333			
12	N	6		0.125667		Variance	
13	Mean	19.78333	0.283333	0.354495		Standard Deviation	
14							

Notice that we have deliberately left cells E12 and E13 blank. We will now use the built-in variance and standard deviation functions of Excel to check our formulas.

The Built-in Statistical Functions of Excel

Click on cell E12, and then type

`=VAR(`

Now click in cell B2, and drag the mouse into cell B7 so that the worksheet appears as follows:

SUM ▾ ✕ ✓ fx =VAR(B2:B7

	A	B	C	D	E	F	G
1		Data	Deviation	Deviation^2			
2		19.4	0.383333	0.146944			
3		19.5	0.283333	0.080278			
4		19.6	0.183333	0.033611			
5		19.8	0.016667	0.000278			
6		20.1	0.316667	0.100278			
7		20.3	0.516667	0.266944			
8			6R x 1C				
9							
10							
11	Total	118.7		0.628333			
12	N	6		0.125667	R(B2:B7	Variance	
13	Mean	19.78333	0.283333	0.354495		Standard Deviation	
14							

Notice that the cell references B2:B7 appear in cell E12 and in the formula bar. Now, let up on the mouse button, and press `[↵]`, and the variance appears in cell E12. If you have performed these operations correctly, the values displayed in cells D12 and E12 are identical.

The active cell should now be E13. If it is not, click on it, and type

=STDEV(

and click and drag to highlight cells B2:B7 as you did previously. Let up on the mouse button, press `[↵]`, and the standard deviation appears in cell E13. The computed values in cells D13 and E13 should be equal. It is important to note that the Excel STDEV and VAR functions calculate the **sample standard deviation** and the **sample variance,** not the corresponding population statistics. These built-in functions are quite convenient, since your sample will generally be sufficiently small that you will want to calculate sample statistics rather than population statistics. Excel also has STDEVP and VARP functions that calculate the standard deviation and variance values for an entire population, respectively, but these should not be used for samples of data.

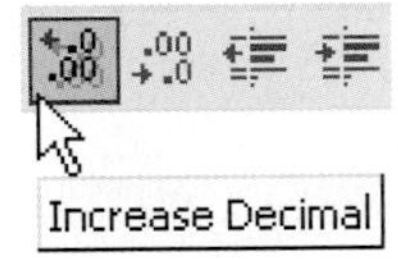

Up to this point, we have paid little attention to the number of decimal places displayed in the cells. To control the number of decimal places in a cell or range of cells, highlight the target cell(s), and click on the Increase Decimal button shown in the margin. Highlight D13:E13 now, and try it. Then click on the Decrease Decimal icon to reverse the process. Excel has no idea how many significant figures to display in a cell; you must control this aspect yourself. Now decrease the number of decimal places until only one significant figure is displayed. Note that Excel conveniently rounds the data for us.

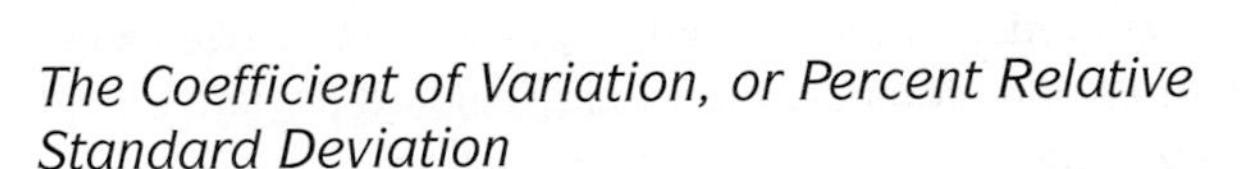

The Coefficient of Variation, or Percent Relative Standard Deviation

Our final goal for this exercise is to calculate the coefficient of variation (CV), also known as the percent relative standard deviation (%RSD) (see Section 6B-5 for an explanation of this term). As shown by Equation 6-9 on page 126, the CV is given by

$$\text{CV} = \frac{s}{x} \times 100\%$$

Click in cell E14, and type

−E13*100/B13[↵]

Then click in cell F13 and type the label `CV, %[↵]`. Your worksheet should now look something like the one that follows. Note that we have multiplied the ratio of E13 to B13 by 100 so that the relative standard deviation is expressed as a percentage. Move the decimal point to indicate only significant figures in the CV.

	A	B	C	D	E	F	G
1		Data	Deviation	Deviation^2			
2		19.4	0.383333	0.146944			
3		19.5	0.283333	0.080278			
4		19.6	0.183333	0.033611			
5		19.8	0.016667	0.000278			
6		20.1	0.316667	0.100278			
7		20.3	0.516667	0.266944			
8							
9							
10							
11	Total	118.7		0.628333			
12	N	6		0.125667	0.125667	Variance	
13	Mean	19.78333	0.283333	0.4	0.4	Standard Deviation	
14					1.791887	CV, %	
15							

We have now constructed a general-purpose spreadsheet that you may use to make basic statistical calculations. To complete this part of the exercise, select a convenient location, construct a formula to display the number of degrees of freedom, and then add a label in an adjacent cell to identify this important variable. Save the file for future use in problems and laboratory calculations. Use the spreadsheet now to check the calculations in Example 6-1. To clear the data from your worksheet, just click and drag to highlight cells B2:B7, and strike **[Delete]**. Alternatively, you may simply click on B2, and begin typing the data. Terminate each piece of data with **[↵]**. Be sure to delete the data in cells B7:D7.

As a final exercise, retrieve the spreadsheet that we created in Chapter 3 for the gravimetric determination of chloride, which we called **grav_chloride.xls**. Enter formulas into cells B12–B14 to compute the mean, standard deviation, and the RSD in parts per thousand of the percent chloride in the samples. In this example, multiply the relative standard deviation by 1000 in cell B14. Adjust the decimal point in the results to display the proper number of significant figures. The worksheet below shows the results. Save your worksheet so that you can use it as a model for making laboratory calculations.

	A	B	C	D
1	Gravimetric Determination of Chloride			
2	Samples	1	2	3
3	Mass of bottle plus sample, g	27.6115	27.2185	26.8105
4	Mass of bottle less sample, g	27.2185	26.8105	26.4517
5	Mass of sample, g	0.3930	0.4080	0.3588
6				
7	Crucible masses, with AgCl, g	21.4296	23.4915	21.8323
8	Crucible masses, empty, g	20.7926	22.8311	21.2483
9	Mass of AgCl, g	0.6370	0.6604	0.5840
10				
11	%Chloride	40.0947	40.0393	40.2625
12	Mean % Chloride	40.1322		
13	Standard Deviation, % Chloride	0.12		
14	RSD, parts per thousand	2.90		
15				

6B-4 Reliability of s as a Measure of Precision

In Chapter 7 we describe several statistical tests that are used to test hypotheses, to produce confidence intervals for results, and to reject outlying data points. Most of these tests are based on sample standard deviations. The probability that these statistical tests provide correct results increases as the reliability of s becomes greater. As N in Equation 6-4 increases, s becomes a better estimator of the population standard deviation, σ. When N is greater than about 20, s is usually a good estimator of σ, and these quantities can be assumed to be identical for most purposes. For example, if the 50 measurements in Table 6-2 (page 108) are divided into 10 subgroups of 5 each, the value of s varies widely from one subgroup to another (0.0023 – 0.0079 mL), even though the average of the computed values of s is that of the entire set (0.0056 mL). In contrast, the computed values of s for two subsets of 25 each are nearly identical (0.0054 and 0.0058 mL).

◀ CHALLENGE: Construct a spreadsheet containing the data in Table 6-2, and show that s better estimates σ as N becomes larger. Also show that s is nearly equal to σ for $N > 20$.

The rapid improvement in the reliability of s with increases in N makes it feasible to obtain a good approximation of σ when the method of measurement is not excessively time consuming and when an adequate supply of sample is available. For example, if the pH of numerous solutions is to be measured in the course of an investigation, it is useful to evaluate s in a series of preliminary experiments. This measurement is simple, requiring only that a pair of rinsed and dried electrodes be immersed in the test solution and the pH read from a scale or a display. To determine s, 20 to 30 portions of a buffer solution of fixed pH can be measured with all steps of the procedure being followed exactly. Normally, it is safe to assume that the random error in this test is the same as that in subsequent measurements. The value of s calculated from Equation 6-4 is a good estimator of the population value, σ.

Spreadsheet Summary In Chapter 2 of *Applications of Microsoft® Excel in Analytical Chemistry,* we introduce the use of Excel's Analysis ToolPak to compute the mean, standard deviation, and other quantities. In addition, the Descriptive Statistics package finds the standard error of the mean, the median, the range, the maximum and minimum values, and parameters that reflect the symmetry of the data set.

Pooling Data to Improve the Reliability of s

If we have several subsets of data, we can get a better estimate of the population standard deviation by pooling (combining) the data than by using only one data set. Again, we must assume the same sources of random error in all the measurements. This assumption is usually valid if the samples have similar compositions and have been analyzed in exactly the same way. We must also assume that the samples are randomly drawn from the same population and thus have a common value of σ.

The pooled estimate of σ, which we call s_{pooled}, is a weighted average of the individual estimates. To calculate s_{pooled}, deviations from the mean for each subset are squared; the squares of the deviations of all subsets are then summed and divided by the appropriate number of degrees of freedom. The pooled s is obtained by taking the square root of the resulting number. One degree of freedom is lost for each subset. Thus, the number of degrees of freedom for the pooled s is equal to the total number of measurements minus the number of subsets. Equation 6-7 in Feature 6-4 gives the full equation for obtaining s_{pooled} for t data sets. Example 6-2 illustrates the application of this type of computation.

FEATURE 6-4

Equation for Calculating the Pooled Standard Deviation

The equation for computing a pooled standard deviation from several sets of data takes the form

$$s_{pooled} = \sqrt{\frac{\sum_{i=1}^{N_1}(x_i - \bar{x}_1)^2 + \sum_{j=1}^{N_2}(x_j - \bar{x}_2)^2 + \sum_{k=1}^{N_3}(x_k - \bar{x}_3)^2 + \cdots}{N_1 + N_2 + N_3 + \cdots - N_t}} \tag{6-7}$$

where N_1 is the number of results in set 1, N_2 is the number in set 2, and so forth. The term N_t is the total number of data sets that are pooled.

EXAMPLE 6-2

Glucose levels are routinely monitored in patients suffering from diabetes. The glucose concentrations in a patient with mildly elevated glucose levels were determined in different months by a spectrophotometric analytical method. The patient was placed on a low-sugar diet to reduce the glucose levels. The following results were obtained during a study to determine the effectiveness of the diet. Calculate a pooled estimate of the standard deviation for the method.

Time	Glucose Concentration, mg/L	Mean Glucose, mg/L	Sum of Squares of Deviation from Mean	Standard Deviation
Month 1	1108, 1122, 1075, 1099, 1115, 1083, 1100	1100.3	1687.43	16.8
Month 2	992, 975, 1022, 1001, 991	996.2	1182.80	17.2
Month 3	788, 805, 779, 822, 800	798.8	1086.80	16.5
Month 4	799, 745, 750, 774, 777, 800, 758	771.9	2950.86	22.2

Total number of measurements = 24   Total sum of squares = 6907.89

For the first month, the sum of the squares in the next to last column was calculated as follows:

$$\begin{aligned}\text{Sum of squares} &= (1108 - 1100.3)^2 + (1122 - 1100.3)^2 \\ &+ (1075 - 1100.3)^2 + (1099 - 1100.3)^2 + (1115 - 1100.3)^2 \\ &+ (1083 - 1100.3)^2 + (1100 - 1100.3)^2 = 1687.43\end{aligned}$$

The other sums of squares were obtained similarly. The pooled standard deviation is then

$$s_{pooled} = \sqrt{\frac{6907.89}{24 - 4}} = 18.58 \approx 19 \text{ mg/ L}$$

Note this pooled value is a better estimate of σ than any of the individual s values in the last column.

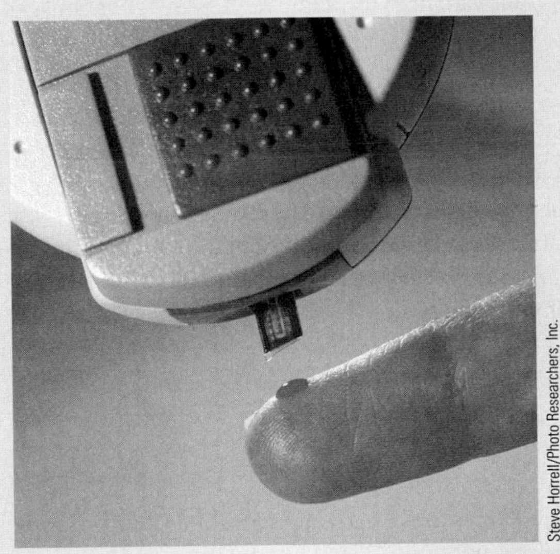

A glucose analyzer.

Note also that one degree of freedom is lost for each of the four data sets. Because 20 degrees of freedom remain, however, the calculated value of s can be considered a good estimate of σ.

Spreadsheet Summary In Chapter 2 of *Applications of Microsoft® Excel in Analytical Chemistry,* we develop a worksheet to calculate the pooled standard deviation of the data from Example 6-2. The Excel function DEVSQ() is introduced to find the sum of the squares of the deviations. As extensions of this exercise, you may use the worksheet to solve some of the pooled standard deviation problems at the end of this chapter. You can also expand the worksheet to accommodate more data points within data sets and larger numbers of sets.

6B-5 Variance and Other Measures of Precision

Chemists ordinarily use the sample standard deviation in reporting the precision of their data. We often encounter three other terms in analytical work.

Variance (s^2)

The **variance** is just the square of the standard deviation. The **sample variance** s^2 is an estimate of the population variance σ^2 and is given by

The sample **variance** s^2 is equal to the square of the sample standard deviation.

$$s^2 = \frac{\sum_{i=1}^{N} (x_i - \bar{x})^2}{N - 1} = \frac{\sum_{i=1}^{N} (d_i)^2}{N - 1} \qquad (6\text{-}8)$$

Note that the standard deviation has the same units as the data, while the variance has the units of the data squared. People who do scientific work tend to use standard deviation rather than variance as a measure of precision. It is easier to relate a measurement and its precision if they both have the same units. The advantage of using variance is that variances are additive in many situations, as we will see later in this chapter.

Relative Standard Deviation (RSD) and Coefficient of Variation (CV)

Scientists frequently quote standard deviations in relative rather than absolute terms. We calculate the relative standard deviation by dividing the standard deviation by the mean value of the data set. The relative standard deviation, RSD, is sometimes given the symbol s_r.

► The International Union of Pure and Applied Chemistry recommends that the symbol s_r be used for relative sample standard deviation and σ_r for relative population standard deviation. In equations where it is cumbersome to use RSD, we will use s_r and σ_r.

$$\text{RSD} = s_r = \frac{s}{\bar{x}}$$

The result is often expressed in parts per thousand (ppt) or in percent by multiplying this ratio by 1000 ppt or by 100%. For example,

$$\text{RSD in ppt} = \frac{s}{\bar{x}} \times 1000 \text{ ppt}$$

The **coefficient of variation,** CV, is the percent relative standard deviation.

The relative standard deviation multiplied by 100% is called the **coefficient of variation** (CV).

$$\text{CV} = \frac{s}{\bar{x}} \times 100\% \qquad (6\text{-}9)$$

Relative standard deviations often give a clearer picture of data quality than do absolute standard deviations. As an example, suppose that a copper determination has a standard deviation of 2 mg. If the sample has a mean value of 50 mg of copper, the CV for this sample is 4% $\left(\frac{2}{50} \times 100\%\right)$. For a sample containing only 10 mg, the CV is 20%.

Spread or Range (w)

The **spread,** or **range,** is another term that is sometimes used to describe the precision of a set of replicate results. It is the difference between the largest value in the set and the smallest. Thus, the spread of the data in Figure 5-1 is $(20.3 - 19.4) = 0.9$ ppm Fe. The spread in the results for month 1 in Example 6-2 is $1122 - 1075 = 47$ mg/L glucose.

EXAMPLE 6-3

For the set of data in Example 6-1, calculate (a) the variance, (b) the relative standard deviation in parts per thousand, (c) the coefficient of variation, and (d) the spread.

In Example 6-1, we found

$$\bar{x} = 0.754 \text{ ppm Pb} \qquad \text{and} \qquad s = 0.0038 \text{ ppm Pb}$$

(a) $s^2 = (0.0038)^2 = 1.4 \times 10^{-5}$

(b) $\text{RSD} = \dfrac{0.0038}{0.754} \times 1000 \text{ ppt} = 5.0 \text{ ppt}$

(c) $\text{CV} = \dfrac{0.0038}{0.754} \times 100\% = 0.50\%$

(d) $w = 0.760 - 0.751 = 0.009$ ppm Pb

6C STANDARD DEVIATION OF CALCULATED RESULTS

We must often estimate the standard deviation of a result that has been calculated from two or more experimental data points, each of which has a known sample standard deviation. As shown in Table 6-4, the way such estimates are made depends on the types of calculations that are involved. The relationships shown in this table are derived in Appendix 9.

6C-1 Standard Deviation of a Sum or Difference

Consider the summation

$$\begin{array}{rl} +\,0.50 & (\pm\,0.02) \\ +\,4.10 & (\pm\,0.03) \\ -\,1.97 & (\pm\,0.05) \\ \hline 2.63 & \end{array}$$

where the numbers in parentheses are absolute standard deviations. If the three individual standard deviations happen by chance to have the same sign, the standard deviation of the sum could be as large as $+0.02 + 0.03 + 0.05 = +0.10$ or $-0.02 - 0.03 - 0.05 = -0.10$. On the other hand, it is possible that the three standard deviations could combine to give an accumulated value of zero: $-0.02 - 0.03 + 0.05 = 0$ or $+0.02 + 0.03 - 0.05 = 0$. More likely, however, the standard deviation of the sum will lie between these two extremes. The variance of a sum or difference is equal to the sum of the individual variances.[3] The most probable value

◀ The variance of a sum or difference is equal to the *sum* of the variances of the numbers making up that sum or difference.

[3]See P. R. Bevington and D. K. Robinson, *Data Reduction and Error Analysis for the Physical Sciences,* 2nd ed., pp. 41–50. New York: McGraw-Hill, 1992.

TABLE 6-4
Error Propagation in Arithmetic Calculations

Type of Calculation	Example*	Standard Deviation of y†	
Addition or subtraction	$y = a + b - c$	$s_y = \sqrt{s_a^2 + s_b^2 + s_c^2}$	(1)
Multiplication or division	$y = a \times b/c$	$\frac{s_y}{y} = \sqrt{\left(\frac{s_a}{a}\right)^2 + \left(\frac{s_b}{b}\right)^2 + \left(\frac{s_c}{c}\right)^2}$	(2)
Exponentiation	$y = a^x$	$\frac{s_y}{y} = x\left(\frac{s_a}{a}\right)$	(3)
Logarithm	$y = \log_{10} a$	$s_y = 0.434 \frac{s_a}{a}$	(4)
Antilogarithm	$y = \text{antilog}_{10}\, a$	$\frac{s_y}{y} = 2.303\, s_a$	(5)

*a, b, and c are experimental variables with standard deviations of s_a, s_b, and s_c, respectively
†These relationships are derived in Appendix 9. The values for s_y/y are absolute values if y is a negative number.

for a standard deviation of a sum or difference can be found by taking the square root of the sum of the squares of the individual absolute standard deviations. So, for the computation

$$y = a(\pm s_a) + b(\pm s_b) - c(\pm s_c)$$

The variance of y, s_y^2, is given by

$$s_y^2 = s_a^2 + s_b^2 + s_c^2$$

Hence, the standard deviation of the result s_y is

For a sum or a difference, the *absolute standard deviation of the answer* is the square root of the sum of the squares of the *absolute standard deviations* of the numbers used to calculate the sum or difference.

$$s_y = \sqrt{s_a^2 + s_b^2 + s_c^2} \tag{6-10}$$

where s_a, s_b, and s_c are the standard deviations of the three terms making up the result. Substituting the standard deviations from the example gives

$$s_y = \sqrt{(0.02)^2 + (0.03)^2 + (0.05)^2} = 0.06$$

and the sum should be reported as 2.64 (±0.06).

6C-2 Standard Deviation of a Product or Quotient

Consider the following computation where the numbers in parentheses are again absolute standard deviations:

$$\frac{4.10(\pm 0.02) \times 0.0050(\pm 0.0001)}{1.97(\pm 0.04)} = 0.010406(\pm ?)$$

In this situation, the standard deviations of two of the numbers in the calculation are larger than the result itself. Evidently, we need a different approach for multiplication and division. As shown in Table 6-4, the *relative standard deviation* of a product or quotient is determined by the *relative standard deviations* of the numbers forming the computed result. For example, in the case of

$$y = \frac{a \times b}{c} \tag{6-11}$$

we obtain the relative standard deviation s_y/y of the result by summing the squares of the relative standard deviations of a, b, and c and extracting the square root of the sum:

$$\frac{s_y}{y} = \sqrt{\left(\frac{s_a}{a}\right)^2 + \left(\frac{s_b}{b}\right)^2 + \left(\frac{s_c}{c}\right)^2} \tag{6-12}$$

For multiplication or division, the *relative standard deviation of the answer* is the square root of the sum of the squares of the *relative standard deviations* of the numbers that are multiplied or divided.

Applying this equation to the numerical example gives

$$\frac{s_y}{y} = \sqrt{\left(\frac{0.02}{4.10}\right)^2 + \left(\frac{0.0001}{0.0050}\right)^2 + \left(\frac{0.04}{1.97}\right)^2}$$

$$= \sqrt{(0.0049)^2 + (0.0200)^2 + (0.0203)^2} = 0.0289$$

To complete the calculation, we must find the absolute standard deviation of the result,

◀ To find the absolute standard deviation in a product or a quotient, first find the relative standard deviation in the result and then multiply it by the result.

$$s_y = y \times (0.0289) = 0.0104 \times (0.0289) = 0.000301$$

and we can write the answer and its uncertainty as 0.0104 (±0.0003). Note that if y is a negative number, we should treat s_y/y as an absolute value.

Example 6-4 demonstrates the calculation of the standard deviation of the result for a more complex calculation.

EXAMPLE 6-4

Calculate the standard deviation of the result of

$$\frac{[14.3(\pm 0.2) - 11.6(\pm 0.2)] \times 0.050(\pm 0.001)}{[820(\pm 10) + 1030(\pm 5)] \times 42.3(\pm 0.4)} = 1.725(\pm ?) \times 10^{-6}$$

First, we must calculate the standard deviation of the sum and the difference. For the difference in the numerator,

$$s_a = \sqrt{(0.2)^2 + (0.2)^2} = 0.283$$

and for the sum in the denominator,

$$s_b = \sqrt{(10)^2 + (5)^2} = 11.2$$

(continued)

We may then rewrite the equation as

$$\frac{2.7(\pm 0.283) \times 0.050(\pm 0.001)}{1850(\pm 11.2) \times 42.3(\pm 0.4)} = 1.725 \times 10^{-6}$$

The equation now contains only products and quotients, and Equation 6-12 applies. Thus,

$$\frac{s_y}{y} = \sqrt{\left(\frac{0.283}{2.7}\right)^2 + \left(\frac{0.001}{0.050}\right)^2 + \left(\frac{11.2}{1850}\right)^2 + \left(\frac{0.4}{42.3}\right)^2} = 0.017$$

To obtain the absolute standard deviation, we write

$$s_y = y \times 0.107 = 1.725 \times 10^{-6} \times 0.107 = 0.185 \times 10^{-6}$$

and round the answer to $1.7(\pm 0.2) \times 10^{-6}$.

6C-3 Standard Deviations in Exponential Calculations

Consider the relationship

$$y = a^x$$

where the exponent x can be considered free of uncertainty. As shown in Table 6-4 and Appendix 9, the relative standard deviation in y resulting from the uncertainty in a is

$$\frac{s_y}{y} = x\left(\frac{s_a}{a}\right) \tag{6-13}$$

Thus, the relative standard deviation of the square of a number is twice the relative standard deviation of the number, the relative standard deviation of the cube root of a number is one-third that of the number, and so forth. Examples 6-5 and 6-6 illustrate these calculations.

EXAMPLE 6-5

The standard deviation in measuring the diameter d of a sphere is ± 0.02 cm. What is the standard deviation in the calculated volume V of the sphere if $d =$ 2.15 cm?

From the equation for the volume of a sphere, we have

$$V = \frac{4}{3}\pi r^3 = \frac{4}{3}\pi\left(\frac{d}{2}\right)^3 = \frac{4}{3}\pi\left(\frac{2.15}{2}\right)^3 = 5.20 \text{ cm}^3$$

Here we may write

$$\frac{s_V}{V} = 3 \times \frac{s_d}{d} = 3 \times \frac{0.02}{2.15} = 0.0279$$

The absolute standard deviation in V is then

$$s_V = 5.20 \times 0.0279 = 0.145$$

Thus,

$$V = 5.2\,(\pm 0.1)\ \text{cm}^3$$

EXAMPLE 6-6

The solubility product K_{sp} for the silver salt AgX is $4.0\,(\pm 0.4) \times 10^{-8}$. The molar solubility of AgX in water is

$$\text{solubility} = (K_{sp})^{1/2} = (4.0 \times 10^{-8})^{1/2} = 2.0 \times 10^{-4}\ \text{M}$$

What is the uncertainty in the calculated solubility of AgX in water? Substituting y = solubility, $a = K_{sp}$, and $x = 1/2$ into Equation 6-13 gives

$$\frac{s_a}{a} = \frac{0.4 \times 10^{-8}}{4.0 \times 10^{-8}}$$

$$\frac{s_y}{y} = \frac{1}{2} \times \frac{0.4}{4.0} = 0.05$$

$$s_y = 2.0 \times 10^{-4} \times 0.05 = 0.1 \times 10^{-4}$$

$$\text{solubility} = 2.0\,(\pm 0.1) \times 10^{-4}\ \text{M}$$

It is important to note that the error propagation in taking a number to a power is different from the error propagation in multiplication. For example, consider the uncertainty in the square of $4.0\,(\pm 0.2)$. Here, the relative error in the result (16.0) is given by Equation 6-13:

$$\frac{s_y}{y} = 2\left(\frac{0.2}{4}\right) = 0.1, \quad \text{or } 10\%$$

The result is then $y = 16\,(\pm 2)$.

Consider now the situation where y is the product of *two independently measured* numbers that by chance happen to have identical values of $a_1 = 4.0\,(\pm 0.2)$

and $a_2 = 4.0\ (\pm 0.2)$. Here, the relative error of the product $a_1a_2 = 16.0$ is given by Equation 6–12:

$$\frac{s_y}{y} = \sqrt{\left(\frac{0.2}{4}\right)^2 + \left(\frac{0.2}{4}\right)^2} = 0.07, \quad \text{or } 7\%$$

► The relative standard deviation of $y = a^3$ is *not* the same as the relative standard deviation of the product of three independent measurements $y = abc$, where $a = b = c$.

The result is now $y = 16\ (\pm 1)$. The reason for the difference between this and the previous result is that with measurements that are independent of one another, the sign associated with one error can be the same as or different from that of the other error. If they happen to be the same, the error is identical to that encountered in the first case, where the signs *must* be the same. On the other hand, if one sign is positive and the other negative, the relative errors tend to cancel. Thus the probable error for the case of independent measurements lies somewhere between the maximum (10%) and zero.

6C-4 Standard Deviations of Logarithms and Antilogarithms

The last two entries in Table 6-4 show that for $y = \log a$

$$s_y = 0.434\frac{s_a}{a} \tag{6-14}$$

and for $y = \text{antilog } a$

$$\frac{s_y}{y} = 2.303 s_a \tag{6-15}$$

Thus, the *absolute* standard deviation of the logarithm of a number is determined by the *relative* standard deviation of the number; conversely, the *relative* standard deviation of the antilogarithm of a number is determined by the *absolute* standard deviation of the number. Example 6-7 illustrates these calculations.

EXAMPLE 6-7

Calculate the absolute standard deviations of the results of the following calculations. The absolute standard deviation for each quantity is given in parentheses.

(a) $y = \log[2.00(\pm 0.02) \times 10^{-4}] = -3.6990 \pm ?$
(b) $y = \text{antilog}[1.200(\pm 0.003)] = 15.849 \pm ?$
(c) $y = \text{antilog}[45.4(\pm 0.3)] = 2.5119 \times 10^{45} \pm ?$

(a) Referring to Equation 6–14, we see that we must multiply the *relative* standard deviation by 0.434:

$$s_y = 0.434 \times \frac{0.02 \times 10^{-4}}{2.00 \times 10^{-4}} = 0.004$$

Thus,

$$y = \log[2.00(\pm 0.02) \times 10^{-4}] = -3.699\ (\pm 0.004)$$

(b) Applying Equation 6-15, we have

$$\frac{s_y}{y} = 2.303 \times (0.003) = 0.0069$$

$$s_y = 0.0069y = 0.0069 \times 15.849 = 0.11$$

Thus,

$$y = \text{antilog}[1.200(\pm 0.003)] = 15.8 \pm 0.1$$

(c) $\dfrac{s_y}{y} = 2.303 \times (0.3) = 0.69$

$$s_y = 0.69y = 0.69 \times 2.5119 \times 10^{45} = 1.7 \times 10^{45}$$

Thus,

$$y = \text{antilog}[45.4(\pm 0.3)] = 2.5(\pm 1.7) \times 10^{45} = 3\ (\pm 2) \times 10^{45}$$

Example 6-7c demonstrates that a large absolute error is associated with the antilogarithm of a number with few digits beyond the decimal point. This large uncertainty is due to the fact that the numbers to the left of the decimal (the *characteristic*) serve only to locate the decimal point. The large error in the antilogarithm results from the relatively large uncertainty in the *mantissa* of the number (that is, 0.4 ± 0.3).

6D REPORTING COMPUTED DATA

A numerical result is worthless to users of the data unless they know something about its quality. Therefore, it is always essential to indicate your best estimate of the reliability of your data. One of the best ways of indicating reliability is to give a confidence interval at the 90% or 95% confidence level, as we describe in Section 7A-2. Another method is to report the absolute standard deviation or the coefficient of variation of the data. In this case, it is a good idea to indicate the number of data points that were used to obtain the standard deviation so that the user has some idea of the reliability of s. A less satisfactory but more common indicator of data quality is the **significant figure convention.**

6D-1 Significant Figures

The **significant figures** in a number are all of the certain digits plus the first uncertain digit.

We often indicate the probable uncertainty associated with an experimental measurement by rounding the result so that it contains only **significant figures.** By definition, the significant figures in a number are all of the digits known with certainty *plus the first uncertain digit.* For example, when you read the 50-mL buret section shown in Figure 6-5, you can easily tell that the liquid level is greater than 30.2 mL and less than 30.3 mL. You can also estimate the position of the liquid between the graduations to about 0.02 mL. So, using the significant figure convention, you should report the volume delivered as 30.24 mL, which has four significant figures. Note that the first three digits are certain and the last digit (4) is uncertain.

► Rules for determining the number of significant figures:

1. Disregard all initial zeros.
2. Disregard all final zeros *unless they follow a decimal point.*
3. All remaining digits, including zeros between nonzero digits, are significant.

A zero may or may not be significant, depending on its location in a number. A zero surrounded by other digits is always significant (such as in 30.24 mL) because it is read directly and with certainty from a scale or instrument readout. On the other hand, zeros that only locate the decimal point for us are not. If we write 30.24 mL as 0.03024 L, the number of significant figures is the same. The only function of the zero before the 3 is to locate the decimal point, so it is not significant. Terminal, or final, zeros may or may not be significant. For example, if the volume of a beaker is expressed as 2.0 L, the presence of the zero tells us that the volume is known to a few tenths of a liter, so both the 2 and the zero are significant figures. If this same volume is reported as 2000 mL, the situation becomes confusing. The last two zeros are not significant because the uncertainty is still a few tenths of a liter, or a few hundred milliliters. To follow the significant figure convention in a case such as this, use scientific notation and report the volume as 2.0×10^3 mL.

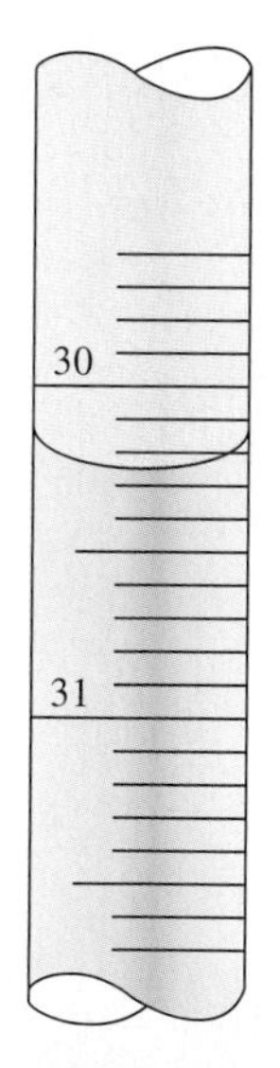

Figure 6-5 Buret section showing the liquid level and meniscus.

6D-2 Significant Figures in Numerical Computations

Care is required to determine the appropriate number of significant figures in the result of an arithmetic combination of two or more numbers.[4]

Sums and Differences

For addition and subtraction, the number of significant figures can be found by visual inspection. For example, in the expression

$$3.4 + 0.020 + 7.31 = 10.730 \quad \text{(round to 10.7)}$$

the second and third decimal places in the answer cannot be significant because 3.4 is uncertain in the first decimal place. Hence, the result should be rounded to 10.7. Note that the result contains three significant digits, even though two of the numbers involved have only two significant figures.

► Express data in scientific notation to avoid confusion in determining whether terminal zeros are significant.

Products and Quotients

A rule of thumb that is sometimes suggested for multiplication and division is to round off the answer so that it contains the same number of significant digits as the original number with the smallest number of significant digits. Unfortunately,

[4]For an extensive discussion of propagation of significant figures, see L. M. Schwartz, *J. Chem. Educ.*, **1985,** *62,* 693.

this procedure often leads to incorrect rounding. For example, consider the two calculations

$$\frac{24 \times 4.52}{100.0} = 1.08 \qquad \text{and} \qquad \frac{24 \times 4.02}{100.0} = 0.965$$

◀ As a rule of thumb, for addition and subtraction, the result should contain the same number of decimal places as the number with the *smallest* number of decimal places.

By the rule of thumb, the first answer would be rounded to 1.1 and the second to 0.96. If, however, we assume a unit uncertainty in the last digit of each number in the first quotient, the relative uncertainties associated with each of these numbers are 1/24, 1/452, and 1/1000. Because the first relative uncertainty is much larger than the other two, the relative uncertainty in the result is also 1/24; the absolute uncertainty is then

$$1.08 \times \frac{1}{24} = 0.045 \approx 0.04$$

◀ When adding and subtracting numbers in scientific notation, express the numbers to the same power of 10. For example,

$$\begin{aligned} 2.432 \times 10^6 &= 2.432 \times 10^6 \\ +6.512 \times 10^4 &= +0.06512 \times 10^6 \\ -1.227 \times 10^5 &= \underline{-0.1227 \times 10^6} \\ &\ \ 2.37442 \times 10^6 \end{aligned}$$

(round to 2.374×10^6)

By the same argument the absolute uncertainty of the second answer is given by

$$0.965 \times \frac{1}{24} = 0.040 \approx 0.04$$

Therefore, the first result should be rounded to three significant figures, or 1.08, but the second should be rounded to only two, that is, 0.96.

◀ The weak link for multiplication and division is the number of *significant figures* in the number with the smallest number of significant figures. *Use this rule of thumb with caution.*

Logarithms and Antilogarithms

Be especially careful in rounding the results of calculations involving logarithms. The following rules apply to most situations.[5] These rules are illustrated in Example 6-8.

1. In a logarithm of a number, keep as many digits to the right of the decimal point as there are significant figures in the original number.
2. In an antilogarithm of a number, keep as many digits as there are digits to the right of the decimal point in the original number.

◀ The number of significant figures in the *mantissa,* or the digits to the right of the decimal point of a logarithm, is the same as the number of significant figures in the original number. Thus, $\log (9.57 \times 10^4) = 4.981$. Since 9.57 has 3 significant figures, there are 3 digits to the right of the decimal point in the result.

EXAMPLE 6-8

Round the following answers so that only significant digits are retained: (a) $\log 4.000 \times 10^{-5} = -4.3979400$ and (b) antilog $12.5 = 3.162277 \times 10^{12}$.

(a) Following rule 1, we retain 4 digits to the right of the decimal point:

$$\log 4.000 \times 10^{-5} = -4.3979$$

(b) Following rule 2, we may retain only 1 digit:

$$\text{antilog } 12.5 = 3 \times 10^{12}$$

[5] D. E. Jones, *J. Chem. Educ.*, **1971,** *49,* 753.

6D-3 Rounding Data

Always round the computed results of a chemical analysis in an appropriate way. For example, consider the following replicate results: 61.60, 61.46, 61.55, and 61.61. The mean of this data set is 61.555, and the standard deviation is 0.069. When we round the mean, do we take 61.55 or 61.56? A good guide to follow when rounding a 5 is always to round to the nearest even number. In this way, we eliminate any tendency to round in a set direction. In other words, there is an equal likelihood that the nearest even number will be the higher or the lower in any given situation. Accordingly, we might choose to report the result as 61.56 ± 0.07. If we had reason to doubt the reliability of the estimated standard deviation, we might report the result as 61.6 ± 0.1.

► In rounding a number ending in 5, always round so that the result ends with an even number. Thus, 0.635 rounds to 0.64 and 0.625 rounds to 0.62.

We should note that *it is seldom justifiable to keep more than one significant figure in the standard deviation* because the standard deviation contains error as well. For certain specialized purposes, such as reporting uncertainties in physical constants in research articles, it may be useful to keep two significant figures, and there is certainly nothing wrong with including a second digit in the standard deviation. However, it is important to recognize that the uncertainty usually lies in the first digit.[6]

6D-4 Expressing Results of Chemical Computations

Two cases are encountered when reporting the results of chemical calculations. If the standard deviations of the values making up the final calculation are known, we then apply the propagation of error methods discussed in Section 6C and round the results to contain significant digits. Often, however, you are asked to perform calculations with data whose precision is indicated only by the significant figure convention. In this second case, common sense assumptions must be made as to the uncertainty in each number. With these assumptions, the uncertainty of the final result is then estimated using the methods presented in Section 6C. Finally, the result is rounded so that it contains only significant digits.

It is especially important to postpone rounding until the calculation is completed. At least one extra digit beyond the significant digits should be carried through all of the computations in order to avoid a *rounding error.* This extra digit is sometimes called a "guard" digit. Modern calculators generally retain several extra digits that are not significant, and the user must be careful to round final results properly so that only significant figures are included. Example 6-9 illustrates this procedure.

EXAMPLE 6-9

A 3.4842-g sample of a solid mixture containing benzoic acid, C_6H_5COOH (122.123 g/mol), was dissolved and titrated with base to a phenolphthalein end point. The acid consumed 41.36 mL of 0.2328 M NaOH. Calculate the percent benzoic acid (HBz) in the sample.

---

[6]For more details on this topic, direct your Web browser to **http://www.chem.uky.edu/courses/che226/download/CI_for_sigma.html.**

As shown in Section 13C-3, the calculation takes the following form:

$$\%\text{HBz} = \frac{41.36 \text{ mL} \times 0.2328 \dfrac{\text{mmol NaOH}}{\text{mL NaOH}} \times \dfrac{1 \text{ mmol HBz}}{\text{mmol NaOH}} \times \dfrac{122.123 \text{ g HBz}}{1000 \text{ mmol HBz}}}{3.842 \text{ g sample}}$$

$$\times 100\%$$

$$= 33.749\%$$

Since all operations are either multiplication or division, the relative uncertainty of the answer is determined by the relative uncertainties of the experimental data. Let us estimate what these uncertainties are.

1. The position of the liquid level in a buret can be estimated to ±0.02 mL (Figure 6-5). Initial and final readings must be made, however, so that the standard deviation of the volume s_V will be

$$s_V = \sqrt{(0.02)^2 + (0.02)^2} = 0.028 \text{ mL} \qquad \text{(Equation 6-10)}$$

The relative uncertainty in volume s_V/V is then

$$\frac{s_V}{V} = \frac{0.028}{41.36} \times 1000 \text{ ppt} = 0.68 \text{ ppt}$$

2. Generally, the absolute uncertainty of a mass obtained with an analytical balance will be on the order of ±0.0001 g. Thus the relative uncertainty of the denominator s_D/D is

$$\frac{0.0001}{3.4842} \times 1000 \text{ ppt} = 0.029 \text{ ppt}$$

3. Usually we can assume that the absolute uncertainty in the molarity of a reagent solution is ±0.0001, and so the relative uncertainty in the molarity of NaOH, s_M/M, is

$$\frac{s_M}{M} = \frac{0.0001}{0.2328} \times 1000 \text{ ppt} = 0.43 \text{ ppt}$$

4. The relative uncertainty in the molar mass of HBz is several orders of magnitude smaller than any of the three experimental data and is of no consequence. Note, however, that we should retain enough digits in the calculation so that the molar mass is given to at least one more digit (the guard digit) than any of the experimental data. Thus, in the calculation, we use 122.123 for the molar mass (here we are carrying two extra digits).
5. No uncertainty is associated with 100% and the 1000 mmol HBz, since these are exact numbers.

(continued)

Substituting the three relative uncertainties into Equation 6-12, we obtain

$$\frac{s_y}{y} = \sqrt{\left(\frac{0.028}{41.36}\right)^2 + \left(\frac{0.0001}{3.4842}\right)^2 + \left(\frac{0.0001}{0.2328}\right)^2}$$
$$= \sqrt{(0.68)^2 + (0.029)^2 + (0.43)^2}$$
$$= 8.02 \times 10^{-4}$$
$$s_y = 8.02 \times 10^{-4} \times y = 8.02 \times 10^{-4} \times 33.749 = 0.027$$

Thus the uncertainty in the calculated result is 0.03% HBz, and we should report the result as 33.75% HBz or, better, 33.75 (± 0.03)% HBz.

► There is no relationship between the number of digits displayed on a computer screen or a calculator and the true number of significant figures.

We must emphasize that rounding decisions are an important part of *every calculation* and that such decisions *cannot* be based on the number of digits displayed on an instrument readout, on the computer screen, or on a calculator display.

WEB WORKS

The National Institute of Standards and Technology (NIST) maintains Web pages of statistical data for testing software. Direct your Web browser to **http://chemistry.brookscole.com/skoogfac/.** From the Chapter Resources menu, choose Web Works, and locate the Chapter 6 section. Here you will find a link to the NIST site. Browse the site to see what kinds of data are available for testing. We use two of the NIST data sets in Problems 6-21 and 6-22. Find the software diagnostics site for the "Healthcare Standards Roadmap Project." Describe why the project is needed and the NIST approach.

QUESTIONS AND PROBLEMS

6-1. Define
  *(a) spread or range.
  (b) coefficient of variation.
  *(c) significant figures.
  (d) Gaussian distribution.

6-2. Differentiate between
  *(a) sample standard deviation and sample variance.
  (b) population mean and sample mean.
  *(c) accuracy and precision.
  (d) random and systematic error.

6-3. Distinguish between
  *(a) the meaning of the word "sample" as it is used in a chemical and in a statistical sense.
  (b) the sample standard deviation and the population standard deviation.

6-4. What is the standard error of a mean? Why is the standard deviation of the mean lower than the standard deviation of the data points in a set?

***6-5.** From the Gaussian error curve, what is the probability that a result from a population lies between 0 and $+1\sigma$ of the mean? What is the probability of a result occurring that is between +1 and $+2\sigma$ of the mean?

6-6. From the normal curve of error, find the probability that a result is outside the limits of $\pm 2\sigma$ from the mean. What is the probability that a result has a more negative deviation from the mean than -2σ?

6-7. Consider the following sets of replicate measurements:

*A	B	*C	D	*E	F
3.5	70.24	0.812	2.7	70.65	0.514
3.1	70.22	0.792	3.0	70.63	0.503
3.1	70.10	0.794	2.6	70.64	0.486
3.3		0.900	2.8	70.21	0.497
2.5			3.2		0.472

For each set, calculate the (a) mean; (b) median; (c) spread, or range; (d) standard deviation; and (e) coefficient of variation.

6-8. The accepted values for the sets of data in Problem 6-7 are as follows: *set A, 3.0; set B, 70.05; *set C, 0.830; set D, 3.4; *set E, 70.05; set F, 0.525. For the mean of each set, calculate (a) the absolute error and (b) the relative error in parts per thousand.

6-9. Estimate the absolute standard deviation and the coefficient of variation for the results of the following calculations. Round each result so that it contains only significant digits. The numbers in parentheses are absolute standard deviations.

*(a) $y = 5.75(\pm 0.03) + 0.833(\pm 0.001) - 8.021(\pm 0.001) = -1.438$

(b) $y = 18.97(\pm 0.04) + 0.0025(\pm 0.0001) + 2.29(\pm 0.08) = 21.2625$

*(c) $y = 66.2(\pm 0.3) \times 1.13(\pm 0.02) \times 10^{-17} = 7.4806 \times 10^{-16}$

(d) $y = 251(\pm 1) \times \dfrac{860(\pm 2)}{1.673(\pm 0.006)} = 129{,}025.70$

*(e) $y = \dfrac{157(\pm 6) - 59(\pm 3)}{1220(\pm 1) + 77(\pm 8)} = 7.5559 \times 10^{-2}$

(f) $y = \dfrac{1.97(\pm 0.01)}{243(\pm 3)} = 8.106996 \times 10^{-3}$

6-10. Estimate the absolute standard deviation and the coefficient of variation for the results of the following calculations. Round each result to include only significant figures. The numbers in parentheses are absolute standard deviations.

*(a) $y = 1.02(\pm 0.02) \times 10^{-8} - 3.54(\pm 0.2) \times 10^{-9}$

(b) $y = 90.31(\pm 0.08) - 89.32(\pm 0.06) + 0.200(\pm 0.004)$

*(c) $y = 0.0020(\pm 0.0005) \times 20.20(\pm 0.02) \times 300(\pm 1)$

(d) $y = \dfrac{163(\pm 0.03) \times 10^{-14}}{1.03(\pm 0.04) \times 10^{-16}}$

*(e) $y = \dfrac{100(\pm 1)}{2(\pm 1)}$

(f) $y = \dfrac{2.45(\pm 0.02) \times 10^{-2} - 5.06(\pm 0.06) \times 10^{-3}}{23.2(\pm 0.7) + 9.11(\pm 0.08)}$

6-11. Calculate the absolute standard deviation and the coefficient of variation for the results of the following calculations. Round each result to include only significant figures. The numbers in parentheses are absolute standard deviations.

*(a) $y = \log[2.00(\pm 0.03) \times 10^{-4}]$

(b) $y = \log[4.42(\pm 0.01) \times 10^{37}]$

*(c) $y = \text{antilog}[1.200(\pm 0.003)]$

(d) $y = \text{antilog}[49.54(\pm 0.04)]$

6-12. Calculate the absolute standard deviation and the coefficient of variation for the results of the following calculations. Round each result to include only significant figures. The numbers in parentheses are absolute standard deviations.

*(a) $y = [4.73(\pm 0.03) \times 10^{-4}]^3$

(b) $y = [2.145(\pm 0.002)]^{1/4}$

***6-13.** The inside diameter of an open cylindrical tank was measured. The results of four replicate measurements were 5.4, 5.2, 5.5, and 5.2 m. Measurements of the height of the tank yielded 9.8, 9.9, and 9.6 m. Calculate the volume in liters of the tank and the standard deviation of the result.

6-14. In a volumetric determination of an analyte A, the data obtained and their standard deviations are as follows:

Initial buret reading	0.23 mL	0.02 mL
Final buret reading	8.76 mL	0.03 mL
Sample weight	50.0 mg	0.2 mg

From the data, find the coefficient of variation of the final result for the % A that is obtained by using the equation below (the equivalent weight can be treated as having no uncertainty).

$$\%\text{ A} = \text{titrant volume} \times \text{equivalent weight} \times 100/\text{sample weight}$$

***6-15.** In Chapter 28, we discuss inductively coupled plasma atomic emission spectrometry. In that method, the number of atoms excited to a particular energy level is a strong function of temperature. For an element of excitation energy E in joules (J), the measured ICP emission signal S can be written

$$S = k'e^{-E/kT}$$

where k' is a constant nearly temperature independent, T is the absolute temperature in Kelvin (K), and k is Boltzmann's constant (1.3807×10^{-23} J K^{-1}). For an ICP of average temperature 6000 K and for Cu with an excitation energy of 6.12×10^{-19} J, how precisely does the ICP temperature need to be controlled for the coefficient of variation in the emission signal to be 1% or less.

6-16. In Chapter 24, we show that quantitative molecular absorption spectrometry is based on Beer's law, which can be written

$$-\log T = \varepsilon b c_X$$

where T is the transmittance of a solution of an analyte X, b is the thickness of the absorbing solution, c_X is the molar concentration of X, and ε is an experimentally determined constant. By measuring a series of standard solutions of X, εb was found to have a value of $2505(\pm 12)\ M^{-1}$, where the number in parentheses is the absolute standard deviation.

An unknown solution of X was measured in a cell identical to the one used to determine εb. The replicate results were $T = 0.273$, 0.276, 0.268, and 0.274. Calculate (a) the molar concentration of the analyte c_X; (b) the absolute standard deviation of the c_X, and (c) the coefficient of variation of c_X.

***6-17.** Analysis of several plant-food preparations for potassium ion yielded the following data:

Sample	Percent K^+
1	5.15, 5.03, 5.04, 5.18, 5.20
2	7.18, 7.17, 6.97
3	4.00, 3.93, 4.15, 3.86
4	4.68, 4.85, 4.79, 4.62
5	6.04, 6.02, 5.82, 6.06, 5.88

The preparations were randomly drawn from the same population.

(a) Find the mean and standard deviation s for each sample.

(b) Obtain the pooled value s_{pooled}.

(c) Why is this a better estimate of σ than the standard deviation from any one sample?

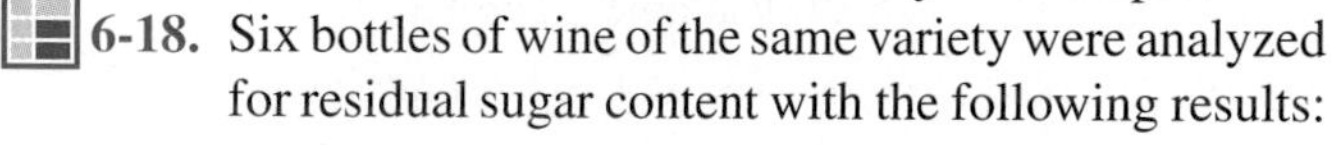

6-18. Six bottles of wine of the same variety were analyzed for residual sugar content with the following results:

Bottle	Percent (w/v) Residual Sugar
1	0.99, 0.84, 1.02
2	1.02, 1.13, 1.17, 1.02
3	1.25, 1.32, 1.13, 1.20, 1.12
4	0.72, 0.77, 0.61, 0.58
5	0.90, 0.92, 0.73
6	0.70, 0.88, 0.72, 0.73

(a) Evaluate the standard deviation s for each set of data.

(b) Pool the data to obtain an absolute standard deviation for the method.

***6-19.** Nine samples of illicit heroin preparations were analyzed in duplicate by a gas chromatographic method. The samples can be assumed to have been drawn randomly from the same population. Pool the following data to establish an estimate of σ for the procedure.

Sample	Heroin, %	Sample	Heroin, %
1	2.24, 2.27	6	1.07, 1.02
2	8.4, 8.7	7	14.4, 14.8
3	7.6, 7.5	8	21.9, 21.1
4	11.9, 12.6	9	8.8, 8.4
5	4.3, 4.2		

6-20. Calculate a pooled estimate of σ from the following spectrophotometric analysis for NTA (nitrilotriacetic acid) in water from the Ohio River:

Sample	NTA, ppb
1	12, 17, 15, 8
2	32, 31, 32
3	25, 29, 23, 29, 26

6-21. Direct your Web browser to **http://chemistry.brookscole.com/skoogfac/.** From the Chapter Resources menu, choose Web Works and locate the Chapter 6 section. Find the link to the NIST page for the speed of light measurements. After reading the page, click on the link labeled Data file *(ASCII Format).* The page that you see contains 100 values for the speed of light as measured by E. N. Dorsey, *Transactions of the American Philosophical Society,* **1944,** *34,* pp. 1–110, Table 22. Once you have the data on the screen, use your mouse to highlight only the 100 values of the speed of light, and click on **Edit/Copy** to place the data on the clipboard. Then start Excel with a clean spreadsheet, and click on **Edit/Paste** to insert the data in column A. Now, find the mean and standard deviation, and compare your values with those presented when you click on Certified Values on the NIST Web page. Be sure to increase the displayed number of digits in your spreadsheet so that you can compare all of the results. Comment on any differences between your results and the certified values. Suggest possible sources for the differences.

6-22. Challenge Problem. Direct your Web browser to **http://chemistry.brookscole.com/skoogfac/.** Locate Web Works, Chapter 6, from the Chapter Resources menu. Find the link to the NIST page containing the atomic mass of silver as presented by L. J. Powell, T. J. Murphy, and J. W. Gramlich, "The Absolute Isotopic Abundance & Atomic Weight of a Reference Sample of Silver," *NBS Journal of Research,* **1982,** *87,* pp. 9–19. The page that you see contains 48 values for the atomic mass of silver: 24 determined by one instrument and 24 determined by another.

(a) We will first import the data. Once you have the data on the screen, click on **File/Save As...,** and Ag_Atomic_Wtt.dat will appear in the File name blank. Click on Save. Then start Excel, click on **File/Open**, and be sure that All Files (*.*) is selected in the Files of type: blank. Find Ag_Atomic_Wtt.dat, highlight the file name, and click on Open. After the Test Import Wizard appears, click on Delimited and then Next. In the next window, be sure that only Space is checked, and scroll down to the bottom of the file to be sure that Excel draws vertical lines to separate the two columns of atomic mass data; then click on Finish. The data should then appear in the spreadsheet. The data in the first 60 rows will look a bit disorganized, but beginning in row 61, the atomic mass data should appear in two columns of the spreadsheet.

(b) Now find the mean and standard deviation of the two sets of data. Also determine the coefficient of variation for each data set.

(c) Next find the pooled standard deviation of the two sets of data and compare your value with that for the certified residual standard deviation presented when you click on Certified Values on the NIST Web page. Be sure to increase the displayed number of digits in your spreadsheet so that you can compare all of the results.

(d) Compare your sum of squares of the deviations from the two means with the NIST value for the certified sum of squares (within instrument). Comment on any differences that you find between your results and the certified values, and suggest possible reasons for the differences.

(e) Compare the mean values of the two sets of data for the atomic mass of silver with the currently accepted value. Assuming the currently accepted value is the true value, determine the absolute error and the percent relative error.

InfoTrac College Edition

For additional readings, go to InfoTrac College Edition, your online research library, at

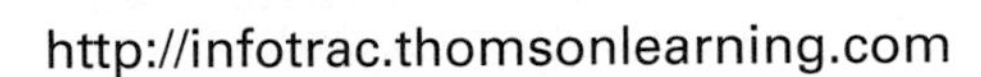

CHAPTER 7

Statistical Data Treatment and Evaluation

American Illustrators Gallery, NYC. © 2003 National Museum of American Illustration, Newport, RI.

The consequences of making errors in statistical tests are often compared with the consequences of errors made in judicial procedures. Shown to the left is Norman Rockwell's *Saturday Evening Post* cover *The Holdout,* from February 14, 1959. One of the 12 jurors does not agree with the others, who are trying to convince her. In the jury room, we can make two types of errors. An innocent person can be convicted or a guilty person can be set free. In our justice system, we consider it a more serious error to convict an innocent person than to acquit a guilty one.

Similarly, in statistical tests to determine whether two quantities are the same, two types of errors can be made. A type I error occurs when we reject the hypothesis that two quantities are the same when they are statistically identical. A type II error occurs when we accept that they are the same when they are not statistically identical. The characteristics of these errors in statistical testing and the ways we can minimize them are among the subjects of this chapter.

Scientists use statistical calculations to sharpen their judgments concerning the quality of experimental measurements. In this chapter, we consider several of the most common applications of statistical tests to the treatment of analytical results. These applications include the following:

1. Defining a numerical interval around the mean of a set of replicate analytical results within which the population mean can be expected to lie with a certain probability. This interval is called the **confidence interval** (CI). The interval is related to the standard deviation of the mean.
2. Determining the number of replicate measurements required to ensure that an experimental mean falls within a certain range with a given level of probability.
3. Estimating the probability that (a) an experimental mean and a true value or (b) two experimental means are different; that is, whether the difference is real or simply the result of random error. This test is particularly important for discovering systematic errors in a method and determining whether two samples come from the same source.
4. Determining at a given probability level whether the precision of two sets of measurements differs.
5. Comparing the means of more than two samples to determine whether differences in the means are real or the result of random error. This process is known as **analysis of variance.**
6. Deciding with a certain probability whether an apparent outlier in a set of replicate measurements is the result of a gross error and can thus be rejected or

whether it is a legitimate part of the population that must be retained in calculating the mean of the set.

7A CONFIDENCE INTERVALS

In most of the situations encountered in chemical analysis, the true value of the mean μ cannot be determined because a huge number of measurements (approaching infinity) would be required. With statistics, however, we can establish an interval surrounding an experimentally determined mean $\bar{x}$ within which the population mean μ is expected to lie with a certain degree of probability. This interval is known as the **confidence interval** and the boundaries are called **confidence limits.** For example, we might say that it is 99% probable that the true population mean for a set of potassium measurements lies in the interval 7.25% $\pm$ 0.15% K. Thus, the mean should lie in the interval from 7.10% to 7.40% K with 99% probability.

The **confidence interval** for the mean is the range of values within which the population mean μ is expected to lie with a certain probability.

The size of the confidence interval, which is computed from the sample standard deviation, depends on how well the sample standard deviation s estimates the population standard deviation σ. If s is a good approximation of σ, the confidence interval can be significantly narrower than if the estimate of σ is based on only a few measurement values.

7A-1 Finding the Confidence Interval When σ Is Known or s Is a Good Estimate of σ

Figure 7-1 shows a series of five normal error curves. In each, the relative frequency is plotted as a function of the quantity z (see Equation 6-2, page 112), which is the deviation from the mean *divided by the population standard deviation.* The shaded areas in each plot lie between the values of $-z$ and $+z$ that are indicated to the left and right of the curves. The numbers within the shaded areas are the percentage of the total area under the curve that is included within these values of z. For example, as shown in curve (a), 50% of the area under any Gaussian curve is located between -0.67σ and $+0.67\sigma$. Proceeding to curves (b) and (c), we see that 80% of the total area lies between -1.28σ and $+1.28\sigma$, and 90% lies between -1.64σ and $+1.64\sigma$. Relationships such as these allow us to define a range of values around a measurement result within which the true mean is likely to lie with a certain probability *provided we have a reasonable estimate of* σ. For example, if we have a result x from a data set with a standard deviation of σ, we may assume that 90 times out of 100, the true mean μ will fall in the interval $x \pm 1.64\sigma$ (see Figure 7-1c). The probability is called the **confidence level** (CL). In the example of Figure 7-1c, the confidence level is 90%, and the confidence interval is from -1.64σ to $+1.64\sigma$. The probability that a result is *outside* the confidence interval is often called the **significance level.**

The **confidence level** is the probability that the true mean lies within a certain interval. It is often expressed as a percentage.

If we make a single measurement x from a distribution of known σ, we can say that the true mean should lie in the interval $x \pm z\sigma$ with a probability dependent on z. This probability is 90% for $z = 1.64$, 95% for $z = 1.96$, and 99% for $z = 2.58$, as shown in Figure 7-1c, d, and e. We find a general expression for the confidence interval of the true mean based on measuring a single value x by rearranging Equation 6-2. (Remember that z can take positive or negative values.) Thus,

$$\text{CI for } \mu = x \pm z\sigma \qquad (7\text{-}1)$$

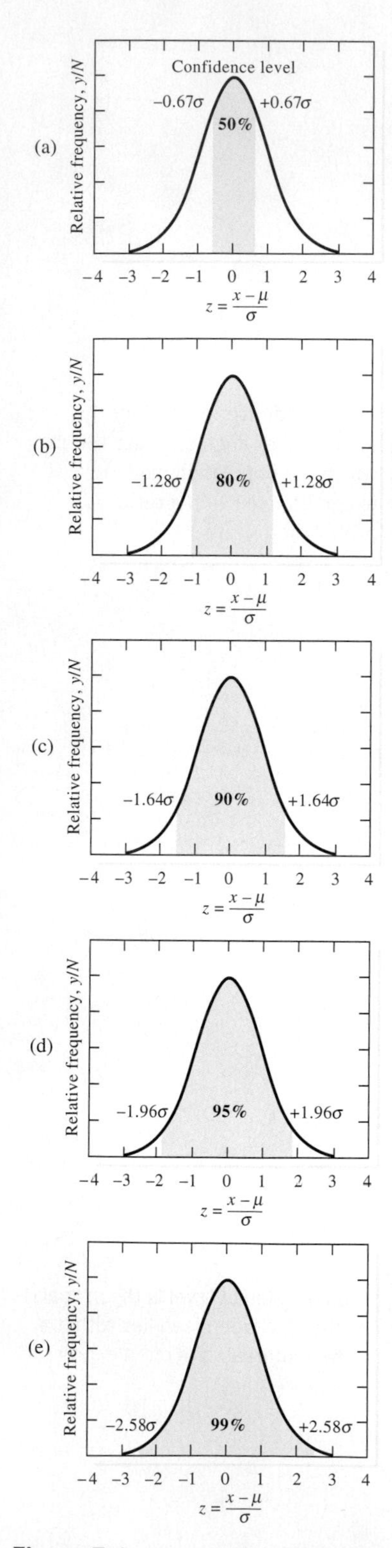

Figure 7-1 Areas under a Gaussian curve for various values of $\pm z$. (a) $z = \pm 0.67$; (b) $z = \pm 1.29$; (c) $z = \pm 1.64$; (d) $z = \pm 1.96$; (e) $z = \pm 2.58$.

TABLE 7-1

Confidence Levels for Various Values of z

Confidence Level, %	z
50	0.67
68	1.00
80	1.28
90	1.64
95	1.96
95.4	2.00
99	2.58
99.7	3.00
99.9	3.29

TABLE 7-2

Size of Confidence Interval as a Function of the Number of Measurements Averaged

Number of Measurements Averaged	Relative Size of Confidence Interval
1	1.00
2	0.71
3	0.58
4	0.50
5	0.45
6	0.41
10	0.32

Rarely do we estimate the true mean from a single measurement, however. Instead, we use the experimental mean $\bar{x}$ of N measurements as a better estimate of μ. In this case, we replace x in Equation 7-1 with $\bar{x}$ and σ with the standard error of the mean, $\sigma/\sqrt{N}$. That is,

$$\text{CI for } \mu = \bar{x} \pm \frac{z\sigma}{\sqrt{N}} \tag{7-2}$$

Values for z at various confidence levels are found in Table 7-1, and the relative size of the confidence interval as a function of N is shown in Table 7-2. Sample calculations of confidence limits and confidence intervals are given in Examples 7-1 and 7-2.

EXAMPLE 7-1

Determine the 80% and 95% confidence intervals for (a) the first entry (1108 mg/L glucose) in Example 6-2 (page 124) and (b) the mean value (1100.3 mg/L) for month 1 in the example. Assume that in each part, $s = 19$ is a good estimate of σ.

(a) From Table 7-1, we see that $z = 1.28$ and 1.96 for the 80% and 95% confidence levels. Substituting into Equation 7-1,

$$80\%\ \text{CI} = 1108 \pm 1.28 \times 19 = 1108 \pm 24.3\ \text{mg/L}$$

$$95\%\ \text{CI} = 1108 \pm 1.96 \times 19 = 1108 \pm 37.2\ \text{mg/L}$$

From these calculations, we conclude that it is 80% probable that μ, the population mean (and, *in the absence of determinate error,* the true value), lies in the interval 1083.7 to 1132.3 mg/L glucose. Furthermore, the probability is 95% that μ lies in the interval between 1070.8 and 1145.2 mg/L.

(b) For the seven measurements,

$$80\%\ \text{CL} = 1100.3 \pm \frac{1.28 \times 19}{\sqrt{7}} = 1100.3 \pm 9.2\ \text{mg/L}$$

$$95\%\ \text{CL} = 1100.3 \pm \frac{1.96 \times 19}{\sqrt{7}} = 1100.3 \pm 14.1\ \text{mg/L}$$

Thus, there is an 80% chance that μ is located in the interval between 1091.1 and 1109.5 mg/L glucose and a 95% chance that it lies between 1086.2 and 1114.4 mg/L glucose.

EXAMPLE 7-2

How many replicate measurements in month 1 in Example 6-2 are needed to decrease the 95% confidence interval to 1100.3 ± 10.0 mg/L of glucose?

Here, we want the term $\pm\frac{z\sigma}{\sqrt{N}}$ to equal ±10.0 mg/L of glucose.

$$\frac{z\sigma}{\sqrt{N}} = \frac{1.96 \times 19}{\sqrt{N}} = 10.0$$

$$\sqrt{N} = \frac{1.96 \times 19}{10.0} = 3.724$$

$$N = (3.724)^2 = 13.9$$

We thus conclude that 14 measurements are needed to provide a slightly better than 95% chance that the population mean will lie within ±14 mg/L of the experimental mean.

Equation 7-2 tells us that the confidence interval for an analysis can be halved by carrying out four measurements. Sixteen measurements will narrow the interval by a factor of 4, and so on. We rapidly reach a point of diminishing returns in acquiring additional data. Ordinarily we take advantage of the relatively large gain attained by averaging two to four measurements but can seldom afford the time or amount of sample required to obtain narrower confidence intervals through additional replicate measurements.

It is essential to keep in mind at all times that confidence intervals based on Equation 7-2 apply *only in the absence of bias and only if we can assume that s is a good approximation of σ*. We indicate that s is a good estimate of σ by using the symbol $s \rightarrow \sigma$ (s approaches σ).

Spreadsheet Summary In Chapter 2 of *Applications of Microsoft® Excel in Analytical Chemistry,* we explore the use of the Excel function CONFIDENCE() to obtain confidence intervals when σ is known. The 80% and 90% confidence intervals are obtained for one result and for seven results.

7A-2 Finding the Confidence Interval When σ Is Unknown

Often, limitations in time or in the amount of available sample prevent us from making enough measurements to assume s is a good estimate of σ. In this case, a single set of replicate measurements must provide not only a mean but also an estimate of precision. As indicated earlier, s calculated from a small set of data may be quite uncertain. Thus, confidence intervals are necessarily broader when we must use a small sample value of s as our estimate of σ.

► The t statistic is often called **Student's** t. Student was the name used by W. S. Gossett when he wrote the classic paper on t that appeared in *Biometrika,* **1908,** *6,* 1. Gossett was employed by the Guinness Brewery to statistically analyze the results of determinations of the alcohol content in their products. As a result of this work, he discovered the now-famous statistical treatment of small sets of data. To avoid the disclosure of any trade secrets of his employer, Gossett published the paper under the name Student.

To account for the variability of s, we use the important statistical parameter t, which is defined in exactly the same way as z (see Equation 6-2) except that s is substituted for σ. For a single measurement with result x, we can define t as

$$t = \frac{x - \mu}{s} \tag{7-3}$$

For the mean of N measurements,

$$t = \frac{\bar{x} - \mu}{s/\sqrt{N}} \tag{7-4}$$

► Remember that the number of degrees of freedom for small data sets is equal to $N - 1$, not N.

Like z in Equation 7-1, t depends on the desired confidence level. But t also depends on the number of degrees of freedom in the calculation of s. Table 7-3 provides values for t for a few degrees of freedom. More extensive tables are found in various mathematical and statistical handbooks. Note that t approaches z as the number of degrees of freedom becomes large.

The confidence interval for the mean $\bar{x}$ of N replicate measurements can be calculated from t by an equation similar to Equation 7-2:

$$\text{CI for } \mu = \bar{x} \pm \frac{ts}{\sqrt{N}} \tag{7-5}$$

The use of the t statistic for confidence intervals is illustrated in Example 7-3.

Spreadsheet Summary In the final exercise in Chapter 2 of *Applications of Microsoft® Excel in Analytical Chemistry,* we use the Analysis ToolPak's Descriptive Statistics package to obtain the confidence interval for cases in which we must use the sample standard deviation as an estimate of σ. The 95% confidence interval is obtained for glucose data of Example 7-1 of this chapter.

TABLE 7-3
Values of *t* for Various Levels of Probability

Degrees of Freedom	80%	90%	95%	99%	99.9%
1	3.08	6.31	12.7	63.7	637
2	1.89	2.92	4.30	9.92	31.6
3	1.64	2.35	3.18	5.84	12.9
4	1.53	2.13	2.78	4.60	8.61
5	1.48	2.02	2.57	4.03	6.87
6	1.44	1.94	2.45	3.71	5.96
7	1.42	1.90	2.36	3.50	5.41
8	1.40	1.86	2.31	3.36	5.04
9	1.38	1.83	2.26	3.25	4.78
10	1.37	1.81	2.23	3.17	4.59
15	1.34	1.75	2.13	2.95	4.07
20	1.32	1.73	2.09	2.84	3.85
40	1.30	1.68	2.02	2.70	3.55
60	1.30	1.67	2.00	2.62	3.46
∞	1.28	1.64	1.96	2.58	3.29

EXAMPLE 7-3

A chemist obtained the following data for the alcohol content of a sample of blood: % C_2H_5OH: 0.084, 0.089, and 0.079. Calculate the 95% confidence interval for the mean assuming (a) the three results obtained are the only indication of the precision of the method and (b) from previous experience on hundreds of samples, we know that the standard deviation of the method s = 0.005% C_2H_5OH and is a good estimate of σ.

(a) $\Sigma x_i = 0.084 + 0.089 + 0.079 = 0.252$

$\Sigma x_i^2 = 0.007056 + 0.007921 + 0.006241 = 0.021218$

$$s = \sqrt{\frac{0.021218 - (0.252)^2/3}{3 - 1}} = 0.0050\%\ C_2H_5OH$$

Here, $\bar{x} = 0.252/4 = 0.084$. Table 7-3 indicates that $t = 4.30$ for two degrees of freedom and the 95% confidence level. Thus,

$$95\%\ \text{CI} = \bar{x} \pm \frac{ts}{\sqrt{N}} = 0.084 \pm \frac{4.30 \times 0.0050}{\sqrt{3}}$$

$$= 0.084 \pm 0.012\%\ C_2H_5OH$$

(b) Because $s = 0.0050\%$ is a good estimate of σ,

$$95\%\ \text{CI} = \bar{x} \pm \frac{z\sigma}{\sqrt{N}} = 0.094 \pm \frac{1.96 \times 0.0050}{\sqrt{3}}$$

$$= 0.084 \pm 0.006\%\ C_2H_5OH$$

Note that a sure knowledge of σ decreases the confidence interval by a significant amount. See Feature 7-1 for a description of alcohol analyzers.

FEATURE 7-1

Breath Alcohol Analyzers

Alcohol determinations at the roadside or at home are typically done with a breath analyzer or "breathalyzer." Because of rapid gas exchange and the vapor pressure of ethanol, the concentration exhaled is directly related to the blood alcohol concentration. The blood alcohol concentration is widely used as a criterion for determining whether a person is under the influence of alcohol. Many states have ruled that a blood alcohol level of 0.1% or greater indicates intoxication.

There are four widely used types of breath alcohol analyzers. In the indicator type, a chemical reaction occurs between ethanol and a reagent to produce a color change that is semiquantitatively related to the ethanol concentration. A second type is based on fuel cell technology. Here the ethanol is electrochemically oxidized to water and CO_2 at a selective platinum anode. The oxidation reaction and the reduction of atmospheric oxygen at the cathode produce a current proportional to the ethanol concentration (see Chapter 23 for principles of voltammetry). Fuel cell devices are small and well suited for handheld instruments. They require no source of electrical power to operate. A third type of analyzer is based on the absorption of infrared radiation (see Chapter 26). A sample of the breath is held in a gas cell, through which infrared radiation is passed. The absorbance at a few wavelengths is used to quantitate the amount of alcohol present. The primary wavelength detects ethanol plus organic contaminants. One or two secondary wavelengths are used to detect interfering substances, which are then used to correct the absorption at the primary wavelength. Such instruments require power and are used in mobile and stationary applications. The newest technology uses a semiconductor sensor. Here the alcohol is adsorbed at the surface of a semiconductor. Usually a change in electrical conductance is monitored and related to blood alcohol level.

These devices are low in cost and simple to use. At present, technical limitations make them unsuitable for accurate quantitative applications. Hence, they are primarily intended for personal home or automobile use. Two of the analyzer types are pictured here.

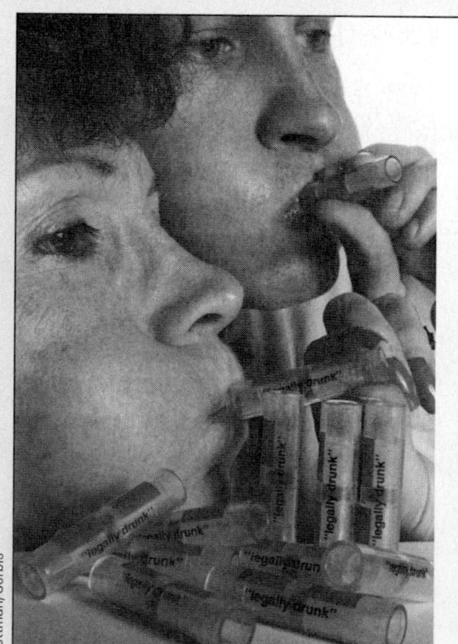

Bettman/Corbis

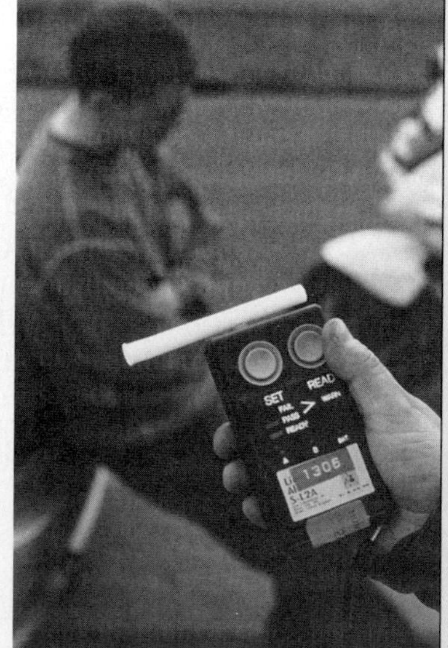

Jim Varney/SPL/Custom Medical Stock Photo

Alcohol breath analyzers.

7B STATISTICAL AIDS TO HYPOTHESIS TESTING

Hypothesis testing is the basis for many decisions made in scientific and engineering work. To explain an observation, a hypothetical model is advanced and is tested experimentally to determine its validity. If the results from these experiments do not support the model, we reject it and seek a new hypothesis. If agreement is found, the hypothetical model serves as the basis for further experiments. When the hypothesis is supported by sufficient experimental data, it becomes recognized as a useful theory until such time as data are obtained that refute it.

Experimental results seldom agree exactly with those predicted from a theoretical model. As a consequence, scientists and engineers frequently must judge whether a numerical difference is a result of the random errors inevitable in all measurements or a result of systematic errors. Certain statistical tests are useful in sharpening these judgments.

Tests of this kind make use of a **null hypothesis,** which assumes that the numerical quantities being compared are, in fact, the same. We then use a probability distribution to calculate the probability that the observed differences are a result of random error. Usually, if the observed difference is greater than or equal to the difference that would occur 5 times in 100 by random chance (a significance level of 0.05), the null hypothesis is considered questionable, and the difference is judged to be significant. Other significance levels, such as 0.01 (1%) or 0.001 (0.1%), may also be adopted, depending on the certainty desired in the judgment. When expressed as a fraction, the significance level is often given the symbol α. The confidence level (CL) is related to α on a percentage basis by CL $= (1 - \alpha) \times 100\%$.

In statistics, a **null hypothesis** postulates that two or more observed quantities are the same.

Specific examples of hypothesis tests that chemists often use include the comparison of (1) the mean of an experimental data set with what is believed to be the true value; (2) the mean to a predicted or cutoff (threshold) value; and (3) the means or the standard deviations from two or more sets of data. The sections that follow consider some of the methods for making these comparisons.

7B-1 Comparing an Experimental Mean with a Known Value

There are many cases in which a scientist or an engineer needs to compare the mean of a data set with a known value. In some cases, the known value is the true or accepted value based on prior knowledge or experience. In other situations, the known value might be a value predicted from theory or it might be a threshold value that we use in making decisions about the presence or absence of a constituent. In all these cases, we use a statistical **hypothesis test** to draw conclusions about the population mean μ and its nearness to the known value, which we call μ_0.

There are two contradictory outcomes that we consider in any hypothesis test. The first, the null hypothesis H_0, states that $\mu = \mu_0$. The second, the alternative hypothesis H_a, can be stated in several ways. We might reject the null hypothesis in favor of H_a if μ is different from μ_0 ($\mu \neq \mu_0$). Other alternative hypotheses are $\mu > \mu_0$ or $\mu < \mu_0$. For example, suppose we are interested in determining whether the concentration of lead in an industrial wastewater discharge exceeds the maximum permissible amount of 0.05 ppm. Our hypothesis test would be written as follows:

$$H_0: \mu = 0.05 \text{ ppm}$$

$$H_a: \mu > 0.05 \text{ ppm}$$

Now suppose instead that experiments over a several year period have determined that the mean lead level is 0.02 ppm. Recently, changes in the industrial process have been made, and we suspect that the mean lead level is now different. Here we do not care whether it is higher or lower than 0.02 ppm. Our hypothesis test would be written as follows:

$$H_0: \mu = 0.02 \text{ ppm}$$

$$H_a: \mu \neq 0.02 \text{ ppm}$$

To carry out the statistical test, a test procedure must be implemented. The crucial elements of a test procedure are the formation of an appropriate test statistic and the identification of a rejection region. The test statistic is formulated from the data on which we will base the decision to accept or reject H_0. The rejection region consists of all the values of the test statistic for which H_0 will be rejected. The null hypothesis is rejected if the test statistic lies within the rejection region. For tests concerning one or two means, the test statistic can be the z statistic if we have a large number of measurements or if we know σ. Alternatively, we must use the t statistic for small numbers with unknown σ. When in doubt, the t statistic should be used.

Large Sample z Test

If a large number of results are available so that s is a good estimate of σ, the z test is appropriate. The procedure that is used is summarized as follows:

1. State the null hypothesis: $H_0: \mu = \mu_0$
2. Form the test statistic: $z = \dfrac{\bar{x} - \mu_0}{\sigma/\sqrt{N}}$
3. State the alternative hypothesis, H_a, and determine the rejection region:
   For $H_a: \mu \neq \mu_0$, reject H_0 if $z \geq z_{crit}$ or if $z \leq -z_{crit}$
   For $H_a: \mu > \mu_0$, reject H_0 if $z \geq z_{crit}$
   For $H_a: \mu < \mu_0$, reject H_0 if $z \leq -z_{crit}$

The rejection regions are illustrated in Figure 7-2 for the 95% confidence level. Note that for $H_a: \mu \neq \mu_0$, we can reject for either a positive value of z or for a negative value of z that exceeds the critical value. This is called a **two-tailed test.** For the 95% confidence level, the probability that z exceeds z_{crit} is 0.025 in each tail or 0.05 total. Hence, there is only a 5% probability that random error will lead to a value of $z \geq z_{crit}$ or $z \leq -z_{crit}$. The significance level overall is $\alpha = 0.05$. From Table 7-1, the critical value is 1.96 for this case.

For $H_a: \mu > \mu_0$, the test is said to be a **one-tailed test.** Here, we can reject only when $z \geq z_{crit}$. Now for the 95% confidence level, we want the probability that z exceeds z_{crit} to be 5% or the total probability in both tails to be 10%. The overall significance level would be $\alpha = 0.10$, and the critical value from Table 7-1 is 1.64. Similarly, if the alternative hypothesis is $\mu < \mu_0$, we can reject only when $z \leq -z_{crit}$. The critical value of z is again 1.64 for this one-tailed test.

Example 7-4 illustrates the use of the z test to determine whether the mean of 35 values agrees with a theoretical value.

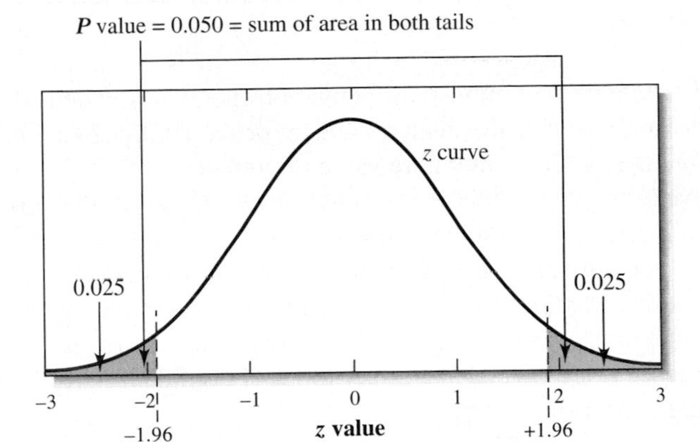

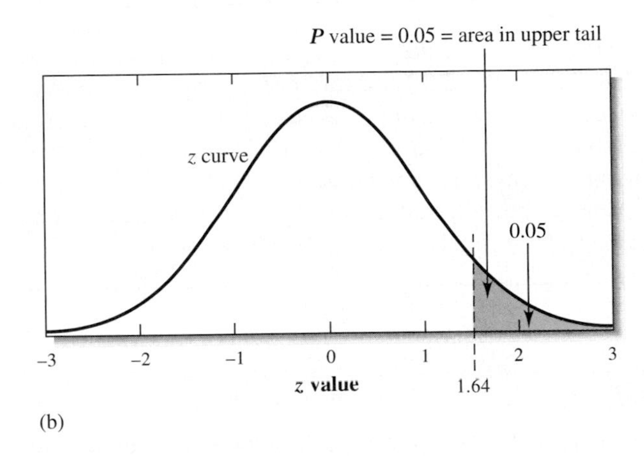

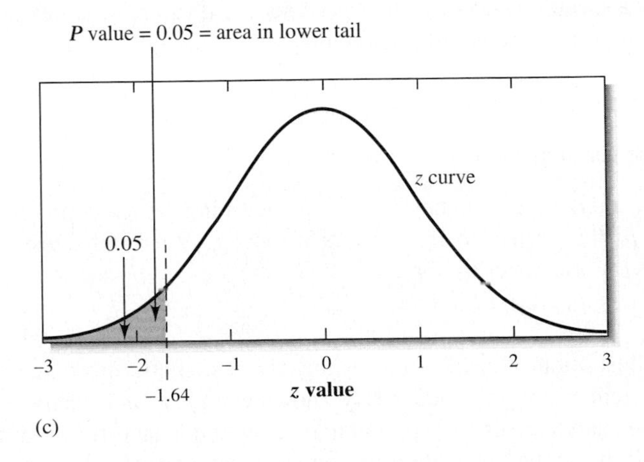

Figure 7-2 Rejection regions for the 95% confidence level. (a) Two-tailed test for H_a: $\mu \neq \mu_0$. Note the critical value of z is 1.96, as in Figure 7-1. (b) One-tailed test for H_a: $\mu > \mu_0$. Here, the critical value of z is 1.64, so that 95% of the area is to the left of z_{crit} and 5% of the area is to the right. (c) One-tailed test for H_a: $\mu < \mu_0$. Here the critical value is again 1.64, so that 5% of the area lies to the left of $-z_{crit}$.

EXAMPLE 7-4

A class of 30 students determined the activation energy of a chemical reaction to be 27.7 kcal/mol (mean value) with a standard deviation of 5.2 kcal/mol. Are the data in agreement with the literature value of 30.8 kcal/mol at (1) the 95% confidence level and (2) the 99% confidence level? Estimate the probability of obtaining a mean equal to the literature value.

We have enough values here so that s should be a good estimate of σ. Our null hypothesis is that $\mu = 30.8$ kcal/mol, and the alternative hypothesis is that $\mu \neq 30.8$ kcal/mol. This is a two-tailed test. From Table 7-1, $z_{crit} = 1.96$ for the 95% confidence level and $z_{crit} = 2.58$ for the 99% confidence level. The test statistic is calculated as follows:

$$z = \frac{\bar{x} - \mu_0}{\sigma/\sqrt{N}} = \frac{27.7 - 30.8}{5.2/\sqrt{30}} = -3.26$$

Since $z \leq -1.96$, we reject the null hypothesis at the 95% confidence level. Note also that since $z \leq -2.58$, we reject H_0 at the 99% confidence level. To estimate the probability of obtaining a mean value $\mu = 30.8$ kcal/mol, we must find the probability of obtaining a z value of 3.26. From Table 7-1, the probability of obtaining a z value this large because of random error is only about 0.2%. All these lead us to conclude that the student mean is actually different from the literature value and not just the result of random error.

Small Sample t Test

For a small number of results, we use a procedure similar to the z test except that the test statistic is the t statistic. Here again, we test the null hypothesis H_0: $\mu = \mu_0$ where μ_0 is a specific value of μ such as an accepted value, a theoretical value, or a threshold value. The procedure is as follows:

1. State the null hypothesis: H_0: $\mu = \mu_0$
2. Form the test statistic: $t = \dfrac{\bar{x} - \mu_0}{s/\sqrt{N}}$
3. State the alternative hypothesis, H_a, and determine the rejection region:
   For H_a: $\mu \neq \mu_0$, reject H_0 if $t \geq t_{crit}$ or if $t \leq -t_{crit}$ (two-tailed test)
   For H_a: $\mu > \mu_0$, reject H_0 if $t \geq t_{crit}$
   For H_a: $\mu < \mu_0$, reject H_0 if $t \leq -t_{crit}$

As an illustration, consider the testing for systematic error in an analytical method wherein a sample of accurately known composition is analyzed. Determination of the analyte gives an experimental mean that is an estimate of the population mean. If the analytical method had no systematic error, or bias, random errors would give the frequency distribution shown by curve A in Figure 7-3. Method B has some systematic error, so that $\bar{x}_B$, which estimates μ_B, differs from the accepted value μ_0. The bias is given by

$$\text{bias} = \mu_B - \mu_0 \tag{7-6}$$

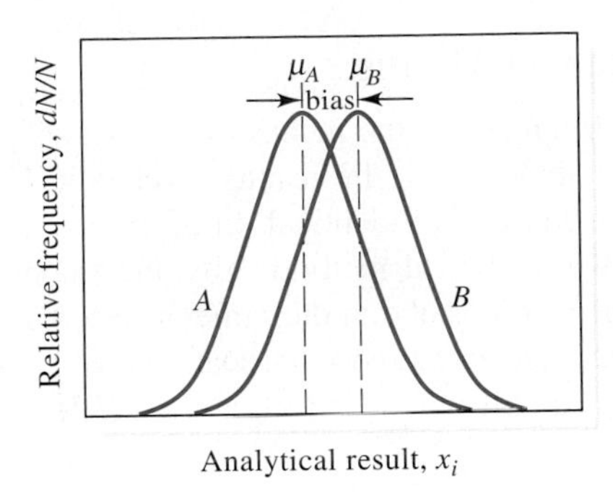

Figure 7-3 Illustration of systematic error in an analytical method. Curve A is the frequency distribution for the accepted value by a method without bias. Curve B illustrates the frequency distribution of results by a method that could have a significant bias.

In testing for bias, we do not know initially whether the difference between the experimental mean and the accepted value is due to random error or to an actual systematic error. The t test is used to determine the significance of the difference. Example 7-5 illustrates the use of the t test to determine whether there is bias in a method.

EXAMPLE 7-5

A new procedure for the rapid determination of the percentage of sulfur in kerosenes was tested on a sample known from its method of preparation to contain 0.123% ($\mu_0 = 0.123\%$) S. The results were % S = 0.112, 0.118, 0.115, and 0.119. Do the data indicate that there is a bias in the method at the 95% confidence level?

The null hypothesis is H_0: $\mu = 0.123\%$ S, and the alternative hypothesis is H_a: $\mu \neq 0.123\%$ S.

$$\Sigma x_i = 0.112 + 0.118 + 0.115 + 0.119 = 0.464$$

$$\bar{x} = 0.464/4 = 0.116\%\ \text{S}$$

$$\Sigma x_i^2 = 0.012544 + 0.013924 + 0.013225 + 0.014161 = 0.53854$$

$$s = \sqrt{\frac{0.053854 - (0.464)^2/4}{4-1}} = \sqrt{\frac{0.000030}{3}} = 0.0032\%\ \text{S}$$

The test statistic can now be calculated as

$$t = \frac{\bar{x} - \mu_0}{s/\sqrt{N}} = \frac{0.116 - 0.123}{0.032/\sqrt{4}} = -4.375$$

From Table 7-3, we find that the critical value of t for 3 degrees of freedom and the 95% confidence level is 3.18. Since $t \leq -3.18$, we conclude that there is a significant difference at the 95% confidence level and thus bias in the method. Note that if we do this test at the 99% confidence level, $t_{crit} = 5.84$ (see Table 7-3). Since $-5.84 < -4.375$, we would accept the null hypothesis at the 99% confidence level and conclude that there is no difference between the experimental and the accepted values. Note in this case that the outcome depends on the confidence level that is used. As we will see, choice of the confidence level depends on our willingness to accept an error in the outcome. The significance level (0.05 or 0.01) is the probability of making an error by rejecting the null hypothesis (see Section 7B-3).

◀ The probability of a difference this large occurring because of only random errors can be obtained from the Excel function TDIST(x,deg_freedom,tails), where x is the test value of t(4.375), deg_freedom is 3 for our case, and tails = 2. The result is TDIST(4.375,3,2) = 0.022. Hence, it is only 2.2% probable to get a value this large because of random error. The critical value of t for a given confidence level can be obtained in Excel from TINV(probability,deg_freedom). In our case, TINV(0.05,3) = 3.1825.

If it were confirmed by further experiments that the method always gives low results, we would say that the method had a **negative bias.**

7B-2 Comparison of Two Experimental Means

Frequently, chemists must judge whether a difference in the means of two sets of data is real or the result of random error. In some cases, the results of chemical analyses are used to determine whether two materials are identical. In other cases, the results are used to determine whether two analytical methods give the same values or whether two analysts using the same methods obtain the same means. An extension of these procedures can be used to analyze paired data. Data are often collected in pairs to eliminate one source of variability by focusing on the differences within each pair.

The t Test for Differences in Means

CD-ROM Simulation: Exploration of the Relationships Between Sample Means, Standard Deviations, and Significant Difference Confidence.

In the case of large numbers of measurements in both sets, the z test, discussed in the previous section, can be modified to take into account a comparison of two sets of data. More often, both sets contain only a few results, and we must use the t test. To illustrate, let us assume that N_1 replicate analyses by analyst 1 yielded a mean value of $\bar{x}_1$ and that N_2 analyses by analyst 2 obtained by the same method gave $\bar{x}_2$. The null hypothesis states that the two means are identical and that any difference is the result of random errors. Thus, we can write H_0: $\mu_1 = \mu_2$. Most often when testing differences in means, the alternative hypothesis is H_a: $\mu_1 \neq \mu_2$, and the test is a two-tailed test. In some situations, however, we could test H_a: $\mu_1 > \mu_2$ or H_a: $\mu_1 < \mu_2$ and use a one-tailed test. We will assume here that a two-tailed test is used.

If the data were collected in the same manner and the analysts were both careful, it would often be safe to assume that the standard deviations of both data sets are similar. Thus, both s_1 and s_2 are estimates of the population standard deviation σ. To get a better estimate of σ than given by s_1 or s_2 alone, we use the pooled standard deviation (see Section 6B-4). From Equation 6-6, the standard deviation of the mean of analyst 1 is given by $s_{m1} = \dfrac{s_1}{\sqrt{N_1}}$. The variance of the mean of analyst 1 is.

$$s_{m1}^2 = \frac{s_1^2}{N_1}$$

Likewise, the variance of the mean of analyst 2 is

$$s_{m2}^2 = \frac{s_2^2}{N_2}$$

In the t test, we are interested in the difference between the means, or $\bar{x}_1 - \bar{x}_2$. The variance of the difference s_d^2 between the means is given by

$$s_d^2 = s_{m1}^2 + s_{m2}^2$$

The standard deviation of the difference between the means is found by taking the square root after substituting the values of s_{m1}^2 and s_{m2}^2 from above.

$$\frac{s_d}{\sqrt{N}} = \sqrt{\frac{s_1^2}{N_1} + \frac{s_2^2}{N_2}}$$

Now if we make the further assumption that the pooled standard deviation s_{pooled} is a better estimate of σ than s_{m1} or s_{m2}, we can write

$$\frac{s_d}{\sqrt{N}} = \sqrt{\frac{s^2_{pooled}}{N_1} + \frac{s^2_{pooled}}{N_2}} = s_{pooled}\sqrt{\frac{N_1 + N_2}{N_1 N_2}}$$

The test statistic t is now found from

$$t = \frac{\bar{x}_1 - \bar{x}_2}{s_{pooled}\sqrt{\dfrac{N_1 + N_2}{N_1 N_2}}} \tag{7-7}$$

The test statistic is then compared with the critical value of t obtained from the table for the particular confidence level desired. The number of degrees of freedom for finding the critical value of t in Table 7-3 is $N_1 + N_2 - 2$. If the absolute value of the test statistic is less than the critical value, the null hypothesis is accepted and no significant difference between the means has been demonstrated. A test value of t greater than the critical value indicates a significant difference between the means. Example 7-6 illustrates the use of the t test to determine if two barrels of wine came from different sources.

EXAMPLE 7-6

Two barrels of wine were analyzed for their alcohol content to determine whether they were from different sources. On the basis of six analyses, the average content of the first barrel was established to be 12.61% ethanol. Four analyses of the second barrel gave a mean of 12.53% alcohol. The 10 analyses yielded a pooled standard deviation s_{pooled} of 0.070%. Do the data indicate a difference between the wines?

The null hypothesis is H_0: $\mu_1 = \mu_2$, and the alternative hypothesis is H_a: $\mu_1 \neq \mu_2$. Here we employ Equation 7-7 to calculate the test statistic t.

$$t = \frac{\bar{x}_1 - \bar{x}_2}{s_{pooled}\sqrt{\dfrac{N_1 + N_2}{N_1 \times N_2}}} = \frac{12.61 - 12.53}{0.07\sqrt{\dfrac{6 + 4}{6 \times 4}}} = 1.771$$

The critical value of t at the 95% confidence level for $10 - 2 = 8$ degrees of freedom is 2.31. Since $1.771 < 2.31$, we accept the null hypothesis at the 95% confidence level and conclude that there is no difference in the alcohol content of the wines. The probability of getting a t value of 1.771 can be calculated using the Excel function TDIST() and is TDIST(1.771,8,2) = 0.11. Thus, there is more than a 10% chance that we could get a value this large due to random error.

In Example 7-5, no significant difference between the two wines was detected at the 95% probability level. This statement is equivalent to saying that μ_1 is equal to μ_2 with a certain degree of confidence. The tests do not prove that the wines come from the same source, however. Indeed, it is conceivable that one wine is red and the other is white. To establish with a reasonable probability that the two wines are from the same source would require extensive testing of other characteristics, such as taste, color, odor, and refractive index, as well as tartaric acid, sugar, and trace element content. If no significant differences are revealed by all these tests and by others, then it might be possible to judge the two wines as having a common origin. In contrast, the finding of *one* significant difference in any test would clearly show that the two wines are different. Thus, the establishment of a significant difference by a single test is much more revealing than the establishment of an absence of difference.

If there is good reason to believe that the standard deviations of the two data sets differ, the **two-sample *t* test** must be employed.[1] However, the significance level for this t test is only approximate, and the number of degrees of freedom is more difficult to calculate.

Spreadsheet Summary In the first exercise in Chapter 3 of *Applications of Microsoft® Excel in Analytical Chemistry,* we use Excel to perform the t test for comparing two means assuming equal variances of the two data sets. We first manually calculate the value of t and compare it with the critical value obtained from Excel's function TINV(). We obtain the probability from Excel's TDIST() function. Then, we use Excel's built-in function TTEST() for the same test. Finally, we employ Excel's Analysis ToolPak to automate the t test with equal variances.

Paired Data

Scientists and engineers often make use of pairs of measurements on the same sample to minimize sources of variability that are not of interest. For example, two methods for determining glucose in blood serum are to be compared. Method A could be performed on samples from five randomly chosen patients and Method B could be performed on samples from five different patients. There would be variability, however, because of the different glucose levels of each patient. A better way to compare the methods would be to use both methods on the same samples and to focus on the differences.

The paired t test uses the same type of procedure as the normal t test except that we analyze pairs of data. The standard deviation is now the standard deviation of the mean difference. Our null hypothesis is H_0: $\mu_d = \Delta_0$, where Δ_0 is a specific value of the difference to be tested, often zero. The test statistic value is

$$t = \frac{\bar{d} - \Delta_0}{s_d/\sqrt{N}}$$

where $\bar{d}$ is the average difference equal to $\Sigma d_i/N$. The alternative hypothesis could be $\mu_d \neq \Delta_0$, $\mu_d > \Delta_0$, or $\mu_d < \Delta_0$. An illustration is given in Example 7-7.

[1]For more information, see J. L. Devore and N. R. Farnum, *Applied Statistics for Engineers and Scientists,* pp 340–344. Pacific Grove, CA: Duxbury Press at Brooks/Cole Publishing Co., 1999.

EXAMPLE 7-7

A new automated procedure for determining glucose in serum (Method A) is to be compared with the established method (Method B). Both methods are performed on serum from the same six patients to eliminate patient-to-patient variability. Do the following results confirm a difference in the two methods at the 95% confidence level?

	Patient 1	Patient 2	Patient 3	Patient 4	Patient 5	Patient 6
Method A glucose, mg/L	1044	720	845	800	957	650
Method B glucose, mg/L	1028	711	820	795	935	639
Difference, mg/L	16	9	25	5	22	11

Let us now test the appropriate hypotheses. If μ_d is the true average difference between the methods, we want to test the null hypothesis H_0: $\mu_d = 0$ and the alternative hypothesis, H_a: $\mu_d \neq 0$. The test statistic is

$$t = \frac{\bar{d} - 0}{s_d/\sqrt{N}}$$

From the table, $N = 6$, $\Sigma d_i = 16 + 9 + 25 + 5 + 22 + 11 = 88$, $\Sigma d_i^2 = 1592$, and $\bar{d} = 14.67$. The standard deviation of the difference s_d is given by

$$s_d = \sqrt{\frac{1592 - \frac{(88)^2}{6}}{6 - 1}} = 7.76$$

and the t statistic is

$$t = \frac{14.67}{7.76/\sqrt{6}} = 4.628$$

From Table 7-3, the critical value of t is 2.57 for the 95% confidence level and 5 degrees of freedom. Since $t > t_{crit}$, we reject the null hypothesis and conclude that the two methods give different results.

Note that if we merely average the results of Method A ($\bar{x}_A = 836.0$ mg/L) and the results of Method B ($\bar{x}_B = 821.3$ mg/L), the large patient-to-patient variation in glucose level gives us large values for s_A (146.5) and s_B (142.7). A comparison of means gives us a t value of 0.176, and we would accept the null hypothesis. Hence, the large patient-to-patient variability masks the method differences that are of interest. Pairing allows us to focus on the differences.

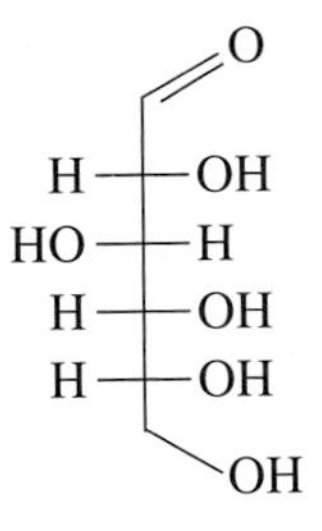

Structural formula of glucose, $C_6H_{12}O_6$.

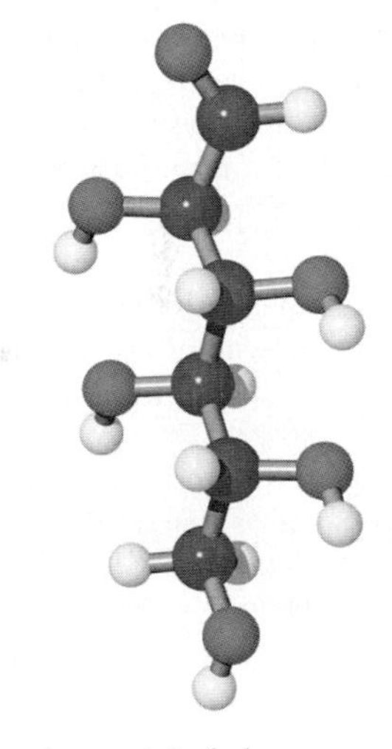

Molecular model of glucose.

Spreadsheet Summary In Chapter 3 of *Applications of Microsoft® Excel in Analytical Chemistry,* we use Excel's Analysis ToolPak to perform the paired t text on the data of Example 7-7. We compare the results obtained with those found without pairing.

7B-3 Errors in Hypothesis Testing

The choice of a rejection region for the null hypothesis is made so that we can readily understand the errors involved. At the 95% confidence level, for example, there is a 5% chance that we will reject the null hypothesis even though it is true. This could happen if an unusual result occurred that put our test statistic z or t into the rejection region. The error that results from rejecting H_0 when it is true is called a **type I error.** The significance level α gives the frequency of rejecting H_0 when it is true.

A **type I error** occurs when H_0 is rejected although it is actually true. In some sciences, a type I error is called a **false negative.** A **type II error** occurs when H_0 is accepted and it is actually false. This is sometimes termed a **false positive.**

The other type of error that is possible is that we accept H_0 when it is false. This is termed a **type II error.** The probability of a type II error is given the symbol β No test procedure can guarantee that we will not commit one error or the other. The error probabilities are the result of using a sample of data to make inferences about the population. At first thought, making α smaller (0.01 instead of 0.05) would appear to make sense to minimize the type I error rate. Decreasing the type I error rate, however, increases the type II error rate because they are inversely related.

It is important when thinking about errors in hypothesis testing to determine the consequences of making a type I or a type II error. If a type I error is much more likely to have serious consequences than a type II error, it is reasonable to choose a small value of α. On the other hand, in some situations a type II error would be quite serious, and so a larger value of α is employed to keep the type II error rate under control. As a general rule of thumb, the largest α that is tolerable for the situation should be used. This ensures the smallest type II error while keeping the type I error within acceptable limits.

► The consequences of making errors in hypothesis testing are often compared with the errors made in judicial procedures. Thus, convicting an innocent person is usually considered a more serious error than setting a guilty person free. If we make it less likely that an innocent person gets convicted, we make it more likely that a guilty person goes free.

7B-4 Comparison of Precision

At times, there is a need to compare the variances (or standard deviations) of two populations. For example, the normal t test requires that the standard deviations of the data sets being compared are equal. A simple statistical test, called the F test, can be used to test this assumption under the provision that the populations follow the normal (Gaussian) distribution. The F test is also used in comparing more than two means (see Section 7C) and in linear regression analysis (see Section 8C-2).

The F test is based on the null hypothesis that the two population variances under consideration are equal, H_0: $\sigma_1^2 = \sigma_2^2$. The test statistic F, which is defined as the ratio of the two sample variances ($F = s_1^2/s_2^2$), is calculated and compared with the critical value of F at the desired significance level. The null hypothesis is rejected if the test statistic differs too much from 1.

Critical values of F at the 0.05 significance level are shown in Table 7-4. Note that two degrees of freedom are given, one associated with the numerator and the other with the denominator. Most mathematical handbooks give more extensive tables of F values at various significance levels.

The F test can be used in either a one-tailed mode or a two-tailed mode. For a one-tailed test, we test the alternative hypothesis that one variance is greater than the other. Hence, the variance of the supposedly more precise procedure is placed in the denominator and that of the less precise procedure is placed in the numerator. The alternative hypothesis is H_a: $\sigma_1^2 > \sigma_2^2$. The critical values of F for the 95% confidence level are given in Table 7-4. For a two-tailed test, we are testing whether the variances are different, H_a: $\sigma_1^2 \neq \sigma_2^2$. For this application, the larger variance always appears in the numerator. This arbitrary placement of the larger variance makes the outcome of the test less certain; thus, the uncertainty level of the F values in Table 7-4 is doubled from 5% to 10%. Example 7-8 illustrates the use of the F test for comparing measurement precision.

TABLE 7-4

Critical Values of *F* at the 5% Probability Level (95% confidence level)

Degrees of Freedom (Denominator)	Degrees of Freedom (Numerator) 2	3	4	5	6	10	12	20	∞
2	19.00	19.16	19.25	19.30	19.33	19.40	19.41	19.45	19.50
3	9.55	9.28	9.12	9.01	8.94	8.79	8.74	8.66	8.53
4	6.94	6.59	6.39	6.26	6.16	5.96	5.91	5.80	5.63
5	5.79	5.41	5.19	5.05	4.95	4.74	4.68	4.56	4.36
6	5.14	4.76	4.53	4.39	4.28	4.06	4.00	3.87	3.67
10	4.10	3.71	3.48	3.33	3.22	2.98	2.91	2.77	2.54
12	3.89	3.49	3.26	3.11	3.00	2.75	2.69	2.54	2.30
20	3.49	3.10	2.87	2.71	2.60	2.35	2.28	2.12	1.84
∞	3.00	2.60	2.37	2.21	2.10	1.83	1.75	1.57	1.00

EXAMPLE 7-8

A standard method for the determination of the carbon monoxide (CO) level in gaseous mixtures is known from many hundreds of measurements to have a standard deviation of 0.21 ppm CO. A modification of the method yields a value for s of 0.15 ppm CO for a pooled data set with 12 degrees of freedom. A second modification, also based on 12 degrees of freedom, has a standard deviation of 0.12 ppm CO. Is either modification significantly more precise than the original?

Here we test the null hypothesis H_0: $\sigma_{std}^2 = \sigma_1^2$, where σ_{std}^2 is the variance of the standard method and σ_1^2 is the variance of the modified method. The alternative hypothesis is one-tailed, H_a: $\sigma_1^2 < \sigma_{std}^2$. Because an improvement is claimed, the variances of the modifications are placed in the denominator. For the first modification,

$$F_1 = \frac{s_{std}^2}{s_1^2} = \frac{(0.21)^2}{(0.15)^2} = 1.96$$

and for the second,

$$F_2 = \frac{(0.21)^2}{(0.12)^2} = 3.06$$

For the standard procedure, s_{std} is a good estimate of σ, and the number of degrees of freedom from the numerator can be taken as infinite. From Table 7-4, the critical value of F at the 95% confidence level is $F_{crit} = 2.30$.

Since F_1 is less than 2.30, we cannot reject the null hypothesis for the first modification. We conclude that there is no improvement in precision. For the second modification, however, $F_2 > 2.30$. Here, we reject the null hypothesis and conclude that the second modification does appear to give better precision at the 95% confidence level.

(continued)

It is interesting to note that if we ask whether the precision of the second modification is significantly better than that of the first, the F test dictates that we must accept the null hypothesis. That is,

$$F = \frac{s_1^2}{s_2^2} = \frac{(0.15)^2}{(0.12)^2} = 1.56$$

In this case, $F_{crit} = 2.69$. Since $F < 2.69$, we must accept H_0 and conclude that the two methods give equivalent precision.

Spreadsheet Summary In Chapter 3 of *Applications of Microsoft® Excel in Analytical Chemistry,* we use two Excel functions to perform the F test. First, we use the built-in function FTEST(), which returns the probability that the variances in two data arrays are not significantly different. Then we use the Analysis ToolPak for the same comparison of variances.

7C ANALYSIS OF VARIANCE

Analysis of variance (ANOVA) is used to test whether a difference exists in the means of more than two populations.

In Section 7B, we introduced methods to compare two sample means or one sample mean to a known value. In this section, we extend these principles to allow comparisons among more than two population means. The methods used for multiple comparisons fall under the general category of analysis of variance, often known by the acronym **ANOVA.** These methods use a single test to determine whether there is or is not a difference among the population means rather than pairwise comparisons, as are done with the t test. After ANOVA indicates a potential difference, **multiple comparison** procedures can be used to identify which specific population means differ from the others. **Experimental design methods** take advantage of ANOVA in planning and performing experiments.

7C-1 ANOVA Concepts

In ANOVA procedures, we detect differences in several population means by comparing the *variances.* For comparing I population means, $\mu_1, \mu_2, \mu_3, \cdots \mu_I$, the null hypothesis H_0 is of the form

$$H_0: \mu_1 = \mu_2 = \mu_3 = \cdots = \mu_I$$

and the alternative hypothesis H_a is

H_a: at least two of the μ_i's are different.

The following are typical applications of ANOVA:

1. Is there a difference in the results of five analysts determining calcium by a volumetric method?
2. Will four different solvent compositions have differing influences on the yield of a chemical synthesis?

3. Are the results of manganese determinations by three different analytical methods different?
4. Are there any differences in the fluorescence of a complex ion at six different values of pH?

In each of these situations, the populations have differing values of a common characteristic called a **factor** or sometimes a **treatment.** In determining calcium by a volumetric method, the factor of interest is the analyst. The different values of the factor of interest are called **levels.** For the calcium example, there are five levels corresponding to analyst 1, analyst 2, analyst 3, analyst 4, and analyst 5. The comparisons among the various populations are made by measuring a **response** for each item sampled. In the case of the calcium determination, the response is the number of mmol Ca determined by each analyst. For the four examples given previously, the factors, levels, and responses are as follows:

Factor	Levels	Response
Analyst	Analyst 1, analyst 2, analyst 3, analyst 4, analyst 5	mmol Ca
Solvent	Composition 1, composition 2, composition 3, composition 4	Synthesis yield, %
Analytical methods	Method 1, method 2, method 3	Concentration Mn, ppm
pH	pH 1, pH 2, pH 3, pH 4, pH 5, pH 6	Fluorescence intensity

The factor can be considered the independent variable, whereas the response is the dependent variable. Figure 7-4 illustrates how to visualize ANOVA data for the five analysts determining Ca in triplicate.

The type of ANOVA shown in Figure 7-4 is known as a single-factor, or one-way, ANOVA. Often, several factors may be involved, such as in an experiment to

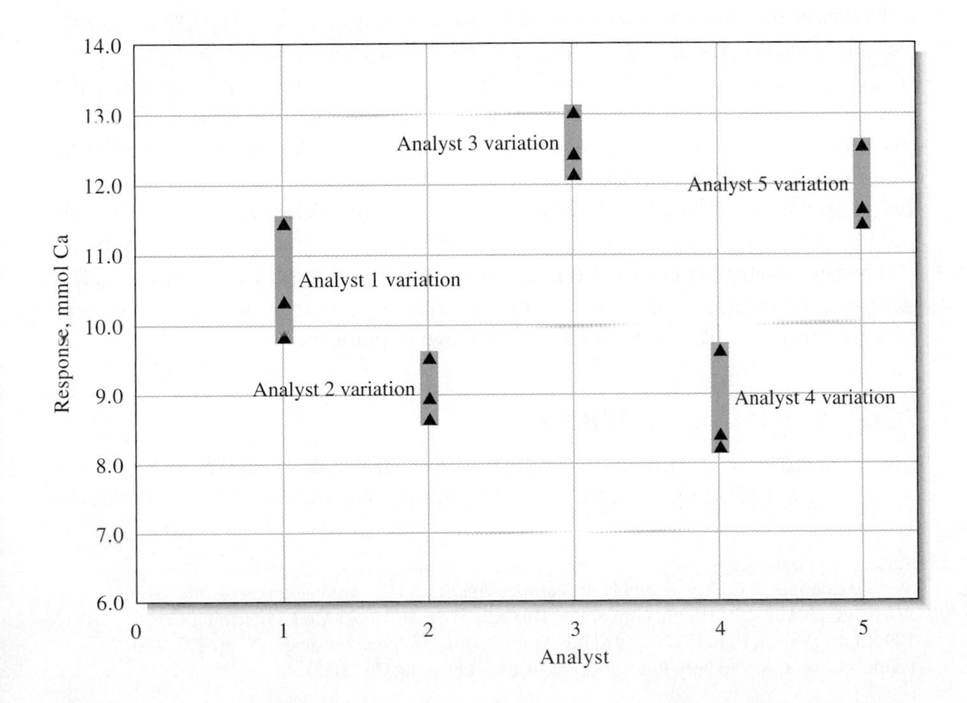

Figure 7-4 Pictorial of the results from the ANOVA study of the determination of calcium by five analysts. Each analyst does the determination in triplicate. Analyst is considered a factor, whereas analyst 1, analyst 2, analyst 3, analyst 4, and analyst 5 are levels of the factor.

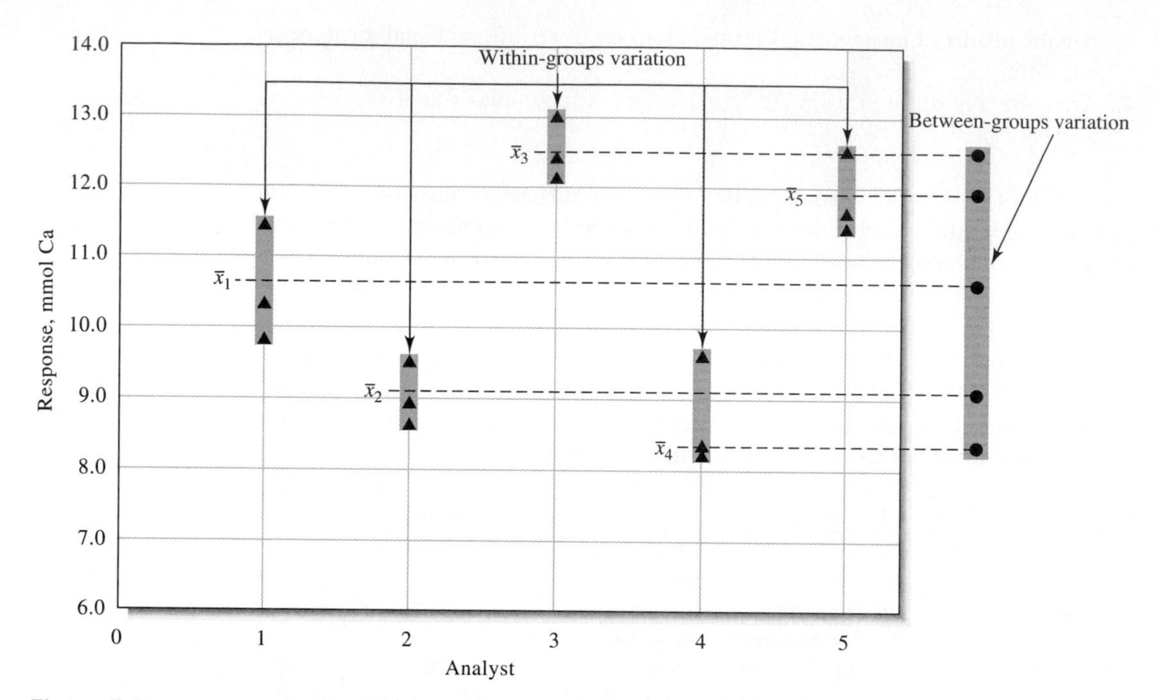

Figure 7-5 Pictorial representation of the ANOVA principle. The results of each analyst are considered a group. The triangles (▲) represent individual results, and the circles (●) represent the means. Here the variation between the group means is compared with that within groups.

determine whether pH and temperature influence the rate of a chemical reaction. Here, the type of ANOVA is known as a two-way ANOVA. Procedures for dealing with multiple factors are given in statistics books.[2] Here, we consider only single-factor ANOVA.

► The basic principle of ANOVA is to compare the variations between the different factor levels (groups) with those within factor levels.

Consider the triplicate results for each analyst in Figure 7-4 to be random samples. In ANOVA, the factor levels are often called groups. The basic principle of ANOVA is to compare the between-groups variation with the within-groups variation. In our specific case, the groups (factor levels) are the different analysts, and this is a comparison of the variation between analysts to the within-analyst variation. Figure 7-5 illustrates this comparison. When H_0 is true, the variation between the group means is close to the variation within groups. When H_0 is false, the variation between group means is large compared with the variation within groups.

The basic statistical test used for ANOVA is the F test, described in Section 7B-4. Here, a large value of F compared with the critical value from the tables may give us reason to reject H_0 in favor of the alternative hypothesis.

7C-2 Single-Factor ANOVA

Several quantities are important in testing the null hypothesis H_0: $\mu_1 = \mu_2 = \mu_3 = \cdots \mu_I$. The sample means of the I populations are $\bar{x}_1, \bar{x}_2, \bar{x}_3, \cdots \bar{x}_I$ and the

[2]See, for example, J. L. Devore and N. R. Farnum, *Applied Statistics for Engineers and Scientists,* pp 411–464. Pacific Grove, CA: Duxbury Press at Brooks/Cole Publishing Co., 1999; J. L. Devore, *Probability and Statistics for Engineering and the Sciences,* pp 433–480. Pacific Grove, CA: Duxbury Press at Brooks/Cole Publishing Co., 2000.

sample variances are $s_1^2, s_2^2, s_3^2 \cdots s_I^2$. These are estimates of the corresponding population values. In addition, we can calculate the grand average $\bar{\bar{x}}$, which is the average of all the data. The grand mean can be calculated as the weighted average of the individual group means, as shown in Equation 7-8:

$$\bar{\bar{x}} = \left(\frac{N_1}{N}\right)\bar{x}_1 + \left(\frac{N_2}{N}\right)\bar{x}_2 + \left(\frac{N_3}{N}\right)\bar{x}_3 + \cdots + \left(\frac{N_I}{N}\right)\bar{x}_I \tag{7-8}$$

where N_1 is the number of measurements in group 1, N_2 is the number in group 2, and so on. The grand average can also be found by summing all the data values and dividing by the total number of measurements N.

To calculate the variance ratio needed in the F test, it is necessary to obtain several other quantities called sums of squares:

1. The sum of the squares due to the factor (SSF):

$$\text{SSF} = N_I(\bar{x}_I - \bar{\bar{x}})^2 + N_2(\bar{x}_2 - \bar{\bar{x}})^2 + N_3(\bar{x}_3 - \bar{\bar{x}})^2 + \cdots + N_I(\bar{x}_I - \bar{\bar{x}})^2 \tag{7-9}$$

2. The sum of the squares due to error (SSE):

$$\text{SSE} = \sum_{j=1}^{N_1}(x_{1j} - \bar{x}_1)^2 + \sum_{j=1}^{N_2}(x_{2j} - \bar{x}_2)^2 + \sum_{j=1}^{N_3}(x_{3j} - \bar{x}_3)^2 + \cdots + \sum_{j=1}^{N_I}(x_{ij} - \bar{x}_I)^2 \tag{7-10}$$

These two sums of squares are used to obtain the between-groups variation and the within-groups variation. The error sum of the squares is related to the individual group variances by

$$\text{SSE} = (N_1 - 1)s_1^2 + (N_2 - 1)s_2^2 + (N_3 - 1)s_3^2 + \cdots + (N_I - 1)s_I^2 \tag{7-11}$$

3. The total sum of the squares (SST) is obtained as the sum of SSF and SSE

$$\text{SST} = \text{SSF} + \text{SSE} \tag{7-12}$$

The total sum of the squares can also be obtained from $(N - 1)s^2$, where s^2 is the sample variance of all the data points.

To apply ANOVA methods, we need to make a few assumptions concerning the populations under study. First, the usual ANOVA methods are based on an equal variance assumption. That is, the variances of the I populations are assumed to be identical. This assumption is sometimes tested (Hartley test) by comparing the maximum and minimum variances in the set with an F test (see Section 7B-4). However, the Hartley test is quite susceptible to departures from the normal distribution. As a rough rule of thumb, the largest s should not be much more than twice the smallest s for equal variances to be assumed.[3] Transforming the data by working with a new variable such as $\sqrt{x}$ or $\log x$ can also

[3] J. L. Devore, *Probability and Statistics for Engineering and the Sciences*, p 406. Pacific Grove, CA: Duxbury Press at Brooks/Cole Publishing Co., 2000.

be used to give populations with more equal variances. Second, each of the I populations is assumed to follow a Gaussian distribution. For cases in which this last assumption is not true, there are distribution-free ANOVA procedures that can be applied.

4. The number of degrees of freedom for each of the sum of squares must be obtained. The total sum of the squares SST has $N - 1$ degrees of freedom. Just as SST is the sum of SSF and SSE, the total number of degrees of freedom $N - 1$ can be decomposed into degrees of freedom associated with SSF and SSE. Since I groups are being compared, SSF has $I - 1$ degrees of freedom. This leaves $N - I$ degrees of freedom for SSE. Or,

$$\begin{array}{ccccc} \text{SST} & = & \text{SSF} & + & \text{SSE} \\ (N-1) & = & (I-1) & + & (N-\text{I}) \end{array}$$

5. By dividing the sums of squares by their corresponding degrees of freedom, we can obtain quantities that are estimates of the between-groups and within-groups variations. These quantities are called **mean square values** and are defined as

$$\text{Mean square due to factor levels} = \text{MSF} = \frac{\text{SSF}}{I-1} \tag{7-13}$$

$$\text{Mean square error} = \text{MSE} = \frac{\text{SSF}}{N-\text{I}} \tag{7-14}$$

The quantity MSE is an estimate of the variance due to error (σ_E^2), while MSF is an estimate of the error variance plus the between-groups variance ($\sigma_E^2 + \sigma_F^2$). If the factor has little effect, the between-groups variance should be small compared with the error variance. Thus, the two mean squares should be nearly identical under these circumstances. If the factor effect is significant, MSF is greater than MSE. The test statistic is the F value, calculated as

$$F = \frac{\text{MSF}}{\text{MSE}} \tag{7-15}$$

To complete the hypothesis test, we compare the value of F calculated above with the critical value from the table at a significance level of α. We reject H_0 if F exceeds the critical value. It is common practice to summarize the results of ANOVA in an **ANOVA table,** as follows:

Source of Variation	**Sum of Squares (SS)**	**Degrees of Freedom (df)**	**Mean Square (MS)**	**Mean Square Estimates**	**F**
Between groups (factor effect)	SSF	$I - 1$	$\text{MSF} = \frac{\text{SSF}}{I-1}$	$\sigma_E^2 + \sigma_F^2$	$\frac{\text{MSF}}{\text{MSE}}$
Within groups (error)	SSE	$N - \text{I}$	$\text{MSE} = \frac{\text{SSE}}{N-I}$	σ_E^0	
Total	SST	$N - 1$			

Example 7-9 shows an application of ANOVA to the determination of calcium by five analysts. The data are those used to construct Figures 7-4 and 7-5.

EXAMPLE 7-9

Five analysts obtained the results (mmol Ca) shown in the following table for determining calcium by a volumetric method. Do the means differ significantly at the 95% confidence level?

Trial no.	Analyst 1	Analyst 2	Analyst 3	Analyst 4	Analyst 5
1	10.3	9.5	12.1	9.6	11.6
2	9.8	8.6	13.0	8.3	12.5
3	11.4	8.9	12.4	8.2	11.4

First, we can obtain the means and standard deviations for each analyst. The mean for analyst one is $\bar{x}_1 = (10.3 + 9.8 + 11.4)/3 = 10.5$ mmol Ca. The remaining means are obtained in the same manner: $\bar{x}_2 = 9.0$ mmol Ca, $\bar{x}_3 = 12.5$ mmol Ca, $\bar{x}_4 = 8.7$ mmol Ca, $\bar{x}_5 = 11.833$ mmol Ca. The standard deviations are obtained as described in Section 6B-3. These results are summarized as follows:

	Analyst 1	Analyst 2	Analyst 3	Analyst 4	Analyst 5
Mean	10.5	9.0	12.5	8.7	11.833
Standard Dev.	0.818535	0.458258	0.458258	0.781025	0.585947

The grand mean can be obtained from

$$\bar{\bar{x}} = \frac{3}{15}(\bar{x}_1 + \bar{x}_2 + \bar{x}_3 + \bar{x}_4 + \bar{x}_5) = 10.507 \text{ mmol Ca}$$

The between-groups sum of the squares is found from Equation 7-9:

$$\begin{aligned} \text{SSF} &= 3(10.5 - 10.507)^2 + 3(9.0 - 10.507)^2 + 3(12.5 - 10.507)^2 \\ &\quad + 3(8.7 - 10.507)^2 + 3(11.833 - 10.507)^2 \\ &= 33.80267 \end{aligned}$$

Note that SSF has associated with it $(5 - 1) = 4$ degrees of freedom.
The error sum of squares is easiest to find from the standard deviations and Equation 7-11:

$$\begin{aligned} \text{SSE} &= 2(0.818535)^2 + 2(0.458258)^2 \\ &\quad + 2(0.458258)^2 + 2(0.781025)^2 + 2(0.585947)^2 \\ &= 4.086667 \end{aligned}$$

The error sum of the squares has $(15 - 5) = 10$ degrees of freedom.

We can now calculate the mean square values, MSF and MSE, from Equations 7-13 and 7-14.

$$\text{MSF} = \frac{33.80267}{4} = 8.450667$$

$$\text{MSE} = \frac{4.086667}{10} = 0.408667$$

(continued)

The F value obtained from Equation 7-15 is

$$F = \frac{8.450667}{0.408667} = 20.68$$

From the F table on page 159, the critical value of F at the 95% confidence level for 4 and 10 degrees of freedom is 3.48. Since F exceeds 3.48, we reject H_0 at the 95% confidence level and conclude that there is a significant difference among the analysts. The ANOVA table is shown here.

Source of Variation	Sum of Squares (SS)	Degrees of Freedom (df)	Mean Square (MS)	F
Between groups	33.80267	4	8.450667	20.68
Within groups	4.086667	10	0.408667	
Total	37.88933	14		

Spreadsheet Summary In Chapter 3 of *Applications of Microsoft® Excel in Analytical Chemistry,* the use of Excel to perform ANOVA procedures is described. There are several ways to do ANOVA with Excel. First, the equations from this section are entered manually into a worksheet, and Excel is invoked to do the calculations. Second, the Analysis ToolPak is used to carry out the entire ANOVA procedure automatically. The results of the five analysts from Example 7-9 are analyzed by both these methods.

7C-3 Determining Which Results Differ

If significant differences are indicated in ANOVA, we are often interested in the cause. Is one mean different from the others? Are all the means different? Are there two distinct groups that the means fall into? There are several methods to determine which means are significantly different. One of the simplest is the **least significant difference** (LSD) method. In this method, a difference is calculated that is judged to be the smallest difference that is significant. The difference between each pair of means is then compared with the least significant difference to determine which means are different.

For an equal number of replicates N_g in each group, the least significant difference is calculated as follows:

$$\text{LSD} = t\sqrt{\frac{2 \times \text{MSE}}{N_g}} \tag{7-16}$$

where MSE is the mean square for error and the value of t should have $N - I$ degrees of freedom. Example 7-10 illustrates the procedure.

EXAMPLE 7-10

For the results of Example 7-9, determine which analysts differ from each other at the 95% confidence level.

First, let us arrange the means in increasing order: 8.7, 9.0, 10.5, 11.833, 12.5. Each analyst did three repetitions, and so we can use Equation 7-16. We obtain a value of t of 2.23 for the 95% confidence level and 10 degrees of freedom. Application of Equation 7-16 gives us

$$\text{LSD} = 2.23\sqrt{\frac{2 \times 0.408667}{3}} = 1.16$$

We now calculate the differences in means and compare them with 1.16. For the various pairs:

$$\bar{x}_{\text{largest}} - \bar{x}_{\text{smallest}} = 12.5 - 8.7 = 3.8 \quad \text{(a significant difference)}$$
$$\bar{x}_{\text{2nd largest}} - \bar{x}_{\text{smallest}} = 11.833 - 8.7 = 3.133 \quad \text{(significant)}$$
$$\bar{x}_{\text{3rd largest}} - \bar{x}_{\text{smallest}} = 10.5 - 8.7 = 1.8 \quad \text{(significant)}$$
$$\bar{x}_{\text{4th largest}} - \bar{x}_{\text{smallest}} = 9.0 - 8.7 = 0.3 \quad \text{(no significant difference)}$$

We then continue to test each pair to determine which are different. From this we conclude that analysts 3, 5, and 1 differ from analyst 4, analysts 3, 5, and 1 differ from analyst 2, analysts 3 and 5 differ from analyst 1, and analyst 3 differs from analyst 5.

7D DETECTION OF GROSS ERRORS

There are times when a set of data contains an outlying result that appears to be outside the range of what random errors in the procedure would give. It is generally considered inappropriate and in some cases unethical to discard data without a reason. However, the **outlier** could be the result of an undetected gross error. Hence, we must develop a criterion to decide whether to retain or reject the outlying data point. The choice of criterion for the rejection of a suspected result has its perils. If our standard is too strict, so that it is quite difficult to reject a questionable result, we run the risk of retaining a spurious value that has an inordinate effect on the mean. If we set a lenient limit and thereby make the rejection of a result easy, we are likely to discard a value that rightfully belongs in the set, thus introducing a bias to the data. Although there is no universal rule to settle the question of retention or rejection, the Q test is generally acknowledged to be an appropriate method for making the decision.[4]

7D-1 The Q Test

The Q test is a simple, widely used statistical test for deciding whether a suspected result should be retained or rejected.[5] In this test, the absolute value of the

[4] J. Mandel, in *Treatise on Analytical Chemistry*, 2nd ed., I. M. Kolthoff and P. J. Elving, Eds., Part I, Vol. 1, pp 282–289. New York: Wiley, 1978.

[5] R. B. Dean and W. J. Dixon, *Anal. Chem.*, **1951,** *23,* 636.

difference between the questionable result x_q and its nearest neighbor x_n is divided by the spread w of the entire set to give the quantity Q:

$$Q = \frac{|x_q - x_n|}{w} \tag{7-17}$$

CD-ROM Tutorial: Detection of Gross Errors Using the Q Test.

This ratio is then compared with critical values Q_{crit} found in Table 7-5. If Q is greater than Q_{crit}, the questionable result can be rejected with the indicated degree of confidence (Figure 7-6).

EXAMPLE 7-11

The analysis of a calcite sample yielded CaO percentages of 55.95, 56.00, 56.04, 56.08, and 56.23. The last value appears anomalous; should it be retained or rejected at the 95% confidence level?

The difference between 56.23 and 56.08 is 0.15%. The spread (56.23 − 55.95) is 0.28%. Thus,

$$Q = \frac{0.15}{0.28} = 0.54$$

For five measurements, Q_{crit} at the 95% confidence level is 0.71. Because $0.54 < 0.71$, we must retain the outlier at the 95% confidence level.

TABLE 7-5
Critical Values for the Rejection Quotient, Q*

Number of Observations	Q_{crit} (Reject if $Q > Q_{crit}$) 90% Confidence	95% Confidence	99% Confidence
3	0.941	0.970	0.994
4	0.765	0.829	0.926
5	0.642	0.710	0.821
6	0.560	0.625	0.740
7	0.507	0.568	0.680
8	0.468	0.526	0.634
9	0.437	0.493	0.598
10	0.412	0.466	0.568

*Reprinted with permission from D. B. Rorabacher, *Anal. Chem.*, **1991,** *63,* 139. Copyright 1991 American Chemical Society.

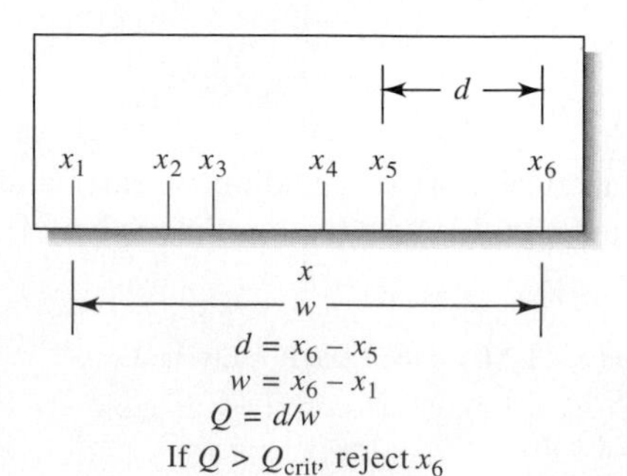

Figure 7-6 The Q test for outliers.

7D-2 Other Statistical Tests

Several other statistical tests have been developed to provide criteria for rejection or retention of outliers. Such tests, like the Q test, assume that the distribution of the population data is normal, or Gaussian. Unfortunately, this condition cannot be proved or disproved for samples that have less than about 50 results. Consequently, statistical rules, which are perfectly reliable for normal distributions of data, should be *used with extreme caution* when applied to samples containing only a few data. J. Mandel, in discussing treatment of small sets of data, writes, "Those who believe that they can discard observations with statistical sanction by using statistical rules for the rejection of outliers are simply deluding themselves."[6] Thus, statistical tests for rejection should be used only as aids to common sense when small samples are involved.

◀ Use extreme caution when rejecting data for any reason.

The blind application of statistical tests to retain or reject a suspect measurement in a small set of data is not likely to be much more fruitful than an arbitrary decision. The application of good judgment based on broad experience with an analytical method is usually a sounder approach. In the end, the only valid reason for rejecting a result from a small set of data is the sure knowledge that a mistake was made in the measurement process. Without this knowledge, *a cautious approach to rejection of an outlier is wise.*

7D-3 Recommendations for Treating Outliers

Recommendations for the treatment of a small set of results that contains a suspect value follow:

1. Reexamine carefully all data relating to the outlying result to see if a gross error could have affected its value. This recommendation demands *a properly kept laboratory notebook containing careful notations of all observations* (see Section 2I).
2. If possible, estimate the precision that can be reasonably expected from the procedure to be sure that the outlying result actually is questionable.
3. Repeat the analysis if sufficient sample and time are available. Agreement between the newly acquired data and those of the original set that appear to be valid will lend weight to the notion that the outlying result should be rejected. Furthermore, if retention is still indicated, the questionable result will have a small effect on the mean of the larger set of data.
4. If more data cannot be secured, apply the Q test to the existing set to see if the doubtful result should be retained or rejected on statistical grounds.
5. If the Q test indicates retention, consider reporting the median of the set rather than the mean. The median has the great virtue of allowing inclusion of all data in a set without undue influence from an outlying value. In addition, the median of a normally distributed set containing three measurements provides a better estimate of the correct value than the mean of the set after the outlying value has been discarded.

[6]J. Mandel, in *Treatise on Analytical Chemistry,* 2nd ed., I. M. Kolthoff and P. J. Elving, Eds., Part I, Vol. 1. p 282. New York: Wiley, 1978.

WEB WORKS

Point your Web browser to **http://chemistry.brookscole.com/skoogfac/.** From the Chapter Resources menu, choose Web Works, and locate the Chapter 7 section. Click on the link to the statistics on-line textbook. Click on the ANOVA/MANOVA button. Read about the partitioning of the sum of squares in ANOVA procedures. Click on the *F*-distribution link in this section. Look at the tail areas for an *F*-distribution with both degrees of freedom equal to 10. Determine the value of *F* for a significance level of 0.10 with both degrees of freedom equal to 10.

QUESTIONS AND PROBLEMS

***7-1.** Describe in your own words why the confidence interval for the mean of five measurements is smaller than that for a single result.

7-2. Assuming a large number of measurements so that s is a good estimate of σ, determine what confidence level was used for each of the following confidence intervals.

(a) $\bar{x} \pm \frac{3.00s}{\sqrt{N}}$ (b) $\bar{x} \pm \frac{1.64s}{\sqrt{N}}$

(c) $\bar{x} \pm \frac{s}{\sqrt{N}}$ (d) $\bar{x} \pm \frac{2.00s}{\sqrt{N}}$

***7-3.** Discuss how the size of the confidence interval for the mean is influenced by the following (all the other factors are constant):
(a) the sample size *N*.
(b) the confidence level.
(c) the standard deviation *s*.

7-4. Consider the following sets of replicate measurements:

*A	B	*C	D	*E	F
3.5	70.24	0.812	2.7	70.65	0.514
3.1	70.22	0.792	3.0	70.63	0.503
3.1	70.10	0.794	2.6	70.64	0.486
3.3		0.900	2.8	70.21	0.497
2.5			3.2		0.472

Calculate the mean and the standard deviation for each of these six data sets. Calculate the 95% confidence interval for each set of data. What does this interval mean?

7-5. Calculate the 95% confidence interval for each set of data in Problem 7-4 if s is a good estimate of σ and has a value of: *set A, 0.20; set B, 0.070; *set C, 0.0090; set D, 0.30; *set E, 0.15; set F, 0.015.

7-6. The last result in each set of data in Problem 7-4 may be an outlier. Apply the Q test (95% confidence level) to determine whether or not there is a statistical basis for rejection.

***7-7.** An atomic absorption method for the determination of the amount of iron present in used jet engine oil was found, from pooling 30 triplicate analyses, to have a standard deviation s = 2.4 μg Fe/mL. If s is a good estimate of σ, calculate the 80% and 95% confidence intervals for the result, 18.5 μg Fe/mL, if it was based on (a) a single analysis, (b) the mean of two analyses, (c) the mean of four analyses.

7-8. An atomic absorption method for determination of copper content in fuels yielded a pooled standard deviation of s_{pooled} = 0.32 μg Cu/mL ($s \rightarrow \sigma$). The analysis of oil from a reciprocating aircraft engine showed a copper content of 8.53 μg Cu/mL. Calculate the 90% and 99% confidence intervals for the result if it was based on (a) a single analysis, (b) the mean of four analyses, (c) the mean of 16 analyses.

***7-9.** How many replicate measurements are needed to decrease the 95% and 99% confidence intervals for the analysis described in Problem 7-7 to ± 1.5 μg Fe/mL?

7-10. How many replicate measurements are necessary to decrease the 95% and 99% confidence intervals for the analysis described in Problem 7-8 to ± 0.2 μg Cu/mL?

***7-11.** A volumetric calcium analysis on triplicate samples of the blood serum of a patient believed to be suffering from a hyperparathyroid condition produced the following data: meq Ca/L = 3.15, 3.25, 3.26. What is the 95% confidence limit for the mean of the data, assuming
(a) no prior information about the precision of the analysis?
(b) $s \rightarrow \sigma$ = 0.056 meq Ca/L?

7-12. A chemist obtained the following data for percent lindane in the triplicate analysis of an insecticide preparation: 7.47, 6.98, 7.27. Calculate the 90%

confidence interval for the mean of the three data, assuming that

(a) the only information about the precision of the method is the precision for the three data items.

(b) on the basis of long experience with the method, it is believed that $s \rightarrow \sigma = 0.28\%$ lindane.

7-13. A standard method for the determination of glucose in serum is reported to have a standard deviation of 0.40 mg/dL. If $s = 0.40$ is a good estimate of σ, how many replicate determinations should be made for the mean for the analysis of a sample to be within

*(a) 0.3 mg/dL of the true mean 99% of the time?

(b) 0.3 mg/dL of the true mean 95% of the time?

(c) 0.2 mg/dL of the true mean 90% of the time?

7-14. A titrimetric method for the determination of calcium in limestone was tested by analysis of a NIST limestone containing 30.15% CaO. The mean result of four analyses was 30.26% CaO, with a standard deviation of 0.085%. By pooling data from several analyses, it was established that $s \rightarrow \sigma = 0.094\%$ CaO.

(a) Do the data indicate the presence of a systematic error at the 95% confidence level?

(b) Would the data indicate the presence of a systematic error at the 95% confidence level if no pooled value for s had been available?

***7-15.** To test the quality of the work of a commercial laboratory, duplicate analyses of a purified benzoic acid (68.8% C, 4.953% H) sample were requested. It is assumed that the relative standard deviation of the method is $s_r \rightarrow \sigma = 4$ ppt for carbon and 6 ppt for hydrogen. The means of the reported results are 68.5% C and 4.882% H. At the 95% confidence level, is there any indication of systematic error in either analysis?

7-16. A prosecuting attorney in a criminal case presented as principal evidence small fragments of glass found imbedded in the coat of the accused. The attorney claimed that the fragments were identical in composition to a rare Belgian stained glass window broken during the crime. The average of triplicate analyses for five elements in the glass are shown here. On the basis of these data, does the defendant have grounds for claiming reasonable doubt as to guilt? Use the 99% confidence level as a criterion for doubt.

	Concentration, ppm		Standard Deviation
Element	From Clothes	From Window	$s \rightarrow \sigma$
As	129	119	9.5
Co	0.53	0.60	0.025
La	3.92	3.52	0.20
Sb	2.75	2.71	0.25
Th	0.61	0.73	0.043

***7-17.** Sewage and industrial pollutants dumped into a body of water can reduce the dissolved oxygen concentration and adversely affect aquatic species. In one study, weekly readings are taken from the same location in a river over a 2-month period.

Week Number	Dissolved O_2, ppm
1	4.9
2	5.1
3	5.6
4	4.3
5	4.7
6	4.9
7	4.5
8	5.1

Some scientists think that 5.0 ppm is a dissolved O_2 level that is marginal for fish to live. Conduct a statistical test to determine whether the mean dissolved O_2 concentration is less than 5.0 ppm at the 95% confidence level. State clearly the null and alternative hypotheses.

7-18. The week 3 measurement in the data set of Problem 7-17 is suspected of being an outlier. Use the Q test to determine if the value can be rejected at the 95% confidence level.

***7-19.** Before agreeing to the purchase of a large order of solvent, a company wants to see conclusive evidence that the mean value of a particular impurity is less than 1.0 ppb. Which hypotheses should be tested? What are the type I and type II errors in this situation?

7-20. The level of a pollutant in a river near a chemical plant is regularly monitored. Over a period of years, the normal level of the pollutant has been established by chemical analyses. Recently, the company has made several changes to the plant that appear to have increased the level of the pollutant. The Environmental Protection Agency (EPA) wants conclusive proof that the pollutant level has not increased. State the relevant null and alternative hypotheses and describe the type I and type II errors that might occur in this situation.

7-21. State quantitatively the null hypothesis H_0 and the alternative hypothesis H_a for the situations given here and describe the type I and type II errors. If these hypotheses were to be tested statistically, comment on whether a one- or two-tailed test would be involved for each case.

*(a) Since this sample gave a concentration lower than the 7.03 ppm level certified by the National Institute of Standards and Technology (NIST), a systematic error must have occurred.

(b) The mean values for Ca determinations by atomic absorption and by titrations differ substantially.

*(c) The atomic absorption results obtained for Cd are less precise than the electrochemical results.

(d) Results show that the batch-to-batch variation in the impurity content of Brand X acetonitrile is lower than Brand Y acetonitrile.

7-22. The homogeneity of the chloride level in a water sample from a lake was tested by analyzing portions drawn from the top and from near the bottom of the lake, with the following results in ppm Cl:

Top	Bottom
26.30	26.22
26.43	26.32
26.28	26.20
26.19	26.11
26.49	26.42

(a) Apply the t test at the 95% confidence level to determine if the means are different.

(b) Now use the paired t test and determine whether there is a significant difference between the top and bottom values at the 95% confidence level.

(c) Why is a different conclusion drawn from using the paired t test than from just pooling the data and using the normal t test for differences in means?

***7-23.** Two different analytical methods were used to determine residual chlorine in sewage effluents. Both methods were used on the same samples, but each sample came from various locations, with differing amounts of contact time with the effluent. The concentration of Cl in mg/L was determined by the two methods, and the following results were obtained:

Sample	Method A	Method B
1	0.39	0.36
2	0.84	1.35
3	1.76	2.56
4	3.35	3.92
5	4.69	5.35
6	7.70	8.33
7	10.52	10.70
8	10.92	10.91

(a) What type of t test should be used to compare the two methods, and why?

(b) Do the two methods give different results? State and test the appropriate hypotheses.

(c) Does the conclusion depend on whether the 90%, 95%, or 99% confidence levels are used?

7-24. Lord Rayleigh prepared nitrogen samples by several different methods. The density of each sample was measured as the mass of gas required to fill a particular flask at a certain temperature and pressure. Masses of nitrogen samples prepared by decomposition of various nitrogen compounds were 2.29280 g, 2.29940 g, 2.29849 g, and 2.30054 g. Masses of "nitrogen" prepared by removing oxygen from air in various ways were 2.31001 g, 2.31163 g, and 2.31028 g. Is the density of nitrogen prepared from nitrogen compounds significantly different from that prepared from air? What are the chances of the conclusion being in error? (Study of this difference led to the discovery of the inert gases by Sir William Ramsey, Lord Rayleigh.)

***7-25.** The phosphorus content was measured for three different soil locations. Five replicate determinations were made on each soil sample. A partial ANOVA table follows:

Variation Source	SS	df	MS	F
Between soils	_____	_____	_____	_____
Within soils	_____	_____	0.0081	
Total	0.374	_____		

(a) Fill in the missing entries in the ANOVA table.

(b) State the null and alternative hypotheses.

(c) Do the three soils differ in phosphorus content at the 95% confidence level?

7-26. The ascorbic acid concentration of five different brands of orange juice was measured. Six replicate samples of each brand were analyzed. The following partial ANOVA table was obtained.

Variation Source	SS	df	MS	F
Between juices	_____	_____	_____	8.45
Within juices	_____	_____	0.913	
Total	_____	_____		

(a) Fill in the missing entries in the table.

(b) State the null and alternative hypotheses.

(c) Is there a difference in the ascorbic acid content of the five juices at the 95% confidence level?

***7-27.** Five different laboratories participated in an interlaboratory study involving determinations of the Fe level in water samples. The following results are replicate determinations of the ppm Fe for laboratories A-E.

Result No.	Lab A	Lab B	Lab C	Lab D	Lab E
1	10.3	9.5	10.1	8.6	10.6
2	11.4	9.9	10.0	9.3	10.5
3	9.8	9.6	10.4	9.2	11.1

(a) State the appropriate hypotheses.
(b) Do the laboratories differ at the 95% confidence level? At the 99% confidence level (F_{crit} = 5.99)? At the 99.9% confidence level (F_{crit} = 11.28)?
(c) Which laboratories are different from each other at the 95% confidence level?

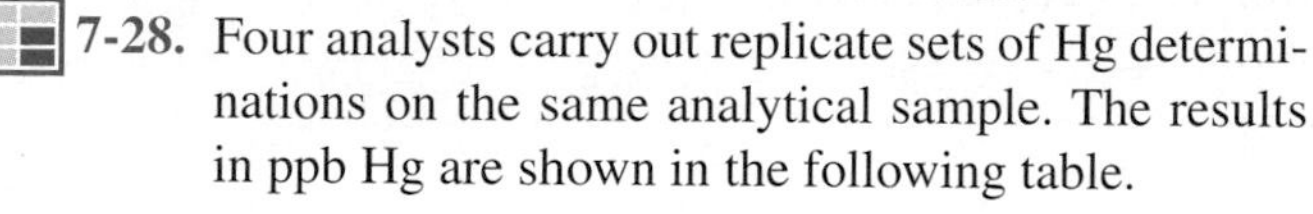

7-28. Four analysts carry out replicate sets of Hg determinations on the same analytical sample. The results in ppb Hg are shown in the following table.

Determination	Analyst 1	Analyst 2	Analyst 3	Analyst 4
1	10.24	10.14	10.19	10.19
2	10.26	10.12	10.11	10.15
3	10.29	10.04	10.15	10.16
4	10.23	10.07	10.12	10.10

(a) State the appropriate hypotheses.
(b) Do the analysts differ at the 95% confidence level? At the 99% confidence level (F_{crit} = 5.95)? At the 99.9% confidence level (F_{crit} = 10.80)?
(c) Which analysts differ from each other at the 95% confidence level?

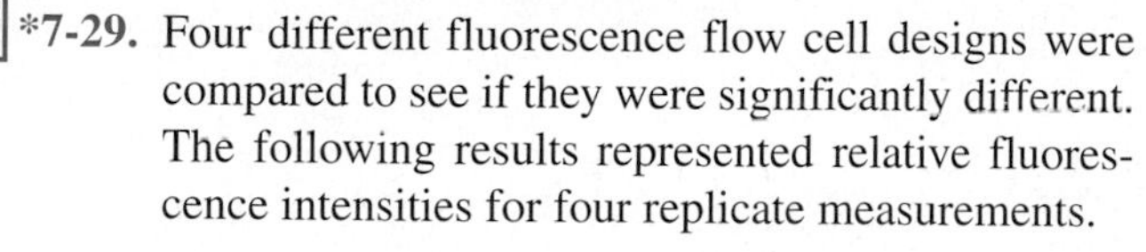

***7-29.** Four different fluorescence flow cell designs were compared to see if they were significantly different. The following results represented relative fluorescence intensities for four replicate measurements.

Measurement No.	Design 1	Design 2	Design 3	Design 4
1	72	93	96	100
2	93	88	95	84
3	76	97	79	91
4	90	74	82	94

(a) State the appropriate hypotheses.
(b) Do the flow cell designs differ at the 95% confidence level?
(c) If differences were detected in part (b), which designs would differ from each other at the 95% confidence level?

7-30. Three different analytical methods are compared for determining Ca. We are interested in knowing whether the methods differ. The results shown here represent ppm Ca determined by colorimetry, EDTA titration, and atomic absorption spectrometry.

Repetition No.	Colorimetry	EDTA Titration	Atomic Absorption
1	3.92	2.99	4.40
2	3.28	2.87	4.92
3	4.18	2.17	3.51
4	3.53	3.40	3.97
5	3.35	3.92	4.59

(a) State the null and alternative hypotheses.
(b) Determine whether there are differences in the three methods at the 95% and 99% confidence levels.
(c) If a difference was found at the 95% confidence level, determine which methods differ from each other.

***7-31.** Apply the Q test to the following data sets to determine whether the outlying result should be retained or rejected at the 95% confidence level.
(a) 41.27, 41.61, 41.84, 41.70
(b) 7.295, 7.284, 7.388, 7.292

7-32. Apply the Q test to the following data sets to determine whether the outlying result should be retained or rejected at the 95% confidence level.
(a) 85.10, 84.62, 84.70
(b) 85.10, 84.62, 84.65, 84.70

***7-33.** The following results were obtained for the determination of the ppm P in blood serum: 4.40, 4.42, 4.60, 4.48, 4.50. Determine whether the 4.60 ppm result is an outlier or should be retained at the 95% confidence level.

7-34. Challenge Problem. The following are three sets of data for the atomic mass of antimony from the work of Willard and McAlpine[7]:

Set 1	Set 2	Set 3
121.771	121.784	121.752
121.787	121.758	121.784
121.803	121.765	121.765
121.781	121.794	

(a) Determine the mean and the standard deviation of each data set.
(b) Determine the 95% confidence intervals for each data set.

[7]H. H. Willard and R. K. McAlpine, *J. Am. Chem. Soc.*, **1921**, *43*, 797.

(c) Determine whether the 121.803 value in the first data set is an outlier for that set at the 95% confidence level.

(d) Use the t test to determine whether the mean of data set 3 is identical to that of data set 1 at the 95% confidence level.

(e) Compare the means of all 3 data sets by ANOVA. State the null hypothesis. Determine whether the means differ at the 95% confidence level.

(f) Pool the data and determine the overall mean and the pooled standard deviation.

(g) Compare the overall mean of the 11 data points with the currently accepted value. Report the absolute error and the percent relative error, assuming the currently accepted value is the true value.

InfoTrac College Edition

For additional readings, go to InfoTrac College Edition, your online research library, at

http://infotrac.thomsonlearning.com

CHAPTER 8

Sampling, Standardization, and Calibration

Sampling is one of the most important operations in a chemical analysis. Chemical analyses use only a small fraction of the available sample. The fractions of the sandy and loam soil samples shown in the photo that are collected for analyses must be representative of the bulk materials. Knowing how much sample to collect and how to further subdivide the collected sample to obtain a laboratory sample is vital in the analytical process. Sampling, standardization, and calibration are the focal points of this chapter. All three steps require a knowledge of statistics.

As discussed in Chapter 1, an analytical procedure consists of several important steps. The specific analytical procedure chosen depends on how much sample is available and, in a broad sense, how much analyte is present. We discuss here a general classification of the types of determinations based on these factors. After selecting the particular method to be used, a representative sample must be acquired. The sampling process involves obtaining a small amount of material that accurately represents the bulk of the material being analyzed. Acquiring a representative sample is a statistical process. Most analytical methods are not absolute and require that results be compared with results on standard materials of accurately known composition. Some methods involve direct comparison with standards, while others involve an indirect calibration procedure. We discuss here in some detail standardization and calibration, including the use of the least-squares method to construct calibration models. This chapter concludes with a discussion of the methods used to compare analytical methods by using various performance criteria, called figures of merit.

8A ANALYTICAL SAMPLES AND METHODS

Many factors are involved in the choice of a specific analytical method, as discussed in Section 1C-1. Among the most important factors are the amount of sample and the concentration of the analyte.

8A-1 Types of Samples and Methods

Analytical methods can be classified in many different ways. Often we distinguish a method of identifying chemical species, a qualitative analysis, from one that determines the amount of a constituent, a quantitative analysis. Quantitative methods, as discussed in Section 1B, are traditionally classified as gravimetric, volumetric, or instrumental. Another way to distinguish methods is based on the size of the sample and the level of the constituents.

Sample Size

The size of the sample is often used to classify the type of analysis performed. As shown in Figure 8-1, the term **macro analysis** is used for samples of mass more than 0.1 g. A **semimicro analysis** is performed on a sample in the range of 0.01 to 0.1g, while the samples for a **micro analysis** are in the range 10^{-4} to 10^{-2} g. For samples whose mass is lower than 10^{-4} g, the term **ultramicro analysis** is sometimes used.

Sample Size	Type of Analysis
>0.1 g	Macro
0.01 to 0.1 g	Semimicro
0.0001 to 0.01 g	Micro
$<10^{-4}$ g	Ultramicro

From the classification in Figure 8-1, we see that the analysis of a 1-g sample of soil for a suspected pollutant would be called a macro analysis, while that of a 5-mg sample of a powder suspected to be an illicit drug would be a micro analysis. A typical analytical laboratory handles samples ranging from the macro size to the micro and even ultramicro size. Techniques for handling very small samples are quite different from those for treating macro samples.

Constituent Types

The constituents determined in an analytical procedure can cover a huge range in concentration. In some cases, analytical methods are used to determine **major constituents.** These constituents are present in the relative weight range of 1% to 100%. Many of the gravimetric and some of the volumetric procedures discussed in Part III are examples of major constituent determinations. As shown in Figure 8-2, species present in the range of 0.01% to 1% are usually termed **minor constituents,** whereas those present in amounts between 100 ppm (0.01%) and 1 ppb are termed **trace constituents.** Components present in amounts less than 1 ppb are usually considered to be **ultratrace constituents.**

Analyte Level	Type of Constituent
1% to 100%	Major
0.01% (100 ppm) to 1%	Minor
1 ppb to 100 ppm	Trace
<1 ppb	Ultratrace

Determination of Hg in the ppb to pspm range in a 1-μL (≈ 1 mg) sample of river water would be a micro analysis of a trace constituent. Determinations of trace and ultratrace constituents are particularly demanding because of potential interferences and contaminations. In extreme cases, determinations must be per-

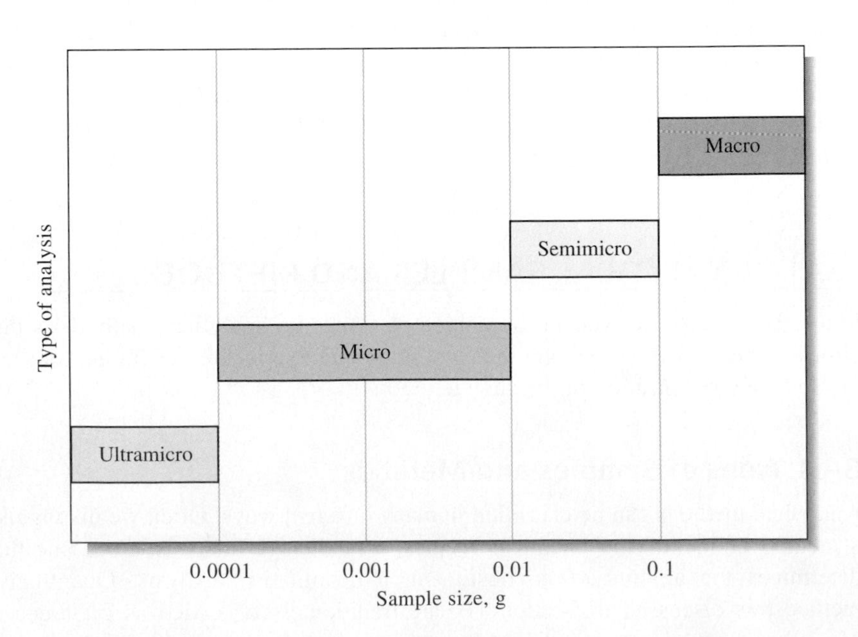

Figure 8-1 Classification of analyses by sample size.

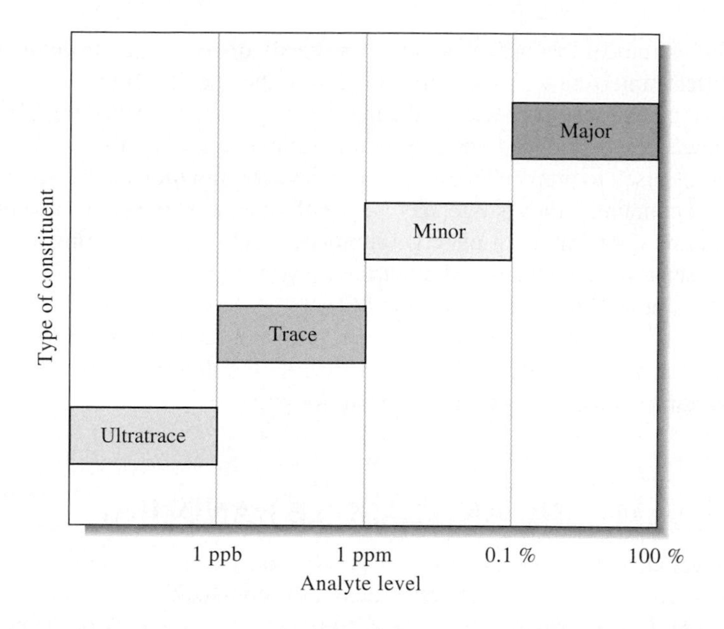

Figure 8-2 Classification of constituent types by analyte level.

formed in special rooms that are kept meticulously clean and free from dust and other contaminants. A general problem in trace procedures is that the reliability of results usually decreases dramatically with a decrease in analyte level. Figure 8-3 shows how the standard deviation between laboratories increases as the level of analyte decreases.

8A-2 Real Samples

The analysis of real samples is complicated by the presence of the sample matrix. This matrix can contain species that have chemical properties similar to the analyte. Such species can react with the same reagents as the analyte or they can

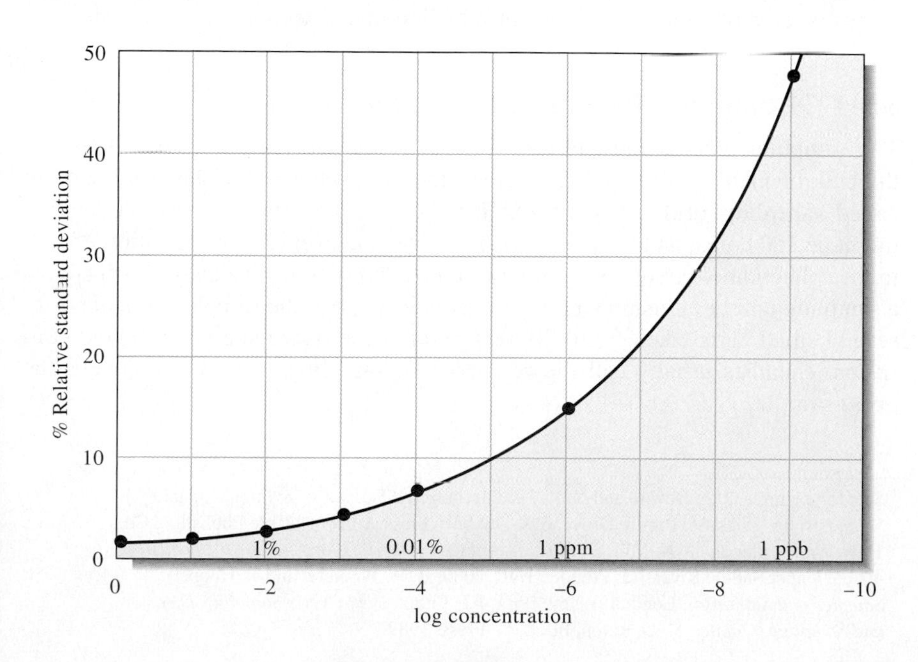

Figure 8-3 Interlaboratory error as a function of analyte concentration. Note that the relative standard deviation dramatically increases as the analyte concentration decreases. In the ultratrace range, the relative standard deviation approaches 100%. (From W. Horowitz, *Anal. Chem.*, **1982,** *54,* 67A–76A.)

cause an instrument response that cannot be easily distinguished from the analyte. These effects interfere with the determination of the analyte. If these interferences are caused by extraneous species in the matrix, they are often called **matrix effects.** Such effects can be induced not only by the sample itself but also by the reagents and solvents used to prepare the samples for the determination. The composition of the matrix containing the analyte may vary with time, as is the case when materials lose water by dehydration or undergo photochemical reactions during storage. We further discuss matrix effects and other interferences in the context of standardization and calibration methods in Section 8C.

► Samples are analyzed, but constituents or concentrations are determined.

As discussed in Section 1C, samples are *analyzed,* but species or concentrations are *determined.* Hence, we can correctly discuss the determination of glucose in blood serum or the analysis of blood serum for glucose.

8B SAMPLING AND SAMPLE HANDLING

A chemical analysis is most often performed on only a small fraction of the material whose composition is of interest. Clearly, the composition of this fraction must reflect as closely as possible the average composition of the bulk of the material if the results are to have value. The process by which a representative fraction is acquired is termed **sampling.** Often, sampling is the most difficult step in the entire analytical process and the step that limits the accuracy of the procedure. This statement is particularly true when the material to be analyzed is a large and inhomogeneous liquid, such as a lake, or an inhomogeneous solid, such as an ore, a soil, or a piece of animal tissue.

► Sampling is often the most difficult aspect of an analysis.

Sampling for a chemical analysis necessarily involves statistics because conclusions will be drawn about a much larger amount of material from the analysis of a small laboratory sample. This is the same process that we discussed in Chapters 6 and 7 of examining a finite number of items drawn from a population. From the observation of the sample, we use statistical tools, such as the mean and standard deviation, to draw conclusions about the population. The literature on sampling is extensive[1]; we provide only a brief introduction in this section.

8B-1 Obtaining A Representative Sample

The sampling process must ensure that the items chosen are representative of the bulk of material or population. Here, the items chosen for analysis are often called **sampling units** or **sampling increments.** For example, our population might be 100 coins, and we might wish to know the average concentration of lead in the collection of coins. Our sample is to be composed of five coins. Each coin is a sampling unit or an increment. In the statistical sense, the sample corresponds to several small parts taken from different parts of the bulk material. To avoid confusion, chemists usually call the collection of sampling units or increments the **gross sample.**

► The composition of the **gross sample** and the **laboratory sample** must closely resemble the average composition of the total mass of material to be analyzed.

[1] See, for example, J. L. Devore and N. R. Farnum, *Applied Statistics for Engineers and Scientists,* pp. 158–166. Pacific Grove, CA: Duxbury Press at Brooks/Cole Publishing Co., 1999; J. C. Miller and J. N. Miller, *Statistics and Chemometrics for Analytical Chemistry,* 4th ed. Upper Saddle River, NJ: Prentice-Hall, 2000; B. W. Woodget and D. Cooper, *Samples and Standards.* London: Wiley, 1987; F.F. Pitard, *Pierre Gy's Sampling Theory and Sampling Practice.* Boca Raton, FL: CRC Press, 1989.

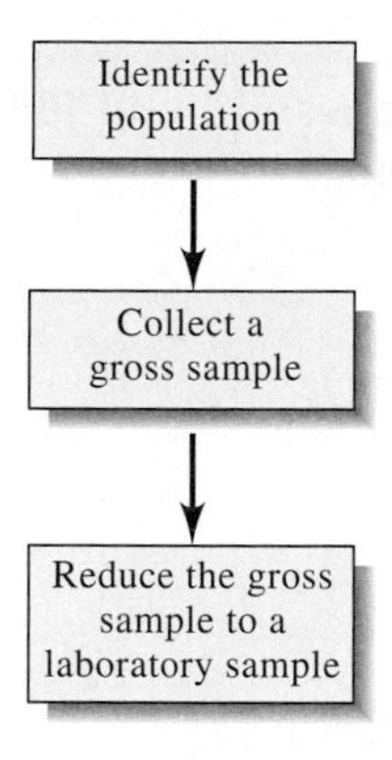

Figure 8-4 Steps in obtaining a laboratory sample. The laboratory sample consists of a few grams to at most a few hundred grams. It may constitute as little as 1 part in 10^7 or 10^8 of the bulk material.

For analysis in the laboratory, the gross sample is usually reduced in size and made homogeneous to become the **laboratory sample.** In some cases, such as sampling powders, liquids, and gases, we do not have obvious discrete items. Such materials may not be homogeneous because they may consist of microscopic particles of different compositions or, in the case of fluids, zones where concentrations differ. With these materials, we can assure a representative sample by taking our sample increments from different regions of the bulk material. Figure 8-4 illustrates the three steps that are commonly involved in obtaining the laboratory sample. Ordinarily, step 1 is straightforward, with the population being as diverse as a carton of bottles containing vitamin tablets, a field of wheat, the brain of a rat, or the mud from a stretch of river bottom. Steps 2 and 3 are seldom simple and may require a good deal of effort and ingenuity.

Sampling is the process by which a sample population is reduced in size to an amount of homogeneous material that can be conveniently handled in the laboratory and whose composition is representative of the population.

Statistically, the goals of the sampling process are

1. To obtain a mean value that is an unbiased estimate of the population mean. This goal can be realized only if all members of the population have an equal probability of being included in the sample.
2. To obtain a variance that is an unbiased estimate of the population variance so that valid confidence limits can be found for the mean, and various hypothesis tests can be applied. This goal can be reached only if every possible sample is equally likely to be drawn.

Both goals require obtaining a **random sample.** Here, the term random sample does not imply that the samples are chosen in a haphazard manner. Instead, a randomization procedure is applied to obtain such a sample. For example, suppose our sample is to consist of 10 pharmaceutical tablets to be drawn from 1000 tablets off a production line. One way to ensure a random sample is to choose the tablets to be tested from a table of random numbers. These can be conveniently generated from a random number table or from a spreadsheet, as shown in Figure 8-5. Here, we would assign each of the tablets a number from 1 to 1000 and use the sorted random numbers in column C of the spreadsheet to pick tablet 37, 71, 171, and so forth for analysis.

8B-2 Sampling Uncertainties

In Chapter 5, we concluded that both systematic and random errors in analytical data can be traced to instrument, method, and personal causes. Most systematic

Figure 8-5 Generating 10 random numbers from 1 to 1000 by use of a spreadsheet. The random number function in Excel [=RAND()] generates random numbers between 0 and 1. The multiplier shown in the documentation ensures that the numbers generated in column B will be between 1 and 1000. To obtain integer numbers, we use the Format/Cells . . . command on the menu bar, choose Number and then 0 decimal places. So that the numbers do not change with every recalculation, the random numbers in column B are copied and then pasted as values into column C using the Edit/Paste Special . . . command on the menu bar. In column C, the numbers were sorted in ascending order using Excel's Data Sort . . . command on the menu bar.

	A	B	C	D	E
1	**Spreadsheet to generate random numbers between 1 and 1000**				
2		Random Numbers	Sorted Numbers		
3		309	37		
4		184	71		
5		71	171		
6		171	184		
7		382	309		
8		933	382		
9		935	881		
10		37	933		
11		881	935		
12		961	961		
13					
14	**Spreadsheet Documentation**				
15	Cell B3=RAND()*(1000-1)+1				

errors can be eliminated by exercising care, by calibration, and by the proper use of standards, blanks, and reference materials. Random errors, which are reflected in the precision of data, can generally be kept at an acceptable level by closely controlling the variables that influence the measurements. Errors due to invalid sampling are unique in the sense that they are not controllable by the use of blanks and standards or by closer control of experimental variables. For this reason, sampling errors are ordinarily treated separately from the other uncertainties associated with an analysis.

For random and independent uncertainties, the overall standard deviation s_o for an analytical measurement is related to the standard deviation of the sampling process s_s and to the standard deviation of the method s_m by the relationship

$$s_o^2 = s_s^2 + s_m^2 \tag{8-1}$$

In many cases, the method variance will be known from replicate measurements of a single laboratory sample. Under this circumstance, s_s can be computed from measurements of s_o for a series of laboratory samples, each of which is obtained from several gross samples. An analysis of variance (see Section 7C) can reveal whether the between-samples variation (sampling plus measurement variance) is significantly greater than the within-samples variation (measurement variance).

► When $s_m < s_s/3$, there is no point in trying to improve the measurement precision. Equation 8-1 shows that s_o is determined predominantly by the sampling uncertainty under these conditions.

Youden has shown that once the measurement uncertainty has been reduced to one third or less of the sampling uncertainty (that is, $s_m < s_s/3$), further improvement in the measurement uncertainty is fruitless.[2] As a consequence, if the sampling uncertainty is large and cannot be improved, it is often a good idea to switch to a less precise but faster method of analysis so that more samples can be

[2] W. J. Youden, *J. Assoc. Off. Anal. Chem.*, **1981,** *50,* 1007.

analyzed in a given time. Since the standard deviation of the mean is lower by a factor of $\sqrt{N}$, taking more samples can improve precision.

8B-3 The Gross Sample

Ideally, the gross sample is a miniature replica of the entire mass of material to be analyzed. It should correspond to the bulk material in chemical composition and, if composed of particles, in particle size distribution.

◀ The gross sample is the collection of individual sampling units. It must be representative of the whole in composition and in particle size distribution.

Size of the Gross Sample

From the standpoint of convenience and economy, it is desirable that the gross sample weigh no more than absolutely necessary. Basically, gross sample weight is determined by (1) the uncertainty that can be tolerated between the composition of the gross sample and that of the whole, (2) the degree of heterogeneity of the whole, and (3) the level of particle size at which heterogeneity begins.[3]

The last point warrants amplification. A well-mixed, homogeneous solution of a gas or liquid is heterogeneous only on the molecular scale, and the weight of the molecules themselves governs the minimum weight of the gross sample. A particulate solid, such as an ore or a soil, represents the opposite situation. In such materials, the individual pieces of solid differ from each other in composition. Here, heterogeneity develops in particles that may have dimensions on the order of a centimeter or more and may weigh several grams. Intermediate between these extremes are colloidal materials and solidified metals. With the former, heterogeneity is first encountered in the range of 10^{-5} cm or less. In an alloy, heterogeneity first occurs in the crystal grains.

To obtain a truly representative gross sample, a certain number N of particles must be taken. The magnitude of this number depends on the uncertainty that can be tolerated and the heterogeneity of the material. The number may range from a few particles to as many as 10^{12} particles. The need for large numbers of particles is of no great concern for homogeneous gases and liquids because heterogeneity among particles first occurs at the molecular level. Thus, even a very small weight of sample will contain more than the requisite number of particles. The individual particles of a particulate solid may weigh a gram or more, however, which sometimes leads to a gross sample that weighs several tons. Sampling of such material is a costly, time-consuming procedure at best. To minimize cost, it is important to determine the smallest weight of material required to provide the desired information.

◀ The number of particles required in a gross sample ranges from a few particles to 10^{12} particles.

The laws of probability govern the composition of a gross sample removed randomly from a bulk of material. Because of this, it is possible to predict the likelihood that a selected fraction is similar to the whole. We can take an idealized case of a two-component mixture as a first example. A pharmaceutical mixture contains just two types of particles, type A particles, containing the active ingredient, and type B particles, containing only an inactive filler material. All particles are the same size. We wish to collect a gross sample that will allow us to determine the percentage of particles containing the active ingredient in the bulk material.

Let us assume that the probability of randomly drawing a type A particle is p and that of randomly drawing a type B particle is $(1 - p)$. If N particles of the

[3]For a paper on sample weight as a function of particle size, see G. H. Fricke, P.G. Mischler, F. P. Staffieri, and C. L. Housmyer, *Anal. Chem.*, **1987,** *59,* 1213.

mixture are taken, the most probable value for the number of type A particles is pN, whereas the most probable number of type B particles is $(1 - p)N$. For such a binary population, the Bernoulli equation[4] can be used to calculate the standard deviation of the number of A particles drawn, σ_A.

$$\sigma_A = \sqrt{Np(1 - p)} \tag{8-2}$$

► We use the symbol σ_r to indicate relative standard deviation in accordance with the International Union of Pure and Applied Chemistry (IUPAC) recommendations (see footnote 5). You should bear in mind that σ_r is a ratio.

The relative standard deviation σ_r[5] of drawing type A particles is σ_A/Np.

$$\sigma_r = \frac{\sigma_A}{Np} = \sqrt{\frac{1 - p}{Np}} \tag{8-3}$$

From Equation 8-3, we can obtain the number of particles needed to achieve a given relative standard deviation, as shown in Equation 8-4.

$$N = \frac{1 - p}{p\sigma_r^2} \tag{8-4}$$

Thus, for example, if 80% of the particles are type A ($p = 0.8$) and the desired relative standard deviation is 1% ($\sigma_r = 0.01$), the number of particles making up the gross sample should be

$$N = \frac{1 - 0.8}{0.8(0.01)^2} = 2500$$

Here, a random sample containing 2500 particles should be collected. A relative standard deviation of 0.1% would require 250,000 particles. Such a large number of particles would, of course, be determined by weighing, not by counting.

Let us now make the problem more realistic and assume that both of the components in the mixture contain the active ingredient (analyte), but in differing percentages. The type A particles contain a higher percentage of analyte, P_A, and the type B particles contain a lesser amount, P_B. Furthermore, the average density d of the particles differs from the densities d_A and d_B of these components. We are now interested in deciding the number of particles and thus the weight that we need to ensure a sample with the overall average percentage of active ingredient P, with a sampling relative standard deviation of σ_r. Equation 8-4 can be extended to include these stipulations:

$$N = p(1 - p)\left(\frac{d_A d_B}{d^2}\right)^2\left(\frac{P_A - P_B}{\sigma_r P}\right)^2 \tag{8-5}$$

From this equation, we see that the demands of precision are costly, in terms of the sample size required, because of the inverse-square relationship between the allow-

[4] A. A. Benedetti Pichler, in *Physical Methods in Chemical Analysis,* W. G. Berl, Ed., Vol. 3, pp. 183–194; New York: Academic Press, 1956, A. A. Beneditti-Pichler, *Essentials of Quantitative Analysis* Chapter 19, New York, Ronald Press, 1956.

[5] Compendium of Analytical Nomenclature: Definitive Rules, 1997, International Union of Pure and Applied Chemistry, prepared by J. Inczedy, T. Lengyel, and A. M. Ure, pp. 2–8. Malden, MA: Blackwell Science, 1998.

able standard deviation and the number of particles taken. Also, we can see that a greater number of particles must be taken as the average percentage P of the active ingredient becomes smaller.

The degree of heterogeneity as measured by $P_A - P_B$ has a large influence on the number of particles required since N increases with the square of the difference in composition of the two components of the mixture.

We can rearrange Equation 8-5 to calculate the relative standard deviation of sampling, σ_r.

$$\sigma_r = \frac{|P_A - P_B|}{P} \times \frac{d_A d_B}{d^2} \sqrt{\frac{p(1-p)}{N}} \tag{8-6}$$

If we make the assumption that the sample mass m is proportional to the number of particles and the other quantities in Equation 8-6 are constant, the product of m and σ_r should be a constant. This constant K_s is called Ingamells sampling constant.[6] Thus,

$$K_s = m \times (\sigma_r \times 100\%)^2 \tag{8-7}$$

where the factor of 100% converts σ_r to the relative standard deviation in percent. Hence, when $\sigma_r = 0.01$, $\sigma_r \times 100\% = 1\%$ and K_s is just equal to m. We can thus interpret the sampling constant K_s to be the minimum sample mass needed to reduce the sampling uncertainty to 1%.

The problem of deciding on the weight of the gross sample for a solid substance is ordinarily even more difficult than this example because most materials not only contain more than two components but also consist of a range of particle sizes. In most instances, the first of these problems can be met by dividing the sample into an imaginary two-component system. Thus, with an actual complex mixture of substances, one component selected might be all the various analyte-containing particles and the other all the residual components containing little or no analyte. After average densities and percentages of analyte are assigned to each part, the system is treated as if it has only two components.

◀ To simplify the problem of defining the weight of a gross sample of a multi-component mixture, assume that the sample is a hypothetical two-component mixture.

The problem of variable particle size can be handled by calculating the number of particles that would be needed if the sample consisted of particles of a single size. The gross sample weight is then determined by taking into account the particle size distribution. One approach is to calculate the needed weight by assuming that all particles are the size of the largest. This procedure is not very efficient, however, for it usually requires removal of a larger weight of material than necessary. Benedetti-Pichler gives alternative methods for computing the weight of gross sample to be chosen.[7]

An interesting conclusion from Equation 8-5 is that the number of particles in the gross sample is independent of particle size. The weight of the sample, of course, increases directly with the volume (or as the cube of the particle diameter), so that reduction in the particle size of a given material has a large effect on the weight required in the gross sample.

---

[6]C. O. Ingamells and P. Switzer, *Talanta,* **1973,** *20,* 547–568.

[7]A. A. Benedetti-Pichler in *Physical Methods in Chemical Analysis,* W. G. Berl, Ed., Vol. 3, p. 192. New York: Academic Press, 1956.

Clearly, a great deal of information must be known about a substance to make use of Equation 8-5. Fortunately, reasonable estimates of the various parameters in the equation can often be made. These estimates can be based on a qualitative analysis of the substance, visual inspection, and information from the literature on substances of similar origin. Crude measurements of the density of the various sample components may also be necessary.

EXAMPLE 8-1

A column-packing material for chromatography consists of a mixture of two types of particles. Assume that the average particle in the batch being sampled is approximately spherical with a radius of about 0.5 mm. Roughly 20% of the particles appear to be pink and are known to have about 30% by weight of a polymeric stationary phase attached (analyte). The pink particles have a density of 0.48 g/cm^3. The remaining particles have a density of about 0.24 g/cm^3 and contain little or no polymeric stationary phase. What mass of the material should the gross sample contain if the sampling uncertainty is to be kept below 0.5% relative?

We first compute values for the average density and percent polymer:

$$d = 0.20 \times 0.48 + 0.80 \times 0.24 = 0.288 \text{ g/cm}^3$$

$$P = \frac{(0.20 \times 0.48 \times 0.30) \text{ g polymer/cm}^3}{0.288 \text{ g sample/cm}^3} \times 100\% = 0.10\%$$

Then, substituting into Equation 8-5 gives

$$N = 0.20(1 - 0.20)\left[\frac{0.48 \times 0.24}{(0.288)^2}\right]^2\left(\frac{30 - 0}{0.005 \times 10.0}\right)^2$$

$$= 1.11 \times 10^5 \text{ particles required}$$

$$\text{weight of sample} = 1.11 \times 10^5 \cancel{\text{particles}} \times \frac{4}{3}\pi(0.05)^3 \frac{\cancel{\text{cm}^3}}{\cancel{\text{particle}}} \times \frac{0.288 \text{ g}}{\cancel{\text{cm}^3}}$$

$$= 5.3 \text{ g}$$

Sampling Homogeneous Solutions of Liquids and Gases

► Well-mixed solutions of liquids and gases require only a very small sample because they are homogeneous down to the molecular level.

For solutions of liquids or gases, the gross sample can be relatively small, since ordinarily nonhomogeneity first occurs at the molecular level, and even small volumes of sample will contain many more particles than the number computed from Equation 8-5. Whenever possible, the liquid or gas to be analyzed should be stirred well before sampling to make sure that the gross sample is homogeneous. With large volumes of solutions, mixing may be impossible; it is then best to sample several portions of the container with a "sample thief," a bottle that can be opened and filled at any desired location in the solution. This type of sampling is important, for example, in determining the constituents of liquids exposed to the atmosphere. Thus, the oxygen content of lake water may vary by a factor of 1000 or more over a depth difference of a few meters.

With the advent of portable sensors, it has become common in recent years to bring the laboratory to the sample instead of bringing the sample back to the laboratory. Most sensors, however, measure only local concentrations and do not average or sense remote concentrations.

In process control and other applications, samples of liquids are collected from flowing streams. Care must be taken so that the sample collected represents a constant fraction of the total flow and that all portions of the stream are sampled.

Gases can be sampled by several methods. In some cases, a sampling bag is simply opened and filled with the gas. In other cases, gases can be *trapped* in a liquid or adsorbed onto the surface of a solid.

Sampling Particulate Solids

It is often difficult to obtain a random sample from a bulky particulate material. Random sampling can best be accomplished while the material is being transferred. Mechanical devices have been developed for handling many types of particulate matter. Details on sampling these materials are beyond the scope of this book.

Sampling Metals and Alloys

Samples of metals and alloys are obtained by sawing, milling, or drilling. In general, it is not safe to assume that chips of the metal removed from the surface are representative of the entire bulk, so solid material from the interior must be sampled as well. With some materials, a representative sample can be obtained by sawing across the piece at random intervals and collecting the "sawdust" as the sample. Alternatively, the specimen may be drilled, again at various randomly spaced intervals, and the drillings collected as the sample; the drill should pass entirely through the block or halfway through from opposite sides. The drillings can be broken up and mixed or melted together in a special crucible made from graphite. A granular sample can often then be produced by pouring the melt into distilled water.

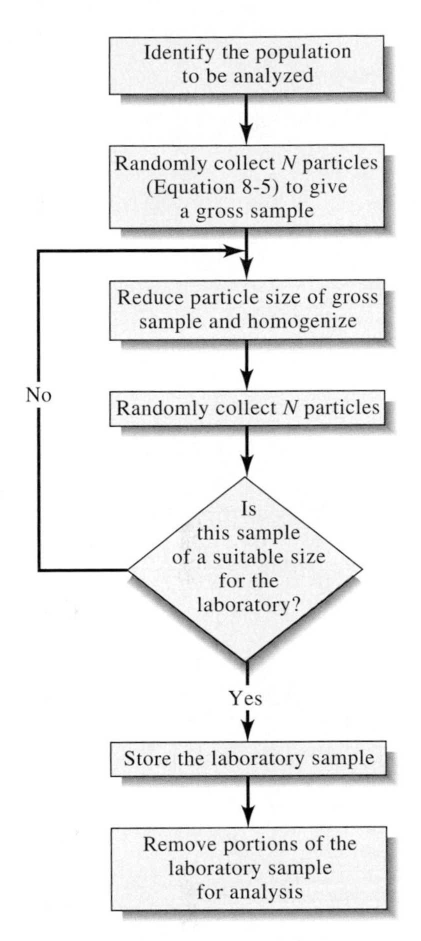

Figure 8-6 Steps in sampling a particulate solid.

8B-4 Preparing A Laboratory Sample

For nonhomogeneous solids, the gross sample may weigh from hundreds of grams to kilograms or more; therefore, reduction of the gross sample to a finely ground and homogeneous laboratory sample, weighing at the most a few hundred grams, is necessary. As shown in Figure 8-6, this process involves a cycle of operations that includes crushing and grinding, sieving, mixing, and dividing the sample (often into halves) to reduce its weight. During each division, a weight of sample that contains the number of particles computed from Equation 8-5 is retained.

◀ The laboratory sample should have the same number of particles as the gross sample.

EXAMPLE 8-2

A carload of lead ore containing galena ($\approx$ 70% Pb) and other particles with little or no lead is to be sampled. From the densities (galena = 7.6 g/cm^3, other particles = 3.5 g/cm^3, average density = 3.7 g/cm^3) and rough percentage of lead, Equation 8-5 indicates that 8.45×10^5 particles are required to keep the sampling error below 0.5% relative. The particles appear spherical with a radius of 5 mm. A calculation of the weight required, similar to that in Example 8-1, shows that the gross sample should weigh about 1.6×10^6 g (1.8 tons). We wish to reduce this gross sample to a laboratory sample that weighs about 100 g. How can this be done? *(continued)*

The laboratory sample should contain the same number of particles as the gross sample, or 8.45×10^5. For each particle,

$$\text{avg wt of particle} = \frac{100\text{ g}}{8.45 \times 10^5 \text{ particles}} = 1.18 \times 10^{-4}\text{ g/particle}$$

The average weight of a particle is related to its radius by the equation

$$\text{avg wt of particle} = \frac{4}{3}\pi\left[r(\text{cm})^3\right] \times \frac{3.7\text{ g}}{\text{cm}^3}$$

If we equate these two relationships and solve for r, we get

$$r = \left(1.18 \times 10^{-4}\text{ g} \times \frac{3}{4\pi} \times \frac{\text{cm}^3}{3.7\text{ g}}\right)^{1/3} = 1.97 \times 10^{-2}\text{ cm, or } 0.2\text{ mm}$$

Thus, the sample should be repeatedly ground, mixed, and divided until the particles are about 0.2 mm in diameter.

Additional information on details of preparing the laboratory sample can be found in Chapter 35 and in the literature.[8]

8B-5 Number of Laboratory Samples

Once the laboratory samples have been prepared, the question that remains is how many samples should be taken for the analysis. If we have reduced the measurement uncertainty such that it is less than one third the sampling uncertainty, the latter will limit the precision of the analysis. The number, of course, depends on what confidence interval we want to report for the mean value and the desired relative standard deviation of the method. If the sampling standard deviation σ_s is known from previous experience, we can use the values of z from the tables (see Section 7A-1).

$$\text{CI for } \mu = \bar{x} + \frac{z\sigma_s}{\sqrt{N}}$$

More often, we use an estimate of σ_s and so must use the t tables (see Section 7A-2).

$$\text{CI for } \mu = \bar{x} + \frac{ts_s}{\sqrt{N}}$$

The last term in this equation represents the absolute uncertainty that we can tolerate at a particular confidence level. If we divide this term by the mean value,

[8] *Standard Methods of Chemical Analysis,* F. J. Welcher, Ed. Vol. 2, Part A, pp. 21–55. Princeton, NJ: Van Nostrand, 1963. An extensive bibliography of specific sampling information has been compiled by C. A. Bicking, in *Treatise on Analytical Chemistry,* I. M. Kolthoff and P. J. Elving, Eds., 2nd ed., Vol. 1, p. 299. New York: Wiley, 1978.

$\bar{x}$, we can calculate the relative uncertainty σ_r that is tolerable at a given confidence level.

$$\sigma_r = \frac{ts_s}{\bar{x}\sqrt{N}} \tag{8-8}$$

If we solve Equation 8-8 for the number of samples N, we obtain

$$N = \frac{t^2 s_s^2}{\bar{x}^2 \sigma_r} \tag{8-9}$$

The use of t instead of z in Equation 8-9 does lead to one complication, since the value of t itself depends on N. Usually, however, we can solve the equation by iteration as shown in Example 8-3 and obtain the desired number of samples.

EXAMPLE 8-3

The determination of copper in a sea water sample gave a mean value of 77.81 μg/L and a standard deviation s_s of 1.74 μg/L. (*Note:* The insignificant figures were retained here because these results are used later in a calculation.) How many samples must be analyzed to obtain a relative standard deviation of 1.7% in the results at the 95% confidence level?

We begin by assuming an infinite number of samples, which gives a value of t of 1.96 at the 95% confidence level. Since $\sigma_r = 0.017$, $s_s = 1.74$, and $\bar{x} = 77.81$, Equation 8-9 gives

$$N = \frac{(1.96)^2 \times (1.74)^2}{(0.017)^2 \times (77.81)^2} = 6.65$$

If we round this result to seven samples, we find the value of t for 6 degrees of freedom is 2.45. A second value for N can then be calculated using this t value, which gives $N = 10.38$. If we use 9 degrees of freedom and $t = 2.26$, the next value is $N = 8.84$. The iterations converge with an N value of approximately 9. Note that it would be good strategy to reduce the sampling uncertainty so that fewer samples would be needed.

8B-6 Automated Sample Handling

Once sampling has been accomplished and the number of samples and replicates chosen, sample processing begins (recall Figure 1-2). Because of their reliability and cost-effectiveness, many laboratories are using automated sample handling methods. In some cases, automated sample handling is used for only a few specific operations, such as dissolving the sample and removing interferences; in other cases all the remaining steps in the analytical procedure are automated. Two different methods of automated sample handling are described here: the **batch,** or **discrete approach,** and the **continuous flow approach.**

◀ Automated sample handling can lead to higher throughput (more analyses per unit time), higher reliability, and lower costs than manual sample handling.

Discrete Methods

Systems that process samples in a discrete manner often mimic the operations that would be performed manually. Laboratory robots are used to process samples when it might be dangerous for humans to be involved or when a large number of routine steps might be involved. Small laboratory robots have been available commercially since the mid-1980s.[9] The robotic system is controlled by a computer programmed by the user. Laboratory robots can be used to dilute, filter, partition, grind, centrifuge, homogenize, extract, and treat samples with reagents. They can also be trained to heat and shake samples, dispense measured volumes of liquids, inject samples into chromatographic columns, weigh samples, and transport them to the appropriate instrument for measurement.

Some discrete sample processors automate only the measurement step of the procedure or a few chemical steps and the measurement step. One type, based on the use of centrifugal force, mixes samples and reagents and transfers them to a photometric instrument for measurement. Another type, based on multilayer film technology, carries out a series of chemical reactions or physical processes in a sequential manner.[10]

Continuous Flow Methods

In continuous flow methods, the sample is inserted into a flowing stream, where a number of operations can be performed before it is transported to a flow-through detector. Hence, these systems behave as automated analyzers in that they can perform not only sample processing operations but also the final measurement step. Such sample processing operations as reagent addition, dilution, incubation, mixing, dialysis, extraction, and many others can be implemented between the point of sample introduction and detection. There are two different types of continuous flow systems: segmented flow analyzers and flow injection analyzers.

► Two types of continuous flow analyzers are the segmented flow analyzer and the flow injection analyzer.

The segmented flow analyzer divides the sample into discrete segments separated by gas bubbles, as shown in Figure 8-7a. As shown in Figure 8-7b, the gas bubbles provide barriers to prevent the sample from spreading out along the tube as a result of dispersion processes. The bubbles thus confine the sample and minimize cross-contamination between different samples. They also enhance mixing between the samples and the reagents. The concentration profiles of the analyte are shown in Figure 8-7c. Samples are introduced at the sampler as plugs (left). Some broadening due to dispersion occurs by the time the samples reach the detector. Hence, the type of signals shown on the right are typically used to obtain quantitative information about the analyte. Samples can be analyzed at a rate of 30 to 120 per hour.

Dispersion is a band-spreading or mixing phenomenon that results from the coupling of fluid flow with molecular diffusion. **Diffusion** is mass transport due to a concentration gradient.

---

[9]For a description of laboratory robots, see G.J. Kost, Ed., *Handbook of Clinical Automation, Robotics and Optimization.* New York: Wiley, 1996; J. R. Strimaltis and G. L. Hawk, *Advances in Laboratory Automation-Robotics.* Hopkinton, MA: Zymark Corp, 1998; V. Berry, *Anal. Chem.,* **1990,** *62,* 337A; J. R. Strimaltis, *J. Chem. Educ.,* **1989,** *66,* A8; **1990,** *67,* A20; W. J. Hurst and J. W. Mortimer, *Laboratory Robotics.* New York: VCH Publishers, 1987.

[10]For a more extensive discussion of automated analyzers based on multilayer films, see D. A. Skoog, F. J. Holler, and T. A. Nieman, *Principles of Instrumental Analysis,* 5th ed., pp. 845–849. Belmont, CA: Brooks/Cole, 1998.

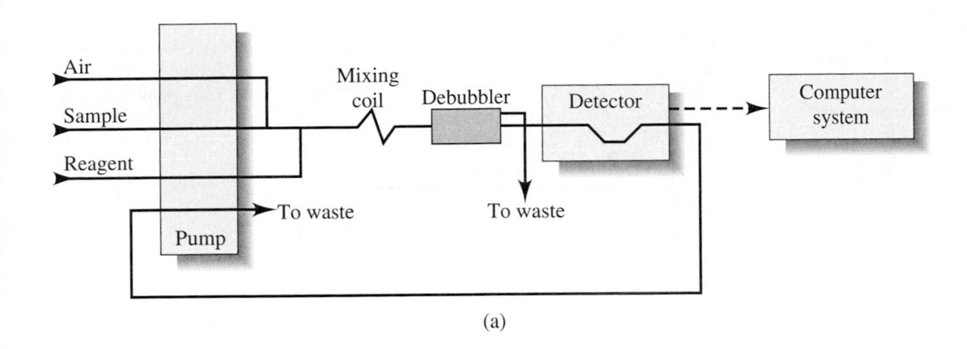

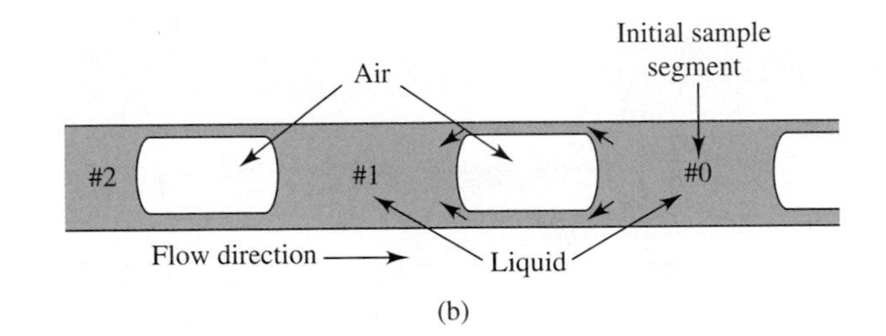

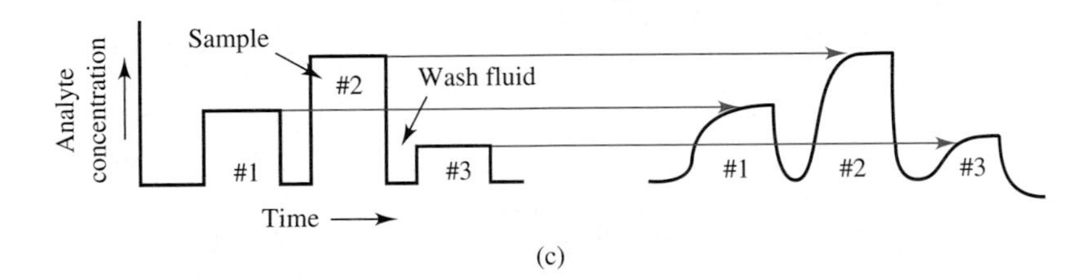

Figure 8-7 Segmented continuous flow analyzer. (a), Samples are aspirated from sample cups in the sampler and pumped into the manifold, where they are mixed with one or more reagents. Air is also injected to segment the samples with bubbles. The bubbles are usually removed by a debubbler before the stream reaches the detector. The segmented sample is shown in more detail in (b). The bubbles minimize dispersion of the sample, which can cause broadening of the zones and cross-contamination from different samples. The analyte concentration profiles at the sampler and at the detector are shown in (c). Normally, the height of a sample peak is related to the concentration of the analyte.

Flow injection analysis (FIA) is a more recent development.[11] In this process, samples are injected from a sample loop into a flowing stream containing one or more reagents, as shown in Figure 8-8a. The sample plug is allowed to disperse in a controlled manner before it reaches the detector, as illustrated in Figure 8-8b. Injection of the sample into a reagent stream yields the type of responses shown on the right. In merging zones FIA, the sample and reagent are both injected into carrier streams and merged at a tee mixer. In either normal or merging zones FIA, sample dispersion is controlled by the sample size, the flow rate, and the length and diameter of the tubing. It is also possible to stop the flow when the sample reaches the detector to allow concentration-time profiles to be measured for kinetic methods (see Chapter 29).

Flow injection systems can also incorporate several sample processing units, such as solvent extraction modules, dialysis modules, heating modules and others.

[11]For more information on FIA see J. Ruzicka and E. H. Hansen, *Flow Injection Analysis*, 2nd ed. New York: Wiley, 1988; M. Valcarcel and M. D. Luque de Castro, *Flow Injection Analysis: Principles and Applications*. Chichester, England: Ellis Horwood, 1987; B. Karlberg and G. E. Pacey, *Flow Injection Analysis: A Practical Guide*. New York: Elsevier, 1989: J. P. Smith and V. Hinson-Smith, *Anal. Chem.*, **2002,** *74,* 385A.

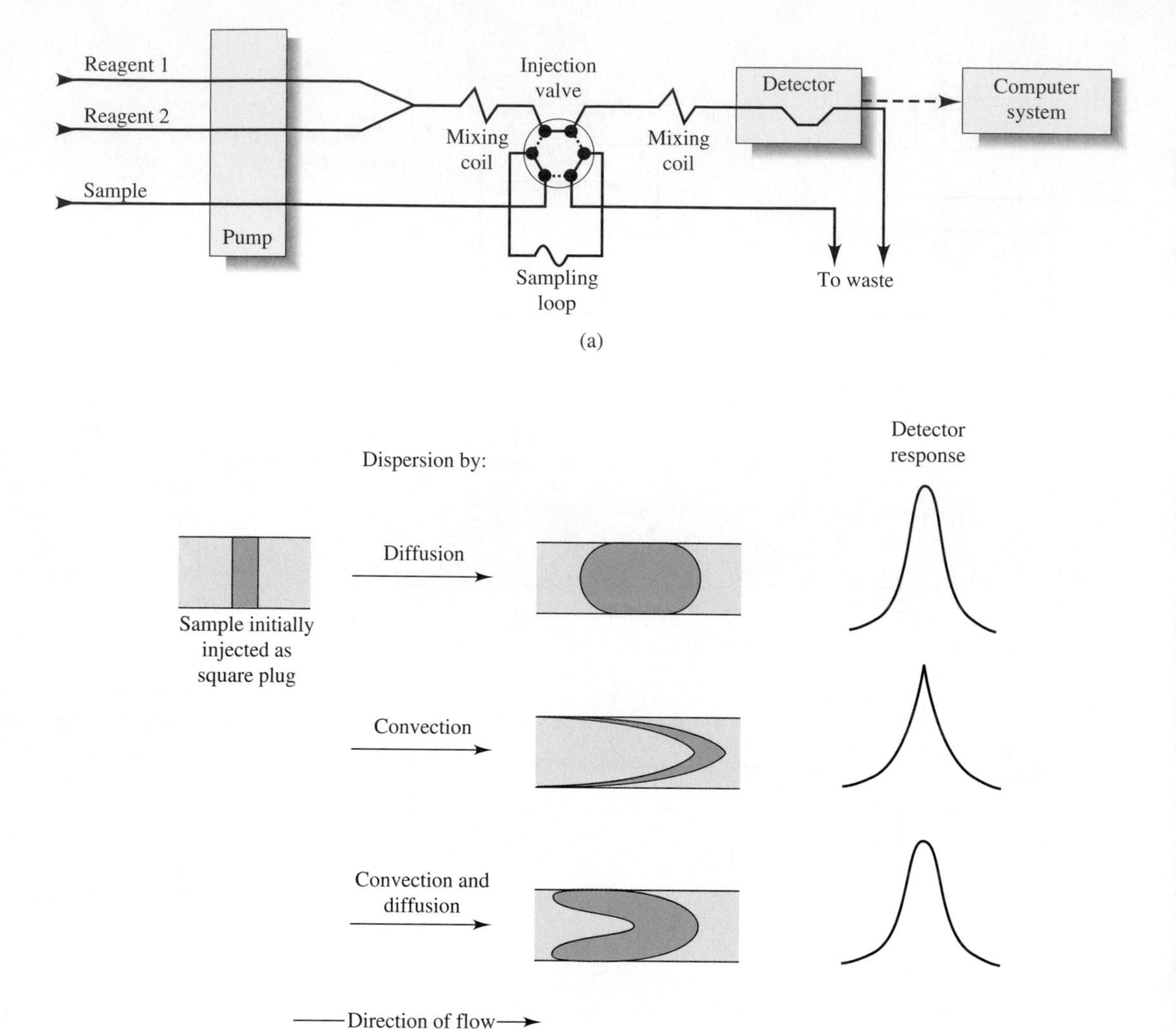

Figure 8-8 Flow injection analyzer. In (a), the sample is loaded from a sampler into the sample loop of a sampling valve. The valve, shown in the load position, also has a second inject position, shown by dotted lines. When switched to the inject position, the stream containing the reagent flows through the sample loop. Sample and reagent are allowed to mix and react in the mixing coil before reaching the detector. In this case, the sample plug is allowed to disperse before reaching the detector (b). The resulting concentration profile (detector response) depends on the degree of dispersion.

Samples can be processed with FIA at rates varying from 60 to 300 per hour. In recent work, FIA systems have been miniaturized to either capillary (inner diameters from 20 to 100 μm) or microchip (see Feature 8-1) dimensions.[12] Such miniature analyzers have the potential to enable manipulations and measurements on such small samples as single cells and to minimize the amount of reagent consumed in an analysis.

[12]Examples of miniature FIA systems can be found in D. M. Spence and S. R. Crouch, *Anal. Chem.*, **1997,** *69,* 165; A. G. Hadd, D. E. Raymond, J. W. Halliwell, S. C. Jacobson, and J. M. Ramsey, *Anal. Chem.*, **1997,** *69,* 3407.

FEATURE 8-1

Lab-on-a-Chip[13]

The concept of a complete laboratory on a chip has evolved over the past several years. Miniaturization of laboratory operations to the chip scale promises to reduce analysis costs by lowering reagent consumption, by automating the procedures, and by increasing the number of analyses that can be done per day. There have been several approaches to implementing the lab-on-a-chip concept. The most successful use the same photolithography technology developed for preparing electronic integrated circuits. This technology is employed to produce the valves, propulsion systems, and reaction chambers needed to perform chemical analyses. The development of microfluidic devices is an active research area involving scientists and engineers from academic and industrial laboratories.[14]

Several different fluid propulsion systems have been investigated, including electroosmosis (see Chapter 33), microfabricated mechanical pumps, and hydrogels that emulate human muscles. Flow injection techniques as well as such separation methods as liquid chromatography (see Chapter 32), capillary electrophoresis, and capillary electrokinetic chromatography (see Chapter 33) have been implemented. Figure 8F-1 shows the layout of a microstructure used for FIA or for FIA combined with capillary electrophoresis. This type of system is often called a miniaturized total chemical analysis system, or a μ-TAS.

The lab-on-a-chip devices have been used in research demonstrations to separate and detect explosives, sequence DNA, determine species of clinical importance, and screen for drugs. These devices should become more important as the technology matures.

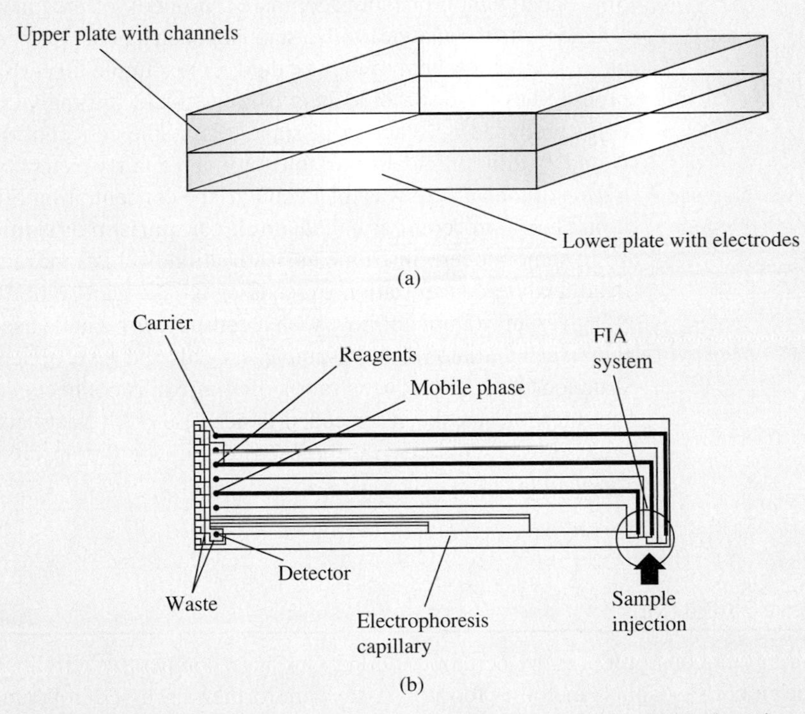

Figure 8F-1 Layout of a microfabricated structure for combining FIA with a capillary electrophoretic separation. (a) Two glass plates are used in a sandwich structure. The upper plate contains the channel structure (30 μm wide by 10 μm deep), and the lower plate contains platinum electrodes for controlling the flow. (b) Samples are injected, mixed with reagents, and carried to a detector. An electrophoretic separation can also be achieved if desired. Detectors have been conductivity, electrochemical, and fluorescence. (Modified from A. Manz, J. C. Fettinger, E. Verpoorte, H. Ludi, H. M. Widmer, and D. J. Harrison, *Trends in Analytical Chemistry (TRAC)*, **1991,** *10,* 144, with permission from Elsevier Science.)

[13]For an overview see D. Figeys, *Anal. Chem.*, **2000,** *72,* 330A.

[14]See N. A. Polson and M. A. Hayes, *Anal. Chem.*, **2001,** *73,* 313A.

8C STANDARDIZATION AND CALIBRATION

A very important part of all analytical procedures is the calibration and standardization process. **Calibration** determines the relationship between the analytical response and the analyte concentration. Usually this is accomplished by the use of **chemical standards.** In the deer kill case study of Feature 1-1, the arsenic concentration was found by calibrating the absorbance scale of a spectrophotometer with solutions of known arsenic concentration. Almost all analytical methods require some type of calibration with chemical standards. Gravimetric methods (see Chapter 12) and some coulometric methods (see Chapter 22) are among the few **absolute** methods that do not rely on calibration with chemical standards. Several types of calibration procedures are described in this section.

8C-1 Comparison with Standards

Two types of comparison methods are described in this section: the direct comparison technique and the titration procedure.

Direct Comparison

Some analytical procedures compare a property of the analyte (or the product of a reaction with the analyte) with a standard such that the property being tested matches or nearly matches that of the standard. For example, in early colorimeters, the color produced as the result of a chemical reaction of the analyte was compared with the color produced by reaction of standards. If the concentration of the standard was varied by dilution, it was possible to obtain a fairly exact color match. The concentration of the analyte was then equal to the concentration of the standard after dilution. Such a procedure is called a **null comparison** or **isomation method.**[15]

In some modern instruments, a variation of this procedure is used to determine if an analyte concentration exceeds or is less than some threshold level. Feature 8-2 gives an example of how such a **comparator** can be used to determine whether the level of aflatoxin in a sample exceeds the level that would indicate a toxic situation. The exact concentration of aflatoxin is not needed; only an indication that the threshold has been exceeded is necessary. Alternatively, a simple comparison with several standards can be used to indicate the approximate concentration of the analyte.

FEATURE 8-2

A Comparison Method for Aflatoxins[16]

Aflatoxins are potential carcinogens produced by certain molds that may be found in corn, peanuts, and other food items. They are colorless, odorless, and tasteless. The toxic nature of aflatoxins was made evident by a large "turkey kill" in England in 1960. One method of detecting aflatoxins is a competitive binding immunoassay. Such assays are further discussed in Feature 11-1.

In the analysis, antibodies specific to the aflatoxin are coated on the base of a plastic compartment or microtiter well in an array on a plate such as that shown in Figure

[15]See, for example, H. V. Malmstadt and J. D. Winefordner, *Anal. Chim. Acta,* **1960,** *20,* 283; L. Ramaley and C. G. Enke, *Anal. Chem.,* **1965,** *37,* 1073.

[16]P. R. Kraus, A. P. Wade, S. R. Crouch, J. F. Holland, and B. M. Miller, *Anal. Chem.,* **1988,** *60,* 1387.

8F-2. The aflatoxin behaves as the antigen. During the analysis, an enzyme reaction causes a blue product to form. As the amount of aflatoxin in the sample increases, the blue color decreases in intensity. The color-measuring instrument is the basic fiber-optic comparator shown in Figure 8F-3. The instrument can be used to compare the color intensity of the sample with that of a reference solution to indicate whether the aflatoxin level exceeds the threshold level. In another mode, a series of increasingly concentrated standards can be placed in the reference well holder. The sample aflatoxin concentration is then between the two standards that are slightly more and slightly less concentrated than the analyte, as indicated by the green and red indicator light-emitting diodes (LEDs).

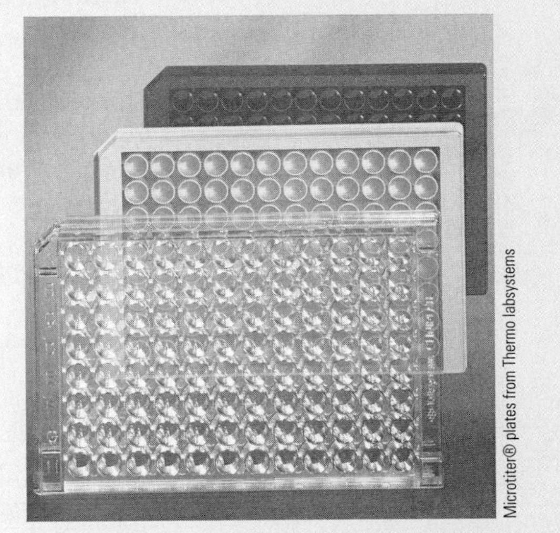

Figure 8F-2 Microtiter plates. Several different sizes and configurations are available commercially. Most are arrays of 24 or 96 wells. Some are strips or can be cut into strips.

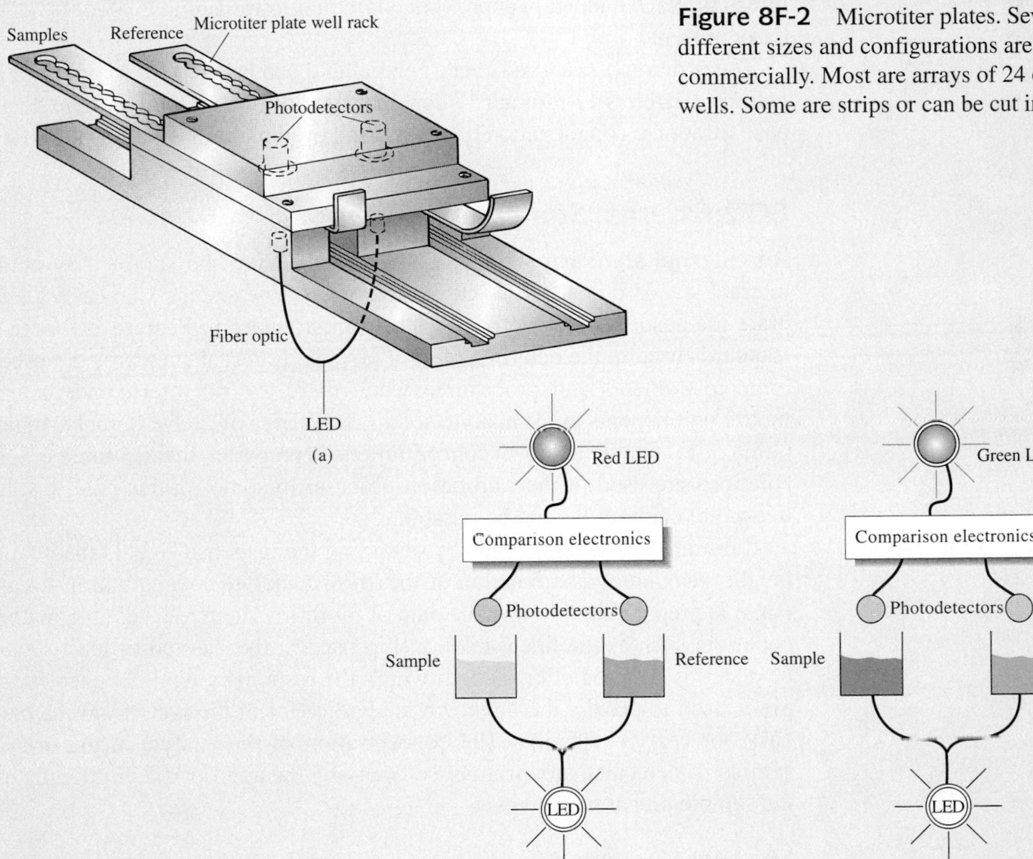

Figure 8F-3 Optical comparator. (a) An optical fiber that splits into two branches carries light from a light-emitting diode (LED) through sample and reference wells in a microtiter plate holder. In the comparison mode, a standard containing the threshold level of analyte (aflatoxin) is placed in one of the reference well holders. The samples containing unknown amounts of the analyte are placed in the sample well holder. If the sample contains more aflatoxin than the standard (b), the sample well absorbs less light at 650 nm than the reference well. An electronic circuit lights a red LED to indicate a dangerous amount of aflatoxin. If the sample contains less aflatoxin than the standard (c), a green LED is lit.

Titrations

Titrations are among the most accurate of all analytical procedures. In a titration, the analyte reacts with a standardized reagent (the titrant) in a reaction of known stoichiometry. Usually the amount of titrant is varied until chemical equivalence is reached, as indicated by the color change of a chemical indicator or by the change in an instrument response. The amount of the standardized reagent needed to achieve chemical equivalence can then be related to the amount of analyte present. The titration is thus a type of chemical comparison.

For example, in the titration of the strong acid HCl with the strong base NaOH, a standardized solution of NaOH is used to determine the amount of HCl present. The reaction is

$$HCl + NaOH \rightarrow NaCl + H_2O$$

The standardized solution of NaOH is added from a buret until an indicator like phenolphthalein changes color. At this point, called the **end point,** the number of moles of NaOH added is approximately equal to the number of moles of HCl initially present.

The titration procedure is very general and can be used for a variety of determinations. Chapters 13 through 17 consider the titration method in more detail. Acid-base titrations, complexation titrations, and precipitation titrations are described.

8C-2 External Standard Calibration

An **external standard** is prepared separately from the sample. By contrast, an internal standard is added to the sample itself. The arsenic standards used to calibrate the absorbance scale of the spectrophotometer in Feature 1-1 were external standards used in the determination of arsenic. External standards are used to calibrate instruments and procedures when there are no interference effects from matrix components in the analyte solution. A series of such external standards containing the analyte in known concentrations is prepared. Ideally, three or more such solutions are used in the calibration process. In some routine analyses, however, two-point calibrations can be reliable.

Calibration is accomplished by obtaining the response signal (absorbance, peak height, peak area) as a function of the known analyte concentration. A calibration curve is prepared by plotting the data or by fitting them to a suitable mathematical equation, such as the linear relationship used in the method of least squares. The next step is the prediction step, in which the response signal obtained for the sample is used to *predict* the unknown analyte concentration, c_u, from the calibration curve or best-fit equation. The concentration of the analyte in the original bulk sample is then calculated from c_u by applying the appropriate dilution factors from the sample preparation steps.

The Least-Squares Method

A typical calibration curve is shown in Figure 8-9 for the determination of isooctane in a hydrocarbon sample. Here a series of isooctane standards were injected into a gas chromatograph, and the area of the isooctane peak was obtained as a function of concentration. The ordinate is the dependent variable (peak area), while the abscissa is the independent variable (mole percent of isooctane). As is typical and usually desirable, the plot approximates a straight line. Note, however, that

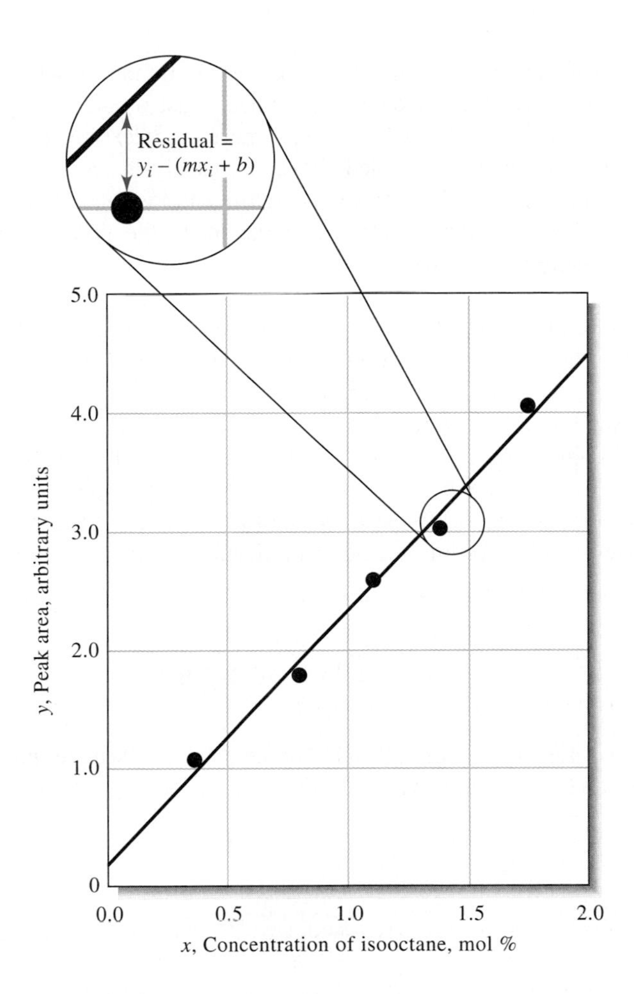

Figure 8-9 Calibration curve for the determination of isooctane in a hydrocarbon mixture.

because of the indeterminate errors in the measurement process, not all the data fall exactly on the line. Thus, the investigator must try to draw the "best" straight line among the data points. **Regression analysis** provides the means for objectively obtaining such a line and also for specifying the uncertainties associated with its subsequent use. We consider here only the basic **method of least squares** for two-dimensional data.

Assumptions of the Least-Squares Method Two assumptions are made in using the method of least squares. The first is that a linear relationship actually exists between the measured response y and the standard analyte concentration x. The mathematical relationship describing this assumption is called the **regression model,** which may be represented as

$$y = mx + b$$

where b is the y intercept (the value of y when x is zero) and m is the slope of the line (Figure 8-10). We also assume that any deviation of the individual points from the straight line arises from error in the *measurement.* That is, we assume that there is no error in the x values of the points (concentrations). Both these assumptions are

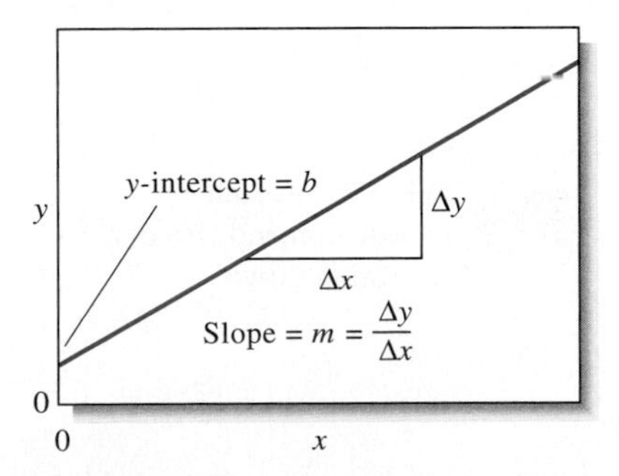

Figure 8-10 The slope-intercept form of a straight line.

► Linear least-squares analysis gives you the equation for the best straight line among a set of x, y data points when the x data contain negligible uncertainty.

appropriate for many analytical methods, but bear in mind that whenever there is significant uncertainty in the x data, basic linear least-squares analysis may not give the best straight line. In such a case, a more complex **correlation analysis** may be necessary. In addition, simple least-squares analysis may not be appropriate when the uncertainties in the y values vary significantly with x. In this case, it may be necessary to apply different weighting factors to the points and perform a **weighted least-squares analysis.**

Finding the Least-Squares Line As illustrated in Figure 8-9, the vertical deviation of each point from the straight line is called a **residual.** The line generated by the least-squares method is the one that minimizes the sum of the squares of the residuals for all the points. In addition to providing the best fit between the experimental points and the straight line, the method gives the standard deviations for m and b.

The least-squares method finds the sum of the squares of the residuals SS_{resid} and minimizes these according to the minimization technique of calculus.[17] The value of SS_{resid} is found from

$$SS_{resid} = \sum_{i=1}^{N} [y_i - (b + mx_i)]^2$$

where N is the number of points used. The calculation of slope and intercept is simplified when three quantities are defined, S_{xx}, S_{yy}, and S_{xy}, as follows:

► The equations for S_{xx} and S_{yy} are the numerators in the equations for the variance in x and the variance in y. Likewise, S_{xy} is the numerator in the covariance of x and y.

$$S_{xx} = \Sigma(x_i - \bar{x})^2 = \Sigma x_i^2 - \frac{(\Sigma x_i)^2}{N} \qquad (8\text{-}10)$$

$$S_{yy} = \Sigma(y_i - \bar{y})^2 = \Sigma y_i^2 - \frac{(\Sigma y_i)^2}{N} \qquad (8\text{-}11)$$

$$S_{xy} = \Sigma(x_i - \bar{x})(y_i - \bar{y}) = \Sigma x_i y_i - \frac{\Sigma x_i \Sigma y_i}{N} \qquad (8\text{-}12)$$

where x_i and y_i are individual pairs of data for x and y, N is the number of pairs, and $\bar{x}$ and $\bar{y}$ are the average values for x and y; that is, $\bar{x} = \frac{\Sigma x_i}{N}$ and $\bar{y} = \frac{\Sigma y_i}{N}$.

Note that S_{xx} and S_{yy} are the sum of the squares of the deviations from the mean for individual values of x and y. The expressions shown on the far right in Equations 8-10 through 8-12 are more convenient when a calculator without a built-in regression function is being used.

Six useful quantities can be derived from S_{xx}, S_{yy}, and S_{xy}, as follows:

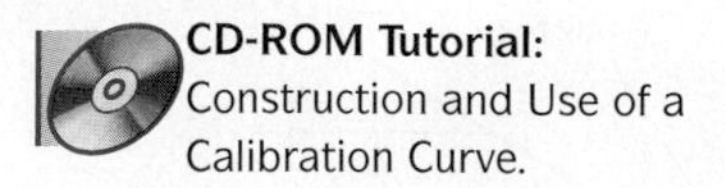
CD-ROM Tutorial: Construction and Use of a Calibration Curve.

1. The slope of the line, m:

$$m = \frac{S_{xy}}{S_{xx}} \qquad (8\text{-}13)$$

[17]The procedure involves differentiating SS_{resid} with respect to first m and then b and setting the derivatives equal to zero. This yields two equations, called normal equations, in the two unknowns m and b. These are then solved to give the least-squares best estimates of these parameters.

2. The intercept, b:

$$b = \bar{y} - m\bar{x} \tag{8-14}$$

3. The standard deviation about regression, s_r:

$$s_r = \sqrt{\frac{S_{yy} - m^2 S_{xx}}{N - 2}} \tag{8-15}$$

4. The standard deviation of the slope, s_m:

$$s_m = \sqrt{\frac{s_i^2}{S_{xx}}} \tag{8-16}$$

5. The standard deviation of the intercept, s_b:

$$s_b = s_r \sqrt{\frac{\Sigma x_i^2}{N\Sigma x_i^2 - (\Sigma x_i)^2}} = s_r \sqrt{\frac{1}{N - (\Sigma x_i)^2/\Sigma x_i^2}} \tag{8-17}$$

6. The standard deviation for results obtained from the calibration curve, s_c:

$$s_c = \frac{s_r}{m}\sqrt{\frac{1}{M} + \frac{1}{N} + \frac{(\bar{y}_c - \bar{y})^2}{m^2 S_{xx}}} \tag{8-18}$$

Equation 8-18 gives us a way to calculate the standard deviation from the mean $\bar{y}_c$ of a set of M replicate analyses of unknowns when a calibration curve that contains N points is used; recall that $\bar{y}$ is the mean value of y for the N calibration points. This equation is only approximate and assumes that slope and intercept are independent parameters, which is not strictly true.

The standard deviation about regression s_r (see Equation 8-15) is the standard deviation for y when the deviations are measured not from the mean of y (as is the usual case) but from the straight line that results from the least-squares prediction. The value of s_r is related to SS_{resid} by

$$s_r = \sqrt{\frac{\sum_{i=1}^{N} [(y_i - (b + mx_i)]^2}{N - 2}} = \sqrt{\frac{SS_{resid}}{N - 2}}$$

In this equation, the number of degrees of freedom is $N - 2$, since one degree of freedom is lost in calculating m and one in determining b. The **standard deviation about regression** is often called the **standard error of the estimate.** It roughly corresponds to the size of a typical deviation from the estimated regression line. Examples 8-4 and 8-5 illustrate how these quantities are calculated and used. The spreadsheet calculation of these quantities is illustrated in the Spreadsheet Exercise later in this section.

The **standard deviation about regression,** also called the **standard error of the estimate** or just the **standard error,** is a rough measure of the magnitude of a typical deviation from the regression line.

EXAMPLE 8-4

Carry out a least-squares analysis of the experimental data provided in the first two columns of Table 8-1 and plotted in Figure 8-9.

TABLE 8-1

Calibration Data for the Chromatographic Determination of Isooctane in a Hydrocarbon Mixture

Mole Percent Isooctane, x_i	Peak Area y_i	x_i^2	y_i^2	$x_i y_i$
0.352	1.09	0.12390	1.1881	0.38368
0.803	1.78	0.64481	3.1684	1.42934
1.08	2.60	1.16640	6.7600	2.80800
1.38	3.03	1.90440	9.1809	4.18140
1.75	4.01	3.06250	16.0801	7.01750
5.365	12.51	6.90201	36.3775	15.81992

Columns 3, 4, and 5 of the table contain computed values for x_i^2, y_i^2, and $x_i y_i$, respectively, with their sums appearing as the last entry in each column. Note that the number of digits carried in the computed values should be the *maximum allowed by the calculator or computer;* that is, *rounding should not be performed until the calculation is complete.*

We now substitute into Equations 8-10, 8-11, and 8-12 and obtain

$$S_{xx} = \Sigma x_i^2 - \frac{(\Sigma x_i)^2}{N} = 6.90201 - \frac{(5.365)^2}{5} = 1.14537$$

$$S_{yy} = \Sigma y_i^2 - \frac{(\Sigma y_i)^2}{N} = 36.3775 - \frac{(12.51)^2}{5} = 5.07748$$

$$S_{xy} = \Sigma x_i y_i - \frac{\Sigma x_i \Sigma y_i}{N} = 15.81992 - \frac{5.365 \times 12.51}{5} = 2.39669$$

Substitution of these quantities into Equations 8-13 and 8-14 yields

$$m = \frac{2.39669}{1.14537} = 2.0925 \approx 2.09$$

$$b = \frac{12.51}{5} - 2.0925 \times \frac{5.365}{5} = 0.2567 \approx 0.26$$

Thus, the equation for the least-squares line is

$$y = 2.09x + 0.26$$

Substitution into Equation 8-15 yields the standard deviation about regression.

$$s_r = \sqrt{\frac{S_{yy} - m^2 S_{xx}}{N - 2}} = \sqrt{\frac{5.07748 - (2.0925)^2 \times 1.14537}{5 - 2}} = 0.1442 \approx 0.14$$

and substitution into Equation 8-16 gives the standard deviation of the slope,

$$s_m = \sqrt{\frac{s_r^2}{S_{xx}}} = \sqrt{\frac{(0.1442)^2}{1.14537}} = 0.13$$

Finally, we find the standard deviation of the intercept from Equation 8-17:

$$s_b = 0.1442\sqrt{\frac{1}{5 - (5.365)^2/6.9021}} = 0.16$$

EXAMPLE 8-5

The calibration curve found in Example 8-4 was used for the chromatographic determination of isooctane in a hydrocarbon mixture. A peak area of 2.65 was obtained. Calculate the mole percent of isooctane in the mixture and the standard deviation if the area was (a) the result of a single measurement and (b) the mean of four measurements.

In either case, the unknown concentration is found from rearranging the least-squares equation for the line, which gives

$$x = \frac{y - b}{m} = \frac{y - 0.2567}{2.0925} = \frac{2.65 - 0.2567}{2.0925} = 1.144 \text{ mol }\%$$

(a) Substituting into Equation 8-18, we obtain

$$s_c = \frac{0.1442}{2.0925}\sqrt{\frac{1}{1} + \frac{1}{5} + \frac{(2.65 - 12.51/5)^2}{(2.0925)^2 \times 1.145}} = 0.076 \text{ mol }\%$$

(b) For the mean of four measurements

$$s_c = \frac{0.1442}{2.0925}\sqrt{\frac{1}{4} + \frac{1}{5} + \frac{(2.65 - 12.51/5)^2}{(2.0925)^2 \times 1.145}} = 0.046 \text{ mol }\%$$

Interpretation of Least-Squares Results The closer the data points are to the line predicted by a least-squares analysis, the smaller are the residuals. The sum of the squares of the residuals, SS_{resid}, measures the variation in the observed values of the dependent variable (y values) that are not explained by the presumed linear relationship between x and y.

$$SS_{resid} = \sum_{i=1}^{N} [y_i - (b + mx_i)]^2 \tag{8-19}$$

We can also define a total sum of the squares, SS_{tot}, as

$$SS_{tot} = S_{yy} = \Sigma(y_i - \bar{y})^2 = \Sigma y_i^2 - \frac{(\Sigma y_i)^2}{N} \tag{8-20}$$

The total sum of the squares is a measure of the total variation in the observed values of y because the deviations are measured from the mean value of y.

An important quantity called the **coefficient of determination** (R^2) measures the fraction of the observed variation in y that is explained by the linear relationship and is given by

The **coefficient of determination** (R^2) is a measure of the fraction of the total variation in y that can be explained by the linear relationship between x and y.

$$R^2 = 1 - \frac{SS_{resid}}{SS_{tot}} \tag{8-21}$$

The closer R^2 is to unity, the better the linear model explains the y variations, as shown in Example 8-6. The difference between SS_{tot} and SS_{resid} is the sum of the squares due to regression, SS_{regr}. In contrast to SS_{resid}, SS_{regr} is a measure of the explained variation. We can write,

$$SS_{regr} = SS_{tot} - SS_{resid} \quad \text{and} \quad R^2 = \frac{SS_{regr}}{SS_{tot}}$$

By dividing the sum of squares by the appropriate number of degrees of freedom, we can obtain the mean square values for regression and for the residuals (error) and then the F value. The F value gives us an indication of the significance of the regression. The F value is used to test the null hypothesis that the total variance in y is equal to the variance due to error. A value of F smaller than the value from the tables at the chosen confidence level indicates that the null hypothesis should be accepted and that the regression is not significant. A large value of F indicates that the null hypothesis should be rejected and that the regression is significant.

► A significant regression is one in which the variation in the y values due to the presumed linear relationship is large compared with that due to error (residuals). When the regression is significant, a large value of F occurs.

EXAMPLE 8-6

Find the coefficient of determination for the chromatographic data of Example 8-4.

For each value of x_i, we can find a predicted value of y_i from the linear relationship. Let us call the predicted values of y_i, $\hat{y}_i$. We can write $\hat{y}_i = b + mx_i$ and make a table of the observed y_i values, the predicted values, $\hat{y}_i$, the residuals, $y_i - \hat{y}_i$, and the squares of the residuals, $(y_i - \hat{y}_i)^2$. By summing the latter values, we obtain SS_{resid}, as shown in Table 8-2.

From Example 8-4, the value of $S_{yy} = 5.07748$. Hence,

$$R^2 = 1 - \frac{SS_{resid}}{SS_{tot}} = 1 - \frac{0.0624}{5.07748} = 0.9877$$

This shows that more than 98% of the variation in peak area can be explained by the linear model.

We can also calculate SS_{regr} as

$$SS_{regr} = SS_{tot} - SS_{resid} = 5.07748 - 0.06240 = 5.01508$$

TABLE 8-2
Finding the Sum of the Squares of the Residuals

x_i	y_i	$\hat{y}_i$	$y_i - \hat{y}_i$	$(y_i - \hat{y}_i)^2$
0.352	1.09	0.99326	0.09674	0.00936
0.803	1.78	1.93698	−0.15698	0.02464
1.08	2.60	2.51660	0.08340	0.00696
1.38	3.03	3.14435	−0.11435	0.01308
1.75	4.01	3.91857	0.09143	0.00836
5.365	12.51			0.06240

Now we can calculate the F value. Five xy pairs were used for the analysis. The total sum of the squares has 4 degrees of freedom associated with it, since one is lost in calculating the mean of the y values. The sum of the squares due to the residuals has 3 degrees of freedom because two parameters m and b are estimated. Hence, SS_{regr} has only 1 degree of freedom because it is the difference between SS_{tot} and SS_{resid}. In our case, we can find F from

$$F = \frac{SS_{regr}/1}{SS_{resid}/3} = \frac{5.01508/1}{0.0624/3} = 241.11$$

This very large value of F has a very small chance of occurring by random chance, and hence we conclude that this is a significant regression.

Transformed Variables Sometimes an alternative to a simple linear model is suggested by a theoretical relationship or by examining residuals from a linear regression. In some cases, linear least-squares analysis can be used after the simple transformations shown in Table 8-3.

Although transforming variables is quite common, there are some warnings that should be heeded. The linear least-squares method will give best estimates of the transformed variables, but these may not be optimal when transformed back to obtain estimates of the original parameters. For the original parameters, nonlinear regression methods may give better estimates. The transformed variables method does not give good estimates if the errors are not normally distributed. The statistics produced by ANOVA after transformation always refer to the transformed variables.

TABLE 8-3
Transformations to Linearize Functions

Function	Transformation to Linearize	Resulting Equation
Exponential: $y = be^{mx}$	$y' = \ln(y)$	$y' = \ln(b) + mx$
Power: $y = bx^m$	$y' = \log(y)$, $x' = \log(x)$	$y' = \log(b) + mx'$
Reciprocal $y = b + m\left(\frac{1}{x}\right)$	$x' = \frac{1}{x}$	$y = b + mx'$

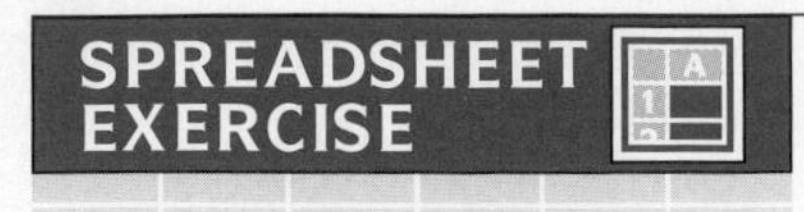

USING EXCEL TO DO LEAST SQUARES

Linear least-squares analysis is quite easy with Excel. This type of analysis can be accomplished in several ways: by using the equations presented in this chapter, by employing the basic built-in functions of Excel, or by using the regression data analysis tool. Because the built-in functions are the easiest of these options, we explore them in detail here and see how they may be used to evaluate analytical data.

The Slope and Intercept

As usual, we begin with a blank worksheet. Enter the data from Table 8-1 into the worksheet so that it appears as follows.

	A	B	C
1		x	y
2		0.352	1.09
3		0.803	1.78
4		1.08	2.6
5		1.38	3.03
6		1.75	4.01
7			
8	slope		
9	intercept		
10			
11			

Now, click on cell B8, and then click on the Insert (Paste) Function icon shown in the margin so that the Insert Function window appears, and then click on Statistical. The window appears as follows.

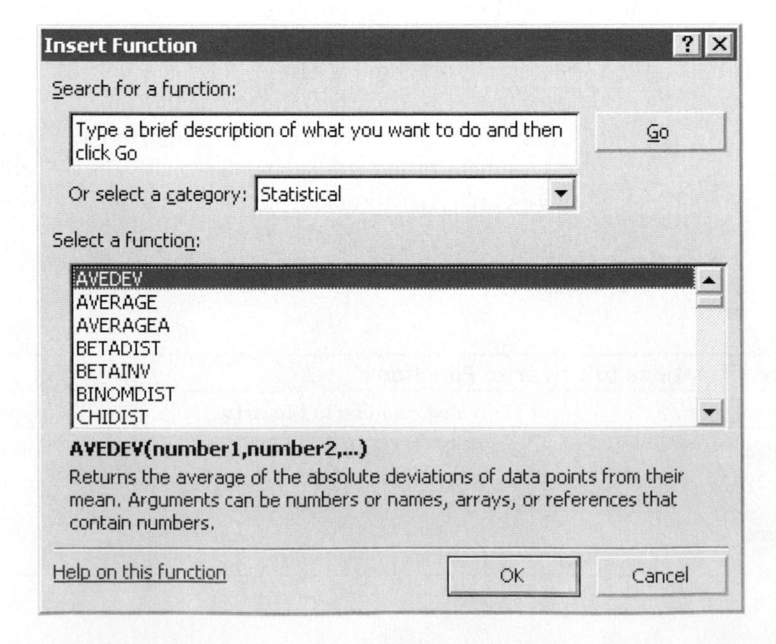

Note that numerous statistical functions appear in the window labeled Select a function:. Use the mouse to scroll down the list of functions until you come to the SLOPE function, and then click on it. The function appears in bold under the left-hand window, and a description of the function appears below it. Read the description of the slope function, and then click OK. The following window appears just below the formula bar.

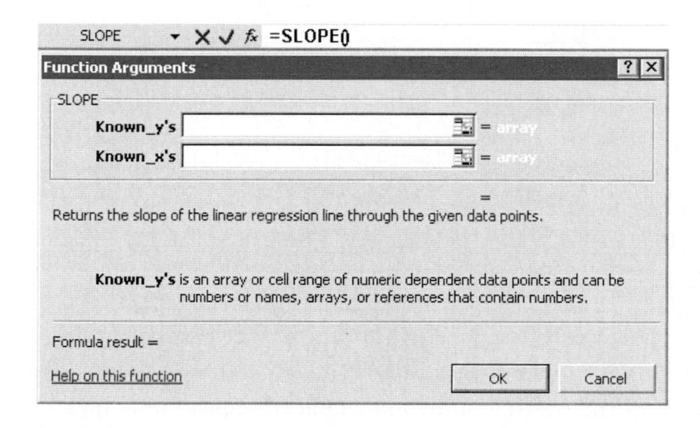

Look at the information that is provided in the window and in the formula bar. The SLOPE() function appears in the formula bar with no arguments, so we must select the data that Excel will use to determine the slope of the line. Now click on the selection button at the right end of the Known_y's box, use the mouse to select cells C2:C6, and type [↵]. Similarly, click on the selection button for the Known_x's box and select cells B2:B8 followed by [↵], which produces the following in the window.

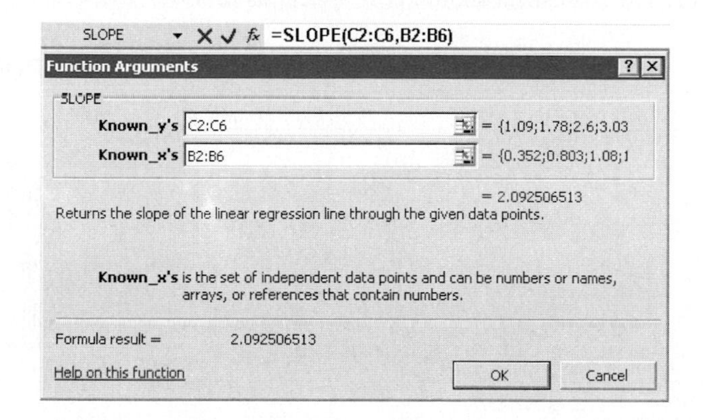

The window shows not only the cell references for the x and y data but also the first few of the data points to the right and displays the result of the slope calculation. Now click on OK, and the slope of the line appears in cell B8.

Click on cell B9 followed by the Insert Function icon, and repeat the process that we just carried out, except that now you should select the INTERCEPT function. When the intercept function window appears, select the Known_y's and the Known_x's as before, and click OK. When you have finished, the worksheet will have the following appearance.

	A	B	C
1		x	y
2		0.352	1.09
3		0.803	1.78
4		1.08	2.6
5		1.38	3.03
6		1.75	4.01
7			
8	slope	2.092507	
9	intercept	0.256741	
10			

At this point, you may wish to compare these results with those obtained for the slope and intercept in Example 8-4. We should note at this point that Excel provides many digits that are not significant. We shall see how many figures are significant after we find the standard deviations of the slope and intercept.

Using LINEST

Now we see how the LINEST function can accomplish many important functions in a single procedure. Begin by using the mouse to select an array of cells two cells wide and five cells high, such as E2:F6. Then click on the Insert Function icon, select STATISTICAL and LINEST in the left and right windows, respectively, and click on OK. Select the Known_y's and Known_x's as before, then click on the box labeled Const and type **`true`**. Also type **`true`** in the box labeled Stats. When you click on each of the latter two boxes, notice that a description of the meaning of these logical variables appears below the box. To activate the LINEST function, you must now type the rather unusual keystroke combination **`Ctrl+Shift+[↵]`**. This keystroke combination must be used whenever you perform a function on an array of cells. The worksheet should now appear as follows.

E2 ▾ *fx* {=LINEST(C2:C6,B2:B6,TRUE,TRUE)}

	A	B	C	D	E	F
1		x	y			
2		0.352	1.09		2.092507	0.256741
3		0.803	1.78		0.134749	0.158318
4		1.08	2.6		0.987712	0.144211
5		1.38	3.03		241.1465	3
6		1.75	4.01		5.015089	0.062391
7						
8	slope	2.092507				
9	intercept	0.256741				

As you can see, cells E2 and F2 contain the slope and intercept of the least-squares line. Cells E3 and F3 are the respective standard deviations of the slope and intercept. Cell E4 contains the coefficient of determination (R^2). The standard deviation about regression (s_r, standard error of the estimate) is located in cell F4. The smaller the s_r value, the better the fit. The square of the standard error of the estimate is the mean square for the residuals (error). The value in cell E5 is the *F* statistic. Cell F5 contains the number of degrees of freedom associated with the error.

Finally, cells E6 and F6 contain the sum of the squares of the regression and the sum of the squares of the residuals, respectively. Note that the F value can be calculated from these latter quantities, as described on page 201.

It is worth noting that the number of significant figures that are kept in a least-squares analysis depend on the use for which the data are intended. If the results are to be used to carry out further spreadsheet computations, wait until final results are computed before rounding to an appropriate number of significant figures. Excel provides 15 digits of numerical precision, and so, in general, spreadsheet computations will not contribute to the uncertainty in the results. Final answers based on the least-squares equation should be rounded to be consistent with the uncertainties reflected in the standard deviations of the slope and intercept and the standard error of the estimate. The standard deviations of the slope and intercept in our example suggest that at most, we should express both the slope and the intercept to only two decimal places. Thus, the least-squares results for the slope and intercept may be expressed as 2.09 ± 0.13 and 0.26 ± 0.16, respectively, or as 2.1 ± 0.1 and 0.3 ± 0.2.

Plotting a Graph of the Data and the Least-Squares Fit

It is customary and useful to plot a graph of the data and the least-squares fitted line similar to Figure 8-9. The built-in Chart Wizard of Excel makes creating such plots relatively easy. There are several ways to display the data points and the predicted line simultaneously. One way is to plot the predicted values $\hat{y}$ and the experimental y values simultaneously. The predicted values are given in Table 8-2. The easiest way is to have Excel add the line, called a TrendLine, itself.

To plot the points, select the xy data (cells B2:D6) from the original worksheet. Click on the Chart Wizard icon shown in the margin. Select XY(Scatter) from the standard types list and click on Next>. When the Step 2 of 4 window appears, click on Next> again. Click on the Gridlines tab, and check Major gridlines under Value (X) axis. Then click on the Titles tab, and enter **x** in the Value (X) axis blank and **y** in the Value (Y) axis blank. Finally, click on Finish to produce the following graph of the data.

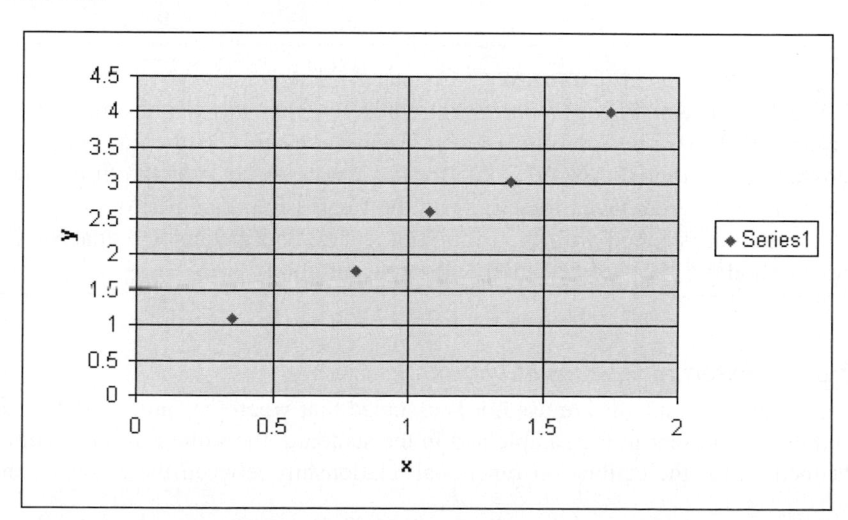

Now right-click on any data point and then click on Add Trendline.... Under Type, select Linear. Under options, check Display equation on chart and Display R-squared value on chart. Then click on OK. The weight of the line can be adjusted by right-clicking on the line and selecting Format Trendline. . . . Under Patterns, select a line of the desired weight. You can also move the equation and R^2 text to a more convenient place, as indicated in the chart that follows.

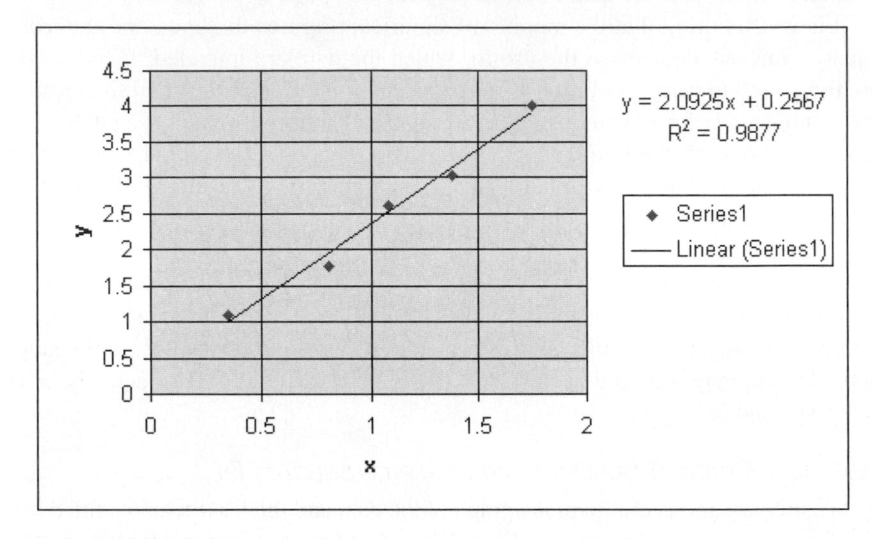

As an extension of this exercise, modify your spreadsheet to include a column of residuals, as shown in Table 8-2. Create a plot of the residuals as a function of x. Residual plots can assist you in detecting any systematic deviations of the experimental points from the least-squares line. Be sure to save your spreadsheet in a file for reference and for use in analyzing laboratory data.

Although we have focused on linear calibration relationships, there are cases in analytical chemistry in which nonlinear calibration is used. Sometimes the relationship between the analytical response and concentration is inherently nonlinear. In other cases, nonlinearities arise because solutions do not behave ideally. In either case, **nonlinear regression** can be used to develop the calibration model.[18]

Spreadsheet Summary Chapter 4 of *Applications of Microsoft® Excel in Analytical Chemistry* introduces another way to perform a least-squares analysis. The Analysis ToolPak Regression tool has the advantage of producing a complete ANOVA table for the results. A chart of the fit and the residuals can be produced directly from the Regression window. An unknown concentration is found with the calibration curve, and a statistical analysis is used to find the standard deviation of the concentration.

Errors in External Standard Calibration

When external standards are used, it is assumed that when the same analyte concentration is present in the sample and in the standard, the same response will be obtained. Thus, the calibration functional relationship between the response and

[18]See D. M. Bates and D. G. Watts, *Nonlinear Regression Analysis and Its Applications.* New York: Wiley, 1988.

the analyte concentration must apply to the sample as well. Usually in a determination, the raw response from the instrument is not used. Instead, the raw analytical response is corrected by measuring a **blank.** The **ideal blank** is identical to the sample, but without the analyte. In practice, with complex samples, it is too time consuming or impossible to prepare an ideal blank, and a compromise must be made. Most often, a real blank is either a **solvent blank,** containing the same solvent in which the sample is dissolved, or a **reagent blank,** containing the solvent plus all the reagents used in sample preparation.

Even with blank corrections, several factors can cause the basic assumption of the external standard method to break down. Matrix effects due to extraneous species in the sample that are not present in the standards or blank can cause the same analyte concentrations in the sample and standards to give different responses. Differences in experimental variables at the times at which blank, sample, and standard are measured can also invalidate the established calibration function. Even when the basic assumption is valid, errors can still occur owing to contamination during the sampling or sample preparation steps.

Systematic errors can also occur during the calibration process. For example, if the standards are prepared incorrectly, an error will result. The accuracy with which the standards are prepared depends on the accuracy of the gravimetric and volumetric techniques and equipment used. The chemical form of the standards must be identical to that of the analyte in the sample; the state of oxidation, isomerization, or complexation of the analyte can alter the response. Once prepared, the concentration of the standards can change owing to decomposition, volatilization, or adsorption onto container walls. Contamination of the standards can also result in higher analyte concentrations than expected. A systematic error can occur if some bias exists in the calibration model. For example, errors can arise if the calibration function is obtained without using enough standards to obtain good statistical estimates of the parameters.

◀ To avoid systematic errors in calibration, the standards must be accurately prepared and their chemical state must be identical to that of the analyte in the sample. The standards should be stable in concentration, at least during the calibration process.

Random errors can also influence the accuracy of results obtained from calibration curves. From Equation 8-18, it can be seen that the standard deviation in the concentration of analyte s_c obtained from a calibration curve is lowest when the response $\hat{y}_c$ is close to the mean value $\hat{y}$. The point $\bar{x}, \bar{y}$ represents the centroid of the regression line. Points close to this value are determined with more certainty than those far away from the centroid. Figure 8-11 shows a calibration curve with confidence limits. Note that measurements made near the center of the curve will give less uncertainty in analyte concentration than those made at the extremes.

8C-3 Minimizing Errors in Analytical Procedures

There are several steps that can be taken to ensure accuracy in analytical procedures.[19] Most of these depend on minimizing or correcting errors that might occur in the measurement step. We should note, however, that the overall accuracy and precision of an analysis might be limited not by the measurement step but by factors such as sampling, sample preparation, and calibration, as discussed earlier in this chapter.

[19]For a more extensive discussion of error minimization, see J. D. Ingle, Jr., and S. R. Crouch, *Spectrochemical Analysis,* pp. 176–183. Upper Saddle River, NJ: Prentice-Hall, 1988.

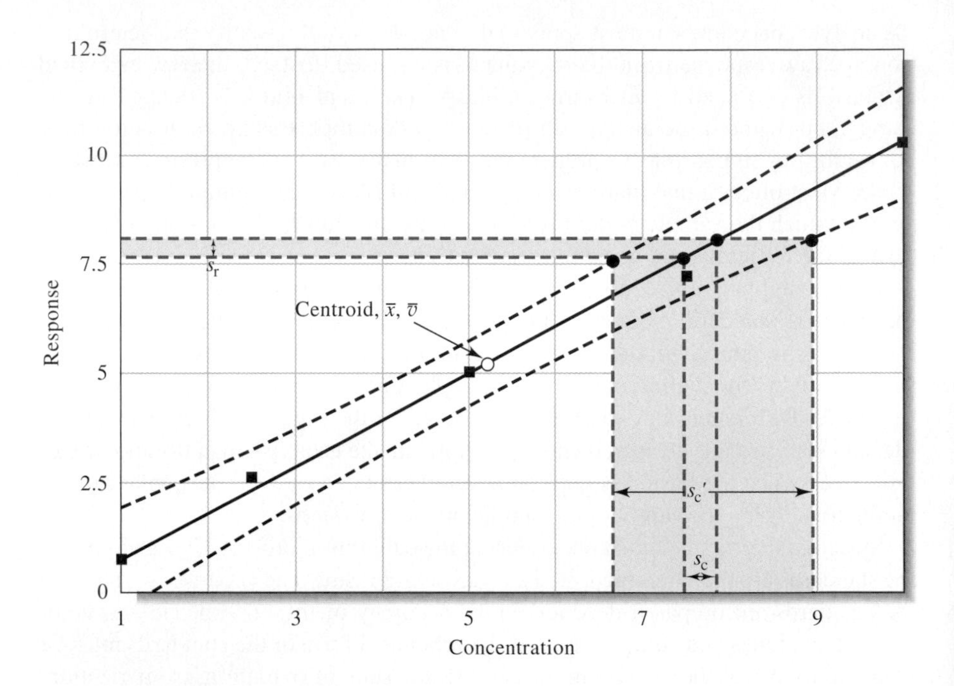

Figure 8-11 Effect of calibration curve uncertainty. The dashed lines show confidence limits for concentrations determined by the regression line. Note that uncertainties increase at the extremities of the plot. Usually, we estimate the uncertainty in analyte concentration only from the standard deviation of the response. Calibration curve uncertainty can significantly increase the uncertainty in the analyte concentration from s_c to s_c'as shown.

FEATURE 8-3

Multivariate Calibration

The least-squares procedure just described is an example of a univariate calibration procedure because only one response is used per sample. The process of relating multiple instrument responses to an analyte or a mixture of analytes is known as **multivariate calibration.** Multivariate calibration methods[20] have become quite popular in recent years as new instruments become available that produce multidimensional responses (absorbance of several samples at multiple wavelengths, mass spectrum of chromatographically separated components, and so forth). Multivariate calibration methods are very powerful. They can be used to determine multiple components in mixtures simultaneously and can provide redundancy in measurements to improve precision. Recall that repeating a measurement N times provides a $\sqrt{N}$ improvement in the precision of the mean value. These methods can also be used to detect the presence of interferences that would not be identified in a univariate calibration.

Multivariate techniques are **inverse calibration methods.** In normal least-squares methods, often called **classical least-squares methods,** the system response is modeled as a function of analyte concentration. In inverse methods, the concentrations are treated as functions of the responses. The latter has some advantages in that concentrations can be accurately predicted even in the presence of chemical and physical sources of interference. In classical methods, all components in the system need to be considered in the mathematical model produced (regression equation).

[20]For a more extensive discussion, see K. R. Beebe, R. J. Pell, and M. B. Seasholtz, *Chemometrics: A Practical Guide,* Chapter 5. New York: Wiley, 1998; H. Martens and T. Naes, *Multivariate Calibration.* New York: Wiley, 1989.

The common multivariate calibration methods are **multiple linear regression, partial least-squares regression,** and **principal components regression.** These differ in the exact way in which variations in the data (responses) are used to predict the concentration. Software for accomplishing multivariate calibration is available from several companies. The use of multivariate statistical methods for quantitative analysis is part of the subdiscipline of chemistry called **chemometrics.**

The multicomponent determination of Ni(II) and Ga(III) in mixtures is an example of the use of multivariate calibration.[21] Both metals react with 4-(2-pyridylazo)-resorcinol (PAR) to form colored products. The absorption spectra of the products are slightly different, and they form at slightly different rates. Advantage can be taken of these small differences to perform simultaneous determinations of the metals in mixtures. In one analysis, 16 standard mixtures containing the two metals were used to obtain the calibration model. A multiwavelength diode array spectrometer (see Chapter 25) collected data for 26 time intervals and 26 wavelengths. Concentrations of the metals in the μM range were determined in unknown mixtures at pH 8.5 with relative errors of less than 10% by partial least-squares regression and principal components regression.

Chemical structure of 4-(2-pyridylazo)-resorcinol.

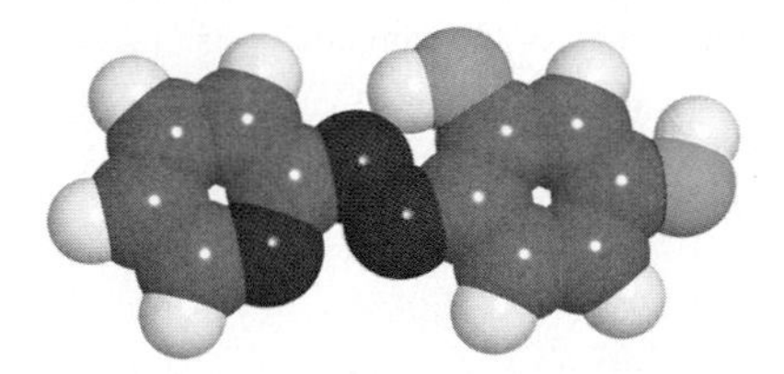

Molecular model of PAR.

Separations

Sample cleanup by separation methods is an important way to minimize errors due to possible interferences in the sample matrix. Techniques such as filtration, precipitation, dialysis, solvent extraction, volatilization, ion exchange, and chromatography are all useful in ridding the sample of potential interfering constituents. Most separation methods are, however, time consuming and may increase the chances that some of the analyte will be lost or that the sample may be contaminated. In many cases, though, separations are the only way to eliminate an interfering species. Some modern instruments include an automated front-end sample delivery system that includes a separation step (flow injection or chromatography).

Saturation, Matrix Modification, and Masking

The **saturation method** involves adding the interfering species to all the samples, standards, and blanks so that the interference effect becomes independent of the original concentration of the interfering species in the sample. This can, however, degrade the sensitivity and detectability of the analyte.

[21] T. F. Cullen and S. R. Crouch, *Anal. Chim. Acta,* **2000,** *407,* 135.

A **matrix modifier** is a species, not itself an interfering species, added to samples, standards, and blanks in sufficient amounts to make the analytical response independent of the concentration of the interfering species. For example, a buffer might be added to keep the pH within limits regardless of the sample pH. Sometimes a **masking agent** is added, which reacts selectively with the interfering species to form a complex that does not interfere.

In both these methods, care must be taken that the added reagents do not contain significant quantities of the analyte or other interfering species.

Dilution and Matrix Matching

The **dilution method** can sometimes be used if the interfering species produces no significant effect below a certain concentration level. With this method, the interference effect is minimized simply by diluting the sample. Dilution may influence our ability to detect the analyte or to measure its response with accuracy and precision, so care is necessary in using this method.

► Errors in procedures can be minimized by saturating with interfering species, by adding matrix modifiers or masking agents, by diluting the sample, or by matching the matrix of the sample.

The **matrix matching method** attempts to duplicate the sample matrix by adding the major matrix constituents to the standard and blank solutions. For example, in the analysis of sea water samples for a trace metal, the standards can be prepared in a synthetic sea water containing Na^+, K^+, Cl^-, Ca^{2+}, Mg^{2+} and other components. The concentrations of these species are well known in sea water and fairly constant. In some cases, the analyte can be removed from the original sample matrix and the remaining components used to prepare standards and blanks. Again, we must be careful that added reagents do not contain the analyte or cause extra interference effects.

Internal Standard Methods

An **internal standard** is a reference species, chemically and physically similar to the analyte, that is added to samples, standards, and blanks. The ratio of the response of the analyte to that of the internal standard is plotted versus the concentration of analyte.

In the **internal standard method,** a known amount of a reference species is added to all the samples, standards, and blanks. The response signal is then not the analyte signal itself but the *ratio* of the analyte signal to the reference species signal. A calibration curve is prepared in which the y-axis is the ratio of responses and the x-axis is the analyte concentration in the standards, as usual. Figure 8-12 illustrates the use of the internal standard method for peak-shaped responses.

The internal standard method can compensate for certain types of errors if these influence both the analyte and the reference species to the same proportional extent. For example, if temperature influences both the analyte and the reference species to the same extent, taking the ratio can compensate for variations in temperature. For compensation to occur, a reference species is chosen that has chemical and physical properties similar to those of the analyte. The use of an internal standard in flame spectrometry is illustrated in Example 8-7.

Standard Addition Methods

In the **method of standard additions,** a known amount of a standard solution of analyte is added to one portion of the sample. The responses before and after the addition are measured and used to obtain the analyte concentration. Alternatively, multiple additions are made to several portions of the sample. The standard additions method assumes a linear response. This should always be confirmed or the **multiple additions method** used to check linearity.

We use the **method of standard additions** when it is difficult or impossible to duplicate the sample matrix. In general, the sample is "spiked" with a known amount or amounts of a standard solution of the analyte. In the single-point standard addition method, two portions of the sample are taken. One portion is measured as usual, but a known amount of standard analyte solution is added to the second portion. The responses for the two portions are then used to calculate the unknown concentration, assuming a linear relationship between response and analyte concentration (see Example 8-8). In the **multiple additions method,** additions of known amounts of standard analyte solution are made to several portions of the sample, and a multiple additions calibration curve is obtained. The multiple additions method gives some

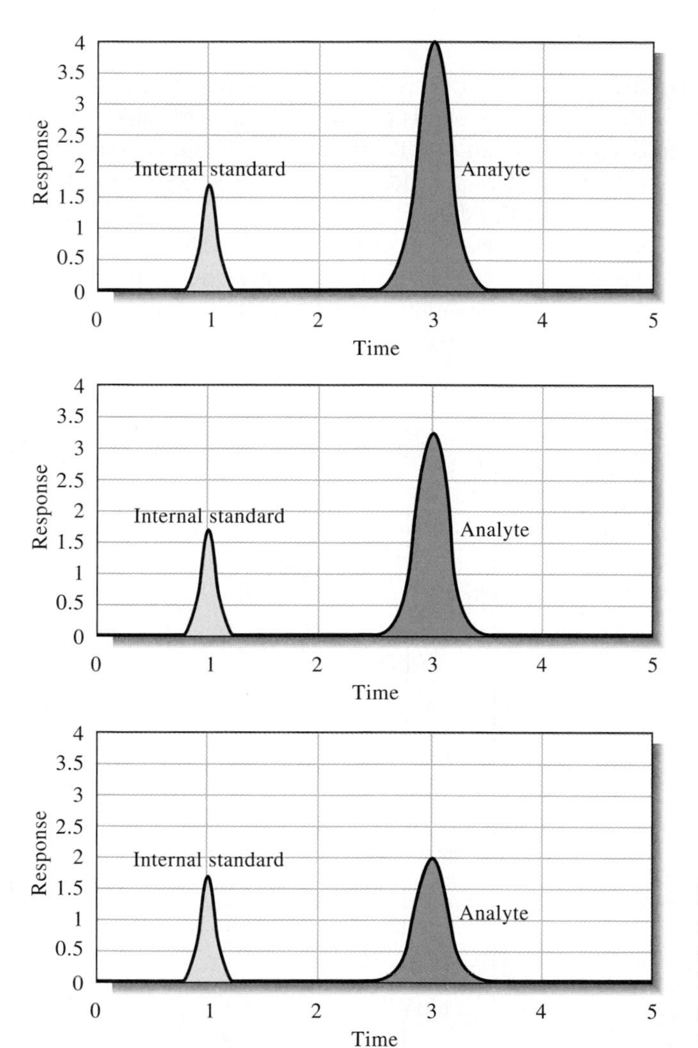

Figure 8-12 Illustration of the internal standard method. A fixed amount of the internal standard species is added to all samples, standards, and blanks. The calibration curve plots the ratio of the analyte signal to the internal standard signal against the concentration of the analyte.

EXAMPLE 8-7

The intensities of flame emission lines can be influenced by a variety of instrumental factors, including flame temperature, flow rate of solution, and nebulizer efficiency. We can compensate for variations in these factors by using the internal standard method. Here, we add the same amount of internal standard to mixtures containing known amounts of the analyte and to the samples of unknown analyte concentration. We then take the ratio of the intensity of the analyte line to that of the internal standard. The internal standard should be absent in the sample to be analyzed.

In the flame emission determination of sodium, lithium is often added as an internal standard. The following data were obtained for solutions containing Na and 1000 ppm Li.

ppm Na	Na emission intensity	Li emission intensity
0.10	0.11	86
0.50	0.52	80
1.00	1.8	128
5.00	5.9	91
10.00	9.5	73
Unknown	4.4	95

(continued)

Construct a spreadsheet to determine the intensity ratio of sodium to lithium and plot this versus the ppm sodium. Also plot the sodium intensity versus ppm sodium. Determine the concentration of the unknown and its standard deviation.

The spreadsheet is shown in Figure 8-13. The data are entered into columns A through C. In cells D4 through D9, the intensity ratio is calculated by the formula shown in documentation cell A22. A plot of the normal calibration curve is shown as the upper plot in the figure. The lower plot is the internal standard calibration curve. Note the improvement in the calibration curve when using the internal standard. The linear regression statistics are calculated in cells B11 through B20 using the same approach as described in Section 8C-2. The statistics are calculated by the formulas in documentation cells A23 through A31. The sodium concentration in the unknown is found to be 3.55 ± 0.05 ppm.

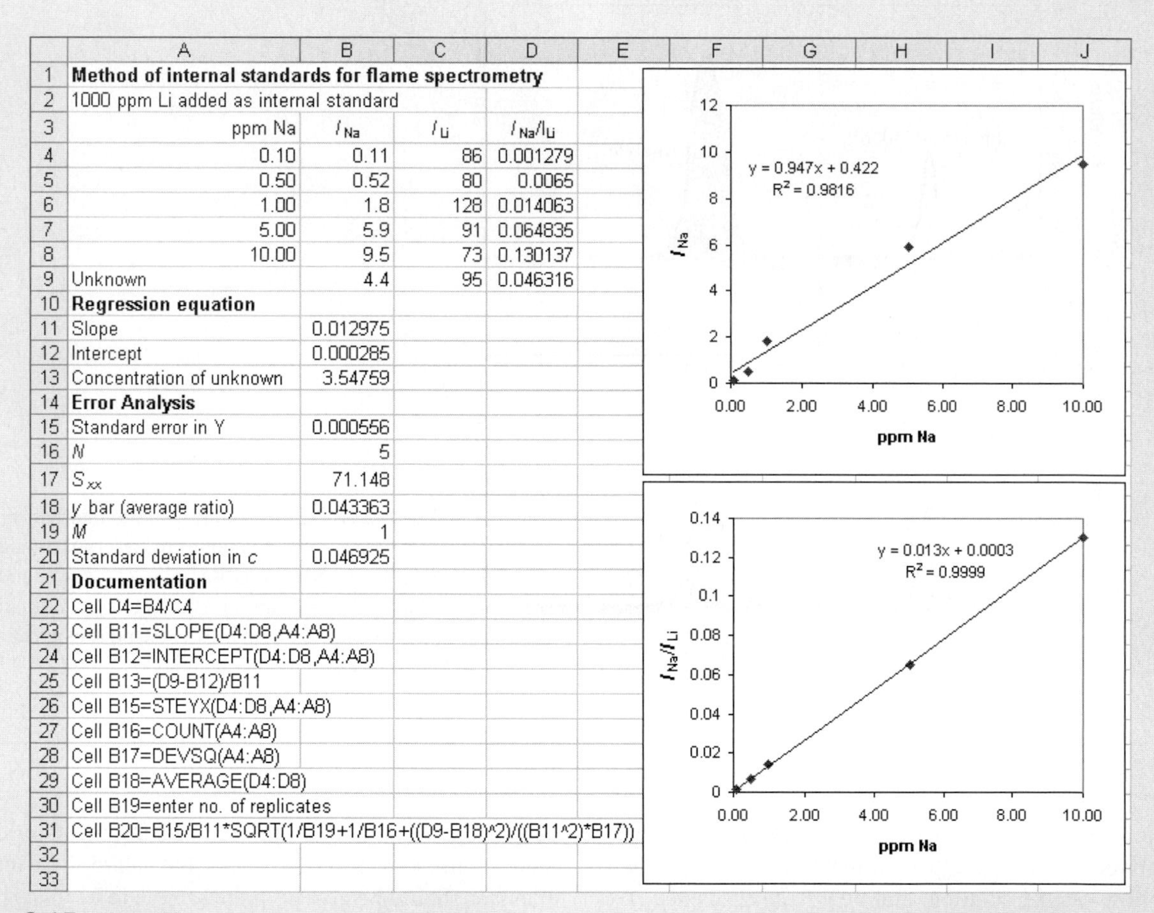

	A	B	C	D
1	**Method of internal standards for flame spectrometry**			
2	1000 ppm Li added as internal standard			
3	ppm Na	I_{Na}	I_{Li}	I_{Na}/I_{Li}
4	0.10	0.11	86	0.001279
5	0.50	0.52	80	0.0065
6	1.00	1.8	128	0.014063
7	5.00	5.9	91	0.064835
8	10.00	9.5	73	0.130137
9	Unknown	4.4	95	0.046316
10	**Regression equation**			
11	Slope	0.012975		
12	Intercept	0.000285		
13	Concentration of unknown	3.54759		
14	**Error Analysis**			
15	Standard error in Y	0.000556		
16	*N*	5		
17	S_{xx}	71.148		
18	*y* bar (average ratio)	0.043363		
19	*M*	1		
20	Standard deviation in *c*	0.046925		
21	**Documentation**			
22	Cell D4=B4/C4			
23	Cell B11=SLOPE(D4:D8,A4:A8)			
24	Cell B12=INTERCEPT(D4:D8,A4:A8)			
25	Cell B13=(D9-B12)/B11			
26	Cell B15=STEYX(D4:D8,A4:A8)			
27	Cell B16=COUNT(A4:A8)			
28	Cell B17=DEVSQ(A4:A8)			
29	Cell B18=AVERAGE(D4:D8)			
30	Cell B19=enter no. of replicates			
31	Cell B20=B15/B11*SQRT(1/B19+1/B16+((D9-B18)^2)/((B11^2)*B17))			
32				
33				

Figure 8-13 Spreadsheet to illustrate the internal standard method for the flame emission determination of sodium.

verification that the linear relationship between response and analyte concentration holds. We will further discuss the multiple additions method in Chapter 26, where it is used in conjunction with molecular absorption spectroscopy (Fig. 26-8).

The method of standard additions is quite powerful when used properly. First, there must be a good blank measurement so that extraneous species do not contribute to the analytical response. Second, the calibration curve for the analyte must be linear in the sample matrix. The multiple additions method provides a check on this assumption. A significant disadvantage of the multiple additions method is the extra time required for making the additions and measurements. The major benefit

EXAMPLE 8-8

The single-point standard addition method was used in the determination of phosphate by the molybdenum blue method. A 2.00-mL urine sample was treated with molybdenum blue reagents to produce a species absorbing at 820 nm, after which the sample was diluted to 100.00 mL. A 25.00-mL aliquot gave an instrument reading (absorbance) of 0.428 (solution 1). Addition of 1.00 mL of a solution containing 0.0500 mg of phosphate to a second 25.0-mL aliquot gave an absorbance of 0.517 (solution 2). Use these data to calculate the concentration of phosphate in milligrams per milliliter of the sample, assuming a linear relationship between absorbance and concentration and a blank measurement has been made.

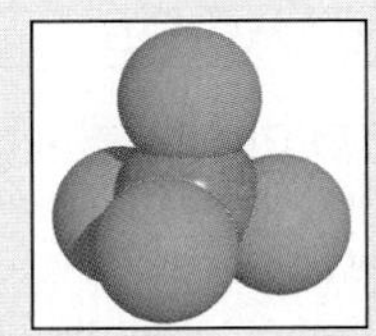

Molecular model of phosphate ion (PO_4^{3-}).

The absorbance of the first solution is given by

$$A_1 = kc_u$$

where c_u is the unknown concentration of phosphate in the first solution and k is a proportionality constant. The absorbance of the second solution is given by

$$A_2 = \frac{kV_u c_u}{V_t} + \frac{kV_s c_s}{V_t}$$

where V_u is the volume of the solution of unknown phosphate concentration (25.00 mL), V_s is the volume of the standard solution of phosphate added (1.00 mL), V_t is the total volume after the addition (26.00 mL), and c_s is the concentration of the standard solution (0.500 mg mL^{-1}). If we solve the first equation for k, substitute the result into the second equation, and solve for c_u, we obtain

$$c_u = \frac{A_1 c_s V_s}{A_2 V_t - A_1 V_u}$$

$$= \frac{0.428 \times 0.0500 \text{ mg mL}^{-1} \times 1.00 \text{ mL}}{0.517 \times 26.00 \text{ mL} - 0.428 \times 25.00 \text{ mL}} = 0.0780 \text{ mg mL}^{-1}$$

This is the concentration of the diluted sample. To obtain the concentration of the original urine sample, we need to multiply by 100.00/2.00. Thus,

$$\text{concentration of phosphate} = 0.00780 \text{ mg mL}^{-1} \times 100.00 \text{ mL}/2.00 \text{ mL}$$
$$= 0.390 \text{ mg mL}^{-1}$$

is the potential compensation for complex interference effects that may be unknown to the user.

Spreadsheet Summary In Chapter 4 of *Applications of Microsoft® Excel in Analytical Chemistry,* a multiple standard additions procedure is illustrated. The determination of strontium in sea water with inductively coupled plasma atomic emission spectrometry is used as an example. The worksheet is prepared, and the standard additions plot is made. The unknown Sr concentration and its standard deviation are obtained.

8D FIGURES OF MERIT FOR ANALYTICAL METHODS

Analytical procedures are characterized by a number of figures of merit such as accuracy, precision, sensitivity, detection limit, and dynamic range. We discussed in Chapter 5 the general concepts of accuracy and precision. Here, we describe those additional figures of merit that are commonly used and discuss the validation and reporting of analytical results.

8D-1 Sensitivity and Detection Limit

The word **sensitivity** is often used in describing an analytical method. Unfortunately, it is occasionally used indiscriminately and incorrectly. The most often-used definition of sensitivity is the **calibration sensitivity,** or the change in the response signal per unit change in analyte concentration. The calibration sensitivity is thus the slope of the calibration curve, as shown in Figure 8-14. If the calibration curve is linear, the sensitivity is constant and independent of concentration. If nonlinear, sensitivity changes with concentration and is not a single value.

The calibration sensitivity does not indicate what concentration differences can be detected. Noise in the response signals must be taken into account to be quantitative about what differences can be detected. For this reason, the term **analytical sensitivity** is sometimes used. The analytical sensitivity is the ratio of the calibration curve slope to the standard deviation of the analytical signal at a given analyte concentration. The analytical sensitivity is usually a strong function of concentration.

The **detection limit** (DL) is the smallest concentration that can be reported with a certain level of confidence. Every analytical technique has a detection limit. For methods that employ a calibration curve, the detection limit is defined as the analyte concentration yielding a response of a confidence factor k higher than the standard deviation of the blank, s_b, as given in Equation 8-22.

$$\mathrm{DL} = \frac{k s_b}{m} \tag{8-22}$$

where m is the calibration sensitivity. Usually, the factor k is chosen to be 2 or 3. A k value of 2 corresponds to a confidence level of 92.1%, while a k value of 3 corresponds to a 98.3% confidence level.[22]

[22]See J. D. Ingle, Jr., and S. R. Crouch, *Spectrochemical Analysis,* p 174. Upper Saddle River, NJ: Prentice-Hall, 1988.

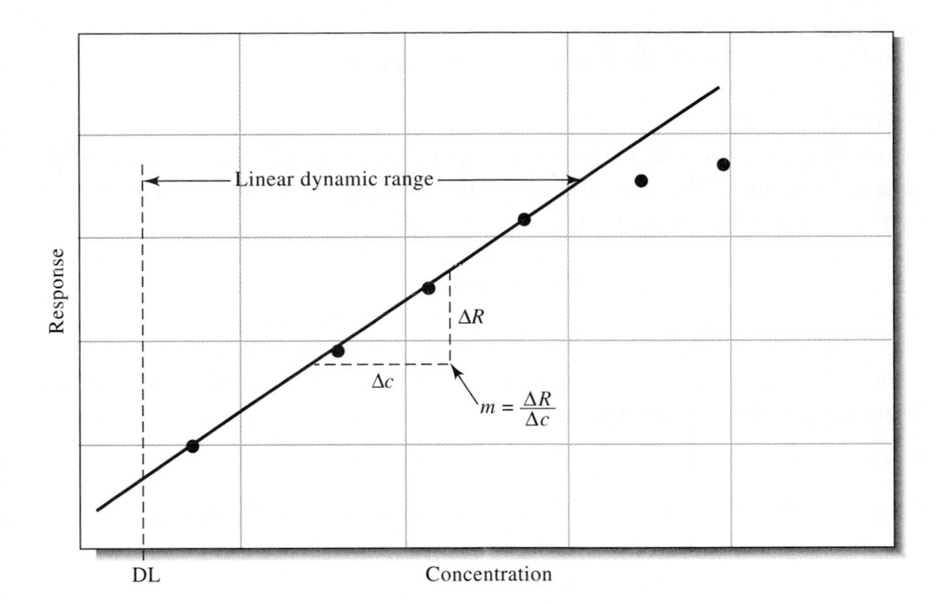

Figure 8-14 Calibration curve of response, R, vs. concentration, c. The slope of the calibration curve is called the calibration sensitivity, m. The detection limit, DL, designates the lowest concentration that can be measured at a specified confidence level.

Detection limits reported by researchers or instrument companies may not apply to real samples. The values reported are usually obtained on ideal standards with optimized instruments. These limits are useful, however, in comparing methods or instruments.

8D-2 Linear Dynamic Range

The **linear dynamic range** of an analytical method most often refers to the concentration range that can be determined with a linear calibration curve. The lower limit of the dynamic range is generally considered the detection limit. The upper end of the range is usually taken as the concentration at which the analytical signal or the slope of the calibration curve deviates by a specified amount from the linear relationship. Usually, a deviation of 5% from linearity is considered the upper limit. Deviations from linearity are common at high concentrations because of nonideal detector responses or chemical effects. Some analytical techniques, such as absorption spectrophotometry, are linear over only 1 to 2 orders of magnitude. Other methods, such as mass spectrometry, may exhibit linearity over 4 to 5 orders of magnitude.

A linear calibration curve is preferred because of its mathematical simplicity and because it makes detecting an abnormal response easy. With linear calibration curves, fewer standards and a linear regression procedure can be used. Nonlinear calibration curves can be employed, but more standards are required to establish the calibration function. A large linear dynamic range is desirable because a wide range of concentrations can be determined without dilution. In some determinations, such as the determination of sodium in blood serum, only a small dynamic range is required because variations of the sodium level in humans are quite small.

8D-3 Quality Assurance of Analytical Results

When analytical methods are applied to real-world problems, the quality of results as well as the performance quality of the tools and instruments used must be evaluated constantly. The major activities involved are quality control, validation of results, and reporting.[23] We briefly describe each of these.

Control Charts

A **control chart** is a sequential plot of some characteristic that is a criterion of quality.

A **control chart** is a sequential plot of some characteristic that is important in quality assurance. The chart also shows the statistical limits of variation that are permissible for the characteristic being measured.

As an example, consider monitoring the performance of a modern analytical balance. Both the accuracy and the precision of the balance can be monitored by periodically determining the mass of a standard. We can then determine whether the measurements on consecutive days are within certain limits of the standard mass. These limits are called the **upper control limit** (UCL) and the **lower control limit** (LCL). They are defined as

$$\text{UCL} = \mu + \frac{3\sigma}{\sqrt{N}}$$

$$\text{LCL} = \mu - \frac{3\sigma}{\sqrt{N}}$$

where μ is the population mean for the mass measurement, σ is the population standard deviation for the measurement, and N is the number of replicates that are obtained for each sample. The population mean and standard deviation for the standard mass must have been estimated from preliminary studies. Note that the UCL and the LCL are three standard deviations on either side of the population mean and form a range within which a measured mass is expected to lie 99.7% of the time.

Figure 8-15 is a typical instrument control chart for an analytical balance. Mass data were collected on 20 consecutive days for a 20.000-g standard mass certified by the National Institute of Standards and Technology. On each day, five replicate determinations were made. From independent experiments, estimates of the population mean and standard deviation were found to be μ = 20.000 g and

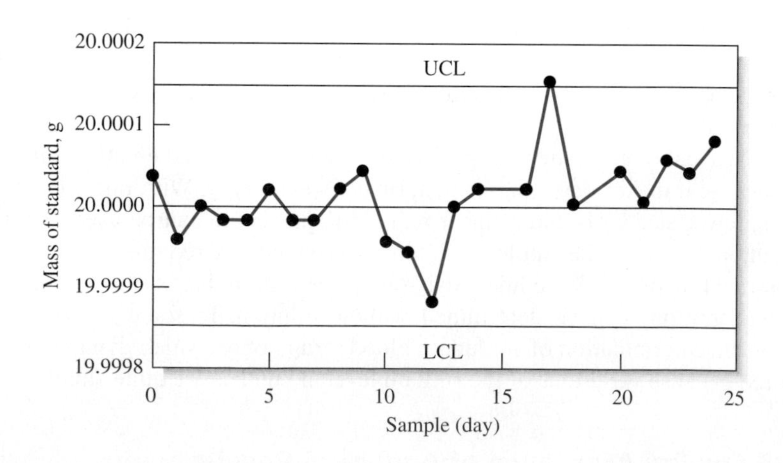

Figure 8-15 A control chart for a modern analytical balance. The results appear to fluctuate normally about the mean except for those obtained on day 17. Investigation led to the conclusion that the questionable value resulted from a dirty balance pan. UCL = upper control limit; LCL = lower control limit.

[23] For more information, see J. K. Taylor, *Quality Assurance of Chemical Measurements.* Chelsea, MI: Lewis Publishers, 1987.

$\sigma = 0.00012$ g, respectively. For the mean of five measurements, $3 \times (0.00012/\sqrt{5}) = 0.00016$. Hence, the UCL value equals 20.00016 g, and the LCL value equals 19.99984 g. With these values and the mean masses for each day, the control chart shown in Figure 8-15 can be constructed. As long as the sample mean mass remains between the LCL and the UCL, the balance is said to be in **statistical control.** On day 17, the balance went out of control, and an investigation was made to find the cause. In this example, the balance was not properly cleaned on day 17, so that there was dust on the balance pan. Systematic deviations from the mean are relatively easy to spot on a control chart.

Structure of benzoyl peroxide.

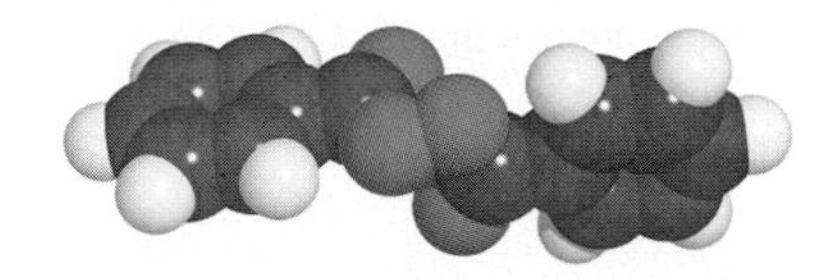

Molecular model of benzoyl peroxide.

In another example, a control chart was used to monitor the production of medications containing benzoyl peroxide, which are used for treating acne. Benzoyl peroxide is a bactericide that is effective when applied to the skin as a cream or gel containing 10% of the active ingredient. These substances are regulated by the Food and Drug Administration (FDA). Concentrations of benzoyl peroxide must, therefore, be monitored and maintained in statistical control. Benzoyl peroxide is an oxidizing agent that can be combined with an excess of iodide to produce iodine, which is titrated with standard sodium thiosulfate to provide a measure of the benzoyl peroxide in the sample.

The control chart of Figure 8-16 shows the results of 89 production runs of a cream containing a nominal 10% benzoyl peroxide measured on consecutive days. Each sample is represented by the mean percent benzoyl peroxide determined from the results of five titrations of different analytical samples of the cream.

The chart shows that until day 83, the manufacturing process was in statistical control, with normal random fluctuations in the amount of benzoyl peroxide. On day 83, the system went out of control, with a dramatic systematic increase above the UCL. This increase caused considerable concern at the manufacturing facility until its source was discovered and corrected. These examples show how control charts are effective for presenting quality control data in a variety of situations.

Validation

Validation determines the suitability of an analysis for providing the desired information. Validation can apply to samples, to methodologies, and to data. Validation is often done by the analyst but can also be done by supervisory personnel.

Validation of samples is often used to accept samples as members of the population being studied, to admit samples for measurement, to establish the authenticity of samples, and to allow for resampling if necessary. In the validation process, samples can be rejected because of questions about the sample identity, questions about sample handling, or knowledge that the method of sample collection was not appropriate or in doubt. For example, contamination of blood samples during collection as evidence in a forensic examination would be reason to reject the samples.

There are several different ways to validate analytical methods. Some of these were discussed in Section 5B-4. The most common methods include analysis

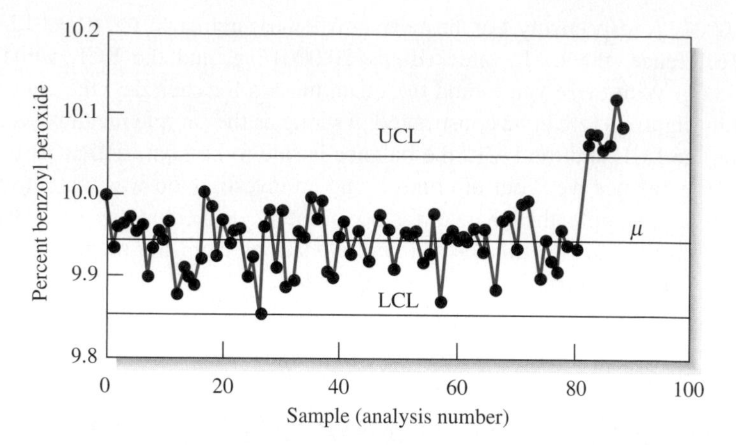

Figure 8-16 A control chart for monitoring the concentration of benzoyl peroxide in a commercial acne preparation. The manufacturing process became out of statistical control with sample 83 and exhibited a systematic change in the mean concentration. UCL = upper control limit; LCL = lower control limit.

of standard reference materials when available, analysis by a different analytical method, analysis of "spiked" samples, and analysis of synthetic samples approximating the chemical composition of the test samples. Individual analysts and laboratories often must periodically demonstrate the validity of the methods and techniques used.

Data validation is the final step before releasing results. This process starts with validating the samples and methods used. Then the data are reported with statistically valid limits of uncertainty after a thorough check has been made to eliminate errors in sampling and sample handling, in performing the analysis, in identifying samples, and in the calculations used.

Reporting Analytical Results

The specific reporting formats and procedures vary from laboratory to laboratory. A few general guidelines can be mentioned here, however. Whenever appropriate, reports should follow the procedure of good laboratory practice (GLP).[24]

Generally, analytical results should be reported as the mean value and the standard deviation. Sometimes, the standard deviation of the mean is given instead of that of the data set. Either of these is acceptable as long as it is clear what is being presented. A confidence interval for the mean should also be reported. Usually, the 95% confidence level is a reasonable compromise between being too inclusive and being too restrictive. Again, the interval and its confidence level should be explicitly stated. The results of various statistical tests on the data should also be included when appropriate, as should the rejection of any outlying results along with the rejection criterion.

Significant figures are important when reporting results. Significant figures should be based on statistical evaluation of the data. The significant figure convention stated in Section 6D-1 should be followed when possible, and rounding of the data should be done with careful attention to the guidelines.

Graphical presentation should include error bars on the data points to indicate their uncertainty when possible. Some graphical software allows the user to choose

[24] J. K. Taylor, *Quality Assurance of Chemical Measurements,* pp. 113–114. Chelsea, MI: Lewis Publishers, 1987.

different error bar limits of $\pm 1s$, $\pm 2s$, and so forth, while other packages automatically choose the size of the error bars. Whenever appropriate, the regression equation and its statistics should also be reported.

Validating and reporting analytical results are not the most glamorous parts of an analysis, but they can be considered one of the most important parts. Validation gives us confidence in the conclusions drawn. The report is often the "public" part of the procedure and may be brought to light during hearings, trials, patent applications, and other events.

WEB WORKS

Go to **http://chemistry.brookscole.com/skoogfac/.** From the Chapter Resources menu, choose Web Works, and locate the Chapter 8 section. Set your browser to the NIST link. Look up information on Standard Reference Materials (SRMs) in the Food and Agriculture area. Find the SRM for rice flour and look up the Certificate of Analysis. For how many elements are there certified values? Which elements?

QUESTIONS AND PROBLEMS

***8-1.** Describe the steps in a sampling operation.

8-2. What is the object of the sampling step in an analysis?

***8-3.** What factors determine the weight of a gross sample?

8-4. The following results were obtained for the determination of calcium in a NIST limestone sample: % CaO = 50.38, 50.20, 50.31, 50.22, and 50.41. Five gross samples were then obtained for a carload of limestone. The average percent CaO values for the gross samples were found to be 49.53, 50.12, 49.60, 49.87, and 50.49. Calculate the relative standard deviation associated with the sampling step.

***8-5.** A coating that weighs at least 3.00 mg is needed to impart adequate shelf life to a pharmaceutical tablet. A random sampling of 250 tablets revealed that 14 failed to meet this requirement.
- (a) Use this information to estimate the relative standard deviation for the measurement.
- (b) What is the 90% confidence interval for the number of unsatisfactory tablets?
- (c) Assuming that the fraction of rejects remains unchanged, how many tablets should be taken to ensure a relative standard deviation of 10% in the measurement?

8-6. Changes in the method used to coat the tablets lowered the percentage of rejects from 5.6% (see Problem 8-5) to 2.0%. How many tablets should be taken for inspection if the permissible relative standard deviation in the measurement is to be
*(a) 25%? (b) 10%? *(c) 5%? (d) 1%?

***8-7.** The mishandling of a shipping container loaded with 750 cases of wine caused some of the bottles to break. An insurance adjuster proposed to settle the claim at 20.8% of the value of the shipment, based on a random 250-bottle sample in which 52 were cracked or broken. Calculate
- (a) the relative standard deviation of the adjuster's evaluation.
- (b) the absolute standard deviation for the 750 cases (12 bottles per case).
- (c) the 90% confidence interval for the total number of bottles.
- (d) the size of a random sampling needed for a relative standard deviation of 5.0%, assuming a breakage rate of about 21%.

8-8. Approximately 15% of the particles in a shipment of silver-bearing ore are judged to be argentite, Ag_2S (d = 7.3 g cm^{-3}, 87% Ag); the remainder are siliceous (d = 2.6 g cm^{-3}) and contain essentially no silver.
- (a) Calculate the number of particles that should be taken for the gross sample if the relative standard deviation due to sampling is to be 1% or less.

(b) Estimate the mass of the gross sample, assuming that the particles are spherical and have an average diameter of 4.0 mm.

(c) The sample taken for analysis is to weigh 0.600 g and contain the same number of particles as the gross sample. To what diameter must the particles be ground to satisfy these criteria?

***8-9.** In the determination of lead in a paint sample, it is known that the sampling variance is 10 ppm and the measurement variance is 4 ppm. Two different sampling schemes are under consideration:

Scheme a: Take five sample increments and blend them. Perform a duplicate analysis of the blended sample.

Scheme b: Take three sample increments and perform a duplicate analysis on each.

Which sampling scheme, if any, should have the lower variance of the mean?

8-10. The following data represent the concentration of glucose in the blood serum of an adult patient. On 4 consecutive days, a blood sample was drawn from the patient and analyzed in triplicate. The variance for a given sample is an estimate of the measurement variance, whereas the day-to-day variance reflects both the measurement variance and the sampling variance.

Day	Glucose Concentration, mg/100 mL		
1	62	60	63
2	58	57	57
3	51	47	48
4	54	59	57

(a) Perform an analysis of variance and see whether the mean concentrations vary significantly from day to day.

(b) Estimate the sampling variance.

(c) What is the best way to lower the overall variance?

***8-11.** The seller of a mining claim took a random ore sample that weighed approximately 5 lb and had an average particle diameter of 5.0 mm. Inspection revealed that about 1% of the sample was argentite (see Problem 8-8), and the remainder had a density of about 2.6 g cm^3 and contained no silver. The prospective buyer insisted on knowing the silver content of the claim with a relative error no greater than 5%. Did the seller provide a sufficiently large sample to permit such an evaluation?

8-12. A method for the determination of the corticosteroid methylprednisolone acetate in solutions obtained from pharmaceutical preparations yielded a mean value of 3.5 mg mL^{-1} with a standard deviation of 0.2 mg mL^{-1}. For quality control purposes, the relative uncertainty in the concentration should be no more than 5%. How many samples of each batch should be analyzed to ensure that the relative standard deviation does not exceed 5% at the 95% confidence level?

8-13. The sulfate ion concentration in natural water can be determined by measuring the turbidity that results when an excess of $BaCl_2$ is added to a measured quantity of the sample. A turbidimeter, the instrument used for this analysis, was calibrated with a series of standard Na_2SO_4 solutions. The following data were obtained in the calibration:

mg SO_4^{2-}/L, C_x	Turbidimeter Reading, R
0.00	0.06
5.00	1.48
10.00	2.28
15.0	3.98
20.0	4.61

Assume that a linear relationship exists between the instrument reading and the concentration.

(a) Plot the data and draw a straight line through the points by eye.

*(b) Compute the least-squares slope and intercept for the best straight line among the points.

(c) Compare the straight line from the relationship determined in (b) with that in (a).

*(d) Do ANOVA and find the R^2 value, the adjusted R^2 value, and the significance of the regression. Comment on the interpretation of these values.

(e) Obtain the concentration of sulfate in a sample yielding a turbidimeter reading of 2.84. Find the absolute standard deviation and the coefficient of variation.

*(f) Repeat the calculations in (e) assuming that the 2.84 was the mean of six turbidimeter readings.

8-14. The following data were obtained in calibrating a calcium ion electrode for the determination of pCa. A linear relationship between the potential and pCa is known to exist.

pCa = $-\log[Ca^{2+}]$	E, mV
5.00	−53.8
4.00	−27.7
3.00	+2.7
2.00	+31.9
1.00	+65.1

(a) Plot the data and draw a line through the points by eye.
(b) Find the least-squares expression for the best straight line through the points. Plot this line.
(c) Do ANOVA and report the statistics given in the ANOVA table. Comment on the meaning of the ANOVA statistics.
(d) Calculate the pCa of a serum solution in which the electrode potential was 10.7 mV. Find the absolute and relative standard deviations for pCa if the result was from a single voltage measurement.
(e) Find the absolute and relative standard deviations for pCa if the millivolt reading in (d) was the mean of two replicate measurements. Repeat the calculation based on the mean of eight measurements.

8-15. The following are relative peak areas for chromatograms of standard solutions of methyl vinyl ketone (MVK).

Concentration MVK, mmol/L	Relative Peak Area
0.500	3.76
1.50	9.16
2.50	15.03
3.50	20.42
4.50	25.33
5.50	31.97

*(a) Determine the coefficients of the best-fit line using the least-squares method.
(b) Construct an ANOVA table.
(c) Plot the least-squares line as well as the experimental points.
*(d) A sample containing MVK yielded relative peak area of 10.3. Calculate the concentration of MVK in the solution.
(e) Assume that the result in (d) represents a single measurement as well as the mean of four measurements. Calculate the respective absolute and relative standard deviations.
*(f) Repeat the calculations in (d) and (e) for a sample that gave a peak area of 22.8.

8-16. The data in this table were obtained during a colorimetric determination of glucose in blood serum.

Glucose Concentration, mM	Absorbance, A
0.0	0.002
2.0	0.150
4.0	0.294
6.0	0.434
8.0	0.570
10.0	0.704

(a) Assuming a linear relationship, find the least-squares estimates of the slope and intercept.
(b) What are the standard deviations of the slope and intercept? What is the standard error of the estimate?
(c) Determine the 95% confidence intervals for the slope and intercept.
(d) A serum sample gave an absorbance of 0.350. Find the 95% confidence interval for glucose in the sample.

8-17. The data in this table represent electrode potential E vs. concentration c.

E, mV	Concentration c in mol L^{-1}
106	0.20000
115	0.07940
121	0.06310
139	0.03160
153	0.02000
158	0.01260
174	0.00794
182	0.00631
187	0.00398
211	0.00200
220	0.00126
226	0.00100

*(a) Transform the data to E vs. $-\log c$ values.
(b) Plot E vs. $-\log c$ and find the least-squares estimate of the slope and intercept. Write the least-squares equation.
*(c) Find the 95% confidence limits for the slope and intercept.
(d) Use the F test to comment on the significance of regression.
*(e) Find the standard error of the estimate, the correlation coefficient, and the multiple correlation coefficient.

8-18. A study was done to determine the activation energy E_A for a chemical reaction. The rate constant k was determined as a function of temperature T, and the data in the following table were obtained.

Temperature, T, K	k, s^{-1}
599	0.00054
629	0.0025
647	0.0052
666	0.014
683	0.025
700	0.064

The data should fit a linear model of the form $\log k = \log A - E_A/(2.303RT)$, where A is a preexponential factor and R is the gas constant.

(a) Fit the data to a straight line of the form $\log k = a - 1000b/T$.

(b) Find the slope, intercept, and standard error of the estimate.

(c) Noting that $E_A = -b \times 2.303R \times 1000$, find the activation energy and its standard deviation. (Use $R = 1.987$ cal mol^{-1} K^{-1}.)

(d) A theoretical prediction gave $E_A = 41.00$ kcal mol^{-1} K^{-1}. Test the null hypothesis that E_A is this value at the 95% confidence level.

***8-19.** Water can be determined in solid samples by infrared spectroscopy. The water content of calcium sulfate hydrates is to be measured using calcium carbonate as an internal standard to compensate for some systematic errors in the procedure. A series of standard solutions containing calcium sulfate dihydrate and a constant known amount of the internal standard are prepared. The solution of unknown water content is also prepared with the same amount of internal standard. The absorbance of the dihydrate is measured at one wavelength (A_{sample}) along with that of the internal standard at another wavelength (A_{std}). The following results were obtained.

A_{sample}	A_{std}	% Water
0.15	0.75	4.0
0.23	0.60	8.0
0.19	0.31	12.0
0.57	0.70	16.0
0.43	0.45	20.0
0.37	0.47	Unknown

(a) Plot the absorbance of the sample (A_{sample}) vs. the % water and determine whether the plot is linear from the regression statistics.

(b) Plot the ratio A_{sample}/A_{std} vs. % water and comment on whether use of the internal standard improves the linearity from that in part (a). If it improves the linearity, why?

(c) Calculate the percentage of water in the unknown using the internal standard data.

8-20. Potassium can be determined by flame emission spectrometry (flame photometry) using a lithium internal standard. The following data were obtained for standard solutions of KCl and an unknown containing a constant known amount of LiCl as the internal standard. All the intensities were corrected for background by subtracting the intensity of a blank.

Concentration of K, ppm	Intensity of K Emission	Intensity of Li Emission
1.0	10.0	10.0
2.0	15.3	7.5
5.0	34.7	6.8
7.5	65.2	8.5
10.0	95.8	10.0
20.0	110.2	5.8
Unknown	47.3	9.1

(a) Plot the K emission intensity vs. the concentration of K and determine the linearity from the regression statistics.

(b) Plot the ratio of the K intensity to the Li intensity vs. the concentration of K and compare the resulting linearity to that in part (a). Why does the internal standard improve linearity?

(c) Calculate the concentration of K in the unknown.

***8-21.** Copper was determined in a river water sample by atomic absorption spectrometry and the method of standard additions. For the addition, 100.0 μL of a 1000.0 μg/mL Cu standard was added to 100.0 mL of solution. The following data were obtained:

Absorbance of reagent blank = 0.020

Absorbance of sample = 0.520

Absorbance of sample plus addition − blank = 1.020

(a) Calculate the copper concentration in the sample.

(b) Later studies showed that the reagent blank used to obtain these data was inadequate and that the actual blank absorbance was 0.100. Find the copper concentration with the appropriate blank and determine the error caused by using an improper blank.

8-22. The method of standard additions was used to determine nitrite in a soil sample. An aliquot of 1.00 mL of the sample was mixed with 24.00 mL of a colorimetric reagent, and the nitrite was converted to a colored product with a blank-corrected absorbance of 0.300. To 50.00 mL of the original sample, 1.00 mL of a standard solution of 1.00×10^{-3} M nitrite was added. The same color-forming procedure was followed, and the new absorbance was 0.530. What was the concentration of nitrite in the original undiluted sample?

***8-23.** The following atomic absorption results were obtained for determinations of Zn in a multivitamin tablet. All absorbance values were corrected for the appropriate reagent blank (c_{Zn} = 0.0 ng/mL). The mean value for the blank was 0.0000 with a standard deviation of 0.0047 absorbance units.

c_{Zn}, ng/mL	A
5.0	0.0519
5.0	0.0463
5.0	0.0485
10.0	0.0980
10.0	0.1033
10.0	0.0925
Tablet sample	0.0672
Tablet sample	0.0614
Tablet sample	0.0661

(a) Find the mean absorbance values for the 5.0 and 10.0 ng/mL standards and for the tablet sample. Find the standard deviations of these values.

(b) Find the least-squares best line through the points at c_{Zn} = 0.0, 5.0, and 10.0 ng/mL. Find the calibration sensitivity and the analytical sensitivity.

(c) Find the detection limit for a k value of 3. To what level of confidence does this correspond?

(d) Find the concentration of Zn in the tablet sample and the standard deviation in the concentration.

8-24. Atomic emission measurements were made to determine sodium in a blood serum sample. The following emission intensities were obtained for standards of 5.0 and 10.0 ng/mL and for the serum sample. All emission intensities were corrected for any blank emission. The mean value for the blank intensity (c_{Na} = 0.0) was 0.000 with a standard deviation of 0.0071 (arbitrary units).

c_{Na}, ng/mL	Emission Intensity
5.0	0.51
5.0	0.49
5.0	0.48
10.0	1.02
10.0	1.00
10.0	0.99
Serum	0.71
Serum	0.77
Serum	0.78

(a) Find the mean emission intensity values for the 5.0 and 10.0 ng/mL standards and for the serum sample. Find the standard deviations of these values.

(b) Find the least-squares best line through the points at c_{Na} = 0.0, 5.0, and 10.0 ng/mL. Find the calibration sensitivity and the analytical sensitivity.

(c) Find the detection limit for k values of 2 and 3. To what level of confidence do these correspond?

(d) Find the concentration of Na in the serum sample and the standard deviation in the concentration.

***8-25.** The following data represent measurements made on a process for 30 days. One measurement was made each day. Assuming that 30 measurements are enough that $\bar{x} \rightarrow \mu$ and $s \rightarrow \sigma$, find the mean of the values, the standard deviation, and the upper and lower control limits. Plot the data points along with the statistical quantities on a chart and determine whether the process was always in statistical control.

Day	Value	Day	Value	Day	Value
1	49.8	11	49.5	21	58.8
2	48.4	12	50.5	22	51.3
3	49.8	13	48.9	23	50.6
4	50.8	14	49.7	24	48.8
5	49.6	15	48.9	25	52.6
6	50.2	16	48.8	26	54.2
7	51.7	17	48.6	27	49.3
8	50.5	18	48.1	28	47.9
9	47.7	19	53.8	29	51.3
10	50.3	20	49.6	30	49.3

8-26. The following table gives the sample means and standard deviations for six measurements each day of the purity of a polymer in a process. The purity is monitored for 24 days. Determine the overall mean and standard deviation of the measurements and construct a control chart with upper and lower control limits. Do any of the means indicate a loss of statistical control?

Day	Mean	SD	Day	Mean	SD
1	96.50	0.80	13	96.64	1.59
2	97.38	0.88	14	96.87	1.52
3	96.85	1.43	15	95.52	1.27
4	96.64	1.59	16	96.08	1.16
5	96.87	1.52	17	96.48	0.79
6	95.52	1.27	18	96.63	1.48
7	96.08	1.16	19	95.47	1.30
8	96.48	0.79	20	96.43	0.75
9	96.63	1.48	21	97.06	1.34
10	95.47	1.30	22	98.34	1.60
11	97.38	0.88	23	96.42	1.22
12	96.85	1.43	24	95.99	1.18

8-27. Challenge Problem. Zwanziger and Sârbu[25] conducted a study to validate analytical methods and instruments. The following data are results obtained in the determination of mercury in solid wastes by atomic absorption spectroscopy using two different sample preparation methods: a microwave digestion method and a traditional digestion method.

x, Mercury Concentration, ppm (traditional)	*y*, Mercury Concentration, ppm (microwave)
7.32	5.48
15.80	13.00
4.60	3.29
9.04	6.84
7.16	6.00
6.80	5.84
9.90	14.30
28.70	18.80

(a) Perform a least-squares analysis on the data in the table, assuming that the traditional method (x) is the independent variable. Determine the slope, the intercept, the R^2 value, the standard error, and any other relevant statistics.

(b) Plot the results obtained in part (a) and give the equation of the regression line.

(c) Now assume that the microwave digestion method (y) is the independent variable, once again perform a regression analysis, and determine the relevant statistics.

(d) Plot the data in part (c), and determine the regression equation.

(e) Compare the regression equation obtained in (b) with the equation from (d). Why are the equations different?

(f) Is there any conflict between the procedure that you have just performed and the assumptions of the least-squares method? What type of statistical analysis would be more appropriate than linear least-squares in dealing with data sets of this type?

(g) Look up the paper in reference 25, and compare your results with those presented in the paper for Example 4 in Table 2. You will note that your results from (d) differ from the authors' results. What is the most probable explanation for this discrepancy?

(h) Download the test data found in Table 1 of reference 25 from our Web site **http://chemistry.brookscole.com/skoogfac/,** perform the same type of analysis for Example 1 and Example 3, and compare your results with those in Table 2 of the paper. Note that in Example 3, you must include all 37 data pairs.

(i) What other methods for dealing with method comparison data are suggested in the paper?

(j) What is implied when we compare two methods by linear regression and the slope is not equal to one? What is implied when the intercept is not zero?

InfoTrac College Edition

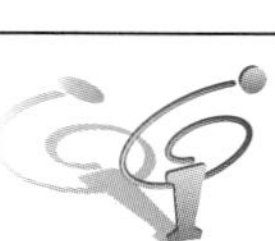

For additional readings, go to InfoTrac College Edition, your online research library, at

http://infotrac.thomsonlearning.com

[25] H.W. Zwanziger and C. Sârbu, *Anal. Chem.,* **1998,** 70, 1277.

PART II

Chemical Equilibria

A conversation with *Sylvia Daunert*

© Lee P. Thomas

Sylvia Daunert lives in Kentucky, but her accent is no Southern drawl: It reflects her cosmopolitan background. She is from Barcelona, Spain, and is of German origin. She attended a German school and spent summers in boarding schools throughout Europe and in the United States. Daunert studied to be a pharmacist at the University of Barcelona; as a Fulbright scholar, she received her M.S. in medicinal chemistry at the University of Michigan. There she met her husband, Leonidas Bachas, who is Greek. After Leonidas accepted a position as a chemistry professor at the University of Kentucky, she traveled back and forth between Lexington and Spain until she received her Ph.D. from the University of Barcelona. She is now also a chemistry professor at the University of Kentucky. She is interested in using recombinant DNA technology to develop new bioanalytical techniques. She is currently devising bioluminescence assays to detect biomolecules and toxic compounds. Soon, the products of her research will be implanted in patients with chronic illnesses or will be used to maintain the health of astronauts on long-term space missions. To add to Daunert's hectic life, she has three children, two adolescents and a baby.

Q: How did you become interested in chemistry?

A: When I was young I liked to mix things, especially in the kitchen. Chemists tend to be cooks. We always had a cook at home, but on weekends I would do the cooking. My parents always told me that being a woman shouldn't limit me. They told me that I could do whatever I want and achieve whatever I want.

Q: Where did you receive your training?

A: As a Fulbright scholar, I could go anywhere. I was interested in the University of Michigan because I wanted to work at the interface between pharmacy and chemistry, and they had the best program. The most important thing that happened to me there—besides meeting my husband—was being exposed to biosensors. I loved it and decided to work in that field.

My husband was 3 years ahead of me. After he got his Ph.D., he got a tenure-track position at the University of Kentucky. I had to decide whether to stay in Michigan to earn my Ph.D. and live apart from him for 3 years or get my master of science degree, go with him, and figure out how to get my Ph.D. Of course, I went with him! I set up a collaboration with a professor in Spain and flew back and forth, doing research both in my advisor's lab in Spain and in my husband's lab in Kentucky. In the meantime, I gave birth to my two older kids—at first I was flying back and forth pregnant, then I was flying back and forth pregnant with a young child! When the second one was 4 months old, I defended my thesis. It was very rewarding because I received an award from the Spanish Royal Academy of Doctors for my dissertation.

I then became a research professor at Kentucky. Because I was successful at bringing in grants, they created a tenure-track position for me. I started in 1994 and got one of the fastest promotions ever in my department to associate professor and one of the fastest to full professor. In 2002, I received the Gill Eminent Professorship in Analytical and Biological Chemistry.

Q: What is your research focus in your laboratory?

A: In my lab, we genetically engineer proteins and cells to do analytical chemistry. We use proteins from a bioluminescent jellyfish found near Seattle. When a predator is nearby or the organism is interested in mating, an internal reaction triggered by calcium occurs that causes the jellyfish to emit a very strong flash of blue light. The bioluminescence comes from a protein contained in certain cells in the umbrella of the jellyfish. When the jellyfish is in deep or very cold water the water is blue, and the blue light can't be seen. Therefore, this light excites another luminescent protein that emits green fluorescent light that can be seen. In our lab, we mimic nature. We genetically engineer the proteins to develop assays for biomolecules—drugs, hormones, neuropeptides—that are difficult to detect because of their very low concentrations. Because the bioluminescent signal is so powerful, we can detect it at extremely low levels, down to single cells. Blood, saliva, and urine are colored, so optical detection methods have backgrounds we need to worry about. With bioluminescence, there is virtually no background, so there is no interference from anything in the sample. Also, the flash emission allows for fast detection, which is advantageous when a quick answer is needed, like in an emergency room situation.

Q: Are there any other potential applications of bioluminescence?

A: We genetically engineer whole cells—bacterial, yeast, or even mammalian—to detect molecules in the environment. In bacteria, we engineer a plasmid to harbor a protein capable of detecting a toxic compound, along with a fluorescent reporter protein. The detector protein recognizes the toxic compound, then allows for the reporter protein to be produced and generate a signal, usually light. Light intensity is directly proportional to the amount of the toxic compound. We can engineer the cells to glow in an array of colors—different colors for different compounds.

Advances in analytical chemistry will be from bringing in techniques from other fields: material science, nanotechnology, microfabrication, microelectronics, and, of course, proteomics and genomics.

We are also working with engineers to microfabricate channels in a disk, like a CD, that we use in a Sony Walkman–like device. The channels are at micron or submicron level; we put biosensing systems in them based on either the genetically engineered bacteria or rationally designed binding proteins. When the disk spins, reagents go through the channels. We have a detection chamber at the end, and the luminescence signal tells us how much of the compound we have. We're interested in developing these detectors for NASA to monitor the health of astronauts and the environment in the spaceship. They are being designed to go to the space station or someday to Mars, where you need to continually monitor biochemicals in the body fluids of astronauts. Eventually, this could be used for detection of organisms on other planets.

Q: What products is your company, ChipRx, working on?

A: I'm one of the founders of ChipRx. We're developing responsive therapeutic systems for individualized therapy for patients. These integrate biosensors with drug delivery technologies to fabricate smart implantable devices. The biosensors are based on different kinds of genetically engineered proteins. When they bind to an analyte, they open and close like a hinge and generate a very specific signal. Since no two patients respond to drugs in the same manner, by sensing the level of a particular molecule, you can administer the right amount of drug. One example is glucose binding protein. We are set to incorporate a biosensor into a device, which we will implant subcutaneously, that continuously senses glucose levels. When they are too high, the biosensor emits a signal that orders the release of the right amount of insulin. The drug is in microfabricated chambers in the device, which are blasted open. The device operates on a tiny battery and works by telemetry, so there are no wires. Diabetic patients who need to test insulin levels many times a day may experience hypoglycemia during the night and enter a diabetic coma. These individuals will really benefit from this device. It will act as an alarm for the onset of hypoglycemia. Other applications we are targeting are in cardiology, pain management, and hormone treatment.

Q: Do you have any advice for students interested in analytical chemistry?

A: Students entering the field of analytical chemistry need to be open minded. If you are a well-trained analytical chemist, you must not be afraid of touching other fields to solve your problem. There are no boundaries! Advances in analytical chemistry will be the result of bringing in techniques from other fields: material science, nanotechnology, microfabrication, microelectronics and, of course, proteomics and genomics. Students need to learn to talk to people in other disciplines.

Q: Have you received any acknowledgment for your work?

A: In 2001, I won the Findeis Award from the American Chemical Society Division of Analytical Chemistry, which is for a young analytical chemist who is within 10 years of receiving the Ph.D. It was special because it was given by my colleagues in the analytical chemistry community. They had a scientific session in my honor, and I chose the speakers I wanted. It was really wonderful! It made me remember the first paper I sent to the journal of *Analytical Chemistry.* One of the reviewers sent me the comment that my paper was not analytical chemistry, but the other said that it was beautiful science. Luckily the editor liked it. He thought that it was the future of analytical chemistry, and he accepted it. At the time, there was hardly anything to do with DNA in the journal; my research was foreign. Now you see so much DNA work when you open an analytical chemistry book!

Q: Is it difficult to balance your career and your family life?

A: My husband has always been very encouraging to me in my career, and he helps a lot with the kids. To get everything done, there's a lot of coordination between us. But still, it's hard. I work at home a lot in the evenings and over the weekend. On vacations, I always take my computer and all my electronic gadgets, and I work in my free time. There's never a time that I'm not working. I enjoy my research tremendously, so working on it doesn't seem like work! ■

CHAPTER 9

Aqueous Solutions and Chemical Equilibria

Most analytical techniques require the state of chemical equilibrium. At equilibrium, the rate of a forward process or reaction and that of the reverse process are equal. The photo at left shows the beautiful natural formation called "Frozen Niagra" in Mammoth Cave National Park in Kentucky. As water seeps over the limestone surface of the cave, calcium carbonate dissolves in the water according to the chemical equilibrium

$$CaCO_3(s) + CO_2(g) + H_2O(l) \rightleftharpoons Ca^{2+}(aq) + 2HCO_3^-(aq)$$

The flowing water becomes saturated with calcium carbonate; as carbon dioxide is swept away, the reverse reaction becomes favored, and limestone is deposited in formations whose shapes are governed by the path of the flowing water. Stalactites and stalagmites are examples of similar formations found where water saturated with calcium carbonate drips from the ceiling to the floor of caves over eons.

This chapter provides a fundamental approach to chemical equilibrium, including calculations of chemical composition and of equilibrium concentrations for monoprotic acid/base systems. Buffer solutions, which are extremely important in many areas of science, are also discussed, and the properties of buffer solutions are described.

9A THE CHEMICAL COMPOSITION OF AQUEOUS SOLUTIONS

Water is the most plentiful solvent available on Earth, is easily purified, and is not toxic. It therefore finds widespread use as a medium for carrying out chemical analyses.

9A-1 Classifying Solutions of Electrolytes

Most of the solutes we will discuss are *electrolytes,* which form ions when dissolved in water (or certain other solvents) and thus produce solutions that conduct electricity. *Strong electrolytes* ionize essentially completely in a solvent, whereas *weak electrolytes* ionize only partially. This means that a solution of a weak electrolyte will not conduct electricity as well as a solution containing an equal concentration of a strong electrolyte. Table 9-1 shows various solutes that act as strong

TABLE 9-1
Classification of Electrolytes

Strong	Weak
1. Inorganic acids such as HNO_3, $HClO_4$, $H_2SO_4^*$, HCl, HI, HBr, $HClO_3$, $HBrO_3$	1. Many inorganic acids, including H_2CO_3, H_3BO_3, H_3PO_4, H_2S, H_2SO_3
2. Alkali and alkaline-earth hydroxides	2. Most organic acids
3. Most salts	3. Ammonia and most organic bases
	4. Halides, cyanides, and thiocyanates of Hg, Zn, and Cd

*H_2SO_4 is completely dissociated into HSO_4^- and H_3O^+ ions and for this reason is classified as a strong electrolyte. It should be noted, however, that the HSO_4^- ion is a weak electrolyte, being only partially dissociated into SO_4^{2-} and H_3O^+.

and weak electrolytes in water. Among the strong electrolytes listed are acids, bases, and **salts.**

A **salt** is produced in the reaction of an acid with a base. Examples include NaCl, Na_2SO_4, and $NaOOCCH_3$ (sodium acetate).

9A-2 Acids and Bases

In 1923, two chemists, J. N. Brønsted in Denmark and J. M. Lowry in England, proposed independently a theory of acid/base behavior that is particularly useful in analytical chemistry. According to the Brønsted-Lowry theory, an **acid** is a proton donor and a **base** is a proton acceptor. For a molecule to behave as an acid, it must encounter a proton acceptor (or base). Likewise, a molecule that can accept a proton behaves as a base if it encounters an acid.

An **acid** donates protons; a **base** accepts protons.

◀ An acid donates protons only in the presence of a proton acceptor (a base). Likewise, a base accepts protons only in the presence of a proton donor (an acid).

Conjugate Acids and Bases

An important feature of the Brønsted-Lowry concept is the idea that the product formed when an acid gives up a proton is a potential proton acceptor and is called the **conjugate base** of the parent acid. For example, when the species acid_1 gives up a proton, the species base_1 is formed, as shown by the reaction

$$\text{acid}_1 \rightleftharpoons \text{base}_1 + \text{proton}$$

Here, acid_1 and base_1 are a conjugate acid/base pair.

Similarly, every base produces a **conjugate acid** as a result of accepting a proton. That is,

$$\text{base}_2 + \text{proton} \rightleftharpoons \text{acid}_2$$

A **conjugate base** is formed when an acid loses a proton. For example, acetate ion is the conjugate base of acetic acid; similarly, ammonium ion is the conjugate acid of the base ammonia.

A **conjugate acid** is formed when a base accepts a proton.

When these two processes are combined, the result is an acid/base, or **neutralization,** reaction:

$$\text{acid}_1 + \text{base}_2 \rightleftharpoons \text{base}_1 + \text{acid}_2$$

◀ A substance acts as an acid only in the presence of a base, and vice versa.

The extent to which this reaction proceeds depends on the relative tendencies of the two bases to accept a proton (or the two acids to donate a proton).

Examples of conjugate acid/base relationships are shown in Equations 9-1 through 9-4.

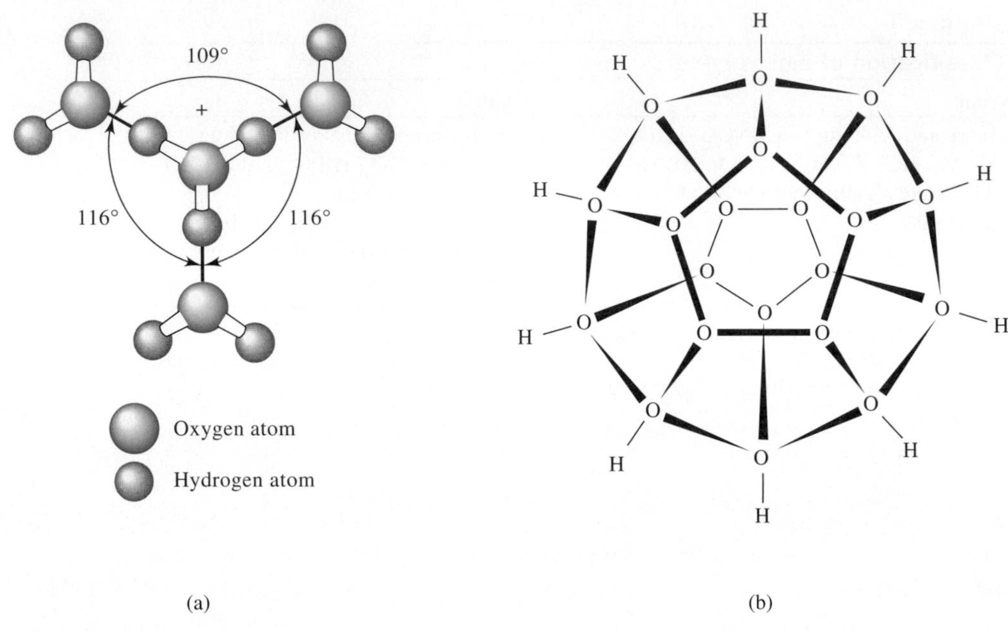

Figure 9-1 Possible structures for the hydronium ion. (a) The species $H_9O_4^+$ has been observed in the solid state and may be an important contributor in aqueous solution. (b) The species $(H_2O)_{21}H^+$ exhibits a dodecahedral caged structure in mixed water-trimethylamine cluster ions. The hydronium ion (not shown) is encased in the hydrogen-bonded cage with 10 non–hydrogen-bonded protons protruding from its surface. S. Wei, Z. Shi, and A. W. Castleman, Jr., *J. Chem. Phys,* **1991,** *94,* 3268. Structure reproduced by courtesy of the American Institute of Physics.

Many solvents are proton donors or proton acceptors and can thus induce basic or acidic behavior in solutes dissolved in them. For example, in an aqueous solution of ammonia, water can donate a proton and thus acts as an acid with respect to the solute:

$$\underset{\text{base}_1}{NH_3} + \underset{\text{acid}_2}{H_2O} \rightleftharpoons \underset{\substack{\text{conjugate}\\ \text{acid}_1}}{NH_4^+} + \underset{\substack{\text{conjugate}\\ \text{base}_2}}{OH^-} \qquad (9\text{-}1)$$

In this reaction, ammonia (base_1) reacts with water, which is labeled acid_2, to give the conjugate acid ammonium ion (acid_1) and hydroxide ion, which is the conjugate base (base_2) of the acid water. In contrast, water acts as a proton acceptor, or base, in an aqueous solution of nitrous acid:

$$\underset{\text{base}_1}{H_2O} + \underset{\text{acid}_2}{HNO_2} \rightleftharpoons \underset{\substack{\text{conjugate}\\ \text{acid}_1}}{H_3O^+} + \underset{\substack{\text{conjugate}\\ \text{base}_2}}{NO_2^-} \qquad (9\text{-}2)$$

The conjugate base of the acid HNO_2 is nitrite ion. The conjugate acid of water is the hydrated proton written as H_3O^+. This species is called the **hydronium ion** and consists of a proton covalently bonded to one water molecule. Higher hydrates such as $H_5O_2^+$, $H_9O_4^+$, and the cage structure shown in Figure 9-1 may also exist in aqueous solutions of protons. For convenience, however, chemists generally use the notation H_3O^+ or, more simply H^+, in writing chemical equations in which the proton is involved.

An acid that has donated a proton becomes a conjugate base capable of accepting a proton to re-form the original acid; the converse holds equally well. Thus, nitrite ion, the species produced by the loss of a proton from nitrous acid, is a potential acceptor of a proton from a suitable donor. It is this reaction that causes an aqueous solution of sodium nitrite to be slightly basic:

$$\underset{\text{base}_1}{NO_2^-} + \underset{\text{acid}_2}{H_2O} \rightleftharpoons \underset{\substack{\text{conjugate}\\ \text{acid}_1}}{HNO_2} + \underset{\substack{\text{conjugate}\\ \text{base}_2}}{OH^-}$$

© Hulton-Deutsch Collection/Corbis

Svante Arrhenius (1859–1927), Swedish chemist, formulated many of the early ideas regarding ionic dissociation in solution. His ideas were not accepted at first; in fact, he was given the lowest possible passing grade for his Ph.D. examination. In 1903, Arrhenius was awarded the Nobel Prize in chemistry for these revolutionary ideas. He was one of the first scientists to suggest the relationship between the amount of carbon dioxide in the atmosphere and global temperature, a phenomenon that has come to be known as the **greenhouse effect.** You may like to read Arrhenius's original paper "On the influence of carbonic acid in the air upon the temperature of the ground," *London Edinburgh Dublin Philos. Mag. J. Sci.,* **1896,** *41,* 237–276.

9A-3 Amphiprotic Species

Species that possess both acidic and basic properties are **amphiprotic.** An example is dihydrogen phosphate ion, $H_2PO_4^-$, which behaves as a base in the presence of a proton donor such as H_3O^+.

$$\underset{\text{base}_1}{H_2PO_4^-} + \underset{\text{acid}_2}{H_3O^+} \rightleftharpoons \underset{\text{acid}_1}{H_3PO_4} + \underset{\text{base}_2}{H_2O}$$

Here, H_3PO_4 is the conjugate acid of the original base. In the presence of a proton acceptor, such as hydroxide ion, however, $H_2PO_4^-$ behaves as an acid and donates a proton to form the conjugate base HPO_4^{2-}.

$$\underset{\text{acid}_1}{H_2PO_4^-} + \underset{\text{base}_2}{OH^-} \rightleftharpoons \underset{\text{base}_1}{HPO_4^{2-}} + \underset{\text{acid}_2}{H_2O}$$

The simple amino acids are an important class of amphiprotic compounds that contain both a weak acid and a weak base functional group. When dissolved in water, an amino acid such as glycine undergoes a kind of internal acid/base reaction to produce a **zwitterion**—a species that bears both a positive and a negative charge. Thus,

$$\underset{\text{glycine}}{NH_2CH_2COOH} \rightleftharpoons \underset{\text{zwitterion}}{NH_3^+CH_2COO^-}$$

This reaction is analogous to the acid/base reaction between a carboxylic acid and an amine:

$$\underset{\text{acid}_1}{R'COOH} + \underset{\text{base}_2}{R''NH_2} \rightleftharpoons \underset{\text{base}_1}{R'COO^-} + \underset{\text{acid}_2}{R''NH_3^+}$$

A **zwitterion** is an ion that bears both a positive and a negative charge.

Water is the classic example of an **amphiprotic solvent**—that is, a solvent that can act either as an acid (Equation 9-1) or as a base (Equation 9-2), depending on the solute. Other common amphiprotic solvents are methanol, ethanol, and anhydrous acetic acid. In methanol, for example, the equilibria analogous to those shown in Equations 9-1 and 9-2 are

◀ Water can act as either an acid or a base.

Amphiprotic solvents behave as acids in the presence of basic solutes and bases in the presence of acidic solutes.

$$\underset{\text{base}_1}{NH_3} + \underset{\text{acid}_2}{CH_3OH} \rightleftharpoons \underset{\substack{\text{conjugate}\\ \text{acid}_1}}{NH_4^+} + \underset{\substack{\text{conjugate}\\ \text{base}_2}}{CH_3O^-} \quad (9\text{-}3)$$

$$\underset{\text{base}_1}{CH_3OH} + \underset{\text{acid}_2}{HNO_2} \rightleftharpoons \underset{\substack{\text{conjugate}\\ \text{acid}_1}}{CH_3OH_2^+} + \underset{\substack{\text{conjugate}\\ \text{base}_2}}{NO_2^-} \quad (9\text{-}4)$$

9A-4 Autoprotolysis

Autoprotolysis (also called autoionization) involves the spontaneous reaction of molecules of a substance to give a pair of ions.

Amphiprotic solvents undergo self-ionization, or **autoprotolysis,** to form a pair of ionic species. Autoprotolysis is yet another example of acid/base behavior, as illustrated by the following equations.

$$\begin{array}{llll} \text{base}_1 & + \text{acid}_2 & \rightleftharpoons \text{acid}_1 & + \text{base}_2 \\ H_2O & + H_2O & \rightleftharpoons H_3O^+ & + OH^- \\ CH_3OH & + CH_3OH & \rightleftharpoons CH_3OH_2^+ & + CH_3O^- \\ HCOOH & + HCOOH & \rightleftharpoons HCOOH_2^+ & + HCOO^- \\ NH_3 & + NH_3 & \rightleftharpoons NH_4^+ & + NH_2^- \end{array}$$

The **hydronium ion** is the hydrated proton formed when water reacts with an acid. It is usually formulated as H_3O^+, although there are several possible higher hydrates, as shown in Figure 9-1.

The extent to which water undergoes autoprotolysis at room temperature is slight. Thus, the hydronium and hydroxide ion concentrations in pure water are only about 10^{-7} M. Despite the small values of these concentrations, this dissociation reaction is of utmost importance in understanding the behavior of aqueous solutions.

► In this text, we use the symbol H_3O^+ in those chapters that deal with acid/base equilibria and acid/base equilibrium calculations. In the remaining chapters, we simplify to the more convenient H^+, with the understanding that this symbol represents the hydrated proton.

► The common strong bases include NaOH, KOH, $Ba(OH)_2$, and the quaternary ammonium hydroxide R_4NOH, where R is an alkyl group such as CH_3 or C_2H_5.

► The common strong acids include HCl, $HClO_4$, HNO_3, the first proton in H_2SO_4, HBr, HI, and the organic sulfonic acid RSO_3H.

9A-5 Strengths of Acids and Bases

Figure 9-2 shows the dissociation reactions of a few common acids in water. The first two are *strong acids* because reaction with the solvent is sufficiently complete that no undissociated solute molecules are left in aqueous solution. The remainder are *weak acids,* which react incompletely with water to give solutions containing significant quantities of both the parent acid and its conjugate base. Note that acids can be cationic, anionic, or electrically neutral. The same holds for bases.

The acids in Figure 9-2 become progressively weaker from top to bottom. Perchloric acid and hydrochloric acid are completely dissociated, but only about 1% of acetic acid ($HC_2H_3O_2$) is dissociated. Ammonium ion is an even weaker acid; only about 0.01% of this ion is dissociated into hydronium ions and ammonia molecules. Another generality illustrated in Figure 9-2 is that the weakest acid forms the strongest conjugate base; that is, ammonia has a much stronger affinity for protons than any base above it. Perchlorate and chloride ions have no affinity for protons.

The tendency of a solvent to accept or donate protons determines the strength of a solute acid or base dissolved in it. For example, perchloric and hydrochloric acids are strong acids in water. If anhydrous acetic acid, a weaker proton acceptor than water, is substituted *as the solvent,* neither of these acids undergoes complete dissociation; instead, equilibria such as the following are established:

$$\underset{\text{base}_1}{CH_3COOH} + \underset{\text{acid}_2}{HClO_4} \rightleftharpoons \underset{\text{acid}_1}{CH_3COOH_2^+} + \underset{\text{base}_2}{ClO_4^-}$$

Perchloric acid is, however, considerably stronger than hydrochloric acid in this solvent, its dissociation being about 5000 times greater. Acetic acid thus acts as a

Strongest acid ↑ Weakest acid | Weakest base ↓ Strongest base

$$\begin{array}{rcl} HClO_4 + H_2O & \longrightarrow & H_3O^+ + ClO_4^- \\ HCl + H_2O & \longrightarrow & H_3O^+ + Cl^- \\ H_3PO_4 + H_2O & \rightleftharpoons & H_3O^+ + H_2PO_4^- \\ Al(H_2O)_6^{3+} + H_2O & \rightleftharpoons & H_3O^+ + AlOH(H_2O)_5^{2+} \\ HC_2H_3O_2 + H_2O & \rightleftharpoons & H_3O^+ + C_2H_3O_2^- \\ H_2PO_4^- + H_2O & \rightleftharpoons & H_3O^+ + HPO_4^{2-} \\ NH_4^+ + H_2O & \rightleftharpoons & H_3O^+ + NH_3 \end{array}$$

Figure 9-2 Dissociation reactions and relative strengths of some common acids and their conjugate bases. Note that HCl and $HClO_4$ are completely dissociated in water.

differentiating solvent toward the two acids by revealing the inherent differences in their acidities. Water, on the other hand, is a **leveling solvent** for perchloric, hydrochloric, and nitric acids because all three are completely ionized in this solvent and show no differences in strength. Differentiating and leveling solvents also exist for bases.

In a **differentiating solvent,** various acids dissociate to different degrees and have different strengths. In a **leveling solvent,** several acids are completely dissociated and show the same strength.

9B CHEMICAL EQUILIBRIUM

◀ Of all the acids listed in the marginal note on page 232 and in Figure 9-2, only perchloric acid is a strong acid in methanol and ethanol. These two alcohols are therefore also differentiating solvents.

The reactions used in analytical chemistry never result in complete conversion of reactants to products. Instead, they proceed to a state of **chemical equilibrium** in which the ratio of concentrations of reactants and products is constant. **Equilibrium-constant expressions** are *algebraic* equations that describe the concentration relationships existing among reactants and products at equilibrium. Among other things, equilibrium-constant expressions permit calculation of the error in an analysis resulting from the quantity of unreacted analyte that remains when equilibrium has been reached.

The discussion that follows deals with the use of equilibrium-constant expressions to gain information about analytical systems in which no more than one or two equilibria are present. Chapter 11 extends these methods to systems containing several simultaneous equilibria. Such complex systems are often encountered in analytical chemistry.

9B-1 The Equilibrium State

Consider the chemical reaction

$$H_3AsO_4 + 3I^- + 2H^+ \rightleftharpoons H_3AsO_3 + I_3^- + H_2O \qquad (9\text{-}5)$$

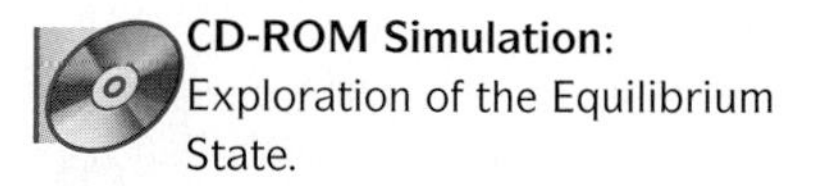

CD-ROM Simulation:
Exploration of the Equilibrium State.

The rate of this reaction and the extent to which it proceeds to the right can be readily judged by observing the orange-red color of the triiodide ion I_3^-. (The other participants in the reaction are colorless.) If, for example, 1 mmol of arsenic acid, H_3AsO_4, is added to 100 mL of a solution containing 3 mmol of potassium iodide, the red color of the triiodide ion appears almost immediately. Within a few seconds, the intensity of the color becomes constant, which shows that the triiodide concentration has become constant (see color plates 1b and 2b).

A solution of identical color intensity (and hence identical triiodide concentration) can also be produced by adding 1 mmol of arsenous acid, H_3AsO_3, to 100 mL of a solution containing 1 mmol of triiodide ion (see color plate 1a). Here, the color intensity is initially greater than in the first solution but rapidly decreases as a result of the reaction

$$H_3AsO_3 + I_3^- + H_2O \rightleftharpoons H_3AsO_4 + 3I^- + 2H^+$$

Ultimately, the color of the two solutions is identical. Many other combinations of the four reactants can be used to yield solutions that are indistinguishable from the two just described.

The results of the experiments shown in color plates 1 and 2 illustrate that the concentration relationship at chemical equilibrium (that is, the *position of equilibrium*) is independent of the route by which the equilibrium state is achieved. This relationship is altered by the application of stress to the system, however. Such stresses include changes in temperature, in pressure (if one of the reactants or

◀ The position of a chemical equilibrium is independent of the route by which equilibrium is reached.

The **Le Châtelier principle** states that the position of an equilibrium always shifts in such a direction as to relieve a stress that is applied to the system.

The **mass-action effect** is a shift in the position of an equilibrium caused by adding one of the reactants or products to a system.

► Chemical reactions do not cease at equilibrium. Instead, the amounts of reactants and products are constant because the rates of the forward and reverse processes are identical.

Thermodynamics is a branch of chemical science that deals with the flow of heat and energy in chemical reactions. The position of a chemical equilibrium is related to these energy changes.

► Equilibrium-constant expressions provide *no* information about whether a chemical reaction is fast enough to be useful in an analytical procedure.

Cato Guldberg (1836–1902) and Peter Waage (1833–1900) were Norwegian chemists whose primary interests were in the field of thermodynamics. In 1864, these workers were the first to propose the law of mass action, which is expressed in Equation 9-7. If you would like to learn more about Guldberg and Waage and read a translation of their original paper on the law of mass action, use your Web browser to connect to **http://chemistry.brookscole.com/skoogfac/.** From the Chapter Resources menu, choose Web Works, find Chapter 9, and click on the link to the paper.

products is a gas), or in total concentration of a reactant or a product. These effects can be predicted qualitatively from the **principle of Le Châtelier,** which states that the position of chemical equilibrium always shifts in a direction that tends to relieve the effect of an applied stress. For example, an increase in temperature alters the concentration relationship in the direction that tends to absorb heat, and an increase in pressure favors those participants that occupy a smaller total volume.

In an analysis, the effect of introducing an additional amount of a participating species to the reaction mixture is particularly important. Here, the resulting stress is relieved by a shift in equilibrium in the direction that partially uses up the added substance. Thus, for the equilibrium we have been considering (Equation 9-5), the addition of arsenic acid (H_3AsO_4) or hydrogen ions causes an increase in color as more triiodide ion and arsenous acid are formed; the addition of arsenous acid has the reverse effect. An equilibrium shift brought about by changing the amount of one of the participating species is called a **mass-action effect.**

Theoretical and experimental studies of reacting systems on the molecular level show that reactions among the participating species continue even after equilibrium is achieved. The constant concentration ratio of reactants and products results from the equality in the rates of the forward and reverse reactions. In other words, chemical equilibrium is a dynamic state in which the rates of the forward and reverse reactions are identical.

9B-2 Equilibrium-Constant Expressions

The influence of concentration (or pressure if the species are gases) on the position of a chemical equilibrium is conveniently described in quantitative terms by means of an *equilibrium-constant expression.* Such expressions are derived from thermodynamics. They are important because they permit the chemist to predict the direction and completeness of a chemical reaction. An equilibrium-constant expression, however, yields no information concerning the *rate* at which equilibrium is approached. In fact, we sometimes encounter reactions that have highly favorable equilibrium constants but are of little analytical use because their rates are low. This limitation can often be overcome by the use of a catalyst, which speeds the attainment of equilibrium without changing its position.

Consider a generalized equation for a chemical equilibrium

$$wW + xX \rightleftharpoons yY + zZ \tag{9-6}$$

where the capital letters represent the formulas of participating chemical species and the lowercase italic letters are the small whole numbers required to balance the equation. Thus, the equation states that w moles of W react with x moles of X to form y moles of Y and z moles of Z. The equilibrium-constant expression for this reaction is

$$K = \frac{[Y]^y[Z]^z}{[W]^w[X]^x} \tag{9-7}$$

where the square-bracketed terms have the following meanings:

1. molar concentration if the species is a dissolved solute.
2. partial pressure in atmospheres if the species is a gas; in fact, we will often replace the square bracketed term (say [Z] in Equation 9-7) with the symbol p_z, which stands for the partial pressure of the gas Z in atmospheres.

If one (or more) of the species in Equation 9-7 is a pure liquid, a pure solid, or the solvent present in excess, no term for this species appears in the equilibrium-constant expression. For example, if Z in Equation 9-6 is the solvent H_2O, the equilibrium-constant expression simplifies to

$$K = \frac{[Y]^y}{[W]^w[X]^x}$$

We discuss the reason for this simplification in the sections that follow.

The constant K in Equation 9-7 is a temperature-dependent numerical quantity called the *equilibrium constant.* By convention, the concentrations of the products, *as the equation is written,* are always placed in the numerator and the concentrations of the reactants in the denominator.

Equation 9-7 is only an approximate form of a thermodynamic equilibrium constant expression. The exact form is given by Equation 9-8 (in the margin). Generally, we use the approximate form of this equation because it is less tedious and time consuming. In Section 10B, we show when the use of Equation 9-7 is likely to lead to serious errors in equilibrium calculations and how Equation 9-8 is modified in these cases.

◀ $[Z]^z$ in Equation 9-7 is replaced with p_z in atmospheres if Z is a gas. No term for Z is included in the equation if this species is a pure solid, a pure liquid, or the solvent of a dilute solution.

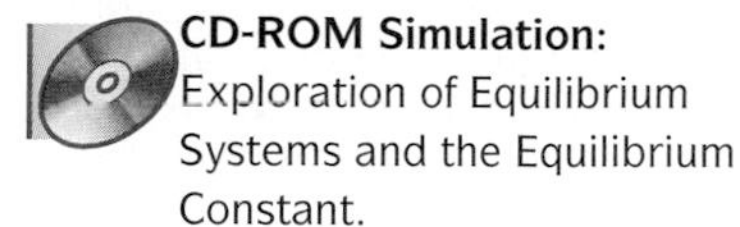

CD-ROM Simulation: Exploration of Equilibrium Systems and the Equilibrium Constant.

◀ Remember: Equation 9-7 is only an approximate form of an equilibrium-constant expression. The exact expression takes the form

$$K = \frac{a_Y^y \times a_Z^z}{a_W^w \times a_X^x} \quad (9\text{-}8)$$

where a_Y, a_Z, a_W, and a_X are the *activities* of species Y, Z, W, and X (see Section 10B).

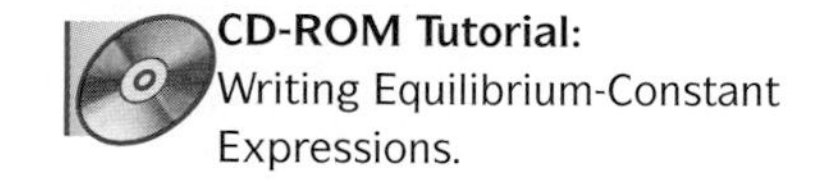

CD-ROM Tutorial: Writing Equilibrium-Constant Expressions.

9B-3 Types of Equilibrium Constants Encountered in Analytical Chemistry

Table 9-2 summarizes the types of chemical equilibria and equilibrium constants that are of importance in analytical chemistry. Simple applications of some of these constants are illustrated in the three sections that follow.

TABLE 9-2

Equilibria and Equilibrium Constants Important in Analytical Chemistry

Type of Equilibrium	Name and Symbol of Equilibrium Constant	Typical Example	Equilibrium-Constant Expression
Dissociation of water	Ion-product constant, K_w	$2\,H_2O \rightleftharpoons H_3O^+ + OH^-$	$K_w = [H_3O^+][OH^-]$
Heterogeneous equilibrium between a slightly soluble substance and its ions in a saturated solution	Solubility product, K_{sp}	$BaSO_4(s) \rightleftharpoons Ba^{2+} + SO_4^{2-}$	$K_{sp} = [Ba^{2+}][SO_4^{2-}]$
Dissociation of a weak acid or base	Dissociation constant, K_a or K_b	$CH_3COOH + H_2O \rightleftharpoons H_3O^+ + CH_3COO^-$	$K_a = \frac{[H_3O^+][CH_3COO^-]}{[CH_3COOH]}$
		$CH_3COO^- + H_2O \rightleftharpoons OH^- + CH_3COOH$	$K_b = \frac{[OH^-][CH_3COOH]}{[CH_3COO^-]}$
Formation of a complex ion	Formation constant, β_n	$Ni^{2+} + 4CN^- \rightleftharpoons Ni(CN)_4^{2-}$	$\beta_4 = \frac{[Ni(CN)_4^{2-}]}{[Ni^{2+}][CN^-]^4}$
Oxidation/reduction equilibrium	K_{redox}	$MnO_4 + 5Fe^{2+} + 8H^+ \rightleftharpoons Mn^{2+} + 5Fe^{3+} + 4H_2O$	$K_{redox} = \frac{[Mn^{2+}][Fe^{3+}]^5}{[MnO_4^-][Fe^{2+}]^5[H^+]^8}$
Distribution equilibrium for a solute between immiscible solvents	K_d	$I_2(aq) \rightleftharpoons I_2(org)$	$K_d = \frac{[I_2]_{org}}{[I_2]_{aq}}$

FEATURE 9-1

Stepwise and Overall Formation Constants for Complex Ions

The formation of $Ni(CN)_4^{2-}$ (Table 9-2) is typical in that it occurs in steps as shown. Note that *stepwise formation* constants are symbolized by K_1, K_2, and so forth.

$$Ni^{2+} + CN^- \rightleftharpoons Ni(CN)^+ \qquad K_1 = \frac{[Ni(CN)^+]}{[Ni^{2+}][CN^-]}$$

$$Ni(CN)^+ + CN^- \rightleftharpoons Ni(CN)_2 \qquad K_2 = \frac{[Ni(CN)_2]}{[Ni(CN)^+][CN^-]}$$

$$Ni(CN)_2 + CN^- \rightleftharpoons Ni(CN)_3^- \qquad K_3 = \frac{[Ni(CN)_3^-]}{[Ni(CN)_2][CN^-]}$$

$$Ni(CN)_3^- + CN^- \rightleftharpoons Ni(CN)_4^2 \qquad K_4 = \frac{[Ni(CN)_4^{2-}]}{[Ni(CN)_3^-][CN^-]}$$

Overall constants are designated by the symbol β_n. Thus,

$$Ni^{2+} + 2CN^- \rightleftharpoons Ni(CN)_2 \qquad \beta_2 = K_1K_2 = \frac{[Ni(CN)_2]}{[Ni^{2+}][CN^-]^2}$$

$$Ni^{2+} + 3CN \rightleftharpoons Ni(CN)_3^- \qquad \beta_3 = K_1K_2K_3 = \frac{[Ni(CN)_3^-]}{[Ni^{2+}][CN^-]^3}$$

$$Ni^{2+} + 4CN^- \rightleftharpoons Ni(CN)_4^{2-} \qquad \beta_4 = K_1K_2K_3K_4 = \frac{[Ni(CN)_4^{2-}]}{[Ni^{2+}][CN^-]^4}$$

CD-ROM Tutorial: Calculations Involving $[H_3O^+]$, $[OH^-]$, pH and pOH.

9B-4 Applying the Ion-Product Constant for Water

Aqueous solutions contain small concentrations of hydronium and hydroxide ions as a consequence of the dissociation reaction

$$2H_2O \rightleftharpoons H_3O^+ + OH^- \qquad (9\text{-}9)$$

CD-ROM Simulation: Exploration of the Relationships Between $[H_3O^+]$, $[OH^-]$, pH and pOH.

An equilibrium constant for this reaction can be formulated as shown in Equation 9-7:

$$K = \frac{[H_3O^+][OH^-]}{[H_2O]^2} \qquad (9\text{-}10)$$

► A useful relationship is obtained by taking the negative logarithm of Equation 9-11.

$$-\log K_w = -\log [H_3O^+] - \log [OH^-]$$

By the definition of p-function, (see Section 4B-1)

$$pK_w = pH + pOH \qquad (9\text{-}12)$$

At 25°C, $pK_w = 14.00$.

The concentration of water in dilute aqueous solutions is enormous, however, when compared with the concentration of hydrogen and hydroxide ions. As a consequence, $[H_2O]^2$ in Equation 9-10 can be taken as constant, and we write

$$K[H_2O]^2 = K_w = [H_3O^+][OH^-] \qquad (9\text{-}11)$$

where the new constant K_w is given a special name, the *ion-product constant for water.*

FEATURE 9-2

Why $[H_2O]$ Does Not Appear in Equilibrium-Constant Expressions for Aqueous Solutions

In a dilute aqueous solution, the molar concentration of water is

$$[H_2O] = \frac{1000 \text{ g } H_2O}{\text{L } H_2O} \times \frac{1 \text{ mol } H_2O}{18.0 \text{ g } H_2O} = 55.6 \text{ M}$$

Suppose we have 0.1 mol of HCl in 1 L of water. The presence of this acid will shift the equilibrium shown in Equation 9-9 to the left. Originally, however, there were only 10^{-7} mol/L OH^- to consume the added protons. Thus, even if all the OH^- ions are converted to H_2O, the water concentration will increase to only

$$[H_2O] = 55.6 \frac{\text{mol } H_2O}{\text{L } H_2O} + 1 \times 10^{-7} \frac{\text{mol } OH^-}{\text{L } H_2O} \times \frac{1 \text{ mol } H_2O}{\text{mol } OH^-} \approx 55.6 \text{ M}$$

The percent change in water concentration is

$$\frac{10^{-7} \text{ M}}{55.6 \text{ M}} \times 100\% = 2 \times 10^{-7}\%$$

which is certainly inconsequential. Thus, $K[H_2O]^2$ in Equation 9-10 is, for all practical purposes, a constant. That is,

$$K(55.6)^2 = K_w = 1.00 \times 10^{-14} \text{ at } 25°\text{C}$$

At 25°C, the ion-product constant for water is 1.008×10^{-14}. For convenience, we use the approximation that at room temperature $K_w \approx 1.00 \times 10^{-14}$. Table 9-3 shows the dependence of this constant on temperature. The ion-product constant for water permits the ready calculation of the hydronium and hydroxide ion concentrations of aqueous solutions.

TABLE 9-3
Variation of K_w with Temperature

Temperature, °C	K_w
0	0.114×10^{-14}
25	1.01×10^{-14}
50	5.47×10^{-14}
100	49×10^{-14}

EXAMPLE 9-1

Calculate the hydronium and hydroxide ion concentrations of pure water at 25°C and 100°C.

Because OH^- and H_3O^+ are formed only from the dissociation of water, their concentrations must be equal:

$$[H_3O^+] = [OH^-]$$

Substitution into Equation 9-11 gives

$$[H_3O^+]^2 = [OH^-]^2 = K_w$$

$$[H_3O^+] = [OH^-] = \sqrt{K_w}$$

(continued)

At 25°C,

$$[H_3O^+] = [OH^-] = \sqrt{1.00 \times 10^{-14}} = 1.00 \times 10^{-7}\ M$$

At 100°C, from Table 9-3,

$$[H_3O^+] = [OH^-] = \sqrt{49 \times 10^{-14}} = 7.0 \times 10^{-7}\ M$$

EXAMPLE 9-2

Calculate the hydronium and hydroxide ion concentrations and the pH and pOH of 0.200 M aqueous NaOH at 25°C.

Sodium hydroxide is a strong electrolyte, and its contribution to the hydroxide ion concentration in this solution is 0.200 mol/L. As in Example 9-1, hydroxide ions and hydronium ions are formed *in equal amounts* from dissociation of water. Therefore, we write

$$[OH^-] = 0.200 + [H_3O^+]$$

where $[H_3O^+]$ accounts for the hydroxide ions contributed by the solvent. The concentration of OH^- from the water is insignificant, however, when compared with 0.200, so we can write

$$[OH^-] \approx 0.200$$

$$pOH = -\log 0.200 = 0.699$$

Equation 9-11 is then used to calculate the hydronium ion concentration:

$$[H_3O^+] = \frac{K_w}{[OH^-]} = \frac{1.00 \times 10^{-14}}{0.200} = 5.00 \times 10^{-14}\ M$$

$$pH = -\log 0.500 \times 10^{-14} = 13.301$$

Note that the approximation

$$[OH^-] = 0.200 + 5.00 \times 10^{-14} \approx 0.200\ M$$

causes no significant error.

9B-5 Applying Solubility-Product Constants

▶ When we say that a sparingly soluble salt is completely dissociated, *we do not imply* that all of the salt dissolves. Instead, the very small amount that *does* go into solution dissociates completely.

Most, but not all, sparingly soluble salts are essentially completely dissociated in saturated aqueous solution. For example, when an excess of barium iodate is equilibrated with water, the dissociation process is adequately described by the equation

$$Ba(IO_3)_2(s) \rightleftharpoons Ba^{2+}(aq) + 2IO_3^-(aq)$$

Using Equation 9-7, we write

$$K = \frac{[Ba^{2+}][IO_3^-]^2}{[Ba(IO_3)_2(s)]}$$

The denominator represents the molar concentration of $Ba(IO_3)_2$ *in the solid,* which is a phase that is separate from but in contact with the saturated solution. The concentration of a compound in its solid state is, however, constant. In other words, the number of moles of $Ba(IO_3)_2$ divided by the *volume* of the solid $Ba(IO_3)_2$ is constant no matter how much excess solid is present. Therefore, the previous equation can be rewritten in the form

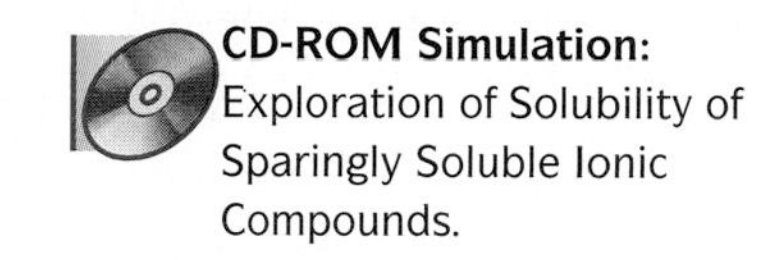

CD-ROM Simulation: Exploration of Solubility of Sparingly Soluble Ionic Compounds.

$$K[Ba(IO_3)_2(s)] = K_{sp} = [Ba^{2+}][IO_3^-]^2 \tag{9-13}$$

◀ For Equation 9-13 to apply, it is necessary only that *some solid be present. You should always keep in mind that in the absence of $Ba(IO_3)(s)$, Equation 9-13 is not applicable.*

where the new constant is called the **solubility-product constant** or the **solubility product.** It is important to appreciate that Equation 9-13 shows that the position of this equilibrium is independent of the *amount* of $Ba(IO_3)_2$ as long as some solid is present; that is, it does not matter whether the amount is a few milligrams or several grams.

A table of solubility-product constants for numerous inorganic salts is found in Appendix 2. The examples that follow demonstrate some typical uses of solubility-product expressions. Further applications are considered in later chapters.

The Solubility of a Precipitate in Pure Water

The solubility-product expression permits the ready calculation of the solubility of a sparingly soluble substance that ionizes completely in water.

EXAMPLE 9-3

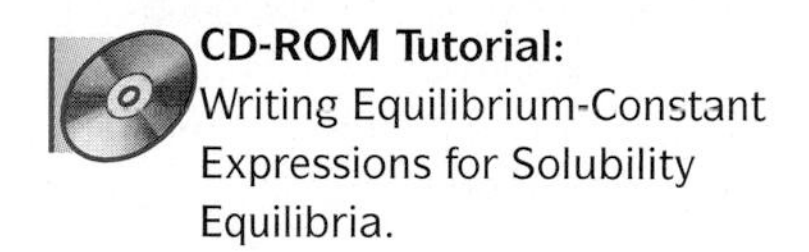

CD-ROM Tutorial: Writing Equilibrium-Constant Expressions for Solubility Equilibria.

How many grams of $Ba(IO_3)_2$ (487 g/mol) can be dissolved in 500 mL of water at 25°C?

The solubility-product constant for $Ba(IO_3)_2$ is 1.57×10^{-9} (see Appendix 2). The equilibrium between the solid and its ions in solution is described by the equation

$$Ba(IO_3)_2(s) \rightleftharpoons Ba^{2+} + 2IO_3^-$$

and so

$$K_{sp} = [Ba^{2+}][IO_3^-]^2 = 1.57 \times 10^{-9}$$

The equation describing the equilibrium reveals that 1 mol of Ba^{2+} is formed for each mole of $Ba(IO_3)_2$ that dissolves. Therefore,

$$\text{molar solubility of } Ba(IO_3)_2 = [Ba^{2+}]$$

(continued)

Since two moles of iodate are produced for each mole of barium ion, the iodate concentration is twice the barium ion concentration:

$$[IO_3^-] = 2[Ba^{2+}]$$

► Note that the molar solubility is equal to $[Ba^{2+}]$ or to $\frac{1}{2}[IO_3^-]$.

Substituting this last equation into the equilibrium-constant expression gives

$$[Ba^{2+}](2[Ba^{2+}])^2 = 4[Ba^{2+}]^3 = 1.57 \times 10^{-9}$$

$$[Ba^{2+}] = \left(\frac{1.57 \times 10^{-9}}{4}\right)^{1/3} = 7.32 \times 10^{-4}\ M$$

Since 1 mol Ba^{2+} is produced for every mole of $Ba(IO_3)_2$,

$$\text{solubility} = 7.32 \times 10^{-4}\ M$$

To compute the number of millimoles of $Ba(IO_3)_2$ dissolved in 500 mL of solution, we write

$$\text{no. mmol } Ba(IO_3)_2 = 7.32 \times 10^{-4} \frac{\text{mmol } Ba(IO_3)_2}{\cancel{\text{mL}}} \times 500\ \cancel{\text{mL}}$$

The mass of $Ba(IO_3)_2$ in 500 mL is given by

$$\text{mass } Ba(IO_3)_2 =$$

$$(7.32 \times 10^{-4} \times 500)\ \cancel{\text{mmol } Ba(IO_3)_2} \times 0.487 \frac{\text{g } Ba(IO_3)_2}{\cancel{\text{mmol } Ba(IO_3)_2}}$$

$$= 0.178\ g$$

CD-ROM Simulation: Exploration of the Relationship between Ionic Compound Solubility and the Presence of Ions Common to the Solubility Equilibrium.

The **common-ion effect** is responsible for the reduction in solubility of an ionic precipitate when a soluble compound combining one of the ions of the precipitate is added to the solution in equilibrium with the precipitate (see color plate 4).

CD-ROM Tutorial: Calculation of the Solubility of an Ionic Compound in the Presence of an Ion Common to the Solubility Equilibrium.

The Effect of a Common Ion on the Solubility of a Precipitate

The **common-ion effect** is a mass-action effect predicted from the Le Châtelier principle and is demonstrated by the following examples.

EXAMPLE 9-4

Calculate the molar solubility of $Ba(IO_3)_2$ in a solution that is 0.0200 M in $Ba(NO_3)_2$.

The solubility is no longer equal to $[Ba^{2+}]$ because $Ba(NO_3)_2$ is also a source of barium ions. We know, however, that the solubility is related to $[IO_3^-]$:

$$\text{molar solubility of } Ba(IO_3)_2 = \frac{1}{2}[IO_3^-]$$

There are two sources of barium ions: $Ba(NO_3)_2$ and $Ba(IO_3)_2$. The contribution from the former is 0.0200 M, and that from the latter is equal to the molar solubility, or $\frac{1}{2}[IO_3^-]$. Thus,

$$[Ba^{2+}] = 0.0200 + \frac{1}{2}[IO_3^-]$$

Substitution of these quantities into the solubility-product expression yields

$$\left(0.0200 + \frac{1}{2}[IO_3^-]\right)[IO_3^-]^2 = 1.57 \times 10^{-9}$$

Since the exact solution for $[IO_3^-]$ requires solving a cubic equation, we seek an approximation that simplifies the algebra. The small numerical value of K_{sp} suggests that the solubility of $Ba(IO_3)_2$ is not large, and this is confirmed by the result obtained in Example 9-3. Moreover, barium ion from $Ba(NO_3)_2$ will further repress the limited solubility of $Ba(IO_3)_2$. Thus, it is reasonable to seek a provisional answer to the problem by assuming that 0.0200 is large with respect to $\frac{1}{2}[IO_3^-]$. That is, $\frac{1}{2}[IO_3^-] \ll 0.0200$, and

$$[Ba^{2+}] = 0.0200 + \frac{1}{2}[IO_3^-] \approx 0.0200\ M$$

The original equation then simplifies to

$$0.0200\,[IO_3^-]^2 = 1.57 \times 10^{-9}$$

$$[IO_3^-] = \sqrt{1.57 \times 10^{-9}/0.0200} = \sqrt{7.85 \times 10^{-8}} = 2.80 \times 10^{-4}\ M$$

The assumption that $(0.0200 + \frac{1}{2} \times 2.80 \times 10^{-4}) \approx 0.0200$ does not appear to cause serious error because the second term, representing the amount of Ba^{2+} arising from the dissociation of $Ba(IO_3)_2$, is only about 0.7% of 0.0200. Ordinarily, we consider an assumption of this type to be satisfactory if the discrepancy is less than 10%.[1] Finally, then,

$$\text{solubility of } Ba(IO_3)_2 = \frac{1}{2}[IO_3^-] = \frac{1}{2} \times 2.80 \times 10^{-4} = 1.40 \times 10^{-4}\ M$$

If we compare this result with the solubility of barium iodate in pure water (Example 9-3), we see that the presence of a small concentration of the common ion has lowered the molar solubility of $Ba(IO_3)_2$ by a factor of about 5.

---

[1]Ten percent error is a somewhat arbitrary cutoff, but since we do not consider activity coefficients in our calculations, which often create errors of at least 10%, our choice is reasonable. Many general chemistry and analytical chemistry texts suggest that 5% error is appropriate, but such decisions should be based on the goal of the calculation. If you require an exact answer, the method of successive approximations presented in Feature 9-4 may be used; a spreadsheet solution may be appropriate for complex examples.

EXAMPLE 9-5

Calculate the solubility of $Ba(IO_3)_2$ in a solution prepared by mixing 200 mL of 0.0100 M $Ba(NO_3)_2$ with 100 mL of 0.100 M $NaIO_3$.

First establish whether either reactant is present in excess at equilibrium. The amounts taken are

$$\text{no. mmol } Ba^{2+} = 200 \cancel{mL} \times 0.0100 \text{ mmol}/\cancel{mL} = 2.00$$

$$\text{no. mmol } IO_3^- = 100 \cancel{mL} \times 0.100 \text{ mmol}/\cancel{mL} = 10.0$$

If the formation of $Ba(IO_3)_2$ is complete,

$$\text{no. mmol excess } NaIO_3 = 10.0 - 2 \times 2.00 = 6.00$$

Thus,

$$[IO_3^-] = \frac{6.00 \text{ mmol}}{200 \text{ mL} + 100 \text{ mL}} = \frac{6.00 \text{ mmol}}{300 \text{ mL}} = 0.0200 \text{ M}$$

► The uncertainty in $[IO_3^-]$ is 0.1 part in 6.0 or 1 part in 60. Thus, $0.0200 \times (1/60) = 0.0003$, and we round to 0.0200 M.

As in Example 9-3,

$$\text{molar solubility of } Ba(IO_3)_2 = [Ba^{2+}]$$

Here, however,

$$[IO_3^-] = 0.0200 + 2[Ba^{2+}]$$

where $2[Ba^{2+}]$ represents the iodate contributed by the sparingly soluble $Ba(IO_3)_2$. We can obtain a provisional answer after making the assumption that $[IO_3^-] \approx 0.0200$; thus

$$\text{solubility of } Ba(IO_3)_2 = [Ba^{2+}] = \frac{K_{sp}}{[IO_3^-]^2} = \frac{1.57 \times 10^{-9}}{(0.0200)^2}$$

$$= 3.93 \times 10^{-6} \text{ mol/L}$$

Since the provisional answer is nearly four orders of magnitude less than 0.02 M, our approximation is justified, and the solution does not need further refinement.

► A 0.02 M excess of Ba^{2+} decreases the solubility of $Ba(IO_3)_2$ by a factor of about 5; this same excess of IO_3^- lowers the solubility by a factor of about 200.

Note that the results from the last two examples demonstrate that an excess of iodate ions is more effective in decreasing the solubility of $Ba(IO_3)_2$ than is the same excess of barium ions.

9B-6 Applying Acid-Base Dissociation Constants

When a weak acid or a weak base is dissolved in water, partial dissociation occurs. Thus, for nitrous acid, we can write

$$HNO_2 + H_2O \rightleftharpoons H_3O^+ + NO_2^- \qquad K_a = \frac{[H_3O^+][NO_2^-]}{[HNO_2]}$$

where K_a is the **acid dissociation constant** for nitrous acid. In an analogous way, the **base dissociation constant** for ammonia is

$$NH_3 + H_2O \rightleftharpoons NH_4^+ + OH^- \qquad K_b = \frac{[NH_4^+][OH^-]}{[NH_3]}$$

Note that $[H_2O]$ does not appear in the denominator of either equation because the concentration of water is so large relative to the concentration of the weak acid or base that the dissociation does not alter $[H_2O]$ appreciably (see Feature 9-2). Just as in the derivation of the ion-product constant for water, $[H_2O]$ is incorporated into the equilibrium constants K_a and K_b. Dissociation constants for weak acids are found in Appendix 3.

Dissociation Constants for Conjugate Acid/Base Pairs

Consider the base dissociation-constant expression for ammonia and the acid dissociation-constant expression for its conjugate acid, ammonium ion:

$$NH_3 + H_2O \rightleftharpoons NH_4^+ + OH^- \qquad K_b = \frac{[NH_4^+][OH^-]}{[NH_3]}$$

$$NH_4^+ + H_2O \rightleftharpoons NH_3 + H_3O^+ \qquad K_a = \frac{[NH_3][H_3O^+]}{[NH_4^+]}$$

Multiplication of one equilibrium-constant expression by the other gives

$$K_aK_b = \frac{\cancel{[NH_3]}[H_3O^+]}{\cancel{[NH_4^+]}} \times \frac{\cancel{[NH_4^+]}[OH^-]}{\cancel{[NH_3]}} = [H_3O^+][OH^-]$$

but

$$K_w = [H_3O^+][OH^-]$$

and therefore

$$K_w = K_aK_b \tag{9-14}$$

This relationship is general for all conjugate acid/base pairs. Many compilations of equilibrium-constant data list only acid dissociation constants, since it is so easy to calculate basic dissociation constants by using Equation 9-14. For example, in Appendix 3, we find no data on the basic dissociation of ammonia (nor for any other bases). Instead, we find the acid dissociation constant for the conjugate acid, ammonium ion. That is,

$$NH_4^+ + H_2O \rightleftharpoons H_3O^+ + NH_3 \qquad K_a = \frac{[H_3O^+][NH_3]}{[NH_4^+]} = 5.70 \times 10^{-10}$$

and we can write

$$NH_3 + H_2O \rightleftharpoons NH_4^+ + OH^-$$

$$K_b = \frac{[NH_4^+][OH^-]}{[NH_3]} = \frac{K_w}{K_a} = \frac{1.00 \times 10^{-14}}{5.70 \times 10^{-10}} = 1.75 \times 10^{-5}$$

◀ To find a dissociation constant for a base at 25°C, we look up the dissociation constant for its conjugate acid and then divide 1.00×10^{-14} by the K_a.

FEATURE 9-3

Relative Strengths of Conjugate Acid/Base Pairs

Equation 9-14 confirms the observation in Figure 9-2 that as the acid of a conjugate acid/base pair becomes weaker, its conjugate base becomes stronger and vice versa. Thus, the conjugate base of an acid with a dissociation constant of 10^{-2} will have a basic dissociation constant of 10^{-12}, whereas an acid with a dissociation constant of 10^{-9} has a conjugate base with a dissociation constant of 10^{-5}.

EXAMPLE 9-6

What is K_b for the equilibrium

$$CN^- + H_2O \rightleftharpoons HCN + OH^-$$

Appendix 3 lists a K_a value of 6.2×10^{-10} for HCN. Thus,

$$K_b = \frac{K_w}{K_a} = \frac{[HCN][OH^-]}{[CN^-]}$$

$$K_b = \frac{1.00 \times 10^{-14}}{6.2 \times 10^{-10}} = 1.61 \times 10^{-5}$$

Hydronium Ion Concentration of Solutions of Weak Acids

When the weak acid HA is dissolved in water, two equilibria are established that yield hydronium ions:

$$HA + H_2O \rightleftharpoons H_3O^+ + A^- \qquad K_a = \frac{[H_3O^+][A^-]}{[HA]}$$

$$2H_2O \rightleftharpoons H_3O^+ + OH^- \qquad K_w = [H_3O^+][OH^-]$$

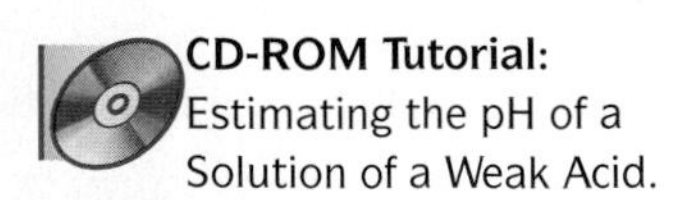
CD-ROM Tutorial: Estimating the pH of a Solution of a Weak Acid.

Ordinarily, the hydronium ions produced from the first reaction suppress the dissociation of water to such an extent that the contribution of hydronium ions from the second equilibrium is negligible. Under these circumstances, one H_3O^+ ion is formed for each A^- ion, and we write

$$[A^-] \approx [H_3O^+] \tag{9-15}$$

Furthermore, the sum of the molar concentrations of the weak acid and its conjugate base must equal the analytical concentration of the acid c_{HA} because the solution contains no other source of A^- ions. Thus,

$$c_{HA} = [A^-] + [HA] \tag{9-16}$$

Substituting $[H_3O^+]$ for $[A^-]$ (see Equation 9-15) into Equation 9-16 yields

$$c_{HA} = [H_3O^+] + [HA]$$

which rearranges to

$$[HA] = c_{HA} - [H_3O^+] \tag{9-17}$$

When $[A^-]$ and [HA] are replaced by their equivalent terms from Equations 9-15 and 9-17, the equilibrium-constant expression becomes

$$K_a = \frac{[H_3O^+]^2}{c_{HA} - [H_3O^+]} \tag{9-18}$$

which rearranges to

$$[H_3O^+]^2 + K_a[H_3O^+] - K_a c_{HA} = 0 \tag{9-19}$$

The positive solution to this quadratic equation is

$$[H_3O^+] = \frac{-K_a + \sqrt{K_a^2 + 4K_a c_{HA}}}{2} \tag{9-20}$$

As an alternative to using Equation 9-20, Equation 9-19 may be solved by successive approximations, as shown in Feature 9-4.

Equation 9-17 can frequently be simplified by making the additional assumption that dissociation does not appreciably decrease the molar concentration of HA. Thus, provided $[H_3O^+] \ll c_{HA}$, $c_{HA} - [H_3O^+] \approx c_{HA}$, and Equation 9-18 reduces to

$$K_a = \frac{[H_3O^+]^2}{c_{HA}} \tag{9-21}$$

and

$$[H_3O^+] = \sqrt{K_a c_{HA}} \tag{9-22}$$

The magnitude of the error introduced by the assumption that $[H_3O^+] \ll c_{HA}$ increases as the molar concentration of acid becomes smaller and its dissociation constant becomes larger. This statement is supported by the data in Table 9-4. Note that the error introduced by the assumption is about 0.5% when the ratio c_{HA}/K_a is 10^4. The error increases to about 1.6% when the ratio is 10^3, to about 5% when it is 10^2, and to about 17% when it is 10. Figure 9-3 illustrates the effect graphically. Notice that the hydronium ion concentration computed with the approximation becomes equal to or greater than the molar concentration of the acid when the ratio is less than or equal to 1, which is clearly a meaningless result.

In general, it is good practice to make the simplifying assumption and obtain a trial value for $[H_3O^+]$ that can be compared with c_{HA} in Equation 9-17. If the trial value alters [HA] by an amount smaller than the allowable error in the calculation, the solution may be considered satisfactory. Otherwise, the quadratic equation must be solved to obtain a better value for $[H_3O^+]$. Alternatively, the method of successive approximations (see Feature 9-4) may be used.

TABLE 9-4

Error Introduced by Assuming H_3O^+ Concentration Is Small Relative to c_{HA} in Equation 9-16

K_a	c_{HA}	$[H_3O^+]$ Using Assumption	$\frac{c_{HA}}{K_a}$	$[H_3O^+]$ Using More Exact Equation	Percent Error
1.00×10^{-2}	1.00×10^{-3}	3.16×10^{-3}	10^{-1}	0.92×10^{-3}	244
	1.00×10^{-2}	1.00×10^{-2}	10^{0}	0.62×10^{-2}	61
	1.00×10^{-1}	3.16×10^{-2}	10^{1}	2.70×10^{-2}	17
1.00×10^{-4}	1.00×10^{-4}	1.00×10^{-4}	10^{0}	0.62×10^{-4}	61
	1.00×10^{-3}	3.16×10^{-4}	10^{1}	2.70×10^{-4}	17
	1.00×10^{-2}	1.00×10^{-3}	10^{2}	0.95×10^{-3}	5.3
	1.00×10^{-1}	3.16×10^{-3}	10^{3}	3.11×10^{-3}	1.6
1.00×10^{-6}	1.00×10^{-5}	3.16×10^{-6}	10^{1}	2.70×10^{-6}	17
	1.00×10^{-4}	1.00×10^{-5}	10^{2}	0.95×10^{-5}	5.3
	1.00×10^{-3}	3.16×10^{-5}	10^{3}	3.11×10^{-5}	1.6
	1.00×10^{-2}	1.00×10^{-4}	10^{4}	9.95×10^{-5}	0.5
	1.00×10^{-1}	3.16×10^{-4}	10^{5}	3.16×10^{-4}	0.0

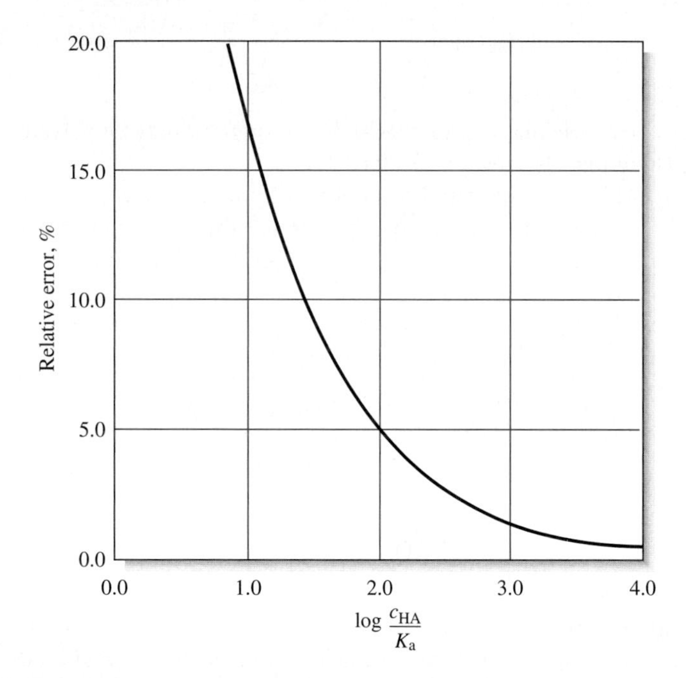

Figure 9-3 Relative error resulting from the assumption that $[H_3O^+] \ll c_{HA}$ in Equation 9-18.

EXAMPLE 9-7

Calculate the hydronium ion concentration in 0.120 M nitrous acid. The principal equilibrium is

$$HNO_2 + H_2O \rightleftharpoons H_3O^+ + NO_2^-$$

for which (see Appendix 2)

$$K_a = 7.1 \times 10^{-4} = \frac{[H_3O^+][NO_2^-]}{[HNO_2]}$$

Substitution into Equations 9-15 and 9-17 gives

$$[NO_2^-] = [H_3O^+]$$
$$[HNO_2] = 0.120 - [H_3O^+]$$

When these relationships are introduced into the expression for K_a, we obtain

$$K_a = \frac{[H_3O^+]^2}{0.120 - [H_3O^+]} = 7.1 \times 10^{-4}$$

If we now assume that $[H_3O^+] \ll 0.120$, we find

$$\frac{[H_3O^+]^2}{0.120} = 7.1 \times 10^{-4}$$

$$[H_3O^+] = \sqrt{0.120 \times 7.1 \times 10^{-4}} = 9.2 \times 10^{-3}\,M$$

We now examine the assumption that $0.120 - 0.0092 \approx 0.120$ and see that the error is about 8%. The relative error in $[H_3O^+]$ is actually smaller than this figure, however, as we can see by calculating $\log(c_{HA}/K_a) = 2.2$, which, from Figure 9-3, suggests an error of about 4%. If a more accurate figure is needed, solution of the quadratic equation yields 8.9×10^{-3} M for the hydronium ion concentration.

EXAMPLE 9-8

Calculate the hydronium ion concentration in a solution that is 2.0×10^{-4} M in aniline hydrochloride, $C_6H_5NH_3Cl$.

In aqueous solution, dissociation of the salt to Cl^- and $C_6H_5NH_3^+$ is complete. The weak acid $C_6H_5NH_3^+$ dissociates as follows:

$$C_6H_5NH_3^+ + H_2O \rightleftharpoons C_6H_5NH_2 + H_3O^+ \qquad K_a = \frac{[H_3O^+][C_6H_5NH_2]}{[C_6H_5NH_3^+]}$$

If we look in Appendix 3, we find that the K_a for $C_6H_5NH_3^+$ is 2.51×10^{-5}. Proceeding as in Example 9-7, we have

$$[H_3O^+] = [C_6H_5NH_2]$$
$$[C_6H_5NH_3^+] = 2.0 \times 10^{-4} - [H_3O^+]$$

Assume that $[H_3O^+] \ll 2.0 \times 10^{-4}$, and substitute the simplified value for $[C_6H_5NH_3^+]$ into the dissociation-constant expression to obtain (see Equation 9-21)

$$\frac{[H_3O^+]^2}{2.0 \times 10^{-4}} = 2.51 \times 10^{-5}$$

$$[H_3O^+] = \sqrt{5.02 \times 10^{-9}} = 7.09 \times 10^{-5}\,M$$

(continued)

Comparison of 7.09×10^{-5} with 2.0×10^{-4} suggests that a significant error has been introduced by the assumption that $[H_3O^+] \ll c_{C_6H_5NH_3^+}$. (Figure 9-3 indicates that this error is about 20%.) Thus, unless only an approximate value for $[H_3O^+]$ is needed, it is necessary to use the more nearly exact expression (Equation 9-19)

$$\frac{[H_3O^+]^2}{2.0 \times 10^{-4} - [H_3O^+]} = 2.51 \times 10^{-5}$$

which rearranges to

$$[H_3O^+]^2 + 2.51 \times 10^{-5}\,[H_3O^+] - 5.02 \times 10^{-9} = 0$$

$$[H_3O^+] = \frac{-2.51 \times 10^{-5} + \sqrt{(2.54 \times 10^{-5})^2 + 4 \times 5.02 \times 10^{-9}}}{2}$$

$$= 5.94 \times 10^{-5}\ \text{M}$$

The quadratic equation can also be solved by the iterative method shown in Feature 9-4.

FEATURE 9-4
The Method of Successive Approximations

For convenience, write the quadratic equation in Example 9-8 in the form

$$x^2 + 2.51 \times 10^{-5}x - 5.02 \times 10^{-9} = 0$$

where $x = [H_3O^+]$.

As a first step, rearrange the equation to the form

$$x = \sqrt{5.02 \times 10^{-9} - 2.51 \times 10^{-5}x}$$

We then assume that x on the right-hand side of the equation is zero and calculate a first value, x_1.

$$x_1 = \sqrt{5.02 \times 10^{-9} - 2.51 \times 10^{-5} \times 0} = 7.09 \times 10^{-5}$$

We then substitute this value into the original equation and derive a second value, x_2. That is,

$$x_2 = \sqrt{5.02 \times 10^{-9} - 2.51 \times 10^{-5} \times 7.09 \times 10^{-5}} = 5.69 \times 10^{-5}$$

Repeating this calculation gives

$$x_3 = \sqrt{5.02 \times 10^{-9} - 2.51 \times 10^{-5} \times 5.69 \times 10^{-5}} = 5.99 \times 10^{-5}$$

Continuing in the same way, we obtain

$$x_4 = 5.93 \times 10^{-5}$$
$$x_5 = 5.94 \times 10^{-5}$$
$$x_6 = 5.94 \times 10^{-5}$$

Note that after three iterations, x_3 is 5.99×10^{-5}, which is within about 0.8% of the final value of 5.94×10^{-5} M.

The method of successive approximations is particularly useful when cubic or higher-power equations need to be solved.

As shown in Chapter 5 of *Applications of Microsoft® Excel in Analytical Chemistry* iterative solutions can be readily performed using a spreadsheet.

Hydronium Ion Concentration of Solutions of Weak Bases

The techniques discussed in previous sections are readily adapted to the calculation of the hydroxide or hydronium ion concentration in solutions of weak bases.

Aqueous ammonia is basic by virtue of the reaction

$$NH_3 + H_2O \rightleftharpoons NH_4^+ + OH^-$$

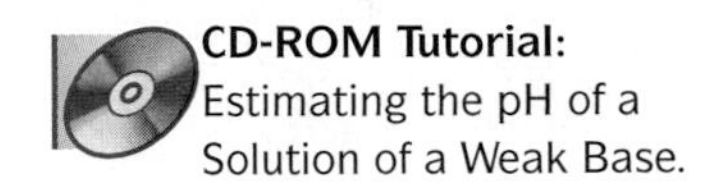

CD-ROM Tutorial:
Estimating the pH of a Solution of a Weak Base.

The predominant species in such solutions has been clearly demonstrated to be NH_3. Nevertheless, solutions of ammonia are still called ammonium hydroxide occasionally because at one time chemists thought that NH_4OH rather than NH_3 was the undissociated form of the base. Application of the mass law to the equilibrium as written yields

$$K_b = \frac{[NH_4^+][OH^-]}{[NH_3]}$$

EXAMPLE 9-9

Calculate the hydroxide ion concentration of a 0.0750 M NH_3 solution. The predominant equilibrium is

$$NH_3 + H_2O \rightleftharpoons NH_4^+ + OH^-$$

As shown on page 243.

$$K_b = \frac{[NH_4^+][OH^-]}{[NH_3]} = \frac{1.00 \times 10^{-14}}{5.70 \times 10^{-10}} = 1.75 \times 10^{-5}$$

The chemical equation shows that

$$[NH_4^+] = [OH^-]$$

Both NH_4^+ and NH_3 come from the 0.0750 M solution. Thus,

$$[NH_4^+] + [NH_3] = c_{NH_3} = 0.0750 \text{ M}$$

If we substitute $[OH^-]$ for $[NH_4^+]$ in the second of these equations and rearrange, we find that

$$[NH_3] = 0.0750 - [OH^-]$$

(continued)

Substituting these quantities into the dissociation-constant expression yields

$$\frac{[OH^-]^2}{7.50 \times 10^{-2} - [OH^-]} = 1.75 \times 10^{-5}$$

which is analogous to Equation 9-17 for weak acids. Provided that $[OH^-] \ll 7.50 \times 10^{-2}$, this equation simplifies to

$$[OH^-]^2 \approx 7.50 \times 10^{-2} \times 1.75 \times 10^{-5}$$

$$[OH^-] = 1.15 \times 10^{-3}\ M$$

Comparing the calculated value for $[OH^-]$ with 7.50×10^{-2}, we see that the error in $[OH^-]$ is less than 2%. If needed, a better value for $[OH^-]$ can be obtained by solving the quadratic equation.

EXAMPLE 9-10

Calculate the hydroxide ion concentration in a 0.0100 M sodium hypochlorite solution.

The equilibrium between OCl^- and water is

$$OCl^- + H_2O \rightleftharpoons HOCl + OH^-$$

for which

$$K_b = \frac{[HOCl][OH^-]}{[OCl^-]}$$

Appendix 3 reveals that the acid dissociation constant for HOCl is 3.0×10^{-8}. Therefore, we rearrange Equation 9-14 and write

$$K_b = \frac{K_w}{K_a} = \frac{1.00 \times 10^{-14}}{3.0 \times 10^{-8}} = 3.33 \times 10^{-7}$$

Proceeding as in Example 9-9, we have

$$[OH^-] = [HOCl]$$

$$[OCl^-] + [HOCl] = 0.0100$$

$$[OCl^-] = 0.0100 - [OH^-] \approx 0.0100$$

Here we have assumed that $[OH^-] \ll 0.0100$. Substitution into the equilibrium-constant expression gives

$$\frac{[OH^-]^2}{0.0100} = 3.33 \times 10^{-7}$$

$$[OH^-] = 5.8 \times 10^{-5}\ M$$

Note that the error resulting from the approximation is small.

Spreadsheet Summary In the first three exercises in Chapter 5 of *Applications of Microsoft® Excel in Analytical Chemistry,* we explore the solution to the types of equations found in chemical equilibria. A general purpose quadratic equation solver is developed and used for equilibrium problems. Then, Excel is used to find iterative solutions by successive approximations. Excel's Solver is next employed to solve quadratic, cubic, and quartic equations of the type encountered in equilibrium calculations.

9C BUFFER SOLUTIONS

By definition, a *buffer solution* resists changes in pH with dilution or with addition of acids or bases. Generally, buffer solutions are prepared from a conjugate acid/base pair, such as acetic acid/sodium acetate or ammonium chloride/ammonia. Chemists use buffers to maintain the pH of solutions at a relatively constant and predetermined level. You will find many references to buffers throughout this text.

A **buffer** is a mixture of a weak acid and its conjugate base or a weak base and its conjugate acid that resists changes in pH of a solution.

◀ Buffers are used in all types of chemistry whenever it is desirable to maintain the pH of a solution at a constant and predetermined level.

9C-1 Calculation of the pH of Buffer Solutions

A solution containing a weak acid, HA, and its conjugate base, A^-, may be acidic, neutral, or basic, depending on the position of two competitive equilibria:

$$HA + H_2O \rightleftharpoons H_3O^+ + A^- \qquad K_a = \frac{[H_3O^+][A^-]}{[HA]} \tag{9-23}$$

$$A^- + H_2O \rightleftharpoons OH^- + HA \qquad K_b = \frac{[OH^-][HA]}{[A^-]} = \frac{K_w}{K_a} \tag{9-24}$$

◀ Buffered aspirin contains buffers to help prevent stomach irritation from the acidity of the carboxylic acid group in aspirin.

If the first equilibrium lies farther to the right than the second, the solution is acidic. If the second equilibrium is more favorable, the solution is basic. These two equilibrium-constant expressions show that the relative concentrations of the hydronium and hydroxide ions depend not only on the magnitudes of K_a and K_b but also on the ratio between the concentrations of the acid and its conjugate base.

To find the pH of a solution containing both an acid, HA, and its conjugate base, NaA, we need to express the equilibrium concentrations of HA and NaA in terms of their analytical concentrations, c_{HA} and c_{NaA}. An examination of the two equilibria reveals that the first reaction decreases the concentration of HA by an amount equal to $[H_3O^+]$, whereas the second increases the HA concentration by an amount equal to $[OH^-]$. Thus, the species concentration of HA is related to its analytical concentration by the equation

$$[HA] = c_{HA} - [H_3O^+] + [OH^-] \tag{9-25}$$

Similarly, the first equilibrium will increase the concentration of A^- by an amount equal to $[H_3O^+]$, and the second will decrease this concentration by the amount $[OH^-]$. Thus, the equilibrium concentration is given by a second equation similar to Equation 9-25.

$$[A^-] = c_{NaA} + [H_3O^+] - [OH^-] \tag{9-26}$$

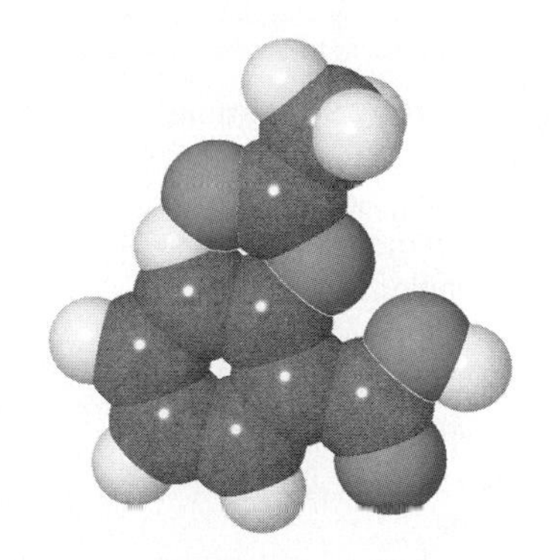

Molecular model and structure of aspirin. The analgesic action is thought to arise because aspirin interferes with the synthesis of prostaglandins, which are hormones involved in the transmission of pain signals.

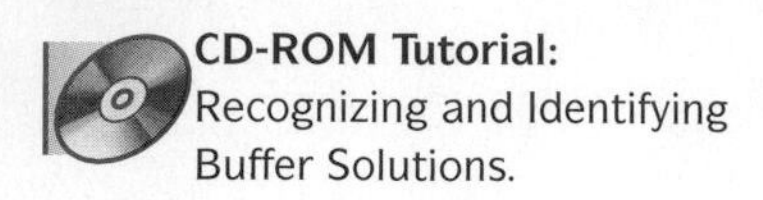
CD-ROM Tutorial: Recognizing and Identifying Buffer Solutions.

Because of the inverse relationship between $[H_3O^+]$ and $[OH^-]$, it is *always* possible to eliminate one or the other from Equations 9-25 and 9-26. Moreover, the *difference* in concentration between these two species is usually so small relative to the molar concentrations of acid and conjugate base that Equations 9-25 and 9-26 simplify to

$$[HA] \approx c_{HA} \tag{9-27}$$

$$[A^-] \approx c_{NaA} \tag{9-28}$$

Substitution of Equations 9-27 and 9-28 into the dissociation-constant expression and rearrangement yields

$$[H_3O^+] = K_a \frac{c_{HA}}{c_{NaA}} \tag{9-29}$$

The assumption leading to Equations 9-27 and 9-28 sometimes breaks down with acids or bases that have dissociation constants greater than about 10^{-3} or when the molar concentration of either the acid or its conjugate base (or both) is very small. In these circumstances, either $[OH^-]$ or $[H_3O^+]$ must be retained in Equations 9-25 and 9-26, depending on whether the solution is acidic or basic. In any case, Equations 9-27 and 9-28 should always be used initially. The provisional values for $[H_3O^+]$ and $[OH^-]$ can then be used to test the assumptions.

Within the limits imposed by the assumptions made in its derivation, Equation 9-29 says that the hydronium ion concentration of a solution containing a weak acid and its conjugate base is dependent only on the *ratio* of the molar concentrations of these two solutes. Furthermore, this ratio is *independent of dilution* because the concentration of each component changes proportionately when the volume changes.

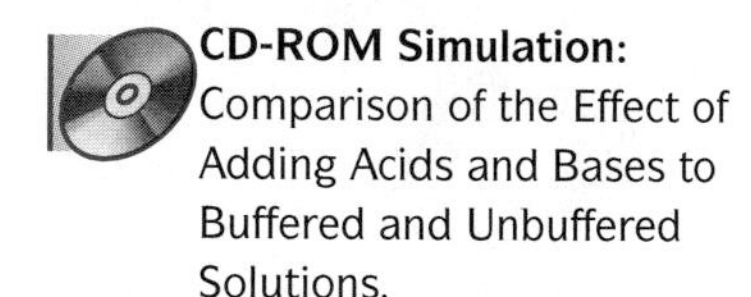
CD-ROM Simulation: Comparison of the Effect of Adding Acids and Bases to Buffered and Unbuffered Solutions.

FEATURE 9-5

The Henderson-Hasselbalch Equation

The Henderson-Hasselbalch equation, which is used to calculate the pH of buffer solutions, is frequently encountered in the biological literature and biochemical texts. It is obtained by expressing each term in Equation 9-29 in the form of its negative logarithm and inverting the concentration ratio to keep all signs positive:

$$-\log [H_3O^+] = -\log K_a + \log \frac{c_{NaA}}{c_{HA}}$$

Therefore,

$$pH = pK_a + \log \frac{c_{NaA}}{c_{HA}} \tag{9-30}$$

If the assumptions leading to Equation 9-28 are not valid, the values for [HA] and $[A^-]$ are given by Equations 9-24 and 9-25, respectively. If we take the negative logarithms of these expressions, we derive extended Henderson-Hasselbalch equations.

EXAMPLE 9-11

What is the pH of a solution that is 0.400 M in formic acid and 1.00 M in sodium formate?

The pH of this solution will be affected by the K_w of formic acid and the K_b of formate ion.

$$HCOOH + H_2O \rightleftharpoons H_3O^+ + HCOO^- \qquad K_a = 1.80 \times 10^{-4}$$

$$HCOO^- + H_2O \rightleftharpoons HCOOH + OH^- \qquad K_b = \frac{K_w}{K_a} = 5.56 \times 10^{-11}$$

Since the K_a for formic acid is orders of magnitude larger than the K_b for formate, the solution will be acidic and K_a will determine the H_3O^+ concentration. We can thus write

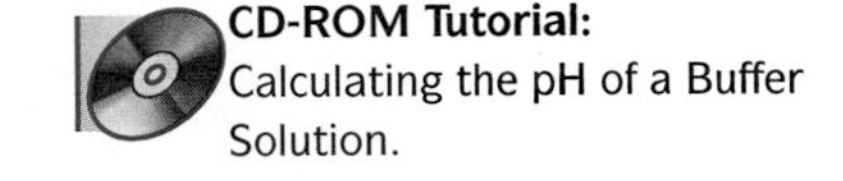
CD-ROM Tutorial:
Calculating the pH of a Buffer Solution.

$$K_a = \frac{[H_3O^+][HCOO^-]}{[HCOOH]} = 1.80 \times 10^{-4}$$

$$[HCOO^-] \approx c_{HCOO^-} = 1.00 \text{ M}$$

$$[HCOOH] \approx c_{HCOOH} = 0.400 \text{ M}$$

Substitution into Equation 9-29 gives, with rearrangement,

$$[H_3O^+] = 1.80 \times 10^{-4} \times \frac{0.400}{1.00} = 7.20 \times 10^{-5} \text{ M}$$

Note that the assumption that $[H_3O^+] \ll c_{HCOOH}$ and that $[H_3O^+] \ll c_{HCOO^-}$ is valid. Thus,

$$pH = -\log(7.20 \times 10^{-5}) = 4.14$$

As shown in Example 9-12, Equations 9-25 and 9-26 also apply to buffer systems consisting of a weak base and its conjugate acid. Furthermore, in most cases it is possible to simplify these equations so that Equation 9-29 can be used.

EXAMPLE 9-12

Calculate the pH of a solution that is 0.200 M in NH_3 and 0.300 M in NH_4Cl. In Appendix 3, we find that the acid dissociation constant K_a for NH_4^+ is 5.70×10^{-10}.

The equilibria we must consider are

$$NH_4^+ + H_2O \rightleftharpoons NH_3 + H_3O^+ \qquad K_a = 5.70 \times 10^{-10}$$

$$NH_3 + H_2O \rightleftharpoons NH_4^+ + OH^- \qquad K_b = \frac{K_w}{K_a} = \frac{1.00 \times 10^{-14}}{5.70 \times 10^{-10}} = 1.75 \times 10^{-5}$$

(continued)

Using the arguments that led to Equations 9-25 and 9-26, we obtain

$$[NH_4^+] = c_{NH_4Cl} + [OH^-] - [H_3O^+] \approx c_{NH_4Cl} + [OH^-]$$

$$[NH_3] = c_{NH_3} + [H_3O^+] - [OH^-] \approx c_{NH_3} - [OH^-]$$

Because K_b is several orders of magnitude larger than K_a, we have assumed that the solution is basic and that $[OH^-]$ is much larger than $[H_3O^+]$. Thus, we have neglected the concentration of H_3O^+ in these approximations.

Also assume that $[OH^-]$ is much smaller than c_{NH_4Cl} and c_{NH_3} so that

$$[NH_4^+] \approx c_{NH_4Cl} = 0.300 \text{ M}$$

$$[NH_3] \approx c_{NH_3} = 0.200 \text{ M}$$

Substituting into the acid dissociation constant for NH_4^+, we obtain a relationship similar to Equation 9-29. That is,

$$[H_3O^+] = \frac{K_a \times [NH_4^+]}{[NH_3]} = \frac{5.70 \times 10^{-10} \times c_{NH_4Cl}}{c_{NH_3}}$$

$$= \frac{5.70 \times 10^{-10} \times 0.300}{0.200} = 8.55 \times 10^{-10} \text{ M}$$

To check the validity of our approximations, we calculate $[OH^-]$. Thus,

$$[OH^-] = \frac{1.00 \times 10^{-14}}{8.55 \times 10^{-10}} = 1.17 \times 10^{-5} \text{ M}$$

which is certainly much smaller than c_{NH_4Cl} or c_{NH_3}. Thus, we may write

$$\text{pH} = -\log(8.55 \times 10^{-10}) = 9.07$$

9C-2 Properties of Buffer Solutions

In this section, we illustrate the resistance of buffers to changes of pH brought about by dilution or addition of strong acids or bases.

The Effect of Dilution

The pH of a buffer solution remains essentially independent of dilution until the concentrations of the species it contains are decreased to the point where the approximations used to develop Equations 9-27 and 9-28 become invalid. Figure 9-4 contrasts the behavior of buffered and unbuffered solutions with dilution. For each, the initial solute concentration is 1.00 M. The resistance of the buffered solution to changes in pH during dilution is clear.

The Effect of Added Acids and Bases

Example 9-13 illustrates a second property of buffer solutions, their resistance to pH change after addition of small amounts of strong acids or bases.

EXAMPLE 9-13

Calculate the pH change that takes place when a 100-mL portion of (a) 0.0500 M NaOH and (b) 0.0500 M HCl is added to 400 mL of the buffer solution that was described in Example 9-12.

(a) Addition of NaOH converts part of the NH_4^+ in the buffer to NH_3:

$$NH_4^+ + OH^- \rightleftharpoons NH_3 + H_2O$$

The analytical concentrations of NH_3 and NH_4Cl then become

$$c_{NH_3} = \frac{400 \times 0.200 + 100 \times 0.0500}{500} = \frac{85.0}{500} = 0.170 \text{ M}$$

$$c_{NH_4Cl} = \frac{400 \times 0.300 - 100 \times 0.0500}{500} = \frac{115}{500} = 0.230 \text{ M}$$

When substituted into the acid dissociation-constant expression for NH_4^+, these values yield

$$[H_3O^+] = 5.70 \times 10^{-10} \times \frac{0.230}{0.170} = 7.71 \times 10^{-10} \text{ M}$$

$$pH = -\log 7.71 \times 10^{-10} = 9.11$$

and the change in pH is

$$\Delta\, pH = 9.11 - 9.07 = 0.04$$

(b) Addition of HCl converts part of the NH_3 to NH_4^+; thus,

$$NH_3 + H_3O^+ \rightleftharpoons NH_4^+ + H_2O$$

$$c_{NH_3} = \frac{400 \times 0.200 - 100 \times 0.0500}{500} = \frac{75}{500} = 0.150 \text{ M}$$

$$c_{NH_4^+} = \frac{400 \times 0.300 + 100 \times 0.0500}{500} = \frac{125}{500} = 0.250 \text{ M}$$

$$[H_3O^+] = 5.70 \times 10^{-10} \times \frac{0.250}{0.150} = 9.50 \times 10^{-10}$$

$$pH = -\log 9.50 \times 10^{-10} = 9.02$$

$$\Delta\, pH = 9.02 - 9.07 = -0.05$$

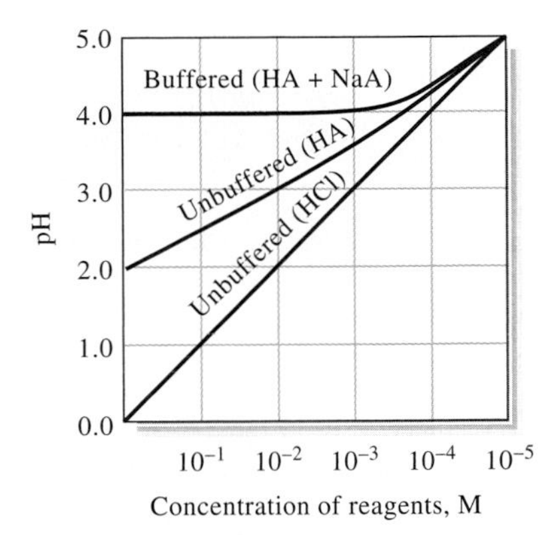

Figure 9-4 The effect of dilution of the pH of buffered and unbuffered solutions. The dissociation constant for HA is 1.00×10^{-4}. Initial solute concentrations are 1.00 M.

◀ Buffers do not maintain pH at an absolutely constant value, but changes in pH are relatively small when small amounts of acid or base are added.

It is interesting to contrast the behavior of an unbuffered solution with a pH of 9.07 to that of the buffer in Example 9-13. It can be readily shown that adding the same quantity of base to the unbuffered solution would increase the pH to 12.00—a pH change of 2.93 units. Adding the acid would decrease the pH by slightly more than 7 units.

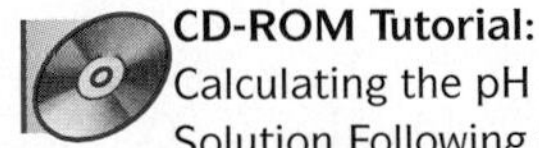

CD-ROM Tutorial: Calculating the pH of a Buffer Solution Following the Addition of a Strong Acid or Strong Base.

The Composition of Buffer Solutions as a Function of pH; Alpha Values

The composition of buffer solutions can be visualized by plotting the *relative* equilibrium concentrations of the two components of a conjugate acid/base as a function of the pH of the solution. These relative concentrations are called *alpha values.* For example, if we let c_T be the sum of the analytical concentrations of acetic acid and sodium acetate in a typical buffer solution, we may write

$$c_T = c_{HOAc} + c_{NaOAc} \tag{9-31}$$

We then define α_0, the fraction of the total concentration of acid that is undissociated, as

$$\alpha_0 = \frac{[HOAc]}{c_T} \tag{9-32}$$

and α_1, the fraction dissociated, as

$$\alpha_1 = \frac{[OAc^-]}{c_T} \tag{9-33}$$

Alpha values are unitless ratios whose sum must equal unity. That is,

$$\alpha_0 + \alpha_1 = 1$$

► Alpha values do not depend on c_T.

Alpha values depend *only* on $[H_3O^+]$ and K_a and are independent of c_T. To obtain expressions for α_0, we rearrange the dissociation-constant expression to

$$[OAc^-] = \frac{K_a[HOAc]}{[H_3O^+]}$$

The total concentration of acetic acid, c_T, is in the form of either HOAc or OAc^-. Thus,

$$c_T = [HOAc] + [OAc^-] \tag{9-34}$$

Substituting Equation 9-34 into Equation 9-35 gives

$$c_T = [HOAc] + \frac{K_a[HOAc]}{[H_3O^+]} = [HOAc]\left(\frac{[H_3O^+] + K_a}{[H_3O^+]}\right)$$

On rearrangement, we obtain

$$\frac{[HOAc]}{c_T} = \frac{[H_3O^+]}{[H_3O^+] + K_a}$$

But by definition, $[HOAc]/c_T = \alpha_0$ (see Equation 9-32), or

$$\alpha_0 = \frac{[HOAc]}{c_T} = \frac{[H_3O^+]}{[H_3O^+] + K_a} \tag{9-35}$$

To obtain an expression for α_1, we rearrange the dissociation-constant expression to

$$[HOAc] = \frac{[H_3O^+][OAc^-]}{K_a}$$

and substitute into Equation 9-35

$$c_T = \frac{[H_3O^+][OAc^-]}{K_a} + [OAc^-] = [OAc^-]\left(\frac{[H_3O^+] + K_a}{K_a}\right)$$

Rearranging this gives α_1 as defined by Equation 9-33

$$\alpha_1 = \frac{[OAc^-]}{c_T} = \frac{K_a}{[H_3O^+] + K_a} \tag{9-36}$$

Note that the denominator is the same in Equations 9-35 and 9-36.

Figure 9-5 illustrates how α_0 and α_1 vary as a function of pH. The data for these plots were calculated from Equations 9-35 and 9-36.

Notice that the two curves cross at the point where pH = pK_{HOAc} = 4.74. At this point, the concentrations of acetic acid and acetate ion are equal, and the fractions of the total analytical concentration of acid both equal one half.

Buffer Capacity

Figure 9-4 and Example 9-13 demonstrate that a solution containing a conjugate acid/base pair possesses remarkable resistance to changes in pH. The ability of a buffer to prevent a significant change in pH is directly related to the total concentration of the buffering species as well as to their concentration ratio. For example, the pH of a 400-mL portion of a buffer formed by diluting the solution described in Example 9-13 by 10 would change by about 0.4 to 0.5 unit when treated with 100 mL of 0.0500 M sodium hydroxide or 0.0500 M hydrochloric acid. We showed in Example 9-13 that the change is only about 0.04 to 0.05 unit for the more concentrated buffer.

The **buffer capacity,** β, of a solution is defined as the number of moles of a strong acid or a strong base that causes 1.00 L of the buffer to undergo a 1.00-unit change in pH. Mathematically, buffer capacity is given by

> The **buffer capacity** of a buffer is the number of moles of strong acid or strong base that 1 L of the buffer can absorb without changing pH by more than 1.

$$\beta = \frac{dc_b}{dpH} = -\frac{dc_a}{dpH}$$

where dc_b is the number of moles per liter of strong base and dc_a is the number of moles per liter of strong acid added to the buffer. Since adding strong acid to a buffer causes the pH to decrease, dc_a/dpH is negative, and *buffer capacity is always positive.*

The capacity of a buffer depends not only on the total concentration of the two buffer components but also on their concentration ratio. Buffer capacity decreases

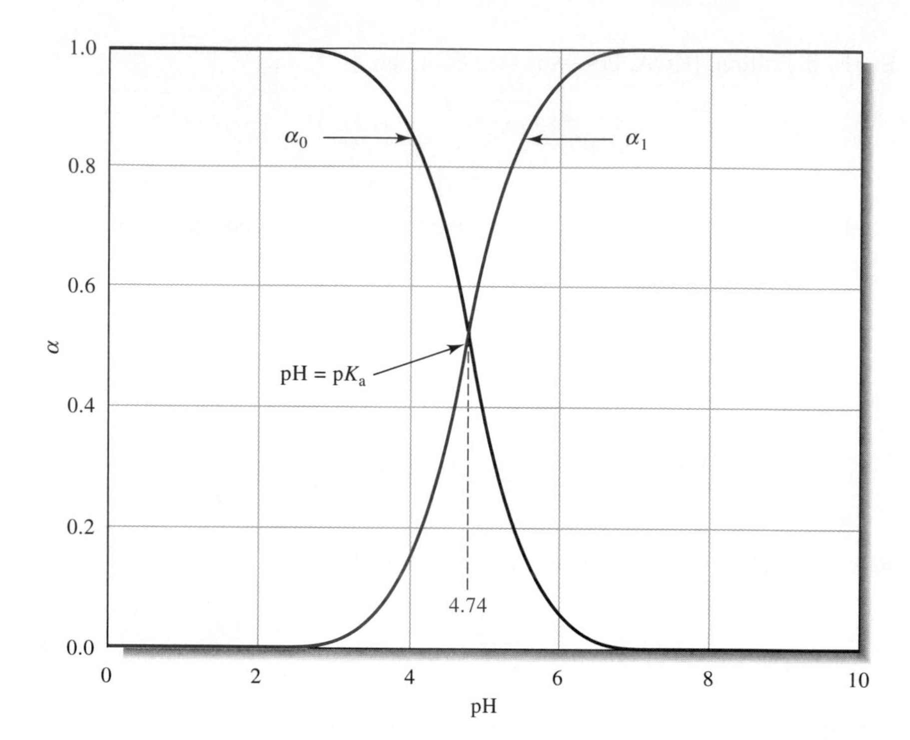

Figure 9-5 Variation in α with pH. Note that most of the transition between α_0 and α_1 occurs within ± 1 pH unit of the crossover point of the two curves. The crossover point where $\alpha_0 = \alpha_1 = 0.5$ occurs when pH = $pK_{HOAc} = 4.74$.

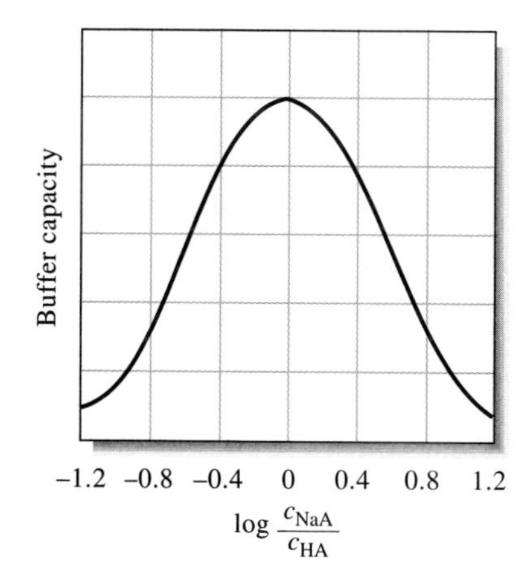

Figure 9-6 Buffer capacity as a function of the logarithm of the ratio c_{NaA}/c_{HA}. The maximum buffer capacity occurs when the concentration of acid and conjugate base are equal; that is, when $\alpha_0 = \alpha_1 = 0.5$.

fairly rapidly as the concentration ratio of acid to conjugate base becomes larger or smaller than unity (Figure 9-6). For this reason, the pK_a of the acid chosen for a given application should lie within ± 1 unit of the desired pH for the buffer to have a reasonable capacity.

Preparation of Buffers

In principle, a buffer solution of any desired pH can be prepared by combining calculated quantities of a suitable conjugate acid/base pair. In practice, however, the pH values of buffers prepared from theoretically generated recipes differ from the predicted values because of uncertainties in the numerical values of many dissociation constants and from the simplifications used in calculations. Because of these uncertainties, we prepare buffers by making up a solution of approximately the desired pH (see Example 9-14) and then adjusting it by adding strong acid or strong base until the required pH is indicated by a pH meter. Alternatively, empirically derived recipes for preparing buffer solutions of known pH are available in chemical handbooks and reference works.[2]

EXAMPLE 9-14

Describe how you might prepare approximately 500.0 mL or a pH 4.5 buffer solution from 1.0 M acetic acid (HOAc) and sodium acetate (NaOAc).

It is reasonable to assume there is little volume change if we add solid sodium acetate to the acetic acid solution. We then calculate the mass of NaOAc to add to 500.0 mL of 1.0 M HOAc. The H_3O^+ concentration should be

[2]See, for example, J. A. Dean, *Analytical Chemistry Handbook*, pp. 14–29 through 14–34. New York: McGraw-Hill, 1995.

$$[H_3O^+] = 10^{-4.5} = 3.16 \times 10^{-5}\,M$$

$$K_a = \frac{[H_3O^+][OAc^-]}{[HOAc]} = 1.75 \times 10^{-5}$$

$$\frac{[OAc^-]}{[HOAc]} = \frac{1.75 \times 10^{-5}}{[H_3O^+]} = \frac{1.75 \times 10^{-5}}{3.16 \times 10^{-5}} = 0.5534$$

The acetate concentration should be

$$[OAc^-] = 0.5534 \times 1.0\,M = 0.5534\,M$$

The mass of NaOAc needed is then

$$\text{mass NaOAc} = \frac{0.5534\ \cancel{\text{mol NaOAc}}}{\cancel{L}} \times 0.500\ \cancel{L} \times \frac{82.034\text{ g NaOAc}}{\cancel{\text{mol NaOAc}}} = 22.7\text{ g}$$

NaOAc

After dissolving this quantity of NaOAc in the acetic acid solution, we would check the pH with a pH meter and, if necessary, adjust the pH slightly by adding a small amount of acid or base.

Buffers are of tremendous importance in biological and biochemical studies in which a low but constant concentration of hydronium ions (10^{-6} to 10^{-10} M) must be maintained throughout experiments. Chemical and biological supply houses offer a variety of such buffers.

FEATURE 9-6

Acid Rain and the Buffer Capacity of Lakes

Acid rain has been the subject of considerable controversy over the past two decades. Acid rain forms when the gaseous oxides of nitrogen and sulfur dissolve in water droplets in the air. These gases form at high temperatures in power plants, automobiles, and other combustion sources. The combustion products pass into the atmosphere, where they react with water to form nitric acid and sulfuric acid as shown by the equations

$$4NO_2(g) + 2H_2O(l) + O_2(g) \rightarrow 4HNO_3(aq)$$

$$SO_3(g) + H_2O(l) \rightarrow H_2SO_4(aq)$$

Eventually, the droplets coalesce with other droplets to form acid rain. The profound effects of acid rain have been highly publicized. Stone buildings and monuments literally dissolve as acid rain flows over their surfaces. Forests are slowly being killed off in some locations. To illustrate the effects on aquatic life, consider the changes in pH that have occurred in the lakes of the Adirondack Mountains area of New York, illustrated in the bar graphs of Figure 9F-1.

The graphs show the distribution of pH in these lakes, which were studied first in the 1930s and then again in 1975.[3] The shift in pH of the lakes over a

(continued)

[3] R. F. Wright and E. T. Gjessing, *Ambio,* **1976,** *5,* 219.

40-year period is dramatic. The average pH of the lakes changed from 6.4 to about 5.1, which represents a 20-fold change in the hydronium ion concentration. Such changes in pH have a profound effect on aquatic life, as shown by a study of the fish population in lakes in the same area.[4] In the graph of Figure 9F-2, the number of lakes is plotted as a function of pH. The darker bars represent lakes containing fish, and lakes having no fish are lighter in color. There is a distinct correlation between pH changes in the lakes and diminished fish population.

Many factors contribute to pH changes in groundwater and lakes in a given geographical area. These include the prevailing wind patterns and weather, types of soils, water sources, nature of the terrain, characteristics of plant life, human activity, and geological characteristics. The susceptibility of natural water to acidification is largely determined by its buffer capacity, and the principal buffer of natural water is a mixture of bicarbonate ion and carbonic acid. Recall that the buffer capacity of a solution is proportional to the concentration of the buffering agent. So the higher the concentration of dissolved bicarbonate, the

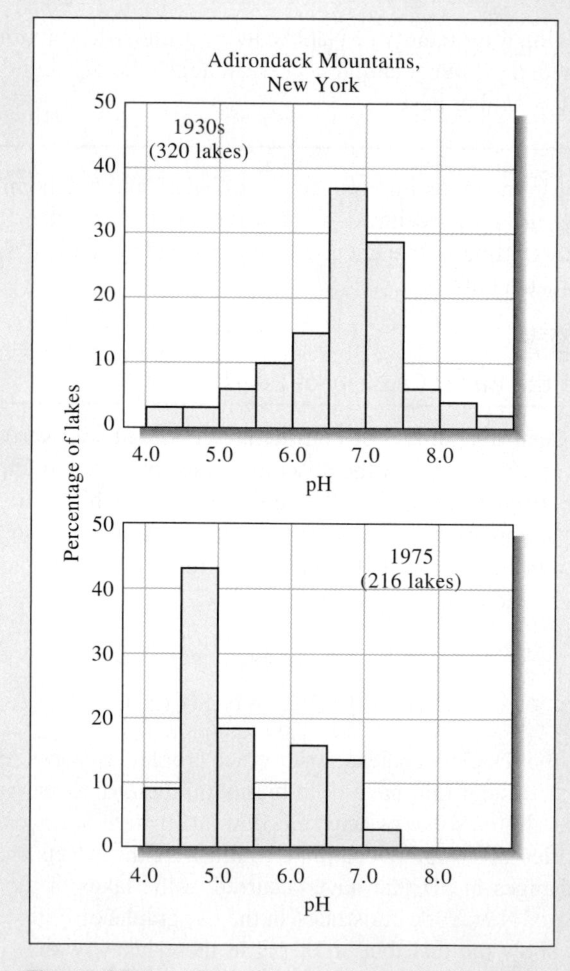

Figure 9F-1 Changes in pH of lakes between 1930 and 1975.

[4]C. L. Schofield, *Ambio,* **1976,** *5,* 228.

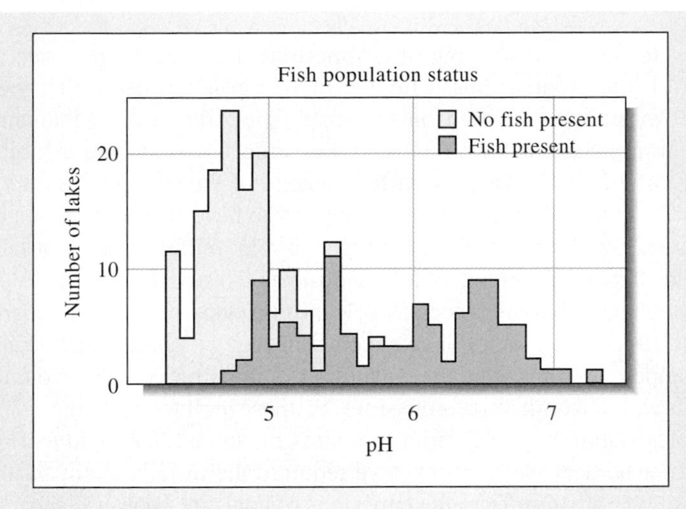

Figure 9F-2 Effect of pH of lakes on their fish population.

greater is the capacity of the water to neutralize acid from acid rain. The most important source of bicarbonate ion in natural water is limestone, or calcium carbonate, which reacts with hydronium ion as shown in the following equation:

$$CaCO_3(s) + H_3O^+(aq) \rightleftharpoons HCO_3^-(aq) + Ca^{2+}(aq) + H_2O(l)$$

Limestone-rich areas have lakes with relatively high concentrations of dissolved bicarbonate and thus low susceptibility to acidification. Granite, sandstone, shale, and other rock containing little or no calcium carbonate are associated with lakes having high susceptibility to acidification.

The map of the United States shown in Figure 9F-3 vividly illustrates the correlation between the absence of limestone-bearing rocks and the acidification of

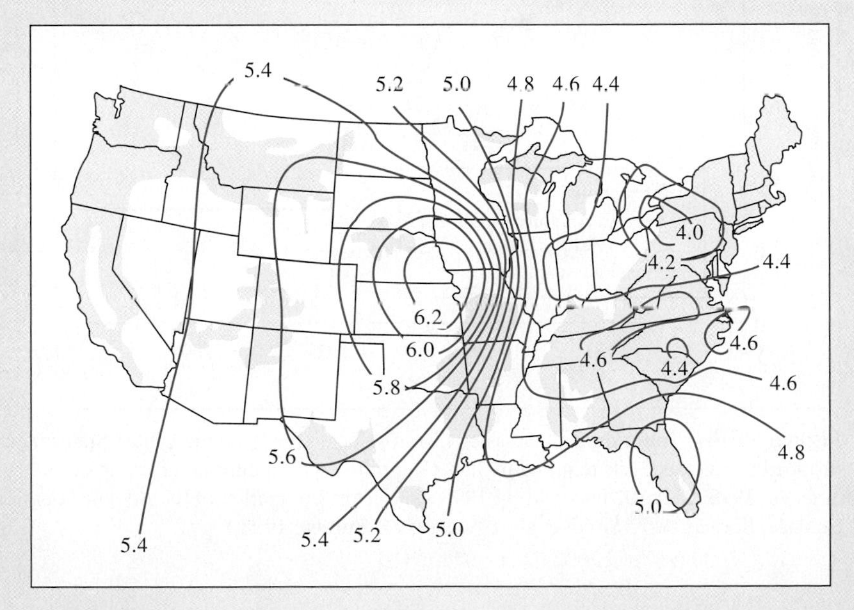

Figure 9F-3 Effect of presence of limestone on pH of lakes in the United States. Shaded areas contain little limestone.

(continued)

groundwater.[5] Areas containing little limestone are shaded; areas rich in limestone are white. Contour lines of equal pH for groundwater during the period 1978–1979 are superimposed on the map. The Adirondack Mountains area, located in northeastern New York, contains little limestone and exhibits pH in the range of 4.2 to 4.4. The low buffer capacity of the lakes in this region combined with the low pH of precipitation appears to have caused the decline of fish populations. Similar correlations among acid rain, buffer capacity of lakes, and wildlife decline occur throughout the industrialized world.

Although natural sources such as volcanos produce sulfur trioxide, and lightning discharges in the atmosphere generate nitrogen dioxide, large quantities of these compounds come from the burning of high-sulfur coal and from automobile emissions. To minimize emissions of these pollutants, some states have enacted legislation imposing strict standards on automobiles sold and operated within their borders. Some states have required the installation of scrubbers to remove oxides of sulfur from the emissions of coal-fired power plants. To minimize the effects of acid rain on lakes, powdered limestone is dumped on their surfaces to increase the buffer capacity of the water. Solutions to these problems require the expenditure of much time, energy, and money. We must sometimes make difficult economic decisions to preserve the quality of our environment and to reverse trends that have operated for many decades.

The 1990 Clean Air Act Amendments provided a dramatic new way to regulate sulfur dioxide. Congress issued specific emission limits to power plant operators, as shown in Figure 9F-4, but no specific methods were proposed for

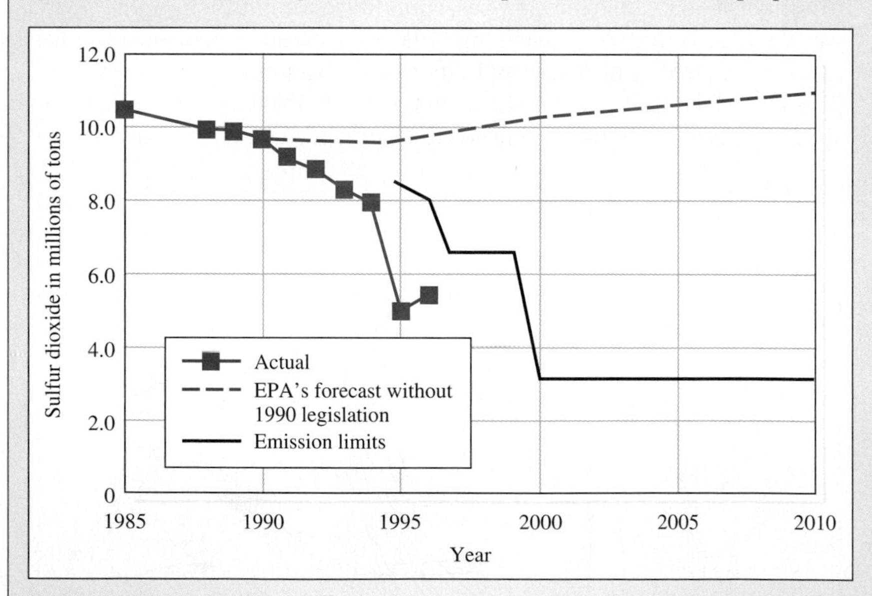

Figure 9F-4 Sulfur dioxide emissions from selected plants in the United States have dropped below the levels required by law. (Reprinted with permission from R. A. Kerr, *Science*, **1998,** *282,* 1024. Copyright 1998 American Association of the Advancement of Science. Source: A. E. Smith et al., 1998, and D. Burtaw, 1998.)

[5]J. Root, et al., cited in *The Effects of Air Pollution and Acid Rain on Fish, Wildlife, and Their Habitats—Introduction.* U.S. Fish and Wildlife Service, Biological Services Program, Eastern Energy and Land Use Team, M. A. Peterson, Ed., p. 63. U.S. Government Publication FWS/OBS-80/40.3.

meeting the standards. In addition, Congress established an emissions trading system by which power plants could buy, sell, and trade rights to pollute. Although detailed scientific and economic analysis of the effects of these congressional measures is still under way, it is clear from the results so far that the Clean Air Act Amendments have had a profound positive effect on the causes and effects of acid rain.[6]

Figure 9F-4 shows that sulfur dioxide emissions have decreased dramatically since 1990 and are well below levels forecasted by the EPA and within the limits set by Congress. The effects of these measures on acid rain are depicted in the map in Figure 9F-5, which shows the percent change in acidity in various regions of the eastern United States from 1983 to 1994. The significant improvement in acid rain shown on the map has been attributed tentatively to the flexibility of the regulatory statutes imposed in 1990. Another surprising result of the statutes is that their implementation has apparently been much less costly than originally projected. Initial estimates of the cost of meeting the emission standards were as high as $10 billion per year, but recent surveys indicate that actual costs may be as low as $1 billion per year.[7]

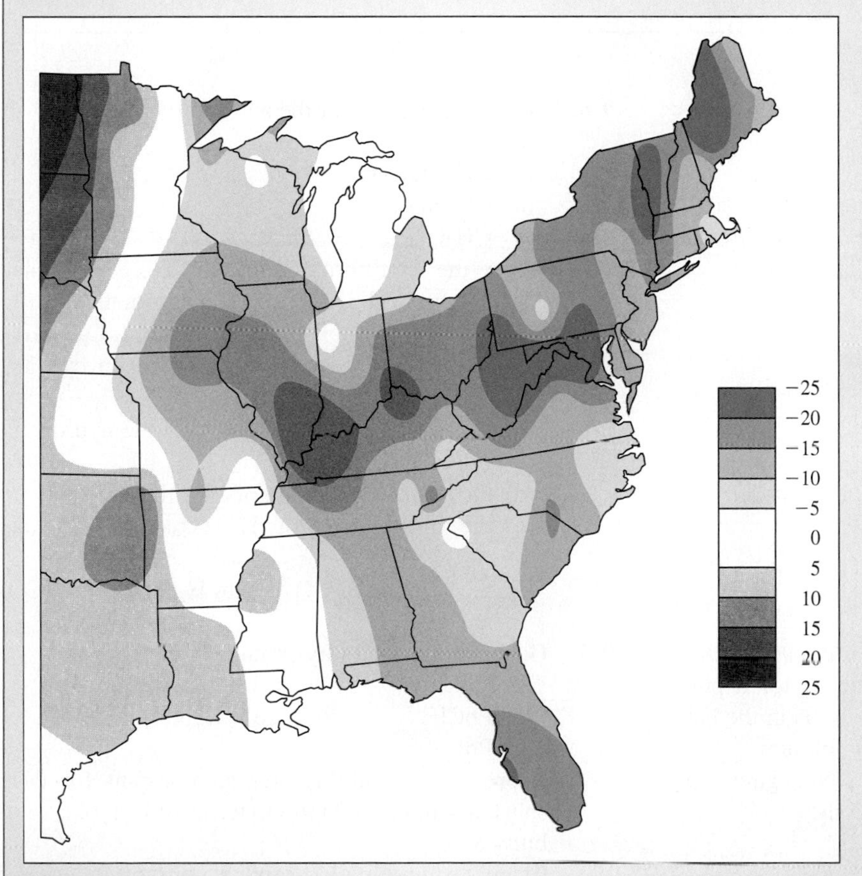

Figure 9F-5 Precipitation over much of the eastern United States has become less acidic, as shown by the percent change from 1983 to 1994. (Reprinted with permission from R. A. Kerr, *Science*, **1998,** *282,* 1024. Copyright 1998 American Association of the Advancement of Science. Source: James A. Lynch/Penn State University.)

[6]R. A. Kerr, *Science,* **1998,** *282,* 1024.

[7]C. C. Park, *Acid Rain.* New York: Methuen, 1987.

WEB WORKS

Use your Web browser to connect to **http://chemistry.brookscole.com/skoogfac/**. From the Chapter Resources menu, choose Web Works. Locate the Chapter 9 section, and click on the link to the Web site of the Swedish Environmental Protection Agency. Click on the Pollutants link on the left side of the home page, and then follow the links to the Acidification and Liming page. Read the article on this page, and answer the following questions. According to the article, where does most of Sweden's pollution come from? The article describes the critical acid load. What does this term mean? Roughly how much has soil pH changed in Sweden over the last few decades? Why has southern Sweden been affected more than northern Sweden? Characterize the effect of liming on the acidification of lakes in Sweden.

For an unusual connection with acid rain, browse to the *Scientific American* Web site and perform a search using the words "acid rain." One of the hits should be an article on the effects of acid rain following the impact of a comet with Earth. How would the effects of such an impact compare with the effects from pollution that we have observed over the past few decades?

QUESTIONS AND PROBLEMS

9-1. Briefly describe or define and give an example of
*(a) a weak electrolyte.
(b) a Brønsted-Lowry acid.
*(c) the conjugate acid of a Brønsted-Lowry base.
(d) neutralization, in terms of the Brønsted-Lowry concept.
*(e) an amphiprotic solvent.
(f) a zwitterion.
*(g) autoprotolysis.
(h) a strong acid.
*(i) the Le Châtelier principle.
(j) the common-ion effect.

9-2. Briefly describe or define and give an example of
*(a) an amphiprotic solute.
(b) a differentiating solvent.
*(c) a leveling solvent.
(d) a mass-action effect.

***9-3.** Briefly explain why there is no term in an equilibrium constant expression for water or for a pure solid, even though one (or both) appears in the balanced net ionic equation for the equilibrium.

9-4. Identify the acid on the left and its conjugate base on the right in the following equations:
*(a) $HOCl + H_2O \rightleftharpoons H_3O^+ + OCl^-$
(b) $HONH_2 + H_2O \rightleftharpoons HONH_3^+ + OH^-$
*(c) $NH_4^+ + H_2O \rightleftharpoons NH_3 + H_3O^+$
(d) $2HCO_3^- \rightleftharpoons H_2CO_3 + CO_3^{2-}$
*(e) $PO_4^{3-} + H_2PO_4^- \rightleftharpoons 2HPO_4^{2-}$

9-5. Identify the base on the left and its conjugate acid on the right in the equations for Problem 9-4.

9-6. Write expressions for the autoprotolysis of
*(a) H_2O.
(b) CH_3COOH.
*(c) CH_3NH_2.
(d) CH_3OH.

9-7. Write the equilibrium-constant expressions and obtain numerical values for each constant in
*(a) the basic dissociation of ethylamine, $C_2H_5NH_2$.
(b) the acidic dissociation of hydrogen cyanide, HCN.
*(c) the acidic dissociation of pyridine hydrochloride, C_5H_5NHCl.
(d) the basic dissociation of NaCN.
*(e) the dissociation of H_3AsO_4 to H_3O^+ and AsO_4^{3-}.
(f) the reaction of CO_3^{2-} with H_2O to give H_2CO_3 and OH^-.

9-8. Generate the solubility-product expression for
*(a) CuI.
*(b) PbClF.
*(c) PbI_2.
(d) BiI_3.
(e) $MgNH_4PO_4$.

9-9. Express the solubility-product constant for each substance in Problem 9-8 in terms of its molar solubility *S*.

9-10. Calculate the solubility-product constant for each of the following substances, given that the molar concentrations of their saturated solutions are as indicated:
(a) $CuSeO_3$ (1.42×10^{-4} M).
*(b) $Pb(IO_3)_2$ (4.3×10^{-5} M).

(c) SrF_2 (8.6×10^{-4} M).
*(d) $Th(OH)_4$ (3.3×10^{-4} M).

9-11. Calculate the solubility of the solutes in Problem 9-10 for solutions in which the cation concentration is 0.050 M.

9-12. Calculate the solubility of the solutes in Problem 9-10 for solutions in which the anion concentration is 0.050 M.

***9-13.** What CrO_4^{2-} concentration is required to
(a) initiate precipitation of Ag_2CrO_4 from a solution that is 3.41×10^{-2} M in Ag^+?
(b) lower the concentration of Ag^+ in a solution to 2.00×10^{-6} M?

9-14. What hydroxide concentration is required to
(a) initiate precipitation of Al^{3+} from a 2.50×10^{-2} M solution of $Al_2(SO_4)_3$?
(b) lower the Al^{3+} concentration in the foregoing solution to 2.00×10^{-7} M?

***9-15.** The solubility-product constant for $Ce(IO_3)_3$ is 3.2×10^{-10}. What is the Ce^{3+} concentration in a solution prepared by mixing 50.0 mL of 0.0250 M Ce^{3+} with 50.00 mL of
(a) water?
(b) 0.040 M IO_3^-?
(c) 0.250 M IO_3^-?
(d) 0.150 M IO_3^-?

9-16. The solubility-product constant for K_2PdCl_6 is 6.0×10^{-6} ($K_2PdCl_6 \rightleftharpoons 2K^+ + PdCl_6^{2-}$). What is the K^+ concentration of a solution prepared by mixing 50.0 mL of 0.200 M KCl with 50.0 mL of
(a) 0.0500 M $PdCl_6^{2-}$?
(b) 0.100 M $PdCl_6^{2-}$?
(c) 0.200 M $PdCl_6^{2-}$?

***9-17.** The solubility products for a series of iodides are

CuI	$K_{sp} = 1 \times 10^{-12}$
AgI	$K_{sp} = 8.3 \times 10^{-17}$
PbI_2	$K_{sp} = 7.1 \times 10^{-9}$
BiI_3	$K_{sp} = 8.1 \times 10^{-19}$

List these four compounds in order of decreasing molar solubility in
(a) water.
(b) 0.10 M NaI.
(c) a 0.010 M solution of the solute cation.

9-18. The solubility products for a series of hydroxides are

BiOOH	$K_{sp} = 4.0 \times 10^{-10} = [BiO^+][OH^-]$
$Be(OH)_2$	$K_{sp} = 7.0 \times 10^{-22}$
$Tm(OH)_3$	$K_{sp} = 3.0 \times 10^{-24}$
$Hf(OH)_4$	$K_{sp} = 4.0 \times 10^{-26}$

Which hydroxide has
(a) the lowest molar solubility in H_2O?
(b) the lowest molar solubility in a solution that is 0.10 M in NaOH?

9-19. Calculate the pH of water at 0°C and 100°C.

9-20. At 25°C, what are the molar H_3O^+ and OH^- concentrations in
*(a) 0.0300 M HOCl?
(b) 0.0600 M butanoic acid?
*(c) 0.100 M ethylamine?
(d) 0.200 M trimethylamine?
*(e) 0.200 M NaOCl?
(f) 0.0860 M CH_3CH_2COONa?
*(g) 0.250 M hydroxylamine hydrochloride?
(h) 0.0250 M ethanolamine hydrochloride?

9-21. At 25°C, what is the hydronium ion concentration in
*(a) 0.100 M chloroacetic acid?
*(b) 0.100 M sodium chloroacetate?
(c) 0.0100 M methylamine?
(d) 0.0100 M methylamine hydrochloride?
*(e) 1.00×10^{-3} M aniline hydrochloride?
(f) 0.200 M HIO_3?

9-22. What is a buffer solution, and what are its properties?

***9-23.** Define buffer capacity.

9-24. Which has the greater buffer capacity: (a) a mixture containing 0.100 mol of NH_3 and 0.200 mol of NH_4Cl or (b) a mixture containing 0.0500 mol of NH_3 and 0.100 mol of NH_4Cl?

***9-25.** Consider solutions prepared by
(a) dissolving 8.00 mmol of NaOAc in 200 mL of 0.100 M HOAc.
(b) adding 100 mL of 0.0500 M NaOH to 100 mL of 0.175 M HOAc.
(c) adding 40.0 mL of 0.1200 M HCl to 160.0 mL of 0.0420 M NaOAc.

In what respects do these solutions resemble one another? How do they differ?

9-26. Consult Appendix 3 and pick out a suitable acid/base pair to prepare a buffer with a pH of
*(a) 3.5. (b) 7.6. *(c) 9.3. (d) 5.1.

***9-27.** What weight of sodium formate must be added to 400.0 mL of 1.00 M formic acid to produce a buffer solution that has a pH of 3.50?

9-28. What weight of sodium glycolate should be added to 300.0 mL of 1.00 M glycolic acid to produce a buffer solution with a pH of 4.00?

***9-29.** What volume of 0.200 M HCl must be added to 250.0 mL of 0.300 M sodium mandelate to produce a buffer solution with a pH of 3.37?

9-30. What volume of 2.00 M NaOH must be added to 300.0 mL of 1.00 M glycolic acid to produce a buffer solution having a pH of 4.00?

9-31. Is the following statement true or false, or both? Define your answer with equations, examples, or graphs. "A buffer maintains the pH of a solution constant."

9-32. Challenge Problem: It can be shown[8] that the buffer capacity is

$$\beta = 2.303\left(\frac{K_w}{[H_3O^+]} + [H_3O^+] + \frac{c_T K_a [H_3O^+]}{(K_a + [H_3O^+])^2}\right)$$

where c_T is the molar analytical concentration of the buffer.

(a) Show that

$$\beta = 2.303\ ([OH^-] + [H_3O^+] + c_T\alpha_0\alpha_1)$$

(b) Use the equation in (a) to explain the shape of Figure 9-6.

(c) Differentiate the equation presented at the beginning of the problem and show that the buffer capacity is at a maximum when $\alpha_0 = \alpha_1 = 0.5$.

(d) Describe the conditions under which these relationships apply.

InfoTrac College Edition

For additional readings, go to InfoTrac College Edition, your online research library, at

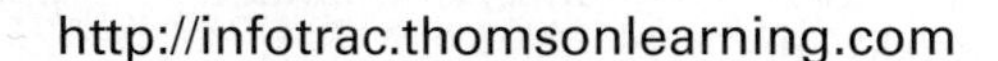

http://infotrac.thomsonlearning.com

[8] J. N. Butler, *Ionic Equilibrium: A Mathematical Approach,* p. 151. Menlo Park, CA: Addison-Wesley, 1964.

CHAPTER 10

Effect of Electrolytes on Chemical Equilibria

Hulton-Deutsch Collection/CORBIS

This calotype of a leaf was taken by the inventor of the process, William Henry Fox Talbot, in 1844. In its earliest form, the photosensitive paper was created by coating the paper with a sodium chloride solution, allowing the paper to dry, and then applying a second coat of silver nitrate, which produced a film of silver chloride. The leaf was then placed on the paper and exposed to light. The silver chloride in the paper was produced by the chemical equilibrium $Ag^+ + Cl^- \rightleftharpoons AgCl(s)$, which is driven by the activities of reactants and products.

In this chapter, we explore the detailed effects of electrolytes on chemical equilibria. The equilibrium constants for chemical reactions should be, strictly speaking, written in terms of the activities of the participating species. The ***activity*** *of a species is related to its concentration by a factor called the* ***activity coefficient.*** *In some cases, the activity of a reactant is essentially equal to its concentration, and we can write the equilibrium constant in terms of the concentrations of the participating species. In the case of ionic equilibria, however, activities and concentrations can be substantially different. Such equilibria are also affected by the concentrations of electrolytes in solution that may not participate directly in the reaction.*

The concentration-based equilibrium constant embodied in Equation 9-7 on page 234 provides only an approximation to real laboratory measurements. In this chapter, we show how the approximate form of the equilibrium constant often leads to significant error. We explore the difference between the activity of a solute and its concentration, calculate activity coefficients, and use them to modify the approximate expression to compute species concentrations that more closely match real laboratory systems at chemical equilibrium.

10A THE EFFECT OF ELECTROLYTES ON CHEMICAL EQUILIBRIA

Experimentally, we find that the position of most solution equilibria depends on the electrolyte concentration of the medium, even when the added electrolyte contains no ion in common with those involved in the equilibrium. For example, consider again the oxidation of iodide ion by arsenic acid that we described in Section 9B-1:

$$H_3AsO_4 + 3I^- + 2H^+ \rightleftharpoons H_3AsO_3 + I_3^- + H_2O$$

If an electrolyte, such as barium nitrate, potassium sulfate, or sodium perchlorate, is added to this solution, the color of the triiodide ion becomes less intense. This decrease in color intensity indicates that the concentration of I_3^- has decreased and that the equilibrium has been shifted to the left by the added electrolyte.

► Concentration-based equilibrium constants are often indicated by adding a prime mark, for example, K'_w, K'_{sp}, K'_a.

► As the electrolyte concentration becomes very small, concentration-based equilibrium constants approach their thermodynamic values: K_w, K_{sp}, K_a.

Figure 10-1 further illustrates the effect of electrolytes. Curve *A* is a plot of the product of the molar hydronium and hydroxide ion *concentrations* ($\times 10^{14}$) as a function of the concentration of sodium chloride. This *concentration*-based ion product is designated K'_w. At low sodium chloride concentrations, K'_w becomes independent of the electrolyte concentration and is equal to 1.00×10^{-14}, which is the *thermodynamic* ion-product constant for water, K_w. A relationship that approaches a constant value as some parameter (here, the electrolyte concentration) approaches zero is called a **limiting law;** the constant numerical value observed at this limit is referred to as a **limiting value.**

The vertical axis for curve *B* in Figure 10-1 is the product of the molar concentrations of barium and sulfate ions ($\times 10^{10}$) in saturated solutions of barium sulfate. This concentration-based solubility product is designated as K'_{sp}. At low electrolyte concentrations, K'_{sp} has a limiting value of 1.1×10^{-10}, which is the accepted thermodynamic value of K_{sp} for barium sulfate.

Curve *C* is a plot of K'_a ($\times 10^5$), the concentration quotient for the equilibrium involving the dissociation of acetic acid, as a function of electrolyte concentration. Here again, the ordinate function approaches a limiting value K_a, which is the thermodynamic acid dissociation constant for acetic acid.

The dashed lines in Figure 10-1 represent ideal behavior of the solutes. Note that departures from ideality can be significant. For example, the product of the molar concentrations of hydrogen and hydroxide ion increases from 1.0×10^{-14} in pure water to about 1.7×10^{-14} in a solution that is 0.1 M in sodium chloride, a 70% increase. The effect is even more pronounced with barium sulfate; here, K'_{sp} in 0.1 M sodium chloride is more than twice that of its limiting value.

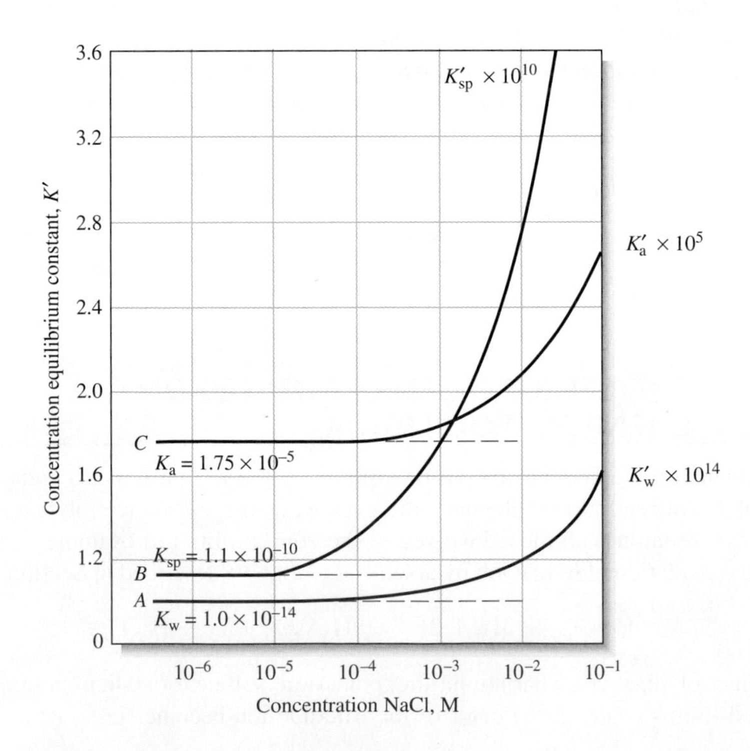

Figure 10-1 Effect of electrolyte concentration on concentration-based equilibrium constants.

The electrolyte effect shown in Figure 10-1 is not peculiar to sodium chloride. Indeed, we would see identical curves if potassium nitrate or sodium perchlorate were substituted for sodium chloride. In each case, the origin of the effect is the electrostatic attraction between the ions of the electrolyte and the ions of reacting species of opposite charge. Since the electrostatic forces associated with all singly charged ions are approximately the same, the three salts exhibit essentially identical effects on equilibria.

Next, we consider how we can take the electrolyte effect into account when we wish to make more accurate equilibrium calculations.

10A-1 The Effect of Ionic Charges on Equilibria

Extensive studies have revealed that the magnitude of the electrolyte effect is highly dependent on the charges of the participants in an equilibrium. When only neutral species are involved, the position of equilibrium is essentially independent of electrolyte concentration. With ionic participants, the magnitude of the electrolyte effect increases with charge. This generality is demonstrated by the three solubility curves in Figure 10-2. Note, for example, that in a 0.02 M solution of potassium nitrate, the solubility of barium sulfate, with its pair of doubly charged ions, is larger than it is in pure water by a factor of 2. This same change in electrolyte concentration increases the solubility of barium iodate by a factor of only 1.25 and that of silver chloride by 1.2. The enhanced effect due to doubly charged ions is also reflected in the greater slope of curve *B* in Figure 10-1.

10A-2 The Effect of Ionic Strength

Systematic studies have shown that the effect of added electrolyte on equilibria is *independent* of the chemical nature of the electrolyte but depends on a property of the solution called the **ionic strength.** This quantity is defined as

$$\text{ionic strength} = \mu = \frac{1}{2}([\text{A}]\,Z_{\text{A}}^2 + [\text{B}]\,Z_{\text{B}}^2 + [\text{C}]\,Z_{\text{C}}^2 + \cdots) \qquad (10\text{-}1)$$

where [A], [B], [C], ... represent the species molar concentrations of ions A, B, C, ... and Z_A, Z_B, Z_C, ... are their charges.

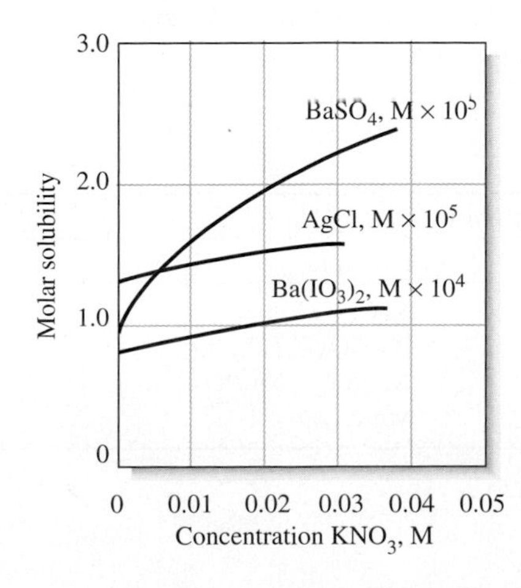

Figure 10-2 Effect of electrolyte concentration on the solubility of some salts.

EXAMPLE 10-1

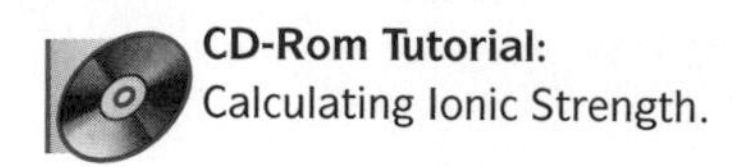

CD-Rom Tutorial: Calculating Ionic Strength.

Calculate the ionic strength of (a) a 0.1 M solution of KNO_3 and (b) a 0.1 M solution of Na_2SO_4.

(a) For the KNO_3 solution, $[K^+]$ and $[NO_3^-]$ are 0.1 M and

$$\mu = \frac{1}{2}(0.1\text{ M} \times 1^2 + 0.1\text{ M} \times 1^2) = 0.1\text{ M}$$

(b) For the Na_2SO_4 solution, $[Na^+] = 0.2$ M and $[SO_4^{2-}] = 0.1$ M. Therefore,

$$\mu = \frac{1}{2}(0.2\text{ M} \times 1^2 + 0.1\text{ M} \times 2^2) = 0.3\text{ M}$$

EXAMPLE 10-2

What is the ionic strength of a solution that is 0.05 M in KNO_3 and 0.1 M in Na_2SO_4?

$$\mu = \frac{1}{2}(0.05\text{ M} \times 1^2 + 0.05\text{ M} \times 1^2 + 0.2\text{ M} \times 1^2 + 0.1\text{ M} \times 2^2) = 0.35\text{ M}$$

These examples show that the ionic strength of a solution of a strong electrolyte consisting solely of singly charged ions is identical to its total molar salt concentration. The ionic strength is greater than the molar-concentration, however, if the solution contains ions with multiple charges (Table 10-1).

CD-Rom Simulation: Exploration of the Effects of Electrolyte Strength on Solution Ionic Strength.

For solutions with ionic strengths of 0.1 M or less, the electrolyte effect is *independent of the kind of ions* and *dependent only on the ionic strength.* Thus, the solubility of barium sulfate is the same in aqueous sodium iodide, potassium nitrate, or aluminum chloride provided the concentrations of these species are such that the ionic strengths are identical. Note that this independence with respect to electrolyte species disappears at high ionic strengths.

TABLE 10-1
Effect of Charge on Ionic Strength

Type Electrolyte	Example	Ionic Strength*
1:1	NaCl	c
1:2	$Ba(NO_3)_2$, NA_2SO_4	$3c$
1:3	$Al(NO_3)_3$, Na_3PO_4	$6c$
2:2	$MgSO_4$	$4c$

*c = molarity of the salt.

10A-3 The Salt Effect

The electrolyte effect (also called the **salt effect**), which we have just described, results from the electrostatic attractive and repulsive forces that exist between the ions of an electrolyte and the ions involved in an equilibrium. These forces cause each ion from the dissociated reactant to be surrounded by a sheath of solution that contains a slight excess of electrolyte ions of opposite charge. For example, when a barium sulfate precipitate is equilibrated with a sodium chloride solution, each dissolved barium ion is surrounded by an ionic atmosphere that (because of electrostatic attraction and repulsion) carries a small net negative charge on the average owing to repulsion of sodium ions and attraction of chloride ions. Similarly, each sulfate ion is surrounded by an ionic atmosphere that tends to be slightly positive. These charged layers make the barium ions seem somewhat less positive and the sulfate ions somewhat less negative than in the absence of electrolyte. The consequence of this shielding effect is a decrease in overall attraction between barium and sulfate ions and an increase in solubility, which becomes greater as the number of electrolyte ions in the solution becomes larger. That is, the *effective concentration* of barium ions and of sulfate ions becomes less as the ionic strength of the medium becomes greater.

10B ACTIVITY COEFFICIENTS

Chemists use a term called activity, a, to account for the effects of electrolytes on chemical equilibria. The activity, or effective concentration, of species X depends on the ionic strength of the medium and is defined by

$$a_X = [X]\,\gamma_X \qquad (10\text{-}2)$$

where a_X is the activity of the species X, [X] is its molar concentration, and γ_X is a dimensionless quantity called the **activity coefficient.** The activity coefficient and thus the activity of X vary with ionic strength such that substitution of a_X for [X] in any equilibrium-constant expression makes the equilibrium constant independent of the ionic strength. To illustrate, if X_mY_n is a precipitate, the thermodynamic solubility product expression is defined by the equation

◀ The activity of a species is a measure of its effective concentration as determined by colligative properties (such as increasing the boiling point or decreasing the freezing point of water), by electrical conductivity, and by the mass action effect.

$$K_{sp} = a_X^m \cdot a_Y^n \qquad (10\text{-}3)$$

Applying Equation 10-2 gives

$$K_{sp} = [X]^m\,[Y]^n \cdot \gamma_X^m \gamma_Y^n = K'_{sp} \cdot \gamma_X^m \gamma_Y^n \qquad (10\text{-}4)$$

Here K'_{sp} is the **concentration solubility product constant** and K_{sp} is the thermodynamic equilibrium constant.[1] The activity coefficients γ_X and γ_Y vary with ionic

[1] In the chapters that follow, we use the prime notation only when it is necessary to distinguish between thermodynamic and concentration equilibrium constants.

strength in such a way as to keep K_{sp} numerically constant and independent of ionic strength (in contrast to the concentration constant, K'_{sp}).

10B-1 Properties of Activity Coefficients

Activity coefficients have the following properties:

1. The activity coefficient of a species is a measure of the effectiveness with which that species influences an equilibrium in which it is a participant. In very dilute solutions, in which the ionic strength is minimal, this effectiveness becomes constant, and the activity coefficient is unity. Under such circumstances, the activity and the molar concentration are identical (as are thermodynamic and concentration equilibrium constants). As the ionic strength increases, however, an ion loses some of its effectiveness, and its activity coefficient decreases. We may summarize this behavior in terms of Equations 10-2 and 10-3. At moderate ionic strengths, $\gamma_X < 1$; as the solution approaches infinite dilution, however, $\gamma_X \rightarrow 1$ and thus $a_X \rightarrow [X]$ and $K'_{sp} \rightarrow K_{sp}$. At high ionic strengths ($\mu > 0.1$ M), activity coefficients often increase and may even become greater than unity. Because interpretation of the behavior of solutions in this region is difficult, we confine our discussion to regions of low or moderate ionic strength (that is, where $\mu \leq 0.1$ M). The variation of typical activity coefficients as a function of ionic strength is shown in Figure 10-3.
2. In solutions that are not too concentrated, the activity coefficient for a given species is independent of the nature of the electrolyte and dependent only on the ionic strength.
3. For a given ionic strength, the activity coefficient of an ion departs farther from unity as the charge carried by the species increases. This effect is shown in Figure 10-3.
4. The activity coefficient of an uncharged molecule is approximately unity, regardless of ionic strength.
5. At any given ionic strength, the activity coefficients of ions of the same charge are approximately equal. The small variations that do exist can be correlated with the effective diameter of the hydrated ions.

► As $\mu \rightarrow 0$, $\gamma_X \rightarrow 1$, $a_X \rightarrow [X]$, and $K'_{sp} \rightarrow K_{sp}$.

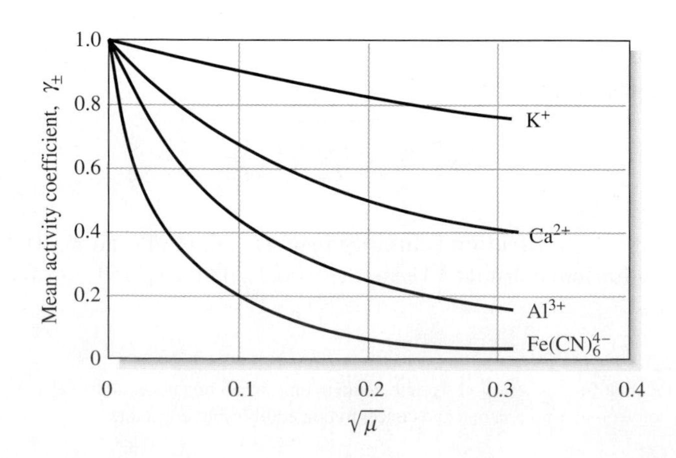

Figure 10-3 Effect of ionic strength on activity coefficients.

6. The activity coefficient of a given ion describes its effective behavior in all equilibria in which it participates. For example, at a given ionic strength, a single activity coefficient for cyanide ion describes the influence of that species on any of the following equilibria:

$$HCN + H_2O \rightleftharpoons H_3O^+ + CN^-$$

$$Ag^+ + CN^- \rightleftharpoons AgCN(s)$$

$$Ni^{2+} + 4CN^- \rightleftharpoons Ni(CN)_4^{2-}$$

Cornell University, courtesy of AIP Emilio Segrè Visual Archives.

Peter Debye (1884–1966) was born and educated in Europe but became Professor of Chemistry at Cornell University in 1940. He was noted for his work in several different areas of chemistry, including electrolyte solutions, X-ray diffraction, and the properties of polar molecules. He received the 1936 Nobel Prize in Chemistry.

10B-2 The Debye-Hückel Equation

In 1923, P. Debye and E. Hückel used the ionic atmosphere model, described in Section 10A-3, to derive an equation that permits the calculation of activity coefficients of ions from their charge and their average size.[2] This equation, which has become known as the **Debye-Hückel equation,** takes the form

$$-\log \gamma_X = \frac{0.51 Z_X^2 \sqrt{\mu}}{1 + 3.3\alpha_X \sqrt{\mu}} \qquad (10\text{-}5)$$

where

γ_X = activity coefficient of the species X
Z_X = charge on the species X
μ = ionic strength of the solution
α_X = effective diameter of the hydrated ion X in nanometers (10^{-9} m)

The constants 0.51 and 3.3 are applicable to aqueous solutions at 25°C; other values must be used at other temperatures.

Unfortunately, considerable uncertainty exists regarding the magnitude of α_X in Equation 10-5. Its value appears to be approximately 0.3 nm for most singly charged ions; for these species, then, the denominator of the Debye-Hückel equation simplifies to approximately $1 + \sqrt{\mu}$. For ions with higher charge, α_X may be as large as 1.0 nm. This increase in size with increase in charge makes good chemical sense. The larger the charge on an ion, the larger the number of polar water molecules that will be held in the solvation shell about the ion. The second term of the denominator is small with respect to the first when the ionic strength is less than 0.01 M. At these ionic strengths, uncertainties in α_A are of little significance in calculating activity coefficients.

◀ When μ is less than 0.01 M, $1 + \sqrt{\mu} \approx 1$, and Equation 10-5 becomes

$$-\log \gamma_X = 0.51\, Z_X^2 \sqrt{\mu}.$$

This equation is referred to as the Debye-Hückel Limiting Law (DHLL). Thus, in solutions of very low ionic strength ($\mu < 0.01$ M), the DHLL can be used to calculate approximate activity coefficients.

Kielland[3] has estimated values of α_X for numerous ions from a variety of experimental data. His best values for effective diameters are given in Table 10-2. Also presented are activity coefficients calculated from Equation 10-5 using these values for the size parameter.

Experimental determination of single-ion activity coefficients such as those shown in Table 10-2 is unfortunately impossible because all experimental methods give only a mean activity coefficient for the positively and negatively charged ions

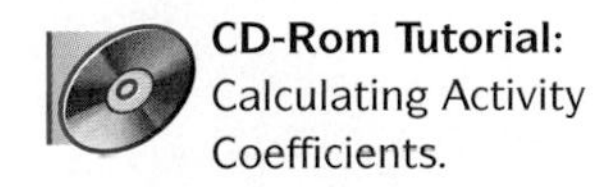

CD-Rom Tutorial: Calculating Activity Coefficients.

[2]P. Debye and E. Hückel, *Physik. Z.*, **1923,** *24,* 185.

[3]J. Kielland, *J. Amer. Chem. Soc.*, **1937,** *59,* 1675.

TABLE 10-2

Activity Coefficients for Ions at 25 °C

Ion	α_X, nm	Activity Coefficient at Indicated Ionic Strength				
		0.001	0.005	0.01	0.05	0.1
H_3O^+	0.9	0.967	0.934	0.913	0.85	0.83
Li^+, $C_6H_5COO^-$	0.6	0.966	0.930	0.907	0.83	0.80
Na^+, IO_3^-, HSO_3^-, HCO_3^-, $H_2PO_4^-$, $H_2AsO_4^-$, OAc^-	0.4–0.45	0.965	0.927	0.902	0.82	0.77
OH^-, F^-, SCN^-, HS^-, ClO_3^-, ClO_4^-, BrO_3^-, IO_3^-, MnO_4^-	0.35	0.965	0.926	0.900	0.81	0.76
K^+, Cl^-, Br^-, I^-, CN^-, NO_2^-, NO_3^-, $HCOO^-$	0.3	0.965	0.925	0.899	0.81	0.75
Rb^+, Cs^+, Tl^+, Ag^+, NH_4^+	0.25	0.965	0.925	0.897	0.80	0.75
Mg^{2+}, Be^{2+}	0.8	0.872	0.756	0.690	0.52	0.44
Ca^{2+}, Cu^{2+}, Zn^{2+}, Sn^{2+}, Mn^{2+}, Fe^{2+}, Ni^{2+}, Co^{2+}, Phthalate^{2-}	0.6	0.870	0.748	0.676	0.48	0.40
Sr^{2+}, Ba^{2+}, Cd^{2+}, Hg^{2+}, S^{2-}	0.5	0.869	0.743	0.668	0.46	0.38
Pb^{2+}, CO_3^{2-}, SO_3^{2-}, $C_2O_4^{2-}$	0.45	0.868	0.741	0.665	0.45	0.36
Hg_2^{2+}, SO_4^{2-}, $S_2O_3^{2-}$, Cr_4^{2-}, HPO_4^{2-}	0.40	0.867	0.738	0.661	0.44	0.35
Al^{3+}, Fe^{3+}, Cr^{3+}, La^{3+}, Ce^{3+}	0.9	0.737	0.540	0.443	0.24	0.18
PO_4^{3-}, $Fe(CN)_6^{3-}$	0.4	0.726	0.505	0.394	0.16	0.095
Th^{4+}, Zr^{4+}, Ce^{4+}, Sn^{4+}	1.1	0.587	0.348	0.252	0.10	0.063
$Fe(CN)_6^{4-}$	0.5	0.569	0.305	0.200	0.047	0.020

Source: Reprinted with permission from J. Kielland, *J. Am. Chem. Soc.*, **1937,** *59,* 1675. Copyright 1937 American Chemical Society.

in a solution. In other words, it is impossible to measure the properties of individual ions in the presence of counter-ions of opposite charge and solvent molecules. We should point out, however, that mean activity coefficients calculated from the data in Table 10-2 agree satisfactorily with the experimental values.

FEATURE 10-1

Mean Activity Coefficients

The mean activity coefficient of the electrolyte A_mB_n is defined as

$$\gamma_{\pm} = \text{mean activity coefficient} = (\gamma_A^m \gamma_B^n)^{1/(m+n)}$$

The mean activity coefficient can be measured in any of several ways, but it is impossible experimentally to resolve this term into the individual activity coefficients for γ_A and γ_B. For example, if

$$K_{sp} = [A]^m [B]^n \cdot \gamma_A^m \gamma_B^n = [A]^m [B]^n (\gamma_{\pm})^{m+n}$$

We can obtain K_{sp} by measuring the solubility of A_mB_n in a solution in which the electrolyte concentration approaches zero (that is, where both γ_A and $\gamma_B \rightarrow 1$). A second solubility measurement at some ionic strength μ_1 gives values for [A] and [B]. These data then permit the calculation of $\gamma_A^m \gamma_B^n = (\gamma_{\pm})^{m+n}$ for ionic strength μ_1.

It is important to understand that this procedure does not provide enough experimental data to permit the calculation of the *individual* quantities γ_A and γ_B and that there appears to be no additional experimental information that would permit evaluation of these quantities. This situation is general, and the *experimental* determination of an individual activity coefficient is impossible.

EXAMPLE 10-3

(a) Use Equation 10-5 to calculate the activity coefficient for Hg^{2+} in a solution that has an ionic strength of 0.085 M. Use 0.5 nm for the effective diameter of the ion. (b) Compare the value obtained in (a) with the activity coefficient obtained by linear interpolation of the data in Table 10-2 for coefficients of the ion at ionic strengths of 0.1 M and 0.05 M.

(a)
$$-\log \gamma_{Hg^{2+}} = \frac{(0.51)(2)^2\sqrt{0.085}}{1 + (3.3)(0.5)\sqrt{0.085}} \approx 0.4016$$

$$\gamma_{Hg^{2+}} = 10^{-0.4016} = 0.397 \approx 0.40$$

◀ Values for activity coefficients at ionic strengths not listed in Table 10-2 can be approximated by interpolation, as shown in Example 10-3(b).

(b) From Table 10-1

μ	$\gamma_{Hg^{2+}}$
0.1 M	0.38
0.05 M	0.46

Thus, when $\Delta\mu = (0.10\text{ M} - 0.05\text{ M}) = 0.05\text{ M}$, $\Delta\gamma_{Hg^{2+}} = 0.46 - 0.38 = 0.08$. At an ionic strength of 0.085 M,

$$\Delta\mu = (0.100\text{ M} - 0.085\text{ M}) = 0.015\text{ M}$$

and

$$\Delta\gamma_{Hg^{2+}} = \frac{0.015}{0.05} \times 0.08 = 0.024$$

Thus,

$$\gamma_{Hg^{2+}} = 0.38 + 0.024 = 0.404 \approx 0.40$$

Based on agreement between calculated and experimental values of mean ionic activity coefficients, we can infer that the Debye-Hückel relationship and the data in Table 10-2 give satisfactory activity coefficients for ionic strengths up to about 0.1 M. Beyond this value, the equation fails, and we must determine mean activity coefficients experimentally.

10B-3 Equilibrium Calculations Using Activity Coefficients

Equilibrium calculations with activities yield results that agree with experimental data more closely than those obtained with molar concentrations. Unless otherwise specified, equilibrium constants found in tables are generally based on activities and are thus thermodynamic. The examples that follow illustrate how activity coefficients from Table 10-2 are applied to such data.

EXAMPLE 10-4

Find the relative error introduced by neglecting activities in calculating the solubility of $Ba(IO_3)_2$ in a 0.033 M solution of $Mg(IO_3)_2$. The thermodynamic solubility product for $Ba(IO_3)_2$ is 1.57×10^{-9} (see Appendix 2).

At the outset, we write the solubility-product expression in terms of activities:

$$a_{Ba^{2+}} \cdot a^2_{IO_3^-} = K_{sp} = 1.57 \times 10^{-9}$$

where $a_{Ba^{2+}}$ and $a_{IO_3^-}$ are the activities of barium and iodate ions. Replacing activities in this equation with activity coefficients and concentrations from Equation 10-2 yields

$$[Ba^{2+}]\,\gamma_{Ba^{2+}} \cdot [IO_3^-]^2\,\gamma^2_{IO_3^-} = K_{sp}$$

where $\gamma_{Ba^{2+}}$ and $\gamma_{IO_3^-}$ are the activity coefficients for the two ions. Rearranging this expression gives

$$K'_{sp} = \frac{K_{sp}}{\gamma_{Ba^{2+}}\gamma^2_{IO_3^-}} = [Ba^{2+}]\,[IO_3^-]^2 \qquad (10\text{-}6)$$

where K'_{sp} is the *concentration-based solubility product.*

The ionic strength of the solution is obtained by substituting into Equation 10-1:

$$\mu = \frac{1}{2}([Mg^{2+}] \times 2^2 + [IO_3^-] \times 1^2)$$

$$= \frac{1}{2}(0.033\text{ M} \times 4 + 0.066\text{ M} \times 1) = 0.099\text{ M} \approx 0.1\text{ M}$$

In calculating μ, we have assumed that the Ba^{2+} and IO_3^- ions from the precipitate do not significantly affect the ionic strength of the solution. This simplification seems justified, considering the low solubility of barium iodate and the relatively high concentration of $Mg(IO_3)_2$. In situations in which it is not possible to make such an assumption, the concentrations of the two ions can be approximated by solubility calculation in which activities and concentrations are assumed to be identical (as in Examples 9-3, 9-4, and 9-5). These concentrations can then be introduced to give a better value for μ.

Turning now to Table 10-2, we find that at an ionic strength of 0.1 M,

$$\gamma_{Ba^{2+}} = 0.38 \qquad \gamma_{IO_3^-} = 0.77$$

If the calculated ionic strength did not match that of one of the columns in the table, $\gamma_{Ba^{2+}}$ and $\gamma_{IO_3^-}$ could be calculated from Equation 10-5.

Substituting into the thermodynamic solubility-product expression gives

$$K'_{sp} = \frac{1.57 \times 10^{-9}}{(0.38)(0.77)^2} = 6.97 \times 10^{-9}$$

$$[Ba^{2+}]\,[IO_3^-]^2 = 6.97 \times 10^{-9}$$

Proceeding now as in earlier solubility calculations,

$$\text{solubility} = [Ba^{2+}]$$

$$[IO_3^-] = 2 \times 0.033\ \text{M} + 2[Ba^{2+}] \approx 0.066\ \text{M}$$

$$[Ba^{2+}]\,(0.066)^2 = 6.97 \times 10^{-9}$$

$$[Ba^{2+}] = \text{solubility} = 1.60 \times 10^{-6}\ \text{M}$$

If we neglect activities, the solubility is,

$$[Ba^{2+}]\,(0.066)^2 = 1.57 \times 10^{-9}$$

$$[Ba^{2+}] = \text{solubility} = 3.60 \times 10^{-7}\ \text{M}$$

$$\text{relative error} = \frac{3.60 \times 10^{-7} - 1.60 \times 10^{-6}}{1.60 \times 10^{-6}} \times 100\% = -77\%$$

EXAMPLE 10-5

Use activities to calculate the hydronium ion concentration in a 0.120 M solution of HNO_2 that is also 0.050 M in NaCl. What is the relative percent error incurred by neglecting activity corrections?

The ionic strength of this solution is

$$\mu = \frac{1}{2}(0.0500\ \text{M} \times 1^2 + 0.0500\ \text{M} \times 1^2) = 0.0500\ \text{M}$$

In Table 10-2, at ionic strength 0.050 M, we find

$$\gamma_{H_3O^+} = 0.85 \qquad \gamma_{NO_2^-} = 0.81$$

Also, from rule 4 (page 272), we can write

$$\gamma_{HNO_2} = 1.0$$

These three values for γ permit the calculation of a concentration-based dissociation constant from the thermodynamic constant of 7.1×10^{-4} (see Appendix 3):

$$K_a' = \frac{[H_3O^+][NO_2^-]}{[HNO_2]} = \frac{K_a \cdot \gamma_{HNO_2}}{\gamma_{H_3O^+}\gamma_{NO_2^-}} = \frac{7.1 \times 10^{-4} \times 1.0}{0.85 \times 0.81} = 1.03 \times 10^{-3}$$

Proceeding as in Example 9-7, we write

$$[H_3O^+] = \sqrt{K_a \times c_a} = \sqrt{1.03 \times 10^{-3} \times 0.120} = 1.11 \times 10^{-2}\ \text{M}$$

(continued)

Note that assuming unit activity coefficients gives $[H_3O^+] = 9.2 \times 10^{-3}$ M.

$$\text{relative error} = \frac{9.2 \times 10^{-3} - 1.11 \times 10^{-2}}{1.11 \times 10^{-2}} \times 100\% = -17\%$$

In this example, we assumed that the contribution of the acid dissociation to the ionic strength was negligible. In addition, we used the approximate solution for calculating the hydronium ion concentration. See Problem 10-18 for a discussion of these approximations.

10B-4 Omitting Activity Coefficients in Equilibrium Calculations

Ordinarily, we neglect activity coefficients and simply use molar concentrations in applications of the equilibrium law. This approximation simplifies the calculations and greatly decreases the amount of data needed. For most purposes, the error introduced by the assumption of unity for the activity coefficient is not large enough to lead to false conclusions. It should be apparent from the preceding examples, however, that disregarding activity coefficients may introduce a significant numerical error in calculations of this kind. Note, for example, that neglecting activities in Example 10-4 resulted in an error of about −77%. Be alert to situations in which the substitution of concentration for activity is likely to lead to the largest error. Significant discrepancies occur when the ionic strength is large (0.01 M or larger) or when the ions involved have multiple charges (Table 10-2). With dilute solutions (ionic strength < 0.01 M) of nonelectrolytes or of singly charged ions, the use of concentrations in a mass-law calculation often provides reasonably accurate results. When, as is often the case, solutions have ionic strengths of more than 0.01 M, activity corrections must be made. Computer applications such as Excel greatly reduce the time and effort required to make these calculations.

It is also important to note that the decrease in solubility resulting from the presence of an ion common to the precipitate is in part counteracted by the larger electrolyte concentration associated with the presence of the salt containing the common ion.

Spreadsheet Summary In Chapter 5 of *Applications of Microsoft® Excel in Analytical Chemistry,* we explore the solubility of a salt in the presence of an electrolyte that changes the ionic strength of the solution. The solubility also changes the ionic strength. An iterative solution is first found, in which the solubility is determined by assuming that activity coefficients are unity. The ionic strength is then calculated and used to find the activity coefficients, which in turn are used to obtain a new solubility. The iteration process is continued until the results reach a steady value. Excel's Solver is then used to find the solubility directly from an equation containing all the variables.

WEB WORKS

It is often interesting and instructive to read the original papers describing important discoveries in your field of interest. Two Web sites, *Selected Classic Papers from the History of Chemistry* and *Classic Papers from the History of Chemistry (and Some Physics too)*, present many original papers or their translations for those who wish to explore pioneering work in chemistry. To learn about early work on the subject of this chapter, use your Web browser to connect to **http://chemistry.brookscole.com/skoogfac/.** From the Chapter Resources Menu, choose Web Works. Locate the Chapter 10 section. Click on the link to one of the Web sites just listed. Locate the link to the famous 1923 paper by Debye and Hückel on the theory of electrolytic solutions and click on it. Read the paper and compare the notation in the paper to the notation in this chapter. What symbol do the authors use for the activity coefficient? What important phenomena do the authors relate to their theory? Note that the mathematical details are missing from the translation of the paper.

QUESTIONS AND PROBLEMS

***10-1.** Make a distinction between
(a) activity and activity coefficient.
(b) thermodynamic and concentration equilibrium constants.

10-2. List general properties of activity coefficients.

***10-3.** Neglecting any effects caused by volume changes, would you expect the ionic strength to (1) increase, (2) decrease, or (3) remain essentially unchanged by the addition of NaOH to a dilute solution of
(a) magnesium chloride [$Mg(OH)_2$ *(s)* forms]?
(b) hydrochloric acid?
(c) acetic acid?

10-4. Neglecting any effects caused by volume changes, would you expect the ionic strength to (1) increase, (2) decrease, or (3) remain essentially unchanged by the addition of iron(III) chloride to
(a) HCl?
(b) NaOH?
(c) $AgNO_3$?

***10-5.** Explain why the initial slope for Ca^{2+} in Figure 10-3 is steeper than that for K^+?

10-6. What is the numerical value of the activity coefficient of aqueous ammonia (NH_3) at an ionic strength of 0.1?

10-7. Calculate the ionic strength of a solution that is
*(a) 0.040 M in $FeSO_4$.
(b) 0.20 M in $(NH_4)_2CrO_4$.
*(c) 0.10 M in $FeCl_3$ and 0.20 M in $FeCl_2$.
(d) 0.060 M in $La(NO_3)_3$ and 0.030 M in $Fe(NO_3)_2$.

10-8. Use Equation 10-5 to calculate the activity coefficient of
*(a) Fe^{3+} at $\mu = 0.075$.
(b) Pb^{2+} at $\mu = 0.012$.
*(c) Ce^{4+} at $\mu = 0.080$.
(d) Sn^{4+} at $\mu = 0.060$.

10-9. Calculate activity coefficients for the species in Problem 10-8 by linear interpolation of the data in Table 10-2.

10-10. For a solution in which μ is 5.0×10^{-2}, calculate K'_{sp} for
*(a) AgSCN.
(b) PbI_2.
*(c) $La(IO_3)_3$.
(d) $MgNH_4PO_4$.

***10-11.** Use activities to calculate the molar solubility of $Zn(OH)_2$ in
(a) 0.0100 M KCl.
(b) 0.0167 M K_2SO_4.
(c) the solution that results when you mix 20.0 mL of 0.250 M KOH with 80.0 mL of 0.0250 M $ZnCl_2$.
(d) the solution that results when you mix 20.0 mL of 0.100 M KOH with 80.0 mL of 0.0250 M $ZnCl_2$.

***10-12.** Calculate the solubilities of the following compounds in a 0.0333 M solution of $Mg(ClO_4)_2$ using (1) activities and (2) molar concentrations:
(a) AgSCN.
(b) PbI_2.
(c) $BaSO_4$.
(d) $Cd_2Fe(CN)_6$.

$$Cd_2Fe(CN)_6(s) \rightleftharpoons 2Cd^{2+} + Fe(CN)_6^{4-} \qquad K_{sp} = 3.2 \times 10^{-17}$$

***10-13.** Calculate the solubilities of the following compounds in a 0.0167 M solution of $Ba(NO_3)_2$ using (1) activities and (2) molar concentrations:
(a) $AgIO_3$.
(b) $Mg(OH)_2$.
(c) $BaSO_4$.
(d) $La(IO_3)_3$.

***10-14.** Calculate the % relative error in solubility by using concentrations instead of activities for the following compounds in 0.05000 M KNO_3 using the thermodynamic solubility products listed in Appendix 2.
*(a) CuCl ($\alpha_{Cu^+} = 0.3$ nm).
(b) $Fe(OH)_2$.
*(c) $Fe(OH)_3$.
(d) $La(IO_3)_3$.
*(e) Ag_3AsO_4 ($\alpha_{AsO_4} = 0.4$ nm).

10-15. Calculate the % relative error in hydronium ion concentration by using concentrations instead of activities in calculating the pH of solution of the following species using the thermodynamic constants found in Appendix 3.
*(a) 0.100 M HOAc and 0.200 M NaOAc.
(b) 0.0500 M NH_3 and 0.200 M NH_4Cl.
(c) 0.0100 M $ClCH_2COOH$ and 0.0600 M $ClCH_3COONa$.

10-16. (a) Repeat the computations of Problem 10–15 using a spreadsheet. Vary the concentration of $Ba(NO_3)_3$ from 0.0001 M to 1 M in a manner similar to that used in the spreadsheet exercise.
(b) Plot p*S* versus p*c*, where p*c* is the negative logarithm of the concentration of $Ba(NO_3)_3$.

10-17. Design and construct a spreadsheet to calculate activity coefficients in a format similar to Table 10-2. Enter values of α_X in cells A3, A4, A5, and so forth, and enter ionic charges in cells B3, B4, B5, and so forth. Enter in cells C2:G2 the same set of values for ionic strength listed in Table 10-2. Enter the formula for the activity coefficients in cells C3:G3. Be sure to use absolute cell references for ionic strength in your formulas for the activity coefficients. Finally, copy the formulas for the activity coefficients into the rows below row C by highlighting C3:G3 and dragging the fill handle downward. Compare the activity coefficients that you calculate to those in Table 10-2. Do you find any discrepancies? If so, explain how they arise.

10-18. Challenge Problem. In example 10-5, we neglected the contribution of nitrous acid to the ionic strength. We also used the simplified solution for the hydronium ion concentration,

$$[H_3O^+] = \sqrt{K_a c_a}$$

(a) Carry out an iterative solution to the problem in which you actually calculate the ionic strength, first without taking into account the dissociation of the acid. Then calculate corresponding activity coefficients for the ions using the Debye-Hückel equation, compute a new K_a, and find a new value for $[H_3O^+]$. Repeat the process, but use the concentrations of $[H_3O^+]$ and $[NO_2^-]$ along with the 0.05 M NaCl to calculate a new ionic strength; once again, find the activity coefficients, K_a, and a new value for $[H_3O^+]$. Iterate until you obtain two consecutive values of $[H_3O^+]$ that are equal to within 0.1%. How many iterations did you need? What is the relative error between your final value and the value obtained in Example 10-5 with no activity correction? What is the relative error between the first value that you calculated and the last one? You may want to use a spreadsheet to assist you in these calculations.
(b) Now perform the same calculation, except this time calculate the hydronium ion concentration using the quadratic equation or the method of successive approximations each time you compute a new ionic strength. How much improvement do you observe over the results that you obtained in (a)?
(c) When are activity corrections like those that you carried out in (a) necessary? What variables must be considered in deciding whether to make such corrections?
(d) When are corrections such as those in (b) necessary? What criteria do you use to decide whether these corrections should be made?
(e) Suppose that you are attempting to determine ion concentrations in a complex matrix such as blood serum or urine. Is it possible to make activity corrections in such a system? Explain your answer.

InfoTrac College Edition

For additional readings, go to InfoTrac College Edition, your online research library, at

http://infotrac.thomsonlearning.com

CHAPTER 11

Solving Equilibrium Problems for Complex Systems

Complex equilibria are extremely important in many areas of science. Such equilibria play important roles in the environment. Rivers and lakes, as shown in the photo, are subject to many sources of pollution that may make the water unsuitable for drinking, swimming, or fishing. One of the most common problems with lakes is the nutrient overload caused by the increased flow of plant nutrients, such as phosphates and nitrates, from sewage treatment plants, fertilizers, detergents, animal wastes, and soil erosion. These nutrients are involved in complex equilibria that cause rooted plants like water hyacinths and algae to undergo population explosions. When the plants die and fall to the bottom of the lake, the decomposing bacteria deplete the lake's lower layers of dissolved oxygen, which may lead food fish to die of oxygen starvation.

The calculations involved in complex equilibria are the major subject of this chapter. The systematic approach to solving multiple-equilibrium problems is described. The calculation of solubility when the equilibrium is influenced by pH and the formation of complexes is also discussed.

Charles D. Winters

Aqueous solutions encountered in the laboratory often contain several species that interact with one another and water to yield two or more equilibria that function simultaneously. For example, when water is saturated with sparingly soluble barium sulfate, three equilibria develop:

$$BaSO_4(s) \rightleftharpoons Ba^{2+} + SO_4^{2-} \qquad (11\text{-}1)$$

$$SO_4^{2-} + H_3O^+ \rightleftharpoons HSO_4^- + H_2O \qquad (11\text{-}2)$$

$$2H_2O \rightleftharpoons H_3O^+ + OH^- \qquad (11\text{-}3)$$

◄ The introduction of a new equilibrium system into a solution does not change the equilibrium constants for any existing equilibria.

If hydronium ions are added to this system, the second equilibrium is shifted to the right by the common-ion effect. The resulting decrease in sulfate concentration causes the first equilibrium to shift to the right as well, which increases the solubility of the barium sulfate.

The solubility of barium sulfate is also increased when acetate ions are added to an aqueous suspension of barium sulfate because acetate ions tend to form a soluble complex with barium ions, as shown by the reaction

$$Ba^{2+} + OAc^- \rightleftharpoons BaOAc^+ \qquad (11\text{-}4)$$

Again, the common-ion effect causes both this equilibrium and the solubility equilibrium to shift to the right; an increase in solubility results.

If we wish to calculate the solubility of barium sulfate in a system containing hydronium and acetate ions, we must take into account not only the solubility equilibrium but also the other three equilibria. We find, however, that using four equilibrium-constant expressions to calculate solubility is much more difficult and complex than the simple procedure illustrated in Examples 9-4, 9-5, and 9-6. To solve this type of problem, the systematic approach described in Section 11A is helpful. We then use this approach to illustrate the effect of pH and complex formation on the solubility of typical analytical precipitates. In later chapters, we use this same systematic method for solution of problems involving multiple equilibria of several types.

11A SOLVING MULTIPLE-EQUILIBRIUM PROBLEMS BY A SYSTEMATIC METHOD

Solution of a multiple-equilibrium problem requires us to develop as many independent equations as there are participants in the system being studied. For example, if we wish to compute the solubility of barium sulfate in a solution of acid, we need to be able to calculate the concentration of all the species present in the solution. There are five species: $[Ba^{2+}]$, $[SO_4^{2-}]$, $[HSO_4^-]$, $[H_3O^+]$, and $[OH^-]$. To calculate the solubility of barium sulfate in this solution rigorously, it is then necessary to develop five independent algebraic equations that can be solved simultaneously to give the five concentrations.

Three types of algebraic equations are used in solving multiple-equilibrium problems: (1) equilibrium-constant expressions, (2) *mass-balance* equations, and (3) a single *charge-balance* equation. We showed in Section 4B how equilibrium-constant expressions are written; we now turn our attention to the development of the other two types of equations.

11A-1 Mass-Balance Equations

▶ The term "mass-balance equation," although widely used, is somewhat misleading because such equations are really based on balancing *concentrations* rather than *masses.* Since all species are in the same volume of solvent, however, equating masses to concentrations does not create a problem.

Mass-balance equations relate the *equilibrium* concentrations of various species in a solution to one another and to the *analytical* concentrations of the various solutes. We can derive such equations from information about how the solution was prepared and from a knowledge of the kinds of equilibria that are present in the solution.

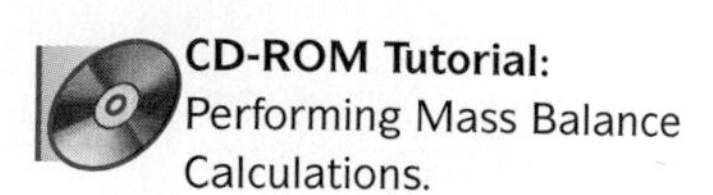
CD-ROM Tutorial:
Performing Mass Balance Calculations.

EXAMPLE 11-1

Write mass-balance expressions for a 0.0100 M solution of HCl that is in equilibrium with an excess of solid $BaSO_4$.

As shown by Equations 11-1, 11-2, and 11-3, three equilibria are present in this solution. That is,

$$BaSO_4(s) \rightleftharpoons Ba^{2+} + SO_4^{2-}$$

$$SO_4^{2-} + H_3O^+ \rightleftharpoons HSO_4^- + H_2O$$

$$2H_2O \rightleftharpoons H_3O^+ + OH^-$$

Because the only source for the two sulfate species is the dissolved $BaSO_4$, the barium ion concentration must equal the total concentration of sulfate-containing species, and a mass-balance equation can be written that expresses this equality. Thus,

$$[Ba^{2+}] = [SO_4^{2-}] + [HSO_4^-]$$

The hydronium ions in the solution can exist either as free H_3O^+ ions or combined with SO_4^{2-} to form HSO_4^-, according to the second reaction above. These hydronium ions have two sources: HCl and the dissociation of water. Thus,

$$[H_3O^+] + [HSO_4^-] = c_{HCl} + [OH^-] = 0.0100 + [OH^-]$$

Since the only source of hydroxide is the dissociation of water, $[OH^-]$ is equal to the hydronium ion concentration from the dissociation of water.

◀ For a slightly soluble salt with a 1:1 stoichiometry, the equilibrium concentration of the cation is equal to the equilibrium concentration of the anion. This equality is the mass-balance expression. For anions that can be protonated, the equilibrium concentration of the cation is equal to the sum of the concentrations of the various forms of the anion.

EXAMPLE 11-2

Write mass-balance expressions for the system formed when a 0.010 M NH_3 solution is saturated with AgBr.

Here, equations for the pertinent equilibria in the solution are

$$AgBr(s) \rightleftharpoons Ag^+ + Br^-$$

$$Ag^+ + NH_3 \rightarrow AgNH_3^+$$

$$Ag(NH_3)^+ + NH_3 \rightleftharpoons Ag(NH_3)_2^+$$

$$NH_3 + H_2O \rightleftharpoons NH_4^+ + OH^-$$

$$2H_2O \rightleftharpoons H_3O^+ + OH^-$$

Because AgBr is the only source of Br^-, Ag^+, $Ag(NH_3)^+$, and $Ag(NH_3)_2^+$ and because silver and bromide ions are present in a 1:1 ratio in that compound, it follows that one mass-balance equation is

$$[Ag^+] + [Ag(NH_3)^+] + [Ag(NH_3)_2^+] = [Br^-]$$

where the bracketed terms are molar species concentrations. Also, we know that the only source of ammonia-containing species is the 0.010 M NH_3. Therefore,

$$c_{NH_3} = [NH_3] + [NH_4^+] + [Ag(NH_3)^+] + 2\,[Ag(NH_3)_2^+] = 0.010$$

From the last two equilibria, we see that one hydroxide ion is formed for each NH_4^+ and each hydronium ion. Therefore,

$$[OH^-] = [NH_4^+] + [H_3O^+]$$

◀ For slightly soluble salts with stoichiometries other than 1:1, the mass-balance expression is obtained by multiplying the concentration of one of the ions by the stoichiometric ratio. For example, in a solution saturated with PbI_2, the iodide ion concentration is twice that of the Pb^{2+}. That is,

$$[I^-] = 2[Pb^{2+}]$$

11A-2 Charge-Balance Equation

Electrolyte solutions are electrically neutral even though they may contain millions of charged ions. Solutions are neutral because the *molar concentration of positive charge* in an electrolyte solution always equals *the molar concentration of negative charge.* That is, for any solution containing electrolytes, we may write

$$\text{no. mol/L positive charge} = \text{no. mol/L negative charge}$$

CD-ROM Tutorial: Performing Charge Balance Calculations.

This equation represents the charge-balance condition and is called the charge-balance equation. To be useful for equilibrium calculations, the equality must be expressed in terms of the molar concentrations of the species that carry a charge in the solution.

How much charge is contributed to a solution by 1 mol of Na^+? How about 1 mol of Mg^{2+} or 1 mol of PO_4^{3-}? The concentration of charge contributed to a solution by an ion is equal to the molar concentration of that ion multiplied by its charge. Thus, the molar concentration of positive charge in a solution due to the presence of sodium ions is the molar sodium ion concentration. That is,

► Always remember that a charge-balance equation is based on the equality in *molar charge concentrations* and that to obtain the charge concentration of an ion, you must multiply the molar concentration of the ion by its charge.

$$\frac{\text{mol positive charge}}{\text{L}} = \frac{1 \text{ mol positive charge}}{\cancel{\text{mol Na}^+}} \times \frac{\cancel{\text{mol Na}^+}}{\text{L}}$$
$$= 1 \times [Na^+]$$

The concentration of positive charge due to magnesium ions is

$$\frac{\text{mol positive charge}}{\text{L}} = \frac{2 \text{ mol positive charge}}{\cancel{\text{mol Mg}^{2+}}} \times \frac{\cancel{\text{mol Mg}^{2+}}}{\text{L}}$$
$$= 2 \times [Mg^{2+}]$$

since each mole of magnesium ion contributes 2 mol of positive charge to the solution. Similarly, we may write for phosphate ion

► In some systems, a useful charge-balance equation cannot be written because not enough information is available or because the charge-balance equation is identical to one of the mass-balance equations.

$$\frac{\text{mol negative charge}}{\text{L}} = \frac{3 \text{ mol negative charge}}{\cancel{\text{mol PO}_4^{3-}}} \times \frac{\cancel{\text{mol PO}_4^{3-}}}{\text{L}}$$
$$= 3 \times [PO_4^{3-}]$$

Now, consider how we would write a charge-balance equation for a 0.100 M solution of sodium chloride. Positive charges in this solution are supplied by Na^+ and H_3O^+ (from dissociation of water). Negative charges come from Cl^- and OH^-. The molarity of positive and negative charges are

$$\text{mol/L positive charge} = [Na^+] + [H_3O^+] = 0.100 + 1 \times 10^{-7}$$
$$\text{mol/L negative charge} = [Cl^-] + [OH^-] = 0.100 + 1 \times 10^{-7}$$

We write the charge-balance equation by equating the concentrations of positive and negative charges. That is,

$$[Na^+] + [H_3O^+] = [Cl^-] + [OH^-] = 0.100 + 1 \times 10^{-7}$$

Now consider a solution that has an analytical concentration of magnesium chloride of 0.100 M. Here, the molarities of positive and negative charge are given by

$$\text{mol/L positive charge} = 2[Mg^{2+}] + [H_3O^+] = 2 \times 0.100 + 1 \times 10^{-7}$$

$$\text{mol/L negative charge} = [Cl^-] + [OH^-] = 2 \times 0.100 + 1 \times 10^{-7}$$

In the first equation, the molar concentration of magnesium ion is multiplied by two (2×0.100) because 1 mol of that ion contributes 2 mol of positive charge to the solution. In the second equation, the molar chloride ion concentration is twice that of the magnesium chloride concentration, or 2×0.100. To obtain the charge-balance equation, we equate the concentration of positive charge with the concentration of negative charge to obtain

$$2[Mg^{2+}] + [H_3O^+] = [Cl^-] + [OH^-] = 0.200 + 1 \times 10^{-7}$$

For a neutral solution, $[H_3O^+]$ and $[OH^-]$ are very small and equal, so that we can ordinarily simplify the charge-balance equation to

$$2[Mg^{2+}] \approx [Cl^-] = 0.200 \text{ M}$$

EXAMPLE 11-3

Write a charge-balance equation for the system in Example 11-2.

$$[Ag^+] + [Ag(NH_3)^+] + [Ag(NH_3)_2^+] + [H_3O^+] + [NH_4^+] = [OH^-] + [Br^-]$$

EXAMPLE 11-4

Write a charge-balance equation for an aqueous solution that contains NaCl, $Ba(ClO_4)_2$, and $Al_2(SO_4)_3$.

$$[Na^+] + [H_3O^+] + 2[Ba^+] + 3[Al^{3+}] =$$
$$[ClO_4^-] + [NO_3^-] + 2[SO_4^{2-}] + [HSO_4^-] + [OH^-]$$

11A-3 Steps for Solving Problems Involving Several Equilibria

Step 1. Write a set of balanced chemical equations for all pertinent equilibria.
Step 2. State the quantity being sought in terms of equilibrium concentrations.
Step 3. Write equilibrium-constant expressions for all equilibria developed in Step 1, and find numerical values for the constants in tables of equilibrium constants.
Step 4. Write mass-balance expressions for the system.
Step 5. If possible, write a charge-balance expression for the system.

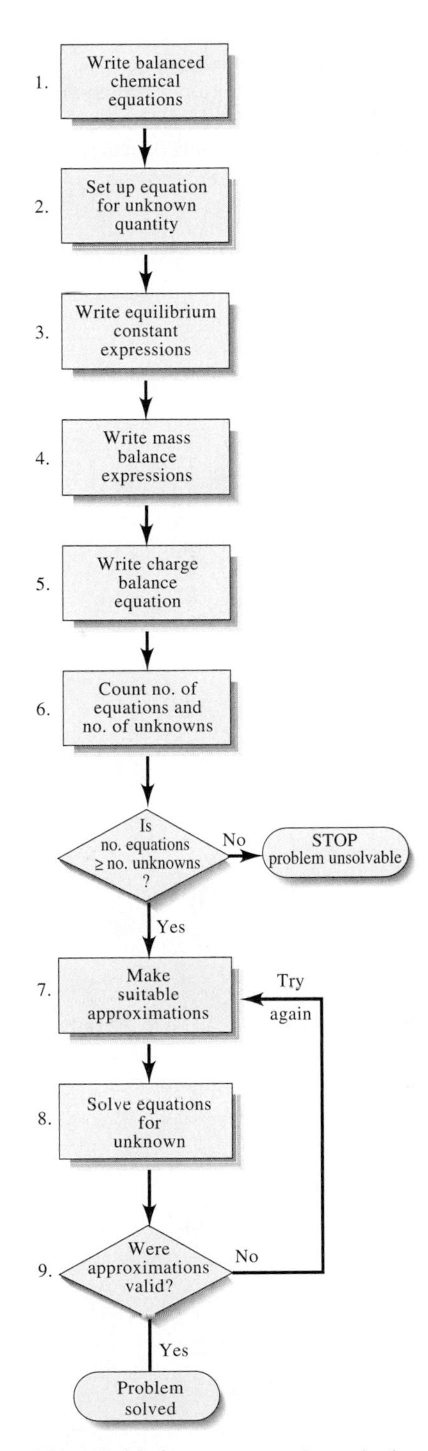

Figure 11-1 A systematic method for solving multiple-equilibrium problems.

► Do not waste time starting the algebra in an equilibrium calculation until you are absolutely sure that you have enough independent equations to make the solution feasible.

Step 6. Count the number of unknown concentrations in the equations developed in Steps 3, 4, and 5, and compare this number with the number of independent equations. Step 6 is critical because it shows whether an exact solution to the problem is possible. If the number of unknowns is identical to the number of equations, the problem has been reduced to one of *algebra* alone. That is, answers can be obtained with sufficient perseverance. On the other hand, if there are not enough equations even after approximations are made, the problem should be abandoned.

If a sufficient number of equations have been developed, proceed to either Step 7a or Step 7b.

Step 7a. Make suitable approximations to reduce the number of unknown equilibrium concentrations and thus the number of equations needed to provide an answer, as defined in Step 2. Proceed to Steps 8 and 9.

Step 7b. Solve the simultaneous equations exactly for the concentrations required by Step 2 by means of a computer program.

Step 8. Solve manually the simplified algebraic equations to give provisional concentrations for the species in the solution.

Step 9. Check the validity of the approximations.

These steps are illustrated in Figure 11-1.

11A-4 Using Approximations to Solve Equilibrium Calculations

When Step 6 of the systematic approach is complete, we have a *mathematical* problem of solving several nonlinear simultaneous equations. This job is often formidable, tedious, and time consuming unless a suitable computer program is available or approximations can be found that decrease the number of unknowns and equations. In this section, we consider in general terms how equations describing equilibrium relationships can be simplified by suitable approximations.

► Approximations can be made only in charge-balance and mass-balance equations—never in equilibrium-constant expressions.

Bear in mind that *only* the mass-balance and charge-balance equations can be simplified because only in these equations do the concentration terms appear as sums or differences rather than as products or quotients. It is always possible to assume that one (or more) of the terms in a sum or difference is so much smaller than the others that it can be ignored without significantly affecting the equality. The assumption that a concentration term in an equilibrium-constant expression is zero makes the expression meaningless.

The assumption that a given term in a mass- or charge-balance equation is sufficiently small that it can be neglected is generally based on a knowledge of the chemistry of the system. For example, in a solution containing a reasonable concentration of an acid, the hydroxide concentration will often be negligible with respect to the other species in the solution, and the term for the hydroxide concentration can usually be neglected in a mass- or charge-balance expression without introducing significant error to such a calculation.

► Never be afraid to make an assumption while attempting to solve an equilibrium problem. If the assumption is not valid, you will know it as soon as you have an approximate answer.

Many students find Step 7a to be troublesome because they fear that making invalid approximations will lead to serious errors in their computed results. *Such fears are groundless.* Experienced scientists are often as puzzled as beginners when making an approximation that simplifies an equilibrium calculation. Nonetheless, they make such approximations without fear because they know that the effects of an invalid assumption will become obvious by the time a computation is completed (see Example 11-6). It is a good idea to try questionable assumptions early during solution of a problem. If the assumption leads to an intolerable error (which is easily recognized), recalculate without the faulty approximation to arrive at a tentative answer. It is usually more efficient to try a questionable assumption at the beginning of a problem than to make a more laborious and time-consuming calculation without the assumption.

11A-5 Use of Computer Programs to Solve Multiple-Equilibrium Problems

So far, we have learned that if we know the chemical equilibria involved in a system, we can write a corresponding system of equations that allows us to solve for the concentrations of all species in the system. Although the systematic method gives us the means to solve equilibrium problems of great complexity, it is sometimes tedious and time consuming, particularly when a system must be solved for various sets of experimental conditions. For example, if we wish to find the solubility of silver chloride as a function of the concentration of added chloride, the system of five equations and five unknowns must be solved repetitively for each different concentration of chloride (see Example 11-9).

A number of powerful general-purpose software applications are available for solving equations. These include Mathcad, Mathematica, MATLAB, TK Solver, and Excel, among many others. Once a system of equations has been set up, they may be solved repetitively for many different sets of conditions. Furthermore, the accuracy of the solutions to the equations may be controlled by choosing appropriate tolerances within the programs. The equation-solving features of these applications coupled with their graphical capabilities enable you to solve complex systems of equations and present the results in graphical form. In this way, you can explore many different types of systems quickly and efficiently and develop your chemical intuition based on the results. A word of caution is in order. Nearly all equation-solving software requires initial estimates of the solutions to solve systems of equations. To provide these estimates, you must think about the chemistry a bit before beginning to solve the equations, and you should check the solutions that you find to be sure that they make good chemical sense.

◀ Several software packages are available for solving multiple nonlinear simultaneous equations rigorously. Three such programs are Mathcad, Mathematica, and Excel.

Computers know no chemistry. They will dutifully find solutions to the equations that you write based on the initial estimates that you give. If you make errors in the equations, software applications can sometimes flag errors based on certain mathematical constraints, but they will not find errors in the chemistry. If a program does not find a solution to a set of equations, it is often because of faulty initial estimates. Always be skeptical of computer results and respectful of software limitations. Used wisely, computer applications can be a marvelous aid in your study of chemical equilibria. For examples of the use of Excel in solving systems of equations such as those found in this chapter, see Chapter 6 of *Applications of Microsoft® Excel in Analytical Chemistry*.

11B CALCULATING SOLUBILITIES BY THE SYSTEMATIC METHOD

The use of the systematic method is illustrated in the sections that follow with examples involving the solubility of precipitates under various conditions. In later chapters, we apply this method to other types of equilibria.

11B-1 The Solubility of Metal Hydroxides

Examples 11-5 and 11-6 involve calculating the solubilities of two metal hydroxides. These examples illustrate how to make approximations and check their validity.

EXAMPLE 11-5

Calculate the molar solubility of $Mg(OH)_2$ in water.

Step 1. Write Equations for the Pertinent Equilibria Two equilibria that need to be considered are

$$Mg(OH)_2(s) \rightleftharpoons Mg^{2+} + 2OH^-$$

$$2H_2O \rightleftharpoons H_3O^+ + OH^-$$

Step 2. Define the Unknown Since 1 mol of Mg^{2+} is formed for each mole of $Mg(OH)_2$ dissolved,

$$\text{solubility } Mg(OH)_2 = [Mg^{2+}]$$

Step 3. Write All Equilibrium-Constant Expressions

$$[Mg^{2+}][OH^-]^2 = 7.1 \times 10^{-12} \quad (11\text{-}5)$$

$$[H_3O^+][OH^-] = 1.00 \times 10^{-14} \quad (11\text{-}6)$$

Step 4. Write Mass-Balance Expressions As shown by the two equilibrium equations, there are two sources of hydroxide ions: $Mg(OH)_2$ and H_2O. The hydroxide ion concentration resulting from dissociation of $Mg(OH)_2$ is twice the magnesium ion concentration, and the hydroxide ion concentration from the dissociation of water is equal to the hydronium ion concentration. Thus,

$$[OH^-] = 2[Mg^{2+}] + [H_3O^+] \quad (11\text{-}7)$$

► To arrive at Equation 11-7, we reasoned that if $[OH^-]_{H_2O}$ and $[OH^-]_{Mg(OH)_2}$ are the concentrations of OH^- produced from H_2O and $Mg(OH)_2$, respectively, then

$$[OH^-]_{H_2O} = [H_3O^+]$$

$$[OH^-]_{Mg(OH)_2} = 2[Mg^{2+}]$$

$$[OH^-]_{total} = [OH^-]_{H_2O} + [OH^-]_{Mg(OH)_2} = [H_3O^+] + 2[Mg^{2+}]$$

Step 5. Write the Charge-Balance Expression

$$[OH^-] = 2[Mg^{2+}] + [H_3O^+]$$

Note that this equation is identical to Equation 11-7. Often, a mass-balance and a charge-balance equation are the same.

Step 6. Count the Number of Independent Equations and Unknowns We have developed three independent algebraic equations (Equations 11-5, 11-6, and 11-7) and have three unknowns ($[Mg^{2+}]$, $[OH^-]$, and $[H_3O^+]$). Therefore, the problem can be solved rigorously.

Step 7a. Make Approximations We can make approximations only in Equation 11-7. Since the solubility-product constant for $Mg(OH)_2$ is relatively large, the solution will be somewhat basic. Therefore, it is reasonable to assume that $[H_3O^+] << [OH^-]$. Equation 11-7 then simplifies to

$$2[Mg^{2+}] \approx [OH^-] \tag{11-8}$$

Step 8. Solve the Equations Substitution of Equation 11-8 into Equation 11-5 gives

$$[Mg^{2+}](2[Mg^{2+}])^2 = 7.1 \times 10^{-12}$$

$$[Mg^{2+}]^3 = \frac{7.1 \times 10^{-12}}{4} = 1.78 \times 10^{-12}$$

$$[Mg^{2+}] = \text{solubility} = (1.78 \times 10^{-12})^{1/3} = 1.21 \times 10^{-4} \text{ or } 1.2 \times 10^{-4}\text{ M}$$

Step 9. Check the Assumptions Substitution into Equation 11-8 yields

$$[OH^-] = 2 \times 1.21 \times 10^{-4} = 2.42 \times 10^{-4}\text{ M}$$

and from Equation 11-6,

$$[H_3O^+] = \frac{1.00 \times 10^{-14}}{2.42 \times 10^{-4}} = 4.1 \times 10^{-11}\text{ M}$$

Thus, our assumption that $[H_3O^+] << [OH^-]$ is certainly valid.

EXAMPLE 11-6

Calculate the solubility of $Fe(OH)_3$ in water. Proceeding by the systematic approach used in Example 11-5, we write.

Step 1. Write Equations for the Pertinent Equilibrium

$$Fe(OH)_3(s) \rightleftharpoons Fe^{3+} + 3OH^-$$

$$2H_2O \rightleftharpoons H_3O^+ + OH^-$$

Step 2. Define the Unknown

$$\text{solubility} = [Fe^{3+}]$$

Step 3. Write All the Equilibrium-Constant Expressions

$$[Fe^{3+}][OH^-]^3 = 2 \times 10^{-39}$$

$$[H_3O^+][OH^-] = 1.00 \times 10^{-14}$$

(continued)

Step 4 and 5. Write Mass-Balance and Charge-Balance Equations As in Example 11-5, the mass-balance equation and the charge-balance equations are identical. That is,

$$[OH^-] = 3[Fe^{3+}] + [H_3O^+]$$

Step 6. Count the Number of Independent Equations and Unknowns We see that we have enough equations to calculate the three unknowns.

Step 7. Make Approximations As in Example 11-5, assume that $[H_3O^+]$ is very small, so that $[H_3O^+] << 3[Fe^{3+}]$, and

$$3[Fe^{3+}] \approx [OH^-]$$

Step 8. Solve the Equations Substituting $[OH^-] = 3[Fe^{3+}]$ into the solubility-product expression gives

$$[Fe^{3+}](3[Fe^{3+}])^3 = 2 \times 10^{-39}$$

$$[Fe^{3+}] = \left(\frac{2 \times 10^{-39}}{27}\right)^{1/4} = 9 \times 10^{-11}$$

$$\text{solubility} = [Fe^{3+}] = 9 \times 10^{-11}\ \text{M}$$

Step 9. Check the Assumption From the assumption made in Step 7, we can calculate a provisional value of $[OH^-]$. That is,

$$[OH^-] \approx 3[Fe^{3+}] = 3 \times 9 \times 10^{-11} = 3 \times 10^{-10}\ \text{M}$$

Then use this value of $[OH^-]$ to compute a *provisional* value for $[H_3O^+]$:

$$[H_3O^+] = \frac{1.00 \times 10^{-14}}{3 \times 10^{-10}} = 3 \times 10^{-5}\ \text{M}$$

But 3×10^{-5} is not much smaller than three times our provisional value of $[Fe^{3+}]$. This discrepancy means that our assumption was invalid and the provisional values for $[Fe^{3+}]$, $[OH^-]$, and $[H_3O^+]$ are all significantly in error. Therefore, go back to Step 7a and assume that

$$3[Fe^{3+}] << [H_3O^+]$$

Now the mass-balance expression becomes

$$[H_3O^+] = [OH^-]$$

Substituting this equality into the expression for K_w gives

$$[H_3O^+] = [OH^-] = 1.00 \times 10^{-7}\ \text{M}$$

Substituting this number into the solubility-product expression developed in Step 3 gives

$$[Fe^{3+}] = \frac{2 \times 10^{-39}}{(1.00 \times 10^{-7})^3} = 2 \times 10^{-18}\,M$$

In this case, we assumed that $3[Fe^{3+}] << [OH^-]$ or $3 \times 2 \times 10^{-18} << 10^{-7}$. Clearly, the assumption is valid, and we may write

$$\text{solubility} = 2 \times 10^{-18}\,M$$

Note the very large error introduced by the invalid assumption.

11B-2 The Effect of pH on Solubility

The solubility of precipitates containing an anion with basic properties, a cation with acidic properties, or both will depend on pH.

◀ All precipitates containing an anion that is the conjugate base of a weak acid are more soluble at low than at high pH.

Solubility Calculations When the pH Is Constant

Analytical precipitations are usually performed in buffered solutions in which the pH is fixed at some predetermined and known value. The calculation of solubility under this circumstance is illustrated by the following example.

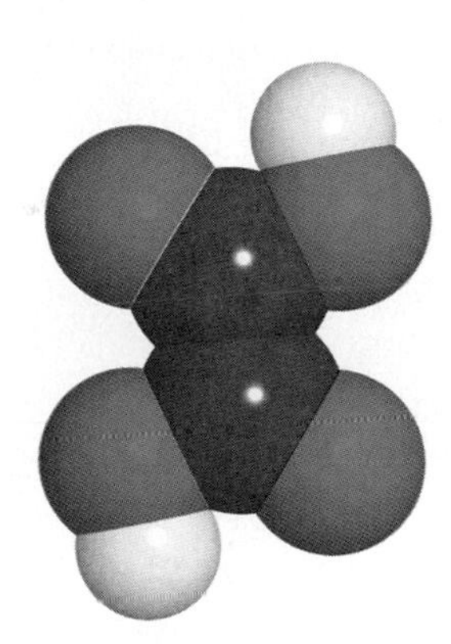

The molecular structure of oxalic acid. Oxalic acid occurs naturally in many plants as the potassium or sodium salt, and molds produce oxalic acid as the calcium salt. The sodium salt is used as a primary standard in redox titrimetry (see Chapter 20). The acid is widely used in the dye industry as a cleaning agent in a variety of applications, including the cleaning and restoration of wood surfaces; in the ceramics industry; in metallurgy; in the paper industry; and in photography. It is poisonous if ingested and may cause severe gastroenteritis or kidney damage. It can be prepared by passing carbon monoxide into concentrated sodium hydroxide.

EXAMPLE 11-7

Calculate the molar solubility of calcium oxalate in a solution that has been buffered so that its pH is constant and equal to 4.00.

Step 1. Write Pertinent Equilibria

$$CaC_2O_4(s) \rightleftharpoons Ca^{2+} + C_2O_4^{2-} \qquad (11\text{-}9)$$

Oxalate ions react with water to form $HC_2O_4^-$ and $H_2C_2O_4$. Thus, there are three other equilibria present in this solution:

$$H_2C_2O_4 + H_2O \rightleftharpoons H_3O^+ + HC_2O_4^- \qquad (11\text{-}10)$$

$$HC_2O_4^- + H_2O \rightleftharpoons H_3O^+ + C_2O_4^{2-} \qquad (11\text{-}11)$$

$$2H_2O \rightleftharpoons H_3O^+ + OH^-$$

Step 2. Define the Unknown Calcium oxalate is a strong electrolyte, so that its molar analytical concentration is equal to the equilibrium calcium ion concentration. That is,

$$\text{solubility} = [Ca^{2+}] \qquad (11\text{-}12)$$

(continued)

Step 3. Write All the Equilibrium-Constant Expressions

$$[Ca^{2+}][C_2O_4^{2-}] = K_{sp} = 1.7 \times 10^{-9} \quad (11\text{-}13)$$

$$\frac{[H_3O^+][HC_2O_4^-]}{[H_2C_2O_4]} = K_1 = 5.60 \times 10^{-2} \quad (11\text{-}14)$$

$$\frac{[H_3O^+][C_2O_4^{2-}]}{[HC_2O_4^-]} = K_2 = 5.42 \times 10^{-5} \quad (11\text{-}15)$$

$$[H_3O^+][OH^-] = K_w = 1.0 \times 10^{-14}$$

Step 4. Mass-Balance Expressions Because CaC_2O_4 is the only source of Ca^{2+} and the three oxalate species.

$$[Ca^{2+}] = [C_2O_4^{2-}] + [HC_2O_4^-] + [H_2C_2O_4] = \text{solubility} \quad (11\text{-}16)$$

Moreover, the problem states that the pH is 4.00. Thus,

$$[H_3O^+] = 1.00 \times 10^{-4} \text{ and } [OH^-] = K_w/[H_3O^+] = 1.00 \times 10^{-10}$$

A **buffer** keeps the pH of a solution nearly constant (see Chapter 9).

Step 5. Write Charge-Balance Expression A buffer is required to maintain the pH at 4.00. The buffer most likely consists of some weak acid HA and its conjugate base, A^-. The nature of the three species and their concentrations have not been specified, however, so we do not have enough information to write a charge-balance equation.

Step 6. Count the Number of Independent Equations and Unknowns We have four unknowns ($[Ca^{2+}]$, $[C_2O_4^{2-}]$, $[HC_2O_4^-]$, and $[H_2C_2O_4]$) as well as four independent algebraic relationships (Equations 11-13, 11-14, 11-15, and 11-16). Therefore, an exact solution can be obtained, and the problem becomes one of algebra.

Step 7a. Make Approximations An exact solution is so readily obtained in this case that we will not bother with approximations.

Step 8. Solve the Equations A convenient way to solve the problem is to substitute Equations 11-14 and 11-15 into 11-16 in such a way as to develop a relationship between $[Ca^{2+}]$, $[C_2O_4^{2-}]$, and $[H_3O^+]$. Thus, we rearrange Equation 11-15 to give

$$[HC_2O_4^-] = \frac{[H_3O^+][C_2O_4^{2-}]}{K_2}$$

Substituting numerical values for $[H_3O^+]$ and K_2 gives

$$[HC_2O_4^-] = \frac{1.00 \times 10^{-4}[C_2O_4^{2-}]}{5.42 \times 10^{-5}} = 1.85\,[C_2O_4^{2-}]$$

Substituting this relationship into Equation 11-14 and rearranging gives

$$[H_2C_2O_4] = \frac{[H_3O^+][C_2O_4^{2-}] \times 1.85}{K_1}$$

Substituting numerical values for $[H_3O^+]$ and K_1 yields

$$[H_2C_2O_4] = \frac{1.85 \times 10^{-4}[C_2O_4^{2-}]}{5.60 \times 10^{-2}} = 3.30 \times 10^{-3}\,[C_2O_4^{2-}]$$

Substituting these expressions for $[HC_2O_4^-]$ and $[H_2C_2O_4]$ into Equation 11-16 gives

$$[Ca^{2+}] = [C_2O_4^{2-} + 1.85\,[C_2O_4^{2-}] + 3.30 \times 10^{-3}\,[C_2O_4^{2-}]$$
$$= 2.85\,[C_2O_4^{2-}]$$

or $$[C_2O_4^{2-}] = [Ca^{2+}]/2.85$$

Substituting into Equation 11-13 gives

$$\frac{[Ca^{2+}][Ca^{2+}]}{2.85} = 1.7 \times 10^{-9}$$

$$[Ca^{2+}] = \text{solubility} = \sqrt{2.85 \times 1.7 \times 10^{-9}} = 7.0 \times 10^{-5}\ \text{M}$$

Solubility Calculations When the pH Is Variable

Computing the solubility of a precipitate such as calcium oxalate in a solution in which the pH is not fixed and known is considerably more complicated than in the example that we just explored. Thus, to determine the solubility of CaC_2O_4 in pure water, we must take into account the change in OH^- and H_3O^+ that accompanies the solution process. In this example, there are four equilibria to consider.

$$CaC_2O_4(s) \rightleftharpoons Ca^{2+} + C_2O_4^{2-}$$

$$C_2O_4^{2-} + H_2O \rightleftharpoons HC_2O_2^- + OH^-$$

$$HC_2O_4^- + H_2O \rightharpoonup H_2C_2O_4 + OH$$

$$2H_2O \rightleftharpoons H_3O^+ + OH^-$$

In contrast to Example 11-7, the hydroxide ion concentration now becomes an unknown, and an additional algebraic equation must therefore be developed to calculate the solubility of calcium oxalate.

It is not difficult to write the six algebraic equations needed to calculate the solubility of calcium oxalate (see Feature 11-1). Solving the six equations manually, however, is somewhat tedious and time consuming.

FEATURE 11-1

Algebraic Expressions Needed to Calculate the Solubility of CaC_2O_4 in Water

Here, as in Example 11-7, the solubility is equal to the cation concentration, $[Ca^{2+}]$.

$$\text{solubility} = [Ca^{2+}] = [C_2O_4^{2-}] + [HC_2O_4^-] + [H_2C_2O_4]$$

In this case, however, we must take into account one additional equilibrium—the dissociation of water. The equilibrium-constant expressions for the four equilibria are then

$$K_{sp} = [Ca^{2+}][C_2O_4^{2-}] = 1.7 \times 10^{-9} \quad (11\text{-}17)$$

$$K_2 = \frac{[H_3O^+][C_2O_4^{2-}]}{[HC_2O_4^-]} = 5.42 \times 10^{-5} \quad (11\text{-}18)$$

$$K_1 = \frac{[H_3O^+][HC_2O_4^-]}{[H_2C_2O_4]} = 5.60 \times 10^{-2} \quad (11\text{-}19)$$

$$K_w = [H_3O^+][OH^-] = 1.00 \times 10^{-14} \quad (11\text{-}20)$$

The mass-balance equation is

$$[Ca^{2+}] = [C_2O_4^{2-}] + [HC_2O_4^-] + [H_2C_2O_4] \quad (11\text{-}21)$$

The charge-balance equation is

$$2[Ca^{2+}] + [H_3O^+] = 2[C_2O_4^{2-}] + [HC_2O_4^-] + [OH^-] \quad (11\text{-}22)$$

We now have six unknowns ($[Ca^{2+}]$, $[C_2O_4^{2-}]$, $[HC_2O_4^-]$, $[H_2C_2O_4]$, $[H_3O^+]$, and $[OH^-]$) and six equations (11-17 through 11-22). Thus, in principle, the problem can be solved exactly.

11B-3 The Effect of Undissociated Solutes on Precipitation Calculations

Thus far, we have considered only solutes that dissociate completely when dissolved in aqueous media. There are some inorganic substances, however, such as calcium sulfate and the silver halides, that act as weak electrolytes and only partially dissociate in water. For example, a saturated solution of silver chloride contains significant amounts of undissociated silver chloride molecules as well as silver and chloride ions. Here, two equilibria are required to describe the system:

$$AgCl(s) \rightleftharpoons AgCl(aq) \quad (11\text{-}23)$$

$$AgCl(aq) \rightleftharpoons Ag^+ + Cl^- \quad (11\text{-}24)$$

The equilibrium constant for the first reaction takes the form

$$\frac{[AgCl(aq)]}{[AgCl(s)]} = K$$

where the numerator is the concentration of the undissociated species in *the solution* and the denominator is the concentration of silver chloride in *the solid phase.* The latter term is a constant, however (page 239), and so the equation can be written

$$[AgCl(aq)] = K[AgCl(s)] = K_s = 3.6 \times 10^{-7} \tag{11-25}$$

where K_s is the constant for the equilibrium shown in Equation 11-23. It is evident from this equation that at a given temperature, the concentration of the undissociated silver chloride is constant and *independent* of the chloride and silver ion concentrations.

The equilibrium constant K_d for the dissociation reaction (Equation 11-24) is

$$\frac{[Ag^+][Cl^-]}{[AgCl(aq)]} = K_d = 5.0 \times 10^{-4} \tag{11-26}$$

The product of these two constants is equal to the solubility product:

$$[Ag^+][Cl^-] = K_d K_s = K_{sp}$$

As shown by Example 11-8, both Reaction 11-23 and Reaction 11-24 contribute to the solubility of silver chloride in water.

EXAMPLE 11-8

Calculate the solubility of AgCl in distilled water.

$$\text{Solubility} = S = [AgCl(aq)] + [Ag^+]$$

$$[Ag^+] = [Cl^-]$$

$$[Ag^+][Cl^-] = K_{sp} = 1.82 \times 10^{-10}$$

$$[Ag^+] = \sqrt{1.82 \times 10^{-10}} = 1.35 \times 10^{-5}$$

Substituting this value and K_s from Equation 11-25 gives

$$S = 1.35 \times 10^{-5} + 3.6 \times 10^{-7} = 1.38 \times 10^{-5}\text{ M}$$

11B-4 The Solubility of Precipitates in The Presence of Complexing Agents

The solubility of a precipitate may increase dramatically in the presence of reagents that form complexes with the anion or the cation of the precipitate. For

▶ The solubility of a precipitate always increases in the presence of a complexing agent that reacts with the cation of the precipitate.

example, fluoride ions prevent the quantitative precipitation of aluminum hydroxide even though the solubility product of this precipitate is remarkably small (2×10^{-32}). The cause of the increase in solubility is shown by the equations

$$Al(OH)_3(s) \rightleftharpoons Al^{3+} + 3OH^-$$
$$+$$
$$6F^-$$
$$\upharpoonleft\downharpoonright$$
$$AlF_6^{3-}$$

The fluoride complex is sufficiently stable to permit fluoride ions to compete successfully with hydroxide ions for aluminum ions.

Many precipitates react with excesses of the precipitating reagent to form soluble complexes. In a gravimetric analysis, this tendency may have the undesirable effect of reducing the recovery of analytes if too large an excess of reagent is used. For example, silver is often determined by precipitation of silver ion by addition of an excess of a potassium chloride solution. The effect of excess reagent is complex, as revealed by the following set of equations that describe the system:

$$AgCl(s) \rightleftharpoons AgCl(aq) \qquad (11\text{-}27)$$

$$AgCl(aq) \rightleftharpoons Ag^+ + Cl^- \qquad (11\text{-}28)$$

$$AgCl(s) + Cl^- \rightleftharpoons AgCl_2^- \qquad (11\text{-}29)$$

$$AgCl_2^- + Cl^- \rightleftharpoons AgCl_3^{2-} \qquad (11\text{-}30)$$

Note that Equilibrium 11-28 and thus Equilibrium 11-27 shift to the left with added chloride ion, whereas Equilibria 11-29 and 11-30 shift to the right under the same circumstance. The consequence of these opposing effects is that a plot of silver chloride solubility as a function of concentration of added chloride exhibits a minimum. Example 11-9 illustrates how this behavior can be described in quantitative terms.

EXAMPLE 11-9

Derive an equation that describes the effect of the analytical concentration of KCl on the solubility of AgCl in an aqueous solution. Calculate the concentration of KCl at which the solubility is a minimum.

Step 1. Pertinent Equilibria Equations 11-27 through 11-30 describe the pertinent equilibria.

Step 2. Definition of Unknown The molar solubility S of AgCl is equal to the sum of the concentrations of the silver-containing species:

$$\text{solubility} = S = [AgCl(aq)] + [Ag^+] + [AgCl_2^-] + [AgCl_3^{2-}] \qquad (11\text{-}31)$$

Step 3. Equilibrium-Constant Expressions Equilibrium constants available in the literature include

$$[Ag^+][Cl^-] = K_{sp} = 1.82 \times 10^{-10} \quad (11\text{-}32)$$

$$\frac{[Ag^+][Cl^-]}{[AgCl(aq)]} = K_d = 3.9 \times 10^{-4} \quad (11\text{-}33)$$

$$\frac{[AgCl_2^-]}{[AgCl(aq)][Cl^-]} = K_2 = 2.0 \times 10^{-5} \quad (11\text{-}34)$$

$$\frac{[AgCl_3^{2-}]}{[AgCl_2^-][Cl^-]} = K_3 = 1 \quad (11\text{-}35)$$

Step 4. Mass-Balance Equation

$$[Cl^-] = c_{KCl} + [Ag^+] - [AgCl_2^-] - 2[AgCl_3^{2-}] \quad (11\text{-}36)$$

The second term on the right-hand side of this equation gives the chloride ion concentration produced by the dissolution of the precipitate, and the next two terms correspond to the *decrease* in chloride ion concentration resulting from the formation of the two chloro complexes from AgCl.

Step 5. Charge-Balance Equation As in some of the earlier examples, the charge-balance equation is identical to the mass-balance equation.

Step 6. Number of Equations and Unknowns We have five equations (11-32 through 11-36) and five unknowns ($[Ag^+]$, $[AgCl(aq)]$, $[AgCl_2^-]$, $[AgCl_3^{2-}]$, and $[Cl^-]$).

Step 7a. Assumptions We assume that over a considerable range of chloride ion concentrations, the solubility of AgCl is so small that Equation 11-36 can be greatly simplified by the assumption that

$$[Ag^+] - [AgCl_2^-] - 2[AgCl_3^{2-}] << c_{KCl}$$

It is not certain that this is a valid assumption, but it is worth trying because it simplifies the problem so much. With this assumption, then, Equation 11-36 reduces to

$$[Cl^-] = c_{KCl} \quad (11\text{-}37)$$

Step 8. Solution of Equations For convenience, we multiply Equations 11-34 and 11-35 to give

$$\frac{[AgCl_3^{2-}]}{[Cl^-]^2} = K_2K_3 = 2.0 \times 10^{-5} \times 1 = 2.0 \times 10^{-5} \quad (11\text{-}38)$$

To calculate [AgCl(*aq*)], we divide Equation 11-32 by Equation 11-33 and rearrange:

$$[\mathrm{AgCl}(aq)] = \frac{K_{sp}}{K_d} = \frac{1.82 \times 10^{-10}}{3.9 \times 10^{-4}} = 4.7 \times 10^{-7} \qquad (11\text{-}39)$$

Note that the concentration of this species is *constant and independent of the chloride concentration.*

Substitution of Equations 11-39, 11-32, 11-33, and 11-38 into Equation 11-31 permits us to express the solubility in terms of the chloride ion concentration and the several constants.

$$S = \frac{K_{sp}}{K_d} + \frac{K_{sp}}{[\mathrm{Cl}^-]} + K_2[\mathrm{Cl}^-] + K_2K_3[\mathrm{Cl}^-]^2 \qquad (11\text{-}40)$$

Substitution of Equation 11-37 yields the desired relationship between the solubility and the analytical concentration of KCl:

$$S = \frac{K_{sp}}{K_d} + \frac{K_{sp}}{c_{KCl}} + K_2c_{KCl} + K_2K_3c_{KCl}^2 \qquad (11\text{-}41)$$

To find the minimum in *S*, we set the derivative of *S* with respect to c_{KCl} equal to zero:

$$\frac{dS}{dc_{KCl}} = 0 = \frac{K_{sp}}{c_{KCl}^2} + K_2 + 2\,K_2K_3c_{KCl}$$

$$2\,K_2K_3c_{KCl}^3 + c_{KCl}^2K_2 - K_{sp} = 0$$

Substituting numerical values gives

$$(4.0 \times 10^{-5})\,c_{KCl}^3 + (2.0 \times 10^{-5})\,c_{KCl}^2 - 1.82 \times 10^{-10} = 0$$

Following the procedure shown in Feature 6-4, we can solve this equation by successive approximations to obtain

$$c_{KCl} = 0.0030 = [\mathrm{Cl}^-]$$

To check the assumption made earlier, we calculate the concentrations of the various species. Substitutions into Equations 11-32, 11-34, and 11-36 yield

$$[\mathrm{Ag}^+] = (1.82 \times 10^{-10})/0.0030 = 6.1 \times 10^{-8}\,\mathrm{M}$$

$$[\mathrm{AgCl}_2^-] = 2.0 \times 10^{-5} \times 0.0030 = 6.0 \times 10^{-8}\,\mathrm{M}$$

$$[\mathrm{AgCl}_3^{2-}] = 2.0 \times 10^{-5} \times (0.0030)^2 = 1.8 \times 10^{-10}\,\mathrm{M}$$

Thus, our assumption that c_{KCl} is much larger than the concentrations of the ions of the precipitate is reasonable. The minimum solubility is obtained by substitution of these concentrations and [AgCl(*aq*)] into Equation 11-31:

$$S = 4.7 \times 10^{-7} + 6.1 \times 10^{-8} + 6.0 \times 10^{-8} + 1.8 \times 10^{-10}$$
$$= 5.9 \times 10^{-7}\,\mathrm{M}$$

The solid curve in Figure 11-2 illustrates the effect of chloride ion concentration on the solubility of silver chloride; data for the curve were obtained by substituting various chloride concentrations into Equation 11-41. Note that at high concentrations of the common ion, the solubility becomes greater than that in pure water. The broken lines represent the equilibrium concentrations of the various silver-containing species as a function of c_{KCl}. Note that at the solubility minimum, undissociated silver chloride, AgCl(*aq*), is the major silver species in the solution, representing about 80% of the total dissolved silver. Its concentration is invariant, as has been demonstrated.

Unfortunately, reliable equilibrium data regarding undissociated species such as AgCl(*aq*) and complex species such as $AgCl_2^-$ are not abundant; consequently, solubility calculations are often, of necessity, based on solubility-product equilibria alone. Example 11-9 shows that under some circumstances, such neglect of other equilibria can lead to serious error.

Spreadsheet Summary In the first exercise in Chapter 6 of *Applications of Microsoft® Excel in Analytical Chemistry,* we explore the use of Excel's Solver to find the concentrations of Mg^{2+}, OH^-, and H_3O^+ in the $Mg(OH)_2$ system of Example 11-5. Solver finds the concentrations from the mass-balance expression, the solubility product of $Mg(OH)_2$, and the ion product of water. Then Excel's built-in tool Goal Seek is used to solve a cubic equation for the same system. The final exercise in Chapter 6 uses Solver to find the solubility of calcium oxalate at a known pH (see Example 11-7) and when the pH is unknown (see Feature 11-1).

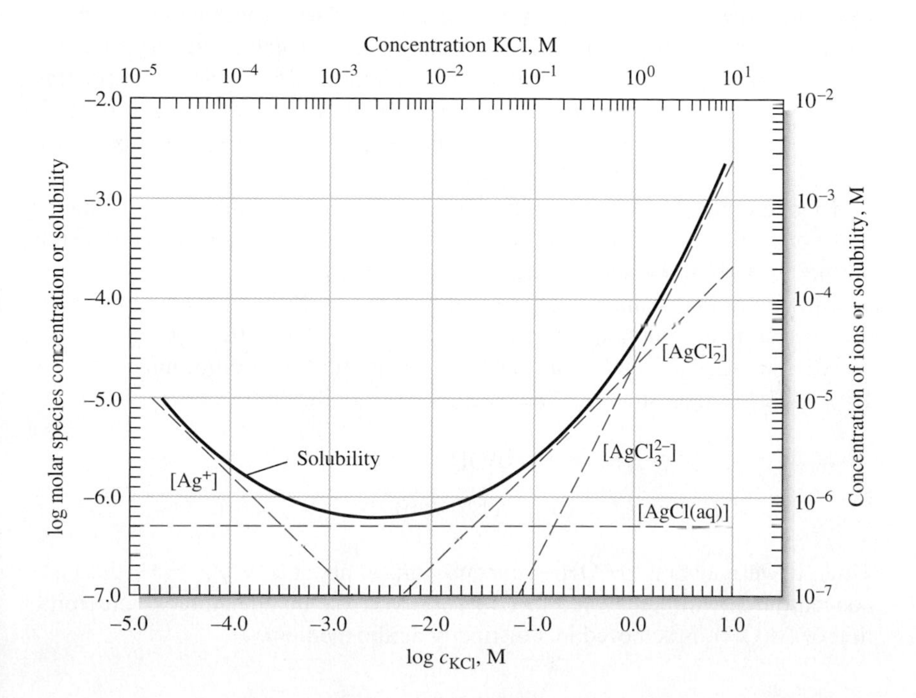

Figure 11-2 The effect of chloride ion concentration on the solubility of AgCl. The solid curve shows the total concentration of dissolved AgCl. The broken lines show the concentrations of the various silver-containing species.

11C SEPARATION OF IONS BY CONTROL OF THE CONCENTRATION OF THE PRECIPITATING AGENT

Several precipitating agents permit separation of ions based on solubility differences. Such separations require close control of the active reagent concentration at a suitable and predetermined level. Most often, such control is achieved by controlling the pH of the solution with suitable buffers. This technique is applicable to anionic reagents in which the anion is the conjugate base of a weak acid. Examples include sulfide ion (the conjugate base of hydrogen sulfide), hydroxide ion (the conjugate base of water), and the anions of several organic weak acids.

11C-1 Calculation of the Feasibility of Separations

The following example illustrates how solubility-product calculations are used to determine the feasibility of separations based on solubility differences.

EXAMPLE 11-10

Can Fe^{3+} and Mg^{2+} be separated quantitatively as hydroxides from a solution that is 0.10 M in each cation? If the separation is possible, what range of OH^- concentrations is permissible? Solubility-product constants for the two precipitates are

$$[Fe^{3+}][OH^-]^3 = 2 \times 10^{-39}$$

$$[Mg^{2+}][OH^-]^2 = 7.1 \times 10^{-12}$$

The K_{sp} for $Fe(OH)_3$ is so much smaller than that for $Mg(OH)_2$ that it appears likely that the former will precipitate at a lower OH^- concentration. We can answer the questions posed in this problem by (1) calculating the OH^- concentration required to achieve quantitative precipitation of Fe^{3+} and (2) computing the OH^- concentration at which $Mg(OH)_2$ just begins to precipitate. If (1) is smaller than (2), a separation is feasible in principle, and the range of permissible OH^- concentrations is defined by the two values.

To determine (1), we must first specify what constitutes a quantitative removal of Fe^{3+} from the solution. The decision here is arbitrary and depends on the purpose of the separation. In this example and the next, we consider a precipitation to be quantitative when all but 1 part in 1000 of the ion has been removed from the solution—that is, when $[Fe^{3+}] << 1 \times 10^{-4}$ M.

We can readily calculate the OH^- concentration in equilibrium with 1×10^{-1} M Fe^{3+} by substituting directly into the solubility-product expression:

$$(1.0 \times 10^{-4})[OH^-]^3 = 2 \times 10^{-39}$$

$$[OH^-] = [(2 \times 10^{-39})/(1.0 \times 10^{-4})]^{1/3} = 3 \times 10^{-12}\ \text{M}$$

Thus, if we maintain the OH^- concentration at about 3×10^{-12} M, the Fe^{3+} concentration will be lowered to 1×10^{-4} M. Note that quantitative precipitation of $Fe(OH)_3$ is achieved in a distinctly acidic medium.

To determine what maximum OH^- concentration can exist in the solution without causing formation of $Mg(OH)_2$, we note that precipitation cannot occur until the product $[Mg^{2+}][OH^-]^2$ exceeds the solubility product, 7.1×10^{-12}. Substitution of 0.1 (the molar Mg^{2+} concentration of the solution) into the solubility-product expression permits the calculation of the *maximum* OH^- concentration that can be tolerated:

$$0.10\,[OH^-]^2 = 7.1 \times 10^{-12}$$

$$[OH^-] = 8.4 \times 10^{-6}\ M$$

When the OH^- concentration exceeds this level, the solution will be supersaturated with respect to $Mg(OH)_2$, and precipitation may begin.

From these calculations, we conclude that quantitative separation of $Fe(OH)_3$ can be achieved if the OH^- concentration is greater than 3×10^{-12} M and that $Mg(OH)_2$ will not precipitate until an OH^- concentration of 8.4×10^{-6} M is reached. Therefore, it is possible, in principle, to separate Fe^{3+} from Mg^{2+} by maintaining the OH^- concentration between these levels. In practice, the concentration of OH^- is kept as low as practical—often about 10^{-10} M.

11C-2 Sulfide Separations

Sulfide ion forms precipitates with heavy metal cations that have solubility products that vary from 10^{-10} to 10^{-90} or smaller. In addition, the concentration of S^{2-} can be varied over a range of about 0.1 M to 10^{-22} M by controlling the pH of a saturated solution of hydrogen sulfide. These two properties make possible a number of useful cation separations. To illustrate the use of hydrogen sulfide to separate cations based on pH control, consider the precipitation of the divalent cation M^{2+} from a solution that is kept saturated with hydrogen sulfide by bubbling the gas continuously through the solution. The important equilibria in this solution are:

$$MS(s) \rightleftharpoons M^{2+} + S^{2-} \qquad [M^{2+}][S^{2-}] = K_{sp}$$

$$H_2S + H_2O \rightleftharpoons H_3O^+ + HS^- \qquad \frac{[H_3O^+][HS^-]}{[H_2S]} = K_1 = 9.6 \times 10^{-8}$$

$$HS^- + H_2O \rightleftharpoons H_3O^+ + S^{2-} \qquad \frac{[H_3O^+][S^{2-}]}{[HS^-]} = K_2 = 1.3 \times 10^{-14}$$

We may also write

$$\text{solubility} = [M^{2+}]$$

The concentration of hydrogen sulfide in a saturated solution of the gas is approximately 0.1 M. Thus, we may write as a mass-balance expression

$$0.1 = [S^{2-}] + [HS^-] + [H_2S]$$

Because we know the hydronium ion concentration, we have four unknowns, the concentration of the metal ion and the three sulfide species.

We can simplify the calculation greatly by assuming that ($[S^{2-}] + [HS^-]$) << $[H_2S]$, so that

$$[H_2S] \approx 0.10 \text{ M}$$

The two dissociation-constant expressions for hydrogen sulfide may be multiplied to give an expression for the overall dissociation of hydrogen sulfide to sulfide ion:

$$H_2S + 2H_2O \rightleftharpoons 2H_3O^+ + S^{2-} \qquad \frac{[H_3O^+]^2[S^{2-}]}{[H_2S]} = K_1K_2 = 1.2 \times 10^{-21}$$

The constant for this overall reaction is simply the product of K_1 and K_2.

Substituting the numerical value for $[H_2S]$ into this equation gives

$$\frac{[H_3O^+]^2[S^{2-}]}{0.10} = 1.2 \times 10^{-21}$$

On rearranging this equation, we obtain

$$[S^{2-}] = \frac{1.2 \times 10^{-22}}{[H_3O^+]^2} \tag{11-42}$$

Thus, we see that the sulfide ion concentration of a saturated hydrogen sulfide solution varies inversely with the square of the hydrogen ion concentration. Figure 11-3, which was obtained with this equation, reveals that the sulfide ion concentra-

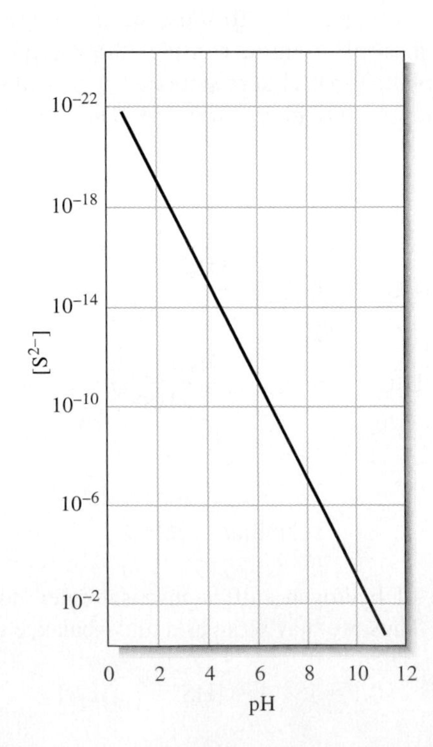

Figure 11-3 Sulfide ion concentration as a function of pH in a saturated H_2S solution.

tion of an aqueous solution can be varied by more than 20 orders of magnitude by varying the pH from 1 to 11.

Substituting Equation 11-42 into the solubility-product expression gives

$$\frac{[M^{2+}] \times 1.2 \times 10^{-22}}{[H_3O^+]^2} = K_{sp}$$

$$[M^{2+}] = \text{solubility} = \frac{[H_3O^+]^2 K_{sp}}{1.2 \times 10^{-22}}$$

Thus, the solubility of a divalent metal sulfide increases with the square of the hydronium ion concentration.

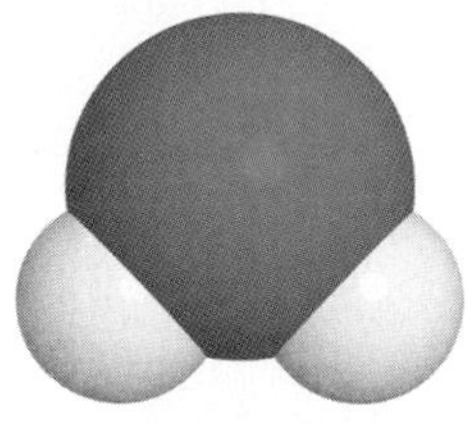

Hydrogen sulfide is a colorless, flammable gas with important chemical and toxicological properties. It is the product of a number of natural processes, including the decay of sulfur-containing material. Its noxious odor of rotten eggs permits its detection at extremely low concentration (0.02 ppm). Because the olfactory sense is dulled by its action, however, higher concentrations may be tolerated, and the lethal concentration of 100 ppm may be exceeded. Aqueous solutions of the gas were used traditionally as a source of sulfide for the precipitation of metals, but because of the toxicity of H_2S, this role has been taken over by other sulfur-containing compounds such as thioacetamide.

EXAMPLE 11-11

Cadmium sulfide is less soluble than thallium(I) sulfide. Find the conditions under which Cd^{2+} and Tl^{+} can, in theory, be separated quantitatively with H_2S from a solution that is 0.1 M in each cation.

The constants for the two solubility equilibria are:

$$CdS(s) \rightleftharpoons Cd^{2+} + S^{2-} \qquad [Cd^{2+}][S^{2-}] = 1 \times 10^{-27}$$

$$Tl_2S(s) \rightleftharpoons 2Tl^{+} + S^{2-} \qquad [Tl^{+}]^2[S^{2-}] = 6 \times 10^{-22}$$

Since CdS precipitates at a lower $[S^{2-}]$ than does Tl_2S, we first compute the sulfide ion concentration necessary for quantitative removal of Cd^{2+} from solution. As in Example 11-10, we arbitrarily specify that separation is quantitative when all but 1 part in 1000 of the Cd^{2+} has been removed; that is, the concentration of the cation has been lowered to 1.00×10^{-4} M. Substituting this value into the solubility-product expression gives

$$10^{-4}[S^{2-}] = 1 \times 10^{-27}$$

$$[S^{2-}] = 1 \times 10^{-23} \text{ M}$$

Thus, if we maintain the sulfide concentration at this level or greater, we may assume that quantitative removal of the cadmium will take place. Next, we compute the $[S^{2-}]$ needed to initiate precipitation of Tl_2S from a 0.1 M solution. Precipitation will begin when the solubility product is just exceeded. Since the solution is 0.1 M in Tl^{+},

$$(0.1)^2[S^{2-}] = 6 \times 10^{-22}$$

$$[S^{2-}] = 6 \times 10^{-20} \text{ M}$$

These two calculations show that quantitative precipitation of Cd^{2+} takes place if $[S^{2-}]$ is made greater than 1×10^{-23} M. No precipitation of Tl^{+} occurs, however, until $[S^{2-}]$ becomes greater than 6×10^{-20} M.

(continued)

Substituting these two values for $[S^{2-}]$ into Equation 11-42 permits calculation of the $[H_3O^+]$ range required for the separation.

$$[H_3O^+]^2 = \frac{1.2 \times 10^{-22}}{1 \times 10^{-23}} = 12$$

$$[H_3O^+] = 3.5\ M$$

and

$$[H_3O^+]^2 = \frac{1.2 \times 10^{-22}}{6 \times 10^{-20}} = 2.0 \times 10^{-3}$$

$$[H_3O^+] = 0.045\ M$$

By maintaining $[H_3O^+]$ between approximately 0.045 and 3.5 M, we can in theory separate Cd^{2+} quantitatively from Tl^+.

FEATURE 11-2

Immunoassay: Equilibria in the Specific Determination of Drugs

The determination of drugs in the human body is a matter of great importance in drug therapy and in the detection and prevention of drug abuse. The diversity of drugs and their typical low levels of concentration in body fluids make it difficult to identify them and measure their concentrations. Fortunately, it is possible to harness one of nature's own mechanisms, the immune response, to determine quantitatively a variety of therapeutic and illicit drugs.

When a foreign substance, or antigen (Ag), shown schematically in Figure 11F-1a, is introduced into the body of a mammal, the immune system synthesizes protein-based molecules (Figure 11F-1b) called antibodies (Ab), which specifically bind to the antigen molecules via electrostatic interactions, hydrogen bonding, and other noncovalent short-range forces. These massive molecules (molar mass ≈ 150,000) form a complex with antigens, as shown in the following reaction and in Figure 11F-1c.

$$\text{Ag} + \text{Ab} \rightleftharpoons \text{AgAb} \qquad K = \frac{[\text{AgAb}]}{[\text{Ag}][\text{Ab}]}$$

The immune system does not recognize relatively small molecules, so we must use a trick to prepare antibodies with binding sites that are specific for a particular drug. As shown in Figure 11F-1d, we attach the drug covalently to an antigenic carrier molecule such as bovine serum albumin (BSA), a protein that is obtained from the blood of cattle.

$$\text{D} + \text{Ag} \rightarrow \text{D-Ag}$$

When the resulting drug-antigen conjugate (D-Ag) is injected into the bloodstream of a rabbit, the immune system of the rabbit synthesizes antibodies with

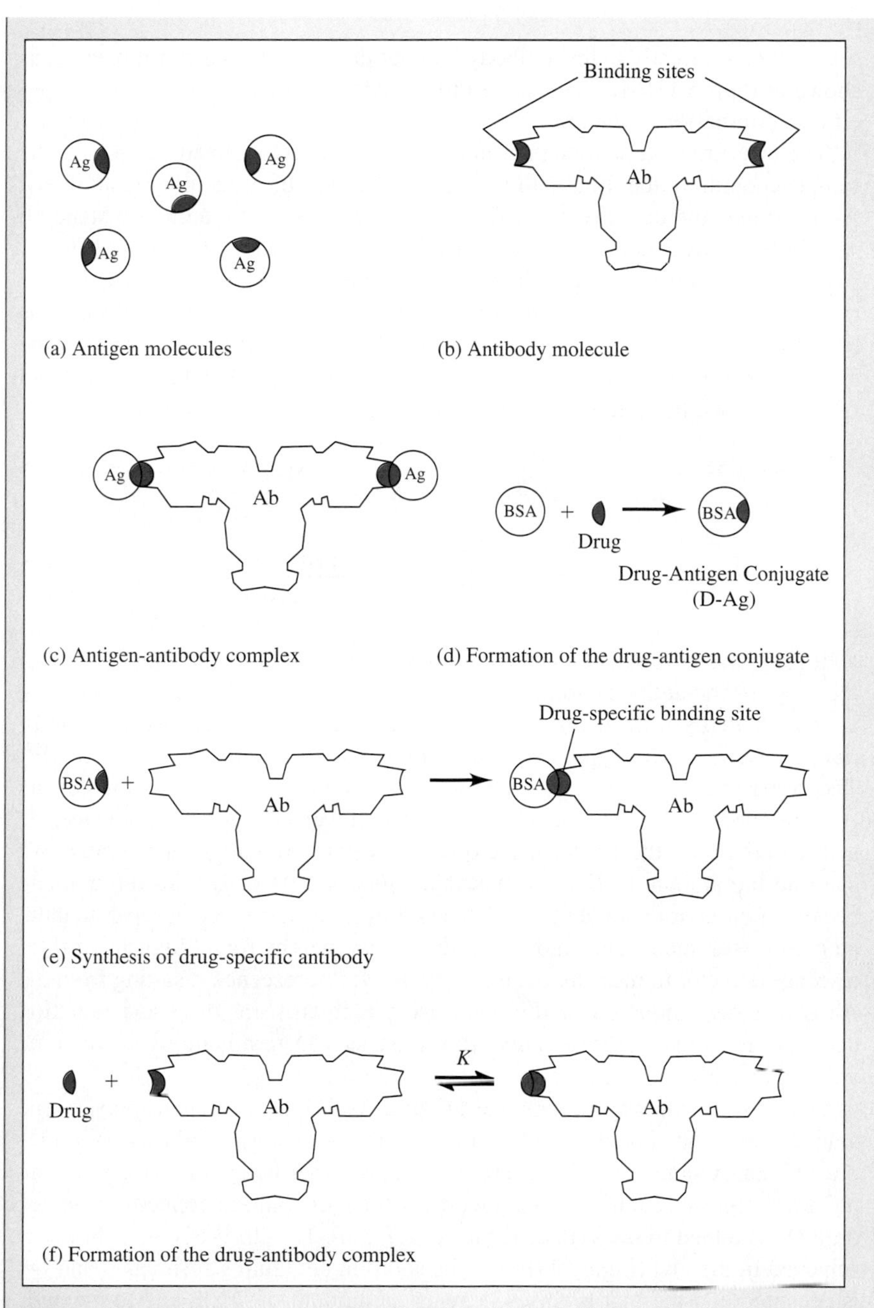

Figure 11F-1 Antibody-antigen interaction.

binding sites that are specific for the drug, as illustrated in Figure 11F-1e. Approximately 3 weeks after injection of the antigen, blood is drawn from the rabbit, the serum is separated from the blood, and the antibodies of interest are separated from the serum and other antibodies, usually by chromatographic methods (see Chapters 31 and 32). It is important to note that once the drug-specific antibody has been synthesized by the immune system of the rabbit, the

(continued)

drug can bind directly to the antibody without the aid of the carrier molecule, as shown in Figure 11F-1f. This direct drug-antibody forms the basis for the specific determination of the drug.

The measurement step of the immunoassay is accomplished by mixing the sample containing the drug with a measured amount of drug-specific antibody. At this point, the quantity of Ab-D must be determined by adding a standard sample of the drug that has been chemically altered to contain a detectable *label.* Typical labels are enzymes, fluorescent or chemiluminescent molecules, or radioactive atoms. For our example, we assume that a fluorescent molecule has been attached to the drug to produce the labeled drug D*.[1] If the amount of the antibody is somewhat less than the sum of the amounts of D and D*, then D and D* compete for the antibody, as shown in the following equilibria.

$$\text{D}^* + \text{Ab} \rightleftharpoons \text{Ab-D}^* \qquad K^* = \frac{[\text{Ab-D}^*]}{[\text{D}^*][\text{Ab}]}$$

$$\text{D} + \text{Ab} \rightleftharpoons \text{Ab-D} \qquad K = \frac{[\text{Ab-D}]}{[\text{D}][\text{Ab}]}$$

It is important to select a label that does not substantially alter the affinity of the drug for the antibody so that the labeled and unlabeled drugs bind with the antibody equally well. If this is true, then $K = K^*$. Typical values for equilibrium constants of this type, called **binding constants,** range from 10^7 to 10^{12}. The larger the concentration of the unknown, unlabeled drug, the smaller the concentration of Ab-D*, and vice versa. This inverse relationship between D and Ab-D* forms the basis for the quantitative determination of the drug. We can find the amount of D if we measure *either* Ab-D* or D*. To differentiate between bound drug and unbound labeled drug, it is necessary to separate them before measurement. The amount of Ab-D* can then be found by using a fluorescence detector to measure the intensity of the fluorescence resulting from the Ab-D*. A determination of this type using a fluorescent drug and radiation detection is called a **fluorescence immunoassay.** Determinations of this type are very sensitive and selective.

One convenient way to separate D* and Ag-D* is to prepare polystyrene vials that are coated on the inside with antibody molecules, as illustrated in Figure 11F-2a. A sample of blood serum, urine, or other body fluid containing an unknown concentration of D along with a volume of solution containing labeled drug D* is added to the vial, as depicted in Figure 11F-2b. After equilibrium is achieved in the vial (Figure 11F-2c), the solution containing residual D and D* is decanted and the vial is rinsed, leaving an amount of D* bound to the antibody that is inversely proportional to the concentration of D in the sample (Figure 11F-2d). Finally, the fluorescence intensity of the bound D* is determined using a fluorometer, as shown in Figure 11F-2e. This procedure is repeated for several standard solutions of D to produce a nonlinear working curve called a **dose-response curve** similar to the curve of Figure 11F-3. The fluorescence intensity for an unknown solution of D is located on the calibration curve, and

[1]For a discussion of molecular fluorescence, see Chapter 27.

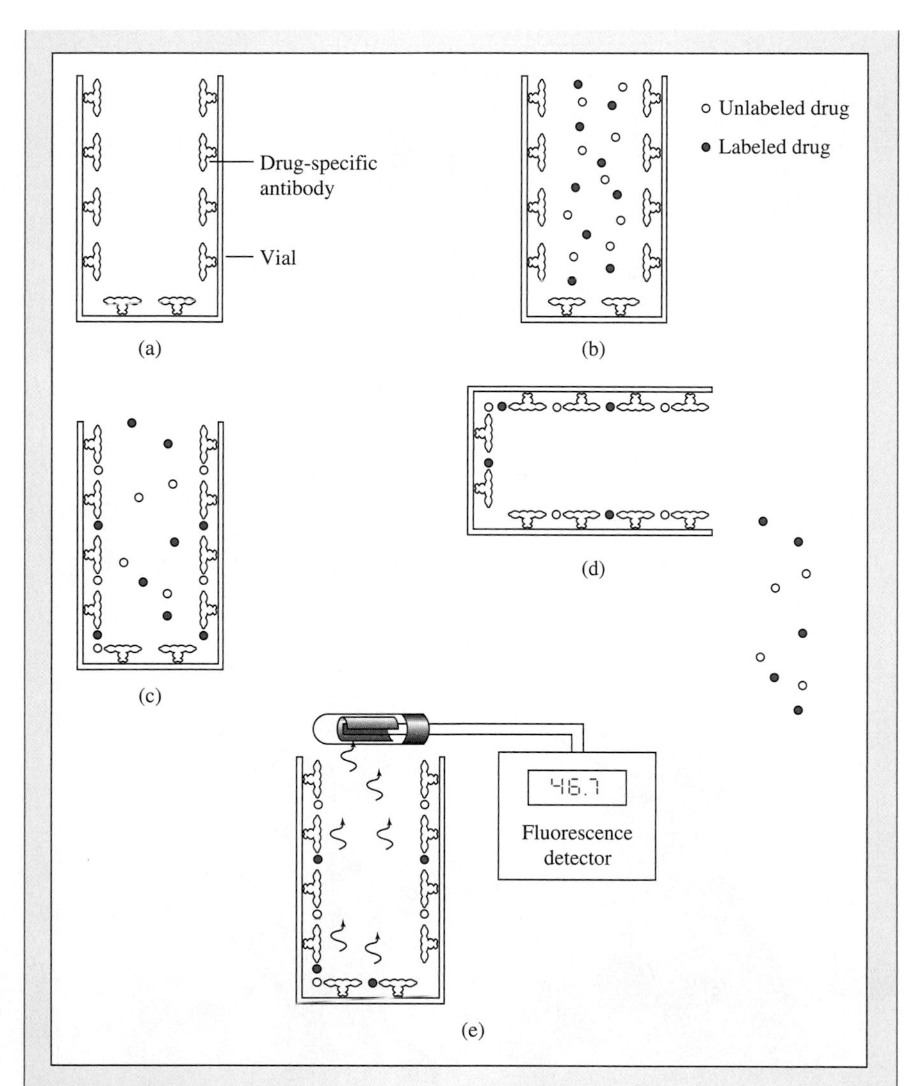

Figure 11F-2 Procedure for determining drugs by immunoassay with fluorescence labeling. (a) Vial is lined with drug-specific antibodies; (b) vial is filled with solution containing both labeled and unlabeled drug; (c) labeled and unlabeled drug binds to antibodies; (d) solution is discarded leaving bound drug behind; (e) fluorescence of the bound, labeled drug is measured. The concentration of drug is determined by using the dose-response curve of Figure 11F-3.

the concentration is read from the concentration axis. Immunoassay is a powerful tool in the clinical laboratory and is one of the most widely used of all analytical techniques. Reagent kits for many different immunoassays are available commercially, as are automated instruments for carrying out fluorescent immunoassays and immunoassays of other types. In addition to concentrations of drugs, vitamins, proteins, growth hormones, pregnancy hormones, cancer and other disease indicators, and pesticide residues in natural waters and food are determined by immunoassay. The structure of an antigen-antibody complex is shown in Figure 11F-4.

(continued)

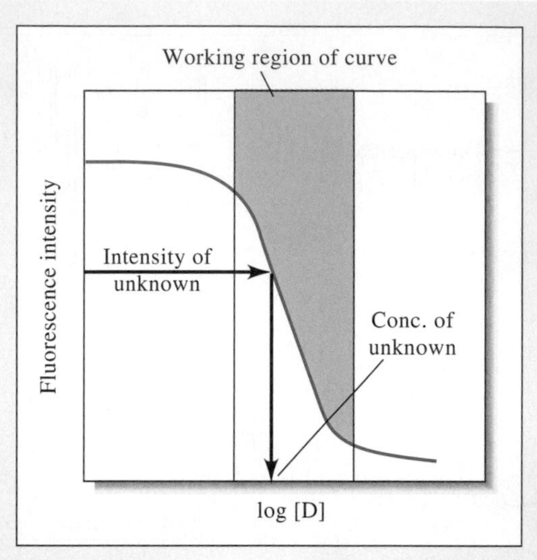

Figure 11F-3 Dose-response curve for determining drugs by fluorescence-based immunoassay.

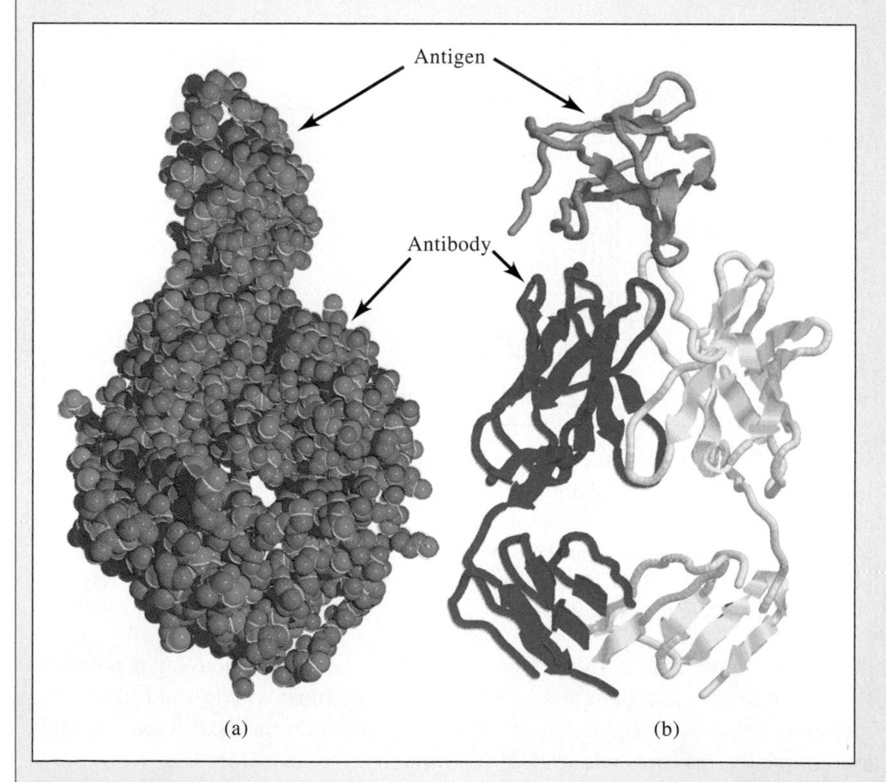

Figure 11F-4 The molecular structure of an antigen-antibody complex. Pictured are two representations of the complex formed between a digestion fragment of intact mouse antibody A6 and genetically engineered human interferon-gamma receptor alpha chain. (a) A space-filling model of the molecular structure of the complex. (b) A ribbon diagram showing the protein chains in the complex. (From the Protein Data Base, Rutgers University, Structure 1JRH, S. Sogabe, F. Stuart, C. Henke, A. Bridges, G. Williams, A. Birch, F. K. Winkler, and J. A. Robinson, 1997; http://www.rcsb.org)

WEB WORKS

The Centers for Disease Control and Prevention (CDC) maintains a Web site to provide information related to AIDS and HIV. Use your Web browser to connect to **http://chemistry.brookscole.com/skoogfac/.** From the Chapter Resource Menu, choose Web Works. Locate the Chapter 11 section, and click on the link to the CDC Web site. Use the search facility at the CDC site to locate pages containing information on testing for HIV. You will find that various types of immunoassay are useful for HIV testing. What are these immunoassay types? Then use Google to search the Web for these types of immunoassay. What physical or chemical properties are used in the immunoassays? What are the chemical principles behind these methods?

QUESTIONS AND PROBLEMS

11-1. Demonstrate how the sulfide ion concentration is related to the hydronium ion concentration of a solution that is kept saturated with hydrogen sulfide.

11-2. Why are simplifying assumptions restricted to relationships that are sums or differences?

***11-3.** Why do molar concentrations of some species appear as multiples in charge-balance equations?

11-4. Write the mass-balance expressions for a solution that is

*(a) 0.20 M in H_3AsO_4.

(b) 0.10 M in Na_2HAsO_4.

*(c) 0.0500 M in HClO and 0.100 M in NaClO.

(d) 0.25 M in NaF and saturated with CaF_2.

*(e) 0.100 M in NaOH and saturated with $Zn(OH)_2$, which undergoes the reaction $Zn(OH)_2 + 2OH^- \rightleftharpoons Zn(OH)_4^{2-}$.

(f) saturated with BaC_2O_4.

*(g) saturated with CaF_2.

11-5. Write the charge-balance equations for the solutions in Problem 11-4.

11-6. Calculate the molar solubility of $Ag_2C_2O_4$ in a solution that has a fixed H_3O^+ concentration of

*(a) 1.0×10^{-6} M.

(b) 1.0×10^{-7} M.

*(c) 1.0×10^{-9} M.

(d) 1.0×10^{-11} M.

11-7. Calculate the molar solubility of $BaSO_4$ in a solution in which $[H_3O^+]$ is

*(a) 2.5 M.

(b) 1.5 M.

*(c) 0.060 M.

(d) 0.200 M.

***11-8.** Calculate the molar solubility of CuS in a solution in which the $[H_3O^+]$ concentration is held constant at (a) 2.0×10^{-1} M and (b) 2.0×10^{-4} M.

11-9. Calculate the concentration of CdS in a solution in which the $[H_3O^+]$ is held constant at (a) 2.0×10^{-1} M and (b) 2.0×10^{-4} M.

11-10. Calculate the molar solubility of MnS (green) in a solution with a constant $[H_3O^+]$ of *(a) 2.00×10^{-5} and (b) 2.00×10^{-7}.

***11-11.** Calculate the molar solubility of $PbCO_3$ in a solution buffered to a pH of 7.00.

11-12. Calculate the molar solubility of Ag_2SO_3 ($K_{sp} = 1.5 \times 10^{-14}$) in a solution buffered to a pH of 8.00.

***11-13.** Dilute NaOH is introduced into a solution that is 0.050 M in Cu^{2+} and 0.040 M in Mn^{2+}.

(a) Which hydroxide precipitates first?

(b) What OH^- concentration is needed to initiate precipitation of the first hydroxide?

(c) What is the concentration of the cation forming the less soluble hydroxide when the more soluble hydroxide begins to form?

11-14. A solution is 0.040 M in Na_2SO_4 and 0.050 M in $NaIO_3$. To this is added a solution containing Ba^{2+}. Assuming that no HSO_4^- is present in the original solution,

(a) which barium salt will precipitate first?

(b) what is the Ba^{2+} concentration as the first precipitate forms?

(c) what is the concentration of the anion that forms the less soluble barium salt when the more soluble precipitate begins to form?

***11-15.** Silver ion is being considered as a reagent for separating I^- from SCN^- in a solution that is 0.060 M in KI and 0.070 M in NaSCN.

(a) What Ag^+ concentration is needed to lower the I^- concentration to 1.0×10^{-6} M?

(b) What is the Ag^+ concentration of the solution when AgSCN begins to precipitate?

(c) What is the ratio of SCN^- to I^- when AgSCN begins to precipitate?
(d) What is the ratio of SCN^- to I^- when the Ag^+ concentration is 1.0×10^{-3} M?

11-16. Using 1.0×10^{-6} M as the criterion for quantitative removal, determine whether it is feasible to use
(a) SO_4^{2-} to separate Ba^{2+} and Sr^{2+} in a solution that is initially 0.050 M in Sr^{2+} and 0.30 M in Ba^{2+}.
(b) SO_4^{2-} to separate Ba^{2+} and Ag^+ in a solution that is initially 0.020 M in each cation. For Ag_2SO_4, $K_{sp} = 1.6 \times 10^{-5}$.
(c) OH^- to separate Be^{3+} and Hf^{4+} in a solution that is initially 0.020 M in Be^{2+} and 0.010 M in Hf^{4+}. For $Be(OH)_3$, $K_{sp} = 7.0 \times 10^{-22}$ and for $Hf(OH)_4$, $K_{sp} = 4.0 \times 10^{-26}$.
(d) IO_3^- to separate In^{3+} and Tl^+ in a solution that is initially 0.20 M in In^{3+} and 0.090 M in Tl^+. For $In(IO_3)_3$, $K_{sp} = 3.3 \times 10^{-11}$; for $TlIO_3$, $K_{sp} = 3.1 \times 10^{-6}$.

***11-17.** What weight of AgBr dissolves in 200 mL of 0.100 M NaCN?

$$Ag^+ + 2CN^- \rightleftharpoons Ag(CN)_2^- \qquad \beta_2 = 1.3 \times 10^{21}$$

11-18. The equilibrium constant for formation of $CuCl_2^-$ is given by

$$Cu^+ + 2Cl^- \rightleftharpoons CuCl_2^-$$

$$\beta_2 = \frac{[CuCl_2^-]}{[Cu^+][Cl^-]^2} = 7.9 \times 10^4$$

What is the solubility of CuCl in solutions having the following analytical NaCl concentrations:
(a) 2.0 M?
(b) 2.0×10^{-1} M?
(c) 2.0×10^{-2} M?
(d) 2.0×10^{-3} M?
(e) 2.0×10^{-4} M?

***11-19.** In contrast to many salts, calcium sulfate is only partially dissociated in aqueous solution:

$$CaSO_4(aq) \rightleftharpoons Ca^{2+} + SO_4^{2-} \qquad K_d = 5.2 \times 10^{-3}$$

The solubility-product constant for $CaSO_4$ is 2.6×10^{-5}. Calculate the solubility of $CaSO_4$ in (a) water and (b) 0.0100 M Na_2SO_4. In addition, calculate the percent of undissociated $CaSO_4$ in each solution.

11-20. Calculate the molar solubility of TIS as a function of pH over the range of pH 10 to pH 1. Find values at every 0.5 pH unit and use the charting function of Excel to plot solubility versus pH.

11-21. Challenge Problem.
(a) The solubility of CdS is ordinarily very low, but can be increased by lowering the solution pH. Calculate the molar solubility of CdS as a function of pH from pH 11 to pH 1. Find values at every 0.5 pH unit and plot solubility versus pH.
(b) A solution contains 1×10^{-4} M of both Fe^{2+} and Cd^{2+}. Sulfide ions are slowly added to this solution to precipitate either FeS or CdS. Determine which ion precipitates first and the range of S^{2-} concentration that will allow a clean separation of the two ions.
(c) The analytical concentration of H_2S in a solution saturated with $H_2S(g)$ is 0.10 M. What pH range is necessary for the clean separation described in part (b)?
(d) If there is no pH control from a buffer, what is the pH of a saturated H_2S solution?
(e) Plot the α_0 and α_1 values for H_2S over the pH range of 10 to 1.
(f) A solution contains H_2S and NH_3. Four Cd^{2+} complexes form with NH_3 in a stepwise fashion: $Cd(NH_3)^{2+}$, $Cd(NH_3)_2^{2+}$, $Cd(NH_3)_3^{2+}$, and $Cd(NH_3)_4^{2+}$. Find the molar solubility of CdS in a solution of 0.1 M NH_3.
(g) For the same solution components as in part (f), buffers are prepared with a total concentration of $NH_3 + NH_4Cl = 0.10$ M. The pH values are 8.0, 8.5, 9.0, 9.5, 10.0, 10.5, and 11.0. Find the molar solubility of CdS in these solutions.
(h) For the solutions in part (g), how could you determine whether the solubility increase with pH is due to complex formation or to an activity effect?

InfoTrac College Edition

For additional readings, go to InfoTrac College Edition, your online research library, at

http://infotrac.thomsonlearning.com

PART III

Classical Methods of Analysis

A conversation with *Larry R. Faulkner*

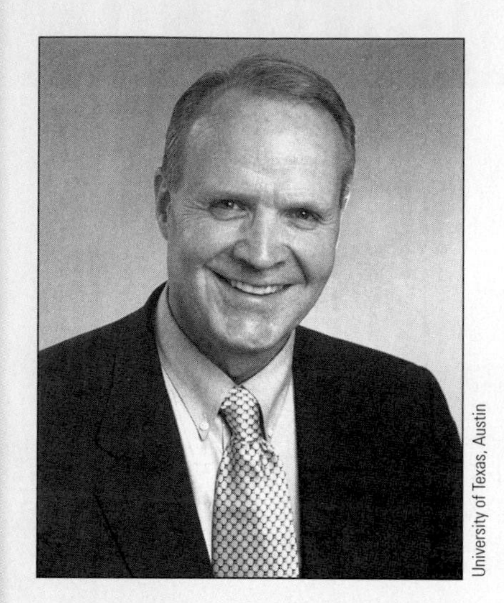
University of Texas, Austin

Larry R. Faulkner once was one of the foremost analytical chemists in the world. The past tense is appropriate because his analytical chemistry career has given way to a second career in university administration. Now Faulkner is the president of the University of Texas at Austin, where he thinks more about improving undergraduate education and helping the university become more of an asset to the Texas economy than about the electrochemical problems that first captured his professional interests.

Faulkner began his career at Southern Methodist University, where he received a B.S. in Chemistry in 1966. He then moved to Austin for the first time to work toward his doctorate in chemistry at the University of Texas. Faulkner's advisor was Allen J. Bard, the subject of an interview in Part IV of this text. After receiving his Ph.D., Faulkner went on to professorships at Harvard University, the University of Illinois, and the University of Texas. He transitioned into administration when he returned to the University of Illinois to head the chemistry department. He was then made dean of the College of Liberal Arts and Sciences and later provost and vice chancellor for academic affairs. He returned to the University of Texas as its president in 1998.

As an analytical chemist, Faulkner published more than 120 papers. He and Bard are co-authors of the textbook Electrochemical Methods: Fundamentals and Applications, now in its second edition. He is also a co-inventor of the cybernetic potentiostat, an instrument for electrochemical research and analysis. Among Faulkner's research awards are the American Chemical Society Award in Analytical Chemistry, the U.S. Department of Energy Award for Outstanding Scientific Achievement in Materials Chemistry, and the Charles N. Reilly Award from the Society for Electroanalytical Chemistry.

Q: What influence did your early education have in your choice of career?

A: As I look back, I was interested in electricity, light, and optics even in junior high school. It's interesting that I've carried those interests through to a career-long engagement with electrochemistry, luminescence, and light-producing reactions. I had two spectacular teachers in introductory chemistry, one in high school and one in college. Both had great rapport with their classes, a real affection for the subject, and a way of conveying it that imparted enthusiasm. As an undergraduate I did research on the magnetic susceptibility of inorganic compounds with a physical chemist. He had a spectacular commitment to science and high standards that I really wanted to be connected to.

Q: What interests did you pursue in your doctoral work?

A: When I came to Texas, Al Bard was a 32-year-old associate professor. He was young and enthusiastic, a marvelous teacher. Al's a brilliant chemist with a very high order of dedication to science—that is, to science understood broadly. I was captivated by his enthusiasm for his subject. You can't be around him without gaining a tremendous respect for him and without his affecting your view of what you want to do.

I came to Texas about 2 years after the discovery of electrogenerated chemiluminescence in Al's lab, where it had been learned that species undergoing electron transfer reactions could produce light. I was one of Al's first grad students in that area, and I stayed with it for two decades at Harvard and Illinois. The work had a lot to say about how electron transfer reactions occurred and how molecules handle the need to dissipate a large amount of energy in very energetic electron transfer reactions. It led us into the theory of electron transfer and all of the chemistry and physics associated with that.

Q: What made you choose a career in academia?

A: I became interested in the academic world as a sophomore in college, when I began to realize the scope of things that happen in a university, the interaction of teaching and research, and the generation of new knowledge. I was fortunate to be around a group of people whose level of dedication and fascination was really attractive to me. I was pointed in the academic direction through grad school, although not exclusively. Even in my last year I thought about taking the industrial path, but I was drawn back to the university because of the independence of intellectual activity.

Analytical chemistry in my era has literally gone from a period of not having confidence in its future to the point where it is a central player in public policy.

Q: What do you consider to be your most satisfying research accomplishment?

A: My research group did a lot to advance the art of electrochemical instrumentation. The early 1980s brought a new concept of coordination of instrumental methods that brought artificial intelligence into machine-operator interactions. Before that time, investigators in electrochemistry had separate apparatuses to carry out each of several experimental methods, or they had extremely complicated multipurpose instruments. We integrated about 40 methods into a single instrument that used a computer to simplify the queries made of the operator, to allow optimization of experimental conditions based on artificial intelligence, and to provide elaborate presentation of results by graphical means. Those things represent the standard now, but when we first brought the prototype out at the Pittsburgh Conference, it was pretty stunning. Having had the imagination in our group to create the concept and to bring it into reality has been very satisfying. The sign of real success is that nearly everything in the world of electrochemical instrumentation—and in the larger instrumental world, too—works that way now. Of course, that's not all due to our contribution, but I do believe that we contributed significantly.

My group was also one of first into nanotechnology—although we didn't call it that in the 1970s when we started. I got into structures based on very thin films on electrodes, into electron transfer in controlled structures on electrodes, and into what could be done to create sophisticated local electrochemical environments.

Q: What are your thoughts on analytical chemistry?

A: Analytical chemistry is a remarkable domain of chemistry. It is an area that has to take the techniques and knowledge of all of the rest of chemistry and bend that knowledge toward the goal of developing methods and techniques that can produce answers to very pragmatic questions in very practical circumstances. I've always had an interest in fundamental science, but I'm also interested in its relationship to the industrial world, to the clinical world, and to the environment. That is, how do we take things out of the lab and bring them into the larger world of human society?

Over the years, I have watched analytical chemistry as it has become central to huge questions of public concern. The global question of how to deliver health care in cost-effective ways rests in large measure on analytical chemistry. The question of how we're going to control terrorism has analytical chemistry as a serious component. Understanding the environment and learning how to protect it is centrally dependent on analytical chemistry. Better quality in manufacturing is strongly dependent on analytical chemistry. Analytical chemistry in my era has literally gone from a period of not having confidence in its future to the point where it is a central player in public policy.

I've been fortunate to live through some of the most stunning advances in analytical chemistry. When I was an undergrad, a large part of analytical practice involved classical methods, such as titration. In my years as a practicing scientist, not only did the electronics revolution occur in analytical chemistry, but also brought into the field were tremendous advances based on surface science and related fields, magnetic resonance, powerful separation methods, . . . a whole host of instrumental approaches that didn't exist in the 1960s came on the scene. It's a tremendous privilege to have been a part of it all.

Q: How did you become interested in being a university president?

A: I was not always interested in university leadership, certainly not at the presidential level. In fact, I never dealt seriously with that possibility until I became a provost. There are the same number of provosts as presidents, so at that point you inevitably have to confront whether you'd want to become a president! I decided that I wanted to do that only for an institution that I cared a lot about, so I didn't enter into many candidacies. At Texas I had been both a student and a professor. My family roots are all in this region, and I was interested in helping to build the future of Texas.

Q: What are the goals of your presidency?

A: In serving as president, my greatest desire is to preserve and extend this tremendous invention that has come out of American society. To achieve that goal, I, and others who lead similar institutions, must be effective at communicating the essential social role of the American research university. It's very important for people to realize that the integration of our nation's basic research capacity with its powerful universities is an American novelty. There are other countries that have followed in that line after the U.S. invented the model, but most countries do not use it. Instead, they separate research into institutes or corporations and leave universities mainly with a teaching role. In this country, we have gained great synergy and have produced both a stronger educational enterprise and stronger research enterprise by bringing these things together. It's a powerful concept, with proven results, that needs to be understood by policymakers and by citizens. ■

CHAPTER 12

Gravimetric Methods of Analysis

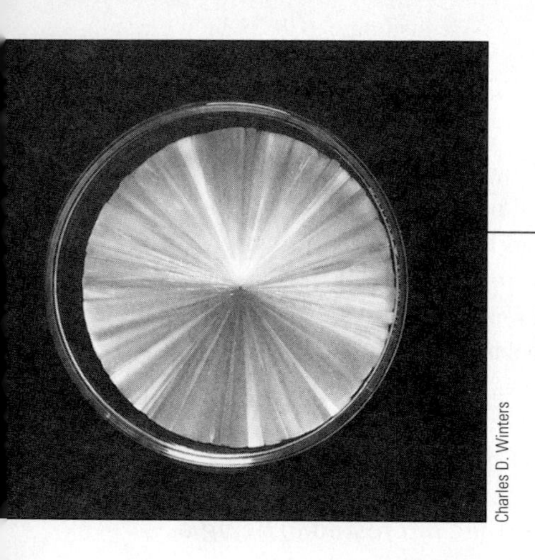
Charles D. Winters

The formation and growth of precipitates and crystals are very important in analytical chemistry and in other areas of science. Shown in the photo is the growth of sodium acetate crystals from a supersaturated solution. Because supersaturation leads to small particles that are difficult to filter, it is desirable in gravimetric analysis to minimize the supersaturation and thus increase the particle size of the solid that is formed.

The properties of precipitates that are used in chemical analysis are described in this chapter. The techniques for obtaining easily filterable precipitates that are free from contaminants are major topics in this chapter. Such precipitates are not only used in gravimetric analysis but also employed in the separation of interferences for other analytical procedures.

Several analytical methods are based on mass measurements. In ***precipitation gravimetry,*** *the analyte is separated from a solution of the sample as a precipitate and is converted to a compound of known composition that can be weighed. In* ***volatilization gravimetry,*** *the analyte is separated from other constituents of a sample by conversion to a gas of known chemical composition. The weight of this gas then serves as a measure of the analyte concentration. These two types of gravimetry are considered in this chapter.*[1] *In* ***electrogravimetry,*** *the analyte is separated by deposition on an electrode by an electrical current. The mass of this product then provides a measure of the analyte concentration. Electrogravimetry is described in Section 22C.*

Two other types of analytical methods are based on mass. In ***gravimetric titrimetry,*** *which is described in Section 13D, the mass of a reagent, of known concentration, required to react completely with the analyte provides the information needed to determine the analyte concentration.* ***Atomic mass spectrometry*** *uses a mass spectrometer to separate the gaseous ions formed from the elements making up a sample of matter. The concentration of the resulting ions is then determined by measuring the electrical current produced when they fall on the surface of an ion detector. This technique is described briefly in Chapter 28.*

Gravimetric methods are quantitative methods that are based on determining the mass of a pure compound to which the analyte is chemically related.

► Gravimetric methods of analysis are based on mass measurements made with an analytical balance, an instrument that yields highly accurate and precise data. In fact, if you perform a gravimetric chloride determination in the laboratory, you may make some of the most accurate and precise measurements of your life.

[1] For an extensive treatment of gravimetric methods, see C. L. Rulfs, in *Treatise on Analytical Chemistry,* I. M. Kolthoff and P. J. Elving, Eds., Part I, Vol. 11, Chapter 13. New York: Wiley, 1975.

12A PRECIPITATION GRAVIMETRY

In precipitation gravimetry, the analyte is converted to a sparingly soluble precipitate. This precipitate is then filtered, washed free of impurities, converted to a product of known composition by suitable heat treatment, and weighed. For example, a precipitation method for determining calcium in natural waters is recommended by the Association of Official Analytical Chemists. Here, an excess of oxalic acid, $H_2C_2O_4$, is added to an aqueous solution of the sample. Ammonia is then added, which neutralizes the acid and causes essentially all of the calcium in the sample to precipitate as calcium oxalate. The reactions are

$$2NH_3 + H_2C_2O_4 \rightarrow 2NH_4^+ + C_2O_4^{2-}$$

$$Ca^{2+}(aq) + C_2O_4^{2-}(aq) \rightarrow CaC_2O_4(s)$$

The precipitate is filtered using a weighed filtering crucible, then dried and ignited. This process converts the precipitate entirely to calcium oxide. The reaction is

$$CaC_2O_4(s) \xrightarrow{\Delta} CaO(s) + CO(g) + CO_2(g)$$

After cooling, the crucible and precipitate are weighed, and the mass of calcium oxide is determined by subtracting the known mass of the crucible. The calcium content of the sample is then computed as shown in Example 12-1, Section 12B.

12A-1 Properties of Precipitates and Precipitating Reagents

Ideally, a gravimetric precipitating agent should react *specifically* or at least *selectively* with the analyte. Specific reagents, which are rare, react only with a single chemical species. Selective reagents, which are more common, react with a limited number of species. In addition to specificity and selectivity, the ideal precipitating reagent would react with the analyte to give a product that is

◀ An example of a selective reagent is $AgNO_3$. The only common ions that it precipitates from acidic solution are Cl^-, Br^-, I^-, and SCN^-. Dimethylglyoxime, which is discussed in Section 12C-3, is a specific reagent that precipitates only Ni^{2+} from alkaline solutions.

1. easily filtered and washed free of contaminants;
2. of sufficiently low solubility that no significant loss of the analyte occurs during filtration and washing;
3. unreactive with constituents of the atmosphere;
4. of known chemical composition after it is dried or, if necessary, ignited.

Few, if any, reagents produce precipitates that possess all these desirable properties.

The variables that influence solubility (the second property in the foregoing list) are discussed in Section 11B. In the sections that follow, we are concerned with methods to obtain easily filtered and pure solids of known composition.[2]

[2]For a more detailed treatment of precipitates, see H. A. Laitinen and W. E. Harris, *Chemical Analysis,* 2nd ed., Chapters 8 and 9. New York: McGraw-Hill, 1975; A. E. Nielsen, in *Treatise on Analytical Chemistry,* 2nd ed., I. M. Kolthoff and P. J. Elving, Eds., Part I, Vol. 3, Chapter 27. New York: Wiley, 1983.

A **colloid** consists of solid particles with diameters that are less than 10^{-4} cm.

In diffuse light, **colloidal suspensions** may be perfectly clear and appear to contain no solid. The presence of the second phase can be detected, however, by shining the beam of a flashlight into the solution. Because particles of colloidal dimensions scatter visible radiation, the path of the beam through the solution can be seen by the eye. This phenomenon is called the **Tyndall effect** (see color plate 6).

► The particles of a colloidal suspension are not easily filtered. To trap these particles, the pore size of the filtering medium must be so small that filtrations take inordinately long. With suitable treatment, however, the individual colloidal particles can be made to stick together, thus giving a filterable mass.

► Equation 12-1 is known as the Von Weimarn equation in recognition of the scientist who proposed it in 1925.

A **supersaturated solution** is an unstable solution that contains a higher solute concentration than a saturated solution. With time, supersaturation is relieved by precipitation of the excess solute (see color plate 5).

► To increase the particle size of a precipitate, minimize the relative supersaturation during precipitate formation.

Nucleation is a process in which a minimum number of atoms, ions, or molecules join together to give a stable solid.

► Precipitates form by nucleation and by particle growth. If nucleation predominates, a large number of very fine particles results; if particle growth predominates, a smaller number of larger particles is obtained.

12A-2 Particle Size and Filterability of Precipitates

Precipitates consisting of large particles are generally desirable for gravimetric work because these particles are easy to filter and wash free of impurities. In addition, precipitates of this type are usually purer than are precipitates made up of fine particles.

Factors That Determine the Particle Size of Precipitates

The particle size of solids formed by precipitation varies enormously. At one extreme are **colloidal suspensions,** whose tiny particles are invisible to the naked eye (10^{-7} to 10^{-4} cm in diameter). Colloidal particles show no tendency to settle from solution and are not easily filtered. At the other extreme are particles with dimensions on the order of tenths of a millimeter or greater. The temporary dispersion of such particles in the liquid phase is called a **crystalline suspension.** The particles of a crystalline suspension tend to settle spontaneously and are easily filtered.

Scientists have studied precipitate formation for many years, but the mechanism of the process is still not fully understood. It is certain, however, that the particle size of a precipitate is influenced by such experimental variables as precipitate solubility, temperature, reactant concentrations, and rate at which reactants are mixed. The net effect of these variables can be accounted for, at least qualitatively, by assuming that the particle size is related to a single property of the system called the **relative supersaturation,** where

$$\text{relative supersaturation} = \frac{Q - S}{S} \tag{12-1}$$

In this equation, Q is the concentration of the solute at any instant and S is its equilibrium solubility.

Generally, precipitation reactions are slow, so that even when a precipitating reagent is added drop by drop to a solution of an analyte, some supersaturation is likely. Experimental evidence indicates that the particle size of a precipitate varies inversely with the average relative supersaturation during the time when the reagent is being introduced. Thus, when $(Q - S)/S$ is large, the precipitate tends to be colloidal; when $(Q - S)/S$ is small, a crystalline solid is more likely.

Mechanism of Precipitate Formation

The effect of relative supersaturation on particle size can be explained if we assume that precipitates form in two ways; by **nucleation** and by **particle growth.** The particle size of a freshly formed precipitate is determined by the mechanism that predominates.

In nucleation, a few ions, atoms, or molecules (perhaps as few as four or five) come together to form a stable solid. Often, these nuclei form on the surface of suspended solid contaminants, such as dust particles. Further precipitation then involves a competition between additional nucleation and growth on existing nuclei (particle growth). If nucleation predominates, a precipitate containing a large number of small particles results; if growth predominates, a smaller number of larger particles is produced.

The rate of nucleation is believed to increase enormously with increasing relative supersaturation. In contrast, the rate of particle growth is only moderately enhanced by high relative supersaturations. Thus, when a precipitate is formed at high relative supersaturation, nucleation is the major precipitation mechanism, and a large number of small particles is formed. At low relative supersaturations, on the other hand, the rate of particle growth tends to predominate, and deposition of solid on existing particles occurs to the exclusion of further nucleation; a crystalline suspension results.

Experimental Control of Particle Size

Experimental variables that minimize supersaturation and thus produce crystalline precipitates include elevated temperatures to increase the solubility of the precipitate (S in Equation 12-1), dilute solutions (to minimize Q), and slow addition of the precipitating agent with good stirring. The last two measures also minimize the concentration of the solute (Q) at any given instant.

Larger particles can also be obtained by controlling pH, provided the solubility of the precipitate depends on pH. For example, large, easily filtered crystals of calcium oxalate are obtained by forming the bulk of the precipitate in a mildly acidic environment in which the salt is moderately soluble. The precipitation is then completed by slowly adding aqueous ammonia until the acidity is sufficiently low for removal of substantially all of the calcium oxalate. The additional precipitate produced during this step deposits on the solid particles formed in the first step.

Unfortunately, many precipitates cannot be formed as crystals under practical laboratory conditions. A colloidal solid is generally encountered when a precipitate has such a low solubility that S in Equation 12-1 always remains negligible relative to Q. The relative supersaturation thus remains enormous throughout precipitate formation, and a colloidal suspension results. For example, under conditions feasible for an analysis, the hydrous oxides of iron(III), aluminum, and chromium(III) and the sulfides of most heavy-metal ions form only as colloids because of their very low solubilities.[3]

◀ Precipitates that have very low solubilities, such as many sulfides and hydrous oxides, generally form as colloids.

12A-3 Colloidal Precipitates

Individual colloidal particles are so small that they are not retained by ordinary filters. Moreover, Brownian motion prevents their settling out of solution under the influence of gravity. Fortunately, however, we can coagulate, or agglomerate, the individual particles of most colloids to give a filterable, amorphous mass that will settle out of solution.

Coagulation of Colloids

Coagulation can be hastened by heating, by stirring, and by adding an electrolyte to the medium. To understand the effectiveness of these measures, we need to look into why colloidal suspensions are stable and do not coagulate spontaneously.

Colloidal suspensions are stable because all of the particles of the colloid are either positively or negatively charged. This charge results from cations or anions that are bound to the surface of the particles. We can easily demonstrate that colloidal particles are charged by observing their migration when placed in an electrical field. The process by which ions are retained *on the surface of a solid* is known as **adsorption.**

Adsorption is a process in which a substance (gas, liquid, or solid) is held *on the surface* of a solid. In contrast, **absorption** involves retention of a substance *within* the pores of a solid.

[3]Silver chloride illustrates that the relative supersaturation concept is imperfect. This compound ordinarily forms as a colloid, yet its molar solubility is not significantly different from that of other compounds, such as $BaSO_4$, which generally form as crystals.

The adsorption of ions on an ionic solid originates from the normal bonding forces that are responsible for crystal growth. For example, a silver ion at the surface of a silver chloride particle has a partially unsatisfied bonding capacity for anions because of its surface location. Negative ions are attracted to this site by the same forces that hold chloride ions in the silver chloride lattice. Chloride ions at the surface of the solid exert an analogous attraction for cations dissolved in the solvent.

The kind of ions retained on the surface of a colloidal particle and their number depend, in a complex way, on several variables. For a suspension produced in the course of a gravimetric analysis, however, the species adsorbed, and hence the charge on the particles, can be easily predicted because lattice ions are generally more strongly held than others. For example, when silver nitrate is first added to a solution containing chloride ion, the colloidal particles of the precipitate are negatively charged as a result of adsorption of some of the excess chloride ions. This charge, however, becomes positive when enough silver nitrate has been added to provide an excess of silver ions. The surface charge is at a minimum when the supernatant liquid contains an excess of neither ion.

► The charge on a colloidal particle formed in a gravimetric analysis is determined by the charge of the lattice ion that is in excess when the precipitation is complete.

The extent of adsorption and thus the charge on a given particle increase rapidly as the concentration of a common ion becomes greater. Eventually, however, the surface of the particles becomes covered with the adsorbed ions, and the charge becomes constant and independent of concentration.

Figure 12-1 shows a colloidal silver chloride particle in a solution that contains an excess of silver nitrate. Attached directly to the solid surface is the **primary**

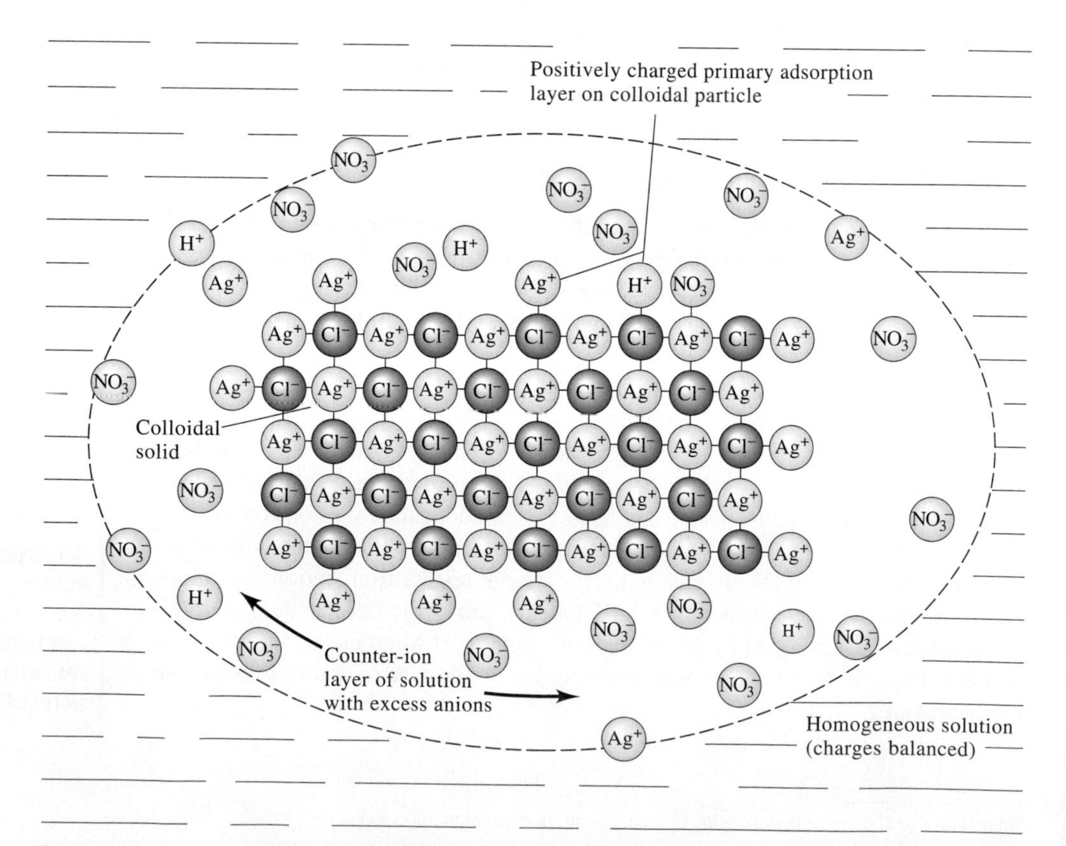

Figure 12-1 A colloidal silver chloride particle suspended in a solution of silver nitrate.

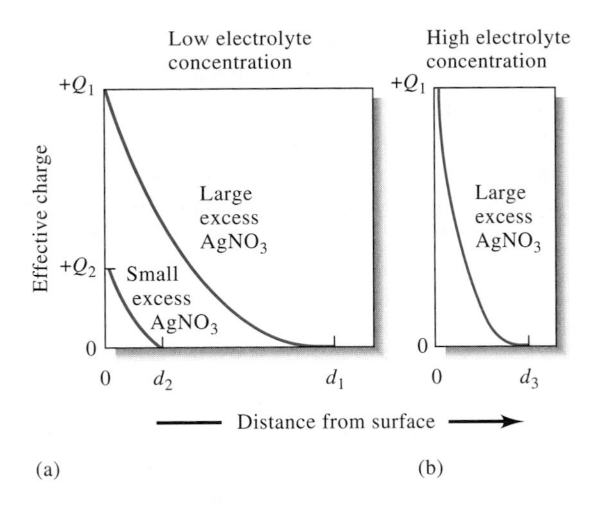

Figure 12-2 Effect of $AgNO_3$ and electrolyte concentration on the thickness of the double layer surrounding a colloidal AgCl particle in a solution containing excess $AgNO_3$.

adsorption layer, which consists mainly of adsorbed silver ions. Surrounding the charged particle is a layer of solution, called the **counter-ion layer,** which contains sufficient excess of negative ions (principally nitrate) to just balance the charge on the surface of the particle. The primarily adsorbed silver ions and the negative counter-ion layer constitute an **electric double layer** that imparts stability to the colloidal suspension. As colloidal particles approach one another, this double layer exerts an electrostatic repulsive force that prevents particles from colliding and adhering.

Figure 12-2a shows the effective charge on two silver chloride particles. The upper curve represents a particle in a solution that contains a reasonably large excess of silver nitrate, whereas the lower curve depicts a particle in a solution that has a much lower silver nitrate content. The effective charge can be thought of as a measure of the repulsive force that the particle exerts on like particles in the solution. Note that the effective charge falls off rapidly as the distance from the surface increases, and it approaches zero at the points d_1 or d_2. These decreases in effective charge (in both cases positive) are caused by the negative charge of the excess counter-ions in the double layer surrounding each particle. At points d_1 and d_2, the number of counter-ions in the layer is approximately equal to the number of primarily adsorbed ions on the surfaces of the particles; therefore, the effective charge of the particles approaches zero at this point.

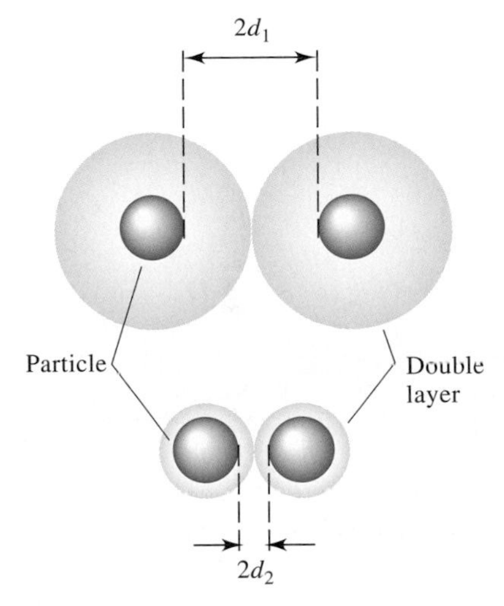

Figure 12-3 The electrical double layer of a colloid consists of a layer of charge adsorbed on the surface of the particle (the primary adsorption layer) and a layer of opposite charge (the counter-ion layer) in the solution surrounding the particle. Increasing the electrolyte concentration has the effect of decreasing the volume of the counter-ion layer, thereby increasing the chances for coagulation.

The upper portion of Figure 12-3 depicts two silver chloride particles and their counter-ion layers as they approach one another in the concentrated silver nitrate just considered. Note that the effective charge on the particles prevents them from approaching one another more closely than about $2d_1$—a distance that is too great for coagulation to occur. As shown in the lower part of Figure 12-3, in the more dilute silver nitrate solution, the two particles can approach within $2d_2$ of one another. Ultimately, as the concentration of silver nitrate is further decreased, the distance between particles becomes small enough for the forces of agglomeration to take effect and a coagulated precipitate to appear.

Coagulation of a colloidal suspension can often be brought about by a short period of heating, particularly if accompanied by stirring. Heating decreases the number of adsorbed ions and thus the thickness, d_i, of the double layer. The particles may also gain enough kinetic energy at the higher temperature to overcome the barrier to close approach posed by the double layer.

◀ Colloidal suspensions can often be coagulated by heating, stirring, and adding an electrolyte.

An even more effective way to coagulate a colloid is to increase the electrolyte concentration of the solution. If we add a suitable ionic compound to a colloidal suspension, the concentration of counter-ions increases in the vicinity of each particle. As a result, the volume of solution that contains sufficient counter-ions to balance the charge of the primary adsorption layer decreases. The net effect of adding an electrolyte is thus a shrinkage of the counter-ion layer, as shown in Figure 12-2b. The particles can then approach one another more closely and agglomerate.

Peptization of Colloids

Peptization is a process by which a coagulated colloid returns to its dispersed state.

Peptization is the process by which a coagulated colloid reverts to its original dispersed state. When a coagulated colloid is washed, some of the electrolyte responsible for its coagulation is leached from the internal liquid in contact with the solid particles. Removal of this electrolyte has the effect of increasing the volume of the counter-ion layer. The repulsive forces responsible for the original colloidal state are then reestablished, and particles detach themselves from the coagulated mass. The washings become cloudy as the freshly dispersed particles pass through the filter.

The chemist is thus faced with a dilemma in working with coagulated colloids. On the one hand, washing is needed to minimize contamination; on the other, there is a risk of losses resulting from peptization if pure water is used. The problem is commonly solved by washing the precipitate with a solution containing an electrolyte that volatilizes when the precipitate is dried or ignited. For example, silver chloride is ordinarily washed with a dilute solution of nitric acid. While the precipitate undoubtedly becomes contaminated with the acid, no harm results, since the nitric acid is volatilized during the ensuing drying step.

Practical Treatment of Colloidal Precipitates

Colloids are best precipitated from hot, stirred solutions containing sufficient electrolyte to ensure coagulation. The filterability of a coagulated colloid frequently improves if it is allowed to stand for an hour or more in contact with the hot solution from which it was formed. During this process, which is known as **digestion,** weakly bound water appears to be lost from the precipitate; the result is a denser mass that is easier to filter.

Digestion is a process in which a precipitate is heated for an hour or more in the solution from which it was formed (the mother liquor).

12A-4 Crystalline Precipitates

Mother liquor is the solution from which a precipitate was formed.

Crystalline precipitates are generally more easily filtered and purified than are coagulated colloids. In addition, the size of individual crystalline particles, and thus their filterability, can be controlled to a degree.

Methods of Improving Particle Size and Filterability

The particle size of crystalline solids can often be improved significantly by minimizing Q or maximizing S, or both, in Equation 12-1. Minimization of Q is generally accomplished by using dilute solutions and adding the precipitating reagent slowly and with good mixing. Often, S is increased by precipitating from hot solution or by adjusting the pH of the precipitation medium.

► Digestion improves the purity and filterability of both colloidal and crystalline precipitates.

Digestion of crystalline precipitates (without stirring) for some time after formation frequently yields a purer, more filterable product. The improvement in filterability undoubtedly results from the dissolution and recrystallization that occur continuously and at an enhanced rate at elevated temperatures. Recrystallization apparently results in bridging between adjacent particles, a process that yields larger and more easily filtered crystalline aggregates. This view is supported by the observation that little improvement in filtering characteristics occurs if the mixture is stirred during digestion.

12A-5 Coprecipitation

Coprecipitation is a phenomenon in which *otherwise soluble* compounds are removed from solution during precipitate formation. It is important to understand that contamination of a precipitate by a second substance whose solubility product has been exceeded *does not constitute coprecipitation.*

Coprecipitation is a process in which *normally soluble* compounds are carried out of solution by a precipitate.

There are four types of coprecipitation: **surface adsorption, mixed-crystal formation, occlusion,** and **mechanical entrapment.**[4] Surface adsorption and mixed-crystal formation are equilibrium processes, whereas occlusion and mechanical entrapment arise from the kinetics of crystal growth.

Surface Adsorption

Adsorption is a common source of coprecipitation and is likely to cause significant contamination of precipitates with large specific surface areas—that is, coagulated colloids (see Feature 12-1 for definition of specific area). Although adsorption does occur in crystalline solids, its effects on purity are usually undetectable because of the relatively small specific surface area of these solids.

◀ Adsorption is often the major source of contamination in coagulated colloids but of no significance in crystalline precipitates.

Coagulation of a colloid does not significantly decrease the amount of adsorption because the coagulated solid still contains large internal surface areas that remain exposed to the solvent (Figure 12-4). The coprecipitated contaminant on the coagulated colloid consists of the lattice ion originally adsorbed on the surface before coagulation plus the counter-ion of opposite charge held in the film of solution immediately adjacent to the particle. *The net effect of surface adsorption is therefore the carrying down of an otherwise soluble compound as a surface contaminant.* For example, the coagulated silver chloride formed in the gravimetric determination of chloride ion is contaminated with primarily adsorbed silver ions with nitrate or other anions in the counter-ion layer. As a consequence, silver nitrate, a normally soluble compound, is coprecipitated with the silver chloride.

◀ In adsorption, a normally soluble compound is carried out of solution on the surface of a coagulated colloid. This compound consists of the primarily adsorbed ion and an ion of opposite charge from the counter-ion layer.

Minimizing Adsorbed Impurities on Colloids The purity of many coagulated colloids is improved by digestion. During this process, water is expelled from the solid to give a denser mass that has a smaller specific surface area for adsorption.

Washing a coagulated colloid with a solution containing a volatile electrolyte may also be helpful because any nonvolatile electrolyte added earlier to cause coagulation is displaced by the volatile species. Washing generally does not remove much of the primarily adsorbed ions because the attraction between these ions and the surface of the solid is too strong. Exchange occurs, however, between existing *counter-ions* and ions in the wash liquid. For example, in the determination of silver by precipitation with chloride ion, the primarily adsorbed species is

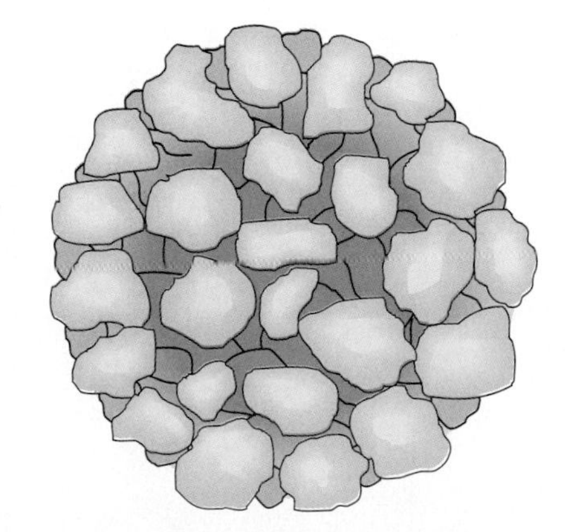

Figure 12-4 A coagulated colloid. This figure suggests that a coagulated colloid continues to expose a large surface area to the solution from which it was formed.

[4]Several systems of classification of coprecipitation phenomena have been suggested. We follow the simple system proposed by A. E. Nielsen, in *Treatise on Analytical Chemistry,* 2nd ed., I. M. Kolthoff and P. J. Elving, Eds., Part I, Vol. 3, p 333. New York: Wiley, 1983.

FEATURE 12-1

Specific Surface Area of Colloids

Specific surface area is defined as the surface area per unit mass of solid and is ordinarily expressed in terms of square centimeters per gram. For a given mass of solid, the specific surface area increases dramatically as particle size decreases, and it becomes enormous for colloids. For example, the solid cube shown in Figure 12F-1, which has dimensions of 1 cm on a side, has a surface area of 6 cm^2. If this cube weighs 2 g, its specific surface area is 6 cm^2/2 g = 3 cm^2/g. This cube could be divided into 1000 cubes, each having an edge length of 0.1 cm. The surface area of each face of these cubes is now 0.1 cm $\times$ 0.1 cm = 0.01 cm^2, and the total area for the six faces of the cube is 0.06 cm^2. Because there are 1000 of these cubes, the total surface area for the 2 g of solid is now 60 cm^2; the specific surface area is 30 cm^2/g. Continuing in this way, we find that the specific surface area becomes 300 cm^2/g when we have 10^6 cubes that are 0.01 cm on a side. The particle size of a typical crystalline suspension lies in the region of 0.01 to 0.1 cm, so a typical crystalline precipitate has a specific surface area between 30 cm^2/g and 300 cm^2/g. Contrast these figures with those for 2 g of a colloid made up of 10^{18} particles, each having an edge of 10^{-6} cm. Here, the specific area is 3×10^6 cm^2/g, which converts to something over 3000 ft^2/g. Based on these calculations, 1 g of a typical colloidal suspension has a surface area that is equivalent to the floor area of a good-sized home.

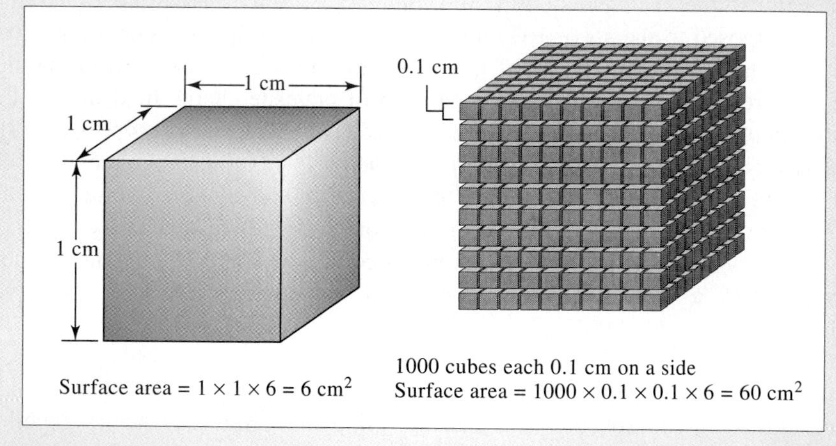

Figure 12F-1 Increase in surface area per unit mass with decrease in particle size.

chloride. Washing with an acidic solution converts the counter-ion layer largely to hydrogen ions so that both chloride and hydrogen ions are retained by the solid. Volatile HCl is then given off when the precipitate is dried.

Regardless of the method of treatment, a coagulated colloid is always contaminated to some degree, even after extensive washing. The error introduced into the analysis from this source can be as low as 1 to 2 ppt, as in the coprecipitation of silver nitrate on silver chloride. In contrast, coprecipitation of heavy-metal hydroxides on the hydrous oxides of trivalent iron or aluminum can result in errors as large as several percent, which is generally intolerable.

Reprecipitation A drastic but effective way to minimize the effects of adsorption is **reprecipitation.** In this process, the filtered solid is redissolved and reprecipitated. The first precipitate ordinarily carries down only a fraction of the contaminant present in the original solvent. Thus, the solution containing the redissolved precipitate has a significantly lower contaminant concentration than the original, and even less adsorption occurs during the second precipitation. Reprecipitation adds substantially to the time required for an analysis but is often necessary for such precipitates as the hydrous oxides of iron(III) and aluminum, which have extraordinary tendencies to adsorb the hydroxides of heavy-metal cations such as zinc, cadmium, and manganese.

Mixed-Crystal Formation

Mixed-crystal formation is a type of coprecipitation in which a contaminant ion replaces an ion in the lattice of a crystal.

In mixed-crystal formation, one of the ions in the crystal lattice of a solid is replaced by an ion of another element. For this exchange to occur, it is necessary that the two ions have the same charge and that their sizes differ by no more than about 5%. Furthermore, the two salts must belong to the same crystal class. For example, barium sulfate formed by adding barium chloride to a solution containing sulfate, lead, and acetate ions is found to be severely contaminated by lead sulfate even though acetate ions normally prevent precipitation of lead sulfate by complexing the lead. Here, lead ions replace some of the barium ions in the barium sulfate crystals. Other examples of coprecipitation by mixed-crystal formation include $MgKPO_4$ in $MgNH_4PO_4$, $SrSO_4$ in $BaSO_4$, and MnS in CdS.

The extent of mixed-crystal contamination is governed by the law of mass action and increases as the ratio of contaminant to analyte concentration increases. Mixed-crystal formation is a particularly troublesome type of coprecipitation because little can be done about it when certain combinations of ions are present in a sample matrix. This problem is encountered with both colloidal suspensions and crystalline precipitates. When mixed-crystal formation occurs, the interfering ion may have to be separated before the final precipitation step. Alternatively, a different precipitating reagent that does not give mixed crystals with the ions in question may be used.

Occlusion and Mechanical Entrapment

Occlusion is a type of coprecipitation in which a compound is trapped within a pocket formed during rapid crystal growth.

When a crystal is growing rapidly during precipitate formation, foreign ions in the counter-ion layer may become trapped, or *occluded,* within the growing crystal. Because supersaturation and thus growth rate decrease as precipitation progresses, the amount of occluded material is greatest in that part of a crystal that forms first.

Mechanical entrapment occurs when crystals lie close together during growth. Several crystals grow together and in so doing trap a portion of the solution in a tiny pocket.

◀ Mixed-crystal formation may occur in both colloidal and crystalline precipitates, whereas occlusion and mechanical entrapment are confined to crystalline precipitates.

Both occlusion and mechanical entrapment are at a minimum when the rate of precipitate formation is low—that is, under conditions of low supersaturation. In addition, digestion is often remarkably helpful in reducing these types of coprecipitation. Undoubtedly, the rapid solution and reprecipitation that go on at the elevated temperature of digestion open up the pockets and allow the impurities to escape into the solution.

Coprecipitation Errors

◀ Coprecipitation can cause either negative or positive errors.

Coprecipitated impurities may cause either negative or positive errors in an analysis. If the contaminant is not a compound of the ion being determined, a positive error will always result. Thus, a positive error is observed whenever colloidal silver chloride adsorbs silver nitrate during a chloride analysis. In contrast, when the contaminant does contain the ion being determined, either positive or negative errors may be observed. For example, in the determination of barium by precipitation as barium sulfate, occlusion of other barium salts occurs. If the occluded contaminant is barium nitrate, a positive error is observed because this compound has a larger molar mass than the barium sulfate that would have formed had no coprecipitation occurred. If barium chloride is the contaminant, the error is negative because its molar mass is less than that of the sulfate salt.

12A-6 Precipitation from Homogeneous Solution

Homogeneous precipitation is a process in which a precipitate is formed by slow generation of a precipitating reagent homogeneously throughout a solution.

► Solids formed by homogeneous precipitation are generally purer and more easily filtered than precipitates generated by direct addition of a reagent to the analyte solution.

Precipitation from homogeneous solution is a technique in which a precipitating agent is generated in a solution of the analyte by a slow chemical reaction.[5] Local reagent excesses do not occur because the precipitating agent appears gradually and homogeneously throughout the solution and reacts immediately with the analyte. As a result, the relative supersaturation is kept low during the entire precipitation. In general, homogeneously formed precipitates, both colloidal and crystalline, are better suited for analysis than a solid formed by direct addition of a precipitating reagent.

Urea is often used for the homogeneous generation of hydroxide ion. The reaction can be expressed by the equation

$$(H_2N)_2CO + 3H_2O \rightarrow CO_2 + 2NH_4^+ + 2OH^-$$

This hydrolysis proceeds slowly at temperatures just below 100°C, and 1 to 2 hours is needed to complete a typical precipitation. Urea is particularly valuable for the precipitation of hydrous oxides or basic salts. For example, hydrous oxides of iron(III) and aluminum, formed by direct addition of base, are bulky and gelatinous masses that are heavily contaminated and difficult to filter. In contrast, when these same products are produced by homogeneous generation of hydroxide ion, they are dense and easily filtered and have considerably higher purity. Figure 12-5 shows hydrous oxide precipitates of aluminum formed by direct addition of base and by homogeneous precipitates with urea. Homogeneous precipitation of crystalline precipitates also results in marked increases in crystal size as well as improvements in purity.

Representative methods based on precipitation by homogeneously generated reagents are given in Table 12-1.

Figure 12-5 Aluminum hydroxide formed by the direct addition of ammonia (left) and the homogeneous production of hydroxide (right).

12A-7 Drying and Ignition of Precipitates

After filtration, a gravimetric precipitate is heated until its mass becomes constant. Heating removes the solvent and any volatile species carried down with the precipitate. Some precipitates are also ignited to decompose the solid and form a compound of known composition. This new compound is often called the *weighing form.*

[5]For a general reference on this technique, see L. Gordon, M. L. Salutsky, and H. H. Willard, *Precipitation from Homogeneous Solution,* New York: Wiley, 1959.

TABLE 12-1

Methods for Homogeneous Generation of Precipitating Agents

Precipitating Agent	Reagent	Generation Reaction	Elements Precipitated
OH^-	Urea	$(NH_2)_2CO + 3H_2O \rightarrow CO_2 + 2NH_4^+ + 2OH^-$	Al, Ga, Th, Bi, Fe, Sn
PO_4^{3-}	Trimethyl phosphate	$(CH_3O)_3PO + 3H_2O \rightarrow 3CH_3OH + H_3PO_4$	Zr, Hf
$C_2O_4^{2-}$	Ethyl oxalate	$(C_2H_5)_2C_2O_4 + 2H_2O \rightarrow 2C_2H_5OH + H_2C_2O_4$	Mg, Zn, Ca
SO_4^{2-}	Dimethyl sulfate	$(CH_3O)_2SO_2 + 4H_2O \rightarrow 2CH_3OH + SO_4^{2-} + 2H_3O^+$	Ba, Ca, Sr, Pb
CO_3^{2-}	Trichloroacetic acid	$Cl_3CCOOH + 2OH^- \rightarrow CHCl_3 + CO_3^{2-} + H_2O$	La, Ba, Ra
H_2S	Thioacetamide*	$CH_3CSNH_2 + H_2O \rightarrow CH_3CONH_2 + H_2S$	Sb, Mo, Cu, Cd
DMG†	Biacetyl + hydroxylamine	$CH_3COCOCH_3 + 2H_2NOH \rightarrow DMG + 2H_2O$	Ni
HOQ‡	8-Acetoxyquinoline§	$CH_3COOQ + H_2O \rightarrow CH_3COOH + HOQ$	Al, U, Mg, Zn

*$CH_3-C(=S)-NH_2$

†DMG = Dimethylglyoxime = $CH_3-C(=N-OH)-C(=N-OH)-CH_3$

‡HOQ = 8-Hydroxyquinoline = [structure]

§$CH_3-C(=O)-O-$[quinolin-8-yl]

The temperature required to produce a suitable weighing form varies from precipitate to precipitate. Figure 12-6 shows mass loss as a function of temperature for several common analytical precipitates. These data were obtained with an automatic thermobalance,[6] an instrument that records the mass of a substance continuously as its temperature is increased at a constant rate (Figure 12-7). Heating three of the precipitates—silver chloride, barium sulfate, and aluminum oxide—simply causes removal of water and perhaps volatile electrolytes. Note the vastly different temperatures required to produce an anhydrous precipitate of constant mass. Moisture is completely removed from silver chloride at temperatures higher than 110°C, but dehydration of aluminum oxide is not complete until a temperature greater than 1000°C is achieved. Aluminum oxide formed homogeneously with urea can be completely dehydrated at about 650°C.

The thermal curve for calcium oxalate is considerably more complex than the others shown in Figure 12-6. Below about 135°C, unbound water is eliminated to give the monohydrate $CaC_2O_4 \cdot H_2O$. This compound is then converted to the

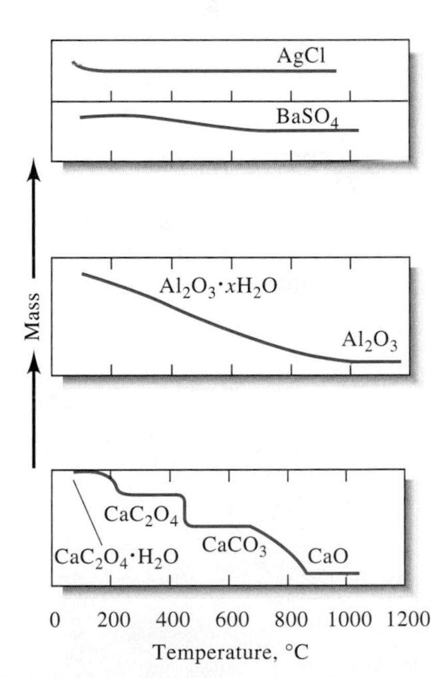

Figure 12-6 Effect of temperature on precipitate mass.

[6]For descriptions of thermobalances, see W. W. Wendlandt, *Thermal Methods of Analysis*, 3rd ed. New York: Wiley, 1985; A. J. Paszto, in *Handbook of Instrumental Techniques for Analytical Chemistry*, F. Settle, Ed., Upper Saddle River, NJ: Prentice Hall, 1997, Ch. 50.

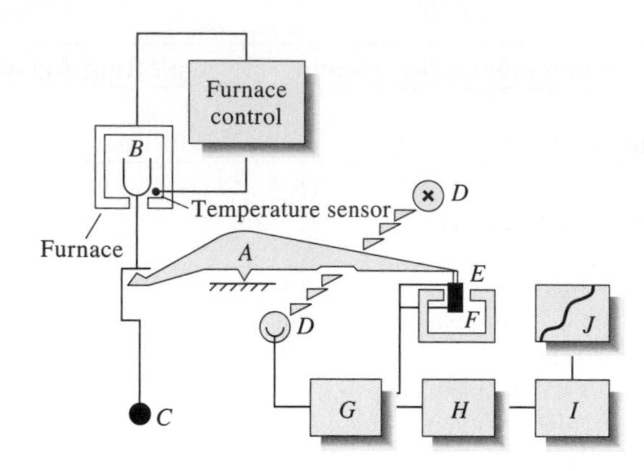

Figure 12-7 Schematic of thermobalance: *A:* beam; *B:* sample cup and holder; *C:* counterweight; *D:* lamp and photodiodes; *E:* coil; *F:* magnet; *G:* control amplifier; *H:* tare calculator; *I:* amplifier; and *J:* recorder. (Courtesy of Mettler Toledo, Inc., Columbus, OH.)

▶ The temperature required to dehydrate a precipitate completely may be as low as 100°C or as high as 1000°C.

anhydrous oxalate CaC_2O_4 at 225°C. The abrupt change in mass at about 450°C signals the decomposition of calcium oxalate to calcium carbonate and carbon monoxide. The final step in the curve depicts the conversion of the carbonate to calcium oxide and carbon dioxide. As can be seen, the compound finally weighed in a gravimetric calcium determination based on oxalate precipitation is highly dependent on the ignition temperature.

Recording thermal decomposition curves is called **thermogravimetric analysis,** and the mass versus temperature curves are termed **thermograms.**

12B CALCULATION OF RESULTS FROM GRAVIMETRIC DATA

The results of a gravimetric analysis are generally computed from two experimental measurements: the mass of sample and the mass of a product of known composition. The examples that follow illustrate how such computations are carried out.

EXAMPLE 12-1

CD-ROM Tutorial: Calculating Percentage of a Compound in a Sample Using Production of a Different Compound.

The calcium in a 200.0-mL sample of a natural water was determined by precipitating the cation as CaC_2O_4. The precipitate was filtered, washed, and ignited in a crucible with an empty mass of 26.6002 g. The mass of the crucible plus CaO (56.077 g/mol) was 26.7134 g. Calculate the concentration of Ca (40.078 g/mol) in water in units of grams per 100 mL of the water.

The mass of CaO is

$$26.7134 \text{ g} - 26.6002 \text{ g} = 0.1132 \text{ g}$$

The number of moles Ca in the sample is equal to the number of moles CaO or

$$\text{amount of Ca} = 0.1132 \text{ g CaO} \times \frac{1 \text{ mol CaO}}{56.077 \text{ g CaO}} \times \frac{1 \text{ mol Ca}}{\text{mol CaO}}$$
$$= 2.0186 \times 10^{-3} \text{ mol Ca}$$

$$\text{conc. Ca} = \frac{2.0186 \times 10^{-3} \text{ mol Ca} \times 40.078 \text{ g Ca/mol Ca}}{200 \text{ mL sample}} \times 100 \text{ mL}$$
$$= 0.04045 \text{ g/100 mL}$$

EXAMPLE 12-2

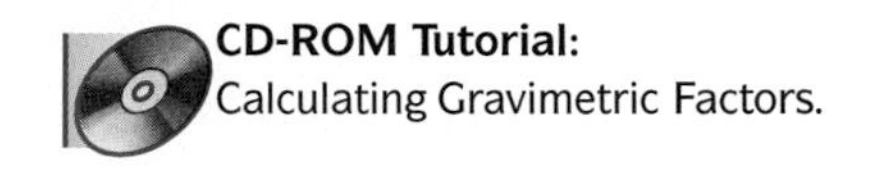
CD-ROM Tutorial: Calculating Gravimetric Factors.

An iron ore was analyzed by dissolving a 1.1324-g sample in concentrated HCl. The resulting solution was diluted with water, and the iron(III) was precipitated as the hydrous oxide $Fe_2O_3 \cdot xH_2O$ by the addition of NH_3. After filtration and washing, the residue was ignited at a high temperature to give 0.5394 g of pure Fe_2O_3 (159.69 g/mol). Calculate (a) the % Fe (55.847 g/mol) and (b) the % Fe_3O_4 (231.54 g/mol) in the sample.

For both parts of this problem, we need to calculate the number of moles of Fe_2O_3. Thus,

$$\text{amount } Fe_2O_3 = 0.5394 \cancel{\text{ g } Fe_2O_3} \times \frac{1 \text{ mol } Fe_2O_3}{159.69 \cancel{\text{ g } Fe_2O_3}}$$

$$= 3.3778 \times 10^{-3} \text{ mol } Fe_2O_3$$

(a) The number of moles of Fe is twice the number of moles of Fe_2O_3, and

$$\text{mass Fe} = 3.3778 \times 10^{-3} \cancel{\text{ mol } Fe_2O_3} \times \frac{2 \cancel{\text{ mol Fe}}}{\text{mol } \cancel{Fe_2O_3}} \times 55.847 \frac{\text{g Fe}}{\cancel{\text{mol Fe}}}$$

$$= 0.37728 \text{ g Fe}$$

$$\% \text{ Fe} = \frac{0.37728 \text{ g Fe}}{1.1324 \text{ g sample}} \times 100\% = 33.32\%$$

(b) As shown by the following balanced equation, 3 mol of Fe_2O_3 are chemically equivalent to 2 mol of Fe_3O_4. That is,

$$3Fe_2O_3 \rightarrow 2Fe_3O_4 + \frac{1}{2}O_2$$

$$\text{mass } Fe_3O_4 = 3.3778 \times 10^{-3} \cancel{\text{ mol } Fe_2O_3} \times \frac{2 \cancel{\text{ mol } Fe_3O_4}}{3 \cancel{\text{ mol } Fe_2O_3}} \times \frac{231.54 \text{ g } Fe_3O_4}{\cancel{\text{mol } Fe_3O_4}}$$

$$= 0.52140 \text{ g } Fe_3O_4$$

$$\% \text{ } Fe_3O_4 = \frac{0.5140 \text{ g } Fe_3O_4}{1.1324 \text{ g sample}} \times 100\% = 46.04\%$$

EXAMPLE 12-3

CD-ROM Tutorial: Using Gravimetric Analysis for Determining Percent in a Mixture.

A 0.2356-g sample containing *only* NaCl (58.44 g/mol) and $BaCl_2$ (208.23 g/mol) yielded 0.4637 g of dried AgCl (143.32 g/mol). Calculate the percent of each halogen compound in the sample.

If we let x be the mass of NaCl in grams and y be the mass of $BaCl_2$ in grams, we can write as a first equation

$$x + y = 0.2356 \text{ g sample}$$

(continued)

To obtain the mass of AgCl from the NaCl, we write an expression for the number of moles of AgCl formed from the NaCl. That is,

$$\text{amount AgCl from NaCl} = x\,\cancel{\text{g NaCl}} \times \frac{1\,\cancel{\text{mol NaCl}}}{58.44\,\cancel{\text{g NaCl}}} \times \frac{1\text{ mol AgCl}}{\cancel{\text{mol NaCl}}}$$
$$= 0.017111x\text{ mol AgCl}$$

The mass of AgCl from this source is

$$\text{mass AgCl from NaCl} = 0.017111x\,\cancel{\text{mol AgCl}} \times 143.32\frac{\text{g AgCl}}{\cancel{\text{mol AgCl}}}$$
$$= 2.4524x\text{ g AgCl}$$

Proceeding in the same way, we can write that the number of moles of AgCl from the $BaCl_2$ is given by

$$\text{amount AgCl from BaCl}_2 = y\,\cancel{\text{g BaCl}_2} \times \frac{1\,\cancel{\text{mol BaCl}_2}}{208.23\,\cancel{\text{g BaCl}_2}} \times \frac{2\text{ mol AgCl}}{\cancel{\text{mol BaCl}_2}}$$
$$= 9.605 \times 10^{-3}y\text{ mol AgCl}$$
$$\text{amount AgCl from BaCl}_2 = 9.605 \times 10^{-3}y\,\cancel{\text{mol AgCl}} \times 143.32\frac{\text{g AgCl}}{\cancel{\text{mol AgCl}}}$$
$$= 1.3766y\text{ g AgCl}$$

Because 0.4637 g of AgCl comes from the two compounds, we can write

$$2.4524x + 1.3766y = 0.4637$$

The first equation can be rewritten as

$$y = 0.2356 - x$$

Substituting into the previous equation gives

$$2.4524x + 1.3766\,(0.2356 - x) = 0.4637$$

which rearranges to

$$1.0758\,x = 0.13942$$

$$x = \text{mass NaCl} = 0.12960\text{ g NaCl}$$

$$\%\,\text{NaCl} = \frac{0.12956\text{ g NaCl}}{0.2356\text{ g sample}} \times 100\% = 55.01\%$$

$$\%\,\text{BaCl}_2 = 100.00\% - 55.01\% = 44.99\%$$

Spreadsheet Summary In some chemical problems, two or more simultaneous equations must be solved to obtain the desired result. Example 12-3 is such a problem. In Chapter 6 of *Applications of Microsoft® Excel in Analytical Chemistry,* the method of determinants and the matrix inversion method are explored for solving such equations. The matrix method is extended to solve a system of 4 equations in 4 unknowns. The matrix method is used to confirm the results of Example 12-3.

◀ Gravimetric methods do not require a calibration or standardization step (as do all other analytical procedures except coulometry) because the results are calculated directly from the experimental data and atomic masses. Thus, when only one or two samples are to be analyzed, a gravimetric procedure may be the method of choice because it requires less time and effort than a procedure that requires preparation of standards and calibration.

12C APPLICATIONS OF GRAVIMETRIC METHODS

Gravimetric methods have been developed for most inorganic anions and cations, as well as for such neutral species as water, sulfur dioxide, carbon dioxide, and iodine. A variety of organic substances can also be easily determined gravimetrically. Examples include lactose in milk products, salicylates in drug preparations, phenolphthalein in laxatives, nicotine in pesticides, cholesterol in cereals, and benzaldehyde in almond extracts. Indeed, gravimetric methods are among the most widely applicable of all analytical procedures.

12C-1 Inorganic Precipitating Agents

Table 12-2 lists common inorganic precipitating agents. These reagents typically form slightly soluble salts or hydrous oxides with the analyte. As you can see from the many entries for each reagent, few inorganic reagents are selective.

12C-2 Reducing Agents

Table 12-3 lists several reagents that convert an analyte to its elemental form for weighing.

12C-3 Organic Precipitating Agents

Numerous organic reagents have been developed for the gravimetric determination of inorganic species. Some of these reagents are significantly more selective in their reactions than are most of the inorganic reagents listed in Table 12-2.

We encounter two types of organic reagents. One forms slightly soluble nonionic products called **coordination compounds;** the other forms products in which the bonding between the inorganic species and the reagent is largely ionic.

Organic reagents that yield sparingly soluble coordination compounds typically contain at least two functional groups. Each of these groups is capable of bonding with a cation by donating a pair of electrons. The functional groups are located in the molecule such that a five- or six-membered ring results from the reaction. Reagents that form compounds of this type are called **chelating agents,** and their products are called **chelates** (see Chapter 17).

Metal chelates are relatively nonpolar and, as a consequence, have solubilities that are low in water but high in organic liquids. Usually, these compounds possess low densities and are often intensely colored. Because they are not wetted by water, coordination compounds are easily freed of moisture at low temperatures. Two widely used chelating reagents are described in the paragraphs that follow.

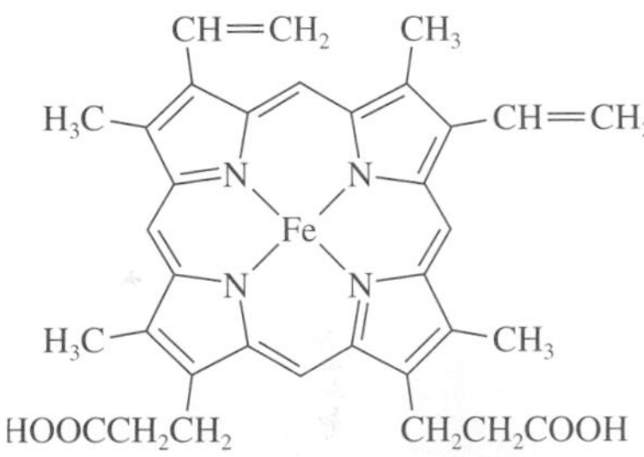

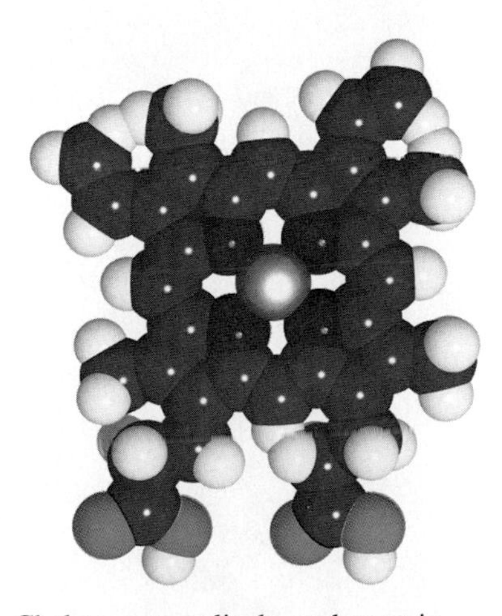

Chelates are cyclical metal-organic compounds in which the metal is a part of one or more five- or six-membered rings. The chelate pictured here is heme, which is a part of hemoglobin, the oxygen-carrying molecule in human blood. Notice the four six-membered rings that are formed with Fe^{2+}.

TABLE 12-3

Some Reducing Agents Employed in Gravimetric Methods

Reducing Agent	Analyte
SO_2	Se, Au
$SO_2 + H_2NOH$	Te
H_2NOH	Se
$H_2C_2O_4$	Au
H_2	Re, Ir
HCOOH	Pt
$NaNO_2$	Au
$SnCl_2$	Hg
Electrolytic reduction	Co, Ni, Cu, Zn Ag, In, Sn, Sb, Cd, Re, Bi

TABLE 12-2

Some Inorganic Precipitating Agents

Precipitating Agent	Element Precipitated*
$NH_3(aq)$	**Be** (BeO), **Al** (Al_2O_3), **Sc** (Sc_2O_3), Cr (Cr_2O_3)†, **Fe** (Fe_2O_3), Ga (Ga_2O_3), Zr (ZrO_2), **In** (In_2O_3), Sn (SnO_2), U (U_3O_8)
H_2S	Cu (CuO)†, **Zn** (ZnO, or $ZnSO_4$), **Ge** (GeO_2), As ($\underline{As_2O_3}$, or As_2O_5), Mo (MoO_3), Sn (SnO_2)†, Sb ($\underline{Sb_2O_3}$), or Sb_2O_5), Bi (Bi_2S_3)
$(NH_4)_2S$	Hg ($\underline{HgS}$), Co (Co_3O_4)
$(NH_4)_2HPO_4$	**Mg** ($Mg_2P_2O_7$), Al ($AlPO_4$), Mn ($Mn_2P_2O_7$), Zn ($Zn_2P_2O_7$), Zr ($Zr_2P_2O_7$), Cd ($Cd_2P_2O_7$), Bi ($BiPO_4$)
H_2SO_4	Li, Mn, **Sr, Cd, Pb, Ba** (all as sulfates)
H_2PtCl_6	K (K_2PtCl_6, or Pt), Rb ($\underline{Rb_2PtCl_6}$), Cs ($\underline{Cs_2PtCl_6}$)
$H_2C_2O_4$	Ca (CaO), Sr (SrO), **Th** (ThO_2)
$(NH_4)_2MoO_4$	Cd ($CdMoO_4$)†, Pb ($\underline{PbMoO_4}$)
HCl	**Ag** (AgCl), Hg (Hg_2Cl_2), Na (as NaCl from butyl alcohol), Si (SiO_2)
$AgNO_3$	**Cl** (AgCl), Br ($\underline{AgBr}$), I($\underline{AgI}$)
$(NH_4)_2CO_3$	**Bi** (Bi_2O_3)
NH_4SCN	Cu [$Cu_2(SCN)_2$]
$NaHCO_3$	Ru, Os, Ir (precipitated as hydrous oxides; reduced with H_2 to metallic state)
HNO_3	Sn (SnO_2)
H_5IO_6	Hg [$Hg_5(IO_6)_2$]
NaCl, $Pb(NO_3)_2$	F (PbClF)
$BaCl_2$	$\mathbf{SO_4^{2-}}$ ($BaSO_4$)
$MgCl_2$, NH_4Cl	$\mathbf{PO_4^{3-}}$ ($Mg_2P_2O_7$)

*Boldface type indicates that gravimetric analysis is the preferred method for the element or ion. The weighed form is indicated in parentheses.

†A dagger indicates that the gravimetric method is seldom used. An underscore indicates the most reliable gravimetric method.

From W. F. Hillebrand, G. E. F. Lundell, H. A. Bright, and J. I. Hoffman, *Applied Inorganic Analysis.* New York: Wiley, 1953.

Magnesium complex with 8-hydroxyquinoline.

8-Hydroxyquinoline (oxine)

Approximately two dozen cations form sparingly soluble chelates with 8-hydroxyquinoline. The structure of magnesium 8-hydroxyquinolate is typical of these chelates.

$$2\ \text{(8-hydroxyquinoline, OH, N)} + Mg^{2+} \rightarrow \text{(O, N}\rightarrow Mg \leftarrow \text{N, O chelate)} + 2H^+$$

The solubilities of metal 8-hydroxyquinolates vary widely from cation to cation and are pH dependent because 8-hydroxyquinoline is always deprotonated during chelation reaction. Therefore, we can achieve a considerable degree of selectivity in the use of 8-hydroxyquinoline by controlling pH.

CD-ROM Tutorial: Performing an Analysis Using an Organic Precipitating Agent.

Dimethylglyoxime

Dimethylglyoxime is an organic precipitating agent of unparalleled specificity. Only nickel(II) is precipitated from a weakly alkaline solution. The reaction is

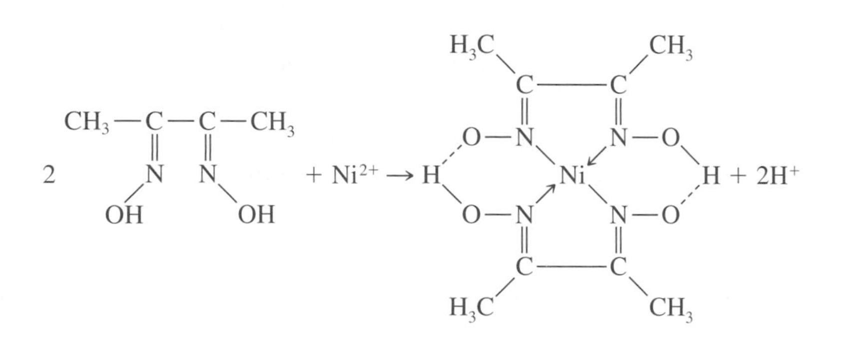

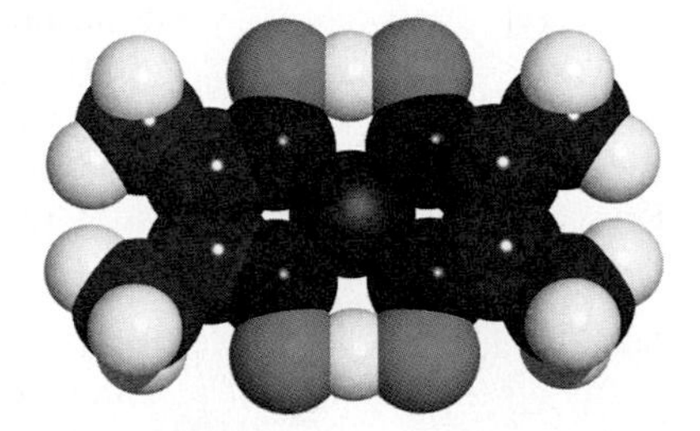

Nickel dimethylglyoxime is spectacular in appearance. As shown in color plate 7, it has a beautiful vivid red color.

This precipitate is so bulky that only small amounts of nickel can be handled conveniently. It also has an exasperating tendency to creep up the sides of the container as it is filtered and washed. The solid is conveniently dried at 110°C and has the composition indicated by its formula.

Sodium Tetraphenylborate

Sodium tetraphenylborate, $(C_6H_5)_4B^-Na^+$, is an important example of an organic precipitating reagent that forms salt-like precipitates. In cold mineral acid solutions, it is a near-specific precipitating agent for potassium and ammonium ions. The precipitates have stoichiometric composition and contain one mole of potassium or ammonium ion for each mole of tetraphenylborate ion; these ionic compounds are easily filtered and can be brought to constant mass at 105°C to 120°C. Only mercury(II), rubidium, and cesium interfere and must be removed by prior treatment.

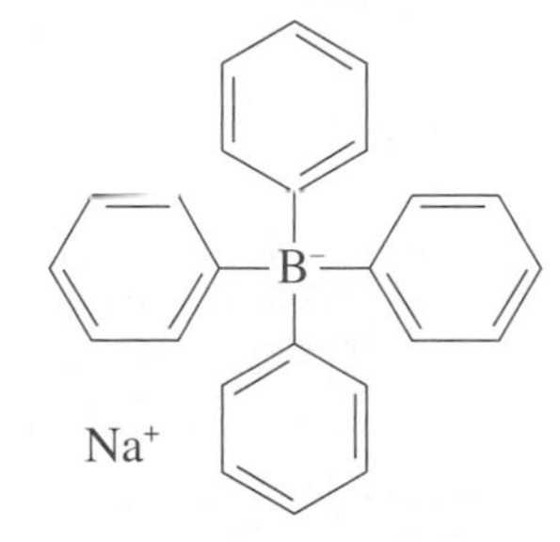

Sodium tetraphenylborate.

12C-4 Organic Functional Group Analysis

Several reagents react selectively with certain organic functional groups and thus can be used for the determination of most compounds containing these groups. A list of gravimetric functional group reagents is given in Table 12-4. Many of the reactions shown can also be used for volumetric and spectrophotometric determinations.

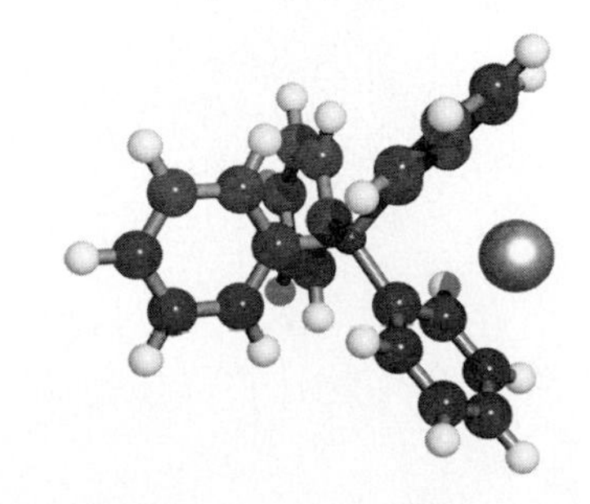

Molecular model of sodium tetraphenylborate.

12C-5 Volatilization Gravimetry

The two most common gravimetric methods based on volatilization are those for determining water and carbon dioxide.

TABLE 12-4
Gravimetric Methods for Organic Functional Groups

Functional Group	Basis for Method	Reaction and Product Weighed*
Carbonyl	Mass of precipitate with 2,4-dinitrophenylhydrazine	$RCHO + H_2NNHC_6H_3(NO_2)_2 \rightarrow \underline{R{-}CH = NNHC_6H_3(NO_2)_2(s)} + H_2O$ (RCOR′ reacts similarly)
Aromatic carbonyl	Mass of CO_2 formed at 230°C in quinoline; CO_2 distilled, absorbed, and weighed	$ArCHO \xrightarrow[CuCO_3]{230°C} Ar + \underline{CO_2(g)}$
Methoxyl and ethoxyl	Mass of AgI formed after distillation and decomposition of CH_3I or C_2H_5I	$\left.\begin{array}{l} ROCH_3 + HI \rightarrow ROH + CH_3I \\ RCOOCH_3 + HI \rightarrow RCOOH + CH_3I \\ ROC_2H_5 + HI \rightarrow ROH + C_2H_5I \end{array}\right\} CH_3I + Ag^+ + H_2O \rightarrow \underline{AgI(s)} + CH_3OH$
Aromatic nitro	Mass loss of Sn	$RNO_2 + \frac{3}{2}\underline{Sn(s)} + 6H^+ \rightarrow RNH_2 + \frac{3}{2}Sn^{4+} + 2H_2O$
Azo	Mass loss of Cu	$RN = NR' + 2\underline{Cu(s)} + 4H^+ \rightarrow RNH_2 + R'NH_2 + 2Cu^{2+}$
Phosphate	Mass of Ba salt	$ROP(=O)(OH)_2 + Ba^{2+} \rightarrow \underline{ROP(=O)O_2Ba(s)} + 2H^+$
Sulfamic acid	Mass of $BaSO_4$ after oxidation with HNO_2	$RNHSO_3H + HNO_2 + Ba^{2+} \rightarrow ROH + \underline{BaSO_4(s)} + N_2 + 2H^+$
Sulfinic acid	Mass of Fe_2O_3 after ignition of Fe(III) sulfinate	$3ROSOH + Fe^{3+} \rightarrow (ROSO)_3Fe(s) + 3H^+$; $(ROSO)_3Fe \xrightarrow[O_2]{} CO_2 + H_2O + SO_2 + \underline{Fe_2O_3(s)}$

*The substance weighed is underlined.

Water is quantitatively distilled from many materials by heating. In direct determination, water vapor is collected on any of several solid desiccants, and its mass is determined from the mass gain of the desiccant. The indirect method, in which the amount of water is determined by the loss of mass of the sample during heating, is less satisfactory because it must be assumed that water is the only component volatilized. This assumption is frequently unjustified, however, because heating of many substances results in their decomposition and a consequent change in mass, irrespective of the presence of water. Nevertheless, the indirect method has found wide use for the determination of water in items of commerce. For example, a semiautomated instrument for the determination of moisture in cereal grains can be purchased. It consists of a platform balance on which a 10-g sample is heated with an infrared lamp. The percent moisture is read directly.

► Automatic instruments for the routine determination of water in various products of agriculture and commerce are marketed by several instrument manufacturers.

An example of a gravimetric procedure involving volatilization of carbon dioxide is the determination of the sodium hydrogen carbonate content of antacid tablets. Here, a weighed sample of the finely ground tablets is treated with dilute sulfuric acid to convert the sodium hydrogen carbonate to carbon dioxide:

$$NaHCO_3(aq) + H_2SO_4(aq) \rightarrow CO_2(g) + H_2O(l) + NaHSO_4(aq)$$

As shown in Figure 12-8, this reaction is carried out in a flask connected to a weighed absorption tube containing the absorbent Ascarite II,[7] which consists of sodium hydroxide absorbed on a nonfibrous silicate. This material retains carbon dioxide by the reaction

$$2NaOH + CO_2 \rightarrow Na_2CO_3 + H_2O$$

[7]Thomas Scientific, Swedesboro, NJ.

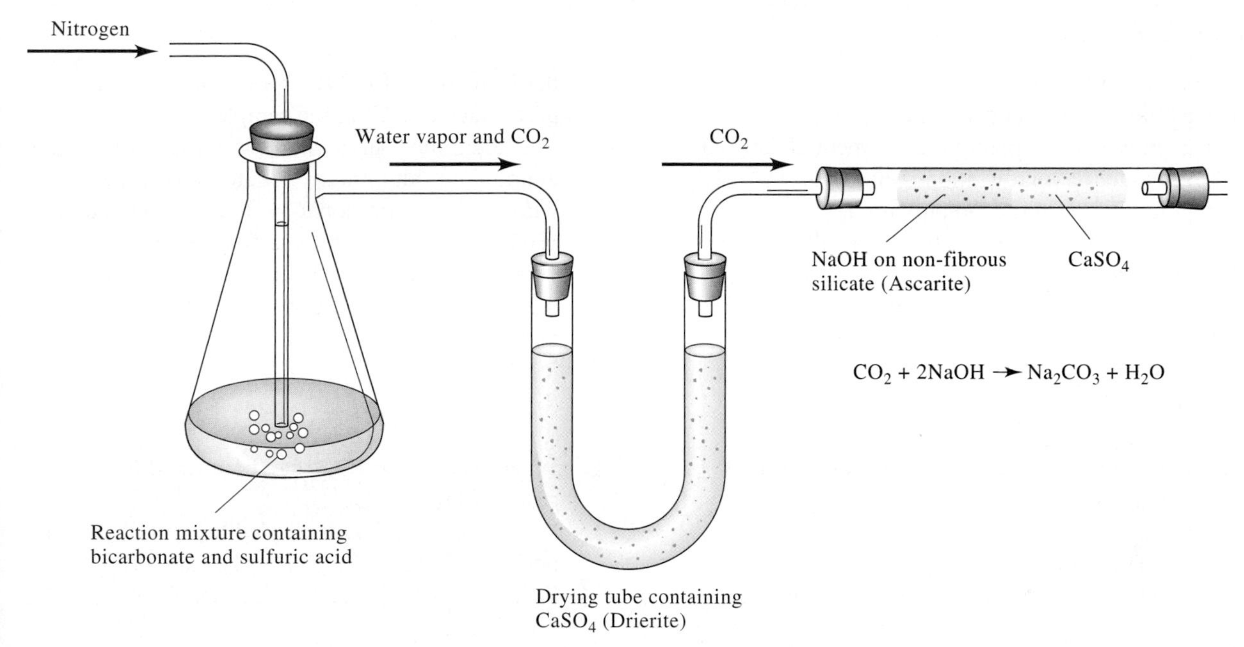

Figure 12-8 Apparatus for determining the sodium hydrogen carbonate content of antacid tablets by a gravimetric volatilization procedure.

The absorption tube must also contain a desiccant to prevent loss of the water produced by the reaction.

Sulfides and sulfites can also be determined by volatilization. Hydrogen sulfide or sulfur dioxide evolved from the sample after treatment with acid is collected in a suitable absorbent.

Finally, the classical method for the determination of carbon and hydrogen in organic compounds is a gravimetric volatilization procedure in which the combustion products (H_2O and CO_2) are collected selectively on weighed absorbents. The increase in mass serves as the analytical parameter.

WEB WORKS

Use your Web browser to connect to **http://chemistry.brookscole.com/skoogfac/.** From the Chapter Resources menu, choose Web Works. Locate the Chapter 12 Section, and click on the link to the articles on classical analysis by C. M. Beck. In these articles, which were originally published in the scientific literature,[8] Beck makes a strong case for the revival of classical analysis. What is Beck's definition of classical analysis? Why does Beck maintain that classical analysis should be cultivated in this age of automated, computerized instrumentation? What solution does he propose for the problem of dwindling numbers of qualified classical analysts? List three reasons why, in Beck's view, a supply of classical analysts must be maintained.

[8]C. M. Beck, *Anal. Chem.*, **1991,** *63*(20), 993A–1003A, and C. M. Beck, *Metrologia,* **1997,** *34*(1), 19–30.

QUESTIONS AND PROBLEMS

12-1. Explain the difference between
*(a) a colloidal and a crystalline precipitate.
(b) a gravimetric precipitation method and a gravimetric volatilization method.
*(c) precipitation and coprecipitation.
(d) peptization and coagulation of a colloid.
*(e) occlusion and mixed-crystal formation.
(f) nucleation and particle growth.

12-2. Define
*(a) digestion.
(b) adsorption.
*(c) reprecipitation.
(d) precipitation from homogeneous solution.
*(e) counter-ion layer.
(f) mother liquor.
*(g) supersaturation.

***12-3.** What are the structural characteristics of a chelating agent?

12-4. How can the relative supersaturation be varied during precipitate formation?

***12-5.** An aqueous solution contains $NaNO_3$ and KBr. The bromide ion is precipitated as AgBr by addition of $AgNO_3$. After an excess of the precipitating reagent has been added,
(a) what is the charge on the surface of the coagulated colloidal particles?
(b) what is the source of the charge?
(c) what ions make up the counter-ion layer?

12-6. Suggest a method by which Ni^{2+} can be precipitated homogeneously as NiS.

***12-7.** What is peptization and how is it avoided?

12-8. Suggest a precipitation method for separation of K^+ from Na^+ and Li^+.

12-9. Write an equation showing how the mass of the substance sought can be converted to the mass of the weighed substance on the right.

Sought	Weighed	Sought	Weighed
*(a) SO_2	$BaSO_4$	(f) $MnCl_2$	Mn_3O_4
(b) Mg	$Mg_2P_2O_7$	(g) Pb_3O_4	PbO_2
*(c) In	In_2O_3	(h) $U_2P_2O_{11}$	P_2O_5
(d) K	K_2PtCl_6	*(i) $Na_2B_4O_7 \cdot 10H_2O$	B_2O_3
*(e) CuO	$Cu_2(SCN)_2$	(j) Na_2O	†

†$NaZn(UO_2)_3(C_2H_3O_2)_9 \cdot 6H_2O$

***12-10.** Treatment of a 0.2500-g sample of impure potassium chloride with an excess of $AgNO_3$ resulted in the formation of 0.2912 g of AgCl. Calculate the percentage of KCl in the sample.

12-11. The aluminum in a 0.910-g sample of impure ammonium aluminum sulfate was precipitated with aqueous ammonia as the hydrous $Al_2O_3 \cdot xH_2O$. The precipitate was filtered and ignited at 1000°C to give anhydrous Al_2O_3, which weighed 0.2001 g. Express the result of this analysis in terms of
(a) % $NH_4Al(SO_4)_2$.
(b) % Al_2O_3.
(c) % Al.

***12-12.** What mass of $Cu(IO_3)_2$ can be formed from 0.500 g of $CuSO_4 \cdot 5H_2O$?

12-13. What mass of KIO_3 is needed to convert the copper in 0.2000 g of $CuSO_4 \cdot 5H_2O$ to $Cu(IO_3)_2$?

***12-14.** What mass of AgI can be produced from a 0.512-g sample that assays 20.1% AlI_3?

12-15. Precipitates used in the gravimetric determination of uranium include $Na_2U_2O_7$ (634.0 g/mol), $(UO_2)_2P_2O_7$ (714.0 g/mol), and $V_2O_5 \cdot 2UO_3$ (753.9 g/mol). Which of these weighing forms provides the greatest mass of precipitate from a given quantity of uranium?

12-16. A 0.8102-g sample of impure $Al_2(CO_3)_3$ decomposed with HCl; the liberated CO_2 was collected on calcium oxide and found to weigh 0.0515 g. Calculate the percentage of aluminum in the sample.

12-17. The hydrogen sulfide in a 75.0-g sample of crude petroleum was removed by distillation and uncollected in a solution of $CdCl_2$. The precipitated CdS was then filtered, washed, and ignited to $CdSO_4$. Calculate the percentage of H_2S in the sample if 0.117 g of $CdSO_4$ was recovered.

***12-18.** A 0.2121-g sample of an organic compound was burned in a stream of oxygen, and the CO_2 produced was collected in a solution of barium hydroxide. Calculate the percentage of carbon in the sample if 0.6006 g of $BaCO_3$ was formed.

12-19. A 5.000-g sample of a pesticide was decomposed with metallic sodium in alcohol, and the liberated chloride ion was precipitated as AgCl. Express the results of this analysis in terms of percent DDT ($C_{14}H_9Cl_5$) based on the recovery of 0.1606 g of AgCl.

***12-20.** The mercury in a 0.8142-g sample was precipitated with an excess of paraperiodic acid, H_5IO_6:

$$5Hg^{2+} + 2H_5IO_6 \rightarrow Hg_5(IO_6)_2 + 10H^+$$

The precipitate was filtered, washed free of precipitating agent, dried, and weighed, and 0.4114 g was recovered. Calculate the percentage of Hg_2Cl_2 in the sample.

12-21. The iodide in a sample that also contained chloride was converted to iodate by treatment with an excess of bromine:

$$3H_2O + 3Br_2 + I^- \rightarrow 6Br^- + IO_3^- + 6H^+$$

The unused bromine was removed by boiling; an excess of barium ion was then added to precipitate the iodate:

$$Ba^{2+} + 2IO_3^- \rightarrow Ba(IO_3)_2$$

In the analysis of a 1.97-g sample, 0.0612 g of barium iodate was recovered. Express the results of this analysis as percent potassium iodide.

***12-22.** Ammoniacal nitrogen can be determined by treatment of the sample with chloroplatinic acid; the product is slightly soluble ammonium chloroplatinate:

$$H_2PtCl_6 + 2NH_4^+ \rightarrow (NH_4)_2PtCl_6 + 2H^+$$

The precipitate decomposes on ignition, yielding metallic platinum and gaseous products:

$$(NH_4)_2PtCl_6 \rightarrow Pt(s) + 2Cl_2(g) + 2NH_3(g) + 2HCl(g)$$

Calculate the percentage of ammonia in a sample if 0.2115 g gave rise to 0.4693 g of platinum.

12-23. A 0.6447-g portion of manganese dioxide was added to an acidic solution in which 1.1402 g of a chloride-containing sample was dissolved. Evolution of chlorine took place as a consequence of the following reaction:

$$MnO_2(s) + 2Cl^- + 4H^+ \rightarrow Mn^{2+} + Cl_2(g) + 2H_2O$$

After the reaction was complete, the excess MnO_2 was collected by filtration, washed, and weighed, and 0.3521 g was recovered. Express the results of this analysis in terms of percent aluminum chloride.

***12-24.** A series of sulfate samples is to be analyzed by precipitation as $BaSO_4$. If it is known that the sulfate content in these samples ranges between 20% and 55%, what minimum sample mass should be taken to ensure that a precipitate mass no smaller than 0.200 g is produced? What is the maximum precipitate weight to be expected if this quantity of sample is taken?

12-25. The addition of dimethylglyoxime, $H_2C_4H_6O_2N_2$, to a solution containing nickel(II) ion gives rise to a precipitate:

$$Ni^{2+} + 2H_2C_4H_6O_2N_2 \rightarrow 2H^+ + Ni(HC_4H_6O_2N_2)_2$$

Nickel dimethylglyoxime is a bulky precipitate that is inconvenient to manipulate in amounts greater than 175 mg. The amount of nickel in a type of permanent-magnet alloy ranges between 24% and 35%. Calculate the sample size that should not be exceeded when analyzing these alloys for nickel.

***12-26.** The efficiency of a particular catalyst is highly dependent on its zirconium content. The starting material for this preparation is received in batches that assay between 68% and 84% $ZrCl_4$. Routine analysis based on precipitation of AgCl is feasible, it having been established that there are no sources of chloride ion other than the $ZrCl_4$ in the sample.

(a) What sample mass should be taken to ensure an AgCl precipitate that weighs at least 0.400 g?

(b) If this sample mass is used, what is the maximum weight of AgCl that can be expected in this analysis?

(c) To simplify calculations, what sample mass should be taken to have the percentage of $ZrCl_4$ exceed the mass of AgCl produced by a factor of 100?

12-27. A 0.8720-g sample of a mixture consisting solely of sodium bromide and potassium bromide yields 1.505 g of silver bromide. What are the percentages of the two salts in the sample?

***12-28.** A 0.6407-g sample containing chloride and iodide ions gave a silver halide precipitate weighing 0.4430 g. This precipitate was then strongly heated in a stream of Cl_2 gas to convert the AgI to AgCl; on completion of this treatment, the precipitate weighed 0.3181 g. Calculate the percentage of chloride and iodide in the sample.

12-29. The phosphorus in a 0.1969-g sample was precipitated as the slightly soluble $(NH_4)_3PO_4 \cdot 12MoO_3$. This precipitate was filtered, washed, and then redissolved in acid. Treatment of the resulting solution with an excess of Pb^{2+} resulted in the formation of 0.2554 g of $PbMoO_4$. Express the results of this analysis in terms of percent P_2O_5.

*12-30. How many grams of CO_2 are evolved from a 1.500-g sample that is 38.0% $MgCO_3$ and 42.0% K_2CO_3 by mass?

12-31. A 6.881-g sample containing magnesium chloride and sodium chloride was dissolved in sufficient water to give 500 mL of solution. Analysis for the chloride content of a 50.0-mL aliquot resulted in the formation of 0.5923 g of AgCl. The magnesium in a second 50.0-mL aliquot was precipitated as $MgNH_4PO_4$; on ignition, 0.1796 g of $Mg_2P_2O_7$ was found. Calculate the percentage of $MgCl_2 \cdot 6H_2O$ and of NaCl in the sample.

*12-32. A 50.0-mL portion of a solution containing 0.200 g of $BaCl_2 \cdot 2H_2O$ is mixed with 50.0 mL of a solution containing 0.300 g of $NaIO_3$. Assume that the solubility of $Ba(IO_3)_2$ in water is negligibly small and calculate

(a) the mass of the precipitated $Ba(IO_3)_2$.

(b) the mass of the unreacted compound that remains in solution.

12-33. When a 100.0-mL portion of a solution containing 0.500 g of $AgNO_3$ is mixed with 100.0 mL of a solution containing 0.300 g of K_2CrO_4, a bright red precipitate of Ag_2CrO_4 forms.

(a) Assuming that the solubility of Ag_2CrO_4 is negligible, calculate the mass of the precipitate.

(b) Calculate the mass of the unreacted component that remains in solution.

12-34. **Challenge Problem.** Stones form in the urinary tract when certain chemicals become too concentrated in urine. By far the most common kidney stones are those formed from calcium and oxalate. Magnesium is known to inhibit the formation of kidney stones.

(a) The solubility of calcium oxalate (CaC_2O_4) in urine is 9×10^{-5} M. What is the solubility product, K_{sp}, of CaC_2O_4 in urine?

(b) The solubility of magnesium oxalate (MgC_2O_4) in urine is 0.0093 M. What is the solubility product, K_{sp}, of MgC_2O_4 in urine.

(c) The concentration of calcium in urine is approximately 5 mM. What is the maximum concentration of oxalate that can be tolerated and not precipitate CaC_2O_4?

(d) The pH of Subject A's urine was 5.9. What fraction of total oxalate, c_T, is present as oxalate ion, $C_2O_4^{2-}$, at pH 5.9? The K_a values for oxalic acid in urine are the same as in water. *Hint:* Find the ratio $[C_2O_4^{2-}]/c_T$ at pH 5.9.

(e) If the total oxalate concentration in Subject A's urine was 15.0 mM, should a calcium oxalate precipitate form?

(f) In actuality, Subject A does not show the presence of calcium oxalate crystals in urine. Give a plausible reason for this observation.

(g) Why would magnesium inhibit the formation of CaC_2O_4 crystals?

(h) Why are patients with CaC_2O_4 kidney stones often advised to drink large amounts of water?

(i) The calcium and magnesium in a urine sample were precipitated as oxalates. A mixed precipitate of CaC_2O_4 and MgC_2O_4 resulted and was analyzed by a thermogravimetric procedure. The precipitate mixture was heated to form $CaCO_3$ and MgO. This second mixture weighed 0.0433 g. After ignition to form CaO and MgO, the resulting solid weighed 0.0285 g. What was the mass of Ca in the original sample?

InfoTrac College Edition

For additional readings, go to InfoTrac College Edition, your online research library, at

http://infotrac.thomsonlearning.com

CHAPTER 13

Titrimetric Methods; Precipitation Titrimetry

Titrations are widely used in analytical chemistry to determine acids, bases, oxidants, reductants, metal ions, proteins, and many other species. Titrations are based on a reaction between the analyte and a standard reagent known as the titrant. The reaction is of known and reproducible stoichiometry. The volume, or the mass, of the titrant needed to react essentially completely with the analyte is determined and used to obtain the quantity of analyte. A volume-based titration is shown in this figure, in which the standard solution is added from a buret, and the reaction occurs in the Erlenmeyer flask. In some titrations, known as coulometric titrations, the quantity of charge needed to completely consume the analyte is obtained. In any titration, the point of chemical equivalence, experimentally called the end point, is signaled by an indicator color change or a change in an instrumental response.

This chapter introduces the titration principle and the calculations involved. Titration curves, which show the progress of the titration, are introduced. The titration process is illustrated by reactions involving the formation of precipitates.

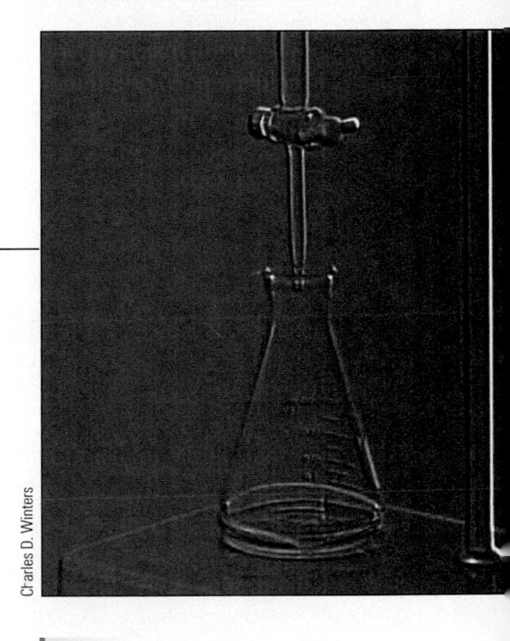

Charles D. Winters

Titrimetric methods include a large and powerful group of quantitative procedures that are based on measuring the amount of a reagent of known concentration that is consumed by an analyte. ***Volumetric titrimetry*** *involves measuring the volume of a solution of known concentration that is needed to react essentially completely with the analyte.* ***Gravimetric titrimetry*** *differs only in that the mass of the reagent is measured instead of its volume. In* ***coulometric titrimetry,*** *the "reagent" is a constant direct electrical current of known magnitude that consumes the analyte; here, the time required (and thus the total charge) to complete the electrochemical reaction is measured.*

This chapter provides introductory material that applies to all types of titrimetric methods of analysis, using precipitation titrimetry to illustrate the various theoretical aspects of the titration process. Chapters 14, 15, and 16 are devoted to the various types of neutralization titrations, in which the analyte and titrants undergo acid/base reactions. Chapter 17 provides information about titrations in which the analytical reactions involve complex formation. These methods are of particular importance for the determination of a variety of cations. Finally, Chapters 18 and 19 are devoted to volumetric methods, in which the analytical reactions involve electron transfer. These methods are often called ***redox titrations.*** *Some additional titration methods are explored in later chapters. These methods include* ***amperometric titration,*** *in Section 23B-4, and* ***spectrophotometric titrations,*** *in Section 26A-4.*

Titrimetry includes a group of analytical methods that are based on determining the quantity of a reagent of known concentration that is required to react completely with the analyte. The reagent may be a standard solution of a chemical or an electric current of known magnitude.

Volumetric titrimetry is a type of titrimetry in which the volume of a standard reagent is the measured quantity.

Coulometric titrimetry is a type of titrimetry in which the quantity of charge in coulombs required to complete a reaction with the analyte is the measured quantity.

13A SOME TERMS USED IN VOLUMETRIC TITRIMETRY[1]

A **standard solution** is a reagent of exactly known concentration that is used in a titrimetric analysis.

Titration is a process in which a standard reagent is added to a solution of an analyte until the reaction between the analyte and reagent is judged to be complete.

Back-titration is a process in which the excess of a standard solution used to consume an analyte is determined by titration with a second standard solution. Back-titrations are often required when the rate of reaction between the analyte and reagent is slow or when the standard solution lacks stability.

The **equivalence point** is the point in a titration when the amount of added standard reagent is exactly equivalent to the amount of analyte.

The **end point** is the point in a titration when a physical change occurs that is associated with the condition of chemical equivalence.

In volumetric methods, the **titration error** E_t is given by

$$E_t = V_{ep} - V_{eq}$$

where V_{ep} is the actual volume of reagent required to reach the end point and V_{eq} is the theoretical volume to reach the equivalence point.

A **standard solution** (or a **standard titrant**) is a reagent of known concentration that is used to carry out a titrimetric analysis. A **titration** is performed by slowly adding a standard solution from a buret or other liquid-dispensing device to a solution of the analyte until the reaction between the two is judged complete. The volume or mass of reagent needed to complete the titration is determined from the difference between the initial and final readings. A volumetric titration process is depicted in Figure 13-1.

It is sometimes necessary to add an excess of the standard titrant and then determine the excess amount by **back-titration** with a second standard titrant. For example, the amount of phosphate in a sample can be determined by adding a measured excess of standard silver nitrate to a solution of the sample, which leads to the formation of insoluble silver phosphate:

$$3Ag^+ + PO_4^{3-} \rightarrow Ag_3PO_4(s)$$

The excess silver nitrate is then back-titrated with a standard solution of potassium thiocyanate:

$$Ag^+ + SCN^- \rightarrow AgSCN(s)$$

Here, the amount of silver nitrate is chemically equivalent to the amount of phosphate ion plus the amount of thiocyanate used for the back-titration.

13A-1 Equivalence Points and End Points

The **equivalence point** in a titration is a theoretical point reached when the amount of added titrant is chemically equivalent to the amount of analyte in the sample. For example, the equivalence point in the titration of sodium chloride with silver nitrate occurs after exactly 1 mol of silver ion has been added for each mole of chloride ion in the sample. The equivalence point in the titration of sulfuric acid with sodium hydroxide is reached after introduction of 2 mol of base for each mole of acid.

We cannot determine the equivalence point of a titration experimentally. Instead, we can only estimate its position by observing some physical change associated with the condition of equivalence. This change is called the **end point** for the titration. Every effort is made to ensure that any volume or mass difference between the equivalence point and the end point is small. Such differences do exist, however, as a result of inadequacies in the physical changes and in our ability to observe them. The difference in volume or mass between the equivalence point and the end point is the **titration error.**

Indicators are often added to the analyte solution to produce an observable physical change (the end point) at or near the equivalence point. Large changes in the relative concentration of analyte or titrant occur in the equivalence-point region. These concentration changes cause the indicator to change in appearance. Typical indicator changes include the appearance or disappearance of a color, a

[1]For a detailed discussion of volumetric methods, see J. I. Watters, in *Treatise on Analytical Chemistry,* I. M. Kolthoff and P. J. Elving, Eds., Part I, Vol. 11, Chapter 114. New York: Wiley, 1975.

Figure 13-1 The titration process.

© Charles D. Winters

Typical setup for carrying out a titration. The apparatus consists of a buret, a buret stand and clamp with a white porcelain base to provide an appropriate background for viewing indicator changes, and a wide-mouth Erlenmeyer flask containing a precisely known volume of the solution to be titrated. The solution is normally delivered into the flask using a pipet, as shown in Figure 2-22.

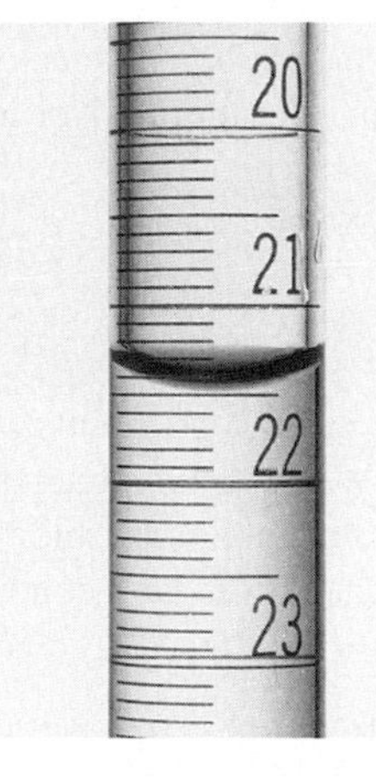

© Charles D. Winters

Detail of the buret graduations. Normally, the buret is filled with titrant solution to within 1 or 2 mL of the zero position at the top. The initial volume of the buret is read to the nearest ±0.01 mL. The reference point on the meniscus and the proper position of the eye for reading are depicted in Figure 2-21.

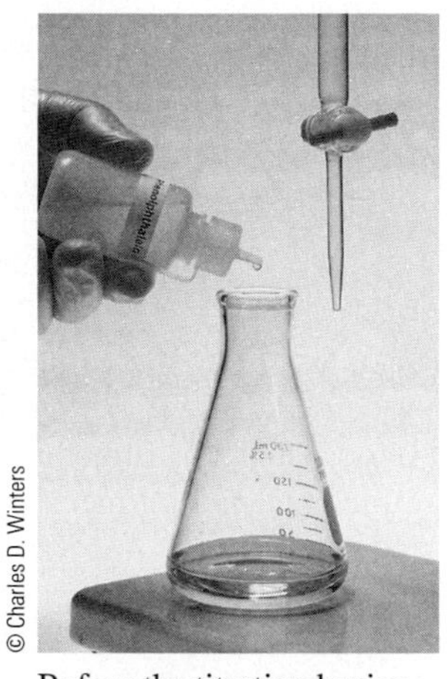

© Charles D. Winters

Before the titration begins. The solution to be titrated, an acid in this example, is placed in the flask and the indicator is added as shown in the photo. The indicator in this case is phenolphthalein, which turns pink in basic solution.

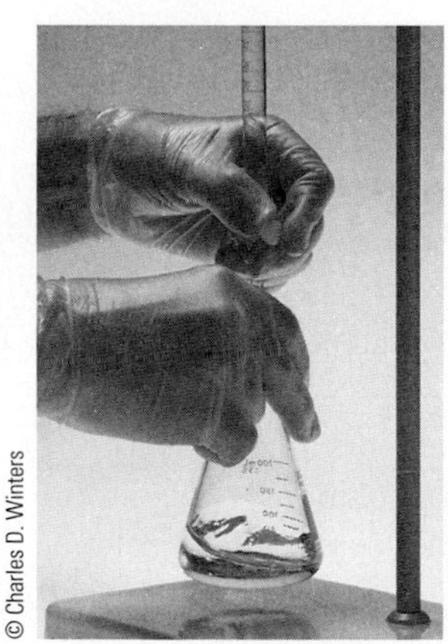

© Charles D. Winters

During titration. The titrant is added to the flask with swirling until the color of the indicator persists. In the initial region of the titration, titrant may be added rather rapidly, but as the end point is approached, increasingly smaller portions are added; at the end point, less than half a drop of titrant should cause the indicator to change color.

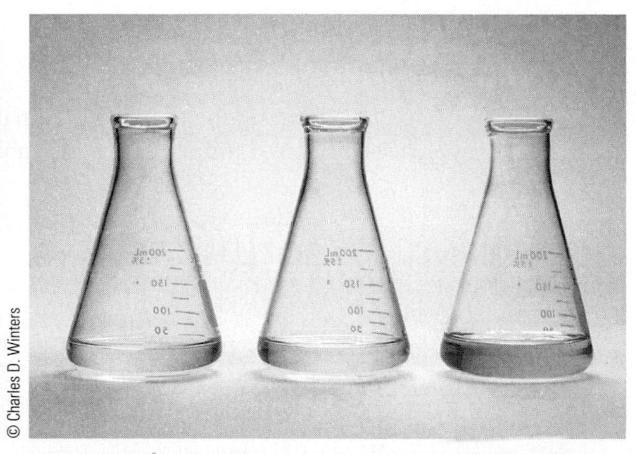

© Charles D. Winters

Titration end point. The end point is achieved when the barely perceptible pink color of phenolphthalein persists. The flask on the left shows the titration less than half a drop prior to the end point; the middle flask shows the end point. The final reading of the buret is made at this point, and the volume of base delivered in the titration is calculated from the difference between the initial and final buret readings. The flask on the right shows what happens when a slight excess of base is added to the titration mixture. The solution turns a deep pink color, and the end point has been exceeded. In color plate 9, the color change at the end point is much easier to see than in this black-and-white version.

change in color, or the appearance or disappearance of turbidity. As an example, the indicator used in the precipitation titration of silver ion with potassium thiocyanate is a small amount of ferric chloride, which reacts with thiocyanate ions to give a red color. The indicator reaction is

$$Fe^{3+} + SCN^- \rightarrow \underset{\text{red}}{FeSCN^{2-}}$$

We often use instruments to detect end points. These instruments respond to properties of the solution that change in a characteristic way during the titration. Among such instruments are colorimeters, turbidimeters, temperature monitors, refractometers, voltmeters, current meters, and conductivity meters.

13A-2 Primary Standards

A **primary standard** is an ultrapure compound that serves as the reference material for a titrimetric method of analysis.

A **primary standard** is a highly purified compound that serves as a reference material in volumetric and mass titrimetric methods. The accuracy of a method is critically dependent on the properties of this compound. Important requirements for a primary standard are the following:

1. High purity. Established methods for confirming purity should be available.
2. Atmospheric stability.
3. Absence of hydrate water so that the composition of the solid does not change with variations in humidity.
4. Modest cost.
5. Reasonable solubility in the titration medium.
6. Reasonably large molar mass so that the relative error associated with weighing the standard is minimized.

Very few compounds meet or even approach these criteria, and only a limited number of primary-standard substances are available commercially. As a consequence, less pure compounds must sometimes be used in place of a primary standard. The purity of such a **secondary standard** must be established by careful analysis.

A **secondary standard** is a compound whose purity has been established by chemical analysis and that serves as the reference material for a titrimetric method of analysis.

13B STANDARD SOLUTIONS

Standard solutions play a central role in all titrimetric methods of analysis. Therefore, we need to consider the desirable properties for such solutions, how they are prepared, and how their concentrations are expressed. The ideal standard solution for a titrimetric method will

1. be sufficiently stable so that it is necessary to determine its concentration only once;
2. react rapidly with the analyte so that the time required between additions of reagent is minimized;
3. react more or less completely with the analyte so that satisfactory end points are realized; and
4. undergo a selective reaction with the analyte that can be described by a balanced equation.

Few reagents meet all these ideals perfectly.

The accuracy of a titrimetric method can be no better than the accuracy of the concentration of the standard solution used in the titration. Two basic methods are used to establish the concentration of such solutions. The first is the **direct method,** in which a carefully weighed quantity of a primary standard is dissolved in a suitable solvent and diluted to an exactly known volume in a volumetric flask. The second is by **standardization,** in which the titrant to be standardized is used to titrate (1) a weighed quantity of a primary standard, (2) a weighed quantity of a secondary standard, or (3) a measured volume of another standard solution. A titrant that is standardized against a secondary standard or against another standard solution is sometimes referred to as a **secondary-standard solution.** The concentration of a secondary-standard solution is subject to a larger uncertainty than is that of a primary-standard solution. If there is a choice, then, solutions are best prepared by the direct method. Many reagents lack the properties required for a primary standard, however, and therefore require standardization.

In a **standardization,** the concentration of a volumetric solution is determined by titrating it against a carefully measured quantity of a primary or secondary standard or an exactly known volume of another standard solution.

13C VOLUMETRIC CALCULATIONS

As we indicated in Section 4B-1, we express the concentration of solutions in several ways. For standard solutions used in titrimetry, either **molarity** c or **normality** c_N is usually employed. The first term gives the number of moles of reagent contained in one liter of solution, and the second gives the number of **equivalents** of reagent in the same volume.

Throughout this text, we base volumetric calculations exclusively on molarity and molar masses. We have also included in Appendix 6 a discussion of how volumetric calculations are performed based on normality and equivalent weights because you may encounter these terms and their uses in the industrial and health science literature.

13C-1 Some Useful Algebraic Relationships

Most volumetric calculations are based on two pairs of simple equations that are derived from definitions of the millimole, the mole, and the molar concentration. For the chemical species A, we may write

◀ $n_A = \frac{m_A}{\mathcal{M}_A}$

where n_A is the amount of A, m_A is the mass of A, and $\mathcal{M}_A$ is the molar mass of A.

$$\text{amount A (mmol)} = \frac{\text{mass A (g)}}{\text{millimolar mass A (g/mmol)}} \tag{13-1}$$

◀ $c_A = \frac{n_A}{V}$ or $n_A = V \times c_A$

$$\text{amount A (mol)} = \frac{\text{mass A (g)}}{\text{molar mass A (g/mol)}} \tag{13-2}$$

The second pair is derived from the definition of molar concentration. That is,

$$\text{amount A (mmol)} = V\,\text{(mL)} \times c_A\,\text{(mmol A/mL)} \tag{13-3}$$

$$\text{amount A (mol)} = V\,\text{(L)} \times c_A\,\text{(mol A/L)} \tag{13-4}$$

where V is the volume of the solution.

Use Equations 13-1 and 13-3 when volumes are measured in milliliters and Equations 13-2 and 13-4 when the units are liters.

◀ Any combination of grams, moles, and liters can be replaced with any analogous combination expressed in milligrams, millimoles, and milliliters. For example, a 0.1 M solution contains 0.1 mol of a species per liter or 0.1 mmol per milliliter. Similarly, the number of moles of a compound is equal to the mass in grams of that compound divided by its molar mass in grams or the mass in milligrams divided by its millimolar mass in milligrams.

13C-2 Calculating the Molarity of Standard Solutions

The following three examples illustrate how the concentrations of volumetric reagents are computed.

EXAMPLE 13-1

Describe the preparation of 2.000 L of 0.0500 M $AgNO_3$ (169.87 g/mol) from the primary-standard–grade solid.

Since the volume is in liters, we base our calculations on the mole rather than the millimole. Thus, to obtain the amount of $AgNO_3$ needed, we write

$$\text{amount } AgNO_3 = V_{soln}(\text{L}) \times c_{AgNO_3}(\text{mol/L})$$

$$= 2.000\ \cancel{\text{L}} \times \frac{0.0500 \text{ mol } Na_2CO_3}{\cancel{\text{L}}} = 0.1000 \text{ mol } AgNO_3$$

To obtain the mass of $AgNO_3$, we rearrange Equation 13-2 to give

$$\text{mass } AgNO_3 = 0.1000\ \cancel{\text{mol AgNO}_3} \times \frac{169.87 \text{ g } AgNO_3}{\cancel{\text{mol AgNO}_3}}$$

$$= 16.98 \text{ g } AgNO_3$$

Therefore, the solution is prepared by dissolving 16.98 g of $AgNO_3$ in water and diluting to exactly 2.000 L.

EXAMPLE 13-2

A standard 0.0100 M solution of Na^+ is required to calibrate a flame photometric method to determine the element. Describe how 500 mL of this solution can be prepared from primary standard Na_2CO_3 (105.99 g/mL).

We wish to compute the mass of reagent required to give a species molarity of 0.0100. Here, we will use millimoles, since the volume is in milliliters. Because Na_2CO_3 dissociates to give two Na^+ ions, we can write that the number of millimoles of Na_2CO_3 needed is

$$\text{amount } Na_2CO_3 = 500\ \cancel{\text{mL}} \times \frac{0.0100\ \cancel{\text{mmol Na}^+}}{\cancel{\text{mL}}} \times \frac{1 \text{ mmol } Na_2CO_3}{2\ \cancel{\text{mmoi Na}^+}}$$

$$= 2.50 \text{ mmol}$$

From the definition of millimole, we write

$$\text{mass } Na_2CO_3 = 2.50\ \cancel{\text{mmol Na}_2\text{CO}_3} \times 0.10599 \frac{\text{g } Na_2CO_3}{\cancel{\text{mmol Na}_2\text{CO}_3}} = 0.265 \text{ g}$$

The solution is therefore prepared by dissolving 0.265 g of Na_2CO_3 in water and diluting to 500 mL.

EXAMPLE 13-3

How would you prepare 50.0-mL portions of standard solutions that are 0.00500 M, 0.00200 M, and 0.00100 M in Na^+ from the solution in Example 13-2?

The number of millimoles of Na^+ taken from the concentrated solution must equal the number in the diluted solutions. Thus,

$$\text{amount } Na^+ \text{ from concd soln} = \text{amount } Na^+ \text{ in dil soln}$$

Recall that the number of millimoles is equal to the number of millimoles per milliliter times the number of milliliters. That is,

$$V_{concd} \times c_{concd} = V_{dil} \times c_{dil}$$

◀ A useful algebraic relationship is $V_{concd} \times c_{concd} = V_{dil} \times c_{dil}$.

where V_{concd} and V_{dil} are the volumes in milliliters of the concentrated and diluted solutions, respectively, and c_{concd} and c_{dil} are their molar Na^+ concentrations. For the 0.00500-M solution, this equation rearranges to

$$V_{concd} = \frac{V_{dil} \times c_{dil}}{c_{concd}} = \frac{50.0 \text{ mL} \times 0.00500 \text{ mmol Na}^+/\text{mL}}{0.0100 \text{ mmol Na}^+/\text{mL}} = 25.0 \text{ mL}$$

Thus, to produce 50.0 mL of 0.00500 M Na^+, 25.0 mL of the concentrated solution should be diluted to exactly 50.0 mL.

Repeat the calculation for the other two molarities to confirm that diluting 10.0 and 5.00 mL of the concentrated solution to 50.0 mL produces the desired solutions.

13C-3 Treating Titration Data

In this section, we describe two types of volumetric calculations. The first involves computing the molarity of solutions that have been standardized against either a primary-standard or another standard solution. The second involves calculating the amount of analyte in a sample from titration data. Both types are based on three algebraic relationships. Two of these are Equations 13-1 and 13-3, both of which are based on millimoles and milliliters. The third relationship is the stoichiometric ratio of the number of millimoles of the analyte to the number of millimoles of titrant.

Calculating Molarities from Standardization Data

Examples 13-4 and 13-5 illustrate how standardization data are treated.

EXAMPLE 13-4

A 50.00-mL portion of an HCl solution required 29.71 mL of 0.01963 M $Ba(OH)_2$ to reach an end point with bromocresol green indicator. Calculate the molarity of the HCl.

(continued)

In the titration, 1 mmol of $Ba(OH)_2$ reacts with 2 mmol of HCl:

$$Ba(OH)_2 + 2HCl \rightarrow BaCl_2 + 2H_2O$$

Thus, the stoichiometric ratio is

$$\text{stoichiometric ratio} = \frac{2 \text{ mmol HCl}}{1 \text{ mmol Ba(OH)}_2}$$

The number of millimoles of the standard is obtained by substituting into Equation 13-3:

$$\text{amount Ba(OH)}_2 = 29.71 \cancel{\text{mL Ba(OH)}_2} \times 0.01963 \frac{\text{mmol Ba(OH)}_2}{\cancel{\text{mL Ba(OH)}_2}}$$

To obtain the number of millimoles of HCl, we multiply this result by the stoichiometric ratio determined initially:

$$\text{amount HCl} = (29.71 \times 0.01963) \cancel{\text{mmol Ba(OH)}_2} \times \frac{2 \text{ mmol HCl}}{1 \cancel{\text{mmol Ba(OH)}_2}}$$

► In determining the number of significant figures to retain in volumetric calculations, the stoichiometric ratio is assumed to be known exactly without uncertainty.

To obtain the number of millimoles of HCl per mL, we divide by the volume of the acid. Thus,

$$c_{HCl} = \frac{(29.71 \times 0.01963 \times 2) \text{ mmol HCl}}{50.0 \text{ mL HCl}}$$

$$= 0.023328 \frac{\text{mmol HCl}}{\text{mL HCl}} = 0.02333 \text{ M}$$

EXAMPLE 13-5

CD-ROM Tutorials: Analyzing Acids and Bases Using Titrations.

Titration of 0.2121 g of pure $Na_2C_2O_4$ (134.00 g/mol) required 43.31 mL of $KMnO_4$. What is the molarity of the $KMnO_4$ solution? The chemical reaction is

$$2MnO_4^- + 5C_2O_4^{2-} + 16H^+ \rightarrow 2Mn^{2+} + 10CO_2 + 8H_2O$$

From this equation, we see that

$$\text{stoichiometric ratio} = \frac{2 \text{ mmol KMnO}_4}{5 \text{ mmol Na}_2\text{C}_2\text{O}_4}$$

The amount of primary-standard $Na_2C_2O_4$ is given by Equation 13-1:

$$\text{amount Na}_2\text{C}_2\text{O}_4 = 0.2121 \cancel{\text{g Na}_2\text{C}_2\text{O}_4} \times \frac{1 \text{ mmol Na}_2\text{C}_2\text{O}_4}{0.13400 \cancel{\text{g Na}_2\text{C}_2\text{O}_4}}$$

To obtain the number of millimoles of $KMnO_4$, we multiply this result by the stoichiometric ratio:

$$\text{amount KMnO}_4 = \frac{0.2121}{0.1340}\,\cancel{\text{mmol Na}_2\text{C}_2\text{O}_4} \times \frac{2\ \text{mmol KMnO}_4}{5\ \cancel{\text{mmol Na}_2\text{C}_2\text{O}_4}}$$

The molarity is then obtained by dividing by the volume of $KMnO_4$ consumed. Thus,

$$c_{\text{KMnO4}} = \frac{\left(\dfrac{0.2121}{0.13400} \times \dfrac{2}{5}\right)\text{mmol KMnO}_4}{43.31\ \text{mL KMnO}_4} = 0.01462\ \text{M}$$

Note that units are carried through all calculations as a check on the correctness of the relationships used in Examples 13-4 and 13-5.

Calculating the Quantity of Analyte from Titration Data

As shown by the examples that follow, the same systematic approach just described is also used to compute analyte concentrations from titration data.

EXAMPLE 13-6

A 0.8040-g sample of an iron ore is dissolved in acid. The iron is then reduced to Fe^{2+} and titrated with 47.22 mL of 0.02242 M $KMnO_4$ solution. Calculate the results of this analysis in terms of (a) % Fe (55.847 g/mol) and (b) % Fe_3O_4 (231.54 g/mol). The reaction of the analyte with the reagent is described by the equation

$$MnO_4^- + 5Fe^{2+} + 8H^+ \rightarrow Mn^{2+} + 5Fe^{3+} + 4H_2O$$

(a)

$$\text{stoichiometric ratio} = \frac{5\ \text{mmol Fe}^{2+}}{1\ \text{mmol KMnO}_4}$$

$$\text{amount KMnO}_4 = 47.22\ \cancel{\text{mL KMnO}_4} \times \frac{0.02242\ \text{mmol KMnO}_4}{\cancel{\text{mL KMnO}_4}}$$

$$\text{amount Fe}^{2+} = (47.22 \times 0.02242)\ \cancel{\text{mmol KMnO}_4} \times \frac{5\ \text{mmol Fe}^{2+}}{1\ \cancel{\text{mmol KMnO}_4}}$$

The mass of Fe^{2+} is then given by

$$\text{mass Fe}^{2+} = (47.22 \times 0.02242 \times 5)\ \cancel{\text{mmol Fe}^{2+}} \times 0.055847\ \frac{\text{g Fe}^{2+}}{\cancel{\text{mmol Fe}^{2+}}}$$

(continued)

The percent Fe^{2+} is

$$\% \; Fe^{2+} = \frac{(47.22 \times 0.02242 \times 5 \times 0.55847) \text{ g } Fe^{2+}}{0.8040 \text{ g sample}} \times 100\% = 36.77\%$$

(b) To determine the correct stoichiometric ratio, we note that

$$5 \; Fe^{2+} \equiv 1 \; MnO_4^-$$

Therefore,

$$5Fe_3O_4 \equiv 15Fe^{2+} \equiv 3MnO_4^-$$

and

$$\text{stoichiometric ratio} = \frac{5 \text{ mmol } Fe_3O_4}{3 \text{ mmol } KMnO_4}$$

As in part (a),

$$\text{amount } KMnO_4 = \frac{47.22 \; \cancel{\text{mL } KMnO_4} \times 0.02242 \text{ mmol } KMnO_4}{\cancel{\text{mL } KMnO_4}}$$

$$\text{amount } Fe_3O_4 = (47.22 \times 0.02242) \; \cancel{\text{mmol } KMnO_4} \times \frac{5 \text{ mmol } Fe_3O_4}{3 \; \cancel{\text{mmol } KMnO_4}}$$

$$\text{mass } Fe_3O_4 = \left(47.22 \times 0.02242 \times \frac{5}{3}\right) \cancel{\text{mmol } Fe_3O_4} \times 0.23154 \frac{\text{g } Fe_3O_4}{\cancel{\text{mmol } Fe_3O_4}}$$

$$\% \; Fe_3O_4 = \frac{\left(47.22 \times 0.02242 \times \frac{5}{3}\right) \times 0.23154 \text{ g } Fe_3O_4}{0.8040 \text{ g sample}} \times 100\% = 50.81\%$$

FEATURE 13-1

Another Approach to Example 13-6(a)

Some people find it easier to write out the solution to a problem in such a way that the units in the denominator of each succeeding term eliminate the units in the numerator of the preceding one until the units of the answer are obtained. For example, the solution to part (a) of Example 13-6 can be written

$$47.22 \; \cancel{\text{mL } KMnO_4} \times \frac{0.02242 \; \cancel{\text{mmol } KMnO_4}}{\cancel{\text{mL } KMnO_4}} \times \frac{5 \; \cancel{\text{mmol Fe}}}{1 \; \cancel{\text{mmol } KMnO_4}}$$

$$\times \frac{0.05585 \text{ g Fe}}{\cancel{\text{mmol Fe}}} \times \frac{1}{0.8040 \text{ g sample}} \times 100\% = 36.77\% \text{Fe}$$

EXAMPLE 13-7

A 100.0-mL sample of brackish water was made ammoniacal, and the sulfide it contained was titrated with 16.47 mL of 0.02310 M $AgNO_3$. The analytical reaction is

$$2Ag^+ + S^{2-} \rightarrow Ag_2S(s)$$

Calculate the concentration of H_2S in the water in parts per million.

At the end point

$$\text{stoichiometric ratio} = \frac{1 \text{ mmol } H_2S}{2 \text{ mmol } AgNO_3}$$

$$\text{amount } AgNO_3 = 16.47 \cancel{\text{mL } AgNO_3} \times 0.02310 \frac{\text{mmol } AgNO_3}{\cancel{\text{mL } AgNO_3}}$$

$$\text{amount } H_2S = (16.47 \times 0.02310) \cancel{\text{mmol } AgNO_3} \times \frac{1 \text{ mmol } H_2S}{2 \cancel{\text{mmol } AgNO_3}}$$

$$\text{mass } H_2S = \left(16.47 \times 0.02310 \times \frac{1}{2}\right) \cancel{\text{mmol } H_2S} \times 0.034802 \frac{\text{g } H_2S}{\cancel{\text{mmol } H_2S}}$$

$$= 6.620 \times 10^{-3} \text{ g } H_2S$$

$$\text{concd } H_2S = \frac{6.620 \times 10^{-3} \text{ g } H_2S}{100.0 \cancel{\text{mL sample}} \times 1.000 \text{ g sample}/\cancel{\text{mL sample}}} \times 10^6 \text{ ppm}$$

$$= 6.62 \text{ ppm } H_2S$$

FEATURE 13-2

Rounding the Answer to Example 13-7

Note that the input data for Example 13-7 all contained four or more significant figures, but the answer was rounded to three. Why?

We can make the rounding decision by doing a couple of rough calculations in our heads. Assume that the input data are uncertain to 1 part in the last significant figure. The largest *relative* error will then be associated with the sample size. Here, the relative uncertainty is 0.1/100.0. Thus, the uncertainty is about 1 part in 1000 (compared with about 1 part in 1647 for the volume of $AgNO_3$ and 1 part in 2300 for the reagent molarity). We then assume that the calculated result is uncertain to about the same amount as the least precise measurement, or 1 part in 1000. The absolute uncertainty of the final result is then 6.62 ppm $\times$ 1/1000 = 0.0066, or about 0.01 ppm, and we round to the second figure to the right of the decimal point. Thus, we report 6.62 ppm.

Practice making this rough type of rounding decision whenever you make a computation.

EXAMPLE 13-8

The phosphorus in a 4.258-g sample of a plant food was converted to PO_4^{3-} and precipitated as Ag_3PO_4 through the addition of 50.00 mL of 0.0820 M $AgNO_3$. The excess $AgNO_3$ was back-titrated with 4.86 mL of 0.0625 M KSCN. Express the results of this analysis in terms of % P_2O_5.

The chemical reactions are

$$P_2O_5 + 9H_2O \rightarrow 2PO_4^{3-} + 6H_3O^+$$

$$2PO_4^{3-} + \underset{\text{excess}}{6Ag^+} \rightarrow 2\,Ag_3PO_4(s)$$

$$Ag^+ + SCN^- \rightarrow AgSCN(s)$$

Thus, the stoichiometric ratios are

$$\frac{1 \text{ mmol } P_2O_5}{6 \text{ mmol } AgNO_3} \quad \text{and} \quad \frac{1 \text{ mmol KSCN}}{1 \text{ mmol } AgNO_3}$$

$$\text{total amount } AgNO_3 = 50.00\ \cancel{\text{mL}} \times 0.0820\ \frac{\text{mmol } AgNO_3}{\cancel{\text{mL}}} = 4.100$$

$$\text{amount } AgNO_3 \text{ consumed by KSCN} = 4.06 \text{ mL} \times 0.0625\ \frac{\cancel{\text{mmol KSCN}}}{\text{mL}} \times \frac{1 \text{ mmol } AgNO_3}{\cancel{\text{mmol KSCN}}} = 0.2538 \text{ mmol}$$

$$\text{amount } P_2O_5 = (4.100 - 0.254)\ \cancel{\text{mmol } AgNO_3} \times \frac{1 \text{ mmol } P_2O_5}{6\ \cancel{\text{mmol } AgNO_3}} = 0.6410 \text{ mmol } P_2O_5$$

$$\% P_2O_5 = \frac{0.6410\ \cancel{\text{mmol}} \times \dfrac{0.1419 \text{ g } P_2O_5}{\cancel{\text{mmol}}}}{4.258 \text{ g sample}} \times 100\% = 2.14\%$$

EXAMPLE 13-9

The CO in a 20.3-L sample of gas was converted to CO_2 by passing the gas over iodine pentoxide heated to 150°C:

$$I_2O_5(s) + 5CO(g) \rightarrow 5CO_2(g) + I_2(g)$$

The iodine was distilled at this temperature and was collected in an absorber containing 8.25 mL of 0.01101 M $Na_2S_2O_3$.

$$I_2(g) + 2S_2O_3^{2-}(aq) \rightarrow 2I^-(aq) + S_4O_6^{2-}(aq)$$

The excess $Na_2S_2O_3$ was back-titrated with 2.16 mL of 0.00947 M I_2 solution. Calculate the concentration in milligrams of CO (28.01 g/mol) per liter of sample.

Based on the two reactions, the stoichiometric ratios are

$$\frac{5 \text{ mmol CO}}{1 \text{ mmol I}_2} \quad \text{and} \quad \frac{2 \text{ mmol Na}_2\text{S}_2\text{O}_3}{1 \text{ mmol I}_2}$$

We divide the first ratio by the second to get a third useful ratio:

$$\frac{5 \text{ mmol CO}}{2 \text{ mmol Na}_2\text{S}_2\text{O}_3}$$

This relationship reveals that 5 mmol of CO are responsible for the consumption of 2 mmol of $Na_2S_2O_3$. The total amount of $Na_2S_2O_3$ is

$$\text{amount Na}_2\text{S}_2\text{O}_3 = 8.25 \cancel{\text{mL Na}_2\text{S}_2\text{O}_3} \times 0.01101 \frac{\text{mmol Na}_2\text{S}_2\text{O}_3}{\cancel{\text{mL Na}_2\text{S}_2\text{O}_3}}$$
$$= 0.09083 \text{ mmol Na}_2\text{S}_2\text{O}_3$$

The amount of $Na_2S_2O_3$ consumed in the back-titration is

$$\text{amount Na}_2\text{S}_2\text{O}_3 = 2.16 \cancel{\text{mL I}_2} \times 0.00947 \frac{\cancel{\text{mmol I}_2}}{\cancel{\text{mL I}_2}} \times \frac{2 \text{ mmol Na}_2\text{S}_2\text{O}_3}{\cancel{\text{mmol I}_2}}$$
$$= 0.04091 \text{ mmol Na}_2\text{S}_2\text{O}_3$$

The number of millimoles of CO can then be obtained by using the third stoichiometric ratio:

$$\text{amount CO} = (0.09083 - 0.04091) \cancel{\text{mmol Na}_2\text{S}_2\text{O}_3} \times \frac{5 \text{ mmol CO}}{2 \cancel{\text{mmol Na}_2\text{S}_2\text{O}_3}}$$
$$= 0.1248 \text{ mmol CO}$$

$$\text{mass CO} = 0.1248 \cancel{\text{mmol CO}} \times \frac{28.01 \text{ mg CO}}{\cancel{\text{mmol CO}}} = 3.4956 \text{ mg}$$

$$\frac{\text{mass CO}}{\text{vol sample}} = \frac{3.4956 \text{ mg CO}}{20.3 \text{ L sample}} = 0.172 \frac{\text{mg CO}}{\text{L sample}}$$

13D GRAVIMETRIC TITRIMETRY

Weight or **gravimetric titrimetry** differs from its volumetric counterpart in that the *mass* of titrant is measured rather than the volume. Thus, in a weight titration, a balance and a weighable solution dispenser are substituted for a buret and its markings. Weight titrimetry actually predates volumetric titrimetry by more than 50 years.[2] With the advent of reliable burets, however, weight titrations were largely supplanted by volumetric methods because the former required relatively elaborate equipment and were tedious and time consuming. The availability of sensitive,

[2] For a brief history of gravimetric and volumetric titrimetry, see B. Kratochvil and C. Maitra, *Amer. Lab.*, **1982** (1), 22.

low-cost, top-loading digital analytical balances and convenient plastic solution dispensers has changed this situation completely, and weight titrations can now be performed more easily and rapidly than volumetric titrations.

13D-1 Calculations Associated with Weight Titrations

The most convenient unit of concentration for weight titrations is **weight molarity,** M_w, which is the number of moles of a reagent in one kilogram of solution or the number of millimoles in one gram of solution. Thus, aqueous 0.1 M_w NaCl contains 0.1 mol of the salt in 1 kg of solution or 0.1 mmol in 1 g of the solution.

The weight molarity $c_{w(A)}$ of a solution of a solute A is computed by means of either of two equations that are analogous to Equation 4-1:

$$\text{weight molarity} = \frac{\text{no. mol A}}{\text{no. kg solution}} = \frac{\text{no. mmol A}}{\text{no. g solution}} \tag{13-5}$$

$$c_{w(A)} = \frac{n_A}{m_{soln}}$$

Weight titration data can then be treated by using the methods illustrated in Sections 13C-2 and 13C-3 after substitution of weight molarity for molarity and grams and kilograms for milliliters and liters.

13D-2 Advantages of Weight Titrations

In addition to greater speed and convenience, weight titrations offer certain other advantages over their volumetric counterparts:

1. Calibration of glassware and tedious cleaning to ensure proper drainage are completely eliminated.
2. Temperature corrections are unnecessary because weight molarity does not change with temperature, in contrast to volume molarity. This advantage is particularly important in nonaqueous titrations because of the high coefficients of expansion of most organic liquids (about ten times that of water).
3. Weight measurements can be made with considerably greater precision and accuracy than can volume measurements. For example, 50 g or 100 g of an aqueous solution can be readily measured to ± 1 mg, which corresponds to ± 0.001 mL. This greater sensitivity makes it possible to choose sample sizes that lead to significantly smaller consumption of standard reagents.
4. Weight titrations are more easily automated than are volumetric titrations.

13E TITRATION CURVES IN TITRIMETRIC METHODS

As noted in Section 13A-1, an end point is an observable physical change that occurs near the equivalence point of a titration. The two most widely used end points involve (1) changes in color due to the reagent, the analyte, or an indicator and (2) a change in potential of an electrode that responds to the concentration of the reagent or the analyte.

To understand the theoretical basis of end points and the sources of titration errors, we calculate the data points necessary to construct **titration curves** for the systems under consideration. Titration curves plot reagent volume on the horizontal axis and some function of the analyte or reagent concentration on the vertical axis.

◀ Titration curves are plots of a concentration-related variable as a function of reagent volume.

13E-1 Types of Titration Curves

Two general types of titration curves (and thus two general types of end points) are encountered in titrimetric methods. In the first type, called a sigmoidal curve, important observations are confined to a small region (typically ±0.1 to ±0.5 mL) surrounding the equivalence point. A **sigmoidal curve,** in which the p-function of analyte (or sometimes the reagent) is plotted as a function of reagent volume, is shown in Figure 13-2a. In the second type of curve, called a **linear segment curve,** measurements are made on both sides of, but well away from, the equivalence point. Measurements near equivalence are avoided. In this type of curve, the vertical axis represents an instrument reading that is directly proportional to the concentration of the analyte or the reagent. A typical linear segment curve is found in Figure 13-2b. The sigmoidal type offers the advantages of speed and convenience. The linear segment type is advantageous for reactions that are complete only in the presence of a considerable excess of the reagent or analyte.

In this chapter and several that follow, we deal exclusively with sigmoidal titration curves. We explore linear-segment curves in Section 26A-5.

◀ The vertical axis in a sigmoidal titration curve is either the p-function of the analyte or reagent or the potential of an analyte- or reagent-sensitive electrode.

◀ The vertical axis of a linear-segment titration curve is an instrument signal that is proportional to the concentration of the analyte or reagent.

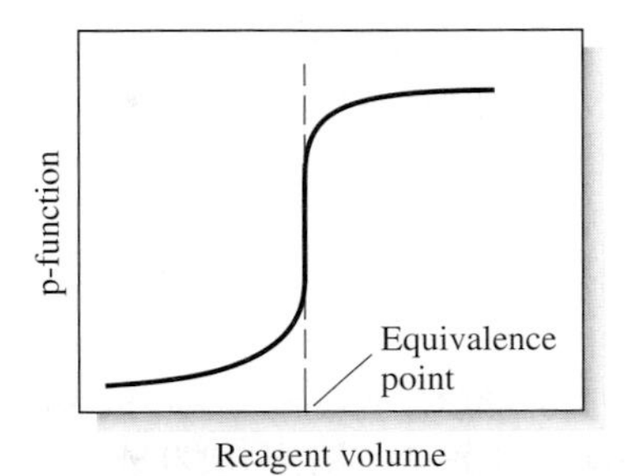

(a) Sigmoidal curve

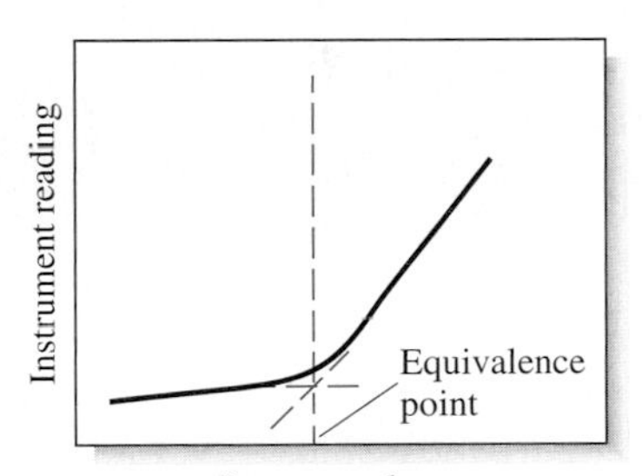

(b) Linear-segment curve

Figure 13-2 Two types of titration curves.

13E-2 Concentration Changes during Titrations

The equivalence point in a titration is characterized by major changes in the *relative* concentrations of reagent and analyte. Table 13-1 illustrates this phenomenon. The data in the second column of the table show the silver ion concentration as a 50.00-mL aliquot of a 0.1000 M solution of silver nitrate acid is titrated with a 0.1000 M solution of potassium thiocyanate. The precipitation reaction is described by the equation

$$Ag^+ + SCN^- \rightarrow AgSCN(s) \tag{13-6}$$

TABLE 13-1

Concentration Change during the Titration of 50.00 mL of 0.1000 M $AgNO_3$ with 0.1000 M KSCN

Volume of 0.1000 M KSCN	$[Ag^+]$ mmol/L	Milliliters KSCN to Cause a Tenfold Decrease in $[Ag^+]$	pAg	pSCN
0.00	1.0×10^{-1}		1.0	
40.91	1.0×10^{-2}	40.91	2.0	10.0
49.01	1.0×10^{-3}	8.10	3.0	9.0
49.90	1.0×10^{-4}	0.89	4.0	8.0
49.99	1.0×10^{-5}	0.09	5.0	7.0
50.00	1.0×10^{-6}	0.01	6.0	6.0
50.01	1.0×10^{-7}	0.01	7.0	5.0
50.10	1.0×10^{-8}	0.09	8.0	4.0
51.01	1.0×10^{-9}	0.91	9.0	3.0
61.11	1.0×10^{-10}	10.10	10.0	2.0

◀ At the beginning of the titration described in Table 13-1, about 41 mL of reagent bring about a 10-fold decrease in the concentration of A; only 0.001 mL is required to cause this same change at the equivalence point.

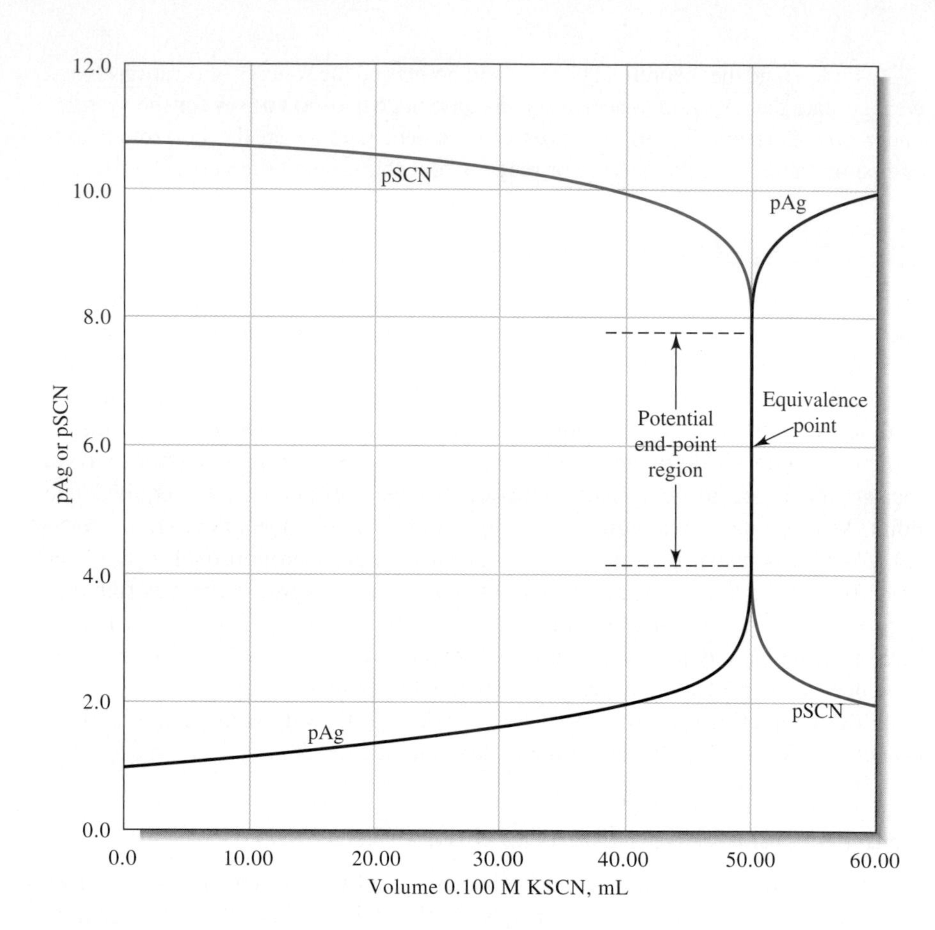

Figure 13-3 Titration curve for the titration of 50.00 mL of 0.1000 M $AgNO_3$ with 0.1000 M KSCN.

To emphasize the changes in *relative* concentration that occur in the equivalence point region, the volume increments computed are those required to cause tenfold decreases in the concentration of Ag^+. Thus, we see in the third column that an addition of 40.91 mL of KSCN is needed to decrease the concentration of the silver ion by one order of magnitude, from 0.10 M to 0.010 M. An addition of only 8.1 mL is required to lower the concentration by another factor of 10 to 0.0010 M; 0.89 mL causes yet another 10-fold decrease. A corresponding increase in thiocyanate ion concentration occurs at the same time. End-point detection, then, is based on this large change in *relative* concentration of the analyte (or the reagent) that occurs at and near the equivalence point for every type of titration.

The large changes in relative concentration that occur in the region of chemical equivalence are shown by plotting the negative logarithm of the analyte or the reagent concentration (the p-function) against reagent volume, as in Figure 13-3. The data for these plots are found in the fourth and fifth columns of Table 13-1. Titration curves for reactions involving complex formation, precipitation, and oxidation/reduction all exhibit the same sharp increase or decrease in p-function as is shown in Figure 13-3 in the equivalence-point region. Titration curves define the properties required of an indicator and allow us to estimate the error associated with titration methods. For example, as shown in Figure 13-3, the equivalence point is found in the center of the steeply rising part of the curve at a pAg of about

6.0. Any end-point signal that occurs at a pAg between about 4.0 and 8.0 will yield a titration error of about ±0.01 mL or less, which corresponds to a relative error of about ±0.02% for an analysis based on this reaction.

13F PRECIPITATION TITRIMETRY

Precipitation titrimetry, which is based on reactions that yield ionic compounds of limited solubility, is one of the oldest analytical techniques, dating back to the mid-1800s. Because of the slow rate of formation of most precipitates, however, there are only a few precipitating agents that can be used in titrimetry. By far the most widely used and most important precipitating reagent is silver nitrate, which is used for the determination of the halides, the halide-like anions (SCN^-, CN^-, CNO^-), mercaptans, fatty acids, and several divalent and trivalent inorganic anions. Titrimetric methods based on silver nitrate are sometimes called **argentometric methods.** In this text, we limit our discussion of precipitation titrimetry to argentometric methods.

◀ Argentometric is derived from the Latin noun *argentum,* which means silver.

13F-1 Precipitation Titration Curves Involving Silver Ion

The most common method of determining the halide ion concentration of aqueous solutions is titration with a standard solution of silver nitrate. The reaction product is solid silver halide. A titration curve for this method usually consists of a plot of pAg versus the volume of silver nitrate added. To construct titration curves, three type of calculations are required, each of which corresponds to a distinct stage in the reaction: (1) preequivalence, (2) equivalence, and (3) postequivalence. Example 13-10 demonstrates how pAg is determined for each of these stages.

EXAMPLE 13-10

Perform calculations needed to generate a titration curve for 50.00 mL of 0.0500 M NaCl with 0.1000 M $AgNO_3$ (for AgCl, $K_{sp} = 1.82 \times 10^{-10}$).

$$\text{Reaction:} \qquad Ag^+(aq) + Cl^-(aq) \rightleftharpoons AgCl(s)$$

(1) **Preequivalence-Point Data**

Here the molar analytical concentration c_{NaCl} is readily computed. For example, when 10.00 mL of $AgNO_3$ has been added,

$$c_{NaCl} = \frac{\text{original number of mmol NaCl} - \text{no. mmol } AgNO_3 \text{ added}}{\text{total volume solution}}$$

(continued)

But

$$\text{original number of mmol NaCl} = 50.00\ \cancel{\text{mL}} \times 0.0500\ \frac{\text{mmol NaCl}}{\cancel{\text{mL}}} = 2.500$$

$$\text{number of mmol AgNO}_3\ \text{added} = 10.00\ \cancel{\text{mL}} \times 0.1000\ \frac{\text{mmol AgNO}_3}{\cancel{\text{mL}}} = 1.000$$

$$\text{no. mmol NaCl remaining} = 1.500$$

$$c_{\text{NaCl}} = \frac{1.500\ \text{mmol NaCl}}{(50.00 + 10.00)\ \text{mL}} = 0.02500\ \frac{\text{mmol NaCl}}{\text{mL}} = 0.02500\ \text{M}$$

$$[\text{Cl}^-] = 0.02500\ \text{M}$$

$$[\text{Ag}^+] = K_{sp}/[\text{Cl}^-] = \frac{1.82 \times 10^{-10}}{0.02500} = 7.28 \times 10^{-9}\ \text{M}$$

$$\text{pAg} = -\log(7.28 \times 10^{-9}) = 8.14$$

Additional points defining the curve in the preequivalence-point region are obtained in the same way. Results of calculations of this kind are shown in the second column of Table 13-2.

(2) **Equivalence Point pAg**

Here,

$$[\text{Ag}^+] = [\text{Cl}^-] \quad \text{and} \quad [\text{Ag}^+][\text{Cl}^-] = 1.82 \times 10^{-10} = [\text{Ag}^+]^2$$

$$[\text{Ag}^+] = 1.349 \times 10^{-5}\ \text{M} \quad \text{and} \quad \text{pAg} = -\log(1.349 \times 10^{-5}) = 4.87$$

(3) **Postequivalence-Point Data**

At 26.00 mL $AgNO_3$ added, Ag^+ is in excess, so

$$[\text{Ag}^+] = c_{\text{AgNO}_3} = \frac{26.00 \times 0.1000 - 50.00 \times 0.0500}{50.00 - 26.00} = 1.316 \times 10^{-3}\ \text{M}$$

$$\text{pAg} = -\log(1.316 \times 10^{-3}) = 2.88$$

Additional postequivalence-point data are obtained in the same way and are shown in Table 13-2.

TABLE 13-2

Changes in pAg in the Titration of Cl^- With Standard $AgNO_3$

	pAg	
Volume of $AgNO_3$	**50.00 mL of 0.0500 M NaCl with 0.1000 M $AgNO_3$**	**50.00 mL of 0.00500 M NaCl with 0.01000 M $AgNO_3$**
10.00	8.14	7.14
20.00	7.59	6.59
24.00	6.87	5.87
25.00	4.87	4.87
26.00	2.88	3.88
30.00	2.20	3.20
40.00	1.78	2.78

The Effect of Concentration on Titration Curves

The effect of reagent and analyte concentrations on titration curves was shown by the two sets of data in Table 13-2 and the two titration curves in Figure 13-4. With 0.1 M $AgNO_3$ (Curve *A*), the change in pAg in the equivalence-point region is large. With the 0.01 M reagent, the change is markedly less but still pronounced. Thus, an indicator for Ag^+ that produces a signal in the 4.0 to 6.0 pAg range should give a minimum error for the stronger solution. For the more dilute chloride solution, the change in pAg in the equivalence-point region would be too small to be detected precisely with a visual indicator.

The Effect of Reaction Completeness on Titration Curves

Figure 13-5 illustrates the effect of solubility product on the sharpness of the end point in titrations with 0.1 M silver nitrate. Clearly, the change in pAg at the equivalence point becomes greater as the solubility products become smaller, that is, as the reaction between the analyte and silver nitrate becomes more complete. By careful choice of indicator—one that changes color in the region of pAg from 4 to 6—titration of chloride ion should be possible with a minimal titration error. Note that ions forming precipitates with solubility products much larger than about 10^{-10} do not yield satisfactory end points.

◀ You can derive a useful relationship by taking the negative logarithm of both sides of a solubility-product expression. Thus, for silver chloride,

$$\begin{aligned} -\log K_{sp} &= -\log([Ag^+][Cl^-]) \\ &= -\log[Ag^+] - \log[Cl^-] \\ pK_{sp} &= pAg + pCl \\ &= -\log(1.82 \times 10^{-10}) \\ &= 9.74 = pAg + pCl \end{aligned}$$

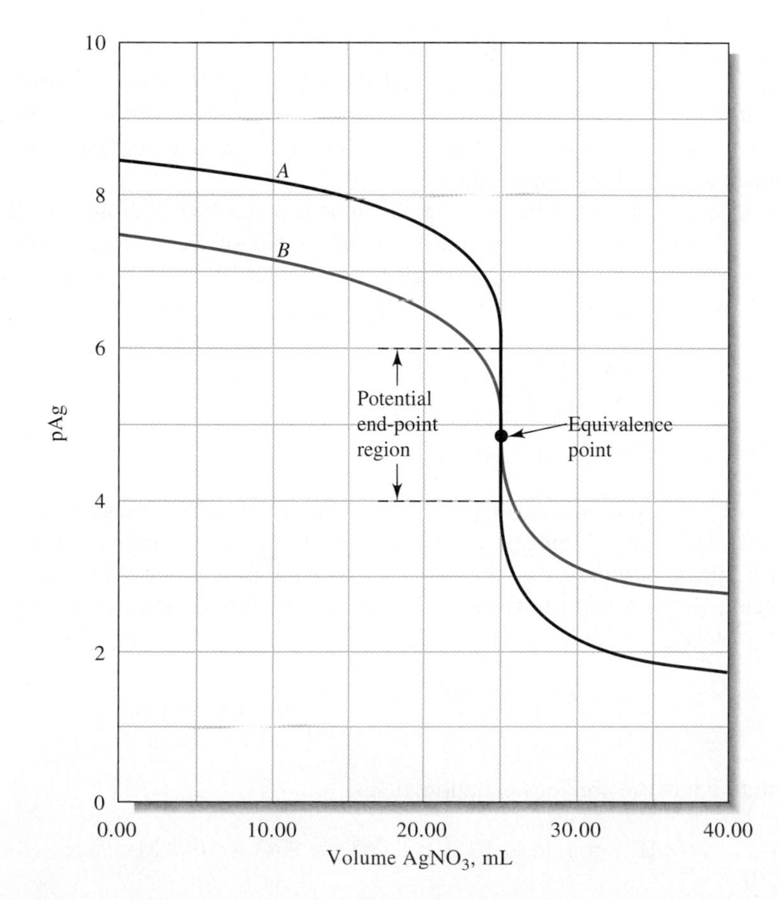

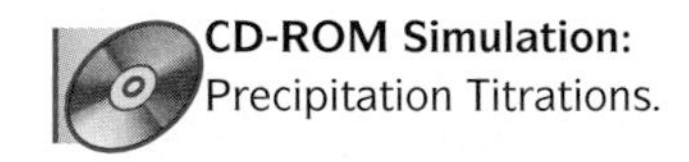

Figure 13-4 Titration curve for *A*, 50.00 mL of 0.0500 M NaCl with 0.1000 M $AgNO_3$, and *B*, 50.00 mL of 0.00500 M NaCl with 0.0100 M $AgNO_3$.

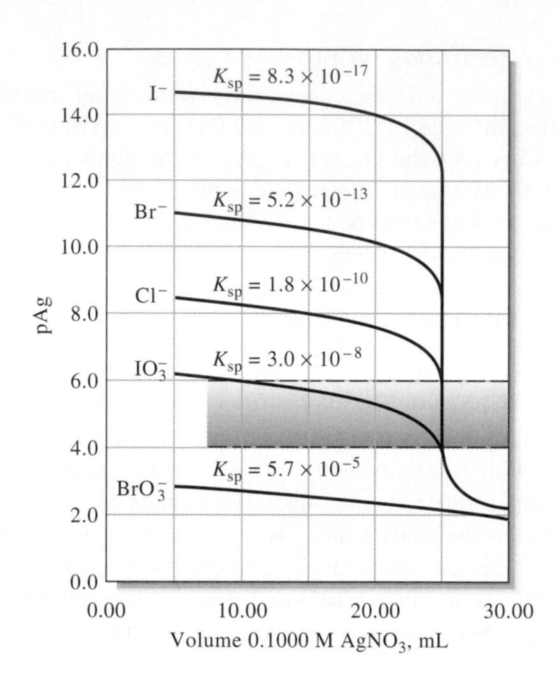

Figure 13-5 Effect of reaction completeness on precipitation titration curves. For each curve, 50.00 mL of a 0.0500 M solution of the anion was titrated with 0.1000 M $AgNO_3$. Note that smaller values of K_{sp} give much sharper breaks at the end point.

13F-2 Titration Curves for Mixtures of Anions

The methods developed in the previous section for deriving titration curves can be extended to mixtures that form precipitates of different solubilities. To illustrate, consider the titration of 50.00 mL of a solution that is 0.0500 M in iodide ion and 0.0800 M in chloride ion with 0.1000 M silver nitrate. The curve for the initial stages of this titration is identical to the curve shown for iodide in Figure 13-5 because silver chloride, with its much larger solubility product, does not begin to precipitate until well into the titration.

It is of interest to determine how much iodide is precipitated before appreciable amounts of silver chloride form. With the appearance of the smallest amount of solid silver chloride, the solubility-product expressions for both precipitates apply, and division of one by the other provides the useful relationship

$$\frac{\cancel{[Ag^+]}[I^-]}{\cancel{[Ag^+]}[Cl^-]} = \frac{8.3 \times 10^{-17}}{1.82 \times 10^{-10}} = 4.56 \times 10^{-7}$$

$$[I^-] = (4.56 \times 10^{-7})\,[Cl^-]$$

From this relationship, we see that the iodide concentration decreases to a tiny fraction of the chloride ion concentration before silver chloride begins to precipitate. So for all practical purposes, silver chloride forms only after 25.00 mL of titrant have been added in this titration. At this point, the chloride ion concentration is approximately

$$c_{Cl} = [Cl^-] = \frac{50.00 \times 0.0800}{50.00 + 25.00} = 0.0533 \text{ M}$$

Substituting into the previous equation yields

$$[I^-] = 4.56 \times 10^{-7} \times 0.0533 = 2.43 \times 10^{-8} \text{ M}$$

The percentage of iodide unprecipitated at this point can be calculated as follows:

$$\text{no. mmol I}^- = (75.00 \text{ mL})(2.43 \times 10^{-8} \text{ mmol I}^-/\text{mL}) = 1.82 \times 10^{-6}$$

$$\text{original no. mmol I}^- = (50.00 \text{ mL})(0.0500 \text{ mmol/mL} = 2.50$$

$$\text{I}^- \text{ unprecipitated} = \frac{1.82 \times 10^{-6}}{2.50} \times 100\% = 7.3 \times 10^{-5}\,\%$$

Thus, to within about 7.3×10^{-5} percent of the equivalence point for iodide, no silver chloride forms; up to this point, the titration curve is indistinguishable from that for the iodide alone (Figure 13-6). The data points for the first part of the titration curve, shown by the solid line in Figure 13-6, were computed on this basis.

As chloride ion begins to precipitate, the rapid decrease in pAg is terminated abruptly at a level that can be calculated from the solubility-product constant for silver chloride and the computed chloride concentration:

$$[\text{Ag}^+] = \frac{1.82 \times 10^{-10}}{0.0533} = 3.41 \times 10^{-9}$$

$$\text{pAg} = -\log(3.41 \times 10^{-9}) = 8.47$$

Further additions of silver nitrate decrease the chloride ion concentration, and the curve then becomes that for the titration of chloride by itself. For example, after 30.00 mL of titrant have been added,

$$c_{\text{Cl}} = [\text{Cl}^-] = \frac{50.00 \times 0.0800 + 50.00 \times 0.0500 - 30.00 \times 0.100}{50.00 + 30.00}$$

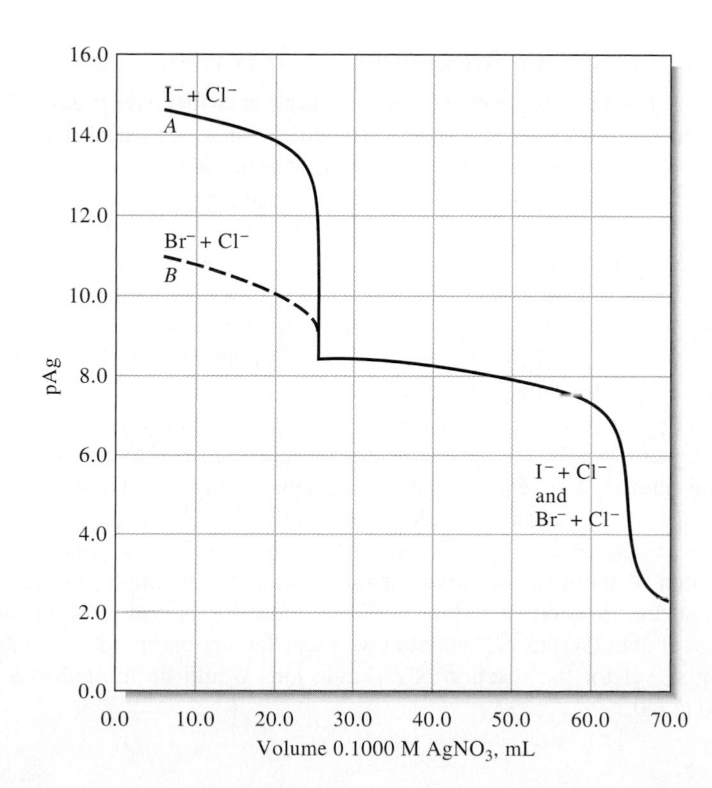

Figure 13-6 Titration curves for 50.00 mL of a solution 0.0800 M in Cl^- and 0.0500 M in I^- or Br^-.

Here, the first two terms in the numerator give the number of millimoles of chloride and iodide, respectively, and the third is the number of millimoles of titrant added. Thus,

$$[Cl^-] = 0.0438 \text{ M}$$

$$[Ag^+] = \frac{1.82 \times 10^{-10}}{0.0438} = 4.16 \times 10^{-9}$$

$$pAg = 8.38$$

The remainder of the data points for this curve can be computed in the same way as for a curve of chloride by itself.

Curve *A* in Figure 13-6, which is the titration curve for the chloride/iodide mixture just considered, is a composite of the individual curves for the two anionic species. Two equivalence points are evident. Curve *B* is the titration curve for a mixture of bromide and chloride ions. Clearly, the change associated with the first equivalence point becomes less distinct as the solubilities of the two precipitates approach one another. In the bromide/chloride titration, the initial pAg values are lower than they are in the iodide/chloride titration because the solubility of silver bromide exceeds that of silver iodide. Beyond the first equivalence point, however, where chloride ion is being titrated, the two titration curves are identical.

As shown in Section 37J-2, curves similar to those in Figure 13-6 can be obtained experimentally by potential measurement of a silver electrode immersed in the analyte solution. These curves can then be employed for the determination of the concentration of each of the ions in mixtures of this kind.

13F-3 Indicators for Argentometric Titrations

Three types of end points are encountered in titrations with silver nitrate: (1) chemical, (2) potentiometric, and (3) amperometric. Three chemical indicators are described in the sections that follow. Potentiometric end points are obtained by measuring the potential between a silver electrode and a reference electrode whose potential is constant and independent of the added reagent. Titration curves similar to those shown in Figures 13-3, 13-4, and 13-5 are obtained. Potentiometric end points are discussed in Section 21C. To obtain an amperometric end point, the current generated between a pair of silver microelectrodes in the solution of the analyte is measured and plotted as a function of reagent volume. Amperometric methods are considered in Section 23B-4.

The end point produced by a chemical indicator usually consists of a color change or, occasionally, the appearance or disappearance of turbidity in the solution being titrated. The requirements for an indicator for a precipitation titration are that (1) the color change should occur over a limited range in p-function of the reagent or the analyte and (2) the color change should take place within the steep portion of the titration curve for the analyte. For example, in Figure 13-5, we see that the titration of iodide with any indicator providing a signal in the pAg range of about 4.0 to 12.0 should give a satisfactory end point. In contrast, the end-point signal for the reaction of chloride ions would be limited to a pAg of about 4.0 to 6.0.

Chromate Ion; The Mohr Method

Sodium chromate can serve as an indicator for the argentometric determination of chloride, bromide, and cyanide ions by reacting with silver ion to form a brick-red silver chromate (Ag_2CrO_4) precipitate in the equivalence-point region. The silver ion concentration at chemical equivalence in the titration of chloride with silver ions is given by

◀ The Mohr method was first described in 1865 by K. F. Mohr, a German pharmaceutical chemist who was a pioneer in the development of titrimetry. Because Cr(VI) has been discovered to be a carcinogen, the Mohr method is now seldom used.

$$[Ag^+] = \sqrt{K_{sp}} = \sqrt{1.82 \times 10^{-10}} = 1.35 \times 10^{-5}\text{ M}$$

◀ Mohr method for chloride.
Titration reaction

$$Ag^+ + Cl^- \rightleftharpoons \underset{\text{white}}{AgCl(s)}$$

Indicator reaction

$$2\,Ag^+ + CrO_4^{2-} \rightleftharpoons \underset{\text{red}}{Ag_2CrO_4(s)}$$

The chromate ion concentration required to initiate formation of silver chromate under this condition can be computed from the solubility constant for silver chromate,

$$[CrO_4^{2-}] = \frac{K_{sp}}{[Ag^+]^2} = \frac{1.2 \times 10^{-12}}{(1.35 \times 10^{-5})^2} = 6.6 \times 10^{-3}\text{ M}$$

In principle, then, chromate should be added in an amount to give this concentration for the red precipitate to appear just after the equivalence point. In fact, however, a chromate ion concentration of 6.6×10^{-3} M imparts such an intense yellow color to the solution that formation of the red silver chromate is not readily detected, and lower concentrations of chromate ion are generally used for this reason. As a consequence, excess silver nitrate is required before precipitation begins. An additional excess of the reagent must also be added to produce enough silver chromate to be seen. These two factors create a positive systematic error in the Mohr method that becomes significant at reagent concentrations lower than about 0.1 M. A correction for this error can readily be made by blank titration of a chloride-free suspension of calcium carbonate. Alternatively, the silver nitrate solution can be standardized against primary-standard–grade sodium chloride using the same conditions as in the analysis. This technique compensates not only for the overconsumption of reagent but also for the acuity of the analyst in detecting the appearance of the color.

The Mohr titration must be carried out at a pH of 7 to 10 because chromate ion is the conjugate base of the weak chromic acid. Consequently, in more acidic solutions, the chromate ion concentration is too low to produce the precipitate near the equivalence point. Normally, a suitable pH is achieved by saturating the analyte solution with sodium hydrogen carbonate.

Adsorption Indicators: The Fajans Method

An **adsorption indicator** is an organic compound that tends to be adsorbed onto the surface of the solid in a precipitation titration. Ideally, the adsorption (or desorption) occurs near the equivalence point and results not only in a color change but also in a transfer of color from the solution to the solid (or the reverse).

◀ Adsorption indicators were first described by K. Fajans, a Polish chemist, in 1926. His name is pronounced Fay′yahns.

Fluorescein is a typical adsorption indicator that is useful for the titration of chloride ion with silver nitrate. In aqueous solution, fluorescein partially dissociates into hydronium ions and negatively charged fluoresceinate ions that are yellow-green. The fluoresceinate ion forms an intensely red silver salt. Whenever this dye is used as an indicator, however, *its concentration is never large enough to precipitate as silver fluoresceinate.*

Fluorescein structural formula and molecular model. This strongly fluorescent dye has many applications. It is widely used to study retinal circulation and various diseases involving the retina. The technique is known as fluorescein angiography. Fluorescein can be bound to DNA and other proteins and its fluorescence used as a probe of these molecules and their interactions. Fluorescein is also used for water tracing to provide information on the contamination of underground wells. In addition, it has been used as a laser dye.

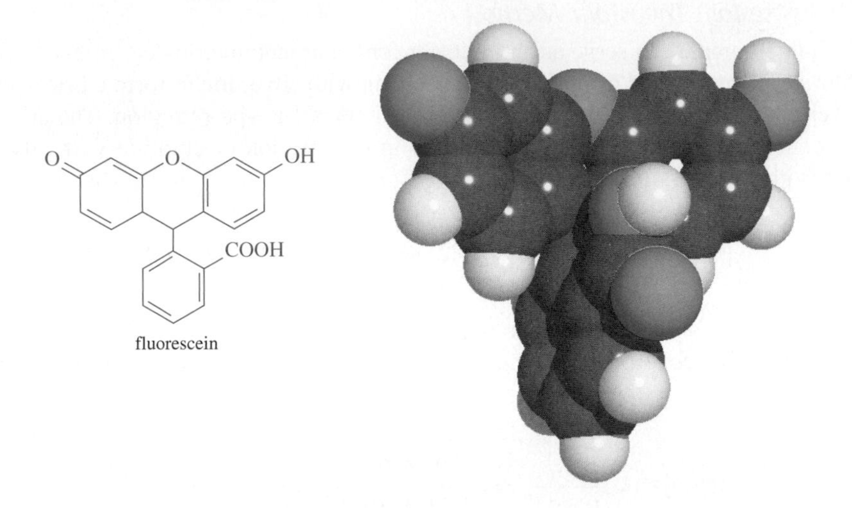

In the early stages of the titration of chloride ion with silver nitrate, the colloidal silver chloride particles are negatively charged because of adsorption of excess chloride ions (see Section 5B-2). The dye anions are repelled from this surface by electrostatic repulsion and impart a yellow-green color to the solution. Beyond the equivalence point, however, the silver chloride particles strongly adsorb silver ions and thereby acquire a positive charge. Fluoresceinate anions are now attracted *into the counter-ion layer* that surrounds each colloidal silver chloride particle. The net result is the appearance of the red color of silver fluoresceinate *in the surface layer of the solution surrounding the solid.* It is important to emphasize that the color change is an *adsorption* (not a precipitation) process, because the solubility product of the silver fluoresceinate is never exceeded. The adsorption is reversible, the dye being desorbed on back-titration with chloride ion.

Titrations involving adsorption indicators are rapid, accurate, and reliable, but their application is limited to the relatively few precipitation reactions in which a colloidal precipitate is formed rapidly.

Iron(III) Ion; The Volhard Method

► The Volhard method was first described by Jacob Volhard, a German chemist, in 1874.

In the Volhard method, silver ions are titrated with a standard solution of thiocyanate ion:

$$Ag^+ + SCN^- \rightleftharpoons AgSCN(s)$$

► Volhard method for chloride.

$$\underset{\text{excess}}{Ag^+} + Cl^- \rightleftharpoons \underset{\text{white}}{AgCl(s)}$$

$$SCN^- + Ag^+ \rightleftharpoons \underset{\text{white}}{AgSCN(s)}$$

$$Fe^{3+} + SCN^- \rightleftharpoons \underset{\text{red}}{Fe(SCN)^{2+}}$$

Iron(III) serves as the indicator. The solution turns red with the first slight excess of thiocyanate ion:

$$Fe^{3+} + SCN^- \rightleftharpoons \underset{\text{red}}{FeSCN^{2+}} \qquad K_f = 1.05 \times 10^3 = \frac{[Fe(SCN)^{2+}]}{[Fe^{3+}][SCN^-]}$$

The titration must be carried out in acidic solution to prevent precipitation of iron(III) as the hydrated oxide.

FEATURE 13-3

Calculating the Concentration of Indicator Solutions

Experiments show that the average observer can just detect the red color of $Fe(SCN)^{2+}$ when its concentration is 6.4×10^{-6} M. In the titration of 50.0 mL of 0.050 M Ag^+ with 0.100 M KSCN, what concentration of Fe^{3+} should be used to lower the titration error to near zero?

For a zero titration error, the $FeSCN^{2+}$ color should appear when the concentration of Ag^+ remaining in the solution is identical to the sum of the two thiocyanate species. That is, at the equivalence point

$$[Ag^+] = [SCN^-] + [Fe(SCN)^{2+}]$$

Substituting the detectable concentration of $FeSCN^{2+}$ gives

$$[Ag^+] = [SCN^-] + 6.4 \times 10^{-6}$$

or

$$[Ag^+] = \frac{K_{sp}}{[SCN^-]} = \frac{1.1 \times 10^{-12}}{[SCN^-]} = [SCN^-] + 6.4 \times 10^{-6}$$

which rearranges to

$$[SCN^-]^2 + 6.4 \times 10^{-6}\,[SCN^-] - 1.1 \times 10^{-12} = 0$$

$$[SCN^-] = 1.7 \times 10^{-7}\text{ M}$$

The formation constant for $FeSCN^{2+}$ is

$$K_f = 1.05 \times 10^3 = \frac{[Fe(SCN)^{2+}]}{[Fe^{3+}][SCN^-]}$$

If we now substitute the $[SCN^-]$ necessary to give a detectable concentration of $FeSCN^{2+}$ at the equivalence point, we obtain

$$1.05 \times 10^3 = \frac{6.4 \times 10^{-6}}{[Fe^{3+}]1.7 \times 10^{-7}}$$

$$[Fe^{3+}] = 0.036\text{ M}$$

The indicator concentration is not critical in the Volhard titration. In fact, calculations similar to those shown in Feature 13-3 demonstrate that a titration error of one part in a thousand or less is possible if the iron(III) concentration is held between 0.002 M and 1.6 M. In practice, an indicator concentration greater than 0.2 M imparts sufficient color to the solution to make detection of the thiocyanate complex difficult because of the yellow color of Fe^{3+}. Therefore, lower concentrations (usually about 0.01 M) of iron(III) ion are employed.

The most important application of the Volhard method is the indirect determination of halide ions. A measured excess of standard silver nitrate solution is added to the sample, and the excess silver is determined by back-titration with a standard thiocyanate solution. The strong acidic environment required for the Volhard procedure represents a distinct advantage over other titrimetric methods of halide analysis because such ions as carbonate, oxalate, and arsenate (which form slightly soluble silver salts in neutral media but not in acidic media) do not interfere.

► The Volhard procedure requires that the analyte solution be distinctly acidic.

Silver chloride is more soluble than silver thiocyanate. As a consequence, in chloride determinations by the Volhard method, the reaction

$$AgCl(s) + SCN^- \rightleftharpoons AgSCN(s) + Cl^-$$

occurs to a significant extent near the end of the back-titration of the excess silver ion. This reaction causes the end point to fade and results in an overconsumption of thiocyanate ion, which in turn leads to low values for the chloride analysis. This error can be circumvented by filtering the silver chloride before undertaking the back-titration. Filtration is not required in the determination of other halides because they all form silver salts that are less soluble than silver thiocyanate.

13F-4 Applications of Standard Silver Nitrate Solutions

Table 13-3 lists some typical applications of precipitation titrations in which silver nitrate is the standard solution. In most of these methods, the analyte is precipitated with a measured excess of silver nitrate, and the excess is determined by a Volhard titration with standard potassium thiocyanate.

Both silver nitrate and potassium thiocyanate are obtainable in primary-standard quality. The latter is, however, somewhat hygroscopic, and thiocyanate solutions are ordinarily standardized against silver nitrate. Both silver nitrate and potassium thiocyanate solutions are stable indefinitely.

TABLE 13-3

Typical Argentometric Precipitation Methods

Substance Being Determined	End Point	Remarks
AsO_4^{3-}, Br^-, I^-, CNO^-, SCN^-	Volhard	Removal of silver salt not required
CO_3^{2-}, CrO_4^{2-}, CN^-, Cl^-, $C_2O_4^{2-}$, PO_4^{3-}, S^{2-}, NCN^{2-}	Volhard	Removal of silver salt required before back-titration of excess Ag^+
BH_4^-	Modified Volhard	Titration of excess Ag^+ following $BH_4^- + 8Ag^+ + 8OH^- \rightarrow 8Ag(s) + H_2BO_3^- + 5H_2O$
Epoxide	Volhard	Titration of excess Cl^- following hydrohalogenation
K^+	Modified Volhard	Precipitation of K^+ with known excess of $B(C_6H_5)_4^-$, addition of excess Ag^+ giving $AgB(C_6H_5)_4(s)$, and back-titration of the excess
Br^-, Cl^-	$2Ag^+ + CrO_4^{2-} \rightarrow Ag_2CrO_4(s)$ **red**	In neutral solution
Br^-, Cl^-, I^-, SeO_3^{2-}	Adsorption indicator	
$V(OH)_4^+$, fatty acids, mercaptans	Electroanalytical	Direct titration with Ag^+
Zn^{2+}	Modified Volhard	Precipitation as $ZnHg(SCN)_4$, filtration, dissolution in acid addition of excess Ag^+, back-titration of excess Ag^+
F^-	Modified Volhard	Precipitation as PbClF, filtration, dissolution in acid, addition of excess Ag^+, back-titration of excess Ag^+

Spreadsheet Summary In the first exercise in Chapter 7 of *Applications of Microsoft® Excel in Analytical Chemistry,* we plot a titration curve for the titration of NaCl with $AgNO_3$ (Example 13-10). A stoichiometric approach is used to calculate the Ag^+ concentration and pAg versus the volume of titrant added. A master equation approach is then explored for the same titration. Finally, the problem is inverted, and the volume needed to achieve a given pAg value is calculated.

WEB WORKS

Use your Web browser to connect to **http://chemistry.brookscole.com/skoogfac/.** From the Chapter Resources menu, choose Web Works. Locate the Chapter 13 section, and click on the link to the Virtual Titrator. Click on the indicated frame to invoke the Virtual Titrator Java applet and display two windows: the Menu Panel and the Virtual Titrator main window. To begin, click on Acids on the main window menu bar, and select the diprotic acid o-phthalic acid. Examine the titration curve that results. Then click on Graphs/Alpha Plot vs. pH and observe the result. Click on Graphs/Alpha Plot vs. mL base. Repeat the process for several monoprotic and polyprotic acids, and note the results.

QUESTIONS AND PROBLEMS

13-1. Write two equations that—along with the stoichiometric factor—form the basis for the calculation of volumetric titrimetry.

13-2. Define
- *(a) millimole.
- (b) titration.
- *(c) stoichiometric ratio.
- (d) titration error.

13-3. Distinguish between
- *(a) the equivalence point and the end point of a titration.
- (b) a primary standard and a secondary standard.

***13-4.** In what respect is the Fajans method superior to the Volhard method for the titration of chloride ion?

13-5. Calculations of volumetric analysis ordinarily consist of transforming the quantity of titrant used (in chemical units) to a chemically equivalent quantity of analyte (also in chemical units) through use of a stoichiometric factor. Use chemical formulas (NO CALCULATIONS REQUIRED) to express this ratio for calculation of the percentage of

*(a) hydrazine in rocket fuel by titration with standard iodine. Reaction:

$$H_2NNH_2 + 2I_2 \rightarrow N_2(g) + 4I^- + 4H^+$$

(b) hydrogen peroxide in a cosmetic preparation by titration with standard permanganate. Reaction:

$$5H_2O_2 + 2MnO_4^- + 6H^+ \rightarrow 2Mn^{2+} + 5O_2(g) + 8H_2O$$

*(c) boron in a sample of borax, $Na_2B_4O_7 \cdot 10\,H_2O$, by titration with standard acid. Reaction:

$$B_4O_7^{2-} + 2H^+ + 5H_2O \rightarrow 4H_3BO_3$$

(d) sulfur in an agricultural spray that was converted to thiocyanate with an unmeasured excess of cyanide. Reaction:

$$S(s) + CN^- \rightarrow SCN^-$$

After removal of the excess cyanide, the thiocyanate was titrated with a standard potassium iodate solution in strong HCl. Reaction:

$$2SCN^- + 3IO_3^- + 2H^+ + 6Cl^- \rightarrow 2SO_4^{2-} + 2CN^- + 3ICl_2^- + H_2O$$

***13-6.** Why does a Volhard determination of iodide ion require fewer steps than a Volhard determination of
(a) carbonate ion?
(b) cyanide ion?

13-7. Why does the charge on the surface of precipitate particles change sign at the equivalence point in a titration?

***13-8.** Describe the preparation of
(a) 500 mL of 0.0750 M $AgNO_3$ from the solid reagent.
(b) 2.00 L of 0.325 M HCl, starting with a 6.00 M solution of the reagent.
(c) 750 mL of a solution that is 0.0900 M in K^+, starting with solid $K_4Fe(CN)_6$.
(d) 600 mL of 2.00% (w/v) aqueous $BaCl_2$ from a 0.500 M $BaCl_2$ solution.
(e) 2.00 L of 0.120 M $HClO_4$ from the commercial reagent [60% $HClO_4$ (w/w), sp gr 1.60].
(f) 9.00 L of a solution that is 60.0 ppm in Na^+, starting with solid Na_2SO_4.

13-9. Describe the preparation of
(a) 1.00 L of 0.150 M $KMnO_4$ from the solid reagent.
(b) 2.50 L of 0.500 M $HClO_4$, starting with a 9.00 M solution of the reagent.
(c) 400 mL of a solution that is 0.0500 M in I^-, starting with MgI_2.
(d) 200 mL of 1.00% (w/v) aqueous $CuSO_4$, from a 0.218 M $CuSO_4$ solution.
(e) 1.50 L of 0.215 M NaOH from the concentrated commercial reagent [50% NaOH (w/w), sp gr 1.525].
(f) 1.50 L of a solution that is 12.0 ppm in K^+, starting with solid $K_4Fe(CN)_6$.

***13-10.** A solution of $HClO_4$ was standardized by dissolving 0.4125 g of primary-standard–grade HgO in a solution of KBr:

$$HgO(s) + 4Br^- + H_2O \rightarrow HgBr_4^{2-} + 2OH^-$$

The liberated OH^- consumed 46.51 mL of the acid. Calculate the molarity of the $HClO_4$.

13-11. A 0.4512-g sample of primary-standard–grade Na_2CO_3 required 36.44 mL of an H_2SO_4 solution to reach the end point in the reaction

$$CO_3^{2-} + 2H^+ \rightarrow H_2O + CO_2(g)$$

What is the molarity of the H_2SO_4?

***13-12.** A 0.4000-g sample that assayed 96.4% Na_2SO_4 required 41.25 mL of a barium chloride solution. Reaction:

$$Ba^{2+} + SO_4^{2-} \rightarrow BaSO_4(s)$$

Calculate the analytical molarity of $BaCl_2$ in the solution.

***13-13.** A 0.3125-g sample of primary standard Na_2CO_3 was treated with 40.00 mL of dilute perchloric acid. The solution was boiled to remove CO_2, following which the excess $HClO_4$ was back-titrated with 10.12 mL of dilute NaOH. In a separate experiment, it was established that 27.43 mL of the $HClO_4$ neutralized the NaOH in a 25.00-mL portion. Calculate the molarities of the $HClO_4$ and NaOH.

13-14. Titration of 50.00 mL of 0.05251 M $Na_2C_2O_4$ required 36.75 mL of a potassium permanganate solution.

$$2MnO_4^- + 5H_2C_2O_4 + 6H^+ \rightarrow 2Mn^{2+} + 10CO_2(g) + 8H_2O$$

Calculate the molarity of the $KMnO_4$ solution.

***13-15.** Titration of the I_2 produced from 0.1045 g of primary standard KIO_3 required 30.72 mL of sodium thiosulfate.

$$IO_3^- + 5I^- + 6H^+ \rightarrow 3I_2 + 3H_2O$$

$$I_2 + 2S_2O_3^{2-} \rightarrow 2I^- + S_4O_6^{2-}$$

Calculate the concentration of the $Na_2S_2O_3$.

***13-16.** The monochloroacetic acid ($ClCH_2COOH$) preservative in 100.0 mL of a carbonated beverage was extracted into diethyl ether and then returned to aqueous solution as $ClCH_3COO^-$ by extraction with 1 M NaOH. This aqueous extract was acidified and treated with 50.00 mL of 0.04521 M $AgNO_3$. The reaction is

$$ClCH_2COOH + Ag^+ + H_2O \rightarrow HOCH_2COOH + H^+ + AgCl(s)$$

After filtration of the AgCl, titration of the filtrate and washings required 10.43 mL of an NH_4SCN solution. Titration of a blank taken through the entire process used 22.98 mL of the NH_4SCN. Calculate the weight (in milligrams) of $ClCH_2COOH$ in the sample.

13-17. An analysis for borohydride ion is based on its reaction with Ag^+:

$$BH_4^- + 8Ag^+ + 8OH^- \rightarrow H_2BO_3^- + 8Ag(s) + 5H_2O$$

The purity of a quantity of KBH_4 to be used in an organic synthesis was established by diluting 3.213 g of the material to exactly 500.0 mL, treating a 100.0-mL aliquot with 50.00 mL of 0.2221 M $AgNO_3$, and titrating the excess silver ion with 3.36 mL of 0.0397 M KSCN. Calculate the percent purity of the KBH_4 (53.941 g/mol).

***13-18.** The arsenic in a 1.010-g sample of a pesticide was converted to H_3AsO_4 by suitable treatment. The acid was then neutralized, and exactly 40.00 mL of 0.06222 M $AgNO_3$ was added to precipitate the arsenic quantitatively as Ag_3AsO_4. The excess Ag^+ in the filtrate and in the washings from the precipitate was titrated with 10.76 mL of 0.1000 M KSCN; the reaction was

$$Ag^+ + SCN^- \rightarrow AgSCN(s)$$

Calculate the percent As_2O_3 in the sample.

***13-19.** The Association of Official Analytical Chemists recommends a Volhard titration for analysis of the insecticide heptachlor, $C_{10}H_5Cl_7$. The percentage of heptachlor is given by

$$\% \text{ heptachlor} = \frac{(mL_{Ag} \times c_{Ag} - mL_{SCN} \times c_{SCN}) \times 37.33}{\text{mass sample}}$$

What does this calculation reveal concerning the stoichiometry of this titration?

13-20. A carbonate fusion was needed to free the Bi from a 0.6423-g sample containing the mineral eulytite ($2Bi_2O_3 \cdot 3SiO_2$). The fused mass was dissolved in dilute acid, following which the Bi^{3+} was titrated with 27.36 mL of 0.03369 M NaH_2PO_4. The reaction is

$$Bi^{3+} + H_2PO_4^- \rightarrow BiPO_4(s) + 2H^+$$

Calculate the percent purity of eulytite (1112 g/mol) in the sample.

***13-21.** A solution of $Ba(OH)_2$ was standardized against 0.1175 g of primary-standard-grade benzoic acid, C_6H_5COOH (122.12 g/mol). An end point was observed after addition of 40.42 mL of base.

(a) Calculate the molarity of the base.
(b) Calculate the standard deviation of the molarity if the standard deviation for weighing was ±0.2 mg and that for the volume measurement was ±0.03 mL.
(c) Assuming an error of −0.3 mg in the weighing, calculate the absolute and relative systematic error in the molarity.

13-22. A 0.1475 M solution of $Ba(OH)_2$ was used to titrate the acetic acid (60.05 g/mol) in a dilute aqueous solution. The following results were obtained.

Sample	Sample Volume, mL	$Ba(OH)_2$ Volume, mL
1	50.00	43.17
2	49.50	42.68
3	25.00	21.47
4	50.00	43.33

(a) Calculate the mean w/v percentage of acetic acid in the sample.
(b) Calculate the standard deviation for the results.
(c) Calculate the 90% confidence interval for the mean.
(d) At the 90% confidence level, could any of the results be discarded?
(e) Assume that the buret used to measure out the acetic acid had a systematic error of −0.05 mL at all volumes delivered. Calculate the systematic error in the mean result.

***13-23.** A 20-tablet sample of soluble saccharin was treated with 20.00 mL of 0.08181 M $AgNO_3$. The reaction is

$$\text{C}_6\text{H}_4(\text{CO})(\text{SO}_2)\text{NNa} + Ag^+ \longrightarrow \text{C}_6\text{H}_4(\text{CO})(\text{SO}_2)\text{NAg}(s) + Na^+$$

After removal of the solid, titration of the filtrate and washings required 2.81 mL of 0.04124 M KSCN. Calculate the average number of milligrams of saccharin (205.17 g/mol) in each tablet.

13-24. (a) A 0.1752-g sample of primary standard $AgNO_3$ was dissolved in 502.3 g of distilled water. Calculate the weight molarity of Ag^+ in this solution.

(b) The standard solution described in part (a) was used to titrate a 25.171-g sample of a KSCN solution. An end point was obtained after adding 23.765 g of the $AgNO_3$ solution. Calculate the weight molarity of the KSCN solution.

(c) The solutions described in parts (a) and (b) were used to determine the $BaCl_2 \cdot 2H_2O$ in a 0.7120-g sample. A 20.102-g sample of the $AgNO_3$ was added to a solution of the sample, and the excess $AgNO_3$ was back-titrated with 7.543 g of the KSCN solution. Calculate the percent $BaCl_2 \cdot 2H_2O$ in the sample.

13-25. A solution was prepared by dissolving 10.12 g of $KCl \cdot MgCl_2 \cdot 6H_2O$ (277.85 g/mol) in sufficient water to give 2.000 L. Calculate

(a) the molar analytical concentration of $KCl \cdot MgCl_2$ in this solution.

(b) the molar concentration of Mg^{2+}.

(c) the molar concentration of Cl^-.

(d) the weight/volume percentage of $KCl \cdot MgCl_2 \cdot 6H_2O$.

(e) the number of millimoles of Cl^- in 25.0 mL of this solution.

(f) ppm K^+.

***13-26.** The formaldehyde in a 5.00-g sample of a seed disinfectant was steam distilled, and the aqueous distillate was collected in a 500-mL volumetric flask. After dilution to volume, a 25.0-mL aliquot was treated with 30.0 mL of 0.121 M KCN solution to convert the formaldehyde to potassium cyanohydrin.

$$K^+ + CH_2O + CN^- \rightarrow KOCH_2CN$$

The excess KCN was then removed by addition of 40.0 mL of 0.100 M $AgNO_3$.

$$2CN^- + 2Ag^+ \rightarrow Ag_2(CN)_2(s)$$

The excess Ag^+ in the filtrate and washings required a 16.1-mL titration with 0.134 M NH_4SCN. Calculate the percent CH_2O in the sample.

***13-27.** The action of an alkaline I_2 solution on the rodenticide warfarin, $C_{19}H_{16}O_4$ (308.34 g/mol), results in the formation of 1 mol of iodoform, CHI_3 (393.73 g/mol), for each mole of the parent compound reacted. Analysis for warfarin can then be based on the reaction between CHI_3 and Ag^+.

$$CHI_3 + 3Ag^+ + H_2O \rightarrow 3AgI(s) + 3H^+ + CO(g)$$

The CHI_3 produced from a 13.96-g sample was treated with 25.00 mL of 0.02979 M $AgNO_3$, and the excess Ag^+ was then titrated with 2.85 mL of 0.05411 M KSCN. Calculate the percentage of warfarin in the sample.

13-28. A 5.00-mL aqueous suspension of elemental selenium was treated with 25.00 mL of ammoniacal 0.0360 M $AgNO_3$. Reaction:

$$6Ag(NH_3)_2^+ + 3Se(s) + 3H_2O \rightarrow 2Ag_2Se(s) + Ag_2SeO_3(s) + 6NH_4^+$$

After this reaction was complete, nitric acid was added to dissolve the Ag_2SO_3 but not the Ag_2Se. The Ag^+ from the dissolved Ag_2SeO_3 and the excess reagent required 16.74 mL of 0.01370 M KSCN in a Volhard titration. How many milligrams of Se were contained in each milliliter of sample?

***13-29.** A 1.998-g sample containing Cl^- and ClO_4^- was dissolved in sufficient water to give 250.0 mL of solution. A 50.00-mL aliquot required 13.97 mL of 0.08551 M $AgNO_3$ to titrate the Cl^-. A second 50.00-mL aliquot was treated with $V_2(SO_4)_3$ to reduce the ClO_4^- to Cl^-:

$$ClO_4^- + 4V_2(SO_4)_3 + 4H_2O \rightarrow Cl^- + 12SO_4^{2-} + 8VO^{2+} + 8H^+$$

Titration of the reduced sample required 40.12 mL of the $AgNO_3$ solution. Calculate the percentages of Cl^- and ClO_4^- in the sample.

13-30. For each of the following precipitation titrations, calculate the cation and anion concentrations at equivalence as well as at reagent volumes corresponding to ±20.00 mL, ±10.00 mL, and ±1.00 mL of equivalence. Construct a titration curve from the data, plotting the p-function of the cation versus reagent volume.

(a) 25.00 mL of 0.05000 M $AgNO_3$ with 0.02500 M NH_4SCN.

(b) 20.00 mL of 0.06000 M $AgNO_3$ with 0.03000 M KI.

(c) 30.00 mL of 0.07500 M $AgNO_3$ with 0.07500 M NaCl.

(d) 35.00 mL of 0.4000 M Na_2SO_4 with 0.2000 M $Pb(NO_3)_2$.

(e) 40.00 mL of 0.02500 M $BaCl_2$ with 0.05000 M Na_2SO_4.

(f) 50.00 mL of 0.2000 M NaI with 0.4000 M $TINO_3$ (K_{sp} for TlI = 6.5×10^{-8})

13-31. Calculate the silver ion concentration after the addition of 5.00*, 15.00, 25.00, 30.00, 35.00, 39.00, 40.00*, 41.00, 45.00*, and 50.00 mL of 0.05000 M $AgNO_3$ to 50.0 mL of 0.0400 M KBr. Construct a titration curve from these data plotting pAg as a function of titrant volume.

13-32. Challenge Problem. The Volhard titration for Ag^+ is being evaluated for use in determining silver in a thiosulfate photographic fixing bath. An independent analysis of the bath solution by atomic absorption spectrometry gives the known silver concentration for comparison. In the Volhard titration, a typical observer can just detect 1×10^{-5} M $Fe(SCN)^{2+}$. The formation constant for $Fe(SCN)^{2+}$ is 1.05×10^3. If a volume of 50.00 mL of the bath solution is withdrawn and titrated with 0.025 M SCN^-, what titration error would result if the known silver concentration was:

*(a) 0.250%.

(b) 0.100%.

*(c) 0.050%.

InfoTrac College Edition

For additional readings, go to InfoTrac College Edition, your online research library, at

http://infotrac.thomsonlearning.com

CHAPTER 14

Principles of Neutralization Titrations

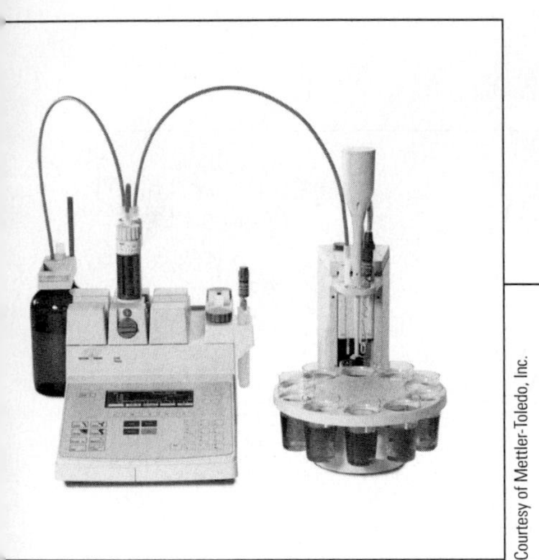

Courtesy of Mettler-Toledo, Inc.

Neutralization titrations are widely employed to determine the amounts of acids and bases. In addition, they can be used to monitor the progress of reactions that produce or consume hydrogen ions. In clinical chemistry, for example, pancreatitis can be diagnosed by measuring the activity of serum lipase. Lipases hydrolyze the long-chain fatty acid triglyceride. The reaction liberates two moles of fatty acid and one mole of β-monoglyceride for each mole of triglyceride present according to the following:

$$\text{triglyceride} \xrightarrow{\text{lipase}} \text{monoglyceride} + 2 \text{ fatty acid}$$

The reaction is allowed to proceed for a certain amount of time, and then the liberated fatty acid is titrated with NaOH using a phenolphthalein indicator or a pH meter. The amount of fatty acid produced in a fixed time is related to the lipase activity (see Chapter 29). The entire procedure can be automated with an automatic titrator such as that shown here.

Acid/base equilibria are ubiquitous in both chemistry and science in general. For example, you will find that the material in this chapter and in Chapter 15 is directly relevant to the acid/base reactions that are so important in biochemistry and the other biological sciences.

Standard solutions of strong acids and strong bases are used extensively for determining analytes that are themselves acids or bases or that can be converted to such species by chemical treatment. This chapter introduces neutralization titrations, deals with the principles of titration, and discusses the common indicators that are used. In addition, titration curves that are plots of pH versus volume of titrant are explored, and several examples of pH calculations are presented. Titration curves for strong and weak acids and bases are described.

14A SOLUTIONS AND INDICATORS FOR ACID/BASE TITRATIONS

Like all titrations, neutralization titrations depend on a chemical reaction between the analyte and a standard reagent. The point of chemical equivalence is indicated by a chemical indicator or an instrumental method. The discussion here focuses on the types of standard solutions and the chemical indicators that are used for neutralization titrations.

14A-1 Standard Solutions

The standard solutions used in neutralization titrations are strong acids or strong bases because these substances react more completely with an analyte than do their weaker counterparts, and they therefore provide sharper end points. Standard solutions of acids are prepared by diluting concentrated hydrochloric, perchloric, or sulfuric acid. Nitric acid is seldom used because its oxidizing properties offer the potential for undesirable side reactions. *Hot concentrated perchloric and sulfuric acids are potent oxidizing agents and are very hazardous.* Fortunately, cold dilute solutions of these reagents are relatively benign and can be used in the analytical laboratory without any special precautions other than eye protection.

◀ The standard reagents used in acid/base titrations are always strong acids or strong bases, most commonly HCl, $HClO_4$, H_2SO_4, NaOH, and KOH. Weak acids and bases are never used as standard reagents because they react incompletely with analytes.

Standard solutions of bases are ordinarily prepared from solid sodium, potassium, and occasionally barium hydroxides. Again, eye protection should always be used when handling dilute solutions of these reagents.

14A-2 Acid/Base Indicators

Many substances, both naturally occurring and synthetic, display colors that depend on the pH of the solutions in which they are dissolved. Some of these substances, which have been used for centuries to indicate the acidity or alkalinity of water, are still employed as acid/base indicators.

◀ For a list of common acid/base indicators and their colors, look inside the front cover of this book. See also color plate 8 for photographs showing the colors and transition ranges of 12 common indicators.

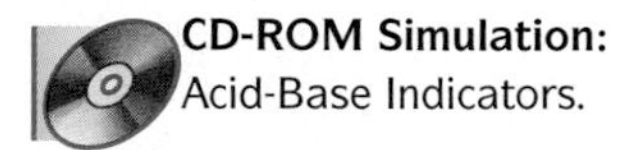
CD-ROM Simulation: Acid-Base Indicators.

An acid/base indicator is a weak organic acid or a weak organic base whose undissociated form differs in color from its conjugate base or its conjugate acid form. For example, the behavior of an acid-type indicator, HIn, is described by the equilibrium

$$\underset{\text{acid color}}{HIn} + H_2O \rightleftharpoons \underset{\text{base color}}{In^-} + H_3O^+$$

Here, internal structural changes accompany dissociation and cause the color change (Figure 14-1). The equilibrium for a base-type indicator, In, is

$$\underset{\text{base color}}{In} + H_2O \rightleftharpoons \underset{\text{acid color}}{InH^+} + OH^-$$

In the paragraphs that follow, we focus on the behavior of acid-type indicators. The principles, however, can be easily extended to base-type indicators as well.

The equilibrium-constant expression for the dissociation of an acid-type indicator takes the form

$$K_a = \frac{[H_3O^+][In^-]}{[HIn]} \tag{14-1}$$

Rearranging leads to

$$[H_3O^+] = K_a\frac{[HIn]}{[In^-]} \tag{14-2}$$

We then see that the hydronium ion concentration determines the ratio of the acid to the conjugate base form of the indicator, which in turn controls the color of the solution.

The human eye is not very sensitive to color differences in a solution containing a mixture of HIn and In^-, particularly when the ratio $[HIn]/[In^-]$ is greater than about 10 or smaller than about 0.1. Consequently, the color change detected by an average observer occurs within a limited range of concentration ratios from about

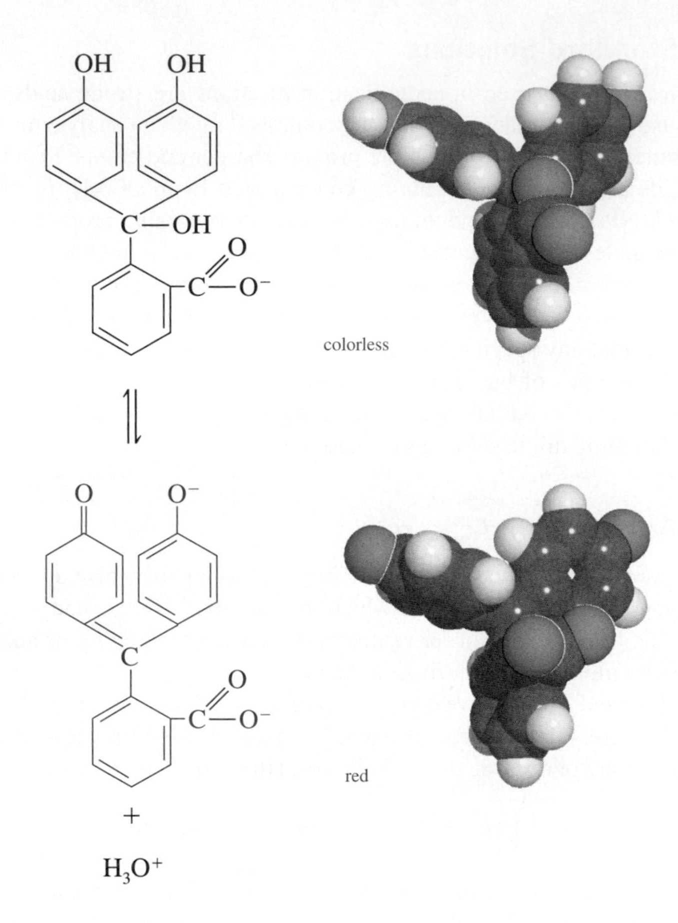

Figure 14-1 Color change and molecular model for phenolphthalein.

10 to about 0.1. At greater or smaller ratios, the color appears essentially constant to the eye and is independent of the ratio. As a result, we can write that the average indicator, HIn, exhibits its pure acid color when

$$\frac{[\mathrm{HIn}]}{[\mathrm{In}^-]} \geq \frac{10}{1}$$

and its base color when

$$\frac{[\mathrm{HIn}]}{[\mathrm{In}^-]} \leq \frac{1}{10}$$

The color appears to be intermediate for ratios between these two values. These ratios vary considerably from indicator to indicator, of course. Furthermore, people differ significantly in their ability to distinguish between colors; indeed a color-blind person may be unable to distinguish any color change at all.

If the two concentration ratios are substituted into Equation 14-2, the range of hydronium ion concentrations needed to change the indicator color can be evaluated. So, for the full acid color,

$$[\mathrm{H_3O^+}] = 10K_\mathrm{a}$$

and in the same way, for the full base color,

$$[H_3O^+] = 0.1K_a$$

To obtain the indicator pH range, we take the negative logarithms of the two expressions:

$$\text{pH(acid color)} = -\log(10K_a) = pK_a + 1$$

$$\text{pH(basic color)} = -\log(0.1K_a) = pK_a - 1$$

◀ The approximate pH transition range of most acid-type indicators is roughly $pK_a \pm 1$.

$$\text{indicator pH range} = pK_a \pm 1 \qquad (14\text{-}3)$$

This expression shows that an indicator with an acid dissociation constant of 1×10^{-5} ($pK_a = 5$) typically shows a complete color change when the pH of the solution in which it is dissolved changes from 4 to 6 (Figure 14-2). With a little more algebra, we can derive a similar relationship for a basic-type indicator.

Titration Errors with Acid/Base Indicators

We find two types of titration errors in acid/base titrations. The first is a determinate error that occurs when the pH at which the indicator changes color differs from the pH at the equivalence point. This type of error can usually be minimized by choosing the indicator carefully or by making a blank correction.

The second type is an indeterminate error that originates from the limited ability of the eye to distinguish reproducibly the intermediate color of the indicator. The magnitude of this error depends on the change in pH per milliliter of reagent at the equivalence point, on the concentration of the indicator, and on the sensitivity of the eye to the two indicator colors. On the average, the visual uncertainty with an acid/base indicator is in the range of ±0.5 to ±1 pH unit. This uncertainty can often be decreased to as little as ±0.1 pH unit by matching the color of the solution being titrated to that of a reference standard containing a similar amount of indicator at the appropriate pH. These uncertainties are of course approximations that vary considerably from indicator to indicator as well as from person to person.

Variables That Influence the Behavior of Indicators

The pH interval over which a given indicator exhibits a color change is influenced by temperature, by the ionic strength of the medium, and by the presence of organic solvents and colloidal particles. Some of these effects, particularly the last two, can cause the transition range to shift by one or more pH units.[1]

The Common Acid/Base Indicators

The list of acid/base indicators is large and includes a number of organic compounds. Indicators are available for almost any desired pH range. A few common indicators and their properties are listed in Table 14-1. Note that the transition ranges vary from 1.1 to 2.2, with the average being about 1.6 units. These indicators and several more are shown along with their transition ranges in the color chart inside the front cover of this book.

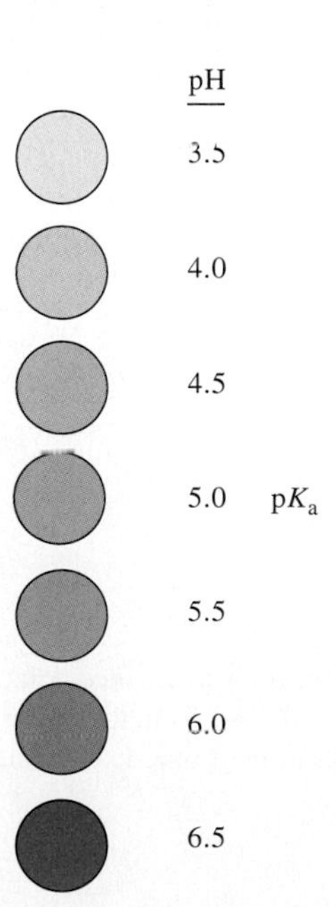

Figure 14-2 Indicator color as a function of pH ($pK_a = 5.0$).

[1]For a discussion of these effects, see H.A. Laitinen and W. E. Harris, *Chemical Analysis,* 2nd ed., pp. 48–51. New York: McGraw-Hill, 1975.

TABLE 14-1
Some Important Acid/Base Indicators

Common Name	Transition Range, pH	pK_a*	Color Change†	Indicator Type‡
Thymol blue	1.2–2.8	1.65§	R–Y	1
	8.0–9.6	8.96§	Y–B	
Methyl yellow	2.9–4.0		R–Y	2
Methyl orange	3.1–4.4	3.46§	R–O	2
Bromocresol green	3.8–5.4	4.66§	Y–B	1
Methyl red	4.2–6.3	5.00§	R–Y	2
Bromocresol purple	5.2–6.8	6.12§	Y–P	1
Bromothymol blue	6.2–7.6	7.10§	Y–B	1
Phenol red	6.8–8.4	7.81§	Y–R	1
Cresol purple	7.6–9.2		Y–P	1
Phenolphthalein	8.3–10.0		C–R	1
Thymolphthalein	9.3–10.5		C–B	1
Alizarin yellow GG	10–12		C–Y	2

*At ionic strength of 0.1.
†B = blue; C = colorless; O = orange; P = purple; R = red; Y = yellow.
‡(1) Acid type: $HIn + H_2O \rightleftharpoons H_3O^+ + In^-$; (2) Base type: $In + H_2O \rightleftharpoons InH^+ + OH^-$.
§For the reaction $InH^+ + H_2O \rightleftharpoons H_3O^+ + In$.

14B TITRATION OF STRONG ACIDS AND STRONG BASES

The hydronium ions in an aqueous solution of a strong acid have two sources: (1) the reaction of the acid with water and (2) the dissociation of water itself. In all but the most dilute solutions, however, the contribution from the strong acid far exceeds that from the solvent. So, for a solution of HCl with a concentration greater than about 10^{-6} M, we can write

► In solutions of a strong acid that are more concentrated than about 1×10^{-6} M, we can assume that the equilibrium concentration of H_3O^+ is equal to the analytical concentration of the acid. The same is true for $[OH^-]$ in solutions of strong bases.

$$[H_3O^+] = c_{HCl} + [OH^-] \approx c_{HCl}$$

where $[OH^-]$ represents the contribution of hydronium ions from the dissociation of water. An analogous relationship applies for a solution of a strong base, such as sodium hydroxide. That is,

$$[OH^-] = c_{NaOH} + [H_3O^+] \approx c_{NaOH}$$

14B-1 Titrating a Strong Acid with a Strong Base

We are interested here, and in the next several chapters, in calculating *hypothetical* titration curves of pH versus volume of titrant. We distinguish between the curves constructed by computing the values of pH and the *experimental* titration curves observed in the laboratory. Three types of calculations must be done to construct the hypothetical curve for titrating a solution of a strong acid with a strong base. Each of these corresponds to a distinct stage in the titration: (1) preequivalence, (2) equivalence, and (3) postequivalence. In the preequivalence stage, we compute the concentration of the acid from its starting concentration and the amount of base

► Before the equivalence point, we calculate the pH from the molar concentration of unreacted acid.

added. At the equivalence point, the hydronium and hydroxide ions are present in equal concentrations, and the hydronium ion concentration is derived directly from the ion-product constant for water. In the postequivalence stage, the analytical concentration of the excess base is computed, and the hydroxide ion concentration is assumed to be equal to or a multiple of the analytical concentration. Our approach is analogous to the method that we used in the silver chloride titration in Example 13-10.

◀ Beyond the equivalence point, we first calculate pOH and then pH. Remember that $pH = pK_w - pOH = 14.00 - pOH$.

A convenient way of converting hydroxide concentration to pH is to take the negative logarithm of both sides of the ion-product constant expression for water. Thus,

$$K_w = [H_3O^+][OH^-]$$

$$-\log K_w = -\log[H_3O^+][OH^-] = -\log[H_3O^+] - \log[OH^-]$$

$$pK_w = pH + pOH$$

$$-\log 10^{-14} = pH + pOH = 14.00$$

EXAMPLE 14-1

Generate the hypothetical titration curve for the titration of 50.00 mL of 0.0500 M HCl with 0.1000 M NaOH.

Initial Point

Before any base is added, the solution is 0.0500 M in H_3O^+, and

$$pH = -\log[H_3O^+] = -\log 0.0500 = 1.30$$

After Addition of 10.00 mL of Reagent

The hydronium ion concentration is decreased as a result of both reaction with the base and dilution. So the analytical concentration of HCl is

$$c_{HCl} = \frac{\text{no. mmol HCl remaining after addition of NaOH}}{\text{total volume soln}}$$

$$= \frac{\text{original no. mmol HCl} - \text{no. mmol NaOH added}}{\text{total volume soln}}$$

$$= \frac{(50.00\text{ mL} \times 0.0500\text{ M}) - (10.00\text{ mL} \times 0.1000\text{ M})}{50.00\text{ mL} + 10.00\text{ mL}}$$

$$= \frac{(2.500\text{ mmol} - 1.000\text{ mmol})}{60.00\text{ mL}} = 2.500 \times 10^{-2}\text{ M}$$

$$[H_3O^+] = 2.500 \times 10^{-2}\text{ M}$$

$$\text{and pH} = -\log[H_3O^+] = -\log(2.500 \times 10^{-2}) = 1.60$$

(continued)

TABLE 14-2

Changes in pH during the Titration of a Strong Acid with a Strong Base

	pH	
Volume of NaOH, mL	50.00 mL of 0.0500 M HCl with 0.100 M NaOH	50.00 mL of 0.000500 M HCl with 0.00100 M NaOH
0.00	1.30	3.30
10.00	1.60	3.60
20.00	2.15	4.15
24.00	2.87	4.87
24.90	3.87	5.87
25.00	7.00	7.00
25.10	10.12	8.12
26.00	11.12	9.12
30.00	11.80	9.80

We calculate additional points defining the curve in the region before the equivalence point in the same way. The results of these calculations are shown in the second column of Table 14-2.

After Addition of 25.00 mL of Reagent: The Equivalence Point

At the equivalence point, neither HCl nor NaOH is in excess, and so the concentrations of hydronium and hydroxide ions must be equal. Substituting this equality into the ion-product constant for water yields

► At the equivalence point, the solution is neutral, and pH = 7.00.

$$[H_3O^+] = \sqrt{K_w} = \sqrt{1.00 \times 10^{-14}} = 1.00 \times 10^{-7}\,M$$

$$pH = -\log(1.00 \times 10^{-7}) = 7.00$$

After Addition of 25.10 mL of Reagent

The solution now contains an excess of NaOH, and we can write

$$c_{NaOH} = \frac{\text{no. mmol NaOH added} - \text{original no. mmol HCl}}{\text{total volume soln}}$$

$$= \frac{25.10 \times 0.100 - 50.00 \times 0.0500}{75.10} = 1.33 \times 10^{-4}\,M$$

and the equilibrium concentration of hydroxide ion is

$$[OH^-] = c_{NaOH} = 1.33 \times 10^{-4}\,M$$

$$pOH = -\log(1.33 \times 10^{-4}) = 3.88$$

and

$$pH = 14.00 - 3.88 = 10.12$$

We compute additional data defining the curve beyond the equivalence point in the same way. The results of these computations are shown in Table 14-2.

FEATURE 14-1

Using the Charge-Balance Equation to Construct Titration Curves

In Example 14-1, we generated an acid/base titration curve from the reaction stoichiometry. We can show that all points on the curve can also be calculated from the charge-balance equation.

For the system treated in Example 14-1, the charge-balance equation is given by

$$[H_3O^+] + [Na^+] = [OH^-] + [Cl^-]$$

where the sodium and chloride ion concentrations are given by

$$[Na^+] = \frac{V_{NaOH}c_{NaOH}}{V_{NaOH} + V_{HCl}}$$

$$[Cl^-] = \frac{V_{HCl}c_{HCl}}{V_{NaOH} + V_{HCl}}$$

We can rewrite the first equation in the form

$$[H_3O^+] = [OH^-] + [Cl^-] - [Na^+]$$

For volumes of NaOH short of the equivalence point, $[OH^-] \ll [Cl^-]$, so

$$[H_3O^+] \approx [Cl^-] - [Na^+]$$

and

$$[H_3O^+] = \frac{V_{HCl}c_{HCl}}{V_{HCl} + V_{NaOH}} - \frac{V_{NaOH}c_{NaOH}}{V_{HCl} + V_{NaOH}} = \frac{V_{HCl}c_{HCl} - V_{NaOH}c_{NaOH}}{V_{HCl} + V_{NaOH}}$$

At the equivalence point, $[Na^+] = [Cl^-]$ and

$$[H_3O^+] = [OH^-]$$

$$[H_3O^+] = \sqrt{K_w}$$

Beyond the equivalence point, $[H_3O^+] \ll [Na^+]$, and the original equation rearranges to

$$[OH^-] \approx [Na^+] - [Cl^-]$$

$$= \frac{V_{NaOH}c_{NaOH}}{V_{NaOH} + V_{HCl}} - \frac{V_{HCl}c_{HCl}}{V_{NaOH} + V_{HCl}} = \frac{V_{NaOH}c_{NaOH} - V_{HCl}c_{HCl}}{V_{NaOH} + V_{HCl}}$$

The Effect of Concentration

The effects of reagent and analyte concentrations on the neutralization titration curves for strong acids are shown by the two sets of data in Table 14-2. and the

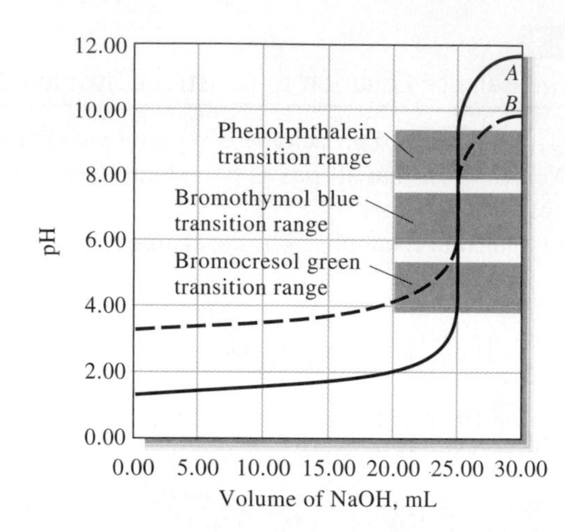

Figure 14-3 Titration curves for HCl with NaOH. Curve *A:* 50.00 mL of 0.0500 M HCl with 0.1000 M NaOH. Curve *B:* 50.00 mL of 0.000500 M HCl with 0.001000 M NaOH.

plots in Figure 14-3. Note that with 0.1 M NaOH as the titrant, the change in pH in the equivalence-point region is large. With 0.001 M NaOH, the change is markedly less but still pronounced.

Choosing an Indicator

Figure 14-3 shows that the selection of an indicator is not critical when the reagent concentration is approximately 0.1 M. Here, the volume differences in titrations with the three indicators shown are of the same magnitude as the uncertainties associated with reading the buret; therefore, they are negligible. Note, however, that bromocresol green is unsuited for a titration involving the 0.001 M reagent because the color change occurs over a 5-mL range well before the equivalence point. The use of phenolphthalein is subject to similar objections. Of the three indicators, then, only bromothymol blue provides a satisfactory end point with a minimal systematic error in the titration of the more dilute solution.

14B-2 Titrating a Strong Base with a Strong Acid

Titration curves for strong bases are derived in an analogous way to those for strong acids. Short of the equivalence point, the solution is highly basic, the hydroxide ion concentration being numerically related to the analytical molarity of the base. The solution is neutral at the equivalence point and becomes acidic in the region beyond the equivalence point; then the hydronium ion concentration is equal to the analytical concentration of the excess strong acid.

EXAMPLE 14-2

Calculate the pH during the titration of 50.00 mL of 0.0500 M NaOH with 0.1000 M HCl after the addition of the following volumes of reagent: (a) 24.50 mL, (b) 25.00 mL, (c) 25.50 mL.

(a) At 24.50 mL added, $[H_3O^+]$ is very small and cannot be computed from stoichiometric considerations but can be obtained from $[OH^-]$

$$[OH^-] = c_{NaOH} = \frac{\text{original no. mmol NaOH} - \text{no. mmol HCl added}}{\text{total volume of solution}}$$

$$= \frac{50.00 \times 0.0500 - 24.50 \times 0.100}{50.00 + 24.50} = 6.71 \times 10^{-4}\ M$$

$$[H_3O^+] = K_w/(6.71 \times 10^{-4}) = 1.00 \times 10^{-14}/(6.71 \times 10^{-4})$$

$$= 1.49 \times 10^{-11}\ M$$

$$pH = -\log(1.49 \times 10^{-11}) = 10.83$$

(b) This is the equivalence point where $[H_3O^+] = [OH^-]$

$$[H_3O^+] = \sqrt{K_w} = \sqrt{1.00 \times 10^{-14}} = 1.00 \times 10^{-7}\ M$$

$$pH = -\log(1.00 \times 10^{-7}) = 7.00$$

(c) At 25.50 mL added,

$$[H_3O^+] = c_{HCl} = \frac{(25.50 \times 0.100 - 50.00 \times 0.0500}{75.50}$$

$$= 6.62 \times 10^{-4}\ M$$

$$pH = -\log(6.62 \times 10^{-4}) = 3.18$$

Curves for the titration of 0.0500 M and 0.00500 M NaOH with 0.1000 M and 0.0100 M HCl are shown in Figure 14-4. Indicator selection is based on the same considerations described for the titration of a strong acid with a strong base.

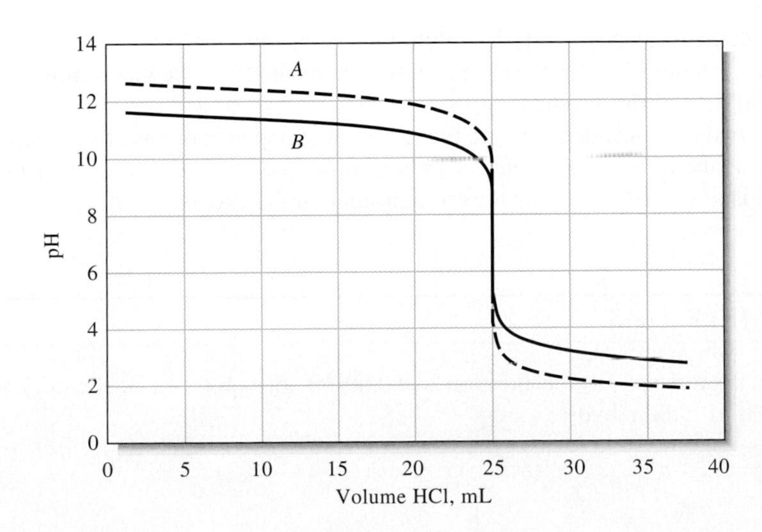

Figure 14-4 Titration curves for NaOH with HCl. Curve *A:* 50.00 mL of 0.0500 M NaOH with 0.1000 M HCl. Curve *B:* 50.00 mL of 0.00500 M NaOH with 0.0100 M HCl.

FEATURE 14-2

How Many Significant Figures Should We Retain in Titration Curve Calculations?

Concentrations calculated in the equivalence-point region of titration curves are generally of low precision because they are based on small differences between large numbers. For example, in the calculation of c_{NaOH} after introduction of 25.10 mL of NaOH in Example 14-1, the numerator (2.510–2.500 = 0.010) is known to only two significant figures. To minimize rounding error, however, three digits were retained in c_{NaOH} (1.33×10^{-4}), and rounding was postponed until pOH and pH were computed.

In rounding the calculated values for p-functions, you should remember (see Section 6D-2) that it is *the mantissa of a logarithm* (that is, the number to the right of the decimal point) *that should be rounded to include only significant figures* because the characteristic (the number to the left of the decimal point) serves merely to locate the decimal point. Fortunately, the large changes in p-functions characteristic of most equivalence points are not obscured by the limited precision of the calculated data. Generally, in deriving data for titration curves, we round p-functions to two places to the right of the decimal point whether or not such rounding is called for.

14C TITRATION CURVES FOR WEAK ACIDS

Four distinctly different types of calculations are needed to derive a titration curve for a weak acid (or a weak base):

1. At the beginning, the solution contains only a weak acid or a weak base, and the pH is calculated from the concentration of that solute and its dissociation constant.
2. After various increments of titrant have been added (in quantities up to, but not including, an equivalent amount), the solution consists of a series of buffers. The pH of each buffer can be calculated from the analytical concentrations of the conjugate base or acid and the residual concentrations of the weak acid or base.
3. At the equivalence point, the solution contains only the conjugate of the weak acid or base being titrated (that is, a salt), and the pH is calculated from the concentration of this product.
4. Beyond the equivalence point, the excess of strong acid or base titrant represses the acidic or basic character of the reaction product to such an extent that the pH is governed largely by the concentration of the excess titrant.

► Titration curves for strong and weak acids are identical just slightly beyond the equivalence point. The same is true for strong and weak bases.

EXAMPLE 14-3

Generate a curve for the titration of 50.00 mL of 0.1000 M acetic acid with 0.1000 M sodium hydroxide.

Initial pH

First, we must calculate the pH of a 0.1000 M solution of HOAc using Equation 9-22.

$$[H_3O^+] = \sqrt{K_a c_{HOAc}} = \sqrt{1.75 \times 10^{-5} \times 0.100} = 1.32 \times 10^{-3}\ M$$

$$pH = -\log(1.32 \times 10^{-3}) = 2.88$$

pH after Addition of 5.00 mL of Reagent

A buffer solution consisting of NaOAc and HOAc has now been produced. The analytical concentrations of the two constituents are

$$c_{HOAc} = \frac{50.00\ mL \times 0.100\ M - 5.00\ mL \times 0.100\ M}{60.00\ mL} = \frac{4.500}{60.00}\ M$$

$$c_{NaOAc} = \frac{5.00\ mL \times 0.100\ M}{60.00\ mL} = \frac{0.500}{60.00}\ M$$

Now for the 5.00-mL volume, we substitute the concentrations of HOAc and OAc^- into the dissociation-constant expression for acetic acid and obtain

$$K_a = \frac{[H_3O^+](0.500/\cancel{60.00})}{4.500/\cancel{60.00}} = 1.75 \times 10^{-5}$$

$$[H_3O^+] = 1.58 \times 10^{-4}\ M$$

$$pH = 3.80$$

Note that the total volume of solution is present in both numerator and denominator and thus cancels in the expression for $[H_3O^+]$. Calculations similar to this provide points on the curve throughout the buffer region. Data from these calculations are presented in column 2 of Table 14-3.

TABLE 14-3

Changes in pH during the Titration of a Weak Acid with a Strong Base

	pH	
Volume of NaOH, mL	50.00 mL of 0.1000 M HOAc with 0.1000 M NaOH	50.00 mL of 0.001000 M HOAc with 0.001000 M NaOH
0.00	2.88	3.91
10.00	4.16	4.30
25.00	4.76	4.80
40.00	5.36	5.38
49.00	6.45	6.46
49.90	7.46	7.47
50.00	8.73	7.73
50.10	10.00	8.09
51.00	11.00	9.00
60.00	11.96	9.96
70.00	12.22	10.25

(continued)

pH after Addition of 25.00 mL of Reagent

As in the previous calculation, the analytical concentrations of the two constituents are

$$c_{\text{HOAc}} = \frac{50.00 \text{ mL} \times 0.100 \text{ M} - 25.00 \text{ mL} \times 0.100 \text{ M}}{60.00 \text{ mL}} = \frac{2.500}{60.00} \text{ M}$$

$$c_{\text{NaOAc}} = \frac{25.00 \text{ mL} \times 0.100 \text{ M}}{60.00 \text{ mL}} = \frac{2.500}{60.00} \text{ M}$$

Now for the 25.00-mL volume, we substitute the concentrations of HOAc and OAc^- into the dissociation-constant expression for acetic acid and obtain

$$K_a = \frac{[H_3O^+](\cancel{2.500/60.00})}{\cancel{2.500/60.00}} = [H_3O^+] = 1.75 \times 10^{-5}$$

$$\text{pH} = \text{p}K_a = 4.76$$

At this point in the titration, both the analytical concentrations of the acid and conjugate base as well as the total volume of solution cancel in the expression for $[H_3O^+]$.

Equivalence Point pH

At the equivalence point, all the acetic acid has been converted to sodium acetate. The solution is therefore similar to one formed by dissolving that salt in water, and the pH calculation is identical to that shown in Example 9-10 (page 250) for a weak base. In the present example, the NaOAc concentration is 0.0500 M. Thus,

$$OAc^- + H_2O \rightleftharpoons HOAc + OH^-$$

$$[OH^-] = [HOAc]$$

$$[OAc^-] = 0.0500 - [OH^-] \approx 0.0500$$

Substituting in the base dissociation-constant expression for OAc^- gives

$$\frac{[OH^-]^2}{0.0500} = \frac{K_w}{K_a} = \frac{1.00 \times 10^{-14}}{1.75 \times 10^{-5}} = 5.71 \times 10^{-10}$$

$$[OH^-] = \sqrt{0.0500 \times 5.71 \times 10^{-10}} = 5.34 \times 10^{-6} \text{ M}$$

$$\text{pH} = 14.00 - (-\log 5.34 \times 10^{-6}) = 8.73$$

► Note that the pH at the equivalence point of this titration is greater than 7. The solution is alkaline.

pH after Addition of 50.01 mL of Base

After the addition of 50.01 mL of NaOH, both the excess base and the acetate ion are sources of hydroxide ion. The contribution from the acetate ion is small, however, because the excess of strong base represses the reaction of acetate with water. This fact becomes evident when we consider that the hydroxide ion concentration is only 5.35×10^{-6} at the equivalence point; once a tiny excess of strong base is added, the contribution from the reaction of the acetate is even smaller. We then have

$$[OH^-] \approx c_{NaOH} = \frac{50.01 \text{ mL} \times 0.1000 \text{ M} - 50.00 \text{ mL} \times 0.1000 \text{ M}}{100.01 \text{ mL}}$$

$$= 1.00 \times 10^{-5} \text{ M}$$

$$\text{pH} = 14.00 - [-\log(1.00 \times 10^{-5})] = 9.00$$

Note that the titration curve for a weak acid with a strong base is identical to that for a strong acid with a strong base in the region slightly beyond the equivalence point.

Table 14-3 and Figure 14-5 compare the pH values calculated in this example with a more dilute titration. The effect of concentration is discussed in Section 14C-1.

Note from Example 14-3 that the analytical concentrations of acid and conjugate base are identical when an acid has been half neutralized (after the addition of exactly 25.00 mL of base in this case). Thus, these terms cancel in the equilibrium-constant expression, and the hydronium ion concentration is numerically equal to the dissociation constant. Likewise, in the titration of a weak base, the hydroxide ion concentration is numerically equal to the dissociation constant of the base at the midpoint in the titration curve. In addition, the buffer capacities of each of the solutions are at a maximum at this point. These points are often called the **half-titration points.**

◀ At the half-titration point in the titration of a weak acid, $[H_3O^+] = K_a$ or $pH = pK_a$.

◀ At the half-titration point in the titration of a weak base, $[OH^-] = K_b$ or $pOH = pK_b$.

FEATURE 14-3

Determining Dissociation Constants for Weak Acids and Bases

The dissociation constants of weak acids or weak bases are often determined by monitoring the pH of the solution while the acid or base is being titrated. A pH meter with a glass pH electrode (see Section 21D-3) is used for the measurements. For an acid, the measured pH when the acid is exactly half neutralized is numerically equal to pK_a. For a weak base, the pH at half titration must be converted to pOH, which is then equal to pK_b.

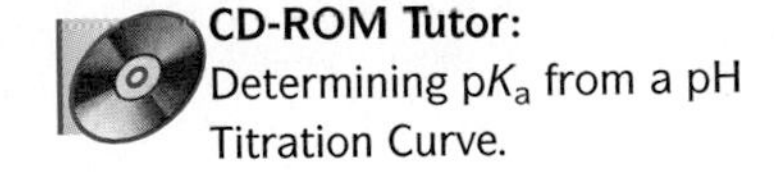
CD-ROM Tutor: Determining pK_a from a pH Titration Curve.

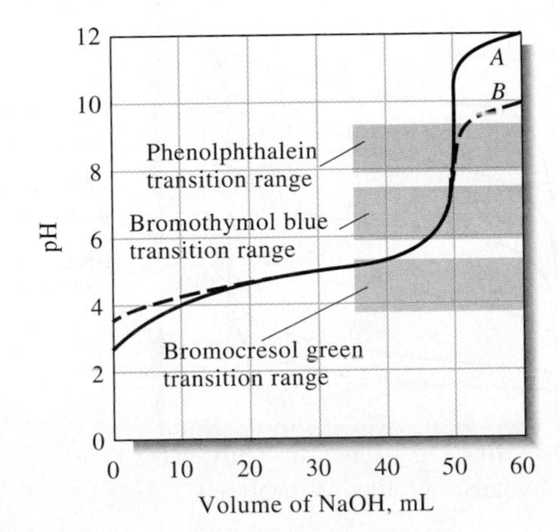

Figure 14-5 Curve for the titration of acetic acid with sodium hydroxide. Curve *A:* 0.1000 M acid with 0.1000 M base. Curve *B:* 0.001000 M acid with 0.001000 M base.

14C-1 The Effect of Concentration

► CHALLENGE: Show that the pH values given in the third column of Table 14-3 are correct.

The second and third columns of Table 14-3 contain pH data for the titration of 0.1000 M and 0.001000 M acetic acid with sodium hydroxide solutions of the same two concentrations. In calculating the values for the more dilute acid, none of the approximations shown in Example 14-3 were valid, and solution of a quadratic equation was necessary until after the equivalence point. In the postequivalence point region, the excess OH^- predominates, and the simple calculation suffices.

Figure 14-5 is a plot of the data in Table 14-3. Note that the initial pH values are higher and the equivalence-point pH is lower for the more dilute solution (curve *B*). At intermediate titrant volumes, however, the pH values differ only slightly because of the buffering action of the acetic acid/sodium acetate system that is present in this region. Figure 14-5 is graphical confirmation that the pH of buffers is largely independent of dilution. Note that the change in OH^- in the vicinity of the equivalence point becomes smaller with lower analyte and reagent concentrations. This effect is analogous to the effect with the titration of a strong acid with a strong base (see Figure 14-3).

14C-2 The Effect of Reaction Completeness

Titration curves for 0.1000 M solutions of acids with different dissociation constants are shown in Figure 14-6. Note that the pH change in the equivalence-point region becomes smaller as the acid becomes weaker—that is, as the reaction between the acid and the base becomes less complete.

14C-3 Choosing an Indicator: The Feasibility of Titration

Figures 14-5 and 14-6 show that the choice of indicator is more limited for the titration of a weak acid than for the titration of a strong acid. For example, Figure 14-5 illustrates that bromocresol green is totally unsuited for titration of 0.1000 M acetic acid. Bromothymol blue does not work either because its full color change occurs over a range of titrant volume from about 47 mL to 50 mL of 0.1000 M base. An indicator exhibiting a color change in the basic region, such as phenolphthalein, however, should provide a sharp end point with a minimal titration error.

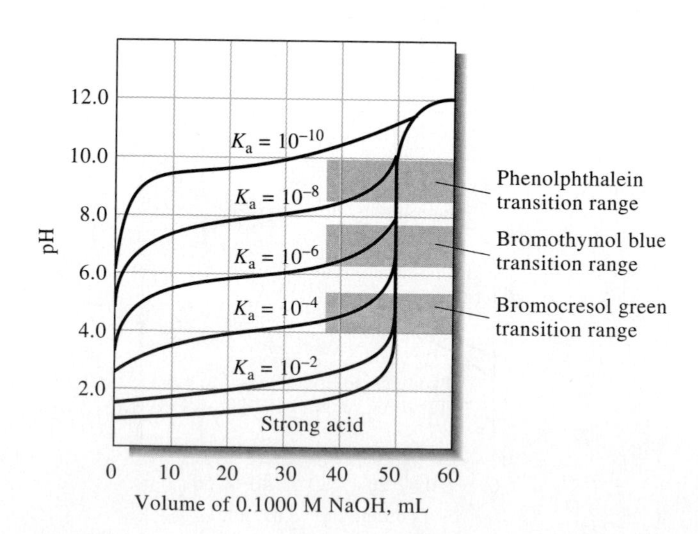

Figure 14-6 The effect of acid strength (dissociation constant) on titration curves. Each curve represents the titration of 50.00 mL of 0.1000 M acid with 0.1000 M base.

The end-point pH change associated with the titration of 0.001000 M acetic acid (curve *B*, Figure 14-5) is so small that a significant titration error is likely to be introduced regardless of indicator. Use of an indicator with a transition range between that of phenolphthalein and that of bromothymol blue in conjunction with a suitable color comparison standard, however, makes it possible to establish the end point in this titration with a reproducibility of a few percent relative.

Figure 14-6 illustrates that similar problems occur as the strength of the acid being titrated decreases. Precision on the order of ±2 ppt can be achieved in the titration of a 0.1000 M acid solution with a dissociation constant of 10^{-8} provided a suitable color comparison standard is available. With more concentrated solutions, somewhat weaker acids can be titrated with reasonable precision.

Spreadsheet Summary In the Acid/Base Titrations section of Chapter 7 of *Applications of Microsoft® Excel in Analytical Chemistry,* a master equation approach is used to carry out the calculations and plot a titration curve for the titration of a weak acid with a strong base. Excel's Goal Seek is used to solve the charge-balance expression for the H_3O^+ concentration and the pH.

14D TITRATION CURVES FOR WEAK BASES

The calculations needed to draw the titration curve for a weak base are analogous to those for a weak acid.

EXAMPLE 14-4

A 50.00-mL aliquot of 0.0500 M NaCN is titrated with 0.1000 M HCl. The reaction is

$$CN^- + H_3O^+ \rightleftharpoons HCN + H_2O$$

Calculate the pH after the addition of (a) 0.00, (b) 10.00, (c) 25.00, and (d) 26.00 mL of acid.

(a) **0.00 mL of Reagent**
The pH of a solution of NaCN can be derived by the method shown in Example 9-10, page 250:

$$CN^- + H_2O \rightleftharpoons HCN + OH^-$$

$$K_b = \frac{[OH^-][HCN]}{[CN^-]} = \frac{K_w}{K_a} = \frac{1.00 \times 10^{-14}}{6.2 \times 10^{-10}} = 1.61 \times 10^{-5}$$

$$[OH^-] = [HCN]$$

$$[CN^-] = c_{NaCN} - [OH^-] \approx c_{NaCN} = 0.050 \text{ M}$$

(continued)

Substitution into the dissociation-constant expression gives, after rearrangement,

$$[OH^-] = \sqrt{K_b c_{NaCN}} = \sqrt{1.61 \times 10^{-5} \times 0.0500} = 8.97 \times 10^{-4}$$

$$pH = 14.00 - (-\log 8.97 \times 10^{-4}) = 10.95$$

(b) **10.00 mL of Reagent**

Addition of acid produces a buffer with a composition given by

$$c_{NaCN} = \frac{50.00 \times 0.0500 - 10.00 \times 0.1000}{60.00} = \frac{1.500}{60.00}\,M$$

$$c_{HCN} = \frac{10.00 \times 0.1000}{60.00} = \frac{1.000}{60.00}\,M$$

► CHALLENGE: Show that the pH of the buffer can be calculated with K_a for HCN, as was done here, or equally well with K_b. We used K_a because it gives $[H_3O^+]$ directly; K_b gives $[OH^-]$.

These values are then substituted into the expression for the acid dissociation constant of HCN to give $[H_3O^+]$ directly (see Margin Note):

$$[H_3O^+] = \frac{6.2 \times 10^{-10} \times (1.000/\cancel{60.00})}{1.500/\cancel{60.00}} = 4.13 \times 10^{-10}$$

$$pH = -\log(4.13 \times 10^{-10}) = 9.38$$

(c) **25.00 mL of Reagent**

This volume corresponds to the equivalence point, where the principal solute species is the weak acid HCN. Thus,

$$c_{HCN} = \frac{25.00 \times 0.1000}{75.00} = 0.03333\,M$$

Applying Equation 9-22 gives

► Since the principal solute species at the equivalence point is HCN, the pH is acidic.

$$[H_3O^+] = \sqrt{K_a c_{HCN}} = \sqrt{6.2 \times 10^{-10} \times 0.03333} = 4.45 \times 10^{-6}\,M$$

$$pH = -\log(4.45 \times 10^{-6}) = 5.34$$

(d) **26.00 mL of Reagent**

The excess of strong acid now present represses the dissociation of the HCN to the point where its contribution to the pH is negligible. Thus,

$$[H_3O^+] = c_{HCl} = \frac{26.00 \times 0.1000 - 50.00 \times 0.0500}{76.00} = 1.32 \times 10^{-3}\,M$$

$$pH = -\log(1.32 \times 10^{-3}) = 2.88$$

► When you titrate a weak base, use an indicator with an acidic transition range. When titrating a weak acid, use an indicator with a basic transition range.

Figure 14-7 shows hypothetical titration curves for a series of weak bases of different strengths. The curves show that indicators with *acidic* transition ranges must be used for weak bases.

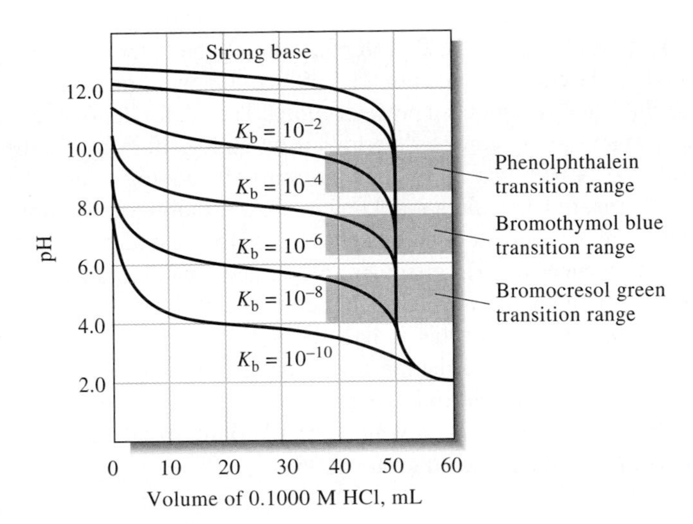

Figure 14-7 The effect of base strength (K_b) on titration curves. Each curve represents the titration of 50.00 mL of 0.1000 M base with 0.1000 M HCl.

FEATURE 14-4

Determining the p*K* Values for Amino Acids

Amino acids contain both an acidic and a basic group. For example, the structure of alanine is represented by Figure 14F-1.

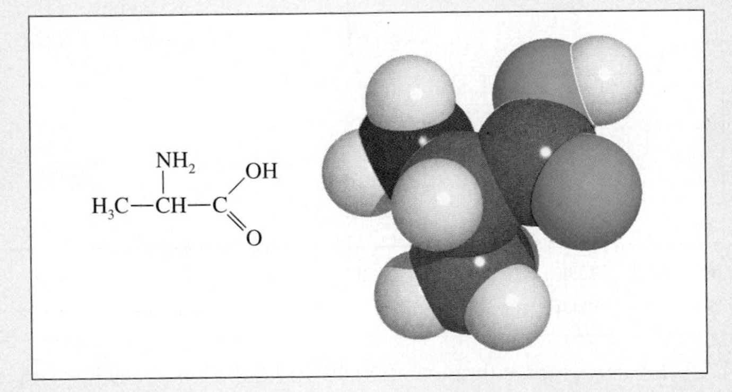

Figure 14F-1 Structure and molecular model of alanine. Alanine is an amino acid. It can exist in two mirror image forms, the left-handed (L) form and the right-handed (D) form. All naturally occurring amino acids are left handed.

The amine group behaves as a base, while the carboxyl group acts as an acid. In aqueous solution, the amino acid is an internally ionized molecule, or "zwitterion," in which the amine group acquires a proton and becomes positively charged while the carboxyl group, having lost a proton, becomes negatively charged.

(continued)

The pK values for amino acids can conveniently be determined by the general procedure described in Feature 14-3. Since the zwitterion has both acidic and basic character, two pKs can be determined. The pK for deprotonation of the protonated amine group can be determined by adding base, while the pK for protonating the carboxyl group can be determined by adding acid. In practice, a solution is prepared containing a known concentration of the amino acid. Hence, we know the amount of base or acid to add to reach halfway to the equivalence point. A curve of pH versus volume of acid or base added is shown in Figure 14F-2. Here the titration starts in the middle of the plot (0.00 mL added) and is only taken to a point that is half the volume required for equivalence. Note that in this example, for alanine, a volume of 20.00 mL of HCl is needed to completely protonate the carboxyl group. By adding acid to the zwitterion, the curve on the left is obtained. At a volume of 10.00 mL of HCl added, the pH is equal to the pK_a for the carboxyl group, 2.35.

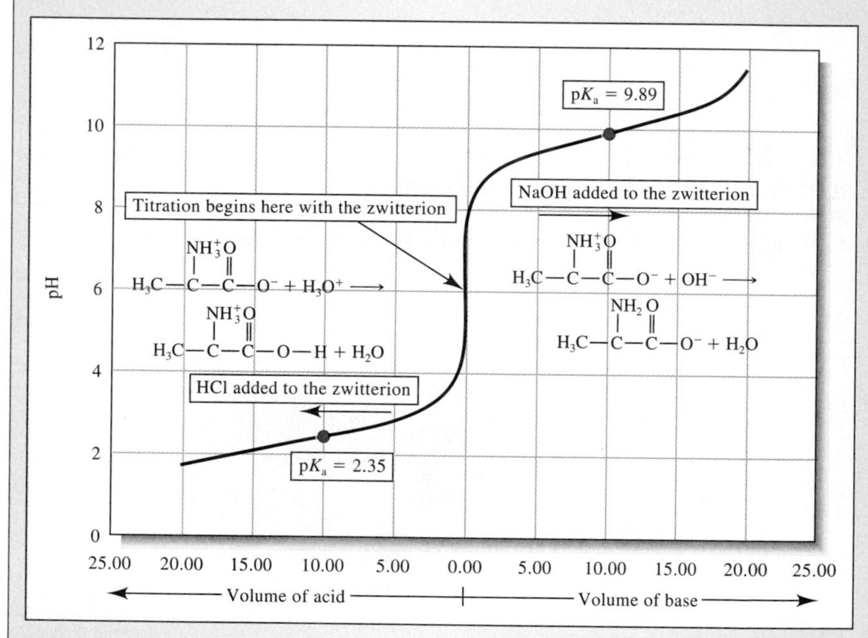

Figure 14F-2 Curves for the titration of 20.00 mL of 0.1000 M alanine with 0.1000 M NaOH and 0.1000 M HCl. Note that the zwitterion is present before any acid or base has been added. Adding acid protonates the carboxylate group with a pK_a of 2.35. Adding base reacts with the protonated amine group with a pK_a of 9.89.

By adding NaOH to the zwitterion, the pK for deprotonating the NH_3^+ group can be determined. Now 20.00 mL of base are required for complete deprotonation. At a volume of 10.00 mL of NaOH added, the pH is equal to the pK_a for the amine group, or 9.89. The pK_a values for other amino acids and more complicated biomolecules such as peptides and proteins can often be obtained in a similar manner. Some amino acids have more than one carboxyl or amine group. Aspartic acid is an example (Fig. 14F-3).

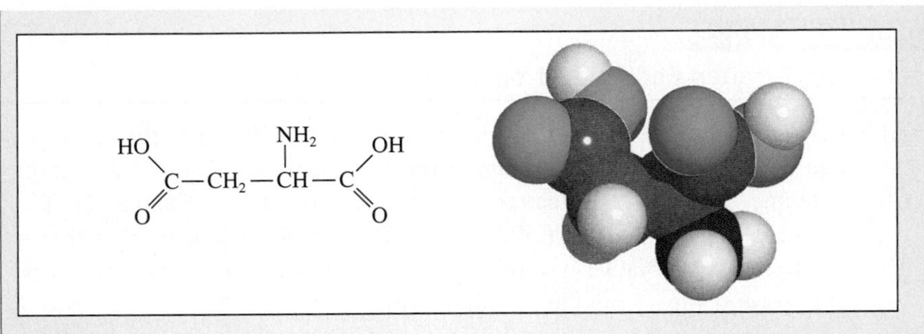

Figure 14F-3 Aspartic acid is an amino acid with two carboxyl groups. It can be combined with phenylalanine to make the artificial sweetener aspartame, which is sweeter and less fattening than ordinary sugar (sucrose).

It is important to note that in general, amino acids cannot be quantitatively determined by direct titration because end points for completely protonating or deprotonating the zwitterion are often indistinct. Amino acids are normally determined by high-performance liquid chromatography (see Chapter 32) or spectroscopic methods (see Part V).

14E THE COMPOSITION OF SOLUTIONS DURING ACID/BASE TITRATIONS

We are often interested in the changes in composition that occur while a solution of a weak acid or a weak base is being titrated. These changes can be visualized by plotting the *relative* equilibrium concentration α_0 of the weak acid as well as the relative equilibrium concentration of the conjugate base α_1 as functions of the pH of the solution.

The solid straight lines labeled α_0 and α_1 in Figure 14-8 were calculated with Equations 9-35 and 9-36 using values for $[H_3O^+]$ shown in column 2 of Table 14-3. The actual titration curve is shown as the curved line in Figure 14-8. Note that at the beginning of the titration, α_0 is nearly 1 (0.987), meaning that 98.7% of the acetate-containing species is present as HOAc and only 1.3% is present as OAc^-. At the equivalence point, α_0 decreases to 1.1×10^{-4} and α_1 approaches 1. Thus, only about 0.011% of the acetate-containing species is HOAc. Notice that at the half-titration point (25.00 mL), α_0 and α_1 are both 0.5.

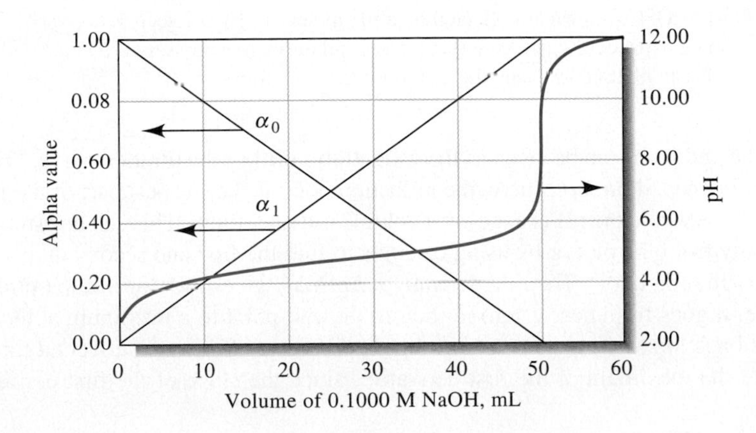

Figure 14-8 Plots of relative amounts of acetic acid and acetate ion during a titration. The straight lines show the change in relative amounts of HOAc (α_0) and OAc^-(α_1) during the titration of 50.00 mL of 0.1000 M acetic acid. The curved line is the titration curve for the system.

FEATURE 14-5

Locating Titration End Points from pH Measurements

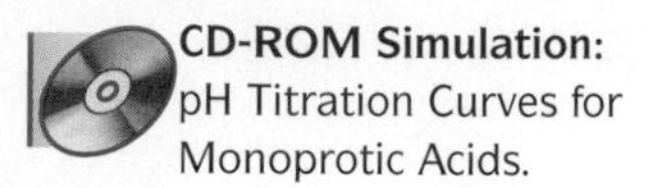

CD-ROM Simulation: pH Titration Curves for Monoprotic Acids.

Although indicators are still widely used in acid/base titrations, the glass pH electrode and pH meter allow the direct measurement of pH as a function of titrant volume. The glass pH electrode is discussed in detail in Chapter 21. The titration curve for the titration of 50.00 mL of 0.1000 M weak acid ($K_a = 1.0 \times 10^{-5}$) with 0.1000 M NaOH is shown in Figure 14F-4a. The end point can be located in several ways from the pH versus volume data.

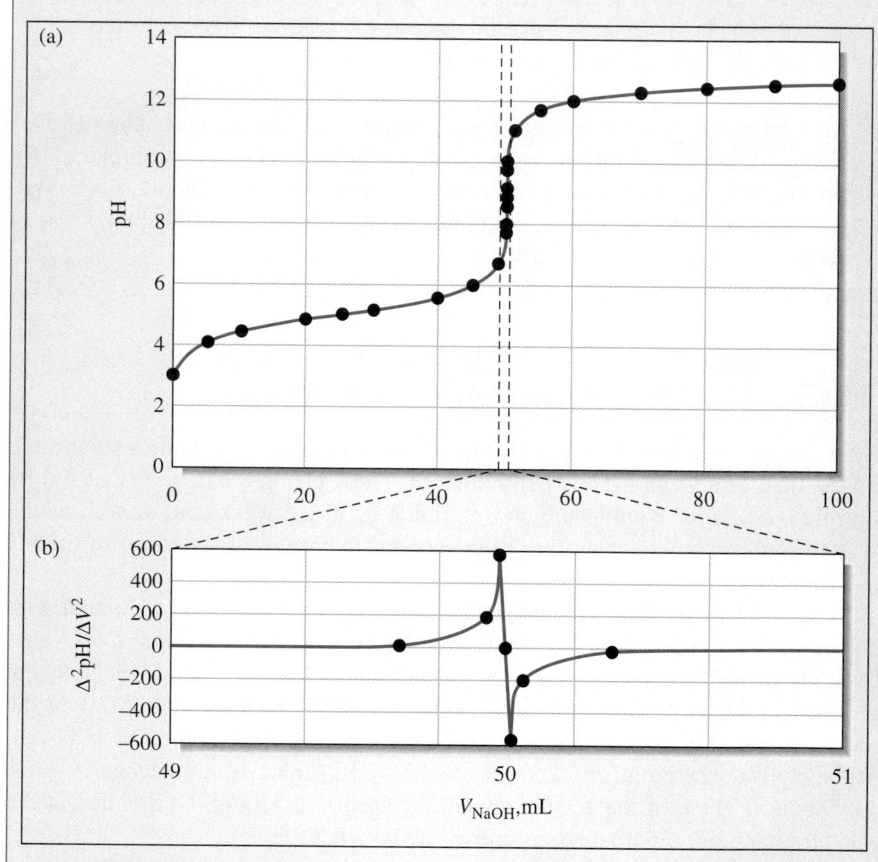

Figure 14F-4 In (a) the titration curve of 50.00 mL of 0.1000 M weak acid with 0.1000 M NaOH is shown as collected by a pH meter. In (b), the second derivative is shown on an expanded scale. Note that the second derivative crosses zero at the end point. This can be used to locate the end point very precisely.

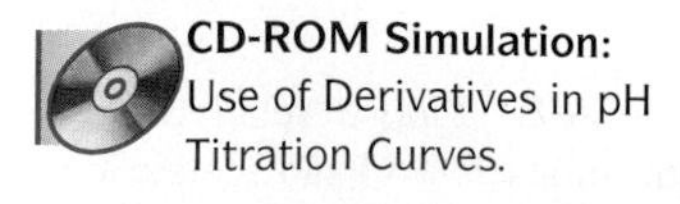

CD-ROM Simulation: Use of Derivatives in pH Titration Curves.

The end point can be taken as the **inflection point** of the titration curve. With a sigmoid-shaped titration curve, the inflection point is the steepest part of the titration curve where the pH change with volume is a maximum. This can be estimated visually from the plot or by using calculus to find the first and second derivatives of the titration curve. The first derivative, ΔpH/ΔV, gives the slope of the titration curve. It goes from nearly zero far before the end point to a maximum at the end point back to zero far beyond the end point. We can differentiate a second time to locate the maximum of the first derivative, since the slope of the first derivative goes from positive to negative as we pass through the maximum. This is the basis for locating the end point by taking the second derivative. The second derivative, $\Delta^2\text{pH}/\Delta V^2$, is zero at the end point as shown in Figure 14F-4b. Note that we have expanded the scale to make it easier to locate the zero crossing of the second derivative. The details of calculating the derivatives are given in Section 21G. The spreadsheet approach for obtaining these derivatives and making the plots is developed in Chapter 7 of *Applications of Microsoft® Excel in Analytical Chemistry.*

The Gran plot is an alternative method for locating the end point in a titration. In this method, a linear plot is produced that can reveal both the acid dissociation constant and the volume of base required to reach the end point. Unlike the normal titration curve and derivative curves, which find the end point only from data located in the end point region, the Gran plot uses data far away from the end point. This can decrease the tedium of taking many measurements after dispensing very small volumes of titrant in the end point region.

Prior to the equivalence point of the titration of a weak acid with a strong base, the concentration of acid remaining, c_{HA}, is given by

$$c_{HA} = \frac{\text{no. mmoles of HA initially present}}{\text{total volume of solution}} - \frac{\text{no. mmoles NaOH added}}{\text{total volume of solution}}$$

or

$$c_{HA} = \frac{c_{HA}^0 V_{HA}}{V_{HA} + V_{NaOH}} - \frac{c_{NaOH} V_{NaOH}}{V_{HA} + V_{NaOH}}$$

where c_{HA}^0 is the initial analytical concentration of HA. The equivalence-point volume of NaOH, V_{eq}, can be found from the stoichiometry, which for a 1:1 reaction is given by

$$c_{HA}^0 V_{HA} = c_{NaOH} V_{eq}$$

Substituting for $c_{HA}^0 V_{HA}$ in the equation for c_{HA}, and rearranging, yields

$$c_{HA} = \frac{c_{NaOH}}{V_{HA} + V_{NaOH}} (V_{eq} - V_{NaOH})$$

If K_a is not too large, the equilibrium concentration of acid in the preequivalence-point region is approximately equal to the analytical concentration (see Equation 9-27). That is,

$$[\text{HA}] \approx c_{HA} = \frac{c_{NaOH}}{V_{HA} + V_{NaOH}} (V_{eq} - V_{NaOH})$$

With moderate dissociation of the acid, the equilibrium concentration of A^- at any point is approximately the number of millimoles of base added divided by the total solution volume.

$$[\text{A}^-] \approx \frac{c_{NaOH} V_{NaOH}}{V_{HA} + V_{NaOH}}$$

(continued)

The concentration of H_3O^+ can be found from the equilibrium constant as

$$[H_3O^+] = \frac{K_a[HA]}{[A^-]} = \frac{K_a(V_{eq} - V_{NaOH})}{V_{NaOH}}$$

Multiplying both sides by V_{NaOH} gives,

$$[H_3O^+]V_{NaOH} = K_aV_{eq} - K_aV_{NaOH}$$

A plot of the left-hand side of this equation versus the volume of titrant V_{NaOH} should yield a straight line with a slope of $-K_a$ and an intercept of K_aV_{eq}. In Figure 14F-5, a Gran plot of the titration of 50.00 mL of 0.1000 M weak acid ($K_a = 1.0 \times 10^{-5}$) with 0.1000 M NaOH is shown along with the least-squares equation. From the intercept value of 0.0005, we calculate an end point volume of 50.00 mL by dividing by the value for K_a. Usually, points in the mid stages of the titration are plotted and used to obtain the slope and intercept values. The Gran plot can exhibit curvature in the early stages if K_a is too large, and it can curve near the equivalence point. The spreadsheet approach to constructing Gran plots is given in Chapter 7 of *Applications of Microsoft® Excel in Analytical Chemistry.*

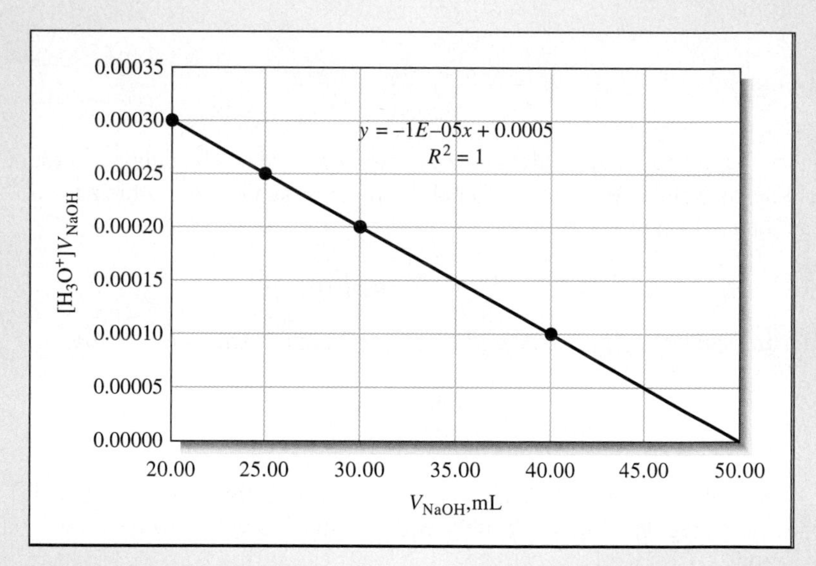

Figure 14F-5 Gran plot for the titration of 50.00 mL of 0.1000 M weak acid ($K_a = 1.0 \times 10^{-5}$) with 0.1000 M NaOH. The least-squares equation for the line is given in the figure.

Spreadsheet Summary In the final three exercises in Chapter 7 of *Applications of Microsoft® Excel in Analytical Chemistry,* we first use Excel to plot a simple distribution of species diagram (α plot) for a weak acid. Then, the first and second derivatives of the titration curve are plotted to better determine the titration end point. A combination plot is produced that simultaneously displays the pH versus volume curve and the second-derivative curve. Finally, a Gran plot is explored for locating the end point by a linear regression procedure.

WEB WORKS

Use the search engine Google to locate the Web document *The Fall of the Proton: Why Acids React with Bases* by Stephen Lower. This document explains acid/base behavior in terms of the concept of proton free energy. How is an acid/base titration described in this view? In a titration of strong acid with strong base, what is the free energy sink? In a complex mixture of weak acid/base systems, such as serum, what happens to protons?

QUESTIONS AND PROBLEMS

In this chapter, round all calculated values for pH and pOH to two figures to the right of the decimal point unless otherwise instructed.

***14-1.** Consider curves for the titration of 0.10 M NaOH and 0.010 M NH_3 with 0.10 M HCl.
(a) Briefly account for the differences between curves for the two titrations.
(b) In what respect will the two curves be indistinguishable?

14-2. What factors affect end-point sharpness in an acid/base titration?

***14-3.** Why does the typical acid/base indicator exhibit its color change over a range of about 2 pH units?

14-4. What variables can cause the pH range of an indicator to shift?

***14-5.** Why are the standard reagents used in neutralization titrations generally strong acids and bases rather than weak acids and bases?

14-6. Which solute would provide the sharper end point in a titration with 0.10 M HCl:
*(a) 0.10 M NaOCl or 0.10 M hydroxylamine?
(b) 0.10 M NH_3 or 0.10 M sodium phenolate?
*(c) 0.10 M methylamine or 0.10 M hydroxylamine?
(d) 0.10 M hydrazine or 0.10 M NaCN?

14-7. Which solute would provide the sharper end point in a titration with 0.10 M NaOH:
*(a) 0.10 M nitrous acid or 0.10 M iodic acid?
(b) 0.10 M anilinium hydrochloride ($C_6H_5NH_3Cl$) or 0.10 M benzoic acid?
*(c) 0.10 M hypochlorous acid or 0.10 M pyruvic acid?
(d) 0.10 M salicylic acid or 0.10 M acetic acid?

14-8. Before glass electrodes and pH meters became so widely used, pH was often determined by measuring the concentration of the acid and base forms of the indicator colorimetrically. If bromothymol blue is introduced into a solution and the concentration ratio of acid to base form is found to be 1.43, what is the pH of the solution?

***14-9.** The procedure described in Problem 14-8 was used to determine pH with methyl orange as the indicator. The concentration ratio of the acid to base form of the indicator was 1.64. Calculate the pH of the solution.

14-10. Values for K_w at 0°C, 50°C, and 100°C are 1.14×10^{-15}, 5.47×10^{-14}, and 4.9×10^{-13}, respectively. Calculate the pH for a neutral solution at each of these temperatures.

14-11. Using the data in Problem 14-10, calculate pK_w at
*(a) 0°C.
(b) 50°C.
(c) 100°C.

14-12. Using the data in Problem 14-10, calculate the pH of a 1.00×10^{-2} M NaOH solution at
*(a) 0°C.
(b) 50°C.
(c) 100°C.

***14-13.** What is the pH of an aqueous solution that is 14.0% HCl by weight and has a density of 1.054 g/mL?

14-14. Calculate the pH of a solution that contains 9.00% (w/w) NaOH and has a density of 1.098 g/mL.

***14-15.** What is the pH of a solution that is 2.00×10^{-8} M in NaOH? (Hint: In such a dilute solution, you must take into account the contribution of H_2O to the hydroxide ion concentration.)

14-16. What is the pH of a 2.00×10^{-8} M HCl solution?

14-17. What is the pH of the solution that results when 0.102 g of $Mg(OH)_2$ is mixed with
(a) 75.0 mL of 0.0600 M HCl?
(b) 15.0 mL of 0.0600 M HCl?
(c) 30.0 mL of 0.0600 M HCl?
(d) 30.0 mL of 0.0600 M $MgCl_2$?

***14-18.** Calculate the pH of the solution that results when mixing 20.0 mL of 0.2000 M HCl with 25.0 mL of
(a) distilled water.
(b) 0.132 M $AgNO_3$.
(c) 0.132 M NaOH.
(d) 0.132 M NH_3.
(e) 0.232 M NaOH.

*14-19. Calculate the hydronium ion concentration and pH of a solution that is 0.0500 M in HCl
(a) neglecting activity corrections.
(b) using activity coefficients.

14-20. Calculate the hydroxide ion concentration and the pH of a 0.0167 M $Ba(OH)_2$ solution
(a) neglecting activity corrections.
(b) using activity coefficients.

*14-21. Calculate the pH of an HOCl solution that is
(a) 1.00×10^{-1} M.
(b) 1.00×10^{-2} M.
(c) 1.00×10^{-4} M.

14-22. Calculate the pH of an NaOCl solution that is
(a) 1.00×10^{-1} M.
(b) 1.00×10^{-2} M.
(c) 1.00×10^{-4} M.

*14-23. Calculate the pH of an ammonia solution that is
(a) 1.00×10^{-1} M.
(b) 1.00×10^{-2} M.
(c) 1.00×10^{-4} M.

14-24. Calculate the pH of an NH_4Cl solution that is
(a) 1.00×10^{-1} M.
(b) 1.00×10^{-2} M.
(c) 1.00×10^{-4} M.

*14-25. Calculate the pH of a solution in which the concentration of the piperidine is
(a) 1.00×10^{-1} M.
(b) 1.00×10^{-2} M.
(c) 1.00×10^{-4} M.

14-26. Calculate the pH of an iodic acid solution that is
(a) 1.00×10^{-1} M.
(b) 1.00×10^{-2} M.
(c) 1.00×10^{-4} M.

*14-27. Calculate the pH of a solution prepared by
(a) dissolving 43.0 g of lactic acid in water and diluting to 500 mL.
(b) diluting 25.0 mL of the solution in (a) to 250 mL.
(c) diluting 10.0 mL of the solution in (b) to 1.00 L.

14-28. Calculate the pH of a solution prepared by
(a) dissolving 1.05 g of picric acid, $(NO_2)_3C_6H_2OH$ (229.11 g/mol), in 100 mL of water.
(b) diluting 10.0 mL of the solution in (a) to 100 mL.
(c) diluting 10.0 mL of the solution in (b) to 1.00 L.

*14-29. Calculate the pH of the solution that results when 20.0 mL of 0.200 M formic acid is
(a) diluted to 45.0 mL with distilled water.
(b) mixed with 25.0 mL of 0.160 M NaOH solution.
(c) mixed with 25.0 mL of 0.200 M NaOH solution.
(d) mixed with 25.0 mL of 0.200 sodium formate solution.

14-30. Calculate the pH of the solution that results when 40.0 mL of 0.100 M NH_3 is
(a) diluted to 20.0 mL with distilled water.
(b) mixed with 20.0 mL of 0.200 M HCl solution.
(c) mixed with 20.0 mL of 0.250 M HCl solution.
(d) mixed with 20.0 mL of 0.200 M NH_4Cl solution.
(e) mixed with 20.0 mL of 0.100 M HCl solution.

14-31. A solution is 0.0500 M in NH_4Cl and 0.0300 M in NH_3. Calculate its OH^-concentration and its pH
(a) neglecting activity corrections.
(b) taking activities coefficients into account.

*14-32. What is the pH of a solution that is
(a) prepared by dissolving 9.20 g of lactic acid (90.08 g/mol) and 11.15 g of sodium lactate (112.06 g/mol) in water and diluting to 1.00 L?
(b) 0.0550 M in acetic acid and 0.0110 M in sodium acetate?
(c) prepared by dissolving 3.00 g of salicylic acid, $C_6H_4(OH)COOH$ (138.12 g/mol), in 50.0 mL of 0.1130 M NaOH and diluting to 500.0 mL?
(d) 0.0100 M in picric acid and 0.100 M in sodium picrate?

14-33. What is the pH of a solution that is
(a) prepared by dissolving 3.30 g of $(NH_4)_2SO_4$ in water, adding 125.0 mL of 0.1011 M NaOH, and diluting to 500.0 mL?
(b) 0.120 M in piperidine and 0.080 M in its chloride salt?
(c) 0.050 M in ethylamine and 0.167 M in its chloride salt?
(d) prepared by dissolving 2.32 g of aniline (93.13 g/mol) in 100 mL of 0.0200 M HCl and diluting to 250.0 mL?

14-34. Calculate the change in pH that occurs in each of the solutions listed below as a result of a tenfold dilution with water. Round calculated values for pH to three figures to the right of the decimal point.
*(a) H_2O.
(b) 0.0500 M HCl.
*(c) 0.0500 M NaOH.
(d) 0.0500 M CH_3COOH.
*(e) 0.0500 M CH_3COONa.
(f) 0.0500 M CH_3COOH + 0.0500 M CH_3COONa.
*(g) 0.500 M CH_3COOH + 0.500 M CH_3COONa.

14-35. Calculate the change in pH that occurs when 1.00 mmol of a strong acid is added to 100 mL of the solutions listed in Problem 14-34.

14-36. Calculate the change in pH that occurs when 1.00 mmol of a strong base is added to 100 mL of the solutions listed in Problem 14-34. Calculate values to three decimal places.

14-37. Calculate the change in pH to three decimal places that occurs when 0.50 mmol of a strong acid is added to 100 mL of

(a) 0.0200 M lactic acid + 0.0800 M sodium lactate.

*(b) 0.0800 M lactic acid + 0.0200 M sodium lactate.

(c) 0.0500 M lactic acid + 0.0500 M sodium lactate.

***14-38.** A 50.00-mL aliquot of 0.1000 M NaOH is titrated with 0.1000 M HCl. Calculate the pH of the solution after the addition of 0.00, 10.00, 25.00, 40.00, 45.00, 49.00, 50.00, 51.00, 55.00, and 60.00 mL of acid, and prepare a titration curve from the data.

***14-39.** In a titration of 50.00 mL of 0.05000 M formic acid with 0.1000 M KOH, the titration error must be smaller than 0.05 mL. What indicator can be chosen to realize this goal?

14-40. In a titration of 50.00 mL of 0.1000 M ethylamine with 0.1000 M $HClO_4$, the titration error must be no more than 0.05 mL. What indicator can be chosen to realize this goal?

14-41. Calculate the pH after addition of 0.00, 5.00, 15.00, 25.00, 40.00, 45.00, 49.00, 50.00, 51.00, 55.00, and 60.00 mL of 0.1000 M NaOH in the titration of 50.00 mL of

*(a) 0.1000 M HNO_2.

(b) 0.1000 M lactic acid.

*(c) 0.1000 M pyridinium chloride.

14-42. Calculate the pH after addition of 0.00, 5.00, 15.00, 25.00, 40.00, 45.00, 49.00, 50.00, 51.00, 55.00, and 60.00 mL of 0.1000 M HCl in the titration of 50.00 mL of

*(a) 0.1000 M ammonia.

(b) 0.1000 M hydrazine.

(c) 0.1000 M sodium cyanide.

14-43. Calculate the pH after addition of 0.00, 5.00, 15.00, 25.00, 40.00, 49.00, 50.00, 51.00, 55.00, and 60.00 mL of reagent in the titration of 50.0 mL of

*(a) 0.1000 M anilinium chloride with 0.1000 M NaOH.

(b) 0.01000 M chloroacetic acid with 0.01000 M NaOH.

*(c) 0.1000 M hypochlorous acid with 0.1000 M NaOH.

(d) 0.1000 M hydroxylamine with 0.1000 M HCl.

Construct titration curves from the data.

14-44. Calculate α_0 and α_1 for

*(a) acetic acid species in a solution with a pH of 5.320.

(b) picric acid species in a solution with a pH of 1.250.

*(c) hypochlorous acid species in a solution with a pH of 7.000.

(d) hydroxylamine acid species in a solution with a pH of 5.120.

*(e) piperidine species in a solution with a pH of 10.080.

***14-45.** Calculate the equilibrium concentration of undissociated HCOOH in a formic acid solution with an analytical formic acid concentration of 0.0850 and a pH of 3.200.

14-46. Calculate the equilibrium concentration of methylammonia in a solution that has a molar analytical CH_3NH_2 concentration of 0.120 and a pH of 11.471.

14-47. Supply the missing data in the table below.

Acid	Molar Analytical Concentration, c_T ($c_T = c_{HA} + c_{A^-}$)	pH	[HA]	$[A^-]$	α_0	α_1
*Lactic	0.120	____	____	____	0.640	____
Iodic	0.200	____	____	____	____	0.765
Butanoic	____	5.00	0.644	____	____	____
Hypochlorous	0.280	7.00	____	____	____	____
Nitrous	____	____	____	0.105	0.413	0.587
Hydrogen cyanide	____	____	0.145	0.221	____	____
*Sulfamic	0.250	1.20	____	____	____	____

14-48. Challenge Problem. This photo shows a buret that has at least two defects on the scale that were created during its fabrication.

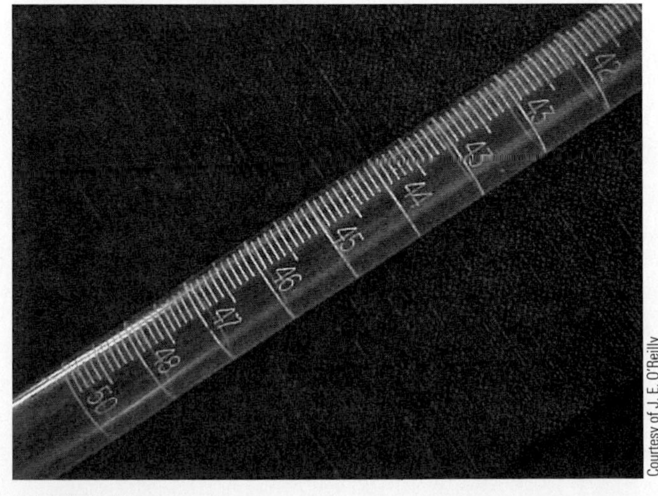

Courtesy of J. E. O'Reilly

Mislabeled buret.

Answer the following questions about the buret, its origin, and its use.

(a) Under what conditions is the buret usable?

(b) Assuming that the user do not notice the defects in the buret, what type of error would occur if the liquid level was between the second 43-mL mark and the 48-mL mark?

(c) Assume that the initial reading in a titration is 0.00 mL (very unlikely), and calculate the relative error in the volume if the final reading is 43.00 mL (upper mark). What is the relative error if the same reading is made on the lower mark? Perform the same calculation for a final reading made at the 48.00-mL mark. What do these calculations demonstrate about the type of error caused by the defect in the buret?

(d) Speculate on the age of the buret. How would you guess that the markings were made on the glass? Is it likely that the same type of defect would appear in a buret manufactured today? Explain your answer.

(e) Modern electronic chemical instruments such as pH meters, balances, titrators, and spectrophotometers are normally assumed to be free of manufacturing defects analogous to the one illustrated in the photo. Comment on the wisdom of making such assumptions.

(f) Burets in automated titrators contain a motor connected to a screw-driven plunger that delivers titrant in much the same way that a hypodermic syringe delivers liquids. The distance of travel of the plunger is proportional to the volume of liquid delivered. What kinds of manufacturer's defects would lead to inaccuracy or imprecision in the volume of liquid dispensed by these devices?

(g) What steps can you take to avoid measurement errors while using modern chemical instruments?

InfoTrac College Edition

For additional readings, go to InfoTrac College Edition, your online research library, at

http://infotrac.thomsonlearning.com

CHAPTER 15

Titration Curves for Complex Acid/Base Systems

Polyfunctional acids and bases play important roles in many chemical and biological systems. The human body contains a complicated system of buffers within cells and within bodily fluids, such as human blood. The photo is a scanning electronmicrograph of red blood cells traveling through an artery. The pH of human blood is controlled to be within the range of 7.35 to 7.45, primarily by the carbonic acid/bicarbonate buffer system.

$$CO_2(g) + H_2O(l) \rightleftharpoons H_2CO_3(aq)$$
$$H_2CO_3 + H_2O \rightleftharpoons H_3O^+ + HCO_3^-.$$

This chapter describes polyfunctional acid and base systems, including buffer solutions. Calculations of pH and titration curves are also described.

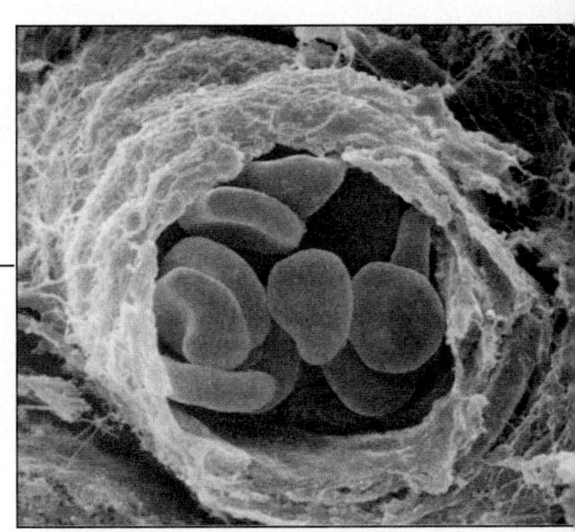

In this chapter, we describe methods for calculating titration curves for complex acid/base systems. For the purpose of this discussion, complex systems are defined as solutions made up of (1) two acids or two bases of different strengths, (2) an acid or a base that has two or more acidic or basic functional groups, or (3) an amphiprotic substance, which is capable of acting as both an acid and a base. Equations for more than one equilibrium are required to describe the characteristics of any of these systems.

15A MIXTURES OF STRONG AND WEAK ACIDS OR STRONG AND WEAK BASES

It is possible to determine each of the components in a mixture containing a strong acid and a weak acid (or a strong base and a weak base) provided that the concentrations of the two are of the same order of magnitude and that the dissociation constant for the weak acid or base is somewhat less than about 10^{-4}. To demonstrate that this statement is true, let us show how a titration curve can be constructed for a solution containing roughly equal concentrations of HCl and HA, where HA is a weak acid with a dissociation constant of 10^{-4}.

EXAMPLE 15-1

Calculate the pH of a mixture that is 0.1200 M in hydrochloric acid and 0.0800 M in the weak acid HA ($K_a = 1.00 \times 10^{-4}$) during its titration with 0.1000 M KOH. Compute results for additions of the following volumes of base: (a) 0.00 mL and (b) 5.00 mL.

(a) 0.00 mL KOH Added

The molar hydronium concentration in this mixture is equal to the concentration of HCl plus the concentration of hydronium ions that results from dissociation of HA and H_2O. In the presence of the two acids, however, we can be certain that the concentration of hydronium ions from the dissociation of water is extremely small. We therefore need to take into account only the other two sources of protons. Thus, we may write

$$[H_3O^+] = c_{HCl} + [A^-] = 0.1200 + [A^-]$$

Note that $[A^-]$ is equal to the concentration of hydronium ions from the dissociation of HA.

Now assume that the presence of the strong acid so represses the dissociation of HA that $[A^-] << 0.1200$ M; then

$$[H_3O^+] \approx 0.1200 \text{ M, and the pH is } 0.92$$

To check this assumption, the provisional value for $[H_3O^+]$ is substituted into the dissociation-constant expression for HA. When this expression is rearranged, we obtain

$$\frac{[A^-]}{[HA]} = \frac{K_a}{[H_3O^+]} = \frac{1.00 \times 10^{-4}}{0.1200} = 8.33 \times 10^{-4}$$

This expression can be rearranged to

$$[HA] = [A^-]/(8.33 \times 10^{-4})$$

From the concentration of the weak acid, we can write the mass-balance expression

$$c_{HA} = [HA] + [A^-] = 0.0800 \text{ M}$$

Substituting the value of [HA] from the previous equation gives

$$[A^-]/(8.33 \times 10^{-4}) + [A^-] \approx (1.20 \times 10^3)[A^-] = 0.0800 \text{ M}$$
$$[A^-] = 6.7 \times 10^{-5} \text{ M}$$

We see that $[A^-]$ is indeed much smaller than 0.1200 M, as assumed.

(b) After Adding 5.00 mL of Base

$$c_{HCl} = \frac{25.00 \times 0.1200 - 5.00 \times 0.1000}{25.00 + 5.00} = 0.0833 \text{ M}$$

and we may write

$$[H_3O^+] = 0.0833 + [A^-] \approx 0.0833 \text{ M}$$
$$\text{pH} = 1.08$$

To determine whether our assumption is still valid, we compute $[A^-]$ as we did in part (a), knowing that the concentration of HA is now $0.0800 \times 25.00/30.00 = 0.0667$, and find

$$[A^-] = 8.0 \times 10^{-5} \text{ M}$$

which is still much smaller than 0.0833.

Example 15-1 demonstrates that hydrochloric acid represses the dissociation of the weak acid in the early stages of the titration to such an extent that we can assume that $[A^-] << c_{HCl}$ and $[H_3O^+] = c_{HCl}$. In other words, the hydronium ion concentration is simply the molar concentration of the strong acid.

The approximation employed in Example 15-1 can be shown to apply until most of the hydrochloric acid has been neutralized by the titrant. Therefore, the curve in this region *is identical to the titration curve for a 0.1200 M solution of a strong acid by itself.*

As shown by Example 15-2, the presence of HA must be taken into account as the first end point in the titration is approached.

EXAMPLE 15-2

Calculate the pH of the solution that results when 29.00 mL of 0.1000 M NaOH is added to 25.00 mL of the solution described in Example 15-1.

Here,

$$c_{HCl} = \frac{25.00 \times 0.1200 - 29.00 \times 0.1000}{54.00} = 1.85 \times 10^{-3} \text{ M}$$

$$c_{HA} = \frac{25.00 \times 0.0800}{54.00} = 3.70 \times 10^{-2} \text{ M}$$

A provisional result based (as in the previous example) on the assumption that $[H_3O^+] = 1.85 \times 10^{-3}$ yields a value of 1.90×10^{-3} for $[A^-]$. Clearly, $[A^-]$ is no longer much smaller than $[H_3O^+]$, and we must write

$$[H_3O^+] = c_{HCl} + [A^-] = 1.85 \times 10^{-3} + [A^-] \quad (15\text{-}1)$$

In addition, from mass-balance considerations, we know that

$$[HA] + [A^-] = c_{HA} = 3.70 \times 10^{-2} \quad (15\text{-}2)$$

(continued)

We rearrange the acid dissociation-constant expression for HA and obtain

$$[HA] = \frac{[H_3O^+][A^-]}{1.00 \times 10^{-4}}$$

Substitution of this expression into Equation 15-2 yields

$$\frac{[H_3O^+][A^-]}{1.00 \times 10^{-4}} + [A^-] = 3.70 \times 10^{-2}$$

$$[A^-] = \frac{3.70 \times 10^{-6}}{[H_3O^+] + 1.00 \times 10^{-4}}$$

Substitution for $[A^-]$ and c_{HCl} in Equation 15-1 yields

$$[H_3O^+] = 1.85 \times 10^{-3} + \frac{3.70 \times 10^{-6}}{[H_3O^+] + 1.00 \times 10^{-4}}$$

$$[H_3O^+]^2 + (1.00 \times 10^{-4})[H_3O^+] = (1.85 \times 10^{-3})[H_3O^+] + 1.85 \times 10^{-7} + 3.7 \times 10^{-6}$$

Collecting terms gives

$$[H_3O^+]^2 - (1.75 \times 10^{-3})[H_3O^+] - 3.885 \times 10^{-6} = 0$$

Solving the quadratic equation gives

$$[H_3O^+] = 3.03 \times 10^{-3}\ M$$
$$pH = 2.52$$

Note that the contributions to the hydronium ion concentration from HCl (1.85×10^{-3} M) and HA (3.03×10^{-3} M $-$ 1.85×10^{-3} M) are of comparable magnitude.

When the amount of base added is equivalent to the amount of hydrochloric acid originally present, the solution is identical in all respects to one prepared by dissolving appropriate quantities of the weak acid and sodium chloride in a suitable volume of water. The sodium chloride, however, has no effect on the pH (neglecting the influence of increased ionic strength); thus, the remainder of the titration curve is identical to that for a dilute solution of HA.

The shape of the curve for a mixture of weak and strong acids, and hence the information obtainable from it, depends in large measure on the strength of the weak acid. Figure 15-1 depicts the pH changes that occur during the titration of mixtures containing hydrochloric acid and several weak acids. Note that the rise in pH at the first equivalence point is small or essentially nonexistent when the weak acid has a relatively large dissociation constant (curves *A* and *B*). For titrations such as these, only the total number of millimoles of weak and strong acid can be determined accurately. Conversely, when the weak acid has a very small dissociation

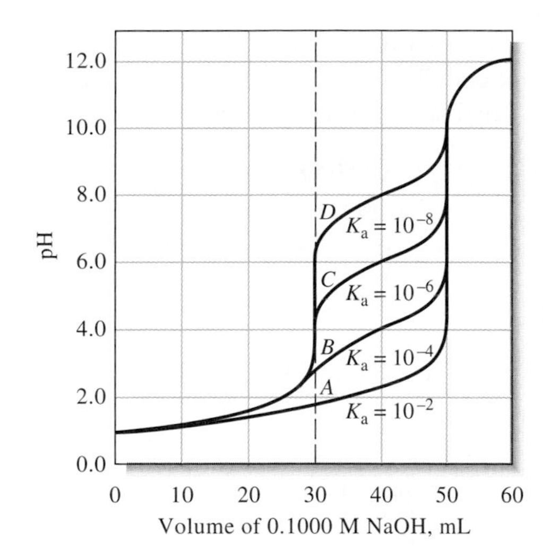

Figure 15-1 Curves for the titration of strong acid/weak acid mixtures with 0.1000 M NaOH. Each titration is on 25.00 mL of a solution that is 0.1200 M in HCl and 0.0800 M in HA.

constant, only the strong acid content can be determined. For weak acids of intermediate strength (K_a somewhat less than 10^{-4} but greater than 10^{-8}), there are usually two useful end points.

Determination of the amount of each component in a mixture that contains a strong base and a weak base is also possible, subject to the constraints just described for the strong acid/weak acid system. The computation of a curve for such a titration is analogous to that for a mixture of acids.

◀ The composition of a mixture of a strong acid and a weak acid can be determined by titration with suitable indicators if the weak acid has a dissociation constant that lies between 10^{-4} and 10^{-8} and the concentrations of the two acids are of the same order of magnitude.

15B POLYFUNCTIONAL ACIDS AND BASES

Several species are encountered in analytical chemistry that have two or more acidic or basic functional groups. Generally, the two groups differ in strength and, as a consequence, exhibit two or more end points in a neutralization titration.

15B-1 The Phosphoric Acid System

Phosphoric acid is a typical polyfunctional acid. In aqueous solution, it undergoes the following three dissociation reactions:

◀ Throughout the remainder of this chapter, we use K_{a1}, K_{a2} to represent the first and second dissociation constants of acids and K_{b1}, K_{b2} to represent the stepwise constants for bases.

$$H_3PO_4 + H_2O \rightleftharpoons H_2PO_4^- + H_3O^+ \qquad K_{a1} = \frac{[H_3O^+][H_2PO_4^-]}{[H_3PO_4]} = 7.11 \times 10^{-3}$$

$$H_2PO_4^- + H_2O \rightleftharpoons HPO_4^{2-} + H_3O^+ \qquad K_{a2} = \frac{[H_3O^+][HPO_4^{2-}]}{[H_2PO_4^-]} = 6.32 \times 10^{-8}$$

$$HPO_4^{2-} + H_2O \rightleftharpoons PO_4^{3-} + H_3O^+ \qquad K_{a3} = \frac{[H_3O^+][PO_4^{3-}]}{[HPO_4^{2-}]} = 4.5 \times 10^{-13}$$

◀ Generally, $K_{a1} > K_{a2}$, often by a factor of 10^4 to 10^5 because of electrostatic forces. That is, the first dissociation involves separating a single positively charged hydronium ion from a singly charged anion. In the second step, a hydronium ion is separated from a doubly charged anion, a process that requires considerably more energy.

With this acid, as with other polyprotic acids, $K_{a1} > K_{a2} > K_{a3}$.

► A second reason that $K_{a1} > K_{a2}$ is a statistical one. In the first step, a proton can be removed from two locations; in the second step, only one can be removed. Thus, the first dissociation is twice as probable as the second.

When we *add* two adjacent stepwise equilibria, we *multiply* the two equilibrium constants to obtain the equilibrium constant for the resulting overall reaction. Thus, for the first two dissociation equilibria for H_3PO_4, we write

$$K_{a1}K_{a2} = \frac{[H_3O^+]^2[HPO_4^{2-}]}{[H_3PO_4]}$$
$$= 7.11 \times 10^{-3} \times 6.32 \times 10^{-8} = 4.49 \times 10^{-10}$$

Similarly, for the reaction

$$H_3PO_4 \rightleftharpoons 3H_3O^+ + PO_4^{3-}$$

we may write

$$K_{a1}K_{a2}K_{a3} = \frac{[H_3O^+]^3[PO_4^{3-}]}{H_3PO_4}$$
$$= 7.11 \times 10^{-3} \times 6.32 \times 10^{-8} \times 4.5 \times 10^{-13} = 2.0 \times 10^{-22}$$

15B-2 The Carbon Dioxide Carbonic Acid System

When carbon dioxide is dissolved in water, a dibasic acid system is formed by the following reactions:

$$CO_2(aq) + H_2O \rightleftharpoons H_2CO_3 \qquad K_{hyd} = \frac{[H_2CO_3]}{[CO_2(aq)]} = 2.8 \times 10^{-3} \qquad (15\text{-}3)$$

$$H_2CO_3 + H_2O \rightleftharpoons H_3O^+ + HCO_3^-$$
$$K_1 = \frac{[H_3O^+][HCO_3^-]}{[H_2CO_3]} = 1.5 \times 10^{-4} \qquad (15\text{-}4)$$

$$HCO_3^- + H_2O \rightleftharpoons H_3O^+ + CO_3^{2-}$$
$$K_2 = \frac{[H_3O^+][CO_3^{2-}]}{[HCO_3^-]} = 4.69 \times 10^{-11} \qquad (15\text{-}5)$$

The first reaction describes the hydration of aqueous CO_2 to form carbonic acid. Note that the magnitude of K_{hyd} indicates that the concentration of $CO_2(aq)$ is much larger than the concentration of H_2CO_3 (that is, $[H_2CO_3]$ is only about 0.3% that of $[CO_2(aq)]$). Thus, a more useful way of discussing the acidity of solutions of carbon dioxide is to combine Equation 15-3 and 15-4 to give

$$CO_2(aq) + 2H_2O \rightleftharpoons H_3O^+ + HCO_3^- \qquad K_{a1} = \frac{[H_3O^+][HCO_3^-]}{[CO_2(aq)]} \qquad (15\text{-}6)$$
$$= 2.8 \times 10^{-3} \times 1.5 \times 10^{-4}$$
$$= 4.2 \times 10^{-7}$$

$$HCO_3^- + H_2O \rightleftharpoons H_3O^+ + CO_3^{2-} \qquad K_{a2} = 4.69 \times 10^{-11} \qquad (15\text{-}7)$$

EXAMPLE 15-3

Calculate the pH of a solution that is 0.02500 M CO_2. From mass-balance considerations,

$$c_{CO_2} = 0.02500 = [CO_2(aq)] + [H_2CO_3] + [HCO_3^-] + [CO_3^{2-}]$$

The small magnitude of K_{hyd}, K_1, and K_2 (see Equations 15-3, 15-4, and 15-5) suggests that

$$([H_2CO_3] + [HCO_3^-] + [CO_3^{2-}]) << [CO_2(aq)]$$

and we may write

$$[CO_2(aq)] \approx c_{CO_2} = 0.02500\ \text{M}$$

From charge-balance considerations,

$$[H_3O^+] = [HCO_3^-] + 2[CO_3^{2-}] + [OH^-]$$

We will then assume

$$2([CO_3^{2-}] + [OH^-]) << [HCO_3^-]$$

Thus,

$$[H_3O^+] \approx [HCO_3^-]$$

Substituting these approximations into Equation 15-6 leads to

$$\frac{[H_3O^+]^2}{0.02500} = K_{a1} = 4.2 \times 10^{-7}$$

$$[H_3O^+] = \sqrt{0.02500 \times 4.2 \times 10^{-7}} = 1.02 \times 10^{-4}\ \text{M}$$

$$\text{pH} = -\log(1.02 \times 10^{-4}) = 3.99$$

Calculated tentative values for $[H_2CO_3]$, $[CO_3^{2-}]$, and $[OH^-]$ indicate that the assumptions were valid.

The pH of polyfunctional systems, such as phosphoric acid or sodium carbonate, can be computed rigorously through use of the systematic approach to multi-equilibrium problems described in Chapter 11. Solution of the several simultaneous equations that are involved is difficult and time consuming, however. Fortunately, simplifying assumptions can be invoked when the successive equilibrium constants for the acid (or base) differ by a factor of about 10^3 (or more). With one exception, these assumptions make it possible to compute pH data for titration curves by the techniques we have discussed in earlier chapters.

◀ CHALLENGE: Write a sufficient number of equations to make it possible to calculate all of the species in a solution containing known molar analytical concentrations of Na_2CO_3 and $NaHCO_3$.

15C BUFFER SOLUTIONS INVOLVING POLYPROTIC ACIDS

Two buffer systems can be prepared from a weak dibasic acid and its salts. The first consists of free acid H_2A and its conjugate base NaHA, and the second makes use of the acid NaHA and its conjugate base Na_2A. The pH of the latter system is higher than that of the former because the acid dissociation constant for HA^- is always less than that for H_2A.

Sufficient independent equations are readily written to permit a rigorous evaluation of the hydronium ion concentration for either of these systems. Ordinarily, however, it is permissible to introduce the simplifying assumption that only one of the equilibria is important in determining the hydronium ion concentration of the solution. Thus, for a buffer prepared from H_2A and NaHA, the dissociation of HA^- to yield A^{2-} is neglected, and the calculation is based only on the first dissociation. With this simplification, the hydronium ion concentration is calculated by the method described in Section 9C-1 for a simple buffer solution. As shown in Example 15-4, it is easy to check the validity of the assumption by calculating an approximate concentration of A^{2-} and comparing this value with the concentrations of H_2A and HA^-.

EXAMPLE 15-4

Calculate the hydronium ion concentration for a buffer solution that is 2.00 M in phosphoric acid and 1.50 M in potassium dihydrogen phosphate.

The principal equilibrium in this solution is the dissociation of H_3PO_4.

$$H_3PO_4 + H_2O \rightleftharpoons H_3O^+ + H_2PO_4^- \qquad \frac{[H_3O^+][H_2PO_4^-]}{[H_3PO_4]} = K_{a1} = 7.11 \times 10^{-3}$$

We assume that the dissociation of $H_2PO_4^-$ is negligible; that is, $[HPO_4^{2-}]$ and $[PO_4^{3-}] << [H_2PO_4^-]$ and $[H_3PO_4]$. Then,

$$[H_3PO_4] \approx c_{H_3PO_4} = 2.00 \text{ M}$$

$$[H_2PO_4^-] \approx c_{KH_2PO_4} = 1.50 \text{ M}$$

$$[H_3O^+] = \frac{7.11 \times 10^{-3} \times 2.00}{1.50} = 9.48 \times 10^{-3} \text{ M}$$

We now use the equilibrium-constant expression for K_{a2} to show that $[HPO_4^{2-}]$ can be neglected:

$$\frac{[H_3O^+][HPO_4^{2-}]}{[H_2PO_4^-]} = \frac{9.48 \times 10^{-3}[HPO_4^{2-}]}{1.50} = K_{a2} = 6.34 \times 10^{-8}$$

$$[HPO_4^{2-}] = 1.00 \times 10^{-5} \text{ M}$$

and our assumption is valid. Note that $[PO_4^{3-}]$ is even smaller than $[HPO_4^{2-}]$.

For a buffer prepared from NaHA and Na_2A, the second dissociation will ordinarily predominate, and the equilibrium

$$HA^- + H_2O \rightleftharpoons H_2A + OH^-$$

is disregarded. The concentration of H_2A is negligible compared with that of HA^- or A^{2-}. The hydronium ion concentration can be calculated from the second dissociation constant, again employing the techniques for a simple buffer solution. To test the assumption, we compare an estimate of the H_2A concentration with the concentrations of HA^- and A^{2-}, as in Example 15-5.

EXAMPLE 15-5

Calculate the hydronium ion concentration of a buffer that is 0.0500 M in potassium hydrogen phthalate (KHP) and 0.150 M in potassium phthalate (K_2P).

$$HP^- + H_2O \rightleftharpoons H_3O^+ + P^{2-} \qquad \frac{[H_3O^+][P^{2-}]}{[HP^-]} = K_{a2} = 3.91 \times 10^{-6}$$

Provided that the concentration of H_2P in this solution is negligible,

$$[HP^-] \approx c_{KHP} \approx 0.0500 \text{ M}$$

$$[P^{2-}] \approx c_{K_2P} = 0.150 \text{ M}$$

$$[H_3O^+] = \frac{3.91 \times 10^{-6} \times 0.0500}{0.150} = 1.30 \times 10^{-6} \text{ M}$$

To check the first assumption, an approximate value for $[H_2P]$ is calculated by substituting numerical values for $[H_3O^+]$ and $[HP^-]$ into the expression for K_{a1}:

$$\frac{(1.30 \times 10^{-6})(0.0500)}{[H_2P]} = K_{a1} = 1.12 \times 10^{-3}$$

$$[H_2P] = 6 \times 10^{-5} \text{ M}$$

This result justifies the assumption that $[H_2P] << [HP^-]$ and $[P^{2-}]$—that is, that the reaction of HP^- as a base can be neglected.

In all but a few situations, the assumption of a single principal equilibrium, as invoked in Examples 15-4 and 15-5, provides a satisfactory estimate of the pH of buffer mixtures derived from polybasic acids. Appreciable errors occur, however, when the concentration of the acid or the salt is very low or when the two dissociation constants are numerically close to one another. A more laborious and rigorous calculation is then required.

15D CALCULATION OF THE pH OF SOLUTIONS OF NaHA

Thus far, we have not considered how to calculate the pH of solutions of salts that have both acidic and basic properties—that is, salts that are *amphiprotic*. Such salts are formed during neutralization titration of polyfunctional acids and bases. For example, when 1 mol of NaOH is added to a solution containing 1 mol of the acid

H_2A, 1 mol of NaHA is formed. The pH of this solution is determined by two equilibria established between HA^- and water:

$$HA^- + H_2O \rightleftharpoons A^{2-} + H_3O^+$$

and

$$HA^- + H_2O \rightleftharpoons H_2A + OH^-$$

One of these reactions produces hydronium ions and the other hydroxide ions. A solution of NaHA will be acidic or basic depending on the relative magnitude of the equilibrium constants for these processes:

$$K_{a2} = \frac{[H_3O^+][A^{2-}]}{[HA^-]} \tag{15-8}$$

$$K_{b2} = \frac{K_w}{K_{a1}} = \frac{[H_2A][OH^-]}{[HA^-]} \tag{15-9}$$

where K_{a1} and K_{a2} are the acid dissociation constants for H_2A and K_{b2} is the *basic* dissociation constant for HA^-. If K_{b2} is greater than K_{a2}, the solution is basic; otherwise, it is acidic.

To derive an expression for the hydronium ion concentration of a solution of HA^-, we employ the systematic approach described in Section 11A. We first write a mass-balance expression. That is,

$$c_{NaHA} = [HA^-] + [H_2A] + [A^{2-}] \tag{15-10}$$

The charge-balance equation takes the form:

$$[Na^+] + [H_3O^+] = [HA^-] + 2[A^{2-}] + [OH^-]$$

Since the sodium ion concentration is equal to the molar analytical concentration of the salt, the last equation can be rewritten as

$$c_{NaHA} + [H_3O^+] = [HA^-] + 2[A^{2-}] + [OH^-] \tag{15-11}$$

We now have four algebraic equations (Equations 15-10 and 15-11 and the two dissociation constant expressions for H_2A) and need one additional expression to solve for the five unknowns. The ion-product constant for water serves this purpose:

$$K_w = [H_3O^+][OH^-]$$

The rigorous computation of the hydronium ion concentration from these five equations is difficult. A reasonable approximation, applicable to solutions of most acid salts, can be obtained as follows, however.

We first subtract the mass-balance equation from the charge-balance equation.

$$c_{NaHA} + [H_3O^+] = [HA^-] + 2[A^{2-}] + [OH^-] \quad \text{charge balance}$$

$$c_{NaHA} = [H_2A] + [HA^-] + [A^{2-}] \quad \text{mass balance}$$

$$[H_3O^+] = [A^{2-}] + [OH^-] - [H_2A] \tag{15-12}$$

We then rearrange the acid-dissociation constant expressions for H_2A to obtain

$$[H_2A] = \frac{[H_3O^+][HA^-]}{K_{a1}}$$

and for HA^- to give

$$[A^{2-}] = \frac{K_{a2}[HA^-]}{[H_3O^+]}$$

Substituting these expressions and the expression for K_w into Equation 15-12 yields

$$[H_3O^+] = \frac{K_{a2}[HA^-]}{[H_3O^+]} + \frac{K_w}{[H_3O^+]} - \frac{[H_3O^+][HA^-]}{K_{a1}}$$

Multiplying through by $[H_3O^+]$ gives

$$[H_3O^+]^2 = K_{a2}[HA^-] + K_w - \frac{[H_3O^+]^2[HA^-]}{K_{a1}}$$

We collect terms to obtain

$$[H_3O^+]^2\left(\frac{[HA^-]}{K_{a1}} + 1\right) = K_{a2}[HA^-] + K_w$$

Finally, this equation rearranges to

$$[H_3O^+] = \sqrt{\frac{K_{a2}[HA^-] + K_w}{1 + [HA^-]/K_{a1}}} \qquad (15\text{-}13)$$

Under most circumstances, we can make the approximation that

$$[HA^-] \approx c_{NaHA} \qquad (15\text{-}14)$$

Introduction of this relationship into Equation 15-13 gives

$$[H_3O^+] = \sqrt{\frac{K_{a2}c_{NaHA} + K_w}{1 + c_{NaHA}/K_{a1}}} \qquad (15\text{-}15)$$

The approximation shown as Equation 15-14 requires that $[HA^-]$ be much larger than any of the other equilibrium concentrations in Equations 15-10 and 15-11. This assumption is not valid for very dilute solutions of NaHA or when K_{a2} or K_w/K_{a1} is relatively large.

Frequently, the ratio c_{NaHA}/K_{a1} is much larger than unity in the denominator of Equation 15-15, and $K_{a2}c_{NaHA}$ is considerably greater than K_w in the numerator. With these assumptions, the equation simplifies to

$$[H_3O^+] \approx \sqrt{K_{a1}K_{a2}} \qquad (15\text{-}16)$$

◀ Always check the assumptions that are inherent in Equation 15-16.

Note that Equation 15-16 does not contain c_{NaHA}, which implies that the pH of solutions of this type remains constant over a considerable range of solute concentrations.

EXAMPLE 15-6

Calculate the hydronium ion concentration of a 1.00×10^{-3} M Na_2HPO_4 solution.

The pertinent dissociation constants are K_{a2} and K_{a3}, which both contain $[HPO_4^{2-}]$. Their values are $K_{a2} = 6.32 \times 10^{-8}$ and $K_{a3} = 4.5 \times 10^{-13}$. Considering again the assumptions that led to Equation 15-16, we find that $(1.0 \times 10^{-3})/(6.32 \times 10^{-8})$ is much larger than 1, so that the denominator can be simplified. The product $K_{a2}c_{Na_2HPO_4}$ is by no means much larger than K_w, however. We therefore use a partially simplified version of Equation 15-15:

$$[H_3O^+] = \sqrt{\frac{4.5 \times 10^{-13} \times 1.00 \times 10^{-3} + 1.00 \times 10^{-14}}{(1.00 \times 10^{-3})/(6.32 \times 10^{-8})}} = 8.1 \times 10^{-10}\text{ M}$$

Use of Equation 15-16 yields a value of 1.7×10^{-10} M.

EXAMPLE 15-7

Find the hydronium ion concentration of a 0.0100 M NaH_2PO_4 solution.

The two dissociation constants of importance (those containing $[H_2PO_4^-]$) are $K_{a1} = 7.11 \times 10^{-3}$ and $K_{a2} = 6.32 \times 10^{-8}$. We see that the denominator of Equation 15-15 cannot be simplified, but the numerator reduces to $K_{a2}c_{NaH_2PO_4}$. Thus, Equation 15-15 becomes

$$[H_3O^+] = \sqrt{\frac{6.32 \times 10^{-8} \times 1.00 \times 10^{-2}}{1.00 + (1.00 \times 10^{-2})/(7.11 \times 10^{-3})}} = 1.62 \times 10^{-5}\text{ M}$$

EXAMPLE 15-8

Calculate the hydronium ion concentration of a 0.100 M $NaHCO_3$ solution.

We assume, as we have earlier (page 400), that $[H_2CO_3] << [CO_2(aq)]$ and that the following equilibria describe the system adequately:

$$CO_2(aq) + 2H_2O \rightleftharpoons H_3O^+ + HCO_3^- \qquad K_{a1} = \frac{[H_3O^+][HCO_3^-]}{[CO_2(aq)]} = 4.2 \times 10^{-7}$$

$$HCO_3^- + H_2O \rightleftharpoons H_3O^+ + CO_3^{2-} \qquad K_{a2} = \frac{[H_3O^+][CO_3^{2-}]}{[HCO_3^-]} = 4.69 \times 10^{-11}$$

Clearly, c_{NaHA}/K_{a1} in the denominator of Equation 15-15 is much larger than unity; in addition, $K_{a2}c_{NaHA}$ has a value of 4.69×10^{-12}, which is substantially greater than K_w. Thus, Equation 15-16 applies, and

$$[H_3O^+] = \sqrt{4.2 \times 10^{-7} \times 4.69 \times 10^{-11}} = 4.4 \times 10^{-9} M$$

15E TITRATION CURVES FOR POLYFUNCTIONAL ACIDS

Compounds with two or more acid functional groups yield multiple end points in a titration provided that the functional groups differ sufficiently in strength as acids. The computational techniques described in Chapter 14 permit construction of reasonably accurate theoretical titration curves for polyprotic acids if the ratio K_{a1}/K_{a2} is somewhat greater than 10^3. If this ratio is smaller, the error becomes excessive, particularly in the region of the first equivalence point, and a more rigorous treatment of the equilibrium relationships is required.

Figure 15-2 shows the titration curve for a diprotic acid H_2A with dissociation constants of $K_{a1} = 1.00 \times 10^{-3}$ and $K_{a2} = 1.00 \times 10^{-7}$. Because the K_{a1}/K_{a2} ratio is significantly greater than 10^3, we can calculate this curve (except for the first equivalence point) using the techniques developed in Chapter 14 for simple monoprotic weak acids. Thus, to obtain the initial pH (point *A*), we treat the system as if it contained a single monoprotic acid with a dissociation constant of $K_{a1} = 1.00 \times 10^{-3}$. In region *B*, we have the equivalent of a simple buffer solution consisting of the weak acid H_2A and its conjugate base NaHA. That is, we assume that the concentration of A^{2-} is negligible with respect to the other two A-containing species and employ

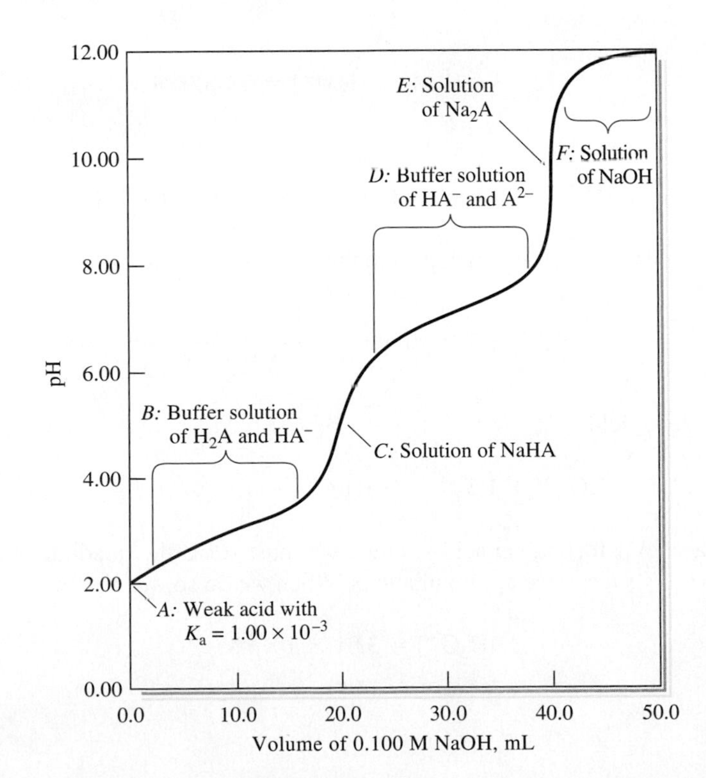

Figure 15-2 Titration of 20.00 mL of 0.1000 M H_2A with 0.1000 M NaOH. For H_2A, $K_{a1} = 1.00 \times 10^{-3}$ and $K_{a2} = 1.00 \times 10^{-7}$. The method of pH calculation is shown for several points and regions on the titration curve.

Equation 9-29 (page 252) to obtain $[H_3O^+]$. At the first equivalence point (point C), we have a solution of an acid salt and use Equation 15-15 or one of its simplifications to compute the hydronium ion concentration. In the region labeled D, we have a second buffer consisting of a weak acid HA^- and its conjugate base Na_2A, and we calculate the pH employing the second dissociation constant, $K_{a2} = 1.00 \times 10^{-7}$. At point E, the solution contains the conjugate base of a weak acid with a dissociation constant of 1.00×10^{-7}. That is, we assume that the hydroxide concentration of the solution is determined solely by the reaction of A^{2-} with water to form HA^- and OH^-. Finally, in the region labeled F, we compute the hydroxide concentration from the molarity of the NaOH and compute the pH from this quantity.

EXAMPLE 15-9

Construct a curve for the titration of 25.00 mL of 0.1000 M maleic acid, HOOC—CH═CH—COOH, with 0.1000 M NaOH.

Symbolizing the acid as H_2M, we can write the two dissociation equilibria as

$$H_2M + H_2O \rightleftharpoons H_3O^+ + HM^- \qquad K_{a1} = 1.3 \times 10^{-2}$$

$$HM^- + H_2O \rightleftharpoons H_3O^+ + M^{2-} \qquad K_{a2} = 5.9 \times 10^{-7}$$

Because the ratio K_{a1}/K_{a2} is large (2×10^4), we proceed as just described.

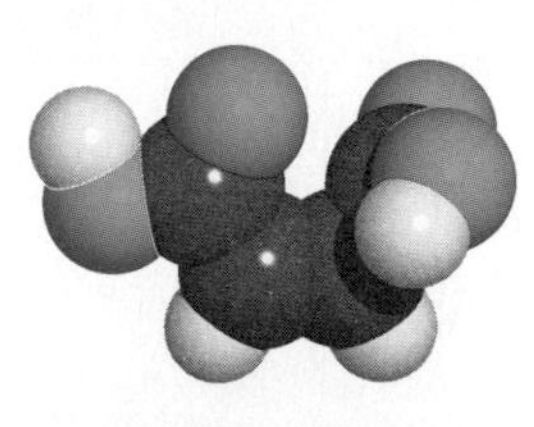

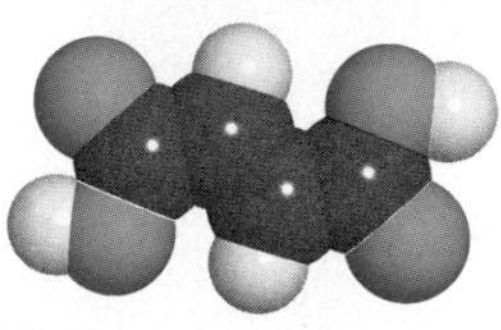

Molecular models of maleic acid, or (Z)-butenedioic acid (top), and fumaric acid, or (E)-butenedioic acid (bottom). These geometric isomers exhibit striking differences in both their physical and their chemical properties. Because the *cis* isomer (maleic acid) has both carboxyl groups on the same side of the molecule, the compound eliminates water to form cyclic maleic anhydride, which is a very reactive precursor widely used in plastics, dyes, pharmaceuticals, and agrichemicals. Fumaric acid, which is essential to animal and vegetable respiration, is used industrially as an antioxidant, to synthesize resins, and to fix colors in dyeing. It is interesting to compare the pK_a values for the two acids; for fumaric acid, $pK_{a1} = 3.05$ and $pK_{a2} = 4.49$; for maleic acid, $pK_{a1} = 1.89$ and $pK_{a2} = 6.23$. Challenge: Explain the differences in the pK_a values based on differences in the molecular structures.

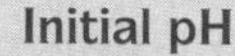

Initial pH

Only the first dissociation makes an appreciable contribution to $[H_3O^+]$; thus,

$$[H_3O^+] \approx [HM^-]$$

Mass balance requires that

$$c_{H_2M} \approx [H_2M] + [HM^-] = 0.1000 \text{ M}$$

or

$$[H_2M] = 0.1000 - [HM^-] = 0.1000 - [H_3O^+]$$

Substituting these relationships into the expression for K_{a1} gives

$$K_{a1} = 1.3 \times 10^{-2} = \frac{[H_3O^+]^2}{0.1000 - [H_3O^+]}$$

Rearranging yields

$$[H_3O^+]^2 + 1.3 \times 10^{-2}[H_3O^+] - 1.3 \times 10^{-3} = 0$$

Because K_{a1} for maleic acid is large, we must solve the quadratic equation exactly or by successive approximations. When we do so, we obtain

$$[H_3O^+] = 3.01 \times 10^{-2} \text{ M}$$

$$\text{pH} = 2 - \log 3.01 = 1.52$$

First Buffer Region

The addition of 5.00 mL of base results in the formation of a buffer consisting of the weak acid H_2M and its conjugate base HM^-. To the extent that dissociation of HM^- to give M^{2-} is negligible, the solution can be treated as a simple buffer system. Thus, applying Equations 9-27 and 9-28 (page 252) gives

$$c_{NaHM} \approx [HM^-] = \frac{5.00 \times 0.1000}{30.00} = 1.67 \times 10^{-2}\text{ M}$$

$$c_{H_2M} \approx [H_2M] = \frac{25.00 \times 0.1000 - 5.00 \times 0.1000}{30.00} = 6.67 \times 10^{-2}\text{ M}$$

Substitution of these values into the equilibrium-constant expression for K_{a1} yields a tentative value of 5.2×10^{-2} M for $[H_3O^+]$. It is clear, however, that the approximation $[H_3O^+] << c_{H_2M}$ or c_{HM^-} is not valid; therefore, Equations 9-25 and 9-26 must be used, and

$$[HM^-] = 1.67 \times 10^{-2} + [H_3O^+] - [OH^-]$$

$$[H_2M] = 6.67 \times 10^{-2} - [H_3O^+] + [OH^-]$$

Because the solution is quite acidic, the approximation that $[OH^-]$ is very small is surely justified. Substitution of these expressions into the dissociation-constant relationship gives

$$\frac{[H_3O^+](1.67 \times 10^{-2} + [H_3O^+])}{6.67 \times 10^{-2} - [H_3O^+]} = 1.3 \times 10^{-2} = K_{a1}$$

$$[H_3O^+]^2 + (2.97 \times 10^{-2})\,[H_3O^+] - 8.67 \times 10^{-4} = 0$$

$$[H_3O^+] = 1.81 \times 10^{-2}\text{ M}$$

$$pH = -\log(1.81 \times 10^{-2}) = 1.74$$

Additional points in the first buffer region can be computed in a similar way.

Just Prior to First Equivalence Point

Just prior to the first equivalence point, the concentration of H_2M is so small that it becomes comparable to the concentration of M^{2-}, and the second equilibrium must also be considered. Within approximately 0.1 mL of the first equivalence point, we have a solution of primarily HM^- with a small amount of H_2M remaining and a small amount of M^{2-} formed. For example, at 24.90 mL of NaOH added,

$$[HM^-] \approx c_{NaHM} = \frac{24.90 \times 0.1000}{49.90} = 4.99 \times 10^{-2}\text{ M}$$

$$c_{H_2M} = \frac{25.00 \times 0.1000}{49.90} - \frac{24.90 \times 0.1000}{49.90} = 2.00 \times 10^{-4}\text{ M}$$

Mass balance gives

$$c_{H_2M} + c_{NaHM} = [H_2M] + [HM^-] + [M^{2-}]$$

(continued)

Charge balance gives

$$[H_3O^+] + [Na^+] = [HM^-] + 2[M^{2-}] + [OH^-]$$

Since the solution consists primarily of the acid HM^- at the first equivalence point, we can neglect $[OH^-]$ in the previous equation and can replace $[Na^+]$ with c_{NaHM}. After rearranging, we obtain

$$c_{NaHM} = [HM^-] + 2[M^{2-}] - [H_3O^+]$$

Substituting this into the mass-balance expression and solving for $[H_3O^+]$ gives

$$[H_3O^+] = c_{H_2M} + [M^{2-}] - [H_2M]$$

If we express $[M^{2-}]$ and $[H_2M]$ in terms of $[HM^-]$ and $[H_3O^+]$, the result is

$$[H_3O^+] = c_{H_2M} + \frac{K_{a2}[HM^-]}{[H_3O^+]} - \frac{[H_3O^+][HM^-]}{K_{a1}}$$

Multiplying through by $[H_3O^+]$ gives, after rearrangement,

$$[H_3O^+]^2\left(1 + \frac{[HM^-]}{K_{a1}}\right) - c_{H_2M}[H_3O^+] - K_{a2}[HM^-] = 0$$

Substituting $[HM^-] \approx 4.99 \times 10^{-2}$, $c_{HM^-} = 2.00 \times 10^{-4}$, and the values for K_{a1} and K_{a2} leads to

$$4.838\,[H_3O^+]^2 - 2.00 \times 10^{-4}\,[H_3O^+] - 2.94 \times 10^{-8}$$

The solution for this equation is

$$[H_3O^+] = 1.014 \times 10^{-4}\text{ M} \qquad \text{or} \qquad \text{pH} = 3.99$$

The same reasoning applies at 24.99 mL of titrant, where we find

$$[H_3O^+] = 8.01 \times 10^{-5}\text{ M}$$
$$\text{pH} = 4.10$$

First Equivalence Point

At the first equivalence point,

$$[HM^-] \approx c_{NaHM} = \frac{25.00 \times 0.1000}{50.00} = 5.00 \times 10^{-2}\text{ M}$$

Our simplification of the numerator in Equation 15-15 is clearly justified. On the other hand, the second term in the denominator is not << 1. So,

$$[H_3O^+] = \sqrt{\frac{K_{a2}c_{HM^-}}{1 + c_{HM^-}/K_{a1}}} = \sqrt{\frac{5.9 \times 10^{-7} \times 5.00 \times 10^{-2}}{1 + (5.00 \times 10^{-2})/(1.3 \times 10^{-2})}}$$
$$= 7.80 \times 10^{-5}\text{ M}$$

$$\text{pH} = -\log(7.80 \times 10^{-5}) = 4.11$$

Just after the First Equivalence Point

Until the second equivalence point, we can obtain the analytical concentration of HM^- and M^{2-} from the titration stoichiometry. At 25.01 mL, the values are calculated as

$$c_{HM^-} = \frac{\text{mmol NaHM formed} - (\text{mmol NaOH added} - \text{mmol NaHM formed})}{\text{total volume of solution}}$$

$$= \frac{25.00 \times 0.1000 - (25.01 - 25.00) \times 0.100}{50.01} = 0.04997$$

$$c_{M^{2-}} = \frac{(\text{mmol NaOH added} - \text{mmol NaHM formed})}{\text{total volume of solution}} = 1.996 \times 10^{-5}$$

In the region of a few tenths of a milliliter beyond the first equivalence point, the solution is primarily HM^- with some M^{2-} formed as a result of the titration. So, the mass balance is

$$c_{Na_2M} + c_{NaHM} = [H_2M] + [HM^-] + [M^{2-}] = 0.04997 + 1.996 \times 10^{-5}$$
$$= 0.049999$$

and the charge balance is

$$[H_3O^+] + [Na^+] = [HM^-] + 2[M^{2-}] + [OH^-]$$

Again, the solution should be acidic, and so we can neglect OH^- as an important species. The Na^+ concentration equals the millimoles of NaOH added divided by the total volume, or

$$[Na^+] = \frac{25.01 \times 0.1000}{50.01} = 0.05001 \text{ M}$$

Subtracting the mass balance from the charge balance and solving for $[H_3O^+]$ gives

$$[H_3O^+] = [M^{2-}] - [H_2M] - (0.05001 - 0.049999)$$

Expressing the $[M^{2-}]$ and $[H_2M]$ in terms of the predominant species HM^- gives

$$[H_3O^+] = \frac{K_{a2}[HM^-]}{[H_3O^+]} - \frac{[H_3O^+][HM^-]}{K_{a1}} - 1.9996 \times 10^{-5}$$

Since $[HM^-] \approx c_{NaHM} = 0.04997$, we can solve for $[H_3O^+]$ as

$$[H_3O^+] =$$
$$\frac{-1.9996 \times 10^{-5} \pm \sqrt{(1.9996 \times 10^{-5})^2 - 4 \times 4.8438 \times (-2.948 \times 10^{-8})}}{2 \times 4.8438}$$
$$= 7.40 \times 10^{-5} \text{ M}$$

$$\text{pH} = 4.13$$

(continued)

Second Buffer Region

Further additions of base to the solution create a new buffer system consisting of HM^- and M^{2-}. When enough base has been added that the reaction of HM^- with water to give OH^- can be neglected (a few tenths of a milliliter beyond the first equivalence point), the pH of the mixture is readily obtained from K_{a2}. When we introduce 25.50 mL of NaOH, for example,

$$[M^{2-}] \approx c_{Na_2M} \approx \frac{(25.50 - 25.00)(0.1000)}{50.50} = \frac{0.050}{50.50}\text{ M}$$

and the molar concentration of NaHM is

$$[HM^-] \approx c_{NaHM} \approx \frac{(25.00 \times 0.1000) - (25.50 - 25.00)(0.1000)}{50.50} = \frac{2.45}{50.50}\text{ M}$$

Substituting these values into the expression for K_{a2} gives

$$\frac{[H_3O^+](0.050/\cancel{50.50})}{2.45/\cancel{50.50}} = 5.9 \times 10^{-7}$$

$$[H_3O^+] = 2.89 \times 10^{-5}\text{ M}$$

The assumption that $[H_3O^+]$ is small relative to c_{HM^-} and $c_{M^{2-}}$ is valid, and pH = 4.54.

Just Prior to Second Equivalence Point

At 49.90 mL and 49.99 mL, the ratio M^{2-}/HM^- becomes large, and the simple buffer equation no longer applies. At 49.90 mL, $c_{HM^-} = 1.335 \times 10^{-4}$ and $c_{M^{2-}} = 0.03324$. The primary equilibrium is now

$$M^{2-} + H_2O \rightleftharpoons HM^- + OH^-$$

We can write the equilibrium constant as

$$K_{b1} = \frac{K_w}{K_{a2}} = \frac{[OH^-][HM^-]}{[M^{2-}]} = \frac{[OH^-](1.335 \times 10^{-4} + [OH^-])}{(0.03324 - [OH^-])}$$

$$= \frac{1.00 \times 10^{-14}}{5.9 \times 10^{-7}} = 1.69 \times 10^{-8}$$

It is easier to solve for $[OH^-]$ than for $[H_3O^+]$. This gives

$$[OH^-]^2 + (1.335 \times 10^{-4} + K_{b1})[OH^-] - 0.03324\,K_{b1} = 0$$

$$[OH^-] = 4.10 \times 10^{-6}\text{ M}$$

$$\text{pOH} = 5.39$$

and

$$\text{pH} = 14 - 5.39 = 8.61$$

This same reasoning is used for 49.99 mL, which leads to $[OH^-] = 1.80 \times 10^{-5}$ M and pH = 9.26.

Second Equivalence Point

After the addition of 50.00 mL of 0.1000 M sodium hydroxide, the solution is 0.0333 M in Na_2M. Reaction of the base M^{2-} with water is the predominant equilibrium in the system and the only one that we need to take into account. Thus,

$$M^{2-} + H_2O \rightleftharpoons OH^- + HM^-$$

$$\frac{[OH^-][HM^-]}{[M^{2-}]} = \frac{K_w}{K_{a2}} = \frac{1.00 \times 10^{-14}}{5.9 \times 10^{-7}} = 1.69 \times 10^{-8} = K_{b1}$$

$$[OH^-] \approx [HM^-]$$

$$[M^{2-}] = 0.0333 - [OH^-] \approx 0.0333 \text{ M}$$

$$\frac{[OH^-]^2}{0.0333} = \frac{1.00 \times 10^{-14}}{5.9 \times 10^{-7}}$$

$$[OH^-] = 2.38 \times 10^{-5} \text{ M}$$

$$\text{pOH} = -\log(2.38 \times 10^{-5}) = 4.62$$

$$\text{pH} = 14.00 - 4.62 = 9.38$$

pH Just Beyond the Second Equivalence Point

In the region just beyond the second equivalence point (50.01 mL, for example), we still need to take into account the reaction of M^{2-} with water. The analytical concentration of M^{2-} is the number of millimoles of M^{2-} produced divided by the total solution volume.

$$c_{M^{2-}} = \frac{25.00 \times 0.1}{75.01} = 0.03333 \text{ M}$$

The $[OH^-]$ now comes from the reaction of M^{2-} with water and from the excess OH^- added as titrant. The excess OH^- is then the number of millimoles of NaOH added minus the number required to reach the second equivalence point divided by the total solution volume. Or,

$$\text{excess OH}^- = \frac{(50.01 - 50.00) \times 0.1}{75.01} = 1.3333 \times 10^{-5} \text{ M}$$

It is now relatively easy to solve for $[HM^-]$ from K_{b1}.

$$[M^{2-}] = c_{M^{2-}} - [HM^-] = 0.0333 - [HM^-]$$

$$[OH^-] = 1.3333 \times 10^{-5} + [HM^-]$$

$$K_{b1} = \frac{[HM^-][OH^-]}{[M^{2-}]} = \frac{[HM^-](1.3333 \times 10^{-5} + [HM^-])}{0.03333 - [HM^-]}$$

(continued)

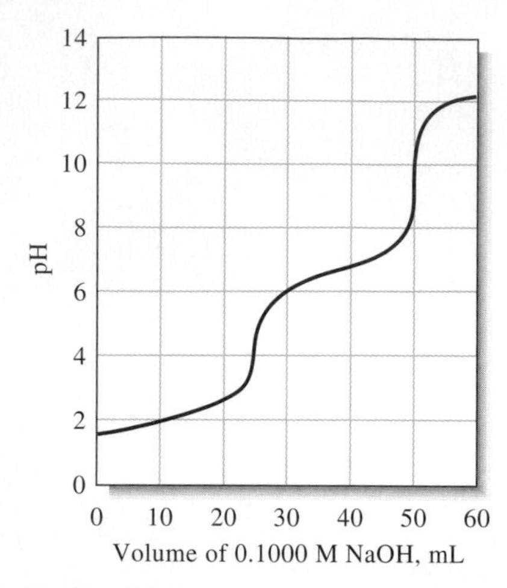

Figure 15-3 Titration curve for 25.00 mL of 0.1000 M maleic acid, H_2M, with 0.1000 M NaOH.

The quadratic formula for $[HM^-]$ is

$$[HM^-]^2 + (1.33 \times 10^{-5} + K_{b1})\,[HM^-] - 0.03333\,K_{b1} = 0$$

$$[HM^-] = 1.807 \times 10^{-5}\ \text{M}$$

$$[OH^-] = 1.333 \times 10^{-5} + 1.807 \times 10^{-5} = 3.14 \times 10^{-5}\ \text{M}$$

$$\text{pOH} = 4.50 \quad \text{and} \quad \text{pH} = 14 - \text{pOH} = 9.50$$

The same reasoning applies to 50.10 mL, where the calculations give

$$\text{pH} = 10.14$$

pH Beyond the Second Equivalence Point

Further additions of sodium hydroxide repress the basic dissociation of M^{2-}. The pH is calculated from the concentration of NaOH added in excess of that required for the complete neutralization of H_2M. Thus, when 51.00 mL of NaOH have been added, we have 1.00 mL excess of 0.1000 M NaOH and

$$[OH^-] = \frac{1.00 \times 0.1000}{76.00} = 1.32 \times 10^{-3}\ \text{M}$$

$$\text{pOH} = -\log(1.32 \times 10^{-3})$$

$$\text{pH} = 14.00 - \text{pOH} = 11.12$$

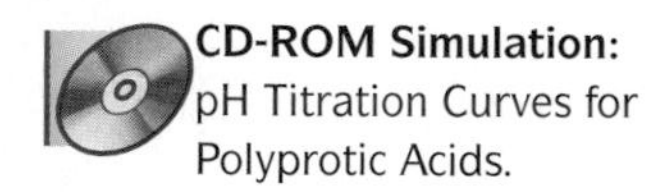

CD-ROM Simulation: pH Titration Curves for Polyprotic Acids.

► In titrating a polyprotic acid or base, two usable end points are obtained if the ratio of dissociation constants is greater than 10^4 and if the weaker acid or base has a dissociation constant greater than 10^{-8}.

Figure 15-3 is the titration curve for 0.1000 M maleic acid constructed as shown in Example 15-9. Two end points are apparent, either of which could in principle be used as a measure of the concentration of the acid. The second end point is clearly more satisfactory, however, inasmuch as the pH change is more pronounced.

Figure 15-4 shows titration curves for three other polyprotic acids. These curves illustrate that a well-defined end point corresponding to the first equivalence point is observed only when the degree of dissociation of the two acids is sufficiently different. The ratio of K_{a1} to K_{a2} for oxalic acid (curve *B*) is approximately 1000. The curve for this titration shows an inflection corresponding to the first equivalence point. The magnitude of the pH change is too small to permit precise location of equivalence with an indicator; however, the second end point provides a means for the accurate determination of oxalic acid.

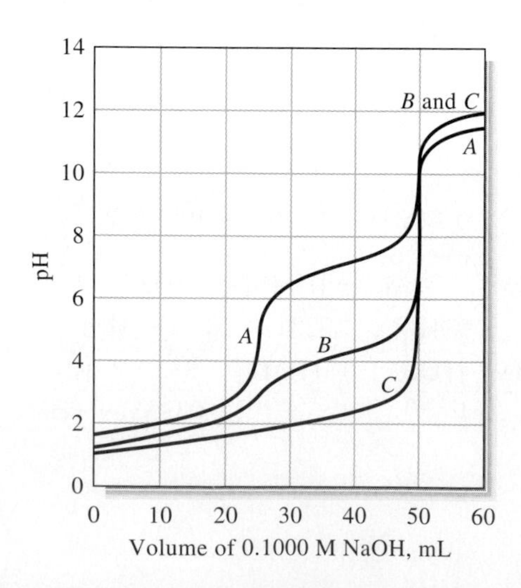

Figure 15-4 Curves for the titration of polyprotic acids. A 0.1000 M NaOH solution is used to titrate 25.00 mL of 0.1000 M H_3PO_4 (curve *A*), 0.1000 M oxalic acid (curve *B*), and 0.1000 M H_2SO_4 (curve *C*).

Curve *A* in Figure 15-4 is the theoretical titration curve for triprotic phosphoric acid. Here, the ratio K_{a1}/K_{a2} is approximately 10^5, as is K_{a2}/K_{a3}. This ratio results in two well-defined end points, either of which is satisfactory for analytical purposes. An acid range indicator will provide a color change when 1 mol of base has been introduced for each mole of acid; a base range indicator will require 2 mol of base per mole of acid. The third hydrogen of phosphoric acid is so slightly dissociated ($K_{a3} = 4.5 \times 10^{-13}$) that no practical end point is associated with its neutralization. The buffering effect of the third dissociation is noticeable, however, and causes the pH for curve *A* to be lower than the pH for the other two curves in the region beyond the second equivalence point.

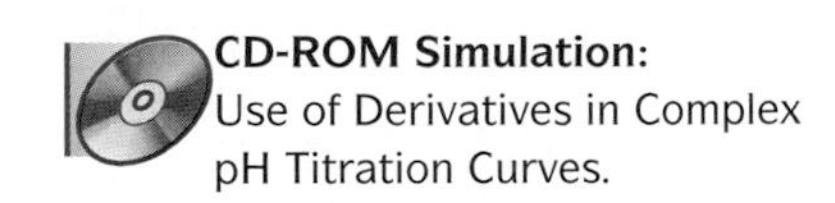
CD-ROM Simulation: Use of Derivatives in Complex pH Titration Curves.

Curve *C* is the titration curve for sulfuric acid, a substance that has one fully dissociated proton and one that is dissociated to a relatively large extent ($K_{a2} = 1.02 \times 10^{-2}$). Because of the similarity in strengths of the two acids, only a single end point, corresponding to the titration of both protons, is observed.

► CHALLENGE: Construct a titration curve for 50.0 mL of 0.0500 M H_2SO_4 with 0.1000 M NaOH.

In general, the titration of acids or bases that have two reactive groups yields individual end points that are of practical value only when the ratio between the two dissociation constants is at least 10^4. If the ratio is much smaller than this, the pH change at the first equivalence point will prove less satisfactory for an analysis.

FEATURE 15-1

The Dissociation of Sulfuric Acid

Sulfuric acid is unusual in that one of its protons behaves as a strong acid and the other as a weak acid ($K_{a2} = 1.02 \times 10^{-2}$). As an example, consider how the hydronium ion concentration of sulfuric acid solutions is computed using a 0.0400 M solution.

We first assume that the dissociation of HSO_4^- is negligible because of the large excess of H_3O^+ resulting from the complete dissociation of H_2SO_4. Therefore,

$$[H_3O^+] \approx [HSO_4^-] \approx 0.0400 \text{ M}$$

An estimate of $[SO_4^{2-}]$ based on this approximation and the expression for K_{a2} reveals that

$$\frac{\cancel{0.0400}[SO_4^{2-}]}{\cancel{0.0400}} = 1.02 \times 10^{-2}$$

Clearly, $[SO_4^{2-}]$ is *not* small relative to $[HSO_4^-]$, and a more rigorous solution is required.

From stoichiometric considerations, it is necessary that

$$[H_3O^+] = 0.0400 + [SO_4^{2-}]$$

The first term on the right is the concentration of H_3O^+ from dissociation of the H_2SO_4 to HSO_4^-. The second term is the contribution of the dissociation of HSO_4^-. Rearrangement yields

$$[SO_4^{2-}] = [H_3O^+] - 0.0400$$

(continued)

Mass-balance considerations require that

$$c_{H_2SO_4} = 0.0400 = [HSO_4^-] + [SO_4^{2-}]$$

Combining the last two equations and rearranging yield

$$[HSO_4^-] = 0.0800 - [H_3O^+]$$

Introduction of these equations for $[SO_4^{2-}]$ and $[HSO_4^-]$ into the expression for K_{a2} yields

$$\frac{[H_3O^+]([H_3O^+] - 0.0400)}{0.0800 - [H_3O^+]} = 1.02 \times 10^{-2}$$

$$[H_3O^+]^2 - (0.0298)[H_3O^+] - 8.16 \times 10^{-4} = 0$$

$$[H_3O^+] = 0.0471 \text{ M}$$

Spreadsheet Summary In Chapter 8 of *Applications of Microsoft® Excel in Analytical Chemistry,* we extend the treatment of neutralization titration curves to polyfunctional acids. Both a stoichiometric approach and a master equation approach are used for the titration of maleic acid with sodium hydroxide.

15F TITRATION CURVES FOR POLYFUNCTIONAL BASES

The construction of a titration curve for a polyfunctional base involves no new principles. To illustrate, consider the titration of a sodium carbonate solution with standard hydrochloric acid. The important equilibrium constants are

$$CO_3^{2-} + H_2O \rightleftharpoons OH^- + HCO_3^- \qquad K_{b1} = \frac{K_w}{K_{a2}} = \frac{1.00 \times 10^{-14}}{4.69 \times 10^{-11}} = 2.13 \times 10^{-4}$$

$$HCO_3^- + H_2O \rightleftharpoons OH^- + CO_2(aq) \qquad K_{b2} = \frac{K_w}{K_{a1}} = \frac{1.00 \times 10^{-14}}{4.2 \times 10^{-7}} = 2.4 \times 10^{-8}$$

The reaction of carbonate ion with water governs the initial pH of the solution, which can be computed by the method shown for the second equivalence point in Example 15-9. With the first additions of acid, a carbonate/hydrogen carbonate buffer is established. In this region, the pH can be calculated from *either* the hydroxide ion concentration calculated from K_{b1} *or* the hydronium ion concentration calculated from K_{a2}. Because we are usually interested in calculating $[H_3O^+]$ and pH, the expression for K_{a2} is easier to use.

► CHALLENGE: Show that either K_{b2} or K_{a1} can be used to calculate the pH of a buffer that is 0.100 M in Na_2CO_3 and 0.100 M in $NaHCO_3$.

Sodium hydrogen carbonate is the principal solute species at the first equivalence point, and Equation 15-16 is used to compute the hydronium ion concentration (see Example 15-8). With the addition of more acid, a new buffer consisting of sodium hydrogen carbonate and carbonic acid is formed. The pH of this buffer is readily obtained from either K_{b2} or K_{a1}.

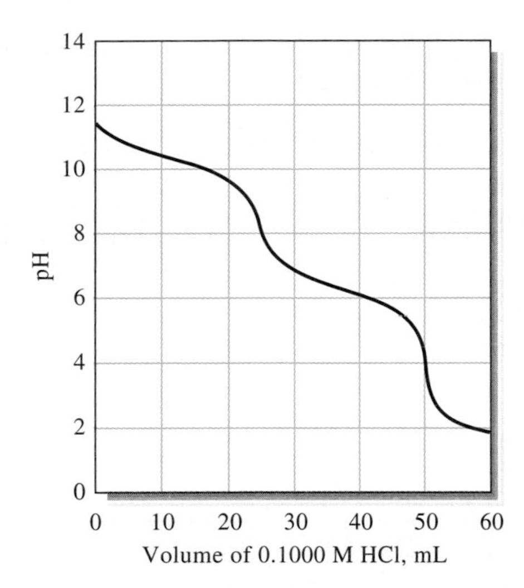

Figure 15-5 Curve for the titration of 25.00 mL of 0.1000 M Na_2CO_3 with 0.1000 M HCl.

At the second equivalence point, the solution consists of carbon dioxide and sodium chloride. The carbon dioxide can be treated as a simple weak acid having a dissociation constant K_{a1}. After excess hydrochloric acid has been introduced, the dissociation of the weak acid is repressed to the point where the hydronium ion concentration is essentially that of the molar concentration of the strong acid.

Figure 15-5 illustrates that two end points are observed in the titration of sodium carbonate, the second being appreciably sharper than the first. It is apparent that the individual components in mixtures of sodium carbonate and sodium hydrogen carbonate can be determined by neutralization methods.

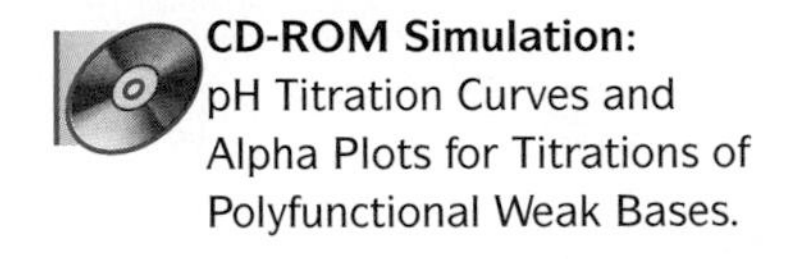

CD-ROM Simulation: pH Titration Curves and Alpha Plots for Titrations of Polyfunctional Weak Bases.

Spreadsheet Summary The titration curve for a difunctional base being titrated with a strong acid is developed in Chapter 8 of *Applications of Microsoft® Excel in Analytical Chemistry.* In the example studied, ethylenediamine is titrated with hydrochloric acid. A master equation approach is explored, and the spreadsheet is used to plot pH versus fraction titrated.

15G TITRATION CURVES FOR AMPHIPROTIC SPECIES

As noted earlier, an amphiprotic substance, when dissolved in a suitable solvent, behaves both as a weak acid and as a weak base. If either its acidic or basic character predominates sufficiently, titration of the species with a strong base or a strong acid may be feasible. For example, in sodium dihydrogen phosphate solution, the principal equilibria are:

$$H_2PO_4^- + H_2O \rightleftharpoons H_3O^+ + HPO_4^{2-} \qquad K_{a2} = 6.32 \times 10^{-8}$$

$$H_2PO_4^- + H_2O \rightleftharpoons OH^- + H_3PO_4 \qquad K_{b3} = \frac{K_w}{K_{a1}} = \frac{1.00 \times 10^{-14}}{7.11 \times 10^{-3}} = 1.41 \times 10^{-12}$$

Note that K_{b3} is much too small to permit titration of $H_2PO_4^-$ with an acid, but K_{a2} is large enough for a successful titration of the ion with a standard base solution.

A different situation prevails in solutions containing disodium hydrogen phosphate, for which the analogous equilibria are:

$$HPO_4^{2-} + H_2O \rightleftharpoons H_3O^+ + PO_4^{3-} \qquad K_{a3} = 4.5 \times 10^{-13}$$

$$HPO_4^{2-} + H_2O \rightleftharpoons OH^- + H_2PO_4^- \qquad K_{b2} = \frac{K_w}{K_{a2}} = \frac{1.00 \times 10^{-14}}{6.32 \times 10^{-8}} = 1.58 \times 10^{-7}$$

The magnitude of the constants indicates that HPO_4^{2-} can be titrated with standard acid but not with standard base.

FEATURE 15-2

Acid/Base Behavior of Amino Acids

▶ Amino acids are amphiprotic.

The simple amino acids are an important class of amphiprotic compounds that contain both a weak acid and a weak base functional group. In an aqueous solution of a typical amino acid, such as glycine, three important equilibria operate:

$$NH_2CH_2COOH \rightleftharpoons NH_3^+CH_2COO^- \qquad (15\text{-}17)$$

$$NH_3^+CH_2COO^- + H_2O \rightleftharpoons NH_2CH_2COO^- + H_3O^+ \qquad K_a = 2 \times 10^{-10} \qquad (15\text{-}18)$$

$$NH_3^+CH_2COO^- + H_2O \rightleftharpoons NH_3^+CH_2COOH + OH^- \qquad K_b = 2 \times 10^{-12} \qquad (15\text{-}19)$$

The first equilibrium constitutes a kind of internal acid/base reaction and is analogous to the reaction one would observe between a carboxylic acid and an amine:

$$R_1NH_2 + R_2COOH \rightleftharpoons R_1NH_3^+ + R_2COO^- \qquad (15\text{-}20)$$

The typical aliphatic amine has a base dissociation constant of 10^{-4} to 10^{-5} (see Appendix 3), while many carboxylic acids have acid dissociation constants of about the same magnitude. The consequence is that both Reaction 15-18 and Reaction 15-19 proceed far to the right, with the product or products being the predominant species in the solution.

A **zwitterion** is an ionic species that has both a positive and a negative charge.

The amino acid species in Equation 15-17, which bears both a positive and a negative charge, is called a **zwitterion.** As shown by Equations 15-18 and 15-19, the zwitterion of glycine is stronger as an acid than as a base. Thus, an aqueous solution of glycine is somewhat acidic.

The **isoelectric point** is the pH at which no net migration of amino acids occurs when they are placed in an electric field.

The zwitterion of an amino acid, containing as it does a positive and a negative charge, has no tendency to migrate in an electric field, whereas the singly charged anionic and cationic species are attracted to electrodes of opposite charge. No *net* migration of the amino acid occurs in an electric field when the pH of the solvent is such that the concentrations of the anionic and cationic forms are identical. The pH at which no net migration occurs is called the **isoelectric point** and is an important physical constant for characterizing amino acids. The isoelectric point is readily related to the ionization constants for the species. Thus, for glycine,

$$K_a = \frac{[H_3O^+][NH_2CH_2COO^-]}{[NH_3^+CH_2COO^-]}$$

$$K_b = \frac{[OH^-][NH_3^+CH_2COOH]}{[NH_3^+CH_2COO^-]}$$

At the isoelectric point,

$$[NH_2CH_2COO^-] = [NH_3^+CH_2COOH]$$

Thus, if we divide K_a by K_b and substitute this relationship, we obtain for the isoelectric point

$$\frac{K_a}{K_b} = \frac{[H_3O^+][NH_2CH_2COO^-]}{[OH^-][NH_3^+CH_2COOH]} = \frac{[H_3O^+]}{[OH^-]}$$

Substitution of $K_w/[H_3O^+]$ for $[OH^-]$ and rearrangement yield

$$[H_3O^+] = \sqrt{\frac{K_aK_w}{K_b}}$$

The isoelectric point for glycine occurs at a pH of 6.0, as shown by the following:

$$[H_3O^+] = \sqrt{\frac{(2 \times 10^{-10})(1 \times 10^{-14})}{2 \times 10^{-12}}} = 1 \times 10^{-6}\text{ M}$$

For simple amino acids, K_a and K_b are generally so small that their determination by direct neutralization is impossible. Addition of formaldehyde removes the amine functional group, however, and leaves the carboxylic acid available for titration with a standard base. For example, with glycine,

$$NH_3^+CH_2COO^- + CH_2O \rightarrow CH_2{=}NCH_2COOH + H_2O$$

The titration curve for the product is that of a typical carboxylic acid.

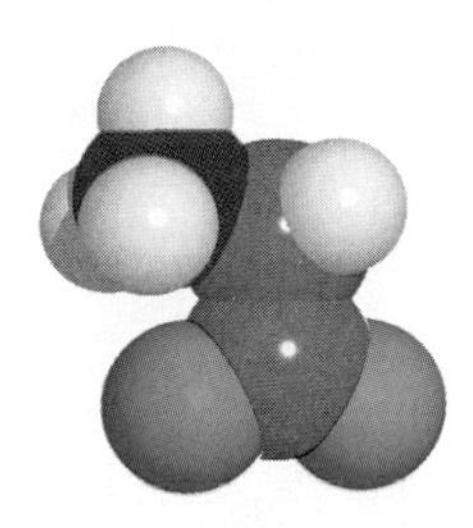

The molecular structure of the glycine zwitterion, $NH_3^+CH_2COO^-$. Glycine is one of the so-called nonessential amino acids; it is nonessential in the sense that it is synthesized in the bodies of mammals and so is not generally essential in the diet. Because of its compact structure, glycine acts as a versatile building block in protein synthesis and in the biosynthesis of hemoglobin. A significant fraction of the collagen—or the fibrous protein constituent of bone, cartilage, tendon, and other connective tissue in the human body—is made up of glycine. Glycine is also an inhibitory *neurotransmitter* and, as a result, has been suggested as a possible therapeutic agent for diseases of the central nervous system such as multiple sclerosis and epilepsy. The calming effects of glycine are currently being investigated to assess its usefulness in the treatment of schizophrenia.

Spreadsheet Summary The final exercise in Chapter 8 of *Applications of Microsoft® Excel in Analytical Chemistry* considers the titration of an amphiprotic species, phenylalanine. A spreadsheet is developed to plot the titration curve of this amino acid, and the isoelectric pH is calculated.

15H THE COMPOSITION OF SOLUTIONS OF A POLYPROTIC ACID AS A FUNCTION OF pH

In Section 14E, we showed how alpha values are useful in visualizing the changes in the concentration of various species that occur in titration of a simple weak acid. Alpha values can also be calculated for polyfunctional acids and bases. For example, if we let c_T be the sum of the molar concentrations of the maleate-containing

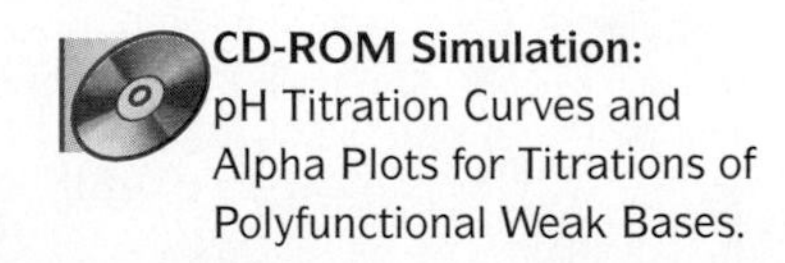

CD-ROM Simulation: pH Titration Curves and Alpha Plots for Titrations of Polyfunctional Weak Bases.

species in the solution throughout the titration described in Example 15-9, the alpha value for the free acid α_0 is defined as

$$\alpha_0 = \frac{[H_2M]}{c_T}$$

where

$$c_T = [H_2M] + [HM^-] + [M^{2-}] \tag{15-21}$$

Alpha values for HM^- and M^{2-} are given by similar equations:

$$\alpha_1 = \frac{[HM^-]}{c_T}$$

$$\alpha_2 = \frac{[M^{2-}]}{c_T}$$

As noted in Section 9C-2, the sum of the alpha values for a system must equal unity:

$$\alpha_0 + \alpha_1 + \alpha_2 = 1$$

Alpha values for the maleic acid system are expressed in terms of $[H_3O^+]$, K_{a1}, and K_{a2}. To obtain such expressions, we follow the method used to derive Equations 9-35 and 9-36 in Section 9C-2, and the following equations result.

$$\alpha_0 = \frac{[H_3O^+]^2}{[H_3O^+]^2 + K_{a1}[H_3O^+] + K_{a1}K_{a2}} \tag{15-22}$$

$$\alpha_1 = \frac{K_{a1}[H_3O^+]}{[H_3O^+]^2 + K_{a1}[H_3O^+] + K_{a1}K_{a2}} \tag{15-23}$$

$$\alpha_2 = \frac{K_{a1}K_{a2}}{[H_3O^+]^2 + K_{a1}[H_3O^+] + K_{a1}K_{a2}} \tag{15-24}$$

► CHALLENGE: Derive Equations 15-22, 15-23, and 15-24.

Note that the denominator is the same for each expression. Note also that the fractional amount of each species is fixed at any pH and is *independent* of the total concentration, c_T.

FEATURE 15-3

A General Expression for Alpha Values

For the weak acid H_nA, the denominator in all alpha value expressions takes the form:

$$[H_3O^+]^n + K_{a1}[H_3O^+]^{(n-1)} + K_{a1}K_{a2}[H_3O^+]^{(n-2)} + \cdots + K_{a1}K_{a2}\cdots K_{an}$$

The numerator for α_0 is the first term in the denominator; for α_1, it is the second term, and so forth. Thus, if we let D be the denominator, $\alpha_0 = [H_3O^+]^n/D$ and $\alpha_1 = K_{a1}[H_3O^+]^{(-1)}/D$.

Alpha values for polyfunctional bases are generated in an analogous way, with the equations being written in terms of base dissociation constants and $[OH^-]$.

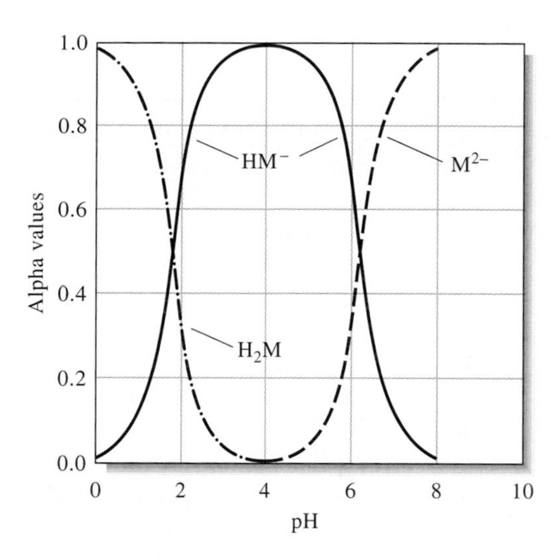

Figure 15-6 Composition of H_2M solutions as a function of pH.

The three curves plotted in Figure 15-6 show the alpha values for each maleate-containing species as a function of pH. The solid curves in Figure 15-7 depict the same alpha values but now plotted as a function of volume of sodium hydroxide as the acid is titrated. The titration curve is also shown by the dashed line in Figure 15-7. Consideration of these curves gives a clear picture of all concentration changes that occur during the titration. For example, Figure 15-7 reveals that before the addition of any base, α_0 for H_2M is roughly 0.7, and α_1 for HM^- is approximately 0.3. For all practical purposes, α_2 is zero. Thus, approximately 70% of the maleic acid exists as H_2M and 30% as HM^-. With addition of base, the pH rises, as does the fraction of HM^-. At the first equivalence point (pH = 4.11), essentially all of the maleate is present as HM^- ($\alpha_1 \rightarrow 1$). Beyond the first equivalence point, HM^- decreases and M^{2-} increases. At the second equivalence point (pH = 9.38) and beyond, essentially all of the maleate exists as M^{2-}.

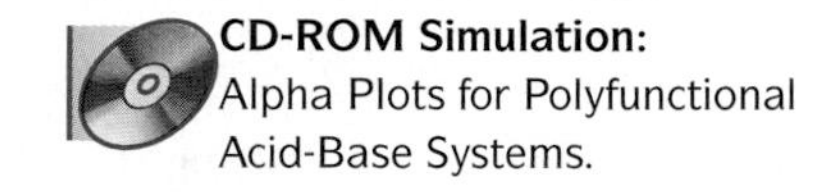

CD-ROM Simulation:
Alpha Plots for Polyfunctional Acid-Base Systems.

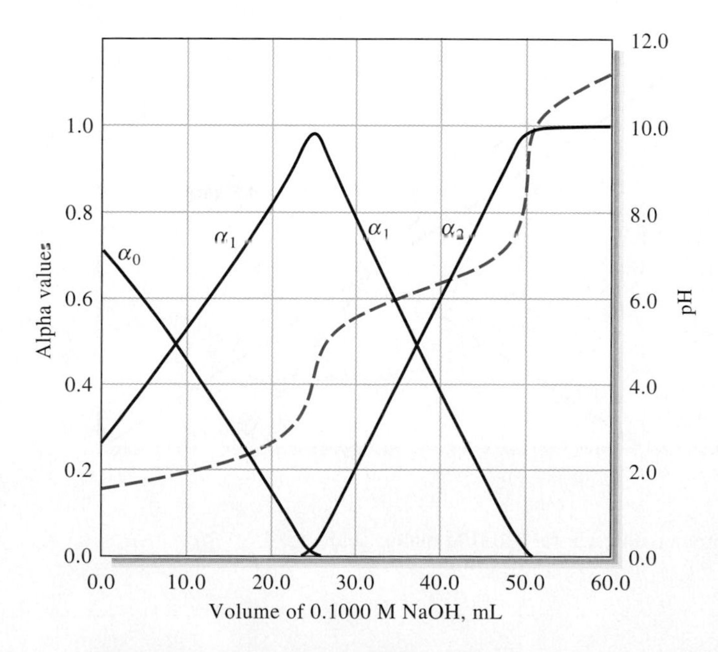

Figure 15-7 Titration of 25.00 mL of 0.1000 M maleic acid with 0.1000 M NaOH. The solid curves are plots of alpha values as a function of volume. The broken curve is a plot of pH as a function of volume.

FEATURE 15-4
Logarithmic Concentration Diagrams

A logarithmic concentration diagram is a plot of log concentration versus a master variable such as pH. Such diagrams are useful because they express the concentrations of all species in a polyprotic acid solution as a function of pH. This allows us to observe at a glance the species that are important at a particular pH. The logarithmic scale is used because the concentrations can vary over many orders of magnitude.

The logarithmic concentration diagram applies only for a specific acid and for a particular initial concentration of acid. Such diagrams can be readily obtained from the distribution diagrams previously discussed. The details of constructing logarithmic concentration diagrams are given in Chapter 8 of *Applications of Microsoft® Excel in Analytical Chemistry.*

Logarithmic concentration diagrams can be obtained from the concentration of acid and the dissociation constants. We use as an example the maleic acid system discussed previously. The diagram shown in Figure 15F-1 is a logarithmic concentration diagram for a maleic acid concentration of 0.10 M (c_T = 0.10 M maleic acid). The diagram expresses the concentrations of all forms of maleic acid, H_2M, HM^-, and M^{2-}, as a function of pH. We usually include the H_3O^+ and OH^- concentrations as well. The diagram is based on the mass-balance condition and the acid dissociation constants. The changes in slope in the diagram for the maleic acid species occur near what are termed **system points.** These are defined by the total acid concentration, 0.10 M in our case, and the pK_a values. For maleic acid, the first system point occurs at log c_T = -1 and pH = pK_{a1} = $-\log (1.30 \times 10^{-2})$ = 1.89, while the second system point is at pH = pK_{a2} = $-\log (5.90 \times 10^{-7})$ = 6.23 and log c_T = -1. Note that when pH = pK_{a1}, the concentrations of H_2M and HM^- are equal, as shown by the crossing of the lines indicating these concentrations. Also, note that at this first system point, $[M^{2-}] << [HM^-]$ and $[M^{2-}] << [H_2M]$. Near this first system point, we can thus neglect the unprotonated maleate ion and express the mass balance as $c_T \approx [H_2M] + [HM^-]$.

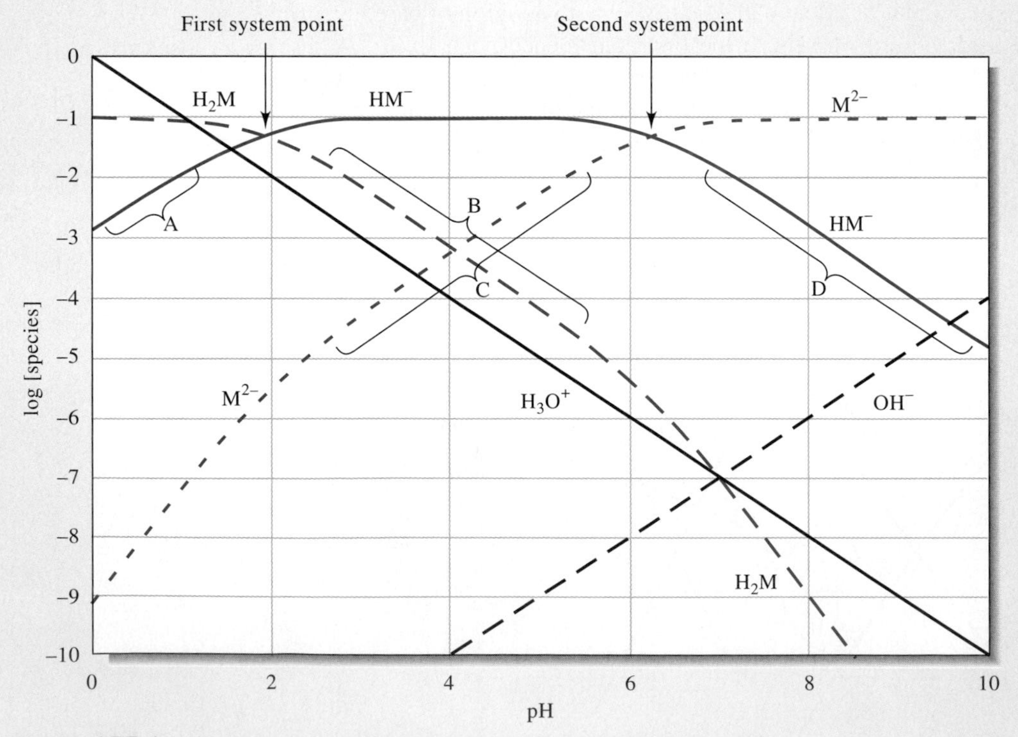

Figure 15F-1 Logarithmic concentration diagram for 0.100 M maleic acid.

To the left of this first system point, $[H_2M] >> [HM^-]$ and so $c_T \approx [H_2M]$. This is indicated on the diagram by the slope of 0 for the H_2M line between pH values of 0 to about 1. In this same region, the HM concentration is steadily increasing with increasing pH since protons are removed from H_2M as the pH increases. From the K_{a1} expression, we can write

$$[HM^-] = \frac{[H_2M]K_{a1}}{[H_3O^+]} \approx \frac{c_T K_{a1}}{[H_3O^+]}$$

Taking the logarithms of both sides of this equation gives

$$\log [HM^-] = \log c_T + \log K_{a1} - \log [H_3O^+]$$
$$= \log c_T + \log K_{a1} + \text{pH}$$

Hence, to the left of the first system point (region A), a plot of $\log [HM^-]$ versus pH is a straight line of slope $+1$.

Using similar reasoning, we conclude that to the right of the first system point, $c_T \approx [HM^-]$, and

$$[H_2M] \approx \frac{c_T[H_3O^+]}{K_{a1}}$$

Taking the logarithms of both sides of this equation reveals that a plot of $\log [H_2M]$ versus pH (region B) should be linear with a slope of -1. This holds until we get near the second system point, which occurs at pH $= pK_{a2} = -\log (5.90 \times 10^{-7}) = 6.23$ and $\log c_T = -1$.

At the second system point, the HM^- and M^{2-} concentrations are equal. Note that to the left of the second system point, $[HM^-] \approx c_T$ and $\log [M^{2-}]$ increases with increasing pH with a slope of $+1$ (region C). To the right of the second system point, $[M^{2-}] \approx c_T$ and $\log [HM^-]$ decreases with increasing pH with a slope of -1 (region D). The H_3O^+ lines and the OH^- lines are easy to draw, since

$$\log [H_3O^+] = -\text{pH} \quad \text{and} \quad \log [OH^-] = \text{pH} - 14$$

We can draw a logarithmic concentration diagram easily by noting the relationships just given. An easier method is to modify the distribution diagram so that it produces the logarithmic concentration diagram. This is the method illustrated in *Applications of Microsoft® Excel in Analytical Chemistry,* Chapter 8. Note that the plot is specific for a total analytical concentration of 0.10 M and for maleic acid, since the acid dissociation constants are included.

Estimating Concentrations at a Given pH Value

The log concentration diagram can be very useful in making more exact calculations and in determining which species are important at a given pH. For example, if we are interested in calculating concentrations at pH 5.7, we can use the diagram in Figure 15F-1 to tell us which species to include in the calculation. At pH 5.7, the concentrations of the maleate-containing species are $[H_2M] \approx 10^{-5}$ M, $[HM^-] \approx 0.07$ M, and $[M^{2-}] \approx 0.02$ M. Hence, the only maleate species of importance at this pH are HM^- and M^{2-}. Since $[OH^-]$ is four orders of magnitude lower than $[H_3O^+]$, we could carry out a more accurate calculation than the previous estimates by considering only three species. If we do so, we find the following concentrations: $[H_2M] \approx 1.18 \times 10^{-5}$ M, $[HM^-] \approx 0.077$ M, and $[M^{2-}] \approx 0.023$ M.

Finding pH Values

If we do not know the pH, the logarithmic concentration diagram can also be used to give an approximate pH value. For example, find the pH of a 0.1 M maleic acid solution. Since the log concentration diagram expresses mass balance and the equilibrium constants, we need only one additional equation such as charge balance to solve the problem exactly. The charge-balance equation for this system is

$$[H_3O^+] = [HM^-] + 2[M^{2-}] + [OH^-]$$

The pH is found by graphically superimposing this equation on the log concentration diagram, as described later. Beginning with a pH of 0, move from left to right along the H_3O^+ line until it intersects a line representing one of the species on the right-hand side of the charge-balance equation. We see that the H_3O^+ line first intersects the HM^- line at a pH of approximately 1.5. At this point, $[H_3O^+] = [HM^-]$. We also see that the concentrations of the other negatively charged species M^{2-} and OH^- are negligible compared with the HM^- concentration. Hence, the pH of a 0.1 M solution of maleic acid is approximately 1.5. A more accurate calculation using the quadratic formula gives pH = 1.52.

We can ask another question: What is the pH of a 0.100 M solution of NaHM? In this case, the charge-balance equation is

$$[H_3O^+] + [Na^+] = [HM^-] + 2[M^{2-}] + [OH^-]$$

(continued)

The Na^+ concentration is the total concentration of maleate-containing species:

$$[Na^+] = c_T = [H_2M] + [HM^-] + [M^{2-}]$$

Substituting this latter equation into the charge-balance equation gives

$$[H_3O^+] + [H_2M] = [M^{2-}] + [OH^-]$$

Now we superimpose this equation on the log concentration diagram. If we again begin on the left at pH 0 and move along either the H_3O^+ line or the H_2M line, we see that at pH values greater than about 2, the concentration of H_2M exceeds the H_3O^+ concentration by about an order of magnitude. Hence, we move along the H_2M line until it intersects either the M^{2-} line or the OH^- line. We see that it intersects the M^{2-} line first at pH $\approx$ 4.1. Thus, $[H_2M] \approx [M^{2-}]$, and the concentrations of the $[H_3O^+]$ and $[OH^-]$ are relatively small compared with H_2M and M^{2-}. We conclude that the pH of a 0.100 M NaHM solution is approximately 4.1. A more exact calculation using the quadratic formula reveals that the pH of this solution is 4.08.

Finally, find the pH of a 0.100 M solution of Na_2M. The charge-balance equation is the same as before:

$$[H_3O^+] + [Na^+] = [HM^-] + 2[M^{2-}] + [OH^-]$$

Now, however, the Na^+ concentration is given by

$$[Na^+] = 2c_T = 2[H_2M] + 2[HM^-] + 2[M^{2-}]$$

Substituting this into the charge-balance equation gives

$$[H_3O^+] + 2[H_2M] + [HM^-] = [OH^-]$$

In this case, it is easier to find the OH^- concentration. Again, we move down the OH^- line, now from right to left, until it intersects the HM^- line at a pH of approximately 9.7. Since $[H_3O^+]$ and $[H_2M]$ are negligibly small at this intersection, $[HM^-] \approx [OH^-]$, we conclude that 9.7 is the approximate pH of a 0.100 M solution of Na_2M. A more exact calculation using the quadratic formula gives the pH as 9.61.

Spreadsheet Summary In the first exercise in Chapter 8 of *Applications of Microsoft® Excel in Analytical Chemistry,* we investigate the calculation of distribution diagrams for polyfunctional acids and bases. The alpha values are plotted as a function of pH. The plots are used to find concentrations at a given pH and to infer which species can be neglected in more extensive calculations. A logarithmic concentration diagram is constructed. These plots are used to estimate concentrations at a given pH and to find the pH for various starting conditions with a weak acid system.

WEB WORKS

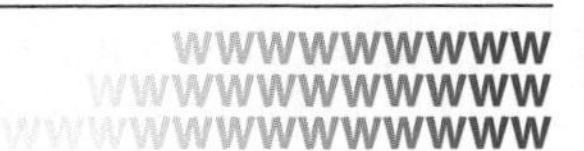

Use your Web browser to connect to **http://chemistry.brookscole.com/skoogfac/.** From the Chapter Resources menu, choose Web Works. Locate the Chapter 15 section, and click on the link to the Virtual Titrator. Click on the indicated frame to invoke the Virtual Titrator Java applet and display two windows: the Menu Panel and the Virtual Titrator main window. To begin, click on Acids on the main window menu bar, and select the diprotic acid *i*-phthalic acid. Examine the titration curve that results. Then click on Graphs/Alpha Plot vs. pH and observe the result. Click on Graphs/Alpha Plot vs. mL base. Repeat the process for several monoprotic and polyprotic acids, and note the results.

QUESTIONS AND PROBLEMS

*15-1. Why is it impossible to titrate all three protons of phosphoric acid in aqueous solution?

15-2. Indicate whether an aqueous solution of the following compounds is acidic, neutral, or basic. Explain your answer.
*(a) NH_4OAc
(b) KNO_2
*(c) KNO_3
(d) KHC_2O_4
*(e) $K_2C_2O_4$
(f) K_2HAsO_4
*(g) KH_2AsO_4
(h) K_3AsO_4

15-3. Suggest an indicator that could be used to provide an end point for the titration of the first proton in H_3PO_4.

*15-4. Suggest an indicator that would give an end point for the titration of the first two protons in H_3PO_4.

15-5. Suggest a method for the determination of the amounts of H_3PO_4 and NaH_2PO_4 in an aqueous solution.

15-6. Suggest a suitable indicator for a titration based on the following reactions; use 0.05 M if an equivalence point concentration is needed.
*(a) $H_3AsO_4 + NaOH \rightarrow NaH_2AsO_4 + H_2O$
(b) $H_2P + 2NaOH \rightarrow Na_2P + 2H_2O$
(H_2P = *o*-phthalic acid)
*(c) $H_2T + 2NaOH \rightarrow Na_2T + 2H_2O$
(H_2T = tartaric acid)
(d) $NH_2C_2H_4NH_2 + HCl \rightarrow NH_2C_2H_4NH_3Cl$
*(e) $NH_2C_2H_4NH_2 + 2HCl \rightarrow ClNH_3C_2H_4NH_3Cl$
(f) $H_2SO_3 + NaOH \rightarrow NaHSO_3 + H_2O$
*(g) $H_2SO_3 + 2NaOH \rightarrow Na_2SO_3 + 2H_2O$

15-7. Calculate the pH of a solution that is 0.0400 M in
*(a) H_3AsO_3.
(b) $C_6H_4(COOH)_2$.
*(c) H_3PO_3.
(d) H_2SO_3.
*(e) H_2S.
(f) $H_2NC_2H_4NH_2$.

15-8. Calculate the pH of a solution that is 0.0400 M in
*(a) NaH_2AsO_4.
(b) $NaHC_2O_4$.
*(c) NaH_2PO_3.
(d) $NaHSO_3$.
*(e) NaHS.
(f) $H_2NC_2H_4NH_3^+Cl^-$.

15-9. Calculate the pH of a solution that is 0.0400 M in
*(a) Na_3AsO_4.
(b) $Na_2C_2O_4$.
*(c) Na_2HPO_3.
(d) Na_2SO_3.
*(e) Na_2S.
(f) $C_2H_4(NH_3^+Cl^-)_2$.

*15-10. Calculate the pH of a solution that is made up to contain the following analytical concentrations:
(a) 0.0500 M in H_3PO_4 and 0.0200 M in NaH_2PO_4.
(b) 0.0300 M in NaH_2AsO_4 and 0.0500 M in Na_2HAsO_4.
(c) 0.0600 M in Na_2CO_3 and 0.0300 M in $NaHCO_3$.
(d) 0.0400 M in H_3PO_4 and 0.0200 M in Na_2HPO_4.
(e) 0.0500 M in $NaHSO_4$ and 0.0400 M in Na_2SO_4.

15-11. Calculate the pH of a solution made up to contain the following analytical concentrations:
(a) 0.240 M in H_3PO_4 and 0.480 M in NaH_2PO_4.
(b) 0.0670 M in Na_2SO_3 and 0.0315 M in $NaHSO_3$.
(c) 0.640 M in $HOC_2H_4NH_2$ and 0.750 M in $HOC_2H_4NH_3Cl$.
(d) 0.0240 in $H_2C_2O_4$ (oxalic acid) and 0.0360 M in $Na_2C_2O_4$.
(e) 0.0100 M in $Na_2C_2O_4$ and 0.0400 M in $NaHC_2O_4$.

*15-12. Calculate the pH of a solution that is
(a) 0.0100 M in HCl and 0.0200 M in picric acid.
(b) 0.0100 M in HCl and 0.0200 M in benzoic acid.
(c) 0.0100 M in NaOH and 0.100 M in Na_2CO_3.
(d) 0.0100 M in NaOH and 0.100 M in NH_3.

15-13. Calculate the pH of a solution that is
(a) 0.0100 M in $HClO_4$ and 0.0300 M in monochloroacetic acid.
(b) 0.0100 M in HCl and 0.0150 M in H_2SO_4.
(c) 0.0100 M in NaOH and 0.0300 M in Na_2S.
(d) 0.0100 M in NaOH and 0.0300 M in sodium acetate.

*15-14. Identify the principal conjugate acid/base pair and calculate the ratio between them in a solution that is buffered to pH 6.00 and contains
(a) H_2SO_3.
(b) citric acid.
(c) malonic acid.
(d) tartaric acid.

15-15. Identify the principal conjugate acid/base pair and calculate the ratio between them in a solution that is buffered to pH 9.00 and contains

(a) H_2S.
(b) ethylenediamine dihydrochloride.
(c) H_3AsO_4.
(d) H_2CO_3.

***15-16.** How many grams of $Na_2HPO_4 \cdot 2H_2O$ must be added to 400 mL of 0.200 M H_3PO_4 to give a buffer of pH 7.30?

15-17. How many grams of dipotassium phthalate must be added to 750 mL of 0.0500 M phthalic acid to give a buffer of pH 5.75?

***15-18.** What is the pH of the buffer formed by mixing 50.0 mL of 0.200 M NaH_2PO_4 with
(a) 50.0 mL of 0.120 M HCl?
(b) 50.0 mL of 0.120 M NaOH?

15-19. What is the pH of the buffer formed by adding 100 mL of 0.150 M potassium hydrogen phthalate to
(a) 100 mL of 0.0800 M NaOH?
(b) 100 mL of 0.0800 M HCl?

***15-20.** How would you prepare 1.00 L of a buffer with a pH of 9.60 from 0.300 M Na_2CO_3 and 0.200 M HCl?

15-21. How would you prepare 1.00 L of a buffer with a pH of 7.00 from 0.200 M H_3PO_4 and 0.160 M NaOH?

***15-22.** How would you prepare 1.00 L of a buffer with a pH of 6.00 from 0.500 M Na_3AsO_4 and 0.400 M HCl?

15-23. Identify by letter the curve you would expect in the titration of a solution containing
(a) disodium maleate, Na_2M, with standard acid.
(b) pyruvic acid, HP, with standard base.
(c) sodium carbonate, Na_2CO_3, with standard acid.

15-24. Describe the composition of a solution that would be expected to yield a curve resembling (see Problem 15-23):
(a) curve *B*.
(b) curve *A*.
(c) curve *E*.

***15-25.** Briefly explain why curve *B cannot* describe the titration of a mixture consisting of H_3PO_4 and NaH_2PO_4.

15-26. Construct a curve for the titration of 50.00 mL of a 0.1000 M solution of compound A with a 0.2000 M solution of compound B in the following list. For each titration, calculate the pH after the addition of 0.00, 12.50, 20.00, 24.00, 25.00, 26.00, 37.50, 45.00, 49.00, 50.00, 51.00, and 60.00 mL of compound B:

	A	B
(a)	H_2SO_3	NaOH
(b)	ethylenediamine	HCl
(c)	H_2SO_4	NaOH

***15-27.** Generate a curve for the titration of 50.00 mL of a solution in which the analytical concentration of NaOH is 0.1000 M and that for hydrazine is 0.0800 M. Calculate the pH after addition of 0.00, 10.00, 20.00, 24.00, 25.00, 26.00, 35.00, 44.00, 45.00, 46.00, and 50.00 mL of 0.2000 M $HClO_4$.

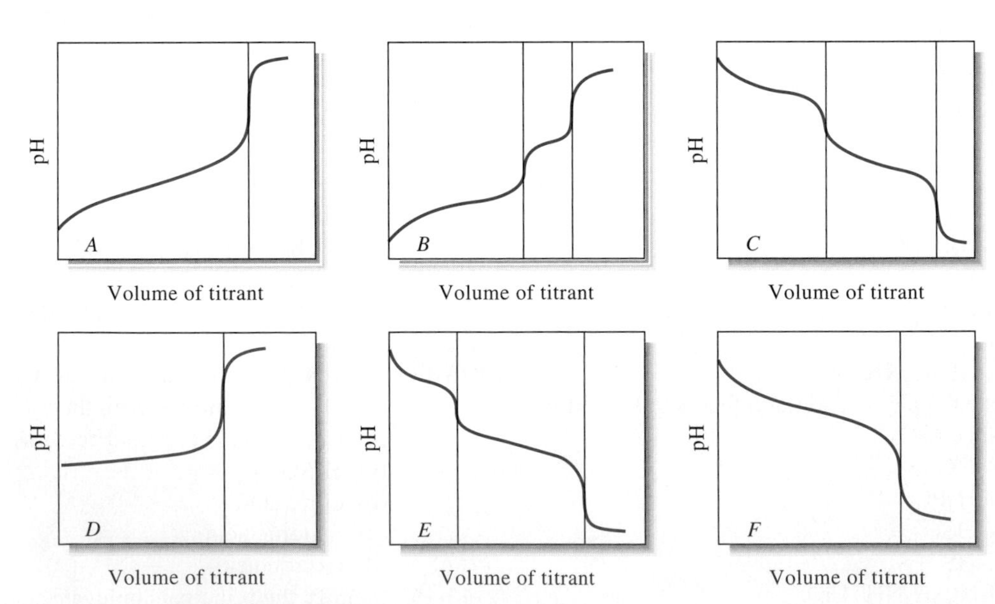

Titration curves for Problem 15–23.

15-28. Generate a curve for the titration of 50.00 mL of a solution in which the analytical concentration of $HClO_4$ is 0.1000 M and that for formic acid is 0.0800 M. Calculate the pH after addition of 0.00, 10.00, 20.00, 24.00, 25.00, 26.00, 35.00, 44.00, 45.00, 46.00, and 50.00 mL of 0.2000 M KOH.

15-29. Formulate equilibrium constants for the following equilibria, giving numerical values for the constants:

*(a) $H_2AsO_4^- + H_2AsO_4^- \rightleftharpoons H_3AsO_4 + HAsO_4^{2-}$

(b) $HAsO_4^{2-} + HAsO_4^{2-} \rightleftharpoons AsO_4^{3-} + H_2AsO_4^-$

***15-30.** Compute the numerical value for the equilibrium constant for the reaction:

$$NH_4^+ + OAc^- \rightleftharpoons NH_3 + HOAc$$

15-31. For pH values of 2.00, 6.00, and 10.00, calculate the alpha value for each species in an aqueous solution of

*(a) phthalic acid.
(b) phosphoric acid.
*(c) citric acid.
(d) arsenic acid.
*(e) phosphorous acid.
(f) oxalic acid.

15-32. Derive equations that define α_0, α_1, α_2, and α_3 for the acid H_3AsO_4.

15-33. Challenge Problem. Malonic acid

$$HOOC—CH_2—COOH, (H_2Ml)$$

is a diprotic acid that undergoes the following acid dissociation reactions:

$$H_2Ml \rightleftharpoons HMl^- + H^+ \qquad pK_{a1} = 2.86$$
$$HMl^- \rightleftharpoons Ml^{2-} + H^+ \qquad pK_{a2} = 5.70$$

(a) Construct a logarithmic concentration diagram for a total malonic acid concentration, c_T, of 0.050 M.
(b) From the logarithmic concentration diagram, determine the approximate concentrations of all species at pH values of 2.00, 3.60, 4.80, and 6.10.
(c) Determine the pH of a solution containing 0.050 M sodium malonate, Na_2Ml.
(d) Find the pH of a solution containing 0.050 M sodium hydrogen malonate, NaHMl.
(e) Discuss how you might modify the logarithmic concentration diagram so that it shows the pH in terms of the hydrogen ion activity a_{H^+} instead of the hydrogen ion concentration (pH = $-\log a_{H^+}$ instead of pH = $-\log c_{H^+}$). Be specific in your discussion and show what the difficulties might be.

InfoTrac College Edition

For additional readings, go to InfoTrac College Edition, your online research library, at

http://infotrac.thomsonlearning.com

CHAPTER 16

Applications of Neutralization Titrations

© Wally McNamee/Corbis

Acids and bases are very important in the environment, in our bodies, and in many other systems. In the environment, acid rain falling on the surface waters of lakes and rivers can cause these waters to become acidic. In the eastern United States, the number of acidic lakes increased from 1930 to 1970 as a result of acid rain. Many lakes in the Midwest have no problem with acidification, however, even though the industrial Midwest is presumed to be a major source of the acids found in acid rain. In the Midwest, the surface rock is mostly limestone (calcium carbonate), which reacts with CO_2 and H_2O to form bicarbonate. Bicarbonate in turn neutralizes acids to maintain the pH relatively constant. This effect is characterized by the **acid neutralizing capacity** of the lake, which is usually quite large in limestone-rich areas. In contrast, many eastern lakes and streams are surrounded by granite, which is a much less reactive rock. These bodies of water have little neutralizing capacity and so are more susceptible to acidification. To combat this problem, limestone is often imported from limestone-rich states into the eastern states and applied to lakes and streams. The photo shows workers dumping pulverized limestone into Cedar Creek, Shenandoah County, Virginia, to neutralize acidic waters that had previously killed many stocked rainbow trout. Acid neutralizing capacity is frequently determined by titration with a standard solution of acid.

Neutralization titrations are widely used to determine the concentration of analytes that are themselves acids or bases or are convertible to such species by suitable treatment.[1] *Water is the usual solvent for neutralization titrations because it is readily available, inexpensive, and nontoxic. Its low temperature coefficient of expansion is an added virtue. Some analytes, however, are not titratable in aqueous media because their solubilities are too low or because their strengths as acids or bases are not sufficiently great to provide satisfactory end points. Such substances can often be titrated in a solvent other than water.*[2] *We shall restrict our discussions to aqueous systems.*

► Nonaqueous solvents, such as methyl and ethyl alcohol, glacial acetic acid, and methyl isobutyl ketone, often make it possible to titrate acids or bases that are too weak to titrate in aqueous solution.

[1]For reviews of applications of neutralization titrations, see J. A. Dean, *Analytical Chemistry Handbook,* Section 3.2, p. 3.28. New York: McGraw-Hill, 1995; D. Rosenthal and P. Zuman, in *Treatise on Analytical Chemistry,* 2nd ed., I. M. Kolthoff and P. J. Elving, Eds., Part I, Vol. 2, Chapter 18. New York: Wiley, 1979.

[2]For reviews of nonaqueous acid/base titrimetry, see J. A. Dean, *Analytical Chemistry Handbook,* Section 3.3, p. 3.48. New York: McGraw-Hill, 1995; *Treatise on Analytical Chemistry,* 2nd ed., I. M. Kolthoff and P. J. Elving, Eds., Part I, Vol. 2, Chapters 19A–19E. New York: Wiley, 1979.

16A REAGENTS FOR NEUTRALIZATION TITRATIONS

In Chapter 14, we noted that strong acids and strong bases cause the most pronounced change in pH at the equivalence point. For this reason, standard solutions for neutralization titrations are always prepared from these reagents.

16A-1 Preparation of Standard Acid Solutions

Hydrochloric acid is widely used for titration of bases. Dilute solutions of the reagent are stable indefinitely and do not cause troublesome precipitation reactions with most cations. It is reported that 0.1 M solutions of HCl can be boiled for as long as an hour without loss of acid, provided that the water lost by evaporation is periodically replaced; 0.5 M solutions can be boiled for at least 10 minutes without significant loss.

Solutions of perchloric and sulfuric acid are also stable and are useful for titrations where chloride ion interferes by forming precipitates. Standard solutions of nitric acid are seldom encountered because of their oxidizing properties.

◀ Solutions of HCl, $HClO_4$, and H_2SO_4 are stable indefinitely. Restandardization is not required unless evaporation occurs.

Standard acid solutions are ordinarily prepared by diluting an approximate volume of the concentrated reagent and subsequently standardizing the diluted solution against a primary-standard base. Less frequently, the composition of the concentrated acid is established through careful density measurement; a weighed quantity is then diluted to a known volume. (Tables relating reagent density to composition are found in most chemistry and chemical engineering handbooks.) A stock solution with an exactly known hydrochloric acid concentration can also be prepared by dilution of a quantity of the concentrated reagent with an equal volume of water followed by distillation. Under controlled conditions, the final quarter of the distillate, which is known as *constant-boiling* HCl, has a fixed and known composition, its acid content being dependent only on atmospheric pressure. For a pressure P between 670 and 780 torr, the mass in air of the distillate that contains exactly one mole of H_3O^+ is[3]

$$\frac{\text{mass of constant-boiling HCl in g}}{\text{mol } H_3O^+} = 164.673 + 0.02039\,P \qquad (16\text{-}1)$$

Standard solutions are prepared by diluting weighed quantities of this acid to accurately known volumes.

16A-2 The Standardization of Acids

Sodium Carbonate

Acids are frequently standardized against weighed quantities of sodium carbonate. Primary-standard grade sodium carbonate is available commercially or can be prepared by heating purified sodium hydrogen carbonate to 270°C to 300°C for 1 hour:

$$2NaHCO_3(s) \rightarrow Na_2CO_3(s) + H_2O(g) + CO_2(g)$$

◀ Sodium carbonate occurs naturally in large deposits as *washing soda*, $Na_2CO_3 \cdot 10H_2O$, and as *trona*, $Na_2CO_3 \cdot NaHCO_3 \cdot 2H_2O$. These minerals find wide use in the glass industry as well as many others. Primary-standard sodium carbonate is manufactured by extensive purification of these minerals.

[3] *Official Methods of Analysis of the AOAC,* 15th ed., p. 692. Washington, D.C.: Association of Official Analytical Chemists, 1990.

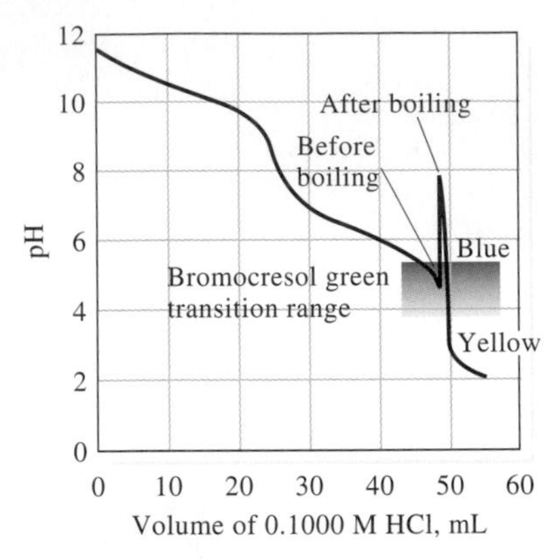

Figure 16-1 Titration of 25.00 mL of 0.1000 M Na_2CO_3 with 0.1000 M HCl. After about 49 mL of HCl have been added, the solution is boiled, causing the increase in pH shown. The change in pH when more HCl is added is much larger.

As shown in Figure 15-5, two end points are observed in the titration of sodium carbonate. The first, corresponding to the conversion of carbonate to hydrogen carbonate, occurs at about pH 8.3; the second, involving the formation of carbon dioxide, is observed at about pH 3.8. The second end point is always used for standardization because the change in pH is greater than that of the first. An even sharper end point can be achieved by boiling the solution briefly to eliminate the reaction product, carbonic acid and carbon dioxide. The sample is titrated to the first appearance of the acid color of the indicator (such as bromocresol green or methyl orange). At this point, the solution contains a large amount of dissolved carbon dioxide and small amounts of carbonic acid and unreacted hydrogen carbonate. Boiling effectively destroys this buffer by eliminating the carbonic acid:

$$H_2CO_3(aq) \rightarrow CO_2(g) + H_2O(l)$$

The solution then becomes alkaline again due to the residual hydrogen carbonate ion. The titration is completed after the solution has cooled. Now, however, a substantially larger decrease in pH occurs during the final additions of acid, thus giving a more abrupt color change (Figure 16-1).

As an alternative, the acid can be introduced in an amount slightly in excess of that needed to convert the sodium carbonate to carbonic acid. The solution is boiled as before to remove carbon dioxide and cooled; the excess acid is then back-titrated with a dilute solution of base. Any indicator suitable for a strong acid/strong base titration is satisfactory. The volume ratio of acid to base must of course be established by an independent titration.

Other Primary Standards for Acids

► A high mass per proton consumed is desirable in a primary standard because a larger mass of reagent must be used, thus decreasing the relative weighing error.

Tris-(hydroxymethyl)aminomethane, $(HOCH_2)_3CNH_2$, known also as TRIS or THAM, is available in primary-standard purity from commercial sources. It possesses the advantage of a substantially greater mass per mole of protons consumed (121.1) than sodium carbonate (53.0) (see Example 16-1). The reaction of TRIS with acids is

$$(HOCH_2)_3CNH_2 + H_3O^+ \rightleftharpoons (HOCH_2)_3CNH_3^+ + H_2O$$

► *Borax*, $Na_2B_4O_7 \cdot 10H_2O$, is a mineral that is mined in the desert and is widely used in cleaning preparations. A highly purified form of borax is used as a primary standard for bases.

Sodium tetraborate decahydrate and mercury(II) oxide have also been recommended as primary standards. The reaction of an acid with the tetraborate is

$$B_4O_7^{2-} + 2H_3O^+ + 3H_2O \rightarrow 4H_3BO_3$$

EXAMPLE 16-1

Use a spreadsheet to compare the masses of (a) TRIS (121 g/mol), (b) Na_2CO_3 (106 g/mol), and (c) $Na_2B_4O_7 \cdot 10H_2O$ (381 g/mol) that should be taken to standardize an approximately 0.020 molar solution of HCl for the following volumes of HCl: 20.00 mL, 30.00 mL, 40.00 mL, and 50.00 mL. If the standard deviation associated with weighing out the primary-standard bases is 0.1 mg, use the spreadsheet to calculate the percent relative standard deviation that this uncertainty would introduce into each of the calculated molarities.

The spreadsheet is shown in Figure 16-2. We enter the molarity of HCl in cell B2 and the molecular weights of the three primary standards in cells B3, B4, and B5. Appropriate labels are entered into columns A and C. The volumes of HCl for which the calculations are desired are entered in cells A8 through A11. We will do a sample calculation for the 20.0-mL volume here and show the spreadsheet entry. In each case, the number of millimoles of HCl is calculated from

$$\text{mmol HCl} = \cancel{\text{mL HCl}} \times 0.020 \frac{\text{mmol HCl}}{\cancel{\text{mL HCl}}}$$

(a) For TRIS,

$$\text{mass TRIS} = \cancel{\text{mmol HCl}} \times \frac{1\ \cancel{\text{mmol TRIS}}}{\cancel{\text{mmol HCl}}} \times \frac{121\text{ g TRIS}/\cancel{\text{mol TRIS}}}{1000\ \cancel{\text{mmol TRIS}}/\cancel{\text{mol TRIS}}}$$

For the 20.00-mL volume of HCl, the appropriate entry in cell B8 is =$B$2*A8*$B$3/1000, as shown in the documentation section of Figure 16-2. The result returned is 0.048 g. The formula in cell B8 is then copied into cells B9 through B11 to complete the column. The relative uncertainty in the molarity due to weighing would be equal to the relative uncertainty in the weighing process. For 0.048 g of TRIS, the percent relative standard deviation (%RSD) is

(continued)

$$\begin{array}{c} CH_2OH \\ | \\ H_2N\text{—}C\text{—}CH_2OH \\ | \\ CH_2OH \end{array}$$

Molecular model and structure of TRIS.

	A	B	C	D	E	F	G
1	**Spreadsheet to compare masses required for various bases in the standardization of 0.020 M HCl**						
2	M HCl	0.020					
3	MW TRIS	121	g/mol	**Note:** All weighings have standard deviations of 0.1 mg (0.0001 g)			
4	MW Na_2CO_3	106	g/mol				
5	MW $Na_2B_4O_7 \bullet H_2O$	381	g/mol				
6							
7	**mL HCl**	**g TRIS**	**RSD TRIS**	**g Na_2CO_3**	**RSD Na_2CO_3**	**g $Na_2B_4O_7 \bullet H_2O$**	**RSD $Na_2B_4O_7 \bullet H_2O$**
8	20.00	0.048	0.21	0.021	0.47	0.08	0.13
9	30.00	0.073	0.14	0.032	0.31	0.11	0.09
10	40.00	0.097	0.10	0.042	0.24	0.15	0.07
11	50.00	0.121	0.08	0.053	0.19	0.19	0.05
12							
13	**Documentation**						
14	Cell B8=$B$2*A8*1*$B$3/1000						
15	Cell C8=(0.0001/B8)*100						
16	Cell D8=$B$2*A8*1/2*$B$4/1000						
17	Cell E8=(0.0001/D8)*100						
18	Cell F8=$B$2*A8*1/2*$B$5/1000						
19	Cell G8=(0.0001/F8)*100						

Figure 16-2 Spreadsheet to compare masses and relative errors associated with using different primary-standard bases to standardize HCl solutions.

(0.0001 g/0.048 g) × 100%, so the entry in cell C8 is as shown in Figure 16-2. This formula in cell C8 is then copied into C9:C11.

(b) For Na_2CO_3,

$$\text{mass } Na_2CO_3 = \cancel{\text{mmol HCl}} \times \frac{1 \text{ mmol } \cancel{Na_2CO_3}}{2 \ \cancel{\text{mmol HCl}}} \times \frac{106 \text{ g } Na_2CO_3/\cancel{\text{mol } Na_2CO_3}}{1000 \ \cancel{\text{mmol } Na_2CO_3}/\cancel{\text{mol } Na_2CO_3}}$$

This result is entered into cell D8 as shown in Figure 16-2 and copied into D9:D11. The percent relative standard deviation in cell E8 is calculated as (0.0001/D8) × 100. The formula in E8 is copied into E9:E11.

(c) Similarly, for $Na_2B_4O_7 \cdot 10\, H_2O$,

$$\text{mass borax} = \cancel{\text{mmol HCl}} \times \frac{1 \ \cancel{\text{mmol borax}}}{2 \ \cancel{\text{mmol HCl}}} \times \frac{381 \text{ g borax}/\cancel{\text{mol borax}}}{1000 \ \cancel{\text{mmol borax}}/\cancel{\text{mol borax}}}$$

Note in Figure 16-2 that the relative standard deviation for TRIS is 0.10% or less if the volume of HCl taken is more than 40.00 mL. For Na_2CO_3, more than 50.00 mL of HCl would be required for this level of uncertainty. For borax, any volume of more than about 26.00 mL would suffice.

16A-3 Preparation of Standard Solutions of Base

Sodium hydroxide is the most common base for preparing standard solutions, although potassium hydroxide and barium hydroxide are also encountered. Because none of these is obtainable in primary-standard purity, standardization of these solutions is required after preparation.

The Effect of Carbon Dioxide on Standard Base Solutions

In solution as well as in the solid state, the hydroxides of sodium, potassium, and barium react rapidly with atmospheric carbon dioxide to produce the corresponding carbonate:

$$CO_2(g) + 2OH^- \rightarrow CO_3^{2-} + H_2O$$

► Absorption of carbon dioxide by a standardized solution of sodium or potassium hydroxide leads to a negative systematic error in analyses in which an indicator with a basic range is used; no systematic error is incurred when an indicator with an acidic range is used.

Although production of each carbonate ion uses up two hydroxide ions, the uptake of carbon dioxide by a solution of base does not necessarily alter its combining capacity for hydronium ions. Thus, at the end point of a titration that requires an acid-range indicator (such as bromocresol green), each carbonate ion produced from sodium or potassium hydroxide will have reacted with two hydronium ions of the acid (see Figure 16-1):

$$CO_3^{2-} + 2H_3O^+ \rightarrow H_2CO_3 + 2H_2O$$

Because the amount of hydronium ion consumed by this reaction is identical to the amount of hydroxide lost during formation of the carbonate ion, no error is incurred.

Unfortunately, most applications of standard base require an indicator with a basic transition range (phenolphthalein, for example). Here, each carbonate ion has reacted with only one hydronium ion when the color change of the indicator is observed:

$$CO_3^{2-} + H_3O^+ \rightarrow HCO_3^- + H_2O$$

The effective concentration of the base is thus diminished by absorption of carbon dioxide, and a systematic error (called a **carbonate error**) results.

EXAMPLE 16-2

A carbonate-free NaOH solution was found to be 0.05118 M immediately after preparation. Exactly 1.000 L of this solution was exposed to air for some time and absorbed 0.1962 g CO_2. Calculate the relative carbonate error that would arise in the determination of acetic acid with the contaminated solution if phenolphthalein were used as an indicator.

$$2NaOH + CO_2 \rightarrow Na_2CO_3 + H_2O$$

$$c_{Na_2CO_3} = 0.1962\ \cancel{g\ CO_2} \times \frac{1\ \cancel{mol\ CO_2}}{44.01\ \cancel{g\ CO_2}} \times \frac{1\ mol\ Na_2CO_3}{\cancel{mol\ CO_2}} \times \frac{1}{1.000\ L\ soln}$$

$$= 4.458 \times 10^{-3}\ M$$

The effective concentration c_{NaOH} of NaOH for acetic acid is then

$$c_{NaOH} = 0.05118\ \frac{mol\ NaOH}{L} - \frac{4.458 \times 10^{-3}\ \cancel{mol\ Na_2CO_3}}{L}$$

$$\times \frac{1\ \cancel{mol\ HCl}}{\cancel{mol\ Na_2CO_3}} \times \frac{1\ mol\ NaOH}{\cancel{mol\ HCl}}$$

$$= 0.04672\ M$$

$$\text{rel error} = \frac{0.04672 - 0.05118}{0.05118} \times 100\ \% = -8.7\%$$

The solid reagents used to prepare standard solutions of base are always contaminated by significant amounts of carbonate ion. The presence of this contaminant does not cause a carbonate error provided that the same indicator is used for both standardization and analysis. It does, however, lead to less sharp end points. Consequently, steps are usually taken to remove carbonate ion before a solution of a base is standardized.

◀ Carbonate ion in standard base solutions is undesirable because it decreases the sharpness of end points.

◀ **WARNING:** Concentrated solutions of NaOH (and KOH) are extremely corrosive to the skin. In making up standard solutions of NaOH, a face shield, rubber gloves, and protective clothing **must be worn at all times.**

The best method for preparation of carbonate-free sodium hydroxide solutions takes advantage of the very low solubility of sodium carbonate in concentrated solutions of the base. An approximately 50% aqueous solution of sodium hydroxide is prepared (or purchased from commercial sources). The solid sodium carbonate is allowed to settle to produce a clear liquid that is decanted and diluted to give the desired concentration. (Alternatively, the solid is removed by vacuum filtration.)

Water that is used to prepare carbonate-free solutions of base must also be free of carbon dioxide. Distilled water, which is sometimes supersaturated with carbon dioxide, should be boiled briefly to eliminate the gas. The water is then allowed to cool to room temperature before the introduction of base, because hot alkali solutions rapidly absorb carbon dioxide. Deionized water ordinarily does not contain significant amounts of carbon dioxide.

► Water that is in equilibrium with atmospheric constituents contains only about 1.5×10^{-5} mol CO_2/L, an amount that has a negligible effect on the strength of most standard bases. As an alternative to boiling to remove CO_2 from supersaturated solutions of CO_2, the excess gas can be removed by bubbling air through the water for several hours. This process is called *sparging* and produces a solution that contains the equilibrium concentration of CO_2.

A tightly capped polyethylene bottle usually provides adequate short-term protection against the uptake of atmospheric carbon dioxide. Before capping, the bottle is squeezed to minimize the interior air space. Care should also be taken to keep the bottle closed except during the brief periods when the contents are being transferred to a buret. Sodium hydroxide solutions will ultimately cause a polyethylene bottle to become brittle.

Sparging is the process of removing a gas from a solution by bubbling an inert gas through the solution.

The concentration of a sodium hydroxide solution will decrease slowly (0.1% to 0.3% per week) if the base is stored in glass bottles. The loss in strength is caused by the reaction of the base with the glass to form sodium silicates. For this reason, standard solutions of base should not be stored for extended periods (longer than 1 or 2 weeks) in glass containers. In addition, bases should never be kept in glass-stoppered containers because the reaction between the base and the stopper may cause the latter to "freeze" after a brief period. Finally, to avoid the same type of freezing, burets with glass stopcocks should be promptly drained and thoroughly rinsed with water after use with standard base solutions. Burets equipped with Teflon stopcocks do not have this problem.

► Solutions of bases are preferably stored in polyethylene bottles rather than glass because of the reaction between bases and glass. Such solutions should never be stored in glass-stoppered bottles; after standing for a period, removal of the stopper often becomes impossible.

16A-4 The Standardization of Bases

Several excellent primary standards are available for the standardization of bases. Most are weak organic acids that require the use of an indicator with a basic transition range.

► Standard solutions of strong bases cannot be prepared directly by mass and must always be standardized against a primary standard with acidic properties.

Potassium Hydrogen Phthalate

Potassium hydrogen phthalate, $KHC_8H_4O_4$, is an ideal primary standard. It is a nonhygroscopic crystalline solid with a high molar mass (204.2 g/mol). For most purposes, the commercial analytical-grade salt can be used without further purification. For the most exacting work, potassium hydrogen phthalate of certified purity is available from the National Institute of Standards and Technology.

Other Primary Standards for Bases

Benzoic acid is obtainable in primary-standard purity and can be used for the standardization of bases. Because its solubility in water is limited, this reagent is ordinarily dissolved in ethanol prior to dilution with water and titration. A blank should always be carried through this standardization because commercial alcohol is sometimes slightly acidic.

Potassium hydrogen iodate, $KH(IO_3)_2$, is an excellent primary standard with a high molecular mass per mole of protons. It is also a strong acid that can be titrated using virtually any indicator with a transition range between a pH of 4 and 10.

► $KH(IO_3)_2$, in contrast to all other primary standards for bases, has the advantage of being a strong acid, thus making the choice of indicator less critical.

16B TYPICAL APPLICATIONS OF NEUTRALIZATION TITRATIONS

Neutralization titrations are used to determine the innumerable inorganic, organic, and biological species that possess inherent acidic or basic properties. Equally important, however, are the many applications that involve conversion of an analyte to an acid or base by suitable chemical treatment followed by titration with a standard strong base or acid.

Two major types of end points find widespread use in neutralization titrations. The first is a visual end point based on indicators such as those described in Section 14A. The second is a *potentiometric* end point, in which the potential of a glass/calomel electrode system is determined with a voltage-measuring device. The measured potential is directly proportional to pH. Potentiometric end points are described in Section 21G.

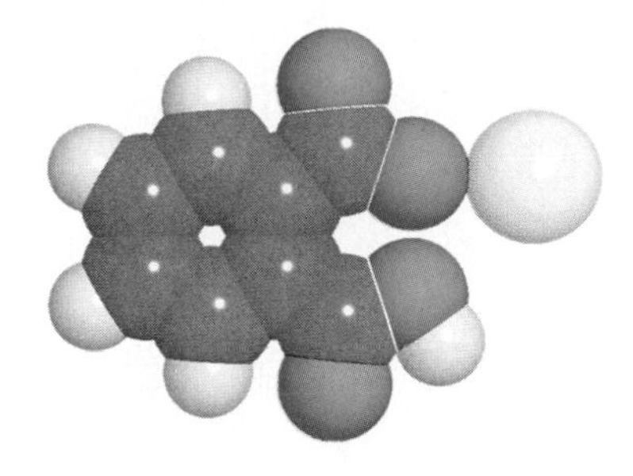

$C_6H_4(COO^-K^+)(COOH)$

Molecular model and structure of potassium hydrogen phthalate.

◀ Neutralization titrations are still among the most widely used analytical methods.

16B-1 Elemental Analysis

Several important elements that occur in organic and biological systems are conveniently determined by methods that involve an acid/base titration as the final step. Generally, the elements susceptible to this type of analysis are nonmetallic and include carbon, nitrogen, chlorine, bromine, and fluorine, as well as a few other less common species. Pretreatment converts the element to an inorganic acid or base that is then titrated. A few examples follow.

Nitrogen

◀ Kjeldahl is pronounced Kyell′dahl. Hundreds of thousands of Kjeldahl nitrogen determinations are performed each year, primarily to provide a measure of the protein content of meats, grains, and animal feeds.

Nitrogen occurs in a wide variety of substances of interest in research, industry, and agriculture. Examples include amino acids, proteins, synthetic drugs, fertilizers, explosives, soils, potable water supplies, and dyes. Thus, analytical methods for the determination of nitrogen, particularly in organic substrates, are of singular importance.

The most common method for determining organic nitrogen is the *Kjeldahl method,* which is based on a neutralization titration. The procedure is straightforward, requires no special equipment, and is readily adapted to the routine analysis of large numbers of samples. It (or one of its modifications) is the standard means of determining the protein content of grains, meats, and other biological materials. Since most proteins contain approximately the same percentage of nitrogen, multiplication of this percentage by a suitable factor (6.25 for meats, 6.38 for dairy products, and 5.70 for cereals) gives the percentage of protein in a sample.

The Kjeldahl method was developed by a Danish chemist who first described it in 1883. (J. Kjeldahl, *Z. Anal. Chem.*, **1883,** *22,* 366.)

FEATURE 16-1

Determining Total Serum Protein

The determination of total serum protein is an important clinical measurement used in diagnosing liver malfunctions. Although the Kjeldahl method is capable of high precision and accuracy, it is too slow and cumbersome to be used routinely in determining total serum protein. The Kjeldahl procedure, however, has historically been the reference method against which other methods are compared. Methods commonly used include the *Biuret method* and the *Lowry*

(continued)

method. In the Biuret method, a reagent containing cupric ions is used, and a violet-colored complex is formed between the Cu^{2+} ions and peptide bonds. The increase in the absorption of visible radiation is used to measure serum protein. This method is readily automated. In the Lowry procedure, the serum sample is pretreated with an alkaline copper solution followed by a phenolic reagent. A color develops because of reduction of phosphotungstic acid and phophomolybdic acid to a blue heteropoly acid. Both the Biuret and Lowry methods use spectrophotometry (see Chapter 26) for quantitative measurements.

FEATURE 16-2

Other Methods For Determining Organic Nitrogen

Several other methods are used to determine the nitrogen content of organic materials. In the *Dumas method,* the sample is mixed with powdered copper(II) oxide and ignited in a combustion tube to give carbon dioxide, water, nitrogen, and small amounts of nitrogen oxides. A stream of carbon dioxide carries these products through a packing of hot copper, which reduces any oxides of nitrogen to elemental nitrogen. The mixture then is passed into a gas buret filled with concentrated potassium hydroxide. The only component not absorbed by the base is nitrogen, and its volume is measured directly.

The newest method for determining organic nitrogen involves combusting the sample at 1100°C for a few minutes to convert the nitrogen to nitric oxide, NO. Ozone is then introduced into the gaseous mixture, which oxidizes the nitric oxide to nitrogen dioxide. This reaction gives off visible radiation *(chemiluminescence),* the intensity of which is measured and is proportional to the nitrogen content of the sample. Chemiluminescence is discussed further in Chapter 27. An instrument for this procedure is available from commercial sources.

In the Kjeldahl method, the sample is decomposed in hot, concentrated sulfuric acid to convert the bound nitrogen to ammonium ion. The resulting solution is then cooled, diluted, and made basic. The liberated ammonia is distilled, collected in an acidic solution, and determined by a neutralization titration.

► $-NO_2$ (nitro group) $-N{=}N-$ (azo group) $-N^+{=}N-$ with O^- on the N (azoxy group)

The critical step in the Kjeldahl method is the decomposition with sulfuric acid, which oxidizes the carbon and hydrogen in the sample to carbon dioxide and water. The fate of the nitrogen, however, depends on its state of combination in the original sample. Amine and amide nitrogens are quantitatively converted to ammonium ion. In contrast, nitro, azo, and azoxy groups are likely to yield the element or its various oxides, all of which are lost from the hot acidic medium. This loss can be avoided by pretreating the sample with a reducing agent to form products that behave as amide or amine nitrogen. In one such prereduction scheme, salicylic acid and sodium thiosulfate are added to the concentrated sulfuric acid solution containing the sample. After a brief period, the digestion is performed in the usual way.

Certain aromatic heterocyclic compounds, such as pyridine and its derivatives, are particularly resistant to complete decomposition by sulfuric acid. Such compounds yield low results as a consequence (Figure 5-3) unless special precautions are taken.

The decomposition step is frequently the most time-consuming aspect of a Kjeldahl determination. Some samples may require heating periods in excess of 1 hour. Numerous modifications of the original procedure have been proposed with the aim of shortening the digestion time. In the most widely used modification, a neutral salt, such as potassium sulfate, is added to increase the boiling point of the sulfuric acid solution and thus the temperature at which the decomposition occurs. In another modification, a solution of hydrogen peroxide is added to the mixture after the digestion has decomposed most of the organic matrix.

Many substances catalyze the decomposition of organic compounds by sulfuric acid. Mercury, copper, and selenium, either combined or in the elemental state, are effective. Mercury(II), if present, must be precipitated with hydrogen sulfide prior to distillation to prevent retention of ammonia as a mercury(II) ammine complex.

EXAMPLE 16-3

A 0.7121 g sample of a wheat flour was analyzed by the Kjeldahl method. The ammonia formed by addition of concentrated base after digestion with H_2SO_4 was distilled into 25.00 mL of 0.04977 M HCl. The excess HCl was then back-titrated with 3.97 mL of 0.04012 M NaOH. Calculate the percent protein in the flour.

$$\text{amount HCl} = 25.00\ \cancel{\text{mL HCl}} \times 0.04977\ \frac{\text{mmol HCl}}{\cancel{\text{mL HCl}}} = 1.2443\ \text{mmol}$$

$$\text{amount NaOH} = 3.97\ \cancel{\text{mL NaOH}} \times 0.04012\ \frac{\text{mmol NaOH}}{\cancel{\text{mL NaOH}}} = 0.1593\ \text{mmol}$$

$$\text{amount N} = 1.0850\ \text{mmol}$$

$$\%\ \text{N} = \frac{1.0850\ \cancel{\text{mmol N}} \times \dfrac{0.014007\ \text{g N}}{\cancel{\text{mmol N}}}}{0.7121\ \text{g sample}} \times 100\ \% = 2.1341$$

$$\%\ \text{protein}\ 2.1341\ \cancel{\%\ \text{N}} \times \frac{5.70\ \%\ \text{protein}}{\cancel{\%\ \text{N}}} = 12.16$$

Sulfur

Sulfur in organic and biological materials is conveniently determined by burning the sample in a stream of oxygen. The sulfur dioxide (as well as the sulfur trioxide) formed during the oxidation is collected by distillation into a dilute solution of hydrogen peroxide:

$$SO_2(g) + H_2O_2 \rightarrow H_2SO_4$$

◀ Sulfur dioxide in the atmosphere is often determined by drawing a sample through a hydrogen peroxide solution and then titrating the sulfuric acid that is produced.

The sulfuric acid is then titrated with standard base.

Other Elements

Table 16-1 lists other elements that can be determined by neutralization methods.

TABLE 16-1

Elemental Analyses Based on Neutralization Titrations

Element	Converted to	Adsorption or Precipitation Products	Titration
N	NH_3	$NH_3(g) + H_3O^+ \rightarrow NH_4^+ + H_2O$	Excess HCl with NaOH
S	SO_2	$SO_2(g) + H_2O_2 \rightarrow H_2SO_4$	NaOH
C	CO_2	$CO_2(g) + Ba(OH)_2 \rightarrow Ba(CO)_3(s) + H_2O$	Excess $Ba(OH)_2$ with HCl
Cl(Br)	HCl	$HCl(g) + H_2O \rightarrow Cl^- + H_3O^+$	NaOH
F	SiF_4	$SiF_4(g) + H_2O \rightarrow H_2SiF_6$	NaOH
P	H_3PO_4	$12H_2MoO_4 + 3NH_4^+ + H_3PO_4 \rightarrow (NH_4)_3PO_4 \cdot 12MoO_3(s) + 12H_2O + 3H^+$ $(NH_4)_3PO_4 \cdot 12MoO_3(s) + 26OH^- \rightarrow HPO_4^{2-} + 12MoO_4^{2-} + 14H_2O + 3NH_3(g)$	Excess NaOH with HCl

16B-2 The Determination of Inorganic Substances

Numerous inorganic species can be determined by titration with strong acids or bases. A few examples follow.

Ammonium Salts

Ammonium salts are conveniently determined by conversion to ammonia with strong base followed by distillation. The ammonia is collected and titrated as in the Kjeldahl method.

Nitrates and Nitrites

The method just described for ammonium salts can be extended to the determination of inorganic nitrate or nitrite. These ions are first reduced to ammonium ion by Devarda's alloy (50% Cu, 45% Al, 5% Zn). Granules of the alloy are introduced into a strongly alkaline solution of the sample in a Kjeldahl flask. The ammonia is distilled after reaction is complete. Arnd's alloy (60% Cu, 40% Mg) has also been used as the reducing agent.

Carbonate and Carbonate Mixtures

The qualitative and quantitative determination of the constituents in a solution containing sodium carbonate, sodium hydrogen carbonate, and sodium hydroxide, either alone or admixed, provides interesting examples of how neutralization titrations can be employed to analyze mixtures. No more than two of these three constituents can exist in appreciable amount in any solution because reaction elim-

TABLE 16-2

Volume Relationships in the Analysis of Mixtures Containing Hydroxide, Carbonate, and Hydrogen Carbonate Ions

Constituents in Sample	Relationship between V_{phth} and V_{beg} in the Titration of an Equal Volume of Sample*
NaOH	$V_{phth} = V_{beg}$
Na_2CO_3	$V_{phth} = \frac{1}{2}V_{beg}$
$NaHCO_3$	$V_{phth} = 0;\ V_{beg} > 0$
NaOH, Na_2CO_3	$V_{phth} > \frac{1}{2}V_{beg}$
Na_2CO_3, $NaHCO_3$	$V_{phth} < \frac{1}{2}V_{beg}$

*V_{phth} = volume of acid needed for a phenolphthalein end point; V_{beg} = volume of acid needed for a bromocresol green end point.

inates the third. Thus, mixing sodium hydroxide with sodium hydrogen carbonate results in the formation of sodium carbonate until one or the other (or both) of the original reactants is exhausted. If the sodium hydroxide is used up, the solution will contain sodium carbonate and sodium hydrogen carbonate; if sodium hydrogen carbonate is depleted, sodium carbonate and sodium hydroxide will remain; if equimolar amounts of sodium hydrogen carbonate and sodium hydroxide are mixed, the principal solute species will be sodium carbonate.

The analysis of such mixtures requires two titrations; one with an alkaline-range indicator, such as phenolphthalein, and the other with an acid-range indicator, such as bromocresol green. The composition of the solution can then be deduced from the relative volumes of acid needed to titrate equal volumes of the sample (Table 16-2 and Figure 16-3). Once the composition of the solution has been established, the volume data can be used to determine the concentration of each component in the sample.

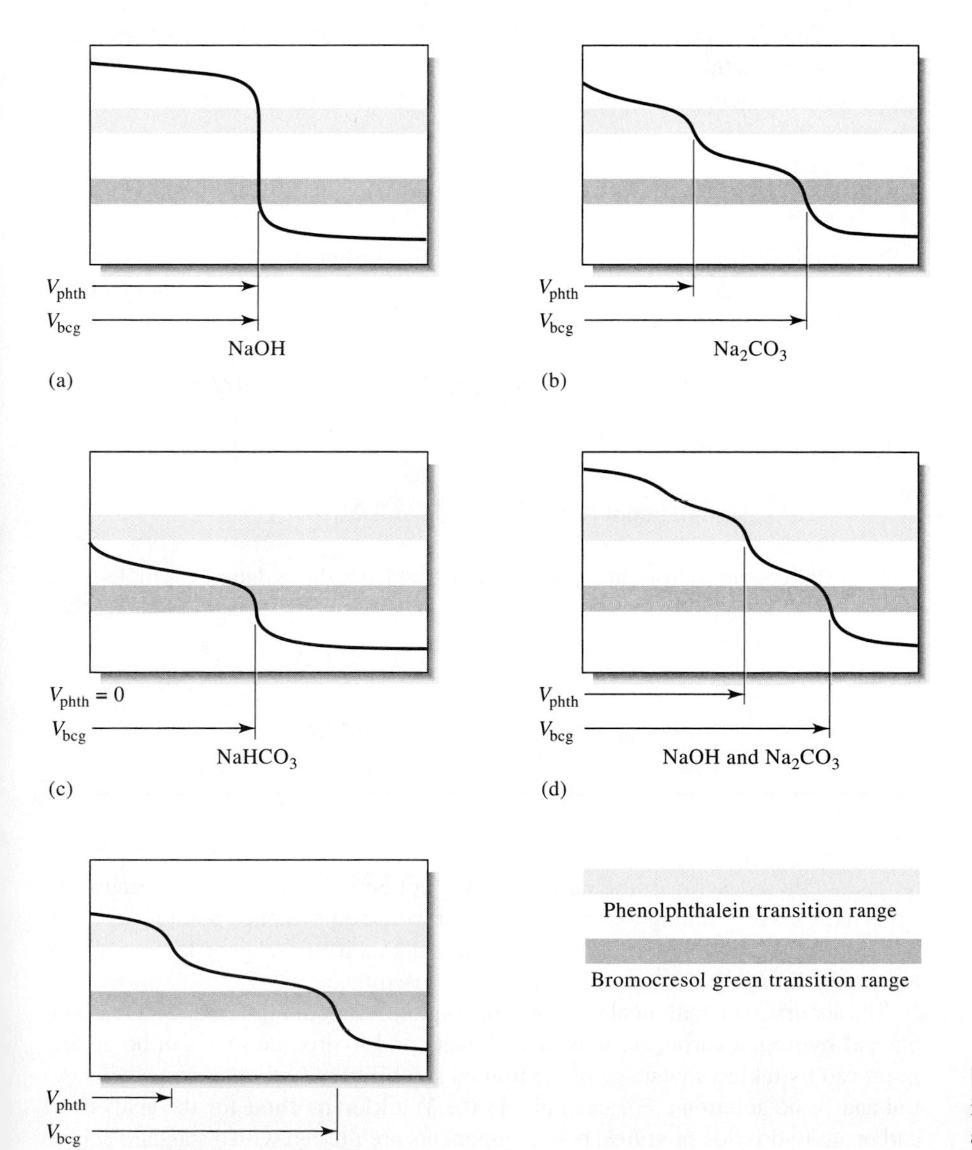

Figure 16-3 Titration curves and indicator transition ranges for the analysis of mixtures containing hydroxide, carbonate, and hydrogen carbonate ions.

► Compatible mixtures containing two of the following can also be analyzed in a similar way: HCl, H_3PO_4, NaH_2PO_4, Na_2HPO_4, Na_3PO_4, and NaOH.

► How could you analyze a mixture of HCl and H_3PO_4? A mixture of Na_3PO_4 and Na_2HPO_4? See Figure 15-4, curve *A*.

EXAMPLE 16-4

A solution contains $NaHCO_3$, Na_2CO_3, and NaOH, either alone or in permissible combination. Titration of a 50.0-mL portion to a phenolphthalein end point requires 22.1 mL of 0.100 M HCl. A second 50.0-mL aliquot requires 48.4 mL of the HCl when titrated to a bromocresol green end point. Deduce the composition, and calculate the molar solute concentrations of the original solution.

If the solution contained only NaOH, the volume of acid required would be the same regardless of indicator (see Figure 16-3a). Similarly, we can rule out the presence of Na_2CO_3 alone because titration of this compound to a bromocresol green end point would consume just twice the volume of acid required to reach the phenolphthalein end point (see Figure 16-3b). In fact, however, the second titration requires 48.4 mL. Because less than half of this amount is involved in the first titration, the solution must contain some $NaHCO_3$ in addition to Na_2CO_3 (see Figure 16-3e). We can now calculate the concentration of the two constituents.

When the phenolphthalein end point is reached, the CO_3^{2-} originally present is converted to HCO_3^-. Thus,

$$\text{no. mmol } Na_2CO_3 = 22.1 \cancel{\text{mL}} \times 0.100 \text{ mmol}/\cancel{\text{mL}} = 2.21$$

The titration from the phenolphthalein to the bromocresol green end point $(48.4 - 22.1 = 26.3 \text{ mL})$ involves both the hydrogen carbonate originally present and that formed by titration of the carbonate. Thus,

$$\text{no. mmol } NaHCO_3 + \text{no. mmol } Na_2CO_3 = 26.3 \times 0.100 = 2.63$$

Hence,

$$\text{no. mmol } NaHCO_3 = 2.63 - 2.21 = 0.42$$

The molar concentrations are readily calculated from these data:

$$c_{Na_2CO_3} = \frac{2.21 \text{ mmol}}{50.0 \text{ mL}} = 0.0442 \text{ M}$$

$$c_{NaHCO_3} = \frac{0.42 \text{ mmol}}{50.0 \text{ mL}} = 0.084 \text{ M}$$

The method described in Example 16-4 is not entirely satisfactory, because the pH change corresponding to the hydrogen carbonate equivalence point is not sufficient to give a sharp color change with a chemical indicator (Figure 15-5). Relative errors of 1% or more must be expected as a consequence.

The accuracy of analytical methods for solutions containing mixtures of carbonate and hydrogen carbonate ions or carbonate and hydroxide ions can be greatly improved by taking advantage of the limited solubility of barium carbonate in neutral and basic solutions. For example, in the **Winkler method** for the analysis of carbonate/hydroxide mixtures, both components are titrated with a standard acid to the end point with an acid-range indicator, such as bromocresol green. (The end point is established after the solution is boiled to remove carbon dioxide.) An unmeasured excess of neutral barium chloride is then added to a second aliquot of the sample solution to precipitate the carbonate ion, after which the hydroxide ion is titrated to a phenolphthalein end point. The presence of the sparingly soluble barium carbonate does not interfere as long as the concentration of barium ion is greater than 0.1 M.

Carbonate and hydrogen carbonate ions can be accurately determined in mixtures by first titrating both ions with standard acid to an end point with an acid-range indicator (with boiling to eliminate carbon dioxide). The hydrogen carbonate in a second aliquot is converted to carbonate by the addition of a known excess of standard base. After a large excess of barium chloride has been introduced, the excess base is titrated with standard acid to a phenolphthalein end point.

The presence of solid barium carbonate does not hamper end point detection in either of these methods.

16B-3 The Determination of Organic Functional Groups

Neutralization titrations provide convenient methods for the direct or indirect determination of several organic functional groups. Brief descriptions of methods for the more common groups follow.

Carboxylic and Sulfonic Acid Groups

Carboxylic and sulfonic acid groups are the two most common structures that impart acidity to organic compounds. Most carboxylic acids have dissociation constants that range between 10^{-4} and 10^{-6}, and thus these compounds are readily titrated. An indicator that changes color in a basic range, such as phenolphthalein, is required.

Many carboxylic acids are not sufficiently soluble in water to permit direct titration in this medium. Where this problem exists, the acid can be dissolved in ethanol and titrated with aqueous base. Alternatively, the acid can be dissolved in an excess of standard base followed by back-titration with standard acid.

Sulfonic acids are generally strong acids and readily dissolve in water. Their titration with a base is therefore straightforward.

Neutralization titrations are often employed to determine the equivalent weight of purified organic acids (Feature 16-3). Equivalent weights serve as an aid in the qualitative identification of organic acids.

The **equivalent weight** of an acid or base is the mass of the acid or base in grams that reacts with or contains one mole of protons. Thus, the equivalent weight of KOH (56.11 g/mol) is its molar mass; for $Ba(OH)_2$, it is the molar mass divided by 2 $\left(\frac{1}{2} \times 171.3 \text{ g/mol}\right)$.

Amine Groups

Aliphatic amines generally have base dissociation constants on the order of 10^{-5} and can thus be titrated directly with a solution of a strong acid. In contrast, aromatic amines such as aniline and its derivatives are usually too weak for titration in aqueous medium ($K_b \approx 10^{-10}$). The same is true for cyclic amines with aromatic character, such as pyridine and its derivatives. Many saturated cyclic amines, such as piperidine, tend to resemble aliphatic amines in their acid/base behavior and thus can be titrated in aqueous media.

Many amines that are too weak to be titrated as bases in water are readily titrated in nonaqueous solvents, such as anhydrous acetic acid, which enhance their basicity.

FEATURE 16-3

Equivalent Weights of Acids and Bases

The equivalent weight of a participant in a neutralization reaction is the weight that reacts with or supplies one mole of protons *in a particular reaction.* For example, the equivalent weight of H_2SO_4 is one half of its formula weight. The equivalent weight of Na_2CO_3 is usually one half of its formula weight because in most applications its reaction is

$$Na_2CO_3 + 2H_3O^+ \rightarrow 3H_2O + CO_2 + 2Na^+$$

When titrated with some indicators, however, it consumes but a single proton:

$$Na_2CO_3 + H_3O^+ \rightarrow NaHCO_3 + Na^+$$

Here, the equivalent weight and the formula weight of Na_2CO_3 are identical. These observations demonstrate that the equivalent weight of a compound cannot be defined without having a particular reaction in mind (see Appendix 7).

Ester Groups

Saponification is the process by which an ester is hydrolyzed in alkaline solution to give an alcohol and a conjugate base. For example,

$$CH_3\overset{\displaystyle O}{\overset{\|}{C}}OCH_3 + OH^- \rightarrow CH_3\overset{\displaystyle O}{\overset{\|}{C}}{-}O^- + CH_3OH$$

Esters are commonly determined by **saponification** with a measured quantity of standard base:

$$R_1COOR_2 + OH^- \rightarrow R_1COO^- + HOR_2$$

The excess base is then titrated with standard acid.

Esters vary widely in their rates of saponification. Some require several hours of heating with a base to complete the process. A few react rapidly enough to permit direct titration with the base. Typically, the ester is refluxed with standard 0.5 M KOH for 1 to 2 hours. After cooling, the excess base is determined with standard acid.

Hydroxyl Groups

Hydroxyl groups in organic compounds can be determined by esterification with various carboxylic acid anhydrides or chlorides; the two most common reagents are acetic anhydride and phthalic anhydride. With acetic anhydride, the reaction is

$$(CH_3CO)_2O + ROH \rightarrow CH_3COOR + CH_3COOH$$

The acetylation is ordinarily carried out by mixing the sample with a carefully measured volume of acetic anhydride in pyridine. After heating, water is added to hydrolyze the unreacted anhydride:

$$(CH_3CO)_2O + H_2O \rightarrow 2CH_3COOH$$

The acetic acid is then titrated with a standard solution of alcoholic sodium or potassium hydroxide. A blank is carried through the analysis to establish the original amount of anhydride.

Amines, if present, are converted quantitatively to amides by acetic anhydride; a correction for this source of interference is frequently possible by direct titration of another portion of the sample with standard acid.

Carbonyl Groups

Many aldehydes and ketones can be determined with a solution of hydroxylamine hydrochloride. The reaction, which produces an oxime, is

$$(R_1)(R_2)C{=}O + NH_2OH \cdot HCl \longrightarrow (R_1)(R_2)C{=}NOH + HCl + H_2O$$

where R_2 may be an atom of hydrogen. The liberated hydrochloric acid is titrated with base. Here again, the conditions necessary for quantitative reaction vary. Typically, 30 minutes suffices for aldehydes. Many ketones require refluxing with the reagent for 1 hour or more.

16B-4 The Determination of Salts

The total salt content of a solution can be accurately and readily determined by an acid/base titration. The salt is converted to an equivalent amount of an acid or a base by passage through a column packed with an ion-exchange resin. (This application is considered in more detail in Section 33D).

Standard acid or base solutions can also be prepared with ion-exchange resins. Here, a solution containing a known mass of a pure compound, such as sodium chloride, is washed through the resin column and diluted to a known volume. The salt liberates an equivalent amount of acid or base from the resin, permitting calculation of the molarity of the reagent in a straightforward way.

WEB WORKS

Use your Web browser to connect to **http://chemistry.brookscole.com/skoogfac/.** From the Chapter Resources menu, choose Web Works. Locate the Chapter 16 section, and click on the link to the executive summary of the Lake Champlain Basin Agricultural Watersheds Project. The report summarizes a project to improve water quality in Lake Champlain in Vermont and New York. Based on your reading of the report, what appears to be the primary general cause of the eutrophication of Lake Champlain? What types of industry are the sources of the pollution? What measures have been taken to reduce the pollution? Briefly describe the experimental design to determine whether these measures have been effective. Total Kjeldahl nitrogen (TKN) determinations were one of the quantities measured in this study; name three others. Explain how TKN provides a measure of pollution in the lake. Based on the TKN measurements and other data in the report, have the pollution abatement measures been effective? What are the final recommendations of the report?

QUESTIONS AND PROBLEMS

***16-1.** The boiling points of HCl and CO_2 are nearly the same (−85°C and −78°C). Explain why CO_2 can be removed from an aqueous solution by boiling briefly while essentially no HCl is lost even after boiling for 1 hour or more.

16-2. Why is HNO_3 seldom used to prepare standard acid solutions?

***16-3.** Describe how Na_2CO_3 of primary-standard grade can be prepared from primary-standard $NaHCO_3$.

16-4. Why is it common practice to boil the solution near the equivalence point in the standardization of Na_2CO_3 with acid?

***16-5.** Give two reasons why $KH(IO_3)_2$ would be preferred over benzoic acid as a primary standard for a 0.010 M NaOH solution.

16-6. Briefly describe the circumstance in which the molarity of a sodium hydroxide solution will apparently be unaffected by the absorption of carbon dioxide.

16-7. What types of organic nitrogen-containing compounds tend to yield low results with the Kjeldahl method unless special precautions are taken?

***16-8.** How would you prepare 2.00 L of
(a) 0.15 M KOH from the solid?
(b) 0.015 M $Ba(OH)_2 \cdot 8H_2O$ from the solid?
(c) 0.200 M HCl from a reagent that has a density of 1.0579 g/mL and is 11.50% HCl (w/w)?

16-9. How would you prepare 500.0 mL of
(a) 0.250 M H_2SO_4 from a reagent that has a density of 1.1539 g/mL and is 21.8% H_2SO_4 (w/w)?
(b) 0.30 M NaOH from the solid?
(c) 0.08000 M Na_2CO_3 from the pure solid?

***16-10.** Standardization of a sodium hydroxide solution against potassium hydrogen phthalate (KHP) yielded the accompanying results.

mass KHP, g	0.7987	0.8365	0.8104	0.8039
volume NaOH, mL	38.29	39.96	38.51	38.29

Calculate
(a) the average molarity of the base.
(b) the standard deviation and the coefficient of variation for the data.
(c) the spread of the data.

16-11. The molarity of a perchloric acid solution was established by titration against primary-standard sodium carbonate (product: CO_2); the following data were obtained.

mass Na_2CO_3, g	0.2068	0.1997	0.2245	0.2137
volume $HClO_4$, mL	36.31	35.11	39.00	37.54

(a) Calculate the average molarity of the acid.
(b) Calculate the standard deviation for the data and the coefficient of variation for the data.
(c) Does statistical justification exist for disregarding the outlying result?

***16-12.** If 1.000 L of 0.1500 M NaOH is unprotected from the air after standardization and absorbs 11.2 mmol of CO_2, what will be its new molarity when it is standardized against a standard solution of HCl using
(a) phenolphthalein?
(b) bromocresol green?

16-13. A NaOH solution was 0.1019 M immediately after standardization. Exactly 500.0 mL of the reagent was left exposed to air for several days and absorbed 0.652 g of CO_2. Calculate the relative carbonate error in the determination of acetic acid with this solution if the titrations were performed with phenolphthalein.

***16-14.** Calculate the molar concentration of a dilute HCl solution if
(a) a 50.00-mL aliquot yielded 0.6010 g of AgCl.
(b) titration of 25.00 mL of 0.04010 M $Ba(OH)_2$ required 19.92 mL of the acid.
(c) titration of 0.2694 g of primary-standard Na_2CO_3 required 38.77 mL of the acid (products: CO_2 and H_2O).

16-15. Calculate the molarity of a dilute $Ba(OH)_2$ solution if
(a) 50.00 mL yielded 0.1684 g of $BaSO_4$.
(b) titration of 0.4815 g of primary-standard potassium hydrogen phthalate (KHP) required 29.41 mL of the base.
(c) addition of 50.00 mL of the base to 0.3614 g of benzoic acid required a 4.13-mL back-titration with 0.05317 M HCl.

16-16. Suggest a range of sample masses for the indicated primary standard if it is desired to use between 35 and 45 mL of titrant:
*(a) 0.150 M $HClO_4$ titrated against Na_2CO_3 (CO_2 product).
(b) 0.075 M HCl titrated against $Na_2C_2O_4$:

$$Na_2C_2O_4 \rightarrow Na_2CO_3 + CO$$
$$CO_3^{2-} + 2H^+ \rightarrow H_2O + CO_2$$

*(c) 0.20 M NaOH titrated against benzoic acid.
(d) 0.030 M $Ba(OH)_2$ titrated against $KH(IO_3)_2$.
*(e) 0.040 M $HClO_4$ titrated against TRIS.

(f) 0.080 M H_2SO_4 titrated against $Na_2B_4O_7 \cdot 10H_2O$. Reaction:

$$B_4O_7^{2-} + 2H_3O^+ + 3H_2O \rightarrow 4H_3BO_3$$

*16-17. Calculate the relative standard deviation in the computed molarity of 0.0200 M HCl if this acid was standardized against the masses derived in Example 16-1 for: (a) TRIS, (b) Na_2CO_3, and (c) $Na_2B_4O_7 \cdot 10H_2O$. Assume that the absolute standard deviation in the mass measurement is 0.0001 g and that this measurement limits the precision of the computed molarity.

16-18. (a) Compare the masses of potassium hydrogen phthalate (204.22 g/mol); potassium hydrogen iodate (389.91 g/mol); and benzoic acid (122.12 g/mol) needed for a 30.00-mL standardization of 0.0400 M NaOH.
(b) What would be the relative standard deviation in the molarity of the base if the standard deviation in the measurement of mass in (a) is 0.002 g and this uncertainty limits the precision of the calculation?

*16-19. A 50.00-mL sample of a white dinner wine required 21.48 mL of 0.03776 M NaOH to achieve a phenolphthalein end point. Express the acidity of the wine in terms of grams of tartaric acid ($H_2C_4H_4O_6$, 150.09 g/mol) per 100 mL. (Assume that both acidic protons of the compound are titrated.)

16-20. A 25.0-mL aliquot of vinegar was diluted to 250 mL in a volumetric flask. Titration of 50.0-mL aliquots of the diluted solution required an average of 34.88 mL of 0.09600 M NaOH. Express the acidity of the vinegar in terms of the percentage (w/v) of acetic acid.

*16-21. Titration of a 0.7439-g sample of impure $Na_2B_4O_7$ required 31.64 mL of 0.1081 M HCl (see Problem 16-16f for reaction). Express the results of this analysis in terms of percent
(a) $Na_2B_4O_7$.
(b) $Na_2B_4O_7 \cdot 10H_2O$.
(c) B_2O_3.
(d) B.

16-22. A 0.6334-g sample of impure mercury(II) oxide was dissolved in an unmeasured excess of potassium iodide. Reaction:

$$HgO(s) + 4I^- + H_2O \rightarrow HgI_4^{2-} + 2OH^-$$

Calculate the percentage of HgO in the sample if titration of the liberated hydroxide required 42.59 mL of 0.1178 M HCl.

*16-23. The formaldehyde content of a pesticide preparation was determined by weighing 0.3124 g of the liquid sample into a flask containing 50.0 mL of 0.0996 M NaOH and 50 mL of 3% H_2O_2. On heating, the following reaction took place:

$$OH^- + HCHO + H_2O_2 \rightarrow HCOO^- + 2H_2O$$

After cooling, the excess base was titrated with 23.3 mL of 0.05250 M H_2SO_4. Calculate the percentage of HCHO (30.026 g/mol) in the sample.

16-24. The benzoic acid extracted from 106.3 g of catsup required a 14.76-mL titration with 0.0514 M NaOH. Express the results of this analysis in terms of percent sodium benzoate (144.10 g/mol).

*16-25. The active ingredient in Antabuse, a drug used for the treatment of chronic alcoholism, is tetraethylthiuram disulfide

$$(C_2H_5)_2N\overset{\displaystyle S}{\overset{\|}{C}}SS\overset{\displaystyle S}{\overset{\|}{C}}N(C_2H_5)_2$$

(296.54 g/mol). The sulfur in a 0.4329-g sample of an Antabuse preparation was oxidized to SO_2, which was absorbed in H_2O_2 to give H_2SO_4. The acid was titrated with 22.13 mL of 0.03736 M base. Calculate the percentage of active ingredient in the preparation.

16-26. A 25.00-mL sample of a household cleaning solution was diluted to 250.0 mL in a volumetric flask. A 50.00-mL aliquot of this solution required 40.38 mL of 0.2506 M HCl to reach a bromocresol green end point. Calculate the weight/volume percentage of NH_3 in the sample. (Assume that all the alkalinity results from the ammonia.)

*16-27. A 0.1401-g sample of a purified carbonate was dissolved in 50.00 mL of 0.1140 M HCl and boiled to eliminate CO_2. Back-titration of the excess HCl required 24.21 mL of 0.09802 M NaOH. Identify the carbonate.

16-28. A dilute solution of an unknown weak acid required a 28.62-mL titration with 0.1084 M NaOH to reach a phenolphthalein end point. The titrated solution was evaporated to dryness. Calculate the equivalent weight (see Margin Note, page 441) of the acid if the sodium salt was found to weigh 0.2110 g.

*16-29. A 3.00-L sample of urban air was bubbled through a solution containing 50.0 mL of 0.0116 M $Ba(OH)_2$, which caused the CO_2 in the sample to precipitate as $BaCO_3$. The excess base was back-titrated to a phenolphthalein end point with

23.6 mL of 0.0108 M HCl. What is the concentration of CO_2 in the air in parts per million (that is, mL $CO_2/10^6$ mL air); use 1.98 g/L for the density of CO_2.

16-30. Air was bubbled at a rate of 30.0 L/min through a trap containing 75 mL of 1% H_2O_2 (H_2O_2 + $SO_2 \rightarrow H_2SO_4$). After 10.0 minutes the H_2SO_4 was titrated with 11.1 mL of 0.00204 M NaOH. Calculate the SO_2 concentration in ppm (that is, mL $SO_2/10^6$ mL air) if the density of SO_2 is 0.00285 g/mL.

***16-31.** The digestion of a 0.1417-g sample of a phosphorus-containing compound in a mixture of HNO_3 and H_2SO_4 resulted in the formation of CO_2, H_2O, and H_3PO_4. Addition of ammonium molybdate yielded a solid having the composition $(NH_4)_3PO_4 \cdot 12MoO_3$ (1876.3 g/mol). This precipitate was filtered, washed, and dissolved in 50.00 mL of 0.2000 M NaOH:

$$(NH_4)_3PO_4 \cdot 12MoO_3(s) + 26OH^- \rightarrow HPO_4^{2-} + 12MoO_4^{2-} + 14H_2O + 3NH_3(g)$$

After the solution was boiled to remove the NH_3, the excess NaOH was titrated with 14.17 mL of 0.1741 M HCl to a phenolphthalein end point. Calculate the percentage of phosphorus in the sample.

***16-32.** A 0.8160-g sample containing dimethylphthalate, $C_6H_4(COOCH_3)_2$ (194.19 g/mol), and unreactive species was refluxed with 50.00 mL of 0.1031 M NaOH to hydrolyze the ester groups (this process is called *saponification*).

$$C_6H_4(COOH_3)_2 + 2OH^- \rightarrow C_6H_4(COO)^{2-} + H_2O$$

After the reaction was complete, the excess NaOH was back-titrated with 32.25 mL of 0.1251 M HCl. Calculate the percentage of dimethylphthalate in the sample.

***16-33.** Neohetramine, $C_{16}H_{21}ON_4$ (285.37 g/mol), is a common antihistamine. A 0.1532-g sample containing this compound was analyzed by the Kjeldahl method. The ammonia produced was collected in H_3BO_3; the resulting $H_2BO_3^-$ was titrated with 36.65 mL of 0.01522 M HCl. Calculate the percentage of neohetramine in the sample.

16-34. The *Merck Index* indicates that 10 mg of guanidine, CH_5N_3, may be administered for each kilogram of body weight in the treatment of myasthenia gravis. The nitrogen in a 4-tablet sample that weighed a total of 7.50 g was converted to ammonia by Kjeldahl digestion, followed by distillation into 100.0 mL of 0.1750 M HCl. The analysis was completed by titrating the excess acid with 11.37 mL of 0.1080 M NaOH. How many of these tablets represent a proper dose for patients who weigh (a) 100 lb, (b) 150 lb, and (c) 275 lb?

***16-35.** A 0.992-g sample of canned tuna was analyzed by the Kjeldahl method; 22.66 mL of 0.1224 M HCl were required to titrate the liberated ammonia. Calculate the percentage of nitrogen in the sample.

16-36. Calculate the mass in grams of protein in a 6.50-oz can of tuna in Problem 16-35.

***16-37.** A 0.5843-g sample of a plant food preparation was analyzed for its N content by the Kjeldahl method, the liberated NH_3 being collected in 50.00 mL of 0.1062 M HCl. The excess acid required an 11.89-mL back-titration with 0.0925 M NaOH. Express the results of this analysis in terms of percent

*(a) N.
(b) urea, H_2NCONH_2.
*(c) $(NH_4)_2SO_4$.
(d) $(NH_4)_3PO_4$.

16-38 A 0.9092-g sample of a wheat flour was analyzed by the Kjeldahl procedure. The ammonia formed was distilled into 50.00 mL of 0.05063 M HCl; a 7.46-mL back-titration with 0.04917 M NaOH was required. Calculate the percentage of protein in the flour.

***16-39.** A 1.219-g sample containing $(NH_4)_2SO_4$, NH_4NO_3, and nonreactive substances was diluted to 200 mL in a volumetric flask. A 50.00-mL aliquot was made basic with strong alkali, and the liberated NH_3 was distilled into 30.00 mL of 0.08421 M HCl. The excess HCl required 10.17 mL of 0.08802 M NaOH. A 25.00-mL aliquot of the sample was made alkaline after the addition of Devarda's alloy, and the NO_3^- was reduced to NH_3. The NH_3 from both NH_4^+ and NO_3^- was then distilled into 30.00 mL of the standard acid and back-titrated with 14.16 mL of the base. Calculate the percentages of $(NH_4)_2SO_4$ and NH_4NO_3 in the sample.

***16-40.** A 1.217-g sample of commercial KOH contaminated by K_2CO_3 was dissolved in water, and the resulting solution was diluted to 500.0 mL. A 50.00-mL aliquot of this solution was treated with 40.00 mL of 0.05304 M HCl and boiled to remove CO_2. The excess acid consumed 4.74 mL of 0.04983 M NaOH (phenolphthalein indicator). An excess of neutral $BaCl_2$ was added to another

50.00-mL aliquot to precipitate the carbonate as $BaCO_3$. The solution was then titrated with 28.56 mL of the acid to a phenolphthalein end point. Calculate the percentage KOH, K_2CO_3, and H_2O in the sample, assuming that these are the only compounds present.

16-41. A 0.5000-g sample containing $NaHCO_3$, Na_2CO_3, and H_2O was dissolved and diluted to 250.0 mL. A 25.00-mL aliquot was then boiled with 50.00 mL of 0.01255 M HCl. After cooling, the excess acid in the solution required 2.34 mL of 0.01063 M NaOH when titrated to a phenolphthalein end point. A second 25.00-mL aliquot was then treated with an excess of $BaCl_2$ and 25.00 mL of the base; precipitation of all the carbonate resulted, and 7.63 mL of the HCl was required to titrate the excess base. Calculate the composition of the mixture.

***16-42.** Calculate the volume of 0.06122 M HCl needed to titrate
(a) 10.00, 15.00, 25.00, and 40.00 mL of 0.05555 M Na_3PO_4 to a thymolphthalein end point.
(b) 10.00, 15.00, 20.00, and 25.00 mL of 0.05555 M Na_3PO_4 to a bromocresol green end point.
(c) 20.00, 25.00, 30.00, and 40.00 mL of a solution that is 0.02102 M in Na_3PO_4 and 0.01655 M in Na_2HPO_4 to a bromocresol green end point.
(d) 15.00, 20.00, 35.00, and 40.00 mL of a solution that is 0.02102 M in Na_3PO_4 and 0.01655 M in NaOH to a thymolphthalein end point.

16-43. Calculate the volume of 0.07731 M NaOH needed to titrate
(a) 25.00 mL of a solution that is 0.03000 M in HCl and 0.01000 M in H_3PO_4 to a bromocresol green end point.
(b) the solution in (a) to a thymolphthalein end point.
(c) 10.00, 20.00, 30.00, and 40.00 mL of 0.06407 M NaH_2PO_4 to a thymolphthalein end point.
(d) 20.00, 25.00, and 30.00 mL of a solution that is 0.02000 M in H_3PO_4 and 0.03000 M in NaH_2PO_4 to a thymolphthalein end point.

***16-44.** A series of solutions containing NaOH, Na_2CO_3, and $NaHCO_3$, alone or in compatible combination, were titrated with 0.1202 M HCl. Tabulated below are the volumes of acid needed to titrate 25.00-mL portions of each solution to a (1) phenolphthalein and (2) bromocresol green end point. Use this information to deduce the composition of the solutions. In addition, calculate the concentration of each solute in milligrams per milliliter of solution.

	(1)	(2)
(a)	22.42	22.44
(b)	15.67	42.13
(c)	29.64	36.42
(d)	16.12	32.23
(e)	0.00	33.333

16-45. A series of solutions containing NaOH, Na_3AsO_4, and Na_2HAsO_4, alone or in compatible combination, were titrated with 0.08601 M HCl. Tabulated below are the volumes of acid needed to titrate 25.00-mL portions of each solution to a (1) phenolphthalein and (2) bromocresol green end point. Use this information to deduce the composition of the solutions. In addition, calculate the concentration of each solute in milligrams per milliliter of solution.

	(1)	(2)
(a)	0.00	18.15
(b)	21.00	28.15
(c)	19.80	39.61
(d)	18.04	18.03
(e)	16.00	37.37

***16-46.** Define the equivalent weight of (a) an acid and (b) a base.

16-47. Calculate the equivalent weight of oxalic acid dihydrate ($H_2C_2O_4 \cdot 2H_2O$, 126.1 g/mol) when it is titrated to (a) a bromocresol green end point and (b) a phenolphthalein end point.

***16-48.** A 10.00-mL sample of vinegar (acetic acid, CH_3COOH) was pipetted into a flask, two drops of phenolphthalein indicator were added, and the acid was titrated with 0.1008 M NaOH.
(a) If 45.62 mL of the base was required for the titration, what was the molar concentration of acetic acid in the sample?
(b) If the density of the pipetted acetic acid solution was 1.004 g/mL, what was the percentage of acetic acid in the sample?

16-49. Challenge Problem
(a) Why are indicators used only in the form of dilute solutions?
(b) Suppose that 0.1% methyl red (molar mass 269 g/mol) is used as the indicator in a titration to determine the acid neutralizing capacity of an Ohio lake. Five drops (0.25 mL) of methyl red solution are added to a 100-mL sample of water, and 4.74 mL of 0.01072 M hydrochloric acid is required to change the indicator to the midpoint in its transition

range. Assuming that there is no indicator error, what is the acid neutralizing capacity of the lake expressed as milligrams of calcium bicarbonate per liter in the sample?

(c) If the indicator was initially in its acid form, what is the indicator error expressed as a percentage of the acid neutralizing capacity?

(d) What is the correct value for the acid neutralizing capacity determination?

(e) List four species other than carbonate or bicarbonate that may contribute to acid neutralizing capacity.

(f) It is normally assumed that species other than carbonate or bicarbonate do not contribute appreciably to acid neutralizing capacity. Suggest circumstances under which this assumption may not be valid.

(g) Particulate matter may make a significant contribution to acid neutralizing capacity. Explain how you would deal with this problem.

(h) Explain how you would determine separately the contribution to the acid neutralizing capacity from particulate matter and the contribution from soluble species.

InfoTrac College Edition

For additional readings, go to InfoTrac College Edition, your online research library, at

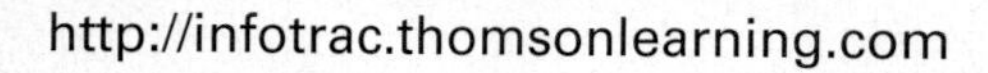

CHAPTER 17

Complexation Reactions and Titrations

Complexation reactions are important in many areas of science and everyday life, such as black-and-white photography. Shown to the right are photomicrographs of a capillary chromatography column at ×1300 (top) and ×4900 (bottom) magnification. Black-and-white film consists of an emulsion of finely divided AgBr coated on a polymer strip. Exposure to light from the scanning electron microscope causes reduction of some of the Ag^+ ions to Ag atoms and corresponding oxidation of Br^- to Br atoms. These atoms remain in the crystal lattice of AgBr as invisible defects, or the so-called latent image. Developing reduces many more Ag^+ ions to Ag atoms in the granules of AgBr containing Ag atoms from the original latent image. This produces a visible negative image, in which dark regions of Ag atoms represent areas where light has exposed the film. The fixing step removes the unexposed AgBr by forming the highly stable silver thiosulfate complex $[Ag(S_2O_3)_2]^{3-}$. The black metallic silver of the negative remains.

$$AgBr(s) + 2S_2O_3^{2-}(aq) \rightarrow [Ag(S_2O_3)_2]^{3-}(aq) + Br^-(aq)$$

After the negative has been fixed, a positive image is produced by projecting light through the negative onto photographic paper. (© American Chemical Society. Courtesy of R. N. Zare, Stanford University, Chemistry Dept.)

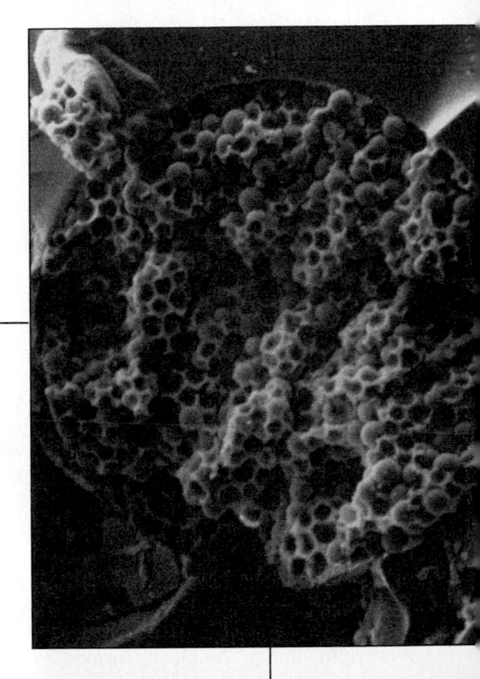

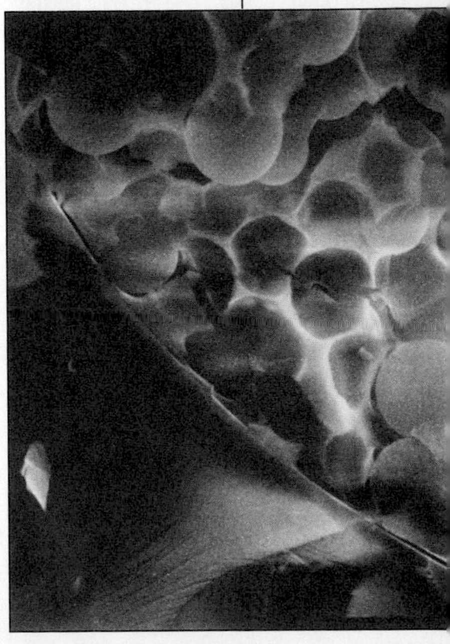

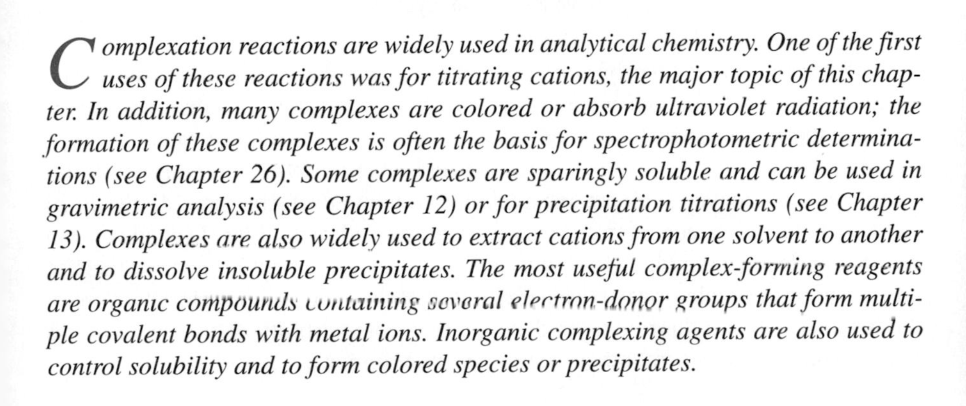

Complexation reactions are widely used in analytical chemistry. One of the first uses of these reactions was for titrating cations, the major topic of this chapter. In addition, many complexes are colored or absorb ultraviolet radiation; the formation of these complexes is often the basis for spectrophotometric determinations (see Chapter 26). Some complexes are sparingly soluble and can be used in gravimetric analysis (see Chapter 12) or for precipitation titrations (see Chapter 13). Complexes are also widely used to extract cations from one solvent to another and to dissolve insoluble precipitates. The most useful complex-forming reagents are organic compounds containing several electron-donor groups that form multiple covalent bonds with metal ions. Inorganic complexing agents are also used to control solubility and to form colored species or precipitates.

17A THE FORMATION OF COMPLEXES

Most metal ions react with electron-pair donors to form coordination compounds or complexes. The donor species, or **ligand,** must have at least one pair of unshared electrons available for bond formation. Water, ammonia, and halide ions are

A **ligand** is an ion or a molecule that forms a covalent bond with a cation or a neutral metal atom by donating a pair of electrons, which are then shared by the two.

common inorganic ligands. In fact, most metal ions in aqueous solution actually exist as aquo complexes. Copper(II) in aqueous solution, for example, is readily complexed by water molecules to form species such as $Cu(H_2O)_4^{2+}$. We often simplify such complexes in chemical equations by writing the metal ion as if it were uncomplexed Cu^{2+}. Remember, however, that such ions are actually aquo complexes in aqueous solution.

The number of covalent bonds that a cation tends to form with electron donors is its coordination number. Typical values for coordination numbers are 2, 4, and 6. The species formed as a result of coordination can be electrically positive, neutral, or negative. For example, copper(II), which has a coordination number of 4, forms a cationic ammine complex, $Cu(NH_3)_4^{2+}$; a neutral complex with glycine, $Cu(NH_2CH_2COO)_2$; and an anionic complex with chloride ion, $CuCl_4^{2-}$.

► **Chelate** is pronounced *kee'late* and is derived from the Greek word for claw. The copper/glycine complexation reaction is shown below.

Cu^{2+} + 2H—C(H)(NH$_2$)—C(=O)—OH ⟶

Glycine

Cu/glycine complex + $2H^+$

Titrimetric methods based on complex formation, sometimes called **complexometric methods,** have been used for more than a century. The truly remarkable growth in their analytical application, based on a particular class of coordination compounds called **chelates,** began in the 1940s. A chelate is produced when a metal ion coordinates with two or more donor groups of a single ligand to form a five- or six-member heterocyclic ring. The copper complex of glycine mentioned in the previous paragraph is an example. Here, copper bonds to both the oxygen of the carboxyl group and the nitrogen of the amine group (see Margin Note).

A ligand that has a single donor group, such as ammonia, is called **unidentate** (single-toothed), whereas one such as glycine, which has two groups available for covalent bonding, is called **bidentate.** Tridentate, tetradentate, pentadentate, and hexadentate chelating agents are also known.

Dentate (Latin) means having tooth-like projections.

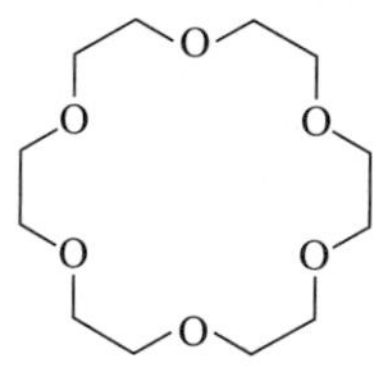

18-crown-6

dibenzo-18-crown-6

cryptand 2,2,2

Crown ethers and cryptands.

► The selectivity of a ligand for one metal ion over another refers to the stability of the complexes formed. The higher the formation constant of the metal-ligand complex, the better the selectivity of the ligand for the metal relative to similar complexes formed with other metals.

Another important type of complex is formed between metal ions and cyclic organic compounds, known as **macrocycles.** These molecules contain nine or more atoms in the cycle and include at least three heteroatoms, usually oxygen, nitrogen, or sulfur. Crown ethers such as 18-crown-6 and dibenzo-18-crown-6 are examples of organic macrocycles. Some macrocyclic compounds form three-dimensional cavities that can just accommodate appropriately sized metal ions. Ligands known as **cryptands** are examples. Selectivity occurs to a large extent because of the size and shape of the cycle or cavity relative to that of the metal, although the nature of the heteroatoms and their electron densities, the compatibility of the donor atoms with the metal, and several other factors also play important roles.

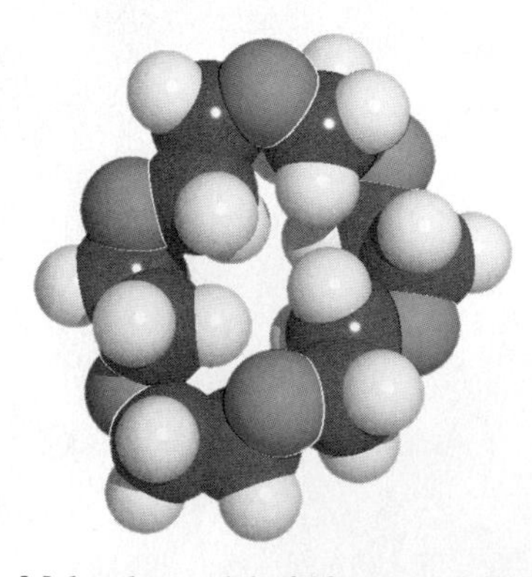

Molecular model of 18-crown-6. This crown ether can form strong complexes with alkali metal ions. The formation constants of the Na^+, K^+ and Rb^+ complexes are in the 10^5 to 10^6 range.

17A-1 Complexation Equilibria

Complexation reactions involve a metal ion M reacting with a ligand L to form a complex ML, as shown in Equation 17-1,

$$M + L \rightleftharpoons ML \qquad (17\text{-}1)$$

where the charges on the ions have been omitted so as to be general. Complexation reactions occur in a stepwise fashion; the reaction in Equation 17-1 is often followed by additional reactions:

$$ML + L \rightleftharpoons ML_2 \tag{17-2}$$

$$ML_2 + L \rightleftharpoons ML_3 \tag{17-3}$$

$$\vdots \qquad \vdots$$

$$ML_{n-1} + L \rightleftharpoons ML_n \tag{17-4}$$

Unidentate ligands invariably add in a series of steps, as shown here. With multidentate ligands, the maximum coordination number of the cation may be satisfied with only one ligand or a few added ligands. For example, Cu(II), with a maximum coordination number of 4, can form complexes with ammonia that have the formulas $Cu(NH_3)^{2+}$, $Cu(NH_3)_2^{2+}$, $Cu(NH_3)_3^{2+}$, and $Cu(NH_3)_4^{2+}$. With the bidentate ligand glycine (gly), the only complexes that form are $Cu(gly)^{2+}$ and $Cu(gly)_2^{2+}$.

The equilibrium constants for complex formation reactions are generally written as formation constants, as discussed in Chapter 9. Thus, each of the Reactions 17-1 through 17-4 is associated with a stepwise formation constant, K_1 through K_4. For example, $K_1 = [ML]/[M][L]$, $K_2 = [ML_2]/[ML][L]$, and so on. We can also write the equilibria as the sum of individual steps. These have overall formation constants designated by the symbol β_n. Thus,

$$M + L \rightleftharpoons ML \qquad \beta_1 = \frac{[ML]}{[M][L]} = K_1 \tag{17-5}$$

$$M + 2L \rightleftharpoons ML_2 \qquad \beta_2 = \frac{[ML_2]}{[M][L]^2} = K_1K_2 \tag{17-6}$$

$$M + 3L \rightleftharpoons ML_3 \qquad \beta_3 = \frac{[ML_3]}{[M][L]^3} = K_1K_2K_3 \tag{17-7}$$

$$\vdots \qquad \vdots$$

$$M + nL \rightleftharpoons ML_n \qquad \beta_n = \frac{[ML_n]}{[M][L]^n} = K_1K_2 \cdots K_n \tag{17-8}$$

Except for the first step, the overall formation constants are products of the stepwise formation constants for the individual steps leading to the product.

For a given species like ML, we can calculate an alpha value, which is the fraction of the total metal concentration existing in that form. Thus, α_M is the fraction of the total metal present at equilibrium as the free metal, α_{ML} is the fraction present as ML, and so on. As derived in Feature 17-1, the alpha values are given by

$$\alpha_M = \frac{1}{1 + \beta_1[L] + \beta_2[L]^2 + \beta_3[L]^3 + \cdots + \beta_n[L]^n} \tag{17-9}$$

$$\alpha_{ML} = \frac{\beta_1[L]}{1 + \beta_1[L] + \beta_2[L]^2 + \beta_3[L]^3 + \cdots + \beta_n[L]^n} \tag{17-10}$$

$$\alpha_{ML_2} = \frac{\beta_2[L]^2}{1 + \beta_1[L] + \beta_2[L]^2 + \beta_3[L]^3 + \cdots + \beta_n[L]^n} \tag{17-11}$$

$$\alpha_{ML_n} = \frac{\beta_n[L]^n}{1 + \beta_1[L] + \beta_2[L]^2 + \beta_3[L]^3 + \cdots + \beta_n[L]^n} \tag{17-12}$$

Note that these expressions are analogous to the α expressions we wrote for polyfunctional acids and bases except that the equations here are written in terms of formation equilibria while those for acids or bases are written in terms of dissociation equilibria. Also, the master variable is the ligand concentration [L] instead of the hydronium ion concentration. The denominators are the same for each α value. Plots of the α values versus p[L] are known as **distribution diagrams.**

Spreadsheet Summary In the first exercise in Chapter 9 of *Applications of Microsoft® Excel in Analytical Chemistry,* α values for the Cu(II)/NH_3 complexes are calculated and used to plot distribution diagrams. The α values for the Cd(II)/Cl^- system are also calculated.

CD-ROM Simulation: Titration Curves and Alpha Plots for Metal Complex-forming Titrations.

FEATURE 17-1

Calculation of Alpha Values for Metal Complexes

The alpha values for metal-ligand complexes can be derived in the same way that we derived values for polyfunctional acids in Section 15H. The alphas are defined as

$$\alpha_M = \frac{[M]}{c_M} \qquad \alpha_{ML_2} = \frac{[ML_2]}{c_M}$$

$$\alpha_{ML} = \frac{[ML]}{c_M} \qquad \alpha_{ML_n} = \frac{[ML_n]}{c_M}$$

The total metal concentration c_M can be written

$$c_M = [M] + [ML] + [ML_2] + \cdots + [ML_n]$$

From the overall formation constants (see Equations 17-5 through 17-8), the concentrations of the complexes can be expressed in terms of the free metal concentration [M], to give

$$\begin{aligned} c_M &= [M] + \beta_1[M][L] + \beta_2[M][L]^2 + \cdots + \beta_n[M][L]^n \\ &= [M]\{1 + \beta_1[L] + \beta_2[L]^2 + \cdots + \beta_n[L]^n\} \end{aligned}$$

Now α_M can be found as

$$\begin{aligned} \alpha_M = \frac{[M]}{c_M} &= \frac{[M]}{[M] + \beta_1[M][L] + \beta_2[M][L]^2 + \cdots + \beta_n[M][L]^n} \\ &= \frac{1}{1 + \beta_1[L] + \beta_2[L]^2 + \beta_3[L]^3 + \cdots + \beta_n[L]^n} \end{aligned}$$

Note that the last form is Equation 17-9. We can find a_{ML} from

$$\alpha_{ML} = \frac{[ML]}{c_M} = \frac{\beta_1[M][L]}{[M] + \beta_1[M][L] + \beta_2[M][L]^2 + \cdots + \beta_n[M][L]^n}$$
$$= \frac{\beta_1[L]}{1 + \beta_1[L] + \beta_2[L]^2 + \beta_3[L]^3 + \cdots + \beta_n[L]^n}$$

This last form is identical to Equation 17-10. The other alpha values in Equations 17-11 and 17-12 can be found in a similar manner.

17A-2 The Formation of Insoluble Species

In the cases discussed in the previous section, the complexes formed are soluble in solution. The addition of ligands to a metal ion, however, may result in insoluble species, such as the familiar nickel-dimethylglyoxime precipitate. In many cases, the intermediate uncharged complexes in the stepwise formation scheme may be sparingly soluble, whereas the addition of more ligand molecules may result in soluble species. For example, adding Cl^- to Ag^+ results in the insoluble AgCl precipitate. Addition of a large excess of Cl^- produces soluble species $AgCl_2^-$, $AgCl_3^{2-}$, and $AgCl_4^{3-}$.

In contrast to complexation equilibria, which are most often treated as formation reactions, solubility equilibria are normally treated as dissociation reactions, as discussed in Chapter 9. In general, for a sparingly soluble salt M_xA_y in a saturated solution, we can write

$$M_xA_y(s) \rightleftharpoons xM^{y+}(aq) + yA^{x-}(aq) \qquad K_{sp} = [M^{y+}]^x[A^{x-}]^y \qquad (17\text{-}13)$$

where K_{sp} is the solubility product. Hence, for BiI_3, the solubility product is written $K_{sp} = [Bi^{3+}][I^-]^3$.

17A-3 Ligands That Can Protonate

Complexation equilibria can be complicated by side reactions involving the metal or the ligand. Such side reactions make it possible to exert some additional control over the complexes that form. Metals can form complexes with ligands other than the one of interest. If these complexes are strong, we can effectively prevent complexation with the ligand of interest. Ligands can also undergo side reactions. One of the most common side reactions is of a ligand that can protonate; that is, the ligand is a weak acid.

Complexation with Protonating Ligands

Consider the formation of soluble complexes between the metal M and the ligand L. Assume that L is the conjugate base of a polyprotic acid and forms HL, H_2L, . . . H_nL, where again the charges have been omitted for generality. Adding acid to a solution containing M and L reduces the concentration of free L available to complex with M and, thus, decreases the effectiveness of L as a complexing agent (Le

Châtelier's principle). For example, ferric ions (Fe^{3+}) form complexes with oxalate ($C_2O_4^{2-}$, abbreviated Ox^{2-}) with formulas $(FeOx)^+$, $(FeOx_2)^-$, and $(FeOx_3)^{3-}$. Oxalate can protonate to form HOx^- and H_2Ox. In basic solution, where most of the oxalate is present as Ox^{2-} before complexation with Fe^{3+}, the ferric/oxalate complexes are very stable. Adding acid, however, protonates the oxalate ion, which in turn causes dissociation of the ferric complexes.

For a diprotic acid like oxalic acid, the fraction of the total oxalate-containing species in any given form (Ox^{2-}, HOx^-, and H_2Ox) is given by an alpha value (recall Section 15H). Since

$$c_T = [H_2Ox] + [HOx^-] + [Ox^{2-}] \tag{17-14}$$

we can write the alpha values α_2, α_1, and α_0 as

$$\alpha_0 = \frac{[H_2Ox]}{c_T} = \frac{[H^+]^2}{[H^+]^2 + K_{a1}[H^+] + K_{a1}K_{a2}} \tag{17-15}$$

$$\alpha_1 = \frac{[HOx^-]}{c_T} = \frac{K_{a1}[H^+]}{[H^+]^2 + K_{a1}[H^+] + K_{a1}K_{a2}} \tag{17-16}$$

$$\alpha_2 = \frac{[Ox^{2-}]}{c_T} = \frac{K_{a1}K_{a2}}{[H^+]^2 + K_{a1}[H^+] + K_{a1}K_{a2}} \tag{17-17}$$

Since we are interested in the free oxalate concentration, we will be most concerned with the highest α value, here α_2. From Equation 17-17, we can write

$$[Ox^{2-}] = c_T\alpha_2 \tag{17-18}$$

Note that as the solution gets more acidic, the first two terms in the denominator of Equation 17-17 dominate, and α_2 and the free oxalate concentration decrease. When the solution is very basic, α_2 becomes nearly unity, and $[Ox^{2-}] \approx c_T$, indicating that nearly all the oxalate is in the Ox^{2-} form in basic solution.

Conditional Formation Constants

To take into account the effect of pH on the free ligand concentration in a complexation reaction, it is useful to introduce a **conditional** or **effective formation constant.** Such constants are pH-dependent equilibrium constants that apply at a single pH only. For the reaction of Fe^{3+} with oxalate, for example, we can write the formation constant K_1 for the first complex as

$$K_1 = \frac{[FeOx^+]}{[Fe^{3+}][Ox^{2-}]} = \frac{[FeOx^+]}{[Fe^{3+}]\alpha_2 c_T} \tag{17-19}$$

At a particular pH value, α_2 is constant, and we can combine K_1 and α_2 to yield a new conditional constant, K_1':

$$K_1' = \alpha_2 K_1 = \frac{[FeOx^+]}{[Fe^{3+}]c_T} \tag{17-20}$$

The use of conditional constants greatly simplifies calculations because c_T is often known or readily computed, whereas the free ligand concentration is not as easily determined. The overall formation constants (β values) for the higher complexes, $(FeOx_2)^-$ and $(FeOx_3)^{3-}$, can also be written as conditional constants.

Spreadsheet Summary Ligands that protonate are treated in Chapter 9 of *Applications of Microsoft® Excel in Analytical Chemistry.* Alpha values and conditional formation constants are calculated.

17B TITRATIONS WITH INORGANIC COMPLEXING AGENTS

Complexation reactions have many uses in analytical chemistry, but their classical application is in **complexometric titrations.** Here, a metal ion reacts with a suitable ligand to form a complex, and the equivalence point is determined by an indicator or an appropriate instrumental method. The formation of soluble inorganic complexes is not widely used for titrations, as discussed later, but the formation of precipitates, particularly with silver nitrate as the titrant, is the basis for many important determinations (see Section 13F).

The progress of a complexometric titration is generally illustrated by a titration curve, which is usually a plot of $pM = -\log[M]$ as a function of the volume of titrant added. Most often, in complexometric titrations the ligand is the titrant and the metal ion the analyte, although occasionally the reverse is true. Many precipitation titrations, as discussed in Section 13F, use the metal ion as the titrant. Most simple inorganic ligands are unidentate, which can lead to low complex stability and indistinct titration end points. As titrants, multidentate ligands, particularly those having four or six donor groups, have two advantages over their unidentate counterparts. First, they generally react more completely with cations and thus provide sharper end points. Second, they ordinarily react with metal ions in a single-step process, whereas complex formation with unidentate ligands usually involves two or more intermediate species (recall Equations 17-1 through 17-4).

◀ Tetradentate or hexadentate ligands are more satisfactory as titrants than ligands with a lesser number of donor groups because their reactions with cations are more complete and because they tend to form 1:1 complexes.

The advantage of a single-step reaction is illustrated by the titration curves shown in Figure 17-1. Each of the titrations involves a reaction that has an overall

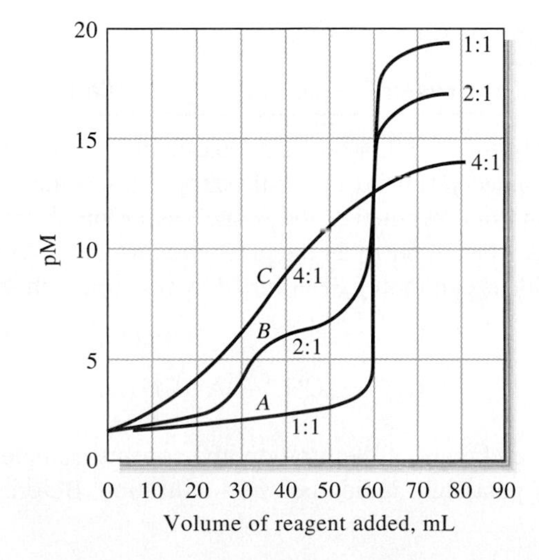

Figure 17-1 Titration curves for complexometric titrations. Titration of 60.0 mL of a solution that is 0.020 M in metal M with (A) a 0.020 M solution of the tetradentate ligand D to give MD as the product; (B) a 0.040 M solution of the bidentate ligand B to give MB_2; and (C) a 0.080 M solution of the unidentate ligand A to give MA_4. The overall formation constant for each product is 10^{20}.

TABLE 17-1

Typical Inorganic Complex–Forming Titrations

Titrant	Analyte	Remarks
$Hg(NO_3)_2$	Br^-, Cl^-, SCN^-, CN^-, thiourea	Products are neutral Hg(II) complexes; various indicators used
$AgNO_3$	CN^-	Product is $Ag(CN)_2^-$; indicator is I^-; titrate to first turbidity of AgI
$NiSO_4$	CN^-	Product is $Ni(CN)_4^{2-}$; indicator is I^-; titrate to first turbidity of AgI
KCN	Cu^{2+}, Hg^{2+}, Ni^{2+}	Products are $Cu(CN)_4^{2-}$, $Hg(CN)_2$, and $Ni(CN)_4^{2-}$; various indicators used

equilibrium constant of 10^{20}. Curve *A* is derived for a reaction in which a metal ion M having a coordination number of 4 reacts with a tetradentate ligand D to form the complex of MD. (We have again omitted the charges on the two reactants for convenience.) Curve *B* is for the reaction of M with a hypothetical bidentate ligand B to give MB_2 in two steps. The formation constant for the first step is 10^{12} and for the second 10^8. Curve *C* involves a unidentate ligand A that forms MA_4 in four steps with successive formation constants of 10^8, 10^6, 10^4, and 10^2. These curves demonstrate that a much sharper end point is obtained with a reaction that takes place in a single step. For this reason, multidentate ligands are ordinarily preferred for complexometric titrations.

The most widely used complexometric titration employing a unidentate ligand is the titration of cyanide with silver nitrate, a method introduced by Liebig in the 1850s. This method involves the formation of the soluble $Ag(CN)_2^-$, as discussed in Feature 17-2. Other common inorganic complexing agents and their applications are listed in Table 17-1.

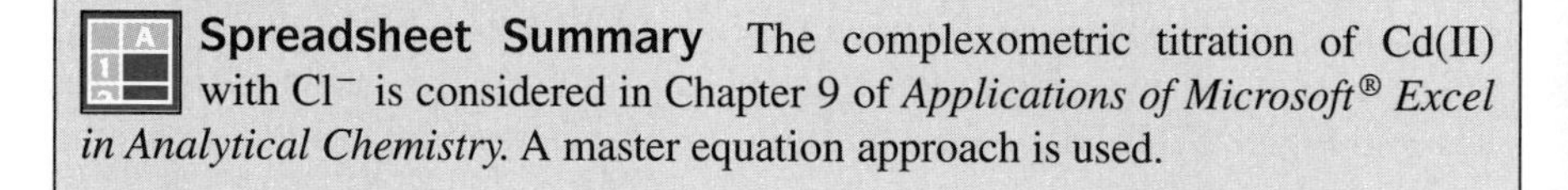
Spreadsheet Summary The complexometric titration of Cd(II) with Cl^- is considered in Chapter 9 of *Applications of Microsoft® Excel in Analytical Chemistry.* A master equation approach is used.

FEATURE 17-2

Determination of Hydrogen Cyanide in Acrylonitrile Plant Streams

Acrylonitrile, $CH_2{=}CH{-}C{\equiv}N$, is an extremely important chemical in the production of polyacrylonitrile. This thermoplastic is drawn out into fine threads and woven into synthetic fabrics such as Orlon, Acrilan, and Creslan. Hydrogen cyanide is an impurity in the plant streams that carry aqueous acrylonitrile. The cyanide is commonly determined by titration with $AgNO_3$. The titration reaction is

$$Ag^+ + 2CN^- \rightarrow Ag(CN)_2^-$$

To determine the end point of the titration, the aqueous sample is mixed with a basic solution of potassium iodide before the titration. Before the equivalence point, cyanide is in excess and all the Ag^+ is complexed. As soon as all the cyanide has been reacted, the first excess of Ag^+ causes a permanent turbidity to appear in the solution because of the formation of the AgI precipitate, according to

$$Ag^+ + I^- \rightarrow AgI(s)$$

17C ORGANIC COMPLEXING AGENTS

Many different organic complexing agents have become important in analytical chemistry because of their inherent sensitivity and potential selectivity in reacting with metal ions. Such reagents are particularly useful in precipitating metals, in binding metals to prevent interferences, in extracting metals from one solvent to another, and in forming complexes that absorb light for spectrophotometric determinations. The most useful organic reagents form chelate complexes with metal ions.

Many organic reagents are used to convert metal ions into forms that can be readily extracted from water into an immiscible organic phase. Extractions are widely employed to separate metals of interest from potential interfering ions and to achieve a concentrating effect by extracting into a phase of smaller volume. Extractions are applicable to much smaller amounts of metals than precipitations, and they avoid problems associated with coprecipitation. Separations by extraction are considered in Section 30C.

Several of the most widely used organic complexing agents for extractions are listed in Table 17-2. Some of these same reagents normally form insoluble species with metal ions in aqueous solution. In extraction applications, however, the solubility of the metal chelate in the organic phase keeps the complex from precipitating in the aqueous phase. In many cases, the pH of the aqueous phase is used to achieve some control over the extraction process, since most of the reactions are pH dependent, as shown in Equation 17-21.

$$n\text{HX}(org) + \text{M}^{n+}(aq) \rightleftharpoons \text{MX}_n(org) + n\text{H}^+(aq) \tag{17-21}$$

TABLE 17-2
Organic Reagents for Extracting Metals

Reagent	Metal Ions Extracted	Solvents
8-Hydroxyquinoline	Zn^{2+}, Cu^{2+}, Ni^{2+}, Al^{3+}, many others	Water → Chloroform ($CHCl_3$)
Diphenylthiocarbazone (dithizone)	Cd^{2+}, Co^{2+}, Cu^{2+}, Pb^{2+}, many others	Water → $CHCl_3$, or CCl_4
Acetylacetone	Fe^{3+}, Cu^{2+}, Zn^{2+}, U(VI), many others	Water → $CHCl_3$, CCl_4, or C_6H_6
Ammonium pyrrolidine dithiocarbamate	Transition metals	Water → Methyl isobutyl ketone
Tenoyltrifluoroacetone	Ca^{2+}, Sr^{2+}, La^{3+}, Pr^{3+}, other rare earths	Water → Benzene
Dibenzo-18-crown-6	Alkali metals, some alkaline earths	Water → Benzene

Another important application of organic complexing agents is the formation of stable complexes that bind a metal and prevent it from interfering in a determination. Such complexing agents are called **masking agents** and are discussed in Section 17D-8. Organic complexing agents are also widely used in spectrophotometric determinations of metal ions (see Chapter 26). Here, the metal-ligand complex either is colored or absorbs ultraviolet radiation. Organic complexing agents are also commonly employed in electrochemical determinations and in molecular fluorescence spectrometry.

17D AMINOCARBOXYLIC ACID TITRATIONS

Tertiary amines that also contain carboxylic acid groups form remarkably stable chelates with many metal ions.[1] Gerold Schwarzenbach first recognized their potential as analytical reagents in 1945. Since this original work, investigators throughout the world have described applications of these compounds to the volumetric determination of most of the metals in the periodic table.

17D-1 Ethylenediaminetetraacetic Acid (EDTA)

► EDTA, a hexadentate ligand, is among the most important and widely used reagents in titrimetry.

Ethylenediaminetetraacetic acid—also called (ethylenedinitrilo)tetraacetic acid—which is commonly shortened to EDTA, is the most widely used complexometric titrant. EDTA has the structural formula

$$\begin{matrix} HOOC-H_2C & & & & CH_2-COOH \\ & \diagdown & & \diagup & \\ & & N-CH_2-CH_2-N & & \\ & \diagup & & \diagdown & \\ HOOC-H_2C & & & & CH_2-COOH \end{matrix}$$

The EDTA molecule has six potential sites for bonding a metal ion: the four carboxyl groups and the two amino groups, each of the latter with an unshared pair of electrons. Thus, EDTA is a hexadentate ligand.

Acidic Properties of EDTA

The dissociation constants for the acidic groups in EDTA are $K_1 = 1.02 \times 10^{-2}$, $K_2 = 2.14 \times 10^{-3}$, $K_3 = 6.92 \times 10^{-7}$, and $K_4 = 5.50 \times 10^{-11}$. It is of interest that the first two constants are of the same order of magnitude, which suggests that the two protons involved dissociate from opposite ends of the rather long molecule. As a consequence of their physical separation, the negative charge created by the first dissociation does not greatly affect the removal of the second proton. The same cannot be said for the dissociation of the other two protons, however, which are much closer to the negatively charged carboxylate ions created by the initial dissociations.

The various EDTA species are often abbreviated H_4Y, H_3Y^-, H_2Y^{2-}, HY^{3-}, and Y^{4-}. Figure 17-2 illustrates how the relative amounts of these five species vary as a

[1]See for example, R. Pribil, *Applied Complexometry.* New York: Pergamon, 1982; A. Ringbom and E. Wanninen, in *Treatise on Analytical Chemistry,* 2nd ed., I. M. Kolthoff and P. J. Elving, Eds., Part I, Vol. 2, Chapter 11. New York: Wiley, 1979.

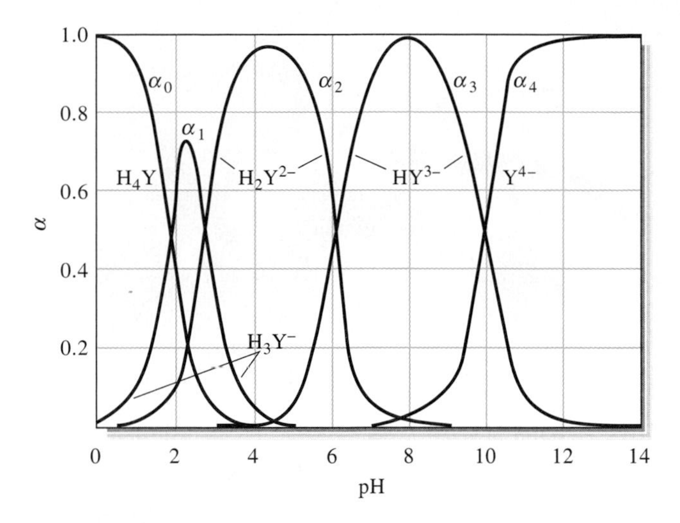

Figure 17-2 Composition of EDTA solutions as a function of pH. Note that the fully protonated form, H_4Y is only a major component in very acidic solutions (pH < 3). Throughout the pH range of 3 to 10, the species H_2Y^{2-} and HY^{3-} are predominant. The fully unprotonated form Y^{4-} is a significant component only in very basic solutions (pH > 10).

function of pH. Note that the species H_2Y^{2-} predominates in moderately acidic medium (pH 3 to 6).

Reagents for EDTA Titrations

The free acid H_4Y and the dihydrate of the sodium salt, $Na_2H_2Y \cdot 2H_2O$, are commercially available in reagent quality. The former can serve as a primary standard after it has been dried for several hours at 130°C to 145°C. It is then dissolved in the minimum amount of base required for complete solution.

◀ Standard EDTA solutions are ordinarily prepared by dissolving weighed quantities of $Na_2H_2Y \cdot 2H_2O$ and diluting to the mark in a volumetric flask.

Under normal atmospheric conditions, the dihydrate, $Na_2H_2Y \cdot 2H_2O$, contains 0.3% moisture in excess of the stoichiometric amount. For all but the most exacting work, this excess is sufficiently reproducible to permit use of a corrected weight of the salt in the direct preparation of a standard solution. If necessary, the pure dihydrate can be prepared by drying at 80°C for several days in an atmosphere of 50% relative humidity.

A number of compounds that are chemically related to EDTA have also been investigated but do not appear to offer significant advantages. We thus limit our discussion here to the properties and applications of EDTA.

◀ Nitrilotriacetic acid (NTA) is the second most common aminopolycarboxylic acid used for titrations. It is a tetradentate chelating agent and has the structure

```
                   H2
COOH—CH2         C—COOH
         \      /
           N
            \
             CH2—COOH
```

Structural formula of NTA.

FEATURE 17-3

Species Present in a Solution of EDTA

When dissolved in water, EDTA behaves like an amino acid, such as glycine (see Features 14-4 and 15-2). Here, however, a double zwitterion forms, which has the structure shown in Figure 17F-1a. Note that the net charge on this species is zero and that it contains four dissociable protons, two associated with two of the carboxyl groups and the other two associated with the two amine groups. For simplicity, we generally formulate the double zwitterion as H_4Y, where Y^{4-} is the fully deprotonated form of Figure 17F-1e. The first and second steps in the dissociation process involve successive loss of protons from the two carboxylic acid groups; the third and fourth steps involve dissociation of the protonated amine groups. The structural formulas of H_3Y^-, H_2Y^{2-}, and HY^{3-} are shown in Figure 17F-1b, c, and d.

(continued)

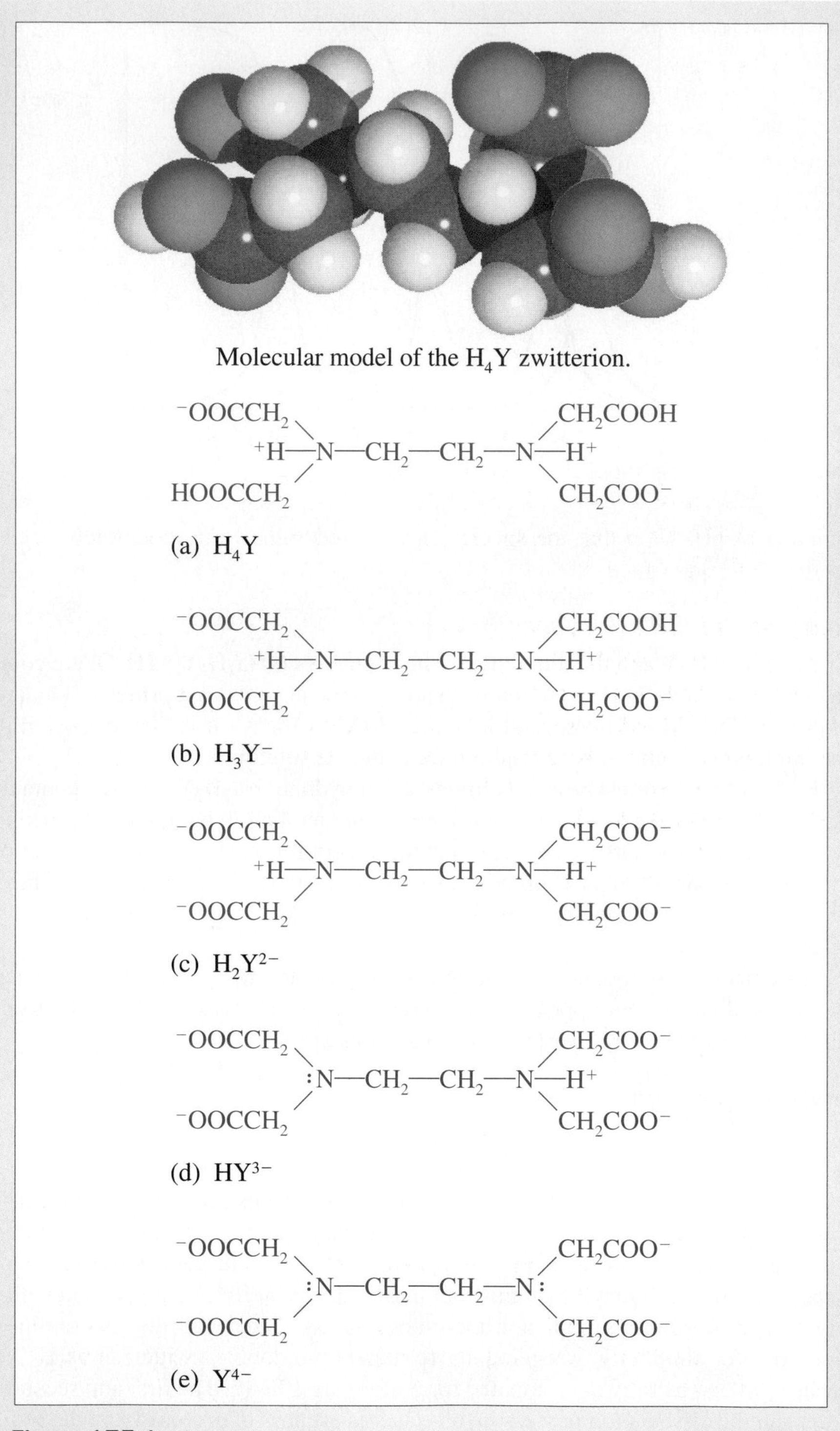

Figure 17F-1 Structure of H_4Y and its dissociation products. Note that the fully protonated species H_4Y exists as the double zwitterion with the amine nitrogens and two of the carboxylic acid groups protonated. The first two protons dissociate from the carboxyl groups, while the last two come from the amine groups.

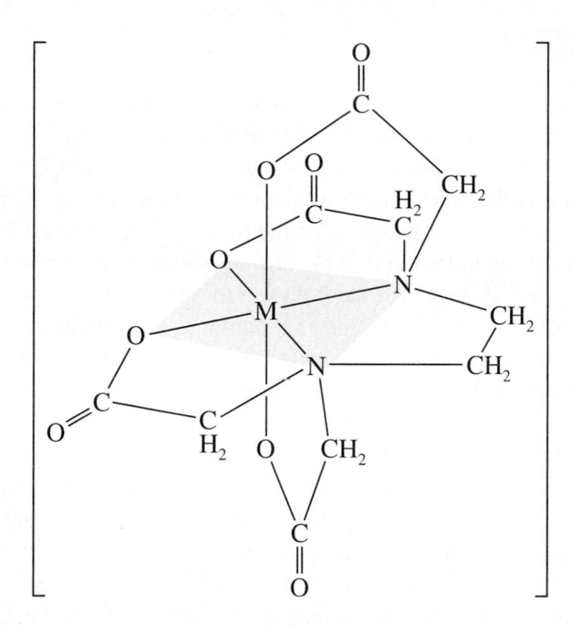

Figure 17-3 Structure of a metal/EDTA complex. Note that EDTA behaves here as a hexadentate ligand in that six donor atoms are involved in bonding the divalent metal cation.

17D-2 Complexes of EDTA and Metal Ions

Solutions of EDTA are particularly valuable as titrants because the reagent *combines with metal ions in a 1:1 ratio regardless of the charge on the cation.* For example, the silver and aluminum complexes are formed by the reactions:

◀ In general, we can write the reaction of the EDTA anion with a metal ion M^{n+} as $M^{n+} + Y^{4-} \rightleftharpoons MY^{(n-4)+}$

$$Ag^{+} + Y^{4-} \rightleftharpoons AgY^{3-}$$

$$Al^{3+} + Y^{4-} \rightleftharpoons AlY^{-}$$

EDTA is a remarkable reagent not only because it forms chelates with all cations except alkali metals but also because most of these chelates are sufficiently stable for titrations. This great stability undoubtedly results from the several complexing sites within the molecule that give rise to a cage-like structure, in which the cation is effectively surrounded by and isolated from solvent molecules. One of the common structures for metal/EDTA complexes is shown in Figure 17-3. The ability of EDTA to complex metals is responsible for its widespread use as a preservative in foods and in biological samples, as discussed in Feature 17-4.

Table 17-3 lists formation constants K_{MY} for common EDTA complexes. Note that the constant refers to the equilibrium involving the fully unprotonated species Y^{4-} with the metal ion:

$$M^{n+} + Y^{4-} \rightleftharpoons MY^{(n-4)+} \qquad K_{MY} = \frac{[MY^{(n-4)+}]}{[M^{n+}][Y^{4-}]} \tag{17-22}$$

17D-3 Equilibrium Calculations Involving EDTA

A titration curve for the reaction of a cation M^{n+} with EDTA consists of a plot of pM versus reagent volume. Values for pM are readily computed in the early stage of a titration by assuming that the equilibrium concentration of M^{n+} is equal to its analytical concentration, which in turn is readily derived from stoichiometric data.

FEATURE 17-4

EDTA as a Preservative

Trace quantities of metal ions can effectively catalyze the air oxidation of many of the compounds present in foods and biological samples (for example, proteins in blood). To prevent such oxidation reactions, it is important to inactivate or remove even trace amounts of metal ions. In processed foods, we can readily get trace quantities of metal ions as a result of contact with various metallic containers (kettles and vats) during the processing stages. EDTA is an excellent preservative for foods and is a common ingredient of such commercial food products as mayonnaise, salad dressings, and oils. When EDTA is added to foods, it so tightly binds most metal ions that they are unable to catalyze the air oxidation reaction. EDTA and other similar chelating agents are often called **sequestering agents** because of their ability to remove or inactivate metal ions. In addition to EDTA, some other common sequestering agents are salts of citric and phosphoric acid. These agents can protect the unsaturated side chains of triglycerides and other components from air oxidation. Such oxidation reactions are responsible for making fats and oils turn rancid. Sequestering agents are also added to prevent oxidation of easily oxidized compounds, such as ascorbic acid.

With biological samples, it is important to add EDTA as a preservative if the sample is to be stored for a long period. As with foods, EDTA tightly complexes metal ions and prevents them from catalyzing air oxidation reactions that can lead to decomposition of proteins and other compounds. During the trial of O. J. Simpson, the use of EDTA as a preservative became an important point. The prosecution team contended that if blood evidence had been planted on the back fence at Nicole Brown Simpson's home, EDTA should have been present, but if the blood were from the criminal, no preservative should be seen. Analytical evidence obtained by using a sophisticated instrumental system (liquid chromatography combined with tandem mass spectrometry) did show traces of EDTA, but the amounts were very small and subject to differing interpretations.

TABLE 17-3

Formation Constants for EDTA Complexes

Cation	K_{MY}*	log K_{MY}	Cation	K_{MY}	log K_{MY}
Ag^+	2.1×10^7	7.32	Cu^{2+}	6.3×10^{18}	18.80
Mg^{2+}	4.9×10^8	8.69	Zn^{2+}	3.2×10^{16}	16.50
Ca^{2+}	5.0×10^{10}	10.70	Cd^{2+}	2.9×10^{16}	16.46
Sr^{2+}	4.3×10^8	8.63	Hg^{2+}	6.3×10^{21}	21.80
Ba^{2+}	5.8×10^7	7.76	Pb^{2+}	1.1×10^{18}	18.04
Mn^{2+}	6.2×10^{13}	13.79	Al^{3+}	1.3×10^{16}	16.13
Fe^{2+}	2.1×10^{14}	14.33	Fe^{3+}	1.3×10^{25}	25.1
Co^{2+}	2.0×10^{16}	16.31	V^{3+}	7.9×10^{25}	25.9
Ni^{2+}	4.2×10^{18}	18.62	Th^{4+}	1.6×10^{23}	23.2

*Constants are valid at 20°C and ionic strength of 0.1.
Data from G. Schwarzenbach, *Complexometric Titrations,* p. 8. London: Chapman and Hall, 1957.

Calculation of M^{n+} at and beyond the equivalence point requires the use of Equation 17-22. The computation in this region is troublesome and time consuming if the pH is unknown and variable because both $[MY^{(n-4)+}]$ and $[M^{n+}]$ are pH dependent. Fortunately, EDTA titrations are always performed in solutions that are buffered to a known pH to avoid interference by other cations or to ensure satisfactory indicator behavior. Calculating $[M^{n+}]$ in a buffered solution containing EDTA is a relatively straightforward procedure provided that the pH is known. In this computation, use is made of the alpha value for H_4Y. Recall from Section 15H that α_4 for H_4Y can be defined as

$$\alpha_4 = \frac{[Y^{4-}]}{c_T} \tag{17-23}$$

where c_T is the total molar concentration of *uncomplexed* EDTA

$$c_T = [Y^{4-}] + [HY^{3-}] + [H_2Y^{2-}] + [H_3Y^-] + [H_4Y]$$

Conditional Formation Constants

To obtain the conditional formation constant for the equilibrium shown in Equation 17-22, we substitute $\alpha_4 c_T$ from Equation 17-23 for $[Y^{4-}]$ in the formation constant expression (see Equation 17-22):

$$M^{n+} + Y^{4-} \rightleftharpoons MY^{(n-4)+} \qquad K_{MY} = \frac{[MY^{(n-4)+}]}{[M^{n+}]\alpha_4 c_T} \tag{17-24}$$

Combining the two constants α_4 and K_{MY} yields the conditional formation constant K'_{MY}

$$K'_{MY} = \alpha_4 K_{MY} = \frac{[MY^{(n-4)+}]}{[M^{n+}]c_T} \tag{17-25}$$

where K'_{MY} is a constant *only at the pH for which α_4 is applicable.*

◀ Conditional formation constants are pH dependent.

Conditional constants are readily computed and provide a simple means by which the equilibrium concentration of the metal ion and the complex can be calculated at the equivalence point and where there is an excess of reactant. Note that replacement of $[Y^{4-}]$ with c_T in the equilibrium-constant expression greatly simplifies calculations because c_T is easily determined from the reaction stoichiometry, whereas $[Y^{4-}]$ is not.

Computing α_4 Values for EDTA Solutions

An expression for calculating α_4 at a given hydrogen ion concentration is derived by the method given in Section 15-H (see Feature 15-3). Thus, α_4 for EDTA is

$$\alpha_4 = \frac{K_1K_2K_3K_4}{[H^+]^4 + K_1[H^+]^3 + K_1K_2[H^+]^2 + K_1K_2K_3[H^+] + K_1K_2K_3K_4} \tag{17-26}$$

$$\alpha_4 = \frac{K_1K_2K_3K_4}{D} \tag{17-27}$$

◀ The alpha values for the other EDTA species are calculated in a similar manner and are found to be

$\alpha_0 = [H^+]^4/D \qquad \alpha_2 = K_1K_2[H^+]^2/D$
$\alpha_1 = K_1[H^+]^3/D \qquad \alpha_3 = K_1K_2K_3[H^+]/D$

Only α_4 is needed in deriving titration curves.

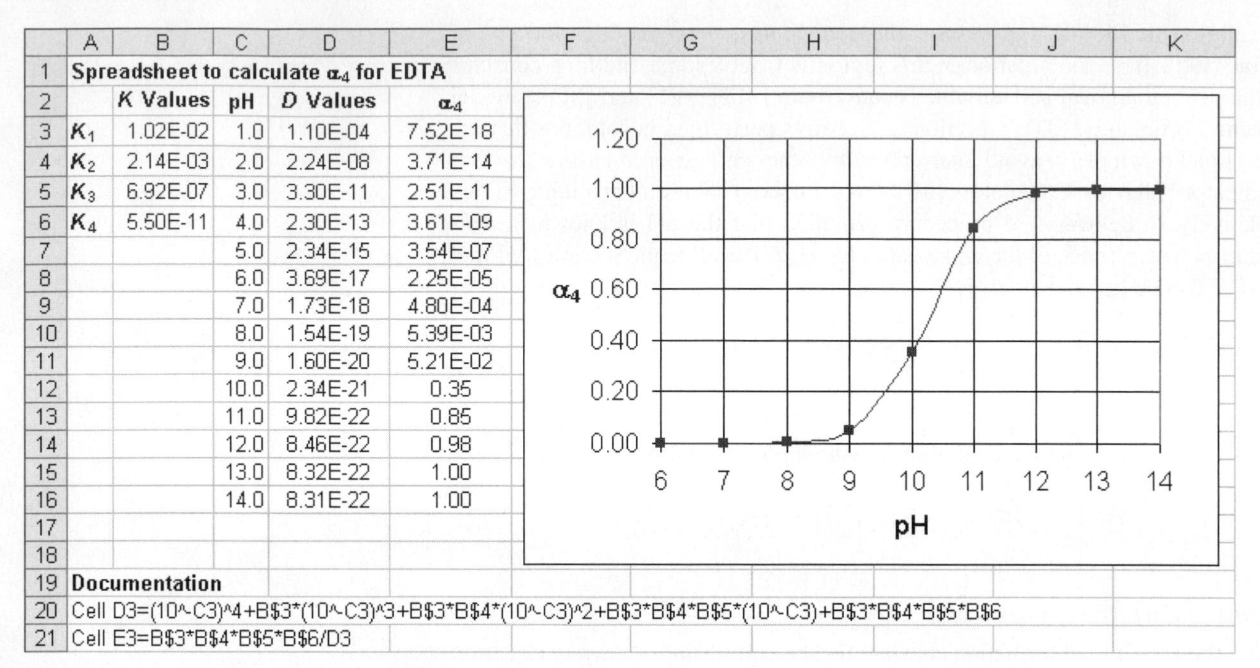

	A	B	C	D	E
1	**Spreadsheet to calculate α_4 for EDTA**				
2		*K* Values	pH	*D* Values	α_4
3	K_1	1.02E-02	1.0	1.10E-04	7.52E-18
4	K_2	2.14E-03	2.0	2.24E-08	3.71E-14
5	K_3	6.92E-07	3.0	3.30E-11	2.51E-11
6	K_4	5.50E-11	4.0	2.30E-13	3.61E-09
7			5.0	2.34E-15	3.54E-07
8			6.0	3.69E-17	2.25E-05
9			7.0	1.73E-18	4.80E-04
10			8.0	1.54E-19	5.39E-03
11			9.0	1.60E-20	5.21E-02
12			10.0	2.34E-21	0.35
13			11.0	9.82E-22	0.85
14			12.0	8.46E-22	0.98
15			13.0	8.32E-22	1.00
16			14.0	8.31E-22	1.00
17					
18					
19	**Documentation**				
20	Cell D3=(10^-C3)^4+B$3*(10^-C3)^3+B$3*B$4*(10^-C3)^2+B$3*B$4*B$5*(10^-C3)+B$3*B$4*B$5*B$6				
21	Cell E3=B$3*B$4*B$5*B$6/D3				

Figure 17-4 Spreadsheet to calculate α_4 for EDTA at selected pH values. Note that the acid dissociation constants for EDTA are entered in column B (labels in column A). Next, the pH values for which the calculations are to be done are entered in column C. The formula for calculating the denominator *D* in Equations 17-26 and 17-27 is placed into cell D3 and copied into D4 through D16. The final column E contains the equation for calculating the α_4 values as given in Equation 17-27. The graph shows a plot of α_4 versus pH.

where K_1, K_2, K_3, and K_4 are the four dissociation constants for H_4Y and *D* is the denominator of Equation 17-26.

Figure 17-4 is an Excel spreadsheet that calculates α_4 at selected pH values according to Equations 17-26 and 17-27. Note from the results that only about 4×10^{-12} percent of EDTA exists as Y^{4-} at pH 2.00. Example 17-1 illustrates how Y^{4-} is calculated for a solution of known pH.

EXAMPLE 17-1

Calculate the molar Y^{4-} concentration in a 0.0200 M EDTA solution buffered to a pH of 10.00.

At pH 10.00, α_4 is 0.35 (see Figure 17-4). Thus,

$$[Y^{4-}] = \alpha_4 c_T = 0.35 \times 0.0200 = 7.00 \times 10^{-3}\ M$$

Calculation of the Cation Concentration in EDTA Solutions

In an EDTA titration, we are interested in finding the cation concentration as a function of the amount of titrant (EDTA) added. Prior to the equivalence point, the cation is in excess, and its concentration can be found from the reaction stoichiometry. At the equivalence point and in the postequivalence-point region, however, the conditional formation constant of the complex must be used to calculate the cation concentration. Example 17-2 demonstrates how the cation concentration can be calculated in a solution of an EDTA complex. Example 17-3 illustrates this calculation when excess EDTA is present.

EXAMPLE 17-2

Calculate the equilibrium concentration of Ni^{2+} in a solution with an analytical NiY^{2-} concentration of 0.0150 M at pH (a) 3.0 and (b) 8.0.

From Table 17-3,

$$Ni^{2+} + Y^{4-} \rightleftharpoons NiY^{2-} \qquad K_{NiY} = \frac{[NiY^{2-}]}{[Ni^{2+}][Y^{4-}]} = 4.2 \times 10^{18}$$

The equilibrium concentration of NiY^{2-} is equal to the analytical concentration of the complex minus the concentration lost by dissociation. The latter is identical to the equilibrium Ni^{2+} concentration. Thus,

$$[NiY^{2-}] = 0.0150 - [Ni^{2+}]$$

If we assume that $[Ni^{2+}] << 0.0150$, an assumption that is almost certainly valid in light of the large formation constant of the complex, this equation simplifies to

$$[NiY^{2-}] \approx 0.0150$$

Since the complex is the only source of both Ni^{2+} and the EDTA species,

$$[Ni^{2+}] = [Y^{4-}] + [HY^{3-}] + [H_2Y^{2-}] + [H_3Y^{-}] + [H_4Y] = c_T$$

Substitution of this equality into Equation 17-25 gives

$$K'_{NiY} = \frac{[NiY^{2-}]}{[Ni^{2+}]c_T} = \frac{[NiY^{2-}]}{[Ni^{2+}]^2} = \alpha_4 K_{NiY}$$

(a) The spreadsheet in Figure 17-4 indicates that α_4 is 2.5×10^{-11} at pH 3.0. If we substitute this value and the concentration of NiY^{2-} into the equation for K'_{MY}, we get

$$\frac{0.0150}{[Ni^{2+}]^2} = 2.5 \times 10^{-11} \times 4.2 \times 10^{18} = 1.05 \times 10^{8}$$

$$[Ni^{2+}] = \sqrt{1.43 \times 10^{-10}} = 1.2 \times 10^{-5}\,M$$

(b) At pH 8.0, the conditional constant is much larger. Thus,

$$K'_{NiY} = 5.4 \times 10^{-3} \times 4.2 \times 10^{18} = 2.27 \times 10^{16}$$

and after we substitute this into the equation for K'_{NiY}, we get

$$[Ni^{2+}] = \sqrt{0.0150/(2.27 \times 10^{16})} = 8.1 \times 10^{-10}\,M$$

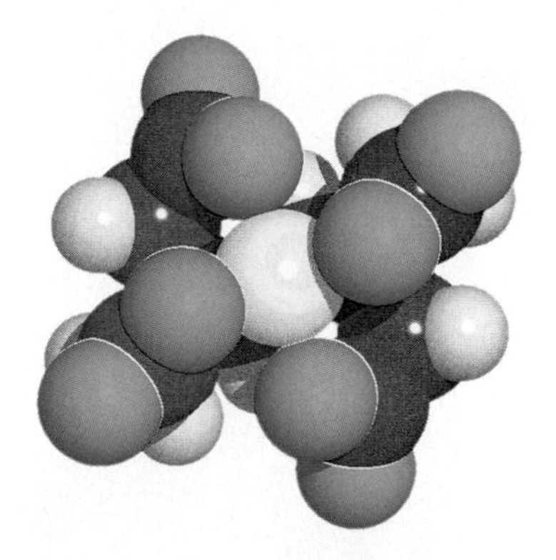

Molecular model of NiY^{2-}. This complex is typical of the strong complexes that EDTA forms with metal ions. The formation constant of the Ni^{2+} complex is 4.2×10^{18}.

◀ Note that for both pH 3.0 and pH 8.0, our assumption that $[Ni^{2+}] <<$ 0.0150 M is valid.

EXAMPLE 17-3

Calculate the concentration of Ni^{2+} in a solution that was prepared by mixing 50.0 mL of 0.0300 M Ni^{2+} with 50.00 mL of 0.0500 M EDTA. The mixture was buffered to a pH of 3.0.

Here, the solution has an excess of EDTA, and the analytical concentration of the complex is determined by the amount of Ni^{2+} originally present. Thus,

$$c_{NiY^{2+}} = 50.00 \text{ mL} \times \frac{0.0300 \text{ M}}{100 \text{ mL}} = 0.0150 \text{ M}$$

$$c_{EDTA} = \frac{(50.0 \times 0.0500) \text{ mmol} - (50.0 \times 0.0300) \text{ mmol}}{100 \text{ mL}} = 0.0100 \text{ M}$$

Again let us assume that $[Ni^{2+}] << [NiY^{2-}]$, so that

$$[NiY^{2-}] = 0.0150 - [Ni^{2+}] \approx 0.0150 \text{ M}$$

At this point, the total concentration of uncomplexed EDTA is given by its molarity:

$$c_T = 0.0100 \text{ M}$$

If we substitute this value in Equation 17-25, we get

$$K'_{NiY} = \frac{0.0150}{[Ni^{2+}]0.0100} = \alpha_4 K_{NiY}$$

$$[Ni^{2+}] = \frac{0.0150}{0.0100 \times 1.05 \times 10^8} = 1.4 \times 10^{-8} \text{ M}$$

Note again that our assumption that $[Ni^{2+}] << [NiY^{2-}]$ is valid.

► The value for K'_{NiY} was found in Example 17-2 to be 1.05×10^8 at pH 3.00.

Spreadsheet Summary The alpha values for EDTA are obtained and used to plot a distribution diagram in Chapter 9 of *Applications of Microsoft® Excel in Analytical Chemistry.* The titration of the tetraprotic acid EDTA with base is also considered.

17D-4 EDTA Titration Curves

The principles illustrated in Examples 17-2 and 17-3 can be used in the derivation of the titration curve for a metal ion with EDTA in a solution of fixed pH. Example 17-4 demonstrates how the titration curve is constructed with a spreadsheet.

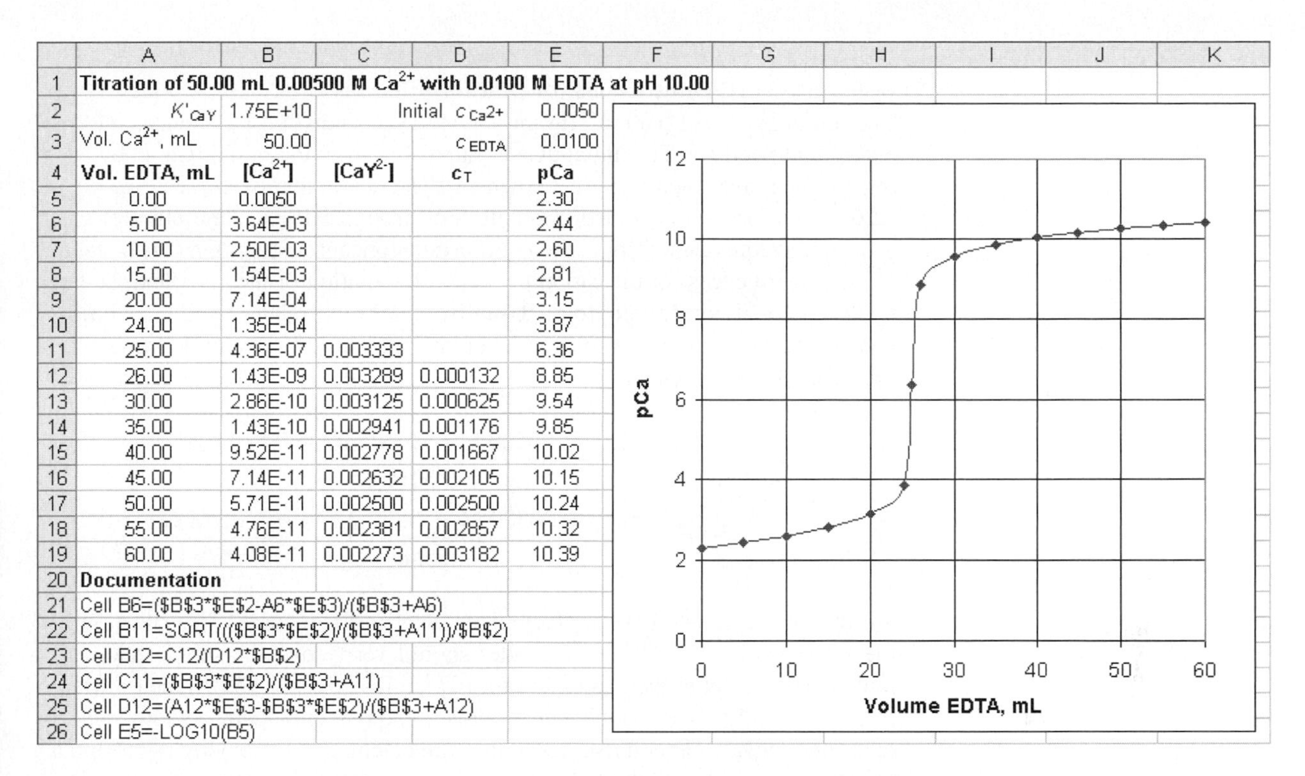

	A	B	C	D	E
1	**Titration of 50.00 mL 0.00500 M Ca^{2+} with 0.0100 M EDTA at pH 10.00**				
2	K'_{CaY}	1.75E+10	Initial	c_{Ca2+}	0.0050
3	Vol. Ca^{2+}, mL	50.00		c_{EDTA}	0.0100
4	**Vol. EDTA, mL**	**$[Ca^{2+}]$**	**$[CaY^{2-}]$**	c_T	**pCa**
5	0.00	0.0050			2.30
6	5.00	3.64E-03			2.44
7	10.00	2.50E-03			2.60
8	15.00	1.54E-03			2.81
9	20.00	7.14E-04			3.15
10	24.00	1.35E-04			3.87
11	25.00	4.36E-07	0.003333		6.36
12	26.00	1.43E-09	0.003289	0.000132	8.85
13	30.00	2.86E-10	0.003125	0.000625	9.54
14	35.00	1.43E-10	0.002941	0.001176	9.85
15	40.00	9.52E-11	0.002778	0.001667	10.02
16	45.00	7.14E-11	0.002632	0.002105	10.15
17	50.00	5.71E-11	0.002500	0.002500	10.24
18	55.00	4.76E-11	0.002381	0.002857	10.32
19	60.00	4.08E-11	0.002273	0.003182	10.39
20	**Documentation**				
21	Cell B6=($B$3*$E$2-A6*$E$3)/($B$3+A6)				
22	Cell B11=SQRT((($B$3*$E$2)/($B$3+A11))/$B$2)				
23	Cell B12=C12/(D12*$B$2)				
24	Cell C11=($B$3*$E$2)/($B$3+A11)				
25	Cell D12=(A12*$E$3-$B$3*$E$2)/($B$3+A12)				
26	Cell E5=-LOG10(B5)				

Figure 17-5 Spreadsheet for the titration of 50.00 mL of 0.00500 M Ca^{2+} with 0.0100 M EDTA in a solution buffered at pH 10.0.

EXAMPLE 17-4

Use a spreadsheet to construct the titration curve of pCa versus volume of EDTA for 50.0 mL of 0.00500 M Ca^{2+} being titrated with 0.0100 M EDTA in a solution buffered to a constant pH of 10.0.

Initial Entries

The spreadsheet is shown in Figure 17-5. The initial volume of Ca^{2+} is entered into cell B3, and the initial Ca^{2+} concentration is entered into E2. The EDTA concentration is entered into cell E3. The volumes for which pCa values are to be calculated are entered into cells A5 through A19.

Calculating the Conditional Constant

The conditional formation constant for the calcium/EDTA complex at pH 10 is obtained from the formation constant of the complex (see Table 17-3) and the α_4 value for EDTA at pH 10 (see Figure 17-4). Thus, if we substitute into Equation 17-25, we get

$$K'_{CaY} = \frac{[CaY^{2-}]}{[CaY^{2+}]c_T} = \alpha_4 K_{CaY}$$

$$= 0.35 \times 5.0 \times 10^{10} = 1.75 \times 10^{10}$$

This value is entered into cell B2.

(continued)

Preequivalence-Point Values for pCa

The initial $[Ca^{2+}]$ at 0.00 mL titrant is just the value in cell E2. Hence, **=E2** is entered into cell B5. The initial pCa is calculated from the initial $[Ca^{2+}]$ by taking the negative logarithm, as shown in the documentation for cell E5 (cell A26). This formula is copied into cells E6 through E19. For the other entries prior to the equivalence point, the equilibrium concentration of Ca^{2+} is equal to the untitrated excess of the cation plus any dissociation of the complex, the latter being equal numerically to c_T. Usually, c_T is small relative to the analytical concentration of the uncomplexed calcium ion. Thus, for example, after 5.00 mL of EDTA has been added,

$$[Ca^{2+}] = \frac{50.0 \text{ mL} \times 0.00500 \text{ M} - 5.00 \text{ mL} \times 0.0100 \text{ M}}{(50 + 5.00) \text{ mL}} + c_T$$

$$\approx \frac{50.0 \text{ mL} \times 0.00500 \text{ M} - 5.00 \text{ mL} \times 0.0100 \text{ M}}{55.00 \text{ mL}}$$

We thus enter into cell B6 the formula shown in the documentation section of the spreadsheet (cell A21). The reader should verify that the spreadsheet formula is equivalent to the expression for $[Ca^{2+}]$ just given. The volume of titrant (A6) is the only value that changes in this preequivalence-point region. Hence, other preequivalence-point values of pCa are calculated by copying the formula in cell B6 into cells B7 through B10.

Equivalence-Point pCa

At the equivalence point (25.00 mL of EDTA), we follow the method shown in Example 17-2 and first compute the analytical concentration of CaY^{2-}:

$$c_{CaY^{2-}} = \frac{(50.0 \times 0.00500) \text{ mmol}}{(50.0 + 25.0) \text{ mL}}$$

The only source of Ca^{2+} ions is the dissociation of the complex. It also follows that the Ca^{2+} concentration must be equal to the sum of the concentrations of the uncomplexed EDTA, c_T. Thus,

$$[Ca^{2+}] = c_T \qquad \text{and} \qquad [CaY^{2-}] = c_{CaY^{2-}} - [Ca^{2+}] \approx c_{CaY^{2-}}$$

The formula for $[CaY^{2-}]$ is thus entered into cell C11, as shown in the documentation in cell A24. The reader should again be able to verify this formula. To obtain $[Ca^{2+}]$, we substitute into the expression for K'_{CaY},

$$K'_{CaY} = \frac{[CaY^{2-}]}{[Ca^{2+}]c_T} \approx \frac{c_{CaY^{2-}}}{[Ca^{2+}]^2}$$

$$[Ca^{2+}] = \sqrt{\frac{c_{CaY^{2-}}}{K'_{CaY}}}$$

We thus enter into cell B11 the formula corresponding to this expression, as shown in cell A22.

Postequivalence-Point pCa

Beyond the equivalence point, analytical concentrations of CaY^{2-} and EDTA are obtained directly from the stoichiometric data. Since there is now excess EDTA, a calculation similar to that in Example 17-3 is performed. Thus, after the addition of 26.0 mL of EDTA, we can write

$$c_{CaY^{2-}} = \frac{(50.0 \times 0.00500) \text{ mmol}}{(50.0 + 26.0) \text{ mL}}$$

$$c_{EDTA} = \frac{(26.0 \times 0.0100) \text{ mL} - (50.0 \times 0.00500) \text{ mL}}{76.0 \text{ mL}}$$

As an approximation,

$$[CaY^{2-}] = c_{CaY^{2-}} - [Ca^{2+}] \approx c_{CaY^{2}} = \frac{(50.0 \times 0.00500) \text{ mmol}}{(50.0 + 26.0) \text{ mL}}$$

Since this expression is the same as that previously entered into cell C11, we copy that equation into cell C12. We also note that $[CaY^{2-}]$ will be given by this same expression (with the volume varied) throughout the remainder of the titration. Hence, the formula in cell C12 is copied into cells C13 through C19. Also, we approximate

$$c_T = c_{EDTA} + [Ca^{2+}] \approx c_{EDTA} = \frac{(26.0 \times 0.0100) \text{ mL} - (50.0 \times 0.00500) \text{ mL}}{76.0 \text{ mL}}$$

We enter this formula into cell D12 as shown in the documentation (cell A25) and copy it into cells D13 through D16.

To calculate $[Ca^{2+}]$, we then substitute into the conditional formation-constant expression and obtain

$$K'_{CaY} = \frac{[CaY^{2-}]}{[Ca^{2+}] \times c_T} \approx \frac{c_{CaY^{2-}}}{[Ca^{2+}] \times c_{EDTA}}$$

$$[Ca^{2+}] = \frac{c_{CaY^{2-}}}{c_{EDTA} \times K'_{CaY}}$$

Hence, the $[Ca^{2+}]$ in cell B12 is computed from the values in cells C12 and D12, as shown in cell A23. We copy this formula into cells B13 through B19 and plot the titration curve shown in Figure 17-5.

Curve *A* in Figure 17-6 is a plot of data for the titration in Example 17-4. Curve *B* is the titration curve for a solution of magnesium ion under identical conditions. The formation constant for the EDTA complex of magnesium is smaller than that of the calcium complex, which results in a smaller change in the p-function in the equivalence-point region.

Figure 17-7 provides titration curves for calcium ion in solutions buffered to various pH levels. Recall that α_4, and hence K'_{CaY}, becomes smaller as the pH

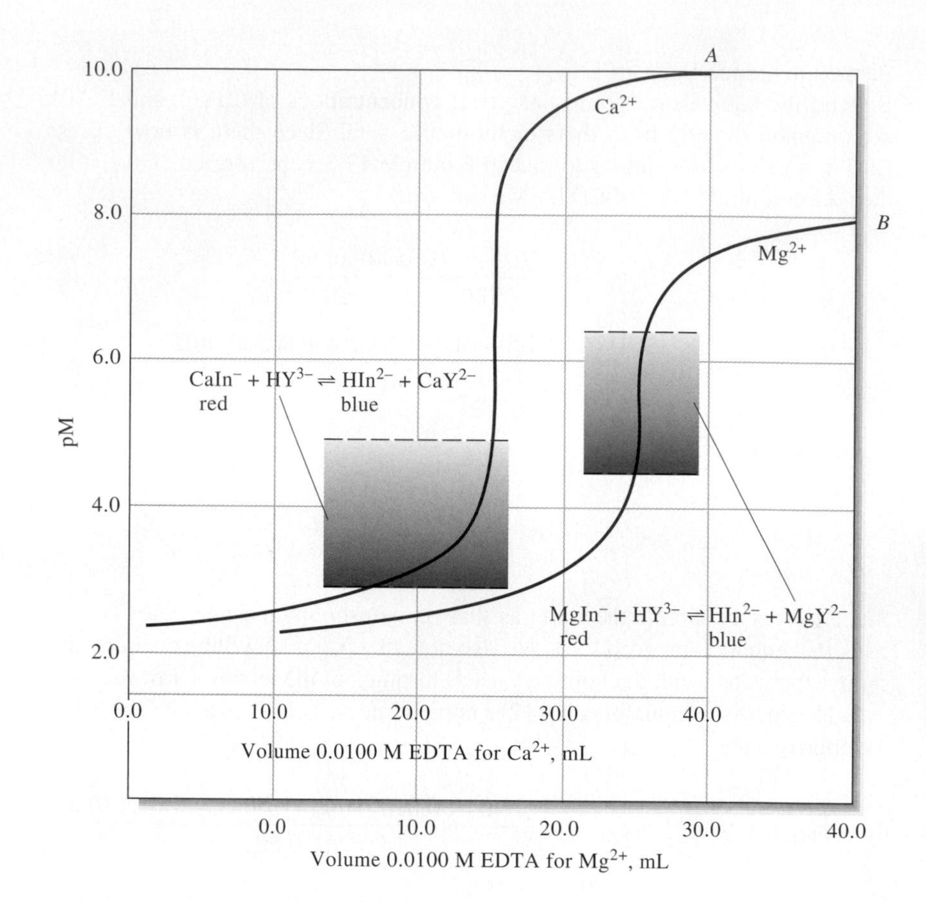

Figure 17-6 EDTA titration curves for 50.0 mL of 0.00500 M Ca^{2+} ($K'_{CaY} = 1.75 \times 10^{10}$) and Mg^{2+} ($K'_{MgY} = 1.72 \times 10^{8}$) at pH 10.0. Note that because of the larger formation constant, the reaction of calcium ion with EDTA is more complete, and a larger change occurs in the equivalence-point region. The shaded areas show the transition range for the indicator Eriochrome Black T.

CD-ROM Simulation: EDTA Titration Curves.

decreases. The less favorable equilibrium constant leads to a smaller change in pCa in the equivalence-point region. It can be seen from Figure 17-7 that an adequate end point in the titration of calcium requires a pH of about 8 or greater. As shown in Figure 17-8, however, cations with larger formation constants provide good end

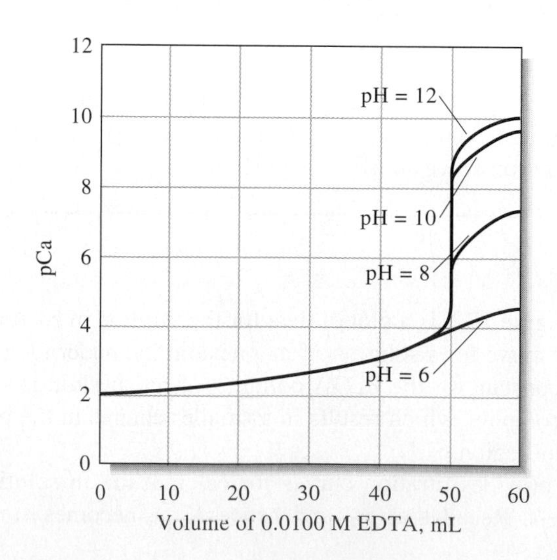

Figure 17-7 Influence of pH on the titration of 0.0100 M Ca^{2+} with 0.0100 M EDTA. Note that the end point becomes less sharp as the pH decreases because the complex formation reaction is less complete under these circumstances.

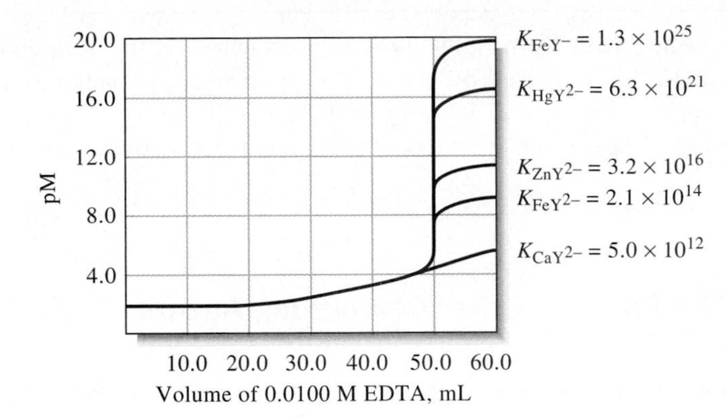

Figure 17-8 Titration curves for 50.0 mL of 0.0100 M solutions of various cations at pH 6.0.

points even in acidic media. Figure 17-9 shows the minimum permissible pH for a satisfactory end point in the titration of various metal ions in the absence of competing complexing agents. Note that a moderately acidic environment is satisfactory for many divalent heavy-metal cations and that a strongly acidic medium can be tolerated in the titration of such ions as iron(III) and indium(III).

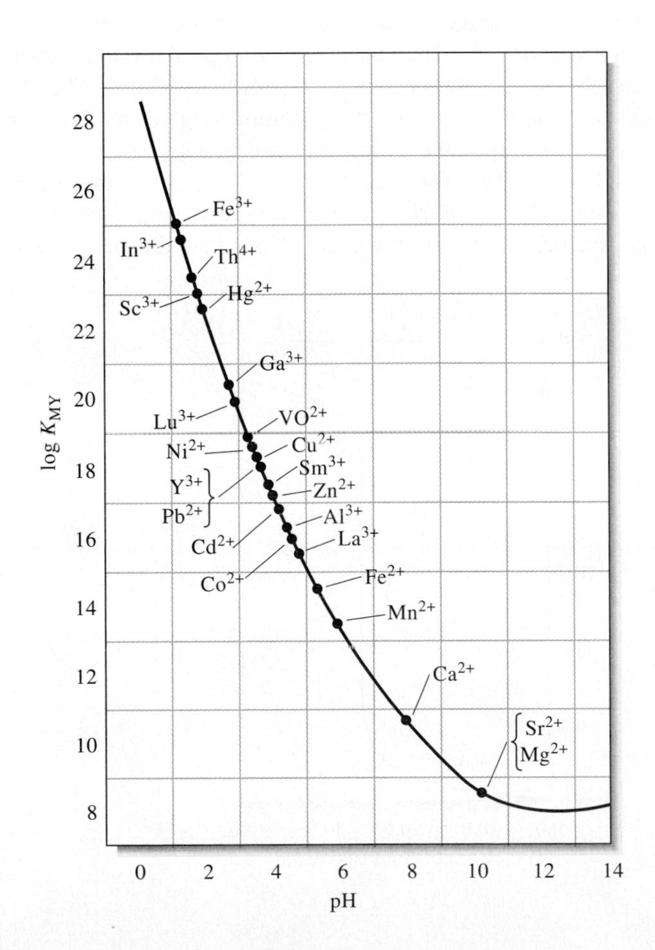

Figure 17-9 Minimum pH needed for satisfactory titration of various cations with EDTA. (From C.N. Reilley and R.W. Schmid, *Anal. Chem.*, **1958,** *30,* 947. Copyright 1958 American Chemical Society. Reprinted with permission of the American Chemical Society.)

Spreadsheet Summary The titration curve for the titration of Ca^{2+} with EDTA is derived by both a stoichiometric approach and a master equation approach in Chapter 9 of *Applications of Microsoft® Excel in Analytical Chemistry.* The effect of pH on the shape and end point of the titration curve is examined.

17D-5 The Effect of Other Complexing Agents on EDTA Titration Curves

► Often, auxiliary complexing agents must be used in EDTA titrations to prevent precipitation of the analyte as a hydrous oxide. Such reagents cause the end points to be less sharp.

Many cations form hydrous oxide precipitates when the pH is raised to the level required for their successful titration with EDTA. When this problem is encountered, an auxiliary complexing agent is needed to keep the cation in solution. For example, zinc(II) is ordinarily titrated in a medium that has fairly high concentrations of ammonia and ammonium chloride. These species buffer the solution to a pH that ensures complete reaction between cation and titrant; in addition, ammonia forms ammine complexes with zinc(II) and prevents formation of the sparingly soluble zinc hydroxide, particularly in the early stages of the titration. A somewhat more realistic description of the reaction is then

$$Zn(NH_3)_4^{2+} + HY^{3-} \rightarrow ZnY^{2-} + 3NH_3 + NH_4^+$$

The solution also contains such other zinc/ammonia species as $Zn(NH_3)_3^{2+}$, $Zn(NH_3)_2^{2+}$, and $Zn(NH_3)^{2+}$. Calculation of pZn in a solution that contains ammonia must take these species into account, as shown in Feature 17-5. Qualitatively, complexation of a cation by an auxiliary complexing reagent causes preequivalence pM values to be larger than in a comparable solution with no such reagent.

Figure 17-10 shows two theoretical curves for the titration of zinc(II) with EDTA at pH 9.00. The equilibrium concentration of ammonia was 0.100 M for one titration and 0.0100 M for the other. Note that the presence of ammonia

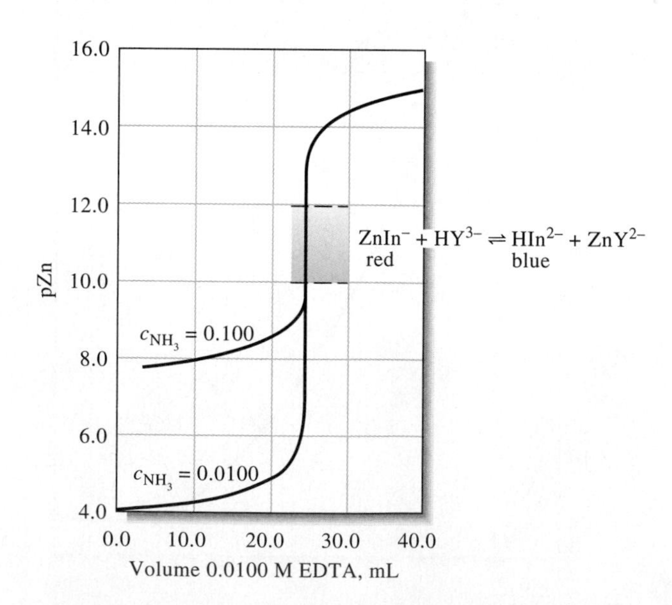

Figure 17-10 Influence of ammonia concentration on the end point for the titration of 50.0 mL of 0.00500 M Zn^{2+}. Solutions are buffered to pH 9.00. The shaded region shows the transition range for Eriochrome Black T. Note that ammonia decreases the change in pZn in the equivalence-point region.

decreases the change in pZn near the equivalence point. For this reason, the concentration of auxiliary complexing reagents should always be the minimum required to prevent precipitation of the analyte. Note that the auxiliary complexing agent does not affect pZn beyond the equivalence point. Also keep in mind that α_4, and thus pH, plays an important role in defining this part of the titration curve (see Figure 17-7).

FEATURE 17-5

EDTA Titration Curves When a Complexing Agent Is Present

A quantitative description of the effects of an auxiliary complexing reagent can be derived by a procedure similar to that used to determine the influence of pH on EDTA titration curves. Here, a quantity α_M is defined that is analogous to α_4.

$$\alpha_M = \frac{[M^{n+}]}{c_M} \tag{17-28}$$

where c_M is the sum of the concentrations of species containing the metal ion *exclusive* of that combined with EDTA. For solutions containing zinc(II) and ammonia, then

$$c_M = [Zn^{2+}] + [Zn(NH_3)^{2+}] + [Zn(NH_3)_2^{2+}] + [Zn(NH_3)_3^{2+}] + [Zn(NH_3)_4^{2+}] \tag{17-29}$$

The value of α_M can be expressed readily in terms of the ammonia concentration and the formation constants for the various ammine complexes, as described for a general metal-ligand reaction in Feature 17-1. The result is an equation analogous to Equation 17-9:

$$\alpha_M = \frac{1}{1 + \beta_1[NH_3] + \beta_2[NH_3]^2 + \beta_3[NH_3]^3 + \beta_4[NH_3]^4} \tag{17-30}$$

Finally, a conditional constant for the equilibrium between EDTA and zinc(II) in an ammonia/ammonium chloride buffer is obtained by substituting Equation 17-28 into Equation 17-25 and rearranging

$$K''_{ZnY} = \alpha_4 \alpha_M K_{ZnY} = \frac{[ZnY^{2-}]}{c_M c_T} \tag{17-31}$$

where K''_{ZnY} is a new conditional constant that applies at a single pH as well as a single concentration of ammonia.

To show how Equations 17-28 to 17-31 can be used to obtain a titration curve, calculate the pZn of solutions prepared by adding 20.0, 25.0, and 30.0 mL of 0.0100 M EDTA to 50.0 mL of 0.00500 M Zn^{2+}. Assume that both the

(continued)

Zn^{2+} and EDTA solutions are 0.100 M in NH_3 and 0.175 M in NH_4Cl to provide a constant pH of 9.0.

In Appendix 4, we find that the logarithms of the stepwise formation constants for the four zinc complexes with ammonia are 2.21, 2.29, 2.36, and 2.03. Thus,

$$\beta_1 = \text{antilog } 2.21 = 1.62 \times 10^2$$

$$\beta_2 = \text{antilog } (2.21 + 2.29) = 3.16 \times 10^4$$

$$\beta_3 = \text{antilog } (2.21 + 2.29 + 2.36) = 7.24 \times 10^6$$

$$\beta_4 = \text{antilog } (2.21 + 2.29 + 2.36 + 2.03) = 7.76 \times 10^8$$

Calculation of a Conditional Constant

A value for α_M can be obtained from Equation 17-30 by assuming that the molar and analytical concentrations of ammonia are essentially the same; thus, for $[NH_3] = 0.100$,

$$\alpha_M = \frac{1}{1 + 16 + 316 + 7.24 \times 10^3 + 7.76 \times 10^4} = 1.17 \times 10^{-5}$$

A value for K_{ZnY} is found in Table 17-3, and α_4 for pH 9.0 is given in Figure 17-4. Substituting into Equation 17-31, we find

$$K''_{ZnY} = 5.21 \times 10^{-2} \times 1.17 \times 10^{-5} \times 3.2 \times 10^{16} = 1.9 \times 10^{10}$$

Calculation of pZn after Addition of 20.0 mL of EDTA

At this point, only part of the zinc has been complexed by EDTA. The remainder is present as Zn^{2+} and the four ammine complexes. By definition, the sum of the concentrations of these five species is c_M. Therefore,

$$c_M = \frac{50.00 \times 0.00500 - 20.0 \times 0.0100}{70.0} = 7.14 \times 10^{-4}\text{ M}$$

Substitution of this value into Equation 17-28 gives

$$[Zn^{2+}] = c_M\alpha_M = (7.14 \times 10^{-4})(1.17 \times 10^{-5}) = 8.35 \times 10^{-9}\text{ M}$$

$$\text{pZn} = 8.08$$

Calculation of pZn after Addition of 25.0 mL of EDTA

At the equivalence point, the analytical concentration of ZnY^{2-} is

$$c_{ZnY^{2-}} = \frac{50.0 \times 0.00500}{20.0 + 25.0} = 3.33 \times 10^{-3}\text{ M}$$

The sum of the concentrations of the various zinc species not combined with EDTA equals the sum of the concentrations of the uncomplexed EDTA species:

$$c_M = c_T$$

and

$$[ZnY^{2-}] = 3.33 \times 10^{-3} - c_M \approx 3.33 \times 10^{-3}\text{ M}$$

Substituting into Equation 17-31, we have

$$K''_{ZnY} = \frac{3.33 \times 10^{-3}}{c_M^2} = 1.9 \times 10^{10}$$

$$c_M = 4.19 \times 10^{-7}\text{ M}$$

With Equation 17-28, we obtain

$$[Zn^{2+}] = c_M\alpha_M = (4.18 \times 10^{-7})(1.17 \times 10^{-5}) = 4.90 \times 10^{-12}\text{ M}$$

$$\text{pZn} = 11.31$$

Calculation of pZn after Addition of 30.0 mL of EDTA
The solution now contains an excess of EDTA; thus,

$$c_{EDTA} = c_T = \frac{30.0 \times 0.0100 - 50.0 \times 0.00500}{80.0} = 6.25 \times 10^{-4}\text{ M}$$

and since essentially all of the original Zn^{2+} is now complexed,

$$c_{ZnY^{2-}} = [ZnY^{2-}] = \frac{50.00 \times 0.00500}{80.0} = 3.12 \times 10^{-3}\text{ M}$$

Rearranging Equation 17-31 gives

$$c_M = \frac{[ZnY^{2-}]}{c_T K''_{ZnY}} = \frac{3.12 \times 10^{-3}}{(6.25 \times 10^{-4})(1.9 \times 10^{10})} = 2.63 \times 10^{-10}\text{ M}$$

and, from Equation 17-28,

$$[Zn^{2+}] = c_M\alpha_M = (2.63 \times 10^{-10})(1.17 \times 10^{-5}) = 3.07 \times 10^{-15}\text{ M}$$

$$\text{pZn} = 14.51$$

17D-6 Indicators for EDTA Titrations

Nearly 200 organic compounds have been investigated as indicators for metal ions in EDTA titrations. The most common indicators are given by Dean.[2] In general, these indicators are organic dyes that form colored chelates with metal ions in a pM range that is characteristic of the particular cation and dye. The complexes are often intensely colored and are discernible to the eye at concentrations in the range of 10^{-6} to 10^{-7} M.

[2] J. A. Dean, *Analytical Chemistry Handbook*, p. 3.95. New York: McGraw-Hill, 1995.

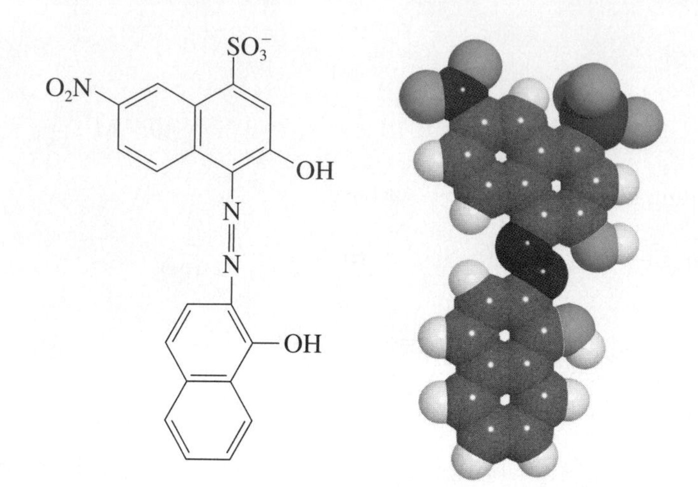

Figure 17-11 Structure and molecular model of Eriochrome Black T. The compound contains a sulfonic acid group that completely dissociates in water and two phenolic groups that only partially dissociate.

Eriochrome Black T is a typical metal ion indicator that is used in the titration of several common cations. The structural formula of Eriochrome Black T is shown in Figure 17-11. Its behavior as a weak acid is described by the equations

$$\underset{}{H_2O} + \underset{\text{red}}{H_2In^-} \rightleftharpoons \underset{\text{blue}}{HIn^{2-}} + H_3O^+ \qquad K_1 = 5 \times 10^{-7}$$

$$H_2O + \underset{\text{blue}}{HIn^{2-}} \rightleftharpoons \underset{\text{orange}}{In^{3-}} + H_3O^+ \qquad K_2 = 2.8 \times 10^{-12}$$

Note that the acids and their conjugate bases have different colors. Thus, Eriochrome Black T behaves as an acid/base indicator as well as a metal ion indicator.

The metal complexes of Eriochrome Black T are generally red, as is H_2In^-. Thus, for metal ion detection, it is necessary to adjust the pH to 7 or above so that the blue form of the species, HIn^{2-}, predominates in the absence of a metal ion. Until the equivalence point in a titration, the indicator complexes the excess metal ion so that the solution is red. With the first slight excess of EDTA, the solution turns blue as a consequence of the reaction

$$\underset{\text{red}}{MIn^-} + HY^{3-} \rightleftharpoons \underset{\text{blue}}{HIn^{2-}} + MY^{2-}$$

Eriochrome Black T forms red complexes with more than two dozen metal ions, but the formation constants of only a few are appropriate for end point detection. As shown in Example 17-5, the applicability of a given indicator for an EDTA titration can be determined from the change in pM in the equivalence-point region, provided that the formation constant for the metal indicator complex is known.[3]

[3]C. N. Reilley and R. W. Schmid, *Anal. Chem.*, **1959,** *31,* 887.

EXAMPLE 17-5

Determine the transition ranges for Eriochrome Black T in titrations of Mg^{2+} and Ca^{2+} at pH 10.0, given that (a) the second acid dissociation constant for the indicator is

$$HIn^{2-} + H_2O \rightleftharpoons In^{3-} + H_3O^+ \qquad K_2 = \frac{[H_3O^+][In^{3-}]}{[HIn^{2-}]} = 2.8 \times 10^{-12}$$

(b) the formation constant for $MgIn^-$ is

$$Mg^{2+} + In^{3-} \rightleftharpoons MgIn^- \qquad K_f = \frac{[MgIn^-]}{[Mg^{2+}][In^{3-}]} = 1.0 \times 10^7$$

and (c) the analogous constant for Ca^{2+} is 2.5×10^5.

We assume, as we did earlier (see Section 14A-1), that a detectable color change requires a 10-fold excess of one or the other of the colored species; that is, a detectable color change is observed when the ratio $[MgIn^-]/[HIn^{2-}]$ changes from 10 to 0.10. Multiplication of K_2 for the indicator by K_f for $MgIn^-$ gives an expression that contains this ratio:

$$\frac{[MgIn^-][H_3O^+]}{[HIn^{2-}][Mg^{2+}]} = 2.8 \times 10^{-12} \times 1.0 \times 10^7 = 2.8 \times 10^{-5}$$

which rearranges to

$$[Mg^{2+}] = \frac{[MgIn^-]}{[HIn^{2-}]} \times \frac{[H_3O^+]}{2.8 \times 10^{-5}}$$

Substitution of 1.0×10^{-10} for $[H_3O^+]$ and 10 and 0.10 for the ratio yields the range of $[Mg^{2+}]$ over which the color change occurs:

$$[Mg^{2+}] = 3.6 \times 10^{-5}\ M \quad \text{to} \quad 3.6 \times 10^{-7}\ M$$

$$pMg = 5.4 \pm 1.0$$

Proceeding in the same way, we find the range for pCa to be 3.8 ± 1.0.

Transition ranges for magnesium and calcium are indicated on the titration curves in Figure 17-6. As can be seen, the indicator is ideal for the titration of magnesium but quite unsatisfactory for calcium. Note that the formation constant for $CaIn^-$ is only about $^1/_{40}$ that for $MgIn^-$. As a consequence, significant conversion of $CaIn^-$ to HIn^{2-} occurs well before equivalence. A similar calculation shows that Eriochrome Black T is also well suited for the titration of zinc with EDTA (see Figure 17-10).

A limitation of Eriochrome Black T is that its solutions decompose slowly with standing. It is claimed that solutions of Calmagite (Figure 17-12), an indicator that

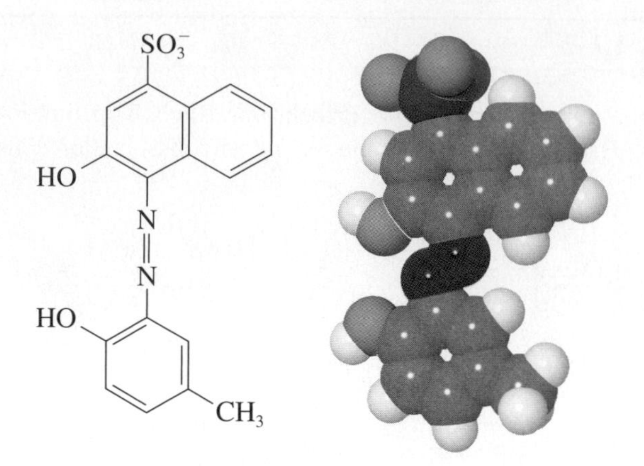

Figure 17-12 Structural formula and molecular model of Calmagite. Note the similarity to Eriochrome Black T (see Figure 17-11).

for all practical purposes is identical in behavior to Eriochrome Black T, do not suffer this disadvantage. Many other metal indicators have been developed for EDTA titrations.[4] In contrast to Eriochrome Black T, some of these indicators can be used in strongly acidic media.

17D-7 Titration Methods Employing EDTA

Several different types of titration methods can be used with EDTA, as described next.

Direct Titration

► Direct titration procedures with a metal ion indicator that responds to the analyte are the easiest and most convenient to use. Methods that use an added metal ion are also widely employed.

Many of the metals in the periodic table can be determined by titration with standard EDTA solutions. Some methods are based on indicators that respond to the analyte itself, while others are based on an added metal ion.

Methods Based on Indicators for the Analyte Dean[5] lists nearly 40 metal ions that can be determined by direct titration with EDTA using metal ion indicators. Indicators that respond to the metal directly cannot be used in all cases, either because no indicator with an appropriate transition range is available or because the reaction between the metal ion and EDTA is so slow as to make titration impractical.

Methods Based on Indicators for an Added Metal Ion When a good, direct indicator for the analyte is unavailable, a small amount of a metal ion for which a good indicator is available can be added. The metal ion must form a complex that is less stable than the analyte complex. For example, indicators for calcium ion are generally less satisfactory than those we have described for magnesium ion. Consequently, a small amount of magnesium chloride is often added to an EDTA solution that is to be used for the determination of calcium. In this case, Eriochrome Black T can be used in the titration. In the initial stages, magnesium

[4]See, for example, J. A. Dean, *Analytical Chemistry Handbook,* pp. 3.94–3.96. New York: McGraw-Hill, 1995.

[5]Reference 4 pp. 3.104–3.109.

ions are displaced from the EDTA complex by calcium ions and are free to combine with the Eriochrome Black T, thus imparting a red color to the solution. When all of the calcium ions have been complexed, however, the liberated magnesium ions again combine with the EDTA until the end point is observed. This procedure requires standardization of the EDTA solution against primary-standard calcium carbonate.

Potentiometric Methods Potential measurements can be used for end point detection in the EDTA titration of those metal ions for which specific ion electrodes are available. Electrodes of this type are described in Section 21D-1. In addition, a mercury electrode can be made sensitive to EDTA ions and used in titrations with this reagent.

Spectrophotometric Methods Measurement of UV/visible absorption can also be used to determine the end points of titrations (see Section 26A-4). In these cases, an instrument responds to the color change in the titration rather than relying on a visual determination of the end point.

Back-titration Methods

Back-titration is useful for the determination of cations that form stable EDTA complexes and for which a satisfactory indicator is not available. The method is also useful for cations such as Cr(III) and Co(III) that react only slowly with EDTA. A measured excess of standard EDTA solution is added to the analyte solution. After the reaction is judged complete, the excess EDTA is back-titrated with a standard magnesium or zinc ion solution to an Eriochrome Black T or Calmagite end point.[6] For this procedure to be successful, it is necessary that the magnesium or zinc ions form an EDTA complex that is less stable than the corresponding analyte complex.

◀ Back-titration procedures are used when no suitable indicator is available, when the reaction between analyte and EDTA is slow, or when the analyte forms precipitates at the pH required for its titration.

Back-titration is also useful for analyzing samples that contain anions that would otherwise form sparingly soluble precipitates with the analyte under the analytical conditions. Here, the excess EDTA prevents precipitate formation.

Displacement Methods

In displacement titrations, an unmeasured excess of a solution containing the magnesium or zinc complex of EDTA is introduced into the analyte solution. If the analyte forms a more stable complex than that of magnesium or zinc, the following displacement reaction occurs:

◀ Displacement titrations are used when no indicator for an analyte is available.

$$MgY^{2-} + M^{2+} \rightarrow MY^{2-} + Mg^{2+}$$

where M^{2+} represents the analyte cation. The liberated Mg^{2+} or, in some cases Zn^{2+}, is then titrated with a standard EDTA solution.

17D-8 The Scope of EDTA Titrations

Complexometric titrations with EDTA have been applied to the determination of virtually every metal cation, with the exception of the alkali metal ions. Because EDTA complexes most cations, the reagent might appear at first glance to be totally

[6]For a discussion of the back-titration procedure, see C. Macca and M. Fiorana, *J. Chem. Educ.*, **1986,** *63,* 121.

lacking in selectivity. In fact, however, considerable control over interferences can be realized by pH regulation. For example, trivalent cations can usually be titrated without interference from divalent species by maintaining the solution at a pH of about 1 (see Figure 17-8). At this pH, the less stable divalent chelates do not form to any significant extent, but the trivalent ions are quantitatively complexed.

Similarly, ions such as cadmium and zinc, which form more stable EDTA chelates than does magnesium, can be determined in the presence of the latter ion by buffering the mixture to a pH of 7 before titration. Eriochrome Black T serves as an indicator for the cadmium or zinc end points without interference from magnesium because the indicator chelate with magnesium is not formed at this pH.

A **masking agent** is a complexing agent that reacts selectively with a component in a solution to prevent that component from interfering in a determination.

Finally, interference from a particular cation can sometimes be eliminated by adding a suitable masking agent, an auxiliary ligand that preferentially forms highly stable complexes with the potential interfering ion.[7] Thus, cyanide ion is often employed as a masking agent to permit the titration of magnesium and calcium ions in the presence of ions such as cadmium, cobalt, copper, nickel, zinc, and palladium. All of the latter form sufficiently stable cyanide complexes to prevent reaction with EDTA. Feature 17-6 illustrates how masking and demasking reagents are used to improve the selectivity of EDTA reactions.

FEATURE 17-6

How Masking and Demasking Agents Can Be Used to Enhance the Selectivity of EDTA Titrations

Lead, magnesium, and zinc can be determined in a single sample by two titrations with standard EDTA and one titration with standard Mg^{2+}. The sample is first treated with an excess of NaCN, which masks Zn^{2+} and prevents it from reacting with EDTA:

$$Zn^{2+} + 4CN^- \rightleftharpoons Zn(CN)_4^{2-}$$

The Pb^{2+} and Mg^{2+} are then titrated with standard EDTA. After the equivalence point has been reached, a solution of the complexing agent BAL (2-3-dimercapto-1-propanol, $CH_2SHCHSHCH_2OH$), which we will write as $R(SH)_2$, is added to the solution. This bidentate ligand reacts selectively to form a complex with Pb^{2+} that is much more stable than PbY^{2-}:

$$PbY^{2-} + 2R(SH)_2 \rightarrow Pb(RS)_2 + 2H^+ + Y^{4-}$$

The liberated Y^{4-} is then titrated with a standard solution of Mg^{2+}. Finally, the zinc is demasked by adding formaldehyde:

$$Zn(CN)_4^{2-} + 4HCHO + 4H_2O \rightarrow Zn^{2+} + 4HOCH_2CN + 4OH^-$$

The liberated Zn^{2+} is then titrated with the standard EDTA solution.

[7] For further information, see D. D. Perrin, *Masking and Demasking of Chemical Reactions.* New York: Wiley-Interscience, 1970; J. A. Dean, *Analytical Chemistry Handbook,* pp. 3.92–3.111. New York: McGraw-Hill, 1995.

Suppose the initial titration of Mg^{2+} and Pb^{2+} required 42.22 mL of 0.02064 M EDTA. Titration of the Y^{4-} liberated by the BAL consumed 19.35 mL of 0.007657 M Mg^{2+}. After addition of formaldehyde, the liberated Zn^{2+} was titrated with 28.63 mL of the EDTA. Calculate the percent of the three elements if a 0.4085-g sample was used.

The initial titration reveals the number of millimoles of Pb^{2+} and Mg^{2+} present. That is,

$$\text{mmol }(Pb^{2} + Mg^{2+}) = 42.22 \times 0.02064 = 0.87142$$

The second titration gives the number of millimoles of Pb^{2+}. Thus,

$$\text{mmol }Pb^{2+} = 19.35 \times 0.007657 = 0.14816$$

$$\text{mmol }Mg^{2+} = 0.87142 - 0.14816 = 0.72326$$

Finally, from the third titration, we obtain

$$\text{mmol }Zn^{2+} = 28.63 \times 0.02064 = 0.59092$$

To obtain the percentages, we write

$$\frac{0.014816\ \cancel{\text{mmol Pb}} \times 0.2075\ \text{g Pb}/\cancel{\text{mmol Pb}}}{0.4085\ \text{g sample}} \times 100\% = 7.515\%\ \text{Pb}$$

$$\frac{0.72326\ \cancel{\text{mmol Mg}} \times 0.024305\ \text{g Mg}/\cancel{\text{mmol Mg}}}{0.4085\ \text{g sample}} \times 100\% = 4.303\%\ \text{Mg}$$

$$\frac{0.59095\ \cancel{\text{mmol Zn}} \times 0.06539\ \text{g Zn}/\cancel{\text{mmol Zn}}}{0.4085\ \text{g sample}} \times 100\% = 9.459\%\ \text{Zn}$$

17D-9 The Determination of Water Hardness

Historically, water "hardness" was defined in terms of the capacity of cations in the water to replace the sodium or potassium ions in soaps and to form sparingly soluble products that cause "scum" in the sink or bathtub. Most multiply charged cations share this undesirable property. In natural waters, however, the concentrations of calcium and magnesium ions generally far exceed those of any other metal ion. Consequently, hardness is now expressed in terms of the concentration of calcium carbonate that is equivalent to the total concentration of all the multivalent cations in the sample.

Hard water contains calcium, magnesium, and heavy-metal ions that form precipitates with soap (but not detergents).

The determination of hardness is a useful analytical test that provides a measure of the quality of water for household and industrial uses. The test is important to industry because hard water, on being heated, precipitates calcium carbonate, which clogs boilers and pipes.

Water hardness is ordinarily determined by an EDTA titration after the sample has been buffered to pH 10. Magnesium, which forms the least stable EDTA complex of all of the common multivalent cations in typical water samples, is not titrated until enough reagent has been added to complex all of the other cations in the sample. Therefore, a magnesium ion indicator, such as Calmagite or Eriochrome Black T, can serve as indicator in water hardness titrations. Often, a small concentration of the magnesium-EDTA chelate is incorporated into the buffer or the titrant to ensure sufficient magnesium ions for satisfactory indicator action.

FEATURE 17-7
Test Kits for Water Hardness

Test kits for determining the hardness of household water are available at stores selling water softeners and plumbing supplies. They usually consist of a vessel calibrated to contain a known volume of water, a packet containing an appropriate amount of a solid buffer mixture, an indicator solution, and a bottle of standard EDTA, which is equipped with a medicine dropper. A typical kit is shown in Figure 17F-2. The drops of standard reagent needed to cause a color change are counted. The concentration of the EDTA solution is ordinarily such that one drop corresponds to one grain (about 0.065 g) of calcium carbonate per gallon of water. Home water softeners that use ion exchange processes to remove hardness are discussed in Feature 30-2.

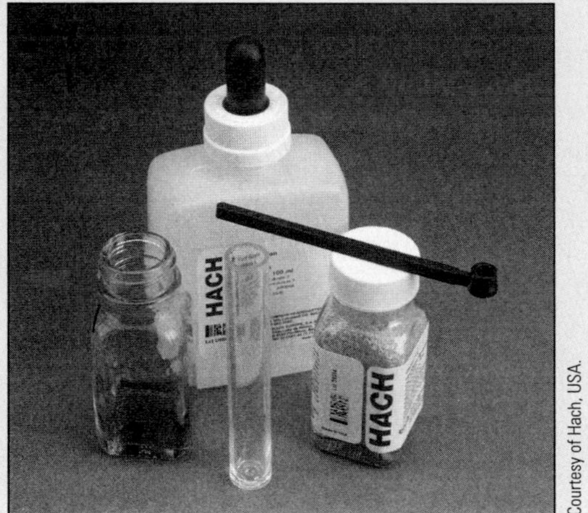

Figure 17F-2 Typical kit for testing for water hardness in household water.

WEB WORKS

Go to **http://chemistry.brookscole.com/skoogfac/.** From the Chapter Resources menu, choose Web Works, and locate the Chapter 17 section, where you will find links to several good educational sites that give additional help on complexation equilibria and complexometric titrations. Several Web sites describe experiments that can be done in the laboratory based on complexation methods. Find the abstract of the paper from the *Journal of Chemical Education* that deals with the determination of zinc by EDTA titration. Find the indicator and the buffer pH used in the titration. There is also a link to additional information on chemistry applied to aquatic systems. Compare some of the complexation equilibria described in these documents to those discussed in this chapter.

QUESTIONS AND PROBLEMS

17-1. Define
*(a) chelate.
(b) tetradentate chelating agent.
*(c) ligand.
(d) coordination number.
*(e) conditional formation constant.
(f) NTA.
*(g) water hardness.
(h) EDTA displacement titration.

17-2. Describe three general methods for performing EDTA titrations. What are the advantages of each?

***17-3.** Why are multidentate ligands preferable to unidentate ligands for complexometric titrations?

17-4. Write chemical equations and equilibrium-constant expressions for the stepwise formation of
*(a) $Ni(CN)_4^{2-}$.
(b) $Cd(SCN)_3^-$.

***17-5.** Write chemical formulas for the following complex ions:
(a) hexamminezinc(II).
(b) dichloroargentate.
(c) disulfatocuprate(II).
(d) trioxalotoferrate(III).
(e) hexacyanoferrate(II).

17-6. Explain how stepwise and overall formation constants are related.

17-7. Write equations in terms of the acid dissociation constants and $[H^+]$ for the highest alpha value for each of the following weak acid ligands:
(a) acetate (α_1).
(b) tartrate (α_2).
(c) phosphate (α_3).

17-8. Write conditional formation constants for 1:1 complexes of Fe(III) with each of the ligands in Problem 17-7. Express these constants in terms of the α value and the formation constant and in terms of concentrations, as in Equation 17-20.

***17-9.** Write a conditional overall formation constant for $Fe(Ox)_3^{3-}$ in terms of α_2 for oxalic acid and the β value for the complex. Also express the conditional constant in terms of concentrations as in Equation 17-20.

17-10. Propose a complexometric method for the determination of the individual components in a solution containing In^{3+}, Zn^{2+}, and Mg^{2+}.

***17-11.** Given an overall complex formation reaction of $M + nL \rightleftharpoons ML_n$ with an overall formation constant of β_n, show that the following relationship holds:

$$\log \beta_n = pM + npL - pML_n$$

17-12. Why is a small amount of MgY^{2-} often added to a water specimen that is to be titrated for hardness?

***17-13.** An EDTA solution was prepared by dissolving 3.156 g of purified and dried $Na_2H_2Y_2 \cdot H_2O$ in sufficient water to give 1.000 L. Calculate the molar concentration, given that the solute contained 0.3% excess moisture (see page 459).

17-14. A solution was prepared by dissolving about 3.0 g of $NaH_2Y \cdot 2H_2O$ in approximately 1 L of water and standardizing against 50.00-mL aliquots of 0.004517 M Mg^{2+}. An average titration of 32.22 mL was required. Calculate the molar concentration of the EDTA.

17-15. Calculate the volume of 0.0500 M EDTA needed to titrate
*(a) 27.16 mL of 0.0741 M $Mg(NO_3)_2$.
(b) the Ca in 0.1973 g of $CaCO_3$.
*(c) the Ca in a 0.5140-g mineral specimen that is 81.4% brushite, $CaHPO_4 \cdot 2H_2O$ (172.09 g/mol).
(d) the Mg in a 0.2222-g sample of the mineral hydromagnesite, $3MgCO_3Mg(OH)_2 \cdot 3H_2O$ (365.3 g/mol).
*(e) the Ca and Mg in a 0.1414-g sample that is 92.5% dolomite, $CaCO_3 \cdot MgCO_3$ (184.4 g/mol).

17-16. A solution contains 1.694 mg of $CoSO_4$ (155.0 g/mol) per milliliter. Calculate
(a) the volume of 0.08640 M EDTA needed to titrate a 25.00-mL aliquot of this solution.
(b) the volume of 0.009450 M Zn^{2+} needed to titrate the excess reagent after addition of 50.00 mL of 0.008640 M EDTA to a 25.00-mL aliquot of this solution.
(c) the volume of 0.008640 M EDTA needed to titrate the Zn^{2+} displaced by Co^{2+} following addition of an unmeasured excess of ZnY^{2-} to a 25.00-mL aliquot of the $CoSO_4$ solution. The reaction is

$$Co^{2+} + ZnY^{2-} \rightarrow CoY^{2-} + Zn^{2+}$$

***17-17.** The Zn in a 0.7162-g sample of foot powder was titrated with 21.27 mL of 0.01645 M EDTA. Calculate the percent Zn in this sample.

17-18. The Cr plating on a surface that measured 3.00 × 4.00 cm was dissolved in HCl. The pH was suitably adjusted, following which 15.00 mL of 0.01768 M EDTA were introduced. The excess reagent required a 4.30-mL back-titration with

0.008120 M Cu^{2+}. Calculate the average weight of Cr on each square centimeter of surface.

***17-19.** The Tl in a 9.76-g sample of rodenticide was oxidized to the trivalent state and treated with an unmeasured excess of Mg/EDTA solution. The reaction is

$$Tl^{3+} + MgY^{2-} \rightarrow TlY^{-} + Mg^{2+}$$

Titration of the liberated Mg^{2+} required 13.34 mL of 0.03560 M EDTA. Calculate the percentage of Tl_2SO_4 (504.8 g/mol) in the sample.

17-20. An EDTA solution was prepared by dissolving approximately 4 g of the disodium salt in approximately 1 L of water. An average of 42.35 mL of this solution was required to titrate 50.00-mL aliquots of a standard that contained 0.7682 g of $MgCO_3$ per liter. Titration of a 25.00-mL sample of mineral water at pH 10 required 18.81 mL of the EDTA solution. A 50.00-mL aliquot of the mineral water was rendered strongly alkaline to precipitate the magnesium at $Mg(OH)_2$. Titration with a calcium-specific indicator required 31.54 mL of the EDTA solution. Calculate

(a) the molarity of the EDTA solution.
(b) the concentration of $CaCO_3$ in the mineral water (ppm).
(c) the concentration of $MgCO_3$ in the mineral water (ppm).

***17-21.** A 50.00-mL aliquot of a solution containing iron(II) and iron(III) required 13.73 mL of 0.01200 M EDTA when titrated at pH 2.0 and 29.62 mL when titrated at pH 6.0. Express the concentration of the solution in terms of the parts per million of each solute.

17-22. A 24-hour urine specimen was diluted to 2.000 L. After the solution was buffered to pH 10, a 10.00-mL aliquot was titrated with 27.32 mL of 0.003960 M EDTA. The calcium in a second 10.00-mL aliquot was isolated as $CaC_2O_4(s)$, redissolved in acid, and titrated with 12.21 mL of the EDTA solution. Assuming that 15 to 300 mg of magnesium and 50 to 400 mg of calcium per day are normal, did this specimen fall within these ranges?

***17-23.** A 1.509-g sample of a Pb/Cd alloy was dissolved in acid and diluted to exactly 250.0 mL in a volumetric flask. A 50.00-mL aliquot of the diluted solution was brought to a pH of 10.0 with an NH_4^+/NH_3 buffer; the subsequent titration involved both cations and required 28.89 mL of 0.06950 M EDTA. A second 50.00-mL aliquot was brought to a pH of 10.0 with an HCN/NaCN buffer, which also served to mask the Cd^{2+}; 11.56 mL of the EDTA solution were needed to titrate the Pb^{2+}. Calculate the percent Pb and Cd in the sample.

17-24. A 0.6004-g sample of Ni/Cu condenser tubing was dissolved in acid and diluted to 100.0 mL in a volumetric flask. Titration of both cations in a 25.00-mL aliquot of this solution required 45.81 mL of 0.05285 M EDTA. Mercaptoacetic acid and NH_3 were then introduced; production of the Cu complex with the former resulted in the release of an equivalent amount of EDTA, which required a 22.85-mL titration with 0.07238 M Mg^{2+}. Calculate the percent Cu and Ni in the alloy.

***17-25.** Calamine, which is used for relief of skin irritations, is a mixture of zinc and iron oxides. A 1.022-g sample of dried calamine was dissolved in acid and diluted to 250.0 mL. Potassium fluoride was added to a 10.00-mL aliquot of the diluted solution to mask the iron; after suitable adjustment of the pH, Zn^{2+} consumed 38.71 mL of 0.01294 M EDTA. A second 50.00-mL aliquot was suitably buffered and titrated with 2.40 mL of 0.002727 M ZnY^{2-}solution:

$$Fe^{3+} + ZnY^{2-} \rightarrow FeY^{-} + Zn^{2+}$$

Calculate the percentages of ZnO and Fe_2O_3 in the sample.

17-26. A 3.650-g sample containing bromate and bromide was dissolved in sufficient water to give 250.0 mL. After acidification, silver nitrate was introduced to a 25.00-mL aliquot to precipitate AgBr, which was filtered, washed, and then redissolved in an ammoniacal solution of potassium tetracyanonickelate(II):

$$Ni(CN)_4^{2-} + 2AgBr(s) \rightarrow 2Ag(CN)_2^{-} + Ni^{2+} + 2Br^{-}$$

The liberated nickel ion required 26.73 mL of 0.02089 M EDTA. The bromate in a 10.00-mL aliquot was reduced to bromide with arsenic(III) prior to the addition of silver nitrate. The same procedure was followed, and the released nickel ion was titrated with 21.94 mL of the EDTA solution. Calculate the percentages of NaBr and $NaBrO_3$ in the sample.

***17-27.** The potassium ion in a 250.0-mL sample of mineral water was precipitated with sodium tetraphenylborate:

$$K^{+} + B(C_6H_5)_4^{-} \rightarrow KB(C_6H_5)(s)$$

The precipitate was filtered, washed, and redissolved in an organic solvent. An excess of the mercury(II)/EDTA chelate was added:

$$4HgY^{2-} + B(C_6H_4)_4^- + 4H_2O \rightarrow$$
$$H_3BO_3 + 4C_6H_5Hg^+ + 4HY^{3-} + OH^-$$

The liberated EDTA was titrated with 29.64 mL of 0.05581 M Mg^{2+}. Calculate the potassium ion concentration in parts per million.

17-28. Chromel is an alloy composed of nickel, iron, and chromium. A 0.6472-g sample was dissolved and diluted to 250.0 mL. When a 50.00-mL aliquot of 0.05182 M EDTA was mixed with an equal volume of the diluted sample, all three ions were chelated, and a 5.11-mL back-titration with 0.06241 M copper(II) was required. The chromium in a second 50.0-mL aliquot was masked through the addition of hexamethylenetetramine; titration of the Fe and Ni required 36.28 mL of 0.05182 M EDTA. Iron and chromium were masked with pyrophosphate in a third 50.0-mL aliquot, and the nickel was titrated with 25.91 mL of the EDTA solution. Calculate the percentages of nickel, chromium, and iron in the alloy.

***17-29.** A 0.3284-g sample of brass (containing lead, zinc, copper, and tin) was dissolved in nitric acid. The sparingly soluble $SnO_2 \cdot 4H_2O$ was removed by filtration, and the combined filtrate and washings were then diluted to 500.0 mL. A 10.00-mL aliquot was suitably buffered; titration of the lead, zinc, and copper in this aliquot required 37.56 mL of 0.002500 M EDTA. The copper in a 25.00-mL aliquot was masked with thiosulfate; the lead and zinc were then titrated with 27.67 mL of the EDTA solution. Cyanide ion was used to mask the copper and zinc in a 100-mL aliquot; 10.80 mL of the EDTA solution was needed to titrate the lead ion. Determine the composition of the brass sample; evaluate the percentage of tin by difference.

17-30. Calculate conditional constants for the formation of the EDTA complex of Fe^{2+} at a pH of (a) 6.0, (b) 8.0, (c) 10.0.

***17-31.** Calculate conditional constants for the formation of the EDTA complex of Ba^{2+} at a pH of (a) 7.0, (b) 9.0, (c) 11.0.

17-32. Construct a titration curve for 50.00 mL of 0.01000 M Sr^{2+} with 0.02000 M EDTA in a solution buffered to pH 11.0. Calculate pSr values after the addition of 0.00, 10.00, 24.00, 24.90, 25.00, 25.10, 26.00, and 30.00 mL of titrant.

17-33. Construct a titration curve for 50.00 mL of 0.0150 M Fe^{2+} with 0.0300 M EDTA in a solution buffered to pH 7.0. Calculate pFe values after the addition of 0.00, 10.00, 24.00, 24.90, 25.00, 25.10, 26.00, and 30.00 mL of titrant.

17-34. Titration of Ca^{2+} and Mg^{2+} in a 50.00-mL sample of hard water required 23.65 mL of 0.01205 M EDTA. A second 50.00-mL aliquot was made strongly basic with NaOH to precipitate Mg^{2+} as $Mg(OH)_2(s)$. The supernatant liquid was titrated with 14.53 mL of the EDTA solution. Calculate
(a) the total hardness of the water sample, expressed as ppm $CaCO_3$.
*(b) the concentration in ppm of $CaCO_3$ in the sample.
(c) the concentration in ppm of $MgCO_3$ in the sample.

17-35. Challenge Problem. Zinc sulfide, ZnS, is sparingly soluble under most situations. With ammonia, Zn^{2+} forms four complexes, $Zn(NH_3)^{2+}$, $Zn(NH_3)_2^{2+}$, $Zn(NH_3)_3^{2+}$, and $Zn(NH_3)_4^{2+}$. Ammonia is, of course, a base, and S^{2-} is the anion of the weak diprotic acid, H_2S. Find the molar solubility of zinc sulfide in:
(a) pH 7.0 water.
(b) a solution containing 0.100 M NH_3.
(c) a pH 9.00 ammonia/ammonium ion buffer with a total NH_3/NH_4^+ concentration of 0.100 M.
(d) the same solution as in part (c) except that it also contains 0.100 M EDTA.
(e) Use a search engine and locate a Materials Safety Data sheet (MSDS) for ZnS. Determine what health hazards ZnS poses.
(f) Determine if there is a phosphorescent pigment containing ZnS. What activates the pigment to "glow in the dark"?
(g) Determine what use ZnS has in making optical components. Why is ZnS useful for these components?

InfoTrac College Edition

For additional readings, go to InfoTrac College Edition, your online research library, at

http://infotrac.thomsonlearning.com

PART IV

Electrochemical Methods

A conversation with *Allen J. Bard*

Allen J. Bard is a New Yorker turned Texan by way of Boston. He received his B.S. from City College of New York, completed his doctorate at Harvard, and has been on the faculty at the University of Texas, Austin since 1958. At Texas, he holds the Norman Hackerman/Welch Regents Chair and is founder and director of the Laboratory of Electrochemistry. The lab develops electroanalytical methods and instruments and applies them to the study of problems in electroorganic chemistry, photoelectrochemistry, and electroanalytical chemistry. Bard and his laboratory hold more than 20 patents. Along with his former graduate student Larry R. Faulkner, he co-authored the important textbook Electrochemical Methods. *In 2002, Bard added the Priestly Medal, the top award from the American Chemical Society, to his many other national and international prizes in chemistry. He recently stepped down as editor-in-chief of the* Journal of the American Chemical Society, *a position he held for 20 years.*

Q: How did you become interested in chemistry?

A: I went to the Bronx High School of Science. I liked chemistry and was good at it. I also liked studying organisms, and I might have gone into biology but didn't see a future in it. I thought of biology as classifying and collecting, which shows how short-sighted I was! Of course, this was before molecular biology came on the scene.

Q: Where did you do your graduate work?

A: I chose Harvard because it was a good school and I wanted to leave New York. I started in inorganic chemistry. I was doing research with Geoff Wilkinson, an assistant professor, who was working on ferrocenes and related compounds. He didn't get tenure my first semester at Harvard, so I had to find something else to do. Later, he went on to get the Nobel Prize for his work on sandwich (organometallic) compounds. I liked instrumentation and I liked electronics, so my next choice was analytical chemistry. James J. Lingane was very prominent in the field, so I decided to work with him.

Q: In your experience, have you seen any changes in how science is done?

A: When I was in graduate school, science was not heavily supported by special funding. My Ph.D. advisor at Harvard never had a federal research grant. There was little federal funding of science before World War II, and during the war there was only focused federal funding. The big change came in 1957, when Sputnik went up and we found ourselves in a science and technology race with Russia. All of a sudden, scientists were highly funded! Big and bigger science really started at about that time. At the start of my career, I bought some of my chemicals out of my own pocket money, but soon I learned the game and got funding early on—from the National Science Foundation and the Welch Foundation, which started funding chemistry in Texas in the 1950s. Getting large grants has become more and more important as the years have gone by. This has greatly increased the scope of the science one can do, but a scientist must spend more time writing and reading proposals and reports than ever before. That's a big drag on time and can affect one's creativity.

Q: Have you also seen any changes in the relationship of academia with industry?

A: The nature of the interaction of academia with industry and small companies has changed a lot over the years. When I got out of graduate school, consulting for industry was not very common. If you consulted for industry and said you were doing it to expand your knowledge, the money would usually go into the department pot, not the consultant's pocket. The idea that faculty would become entrepreneurs and start their own companies was almost unheard of. The university also wasn't as hungry for patents as they are today. For example, initially the Welch Foundation said that a discovery should be dedicated to mankind and not patented. They soon realized, however, that if you don't patent something, nobody's going to do anything with it! It's good to have industrial interactions to broaden your knowledge, to expand your view of science and what's important, and to get to know more people. It can also have bad effects, however—for example, by encouraging more applied research. If you get seriously into entrepreneurial endeavors, it takes away from the time you have to spend on other university functions, such as interacting with students.

Q: What advances have you made in the field of scanning electrochemical microscopy?

A: For the past 10 years, we've been developing the technique of scanning electrochemical microscopy (SECM), which uses very small electrodes. For some applications, the smaller the better. The biggest ones are 10 micrometers, and they go down into the 50-nanometer range. We can bring these electrodes into close proximity with a surface containing a system of interest, such as a cell or a piece of material that's undergoing a chemical change, and with very high resolution examine the chemistry on the surface. We can apply the technique to biological systems to learn how things are transported through membranes—i.e., to observe the flux of material through membranes—and to see enzymes and how they're distributed on the outside of a cell. We're now trying to combine that technique with optical microscopy in a form called near-field scanning optical microscopy (NSOM), which is not limited by the wavelength of light. In this technique, you pull down a tip of glass or quartz fiber to a small point, much smaller than the wavelength of light, then shine a laser down the tip. The resolution is determined by the size of the tip. We're trying to combine this with scanning electrochemical microscopy by putting an electrode around the tip. We can then make simultaneous optical and electrical measurements of the system.

One of the driving forces in all scanning probe techniques is to examine things at very high spatial and temporal resolutions. The goal is to do analyses not of a bulk sample, but of little spots or areas on cells or surfaces, a semiconductor chip, or anything else.

The goal [of scanning probe techniques] is to do analyses not of a bulk sample, but of little spots or areas on cells or surfaces, a semiconductor chip, or anything else.

Q: What work are you doing in regard to electrogenerated chemiluminescence?

A: Another area we really like that we've been in since the 1960s—and it has really blossomed—is electrogenerated chemiluminescence (ECL). ECL is the generation of light through electrochemical reactions. We take an electrode and choose two reactants that undergo an electron transfer reaction at the electrode. The selected reaction is so energetic that it doesn't form ground state products but rather an excited state that gives out light. It's a little like fluorescence, but instead of inserting a photon to make an emitting excited state, you do it by an electron transfer reaction. You can measure light with very high sensitivity. Since light is coming out of the system but no light is going into it, there are no problems with scattered light or impurities. It's selective for the molecules that are capable of ECL, and it's very sensitive. It's been picked up by companies that have developed ECL-based tags to form labeled molecules for use in an immunoassay or as a DNA probe. We're now trying to get new tags and new analytical applications. Our dream is to look at a single molecule on a surface by this technique, but we're not nearly there yet.

Q: How do you prefer to work?

A: There are all kinds of scientists who get their kicks in different ways. There are the scientists who are like me, who like to be by themselves in a field for a while, to try new things and probe the frontier. I tend to get out of a field once it gets hot. Now there's a huge tendency to be in fashionable areas. Congress and the funding agencies jump on bandwagons—right now it's nanoscience—and they drop a whole lot of money into those fields, so younger scientists tend to gravitate to these. I'd rather be at my own frontier.

Q: How do you feel about receiving awards and honors for your work?

A: Most of the awards I've won recognize a body of work. I'm proud to be the recipient of the 2002 Priestly medal. I think that in your lifetime, you're either underappreciated or overappreciated. When I was younger I was completely underappreciated; now I'm sure I'm completely overappreciated!

Q: What do you enjoy most about working with graduate students?

A: I like watching students develop. The greatest pleasure in working with graduate students is that you can see them develop from students who don't know very much about what they're doing—who don't have any idea what science is about—and in three or four years they become mature, valuable scientists who you hate to see leave the lab. In teaching courses you see it too, although you don't see quite that level of development. It's fascinating to watch students when they suddenly understand an idea or concept.

Q: Do you have any advice for young people considering a career in academic science?

A: The great thing about a career in science is that you're probably not going to make a lot of money but usually you interact with a lot of very good people. You're doing interesting things, and if you're in academia, your life is your own. You're the master of your fate! To me these things are worth a heck of a lot. That to me is the nice part about science. ■

CHAPTER 18

Introduction to Electrochemistry

© Jeffrey L. Rotman/Corbis

From the earliest days of experimental science, workers such as Galvani, Volta, and Cavendish realized that electricity interacts in interesting and important ways with animal tissues. Electrical charge causes muscles to contract, for example. Perhaps more surprising is that a few animals such as the *torpedo* (shown in the photo) *produce* charge by physiological means. More than 50 billion nerve terminals in the torpedo's flat "wings" on its left and right sides rapidly emit acetylcholine on the bottom side of membranes housed in the wings. The acetylcholine causes sodium ions to surge through the membranes, which produces a rapid separation of charge and a corresponding potential difference, or voltage, across the membrane.[1] The potential difference then produces an electric current of several amperes in the surrounding sea water that may be used to stun or kill prey, detect and ward off enemies, or navigate. Natural devices for separating charge and creating electrical potential difference are relatively rare, but humans have learned to separate charge mechanically, metallurgically, and chemically to create cells, batteries, and other useful charge storage devices.

We now turn our attention to several analytical methods that are based on oxidation/reduction reactions. These methods, which are described in Chapters 18 through 23, include oxidation/reduction titrimetry, potentiometry, coulometry, electrogravimetry, and voltammetry. Fundamentals of electrochemistry that are necessary for understanding the principles of these procedures are presented in this chapter.

18A CHARACTERIZING OXIDATION/REDUCTION REACTIONS

Oxidation/reduction reactions are sometimes called **redox** reactions.

In an **oxidation/reduction reaction,** electrons are transferred from one reactant to another. An example is the oxidation of iron(II) ions by cerium(IV) ions. The reaction is described by the equation

$$Ce^{4+} + Fe^{2+} \rightleftharpoons Ce^{3+} + Fe^{3+} \qquad (18\text{-}1)$$

A **reducing agent** is an electron donor. An **oxidizing agent** is an electron acceptor.

In this reaction, an electron is transferred from Fe^{2+} to Ce^{4+} to form Ce^{3+} and Fe^{3+} ions. A substance that has a strong affinity for electrons, such as Ce^{4+}, is called an **oxidizing agent,** or an **oxidant.** A **reducing agent,** or **reductant,** is a species, such

[1] Y. Dunant and M. Israel, *Sci. Am.* **1985,** *252,* 58.

as Fe^{2+}, that easily donates electrons to another species. To describe the chemical behavior represented by Equation 18-1, we say that Fe^{2+} is oxidized by Ce^{4+}; similarly, Ce^{4+} is reduced by Fe^{2+}.

We can split any oxidation/reduction equation into two half-reactions that show which species gains electrons and which loses them. For example, Equation 18-1 is the sum of the two half-reactions

$$Ce^{4+} + e^- \rightleftharpoons Ce^{3+} \qquad \text{(reduction of } Ce^{4+}\text{)}$$

$$Fe^{2+} \rightleftharpoons Fe^{3+} + e^- \qquad \text{(oxidation of } Fe^{2+}\text{)}$$

◀ It is important to understand that while we can easily write an equation for a half-reaction in which electrons are consumed or generated, we cannot observe an isolated half-reaction experimentally because there must always be a second half-reaction that serves as a source of electrons or a recipient of electrons—that is, an individual half-reaction is a theoretical concept.

The rules for balancing half-reactions (see Feature 18-1) are the same as those for other reaction types; that is, the number of atoms of each element as well as the net charge on each side of the equation must be the same. Thus, for the oxidation of Fe^{2+} by MnO_4^-, the half-reactions are

$$MnO_4^- + 5e^- + 8H^+ \rightleftharpoons Mn^{2+} + 4H_2O$$

$$5Fe^{2+} \rightleftharpoons 5Fe^{3+} + 5e^-$$

In the first half-reaction, the net charge on the left side is $(-1 - 5 + 8) = +2$, which is the same as the charge on the right. Note also that we have multiplied the second half-reaction by 5 so that the number of electrons lost by Fe^{2+} equals the number gained by MnO_4^-. We can then write a balanced net-ionic equation for the overall reaction by adding the two half-reactions

$$MnO_4^- + 5Fe^{2+} + 8H^+ \rightleftharpoons Mn^{2+} + 5Fe^{3+} + 4H_2O$$

18A-1 Comparing Redox Reactions to Acid/Base Reactions

Oxidation/reduction reactions can be viewed in a way that is analogous to the Brønsted-Lowry concept of acid/base reactions (see Section 9A-2). Both involve

◀ Recall that in the Brønsted/Lowry concept, an acid/base reaction, is described by the equation

$$\text{acid}_1 + \text{base}_2 \rightleftharpoons \text{base}_1 + \text{acid}_2$$

B.C. by johnny hart

WHAT'S OXIDATION, ANYWAY?

GEE, I DON'T KNOW... MY SCIENCE IS A LITTLE RUSTY.

FEATURE 18-1
Balancing Redox Equations

Knowing how to balance oxidation/reduction reactions is essential to understanding all the concepts covered in this chapter. Although you probably remember this technique from your general chemistry course, we present a quick review here to remind you of how the process works. For practice, complete and balance the following equation after adding H^+, OH^-, or H_2O as needed.

$$MnO_4^- + NO_2^- \rightleftharpoons Mn^{2+} + NO_3^-$$

First, we write and balance the two half-reactions involved. For MnO_4^-, we write

$$MnO_4^- \rightleftharpoons Mn^{2+}$$

To account for the 4 oxygen atoms on the left-hand side of the equation we add $4H_2O$ on the right-hand side of the equation, which means that we must provide $8H^+$ on the left:

$$MnO_4^- + 8H^+ \rightleftharpoons Mn^{2+} + 4H_2O$$

To balance the charge, we need to add 5 electrons to the left side of the equation. Thus,

$$MnO_4^- + 8H^+ + 5e^- \rightleftharpoons Mn^{2+} + 4H_2O$$

For the other half-reaction,

$$NO_2^- \rightleftharpoons NO_3^-$$

we add one H_2O to the left side of the equation to supply the needed oxygen and $2H^+$ on the right to balance hydrogen:

$$NO_2^- + H_2O \rightleftharpoons NO_3^- + 2H^+$$

Then we add two electrons to the right-hand side to balance the charge:

$$NO_2^- + H_2O \rightleftharpoons NO_3^- + 2H^+ + 2e^-$$

Before combining the two equations, we must multiply the first by 2 and the second by 5 so that the number of electrons lost will be equal to the number of electrons gained. We then add the two half-reactions to obtain

$$2MnO_4^- + 16H^+ + 10e^- + 5NO_2^- + 5H_2O \rightleftharpoons$$
$$2Mn^{2+} + 8H_2O + 5NO_3^- + 10H^+ + 10e^-$$

which then rearranges to the balanced equation

$$2MnO_4^- + 6H^+ + 5NO_2^- \rightleftharpoons 2Mn^{2+} + 5NO_3^- + 3H_2O$$

the transfer of one or more charged particles from a donor to an acceptor—the particles being electrons in oxidation/reduction and protons in neutralization. When an acid donates a proton, it becomes a conjugate base that is capable of accepting a proton. By analogy, when a reducing agent donates an electron, it becomes an oxidizing agent that can then accept an electron. This product could be called a conjugate oxidant, but that terminology is seldom, if ever, used. With this idea in mind, we can write a generalized equation for a redox reaction as

$$A_{red} + B_{ox} \rightleftharpoons A_{ox} + B_{red} \quad (18\text{-}2)$$

Here, B_{ox}, the oxidized form of species B, accepts electrons from A_{red} to form the new reductant, B_{red}. At the same time, reductant A_{red}, having given up electrons, becomes an oxidizing agent, A_{ox}. If we know from chemical evidence that the equilibrium in Equation 18-2 lies to the right, we can state that B_{ox} is a better electron acceptor (stronger oxidant) than A_{ox}. Furthermore, A_{red} is a more effective electron donor (better reductant) than B_{red}.

C. D. Winters

Figure 18-1 Photograph of a "silver tree."

EXAMPLE 18-1

The following reactions are spontaneous and thus proceed to the right, as written

$$2H^+ + Cd(s) \rightleftharpoons H_2 + Cd^{2+}$$

$$2Ag^+ + H_2(g) \rightleftharpoons 2Ag(s) + 2H^+$$

$$Cd^{2+} + Zn(s) \rightleftharpoons Cd(s) + Zn^{2+}$$

What can we deduce regarding the strengths of H^+, Ag^+, Cd^{2+}, and Zn^{2+} as electron acceptors (or oxidizing agents)?

The second reaction establishes that Ag^+ is a more effective electron acceptor than H^+; the first reaction demonstrates that H^+ is more effective than Cd^{2+}. Finally, the third equation shows that Cd^{2+} is more effective than Zn^{2+}. Thus, the order of oxidizing strength is $Ag^+ > H^+ > Cd^{2+} > Zn^{2+}$.

18A-2 Oxidation/Reduction Reactions in Electrochemical Cells

Many oxidation/reduction reactions can be carried out in either of two ways that are physically quite different. In one, the reaction is performed by bringing the oxidant and the reductant into direct contact in a suitable container. In the second, the reaction is carried out in an electrochemical cell in which the reactants do not come in direct contact with one another. A marvelous example of direct contact is the famous "silver tree" experiment, in which a piece of copper is immersed in a silver nitrate solution (Figure 18-1). Silver ions migrate to the metal and are reduced:

◀ For an interesting illustration of this reaction, immerse a piece of copper in a solution of silver nitrate. The result is the deposition of silver on the copper in the form of a "silver tree." See Figure 18-1 and color plate 9.

$$Ag^+ + e^- \rightleftharpoons Ag(s)$$

At the same time, an equivalent quantity of copper is oxidized:

$$Cu(s) \rightleftharpoons Cu^{2+} + 2e^-$$

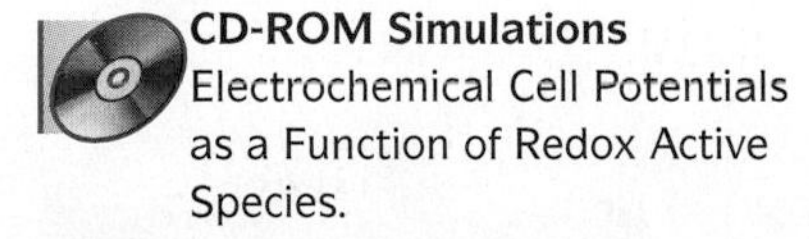

CD-ROM Simulations
Electrochemical Cell Potentials as a Function of Redox Active Species.

By multiplying the silver half-reaction by two and adding the reactions, we obtain a net ionic equation for the overall process.

$$2Ag^+ + Cu(s) \rightleftharpoons 2Ag(s) + Cu^{2+} \qquad (18\text{-}3)$$

A unique aspect of oxidation/reduction reactions is that the transfer of electrons—and thus an identical net reaction—can often be brought about in an **electrochemical cell,** in which the oxidizing agent and the reducing agent are physically separated from one another. Figure 18-2a shows such an arrangement. Note that a **salt bridge** isolates the reactants but maintains electrical contact between the two halves of the cell. When a voltmeter of high internal resistance is connected as shown or the electrodes are not connected externally, the cell is said to be at **open circuit** and delivers the full cell potential. When the circuit is open, no net reaction occurs in the cell, although we will show that the cell has the **potential** for doing work. The voltmeter measures the potential difference, or **voltage,** between the two electrodes at any instant. This voltage is a measure of the tendency of the cell reaction to proceed toward equilibrium.

► Salt bridges are widely used in electrochemistry to prevent mixing of the contents of the two electrolyte solutions making up electrochemical cells. Ordinarily, the two ends of the bridge are fitted with sintered glass disks or other porous materials to prevent liquid from siphoning from one part of the cell to the other.

In Figure 18-2b, the cell is connected so that electrons can pass through a low-resistance external circuit. The potential energy of the cell is now converted to electrical energy to light a lamp, run a motor, or do some other type of electrical work. In the cell in Figure 18-2b, metallic copper is oxidized at the left-hand electrode, silver ions are reduced at the right-hand electrode, and electrons flow through the external circuit to the silver electrode. As the reaction goes on, the cell potential, initially 0.412 V when the circuit is open, decreases continuously and approaches zero as the overall reaction approaches equilibrium. When the cell is at equilibrium, both cell half-reactions occur at the same rate, and the cell voltage is zero. A cell with zero voltage does not perform work, as anyone who has found a "dead" battery in a flashlight or in a laptop computer can attest.

► When the $CuSO_4$ and $AgNO_3$ solutions are 0.0200 M, the cell develops a potential of 0.412 V, as shown in Figure 18-2a.

► The equilibrium-constant expression for the reaction shown in Equation 18-3 is

$$K_{eq} = \frac{[Cu^{2+}]}{[Ag^+]} = 4.1 \times 10^{15} \qquad (18\text{-}4)$$

This expression applies whether the reaction occurs directly between reactants or within an electrochemical cell.

When zero voltage is reached in the cell in Figure 18-2b, the concentrations of Cu(II) and Ag(I) ions will have values that satisfy the equilibrium-constant expression shown in Equation 18-4. At this point, no further net flow of electrons will occur. *It is important to recognize that the overall reaction and its position of equilibrium are totally independent of the way the reaction is carried out,* whether it is by direct reaction in a solution or by indirect reaction in an electrochemical cell.

► At equilibrium, the two half-reactions in a cell continue, but their rates are equal.

18B ELECTROCHEMICAL CELLS

We can study oxidation/reduction equilibria conveniently by measuring the potentials of electrochemical cells in which the two half-reactions making up the equilibrium are participants. For this reason, we must consider some characteristics of electrochemical cells.

An electrochemical cell consists of two conductors called **electrodes,** each of which is immersed in an electrolyte solution. In most of the cells that will be of interest to us, the solutions surrounding the two electrodes are different and must be separated to avoid direct reaction between the reactants. The most common way of avoiding mixing is to insert a salt bridge, such as that shown in Figure 18-2, between the solutions. Conduction of electricity from one electrolyte solution to the other then occurs by migration of potassium ions in the bridge in one direction and chloride ions in the other. However, direct contact between copper metal and silver ions is prevented.

The electrodes in some cells share a common electrolyte; these are known as **cells without liquid junction.** For an example of such a cell, see Figure 19-2 and Example 19-7.

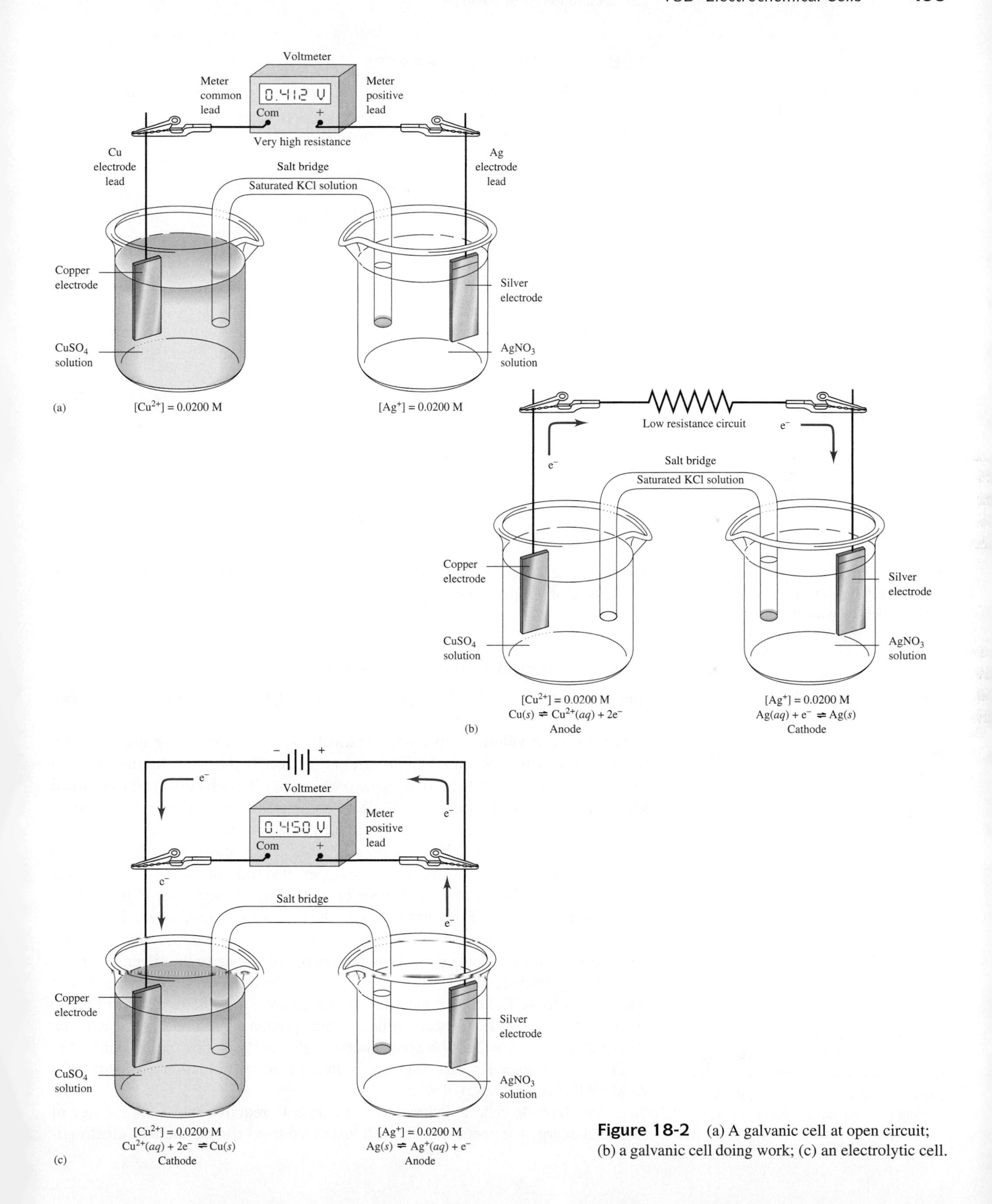

Figure 18-2 (a) A galvanic cell at open circuit; (b) a galvanic cell doing work; (c) an electrolytic cell.

18B-1 Cathodes and Anodes

A **cathode** is an electrode where reduction occurs. An **anode** is an electrode where oxidation occurs.

The **cathode** in an electrochemical cell is the electrode at which reduction occurs. The **anode** is the electrode at which oxidation takes place.

Examples of typical cathodic reactions include

$$Ag^+ + e^- \rightleftharpoons Ag(s)$$
$$Fe^{3+} + e^- \rightleftharpoons Fe^{2+}$$
$$NO_3^- + 10H^+ + 8e^- \rightleftharpoons NH_4^+ + 3H_2O$$

► The reaction $2H^+ + 2e^- \rightleftharpoons H_2(g)$ occurs at a cathode when an aqueous solution contains no easily reduced species.

We can force a desired reaction to occur by applying a suitable potential to an electrode made of an unreactive material such as platinum. Note that the reduction of NO_3^- in the third reaction reveals that anions can migrate to a cathode and be reduced.

Typical anodic reactions include

$$Cu(s) \rightleftharpoons Cu^{2+} + 2e^-$$
$$2\,Cl^- \rightleftharpoons Cl_2(g) + 2e^-$$
$$Fe^{2+} \rightleftharpoons Fe^{3+} + e^-$$

► The Fe^{2+}/Fe^{3+} half-reaction may seem somewhat unusual because a cation rather than an anion migrates to the anode and gives up an electron. Oxidation of a cation at an anode or reduction of an anion at a cathode is a relatively common process.

The first reaction requires a copper anode, but the other two can be carried out at the surface of an inert platinum electrode.

18B-2 Types of Electrochemical Cells

► The reaction $2H_2O \rightleftharpoons O_2(g) + 4H^+ + 4e^-$ occurs at an anode when an aqueous solution contains no other easily oxidized species.

► For both galvanic and electrolytic cells, remember that (1) reduction always takes place at the cathode and (2) oxidation always takes place at the anode. The cathode in a galvanic cell becomes the anode, however, when the cell is operated electrolytically.

Electrochemical cells are either galvanic or electrolytic. They can also be classified as reversible or irreversible.

Galvanic, or **voltaic, cells** store electrical energy. **Batteries** are usually made from several such cells connected in series to produce higher voltages than a single cell can produce. The reactions at the two electrodes in such cells tend to proceed spontaneously and produce a flow of electrons from the anode to the cathode via an external conductor. The cell shown in Figure 18-2a is a galvanic cell that develops a potential of about 0.412 V when no current is being drawn from it. The silver electrode is positive with respect to the copper electrode in this cell. The copper electrode, which is negative with respect to the silver electrode, is a potential source of electrons to the external circuit when the cell is discharged. The cell in Figure 18-2b is the same galvanic cell, but now under discharge so that electrons move through the external circuit from the copper electrode to the silver electrode. While being discharged, the silver electrode is the *cathode,* since the reduction of Ag^+ occurs here. The copper electrode is the *anode,* since the oxidation of $Cu(s)$ occurs at this electrode. Galvanic cells operate spontaneously, and the net reaction during discharge is called the **spontaneous cell reaction.** For the cell in Figure 18-2b, the spontaneous cell reaction is that given by Equation 18-3—that is, $2Ag^+ + Cu(s) \rightleftharpoons 2Ag(s) + Cu^{2+}$.

An **electrolytic cell,** in contrast to a voltaic cell, requires an external source of electrical energy for operation. The cell just considered can be operated electrolyt-

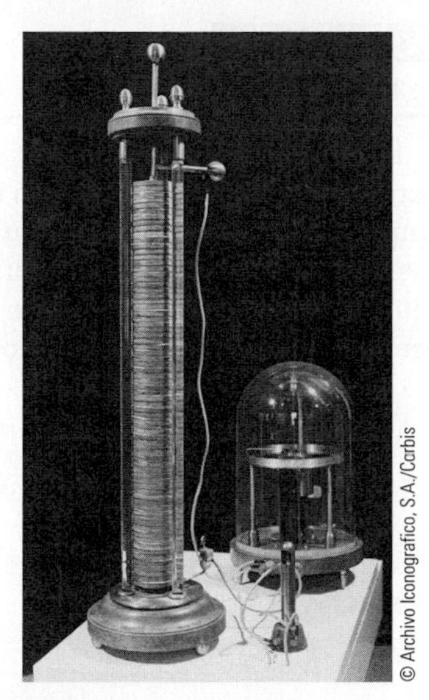

Alessandro Volta (1745–1827), Italian physicist, was the inventor of the first battery, the so-called voltaic pile (shown on the right). It consisted of alternating disks of copper and zinc separated by disks of cardboard soaked with salt solution. In honor of his many contributions to electrical science, the unit of potential difference, the volt, is named for Volta. In fact, in modern usage we often call the quantity the voltage instead of potential difference.

ically by connecting the positive terminal of an external voltage source with a potential somewhat greater than 0.412 V to the silver electrode and the negative terminal of the source to the copper electrode, as shown in Figure 18-2c. Since the negative terminal of the external voltage source is electron rich, electrons flow from this terminal to the copper electrode, where reduction of Cu^{2+} to Cu(*s*) occurs. The current is sustained by the oxidation of Ag(*s*) to Ag^+ at the right-hand electrode, producing electrons that flow to the positive terminal of the voltage source. Note that in the electrolytic cell, the direction of the current is the reverse of that in the galvanic cell in Figure 18-2b, and the reactions at the electrodes are reversed as well. The silver electrode is forced to become the *anode,* while the copper electrode is forced to become the *cathode.* The net reaction that occurs when a voltage higher than the galvanic cell voltage is applied is the opposite of the spontaneous cell reaction. That is,

$$2Ag(s) + Cu^{2+} \rightleftharpoons 2Ag^+ + Cu(s)$$

The cell in Figure 18-2 is an example of a reversible cell, in which the direction of the electrochemical reaction is reversed when the direction of electron flow is changed. In an irreversible cell, changing the direction of current causes entirely different half-reactions to occur at one or both electrodes. The lead-acid storage battery in an automobile is a common example of a series of reversible cells. When the battery is being charged by the generator or an external charger, its cells are electrolytic. When it is used to operate the headlights, the radio, or the ignition, its cells are galvanic.

In a **reversible cell**, reversing the current reverses the cell reaction. In an **irreversible cell,** reversing the current causes a different half-reaction to occur at one or both of the electrodes.

► A modern Daniell cell is shown in color plate 10.

FEATURE 18-2

The Daniell Gravity Cell

The Daniell gravity cell was one of the earliest galvanic cells to find widespread practical application. It was used in the mid-1800s to power telegraphic communication systems. As shown in Figure 18F-1, the cathode was a piece of copper immersed in a saturated solution of copper sulfate. A much less dense solution of dilute zinc sulfate was layered on top of the copper sulfate, and a massive zinc electrode was located in this solution. The electrode reactions were

$$Zn(s) \rightleftharpoons Zn^{2+} + 2e^-$$

$$Cu^{2+} + 2e^- \rightleftharpoons Cu(s)$$

This cell develops an initial voltage of 1.18 V, which gradually decreases as the cell discharges.

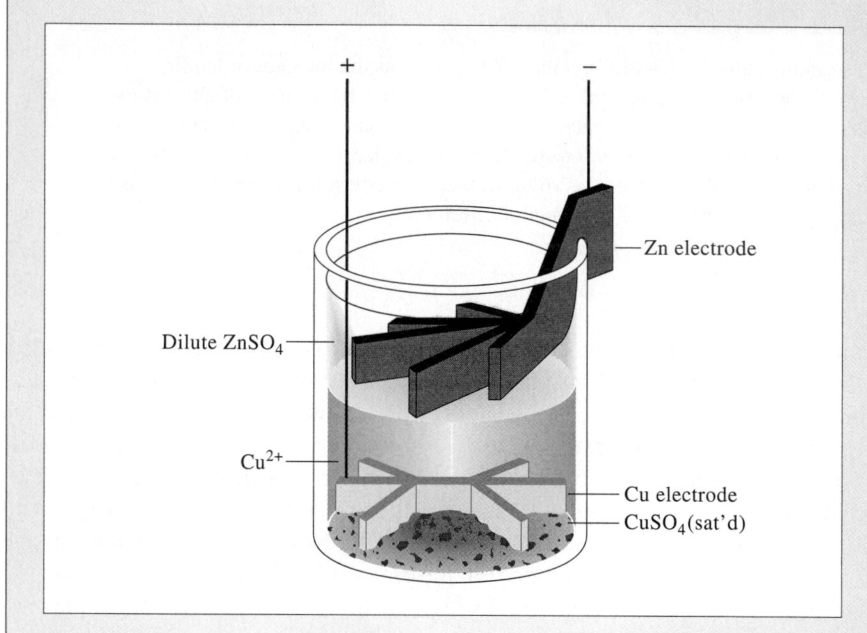

Figure 18F-1 A Daniell gravity cell.

18B-3 Representing Cells Schematically

Chemists frequently use a shorthand notation to describe electrochemical cells. The cell in Figure 18-2a, for example, is described by

$$Cu\,|\,Cu^{2+}(0.0200\ M)\,||\,Ag^{+}(0.0200\ M)\,|\,Ag \tag{18-5}$$

By convention, a single vertical line indicates a phase boundary, or interface, at which a potential develops. For example, the first vertical line in this schematic indicates that a potential develops at the phase boundary between the copper electrode and the copper sulfate solution. The double vertical line represents two phase boundaries, one at each end of the salt bridge. A **liquid-junction potential** develops at each of these interfaces. The junction potential results from differences in the rates at which the ions in the cell compartments and the salt bridge migrate across the interfaces. A liquid-junction potential can amount to as much as several hundredths of a volt but can be negligibly small if the electrolyte in the salt bridge has an anion and a cation that migrate at nearly the same rate. A saturated solution of potassium chloride, KCl, is the electrolyte that is most widely used; it can reduce the junction potential to a few millivolts or less. For our purposes, we will neglect the contribution of liquid-junction potentials to the total potential of the cell. There are also several examples of cells that are without liquid junction and therefore do not require a salt bridge.

An alternative way of writing the cell shown in Figure 18-2a is

$$\mathrm{Cu}\,|\,\mathrm{CuSO_4}(0.0200\ \mathrm{M})\,||\,\mathrm{AgNO_3}(0.0200\ \mathrm{M})\,|\,\mathrm{Ag}$$

Here, the compounds used to prepare the cell are indicated rather than the active participants in the cell half-reactions.

18B-4 Currents in Electrochemical Cells

Figure 18-3 shows the movement of various charge carriers in a galvanic cell during discharge. The electrodes are connected with a wire so that the spontaneous cell reaction occurs. Charge is transported through such an electrochemical cell by three mechanisms:

1. Electrons carry the charge within the electrodes as well as the external conductor. Notice that by convention, current, which is normally indicated by the symbol I, is opposite in direction to electron flow.
2. Anions and cations are the charge carriers within the cell. At the left-hand electrode, copper is oxidized to copper ions, giving up electrons to the electrode. As shown in Figure 18-3, the copper ions formed move away from the copper electrode into the bulk of solution, while anions, such as sulfate and hydrogen sulfate ions, migrate toward the copper anode. Within the salt bridge, chloride ions migrate toward and into the copper compartment, and potassium ions move in the opposite direction. In the right-hand compartment, silver ions move toward the silver electrode, where they are reduced to silver metal, and the nitrate ions move away from the electrode into the bulk of solution.
3. The ionic conduction of the solution is coupled to the electronic conduction in the electrodes by the reduction reaction at the cathode and the oxidation reaction at the anode.

◀ In a cell, electricity is carried by the movement of ions. Both anions and cations contribute.

The phase boundary between an electrode and its solution is called an **interface.**

18C ELECTRODE POTENTIALS

The potential difference that develops between the electrodes of the cell in Figure 18-4a is a measure of the tendency for the reaction

$$2\mathrm{Ag}(s) + \mathrm{Cu}^{2+} \rightleftharpoons 2\mathrm{Ag}^{+} + \mathrm{Cu}(s)$$

to proceed from a nonequilibrium state to the condition of equilibrium. The cell potential E_{cell} is related to the free energy of the reaction ΔG by

$$\Delta G = -nFE_{\mathrm{cell}} \qquad (18\text{-}6)$$

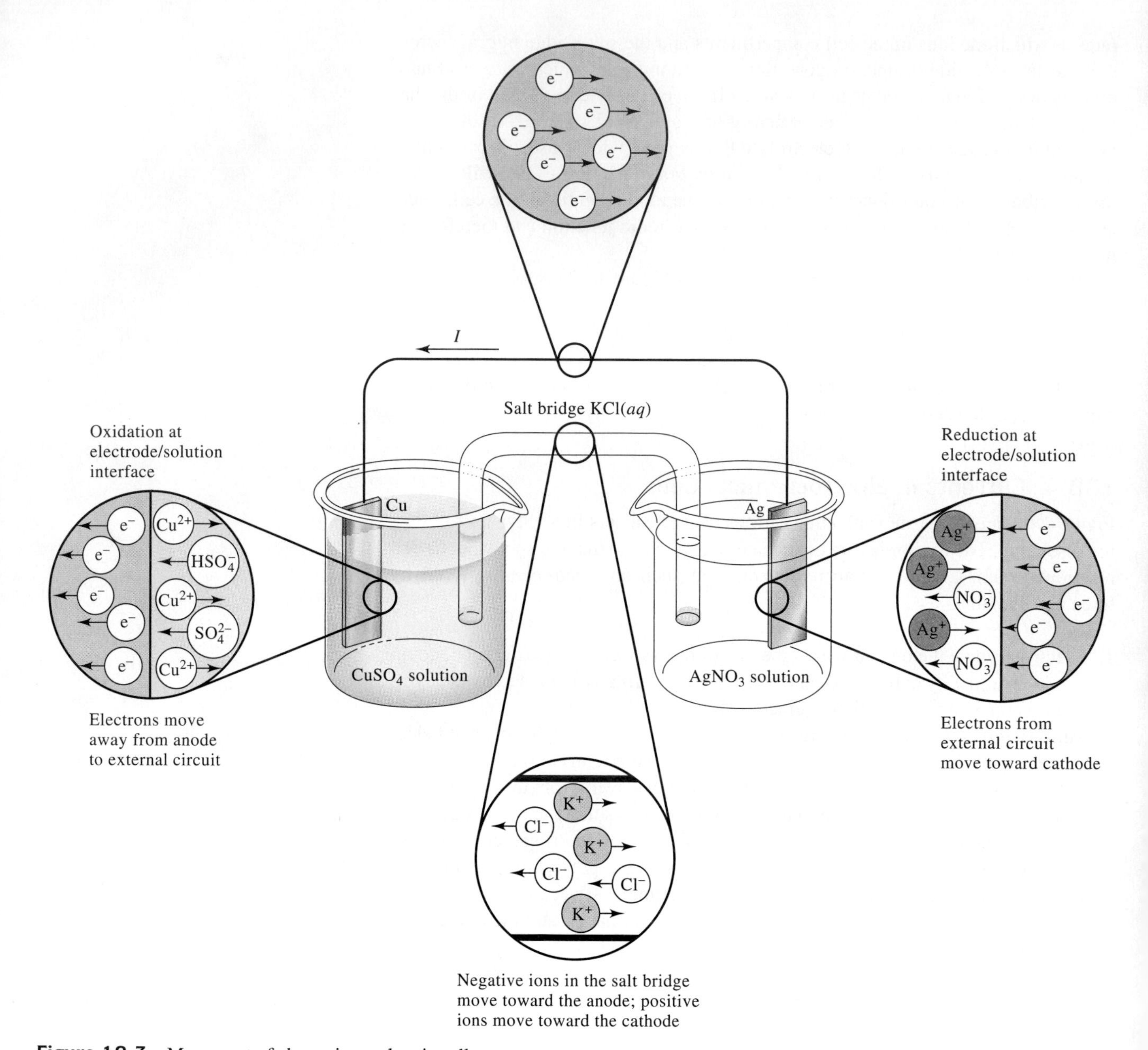

Figure 18-3 Movement of charge in a galvanic cell.

If the reactants and products are in their **standard states,** the resulting cell potential is called the **standard cell potential.** This latter quantity is related to the standard free-energy change for the reaction and thus to the equilibrium constant by

$$\Delta G^0 = -nFE^0_{\text{cell}} = -RT \ln K_{\text{eq}} \tag{18-7}$$

where R is the gas constant and T is the absolute temperature.

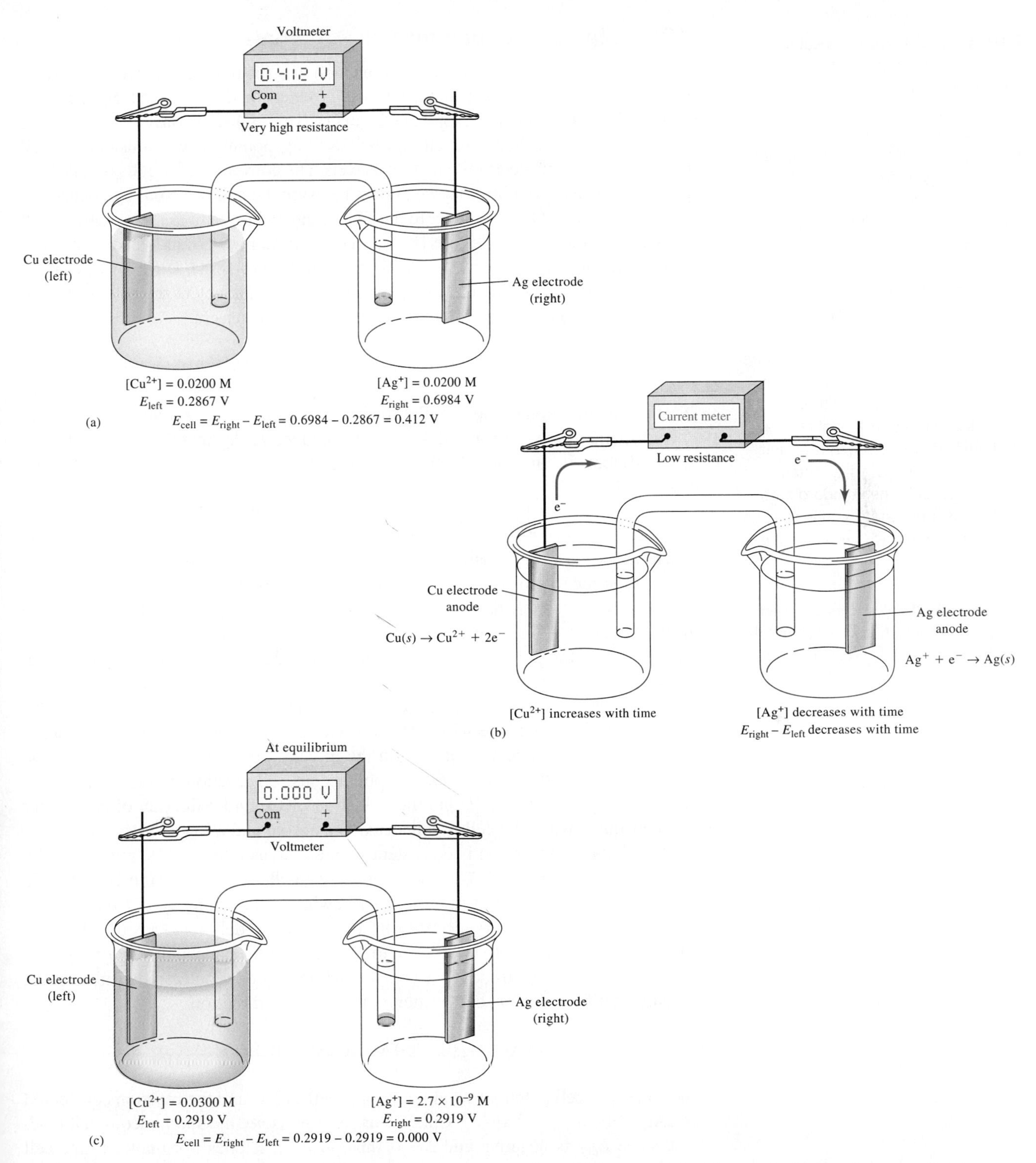

Figure 18-4 Change in cell potential after passage of current until equilibrium is reached. In (a), the high-resistance voltmeter prevents any significant electron flow, and the full open circuit cell potential is measured. For the concentrations shown this is +0.412 V. In (b), the voltmeter is replaced with a low-resistance current meter, and the cell discharges with time until eventually equilibrium is reached. In (c), after equilibrium is reached, the cell potential is again measured with a voltmeter and is found to be 0.000 V. The concentrations in the cell are now those at equilibrium as shown.

The **standard state** of a substance is a reference state that allows us to obtain relative values of such thermodynamic quantities as free energy, activity, enthalpy, and entropy. All substances are assigned unit activity in their standard state. For gases, the standard state has the properties of an ideal gas, but at one atmosphere pressure. It is thus said to be a *hypothetical* state. For pure liquids and solvents, the standard states are *real* states and are the pure substances at a specified temperature and pressure. For solutes in dilute solution, the standard state is a hypothetical state that has the properties of an infinitely dilute solute, but at unit concentration (molarity, molality, or mole fraction). The standard state of a solid is a real state and is the pure solid in its most stable crystalline form.

► The leads of voltmeters are color coded. The positive lead is red and the common, or ground, lead is black.

18C-1 Sign Convention for Cell Potentials

When we consider a normal chemical reaction, we speak of the reaction occurring from reactants on the left side of the arrow to products on the right side. By the International Union of Pure and Applied Chemistry (IUPAC) sign convention, when we consider an electrochemical cell and its resulting potential, we consider the cell reaction to occur in a certain direction as well. The convention for cells is called the **plus right rule;** it implies that we always measure the cell potential by connecting the positive lead of the voltmeter to the right-hand electrode in the schematic or cell drawing (Ag electrode in Figure 18-4) and the common, or ground, lead of the voltmeter to the left-hand electrode (Cu electrode in Figure 18-4). If we always follow this convention, the value of E_{cell} is a measure of the tendency of the cell reaction to occur spontaneously in the direction written from left to right.

$$\text{Cu}\,|\,\text{Cu}^{2+}(0.0200\text{ M})\,\|\,\text{Ag}^{+}(0.0200\text{ M})\,|\,\text{Ag}$$

That is, the direction of the overall process has Cu metal being oxidized to Cu^{2+} in the left-hand compartment and Ag^+ being reduced to Ag metal in the right-hand compartment. In other words, the reaction being considered is $Cu(s) + 2Ag^+ \rightleftharpoons Cu^{2+} + 2Ag(s)$.

Implications of the IUPAC Convention

There are several implications of the sign convention that may not be obvious. First, if the measured value of E_{cell} is positive, the right-hand electrode is positive with respect to the left-hand electrode, and the free energy change for the reaction in the direction being considered is negative according to Equation 18-6. Hence, the reaction in the direction being considered would occur spontaneously if the cell were short-circuited or connected to some device to perform work (e.g., light a lamp, power a radio, start a car). On the other hand, if E_{cell} is negative, the right-hand electrode is negative with respect to the left-hand electrode, the free energy change is positive, and the reaction in the direction considered (oxidation on the left, reduction on the right) is *not* the spontaneous cell reaction. For our cell in Figure 18-4, $E_{cell} = +0.412$ V, and the oxidation of Cu and reduction of Ag^+ occur spontaneously when the cell is connected to a device and allowed to do so.

The IUPAC convention is consistent with the signs that the electrodes actually develop in a galvanic cell. That is, in the Cu/Ag cell shown in Figure 18-4, the Cu electrode becomes electron rich (negative) owing to the tendency of Cu to be oxidized to Cu^{2+}, while the Ag electrode becomes electron deficient (positive) because of the tendency for Ag^+ to be reduced to Ag. As the galvanic cell discharges spontaneously, the silver electrode is the cathode, while the copper electrode is the anode.

Note that for the same cell written in the opposite direction

$$\text{Ag}\,|\,\text{AgNO}_3\ (0.0200\text{ M})\,\|\,\text{CuSO}_4\ (0.0200\text{ M})\,|\,\text{Cu}$$

the measured cell potential would be $E_{cell} = -0.412$ V, and the reaction considered is $2Ag(s) + Cu^{2+} \rightleftharpoons 2Ag^+ + Cu(s)$. This reaction is *not* the spontaneous cell reaction since E_{cell} is negative and ΔG is thus positive. It does not matter to the cell which electrode is written in the schematic on the right and which is written on the left. The spontaneous cell reaction is always $Cu(s) + 2Ag^+ \rightleftharpoons Cu^{2+} + 2Ag(s)$. By convention, we just measure the cell in a standard manner and consider the cell reaction in a standard direction. Finally, we must emphasize that no matter how we may write the cell schematic or arrange the cell in the laboratory, if we connect a wire or a low-resistance circuit to the cell, *the spontaneous cell reaction will occur.* The only way to achieve the reverse reaction is to connect an external voltage source and force the electrolytic reaction $2Ag(s) + Cu^{2+} \rightleftharpoons 2Ag^+ + Cu(s)$ to occur.

Half-Cell Potentials

The potential of a cell such as that shown in Figure 18-4a is the difference between two half-cell or single-electrode potentials, one associated with the half-reaction at the right-hand electrode (E_{right}), the other associated with the half-reaction at the left-hand electrode (E_{left}). According to the IUPAC sign convention, as long as the liquid-junction potential is negligible or there is no liquid junction, we may write the cell potential E_{cell} as

$$E_{cell} = E_{right} - E_{left} \qquad (18\text{-}8)$$

Although we cannot determine absolute potentials of electrodes such as these (see Feature 18-3), we can easily determine relative electrode potentials. For example, if we replace the copper electrode in the cell in Figure 18-2 with a cadmium electrode immersed in a cadmium sulfate solution, the voltmeter reads about 0.7 V more positive than the original cell. Since the right-hand compartment remains unaltered, we conclude that the half-cell potential for cadmium is about 0.7 V less than that for copper (that is, cadmium is a stronger reductant than is copper). Substituting other electrodes while keeping one of the electrodes unchanged allows us to construct a table of relative electrode potentials, as discussed in Section 18-C3.

Discharging a Galvanic Cell

The galvanic cell in Figure 18-4a is in a nonequilibrium state because the very high resistance of the voltmeter prevents the cell from discharging significantly. So when we measure the cell potential, no reaction occurs, and what we measure is the tendency of the reaction to occur should we allow it to proceed. For the Cu/Ag cell with the concentrations shown, the cell potential measured under open circuit conditions is + 0.412 V, as previously noted. If we now allow the cell to discharge by replacing the voltmeter with a low-resistance current meter, as shown in Figure 18-4b, the spontaneous cell reaction occurs. The current, initially high, decreases exponentially with time (Figure 18-5). As shown in Figure 18-4c, when equilibrium is reached, there is no net current in the cell, and the cell potential is 0.000 V. The copper ion concentration at equilibrium is then 0.0300 M, while the silver ion concentration falls to 2.7×10^{-9} M.

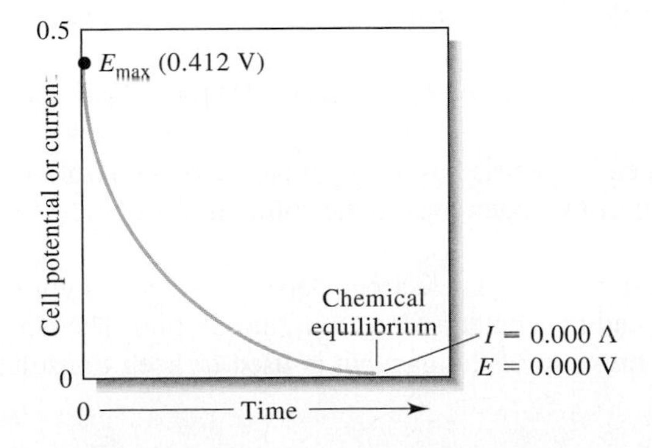

Figure 18-5 Cell potential in the galvanic cell of Figure 18-4b as a function of time. The cell current, which is directly related to the cell potential, also decreases with the same time behavior.

FEATURE 18-3

Why We Cannot Measure Absolute Electrode Potentials

Although it is not difficult to measure *relative* half-cell potentials, it is impossible to determine absolute half-cell potentials because all voltage-measuring devices measure only *differences* in potential. To measure the potential of an electrode, one contact of a voltmeter is connected to the electrode in question. The other contact from the meter must then be brought into electrical contact with the solution in the electrode compartment via another conductor. This second contact, however, inevitably involves a solid/solution interface that acts as a second half-cell when the potential is measured. Thus, an absolute half-cell potential is not obtained. What we do obtain is the difference between the half-cell potential of interest and a half-cell made up of the second contact and the solution.

Our inability to measure absolute half-cell potentials presents no real obstacle because relative half-cell potentials are just as useful provided that they are all measured against the same reference half-cell. Relative potentials can be combined to give cell potentials. We can also use them to calculate equilibrium constants and generate titration curves.

The standard hydrogen electrode is sometimes called the **normal hydrogen electrode (NHE).**

► SHE is the abbreviation for standard hydrogen electrode.

► Platinum black is a layer of finely divided platinum that is formed on the surface of a smooth platinum electrode by electrolytic deposition of the metal from a solution of chloroplatinic acid, H_2PtCl_6. The platinum black provides a large specific surface area of platinum at which the H^+/H_2 reaction can occur. Platinum black catalyzes the reaction shown in Equation 18-9. Remember that catalysts do not change the position of equilibrium but simply shorten the time it takes to reach equilibrium.

► The reaction shown as Equation 18-9 involves two equilibria:

$$2H^+ + 2e^- \rightleftharpoons H_2(aq)$$

$$H_2(aq) \rightleftharpoons H_2(g)$$

The continuous stream of gas at constant pressure provides the solution with a constant molecular hydrogen concentration.

18C-2 The Standard Hydrogen Reference Electrode

For relative electrode potential data to be widely applicable and useful, we must have a generally agreed-upon reference half-cell against which all others are compared. Such an electrode must be easy to construct, reversible, and highly reproducible in its behavior. The **standard hydrogen electrode (SHE)** meets these specifications and has been used throughout the world for many years as a universal reference electrode. It is a typical **gas electrode.**

Figure 18-6 shows how a hydrogen electrode is constructed. The metal conductor is a piece of platinum that has been coated, or **platinized,** with finely divided platinum (platinum black) to increase its specific surface area. This electrode is immersed in an aqueous acid solution of known, constant hydrogen ion activity. The solution is kept saturated with hydrogen by bubbling the gas at constant pressure over the surface of the electrode. The platinum does not take part in the electrochemical reaction and serves only as the site where electrons are transferred. The half-reaction responsible for the potential that develops at this electrode is

$$2H^+(aq) + 2e^- \rightleftharpoons H_2(g) \tag{18-9}$$

The hydrogen electrode shown in Figure 18-6 can be represented schematically as

$$\text{Pt, } H_2(p = 1.00 \text{ atm}) \,|\, ([H^+] = x \text{ M}) \,||$$

Here, the hydrogen is specified as having a partial pressure of one atmosphere and the concentration of hydrogen ions in the solution is x M. The hydrogen electrode is reversible.

The potential of a hydrogen electrode depends on temperature and the activities of hydrogen ion and molecular hydrogen in the solution. The latter, in turn, is proportional to the pressure of the gas that is used to keep the solution saturated in

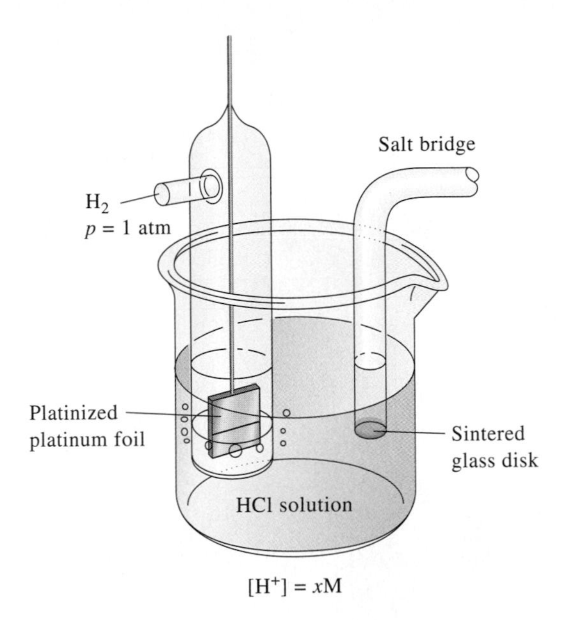

Figure 18-6 The hydrogen gas electrode.

hydrogen. For the SHE, the activity of hydrogen ions is specified as unity and the partial pressure of the gas is specified as one atmosphere. *By convention, the potential of the standard hydrogen electrode is assigned a value of 0.000 V at all temperatures.* As a consequence of this definition, any potential developed in a galvanic cell consisting of a standard hydrogen electrode and some other electrode is attributed entirely to the other electrode.

◀ At p_{H_2} = 1.00 and a_{H^+} = 1.00, the potential of the hydrogen electrode is assigned a value of exactly 0.000 V at all temperatures.

Several other reference electrodes that are more convenient for routine measurements have been developed. Some of these are described in Section 21B.

18C-3 Electrode Potential and Standard Electrode Potential

An **electrode potential** is defined as the potential of a cell in which the electrode in question is the right-hand electrode and the standard hydrogen electrode is the left-hand electrode. So if we want to obtain the potential of a silver electrode in contact with a solution of Ag^+, we would construct a cell as shown in Figure 18-7. In this cell, the half-cell on the right consists of a strip of pure silver in contact with a solution containing silver ions; the electrode on the left is the standard hydrogen electrode. The cell potential is defined as in Equation 18-8. Because the left-hand electrode is the standard hydrogen electrode with a potential that has been assigned a value of 0.000 V, we can write

◀ An electrode potential is the potential of a cell that has a standard hydrogen electrode as the left electrode (reference).

$$E_{cell} = E_{right} - E_{left} = E_{Ag} - E_{SHE} = E_{Ag} - 0.000 = E_{Ag}$$

where E_{Ag} is the potential of the silver electrode. Despite its name, an electrode potential is in fact the potential of an electrochemical cell involving a carefully defined reference electrode. Often, the potential of an electrode, such as the silver electrode in Figure 18-7, is referred to as E_{Ag} versus SHE to emphasize that it is the potential of a complete cell measured against the standard hydrogen electrode as a reference.

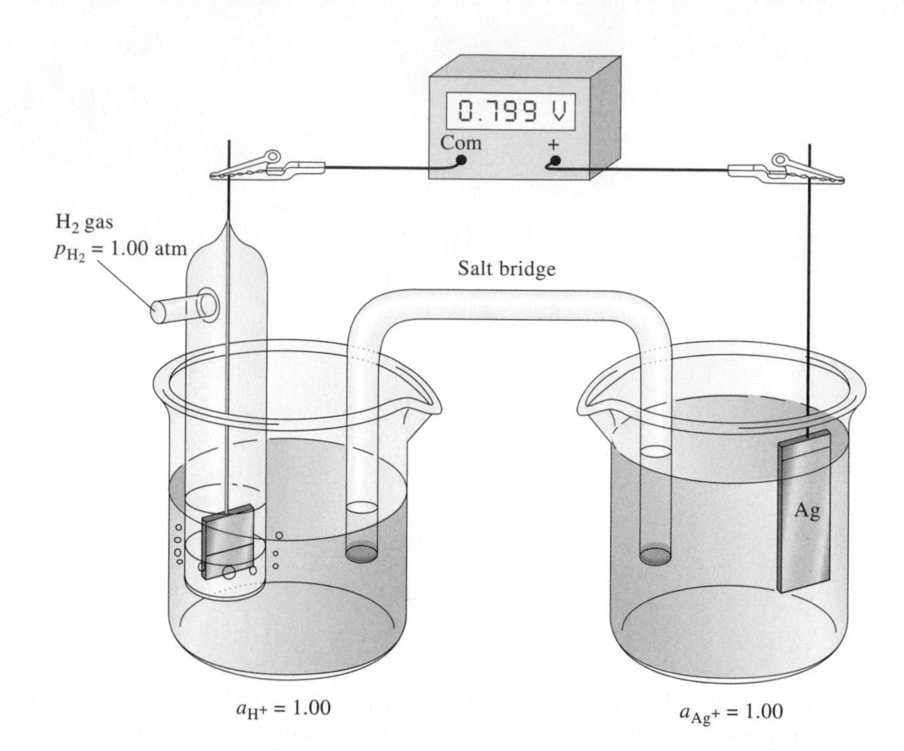

Figure 18-7 Measurement of the electrode potential for an Ag electrode. If the silver ion activity in the right-hand compartment is 1.00, the cell potential is the standard electrode potential of the Ag^+/Ag half-reaction.

The **standard electrode potential, E^0,** of a half-reaction is defined as its electrode potential when the activities of the reactants and products are all unity. For the cell in Figure 18-7, the E^0 value for the half-reaction

$$Ag^+ + e^- \rightleftharpoons Ag(s)$$

can be obtained by measuring E_{cell} with the activity of Ag^+ equal to 1.00. In this case, the cell shown in Figure 18-7 can be represented schematically as

$$Pt, H_2(p = 1.00 \text{ atm}) \,|\, H^+(a_{H^+} = 1.00) \,\|\, Ag^+(a_{Ag^+} = 1.00) \,|\, Ag$$

or alternatively as

$$\text{SHE} \,\|\, Ag^+(a_{Ag^+} = 1.00) \,|\, Ag$$

This galvanic cell develops a potential of + 0.799 V with the silver electrode on the right; that is, the spontaneous cell reaction is oxidation in the left-hand compartment and reduction in the right-hand compartment:

$$2Ag^+ + H_2(g) \rightleftharpoons 2Ag(s) + 2H^+$$

A half-cell is sometimes called a **couple.**

Because the silver electrode is on the right, the measured potential is, by definition, the standard electrode potential for the silver half-reaction, or the **silver couple.** Note that the silver electrode is positive with respect to the standard hydrogen electrode. Therefore, the standard electrode potential is given a positive sign, and we write

$$Ag^+ + e^- \rightleftharpoons Ag(s) \qquad E^0_{Ag^+/Ag} = +0.799 \text{ V}$$

Figure 18-8 illustrates a cell used to measure the standard electrode potential for the half-reaction

$$Cd^{2+} + 2e^- \rightleftharpoons Cd(s)$$

In contrast to the silver electrode, the cadmium electrode is negative with respect to the standard hydrogen electrode. Consequently, the standard electrode potential of the Cd/Cd^{2+} couple is *by convention* given a negative sign, and $E^0_{Cd^{2+}/Cd} = -0.403$ V. Because the cell potential is negative, the spontaneous cell reaction is not the reaction as written (that is, oxidation on the left and reduction on the right). Rather, the spontaneous reaction is in the opposite direction.

$$Cd(s) + 2H^+ \rightleftharpoons Cd^{2+} + H_2(g)$$

A zinc electrode immersed in a solution having a zinc ion activity of unity develops a potential of − 0.763 V when it is the right-hand electrode paired with a standard hydrogen electrode on the left. Thus, we can write $E^0_{Zn^{2+}/Zn} = -0.763$ V.

The standard electrode potentials for the four half-cells just described can be arranged in the following order:

Half-Reaction	Standard Electrode Potential, V
$Ag^+ + e^- \rightleftharpoons Ag(s)$	+ 0.799
$2H^+ + 2e^- \rightleftharpoons H_2(g)$	0.000
$Cd^{2+} + 2e^- \rightleftharpoons Cd(s)$	− 0.403
$Zn^{2+} + 2e^- \rightleftharpoons Zn(s)$	− 0.763

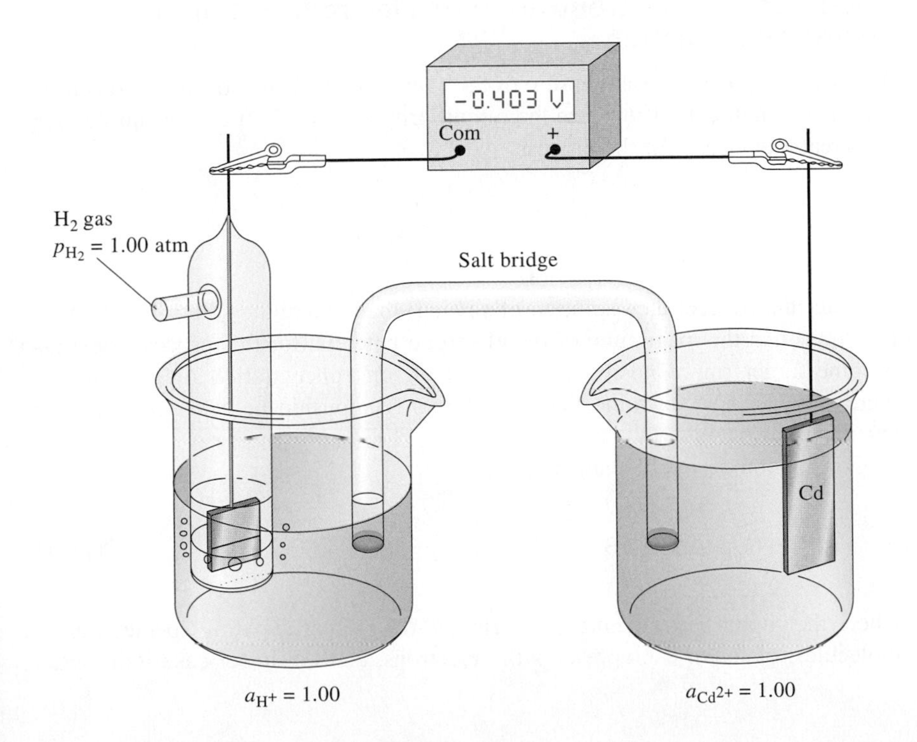

Figure 18-8 Measurement of the standard electrode potential for $Cd^{2+} + 2e^- \rightleftharpoons Cd(s)$.

The magnitudes of these electrode potentials indicate the relative strength of the four ionic species as electron acceptors (oxidizing agents); that is, in decreasing strength, $Ag^+ > H^+ > Cd^{2+} > Zn^{2+}$.

18C-4 Additional Implications of the IUPAC Sign Convention

The sign convention described in the previous section was adopted at the IUPAC meeting in Stockholm in 1953 and is now accepted internationally. Prior to this agreement, chemists did not always use the same convention, and this was the cause of controversy and confusion in the development and routine use of electrochemistry.

► An electrode potential is, by definition, a reduction potential. An oxidation potential is the potential for the half-reaction written in the opposite way. The sign of an oxidation potential is, therefore, opposite that of a reduction potential, but the magnitude is the same.

Any sign convention must be based on expressing half-cell processes in a single way—that is, either as oxidations or as reductions. According to the IUPAC convention, the term "electrode potential" (or, more exactly, "relative electrode potential") *is reserved exclusively to describe half-reactions written as reductions.* There is no objection to the use of the term "oxidation potential" to indicate a process written in the opposite sense, but it is not proper to refer to such a potential as an electrode potential.

► The IUPAC sign convention is based on the actual sign of the half-cell of interest when it is part of a cell containing the standard hydrogen electrode as the other half-cell.

The sign of an electrode potential is determined by the sign of the half-cell in question when it is coupled to a standard hydrogen electrode. When the half-cell of interest exhibits a positive potential versus the SHE (see Figure 18-7), it will behave spontaneously as the cathode when the cell is discharging. When the half-cell of interest is negative versus the SHE (see Figure 18-8), it will behave spontaneously as the anode when the cell is discharging.

18C-5 Effect of Concentration on Electrode Potentials: The Nernst Equation

An electrode potential is a measure of the extent to which the concentrations of the species in a half-cell differ from their equilibrium values. Thus, for example, there is a greater tendency for the process

$$Ag^+ + e^- \rightleftharpoons Ag(s)$$

to occur in a concentrated solution of silver(I) than in a dilute solution of that ion. It follows that the magnitude of the electrode potential for this process must also become larger (more positive) as the silver ion concentration of a solution is increased. We now examine the quantitative relationship between concentration and electrode potential.

Consider the reversible half-reaction

$$aA + bB + \cdots + ne^- \rightleftharpoons cC + dD + \cdots \qquad (18\text{-}10)$$

where the capital letters represent formulas for the participating species (atoms, molecules, or ions), e^- represents the electrons, and the lower-case italic letters indicate the number of moles of each species appearing in the half-reaction as it has been written. The electrode potential for this process is given by the equation

$$E = E^0 - \frac{RT}{nF} \ln \frac{[C]^c[D]^d \cdots}{[A]^a[B]^b \cdots} \quad (18\text{-}11)$$

◀ The meanings of the bracketed terms in Equations 18-11 and 18-12 are,

for a solute A, [A] = molar concentration
for a gas B, [B] = p_B = partial pressure in atmospheres.
If one or more of the species appearing in Equation 18-11 is a pure liquid, a pure solid, or the solvent present in excess, then no bracketed term for this species appears in the quotient because the activities of these are unity.

where

E^0 = the *standard electrode potential*, which is characteristic for each half-reaction
R = the ideal gas constant, 8.314 J K^{-1} mol^{-1}
T = temperature, K
n = number of moles of electrons that appears in the half-reaction for the electrode process as written
F = the faraday = 96,485 C (coulombs) per mole of electrons
ln = natural logarithm = 2.303 log

If we substitute numerical values for the constants, convert to base 10 logarithms, and specify 25°C for the temperature, we get

$$E = E^0 - \frac{0.0592}{n} \log \frac{[C]^c[D]^d \cdots}{[A]^a[B]^b \cdots} \quad (18\text{-}12)$$

Strictly speaking, the letters in brackets represent activities, but we will usually follow the practice of substituting molar concentrations for activities in most calculations. Thus, if some participating species A is a solute, [A] is the concentration of A in moles per liter. If A is a gas, [A] in Equation 18-12 is replaced by p_A, the partial pressure of A in atmospheres. If A is a pure liquid, a pure solid, or the solvent, its activity is unity, and no term for A is included in the equation. The rationale for these assumptions is the same as that described in Section 9B-2, which deals with equilibrium-constant expressions.

Equation 18-12 is known as the Nernst equation in honor of the German chemist Walther Nernst, who was responsible for its development.

Emilio Segrè Visual Archives/AIP

Walther Nernst (1864–1941) received the 1920 Nobel Prize in chemistry for his numerous contributions to the field of chemical thermodynamics. Nernst (far left) is seen here with Albert Einstein, Max Planck, Robert A. Millikan, and Max von Laue in 1928.

EXAMPLE 18-2

Typical half-cell reactions and their corresponding Nernst expressions follow.

(1) $Zn^{2+} + 2e^- \rightleftharpoons Zn(s)$ $\quad E = E^0 - \frac{0.0592}{2} \log \frac{1}{[Zn^{2+}]}$

No term for elemental zinc is included in the logarithmic term because it is a pure second phase (solid). Thus, the electrode potential varies linearly with the logarithm of the reciprocal of the zinc ion concentration.

(2) $Fe^{3+} + e^- \rightleftharpoons Fe^{2+}(s)$ $\quad E = E^0 - \frac{0.0592}{1} \log \frac{[Fe^{2+}]}{[Fe^{3+}]}$

The potential for this couple can be measured with an inert metallic electrode immersed in a solution containing both iron species. The potential depends on the logarithm of the ratio between the molar concentrations of these ions.

(3) $2H^+ + 2e^- \rightleftharpoons H_2(g)$ $\quad E = E^0 - \frac{0.0592}{2} \log \frac{p_{H_2}}{[H^+]^2}$

In this example, p_{H_2} is the partial pressure of hydrogen (in atmospheres) at the surface of the electrode. Usually, its value will be the same as atmospheric pressure.

(4) $MnO_4^- + 5e^- + 8H^+ \rightleftharpoons Mn^{2+} + 4H_2O$

$$E = E^0 - \frac{0.0592}{5} \log \frac{[Mn^{2+}]}{[MnO_4^-][H^+]^8}$$

In this situation, the potential depends not only on the concentrations of the manganese species but also on the pH of the solution.

▶ The Nernst expression in part (5) of Example 18-2 requires an excess of solid AgCl so that the solution is saturated with the compound at all times.

(5) $AgCl(s) + e^- \rightleftharpoons Ag(s) + Cl^-$ $\quad E = E^0 - \frac{0.0592}{1} \log [Cl^-]$

This half-reaction describes the behavior of a silver electrode immersed in a chloride solution that is *saturated* with AgCl. To ensure this condition, an excess of the solid AgCl must always be present. Note that this electrode reaction is the sum of the following two reactions:

$$AgCl(s) \rightleftharpoons Ag^+ + Cl^-$$

$$Ag^+ + e^- \rightleftharpoons Ag(s)$$

Note also that the electrode potential is independent of the amount of AgCl present as long as there is at least some present to keep the solution saturated.

18C-6 The Standard Electrode Potential, E^0

When we look carefully at Equations 18-11 and 18-12, we see that the constant E^0 is the electrode potential whenever the concentration quotient (actually, the activity quotient) has a value of 1. This constant is by definition the standard electrode potential for the half-reaction. Note that the quotient is always equal to 1 when the activities of the reactants and products of a half-reaction are unity.

The **standard electrode potential** for a half-reaction, E^0, is defined as the electrode potential when all reactants and products of a half-reaction are at unit activity.

The standard electrode potential is an important physical constant that provides quantitative information regarding the driving force for a half-cell reaction.[2] The important characteristics of these constants are the following:

1. The standard electrode potential is a relative quantity in the sense that it is the potential of an electrochemical cell in which the reference electrode (left-hand electrode) is the standard hydrogen electrode, whose potential has been assigned a value of 0.000 V.
2. The standard electrode potential for a half-reaction refers exclusively to a reduction reaction; that is, it is a relative reduction potential.
3. The standard electrode potential measures the relative force tending to drive the half-reaction from a state in which the reactants and products are at unit activity to a state in which the reactants and products are at their equilibrium activities relative to the standard hydrogen electrode.
4. The standard electrode potential is independent of the number of moles of reactant and product shown in the balanced half-reaction. Thus, the standard electrode potential for the half-reaction

$$Fe^{3+} + e^- \rightleftharpoons Fe^{2+} \qquad E^0 = +0.771 \text{ V}$$

does not change if we choose to write the reaction as

$$5Fe^{3+} + 5e^- \rightleftharpoons 5Fe^{2+} \qquad E^0 = +0.771 \text{ V}$$

Note, however, that the Nernst equation must be consistent with the half-reaction as written. For the first case, it will be

$$E = 0.771 - \frac{0.0592}{1} \log \frac{[Fe^{2+}]}{[Fe^{3+}]}$$

and for the second

$$E = 0.771 - \frac{0.0592}{5} \log \frac{[Fe^{2+}]^5}{[Fe^{3+}]^5} = 0.771 - \frac{0.0592}{5} \log \left(\frac{[Fe^{2+}]}{[Fe^{3+}]}\right)^5$$

$$= 0.771 - \frac{\cancel{5} \times 0.0592}{\cancel{5}} \log \frac{[Fe^{2+}]}{[Fe^{3+}]}$$

◀ Note that the two log terms have identical values. That is,

$$\frac{0.0592}{1} \log \frac{[Fe^{2+}]}{[Fe^{3+}]} = \frac{0.0592}{\cancel{5}} \log \frac{[Fe^{2+}]^{\cancel{5}}}{[Fe^{3+}]^{\cancel{5}}}$$

5. A positive electrode potential indicates that the half-reaction in question is spontaneous with respect to the standard hydrogen electrode half-reaction. That

[2]For further reading on standard electrode potentials, see R. G. Bates, in *Treatise on Analytical Chemistry,* 2nd ed., I. M. Kolthoff and P. J. Elving, Eds., Part I, Vol. 1, Chapter 13, New York: Wiley, 1978.

is, the oxidant in the half-reaction is a stronger oxidant than is hydrogen ion. A negative sign indicates just the opposite.

6. The standard electrode potential for a half-reaction is temperature dependent.

Standard electrode potential data are available for an enormous number of half-reactions. Many have been determined directly from electrochemical measurements. Others have been computed from equilibrium studies of oxidation/reduction systems and from thermochemical data associated with such reactions. Table 18-1 contains standard electrode potential data for several half-reactions that we will be considering in the pages that follow. A more extensive listing is found in Appendix 5.[3]

CD-ROM Tutorial: Calculating Standard Cell Potentials Using Tables of Standard Reduction Half-Cell Potentials.

Table 18-1 and Appendix 5 illustrate the two common ways for tabulating standard potential data. In Table 18-1, potentials are listed in decreasing numerical order. As a consequence, the species in the upper left part are the most effective electron acceptors, as evidenced by their large positive values. They are therefore the strongest oxidizing agents. As we proceed down the left side of such a table, each succeeding species is less effective as an electron acceptor than the one above it. The half-cell reactions at the bottom of the table have little or no tendency to take place as they are written. On the other hand, they do tend to occur in the opposite sense. The most effective reducing agents, then, are those species that appear in the lower right portion of the table.

► Based on the E^0 values in Table 18-1 for Fe^{3+} and I_3^-, which species would you expect to predominate in a solution produced by mixing iron(III) and iodide ions? See color plate 11.

TABLE 18-1
Standard Electrode Potentials*

Reaction	E^0 at 25°C, V
$Cl_2(g) + 2e^- \rightleftharpoons 2Cl^-$	+1.359
$O_2(g) + 4H^+ + 4e^- \rightleftharpoons 2H_2O$	+1.229
$Br_2(aq) + 2e^- \rightleftharpoons 2Br^-$	+1.087
$Br_2(l) + 2e^- \rightleftharpoons 2Br^-$	+1.065
$Ag^+ + e^- \rightleftharpoons Ag(s)$	+ 0.799
$Fe^{3+} + e^- \rightleftharpoons Fe^{2+}$	+ 0.771
$I_3^- + 2e^- \rightleftharpoons 3I^-$	+ 0.536
$Cu^{2+} + 2e^- \rightleftharpoons Cu(s)$	+ 0.337
$UO_2^{2+} + 4H^+ + 2e^- \rightleftharpoons U^{4+} + 2H_2O$	+ 0.334
$Hg_2Cl_2(s) + 2e^- \rightleftharpoons 2Hg(l) + 2Cl^-$	+ 0.268
$AgCl(s) + e^- \rightleftharpoons Ag(s) + Cl^-$	+ 0.222
$Ag(S_2O_3)_2^{3-} + e^- \rightleftharpoons Ag(s) + 2S_2O_3^{2-}$	+ 0.017
$\mathbf{2H^+ + 2e^- \rightleftharpoons H_2(g)}$	**0.000**
$AgI(s) + e^- \rightleftharpoons Ag(s) + I^-$	− 0.151
$PbSO_4 + 2e^- \rightleftharpoons Pb(s) + SO_4^{2-}$	− 0.350
$Cd^{2+} + 2e^- \rightleftharpoons Cd(s)$	− 0.403
$Zn^{2+} + 2e^- \rightleftharpoons Zn(s)$	− 0.763

*See Appendix 5 for a more extensive list.

[3]Comprehensive sources for standard electrode potentials include *Standard Potentials in Aqueous Solution,* A. J. Bard, R. Parsons, and J. Jordan, Eds. New York: Marcel Dekker, 1985; G. Milazzo and S. Caroli, *Tables of Standard Electrode Potentials.* New York: Wiley-Interscience, 1977; M. S. Antelman with F. J. Harris, Jr., *The Encyclopedia of Chemical Electrode Potentials.* New York: Plenum Press, 1982. Some compilations are arranged alphabetically by element; others are tabulated according to the numerical value of E^0.

FEATURE 18-4

Sign Conventions in the Older Literature

Reference works, particularly those published before 1953, often contain tabulations of electrode potentials that are not in accord with the IUPAC recommendations. For example, in a classic source of standard-potential data complied by Latimer,[4] one finds

$$Zn(s) \rightleftharpoons Zn^{2+} + 2e^- \qquad E = +0.76 \text{ V}$$

$$Cu(s) \rightleftharpoons Cu^{2+} + 2e^- \qquad E = +0.34 \text{ V}$$

To convert these oxidation potentials to electrode potentials as defined by the IUPAC convention, one must mentally (1) express the half-reactions as reductions and (2) change the signs of the potentials.

The sign convention used in a tabulation of electrode potentials may not be explicitly stated. This information can be readily deduced, however, by noting the direction and sign of the potential for a half-reaction with which one is familiar. If the sign agrees with the IUPAC convention, the table can be used as is; if not, the signs of all of the data must be reversed. For example, the reaction

$$O_2(g) + 4H^+ + 4e^- \rightleftharpoons 2H_2O \qquad E = +1.229 \text{ V}$$

occurs spontaneously with respect to the standard hydrogen electrode and thus carries a positive sign. If the potential for this half-reaction is negative in a tabulation, it and all the other potentials should be multiplied by -1.

Compilations of electrode-potential data, such as that shown in Table 18-1, provide chemists with qualitative insights into the extent and direction of electron-transfer reactions. For example, the standard potential for silver(I) (+ 0.799 V) is more positive than that for copper(II) (+ 0.337 V). We therefore conclude that a piece of copper immersed in a silver(I) solution will cause the reduction of that ion and the oxidation of the copper. On the other hand, we would expect no reaction if we place a piece of silver in a copper(II) solution.

In contrast to the data in Table 18-1, standard potentials in Appendix 5 are arranged alphabetically by element to make it easier to locate data for a given electrode reaction.

Systems Involving Precipitates or Complex Ions

In Table 18-1, we find several entries involving Ag(I), including

$$Ag^+ + e^- \rightleftharpoons Ag(s) \qquad E^0_{Ag^+/Ag} = +0.799 \text{ V}$$

$$AgCl(s) + e^- \rightleftharpoons Ag(s) + Cl^- \qquad E^0_{AgCl/Ag} = +0.222 \text{ V}$$

$$Ag(S_2O_3)_2^{3-} + e^- \rightleftharpoons Ag(s) + 2S_2O_3^{2-} \qquad E^0_{Ag(S_2O_3)_2^{3-}/Ag} = +0.017 \text{ V}$$

[4]W. M. Latimer, *The Oxidation States of the Elements and Their Potentials in Aqueous Solutions,* 2nd ed. Englewood Cliffs, NJ: Prentice-Hall, 1952.

Each gives the potential of a silver electrode in a different environment. Let us see how the three potentials are related.

The Nernst expression for the first half-reaction is

$$E = E^0_{Ag^+/Ag} - \frac{0.0592}{1} \log \frac{1}{[Ag^+]}$$

If we replace $[Ag^+]$ with $K_{sp}/[Cl^-]$, we obtain

$$E = E^0_{Ag^+/Ag} - \frac{0.0592}{1} \log \frac{[Cl^-]}{K_{sp}} = E^0_{Ag^+/Ag} + 0.0592 \log K_{sp} - 0.0592 \log [Cl^-]$$

By definition, the standard potential for the second half-reaction is the potential where $[Cl^-] = 1.00$. That is, when $[Cl^-] = 1.00$, $E = E^0_{AgCl/Ag}$. Substituting these values gives

$$E^0_{AgCl/Ag} = E^0_{Ag^+/Ag} - 0.0592 \log 1.82 \times 10^{-10} - 0.0592 \log (1.00)$$
$$= 0.799 + (-0.577) - 0.000 = 0.222 \text{ V}$$

Figure 18-9 illustrates measurement of the standard electrode potential for the Ag/AgCl electrode.

If we proceed in the same way, we can obtain an expression for the standard electrode potential for the reduction of the thiosulfate complex of silver ion depicted in the third equilibrium shown at the start of this section. Here the standard potential is given by

► CHALLENGE: Derive Equation 18-13.

$$E^0_{Ag(S_2O_3)_2^{3-}/Ag} = E^0_{Ag^+/Ag} - 0.0592 \log \beta_2 \qquad (18\text{-}13)$$

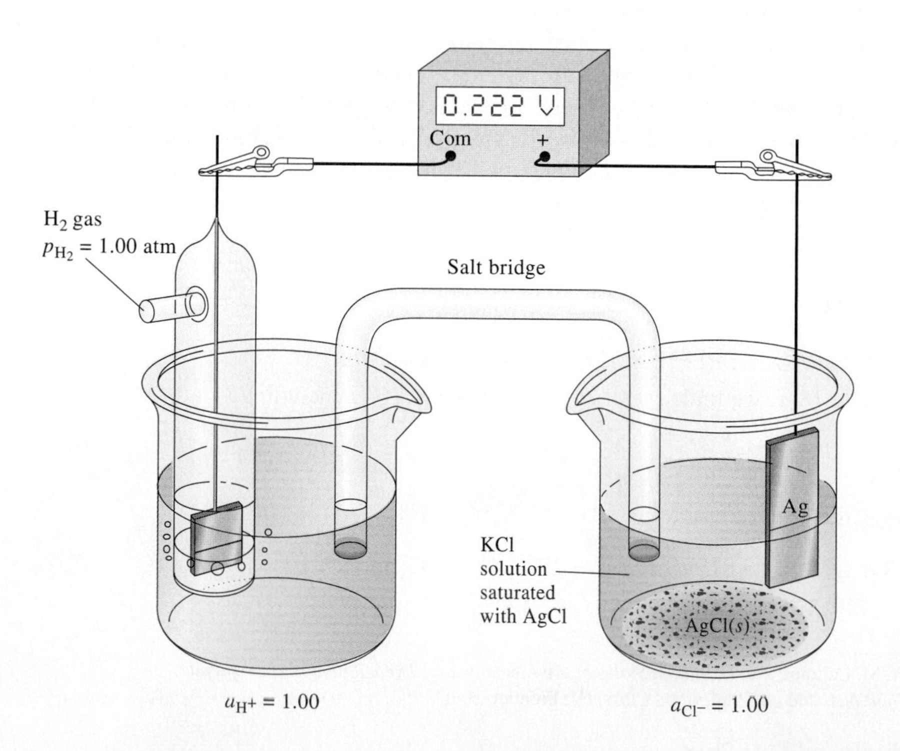

Figure 18-9 Measurement of the standard electrode potential for an Ag/AgCl electrode.

where β_2 is the formation constant for the complex. That is,

$$\beta_2 = \frac{[\mathrm{Ag(S_2O_3)_2^{3-}}]}{[\mathrm{Ag^+}][\mathrm{S_2O_3^{2-}}]^2}$$

EXAMPLE 18-3

Calculate the electrode potential of a silver electrode immersed in a 0.0500 M solution of NaCl using (a) $E^0_{\mathrm{Ag^+/Ag}} = 0.799$ V and (b) $E^0_{\mathrm{AgCl/Ag}} = 0.222$ V.

(a) $\mathrm{Ag^+} + \mathrm{e^-} \rightleftharpoons \mathrm{Ag}(s) \quad E^0_{\mathrm{Ag^+/Ag}} = +0.799$ V

The Ag^+ concentration of this solution is given by

$$[\mathrm{Ag^+}] = \frac{K_{sp}}{[\mathrm{Cl^-}]} = \frac{1.82 \times 10^{-10}}{0.0500} = 3.64 \times 10^{-9}\ \mathrm{M}$$

Substituting into the Nernst expression gives

$$E = 0.799 - 0.0592 \log \frac{1}{3.64 \times 10^{-9}} = 0.299\ \mathrm{V}$$

(b) Here we may write

$$E = 0.222 - 0.0592 \log [\mathrm{Cl^-}] = 0.222 - 0.0592 \log 0.0500$$
$$= 0.299\ \mathrm{V}$$

FEATURE 18-5

Why Are There Two Electrode Potentials for Br_2 In Table 18-1?

In Table 18-1, we find the following data for Br_2:

$$\mathrm{Br_2}(aq) + 2\mathrm{e^-} \rightleftharpoons 2\mathrm{Br^-} \qquad E^0 = +1.087\ \mathrm{V}$$

$$\mathrm{Br_2}(l) + 2\mathrm{e^-} \rightleftharpoons 2\mathrm{Br^-} \qquad E^0 = +1.065\ \mathrm{V}$$

The second standard potential applies only to a solution that is saturated with Br_2 and not to undersaturated solutions. You should use 1.065 V to calculate the electrode potential of a 0.0100 M solution of KBr that is saturated with Br_2 and in contact with an excess of the liquid. In such a case,

$$E = 1.065 - \frac{0.0592}{2} \log [\mathrm{Br^-}]^2 = 1.065 - \frac{0.0592}{2} \log (0.0100)^2$$

$$= 1.065 - \frac{0.0592}{2} \times (-4.00) = 1.183\ \mathrm{V}$$

(continued)

In this calculation, no term for Br_2 appears in the logarithmic term because it is a pure liquid present in excess (unit activity).

The standard electrode potential shown in the first entry for $Br_2(aq)$ is hypothetical because the solubility of Br_2 at 25°C is only about 0.18 M. Thus, the recorded value of 1.087 V is based on a system that—in terms of our definition of E^0—cannot be realized experimentally. Nevertheless, the hypothetical potential does permit us to calculate electrode potentials for solutions that are undersaturated in Br_2. For example, if we wish to calculate the electrode potential for a solution that was 0.0100 M in KBr and 0.00100 M in Br_2, we would write

$$E = 1.087 - \frac{0.0592}{2}\log\frac{[Br^-]^2}{[Br_2(aq)]} = 1.087 - \frac{0.0592}{2}\log\frac{(0.0100)^2}{0.00100}$$

$$= 1.087 - \frac{0.0592}{2}\log 0.100 = 1.117 \text{ V}$$

18C-7 Limitations to the Use of Standard Electrode Potentials

We will use standard electrode potentials throughout the rest of this text to calculate cell potentials and equilibrium constants for redox reactions as well as to calculate data for redox titration curves. You should be aware that such calculations sometimes lead to results that are significantly different from those you would obtain in the laboratory. There are two main sources of these differences: (1) the necessity of using concentrations in place of activities in the Nernst equation and (2) failure to take into account other equilibria such as dissociation, association, complex formation, and solvolysis. Measurement of electrode potentials can allow us to investigate these equilibria and determine their equilibrium constants, however.

Use of Concentrations Instead of Activities

Most analytical oxidation/reduction reactions are carried out in solutions that have such high ionic strengths that activity coefficients cannot be obtained via the Debye-Hückel equation (see Equation 10-1, Section 10B-2). Significant errors may result, however, if concentrations are used in the Nernst equation rather than activities. For example, the standard potential for the half-reaction

$$Fe^{3+} + e^- \rightleftharpoons Fe^{2+} \qquad E^0 = +0.771 \text{ V}$$

is + 0.771 V. When the potential of a platinum electrode immersed in a solution that is 10^{-4} M in iron(III) ion, iron(II) ion, and perchloric acid is measured against a standard hydrogen electrode, a reading of close to + 0.77 V is obtained, as predicted by theory. If, however perchloric acid is added to this mixture until the acid concentration is 0.1 M, the potential is found to decrease to about + 0.75 V. This difference is attributable to the fact that the activity coefficient of iron(III) is considerably smaller than that of iron(II) (0.4 versus 0.18) at the high ionic strength of the 0.1 M perchloric acid medium (see Table 10-1, page 270). As a consequence, the ratio of activities of the two species ($[Fe^{2+}]/[Fe^{3+}]$) in the Nernst equation is greater than unity, a condition that leads to a decrease in the electrode potential. In 1 M $HClO_4$, the electrode potential is still smaller ($\approx$ 0.73 V).

Effect of Other Equilibria

The application of standard electrode potential data to many systems of interest in analytical chemistry is further complicated by association, dissociation, complex formation, and solvolysis equilibria involving the species that appear in the Nernst equation. These phenomena can be taken into account only if their existence is known and appropriate equilibrium constants are available. More often than not, neither of these requirements is met and significant discrepancies arise as a consequence. For example, the presence of 1 M hydrochloric acid in the iron(II)/iron(III) mixture we have just discussed leads to a measured potential of + 0.70 V; in 1 M sulfuric acid, a potential of + 0.68 V is observed; and in 2 M phosphoric acid, the potential is + 0.46 V. In each of these cases, the iron(II)/iron(III) activity ratio is larger because the complexes of iron(III) with chloride, sulfate, and phosphate ions are more stable than those of iron(II); thus, the ratio of the species concentrations, $[Fe^{2+}]/[Fe^{3+}]$, in the Nernst equation is greater than unity and the measured potential is less than the standard potential. If formation constants for these complexes were available, it would be possible to make appropriate corrections. Unfortunately, such data are often not available, or, if they are, they are not very reliable.

Formal Potentials

Formal potentials are empirically derived potentials that compensate for the types of activity and competing equilibria effects that we have just described. The formal potential $E^{0'}$ of a system is the potential of the half-cell with respect to the standard hydrogen electrode measured under conditions such that the ratio of analytical concentrations of reactants and products as they appear in the Nernst equation is exactly unity and the concentrations of other species in the system are all carefully specified. For example, the formal potential for the half-reaction

A **formal potential** is the electrode potential when the ratio of **analytical concentrations** of reactants and products of a half-reaction is exactly 1.00 and the molar concentrations of any other solutes are specified.

$$Ag^+ + e^- \rightleftharpoons Ag(s) \qquad E^{0'} = 0.792 \text{ V in 1 M } HClO_4$$

could be obtained by measuring the potential of the cell shown in Figure 18-10. Here, the right-hand electrode is a silver electrode immersed in a solution that is 1.00 M in $AgNO_3$ and 1.00 M in $HClO_4$, the reference electrode on the left is a standard hydrogen electrode. This cell develops a potential of + 0.792 V, which is the formal potential of the Ag^+/Ag couple in 1.00 M $HClO_4$. Note that the standard potential for this couple is + 0.799 V.

Formal potentials for many half-reactions are listed in Appendix 5. Note that large differences exist between the formal and standard potentials for some half-reactions. For example, the formal potential for

$$Fe(CN)_6^{3-} + e^- \rightleftharpoons Fe(CN)_6^{4-} \qquad E^0 = +\,0.36 \text{ V}$$

is 0.72 V in 1 M perchloric or sulfuric acids, which is 0.36 V greater than the standard electrode potential for the half-reaction. The reason for this difference is that in the presence of high concentrations of hydrogen ion, hexacyanoferrate(II) ions ($Fe(CN)_6^{4-}$) and hexacyanoferrate(III) ions ($Fe(CN)_6^{3-}$) combine with one or more protons to form hydrogen hexacyanoferrate(II) and hydrogen hexacyanoferrate(III) acid species. Because $H_4Fe(CN)_6$ is a weaker acid than $H_3Fe(CN)_6$, the ratio of the species concentrations, $[Fe(CN)_6^{4-}]/[Fe(CN)_6^{3-}]$, in the Nernst equation is less than 1, and so the observed potentials are greater.

Substitution of formal potentials for standard electrode potentials in the Nernst equation yields better agreement between calculated and experimental results—

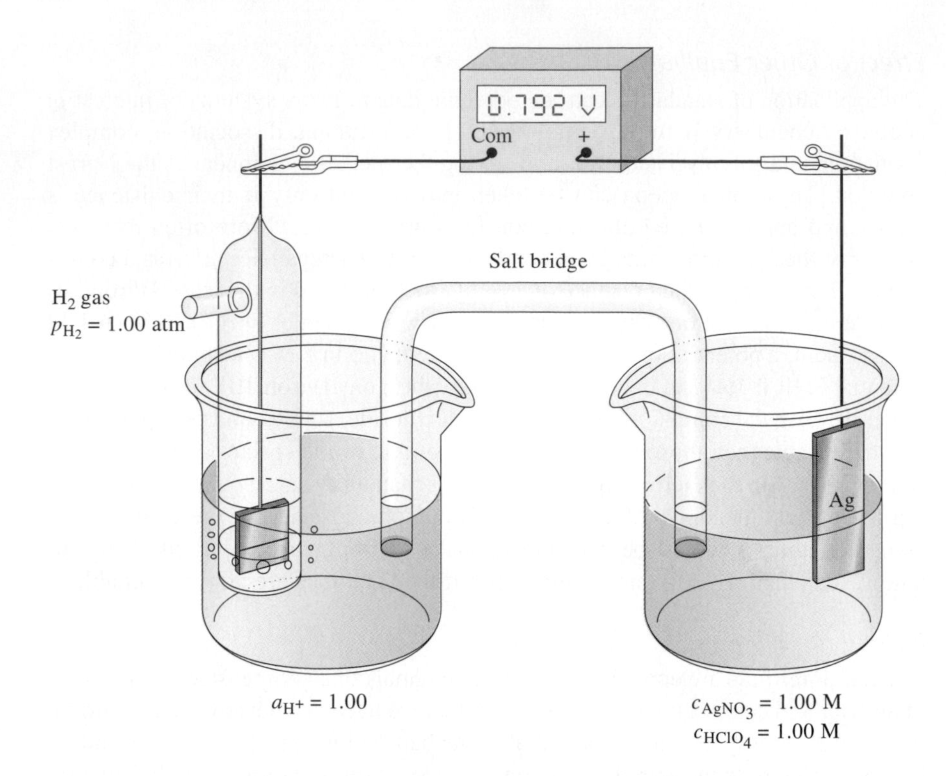

Figure 18-10 Measurement of the formal potential of the Ag^+/Ag couple in 1 M $HClO_4$.

provided, of course, that the electrolyte concentration of the solution approximates that for which the formal potential is applicable. Not surprisingly, attempts to apply formal potentials to systems that differ substantially in type and in concentration of electrolyte can result in errors that are larger than those associated with the use of standard electrode potentials. In this text, we use whichever is the more appropriate.

Spreadsheet Summary In the first exercise in Chapter 10 of *Applications of Microsoft® Excel in Analytical Chemistry,* a spreadsheet is developed to calculate the electrode potentials as a function of the ratio of reductant-to-oxidant concentration ([R]/[O]) for the case of two soluble species. Plots of E versus [R]/[O] and E versus log([R]/[O]) are made, and the slopes and intercepts are determined. The spreadsheet is modified for metal/metal ion systems.

WEB WORKS

Fuel cells have been used to provide electrical power for spacecraft since the 1960s. In recent years, fuel cell technology has begun to mature, and batteries made up of fuel cells will soon be or are now available for small-scale power generation and electric automobiles. Use your Web browser to connect to **http://chemistry.brookscole.com/skoogfac/.** From the Chapter Resources menu, choose Web Works. Locate the Chapter 18 section and click on the link to the *Scientific American* Web site article on fuel cells. Describe a proton-exchange membrane from the information and links given.

QUESTIONS AND PROBLEMS

Note: Numerical data are molar analytical concentrations where the full formula of a species is provided. Molar equilibrium concentrations are supplied for species displayed as ions.

18-1. Briefly describe or define
*(a) oxidation.
(b) oxidizing agent.
*(c) salt bridge.
(d) liquid junction.
*(e) Nernst equation.

18-2. Briefly describe or define
*(a) electrode potential.
(b) formal potential.
*(c) standard electrode potential.
(d) liquid-junction potential.
(e) oxidation potential.

18-3. Make a clear distinction between
*(a) reduction and reducing agent.
(b) a galvanic cell and an electrolytic cell.
*(c) the anode and the cathode of an electrochemical cell.
(d) a reversible electrochemical cell and an irreversible electrochemical cell.
*(e) standard electrode potential and formal potential.

***18-4.** The following entries are found in a table of standard electrode potentials:

$$I_2(s) + 2e^- \rightleftharpoons 2I^- \qquad E^0 = 0.5355\text{ V}$$

$$I_2(aq) + 2e^- \rightleftharpoons 2I^- \qquad E^0 = 0.615\text{ V}$$

What is the significance of the difference between these two standard potentials?

***18-5.** Why is it necessary to bubble hydrogen through the electrolyte in a hydrogen electrode?

18-6. The standard electrode potential for the reduction of Ni^{2+} to Ni is − 0.25V. Would the potential of a nickel electrode immersed in a 1.00 M NaOH solution saturated with $Ni(OH)_2$ be more negative than $E^0_{Ni^{2+}/Ni}$ or less? Explain.

***18-7.** Write balanced net ionic equations for the following reactions. Supply H^+ and/or H_2O as needed to obtain balance.
*(a) $Fe^{3+} + Sn^{2+} \rightarrow Fe^{2+} + Sn^{4+}$
(b) $Cr(s) + Ag^+ \rightarrow Cr^{3+} + Ag(s)$
*(c) $NO_3^- + Cu(s) \rightarrow NO_2(g) + Cu^{2+}$
(d) $MnO_4^{2-} + H_2SO_3 \rightarrow Mn^{2+} + SO_4^{2-}$
*(e) $Ti^{3+} + Fe(CN)_6^{3-} \rightarrow TiO^{2+} + Fe(CN)_6^{4-}$
(f) $H_2O_2 + Ce^{4+} \rightarrow O_2(g) + Ce^{3+}$
*(g) $Ag(s) + I^- + Sn^{4+} \rightarrow AgI(s) + Sn^{2+}$
(h) $UO_2^{2+} + Zn(s) \rightarrow U^{4+} + Zn^{2+}$
*(i) $HNO_2 + MnO_4^- \rightarrow NO_3^- + Mn^{2+}$
(j) $HN_2NNH_2 + IO_3^- + Cl^- \rightarrow N_2(g) + ICl_2^-$

***18-8.** Identify the oxidizing agent and the reducing agent on the left side of each equation in Problem 18-7; write a balanced equation for each half-reaction.

18-9. Write balanced net ionic equations for the following reactions. Supply H^+ and/or H_2O as needed to obtain balance.
*(a) $MnO_4^- + VO^{2+} \rightarrow Mn^{2+} + V(OH)_4^+$
(b) $I_2 + H_2S(g) \rightarrow I^- + S(s)$
*(c) $Cr_2O_7^{2-} + U^{4+} \rightarrow Cr^{3+} + UO_2^{2+}$
(d) $Cl^- + MnO_2(s) \rightarrow Cl_2(g) + Mn^{2+}$
*(e) $IO_3^- + I^- \rightarrow I_2(aq)$
(f) $IO_3^- + I^- + Cl^- \rightarrow ICl_2^-$
*(g) $HPO_3^{2-} + MnO_4^- + OH^- \rightarrow PO_4^{3-} + MnO_4^{2-}$
(h) $SCN^- + BrO_3^- \rightarrow Br^- + SO_4^{2-} + HCN$
*(i) $V^{2+} + V(OH)_4^+ \rightarrow VO^{2+}$
(j) $MnO_4^- + Mn^{2+} + OH^- \rightarrow MnO_2(s)$

18-10. Identify the oxidizing agent and the reducing agent on the left side of each equation in Problem 18-9; write a balanced equation for each half-reaction.

***18-11.** Consider the following oxidation/reduction reactions:

$$AgBr(s) + V^{2+} \rightarrow Ag(s) + V^{3+} + Br^-$$

$$Tl^{3+} + 2Fe(CN)_6^{4-} \rightarrow Tl^+ + 2Fe(CN)_6^{3-}$$

$$2V^{3+} + Zn(s) \rightarrow 2V^{2+} + Zn^{2+}$$

$$Fe(CN)_6^{3-} + Ag(s) + Br^- \rightarrow Fe(CN)_6^{4-} + AgBr(s)$$

$$S_2O_8^{2-} + Tl^+ \rightarrow 2SO_4^{2-} + Tl^{3+}$$

(a) Write each net process in terms of two balanced half-reactions.
(b) Express each half-reaction as a reduction.
(c) Arrange the half-reactions in (b) in order of decreasing effectiveness as electron acceptors.

18-12. Consider the following oxidation/reduction reactions:

$$2H^+ + Sn(s) \rightarrow H_2(g) + Sn^{2+}$$

$$Ag^+ + Fe^{2+} \rightarrow Ag(s) + Fe^{3+}$$

$$Sn^{4+} + H_2(g) \rightarrow Sn^{2+} + 2H^+$$

$$2Fe^{3+} + Sn^{2+} \rightarrow 2Fe^{2+} + Sn^{4+}$$

$$Sn^{2+} + Co(s) \rightarrow Sn(s) + Co^{2+}$$

(a) Write each net process in terms of two balanced half-reactions.
(b) Express each half-reaction as a reduction.
(c) Arrange the half-reactions in (b) in order of decreasing effectiveness as electron acceptors.

***18-13.** Calculate the potential of a copper electrode immersed in
(a) 0.0440 M $Cu(NO_3)_2$.
(b) 0.0750 M in NaCl and saturated with CuCl.
(c) 0.0400 M in NaOH and saturated with $Cu(OH)_2$.
(d) 0.0250 M in $Cu(NH_3)_4^{2+}$ and 0.128 M in NH_3. β_4 for $Cu(NH_3)_4^{2+}$ is 5.62×10^{11}.
(e) a solution in which the molar analytical concentration of $Cu(NO_3)_2$ is 4.00×10^{-3} M, that for H_2Y^{2-} is 2.90×10^{-2} M (Y = EDTA), and the pH is fixed at 4.00.

18-14. Calculate the potential of a zinc electrode immersed in
(a) 0.0600 M $Zn(NO_3)_2$.
(b) 0.01000 M in NaOH and saturated with $Zn(OH)_2$.
(c) 0.0100 M in $Zn(NH_3)_4^{2+}$ and 0.250 M in NH_3. β_4 for $Zn(NH_3)_4^{2+}$ is 7.76×10^8.
(d) a solution in which the molar analytical concentration of $Zn(NO_3)_2$ is 5.00×10^{-3}, that for H_2Y^{2-} is 0.0445 M, and the pH is fixed at 9.00.

18-15. Use activities to calculate the electrode potential of a hydrogen electrode in which the electrolyte is 0.0100 M HCl and the activity of H_2 is 1.00 atm.

***18-16.** Calculate the potential of a platinum electrode immersed in a solution that is
(a) 0.0263 M in K_2PtCl_4 and 0.1492 M in KCl.
(b) 0.0750 M in $Sn(SO_4)_2$ and 2.5×10^{-3} M in $SnSO_4$.
(c) buffered to a pH of 6.00 and saturated with $H_2(g)$ at 1.00 atm.
(d) 0.0353 M in $VOSO_4$, 0.0586 M in $V_2(SO_4)_3$, and 0.100 M in $HClO_4$.
(e) prepared by mixing 25.00 mL of 0.0918 M $SnCl_2$ with an equal volume of 0.1568 M $FeCl_3$.
(f) prepared by mixing 25.00 mL of 0.0832 M with 50.00 mL of 0.01087 M $V_2(SO_4)_3$ and has a pH of 1.00.

18-17. Calculate the potential of a platinum electrode immersed in a solution that is
(a) 0.0813 M in $K_4Fe(CN)_6$ and 0.00566 M in $K_3Fe(CN)_6$.
(b) 0.0400 M in $FeSO_4$ and 0.00845 M in $Fe_2(SO_4)_3$.
(c) buffered to a pH of 5.55 and saturated with H_2 at 1.00 atm.
(d) 0.1996 M in $V(OH)_4^+$, 0.0789 M in VO^{2+}, and 0.0800 M in $HClO_4$.
(e) prepared by mixing 50.00 mL of 0.0607 M $Ce(SO_4)_2$ with an equal volume of 0.100 M $FeCl_2$. Assume solutions were 1.00 M in H_2SO_4 and use formal potentials.
(f) prepared by mixing 25.00 mL of 0.0832 M $V_2(SO_4)_3$ with 50.00 mL of 0.00628 M $V(OH)_4^+$ and has a pH of 1.00.

***18-18.** If the following half-cells are the right-hand electrode in a galvanic cell with a standard hydrogen electrode on the left, calculate the cell potential. If the cell were shorted, indicate whether the electrodes shown would act as an anode or a cathode.
(a) Ni | Ni^{2+}(0.0943 M)
(b) Ag | AgI(sat'd), KI(0.0922 M)
(c) Pt, O_2(780 torr), HCl(1.50×10^{-4} M)
(d) Pt | Sn^{2+}(0.0944 M), Sn^{4+}(0.350 M)
(e) Ag | $Ag(S_2O_3)_2^{3-}$ (0.00753 M), $Na_2S_2O_3$ (0.1439 M)

18-19. The following half-cells are on the left and coupled with the standard hydrogen electrode on the right to form a galvanic cell. Calculate the cell potential. Indicate which electrode would be the cathode if each cell were short circuited.
(a) Cu | Cu^{2+}(0.0897 M)
(b) Cu | CuI(sat'd), KI(0.1214 M)
(c) Pt, H_2(0.984 atm) | HCl(1.00×10^{-4} M)
(d) Pt | Fe^{3+}(0.0906 M), Fe^{2+}(0.1628 M)
(e) Ag | $Ag(CN)_2^-$ (0.0827 M), KCN(0.0699 M)

***18-20.** The solubility-product constant for Ag_2SO_3 is 1.5×10^{-14}. Calculate E^0 for the process

$$Ag_2SO_3(s) + 2e^- \rightleftharpoons 2Ag + SO_3^{2-}$$

18-21. The solubility-product constant for $Ni_2P_2O_7$ is 1.7×10^{-13}. Calculate E^0 for the process

$$Ni_2P_2O_7(s) + 4e^- \rightleftharpoons 2Ni(s) + P_2O_7^{4-}$$

***18-22.** The solubility-product constant for Tl_2S is 6×10^{-22}. Calculate E^0 for the reaction

$$Tl_2S(s) + 2e^- \rightleftharpoons 2Tl(s) + S^{2-}$$

18-23. The solubility product for $Pb_3(AsO_4)_2$ is 4.1×10^{-36}. Calculate E^0 for the reaction

$$_{W}Pb_3(AsO_4)_2(s) + 6e^- \rightleftharpoons 3Pb(s) + 2AsO_4^{3-}$$

***18-24.** Compute E^0 for the process

$$ZnY^{2-} + 2e^- \rightleftharpoons Zn(s) + Y^{4-}$$

where Y^{4-} is the completely deprotonated anion of EDTA. The formation constant for ZnY^{2-} is 3.2×10^{16}.

***18-25.** Given the formation constants

$$Fe^{3+} + Y^{4-} \rightleftharpoons FeY^- \qquad K_f = 1.3 \times 10^{25}$$

$$Fe^{2+} + Y^{4+} \rightleftharpoons FeY^{2-} \qquad K_f = 2.1 \times 10^{14}$$

calculate E^0 for the process

$$FeY^- + e^- \rightleftharpoons FeY^{2-}$$

18-26. Calculate E^0 for the process

$$Cu(NH_3)_4^{2+} + e^- \rightleftharpoons Cu(NH_3)_2^+ + 2\,NH_3$$

given that

$$Cu^+ + 2NH_3 \rightleftharpoons Cu(NH_3)_2^+ \qquad \beta_2 = 7.2 \times 10^{10}$$

$$Cu^{2+} + 4NH_3 \rightleftharpoons Cu(NH_3)_4^{2+} \qquad \beta_4 = 5.62 \times 10^{11}$$

18-27. For a $Pt|Fe^{3+}, Fe^{2+}$ half-cell, find the potential for the following ratios of $[Fe^{3+}]/[Fe^{2+}]$: 0.001, 0.0025, 0.005, 0.0075, 0.010, 0.025, 0.050, 0.075, 0.100, 0.250, 0.500, 0.750, 1.00, 1.250, 1.50, 1.75, 2.50, 5.00, 10.00, 25.00, 75.00, 100.00.

18-28. For a $Pt\,|\,Ce^{4+}, Ce^{3+}$ half-cell, find the potential for the same ratios of $[Ce^{4+}]/[Ce^{3+}]$ as given in Problem 18-27 for $[Fe^{3+}]/[Fe^{2+}]$.

18-29. Plot the half-cell potential versus concentration ratio for the half-cells of Problems 18-27 and 18-28. How would the plot look if potential were plotted against log(concentration ratio)?

18-30. Challenge Problem. At one time, the standard hydrogen electrode was used for measuring pH.

(a) Sketch a diagram of an electrochemical cell that could be used to measure pH and label all parts of the diagram. Use the SHE for both half-cells.

(b) Derive an equation that gives the potential of the cell in terms of the hydronium ion concentration $[H_3O^+]$ in both half-cells.

(c) One half-cell should contain a solution of known hydronium ion concentration and the other should contain the unknown solution. Solve the equation in (b) for the pH of the solution in the unknown half-cell.

(d) Modify your resulting equation to account for activity coefficients, and express the result in terms of $pa_H = -\log a_H$, the negative logarithm of the hydronium ion activity.

(e) Describe the circumstances under which you would expect the cell to provide accurate measurements of pa_H.

(f) Could your cell be used to make practical absolute measurements of pa_H or would you have to calibrate your cell with solutions of known pa_H? Explain your answer in detail.

(g) How (or where) could you obtain solutions of known pa_H?

(h) Discuss the practical problems that you might encounter in using your cell for making pH measurements.

(i) Klopsteg[5] discusses how to make hydrogen electrode measurements. In Figure 2 of his paper, he suggests using a slide rule, a segment of which is shown here, to convert hydronium ion concentrations to pH and vice versa.

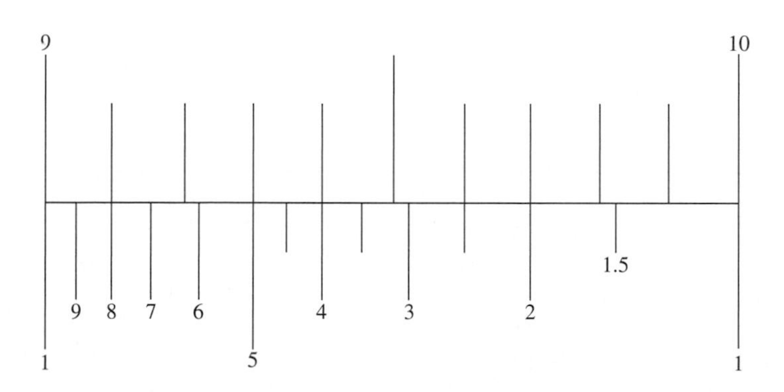

[5] P. E. Klopsteg, *Ind. Eng. Chem,* **1922,** *14,* 399.

Explain the principles of operation of this slide rule and describe how it works. What reading would you obtain from the slide rule for a hydronium ion concentration of 3.56×10^{-10} M? How many significant figures are there in the resulting pH? What is the hydronium ion concentration of a solution of pH = 9.85?

InfoTrac College Edition

For additional readings, go to InfoTrac College Edition, your online research library, at

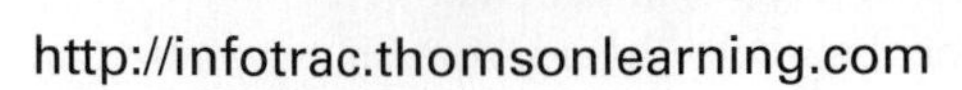

http://infotrac.thomsonlearning.com

CHAPTER 19

Applications of Standard Electrode Potentials

This composite satellite image displays areas on the surface of the Earth where chlorophyll-bearing plants are located. Chlorophyll, which is one of nature's most important biomolecules, is a member of a class of compounds called **porphyrins.** This class also includes hemoglobin and cytochrome c, which is discussed in Feature 19-1. Many analytical techniques have been used to measure the chemical and physical properties of chlorophyll to explore its role in **photosynthesis.** The redox titration of chlorophyll with other standard redox couples reveals the oxidation/reduction properties of the molecule that help explain the photophysics of the complex process that green plants use to oxidize water to molecular oxygen.

Roger Ressmeyer/Corbis

In this chapter, we show how standard electrode potentials can be used for (1) calculating thermodynamic cell potentials, (2) calculating equilibrium constants for redox reactions, and (3) constructing redox titration curves.

19A CALCULATING POTENTIALS OF ELECTROCHEMICAL CELLS

We can use standard electrode potentials and the Nernst equation to calculate the potential obtainable from a galvanic cell or the potential required to operate an electrolytic cell. The calculated potentials (sometimes called thermodynamic potentials) are theoretical in the sense that they refer to cells in which there is no current. As we show in Chapter 22, additional factors must be taken into account if a current is involved.

The thermodynamic potential of an electrochemical cell is the difference between the electrode potential of the right-hand electrode and the electrode potential of the left-hand electrode. That is,

$$E_{cell} = E_{right} - E_{left} \tag{19-1}$$

◀ It is important to note that E_{right} and E_{left} in Equation 19-1 are both *electrode potentials,* as defined at the beginning of Section 18C-3.

where E_{right} and E_{left} are the electrode potentials of the right-hand and left-hand electrodes, respectively.

Gustav Robert Kirchhoff (1824–1877) was a German physicist who made many important contributions to physics and chemistry. In addition to his work in spectroscopy, he is known for Kirchhoff's laws of current and voltage in electrical circuits. These laws can be summarized by the following equations: $\Sigma I = 0$ and $\Sigma E = 0$. These equations state that the sum of the currents into any circuit point (node) is zero and the sum of the potential differences around any circuit loop is zero.

EXAMPLE 19-1

Calculate the thermodynamic potential of the following cell and the free energy change associated with the cell reaction.

$$\text{Cu}|\text{Cu}^{2+}(0.0200\text{ M})||\text{Ag}^{+}(0.0200\text{ M})|\text{Ag}$$

Note that this cell is the galvanic cell shown in Figure 18-2a.

The two half-reactions and standard potentials are

$$\text{Ag}^{+} + e^{-} \rightleftharpoons \text{Ag}(s) \qquad E^0 = 0.799\text{ V} \qquad (19\text{-}2)$$

$$\text{Cu}^{2+} + 2e^{-} \rightleftharpoons \text{Cu}(s) \qquad E^0 = 0.337\text{ V} \qquad (19\text{-}3)$$

The electrode potentials are

$$E_{\text{Ag}^+/\text{Ag}} = 0.799 - 0.0592 \log \frac{1}{0.0200} = 0.6984\text{ V}$$

$$E_{\text{Cu}^{2+}/\text{Cu}} = 0.337 - \frac{0.0592}{2} \log \frac{1}{0.0200} = 0.2867\text{ V}$$

We see from the cell diagram that the silver electrode is the right-hand electrode and the copper electrode is the left-hand electrode. Therefore, application of Equation 19-1 gives

$$E_{\text{cell}} = E_{\text{right}} - E_{\text{left}} = E_{\text{Ag}^+/\text{Ag}} - E_{\text{Cu}^{2+}/\text{Cu}} = 0.6984 - 0.2867 = +0.412\text{ V}$$

The free energy change ΔG for the reaction $\text{Cu}(s) + 2\text{Ag}^{+} \rightleftharpoons \text{Cu}^{2+} + \text{Ag}(s)$ is found from

$$\Delta G = -nFE_{\text{cell}} = -2 \times 96485\text{ C} \times 0.412\text{ V} = -79{,}503\text{ J }(18.99\text{ kcal})$$

EXAMPLE 19-2

Calculate the potential for the cell

$$\text{Ag}|\text{Ag}^{+}(0.0200\text{ M})||\text{Cu}^{2+}(0.0200\text{ M})|\text{Cu}$$

The electrode potentials for the two half-reactions are identical to the electrode potentials calculated in Example 19-1. That is,

$$E_{\text{Ag}^+/\text{Ag}} = 0.6984\text{ V} \qquad \text{and} \qquad E_{\text{Cu}^{2+}/\text{Cu}} = 0.2867\text{ V}$$

In contrast to the previous example, however, the silver electrode is on the left and the copper electrode is on the right. Substituting these electrode potentials into Equation 19-1 gives

$$E_{\text{cell}} = E_{\text{right}} - E_{\text{left}} = E_{\text{Cu}^{2+}/\text{Cu}} - E_{\text{Ag}^+/\text{Ag}} = 0.2867 - 0.6984 = -0.412\text{ V}$$

Examples 19-1 and 19-2 illustrate an important fact. The magnitude of the potential difference between the two electrodes is 0.412 V independent of which electrode is considered the left or reference electrode. If the Ag electrode is the left electrode, as in Example 19-2, the cell potential has a negative sign, but if the Cu electrode is the reference, as in Example 19-2, the cell potential has a positive sign. No matter how the cell is arranged, however, the spontaneous cell reaction is oxidation of Cu and reduction of Ag^+, and the free energy change is 79,503 J. Examples 19-3 and 19-4 illustrate other types of electrode reactions.

EXAMPLE 19-3

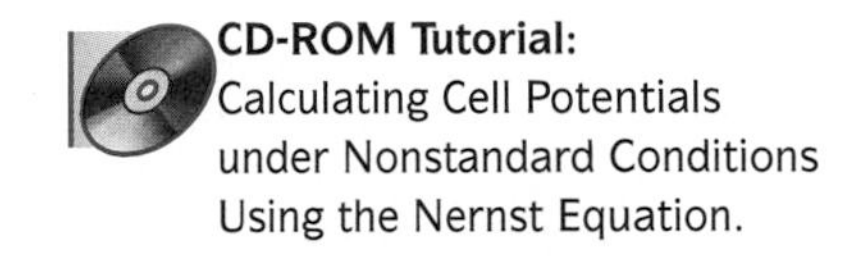
CD-ROM Tutorial: Calculating Cell Potentials under Nonstandard Conditions Using the Nernst Equation.

Calculate the potential of the following cell and indicate the reaction that would occur spontaneously if the cell were short circuited (Figure 19-1).

$$Pt|U^{4+}(0.200\ M),\ UO_2^{2+}\ (0.0150\ M),\ H^+(0.0300\ M)||$$
$$Fe^{2+}(0.0100\ M),\ Fe^{3+}(0.0250\ M)|Pt$$

The two half-reactions are

$$Fe^{3+} + e^- \rightleftharpoons Fe^{2+} \qquad E^0 = +0.771\ V$$

$$UO_2^{2+} + 4H^+ + 2e^- \rightleftharpoons U^{4+} + 2H_2O \qquad E^0 = +0.334\ V$$

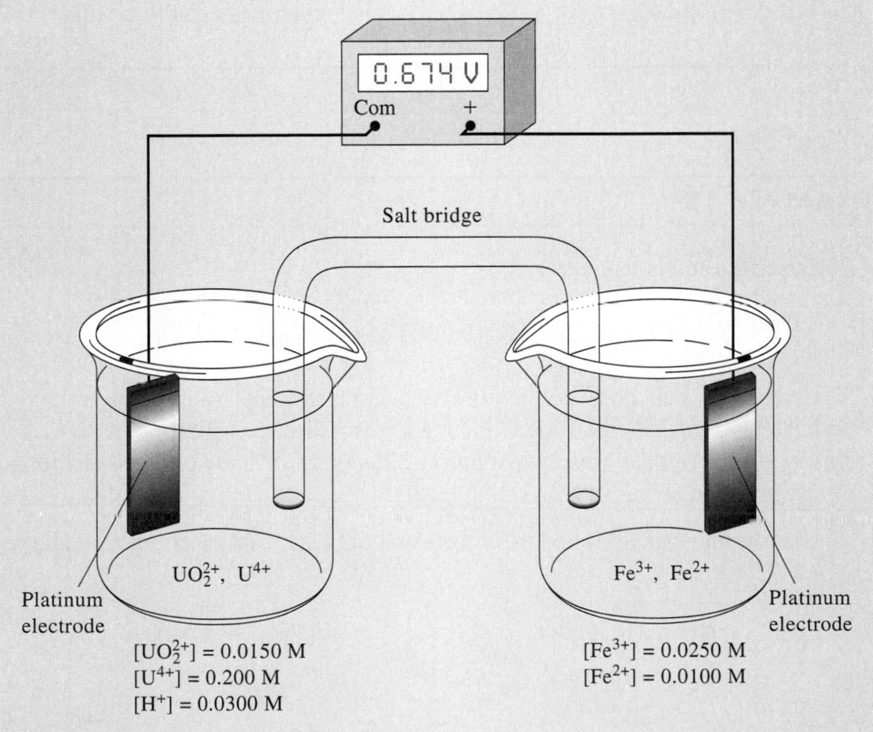

Figure 19-1 Cell for Example 19-3.

(continued)

The electrode potential for the right-hand electrode is

$$E_{right} = 0.771 - 0.0592 \log \frac{[Fe^{2+}]}{[Fe^{3+}]}$$

$$= 0.771 - 0.0592 \log \frac{0.0100}{0.0250} = 0.771 - (-0.0236)$$

$$= 0.7946 \text{ V}$$

The electrode potential for the left-hand electrode is

$$E_{left} = 0.334 - \frac{0.0592}{2} \log \frac{[U^{4+}]}{[UO_2^{2+}][H^+]^4}$$

$$= 0.334 - \frac{0.0592}{2} \log \frac{0.200}{(0.0150)(0.0300)^4}$$

$$= 0.334 - 0.2136 = 0.1204 \text{ V}$$

and

$$E_{cell} = E_{right} - E_{left} = 0.7946 - 0.2136 = 0.674 \text{ V}$$

The positive sign means that the spontaneous reaction is the oxidation of U^{4+} on the left and the reduction of Fe^{3+} on the right, or

$$U^{4+} + 2Fe^{3+} + 2H_2O \rightleftharpoons UO_2^{2+} + 2Fe^{2+} + 4H^+$$

EXAMPLE 19-4

Calculate the cell potential for

$$Ag|AgCl(\text{sat'd}), HCl(0.0200 \text{ M})|H_2(0.800 \text{ atm}), Pt$$

Note that this cell does not require two compartments (nor a salt bridge) because molecular H_2 has little tendency to react directly with the low concentration of Ag^+ in the electrolyte solution. This is an example of a **cell without liquid junction** (Figure 19-2).

The two half-reactions and their corresponding standard electrode potentials are (see Table 18-1)

$$2H^+ + 2e^- \rightleftharpoons H_2(g) \qquad E^0_{H^+/H_2} = 0.000 \text{ V}$$

$$AgCl(s) + e^- \rightleftharpoons Ag(s) + Cl^- \qquad E^0_{AgCl/Ag} = 0.222 \text{ V}$$

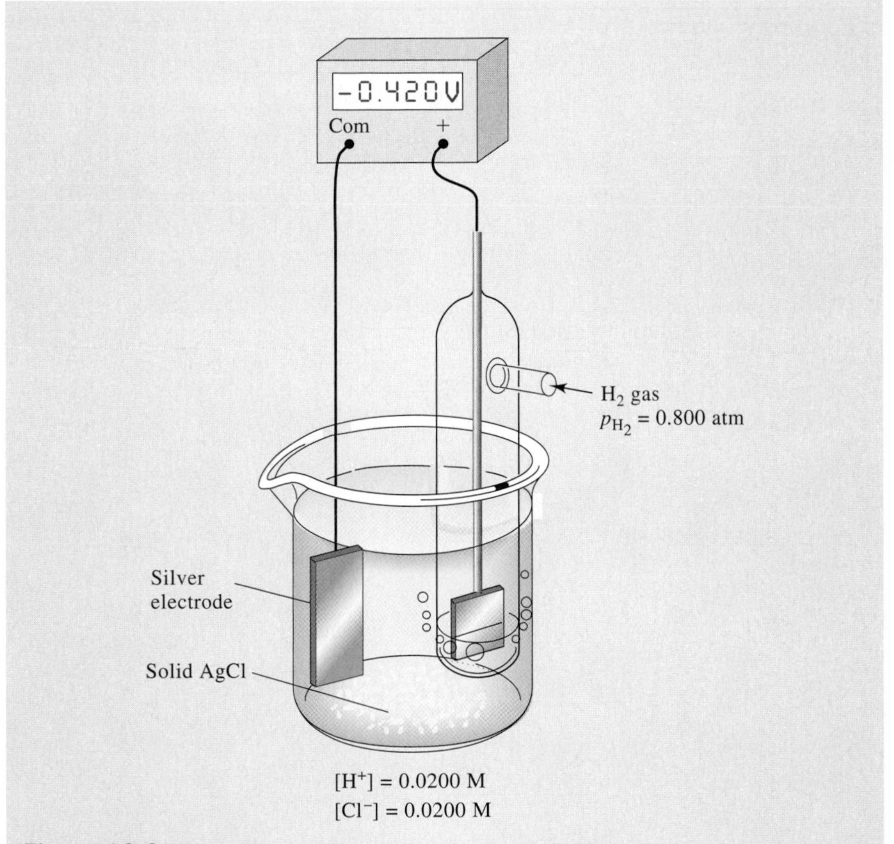

Figure 19-2 Cell without liquid junction for Example 19-4.

The two electrode potentials are

$$E_{right} = 0.000 - \frac{0.0592}{2}\log\frac{p_{H_2}}{[H^+]^2} = -\frac{0.0592}{2}\log\frac{0.800}{(0.0200)^3}$$
$$= -0.0977 \text{ V}$$
$$E_{left} = 0.222 - 0.0592\log[Cl^-] = 0.222 - 0.0592\log 0.0200$$
$$= 0.3226 \text{ V}$$

The cell potential is thus

$$E_{cell} = E_{right} - E_{left} = -0.0977 - 0.3226 = -0.420 \text{ V}$$

The negative sign indicates that the cell reaction as considered

$$2H^+ + 2Ag(s) \rightleftharpoons H_2(g) + 2AgCl(s)$$

is nonspontaneous. To get this reaction to occur, we would have to apply an external voltage and construct an electrolytic cell.

EXAMPLE 19-5

Calculate the potential for the following cell using (a) concentrations and (b) activities:

$$Zn|ZnSO_4(x\ M),\ PbSO_4(sat'd)|Pb$$

where $x = 5.00 \times 10^{-4}$, 2.00×10^{-3}, 1.00×10^{-2}, and 5.00×10^{-2}.

(a) In a neutral solution, little HSO_4^- is formed, and we can assume that

$$[SO_4^{2-}] = c_{ZnSO_4} = x = 5.00 \times 10^{-4}\ M$$

The half-reactions and standard electrode potentials are (see Table 18-1)

$$PbSO_4(s) + 2e^- \rightleftharpoons Pb(s) + SO_4^{2-} \qquad E^0_{PbSO_4/Pb} = -0.350\ V$$

$$Zn^{2+} + 2e^- \rightleftharpoons Zn(s) \qquad E^0_{Zn^{2+}/Zn} = -0.763\ V$$

The lead electrode potential is

$$E_{PbSO_4/Pb} = E^0_{PbSO_4/Pb} - \frac{0.0592}{2}\log[SO_4^{2-}]$$

$$= -0.350 - \frac{0.0592}{2}\log(5.00 \times 10^{-4}) = -0.252\ V$$

The zinc electrode potential is

$$E_{Zn^{2+}/Zn} = E^0_{Zn^{2+}/Zn} - \frac{0.0592}{2}\log\frac{1}{[Zn^{2+}]}$$

$$= -0.763 - \frac{0.0592}{2}\log\frac{1}{5.00 \times 10^{-4}} = -0.860\ V$$

The cell potential is thus

$$E_{cell} = E_{right} - E_{left} = E_{PbSO_4/Pb} - E_{Zn^{2+}/Zn} = -0.252 - (-0.860) = 0.608\ V$$

Cell potentials at the other concentrations can be derived in the same way. Their values are given in Table 19-1.

(b) To calculate activity coefficients for Zn^{2+} and SO_4^{2-}, we must first find the ionic strength of the solution using Equation 10-1:

$$\mu = \frac{1}{2}[5.00 \times 10^{-4} \times (2)^2 + 5.00 \times 10^{-4} \times (2)^2] = 2.00 \times 10^{-3}$$

In Table 10-1, we find that $\alpha_{SO_4^{2-}} = 0.4$ nm and $\alpha_{Zn^{2+}} = 0.4$ nm. If we substitute these values into Equation 10-5, we find that

$$-\log\gamma_{SO_4^{2-}} = \frac{0.51 \times (2)^2\sqrt{2.00 \times 10^{-3}}}{1 + 3.3 \times 0.4\sqrt{2.00 \times 10^{-3}}} = 8.61 \times 10^{-2}$$

$$\gamma_{SO_4^{2-}} = 0.820$$

Repeating the calculations for Zn^{2+}, we find that

$$\gamma_{Zn^{2+}} = 0.825$$

The Nernst equation for the lead electrode is now

$$E_{PbSO_4/Pb} = E^0_{PbSO_4/Pb} - \frac{0.0592}{2}\log(\gamma_{SO_4^{2-}})(c_{SO_4^{2-}})$$

$$= -0.350 - \frac{0.0592}{2}\log(0.820 \times 5.00 \times 10^{-4}) = -0.250 \text{ V}$$

and for the zinc electrode, we have

$$E_{Zn^{2+}/Zn} = E^0_{Zn^{2+}/Zn} - \frac{0.0592}{2}\log\frac{1}{(\gamma_{Zn^{2+}})(c_{Zn^{2+}})}$$

$$= -0.763 - \frac{0.0592}{2}\log\frac{1}{0.825 \times 5.00 \times 10^{-4}} = -0.863 \text{ V}$$

Finally, we find the cell potential from

$$E_{cell} = E_{right} - E_{left} = E_{PbSO_4/Pb} - E_{Zn^{2+}/Zn} = -0.250 - (-0.863) = 0.613 \text{ V}$$

Values for other concentrations and experimentally determined potentials for the cell are found in Table 19-1.

Table 19-1 shows that cell potentials calculated without activity coefficient corrections exhibit significant error. It is also clear from the data in the fifth column of the table that potentials computed with activities agree reasonably well with experiment.

TABLE 19-1

Effect of Ionic Strength on the Potential of a Galvanic Cell*

Concentration $ZnSO_4$, M	Ionic Strength, μ	E, V, based on Concentrations	E, V, based on Activities	E, V, Experimental Values†
5.00×10^{-4}	2.00×10^{-3}	0.608	0.613	0.611
2.00×10^{-3}	8.00×10^{-3}	0.573	0.582	0.583
1.00×10^{-2}	4.00×10^{-2}	0.531	0.550	0.553
2.00×10^{-2}	8.00×10^{-2}	0.513	0.537	0.542
5.00×10^{-2}	2.00×10^{-1}	0.490	0.521	0.529

*Cell described in Example 19-5.

†Experimental data from I. A. Cowperthwaite and V. K. LaMer, *J. Amer. Chem. Soc.*, **1931,** *53,* 4333.

EXAMPLE 19-6

Calculate the potential required to initiate deposition of copper from a solution that is 0.010 M in $CuSO_4$ and contains sufficient H_2SO_4 to give a pH of 4.00.

The deposition of copper necessarily occurs at the cathode. Since there is no more easily oxidizable species than water in the system, O_2 will evolve at the anode. The two half-reactions and their corresponding standard electrode potentials are (see Table 18-1)

$$O_2(g) + 4H^+ + 4e^- \rightleftharpoons 2H_2O \qquad E^0_{O_2/H_2O} = +1.229 \text{ V}$$

$$Cu^{2+} + e^- \rightleftharpoons Cu(s) \qquad E^0_{AgCl/Ag} = +0.337 \text{ V}$$

The electrode potential for the Cu electrode is

$$E_{Cu^{2+}/Cu} = +0.337 - \frac{0.0592}{2}\log\frac{1}{0.010} = +0.278 \text{ V}$$

If O_2 is evolved at 1.00 atm, the electrode potential for the oxygen electrode is

$$E_{O_2/H_2O} = +1.229 - \frac{0.0592}{4}\log\frac{1}{p_{O_2} \times [H^+]^4}$$

$$= +1.229 - \frac{0.0592}{4}\log\frac{1}{(1 \text{ atm})(1.00 \times 10^{-4})} = +0.992 \text{ V}$$

and the cell potential is thus

$$E_{cell} = E_{right} - E_{left} = E_{Cu^{2+}/Cu} - E_{O_2/H_2O} = +0.278 - 0.992 = -0.714 \text{ V}$$

The negative sign shows that the cell reaction

$$2Cu^{2+} + 2H_2O \rightleftharpoons O_2(g) + 4H^+ + 2Cu(s)$$

is nonspontaneous and that to cause copper to be deposited, we must apply a cathode potential more negative than −0.714 V.

Spreadsheet Summary In the first exercise in Chapter 10 of *Applications of Microsoft® Excel in Analytical Chemistry,* a spreadsheet is developed for use in calculating electrode potentials for simple half-reactions. Plots are made of the potential versus the ratio of the reduced species to the oxidized species and of the potential versus the logarithm of this ratio.

19B DETERMINING STANDARD POTENTIALS EXPERIMENTALLY

Although it is easy to look up standard electrode potentials for hundreds of half-reactions in compilations of electrochemical data, it is important to realize that none of these potentials, including the potential of the standard hydrogen electrode, can be measured directly in the laboratory. The standard hydrogen electrode is a hypothetical electrode, as is any electrode system in which the reactants and products are at unit activity or pressure. Such electrode systems cannot be prepared in the laboratory because there is no way to prepare solutions containing ions whose activities are exactly 1. In other words, no theory is available that permits the calculation of the concentration of solute that must be dissolved to produce a solution of exactly unit activity. At high ionic strengths, the Debye-Hückel relationship (see Section 10B-2) and other extended forms of the equation do a relatively poor job of calculating activity coefficients, and there is no independent experimental method for determining activity coefficients in such solutions. Thus, for example, it is impossible to calculate the concentration of HCl or other acids that will produce a solution in which $a_{H^+} = 1$, and it is impossible to determine the activity experimentally. In spite of this difficulty, data collected in solutions of low ionic strength can be extrapolated to give valid estimates of theoretically defined standard electrode potentials. The following example shows how such hypothetical electrode potentials may be determined experimentally.

EXAMPLE 19-7

D. A. MacInnes[1] found that a cell similar to that shown in Figure 19-2 had a potential of 0.52053 V. The cell is described by the following notation.

$$\text{Pt,H}_2(1.00\text{ atm})|\text{HCl}(3.215 \times 10^{-3}\text{ M}),\ \text{AgCl(sat'd)}|\text{Ag}$$

Calculate the standard electrode potential for the half-reaction

$$\text{AgCl}(s) + e^- \rightleftharpoons \text{Ag}(s) + \text{Cl}^-$$

Here, the electrode potential for the right-hand electrode is

$$E_{\text{right}} = E^0_{\text{AgCl}} - 0.0592 \log (\gamma_{\text{Cl}^-})(c_{\text{HCl}})$$

where γ_{Cl^-} is the activity coefficient of Cl^-. The second half-cell reaction is

$$\text{H}^+ + e^- \rightleftharpoons \frac{1}{2}\text{H}_2(g)$$

and

$$E_{\text{left}} = E^0_{\text{H}^+/\text{H}_2} - \frac{0.0592}{1} \log \frac{p_{\text{H}_2}^{1/2}}{(\gamma_{\text{H}^+})(c_{\text{HCl}})}$$

The cell potential is then the difference between these two potentials

$$E_{\text{cell}} = E_{\text{right}} - E_{\text{left}}$$

$$= [E^0_{\text{AgCl}} - 0.0592 \log(\gamma_{\text{Cl}^-})(c_{\text{HCl}})] - \left[E^0_{\text{H}^+/\text{H}_2} - 0.0592 \log \frac{p_{\text{H}_2}^{1/2}}{(\gamma_{\text{H}^+})(c_{\text{HCl}})}\right]$$

$$= E^0_{\text{AgCl}} - 0.0592 \log (\gamma_{\text{Cl}^-})(c_{\text{HCl}}) - 0.000 - 0.0592 \log \frac{(\gamma_{\text{H}^+})(c_{\text{HCl}})}{p_{\text{H}_2}^{1/2}}$$

(continued)

[1] D. A. MacInnes, *The Principles of Electrochemistry*, p. 187. New York: Reinhold, 1939.

Notice that we have inverted the terms in the second logarithmic relationship. We now combine the two logarithmic terms to find that

$$E_{cell} = 0.52053 = E^0_{AgCl} - 0.0592 \log \frac{(\gamma_{H^+})(\gamma_{Cl^-})(c^2_{HCl})}{p^{1/2}_{H_2}}$$

The activity coefficients for H^+ and Cl^- can be calculated from Equation 10-5 using 3.215×10^{-3} M for the ionic strength μ. These values are 0.945 and 0.939, respectively. If we substitute these values and the experimental data into the previous equation and rearrange it, we obtain

$$E^0_{AgCl} = 0.52053 + 0.0592 \log \frac{(0.945)(0.939)(3.215 \times 10^{-3})^2}{1.00^{1/2}}$$

$$= 0.2223 \approx 0.222 \text{ V}$$

The mean for this and similar measurements at other concentrations is 0.222 V.

19C CALCULATING REDOX EQUILIBRIUM CONSTANTS

Again consider the equilibrium that is established when a piece of copper is immersed into a solution containing a dilute solution of silver nitrate:

$$Cu(s) + 2Ag^+ \rightleftharpoons Cu^{2+} + 2Ag(s) \qquad (19\text{-}4)$$

FEATURE 19-1

Biological Redox Systems

There are many redox systems of importance in biology and biochemistry. The cytochromes are excellent examples of such systems. Cytochromes are iron-heme proteins in which a porphyrin ring is coordinated through nitrogen atoms to an iron atom. They undergo one-electron redox reactions, and their physiological function is to facilitate electron transport. In the respiratory chain, the cytochromes are intimately involved in the formation of water from H_2. Reduced pyridine nucleotides deliver hydrogen to flavoproteins. The reduced flavoproteins are reoxidized by the Fe^{3+} of cytochrome b or c. The result is the formation of H^+ and the transport of electrons. The chain is completed when cytochrome oxidase transfers electrons to oxygen. The resulting oxide ion (O^{2-}) is unstable and immediately picks up two H^+ ions to produce H_2O. The scheme is illustrated in Figure 19F-1.

Most biological redox systems are pH dependent. It has become standard practice to compile electrode potentials of these systems at pH 7.0 to make comparisons of oxidizing or reducing powers. The compiled values are typically formal potentials at pH 7.0 and are sometimes symbolized as $E^{0'}_7$.

Other redox systems of importance in biochemistry include the NADH/NAD system, the flavins, the pyruvate/lactate system, the oxalacetate/malate system, and the quinone/hydroquinone system.

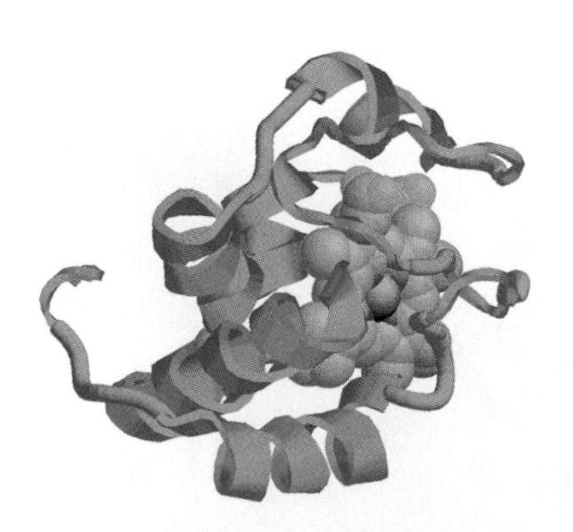

Molecular model of cytochrome c.

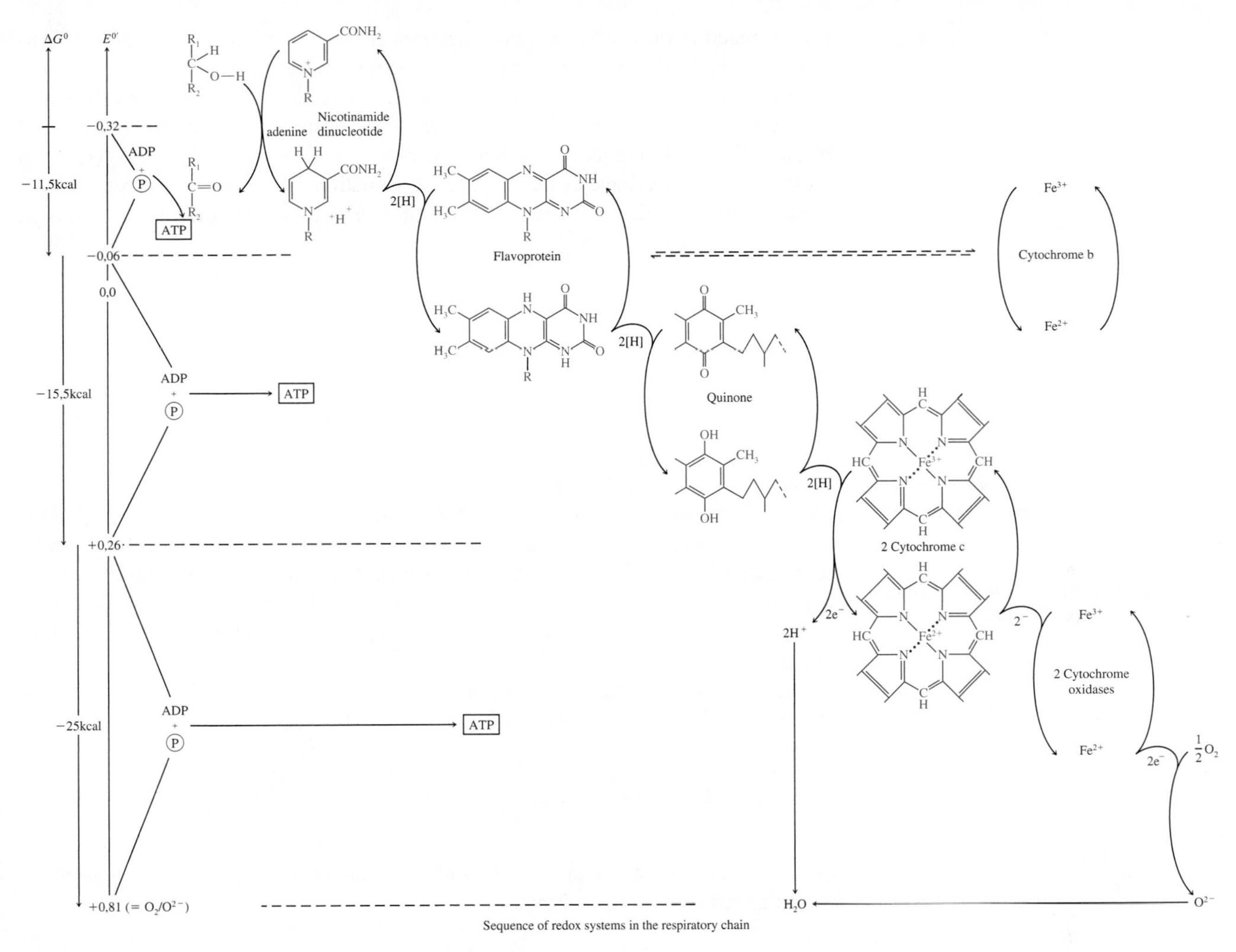

Figure 19F-1 Redox systems in the respiratory chain. P = phosphate ion. (From P. Karlson, *Introduction to Modern Biochemistry.* New York: Academic Press, 1963, with permission.)

The equilibrium constant for this reaction is

$$K_{eq} = \frac{[Cu^{2+}]}{[Ag^{+}]^2} \tag{19-5}$$

As we described in Example 19-1, this reaction can be carried out in the galvanic cell

$$Cu|Cu^{2+}(x\text{ M})||Ag^{+}(y\text{ M})|Ag$$

A sketch of a cell similar to this is shown in Figure 18-2a. Its cell potential at any instant is given by Equation 19-1:

$$E_{cell} = E_{right} - E_{left} = E_{Ag^+/Ag} - E_{Cu^{2+}/Cu}$$

◀ For simplicity in many texts and in the electrochemical literature, the potential of the right-hand electrode in Example 19-1 is symbolized as E_{Ag} and that of the left-hand electrode as E_{Cu}. A completely unambiguous way of describing the redox couple that determines the potential of these electrodes is to symbolize the potentials as $E_{Ag^+/Ag}$ and $E_{Cu^{2+}/Cu}$. Throughout this text, the more unambiguous description is used except for some simple metal ion/metal couples, such Ag^+/Ag and Cu^{2+}/Cu, when the redox couple is clear from the context or the cell schematic.

As the reaction proceeds, the concentration of Cu(II) ions increases and the concentration of Ag(I) ions decreases. These changes make the potential of the copper electrode more positive and that of the silver electrode less positive. As shown in Figure 18-6, the net effect of these changes is a continuous decrease in the potential of the cell as it discharges. Ultimately, the concentrations of Cu(II) and Ag(I) attain their equilibrium values, as determined by Equation 19-5, and the current ceases. Under these conditions, *the potential of the cell becomes zero.* Thus, *at chemical equilibrium,* we may write

$$E_{cell} = 0 = E_{right} - E_{left} = E_{Ag} - E_{Cu}$$

or

$$E_{right} = E_{left} = E_{Ag} = E_{Cu} \quad (19\text{-}6)$$

► Remember that *when redox systems are at equilibrium, the electrode potentials of all redox couples that are present are identical.* This generality applies whether the reactions take place directly in solution or indirectly in a galvanic cell.

We can generalize Equation 19-6 by stating that *at equilibrium, the electrode potentials for all half-reactions in an oxidation/reduction system are equal.* This generalization applies regardless of the number of half-reactions present in the system because interactions among all must take place until the electrode potentials are identical. For example, if we have four oxidation/reduction systems in a solution, interaction among all four takes place until the potentials of all four redox couples are equal.

Returning to the reaction shown in Equation 19-4, substitute Nernst expressions for the two electrode potentials in Equation 19-6 to give

$$E^0_{Ag} - \frac{0.0592}{2}\log\frac{1}{[Ag^+]^2} = E^0_{Cu} - \frac{0.0592}{2}\log\frac{1}{[Cu^{2+}]} \quad (19\text{-}7)$$

Note that the Nernst equation is applied to the silver half-reaction as it appears in the balanced equation (see Equation 19-4):

$$2Ag^+ + 2e^- \rightleftharpoons 2Ag(s) \qquad E^0 = 0.799\text{ V}$$

Rearrangement of Equation 19-7 gives

$$E^0_{Ag} - E^0_{Cu} = \frac{0.0592}{2}\log\frac{1}{[Ag^+]^2} - \frac{0.0592}{2}\log\frac{1}{[Cu^{2+}]}$$

If we invert the ratio in the second logarithmic term, we must change the sign of the term. This gives

$$E^0_{Ag} - E^0_{Cu} = \frac{0.0592}{2}\log\frac{1}{[Ag^+]^2} + \frac{0.0592}{2}\log\frac{[Cu^{2+}]}{1}$$

Finally, combining the logarithmic terms and rearranging gives

$$\frac{2(E^0_{Ag} - E^0_{Cu})}{0.0592} = \log\frac{[Cu^{2+}]}{[Ag^+]^2} = \log K_{eq} \quad (19\text{-}8)$$

The concentration terms in Equation 19-8 are *equilibrium concentrations;* the ratio $[Cu^{2+}]/[Ag^+]^2$ in the logarithmic term is therefore *the equilibrium constant for the reaction.* Note that the term in parenthesis in Equation 19-8 is the standard cell potential E^0_{cell}, which in general is given by

$$E^0_{cell} = E^0_{right} - E^0_{left}$$

We can also obtain Equation 19-8 from the free energy change for the reaction, as was given in Equation 18-7. Rearrangement of this equation gives

$$\ln K_{eq} = -\frac{\Delta G^0}{RT} = \frac{nFE^0_{cell}}{RT} \quad (19\text{-}9)$$

At 25°C after conversion to base 10 logarithms, we can write

$$\log K_{eq} = \frac{nE^0_{cell}}{0.0592} = \frac{n(E^0_{right} - E^0_{left})}{0.0592}$$

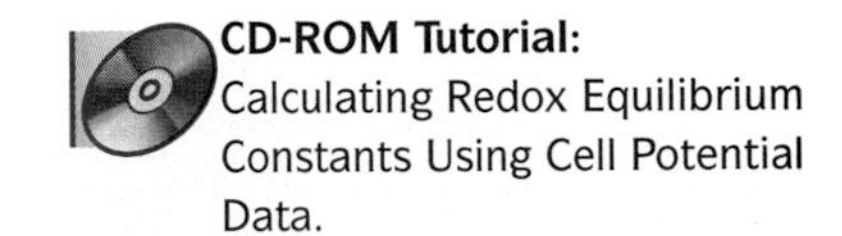
CD-ROM Tutorial: Calculating Redox Equilibrium Constants Using Cell Potential Data.

For the reaction given in Equation 19-4, substituting $E^0_{Ag^+/Ag}$ for E^0_{right} and $E^0_{Cu^{2+}/Cu}$ for E^0_{left} gives Equation 19-8.

EXAMPLE 19-8

Calculate the equilibrium constant for the reaction shown in Equation 19-4 at 25°C.

Substituting numerical values into Equation 19-8 yields

$$\log K_{eq} = \log\frac{[Cu^{2+}]}{[Ag^+]^2} = \frac{2(0.799 - 0.337)}{0.0592} = 15.61$$

$$K_{eq} = \text{antilog } 15.61 = 4.1 \times 10^{15}$$

◀ In making calculations of the sort shown in Example 19-8, follow the rounding rule for antilogs given on page 135.

EXAMPLE 19-9

Calculate the equilibrium constant for the reaction

$$2Fe^{3+} + 3I^- \rightleftharpoons 2Fe^{2+} + I_3^-$$

In Appendix 5, we find

$$2Fe^{3+} + 2e^- \rightleftharpoons 2Fe^{2+} \qquad E^0 = 0.771 \text{ V}$$

$$I_3^- + 2e^- \rightleftharpoons 3I^- \qquad E^0 = 0.536 \text{ V}$$

We have multiplied the first half-reaction by 2 so that the number of moles of Fe^{3+} and Fe^{2+} will be the same as in the balanced overall equation. We write the

(continued)

Nernst equation for Fe^{3+} based on the half-reaction for a two-electron transfer. That is,

$$E_{Fe^{3+}/Fe^{2+}} = E^0_{Fe^{3+}/Fe^{2+}} - \frac{0.0592}{2} \log \frac{[Fe^{2+}]^2}{[Fe^{3+}]^2}$$

and

$$E_{I_3^-/I^-} = E^0_{I_3^-/I^-} - \frac{0.0592}{2} \log \frac{[I^-]^3}{[I_3^-]}$$

At equilibrium, the electrode potentials are equal, and

$$E_{Fe^{3+}/Fe^{2+}} = E_{I_3^-/I^-}$$

$$E^0_{Fe^{3+}/Fe^{2+}} - \frac{0.0592}{2} \log \frac{[Fe^{2+}]^2}{[Fe^{3+}]^2} = E^0_{I_3^-/I^-} - \frac{0.0592}{2} \log \frac{[I^-]^3}{[I_3^-]}$$

This equation rearranges to

$$\frac{2(E^0_{Fe^{3+}/Fe^{2+}} - E^0_{I_3^-/I^-})}{0.0592} = \log \frac{[Fe^{2+}]^2}{[Fe^{3+}]^2} - \log \frac{[I^-]^3}{[I_3^-]}$$

$$= \log \frac{[Fe^{2+}]^2}{[Fe^{3+}]^2} + \log \frac{[I_3^-]}{[I^-]^3}$$

$$= \log \frac{[Fe^{2+}]^2[I_3^-]}{[Fe^{3+}]^2[I^-]^3}$$

Notice that we have changed the sign of the second logarithmic term by inverting the fraction. Further arrangement gives

$$\log \frac{[Fe^{2+}]^2[I_3^-]}{[Fe^{3+}]^2[I^-]^3} = \frac{2(E^0_{Fe^{3+}/Fe^{2+}} - E^0_{I_3^-/I^-})}{0.0592}$$

Recall, however, that here the concentration terms are *equilibrium concentrations,* and

$$\log K_{eq} = \frac{2(E^0_{Fe^{3+}/Fe^{2+}} - E^0_{I_3^-/I^-})}{0.0592} = \frac{2(0.771 - 0.536)}{0.0592} = 7.94$$

$$K_{eq} = \text{antilog } 7.94 = 8.7 \times 10^7$$

We round the answer to two figures because $\log K_{eq}$ contains only two significant figures (the two to the right of the decimal point).

FEATURE 19-2

A General Expression for Calculating Equilibrium Constants from Standard Potentials

To derive a general relationship for computing equilibrium constants from standard-potential data, consider a reaction in which a species A_{red} reacts with a species B_{ox} to yield A_{ox} and B_{red}. The two electrode reactions are

$$A_{ox} + ae^- \rightleftharpoons A_{red}$$

$$B_{ox} + be^- \rightleftharpoons B_{red}$$

We obtain a balanced equation for the desired reaction by multiplying the first equation by b and the second by a to give

$$bA_{ox} + bae^- \rightleftharpoons bA_{red}$$

$$aB_{ox} + bae^- \rightleftharpoons aB_{red}$$

We then subtract the first equation from the second to obtain a balanced equation for the redox reaction

$$bA_{red} + aB_{ox} \rightleftharpoons bA_{ox} + aB_{red}$$

When this system is at equilibrium, the two electrode potentials E_A and E_B are equal; that is,

$$E_A = E_B$$

If we substitute the Nernst expression for each couple into this equation, we find that *at equilibrium*

$$E_A^0 - \frac{0.0592}{ab} \log \frac{[A_{red}]^b}{[A_{ox}]^b} = E_B^0 - \frac{0.0592}{ab} \log \frac{[B_{red}]^a}{[B_{ox}]^a}$$

which rearranges to

$$E_B^0 - E_A^0 = \frac{0.0592}{ab} \log \frac{[A_{ox}]^b[B_{red}]^a}{[A_{red}]^b[B_{ox}]^a} = \frac{0.0592}{ab} \log K_{eq}$$

Finally, then,

$$\log K_{eq} = \frac{ab(E_B^0 - E_A^0)}{0.0592} \quad (19\text{-}10)$$

Note that many reactions are more complex than that shown here in that H^+, OH^-, or other species may be involved.

◀ Note that the product ab is the total number of electrons gained in the reduction (and lost in the oxidation) represented by the balanced redox equation. Thus, if $a = b$, it is not necessary to multiply the half-reactions by a and b. If $a = b = n$, the equilibrium constant is determined from

$$\log K_{eq} = \frac{n(E_B^0 - E_A^0)}{0.0592}$$

EXAMPLE 19-10

Calculate the equilibrium constant for the reaction

$$2MnO_4^- + 3Mn^{2+} + 2H_2O \rightleftharpoons 5MnO_2(s) + 4H^+$$

In Appendix 5, we find

$$2MnO_4^- + 8H^+ + 6e^- \rightleftharpoons MnO_2(s) + 4H_2O \qquad E^0 = +1.695 \text{ V}$$

$$3MnO_2(s) + 12H^+ + 6e^- \rightleftharpoons 3Mn^{2+} + 6H_2O \qquad E^0 = +1.23 \text{ V}$$

Again we have multiplied both equations by integers so that the numbers of electrons are equal. When this system is at equilibrium.

$$E_{MnO_4^-/MnO_2} = E_{MnO_2/Mn^{2+}}$$

$$1.695 - \frac{0.0592}{6} \log \frac{1}{[MnO_4^-]^2[H^+]^8} = 1.23 - \frac{0.0592}{6} \log \frac{[Mn^{2+}]^3}{[H^+]^{12}}$$

If we invert the logarithmic term on the right and rearrange, we obtain

$$\frac{6(1.695 - 1.23)}{0.0592} = \log \frac{1}{[MnO_4^-]^2[H^+]^8} + \log \frac{[H^+]^{12}}{[Mn^{2+}]^3}$$

Adding the two logarithmic terms gives

$$\frac{6(1.695 - 1.23)}{0.0592} = \log \frac{[H^+]^{12}}{[MnO_4^-]^2[Mn^{2+}]^3[H^+]^8}$$

$$47.1 = \log \frac{[H^+]^4}{[MnO_4^-]^2[Mn^{2+}]^3} = \log K_{eq}$$

$$K_{eq} = \text{antilog } 47.1 = 1 \times 10^{47}$$

Note that the final result has only one significant figure.

Spreadsheet Summary In the second exercise in Chapter 10 of *Applications of Microsoft® Excel in Analytical Chemistry,* cell potentials and equilibrium constants are calculated. A spreadsheet is developed for simple reactions to calculate complete cell potentials and equilibrium constants. The spreadsheet calculates E_{left}, E_{right}, E_{cell}, E^0_{cell}, log K_{eq}, and K_{eq}.

19D CONSTRUCTING REDOX TITRATION CURVES

Because most redox indicators respond to changes in electrode potential, the vertical axis in oxidation/reduction titration curves is generally an electrode potential instead of the logarithmic p-functions that were used for complex formation and neutralization titration curves. We saw in Chapter 18 that there is a logarithmic relationship between electrode potential and concentration of the analyte or titrant; as a result, the redox titration curves are similar in appearance to those for other types of titrations in which a p-function is plotted as the ordinate.

19D-1 Electrode Potentials during Redox Titrations

Consider the redox titration of iron(II) with a standard solution of cerium(IV). This reaction is widely used for the determination of iron in various kinds of samples. The titration reaction is

$$Fe^{2+} + Ce^{4+} \rightleftharpoons Fe^{3+} + Ce^{3+}$$

This reaction is rapid and reversible, so that the system is at equilibrium at all times throughout the titration. Consequently, the electrode potentials for the two half-reactions are always identical (see Equation 19-6); that is,

$$E_{Ce^{4+}/Ce^{3+}} = E_{Fe^{3+}/Fe^{2+}} = E_{system}$$

◀ Remember that when redox systems are at equilibrium, *the electrode potentials of all half-reactions are identical.* This generality applies whether the reactions take place directly in solution or indirectly in a galvanic cell.

where we have termed E_{system} as the **potential of the system.** If a redox indicator has been added to this solution, the ratio of the concentrations of its oxidized and reduced forms must adjust so that the electrode potential for the indicator, E_{In}, is also equal to the system potential; thus, using Equation 19-6, we may write

$$E_{In} = E_{Ce^{4+}/Ce^{3+}} = E_{Fe^{3+}/Fe^{2+}} = E_{system}$$

We can calculate the electrode potential of a system from standard potential data. Thus, for the reaction under consideration, the titration mixture is treated as if it were part of the hypothetical cell

$$SHE \| Ce^{4+}, Ce^{3+}, Fe^{3+}, Fe^{2+} | Pt$$

where SHE symbolizes the standard hydrogen electrode. The potential of the platinum electrode with respect to the standard hydrogen electrode is determined by the tendencies of iron(III) and cerium(IV) to accept electrons—that is, by the tendencies of the following half-reactions to occur:

$$Fe^{3+} + e^- \rightleftharpoons Fe^{2+}$$

$$Ce^{4+} + e^- \rightleftharpoons Ce^{3+}$$

At equilibrium, the concentration ratios of the oxidized and reduced forms of the two species are such that their attraction for electrons (and thus their electrode potentials) are identical. Note that these concentration ratios vary continuously throughout the titration, as must E_{system}. End points are determined from the characteristic variation in E_{system} that occurs during the titration.

◀ Most end points in oxidation/reduction titrations are based on the rapid changes in E_{system} that occur at or near chemical equivalence.

Because $E_{Ce^{4+}/Ce^{3+}} = E_{Fe^{3+}/Fe^{2+}} = E_{system}$, data for a titration curve can be obtained by applying the Nernst equation for *either* the cerium(IV) half-reaction *or* the iron(III) half-reaction. It turns out, however, that one or the other will be more convenient, depending on the stage of the titration. Prior to the equivalence point, the analytical concentrations of Fe(II), Fe(III), and Ce(III) are immediately available from the volumetric data and reaction stoichiometry, while the very small amount

◀ Before the equivalence point, E_{system} calculations are easiest to make using the Nernst equation for the analyte. After the equivalence point, the Nernst equation for the titrant is used.

of Ce(IV) can be obtained only by calculations based on the equilibrium constant. Beyond the equivalence point, a different situation prevails; here we can evaluate concentrations of Ce(III), Ce(IV), and Fe(III) directly from the volumetric data, while the Fe(II) concentration is small and more difficult to calculate. In this region, then, the Nernst equation for the cerium couple becomes the more convenient to use. At the equivalence point, yet another situation exists; we can evaluate the concentrations for Fe(III) and Ce(III) from the stoichiometry, but the concentrations of both Fe(II) and Ce(IV) will necessarily be quite small. A method for calculating the equivalence-point potential is given in the next section.

Equivalence-Point Potentials

At the equivalence point, the concentrations of cerium(IV) and iron(II) are minute and cannot be obtained from the stoichiometry of the reaction. Fortunately, equivalence-point potentials are easily obtained by taking advantage of the fact that the two reactant species and the two product species have known concentration ratios at chemical equivalence.

At the equivalence point in the titration of iron(II) with cerium(IV), the potential of the system is given by both

$$E_{eq} = E^0_{Ce^{4+}/Ce^{3+}} - \frac{0.0592}{1} \log \frac{[Ce^{3+}]}{[Ce^{4+}]}$$

and

$$E_{eq} = E^0_{Fe^{3+}/Fe^{2+}} - \frac{0.0592}{1} \log \frac{[Fe^{2+}]}{[Fe^{3+}]}$$

Adding these two expressions gives

$$2E_{eq} = E^0_{Fe^{3+}/Fe^{2+}} + E^0_{Ce^{4+}/Ce^{3+}} - \frac{0.0592}{1} \log \frac{[Ce^{3+}][Fe^{2+}]}{[Ce^{4+}][Fe^{3+}]} \qquad (19\text{-}11)$$

► The concentration quotient in Equation 19-11 is *not* the usual ratio of product concentrations and reactant concentrations that appear in equilibrium-constant expressions.

The definition of equivalence point requires that

$$[Fe^{3+}] = [Ce^{3+}]$$

$$[Fe^{2+}] = [Ce^{4+}]$$

Substitution of these equalities into Equation 19-11 results in the concentration quotient becoming unity and the logarithmic term becoming zero:

$$2E_{eq} = E^0_{Fe^{3+}/Fe^{2+}} + E^0_{Ce^{4+}/Ce^{3+}} - \frac{0.0592}{1} \log \frac{\cancel{[Ce^{3+}]}\cancel{[Ce^{4+}]}}{\cancel{[Ce^{4+}]}\cancel{[Ce^{3+}]}} = E^0_{Fe^{3+}/Fe^{2+}} + E^0_{Ce^{4+}/Ce^{3+}}$$

$$E_{eq} = \frac{E^0_{Fe^{3+}/Fe^{2+}} + E^0_{Ce^{4+}/Ce^{3+}}}{2} \qquad (19\text{-}12)$$

Example 19-11 illustrates how the equivalence-point potential is derived for a more complex reaction.

EXAMPLE 19-11

Obtain an expression for the equivalence-point potential in the titration of 0.0500 M U^{4+} with 0.1000 M Ce^{4+}. Assume both solutions are 1.0 M in H_2SO_4.

$$U^{4+} + 2Ce^{4+} + 2H_2O \rightleftharpoons UO_2^{2+} + 2Ce^{3+} + 4H^+$$

In Appendix 5, we find

$$UO_2^{2+} + 4H^+ + 2e^- \rightarrow U^{4+} + 2H_2O \qquad E^0 = 0.334\ V$$

$$Ce^{4+} + e^- \rightleftharpoons Ce^{3+} \qquad E^{0'} = 1.44\ V$$

Here we use the formal potential for Ce^{4+} in 1.0 M H_2SO_4.

Proceeding as in the cerium(IV)/iron(II) equivalence-point calculation, we write

◀ Recall that we use the prime notation to indicate formal potentials. Thus, the formal potential for Ce^{4+}/Ce^{3+} in 1.0 M H_2SO_4 is symbolized by $E^{0'}$.

$$E_{eq} = E^0_{UO_2^{2+}/U^{4+}} - \frac{0.0592}{2} \log \frac{[U^{4+}]}{[UO_2^{2+}][H^+]^4}$$

$$E_{eq} = E^{0'}_{Ce^{4+}/Ce^{3+}} - \frac{0.0592}{1} \log \frac{[Ce^{3+}]}{[Ce^{4+}]}$$

To combine the logarithmic terms, we must multiply the first equation by 2 to give

$$2E_{eq} = 2E^0_{UO_2^{2+}/U^{4+}} - 0.0592 \log \frac{[U^{4+}]}{[UO_2^{2+}][H^+]^4}$$

Adding this to the previous equation leads to

$$3E_{eq} = 2E^0_{UO_2^{2+}/U^{4+}} + E^{0'}_{Ce^{4+}/Ce^{3+}} - 0.0592 \log \frac{[U^{4+}][Ce^{3+}]}{[UO_2^{2+}][Ce^{4+}][H^+]^4}$$

But at equivalence

$$[U^{4+}] = [Ce^{4+}]/2$$

and

$$[UO_2^{2+}] = [Ce^{3+}]/2$$

Substituting these equations gives, on rearranging,

$$E_{eq} = \frac{2E^0_{UO_2^{2+}/U^{4+}} + E^{0'}_{Ce^{4+}/Ce^{3+}}}{3} - \frac{0.0592}{3} \log \frac{2\cancel{[Ce^{4+}]}\cancel{[Ce^{3+}]}}{2\cancel{[Ce^{3+}]}\cancel{[Ce^{4+}]}[H^+]^4}$$

$$= \frac{2E^0_{UO_2^{2+}/U^{4+}} + E^{0'}_{Ce^{4+}/Ce^{3+}}}{3} - \frac{0.0592}{3} \log \frac{1}{[H^+]^4}$$

We see that in this titration, the equivalence-point potential is pH dependent.

19D-2 The Titration Curve

Consider the titration of 50.00 mL of 0.0500 M Fe^{2+} with 0.1000 M Ce^{4+} in a medium that is 1.0 M in H_2SO_4 at all times. Formal potential data for both half-cell processes are available in Appendix 4 and are used for these calculations. That is,

$$Ce^{4+} + e^- \rightleftharpoons Ce^{3+} \qquad E^{0\prime} = 1.44 \text{ V } (1 \text{ M } H_2SO_4)$$

$$Fe^{3+} + e^- \rightleftharpoons Fe^{2+} \qquad E^{0\prime} = 0.68 \text{ V } (1 \text{ M } H_2SO_4)$$

Initial Potential

The solution contains no cerium species before we add titrant. It is more than likely that there is a small but unknown amount of Fe^{3+} present owing to air oxidation of Fe^{2+}. In any case, we do not have enough information to calculate an initial potential.

Potential after the Addition of 5.00 mL of Cerium(IV)

► Remember, the equation for this reaction is

$$Fe^{3+} + Ce^{4+} \rightleftharpoons Fe^{3+} + Ce^{3+}$$

When oxidant is added, Ce^{3+} and Fe^{3+} are formed, and the solution contains appreciable and easily calculated concentrations of three of the participants; that of the fourth, Ce^{4+}, is vanishingly small. Therefore, it is more convenient to use the concentrations of the two iron species to calculate the electrode potential of the system.

The equilibrium concentration of Fe(III) is equal to its analytical concentration less the equilibrium concentration of the unreacted Ce(IV):

$$[Fe^{3+}] = \frac{5.00 \times 0.1000}{50.00 + 5.00} - [Ce^{4+}] = \frac{0.500}{55.00} - [Ce^{4+}]$$

Similarly, the Fe^{2+} concentration is given by its molarity plus the equilibrium concentration of unreacted $[Ce^{4+}]$:

$$[Fe^{2+}] = \frac{50.00 \times 0.0500 - 5.00 \times 0.1000}{55.00} + [Ce^{4+}] = \frac{2.00}{55.00} + [Ce^{4+}]$$

Generally, redox reactions used in titrimetry are sufficiently complete that the equilibrium concentration of one of the species (in this case $[Ce^{4+}]$) is minuscule with respect to the other species present in the solution. Thus, the foregoing two equations can be simplified to

► Strictly speaking, the concentrations of Fe^{2+} and Fe^{3+} should be corrected for the concentration of unreacted Ce^{4+}. This correction would increase $[Fe^{2+}]$ and decrease $[Fe^{3+}]$. The amount of unreacted Ce^{4+} is usually so small that we can neglect the correction in both cases.

$$[Fe^{3+}] = \frac{0.500}{55.00} \quad \text{and} \quad [Fe^{2+}] = \frac{2.00}{55.00}$$

Substitution for $[Fe^{2+}]$ and $[Fe^{3+}]$ in the Nernst equation gives

$$E_{\text{system}} = +0.68 - \frac{0.0592}{1} \log \frac{2.00/\cancel{55.00}}{0.20/\cancel{55.00}} = 0.64 \text{ V}$$

Note that the volumes in the numerator and denominator cancel, which indicates that the potential is independent of dilution. This independence persists until the solution becomes so dilute that the two assumptions made in the calculation become invalid.

TABLE 19-2

Electrode Potential Versus SHE in Titrations with 0.100 M Ce^{4+}

	Potential, V vs. SHE*		
Reagent Volume, mL	**50.00 mL of 0.0500 M Fe^{2+}**		**50.00 mL of 0.02500 M U^{4+}**
5.00	0.64		0.316
15.00	0.69		0.339
20.00	0.72		0.352
24.00	0.76		0.375
24.90	0.82		0.405
25.00	**1.06**	← **Equivalence point** →	**0.703**
25.10	1.30		1.30
26.00	1.36		1.36
30.00	1.40		1.40

*H_2SO_4 concentration is such that $[H^+] = 1.0$ throughout in both titrations.

It is worth emphasizing again that the use of the Nernst equation for the Ce(IV)/Ce(III) system would yield the same value for E_{system}, but to do so would require computing $[Ce^{4+}]$ by means of the equilibrium constant for the reaction.

Additional potentials needed to define the titration curve short of the equivalence point can be obtained similarly. Such data are given in Table 19-2. You may want to confirm one or two of these values.

Equivalence-Point Potential

Substitution of the two formal potentials into Equation 19-12 yields

$$E_{eq} = \frac{E^{0'}_{Ce^{4+}/Ce^{3+}} + E^{0'}_{Fe^{3+}/Fe^{2+}}}{2} = \frac{1.44 + 0.68}{2} = 1.06 \text{ V}$$

Potential after the Addition of 25.10 mL of Cerium(IV)

The molar concentrations of Ce(III), Ce(IV), and Fe(III) are easily computed at this point, but that for Fe(II) is not. Therefore, E_{system} computations based on the cerium half-reaction are more convenient. The concentrations of the two cerium ion species are

$$[Ce^{3+}] = \frac{25.00 \times 0.1000}{75.10} - [Fe^{2+}] \approx \frac{2.500}{75.10}$$

$$[Ce^{4+}] = \frac{25.10 \times 0.1000 - 50.00 \times 0.0500}{75.10} + [Fe^{2+}] \approx \frac{0.010}{75.10}$$

Here, the iron(II) concentration is negligible with respect to the analytical concentrations of the two cerium species. Substitution into the Nernst equation for the cerium couple gives

$$E = +1.44 - \frac{0.0592}{1} \log \frac{[Ce^{3+}]}{[Ce^{4+}]} = +1.44 - \frac{0.0592}{1} \log \frac{2.500/\cancel{75.10}}{0.010/\cancel{75.10}}$$

$$= +1.30 \text{ V}$$

► In contrast to other titration curves we have encountered, oxidation/reduction curves are *independent* of reactant concentration except for very dilute solutions.

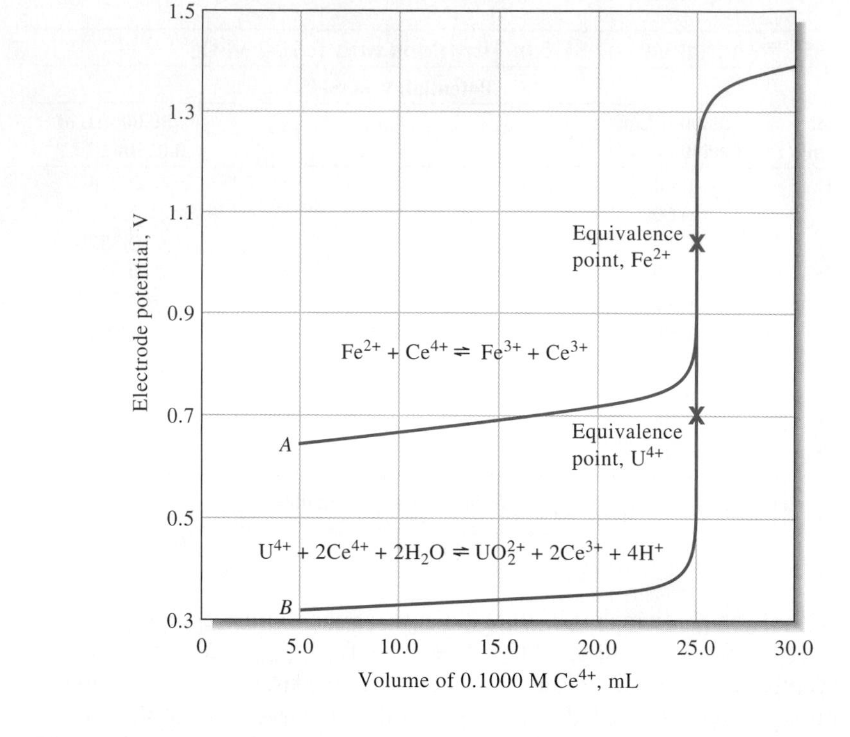

Figure 19-3 Titration curves for 0.1000 M Ce^{4+} titration. Curve *A:* Titration of 50.00 mL of 0.05000 M Fe^{2+}. Curve *B:* Titration of 50.00 mL of 0.02500 M U^{4+}

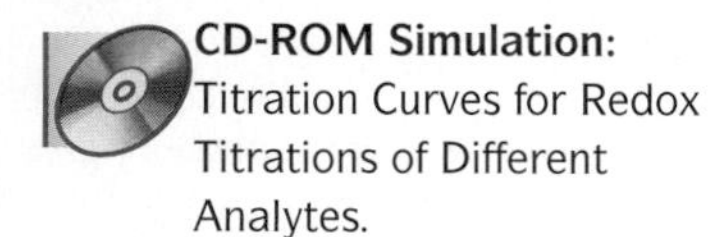

CD-ROM Simulation: Titration Curves for Redox Titrations of Different Analytes.

► Why is it impossible to calculate the potential of the system before titrant is added?

► Redox titration curves are symmetric when the reactants combine in a 1:1 ratio. Otherwise, they are asymmetric.

The additional postequivalence potentials in Table 19-2 were derived in a similar fashion.

The titration curve of iron(II) with cerium(IV) appears as *A* in Figure 19-3. This plot resembles closely the curves encountered in neutralization, precipitation, and complex-formation titrations, with the equivalence point being signaled by a rapid change in the ordinate function. A titration involving 0.00500 M iron(II) and 0.01000 M cerium(IV) yields a curve that for all practical purposes is identical to the one we have derived, since the electrode potential of the system is independent of dilution. A spreadsheet to calculate E_{system} as a function of the volume of Ce(IV) added is shown in Figure 19-4.

The data in the third column of Table 19-2 are plotted as curve *B* in Figure 19-3 to compare the two titrations. The two curves are identical for volumes greater than 25.10 mL because the concentrations of the two cerium species are identical in this region. It is also interesting that the curve for iron(II) is symmetric around the equivalence point, but the curve for uranium(IV) is not. In general, redox titration curves are symmetric when the analyte and titrant react in a 1:1 molar ratio.

EXAMPLE 19-12

Calculate data and construct a titration curve for the reaction of 50.00 mL of 0.02500 M U^{4+} with 0.1000 M Ce^{4+}. The solution is 1.0 M in H_2SO_4 throughout the titration. (For the sake of simplicity, assume that $[H^+]$ for this solution is also about 1.0 M.)

The analytical reaction is

$$U^{4+} + 2Ce^{4+} + 2H_2O \rightleftharpoons UO_2^{2+} + 2Ce^{3+} + 4H^+$$

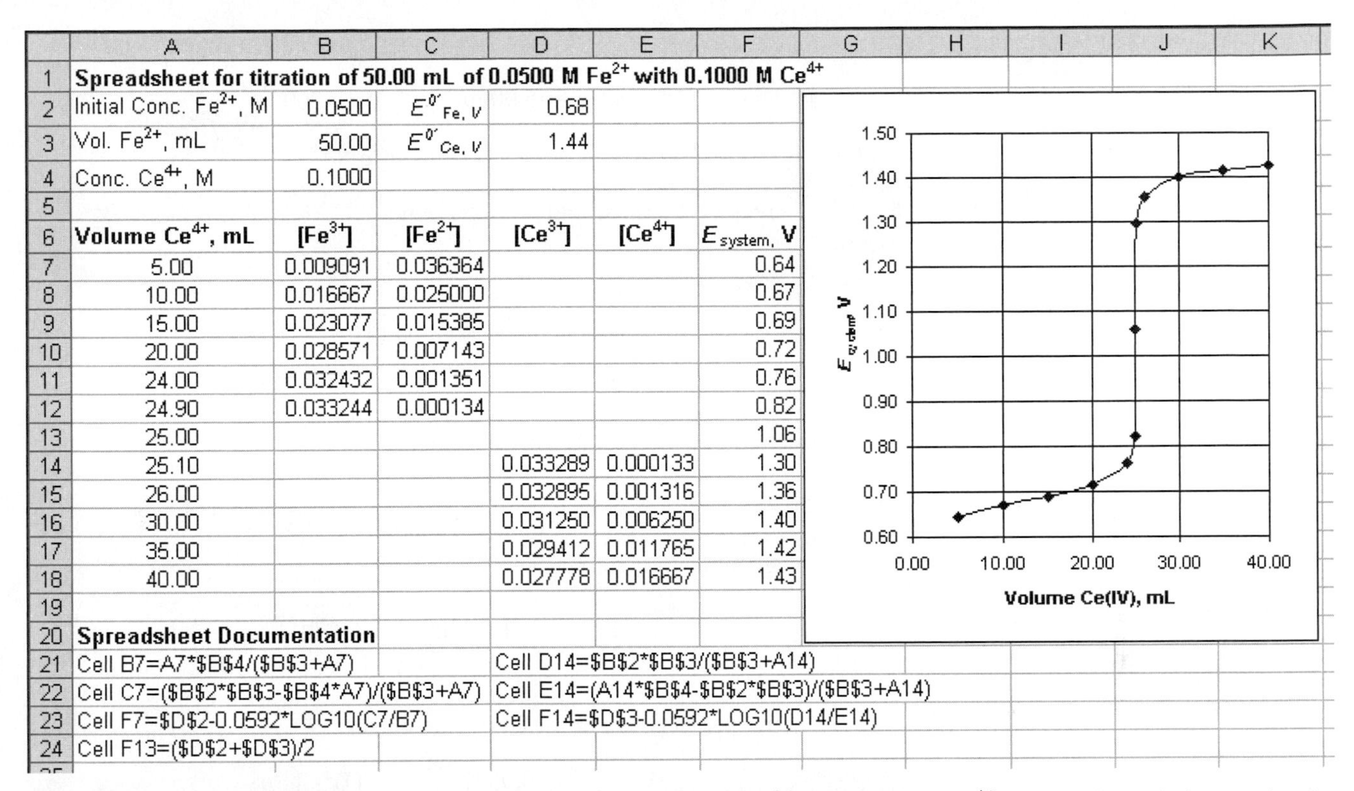

	A	B	C	D	E	F
1	**Spreadsheet for titration of 50.00 mL of 0.0500 M Fe^{2+} with 0.1000 M Ce^{4+}**					
2	Initial Conc. Fe^{2+}, M	0.0500	$E^{0'}_{Fe}$, V	0.68		
3	Vol. Fe^{2+}, mL	50.00	$E^{0'}_{Ce}$, V	1.44		
4	Conc. Ce^{4+}, M	0.1000				
5						
6	**Volume Ce^{4+}, mL**	**$[Fe^{3+}]$**	**$[Fe^{2+}]$**	**$[Ce^{3+}]$**	**$[Ce^{4+}]$**	E_{system}, **V**
7	5.00	0.009091	0.036364			0.64
8	10.00	0.016667	0.025000			0.67
9	15.00	0.023077	0.015385			0.69
10	20.00	0.028571	0.007143			0.72
11	24.00	0.032432	0.001351			0.76
12	24.90	0.033244	0.000134			0.82
13	25.00					1.06
14	25.10			0.033289	0.000133	1.30
15	26.00			0.032895	0.001316	1.36
16	30.00			0.031250	0.006250	1.40
17	35.00			0.029412	0.011765	1.42
18	40.00			0.027778	0.016667	1.43
19						
20	**Spreadsheet Documentation**					
21	Cell B7=A7*$B$4/($B$3+A7)			Cell D14=$B$2*$B$3/($B$3+A14)		
22	Cell C7=($B$2*$B$3-$B$4*A7)/($B$3+A7)			Cell E14=(A14*$B$4-$B$2*$B$3)/($B$3+A14)		
23	Cell F7=$D$2-0.0592*LOG10(C7/B7)			Cell F14=$D$3-0.0592*LOG10(D14/E14)		
24	Cell F13=($D$2+$D$3)/2					

Figure 19-4 Spreadsheet and plot for titration of 50.00 mL of 0.0500 M Fe^{2+} with 0.1000 M Ce^{4+}. Prior to the equivalence point, the system potential is calculated from the Fe^{3+} and Fe^{2+} concentrations. After the equivalence point, the Ce^{4+} and Ce^{3+} concentrations are used in the Nernst equation. The Fe^{3+} concentration in cell B7 is calculated from the number of millimoles of Ce^{4+} added, divided by the total volume of solution. The formula used for the first volume is shown in documentation cell A21. In cell C7, $[Fe^{2+}]$ is calculated as the initial number of millimoles of Fe^{2+} present, minus the number of millimoles of Fe^{3+} formed, divided by the total solution volume. Documentation cell A22 gives the formula for the 5.00-mL volume. The system potential prior to the equivalence point is calculated in cells F7:F12 by using the Nernst equation, expressed for the first volume by the formula shown in documentation cell A23. In cell F13, the equivalence-point potential is found from the average of the two formal potentials, as shown in documentation cell A24. After the equivalence point, the Ce(III) concentration (cell D14) is found from the number of millimoles of Fe^{2+} initially present divided by the total solution volume, as shown for the 25.10-mL volume by the formula in documentation cell D21. The Ce(IV) concentration (E14) is found from the total number of millimoles of Ce(IV) added, minus the number of millimoles of Fe^{2+} initially present, divided by the total solution volume, as shown in documentation cell D22. The system potential in cell F14 is found from the Nernst equation as shown in documentation cell D23. The chart is then the resulting titration curve.

and in Appendix 5 we find

$$UO_2^{2+} + 4H^+ + 2e^- \rightarrow U^{4+} + 2H_2O \qquad E^0 = 0.334\text{ V}$$

$$Ce^{4+} + e^- \rightleftharpoons Ce^{3+} \qquad E^{0'} = 0.144\text{ V}$$

Potential after Adding 5.00 mL of Ce^{4+}

$$\text{original amount } U^{4+} = 50.00 \cancel{\text{mL } U^{4+}} \times 0.02500 \frac{\text{mmol } U^{4+}}{\cancel{\text{mL } U^{4+}}}$$

$$= 1.250 \text{ mmol } U^{4+}$$

(continued)

$$\text{amount Ce}^{4+}\text{ added} = 5.00 \cancel{\text{ml Ce}^{4+}} \times 0.1000 \frac{\text{mmol Ce}^{4+}}{\cancel{\text{mL Ce}^{4+}}}$$

$$= 0.5000 \text{ mmol Ce}^{4+}$$

$$\text{amount U}^{4+}\text{ remaining} = 1.250 \text{ mmol U}^{4+} - 0.2500 \cancel{\text{mmol UO}_2^{2+}} \times \frac{1 \text{ mmol U}^{4+}}{1 \cancel{\text{mmol UO}_2^{2+}}}$$

$$= 1.000 \text{ mmol U}^{4+}$$

$$\text{total volume of solution} = (50.00 + 5.00) \text{ mL} = 55.00 \text{ mL}$$

$$\text{concentration U}^{4+}\text{ remaining} = \frac{1.000 \text{ mmol U}^{4+}}{55.00 \text{ mL}}$$

$$\text{concentration UO}_2^{2+}\text{ formed} = \frac{0.5000 \cancel{\text{mmol Ce}^{4+}} \times \dfrac{1 \text{ mmol UO}_2^{2+}}{2 \cancel{\text{mmol Ce}^{4+}}}}{55.00 \text{ mL}}$$

$$= \frac{0.2500 \text{ mmol UO}_2^{2+}}{55.00 \text{ mL}}$$

Applying the Nernst equation for UO_2^{2+}, we obtain

$$E = 0.334 - \frac{0.0592}{2} \log \frac{[\text{U}^{4+}]}{[\text{UO}_2^{2+}][\text{H}^+]^4}$$

$$= 0.334 - \frac{0.0592}{2} \log \frac{[\text{U}^{4+}]}{[\text{UO}_2^{2+}](1.00)^4}$$

Substituting concentrations of the two uranium species gives

$$E = 0.334 - \frac{0.0592}{2} \log \frac{1.000 \text{ mmol U}^{4+}/\cancel{55.00 \text{ mL}}}{0.2500 \text{ mmol UO}_2^{2+}/\cancel{55.00 \text{ mL}}}$$

$$= 0.316 \text{ V}$$

Other preequivalence-point data, calculated in the same way, are given in the third column in Table 19-2.

Equivalence-Point Potential

Following the procedure shown in Example 19-11, we obtain

$$E_{eq} = \frac{(2E^0_{\text{UO}_2^{2+}/\text{U}^{4+}} + E^{0'}_{\text{Ce}^{4+}/\text{Ce}^{3+}})}{3} - \frac{0.0592}{3} \log \frac{1}{[\text{H}^+]^4}$$

Substituting gives

$$E_{eq} = \frac{2 \times 0.334 + 1.44}{3} - \frac{0.0592}{3} \log \frac{1}{(1.00)^4}$$

$$= \frac{2 \times 0.334 + 1.44}{3} = 0.703 \text{ V}$$

Potential after Adding 25.10 mL of Ce^{4+}

$$\text{total volume of solution} = 75.10 \text{ mL}$$

$$\text{original amount } U^{4+} = 50.00 \cancel{\text{mL } U^{4+}} \times 0.02500 \frac{\text{mmol } U^{4+}}{\cancel{\text{mL } U^{4+}}}$$

$$= 1.250 \text{ mmol } U^{4+}$$

$$\text{amount } Ce^{4+} \text{ added} = 25.10 \cancel{\text{mL } Ce^{4+}} \times 0.1000 \frac{\text{mmol } Ce^{4+}}{\cancel{\text{mL } Ce^{4+}}}$$

$$= 2.510 \text{ mmol } Ce^{4+}$$

$$\text{concentration of } Ce^{3+} \text{ formed} = \frac{1.250 \cancel{\text{mmol } U^{4+}} \times \dfrac{2 \text{ mmol } Ce^{3+}}{\cancel{\text{mmol } U^{4+}}}}{75.10 \text{ mL}}$$

concentration of Ce^{4+} remaining

$$= \frac{2.510 \text{ mmol } Ce^{4+} - 2.500 \cancel{\text{mmol } Ce^{3+}} \times \dfrac{1 \text{ mmol } Ce^{4+}}{\cancel{\text{mmol } Ce^{3+}}}}{75.10 \text{ mL}}$$

Substituting into the expression for the formal potential gives

$$E = 1.44 - 0.0592 \log \frac{2.500/\cancel{75.10}}{0.010/\cancel{75.10}} = 1.30 \text{ V}$$

Table 19-2 contains other postequivalence-point data obtained in this same way.

FEATURE 19-3

The Inverse Master Equation Approach for Redox Titration Curves

α Values for Redox Species

The α values that we used for acid/base and complexation equilibria are also useful in redox equilibria. To calculate redox α values, we must solve the Nernst equation for the ratio of the concentration of the reduced species to the oxidized species. We use an approach similar to that of de Levie.[2] Since

$$E = E^0 - \frac{2.303RT}{nF} \log \frac{[\text{R}]}{[\text{O}]}$$

(continued)

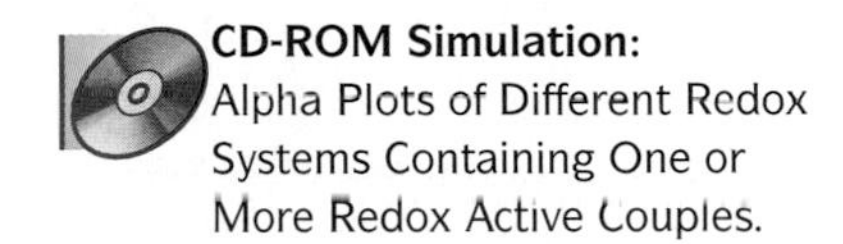

CD-ROM Simulation: Alpha Plots of Different Redox Systems Containing One or More Redox Active Couples.

[2] R. de Levie, *J. Electroanal. Chem.*, **1992,** *323,* 347–355.

we can write

$$\frac{[\mathrm{R}]}{[\mathrm{O}]} = 10^{-\frac{nF(E-E^0)}{2.303RT}} = 10^{-nf(E-E^0)}$$

where at 25°C,

$$f = \frac{F}{2.303RT} = \frac{1}{0.0592}$$

Now we can find the fractions α of the total [R] + [O] as follows:

$$\alpha_\mathrm{R} = \frac{[\mathrm{R}]}{[\mathrm{R}] + [\mathrm{O}]} = \frac{[\mathrm{R}]/[\mathrm{O}]}{[\mathrm{R}]/[\mathrm{O}] + 1} = \frac{10^{-nf(E-E^0)}}{10^{-nf(E-E^0)} + 1}$$

As an exercise, you can show that

$$\alpha_\mathrm{R} = \frac{1}{10^{-nf(E^0-E)} + 1}$$

and that

$$\alpha_\mathrm{O} = 1 - \alpha_\mathrm{R} = \frac{1}{10^{-nf(E-E^0)} + 1}$$

Furthermore, you can rearrange the equations as follows:

$$\alpha_\mathrm{R} = \frac{10^{-nfE}}{10^{-nfE} + 10^{-nfE^0}} \qquad \alpha_\mathrm{O} = \frac{10^{-nfE^0}}{10^{-nfE} + 10^{-nfE^0}}$$

We express α values in this way so that they are in a similar form to those for a weak monoprotic acid (presented in Chapter 14).

$$\alpha_0 = \frac{[\mathrm{H_3O^+}]}{[\mathrm{H_3O^+}] + K_\mathrm{a}} \qquad \alpha_1 = \frac{K_\mathrm{a}}{[\mathrm{H_3O^+}] + K_\mathrm{a}}$$

or, alternatively,

$$\alpha_0 = \frac{10^{-\mathrm{pH}}}{10^{-\mathrm{pH}} + 10^{-\mathrm{p}K_\mathrm{a}}} \qquad \alpha_1 = \frac{10^{-\mathrm{p}K_\mathrm{a}}}{10^{-\mathrm{pH}} + 10^{-\mathrm{p}K_\mathrm{a}}}$$

Notice the very similar forms of α values for redox species and for the weak monoprotic acid. The term 10^{-nfE} in the redox expression is analogous to $10^{-\mathrm{pH}}$ in the acid/base case, and the term 10^{-nfE^0} is analogous to $10^{-\mathrm{p}K_\mathrm{a}}$. These analogies will become more apparent when we plot α_O and α_R versus E in the same way that we plotted α_0 and α_1 versus pH. It is important to recognize that we obtain these relatively straightforward expressions for the redox alphas only for redox half-reactions that have 1:1 stoichiometry. For other stoichiometries, which we will not consider here, the expressions become considerably more complex. For simple cases, these equations provide us with a nice way to visualize redox chemistry and to calculate the data for redox titration curves. If we have formal potential data in a constant ionic strength medium, we can use the $E^{0'}$ values in place of the E^0 values in the α expressions.

Now let us examine graphically the dependence of the redox α-values on the potential E. We shall determine this dependence for both the Fe^{3+}/Fe^{2+} and the Ce^{4+}/Ce^{3+} couples in 1 M H_2SO_4, where the formal potentials are known. For these two couples, the α expressions are given by

$$\alpha_{Fe^{2+}} = \frac{10^{-fE}}{10^{-fE} + 10^{-fE^{0'}_{Fe}}} \qquad \alpha_{Fe^{3+}} = \frac{10^{-fE^{0'}_{Fe}}}{10^{-fE} + 10^{-fE^{0'}_{Fe}}}$$

$$\alpha_{Ce^{3+}} = \frac{10^{-fE}}{10^{-fE} + 10^{-fE^{0'}_{Ce}}} \qquad \alpha_{Ce^{4+}} = \frac{10^{-fE^{0'}_{Ce}}}{10^{-fE} + 10^{-fE^{0'}_{Ce}}}$$

Note that the *only* difference in the expressions for the two sets of α values is the two different formal potentials, $E^{0'}_{Fe} = 0.68$ V and $E^{0'}_{Ce} = 1.44$ V in 1 M H_2SO_4. The effect of this difference will be apparent in the α-plots. Since $n = 1$ for both couples, it does not appear in these equations for α.

The plot of α values is shown in Figure 19F-2. We have calculated the α values every 0.05 V from 0.50 V to 1.75 V. The shapes of the α-plots are identical to those for acid/base systems (treated in Chapters 14 and 15), as you might expect from the form of the analogous expressions.

It is worth mentioning that we normally think of calculating the potential of an electrode for a redox system in terms of concentration rather than the other way around. Just as pH is the independent variable in our α calculations with acid/base systems, potential is the independent variable in redox calculations. It is much easier to calculate α for a series of potential values than to solve the expressions for potential given various values of α.

Inverse Master Equation Approach

At all points during the titration, the concentrations of Fe^{3+} and Ce^{3+} are equal from the stoichiometry. Or

$$[Fe^{3+}] = [Ce^{3+}]$$

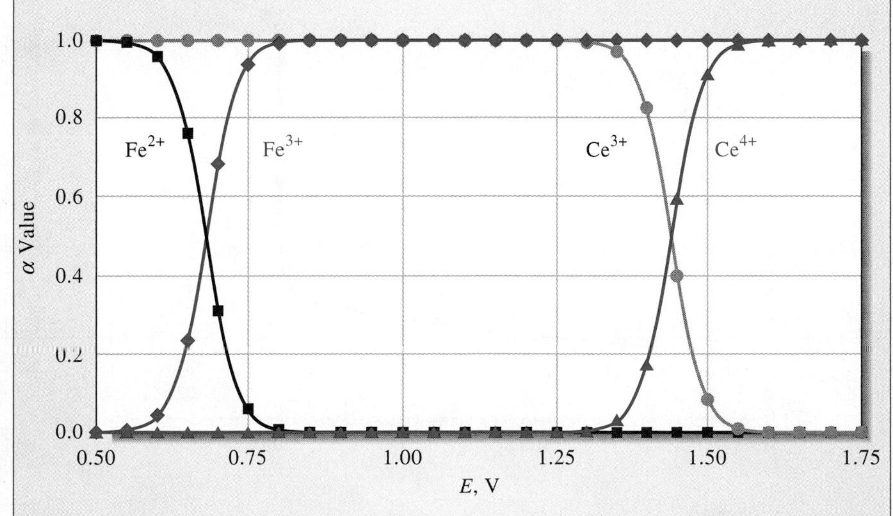

Figure 19F-2 Alpha plot for the Fe^{2+}/Ce^{4+} system.

(continued)

From the α values and the concentrations and volumes of the reagents, we can write

$$\alpha_{Fe^{3+}} \frac{V_{Fe}c_{Fe}}{V_{Fe} + V_{Ce}} = \alpha_{Ce^{3+}} \frac{V_{Ce}c_{Ce}}{V_{Fe} + V_{Ce}}$$

where V_{Fe} and c_{Fe} are the initial volume and concentration, respectively, of Fe^{2+} present, and V_{Ce} and c_{Ce} are the volume and concentration, respectively, of the titrant. By multiplying both sides of the equation by $V_{Fe} + V_{Ce}$ and dividing both sides by $V_{Fe}c_{Fe}\alpha_{Ce^{3+}}$, we find that

$$\phi = \frac{V_{Ce}c_{Ce}}{V_{Fe}c_{Fe}} = \frac{\alpha_{Fe^{3+}}}{\alpha_{Ce^{3+}}}$$

where ϕ is the extent of the titration (fraction titrated). We then substitute the expressions previously derived for the α values and obtain

$$\phi = \frac{\alpha_{Fe^{3+}}}{\alpha_{Ce^{3+}}} = \frac{1 + 10^{-f(E^{0'}_{Ce} - E)}}{1 + 10^{-f(E - E^{0'}_{Fe})}}$$

where E is now the system potential. We then substitute values of E in 0.5-V increments from 0.5 to 1.40 V into this equation to calculate ϕ and plot the resulting data, as shown in Figure 19F-3. An additional point at 1.42 V was added, since 1.45 V gave a ϕ value of more than 2. Compare this graph with Figure 19-4, which was generated using the traditional stoichiometric approach.

At this point, we should mention that some redox titration expressions are more complex than those presented here for a basic 1:1 situation. If you are interested in exploring the master equation approach for pH-dependent redox titrations or other situations, consult the paper by de Levie.[2] You can find the details of the calculations for the two plots in this feature in Chapter 10 of *Applications of Microsoft® Excel in Analytical Chemistry.*

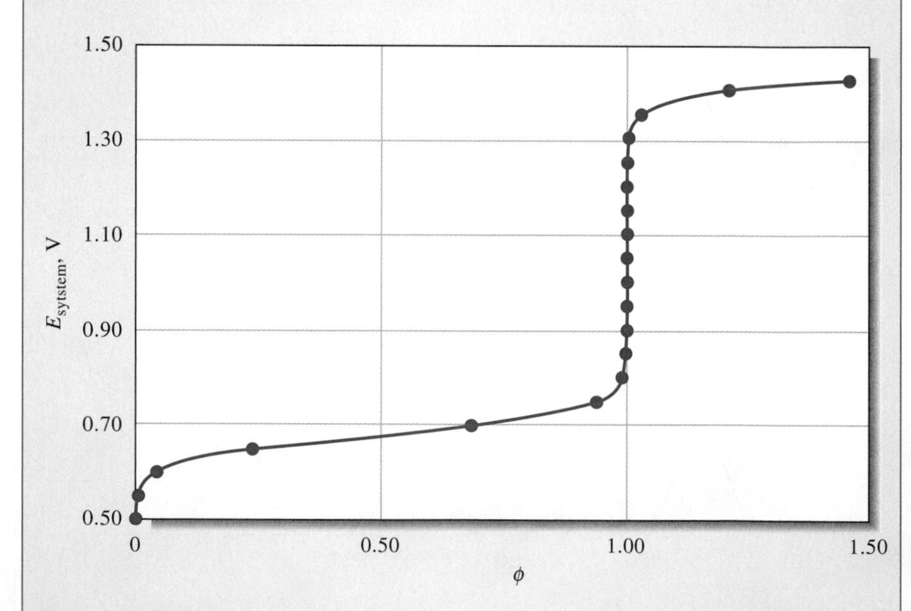

Figure 19F-3 Titration curve calculated using the inverse master equation approach. The extent of titration ϕ is calculated for various values of the system potential E_{system}, but the graph is plotted as E_{system} vs. ϕ.

19D-3 Effect of Variables on Redox Titration Curves

In earlier chapters, we considered the effects of reactant concentrations and completeness of the reaction on titration curves. Here, we describe the effects of these variables on oxidation/reduction titration curves.

Reactant Concentration

As we have just seen, E_{system} for an oxidation/reduction titration is usually independent of dilution. Consequently, titration curves for oxidation/reduction reactions are usually independent of analyte and reagent concentrations. This characteristic is in distinct contrast to that observed in the other types of titration curves we have encountered.

Completeness of the Reaction

The change in potential in the equivalence-point region of an oxidation/reduction titration becomes larger as the reaction becomes more complete. This effect is demonstrated by the two curves in Figure 19-3. The equilibrium constant for the reaction of cerium(IV) with iron(II) is 7×10^{12} while that for U(IV) is 2×10^{37}. The effect of reaction completeness is further demonstrated in Figure 19-5, which shows curves for the titration of a hypothetical reductant having a standard electrode potential of 0.20 V with several hypothetical oxidants with standard potentials ranging from 0.40 to 1.20 V; the corresponding equilibrium constants lie between about 2×10^3 and 8×10^{16}. Clearly, the greatest change in potential of the system is associated with the reaction that is most complete. In this respect, then, oxidation/reduction titration curves are similar to those involving other types of reactions.

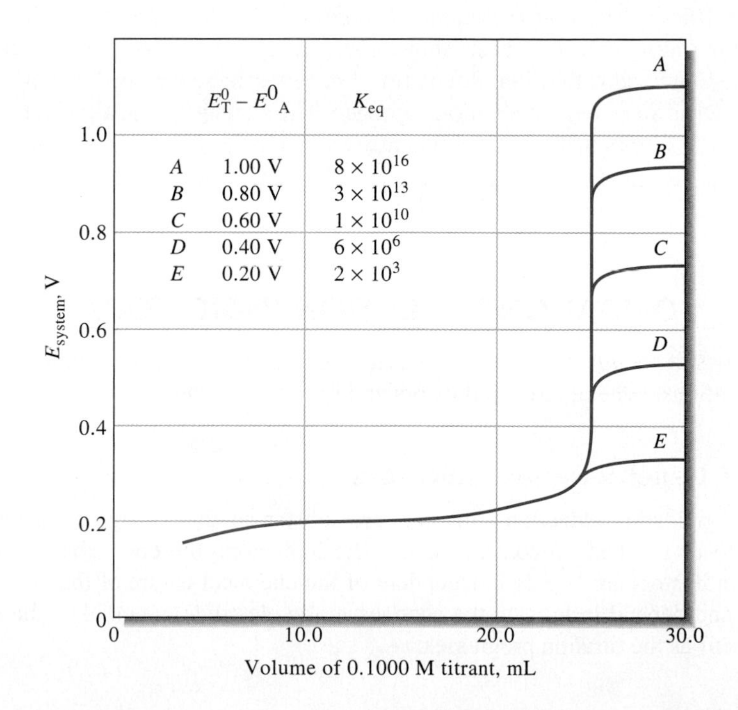

Figure 19-5 Effect of titrant electrode potential on reaction completeness. The standard electrode potential for the analyte (E^0_A) is 0.200 V; starting with curve *A*, standard electrode potentials for the titrant (E^0_T) are 1.20, 1.00, 0.80, 0.60, and 0.40, respectively. Both analyte and titrant undergo a one-electron change.

FEATURE 19-4

Reaction Rates And Electrode Potentials

Standard potentials reveal whether a reaction proceeds far enough toward completion to be useful in a particular analytical problem, but they provide no information about the rate at which the equilibrium state is approached. Consequently, a reaction that appears extremely favorable thermodynamically may be totally unacceptable from the kinetic standpoint. The oxidation of arsenic(III) with cerium(IV) in dilute sulfuric acid is a typical example. The reaction is

$$H_3AsO_3 + 2Ce^{4+} + H_2O \rightleftharpoons H_3AsO_4 + 2Ce^{3+} + 2H^+$$

The formal potentials $E^{0\prime}$ for these two systems are

$$Ce^{4+} + e^- \rightleftharpoons Ce^{3+} \qquad E^{0\prime} = +1.3\text{ V}$$

$$H_3AsO_4 + 2H^+ + 2e^- \rightleftharpoons H_3AsO_3 + H_2O \qquad E^{0\prime} = +1.00\text{ V}$$

and an equilibrium constant of about 10^{28} can be calculated from these data. Even though this equilibrium lies far to the right, titration of arsenic(III) with cerium(IV) is impossible without a catalyst because several hours are required to achieve equilibrium. Fortunately, several substances catalyze the reaction and thus make the titration feasible.

Spreadsheet Summary In Chapter 10 of *Applications of Microsoft® Excel in Analytical Chemistry,* Excel is used to obtain α values for redox species. These show how the species concentrations change throughout a redox titration. Redox titration curves are developed by both a stoichiometric and a master equation approach. The stoichiometric approach is also used for a system that is pH dependent.

19E OXIDATION/REDUCTION INDICATORS

Two types of chemical indicators are used to obtain end points for oxidation/reduction titrations: general redox indicators and specific indicators.

19E-1 General Redox Indicators

► Color changes for general redox indicators depend only on the potential of the system.

General oxidation/reduction indicators are substances that change color on being oxidized or reduced. In contrast to specific indicators, the color changes of true redox indicators are largely independent of the chemical nature of the analyte and titrant and depend instead on the changes in the electrode potential of the system that occur as the titration progresses.

The half-reaction responsible for color change in a typical general oxidation/reduction indicator can be written as

$$\text{In}_{ox} + ne^- \rightleftharpoons \text{In}_{red}$$

If the indicator reaction is reversible, we can write

$$E = E^0_{\text{In}_{ox}/\text{In}_{red}} - \frac{0.0592}{n} \log \frac{[\text{In}_{red}]}{[\text{In}_{ox}]} \quad (19\text{-}13)$$

Typically, a change from the color of the oxidized form of the indicator to the color of the reduced form requires a change of about 100 in the ratio of reactant concentrations; that is, a color change is seen when

$$\frac{[\text{In}_{red}]}{[\text{In}_{ox}]} \leq \frac{1}{10}$$

changes to

$$\frac{[\text{In}_{red}]}{[\text{In}_{ox}]} \geq 10$$

The potential change required to produce the full color change of a typical general indicator can be found by substituting these two values into Equation 19-13, which gives

$$E = E^0_{\text{In}} \pm \frac{0.0592}{n}$$

This equation shows that a typical general indicator exhibits a detectable color change when a titrant causes the system potential to shift from $E^0_{\text{In}} + 0.0592/n$ to $E^0_{\text{In}} - 0.0592/n$, or about $(0.118/n)$ V. For many indicators, $n = 2$, and a change of 0.059 V is thus sufficient.

◀ Protons are involved in the reduction of many indicators. Thus, the range of potentials over which a color change occurs (the *transition potential*) is often pH dependent.

Table 19-3 lists transition potentials for several redox indicators. Note that indicators functioning in any desired potential range up to about +1.25 V are available. Structures for and reactions of a few of the indicators listed in the table are considered in the paragraphs that follow.

Iron(II) Complexes of Orthophenanthrolines

A class of organic compounds known as 1,10-phenanthrolines, or orthophenanthrolines, form stable complexes with iron(II) and certain other ions. The parent compound has a pair of nitrogen atoms located in such positions that each can form a covalent bond with the iron(II) ion.

The compound 1,10-phenanthroline is an excellent complexing agent for Fe(II).

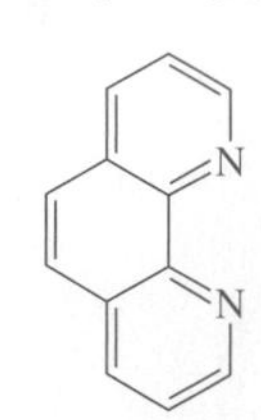

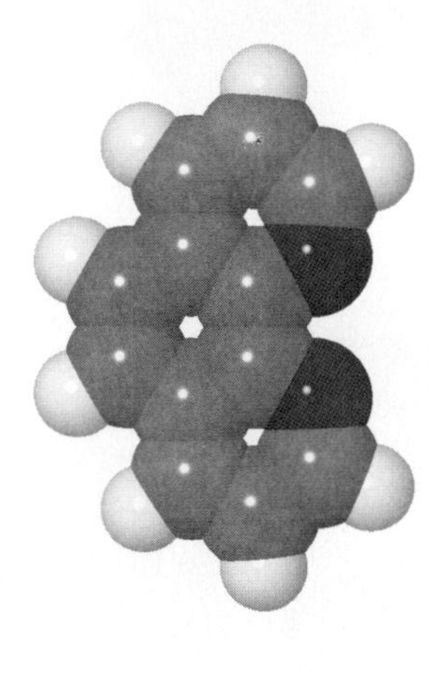

TABLE 19-3*
Selected Oxidation/Reduction Indicators*

Indicator	Color: Oxidized	Color: Reduced	Transition Potential, V	Conditions
5-Nitro-1,10-phenanthroline iron(II) complex	Pale blue	Red-violet	+1.25	1 M H_2SO_4
2,3′-Diphenylamine dicarboxylic acid	Blue-violet	Colorless	+1.12	7–10 M H_2SO_4
1,10-Phenanthroline iron(II) complex	Pale blue	Red	+1.11	1 M H_2SO_4
5-Methyl I,10-phenanthroline iron(II) complex	Pale blue	Red	+1.02	1 M H_2SO_4
Erioglaucin A	Blue-red	Yellow-green	+0.98	0.5 M H_2SO_4
Diphenylamine sulfonic acid	Red-violet	Colorless	+0.85	Dilute acid
Diphenylamine	Violet	Colorless	+0.76	Dilute acid
p-Ethoxychrysoidine	Yellow	Red	+0.76	Dilute acid
Methylene blue	Blue	Colorless	+0.53	1 M acid
Indigo tetrasulfonate	Blue	Colorless	+0.36	1 M acid
Phenosafranine	Red	Colorless	+0.28	1 M acid

*Data in part from I. M. Kolthoff and V. A. Stenger, *Volumetric Analysis,* 2nd ed., Vol. 1, p. 140. New York: Interscience, 1942.

Three orthophenanthroline molecules combine with each iron ion to yield a complex with the structure shown in the margin.

This complex, which is sometimes called "ferroin," is conveniently formulated as $(phen)_3Fe^{2+}$.

The complexed iron in the ferroin undergoes a reversible oxidation/reduction reaction that can be written

$$\underset{\text{pale blue}}{(phen)_3Fe^{3+}} + e^- \rightleftharpoons \underset{\text{red}}{(phen)_3Fe^{2+}}$$

ferroin $(phen)_3Fe^{2+}$

In practice, the color of the oxidized form is so slight as to go undetected, and the color change associated with this reduction is thus from nearly colorless to red. Because of the difference in color intensity, the end point is usually taken when only about 10% of the indicator is in the iron(II) form. The transition potential is thus approximately +1.11 V in 1 M sulfuric acid.

Of all the oxidation/reduction indicators, ferroin approaches most closely the ideal substance. It reacts rapidly and reversibly, its color change is pronounced, and its solutions are stable and easily prepared. In contrast to many indicators, the oxidized form of ferroin is remarkably inert toward strong oxidizing agents. At temperatures above 60°C, ferroin decomposes.

A number of substituted phenanthrolines have been investigated for their indicator properties, and some have proved to be as useful as the parent compound. Among these, the 5-nitro and 5-methyl derivatives are noteworthy, with transition potentials of +1.25 V and +1.02 V, respectively.

5-nitro-1, 10-phenanthroline

5-methyl-1, 10-phenanthroline

Starch/Iodine Solutions

Starch, which forms a blue complex with triiodide ion, is a widely used specific indicator in oxidation/reduction reactions involving iodine as an oxidant or iodide ion as a reductant. A starch solution containing a little triiodide or iodide ion can also function as a true redox indicator. In the presence of excess oxidizing agent, the concentration ratio of iodine to iodide is high, giving a blue color to the solution. With excess reducing agent, on the other hand, iodide ion predominates, and the blue color is absent. Thus, the indicator system changes from colorless to blue in the titration of many reducing agents with various oxidizing agents. This color change is independent of the chemical composition of the reactants, depending only on the potential of the system at the equivalence point.

The Choice of Redox Indicator

Figure 19-5 demonstrates that all the indicators in Table 19-3 except for the first and the last could be used with titrant *A*. In contrast, with titrant *D*, only indigo tetrasulfonate could be used. The change in potential with titrant *E* is too small to be satisfactorily detected by an indicator.

19E-2 Specific Indicators

Perhaps the best-known specific indicator is starch, which forms a dark blue complex with triiodide ion. This complex signals the end point in titrations in which iodine is either produced or consumed.

Another specific indicator is potassium thiocyanate, which may be used, for example, in the titration of iron(III) with solutions of titanium(III) sulfate. The end point involves the disappearance of the red color of the iron(III)/thiocyanate complex as a result of the marked decrease in the iron(III) concentration at the equivalence point.

19F POTENTIOMETRIC END POINTS

We can observe end points for many oxidation/reduction titrations by making the solution of the analyte part of the cell

$$\text{reference electrode} \| \text{analyte solution} | \text{Pt}$$

By measuring the potential of this cell during a titration, data for curves analogous to those shown in Figures 19-3 and 19-5 can be generated. End points are easily estimated from such curves. We consider potentiometric end points in detail in Chapter 21.

WEB WORKS

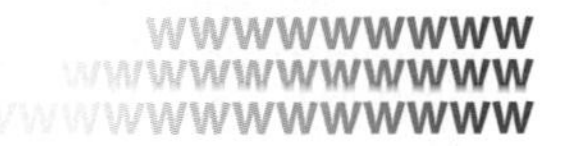

The Electroanalytical Chemistry Commission, which is a subdivision of the Analytical Chemistry Division of the International Union of Pure and Applied Chemistry (IUPAC), provides authoritative guidance on nomenclature, terminologies, symbols, units, and procedures used in electroanalytical chemistry. Use your Web browser to connect to **http://chemistry.brookscole.com/skoogfac/.** From the Chapter Resources menu, choose Web Works. Locate the Chapter 19 section, and click on the link to the Commission for Electroanalytical Chemistry Web site. Then click on the link to projects, choose one of the published reports, and view the Adobe Acrobat PDF file containing the report. Write a short summary of the report and describe its purpose.

QUESTIONS AND PROBLEMS

***19-1.** Briefly define the electrode potential of a system that contains two or more redox couples.

19-2. For an oxidation/reduction reaction, briefly distinguish between
(a) equilibrium and equivalence.
(b) a true oxidation/reduction indicator and a specific indicator.

19-3. What is unique about the condition of equilibrium in an oxidation/reduction reaction?

***19-4.** How is an oxidation/reduction titration curve generated through the use of standard electrode potentials for the analyte species and the volumetric titrant?

19-5. How does calculation of the electrode potential of the system at the equivalence point differ from that for any other point of an oxidation/reduction titration?

***19-6.** Under what circumstance is the curve for an oxidation/reduction titration asymmetric about the equivalence point?

19-7. Calculate the potentials of the following cells. Indicate whether the reaction will proceed spontaneously in the direction considered (oxidation on the left, reduction on the right) or an external voltage source is needed to force this reaction to occur.
(a) $Pb|Pb^{2+}$ (0.1393 M)$||Cd^{2+}$(0.0511)$|Cd$
(b) $Zn|Zn^{2+}$(0.0364 M)$||Tl^{3+}$(9.06×10^{-3} M), Tl^{+}(0.0620 M)$|Pt$
(c) Pt, H_2(765 torr)$|HCl$(1.00×10^{-4} M)$||$ Ni^{2+}(0.0214 M)$|Ni$
(d) $Pb|PbI_2$(sat'd), I^-(0.0120 M)$||Hg^{2+}$(4.59×10^{-3} M)$|Hg$
(e) Pt, H_2(1.00 atm)$|NH_3$(0.438 M), NH_4^+(0.379 M)$||$SHE
(f) $Pt|TiO^{2+}$(0.0790 M), Ti^{3+}(0.00918 M), H^+(1.47×10^{-2} M)$||VO^{2+}$(0.1340 M), V^{3+}(0.0784 M), H^+(0.0538 M)$|Pt$

***19-8.** Calculate the potentials of the following cells. If the cell is short circuited, indicate the direction of the spontaneous cell reaction.
(a) $Zn|Zn^{2+}$ (0.0955 M)$||Co^{2+}$(6.78×10^{-3} M)$|$ Co
(b) $Pt|Fe^{3+}$(0.1310 M), Fe^{2+}(0.0681 M)$||Hg^{2+}$ (0.0671 M)$|Hg$
(c) $Ag|Ag^+$(0.1544 M)$|H^+$(0.0794 M)$|O_2$(1.12 atm), Pt
(d) $Cu|Cu^{2+}$(0.0601 M)$||$(0.1350 M), AgI(sat'd)$|$ Ag
(e) SHE$||$HCOOH(0.1302 M), $HCOO^-$(0.0764 M)$|H_2$(1.00 atm), Pt
(f) $Pt|UO_2^{2+}$(7.93×10^{-3} M), U^{4+}(6.37×10^{-2} M), H^+(1.16×10^{-3} M)$||$ Fe^{3+}(0.003876 M), Fe^{2+}(0.1134 M)$|Pt$

19-9. Calculate the potential of the following two half-cells that are connected by a salt bridge:
*(a) a galvanic cell consisting of a lead electrode on the left immersed in 0.0848 M Pb^{2+} and a zinc electrode on the right in contact with 0.1364 M Zn^{2+}.
(b) a galvanic cell with two platinum electrodes, the one on the left immersed in a solution that is 0.0301 M in Fe^{3+} and 0.0760 M in Fe^{2+}, the one on the right in a solution that is 0.00309 M in $Fe(CN)_6^{4-}$ and 0.1564 M in $Fe(CN)_6^{3-}$.
*(c) a galvanic cell consisting of a standard hydrogen electrode on the left and a platinum electrode immersed in a solution that is 1.46×10^{-3} M in TiO^{2+}, 0.02723 M in Ti^{3+}, and buffered to a pH of 3.00.

19-10. Use the shorthand notation (page 498) to describe the cells in Problem 19-9. Each cell is supplied with a salt bridge to provide electrical contact between the solutions in the two cell compartments.

19-11. Generate equilibrium-constant expressions for the following reactions. Calculate numerical values for K_{eq}.
*(a) $Fe^{3+} + V^{2+} \rightleftharpoons Fe^{2+} + V^{3+}$
(b) $Fe(CN)_6^{3-} + Cr^{2+} \rightleftharpoons Fe(CN)_6^{4-} + Cr^{3+}$
*(c) $2V(OH)_4^+ + U^{4+} \rightleftharpoons 2VO^{2+} + UO_2^{2+} + 4H_2O$
(d) $Tl^{3+} + 2Fe^{2+} \rightleftharpoons Tl^+ + 2Fe^{3+}$
*(e) $2Ce^{4+} + H_3AsO_3 + H_2O \rightleftharpoons 2Ce^{3+} + H_3AsO_4 + 2H^+$ (1 M $HClO_4$)
(f) $2V(OH)_4^+ + H_2SO_3 \rightleftharpoons SO_4^{2-} + 2VO^{2+} + 5H_2O$
*(g) $VO^{2+} + V^{2+} + 2H^+ \rightleftharpoons 2V^{3+} + H_2O$
(h) $TiO^{2+} + Ti^{2+} + 2H^+ \rightleftharpoons 2Ti^{3+} + H_2O$

19-12. Calculate the electrode potential of the system at the equivalence point for each of the reactions in Problem 19-11. Use 0.100 M where a value for $[H^+]$ is needed and is not otherwise specified.

19-13. If you start with 0.1000 M solutions and the first-named species is the titrant, what will be the concentration of each reactant and product at the equivalence point of the titrations in Problem 19-11? Assume that there is no change in $[H^+]$ during the titration.

*19-14. Select an indicator from Table 19-3 that might be suitable for each of the titrations in Problem 19-11. Write NONE if no indicator listed in Table 19-3 is suitable.

19-15. Use a spreadsheet and construct curves for the following titrations. Calculate potentials after the addition of 10.00, 25.00, 49.00, 49.90, 50.00, 50.10, 51.00, and 60.00 mL of the reagent. Where necessary, assume that $[H^+] = 1.00$ throughout.

*(a) 50.00 mL of 0.1000 M V^{2+} with 0.05000 M Sn^{4+}.

(b) 50.00 mL of 0.1000 M $Fe(CN)_6^{3-}$ with 0.1000 M Cr^{2+}.

*(c) 50.00 mL of 0.1000 M $Fe(CN)_6^{4-}$ with 0.05000 M Tl^{3+}.

(d) 50.00 mL of 0.1000 M Fe^{3+} with 0.05000 M Sn^{2+}.

*(e) 50.00 mL of 0.05000 M U^{4+} with 0.02000 M MnO_4^-.

19-16. **Challenge Problem.** As part of a study measuring the dissociation constant of acetic acid, Harned and Ehlers[3] needed to measure E^0 for the following cell.

$$Pt,H_2(1\ atm)|HCl(m),AgCl(sat'd)|Ag$$

(a) Write an expression for the potential of the cell.

(b) Show that the expression can be written as

$$E = E^0 - \frac{RT}{F}\ln \gamma_{H_3O^+}\gamma_{Cl^-}m_{H_3O^+}m_{Cl^-}$$

where $\gamma_{H_3O^+}$ and γ_{Cl^-} are the activity coefficients of hydronium ion and chloride ion, respectively, and $m_{H_3O^+}$ and m_{Cl^-} are their respective molal (mole solute/kg solvent) concentrations.

(c) Under what circumstances is this expression valid?

(d) Show that the expression in (b) may be written

$E + 2k \log m = E^0 - 2k \log\gamma$, where $k = \ln 10RT/F$. What are m and γ?

(e) A considerably simplified version of the Debye-Hückel expression that is valid for very dilute solutions is $\log \gamma = -0.5\sqrt{m} + cm$, where c is a constant. Show that the expression for the cell potential in (d) may be written as $E + 2k \log m - k\sqrt{m} = E^0 - 2kcm$.

(f) The previous expression is a "limiting law" that becomes linear as the concentration of the electrolyte approaches zero. The equation assumes the form $y = ax + b$, where $y = E + 2k \log m - k\sqrt{m}$, $x = m$, the slope $a = -2kc$, and the y-intercept $b = E^0$. Harned and Ehlers very accurately measured the potential of the cell without liquid junction presented at the beginning of the problem as a function of

Potential Measurements of Cell Pt,H_2(1 atm) |HCl(m),AgCl(sat'd)| Ag without Liquid Junction as a Function of Concentration (molality) and Temperature (°C)

	E_T, volts							
m, molal	E_0	E_5	E_{10}	E_{15}	E_{20}	E_{25}	E_{30}	E_{35}
0.005	0.48916	0.49138	0.49338	0.49521	0.44690	0.49844	0.49983	0.50109
0.006	0.48089	0.48295	0.48480	0.48647	0.48800	0.48940	0.49065	0.49176
0.007	0.4739	0.47584	0.47756	0.47910	0.48050	0.48178	0.48289	0.48389
0.008	0.46785	0.46968	0.47128	0.47270	0.47399	0.47518	0.47617	0.47704
0.009	0.46254	0.46426	0.46576	0.46708	0.46828	0.46937	0.47026	0.47103
0.01	0.4578	0.45943	0.46084	0.46207	0.46319	0.46419	0.46499	0.46565
0.02	0.42669	0.42776	0.42802	0.42925	0.42978	0.43022	0.43049	0.43058
0.03	0.40859	0.40931	0.40993	0.41021	0.41041	0.41056	0.41050	0.41028
0.04	0.39577	0.39624	0.39668	0.39673	0.39673	0.39666	0.39638	0.39595
0.05	0.38586	0.38616	0.38641	0.38631	0.38614	0.38589	0.38543	0.38484
0.06	0.37777	0.37793	0.37802	0.37780	0.37749	0.37709	0.37648	0.37578
0.07	0.37093	0.37098	0.37092	0.37061	0.37017	0.36965	0.36890	0.36808
0.08	0.36497	0.36495	0.36479	0.36438	0.36382	0.36320	0.36285	0.36143
0.09	0.35976	0.35963	0.35937	0.35888	0.35823	0.35751	0.35658	0.35556
0.1	0.35507	0.35487	0.33451	0.35394	0.35321	0.35240	0.35140	0.35031
E^0	0.23627	0.23386	0.23126	0.22847	0.22550	0.22239	0.21918	0.21591

[3] H. S. Harned, R. W. Ehlers, *J. Am. Chem. Soc.*, **1932,** *54*(4), 1350–1357.

concentration of HCl (molal) and temperature and obtained the data in the table on page 557. For example, they measured the potential of the cell at 25°C with an HCl concentration of 0.01 m and obtained a value of 0.46419 volts. Construct a plot of $E + 2k \log m - k\sqrt{m}$ versus m, and note that the plot is quite linear at low concentration. Extrapolate the line to the y-intercept, and estimate a value for E^0. Compare your value with the value of Harned and Ehlers, and explain any difference. Also compare the value to the one shown in Table 18-1. The simplest way to carry out this exercise is to place the data in a spreadsheet, and use the Excel function INTERCEPT (known_y's, known_x's) to determine the extrapolated value for E^0. Use only the data from 0.005 to 0.01 m to find the intercept.

(g) If you have used a spreadsheet to carry out the data analysis in (f), enter the data for all temperatures into the spreadsheet and determine values for E^0 at all temperatures from 5°C to 35°C. Alternatively, you may download an Excel spreadsheet containing the entire data table. Use your Web browser to connect to **http://chemistry.brookscole.com/skoogfac/.** From the Chapter Resources menu, choose Web Works. Locate the Chapter 19 section, find the links for Chapter 19, and click on the spreadsheet link for this problem.

(h) There are two typographical errors in the previous table that appeared in the original published paper. Find the errors, and correct them. How can you justify these corrections? What statistical criteria can you apply to justify your action? In your judgment, is it likely that these errors have been detected previously? Explain your answer.

(i) Why do you think that these workers used molality in their studies rather than molarity or weight molarity? Explain whether it matters which of these concentration units is used.

19-17. Challenge Problem. As we saw in Problem 19-16, as a preliminary experiment in their effort to measure the dissociation constant of acetic acid, Harned and Ehlers[3] measured E^0 for the cell without liquid junction shown. To complete the study and determine the dissociation constant, these workers also measured the potential of the following cell.

$$\text{Pt,H}_2(1\text{ atm})\,|\,\text{HOAc}(m_1),\ \text{NaOAc}(m_2),\ \text{NaCl}(m_3),\ \text{AgCl(sat'd)}\,|\,\text{Ag}$$

(a) Show that the potential of this cell is given by

$$E = E^0 - \frac{RT}{F}\ln\,(\gamma_{H_3O^+})\,(\gamma_{Cl^-})\,m_{H_3O^+}\,m_{Cl^-}$$

where $\gamma_{H_3O^+}$ and γ_{Cl^-} are the activity coefficients of hydronium ion and chloride ion, respectively, and $m_{H_3O^+}$ and m_{Cl^+} are their respective molal (mole solute/kg solvent) concentrations.

(b) The dissociation constant for acetic acid is given by

$$K = \frac{(\gamma_{H_3O^+})\,(\gamma_{OAc^-})}{\gamma_{HOAc}}\,\frac{m_{H_3O^+}\,m_{OAc^-}}{m_{HOAc}}$$

where γ_{OAo^-} and γ_{HOAc} are the activity coefficients of acetate ion and acetic acid, respectively, and m_{OAc^-} and m_{HOAc} are their respective equilibrium molal (mole solute/kg solvent) concentrations. Show that the potential of the cell in part (a) is given by

$$E = E^0 + \frac{RT}{F}\ln\frac{m_{HOAc}m_{Cl^-}}{m_{OAc^-}}$$

$$= -\frac{RT}{F}\ln\frac{(\gamma_{H_3O^+})(\gamma_{Cl^-})(\gamma_{HOAc})}{(\gamma_{H_3O^+})(\gamma_{OAc^-})} - \frac{RT}{F}\ln K$$

(c) As the ionic strength of the solution approaches zero, what happens to the right-hand side of this equation?

(d) As a result of the answer to part (c), we can write the right-hand side of the equation as $-(RT/F)\ln K'$. Show that

$$K' = \exp\left[-\frac{(E - E^0)F}{RT}\ln\left(\frac{m_{HOAc}m_{Cl^-}}{m_{OAc^-}}\right)\right]$$

(e) The ionic strength of the solution in the cell without liquid junction calculated by Harned and Ehlers is

$$\mu = m_2 + m_3 + m_{H^+}$$

Show that this expression is correct.

(f) These workers prepared solutions of various molal analytical concentrations of acetic acid, sodium acetate, and sodium chloride and measured the potential of the cell presented at the beginning of this problem. Their results are shown in the following table.

Potential Measurements of Cell Pt,H_2(1 atm) | HOAc(c_{HOAc}),NaOAc(c_{NaOAc}),NaCl(c_{NaCl}),AgCl (sat'd) | Ag without Liquid Junction as a Function of Ionic Strength (molality) and Temperature (°C)

c_{HOAc}, m	c_{NaOAc}, m	c_{NaCl}, m	E_0	E_5	E_{10}	E_{15}	E_{20}	E_{25}	E_{30}	E_{35}
0.004779	0.004599	0.004896	0.61995	0.62392	0.62789	0.63183	0.63580	0.63959	0.64335	0.64722
0.012035	0.011582	0.012326	0.59826	0.60183	0.60538	0.60890	0.61241	0.61583	0.61922	0.62264
0.021006	0.020216	0.021516	0.58528	0.58855	0.59186	0.59508	0.59840	0.60154	0.60470	0.60792
0.04922	0.04737	0.05042	0.56546	0.56833	0.57128	0.57413	0.57699	0.57977	0.58257	0.58529
0.08101	0.07796	0.08297	0.55388	0.55667	0.55928	0.56189	0.56456	0.56712	0.56964	0.57213
0.09056	0.08716	0.09276	0.55128	0.55397	0.55661	0.55912	0.56171	0.56423	0.56672	0.56917

Notice that the notation for molal concentration to this point in our discussion of the Harned and Ehlers paper has been in terms of the variables m_x, where x is the species of interest. Do these symbols represent molal analytical concentrations, species, concentrations, or both? Explain. Note that the symbols for concentration in the table adhere to the convention that we have used throughout this book, not the notation of Harned and Ehlers.

(g) Calculate the ionic strength of each of the solutions using the expression for the K_a of acetic acid to calculate $[H_3O^+]$, $[OAc^-]$, and [HOAc] with the usual suitable approximations and a provisional value of $K_a = 1.8 \times 10^{-5}$. Use the potentials in the table for 25°C to calculate values for K' with the expression in part (d). Construct a plot of K' versus μ, and extrapolate the graph to infinite dilution ($\mu = 0$) to find a value for K_a at 25°C. Compare the extrapolated value to the provisional value used to calculate μ. What effect does the provisional value of K_a have on the extrapolated value of K_a? You can perform these calculations most easily using a spreadsheet.

(h) If you have made these computations using a spreadsheet, determine the dissociation constant for acetic acid at all other temperatures for which data are available. How does K_a vary with temperature? At what temperature does the maximum in K_a occur?

InfoTrac College Edition

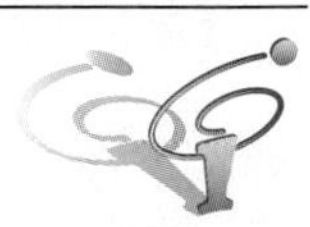

For additional readings, go to InfoTrac College Edition, your online research library, at

http://infotrac.thomsonlearning.com.

CHAPTER 20

Applications of Oxidation/Reduction Titrations

Roger Ressmeyer/CORBIS

Linus Pauling (1901–1994) was one of the most influential and famous chemists of the 20th century. His work in chemical bonding, X-ray crystallography, and related areas had a tremendous impact on chemistry, physics, and biology, spanned eight decades, and led to nearly every award available to chemists. He is the only person to receive two unshared Nobel prizes: for chemistry (1954) and, for his efforts to ban nuclear weapons, the peace prize (1962). In his last years, Pauling devoted his immense intellect and energy to the study of various diseases and their cures. He became convinced that vitamin C, or ascorbic acid, was a panacea. His many books and articles on the subject fueled the popularity of alternative therapies and especially the wide use of vitamin C for preventative maintenance of health. This photo of Pauling tossing an orange into the air is symbolic of this work and the importance of being able to determine concentrations of ascorbic acid at all levels in fruits, vegetables, and commercial vitamin preparations. Redox titrations with iodine are widely used to determine ascorbic acid.

In this chapter, we describe the preparation of standard solutions of oxidants and reductants and their applications in analytical chemistry. In addition, auxiliary reagents that convert an analyte to a single oxidation state are discussed.[1]

20A AUXILIARY OXIDIZING AND REDUCING REAGENTS

The analyte in an oxidation/reduction titration must be in a single oxidation state at the outset. Often, however, the steps that precede the titration, such as dissolving the sample and separating interferences, convert the analyte to a mixture of oxidation states. For example, when a sample containing iron is dissolved, the resulting solution usually contains a mixture of iron(II) and iron(III) ions. If we choose to use a standard oxidant to determine iron, we must first treat the sample solution with an auxiliary reducing agent to convert all of the iron to iron(II). If we plan to titrate with a standard reductant, however, pretreatment with an auxiliary oxidizing reagent is needed.[2]

[1]For further reading on redox titrimetry, see J. A. Dean, *Analytical Chemistry Handbook,* Section 3, pp. 3.65–3.75. New York: McGraw-Hill, 1995.

[2]For a brief summary of auxiliary reagents, see J. A. Goldman and V. A. Stenger, in *Treatise on Analytical Chemistry,* I. M. Kolthoff and P. J. Elving, Eds., Part I, Vol. 11, pp. 7204–7206. New York: Wiley, 1975.

To be useful as a preoxidant or a prereductant, a reagent must react quantitatively with the analyte. In addition, any reagent excess must be easily removable because the excess reagent usually interferes with the titration by reacting with the standard solution.

20A-1 Auxiliary Reducing Reagents

A number of metals are good reducing agents and have been used for the prereduction of analytes. Included among these are zinc, aluminum, cadmium, lead, nickel, copper, and silver (in the presence of chloride ion). Sticks or coils of the metal can be immersed directly in the analyte solution. After reduction is judged complete, the solid is removed manually and rinsed with water. The analyte solution must be filtered to remove granular or powdered forms of the metal. An alternative to filtration is the use of a **reductor,** such as that shown in Figure 20-1.[3] Here, the finely divided metal is held in a vertical glass tube through which the solution is drawn under a mild vacuum. The metal in a reductor is ordinarily sufficient for hundreds of reductions.

A typical **Jones reductor** has a diameter of about 2 cm and holds a 40- to 50-cm column of amalgamated zinc. Amalgamation is accomplished by allowing zinc granules to stand briefly in a solution of mercury(II) chloride, where the following reaction occurs:

$$2Zn(s) + Hg^{2+} \rightarrow Zn^{2+} + Zn(Hg)(s)$$

Zinc amalgam is nearly as effective for reductions as the pure metal and has the important virtue of inhibiting the reduction of hydrogen ions by zinc. This side reaction needlessly uses up the reducing agent and also contaminates the sample solution with a large amount of zinc(II) ions. Solutions that are quite acidic can be passed through a Jones reductor without significant hydrogen formation.

Table 20-1 lists the principal applications of the Jones reductor. Also listed in this table are reductions that can be accomplished with a **Walden reductor,** in

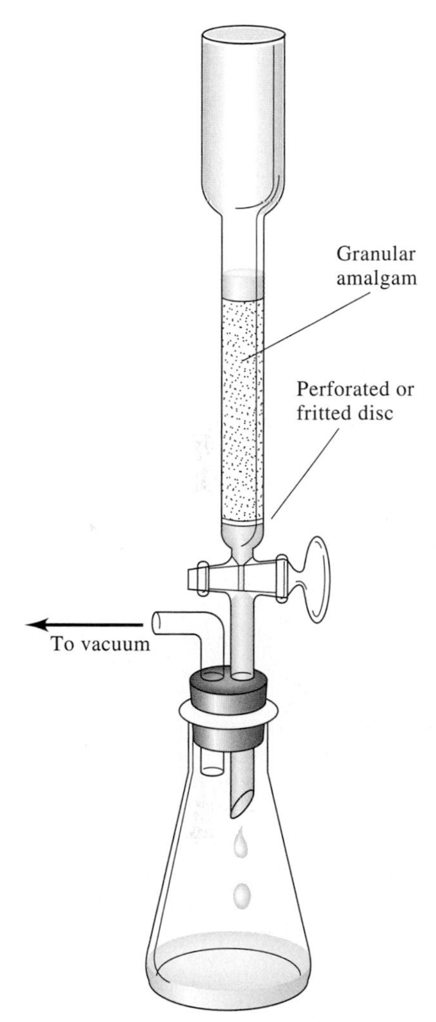

Figure 20-1 A Jones reductor.

TABLE 20-1
Uses of the Walden Reductor and the Jones Reductor*

Walden	Jones
$Ag(s) + Cl^- \rightarrow AgCl(s) + e^-$	**$Zn(Hg)(s) \rightarrow Zn^{2+} + Hg + 2e^-$**
$Fe^{3+} + e^- \rightarrow Fe^{2+}$	$Fe^{3+} + e^- \rightleftharpoons Fe^{2+}$
$Cu^{2+} + e^- \rightarrow Cu^+$	$Cu^{2+} + 2e^- \rightleftharpoons Cu(s)$
$H_2MoO_4 + 2H^+ + e^- \rightarrow MoO_2^+ + 2H_2O$	$H_2MoO_4 + 6H^+ + 3e^- \rightleftharpoons Mo^{3+} + 3H_2O$
$UO_2^{2+} + 4H^+ + 2e^- \rightarrow U^{4+} + 2H_2O$	$UO_2^{2+} + 4H^+ + 2e^- \rightleftharpoons U^{4+} + 2H_2O$
	$UO_2^{2+} + 4H^+ + 3e^- \rightleftharpoons U^{3+} + 2H_2O^\dagger$
$V(OH)_4^+ + 2H^+ + e^- \rightarrow VO^{2+} + 3H_2O$	$V(OH)_4^+ + 4H^+ + 3e^- \rightleftharpoons V^{2+} + 4H_2O$
TiO^{2+} not reduced	$TiO^{2+} + 2H^+ + e^- \rightleftharpoons Ti^{2+} + H_2O$
Cr^{3+} not reduced	$Cr^{3+} + e^- \rightleftharpoons Cr^{2+}$

*From I. M. Kolthoff and R. Belcher, *Volumetric Analysis,* Vol 3, p. 12. New York: Interscience, 1957. This material is used by permission of John Wiley & Sons, Inc.
†A mixture of oxidation states is obtained. The Jones reductor may still be used for the determination of uranium, however, because any U^{2+} formed can be converted to U^{4+} by shaking the solution with air for a few minutes.

[3]For a discussion of reductors, see F. Hecht, in *Treatise on Analytical Chemistry,* I. M. Kolthoff and P. J. Elving, Eds., Part I, Vol. 11, pp. 6703–6707. New York: Wiley, 1975.

which granular metallic silver held in a narrow glass column is the reductant. Silver is not a good reducing agent unless chloride or some other ion that forms a silver salt of low solubility is present. For this reason, prereductions with a Walden reductor are generally carried out from hydrochloric acid solutions of the analyte. The coating of silver chloride produced on the metal is removed periodically by dipping a zinc rod into the solution that covers the packing.

Table 20-1 suggests that the Walden reductor is somewhat more selective in its action than is the Jones reductor.

20A-2 Auxiliary Oxidizing Reagents

Sodium Bismuthate

Sodium bismuthate is a powerful oxidizing agent capable, for example, of converting manganese(II) quantitatively to permanganate ion. This bismuth salt is a sparingly soluble solid with a formula that is usually written as $NaBiO_3$, although its exact composition is somewhat uncertain. Oxidations are performed by suspending the bismuthate in the analyte solution and boiling for a brief period. The unused reagent is then removed by filtration. The half-reaction for the reduction of sodium bismuthate can be written as

$$NaBiO_3(s) + 4H^+ + 2e^- \rightleftharpoons BiO^+ + Na^+ + 2H_2O$$

Ammonium Peroxydisulfate

Ammonium peroxydisulfate, $(NH_4)_2S_2O_8$, is also a powerful oxidizing agent. In acidic solution, it converts chromium(III) to dichromate, cerium(III) to cerium(IV), and manganese(II) to permanganate. The half-reaction is

$$S_2O_8^{2-} + 2e^- \rightleftharpoons 2SO_4^{2-}$$

The oxidations are catalyzed by traces of silver ion. The excess reagent is easily decomposed by a brief period of boiling:

$$2S_2O_8^{2-} + 2H_2O \rightarrow 4SO_4^{2-} + O_2(g) + 4H^+$$

Sodium Peroxide and Hydrogen Peroxide

Peroxide is a convenient oxidizing agent either as the solid sodium salt or as a dilute solution of the acid. The half-reaction for hydrogen peroxide in acidic solution is

$$H_2O_2 + 2H^+ + 2e^- \rightleftharpoons 2H_2O \qquad E^0 = 1.78\ \text{V}$$

After oxidation is complete, the solution is freed of excess reagent by boiling:

$$2H_2O_2 \rightarrow 2H_2O + O_2(g)$$

20B APPLYING STANDARD REDUCING AGENTS

Standard solutions of most reductants tend to react with atmospheric oxygen. For this reason, reductants are seldom used for the direct titration of oxidizing analytes; indirect methods are used instead. The two most common reductants, iron(II) and thiosulfate ions, are discussed in the paragraphs that follow.

20B-1 Iron(II) Solutions

Solutions of iron(II) are easily prepared from iron(II) ammonium sulfate, $Fe(NH_4)_2(SO_4)_2 \cdot 6H_2O$ (Mohr's salt), or from the closely related iron(II) ethylenediamine sulfate, $FeC_2H_4(NH_3)_2(SO_4)_2 \cdot 4H_2O$ (Oesper's salt). Air oxidation of iron(II) takes place rapidly in neutral solutions but is inhibited in the presence of acids, with the most stable preparations being about 0.5 M in H_2SO_4. Such solutions are stable for no more than one day, if that long. Numerous oxidizing agents are conveniently determined by treatment of the analyte solution with a measured excess of standard iron(II) followed by immediate titration of the excess with a standard solution of potassium dichromate or cerium(IV) (see Sections 20C-1 and 20C-2). Just before or just after the analyte is titrated, the volumetric ratio between the standard oxidant and the iron(II) solution is established by titrating two or three aliquots of the latter with the former.

This procedure has been applied to the determination of organic peroxides; hydroxylamine; chromium(VI); cerium(IV); molybdenum(VI); nitrate, chlorate, and perchlorate ions; and numerous other oxidants (see, for example, Problems 20-37 and 20-39).

Molecular model of thiosulfate ion. Sodium thiosulfate, formerly called sodium hyposulfite, or *hypo,* is used to "fix" photographic images, to extract silver from ore, as an antidote in cyanide poisoning, as a mordant in the dye industry, as a bleaching agent in a variety of applications, as the solute in the supersaturated solution of hot packs, and, of course, as an analytical reducing agent. The action of thiosulfate as a photographic fixer is based on its capacity to form complexes with silver and thus dissolve unexposed silver bromide from the surface of photographic film and paper. Thiosulfate is often used as a dechlorinating agent to make aquarium water safe for fish and other aquatic life.

20B-2 Sodium Thiosulfate

Thiosulfate ion ($S_2O_3^{2-}$) is a moderately strong reducing agent that has been widely used to determine oxidizing agents by an indirect procedure that involves iodine as an intermediate. With iodine, thiosulfate ion is oxidized quantitatively to tetrathionate ion ($S_4O_6^{2-}$) according to the half-reaction

$$2S_2O_3^{2-} \rightleftharpoons S_4O_6^{2-} + 2e^-$$

◀ In its reaction with iodine, each thiosulfate ion loses one electron.

The quantitative reaction with iodine is unique. Other oxidants can oxidize the tetrathionate ion to sulfate ion.

The scheme used to determine oxidizing agents involves adding an unmeasured excess of potassium iodide to a slightly acidic solution of the analyte. Reduction of the analyte produces a stoichiometrically equivalent amount of iodine. The liberated iodine is then titrated with a standard solution of sodium thiosulfate, $Na_2S_2O_3$, one of the few reducing agents that is stable toward air oxidation. An example of this procedure is the determination of sodium hypochlorite in bleaches. The reactions are

◀ Sodium thiosulfate is one of the few reducing agents not oxidized by air.

$$OCl^- + 2I^- + 2H^+ \rightarrow Cl^- + I_2 + H_2O \qquad \text{(unmeasured excess KI)}$$

$$I_2 + S_2O_3^{2-} \rightarrow 2I^- + S_4O_6^{2-} \qquad (20\text{-}1)$$

The quantitative conversion of thiosulfate ion to tetrathionate ion shown in Equation 20-1 requires a pH smaller than 7. If strongly acidic solutions must be titrated, air oxidation of the excess iodide must be prevented by blanketing the solution with an inert gas, such as carbon dioxide or nitrogen.

Detecting End Points in Iodine/Thiosulfate Titrations

A solution that is about 5×10^{-6} M in I_2 has a discernible color, which corresponds to less than one drop of a 0.05 M iodine solution in 100 mL. Thus, provided that the analyte solution is colorless, the disappearance of the iodine color can serve as the indicator in titrations with sodium thiosulfate.

More commonly, titrations involving iodine are performed with a suspension of starch as an indicator. The deep blue color that develops in the presence of iodine is believed to arise from the absorption of iodine into the helical chain of β-amylose (see Figure 20-2), a macromolecular component of most starches. The closely related α-amylose forms a red adduct with iodine. This reaction is not easily reversible and is thus undesirable. In commercially available *soluble starch,* the alpha fraction has been removed to leave principally β-amylose; indicator solutions are easily prepared from this product.

Aqueous starch suspensions decompose within a few days, primarily because of bacterial action. The decomposition products tend to interfere with the indicator properties of the preparation and may also be oxidized by iodine. The rate of decomposition can be inhibited by preparing and storing the indicator under sterile conditions and by adding mercury(II) iodide or chloroform as a bacteriostat. Perhaps the simplest alternative is to prepare a fresh suspension of the indicator, which requires only a few minutes, on the day it is to be used.

► Starch undergoes decomposition in solutions with high I_2 concentrations. In titrations of excess I_2 with $Na_2S_2O_3$, addition of the indicator must be deferred until most of the I_2 has been reduced.

Starch decomposes irreversibly in solutions containing large concentrations of iodine. Therefore, in titrating solutions of iodine with thiosulfate ion, as in the indirect determination of oxidants, addition of the indicator is delayed until the color of the solution changes from red-brown to yellow; at this point, the titration is nearly complete. The indicator can be introduced at the outset when thiosulfate solutions are being titrated directly with iodine.

Stability of Sodium Thiosulfate Solutions

► When sodium thiosulfate is added to a strongly acidic medium, a cloudiness develops almost immediately as a consequence of the precipitation of elemental sulfur. Even in neutral solution, this reaction proceeds at such a rate that standard sodium thiosulfate must be restandardized periodically.

Although sodium thiosulfate solutions are resistant to air oxidation, they do tend to decompose to give sulfur and hydrogen sulfite ion:

$$S_2O_3^{2-} + H^+ \rightleftharpoons HSO_3^- + S(s)$$

Variables that influence the rate of this reaction include pH, the presence of microorganisms, the concentration of the solution, the presence of copper(II) ions, and exposure to sunlight. These variables may cause the concentration of a thiosulfate solution to change by several percentage points over a period of a few weeks. Proper attention to detail will yield solutions that need only occasional restandardization. The rate of the decomposition reaction increases markedly as the solution becomes acidic.

The most important single cause of the instability of neutral or slightly basic thiosulfate solutions is bacteria that metabolize thiosulfate ion to sulfite and sulfate ions as well as to elemental sulfur. To minimize this problem, standard solutions of the reagent are prepared under reasonably sterile conditions. Bacterial activity appears to be at a minimum at a pH between 9 and 10, which accounts, at least in part, for the reagent's greater stability in slightly basic solutions. The presence of a bactericide, such as chloroform, sodium benzoate, or mercury(II) iodide, also slows decomposition.

Standardizing Thiosulfate Solutions

Potassium iodate is an excellent primary standard for thiosulfate solutions. In this application, weighed amounts of primary-standard grade reagent are dissolved in water containing an excess of potassium iodide. When this mixture is acidified with a strong acid, the reaction

$$IO_3^- + 5I^- + 6H^+ \rightleftharpoons 3I_2 + 2H_2O$$

occurs instantaneously. The liberated iodine is then titrated with the thiosulfate solution. The stoichiometry of the reaction is

$$1 \text{ mol } IO_3^- = 3 \text{ mol } I_2 = 6 \text{ mol } S_2O_3^{2-}$$

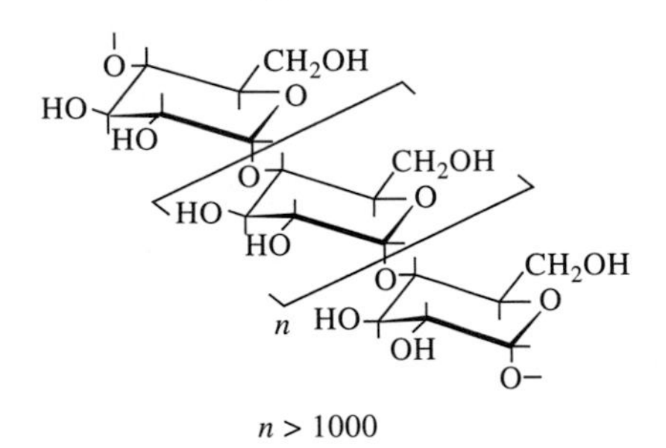

EXAMPLE 20-1

A solution of sodium thiosulfate was standardized by dissolving 0.1210 g KIO_3 (214.00 g/mol) in water, adding a large excess of KI, and acidifying with HCl. The liberated iodine required 41.64 mL of the thiosulfate solution to decolorize the blue starch/iodine complex. Calculate the molarity of the $Na_2S_2O_3$.

$$\text{amount } Na_2S_2O_3 = 0.1210 \cancel{\text{ g } KIO_3} \times \frac{1 \cancel{\text{ mmol } KIO_3}}{0.21400 \cancel{\text{ g } KIO_3}} \times \frac{6 \text{ mmol } Na_2S_2O_3}{\cancel{\text{mmol } KIO_3}}$$
$$= 3.3925 \text{ mmol } Na_2S_2O_3$$

$$c_{Na_2S_2O_3} = \frac{3.3925 \text{ mmol } Na_2S_2O_3}{41.64 \text{ mL } Na_2S_2O_3} = 0.08147 \text{ M}$$

Other primary standards for sodium thiosulfate are potassium dichromate, potassium bromate, potassium hydrogen iodate, potassium hexacyanoferrate(III), and metallic copper. All these compounds liberate stoichiometric amounts of iodine when treated with excess potassium iodide.

Applications of Sodium Thiosulfate Solutions

Numerous substances can be determined by the indirect method involving titration with sodium thiosulfate; typical applications are summarized in Table 20-2.

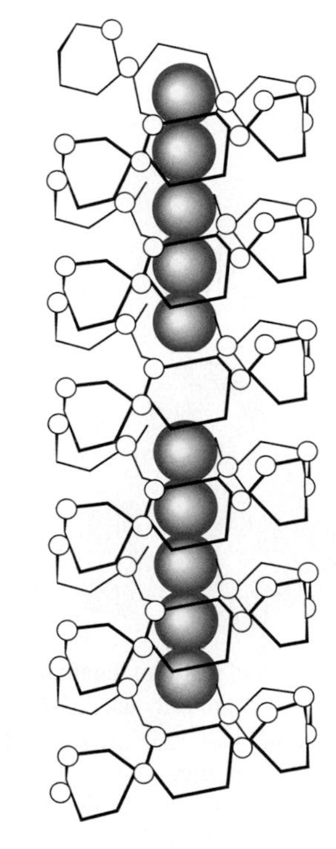

Figure 20-2 Thousands of glucose molecules polymerize to form huge molecules of β-amylose, as shown schematically in (a). Molecules of β-amylose tend to assume a helical structure. The iodine species I_3^- as shown in (b) is incorporated into the amylose helix. For further details, see R. C. Teitelbaum, S. L. Ruby, and T. J. Marks, *J. Amer. Chem. Soc.*, **1980,** *102,* 3322.

TABLE 20-2
Some Applications of Sodium Thiosulfate as a Reductant

Analyte	Half-Reaction	Special Conditions
IO_4^-	$IO_4^- + 8H^+ + 7e^- \rightleftharpoons \frac{1}{2} I_2 + 4H_2O$	Acidic solution
	$IO_4^- + 2H^+ + 2e^- \rightleftharpoons IO_3^- + H_2O$	Neutral solution
IO_3^-	$IO_3^- + 6H^+ + 5e^- \rightleftharpoons \frac{1}{2} I_2 + 3H_2O$	Strong acid
BrO_3^-, ClO_3^-	$XO_3^- + 6H^+ + 6e^- \rightleftharpoons X^- + 3H_2O$	Strong acid
Br_2, Cl_2	$X_2 + 2I^- \rightleftharpoons I_2 + 2X^-$	
NO_2^-	$HNO_2 + H^+ + e^- \rightleftharpoons NO(g) + H_2O$	
Cu^{2+}	$Cu^{2+} + I^- + e^- \rightleftharpoons CuI(s)$	
O_2	$O_2 + 4Mn(OH)_2(s) + 2H_2O \rightleftharpoons Mn(OH)_3(s)$	Basic solution
	$Mn(OH)_3(s) + 3H^+ + e^- \rightleftharpoons Mn^{2+} + 3H_2O$	Acidic solution
O_3	$O_3(g) + 2H^+ + 2e^- \rightleftharpoons O_2(g) + H_2O$	
Organic peroxide	$ROOH + 2H^+ + 2e^- \rightleftharpoons ROH + H_2O$	

20C APPLYING STANDARD OXIDIZING AGENTS

Table 20-3 summarizes the properties of five of the most widely used volumetric oxidizing reagents. Note that the standard potentials for these reagents vary from 0.5 to 1.5 V. The choice among them depends on the strength of the analyte as a reducing agent, the rate of reaction between oxidant and analyte, the stability of the standard oxidant solutions, the cost, and the availability of a satisfactory indicator.

20C-1 The Strong Oxidants—Potassium Permanganate and Cerium(IV)

Solutions of permanganate ion and cerium(IV) ion are strong oxidizing reagents whose applications closely parallel one another. Half-reactions for the two are

$$MnO_4^- + 8H^+ + 5e^- \rightleftharpoons Mn^{2+} + 4H_2O \qquad E^0 = 1.51\ V$$

$$Ce^{4+} + e^- \rightleftharpoons Ce^{3+} \qquad E^{0'} = 1.44\ V (1\ M\ H_2SO_4)$$

The formal potential shown for the reduction of cerium(IV) is for solutions that are 1 M in sulfuric acid. In 1 M perchloric acid and 1 M nitric acid, the potentials are 1.70 V and 1.61 V, respectively. Solutions of cerium(IV) in the latter two acids are not very stable and thus find limited application.

The half-reaction shown for permanganate ion occurs only in solutions that are 0.1 M or greater in strong acid. In less acidic media, the product may be Mn(III), Mn(IV), or Mn(VI), depending on conditions.

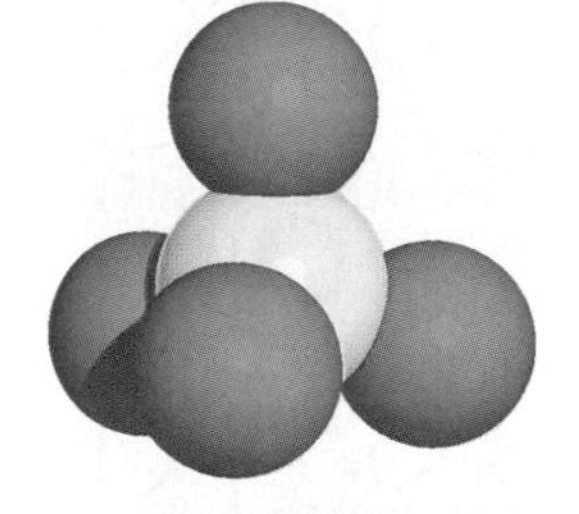

Molecular model of permanganate ion, MnO_4^-. In addition to its use as an analytical reagent, usually in the form of its potassium salt, permanganate is very useful as an oxidizing agent in synthetic organic chemistry. It is used as a bleaching agent with fats, oils, cotton, silk, and other fibers. It has also been used as an antiseptic and anti-infective, as a component in outdoor survival kits, for destroying organic matter in fish ponds, in manufacturing printed wiring boards, for neutralizing the effects of the pesticide rotenone, and for scrubbing flue gases in the determination of mercury. Solid potassium permanganate reacts violently with organic matter, and this effect is often used as a demonstration in general chemistry courses. To further explore these and other uses of permanganate, go to http://www.google.com/. Use "permanganate uses" as your search term.

Comparing the Two Reagents

For all practical purposes, the oxidizing strengths of permanganate and cerium(IV) solutions are comparable. Solutions of cerium(IV) in sulfuric acid,

TABLE 20-3
Some Common Oxidants Used as Standard Solutions

Reagent and Formula	Reduction Product	Standard Potential, V	Standardized with	Indicator*	Stability†
Potassium permanganate, $KMnO_4$	Mn^{2+}	1.51‡	$Na_2C_2O_4$, Fe, As_2O_3	MnO_4^-	(b)
Potassium bromate, $KBrO_3$	Br^-	1.44‡	$KBrO_3$	(1)	(a)
Cerium(IV), Ce^{4+}	Ce^{3+}	1.44‡	$Na_2C_2O_4$, Fe, As_2O_3	(2)	(a)
Potassium dichromate, $K_2Cr_2O_7$	Cr^{3+}	1.33‡	$K_2Cr_2O_7$, Fe,	(3)	(a)
Iodine, I_2	I^-	0.536‡	$BaS_2O_3 \cdot H_2O$, $Na_2S_2O_3$	starch	(c)

*(1) α-Napthoflavone; (2) 1,10-phenanthroline iron(II) complex (ferroin); (3) diphenylamine sulfonic acid.
†(a) Indefinitely stable; (b) moderately stable, requires periodic standardization; (c) somewhat unstable, requires frequent standardization.
‡$E^{0'}$ in 1 M H_2SO_4.

however, are stable indefinitely, whereas permanganate solutions decompose slowly and thus require occasional restandardization. Furthermore, cerium(IV) solutions in sulfuric acid do not oxidize chloride ion and can be used to titrate hydrochloric acid solutions of analytes; in contrast, permanganate ion cannot be used with hydrochloric acid solutions unless special precautions are taken to prevent the slow oxidation of chloride ion that leads to overconsumption of the standard reagent. A further advantage of cerium(IV) is that a primary-standard grade salt of the reagent is available, thus making possible the direct preparation of standard solutions.

Despite these advantages of cerium solutions over permanganate solutions, the latter are more widely used. One reason is the color of permanganate solutions, which is intense enough to serve as an indicator in titrations. A second reason for the popularity of permanganate solutions is their modest cost. The cost of 1 L of 0.02 M $KMnO_4$ solution is about $0.08, whereas 1 L of a comparable strength Ce(IV) solution costs about $2.20 ($4.40 if reagent of primary-standard grade is used). Another disadvantage of cerium(IV) solutions is their tendency to form precipitates of basic salts in solutions that are less than 0.1 M in strong acid.

Detecting the End Points

A useful property of a potassium permanganate solution is its intense purple color, which is sufficient to serve as an indicator for most titrations. If you add as little as 0.01 to 0.02 mL of a 0.02 M solution of permanganate to 100 mL of water, you can perceive the purple color of the resulting solution. If the solution is very dilute, diphenylamine sulfonic acid or the 1,10-phenanthroline complex of iron(II) (see Table 19-2) provides a sharper end point.

The permanganate end point is not permanent because excess permanganate ions react slowly with the relatively large concentration of manganese(II) ions present at the end point, according to the reaction

$$2MnO_4^- + 3Mn^{2+} + 2H_2O \rightleftharpoons 5MnO_2(s) + 4H^+$$

The equilibrium constant for this reaction is about 10^{47}, which indicates that the equilibrium concentration of permanganate ion is incredibly small even in highly acidic media. Fortunately, the rate at which this equilibrium is approached is so slow that the end point fades only gradually over a period of perhaps 30 seconds.

Solutions of cerium(IV) are yellow-orange, but the color is not intense enough to act as an indicator in titrations. Several oxidation/reduction indicators are available for titrations with standard solutions of cerium(IV). The most widely used of these is the iron(II) complex of 1,10-phenanthroline or one of its substituted derivatives (see Table 19-2).

The Preparation and Stability of Standard Solutions

Aqueous solutions of permanganate are not entirely stable because of water oxidation:

$$4MnO_4^- + 2H_2O \rightarrow 4MnO_2(s) + 3O_2(g) + 4OH^-$$

Although the equilibrium constant for this reaction indicates that the products are favored, permanganate solutions, when properly prepared, are reasonably stable because the decomposition reaction is slow. It is catalyzed by light, heat, acids, bases, manganese(II), and manganese dioxide.

► Permanganate solutions are moderately stable provided that they are free of manganese dioxide and are stored in a dark container.

Moderately stable solutions of permanganate ion can be prepared if the effects of these catalysts, particularly manganese dioxide, are minimized. Manganese dioxide is a contaminant in even the best grade of solid potassium permanganate. Furthermore, this compound forms in freshly prepared solutions of the reagent as a consequence of the reaction of permanganate ion with organic matter and dust present in the water used to prepare the solution. Removal of manganese dioxide by filtration before standardization markedly improves the stability of standard permanganate solutions. Before filtration, the reagent solution is allowed to stand for about 24 hours or is heated for a brief period to hasten oxidation of the organic species generally present in small amounts in distilled and deionized water. Paper cannot be used for filtering because permanganate ion reacts with it to form additional manganese dioxide.

Standardized permanganate solutions should be stored in the dark. Filtration and restandardization are required if any solid is detected in the solution or on the walls of the storage bottle. In any event, restandardization every 1 or 2 weeks is a good precautionary measure.

Solutions containing excess standard permanganate should never be heated because they decompose by oxidizing water. This decomposition cannot be compensated for with a blank. It is possible to titrate hot, acidic solutions of reductants with permanganate without error, however, provided that the reagent is added slowly enough so that large excesses do not accumulate.

FEATURE 20-1

Determination of Chromium Species in Water Samples

Chromium is an important metal to monitor in environmental samples. Not only is the total amount of chromium of interest, but also the oxidation state in which the chromium is found is quite important. In water, chromium can exist as Cr(III) or as Cr(VI) species. Chromium(III) is an essential nutrient and is nontoxic. Chromium(VI), however, is a known carcinogen. Hence, the determination of the amount of chromium in each of these oxidation states is often of more interest than the total amount of chromium. There are several good methods available for determining Cr(VI) selectively. One of the most popular involves the oxidation of the reagent 1,5-diphenylcarbohydrazide (diphenylcarbazide) by Cr(VI) in acid solution. The reaction produces a red-purple chelate of Cr(III) and diphenylcarbazide that can be monitored colorimetrically (see Section 26A-3). The direct reaction of Cr(III) itself and the reagent is so slow that essentially only the Cr(VI) is measured. To determine Cr(III), the sample is oxidized with excess permanganate in alkaline solution to convert all the Cr(III) to Cr(VI). The excess oxidant is destroyed with sodium azide. A new colorimetric measurement is made that now determines total chromium [the original Cr(VI) plus that formed by oxidation of Cr(III)]. The amount of Cr(III) present is then obtained by subtracting the amount of Cr(VI) obtained in the original measurement from the amount of total chromium obtained after permanganate oxidation. Note that here permanganate is being used as an auxiliary oxidizing agent.

© Don Conrard/Corbis

Chromium has long been prized for its beauty as a polished coating on metals (see photo) and for its anticorrosive properties in stainless steel and other alloys. In trace amounts, chromimum (III) is an essential nutrient. Chromium (VI) in the form of sodium dichromate is widely used in aqueous solution as a corrosion inhibitor in large-scale industrial processes. See margin note on page 574 for more details on chromium.

EXAMPLE 20-2

Describe how you would prepare 2.0 L of an approximately 0.010 M solution of $KMnO_4$ (158.03 g/mol).

$$\text{mass } KMnO_4 \text{ needed} = 2.0\text{ L} \times 0.010 \frac{\cancel{\text{mol } KMnO_4}}{\text{L}} \times 158.03 \frac{\text{g } KMnO_4}{\cancel{\text{mol } KMnO_4}}$$

$$= 3.16 \text{ g } KMnO_4$$

Dissolve about 3.2 g of $KMnO_4$ in a little water. After solution is complete, add water to bring the volume to about 2.0 L. Heat the solution to boiling for a brief period, and let stand until it is cool. Filter through a glass-filtering crucible and store in a clean dark bottle.

The most widely used compounds for the preparation of solutions of cerium(IV) are listed in Table 20-4. Primary-standard grade cerium ammonium nitrate is available commercially and can be used to prepare standard solutions of the cation directly by weight. More commonly, less expensive reagent-grade cerium(IV)

TABLE 20-4
Analytically Useful Cerium(IV) Compounds

Name	Formula	Molar Mass
Cerium (IV) ammonium nitrate	$Ce(NO_3)_4 \cdot 2NH_4NO_3$	548.2
Cerium(IV) ammonium sulfate	$Ce(SO_4)_2 \cdot 2(NH_4)_2SO_4 \cdot 2H_2O$	632.6
Cerium(IV) hydroxide	$Ce(OH)_4$	208.1
Ce(IV) hydrogen sulfate	$Ce(HSO_4)_4$	528.4

[4]W. J. Blot et al., *J. Occup. Environ. Med.*, **2000,** *423*(7), 194–199; J. P. Fryzek et al., *J. Occup. Environ. Med.*, **2001,** *43*(7), 635–640.

ammonium nitrate or ceric hydroxide is used to prepare solutions that are subsequently standardized. In either case, the reagent is dissolved in a solution that is at least 0.1 M in sulfuric acid to prevent the precipitation of basic salts.

Sulfuric acid solutions of cerium(IV) are remarkably stable and can be stored for months or heated at 100°C for prolonged periods without a change in concentration.

Standardizing Permanganate and Ce(IV) Solutions

Sodium oxalate is a widely used primary standard. In acidic solutions, the oxalate ion is converted to the undissociated acid. Thus, its reaction with permanganate can be described by

$$2MnO_4^- + 5H_2C_2O_4 + 6H^+ \rightarrow 2Mn^{2+} + 10CO_2(g) + 8H_2O$$

Autocatalysis is a type of catalysis in which the product of a reaction catalyses the reaction. This phenomenon causes the rate of the reaction to increase as the reaction proceeds.

The reaction between permanganate ion and oxalic acid is complex and proceeds slowly even at elevated temperature unless manganese(II) is present as a catalyst. Thus, when the first few milliliters of standard permanganate are added to a hot solution of oxalic acid, several seconds are required before the color of the permanganate ion disappears. As the concentration of manganese(II) builds up, however, the reaction proceeds more and more rapidly as a result of autocatalysis.

It has been found that when solutions of sodium oxalate are titrated at 60°C to 90°C, the consumption of permanganate is from 0.1% to 0.4% less than theoretical, probably because of the air oxidation of a fraction of the oxalic acid. This small error can be avoided by adding 90% to 95% of the required permanganate to a cool solution of the oxalate. After the added permanganate is completely consumed (as indicated by the disappearance of color), the solution is heated to about 60°C and titrated to a pink color that persists for about 30 seconds. The disadvantage of this procedure is that it requires prior knowledge of the approximate concentration of the permanganate solution so that a proper initial volume can be added: For most purposes, direct titration of the hot oxalic acid solution is adequate (usually results are 0.2% to 0.3% high). If greater accuracy is required, a direct titration of the hot solution of one portion of the primary standard can be followed by titration of two or three portions in which the solution is not heated until the end.

► Solutions of $KMnO_4$ and Ce^{4+} can also be standardized with electrolytic iron wire or with potassium iodide.

Sodium oxalate is also widely used to standardize Ce(IV) solutions. The reaction between Ce^{4+} and $H_2C_2O_4$ is

$$2Ce^{4+} + H_2C_2O_4 \rightarrow 2Ce^{3+} + 2CO_2(g) + 2H^+$$

Cerium(IV) standardizations against sodium oxalate are usually performed at 50°C in a hydrochloric acid solution containing iodine monochloride as a catalyst.

EXAMPLE 20-3

You wish to standardize the solution in Example 20-2 against primary standard $Na_2C_2O_4$ (134.00 g/mol). If you want to use between 30 and 45 mL of the reagent for the standardization, what range of masses of the primary standard should you weigh out?

For a 30-mL titration:

$$\text{amount KMnO}_4 = 30\ \cancel{\text{mL KMnO}_4} \times 0.010 \frac{\text{mmol KMnO}_4}{\cancel{\text{mL KMnO}_4}}$$

$$= 0.30\ \text{mmol KMnO}_4$$

$$\text{mass Na}_2\text{C}_2\text{O}_4 = 0.30\ \cancel{\text{mmol KMnO}_4} \times \frac{5\ \cancel{\text{mmol Na}_2\text{C}_2\text{O}_4}}{2\ \cancel{\text{mmol KMnO}_4}}$$

$$\times\ 0.134 \frac{\text{g Na}_2\text{C}_2\text{O}_4}{\cancel{\text{mmol Na}_2\text{C}_2\text{O}_4}}$$

$$= 0.101\ \text{g Na}_2\text{C}_2\text{O}_4$$

Proceeding in the same way, we find for a 45-mL titration:

$$\text{mass Na}_2\text{C}_2\text{O}_4 = 45 \times 0.010 \times \frac{5}{2} \times 0.134 = 0.151\ \text{g Na}_2\text{C}_2\text{O}_4$$

Thus, you should weigh 0.10- to 0.15-g samples of the primary standard.

EXAMPLE 20-4

A 0.1278-g sample of primary-standard $Na_2C_2O_4$ required exactly 33.31 mL of the permanganate solution in Example 20-2 to reach the end point. What was the molarity of the $KMnO_4$ reagent?

$$\text{amount Na}_2\text{C}_2\text{O}_4 = 0.1278\ \cancel{\text{g Na}_2\text{C}_2\text{O}_4} \times \frac{1\ \text{mmol Na}_2\text{C}_2\text{O}_4}{0.13400\ \cancel{\text{g Na}_2\text{C}_2\text{O}_4}}$$

$$= 0.95373\ \text{mmol Na}_2\text{C}_2\text{O}_4$$

$$c_{\text{KMnO}_4} = 0.95373\ \cancel{\text{mmol Na}_2\text{C}_2\text{O}_4} \times \frac{2\ \text{mmol KMnO}_4}{5\ \cancel{\text{mmol Na}_2\text{C}_2\text{O}_4}} \times \frac{1}{33.31\ \text{mL KMnO}_4}$$

$$= 0.01145\ \text{M}$$

Using Potassium Permanganate and Cerium(IV) Solutions

Table 20-5 lists some of the many applications of permanganate and cerium(IV) solutions to the volumetric determination of inorganic species. Both reagents have also been applied to the determination of organic compounds with oxidizable functional groups.

TABLE 20-5

Some Applications of Potassium Permanganate and Cerium(IV) Solutions

Substance Sought	Half-Reaction	Conditions
Sn	$Sn^{2+} \rightleftharpoons Sn^{4+} + 2e^-$	Prereduction with Zn
H_2O_2	$H_2O_2 \rightleftharpoons O_2(g) + 2H^+ + 2e^-$	
Fe	$Fe^{2+} \rightleftharpoons Fe^{3+} + e^-$	Prereduction with $SnCl_2$ or with Jones or Walden reductor
$Fe(CN)_6^{4-}$	$Fe(CN)_6^{4-} \rightleftharpoons Fe(CN)_6^{3-} + e^-$	
V	$VO^{2+} + 3H_2O \rightleftharpoons V(OH)_4^{2+} + e^-$	Prereduction with Bi amalgam or SO_2
Mo	$Mo^{3+} + 4H_2O \rightleftharpoons MoO_4^{2-} + 8H^+ + 3e^-$	Prereduction with Jones reductor
W	$W^{3+} + 4H_2O \rightleftharpoons WO_4^{2-} + 8H^+ + 3e^-$	Prereduction with Zn or Cd
U	$U^{4+} + 2H_2O \rightleftharpoons UO_2^{2+} + 4H^+ + 2e^-$	Prereduction with Jones reductor
Ti	$Ti^{3+} + H_2O \rightleftharpoons TiO^{2+} + 2H^+ + e^-$	Prereduction with Jones reductor
$H_2C_2O_4$	$H_2C_2O_4 \rightleftharpoons 2CO_2 + 2H^+ + 2e^-$	
Mg, Ca, Zn, Co, Pb, Ag	$H_2C_2O_4 \rightleftharpoons 2CO_2 + 2H^+ + 2e^-$	Sparingly soluble metal oxalates filtered, washed, and dissolved in acid; liberated oxalic acid titrated
HNO_2	$HNO_2 + H_2O \rightleftharpoons NO_3^- + 3H^+ + 2e^-$	15-min reaction time; excess $KMnO_4$ back titrated
K	$K_2NaCo(NO_2)_6 + 6H_2O \rightleftharpoons Co^{2+} + 6NO_3^- + 12H^+ + 2K^+ + Na^+ + 11e^-$	Precipitated as $K_2NaCo(NO_2)_6$; filtered and dissolved in $KMnO_4$; excess $KMnO_4$ back titrated
Na	$U^{4+} + 2H_2O \rightleftharpoons UO_2^{2+} + 4H^+ + 2e^-$	Precipitated as $NaZn(UO_2)_2(OAc)_9$; filtered washed, dissolved; U determined as above

EXAMPLE 20-5

Aqueous solutions containing approximately 3% (w/w) H_2O_2 are sold in drug stores as disinfectants. Propose a method for determining the peroxide content of such preparation using the standard solution described in Examples 20-3 and 20-4. Assume that you wish to use between 30 and 45 mL of the reagent for a titration. The reaction is

$$5H_2O_2 + 2MnO_4^- + 6H^+ \rightarrow 5O_2 + 2\,Mn^{2+} + 8H_2O$$

The amount of $KMnO_4$ in 35 to 45 mL of the reagent is between

$$\text{amount KMnO}_4 = 35\ \cancel{\text{mL KMnO}_4} \times 0.01145\ \frac{\text{mmol KMnO}_4}{\cancel{\text{mL KMnO}_4}}$$
$$= 0.401\ \text{mmol KMnO}_4$$

and

$$\text{amount KMnO}_4 = 45 \times 0.01145 = 0.515\ \text{mmol KMnO}_4$$

The amount of H_2O_2 consumed by 0.401 mmol of $KMnO_4$ is

$$\text{amount H}_2\text{O}_2 = 0.401\ \cancel{\text{mmol KMnO}_4} \times \frac{5\ \text{mmol H}_2\text{O}_2}{2\ \cancel{\text{mmol KMnO}_4}} = 1.00\ \text{mmol H}_2\text{O}_2$$

and

$$\text{amount H}_2\text{O}_2 = 0.515 \times \frac{5}{2} = 1.29\ \text{mmol H}_2\text{O}_2$$

We, therefore, need to take samples that contain from 1.00 to 1.29 mmol H_2O_2.

$$\text{mass sample} = 1.00 \cancel{\text{mmol } H_2O_2} \times 0.03401 \frac{\text{g } H_2O_2}{\cancel{\text{mmol } H_2O_2}} \times \frac{100 \text{ g sample}}{3 \text{ g } H_2O_2}$$
$$= 1.1 \text{ g sample}$$

to

$$\text{mass sample} = 1.29 \times 0.03401 \times \frac{100}{3} = 1.5 \text{ g sample}$$

Thus, our samples should weigh between 1.1 and 1.5 g. These should be diluted to perhaps 75 to 100 mL with water and made slightly acidic with dilute H_2SO_4 before titration.

FEATURE 20-2

Antioxidants[5]

Oxidation can have deleterious effects on the cells and tissues of the human body. There is a considerable body of evidence that reactive oxygen and nitrogen species, such as superoxide ion O_2^-, hydroxyl radical OH·, peroxyl radicals RO_2·, alkoxyl radicals RO·, nitric oxide NO·, and nitrogen dioxide NO_2·, damage cells and other body components. A group of compounds known as **antioxidants** can help counteract the influence of reactive oxygen and nitrogen species. Antioxidants are reducing agents that are so easily oxidized that they can protect other compounds in the body from oxidation. Typical antioxidants include vitamins A, C, and E; minerals such as selenium; and herbs such as ginkgo, rosemary, and milk thistle.

Several mechanisms of antioxidant action have been proposed. The presence of antioxidants may result in the decreased formation of the reactive oxygen and nitrogen species in the first place. Antioxidants may also scavenge the reactive species or their precursors. Vitamin E is an example of this latter behavior in its inhibition of lipid oxidation by reaction with radical intermediates generated from polyunsaturated fatty acids. Some antioxidants can bind the metal ions needed to catalyze the formation of the reactive oxidants. Other antioxidants can repair oxidative damage to biomolecules or can influence enzymes that catalyze repair mechanisms.

Vitamin E, or α-tocopherol, is thought to deter atherosclerosis, accelerate wound healing, and protect lung tissue from inhaled pollutants. It may also reduce the risk of heart disease and prevent premature skin aging. Researchers suspect that vitamin E has several other beneficial effects, ranging from alleviating rheumatoid arthritis to preventing cataracts. Most of us get enough vitamin E through our diet and do not require supplements. Dark-green leafy vegetables, nuts, vegetable oils, seafood, eggs, and avocados are food sources rich in vitamin E.

(continued)

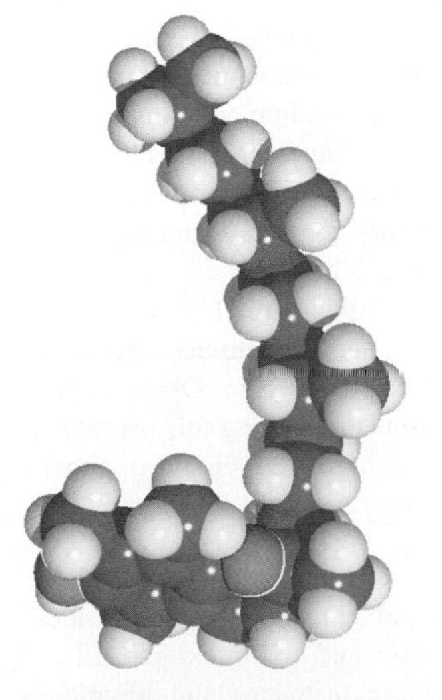

Molecular model of vitamin E.

[5]See B. Halliwell, *Nutr. Rev.*, **1997**, *55*, S44.

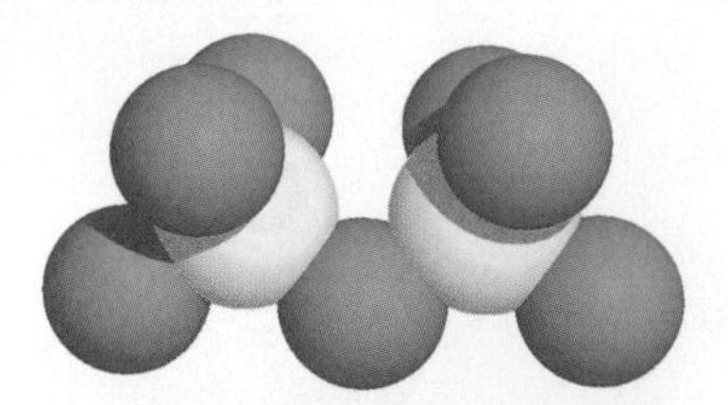

Molecular model of dichromate ion. For many years, dichromate in the form of its ammonium, potassium, or sodium salts was used in nearly all areas of chemistry as a powerful oxidizing agent. In addition to its use as a primary standard in analytical chemistry, it has been used as an oxidizing agent in synthetic organic chemistry; as a pigment in the paint, dye, and photographic industries; as a bleaching agent; and as a corrosion inhibitor. Chromic acid solution made from sodium dichromate and sulfuric acid was once the reagent of choice for thorough cleaning of glassware. Dichromate has been used as the analytical reagent in the alcohol Breathalyzer®, but in recent years these devices have largely been replaced by analyzers based on the absorption of infrared radiation. Early color photography used the colors produced by chromium compounds in the so-called gum bichromate process, but this has been replaced by silver bromide–based processes. The use of chromium compounds in general and dichromate in particular has decreased over the past decade because of the discovery that chromium compounds are carcinogenic. In spite of this, many millions of pounds of chromium compounds are manufactured and consumed by industry each year. Before using dichromate in laboratory work, read the MSDS for potassium dichromate (http://msds.pdc.cornell.edu/) or explore its chemical, toxicological, and carcinogenic properties. Observe all precautions in handling this useful but potentially hazardous chemical either in the solid form or in solution.

► Standard solutions of $K_2Cr_2O_7$ have the great advantage that they are indefinitely stable and do not oxidize HCl. Furthermore, primary-standard grade is inexpensive and readily available commercially.

Selenium has antioxidant effects that complement those of vitamin E. Selenium is a required constituent of several enzymes that remove reactive oxidants. The metal may support the immune function and neutralize some heavy metal poisons. It may also aid in deterring heart disease and some cancers. Good sources of selenium in the diet are whole grains, asparagus, garlic, eggs, mushrooms, lean meats, and seafood. Usually, diet alone provides sufficient selenium for good health. Supplements should be taken only if prescribed by a doctor because high doses can be toxic.

20C-2 Potassium Dichromate

In its analytical applications, dichromate ion is reduced to green chromium(III) ion:

$$Cr_2O_7^{2-} + 14\,H^+ + 6e^- \rightleftharpoons 2\,Cr^{3+} + 7H_2O \qquad E^0 = 1.33\ V$$

Dichromate titrations are generally carried out in solutions that are about 1 M in hydrochloric or sulfuric acid. In these media, the formal potential for the half-reaction is 1.0 to 1.1 V.

Potassium dichromate solutions are indefinitely stable, can be boiled without decomposition, and do not react with hydrochloric acid. Moreover, primary-standard reagent is available commercially and at a modest cost. The disadvantages of potassium dichromate compared with cerium(IV) and permanganate ion are its lower electrode potential and the slowness of its reaction with certain reducing agents.

Preparing Dichromate Solutions

For most purposes, reagent-grade potassium dichromate is sufficiently pure to permit the direct preparation of standard solutions; the solid is simply dried at 150°C to 200°C before being weighed.

The orange color of a dichromate solution is not intense enough for use in end point detection. Diphenylamine sulfonic acid (see Table 19-2) is an excellent indicator for titrations with this reagent, however. The oxidized form of the indicator is violet, and its reduced form is essentially colorless; thus, the color change observed in a direct titration is from the green of chromium(III) to violet.

Applying Potassium Dichromate Solutions

The principal use of dichromate is the volumetric titration of iron(II) based on the reaction

$$Cr_2O_7^{2-} + 6\,Fe^{2+} + 14\,H^+ \rightarrow 2Cr^{3+} + 6\,Fe^{3+} + 7\,H_2O$$

Often, this titration is performed in the presence of moderate concentrations of hydrochloric acid.

The reaction of dichromate with iron(II) has been widely used for the indirect determination of a variety of oxidizing agents. In these applications, a measured excess of an iron(II) solution is added to an acidic solution of the analyte. The excess iron(II) is then back-titrated with standard potassium dichromate (see Section 20B-1). Standardization of the iron(II) solution by titration with the dichromate is performed concurrently with the determination because solutions of iron(II) tend to be air oxidized. This method has been applied to the determination of nitrate, chlorate, permanganate, and dichromate ions as well as organic peroxides and several other oxidizing agents.

EXAMPLE 20-6

A 5.00-mL sample of brandy was diluted to 1.000 L in a volumetric flask. The ethanol (C_2H_5OH) in a 25.00-mL aliquot of the diluted solution was distilled into 50.00 mL of 0.02000 M $K_2Cr_2O_7$ and oxidized to acetic acid with heating. The reaction is

$$3\,C_2H_5OH + 2\,Cr_2O_7^{2-} + 16H^+ \rightarrow 4Cr^{3+} + 3CH_3COOH + 11H_2O$$

After cooling, 20.00 mL of 0.1253 M Fe^{2+} was pipetted into the flask. The excess Fe^{2+} was then titrated with 7.46 mL of the standard $K_2Cr_2O_7$ to a diphenylamine sulfonic acid end point. Calculate the percent (w/v) C_2H_5OH (46.07 g/mol) in the brandy.

total amount $K_2Cr_2O_7$

$$= (50.00 + 7.46)\,\cancel{\text{mL K}_2\text{Cr}_2\text{O}_7} \times 0.02000\,\frac{\text{mmol K}_2\text{Cr}_2\text{O}_7}{\cancel{\text{mL K}_2\text{Cr}_2\text{O}_7}}$$

$$= 1.1492\ \text{mmol K}_2\text{Cr}_2\text{O}_7$$

amount $K_2Cr_2O_7$ consumed by Fe^{2+}

$$= 20.00\,\cancel{\text{mL Fe}^{2+}} \times 0.1253\,\frac{\text{mmol Fe}^{2+}}{\cancel{\text{mL Fe}^{2+}}} \times \frac{1\ \text{mmol K}_2\text{Cr}_2\text{O}_7}{6\ \text{mmol Fe}^{2+}}$$

$$= 0.41767\ \text{mmol K}_2\text{Cr}_2\text{O}_7$$

$$\text{amount K}_2\text{Cr}_2\text{O}_7\ \text{consumed by C}_2\text{H}_5\text{OH} = (1.1492 - 0.41767)\ \text{mmol K}_2\text{Cr}_2\text{O}_7$$

$$= 0.73153\ \text{mmol K}_2\text{Cr}_2\text{O}_7$$

mass C_2H_5OH

$$= 0.73153\,\cancel{\text{mmol K}_2\text{Cr}_2\text{O}_7} \times \frac{3\,\cancel{\text{mmol C}_2\text{H}_5\text{OH}}}{2\,\cancel{\text{mmol K}_2\text{Cr}_2\text{O}_7}} = 0.04607\,\frac{\text{g C}_2\text{H}_5\text{OH}}{\cancel{\text{mmol C}_2\text{H}_5\text{OH}}}$$

$$= 0.050552\ \text{g C}_2\text{H}_5\text{OH}$$

$$\text{percent C}_2\text{H}_5\text{OH} = \frac{0.050552\ \text{g C}_2\text{H}_5\text{OH}}{5.00\ \text{mL sample} \times 25.00\,\cancel{\text{mL}}/1000\,\cancel{\text{mL}}} \times 100\%$$

$$= 40.4\%\ \text{C}_2\text{H}_5\text{OH}$$

20C-3 Iodine

Iodine is a weak oxidizing agent used primarily for the determination of strong reductants. The most accurate description of the half-reaction for iodine in these applications is

$$I_3^- + 2e^- \rightleftharpoons 3I^- \qquad E^0 = 0.536 \text{ V}$$

where I_3^- is the triiodide ion.

Standard iodine solutions have relatively limited application compared with the other oxidants we have described because of their significantly smaller electrode potential. Occasionally, however, this low potential is advantageous because it imparts a degree of selectivity that makes possible the determination of strong reducing agents in the presence of weak ones. An important advantage of iodine is the availability of a sensitive and reversible indicator for the titrations. Iodine solutions lack stability, however, and must be restandardized regularly.

Properties of Iodine Solutions

► Solutions prepared by dissolving iodine in a concentrated solution of potassium iodide are properly called *triiodide solutions.* In practice, however, they are often termed *iodine solutions* because this terminology accounts for the stoichiometric behavior of these solutions ($I_2 + 2e^- \rightarrow 2I^-$).

Iodine is not very soluble in water (0.001 M). To obtain solutions having analytically useful concentrations of the element, iodine is ordinarily dissolved in moderately concentrated solutions of potassium iodide. In this medium, iodine is reasonably soluble as a consequence of the reaction

$$I_2(s) + I^- \rightleftharpoons I_3^- \qquad K = 7.1 \times 10^2$$

Iodine dissolves only very slowly in solutions of potassium iodide, particularly if the iodide concentration is low. To ensure complete solution, the iodine is always dissolved in a small volume of concentrated potassium iodide, care being taken to avoid dilution of the concentrated solution until the last trace of solid iodine has disappeared. Otherwise, the molarity of the diluted solution gradually increases with time. This problem can be avoided by filtering the solution through a sintered glass crucible before standardization.

Iodine solutions lack stability for several reasons, one being the volatility of the solute. Losses of iodine from an open vessel occur in a relatively short time even in the presence of an excess of iodide ion. In addition, iodine slowly attacks most organic materials. Consequently, cork or rubber stoppers are never used to close containers of the reagent, and precautions must be taken to protect standard solutions from contact with organic dusts and fumes.

Air oxidation of iodide ion also causes changes in the molarity of an iodine solution:

$$4I^- + O_2(g) + 4H^+ \rightarrow 2I_2 + 2H_2O$$

In contrast to the other effects, this reaction causes the molarity of the iodine to increase. Air oxidation is promoted by acids, heat, and light.

Standardizing and Applying Iodine Solutions

Iodine solutions can be standardized against anhydrous sodium thiosulfate or barium thiosulfate monohydrate, both of which are available commercially. The reaction between iodine and sodium thiosulfate is discussed in detail in Section 20B-2. Often, solutions of iodine are standardized against solutions of sodium thiosulfate that have in turn been standardized against potassium iodate or potassium dichromate (see Section 20B-2). Table 20-6 summarizes methods that use iodine as an oxidizing agent.

TABLE 20-6

Some Applications of Iodine Solutions

Substance Determined	Half-Reaction
As	$H_3AsO_3 + H_2O \rightleftharpoons H_3AsO_4 + 2H^+ + 2e^-$
Sb	$H_3SbO_3 + H_2O \rightleftharpoons H_3SbO_4 + 2H^+ + 2e^-$
Sn	$Sn^{2+} \rightleftharpoons Sn^{4+} + 2e^-$
H_2S	$H_2S \rightleftharpoons S(s) + 2H^+ + 2e^-$
SO_2	$SO_3^{2-} + H_2O \rightleftharpoons SO_4^{2-} + 2H^+ + 2e^-$
$S_2O_3^{2-}$	$2S_2O_3^{2-} \rightleftharpoons S_4O_6^{2-} + 2e^-$
N_2H_4	$N_2H_4 \rightleftharpoons N_2(g) + 4H^+ + 2e^-$
Ascorbic acid	$C_6H_8O_6 \rightleftharpoons C_6H_6O_6 + 2H^+ + 2e^-$

20C-4 Potassium Bromate as a Source of Bromine

Primary-standard potassium bromate is available from commercial sources and can be used directly to prepare standard solutions that are stable indefinitely. Direct titrations with potassium bromate are relatively few. Instead, the reagent is a convenient and widely used stable source of bromine.[6] In this application, an unmeasured excess of potassium bromide is added to an acidic solution of the analyte. On introduction of a measured volume of standard potassium bromate, a stoichiometric quantity of bromine is produced.

$$\underset{\text{standard solution}}{BrO_3^-} + \underset{\text{excess}}{5Br^-} + 6H^+ \rightarrow 3Br_2 + 3H_2O$$

◀ 1 mol $KBrO_3$ = 3 mol Br_2.

This indirect generation circumvents the problems associated with the use of standard bromine solutions, which lack stability.

The primary use of standard potassium bromate is the determination of organic compounds that react with bromine. Few of these reactions are rapid enough to make direct titration feasible. Instead, a measured excess of standard bromate is added to the solution that contains the sample plus an excess of potassium bromide. After acidification, the mixture is allowed to stand in a glass-stoppered vessel until the bromine/analyte reaction is judged complete. To determine the excess bromine, an excess of potassium iodide is introduced so that the following reaction occurs:

$$2I^- + Br_2 \rightarrow I_2 + 2Br^-$$

The liberated iodine is then titrated with standard sodium thiosulfate (see Equation 20-1).

Substitution Reactions

Bromine is incorporated into an organic molecule either by substitution or by addition. Halogen substitution involves the replacement of hydrogen in an aromatic ring with a halogen. Substitution methods have been successfully applied to the determination of aromatic compounds that contain strong ortho-para-directing groups, particularly amines and phenols.

[6]For a discussion of bromate solutions and their applications, see M. R. F. Ashworth, *Titrimetric Organic Analysis,* Part I, pp. 118–130. New York: Interscience, 1964.

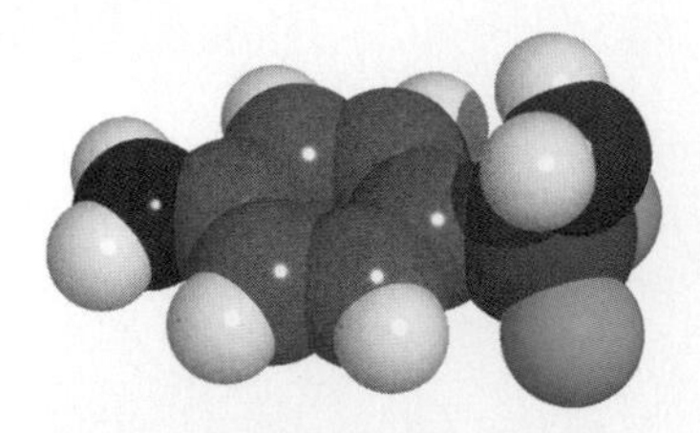

Molecular model of sulfanilamide. In the 1930s, sulfanilamide was found to be an effective antibacterial agent. In an effort to provide a solution of the drug that could be conveniently administered to patients, drug companies distributed sulfanilamide elixir containing a high concentration of ethylene glycol, which is toxic to the kidneys. Consequently, more than 100 people died from the effects of the solvent. This event led to the rapid passage of the 1938 Federal Food, Drug, and Cosmetic Act, which required toxicity testing prior to marketing and listing of active ingredients on product labels. For more information on the history of drug laws, see http://www.fda.gov/fdac/special/newdrug/benlaw.html.

EXAMPLE 20-7

A 0.2981-g sample of an antibiotic powder was dissolved in HCl and the solution diluted to 100.0 mL. A 20.00-mL aliquot was transferred to a flask, followed by 25.00 mL of 0.01767 M $KBrO_3$. An excess of KBr was added to form Br_2, and the flask was stoppered. After 10 min, during which time the Br_2 brominated the sulfanilamide, an excess of KI was added. The liberated iodine titrated with 12.92 mL of 0.1215 M sodium thiosulfate. The reactions are

$$BrO_3^- + 5Br^- + 6H^+ \rightarrow 3Br_2 + 3H_2O$$

NH_2 | ring | SO_2NH_2 (sulfanilamide) $+ 2Br_2 \longrightarrow$ NH_2 | Br, ring, Br | SO_2NH_2 $+ 2H^+ + 2Br^-$

sulfanilamide

$$Br_2 + 2I^- \rightarrow 2Br^- + I_2 \qquad \text{(excess KI)}$$

$$I_2 + 2S_2O_3^{2-} \rightarrow 2S_4O_6^{2-} + 2I^-$$

Calculate the percent sulfanilamide ($NH_2C_6H_4SO_2NH_2$, 172.21 g/mol) in the powder.

$$\text{total amount } Br_2 = 25.00\ \cancel{\text{mL } KBrO_3} \times 0.01767\ \frac{\cancel{\text{mmol } KBrO_3}}{\cancel{\text{mL } KBrO_3}} \times \frac{3 \text{ mmol } Br_2}{\cancel{\text{mmol } KBrO_3}}$$
$$= 1.32525 \text{ mmol } Br_2$$

We next calculate how much Br_2 was in excess over that required to brominate the analyte:

$$\text{amount excess } Br_2 = \text{amount } I_2$$
$$= 12.92\ \cancel{\text{mL } Na_2S_2O_3} \times 0.1215\ \frac{\cancel{\text{mmol } Na_2S_2O_3}}{\cancel{\text{mL } Na_2S_2O_3}} \times \frac{1 \text{ mmol } I_2}{2\ \cancel{\text{mmol } Na_2S_2O_3}}$$
$$= 0.78489 \text{ mmol } Br_2$$

The amount of Br_2 consumed by the sample is given by

$$\text{amount } Br_2 = 1.32525 - 0.78489 = 0.54036 \text{ mmol } Br_2$$

$$\text{mass analyte} = 0.54036\ \cancel{\text{mmol Br}_2} \times \frac{1\ \cancel{\text{mmol analyte}}}{2\ \cancel{\text{mmol Br}_2}} \times 0.17221\ \frac{\text{g analyte}}{\cancel{\text{mmol analyte}}}$$

$$= 0.046528\ \text{g analyte}$$

$$\text{percent analyte} = \frac{0.046528\ \cancel{\text{g}}\ \text{analyte}}{0.2891\ \cancel{\text{g}}\ \text{sample} \times 20.00\ \cancel{\text{mL}}/100\ \cancel{\text{mL}}} \times 100\%$$

$$= 80.47\%\ \text{sulfanilamide}$$

An important example of the use of a bromine substitution reaction is the determination of 8-hydroxyquinoline:

$$\underset{\text{(N, OH)}}{\text{8-hydroxyquinoline}} + Br_2 \longrightarrow \underset{\text{(N, OH, Br, Br)}}{\text{product}} + 2HBr$$

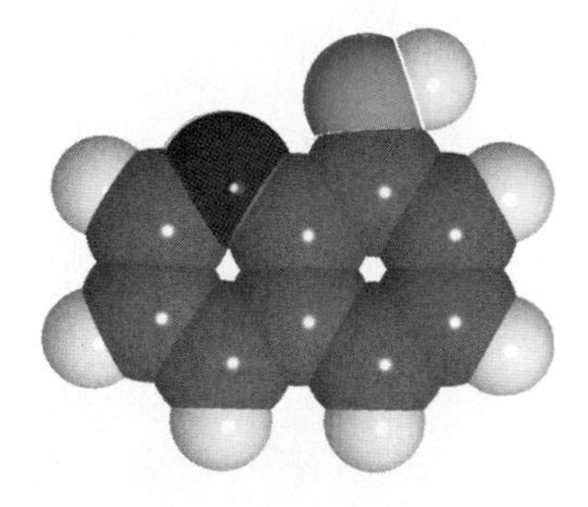

Molecular model of 8-hydroxyquinoline.

In contrast to most bromine substitutions, this reaction takes place rapidly enough in hydrochloric acid solution to make direct titration feasible. The titration of 8-hydroxyquinoline with bromine is particularly significant because the former is an excellent precipitating reagent for cations (see Section 12C-3). For example, aluminum can be determined according to the sequence

$$Al^{3+} + 3HOC_9H_6N \xrightarrow{\text{pH 4–9}} Al(OC_9H_6N)_3(s) + 3H^+$$

$$Al(OC_9H_6N)_3(s) \xrightarrow{\text{hot 4 M HCl}} 3HOC_9H_6N + Al^{3+}$$

$$3HOC_9H_6N + 6Br_2 \longrightarrow 3HOC_9H_4NBr_2 + 6HBr$$

The stoichiometric relationships in this case are

$$1\ \text{mol}\ Al^{3+} = 3\ \text{mol}\ HOC_9H_6N = 6\ \text{mol}\ Br_2 = 2\ \text{mol}\ KBrO_3$$

Addition Reactions

Addition reactions involve the opening of an olefinic double bond. For example, 1 mol of ethylene reacts with 1 mol of bromine in the reaction

$$H\text{—}\overset{H}{C}\text{=}\overset{H}{C}\text{—}H + Br_2 \longrightarrow H\text{—}\underset{Br}{\overset{H}{C}}\text{—}\underset{Br}{\overset{H}{C}}\text{—}H$$

The literature contains numerous references to the use of bromine for the estimation of olefinic unsaturation in fats, oils, and petroleum products. A method for the determination of ascorbic acid in vitamin C tablets is given in Section 37I-3.

20C-5 Determining Water with the Karl Fischer Reagent

In industry and commerce, one of the most widely used analytical methods is the Karl Fischer titration procedure for the determination of water in various types of solids and organic liquids. This important titrimetric method is based on an oxidation/reduction that is relatively specific for water.[7]

Describing the Reaction Stoichiometry

The Karl Fischer reaction is based on the oxidation of sulfur dioxide by iodine. In a solvent that is neither acidic nor basic—an aprotic solvent—the reaction can be summarized by

$$I_2 + SO_2 + 2H_2O \rightarrow 2HI + H_2SO_4$$

In this reaction, two moles of water are consumed for each mole of iodine. The stoichiometry, however, can vary from 2:1 to 1:1 depending on the presence of acids and bases in the solution.

Classical Chemistry To stabilize the stoichiometry and shift the equilibrium further to the right, Fischer added pyridine (C_5H_5N) and used anhydrous methanol as the solvent. A large excess of pyridine was used to complex the I_2 and SO_2. The classical reaction has been shown to occur in two steps. In the first step, I_2 and SO_2 react in the presence of pyridine and water to form pyridinium sulfite and pyridinium iodide.

$$C_5H_5N \cdot I_2 + C_5H_5N \cdot SO_2 + C_5H_5N + H_2O \rightarrow 2C_5H_5N \cdot HI + C_5H_5N \cdot SO_3 \qquad (20\text{-}2)$$

$$C_5H_5N^+ \cdot SO_3^- + CH_3OH \rightarrow C_5H_5N(H)SO_4CH_3 \qquad (20\text{-}3)$$

where I_2, SO_2, and SO_3 are shown as complexed by the pyridine. This second step is important because the pyridinium sulfite can also consume water.

$$C_5H_5N^+ \cdot SO_3^- + H_2O \rightarrow C_5H_5NH^+SO_4H^- \qquad (20\text{-}4)$$

This last reaction is undesirable because it is not as specific for water. It can be prevented completely by having a large excess of methanol present. Note that the stoichiometry is one mole of I_2 per mole of H_2O present.

For volumetric analysis, the classical Karl Fischer reagent consists of I_2, SO_2, pyridine, and anhydrous methanol or another suitable solvent. The reagent decomposes on standing and must be standardized often. Stabilized Karl Fischer reagents are available commercially from several suppliers. For ketones and aldehydes, spe-

[7]For a review of the composition and uses of the Karl Fischer reagent, see S. K. MacLeod, *Anal. Chem.*, **1991,** *63,* 557A; J. D. Mitchell Jr. and D. M. Smith, *Aquametry,* 2nd ed., Vol. 3. New York: Wiley, 1977.

cially formulated reagents are available from commercial sources. For coulometric methods (see Chapter 22), the Karl Fischer reagent contains KI instead of I_2 since, as we will see, the I_2 is generated electrochemically.

Pyridine-Free Chemistry In recent years, pyridine, and its objectionable odor, have been replaced in the Karl Fischer reagent by other amines, particularly imidazole, shown in the margin. These pyridine-free reagents are available commercially for both volumetric and coulometric Karl Fischer procedures. More detailed studies of the reaction have been reported.[8] The reaction is now thought to occur as follows:

(1) Solvolysis $2ROH + SO_2 \rightleftharpoons RSO_3^- + ROH_2^+$

(2) Buffering $B + RSO_3^- + ROH_2^+ \rightleftharpoons BH^+SO_3R^- + ROH$

(3) Redox $B \cdot I_2 + BH^+SO_3R^- + B + H_2O \rightleftharpoons BH^+SO_4R^- + 2BH^+I^-$

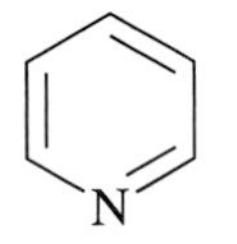

pyridine

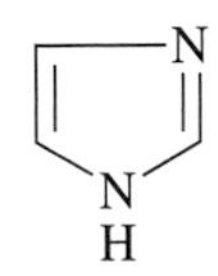

imidazole

Note that the stoichiometry is again one mole of I_2 consumed for each mole of H_2O present in the sample.

Interfering Reactions Several reactions that cause interferences in the Karl Fischer titration can occur. These undesired reactions can cause results to be too high, too low, or just imprecise. Oxidation of iodide in the coulometric reagent by reducing agents such as Cu(II), Fe(III), nitrite, Br_2, Cl_2, or quinones produces I_2, which can react with H_2O and cause low results because not as much generated I_2 is needed. The carbonyl groups on aldehydes and ketones can react with SO_2 and H_2O to form bisulfite complexes. Since this reaction consumes water, the titration results are again too low. Substitution of a weaker base like pyridine for imidazole can lessen the problem.

The iodine generated coulometrically or present in the reagent can be reduced by oxidizable species such as ascorbic acid, ammonia, thiols, Tl^+, Sn^{2+}, In^+, hydroxyl amines, and thiosulfite. This results in consumption of I_2 and water determinations that are too high. Phenolic derivatives and bicarbonates also cause reduction of I_2.

Some interfering compounds react to produce water, which causes the water results to be too high. Carboxylic acids can react with alcohols to produce an ester and water. To minimize this problem, the alcohol can be eliminated in the reagent, or an alcohol that reacts at a slower rate than methanol can be used. The pH of the reagent can be increased because the formation of esters is usually acid catalyzed. Ketones and aldehydes can react with alcoholic solvents to form ketals and acetals with the production of water according to:

$$R_2C{=}O + 2CH_3OH \rightarrow R_2C(OCH_3)_2 + H_2O$$

Aromatic ketones are less reactive than aliphatic ketones; aldehydes are much more reactive than ketones. Some commercial reagent preparations have been formulated to minimize this problem by using alcohols that react slowly and by employing higher pH.

Silanols and cyclic siloxanes also can react with alcohols to produce ethers and water. Some metal oxides, hydroxides, and carbonates can react with HI to produce

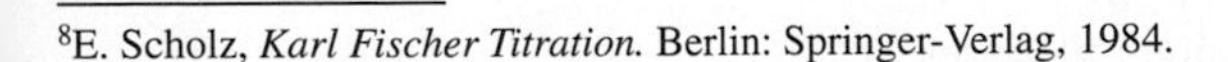
[8] E. Scholz, *Karl Fischer Titration.* Berlin: Springer-Verlag, 1984.

water. All of these increase the amount of I_2 consumed and produce results that are too high.

Detecting the End Point

An end point in a Karl Fischer titration can be observed visually based on the brown color of the excess reagent. More commonly, however, end points are obtained from electroanalytical measurements. Several instrument manufacturers offer automatic or semiautomatic instruments for performing Karl Fischer titrations. All of these are based on electrometric end point detection. The details of operation of Karl Fischer titrators are discussed in Chapter 22.

Reagent Properties

Karl Fischer reagent decomposes on standing. Because decomposition is particularly rapid immediately after preparation, it is common practice to prepare the reagent a day or two before it is to be used. Ordinarily, its strength must be established at least daily against a standard solution of water in methanol. A proprietary commercial Karl Fischer reagent reported to require only occasional restandardization is now available.

It is obvious that great care must be exercised to keep atmospheric moisture from contaminating the Karl Fischer reagent and the sample. All glassware must be carefully dried before use, and the standard solution must be stored out of contact with air. It is also necessary to minimize contact between the atmosphere and the solution during the titration.

Applications

Karl Fischer reagent has been applied to the determination of water in numerous types of samples. There are several variations of the basic technique, depending on the solubility of the material, the state in which the water is retained, and the physical state of the sample. If the sample can be dissolved completely in methanol, a direct and rapid titration is usually feasible. This method has been applied to the determination of water in many organic acids, alcohols, esters, ethers, anhydrides, and halides. The hydrated salts of most organic acids, as well as the hydrates of a number of inorganic salts that are soluble in methanol, can also be determined by direct titration.

Direct titration of samples that are only partially dissolved in the reagent usually leads to incomplete recovery of the water. Satisfactory results with this type of sample are often obtained, however, by the addition of excess reagent and back-titration with a standard solution of water in methanol after a suitable reaction time. An effective alternative is to extract the water from the sample by refluxing with anhydrous methanol or other organic solvents. The resulting solution is then titrated directly with the Karl Fischer solution.

WEB WORKS

Go to **http://chemistry.brookscole.com/skoogfac/.** From the Chapter Resources menu, choose Web Works. Locate the Chapter 20 section, and click on the link to the Cornell University MSDS Web site. Locate and read the MSDS for potassium dichromate and explore its chemical, toxicological, and carcinogenic properties. What are the usual signs and symptoms of overexposure? What first aid procedures are suggested?

QUESTIONS AND PROBLEMS

***20-1.** Write balanced net ionic equations to describe
(a) the oxidation of Mn^{2+} to MnO_4^- by ammonium peroxydisulfate.
(b) the oxidation of Ce^{3+} to Ce^{4+} by sodium bismuthate.
(c) the oxidation of U^{4+} to UO_2^{2+} by H_2O_2.
(d) the reaction of $V(OH)_4^+$ in a Walden reductor.
(e) the titration of H_2O_2 with $KMnO_4$.
(f) the reaction between KI and ClO_3^- in acidic solution.

20-2. Write balanced net ionic equations to describe
(a) the reduction of Fe^{3+} to Fe^{2+} by SO_2.
(b) the reaction of H_2MoO_4 in a Jones reductor.
(c) the oxidation of HNO_2 by a solution of MnO_4^-.
(d) the reaction of aniline ($C_6H_4NH_2$) with a mixture of $KBrO_3$ and KBr in acidic solution.
(e) the air oxidation of $HAsO_3^{2-}$ to $HAsO_4^{2-}$.
(f) the reaction of KI with HNO_2 in acidic solution.

***20-3.** Why is a Walden reductor always used with solutions that contain appreciable concentrations of HCl?

20-4. Why is zinc amalgam preferable to pure zinc in a Jones reductor?

***20-5.** Write a balanced net ionic equation for the reduction of UO_2^{2+} in a Walden reductor.

20-6. Write a balanced net ionic equation for the reduction of TiO^{2+} in a Jones reductor.

***20-7.** Why are standard solutions of reductants less often used for titrations than standard solutions of oxidants?

***20-8.** Why are standard $KMnO_4$ solutions seldom used for the titration of solutions containing HCl?

20-9. Why are Ce^{4+} solutions never used for the titration of reductants in basic solutions?

***20-10.** Write a balanced net ionic equation showing why $KMnO_4$ end points fade.

20-11. Why are $KMnO_4$ solutions filtered before they are standardized?

20-12. Why are solutions of $KMnO_4$ and $Na_2S_2O_3$ generally stored in dark reagent bottles?

***20-13.** When a solution of $KMnO_4$ was left standing in a buret for 3 hours, a brownish ring formed at the surface of the liquid. Write a balanced net ionic equation to account for this observation.

20-14. What is the primary use of standard $K_2Cr_2O_7$ solutions?

***20-15.** Why are iodine solutions prepared by dissolving I_2 in concentrated KI?

20-16. A standard solution of I_2 increased in molarity with standing. Write a balanced net ionic equation that accounts for the increase.

***20-17.** When a solution of $Na_2S_2O_3$ is introduced into a solution of HCl, a cloudiness develops almost immediately. Write a balanced net ionic equation explaining this phenomenon.

20-18. Suggest a way in which a solution of KIO_3 could be used as a source of known quantities of I_2.

***20-19.** Write balanced equations showing how $KBrO_3$ could be used as a primary standard for solutions of $Na_2S_2O_3$.

20-20. Write balanced equations showing how $K_2Cr_2O_7$ could be used as a primary standard for solutions of $Na_2S_2O_3$.

***20-21.** Write a balanced net ionic equation describing the titration of hydrazine (N_2H_4) with standard iodine.

20-22. In the titration of I_2 solutions with $Na_2S_2O_3$, starch indicator is never added until just before chemical equivalence. Why?

20-23. A solution prepared by dissolving a 0.2256-g sample of electrolytic iron wire in acid was passed through a Jones reductor. The iron(II) in the resulting solution required a 35.37-mL titration. Calculate the molar oxidant concentration if the titrant used was
*(a) Ce^{4+} (product: Ce^{3+}).
(b) $Cr_2O_7^{2-}$ (product: Cr^{3+}).
*(c) MnO_4^- (product: Mn^{2+}).
(d) $V(OH)_4^+$ (product: VO^{2+}).
*(e) IO_3^- (product: ICl_2^-).

***20-24.** How would you prepare 500.0 mL of 0.02500 M $K_2Cr_2O_7$?

20-25. How would you prepare 2.000 L of 0.02500 M $KBrO_3$?

***20-26.** How would you prepare 2.0 L of approximately 0.0500 M $KMnO_4$?

20-27. How would you prepare 2.0 L of approximately 0.05 M I_3^-?

***20-28.** Titration of 0.1756 g of primary-standard $Na_2C_2O_4$ required 32.04 mL of a potassium permanganate solution. Calculate the molar concentration of $KMnO_4$ in this solution.

20-29. A 0.1809-g sample of pure iron wire was dissolved in acid, reduced to the +2 state, and titrated with 31.33 mL of cerium(IV). Calculate the molar concentration of the Ce^{4+} solution.

***20-30.** The iodine produced when an excess of KI was added to a solution containing 0.1259 g of $K_2Cr_2O_7$ required a 41.26-mL titration with

$Na_2S_2O_3$. Calculate the molar concentration of the thiosulfate solution.

20-31. A 0.1017-g sample of $KBrO_3$ was dissolved in dilute HCl and treated with an unmeasured excess of KI. The liberated iodine required 39.75 mL of a sodium thiosulfate solution. Calculate the molar concentration of the $Na_2S_2O_3$.

***20-32.** The Sb(III) in a 0.978-g ore sample required a 44.87-mL titration with 0.02870 M I_2 [reaction product: Sb(V)]. Express the results of this analysis in terms of (a) percentage of Sb and (b) percentage of stibnite (Sb_2S_3).

20-33. Calculate the percentage of MnO_2 in a mineral specimen if the I_2 liberated by a 0.1344-g sample in the net reaction

$$MnO_2(s) + 4H^+ + 2I^- \rightarrow Mn^{2+} + I_2 + 2H_2O$$

required 32.30 mL of 0.07220 M $Na_2S_2O_3$.

***20-34.** Under suitable conditions, thiourea is oxidized to sulfate by solutions of bromate

$$3CS(NH_2)_2 + 4BrO_3^- + 3H_2O \rightleftharpoons 3CO(NH_2)_2 + 3SO_4^{2-} + 4Br^- + 6H^+$$

A 0.0715-g sample of a material was found to consume 14.1 mL of 0.00833 M $KBrO_3$. What was the percent purity of the thiourea sample?

***20-35.** A 0.7120-g specimen of iron ore was brought into solution and passed through a Jones reductor. Titration of the Fe(II) produced required 39.21 mL of 0.02086 M $KMnO_4$. Express the results of this analysis in terms of (a) percent Fe and (b) percent Fe_2O_3.

20-36. The Sn in a 0.4352-g mineral specimen was reduced to the +2 state with Pb and titrated with 29.77 mL of 0.01735 M $K_2Cr_2O_7$. Calculate the results of this analysis in terms of (a) percent Sn and (b) percent SnO_2.

***20-37.** Treatment of hydroxylamine (H_2NOH) with an excess of Fe(II) results in the formation of N_2O and an equivalent amount of Fe(II):

$$2H_2NOH + 4Fe^{3+} \rightarrow N_2O(g) + 4Fe^{2+} + 4H^+ + H_2O$$

Calculate the molar concentration of an H_2NOH solution if the Fe(II) produced by treatment of a 50.00-mL aliquot required 19.83 mL of 0.0325 M $K_2Cr_2O_7$.

20-38. The organic matter in a 0.9280-g sample of burn ointment was eliminated by ashing, following which the solid residue of ZnO was dissolved in acid. Treatment with $(NH_4)_2C_2O_4$ resulted in the formation of the sparingly soluble ZnC_2O_4. The solid was filtered, washed, and then redissolved in dilute acid. The liberated $H_2C_2O_4$ required 37.81 mL of 0.01508 M $KMnO_4$. Calculate the percentage of ZnO in the medication.

***20-39.** The $KClO_3$ in a 0.1279-g sample of an explosive was determined by reaction with 50.00 mL of 0.08930 M Fe^{2+}:

$$ClO_3^- + 6Fe^{2+} + 6H^+ \rightarrow Cl^- + 3H_2O + 6Fe^{3+}$$

When the reaction was complete, the excess Fe^{2+} was back-titrated with 14.93 mL of 0.083610 M Ce^{4+}. Calculate the percentage of $KClO_3$ in the sample.

20-40. The tetraethyl lead [$Pb(C_2H_5)_4$] in a 25.00-mL sample of aviation gasoline was shaken with 15.00 mL of 0.02095 M I_2. The reaction is

$$Pb(C_2H_5)_4 + I_2 \rightarrow Pb(C_2H_5)_3I + C_2H_5I$$

After the reaction was complete, the unused I_2 was titrated with 6.09 mL of 0.03465 M $Na_2S_2O_3$. Calculate the weight (in milligrams) of $Pb(C_2H_5)_4$ (323.4 g/mol) in each liter of the gasoline.

***20-41.** A 7.41-g sample of an ant-control preparation was decomposed by wet-ashing with H_2SO_4 and HNO_3. The As in the residue was reduced to the trivalent state with hydrazine. After removal of the excess reducing agent, the As(III) required a 24.56-mL titration with 0.01985 M I_2 in a faintly alkaline medium. Express the results of this analysis in terms of percentage of As_2O_3 in the original sample.

20-42. A sample of alkali metal chlorides was analyzed for sodium by dissolving a 0.800-g sample in water and diluting to exactly 500 mL. A 25.0-mL aliquot of this was treated in such a way as to precipitate the sodium as $NaZn(UO_2)_3(OAc)_9 \cdot 6H_2O$. The precipitate was filtered, dissolved in acid, and passed through a lead reductor, which converted the uranium to U^{4+}. Oxidation of this to UO_2^{2+} required 19.9 mL of 0.100 M $K_2Cr_2O_7$. Calculate the percent NaCl in the sample.

***20-43.** The ethyl mercaptan concentration in a mixture was determined by shaking a 1.534-g sample with 50.0 mL of 0.01293 M I_2 in a tightly stoppered flask:

$$2C_2H_5SH + I_2 \rightarrow C_2H_5SSC_2H_5 + 2I^- + 2H^+$$

The excess I_2 was back-titrated with 15.72 mL of 0.01425 M $Na_2S_2O_3$. Calculate the percentage of C_2H_5SH (62.13 g/mol).

20-44. A 4.971-g sample containing the mineral tellurite was dissolved and then treated with 50.00 mL of 0.03114 M $K_2Cr_2O_7$:

$$3TeO_2 + Cr_2O_7^{2-} + 8H^+ \rightarrow 3H_2TeO_4 + 2Cr^{3+} + H_2O$$

When the reaction is complete, the excess $Cr_2O_7^{2-}$ required a 10.05-mL back-titration with 0.1135 M Fe^{2+}. Calculate the percentage of TeO_2 in the sample.

***20-45.** A sensitive method for I^- in the presence of Cl^- and Br^- entails oxidation of the I^- to IO_3^- with Br_2. The excess Br_2 is then removed by boiling or by reduction with formate ion. The IO_3^- produced is determined by addition of excess I^- and titration of the resulting I_2. A 1.309-g sample of mixed halides was dissolved and analyzed by the foregoing procedure; 19.96 mL of 0.05982 M thiosulfate was required. Calculate the percentage of KI in the sample.

***20-46.** A 1.065-g sample of stainless steel was dissolved in HCl (this treatment converts the Cr present to Cr^{3+}) and diluted to 500.0 mL in a volumetric flask. One 50.00-mL aliquot was passed through a Walden reductor and then titrated with 13.72 mL of 0.01920 M $KMnO_4$. A 100.0-mL aliquot was passed through a Jones reductor into 50 mL of 0.10 M Fe^{3+}. Titration of the resulting solution required 36.43 mL of the $KMnO_4$ solution. Calculate the percentages of Fe and Cr in the alloy.

20-47. A 2.559-g sample containing both Fe and V was dissolved under conditions that converted the elements to Fe(III) and V(V). The solution was diluted to 500.0 mL, and a 50.00-mL aliquot was passed through a Walden reductor and titrated with 17.74 mL of 0.1000 M Ce^{4+}. A second 50.00-mL aliquot was passed through a Jones reductor and required 44.67 mL of the same Ce^{4+} solution to reach an end point. Calculate the percentage of Fe_2O_3 and V_2O_5 in the sample.

***20-48.** A 25.0-mL aliquot of a solution containing Tl(I) ion was treated with K_2CrO_4. The Tl_2CrO_4 was filtered, washed free of excess precipitating agent, and dissolved in dilute H_2SO_4. The $Cr_2O_7^{2-}$ produced was titrated with 39.52 mL of 0.1044 M Fe^{2+} solution. What was the mass of Tl in the sample? The reactions are

$$2Tl^+ + CrO_4^{2-} \rightarrow Tl_2CrO_4(s)$$

$$2Tl_2CrO_4(s) + 2H^+ \rightarrow 4Tl^+ + Cr_2O_7^{2-} + H_2O$$

$$Cr_2O_7^{2-} + 6Fe^{2+} + 14H^+ \rightarrow 6Fe^{3+} + 2Cr^{3+} + 7H_2O$$

***20-49.** A gas mixture was passed at the rate of 2.50 L/min through a solution of sodium hydroxide for a total of 64.00 min. The SO_2 in the mixture was retained as sulfite ion

$$SO_2(g) + 2OH^- \rightarrow SO_3^{2-} + H_2O$$

After acidification with HCl, the sulfite was titrated with 4.98 mL of 0.003125 M KIO_3:

$$IO_3^- + 2H_2SO_3 + 2Cl^- \rightarrow ICl_2^- + SO_4^{2-} + 2H^+$$

Use 1.20 g/L for the density of the mixture and calculate the concentration of SO_2 in ppm.

20-50. A 24.7-L sample of air drawn from the vicinity of a coking oven was passed over iodine pentoxide at 150°C, where CO was converted to CO_2 and a chemically equivalent quantity of I_2 was produced:

$$I_2O_5(s) + 5CO(g) \rightarrow 5CO_2(g) + I_2(g)$$

The I_2 distilled at this temperature and was collected in a solution of KI. The I_3^- produced was titrated with 7.76 mL of 0.00221 M $Na_2S_2O_3$. Does the air in this space comply with federal regulations that mandate a maximum CO level no greater than 50 ppm?

***20-51.** A 30.00-L air sample was passed through an absorption tower containing a solution of Cd^{2+}, where H_2S was retained as CdS. The mixture was acidified and treated with 10.00 mL of 0.01070 M I_2. After the reaction

$$S^{2-} + I_2 \rightarrow S(s) + 2I^-$$

was complete, the excess iodine was titrated with 12.85 mL of 0.01344 M thiosulfate. Calculate the concentration of H_2S in ppm; use 1.20 g/L for the density of the gas stream.

20-52. A square of photographic film 2.0 cm on an edge was suspended in a 5% solution of $Na_2S_2O_3$ to dissolve the silver halides. After removal and washing of the film, the solution was treated with an excess of Br_2 to oxidize the iodide present to IO_3^- and destroy the excess thiosulfate ion. The solution was boiled to remove the bromine, and an excess of iodide was added. The liberated iodine was titrated with 13.7 mL of 0.0352 M thiosulfate solution.

(a) Write balanced equations for the reactions involved in the method.
(b) Calculate the mass in milligrams of AgI per square centimeter of film.

***20-53.** The Winkler method for dissolved oxygen in water is based on the rapid oxidation of solid $Mn(OH)_2$ to $Mn(OH)_3$ in alkaline medium. When acidified, the Mn(III) readily releases iodine from iodide. A 150-mL water sample, in a stoppered vessel, was treated with 1.00 mL of a concentrated solution of NaI and NaOH and 1.00 mL of a manganese(II) solution. Oxidation of the $Mn(OH)_2$ was complete in about 1 min. The precipitates were then dissolved by addition of 2.00 mL of concentrated H_2SO_4, whereupon an amount of iodine equivalent to the $Mn(OH)_3$ (and hence to the dissolved O_2) was liberated. A 25.0-mL aliquot (of the 254 mL) was titrated with 13.67 mL of 0.00942 M thiosulfate. Calculate the mass in milligrams O_2 per milliliter sample. (Assume that the concentrated reagents are O_2 free, and take their dilutions of the sample into account.)

20-54. Use a spreadsheet to do the calculations and plot the titration curves for the following titrations. Calculate potentials after the addition of titrant corresponding to 10%, 20%, 30%, 40%, 50%, 60%, 70%, 80%, 90%, 95%, 99%, 99.9%, 100%, 101%, 105%, 110%, and 120% of the equivalence-point volume.

(a) 25.00 mL of 0.025 M $SnCl_2$ with 0.050 M $FeCl_3$.
(b) 25.00 mL of 0.08467 M $Na_2S_2O_3$ with 0.10235 M I_2.
(c) 0.1250 g of primary-standard grade $Na_2C_2O_4$ with 0.01035 M $KMnO_4$.
(d) 20.00 mL of 0.1034 M Fe^{2+} with 0.01500 M $K_2Cr_2O_7$.
(e) 35.00 mL of 0.0578 M IO_3^- with 0.05362 M $Na_2S_2O_3$.

20-55. Challenge Problem. Verdini and Lagier[9] developed an iodimetric titration procedure for determining ascorbic acid in vegetables and fruits. They compared the results of their titration experiments with similar results from an HPLC method (see Chapter 32). The results of their comparison are shown in the following table.

Method Comparison*

Sample	HPLC, mg/100 g	Voltammetry, mg/100 g
1	138.6	140.0
2	126.6	120.6
3	138.3	140.9
4	126.2	123.7

*Ascorbic acid content determined in kiwi fruit samples by means of HPLC with UV detection and voltammetric titration.

(a) Find the mean and standard deviation of each set of data.
(b) Determine whether there is a difference in the variances of the two data sets at the 95% level.
(c) Determine whether the difference in the means is significant at the 95% level.

These workers also carried out a recovery study in which they determined ascorbic acid in samples, then spiked the samples with additional ascorbic acid and redetermined the mass of the analyte. Their results are shown in the following table.

Recovery Study

Sample	1	2	3	4
	Kiwi fruit			
Amounts				
Initial, mg	9.32	7.29	7.66	7.00
Added, mg	6.88	7.78	8.56	6.68
Found, mg	15.66	14.77	15.84	13.79
	Spinach			
Initial, mg	6.45	7.72	5.58	5.21
Added, mg	4.07	4.32	4.28	4.40
Found, mg	10.20	11.96	9.54	9.36

(d) Calculate the percent recovery for total ascorbic acid in each sample.
(e) Find the mean and standard deviation of the percent recovery, first for the kiwi fruit and then for the spinach.
(f) Determine whether the variances of the percent recovery between the kiwi fruit and the spinach are different at the 95% confidence level.
(g) Determine whether the difference in the percent recovery of ascorbic acid is significant at the 95% confidence level.
(h) Discuss how you would apply the iodimetric method for the determination of ascorbic acid

[9]R. A. Verdini, and C. M. Lagier, *J. Agric. Food Chem.*, **2000,** *48,* 2812.

to various samples of fruits and vegetables. In particular, comment on how you would apply the results of your analysis of the data to the analysis of new samples.

(i) References to several papers are listed on determining ascorbic acid using different analytical techniques. If the papers are available in your library, examine them and briefly describe the methods used in each.

(j) Comment on how each of the methods in (i) might be used and under what circumstances they might be chosen rather than iodimetry. For each method, including iodimetry, compare such factors as speed, convenience, cost of analysis, and quality of the resulting data.

References

A. Campiglio, *Analyst,* **1993,** *118,* 545.

L. Cassella, M. Gulloti, A. Marchesini, and M. Petrarulo, *J. Food Sci.,* **1989,** *54,* 374.

Z. Gao, A. Ivaska, T. Zha, G. Wang, P. Li, and Z. Zhao, *Talanta,* **1993,** *40,* 399.

O. W. Lau, K. K. Shiu, and S. T. Chang, *J. Sci. Food Agric.,* **1985,** *36,* 733.

A. Marchesini, F. Montuori, D. Muffato, and D. Maestri, *J. Food Sci.,* **1974,** *39,* 568.

T. Moeslinger, M. Brunner, I. Volf, and P. G. Spieckermann, *Gen. Clin. Chem.,* **1995,** *41,* 1177.

L. A. Pachla and P. T. Kissinger, *Anal. Chem.,* **1976,** *48,* 364.

InfoTrac College Edition

For additional readings, go to InfoTrac College Edition, your online research library, at

http://infotrac.thomsonlearning.com

CHAPTER 21

Potentiometry

The research vessel *Meteor*, shown in the photo, is owned by the Federal Republic of Germany through the Ministry of Research and Technology and is operated by the German Research Foundation. It is often used by a multinational group of chemical oceanographers to collect data in an effort to better understand the changing chemical composition of the earth's atmosphere and oceans. For example, during December 1992 and January 1993, Meteor sailed from Rio de Janeiro to Capetown, South Africa, while monitoring carbon dioxide and other important ocean concentrations, including total alkalinity of sea water. These scientists made shipboard measurements of total alkalinity by potentiometric titration, which is discussed in this chapter.

***Potentiometric methods** of analysis are based on measuring the potential of electrochemical cells without drawing appreciable current. For nearly a century, potentiometric techniques have been used to locate end points in titrations. In more recent methods, ion concentrations are measured directly from the potential of ion-selective membrane electrodes. These electrodes are relatively free from interferences and provide a rapid, convenient, and nondestructive means of quantitatively determining numerous important anions and cations.*[1]

Analysts make more potentiometric measurements than perhaps any other type of chemical instrumental measurement. The number of potentiometric measurements made on a daily basis is staggering. Manufacturers measure the pH of many consumer products; clinical laboratories determine blood gases as important indicators of disease states; industrial and municipal effluents are monitored continuously to determine pH and concentrations of pollutants; and oceanographers determine carbon dioxide and other related variables in sea water. Potentiometric measurements are also used in fundamental studies to determine thermodynamic equilibrium constants such as K_a, K_b, *and* K_{sp}. *These examples are but a few of the many thousands of applications of potentiometric measurements.*

The equipment for potentiometric methods is simple and inexpensive and includes a reference electrode, an indicator electrode, and a potential-measuring device. The principles of operation and design of each of these components are described in the initial sections of this chapter. Following these discussions, we investigate analytical applications of potentiometric measurements.

[1]R. S. Hutchins and L. G. Bachas, in *Handbook of Instrumental Techniques for Analytical Chemistry,* F. A. Settle, Ed., Chapter 38, pp. 727–748. Upper Saddle River, NJ: Prentice-Hall, 1997.

21A GENERAL PRINCIPLES

In Feature 18-3, we showed that absolute values for individual half-cell potentials cannot be determined in the laboratory. That is, only relative cell potentials can be measured experimentally. Figure 21-1 shows a typical cell for potentiometric analysis. This cell can be represented as

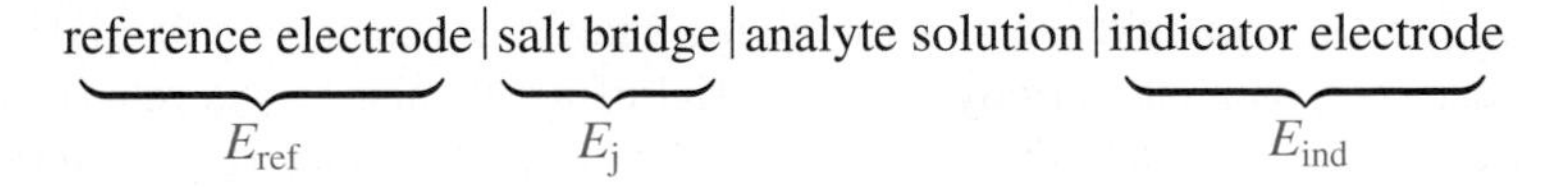

reference electrode | salt bridge | analyte solution | indicator electrode
E_{ref}
E_j
E_{ind}

The **reference electrode** in this diagram is a half-cell with an accurately known electrode potential, E_{ref}, that is independent of the concentration of the analyte or any other ions in the solution under study. It can be a standard hydrogen electrode but seldom is because a standard hydrogen electrode is somewhat troublesome to maintain and use. By convention, the reference electrode is always treated as the left-hand electrode in potentiometric measurements. The **indicator electrode,** which is immersed in a solution of the analyte, develops a potential, E_{ind}, that depends on the activity of the analyte. Most indicator electrodes used in potentiometry are selective in their responses. The third component of a potentiometric cell is a salt bridge that prevents the components of the analyte solution from mixing with those of the reference electrode. As noted in Chapter 18, a potential develops across the liquid junctions at each end of the salt bridge. These two potentials tend to cancel one another if the mobilities of the cation and the anion in the bridge solution are approximately the same. Potassium chloride is a nearly ideal electrolyte for the salt bridge because the mobilities of the K^+ ion and the Cl^- ion are nearly equal. The net potential across the salt bridge E_j is thereby reduced to a few millivolts or less. For most electroanalytical methods, the junction potential is small enough to be neglected. In the potentiometric methods discussed in this chapter, however, the junction potential and its uncertainty can be factors that limit the measurement accuracy and precision.

A **reference electrode** is a half-cell having a known electrode potential that remains constant at constant temperature and is independent of the composition of the analyte solution.

◀ Reference electrodes are *always* treated as the left-hand electrode in this text.

An **indicator electrode** has a potential that varies in a known way with variations in the concentration of an analyte.

◀ A hydrogen electrode is seldom used as a reference electrode for day-to-day potentiometric measurements because it is somewhat inconvenient to use and maintain and is also a fire hazard.

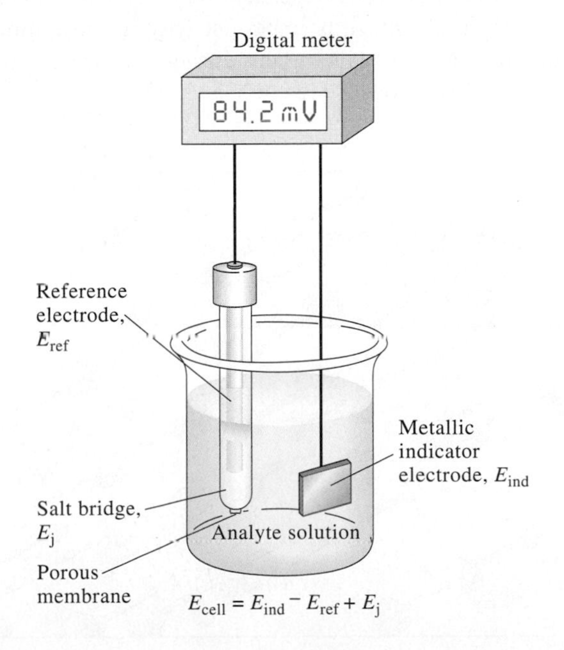

Figure 21-1 A cell for potentiometric determinations.

The potential of the cell we have just considered is given by the equation

$$E_{cell} = E_{ind} - E_{ref} + E_j \quad (21\text{-}1)$$

The first term in this equation, E_{ind}, contains the information that we are looking for—the concentration of the analyte. To make a potentiometric determination of an analyte, then, we must measure a cell potential, correct this potential for the reference and junction potentials, and compute the analyte concentration from the indicator electrode potential. Strictly, the potential of a galvanic cell is related to the activity of the analyte. Only through proper calibration of the electrode system with solutions of known concentration can we determine the concentration of the analyte.

In the sections that follow, we discuss the nature and origin of the three potentials shown on the right side of Equation 21-1.

21B REFERENCE ELECTRODES

The ideal reference electrode has a potential that is accurately known, constant, and completely insensitive to the composition of the analyte solution. In addition, this electrode should be rugged and easy to assemble and should maintain a constant potential while passing minimal currents.

21B-1 Calomel Reference Electrodes

A calomel electrode can be represented schematically as

$$Hg\,|\,Hg_2Cl_2(\text{sat'd}),\ KCl(x\ \text{M})\,\|$$

where x represents the molar concentration of potassium chloride in the solution. Concentrations of potassium chloride that are commonly used in calomel reference electrodes are 0.1 M, 1 M, and saturated (about 4.6 M). The saturated calomel electrode (SCE) is the most widely used because it is easily prepared. Its main disadvantage is that it is somewhat more dependent on temperature than the 0.1 M and 1 M electrodes. This disadvantage is important only in those rare circumstances when substantial temperature changes occur during a measurement. The potential of the saturated calomel electrode is 0.2444 V at 25°C.

► The "saturated" in a saturated calomel electrode refers to the KCl concentration and not the calomel concentration. All calomel electrodes are saturated with Hg_2Cl_2 (calomel).

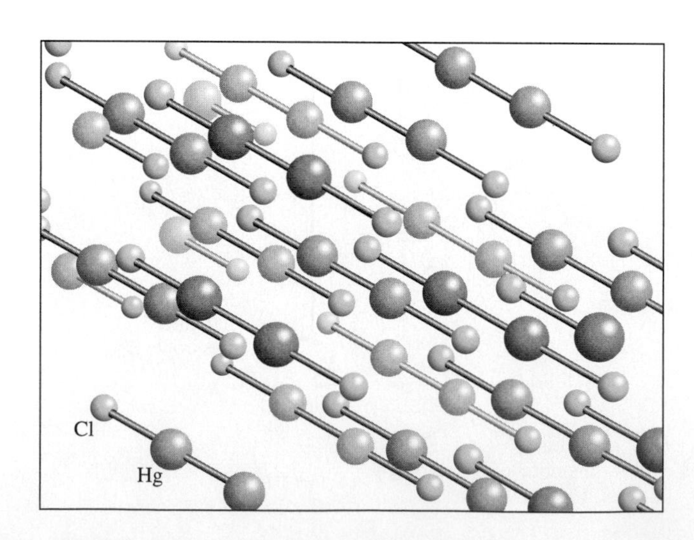

The crystal structure of calomel, Hg_2Cl_2, which has limited solubility in water ($K_{sp} = 1.8 \times 10^{-18}$ at 25°C). Notice the Hg-Hg bond in the structure. There is considerable evidence that a similar type of bonding occurs in aqueous solution, and so mercury(I) is represented as Hg_2^{2+}.

TABLE 21-1

Formal Electrode Potentials for Reference Electrodes as a Function of Composition and Temperature

	Potential vs. SHE, V				
Temperature, °C	0.1M Calomel*	3.5 M Calomel†	Sat'd Calomel*	3.5 M Ag/AgCl†	Sat'd Ag/AgCl†
12	0.3362		0.2528		
15	0.3362	0.254	0.2511	0.212	0.209
20	0.3359	0.252	0.2479	0.208	0.204
25	0.3356	0.250	0.2444	0.205	0.199
30	0.3351	0.248	0.2411	0.201	0.194
35	0.3344	0.246	0.2376	0.197	0.189

*From R. G. Bates, in *Treatise on Analytical Chemistry*, 2nd ed., I. M. Kolthoff and P. J. Elving, Eds., Part I, Vol. 1, p. 793. New York: Wiley, 1978.
†From D. T. Sawyer, A. Sobkowiak, and J. L. Roberts, Jr., *Experimental Electrochemistry for Chemists,* 2nd ed., p. 192. New York: Wiley, 1995.

The electrode reaction in calomel half-cells is

$$Hg_2Cl_2(s) + 2e^- \rightleftharpoons 2Hg(l) + 2Cl^-(aq)$$

Table 21-1 lists the compositions and formal electrode potentials for the three most common calomel electrodes. Note that the electrodes differ only in their potassium chloride concentrations; all are saturated with calomel (Hg_2Cl_2). Figure 21-2 illustrates a typical commercial saturated calomel electrode. It consists of a 5- to 15-cm-long tube that is 0.5 to 1.0 cm in diameter. A mercury/mercury(I) chloride paste in saturated potassium chloride is contained in an inner tube and connected to the saturated potassium chloride solution in the outer tube through a small opening. An inert metal electrode is immersed in the paste. Contact with the analyte solution is made through a fritted disk, a porous fiber, or a piece of porous Vycor ("thirsty glass") sealed in the end of the outer tubing.

Figure 21-3 shows a saturated calomel electrode that one can easily construct from materials available in most laboratories. A salt bridge (see Section 18B-2) provides electrical contact with the analyte solution.

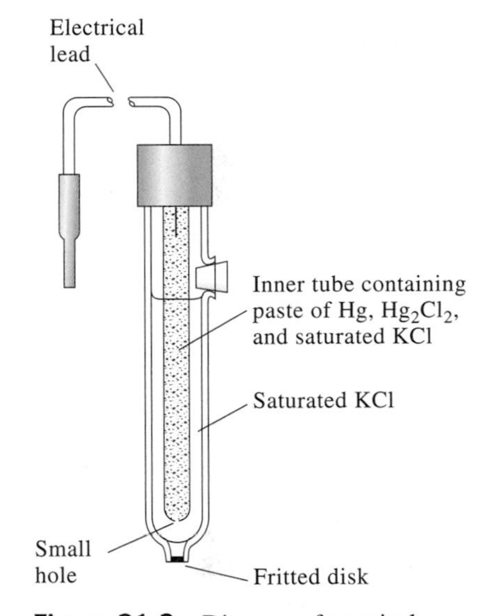

Figure 21-2 Diagram of a typical commercial saturated calomel electrode.

◀ A salt bridge is easily constructed by filling a U-tube with a conducting gel prepared by heating about 5 g of agar in 100 mL of an aqueous solution containing about 35 g of potassium chloride. When the liquid cools, it sets into a gel that is a good conductor but prevents the two solutions at the ends of the tube from mixing. If either of the ions in potassium chloride interfere with the measurement process, ammonium nitrate may be used as the electrolyte in salt bridges.

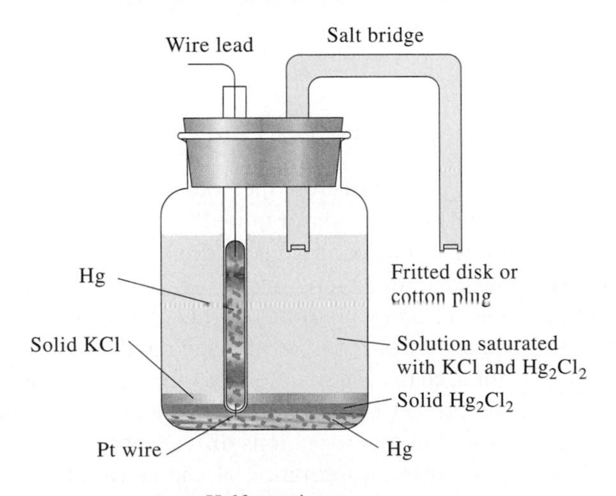

Figure 21-3 A saturated calomel electrode made from materials readily available in any laboratory.

◀ **Agar,** which is available as translucent flakes, is a heteropolysaccharide that is extracted from certain East Indian seaweed. Solutions of agar in hot water set to a gel when they are cooled.

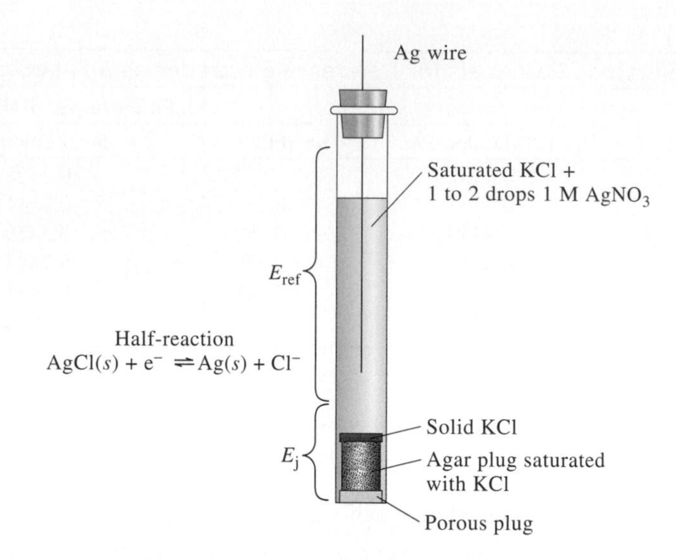

Figure 21-4 Diagram of a silver/silver chloride electrode showing the parts of the electrode that produce the reference electrode potential E_{ref} and the junction potential E_j.

21B-2 Silver/Silver Chloride Reference Electrodes

A system analogous to the use of a saturated calomel electrode employs a silver electrode immersed in a solution that is saturated in both potassium chloride and silver chloride:

► At 25°C, the potential of the saturated calomel electrode versus the standard hydrogen electrode is 0.244 V; for the saturated silver/silver chloride electrode, it is 0.199 V.

$$Ag|AgCl(sat'd), KCl(sat'd)||$$

The half-reaction is

$$AgCl(s) + e^- \rightleftharpoons Ag(s) + Cl^-$$

The potential of this electrode is 0.199 V at 25°C.

Silver/silver chloride electrodes of various sizes and shapes are on the market. A simple and easily constructed electrode of this type is shown in Figure 21-4. The potential characteristics of silver/silver chloride reference electrodes are listed in Table 21-1.

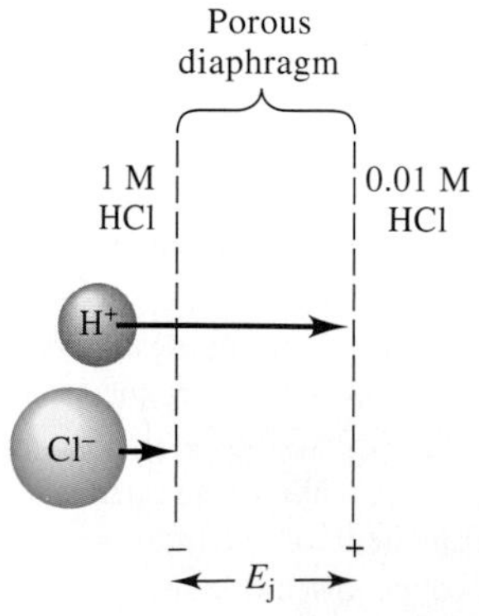

Porous diaphragm

1 M HCl

0.01 M HCl

H^+

Cl^-

$-$ $\leftarrow E_j \rightarrow$ $+$

Figure 21-5 Schematic representation of a liquid junction showing the source of the junction potential, E_j. The length of the arrows corresponds to the relative mobilities of the ions.

21C LIQUID-JUNCTION POTENTIALS

A liquid-junction potential develops across the boundary between two electrolyte solutions that have different compositions. Figure 21-5 shows a very simple liquid junction consisting of a 1 M hydrochloric acid solution that is in contact with a solution that is 0.01 M in that acid. An inert porous barrier, such as a fritted glass plate, prevents the two solutions from mixing. Both hydrogen ions and chloride ions tend to diffuse across this boundary from the more concentrated to the more dilute solution. The driving force for each ion is proportional to the activity difference between the two solutions. In the present example, hydrogen ions are substantially more mobile than chloride ions. Thus, hydrogen ions diffuse more rapidly than chloride ions, and, as shown in the figure, a separation of charge results. The more dilute side of the boundary becomes positively charged because of the more rapid diffusion of hydrogen ions. The concentrated side therefore acquires a negative charge from the excess of slower moving chloride ions. The charge developed tends to counteract the differences in diffusion rates of the two ions so that a steady-state condition is attained rapidly. The potential difference resulting from this charge separation is the junction potential and may be several hundredths of a volt.

The magnitude of the liquid-junction potential can be minimized by placing a salt bridge between the two solutions. The salt bridge is most effective if the mobilities of the negative and positive ions in the bridge are nearly equal and if their concentrations are large. A saturated solution of potassium chloride is good from both standpoints. The net junction potential with such a bridge is typically a few millivolts.

◀ The net junction potential across a typical salt bridge is a few millivolts.

21D INDICATOR ELECTRODES

An ideal indicator electrode responds rapidly and reproducibly to changes in the concentration of an analyte ion (or group of analyte ions). Although no indicator electrode is absolutely specific in its response, a few are now available that are remarkably selective. Indicator electrodes are of three types: metallic, membrane, and ion-sensitive field effect transistors.

◀ The results of potentiometric determinations are the activities of analytes, in contrast to most analytical methods, which give the concentrations of analytes. Recall that the activity of a species a_X is related to the molar concentration of X by Equation 10-2

$$a_X = \gamma_X[X]$$

where γ_X is the activity coefficient of X, a parameter that varies with the ionic strength of the solution. Because potentiometric data are dependent on activities, we will not in most cases make the usual approximation that $a_X \approx [X]$ in this chapter.

21D-1 Metallic Indicator Electrodes

It is convenient to classify metallic indicator electrodes as **electrodes of the first kind, electrodes of the second kind,** or **inert redox electrodes.**

Electrodes of the First Kind

An electrode of the first kind is a pure metal electrode that is in direct equilibrium with its cation in the solution. A single reaction is involved. For example, the equilibrium between a metal X and its cation X^{n+} is

$$X^{n+}(aq) + ne^- \rightleftharpoons X(s)$$

for which

$$E_{ind} = E^0_{X^{n+}/X} - \frac{0.0592}{n}\log\frac{1}{a_{X^{n+}}} = E^0_{X^{n+}/X} + \frac{0.0592}{n}\log a_{X^{n+}} \qquad (21\text{-}2)$$

where E_{ind} is the electrode potential of the metal electrode and $a_{X^{n+}}$ is the activity of the ion (or, in dilute solution, approximately its molar concentration, $[X^{n+}]$).

We often express the electrode potential of the indicator electrode in terms of the p-function of the cation ($pX = -\log a_{X^{n+}}$). Thus, substituting this definition of pX into Equation 21-2 gives

$$E_{ind} = E^0_{X^{n+}/X} + \frac{0.0592}{n}\log a_{X^{n+}} = E^0_{X^{n+}/X} - \frac{0.0592}{n}\text{pX} \qquad (21\text{-}3)$$

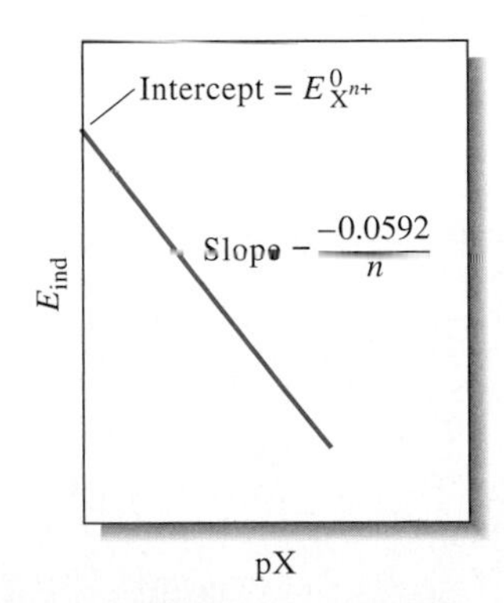

Figure 21-6 A plot of Equation 21-3 for an electrode of the first kind.

This function is plotted in Figure 21-6.

Electrode systems of the first kind are not widely used for potentiometric determinations for several reasons. For one, metallic indicator electrodes are not very selective and respond not only to their own cations but also to other more easily reduced cations. For example, a copper electrode cannot be used for the determination of copper(II) ions in the presence of silver(I) ions because the electrode potential is also a function of the Ag^+ concentration. In addition, many metal electrodes, such as zinc and cadmium, can be used only in neutral or basic solutions because they dissolve in the presence of acids. Third, other metals are so easily oxidized that they can be used only when analyte solutions are deaerated to remove oxygen. Finally, certain harder metals, such as iron, chromium, cobalt, and nickel, do not provide reproducible potentials. Moreover, for these electrodes, plots of pX versus activity yield slopes that differ significantly and irregularly from the theoretical $(-0.0592/n)$. For these reasons, the only electrode systems of the first kind that have been used in potentiometry are Ag/Ag^+ and Hg/Hg^{2+} in neutral solutions and Cu/Cu^{2+}, Zn/Zn^{2+}, Cd/Cd^{2+}, Bi/Bi^{3+}, Tl/Tl^+, and Pb/Pb^{2+} in deaerated solutions.

Electrodes of the Second Kind

Metals not only serve as indicator electrodes for their own cations but also respond to the activities of anions that form sparingly soluble precipitates or stable complexes with such cations. The potential of a silver electrode, for example, correlates reproducibly with the activity of chloride ion in a solution saturated with silver chloride. Here, the electrode reaction can be written as

$$AgCl(s) + e^- \rightleftharpoons Ag(s) + Cl^-(aq) \qquad E^0_{AgCl/Ag} = 0.222\text{ V}$$

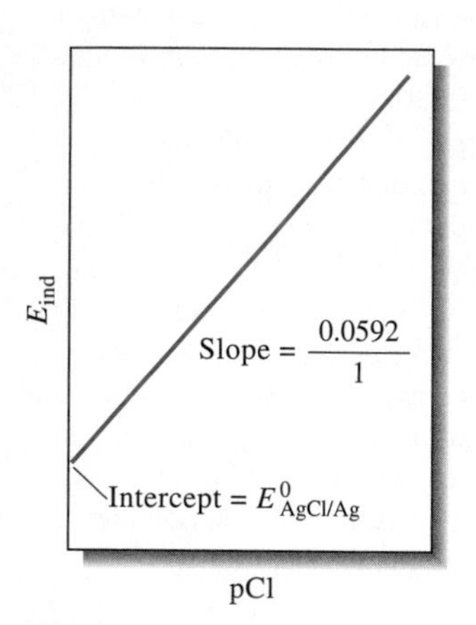

Figure 21-7 A plot of Equation 21-4 for an electrode of the second kind for Cl^-.

The Nernst expression for this process at 25°C is

$$E_{ind} = E^0_{AgCl/Ag} - 0.0592 \log a_{Cl^-} = E^0_{AgCl/Ag} + 0.0592\text{ pCl} \qquad (21\text{-}4)$$

Equation 21-4 shows that the potential of a silver electrode is proportional to pCl, the negative logarithm of the chloride ion activity. Thus, in a solution saturated with silver chloride, a silver electrode can serve as an indicator electrode of the second kind for chloride ion. Note that the sign of the logarithmic term for an electrode of this type is opposite that for an electrode of the first kind (see Equation 21-3). A plot of the potential of the silver electrode versus pCl is shown in Figure 21-7.

Mercury serves as an indicator electrode of the second kind for the EDTA anion Y^{4-}. For example, when a small amount of HgY^{2-} is added to a solution containing Y^{4-}, the half-reaction at a mercury electrode is

$$HgY^{2-} + 2e^- \rightleftharpoons Hg(l) + Y^{4-} \qquad E^0 = 0.21\text{ V}$$

for which

$$E_{ind} = 0.21 - \frac{0.0592}{2} \log \frac{a_{Y^{4-}}}{a_{HgY^{2-}}}$$

The formation constant for HgY^{2-} is very large (6.3×10^{21}), and so the concentration of the complex remains essentially constant over a large range of Y^{4-} concentrations. The Nernst equation for the process can therefore be written as

$$E = K - \frac{0.0592}{2} \log a_{Y^{4-}} = K + \frac{0.0592}{2}\text{ pY} \qquad (21\text{-}5)$$

where

$$K = 0.21 - \frac{0.0592}{2} \log \frac{1}{a_{HgY^{2-}}}$$

The mercury electrode is thus a valuable electrode of the second kind for EDTA titrations, as discussed in Section 21G-2.

Inert Metallic Electrodes for Redox Systems

As noted in Chapter 18, several inert conductors respond to redox systems. Such materials as platinum, gold, palladium, and carbon can be used to monitor redox systems. For example, the potential of a platinum electrode immersed in a solution containing cerium(III) and cerium(IV) is

$$E_{\text{ind}} = E^0_{\text{Ce}^{4+}/\text{Ce}^{3+}} - 0.0592 \log \frac{a_{\text{Ce}^{3+}}}{a_{\text{Ce}^{4+}}}$$

A platinum electrode is a convenient indicator electrode for titrations involving standard cerium(IV) solutions.

21D-2 Membrane Electrodes[2]

For many years, the most convenient method to determine pH has involved measurement of the potential that appears across a thin glass membrane that separates two solutions with different hydrogen ion concentrations. A diagram of the first practical **glass electrode** is shown in Figure 21-8. The phenomenon on which the measurement is based was first reported in 1906 and by now has been extensively studied by many investigators. As a result, the sensitivity and selectivity of glass membranes toward hydrogen ions are reasonably well understood. Furthermore, this understanding has led to the development of other types of membranes that respond selectively to many other ions.

Membrane electrodes are sometimes called **p-ion electrodes** because the data obtained from them are usually presented as p-functions, such as pH, pCa, or pNO_3. In this section, we consider several types of p-ion membranes.

It is important to note at the outset of this discussion that membrane electrodes are fundamentally different from metal electrodes both in design and in principle. We shall use the glass electrode for pH measurements to illustrate these differences.

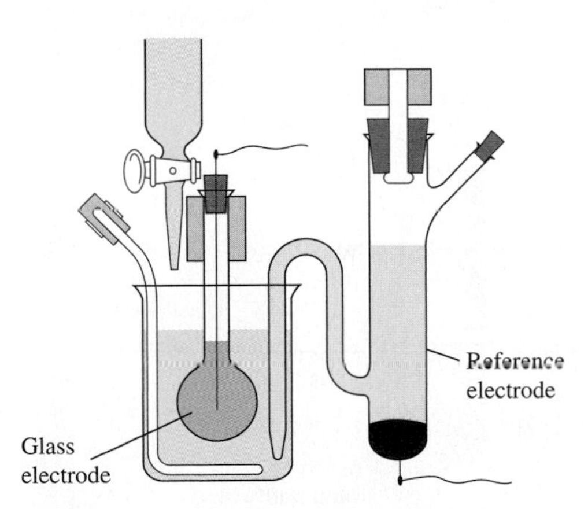

Figure 21-8 The first practical glass electrode. (From Haber and Klemensiewicz, *Z. Phys. Chem.*, **1909,** *65,* 385.)

[2]Some suggested sources for additional information on this topic are A. Evans, *Potentiometry and Ion-Selective Electrodes.* New York: Wiley, 1987; J. Koryta, *Ions, Electrodes, and Membranes,* 2nd ed. New York: Wiley, 1991; R.S. Hutchins and L. G. Bachas, in *Handbook of Instrumental Techniques for Analytical Chemistry,* F. A. Settle, Ed. Upper Saddle River, NJ: Prentice-Hall, 1997.

21D-3 The Glass Electrode for Measuring pH

Figure 21-9a shows a typical *cell* for measuring pH. The cell consists of a glass indicator electrode and a saturated calomel reference electrode immersed in the solution of unknown pH. The indicator electrode consists of a thin, pH-sensitive glass membrane sealed onto one end of a heavy-walled glass or plastic tube. A small volume of dilute hydrochloric acid saturated with silver chloride is contained in the tube. (The inner solution in some electrodes is a buffer containing chloride ion.) A silver wire in this solution forms a silver/silver chloride reference electrode, which is connected to one of the terminals of a potential-measuring device. The calomel electrode is connected to the other terminal.

► The membrane of a typical glass electrode (with a thickness of 0.03 to 0.1 mm) has an electrical resistance of 50 to 500 MΩ.

Figure 21-9a and the representation of this cell in Figure 21-10 show that a glass electrode system contains two reference electrodes: the external calomel electrode and the internal silver/silver chloride electrode. While the internal reference electrode is a part of the glass electrode, it is not the pH-sensing element. Instead, *it is the thin glass membrane bulb at the tip of the electrode that responds to pH.* At first, it may seem unusual that an insulator like glass (see margin note) can be used to detect ions, but keep in mind that whenever there is a charge imbalance across any material, there is an electrical potential difference across the material. In the case of the glass electrode, the concentration of protons inside the membrane is constant, and the concentration outside is determined by the concentration, or activity, of the protons in the analyte solution. This concentration difference produces the potential difference that we measure with a pH meter. Notice that the internal and external reference electrodes are just the means of making electrical contact with the two sides of the glass membrane, and their potentials are essentially constant except for the junction potential, which depends to a small extent on the composition of the analyte solution. The potentials of the two reference electrodes depend on the electrochemical characteristics of their respective redox couples, but the potential across the glass membrane depends on the physicochemical

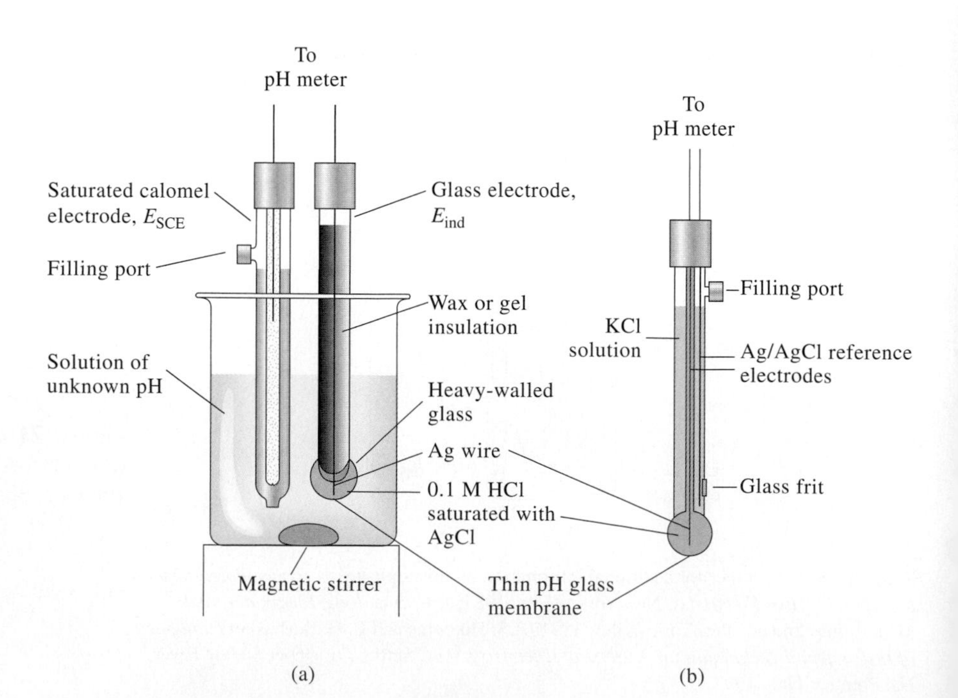

Figure 21-9 Typical electrode system for measuring pH. (a) Glass electrode (indicator) and saturated calomel electrode (reference) immersed in a solution of unknown pH. (b) Combination probe consisting of both an indicator glass electrode and a silver/silver chloride reference. A second silver/silver chloride electrode serves as the internal reference for the glass electrode. The two electrodes are arranged concentrically with the internal reference in the center and the external reference outside. The reference makes contact with the analyte solution through the glass frit or other suitable porous medium. Combination probes are the most common configuration of glass electrode and reference for measuring pH.

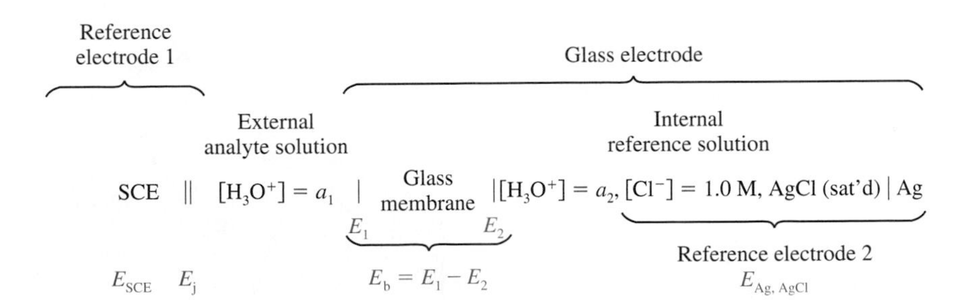

Figure 21-10 Diagram of glass/calomel cell for the measurement of pH. E_{SCE} is the potential of the reference electrode, E_j is the junction potential, a_1 is the activity of hydronium ions in the analyte solution, E_1 and E_2 are the potentials on either side of the glass membrane, E_b is the boundary potential, and a_2 is the activity of hydronium ion in the internal reference solution.

characteristics of the glass and its response to ionic concentrations on both sides of the membrane. To understand how the glass electrode works, we must explore the mechanism of the creation of the charge differential across the membrane that produces the membrane potential. In the next few sections, we investigate this mechanism and the important characteristics of these membranes.

In Figure 21-9b, we see the most common configuration for measuring pH with a glass electrode. In this arrangement, the glass electrode and its Ag/AgCl internal reference electrode are positioned in the center of a cylindrical probe. Surrounding the glass electrode is the external reference electrode, which is most often of the Ag/AgCl type. The presence of the external reference electrode is not as obvious as in the dual-probe arrangement of Figure 21-9a, but the single-probe variety is considerably more convenient and can be made much smaller than the dual system. The pH-sensitive glass membrane is attached to the tip of the probe. These probes are manufactured in many different physical shapes and sizes (5 cm to 5 μm) to suit a broad range of laboratory and industrial applications.

The Composition and Structure of Glass Membranes

There has been a good deal of research devoted to the effects of glass composition on the sensitivity of membranes to protons and other cations, and a number of formulations are now used for the manufacture of electrodes. Corning 015 glass, which has been widely used for membranes, consists of approximately 22% Na_2O, 6% CaO, and 72% SiO_2. This membrane shows excellent specificity toward hydrogen ions up to a pH of about 9. At higher pH values, however, the glass becomes somewhat responsive to sodium, as well as to other singly charged cations. Other glass formulations are now in use in which sodium and calcium ions are replaced to various degrees by barium and lithium ions. These membranes have superior selectivity and lifetimes.

As shown in Figure 21-11 a silicate glass used for membranes consists of an infinite three-dimensional network of groups in which each silicon is bonded to four oxygens and each oxygen is shared by two silicons. Within the empty spaces (interstices) inside this structure are sufficient cations to balance the negative charge of the silicate groups. Singly charged cations, such as sodium and lithium, are mobile in the lattice and are responsible for electrical conduction within the membrane.

The two surfaces of a glass membrane must be hydrated before it will function as a pH electrode. Nonhygroscopic glasses show no pH function. Even hygroscopic glasses lose their pH sensitivity after dehydration by storage over a desiccant. The effect is reversible, however, and the response of a glass electrode can be restored by soaking it in water.

Glasses that absorb water are said to be **hygroscopic**.

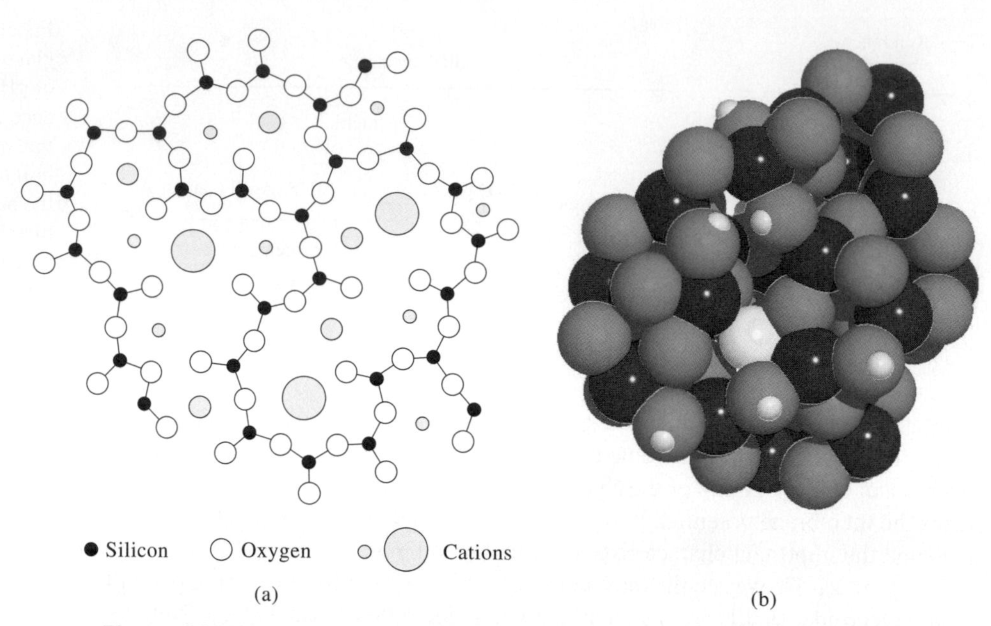

Figure 21-11 (a) Cross-sectional view of a silicate glass structure. In addition to the three Si|O bonds shown, each silicon is bonded to an additional oxygen atom, either above or below the plane of the paper. (Adapted with permission from G. A. Perley, *Anal. Chem.*, **1949,** *21,* 395. Copyright 1949 American Chemical Society.) (b) Model showing three-dimensional structure of amorphous silica with Na^+ ion (large dark blue) and several H^+ ions small dark blue incorporated. Note that the Na^+ ion is surrounded by a cage of oxygen atoms and that each proton in the amorphous lattice is attached to an oxygen. The cavities in the structure, the small size, and the high mobility of the proton ensure that protons can migrate deep into the surface of the silica. Other cations and water molecules may be incorporated into the interstices of the structure as well.

The hydration of a pH-sensitive glass membrane involves an ion-exchange reaction between singly charged cations in the interstices of the glass lattice and protons from the solution. The process involves +1 cations exclusively because +2 and +3 cations are too strongly held within the silicate structure to exchange with ions in the solution. The ion-exchange reaction can then be written as

$$\underset{\text{soln}}{H^+} + \underset{\text{glass}}{Na^+Gl^-} \rightleftharpoons \underset{\text{soln}}{Na^+} + \underset{\text{glass}}{H^+Gl^-} \tag{21-6}$$

Oxygen atoms attached to only one silicon atom are the negatively charged Gl^- sites shown in this equation. The equilibrium constant for this process is so large that the surfaces of a hydrated glass membrane ordinarily consist entirely of silicic acid (H^+Gl^-). An exception to this situation exists in highly alkaline media, where the hydrogen ion concentration is extremely small and the sodium ion concentration is large; here, a significant fraction of the sites are occupied by sodium ions.

Membrane Potentials

The lower part of Figure 21-10 shows four potentials that develop in a cell when pH is being determined with a glass electrode. Two of these, $E_{Ag,AgCl}$ and E_{SCE}, are reference electrode potentials that are constant. A third potential is the junction potential E_j across the salt bridge that separates the calomel electrode from the analyte solution. The fourth and most important potential shown in Figure 21-10 is the **boundary potential,** E_b, *which varies with the pH of the analyte solution.* The two reference electrodes simply provide electrical contacts with the solutions so that changes in the boundary potential can be measured.

The Boundary Potential

Figure 21-10 shows that the boundary potential is determined by potentials, E_1 and E_2, which appear at the two *surfaces* of the glass membrane. The source of these two potentials is the charge that accumulates as a consequence of the reactions

$$\underset{\text{glass}_1}{H^+Gl^-(s)} \rightleftharpoons \underset{\text{soln}_1}{H^+(aq)} + \underset{\text{glass}_1}{Gl^-(s)} \tag{21-7}$$

$$\underset{\text{glass}_2}{H^+Gl^-(s)} \rightleftharpoons \underset{\text{soln}_2}{H^+(aq)} + \underset{\text{glass}_2}{Gl^-(s)} \tag{21-8}$$

where subscript 1 refers to the interface between the exterior of the glass and the analyte solution and subscript 2 refers to the interface between the internal solution and the interior of the glass. These two reactions cause the two glass surfaces to be negatively charged with respect to the solutions with which they are in contact. These negative charges at the surfaces produce the two potentials E_1 and E_2 shown in Figure 21-10. The hydrogen ion concentrations in the solutions on the two sides of the membrane control the positions of the equilibria of Equations 21-7 and 21-8 that in turn determine E_1 and E_2. When the positions of the two equilibria differ, the surface where the greater dissociation has occurred is negative with respect to the other surface. The resulting difference in potential between the two surfaces of the glass is the boundary potential, which is related to the activities of hydrogen ion in each of the solutions by the Nernst-like equation

$$E_b = E_1 - E_2 = 0.0592 \log \frac{a_1}{a_2} \tag{21-9}$$

where a_1 is the activity of the analyte solution and a_2 is that of the internal solution. For a glass pH electrode, the hydrogen ion activity of the internal solution is held constant, so that Equation 21-9 simplifies to

$$E_b = L' + 0.0592 \log a_1 = L' - 0.0592 \text{ pH} \tag{21-10}$$

where

$$L' = -0.0592 \log a_2$$

The boundary potential is then a measure of the hydrogen ion activity of the external solution.

The significance of the potentials and the potential differences shown in Equation 21-10 is illustrated by the potential profiles shown in Figure 21-12. The profiles are plotted across the membrane from the analyte solution on the left through the membrane to the internal solution on the right. The important thing to note about these profiles is that regardless of the absolute potential inside the hygroscopic layers or the glass, the boundary potential is determined by the *difference* in potential on either side of the glass membrane, which is in turn determined by the proton activity on each side of the membrane.

The Asymmetry Potential

When identical solutions and reference electrodes are placed on the two sides of a glass membrane, the boundary potential should in principle be zero. In fact,

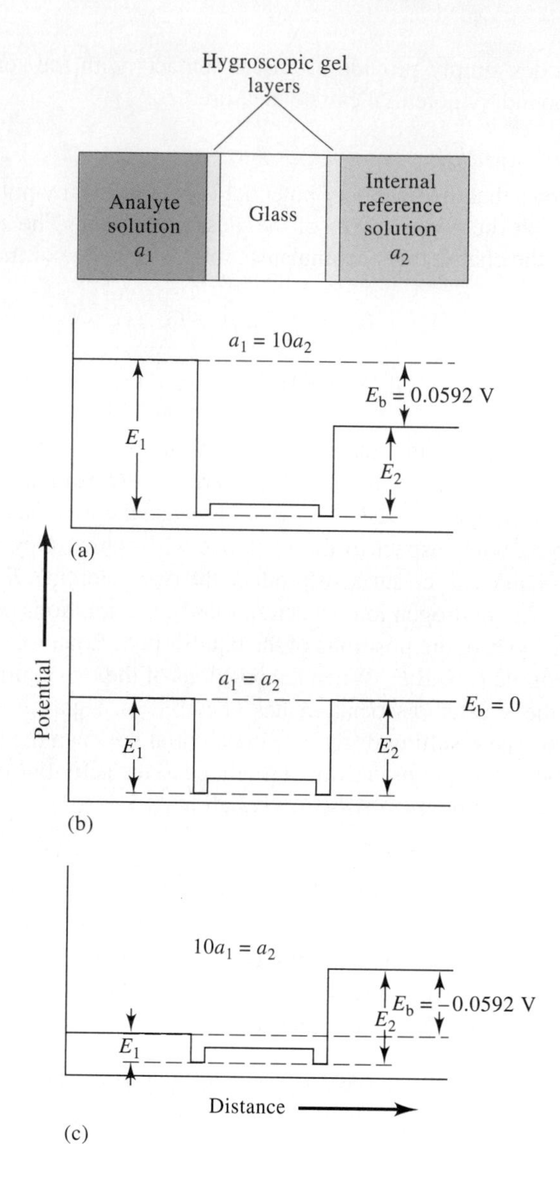

Figure 21-12 Potential profile across a glass membrane from the analyte solution to the internal reference solution. The reference electrode potentials are not shown.

however, we frequently encounter a small asymmetry potential that changes gradually with time.

The sources of the asymmetry potential are obscure but undoubtedly include such causes as differences in strain on the two surfaces of the membrane imparted during manufacture, mechanical abrasion on the outer surface during use, and chemical etching of the outer surface. To eliminate the bias caused by the asymmetry potential, all membrane electrodes must be calibrated against one or more standard analyte solutions. Such calibrations should be carried out at least daily, and more often when the electrode receives heavy use.

The Glass Electrode Potential

The potential of a glass indicator electrode E_{ind} has three components: (1) the boundary potential, given by Equation 21-9; (2) the potential of the internal

Ag/AgCl reference electrode; and (3) the small asymmetry potential, E_{asy}, which changes slowly with time. In equation form, we may write

$$E_{ind} = E_b + E_{Ag/AgCl} + E_{asy}$$

Substitution of Equation 21-10 for E_b gives

$$E_{ind} = L' + 0.0592 \log a_1 + E_{Ag/AgCl} + E_{asy}$$

or

$$E_b = L + 0.0592 \log a_1 = L - 0.0592 \text{ pH} \quad (21\text{-}11)$$

where L is a combination of the three constant terms. Compare Equations 21-11 and 21-3. Although these two equations are similar in form and both potentials are produced by separation of charge, remember that *the mechanisms of charge separation that result in these expressions are considerably different.*

The Alkaline Error

In basic solutions, glass electrodes respond to the concentration of both hydrogen ion and alkali metal ions. The magnitude of the resulting alkaline error for four different glass membranes is shown in Figure 21-13 (curves C to F). These curves refer to solutions in which the sodium ion concentration was held constant at 1 M while the pH was varied. Note that the error is negative (that is, the measured pH values are lower than the true values), which suggests that the electrode is responding to sodium ions as well as to protons. This observation is confirmed by data obtained for solutions containing different sodium ion concentrations. Thus, at pH 12, the electrode with a Corning 015 membrane (curve C in Figure 21-13) registered a pH of 11.3 when immersed in a solution having a sodium ion concentration of 1 M but 11.7 in a solution that was 0.1 M in this ion. All singly charged cations induce an alkaline error whose magnitude depends on both the cation in question and the composition of the glass membrane.

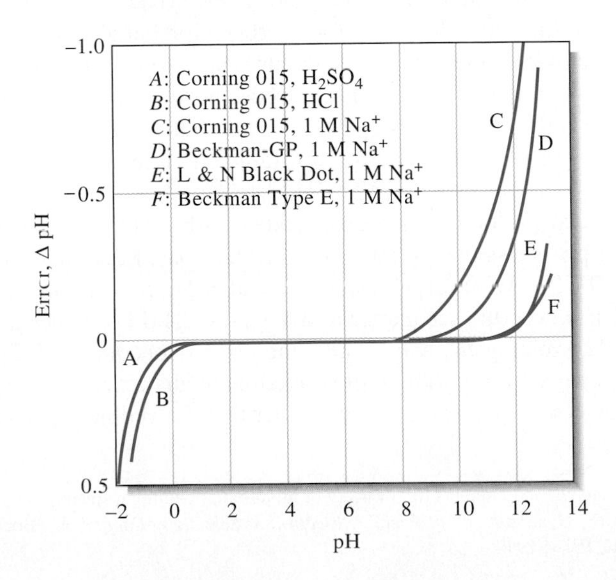

Figure 21-13 Acid and alkaline errors for selected glass electrodes at 25°C. (From R. G. Bates, *Determination of pH,* 2nd ed., p. 365. New York: Wiley, 1973.)

The alkaline error can be satisfactorily explained by assuming an exchange equilibrium between the hydrogen ions on the glass surface and the cations in solution. This process is simply the reverse of that shown in Equation 21-6:

$$\underset{\text{glass}}{H^+Gl^-} + \underset{\text{soln}}{B^+} \rightleftharpoons \underset{\text{glass}}{B^+Gl^-} + \underset{\text{soln}}{H^+}$$

where B^+ represents some singly charged cation, such as sodium ion.

The equilibrium constant for this reaction is

▶ In Equation 21-12, b_1 represents the activity of some singly charged cation such as Na^+ or K^+.

$$K_{ex} = \frac{a_1 b_1'}{a_1' b_1} \tag{21-12}$$

where a_1 and b_1 represent the activities of H^+ and B^+ in solution and a_1' and b_1' are the activities of these ions on the glass surface. Equation 21-12 can be rearranged to give the ratio of the activities B^+ to H^+ on the glass surface:

$$\frac{b_1'}{a_1'} = K_{ex}\frac{b_1}{a_1}$$

For the glasses used for pH electrodes, K_{ex} is so small that the activity ratio b_1'/a_1' is ordinarily minuscule. The situation differs in strongly alkaline media, however. For example, b_1'/a_1' for an electrode immersed in a pH 11 solution that is 1 M in sodium ions (see Figure 21-13) is $10^{11} \times K_{ex}$. Here, the activity of the sodium ions relative to that of the hydrogen ions becomes so large that the electrode responds to both species.

Describing Selectivity

The effect of an alkali metal ion on the potential across a membrane can be accounted for by inserting an additional term in Equation 21-10 to give

$$E_b = L' + 0.0592 \log (a_1 + k_{H,B}b_1) \tag{21-13}$$

The **selectivity coefficient** is a measure of the response of an ion-selective electrode to other ions.

where $k_{H,B}$ is the **selectivity coefficient** for the electrode. Equation 21-13 applies not only to glass indicator electrodes for hydrogen ion but also to all other types of membrane electrodes. Selectivity coefficients range from zero (no interference) to values greater than unity. Thus, if an electrode for ion A responds 20 times more strongly to ion B than to ion A, $k_{H,B}$ has a value of 20. If the response of the electrode to ion C is 0.001 of its response to A (a much more desirable situation), $k_{H,B}$ is 0.001.[3]

The product $k_{H,B}b_1$ for a glass pH electrode is ordinarily small relative to a_1 provided that the pH is less than 9; under these conditions, Equation 21-13 simplifies to Equation 21-10. At high pH values and at high concentrations of a singly charged ion, however, the second term in Equation 21-13 assumes a more important role in determining E_b, and an alkaline error is encountered. For electrodes specifically designed for work in highly alkaline media (curve *E* in Figure 21-13), the magnitude of $k_{H,B}b_1$ is appreciably smaller than for ordinary glass electrodes.

[3]For tables of selectivity coefficients for a variety of membranes and ionic species, see Y. Umezawa, *CRC Handbook of Ion Selective Electrodes: Selectivity Coefficients.* Boca Raton, FL: CRC Press, 1990.

The Acid Error

As shown in Figure 21-13, the typical glass electrode exhibits an error, opposite in sign to the alkaline error, in a solution of pH less than about 0.5; pH readings tend to be too high in this region. The magnitude of the error depends on a variety of factors and is generally not very reproducible. All the causes of the acid error are not well understood, but one source is a saturation effect that occurs when all the surface sites on the glass are occupied with H^+ ions. Under these conditions, the electrode no longer responds to further increases in the H^+ concentration and the pH readings are too high.

21D-4 Glass Electrodes for Other Cations

The alkaline error in early glass electrodes led to investigations concerning the effect of glass composition on the magnitude of this error. One consequence has been the development of glasses for which the alkaline error is negligible below about pH 12 (see curves *E* and *F*, Figure 21-13). Other studies have discovered glass compositions that permit the determination of cations other than hydrogen. Incorporation of Al_2O_3 or B_2O_3 into the glass has the desired effect. Glass electrodes that permit the direct potentiometric measurement of such singly charged species as Na^+, K^+, NH_4^+, Rb^+, Cs^+, Li^+, and Ag^+ have been developed. Some of these glasses are reasonably selective toward particular singly charged cations. Glass electrodes for Na^+, Li^+, NH_4^+, and total concentration of univalent cations are now available from commercial sources.

21D-5 Liquid-Membrane Electrodes

The potential of liquid-membrane electrodes develops across the interface between the solution containing the analyte and a liquid-ion exchanger that selectively bonds with the analyte ion. These electrodes have been developed for the direct potentiometric measurement of numerous polyvalent cations as well as certain anions.

Figure 21-14 is a schematic of a liquid-membrane electrode for calcium. It consists of a conducting membrane that selectively binds calcium ions, an internal solution containing a fixed concentration of calcium chloride, and a silver electrode

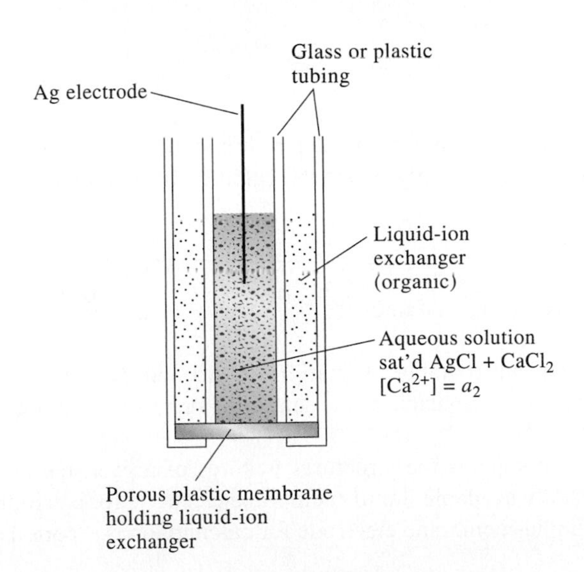

Figure 21-14 Diagram of a liquid-membrane electrode for Ca^{2+}.

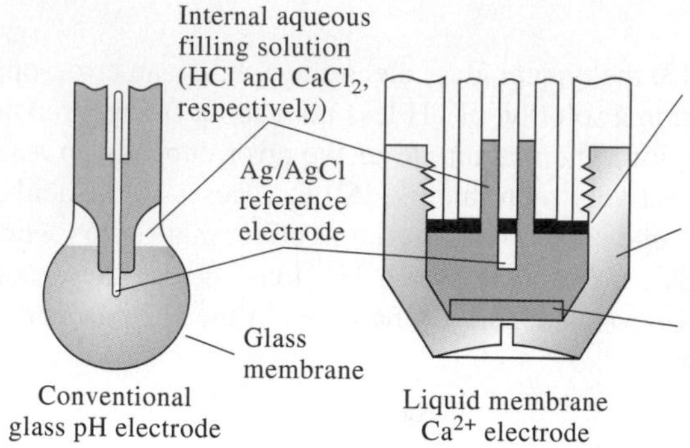

Figure 21-15 Comparison of a liquid-membrane calcium ion electrode with a glass pH electrode. (Courtesy of Thermo Orion, Beverly, MA.)

that is coated with silver chloride to form an internal reference electrode. Notice the similarities between the liquid membrane electrode and the glass electrode, as shown in Figure 21-15. The active membrane ingredient is an ion exchanger consisting of a calcium dialkyl phosphate that is nearly insoluble in water. In the electrode shown in Figures 21-14 and 21-15, the ion exchanger is dissolved in an immiscible organic liquid that is forced by gravity into the pores of a hydrophobic porous disk. This disk then serves as the membrane that separates the internal solution from the analyte solution. In a more recent design, the ion exchanger is immobilized in a tough polyvinyl chloride gel cemented to the end of a tube that holds the internal solution and reference electrode. In either design, a dissociation equilibrium develops at each membrane interface that is analogous to Equations 21-7 and 21-8:

Hydrophobia means fear of water. The hydrophobic disk is porous toward organic liquids but repels water.

$$\underset{\text{organic}}{[(RO)_2POO]_2Ca} \rightleftharpoons \underset{\text{organic}}{2(RO)_2POO^-} + \underset{\text{aqueous}}{Ca^{2+}}$$

where R is a high-molecular-weight aliphatic group. As with the glass electrode, a potential develops across the membrane when the extent of the ion exchanger dissociation at one surface differs from that at the other surface. This potential is a result of differences in the calcium ion activity of the internal and external solutions. The relationship between the membrane potential and the calcium ion activities is given by an equation that is similar to Equation 21-9:

$$E_b = E_1 - E_2 = \frac{0.0592}{2} \log \frac{a_1}{a_2} \qquad (21\text{-}14)$$

where a_1 and a_2 are the activities of calcium ion in the external analyte and internal standard solutions, respectively. Since the calcium ion activity of the internal solution is constant.

$$E_b = N + \frac{0.0592}{2} \log a_1 = N - \frac{0.0592}{2} \text{pCa} \qquad (21\text{-}15)$$

where N is a constant (compare Equations 21-15 and 21-10). Note that because calcium is divalent, a_2 appears in the denominator of the coefficient of the logarithmic term.

Figure 21-15 compares the structural features of a glass-membrane electrode and a commercially available liquid-membrane electrode for calcium ion. The sensitivity of the liquid-membrane electrode for calcium ion is reported to be 50 times

TABLE 21-2
Characteristics of Liquid-Membrane Electrodes*

Analyte Ion	Concentration Range, M	Major Interferences
Ca^{2+}	10^0 to 5×10^{-7}	Pb^{2+}, Fe^{2+}, Ni^{2+}, Hg^{2+}, Si^{2+}
Cl^-	10^0 to 5×10^{-6}	I^-, OH^-, SO_4^{2-}
NO_3^-	10^0 to 7×10^{-6}	ClO_4^-, I^-, ClO_3^-, CN^-, Br^-
ClO_4^-	10^0 to 7×10^{-6}	I^-, ClO_3^-, CN^-, Br^-
K^+	10^0 to 1×10^{-6}	Cs^+, NH_4^+, Tl^+
Water hardness ($Ca^{2+} + Mg^{2+}$)	10^0 to 6×10^{-6}	Cu^{2+}, Zn^{2+}, Ni^{2+}, Fe^{2+}, Sr^{2+}, Ba^{2+}

*From *Orion Guide to Ion Analysis.* Boston, MA: Thermo Electron Corp, 1992.

greater than for magnesium ion and 1000 times greater than for sodium or potassium ions. Calcium ion activities as low as 5×10^{-7} M can be measured. Performance of the electrode is independent of pH in the range between 5.5 and 11. At lower pH levels, hydrogen ions undoubtedly replace some of the calcium ions on the exchanger; the electrode then becomes sensitive to pH as well as to pCa.

The calcium ion liquid-membrane electrode is a valuable tool for physiological investigations because this ion plays important roles in such processes as nerve conduction, bone formation, muscle contraction, cardiac expansion and contraction, renal tubular function, and perhaps hypertension. Most of these processes are influenced more by the activity than the concentration of the calcium ion; activity, of course, is the parameter measured by the membrane electrode. Thus, the calcium ion electrode (and the potassium ion electrode and others) is an important tool in studying physiological processes.

◀ Ion-selective microelectrodes can be used to measure ion activities within a living organism.

A liquid-membrane electrode specific for potassium ion is also of great value for physiologists because the transport of neural signals appears to involve movement of this ion across nerve membranes. Investigation of this process requires an electrode that can detect small concentrations of potassium ion in media that contain much larger concentrations of sodium ion. Several liquid-membrane electrodes show promise in meeting this requirement. One is based on the antibiotic valinomycin, a cyclic ether that has a strong affinity for potassium ion. Of equal importance is the observation that a liquid membrane consisting of valinomycin in diphenyl ether is about 10^4 times as responsive to potassium ion as to sodium ion.[4] Figure 21-16 is a photomicrograph of a tiny electrode used to determine the potassium content of a single cell.

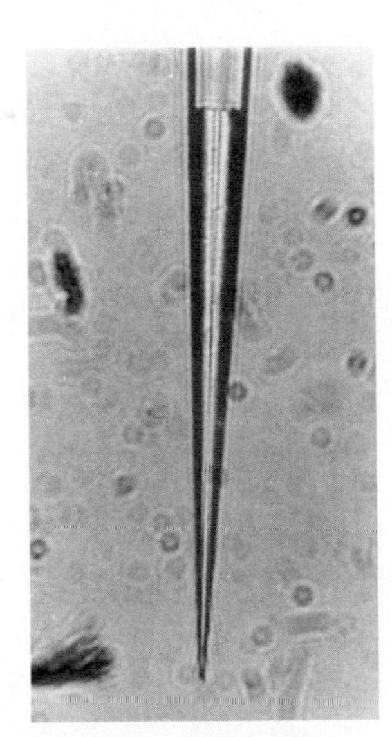

Figure 21-16 Photograph of a potassium liquid-ion exchanger microelectrode with 125 μm of ion exchanger inside the tip. The magnification of the original photo was 400×. (From J. L. Walker, *Anal. Chem.*, **1971,** *43*[3] 91A. Reproduced by permission of the American Chemical Society.)

Table 21-2 lists some liquid-membrane electrodes available from commercial sources. The anion-sensitive electrodes shown make use of a solution containing an anion-exchange resin in an organic solvent. Liquid-membrane electrodes in which the exchange liquid is held in a polyvinyl chloride gel have been developed for Ca^{2+}, K^+, NO_3^-, and BF_4^-. These have the appearance of crystalline electrodes, which are considered in the following section. A homemade liquid-membrane ion-selective electrode is described in Feature 21-1.

[4]M. S. Frant and J. W. Ross, Jr., *Science,* **1970,** *167,* 987.

FEATURE 21-1

An Easily Constructed Liquid-Membrane Ion-Selective Electrode

You can make a liquid-membrane ion-selective electrode with glassware and chemicals available in most laboratories.[5] All you need are a pH meter, a pair of reference electrodes, a fritted-glass filter crucible or tube, trimethylchlorosilane, and a liquid-ion exchanger.

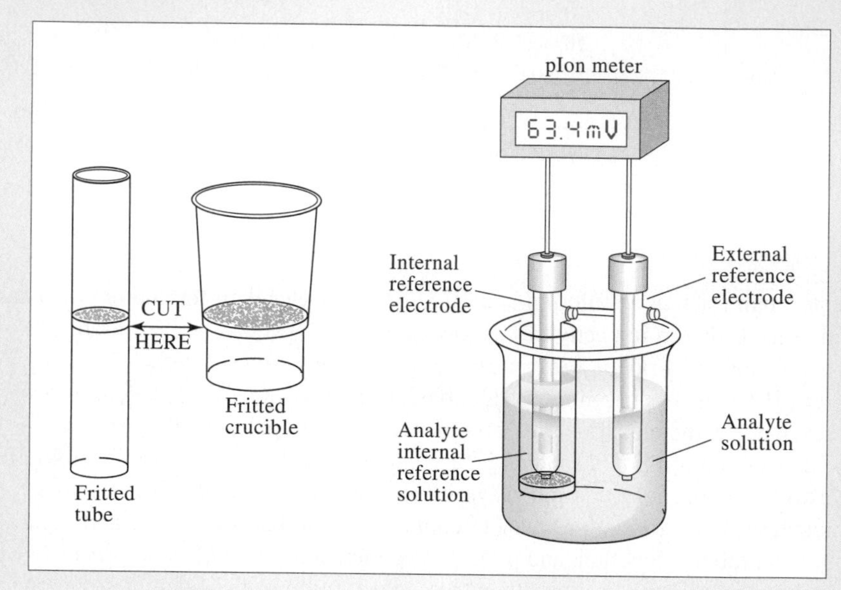

Figure 21F-1 A homemade liquid-membrane electrode.

First, cut the filter crucible (or, alternatively, a fritted tube) as shown in Figure 21F-1. Carefully clean and dry the crucible, and then draw a small amount of trimethylchlorosilane into the frit. This coating makes the glass in the frit hydrophobic. Rinse the frit with water, dry it, and apply a commercial liquid-ion exchanger to it. After a minute, remove the excess exchanger. Add a few milliliters of a 10^{-2} M solution of the ion of interest to the crucible, insert a reference electrode into the solution, and voilá, you have a very nice ion-selective electrode. The exact details of washing, drying, and preparing the electrode are provided in the original article.

Connect the ion-selective electrode and the second reference electrode to the pH meter as shown in Figure 21F-1. Prepare a series of standard solutions of the ion of interest, measure the cell potential for each concentration, plot a working curve of E_{cell} versus log c, and perform a least-squares analysis on the data (see Chapter 8). Compare the slope of the line with the theoretical slope of $(0.0592\ V)/n$. Measure the potential for an unknown solution of the ion and calculate the concentration from the least-squares parameters.

21D-6 Crystalline-Membrane Electrodes

Considerable work has been devoted to the development of solid membranes that are selective toward anions in the same way that some glasses respond to cations. We have seen that anionic sites on a glass surface account for the selectivity of a

[5]See T. K. Christopoulus and E. P. Diamandis, *J. Chem. Educ.*, **1988,** *65,* 648.

TABLE 21-3

Characteristics of Solid-State Crystalline Electrodes*

Analyte Ion	Concentration Range, M	Major Interferences
Br^-	10^0 to 5×10^{-6}	CN^-, I^-, S^{2-}
Cd^{2+}	10^{-1} to 1×10^{-7}	Fe^{2+}, Pb^{2+}, Hg^{2+}, Ag^+, Cu^{2+}
Cl^-	10^0 to 5×10^{-5}	CN^-, I^-, Br^-, S^{2-}, OH^-, NH_3
Cu^{2+}	10^{-1} to 1×10^{-8}	Hg^{2+}, Ag^+, Cd^{2+}
CN^-	10^{-2} to 1×10^{-6}	S^{2-}, I^-
F^-	Sat'd to 1×10^{-6}	OH^-
I^-	10^0 to 5×10^{-8}	CN^-
Pb^{2+}	10^{-1} to 1×10^{-6}	Hg^{2+}, Ag^+, Cu^{2+}
Ag^+/S^{2-}	Ag^+: 10^0 to 1×10^{-7} S^{2-}: 10^0 to 1×10^{-7}	Hg^{2+}
SCN^-	10^0 to 5×10^{-6}	I^-, Bi^-, CN^-, S^{2-}

*From *Orion Guide to Ion Analysis.* Boston, MA: Thermo Electron Corp., 1992.

membrane toward certain cations. By analogy, a membrane with cationic sites might be expected to respond selectively toward anions.

Membranes prepared from cast pellets of silver halides have been used successfully in electrodes for the selective determination of chloride, bromide, and iodide ions. In addition, an electrode based on a polycrystalline Ag_2S membrane is offered by one manufacturer for the determination of sulfide ion. In both types of membranes, silver ions are sufficiently mobile to conduct electricity through the solid medium. Mixtures of PbS, CdS, and CuS with Ag_2S provide membranes that are selective for Pb^{2+}, Cd^{2+}, and Cu^{2+}, respectively. Silver ion must be present in these membranes to conduct electricity because divalent ions are immobile in crystals. The potential that develops across crystalline solid-state electrodes is described by a relationship similar to Equation 21-10.

A crystalline electrode for fluoride ion is available from commercial sources. The membrane consists of a slice of a single crystal of lanthanum fluoride that has been doped with europium(II) fluoride to improve its conductivity. The membrane, supported between a reference solution and the solution to be measured, shows a theoretical response to changes in fluoride ion activity from 10^0 to 10^{-6} M. The electrode is selective for fluoride ion over other common anions by several orders of magnitude; only hydroxide ion appears to offer serious interference.

Some solid-state electrodes available from commercial sources are listed in Table 21-3.

21D-7 Ion-Sensitive Field Effect Transistors (ISFETS)

The **field effect transistor,** or the **metal oxide field effect transistor (MOSFET),** is a tiny solid-state semiconductor device that is widely used in computers and other electronic circuits as a switch to control current in circuits. One of the problems in using this type of device in electronic circuits has been its pronounced sensitivity to ionic surface impurities; a great deal of money and effort has been expended by the electronics industry to minimize or eliminate this sensitivity to produce stable transistors.

Scientists have exploited the sensitivities of MOSFETs to surface ionic impurities for the selective potentiometric determination of various ions. These studies have led to the development of a number of different **ion-selective field effect**

ISFETs stands for ion-sensitive field effect transistors.

transistors, termed **ISFETs.** The theory of their selective ion sensitivity is well understood and is described in Feature 21-2.[6]

ISFETs offer a number of significant advantages over membrane electrodes, including ruggedness, small size, inertness toward harsh environments, rapid response, and low electrical impedance. In contrast to membrane electrodes,

FEATURE 21-2

The Structure and Performance of Ion-Selective Field Effect Transistors

The metal oxide field effect transistor (MOSFET) is a solid-state semiconductor device that is widely used to switch signals in computers and many other types of electronic circuits. Figure 21F-2 shows a cross-sectional diagram (a) and a circuit symbol (b) of an *n*-channel enhancement-mode MOSFET. Modern semiconductor fabrication techniques are used to construct the MOSFET on the surface of a piece of *p*-type semiconductor called the substrate. For a discussion of the characteristics of *p*-type and *n*-type semiconductors, refer to the paragraphs on silicon photodiodes in Section 25A-4. As shown in Figure 21F-2a, two islands of *n*-type semiconductors are formed on the surface of the *p*-type substrate, and the surface is then covered by insulating SiO_2. The last step in the fabrication process is the deposition of metallic conductors that are used to connect the MOSFET to external circuits. There are a total of four such connections to the drain, the gate, the source, and the substrate, as shown in the figure.

The area on the surface of the *p*-type material between the drain and the source is called the channel (see the dark shaded area in Figure 21F-2a). Note that the channel is separated from the gate connection by an insulating layer of SiO_2. When an electrical potential is applied between the gate and the source,

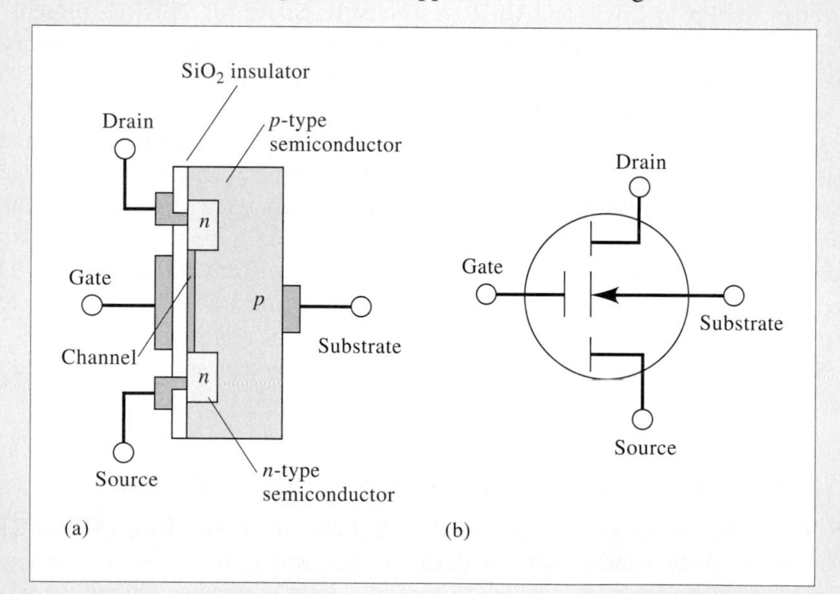

Figure 21F-2 A metal oxide field effect transistor (MOSFET). (a) Cross-sectional diagram; (b) circuit symbol.

[6]For a detailed explanation of the theory of ISFETs, see J. Janata, *Principles of Chemical Sensors,* pp. 125–141. New York: Plenum, 1989.

the electrical conductivity of the channel is enhanced by a factor that is related to the size of the applied potential.

The *ion-selective field effect transistor,* or *ISFET,* is very similar in construction and function to an *n*-channel enhancement mode MOSFET. The ISFET differs only in that variation in the concentration of the ions of interest provides the variable gate voltage to control the conductivity of the channel. As shown in Figure 21F-3, instead of the usual metallic contact, the face of the ISFET is covered with an insulating layer of silicon nitride. The analytical solution, containing hydronium ions in this example, is in contact with this insulating layer and with a reference electrode. The surface of the gate insulator functions very much like the surface of a glass electrode. Protons from the hydronium ions in the test solution are absorbed by available microscopic sites on the silicon nitride. Any change in the hydronium ion concentration (or activity) of the solution results in a change in the concentration of adsorbed protons. The change in concentration of adsorbed protons then gives rise to a changing electrochemical potential between the gate and the source, which in turn changes the conductivity of the channel of the ISFET. The conductivity of the channel can be monitored electronically to provide a signal that is proportional to the logarithm of the activity of hydronium ion in the solution. Note that the entire ISFET except the gate insulator is coated with a polymeric encapsulant to insulate all electrical connections from the analyte solution.

The ion-sensitive surface of the ISFET is naturally sensitive to pH changes, but the device may be modified so that it becomes sensitive to other species by coating the silicon nitride gate insulator with a polymer containing molecules that tend to form complexes with species other than hydronium ion. Furthermore, several ISFETs may be fabricated on the same substrate so that multiple measurements may be made at the same time. All the ISFETs may detect the same species to enhance accuracy and reliability, or each ISFET may be coated with a different polymer so that measurements of several different species may be made. Their small size (about 1 to 2 mm^2), rapid response time relative to glass electrodes, and ruggedness suggest that ISFETs may be the ion detectors of the future for many applications.

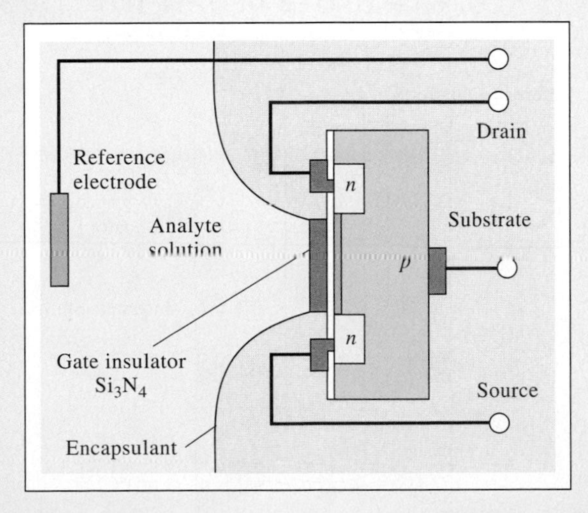

Figure 21F-3 An ion-selective field effect transistor (ISFET) for measuring pH.

ISFETs do not require hydration before use and can be stored indefinitely in the dry state. Despite these many advantages, no ISFET-specific ion electrode appeared on the market until the early 1990s, more than 20 years after its invention. The reason for this delay is that manufacturers were unable to develop the technology to encapsulate the devices to give a product that did not exhibit drift and instability. Several companies now produce ISFETs for the determination of pH, but they are certainly not as routinely used as the glass pH electrode at this time.

21D-8 Gas-Sensing Probes

A **gas-sensing probe** is a galvanic *cell* whose potential is related to the concentration of a gas in a solution. Often, these devices are called gas-sensing electrodes in instrument brochures, which is a misnomer.

Figure 21-17 illustrates the essential features of a potentiometric gas-sensing probe, which consists of a tube containing a reference electrode, a selective ion electrode, and an electrolyte solution. A thin, replaceable, gas-permeable membrane attached to one end of the tube serves as a barrier between the internal and analyte solutions. As can be seen from Figure 21-17, this device is a complete electrochemical cell and is more properly referred to as a probe rather than an electrode, a term that is frequently encountered in advertisements by instrument manufacturers. Gas-sensing probes have found widespread use in the determination of dissolved gases in water and other solvents.

Membrane Composition

A **microporous membrane** is fabricated from a hydrophobic polymer. As the name implies, the membrane is highly porous (the average pore size is less than 1 μm) and allows the free passage of gases; at the same time, the water-repellent polymer prevents water and solute ions from entering the pores. The thickness of the membrane is about 0.1 mm.

The Mechanism of Response

Using carbon dioxide as an example, we can represent the transfer of gas to the internal solution in Figure 21-17 by the following set of equations:

$$\underset{\text{analyte solution}}{CO_2(aq)} \rightleftharpoons \underset{\text{membrane pores}}{CO_2(g)}$$

$$\underset{\text{membrane pores}}{CO_2(g)} \rightleftharpoons \underset{\text{internal solution}}{CO_2(aq)}$$

$$\underset{\text{internal solution}}{CO_2(aq) + 2H_2O} \rightleftharpoons \underset{\text{internal solution}}{HCO_3^- + H_3O^+}$$

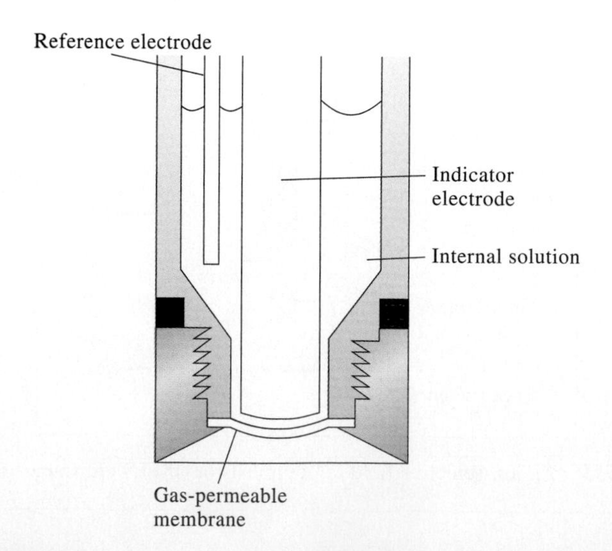

Figure 21-17 Diagram of a gas-sensing probe.

The last equilibrium causes the pH of the internal surface film to change. This change is then detected by the internal glass/calomel electrode system. A description of the overall process is obtained by adding the equations for the three equilibria to give

$$\underset{\text{analyte solution}}{CO_2(aq)} + 2H_2O \rightleftharpoons \underset{\text{internal solution}}{HCO_3^- + H_3O^+}$$

The thermodynamic equilibrium constant K for this overall reaction is

$$K = \frac{(a_{H_3O^+})_{int}(a_{HCO_3^-})_{int}}{(a_{CO_2})_{ext}}$$

For a neutral species such as CO_2, $a_{CO_2} = [CO_2(aq)]$, so

$$K = \frac{(a_{H_3O^+})_{int}(a_{HCO_3^-})_{int}}{[CO_2(aq)]_{ext}}$$

where $[CO_2(aq)]_{ext}$ is the molar concentration of the gas in the analyte solution. For the measured cell potential to vary linearly with the logarithm of the carbon dioxide concentration of the external solution, the hydrogen carbonate activity of the internal solution must be sufficiently large that it is not altered significantly by the carbon dioxide entering from the external solution. Assuming then that $(a_{HCO_3^-})_{int}$ is constant, we can rearrange the previous equations to

$$\frac{(a_{H_3O^+})_{int}}{[CO_2(aq)]_{ext}} = \frac{K}{(a_{HCO_3^-})_{int}} = K_g$$

If we allow a_1 to be the hydrogen ion activity of the internal solution, we rearrange this equation to give

$$(a_{H_3O^+})_{int} = a_1 = K_g[CO_2(aq)]_{ext} \qquad (21\text{-}16)$$

By substituting Equation 21-16 into Equation 21-11, we find

$$E_{ind} = L + 0.0592 \log a_1 = L + 0.0592 \log K_g[CO_2(aq)]_{ext}$$
$$= L + 0.0592 \log K_g + 0.0592 \log[CO_2(aq)]_{ext}$$

Combining the two constant terms to give a new constant L' leads to

$$E_{ind} = L' + 0.0592 \log [CO_2(aq)]_{ext} \qquad (21\text{-}17)$$

Finally, since

$$E_{cell} = E_{ind} - E_{ref}$$

then

$$E_{cell} = L' + 0.0592 \log [CO_2(aq)]_{ext} - E_{ref} \qquad (21\text{-}18)$$

or

$$E_{cell} = L'' + 0.0592 \log [CO_2(aq)]_{ext}$$

where

$$L'' = L + 0.0592 \log K_g - E_{ref}$$

Thus, the potential between the glass electrode and the reference electrode in the internal solution is determined by the CO_2 concentration in the external solution. Note that no electrode comes in direct contact with the analyte solution. Therefore, these devices are gas-sensing cells, or probes, rather than gas-sensing electrodes. Nevertheless, they continue to be called electrodes in some literature and many advertising brochures.

► Although sold as gas-sensing electrodes, these devices are complete electrochemical cells and should be called gas-sensing probes.

The only species that interfere are other dissolved gases that permeate the membrane and then affect the pH of the internal solution. The specificity of gas probes depends only on the permeability of the gas membrane. Gas-sensing probes for CO_2, NO_2, H_2S, SO_2, HF, HCN, and NH_3 are now available from commercial sources.

FEATURE 21-3

Point-of-Care Testing: Blood Gases and Blood Electrolytes with Portable Instrumentation

Modern medicine relies heavily on analytical measurements for diagnosis and treatment in emergency rooms, operating rooms, and intensive care units. Prompt reporting of blood gas values, blood electrolyte concentrations, and other variables is especially important to physicians in these areas. In critical life-and-death situations, there is seldom sufficient time to transport blood samples to the clinical laboratory, perform required analyses, and transmit the results back to the bedside. In this feature, we describe an automated blood gas and electrolyte monitor designed specifically to analyze blood samples at the bedside.[7] The i-STAT Portable Clinical Analyzer, shown in Figure 21F-4, is a handheld device that can measure a variety of important clinical analytes such as potassium, sodium, pH, pCO_2, pO_2, and hematocrit (see margin note). In addition, the computer-based analyzer calculates bicarbonate, total carbon dioxide, base excess, O_2 saturation, and hemoglobin in whole blood. In a study of the performance of the i-STAT system in a neonatal and pediatric intensive care unit, the results shown in the following table were obtained.[8] The results were judged to be sufficiently reliable and cost effective to substitute for similar measurements made in a traditional remote clinical laboratory.

► Hematocrit (Hct) is the ratio of the volume of red blood cells to the total volume of a blood sample expressed as a percentage.

Analyte	Range	Precision, %RSD	Resolution
pO_2	5–800 mm Hg	3.5	1 mm Hg
pCO_2	5–130 mm Hg	1.5	0.1 mm Hg
Na^+	100–180 mmol/L	0.4	1 mmol/L
K^+	2.0–9.0 mmol/L	1.2	0.1 mmol/L
Ca^{2+}	0.25–2.50 mmol/L	1.1	0.01 mmol/L
pH	6.5–8.0	0.07	0.001

[7] i-STAT Corporation, East Windsor, NJ.

[8] J. N. Murthy, J. M. Hicks, and S. J. Soldin, *Clin. Biochem.*, **1997,** *30,* 385.

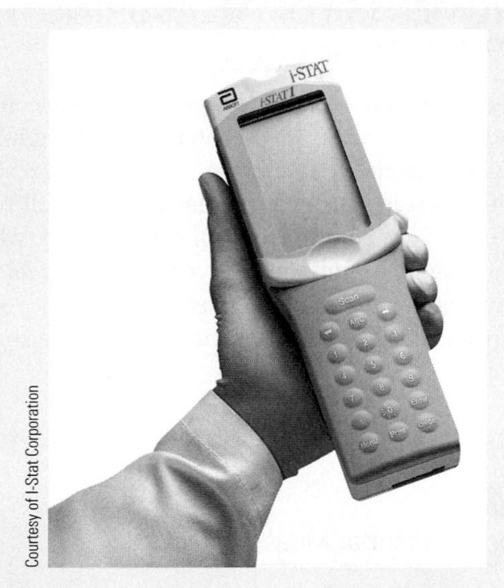

Figure 21F-4 Photograph of i-STAT 1 portable clinical analyzer. (i-STAT Corporation, East Windsor, NJ.)

Most of the analytes (pCO_2, Na^+, K^+, Ca^{2+}, and pH) are determined by potentiometric measurements using membrane-based ion-selective electrode technology. The hematocrit is measured by electrolytic conductivity detection, and pO_2 is determined with a Clark voltammetric sensor (see Section 23B-4). Other results are calculated from these data.

The central component of the monitor is the single-use disposable electrochemical i-STAT sensor array depicted in Figure 21F-5. The individual microfabricated sensor electrodes are located on chips along a narrow flow channel, as shown in the figure. Each new sensor array is automatically calibrated prior to the measurement step. A blood sample withdrawn from the patient is deposited into the sample entry well, and the cartridge is inserted into the i-STAT analyzer. The calibrant pouch, which contains a standard buffered solution of the analytes, is punctured by the i-STAT analyzer and is compressed to force the calibrant through the flow channel across the surface of the sensor array. When the calibration step is complete, the analyzer compresses the air bladder, which forces the blood sample through the flow channel to expel the calibrant solution to waste and to bring the blood into contact with the sensor array. Electrochemical measurements are then made, results are calculated, and the data are presented on the liquid crystal display of the analyzer. The results are stored in the memory of the analyzer and may be transmitted to the hospital laboratory data management system for permanent storage and retrieval.

This feature shows how modern ion-selective electrode technology coupled with computer control of the measurement process and data reporting can be used to provide rapid, essential measurements of analyte concentrations in whole blood at the patient's bedside.

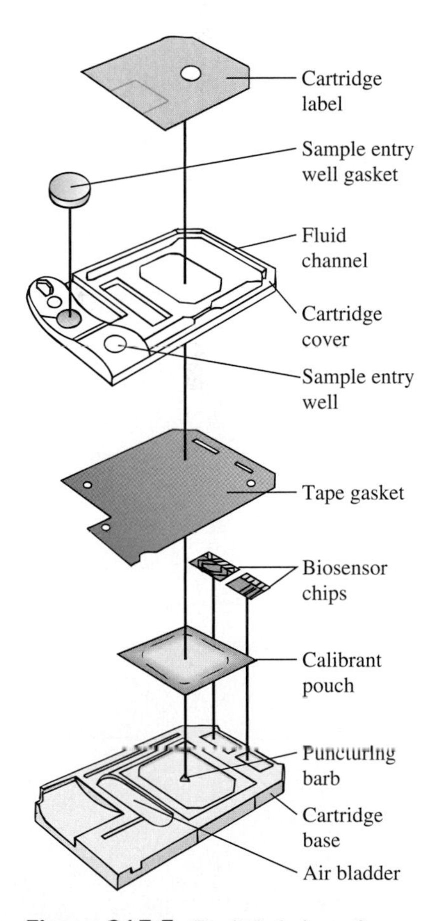

Figure 21F-5 Exploded view of i-STAT sensor array cartridge. (i-STAT Corporation, East Windsor, NJ.)

21E INSTRUMENTS FOR MEASURING CELL POTENTIAL

Most cells containing a membrane electrode have very high electrical resistance (as much as 10^8 ohms or more). To measure potentials of such high-resistance circuits accurately, it is necessary that the voltmeter have an electrical resistance that is several orders of magnitude greater than the resistance of the cell being measured. If the meter resistance is too low, current is drawn from the cell, which has the effect of lowering its output potential, thus creating a negative *loading error.* When the meter and the cell have the same resistance, a relative error of −50% results. When this ratio is 10, the error is about −9%. When it is 1000, the error is less than 0.1% relative.

FEATURE 21-4

The Loading Error in Potential Measurements

When we measure voltages in electrical circuits, the meter becomes a part of the circuit, perturbs the measurement process, and produces a **loading error** in the measurement. This situation is not unique to potential measurements. In fact, it is a basic example of a general limitation to any physical measurement. That is, the process of measurement inevitably disturbs the system of interest so that the quantity actually measured differs from its value prior to the measurement. This type of error can never be completely eliminated, but it can often be reduced to an insignificant level.

The size of the loading error in potential measurements depends on the ratio of the internal resistance of the meter to the resistance of the circuit being studied. The percent relative loading error E_r associated with the measured potential V_M in Figure 21F-6 is given by

$$E_r = \frac{V_M - V_x}{V_x} \times 100\%$$

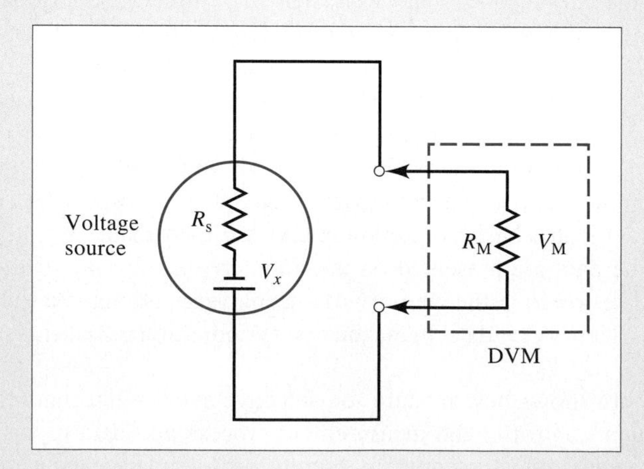

Figure 21F-6 Measurement of output V_x from a potential source with a digital voltmeter.

where V_x is the true voltage of the power source. The voltage drop across the resistance of the meter is given by

$$V_M = V_x \frac{R_M}{R_M + R_s}$$

Substituting this equation into the previous one and rearranging gives

$$E_r = \frac{-R_s}{R_M + R_s} \times 100\%$$

Note in this equation that the relative loading error becomes smaller as the meter resistance R_M becomes larger relative to the source resistance R_s. Table 21F-1 illustrates this effect. Digital voltmeters offer the great advantage of having huge internal resistances (10^{11} to 10^{12} ohms), thus avoiding loading errors except in circuits having load resistances greater than about 10^9 ohms.

TABLE 21F-1

Effect of Meter Resistance on the Accuracy of Potential Measurements

Meter Resistance R_M Ω	Resistance of Source R_s, Ω	R_M/R_s	Relative Error, %
10	20	0.50	−67
50	20	2.5	−29
500	20	25	−3.8
1.0×10^3	20	50	−2.0
1.0×10^4	20	500	−0.2

Numerous high-resistance, direct-reading digital voltmeters with internal resistances of $>10^{11}$ ohms are now on the market. These meters are commonly called **pH meters** but could more properly be referred to as **pIon meters** or **ion meters,** since they are frequently used for the measurement of concentrations of other ions as well. A photograph of a typical pH meter is shown in Figure 21-18.

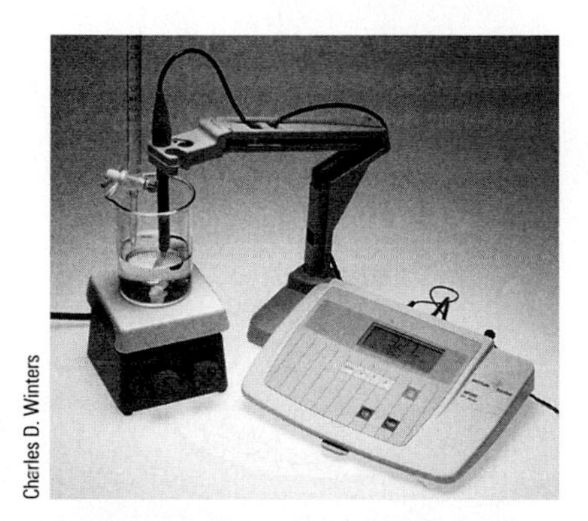

Charles D. Winters

Figure 21-18 Photograph of a typical benchtop pH meter arranged for a potentiometric titration. (Courtesy of Mettler Toledo, Inc., Columbus, OH.)

FEATURE 21-5

Operational Amplifier Voltage Measurements

One of the most important developments in chemical instrumentation over the past three decades has been the advent of compact, inexpensive, versatile integrated-circuit amplifiers (op amps).[9] These devices allow us to make potential measurements on high-resistance cells, such as those that contain a glass electrode, without drawing appreciable current. Even a small current (10^{-7} to 10^{-10} A)

(continued)

[9]For a detailed description of op amp circuits, see H. V. Malmstadt, C. G. Enke, and S. R. Crouch, *Microcomputers and Electronic Instrumentation: Making the Right Connections,* Ch. 5. Washington, DC: American Chemical Society, 1994.

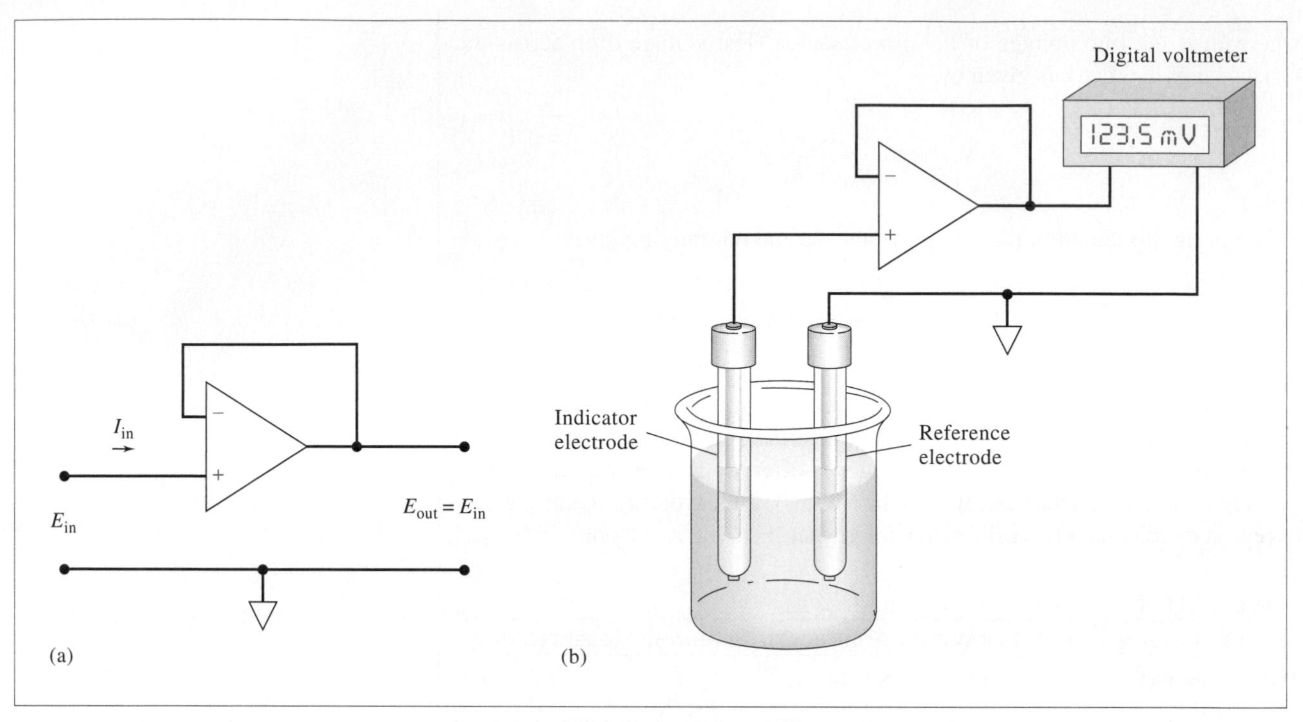

Figure 21F-7 (a) A voltage-follower operational amplifier. (b) Typical arrangement for potentiometric measurements with a membrane electrode.

in a glass electrode produces a large error in the measured voltage due to loading (see Feature 21-4) and electrode polarization (see Chapter 22). One of the most important uses for op amps is the isolation of voltage sources from their measurement circuits. The basic **voltage follower,** which permits this type of measurement, is shown in Figure 21F-7a. This circuit has two important characteristics. The output voltage E_{out} is equal to the input voltage E_{in}, and the input current I_{in} is essentially zero (10^{-9} to 10^{-15} A).

A practical application of this circuit is the measurement of cell potentials. We simply connect the cell to the op amp input as shown in Figure 21F-7b, and we connect the output of the op amp to a digital voltmeter to measure the voltage. Modern op amps are nearly ideal voltage-measurement devices and are incorporated into most ion meters and pH meters to monitor high-resistance indicator electrodes with minimal error.

Modern ion meters are digital, and some are capable of a precision on the order of 0.001 to 0.005 pH unit. Seldom is it possible to measure pH with a comparable degree of *accuracy.* Inaccuracies of ±0.02 to ±0.03 pH unit are typical.

21F DIRECT POTENTIOMETRY

Direct potentiometric measurements provide a rapid and convenient method to determine the activity of a variety of cations and anions. The technique requires only a comparison of the potential developed in a cell containing the indicator electrode in the analyte solution with its potential when immersed in one or more standard solutions of known analyte concentration. If the response of the electrode is specific for the analyte, as it often is, no preliminary separation steps are required. Direct potentiometric measurements are also readily adapted to applications requiring continuous and automatic recording of analytical data.

21F-1 Equations Governing Direct Potentiometry

The sign convention for potentiometry is consistent with the convention described in Chapter 18 for standard electrode potential.[10] In this convention, the indicator electrode is always treated as the right-hand electrode and the reference electrode as the left-hand electrode. For direct potentiometric measurements, the potential of a cell can then be expressed in terms of the potentials developed by the indicator electrode, the reference electrode, and a junction potential, as described in Section 21A:

$$E_{cell} = E_{ind} - E_{ref} + E_j \tag{21-19}$$

In Section 21D, we described the response of various types of indicator electrodes to analyte activities. For the cation X^{n+} at 25°C, the electrode response takes the general *Nernstian* form

$$E_{ind} = L - \frac{0.0592}{n}\text{pX} = L + \frac{0.0592}{n}\log a_X \tag{21-20}$$

where L is a constant and a_X is the activity of the cation. For metallic indicator electrodes, L is ordinarily the standard electrode potential; for membrane electrodes, L is the summation of several constants, including the time-dependent asymmetry potential of uncertain magnitude.

Substitution of Equation 21-20 into Equation 21-19 yields, with rearrangement,

$$\text{pX} = -\log a_X = \frac{E_{cell} - (E_j - E_{ref} + L)}{0.0592/n} \tag{21-21}$$

The constant terms in parentheses can be combined to give a new constant K.

$$\text{pX} = -\log a_X = -\frac{(E_{cell} - K)}{0.0592/n} = -\frac{n(E_{cell} - K)}{0.0592} \tag{21-22}$$

For an anion A^{n-}, the sign of Equation 21-22 is reversed:

$$\text{pA} = \frac{(E_{cell} - K)}{0.0592/n} = \frac{n(E_{cell} - K)}{0.0592} \tag{21-23}$$

All direct potentiometric methods are based on Equation 21-22 or 21-23. The difference in sign in the two equations has a subtle but important consequence in

[10]According to Bates, the convention being described here has been endorsed by standardizing groups in the United States and Great Britain as well as IUPAC. See R. G. Bates, in *Treatise on Analytical Chemistry,* 2nd ed., I. M. Kolthoff and P. J. Elving, Eds., Part I, Vol. 1, pp. 831–832. New York: Wiley, 1978.

the way that ion-selective electrodes are connected to pH meters and pIon meters. When the two equations are solved for E_{cell}, we find that for cations

$$E_{cell} = K - \frac{0.0592}{n} \text{pX} \tag{21-24}$$

and for anions

$$E_{cell} = K + \frac{0.0592}{n} \text{pA} \tag{21-25}$$

Equation 21-24 shows that for a cation-selective electrode, an increase in pX results in a *decrease* in E_{cell}. Thus, when a high-resistance voltmeter is connected to the cell in the usual way, with the indicator electrode attached to the positive terminal, the meter reading decreases as pX increases. Another way of saying this is that as the concentration (and activity) of the cation X increases, pX $= -\log$ [X] decreases, and E_{cell} increases. Notice that the sense of these changes is exactly the opposite of our sense of how pH meter readings change with increasing hydronium ion concentration. To eliminate this reversal from our sense of the pH scale, instrument manufacturers generally reverse the leads so that cation-sensitive electrodes such as glass electrodes are connected to the negative terminal of the voltage-measuring device. Meter readings then increase with increases in pX, and, as a result, they decrease with increasing concentration of the cation. Anion-selective electrodes, however, are connected to the positive terminal of the meter, so that increases in pA also yield larger readings. This sign-reversal conundrum is often confusing, so it is always a good idea to look carefully at the consequences of Equations 21-24 and 21-25 to rationalize the output of the instrument with changes in concentration of the analyte anion or cation and corresponding changes in pX or pA.

21F-2 The Electrode-Calibration Method

► The electrode-calibration method is also referred to as the method of external standards, which is described in some detail in Section 8C-2.

As we have seen from our discussions in Section 18D, the constant K in Equations 21-22 and 21-23 is made up of several constants, of which at least one, the junction potential, cannot be measured directly or calculated from theory without assumptions. Thus, before these equations can be used for the determination of pX or pA, K must be evaluated experimentally with a standard solution of the analyte.

In the electrode-calibration method, K in Equations 21-22 and 21-23 is determined by measuring E_{cell} for one or more standard solutions of known pX or pA. The assumption is then made that K is unchanged when the standard is replaced by the analyte solution. The calibration is ordinarily performed at the time that pX or pA for the unknown is determined. With membrane electrodes, recalibration may be required if measurements extend over several hours because of slow changes in the asymmetry potential.

The electrode-calibration method offers the advantages of simplicity, speed, and applicability to the continuous monitoring of pX or pA. It suffers, however, from a somewhat limited accuracy because of uncertainties in junction potentials.

Inherent Error in the Electrode-Calibration Procedure

A serious disadvantage of the electrode-calibration method is the inherent error that results from the assumption that K in Equations 21-22 and 21-23 remains constant after calibration. This assumption can seldom, if ever, be exactly true because the electrolyte composition of the unknown almost inevitably differs from that of the solution used for calibration. The junction potential term contained in K varies slightly as a consequence, even when a salt bridge is used. This error is frequently on the order of 1 mV or more. Unfortunately, because of the nature of the potential/activity relationship, such an uncertainty has an amplified effect on the inherent accuracy of the analysis.

The magnitude of the error in analyte concentration can be estimated by differentiating Equation 21-22 while assuming E_{cell} to be constant.

$$-\log_{10} e \frac{da_1}{a_1} = -0.434 \frac{da_1}{a_1} = -\frac{dK}{0.0592/n}$$

$$\frac{da_1}{a_1} = \frac{ndK}{0.0257} = 38.9\ ndK$$

When we replace da_1 and dK with finite increments and multiply both sides of the equation by 100%, we obtain

$$\text{percent relative error} = \frac{\Delta a_1}{a_1} \times 100\% = 38.9n\Delta K \times 100\%$$
$$= 3.89 \times 10^3 n\Delta K\% \approx 4000n\Delta K\%$$

The quantity $\Delta a_1/a_1$ is the relative error in a_1 associated with an absolute uncertainty ΔK in K. If, for example, ΔK is ± 0.001 V, a relative error in activity of about $\pm 4n\%$ can be expected. *It is important to appreciate that this error is characteristic of all measurements involving cells that contain a salt bridge and that this error cannot be eliminated by even the most careful measurements of cell potentials or the most sensitive and precise measuring devices.*

Activity versus Concentration

Electrode response is related to analyte activity rather than analyte concentration. We are usually interested in concentration, however, and the determination of this quantity from a potentiometric measurement requires activity coefficient data. Activity coefficients are seldom available because the ionic strength of the solution either is unknown or else is so large that the Debye-Hückel equation is not applicable.

The difference between activity and concentration is illustrated by Figure 21-19, in which the response of a calcium ion electrode is plotted against a logarithmic

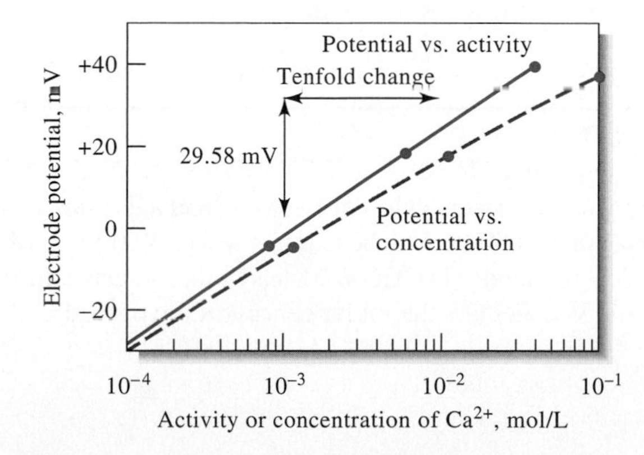

Figure 21-19 Response of a liquid-membrane electrode to variations in the concentration and activity of calcium ion. (Courtesy of Thermo Electron Corp., Beverly, MA.)

function of calcium chloride concentration. The nonlinearity is due to the increase in ionic strength with increasing electrolyte concentration and the consequent decrease in the activity of calcium ion. The upper curve is obtained when these concentrations are converted to activities. This straight line has the theoretical slope of 0.0296 (0.0592/2).

Activity coefficients for singly charged species are less affected by changes in ionic strength than are the coefficients for ions with multiple charges. Thus, the effect shown in Figure 21-19 is less pronounced for electrodes that respond to H^+, Na^+, and other univalent ions.

► Many chemical reactions of physiological importance depend on the activity of metal ions rather than their concentration.

In potentiometric pH measurements, the pH of the standard buffer used for calibration is generally based on the activity of hydrogen ions. Thus, the results are also on an activity scale. If the unknown sample has a high ionic strength, the hydrogen ion *concentration* will differ appreciably from the activity measured.

An obvious way to convert potentiometric measurements from activity to concentration is to make use of an empirical calibration curve, such as the lower plot in Figure 21-19. For this approach to be successful, it is necessary to make the ionic composition of the standards essentially the same as that of the analyte solution. Matching the ionic strength of standards to that of samples is often difficult, particularly for samples that are chemically complex.

A **total ionic strength adjustment buffer (TISAB)** is used to control the ionic strength and the pH of samples and standards in ion-selective electrode measurements.

When electrolyte concentrations are not too great, it is often useful to swamp both samples and standards with a measured excess of an inert electrolyte. The added effect of the electrolyte from the sample matrix becomes negligible under these circumstances, and the empirical calibration curve yields results in terms of concentration. This approach has been used, for example, in the potentiometric determination of fluoride ion in drinking water. Both samples and standards are diluted with a solution that contains sodium chloride, an acetate buffer, and a citrate buffer; the diluent is sufficiently concentrated so that the samples and standards have essentially identical ionic strengths. This method provides a rapid means of measuring fluoride concentrations in the part-per-million range with an accuracy of about 5% relative.

21F-3 The Standard-Addition Method

The standard-addition method (see Section 8C-3) involves determining the potential of the electrode system before and after a measured volume of a standard has been added to a known volume of the analyte solution. Multiple additions can also be made. Often, an excess of an electrolyte is incorporated into the analyte solution at the outset to prevent any major shift in ionic strength that might accompany the addition of standard. It is also necessary to assume that the junction potential remains constant during the measurements.

EXAMPLE 21-1

A cell consisting of a saturated calomel electrode and a lead ion electrode developed a potential of − 0.4706 V when immersed in 50.00 mL of a sample. A 5.00-mL addition of standard 0.02000 M lead solution caused the potential to shift to − 0.4490 V. Calculate the molar concentration of lead in the sample.

We shall assume that the activity of Pb^{2+} is approximately equal to $[Pb^{2+}]$ and apply Equation 21-22. Thus,

$$\text{pPb} = -\log [Pb^{2+}] = -\frac{E'_{\text{cell}} - K}{0.0592/2}$$

where E'_{cell} is the initial measured potential (− 0.4706 V).

After the standard solution is added, the potential becomes E''_{cell} (− 0.4490 V), and

$$-\log \frac{50.00 \times [Pb^{2+}] + 5.00 \times 0.0200}{50.00 + 5.00} = -\frac{E''_{\text{cell}} - K}{0.0592/2}$$

$$-\log(0.9091[Pb^{2+}] + 1.818 \times 10^{-3} = -\frac{E''_{\text{cell}} - K}{0.0592/2}$$

Subtracting this equation from the first leads to

$$-\log \frac{[Pb^{2+}]}{0.09091[Pb^{2+}] + 1.818 \times 10^{-3}} = \frac{2(E''_{\text{cell}} - E'_{\text{cell}})}{0.0592}$$

$$= \frac{2[-0.4490 - (-0.4706)]}{0.0592}$$

$$= 0.7297$$

$$\frac{[Pb^{2+}]}{0.09091[Pb^{2+}] + 1.818 \times 10^{-3}} = \text{antilog}(-0.7297) = 0.1863$$

$$[Pb^{2+}] = 4.08 \times 10^{-4} \text{M}$$

21F-4 Potentiometric pH Measurement with the Glass Electrode[11]

The glass electrode is unquestionably the most important indicator electrode for hydrogen ion. It is convenient to use and subject to few of the interferences that affect other pH-sensing electrodes.

The glass/calomel electrode system is a remarkably versatile tool for the measurement of pH under many conditions. It can be used without interference in solutions containing strong oxidants, strong reductants, proteins, and gases; the pH of viscous or even semisolid fluids can be determined. Electrodes for special applications are available. Included among these are small electrodes for pH measurements in one drop (or less) of solution, in a tooth cavity, or in sweat on the skin; microelectrodes that permit the measurement of pH inside a living cell; rugged electrodes for insertion in a flowing liquid stream to provide continuous monitoring of pH; and small electrodes that can be swallowed to measure the acidity of the stomach contents. (The calomel electrode is kept in the mouth.)

[11]For a detailed discussion of potentiometric pH measurements, see R. G. Bates, *Determination of pH,* 2nd ed. New York: Wiley, 1973.

Errors Affecting pH Measurements

The ubiquity of the pH meter and the general applicability of the glass electrode tend to lull the chemist into the attitude that any measurement obtained with such equipment is surely correct. The reader must be alert to the fact that there are distinct limitations to the electrode, some of which were discussed in earlier sections:

1. *The alkaline error.* The ordinary glass electrode becomes somewhat sensitive to alkali metal ions and gives low readings at pH values greater than 9.
2. *The acid error.* Values registered by the glass electrode tend to be somewhat high when the pH is less than about 0.5.
3. *Dehydration.* Dehydration may cause erratic electrode performance.
4. *Errors in low ionic strength solutions.* It has been found that significant errors (as much as 1 or 2 pH units) may occur when the pH of samples of low ionic strength, such as lake or stream water, is measured with a glass/calomel electrode system.[12] The prime source of such errors has been shown to be nonreproducible junction potentials, which apparently result from partial clogging of the fritted plug or porous fiber that is used to restrict the flow of liquid from the salt bridge into the analyte solution. To overcome this problem, free diffusion junctions of various types have been designed, and one is produced commercially.
5. *Variation in junction potential.* A fundamental source of uncertainty for which a correction cannot be applied is the junction-potential variation resulting from differences in the composition of the standard and the unknown solution.
6. *Error in the pH of the standard buffer.* Any inaccuracies in the preparation of the buffer used for calibration or any changes in its composition during storage cause an error in subsequent pH measurements. The action of bacteria on organic buffer components is a common cause of deterioration.

► Particular care must be taken in measuring the pH of approximately neutral unbuffered solutions, such as samples from lakes and streams.

The Operational Definition of pH

The usefulness of pH as a measure of the acidity and alkalinity of aqueous media, the wide availability of commercial glass electrodes, and the relatively recent proliferation of inexpensive solid-state pH meters have made the potentiometric measurement of pH perhaps the most common analytical technique in all of science. It is thus extremely important that pH be defined in a manner that is easily duplicated at various times and in various laboratories throughout the world. To meet this requirement, it is necessary to define pH in operational terms—that is, by the way the measurement is made. Only then will the pH measured by one worker be the same as that measured by another.

► Perhaps the most common analytical instrumental technique is the measurement of pH.

The operational definition of pH endorsed by the National Institute of Standards and Technology (NIST), similar organizations in other countries, and the IUPAC is based on the direct calibration of the meter with carefully prescribed standard buffers followed by potentiometric determination of the pH of unknown solutions.

► By definition, pH is what you measure with a glass electrode and a pH meter. It is approximately equal to the theoretical definition of $\text{pH} = -\log a_{\text{H}^+}$.

Consider, for example, one of the glass/reference electrode pairs of Figure 21-9. When these electrodes are immersed in a standard buffer, Equation 21-22 applies, and we can write

$$\text{pH}_S = \frac{E_S - K}{0.0592}$$

[12]See W. Davison and C. Woof, *Anal. Chem.*, **1985,** *57,* 2567; T. R. Harbinson and W. Davison, *Anal. Chem.*, **1987,** *59,* 2450; A. Kopelove, S. Franklin, and G. M. Miller, *Amer. Lab.*, **1989,** *21*(6), 40.

where E_S is the cell potential when the electrodes are immersed in the buffer. Similarly, if the cell potential is E_U when the electrodes are immersed in a solution of unknown pH, we have

$$pH_U = \frac{E_U - K}{0.0592}$$

By subtracting the first equation from the second and solving for pH_U, we find

$$pH_U = pH_S - \frac{(E_U - E_S)}{0.0592} \qquad (21\text{-}26)$$

Equation 21-26 has been adopted throughout the world as the *operational definition of pH.*

◀ An operational definition of a quantity defines the quantity in terms of how it is measured.

Workers at NIST and elsewhere have used cells without liquid junctions to study primary-standard buffers extensively. Some of the properties of these buffers are discussed in detail elsewhere.[13] Note that the NIST buffers are described by their molal concentrations (mol solute/kg solvent) for accuracy and precision of preparation. For general use, the buffers can be prepared from relatively inexpensive laboratory reagents; for careful work, however, certified buffers can be purchased from the NIST.

It should be emphasized that the strength of the operational definition of pH is that it provides a coherent scale for the determination of acidity or alkalinity. Measured pH values cannot be expected to yield a detailed picture of solution composition that is entirely consistent with solution theory, however. This uncertainty stems from our fundamental inability to measure single-ion activities. That is, the operational definition of pH does not yield the exact pH as defined by the equation

$$pH = -\log \gamma_{H^+}[H^+]$$

21G POTENTIOMETRIC TITRATIONS

A **potentiometric titration** involves measurement of the potential of a suitable indicator electrode as a function of titrant volume. The information provided by a potentiometric titration is not the same as that obtained from a direct potentiometric measurement. For example, the direct measurement of 0.100 M solutions of hydrochloric and acetic acids would yield two substantially different hydrogen ion concentrations because the latter is only partially dissociated. In contrast, the potentiometric titration of equal volumes of the two acids would require the same amount of standard base because both solutes have the same number of titratable protons.

Potentiometric titrations provide data that are more reliable than data from titrations that use chemical indicators, and they are particularly useful with colored or turbid solutions and for detecting the presence of unsuspected species. Potentiometric titrations have been automated in a variety of different ways, and commercial titrators are available from a number of manufacturers. Manual potentiometric titrations, however, suffer from the disadvantage of being more time consuming than those involving indicators.

[13] R. G. Bates, *Determination of pH,* 2nd ed., Ch. 4. New York: Wiley, 1973.

► Automatic *titrators* for carrying out potentiometric titrations are available from several manufacturers. The operator of the instrument simply adds the sample to the titration vessel and pushes a button to initiate the titration. The instrument adds titrant, records the potential versus volume data, and analyzes the data to determine the concentration of the unknown solution. A photograph of such a device is shown on the opening page of Chapter 14.

Potentiometric titrations offer additional advantages over direct potentiometry. Because the measurement is based on the titrant volume that causes a rapid *change* in potential near the equivalence point, potentiometric titrations are not dependent on measuring absolute values of E_{cell}. This makes the titration relatively free from junction potential uncertainties because the junction potential remains approximately constant during the titration. Titration results instead depend most heavily on having a titrant of accurately known concentration. The potentiometric instrument merely signals the end point and thus behaves in an identical fashion to a chemical indicator. Problems with electrodes fouling or not displaying Nernstian response are not nearly as serious when the electrode system is used to monitor a titration. Likewise, the reference electrode potential does not need to be known accurately in a potentiometric titration. Another advantage of a titration is that the result is analyte concentration even though the electrode responds to activity. For this reason, ionic strength effects are not important in the titration procedure.

Figures 21-18 and 21-20 illustrate a typical apparatus for performing a manual potentiometric titration. Its use involves measuring and recording the cell potential (in units of millivolts or pH, as appropriate) after each addition of reagent. The titrant is added in large increments early in the titration and in smaller and smaller increments as the end point is approached (as indicated by larger changes in response per unit volume).

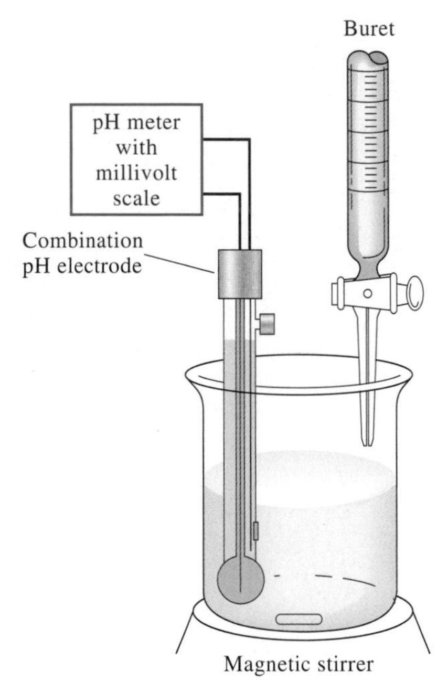

Figure 21-20 Apparatus for a potentiometric titration.

21G-1 Detecting the End Point

Several methods can be used to determine the end point of a potentiometric titration. The most straightforward involves a direct plot of potential as a function of reagent volume. In Figure 21-21a, we plot the data of Table 21-4, visually estimate

TABLE 21-4

Potentiometric Titration Data for 2.433 mmol of Chloride with 0.1000 M Silver Nitrate

Volume $AgNO_3$, mL	E vs. SCE, V	$\Delta E/\Delta V$, V/mL	$\Delta^2 E/\Delta V^2$, V²/mL²
5.00	0.062		
15.00	0.085	0.002	
20.00	0.107	0.004	
22.00	0.123	0.008	
23.00	0.138	0.015	
23.50	0.146	0.016	
23.80	0.161	0.050	
24.00	0.174	0.065	
24.10	0.183	0.09	
24.20	0.194	0.11	2.8
24.30	0.233	0.39	4.4
24.40	0.316	0.83	−5.9
24.50	0.340	0.24	−1.3
24.60	0.351	0.11	−0.4
24.70	0.358	0.07	
25.00	0.373	0.050	
25.50	0.385	0.024	
26.00	0.396	0.022	
28.00	0.426	0.015	

the inflection point in the steeply rising portion of the curve, and take it as the end point.

A second approach to end point detection is to calculate the change in potential per unit volume of titrant (that is, $\Delta E/\Delta V$); that is, we estimate the numerical first derivative of the titration curve. A plot of the first-derivative data (Table 21-4, column 3) as a function of the average volume V produces a curve with a maximum that corresponds to the point of inflection, as shown in Figure 21-21b. Alternatively, this ratio can be evaluated during the titration and recorded rather than the potential. From the plot, it can be seen that the maximum occurs at a titrant volume of about 24.30 mL. If the titration curve is symmetrical, the point of maximum slope coincides with the equivalence point. For the asymmetrical titration curves that are observed when the titrant and analyte half-reactions involve different numbers of electrons, a small titration error occurs if the point of maximum slope is used.

Figure 21-21c shows that the second derivative for the data changes sign at the point of inflection. This change is used as the analytical signal in some automatic titrators. The point at which the second derivative crosses zero is the inflection point, which is taken as the end point of the titration; this point can be located quite precisely.

All the methods of end point detection discussed in the previous paragraphs are based on the assumption that the titration curve is symmetrical about the equivalence point and that the inflection in the curve corresponds to this point. This assumption is valid if the titrant and analyte react in a 1:1 ratio and if the electrode reaction is reversible. Many oxidation/reduction reactions, such as the reaction of iron(II) with permanganate, do not occur in equimolar fashion. Even so, such titration curves are often so steep at the end point that very little error is introduced by assuming that the curves are symmetrical.

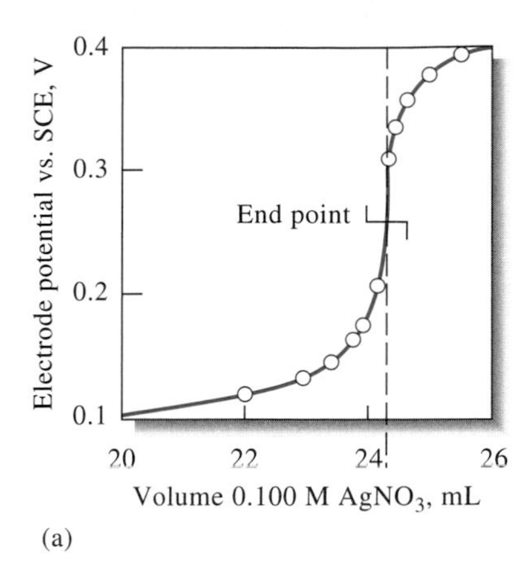

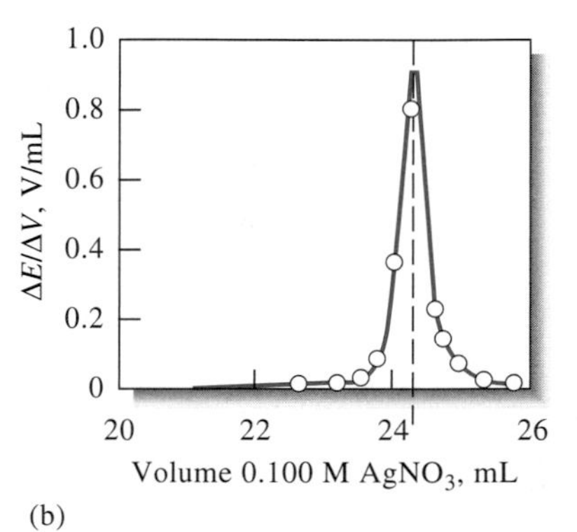

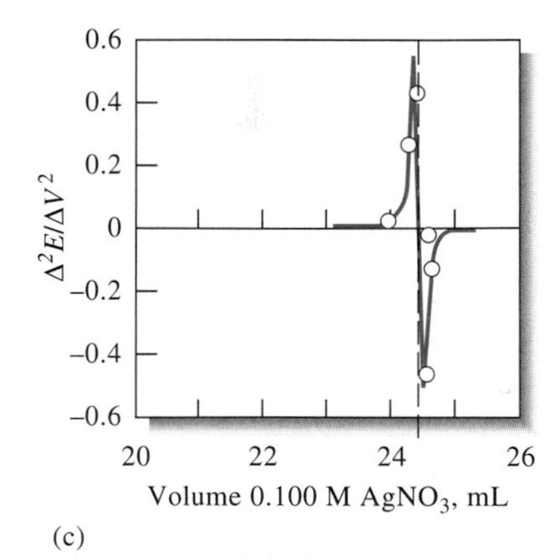

Figure 21-21 Titration of 2.433 mmol of chloride ion with 0.1000 M silver nitrate. (a) Titration curve. (b) First-derivative curve. (c) Second-derivative curve.

Spreadsheet Summary In Chapter 7 of *Applications of Microsoft® Excel in Analytical Chemistry,* the first and second derivatives of an acid/base titration curve are plotted in order to better determine the titration end point. A combination plot is produced that simultaneously displays the pH versus volume curve and the second-derivative curve. Finally, an alternative plotting method, known as a Gran plot, is explored for locating the end point by a linear regression procedure.

21G-2 Complex-Formation Titrations

Both metallic and membrane electrodes have been used to detect end points in potentiometric titrations involving complex formation. Mercury electrodes are useful for EDTA titrations of cations that form complexes that are less stable than HgY^{2-}. See Section 21D-1 for the half-reactions involved and Equation 21-5 for the Nernst expression describing the behavior of the electrode. Hanging mercury drop and thin mercury film electrodes appropriate for EDTA titrations are available from a number of manufacturers. As always, whenever mercury is used in experiments like these, we must take every precaution to avoid spilling it, and it must be stored in a well-ventilated hood or a special cabinet to remove the toxic vapors of the liquid metal. Before working with mercury, be sure to read its Materials Safety Data Sheet (MSDS), and follow all appropriate safety procedures.

21G-3 Neutralization Titrations

Experimental neutralization curves closely approximate the theoretical curves described in Chapters 14 and 15. Usually, the experimental curves are somewhat displaced from the theoretical curves along the pH axis because concentrations rather than activities are used in their derivation. This displacement has little effect on determining end points, and so potentiometric neutralization titrations are quite useful for analyzing mixtures of acids or polyprotic acids. The same is true of bases.

Determining Dissociation Constants

An approximate numerical value for the dissociation constant of a weak acid or base can be estimated from potentiometric titration curves. This quantity can be computed from the pH at any point along the curve, but a very convenient point is the half-titration point. At this point on the curve

$$[HA] \approx [A^-]$$

Therefore,

$$K_a = \frac{[H_3O^+]\cancel{[A^-]}}{\cancel{[HA]}} = [H_3O^+]$$

$$pK_a = pH$$

It is important to note that the use of concentrations instead of activities may cause the value of K_a to differ from its published value by a factor of 2 or more. A more correct form of the dissociation constant for HA is

$$K_a = \frac{a_{H_3O^+}a_{A^-}}{a_{HA}} = \frac{a_{H_3O^+}\gamma_{A^-}\cancel{[A^-]}}{\gamma_{HA}\cancel{[HA]}} = \frac{a_{H_3O^+}\gamma_{A^-}}{\gamma_{HA}} \tag{21-27}$$

Since the glass electrode provides a good approximation of $a_{H_3O^+}$, the measured value of K_a differs from the thermodynamic value by the ratio of the two activity coefficients. The activity coefficient in the denominator of Equation 21-27 does not change significantly as ionic strength increases because HA is a neutral species. The activity coefficient for A^-, on the other hand, decreases as the electrolyte concentration increases. This means that the observed hydrogen ion activity must be numerically larger than the thermodynamic dissociation constant.

EXAMPLE 21-2

To determine K_1 and K_2 for H_3PO_4 from titration data, careful pH measurements are made after 0.5 and 1.5 mol of base are added for each mole of acid. It is then assumed that the hydrogen ion activities computed from these data are identical to the desired dissociation constants. Calculate the relative error incurred by the assumption if the ionic strength is 0.1 at the time of each measurement.

(From Appendix 3, K_1 and K_2 for H_3PO_4 are 7.11×10^{-3} and 6.34×10^{-8}, respectively.)

If we rearrange Equation 21-27, we find that

$$K_a(\text{exptl}) = a_{H_3O^+} = K\left(\frac{\gamma_{HA}}{\gamma_{A^-}}\right)$$

The activity coefficient for H_3PO_4 is approximately equal to 1, since the free acid has no charge. In Table 10-1, we find that the activity coefficient for $H_2PO_4^-$ is 0.78 and that for HPO_4^{2-} is 0.36. When we substitute these values into the equations for K_1 and K_2, we find that

$$K_1(\text{exptl}) = 7.11 \times 10^{-3}\left(\frac{1.00}{0.78}\right) = 9.1 \times 10^{-3}$$

$$\text{error} = \frac{9.1 \times 10^{-3} - 7.11 \times 10^{-3}}{7.11 \times 10^{-3}} \times 100\% = 28\%$$

$$K_2(\text{exptl}) = 6.34 \times 10^{-8}\left(\frac{0.78}{0.36}\right) = 1.37 \times 10^{-7}$$

$$\text{error} = \frac{1.37 \times 10^{-7} - 6.34 \times 10^{-8}}{6.34 \times 10^{-8}} \times 100\% = 116\%$$

It is possible to identify an unknown pure acid by performing a single titration to determine its equivalent mass (molar mass if the acid is monoprotic) and its dissociation constant.

21G-4 Oxidation/Reduction Titrations

An inert indicator electrode constructed of platinum is ordinarily used to detect end points in oxidation/reduction titrations. Occasionally, other inert metals such as silver, palladium, gold, and mercury are used instead. Titration curves similar to those constructed in Section 19D are usually obtained, although they may be displaced along the potential (vertical) axis as a consequence of the high ionic strengths. End points are determined by the methods described earlier in this chapter.

21H POTENTIOMETRIC DETERMINATION OF EQUILIBRIUM CONSTANTS

Numerical values for solubility-product constants, dissociation constants, and formation constants are conveniently evaluated through the measurement of cell potentials. One important virtue of this technique is that the measurement can be made without appreciably affecting any equilibria that may be present in the solution. For example, the potential of a silver electrode in a solution containing silver ion, cyanide ion, and the complex formed between them depends on the activities of the three species. It is possible to measure this potential with negligible current.

Since the activities of the participants are not altered during the measurement, the position of the equilibrium

$$Ag^+ + 2CN^- \rightleftharpoons Ag(CN)_2^-$$

is likewise undisturbed.

EXAMPLE 21-3

Calculate the formation constant K_f for $Ag(CN)_2^-$:

$$Ag^+ + 2CN^- \rightleftharpoons Ag(CN)_2^-$$

if the cell

$$\text{SCE} \,\|\, Ag(CN)_2^-(7.50 \times 10^{-3}\ \text{M}),\ CN^-(0.0250\ \text{M}) \,|\, Ag$$

develops a potential of -0.625 V.

Proceeding as in the earlier examples, we have

$$Ag^+ + e^- \rightleftharpoons Ag(s) \qquad E^0 = +0.799\ \text{V}$$

$$-0.625 = E_{\text{right}} - E_{\text{left}} = E_{Ag^+} - 0.244$$

$$E_{Ag^+} = -0.625 + 0.244 = -0.381\ \text{V}$$

We then apply the Nernst equation for the silver electrode to find that

$$-0.381 = 0.799 - \frac{0.0592}{1} \log \frac{1}{[Ag^+]}$$

$$\log[Ag^+] = \frac{-0.381 - 0.799}{0.0592} = -19.93$$

$$[Ag^+] = 1.2 \times 10^{-20}$$

$$K_f = \frac{[Ag(CN)_2^-]}{[Ag^+][CN^-]^2} = \frac{7.50 \times 10^{-3}}{(1.2 \times 10^{-20})(2.5 \times 10^{-2})^2}$$

$$= 1.0 \times 10^{21} \approx 1 \times 10^{21}$$

In theory, any electrode system in which hydrogen ions are participants can be used to evaluate dissociation constants for acids and bases.

EXAMPLE 21-4

Calculate the dissociation constant K_{HP} for the weak acid HP if the cell

$$\text{SCE} \,\|\, HP(0.010\ \text{M}),\ NaP(0.040\ \text{M}) \,|\, Pt,\ H_2\ (1.00\ \text{atm})$$

develops a potential of -0.591 V.

The diagram for this cell indicates that the saturated calomel electrode is the left-hand electrode. Thus,

$$E_{cell} = E_{right} - E_{left} = E_{right} - 0.244 = -0.591 \text{ V}$$

$$E_{right} = -0.591 + 0.244 = -0.347 \text{ V}$$

We then apply the Nernst equation for the hydrogen electrode to find that

$$-0.347 = 0.000 - \frac{0.0592}{2} \log - \frac{1.00}{[H_3O^+]^2}$$

$$= 0.000 - \frac{2 \times 0.0592}{2} \log[H_3O^+]$$

$$\log[H_3O^+] = \frac{-0.347 - 0.000}{0.0592} = -5.86$$

$$[H_3O^+] = 1.38 \times 10^{-6}$$

By substituting this value of the hydronium ion concentration as well as the concentrations of the weak acid and its conjugate base into the dissociation-constant expression, we obtain

$$K_{HP} = \frac{[H_3O^+][P^-]}{HP} = \frac{(1.38 \times 10^{-6})(0.040)}{0.010} = 5.5 \times 10^{-6}$$

WEB WORKS

Use a Web search engine such as Google to find sites dealing with potentiometric titrators. This search should turn up such companies as Spectralab, Analyticon, Fox Scientific, Brinkmann, Metrohm, Mettler-Toledo, and Thermo Electron. Set your browser to one or two of these and explore the types of titrators that are commercially available. At the sites of two different manufacturers, find application notes or bulletins for determining two analytes by potentiometric titration. For each, list the analyte, the instruments and the reagents that are necessary for the determination, and the expected accuracy and precision of the results. Describe the detailed chemistry behind each determination and the experimental procedure.

QUESTIONS AND PROBLEMS

21-1. Briefly describe or define
- *(a) indicator electrode.
- (b) reference electrode.
- *(c) electrode of the first kind.
- (d) electrode of the second kind.

21-2. Briefly describe or define
- *(a) liquid-junction potential.
- (b) boundary potential.
- (c) asymmetry potential.

***21-3.** Describe how a mercury electrode could function as
- (a) an electrode of the first kind for Hg(II).
- (b) an electrode of the second kind for EDTA.

21-4. What is meant by Nernstian behavior in an indicator electrode?

***21-5.** Describe the source of pH dependence in a glass membrane electrode.

21-6. Why is it necessary for the glass in the membrane of a pH-sensitive electrode to be appreciably hygroscopic?

***21-7.** List several sources of uncertainty in pH measurements with a glass/calomel electrode system.

21-8. What experimental factor places a limit on the number of significant figures in the response of a membrane electrode?

***21-9.** Describe the alkaline error in the measurement of pH. Under what circumstances is this error appreciable? How are pH data affected by alkaline error?

21-10. How does a gas-sensing probe differ from other membrane electrodes?

21-11. What is the source of
*(a) the asymmetry potential in a membrane electrode?
(b) the boundary potential in a membrane electrode?
*(c) a junction potential in a glass/calomel electrode system?
(d) the potential of a crystalline membrane electrode used to determine the concentration of F^-?

***21-12.** How does information supplied by a direct potentiometric measurement of pH differ from that obtained from a potentiometric acid/base titration?

21-13. Give several advantages of a potentiometric titration over a direct potentiometric measurement.

21-14. What is the "operational definition of pH?" Why is it used?

***21-15.** (a) Calculate E^0 for the process

$$AgIO_3(s) + e^- \rightleftharpoons Ag(s) + IO_3^-$$

(b) Use the shorthand notation to describe a cell consisting of a saturated calomel reference electrode and a silver indicator electrode that could be used to measure pIO_3.
(c) Develop an equation that relates the potential of the cell in (b) to pIO_3.
(d) Calculate pIO_3 if the cell in (b) has a potential of 0.294 V.

21-16. (a) Calculate E^0 for the process

$$PbI_2(s) + e^- \rightleftharpoons Pb(s) + 2I^-$$

(b) Use the shorthand notation to describe a cell consisting of a saturated calomel reference electrode, and a lead indicator electrode that could be used for the measurement of pI.
(c) Generate an equation that relates the potential of this cell to pI.
(d) Calculate pI if this cell has a potential of −0.348 V.

21-17. Use the shorthand notation to describe a cell consisting of a saturated calomel reference electrode and a silver indicator electrode for the measurement of
(a) pSCN.
*(b) pI.
(c) pSO_3.
*(d) pPO_4.

21-18. Generate an equation that relates pAnion to E_{cell} for each of the cells in Problem 21-17. (For Ag_2SO_3, $K_{sp} = 1.5 \times 10^{-14}$; for Ag_3PO_4, $K_{sp} = 1.3 \times 10^{-20}$.)

21-19. Calculate
*(a) pSCN if the cell in Problem 21-17(a) has a potential of 0.194 V.
(b) pI if the cell in Problem 21-17(b) has a potential of −211 mV.
*(c) pSO_3 if the cell in Problem 21-17(c) has a potential of 267 mV.
(d) pPO_4 if the cell in Problem 21-17(d) has a potential of 0.244 V.

***21-20.** The cell

$$SCE \| Ag_2CrO_4(sat'd), CrO_4^{2-}(x\ M) | Ag$$

is used for the determination of $pCrO_4$. Calculate $pCrO_4$ when the cell potential is 0.336 V.

***21-21.** The cell

$$SCE \| H^+\ (a = x) | \text{glass electrode}$$

has a potential of 0.2094 V when the solution in the right-hand compartment is a buffer of pH 4.006. The following potentials are obtained when the buffer is replaced with unknowns: (a) −0.2910 V and (b) +0.2011 V. Calculate the pH and the hydrogen ion activity of each unknown. (c) Assuming an uncertainty of 0.002 V in the junction potential, what is the range of hydrogen ion activities within which the true value might be expected to lie?

***21-22.** A 0.5788-g sample of a purified organic acid was dissolved in water and titrated potentiometrically. A plot of the data revealed a single end point after 23.29 mL of 0.0994 M NaOH had been introduced. Calculate the molecular mass of the acid.

21-23. Calculate the potential of a silver indicator electrode versus the standard calomel electrode after the addition of 5.00, 15.00, 25.00, 30.00, 35.00, 39.00, 39.50, 36.60, 39.70, 39.80, 39.90, 39.95, 39.99, 40.00, 40.01, 40.05, 40.10, 40.20, 40.30, 40.40, 40.50, 41.00, 45.00, 50.00, 55.00, and 70.00 mL of 0.1000 M $AgNO_3$ to 50.00 mL of 0.0800 M KSeCN. Construct a titration curve and a first- and second-derivative plot from these data. (K_{sp} for AgSeCN = 4.20×10^{-16}.)

21-24. A 40.00-mL aliquot of 0.05000 M HNO_2 is diluted to 75.00 mL and titrated with 0.0800 M Ce^{4+}. The pH of the solution is maintained at 1.00 throughout the titration; the formal potential of the cerium system is 1.44 V.

*(a) Calculate the potential of the indicator electrode with respect to a saturated calomel reference electrode after the addition of 5.00, 10.00, 15.00, 25.00, 40.00, 49.00, 49.50, 49.60, 49.70, 49.80, 49.90, 49.95, 49.99, 50.00, 50.01, 50.05, 50.10, 50.20, 50.30, 50.40, 50.50, 51.00, 60.00, 75.00, and 90.00 mL of cerium(IV).

(b) Draw a titration curve for these data.

(c) Generate a first- and second-derivative curve for these data. Does the volume at which the second-derivative curve crosses zero correspond to the theoretical equivalence point? Why or why not?

21-25. The titration of Fe(II) with permanganate yields a particularly asymmetrical titration curve because of the different number of electrons involved in the two half-reactions. Consider the titration of 25.00 mL of 0.1 M Fe(II) with 0.1 M MnO_4^-. The H^+ concentration is maintained at 1.0 M throughout the titration. Use a spreadsheet to generate a theoretical titration curve and a first- and second-derivative plot. Do the inflection points obtained from the maximum of the first-derivative plot or the zero crossing of the second-derivative plot correspond to the equivalence point? Explain why or why not.

***21-26.** The Na^+ concentration of a solution was determined by measurement with a sodium ion–selective electrode. The electrode system developed a potential of 0.2331 V when immersed in 10.0 mL of the solution of unknown concentration. After addition of 1.00 mL of 2.00×10^{-2} M NaCl, the potential changed to − 0.1846 V. Calculate the Na^+ concentration of the original solution.

21-27. The F^- concentration of a solution was determined by measurement with a liquid-membrane electrode. The electrode system developed a potential of 0.4965 V when immersed in 25.00 mL of the sample and 0.4117 V after the addition of 2.00 mL of 5.45×10^{-2} M NaF. Calculate pF for the sample.

21-28. A lithium ion–selective electrode gave the potentials given in the table for the following standard solutions of LiCl and three samples of unknown concentration.

Solution (a_{Li})	Potential vs. SCE, mV
0.100 M	+1.0
0.050 M	−30.0
0.010 M	−60.0
0.001 M	−138.0
Unknown 1	−48.5
Unknown 2	−75.3

(a) Draw a calibration curve of electrode potential versus log a_{Li^+} and determine if the electrode follows the Nernst equation.

(b) Use a linear least-squares procedure to determine the concentrations of the two unknowns.

21-29. A fluoride electrode was used to determine the amount of fluoride in drinking water samples. The results given in the table were obtained for four standards and two unknowns. Constant ionic strength and pH conditions were used.

Solution Containing F^-	Potential vs. SCE, mV
5.00×10^{-4} M	0.02
1.00×10^{-4} M	41.4
5.00×10^{-5} M	61.5
1.00×10^{-5} M	100.2
Unknown 1	38.9
Unknown 2	55.3

(a) Plot a calibration curve of potential versus log[F^-]. Determine whether the electrode system shows Nernstian response.

(b) Determine the concentration of F^- in the two unknown samples by a linear least-squares procedure.

21-30. Challenge Problem. In recent work Ceresa, Pretsch, and Bakker[14] investigated three ion-selective electrodes (ISEs) for determining calcium concentrations. All three electrodes used the same membrane, but differed in the composition of the inner solution. Electrode 1 was a conventional ISE with an inner solution of 1.00×10^{-3} M $CaCl_2$ and 0.10 M NaCl. Electrode 2 (low activity of Ca^{2+}) had an inner solution containing the same analytical concentration of $CaCl_2$, but with 5.0×10^{-2} M EDTA adjusted to a pH of 9.0 with 6.0 ×

[14] A. Ceresa, E. Pretsch, and E. Bakker, *Anal. Chem.*, **2000,** *72,* 2054.

10^{-2} M NaOH. Electrode 3 (high Ca^{2+} activity) had an inner solution of 1.00 M $Ca(NO_3)_2$.

(a) Determine the Ca^{2+} concentration in the inner solution of Electrode 2.

(b) Determine the ionic strength of the solution in Electrode 2.

(c) Use the Debye-Hückel equation and determine the activity of Ca^{2+} in Electrode 2. Use 0.6 nm for the α_X value for Ca^{2+}.

(d) Electrode 1 was used in a cell with a calomel reference electrode to measure standard calcium solutions with activities ranging from 0.001 M to 1.00×10^{-9} M. The following data were obtained.

Activity of Ca^{2+}, M	Cell Potential, mV
1.0×10^{-3}	93
1.0×10^{-4}	73
1.0×10^{-5}	37
1.0×10^{-6}	2
1.0×10^{-7}	−23
1.0×10^{-8}	−51
1.0×10^{-9}	−55

Plot the cell potential versus the pCa and determine the pCa value where the plot deviates significantly from linearity. For the linear portion, determine the slope and intercept of the plot. Does the plot obey the expected Equation 21-24?

(e) For Electrode 2, the following results were obtained.

Activity of Ca^{2+}	Cell Potential, V
1.0×10^{-3}	228
1.0×10^{-4}	190
1.0×10^{-5}	165
1.0×10^{-6}	139
5.6×10^{-7}	105
3.2×10^{-7}	63
1.8×10^{-7}	36
1.0×10^{-7}	23
1.0×10^{-8}	18
1.0×10^{-9}	17

Again plot cell potential versus pCa and determine the range of linearity for Electrode 2. Determine the slope and intercept for the linear portion. Does this electrode obey Equation 21-24 for the higher Ca^{2+} activities?

(f) Electrode 2 is said to be super-Nernstian for concentrations from 10^{-7} M to 10^{-6} M. Why is this term used? If you have access to a library that subscribes to *Analytical Chemistry* or has Web access to the journal, read the article. This electrode is said to have Ca^{2+} uptake. What does this mean and how might it explain the response?

(g) Electrode 3 gave the following results.

Activity of Ca^{2+}, M	Cell Potential, mV
1.0×10^{-3}	175
1.0×10^{-4}	150
1.0×10^{-5}	123
1.0×10^{-6}	88
1.0×10^{-7}	75
1.0×10^{-8}	72
1.0×10^{-9}	71

Plot the cell potential versus pCa and determine the range of linearity. Again determine the slope and intercept. Does this electrode obey Equation 21-24?

(h) Electrode 3 is said to have Ca^{2+} release. Explain this term from the article and describe how it might explain the response.

(i) Does the article give any alternative explanations for the experimental results? If so, describe these alternatives.

InfoTrac College Edition

For additional readings, go to InfoTrac College Edition, your online research library, at

http://infotrac.thomsonlearning.com

CHAPTER 22

Bulk Electrolysis: Electrogravimetry and Coulometry

Electrolysis is widely used commercially to provide attractive metal coverings to objects such as truck bumpers, which are chromium plated; silverware, which is often silver plated; and jewelry, which may be electroplated with various precious metals. Another example of an electroplated object is the Oscar, shown in the photo, which is given to recipients of Academy Awards. Each Oscar stands 13.5 inches tall, not including the base, and weighs 8.5 pounds. The statuette is hand cast in brittanium, an alloy of tin, copper, and antimony, in a steel mold. The cast is then electroplated with copper. Nickel electroplating is applied to seal the pores of the metal. The statuette is then washed in a silver-plate, which adheres well to gold. Finally, after polishing, the statuette is electroplated with 24-karat gold and finished in a baked lacquer. The amount of gold deposited on Oscar could be determined by weighing the statuette before and after the final electrolysis step. This technique, called electrogravimetry, is one of the subjects of this chapter. Alternatively, the current during the electroplating process could be integrated to find the total amount of charge required to electroplate Oscar. The number of moles of electrons needed could then be used to calculate the mass of gold deposited. This method, known as coulometry, is also a subject of this chapter.

Sue Ogrocki/Reuters/Corbis

In this chapter we describe two bulk electroanalytical methods: electrogravimetry and coulometry.[1] *In contrast to the potentiometric methods described in Chapters 18 to 21, the methods described here are electrolytic, with a net current and a net cell reaction. Electrogravimetry and coulometry are related methods in which electrolysis is carried out for a sufficient length of time to ensure complete oxidation or reduction of the analyte to a product of known composition. In electrogravimetry, the goal is to determine the amount of analyte present by converting it electrolytically to a product that is weighed as a deposit on one of the electrodes. In coulometric procedures, we determine the amount of analyte by measuring the quantity of electrical charge needed to completely convert it to a product.*

Electrogravimetry and coulometry are moderately sensitive and among the most accurate and precise techniques available to the chemist. Like the gravimetric techniques discussed in Chapter 12, electrogravimetry requires no preliminary calibration against chemical standards because the functional relationship between

◀ Electrogravimetry and coulometry can often exhibit accuracies of a few parts per thousand.

[1]For further information concerning the methods in this chapter, see A. J. Bard and L. R. Faulkner, *Electrochemical Methods,* 2nd ed., Ch. 11 New York: Wiley, 2001; J. A. Dean, *Analytical Chemistry Handbook,* Section 14, pp. 14.93–14.133. New York: McGraw-Hill, 1995.

André Marie Ampère (1775–1836), French mathematician and physicist, was the first to apply mathematics to the study of electrical current. Consistent with Benjamin Franklin's definitions of positive and negative charge, Ampère defined a positive current to be the direction of flow of positive charge. Although we now know that negative electrons carry current in metals, Ampère's definition has survived to the present. The unit of current, the ampere, is named in his honor.

the quantity measured and the analyte concentration can be derived from theory and atomic mass data.

Since we have not previously considered, what happens when current is present in an electrochemical cell, we begin with a discussion of this. Then bulk electrolysis methods are discussed in some detail. The voltammetric methods described in Chapter 23 also require a net current in the cell but use such small electrode areas that no appreciable changes in bulk concentrations occur.

22A THE EFFECT OF CURRENT ON CELL POTENTIAL

When there is a net current in an electrochemical cell, the measured potential across the two electrodes is no longer simply the difference between the two electrode potentials as calculated from the Nernst equation. Two additional phenomena, ***IR* drop** and **polarization,** must be considered when current is present. Because of these phenomena, potentials larger than the thermodynamic potential are needed to operate an electrolytic cell. When present in a galvanic cell, *IR* drop and polarization result in the development of potentials smaller than predicted.

Let us now examine these two phenomena in detail. As an example, consider the following electrolytic cell for the determination of cadmium(II) in hydrochloric acid solutions by electrogravimetry or coulometry:

$$\text{Ag}|\text{AgCl}(s),\text{Cl}^-(0.200\text{ M}),\text{Cd}^{2+}(0.00500\text{ M})|\text{Cd}$$

Current is the rate of charge flow in a circuit or solution. One ampere of current is a charge flow rate of one coulomb per second (1 A = 1 C/s). **Voltage,** the electrical potential difference, is the potential energy that results from the separation of charges. One volt of electrical potential results when one joule of potential energy is required to separate one coulomb of charge (1 V = 1 J/C).

Similar cells can be used to determine Cu(II) and Zn(II) in acid solution. In this cell, the right-hand electrode is a metal electrode that has been coated with a layer of cadmium. Because this is the electrode at which the reduction of Cd^{2+} ions occurs, this **working electrode** operates as a cathode. On the left is a silver/silver chloride electrode whose electrode potential remains more-or-less constant during the analysis. The left-hand electrode is thus the **reference electrode.** Note that this is an example of a cell without liquid junction. As shown in Example 22-1, this cell, as written, has a thermodynamic potential of −0.734 V. Here the negative sign for the cell potential indicates that the spontaneous reaction is *not* the reduction of Cd^{2+} on the right and the oxidation of Ag on the left. To reduce Cd^{2+} to Cd, we must construct an electrolytic cell and *apply* a potential somewhat more negative than −0.734 V. Such a cell is shown in Figure 22-1a. With this cell, we force the Cd electrode to become the cathode so that the net reaction shown in Equation 22-1 occurs in the left-to-right direction.

$$\text{Cd}^{2+} + 2\text{Ag}(s) + 2\text{Cl}^- \rightleftharpoons \text{Cd}(s) + 2\text{AgCl}(s) \qquad (22\text{-}1)$$

Note that this cell is reversible, so that in the absence of the external voltage source shown in the figure, the spontaneous cell reaction is in the right-to-left direction toward oxidation of Cd(*s*) to Cd^{2+}. If the spontaneous reaction is allowed to occur by short-circuiting the galvanic cell, the Cd electrode becomes the anode.

22A-1 Ohmic Potential; *IR* Drop

► Ohm's law: $E = IR$ or $I = E/R$. The units of resistance are ohms (Ω). One ohm equals one volt per ampere. Thus, the product IR has the units of amperes × volts/ampere = volts.

Electrochemical cells, like metallic conductors, resist the flow of charge. Ohm's law describes the effect of this resistance on the magnitude of the current in the cell. The product of the resistance *R* of a cell in ohms (Ω) and the current *I* in amperes (A) is called the ohmic potential or the *IR* drop of the cell. In Figure 22-1b,

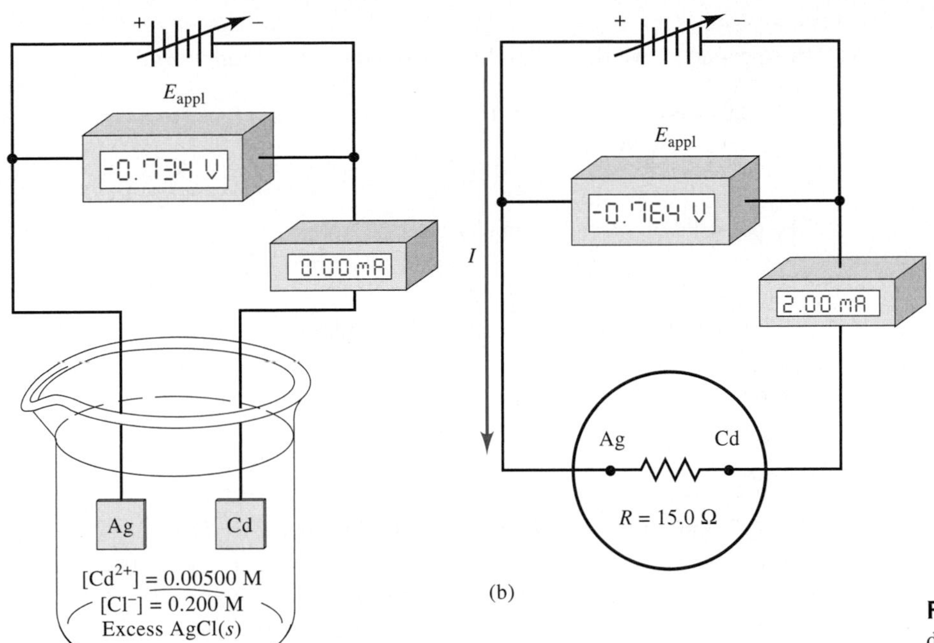

Figure 22-1 An electrolytic cell for determining Cd^{2+}. (a) Current = 0.00 mA. (b) Schematic of cell in (a) with internal resistance of cell represented by a 15.0 Ω resistor and $E_{applied}$ increased to give a current of 2.00 mA.

we have used a resistor R to represent the cell resistance in Figure 22-1a. To generate a current of I amperes in this cell, we must apply a potential that is IR volts more negative than the thermodynamic cell potential, $E_{cell} = E_{right} - E_{left}$. That is,

$$E_{applied} = E_{cell} - IR \qquad (22\text{-}2)$$

Usually we try to minimize the IR drop in the cell by having a very small cell resistance (high ionic strength) or by using a special **three-electrode cell** (see Section 22C-2), in which the current passes between the working electrode and an **auxiliary electrode,** or **counter electrode.** With this arrangement, only a very small current passes between the working electrode and the reference electrode, which minimizes the IR drop.

Direct current (dc) is current that is always in one direction; it is unidirectional. The direction of **alternating current (ac)** reverses periodically. We can also speak of voltage sources that are unidirectional (dc) or of alternating polarity (ac). The terms ac and dc are also used to describe power supplies, circuits, and components designed for alternating or unipolar operation, respectively. DC voltage sources are often given the battery symbol with + and − polarities indicated, as shown in Figure 22-1. An arrow through the battery indicates that the source voltage is variable and can be changed to another dc value.

EXAMPLE 22-1

The following cell has been used for the determination of cadmium in the presence of chloride ions by both electrogravimetry and coulometry.

$$Ag|AgCl(s),Cl^-(0.200\ M),Cd^{2+}(0.00500\ M)|Cd$$

Calculate the potential that (a) must be applied to prevent a current from developing in the cell when the two electrodes are connected and (b) must be applied to cause an electrolytic current of 2.00 mA to develop. Assume that the internal resistance of the cell is 15.0 Ω.

(continued)

(a) In Appendix 5, we find the following standard reduction potentials:

$$Cd^{2+} + 2e^- \rightleftharpoons Cd(s) \qquad E^0 = -0.403 \text{ V}$$

$$AgCl(s) + e^- \rightleftharpoons Ag(s) + Cl^- \qquad E^0 = 0.222 \text{ V}$$

The potential of the cadmium electrode is

$$E_{right} = -0.403 - \frac{0.0592}{2} \log \frac{1}{0.00500} = -0.471 \text{ V}$$

and that of the silver electrode is

$$E_{left} = 0.222 - 0.0592 \log (0.200) = 0.263 \text{ V}$$

Since the current is to be 0.00 mA, we find from Equation 22-2,

$$\begin{aligned} E_{applied} &= E_{cell} = E_{right} - E_{left} \\ &= -0.471 - 0.263 = -0.734 \text{ V} \end{aligned}$$

Hence, to prevent the passage of current in this cell, we would need to apply a voltage of −0.734 V, as shown in Figure 22-1a. Note that to obtain a current of 0.00 mA, the applied voltage must exactly match the galvanic cell potential. This is the basis for a very precise null comparison measurement of the galvanic cell potential. We use a variable, standard voltage source as the applied voltage and adjust its output until a current of 0.00 mA is obtained, as indicated by a very sensitive current meter called a galvanometer. At this *null point,* the standard voltage is read on a voltmeter to obtain the value of E_{cell}. Since there is no current at the null point, this type of voltage measurement prevents the loading error discussed in Section 21E.

(b) To calculate the applied potential needed to develop a current of 2.00 mA, or 2.00×10^{-3} A, we substitute into Equation 22-2 to give

$$\begin{aligned} E_{applied} &= E_{cell} - IR \\ &= -0.734 - 2.00 \times 10^{-3} \text{ A} \times 15 \ \Omega \\ &= -0.734 - 0.030 = -0.764 \text{ V} \end{aligned}$$

Thus, to obtain a 2.00-mA current as in Figure 22-1b, an applied potential of −0.764 V is required.

22A-2 Polarization Effects

If we solve Equation 22-2 for current *I,* we obtain

$$I = \frac{E_{cell} - E_{applied}}{R} = -\frac{E_{applied}}{R} + \frac{E_{cell}}{R} \tag{22-3}$$

Note that a plot of current in an electrolytic cell versus applied potential should be a straight line with a slope equal to the negative reciprocal of the resistance, $-1/R$, and an intercept equal to E_{cell}/R. As can be seen in Figure 22-2, the plot is indeed linear for small currents. In this experiment, the measurements were made in a brief enough time so that neither electrode potential changed significantly as a consequence of the electrolytic reaction. As the applied voltage increases, the current ultimately begins to deviate from linearity.

The term polarization refers to the deviation of the electrode potential from the value predicted by the Nernst equation on the passage of current. Cells that exhibit nonlinear behavior at higher currents exhibit polarization, and the degree of polarization is given by an **overvoltage,** or **overpotential,** which is symbolized by Π in the figure. Note that polarization requires the application of a potential greater than the theoretical value to give a current of the expected magnitude. Thus, the overpotential required to achieve a current of 7.00 mA in the electrolytic cell in Figure 22-2 is about −0.23 V. For an electrolytic cell affected by overvoltage, Equation 22-2 then becomes

Polarization is the departure of the electrode potential from its theoretical Nernst equation value on the passage of current. **Overvoltage** is the potential difference between the theoretical cell potential from Equation 22-2 and the actual cell potential at a given level of current.

$$E_{applied} = E_{cell} - IR - \Pi \qquad (22\text{-}4)$$

Polarization is an electrode phenomenon that may affect either or both of the electrodes in a cell. The degree of polarization of an electrode varies widely. In some instances, it approaches zero, but in others it can be so large that the current in the cell becomes independent of potential. Under this circumstance, polarization is said to be complete. Polarization phenomena can be divided into two categories: **concentration polarization** and **kinetic polarization.**

◀ Factors that influence polarization include (1) electrode size, shape, and composition; (2) composition of the electrolyte solution; (3) temperature and stirring rate; (4) current level; and (5) physical state of species involved in the cell reaction.

Concentration Polarization

Concentration polarization occurs because of the finite rate of mass transfer from the solution to the electrode surface. Electron transfer between a reactive species in a solution and an electrode can take place only from the interfacial region located immediately adjacent to the surface of the electrode; this region is only a fraction of a nanometer in thickness and contains a limited number of reactive ions or molecules. For there to be a steady current in a cell, the interfacial region must be continuously replenished with reactant from the bulk of the solution. That is, as

Mass transfer is the movement of material, such as ions, from one location to another.

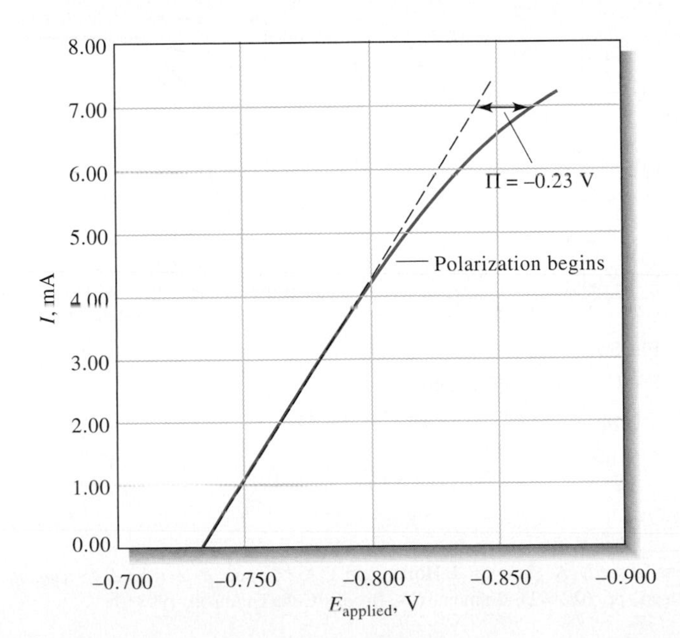

Figure 22-2 Experimental current/voltage curve for operation of the cell shown in Figure 22-1. Dashed line is the theoretical curve assuming no polarization. Overvoltage Π is the potential difference between the theoretical curve and the experimental.

reactant ions or molecules are consumed by the electrochemical reaction, more must be transported into the surface layer at a rate that is sufficient to maintain the current. For example, to have a current of 2.0 mA in the cell described in Figure 22-1b, it is necessary to transport cadmium ions to the cathode surface at a rate of about 1×10^{-8} mol/s or 6×10^{15} cadmium ions per second. Similarly, silver ions must be removed from the anode surface at a rate of 2×10^{-8} mol/s.[2]

Concentration polarization occurs when reactant species do not arrive at the surface of the electrode or when product species do not leave the surface of the electrode fast enough to maintain the desired current. When this happens, the current is limited to values less than that predicted by Equation 22-2.

► Reactants are transported to and products away from an electrode by (1) diffusion, (2) migration, and (3) convection.

Reactants are transported to the surface of an electrode by three mechanisms: (1) *diffusion,* (2) *migration,* and (3) *convection.* Products are removed from electrode surfaces in the same ways.

Diffusion is the movement of a species under the influence of a concentration gradient. It is the process that causes ions or molecules to move from a more concentrated part of a solution to a more dilute region.

CD-ROM Simulation: Diffusion of Redox Active Species as a Function of Distance from an Electrode Surface.

Diffusion When there is a concentration difference between two regions of a solution, ions or molecules move from the more concentrated region to the more dilute. This process is called **diffusion** and ultimately leads to a disappearance of the concentration gradient. The rate of diffusion is directly proportional to the concentration difference. For example, when cadmium ions are deposited at a cadmium electrode, as illustrated in Figure 22-3a, the concentration of Cd^{2+} at the electrode surface $[Cd^{2+}]_0$ becomes lower than the bulk concentration. The difference between the concentration at the surface and the concentration in the bulk solution $[Cd^{2+}]$ creates a concentration *gradient* that causes cadmium ions to diffuse from the bulk of the solution to the surface layer near the electrode (see Figure 22-3b).

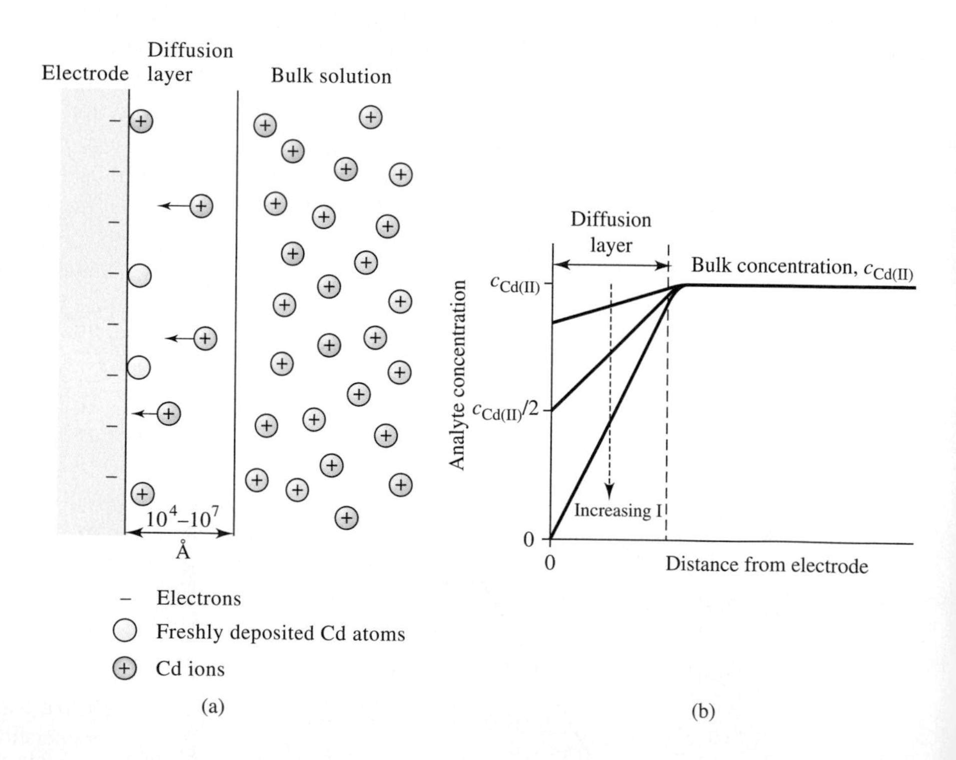

Figure 22-3 Pictorial diagram (a) and concentration vs. distance plot (b) showing concentration changes at the surface of a cadmium electrode. As Cd^{2+} ions are reduced to Cd atoms at the electrode surface, the concentration of Cd^{2+} at the surface becomes smaller than the bulk concentration. Ions then diffuse from the bulk of the solution to the surface as a result of the concentration gradient. The higher the current, the larger the concentration gradient until the surface concentration falls to zero, its lowest possible value. At this point, the maximum possible current, called the limiting current, is obtained.

[2]For more details, see D. A. Skoog, F. J. Holler, and T. A. Nieman, *Principles of Instrumental Analysis,* 5th ed., pp. 622–623. Belmont, CA, Brooks/Cole/Thomson, 1998.

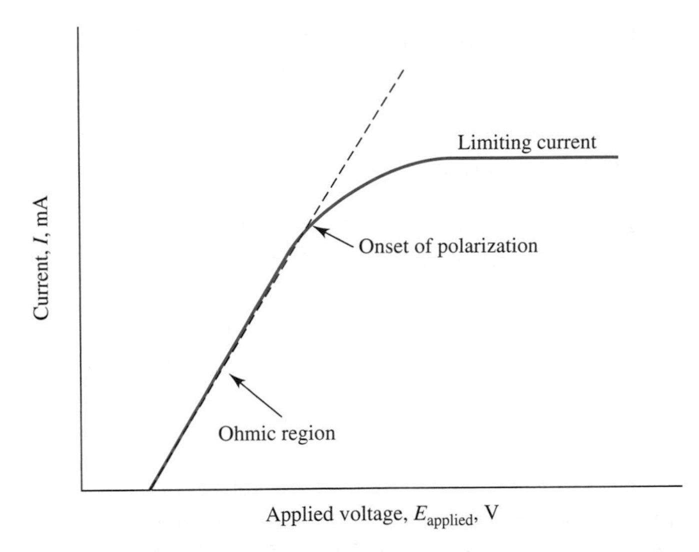

Figure 22-4 Current-potential curve for electrolysis showing the linear or ohmic region, the onset of polarization, and the limiting current plateau. In the limiting current region, the electrode is said to be completely polarized, since its potential can be changed widely without affecting the current.

The rate of diffusion is given by

$$\text{rate of diffusion to cathode surface} = k'([Cd^{2+}] - [Cd^{2+}]_0) \qquad (22\text{-}5)$$

where $[Cd^{2+}]$ is the reactant concentration in the bulk of the solution, $[Cd^{2+}]_0$ is its equilibrium concentration at the electrode surface, and k' is a proportionality or rate constant. The value of $[Cd^{2+}]_0$ at any instant is fixed by the potential of the electrode and can be calculated from the Nernst equation. In the present example, we find the surface cadmium ion concentration from the relationship

$$E_{cathode} = E^0_{Cd^{2+}/Cd} - \frac{0.0592}{2}\log\frac{1}{[Cd^{2+}]_0}$$

where $E_{cathode}$ is the potential applied to the cathode. As the applied potential becomes more and more negative, $[Cd^{2+}]_0$ becomes smaller and smaller. The result is that the rate of diffusion and the current become correspondingly larger until the surface concentration falls to zero and the maximum or **limiting current** is reached, as illustrated in Figure 22-4.

Migration involves the movement of ions through a solution as a result of electrostatic attraction between the ions and the electrodes.

Migration The electrostatic process by which ions move under the influence of an electric field is called **migration.** This process, shown schematically in Figure 22-5, is the primary cause of mass transfer in the bulk of the solution in a cell. The rate at which ions migrate to or away from an electrode surface generally increases as the electrode potential increases. This charge movement constitutes a current, which also increases with potential. Migration causes anions to be attracted to the positive electrode and cations to the negative electrode. Migration of analyte species is undesirable in most types of electrochemistry. We want to reduce anions as well as cations at an electrode of negative polarity and oxidize cations as well as anions at a positive electrode. Migration of analyte species can be minimized by having a high concentration of an inert electrolyte, called a **supporting electrolyte,** present in the cell. The current in the cell is then primarily due to charges carried by ions from the supporting electrolyte. The supporting electrolyte also serves to reduce the resistance of the cell, which decreases the *IR* drop.

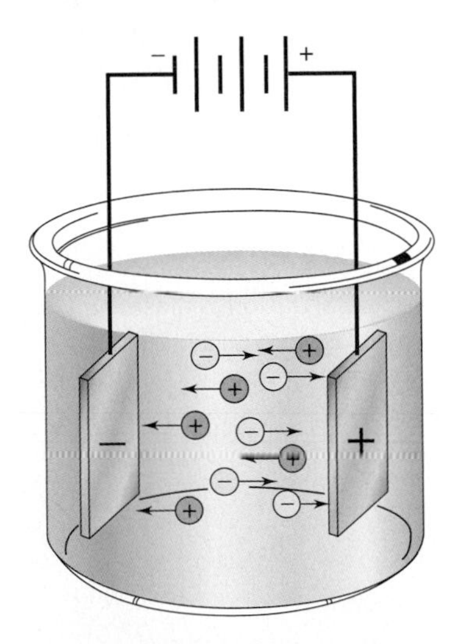

Figure 22-5
The motion of ions through a solution because of the electrostatic attraction between the ions and the electrodes is called migration.

Convection is the transport of ions or molecules through a solution as a result of stirring, vibration, or temperature gradients.

Convection Reactants can also be transferred to or from an electrode by mechanical means. **Forced convection,** such as stirring or agitation, tends to decrease the thickness of the diffusion layer at the surface of an electrode and thus decrease concentration polarization. **Natural convection** resulting from temperature or density differences also contributes to the transport of molecules and ions to and from an electrode.

► The experimental variables that influence the degree of concentration polarization are (1) reactant concentration, (2) total electrolyte concentration, (3) mechanical agitation, and (4) electrode size.

The Importance of Concentration Polarization As noted earlier, concentration polarization occurs when the effects of diffusion, migration, and convection are insufficient to transport a reactant to or from an electrode surface at a rate that produces a current of the magnitude given by Equation 22-2. Concentration polarization requires applied potentials that are larger than calculated from Equation 22-2 to maintain a given current in an electrolytic cell (see Figure 22-2). Similarly, the phenomenon causes a galvanic cell potential to be smaller than the value predicted on the basis of the theoretical potential and the *IR* drop.

Concentration polarization is important in several electroanalytical methods. In some applications, its effects are undesirable, and steps are taken to eliminate it. In others, it is essential to the analytical method, and conditions are adjusted to ensure that it occurs.

Kinetic Polarization

► The current in a kinetically polarized cell is governed by the rate of electron transfer rather than the rate of mass transfer.

In kinetic polarization, the magnitude of the current is limited by the rate of one or both of the electrode reactions—that is, the rate of electron transfer between the reactants and the electrodes. To offset kinetic polarization, an additional potential, or overvoltage, is required to overcome the activation energy of the half-reaction.

Kinetic polarization is most pronounced for electrode processes that yield gaseous products because the kinetics of the gas evolution process are involved and often slow. Kinetic polarization can be negligible for reactions that involve the deposition or solution of such metals as Cu, Ag, Zn, Cd, and Hg. Kinetic polarization can be significant, however, for reactions involving transition metals, such as Fe, Cr, Ni, and Co. Kinetic effects usually decrease with increasing temperature and decreasing current density. These effects also depend on the composition of the electrode and are most pronounced with softer metals, such as lead, zinc, and particularly mercury. The magnitude of overvoltage effects cannot be predicted from present theory and can only be estimated from empirical information in the literature.[3] Just as with *IR* drop, overvoltage effects require the application of voltages larger than calculated to operate an electrolytic cell at a desired current. Kinetic polarization also causes the potential of a galvanic cell to be smaller than calculated from the Nernst equation and the *IR* drop (see Equation 22-2).

Current density is the current per unit surface area of the electrode (A/cm^2).

Kinetic polarization is most commonly encountered when the reactant or product in an electrochemical cell is gas.

The overvoltages associated with the formation of hydrogen and oxygen are often 1 V or more and are quite important because these molecules are frequently produced by electrochemical reactions. The high overvoltage of hydrogen on such metals as copper, zinc, lead, and mercury is particularly interesting. These metals and several others can, therefore, be deposited without interference from hydrogen evolution. In theory, it is not possible to deposit zinc from a neutral aqueous solution because hydrogen forms at a potential that is considerably less than that required for zinc deposition. In fact, zinc can be deposited on a copper electrode with no significant hydrogen formation because the rate at which the gas forms on

[3]Overvoltage data for various gaseous species on different electrode surfaces have been compiled in J.A. Dean, *Analytical Chemistry Handbook,* Section 14, pp. 14.96–14.97. New York: McGraw-Hill, 1995.

FEATURE 22-1

Overvoltage and the Lead/Acid Battery

If it were not for the high overvoltage of hydrogen on lead and lead oxide electrodes, the lead/acid storage batteries found in automobiles and trucks (Figure 22F-1) would not operate because of hydrogen formation at the cathode during both charging and use. Certain trace metals in the system lower this overvoltage and eventually lead to gassing, or hydrogen formation, which limits the lifetime of the battery. The basic difference between a battery with a 48-month warranty and one with a 72-month warranty is the concentration of these trace metals in the system. The overall cell reaction when the cell is discharging is

$$Pb(s) + PbO_2(s) + 2HSO_4^- + 2H^+ \rightarrow 2PbSO_4(s) + 2H_2O$$

The lead/acid storage battery behaves as a galvanic cell during discharge and as an electrolytic cell when it is being charged. Such batteries acting as galvanic cells were once used as voltage sources for electrolysis. They have been supplanted for this purpose by modern line-operated power supplies.

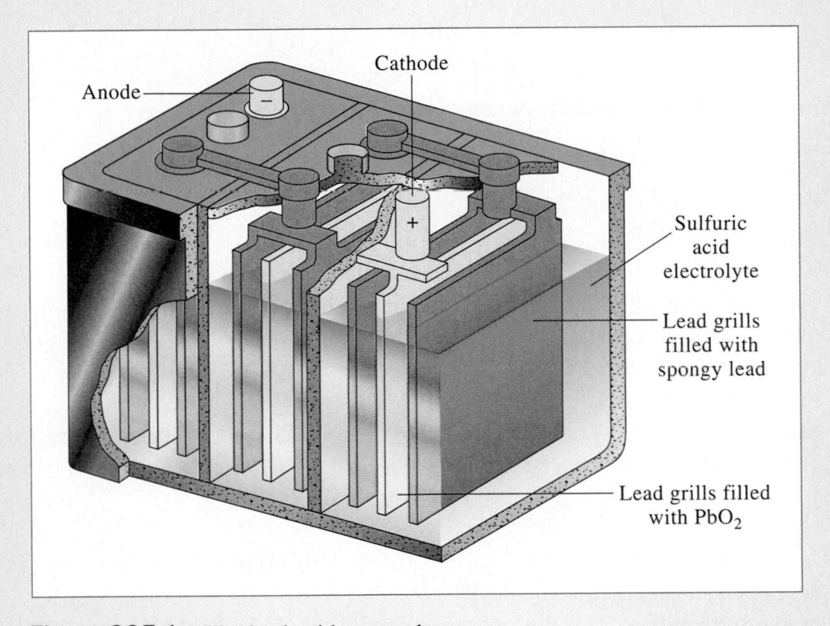

Figure 22F-1 The lead-acid storage battery.

both zinc and copper is negligible, as shown by the high hydrogen overvoltage associated with these metals.

22B THE SELECTIVITY OF ELECTROLYTIC METHODS

In principle, electrolytic methods offer a reasonably selective means of separating and determining a number of ions. The feasibility of and theoretical conditions for accomplishing a given separation can be derived from the standard electrode potentials of the species of interest, as illustrated in Example 22-2.

EXAMPLE 22-2

Is a quantitative separation of Cu^{2+} and Pb^{2+} by electrolytic deposition feasible in principle? If so, what range of cathode potentials versus the saturated calomel electrode (SCE) can be used? Assume that the sample solution is initially 0.1000 M in each ion and that quantitative removal of an ion is realized when only 1 part in 10,000 remains undeposited.

In Appendix 5, we find

$$Cu^{2+} + 2e^- \rightleftharpoons Cu(s) \qquad E^0 = 0.337\ \text{V}$$

$$Pb^{2+} + 2e^- \rightleftharpoons Pb(s) \qquad E^0 = -0.126\ \text{V}$$

Note that based on the standard potentials, copper will begin to deposit at more positive applied voltages than lead. First calculate the potential required to reduce the Cu^{2+} concentration to 10^{-4} of its original concentration (that is, to 1.00×10^{-5} M). Substituting into the Nernst equation, we obtain

$$E = 0.337 - \frac{0.0592}{2} \log \frac{1}{1.00 \times 10^{-5}} = 0.189\ \text{V}$$

Similarly, we can derive the potential at which lead begins to deposit:

$$E = -0.126 - \frac{0.0592}{2} \log \frac{1}{0.1000} = -0.156\ \text{V}$$

Therefore, if the cathode potential is maintained between 0.189 V and −0.156 V versus the standard hydrogen electrode (SHE), we should get a quantitative separation. Now we can convert these to potentials versus the saturated calomel electrode by subtracting E_{SCE}:

$$E_{cell} = E_{cathode} - E_{SCE} = 0.189 - 0.244 = -0.055\ \text{V} \qquad \text{for depositing Cu}$$

and

$$E_{cell} = E_{cathode} - E_{SCE} = -0.156 - 0.244 = -0.400\ \text{V} \qquad \text{for depositing Pb}$$

These results indicate that the cathode potential should be kept between − 0.055 V and − 0.400 V versus the SCE to deposit Cu without depositing any appreciable amounts of Pb.

Calculations such as those in Example 22-2 permit us to find the differences in standard electrode potentials theoretically needed to determine one ion without interference from another. These differences range from about 0.04 V for triply charged ions to approximately 0.24 V for singly charged species.

These theoretical separation limits can be approached only by maintaining the potential of the working electrode (usually the cathode, at which a metal deposits) at the level required. The potential of this electrode can be controlled only by varying the potential applied to the cell, however. Equation 22-4 indicates that variations in $E_{applied}$ affect not only the cathode potential but also the anode potential, the

IR drop, and the overpotential. As a consequence, the only practical way of achieving separation of species whose electrode potentials differ by a few tenths of a volt is to measure the cathode potential continuously against a reference electrode whose potential is known. The applied cell potential can then be adjusted to maintain the cathode potential at the desired level. An analysis performed in this way is called a **controlled-potential electrolysis.** Controlled-potential methods are discussed in Sections 22C-2 and 22D-4.

22C ELECTROGRAVIMETRIC METHODS

Electrolytic deposition has been used for more than a century for the gravimetric determination of metals. In most applications, the metal is deposited on a weighed platinum cathode, and the increase in mass is determined. Some methods use anodic deposition such as the determination of lead as lead dioxide on platinum and of chloride as silver chloride on silver.

There are two general types of electrogravimetric methods. In one, no control of the potential of the working electrode is exercised, and the applied cell potential is held at a more or less constant level that provides a large enough current to complete the electrolysis in a reasonable length of time. The second type of electrogravimetric method is the controlled-potential or **potentiostatic method.**

A **working electrode** is the electrode at which the analytical reaction occurs.

A **potentiostatic method** is an electrolytic procedure in which the potential of the working electrode is maintained at a constant level versus a reference electrode, such as the SCE.

22C-1 Electrogravimetry without Potential Control

Electrolytic procedures in which no effort is made to control the potential of the working electrode use simple and inexpensive equipment and require little operator attention. In these procedures, the potential applied across the cell is maintained at a more or less constant level throughout the electrolysis.

CD-ROM Simulation: Controlled Current Electrolysis Simulation Exploring the Electrogravimetric Method.

Instrumentation

As shown in Figure 22-6, the apparatus for an analytical electrodeposition without cathode potential control consists of a suitable cell and a 6- to 12-V direct-current

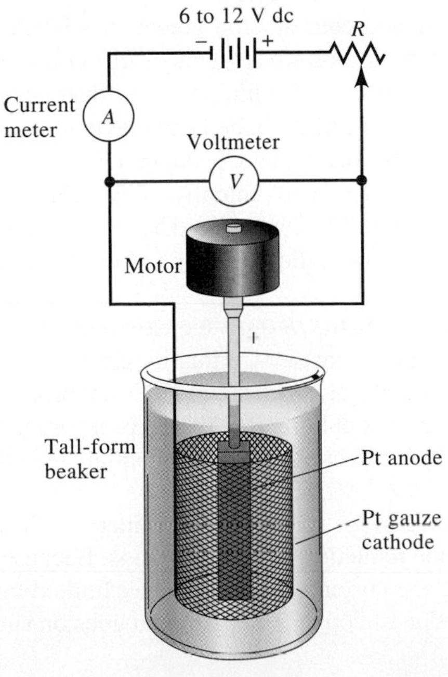

6 to 12 V dc
R
Current meter
A
Voltmeter
V
Motor
Tall-form beaker
Pt anode
Pt gauze cathode

Figure 22-6 Apparatus for electrodeposition of metals without cathode-potential control. Note that this is a two-electrode cell.

power supply. The voltage applied to the cell is controlled by the variable resistor, *R*. A current meter and a voltmeter indicate the approximate current and applied voltage. To perform analytical electrolyses with this apparatus, the applied voltage is adjusted with potentiometer *R* to give a current of several tenths of an ampere. The voltage is then maintained at about the initial level until the deposition is judged to be complete.

Electrolysis Cells

Figure 22-6 shows a typical cell for the deposition of a metal on a solid electrode. Ordinarily, the working electrode is a large-surface-area platinum gauze cylinder 2 or 3 cm in diameter and perhaps 6 cm in length. Copper gauze cathodes and various alloys have also been used. Often, as shown, the anode takes the form of a solid platinum stirring paddle that is located inside and connected to the cathode through the external circuit.

Physical Properties of Electrolytic Precipitates

Ideally, an electrolytically deposited metal should be strongly adherent, dense, and smooth so that it can be washed, dried, and weighed without mechanical loss or reaction with the atmosphere. Good metallic deposits are fine grained and have a metallic luster. Spongy, powdery, or flaky precipitates are usually less pure and less adherent than fine-grained deposits.

The principal factors that influence the physical characteristics of deposits are current density, temperature, and the presence of complexing agents. Ordinarily, the best deposits are formed at low current densities, typically less than 0.1 A/cm^2. Gentle stirring usually improves the quality of a deposit. The effects of temperature are unpredictable and must be determined empirically.

Often, when metals are deposited from solutions of metal complexes, they form smoother and more adherent films than when deposited from the simple ions. Cyanide and ammonia complexes often provide the best deposits. The reasons for this effect are not obvious.

Applications of Electrogravimetric Methods

In practice, electrolysis at a constant cell potential is limited to the separation of easily reduced cations from those that are more difficult to reduce than hydrogen ion or nitrate ion. The reason for this limitation is illustrated in Figure 22-7, which shows the changes of current, *IR* drop, and cathode potential during electrolysis in the cell in Figure 22-6. The analyte here is copper(II) ions in a solution containing an excess of sulfuric or nitric acid. Initially, *R* is adjusted so that the potential applied to the cell is about − 2.5 V, which, as shown in Figure 22-7a, leads to a current of about 1.5 A. The electrolytic deposition of copper is then completed at this applied potential.

As shown in Figure 22-7b, the *IR* drop decreases continually as the reaction proceeds. The reason for this decrease is primarily concentration polarization at the cathode, which limits the rate at which copper ions are brought to the electrode surface and thus the current. From Equation 22-4, it is apparent that the decrease in *IR* must be offset by an increase in the cathode potential since the applied cell potential is constant.

Ultimately, the decrease in current and the increase in cathode potential are slowed at point *B* by the reduction of hydrogen ions. Because the solution contains a large excess of acid, the current is now no longer limited by concentration polarization, and codeposition of copper and hydrogen goes on simultaneously until the

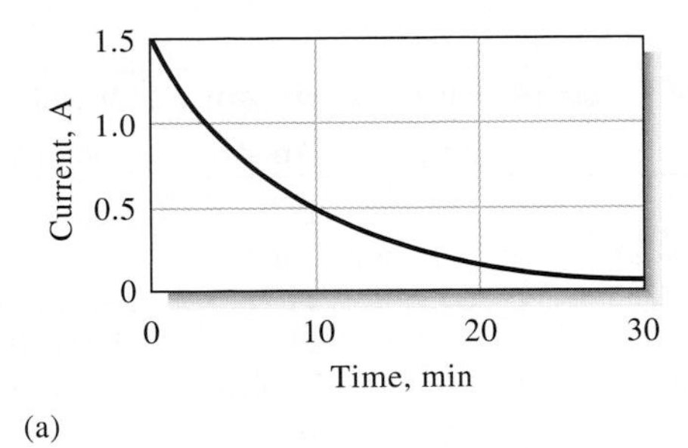

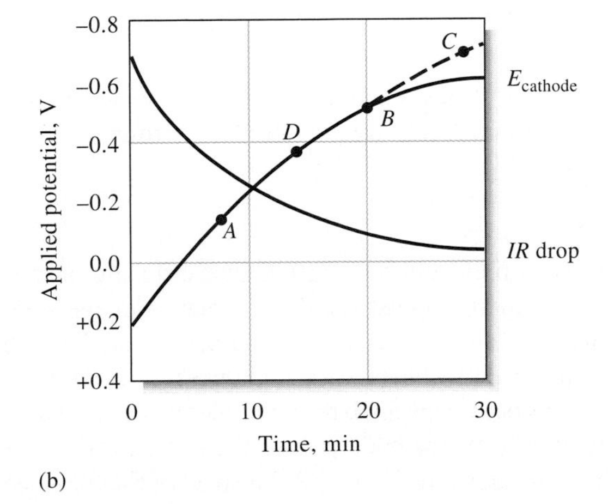

Figure 22-7 (a) Current; (b) *IR* drop and cathode potential change during electrolytic deposition of copper at a constant applied cell potential. The current (a) and *IR* drop (b) decrease steadily with time. The cathode potential shifts negatively to offset the decrease in *IR* drop (b). At point *B*, the cathode becomes depolarized by the reduction of hydrogen ions. Metals that deposit at points *A* or *D* interfere with copper because of codeposition. A metal that deposits at point *C* does not interfere.

remainder of the copper ions are deposited. Under these conditions, the cathode is said to be depolarized by hydrogen ions.

Consider now the fate of some metal ion, such as lead(II), that begins to deposit at point *A* on the cathode potential curve. Lead(II) would codeposit well before copper deposition was complete and would therefore interfere with the determination of copper. In contrast, a metal ion, such as cobalt(II), that reacts at a cathode potential corresponding to point *C* on the curve would not interfere because depolarization by hydrogen gas formation prevents the cathode from reaching this potential.

Codeposition of hydrogen during electrolysis often leads to formation of deposits that do not adhere well. These are usually unsatisfactory for analytical purposes. This problem can be resolved by introducing another species that is reduced at a less negative potential than hydrogen ion and does not adversely affect the physical properties of the deposit. One such cathode **depolarizer** is nitrate ion. Hydrazine and hydroxylamine are also commonly used.

A **depolarizer** is a chemical that is easily reduced (or oxidized). It helps maintain the potential of the working electrode at a relatively small and constant value and prevents interfering reactions that would occur under more reducing or oxidizing conditions.

Electrolytic methods performed without electrode-potential control, while limited by their lack of selectivity, do have several applications of practical importance. Table 22-1 lists the common elements that are often determined by this procedure.

22C-2 Controlled-Potential Electrogravimetry

In the discussion that follows, we assume that the working electrode is a cathode where the analyte is deposited as a metal. The principles can be extended, however,

TABLE 22-1
Some Applications of Electrogravimetry without Potential Control

Analyte	Weighed as	Cathode	Anode	Conditions
Ag^+	Ag	Pt	Pt	Alkaline CN^- solution
Br^-	AgBr (on anode)	Pt	Ag	
Cd^{2+}	Cd	Cu on Pt	Pt	Alkaline CN^- solution
Cu^{2+}	Cu	Pt	Pt	H_2SO_4/HNO_3 solution
Mn^{2+}	MnO_2 (on anode)	Pt	Pt dish	HCOOH/HCOONa solution
Ni^{2+}	Ni	Cu on Pt	Pt	Ammoniacal solution
Pb^{2+}	PbO_2 (on anode)	Pt	Pt	HNO_3 solution
Zn^{2+}	Zn	Cu on Pt	Pt	Acidic citrate solution

to an anodic working electrode where nonmetallic deposits are formed. The determination of Br^- by forming AgBr and of Mn^{2+} by forming MnO_2 are examples of anodic depositions.

Instrumentation

To separate species with electrode potentials that differ by only a few tenths of a volt, we must use a more sophisticated approach than the one just described. Unless something is done, concentration polarization at the cathode causes the potential of that electrode to become so negative that codeposition of the other species present begins before the analyte is completely deposited (see Figure 22-7). A large negative drift in the cathode potential can be avoided by employing the three-electrode system shown in Figure 22-8 instead of the two-electrode system of Figure 22-6.

The controlled-potential apparatus shown in Figure 22-8 is made up of two independent electrical circuits that share a common electrode, the working elec-

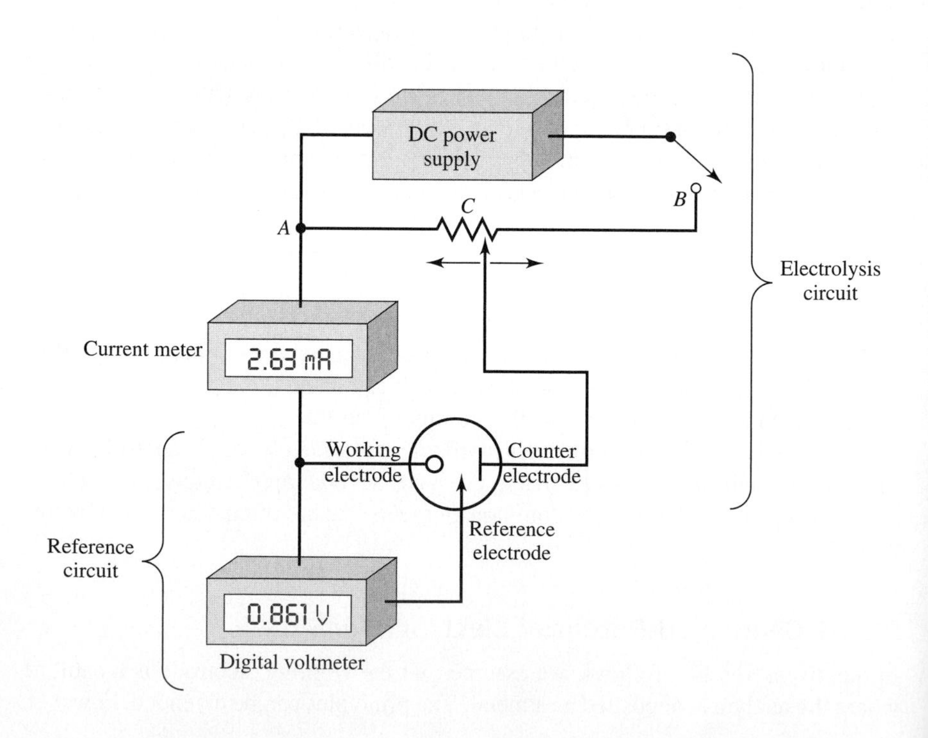

Figure 22-8 Apparatus for controlled-potential electrolysis. The digital voltmeter monitors the potential between the working and the reference electrode. The voltage applied between the working and the counter electrode is varied by adjusting contact *C* on the potentiometer to maintain the working electrode (cathode in this example) at a constant potential versus a reference electrode. The current in the reference electrode is essentially zero at all times. Modern potentiostats are fully automatic and often computer controlled. The electrode symbols shown (—○ Working, → Reference, —| Counter) are the currently accepted notation.

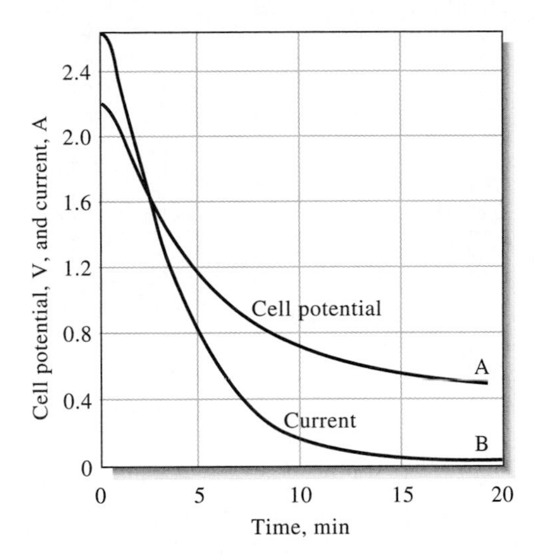

Figure 22-9 Changes in cell potential (A) and current (B) during a controlled-potential deposition of copper. The cathode is maintained at −0.36 V (vs. SCE) throughout the experiment. (Data from J. J. Lingane, *Anal. Chim. Acta,* **1948,** *2,* 590.)

trode where the analyte is deposited. The electrolysis circuit consists of a dc source, a potentiometer (ACB) that permits the voltage applied between the working electrode and a counter electrode to be continuously varied, and a current meter. The control circuit is made up of a reference electrode (often an SCE), a high-resistance digital voltmeter, and the working electrode. The electrical resistance of the control circuit is so large that the electrolysis circuit supplies essentially all of the current for the electrolysis. The control circuit continuously monitors the voltage between the working electrode and the reference electrode and maintains it at a controlled value.

◀ The electrolysis current passes between the working electrode and a **counter electrode.** The counter electrode has no effect on the reaction at the working electrode.

A **potentiostat** maintains the working electrode potential at a constant value relative to a reference electrode.

The current and the cell potential changes that occur in a typical constant-potential electrolysis are illustrated in Figure 22-9. Note that the applied cell potential has to be decreased continuously throughout the electrolysis. Manual adjustment of the potential is tedious (particularly at the outset) and, above all, time consuming. Modern controlled-potential electrolyses are performed with instruments called **potentiostats,** which automatically maintain the working electrode potential at a controlled value versus the reference electrode.

◀ Would Pb^{2+} be expected to interfere with the electrolysis shown in Figure 22-9? Why or why not?

Electrolysis Cells

Electrolysis cells are similar to those shown in Figure 22-6. Tall-form beakers are often used, and solutions are usually mechanically stirred to minimize concentration polarization; frequently, the anode is rotated to act as a mechanical stirrer.

The working electrode is usually a metallic gauze cylinder, as shown in Figure 22-6. Electrodes are often constructed of platinum, although copper, brass, and other metals find occasional use. Some metals, such as bismuth, zinc, and gallium, cannot be desposited directly onto platinum without causing permanent damage to the electrode. Because of this incompatibility, a protective coating of copper is deposited on the platinum electrode before electrolyzing these metals.

The Mercury Cathode

A mercury cathode, such as that shown in Figure 22-10, is particularly useful for removing easily reduced elements as a preliminary step in an analysis. For example, copper, nickel, cobalt, silver, and cadmium are readily separated at this

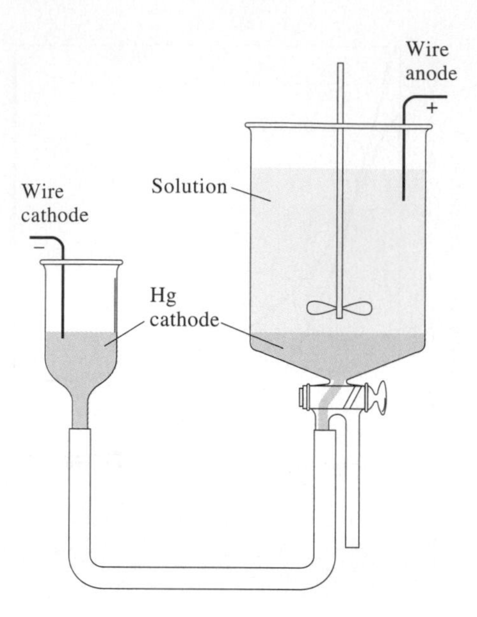

Figure 22-10 A mercury cathode for the electrolytic removal of metal ions from solution.

electrode from such ions as aluminum, titanium, the alkali metals, sulfates, and phosphates. The deposited metals dissolve in the mercury with little hydrogen evolution because even at high applied potentials, formation of the gas is prevented by the high overvoltage on mercury. The metals dissolve in the mercury to form amalgams that are important in several forms of voltammetry (see Section 23B-2). Ordinarily, the deposited metals are not determined after electrolysis but are merely removed from the analyte solution.

Applications of Controlled-Potential Electrogravimetry

The controlled-potential method is a potent tool for separating and determining metallic species having standard potentials that differ by only a few tenths of a volt. For example, copper, bismuth, lead, cadmium, zinc, and tin can be determined in mixtures by successive deposition of the metals on a weighed platinum cathode. The first three elements are deposited from a nearly neutral solution containing tartrate ion to complex the tin(IV) and prevent its deposition. Copper is first reduced quantitatively by maintaining the cathode potential at −0.2 V with respect to the SCE. After being weighed, the copper-plated cathode is returned to the solution, and bismuth is removed at a potential of −0.4 V. Lead is then deposited quantitatively by increasing the cathode potential to −0.6 V. When lead deposition is complete, the solution is made strongly ammoniacal, and cadmium and zinc are deposited successively at −1.2 and −1.5 V. Finally, the solution is acidified to decompose the tin/tartrate complex by the formation of undissociated tartaric acid. Tin is then deposited at a cathode potential of −0.65 V. A fresh cathode must be used here because the zinc redissolves under these conditions. A procedure such as this is particularly attractive for use with computer-controlled potentiostats because little operator time is required for the complete analysis.

Table 22-2 lists some other separations performed by controlled-potential electrolysis. Because of limited sensitivity and the time required for washing, drying, and weighing the electrodes, many electrogravimetric methods have been replaced by the coulometric methods discussed in the next section.

TABLE 22-2
Some Applications of Controlled-Potential Electrolysis*

Metal	Potential vs. SCE	Electrolyte	Other Elements That Can Be Present
Ag	+ 0.10	Acetic acid/acetate buffer	Cu and heavy metals
Cu	− 0.30	Tartrate + hydrazine + Cl^-	Bi, Sb, Pb, Sn, Ni, Cd, Zn
Bi	− 0.40	Tartrate + hydrazine + Cl^-	Pb, Zn, Sb, Cd, Sn
Sb	− 0.35	HCl + hydrazine at 70°C	Pb, Sn
Sn	− 0.60	HCl + hydroxylamine	Cd, Zn, Mn, Fe
Pb	− 0.60	Tartrate + hydrazine	Cd, Sn, Ni, Zn, Mn, Al, Fe
Cd	− 0.80	HCl + hydroxylamine	Zn
Ni	− 1.10	Ammoniacal tartrate + sodium sulfite	Zn, Al, Fe

*From J. J. Lingane, *Electroanalytical Chemistry,* 2nd ed., p. 413. New York: Interscience, 1958. This material is used by permission of John Wiley & Sons, Inc.

22D COULOMETRIC METHODS

Coulometric methods are performed by measuring the quantity of electrical charge required to convert a sample of an analyte quantitatively to a different oxidation state. Coulometric and gravimetric methods share the common advantage that the proportionality constant between the quantity measured and the analyte mass is derived from accurately known physical constants, which can eliminate the need for calibration with chemical standards. In contrast to gravimetric methods, coulometric procedures are usually rapid and do not require the product of the electrochemical reaction to be a weighable solid. Coulometric methods are as accurate as conventional gravimetric and volumetric procedures and in addition are readily automated.[4]

22D-1 Determining the Electrical Charge

Electrical charge is the basis of the other electrical quantities—current, voltage, and power. The charge on an electron (and proton) is defined as 1.6022×10^{-19} coulombs (C). A rate of charge flow equal to one coulomb per second is the definition of one ampere (A) of current. Thus, a coulomb can be considered as that charge carried by a constant current of one ampere for one second. The charge Q that results from a constant current of I amperes operated for t seconds is

The coulomb (C) is the amount of charge required to produce 0.00111800 g of silver metal from silver ions. One coulomb = 1 ampere × 1 s = 1 A s.

$$Q = It \tag{22-6}$$

◀ In describing electric current, it is common to use the upper case letter I for a static or direct current (dc). A variable or alternating current (ac) is commonly indicated by the lower case letter i. Likewise, dc and ac voltages are given the letters E and e, respectively.

For a variable current i, the charge is given by the integral

$$Q = \int_0^t i\,dt \tag{22-7}$$

[4]For additional information about coulometric methods, see J. A. Dean, *Analytical Chemistry Handbook,* Section 14, pp. 14.118–14.133. New York: McGraw-Hill, 1995; D. J. Curran, in *Laboratory Techniques in Electroanalytical Chemistry,* 2nd ed., P. T. Kissinger and W. R. Heinemann, Eds., pp. 739–768. New York: Marcel Dekker, 1996; J. A. Plambeck, *Electroanalytical Chemistry,* Chapter 12. New York: Wiley, 1982.

► The full constants for fundamental quantities are available from the National Institute of Standards and Technology on their Web site, at http://physics.nist.gov/cuu/Constants/index.html. The 1998 value for the faraday is 96485.3415 C mol^{-1} with a standard uncertainty of 0.0039 C mol^{-1}. The value for the electron charge is $1.602176462 \times 10^{-19}$ C with a standard uncertainty of $0.000,000,063 \times 10^{-19}$ C. A detailed description of the data and the analysis that led to the values can be found in P. J. Mohr and B. N. Taylor, *Rev. Mod. Phys.*, **2000,** *72,* 351.

CD-ROM Tutorial: Calculating the Concentration of an Analyte Using a Measurement of Charge Passed.

The faraday (F) is the quantity of charge that corresponds to one mole or 6.022×10^{23} electrons. Since each electron has a charge of 1.6022×10^{-19} C, the faraday also equals 96,485 C.

Faraday's law relates the number of moles of the analyte n_A to the charge

$$n_A = \frac{Q}{nF} \qquad (22\text{-}8)$$

where n is the number of moles of electrons in the analyte half-reaction. As shown in Example 22-3, we can use these definitions to calculate the mass of a chemical species that is formed at an electrode by a current of known magnitude.

EXAMPLE 22-3

A constant current of 0.800 A is used to deposit copper at the cathode and oxygen at the anode of an electrolytic cell. Calculate the number of grams of each product formed in 15.2 min, assuming no other redox reaction.

The two half-reactions are

$$Cu^{2+} + 2e^- \rightarrow Cu(s)$$

$$2H_2O \rightarrow 4\,e^- + O_2(g) + 4H^+$$

Thus, 1 mol of copper is equivalent to 2 mol of electrons and 1 mol of oxygen corresponds to 4 mol of electrons.

Substituting into Equation 22-6 yields

$$Q = 0.800\text{ A} \times 15.2\text{ min} \times 60\text{ s/min} = 729.6\text{ A·s} = 729.6\text{ C}$$

We can find the number of moles of Cu and O_2 from Equation 22-8:

$$n_{Cu} = \frac{729.6\,\cancel{C}}{2\,\cancel{mol\,e^-}/\text{mol Cu} \times 96{,}485\,\cancel{C}/\cancel{mol\,e^-}} = 3.781 \times 10^{-3}\text{ mol Cu}$$

$$n_{O_2} = \frac{729.6\,\cancel{C}}{4\,\cancel{mol\,e^-}/\text{mol }O_2 \times 96{,}485\,\cancel{C}/\cancel{mol\,e^-}} = 1.890 \times 10^{-3}\text{ mol }O_2$$

The masses of Cu and O_2 are given by

$$\text{mass Cu} = 3.781 \times 10^{-3}\,\cancel{mol} \times \frac{63.55\text{ g Cu}}{\cancel{mol}} = 0.240\text{ g Cu}$$

$$\text{mass }O_2 = 1.890 \times 10^{-3}\,\cancel{mol} \times \frac{32.00\text{ g }O_2}{\cancel{mol}} = 0.0605\text{ g }O_2$$

Michael Faraday (1791–1867) was one of the foremost chemists and physicists of his time. Among his most important discoveries was Faraday's law of electrolysis. Faraday, a simple man who lacked mathematical sophistication, was a superb experimentalist and an inspiring teacher and lecturer. The quantity of charge equal to a mole of electrons is named in his honor.

22D-2 Characterizing Coulometric Methods

Two methods have been developed that are based on measuring the quantity of charge: **controlled-potential (potentiostatic) coulometry** and **controlled-current coulometry,** often called **coulometric titrimetry.** Potentiostatic methods are performed in much the same way as controlled-potential gravimetric methods, with

► **Constant-current coulometry** is also called coulometric titrimetry.

the potential of the working electrode being maintained at a constant value relative to a reference electrode throughout the electrolysis. In controlled-potential coulometry, however, the electrolysis current is recorded as a function of time to give a curve similar to curve B in Figure 22-9. The analysis is then completed by integrating the current-time curve (see Equation 22-7) to obtain the charge and, from Faraday's law, the amount of analyte (see Equation 22-8).

Coulometric titrations are similar to other titrimetric methods in that analyses are based on measuring the combining capacity of the analyte with a standard reagent. In the coulometric procedure, the reagent consists of electrons, and the standard solution is a constant current of known magnitude. Electrons are added to the analyte (via the direct current) or to some species that immediately reacts with the analyte until an end point is reached. At that point, the electrolysis is discontinued. The amount of analyte is determined from the magnitude of the current and the time required to complete the titration. The magnitude of the current in amperes is analogous to the molarity of a standard solution, and the time measurement is analogous to the volume measurement in conventional titrimetry.

◀ Electrons are the reagent in a coulometric titration.

CD-ROM Tutorial: Calculating the Time Needed to Deposit a Given Mass of Metal at an Electrode Surface Under a Constant Current.

22D-3 Current-Efficiency Requirements

A fundamental requirement for all coulometric methods is 100% current efficiency; that is, each faraday of electricity must bring about chemical change in the analyte equivalent to one mole of electrons. Note that 100% current efficiency can be achieved without direct participation of the analyte in electron transfer at an electrode. For example, chloride ion may be determined quite easily using potentiostatic coulometry or using coulometric titrations with silver ion at a silver anode. Silver ion then reacts with chloride to form a precipitate or deposit of silver chloride. The quantity of electricity required to complete the silver chloride formation serves as the analytical variable. In this instance, 100% current efficiency is realized because the number of moles of electrons is essentially equal to the number of moles of chloride ion in the sample despite the fact that these ions do not react directly at the electrode surface.

One equivalent of chemical change is the change brought about by 1 mol of electrons. Thus, for the two half-reactions in Example 22-3, one equivalent of chemical change involves production of $\frac{1}{2}$ mol of Cu or $\frac{1}{4}$ mol of O_2.

22D-4 Controlled-Potential Coulometry

In controlled-potential coulometry, the potential of the working electrode is maintained at a constant level such that only the analyte is responsible for conducting charge across the electrode/solution interface. The charge required to convert the analyte to its reaction product is then determined by recording and integrating the current-versus-time curve during the electrolysis.

Instrumentation

The instrumentation for potentiostatic coulometry consists of an electrolysis cell, a potentiostat, and a device for determining the charge consumed by the analyte.

Cells Figure 22-11 illustrates two types of cells that are used for potentiostatic coulometry. The first consists of a platinum-gauze working electrode, a platinum-wire counter electrode, and a saturated calomel reference electrode. The counter electrode is separated from the analyte solution by a salt bridge, which usually contains the same electrolyte as the solution being analyzed. This bridge is needed to prevent the reaction products formed at the counter electrode from diffusing into the analyte solution and interfering. For example, hydrogen gas is a common product at a cathodic counter electrode. Unless this species is physically isolated from

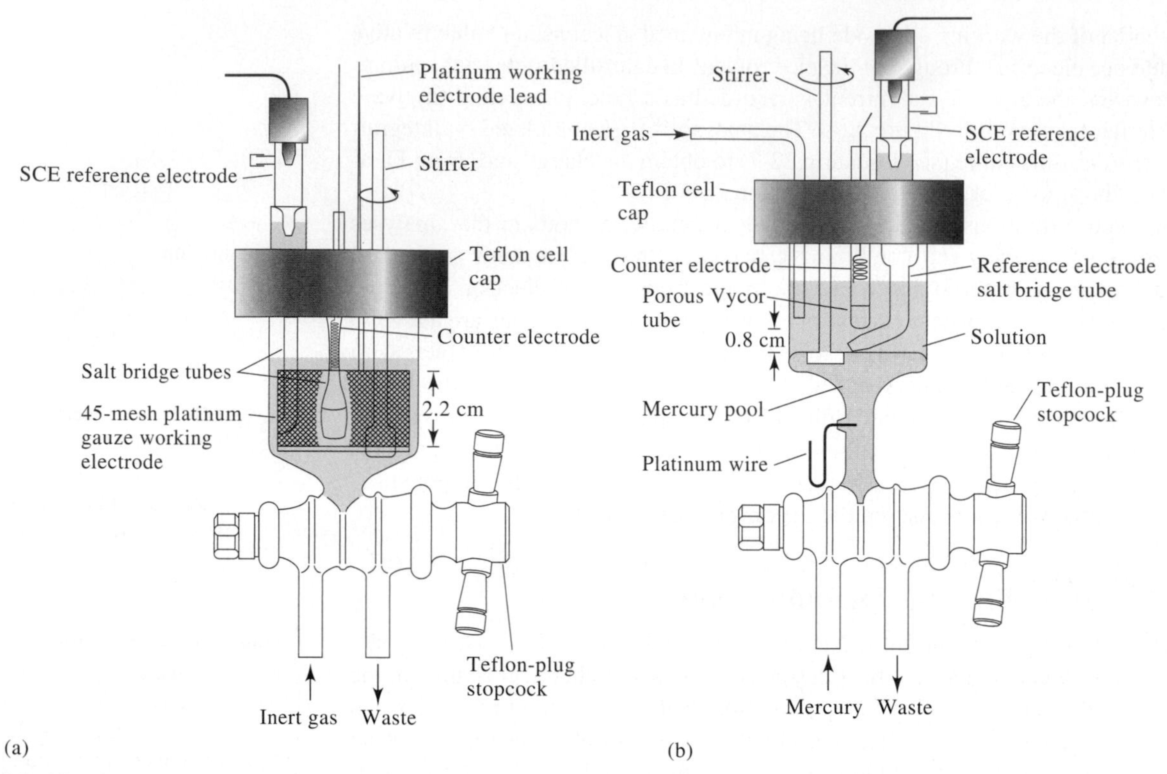

Figure 22-11 Electrolysis cells for potentiostatic coulometry. Working electrode: (a) platinum gauze, (b) mercury pool. (Reprinted with permission from J. E. Harrar and C. L. Pomernacki, *Anal. Chem.*, **1973,** *45,* 57. Copyright 1973 American Chemical Society.)

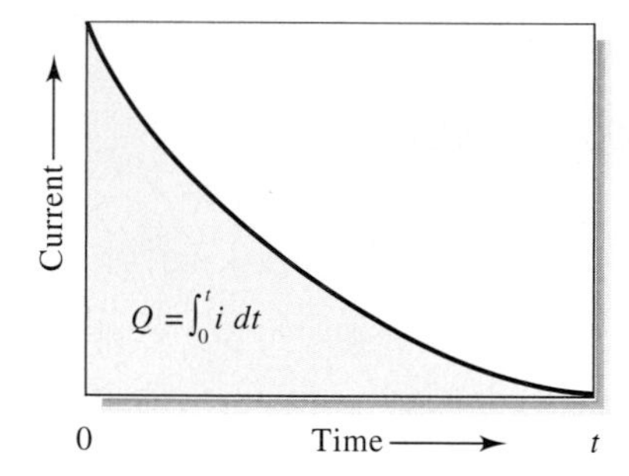

Figure 22-12 For a current that varies with time, the quantity of charge Q in a time t is the shaded area under the curve, obtained by integration of the current-time curve.

the analyte solution by the bridge, it will react directly with many of the analytes that are determined by oxidation at the working anode.

The second cell, shown in Figure 22-11b, is a mercury-pool type. A mercury cathode is particularly useful for separating easily reduced elements as a preliminary step in an analysis. In addition, however, it has found considerable use in the coulometric determination of several metallic cations that form metals that are soluble in mercury. In these applications, little or no hydrogen evolution occurs even at high applied potentials because of the large overvoltage of hydrogen on mercury. A coulometric cell such as that shown in Figure 22-11b is also useful for the coulometric determination of certain types of organic compounds.

Potentiostats and Coulometers For controlled-potential coulometry, we use a potentiostat similar in design to that shown in Figure 22-8. Generally, however, the potentiostat is automated and equipped with a computer or an electronic current integrator that gives the charge in coulombs necessary to complete the reaction, as shown in Figure 22-12.

EXAMPLE 22-4

The Fe(III) in a 0.8202-g sample was determined by coulometric reduction to Fe(II) at a platinum cathode. Calculate the percentage of $Fe_2(SO_4)_3$ ($\mathcal{M}$ = 399.88 g/mol) in the sample if 103.2775 C were required for the reduction.

Since 1 mol of $Fe_2(SO_4)_3$ consumes 2 mol of electrons, we may write from Equation 22-8

$$n_{Fe_2(SO_4)_3} = \frac{103.2775\ \cancel{C}}{2\ \cancel{mol\ e^-}/\text{mol}\ Fe_2(SO_4)_3 \times 96{,}485\ \cancel{C}/\cancel{mol\ e^-}}$$

$$= 5.3520 \times 10^{-4}\ \text{mol}\ Fe_2(SO_4)_3$$

$$\text{mass}\ Fe_2\,(SO_4)_3 = 5.3520 \times 10^{-4}\ \cancel{\text{mol}\ Fe_2(SO_4)_3} \times \frac{399.88\ \text{g}\ Fe_2(SO_4)_3}{\cancel{\text{mol}\ Fe_2(SO_4)_3}}$$

$$= 0.21401\ \text{g}\ Fe_2(SO_4)_3$$

$$\%Fe_2(SO_4)_3 = \frac{0.21401\ \text{g}\ Fe_2(SO_4)_3}{0.8202\ \text{g sample}} \times 100\% = 26.09\%$$

Applications of Controlled-Potential Coulometry

Controlled-potential coulometric methods have been used to determine more than 55 elements in inorganic compounds.[5] Mercury is a popular cathode; methods have been described for the deposition of more than two dozen metals at this electrode. The method has found use in the nuclear-energy field for the relatively interference-free determination of uranium and plutonium.

Controlled-potential coulometry also offers possibilities for the electrolytic determination (and synthesis) of organic compounds. For example, trichloroacetic acid and picric acid are quantitatively reduced at a mercury cathode whose potential is suitably controlled:

$$Cl_3CCOO^- + H^+ + 2e^- \rightarrow Cl_2HCCOO^- + Cl^-$$

$$\text{(OH; } O_2N\text{, } NO_2\text{, } NO_2 \text{ on benzene ring)} + 18H^+ + 18e^- \longrightarrow \text{(OH; } H_2N\text{, } NH_2\text{, } NH_2 \text{ on benzene ring)} + 6H_2O$$

Coulometric measurements permit the determination of these compounds with a relative error of a few tenths of a percent.

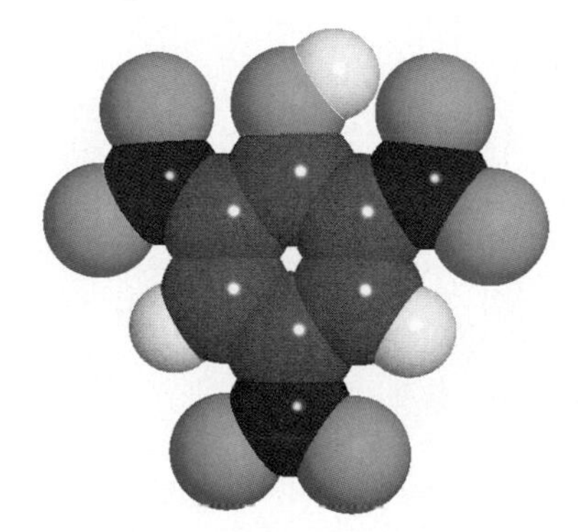

Molecular model of picric acid. Picric acid (2,4,6-trinitrophenol) is a close relative of trinitrotoluene (TNT). It is an explosive compound and has military applications. Picric acid has also been used as a yellow dye and staining agent and as an antiseptic.

Spreadsheet Summary In the first experiment in Chapter 11 of *Applications of Microsoft® Excel in Analytical Chemistry,* numerical integration methods are investigated. These methods are used to determine the charge required to electrolyze a reagent in a controlled-potential coulometric determination. A trapezoidal method and a Simpson's rule method are studied. From the charge, Faraday's law is used to determine the amount of analyte.

[5]For a summary of applications, see J. A. Dean, *Analytical Chemistry Handbook,* Section 14, pp. 14.119–14.123. New York: McGraw-Hill, 1995; A. J. Bard and L. R. Faulkner, *Electrochemical Methods,* 2nd ed., pp. 427–431. New York: Wiley, 2001.

22D-5 Coulometric Titrations[6]

Constant-current generators are sometimes called **galvanostats.**

► Auxiliary reagents are essential in coulometric titrations.

Coulometric titrations are carried out with a constant-current source, sometimes called a **galvanostat,** which senses decreases in current in a cell and responds by increasing the potential applied to the cell until the current is restored to its original level. Because of the effects of concentration polarization, 100% current efficiency with respect to the analyte can be maintained only by having present in large excess an auxiliary reagent that is oxidized or reduced at the electrode to give a product that reacts with the analyte. As an example, consider the coulometric titration of iron(II) at a platinum anode. At the beginning of the titration, the primary anodic reaction directly consumes Fe^{2+} and is

$$Fe^{2+} \rightarrow Fe^{3+} + e^-$$

As the concentration of iron(II) decreases, however, the requirement for a constant current results in an increase in the applied cell potential. Because of concentration polarization, this increase in potential causes the anode potential to increase to the point where the decomposition of water becomes a competing process:

$$2H_2O \rightarrow O_2(g) + 4H^+ + 4e^-$$

The quantity of electricity required to complete the oxidation of iron(II) then exceeds that demanded by theory, and the current efficiency is less than 100%. The lowered current efficiency is prevented, however, by introducing at the outset an unmeasured excess of cerium(III), which is oxidized at a lower potential than is water:

$$Ce^{3+} \rightarrow Ce^{4+} + e^-$$

With stirring, the cerium(IV) produced is rapidly transported from the surface of the electrode to the bulk of the solution, where it oxidizes an equivalent amount of iron(II):

$$Ce^{4+} + Fe^{2+} \rightarrow Ce^{3+} + Fe^{3+}$$

The net effect is an electrochemical oxidation of iron(II) with 100% current efficiency, even though only a fraction of that species is *directly* oxidized at the electrode surface.

Detecting the End Point

Coulometric titrations, like their volumetric counterparts, require a means for determining when the reaction between analyte and reagent is complete. Generally, the end points described in the chapters on volumetric methods are applicable to coulometric titrations as well. Thus, for the titration of iron(II) just described, an oxidation/reduction indicator, such as 1,10-phenanthroline, can be used; as an alternative, the end point can be determined potentiometrically. Potentiometric or

[6]For further details on this technique, see D. J. Curran, in *Laboratory Techniques in Electroanalytical Chemistry,* 2nd ed., P. T. Kissinger and W. R. Heineman, Eds., pp. 750–768. New York: Marcel Dekker, 1996.

amperometric (see Section 23B-4) end points are used in Karl Fischer titrators. Some coulometric titrations utilize a photometric end point (see Section 26A-4).

Instrumentation

As shown in Figure 22-13, the equipment required for a coulometric titration includes a source of constant current from one to several hundred milliamperes, a titration cell, a switch, a timer, and a device for monitoring current. Moving the switch to position 1 simultaneously starts the timer and initiates a current in the titration cell. When the switch is moved to position 2, the electrolysis and the timing are discontinued. With the switch in this position, however, current continues to be drawn from the source and passes through a dummy resistor R_D that has about the same electrical resistance as the cell. This arrangement ensures continuous operation of the source, which aids in maintaining a constant current.

Current Sources The constant-current source for a coulometric titration is an electronic device capable of maintaining a current of 200 mA or more that is constant to a few hundredths of a percent. Such constant-current sources are available from several instrument manufacturers. The electrolysis time can be measured very accurately with a digital timer or a computer-based timing system.

Cells for Coulometric Titrations Figure 22-14 shows a typical coulometric titration cell consisting of a working electrode at which the reagent is produced and a counter (auxiliary) electrode to complete the circuit. The working electrode used to generate reactants in situ is often referred to as the generator electrode. It is usually a platinum rectangle, a coil of wire, or a gauze cylinder with a relatively large surface area to minimize polarization effects. Ordinarily, the counter electrode is isolated from the reaction medium by a sintered disk or some other porous medium to prevent interference by the reaction products from this electrode. For example, hydrogen is sometimes evolved at this electrode. Since hydrogen is a reducing agent, a positive systematic error can occur unless the gas is produced in a separate compartment.

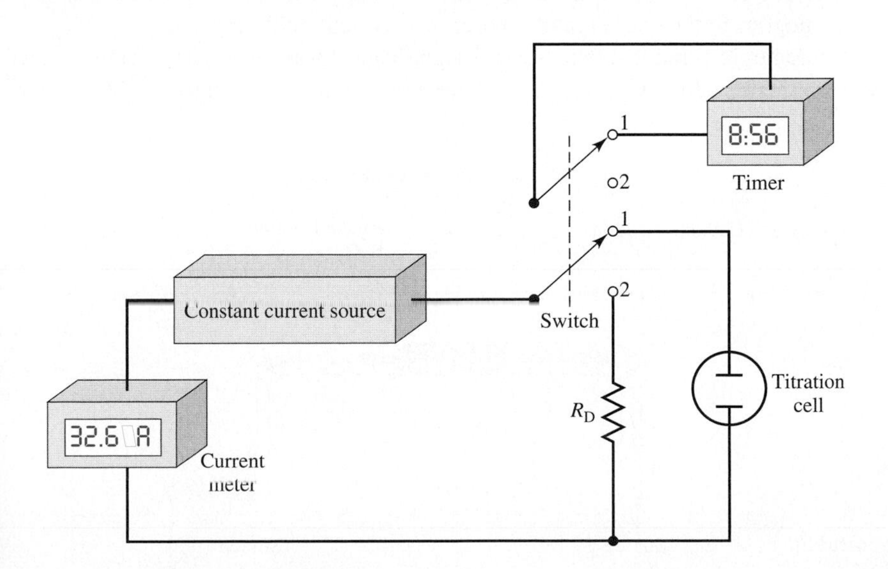

Figure 22-13 Conceptual diagram of a coulometric titration apparatus. Commercial coulometric titrators are totally electronic and usually computer controlled.

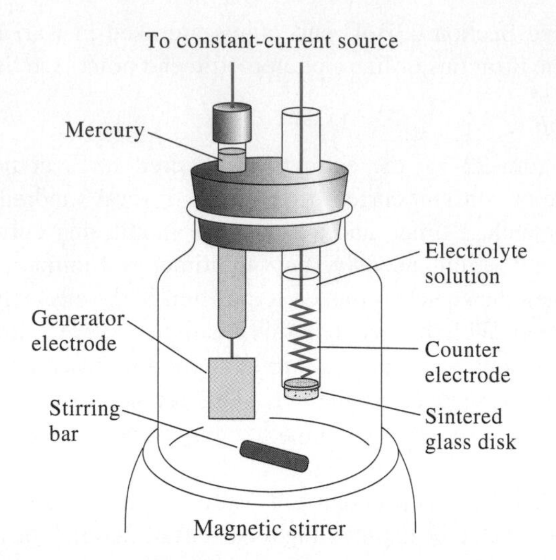

Figure 22-14 A typical coulometric titration cell.

An alternative to isolation of the counter electrode is generation of the reagent externally with a device similar to that shown in Figure 22-15. The external generation cell is arranged so that electrolyte flow continues briefly after the current is switched off, which flushes the residual reagent into the titration vessel. Note that the generation device shown in Figure 22-15 provides either hydrogen or hydroxide ions, depending on which arm is used. External generation cells have also been used for the generation of other reagents such as iodine.

Comparing Coulometric and Conventional Titrations

The various components of the titrator in Figure 22-13 have their counterparts in the reagents and apparatus required for a volumetric titration. The constant-current source of known magnitude serves the same function as the standard solution in a volumetric method. The digital timer and switch correspond to the buret and stopcock, respectively. Electricity is passed through the cell for relatively long periods of time at the outset of a coulometric titration, but the time intervals are made smaller and smaller as chemical equivalence is approached. Note that these steps are analogous to the operation of a buret in a conventional titration.

► Coulometric methods are as accurate and precise as comparable volumetric methods.

Coulometric titration offers several significant advantages over a conventional volumetric procedure. Principal among these is the elimination of the problems asso-

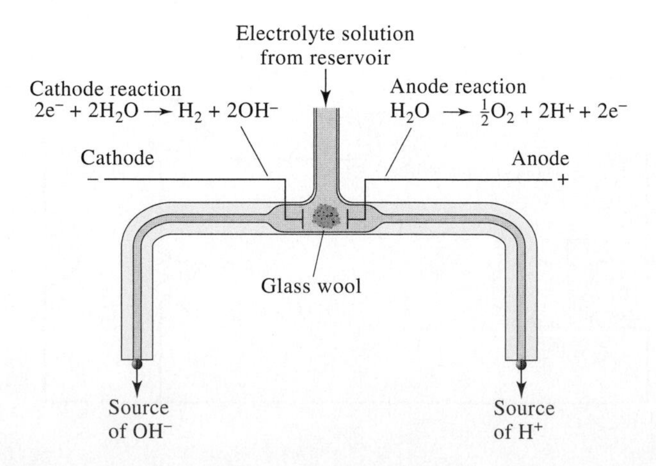

Figure 22-15 A cell for the external coulometric generation of acid and base.

ciated with the preparation, standardization, and storage of standard solutions. This advantage is particularly significant with reagents such as chlorine, bromine, and titanium(III) ion, which are sufficiently unstable in aqueous solution to seriously limit their value as volumetric reagents. Their use in a coulometric determination is, however, straightforward because they are consumed as soon as they are generated.

Coulometric methods also excel when small amounts of analyte have to be titrated because tiny quantities of reagent are generated with ease and accuracy through the proper choice of current. With conventional titrations, it is inconvenient and often inaccurate to use very dilute solutions and small volumes.

A further advantage of the coulometric procedure is that a single constant-current source provides reagents for precipitation, complex formation, neutralization, or oxidation/reduction titrations. Finally, coulometric titrations are more readily automated, since it is easier to control electrical current than liquid flow.

The current-time measurements required for a coulometric titration are inherently as accurate as or more accurate than the comparable volume/molarity measurements of a conventional volumetric method, particularly when small quantities of reagent are involved. When the accuracy of a titration is limited by the sensitivity of the end point, the two titration methods have comparable accuracies.

Applications of Coulometric Titrations

Coulometric titrations have been developed for all types of volumetric reactions.[7] Selected applications are described in this section.

Neutralization Titrations Hydroxide ion can be generated at the surface of a platinum cathode immersed in a solution containing the analyte acid:

$$2H_2O + 2e^- \rightarrow 2OH^- + H_2(g)$$

The platinum anode must be isolated by a diaphragm to eliminate potential interference from the hydrogen ions produced by anodic oxidation of water. As a convenient alternative, a silver wire can be substituted for the platinum anode, provided that chloride or bromide ions are added to the analyte solution. The anode reaction then becomes

$$Ag(s) + Br^- \rightarrow AgBr(s) + e^-$$

Silver bromide does not interfere with the neutralization reaction.

Coulometric titrations of acids are much less susceptible to the carbonate error encountered in volumetric methods (see Section 16A-3). The error can be avoided if carbon dioxide is removed from the solvent by boiling it or by bubbling an inert gas, such as nitrogen, through the solution for a brief period.

Hydrogen ions generated at the surface of a platinum anode can be used for the coulometric titration of strong as well as weak bases:

$$2H_2O \rightarrow O_2 + 4H^+ + 4e^-$$

Here, the cathode must be isolated from the analyte solution to prevent interference from hydroxide ion.

[7] For a summary of applications, see J. A. Dean, *Analytical Chemistry Handbook,* Section 14, pp. 14.127–14.133. New York: McGraw-Hill, 1995.

TABLE 22-3

Summary of Coulometric Titrations Involving Neutralization, Precipitation, and Complex-Formation Reactions

Species Determined	Generator Electrode Reaction	Secondary Analytical Reaction
Acids	$2H_2O + 2e^- \rightleftharpoons 2OH^- + H_2$	$OH^- + H^+ \rightleftharpoons H_2O$
Bases	$H_2O \rightleftharpoons 2H^+ + \frac{1}{2}O_2 + 2e^-$	$H^+ + OH^- \rightleftharpoons H_2O$
Cl^-, Br^-, I^-	$Ag \rightleftharpoons Ag^+ + e^-$	$Ag^+ + X^- \rightleftharpoons AgX(s)$
Mercaptans (RSH)	$Ag \rightleftharpoons Ag^+ + e^-$	$Ag^+ + RSH \rightleftharpoons AgSR(s) + H^+$
Cl^-, Br^-, I^-	$2Hg \rightleftharpoons Hg_2^{2+} + 2e^-$	$Hg_2^{2+} + 2X^- \rightleftharpoons Hg_2X_2(s)$
Zn^{2+}	$Fe(CN)_6^{3-} + e^- \rightleftharpoons Fe(CN)_6^{4-}$	$3Zn^{2+} + 2K^+ + 2Fe(CN)_6^{4-} \rightleftharpoons K_2Zn_3[Fe(CN)_6]_2(s)$
$Ca^{2+}, Cu^{2+}, Zn^{2+}, Pb^{2+}$	See Equation 22-9	$HY^{3-} + Ca^{2+} \rightleftharpoons CaY^{2-} + H^+$, etc.

Precipitation and Complex-Formation Reactions Coulometric titrations with EDTA are carried out by reduction of the ammine mercury(II) EDTA chelate at a mercury cathode:

$$HgNH_3Y^{2-} + NH_4^+ + 2e^- \rightarrow Hg(l) + 2NH_3 + HY^{3-} \tag{22-9}$$

Because the mercury chelate is more stable than the corresponding complexes of such cations as calcium, zinc, lead, or copper, complexation of these ions occurs only after the ligand has been freed by the electrode process.

As shown in Table 22-3, several precipitating reagents can be generated coulometrically. The most widely used of these is silver ion, which is generated at a silver anode, as discussed in Feature 22-2.

Oxidation/Reduction Titrations Coulometric titrations have been developed for many, but not all, redox titrations. Table 22-4 reveals that a variety of redox

FEATURE 22-2

Coulometric Titration of Chloride in Biological Fluids

The accepted reference method for determining chloride in blood serum, plasma, urine, sweat, and other body fluids is the coulometric titration procedure.[8] In this technique, silver ions are generated coulometrically. The silver ions then react with chloride ions to form insoluble silver chloride. The end point is usually detected by amperometry (see Section 23B-4) when a sudden increase in current occurs on the generation of a slight excess of Ag^+. In principle, the absolute amount of Ag^+ needed to react quantitatively with Cl^- can be obtained from application of Faraday's law. In practice, calibration is used. First, the time t_s required to titrate a chloride standard solution with a known number of moles of chloride $(n_{Cl^-})_s$ using a constant current I is measured. The same constant current is next used in the titration of the unknown solution, and the time t_u is measured. The number of moles of chloride in the unknown $(n_{Cl^-})_u$ is then obtained as follows:

$$(n_{Cl^-})_u = \frac{t_u}{t_s} \times (n_{Cl^-})_s$$

► CHALLENGE: Derive the equation shown in Feature 22-2 for the number of moles of chloride ion in the unknown. Begin with Faraday's law.

[8] L. A. Kaplan and A. J. Pesce, *Clinical Chemistry: Theory, Analysis, and Correlation*, p. 1060. St. Louis: C. V. Mosby, 1984.

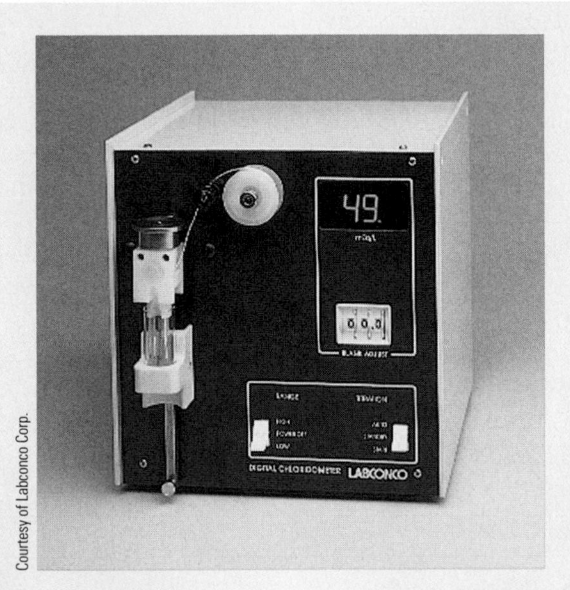

Courtesy of Labconco Corp.

Figure 22F-2 A commercial digital chloridometer. This coulometric titrator is designed to determine chloride ion in such clinical samples as serum, urine, and sweat. It is used in the diagnosis of cystic fibrosis. The chloridometer is also used in food and environmental laboratories. (Courtesy of Labconco Corp., Kansas City, MO.)

If the volumes of the standard solution and the unknown solution are the same, concentrations can be substituted for number of moles in this equation. A commercial coulometric titrator called a chloridometer is shown in Figure 22F-2.

Other popular methods to determine chloride are ion-selective electrodes (see Section 21D), photometric titrations (see Section 26A-4), and isotope dilution mass spectrometry.

reagents can be generated coulometrically. For example, the coulometric generation of bromine forms the basis for a large number of coulometric methods. Of interest as well are reagents such as silver(II), manganese(III), and the chloride complex of copper(I) that are too unstable to be used in conventional volumetric analysis.

TABLE 22-4

Summary of Coulometric Titrations Involving Oxidation/Reduction Reactions

Reagent	Generator Electrode Reaction	Substance Determined
Br_2	$2Br^- \rightleftharpoons Br_2 + 2e^-$	As(III), Sb(III), U(IV), Tl(I), I^-, SCN^-, NH_3, N_2H_4, NH_2OH, phenol, aniline, mustard gas, mercaptans, 8-hydroxyquinoline, olefins
Cl_2	$2Cl^- \rightleftharpoons Cl_2 + 2e^-$	As(III), I^-, styrene, fatty acids
I_2	$2I^- \rightleftharpoons I_2 + 2e^-$	As(III), Sb(III), $S_2O_3^{2-}$, H_2S, ascorbic acid
Ce^{4+}	$Ce^{3+} \rightleftharpoons Ce^{4+} + e^-$	Fe(II), Ti(III), U(IV), As(III), I^-, $Fe(CN)_6^{4-}$
Mn^{3+}	$Mn^{2+} \rightleftharpoons Mn^{3+} + e^-$	$H_2C_2O_4$, Fe(II), As(III)
Ag^{2+}	$Ag^+ \rightleftharpoons Ag^{2+} + e^-$	Ce(III), V(IV), $H_2C_2O_4$, As(III)
Fe^{2+}	$Fe^{3+} + e^- \rightleftharpoons Fe^{2+}$	Cr(VI), Mn(VII), V(V), Ce(IV)
Ti^{3+}	$TiO^{2+} + 2H^+ + e^- \rightleftharpoons Ti^{3+} + H_2O$	Fe(III), V(V), Ce(IV), U(VI)
$CuCl_3^{2-}$	$Cu^{2+} + 3Cl^- + e^- \rightleftharpoons CuCl_3^{2-}$	V(V), Cr(VI), IO_3^-
U^{4+}	$UO_2^{2+} + 4H^+ + 2e^- \rightleftharpoons U^{4+} + 2H_2O$	Cr(VI), Ce(IV)

Automatic Coulometric Titrations

A number of instrument manufacturers offer automatic coulometric titrators, most of which employ a potentiometric end point. Some of these instruments are multipurpose and can be used for the determination of a variety of species. Others are designed for a single type of analysis. Examples of the latter are chloride titrators, in which silver ion is generated coulometrically; sulfur dioxide monitors, where anodically generated bromine oxidizes the analyte to sulfate ions; carbon dioxide monitors, in which the gas, absorbed in monoethanolamine, is titrated with coulometrically generated base; and water titrators, in which Karl Fischer reagent (see Section 20C-5) is generated electrolytically.

Spreadsheet Summary In the second experiment in Chapter 11 of *Applications of Microsoft® Excel in Analytical Chemistry,* a spreadsheet is developed to plot a coulometric titration curve. The end point is located by first- and second-derivative methods.

WEB WORKS

Go to **http://chemistry.brookscole.com/skoogfac/.** From the Chapter Resources menu, choose Web Works. Locate the Chapter 22 section, and click on the link to Bioanalytical Systems. Investigate the electrochemical instruments produced by this instrument company. In particular, describe the features and specifications of the cell for bulk electrolysis. Use the Google search engine to find companies that make coulometers. Compare the features of two instruments from two different instrument companies.

QUESTIONS AND PROBLEMS

Note: Numerical data are molar analytical concentrations if the full formula of a species is provided. Molar equilibrium concentrations are supplied for species displayed as ions.

22-1. Briefly distinguish between
- *(a) concentration polarization and kinetic polarization.
- (b) a galvanostat and a potentiostat.
- *(c) a coulomb and a faraday.
- (d) a working electrode and a counter electrode.
- (e) the electrolysis circuit and the control circuit for controlled-potential methods.

22-2. Briefly define
- *(a) current density.
- (b) ohmic potential.
- *(c) coulometric titration.
- (d) controlled-potential electrolysis.
- *(e) current efficiency.
- (f) an electrochemical equivalent.

***22-3.** Describe three mechanisms responsible for the transport of dissolved species to and from an electrode surface.

22-4. How does the existence of a current affect the potential of an electrochemical cell?

***22-5.** How do concentration polarization and kinetic polarization resemble one another? How do they differ?

22-6. What experimental variables affect concentration polarization in an electrochemical cell?

22-7. What is a supporting electrolyte and what is its role in electrochemistry?

***22-8.** Describe conditions that favor kinetic polarization in an electrochemical cell.

22-9. How do electrogravimetric and coulometric methods differ from potentiometric methods? Consider currents, voltages, and instrumentation in your answer.

***22-10.** Identify three factors that influence the physical characteristics of an electrolytic deposit.

22-11. What is the purpose of a depolarizer?

***22-12.** What is the function of (a) a galvanostat and (b) a potentiostat?

***22-13.** Differentiate between controlled-potential coulometry and constant-current coulometry.

***22-14.** Why is it ordinarily necessary to isolate the working electrode from the counter electrode in a controlled-potential coulometric analysis?

22-15. Why is an auxiliary reagent always required in a coulometric titration?

22-16. Determine the number of ions involved at the surface of an electrode during each second that an electrochemical cell is operated at 0.020 A at 100% current efficiency, and the ions involved are
 (a) univalent.
 *(b) divalent.
 (c) trivalent.

22-17. Calculate the theoretical potential at 25°C needed to initiate the deposition of
 *(a) copper from a solution that is 0.150 M in Cu^{2+} and buffered to a pH of 3.00. Oxygen is evolved at the anode at 1.00 atm.
 (b) tin from a solution that is 0.120 M in Sn^{2+} and buffered to a pH of 4.00. Oxygen is evolved at the anode at 770 torr.
 *(c) silver bromide on a silver anode from a solution that is 0.0864 M in Br^- and buffered to a pH of 3.40. Hydrogen is evolved at the cathode at 765 torr.
 (d) Tl_2O_3 from a solution that is 4.00×10^{-3} M in Tl^+ and buffered to a pH of 8.00. The solution is also made 0.010 M in Cu^{2+}, which acts as a cathode depolarizer for the process

$$Tl_2O_3 + 3H_2O + 4e^- \rightleftharpoons 2Tl^+ + 6OH^- \qquad E^0 = 0.020\text{ V}$$

***22-18.** Calculate the initial potential needed for a current of 0.078 A in the cell

$$Co|Co^{2+}(6.40 \times 10^{-3}\text{ M})\|Zn^{2+}(3.75 \times 10^{-3}\text{ M})|Zn$$

if this cell has a resistance of 5.00 Ω.

22-19. The cell

$$Sn|Sn^{2+}(8.22 \times 10^{-4}\text{ M})\|Cd^{2+}(7.50 \times 10^{-2}\text{ M})|Cd$$

has a resistance of 3.95 Ω. Calculate the initial potential that will be needed for a current of 0.072 A in this cell.

***22-20.** Copper is to be deposited from a solution that is 0.200 M in Cu(II) and is buffered to a pH of 4.00. Oxygen is evolved from the anode at a partial pressure of 740 torr. The cell has a resistance of 3.60 Ω; the temperature is 25°C. Calculate
 (a) the theoretical potential needed to initiate deposition of copper from this solution.
 (b) the *IR* drop associated with a current of 0.10 A in this cell.
 (c) the initial potential, given that the overvoltage of oxygen is 0.50 V under these conditions.
 (d) the potential of the cell when $[Cu^{2+}]$ is 8.00×10^{-6} M, assuming that *IR* drop and O_2 overvoltage remain unchanged.

22-21. Nickel is to be deposited on a platinum cathode (area = 120 cm^2) from a solution that is 0.200 M in Ni^{2+} and buffered to a pH of 2.00. Oxygen is evolved at a partial pressure of 1.00 atm at a platinum anode with an area of 80 cm^2. The cell has a resistance of 3.15 Ω; the temperature is 25°C. Calculate
 (a) the thermodynamic potential needed to initiate the deposition of nickel.
 (b) the *IR* drop for a current of 1.10 A.
 (c) the current density at the anode and the cathode.
 (d) he initial applied potential, given that the overvoltage of oxygen on platinum is approximately 0.52 V under these conditions.
 (e) the applied potential when the nickel concentration has decreased to 2.00×10^{-4} M. (Assume that all variables other than $[Ni^{2+}]$ remain constant.)

***22-22.** Silver is to be deposited from a solution that is 0.150 M in $Ag(CN)_2^-$, 0.320 M in KCN and buffered to a pH of 10.00. Oxygen is evolved at the anode at a partial pressure of 1.00 atm. The cell has a resistance of 2.90 Ω; the temperature is 25°C. Calculate
 (a) the theoretical potential needed to initiate deposition of silver from this solution.
 (b) the *IR* drop associated with a current of 0.12 A.
 (c) the initial applied potential, given that the O_2 overvoltage is 0.80 V.
 (d) the applied potential when $[Ag(CN)_2^-]$ is 1.00×10^{-5} M, assuming no changes in *IR* drop and O_2 overvoltage.

22-23. A solution is 0.150 M in Co^{2+} and 0.0750 M in Cd^{2+}.
 (a) Calculate the Co^{2+} concentration in the solution as the first cadmium starts to deposit.
 (b) Calculate the cathode potential needed to lower the Co^{2+} concentration to 1.00×10^{-5} M.
 (c) Based on (a) and (b) above, can Co^{2+} be quantitatively separated from Cd^{2+}?

***22-24.** A solution is 0.0500 M in BiO^+ and 0.0400 M in Co^{2+} and has a pH of 2.50.
 (a) What is the concentration of the more readily reduced cation at the onset of deposition of the less reducible one?
 (b) What is the potential of the cathode when the concentration of the more easily reduced species is 1.00×10^{-6} M?

(c) Can we achieve a quantitative separation based on your results in (a) and (b) above?

22-25. Electrogravimetric analysis involving control of the cathode potential is proposed as a means to separate Bi^{3+} and Sn^{2+} in a solution that is 0.200 M in each ion and buffered to pH 1.50.

(a) Calculate the theoretical cathode potential at the start of deposition of the more readily reduced ion.

(b) Calculate the residual concentration of the more readily reduced species at the outset of the deposition of the less readily reduced species.

(c) Propose a range (vs. SCE), if such exists, within which the cathode potential should be maintained; consider a residual concentration less than 10^{-6} M as constituting quantitative removal.

*22-26. Halide ions can be deposited on a silver anode via the reaction

$$Ag(s) + X^- \rightarrow AgX(s) + e^-$$

(a) If 1.00×10^{-5} M is used as the criterion for quantitative removal, is it theoretically feasible to separate Br^- from I^- through control of the anode potential in a solution that is initially 0.250 M in each ion?

(b) Is separation of Cl^- and I^- theoretically feasible in a solution that is initially 0.250 M in each ion?

(c) If separation is feasible in either (a) or (b), what range of anode potential (vs. SCE) should be used?

22-27. A solution is 0.100 M in each of two reducible cations, A and B. Removal of the more reducible species (A) is deemed complete when [A] has been decreased to 1.00×10^{-5} M. What minimum difference in standard electrode potentials will permit the isolation of A without interference from B when

A is	B is
(a) univalent	univalent
(b) divalent	univalent
(c) trivalent	univalent
(d) univalent	divalent
(e) divalent	divalent
(f) trivalent	divalent
(g) univalent	trivalent
(h) divalent	trivalent
(i) trivalent	trivalent

*22-28. Calculate the time needed for a constant current of 0.852 A to deposit 0.250 g of Co(II) as

(a) elemental cobalt on the surface of a cathode.

(b) Co_3O_4 on an anode.

Assume 100% current efficiency for both gases.

22-29. Calculate the time needed for a constant current of 1.20 A to deposit 0.500 g of

(a) Tl(III) as the element on a cathode.

(b) Tl(I) as Tl_2O_3 on an anode.

(c) Tl(I) as the element on a cathode.

*22-30. A 0.2416-g sample of a purified organic acid was neutralized by the hydroxide ion produced in 5 min and 24 s by a constant current of 367 mA. Calculate the gram equivalent weight of the acid.

22-31. The CN^- concentration of 10.0 mL of a plating solution was determined by titration with electrogenerated hydrogen ion to a methyl orange end point. A color change occurred after 3 min and 22 s with a current of 43.4 mA. Calculate the number of grams of NaCN per liter of solution. Also calculate the number of ppm of NaCN in the solution.

*22-32. An excess of $HgNH_3Y^{2-}$ was introduced to 25.00 mL of well water. Express the hardness of the water in terms of ppm $CaCO_3$ if the EDTA needed for the titration was generated at a mercury cathode (see Equation 22-9) in 1.05 min by a constant current of 52.7 mA. Assume 100% current efficiency.

22-33. Electrolytically generated I_2 was used to determine the amount of H_2S in 100.0 mL of brackish water. Following addition of excess KI, titration required a constant current of 36.32 mA for 10.12 min. The reaction was

$$H_2S + I_2 \rightarrow S(s) + 2H^+ + 2I^-$$

Express the results of the analysis in terms of ppm H_2S.

*22-34. The nitrobenzene in 194 mg of an organic mixture was reduced to phenylhydroxylamine at a constant potential of − 0.96 V (vs. SCE) applied to a mercury cathode:

$$C_6H_5NO_2 + 4H^+ + 4e^- \rightarrow C_6H_5NHOH + H_2O$$

The sample was dissolved in 100 mL of methanol; after electrolysis for 30 min, the reaction was judged complete. An electronic coulometer in series with the cell indicated that the reduction required 31.23 C. Calculate the percentage of $C_6H_5NO_2$ in the sample.

*22-35. The phenol content of water downstream from a coking furnace was determined by coulometric analysis. A 100-mL sample was rendered slightly acidic, and an excess of KBr was introduced. To produce Br_2 for the reaction

$$C_6H_5OH + 3Br_2 \rightarrow Br_3C_6H_2OH(s) + 2HBr$$

a steady current of 0.0313 A for 7 min and 33 s was required. Express the results of this analysis in terms of parts of C_6H_5OH per million parts of water. (Assume that the density of water is 1.00 g/mL.)

22-36. At a potential of −1.0 V (vs. SCE), CCl_4 in methanol is reduced to $CHCl_3$ at a mercury cathode:

$$2CCl_4 + 2H^+ + 2e^- + 2Hg(l) \rightarrow 2CHCl_3 + Hg_2Cl_2(s)$$

At −1.80 V, the $CHCl_3$ further reacts to give CH_4:

$$2CHCl_3 + 6H^+ + 6e^- + 6Hg(l) \rightarrow 2CH_4 + 3Hg_2Cl_2(s)$$

Several different 0.750-g samples containing CCl_4, $CHCl_3$, and inert organic species were dissolved in methanol and electrolyzed at −1.0 V until the current approached zero. A coulometer indicated the charge required to complete the reaction, as given in the middle column of the following table. The potential of the cathode was then adjusted to −1.8 V. The additional charge given in the last column of the table was required at this potential.

Sample No.	Charge Required at −1.0 V, C	Charge Required at −1.8 V, C
1	11.63	68.60
2	21.52	85.33
3	6.22	45.98
4	12.92	55.31

Calculate the percentage of CCl_4 and $CHCl_3$ in each mixture.

22-37. A single mixture containing only $CHCl_3$ and CH_2Cl_2 was divided into five parts to obtain samples for replicate determinations. Each sample was dissolved in methanol and electrolyzed in a cell containing a mercury cathode; the potential of the cathode was held constant at −1.80 V (vs. SCE). Both compounds were reduced to CH_4 (see Problem 22-36 for the reaction). Calculate the mean value of the percentages of $CHCl_3$ and CH_2Cl_2 in the mixture. Find the standard deviations and the relative standard deviations.

Sample	Mass of Sample, g	Charge Required, C
1	0.1309	306.72
2	0.1522	356.64
3	0.1001	234.54
4	0.0755	176.91
5	0.0922	216.05

22-38. Construct a coulometric titration curve of 100.0 mL of a 1 M H_2SO_4 solution containing Fe(II) titrated with Ce(IV) generated from 0.075 M Ce(III). The titration is monitored by potentiometry. The initial amount of Fe(II) present is 0.05182 mmol. A constant current of 20.0 mA is used. Find the time corresponding to the equivalence point. Then, for about 10 values of time before the equivalence point, use the stoichiometry of the reaction to calculate the amount of Fe^{3+} produced and the amount of Fe^{2+} remaining. Use the Nernst equation to find the system potential. Find the equivalence point potential in the usual manner for a redox titration. For about 10 times after the equivalence point, calculate the amount of Ce^{4+} produced from the electrolysis and the amount of Ce^{3+} remaining. Plot the curve of system potential versus electrolysis time.

*22-39. Traces of aniline, $C_6H_5NH_2$, in drinking water can be determined by reaction with an excess of electrolytically generated Br_2:

$3Br_2$ + [benzene ring with NH_2] ⇌ [benzene ring with NH_2, Br, Br, Br] $+ 3H^+ + 3Br^-$

The polarity of the working electrode is then reversed, and the excess Br_2 is determined by a coulometric titration involving the generation of Cu(I):

$$Br_2 + 2Cu^+ \rightarrow 2Br^- + 2Cu^{2+}$$

Suitable quantities of KBr and $CuSO_4$ were added to a 25.0-mL sample containing aniline. Calculate the mass of $C_6H_5NH_2$ (in micrograms) in the sample from the data:

Working Electrode Functioning as	Generation Time with Constant Current of 1.51 mA, min
Anode	3.76
Cathode	0.270

***22-40.** Quinone can be reduced to hydroquinone with an excess of electrolytically generated Sn(II):

$$C_6H_4O_2 \text{ (quinone)} + Sn^{2+} + 2H^+ \rightleftharpoons C_6H_4(OH)_2 \text{ (hydroquinone)} + Sn^{4+}$$

The polarity of the working electrode is then reversed, and the excess Sn(II) is oxidized with Br_2 generated in a coulometric titration:

$$Sn^{2+} + Br_2 \rightarrow Sn^{4+} + 2\,Br^-$$

Appropriate quantities of $SnCl_4$ and KBr were added to a 50.0-mL sample. Calculate the weight of $C_6H_4O_2$ in the sample from the data:

Working Electrode Functioning as	Generation Time with Constant Current of 1.062 mA, min
Cathode	8.34
Anode	0.691

22-41. Challenge Problem. Sulfide ion (S^{2-}) is formed in wastewater by the action of anaerobic bacteria on organic matter. Sulfide can be readily protonated to form volatile, toxic H_2S. In addition to the toxicity and noxious odor, sulfide and H_2S cause corrosion problems because they can be easily converted to sulfuric acid when conditions change to aerobic. One common method to determine sulfide is by coulometric titration with generated silver ion. At the generator electrode, the reaction is $Ag \rightarrow Ag^+ + e^-$. The titration reaction is $S^{2-} + 2Ag^+ \rightarrow Ag_2S(s)$.

(a) A digital chloridometer was used to determine the mass of sulfide in a wastewater sample. The chloridometer reads out directly in ng Cl^-. In chloride determinations, the same generator reaction is used, but the titration reaction is $Cl^- + Ag^+ \rightarrow AgCl(s)$. Derive an equation that relates the desired quantity, ng S^{2-}, to the chloridometer readout in ng Cl^-.

(b) A particular wastewater standard gave a reading of 1689.6 ng Cl^-. What total charge in coulombs was required to generate the Ag^+ needed to precipitate the sulfide in this standard?

(c) The following results were obtained on 20.00-mL samples containing known amounts of sulfide (D. T. Pierce, M. S. Applebee, C. Lacher, and J. Bessie, *Environ. Sci. Technol.*, **1998,** *32,* 1734). Each standard was analyzed in triplicate and the mass of chloride recorded. Convert each of the chloride results to ng S^{2-}.

Known mass S^{2-}, ng	Mass Cl^- determined, ng		
6365	10447.0	10918.1	10654.9
4773	8416.9	8366.0	8416.9
3580	6528.3	6320.4	6638.9
1989	3779.4	3763.9	3936.4
796	1682.9	1713.9	1669.7
699	1127.9	1180.9	1174.3
466	705.5	736.4	707.7
373	506.4	521.9	508.6
233	278.6	278.6	247.7
0	−22.1	−19.9	−17.7

(d) Determine the average mass of S^- in ng, the standard deviation, and the % RSD of each standard.

(e) Prepare a plot of the average mass of S^{2-} determined (ng) versus the actual mass (ng). Determine the slope, the intercept, the standard error, and the R^2 value. Comment on the fit of the data to a linear model.

(f) Determine the detection limit (DL) in ng and in ppm using a k factor of 2 (see Equation 8-22).

(g) An unknown wastewater sample gave an average reading of 893.2 ng Cl^{2-}. What is the mass of sulfide in ng. If 20.00 mL of the wastewater sample was introduced into the titration vessel, what is the concentration of S^{2-} in ppm?

InfoTrac College Edition

For additional readings, go to InfoTrac College Edition, your online research library, at

http://infotrac.thomsonlearning.com

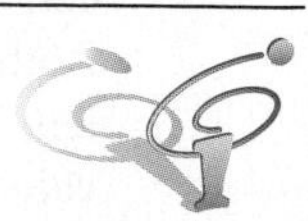

CHAPTER 23

Voltammetry

Lead poisoning in children can cause anorexia, vomiting, convulsions and permanent brain damage. Lead can enter drinking water by being leached from the solder used to join copper pipes and tubes. Anodic stripping voltammetry, discussed in this chapter, is one of the most sensitive analytical methods for determining heavy metals like lead. Shown in the photo is a three-electrode cell used for anodic stripping voltammetry. The working electrode is a glassy carbon electrode on which a thin mercury film has been deposited. An electrolysis step is used to deposit lead into the mercury film as an amalgam. After the electrolysis step, the potential is scanned anodically toward positive values to oxidize (strip) the metal from the film. Levels as low as a few parts per billion can be detected.

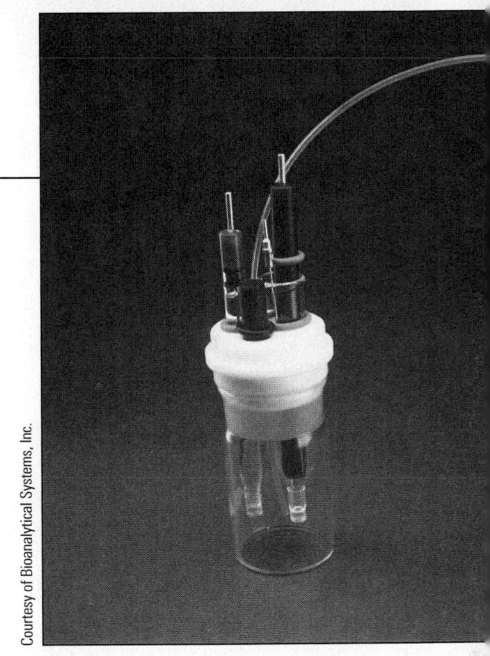
Courtesy of Bioanalytical Systems, Inc.

Electroanalytical methods that depend on the measurement of current as a function of applied potential are called ***voltammetric methods.*** *They employ conditions that encourage polarization of the indicator or working electrode. Generally, to enhance polarization, the working electrodes in voltammetry are relatively small, with surface areas of a few square millimeters at the most and, in some applications, only a few square micrometers.*

Voltammetry is based on the measurement of current in an electrochemical cell under conditions of complete concentration polarization in which the rate of oxidation or reduction of the analyte is limited by the rate of mass transfer of the analyte to the electrode surface. Voltammetry differs from electrogravimetry and coulometry in that in the latter two methods, measures are taken to minimize or compensate for the effects of concentration polarization. Furthermore, in voltammetry a minimal consumption of analyte takes place, whereas in electrogravimetry and coulometry essentially all of the analyte is converted to product.

The field of voltammetry developed from ***polarography,*** *a type of voltammetry that was discovered by the Czechoslovakian chemist Jaroslav Heyrovsky in the early 1920s.*[1] *Polarography, which is still an important branch of voltammetry, differs from other types of voltammetry in that a* ***dropping mercury electrode*** *(DME) is used as the working electrode. The construction and unique properties of this electrode are discussed in Section 23B-5.*[2]

Voltammetry is widely used by analytical, inorganic, physical, and biological chemists for fundamental studies of (1) oxidation and reduction processes in various media, (2) adsorption processes on surfaces, and (3) electron transfer mechanisms at chemically modified electrode surfaces. For analytical purposes, several forms of

Historically, working electrodes with surface areas smaller than a few square millimeters were called **microelectrodes**. Recently, this term has come to signify electrodes with areas on the micrometer scale. In the older literature, micrometer-sized electrodes were often called **ultramicroelectrodes**.

Voltammetric methods are based on measurement of current as a function of the potential applied to a small electrode.

Polarography is voltammetry at the dropping mercury electrode.

[1]J. Heyrovsky, *Chem. Listy,* **1922,** *16,* 256.

[2]For a retrospective on polarography and voltammetry, see A. J. Bard and C. G. Zoski, *Anal. Chem.,* **2000,** *72,* 346A.

Jaroslav Heyrovsky was born in Prague in 1890. He was awarded the 1959 Nobel Prize in chemistry for his discovery and development of polarography. His invention of the polarographic method dates from 1922, and he concentrated the remainder of his career on the development of this new branch of electrochemistry. He died in 1967.

voltammetry are in current use.[3] *Stripping voltammetry is now a significant trace analytical method, particularly for the determination of metals in the environment. Differential pulse polarography and rapid-scan voltammetry are important for the determination of species of pharmaceutical interest. Voltammetric and other electrochemical detectors are frequently employed in high-performance liquid chromatography (HPLC) and capillary electrophoresis (see Sections 32A and 33C). Amperometric techniques are widely used in sensor technology and in monitoring titrations and reactions of biological interest. Modern voltammetric methods continue to be potent tools used by several different kinds of chemists interested in studying and using oxidation, reduction, and adsorption processes.*

23A EXCITATION SIGNALS

In voltammetry, the voltage of the working electrode is varied systematically while the current response is measured. Several different voltage-time functions, called **excitation signals,** can be applied to the electrode. The simplest of these is a linear scan, in which the potential of the working electrode is changed linearly with time. Typically, the potential of the working electrode is varied over a 1- or 2-V range. Other waveforms that can be applied are pulsed waveforms and triangular waveforms. The waveforms of four of the most common excitation signals used in voltammetry are shown in Figure 23-1. The classical voltammetric excitation signal is the linear scan shown in Figure 23-1a, in which the dc voltage applied to the

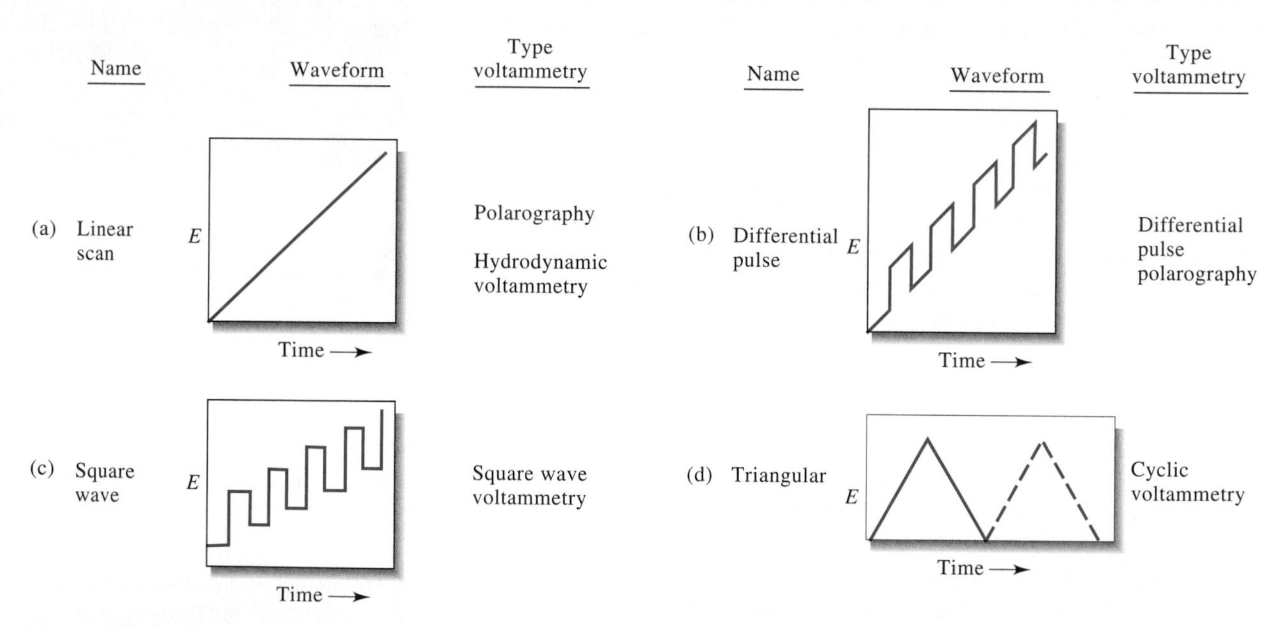

Figure 23-1 Voltage versus time excitation signals used in voltammetry.

[3]For further information on voltammetric methods, see J. A. Dean, *Analytical Chemistry Handbook,* Section 14, pp. 14.57–14.93. New York: McGraw–Hill, 1995; *Analytical Voltammetry,* M. R. Smyth and F. G. Vos, Eds. New York: Elsevier, 1992; A. J. Bard and L. R. Faulkner, *Electrochemical Methods,* 2nd ed. New York: Wiley, 2001; *Laboratory Techniques in Electroanalytical Chemistry,* 2nd ed., P. T. Kissinger and W. R. Heinemann, Eds. New York: Marcel Dekker, 1996.

cell increases linearly as a function of time. The current that develops in the cell is then measured as a function of the applied voltage.

Two pulse-type excitation signals are shown in Figures 23-1b and 1c. Currents are measured at various times during the lifetimes of these pulses, as discussed in Section 23C. With the triangular waveform shown in Figure 23-1d, the potential is varied linearly between a maximum and a minimum value. This process may be repeated numerous times while the current is recorded as a function of potential.

The types of voltammetry that use the various excitation signals are also listed in Figure 23-1. The first three of these techniques in parts a–c of Figure 23-1 are discussed in detail in the sections that follow. Cyclic voltammetry has found considerable application as a diagnostic tool that provides information about the mechanisms of oxidation/reduction reactions under various conditions. Cyclic voltammetry is discussed in Section 23D.

23B LINEAR-SWEEP VOLTAMMETRY

In the first and simplest voltammetric method, the potential of the working electrode is increased or decreased at a typical rate of 2 to 5 mV/s. The current, which is usually in the microampere range, is then recorded to give a **voltammogram,** which is a plot of current as a function of applied potential.

A **supporting electrolyte** is a salt added in excess to the analyte solution. Most commonly, it is an alkali metal salt that does not react at the working electrode at the potentials being used. The salt reduces the effects of migration and lowers the resistance of the solution.

23B-1 Voltammetric Instruments

Figure 23-2 shows the components of a simple apparatus for carrying out linear-sweep voltammetric measurements. The cell is made up of three electrodes immersed in a solution containing the analyte and also an excess of a nonreactive electrolyte called a **supporting electrolyte.** (Note the similarity of this cell to the one for controlled-potential electrolysis shown in Figure 22-7.) One of the three electrodes is the **working electrode,** whose potential versus a **reference electrode** is varied linearly with time. The dimensions of the working electrode are kept small to enhance its tendency to become polarized. The reference electrode has a potential that remains constant throughout the experiment. The third electrode is a

The **working electrode** is the electrode at which the analyte is oxidized or reduced. The potential between the working electrode and the **reference electrode** is controlled. Electrolysis current passes between the working electrode and a **counter electrode.**

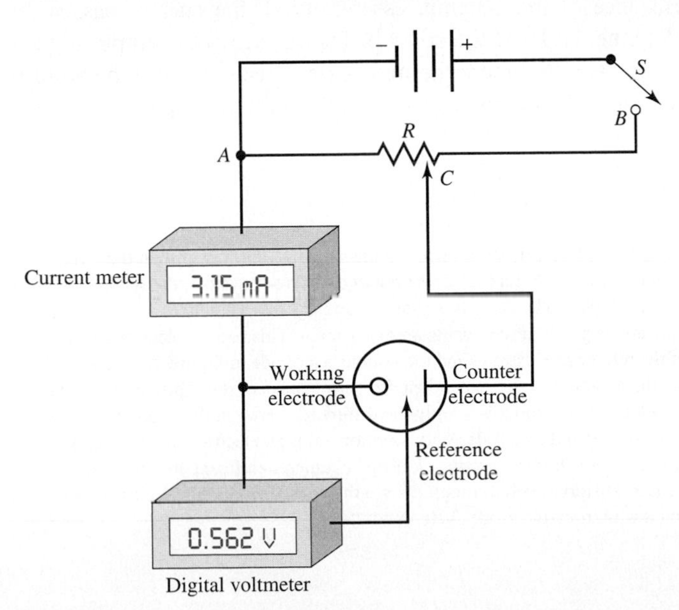

Figure 23-2 A manual potentiostat for voltammetry.

counter electrode, which is often a coil of platinum wire or a pool of mercury. The current in the cell passes between the working electrode and the counter electrode.[4] The signal source is a variable dc voltage source consisting of a battery in series with a variable resistor *R*. The desired potential is selected by moving the contact *C* to the proper position on the resistor. The digital voltmeter has such a high electrical resistance ($>10^{11}\ \Omega$) that there is essentially no current in the circuit containing the meter and the reference electrode. Thus, virtually all the current from the source passes between the counter electrode and the working electrode. A voltammogram is recorded by moving the contact *C* in Figure 23-2 and recording the resulting current as a function of the potential between the working electrode and the reference electrode.

In principle, the manual potentiostat of Figure 23-2 could be used to generate a linear-sweep voltammogram. In such an experiment, contact *C* is moved at a constant rate from *A* to *B* to produce the excitation signal shown in Figure 23-1a. The current and voltage are then recorded at consecutive equal time intervals during the voltage (or time) scan. In modern voltammetric instruments, however, the excitation signals shown in Figure 23-1 are generated electronically. These instruments vary the potential in a systematic way with respect to the reference electrode and record the resulting current. The independent variable in this experiment is the potential of the working electrode versus the reference electrode and not the potential between the working electrode and the counter electrode. A potentiostat that is designed for linear-sweep voltammetry is described in Feature 23-1.

FEATURE 23-1

Voltammetric Instruments Based on Operational Amplifiers

In Feature 21-5, we described the use of operational amplifiers to measure the potential of electrochemical cells. "Op amps" also can be used to measure currents and carry out a variety of other control and measurement tasks. Consider the measurement of current, as illustrated in Figure 23F-1.

In this circuit, a voltage source *E* is attached to one electrode of an electrochemical cell, which produces a current *I* in the cell. Because of the very high input resistance of the op amp, essentially all the current passes through the resistor *R* to the output of the op amp. The voltage at the output of the op amp is given by $E_{out} = -IR$, where the minus sign arises because the amplifier output voltage E_{out} must be opposite in sign to the voltage drop across resistance *R* for

[4]Early voltammetry was performed with a two-electrode system rather than the three-electrode system shown in Figure 23-2. With a two-electrode system, the second electrode is either a large metal electrode, such as a pool of mercury, or a reference electrode large enough to prevent its polarization during an experiment. This second electrode combines the functions of the reference electrode and the counter electrode in Figure 23-2. Here, it is assumed that the potential of this second electrode remains constant throughout a scan, so that the microelectrode potential is simply the difference between the applied potential and the potential of the second electrode. With solutions of high electrical resistance, however, this assumption is not valid because the *IR* drop becomes significant and increases as the current increases. Distorted voltammograms are the consequence. Almost all voltammetry is now performed with three-electrode systems.

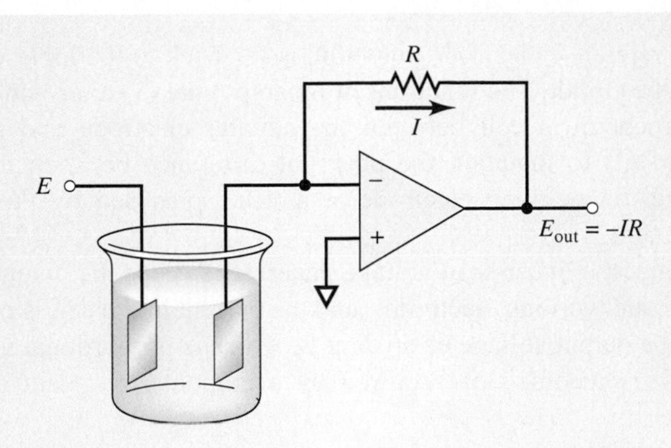

Figure 23F-1 An op amp circuit for measuring voltammetric current.

the potential difference between the op amp inputs to be close to zero volts. By solving this equation for I, we have

$$I = \frac{-E_{out}}{R}$$

In other words, the current in the electrochemical cell is proportional to the voltage output of the op amp. The value of the current can then be calculated from the measured values of E_{out} and the resistance R. The circuit is called a **current-to-voltage converter.**

Op amps can be used to construct an automatic three-electrode potentiostat, as illustrated in Figure 23F-2. Notice that the current-measuring circuit of Figure 23F-1 is connected to the working electrode of the cell (op amp C). The reference electrode is attached to a voltage follower (op amp B). As discussed in Feature 21-4, the voltage follower monitors the potential of the reference electrode without drawing any current from the cell. The output of op amp B,

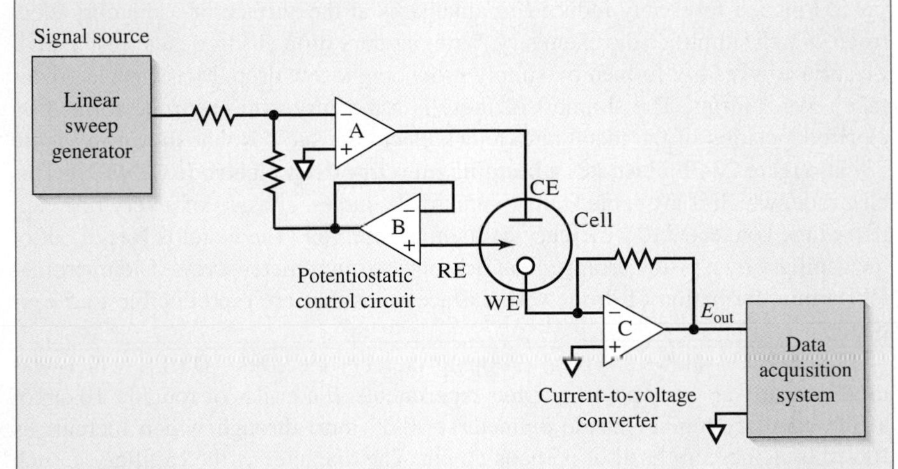

Figure 23F-2 An op amp potentiostat. The three-electrode cell has a working electrode (WE), reference electrode (RE), and a counter electrode (CE).

(continued)

which is the reference electrode potential, is fed back to the input of op amp A to complete the circuit. The functions of op amp A are (1) to provide the current in the electrochemical cell between the counter electrode and the working electrode and (2) to maintain the potential difference between the reference electrode and the working electrode at a value provided by the linear-scan voltage generator.

In operation, the linear-scan voltage generator sweeps the potential between the reference and working electrodes, and the current in the cell is monitored by op amp C. The output voltage of op amp B, which is proportional to the current I in the cell, is recorded or acquired by a computer for data analysis and presentation.[5]

23B-2 Voltammetric Electrodes

The electrodes used in voltammetry take a variety of shapes and forms. Often, as Figure 23-3a shows, they are small flat disks of a conductor that are press fitted into a rod of an inert material, such as Teflon or Kel-F. A wire contact is imbedded in the material. The conductor may be an inert metal, such as platinum or gold; pyrolytic graphite or glassy (vitreous) carbon; a semiconductor, such as tin or indium oxide; or a metal coated with a film of mercury. As shown in Figure 23-4, the range of potentials that can be used with these electrodes in aqueous solutions varies and depends not only on electrode material but also on the composition of the solution in which it is immersed. Generally, the positive potential limitations are caused by the large currents that are due to oxidation of the water to give molecular oxygen. The negative limits arise from the reduction of water, giving hydrogen. Note that relatively large negative potentials can be tolerated with mercury electrodes owing to the high overvoltage of hydrogen on this metal.

► Large negative potentials can be used with mercury electrodes.

Mercury electrodes have been widely used in voltammetry for several reasons. One is the relatively large negative potential range just described. In addition, many metal ions are reversibly reduced to amalgams at the surface of a mercury electrode, which simplifies the chemistry. With mercury drop electrodes, a fresh metallic surface is readily formed by simply producing a new drop. Mercury electrodes take several forms. The simplest of these is a mercury film electrode formed by electrodeposition of the metal onto a disk electrode, such as that shown in Figure 23-3a. Figure 23-3b illustrates a **hanging mercury drop electrode** (HMDE). This electrode, which is available from commercial sources, consists of a very fine capillary tube connected to a mercury-containing reservoir. The metal is forced out of the capillary by a piston arrangement driven by a micrometer screw. The micrometer permits formation of drops with surface areas that are reproducible to 5% or better.

Metals that are soluble in mercury form liquid alloys known as **amalgams.**

Figure 23-3c shows a typical dropping mercury electrode (DME), which was used in nearly all early polarographic experiments. It consists of roughly 10 cm of a fine capillary tubing (inside diameter = 0.05 mm) through which mercury is forced by a mercury head of perhaps 50 cm. The diameter of the capillary is such that a new drop forms and breaks every 2 to 6 s. The diameter of the drop is 0.5 to 1 mm and is highly reproducible. In some applications, the drop time is controlled

[5] For a complete discussion of op amp three-electrode potentiostats, see P. T. Kissinger, in *Laboratory Techniques in Electroanalytical Chemistry,* P. T. Kissinger and W. R. Heineman, Eds., pp. 165–194. New York: Marcel Dekker, 1996.

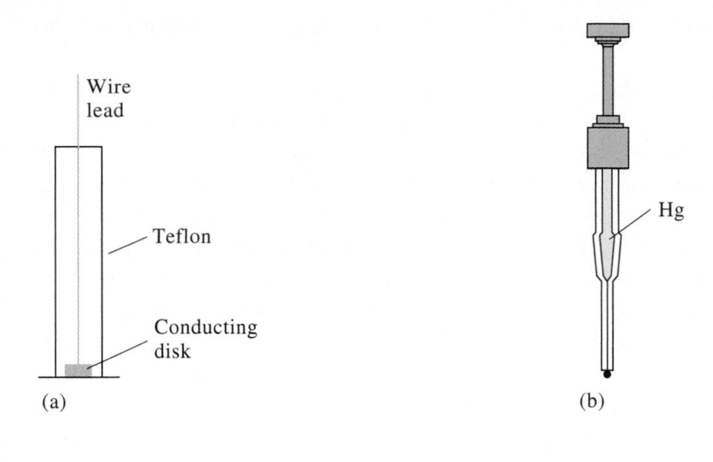

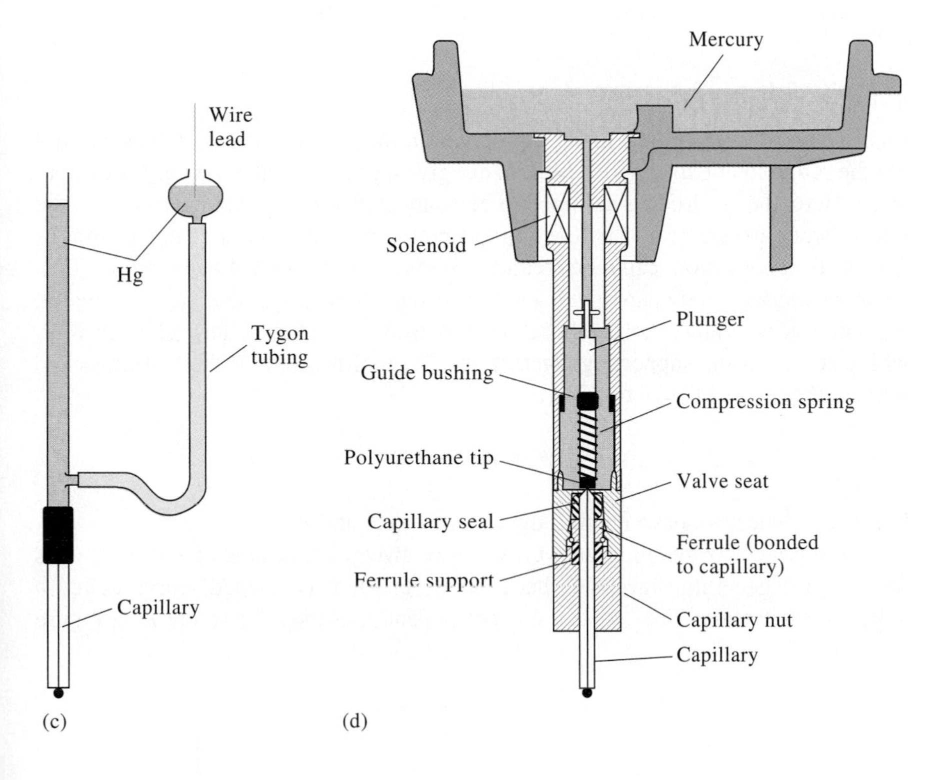

Figure 23-3 Some common types of voltammetric electrodes: (a) a disk electrode; (b) a mercury hanging drop electrode; (c) a dropping mercury electrode; (d) a static mercury dropping electrode.

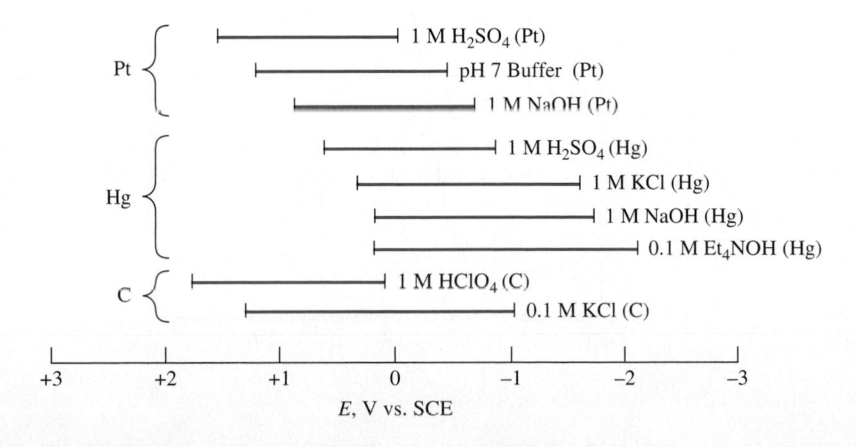

Figure 23-4 Potential ranges for three types of electrodes in various supporting electrolytes. (Adapted from A. J. Bard and L. R. Faulkner, *Electrochemical Methods,* 2nd ed. New York: Wiley, 2001, back cover. This material is used by permission of John Wiley & Sons, Inc.)

by a mechanical knocker that dislodges the drop at a fixed time after it begins to form.

Figure 23-3d shows a commercially available mercury electrode that can be operated as a dropping mercury electrode or a hanging drop electrode. The mercury is contained in a plastic-lined reservoir about 10 inches above the upper end of the capillary. A compression spring forces the polyurethane-tipped plunger against the head of the capillary, thus preventing a flow of mercury. This plunger is lifted on activation of the solenoid by a signal from the control system. The capillary is much larger in diameter (0.15 mm) than the typical one. As a result, the formation of the drop is extremely rapid. After 50, 100, or 200 ms, the valve is closed, leaving a full-sized drop in place until it is dislodged by a mechanical knocker that is built into the electrode support block. This system has the advantage that the full-sized drop forms quickly and current measurements can be delayed until the surface area is stable and constant. This procedure largely eliminates the large current fluctuations that are encountered with the classical dropping electrode.

23B-3 Voltammograms

► The American sign convention for voltammetry considers cathodic currents to be positive and anodic currents to be negative. Voltammograms are plotted with positive current in the top hemisphere and negative currents in the bottom. For mostly historical reasons, the potential axis is arranged such that potentials become less positive (more negative) going from left to right.

Figure 23-5 shows a typical linear-sweep voltammogram for an electrolysis involving the reduction of an analyte species A to give a product P at a mercury film electrode. Here, the electrode is assumed to be connected to the negative terminal of the linear-sweep generator so that the applied potentials are given a negative sign, as shown. By convention, cathodic (reduction) currents are treated as being positive, whereas anodic currents are given a negative sign. In this hypothetical experiment, the solution is assumed to be about 10^{-4} M in A, 0.0 M in P, and 0.1 M in KCl, which serves as the supporting electrolyte. The half-reaction at the working electrode is the reversible reaction

$$A + ne^- \rightleftharpoons P$$

For convenience, we have ignored the charges on A and P.

A **voltammetric wave** is an ʃ-shaped wave obtained in current-voltage plots in voltammetry.

Linear-sweep voltammograms under slow sweep conditions (a few millivolts per second) generally have the shape of a sigmoidal (ʃ-shaped) curve called a **voltammetric wave.** The constant current beyond the steep rise (point Z on Figure

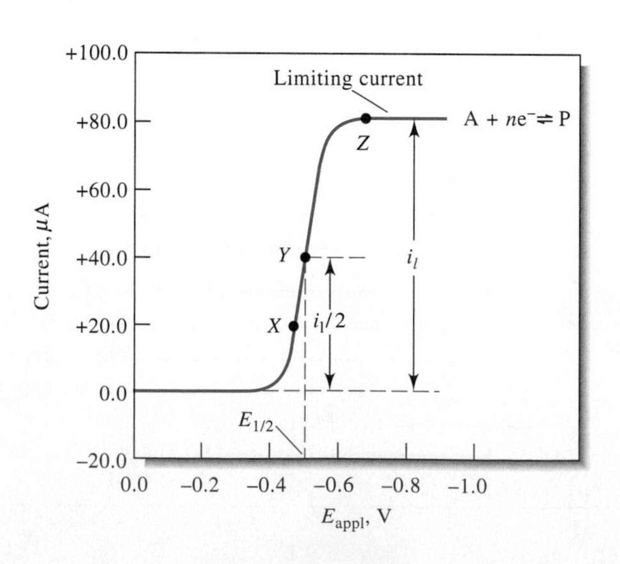

Figure 23-5 Linear-sweep voltammogram for the reduction of a hypothetical species A to give a product P. The limiting current *il* is proportional to the analyte concentration and is used for quantitative analysis. The half-wave potential $E_{1/2}$ is related to the standard potential for the half-reaction and is often used for qualitative identification of species. The half-wave potential is the applied potential at which the current i is $i_l/2$.

23-5) is called the **limiting current** i_l, because it is limited by the rate at which the reactant can be brought to the electrode surface by mass transport processes. Limiting currents are generally directly proportional to reactant concentration. Thus, we may write

The **limiting current** in voltammetry is the current plateau that is observed at the top of the voltammetric wave. It occurs because the surface concentration of the analyte falls to zero. At this point, the mass transfer rate is its maximum value. The limiting current plateau is an example of complete concentration polarization.

$$i_l = kc_A$$

where c_A is the analyte concentration and k is a constant. Quantitative linear-sweep voltammetry is based on this relationship.

The potential at which the current is equal to one half of the limiting current is called the **half-wave potential** and is given the symbol $E_{1/2}$. The half-wave potential is closely related to the standard potential for the half-reaction but is usually not identical to that constant. Half-wave potentials are sometimes useful for identification of the components of a solution.

The **half-wave potential** occurs when the current is equal to one half of the limiting value.

To obtain reproducible limiting currents rapidly, either (1) the solution or the electrode must be in continuous and reproducible motion or (2) a dropping mercury electrode must be used. Linear-sweep voltammetry in which the solution is stirred or the electrode is rotated is called **hydrodynamic voltammetry.** Voltammetry with the dropping mercury electrode is called **polarography.**

In the type of linear-sweep voltammetry discussed thus far, the potential is changed slowly enough and mass transfer is rapid enough that a steady state is reached at the electrode surface. Hence, the mass transport rate of analyte A to the electrode just balances its reduction rate at the electrode. Likewise, the mass transport of P away from the electrode is just equal to its production rate at the electrode surface. There is another type of linear-sweep voltammetry in which fast scan rates (1 V/s or greater) are used with unstirred solutions. In this type of voltammetry, a peak-shaped current-time signal is obtained because of depletion of the analyte in the solution near the electrode. Cyclic voltammetry (see Section 23D) is an example of a process in which forward and reverse linear scans are applied. With cyclic voltammetry, products formed on the forward scan can be detected on the reverse scan if they have not moved away from the electrode or been altered by a chemical reaction.

23B-4 Hydrodynamic Voltammetry

Hydrodynamic voltammetry is performed in several ways. One method involves stirring the solution vigorously while it is in contact with a fixed electrode. Alternatively, the electrode is rotated at a constant high speed in the solution, thus providing the stirring action. Still another way of carrying out hydrodynamic voltammetry is to allow the analyte solution to flow through a tube in which the working electrode is mounted. This last technique is becoming widely used in the detection of oxidizable or reducible analytes as they exit from a liquid chromatographic column (see Section 32A-5).

Hydrodynamic voltammetry is a type of voltammetry in which the analyte solution is kept in continuous motion.

As described in Section 22A-2, during electrolysis, reactant is carried to the surface of an electrode by three mechanisms: (1) migration under the influence of an electric field, (2) convection resulting from stirring or vibration, and (3) diffusion due to any concentration difference between the electrode surface and the bulk of the solution. In voltammetry, every effort is made to minimize the effect of migration by introducing an excess of an inactive supporting electrolyte. When the concentration of supporting electrolyte exceeds that of the analyte by 50- to 100-fold, the fraction of the total current carried by the analyte approaches zero. As a result,

◀ Mass transport processes include diffusion, migration, and convection.

the rate of migration of the analyte toward the electrode of opposite charge becomes essentially independent of applied potential.

Concentration Profiles at Voltammetric Electrode Surfaces during Electrolysis

Throughout this discussion, we consider the electrode reaction $A + ne^- \rightleftharpoons P$ to take place at a mercury-coated electrode in a solution of A that also contains an excess of a supporting electrolyte. We have again left the charges off of A and P for clarity. We further assume that the initial concentration of A is c_A while that of P is zero and that P is not soluble in the mercury. Finally, we assume that the reduction reaction is rapid and reversible so that the concentrations of A and P in the layer of solution immediately adjacent to the electrode are given at any instant by the Nernst equation:

$$E_{appl} = E_A^0 - \frac{0.0592}{n} \log \frac{c_P^0}{c_A^0} - E_{ref} \tag{23-1}$$

► Electrolysis at a small voltammetric electrode does not significantly change the bulk concentration of the analyte solution during the course of a voltammetric experiment.

where E_{appl} is the potential difference between the working electrode and the reference electrode, E_A^0 is the standard electrode potential for the half-reaction, and c_P^0 and c_A^0 are the molar concentrations of P and A *in a thin layer of solution very near the electrode surface.* We also assume that because the electrode is very small, the electrolysis, over short periods of time, does not appreciably change the bulk concentration of the solution. As a result, the concentration of A in the bulk of the solution c_A is essentially constant. Also, the concentration of P in the bulk of the solution c_P remains, for all practical purpose, zero ($c_P \approx 0$).

Voltammetric currents depend on the concentration gradient that is established very near the electrode during electrolysis. To visualize these gradients, consider concentration/distance profiles when the reduction described in the previous section is performed at a planar electrode immersed in a solution that is stirred vigorously. To understand these profiles, we first consider the different types of liquid flow patterns that can exist in a stirred solution. We can identify two types of flow depending on the average flow velocity, as shown in Figure 23-6. Laminar flow occurs at low flow velocities and has smooth and regular motion. Turbulent flow, on the other hand, happens at high velocities and has irregular, fluctuating motion. In a stirred electrochemical cell, we have a region of turbulent flow in the bulk of solution far from the electrode and a region of laminar flow as the electrode is approached. These regions are illustrated in Figure 23-7. In the laminar flow

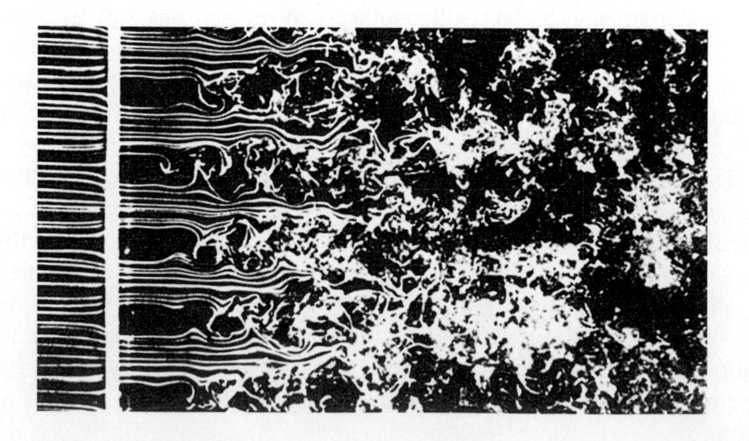

Figure 23-6 Visualization of flow patterns in a flowing stream. Laminar flow, shown on the left, becomes turbulent flow as the average velocity increases. In turbulent flow, the molecules move in an irregular, zigzag fashion, and there are swirls and eddies in the movement. In laminar flow, the streamlines are steady as layers of liquid slide by each other in a regular manner. (From *An Album of Fluid Motion,* assembled by Milton Van Dyke, No. 152, photograph by Thomas Corke and Hassan Nagib, Parabolic Press, Stanford, California, 1982.)

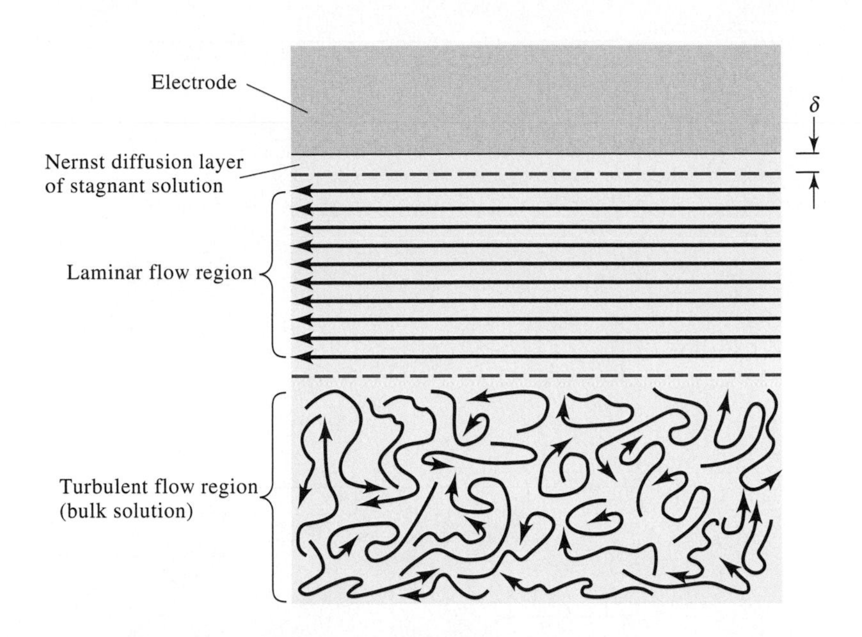

Figure 23-7 Flow patterns and regions of interest near the working electrode in hydrodynamic voltammetry.

region, the layers of liquid slide by one another in a direction parallel to the electrode surface. Very near the electrode, at a distance δ cm from the surface, frictional forces give rise to a region where the flow velocity is essentially zero. The thin layer of solution in this region is a stagnant layer, called the **Nernst diffusion layer.** It is only within the stagnant Nernst diffusion layer that the concentrations of reactant and product vary as a function of distance from the electrode surface and that concentration gradients exist. That is, throughout the laminar flow and turbulent flow regions, convection maintains the concentration of A at its original value and the concentration of P at a very small level.

Figure 23-8a shows the concentration profiles for A at the three potentials labeled *X, Y,* and *Z* in Figure 23-5. The solution is divided into two regions. One region is the bulk of the solution, where mass transport takes place by mechanical convection brought about by stirring. The concentration of A throughout this region is c_A. The second region is the Nernst diffusion layer, which is immediately adjacent to the electrode surface and has a thickness of δ cm. Typically, δ ranges from 0.01 to 0.001 cm depending on the efficiency of the stirring and the viscosity of the liquid. Within the diffusion layer, mass transport takes place by diffusion alone, as would be the case with the unstirred solution. With stirring, however, diffusion is limited to a narrow layer of liquid and cannot extend out indefinitely into the solution. As a consequence, steady, diffusion-controlled currents are realized shortly after the potential is applied.

Figure 23-8b gives concentration profiles for P at the three potentials *X, Y,* and *Z.* In the Nernst diffusion region, the concentration of P decreases linearly with distance from the electrode surface and approaches zero at δ.

Note in the figures that at potential *X,* the equilibrium concentration of A at the electrode surface has been reduced to about 80% of its original value, while the equilibrium concentration P has increased by an equivalent amount (that is, $c_P^0 = c_A - c_A^0$). At potential *Y,* which is the half-wave potential, the equilibrium concentrations of the two species at the surface are approximately the same and are equal to $c_A/2$. Finally, at potential *Z* and beyond, the surface concentration of A approaches zero, whereas that of P approaches the original concentration of A, c_A. At potentials more negative than *Z,* essentially all A ions approaching the electrode

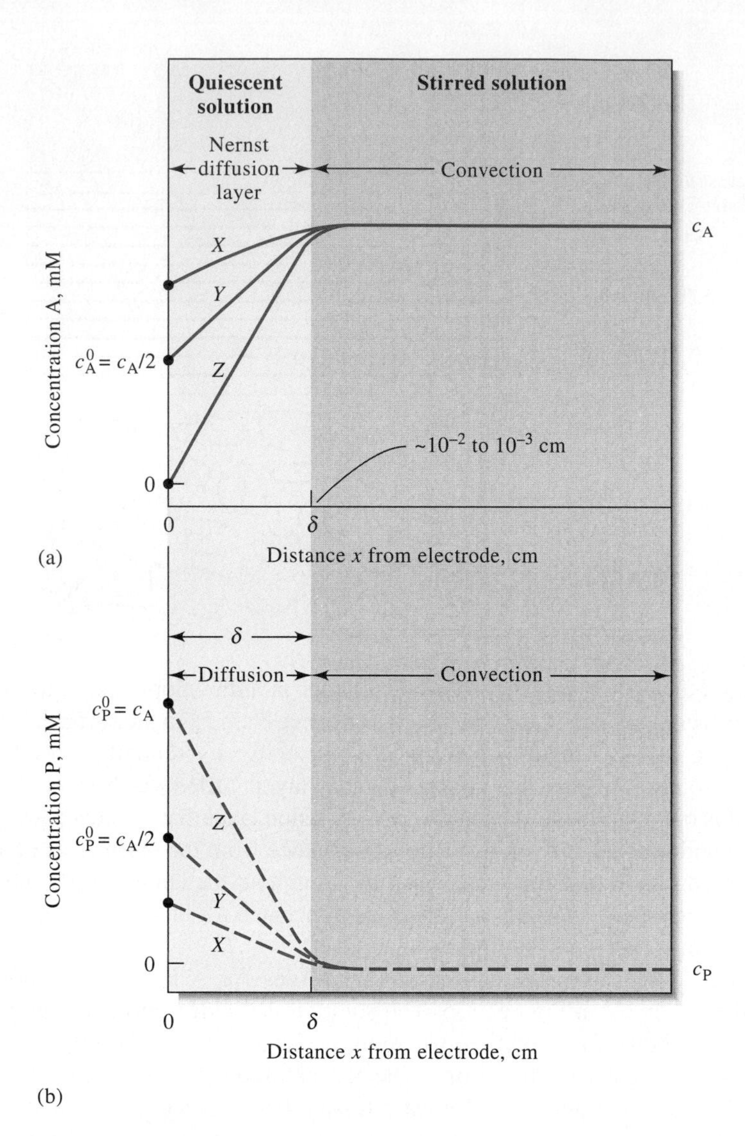

Figure 23-8 Concentration profiles at an electrode/solution interface during the electrolysis $A + ne^- \rightarrow P$ from a stirred solution of A. See Figure 23-5 for potentials corresponding to curves *X, Y,* and *Z.*

surface are immediately reduced to P. The P ions formed in this way rapidly diffuse into the bulk of the solution so that the concentration of P in the surface layer remains constant at c_A.

Voltammetric Currents

The current at any point in the voltammetry experiment described in Figure 23-5 is determined by a combination of (1) the rate of mass transport of A to the edge of the Nernst diffusion layer by convection and (2) the rate of transport of A from the outer edge of the diffusion layer to the electrode surface. Because the product of the electrolysis P diffuses away from the surface and is ultimately swept away by convection, a continuous current is required to maintain the surface concentrations demanded by the Nernst equation. Convection, however, maintains a constant supply of A at the outer edge of the diffusion layer. Thus, a steady-state current results that is determined by the applied potential.

The current in this voltammetry experiment is a quantitative measure of how fast A is being brought to the surface of the electrode; this rate is given by $\partial c_A/\partial x$, where x is the distance in centimeters from the electrode surface. For a planar electrode, it can be shown that the current is given by the expression

$$i = nFAD_A\left(\frac{\partial c_A}{\partial x}\right) \quad (23\text{-}2)$$

where i is the current in amperes, n is the number of moles of electrons per mole of analyte reduced, F is the faraday, A is the electrode surface area in cm^2, D_A is the diffusion coefficient for A in cm^2/s, and c_A is the concentration of A in mol/cm^3. Note that $\partial c_A/\partial x$ is the slope of the initial part of the concentration profiles shown in Figure 23-8a; these slopes can be approximated by $(c_A - c_A^0)/\delta$. Therefore, Equation 23-2 reduces to

$$i = \frac{nFAD_A}{\delta}(c_A - c_A^0) = k_A(c_A - c_A^0) \quad (23\text{-}3)$$

where the constant k_A is equal to $nFAD_A/\delta$.

Equation 23-3 shows that as c_A^0 becomes smaller as a result of a larger applied potential, the current increases until the surface concentration approaches zero, at which point the current becomes constant and independent of the applied potential. Thus, when $c_A^0 \rightarrow 0$, the current becomes the limiting current i_l (see Figure 23-5) and

$$i_l = \frac{nFAD_A}{\delta}c_A = k_A c_A \quad (23\text{-}4)$$

◀ CHALLENGE: Show that the units of Equation 23-4 are amperes if the units of the quantities in the equation are as follows:

Quantity	Units
n	mol electrons/mol analyte
F	coulomb/mol electrons
A	cm^2
D_A	$cm^2\ s^{-1}$
c_A	mol analyte/cm^3
δ	cm

This derivation is based on an oversimplified picture of the diffusion layer, in that the interface between the moving and stationary layers is viewed as a sharply defined edge where transport by convection ceases and transport by diffusion begins. This simplified model does provide, however, a reasonable approximation of the relationship between current and the variables that affect it.[6]

◀ Although our model is oversimplified, it provides a reasonably accurate picture of the processes occurring at the electrode/solution interface.

Current/Voltage Relationships for Reversible Reactions

To develop an equation for the sigmoidal curve shown in Figure 23-5, we can substitute i_l from Equation 23-4 for $k_A c_A$ in Equation 23-3 and rearrange, which gives

$$c_A^0 = \frac{i_l - i}{k_A} \quad (23\text{-}5)$$

The surface concentration of P can also be expressed in terms of the current by employing a relationship similar to Equation 23-3. That is,

$$i = -\frac{nFAD_P}{\delta}(c_P - c_P^0) \quad (23\text{-}6)$$

where the minus sign results from the negative slope of the concentration profile for P. Note that D_P is now the diffusion coefficient of P. We said earlier, however,

[6]For a more rigorous treatment, see A. J. Bard and L. R. Faulkner, *Electrochemical Methods,* 2nd ed., pp. 137–153. New York: Wiley, 2001.

that throughout the electrolysis the concentration of P approaches zero in the bulk of the solution and, therefore, when $c_P \approx 0$,

$$i = -\frac{nFAD_P}{\delta} c_P^0 = k_P c_P^0 \quad (23\text{-}7)$$

where $k_P = -nFAD_P/\delta$. Rearranging this last equation gives

$$c_P^0 = \frac{i}{k_P} \quad (23\text{-}8)$$

If we now substitute Equations 23-5 and 23-8 into Equation 23-1 and rearrange, we obtain

$$E_{appl} = E_A^0 - \frac{0.0592}{n} \log \frac{k_A}{k_P} - \frac{0.0592}{n} \log \frac{i}{i_l - i} - E_{ref} \quad (23\text{-}9)$$

► The half-wave potential is an identifier for the redox couple and is closely related to the standard reduction potential.

The half-wave potential, $E_{1/2}$, is defined as the applied potential when the current i is one half of the limiting current. We can see from Equation 23-9 that when $i = i_l/2$, the third term on the right side of this equation becomes equal to zero. At this point, $E_{appl} = E_{1/2}$ and

$$E_{1/2} = E_A^0 - \frac{0.0592}{n} \log \frac{k_A}{k_P} - E_{ref} \quad (23\text{-}10)$$

If we now substitute this expression into Equation 23-9, we obtain an equation for the entire voltammogram shown in Figure 23-5;

$$E_{appl} = E_{1/2} - \frac{0.0592}{n} \log \frac{i}{i_l - i} \quad (23\text{-}11)$$

Often, the ratio k_A/k_P in Equation 23-10 is nearly unity, so that we may write for species A

$$E_{1/2} \approx E_A^0 - E_{ref} \quad (23\text{-}12)$$

An electrochemical process such as $A + ne^- \rightleftharpoons P$ is said to be *reversible* if the Nernst equation is obeyed under the conditions of the experiment. In a **totally irreversible system,** either the forward or the reverse reaction is so slow as to be completely negligible. In a **partially reversible system,** the reaction in one direction is much slower than the other, but not totally insignificant. A process that appears reversible on a slow time scale may show signs of irreversibility when the time scale of the experiment is increased.

Current/Voltage Relationships for Irreversible Reactions

Many voltammetric electrode processes, particularly those associated with organic systems, are partially or totally irreversible, which leads to drawn-out and less well defined waves. The quantitative description of such waves requires an additional term (involving the activation energy of the reaction) in Equation 23-11 to account for the kinetics of the electrode process. Although half-wave potentials for irreversible reactions ordinarily show some dependence on concentration, diffusion currents are usually still linearly related to concentration; many irreversible processes can, therefore, be adapted to quantitative analysis.

Voltammograms for Mixtures

Ordinarily, the electroactive species in a mixture behave independently of one another at a voltammetric electrode; a voltammogram for a mixture is thus simply the summation of the waves for the individual components. Figure 23-9 shows the voltammograms for a pair of two-component mixtures. The half-wave potentials of the two reactants differ by about 0.1 V in curve *A* and by about 0.2 V in curve *B*. Note that a single voltammogram may permit the quantitative determination of two or more species provided that there is sufficient difference between succeeding half-wave potentials to permit evaluation of individual diffusion currents. Generally, a few tenths of a volt difference is required to resolve different species.

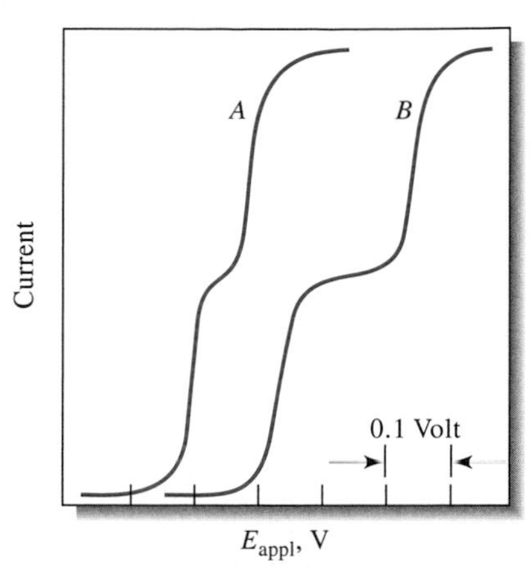

Figure 23-9 Voltammograms for two-component mixtures. Half-wave potentials differ by 0.1 V in curve *A* and 0.2 V in curve *B*.

Anodic and Mixed Anodic/Cathodic Voltammograms

Anodic waves as well as cathodic waves are encountered in voltammetry. An example of an anodic wave is illustrated in curve *A* of Figure 23-10, where the electrode reaction involves the oxidation of iron(II) to iron(III) in the presence of citrate ion. Note that by convention, anodic current is given a negative value. A limiting current is obtained at about +0.1 V, which is due to the half-reaction

$$Fe^{2+} \rightleftharpoons Fe^{3+} + e^-$$

As the potential is made more negative, a decrease in the anodic current occurs; at about −0.02 V, the current becomes zero because the oxidation of iron(II) ion has ended.

Curve *C* represents the voltammogram for a solution of iron(III) in the same medium. Here, a cathodic wave results from reduction of the iron(III) to the divalent state. The half-wave potential is identical to that for the anodic wave, indicating that the oxidation and reduction of the two iron species are perfectly reversible at the working electrode.

Curve *B* is the voltammogram of an equimolar mixture of iron(II) and iron(III). The portion of the curve below the zero-current line corresponds to the oxidation of the iron(II); this reaction ceases at an applied potential equal to the half-wave potential. The upper portion of the curve is due to the reduction of iron(III).

Oxygen Waves

Dissolved oxygen is readily reduced at various electrodes; an aqueous solution saturated with air exhibits two distinct oxygen waves as shown in Figure 23-11. The first results from the reduction of oxygen to peroxide

$$O_2(g) + 2H^+ + 2e^- \rightleftharpoons H_2O_2$$

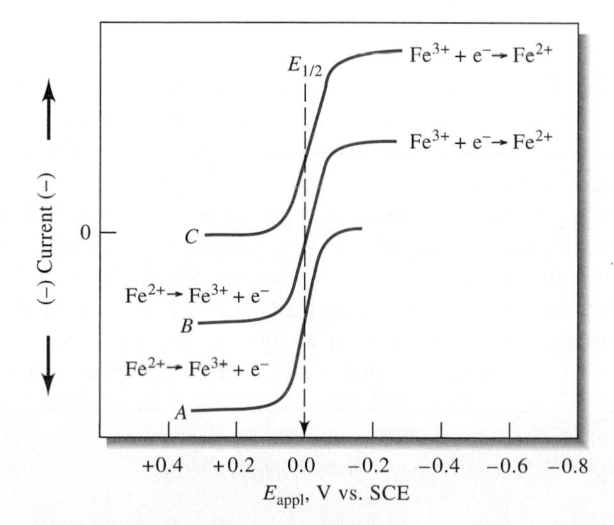

Figure 23-10 Voltammetric behavior of iron(II) and iron(III) in a citrate medium. Curve *A:* anodic wave for a solution in which $c_{Fe^{2+}} = 1 \times 10^{-4}$ M. Curve *B:* anodic/cathodic wave for a solution in which $c_{Fe^{2+}} = c_{Fe^{3+}} = 0.5 \times 10^{-4}$ M. Curve *C:* cathodic wave for a solution in which $c_{Fe^{3+}} = 1 \times 10^{-4}$ M.

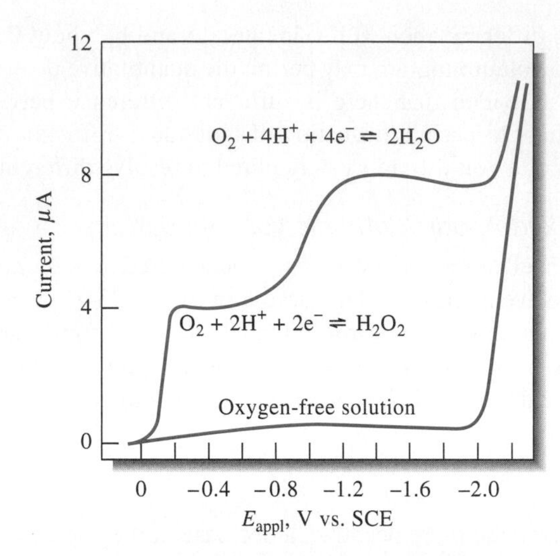

Figure 23-11 Voltammogram for the reduction of oxygen in an air-saturated 0.1 M KCl solution. The lower curve is for a 0.1 M KCl solution in which the oxygen is removed by bubbling nitrogen through the solution.

The second corresponds to the further reduction of the hydrogen peroxide

$$H_2O_2 + 2H^+ + 2e^- \rightleftharpoons 2H_2O$$

As would be expected, the two waves are of equal height. Figure 23-11 shows the sum of the two processes near the second wave.

Voltammetric measurements offer a convenient and widely used method to determine dissolved oxygen in solutions. For determining other species, however, oxygen often interferes with the measurements. Thus, oxygen removal is ordinarily the first step in most voltammetric procedures. Usually, the solution is deaerated for several minutes by bubbling a high-purity inert gas through it (**sparging**). During the analysis, a stream of the same gas, usually nitrogen, is passed over the surface to prevent oxygen from re-entering the solution.

Sparging is a process in which dissolved gases are swept out of a solvent by bubbling an inert gas, such as nitrogen, argon, or helium, through the solution.

Applications of Hydrodynamic Voltammetry

Currently, the most important uses of hydrodynamic voltammetry include (1) detection and determination of chemical species as they exit from chromatographic columns or a continuous-flow apparatus; (2) routine determination of oxygen and certain species of biochemical interest, such as glucose, lactate, and sucrose; (3) detection of end points in coulometric and volumetric titrations; and (4) fundamental studies of electrochemical processes.

Voltammetric Detectors Hydrodynamic voltammetry is becoming widely used for detection and determination of oxidizable or reducible compounds or ions in flowing streams. Compounds that have been separated by liquid chromatography (see Chapter 32) or processed by flow-injection analyzers are typical examples.[7] In these applications, a thin-layer cell such as that shown in Figure 23-12 is used. In such cells, the working electrode is typically imbedded in the wall of an insulating block that is separated from a counter electrode by a thin spacer, as shown. The volume of such a cell is typically 0.1 to 1 μL. A potential corresponding to the limiting current region for analytes is applied between the metal or glassy carbon working electrode and a silver/silver chloride reference electrode

[7]For a recent description of commercially available electrochemical detectors for liquid chromatography, see B. E. Erickson, *Anal. Chem.*, **2000,** *72,* 353A.

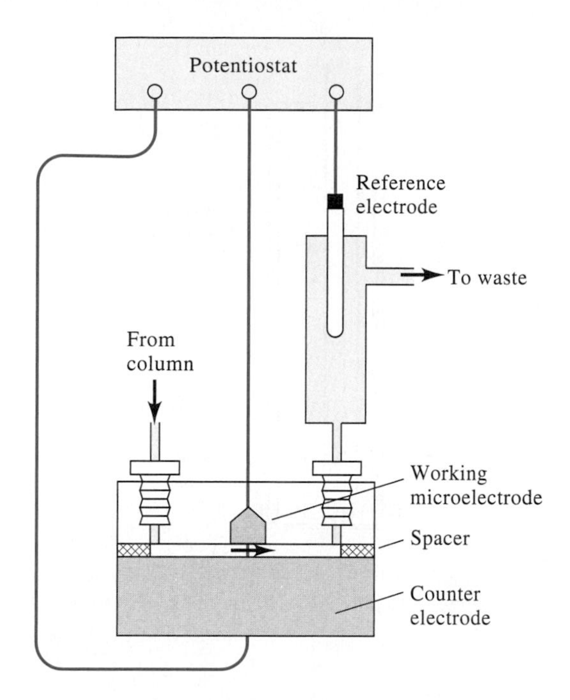

Figure 23-12 A voltammetric system for detecting electroactive species as they elute from a column. The cell volume is 1 μL.

that is located downstream from the detector. In this type of application, analyte detection limits as low as 10^{-9} to 10^{-10} M have been obtained.

Amperometric Sensors A number of voltammetric systems are produced commercially for the determination of specific species of interest in industry and research. These are usually based on measuring the limiting current at a constant applied potential and relating the measured current to concentration. This technique is often called **amperometry.** Amperometric devices are sometimes called electrodes but are, in fact, complete voltammetric cells and are better referred to as sensors. Two of these devices are described here.

The determination of dissolved oxygen in a variety of aqueous environments, such as sea water, blood, sewage, effluents from chemical plants, and soils, is of tremendous importance. One of the most common and convenient devices for making such measurements is the **Clark oxygen sensor,** which was patented by L. C. Clark, Jr., in 1956.[8] A schematic of the Clark oxygen sensor is shown in Figure 23-13. The cell consists of a cathodic platinum-disk working electrode imbedded in a centrally located cylindrical insulator. Surrounding the lower end of this insulator is a ring-shaped silver anode. The tubular insulator and electrodes are mounted inside a second cylinder that contains a buffered solution of potassium chloride. A thin replaceable oxygen-permeable membrane of Teflon or polyethylene is held in place at the bottom end of the tube by an O-ring. The thickness of the electrolyte solution between the cathode and the membrane is approximately 10 μm.

◀ The Clark oxygen sensor is widely used in clinical laboratories for the determination of dissolved O_2 in blood and other body fluids.

When the oxygen sensor is immersed in a flowing or stirred solution of the analyte, oxygen diffuses through the membrane into the thin layer of electrolyte immediately adjacent to the disk cathode, where it diffuses to the electrode and is immediately reduced to water. Two diffusion processes are involved—one through the membrane and the other through the solution between the membrane and the

[8]For a detailed discussion of the Clark oxygen sensor, see M. L. Hitchman, *Measurement of Dissolved Oxygen,* Chapters 3–5. New York: Wiley, 1978.

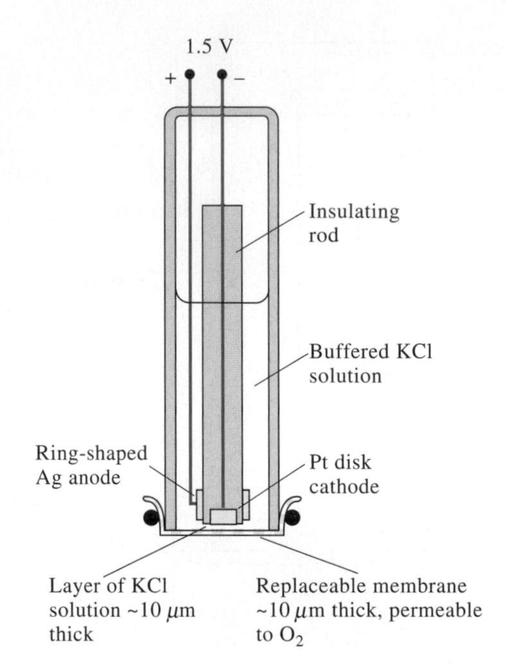

Figure 23-13 The Clark voltammetric oxygen sensor. Cathodic reaction: $O_2 + 4H^+ + 4e^- \rightleftharpoons 2H_2O$. Anodic reaction: $Ag(s) + Cl^- \rightleftharpoons AgCl(s) + e^-$.

electrode surface. For a steady-state condition to be reached in a reasonable time (10 to 20 s), the thickness of the membrane and the electrolyte film must be 20 μm or less. Under these conditions, it is the rate of equilibration of oxygen transfer across the membrane that determines the steady-state current that is reached. This rate is directly proportional to the dissolved oxygen concentration in solution.

► Enzyme-based sensors can be based on detecting hydrogen peroxide, oxygen, or H^+, depending on the analyte and enzyme. Voltammetric sensors are used for H_2O_2 and O_2, while a potentiometric pH electrode is used for H^+.

A number of enzyme-based amperometric sensors are available commercially. An example is a glucose sensor that is widely used in clinical laboratories. This device is similar in construction to the oxygen electrode of Figure 23-13. The membrane in this case is more complex and consists of three layers. The outer layer is a polycarbonate film that is permeable to glucose but impermeable to proteins and other constituents of blood. The middle layer is an immobilized enzyme—here, glucose oxidase. The inner layer is a cellulose acetate membrane, which is permeable to small molecules such as hydrogen peroxide. When this device is immersed in a glucose-containing solution, glucose diffuses through the outer membrane into the immobilized enzyme, where the following catalytic reaction occurs

$$\text{glucose} + O_2 \xrightarrow{\text{glucose oxidase}} \text{gluconic acid} + H_2O_2$$

Molecular model of hydrogen peroxide. Hydrogen peroxide is a strong oxidizing agent that plays an important role in biological and environmental processes. Hydrogen peroxide is produced in enzyme reactions involving the oxidation of sugar molecules. Peroxide radicals can damage cells and body tissues (see Feature 20-2). Peroxide radicals occur in smog and can attack unburned fuel molecules in the environment.

The hydrogen peroxide then diffuses through the inner layer of membrane and to the electrode surface, where it is oxidized to give oxygen. That is,

$$H_2O_2 + 2OH^- \rightarrow O_2 + H_2O + 2e^-$$

The resulting current is directly proportional to the glucose concentration of the analyte solution.

Several other sensors are available that are based on the amperometric measurement of hydrogen peroxide produced by enzymatic reactions. The analytes measured include sucrose, lactose, ethanol, and L-lactate. A different enzyme is, of course, required for each species. In some cases, enzyme electrodes can be based on measuring oxygen or on measuring pH.

Amperometric Titrations Hydrodynamic voltammetry can be employed to estimate the equivalence point of titrations, provided that at least one of the participants or products of the reaction involved is oxidized or reduced at an electrode. The current at some fixed potential in the limiting current region is measured as a function of the reagent volume (or of time if the reagent is generated by a constant-current coulometric process). Plots of the data on either side of the equivalence point are straight lines with differing slopes. We usually determine the end point by extrapolation to the intersection of these straight-line portions.

Amperometric titration curves typically take one of the forms shown in Figure 23-14. The curve in part a represents a titration in which the analyte reacts at the electrode while the titrant does not. Figure 23-14b is typical of a titration in which the reagent reacts at the electrode and the analyte does not. Figure 23-14c corresponds to a titration in which both the analyte and the titrant react at the working electrode.

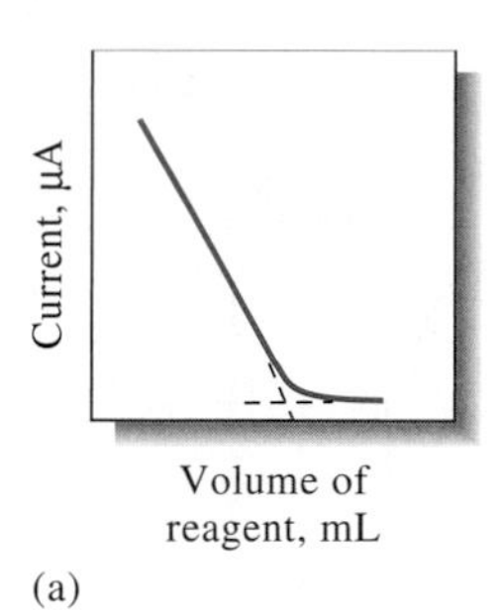

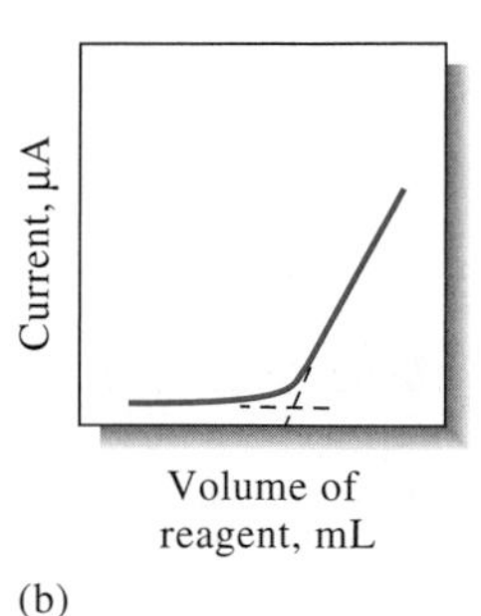

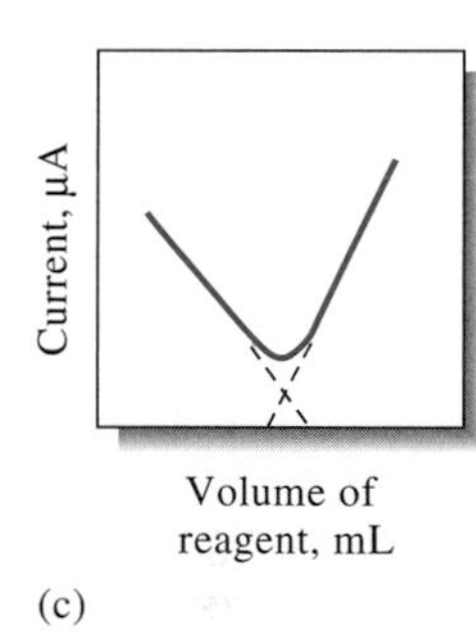

Figure 23-14 Typical amperometric titration curves: (a) analyte is reduced, reagent is not; (b) reagent is reduced, analyte is not; (c) both reagent and analyte are reduced.

Two types of amperometric electrode systems are encountered. One employs a single polarizable working electrode coupled to a reference; the other uses a pair of identical solid-state electrodes immersed in a stirred solution. For the first, the electrode is often a rotating platinum disk connected to a stirring motor, as shown in Figure 23-15. A platinum wire electrode, constructed by sealing a platinum wire into the side of a glass tube, is also used. The dropping mercury electrode is also occasionally employed for amperometric titrations.

Amperometric titrations with one indicator electrode have, with one notable exception, been confined to those cases in which a precipitate or a stable complex is the product. Precipitating reagents include silver nitrate for halide ions, lead(II) nitrate for sulfate ion, and several organic reagents, such as 8-hydroxyquinoline, dimethylglyoxime, and cupferron, for various metallic ions that are reducible at voltammetric electrodes. Several metal ions have also been determined by titration with standard solutions of EDTA. The exception just noted involves titrations of organic compounds, such as certain phenols, aromatic amines, and olefins; hydrazine; and arsenic(III) and antimony(III) with bromine. The bromine is often generated coulometrically. Bromine has also been formed by adding a standard solution of potassium bromate to an acidic solution of the analyte that also contains an excess of potassium bromide by the reaction

$$BrO_3^- + 5Br^- + 6H^+ \rightarrow 3Br_2 + 3H_2O$$

This type of titration has been carried out with a rotating platinum electrode or twin platinum electrodes. No current is observed prior to the equivalence point. After chemical equivalence, a rapid increase in current takes place due to electrochemical reduction of the excess bromine.

The use of a pair of identical metallic electrodes to establish the equivalence point in amperometric titrations offers the advantages of simplicity of equipment and elimination of the need to prepare and maintain a reference electrode. This type of system has been incorporated into equipment designed for the routine automatic determination of a single species, usually with a coulometric generated reagent. An example of this type of system is an instrument for the automatic determination of chloride in samples of serum, sweat, tissue extracts, pesticides, and food products. Here, the reagent is silver ion coulometrically generated from a silver anode. The indicator system consists of a pair of twin silver electrodes that are maintained at a potential of perhaps 0.1 V. Before the equivalence point in the titration of chloride ion, there is essentially no current because no easily reduced species is present in the solution. Consequently, electron transfer at the cathode is precluded and that

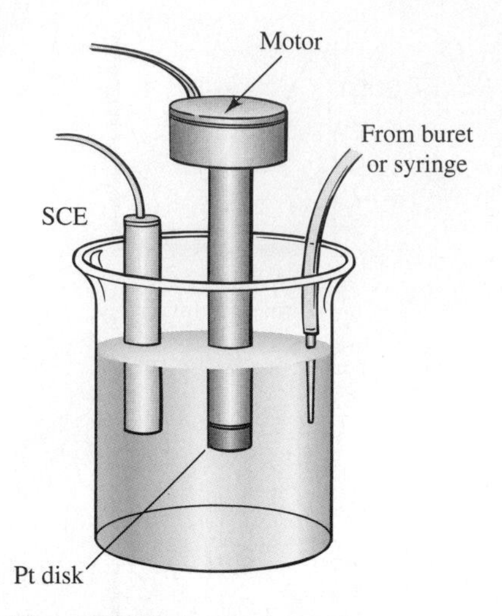

Figure 23-15 Typical cell arrangement for amperometric titrations with a rotating platinum disk electrode.

electrode is completely polarized. Note that the anode is not polarized because the reaction

$$Ag \rightleftharpoons Ag^+ + e^-$$

occurs in the presence of a suitable cathodic reactant or depolarizer.

After the equivalence point has been passed, the cathode becomes depolarized owing to the presence of a significant amount of silver ions, which can react to give silver. That is,

$$Ag^+ + e^- \rightleftharpoons Ag$$

A current develops as a result of this half-reaction and the corresponding oxidation of silver at the anode. The magnitude of the current is, as in other amperometric methods, directly proportional to the concentration of the excess reagent. Thus, the titration curve is similar to that shown in Figure 23-14b. In the automatic titrator just mentioned, the amperometric current signal causes the coulometric generator current to cease; the chloride concentration is then computed from the magnitude of the current and the generation time. The instrument is said to have a range of 1 to 999.9 mM, a precision of 0.1% relative, and an accuracy of 0.5% relative. Typical titration time is 20 s.

The most common end-point detection method for the Karl Fischer titration for determining water (see Section 20C-5) is the amperometric method with dual polarized electrodes. Several manufacturers offer fully automated instruments for use in performing these titrations. A closely related end-point detection method for Karl Fischer titrations measures the potential difference between two identical electrodes through which a small constant current is passed.

Spreadsheet Summary Amperometric titrations are the subject of the final exercise in Chapter 11 of *Applications of Microsoft® Excel in Analytical Chemistry.* An amperometric titration to determine gold in an ore sample is used as an example. Titration curves consisting of two linear segments are extrapolated to find the end point.

23B-5 Polarography

► Polarographic currents are controlled by diffusion alone, not by convection.

Linear-scan polarography was the first type of voltammetry to be discovered and used. It differs from hydrodynamic voltammetry in two regards. First, there is essentially no convection or migration, and, second, a dropping mercury electrode (DME), such as that shown in Figure 23-3c, is used as the working electrode. Because there is no convection, *diffusion alone* controls polarographic limiting currents. Compared with hydrodynamic voltammetry, however, polarographic limiting currents are an order of magnitude or more smaller, since convection is absent in polarography.[9]

[9]References dealing with polarography include A. J. Bard and L. R. Faulkner, *Electrochemical Methods,* 2nd ed., Ch. 7. New York: Wiley, 2001; R. C. Kapoor and B. S. Aggarwal, *Principles of Polarography.* New York: Wiley, 1991; T. Riley and A. Watson, *Polarography and Other Voltammetric Methods.* New York: Wiley, 1987; A. M. Bond, *Modern Polarographic Methods in Analytical Chemistry.* New York: Dekker, 1980; I. M. Kolthoff and J. J. Lingane, *Polarography,* 2nd ed. New York: Wiley, 1952.

Charles D. Winters

$H_2O + I_3^-(aq) + H_3AsO_3(aq) \rightarrow$ (a)

$\leftarrow 3I\ (aq) + H_3AsO_4(aq) + H^+(aq)$ (b)

Color plate 1 Chemical Equilibrium 1: Reaction between iodine and arsenic(III) at pH 1. (a) One mmol I_3^- added to 1 mmol H_3AsO_3. (b) Three mmol I^- added to 1 mmol H_3AsO_4. Both combinations of solutions produce the same final equilibrium state (Section 9B-1, page 233).

Charles D. Winters

$H_2O + I_3^-(aq) + H_3AsO_3(aq) \rightarrow$ (a)

$\leftarrow 3I^-(aq) + H_3AsO_4(aq) + H^+(aq)$ (b)

Color plate 2 Chemical Equilibrium 2: The same reaction as in color plate 1 carried out at pH 7, producing a different equilibrium state than that produced in color plate 1; but similar to the situation in color plate 1, the same state is produced from either the forward (a) or the reverse (b) direction (Section 9B-1, page 233).

Charles D. Winters

$I_3^-(aq) + 2Fe(CN)_6^{4-}(aq) \rightarrow$ (a)

$\leftarrow 3I^-(aq) + 2Fe(CN)_6^{3-}(aq)$ (b)

Color plate 3 Chemical Equilibrium 3: Reaction between iodine and ferrocyanide. One mmol I_3^- added to 2 mmol $Fe(CN)_6^{4-}$. (b) Three mmol I^- added to 2 mmol $Fe(CN)_6^{3-}$ produces the same equilibrium state (Section 9B-1, page 233).

Charles D. Winters

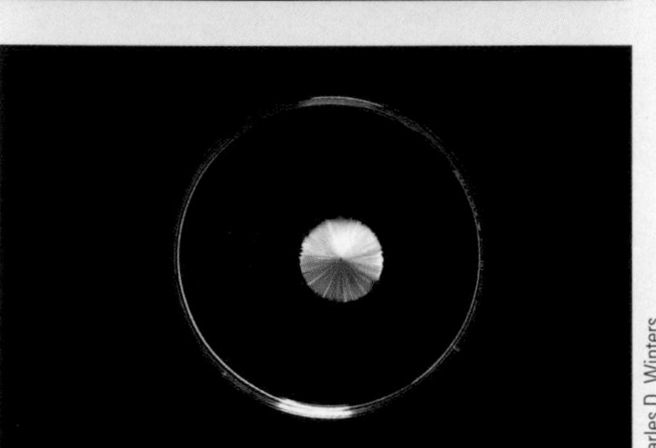
Charles D. Winters

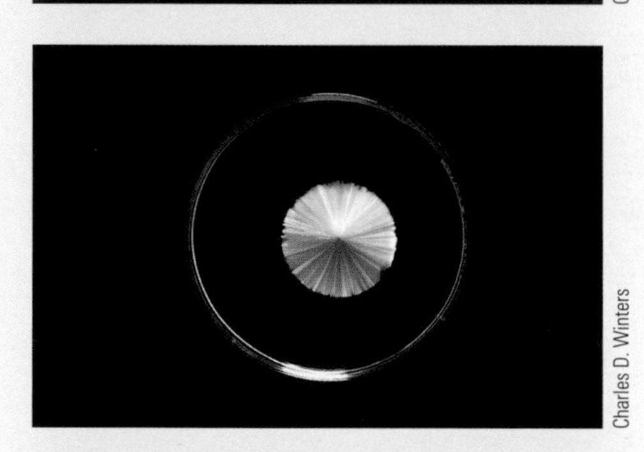
Charles D. Winters

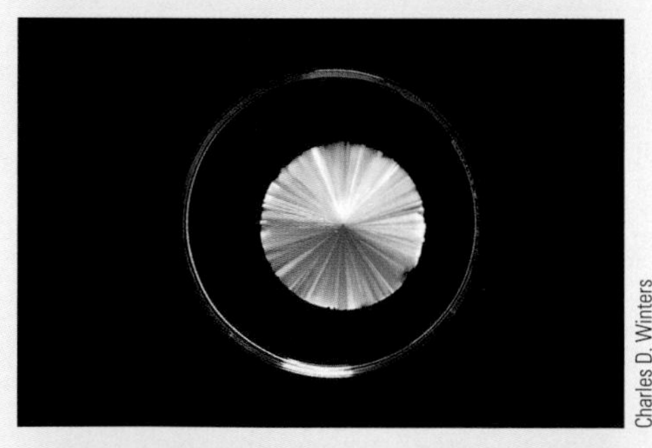
Charles D. Winters

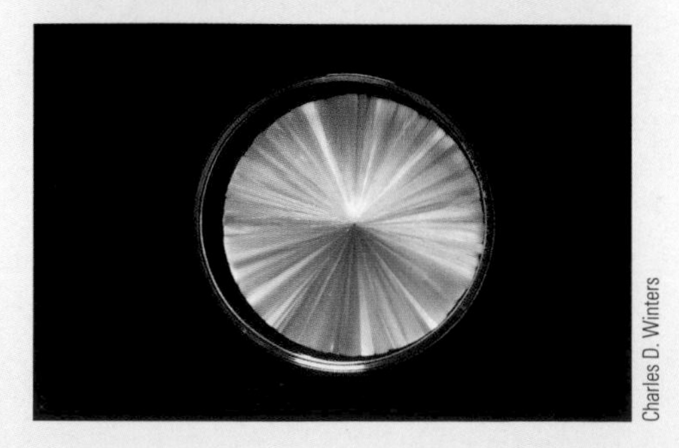
Charles D. Winters

Color plate 5 Crystallization of sodium acetate from a supersaturated solution (Section 12A-2). A tiny seed crystal is dropped into the center of a petri dish containing a supersaturated solution of the compound. The time sequence of photos taken about once per second shows the growth of the beautiful crystals of sodium acetate.

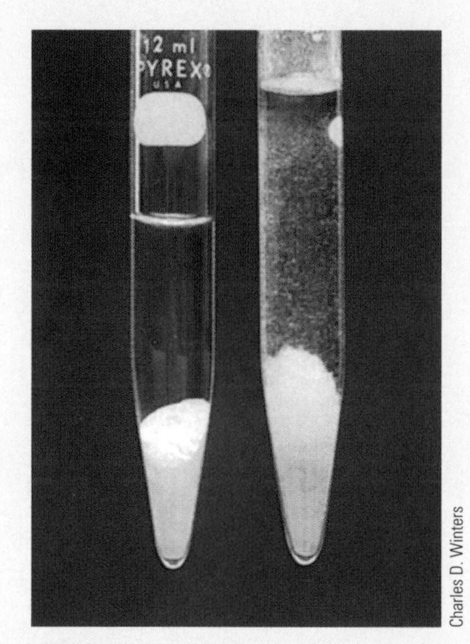

Charles D. Winters

Color plate 4 The common ion effect. The test tube on the left contains a saturated solution of silver acetate, AgOAc. The following equilibrium is established in the test tube:

$$\text{AgOAc}(s) \rightleftharpoons \text{Ag}^+(aq) + \text{OAc}^-(aq)$$

When $AgNO_3$ is added to the test tube, the equilibrium shifts to the left to form more AgOAc, as shown in the test tube on the right (Section 9B-5, page 240).

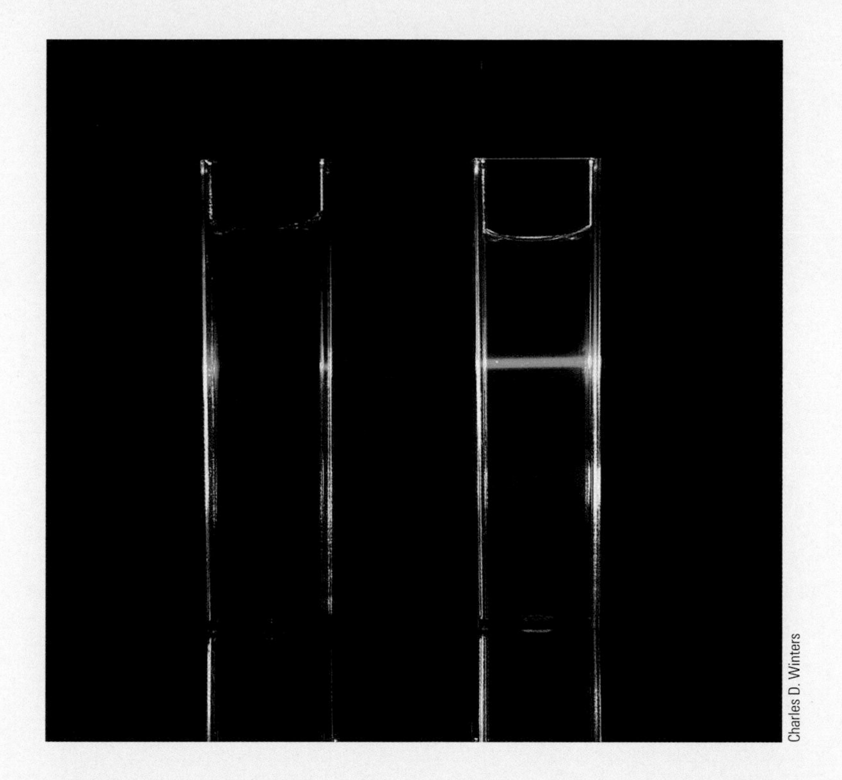

Color plate 6 The Tyndall effect. The photo shows two cuvettes: The one on the left contains only water and the one on the right contains a solution of starch. As red and green laser beams pass through the water in the left cuvette, they are invisible. Colloidal particles in the starch solution in the right cuvette scatter the light from the two lasers, so the beams become visible (see Section 12A-2, margin note, page 316).

Color plate 7 When dimethylglyoxime is added to a slightly basic solution of $Ni^{2+}(aq)$, shown on the left, a bright red precipitate of $Ni(C_4H_7N_2O_2)_2$ is formed, as seen in the beaker on the right (Section 12C-3, page 331).

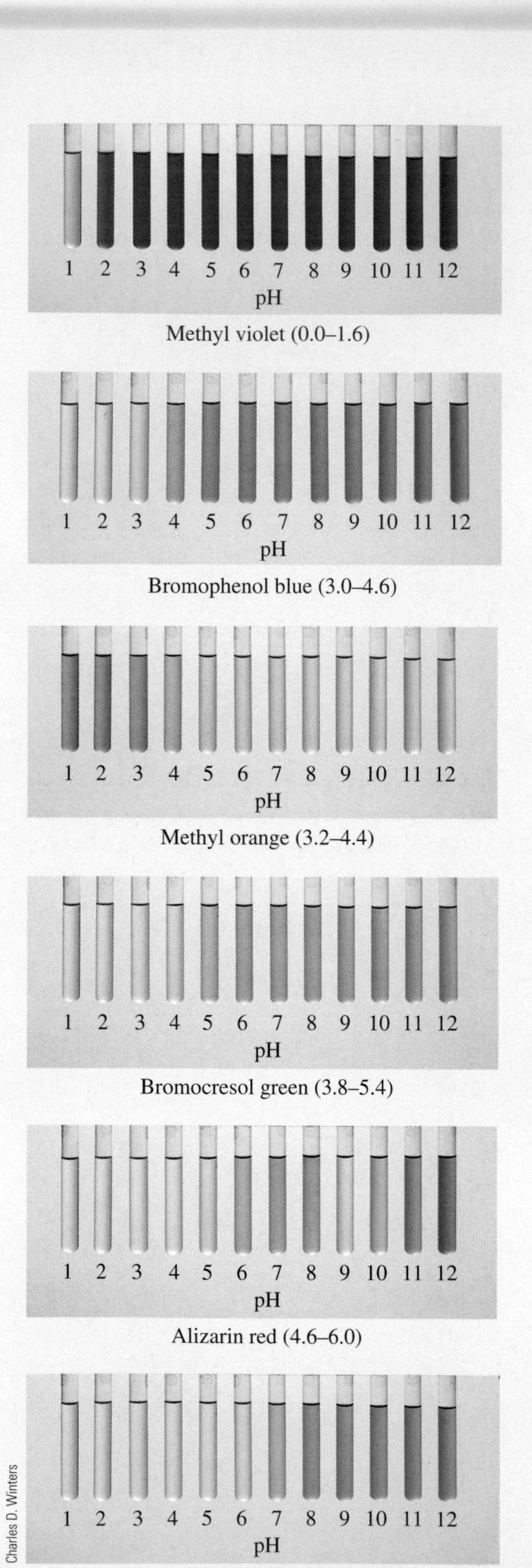

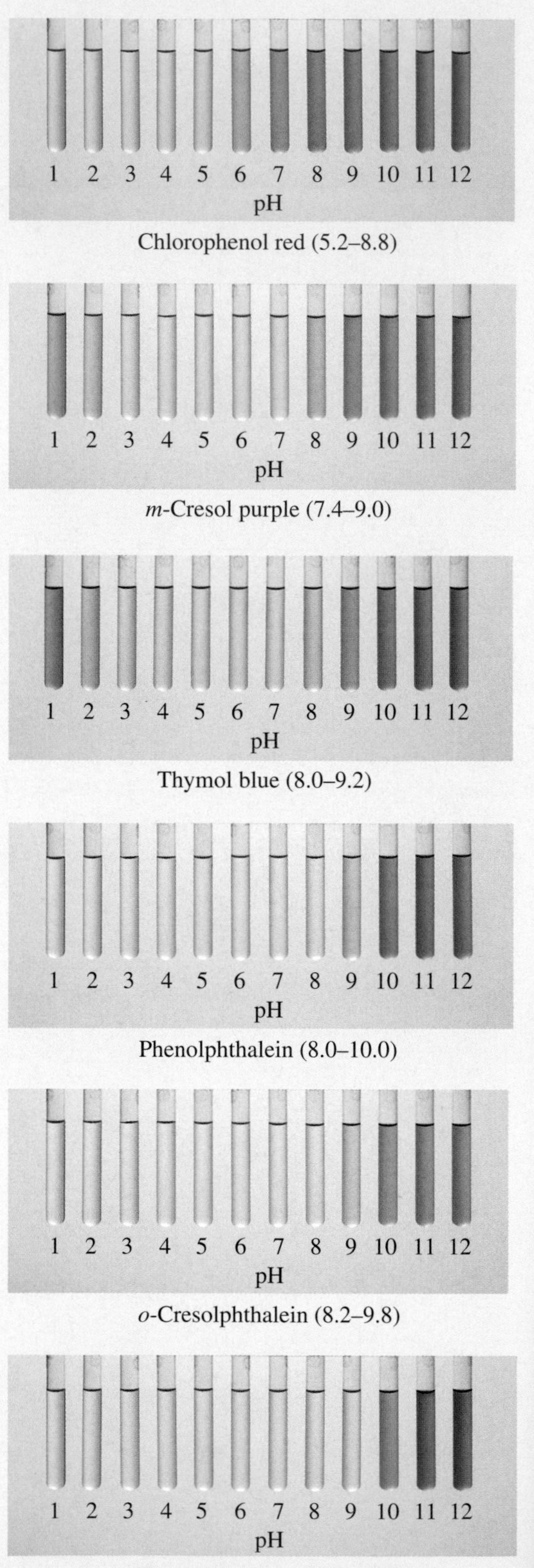

Color plate 8 Acid-base indicators and their transition pH ranges (Section 14A-2).

Color plate 9 Reduction of silver(I) by direct reaction with copper: the "silver tree" (Section 18A-2, page 493).

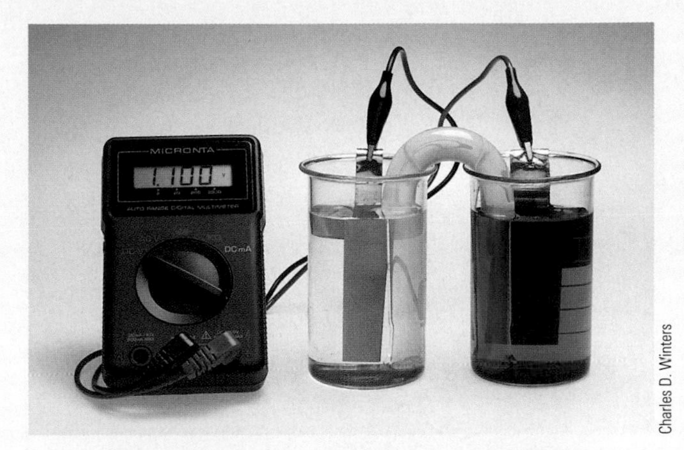

Color plate 10 A modern version of the Daniell cell (Feature 18-2, page 498).

Color plate 11 Reaction between iron(III) and iodide. The species in each beaker are indicated by the colors of the solutions. Iron(III) is pale yellow, iodide is colorless, and triiodide is intense red-orange (see margin note, Section 18C-6, page 512).

Color plate 12 The time dependence of the reaction between permanganate and oxalate (Section 20C-1, page 570).

(a)

(b)

Color plate 13 Cell for polarographic measurements with the dropping mercury electrode. (a) Arrangement of the cell with the reference electrode on the left, the capillary working electrode at the top center, and the counter electrode of the right. The tube in the rear is for purging the cell with nitrogen prior to measurement. (b) Close-up view of the capillary with a mercury drop forming on the tip (Section 23B-5, page 684).

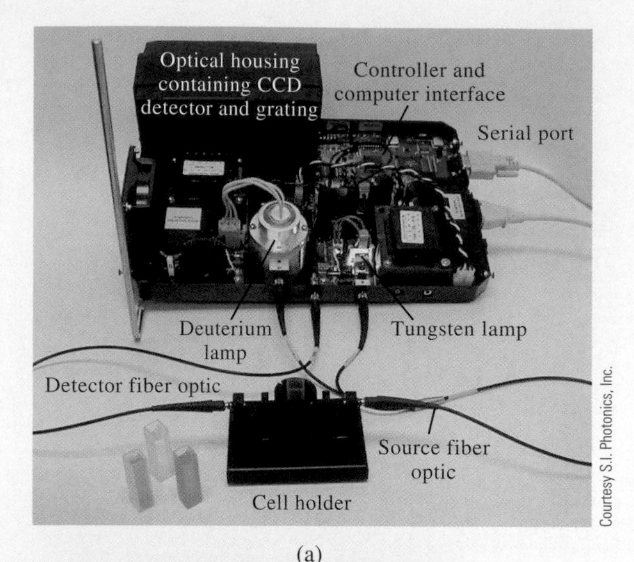

Courtesy S.I. Photonics, Inc.

(a)

Courtesy S.I. Photonics, Inc.

(b)

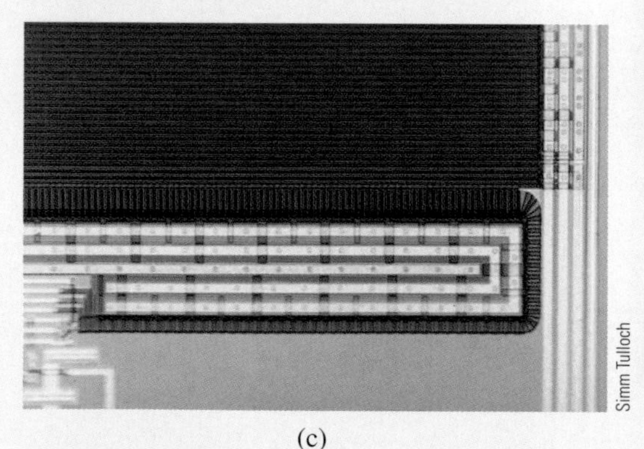

Simm Tulloch

(c)

Astronomical Instrumentation

(d)

Color plate 14 (a) CCD and fiber-optic–based spectrophotometer with cover removed (see Chapter 25). The instrument has a spectral range of 190–980 nm with a bandwidth of <1.0 nm. It has a photometric range of 0.002–3.2 absorbance units with accuracy of 0.005 absorbance units. The optical housing, which is thermostatted at 90°C, contains a 3648-element linear CCD with a holographic grating ruled with 300 lines/mm. Data are acquired and analyzed using LabView® software with routines provided by the manufacturer or written by the user for specific experiments. Both deuterium and tungsten light sources provide radiation that is conveyed to the cell via the fiber optics shown in the photo. Spectra are recorded in about 1 second, and so the unit may be used as a detector for kinetics studies as well as standard spectrophotometric experiments. (b) Typical linear CCD arrays for spectrophotometers. The two arrays arranged vertically on the right have 2048 pixels, the dimensions of which are either 8 μm × 56 μm or 14 μm × 200 μm. The center array is coated with a yellow fluorescent phosphor to increase the range of sensitivity of the transducer. In regions of the spectrum where the transducer is not sensitive, light striking the phosphor produces light in the visible which is then detected by the array. (c) Photomicrograph of a section of a two-dimensional CCD array that is used for imaging and spectroscopy. Light falling on the millions of pixels in the upper left of the photo creates charge that is transferred to the vertical channels at the bottom of the photo and shifted from left to right along the string of channels until it reaches the output amplifier section shown in (d). The amplifier provides a voltage proportional to the charge accumulated on each pixel, which is in turn proportional to the intensity of light falling on the pixel.

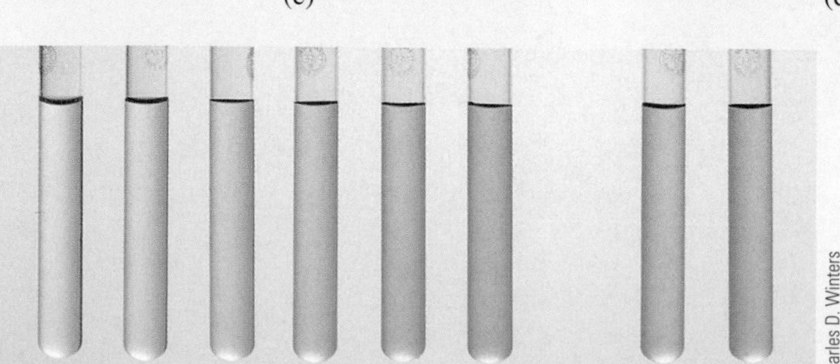

Charles D. Winters

Color plate 15 Series of standards (left) and two unknowns (right) for the spectrophotometric determination of Fe(II) using 1,10-phenanthroline as reagent (See Section 26A-3 and problem 26-26). The color is due to the complex $Fe(phen)_3^{2+}$. The absorbance of the standards is measured, and a working curve is analyzed using linear least-squares (See Section 8C-2). The equation for the line is then used to determine the concentrations of the unknown solutions from their measured absorbances.

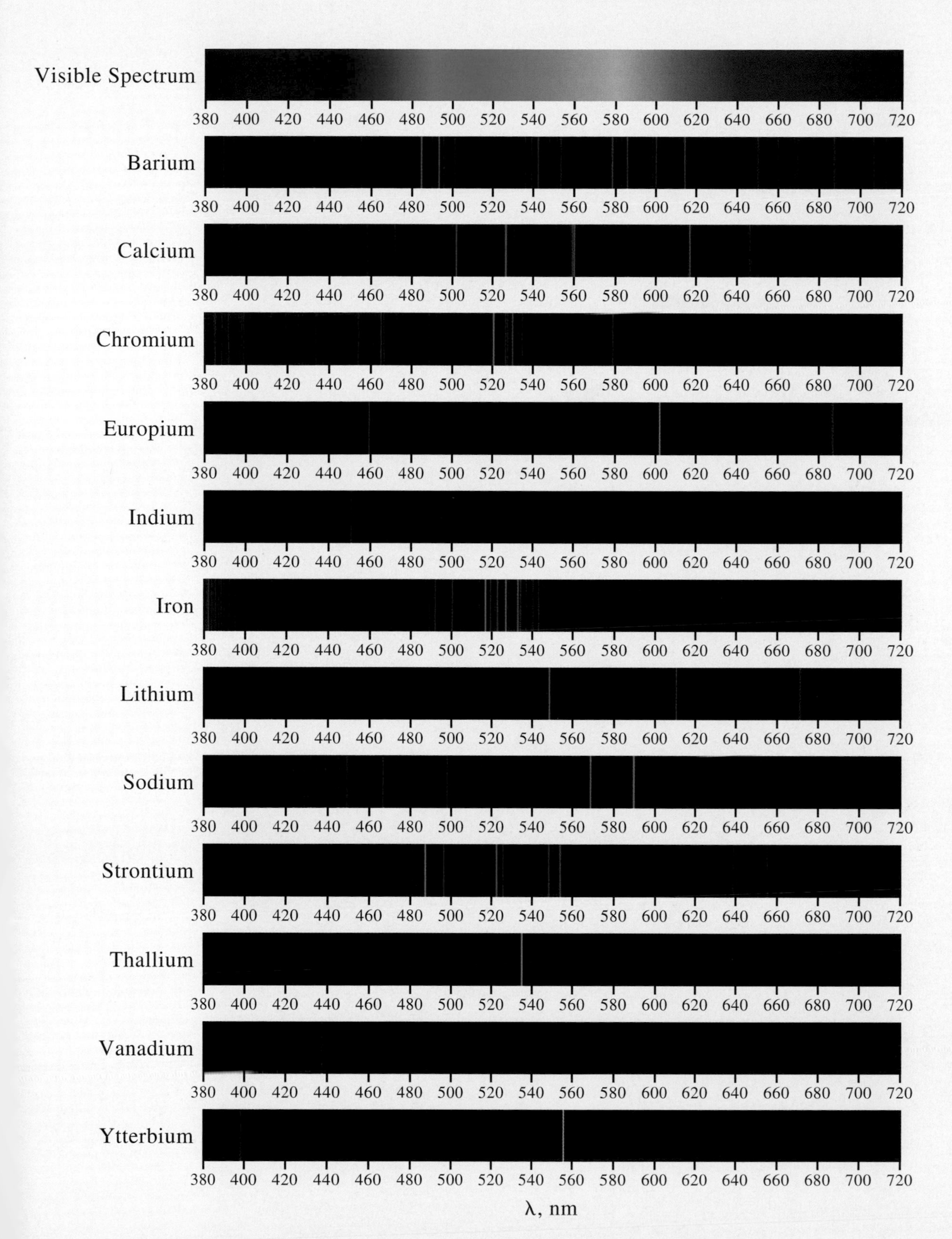

Color plate 16 Spectrum of white light and emission spectra of selected elements (see Chapter 28).

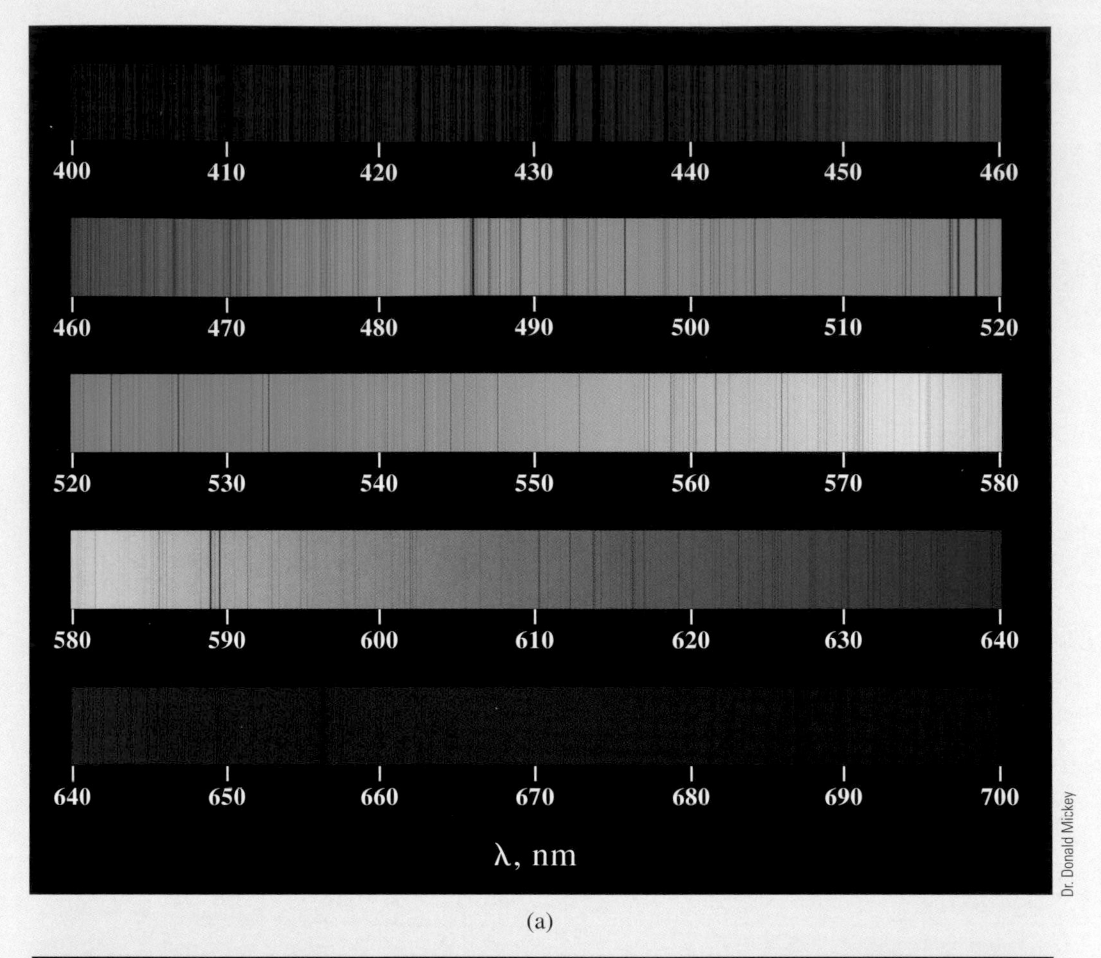

(a)

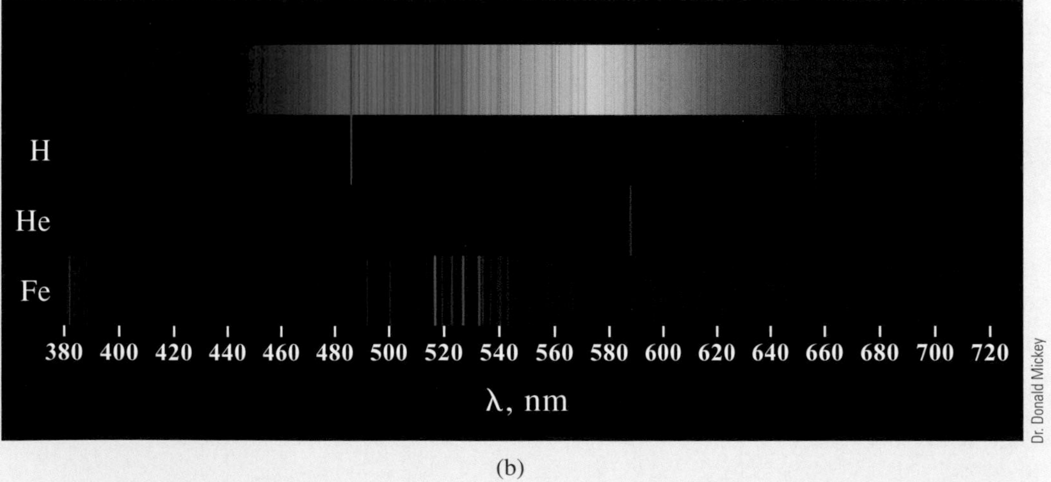

(b)

Color plate 17 The solar spectrum. (a) Expanded color version of the solar spectrum shown in black and white in Feature 24-1 (Figure 24F-1). The huge number of dark absorption lines are produced by all of the elements in the sun. See if you can spot some prominent lines like the famous sodium doublet. (b) Compact version of the solar spectrum in (a) compared with the emission spectra of hydrogen, helium, and iron. It is relatively easy to spot lines in the emission spectra of hydrogen and iron that correspond to absorption lines in the solar spectrum, but the lines of helium are quite obscure. In spite of this, helium was discovered when these lines were observed in the solar spectrum.

(a)

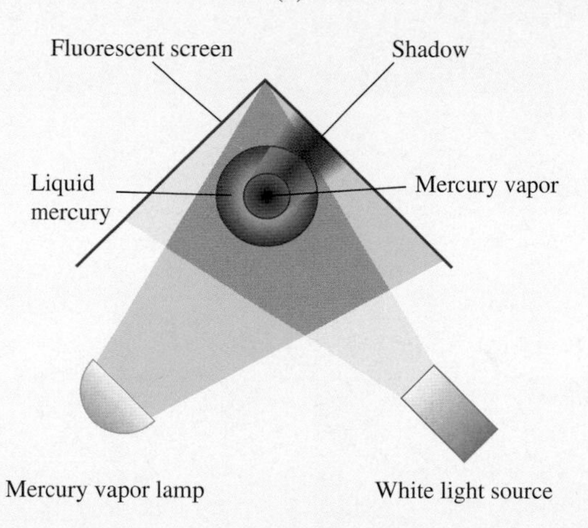

(b)

Color plate 18 (a) Demonstration of atomic absorption by mercury vapor. (b) White light from the source on the right passes through the mercury vapor above the flask, and no shadow appears on the fluorescent screen on the left. Light from the mercury lamp on the left containing the characteristic UV lines of the element is absorbed by the vapor above the flask, which casts a shadow on the screen on the right of the plume of mercury vapor (see Section 28D).

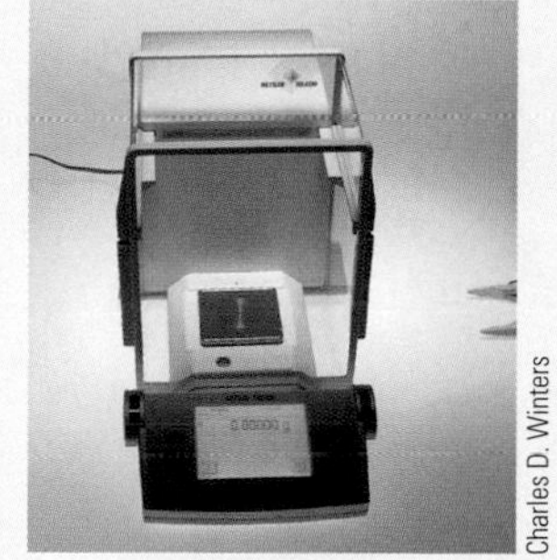

(a)

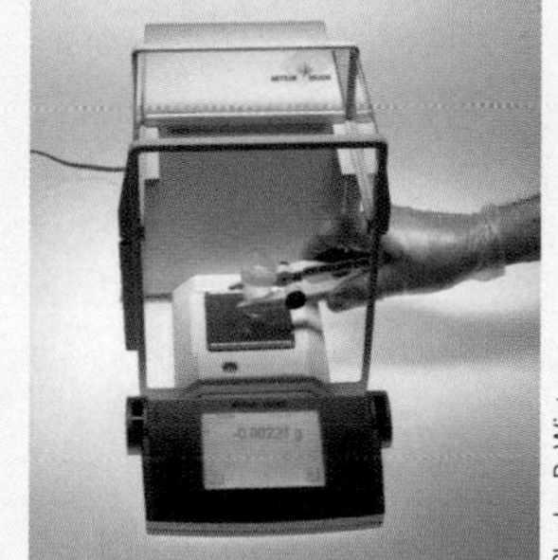

(b)

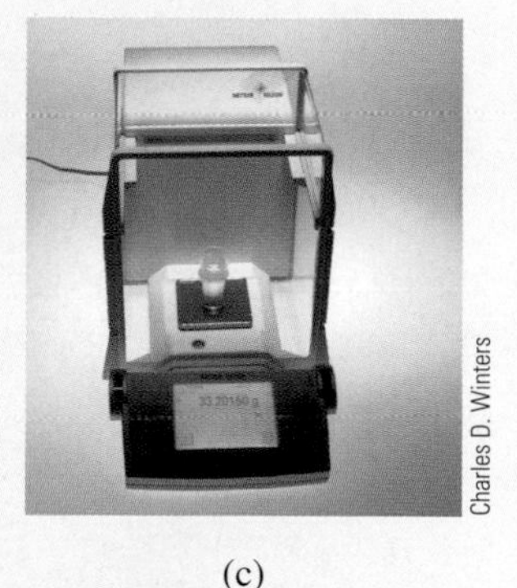

(c)

(d)

Charles D. Winters

(e)

Color plate 19 Weighing by difference the old-fashioned way. (a) Zero the balance. (b) Place a weighing bottle containing the solute on the balance pan. (c) Read the mass (33.2015 g). (d) Transfer the desired amount of solute to a flask. (e) Replace the weighing bottle on the pan, and read the mass (33.0832 g). Finally, calculate the mass of the solute transferred to the flask: 33.2015 g − 33.0832 g = 0.1131 g. (Electronic balance provided by Mettler-Toledo, Inc.)

Color plate 20 Weighing by difference the modern way. Place a weighing bottle containing the solute on the balance pan, and (a) depress the tare or zero button. The balance should then read 0.0000 g, as shown in (b). (c) Transfer the desired amount of solute to a flask. Replace the weighing bottle on the pan, and the balance reads the decrease in mass directly as -0.1070 g. (d) Many modern balances have built-in computers with programs to perform a variety of weighing tasks; for example, it is possible to dispense many consecutive quantities of a substance and automatically read out the loss in mass following each dispensing. Many balances also have computer interfaces so that reading may be logged directly to programs running on the computer. (Electronic balance provided by Mettler-Toledo, Inc.)

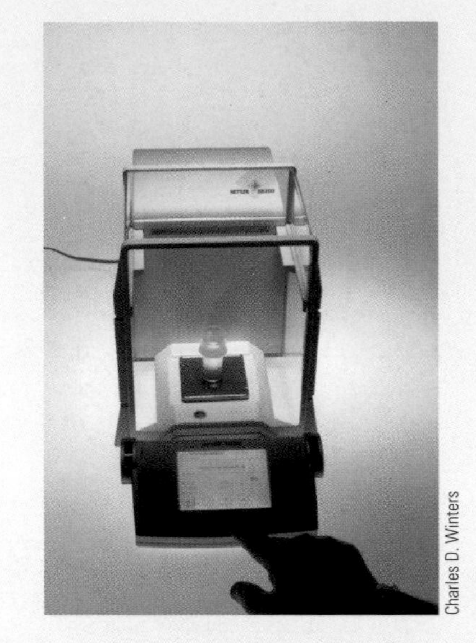

(a)

(b)

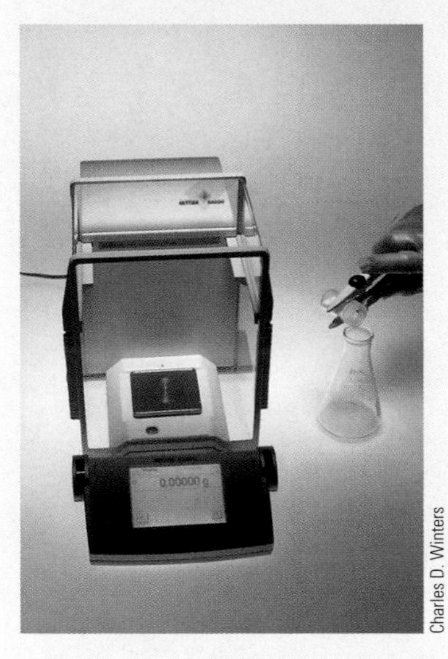

(c)

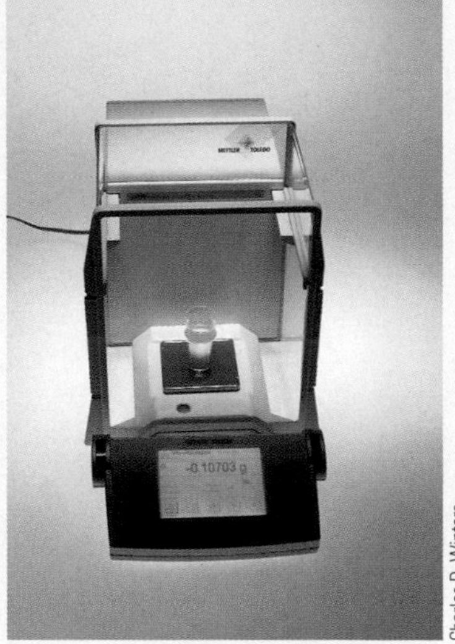

(d)

Polarographic Currents

The current in a cell containing a dropping mercury electrode undergoes periodic fluctuations corresponding in frequency to the drop rate. As a drop dislodges from the capillary, the current falls toward zero, as shown in Figure 23-16. As the surface area of a new drop increases, so does the current. The diffusion current is usually taken at the maximum of the current fluctuations. In the older literature, the *average current* was measured because instruments responded slowly and damped the oscillations. As shown in curve *A* of Figure 23-16, some modern polarographs have electronic filtering that allows either the maximum or the average current to be determined, provided that the drop rate *t* is reproducible. Note the effect of irregular drops in the upper part of curve *A*, probably caused by vibration of the apparatus.

Polarograms

Figure 23-16 shows two polarograms—one for a solution that is 1.0 M in hydrochloric acid and 5.0×10^{-4} M in cadmium ion (curve *A*), and the second for the 1.0 M acid alone (curve *B*). The polarographic wave in curve *A* arises from

$$Cd^{2+} + 2e^- + Hg \rightleftharpoons Cd(Hg)$$

where Cd(Hg) represents elemental cadmium dissolved in mercury to form an amalgam. The sharp increase in current at about -1 V in both polarograms is caused by the reduction of hydrogen ions to give hydrogen. Examination of the polarogram for the supporting electrolyte alone reveals that a small current, called the **residual current,** is present in the cell even in the absence of cadmium ions.

The **residual current** in polarography is the small current observed in the absence of an electroactive species.

As in hydrodynamic voltammetry, limiting currents are observed when the magnitude of the current is limited by the rate at which analyte can be brought up to the electrode surface. In polarography, however, the only mechanism of mass transport is diffusion. For this reason, polarographic limiting currents are usually termed **diffusion currents** and given the symbol i_d. As shown in Figure 23-16, the diffusion current is the difference between the maximum (or average) limiting current and

Diffusion current is the limiting current observed in polarography when the current is limited only by the rate of diffusion to the dropping mercury electrode surface.

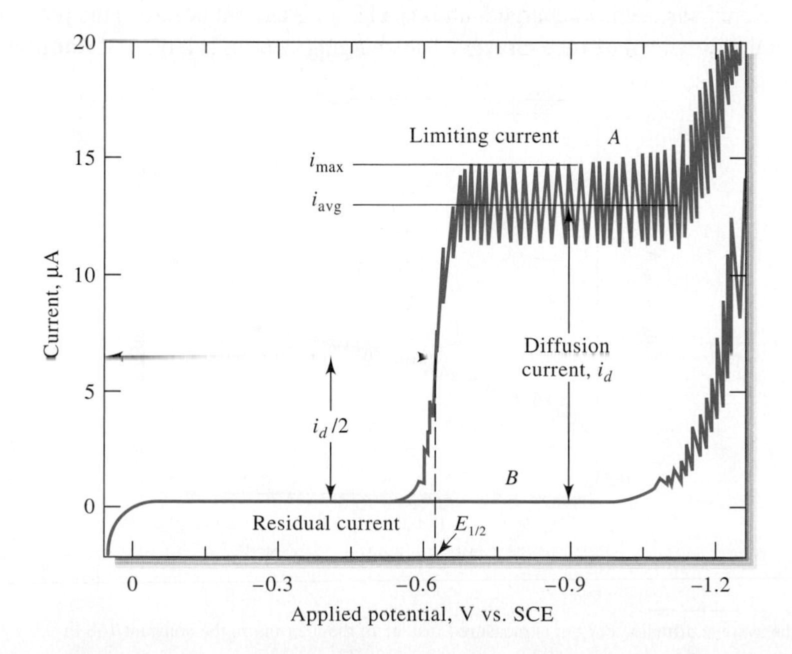

Figure 23-16 Polarograms for *A*, a 1 M solution of HCl that is 5×10^{-4} M in Cd^{2+} and *B*, a 1 M solution of HCl. (From D. T. Sawyer, A. Sobkowiak, and J. L. Roberts, Jr., *Experimental Electrochemistry for Chemists,* 2nd ed., p. 59. New York: Wiley, 1995. This material is used by permission of John Wiley & Sons, Inc.)

► The diffusion current in polarography is proportional to the concentration of analyte.

the residual current. The diffusion current is directly proportional to analyte concentration in the bulk of solution, as shown next.

Diffusion Current at the Dropping Mercury Electrode

To derive an equation for polarographic diffusion currents, we must take into account the rate of growth of the spherical electrode, which is related to the drop time in seconds, *t;* the rate of flow of mercury through the capillary *m* in mg/s; and the diffusion coefficient of the analyte *D* in cm^2/s. These variables are taken into account in the Ilkovic equation:

► In polarography, currents are usually recorded in microamperes. The constant 708 in Equation 23-13 carries units such that the concentration *c* is in millimoles per liter when (i_d) is in microamperes, *D* is in cm^2/s, *m* is in mg/s, and *t* is in s.

$$(i_d)_{max} = 708\ nD^{1/2}m^{2/3}t^{1/6}c \qquad (23\text{-}13)$$ [10]

where $(i_d)_{max}$ is the maximum diffusion current in μA and *c* is the analyte concentration in mM.

Residual Currents

Figure 23-17 shows a residual current curve (obtained at high sensitivity) for a 0.1 M solution of HCl. This current has two sources. The first is the reduction of trace impurities that are almost inevitably present in the blank solution. The contributors here include small amounts of dissolved oxygen, heavy metal ions from the distilled water, and impurities present in the salt used as the supporting electrolyte.

The second component of the residual current is the so-called **charging** or **capacitive current** resulting from a flow of electrons that charge the mercury droplets with respect to the solution; this current may be either negative or positive. At potentials more negative than about −0.4 V, an excess of electrons from the dc source provides the surface of each droplet with a negative charge. These excess electrons are carried down with the drop as it breaks; since each new drop is charged as it forms, a small but continuous current results. At applied potentials less negative than about −0.4 V, the mercury tends to be positive with respect to the solution. Thus, as each drop is formed, electrons are repelled from the surface toward the bulk of mercury, and a negative current is the result. At about −0.4 V, the mercury surface is uncharged, and the charging current is zero. This potential is called the **potential of zero charge.** The charging current is a type of **nonfaradaic**

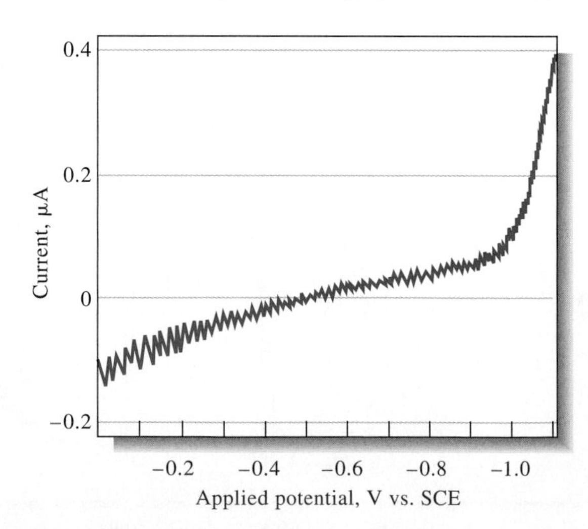

Figure 23-17 Residual current for a 0.1 M solution of HCl.

[10] If the average diffusion current is measured instead of the maximum, the constant 708 in the Ilkovic equation becomes 607 because $(i_d)_{avg} = 6/7\ (i_d)_{max}$.

current in the sense that charge is carried across an electrode/solution interface without an accompanying oxidation/reduction process.

Ultimately, the accuracy and sensitivity of the polarographic method depend on the magnitude of the nonfaradaic residual current and the accuracy with which a correction for its effect can be determined.

A **faradaic current** in an electrochemical cell is the current that results from an oxidation/reduction process. A **nonfaradaic current** is a charging current that results because the mercury drop is expanding and must be charged to the electrode potential. The charging of the double layer is similar to charging a capacitor.

Comparison of Currents from the Dropping Mercury and Stationary Planar Electrodes

Constant currents are not obtained in reasonable periods of time with a planar electrode in an unstirred solution because concentration gradients out from the electrode surface are constantly changing with time. In contrast, the DME exhibits constant reproducible currents nearly instantaneously after an applied voltage adjustment. This behavior represents an advantage of the DME that accounted for its widespread use in the early years of voltammetry.

The rapid achievement of constant currents arises from the highly reproducible nature of the drop formation process and, equally important, the fact that the solution in the electrode area becomes homogenized each time a drop breaks from the capillary. Thus, a concentration gradient is developed only during the brief lifetime of the drop. As we have noted, current changes due to an increase in surface area occur during each lifetime. Changes in the concentration gradient dc/dx also occur during this period, but these changes are highly reproducible, leading to currents that are also very reproducible.

Effect of Complexation on Polarographic Waves

We have already seen (see Section 18C-6) that the potential for the oxidation or reduction of a metallic ion is greatly affected by the presence of species that form complexes with that ion. It is not surprising to find that similar effects are observed with polarographic half-wave potentials. The data in Table 23-1 show clearly that the half-wave potential for the reduction of a metal complex is generally more negative than that for reduction of the corresponding simple metal ion. In fact, this negative shift in potential permits us to determine the composition of the complex ion and its formation constant *provided that the electrode reaction is reversible.* Thus, for the reactions

$$M^{n+} + ne^- + Hg \rightleftharpoons M(Hg)$$

and

$$M^{n+} + xA^- \rightleftharpoons MA^{(n-x)+}$$

TABLE 23-1

Effect of Complexing Agents on Polarographic Half-Wave Potentials ($E_{1/2}$, V)

Ion	Noncomplexing Media	1 M KCN	1 M KCl	1 M NH_3, 1 M NH_4Cl
Cd^{2+}	−0.59	−1.18	−0.64	−0.81
Zn^{2+}	−1.00	NR*	−1.00	−1.35
Pb^{2+}	−0.40	−0.72	−0.44	−0.67
Ni^{2+}	−1.01	−1.36	−1.20	−1.10
Co^{2+}	—	−1.45	−1.20	−1.29
Cu^{2+}	+0.02	NR*	+0.04 and −0.22†	−0.24 and −0.51†

*No reduction occurs before involvement of the supporting electrolyte.
†Reduction occurs in two steps having different electrode potentials.

Lingane[11] derived the following relationship between the molar concentrations of the ligand c_L and the shift in half-wave potential brought about by its presence:

$$(E_{1/2})_c - E_{1/2} = -\frac{0.0592}{n} \log K_f - \frac{0.0592x}{n} \log c_L \qquad (23\text{-}14)$$

where $(E_{1/2})_c$ and $E_{1/2}$ are the half-wave potentials for the complexed and uncomplexed cations, respectively, K_f is the formation constant of the complex, and x is the number of moles of ligand combining with each mole of metal ion.

Equation 23-14 makes it possible to evaluate the formula for the complex. Thus, a plot of the half-wave potential against log c_L for several ligand concentrations gives a straight line with a slope of $0.0592x/n$. If n is known, the combining ratio of ligand to metal ion x is readily calculated. Equation 23-14 can then be employed to calculate K_f.

Effect of pH on Polarograms

Most organic electrode processes and a few inorganic ones involve hydrogen ions. We can represent the typical reaction as

$$R + nH^+ + ne^- \rightleftharpoons RH_n$$

where R and RH_n are the oxidized and reduced forms, respectively, of the reactant species. Half-wave potentials for compounds of this type are therefore markedly pH dependent. Furthermore, changing the pH may result in a different reaction product.

An electrode process that consumes or produces hydrogen ions will alter the pH of the solution *at the electrode surface,* often drastically, unless the solution is well buffered. These changes affect the reduction potential of the reaction and cause drawn-out, poorly defined waves. Moreover, where the electrode process is altered by pH, nonlinearity in the diffusion current/concentration relationship will also be encountered. Thus, good buffering is generally vital for the generation of reproducible half-wave potentials and diffusion currents for organic polarography.

Advantages and Disadvantages of the Dropping Mercury Electrode

► The DME has a high overvoltage for the reduction of H^+ and a renewable metal surface with each droplet. Reproducible currents are obtained very rapidly with the DME.

In the past, the dropping mercury electrode was the most widely used electrode for voltammetry because of several unique features. The first is the unusually high overvoltage associated with the reduction of hydrogen ions. As a consequence, metal ions such as zinc and cadmium can be deposited from acidic solution even though their thermodynamic potentials suggest that deposition of these metals without hydrogen formation is impossible. A second advantage is that a new metal surface is generated continuously, which makes the behavior of the electrode independent of its past history. In contrast, solid metal electrodes are notorious for their irregular behavior, which is related to adsorbed or deposited impurities. A third unusual feature of the DME, which has already been described, is that reproducible currents are *immediately* realized at any given potential whether this potential is approached from lower or higher values.

One serious limitation of the DME is the ease with which mercury is oxidized; this property severely limits the range of anodic potentials that can be used. At potentials greater than about +0.4 V, formation of mercury(I) occurs, giving a

[11] J. J. Lingane, *Chem. Rev.*, **1941,** *29,* 1.

wave that masks the curves of other oxidizable species. In the presence of ions that form precipitates or complexes with mercury(I), this behavior occurs at even lower potentials. For example, in Figure 23-17, the beginning of an anodic wave can be seen at 0 V owing to the reaction:

$$2Hg + 2Cl^- \rightarrow Hg_2Cl_2(s) + 2e^-$$

This anodic wave has been used, however, for the determination of chloride ion.

Another important disadvantage of the DME is the nonfaradaic residual or charging current, which limits the sensitivity of the classical method to concentrations of about 10^{-5} M. At low concentrations, the residual current can be greater than the diffusion current, which prohibits accurate diffusion current measurement. As will be shown in the next sections, methods are now available for enhancing detection limits by one to two orders of magnitude.

◀ The detection limit for classical polarography is about 10^{-5} M. Routine determinations usually involve concentrations in the mM range.

The dropping mercury electrode is also cumbersome to use and tends to malfunction as a result of clogging. An additional problem with classical polarography is the existence of peaks on the current-voltage curves, termed **polarographic maxima.** Although not entirely understood, polarographic maxima are thought to arise from convection around the expanding mercury drop. Generally, the addition of small amounts of surfactants, such as gelatin or Triton X-100, will eliminate maxima. Care must be taken to avoid adding large amounts of these **maxima suppressors** because they can change the solution viscosity and reduce the magnitude of the diffusion current. These limitations, along with the toxicity of mercury, have been responsible for the increasing popularity of solid electrodes versus the DME in voltammetry.

Spreadsheet Summary Polarography is considered in the voltammetry exercise in Chapter 11 of *Applications of Microsoft® Excel in Analytical Chemistry.* A polarographic calibration curve is constructed first; then an accurate determination of half-wave potential is made. Finally, the formation constant and formula of a complex are determined from polarographic data.

23C PULSE POLAROGRAPHIC AND VOLTAMMETRIC METHODS

By the 1960s, linear-scan polarography ceased to be an important analytical tool in most laboratories. The reason for the decline in use of this once-popular technique was not only the appearance of several more convenient spectroscopic methods but also the inherent disadvantages of the method, including slowness, inconvenient apparatus, and, particularly, poor detection limits. These limitations were largely overcome by pulse methods and the development of electrodes such as those shown in Figure 23-3d. Here we discuss the two most important pulse techniques, **differential pulse polarography** and **square-wave polarography.** Both methods have also been applied with electrodes other than the dropping mercury electrode, in which case the procedures are called differential and square-wave voltammetry.[12]

[12]For reviews on pulse and square-wave voltammetry, see G. N. Eccles, *Crit. Rev. Anal. Chem.*, **1991,** *22,* 345; J. Osteryoung, *Acc. Chem. Res.*, **1993,** *26,* 77. See also A. J. Bard and L. R. Faulkner, *Electrochemical Methods,* 2nd ed., pp. 275–301. New York: Wiley, 2001.

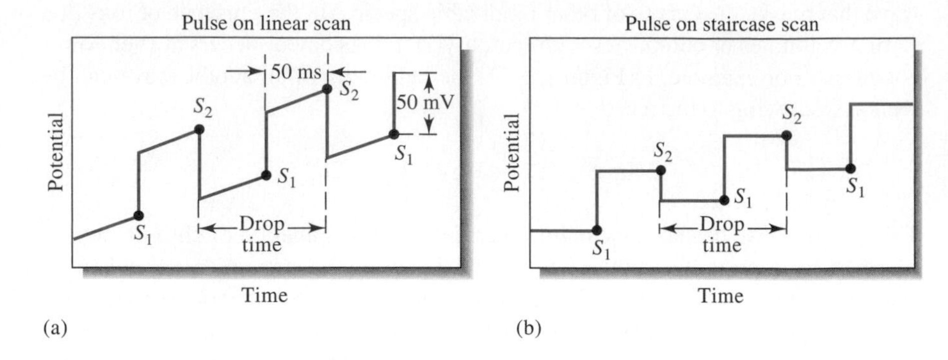

Figure 23-18 Excitation signals for differential pulse polarography.

23C-1 Differential Pulse Polarography

Figure 23-18 shows the two most common excitation signals employed in commercial instruments for differential pulse polarography. Analog instruments use the waveform shown in Figure 23-18a, which is obtained by superimposing a periodic pulse on a linear scan. Digital instruments usually use the waveform shown in Figure 23-18b, which involves the combination of a pulse output and a staircase signal. In either case, a small pulse, typically 50 mV, is applied during the last 50 ms of the lifetime of the mercury drop. Here again, to synchronize the pulse with the drop, the drop is detached at an appropriate time by an electromechanical device.

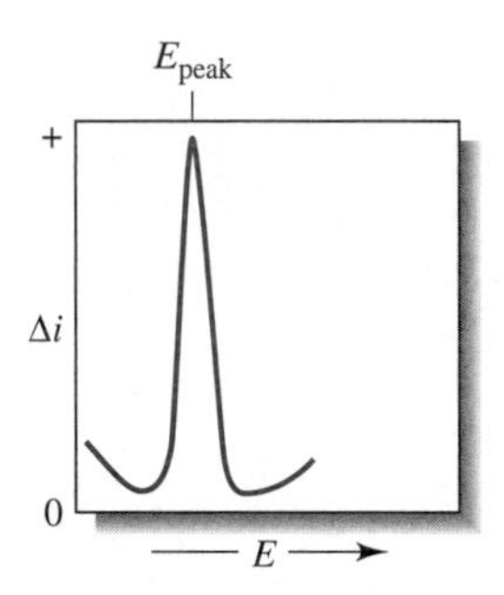

Figure 23-19 Voltammogram for a differential pulse polarography experiment. Here $\Delta i = i_{S_2} - i_{S_1}$ (see Figure 23-18). The peak potential, E_{peak}, is closely related to the polarographic half-wave potential.

As shown in Figure 23-18, two current measurements are made alternately—one at S_1 just prior to the pulse, and one at S_2 just before the end of the pulse. The *difference in current per pulse* Δi is recorded as a function of the linearly increasing voltage. A differential curve with a peak results, as shown in Figure 23-19. The height of the peak is directly proportional to concentration. For a reversible reaction, the peak potential is approximately equal to the standard potential for the half-reaction.

► Derivative polarograms yield peaks that are convenient for qualitative identification of analytes based on the peak potential, E_{peak}.

One advantage of the derivative-type polarogram is that individual peak maxima can be observed for substances with half-wave potentials differing by as little as 0.04 to 0.05 V; in contrast, classical polarography requires a potential difference of about 0.2 V for resolution of waves.

Another advantage is that differential pulse polarography generally has increased sensitivity over normal polarography and significantly lower detection limits. This enhancement is illustrated in Figure 23-20. Note that a classical polarogram for a solution containing 180 ppm of the antibiotic tetracycline gives two barely distinguishable waves; differential pulse polarography, however, provides well-defined peaks at a concentration level that is 500 times lower than that for the classic wave. Note also that the current scale for Δi is in nanoamperes. Generally, detection limits with differential pulse polarography are two to three orders of magnitude lower than those for classical polarography and lie in the range of 10^{-7} to 10^{-8} M.

► The detection limits for differential pulse polarography are two to three orders of magnitude lower than for classical polarography.

The greater sensitivity of differential pulse polarography can be attributed to two sources: an enhancement of the faradaic current or a decrease in the nonfaradaic charging current. To account for the enhancement, consider the events that must occur in the surface layer around an electrode as the potential is suddenly increased by 50 mV. If a reactive species is present in this layer, there will be a surge of current that lowers the reactant concentration to that demanded by the new potential. As the equilibrium concentration for that potential is approached, however, the current decays to a level just sufficient to counteract diffusion (that is, to the diffusion-controlled current). In classical polarography, the initial surge of cur-

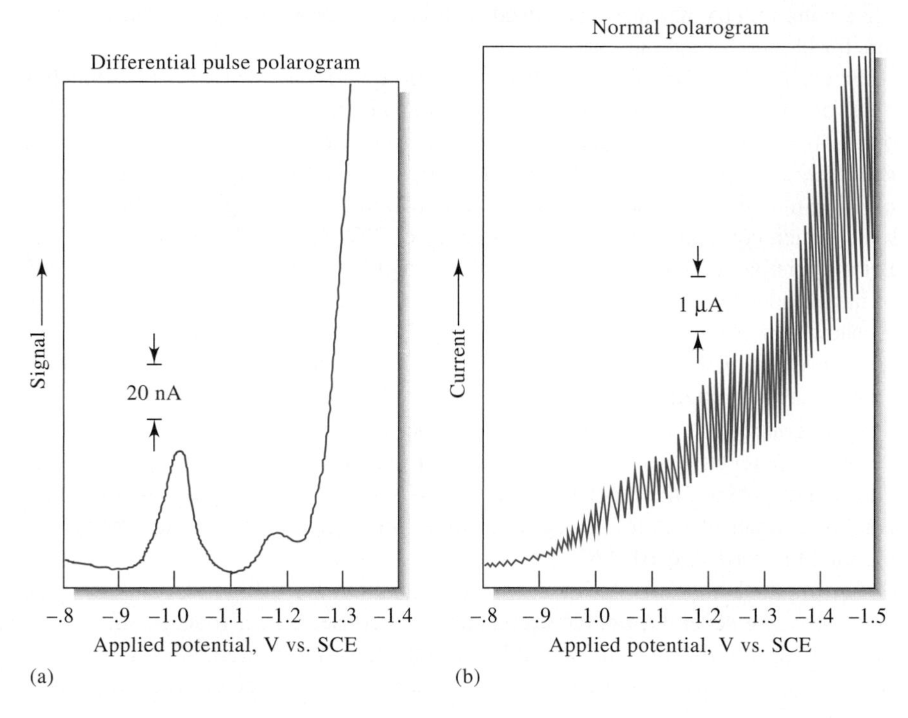

Figure 23-20 (a) Differential pulse polarogram; 0.36 ppm tetracycline·HCl in 0.1 M acetate buffer, pH 4, PAR Model 174 polarographic analyzer, DME, 50-mV pulse amplitude, 1-s drop. (b) DC polarogram: 180 ppm tetracycline·HCl in 0.1 M acetate buffer, pH 4, similar conditions. (Reprinted with permission from J. B. Flato, *Anal. Chem.*, **1972,** *44,* 75A. Published 1972, American Chemical Society.)

rent is not observed because the time scale of the measurement is long relative to the lifetime of the momentary current. In pulse polarography, however, the current measurement is made before the current spike has completely decayed. Thus, the current measured contains both a diffusion-controlled component and a component that has to do with reducing the surface layer to the concentration demanded by the Nernst expression; the total current is typically several times larger than the diffusion current. When the drop is detached, the solution again becomes homogeneous with respect to the analyte. Thus, at any given voltage, an identical current spike accompanies each voltage pulse.

When the potential pulse is first applied to the electrode, a surge in the nonfaradaic current also occurs as the charge on the drop increases. This current, however, decays exponentially with time and approaches zero near the end of the life of a drop when its surface area is changing only slightly. Thus, by measuring currents at this time only, the nonfaradaic residual current is greatly reduced, and the signal-to-noise ratio is larger. Enhanced sensitivity results.

Reliable instruments for differential pulse polarography are available commercially at reasonable cost. The method has thus become the most widely used polarographic procedure.

23C-2 Square-Wave Polarography and Voltammetry[13]

Square-wave polarography is a type of pulse polarography that offers the advantage of great speed and high sensitivity. An entire voltammogram is obtained in less than 10 ms. With a DME, the scan is performed during the last few milliseconds of the life of a single drop when the charging current is essentially constant. Square-

[13]For additional information on square-wave voltammetry, see J. Osteryoung, *Accts. Chem. Res.*, **1993,** *26,* 77; J. Osteryoung and J. J. O'Dea, *Electroanal. Chem.*, **1986,** *14,* 209; J. Osteryoung and R. A. Osteryoung, *Anal. Chem.*, **1985,** *57,* 101A.

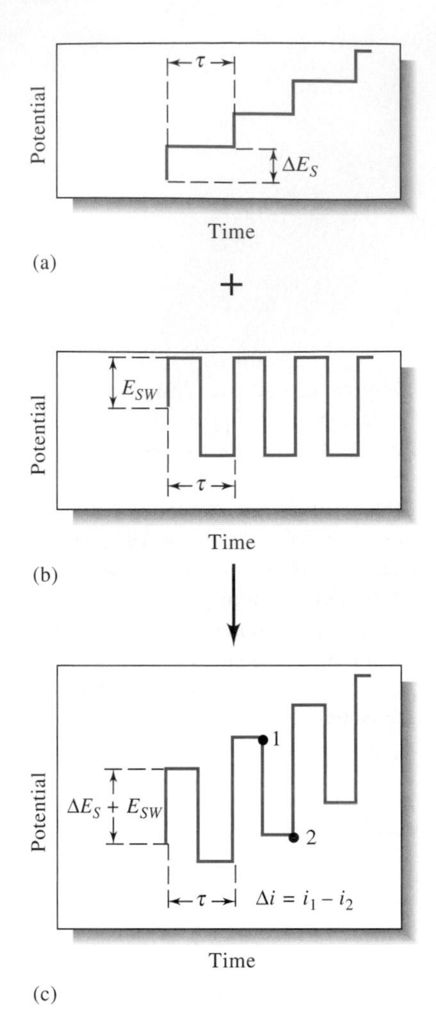

Figure 23-21 Generation of a square-wave voltammetry excitation signal. The staircase signal in (a) is added to the pulse train in (b) to give the square-wave excitation signal in (c). The current response Δi is equal to the current at potential 1 minus that at potential 2.

► Multiple scans from multiple drops can be summed to improve the signal-to-noise ratio of a square-wave voltammogram.

► Detection limits for both differential pulse polarography and square-wave voltammetry are 10^{-7} to 10^{-8} M.

wave voltammetry has also been used with hanging drop electrodes and in detectors for liquid chromatography.

Figure 23-21c shows the excitation signal in square-wave voltammetry, which is obtained by superimposing the pulse train shown in part b on the staircase signal in part a. The length of each step of the staircase and the period of the pulses (τ) are identical, usually about 5 ms. The potential step of the staircase ΔE_S is typically 10 mV. The magnitude of the pulse $2E_{SW}$ is often 50 mV. Operating under these conditions, which correspond to a pulse frequency of 200 Hz, a 1-V scan requires 0.5 s. For a reversible reduction reaction, the size of a pulse is great enough that oxidation of the product formed on the forward pulse occurs during the reverse pulse. Thus, as shown in Figure 23-22, the forward pulse produces a cathodic current i_1, whereas the reverse pulse gives an anodic current i_2. Usually, the difference in these currents Δi is plotted to give voltammograms. This difference is directly proportional to concentration. The potential at the peak corresponds to the polarographic half-wave potential. Because of the speed of the measurement, it is possible and practical to increase the precision of analyses by signal averaging results from several voltammetric scans. Detection limits for square-wave voltammetry are reported to be 10^{-7} to 10^{-8} M.

Commercial instruments for square-wave voltammetry have recently become available from several manufacturers, and, as a consequence, it seems likely that this technique will gain considerable use for analysis of inorganic and organic species. Square-wave voltammetry has also been used in detectors for liquid chromatography.[14]

23C-3 Applications of Pulse Polarography

In the past, linear-scan polarography was used for the quantitative determination of a wide variety of inorganic and organic species, including molecules of biological and biochemical interest. Currently, pulse methods have supplanted the classical method almost completely because of their greater sensitivity, convenience, and selectivity. Generally, quantitative applications are based on calibration curves in which peak heights or areas are plotted as a function of analyte concentration. In some instances the standard addition method (see Section 8C-3) is employed in lieu of calibration curves. In either case, it is essential for the composition of standards to resemble as closely as possible the composition of the sample in regard to both electrolyte concentrations and pH. When this is done, relative standard deviations and accuracies in the 1% to 3% range can often be realized.

Inorganic Applications

The polarographic method is widely applicable to the analysis of inorganic substances. Most metallic cations, for example, are reduced at the DME. Even the alkali and alkaline-earth metals are reducible, provided that the supporting electrolyte does not react at the high potentials required; here, the tetraalkyl ammonium halides are useful electrolytes because of their high reduction potentials.

The successful polarographic determination of cations frequently depends on the supporting electrolyte that is used. To aid in this selection, tabular compilations

[14]See, for example, W. LaCourse, *Pulsed Electrochemical Detection in High-Performance Liquid Chromatography.* New York: Wiley, 1997; S. M. Lunte, C. E. Lunte, and P. T. Kissinger, in *Laboratory Techniques in Electroanalytical Chemistry,* 2nd ed., P. T. Kissinger and W. R. Heinemann, Eds., Ch. 27. New York: Marcel Dekker, 1996.

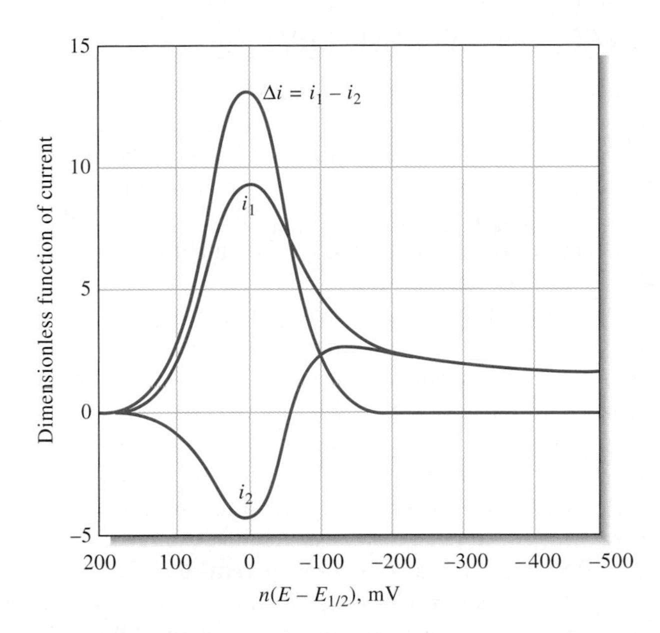

Figure 23-22 Current response for a reversible reaction to excitation signal in Figure 23-21c. This theoretical response plots a dimensionless function of current versus a function of potential, $n(E - E_{1/2})$ in mV. Here i_1 = forward current; i_2 = reverse current; $i_1 - i_2$ = current difference. (From J. J. O'Dea, J. Osteryoung, and R. A. Osteryoung, *Anal. Chem.*, **1981,** *53,* 695. Copyright 1981, American Chemical Society.)

of half-wave potential data are available.[15] The judicious choice of anion often enhances the selectivity of the method. For example, with potassium chloride as a supporting electrolyte, the waves for iron(III) and copper(II) interfere with one another; in a fluoride medium, however, the half-wave potential of the former is shifted by about −0.5 V, while that for the latter is altered by only a few hundredths of a volt. The presence of fluoride thus results in the appearance of well-separated waves for the two ions.

Pulse polarography is also applicable to the analysis of such inorganic anions as bromate, iodate, dichromate, vanadate, selenite, and nitrite. In general, polarograms for these substances are affected by the pH of the solution because the hydrogen ion is a participant in their reduction. As a consequence, strong buffering to some fixed pH is necessary to obtain reproducible data.

Organic Polarographic Analysis

Almost from its inception, polarography has been used for the study and analysis of organic compounds. Several common functional groups are reduced at the dropping electrode, thus making possible the determination of a wide variety of organic compounds.[16]

◀ The following organic functional groups produce polarographic waves:

1. Carbonyl groups
2. Certain carboxylic acids
3. Most peroxides and epoxides
4. Nitro, nitroso, amine oxide, and azo groups
5. Most organic halogen groups
6. Carbon/carbon double bonds
7. Hydroquinones and mercaptans.

In general, the reactions of organic compounds at a voltammetric electrode are slower and more complex than those for inorganic species. Consequently, theoretical interpretation of the data is often more difficult or impossible. Generally, a much stricter adherence to detail is required for quantitative work. Despite these handicaps, organic polarography has proved fruitful for the determination of structure, the quantitative analysis of mixtures, and occasionally the qualitative identification of compounds.

[15]For example, see J. A. Dean, *Analytical Chemistry Handbook* pp. 14.66–14.70. New York: McGraw-Hill, 1995, D. T. Sawyer, A. Sobkowiak, and J. L. Roberts, Jr., *Experimental Electrochemistry for Chemists,* 2nd ed., pp. 100–130. New York: Wiley, 1995.

[16]For a detailed discussion of organic polarographic analysis, see P. Zuman, *Organic Polarographic Analysis.* Oxford: Pergamon Press, 1964; W. F. Smyth, *Polarography of Molecules of Biological Significance.* New York: Academic Press, 1979.

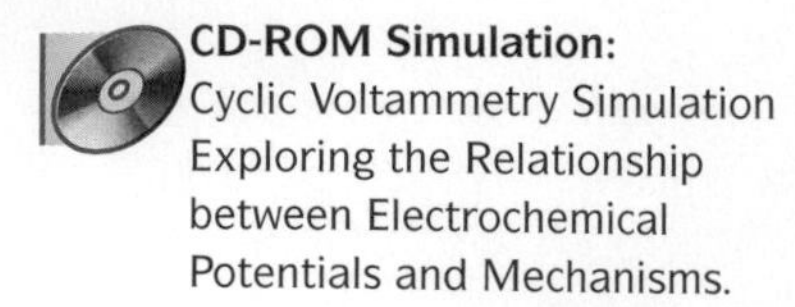

23D CYCLIC VOLTAMMETRY[17]

Cyclic voltammetry (CV) is an important and widely used electroanalytical technique. Although CV is infrequently used for quantitative analysis, it finds wide applicability in the study of oxidation/reduction reactions, the detection of reaction intermediates, and the observation of follow-up reactions of products formed at electrodes. In CV, the applied potential is swept in first one direction and then the other while the current is measured. A CV experiment may use one full cycle, a partial cycle, or several cycles.

During the CV experiment, the current response of a small stationary electrode in an unstirred solution is excited by a triangular potential waveform, such as that shown in Figure 23-23. The triangular waveform produces the forward and then the reverse scan. In the example of Figure 23-23, the potential is first varied linearly from +0.8 V to −0.15 V versus a saturated calomel electrode, at which point the scan direction is reversed and the potential is returned to its original value of +0.8 V. The scan rate in either direction is 50 mV/s in this example. The cycle is often repeated several times. The potentials at which reversal takes place (in this case, −0.15 V and +0.8 V) are called **switching potentials.** For a given experiment, switching potentials are chosen so that we can observe a diffusion-controlled oxidation or reduction of one or more species. The direction of the initial scan may be either negative, as shown, or positive, depending on the composition of the sample. A scan in the direction of more negative potentials is termed a **forward scan,** while one in the opposite direction is called a **reverse scan.** Generally, cycle times range from 1 ms or less to 100 s or more. In this example, the cycle time is 40 s.

Figure 23-24 shows the current response when a solution that is 6 mM in $K_3Fe(CN)_6$ and 1 M in KNO_3 is subjected to the cyclic excitation signal shown in Figure 23-23. The working electrode was a carefully polished stationary platinum electrode, and the reference electrode was a saturated calomel electrode. At the initial potential of +0.8 V, a tiny anodic current is observed, which immediately

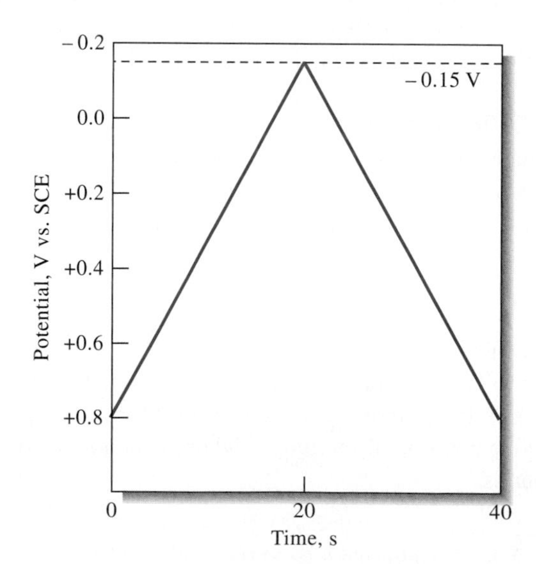

Figure 23-23 Cyclic voltammetric excitation signal.

[17]For further discussion, see A. J. Bard and L. R. Faulkner, *Electrochemical Methods,* 2nd ed., pp. 239–246. New York: Wiley, 2001; P.T. Kissinger and W. R. Heineman, *J. Chem. Educ.,* **1983,** *60,* 702.

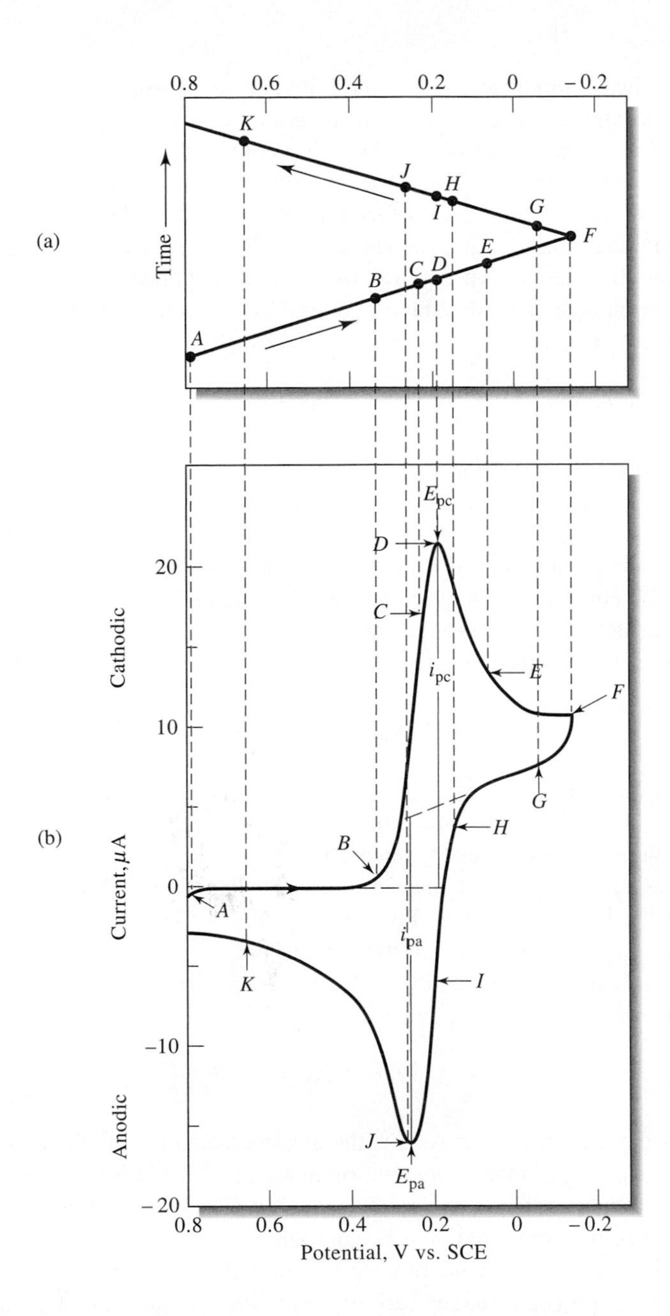

Figure 23-24 (a) Potential vs. time waveform and (b) cyclic voltammogram for a solution that is 6.0 mM in $K_3Fe(CN)_6$ and 1.0 M in KNO_3. (Used with permission from P. T. Kissinger and W. H. Heineman, *J. Chem. Educ.*, **1983,** *60,* 702. Copyright © 1983; Division of Chemical Education, Inc.)

decreases to zero as the scan is continued. This initial negative current arises from the oxidation of water to give oxygen. (At more positive potentials, this current increases rapidly and becomes quite large at about +0.9 V.) No current is observed between a potential of +0.7 and +0.4 V because no reducible or oxidizable species is present in this potential range. When the potential becomes less positive than approximately +0.4 V, a cathodic current begins to develop (point *B*) owing to the reduction of the ferricyanide ion to ferrocyanide ion. The cathodic reaction is then

CD-ROM Simulation: Exploration of Concentration Profiles of Solution Species for Complex Electrochemical Experiments.

$$Fe(CN)_6^{3-} + e^- \rightleftharpoons Fe(CN)_6^{4-}$$

A rapid increase in the current occurs in the region of *B* to *D* as the surface concentration of $Fe(CN)_6^{3-}$ becomes smaller and smaller. The current at the peak is made up of two components. One is the initial current surge required to adjust the surface concentration of the reactant to its equilibrium value given by the Nernst equation. The second is the normal diffusion-controlled current. The first current then decays rapidly (points D to F) as the diffusion layer is extended farther and farther away from the electrode surface (see also Figure 23-7a). At point F (-0.15 V), the scan direction is switched. The current, however, continues to be cathodic even though the scan is toward more positive potentials because the potentials are still negative enough to cause reduction of $Fe(CN)_6^{3-}$. As the potential sweeps in the positive direction, eventually reduction of $Fe(CN)_6^{3-}$ no longer occurs, and the current goes to zero and then becomes anodic. The anodic current results from the reoxidation of $Fe(CN)_6^{4-}$ that has accumulated near the surface during the forward scan. This anodic current peaks and then decreases as the accumulated $Fe(CN)_6^{4-}$ is used up by the anodic reaction.

Important parameters in a cyclic voltammogram are the cathodic peak potential E_{pc}, the anodic peak potential E_{pa}, the cathodic peak current i_{pc}, and the anodic peak current i_{pa}. The definitions and measurements of these parameters are illustrated in Figure 23-24. For a reversible electrode reaction, anodic and cathodic peak currents are approximately equal in absolute value but opposite in sign. For a reversible electrode reaction at 25°C, the difference in peak potentials, ΔE_p, is expected to be

$$\Delta E_p = |E_{pa} - E_{pc}| = 0.059/n \quad (23\text{-}15)$$

where n is the number of electrons involved in the half-reaction. Irreversibility because of slow electron transfer kinetics results in ΔE_p exceeding the expected value. While an electron transfer reaction may appear reversible at a slow sweep rate, increasing the sweep rate may lead to increasing values of ΔE_p, a sure sign of irreversibility. Hence, to detect slow electron transfer kinetics and to obtain rate constants, ΔE_p is measured for different sweep rates.

Quantitative information is obtained from the Randles-Sevcik equation, which at 25°C is

$$i_p = 2.686 \times 10^5 n^{3/2} AcD^{1/2}v^{1/2} \quad (23\text{-}16)$$

where i_p is the peak current in A, A is the electrode area in cm^2, D is the diffusion coefficient in cm^2/s, c is the concentration in mol/cm^3, and v is the scan rate in V/s. Cyclic voltammetry offers a way of determining diffusion coefficients if the concentration, electrode area, and scan rate are known.

The major use of cyclic voltammetry is to provide qualitative information about electrochemical processes under various conditions. As an example, consider the cyclic voltammogram for the agricultural insecticide parathion that is shown in Figure 23-25. Here, the switching potentials were about -1.2 V and $+0.3$ V. The initial forward scan was, however, started at 0.0 V and not $+0.3$ V. Three peaks are observed. The first cathodic peak (A) results from a four-electron reduction of the parathion to give a hydroxylamine derivative

$$\phi NO_2 + 4e^- + 4H^+ \rightarrow \phi NHOH + H_2O$$

The anodic peak at B arises from the oxidation of the hydroxylamine to a nitroso derivative during the reverse scan. The electrode reaction is

$$\phi NHOH \rightarrow \phi NO + 2H^+ + 2e^-$$

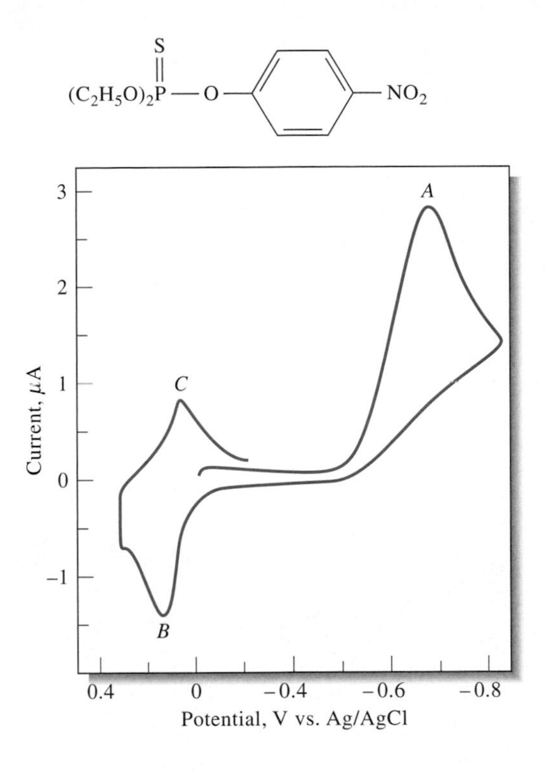

Figure 23-25 Cyclic voltammogram of the insecticide parathion in 0.5 M pH 5 sodium acetate buffer in 50% ethanol. Hanging mercury drop electrode. Scan rate: 200 mV/s. (From W. R. Heineman and P. T. Kissinger, *Amer. Lab.*, **1982** (*11*), 34. Copyright 1982 by International Scientific Communications, Inc. Reprinted with permission.)

The cathodic peak at *C* results from the reduction of the nitroso compound to the hydroxylamine as shown by the equation

$$\Phi NO + 2H^{+} + 2e^{-} \rightarrow \Phi NHOH$$

Cyclic voltammograms for authentic samples of the two intermediates confirm the identities of the compounds responsible for peaks *B* and *C*.

Cyclic voltammetry is widely employed in organic and inorganic chemistry. It is often the first technique selected for investigation of a system with electroactive species. Often, cyclic voltammograms will reveal the presence of intermediates in oxidation/reduction reactions (see Figure 23-25, for example). Platinum electrodes are often used in cyclic voltammetry. For negative potentials, mercury film electrodes can be used. Other popular working electrodes include glassy carbon, gold, graphite, and carbon paste. Chemically modified electrodes are discussed in Feature 23-2.

FEATURE 23-2

Modified Electrodes[18]

An active area of research in electrochemistry is the development of electrodes that are produced by chemical modification of various conductive substrates. Such electrodes can, in principle, be tailored to accomplish various functions. Modifications include the presence of irreversibly adsorbing substances with

(continued)

[18]For more information, see R. W. Murray, "Molecular Design of Electrode Surfaces," in *Techniques in Chemistry, Vol XXII*, W. Weissberger, Founding Ed. New York: Wiley, 1992; A. J. Bard, *Integrated Chemical Systems*. New York: Wiley, 1994.

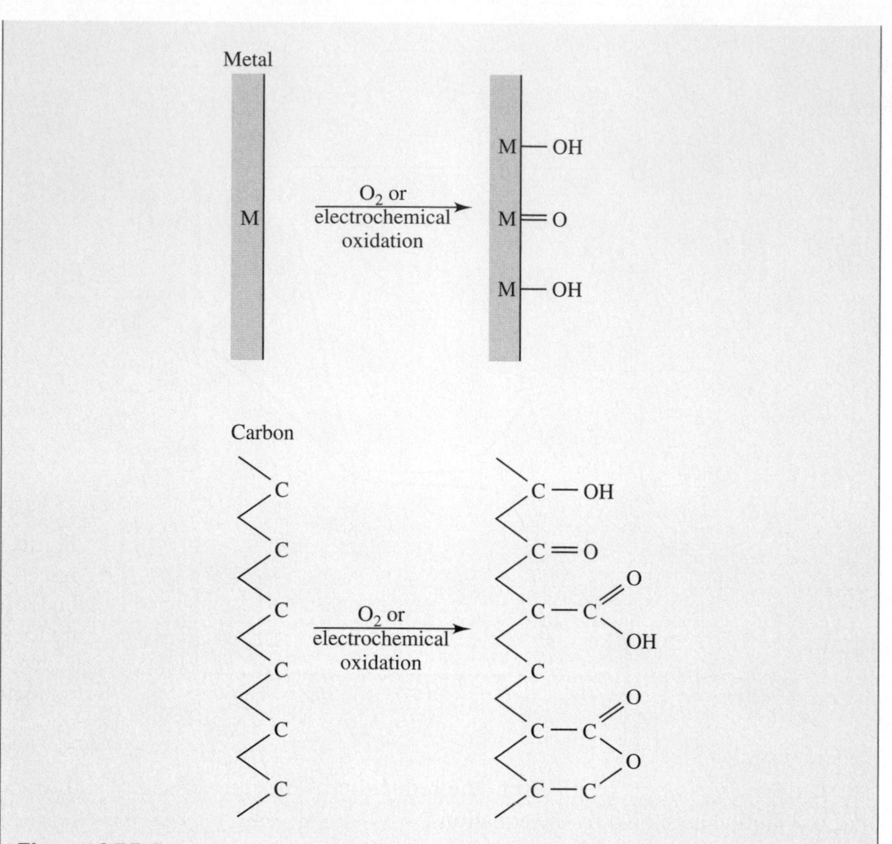

Figure 23F-3 Functional groups formed on a metal or carbon surface by oxidation. Often, linking agents such as orgaonsilanes are bonded to the functionalized surface. Reactive components, such as ferrocenes, viologens, and metal bipyridine complexes, are then attached to form the modified surfaces shown in Figure 23F-4. (From A. J. Bard, *Integrated Chemical Systems.* New York: Wiley, 1994. This material is used by permission of John Wiley & Sons, Inc.)

desired functionalities, covalent bonding of components to the surface, and coating of the electrode with polymer films or films of other substances. The covalent attachment process is shown in Figures 23F-3 and 23F-4. Linking agents such as organosilanes and amines are attached to the surface prior to attaching the group of interest. Polymer films can be prepared from dissolved polymers by dip coating, spin coating, electrodeposition, or covalent attachment. They can also be produced from the monomer by thermal, plasma, photochemical, or electrochemical polymerization methods. Immobilized enzyme biosensors, such as the amperometric sensors described in Section 23B-4, are a type of modified electrode. These can be prepared by covalent attachment, adsorption, or gel entrapment.

Modifed electrodes have many potential applications. A primary interest has been in the area of electrocatalysis. Here, electrodes capable of reducing oxygen to water have been sought for use in fuel cells and batteries. Another potential application is in the production of electrochromic devices that change color on oxidation and reduction. Such devices could be used in displays or *smart win-*

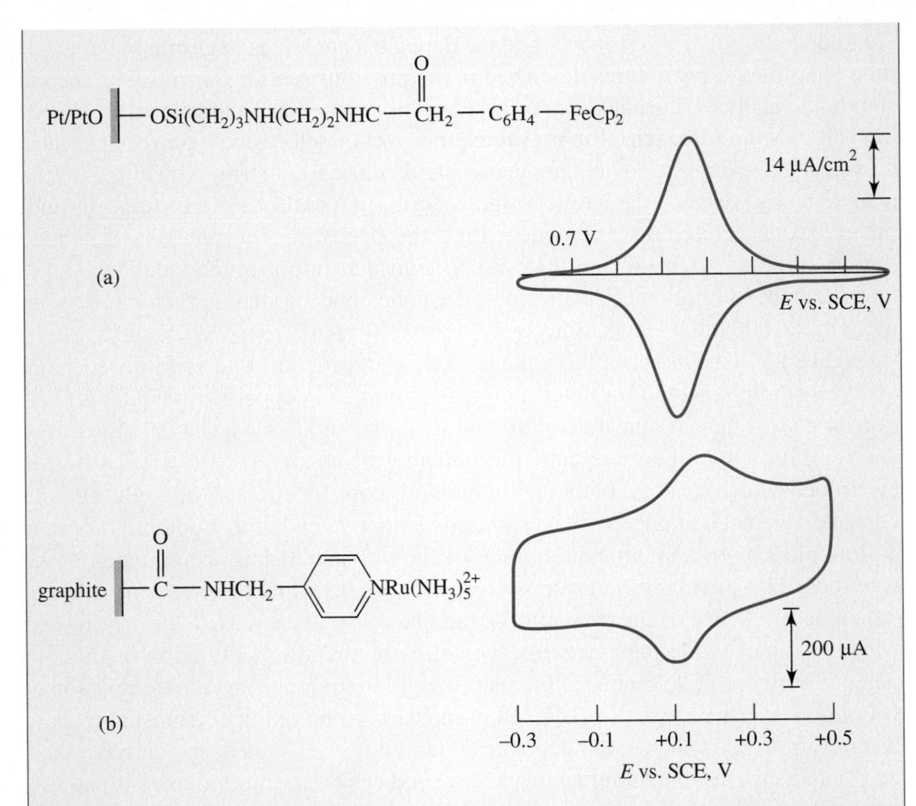

Figure 23F-4 Electrodes modified by covalent attachment of various components. To the right, the cyclic voltammograms are shown. In (a), a Pt electrode is shown with a ferrocene attached. (Reprinted with permission from J. R. Lenhard and R. W. Murray, *J. Am. Chem. Soc.*, **1978,** *100,* 7870. Copyright 1978 American Chemical Society.) In (b), a graphite electrode is shown with attached py-$Ru(NH_3)_5^{2+}$. (Reprinted with permission from C. A. Koval and F. C. Anson, *Anal. Chem.*, **1978,** *50,* 223. Copyright 1978 American Chemical Society.)

dows and *mirrors.* Electrochemical devices that could serve as molecular electronic devices, such as diodes and transistors, are also under intense study. Finally, the most important analytical use for such electrodes is as analytical sensors that are prepared to be selective for a particular species or functional group.

23E STRIPPING METHODS

Stripping methods encompass a variety of electrochemical procedures that include a bulk electrolysis preconcentration step followed by a voltammetric step.[19] In all these procedures, the analyte is first deposited into a small volume of mercury, usually from a stirred solution. A hanging mercury drop or a thin mercury film is most often used. After an accurately measured deposition time, the electrolysis is dis-

[19]For further discussion of stripping methods, see A. J. Bard and L. R. Faulkner, *Electrochemical Methods,* 2nd ed., pp. 458–464. New York: Wiley, 2001; J. Wang, *Stripping Analysis.* Deerfield Beach, FL: VCH Publishers, 1985; A. M. Bond, *Modern Polarographic Methods in Analytical Chemistry,* Ch. 9. New York: Marcel Dekker, 1980.

In **anodic stripping methods**, the analyte is deposited by reduction and then analyzed by oxidation from the small-volume mercury film or drop.

In **cathodic stripping methods**, the analyte is electrolyzed into a small volume of mercury by oxidation and then stripped by reduction.

► A major advantage of stripping analysis is the capability for electrochemical preconcentration of the analyte prior to the measurement step.

continued, the stirring is stopped, and the deposited analyte is determined by one of the voltammetric procedures described in the previous section. During this second step in the analysis, the analyte is redissolved or stripped from the electrode; hence the name. In **anodic stripping methods,** the working electrode behaves as a cathode during the deposition step and as an anode during the stripping step, with the analyte being oxidized back to its original form. In a **cathodic stripping method,** the electrode behaves as an anode during the deposition step and as a cathode during stripping. Because the material is deposited into a much smaller volume than the bulk solution volume, the analyte can be concentrated by factors of 100 to more than 1000 in the deposition step.

Figure 23-26a illustrates the voltage excitation program that is followed in an anodic stripping method for determining cadmium and copper in an aqueous solution of these ions. A linear-scan voltammetric method is used to complete the analysis. Initially, a constant cathodic potential of about −1 V is applied to the electrode, which causes both cadmium and copper ions to be reduced and deposited as metal amalgams. The electrode is maintained at this potential for several minutes, until a significant amount of the two metals has accumulated at the electrode. The stirring is then stopped for perhaps 30 s while the electrode is maintained at −1 V. Next, the potential of the electrode is decreased linearly to less negative values while the current in the cell is recorded as a function of time, or potential. Figure 23-26b shows the resulting voltammogram. At a potential somewhat more negative than −0.6 V, cadmium starts to be oxidized, causing a sharp increase in the current. As the deposited cadmium is consumed, the current peaks and then decreases to its original level. A second peak representing oxidation of the

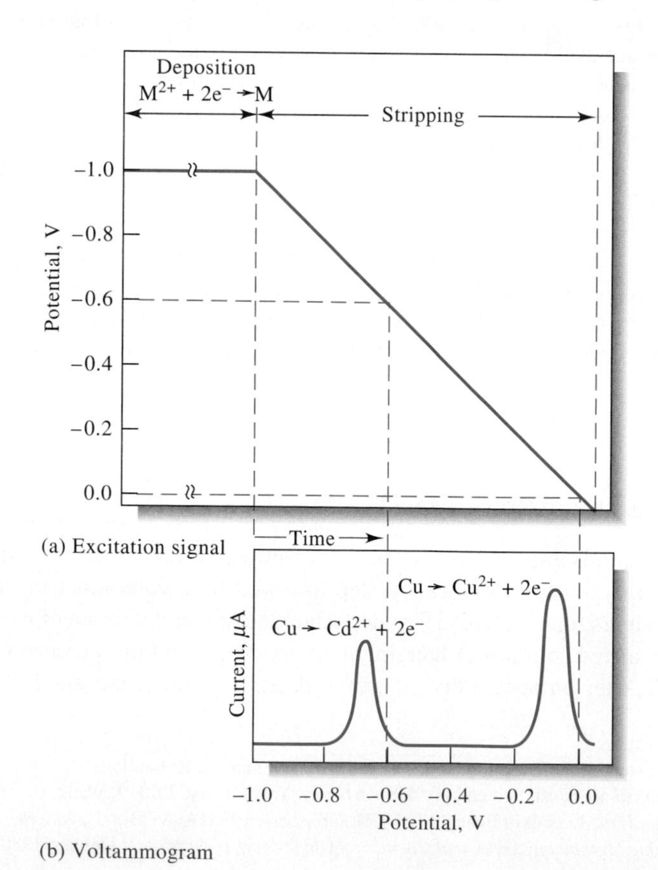

Figure 23-26 (a) Excitation signal for stripping determination of Cd^{2+} and Cu^{2+}. (b) Anodic stripping voltammogram.

copper is observed when the potential has decreased to approximately -0.1 V. The heights of the two peaks are proportional to the mass of the deposited metals.

Stripping methods are of prime importance in trace analysis because the electrodeposition step concentrates the analyte and allows minute amounts to be determined with reasonable accuracy. Thus, analytes in the 10^{-6} to 10^{-9} M range can be determined with stripping methods that use simple and rapid procedures.

23E-1 Electrodeposition Step

Ordinarily, only a fraction of the analyte is deposited during the electrodeposition step; hence, quantitative results depend not only on control of electrode potential but also on such factors as electrode size, deposition time, and stirring rate for both the sample and standard solutions employed for calibration.

Electrodes for stripping methods have been formed from a variety of materials, including mercury, gold, silver, platinum, and carbon in various forms. The most popular electrode is the **hanging mercury drop electrode** (HMDE), which consists of a single drop of mercury in contact with a platinum wire. Hanging drop electrodes are available from several commercial sources. These electrodes often consist of a microsyringe with a micrometer for exact control of drop size. The drop is then formed at the tip of a capillary by displacement of the mercury in the syringe-controlled delivery system (see Figure 23-1b). The system shown in Figure 23-1d is also capable of producing a hanging drop electrode.

To carry out the determination of a metal ion by means of anodic stripping, a fresh hanging drop is formed, gentle stirring is begun, and a potential is applied that is a few tenths of a volt more negative than the half-wave potential for the ion of interest. Deposition is allowed to occur for a carefully measured period, which can vary from a minute or less for 10^{-7} M solutions to 30 min or longer for 10^{-9} M solutions. These times seldom result in complete removal of the analyte. The electrolysis period is determined by the sensitivity of the method ultimately employed for completion of the analysis.

Another electrode that is widely used is a mercury film electrode, where the film is deposited onto a glassy carbon or wax-impregnated disk. Such films are often less than 10 nm in thickness. Mercury film electrodes have smaller volumes than a conventional HMDE and thus permit higher sensitivity determinations. Solid electrodes have been used, but much less frequently than mercury electrodes.

23E-2 Voltammetric Completion Step

The analyte collected in the HMDE or mercury film can be determined by any of several voltammetric procedures. For example, in a linear anodic scan procedure, such as described at the beginning of this section, stirring is discontinued for perhaps 30 s after termination of the deposition. The voltage is then decreased at a linear fixed rate from its original cathodic value, and the resulting anodic current is recorded as a function of the applied voltage. This linear scan produces a curve of the type shown in Figure 23-26b. Analyses of this type are generally based on calibration with standard solutions of the cations of interest. With reasonable care, relative standard deviations of about 2% can be obtained.

Most of the other voltammetric procedures described in the previous section have also been applied to the stripping step. The most widely used of these appears to be an anodic differential pulse technique. Often, narrower peaks are produced by this procedure, which is desirable when mixtures are to be analyzed. Use of the mercury

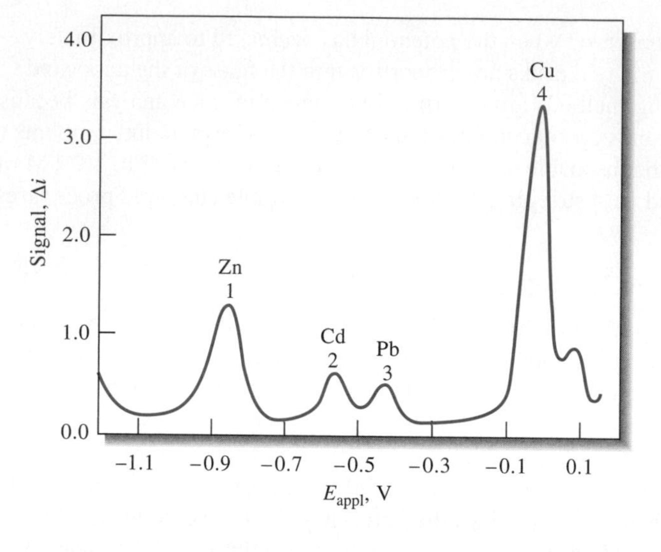

Figure 23-27 Differential pulse anodic stripping voltammogram of 25 ppm zinc, cadmium, lead, and copper. (From W. M. Peterson and R. V. Wong, *Amer. Lab.*, **1981,** *13*(11), 116. Copyright 1981 by International Scientific Communications, Inc. Reprinted with permission.)

film electrode also produces narrower peaks. Because the average diffusion-path length from the film to the solution interface is much shorter than that in a drop of mercury, the analyte can rapidly escape, which leads to narrower and larger voltammetric peaks. The HMDE appears to give more reproducible results, however, especially at higher analyte concentrations. Figure 23-27 is a differential-pulse anodic stripping polarogram for a mixture of cations present at concentrations of 25 ppb showing good resolution and adequate sensitivity for many purposes.

Many other variations of the stripping technique have been developed. For example, a number of cations have been determined by electrodeposition on a platinum cathode. The quantity of electricity required to remove the deposit is then measured coulometrically. Here again, the method is particularly advantageous for trace analyses. Cathodic stripping methods for the halides have also been developed in which the halide ions are first deposited as mercury(I) salts on a mercury anode. Stripping is then performed by a cathodic current.

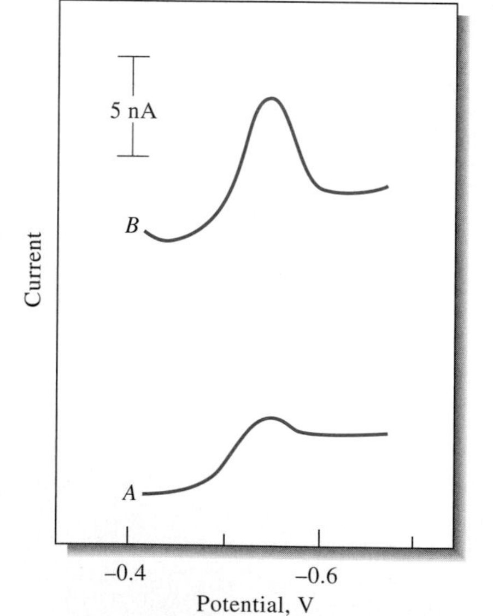

Figure 23-28 Differential pulse voltammogram for 5×10^{-10} M riboflavin. Adsorptive preconcentration for 5 min (*A*) and 30 min (*B*) at −0.2 V. (From J. Wang, *Amer. Lab.*, **1985** (5), 43. Copyright 1985 by International Scientific Communications, Inc. Reprinted with permission.)

23E-3 Adsorption Stripping Methods

Adsorption stripping methods are quite similar to the anodic and cathodic stripping methods we have just considered. Here, a small electrode, most commonly a hanging mercury drop electrode, is immersed in a stirred solution of the analyte for several minutes. Deposition of the analyte then occurs by physical adsorption on the electrode surface rather than by electrolytic deposition. After sufficient analyte has accumulated, the stirring is discontinued, and the deposited material is determined by linear-scan or pulsed voltammetric measurements. Quantitative information is based on calibration with standard solutions that are treated in the same way as samples.

Many organic molecules of clinical and pharmaceutical interest have a strong tendency to be adsorbed from aqueous solutions onto a mercury surface, particularly if the surface is maintained at about −0.4 V (vs. SCE) when the charge on the mercury is zero (see page 686). With good stirring, adsorption is rapid, and only 1 to 5 min are required to accumulate sufficient analyte for analysis from 10^{-7} M solutions—10 to 20 min for 10^{-9} M solutions. Figure 23-28 illustrates the sensitivity of differential-pulse adsorptive stripping voltammetry when it is applied to the determination of riboflavin in a 5×10^{-10} M solution. Many other examples of this type can be found in the recent literature.

Adsorptive stripping voltammetry has also been applied to the determination of a variety of inorganic cations at very low concentrations. In these applications, the cations are generally complexed with surface active complexing agents, such as dimethylglyoxime, catechol, and bipyridine. Detection limits in the 10^{-10} to 10^{-11} M range have been reported.

23F VOLTAMMETRY WITH MICROELECTRODES

Over the last two decades, a number of voltammetric studies have been carried out with microelectrodes that have dimensions that are smaller by an order of magnitude or more than the electrodes we have described so far. The electrochemical behavior of these tiny electrodes is significantly different from classical electrodes and appears to offer advantages in certain analytical applications.[20] Such electrodes have been sometimes called **microscopic electrodes,** or *ultramicroelectrodes,* to distinguish them from classical voltammetric electrodes. The dimensions of such electrodes are generally smaller than about 20 μm and may be as small as a few tenths of a micrometer. These miniature microelectrodes take several forms. The most common is a planar electrode, formed by sealing a carbon fiber with a radius of 5 μm or a gold or platinum wire having dimensions from 0.3 to 20 μm into a fine capillary tube; the fiber or wires are than cut flush with the ends of the tubes. Cylindrical electrodes are also used, in which a small portion of the wire extends from the end of the tube. There are several other forms of these electrodes.

Generally, the instrumentation used with microelectrodes is simpler than that shown in Figure 23-2 because there is no need to employ a three-electrode system. The reason that the reference electrode can be eliminated is that the currents are so small (in the picoampere to nanoampere range) that the *IR* drop does not distort the voltammetric waves the way microampere currents do.

One of the reasons for the early interest in microscopic microelectrodes was the desire to study chemical processes in single cells (Figure 23-29) or processes inside organs of living species, such as in mammalian brains. One approach to this

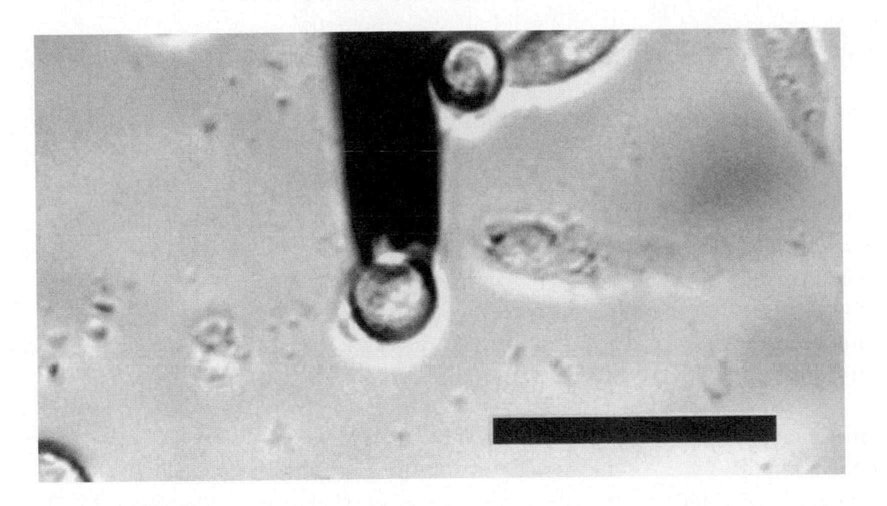

Figure 23-29 Optical image using brightfield microscopy showing a carbon fiber microelectrode adjacent to a bovine chromaffin cell from the adrenal medulla. The extracellular solution was 10 mM TRIS buffer containing 150 mM NaCl, 2 mM $CaCl_2$, 1.2 mM $MgCl_2$, and 5 mM glucose. The black scale bar is 50 μm. (From L. Buhler and R. M. Wightman, unpublished work. With permission.)

[20]See A. C. Michael and R. M. Wightman, in *Laboratory Techniques in Electroanalytical Chemistry,* 2nd ed., P. T. Kissinger and W. R. Heinemann, Eds., Ch. 12. New York: Marcel Dekker, 1996; C. G. Zoski, in *Modern Techniques in Electroanalysis,* P. Vanysek, Ed., Ch. 6. New York: Wiley, 1996; R. M. Wightman, *Science,* **1988,** *240,* 415; S. Pons and M. Fleischmann, *Anal. Chem.,* **1987,** *59,* 1391A.

problem is to use electrodes that are small enough not to cause significant alteration in the function of the organ. It was also realized that microelectrodes have certain advantages that justify their application to other kinds of analytical problems. Among these advantages are the very small *IR* drops, which make them applicable to solvents having low dielectric constants, such as toluene. Second, capacitive charging currents, which often limit detection with ordinary voltammetric electrodes, are reduced to insignificant proportions as the electrode size is diminished. Third, the rate of mass transport to and from an electrode increases as the size of an electrode decreases; as a result, steady-state currents are established in unstirred solutions in less than a microsecond rather than in a millisecond or more, as is the case with classical electrodes. Such high-speed measurements permit the study of intermediates in rapid electrochemical reactions. Undoubtedly, the future will see many more applications of microelectrodes.

WEB WORKS

Use the Google search engine to find companies that make anodic stripping voltammetry (ASV) instruments. You should discover in your search links to companies such as ESA, Inc., Cypress Systems, Inc., and Bioanalytical Systems. For two instrument manufacturers, compare the working electrodes used for anodic stripping voltammetry. Consider the types of electrodes (thin-film, hanging mercury drop, etc.), whether they are rotating electrodes, and whether they pose any health risks. Also, compare the specifications of two instruments from two different manufacturers. Consider in your comparison the deposition potential ranges, the available deposition times, the scanning potential ranges, the scanning sweep rates, and the prices.

QUESTIONS AND PROBLEMS

23-1. Distinguish between
- *(a) voltammetry and polarography.
- (b) linear-scan polarography and pulse polarography.
- *(c) differential pulse polarography and square-wave polarography.
- (d) a hanging drop mercury electrode and a dropping mercury electrode.
- *(e) a limiting current and a residual current.
- (f) a limiting current and a diffusion current.
- *(g) laminar flow and turbulent flow.
- (h) the standard electrode potential and the half-wave potential for a reversible reaction at a voltammetric electrode.

23-2. Why is a high supporting electrolyte concentration used in most electroanalytical procedures?

***23-3.** Why is the reference electrode placed near the working electrode in a three-electrode cell?

23-4. Define
- (a) voltammograms.
- (b) hydrodynamic voltammetry.
- (c) Nernst diffusion layer.
- (d) a mercury film electrode.
- (e) half-wave potential.

***23-5.** Why is it necessary to buffer solutions in organic voltammetry?

23-6. List the advantages and disadvantages of the dropping mercury electrode compared with platinum or carbon electrodes.

***23-7.** Suggest how Equation 23-11 could be employed to determine the number of electrons involved in a reversible reaction at a voltammetric electrode.

23-8. Quinone (Q) undergoes a reversible reduction to hydroquinone (H_2Q) at a dropping mercury electrode. The reaction is

$$Q + 2H^+ + 2e^- \rightleftharpoons H_2Q$$

*(a) Assume that the diffusion coefficient for quinone and hydroquinone are approximately the same, and calculate the approximate half-wave potential (vs. SCE) for the reduction of hydroquinone at a planar electrode from a stirred solution buffered to a pH of 7.0.

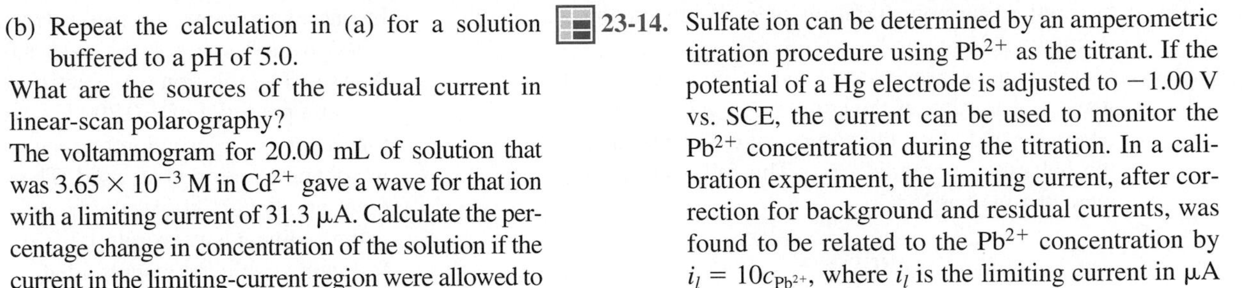

(b) Repeat the calculation in (a) for a solution buffered to a pH of 5.0.

23-9. What are the sources of the residual current in linear-scan polarography?

***23-10.** The voltammogram for 20.00 mL of solution that was 3.65×10^{-3} M in Cd^{2+} gave a wave for that ion with a limiting current of 31.3 μA. Calculate the percentage change in concentration of the solution if the current in the limiting-current region were allowed to continue for (a) 5 min; (b) 10 min; (c) 30 min.

23-11. Calculate the concentration of cadmium in milligrams per milliliter of sample, based on the following data (corrected for residual current):

	Volumes Used, mL				
Solution	Sample	0.400 M KCl	2.00×10^{-3} M Cd^{2+}	H_2O	Current, μA
*(a)	15.0	20.0	0.00	15.0	79.7
	15.0	20.0	5.00	10.0	95.9
(b)	10.0	20.0	0.00	20.0	49.9
	10.0	20.0	10.0	10.0	82.3
*(c)	20.0	20.0	0.00	10.0	41.4
	20.0	20.0	5.00	5.00	57.6
(d)	15.0	20.0	0.00	15.0	67.9
	15.0	20.0	10.0	5.00	100.3

23-12. The following reaction is reversible and has a half-wave potential of − 0.349 V when carried out at a dropping mercury electrode from a solution buffered to pH 2.5.

$$O + 4H^+ + 4e^- \rightleftharpoons R$$

Predict the half-wave potential at pH: (a) 1.0; (b) 3.5; (c) 7.0.

***23-13.** Shown below is the polarogram for a solution that was 1×10^{-4} M in KBr and 0.1 M in KNO_3. Offer an explanation of the wave that occurs at +0.12 V and the rapid change in current that starts at about +0.48 V. Would the wave at 0.12 V have any analytical applications? Explain.

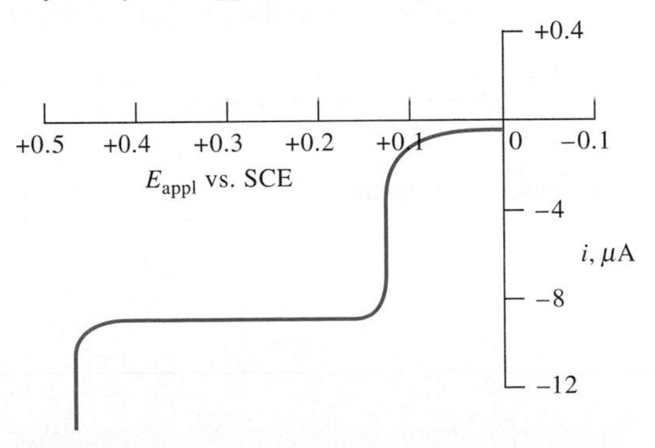

Polarogram.

23-14. Sulfate ion can be determined by an amperometric titration procedure using Pb^{2+} as the titrant. If the potential of a Hg electrode is adjusted to −1.00 V vs. SCE, the current can be used to monitor the Pb^{2+} concentration during the titration. In a calibration experiment, the limiting current, after correction for background and residual currents, was found to be related to the Pb^{2+} concentration by $i_l = 10c_{Pb^{2+}}$, where i_l is the limiting current in μA and $c_{Pb^{2+}}$ is the Pb^{2+} concentration in mM. The titration reaction is

$$SO_4^{2-} + Pb^{2+} \rightleftharpoons PbSO_4(s) \qquad K_{sp} = 1.6 \times 10^{-8}$$

If 25 mL of 0.025 M Na_2SO_4 is titrated with 0.040 M $Pb(NO_3)_2$, develop the titration curve in spreadsheet format and plot the limiting current versus the volume of titrant.

***23-15.** Lead was determined polarographically at the dropping mercury electrode by measurements in 1 M HNO_3. The limiting current on the Pb(II) wave was measured at −0.600 V versus SCE. At this potential, the residual current was 0.12 μA. The method of external standards was used and the following results obtained:

Concentration of Pb(II), mM	Limiting Current, μA
0.50	4.37
1.00	8.67
2.00	17.49
3.00	25.75
4.00	34.35
5.50	47.10
6.50	55.70
Unknown	12.35

Determine the concentration of lead in the unknown solution and its standard deviation.

23-16. Cadmium was determined by polarography at the dropping mercury electrode in solutions that were 1 M in HCl. The limiting current was measured at −0.750 V versus SCE. The residual current at this potential was 0.21 μA. The method of external standards was used and the following results obtained:

Concentration of Cd(II), mM	Limiting current, μA
1.00	4.37
2.00	8.67
3.00	12.87
5.00	21.54
8.00	34.35
12.00	51.25
Unknown	28.53

Determine the concentration of cadmium in the unknown solution and its standard deviation.

23-17. Measurements were made on a polarographic wave to determine if the couple $O + ne^- \rightleftharpoons R$ is reversible and, if so, the number of electrons n and the half-wave potential, $E_{1/2}$. The following data were obtained at 25°C:

$E_{applied}$ vs. SCE, V	i, mA
−0.395	0.49
−0.406	0.96
−0.415	1.48
−0.422	1.95
−0.431	2.42
−0.445	2.95

Determine if the couple behaves reversibly. Find n and $E_{1/2}$.

*23-18. Why are stripping methods more sensitive than other voltammetric procedures?

23-19. What is the purpose of the electrodeposition step in stripping analysis?

23-20. **Challenge Problem.** A new method for determining ultrasmall (nL) volumes by anodic stripping voltammetry has been proposed (W. R. Vandaveer and I. Fritsch, *Anal. Chem.*, **2002,** *74,* 3575). In this method, a metal is exhaustively deposited from the small volume to be measured onto an electrode, from which it is later stripped. The solution volume V_s is related to the total charge Q required to strip the metal by

$$V_s = \frac{Q}{nFC}$$

where n is the number of moles of electrons per mole of analyte, F is the Faraday, and C is the molar concentration of the metal ion before electrolysis.

(a) Beginning with Faraday's law (see Equation 22-8), derive the above equation for V_s.

(b) In one experiment, the metal deposited was Ag(s) from a solution that was 8.00 mM in $AgNO_3$. The solution was electrolyzed for 30 min at a potential of −0.700 V versus a gold top layer as a pseudoreference. A tubular nanoband electrode was used. The silver was then anodically stripped off the electrode using a linear sweep rate of 0.10 V/s. The following table represents idealized anodic stripping results. By integration, determine the total charge required to strip the silver from the tubular electrode. You can do a manual Simpson's rule integration or refer to *Applications of Microsoft® Excel in Analytical Chemistry,* Chapter 11, to do the integration with Excel. From the charge, determine the volume of the solution from which the silver was deposited.

Potential, V	Current, nA	Potential, V	Current, nA
−0.50	0.000	−0.123	−1.10
−0.45	−0.001	−0.115	−1.00
−0.40	−0.02	−0.10	−0.80
−0.30	−0.10	−0.09	−0.65
−0.25	−0.20	−0.08	−0.52
−0.22	−0.30	−0.065	−0.37
−0.20	−0.44	−0.05	−0.22
−0.18	−0.67	−0.025	−0.12
−0.175	−0.80	0.00	−0.05
−0.168	−1.00	0.05	−0.03
−0.16	−1.18	0.10	−0.02
−0.15	−1.34	0.15	−0.005
−0.135	−1.28		

(c) Can you suggest experiments to show whether all the Ag^+ was reduced to Ag(s) in the deposition step?

(d) Would it matter if the droplet were not a hemispherical shape? Why or why not?

(e) Describe an alternative method against which you might test the proposed method.

InfoTrac College Edition

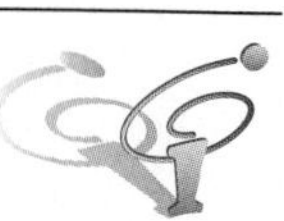

For additional readings, go to InfoTrac College Edition, your online research library, at

http://infotrac.thomsonlearning.com

PART V

Spectrochemical Analysis

A conversation with *Gary M. Hieftje*

Photo by Olan Mills. Courtesy of G. M. Hieftje

Gary Hieftje has many stories to tell: about playing with dangerous compounds as a young boy, setting up gas lines in the family's basement, and driving people out of buildings when chemical reactions were left unattended. He says that his gift of gab stems from his college days when he sold shoes to support his family. He also says that he has never had a life plan—but the places he has ended up are so well suited to him, it seems that he is living a lifelong dream. The common theme is that he has worked very hard, but on problems he finds fascinating.

Hieftje has been a professor in the chemistry department at Indiana University since 1969, where he is now Distinguished Professor and Mann Chair. As a researcher, his goal is to make techniques and instruments better. He investigates basic mechanisms in atomic emission, absorption, fluorescence, and mass spectrometric analysis and is continuing to develop atomic methods of analysis. He is interested in devising a method of on-line computer control of chemical instrumentation and experiments; using time-resolved luminescence processes for analysis; applying information theory to analytical chemistry, analytical mass spectrometry, and near-infrared reflectance analysis; and using stochastic processes to extract basic and kinetic chemical information. His many awards include being named a Fellow by the American Association for the Advancement of Science and receiving the American Chemical Society (ACS) Award in Analytical Chemistry and the ACS Analytical Division Excellence in Teaching Award.

Q: What initially attracted you to the field of chemistry?

A: My life was deeply affected by Marvin Overway, a high school chemistry teacher who lived across the street. It was he who kindled my interest in science. Most chemists liked bright lights, colors, flashes, and explosions as kids. I played with a Gilbert chemistry set and soon learned how to make gunpowder. My mother was deeply offended by the terrible stink the sulfur dioxide created so I substituted cinnamon for the sulfur, which gave off a pleasant fragrance but didn't burn as well. Then the chemistry teacher taught me the nature of oxidizing agents. So I made my own gunpowder, using potassium perchlorate—which is extremely dangerous—instead of potassium nitrate, cinnamon instead of sulfur, and powdered charcoal. This was the way to go. I could shoot a marble down the block!

When I was about 13 I wanted to do glass blowing with my chemistry set, but the alcohol flame was too cool for Pyrex. So I decided I needed a Bunsen burner, and the teacher across the street gave me one. My uncle was a plumber, so I got pipe from him. My dad went crazy 6 months later when he found out that I had put in a gas line! My uncle checked it, and it was all put in right. My parents always encouraged my interests in science, even in doing stupid things like making gunpowder.

Q: What were your undergraduate years like?

A: In college, I already had a family. To support them, I worked the midnight shift as a lab technician in a local chemical manufacturing plant, and I also sold shoes. I lived near Hope College in Holland, Michigan. Hope is a small liberal arts college that had three strengths: Chemistry was one, and that's how I chose my major. I started doing research with the chairperson of the chemistry department, who was a synthetic organic chemist. One time, I was doing the Grignard reaction by combining thiophene with different unsaturated aldehydes. I let the reaction go while I was at calculus class, but someone turned off the water. The result emptied out the entire chemistry building!

Q: How about your experience in graduate school?

A: After college, I intended to get a job in organic chemistry, but the department chairperson convinced me to apply to graduate school. I chose the University of Illinois, but the cost of living was five times as high as near Holland. So for a year I worked as a physical chemist at the State Geological Survey, and at night I sold shoes. During that year, I met Howard Malmstadt. He had an enormous influence on me and still does. He convinced me that I was really an analytical chemist. It wasn't the physical phenomena that intrigued me as much as how to go about measuring things correctly.

I started graduate school in 1965 and supported my family with a National Science Foundation fellowship. I worked in Malmstadt's group, which was an incredibly productive and stimulating environment. Everyone worked hard because we didn't want to look bad in Malmstadt's or in each other's eyes.

I keep following interesting tributaries, which sometimes become more important than the original river. The interesting thing is the more of these tributaries you follow, the more you see how they tie together.

Q: How did you finally choose a career in academia?

A: Malmstadt encouraged me to look for an academic job. In the end, I took the academic position, but I assumed from the outset that I wouldn't receive tenure. I figured that after 5 years of my having fun, they would kick me out, and I'd go into industry and earn twice as much money! Surprisingly, I received tenure and have been at Indiana University ever since.

Q: Are you more interested in fundamentals or in applications?

A: I've always been more interested in fundamental things than in applications. To me, it's more exciting to discover why things happen, how to make techniques and instruments better, and how to make improved measurements. If I see a new area that is interesting, we play around in it for a few weeks. If it succeeds, we write a proposal to work on it and go at it. I keep following interesting tributaries, which sometimes become more important than the original river. The interesting thing is the more of these tributaries you follow, the more you see how they tie together.

Q: What work have you done in understanding plasmas?

A: We use fundamental principles in plasma physics to understand the mechanisms of interference in atomic spectrometry. One project is to study plasmas, such as inductively coupled plasmas (ICPs), in more detail than has ever been possible before. In plasmas, electrons zoom around at enormous velocities. We blast a laser into a plasma to measure the Doppler shift in light scattered from electrons. This tells us how many electrons are there—the more there are, the greater the scatter—and the electron energy distribution (that is, their velocities). We can get this information on a spatially and time-resolved basis because of the pulsed laser; it has a few-nanosecond pulse so that we can measure on a nanosecond time scale. Using Rayleigh scattering, we also can measure concentrations of argon in the plasma, from which we can get the gas-kinetic temperature. We know from the ideal gas law that if the temperature is higher, there are fewer species in the volume, and the scattering is weaker. The interesting thing is that the gas temperature and the electron temperature are different. That tells us that the ICP is controlled by kinetics, not thermodynamics. That observation leads us in all kinds of other directions!

Q: What other topics are you studying?

A: We also have a new light source with interesting characteristics. It's only about 20 microns in size and produces light pulses as short as 10 picoseconds, with a repetition rate of hundreds of millions per second. The beam is incredibly stable. It doesn't need power because it uses a radionuclide, a self-contained energy source, and some form of light-conversion medium to change beta or alpha pulses into showers of photons. We use this to study ultrafast events, such as the rapid kinetic features of various chemical and physical processes.

We developed an ICP time-of-flight mass spectrometer, which is now a commercial instrument. We also have a new device, a double-focusing mass spectrometer with a detector array, to look at many different elements at the same time. The third new type of mass spectrometer geometry is a time-of-flight instrument that uses two ion sources at the same time. One third of proteins contain metal atoms, and we hope to separate the proteins by capillary electrophoresis, then use this spectrometer to characterize the proteins and measure their metal atoms at the same time.

Q: What are your thoughts on teaching?

A: There are two important aspects of teaching for both undergraduate and graduate students: One is in the classroom, and the other is in the research lab. In the lab, you learn the nature of science and the nature of analytical chemistry. There's the incredible excitement of coming up with something that no one has ever known before! There aren't too many things that are more satisfying. The only thing that comes close is seeing the lights come on in students' eyes when they learn something they didn't know before.

To do original research, a person must focus, but there's a big danger in becoming too focused. To become a good problem solver, every scientist has to have a broad range of experiences. Most discoveries are made by people who link things together. In my research group, students have a wide range of activities, and they work side by side. Each scientist makes progress in a narrow area to get a degree but at the same time learns things from the people around him. Students can do so much now because of the sophistication of the instrumentation; as a result, they're better trained. ■

CHAPTER 24

Introduction to Spectrochemical Methods

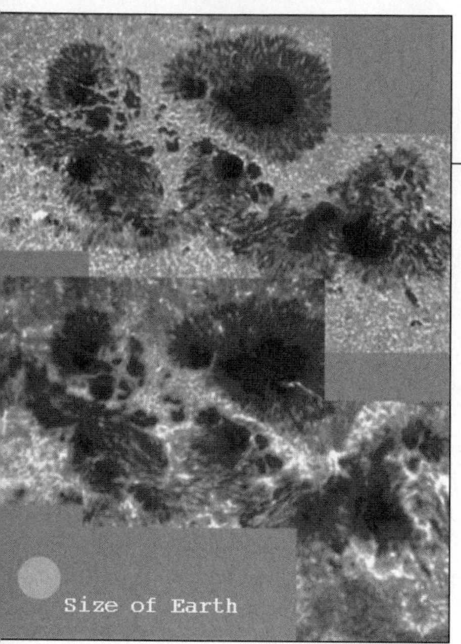

M. Sigwarth, J. Elrod, K.S. Balasubramaniam, S. Fletcher / NSO / AURA / NSF

This composite image of a sunspot group was collected with the Dunn solar telescope at the Sacramento Peak Observatory in New Mexico on March 29, 2001. The lower portion, consisting of four frames, was collected at a wavelength of 393.4 nm, and the upper portion was collected at 430.4 nm. The lower image represents calcium ion concentration, with the intensity of the color proportional to the amount of the ion in the sunspot. The upper image shows the presence of the CH molecule. Using data like these, it is possible to determine the location and abundance of virtually any chemical species in the observable universe. Note that the Earth could fit into the large black core sunspot in the upper left of each of the composite images.

Measurements based on light and other forms of electromagnetic radiation are widely used throughout analytical chemistry. The interactions of radiation and matter are the subject of the science called **spectroscopy.** *Spectroscopic analytical methods are based on measuring the amount of radiation produced or absorbed by molecular or atomic species of interest.*[1] *We can classify spectroscopic methods according to the region of the electromagnetic spectrum involved in the measurement. The regions of the spectrum that have been used include γ-ray, X-ray, ultraviolet (UV), visible, infrared (IR), microwave, and radio-frequency (RF). Indeed, current usage extends the meaning of spectroscopy further to include techniques that do not even involve electromagnetic radiation, such as acoustic, mass, and electron spectroscopy.*

► Other analytically useful types of electromagnetic radiation include γ-ray, X-ray, microwave, and RF radiation. Optical spectroscopic methods involve UV, visible, or IR radiation.

Spectroscopy has played a vital role in the development of modern atomic theory. In addition, **spectrochemical methods** *have provided perhaps the most widely used tools for the elucidation of molecular structure as well as the quantitative and qualitative determination of both inorganic and organic compounds.*

In this chapter, we discuss the basic principles that are necessary to understand measurements made with electromagnetic radiation, particularly those dealing with the absorption of UV, visible, and IR radiation. The nature of electromagnetic radiation and its interactions with matter are stressed. The next four chapters are devoted to spectroscopic instruments (Chapter 25), molecular absorption spectroscopy (Chapter 26), molecular fluorescence spectroscopy (Chapter 27), and atomic spectroscopy (Chapter 28).

[1]For further study, see F. Settle, Ed., *Handbook of Instrumental Techniques for Analytical Chemistry,* Sections III and IV. Upper Saddle River, NJ: Prentice-Hall, 1997; J. D. Ingle, Jr., and S. R. Crouch, *Spectrochemical Analysis.* Upper Saddle River, NJ: Prentice-Hall, 1988; E. J. Meehan, in *Treatise on Analytical Chemistry,* 2nd ed., P. J. Elving, E. J. Meehan, and I. M. Kolthoff, Eds., Part I, Vol. 7, Chapters 1–3. New York: Wiley, 1981; J. E. Crooks, *The Spectrum in Chemistry.* New York: Academic Press, 1978.

24A PROPERTIES OF ELECTROMAGNETIC RADIATION

Electromagnetic radiation is a form of energy that is transmitted through space at enormous velocities. We call electromagnetic radiation in the UV/visible and sometimes in the IR region **light,** although strictly speaking the term refers only to visible radiation. Electromagnetic radiation can be described as a wave with properties of wavelength, frequency, velocity, and amplitude. In contrast to sound waves, light requires no supporting medium for its transmission; thus, it easily passes through a vacuum. Light also travels nearly a million times faster than sound.

The wave model fails to account for phenomena associated with the absorption and emission of radiant energy. For these processes, electromagnetic radiation can be treated as discrete packets of energy or particles called **photons** or **quanta.** These dual views of radiation as particles and waves are not mutually exclusive but are complementary. In fact, the energy of a photon is directly proportional to its frequency, as we shall see. Similarly, this duality applies to streams of electrons, protons, and other elementary particles, which can produce interference and diffraction effects that are typically associated with wave behavior.

Courtesy of the Archives, California Institute of Technology

Richard P. Feynman (1918–1988) was one of the most well-known and renowned scientists of the 20th century. For his role in the development of quantum electrodynamics, he was awarded the Nobel Prize in Physics in 1965. In addition to his many and varied scientific contributions, he was a skilled teacher, and his lectures and books had a major influence on physics education and science education in general.

24A-1 Wave Properties

In dealing with phenomena such as reflection, refraction, interference, and diffraction, electromagnetic radiation is conveniently modeled as waves consisting of perpendicularly oscillating electric and magnetic fields, as shown in Figure 24-1a. The electric field for a single-frequency wave oscillates sinusoidally in space and time, as shown in Figure 24-1b. Here, the electric field is represented as a vector whose length is proportional to the field strength. The x-axis in this plot is either time as the radiation passes a fixed point in space or distance at a fixed time. Note that the direction in which the field oscillates is perpendicular to the direction in which the radiation propagates.

◀ *Now we know how the electrons and photons behave. But what can I call it? If I say they behave like particles I give the wrong impression; also if I say they behave like waves. They behave in their own inimitable way, which technically could be called a quantum mechanical way. They behave in a way that is like nothing that you have ever seen before.*—R. P. Feynman[2]

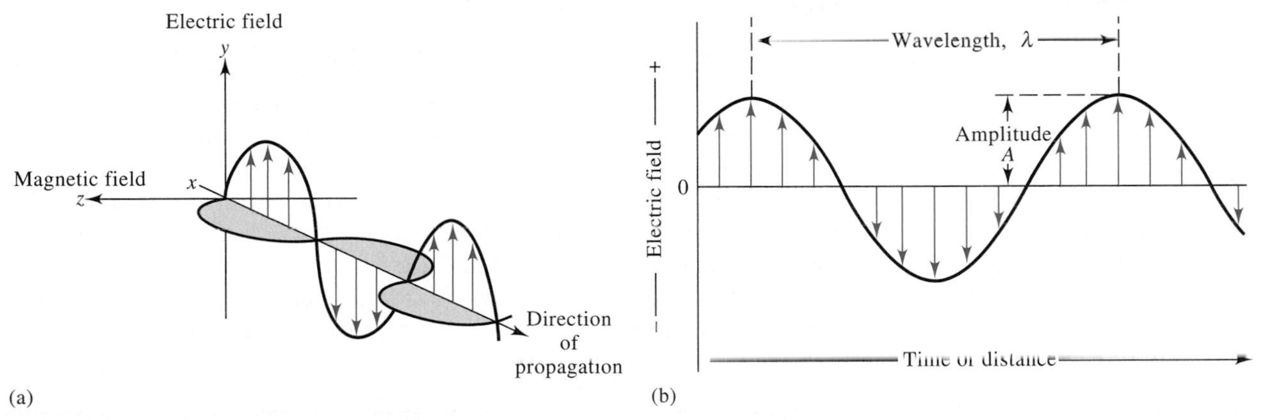

Figure 24-1 Wave nature of a beam of single-frequency electromagnetic radiation. In (a), a plane-polarized wave is shown propagating along the x-axis. The electric field oscillates in a plane perpendicular to the magnetic field. If the radiation were unpolarized, a component of the electric field would be seen in all planes. In (b), only the electric field oscillations are shown. The amplitude of the wave is the length of the electric field vector at the wave maximum, while the wavelength is the distance between successive maxima.

[2]R. P. Feynman, *The Character of Physical Law,* p. 122. New York: Random House, 1994.

The **amplitude** of an electromagnetic wave is a vector quantity that provides a measure of the electric or magnetic field strength at a maximum point in the wave.

The **period** of an electromagnetic wave is the time in seconds that it takes for successive maxima or minima to pass a point in space.

The **frequency** of an electromagnetic wave is the number of oscillations that occur in 1 second.

The unit of frequency is the **hertz** (Hz), which corresponds to one cycle per second. That is, 1 Hz = 1 s^{-1}. The frequency of a beam of electromagnetic radiation does not change as it passes through different media.

► Radiation velocity and wavelength both decrease as the radiation passes from a vacuum or from air to a denser medium. Frequency remains constant.

► Note that in Equation 24-1, v (distance/time) = ν (waves/time) × λ (distance/wave).

► To three significant figures, Equation 24-2 is equally applicable in air or a vacuum.

The **refractive index** η of a medium measures the extent of interaction between electromagnetic radiation and the medium through which it passes. It is defined by $\eta = c/v$. For example, the refractive index of water at room temperature is 1.33, which means that radiation passes through water at a rate of $c/1.33$ or 2.26×10^{10} cm s^{-1}. In other words, light travels 1.33 times slower in water than it does in a vacuum. The velocity and wavelength of radiation become proportionally smaller as the radiation passes from a vacuum or from air to a denser medium, while the frequency remains constant.

TABLE 24-1
Wavelength Units for Various Spectral Regions

Region	Unit	Definition
X-ray	Angstrom unit, Å	10^{-10} m
Ultraviolet/visible	Nanometer, nm	10^{-9} m
Infrared	Micrometer, μm	10^{-6} m

Wave Characteristics

In Figure 24-1b, the **amplitude** of the sine wave is shown, and the wavelength is defined. The time in seconds required for the passage of successive maxima or minima through a fixed point in space is called the **period,** p, of the radiation. The **frequency,** ν, is the number of oscillations of the electric field vector per unit time and is equal to $1/p$.

The frequency of a light wave, or any wave of electromagnetic radiation, is determined by the source that emits it and remains constant regardless of the medium traversed. In contrast, the **velocity,** v, of the wave front through a medium depends on both the medium and the frequency. The **wavelength,** λ, is the linear distance between successive maxima or minima of a wave, as shown in Figure 24-1b. Multiplication of the frequency (in waves per unit time) by the wavelength (in distance per wave) gives the velocity of the wave, in distance per unit time (cm s^{-1} or m s^{-1}), as shown in Equation 24-1. Note that both the velocity and the wavelength depend on the medium.

$$v = \nu\lambda \tag{24-1}$$

Table 24-1 gives the wavelength units for several spectral regions.

The Speed of Light

In a vacuum, light travels at its maximum velocity. This velocity, which is given the special symbol c, is 2.99792×10^8 m s^{-1}. The velocity of light in air is only about 0.03% less than its velocity in a vacuum. Thus, for a vacuum or for air, Equation 24-1 conveniently gives the velocity of light.

$$c = \nu\lambda = 3.00 \times 10^8 \text{ m s}^{-1} = 3.00 \times 10^{10} \text{ cm s}^{-1} \tag{24-2}$$

In a medium containing matter, light travels with a velocity less than c because of interaction between the electromagnetic field and electrons in the atoms or molecules of the medium. Since the frequency of the radiation is constant, the wavelength must decrease as the light passes from a vacuum to a medium containing matter (see Equation 24-2). This effect is illustrated in Figure 24-2 for a beam of visible radiation. Note that the effect can be quite large.

The **wavenumber** $\bar{\nu}$ is another way to describe electromagnetic radiation. It is defined as the number of waves per centimeter and is equal to $1/\lambda$. By definition, $\bar{\nu}$ has the units of cm^{-1}.

EXAMPLE 24-1

Calculate the wavenumber of a beam of infrared radiation with a wavelength of 5.00 μm.

$$\bar{\nu} = \frac{1}{5.00\ \mu\text{m} \times 10^{-4}\ \text{cm}/\mu\text{m}} = 2000\ \text{cm}^{-1}$$

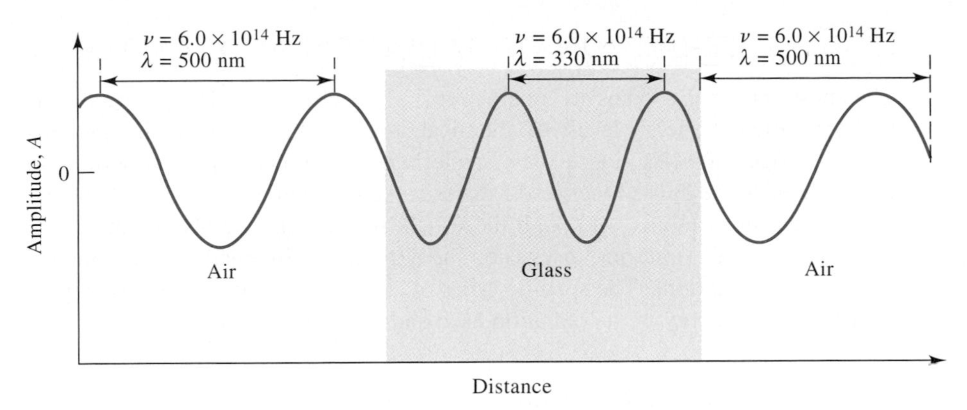

Figure 24-2 Change in wavelength as radiation passes from air into a dense glass and back to air. Note that the wavelength shortens by nearly 200 nm, or more than 30%, as it passes into glass; a reverse change occurs as the radiation again enters air.

Radiant Power and Intensity

The **radiant power** P in watts (W) is the energy of a beam that reaches a given area per unit time. The **intensity** is the radiant power–per-unit solid angle.[3] Both quantities are proportional to the square of the amplitude of the electric field (see Figure 24-1b). Although it is not strictly correct, "radiant power" and "intensity" are frequently used interchangeably.

The **wavenumber** $\bar{\nu}$ in cm^{-1} (Kayser) is most often used to describe radiation in the infrared region. The most useful part of the infrared spectrum for the detection and determination of organic species is 2.5 to 15 μm, which corresponds to a wavenumber range of 4000 to 667 cm^{-1}. The wavenumber of a beam of electromagnetic radiation is directly proportional to its energy and thus its frequency.

24A-2 The Particle Nature of Light: Photons

In many radiation/matter interactions, it is most useful to consider light as consisting of photons or quanta. We can relate the energy of a photon to its wavelength, frequency, and wavenumber by

$$E = h\nu = \frac{hc}{\lambda} = hc\bar{\nu} \tag{24-3}$$

A **photon** is a particle of electromagnetic radiation having zero mass and energy of $h\nu$.

where h is Planck's constant (6.63×10^{-34} J s). Note that the wavenumber and frequency, in contrast to the wavelength, are directly proportional to the photon energy. Wavelength is inversely proportional to energy. The radiant power of a beam of radiation is directly proportional to the number of photons per second.

Equation 24-3 gives the energy of radiation in SI units of **joules**, where one joule (J) is the work done by a force of one newton (N) acting over a distance of one meter.

EXAMPLE 24-2

Calculate the energy in joules of one photon of the radiation described in Example 24-1. Applying Equation 24-3, we write

$$E = hc\bar{\nu} = 6.63 \times 10^{-34}\,\text{J}\cdot\cancel{\text{s}} \times 3.00 \times 10^{10}\,\frac{\cancel{\text{cm}}}{\cancel{\text{s}}} \times 2000\,\cancel{\text{cm}^{-1}}$$

$$= 3.98 \times 10^{-20}\,\text{J}$$

◀ Both frequency and wavenumber are proportional to the energy of a photon.

◀ We sometimes speak of "a mole of photons," meaning 6.022×10^{23} packets of radiation of a given wavelength. The energy of a mole of photons with a wavelength of 5.00 μm is then 6.022×10^{23} photons/mol photons $\times 3.98 \times 10^{-20}$ J/photon = 24.0 kJ/mol photons.

[3]Solid angle is the three-dimensional spread at the vertex of a cone measured as the area intercepted by the cone on a unit sphere whose center is at the vertex. The angle is measured in stereradians (sr).

24B INTERACTION OF RADIATION AND MATTER

The most interesting types of interactions in spectroscopy involve transitions between different energy levels of chemical species. Other types of interactions, such as reflection, refraction, elastic scattering, interference, and diffraction, are often related to the bulk properties of materials rather than to energy levels of specific molecules or atoms. Although these bulk interactions are also of interest in spectroscopy, we limit our discussion here to those interactions that involve energy-level transitions. The specific types of interactions that we observe depend strongly on the energy of the radiation used and the mode of detection.

TABLE 24-2
Regions of the UV, Visible, and IR Spectrum

Region	Wavelength Range
UV	180–380 nm
Visible	380–780 nm
Near-IR	0.78–2.5 μm
Mid-IR	2.5–50 μm

► You can recall the order of the colors in the spectrum by the mnemonic **ROY G BIV**, which is short for **R**ed, **O**range, **Y**ellow, **G**reen, **B**lue, **I**ndigo, and **V**iolet.

The **visible region** of the spectrum extends from about 400 nm to 700 nm.

24B-1 The Electromagnetic Spectrum

The electromagnetic spectrum covers an enormous range of energies (frequencies) and thus wavelengths (Table 24-2). Useful frequencies vary from $>10^{19}$ Hz (γ-ray) to 10^3 Hz (radio waves). An X-ray photon ($\nu \approx 3 \times 10^{18}$ Hz, $\lambda \approx 10^{-10}$ m), for example, is approximately 10,000 times as energetic as a photon emitted by an ordinary light bulb ($\nu \approx 3 \times 10^{14}$ Hz, $\lambda \approx 10^{-6}$ m) and 10^{15} times as energetic as a radio-frequency photon ($\nu \approx 3 \times 10^3$ Hz, $\lambda \approx 10^5$ m).

The major divisions of the spectrum are shown in color in the inside front cover of this book. Note that the visible portion, to which our eyes respond, is only a minute portion of the entire spectrum. Such different types of radiation as gamma (γ) rays or radio waves differ from visible light only in the energy (frequency) of their photons.

Figure 24-3 shows the regions of the electromagnetic spectrum that are used for spectroscopic analyses. Also shown are the types of atomic and molecular transi-

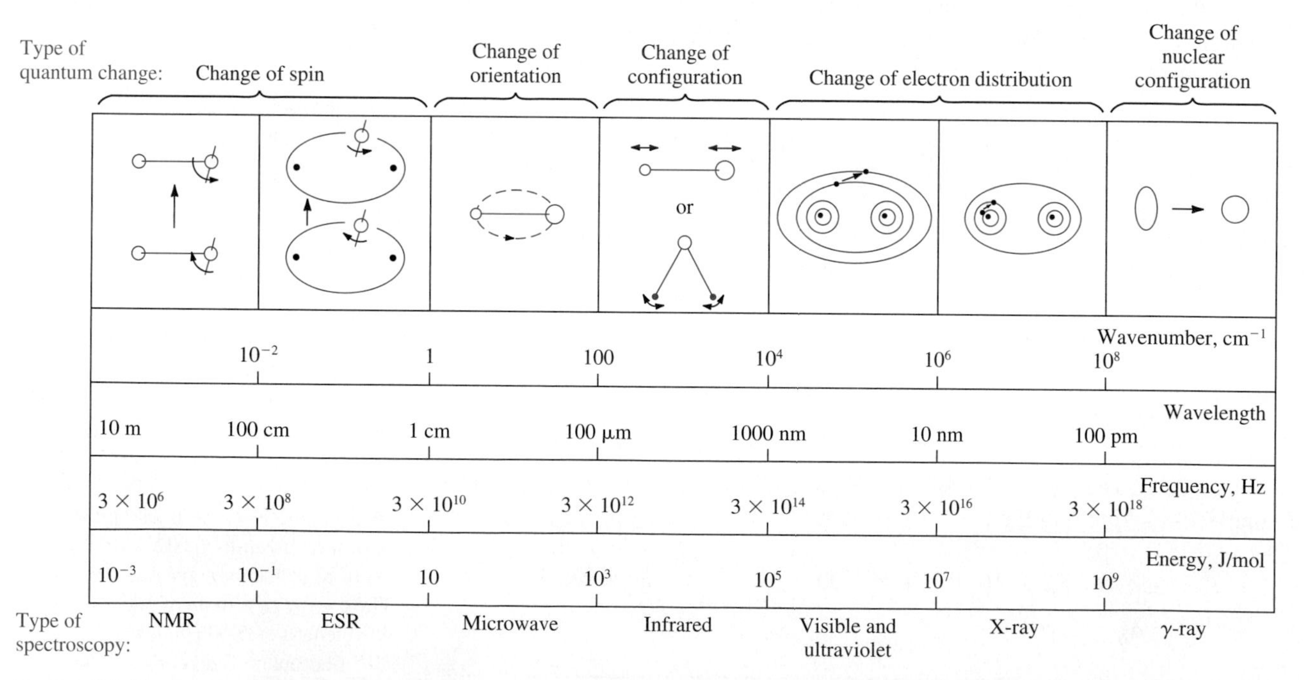

Figure 24-3 The regions of the electromagnetic spectrum. Interaction of an analyte with electromagnetic radiation can result in the types of changes shown. Note that changes in electron distributions occur in the UV/visible region. The wavenumber, wavelength, frequency, and energy are characteristics that describe electromagnetic radiation. (From C.N. Banwell, *Fundamentals of Molecular Spectroscopy,* 3rd ed., p. 7. New York; McGraw-Hill, 1983.)

tions that result from interactions of the radiation with a sample. Note that the low-energy radiation used in nuclear magnetic resonance (NMR) and electron spin resonance (ESR) spectroscopy causes subtle changes, such as changes in spin; the high-energy radiation used in γ-ray spectroscopy can produce much more dramatic effects, such as nuclear configuration changes.

Note that spectrochemical methods that use not only visible but also ultraviolet and infrared radiation are often called **optical methods** in spite of the fact that the human eye is sensitive to neither of the latter two types of radiation. This somewhat ambiguous terminology arises as a result of both the many common features of instruments for the three spectral regions and the similarities in the way in which we view the interactions of the three types of radiation with matter.

Optical methods are spectroscopic methods based on ultraviolet, visible, and infrared radiation.

24B-2 Spectroscopic Measurements

Spectroscopists use the interactions of radiation with matter to obtain information about a sample. Several of the chemical elements were discovered by spectroscopy (see Feature 24-1). The sample is usually stimulated in some way by applying energy in the form of heat, electrical energy, light, particles, or a chemical reaction. Prior to applying the stimulus, the analyte is predominantly in its lowest energy state, or **ground state.** The stimulus then causes some of the analyte species to undergo a transition to a higher energy or **excited state.** We acquire information about the analyte by measuring the electromagnetic radiation emitted as it returns to the ground state or by measuring the amount of electromagnetic radiation absorbed as a result of excitation.

A familiar example of **chemiluminescence** is the light emitted by a firefly. In the firefly reaction, an enzyme, luciferase, catalyzes the oxidative phosphorylation reaction of luciferin with adenosine triphosphate to produce oxyluciferin, carbon dioxide, adenosine monophosphate, and light. Chemiluminescence involving a biological or enzyme reaction is often termed **bioluminescence.** The popular light stick is another familiar example of chemiluminescence.

Figure 24-4 illustrates the processes involved in emission and chemiluminescence spectroscopy. Here, the analyte is stimulated by heat or electrical energy or by a chemical reaction. **Emission spectroscopy** usually involves methods in which the stimulus is heat or electrical energy, while **chemiluminescence spectroscopy** refers to excitation of the analyte by a chemical reaction. In both cases, measurement of the radiant power emitted as the analyte returns to the ground state can give information about its identity and concentration. The results of such a measurement are often expressed graphically by a **spectrum,** which is a plot of the emitted radiation as a function of frequency or wavelength.

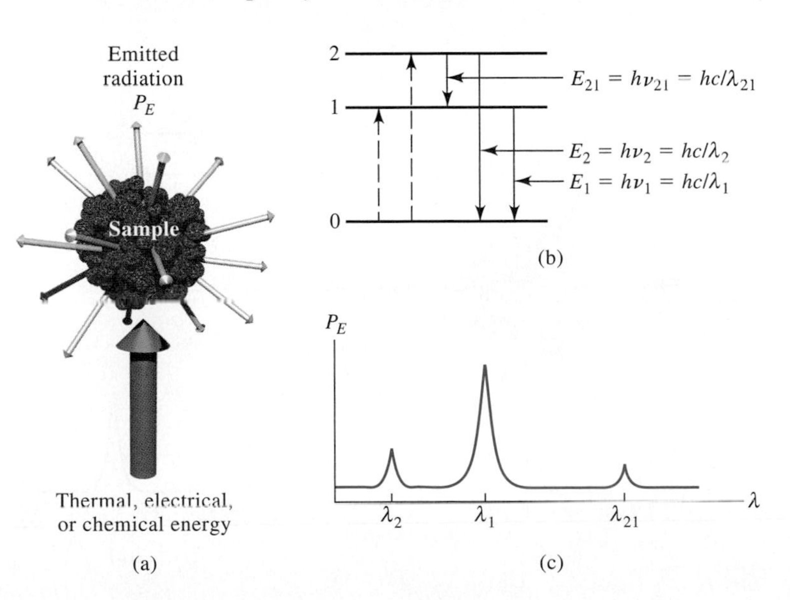

Figure 24-4 Emission or chemiluminescence processes. In (a), the sample is excited by the application of thermal, electrical, or chemical energy. These processes do not involve radiant energy and are hence called nonradiative processes. In the energy-level diagram (b), the dashed lines with upward-pointing arrows symbolize these nonradiative excitation processes, while the solid lines with downward pointing arrows indicate that the analyte loses its energy by emission of a photon. In (c), the resulting spectrum is shown as a measurement of the radiant power emitted P_E as a function of wavelength, λ.

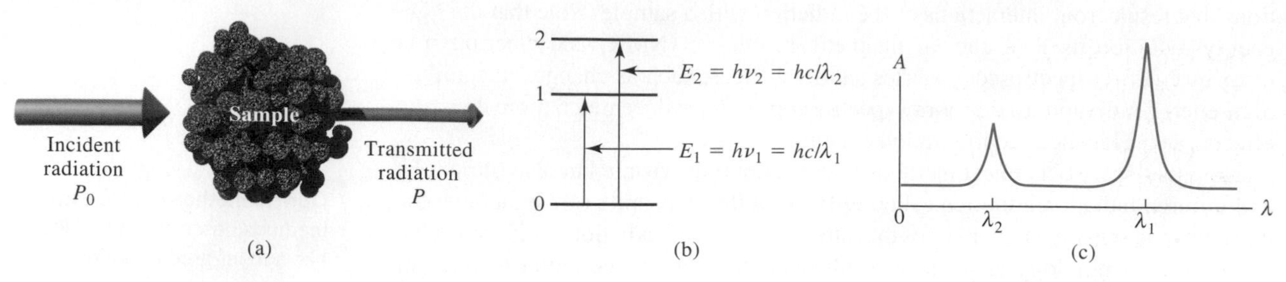

Figure 24-5 Absorption methods. Radiation of incident radiant power P_0 can be absorbed by the analyte, resulting in a transmitted beam of lower radiant power P. For absorption to occur, the energy of the incident beam must correspond to one of the energy differences shown in (b). The resulting absorption spectrum is shown in (c).

When the sample is stimulated by application of an external electromagnetic radiation source, several processes are possible. For example, the radiation can be scattered or reflected. What is important to us is that some of the incident radiation can be absorbed and thus promote some of the analyte species to an excited state, as shown in Figure 24-5. In **absorption spectroscopy,** we measure the amount of light absorbed as a function of wavelength. This can give both qualitative and quantitative information about the sample. In **photoluminescence spectroscopy** (Figure 24-6), the emission of photons is measured after absorption. The most important forms of photoluminescence for analytical purposes are **fluorescence** and **phosphorescence spectroscopy.**

We focus here on absorption spectroscopy in the UV/visible region of the spectrum because it is so widely used in chemistry, biology, forensic science, engineering, agriculture, clinical chemistry, and many other fields. Note that the processes shown in Figures 24-4 through 24-6 can occur in any region of the electromagnetic spectrum; the different energy levels can be nuclear levels, electronic levels, vibrational levels, or spin levels.

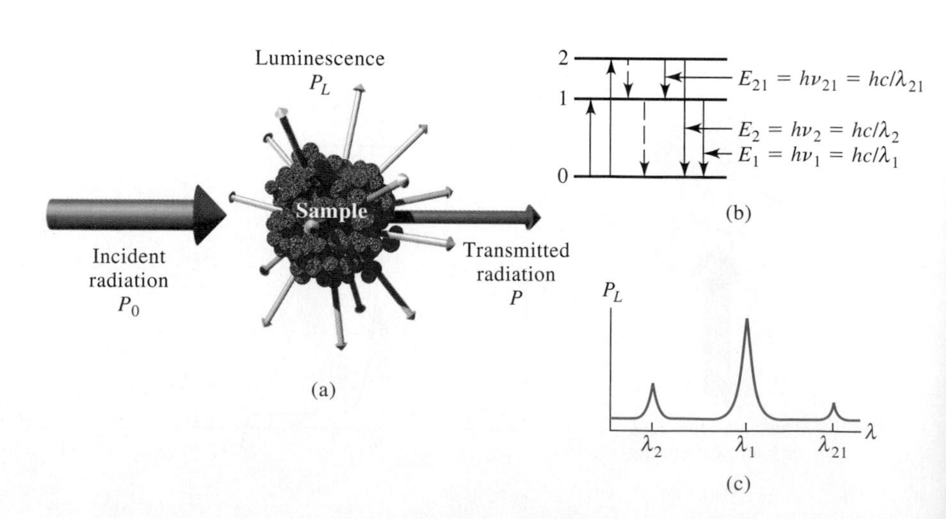

Figure 24-6 Photoluminescence methods (fluorescence and phosphorescence). Fluorescence and phosphorescence result from absorption of electromagnetic radiation and then dissipation of the energy by emission of radiation (a). In (b), the absorption can cause excitation of the analyte to state 1 or state 2. Once excited, the excess energy can be lost by emission of a photon (luminescence, shown as solid line) or by nonradiative processes (dashed lines). The emission occurs over all angles, and the wavelengths emitted (c) correspond to energy differences between levels. The major distinction between fluorescence and phosphorescence is the time scale of emission, with fluorescence being prompt and phosphorescence being delayed.

FEATURE 24-1

Spectroscopy and the Discovery of Elements

The modern era of spectroscopy began with the observation of the spectrum of the sun by Sir Isaac Newton in 1672. In Newton's experiment, rays from the sun passed through a small opening into a dark room, where they struck a prism and dispersed into the colors of the spectrum. The first description of spectral features beyond the simple observation of colors was attributed to Wollaston in 1802, who noticed dark lines on a photographic image of the solar spectrum. These lines, along with more than 500 others—which are shown in the solar spectrum of Figure 24F-1—were later described in detail by Fraunhofer. Based on his observations, the first of which was made in 1817, Fraunhofer assigned the prominent lines letters, starting with "A" at the red end of the spectrum.

It remained, however, for Gustav Kirchhoff and Robert Wilhelm Bunsen in 1859 and 1860 to explain the origin of the Fraunhofer lines. Bunsen had invented his famous burner (Figure 24F-2) a few years earlier, which made possible spectral observations of emission and absorption phenomena in a nearly transparent flame. Kirchhoff concluded that the Fraunhofer "D" lines were due to sodium in the sun's atmosphere and the "A" and "B" lines were due to potassium. We still call the emission lines of sodium the sodium "D" lines. These lines are responsible for the familiar yellow color seen in flames containing sodium or in sodium vapor lamps. The absence of lithium in the sun's spectrum led Kirchhoff to conclude that there was little lithium present in the sun. During these studies, Kirchhoff also developed his famous laws relating the absorption and emission of light from bodies and at interfaces. Together with Bunsen, Kirchhoff observed that different elements could impart different colors to flames and produce spectra exhibiting differently colored bands or lines. Kirchhoff and Bunsen are thus credited with discovering the use of spectroscopy for chemical analysis. The method was soon put to many practical uses, including the discovery of new elements. In 1860, the elements cesium and rubidium were discovered, followed in 1861 by thallium and in 1864 by indium. The age of spectroscopic analysis had clearly begun.

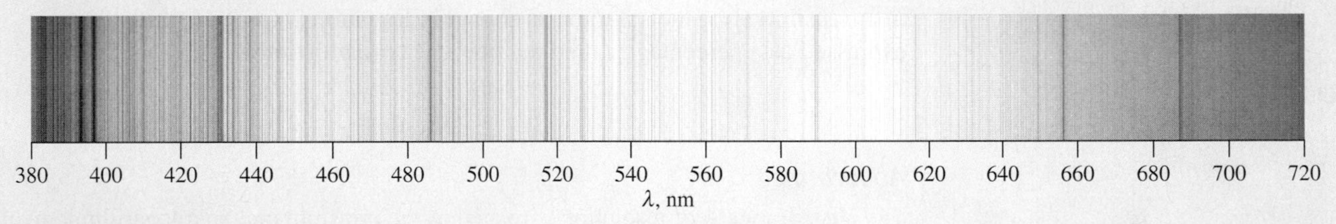

Figure 24F-1 The solar spectrum. The dark vertical lines are the Fraunhofer lines. See color plate 18 for a full-color version of the spectrum. Images created by Dr. Donald Mickey, University of Hawaii Institute for Astronomy, from National Solar Observatory spectral data. NSOS/Kitt Peak FTS data used here were produced by NSF/NOAO.

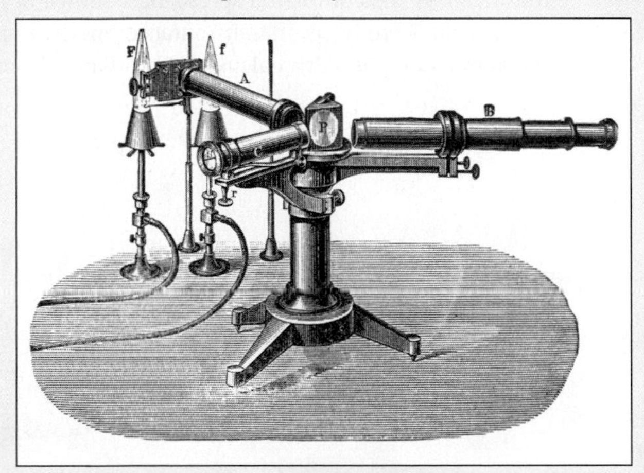

Figure 24F-2 Bunsen burner of the type used in early spectroscopic studies with a prism spectroscope of the type used by Kirchhoff. (From H. Kayser, *Handbuch der Spectroscopie.* Stuttgart, Germany: S. Hirzel Verlag GmbH & Co., 1900.)

24C RADIATION ABSORPTION

Every molecular species is capable of absorbing its own characteristic frequencies of electromagnetic radiation, as described in Figure 24-5. This process transfers energy to the molecule and results in a decrease in the intensity of the incident electromagnetic radiation. Absorption of the radiation thus **attenuates** the beam in accordance with the absorption law described later.

In spectroscopy, to **attenuate** means to decrease the energy per unit area of a beam of radiation. In terms of the photon model, to attenuate means to decrease the number of photons per second in the beam.

24C-1 The Absorption Process

The absorption law, also known as the **Beer-Lambert law** or just **Beer's law,** tells us quantitatively how the amount of attenuation depends on the concentration of the absorbing molecules and the path length over which absorption occurs. As light traverses a medium containing an absorbing analyte, decreases in intensity occur as the analyte becomes excited. For an analyte solution of a given concentration, the longer the length of the medium through which the light passes (path length of light), the more absorbers are in the path, and the greater the attenuation. Also, for a given path length of the light, the higher the concentration of absorbers, the stronger the attenuation.

Figure 24-7 depicts the attenuation of a parallel beam of **monochromatic radiation** as it passes through an absorbing solution of thickness b centimeters and concentration c moles per liter. Because of interactions between the photons and absorbing particles (recall Figure 24-5), the radiant power of the beam decreases from P_0 to P. The **transmittance** T of the solution is the fraction of incident radiation transmitted by the solution, as shown in Equation 24-4. Transmittance is often expressed as a percentage called the **percent transmittance.**

The term **monochromatic radiation** refers to radiation of a single color; that is, a single wavelength or frequency. In practice, it is virtually impossible to produce a single color of light. We discuss the practical problems associated with producing monochromatic radiation in Chapter 25.

$$T = P/P_0 \tag{24-4}$$

► Percent transmittance $= \%T = \frac{P}{P_0} \times 100\%$.

Absorbance

The **absorbance** A of a solution is related to the transmittance in a logarithmic manner, as shown in Equation 24-5. Notice that as the absorbance of a solution increases, the transmittance decreases. The relationship between transmittance and absorbance is illustrated by the conversion spreadsheet shown in Figure 24-8. The scales on earlier instruments were linear in transmittance; modern instruments have linear absorbance scales or a computer that calculates absorbance from measured quantities.

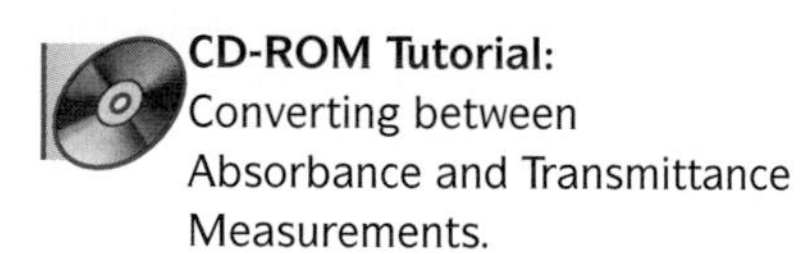

CD-ROM Tutorial:
Converting between Absorbance and Transmittance Measurements.

$$A = -\log T = \log \frac{P_0}{P} \tag{24-5}$$

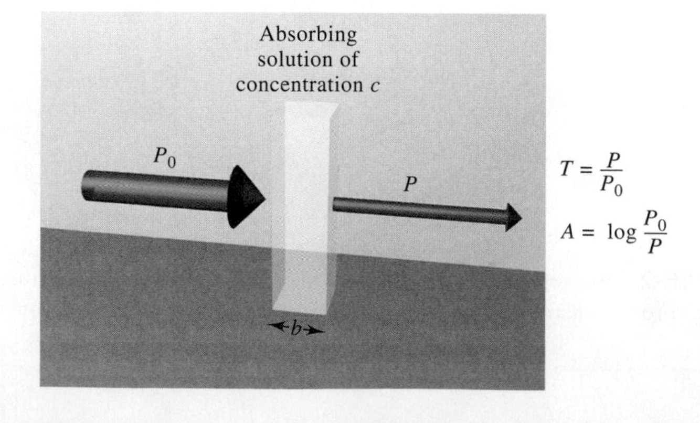

Figure 24-7 Attenuation of a beam of radiation by an absorbing solution. The larger arrow on the incident beam signifies a higher radiant power than is transmitted by the solution. The path length of the absorbing solution is b, and the concentration is c.

	A	B	C	D	E
1	**Calculation of absorbance from transmittance**				
2	T	%T	A	A	
3	0.001	0.1	3.000	3.000	
4	0.010	1.0	2.000	2.000	
5	0.050	5.0	1.301	1.301	
6	0.075	7.5	1.125	1.125	
7	0.100	10.0	1.000	1.000	
8	0.200	20.0	0.699	0.699	
9	0.300	30.0	0.523	0.523	
10	0.400	40.0	0.398	0.398	
11	0.500	50.0	0.301	0.301	
12	0.600	60.0	0.222	0.222	
13	0.700	70.0	0.155	0.155	
14	0.800	80.0	0.097	0.097	
15	0.900	90.0	0.046	0.046	
16	1.000	100.0	0.000	0.000	
17					
18	**Spreadsheet Documentation**				
19	Cell B3=100*A3				
20	Cell C3=-LOG10(A3)				
21	Cell D3=2-LOG10(B3)				

Figure 24-8 Conversion spreadsheet relating transmittance T, percent transmittance $\%T$, and absorbance A. The transmittance data to be converted are entered into cells A3 to A16. The percent transmittance is calculated in cell B3 by the formula shown in the documentation section, cell A19. This formula is copied into cells B4 through B16. The absorbance is calculated from $-\log T$ in cells C3 through C16 and from $2 - \log \%T$ in cells D3 through D16. The formulas for the first cell in the C and D columns are shown in cells A20 and A21.

Measuring Transmittance and Absorbance

Ordinarily, transmittance and absorbance, as defined by Equations 24-4 and 24-5 and depicted in Figure 24-7, cannot be measured as shown because the solution to be studied must be held in some sort of container (cell or cuvette). Reflection and scattering losses can occur at the cell walls, as shown in Figure 24-9. These losses can be substantial. For example, about 8.5% of a beam of yellow light is lost by reflection when it passes through a glass cell. Light can also be scattered in all directions from the surface of large molecules or particles (such as dust) in the solvent, and this scattering can cause further attenuation of the beam as it passes through the solution.

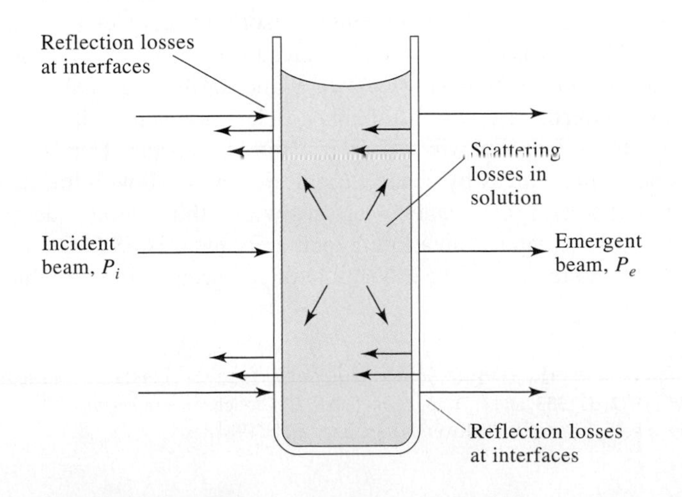

Figure 24-9 Reflection and scattering losses with a solution contained in a typical glass cell. Losses by reflection can occur at all the boundaries that separate the different materials. In this example, the light passes through the following boundaries, called interfaces: air-glass, glass-solution, solution-glass, and glass-air.

To compensate for these effects, the power of the beam transmitted through a cell containing the analyte solution is compared with one that traverses either an identical cell containing only the solvent or a reagent blank. An experimental absorbance that closely approximates the true absorbance for the solution is thus obtained; that is,

$$A = \log \frac{P_0}{P} \approx \log \frac{P_{solvent}}{P_{solution}} \tag{24-6}$$

The terms P_0 and P will henceforth refer to the power of a beam that has passed through cells containing the blank (solvent) and the analyte, respectively.

Beer's Law

According to Beer's law, absorbance is directly proportional to the concentration of the absorbing species c and to the path length b of the absorbing medium, as expressed by Equation 24-7.

$$A = \log (P_0/P) = abc \tag{24-7}$$

► The molar absorptivity of a species at an absorption maximum is characteristic of that species. Peak molar absorptivities for many organic compounds range from 10 or less to 10,000 or more. Some transition metal complexes have molar absorptivities of 10,000 to 50,000. High molar absorptivities are desirable for quantitative analysis because they lead to high analytical sensitivity.

Here, a is a proportionality constant called the **absorptivity.** Because absorbance is a unitless quantity, the absorptivity must have units that cancel the units of b and c. If, for example, c has the units of g L^{-1} and b has the units of cm, absorptivity has the units of L g^{-1} cm^{-1}.

When we express the concentration in Equation 24-7 in moles per liter and b in centimeters, the proportionality constant is called the **molar absorptivity** and is given the special symbol, ε. Thus,

$$A = \varepsilon bc \tag{24-8}$$

where ε has the units of L mol^{-1} cm^{-1}.

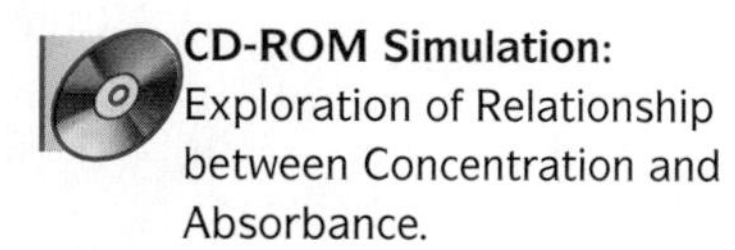

CD-ROM Simulation: Exploration of Relationship between Concentration and Absorbance.

FEATURE 24-2

Deriving Beer's Law[4]

To derive Beer's law, we consider the block of absorbing matter (solid, liquid, or gas) shown in Figure 24F-3. A beam of parallel monochromatic radiation with power P_0 strikes the block perpendicular to a surface; after passing through a length b of the material, which contains n absorbing particles (atoms, ions, or molecules), its power is decreased to P as a result of absorption. Consider now a cross-section of the block having an area S and an infinitesimal thickness dx. Within this section there are dn absorbing particles; associated with each particle, we can imagine a surface at which photon capture will occur. That is, if a photon reaches one of these areas by chance, absorption will follow immediately. The total projected area of these capture surfaces within the section is designated as dS; the ratio of the capture area to the total area, then, is dS/S. On a statistical average, this ratio represents the probability of capturing photons within the sec-

[4]For derivations of Beer's law, see F. C. Strong, *Anal. Chem,* **1952,** *24,* 338; D. J. Swinehart, *J. Chem. Ed.,* **1972,** *32,* 333; and J. D. Ingle, Jr., and S. R. Crouch, *Spectrochemical Analysis,* pp. 34–35. Upper Saddle River, NJ: Prentice-Hall, 1988.

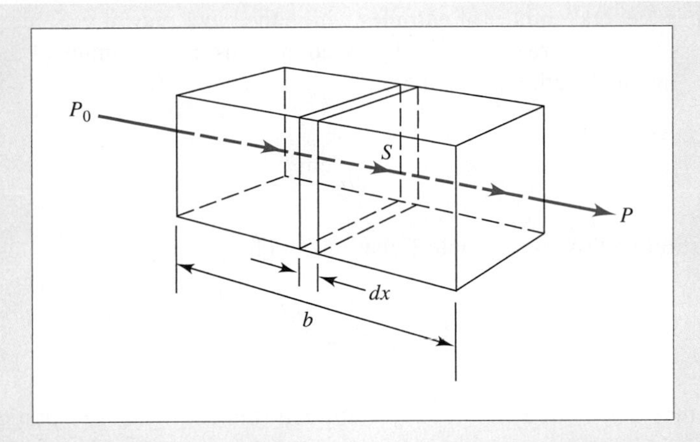

Figure 24F-3 Attenuation with initial power P_0 by a solution containing c mol/L of absorbing solute and a path length of b cm ($P < P_0$).

tion. The power of the beam entering the section, P_x, is proportional to the number of photons per square centimeter per second, and dP_x represents the quantity removed per second within the section; the fraction absorbed is then $-dP_x/P_x$, and this ratio also equals the average probability of capture. The term is given a minus sign to indicate that P undergoes a decrease. Thus,

$$-\frac{dP_x}{P_x} = \frac{dS}{S} \tag{24-9}$$

Recall, that dS is the sum of the capture areas for particles within the section; it must therefore be proportional to the number of particles, or

$$dS = adn \tag{24-10}$$

where dn is the number of particles and a is a proportionality constant, which is called the *capture cross-section.* By combining Equations 24-9 and 24-10 and integrating over the interval between 0 and n, we obtain

$$-\int_{P_0}^{P} \frac{dP_x}{P_x} = \int_0^n \frac{adn}{S}$$

which, when integrated, gives

$$-\ln \frac{P}{P_0} = \frac{an}{S}$$

We then convert to base 10 logarithms, invert the fraction to change the sign, and obtain

$$\log \frac{P_0}{P} = \frac{an}{2.303\, S} \tag{24-11}$$

(continued)

where n is the total number of particles within the block shown in Figure 24F-3. The cross-sectional area S can be expressed in terms of the volume of the block V in cm^3 and its length b in cm. Thus,

$$S = \frac{V}{b}\ \text{cm}^2$$

By substituting this quantity into Equation 24-11, we find

$$\log \frac{P_0}{P} = \frac{anb}{2.303\ V} \tag{24-12}$$

Notice that n/V has the units of concentration (that is, number of particles per cubic centimeter); we can easily convert n/V to moles per liter. Thus, the number of moles is given by

$$\text{number mol} = \frac{n \text{ particles}}{6.022 \times 10^{23} \text{ particles/mol}}$$

and c in mol/L is given by

$$c = \frac{n}{6.022 \times 10^{23}}\ \text{mol} \times \frac{1000\ \cancel{\text{cm}^3}/\text{L}}{V\ \cancel{\text{cm}^3}}$$

$$= \frac{1000\ n}{6.022 \times 10^{23}\ V}\ \text{mol/L}$$

By combining this relationship with Equation 24-12, we have

$$\log \frac{P_0}{P} = \frac{6.022 \times 10^{23}\ abc}{2.303 \times 1000}$$

Finally, the constants in this equation can be collected into a single term ε to give

$$\log \frac{P_0}{P} = \varepsilon bc = A \tag{24-13}$$

which is Beer's law.

Terms Used in Absorption Spectrometry

In addition to the terms we have introduced to describe absorption of radiant energy, you may encounter other terms in the literature or with older instruments. The terms, symbols, and definitions given in Table 24-3 are recommended by the American Society for Testing Materials as well as by the American Chemical Society. The third column contains the older names and symbols. Because a standard nomenclature is highly desirable to avoid ambiguities, we urge you to learn and use the recommended terms and symbols and to avoid the older terms.

TABLE 24-3

Important Terms and Symbols Used in Absorption Measurements

Term and Symbol*	Definition	Alternative Name and Symbol
Incident radiant power, P_0	Radiant power in watts incident on sample	Incident intensity, I_0
Transmitted radiant power, P	Radiant power transmitted by sample	Transmitted intensity, I
Absorbance, A	$\log(P_0/P)$	Optical density, D; extinction, E
Transmittance, T	P/P_0	Transmission, T
Path length of sample, b	Length over which attenuation occurs	l, d
Absorptivity,† a	$A/(bc)$	Extinction coefficient, k
Molar absorptivity,‡ ε	$A/(bc)$	Molar extinction coefficient

*Terminology recommended by the American Chemical Society (*Anal. Chem.*, **1990,** *62,* 91).
†c may be expressed in g L^{-1} or in other specified concentration units; b may be expressed in cm or other units of length.
‡c is expressed in mol L^{-1}; b is expressed in cm.

Using Beer's Law

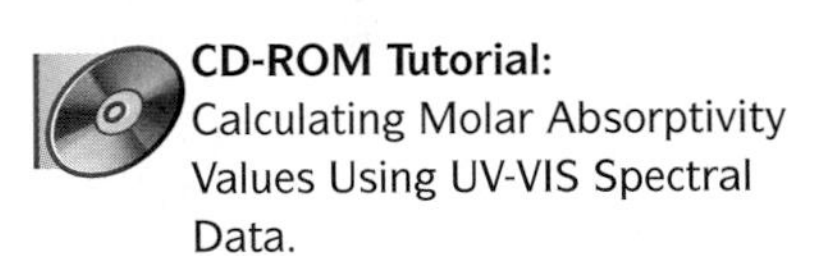
CD-ROM Tutorial: Calculating Molar Absorptivity Values Using UV-VIS Spectral Data.

Beer's law, as expressed in Equations 24-6 and 24-8, can be used in several ways. We can calculate molar absorptivities of species if the concentration is known, as shown in Example 24-3. We can use the measured value of absorbance to obtain concentration if absorptivity and path length are known. Absorptivities, however, are functions of such variables as solvent, solution composition, and temperature. Because of variations in absorptivity with conditions, it is never wise to depend on literature values for quantitative work. Hence, a standard solution of the analyte in the same solvent and at a similar temperature is used to obtain the absorptivity at the time of the analysis. Most often, we use a series of standard solutions of the analyte to construct a calibration curve, or working curve, of A versus c (see Chapter 26, Figure 23-6) or to obtain a linear regression equation (see Chapter 8). It may also be necessary to duplicate closely the overall composition of the analyte solution to compensate for matrix effects. Alternatively, the method of standard additions (see Sections 8C-3 and 26A-4) is used for the same purpose.

EXAMPLE 24-3

A 7.25×10^{-5} M solution of potassium permanganate has a transmittance of 44.1% when measured in a 2.10-cm cell at a wavelength of 525 nm. Calculate (a) the absorbance of this solution; (b) the molar absorptivity of $KMnO_4$.

(a) $A = -\log T = -\log 0.441 = -(-0.3554) = 0.355$
(b) From Equation 24-8,

$$\varepsilon = A/bc = 0.3554/(2.10 \text{ cm} \times 7.25 \times 10^{-5} \text{mol L}^{-1})$$
$$= 2.33 \times 10^{3} \text{ L mol}^{-1} \text{ cm}^{-1}$$

Spreadsheet Summary In the first exercise in Chapter 12 of *Applications of Microsoft® Excel in Analytical Chemistry,* a spreadsheet is developed to calculate the molar absorptivity of permanganate ion. A plot of absorbance versus permanganate concentration is constructed, and least-squares analysis of the linear plot is carried out. The data are analyzed statistically to determine the uncertainty of the molar absorptivity. In addition, other spreadsheets are presented for calibration in quantitative spectrophotometric experiments and for calculation of concentrations of unknown solutions.

Applying Beer's Law to Mixtures

Beer's law also applies to solutions containing more than one kind of absorbing substance. Provided that there is no interaction among the various species, the total absorbance for a multicomponent system at a single wavelength is the sum of the individual absorbances. In other words,

► Absorbances are additive if the absorbing species do not interact.

$$A_{\text{total}} = A_1 + A_2 + \cdots + A_n = \varepsilon_1 bc_1 + \varepsilon_2 bc_2 + \cdots + \varepsilon_n bc_n \quad (24\text{-}14)$$

where the subscripts refer to absorbing components 1, 2, . . . , *n*.

24C-2 Absorption Spectra

► A bit of Latin. One plot of absorbance versus wavelength is called a spectr**um;** two or more plots are called spectr**a.**

An **absorption spectrum** is a plot of absorbance versus wavelength, as illustrated in Figure 24-10. Absorbance could also be plotted against wavenumber or frequency. Most modern scanning spectrophotometers produce such an absorption spectrum directly. Older instruments sometimes displayed transmittance and produced plots of T or $\%T$ versus wavelength. Occasionally, plots with log A as the ordinate are used. The logarithmic axis leads to a loss of spectral detail but is convenient for comparing solutions of widely different concentrations. A plot of molar absorptivity ε as a function of wavelength is independent of concentration. This

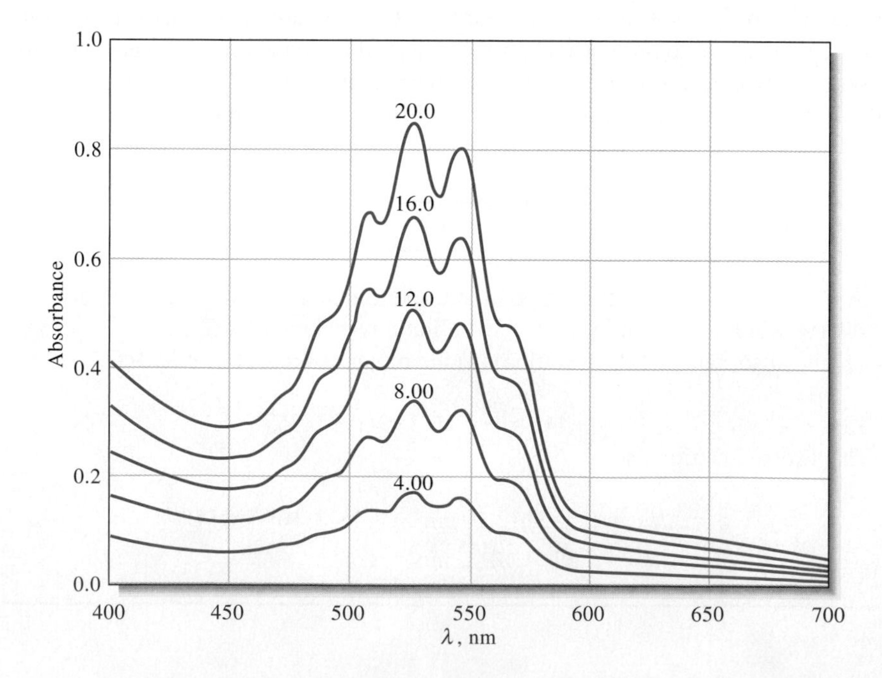

Figure 24-10 Typical absorption spectra of potassium permanganate at five different concentrations. The numbers adjacent to the curves indicate the concentration of manganese in ppm. The absorbing species is permanganate ion, MnO_4^-; the cell path length b is 1.00 cm. A plot of absorbance at the peak wavelength at 525 nm versus concentration of permanganate is linear, and thus the absorber obeys Beer's law.

FEATURE 24-3

Why Is a Red Solution Red?

A solution such as $Fe(SCN)^{2+}$ is red not because the complex adds red radiation to the solvent, but because it absorbs green from the incoming white radiation and transmits the red component unaltered (Figure 24F-4). Thus, in a colorimetric determination of iron based on its thiocyanate complex, the maximum change in absorbance with concentration occurs with green radiation; the absorbance change with red radiation is negligible. In general, the radiation used for a colorimetric analysis should be the complementary color of the analyte solution. The following table shows this relationship for various parts of the visible spectrum.

The Visible Spectrum

Wavelength Region Absorbed, nm	Color of Light Absorbed	Complementary Color Transmitted
400–435	Violet	Yellow-green
435–480	Blue	Yellow
480–490	Blue-green	Orange
490–500	Green-blue	Red
500–560	Green	Purple
560–580	Yellow-green	Violet
580–595	Yellow	Blue
595–650	Orange	Blue-green
650–750	Red	Green-blue

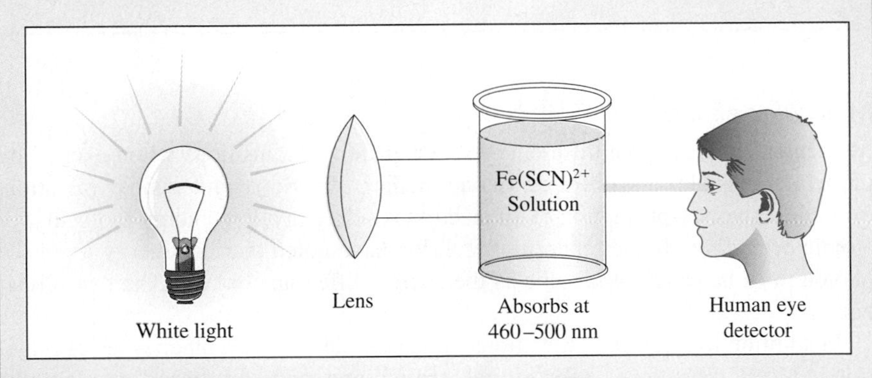

Figure 24F-4 Color of a solution. White light from a lamp or the sun strikes the solution of $Fe(SCN)^{2+}$. The fairly broad absorption spectrum shows a maximum absorbance in the 460- to 500-nm range. The complementary red color is transmitted.

type of spectral plot is characteristic for a given molecule and is sometimes used to aid in identifying or confirming the identity of a particular species. The color of a solution is related to its absorption spectrum (see Feature 24-3).

Atomic Absorption

When a beam of polychromatic ultraviolet or visible radiation passes through a medium containing gaseous atoms, only a few frequencies are attenuated by absorption. When recorded on a very high resolution spectrometer, the spectrum consists of a number of very narrow absorption lines.

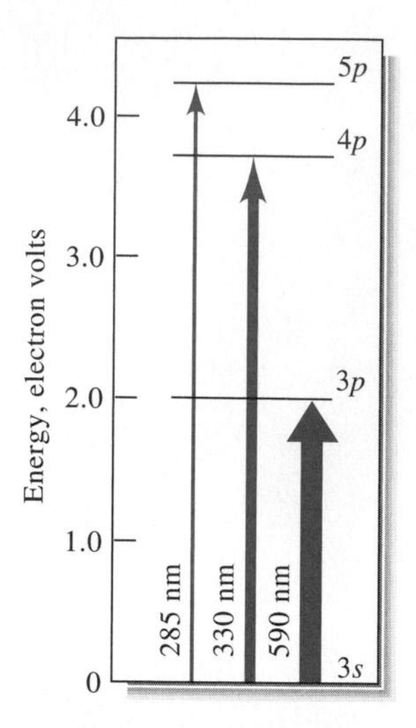

Figure 24-11 Partial energy-level diagram for sodium, showing the transitions resulting from absorption at 590, 330, and 285 nm.

Figure 24-11 is a partial energy-level diagram for sodium that shows the major atomic absorption transitions. The transitions, shown as blue arrows between levels, involve excitation of the single outer electron of sodium from its room-temperature or ground-state 3*s* orbital to the 3*p*, 4*p*, and 5*p* orbitals. These excitations are brought on by absorption of photons of radiation whose energies exactly match the differences in energies between the excited states and the 3*s* ground state. Transitions between two different orbitals are termed **electronic transitions.** Atomic absorption spectra are not ordinarily recorded because of instrumental difficulties. Instead, atomic absorption is measured at a single wavelength using a very narrow, nearly monochromatic source (see Section 28D).

EXAMPLE 24-4

The energy difference between the 3*p* and the 3*s* orbitals in Figure 24-11b is 2.107 eV. Calculate the wavelength of radiation that would be absorbed in exciting the 3*s* electron to the 3*p* state (1 eV = 1.60×10^{-19} J). Rearranging Equation 24-3 gives

$$\lambda = \frac{hc}{E}$$

$$= \frac{6.63 \times 10^{-34}\ \text{J}\ \cancel{\text{s}} \times 3.00 \times 10^{10}\ \cancel{\text{cm}}/\cancel{\text{s}} \times 10^{7}\ \text{nm}/\cancel{\text{cm}}}{2.107\ \cancel{\text{eV}} \times 1.60 \times 10^{-19}\ \text{J}/\cancel{\text{eV}}}$$

$$= 590\ \text{nm}$$

The **electron volt** (eV) is a unit of energy. When an electron with charge $q = 1.60 \times 10^{-19}$ coulombs is moved through a potential difference of 1 volt = 1 joule/coulomb, the energy expended (or released) is then equal to $E = qV = (1.60 \times 10^{-19}\ \text{coulombs})(1\ \text{joule/coulomb}) = 1.60 \times 10^{-19}\ \text{joule} = 1\ \text{eV}$.

$$1\ \text{eV} = 1.60 \times 10^{-19}\ \text{J}$$
$$= 3.83 \times 10^{-20}\ \text{calories}$$
$$= 1.58 \times 10^{-21}\ \text{L atm}$$

An **electronic transition** involves transfer of an electron from one orbital to another. Both atoms (atomic orbitals) and molecules (molecular orbitals) can undergo this type of transition.

► Vibrational and rotational transitions occur with polyatomic species because only this type of species has vibrational and rotational states with different energies.

► The ground state of an atom or a molecular species is the minimum energy state of the species. At room temperature, most species are in their ground state.

Molecular Absorption

Molecules undergo three different types of quantized transitions when excited by ultraviolet, visible, and infrared radiation. For ultraviolet and visible radiation, excitation involves promotion of an electron residing in a low-energy molecular or atomic orbital to a higher energy orbital. We have noted that the energy $h\nu$ of the photon must be exactly the same as the energy difference between the two orbital energies.

In addition to electronic transitions, molecules exhibit two other types of radiation-induced transitions: **vibrational transitions** and **rotational transitions.** Vibrational transitions occur because a molecule has a multitude of quantized energy levels (or vibrational states) associated with the bonds that hold the molecule together.

Figure 24-12 is a partial energy-level diagram that depicts some of the processes that occur when a polyatomic species absorbs infrared, visible, and ultraviolet radiation. The energies E_1 and E_2, two of the several electronically excited states of a molecule, are shown relative to the energy of the ground state E_0. In addition, the relative energies of a few of the many vibrational states associated with each electronic state are indicated by the lighter horizontal lines.

You can get an idea of the nature of vibrational states by picturing a bond in a molecule as a vibrating spring with atoms attached to both ends. In Figure 24-13a, two types of stretching vibrations are shown. With each vibration, atoms first approach and then move away from one another. The potential energy of such a system at any instant depends on the extent to which the spring is stretched or compressed. For an ordinary spring, the energy of the system varies continuously and reaches a maximum when the spring is fully stretched or fully compressed. In contrast, the energy of a spring system of atomic dimensions can assume only certain discrete energies called vibrational energy levels.

Figure 24-13b shows four other types of molecular vibrations. The energies associated with these vibrational states usually differ from one another and from the energies associated with stretching vibrations. Some of the vibrational energy levels associated with each of the electronic states of a molecule are depicted by the lines labeled 1, 2, 3, and 4 in Figure 24-12. (The lowest vibrational levels are labeled 0.) Note that the differences in energy among the vibrational states are significantly smaller than among energy levels of the electronic states (typically, an order of magnitude smaller). Although they are not shown, a molecule has a host of quantized rotational states that are associated with the rotational motion of a molecule around its center of gravity. These rotational energy states are superimposed on each of the vibrational states shown in the energy diagram. The energy differences among these states are smaller than those among vibrational states by an order of magnitude. The total energy E associated with a molecule is then given by

$$E = E_{\text{electronic}} + E_{\text{vibrational}} + E_{\text{rotational}} \qquad (24\text{-}15)$$

where $E_{\text{electronic}}$ is the energy associated with the electrons in the various outer orbitals of the molecule, $E_{\text{vibrational}}$ is the energy of the molecule as a whole due to interatomic vibrations, and $E_{\text{rotational}}$ accounts for the energy associated with rotation of the molecule about its center of gravity.

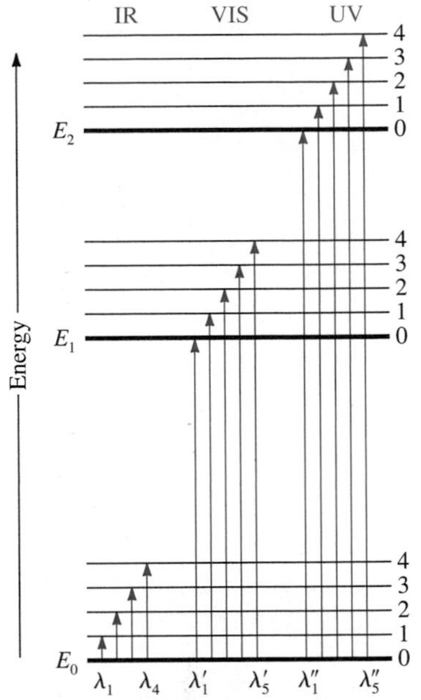

Figure 24-12 Energy-level diagram showing some of the energy changes that occur during absorption of infrared (IR), visible (VIS), and ultraviolet (UV) radiation by a molecular species. Note that with some molecules, a transition from E_0 to E_1 may require UV radiation instead of visible radiation. With other molecules, the transition from E_0 to E_2 may occur with visible radiation instead of UV radiation.

Infrared Absorption Infrared radiation generally is not sufficiently energetic to cause electronic transitions, but it can induce transitions in the vibrational and rotational states associated with the ground electronic state of the molecule. Four of

◀ Infrared radiation is not sufficiently energetic to cause electronic transitions.

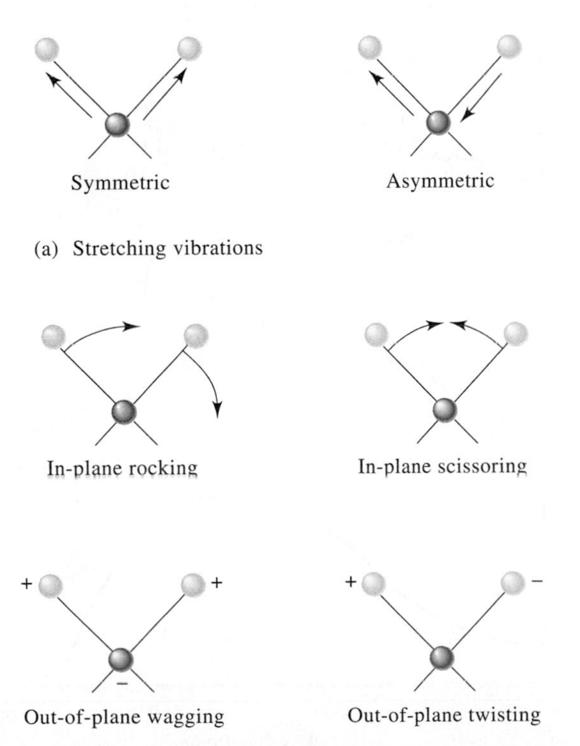

Figure 24-13 Types of molecular vibrations. The plus sign indicates motion from the page toward the reader; the minus sign indicates motion in the opposite direction.

these transitions are depicted in the lower left part of Figure 24-12 (λ_1 to λ_4). For absorption to occur, the source has to emit radiation of frequencies corresponding exactly to the energies indicated by the lengths of the four arrows.

Absorption of Ultraviolet and Visible Radiation The center arrows in Figure 24-12 suggest that the molecules under consideration absorb visible radiation of five wavelengths, thereby promoting electrons to the five vibrational levels of the excited electronic level E_1. Ultraviolet photons that are more energetic are required to produce the absorption indicated by the five arrows to the right.

As suggested by Figure 24-12, molecular absorption in the ultraviolet and visible regions consists of **absorption bands** made up of closely spaced lines. A real molecule has many more energy levels than are shown here; thus, the typical absorption band consists of a very large number of lines. In a solution, the absorbing species are surrounded by solvent, and the band nature of molecular absorption often becomes blurred because collisions tend to spread the energies of the quantum states, giving smooth and continuous absorption peaks.

Figure 21-14 shows visible spectra for 1,2,4,5-tetrazine that were obtained under three different conditions: gas phase, liquid phase, and aqueous solution. Notice that in the gas phase, the individual tetrazine molecules are sufficiently separated from one another to vibrate and rotate freely, so many individual absorption peaks resulting from transitions among the various vibrational and rotational states appear in the spectrum. In the liquid state and in solution, however, tetrazine molecules are unable to rotate freely, so we see no fine structure in the spectrum. Furthermore, because frequent collisions and interactions between tetrazine and water molecules cause the vibrational levels to be modified energetically in an irregular way, the spectrum appears as a single broad peak. The trends shown in the spectra

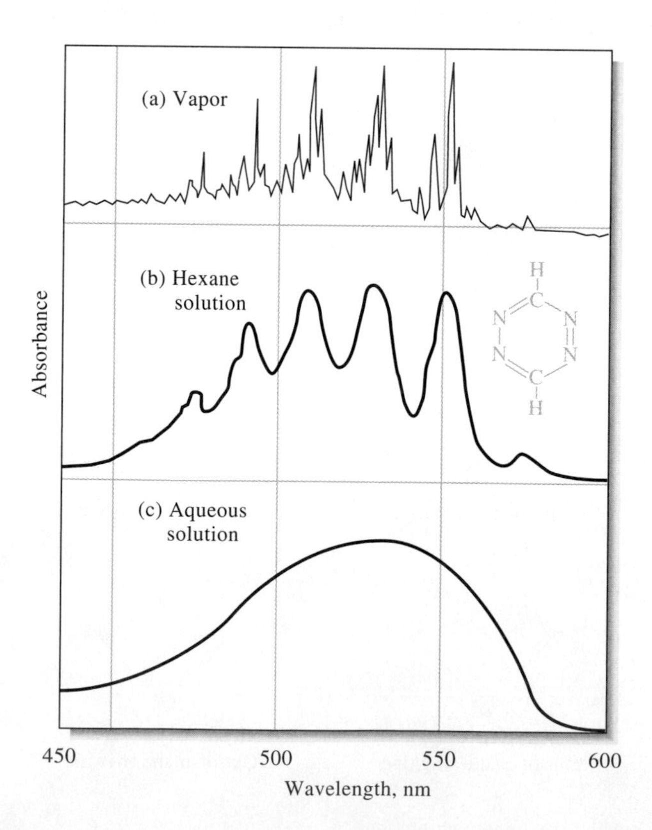

Figure 24-14 Typical ultraviolet absorption spectra. The compound is 1,2,4,5-tetrazine. In (a), the spectrum is shown in the gas phase, where many lines due to electronic, vibrational, and rotational transitions are seen. In a nonpolar solvent (b), the electronic transitions can be observed, but the vibrational and rotational structure has been lost. In a polar solvent (c), the strong intermolecular forces have caused the electronic peaks to blend together to give only a single smooth absorption peak. (From S. F. Mason, *J. Chem. Soc.*, **1959**, 1265.)

of tetrazine in this figure are typical of spectra of other molecules recorded under similar conditions.

24C-3 Limits to Beer's Law

There are few exceptions to the linear relationship between absorbance and path length at a fixed concentration. We frequently observe deviations from the direct proportionality between absorbance and concentration, however, when the path length b is a constant. Some of these deviations, called **real deviations,** are fundamental and represent real limitations to the law. Others are a result of the method that we use to make absorbance measurements **(instrumental deviations)** or a result of chemical changes that occur when the concentration changes **(chemical deviations).**

Real Limitations to Beer's Law

Beer's law describes the absorption behavior only of dilute solutions and in this sense is a **limiting law.** At concentrations exceeding about 0.01 M, the average distances between ions or molecules of the absorbing species are diminished to the point where each particle affects the charge distribution, and thus the extent of absorption, of its neighbors. Because the extent of interaction depends on concentration, the occurrence of this phenomenon causes deviations from the linear relationship between absorbance and concentration. A similar effect sometimes occurs in dilute solutions of absorbers that contain high concentrations of other species, particularly electrolytes. When ions are very close to one another, the molar absorptivity of the analyte can be altered because of electrostatic interactions, and this can lead to departures from Beer's law.

◀ Limiting laws in science are those that hold under limiting conditions such as dilute solutions. In addition to Beer's law, other limiting laws in chemistry include the Debye-Hückel law (see Chapter 10) and the law of independent migration, which describes the conductance of electricity by ions.

Chemical Deviations

As shown in Example 24-5, deviations from Beer's law appear when the absorbing species undergoes association, dissociation, or reaction with the solvent to give products that absorb differently from the analyte. The extent of such departures can be predicted from the molar absorptivities of the absorbing species and the equilibrium constants for the equilibria involved. Unfortunately, since we are usually unaware that such processes are affecting the analyte, there is often no opportunity to correct the measurement. Typical equilibria that give rise to this effect include monomer-dimer equilibria, metal complexation equilibria when more than one complex is present, acid-base equilibria, and solvent-analyte association equilibria.

EXAMPLE 24-5

Solutions containing various concentrations of the acidic indicator HIn ($K_a = 1.42 \times 10^{-5}$) were prepared in 0.1 M HCl and 0.1 M NaOH. In both media, plots of absorbance at either 430 nm or 570 nm versus the total indicator concentration are nonlinear; however, Beer's law is obeyed at both 430 nm and 570 nm for the individual species HIn and In^-. Hence, if we know the equilibrium concentrations of HIn and In^-, we could compensate for the fact that dissociation of HIn occurs. Usually, though the individual concentrations are unknown and only the total concentration, $c_{total} = [HIn] + [In^-]$, is known.

(continued)

Calculate the absorbance for a solution with $c_{total} = 2.00 \times 10^{-5}$ M. The magnitude of the acid dissociation constant suggests that for all practical purposes, the indicator is entirely in the undissociated form (HIn) in the HCl solution and is completely dissociated as In^- in NaOH. The molar absorptivities at the two wavelengths are found to be

	ε_{430}	ε_{570}
HIn (HCl solution)	6.30×10^2	7.12×10^3
In^- (NaOH solution)	2.06×10^4	9.60×10^2

We would now like to find the absorbances (1.00-cm cell) of unbuffered solutions of the indicator ranging in concentration from 2.00×10^{-5} M to 16.00×10^{-5} M. First, find the concentration of HIn and In^- in the unbuffered 2×10^{-5} M solution. From the equation for the dissociation reaction, we know that $[H^+] = [In^-]$. Furthermore, the mass-balance expression for the indicator tells us that $[In^-] + [HIn] = 2.00 \times 10^{-5}$ M. Substitution of these relationships into the K_a expression yields

$$\frac{[In^-]^2}{2.00 \times 10^{-5} - [In^-]} = 1.42 \times 10^{-5}$$

which can be solved to give $[In^-] = 1.12 \times 10^{-5}$ M and $[HIn] = 0.88 \times 10^{-5}$ M. The absorbances at the two wavelengths are found by substituting the values for ε, b, and c into Equation 24-13. The result is that $A_{430} = 0.236$ and $A_{570} = 0.073$. We could similarly calculate A for several other values of c_{total}. Additional data, obtained in the same way, are shown in Table 24-4. Figure 24-15 shows plots at the two wavelengths that were constructed from data obtained in a similar manner.

CHALLENGE: Perform calculations to confirm that $A_{430} = 0.596$ and that $A_{570} = 0.401$ for a solution in which the analytical concentration of HIn is 8.00×10^{-5} M.

The plots of Figure 24-15 illustrate the kinds of departures from Beer's law that occur when the absorbing system undergoes dissociation or association. Notice that the direction of curvature is opposite at the two wavelengths.

TABLE 24-4

Absorbance Data for Various Concentrations of the Indicators in Example 24-5

c_{HIn}, M	[HIn]	$[In^-]$	A_{430}	A_{570}
2.00×10^{-5}	0.88×10^{-5}	1.12×10^{-5}	0.236	0.073
4.00×10^{-5}	2.22×10^{-5}	1.78×10^{-5}	0.381	0.175
8.00×10^{-5}	5.27×10^{-5}	2.73×10^{-5}	0.596	0.401
12.0×10^{-5}	8.52×10^{-5}	3.48×10^{-5}	0.771	0.640
16.0×10^{-5}	11.9×10^{-5}	4.11×10^{-5}	0.922	0.887

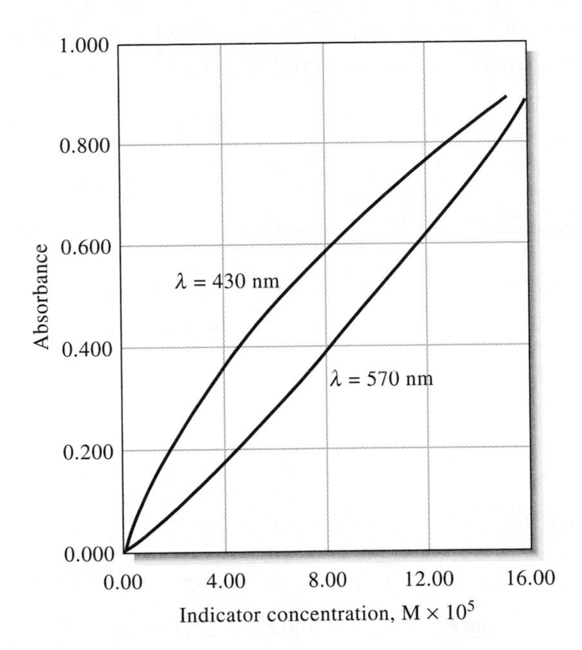

Figure 24-15 Chemical deviations from Beer's law for unbuffered solutions of the indicator HIn. The absorbance values were calculated at various indicator concentrations, as shown in Example 24-5. Note that there are positive deviations at 430 nm and negative deviations at 570 nm. At 430 nm, the absorbance is primarily due to the ionized In^- form of the indicator and is in fact proportional to the fraction ionized. The fraction ionized varies nonlinearly with total concentration. At lower total concentrations ([HIn] + $[In^-]$), the fraction ionized is larger than at high total concentrations, and a positive error occurs. At 570 nm, the absorbance is due principally to the undissociated acid HIn. The fraction in this form begins as a low amount and increases nonlinearly with the total concentration, giving rise to the negative deviation shown.

Instrumental Deviations: Polychromatic Radiation

Beer's law strictly applies only when measurements are made with monochromatic radiation. In practice, polychromatic sources that have a continuous distribution of wavelengths are used in conjunction with a grating or with a filter to isolate a nearly symmetric band of wavelengths around the wavelength to be employed (see Section 25A-3).

◀ Deviations from Beer's law often occur when polychromatic radiation is used to measure absorbance.

The following derivation shows the effect of polychromatic radiation on Beer's law. Consider a beam of radiation consisting of just two wavelengths, λ' and λ''. Assuming that Beer's law applies strictly for each wavelength, we may write for λ'

$$A' = \log \frac{P'_0}{P'} = \varepsilon' bc$$

or

$$\frac{P'_0}{P'} = 10^{\varepsilon' bc}$$

where P'_0 is the incident power and P' is the resultant power at λ'. The symbols b and c are the path length and concentration, respectively, of the absorber, and ε' is the molar absorptivity at λ'. Then

$$P' = P'_0 10^{-\varepsilon' bc}$$

Similarly, for λ''

$$P'' = P''_0 10^{-\varepsilon'' bc}$$

◀ Generally, the better the instrument, the less likely are deviations from Beer's law due to polychromatic radiation.

When an absorbance measurement is made with radiation composed of both wavelengths, the power of the beam emerging from the solution is the sum of the powers emerging at the two wavelengths $P' + P''$. Likewise, the total incident power is the sum $P_0' + P_0''$. Therefore, the measured absorbance A_m is

$$A_m = \log\left(\frac{P_0' + P_0''}{P' + P''}\right)$$

We then substitute for P' and P'' and find that

$$A_m = \log\left(\frac{P_0' + P_0''}{P_0'10^{-\varepsilon' bc} + P_0''10^{-\varepsilon'' bc}}\right)$$

or

$$A_m = \log(P_0' + P_0'') - \log(P_0'10^{-\varepsilon' bc} + P_0''10^{-\varepsilon'' bc})$$

We see that when $\varepsilon' = \varepsilon''$, this equation simplifies to

$$\begin{aligned} A_m &= \log(P_0' + P_0'') - \log[(P_0' + P_0'')(10^{-\varepsilon' bc})] \\ &= \log(P_0' + P_0'') - \log(P_0' + P_0'') - \log(10^{-\varepsilon' bc}) \\ &= \varepsilon' bc = \varepsilon'' bc \end{aligned}$$

and Beer's law is followed. As shown in Figure 24-16, however, the relationship between A_m and concentration is no longer linear when the molar absorptivities differ. In addition, as the difference between ε' and ε'' increases, the deviation from linearity increases. This derivation can be expanded to include additional wavelengths; the effect remains the same.

Polychromatic light, literally multicolored light, is light of many wavelengths, such as that from a tungsten light bulb. Monochromatic light can be produced by filtering, diffracting, or refracting polychromatic light (see Chapter 25, Section 25A-3).

If the band of wavelengths selected for spectrophotometric measurements corresponds to a region of the absorption spectrum in which the molar absorptivity of the analyte is essentially constant, departures from Beer's law will be minimal. Many molecular bands in the UV/visible region fit this description. For these, Beer's law is obeyed, as demonstrated by Band A in Figure 24-17. Some absorption bands in the UV/visible region and many in the infrared region are very narrow, however, and departures from Beer's law are common, as illustrated for Band B in

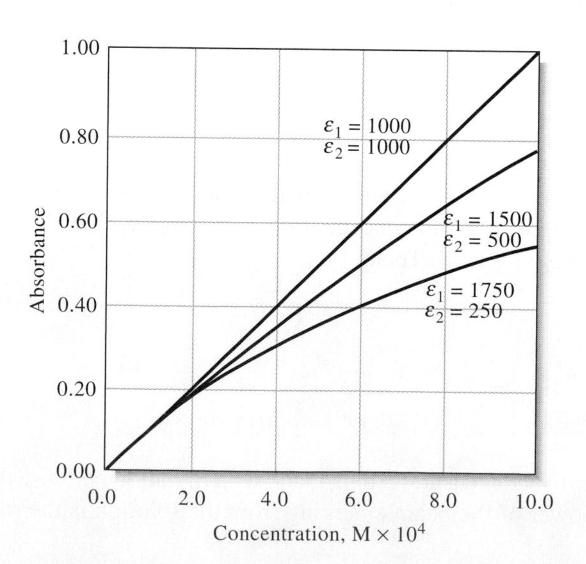

Figure 24-16 Deviations from Beer's law with polychromatic radiation. The absorber has the indicated molar absorptivities at the two wavelengths λ' and λ''.

Figure 24-17. Hence, to avoid deviations, it is advisable to select a wavelength band near the wavelength of maximum absorption, where the analyte absorptivity changes little with wavelength. Atomic absorption lines are so narrow that they require special sources to obtain adherence to Beer's law, as discussed in Chapter 25, Section 25A-2.

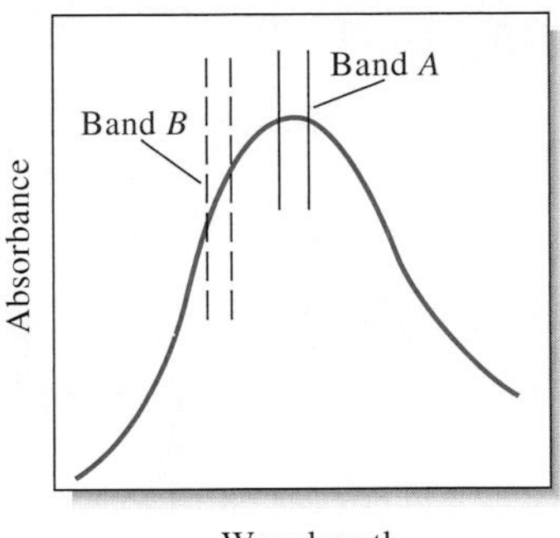

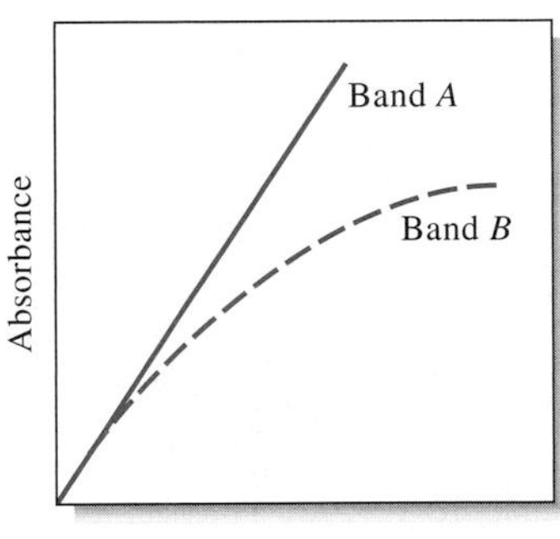

Figure 24-17 The effect of polychromatic radiation on Beer's law. In the absorption spectrum at the top, the absorptivity of the analyte is seen to be nearly constant over Band *A* from the source. Note in the Beer's law plot at the bottom that using Band *A* gives a linear relationship. In the spectrum, Band *B* coincides with a region of the spectrum over which the absorptivity of the analyte changes. Note the dramatic deviation from Beer's law that results in the lower plot.

Instrumental Deviations: Stray Light

Stray radiation, commonly called **stray light,** is defined as radiation from the instrument that is outside the nominal wavelength band chosen for the determination. This stray radiation often is the result of scattering and reflection off the surfaces of gratings, lenses or mirrors, filters, and windows. When measurements are made in the presence of stray light, the observed absorbance is given by

$$A' = \log \frac{P_0 + P_s}{P + P_s}$$

where P_s is the radiant power of the stray light. Figure 24-18 shows a plot of the apparent absorbance A' versus concentration for various levels of P_s relative to P_0. Stray light always causes the apparent absorbance to be lower than the true absorbance. The deviations due to stray light are most significant at high absorbance values. Because stray radiation levels can be as high as 0.5% in modern instruments, absorbance levels greater than 2.0 are rarely measured unless special precautions are taken or special instruments with extremely low stray light levels are used. Some inexpensive filter instruments can exhibit deviations from Beer's law at absorbances as low as 1.0 because of high stray light levels or the presence of polychromatic light.

Mismatched Cells

Another almost trivial but important deviation from adherence to Beer's law is caused by mismatched cells. If the cells holding the analyte and blank solutions are not of equal path length and are not equivalent in optical characteristics, an inter-

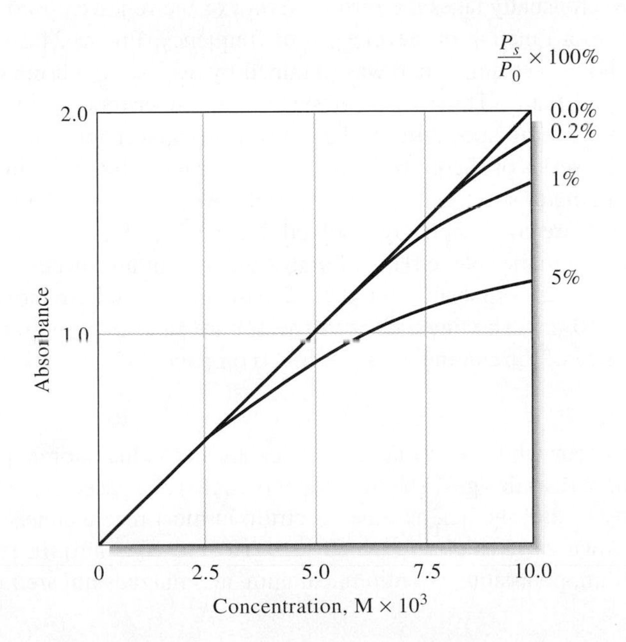

Figure 24-18 Deviation from Beer's law caused by various levels of stray light. Note that absorbance begins to level off with concentration at high stray light levels. Stray light always limits the maximum absorbance that can be obtained because when the absorbance is high, the radiant power transmitted through the sample can become comparable to or lower than the stray light level.

cept will occur in the calibration curve, and $A = \varepsilon bc + k$ will be the actual equation instead of Equation 24-8. This error can be avoided by using either carefully matched cells or a linear regression procedure to calculate both the slope and the intercept of the calibration curve. In most cases, this is the best strategy because an intercept can also occur if the blank solution does not totally compensate for interferences. Another way to avoid the mismatched-cell problem with single-beam instruments is to use only one cell and keep it in the same position for both blank and analyte measurements. After obtaining the blank reading, the cell is emptied by aspiration, washed, and filled with analyte solution.

Spreadsheet Summary In Chapter 12 of *Applications of Microsoft® Excel in Analytical Chemistry,* spreadsheets are presented for modeling the effects of chemical equilibria and stray light on absorption measurements. Chemical and physical variables may be changed to observe their effects on instrument readouts.

24D EMISSION OF ELECTROMAGNETIC RADIATION

► Chemical species can be caused to emit light by (1) bombardment with electrons; (2) heating in a plasma, a flame, or an electric arc; or (3) irradiation with a beam of light.

Atoms, ions, and molecules can be excited to one or more higher energy levels by any of several processes, including bombardment with electrons or other elementary particles; exposure to a high-temperature plasma, flame, or electric arc; or exposure to a source of electromagnetic radiation. The lifetime of an excited species is generally transitory (10^{-9} to 10^{-6} s), and relaxation to a lower energy level or the ground state takes place with a release of the excess energy in the form of electromagnetic radiation, heat, or perhaps both.

24D-1 Emission Spectra

Radiation from a source is conveniently characterized by means of an emission spectrum, which usually takes the form of a plot of the relative power of the emitted radiation as a function of wavelength or frequency. Figure 24-19 illustrates a typical emission spectrum, which was obtained by aspirating a brine solution into an oxyhydrogen flame. Three types of spectra are superimposed in the figure: a **line spectrum,** a **band spectrum,** and a **continuum spectrum.** The line spectrum is made up of a series of sharp, well-defined peaks caused by excitation of individual atoms. The band spectrum is composed of several groups of lines so closely spaced that they are not completely resolved. The source of the bands is small molecules or radicals in the source flame. Finally, the continuum spectrum, shown as a blue dashed line, is responsible for the increase in the background that appears above about 350 nm. The line and band spectra are superimposed on this continuum. The source of the continuum is described on page 737.

Line Spectra

► The line widths of atoms in a medium such as a flame or plasma are about 0.1 to 0.01 Å. The wavelengths of atomic lines are unique for each element and are often used for qualitative analysis.

Line spectra occur when the radiating species are individual atomic particles that are well separated, as in a gas. The individual particles in a gaseous medium behave independently of one another, and the spectrum in most media consists of a series of sharp lines with widths of 10^{-1} to 10^{-2} Å (10^{-2} to 10^{-3} nm). In Figure 24-19, lines for sodium, potassium, strontium, calcium, and magnesium are identified.

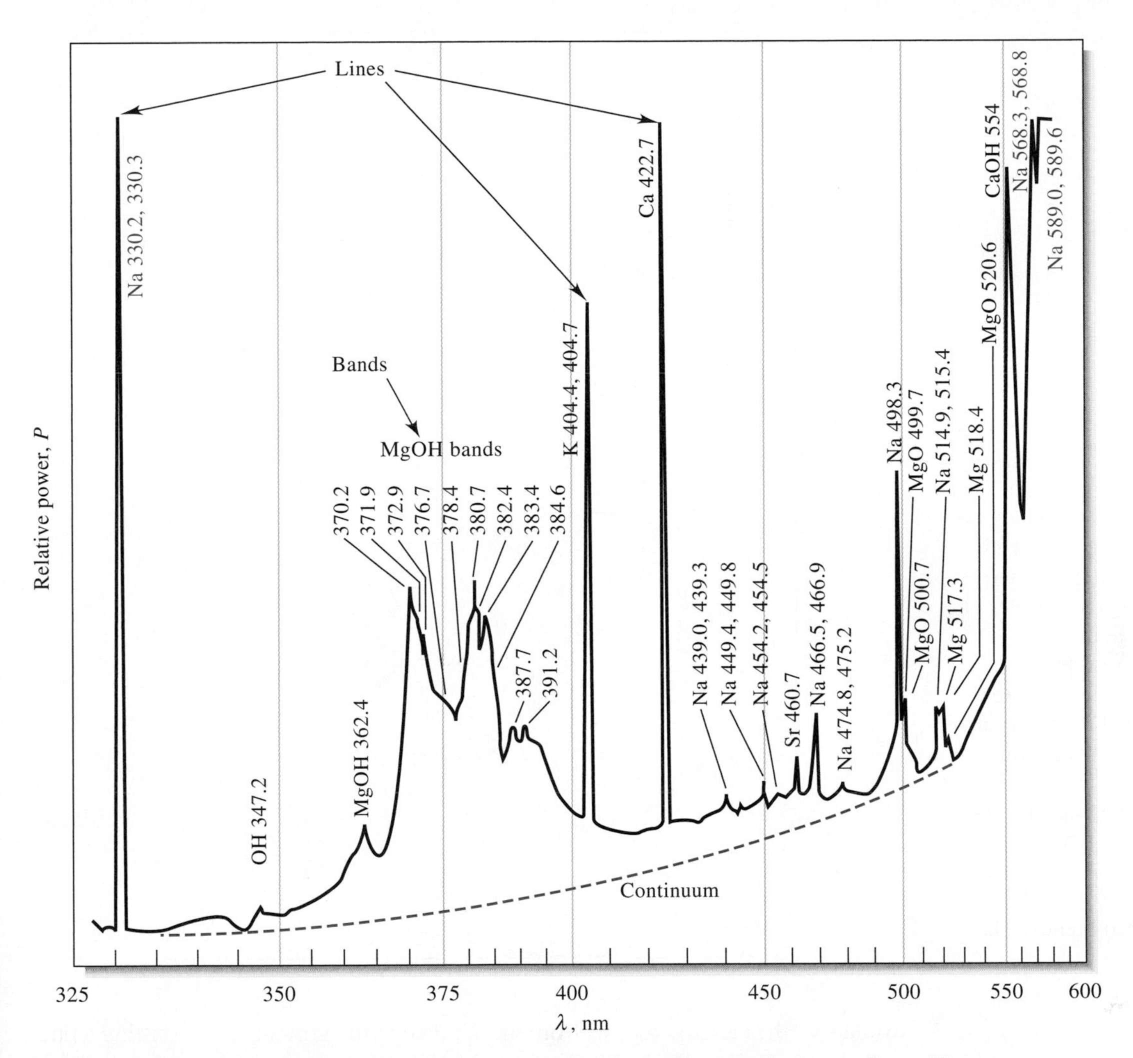

Figure 24-19 Emission spectrum of a brine sample obtained with an oxyhydrogen flame. The spectrum consists of the superimposed line, band, and continuum spectra of the constituents of the sample. The characteristic wavelengths of the species contributing to the spectrum are listed beside each feature. (R. Hermann and C. T. J. Alkemade, *Chemical Analysis by Flame Photometry,* 2nd ed., p. 484. New York: Interscience, 1979.)

The energy-level diagram in Figure 24-20 shows the source of three of the lines that appear in the emission spectrum of Figure 24-19. The horizontal line, labeled $3s$ in Figure 24-20, corresponds to the lowest, or ground-state, energy of the atom E_0. The horizontal lines labeled $3p$, $4p$, and $4d$ are three higher energy electronic levels of sodium. Note that each of the p and d states is split into two closely spaced energy levels as a result of electron spin. The single outer-shell electron in the ground state $3s$ orbital of a sodium atom can be excited into either of these levels by absorption of thermal, electrical, or radiant energy. Energy levels E_{3p} and E'_{3p} then represent the energies of the atom when this electron has been promoted to the two $3p$ states by absorption. The promotion to these states is depicted by the blue line between the $3s$ and the two $3p$ levels in Figure 24-20. A few nanoseconds after

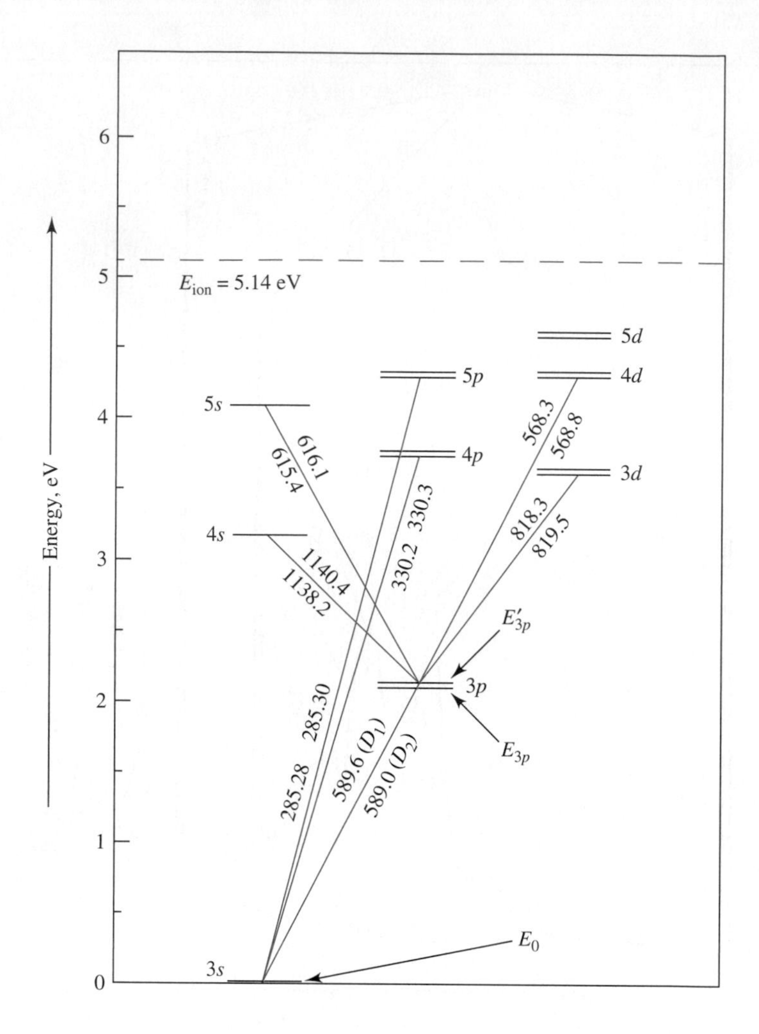

Figure 24-20 Energy-level diagram for sodium in which the horizontal lines represent the atomic orbitals, which are identified with their respective labels. The vertical scale is orbital energy in electron volts (eV), and the energies of excited states relative to the ground-state *3s* orbital can be read from the vertical axis. The blue lines show the allowed transitions resulting from emission of various wavelengths (in nm), indicated adjacent to the lines. The horizontal dashed line represents the ionization energy of sodium. (Adapted from J. D. Ingle, Jr., and S. R. Crouch, *Spectrochemical Analysis,* p. 206. Upper Saddle River, NJ: Prentice-Hall, 1988.)

excitation, the electron returns from the $3p$ state to the ground state, emitting a photon whose wavelength is given by Equation 24-5.

$$\lambda_1 = \frac{hc}{(E_{3p} - E_0)} = 589.6 \text{ nm}$$

In a similar way, relaxation from the $3p'$ state to the ground state yields a photon with $\lambda_2 = 589.0$ nm. This emission process is once again shown by the blue line between the $3s$ and $3p$ levels in Figure 24-20. The result is that the emission process from the two closely spaced $3p$ levels produces two corresponding closely spaced lines in the emission spectrum, called a **doublet.** These lines, indicated by the transitions labeled D_1 and D_2 in Figure 24-20, are the famous Fraunhofer "D" lines discussed in Feature 24-1. They are so intense that they are completely off scale in the upper right corner of the emission spectrum in Figure 24-19.

The transition from the more energetic $4p$ state to the ground state (see Figure 24-20) produces a second doublet at a shorter wavelength. The line appearing at about 330 nm in Figure 24-19 results from these transitions. The $4d$-to-$3p$ transition provides a third doublet at about 568 nm. Notice that all three of these doublets appear in the emission spectrum of Figure 24-19 as single lines. This is a result of the limited resolution of the spectrometer used to produce the spectrum, as discussed in Sections 25A-3 and 28A-1. It is important to note that the emitted wavelengths shown in Figure 24-20 are identical to the wavelengths of the absorption peaks for sodium (see Figure 24-11) because the transitions involved are between the same pairs of states.

At first glance, it may appear that radiation could be absorbed and emitted by atoms between any pair of the states shown in Figure 24-20, but in fact only certain transitions are allowed, while others are forbidden. The transitions that are allowed and forbidden to produce lines in the atomic spectra of the elements are determined by the laws of quantum mechanics in what are called **selection rules.** These rules are beyond the scope of our discussion.[5]

Band Spectra

Band spectra are often produced in spectral sources because of the presence of gaseous radicals or small molecules. For example, in Figure 24-19, bands for OH, MgOH, and MgO are labeled and consist of a series of closely spaced lines that are not fully resolved by the instrument used to obtain the spectrum. Bands arise from the numerous quantized vibrational levels that are superimposed on the ground-state electronic energy level of a molecule. For further discussion of band spectra, see Section 28B-3.

◀ An emission band spectrum is made up of many closely spaced lines that are difficult to resolve.

Continuum Spectra

As shown in Figure 24-21, truly continuous radiation is produced when solids such as carbon and tungsten are heated to incandescence. Thermal radiation of this kind, which is called **blackbody radiation,** is more characteristic of the temperature of the emitting surface than of its material. Blackbody radiation is produced by the innumerable atomic and molecular oscillations excited in the condensed solid by the thermal energy. Note that the energy peaks in Figure 24-21 shift to shorter wavelengths with increasing temperature. As the figure shows, very high temperatures are required to cause a thermally excited source to emit a substantial fraction of its energy as ultraviolet radiation.

◀ Continuum emission spectra have no line character and are generally produced by heating solids to a high temperature.

Part of the continuum background radiation in the flame spectrum shown in Figure 24-19 is probably thermal emission from incandescent particles in the flame. Note that this background decreases rapidly as the wavelength approaches the ultraviolet region of the spectrum.

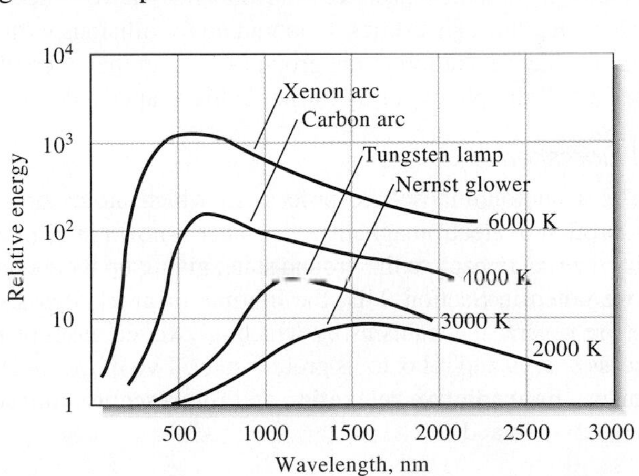

Figure 24-21 Blackbody radiation curves for various light sources. Note the shift in the peaks as the temperature of the sources changes.

[5] J. D. Ingle, Jr., and S. R. Crouch, *Spectrochemical Analysis,* p. 205. Upper Saddle River, NJ: Prentice-Hall, 1988.

In 1900, Max Planck (1858–1947) discovered a formula (now often called the Planck radiation law) that modeled curves like those shown in Figure 24-21 nearly perfectly. He followed this discovery by developing a theory that made two bold assumptions regarding the oscillating atoms or molecules in blackbody radiators. He assumed (1) that these species could have only discrete energies and (2) that they could absorb or emit energy in discrete units, or quanta. These assumptions, which are implicit in Equation 24-3, laid the foundation for the development of quantum theory and eventually won him the Nobel Prize in Physics in 1918.

Resonance fluorescence is radiation that is identical in wavelength to the radiation that excited the fluorescence.

Heated solids are important sources of infrared, visible, and longer wavelength ultraviolet radiation for analytical instruments, as we will see in Chapter 25.

Effect of Concentration on Line and Band Spectra

The radiant power P of a line or a band depends directly on the number of excited atoms or molecules, which in turn is proportional to the total concentration c of the species present in the source. Thus, we can write

$$P = kc \tag{24-16}$$

where k is a proportionality constant. This relationship is the basis of quantitative emission spectroscopy, which is described in some detail in Section 28C.

24D-2 Emission by Fluorescence and Phosphorescence

Fluorescence and phosphorescence are analytically important emission processes in which atoms or molecules are excited by the absorption of a beam of electromagnetic radiation. The excited species then relax to the ground state, giving up their excess energy as photons. Fluorescence takes place much more rapidly than phosphorescence and is generally complete in about 10^{-5} s (or less) from the time of excitation. Phosphorescence emission may extend for minutes or even hours after irradiation has ceased. Because fluorescence is considerably more important than phosphorescence in analytical chemistry, our discussions focus mostly on fluorescence.

Atomic Fluorescence

Gaseous atoms fluoresce when they are exposed to radiation with a wavelength that exactly matches that of one of the absorption (or emission) lines of the element in question. For example, gaseous sodium atoms are promoted to the excited energy state E_{3p} shown in Figure 24-20 through absorption of 589-nm radiation. Relaxation may then take place by reemission of fluorescent radiation of the identical wavelength. When excitation and emission wavelengths are the same, the resulting emission is called **resonance fluorescence.** Sodium atoms could also exhibit resonance fluorescence when exposed to 330-nm or 285-nm radiation. In addition, however, the element could also produce nonresonance fluorescence by first relaxing to energy level E_{3p} through a series of nonradiative collisions with other species in the medium. Further relaxation to the ground state can then take place either by the emission of a 589-nm photon or by further collisional deactivation.

Molecular Fluorescence

Fluorescence is a photoluminescence process in which atoms or molecules are excited by absorption of electromagnetic radiation, as shown in Figure 24-22a. The excited species then relax back to the ground state, giving up their excess energy as photons. As we noted in Section 24D, the lifetime of an excited species is brief because there are several mechanisms by which an excited atom or molecule can give up its excess energy and relax to its ground state. Two of the most important of these mechanisms, **nonradiative relaxation** and **fluorescence emission,** are illustrated in Figures 24-22b and c.

Nonradiative Relaxation Two types of nonradiative relaxation are shown in Figure 24-22b. **Vibrational deactivation,** or **relaxation,** depicted by the short wavy arrows between vibrational energy levels, takes place during collisions between excited molecules and molecules of the solvent. During the collisions, the excess

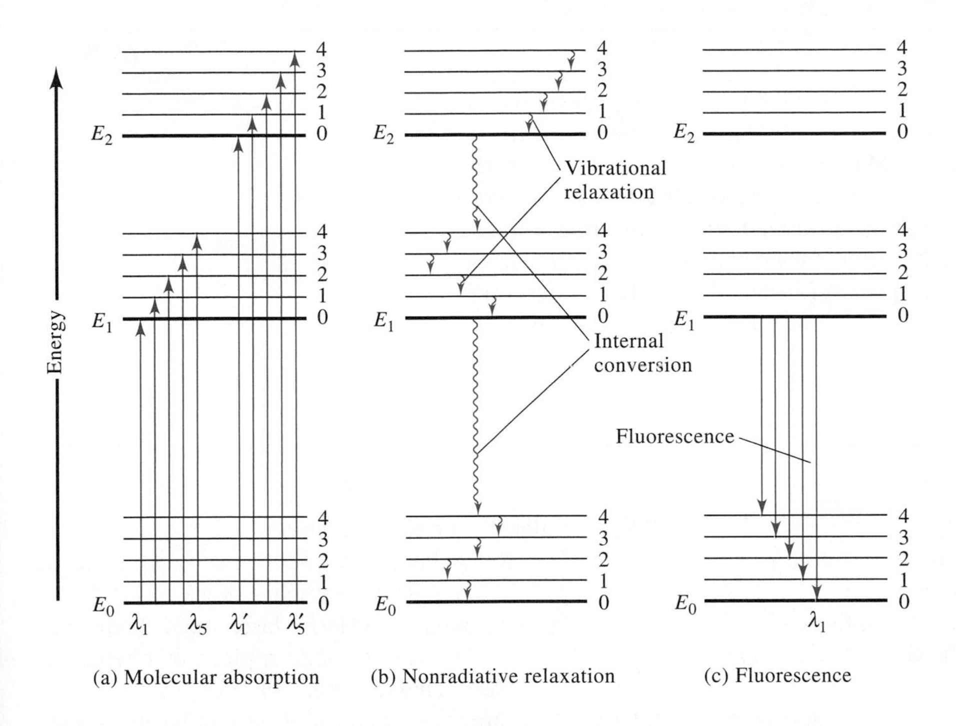

Figure 24-22 Energy-level diagram showing some of the energy changes that occur during absorption, nonradiative relaxation, and fluorescence by a molecular species.

vibrational energy is transferred to solvent molecules in a series of steps, as indicated in the figure. The gain in vibrational energy of the solvent is reflected in a tiny increase in the temperature of the medium. Vibrational relaxation is such an efficient process that the average lifetime of an excited vibrational state is only about 10^{-15} s. Nonradiative relaxation between the lowest vibrational level of an excited electronic state and the upper vibrational level of another electronic state can also occur. This type of relaxation, which is called **internal conversion,** is depicted by the two longer, wavy arrows in Figure 24-22b and is much less efficient than vibrational relaxation, so that the average lifetime of an electronic excited state is between 10^{-9} and 10^{-6} s. The mechanisms by which this type of relaxation occurs are not fully understood, but the net effect is again a rise in the temperature of the medium.

Fluorescence The relative number of molecules that fluoresce is small because fluorescence requires structural features that slow the rate of the nonradiative relaxation processes illustrated in Figure 24-22b and enhance the rate of fluorescence relaxation shown in Figure 24-22c. Most molecules lack these features and undergo nonradiative relaxation at a rate that is significantly greater than the radiative relaxation rate; therefore, fluorescence does not occur. As shown in Figure 24-22c, bands of radiation are produced when molecules relax from the lowest lying vibrational state of an excited state E_1 to the many vibrational levels of the ground state E_0. Like molecular absorption bands, molecular fluorescence bands are made up of a large number of closely spaced lines that are usually difficult to resolve. Notice that the transition from E_1 to the lowest lying vibrational state of the ground state (λ_1) has the highest energy of all the transitions in the band. As a result, all the other lines that terminate in higher vibrational levels of the ground state are lower in energy and produce fluorescence emission at longer wavelengths than λ_1. That is, molecular fluorescence bands consist largely of lines that are longer in wavelength than the band of absorbed radiation responsible for their excitation. This shift in wavelength is sometimes called the **Stokes shift.** A more detailed discussion of molecular fluorescence is given in Chapter 27.

The **Stokes shift** refers to fluorescence radiation that occurs at wavelengths that are longer than the wavelength of radiation used to excite the fluorescence.

WEB WORKS

To learn more about Beer's law, use Google to find the IUPAC Glossary of Terms Used in Photochemistry. Find how the molar absorptivity of a compound (ε) is related to the absorption cross section (σ). Multiply the absorption cross section by Avogadro's number and note the result. How would the result change if absorbance were expressed as $A = -\ln (P/P_0)$ rather than the usual definition in terms of base-10 logarithms? What are the units of σ? Which of the quantities ε or σ is a macroscopic quantity? Which is a microscopic quantity? Notice that the IUPAC term for molar absorptivity is **molar absorption coefficient.** Which of these terms is most descriptive? Explain and justify your answer.

QUESTIONS AND PROBLEMS

***24-1.** Why is a solution of $Cu(NH_3)_4^{2+}$ blue?

24-2. What is the relationship between
- *(a) absorbance and transmittance?
- (b) absorptivity a and molar absorptivity ε?

***24-3.** Identify factors that cause the Beer's law relationship to depart from linearity.

24-4. Describe the difference between "real" deviations from Beer's law and those due to instrumental or chemical factors.

24-5. How does an electronic transition resemble a vibrational transition? How do they differ?

24-6. Calculate the frequency in hertz of
- *(a) an X-ray beam with a wavelength of 2.97 Å.
- (b) an emission line for copper at 324.7 nm.
- *(c) the line at 632.8 nm produced by a He-Ne laser.
- (d) the output of a CO_2 laser at 10.6 μm.
- *(e) an infrared absorption peak at 3.75 μm.
- (f) a microwave beam at 1.86 cm.

24-7. Calculate the wavelength in centimeters of
- *(a) an airport tower transmitting at 118.6 MHz.
- (b) a VOR (radio navigation aid) transmitting at 114.10 kHz.
- *(c) an NMR signal at 135 MHz.
- (d) an infrared absorption peak with a wave number of 1375 cm^{-1}.

24-8. A typical simple infrared spectrophotometer covers a wavelength range from 3 to 15 μm. Express its range (a) in wavenumbers and (b) in hertz.

***24-9.** A sophisticated ultraviolet/visible/near-IR instrument has a wavelength range of 185 to 3000 nm. What are its wavenumber and frequency ranges?

***24-10.** Calculate the frequency in hertz and the energy in joules of an X-ray photon with a wavelength of 2.35 Å.

24-11. Calculate the wavelength and the energy in joules associated with a signal at 220 MHz.

24-12. Calculate the wavelength of
- *(a) the sodium line at 589 nm in an aqueous solution with a refractive index of 1.27.
- (b) the output of a He-Ne laser at 632.8 nm when it is passing through a piece of quartz that has a refractive index of 1.55.

24-13. What are the units for absorptivity when the path length is given in centimeters and the concentration is expressed in
- *(a) parts per million?
- (b) micrograms per liter?
- *(c) weight-volume percent?
- (d) grams per liter?

24-14. Express the following absorbances in terms of percent transmittance:
- *(a) 0.0350
- (b) 0.936
- *(c) 0.310
- (d) 0.232
- *(e) 0.494
- (f) 0.104

24-15. Convert the accompanying transmittance data to absorbances:
- *(a) 22.7%
- (b) 0.567
- *(c) 31.5%
- (d) 7.93%
- *(e) 0.103
- (f) 58.2%

24-16. Calculate the percent transmittance of solutions that have twice the absorbance of the solutions in Problem 24-14.

24-17. Calculate the absorbances of solutions with half the transmittance of those in Problem 24-15.

24-18. Evaluate the missing quantities in the accompanying table. Where needed, use 200 for the molar mass of the analyte.

	A	$\%T$	ε	a	b	c	
			L mol^{-1} cm^{-1}	cm^{-1} ppm^{-1}	cm	M	ppm
*(a)	0.172		4.23×10^3		1.00		
(b)		44.9		0.0258		1.35×10^{-4}	
*(c)	0.520		7.95×10^3		1.00		
(d)		39.6		0.0912			1.76
*(e)			3.73×10^3		0.100	1.71×10^{-3}	
(f)		83.6			1.00	8.07×10^{-6}	
*(g)	0.798				1.50		33.6
(h)		11.1	1.35×10^4			7.07×10^{-5}	
*(i)		5.23	9.78×10^3				5.24
(j)	0.179				1.00	7.19×10^{-5}	

***24-19.** A solution containing 8.75 ppm $KMnO_4$ has a transmittance of 0.743 in a 1.00-cm cell at 520 nm. Calculate the molar absorptivity of $KMnO_4$.

24-20. Beryllium(II) forms a complex with acetylacetone (166.2 g/mol). Calculate the molar absorptivity of the complex, given that a 1.34-ppm solution has a transmittance of 55.7% when measured in a 1.00-cm cell at 295 nm, the wavelength of maximum absorption.

***24-21.** At 580 nm, the wavelength of its maximum absorption, the complex $FeSCN^{2+}$ has a molar absorptivity of 7.00×10^3 L cm^{-1} mol^{-1}. Calculate

(a) the absorbance of a 3.75×10^{-5} M solution of the complex at 580 nm in a 1.00-cm cell.

(b) the absorbance of a solution in which the concentration of the complex is twice that in (a).

(c) the transmittance of the solutions described in (a) and (b).

(d) the absorbance of a solution that has half the transmittance of that described in (a).

***24-22.** A 5.00-mL aliquot of a solution that contains 5.94 ppm iron(III) is treated with an appropriate excess of KSCN and is diluted to 50.0 mL. What is the absorbance of the resulting solution at 580 nm in a 2.50-cm cell? See Problem 24-21 for absorptivity data.

24-23. A solution containing the complex formed between Bi(III) and thiourea has a molar absorptivity of 9.32×10^3 L cm^{-1} mol^{-1} at 470 nm.

(a) What is the absorbance of a 6.24×10^{-5} M solution of the complex at 470 nm in a 1.00-cm cell?

(b) What is the percent transmittance of the solution described in (a)?

(c) What is the molar concentration of the complex in a solution that has the absorbance described in (a) when measured at 470 nm in a 5.00-cm cell?

***24-24.** The complex formed between Cu(I) and 1,10-phenanthroline has a molar absorptivity of 7000 L cm^{-1} mol^{-1} at 435 nm, the wavelength of maximum absorption. Calculate

(a) the absorbance of a 6.77×10^{-5} M solution of the complex when measured in a 1.00-cm cell at 435 nm.

(b) the percent transmittance of the solution in (a).

(c) the concentration of a solution that in a 5.00-cm cell has the same absorbance as the solution in (a).

(d) the path length through a 3.40×10^{-5} M solution of the complex that is needed for an absorbance that is the same as the solution in (a).

***24-25.** A solution with a "true" absorbance [$A = -\log(P/P_0)$] of 2.10 was placed in a spectrophotometer with a stray light percentage (P_s/P_0) of 0.75. What absorbance A' would be measured? What percentage error would result?

24-26. A compound X is to be determined by UV/visible spectrophotometry. A calibration curve is constructed from standard solutions of X with the following results: 0.50 ppm, $A = 0.24$; 1.5 ppm, $A = 0.36$; 2.5 ppm, $A = 0.44$; 3.5 ppm, $A = 0.59$; 4.5 ppm, $A = 0.70$. Find the slope and intercept of the calibration curve, the standard error in y, the concentration of the solution of unknown X concentration, and the standard deviation in the concentration of X. Construct a plot of the calibration curve and determine the unknown concentration by hand from the plot.

***24-27.** One common way to determine phosphorus in urine is to treat the sample, after removing the protein, with molybdenum(VI) and then reduce the resulting 12-molybdophosphate complex with ascorbic acid to give an intense blue-colored species called molybdenum blue. The absorbance of molybdenum blue can be measured at 650 nm. A patient produced 1122 mL of urine in 24 hours. A 1.00-mL aliquot of the sample was treated with Mo(VI) and ascorbic acid and was diluted to a volume of 50.00 mL. A calibration curve was prepared by treating 1.00-mL aliquots of phosphate standard solutions in the same manner as the urine sample. The absorbances of the standards and the urine sample were obtained at 650 nm, and the following results were obtained:

Solution	Absorbance at 650 nm
1.00 ppm P	0.230
2.00 ppm P	0.436
3.00 ppm P	0.638
4.00 ppm P	0.848
Urine sample	0.518

(a) Find the slope, intercept, and standard error in y of the calibration curve. Construct a plot of the calibration curve. Determine the number of ppm of P in the urine sample and its standard deviation from the least-squares equation of the line. Compare the unknown concentration with that obtained manually from a calibration curve.
(b) What mass in grams of phosphorus was eliminated per day by the patient?
(c) What is the phosphate concentration in urine in mM?

24-28. Nitrite is commonly determined by a colorimetric procedure using a reaction called the Griess reaction. In this reaction, the sample containing nitrite is reacted with sulfanilimide and N-(1-Napthyl) ethylenediamine to form a colored species that absorbs at 550 nm. By using an automated flow analysis instrument, the following results were obtained for standard solutions of nitrite and for a sample containing an unknown amount:

Solution	Absorbance at 550 nm
2.00 μM	0.065
6.00 μM	0.205
10.00 μM	0.338
14.00 μM	0.474
18.00 μM	0.598
Unknown	0.402

(a) Find the slope, intercept, and standard deviation of the calibration curve.
(b) Construct a plot of the calibration curve.
(c) Determine the concentration of nitrite in the sample and its standard deviation.

24-29. The equilibrium constant for the reaction

$$2CrO_4^{2-} + 2H^+ \rightleftharpoons Cr_2O_7^{2-} + H_2O$$

is 4.2×10^{14}. The molar absorptivities for the two principal species in a solution of $K_2Cr_2O_7$ are

λ, nm	$\varepsilon_1(CrO_4^{2-})$	$\varepsilon_2(Cr_2O_7^{2-})$
345	1.84×10^3	10.7×10^2
370	4.81×10^3	7.28×10^2
400	1.88×10^3	1.89×10^2

Four solutions were prepared by dissolving 4.00×10^{-4}, 3.00×10^{-4}, 2.00×10^{-4}, and 1.00×10^{-4} moles of $K_2Cr_2O_7$ in water and diluting to 1.00 L with a pH 5.60 buffer. Calculate theoretical absorbance values (1.00-cm cells) for each solution and plot the data for (a) 345 nm; (b) 370 nm; (c) 400 nm.

24-30. Challenge Problem. NIST maintains a database of the spectra of the elements at *http://physlab2.nist.gov/*. The following energy levels for neutral lithium were obtained from this database:

Electronic Configuration	Level, eV
$1s^2 2s^1$	0.00000
$1s^2 2p^1$	1.847819 1.847861
$1s^2 3s^1$	3.373130
$1s^2 3p^1$	3.834260 3.834260
$1s^2 3d^1$	3.878609 3.878614
$1s^2 4s^1$	4.340944
$1s^2 4p^1$	4.521650 4.521650
$1s^2 4d^1$	4.540722 4.540725

(a) Construct a partial energy-level diagram similar to the one in Figure 24-20. Label each energy level with its corresponding orbital.

Look up the first ionization energy for lithium on the NIST site, and indicate it with a horizontal line on your diagram.

(b) Browse to the NIST Web site, and click on the Physical Reference Data link. Locate and click on the link for the Atomic Spectral Database, and click on the Lines icon. Use the form to retrieve the spectral lines for Li I between 300 nm and 700 nm, including energy level information. Note that the retrieved table contains wavelength, relative intensity, and changes in electron configuration for the transitions that give rise to each line. Add connecting lines to the partial energy-level diagram from (a) to illustrate the transitions, and label each line with the wavelength of the emission. Which of the transitions in your diagram are doublets?

(c) Use the intensity vs. wavelength data that you retrieved in (b) to sketch an emission spectrum for lithium. If you placed a sample of $LiCO_3$ in a flame, what color would the flame be?

(d) Describe how the flame spectrum of an ionic lithium compound such as $LiCO_3$ displays the spectrum of neutral lithium atoms.

(e) There appear to be no emission lines for lithium between 544 nm and 610 nm. Why is this?

(f) Describe how the information obtained in this problem could be used to detect the presence of lithium in urine. How would you determine the amount of lithium quantitatively?

InfoTrac College Edition

For additional readings, go to InfoTrac College Edition, your online research library, at

http://infotrac.thomsonlearning.com

CHAPTER 25

Instruments for Optical Spectrometry

The bright star in the middle of the photograph is Supernova 1987a, which was the first supernova visible to the naked eye to appear in more than 400 years. The black dots over the star's image were produced by superimposing a negative of a photo taken 2 years before the supernova appeared on February 23, 1987. Coincident with the supernova was an unusual burst of neutrinos, which was detected by the newly refurbished Irvine-Michigan-Brookhaven underground detector. This detector consists of a 6800 cubic meter volume of water surrounded by 2048 high-sensitivity, large-area photomultiplier tubes and housed in a salt mine under Lake Erie. When at least 20 of the photomultipliers detect a pulse of blue Cherenkov radiation from the impact of neutrinos with water molecules in the detector within a time window of 55 ns, a neutrino is judged to have occurred. This detector and others like it were built in an effort to detect the spontaneous decay of protons in the water molecules. These experiments are very long term, and data from the detector are recorded continuously. As a result, the detector was poised to monitor the neutrino burst from Supernova 1987a. The photomultiplier is one of the radiation detectors described in this chapter.

We often call the UV/visible and IR regions of the spectrum the optical region. Even though the optic nerve is responsive only to visible radiation, the other regions are included because the lenses, mirrors, prisms, and gratings used are similar and function in a comparable manner. Spectroscopy in the UV/visible and IR regions is, therefore, often called **optical spectroscopy.**

The basic components of analytical instruments for absorption spectroscopy, as well as for emission and fluorescence spectroscopy, are remarkably alike in function and in general performance requirements whether the instruments are designed for ultraviolet (UV), visible, or infrared (IR) radiation. Because of the similarities, such instruments are frequently referred to as ***optical instruments*** *even though the eye is sensitive only to the visible region. In this chapter, we first examine the characteristics of the components common to all optical instruments. We then consider the characteristics of typical instruments designed for UV, visible, and IR absorption spectroscopy.*

25A INSTRUMENT COMPONENTS

Most spectroscopic instruments for use in the UV/visible and IR regions are made up of five components: (1) a stable source of radiant energy; (2) a wavelength selector that isolates a limited region of the spectrum for measurement; (3) one or more sample containers; (4) a radiation detector, which converts radiant energy to a measurable electrical signal; and (5) a signal processing and readout unit, usually consisting of electronic hardware and, in modern instruments, a computer. Figure 25-1 illustrates the three configurations of these components for carrying out optical spectroscopic measurements. As can be seen in the figure, components (3), (4), and (5) have similar configurations for each type of measurement.

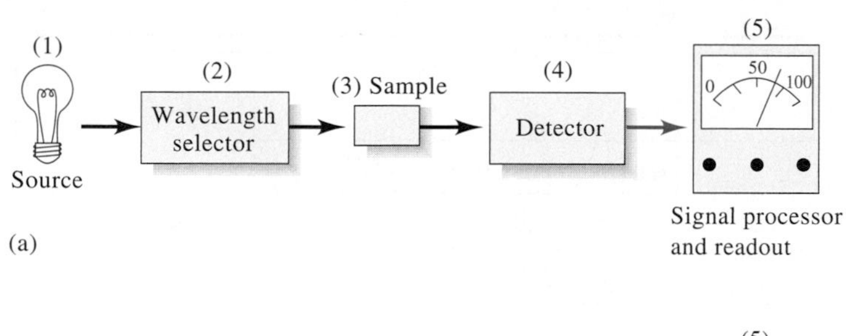

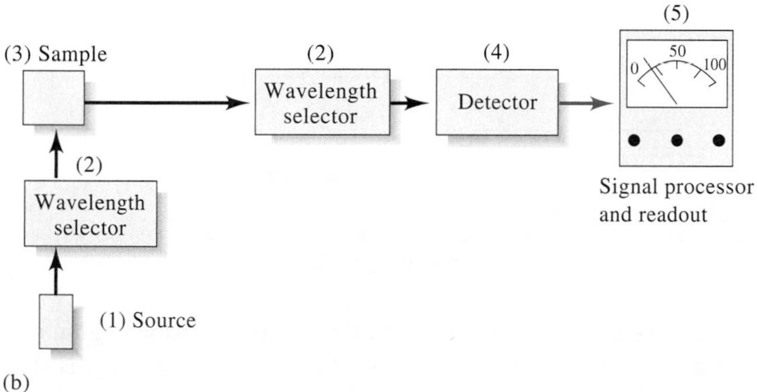

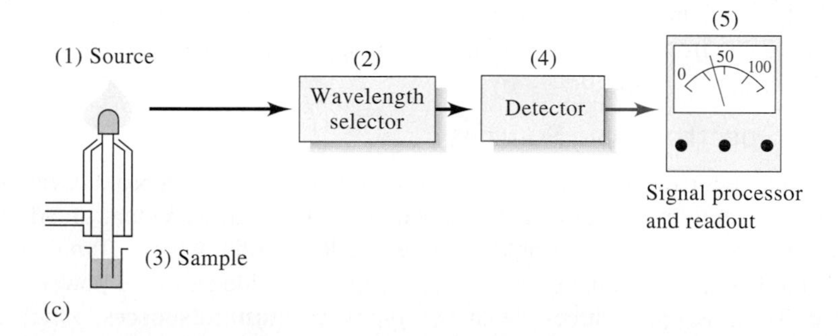

Figure 25-1 Components of various types of instruments for optical spectroscopy. In (a), the arrangement for absorption measurements is shown. Note that source radiation of the selected wavelength is sent through the sample, and the transmitted radiation is measured by the detector/signal processing/readout unit. With some instruments, the position of the sample and wavelength selector is reversed. In (b), the configuration for fluorescence measurements is shown. Here, two wavelength selectors are needed to select the excitation and emission wavelengths. The selected source radiation is incident on the sample and the radiation emitted is measured, usually at right angles to avoid scattering. In (c), the configuration for emission spectroscopy is shown. Here, a source of thermal energy, such as a flame or plasma, produces an analyte vapor that emits radiation isolated by the wavelength selector and converted to an electrical signal by the detector.

The first two designs, for absorption and fluorescence spectroscopy, require an external source of radiation. In absorption measurements (see Figure 25-1a), the attenuation of the source radiation at the selected wavelength is measured. In fluorescence measurements (see Figure 25-1b), the source excites the analyte and causes the emission of characteristic radiation, which is usually measured at a 90-degree angle with respect to the incident source beam. In emission spectroscopy (see Figure 25-1c), the sample itself is the emitter and no external radiation source is needed. In emission methods, the sample is usually fed into a plasma or a flame, which provides enough thermal energy to cause the analyte to emit characteristic radiation. Fluorescence and emission methods are described in more detail in Chapters 27 and 28, respectively.

25A-1 Optical Materials

The cells, windows, lenses, mirrors, and wavelength-selecting elements in an optical spectroscopic instrument must transmit radiation in the wavelength region being investigated. Figure 25-2 shows the usable wavelength range for several optical materials that are used in the UV, visible, and IR regions of the spectrum. Ordi-

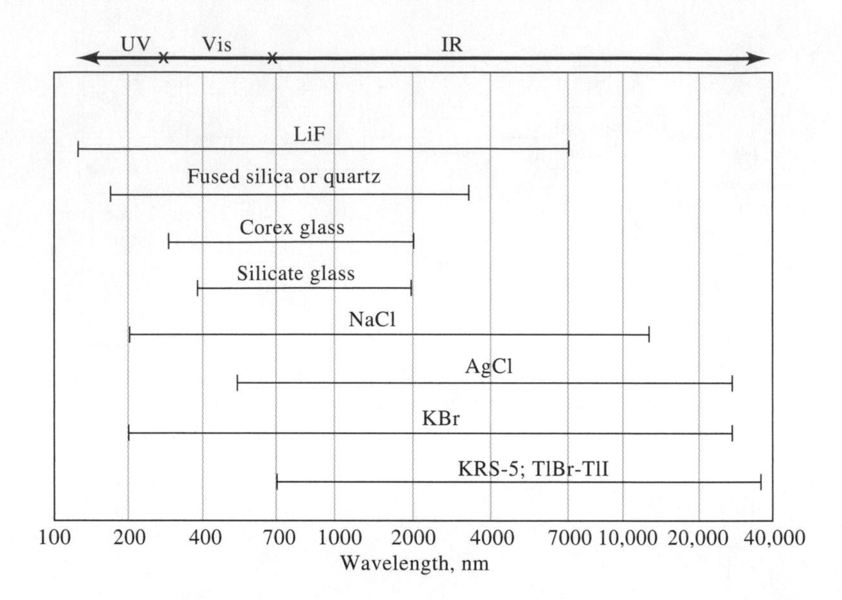

Figure 25-2 Transmittance ranges for various optical materials. Simple glasses are fine in the visible region, while fused silica or quartz is necessary in the UV region (<380 nm). Halide salts (KBr, NaCl, AgCl) are often used in the IR region but have the disadvantages of being expensive and somewhat water soluble.

nary silicate glass is completely adequate for use in the visible region and has the considerable advantage of low cost. In the UV region at wavelengths shorter than about 380 nm, glass begins to absorb, and fused silica or quartz must be substituted. Also, in the IR region, glass, quartz, and fused silica all absorb at wavelengths longer than about 2.5 μm. Hence, optical elements for IR spectrometry are typically made from halide salts or, in some cases, polymeric materials.

25A-2 Spectroscopic Sources

A continuum source provides a broad distribution of wavelengths within a particular spectral range. This distribution is known as a **spectral continuum.**

To be suitable for spectroscopic studies, a source must generate a beam of radiation that is sufficiently powerful to allow easy detection and measurement. In addition, its output power should be stable for reasonable periods of time. Typically, for good stability, a well-regulated power supply must provide electrical power for the source. Spectroscopic sources are of two types: **continuum sources,** which emit radiation that changes in intensity only slowly as a function of wavelength, and **line sources,** which emit a limited number of spectral lines, each of which spans a very limited wavelength range. The distinction between these sources is illustrated in Figure 25-3. Sources can also be classified as **continuous sources,** which emit radiation continuously with time, or **pulsed sources,** which emit radiation in bursts.

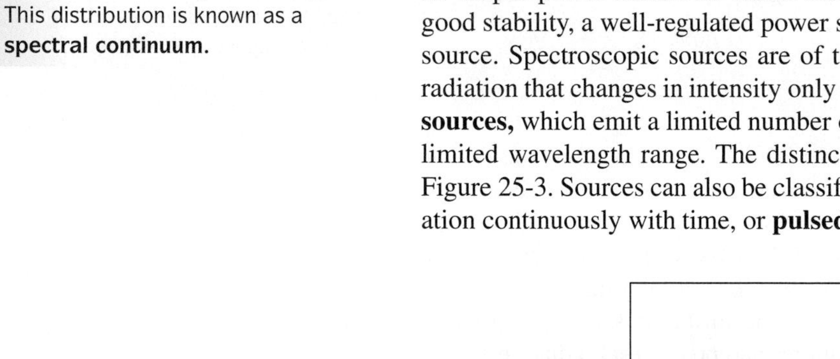

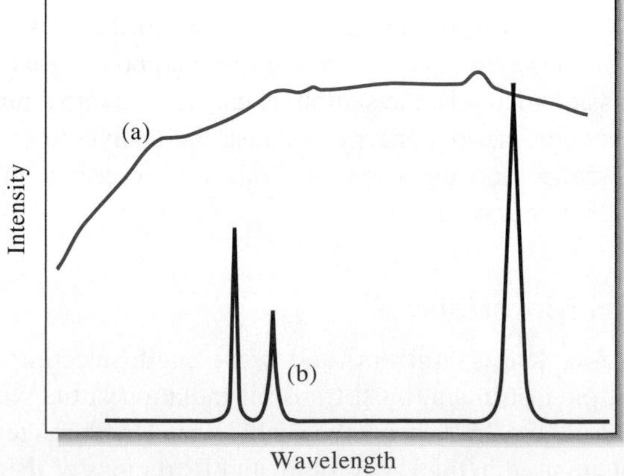

Figure 25-3 Spectral source types. The spectrum of a continuum source (a) is much broader than that of a line source (b).

TABLE 25-1
Continuum Sources for Optical Spectroscopy

Source	Wavelength Region, nm	Type of Spectroscopy
Xenon arc lamp	250–600	Molecular fluorescence
H_2 and D_2 lamps	160–380	UV molecular absorption
Tungsten/halogen lamp	240–2500	UV/visible/near-IR molecular absorption
Tungsten lamp	350–2200	Visible/near-IR molecular absorption
Nernst glower	400–20,000	IR molecular absorption
Nichrome wire	750–20,000	IR molecular absorption
Globar	1200–40,000	IR molecular absorption

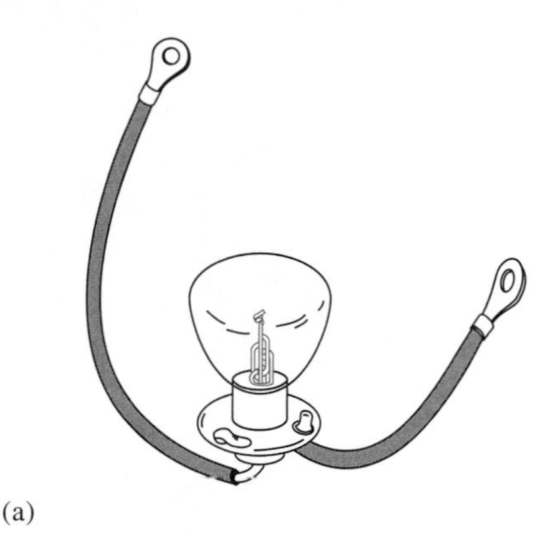
(a)

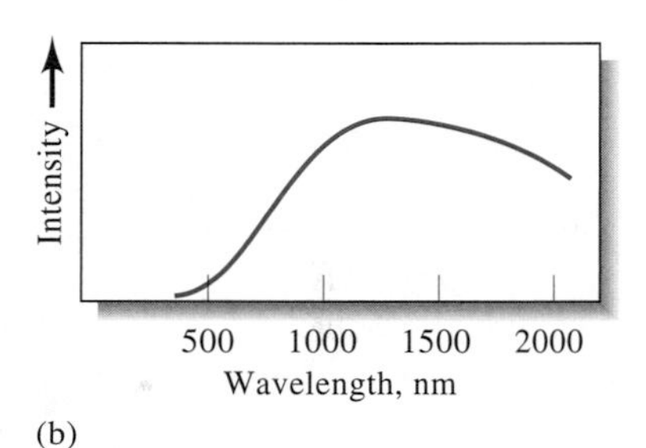

(b)

Figure 25-4 (a) A tungsten lamp of the type used in spectroscopy and its spectrum (b). Intensity of the tungsten source is usually quite low at wavelengths shorter than about 350 nm. Note that the intensity reaches a maximum in the near-IR region of the spectrum (~1200 nm in this case).

Continuum Sources in the Ultraviolet/Visible Region

The most widely used continuum sources in the UV/visible range are listed in Table 25-1. An ordinary tungsten filament lamp provides a distribution of wavelengths from 320 to 2500 nm (Figure 25-4). Generally, these lamps are operated at a temperature of around 2900 K, which produces useful radiation from about 350 to 2200 nm.

Tungsten/halogen lamps, also called quartz/halogen lamps, contain a small amount of iodine within the quartz envelope that houses the filament. Quartz allows the filament to be operated at a temperature of about 3500 K, which leads to higher intensities and extends the range of the lamp well into the UV region. The lifetime of a tungsten/halogen lamp is more than double that of an ordinary tungsten lamp because the life of the latter is limited by sublimation of tungsten from the filament. In the presence of iodine, the sublimed tungsten reacts to give gaseous WI_2 molecules, which then diffuse back to the hot filament, where they decompose and redeposit as W atoms. These lamps are finding ever-increasing use in modern spectroscopic instruments because of their extended wavelength range, greater intensity, and longer life.

Deuterium (and also hydrogen) lamps are most often used to provide continuum radiation in the UV region. A deuterium lamp consists of a cylindrical tube containing deuterium at low pressure, with a quartz window from which the radiation exits (Figure 25-5). The mechanism by which a continuum is produced by this source involves the formation of an excited molecule D_2^* (or H_2^*) by absorption of electrical energy. This species then dissociates to give two hydrogen or deuterium atoms plus an ultraviolet photon. The reactions for hydrogen are

$$H_2 + E_e \rightarrow H_2^* \rightarrow H' + H'' + h\nu$$

where E_e is the electrical energy absorbed by the molecule. The energy for the overall process is

$$E_e = E_{H_2^*} = E_{H'} + E_{H''} + h\nu$$

where $E_{H_2^*}$ is the fixed quantized energy of H_2^*, and H′ and H″ are the kinetic energies of the two hydrogen atoms. The sum of the latter two energies can vary from zero to $E_{H_2^*}$. Thus, the energy and the frequency of the photon can also vary within this range of energies. That is, when the two kinetic energies are by chance small, $h\nu$ is large, and when the two energies are large, $h\nu$ is small. As a result, hydrogen lamps produce a true continuum spectrum from about 160 nm to the beginning of

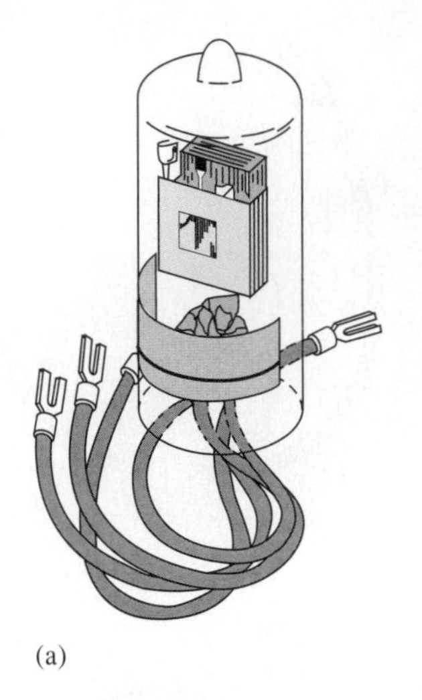

(a)

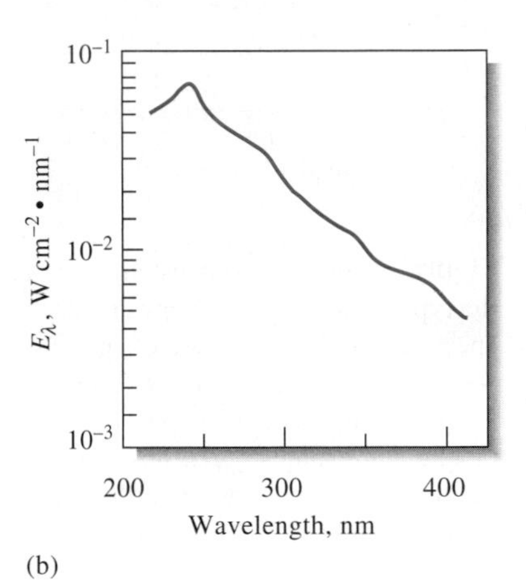

(b)

Figure 25-5 (a) A deuterium lamp of the type used in spectrophotometers and (b) its spectrum. Note that the maximum intensity occurs at ~225 nm. Typically, instruments switch from deuterium to tungsten at ~350 nm.

the visible region. Most modern lamps used to generate ultraviolet radiation contain deuterium and are low voltage; in them, an arc is formed between a heated, oxide-coated filament and a metal electrode (see Figure 25-5a). The heated filament provides electrons to maintain a direct current at a potential of about 40 V; a regulated power supply is required for constant intensities. Both deuterium and hydrogen lamps provide a useful continuum spectrum in the region from 160 to 375 nm, as shown in Figure 25-5b. The deuterium lamp is more widely used than the hydrogen lamp, however, because the deuterium lamp is more intense. At longer wavelengths (>360 nm), the lamps generate emission lines, which are superimposed on the continuum. For many applications, these lines are a nuisance, but they are useful for wavelength calibration of absorption instruments.

Other Ultraviolet/Visible Sources

In addition to the continuum sources just discussed, line sources are also important for use in the UV/visible region. Low-pressure mercury arc lamps are very common sources that are used in liquid chromatography detectors. The dominant line emitted by these sources is the 253.7-nm Hg line. Hollow-cathode lamps are also common line sources that are specifically used for atomic absorption spectroscopy, as discussed in Chapter 28. Lasers (see Feature 25-1) have also been used in molecular and atomic spectroscopy, both for single-wavelength and for scanning applications. Tunable dye lasers can be scanned over wavelength ranges of several hundred nanometers when more than one dye is used.

FEATURE 25-1
Laser Sources: The Light Fantastic

Lasers have become useful as sources in certain types of analytical spectroscopy. To understand how a laser works, consider an assembly of atoms or molecules interacting with an electromagnetic wave. For simplicity, we consider the atoms or molecules to have two energy levels: an upper level 2 with energy E_2 and a lower level 1 with energy E_1. If the electromagnetic wave is of a frequency corresponding to the energy difference between the two levels, excited species in level 2 can be stimulated to emit radiation of the same frequency and phase as the original electromagnetic wave. Each **stimulated emission** generates a photon, while each absorption removes a photon. The number of photons per second, called the **radiant flux Φ,** changes with distance as the radiation interacts with the assembly of atoms or molecules. The change in flux, $d\Phi$, is proportional to the flux itself, to the difference in the populations of the levels, $n_2 - n_1$, and to the path length of the interaction, dz, according to

$$d\Phi = k\Phi(n_2 - n_1)\,dz$$

where k is a proportionality constant related to the absorptivity of the absorbing species. If the upper-level population can be made to exceed that of the lower level, there will be a net gain in flux, and the system will behave as an amplifier. If $n_2 > n_1$, the atomic or molecular system is said to be an **active medium** and to have undergone **population inversion.** The resulting amplifier is called a **laser,** which stands for **l**ight **a**mplification by **s**timulated **e**mission of **r**adiation.

The optical amplifier can be converted into an oscillator by placing the active medium inside a resonant cavity made from two mirrors, as shown in Figure

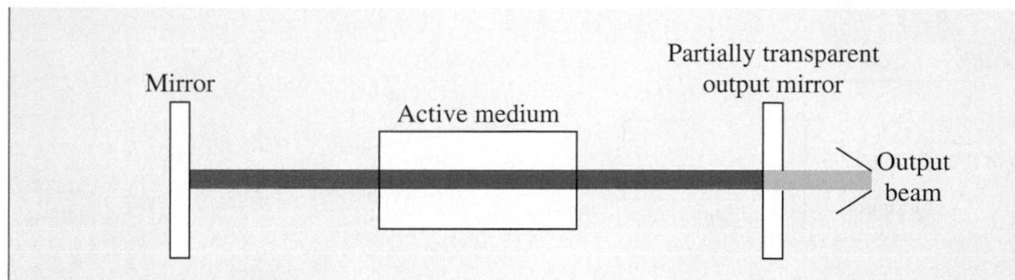

Figure 25F-1 Laser cavity. The electromagnetic wave travels back and forth between the mirrors, and the wave is amplified with each pass. The output mirror is partially transparent to allow only a fraction of the beam to pass out of the cavity.

25F-1. When the gain of the active medium equals the losses in the system, laser oscillation begins.

Population inversion is often achieved by a multi-level atomic or molecular system in which the excitation process, called **pumping,** is accomplished by electrical means, by optical methods, or by chemical reactions. In some cases, the population inversion can be sustained to produce a **continuous wave** (CW) output beam that is continuous with respect to time. In other cases, the lasing action is **self terminating,** so that the laser is operated in a pulsed mode to produce a repetitive pulse train or a single-shot action.[1]

There are many types of lasers. The first operating lasers were **solid-state lasers,** in which the active medium was a ruby crystal. In addition to the ruby laser, there are many other solid-state varieties. A widely used material contains a small concentration of Nd^{3+} embedded in a yttrium-aluminum-garnet (YAG) host. The active material is shaped into a rod and pumped optically by a flashlamp, as illustrated in Figure 25F-2a. The transitions involved are shown in Figure 25F-2b. The Nd:YAG laser generates nanosecond pulses with a very high output power at a wavelength of 1.06 μm. The Nd:YAG laser is very popular as a pumping source for tunable dye lasers.

The very common helium-neon (He-Ne) laser is a **gas laser** that operates in a CW mode. It is widely used as an optical alignment aid and as a source for some types of spectroscopy. The nitrogen laser lases on a transition of the nitrogen molecule at 337.1 nm. It is a self-terminating pulsed laser that requires a very short electrical pulse to pump the appropriate transitions. The N_2 laser is also very popular for pumping tunable dye lasers, as discussed later. **Excimer** (excited dimer or trimer) lasers are among the newest gas lasers. Rare-gas halide excimer lasers were first demonstrated in 1975. In one popular type, a gas mixture of Ar, F_2, and He produces ArF excimers when subjected to an electrical discharge. The excimer laser is an important UV source for photochemical studies, for fluorescence applications, and for pumping tunable dye lasers.

Dye lasers are liquid lasers containing a fluorescent dye such as one of the rhodamines, a coumarin, or a fluorescein. These have been made to lase at wavelengths from the IR region to the UV region. Lasing typically occurs between the first excited singlet state and the ground state. The lasers can be pumped by flashlamps or by another laser, such as those discussed previously.

(continued)

[1]For additional information, see J. D. Ingle, Jr., and S. R. Crouch, *Spectrochemical Analysis.* Upper Saddle River, NJ: Prentice-Hall, 1988.

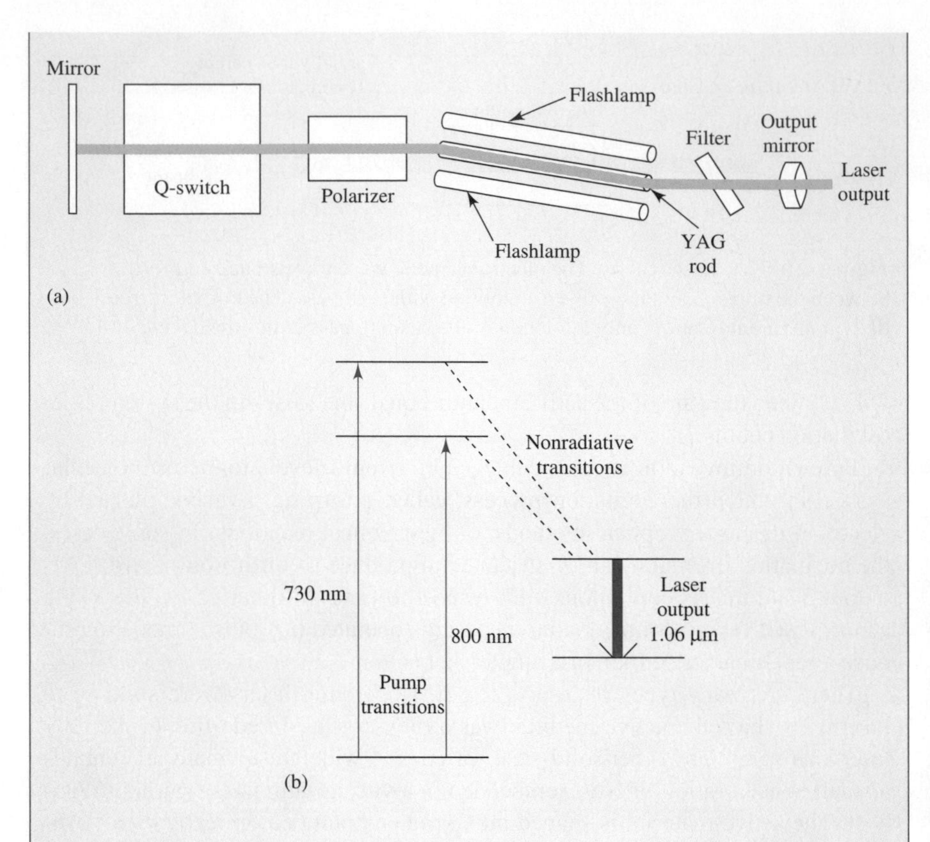

Figure 25F-2 Schematic of a Nd:YAG laser (a) and energy levels (b). The pump transitions are in the red region of the spectrum, and the laser output is in the near-infrared. The laser is flashlamp pumped. The region between the two mirrors is the laser cavity.

Lasing can be sustained over a continuous range of wavelengths on the order of 40 to 50 nm. The broad band over which lasing occurs makes the dye laser suitable for tuning by inserting a grating, a filter, a prism, or an interferometric element into the laser cavity. Dye lasers are very useful for molecular fluorescence spectroscopy and many other applications.

Semiconductor lasers, also known as **diode lasers,** obtain population inversion between the conduction band and the valence band of a *pn*-junction diode. Various compositions of the semiconductor material can be used to give different output wavelengths. Diode lasers can be tuned over small wavelength intervals. Such lasers produce outputs in the IR region of the spectrum. They have become extremely useful in CD players, CD-ROM drives, laser printers, and spectroscopic applications, such as Raman spectroscopy.

Laser radiation is highly directional, spectrally pure, coherent, and of high intensity. These properties have made possible many unique research applications that cannot easily be achieved with conventional sources. Despite the many advances in laser science and technology, only recently have lasers become routinely useful in analytical instruments. Even today, the high-powered Nd:YAG and excimer lasers are difficult to align and use. We should see many new developments in laser technology in the near future.

Continuum Sources in the Infrared Region

The continuum sources for IR radiation are normally heated inert solids. A **Globar** source is a silicon carbide rod; infrared radiation is emitted when the Globar is heated to about 1500°C by the passage of electricity. Table 25-1 gives the wavelength range of these sources.

A **Nernst glower** is a cylinder of zirconium and yttrium oxides that emits IR radiation when heated to a high temperature by an electric current. Electrically heated spirals of nichrome wire also serve as inexpensive IR sources.

25A-3 Wavelength Selectors

Spectroscopic instruments in the UV and visible regions are usually equipped with one or more devices to restrict the radiation being measured to a narrow band that is absorbed or emitted by the analyte. Such devices greatly enhance both the selectivity and the sensitivity of an instrument. In addition, for absorption measurements—as we saw in Section 24C-2—narrow bands of radiation greatly diminish the chance of Beer's law deviations due to polychromatic radiation. Many instruments use a **monochromator** or **filter** to isolate the desired wavelength band so that only the band of interest is detected and measured. Others use a **spectrograph** to spread out, or disperse, the wavelengths so that they can be detected with a multichannel detector.

Monochromators and Polychromators

Monochromators generally have a diffraction grating (see Feature 25-3) to disperse the radiation into its component wavelengths, as shown in Figure 25-6a. By rotat-

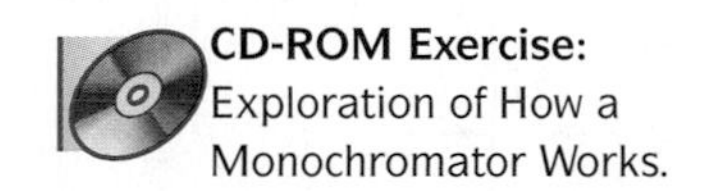
CD-ROM Exercise: Exploration of How a Monochromator Works.

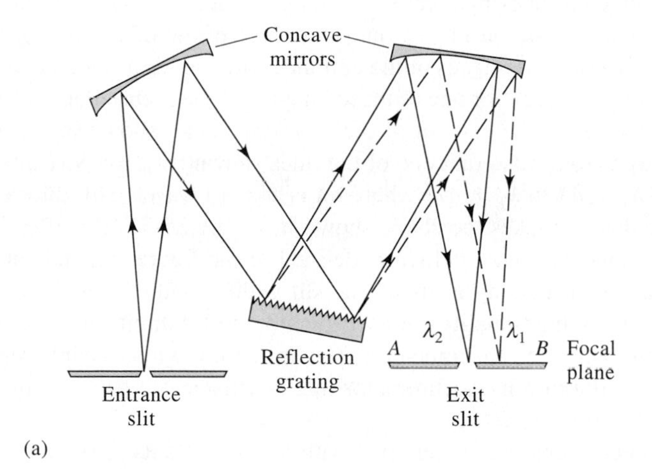

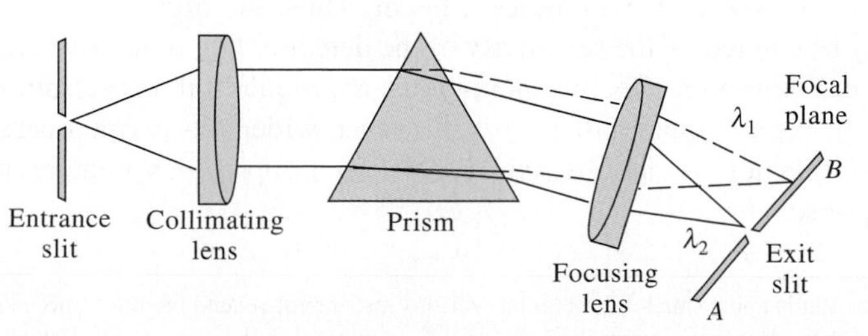

Figure 25-6 Types of monochromators: (a) grating monochromator; (b) prism monochromator. The monochromator design in (a) is a Czerny-Turner design, while the prism monochromator in (b) is a Bunsen design. In both cases, $\lambda_1 > \lambda_2$.

A **spectrograph** is a device that uses a grating to disperse a spectrum. It contains an entrance slit to define the area of the source to be viewed. A large opening at its exit allows a wide range of wavelengths to strike a multi-wavelength detector. A **monochromator** is a device that contains an entrance slit and an exit slit. The latter is used to isolate a small band of wavelengths. One band at a time is isolated, and different bands can be transmitted sequentially by rotating the grating. A **polychromator** contains multiple exit slits so that several wavelength bands can be isolated simultaneously.

ing the grating, different wavelengths can be made to pass through an exit slit. Older instruments used prisms for this purpose (see Figure 25-6b). The output wavelength of a monochromator is thus continuously variable over a considerable spectral range. The wavelength range passed by a monochromator, called the **spectral band-pass** or **effective bandwidth,** can be less than 1 nm for moderately expensive instruments to greater than 20 nm for inexpensive systems. Because of the ease with which the wavelength can be changed with a monochromator-based instrument, these systems are widely used for spectral scanning applications as well as applications requiring a fixed wavelength. With an instrument containing a spectrograph, the sample and wavelength selector are reversed from the configuration shown in Figure 25-1. Like the monochromator, the spectrograph contains a diffraction grating to disperse the spectrum. The spectrograph has no exit slit, however, and this allows the dispersed spectrum to strike a multi-wavelength detector. Still other instruments used for emission spectroscopy contain a device called a **polychromator,** which contains multiple exit slits and multiple detectors. This allows many discrete wavelengths to be measured simultaneously.

Figure 25-6a shows the design of a typical grating monochromator. Radiation from a source enters the monochromator via a narrow rectangular opening, or slit. The radiation is then collimated by a concave mirror, which produces a parallel beam that strikes the surface of a reflection grating. Angular dispersion results from diffraction, which occurs at the reflective surface. For illustrative purposes, the radiation entering the monochromator is shown as consisting of just two wavelengths, λ_1 and λ_2, where λ_1 is longer than λ_2. The pathway of the longer radiation after it is reflected from the grating is shown by the dashed lines; the solid lines show the path of the shorter wavelength. Note that the shorter wavelength radiation λ_2 is reflected off the grating at a sharper angle than is λ_1. That is, **angular dispersion** of the radiation takes place at the grating surface. The two wavelengths are focused by another concave mirror onto the **focal plane** of the monochromator, where they appear as two images of the entrance slit, one for λ_1 and the other as λ_2. By rotating the grating, either one of these images can be focused on the exit slit. If a detector is located at the exit slit of the monochromator shown in Figure 25-6a, and the grating is rotated so that one of the lines shown (say, λ_1) is scanned across the slit from $\lambda_1 - \delta\lambda$ to $\lambda_1 + \delta\lambda$ (where $\delta\lambda$ is a small wavelength difference), the output of the detector takes the shape shown in Figure 25-7.[2] The effective bandwidth of the monochromator, which is defined in the figure, depends on the size and quality of the dispersing element, the slit widths, and the focal length of the monochromator. A high-quality monochromator will exhibit an effective bandwidth of a few tenths of a nanometer or less in the ultraviolet/visible region. The effective bandwidth of a monochromator that is satisfactory for most quantitative applications is about 1 to 20 nm.

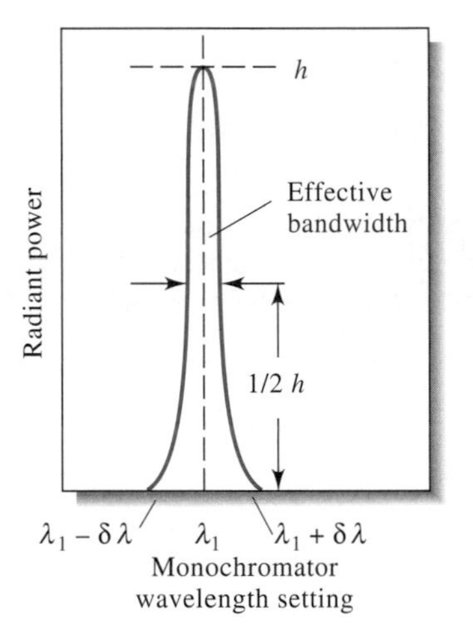

Figure 25-7 Output of an exit slit as the monochromator is scanned from $\lambda_1 - \delta\lambda$ to $\lambda_1 + \delta\lambda$.

The **effective bandwidth** of a wavelength selector is the width of the band of radiation in wavelength units at half-peak height.

Many monochromators are equipped with adjustable slits to permit some control over the bandwidth. A narrow slit decreases the effective bandwidth but also diminishes the power of the emergent beam. Thus, the minimum practical bandwidth may be limited by the sensitivity of the detector. For qualitative analysis, narrow slits and minimum effective bandwidths are required if a spectrum is made up of narrow peaks. For quantitative work, however, wider slits permit operation of the detector system at lower amplification, which in turn provides greater reproducibility of response.

[2]The slit function is approximately triangular. Various instrument-related factors combine to produce the shape shown in Figure 25-7.

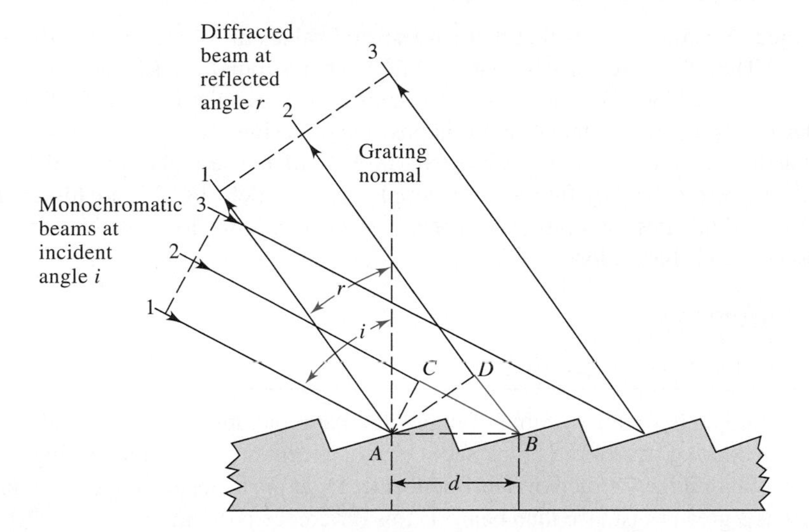

Figure 25-8 Mechanism of diffraction from an echellette-type grating. Angle i from the grating normal is the angle of the incident beam; angle r is the angle of the reflected beam. The distance between successive rulings is d.

Gratings

Most gratings in modern monochromators are replica gratings, which are obtained by making castings of a master grating. The latter consists of a hard, optically flat, polished surface on which a suitably shaped diamond tool has created a large number of parallel and closely spaced grooves. A magnified cross-sectional view of a few typical grooves is shown in Figure 25-8. A grating for the ultraviolet and visible region will typically contain 300 to 2000 grooves/mm, with 1200 to 1400 being most common. The construction of a good master grating is tedious, time-consuming, and expensive because the grooves must be identical in size, exactly parallel, and equally spaced over the length of the grating (3 to 10 cm). Replica gratings are formed from a master grating by a liquid resin casting process that preserves virtually perfectly the optical accuracy of the original master grating on a clear resin surface. This surface is ordinarily made reflective by a coating of aluminum or, sometimes, gold or platinum.

The Echellette Grating One of the most common types of reflection gratings is the echellette grating. Figure 25-8 is a schematic representation of this type of grating, which is grooved or blazed so that it has relatively broad faces from which reflection occurs as well as narrow unused faces.[3] This geometry provides highly efficient diffraction of radiation. In the figure, a parallel beam of monochromatic radiation is approaching the grating surface and an angle i relative to the grating normal. The incident beam is made up of three parallel beams that create a wave front labeled 1, 2, 3. The diffracted beam is reflected at the angle r, which depends on the wavelength of the radiation. In Feature 25-2, we demonstrate that the angle of reflection r is related to the wavelength of the incoming radiation by the equation

$$\mathbf{n}\lambda = d(\sin i + \sin r) \tag{25-1}$$

[3]The echellette grating is blazed for use in relatively low orders, but an **echelle grating** is used in high orders (>10). The echelle grating is often used with a second dispersive element such as a prism to sort out overlapping orders and to provide cross-dispersion. For more on echelle gratings and how they are used, see D. A. Skoog, F. J. Holler, and T. A. Nieman, *Principles of Instrumental Analysis,* 5th ed., Section 10A-3. Belmont, CA: Brooks/Cole, 1998; and J. D. Ingle, Jr., and S. R. Crouch, *Spectrochemical Analysis,* Section 3–5. Upper Saddle River, NJ: Prentice-Hall, 1988.

Equation 25-1 suggests that there are several values of λ for a given diffraction angle *r*. Thus, if a first-order line ($\mathbf{n} = 1$) of 900 nm is found at *r*, second-order (450 nm) and third-order (300 nm) lines also appear at this angle. Ordinarily, the first-order line is the most intense, and it is possible to design gratings that concentrate as much as 90% of the incident intensity in this order. The higher order lines can generally be removed by filters. For example, glass, which absorbs radiation below 350 nm, eliminates the high-order spectra associated with first-order radiation in most of the visible region.

FEATURE 25-2

Derivation of Equation 25-1

In Figure 25-8, parallel beams of monochromatic radiation labeled 1 and 2 are shown striking two of the broad faces at an incident angle *i* to the grating normal. Maximum constructive interference occurs at the reflected angle *r*. Beam 2 travels a greater distance than beam 1; this difference is equal to $\overline{CB} + \overline{BD}$. For constructive interference to occur, this difference must equal $\mathbf{n}\lambda$:

$$\mathbf{n}\lambda = \overline{CB} + \overline{BD}$$

where **n,** a small whole number, is called the **diffraction order.** Note, however, that angle *CAB* is equal to angle *i* and that angle *DAB* is identical to angle *r*. Therefore, from trigonometry,

$$\overline{CB} = d \sin i$$

where *d* is the spacing between the reflecting surfaces. We also see that

$$\overline{BD} = d \sin r$$

Substituting the last two expressions into the first gives Equation 25-1. That is,

$$\mathbf{n}\lambda = d(\sin i + \sin r)$$

Note that when diffraction occurs to the left of the grating normal, values of **n** are positive, and when diffraction occurs to the right of the grating normal, **n** is negative. Thus, $\mathbf{n} = \pm 1, \pm 2, \pm 3$, and so forth.

One of the advantages of a monochromator with an echellette grating is that in contrast to a prism monochromator, the dispersion of radiation along the focal plane is, for all practical purposes, linear. Figure 25-9 demonstrates this property. The linear dispersion of a grating greatly simplifies the design of monochromators.

Concave Gratings Gratings can be formed on a concave surface in much the same way as on a plane surface. A concave grating permits a monochromator design without auxiliary collimating and focusing mirrors or lenses because the concave surface both disperses the radiation and focuses it on the exit slit. Such an arrangement is advantageous in terms of cost; in addition, the reduction in number of optical surfaces increases the energy throughput of a monochromator containing a concave grating.

EXAMPLE 25-1

An echellette grating containing 1450 blazes per millimeter was irradiated with a polychromatic beam at an incident angle 48 degrees to the grating normal. Calculate the wavelengths of radiation that would appear at an angle of reflection of +20, +10, and 0 degrees (angle *r*, Figure 25-8). To obtain *d* in Equation 25-1, we write

$$d = \frac{1 \text{ mm}}{1450 \text{ blazes}} \times 10^6 \frac{\text{nm}}{\text{mm}} = 689.7 \frac{\text{nm}}{\text{blaze}}$$

When *r* in Figure 25-8 equals +20 deg, λ can be obtained by substituting into Equation 25-1. Thus,

$$\lambda = \frac{689.7}{\mathbf{n}} \text{nm} (\sin 48 + \sin 20) = \frac{748.4}{\mathbf{n}} \text{nm}$$

and the wavelengths for the first-, second-, and third-order reflections are 748, 374, and 249 nm, respectively. Further calculations of a similar kind yield the following data:

	Wavelength (nm) for		
***r*, deg**	**n = 1**	**n = 2**	**n = 3**
20	748	374	249
10	632	316	211
0	513	256	171

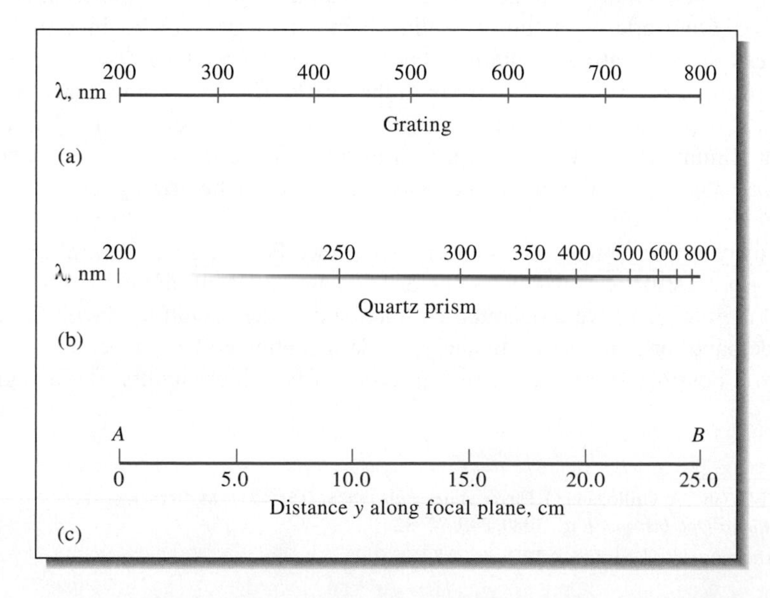

Figure 25-9 Dispersion of radiation along the focal plane *AB* of a typical prism (a) and echellette grating (b). The positions of *A* and *B* in the scale in (c) are shown in Figure 25-6.

Holographic Grating[4] One of the products emerging from laser technology is an optical (rather than mechanical) technique for forming gratings on plane or concave glass surfaces. Holographic gratings produced in this way are appearing in ever-increasing numbers in modern optical instruments, even some of the less expensive ones. Holographic gratings, because of their greater perfection with respect to line shape and dimensions, provide spectra that are freer from stray radiation and ghosts (double images).[5] See Feature 25-3 for a description of the ruling process for both mechanically ruled and holographically ruled gratings.

FEATURE 25-3

Ruling Gratings

Dispersion of UV/visible radiation can be brought about by directing a polychromatic beam through a **transmission grating** or onto the surface of a **reflection grating.** The reflection grating is by far more common. **Replica gratings,** which are used in many monochromators, are manufactured from a **master grating.** The master grating consists of a large number of parallel and closely spaced grooves ruled on a hard, polished surface with a suitably shaped diamond tool. For the UV/visible region, a grating will contain 50 to 6000 grooves mm^{-1}, with 1200 to 2400 being most common. Master gratings are ruled by a diamond tool that is operated by a ruling engine. The construction of a good master grating is tedious, time-consuming, and expensive, because the grooves must be identical in size, exactly parallel, and equally spaced over the typical 3- to 10-cm length of the grating. Because of the difficulty in construction, few master gratings are produced.

The modern era of gratings dates to the 1880s, when Rowland constructed an engine capable of ruling gratings up to 6 inches in width with more than 100,000 grooves. A simplified drawing of the Rowland engine is shown in Figure 25F-3. With this machine, a high-precision screw moves the grating carriage, while a diamond stylus cuts the tiny parallel grooves. Imagine manually ruling a grating with 100,000 grooves in a 6-inch width! The engine required around 5 hours just to warm up to a nearly uniform temperature. After this, nearly 15 hours more were needed to obtain a uniform layer of lubricant on the surface. Only after this time was the diamond lowered to begin the ruling process. Large gratings required almost a week to produce. Two important improvements were made by Strong in the 1930s. The most significant was the vacuum deposition of aluminum onto glass blanks as a medium. The thin layer of aluminum gave a much smoother surface and reduced wear on the diamond tool. Strong's second improvement was to reciprocate the grating blank instead of the diamond tool.

Today, ruling engines use interferometric (see Feature 25-7) control over the ruling process. Fewer than 50 ruling engines are in use around the world. Even if all these engines were operated 24 hours a day, they could not begin to meet the demand for gratings. Fortunately, modern coating and resin technology has made it possible to produce replica gratings of very high quality. Replica grat-

---

[4]See J. Flamand, A. Grillo, and G. Hayat, *Amer. Lab.*, **1975,** *7*(5), 47; J. M. Lerner, et al., *Proc. Photo-Opt. Instrum. Eng.*, **1980,** *240,* 72, 82.

[5]I. R. Altelmose, *J. Chem. Educ.*, **1986,** *63,* A221.

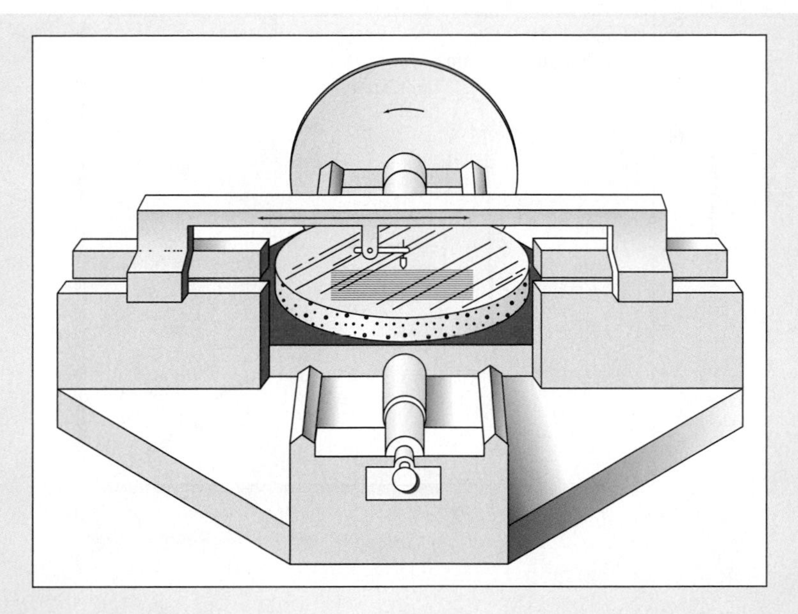

Figure 25F-3 Simplified diagram of the Rowland ruling engine. A single high-precision screw moves the grating carriage. A diamond point then travels over the grating, which is ruled on a concave mirror surface. Machines of this type were the models for many of the ruling engines constructed since Rowland's time. Ruling engines are among the most sensitive and precise macroscopic mechanical devices ever made. The resulting gratings have played an integral role in many of the most important advances in science over the past century.

ings are formed from the master grating by vacuum deposition of aluminum onto a ruled master grating. The aluminum layer is subsequently coated by an epoxy-type material. The material is then polymerized, and the replica is separated from the master. The replica gratings of today are superior to the master gratings produced in the past.

Another way of making gratings is a result of laser technology. These **holographic gratings** are made by coating a flat glass plate with a material that is photosensitive. Beams from a pair of identical lasers then strike the coated glass surface. The resulting interference fringes (see Feature 25-7) from the two beams sensitize the photoresist, producing areas that can be dissolved away, leaving a grooved structure. Aluminum is then vacuum deposited to produce a reflection grating. The spacing of the grooves can be changed by changing the angle of the two laser beams with respect to one another. Nearly perfect gratings with as many as 6000 lines per mm can be manufactured in this way at a relatively low cost. Holographic gratings are not quite as efficient in terms of their light output as ruled gratings; however, they can eliminate false lines, called **grating ghosts,** and reduce scattered light that results from ruling errors.

Radiation Filters

Filters operate by absorbing all but a restricted band of radiation from a continuum source. As shown in Figure 25-10, two types of filters are used in spectroscopy: **interference filters** and **absorption filters.** Interference filters are typically used

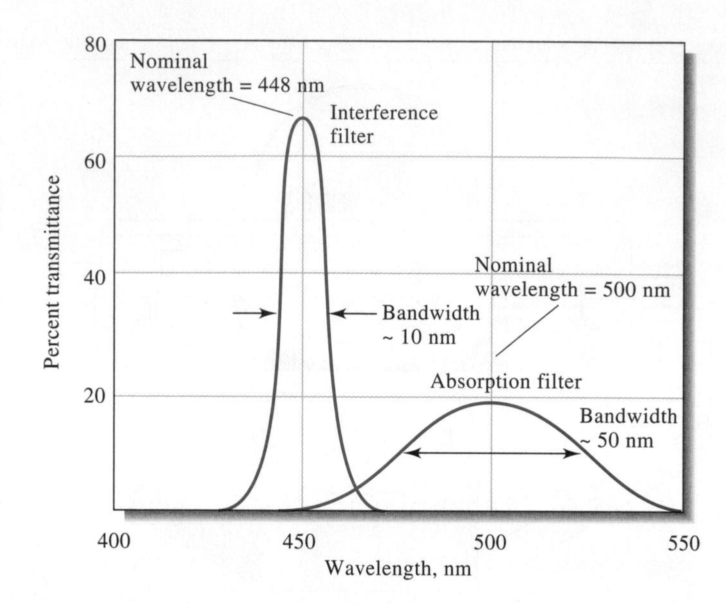

Figure 25-10 Bandwidths for two types of filters.

for absorption measurements, and they generally transmit a much greater fraction of radiation at their nominal wavelengths than do absorption filters.

Interference Filters Interference filters are used with ultraviolet and visible radiation, as well as with wavelengths up to about 14 μm in the infrared region. As the name implies, an interference filter relies on optical interference to provide a relatively narrow band of radiation, typically 5 to 20 nm in width. As shown in Figure 25-11a, an interference filter consists of a very thin layer of a transparent

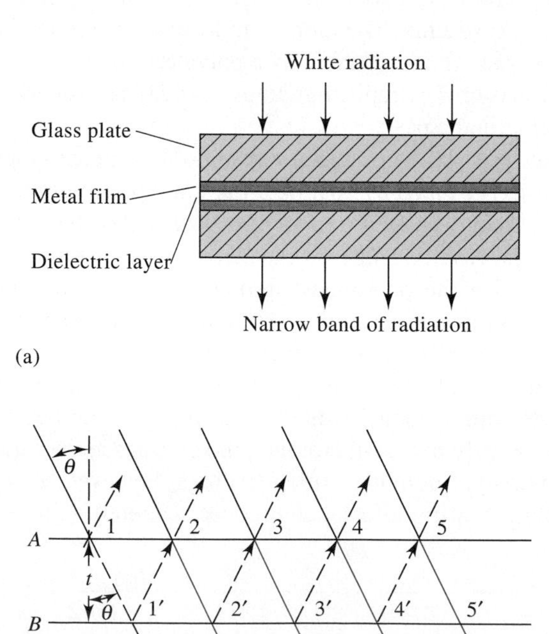

Figure 25-11 (a) Schematic cross section of an interference filter. Note that the drawing is not to scale, and the three central bands are much narrower than shown. (b) Schematic to show the conditions for constructive interference.

dielectric material (frequently calcium fluoride or magnesium fluoride) coated on both sides with a film of metal that is thin enough to transmit approximately half the radiation striking it and to reflect the other half. This array is sandwiched between two glass plates that protect it from the atmosphere. When radiation strikes the central array at a 90-degree angle, approximately half is transmitted by the first metallic layer and the other half reflected. The transmitted radiation undergoes a similar partition when it reaches the second layer of metal. If the reflected portion from the second layer is of the proper wavelength, it is partially reflected from the inner portion of the first layer in phase with the incoming light of the same wavelength. The result is constructive interference of the radiation of this wavelength and destructive removal of most other wavelengths. As shown in Feature 25-4, the nominal wavelength for an interference filter λ_{max} is given by the equation

$$\lambda_{max} = \frac{2t\eta}{\mathbf{n}} \quad (25\text{-}2)$$

where t is the thickness of the central fluoride layer, η is its refractive index, and **n** is an integer called the interference order. The glass layers of the filter are often selected to absorb all but one of the wavelengths transmitted by the central layer, thus restricting the transmission of the filter to a single order. A dielectric is a nonconducting substance or insulator. Such materials are usually optically transparent.

A **dielectric** is a nonconducting substance or insulator. Such materials are usually optically transparent.

FEATURE 25-4

Deriving Equation 25-2

The relationship between the thickness of the dielectric layer t and the transmitted wavelength λ can be found with the aid of Figure 25-11b. For clarity, the incident beam is shown as arriving at an angle θ from the perpendicular. At point 1, the radiation is partially reflected and partially transmitted to point 1′, where partial reflection and transmission again take place. The same process occurs at 2, 2′, and so forth. For reinforcement to occur at point 2, the distance traveled by the beam reflected at 1′ must be some multiple of its wavelength in the medium λ'. Since the path length between surfaces can be expressed as $t/\cos\theta$, the condition for reinforcement is that $\mathbf{n}\lambda' = 2t/\cos\theta$ where **n** is a small whole number.

In ordinary use, θ approaches zero and $\cos\theta$ approaches unity, so that the equation derived from Figure 25-11 simplifies to

$$\mathbf{n}\lambda' = 2t$$

where λ' is the wavelength of radiation *in the dielectric* and t is the thickness of the dielectric. The corresponding wavelength in air is given by

$$\lambda = \lambda'\eta$$

where η is the refractive index of the dielectric medium. Thus, the wavelengths of radiation transmitted by the filter are

$$\lambda = \frac{2t\eta}{\mathbf{n}}$$

Figure 25-10 illustrates the performance characteristics of a typical interference filter. Most filters of this type have bandwidths of less than 1.5% of the nominal wavelength, although this figure is lowered to 0.15% in some narrow-band filters; the latter have a maximum transmittance of about 10%.

Absorption Filters Absorption filters, which are generally less expensive and more rugged than interference filters, are limited in use to the visible region. This type of filter usually consists of a colored glass plate that removes part of the incident radiation by absorption. Absorption filters have effective bandwidths that range from perhaps 30 to 250 nm. Filters that provide the narrowest bandwidths also absorb a significant fraction of the desired radiation and may have a transmittance of 1% or less at their band peaks. Figure 25-10 contrasts the performance characteristics of a typical absorption filter with its interference counterpart. Glass filters with transmittance maxima throughout the entire visible region are available from commercial sources. While their performance characteristics are distinctly inferior to those of interference filters, their cost is appreciably less, and they are perfectly adequate for many routine applications.

Filters have the advantages of simplicity, ruggedness, and low cost. Since one filter can only isolate a single band of wavelengths; a new filter must be used for a different selection. Therefore, filter instruments are used only when measurements are made at a fixed wavelength or when the wavelength is changed infrequently.

In the IR region of the spectrum, most modern instruments do not disperse the spectrum at all, although this was common with older instruments. Instead, an **interferometer** is used, and the constructive and destructive interference of electromagnetic waves is used to obtain spectral information through a technique called Fourier transformation. These IR instruments are further discussed in Feature 25-7 and in Section 26C-2.

25A-4 Detecting and Measuring Radiant Energy

To obtain spectroscopic information, the radiant power transmitted, fluoresced, or emitted must be detected in some manner and converted into a measurable quantity. A **detector** is a device that indicates the existence of some physical phenomenon. Familiar examples of detectors include photographic film (for indicating the presence of electromagnetic or radioactive radiation), the pointer of a balance (for indicating mass differences), and the mercury level in a thermometer (for indicating temperature). The human eye is also a detector; it converts visible radiation into an electrical signal that is passed to the brain via a chain of neurons in the optic nerve and produces vision.

A **transducer** is a type of detector that converts various types of chemical and physical quantities into electrical signals such as electrical charge, current, or voltage.

Invariably in modern instruments, the information of interest is encoded and processed as an electrical signal. The term **transducer** is used to indicate the type of detector that converts quantities, such as light intensity, pH, mass, and temperature, into **electrical signals** that can be subsequently amplified, manipulated, and finally converted into numbers proportional to the magnitude of the original quantity. All the detectors discussed here are radiation transducers.

► Common noise sources include vibration, pickup from 60-Hz lines, temperature variations, frequency or voltage fluctuations in the power supply, and the random arrival of photons at the detector.

Properties of Radiation Transducers

The ideal transducer for electromagnetic radiation responds rapidly to low levels of radiant energy over a broad wavelength range. In addition, it produces an electrical signal that is easily amplified and has a low electrical noise level. Finally, it is

TABLE 25-2
Common Detectors for Absorption Spectroscopy

Type	Wavelength Range, nm
Photon Detectors	
Phototubes	150–1000
Photomultiplier tubes	150–1000
Silicon photodiodes	350–1100
Photoconductive cells	1000–50,000
Thermal Detectors	
Thermocouples	600–20,000
Bolometers	600–20,000
Pneumatic cells	600–40,000
Pyroelectric cells	1000–20,000

essential that the electrical signal produced by the transducer be directly proportional to the radiant power P of the beam, as shown in Equation 25-3:

$$G = KP + K' \quad (25\text{-}3)$$

where G is the electrical response of the detector in units of current, voltage, or charge. The proportionality constant K measures the sensitivity of the detector in terms of electrical response per unit of radiant power input.

Many detectors exhibit a small constant response K', known as a **dark current,** even when no radiation strikes their surfaces. Instruments with detectors that have a significant dark-current response are ordinarily capable of compensation so that the dark current is automatically subtracted. Thus, under ordinary circumstances, we can simplify Equation 25-3 to

Dark current is a current produced by a photoelectric detector when no light is falling on the detector.

$$G = KP \quad (25\text{-}4)$$

Types of Transducers

As shown in Table 25-2, there are two general types of transducers: one type responds to photons, the other to heat. All photon detectors are based on the interaction of radiation with a reactive surface either to produce electrons (**photoemission**) or to promote electrons to energy states in which they can conduct electricity (**photoconduction**). Only UV, visible, and near-IR radiation possess enough energy to cause photoemission to occur; thus, photoemissive detectors are limited to wavelengths shorter than about 2 μm (2000 nm). Photoconductors can be used in the near-, mid-, and far-IR regions of the spectrum.

FEATURE 25-5

Signals, Noise, and the Signal-To-Noise Ratio

The output of an analytical instrument fluctuates in a random way. These fluctuations limit the precision of the instrument and are the net result of a large number of uncontrolled random variables in the instrument and in the chemical system under study. An example of such a variable is the random arrival of photons at the photocathode of a photomultiplier tube. The term *noise* is used to describe these fluctuations, and each uncontrolled variable is a noise source.

(continued)

◀ Generally, the output from analytical instruments fluctuates in a random way as a consequence of the operation of a large number of uncontrolled variables. These fluctuations, which limit the sensitivity of an instrument, are called noise. The terminology is derived from radio engineering, where the presence of unwanted signal fluctuations is audible as static, or noise.

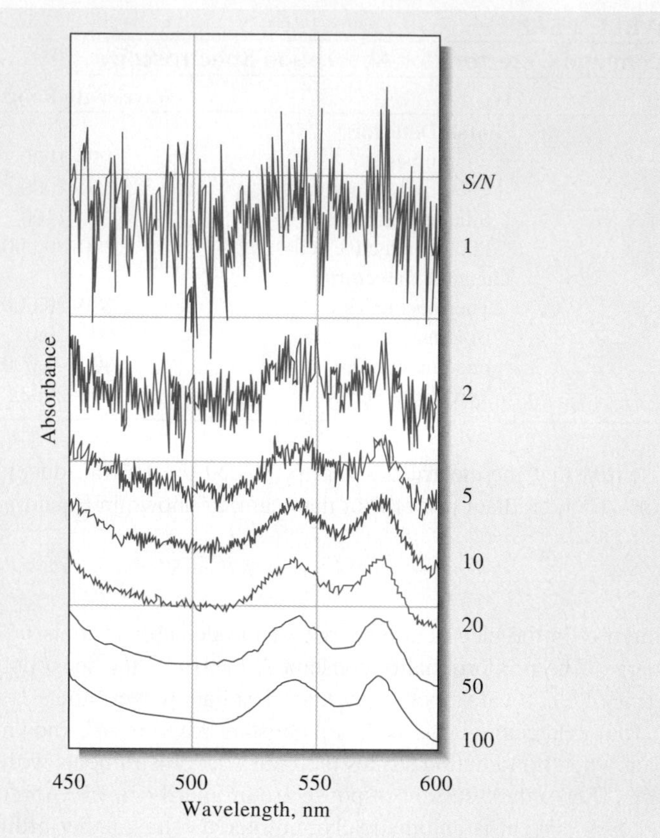

Figure 25F-4 Absorption spectra of hemoglobin with identical signal levels but different amounts of noise. Note that the curves have been offset on the absorbance axis for clarity.

The term comes from audio and electronic engineering, where undesirable signal fluctuations appear to the ear as static, or noise. The average value of the output of an electronic device is called the *signal,* and the standard deviation of the signal is a measure of the noise.

An important figure of merit for analytical instruments, stereos, compact-disk players, and many other types of electronic devices is the signal-to-noise ratio (S/N). The **signal-to-noise ratio** is usually defined as the ratio of the average value of the output signal to its standard deviation. The signal-to-noise behavior of an absorption spectrophotometer is illustrated in the spectra of hemoglobin shown in Figure 25F-4. The spectrum at the bottom of the figure has $S/N = 100$, and you can easily pick out the peaks at 540 nm and 580 nm. As the S/N degrades to about two in the second spectrum from the top of the figure, the peaks are barely visible. Somewhere between $S/N = 2$ and $S/N = 1$, the peaks disappear altogether into the noise and are impossible to identify. As modern instruments have become more computerized and controlled by sophisticated electronic circuits, various methods have been developed to increase the signal-to-noise ratio of instrument outputs. These methods include analog filtering, lock-in amplification, boxcar averaging, smoothing, and Fourier transformation.[6]

[6]D. A. Skoog, F. J. Holler, and T. A. Nieman, *Principles of Instrumental Analysis,* 5th ed., Chapter 5. Belmont, CA: Brooks/Cole, 1998.

Generally, we detect IR radiation by measuring the temperature rise of a blackened material located in the path of the beam or by measuring the increase in electrical conductivity of a photoconducting material when it absorbs IR radiation. Because the temperature changes resulting from the absorption of the IR energy are minute, close control of the ambient temperature is required if large errors are to be avoided. It is usually the detector system that limits the sensitivity and precision of an IR instrument.

Photon Detectors

Widely used types of photon detectors include phototubes, photomultiplier tubes, silicon photodiodes, and photodiode arrays.

Phototubes and Photomultiplier Tubes The response of a phototube or a photomultiplier tube is based on the photoelectric effect. As shown in Figure 25-12, a phototube consists of a semicylindrical photocathode and a wire anode sealed inside an evacuated transparent glass or quartz envelope. The concave surface of the cathode supports a layer of photoemissive material, such as an alkali metal or a metal oxide, that emits electrons when irradiated with light of the appropriate energy. When a voltage is applied across the electrodes, the emitted **photoelectrons** are attracted to the positively charged wire anode. In the complete circuit shown in Figure 25-12, a **photocurrent** then results that is easily amplified and measured. The number of photoelectrons ejected from the photocathode per unit time is directly proportional to the radiant power of the beam striking the surface. With an applied voltage of about 90 V or more, all these photoelectrons are collected at the anode to give a photocurrent that is also proportional to the radiant power of the beam.

The **photomultiplier tube** (PMT) is similar in construction to the phototube but is significantly more sensitive. Its photocathode is similar to that of the phototube, with electrons being emitted on exposure to radiation. In place of a single wire anode, however, the PMT has a series of electrodes called **dynodes,** as shown in Figure 25-13. The electrons emitted from the cathode are accelerated toward the first dynode, which is maintained 90 to 100 V positive with respect to the cathode. Each accelerated photoelectron that strikes the dynode surface produces several electrons, called secondary electrons, that are then accelerated to dynode 2, which

Photoelectrons are electrons that are ejected from a photosensitive surface by electromagnetic radiation. A photocurrent is the current in an external circuit that is limited by the rate of ejection of photoelectrons.

◀ One of the major advantages of photomultipliers is their automatic internal amplification. About 10^6 to 10^7 electrons are produced at the anode for each photon that strikes the photocathode of a photomultiplier tube.

◀ Photomultiplier tubes are among the most widely used types of transducers for detecting ultraviolet/visible radiation.

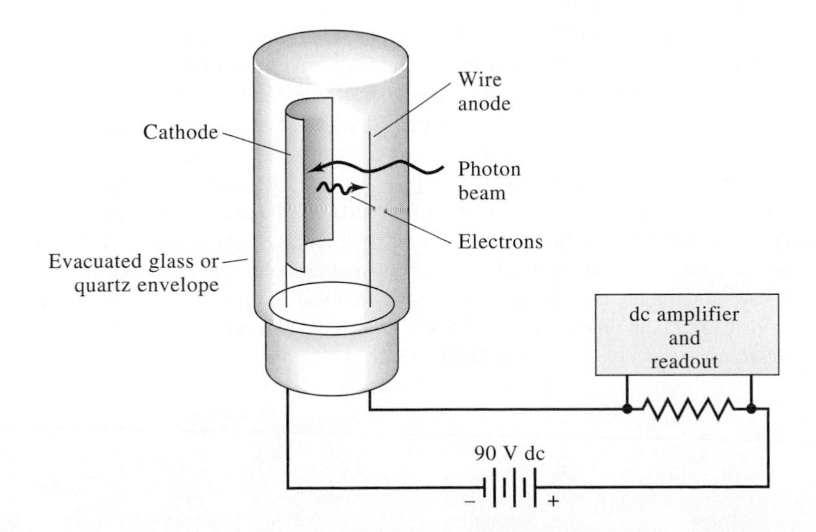

Figure 25-12 A phototube and accompanying circuit. The photocurrent induced by the radiation causes a voltage across the measuring resistor; this voltage is then amplified and measured.

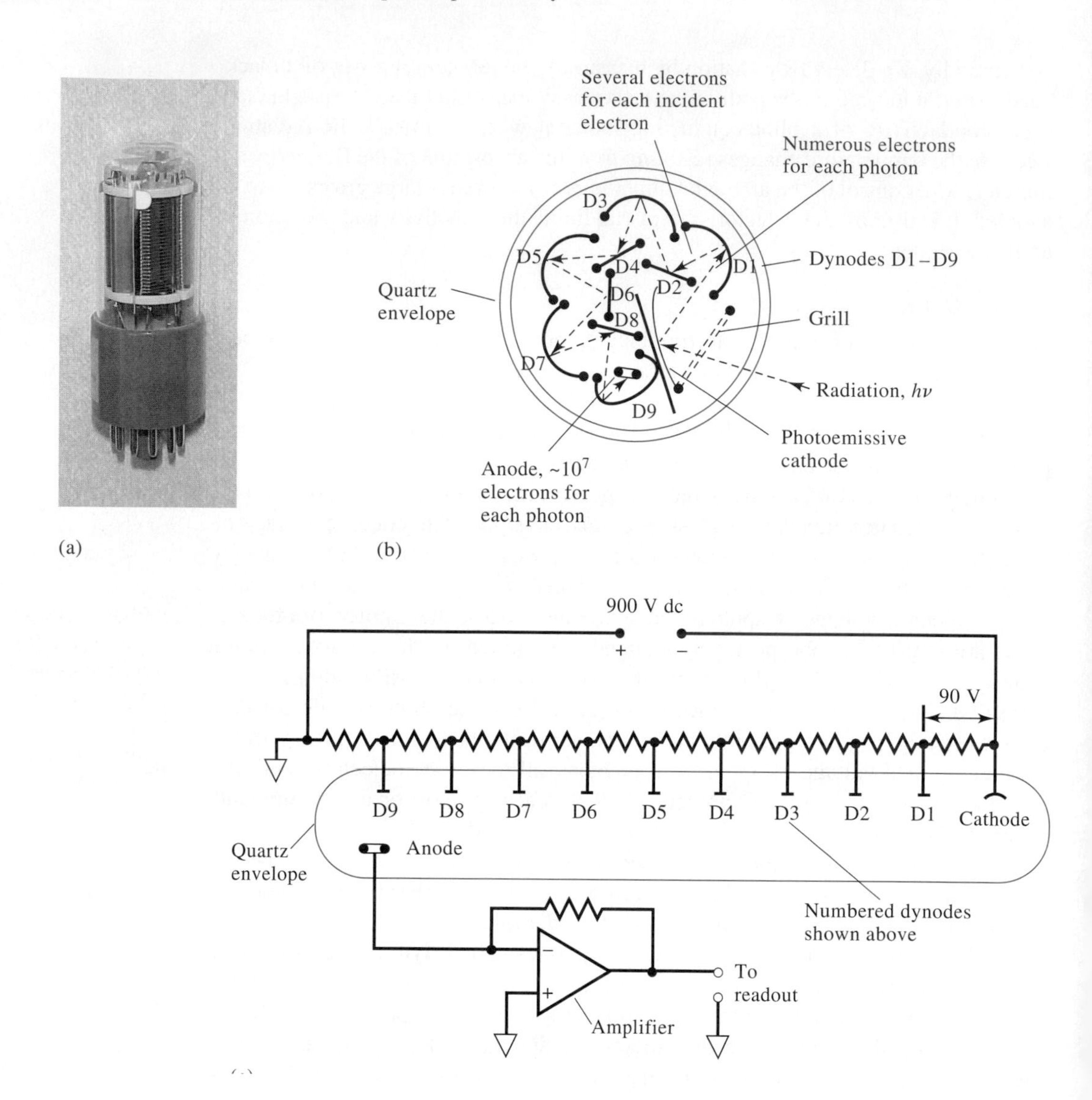

Figure 25-13 Diagram of a photomultiplier tube: (a), photograph; (b), cross-sectional view; (c), electrical diagram illustrating dynode polarization and photocurrent measurement. Radiation striking the photosensitive cathode (b) gives rise to photoelectrons by the photoelectric effect. Dynode D1 is held at a positive voltage with respect to the photocathode. Electrons emitted by the cathode are attracted to the first dynode and accelerated in the field. Each electron striking dynode D1 thus gives rise to 2 to 4 secondary electrons. These are attracted to dynode D2, which is again positive with respect to dynode D1. The resulting amplification at the anode can be 10^6 or greater. The exact amplification factor depends on the number of dynodes and the voltage difference between each. This automatic internal amplification is one of the major advantages of photomultiplier tubes. With modern instrumentation, the arrival of individual photocurrent pulses can be detected and counted instead of being measured as an average current. This technique, called *photon counting,* is advantageous at very low light levels.

is held 90 to 100 V more positive than dynode 1. Again, electron amplification results. By the time this process has been repeated at each of the dynodes, 10^5 to 10^7 electrons have been produced for each incident photon. This cascade of electrons is finally collected at the anode to provide an average current that is further amplified electronically and measured.

◀ With modern electronic instrumentation, it is possible to detect the electron pulses resulting from the arrival of individual photons at the photocathode of a PMT. The pulses are counted, and the accumulated count is a measure of the intensity of the electromagnetic radiation impinging on the PMT. Photon counting is advantageous when light intensity, or the frequency of arrival of photons at the photocathode, is low.

Photoconductive Cells Photoconductive transducers consist of a thin film of a semiconductor material, such as lead sulfide, mercury cadmium telluride (MCT), or indium antimonide, deposited often on a nonconducting glass surface and sealed in an evacuated envelope. Absorption of radiation by these materials promotes nonconducting valence electrons to a higher energy state, which decreases the electrical resistance of the semiconductor. Typically, a photoconductor is placed in series with a voltage source and a load resistor, and the voltage drop across the load resistor serves as a measure of the radiant power of the beam of radiation. The PbS and InSb detectors are quite popular in the near-IR region of the spectrum. The MCT detector is useful in the mid- and far-IR regions when cooled with liquid N_2 to minimize thermal noise.

A **semiconductor** is a substance having a conductivity that lies between that of a metal and that of a dielectric (an insulator).

Silicon Photodiodes and Photodiode Arrays Crystalline silicon is a semiconductor, a material whose electrical conductivity is less than that of a metal but greater than that of an electrical insulator. Silicon is a Group IV element and thus has four valence electrons. In a silicon crystal, each of these electrons is combined with electrons from four other silicon atoms to form four covalent bonds. At room temperature, sufficient thermal agitation occurs in this structure to liberate an occasional electron from its bonded state, leaving it free to move throughout the crystal. Thermal excitation of an electron leaves behind a positively charged region termed a hole, which, like the electron, is also mobile. The mechanism of hole movement is stepwise, with a bound electron from a neighboring silicon atom jumping into the electron-deficient region (the hole) and thereby creating another positive hole in its wake. Conduction in a semiconductor involves the movement of electrons and holes in opposite directions.

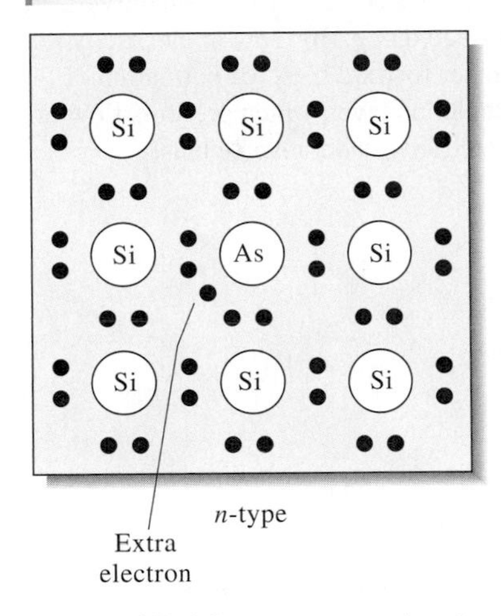

Figure 25-14 Two-dimensional representation of *n*-type silicon showing “impurity” atom.

The conductivity of silicon can be greatly enhanced by doping, a process in which a tiny, controlled amount (approximately 1 ppm) of a Group V or Group III element is distributed homogeneously throughout a silicon crystal. For example, when a crystal is doped with a Group V element, such as arsenic, four out of five of the valence electrons of the dopant form covalent bonds with four silicon atoms, leaving one electron free to conduct (Figure 25-14). When the silicon is doped with a Group III element, such as gallium, which has but three valence electrons, an excess of holes develops, which also enhances conductivity (Figure 25-15). A semiconductor containing unbonded electrons (negative charges) is termed an *n*-type semiconductor, and one containing an excess of holes (positive charges) is a *p*-type. In an *n*-type semiconductor, electrons are the majority carrier; in a *p*-type, holes are the majority carrier.

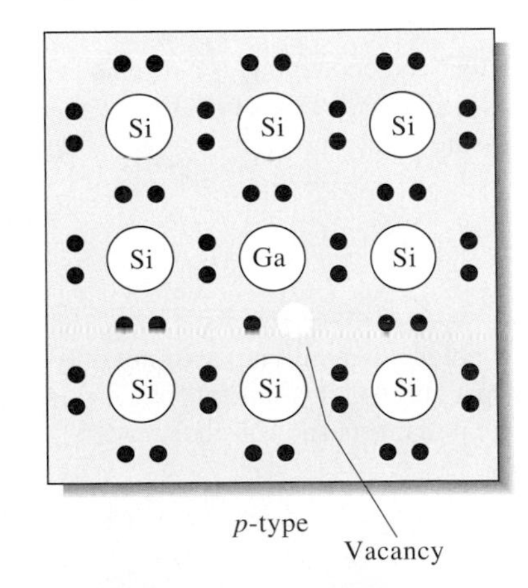

Figure 25-15 Two-dimensional representation of *p*-type silicon showing “impurity” atom.

Present silicon technology makes it possible to fabricate what is called a *pn* junction or a *pn* diode, which is conductive in one direction and not in the other. Figure 25-16a is a schematic of a silicon diode. The *pn* junction is shown as a dashed line through the middle of the crystal. Electrical wires are attached to both ends of the device. Figure 25-16b shows the junction in its conduction mode, wherein the positive terminal of a dc source is connected to the *p* region and the negative terminal to the *n* region. (The diode is said to be **forward biased** under these conditions.) The mobile electrons in the *n* region and the positive holes in the

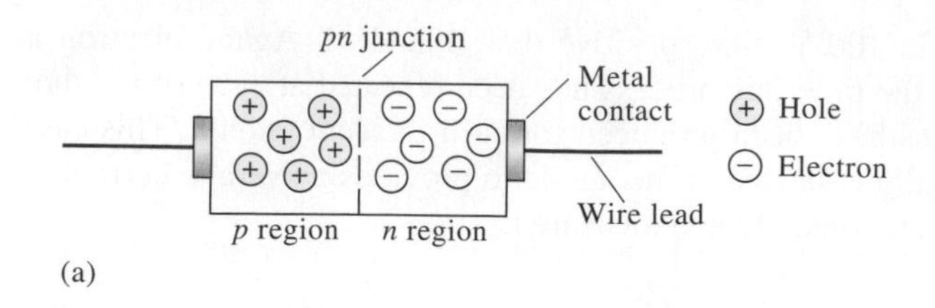

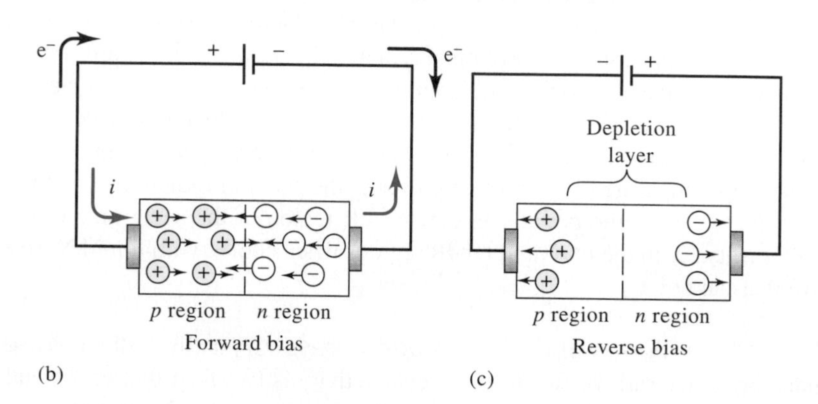

Figure 25-16 (a) Schematic of a silicon diode. (b) Flow of electricity under forward bias. (c) Formation of depletion layer, which prevents flow of electricity under reverse bias.

p region move toward the junction, where they combine and annihilate each other. The negative terminal of the source injects new electrons into the *n* region, which can continue the conduction process. The positive terminal extracts electrons from the *p* region, thus creating new holes that are free to migrate toward the *pn* junction.

Photodiodes are semiconductor *pn*-junction devices that respond to incident light by forming electron-hole pairs. (A *hole* is a mobile positive charge in a semiconductor.) When a voltage is applied to the *pn* diode such that the *p*-type semiconductor is negative with respect to the *n*-type semiconductor, the diode is said to be **reverse biased.** Figure 25-16c illustrates the behavior of a silicon diode under reverse biasing. Here, the majority carriers are drawn away from the junction, leaving a nonconductive **depletion layer.** The conductance under reverse bias is only about 10^{-6} to 10^{-8} of that under forward biasing; thus, a silicon diode is a current rectifier.

In electronics, a **bias** is a dc voltage that is inserted in series with a circuit element.

A reverse-biased silicon diode can serve as a radiation detector because ultraviolet and visible photons are sufficiently energetic to create additional electrons and holes when they strike the depletion layer of a *pn* junction. The resulting increase in conductivity is easily measured and is directly proportional to radiant power. A silicon-diode detector is more sensitive than a simple vacuum phototube but less sensitive than a photomultiplier tube.

► A silicon photodiode is a reverse-biased silicon diode that is used to measure radiant power.

► Photodiode arrays are used not only in spectroscopic instruments but also in optical scanners and bar-code readers.

Diode-Array Detectors Silicon photodiodes have become important recently because 1000 or more can be fabricated side by side on a single small silicon chip. (The width of individual diodes is about 0.02 mm.) With one or two of the diode-array detectors placed along the length of the focal plane of a monochromator, all wavelengths can be monitored simultaneously, thus making high-speed spectroscopy possible. If the number of light-induced charges per unit time is large compared with thermally produced charge carriers, the current in an external circuit, under reverse bias conditions, is directly related to the incident radiant power. Silicon photodiode detectors respond extremely rapidly, usually in nanoseconds.

They are more sensitive than a vacuum phototube but considerably less sensitive than a photomultiplier tube. Diode arrays can also be obtained commercially with front-end devices called **image intensifiers** to provide gain and allow the detection of low light levels.

Charge Transfer Devices Photodiode arrays cannot match the performance of photomultiplier tubes in terms of sensitivity, dynamic range, and signal-to-noise ratio. Thus, their use has been limited to situations in which the multichannel advantage outweighs their other shortcomings. In contrast, performance characteristics of **charge transfer device** (CTD) detectors appear to approach those of photomultiplier tubes in addition to having the multichannel advantage. As a consequence, this type of detector is now appearing in ever-increasing numbers in modern spectroscopic instruments.[7] A further advantage of charge transfer detectors is that they are two dimensional in the sense that individual detector elements are arranged in rows and columns. For example, one detector that we describe in the next section consists of 244 rows of detector elements. Each row is made up of 388 detector elements, giving a two-dimensional array of 19,672 individual detectors, or pixels, contained on a silicon chip having dimensions of 6.5 mm by 8.7 mm. With this device, it becomes possible to record an entire two-dimensional spectrum.

Charge transfer detectors operate much like a photographic film in the sense that they integrate signal information as radiation strikes them. Figure 25-17 is a cross-sectional depiction of one of the pixels making up a charge transfer array. In this case, the pixel consists of two conductive electrodes overlying an insulating layer of silica. (A pixel in some charge transfer devices is made up of more than two electrodes.) This silica layer separates the electrodes from a region of *n*-doped silicon. This assembly constitutes a metal oxide semiconductor capacitor, which stores the charges formed when radiation strikes the doped silicon. When, as shown, a negative charge is applied to the electrodes, a charge inversion region is created under the electrodes, which is energetically favorable for the storage of positive holes. The mobile holes created by the absorption of photons by the silicon then migrate and collect in this region. (Typically, this region, which is called a potential well, is capable of holding as many as 10^5 to 10^6 charges before overflowing into an adjacent pixel.) In the figure, one electrode is shown as being more negative than the other, making the accumulation of charge under this electrode more favorable. The amount of charge generated during exposure to radiation is measured in either of two ways. In a **charge injection device** (CID) detector,

◀ Silica is silicon dioxide, SiO_2, which is an electrical insulator.

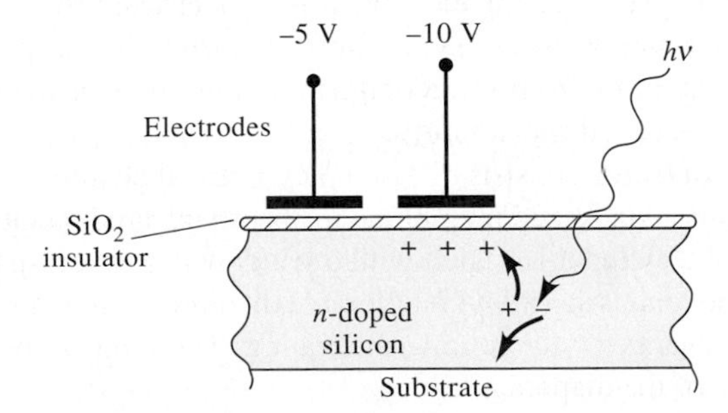

Figure 25-17 Cross section of one of the pixels of a charge transfer device. The positive hole produced by the photon $h\nu$ is collected under the negative electrode.

[7]For details on charge transfer devices, see J. V. Sweedler, *Crit. Rev. Anal. Chem.*, **1993,** *24,* 59; J. V. Sweedler, R. B. Bilhorn, P. M. Epperson, G. R. Sims, and M. B. Denton, *Anal. Chem.*, **1988,** *60,* 282A, 327A.

the voltage change arising from movement of the charge from the region under one electrode to the region under the other is measured. In a **charge coupled device** (CCD) detector (see color plate 14), the charge is moved to a charge-sensing amplifier for measurement.

CCDs and CIDs are appearing in ever-increasing numbers in modern spectroscopic instruments. In spectroscopic applications, charge transfer devices are used in conjunction with multichannel instruments, as discussed in Section 26B-3. In addition to spectroscopic applications, charge transfer devices find widespread applications in solid-state television cameras and microscopy.

Thermal Detectors

The convenient photon detectors discussed in the previous section cannot be used to measure infrared radiation because photons of these frequencies lack the energy to cause photoemission of electrons; as a consequence, thermal detectors must be used. Unfortunately, the performance characteristics of thermal detectors are much inferior to those of phototubes, photomultiplier tubes, silicon diodes, and photovoltaic cells.

A thermal detector has a tiny blackened surface that absorbs infrared radiation and increases in temperature as a consequence. The temperature rise is converted to an electrical signal that is amplified and measured. Under the best of circumstances, the temperature changes involved are minuscule, amounting to a few thousandths of a degree Celsius. The difficulty of measurement is compounded by thermal radiation from the surroundings, which is always a potential source of uncertainty. To minimize the effects of this background radiation, or noise, thermal detectors are housed in a vacuum and are carefully shielded from their surroundings. To further minimize the effects of this external noise, the beam from the source is chopped by a rotating disk inserted between source and detector. Chopping produces a beam that fluctuates regularly from zero intensity to a maximum. The transducer converts this periodic radiation signal to an alternating electric current that can be amplified and separated from the dc signal resulting from the background radiation. Despite all these measures, infrared measurements are significantly less precise than measurements of ultraviolet and visible radiation.

As shown in Table 25-2 (p. 761), four types of thermal detectors are used for infrared spectroscopy. The most widely used is a tiny thermocouple or a group of thermocouples called a **thermopile.** These devices consist of one or more pairs of dissimilar metal junctions that develop a potential difference when their temperatures differ. The magnitude of the potential depends on the temperature difference.

A **bolometer** has a conducting element whose electrical resistance changes as a function of temperature. Bolometers are fabricated from thin strips of metals, such as nickel or platinum, or from semiconductors consisting of oxides of nickel or cobalt; the latter are called **thermistors.**

A **pneumatic detector** consists of a small cylindrical chamber that is filled with xenon and contains a blackened membrane to absorb infrared radiation and heat the gas. One end of the cylinder is sealed with a window that is transparent to infrared radiation; the other end is sealed with a flexible diaphragm that moves in and out as the gas pressure changes with cooling or heating. The temperature is determined from the position of the diaphragm.

Pyroelectric detectors are manufactured from crystals of a pyroelectric material, such as barium titanate or triglycine sulfate. When a crystal of either of these compounds is sandwiched between a pair of electrodes (one of which is transparent to infrared radiation), a temperature-dependent voltage develops that can be amplified and measured.

25A-5 Signal Processors and Readouts

A signal processor is ordinarily an electronic device that amplifies the electrical signal from the detector; in addition, it may alter the signal from dc to ac (or the reverse), change the phase of the signal, and filter it to remove unwanted components. The signal processor may also be called on to perform such mathematical operations on the signal as differentiation, integration, or conversion to a logarithm. Several types of readout devices are found in modern instruments. Digital meters, scales of potentiometers, recorders, cathode-ray tubes, and monitors of microcomputers are some examples.

FEATURE 25-6

Measuring Photocurrents with Operational Amplifiers

The current produced by a reverse-biased silicon photodiode is typically 0.1 μA to 100 μA. Currents produced by these devices, as well as those generated by photomultipliers and phototubes, are so small that they must be converted to a voltage that is large enough to be measured with a digital voltmeter or other voltage-measuring device. We can perform such a conversion with the op amp circuit shown in Figure 25F-5. Light striking the reverse-biased photodiode causes a current I in the circuit. Because the op amp has a very large input resistance, essentially no current enters the op amp input designated by the minus sign. Thus, current in the photodiode must also pass through the resistor R. The current is conveniently calculated from Ohm's law: $E_{out} = -IR$. Since the current is proportional to the radiant power of the light striking the photodiode, $I = kP$, where k is a constant and $E_{out} = -IR = -kPR = K'P$. A voltmeter is connected to the output of the op amp to give a direct readout, which is proportional to the radiant power of the light falling on the photodiode. This same circuit can also be used with vacuum photodiodes or photomultipliers.

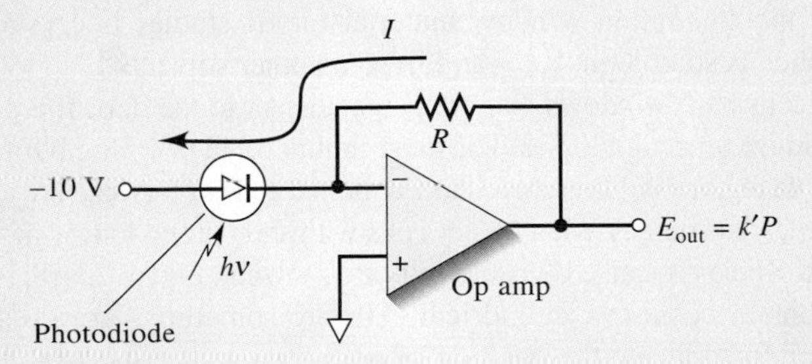

Figure 25F-5 An operational amplifier current-to-voltage converter used to monitor the current in a solid-state photodiode.

25A-6 Sample Containers

Sample containers, which are usually called **cells** or **cuvettes,** must have windows that are transparent in the spectral region of interest. Thus, as shown in Figure 25-2, quartz or fused silica is required for the UV region (wavelengths less than 350

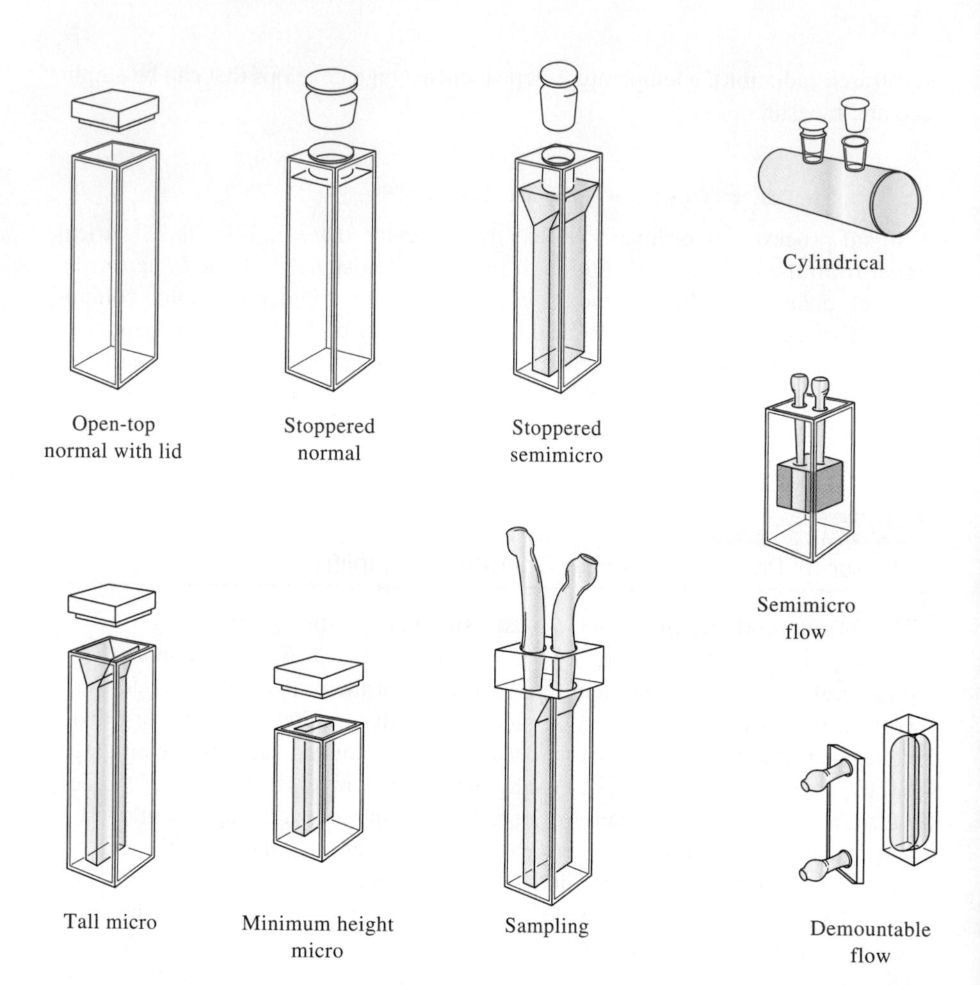

Figure 25-18 Typical examples of commercially available cells for the UV/visible region.

nm) and may be used in the visible region and out to about 3000 nm (3 μm) in the IR region. Silicate glass is ordinarily used for the 375 to 2000-nm region because of its low cost compared with quartz. Plastic cells are also used in the visible region. The most common window material for IR studies is crystalline sodium chloride, which is soluble in water and in some other solvents.

The best cells have windows that are perpendicular to the direction of the beam in order to minimize reflection losses. The most common cell path length for studies in the UV and visible regions is 1 cm; matched, calibrated cells of this size are available from several commercial sources. Many other cells with shorter and longer path lengths can be purchased. Some typical UV-visible cells are shown in Figure 25-18.

For reasons of economy, cylindrical cells are sometimes used. Particular care must be taken to duplicate the position of such cells with respect to the beam; otherwise, variations in path length and reflection losses at the curved surfaces can cause significant error, as discussed in Section 24C-2.

The quality of spectroscopic data is critically dependent on the way that cells are used and maintained. Fingerprints, grease, or other deposits on the walls markedly alter the transmission characteristics of a cell. Thus, thorough cleaning before and after use is imperative, and care must be taken to avoid touching the windows after cleaning is complete. Matched cells should never be dried by heating in an oven or over a flame because this may cause physical damage or a change in path length. Matched cells should be calibrated against each other regularly with an absorbing solution.

25B ULTRAVIOLET-VISIBLE PHOTOMETERS AND SPECTROPHOTOMETERS

The optical components described in Figure 25-1 have been combined in various ways to produce two types of instruments for absorption measurements. Several common terms are used to describe complete instruments. Thus, a **spectrometer** is a spectroscopic instrument that uses a monochromator or a polychromator in conjunction with a transducer to convert the radiant intensities into electrical signals. **Spectrophotometers** are spectrometers that allow measurement of the ratio of the radiant powers of two beams, a requirement to measure absorbance. (Recall from Equation 24-6 on page 720, that $A = \log P_0/P \approx \log P_{solvent}/P_{solution}$). **Photometers** use a filter for wavelength selection in conjunction with a suitable radiation transducer. Spectrophotometers offer the considerable advantage that the wavelength can be varied continuously, thus making it possible to record absorption spectra. Photometers have the advantages of simplicity, ruggedness, and low cost. Several dozen models of spectrophotometers are available commercially. Most spectrophotometers cover the UV/visible and occasionally the near-infrared region, while photometers are most often used for the visible region. Photometers find considerable use as detectors for chromatography, electrophoresis, immunoassays, or continuous-flow analysis. Both photometers and spectrophotometers can be obtained in single- and double-beam varieties.

CD-ROM Exercise: Exploration of How a Spectroscopic-20 Spectrophotometer Works.

25B-1 Single-Beam Instruments

Figure 25-19 shows the design of a simple and inexpensive spectrophotometer, the Spectronic 20, which is designed for use in the visible region of the spectrum. This instrument first appeared on the market in the mid-1950s, and the modified version shown in the figure is still being manufactured and widely sold. More of these instruments are currently in use throughout the world than any other single spectrophotometer model.

The Spectronic 20 reads out in transmittance or in absorbance on a **light-emitting diode display (LED).** The instrument is equipped with an **occluder,** which is a vane that automatically falls between the beam and the detector whenever the cylindrical cell is removed from its holder. The light-control device is a V-shaped aperture that is moved in and out of the beam to control the amount of light reaching the exit slit.

To obtain a percent transmittance reading, the digital readout is first zeroed with the sample compartment empty so that the occluder blocks the beam and no radiation reaches the detector. This process is called the **0% *T* calibration,** or **adjustment.** A cell containing the blank (often the solvent) is then inserted into the cell holder, and the pointer is brought to the 100% *T* mark by adjusting the position of the light-control aperture and thus the amount of light reaching the detector. This adjustment is called the **100% *T* calibration** or **adjustment.** Finally, the sample is placed in the cell compartment, and the percent transmittance or the absorbance is read directly from the LED readout.

◀ The 0% *T* and 100% *T* adjustments should be made immediately before each transmittance or absorbance measurement. To obtain reproducible transmittance measurements, it is essential that the radiant power of the source remain constant while the 100% *T* adjustment is made and the % *T* is read from the meter.

The spectral range of the Spectronic 20 is 340 to 950 nm. Other specifications include a spectral band-pass of 20 nm, a wavelength accuracy of ±2.5 nm, and a photometric accuracy of ±2% *T*. The instrument may be interfaced to a computer for data storage and analysis if this option is available.

Single-beam instruments of the type described are well suited for quantitative absorption measurements at a single wavelength. Here, simplicity of instru-

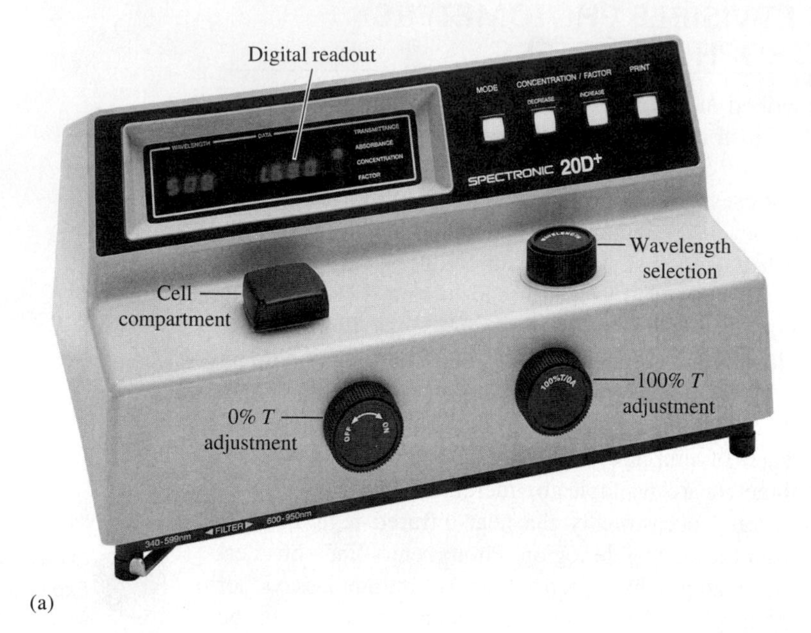

(a)

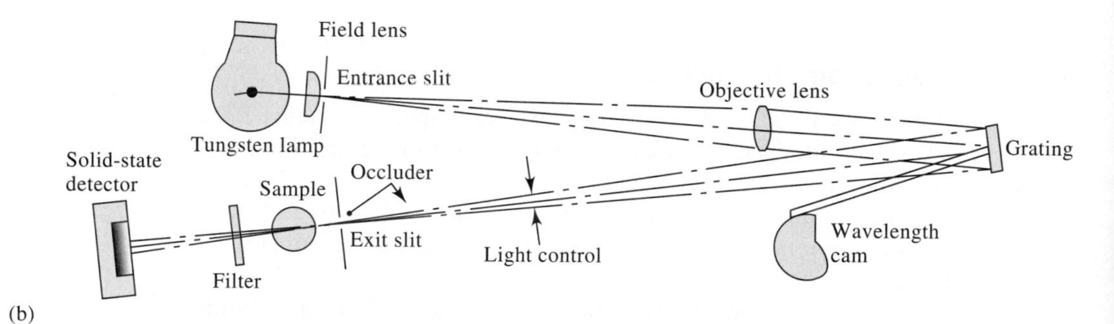

(b)

Figure 25-19 The Spectronic 20 spectrophotometer. A photograph of the instrument is shown in (a), while the optical diagram is seen in (b). Radiation from the tungsten filament source passes through an entrance slit into the monochromator. A reflection grating diffracts the radiation, and the selected wavelength band passes through the exit slit into the sample chamber. A solid-state detector converts the light intensity into a related electrical signal that is amplified and displayed on a digital readout. (Courtesy of Thermo Electron Corp., Madison, WI.)

mentation, low cost, and ease of maintenance offer distinct advantages. Several instrument manufacturers offer single-beam spectrophotometers and photometers of the single-wavelength type. Prices for these instruments are in the range of a thousand to a few thousand dollars. In addition, simple single-beam multichannel instruments based on array detectors are widely available, as discussed in the next section.

25B-2 Double-Beam Instruments

Many modern photometers and spectrophotometers are based on a double-beam design. Figure 25-20 shows two double-beam designs (b and c) compared with a

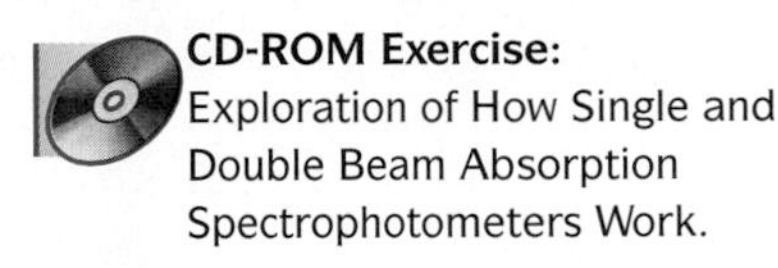

CD-ROM Exercise: Exploration of How Single and Double Beam Absorption Spectrophotometers Work.

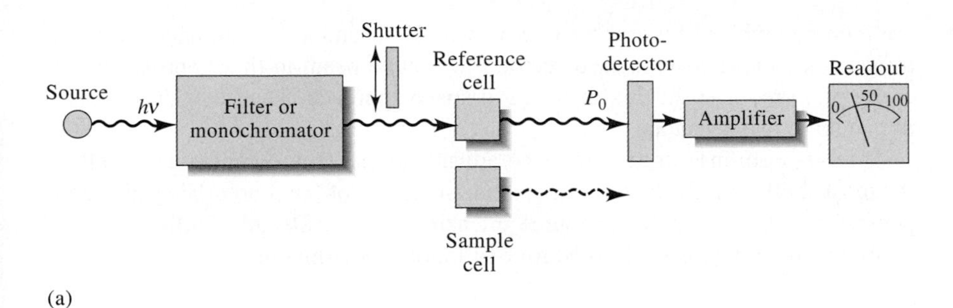

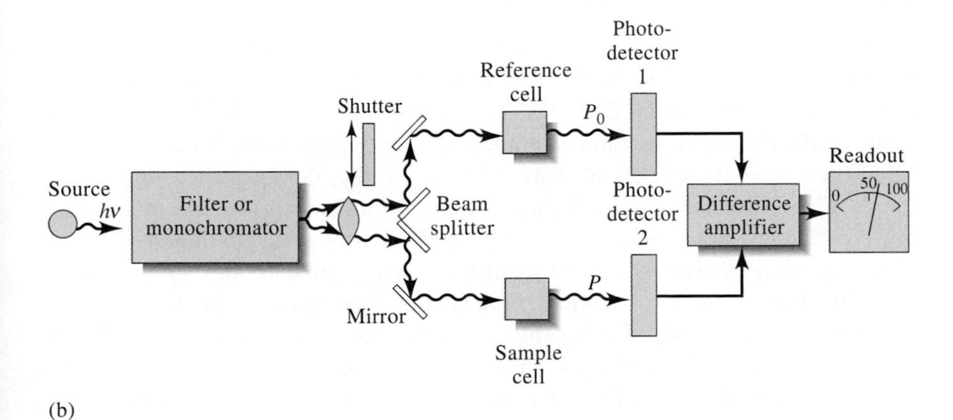

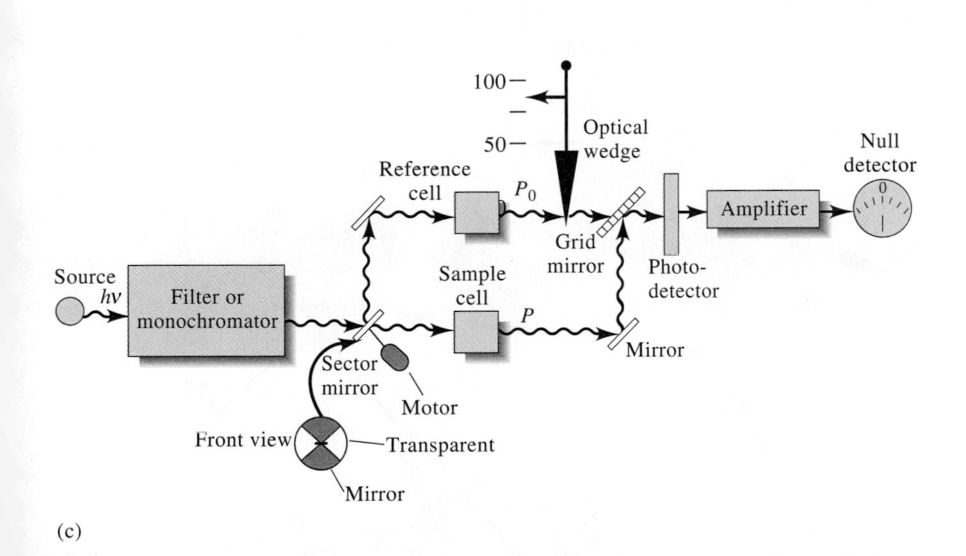

Figure 25-20 Instrument designs for UV/visible photometers or spectrophotometers. In (a), a single-beam instrument is shown. Radiation from the filter or monochromator passes through either the reference cell or the sample cell before striking the photodetector. In (b), a double-beam-in-space instrument is shown. Here, radiation from the filter or monochromator is split into two beams that simultaneously pass through the reference and sample cells before striking two matched photodetectors. In the double-beam-in-time instrument (c), the beam is alternately sent through reference and sample cells before striking a single photodetector. Only a matter of milliseconds separates the beams as they pass through the two cells.

single-beam system (a). Figure 25-20b illustrates a double-beam-in-space instrument in which two beams are formed by a V-shaped mirror called a **beam splitter.** One beam passes through the reference solution to a photodetector, and the second simultaneously passes through the sample to a second, matched photodetector. The two outputs are amplified, and their ratio, or the log of their ratio, is obtained electronically or computed and displayed on the output device.

Figure 25-20c illustrates a double-beam-in-time spectrophotometer. Here the beams are separated in time by a rotating sector mirror that directs the entire beam through the reference cell and then through the sample cell. The pulses of radiation are then recombined by another mirror, which transmits the reference beam and reflects the sample beam to the detector. The double-beam-in-time approach is generally preferred over the double-beam-in-space approach because of the difficulty in matching two detectors.

Double-beam instruments offer the advantage that they compensate for all but the most short-term fluctuations in the radiant output of the source. They also compensate for wide variations of source intensity with wavelength. Furthermore, the double-beam design is well suited for continuous recording of absorption spectra.

25B-3 Multichannel Instruments

Photodiode arrays and charge transfer devices, discussed in Section 25A-4, are the basis of multichannel instruments for UV/visible absorption. These instruments are usually of the single-beam design illustrated in Figure 25-21. With multichannel systems, the dispersive system is a grating spectrograph placed after the sample or reference cell. The photodiode array is placed in the focal plane of the spectrograph. These detectors allow the measurement of an entire spectrum in less than 1 s. With single-beam designs, the array dark current is acquired and stored in computer memory. Next, the spectrum of the source is obtained and stored in memory after dark current subtraction. Finally, the raw spectrum of the sample is obtained, and, after dark current subtraction, the sample values are divided by the source values at each wavelength, and the absorbances calculated. Multichannel instruments can also be configured as double-beam-in-time spectrophotometers.

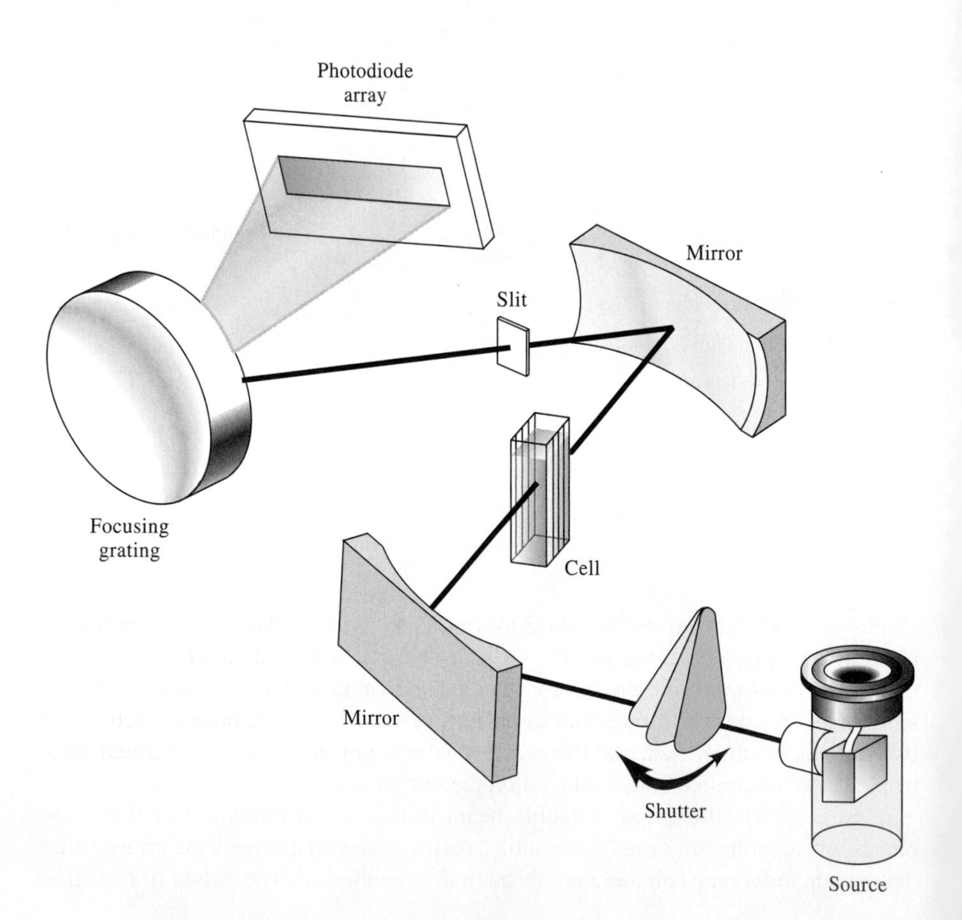

Figure 25-21 Diagram of a multichannel spectrometer based on a grating spectrograph with a photodiode array detector.

The spectrophotometer shown in Figure 25-21 can be controlled by most personal computers. The instrument (without the computer) can be purchased for about $10,000. Several instrument companies are combining array detector systems with fiber-optic probes that transport the light to and from the sample. These instruments allow measurements in convenient locations that are remote to the spectrometer. We are beginning to see more CCD and CID detectors used in multichannel systems, particularly when advantage can be taken of the two-dimensional nature of these detectors for imaging purposes (see color plate 14).

Science Photo Library/Photo Researchers, Inc.

Albert Abraham Michelson (1852–1931) was one of the most gifted and inventive experimentalists of all time. He was a graduate of the United States Naval Academy and eventually became Professor of Physics at The University of Chicago. Michelson spent most of his professional life studying the properties of light and performed several experiments that laid the foundation for our modern view of the universe. He invented the interferometer described in Feature 25-7 to determine the effect of the Earth's motion on the velocity of light. For his many inventions and their application to the study of light, Michelson was awarded the 1907 Nobel Prize in Physics. At the time of his death, Michelson and his collaborators were attempting to measure the speed of light in a mile-long vacuum tube located in what is now Irvine, California.

25C INFRARED SPECTROPHOTOMETERS

Two types of spectrometers are used in IR spectroscopy: the dispersive type and the Fourier transform variety.

25C-1 Dispersive Infrared Instruments

The older IR instruments were invariably dispersive double-beam designs. These were often of the double-beam-in-time variety shown in Figure 25-20c except that the location of the cell compartment with respect to the monochromator was reversed. In most UV/visible instruments, the cell is located between the monochromator and the detector in order to avoid photodecomposition of the sample, which may occur if samples are exposed to the full power of the source. Note that photodiode array instruments avoid this problem because of the short exposure time of the sample to the beam. Infrared radiation, in contrast, is not sufficiently energetic to bring about photodecomposition. Also, most samples are good emitters of IR radiation. Because of this, the cell compartment is usually located between the source and the monochromator in an IR instrument.

As discussed earlier in this section, the components of IR instruments differ significantly from those of UV/visible instruments. Thus, IR sources are heated solids, and IR detectors respond to heat rather than to photons. Furthermore, the optical components of IR instruments are constructed from polished salts, such as sodium chloride or potassium bromide.

25C-2 Fourier Transform Instruments

When Fourier transform infrared (FTIR) spectrometers first appeared on the market in the early 1970s, they were bulky and expensive (more than $100,000) and required frequent mechanical adjustments. For these reasons, their use was limited to special applications in which their unique characteristics (great speed, high resolution, high sensitivity, and excellent wavelength precision and accuracy) were essential. Currently, however, FTIR spectrometers have been reduced to benchtop size and have become very reliable and easy to maintain. Furthermore, the simple models are now priced similarly to simple dispersive spectrometers. Hence, FTIR spectrometers are largely displacing dispersive instruments in most laboratories.

◀ Fourier transform spectrometers detect all the wavelengths all the time. They have greater light-gathering power than dispersive instruments and consequently better precision. Although computation of the Fourier transform is somewhat complex, it is easily accomplished with modern, inexpensive, high-speed personal computers.

Fourier transform IR instruments contain no dispersing element, and all wavelengths are detected and measured simultaneously. Instead of a monochromator, an interferometer is used to produce interference patterns that contain the infrared spectral information. The same types of sources used in dispersive instruments are used in FTIR spectrometers. Transducers are typically triglycine sulfate—a pyroelectric transducer—or mercury cadmium telluride—a photoconductive transducer. To obtain radiant power as a function of wavelength, the interferometer modulates the source signal in such a way that it can be decoded by the mathematical technique of Fourier transformation. This operation requires a high-speed computer to do the necessary calculations. The theory of Fourier transform measurements is discussed in Feature 25-7.[8]

Most benchtop FTIR spectrometers are of the single-beam type. To obtain the spectrum of a sample, the background spectrum is first obtained by Fourier transformation of the interferogram from the background (solvent, ambient water, and carbon dioxide). Next, the sample spectrum is obtained. Finally, the ratio of the single-beam sample spectrum to that of the background spectrum is calculated, and absorbance or transmittance versus wavelength or wavenumber is plotted. Often, benchtop instruments purge the spectrometer with an inert gas or dry, CO_2-free air to reduce the background absorption from water vapor and CO_2.

The major advantages of FTIR instruments over dispersive spectrometers include better speed and sensitivity, better light-gathering power, more accurate wavelength calibration, simpler mechanical design, and the virtual elimination of the problems of stray light and IR emission. Because of these advantages, nearly all the new IR instruments are FTIR systems.

FEATURE 25-7

How Does a Fourier Transform Infrared Spectrometer Work?

Fourier transform infrared spectrometers utilize an ingenious device called a **Michelson interferometer,** which was developed many years ago by A. A. Michelson for making precise measurements of the wavelengths of electromagnetic radiation and for making incredibly accurate distance measurements. The principles of interferometry are utilized in many areas of science including chemistry, physics, astronomy, and metrology and are applicable in many regions of the electromagnetic spectrum.

A diagram of a Michelson interferometer is shown in Figure 25F-6. It consists of a collimated light source (shown on the left of the diagram), a stationary mirror at the top, a movable mirror at the right, a beam splitter, and a detector. The light source may be a continuum source, as in FTIR spectroscopy, or it may be a monochromatic source, such as a laser or a sodium arc lamp for other uses—for example, measuring distances. The mirrors are precision-polished ultra-flat glass with a reflective coating vapor deposited on their surfaces. The movable mirror is usually mounted on a very precise linear bearing that allows it to move along the direction of the light beam while remaining perpendicular to it, as shown in the diagram.

The key to the operation of the interferometer is the *beam splitter,* which is usually a partially silvered mirror similar to the "one-way" mirrors often seen in retail stores and police interrogation rooms. The beam splitter allows a fraction of the light falling on it to pass through the mirror, and another fraction is reflected. This device works in both directions, so that light falling on either side of the beam splitter is partially reflected and partially transmitted.

For simplicity, we will use as our light source the blue line of an argon-ion laser. Beam *A* from the source impinges on the beam splitter, which is tilted at 45° to the incoming beam. Our beam splitter is coated on the right side, so beam *A* enters the glass and is partially reflected off the back side of the coating. It emerges from the beam splitter as beam A' and moves up toward the stationary mirror, where it is reflected back down toward the beam splitter. Part of the beam is then transmitted down through the beam splitter toward the detector. Although the beam loses some intensity with each interaction with the stationary mirror and the beam splitter, the net effect is that a fraction (beam A') of incident beam *A* ends up at the detector.

In its first interaction with the beam splitter, the fraction of beam *A* that is transmitted emerges to the right toward the movable mirror as beam *B*. It then is reflected back to the left to the beam splitter, where it is reflected down toward the detector. With careful alignment, both

[8]See also J. D. Ingle, Jr., and S. R. Crouch, *Spectrochemical Analysis.* Upper Saddle River, NJ: Prentice-Hall, 1988; D. A. Skoog, F. J. Holler, and T. A. Nieman, *Principles of Instrumental Analysis,* 5th ed. Belmont, CA: Brooks/Cole, 1998.

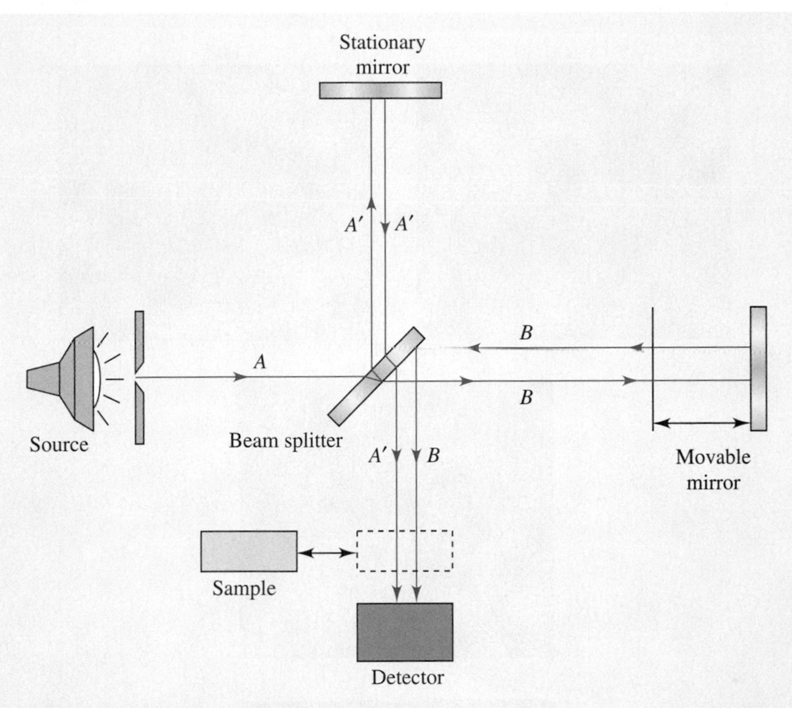

Figure 25F-6 Diagram of a Michelson interferometer. A beam from the light source at left is split into two beams by the beam splitter. The two beams travel two separate paths and converge on the detector. The two beams A' and B converge in the same region of space and form an interference pattern. As the movable mirror on the right is moved, the interference pattern shifts across the detector and modulates the optical signal. The resulting reference interferogram is recorded and used as a measure of the power of the incident beam at all wavelengths. An absorbing sample is then inserted into the beam, and a sample interferogram is recorded. The two interferograms are used to compute the absorption spectrum of the sample.

beam A' and beam B are collinear and impinge on the detector at the same spot.

The overall purpose of the interferometer optics is to split the incident beam into two beams that move through space along separate paths and then recombine at the detector. It is in this region that the two beams, or wavefronts, interact to form an **interference pattern.** The origin of the interference pattern is illustrated in Figure 25F-7, which is a two-dimensional representation of the interaction of the two spherical wavefronts. Beam A' and beam B converge and interact as two point sources of light, represented in the upper portion of the figure. When the two beams interfere, they form a pattern similar to the one shown. In regions where the waves interfere constructively, bright bands appear, and where destructive interference occurs, dark bands form. The alternating light and dark bands are called **interference fringes.** These fringes appear at the detector as the output image shown at the bottom of the figure. In the earliest versions of the Michelson interferometer, the detector was the human eye aided by a telescope. The fringes could be counted or measured through the telescope.

When the movable mirror is moved to the left at constant velocity, the interference pattern gradually sweeps past the detector as the path that beam B follows is gradually shortened. The form of the interference pattern remains the same, but the positions of constructive and destructive interference are shifted as the path difference changes. For example, if the wavelength of our laser source is λ, as we move the mirror a distance of $\lambda/4$, the path difference between the two beams changes by $\lambda/2$, and where we had constructive interference, we now have destructive interference. If we move the mirror another $\lambda/4$, the path difference changes again by $\lambda/2$, and we again return to constructive interference. As the mirror moves, the two wavefronts are shifted in space relative to one another, and alternate light and dark fringes sweep across the detector, as illustrated in Figure 25F-8a. At the detector, we find the

(continued)

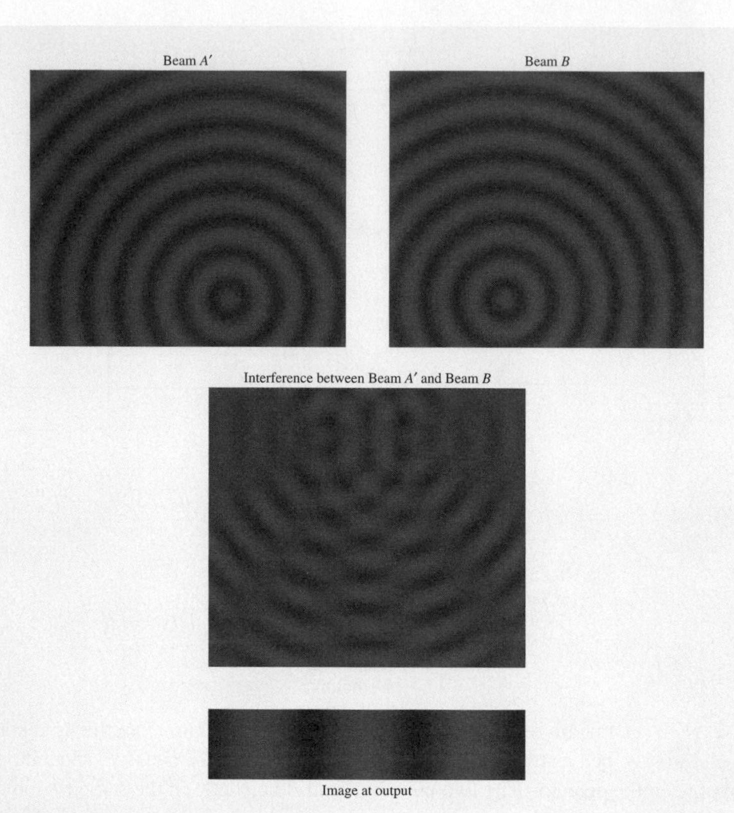

Figure 25F-7 A two-dimensional representation of the interference of two monochromatic wavefronts of the same frequency. Beam A' and beam B at the top form the interference pattern in the middle, and the two wavefronts constructively and destructively interfere. The image shown at the bottom would appear at the output of the Michelson interferometer perpendicular to the plane of the two-dimensional interference pattern.

sinusoidal intensity profile shown in Figure 25F-8b. This profile is called an **interferogram.** The net effect of the constant uniform motion of the mirror is that the light intensity at the output of the interferometer is **modulated,** or systematically varied, in a precisely controlled way, as shown in the figure. In practice, it turns out not to be very easy to move the interferometer mirror at a constant, precisely controlled velocity. There is a better and much more precise way to monitor the mirror motion using a second parallel interferometer.[9] Here, we assume that we can measure or monitor the progress of the mirror and compensate for any nonuniform motion computationally.

We have established that a Michelson interferometer with a monochromatic light source produces a sinusoidally varying signal at the detector when the mirror is moved at constant velocity. Now, we must investigate what happens to the signal once it is recorded. Although the characteristics of Michelson interferometers have been well known for more than a century, and the mathematical apparatus for dealing with the data has been in place for nearly two centuries, the device could not be used routinely for spectroscopy until two developments occurred: (1) High-speed, inexpensive computers had to become available, and (2) appropriate computational methods had to be invented to handle the huge number of rather simple calculations that must be applied to the data acquired in interferometric experiments. Briefly, the principles of Fourier synthesis and analysis tell us that any waveform can be represented as a series of sinusoidal waveforms, and, correspondingly, any combination of sinusoidal waveforms can be broken down into a series of sinusoids of known frequency. We can apply this idea to the sinusoidal signal detected at the output of the Michelson interferometer shown in Figure 25F-8b.

If we subject the signal in the figure to Fourier analysis via a computer algorithm called the fast Fourier transform

[9]D. A. Skoog, F. J. Holler, and T. A. Nieman, *Principles of Instrumental Analysis,* 5th ed., Chapter 5, p. 393. Belmont, CA: Brooks/Cole, 1998.

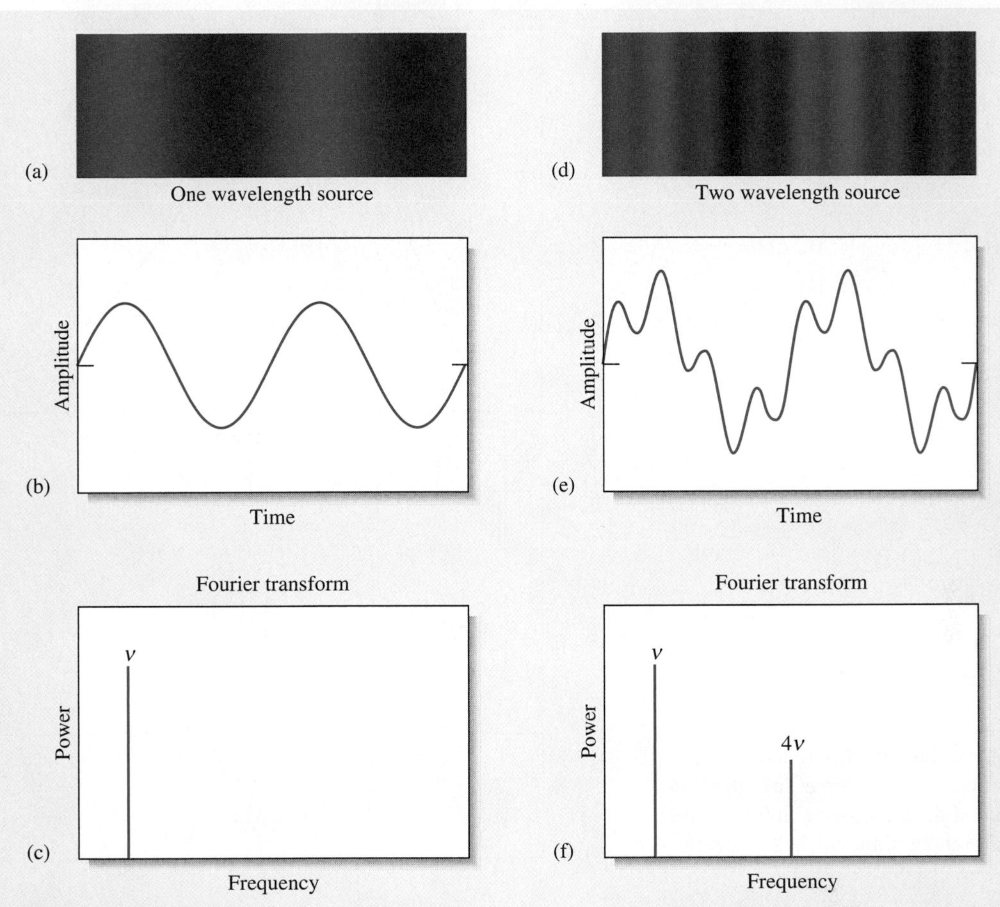

Figure 25F-8 Formation of interferograms at the output of the Michelson interferometer. (a) Interference pattern at the output of the interferometer resulting from a monochromatic source. (b) Sinusoidally varying signal produced at the detector by the pattern in (a). (c) Frequency spectrum of the monochromatic light source resulting from the Fourier transformation of the signal in (b). (d) Interference pattern at the output of the interferometer resulting from a two-color source. (e) Complex signal produced by the interference pattern of (d) as it falls on the detector. (f) Frequency spectrum of the two-color source.

(FFT), we obtain the frequency spectrum illustrated in Figure 25F-8c. Notice that the original waveform in Figure 25F-8b is a time-dependent signal; the resultant output from the FFT is a frequency-dependent signal. In other words, the FFT takes amplitude signals in the **time domain** and converts them to power in the **frequency domain.** Since the output of the interferometer is a sine wave of a single frequency, the frequency spectrum shows a single spike of frequency ν, the frequency of the original sine wave. This frequency is proportional to the optical frequency emitted by the laser source but of much lower value, so that it can be measured and manipulated with modern electronics. We now modify the interferometer so that we can obtain a second sine wave at the output. One way to do this is simply to add a second wavelength to our light source. Experimentally, a second laser or another monochromatic light source at the input of the interferometer gives us a beam that contains just two wavelengths.

For example, assume that the second wavelength is one quarter of the first one; that is, the second frequency is 4ν. Further assume that its intensity is one-half the intensity of the original source. As a result, the signal appearing at the output of the interferometer would exhibit a pattern somewhat more complex than in the single-wavelength example, as shown in Figure 25F-8d. The detector signal plot appears as the sum of two sine waves (Figure 25F-8e). We then apply the FFT to the complex sinusoidal signal to produce the frequency spectrum of Figure 25F-8f.

(continued)

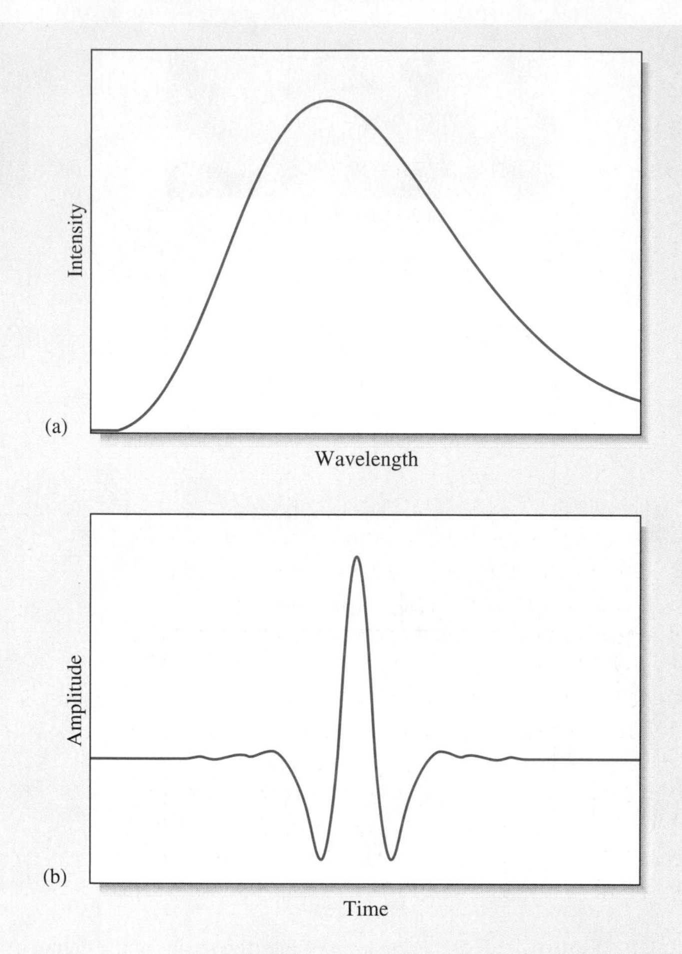

Figure 25F-9 (a) Spectrum of a continuum light source. (b) Interferogram of the light source in (a) produced at the output of the Michelson interferometer.

This spectrum reveals just two frequencies at ν and 4ν, and the relative magnitudes of the two frequency spikes are proportional to the amplitudes of the two sine waves composing the original signal. The two frequencies correspond to the two wavelengths in our interferometer light source, and the FFT has revealed the intensities of the source at those two wavelengths.

To illustrate how the Michelson interferometer is used in practical experiments, we place a continuum infrared light source (Figure 25F-9a) containing a huge number of wavelengths at the input of the interferometer. As the mirror moves along its path, all wavelengths are modulated simultaneously, which produces the very interesting interferogram shown in Figure 25F-9b. This interferogram contains all the information that we require in a spectroscopy experiment regarding the intensity of the light source at all of its component wavelengths.

As suggested in the previous section, there are a number of advantages to acquiring intensity information in this way rather than using a scanning spectrometer.[10] First, there is the advantage of speed. The mirror can be moved in a matter of seconds, and a computer attached to the detector can collect all necessary data during the course of the mirror scan. In just a few more seconds, the computer can perform the FFT and produce the frequency spectrum containing all the intensity information. Next is Fellgett's advantage, which suggests that Michelson interferometers are capable of producing higher signal-to-noise ratios in a shorter time than can equivalent dispersive spectrometers. Finally, we have the throughput, or Jacquinot's advantage, which permits 10 to 200 times more radiation to pass through a sample than do standard dispersive spectrometers. These advantages are often offset by the lower sensitivity of detectors that are used in FTIR spectrometers.

[10] J. D. Ingle, Jr., and S. R. Crouch, *Spectrochemical Analysis,* pp. 425–426. Upper Saddle River, NJ: Prentice-Hall, 1988.

Under these circumstances, the speed of the measurement process and the simplicity and reliability of FTIR spectrometers become primary considerations. We discuss some of these issues further in Chapter 26.

Up to this point in our discussion of the FTIR spectrometer, we have just shown how the Michelson interferometer can provide intensity information for a light source as a function of wavelength. The spectrum of a sample can be obtained by first obtaining a reference interferogram of the source with no sample in the light path, as shown in Figure 25F-6. Then the sample is placed in the path indicated by the arrow and dashed box in the figure, and, once again, we scan the mirror and acquire a second interferogram. In FTIR spectrometry, the sample absorbs infrared radiation, which attenuates the beams in the interferometer. The difference between the second (sample) interferogram and the reference interferogram is then computed. Since the difference interferogram depends only on the absorption of radiation by the sample, the FFT is performed on the resulting data, which produces the IR spectrum of the sample. We will discuss a specific example of this process in Chapter 26. Finally, we should note that the FFT can be accomplished using the most basic modern personal computer equipped with the proper software. Many software packages such as Mathcad, Mathematica, Matlab, and even the Data Analysis Toolpak of Excel have Fourier analysis functions built in. These tools are widely used in science and engineering for a broad range of signal processing tasks.

WEB WORKS

Use your favorite search engine to find companies that manufacture monochromators. Navigate to several Web sites of these companies and find a UV/visible monochromator of the Czerny-Turner design that has better than 0.1-nm resolution. List several other important specifications of monochromators, and describe what they mean and how they affect the quality of analytical spectroscopic measurements. From the specifications and, if available, the prices, determine the factors that have the most significant effect on the cost of the monochromators.

QUESTIONS AND PROBLEMS

25-1. Describe the differences between the following and list any particular advantages possessed by one over the other:
- *(a) filters and monochromators as wavelength selectors
- (b) solid-state photodiodes and phototubes as detectors of electromagnetic radiation
- *(c) phototubes and photomultiplier tubes
- (d) conventional and diode-array spectrophotometers

25-2. Define the term *effective bandwidth of a filter.*

***25-3.** Why are photomultiplier tubes unsuited for the detection of infrared radiation?

25-4. Why do quantitative and qualitative analyses often require different monochromator slit widths?

***25-5.** Why is iodine sometimes introduced into a tungsten lamp?

25-6. Describe the differences between the following and list any particular advantages possessed by one over the other:
- (a) spectrophotometers and photometers
- (b) spectrographs and polychromators
- (c) monochromators and polychromators
- (d) single-beam and double-beam instruments for absorbance measurements
- (e) conventional and diode-array spectrophotometers

25-7. The Wien displacement law states that the wavelength maximum in micrometers for blackbody radiation is

$$\lambda_{max} T = 2.90 \times 10^3$$

where T is the temperature in kelvins. Calculate the wavelength maximum for a blackbody that has been heated to *(a) 4000 K, (b) 3000 K, *(c) 2000 K, and (d) 1000 K.

25-8. Stefan's law states that the total energy emitted by a blackbody per unit time and per unit area is

$$E_t = \alpha T^4$$

where α is 5.69×10^{-8} W/m^{2}K^4. Calculate the total energy output in W/m^2 for the blackbodies described in Problem 25-7.

***25-9.** The relationships described in Problems 23-7 and 23-8 may be of help in solving the following.
(a) Calculate the wavelength of maximum emission of a tungsten-filament bulb operated at 2870 K and at 3000 K.
(b) Calculate the total energy output of the bulb in W/cm^2.

25-10. What minimum requirement is needed to obtain reproducible results with a single-beam spectrophotometer?

***25-11.** What is the purpose of (a) the 0% T adjustment and (b) the 100% T adjustment of a spectrophotometer?

25-12. What experimental variables must be controlled to assure reproducible absorbance data?

***25-13.** What are the major advantages of Fourier transform IR instruments over dispersive IR instruments?

25-14. A photometer with a linear response to radiation gave a reading of 595 mV with a blank in the light path and 139 mV when the blank was replaced by an absorbing solution. Calculate
*(a) the percent transmittance and absorbance of the absorbing solution.
(b) the expected transmittance if the concentration of absorber is one-half that of the original solution.
*(c) the transmittance to be expected if the light path through the original solution is doubled.

25-15. A portable photometer with a linear response to radiation registered 83.2 μA with a blank solution in the light path. Replacement of the blank with an absorbing solution yielded a response of 45.1 μA. Calculate
(a) the percent transmittance of the sample solution.
*(b) the absorbance of the sample solution.
(c) the transmittance to be expected for a solution in which the concentration of the absorber is one-third that of the original sample solution.
*(d) the transmittance to be expected for a solution that has twice the concentration of the sample solution.

25-16. Why does a deuterium lamp produce a continuum rather than a line spectrum in the ultraviolet range?

***25-17.** What are the differences between a photon detector and a heat detector?

25-18. Describe how an absorption photometer and a fluorescence photometer differ.

***25-19.** Describe the basic design difference between a spectrometer for absorption measurements and one for emission studies.

25-20. What data are needed to describe the performance characteristics of an interference filter?

25-21. Define
*(a) dark current.
(b) transducer.
*(c) scattered radiation (in a monochromator).
(d) n-type semiconductor.
*(e) majority carrier.
(f) depletion layer.

***25-22.** An interference filter is to be constructed for isolation of the CS_2 absorption band at 4.54 m.
(a) If the determination is to be based on first-order interference, how thick should the dielectric layer be (refractive index 1.34)?
(b) What other wavelengths will be transmitted?

25-23. The following data were taken from a diode array spectrophotometer in an experiment to measure the spectrum of the Co(II)-EDTA complex. The column labeled $P_{solution}$ is the relative signal obtained with sample solution in the cell after subtraction of the dark signal. The column labeled $P_{solvent}$ is the reference signal obtained with only solvent in the cell after subtraction of the dark signal. Find the transmittance at each wavelength and the absorbance at each wavelength. Plot the spectrum of the compound.

Wavelength, nm	$P_{solvent}$	$P_{solution}$
350	0.002689	0.002560
375	0.006326	0.005995
400	0.016975	0.015143
425	0.035517	0.031648
450	0.062425	0.024978
475	0.095374	0.019073
500	0.140567	0.023275
525	0.188984	0.037448
550	0.263103	0.088537
575	0.318361	0.200872
600	0.394600	0.278072
625	0.477018	0.363525
650	0.564295	0.468281
675	0.655066	0.611062
700	0.739180	0.704126
725	0.813694	0.777466
750	0.885979	0.863224
775	0.945083	0.921446
800	1.000000	0.977237

25-24. **Challenge Problem:** Horlick has described the mathematical principles of the Fourier transform, interpreted them graphically, and described how they may be used in analytical spectroscopy.[11] Read the article, and answer the following questions.

(a) Define *time domain* and *frequency domain.*
(b) Write the equations for the Fourier integral and its transformation, and define each of the terms in the equations.
(c) The paper shows the time-domain signals for a 32-cycle cosine wave, a 21-cycle cosine wave, and a 10-cycle cosine wave as well as the Fourier transforms of these signals. How does the shape of the frequency-domain signal change as the number of cycles in the original waveform changes?
(d) The author describes the phenomenon of *damping*. What effect does damping have on the original cosine waves? What effect does it have on the resulting Fourier transformations?
(e) What is a resolution function?
(f) What is the process of convolution?
(g) Discuss how the choice of the resolution function can affect the appearance of a spectrum.
(h) Convolution may be used to decrease the amount of noise in a noisy spectrum. Consider the following plots of time-domain and frequency-domain signals. Label the axes of the five plots. For example, plot B should be labeled as amplitude vs. time. Characterize each plot as either a time-domain or a frequency-domain signal.
(i) Describe the mathematical relationships among the plots. For example, how could you arrive at plot A from plots D and E?
(j) Discuss the practical importance of being able to reduce noise in spectroscopic signals.

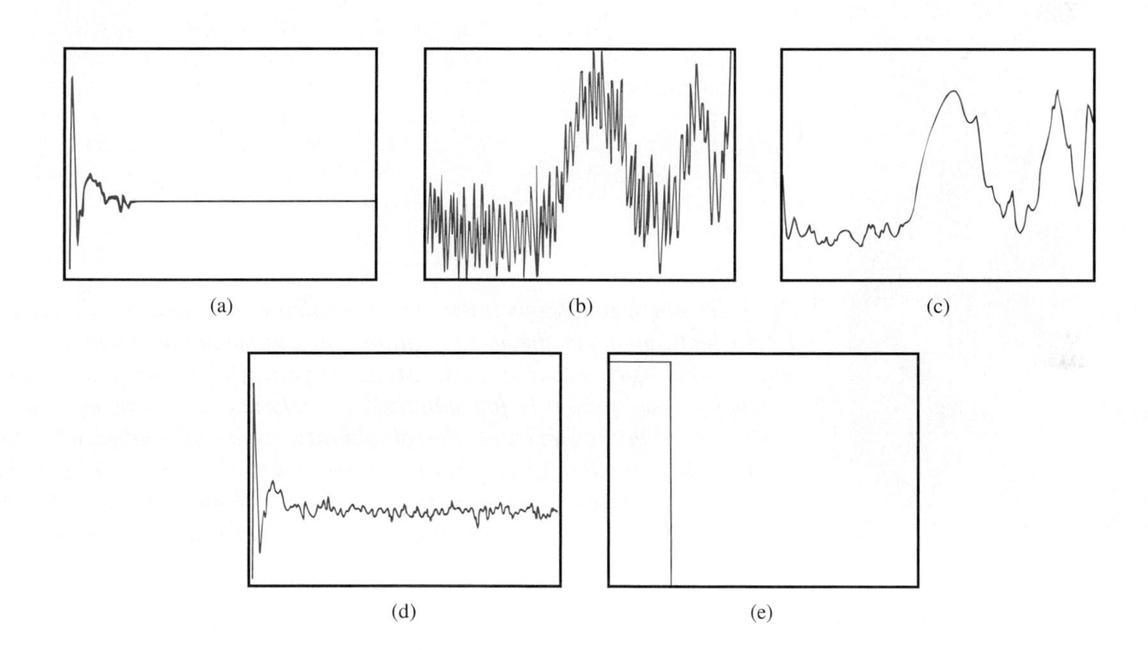

InfoTrac College Edition

For additional readings, go to InfoTrac College Edition, your online research library, at

http://infotrac.thomsonlearning.com

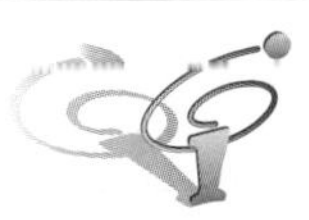

[11]G. Horlick, *Anal. Chem.*, **1971,** *43*(8), 61A–66A.

CHAPTER 26

Molecular Absorption Spectrometry

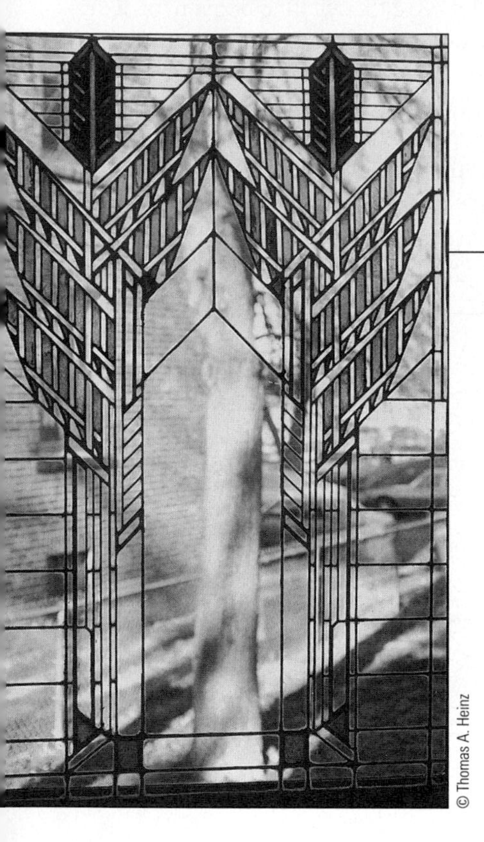

Glassmaking is among the oldest technologies, dating from the Neolithic period nearly 10,000 years ago. Ordinary glass is transparent because valence electrons in the silicate structure do not receive sufficient energy from visible light to be excited from their ground states in the valence band of the silicate structure to the conduction band. Beginning with the Egyptians in the second millennium B.C.E., glassmakers learned to add a variety of compounds to glasses to produce colored glass. These additives often contain transition metals to provide accessible energy levels so that absorption of light occurs, and the resulting glass is colored. Colored glass is used widely in art and architecture as, for example, in the stained glass window shown here. Optical spectroscopy is used to characterize colored glasses by recording their absorption spectra. This information may be used in diverse fields such as art history to characterize, identify, and trace the origin and development of works of art, archaeology to explore the origins of humankind, and forensics to correlate evidence in crime investigations.

Molecular spectroscopy based on ultraviolet, visible, and infrared radiation is widely used for the identification and determination of many inorganic, organic, and biochemical species.[1] Molecular ultraviolet/visible absorption spectroscopy is used primarily for quantitative analysis and is probably more extensively applied in chemical and clinical laboratories throughout the world than any other single method. Infrared absorption spectroscopy is a powerful tool for determining the structure of both inorganic and organic compounds. In addition, it now plays an important role in quantitative analysis, particularly in the area of environmental pollution.

26A ULTRAVIOLET AND VISIBLE MOLECULAR ABSORPTION SPECTROSCOPY

In this section, we first consider types of molecular species that absorb ultraviolet and visible radiation. We then describe qualitative and quantitative applications of

[1]For more detailed treatment of absorption spectroscopy, see E. J. Meehan, in *Treatise on Analytical Chemistry,* 2nd ed., Part I, Vol. 7, Ch. 2, P. J. Elving, E. J. Meehan, and I. M. Kolthoff, Eds. New York: Wiley, 1981; *Techniques in Visible and Ultraviolet Spectrometry,* Vol. 1, C. Burgess and A. Knowles, Eds., New York: Chapman and Hall, 1981; J. D. Ingle, Jr., and S. R. Crouch, *Spectrochemical Analysis,* Ch. 12–14. Upper Saddle River, NJ: Prentice-Hall, 1988.

spectroscopy in the UV-visible region of the spectrum. The section concludes with a discussion of spectrophotometric titrations and studies of the composition of complex ions.

26A-1 Absorbing Species

As noted in Section 24C-2, absorption of ultraviolet and visible radiation by molecules generally occurs in one or more electronic absorption bands, each of which is made up of many closely packed but discrete lines. Each line arises from the transition of an electron from the ground state to one of the many vibrational and rotational energy states associated with each excited electronic energy state. Because there are so many of these vibrational and rotational states and because their energies differ only slightly, the number of lines contained in the typical band is quite large and their displacement from one very small.

◀ A band consists of a large number of closely spaced vibrational and rotational lines. The energies associated with the lines differ little from one another.

As we noted previously in Figure 24-14a (p. 728), the visible absorption spectrum for 1,2,3,4-tetrazine vapor shows the fine structure that is due to the numerous rotational and vibrational levels associated with the excited electronic states of this aromatic molecule. In the gaseous state, the individual tetrazine molecules are sufficiently separated from one another to vibrate and rotate freely, and the many individual absorption lines appear as a result of the large number of vibrational and rotational energy states. In the condensed state or in solution, however, the tetrazine molecules have little freedom to rotate, so lines due to differences in rotational energy levels disappear. Furthermore, when solvent molecules surround the tetrazine molecules, energies of the various vibrational levels are modified in a nonuniform way, and the energy of a given state in a sample of solute molecules appears as a single, broad peak. This effect is more pronounced in polar solvents, such as water, than in nonpolar hydrocarbon media. This solvent effect is illustrated in Figures 24-14b and 24-14c.

Absorption by Organic Compounds

Absorption of radiation by organic molecules in the wavelength region between 180 and 780 nm results from interactions between photons and electrons that either participate directly in bond formation (and are thus associated with more than one atom) or are localized about such atoms as oxygen, sulfur, nitrogen, and the halogens.

The wavelength at which an organic molecule absorbs depends on how tightly its electrons are bound. The shared electrons in carbon/carbon or carbon/hydrogen single bonds are so firmly held that their excitation requires energies corresponding to wavelengths in the vacuum ultraviolet region below 180 nm. Single-bond spectra have not been widely exploited for analytical purposes because of the experimental difficulties of working in this region. These difficulties occur because both quartz and atmospheric components absorb in this region, which requires that evacuated spectrophotometers with lithium fluoride optics be used.

Electrons involved in double and triple bonds of organic molecules are not as strongly held and are therefore more easily excited by radiation; thus, species with unsaturated bonds generally exhibit useful absorption peaks. Unsaturated organic functional groups that absorb in the ultraviolet or visible regions are known as **chromophores.** Table 26-1 lists common chromophores and the approximate wavelengths at which they absorb. The data for position and peak intensity can serve only as a rough guide for identification purposes, since both are influenced by solvent effects as well as other structural details of the molecule. In addition,

Chromophores are unsaturated organic functional groups that absorb in the ultraviolet or visible region.

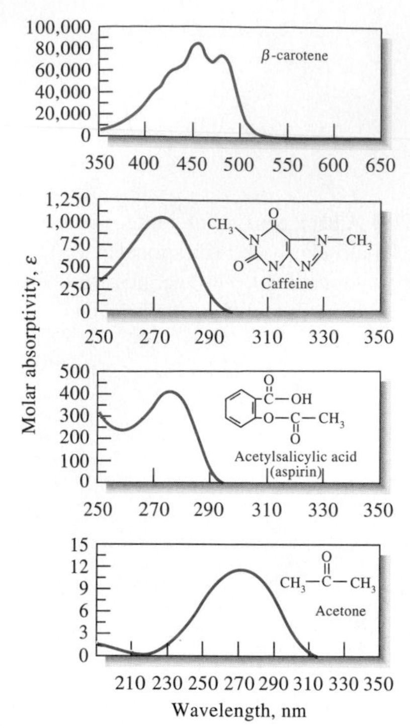

Figure 26-1 Absorption spectra for typical organic compounds.

TABLE 26-1

Absorption Characteristics of Some Common Organic Chromophores

Chromophore	Example	Solvent	λ_{max}, nm	ε_{max}
Alkene	$C_6H_{13}CH{=}CH_2$	*n*-Heptane	177	13,000
Conjugated alkene	$CH_2{=}CHCH{=}CH_2$	*n*-Heptane	217	21,000
Alkyne	$C_5H_{11}C{\equiv}C{-}CH_3$	*n*-Heptane	178	10,000
			196	2,000
			225	160
Carbonyl	$CH_3\overset{O}{\overset{\|}{C}}CH_3$	*n*-Hexane	186	1,000
			280	16
	$CH_3\overset{O}{\overset{\|}{C}}H$	*n*-Hexane	180	Large
			293	12
Carboxyl	$CH_3\overset{O}{\overset{\|}{C}}OH$	Ethanol	204	41
Amido	$CH_3\overset{O}{\overset{\|}{C}}NH_2$	Water	214	60
Azo	$CH_3N{=}NCH_3$	Ethanol	339	5
Nitro	CH_3NO_2	Isooctane	280	22
Nitroso	C_4H_9NO	Ethyl ether	300	100
			665	20
Nitrate	$C_2H_5ONO_2$	Dioxane	270	12
Aromatic	Benzene	*n*-Hexane	204	7,900
			256	200

conjugation between two or more chromophores tends to cause shifts in peak maxima to longer wavelengths. Finally, vibrational effects broaden absorption peaks in the ultraviolet and visible regions, which often makes precise determination of an absorption maximum difficult. Typical spectra for organic compounds are shown in Figure 26-1.

Saturated organic compounds containing such heteroatoms as oxygen, nitrogen, sulfur, or halogens have nonbonding electrons that can be excited by radiation in the 170- to 250-nm range. Table 26-2 lists a few examples of such compounds. Some of these compounds, such as alcohols and ethers, are common solvents, so their absorption in this region prevents measuring absorption of analytes dissolved in these compounds at wavelengths shorter than 180 to 200 nm. Occasionally, absorption in this region is used for determining halogen- and sulfur-bearing compounds.

TABLE 26-2

Absorption by Organic Compounds Containing Unsaturated Heteroatoms

Compound	λ_{max}, nm	ε_{max}
CH_3OH	167	1480
$(CH_3)_2O$	184	2520
CH_3Cl	173	200
CH_3I	258	365
$(CH_3)_2S$	229	140
$CH_3)NH_2$	215	600
$(CH_3)_3N$	227	900

Absorption by Inorganic Species

In general, the ions and complexes of elements in the first two transition series absorb broad bands of visible radiation in at least one of their oxidation states and are, as a result, colored (see, for example, Figure 26-2). Here, absorption involves transitions between filled and unfilled *d*-orbitals with energies that depend on the ligands bonded to the metal ions. The energy differences between these *d*-orbitals (and thus the position of the corresponding absorption peak) depend on the position of the element in the periodic table, its oxidation state, and the nature of the ligand bonded to it.

Absorption spectra of ions of the lanthanide and actinide transitions series differ substantially from those shown in Figure 26-3. The electrons responsible for absorption by these elements (*4f* and *5f,* respectively) are shielded from external influences by electrons that occupy orbitals with larger principal quantum numbers. As a result, the bands tend to be narrow and relatively unaffected by the species bonded by the outer electrons (see Figure 26-3).

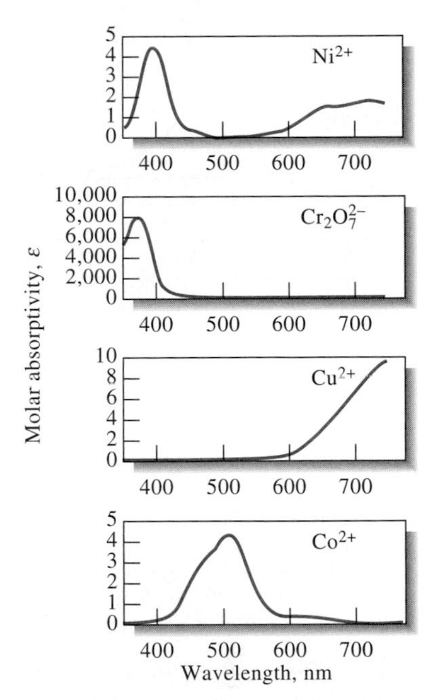

Figure 26-2 Absorption spectra of aqueous solutions of transition metal ions.

Charge-Transfer Absorption

For quantitative purposes, charge-transfer absorption is particularly important because molar absorptivities are unusually large ($\varepsilon > 10{,}000$), which leads to high sensitivity. Many inorganic and organic complexes exhibit this type of absorption and are therefore called charge-transfer complexes.

A **charge-transfer complex** consists of an electron-donor group bonded to an electron acceptor. When this product absorbs radiation, an electron from the donor is transferred to an orbital that is largely associated with the acceptor. The excited state is thus the product of a kind of internal oxidation/reduction process. This behavior differs from that of an organic chromophore, in which the excited electron is in a molecular orbital that is shared by two or more atoms.

A **charge-transfer complex** is a strongly absorbing species that is made up of an electron-donating species bonded to an electron-accepting species.

Familiar examples of charge-transfer complexes include the phenolic complex of iron(III), the 1,10-phenanthroline complex of iron(II), the iodide complex of molecular iodine, and the ferro/ferricyanide complex responsible for the color of Prussian blue. The red color of the iron(III)/thiocyanate complex is a further example of charge-transfer absorption. Absorption of a photon results in the transfer of an electron from the thiocyanate ion to an orbital that is largely associated with the iron(III) ion. The product is an excited species involving predominantly iron(II) and the thiocyanate radical SCN. As with other types of electronic excitation, the electron in this complex ordinarily returns to its original state after a brief period. Occasionally, however, an excited complex may dissociate and produce photochemical oxidation/reduction products. Three spectra of charge-transfer complexes are shown in Figure 26-4.

In most charge-transfer complexes involving a metal ion, the metal serves as the electron acceptor. Exceptions are the 1,10-phenanthroline complexes of iron(II) (Section 37N-2) and copper(I), in which the ligand is the acceptor and the metal ion the donor. A few other examples of this type of complex are known.

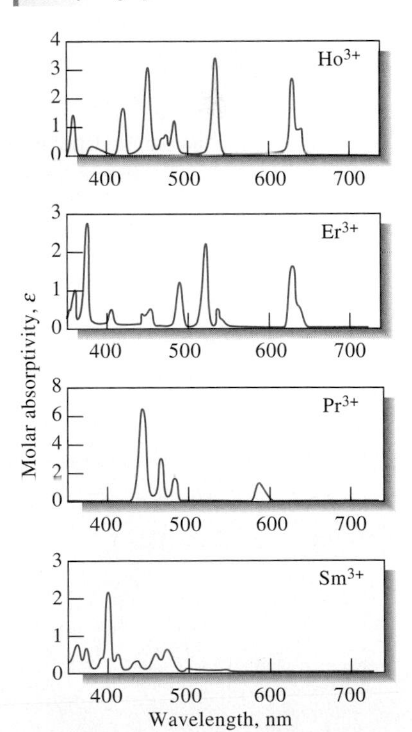

Figure 26-3 Absorption spectra of aqueous solutions of rare earth ions.

26A-2 Qualitative Applications of Ultraviolet/Visible Spectroscopy

Spectrophotometric measurements with ultraviolet radiation are useful for detecting chromophoric groups, such as those shown in Table 26-1.[2] Because large parts of even the most complex organic molecules are transparent to radiation longer than 180 nm, the appearance of one or more peaks in the region from 200 to 400 nm is clear indication of the presence of unsaturated groups or of atoms such as sulfur or halogens. Often, you can get an idea as to the identity of the absorbing groups by comparing the spectrum of an analyte with those of simple molecules

[2]For a detailed discussion of ultraviolet absorption spectroscopy in the identification of organic functional groups, see R. M. Silverstein and F. X. Webster, *Spectrometric Identification of Organic Compounds,* 6th ed., Ch. 7. New York: Wiley, 1997.

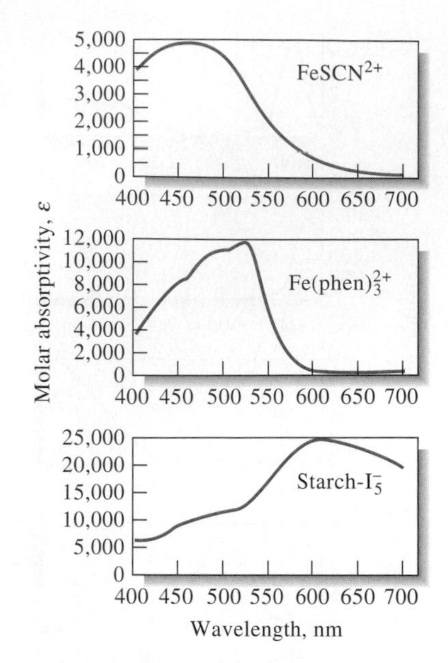

Figure 26-4 Absorption spectra of aqueous charge-transfer complexes.

containing various chromophoric groups.[3] Usually, however, ultraviolet spectra do not have sufficient fine structure to permit an analyte to be identified unambiguously. Thus, ultraviolet qualitative data must be supplemented with other physical or chemical evidence such as infrared, nuclear magnetic resonance, and mass spectra as well as solubility and melting- and boiling-point information.

Solvents

Ultraviolet spectra for qualitative analysis are usually measured using dilute solutions of the analyte. For volatile compounds, however, gas-phase spectra are often more useful than liquid-phase or solution spectra (for example, compare Figure 24-14a and 24-14b). Gas-phase spectra can often be obtained by allowing a drop or two of the pure liquid to evaporate and equilibrate with the atmosphere in a stoppered cuvette.

A solvent for ultraviolet/visible spectroscopy must be transparent in the region of the spectrum where the solute absorbs and should dissolve a sufficient quantity of the sample to give a well-defined analyte spectrum. In addition, we must consider possible interactions of the solvent with the absorbing species. For example, polar solvents, such as water, alcohols, esters, and ketones, tend to obliterate vibration spectra and should thus be avoided to preserve spectral detail. Nonpolar solvents, such as cyclohexane, often provide spectra that more closely approach that of a gas (compare, for example, the three spectra in Figure 24-14). In addition, the polarity of the solvent often influences the position of absorption maxima. For qualitative analysis, it is therefore important to compare analyte spectra with spectra of known compounds measured in the same solvent.

Table 26-3 lists common solvents for studies in the ultraviolet and visible regions and their approximate lower wavelength limits. These limits strongly depend on the purity of the solvent. For example, ethanol and the hydrocarbon solvents are frequently contaminated with benzene, which absorbs below 280 nm.[4]

The Effect of Slit Width

► Use small slit width for qualitative studies to preserve maximum detail in spectra.

The effect of variation in slit width, and hence effective bandwidth, is illustrated by the spectra in Figure 26-5. Clearly, peak heights and separation are distorted at wider bandwidths. Because of this, spectra for qualitative applications should be measured with minimum slit widths.

The Effect of Scattered Radiation at the Wavelength Extremes of a Spectrophotometer

Earlier we demonstrated that scattered radiation may lead to instrumental deviations from Beer's law (p. 733). Another undesirable effect of this type of radiation is that it occasionally causes false peaks to appear when a spectrophotometer is

[3]H. H. Perkampus, *UV-VIS Atlas of Organic Compounds,* 2nd ed. Weinheim: Wiley-VCH, 1992. In addition, in the past, several organizations have published catalogs of spectra that may still be useful, including American Petroleum Institute, Ultraviolet Spectral Data, *A.P.I. Research Project 44*. Pittsburgh: Carnegie Institute of Technology; *Sadtler Handbook of Ultraviolet Spectra.* Philadelphia: Sadtler Research Laboratories; American Society for Testing Materials, Committee E-13, Philadelphia.

[4]Most major suppliers of reagent chemicals in the United States offer spectrochemical grades of solvents. Spectral-grade solvents have been treated to remove absorbing impurities, and they meet or exceed the requirements set forth in *Reagent Chemicals, American Chemical Society Specifications,* 9th ed. Washington, D.C.: American Chemical Society, 2000.

TABLE 26-3
Solvents for the Ultraviolet and Visible Regions

Solvent	Lower Wavelength Limit, nm	Solvent	Lower Wavelength Limit, nm
Water	180	Carbon tetracholoride	260
Ethanol	220	Diethyl ether	210
Hexane	200	Acetone	330
Cyclohexane	200	Dioxane	320
		Cellosolve	320

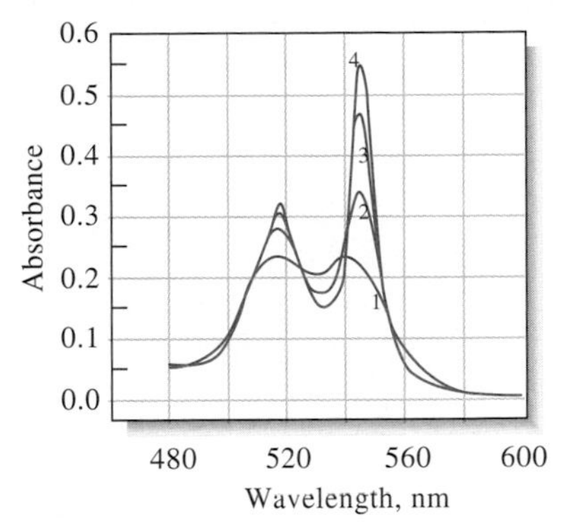

Figure 26-5 Spectra for reduced cytochrome *c* obtained with four spectral bandwidths: (1) 20 nm, (2) 10 nm, (3) 5 nm, and (4) 1 nm. At bandwidths < 1 nm, peak noise became pronounced. (Courtesy of Varian, Inc., Palo Alto, CA.)

being operated at its wavelength extremes. Figure 26-6 shows an example of such behavior. Curve *B* is the true spectrum for a solution of cerium(IV) produced with a research-quality spectrophotometer responsive down to 200 nm or less. Curve *A* was obtained for the same solution with an inexpensive instrument operated with a tungsten source designed for work in the visible region only. The false peak at about 360 nm is directly attributable to scattered radiation, which was not absorbed because it was made up of wavelengths longer than 400 nm. Under most circumstances, such stray radiation has a negligible effect because its power is only a tiny fraction of the total power of the beam exiting from the monochromator. At wavelength settings below 380 nm, however, radiation from the monochromator is greatly attenuated as a result of absorption by glass optical components and cuvettes. In addition, both the output of the source and the photocell sensitivity fall off dramatically below 380 nm. These factors combine to cause a substantial fraction of the measured absorbance to be due to the scattered radiation of wavelengths to which cerium(IV) is transparent. A false peak results.

This same effect is sometimes observed with ultraviolet/visible instruments when attempts are made to measure absorbances at wavelengths lower than about 190 nm.

26A-3 Quantitative Applications

Absorption spectroscopy based on ultraviolet and visible radiation is one of the most useful tools available to the chemist for quantitative analysis. The important characteristics of spectrophotometric and photometric methods are

1. *Wide applicability.* Enormous numbers of inorganic, organic, and biochemical species absorb ultraviolet or visible radiation and are thus amenable to direct quantitative determination. Many nonabsorbing species can also be determined after chemical conversion to absorbing derivatives. It has been estimated that more than 90% of the analyses performed in clinical laboratories are based on ultraviolet and visible absorption spectroscopy.
2. *High sensitivity.* Typical detection limits for absorption spectroscopy range from 10^{-4} to 10^{-5} M. With certain procedural modifications, this range can often be extended to 10^{-6} or even 10^{-7} M.
3. *Moderate to high selectivity.* Often a wavelength can be found at which the analyte alone absorbs, thus making preliminary separations unnecessary. Furthermore, where overlapping absorption bands do occur, corrections based on additional measurements at other wavelengths sometimes eliminate the need for a separation step.

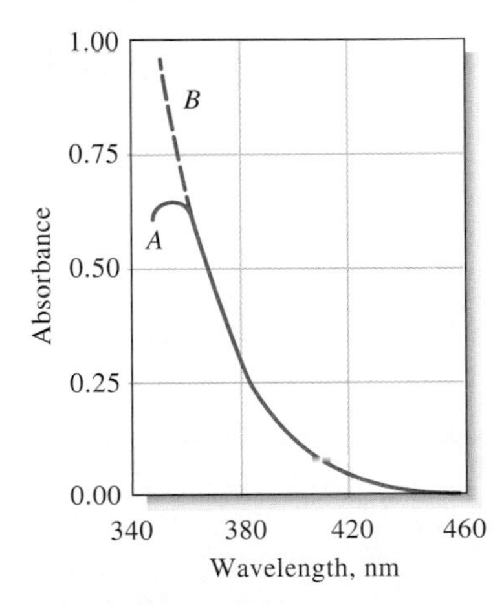

Figure 26-6 Spectra of cerium(IV) obtained with a spectrophotometer having glass optics (*A*) and quartz optics (*B*). The false peak in *A* occurs when stray radiation is transmitted at long wavelengths.

4. *Good accuracy.* The relative errors in concentration encountered with a typical spectrophotometric or photometric procedure using ultraviolet and visible radiation lie in the range from 1% to 5%. With special precautions, such errors can often be decreased to a few tenths of a percent.
5. *Ease and convenience.* Spectrophotometric and photometric measurements are easily and rapidly performed with modern instruments. In addition, the methods readily lend themselves to automation.

Scope

The applications of absorption analysis are not only numerous but also touch on every area in which quantitative information is sought. The reader can obtain a notion of the scope of spectrophotometry by consulting review articles published in *Analytical Chemistry*[5] as well as monographs on the subject.[6]

Application to Absorbing Species Table 26-1 (p. 786) lists many common organic chromophores. Spectrophotometric determination of organic compounds containing one or more of these groups is thus potentially feasible; many such applications can be found in the literature.

A number of inorganic species also absorb. We have noted that many ions of the transition metals are colored in solution and can thus be determined by spectrophotometric measurement. In addition, a number of other species show characteristic absorption peaks, including nitrite, nitrate, and chromate ions, the oxides of nitrogen, the elemental halogens, and ozone.

Applications to Nonabsorbing Species Many nonabsorbing analytes can be determined photometrically by causing them to react with chromophoric reagents to give products that absorb strongly in the ultraviolet and visible regions. The successful application of these color-forming reagents usually requires that their reaction with the analyte be forced to near completion.

Typical inorganic reagents include the following: thiocyanate ion for iron, cobalt, and molybdenum; the anion of hydrogen peroxide for titanium, vanadium, and chromium; and iodide ion for bismuth, palladium, and tellurium. Of even greater importance are organic chelating reagents that form stable colored complexes with cations. Common examples include diethyldithiocarbamate for the determination of copper, diphenylthiocarbazone for lead, 1,10-phenanthroline for iron, and dimethylglyoxime for nickel; Figure 26-7 shows the color-forming reaction for the first two of these reagents. The structure of the 1,10-phenanthroline complex of iron(II) is shown on page 554, and the reaction of nickel with dimethylglyoxime to form a red precipitate is described on page 331. In the application of the last reaction to the photometric determination of nickel, an aqueous solution of the cation is extracted with a solution of the chelating agent in an immiscible organic liquid. The absorbance of the resulting bright red organic layer serves as a measure of the concentration of the metal.

---

[5]L. G. Hargis, J. A. Howell, and R. E. Sutton, *Anal. Chem.* (Review), **1996,** *68,* 169R; J. A. Howell and R. E. Sutton, *Anal. Chem.* (Review), **1998,** *70,* 107R.

[6]H. Onishi, *Photometric Determination of Traces of Metals,* 4th ed. Part IIA, Part IIB. New York: Wiley, 1986, 1989; *Colorimetric Determination of Nonmetals,* 2nd ed., D. F. Boltz, Ed. New York: Interscience, 1978; E. B. Sandell and H. Onishi, *Photometric Determination of Traces of Metals,* 4th ed. New York: Wiley, 1978. F. D. Snell, *Photometric and Fluorometric Methods of Analysis.* New York: Wiley, 1978.

$$2\ (C_2H_5)_2N{-}C(=S){-}SNa + M^{2+} \rightleftharpoons (C_2H_5)_2N{-}C(S)_2M(S)_2C{-}N(C_2H_5)_2 + 2Na^+$$

(a)

$$2\ C_6H_5{-}NH{-}N{=}C(SH){-}N{=}N{-}C_6H_5 + Pb^{2+} \rightleftharpoons C_6H_5{-}NH{-}N{=}C(S)(N{=}N{-}C_6H_5)Pb(S)(N{=}N{-}C_6H_5)C{=}N{-}NH{-}C_6H_5 + 2H^+$$

(b)

Figure 26-7 Typical chelating reagents for absorption. (a) Diethyldithiocarbamate. (b) Diphenylthiocarbazone.

Other reagents are available that react with organic functional groups to produce colors that are useful for quantitative analysis. For example, the red color of the 1:1 complexes that form between low-molecular-weight aliphatic alcohols and cerium(IV) can be used for the quantitative estimation of such alcohols.

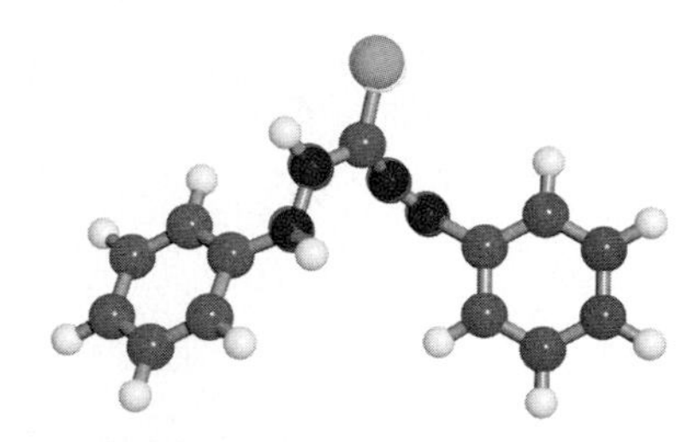

Molecular model of diphenylthiocarbazone.

Procedural Details

A first step in any photometric or spectrophotometric analysis is the development of conditions that yield a reproducible relationship (preferably linear) between absorbance and analyte concentration.

Wavelength Selection For maximum sensitivity, spectrophotometric absorbance measurements are ordinarily made at a wavelength corresponding to an absorption maximum because the change in absorbance per unit of concentration is greatest at this point. In addition, the absorption curve is often flat at a maximum, which leads to good adherence to Beer's law (see Figure 24-17) and less uncertainty from failure to reproduce precisely the wavelength setting of the instrument.

Variables That Influence Absorbance Common variables that influence the absorption spectrum of a substance include the nature of the solvent, the pH of the solution, the temperature, high electrolyte concentrations, and the presence of interfering substances. The effects of these variables must be known and conditions for the analysis chosen such that the absorbance will not be materially affected by small, uncontrolled variations in their magnitudes.

◀ Absorption spectra are affected by such variables as temperature, electrolyte concentration, and the presence of interferences.

Determination of the Relationship between Absorbance and Concentration The calibration standards for a photometric or a spectrophotometric analysis should approximate as closely as possible the overall composition of the actual samples, and should encompass a reasonable range of analyte concentrations. It is seldom safe to assume that Beer's law holds and to use only a single standard to determine the molar absorptivity. It is almost never a good idea to base the results of an analysis solely on a literature value for the molar absorptivity.

The Standard Addition Method Ideally, calibration standards should approximate the composition of the samples to be analyzed with respect to not only the analyte concentration but also the concentrations of the other species in the sample matrix in order to minimize the effects of various components of the sample on the measured absorbance. For example, the absorbance of many colored complexes of metal ions is decreased to a varying degree in the presence of sulfate and phosphate ions as a consequence of the tendency of these anions to form colorless complexes with metal ions. The color formation reaction is often less complete as a consequence, and lowered absorbances are the result. The matrix effect of sulfate and phosphate can often be counteracted by introducing into the standards amounts of the two species that approximate the amounts found in the samples. Unfortunately, when complex materials such as soils, minerals, and plant ash are being analyzed, preparation of standards that match the samples is often impossible or extremely difficult. When this is the case, the standard addition method is often helpful in counteracting matrix effects.

The standard addition method can take several forms as discussed in Section 8C-3; the single-point method was described in Example 8-8.[7] The multiple additions method is often chosen for photometric or spectrophotometric analyses, and this method will be described here. This technique involves adding several increments of a standard solution to sample aliquots of the same size. Each solution is then diluted to a fixed volume before measuring its absorbance. When the amount of sample is limited, standard additions can be carried out by successive addition of increments of the standard to a single measured aliquot of the unknown. The measurements are made on the original solution and after each addition of standard analyte. This procedure is often more convenient for voltammetry.

Assume that several identical aliquots V_x of the unknown solution with a concentration c_x are transferred to volumetric flasks having a volume V_t. To each of these flasks is added a variable volume V_s mL of a standard solution of the analyte having a known concentration c_s. The color development reagents are then added, and each solution is diluted to volume. If the chemical system follows Beer's law, the absorbance of the solutions is described by

$$\begin{aligned} A_s &= \frac{\varepsilon b V_s c_s}{V_t} + \frac{\varepsilon b V_x c_x}{V_t} \\ &= kV_s c_s + kV_x c_x \end{aligned} \tag{26-1}$$

where k is a constant equal to $\varepsilon b/V_t$. A plot of A_s as a function of V_s should yield a straight line of the form

$$A_s = mV_s + b$$

where the slope m and the intercept b are given by

$$m = kc_s$$

and

$$b = kV_x c_x$$

[7]See M. Bader, *J. Chem. Educ.*, **1980,** *57,* 703.

Least-squares analysis (Section 8C-2) of the data can be used to determine m and b; c_x can then be calculated from the ratio of these two quantities and the known values of V_x and V_s. Thus,

$$\frac{m}{b} = \frac{kc_s}{kV_x c_x}$$

which rearranges to

$$c_x = \frac{bc_s}{mV_x} \tag{26-2}$$

An approximate value for the standard deviation in c_x can then be obtained by assuming that the uncertainties in $c_s V_s$, and V_t are negligible with respect to those in m and b. Then, the relative variance of the result $(s_c/c_x)^2$ is assumed to be the sum of the relative variances of m and b. That is,

$$\left(\frac{s_c}{c_x}\right)^2 = \left(\frac{s_m}{m}\right)^2 + \left(\frac{s_b}{b}\right)^2$$

where s_m and s_b are the standard deviations of the slope and intercept, respectively. By taking the square root of this equation, we find the standard deviation of the measured concentration s_c.

$$s_c = c_x\sqrt{\left(\frac{s_m}{m}\right)^2 + \left(\frac{s_b}{b}\right)^2} \tag{26-3}$$

EXAMPLE 26-1

Ten-millimeter aliquots of a natural water sample were pipetted into 50.00-mL volumetric flasks. Exactly 0.00, 5.00, 10.00, 15.00, and 20.00 mL of a standard solution containing 11.1 ppm of Fe^{3+} were added to each, followed by an excess of thiocyanate ion to give the red complex $Fe(SCN)^{2+}$. After dilution to volume, absorbances for the five solutions, measured with a photometer equipped with a green filter, were found to be 0.240, 0.437, 0.621, 0.809, and 1.009, respectively (0.982-cm cells). (a) What was the concentration of Fe^{3+} in the water sample? (b) Calculate the standard deviation of the slope, the intercept, and the concentration of Fe.

(a) In this problem, c_s = 11.1 ppm, V_s = 10.00 mL, and V_t = 50.00 mL. A plot of the data, shown in Figure 26-8, demonstrates that Beer's law is obeyed. To obtain the equation for the line in Figure 26-8, the procedure illustrated in Example 8-4 (p. 198) is followed. The result is m = 0.03820 and b = 0.2412, and thus

$$A_s = 0.03820V_s + 0.2412$$

(continued)

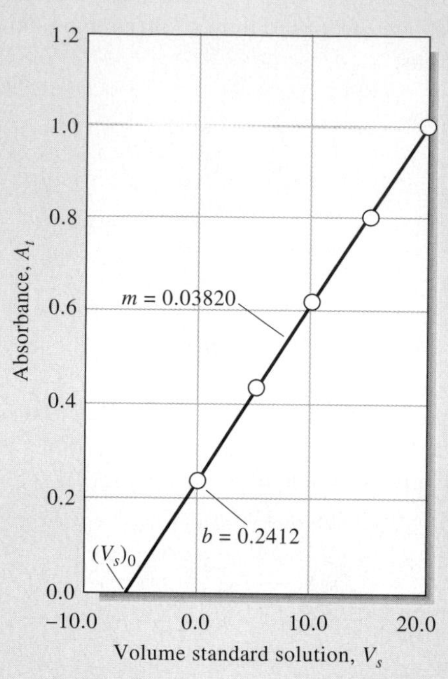

Figure 26-8 Data for standard addition method for the determination of Fe^{3+} as the $Fe(SCN)^{2+}$ complex.

Substituting into Equation 26-2 gives

$$c_x = \frac{(0.2412)(11.1 \text{ ppm } Fe^{3+})}{(0.03820 \text{ mL}^{-1})(10.00 \text{ mL})} = 7.01 \text{ ppm } Fe^{3+}$$

(b) Equations 8-16 and 8-17 give the standard deviation of the slope and the intercept. That is, $s_m = 3.07 \times 10^{-4}$ and $s_b = 3.76 \times 10^{-3}$. Substituting into Equation 26-3 gives

$$s_c = 7.01 \text{ ppm } Fe^{3+} \sqrt{\left(\frac{3.07 \times 10^{-4}}{0.03820}\right)^2 + \left(\frac{3.76 \times 10^{-3}}{0.2412}\right)^2}$$

$$= 0.12 \text{ ppm } Fe^{3+}$$

In the interest of saving time or sample, it is possible to perform a standard addition analysis using only two increments of sample. Here, a single addition of V_s mL of standard would be added to one of the two samples and we can write

$$A_1 = \frac{\varepsilon b V_x c_x}{V_t}$$

$$A_2 = \frac{\varepsilon b V_x c_x}{V_t} + \frac{\varepsilon b V_s c_s}{V_t}$$

where A_1 and A_2 are absorbances of the diluted sample and the diluted sample plus standard, respectively. If we solve the first equation for εb, substitute the result into the second equation, and solve for c_x, we find that

$$c_x = \frac{A_1 c_s V_s}{A_2 V_t - A_1 V_x} \tag{26-4}$$

Single-point standard addition methods are inherently more risky than multiple-point methods. There is no check on linearity with single-point methods, and results depend strongly on the reliability of one measurement.

Spreadsheet Summary In Chapter 12 of *Applications of Microsoft® Excel in Analytical Chemistry,* we investigate the multiple standard additions method for determining solution concentration. A least-squares analysis of the data leads to the determination of the concentration of the analyte as well as the uncertainty of the measured concentration.

EXAMPLE 26-2

The single-point standard addition method was used in the determination of phosphate by the molybdenum blue method. A 2.00-mL urine sample was treated with molybdenum blue reagents to produce a species absorbing at 820 nm, after which the sample was diluted to 100 mL. A 25.00-mL aliquot of this solution gave an absorbance of 0.428 (solution 1). Addition of 1.00 mL of a solution containing 0.0500 mg of phosphate to a second 25.0-mL aliquot gave an absorbance of 0.517 (solution 2). Use these data to calculate the concentration of phosphate in milligrams per milliliter of the specimen.

Here we substitute into Equation 26-4 and obtain

$$c_x = \frac{A_1 c_s V_s}{A_2 V_t - A_1 V_x} = \frac{(0.428)(0.0500 \text{ mg } PO_4^{3-} / \cancel{\text{mL}})(1.00 \cancel{\text{mL}})}{(0.517)(25.00 \text{ mL}) - (0.428)(25.00 \text{ mL})}$$
$$= 0.00780 \text{ mg } PO_4^{3-} \text{ mL}$$

This is the concentration of the diluted sample. To obtain the concentration of the original urine sample, we need to multiply by 100.00/2.00. Thus,

$$\text{concentration of phosphate} = 0.00780 \frac{\text{mg}}{\text{mL}} \times \frac{100.00 \cancel{\text{mL}}}{2.00 \cancel{\text{mL}}}$$
$$= 0.390 \text{ mg/mL}$$

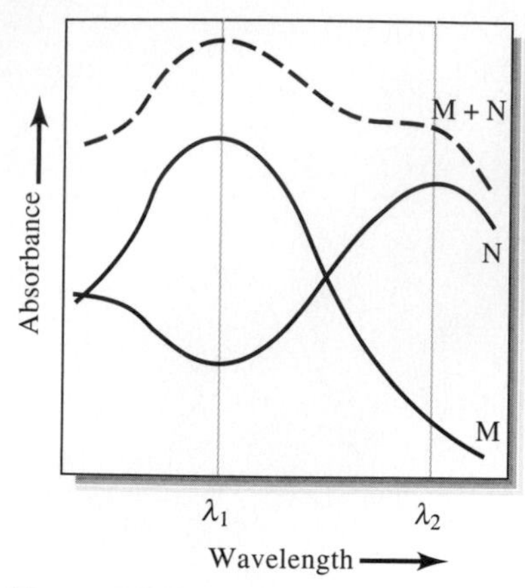

Figure 26-9 Absorption spectrum of a two-component mixture (M + N), with spectra of the individual components.

Analysis of Mixtures The total absorbance of a solution at any given wavelength is equal to the sum of the absorbances of the individual components in the solution (Equation 24-14). This relationship makes it possible in principle to determine the concentrations of the individual components of a mixture even if their spectra overlap completely. For example, Figure 26-9 shows the spectrum of a solution containing a mixture of species M and species N as well as absorption spectra for the individual components. We see that there is no wavelength at which the absorbance is due to just one of these components. To analyze the mixture, molar absorptivities for M and N are first determined at wavelengths λ_1 and λ_2 with sufficient concentrations of the two standard solutions to be sure that Beer's law is obeyed over an absorbance range that encompasses the absorbance of the sample. Note that the wavelengths selected are ones at which the molar absorptivities of the two components differ significantly. Thus, at λ_1, the molar absorptivity of component M is much larger than that for component N. The reverse is true for λ_2. To complete the analysis, the absorbance of the mixture is determined at the same two wavelengths. From the known molar absorptivities and path length, the following equations hold:

$$A_1 = \varepsilon_{M_1} b c_M + \varepsilon_{N_1} b c_N \tag{26-5}$$

$$A_2 = \varepsilon_{M_2} b c_M + \varepsilon_{N_2} b c_N \tag{26-6}$$

where the subscript 1 indicates measurement at λ_1 and the subscript 2 indicates measurement at λ_2. With the known values of ε and b, Equations 26-5 and 26-6 represent two equations in two unknowns (c_M and c_N), which can be readily solved as demonstrated in Example 26-3.

EXAMPLE 26-3

Palladium(II) and gold(III) can be determined simultaneously by reaction with methiomeprazine ($C_{19}H_{24}N_2S_2$). The absorption maximum for the Pd complex occurs at 480 nm, while that for the Au complex is at 635 nm. Molar absorptivity data at these wavelengths are as follows:

	Molar Absorptivity, ε	
	480 nm	635 nm
Pd complex	3.55×10^3	5.64×10^2
Au complex	2.96×10^3	1.45×10^4

A 25.0-mL sample was treated with an excess of methiomeprazine and subsequently diluted to 50.0 mL. Calculate the molar concentrations of Pd(II), c_{Pd}, and Au(III), c_{Au}, in the sample if the diluted solution had an absorbance of 0.533 at 480 nm and 0.590 at 635 nm when measured in a 1.00-cm cell.

At 480 nm from Equation 26-5

$$A_{480} = \varepsilon_{Pd(480)} b c_{Pd} + \varepsilon_{Au(480)} b c_{Au}$$

$$0.533 = (3.55 \times 10^3\ M^{-1}\ \cancel{cm^{-1}})\ (1.00\ \cancel{cm})\ c_{Pd} + (2.96 \times 10^3\ M^{-1}\ \cancel{cm^{-1}})\ (1.00\ \cancel{cm}) c_{Au}$$

or

$$c_{Pd} = \frac{0.533 - 2.96 \times 10^3\ M^{-1}\, c_{Au}}{3.55 \times 10^3\ M^{-1}}$$

At 635 nm from Equation 26-6

$$A_{635} = \varepsilon_{Pd(635)}\, bc_{Pd} + \varepsilon_{Au(635)}\, bc_{Au}$$

$$0.590 = (5.64 \times 10^2\ M^{-1}\ \cancel{cm^{-1}})\,(1.00\ \cancel{cm})\, c_{Pd} + (1.45 \times 10^4\ M^{-1}\ \cancel{cm^{-1}})\,(1.00\ \cancel{cm})c_{Au}$$

Substitution for c_{Pd} in this expression gives

$$0.590 = \frac{(5.64 \times 10^2\ M^{-1})(0.533 - 2.96 \times 10^3\ M^{-1}\, c_{Au})}{3.55 \times 10^3\ M^{-1}} + (1.45 \times 10^4\ M^{-1})\, c_{Au}$$

$$= 0.0847 - (4.70 \times 10^2\ M^{-1})c_{Au} + (1.45 \times 10^4\ M^{-1})c_{Au}$$

$$c_{Au} = \frac{(0.590 - 0.0847)}{(1.45 \times 10^4\ M^{-1} - 4.70 \times 10^2\ M^{-1})} = 3.60 \times 10^{-5}\ M$$

and

$$c_{Pd} = \frac{0.533 - (2.96 \times 10^3\ \cancel{M^{-1}})(3.60 \times 10^{-5}\ \cancel{M})}{3.55 \times 10^3\ M^{-1}} = 1.20 \times 10^{-4}\ M$$

Since the analysis involved a twofold dilution, the concentrations of Pd(II) and Au(III) in the original sample were 7.20×10^{-5} and 2.40×10^{-4} M, respectively.

Mixtures containing more than two absorbing species can be analyzed, in principle at least, if one additional absorbance measurement is made for each added component. The uncertainties in the resulting data become greater, however, as the number of measurements increases. Some computerized spectrophotometers are capable of minimizing these uncertainties by overdetermining the system; that is, these instruments use many more data points than unknowns and effectively match the entire spectrum of the unknown as closely as possible by calculating synthetic spectra for various concentrations of the components. The calculated spectra are then added, and the sum is compared with the spectrum of the analyte solution until a close match is found. The spectra for standard solutions of each component of the mixture are acquired and stored in computer memory prior to making measurements on the analyte mixture.

Spreadsheet Summary In Chapter 12 of *Applications of Microsoft® Excel in Analytical Chemistry,* we use spreadsheet methods to determine concentrations of mixtures of analytes. Solutions to sets of simultaneous equations are evaluated using iterative techniques, the method of determinants, and matrix manipulations.

The Effect of Instrumental Uncertainties[8]

In the context of this discussion, **noise** refers to random variations in the instrument output due not only to electrical fluctuations but also to such other variables as the way the operator reads the meter, the position of the cell in the light beam, the temperature of the solution, and the output of the source.

The accuracy and precision of spectrophotometric analyses are often limited by the indeterminate error, or noise, associated with the instrument. As pointed out in Chapter 25, a spectrophotometric absorbance measurement entails three steps: a 0% T adjustment, a 100% T adjustment, and a measurement of % T. The random errors associated with each of these steps combine to give a net random error for the final value obtained for T. The relationship between the noise encountered in the measurement of T and the resulting *concentration uncertainty* can be derived by writing Beer's law in the form

$$c = -\frac{1}{\varepsilon b}\log T = \frac{-0.434}{\varepsilon b}\ln T$$

Taking the partial derivative of this equation while holding εb constant leads to the expression

$$\partial c = \frac{-0.434}{\varepsilon b T}\partial T$$

where ∂c can be interpreted as the uncertainty in c that results from the noise (or uncertainty) in T. Dividing this equation by the previous one gives

$$\frac{\partial c}{c} = \frac{0.434}{\log T}\left(\frac{\partial T}{T}\right) \qquad (26\text{-}7)$$

where ∂T is the relative random error in T attributable to the noise in the three measurement steps, and is the resulting relative random concentration error.

The best and most useful measure of the random error ∂T is the standard deviation σ_T, which is easily measured for a given instrument by making 20 or more replicate transmittance measurements of an absorbing solution. Substituting σ_T and σ_c for the corresponding differential quantities in Equation 26-5 leads to

$$\frac{\sigma_c}{c} = \frac{0.434}{\log T}\left(\frac{\sigma_T}{T}\right) \qquad (26\text{-}8)$$

where σ_T/T and σ_c/c are relative standard deviations.

Equation 26-6 shows that the uncertainty in a photometric concentration measurement varies in a complex way with the magnitude of the transmittance. The situation is even more complicated than suggested by the equation, however, because the uncertainty σ_T is, under many circumstances, also dependent on T. In a detailed theoretical and experimental study, Rothman, Crouch, and Ingle[9] described several sources of instrumental random errors and showed the net effect of these errors on the precision of concentration measurements. The errors fall into three categories: those for which the magnitude of σ_T is (1) independent of T, (2) proportional to $\sqrt{T^2 + T}$, and (3) proportional to T. Table 26-4 summarizes information about

► Uncertainties in spectrophotometric concentration measurements depend on the magnitude of the transmittance (absorbance) in a complex way. The uncertainties can be independent of T, proportional to $\sqrt{T^2 + T}$, or proportional to T.

[8]For further reading, see J. D. Ingle, Jr., and S. R. Crouch, *Spectrochemical Analysis,* Ch. 5. Upper Saddle River, NJ: Prentice Hall, 1988.

[9]L. D. Rothman, S. R. Crouch, and J. D. Ingle, Jr., *Anal. Chem.,* **1975,** *47,* 1226.

TABLE 26-4

Categories of Instrumental Indeterminate Errors in Transmittance Measurements

Category	Sources	Effect of T on Relative Standard Deviation of Concentration
$\sigma_T = k_1$	Readout resolution; thermal detector noise; dark current and amplifier noise	$\frac{\sigma_c}{c} = \frac{0.434}{\log T} \times \frac{k_1}{T}$ (26-9)
$\sigma_T = k_2\sqrt{T^2 + T}$	Photon detector shot noise	$\frac{\sigma_c}{c} = \frac{0.434}{\log T} \times k_2\sqrt{1 + \frac{1}{T}}$ (26-10)
$\sigma_T = k_3 T$	Cell positioning uncertainty; fluctuation in source intensity	$\frac{\sigma_c}{c} = \frac{0.434}{\log T} \times k_3$ (26-11)

Note: σ_T is the standard deviation of the transmittance measurements; σ_c/c is the relative standard deviation of the concentration measurements; T is transmittance; and k_1, k_2, and k_3 are constants for a given instrument.

these sources of uncertainty. When the three relationships for σ_T in the first column are substituted into Equation 26-8, we obtain three equations for the relative standard deviation in the concentration σ_c/c. These derived equations are shown in the third column of Table 26-4.

Concentration Errors When $\sigma_T = k_1$ For many photometers and spectrophotometers, the standard deviation in the measurement of T is constant and independent of the magnitude of T. We often see this type of random error in direct-reading instruments with analog meter readouts, which have somewhat limited resolution. The size of a typical scale is such that a reading cannot be reproduced to better than a few tenths of a percent of the full-scale reading, and the magnitude of this uncertainty is the same from one end of the scale to the other. For typical inexpensive instruments, we find standard deviations in transmittance of about 0.003 ($\sigma_T = \pm 0.003$).

EXAMPLE 26-4

A spectrophotometric analysis was performed with a manual instrument that exhibited an absolute standard deviation in transmittance of ± 0.003 throughout its transmittance range. Calculate the relative standard deviation in concentration that results from this uncertainty when the analyte solution has an absorbance of (a) 1.000 and (b) 2.000.

(a) To convert absorbance to transmittance, we write

$$\log T = -A = -1.000$$

$$T = \text{antilog}(-1.000) = 0.100$$

(continued)

For this instrument, $\sigma_T = k_1 = \pm 0.003$ (see first entry in Table 26-4). Substituting this value and $T = 0.100$ into Equation 26-8 yields

$$\frac{\sigma_c}{c} = \frac{0.434}{\log 0.100}\left(\frac{\pm 0.003}{0.100}\right) = \pm 0.013 = \pm 1.3\%$$

(b) At $A = 2.000$, $T = \text{antilog}(-2.000) = 0.010$

$$\frac{\sigma_c}{c} = \frac{0.434}{\log 0.010}\left(\frac{\pm 0.003}{0.010}\right) = \pm 0.065 = \pm 6.5\%$$

The data plotted as curve *A* in Figure 26-10 were obtained from calculations similar to those in Example 26-4. Note that the relative standard deviation in the concentration passes through a minimum at an absorbance of about 0.5 and rises rapidly when the absorbance is less than about 0.1 or greater than approximately 1.5.

Figure 26-11a is a plot of the relative standard deviation for experimentally determined concentrations as a function of absorbance. It was obtained with a spectrophotometer similar to the one shown in Figure 25-19. The striking similarity between this curve and curve *A* in Figure 26-10 indicates that the instrument studied is affected by an absolute indeterminate error in transmittance of about ± 0.003 and that this error is independent of transmittance. The source of this uncertainty is probably the limited resolution of the transmittance scale.

Infrared spectrophotometers also exhibit an indeterminate error that is independent of transmittance. The source of the error in these instruments lies in the

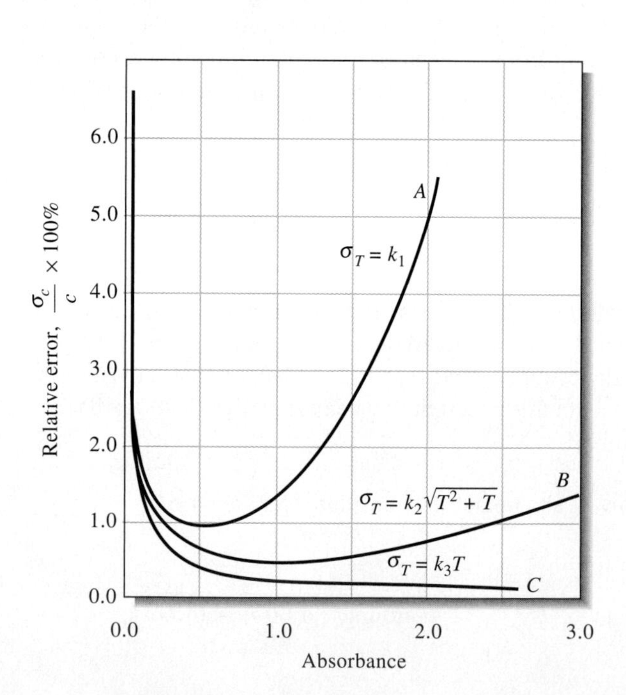

Figure 26-10 Error curves for various categories of instrumental uncertainties.

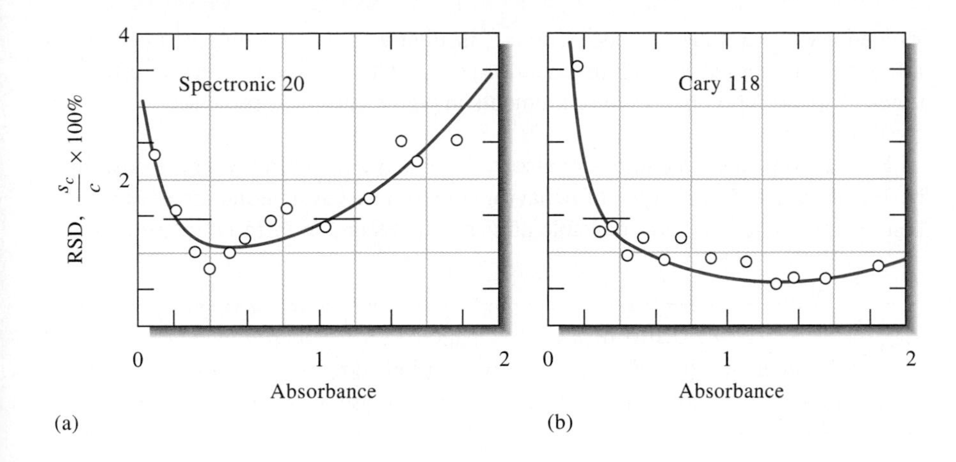

Figure 26-11 Experimental curves relating relative concentration uncertainties to absorbance for two spectrophotometers. Data obtained with (a) a Spectronic 20, a low-cost instrument (Figure 25-19), and (b) a Cary 118, a research-quality instrument. (From W. E. Harris and B. Kratochvil, *An Introduction to Chemical Analysis,* p. 384. Philadelphia: Saunders College Publishing, 1981. With permission.).

thermal detector. Fluctuations in the output of this type of transducer are independent of the output; indeed, fluctuations are observed even in the absence of radiation. An experimental plot of data from an infrared spectrophotometer is similar in appearance to Figure 26-11a. The curve is displaced upward, however, because of the greater standard deviation associated with infrared measurements.

Concentration Errors When $\sigma_T = k_2\sqrt{T^2 + T}$ This type of random uncertainty is characteristic of the highest quality spectrophotometers. It has its origin in the so-called shot noise that causes the output of photomultipliers and phototubes to fluctuate randomly about a mean value. Equation 26-10 in Table 26-4 describes the effect of shot noise on the relative standard deviation of concentration measurements. A plot of this relationship appears as curve *B* in Figure 26-10. We calculated these data assuming that $k_2 = \pm 0.003$, a value that is typical for high-quality spectrophotometers.

Figure 26-11b shows an analogous plot of experimental data obtained with a high-quality research-type ultraviolet/visible spectrophotometer. Note that, in contrast to the less expensive instrument, absorbances of 2.0 or greater can be measured here without serious deterioration in the concentration uncertainty.

Concentration Errors When $\sigma_T = k_3T$ Substituting $\sigma_T = k_3T$ into Equation 26-8 reveals that the relative standard deviation in concentration from this type of uncertainty is inversely proportional to the logarithm of the transmittance (Equation 26-11 in Table 26-4). Curve *C* in Figure 26-10, which is a plot of Equation 26-11, shows that this type of uncertainty is important at low absorbances (high transmittances) but approaches zero at high absorbances.

At low absorbances, the precision obtained with high-quality double-beam instruments is often described by Equation 26-11. The source of this behavior is failure to position cells reproducibly with respect to the beam during replicate measurements. This position dependence is probably the result of small imperfections in the cell windows, which cause reflective losses and transparency to differ from one area of the window to another.

It is possible to evaluate Equation 26-11 by comparing the precision of absorbance measurements made in the usual way with measurements in which the cells are left undisturbed at all times, with replicate solutions being introduced with a syringe. Experiments of this kind with a high-quality spectrophotometer yielded a value of 0.013 for k_3.[9] Curve C in Figure 26-10 was obtained by substituting this numerical value into Equation 26-11. Cell positioning errors affect all types of spectrophotometric measurements in which cells are repositioned between measurements.

Fluctuations in source intensity also yield standard deviations that are described by Equation 26-11. This type of behavior sometimes occurs in inexpensive single-beam instruments that have unstable power supplies and in infrared instruments.

Spreadsheet Summary In Chapter 12 of *Applications of Microsoft® Excel in Analytical Chemistry,* we explore errors in spectrophotometric measurements by simulating error curves such as those shown in Figure 26-11.

26A-4 Photometric and Spectrophotometric Titrations

CD-ROM Simulation: Exploration of Spectrophotometric Titrations with an Emphasis on Wavelength Selection.

Photometric and spectrophotometric measurements are useful for locating the equivalence points of titrations.[10] This application of absorption measurements requires that one or more of the reactants or products absorb radiation or that an absorbing indicator be added to the analyte solution.

Titration Curves

A photometric titration curve is a plot of absorbance (corrected for volume change) as a function of titrant volume. If conditions are chosen properly, the curve consists of two straight-line regions with different slopes, one occurring prior to the equivalence point of the titration and the other located well beyond the equivalence-point region; the end point is taken as the intersection of extrapolated linear portions of the two lines.

Figure 26-12 shows typical photometric titration curves. Figure 26-12a is the curve for the titration of a nonabsorbing species with an absorbing titrant that reacts with the analyte to form a nonabsorbing product. An example is the titration of thiosulfate ion with triiodide ion. The titration curve for the formation of an absorbing product from colorless reactants is shown in Figure 26-12b; an example is the titration iodide ion with a standard solution of iodate ion to form triiodide. The remaining figures illustrate the curves obtained with various combinations of absorbing analytes, titrants, and products.

To obtain titration curves with linear portions that can be extrapolated, the absorbing system(s) must obey Beer's law. In addition, absorbances must be corrected for volume changes by multiplying the observed absorbance by $(V + v)/V$, where V is the original volume of the solution and v is the volume of added titrant.

Instrumentation

Photometric titrations are usually performed with a spectrophotometer or a photometer that has been modified so that the titration vessel is held stationary in the light path. After the instrument is set to a suitable wavelength or an appropriate filter is inserted, the 0%T adjustment is made in the usual way. With radiation passing through the analyte solution to the detector, the instrument is then adjusted to a

[10]For further information, see J. B. Headridge, *Photometric Titrations.* New York: Pergamon Press, 1961.

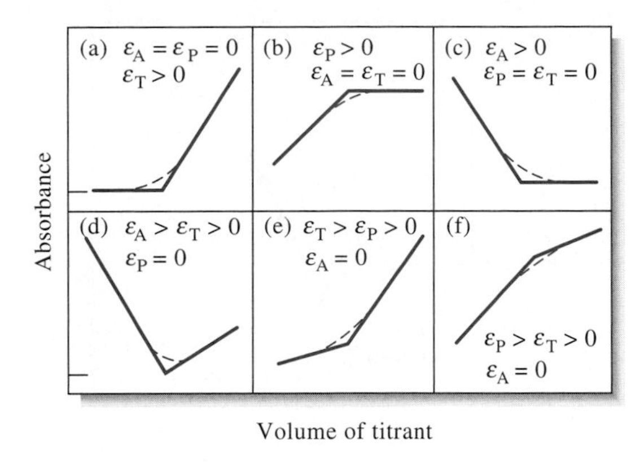

Figure 26-12 Typical photometric titration curves. Molar absorptivities of the analyte titrated, the product, and the titrant are ε_A, ε_P, and ε_T, respectively.

convenient absorbance, reading by varying the source intensity or the detector sensitivity. It is not usually necessary to measure the true absorbance, since relative values are perfectly adequate for end-point detection. Titration data are then collected without changing the instrument settings. The power of the radiation source and the response of the detector must remain constant during a photometric titration. Cylindrical containers are often used in photometric titrations, and it is important to avoid moving the cell so that the pathlength remains constant.

Both filter photometers and spectrophotometers have been used for photometric titrations. To ensure adherence to Beer's law, spectrophotometers are usually used because their bandwidths are narrower than those of photometers.

Applications of Photometric Titrations

Photometric titrations often provide more accurate results than a direct photometric determination because the data from several measurements are used to determine the end point. Furthermore, the presence of other absorbing species may not interfere, since only a change in absorbance is being measured.

◀ Photometric titrations are often more accurate than direct photometric determinations.

An advantage of end points determined from linear-segment photometric titration curves is that the experimental data are collected well away from the equivalence-point region where the absorbance changes gradually. Consequently, the equilibrium constant for the reaction need not be as large as that required for a sigmoidal titration curve that depends on observations near the equivalence point (for example, potentiometric or indicator end points). For the same reason, more dilute solutions may be titrated using photometric detection.

The photometric end point has been applied to many types of reactions. For example, most standard oxidizing agents have characteristic absorption spectra and thus produce photometrically detectable end points. Although standard acids or bases do not absorb, the introduction of acid/base indicators permits photometric neutralization titrations. The photometric end point has also been used to great advantage in titrations with EDTA and other complexing agents. Figure 26-13 illustrates the application of this technique to the successive titration of bismuth(III) and copper(II). At 745 nm, the cations, the reagent, and the bismuth complex formed do not absorb but the copper complex does. Thus, during the first segment of the titration when the bismuth-EDTA complex is being formed ($K_f = 6.3 \times 10^{22}$), the solution exhibits no absorbance until essentially all the bismuth has been titrated. With the first formation of the copper complex ($K_f = 6.3 \times 10^{18}$), an increase in absorbance occurs. The increase continues until the copper equiva-

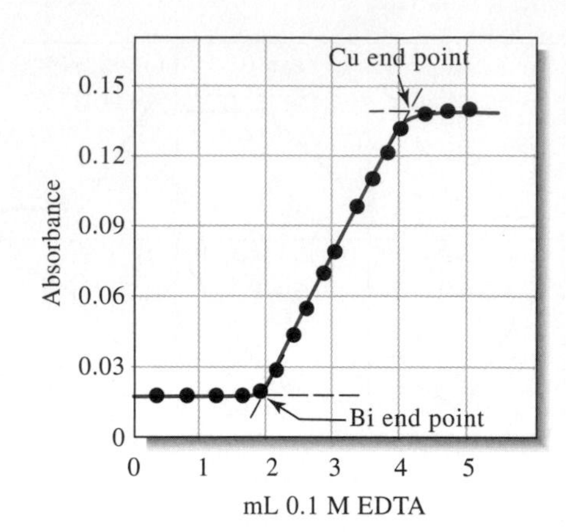

Figure 26-13 Photometric titration curve at 745 nm for 100 mL of a solution that was 2.0×10^{-3} M in Bi^{3+} and Cu^{2+}. (Reprinted with permission from A. L. Underwood, *Anal. Chem.*, **1954,** *26,* 1322. Copyright 1954 the American Chemical Society.)

lence point is reached. Further reagent additions cause no further absorbance change. Two well-defined end points result as shown in Figure 26-13.

The photometric end point has also been adapted to precipitation titrations. The suspended solid product causes a decrease in the radiant power of the light source by scattering from the particles of the precipitate. The equivalence point occurs when the precipitate stops forming, and the amount of light reaching the detector becomes constant. This type of end point detection is called **turbidimetry** because the amount of light reaching the detector is a measure of the **turbidity** of the solution.

Spreadsheet Summary In Chapter 12 of *Applications of Microsoft® Excel in Analytical Chemistry,* methods for treating data from spectrophotometric titrations are explored. We analyze titration data using least-squares procedures and use the resulting parameters to compute the concentration of the analyte.

26A-5 Spectrophotometric Studies of Complex Ions

► The composition of a complex in solution can be determined without actually isolating the complex as a pure compound.

Spectrophotometry is a valuable tool for determining the composition of complex ions in solution and for determining their formation constants. The power of the technique lies in the fact that quantitative absorption measurements can be performed without disturbing the equilibria under consideration. Although many spectrophotometric studies of complexes involve systems in which a reactant or a product absorbs, nonabsorbing systems can also be investigated successfully. For example, the composition and formation constant for a complex of iron(II) and a nonabsorbing ligand may often be determined by measuring the absorbance decreases that occur when solutions of the absorbing iron(II) complex of 1,10-phenanthroline are mixed with various amounts of the nonabsorbing ligand. The success of this approach depends on the well-known values of the formation constant ($K_f = 2 \times 10^{21}$) and the composition of the 1,10-phenanthroline (3:1) complex of iron(II).

The three most common techniques used for complex-ion studies are (1) the method of continuous variations, (2) the mole-ratio method, and (3) the slope-ratio method.

The Method of Continuous Variations

In the method of continuous variations, cation and ligand solutions with identical analytical concentrations are mixed in such a way that the total volume and the total moles of reactants in each mixture is constant but the mole ratio of reactants varies systematically (for example, 9:1, 8:2, 7:3, and so forth). The absorbance of each solution is then measured at a suitable wavelength and corrected for any absorbance the mixture might exhibit if no reaction had occurred. The corrected absorbance is plotted against the volume fraction of one reactant, that is, $V_M/(V_M + V_L)$, where V_M is the volume of the cation solution and V_L is the volume of the ligand solution. A typical continuous-variations plot is shown in Figure 26-14. A maximum (or minimum if the complex absorbs less than the reactants) occurs at a volume ratio V_M/V_L corresponding to the combining ratio of cation and ligand in the complex. In Figure 26-14, $V_M/(V_M + V_L)$ is 0.33 and $V_L/(V_M + V_L)$ is 0.66; thus, V_M/V_L is 0.33/0.66, which suggests that the complex has the formula ML_2.

The curvature of the experimental lines in Figure 26-14 is the result of incompleteness of the complex-formation reaction. A formation constant for the complex can be evaluated from measurements of the deviations from the theoretical straight lines, which represent the curve that would result if the reaction between the ligand and the metal proceeded to completion.

The Mole-Ratio Method

In the mole-ratio method, a series of solutions is prepared in which the analytical concentration of one reactant (usually the cation) is held constant while that of the other is varied. A plot of absorbance versus mole ratio of the reactants is then made. If the formation constant is reasonably favorable, we obtain two straight lines of different slopes that intersect at a mole ratio corresponding to the combining ratio in the complex. Typical mole-ratio plots are shown in Figure 26-15. Notice that the ligand of the 1:2 complex absorbs at the wavelength selected so that the slope beyond the equivalence point is greater than zero. We deduce that the uncomplexed

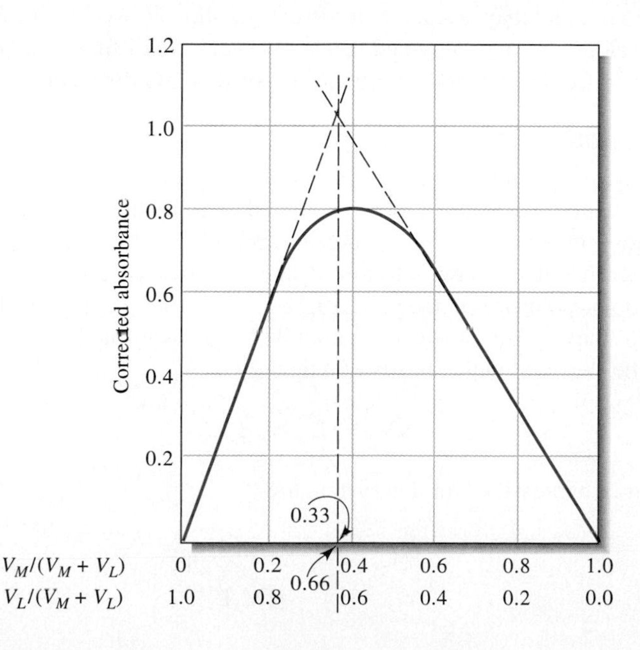

Figure 26-14 Continuous-variation plot for the 1:2 complex ML_2.

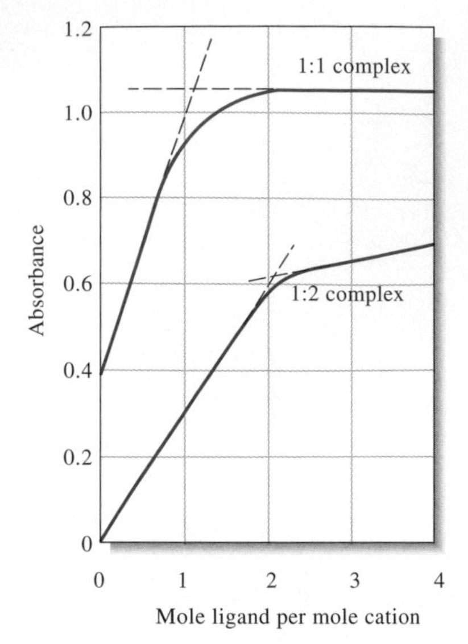

Figure 26-15 Mole-ratio plots for a 1:1 and a 1:2 complex. The 1:2 complex is the more stable of the two complexes as indicated by closeness of the experimental curve to the extrapolated lines. The closer the curve is to the extrapolated lines, the larger the formation constant of the complex; the larger the deviation from the straight lines, the smaller the formation constant of the complex.

cation involved in the 1:1 complex absorbs, because the initial point has an absorbance greater than zero.

Formation constants can be evaluated from the data in the curved portion of mole-ratio plots where the reaction is least complete.

EXAMPLE 26-5

Derive equations to calculate the equilibrium concentrations of all the species involved in the 1:2 complex-formation reaction illustrated in Figure 26-15.

Two mass-balance expressions based on the preparatory data can be written. Thus, for the reaction

$$M + 2L \rightleftharpoons ML_2$$

we can write

$$c_M = [M] + [ML_2]$$

$$c_L = [L] + 2[ML_2]$$

where c_M and c_L are the molar concentrations of M and L before reaction occurs. For 1-cm cells, the absorbance of the solution is

$$A = \varepsilon_M[M] + \varepsilon_L[L] + \varepsilon_{ML_2}[ML_2]$$

From the mole-ratio plot, we see that $\varepsilon_M = 0$. Values for ε_{ML} and ε_{ML_2} can be obtained from the two straight-line portions of the curve. With one or more measurements of A in the curved region of the plot, sufficient data are available to calculate the three equilibrium concentrations and thus the formation constant.

A mole-ratio plot may reveal the stepwise formation of two or more complexes as successive slope changes, provided the complexes have different molar absorptivities and provided the formation constants are sufficiently different from each other.

The Slope-Ratio Method

This approach is particularly useful for weak complexes but is applicable only to systems in which a single complex is formed. The method assumes (1) that the complex-formation reaction can be forced to completion by a large excess of either reactant, (2) that Beer's law is followed under these circumstances, and (3) that only the complex absorbs at the wavelength chosen for the experiment.

Consider the reaction in which the complex M_xL_y is formed by the reaction of x moles of the cation M with y moles of a ligand L:

$$xM + yL \rightleftharpoons M_xL_y$$

Mass-balance expressions for this system are

$$c_M = [M] + x[M_xL_y]$$

$$c_L = [L] + y[M_xL_y]$$

where c_M and c_L are the molar analytical concentrations of the two reactants. We now assume that at very high analytical concentrations of L, the equilibrium is shifted far to the right and $[M] \ll x[M_xL_y]$. Under this condition, the first mass-balance expression simplifies to

$$c_M = x[M_xL_y]$$

If the system obeys Beer's law,

$$A_1 = \varepsilon b[M_xL_y] = \varepsilon bc_M/x$$

where ε is the molar absorptivity of M_xL_y and b is the path length. A plot of absorbance as a function of c_M is linear when there is sufficient L present to justify the assumption that $[M] \ll x[M_xL_y]$. The slope of this plot is $\varepsilon b/x$.

When c_M is made very large, we assume that $[L] \ll y[M_xL_y]$, and the second mass-balance equation reduces to

$$c_L = y[M_xL_y]$$

and

$$A_2 = \varepsilon b[M_xL_y] = \varepsilon bc_L/y$$

Again, if our assumptions are valid, we find that a plot of A versus c_L is linear at high concentrations of M. The slope of this line is $\varepsilon b/y$.

The ratio of the slopes of the two straight lines gives the combining ratio between M and L:

$$\frac{\varepsilon b/x}{\varepsilon b/y} = \frac{y}{x}$$

Spreadsheet Summary In Chapter 12 of *Applications of Microsoft® Excel in Analytical Chemistry,* we investigate the method of continuous variations using the slope and intercept functions and we learn how to produce inset plots.

26B AUTOMATED PHOTOMETRIC AND SPECTROPHOTOMETRIC METHODS

The first fully automated instrument for chemical analysis (the Technicon Auto Analyzer®) appeared on the market in 1957. This instrument was designed to fulfill the needs of clinical laboratories, where blood and urine samples are routinely analyzed for a dozen or more chemical species. The number of such analyses demanded by modern medicine is enormous, so it is necessary to keep their cost at a reasonable level. These two considerations motivated the development of analytical systems that perform several analyses simultaneously with a minimum input of human labor. The use of automatic instruments has spread from clinical laboratories to laboratories for the control of industrial processes and the routine determination of a wide spectrum of species in air, water, soils, and pharmaceutical and agricultural products. In the majority of these applications, the measurement step in the analyses is accomplished by photometry, spectrophotometry, or fluorometry.

In Section 8B-6 we described various automated sample handling techniques including discrete and continuous flow methods. In this section, we explore the instrumentation and two applications of flow-injection analysis with photometric detection.

26B-1 Instrumentation

Figure 26-16a is a flow diagram of the simplest of all flow-injection systems. Here, a colorimetric reagent for chloride ion is pumped by a peristaltic pump directly into a valve that permits injection of samples into the flowing stream. The sample and reagent then pass through a 50-cm reactor coil where the reagent diffuses into the sample plug and produces a colored product by the sequence of reactions

$$Hg(SCN)_2(aq) + 2Cl^- \rightleftharpoons HgCl_2(aq) + 2SCN^-$$

$$Fe^{3+} + SCN^- \rightleftharpoons \underset{\text{red}}{Fe(SCN)^{2+}}$$

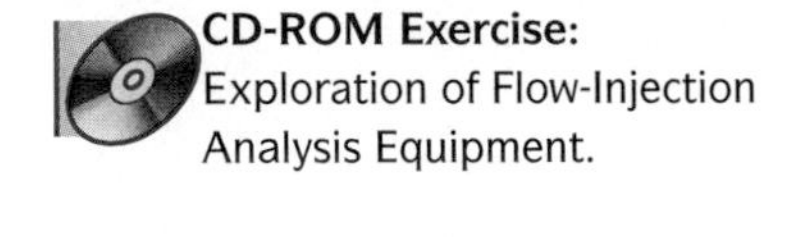

CD-ROM Exercise: Exploration of Flow-Injection Analysis Equipment.

From the reactor coil, the solution passes into a flow-through photometer equipped with a 480-nm interference filter.

The signal output from this system for a series of standards containing from 5 to 75 ppm of chloride is shown on the left of Figure 26-16b. Notice that four injections of each standard were made to demonstrate the reproducibility of the system. The two curves to the right are high-speed recorder scans of one of the samples containing 30 ppm (R_{30}) and another containing 75 ppm (R_{75}) chloride. These curves demonstrate that cross-contamination is minimal in an unsegmented stream. Thus, less than 1% of the first analyte is present in the flow cell after 28 s, the time of the next injection (S_2). This system has been successfully used for the routine determination of chloride ion in brackish and waste waters as well as in serum samples.

Sample and Reagent Transport System

Ordinarily, the solution in a flow-injection analysis is pumped through flexible tubing in the system by a peristaltic pump, a device in which a fluid (liquid or gas) is squeezed through plastic tubing by rollers. Figure 26-17 illustrates the operating principle of the peristaltic pump. The spring-loaded cam, or band, pinches the tubing against two or more of the rollers at all times, thus forcing a continuous flow of

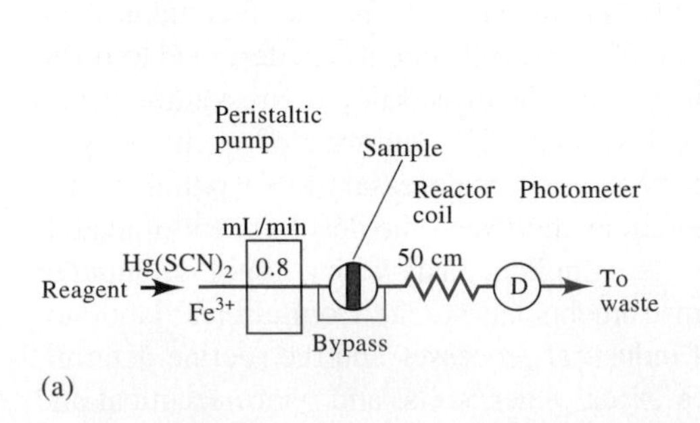

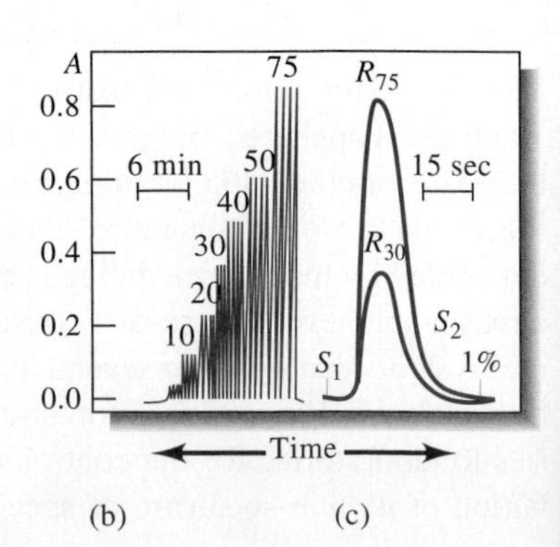

Figure 26-16 Flow-injection determination of chloride: (a) flow diagram; (b) recorder readout for quadruplicate runs on standards containing 5 to 75 ppm of chloride ion; (c) fast scan of two of the standards to demonstrate the low analyte carryover (less than 1%) from run to run. Notice that the point marked 1% corresponds to where the response would just begin for a sample injected at time S_2. (From J. Ruzicka and E. H. Hansen, *Flow Injection Methods,* 2nd ed., p. 16. New York: Wiley, 1988. This material is used by permission of John Wiley & Sons, Inc.)

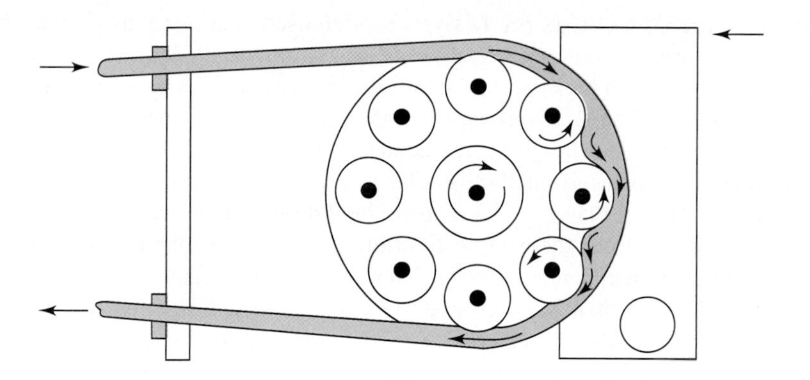

Figure 26-17 Diagram showing one channel of a peristaltic pump. Several additional tubes may be located under the one shown (below the plane of the diagram) to carry multiple channels of reagent or sample. (From B. Karlberg and G. E. Pacey, *Flow Injection Analysis. A Practical Guide,* p. 34. New York: Elsevier, 1989. With permission of Elsevier Science Publishers.)

fluid through the tubing. Modern pumps generally have 8 to 10 rollers, arranged in a circular configuration so that half are squeezing the tube at any instant. This design leads to a flow that is relatively pulse free. The flow rate is controlled by the speed of the motor, which should be greater than 30 rpm, and the inside diameter of the tube. A wide variety of tube sizes (i.d. = 0.25 to 4 mm) are available commercially that permit flow rates as small as 0.0005 mL/min and as great as 40 mL/min. Flow injection has been miniaturized through the use of fused silica capillaries (i.d. = 25–100 μm) or through **lab-on-a-chip** technology (See Feature 8-1). The rollers of typical commercial peristaltic pumps are long enough so that several reagent and sample streams can be pumped simultaneously. Syringe pumps and electroosmosis are also used to induce flow in flow injection systems.

As shown in Figure 26-16a, flow-injection systems often contain a coiled section of tubing (typical coil diameters are about 1 cm or less) whose purpose it is to enhance axial dispersion and to increase radial mixing of the sample and reagent, both of which lead to more symmetric peaks.

Sample Injectors and Detectors

Sample sizes for flow-injection analysis range from less than 1 μL to 200 μL, with 10 to 30 μL being typical for many applications. For a successful analysis, it is important to inject the sample solution rapidly as a plug, or pulse, of liquid; in addition, the injections must not disturb the flow of the carrier stream. The most useful and convenient injector systems are based on sampling loops similar to those used in chromatography (see, for example, Figure 30-4). The method of operation of a sampling loop is illustrated in Figure 26-16a. With the valve of the loop in the position shown, reagents flow through the bypass. When a sample has been injected into the loop and the valve turned 90 deg, the sample enters the flow as a single, well-defined zone. For all practical purposes, flow through the bypass ceases with the valve in this position because the diameter of the sample loop is significantly greater than that of the bypass tubing.

The most common detectors in flow-injection analysis are spectrophotometers, photometers, and fluorometers. Electrochemical systems, refractometers, atomic emission and atomic absorption spectrometers have also been used.

◀ Flow injection analyzers can be fairly simple, consisting of a pump, an injection valve, plastic tubing, and a detector. Filter photometers and spectrophotometers are the most common detectors.

Separations in Flow-Injection Analysis

Separations by dialysis, by liquid/liquid extraction, and by gaseous diffusion are readily carried out automatically with flow-injection systems.

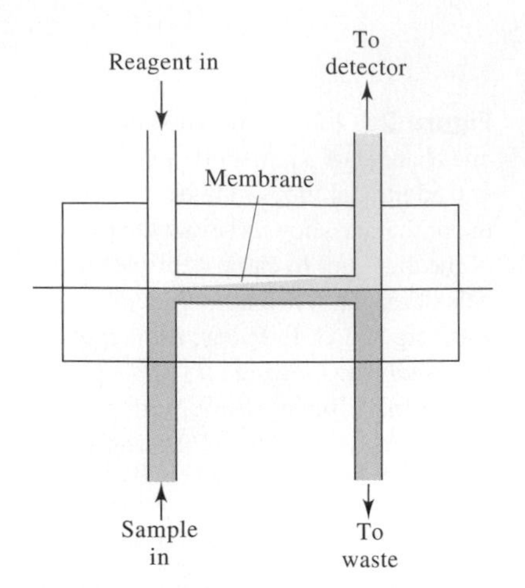

Figure 26-18 A dialysis flow module. The membrane is supported between two grooved Teflon blocks.

Dialysis and Gas Diffusion Dialysis is often used in continuous-flow methods to separate inorganic ions, such as chloride or sodium, or small organic molecules, such as glucose, from high-molecular-weight species, such as proteins. Small ions and molecules diffuse relatively rapidly through hydrophilic membranes of cellulose acetate or nitrate while large molecules do not. Dialysis usually precedes the determination of ions and small molecules in whole blood or serum.

Figure 26-18 is a diagram of a dialysis module in which analyte ions or small molecules diffuse from the sample solution through a membrane into a reagent stream, which often contains a species that reacts with the analyte to form a colored product, which can then be determined photometrically. Large molecules, which interfere in the determination, remain in the original stream and are carried to waste. The membrane is supported between two Teflon plates in which complementary channels have been cut to accommodate the two stream flows on opposite sides of the membrane. The transfer of smaller species through the membrane is usually incomplete (often less than 50%). Thus, successful quantitative analysis requires close control of temperature and flow rates for both samples and standards. Such control is easily accomplished in automated flow-injection systems.

Gas diffusion from a donor stream containing a gaseous analyte to an acceptor stream containing a reagent that permits its determination is a highly selective technique that has found considerable use in flow-injection analysis. The separations are carried out in a module similar to that shown in Figure 26-18. In this application, however, the membrane is usually a hydrophobic microporous material, such as Teflon or isotactic polypropylene. An example of this type of separation technique is found in a method for determining total carbonate in an aqueous solution. Here the sample is injected into a carrier stream of dilute sulfuric acid, which is then directed into a gas-diffusion module, where the liberated carbon dioxide diffuses into an acceptor stream containing an acid/base indicator. This stream then passes through a photometric detector, which yields a signal that is proportional to the carbonate content of the sample.

Solvent Extraction Solvent extraction (see Chapter 30) is another separation method that is readily performed in a flow-injection apparatus. Most commonly an aqueous solution of the analyte is mixed with an immiscible organic solvent, such as hexane or chloroform, which results in transfer of the analyte (or the interferents) into the organic layer. After passing the mixture through a coil of tubing in which the extraction is given time to occur, the more dense liquid is separated from the less dense and one or the other of the phases is passed into a detector for completion of the analysis. Figure 26-19 is a diagram of an apparatus in which the analyte is separated by extraction with chloroform.

26B-2 A Typical Application of Flow-Injection Analysis

Figure 26-19 illustrates a flow-injection system designed for the automatic spectrophotometric determination of caffeine in acetylsalicylic acid drug preparations after extraction of the caffeine into chloroform. The chloroform solvent, after cooling in an ice bath to minimize evaporation, is mixed with the alkaline sample stream in a T-tube (see lower insert). After passing through the 2-m extraction coil, the mixture enters a T-tube separator, which is differentially pumped so that about 35% of the organic phase containing the caffeine passes into the flow cell; the other 65% accompanies the aqueous solution containing the rest of the sample to waste. To avoid contaminating the flow cell with water, Teflon fibers, which are not wet-

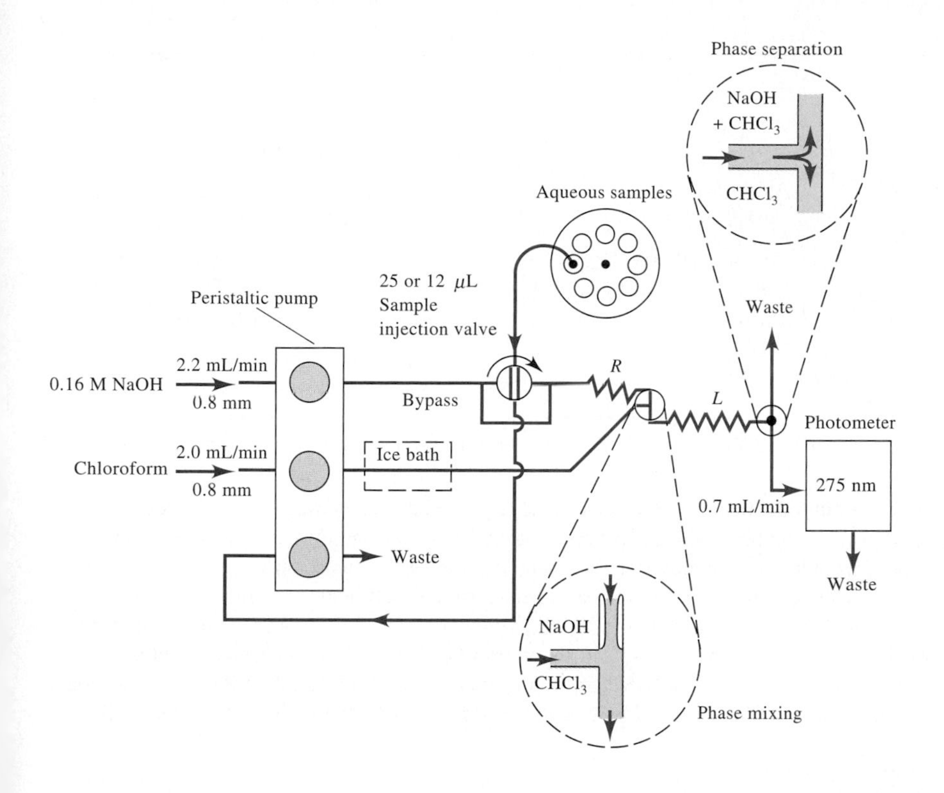

Figure 26-19 Flow-injection apparatus for the determination of caffeine in acetylsalicylic acid preparations. With the valve rotated at 90 deg, the flow in the bypass is essentially zero because of its small diameter. *R* and *L* are Teflon coils with 0.8-mm inside diameters; *L* has a length of 2 m, and the distance from the injection point through *P* to the mixing point is 0.15 m. (Adapted from B. Karlberg and S. Thelander, *Anal. Chim. Acta,* **1978,** *98,* 2. Reprinted with permission from Elsevier.)

ted by water, were twisted into a thread and inserted in the inlet to the T-tube in such a way as to form a smooth downward bend. The chloroform flow then follows this bend to the photometer cell where the caffeine concentration is determined based on its absorption peak at 275 nm. The output of the photometer is similar in appearance to that shown in Figure 26-16b.

26C INFRARED ABSORPTION SPECTROSCOPY

Infrared spectrophotometry is a powerful tool for identifying pure organic and inorganic compounds because, with the exception of a few homonuclear molecules such as O_2, N_2, and Cl_2, all molecular species absorb infrared radiation. In addition, with the exception of chiral molecules in the crystalline state, each molecular species has a unique infrared absorption spectrum. Thus, an exact match between the spectrum of a compound of known structure and the spectrum of an analyte unambiguously identifies the analyte.

Infrared spectroscopy is a less satisfactory tool for quantitative analyses than its ultraviolet and visible counterparts because of lower sensitivity and frequent deviations from Beer's law. Additionally, infrared absorbance measurements are considerably less precise. Nevertheless, in instances where modest precision is adequate, the unique nature of infrared spectra provides a degree of selectivity in a quantitative measurement that may offset these undesirable characteristics.[11]

[11]For a detailed discussion of infrared spectroscopy, see N. B. Colthup, L. H. Daly, and S. E. Wiberley, *Introduction to Infrared and Raman Spectroscopy,* 3rd ed. New York: Academic Press, 1990.

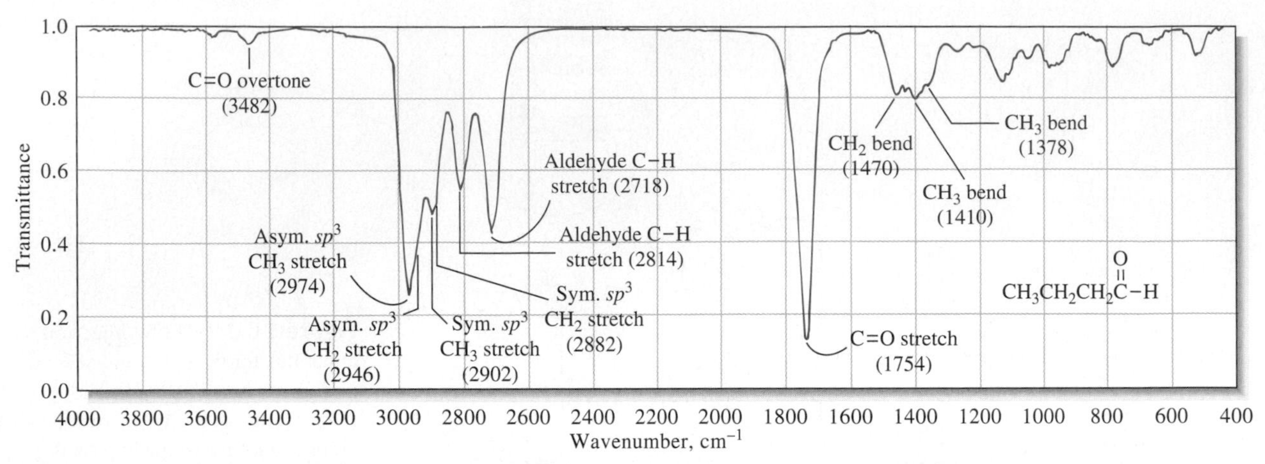

Figure 26-20 Infrared spectrum for *n*-butanal (*n*-butyraldehyde). The vertical scale is plotted as transmittance, as has been common practice in the past. The horizontal scale is linear in wavenumbers, which is proportional to frequency and thus energy. Most modern IR spectrometers are capable of providing data plotted as either transmittance or absorbance on the vertical axis and wavenumber or wavelength on the horizontal axis. IR spectra are usually plotted with frequency increasing from right to left, which is a historical artifact. Early IR spectrometers produced spectra with wavelength increasing from left to right, which led to an auxiliary frequency scale from right to left. Note that several of the bands have been labeled with assignments of the vibrations that produce the bands. (Data from NIST Mass Spec Data Center, S.E. Stein, director, "Infrared Spectra" in NIST Chemistry WebBook, NIST Standard Reference Database Number 69, P .J. Linstrom and W. G. Mallard, Eds. March 2003, National Institute of Standards and Technology, Gaithersburg, MD 20899 [http://webbook.nist.gov].)

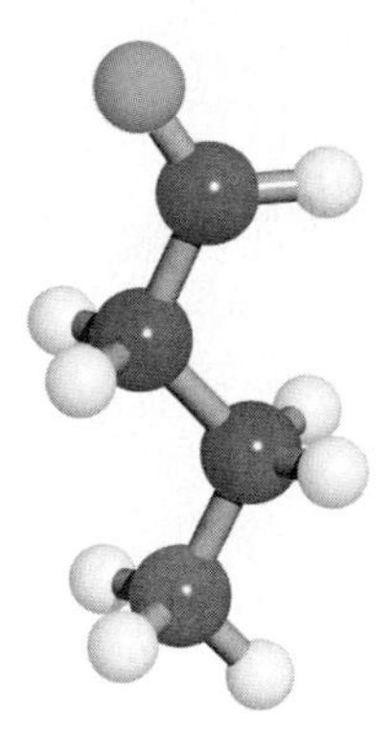

Molecular model of *n*-butanal.

26C-1 Infrared Absorption Spectra

The energy of infrared radiation can excite vibrational and rotational transitions, but it is insufficient to excite electronic transitions. As shown in Figure 26-20, infrared spectra exhibit narrow, closely spaced absorption bands resulting from transitions among the various vibrational quantum levels. Variations in rotational levels may also give rise to a series of peaks for each vibrational state; with liquid or solid samples, however, rotation is often hindered or prevented, and the effects of these small energy differences are not detected. Thus, a typical infrared spectrum for a liquid, such as that in Figure 26-20, consists of a series of vibrational bands.

The number of ways a molecule can vibrate is related to the number of atoms, and thus the number of bonds, it contains. For even a simple molecule, the number of possible vibrations is large. For example, *n*-butanal ($CH_3CH_2CH_2CHO$) has 33 vibrational modes, most differing from each other in energy. Not all of these vibrations produce infrared bands, but, as shown in Figure 26-20, the spectrum for *n*-butanal is relatively complex.

Infrared absorption occurs not only with organic molecules but also with covalently bonded metal complexes, which are generally active in the longer-wavelength infrared region. Infrared spectrophotometric studies have thus provided much useful information about complex metal ions.

26C-2 Instruments for Infrared Spectroscopy

Three types of infrared instruments are found in modern laboratories: dispersive spectrometers (or spectrophotometers), Fourier-transform (FTIR) spectrometers, and filter photometers. The first two are used for obtaining complete spectra for qualitative identification, while filter photometers are designed for quantitative work. Fourier-transform and filter instruments are nondispersive in the sense that neither uses a grating or prism to disperse radiation into its component wavelengths.[12]

Dispersive Instruments

With one difference, dispersive infrared instruments are similar in general design to the double-beam (in time) spectrophotometers shown in Figure 25-20c. The difference lies in the location of the cell compartment with respect to the monochromator. In ultraviolet/visible instruments, cells are always located between the monochromator and the detector in order to avoid photochemical decomposition, which may occur if samples are exposed to the full power of an ultraviolet or visible source. Infrared radiation, in contrast, is not sufficiently energetic to bring about photodecomposition; thus the cell compartment can be located between the source and the monochromator. This arrangement is advantageous because any scattered or emitted radiation generated in the cell compartment is largely removed by the monochromator.

As shown in Section 25A, the components of infrared instruments differ considerably in detail from those in ultraviolet and visible instruments. Thus, infrared sources are heated solids rather than deuterium or tungsten lamps, infrared gratings are much coarser than those required for ultraviolet/visible radiation, and infrared detectors respond to heat rather than photons. In addition, the optical components of infrared instruments are constructed from polished solids, such as sodium chloride or potassium bromide.

Fourier-Transform Spectrometers

Fourier-transform infrared (FTIR) spectrometers offer the advantages of unusually high sensitivity, resolution, and speed of data acquisition (data for an entire spectrum can be obtained in 1 s or less). In the early days of FTIR, instruments were large, intricate, expensive devices controlled by expensive laboratory computers. As the instrumentation evolved and the price of computers dropped dramatically while their power, speed, and ease of use improved by orders of magnitude, FTIR spectrometers have become commonplace in the laboratory.

◀ The FTIR spectrometer is now the most common infrared spectrometer. The great majority of commercial infrared instruments are FTIR systems.

Fourier-transform instruments contain no dispersing element, and all wavelengths are detected and measured simultaneously using a Michelson interferometer as we described in Feature 25-7. To separate wavelengths, it is necessary to modulate the source signal and pass it through the sample in such a way that it can be recorded as an **interferogram.** The interferogram is subsequently decoded by a Fourier transformation, a mathematical operation conveniently carried out by the computer, which is now an integral part of nearly all spectrometers. Although the detailed mathematical theory of Fourier-transform measurements is beyond the scope of this book, the qualitative treatment presented in Feature 25-7 and in Feature 26-1 should give you an idea of how the IR signal is collected and how spectra are extracted from the data.

An **interferogram** is a recording of the signal produced by a Michelson interferometer. The signal is processed by a mathematical process known as the Fourier transform to produce an IR spectrum.

Figure 26-21 shows a typical student-grade benchtop FTIR spectrometer that has a built-in computer for data acquisition, analysis, and presentation. The instrument is relatively inexpensive (approximately $11,000), has a resolution of 4 cm^{-1},

[12]For a discussion of the principles of Fourier-transform spectroscopy, see D. A. Skoog, F. J. Holler, and T. A. Nieman, *Principles of Instrumental Analysis,* 5th ed., pp. 392–396. Belmont, CA: Brooks/Cole, 1998.

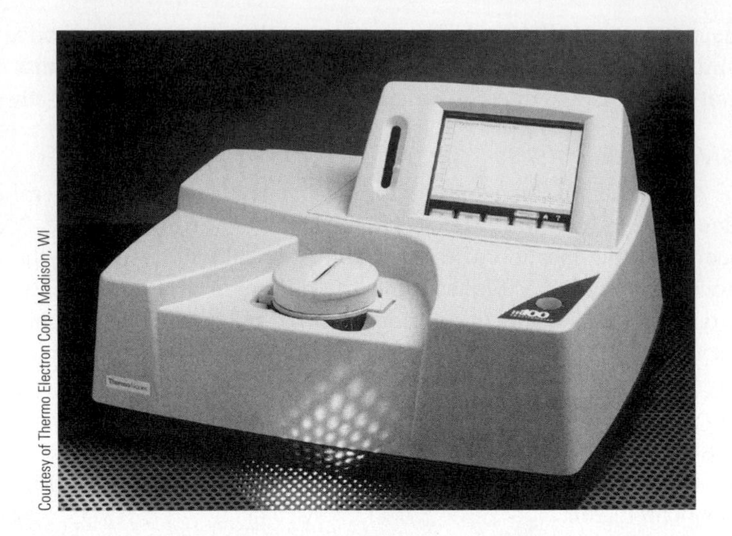

Figure 26-21 Photo of a basic student-grade benchtop FTIR spectrometer. Spectra are recorded in a few seconds and displayed on the LCD panel for viewing and interpretation. The spectra may be stored in a memory card for later retrieval and analysis, or optionally they may be printed.

and achieves a signal-to-noise ratio of 5000 for a 1-min measurement. The measured spectrum appears on an LCD panel where it can be read and interpreted, or it may be printed or stored in a flash memory card for record keeping.

A research-quality instrument may cost \$60,000 or more, will have a resolution of 0.125 cm^{-1} or better, and will exhibit a signal-to-noise ratio of 33,000 or greater for a 1-min measurement period. Typically, a research-grade instrument will be interfaced to a separate computer, which has both advantages and disadvantages. With a separate computer, when the processor becomes obsolete, which would be expected to occur every 2 or 3 years, a newer, more powerful computer can be purchased and attached to the spectrometer with little difficulty. In addition, a separate computer typically has considerably more power for data processing and storage than does a computer integrated within the spectrometer. An additional advantage of a separate generic computer is that software and databases of spectra for performing can be easily installed and used to process spectral data and to match measured spectra with known spectra in the database. In addition, a separate computer provides considerable flexibility for archiving data on CDs or DVDs, and if the computer is connected to a local area network, spectra may be transmitted to colleagues or coworkers, and software or firmware upgrades may be easily downloaded and installed on the computer or in the spectrometer.

Filter Photometers

Infrared photometers designed to monitor the concentration of air pollutants, such as carbon monoxide, nitrobenzene, vinyl chloride, hydrogen cyanide, and pyridine, are often used to ensure compliance with regulations established by the Occupational Safety and Health Administration (OSHA). Interference filters, each designed for the determination of a specific pollutant, are available. These transmit narrow bands of radiation in the range of 3 to 14 μm.

26C-3 Qualitative Applications of Infrared Spectrophotometry

An infrared absorption spectrum, even one for a relatively simple compound, often contains a bewildering array of sharp bands and minima. Absorption bands useful

FEATURE 26-1

Producing Spectra with an FTIR Spectrometer

In Feature 25-7 we described the basic operating principles of the Michelson interferometer and the function of the Fourier transform to produce a frequency spectrum from a measured interferogram. Figure 26F-1 shows an optical diagram for a Michelson interferometer similar to the one in the spectrometer depicted in Figure 26-21.

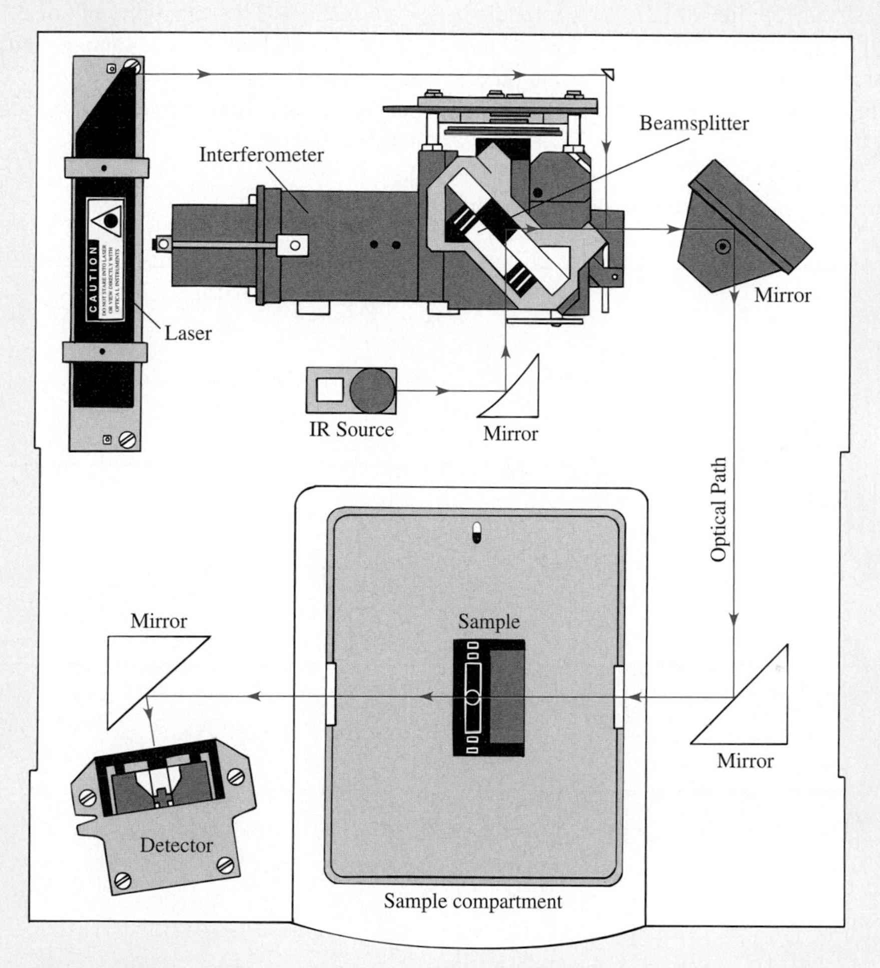

Figure 26-F1 Instrument diagram for a basic FTIR spectrometer. Radiation of all frequencies from the IR source is reflected into the interferometer where it is modulated by the moving mirror on the left. The modulated radiation is then reflected from the two mirrors on the right through the sample in the compartment at the bottom. After passing through the sample, the radiation falls on the detector. A data acquisition system attached to the detector records the signal and stores it in the memory of a computer as an interferogram. (Courtesy of Thermo Electron Corp., Madison, WI. With permission).

The interferometer is actually two parallel interferometers, one to modulate the IR radiation from the source before it passes through the sample and a second to modulate the red light from the He-Ne laser to provide a reference signal for acquiring data from the IR detector. The output of the detector is digitized and stored in the memory of the instrument computer.

(continued)

The first step in producing an IR spectrum is to collect and store a reference interferogram with no sample in the sample cell. Then the sample is placed in the cell and a second interferogram is collected. Figure 26F-2a shows an interferogram collected using an FTIR spectrometer with methylene chloride, CH_2Cl_2, in the sample cell. The Fourier transform is then applied to the two interferograms to compute the IR spectra of the reference and the sample. The ratio of the two spectra can then be computed to produce an IR spectrum of the analyte such as the one illustrated in Figure 26F-2b.

Notice that the methylene chloride IR spectrum exhibits little noise. Since a single interferogram can be scanned in only a second or two, many interferograms can be scanned in a relatively short time and summed in the memory of the computer. This process, which is often called **signal averaging,** reduces the noise on the resulting signal and improves the signal-to-noise ratio of the spectrum as described in Feature 25-5 and illustrated in Figure 25F-4. This capability of noise reduction and speed coupled with Fellgett's advantage and Jacquinot's advantage (see Feature 25-7) make the FTIR spectrometer a marvelous tool for a broad range of qualitative and quantitative analyses.

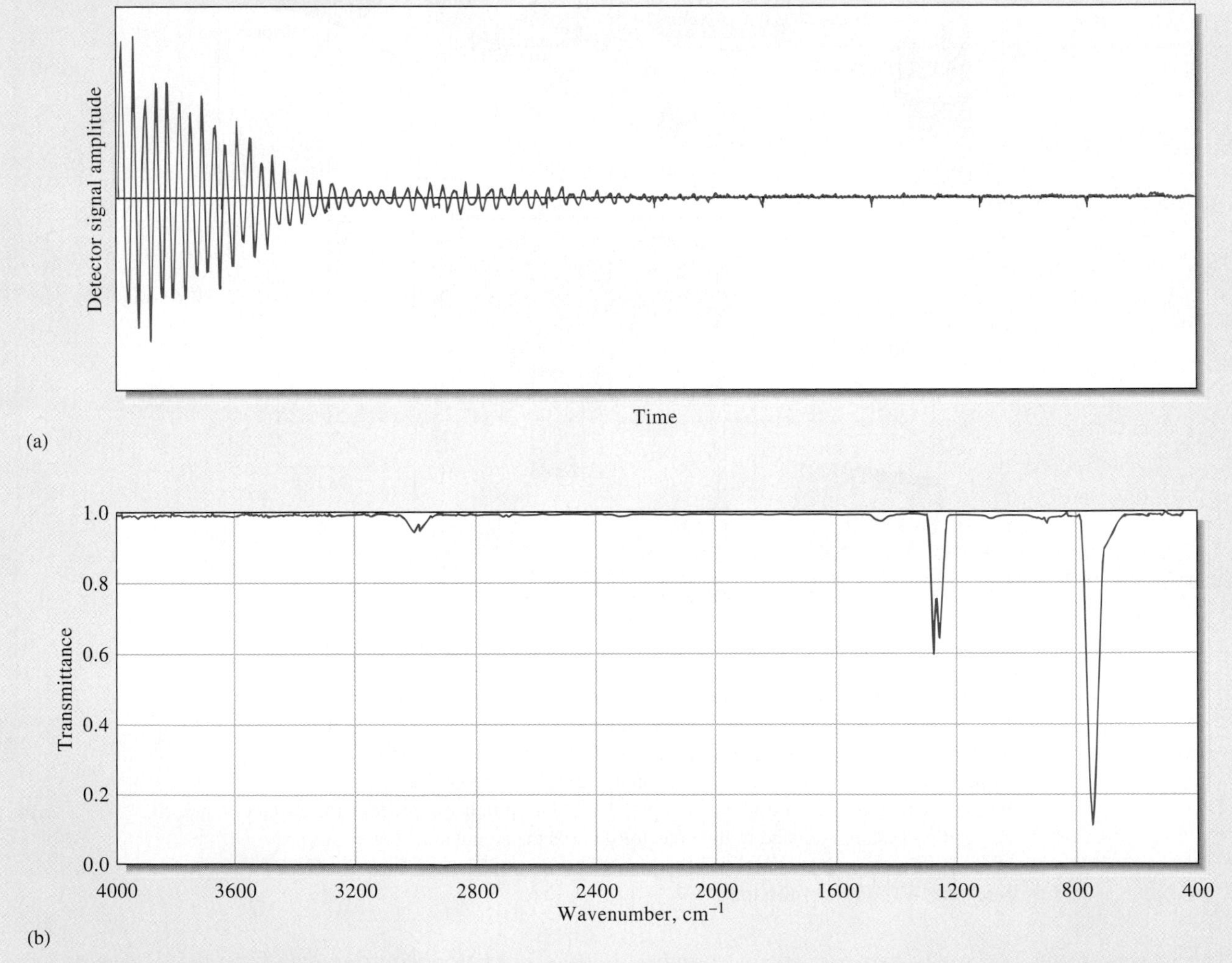

Figure 26-F2 (a) Interferogram obtained from a typical FTIR spectrometer for methylene chloride. The plot shows detector signal output as a function of time, or displacement of the moving mirror of the interferometer. (b) IR spectrum of methylene chloride produced by the Fourier transformation of the data in (a). Note that the Fourier transform takes signal intensity collected as a function of time and produces transmittance as a function of frequency after subtraction of a background interferogram and proper scaling.

TABLE 26-5

Some Characteristic Infrared Absorption Peaks

		Absorption Peaks	
	Functional Group	Wavenumber, cm^{-1}	Wavelength, μm
O—H	Aliphatic and aromatic	3600–3000	2.8–3.3
NH_2	Also secondary and tertiary	3600–3100	2.8–3.2
C—H	Aromatic	3150–3000	3.2–3.3
C—H	Aliphatic	3000–2850	3.3–3.5
C≡N	Nitrile	2400–2200	4.2–4.6
C≡C—	Alkyne	2260–2100	4.4–4.8
COOR	Ester	1750–1700	5.7–5.9
COOH	Carboxylic acid	1740–1670	5.7–6.0
C=O	Aldehydes and ketones	1740–1660	5.7–6.0
$CONH_2$	Amides	1720–1640	5.8–6.1
C=C—	Alkene	1670–1610	6.0–6.2
ϕ—O—R	Aromatic	1300–1180	7.7–8.5
R—O—R	Aliphatic	1160–1060	8.6–9.4

for the identification of functional groups are located in the shorter-wavelength region of the infrared (from about 2.5 to 8.5 μm), where the positions of the absorption maxima are only slightly affected by the carbon skeleton to which the groups are attached. Investigation of this region of the spectrum thus provides considerable information regarding the overall constitution of the molecule under investigation. Table 26-5 gives the positions of characteristic absorption maxima for some common functional groups.[13]

Identifying functional groups in a molecule is seldom sufficient to positively identify the compound, and the entire spectrum from 2.5 to 15 μm must be compared with that of known compounds. Collections of spectra are available for this purpose.[14]

26C-4 Quantitative Infrared Photometry and Spectrophotometry

Quantitative infrared absorption methods differ somewhat from their ultraviolet and visible counterparts because of the greater complexity of the spectra, the narrowness of the absorption bands, and the capabilities of the instruments available for measurements in this spectral region.[15]

[13]For more detailed information, see R. M. Silverstein and F. X. Webster, *Spectrometric Identification of Organic Compounds,* 6th ed., Ch. 3. New York: Wiley, 1997.

[14]Sadtler Standard Spectra. Philadelphia: Sadtler Research Laboratories; C. J. Pouchert, *The Aldrich Library of Infrared Spectra,* 3rd ed. Aldrich Chemical Co., Milwaukee, WI, 1981.; P. J. Linstrom and W. G. Mallard, Eds., NIST Chemistry WebBook, NIST Standard Reference Database Number 69, March 2003, National Institute of Standards and Technology, Gaithersburg, MD 20899 (http://webbook.nist.gov); Thermo Galactic, *Spectra Online,* (http://spectra.galactic.com/).

[15]For an extensive discussion of quantitative infrared analysis, see A. L. Smith, in *Treatise on Analytical Chemistry,* 2nd ed., Part I, Vol. 7, pp. 415–456, P. J. Elving, E. J. Meehan, and I. M. Kolthoff, Eds. New York: Wiley, 1981.

Absorbance Measurements

Using matched cuvettes for solvent and analyte is seldom practical for infrared measurements because it is difficult to obtain cells with identical transmission characteristics. Part of this difficulty results from degradation of the transparency of infrared cell windows (typically polished sodium chloride) with use due to attack by traces of moisture in the atmosphere and in samples. In addition, pathlengths are hard to reproduce because infrared cells are often less than 1 mm thick. Such narrow cells are required to permit the transmission of measurable intensities of infrared radiation through pure samples or through very concentrated solutions of the analyte. Measurements on dilute analyte solutions, as is done in ultraviolet or visible spectroscopy, are usually difficult because there are few good solvents that transmit over appreciable regions of the infrared spectrum.

For these reasons, a reference absorber is often dispensed with entirely in qualitative infrared work, and the intensity of the radiation passing through the sample is simply compared with that of the unobstructed beam; alternatively, a salt plate may be used as a reference. Either way, the resulting transmittance is often less than 1, even in regions of the spectrum where the sample is totally transparent.

Applications of Quantitative Infrared Spectroscopy

Infrared spectrophotometry offers the potential for determining an unusually large number of substances because nearly all molecular species absorb in the infrared region. Moreover, the uniqueness of an infrared spectrum provides a degree of specificity that is matched or exceeded by relatively few other analytical methods. This specificity has particular application to the analysis of mixtures of closely related organic compounds.

The recent proliferation of government regulations on atmospheric contaminants has demanded the development of sensitive, rapid, and highly specific methods for a variety of chemical compounds. Infrared absorption procedures appear to meet this need better than any other single analytical tool.

Table 26-6 illustrates the variety of atmospheric pollutants that can be determined with a simple, portable filter photometer equipped with a separate inter-

TABLE 26-6

Examples of Infrared Vapor Analysis for OSHA Compliance*

Compound	Allowable Exposure, ppm†	Wavelength, μm	Minimum Detectable Concentration, ppm‡
Carbon disulfide	4	4.54	0.5
Chloroprene	10	11.4	4
Diborane	0.1	3.9	0.05
Ethylenediamine	10	13.0	0.4
Hydrogen cyanide	4.7§	3.04	0.4
Methyl mercaptan	0.5	3.38	0.4
Nitrobenzene	1	11.8	0.2
Pyridine	5	14.2	0.2
Sulfur dioxide	2	8.6	0.5
Vinyl chloride	1	10.9	0.3

*Courtesy of The Foxboro Company, Foxboro, MA 02035.

†1992–1993 OSHA exposure limits for 8-hr weighted average.

‡For 20.25-m cell.

§Short-term exposure limit: 15-min time-weighted average that shall not be exceeded at any time during the work day.

ference filter for each analyte species. Of the more than 400 chemicals for which maximum tolerable limits have been set by OSHA, half or more have absorption characteristics that make them amenable to determination by infrared photometry or spectrophotometry. With so many compounds absorbing, overlapping peaks are quite common. In spite of this potential disadvantage, the method provides a moderately high degree of selectivity.

WEB WORKS

Use Google to locate the Thermo Galactic Web site. Navigate to the Spectra Online section of the Galactic site, register as a new user, and search the spectral database to find the IR spectrum of benzonitrile. With the spectrum on the screen, identify the peak that corresponds to the stretching frequency of the cyano group $-C{\equiv}N$. Now identify five other prominent peaks in the spectrum and assign the peaks to vibrational modes in benzonitrile. What other types of spectra can be found in the Galactic database? Find the Spectra Online Collections section of the Galactic site, and read about the NIST Chemistry WebBook and the EPA Vapor Phase FTIR Library. Locate the NIST Chemistry WebBook on the Web, and perform a search for benzonitrile. What data are available for this compound on the NIST site? Click on the link to the IR spectrum, and notice that there are several versions of the spectrum. How are they alike, and how do they differ? Where did the spectra originate? How could you use this database?

QUESTIONS AND PROBLEMS

26-1. Describe the differences between the following and list any particular advantages possessed by one over the other:
- *(a) spectrophotometers and photometers.
- (b) single-beam and double-beam instruments for absorbance measurements.
- *(c) conventional and diode-array spectrophotometers.

26-2. What minimum requirement is needed to obtain reproducible results with a single-beam spectrophotometer?

***26-3.** What experimental variables must be controlled to ensure reproducible absorbance data?

26-4. What is(are) advantage(s) of the standard addition method? What minimum condition is needed for the successful application of this method?

***26-5.** The molar absorptivity for the complex formed between bismuth(III) and thiourea is 9.32×10^3 L cm^{-1} mol^{-1} at 470 nm. Calculate the range of permissible concentrations for the complex if the absorbance is to be no less than 0.10 nor greater than 0.90 when the measurements are made in 1.00-cm cells.

26-6. The molar absorptivity for aqueous solutions of phenol at 211 nm is 6.17×10^3 L cm^{-1} mol^{-1}. Calculate the permissible range of phenol concentrations that can be used if the transmittance is to be less than 80% and greater than 5% when the measurements are made in 1.00-cm cells.

***26-7.** The logarithm of the molar absorptivity for acetone in ethanol is 2.75 at 366 nm. Calculate the range of acetone concentrations that can be used if the absorbance is to be greater than 0.100 and less than 2.000 with a 1.25-cm cell.

26-8. The logarithm of the molar absorptivity of phenol in aqueous solution is 3.812 at 211 nm. Calculate the range of phenol concentrations that can be used if the absorbance is to be greater than 0.100 and less than 2.000 with a 1.25-cm cell.

26-9. A photometer with a linear response to radiation gave a reading of 837 mV with a blank in the light path and 333 mV when the blank was replaced by an absorbing solution. Calculate
- *(a) the transmittance and absorbance of the absorbing solution.
- (b) the expected transmittance if the concentration of absorber is one-half that of the original solution.
- *(c) the transmittance to be expected if the light path through the original solution is doubled.

26-10. A portable photometer with a linear response to radiation registered 73.6 μA with a blank solution in the light path. Replacement of the blank with an absorbing solution yielded a response of 24.9 μA. Calculate
(a) the percent transmittance of the sample solution.
*(b) the absorbance of the sample solution.
(c) the transmittance to be expected for a solution in which the concentration of the absorber is one-third that of the original sample solution.
*(d) the transmittance to be expected for a solution that has twice the concentration of the sample solution.

26-11. Sketch a photometric titration curve for the titration of Sn^{2+} with MnO_4^-. What color radiation should be used for this titration? Explain.

26-12. Iron(III) reacts with thiocyanate ion to form the red complex, $Fe(SCN)^{2+}$. Sketch a photometric titration curve for Fe(III) with thiocyanate ion when a photometer with a green filter is used to collect data. Why is a green filter used?

***26-13.** Ethylenediaminetetraacetic acid abstracts bismuth(III) from its thiourea complex:

$$Bi(tu)_6^{3+} + H_2Y^{2-} \rightarrow BiY^- + 6tu + 2H^+$$

where tu is the thiourea molecule, $(NH_2)_2CS$. Predict the shape of a photometric titration curve based on this process, given that the Bi(III)/thiourea complex is the only species in the system that absorbs at 465 nm, the wavelength selected for the analysis.

26-14. The accompanying data (1.00-cm cells) were obtained for the spectrophotometric titration of 10.00 mL of Pd(II) with 2.44×10^{-4} M Nitroso R (O. W. Rollins and M. M. Oldham, *Anal. Chem.*, **1971,** *43,* 262).

Table for Problem 26–14

Volume of Nitroso R, mL	A_{500}
0	0
1.00	0.147
2.00	0.271
3.00	0.375
4.00	0.371
5.00	0.347
6.00	0.325
7.00	0.306
8.00	0.289

Calculate the concentration of the Pd(II) solution, given that the ligand-to-cation ratio in the colored product is 2:1

26-15. A 4.97-g petroleum specimen was decomposed by wet-ashing and subsequently diluted to 500 mL in a volumetric flask. Cobalt was determined by treating 25.00-mL aliquots of this diluted solution as follows:

Reagent Volume			
Co(II), 3.00 ppm	Ligand	H_2O	**Absorbance**
0.00	20.00	5.00	0.398
5.00	20.00	0.00	0.510

Assume that the Co(II)/ligand chelate obeys Beer's law, and calculate the percentage of cobalt in the original sample.

***26-16.** Iron(III) forms a complex with thiocyanate ion that has the formula $Fe(SCN)^{2+}$. The complex has an absorption maximum at 580 nm. A sample of well water was assayed according to the scheme shown in the table below. Calculate the concentration of iron in parts per million in the well water.

Table for Problem 26–16

	Volumes, mL					
Sample	**Sample Volume**	**Oxidizing Reagent**	**Fe(II) 2.75 ppm**	**KSCN 0.050 M**	**H_2O**	**Absorbance, 580 nm (1.00-cm cells)**
1	50.00	5.00	5.00	20.00	20.00	0.549
2	50.00	5.00	0.00	20.00	25.00	0.231

26-17. A. J. Mukhedkar and N. V. Deshpande (*Anal. Chem.*, **1963,** *35,* 47) report on a simultaneous determination for cobalt and nickel based on absorption by their 8-quinolinol complexes. Molar absorptivities are $\varepsilon_{Co} = 3529$ and $\varepsilon_{Ni} = 3228$ at 365 nm and $\varepsilon_{Co} = 428.9$ and $\varepsilon_{Ni} = 0$ at 700 nm. Calculate the concentration of nickel and cobalt in each of the following solutions (1.00-cm cells):

Solution	A_{365}	A_{700}
1	0.0235	0.617
2	0.0714	0.755
3	0.0945	0.920
4	0.0147	0.592
5	0.0540	0.685

***26-18.** Molar absorptivity data for the cobalt and nickel complexes with 2,3-quinoxalinedithiol are $\varepsilon_{Co} =$ 36,400 and $\varepsilon_{Ni} = 5520$ at 510 nm and $\varepsilon_{Co} = 1240$ and $\varepsilon_{Ni} = 17{,}500$ at 656 nm. A 0.519-g sample was dissolved and diluted to 50.0 mL. A 25.0-mL aliquot was treated to eliminate interferences; after addition of 2,3-quinoxalinedithiol, the volume was adjusted to 50.0 mL. This solution had an absorbance of 0.477 at 510 nm and 0.219 at 656 nm in a 1.00-cm cell. Calculate the concentration in parts per million of cobalt and nickel in the sample.

26-19. The indicator HIn has an acid dissociation constant of 4.80×10^{-6} at ordinary temperatures. The accompanying absorbance data are for 8.00×10^{-5} M solutions of the indicator measured in 1.00-cm cells in strongly acidic and strongly alkaline media.

λ, nm	Absorbance pH 1.00	Absorbance pH 13.00
420	0.535	0.050
445	0.657	0.068
450	0.658	0.076
455	0.656	0.085
470	0.614	0.116
510	0.353	0.223
550	0.119	0.324
570	0.068	0.352
585	0.044	0.360
595	0.032	0.361
610	0.019	0.355
650	0.014	0.284

Estimate the wavelength at which absorption by the indicator becomes independent of pH (the so-called isosbestic point).

26-20. Calculate the absorbance (1.00-cm cells) at 450 nm of a solution in which the total molar concentration of the indicator described in Problem 26-19 is 8.00×10^{-5} and the pH is *(a) 4.92, (b) 5.46, *(c) 5.93, (d) 6.16.

***26-21.** What is the absorbance at 595 nm (1.00-cm cells) of a solution that is 1.25×10^{-4} M in the indicator of Problem 26-19 and has a pH of (a) 5.30, (b) 5.70, (c) 6.10?

26-22. Several buffer solutions were made 1.00×10^{-4} M in the indicator of Problem 26-19. Absorbance data (1.00-cm cells) are

Solution	A_{450}	A_{595}
*A	0.344	0.310
B	0.508	0.212
*C	0.653	0.136
D	0.220	0.380

Calculate the pH of each solution.

26-23. Construct an absorption spectrum for a 7.00×10^{-5} M solution of the indicator of Problem 26-20 when measurements are made with 1.00-cm cells and

(a) $\dfrac{[HIn]}{[In^-]} = 3$

(b) $\dfrac{[HIn]}{[In^-]} = 1$

(c) $\dfrac{[HIn]}{[In^-]} = \dfrac{1}{3}$

26-24. Solutions of P and Q individually obey Beer's law over a large concentration range. Spectral data for these species in 1.00-cm cells are

λ, nm	Absorbance 8.55×10^{-5} M P	Absorbance 2.37×10^{-4} M Q
400	0.078	0.500
420	0.087	0.592
440	0.096	0.599
460	0.102	0.590
480	0.106	0.564
500	0.110	0.515
520	0.113	0.433
540	0.116	0.343
580	0.170	0.170
600	0.264	0.100
620	0.326	0.055
640	0.359	0.030
660	0.373	0.030
680	0.370	0.035
700	0.346	0.063

(a) Plot an absorption spectrum for a solution that is 6.45×10^{-5} M in P and 3.21×10^{-4} M in Q.
(b) Calculate the absorbance (1.00-cm cells) at 440 nm of a solution that is 3.86×10^{-5} M in P and 5.37×10^{-4} M in Q.
(c) Calculate the absorbance (1.00-cm cells) at 620 nm of a solution that is 1.89×10^{-4} M in P and 6.84×10^{-4} M in Q.

26-25. Use the data in Problem 26-22 to calculate the molar concentration of P and Q in each of the following solutions:

	A_{440}	A_{620}
*(a)	0.357	0.803
(b)	0.830	0.448
*(c)	0.248	0.333
(d)	0.910	0.338
*(e)	0.480	0.825
(f)	0.194	0.315

26-26. A standard solution was put through appropriate dilutions to give the concentrations of iron shown below. The iron(II)-1,10,phenanthroline complex was then formed in 25.0-mL aliquots of these solutions, following which each was diluted to 50.0 mL. The following absorbances (1.00-cm cells) were recorded at 510 nm:

Fe(II) Concentration in Original Solutions, ppm	A_{510}
4.00	0.160
10.0	0.390
16.0	0.630
24.0	0.950
32.0	1.260
40.0	1.580

(a) Plot a calibration curve from these data.
*(b) Use the method of least squares to find an equation relating absorbance and the concentration of iron(II).
*(c) Calculate the standard deviation of the slope and intercept.

26-27. The method developed in Problem 26-26 was used for the routine determination of iron in 25.0-mL aliquots of ground water. Express the concentration (as ppm Fe) in samples that yielded the accompanying absorbance data (1.00-cm cell). Calculate the relative standard deviation of the result. Assuming the absorbance data are means of three measurements, repeat the calculation.
*(a) 0.143
(b) 0.675
*(c) 0.068
(d) 1.009
*(e) 1.512
(f) 0.546

***26-28.** The sodium salt of 2-quinizarinsulfonic acid (NaQ) forms a complex with Al^{3+} that absorbs strongly at 560 nm.[16] (a) Use the data from this paper to find the formula of the complex. In all solutions, $c_{Al} = 3.7 \times 10^{-5}$ M, and all measurements were made in 1.00-cm cells. (b) Find the molar absorptivity of the complex and its uncertainty.

c_Q, M	A_{560}
1.00×10^{-5}	0.131
2.00×10^{-5}	0.265
3.00×10^{-5}	0.396
4.00×10^{-5}	0.468
5.00×10^{-5}	0.487
6.00×10^{-5}	0.498
8.00×10^{-5}	0.499
1.00×10^{-4}	0.500

26-29. The accompanying data were obtained in a slope-ratio investigation of the complex formed between Ni^{2+} and 1-cyclopentene-1-dithiocarboxylic acid (CDA). The measurements were made at 530 nm in 1.00-cm cells.

$c_{CDA} = 1.00 \times 10^{-3}$ M		$c_{Ni} = 1.00 \times 10^{-3}$ M	
c_{Ni}, M	A_{530}	c_{CDA}, M	A_{530}
5.00×10^{-6}	0.051	9.00×10^{-6}	0.031
1.20×10^{-5}	0.123	1.50×10^{-5}	0.051
3.50×10^{-5}	0.359	2.70×10^{-5}	0.092
5.00×10^{-5}	0.514	4.00×10^{-5}	0.137
6.00×10^{-5}	0.616	6.00×10^{-5}	0.205
7.00×10^{-5}	0.719	7.00×10^{-5}	0.240

(a) Determine the formula of the complex. Use linear least-squares to analyze the data.
(b) Find the molar absorptivity of the complex and its uncertainty.

[16]E. G. Owens and J. H. Yoe, *Anal. Chem.*, **1959,** *31,* 385.

26-30. The accompanying absorption data were recorded at 390 nm in 1.00-cm cells for a continuous-variation study of the colored product formed between Cd^{2+} and the complexing reagent R.

	Reagent Volumes, mL		
Solution	$c_{Cd} = 1.25 \times 10^{-4}$ M	$c_R = 1.25 \times 10^{-4}$ M	A_{390}
0	10.00	0.00	0.000
1	9.00	1.00	0.174
2	8.00	2.00	0.353
3	7.00	3.00	0.530
4	6.00	4.00	0.672
5	5.00	5.00	0.723
6	4.00	6.00	0.673
7	3.00	7.00	0.537
8	2.00	8.00	0.358
9	1.00	9.00	0.180
10	0.00	10.00	0.000

*(a) Find the ligand-to-metal ratio in the product.
(b) Calculate an average value for the molar absorptivity of the complex and its uncertainty. Assume that in the linear portions of the plot the metal is completely complexed.
(c) Calculate K_f for the complex using the stoichiometric ratio determined in (a) and the absorption data at the point of intersection of the two extrapolated lines.

26-31. Palladium(II) forms an intensely colored complex at pH 3.5 with arsenazo III at 660 nm.[17] A meteorite was pulverized in a ball mill, and the resulting powder was digested with various strong mineral acids. The resulting solution was evaporated to dryness, dissolved in dilute hydrochloric acid, and separated from interferents by ion-exchange chromatography (see Section 32D). The resulting solution containing an unknown amount of Pd(II) was then diluted to 50.00 mL with pH 3.5 buffer. Ten-milliliter aliquots of this analyte solution were then transfered to six 50-mL volumetric flasks. A standard solution was then prepared that was 1.00×10^{-5} M in Pd(II). Volumes of the standard solution shown in the table were then pipetted into the volumetric flasks along with 10.00 mL of 0.01 M arsenazo III. Each solution was then diluted to 50.00 mL, and the absorbance of each solution was measured at 660 nm in 1.00-cm cells.

Volume Standard Solution, mL	A_{660}
0.00	0.216
5.00	0.338
10.00	0.471
15.00	0.596
20.00	0.764
25.00	0.850

(a) Enter the data into a spreadsheet, and construct a standard-additions plot of the data.
(b) Determine the slope and intercept of the line.
(c) Determine the standard deviation of the slope and of the intercept.
(d) Calculate the concentration of Pd(II) in the analyte solution.
(e) Find the standard deviation of the measured concentration.

26-32.  Mercury(II) forms a 1:1 complex with triphenyltetrazolium chloride (TTC) that exhibits an absorption maximum at 255 nm.[18] The mercury (II) in a soil sample was extracted into an organic solvent containing an excess of TTC, and the resulting solution was diluted to 100.0 mL in a volumetric flask. Five-milliliter aliquots of the analyte solution were then transfered to six 25-mL volumetric flasks. A standard solution was then prepared that was 5.00×10^{-6} M in Hg(II). Volumes of the standard solution shown in the table were then pipetted into the volumetric flasks, and each solution was then diluted to 25.00 mL. The absorbance of each solution was measured at 255 nm in 1.00-cm quartz cells.

Volume Standard Solution, mL	A_{255}
0.00	0.582
2.00	0.689
4.00	0.767
6.00	0.869
8.00	1.009
10.00	1.127

(a) Enter the data into a spreadsheet, and construct a standard-additions plot of the data.
*(b) Determine the slope and intercept of the line.
(c) Determine the standard deviation of the slope and of the intercept.
*(d) Calculate the concentration of Hg(II) in the analyte solution.

[17]J. G. Sen Gupta, *Anal. Chem.*, **1967,** *39,* 18.

[18]M. Kamburova, *Talanta.*, **1993,** *40*(5), 719.

(e) Find the standard deviation of the measured concentration.

26-33. Estimate the frequencies of the peaks in the IR spectrum of methylene chloride shown in Figure 26F-2. From these frequencies, assign molecular vibrations of methylene chloride to each of the peaks. Notice that some of the group frequencies that you will need are not listed in Table 26-5, so you will have to look elsewhere.

26-34. Challenge Problem. (a) Show that the overall formation constant for the complex ML_n is

$$K_f = \frac{\left(\frac{A}{A_{extr}}\right)c}{\left[c_M - \left(\frac{A}{A_{extr}}\right)c\right]\left[c_L - n\left(\frac{A}{A_{extr}}\right)c\right]^n}$$

where A is the experimental absorbance at a given value on the x-axis in a continuous-variations plot, A_{extr} is the absorbance determined from the extrapolated lines corresponding to the same point on the x-axis, c_M is the molar analytical concentration of the ligand, c_L is the molar analytical concentration of the metal, and n is the ligand-to-metal ratio in the complex.[19]

(b) Under what assumptions is the equation valid?

(c) What is c?

(d) Discuss the implications of the occurrence of the maximum in a continuous variations plot at a value of less than 0.5.

(e) Using the method of continuous variations, Calabrese and Khan[20] characterized the complex formed between I_2 and I^-. They combined 2.60×10^{-4} M solutions of I_2 and I^- in the usual way to obtain the following data set. Use the data to find the composition of the I_2/I^- complex.

$V(I_2$ soln), mL	A_{350}
0.00	0.002
1.00	0.121
2.00	0.214
3.00	0.279
4.00	0.312
5.00	0.325
6.00	0.301
7.00	0.258
8.00	0.188
9.00	0.100
10.00	0.001

(f) The continuous-variations plot appears to be asymmetrical. Consult the paper by Calabrese and Khan and explain this asymmetry.

(g) Use the equation in part (a) to determine the formation constant of the complex for each of the three central points on the continuous-variations plot.

(h) Explain any trend in the three values of the formation constant in terms of the asymmetry of the plot.

(i) Find the uncertainty in the formation constant determined by this method.

(j) What effect, if any, does the formation constant have on the ability to determine the composition of the complex, using the method of continuous variations?

(k) Discuss the various advantages and potential pitfalls of using the method of continuous variations as a general method for determining the composition and formation constant of a complex compound.

[19] J. Inczédy, *Analytical Applications of Complex Equilibria.* New York: Wiley, 1976.

[20] V. T. Calabrese and A. Khan, *J. Phys. Chem. A*, **2000,** *104,* 1287.

CHAPTER 27

Molecular Fluorescence Spectroscopy

Immunofluorescent light micrograph of HeLa cancer cells. The cell in the center of the photo is in the prophase stage of mitotic cell division. The chromosomes have condensed before dividing to form two nuclei. The cells are stained to reveal actin microfilaments and microtubules of the cytoskeleton, which appear as the filamentary structures surrounding the cell nuclei. The nuclei of the cells are visualized by exposing the cells to structure-specific fluorescent antibodies, prepared by covalently attaching ordinary antibodies to fluorescent molecules. The antibodies collect in the nuclei so that when they are exposed to UV radiation, they glow as shown in the photo. Similar chemistry is used in the fluorescence immunoassay described in Feature 11-2.

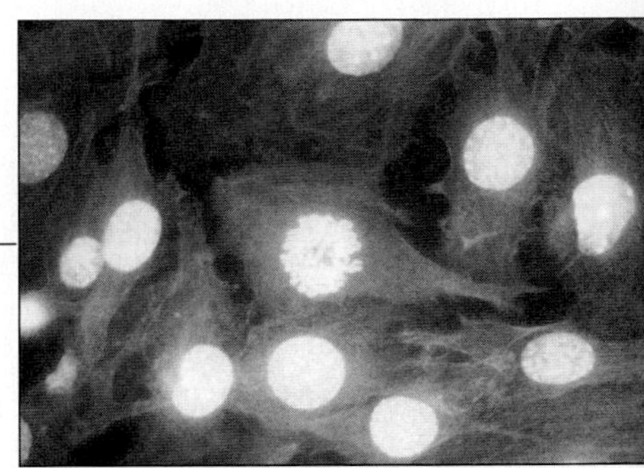

Dr. Gopal Murti/Science Photo Library/Photo Researchers, Inc.

Fluorescence is a photoluminescence process in which atoms or molecules are excited by absorption of electromagnetic radiation (recall Figure 24-6). The excited species then relax to the ground state, giving up their excess energy as photons. One of the most attractive features of molecular fluorescence is its inherent sensitivity, which is often one to three orders of magnitude better than absorption spectroscopy. In fact, for selected species under controlled conditions, single molecules have been detected by fluorescence spectroscopy. Another advantage is the large linear concentration range of fluorescence methods, which is significantly greater than those encountered in absorption spectroscopy. Fluorescence methods are, however, much less widely applicable than absorption methods because of the relatively limited number of chemical systems that show appreciable fluorescence. Fluorescence is also subject to many more environmental interference effects than absorption methods. We consider here some of the most important aspects of molecular fluorescence methods.

27A THEORY OF MOLECULAR FLUORESCENCE

Molecular fluorescence is measured by exciting the sample at the absorption wavelength, also called the excitation wavelength, and measuring the emission at a longer wavelength called the emission or fluorescence wavelength. For example, the reduced form of the coenzyme nicotinamide adenine dinucleotide (NADH) can absorb radiation at 340 nm. The molecule exhibits fluorescence with an emission maximum at 465 nm. Usually, fluorescence emission is measured at right angles to the incident beam, to avoid measuring the incident radiation (recall Figure 25-1b). The short-lived emission that occurs is called **fluorescence,** while luminescence that is much longer lasting is called **phosphorescence.**

◀ Fluorescence emission occurs in 10^{-5} s or less. In contrast, phosphorescence may last for several minutes or even hours. Fluorescence is much more widely used for chemical analysis than phosphorescence.

27A-1 Relaxation Processes

Figure 27-1 shows a partial energy level diagram for a hypothetical molecular species. Three electronic energy states are shown, E_0, E_1, and E_2; E_0 is the ground state, and E_1 and E_2 are excited states. Each of the electronic states is shown as having four excited vibrational levels. Irradiation of this species with a band of radiation made up of wavelengths λ_1 to λ_5 (Figure 23-12a) results in the momentary population of the five vibrational levels of the first excited electronic state, E_1. Similarly, when the molecules are irradiated with a more energetic band of radiation made up of shorter wavelengths λ_1' to λ_5', the five vibrational levels of the higher energy electronic state E_2 become populated briefly.

Once the molecule is excited to E_1 or E_2, several processes can occur that cause the molecule to lose its excess energy. Two of the most important of these mechanisms, **nonradiative relaxation** and **fluorescence emission,** are illustrated in Figure 27-1b and c.

The two most important nonradiative relaxation methods that compete with fluorescence are illustrated in Figure 27-1b. **Vibrational relaxation,** depicted by the short wavy arrows between vibrational energy levels, takes place during collisions between excited molecules and molecules of the solvent. Nonradiative relaxation between the lower vibrational levels of an excited electronic state and the higher vibrational levels of another electronic state can also occur. This type of relaxation, sometimes called **internal conversion,** is depicted by the two longer wavy arrows in Figure 27-1b. Internal conversion is much less efficient than vibrational relaxation, so that the average lifetime of an electronic excited state is between 10^{-9} and 10^{-6} s. The exact mechanism by which these two relaxational processes occur is currently under study, but the net result is a tiny increase in the temperature of the medium.

Figure 27-1c illustrates the relaxation process that is desired: the fluorescence process. Almost always, fluorescence is observed from the lowest-lying excited

Vibrational relaxation involves transfer of the excess energy of a vibrationally excited species to molecules of the solvent. This process takes place in less than 10^{-15} s and leaves the molecules in the lowest vibrational state of an electronic excited state.

Internal conversion is a type of relaxation that involves transfer of the excess energy of a species in the lowest vibrational level of an excited electronic state to solvent molecules and conversion of the excited species to a lower electronic state.

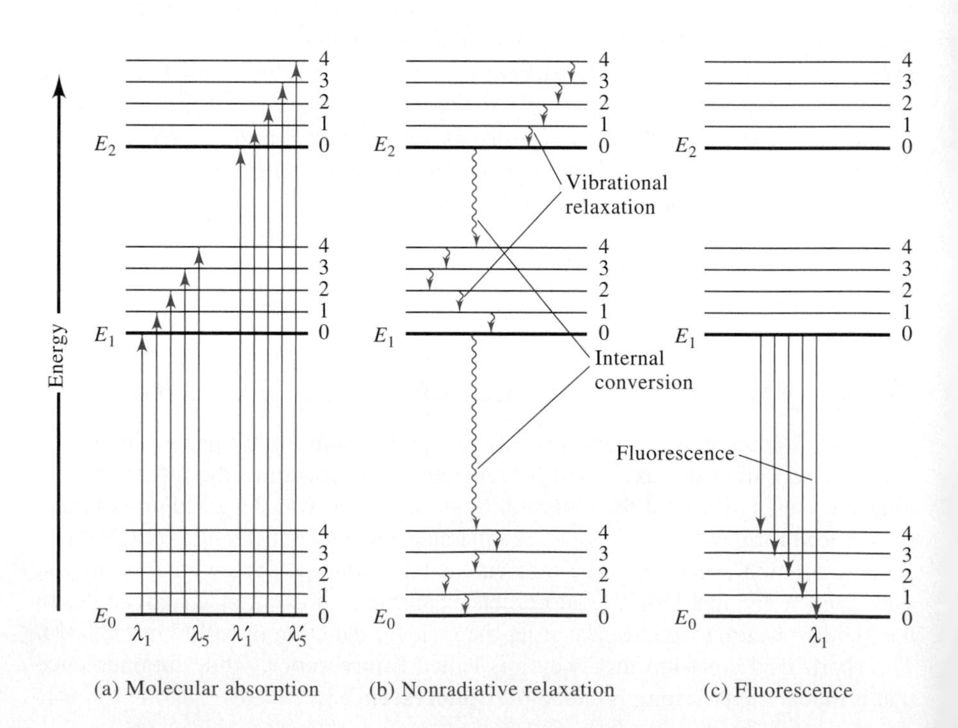

Figure 27-1 Energy-level diagram shows some of the processes that occur during (a) absorption of incident radiation, (b) nonradiative relaxation, and (c) fluorescence emission by a molecular species. Absorption typically occurs in 10^{-15} s, while vibrational relaxation occurs in the 10^{-11} to 10^{-10} s time scale. Internal conversion between different electronic states is also very rapid (10^{-12} s), while fluorescence lifetimes are typically 10^{-10} to 10^{-5} s.

electronic state E_1 to the ground state, E_0. Also, the fluorescence usually occurs only from the lowest vibrational level of E_1 to various vibrational levels of E_0. This is because the internal conversion and vibrational relaxation processes are very rapid compared with fluorescence. Hence, a fluorescence spectrum usually consists of only one band with many closely spaced lines that represent transitions from the lowest vibrational level of E_1 to the many different vibrational levels of E_0.

◀ Fluorescence bands consist of a large number of closely spaced lines.

The line in Figure 27-1c that terminates the fluorescence band on the short-wavelength, or high-energy, side (λ_1) is identical in energy to the line labeled λ_1 in the absorption diagram in Figure 27-1a. Since fluorescence lines in this band originate in the lowest vibrational state of E_1, all of the other lines in the band are of lower energy, or longer wavelength, than the line corresponding to λ_1. Molecular fluorescence bands are mostly made up of lines that are longer in wavelength, lower in frequency, and thus lower in energy, than the band of absorbed radiation responsible for their excitation. This shift to longer wavelength is called the **Stokes shift.**

◀ **Stokes-shifted fluorescence** is longer in wavelength than the radiation that caused the excitation.

Relationship between Excitation Spectra and Fluorescence Spectra

Because the energy differences between vibrational states is about the same for both ground and excited states, the absorption spectrum, or **excitation spectrum,** and the fluorescence spectrum for a compound often appear as approximate mirror images of one another with overlap occurring near the origin transition (0 vibrational level of E_1 to 0 vibrational level of E_0). This effect is demonstrated by the spectra for anthracene shown in Figure 27-2. There are many exceptions to this mirror-image rule, particularly when the excited and ground states have different

CD-ROM Simulation: Exploration of Luminescence Spectral Shape as a Function of Excited State Energy, Vibrational Modes, and Molecular Distortions.

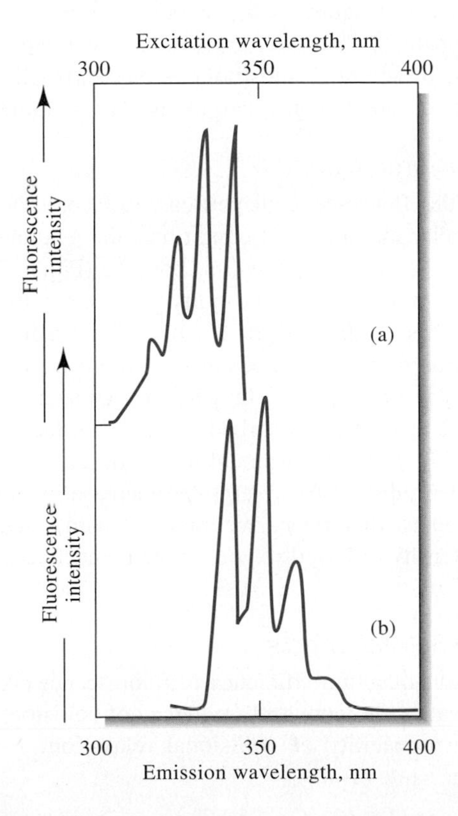

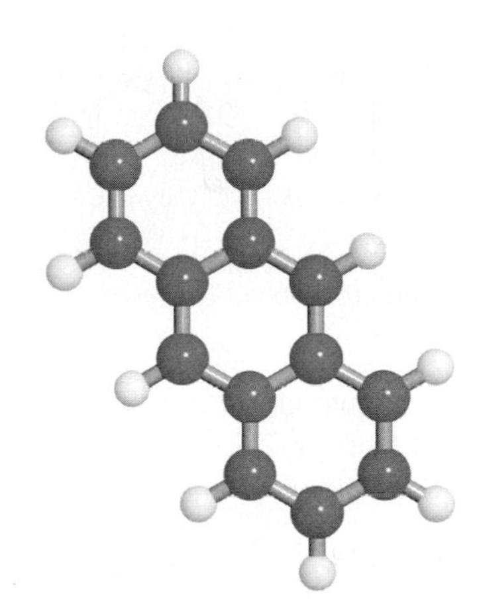

Molecular model of anthracene.

Figure 27-2 Fluorescence spectra for 1 ppm anthracene in alcohol: (a) excitation spectrum; (b) emission spectrum.

molecular geometries or when different fluorescence bands originate from different parts of the molecule.

27A-2 Fluorescent Species

Quantum efficiency is described by the **quantum yield of fluorescence,** Φ_F.

$$\Phi_F = \frac{k_F}{k_F + k_{nr}}$$

where k_F is the first-order rate constant for fluorescence relaxation and k_{nr} is the rate constant for radiationless relaxation. See Chapter 29 for a discussion of rate constants.

As shown in Figure 27-1, fluorescence is one of several mechanisms by which a molecule returns to the ground state after it has been excited by absorption of radiation. All absorbing molecules have the potential to fluoresce, but most compounds do not because their structure provides radiationless pathways for relaxation to occur *at a greater rate* than fluorescence emission. The **quantum yield** of molecular fluorescence is simply the ratio of the number of molecules that fluoresce to the total number of excited molecules, or the ratio of photons emitted to photons absorbed. Highly fluorescent molecules, such as fluorescein, have quantum efficiencies that approach unity under some conditions. Nonfluorescent species have efficiencies that are essentially zero.

Fluorescence and Structure

► Many unsubstituted aromatic compounds fluoresce.

Compounds containing aromatic rings give the most intense and most useful molecular fluorescence emission. While certain aliphatic and alicyclic carbonyl compounds as well as highly conjugate double-bonded structures also fluoresce, there are very few of these compared with the number of fluorescent compounds containing aromatic systems.

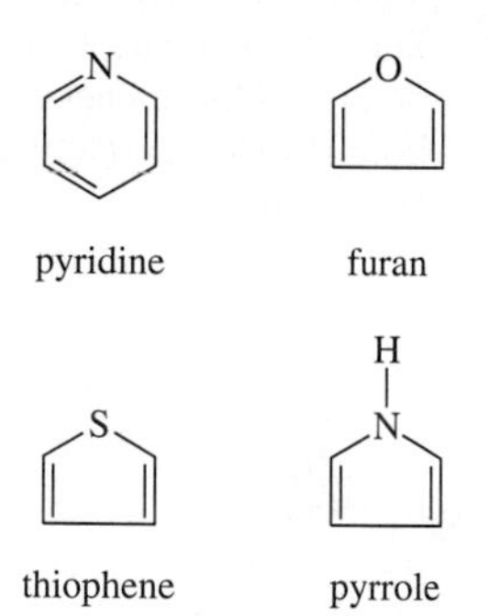

Figure 27-3 Typical aromatic molecules that do not fluoresce.

Most unsubstituted aromatic hydrocarbons fluoresce in solution, with the quantum efficiency increasing with the number of rings and their degree of condensation. The simplest heterocyclics, such as pyridine, furan, thiophene, and pyrrole, do not exhibit molecular fluorescence (Figure 27-3), but fused-ring structures containing these rings often do (Figure 27-4). Substitution on an aromatic ring causes shifts in the wavelength of absorption maxima and corresponding changes in the fluorescence peaks. In addition, substitution frequently affects the fluorescence efficiency. These effects are demonstrated by the data in Table 27-1.

The Effect of Structural Rigidity

► Rigid molecules or complexes tend to fluoresce.

Experiments show that fluorescence is particularly favored in rigid molecules. For example, under similar measurement conditions, the quantum efficiency of fluorene is nearly 1.0, while that of biphenyl is about 0.2 (Figure 27-5). The difference in behavior is a result of the increased rigidity provided by the bridging methylene group in fluorene. This rigidity lowers the rate of nonradiative relaxation to the point where relaxation by fluorescence has time to occur. There are many similar examples of this type of behavior. In addition, enhanced emission frequently results when fluorescing dyes are adsorbed on a solid surface; here again, the added rigidity provided by the solid may account for the observed effect.

The influence of rigidity also explains the increase in fluorescence of certain organic chelating agents when they are complexed with a metal ion. For example, the fluorescence intensity of 8-hydroxyquinoline is much less than that of the zinc complex (Figure 27-6).

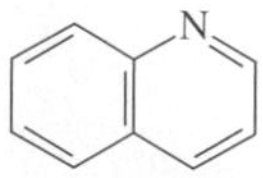

quinoline

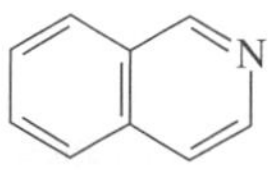

isoquinoline

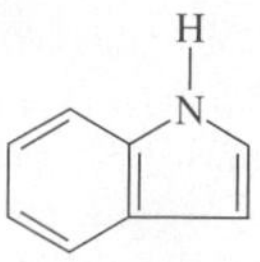

indole

Figure 27-4 Typical aromatic compounds that fluoresce.

Temperature and Solvent Effects

In most molecules, the quantum efficiency of fluorescence decreases with increasing temperature because the increased frequency of collision at elevated temperatures increases the probability of collisional relaxation. A decrease in solvent viscosity leads to the same result.

27B EFFECT OF CONCENTRATION ON FLUORESCENCE INTENSITY

The power of fluorescence radiation F is proportional to the radiant power of the excitation beam absorbed by the system:

$$F = K'(P_0 - P) \tag{27-1}$$

where P_0 is the power of the beam incident on the solution and P is its power after it passes through a length b of the medium. The constant K' depends on the quantum efficiency of the fluorescence. To relate F to the concentration c of the fluorescing particle, we write Beer's law in the form

$$\frac{P}{P_0} = 10^{-\varepsilon bc} \tag{27-2}$$

where ε is the molar absorptivity of the fluorescing species and εbc is the absorbance A. By substituting Equation 27-2 into Equation 27-1, we obtain

$$F = K'P_0(1 - 10^{-\varepsilon bc}) \tag{27-3}$$

Expansion of the exponential term in Equation 27-3 leads to

$$F = K'P_0\left[2.3\varepsilon bc - \frac{(-2.3\varepsilon bc)^2}{2!} - \frac{(-2.3\varepsilon bc)^3}{3!} - \cdots\right] \tag{27-4}$$

When $\varepsilon bc = A < 0.05$, the first term inside the brackets, $2.3\varepsilon bc$, is much larger than subsequent terms, and we can write

$$F = 2.3K'\varepsilon bcP_0 \tag{27-5}$$

or, when the incident power P_0 is constant,

$$F = Kc \tag{27-6}$$

Thus, a plot of the fluorescence power of a solution versus the concentration of the emitting species should be linear at low concentrations. When c becomes large enough that the absorbance is larger than about 0.05 (or the transmittance is smaller than about 0.9), the relationship represented by Equation 27-6 becomes nonlinear and F lies below an extrapolation of the linear plot. This effect is a result of **primary absorption** in which the incident beam is absorbed so strongly that fluorescence is no longer proportional to concentration as shown in the more complete Equation 27-4. At very high concentrations, F reaches a maximum and may even begin to decrease with increasing concentration because of **secondary absorption.** This phenomenon occurs because of absorption of the emitted radiation by other analyte molecules. A typical plot of F versus concentration is shown in Figure 27-7. Note that primary and secondary absorption effects, sometimes called **inner-filter effects,** can also occur because of absorption by molecules in the sample matrix.

TABLE 27-1
Effect of Substitution on the Fluorescence of Benzene Derivatives*

Compound	Relative Intensity of Fluorescence
Benzene	10
Toluene	17
Propylbenzene	17
Fluorobenzene	10
Chlorobenzene	7
Bromobenzene	5
Iodobenzene	0
Phenol	18
Phenolate ion	10
Anisole	20
Aniline	20
Anilinium ion	0
Benzoic acid	3
Benzonitrile	20
Nitrobenzene	0

*In ethanol solution. Taken from W. West, *Chemical Applications of Spectroscopy* (*Techniques of Organic Chemistry.* Vol. IX, p. 730). New York: Interscience, 1956.

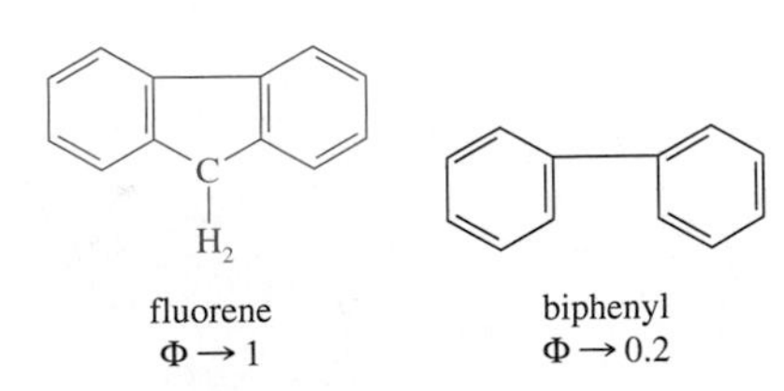

Figure 27-5 Effect of molecular rigidity on quantum yield. The fluorene molecule is held rigid by the central ring, two benzene rings in biphenyl can rotate relative to one another

OH N O Zn N 2

nonfluorescing fluorescing

Figure 27-6 Effect of rigidity on quantum yield in complexes. Free 8-hydroxyquinoline molecules in solution are easily deactivated through collisions with solvent molecules and do not fluoresce. The rigidity of the Zn-8-hydroxyquinoline complex enhances fluorescence.

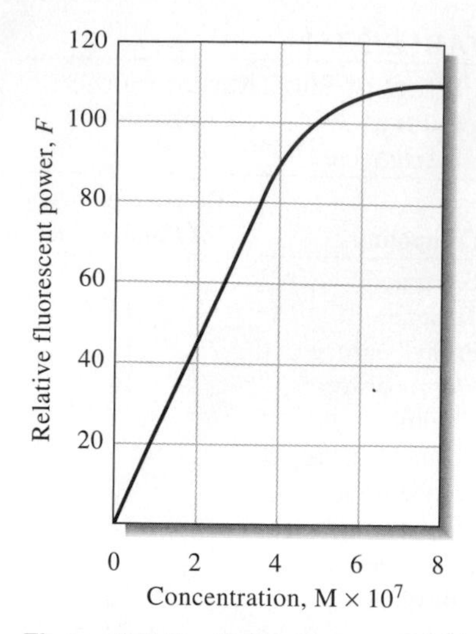

Figure 27-7 Calibration curve for the spectrofluorometric determination of tryptophan in soluble proteins from the lens of a mammalian eye.

27C FLUORESCENCE INSTRUMENTS

There are several different types of fluorescence instruments. All follow the general block diagram of Figure 25-1b. Optical diagrams of typical instruments are shown in Figure 27-8. If the two wavelength selectors are both filters, the instrument is called a fluorometer. If both wavelength selectors are monochromators, the instrument is a spectrofluorometer. Some instruments are hybrids and use an excitation filter along with an emission monochromator. Fluorescence instruments can incorporate a double-beam design to compensate for changes in the source radiant power with time and wavelength. Instruments that correct for the source spectral distribution are called corrected spectrofluorometers.

Sources for fluorescence are usually more powerful than typical absorption sources. In fluorescence, the radiant power emitted is directly proportional to the

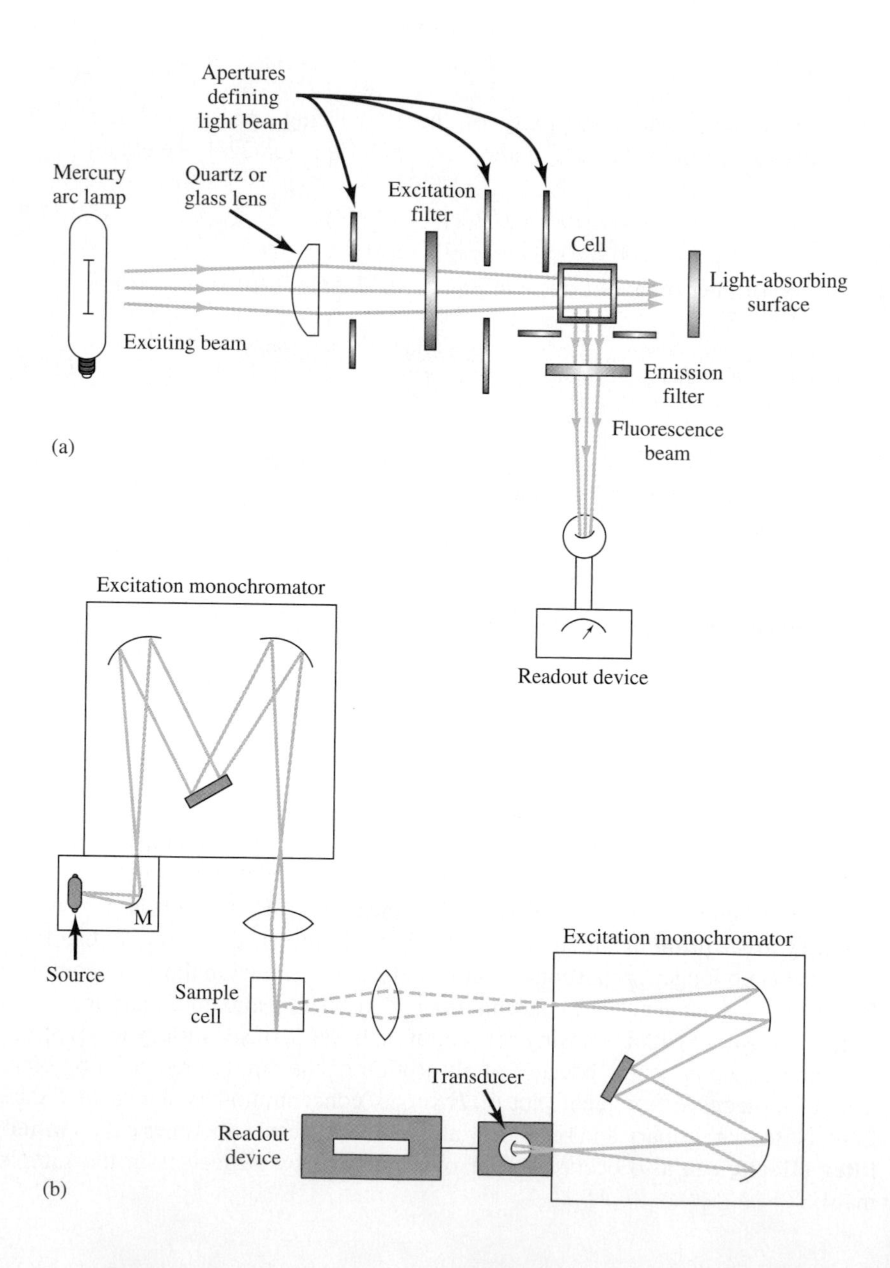

Figure 27-8 Typical fluorescence instruments. A filter fluorometer is shown in (a). Note that the emission is measured at right angles to the mercury arc lamp source. Fluorescence radiation is emitted in all directions, and the 90-deg geometry avoids the detector viewing the source. The spectrofluorometer (b) uses two grating monochromators and also views the emission at right angles. The two monochromators allow the scanning of excitation spectra (excitation wavelength scanned at a fixed emission wavelength), emission spectra (emission wavelength scanned at a fixed excitation wavelength), or synchronous spectra (both wavelengths scanned with a fixed wavelength offset between the two monochromators).

source intensity (Equation 27-5), but absorbance, because it is related to the ratio of radiant powers as shown in Equation 27-7, is essentially independent of source intensity.

$$c = kA = k \log \left(\frac{P_0}{P} \right) \qquad (27\text{-}7)$$

As a result of these differences in dependence on source intensity, fluorescence methods are generally one to three orders of magnitude more sensitive than methods based on absorption. Mercury arc lamps, xenon arc lamps, xenon-mercury arc lamps, and lasers are typical fluorescence sources. Monochromators and transducers are typically similar to those used in absorption spectrophotometers, except that photomultipliers are invariably used in high-sensitivity spectrofluorometers. Fluorometers and spectrofluorometers vary widely in sophistication, performance characteristics, and cost, as do absorption spectrophotometers. Generally, fluorescence instruments are more expensive than absorption instruments of corresponding quality.

◀ Fluorescence methods are 10 to 1000 times more sensitive than absorption methods.

27D APPLICATIONS OF FLUORESCENCE METHODS

Fluorescence spectroscopy is not considered a major structural or qualitative analysis tool, because molecules with subtle structural differences often have similar fluorescence spectra. Also, fluorescence bands in solution are relatively broad at room temperature. However, fluorescence has proved to be a valuable tool in oil spill identification. The source of an oil spill can often be identified by comparing the fluorescence emission spectrum of the spill sample with that of a suspected source. The vibrational structure of polycyclic hydrocarbons present in the oil makes this type of identification possible.

Fluorescence methods are used to study chemical equilibria and kinetics in much the same way as absorption spectrophotometry is used. Often it is possible to study chemical reactions at lower concentrations because of the higher sensitivity of fluorescence methods. In many cases where fluorescence monitoring is ordinarily not feasible, fluorescent probes or tags can be bound covalently to specific sites in molecules such as proteins, thus making them detectable via fluorescence. These tags can be used to provide information about energy transfer processes, the polarity of the protein, and the distances between reactive sites (see, for example, Feature 27-1).

Quantitative fluorescence methods have been developed for inorganic, organic, and biochemical species. Inorganic fluorescence methods can be divided into two classes: direct methods that are based on the reaction of the analyte with a complexing agent to form a fluorescent complex, and indirect methods that depend on the decrease in fluorescence, also called **quenching,** as a result of interaction of the analyte with a fluorescent reagent. Quenching methods are primarily used for the determination of anions and dissolved oxygen. Some fluorescence reagents for cations are shown in Figure 27-9.

Nonradiative relaxation of transition-metal chelates is so efficient that these species seldom fluoresce. It is worth noting that most transition metals absorb in the UV or visible region, while nontransition metal ions do not. For this reason, fluorescence is often considered complementary to absorption for the determination of cations.

The number of applications of fluorescence methods to organic and biochemical problems is impressive. Among the compound types that can be determined by

HO N

8-hydroxyquinoline
(reagent for Al, Be, and other metal ions)

OH HO
HO—N=N—
SO_3Na

alizarin garnet R
(reagent for Al, F^-)

O OH O

flavanol
(reagent for Zr and Sn)

O OH
C—C
H

benzoin
(reagent for B, Zn, Ge, and Si)

Figure 27-9 Some fluorometric chelating agents for metal cations. Alizarin garnet R can detect Al^{3+} at levels as low as 0.007 μg/mL. Detection of F^- with alizarin garnet R is based on quenching of the fluorescence of the Al^{3+} complex. Flavanol can detect Sn^{4+} at the 0.1 μg/mL level.

FEATURE 27-1

Use of Fluorescence Probes in Neurobiology: Probing the Enlightened Mind

Fluorescent indicators have been widely used to probe biological events in individual cells. A particularly interesting probe is the so-called ion probe that changes its excitation or emission spectrum when it binds to specific ions such as Ca^{2+} or Na^{+}. These indicators can be used to record events that take place in different parts of individual neurons or to monitor simultaneously the activity of a collection of neurons. In neurobiology, for example, the dye Fura-2 has been used to monitor the free intracellular calcium concentration following some pharmacological or electrical stimulation. By following the fluorescence changes with time at specific sites in the neuron, researchers can determine when and where a calcium-dependent electrical event took place. One cell that has been studied is the Purkinje neuron in the cerebellum, which is one of the largest in the central nervous system. When this cell is loaded with the Fura-2 fluorescent indicator, sharp changes in fluorescence can be measured that correspond to individual calcium action potentials. The changes are correlated to specific sites in the cell by means of fluorescence imaging techniques. Figure 27F-1 shows the fluorescent image on the right along with fluorescence transients, recorded as the change in fluorescence relative to the steady fluorescence $\Delta F/F$, correlated with sodium action potential spikes. The interpretation of these kinds of patterns can have important implications in understanding the details of synaptic activity.

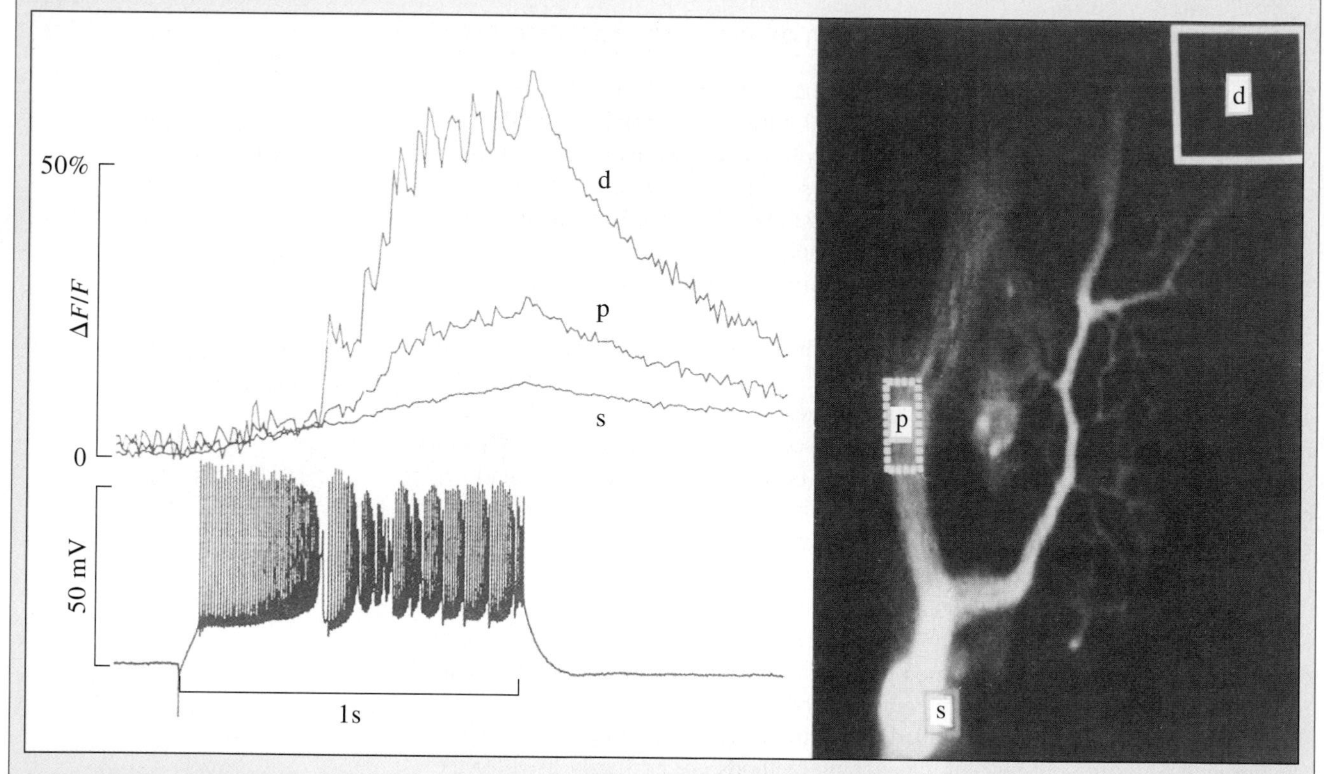

Figure 27F-1 Calcium transients in a cerebellar Purkinje cell. The image on the right is of the cell filled with a fluorescent dye that responds to the calcium concentration. Fluorescent transients are shown on the top left recorded at areas d, p, and s in the cell. The transients in region d correspond to the dendrite region of the cell. Specific calcium signals can be correlated to the action potentials shown on the bottom left. (From V. Lev-Ram, H. Mikayawa, N. Lasser-Ross, W. N. Ross, *J. Neurophysiol.* **1992,** *68,* 1170. With permission of the American Physiological Society.)

fluorescence are amino acids, proteins, coenzymes, vitamins, nucleic acids, alkaloids, porphyrins, steroids, flavonoids, and many metabolites.[1] Because of its sensitivity, fluorescence is widely used as a detection technique for liquid chromatographic methods (see Chapter 32), for flow analysis methods, and for electrophoresis. In addition to methods that are based on measurements of fluorescence intensity, there are many methods based on measurements of fluorescence lifetimes. Several instruments have been developed that provide microscopic images of specific species based on fluorescence lifetimes.[2]

◀ Some typical polycyclic aromatic hydrocarbons found in oil spills are chrysene, perylene, pyrene, fluorene, and 1,2-benzofluorene. Most of these compounds are carcinogenic.

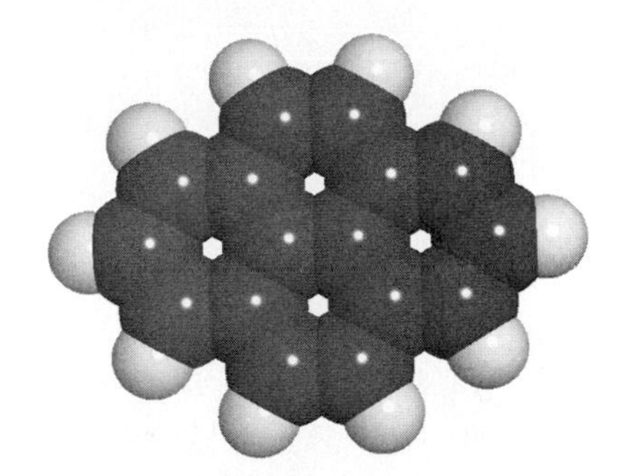

Molecular model of pyrene.

27D-1 Methods for Inorganic Species

The most successful fluorometric reagents for the determination of cations are aromatic compounds having two or more donor functional groups that form chelates with the metal ion. A typical example is 8-hydroxyquinoline, the structure of which is given in Section 12D-3. A few other fluorometric reagents and their applications are found in Table 27-2. With most of these reagents, the cation is extracted into a solution of the reagent in an immiscible organic solvent, such as chloroform. The fluorescence of the organic solution is then measured. For a more complete summary of fluorometric methods for inorganic substances, see the handbook by Dean.[3]

Nonradiative relaxation of transition-metal chelates is so efficient that these species seldom fluoresce. Note that most transition metals absorb in the ultraviolet or visible region, but nontransition metal ions do not. For this reason, fluorometry often complements spectrophotometry as a method for the determination of cations.

TABLE 27-2
Selected Fluorometric Methods for Inorganic Species*

		Wavelength, nm		Sensitivity,	
Ion	**Reagent**	**Absorption**	**Fluorescence**	**μg/mL**	**Interference**
Al^{3+}	Alizarin garnet R	470	500	10.007	Be, Co, Cr, Cu, F^-, NO_3^-, Ni, PO_4^{3-}, Th, Zr
F^-	Al complex of Alizarin garnet R (quenching)	470	500	0.001	Be, Co, Cr, Cu, Fe, Ni, PO_4^{3-}, Th, Zr
$B_4O_7^{2-}$	Benzoin	370	450	0.04	Be, Sb
Cd^{2+}	2-(*o*-Hydroxyphenyl)-benzoxazole	365	Blue	2	NH_3
Li^+	8-Hydroxyquinoline	370	580	0.2	Mg
Sn^{4+}	Flavanol	400	470	0.1	F^-, PO_4^{3-}, Zr
Zn^{2+}	Benzoin	—	Green	10	B, Be, Sb, colored ions

*From J. A. Dean, *Analytical Chemistry Handbook,* New York, McGraw-Hill, 1995, pp. 5.60–5.62.

[1]See for example, O. S. Wolfbeis, in *Molecular Luminescence Spectroscopy: Methods & Applications—Part I,* Ch. 3, S. G. Schulman, Ed. New York: Wiley-Interscience, 1985.

[2]See J. R. Lakowicz, H. Szmacinski, K. Nowacyzk, K. Berndt, and M. L. Johnson, in *Fluorescence Spectroscopy: New Methods and Applications,* Ch. 10, O. S. Wolfbeis, Ed., Berlin: Springer-Verlag, 1993.

[3]J. A. Dean, *Analytical Chemistry Handbook,* pp. 5.60–5.62. New York: McGraw-Hill, 1995.

27D-2 Methods for Organic and Biochemical Species

The number of applications of fluorometric methods to organic problems is impressive. Dean summarizes the most important of these in a table.[4] More than 200 entries are found under the heading "Fluorescence Spectroscopy of Some Organic Compounds," including such diverse compounds as adenine, anthranilic acid, aromatic polycyclic hydrocarbons, cysteine, guanine, isoniazid, naphthols, nerve gases sarin and tabun, proteins, salicylic acid, skatole, tryptophan, uric acid, and warfarin (Coumadin). Many medicinal agents that can be determined fluorometrically are listed, including adrenaline, morphine, penicillin, phenobarbital, procaine, reserpine, and lysergic acid diethylamide (LSD). Without question, the most important application of fluorometry is in the analysis of food products, pharmaceuticals, clinical samples, and natural products. The sensitivity and selectivity of the method make it a particularly valuable tool in these fields. Numerous physiologically important compounds fluoresce.

27E MOLECULAR PHOSPHORESCENCE SPECTROSCOPY

Phosphorescence is a photoluminescence phenomenon that is quite similar to fluorescence. Understanding the distinction between these two phenomena requires an understanding of electron spins and the difference between a **singlet state** and a **triplet state.** Ordinary molecules that are not free radicals exist in the ground state with their electron spins paired. A molecular electronic state in which all electron spins are paired is said to be a *singlet state.* The ground state of a free radical, on the other hand, is a **doublet state,** because the odd electron can assume two orientations in a magnetic field.

When one of a pair of electrons in a molecule is excited to a higher energy level, a singlet or a *triplet* state can be produced. In the excited singlet state, the spin of the promoted electron is still opposite that of the remaining electron. In the triplet state, however, the spins of the two electrons become unpaired and are thus parallel. These states can be represented as illustrated in Figure 27-10. The excited triplet state is less energetic than the corresponding excited singlet state.

Fluorescence of molecules involves a transition from an excited singlet state to the ground singlet state. This transition is highly probable, and thus the lifetime of an excited singlet state is very short (10^{-5} s or less). Molecular phosphorescence, on the other hand, involves a transition from an excited triplet state to the ground singlet state. Because this transition produces a change in electron spin, it is much less probable. Hence, the triplet state has a much longer lifetime (typically 10^{-4} to 10^{4} s). Solid-state phosphors coated on the screen of a cathode-ray tube are responsible for being able to observe the action of electron beams in many oscilloscopes, television sets, and computer monitors.

The long lifetime of phosphorescence is also one of its drawbacks. Because of this long lifetime, nonradiational processes can compete with phosphorescence to deactivate the excited state. Thus, the efficiency of the phosphorescence process, and the corresponding phosphorescence intensity, is relatively low. To increase this efficiency, phosphorescence is commonly observed at low temperatures in rigid media, such as glasses. In recent years, room temperature phosphorescence has

► In room temperature phosphorescence, the triplet state of the analyte can be protected by being incorporated into a surfactant aggregate called a micelle. In aqueous solutions the aggregate has a nonpolar core due to repulsion of the polar head groups. The opposite occurs in nonpolar solvents.

[4] J. A. Dean, *Analytical Chemistry Handbook,* pp. 5.63–5.69. New York: McGraw-Hill, 1995.

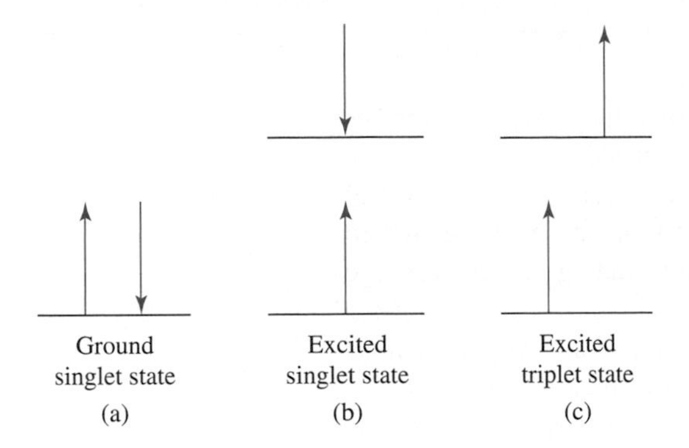

Figure 27-10 Electronic spin states of molecules. In (a) the ground electronic state is shown. In the lowest energy or ground state, the spins are always paired, and the state is said to be a singlet state. In (b) and (c), excited electronic states are shown. If the spins remain paired in the excited state, the molecule is in an excited singlet state (b). If the spins become unpaired, the molecule is in an excited triplet state (c).

become popular. In this technique the molecule is either adsorbed on a solid surface or enclosed in a molecular cavity (micelle or cyclodextrin cavity), which protects the fragile triplet state.

Because of its weak intensity, phosphorescence is much less widely applicable than fluorescence. However, phosphorimetry has been used for the determination of a variety of organic and biochemical species including nucleic acids, amino acids, pyrine and pyrimidine, enzymes, polycyclic hydrocarbons, and pesticides. Many pharmaceutical compounds exhibit measurable phosphorescence signals. The instrumentation for phosphorescence is also somewhat more complex than that for fluorescence. Usually, the phosphorescence instrument allows discrimination of phosphorescence from fluorescence by delaying the phosphorescence measurement until the fluorescence has decayed to nearly zero. Many fluorescence instruments have attachments, called **phosphoroscopes,** that allow the same instrument to be used for phosphorescence measurements.

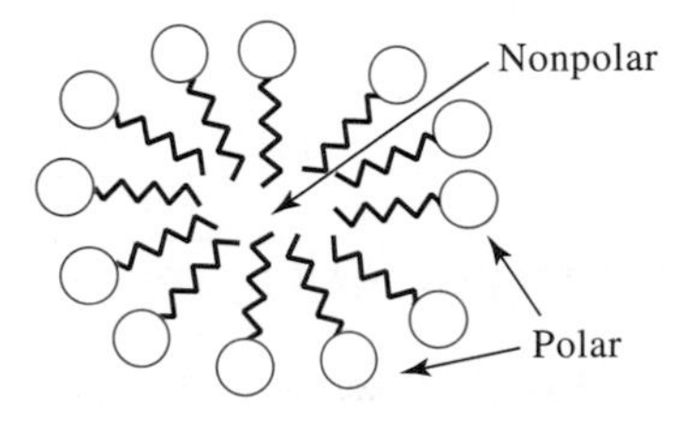

Micelle in aqueous solvent

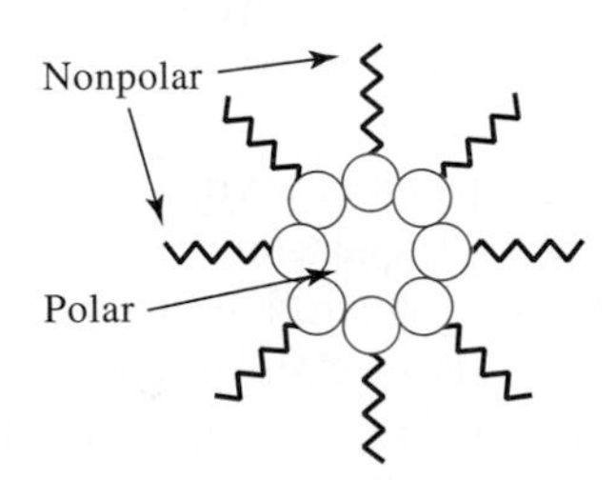

Micelle in nonaqueous solvent

Structure of micelles.

27F CHEMILUMINESCENCE METHODS

Chemiluminescence is produced when a chemical reaction yields an electronically excited molecule, which emits light as it returns to the ground state. Chemiluminescence reactions are encountered in a number of biological systems, where the process is often termed **bioluminescence.** Examples of species exhibiting bioluminescence include the firefly, the sea pansy, certain jellyfish, bacteria, protozoa, and crustacea.

One attractive feature of chemiluminescence for analytical uses is the simple instrumentation. Since no external source of radiation is needed for excitation, the instrument may consist of only a reaction vessel and a photomultiplier tube. Generally, no wavelength selection device is needed because the only source of radiation is the chemical reaction.

Chemiluminescence methods are known for their high sensitivities. Typical detection limits range from parts per million to parts per billion or lower. Applications include the determination of gases, such as oxides of nitrogen, ozone, and sulfur compounds, determination of inorganic species such as hydrogen peroxide and some metal ions, immunoassay techniques, DNA probe assays, and polymerase chain reaction methods.[5]

◀ The firefly produces light by the phenomenon of **bioluminescence.** Different species of fireflies flash with different on-off cycle times. Fireflies mate only with their own species. The familiar bioluminescence reaction occurs when the firefly is looking for a mate.

◀ Several commercial analyzers for the determination of gases are based on chemiluminescence. Nitric oxide (NO) can be determined by reaction with ozone (O_3). The reaction converts the NO to excited NO_2, with the subsequent emission of light.

[5]See, for example, T. A. Nieman, in *Handbook of Instrumental Techniques for Analytical Chemistry,* Chapter 27, F. A. Settle, Ed. Upper Saddle River, NJ: Prentice Hall, 1997.

WEB WORKS

Use your Web browser to connect to **http://chemistry.brookscole.com/skoogfac/.** From the Chapter Resources menu, choose Web Works. Locate the Chapter 27 section, and click on the link to the UK National Physical Laboratory's National Reference Spectrofluorimeter. This instrument uses a scanning monochromator for excitation. What does it use for emission? What is the advantage of this arrangement? At what angle with respect to the incident radiation is the emission collected from the sample? What is the resolution of the detector in nanometers per element? For what purposes does the UK NPL use this spectrofluorimeter? Use Google to find other spectrofluorimeters on the Web, and compare their specifications and characteristics with those of the UK NPL instrument.

QUESTIONS AND PROBLEMS

27-1. Briefly describe or define
- *(a) resonance fluorescence.
- (b) vibrational relaxation.
- *(c) internal conversion.
- (d) fluorescence.
- *(e) Stokes shift.
- (f) quantum yield.
- *(g) self-quenching.

27-2. Why is spectrofluorometry potentially more sensitive than spectrophotometry?

27-3. Which compound in each of the pairs below would you expect to have a greater fluorescence quantum yield? Explain.

*(a)

phenolphthalein

fluorescein

(b)

o,o′-dihydroxyazobenzene

bis(*o*-hydroxyphenyl) hydrazine

27-4. Why do some absorbing compounds fluoresce while others do not?

***27-5.** Describe the characteristics of organic compounds that fluoresce.

27-6. Explain why molecular fluorescence often occurs at a longer wavelength than the exciting radiation.

27-7. Describe the components of a fluorometer.

***27-8.** Why are most fluorescence instruments double beam in design?

27-9. Why are fluorometers often more useful than spectrofluorometers for quantitative analysis?

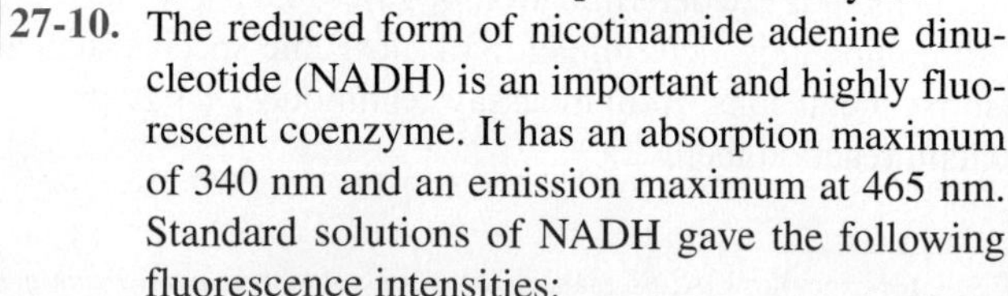

27-10. The reduced form of nicotinamide adenine dinucleotide (NADH) is an important and highly fluorescent coenzyme. It has an absorption maximum of 340 nm and an emission maximum at 465 nm. Standard solutions of NADH gave the following fluorescence intensities:

Concn NADH, μmol/L	Relative Intensity
0.100	2.24
0.200	4.52
0.300	6.63
0.400	9.01
0.500	10.94
0.600	13.71
0.700	15.49
0.800	17.91

(a) Construct a spreadsheet and use it to draw a calibration curve for NADH.

*(b) Find the least-squares slope and intercept for the plot in (a).

(c) Calculate the standard deviation of the slope and the standard deviation about regression for the curve.

*(d) An unknown exhibits a relative fluorescence of 12.16. Use the spreadsheet to calculate the concentration of NADH.

*(e) Calculate the relative standard deviation for the result in part (d).

(f) Calculate the relative standard deviation for the result in part (d) if the reading of 12.16 was the mean of three measurements.

27-11. The following volumes of a solution containing 1.10 ppm of Zn^{2+} were pipetted into separatory funnels each containing 5.00 mL of an unknown zinc solution: 0.00, 1.00, 4.00, 7.00, and 11.00. Each was extracted with three 5-mL aliquots of CCl_4 containing an excess of 8-hydroxyquinoline. The extracts were then diluted to 25.0 mL and their fluorescence measured with a fluorometer. The results were as follows:

Volume Std Zn^{2+}, mL	Fluorometer Reading
0.000	6.12
4.00	11.16
8.00	15.68
12.00	20.64

(a) Construct a working curve from the data.

(b) Calculate a linear least-squares equation for the data.

(c) Calculate the standard deviation of the slope and the standard deviation about regression.

(d) Calculate the concentration of zinc in the sample.

(e) Calculate a standard deviation for the result in part (d).

***27-12.** Quinine in a 1.664-g antimalarial tablet was dissolved in sufficient 0.10 M HCl to give 500 mL of solution. A 15.00-mL aliquot was then diluted to 100.0 mL with the acid. The fluorescence intensity for the diluted sample at 347.5 nm provided a reading of 288 on an arbitrary scale. A standard 100-ppm quinine solution registered 180 when measured under conditions identical to those for the diluted sample. Calculate the mass in milligrams of quinine in the tablet.

27-13. The determination in Problem 27-12 was modified to use the standard addition method. As before, a 2.196-g tablet was dissolved in sufficient 0.10 M HCl to give 1.000 L. Dilution of a 20.00-mL aliquot to 100 mL gave a solution with a reading of 540 at 347.5 nm. A second 20.00-mL aliquot was mixed with 10.0 mL of 50 ppm quinine solution before dilution to 100 mL. The fluorescence intensity of this solution was 600. Calculate the concentration in parts per million of quinine in the tablet.

27-14. Challenge Problem. The following volumes of a standard 10.0-ppb F^- solution were added to four 10.00-mL aliquots of a water sample: 0.00, 1.00, 2.00, and 3.00 mL. Precisely 5.00 mL of a solution containing an excess of the strongly absorbing Al-acid Alizarin Garnet R complex was added to each of the four solutions, and they were each diluted to 50.0 mL. The fluorescence intensity of the four solutions were as follows:

V_s, mL	Meter reading
0.00	68.2
1.00	55.3
2.00	41.3
3.00	28.8

(a) Explain the chemistry of the analytical method.

(b) Construct a plot of the data.

(c) Use the fact that the fluorescence decreases with increasing amounts of the F^- standard to derive a relationship like Equation 26-1 for multiple standard additions. Use that relationship further to obtain an equation for the unknown concentration c_x in terms of the slope and intercept of the standard additions plot, similar to Equation 26-2.

(d) Use linear least-squares to find the equation for the line representing the decrease in fluorescence to the volume of standard fluoride V_s.
(e) Calculate the standard deviation of the slope and intercept.
(f) Calculate the concentration of F^- in the sample in ppb.
(g) Calculate the standard deviation of the result in (e).

InfoTrac College Edition

For additional readings, go to InfoTrac College Edition, your online research library, at

http://infotrac.thomsonlearning.com

CHAPTER 28

Atomic Spectroscopy

Water pollution remains a serious problem in the United States and in other industrial countries. The photo shows land left over after strip mining in Belmont County, Ohio. The various water pools shown are contaminated with waste chemicals. The large pool to the right of center contains sulfuric acid. The smaller pools contain manganese and cadmium. Trace metals in contaminated water samples are often determined by a multielement technique such as inductively coupled plasma mass spectrometry or inductively coupled plasma atomic emission spectroscopy. Both these methods are discussed in this chapter.

Atomic spectroscopic methods are used for the qualitative and quantitative determination of more than 70 elements. Typically, these methods can detect parts-per-million to parts-per-billion amounts, and, in some cases, even smaller concentrations. Atomic spectroscopic methods are, in addition, rapid, convenient, and usually of high selectivity. They can be divided into two groups: ***optical atomic spectrometry***[1] *and* ***atomic mass spectrometry.***[2]

Spectroscopic determination of atomic species can only be performed on a gaseous medium in which the individual atoms or elementary ions, such as Fe^+, Mg^+, or Al^+, are well separated from one another. Consequently, the first step in all atomic spectroscopic procedures is ***atomization,*** *a process in which a sample is volatilized and decomposed in such a way as to produce gas-phase atoms and ions. The efficiency and reproducibility of the atomization step can have a large influence on the sensitivity, precision, and accuracy of the method. In short, atomization is a critical step in atomic spectroscopy.*

Atomization is a process in which a sample is converted into gas-phase atoms or elementary ions.

As shown in Table 28-1, several methods are used to atomize samples for atomic spectroscopic studies. Inductively coupled plasmas, flames, and electrothermal atomizers are the most widely used atomization methods; we consider these three methods as well as direct current plasmas in this chapter. Flames and electrothermal atomizers are widely used in atomic absorption spectrometry, while the inductively coupled plasma is employed in optical emission and in atomic mass spectrometry.

[1]References that deal with the theory and applications of optical atomic spectroscopy include Jose A. C. Broekaert, *Analytical Atomic Spectrometry with Flames and Plasmas.* Weinheim: Cambridge: Wiley-VCH, 2002; L. H. J. Lajunen, *Spectrochemical Analysis by Atomic Absorption and Emission.* Cambridge: Royal Society of Chemistry, 1992; J. D. Ingle Jr., and S. R. Crouch, *Spectrochemical Analysis,* Chapters 7–11. Upper Saddle River, NJ, 1988.

[2]References that deal with atomic mass spectrometry include *Inductively Coupled Plasma Mass Spectrometry,* A. Montaser, Ed. New York: Wiley, 1998; H. E. Taylor, *Inductively Coupled Plasma-Mass Spectrometry: Practices and Techniques.* San Diego: Academic Press, 2000.

TABLE 28-1
Classification of Atomic Spectroscopic Methods

Atomization Method	Typical Atomization Temperature, °C,	Types of Spectroscopy	Common Name and Abbreviation
Inductively coupled plasma	6000–8000	Emission	Inductively coupled plasma atomic emission spectroscopy, ICPAES
		Mass	Inductively coupled plasma mass spectrometry, ICP-MS
Flame	1700–3150	Absorption	Atomic absorption spectroscopy, AAS
		Emission	Atomic emission spectroscopy, AES
		Fluorescence	Atomic fluorescence spectroscopy, AFS
Electrothermal	1200–3000	Absorption	Electrothermal AAS
		Fluorescence	Electrothermal AFS
Direct-current plasma	5000–10,000	Emission	DC plasma spectroscopy, DCP
Electric arc	3000–8000	Emission	Arc-source emission spectroscopy
Electric spark	Varies with time and position	Emission	Spark-source emission spectroscopy
		Mass	Spark-source mass spectroscopy

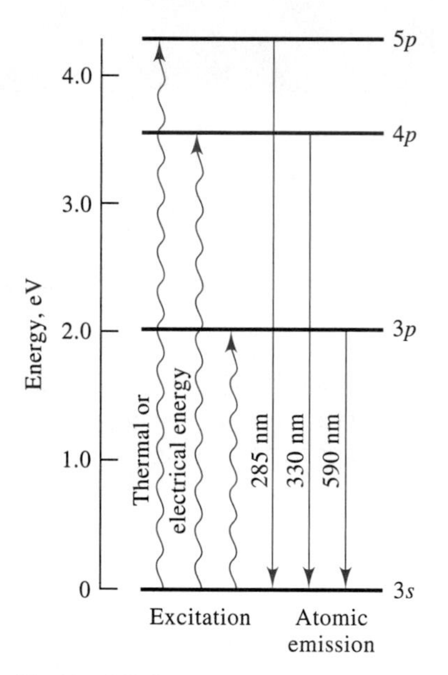

Figure 28-1 Origin of three sodium emission lines.

► Atomic *p* orbitals are in fact split into two energy levels that differ only slightly in energy. The energy difference between the two levels is so small that the emission appears to be a single line, as suggested by Figure 28-1. With a very-high-resolution spectrometer, each of the lines appears as two closely spaced lines known as a **doublet.**

28A ORIGINS OF ATOMIC SPECTRA

Once the sample has been converted into gaseous atoms or elementary ions, various types of spectroscopy can be performed. We consider here optical and mass spectrometric methods.

28A-1 Origins of Optical Spectra

With gas-phase atoms or ions, there are no vibrational or rotational energy states. This means that only electronic transitions occur. Thus, atomic emission, absorption, and fluorescence spectra are made up of a limited number of narrow **spectral lines.**

Emission Spectra

In atomic emission spectroscopy, analyte atoms are excited by external energy in the form of heat or electrical energy, as illustrated in Figure 24-4. The energy typically is supplied by a plasma, a flame, a low-pressure discharge, or a high-powered laser. Figure 28-1 is a partial energy-level diagram for atomic sodium showing the source of three of the most prominent emission lines. Before the external energy source is applied, the sodium atoms are usually in their lowest energy or **ground state.** The applied energy then causes the atoms to be momentarily in a higher energy or **excited state.** With sodium atoms in the ground state, the single valence electrons are in the 3*s* orbital. External energy promotes the outer electrons from their ground-state 3*s* orbitals to 3*p*, 4*p*, or 5*p* excited-state orbitals. After a few nanoseconds, the excited atoms relax to the ground state, giving up their energy as photons of visible or ultraviolet radiation. As shown at the right of the figure, the wavelength of the emitted radiation is 590, 330, and 285 nm. A transition to or from the ground state is called a **resonance transition,** and the resulting spectral line is called a **resonance line.**

Absorption Spectra

In atomic absorption spectroscopy, an external source of radiation impinges on the analyte vapor, as illustrated in Figure 24-5. If the external source radiation is of the appropriate frequency (wavelength), it can be absorbed by the analyte atoms and promote them to excited states. Figure 28-2a shows three of several absorption lines for sodium vapor. The source of these spectral lines is indicated in the partial energy diagram shown in Figure 28-2b. Here, absorption of radiation of 285, 330, and 590 nm excites the single outer electron of sodium from its ground-state 3*s* energy level to the excited 3*p*, 4*p*, and 5*p* orbitals, respectively. After a few nanoseconds, the excited atoms relax to their ground state by transferring their excess energy to other atoms or molecules in the medium.

◀ Note that the wavelengths of the absorption and emission lines for sodium are identical.

The absorption and emission spectra for sodium are fairly simple and consist of relatively few lines. For elements that have several outer electrons that can be excited, absorption and emission spectra may be much more complex.

Fluorescence Spectra

In atomic fluorescence spectroscopy, an external source is used just as in atomic absorption, as shown in Figure 24-6. Instead of measuring the attenuated source radiant power, however, the radiant power of fluorescence, P_F, is measured, usually at right angles to the source beam. In such experiments, we must avoid or discriminate against scattered source radiation. Atomic fluorescence is often measured at the same wavelength as the source radiation, in which case it is called **resonance fluorescence.**

Widths of Atomic Spectral Lines

Atomic spectral lines have finite widths. With ordinary measuring spectrometers, the observed line widths are determined not by the atomic system but by the spectrometer properties. With very-high-resolution spectrometers or with interferometers, the actual widths of spectral lines can be measured. Several factors contribute to atomic spectral line widths.

Natural Broadening The natural line width of an atomic spectral line is determined by the lifetime of the excited state and Heisenberg's uncertainty principle. The shorter the lifetime, the broader the line, and vice versa. Typical radiative lifetimes of atoms are on the order of 10^{-8} s, which leads to natural line widths on the order of 10^{-5} nm.

Collisional Broadening Collisions between atoms and molecules in the gas phase lead to deactivation of the excited state and thus broadening of the spectral line. The amount of broadening increases with the concentration (pressure) of the collision partners. As a result, collisional broadening is sometimes called **pressure broadening.** Pressure broadening increases with increasing temperature. Collisional broadening is highly dependent on the gaseous medium. For Na atoms in flames, such broadening can be as large as 3×10^{-3} nm. In energetic media, collisional broadening greatly exceeds natural broadening.

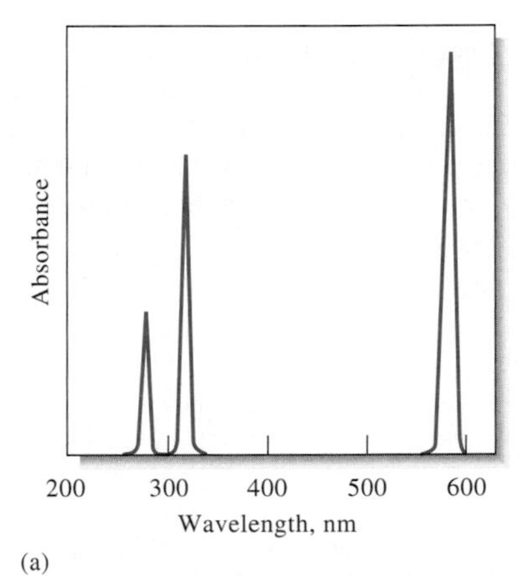

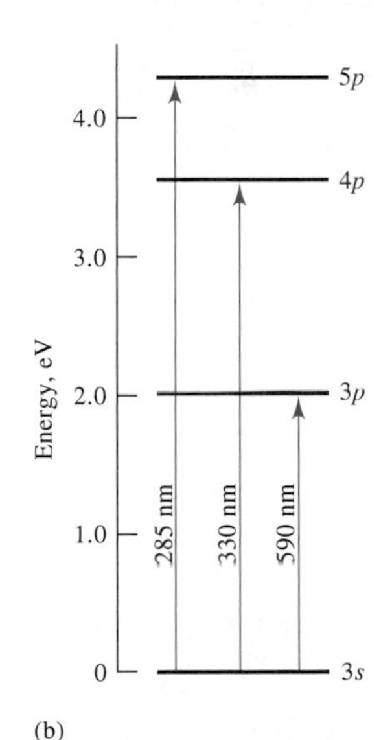

Figure 28-2 (a) Partial absorption spectrum for sodium vapor. (b) Electronic transitions responsible for the absorption lines in (a).

Doppler Broadening Doppler broadening results from the rapid motion of atoms as they emit or absorb radiation. Atoms moving toward the detector emit wavelengths that are slightly shorter than the wavelengths emitted by atoms moving at right angles to the detector. This difference is a manifestation of the

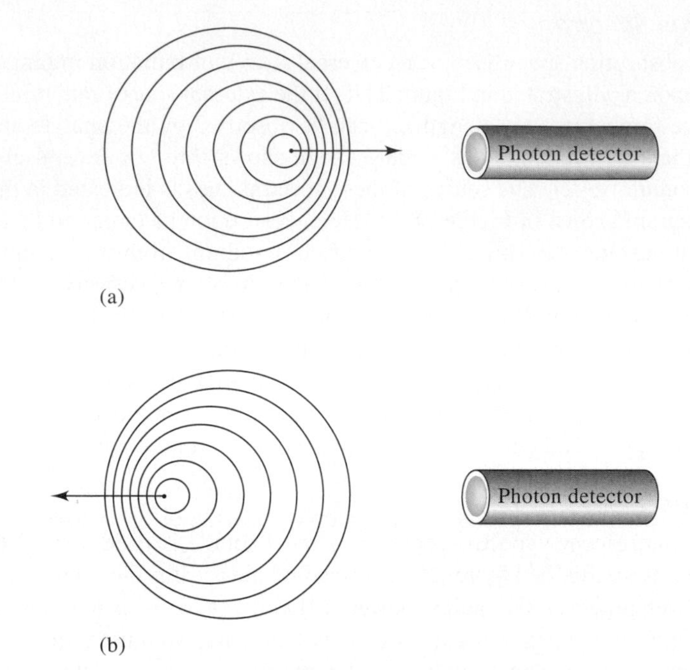

Figure 28-3 Cause of Doppler broadening. (a) When an atom moves toward a photon detector and emits radiation, the detector sees wave crests more often and detects radiation of higher frequency. (b) When an atom moves away from a photon detector and emits radiation, the detector sees crests less frequently and detects radiation of lower frequency. The result in an energetic medium is a statistical distribution of frequencies and thus a broadening of the spectral lines.

well-known Doppler shift; the effect is reversed for atoms moving away from the detector. The net effect is an increase in the width of the emission line, as shown in Figure 28-3. For precisely the same reason, the Doppler effect also causes broadening of absorption lines. This type of broadening becomes more pronounced as the flame temperature increases because of the increased velocity of the atoms. Doppler broadening can be a major contributor to overall line widths. For Na, in flames, the Doppler line widths are on the order of 4×10^{-3} to 5×10^{-3} nm.

▶ Both Doppler broadening and pressure broadening are temperature dependent.

28A-2 Mass Spectra

In atomic mass spectrometry, also called elemental mass spectrometry, it is desirable for the sample to be converted to gas-phase ions rather than gas-phase atoms. With energetic atomization sources, such as plasmas, a substantial fraction of the atoms produced are ionized, usually as singly charged positive ions. The ions of different atomic masses are then separated in a device called a **mass analyzer** to produce the mass spectrum. The separation is on the basis of the **mass-to-charge ratio** of the ionic species. Because the ions produced in atomic mass spectrometry are generally singly charged, the mass-to-charge ratio is sometimes shortened to the more convenient term mass. The atomic masses are usually expressed in terms of **atomic mass units** (amu), or daltons (Da).[3] Some ionization sources, particularly those used in molecular mass spectrometry, produce more highly charged species, in which case referring to the separation as mass-based is incorrect. The mass spectrum is a plot of the number of ions produced versus the mass-to-charge ratio or, for singly charged ions, versus mass, as shown in Figure 28-4.

[3]The amu or Da is defined as $^{1}/_{12}$ the mass of one neutral $^{12}_{6}C$ atom.

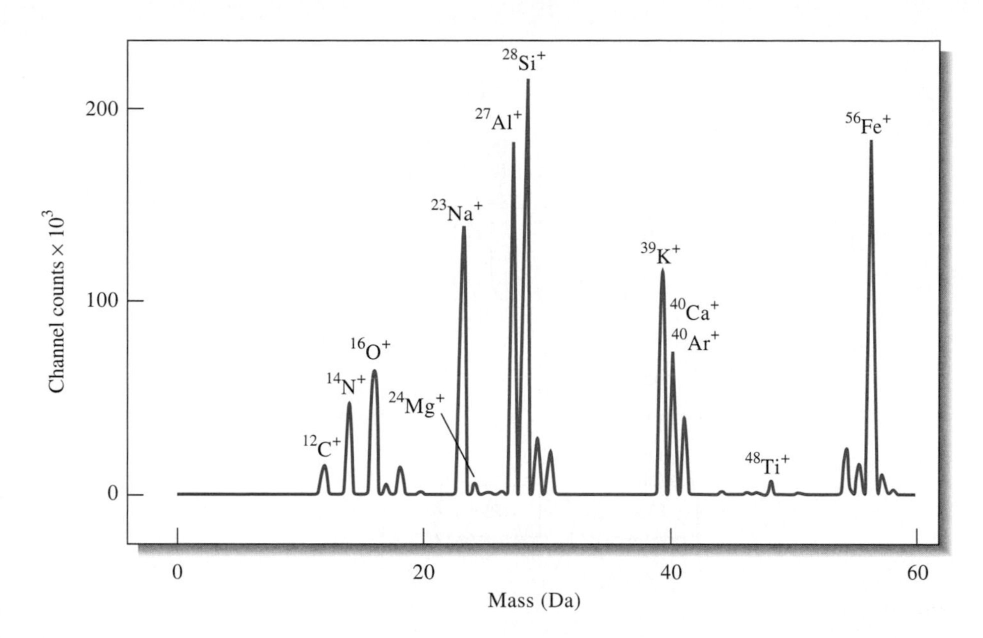

Figure 28-4 The mass spectrum of a standard rock sample obtained by laser ablation/ICP-MS. Major components (%): Na, 5.2; Mg, 0.21; Al, 6.1; Si, 26.3; K, 5.3; Cu, 1.4; Ti, 0.18; and Fe, 4.6. (From *Inorganic Mass Spectrometry,* F. Adams, R. Gijbek, and R. Van Grieken, Eds., p. 297. New York: Wiley, 1988. This material is used by permission of Wiley-Liss, Inc., a subsidiary of John Wiley & Sons, Inc.)

28B PRODUCTION OF ATOMS AND IONS

In all atomic spectroscopic techniques, we must atomize the sample, converting it into gas-phase atoms and ions. Samples are most commonly presented to the atomizer in solution form, although we sometimes introduce gases and solids. Hence, the atomization device must perform the complex task of converting analyte species in solution into gas-phase free atoms or elementary ions, or both.

28B-1 Sample Introduction Systems

Atomization devices fall into two classes; **continuous atomizers** and **discrete atomizers.** With continuous atomizers, such as plasmas and flames, samples are introduced in a continuous manner. With discrete atomizers, samples are introduced in a discrete manner with a device such as a syringe or an autosampler. The most common discrete atomizer is the **electrothermal atomizer.**

The general methods of introducing solution samples into plasma and flames are illustrated in Figure 28-5. Direct **nebulization** is most often used. In this case, the **nebulizer** constantly introduces the sample in the form of a fine spray of droplets, called an **aerosol.** With such continuous sample introduction into a flame or plasma, a steady-state population of atoms, molecules, and ions is produced. When flow injection or liquid chromatography is used, a time-varying plug of sample is nebulized, producing a time-dependent vapor population. The complex processes that must occur to produce free atoms or elementary ions are illustrated in Figure 28-6.

To **nebulize** means to convert a liquid into a fine spray or mist.

An **aerosol** is a suspension of finely divided liquid or solid particles in a gas.

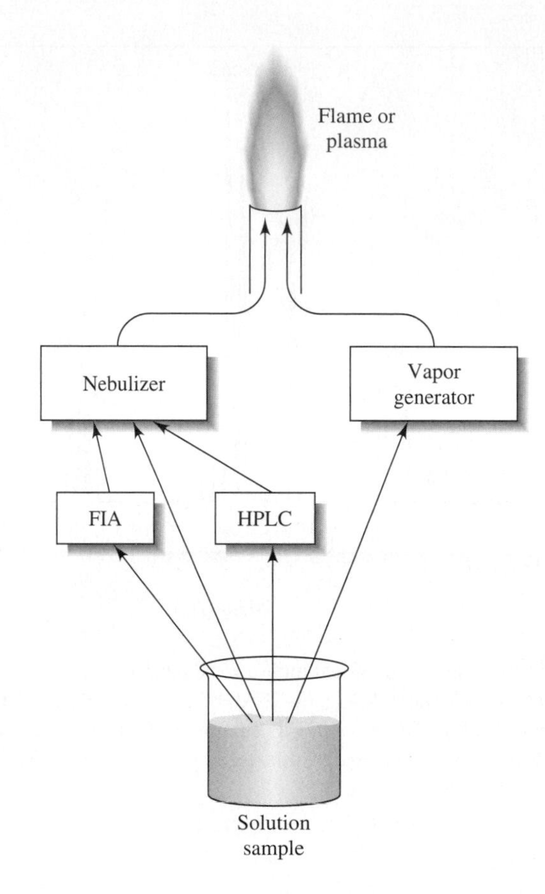

Figure 28-5 Continuous sample introduction methods. Samples are frequently introduced into plasmas or flames by means of a nebulizer, which produces a mist or spray. Samples can be introduced directly to the nebulizer or by means of flow injection (FIA) or high-performance liquid chromatography (HPLC). In some cases, samples are separately converted to a vapor by a vapor generator, such as a hydride generator or an electrothermal vaporizer.

Discrete solution samples are introduced by transferring an aliquot of the sample to the atomizer. The vapor cloud produced with electrothermal atomizers is transient because of the limited amount of sample available.

Solid samples can be introduced into plasmas by vaporizing them with an electrical spark or with a laser beam. Laser volatilization, often called **laser ablation,** has become a popular method to introduce samples into inductively coupled plasmas. Here a high-powered laser beam, usually a Nd:YAG or excimer laser, is directed onto a portion of the solid sample. The sample is then vaporized by radiative heating. The plume of vapor produced is swept into the plasma by means of a carrier gas.

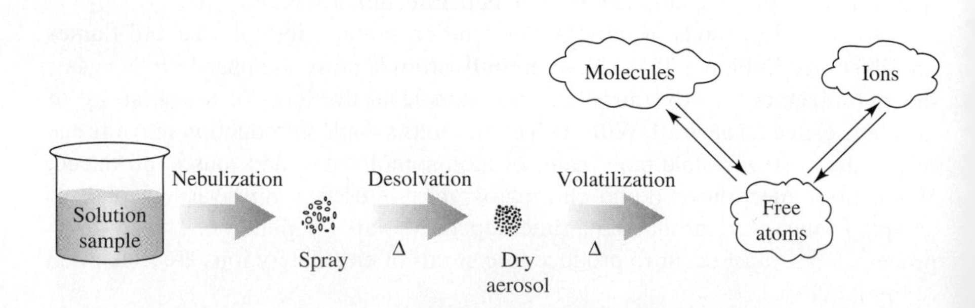

Figure 28-6 Processes leading to atoms, molecules, and ions with continuous sample introduction into a plasma or flame. The solution sample is converted into a spray by the nebulizer. The high temperature of the flame or plasma causes the solvent to evaporate, leaving dry aerosol particles. Further heating volatilizes the particles, producing atomic, molecular, and ionic species. These species are often in equilibrium, at least in localized regions.

28B-2 Plasma Sources

Plasma atomizers, which became available commercially in the mid-1970s, offer several advantages for analytical atomic spectroscopy.[4] Plasma atomization has been used for atomic emission, atomic fluorescence, and atomic mass spectrometry.

By definition, a **plasma** is a conducting gaseous mixture containing a significant concentration of cations and electrons. In the argon plasma used for atomic spectroscopy, argon ions and electrons are the principal conducting species, although cations from the sample also contribute. Argon ions, once formed in a plasma, are capable of absorbing sufficient power from an external source to maintain the temperature at a level at which further ionization sustains the plasma indefinitely; temperatures as great as 10,000 K are encountered.

A **plasma** is a hot, partially ionized gas. It contains relatively high concentrations of ions and electrons.

Three power sources have been employed in argon plasma spectroscopy. One is a dc electrical source capable of maintaining a current of several amperes between electrodes immersed in the argon plasma. The second and third are powerful radio-frequency and microwave-frequency generators through which the argon flows. Of the three, the radio-frequency, or **inductively coupled plasma** (ICP), source offers the greatest advantage in terms of sensitivity and freedom from interference. It is commercially available from a number of instrument companies for use in optical emission and mass spectroscopy. A second source, the **dc plasma source** (DCP), has seen some commercial success and has the virtues of simplicity and lower cost.

Inductively Coupled Plasmas

Figure 28-7 is a schematic drawing of an inductively coupled plasma (ICP) source. It consists of three concentric quartz tubes through which streams of argon flow at a total rate of 11 to 17 L/min. The diameter of the largest tube is about 2.5 cm. Surrounding the top of this tube is a water-cooled induction coil powered by a radio-frequency generator capable of producing about 2 kW of energy at either 27 MHz or 40 MHz. Ionization of the flowing argon is initiated by a spark from a Tesla coil. The resulting ions and their associated electrons then interact with the fluctuating magnetic field (labeled H in Figure 28-7) produced by the induction coil I. This interaction causes the ions and electrons within the coil to flow in the closed annular paths depicted in the figure; ohmic heating is the consequence of their resistance to this movement.

The temperature of the ICP is high enough that it must be thermally isolated from the quartz cylinder. Isolation is achieved by flowing argon tangentially around the walls of the tube, as indicated by the arrows in Figure 28-7. The tangential flow cools the inside walls of the central tube and centers the plasma radially.

Viewing the plasma at right angles, as shown in Figure 28-8a, is called the **radial viewing geometry.** Recent ICP instruments have incorporated an **axial viewing geometry,** shown in Figure 28-8b. Here the torch is turned 90°. The axial geometry was originally popular for torches used as ionization sources for mass

[4]For a detailed discussion of the various plasma sources, see S. J. Hill, *Inductively Coupled Plasma Spectrometry and Its Applications.* Boca Raton, FL: CRC Press, 1999. *Inductively Coupled Plasmas in Analytical Atomic Spectroscopy,* 2nd ed. A. Montaser and D. W. Golightly, Eds. New York: Wiley-VCH Publishers, 1992; *Inductively Coupled Plasma Mass Spectrometry,* A. Montaser, Ed. New York: Wiley, 1998; *Inductively Coupled Plasma Emission Spectroscopy,* Parts 1 and 2, P. W. J. M. Boumans, Ed. New York: Wiley, 1987.

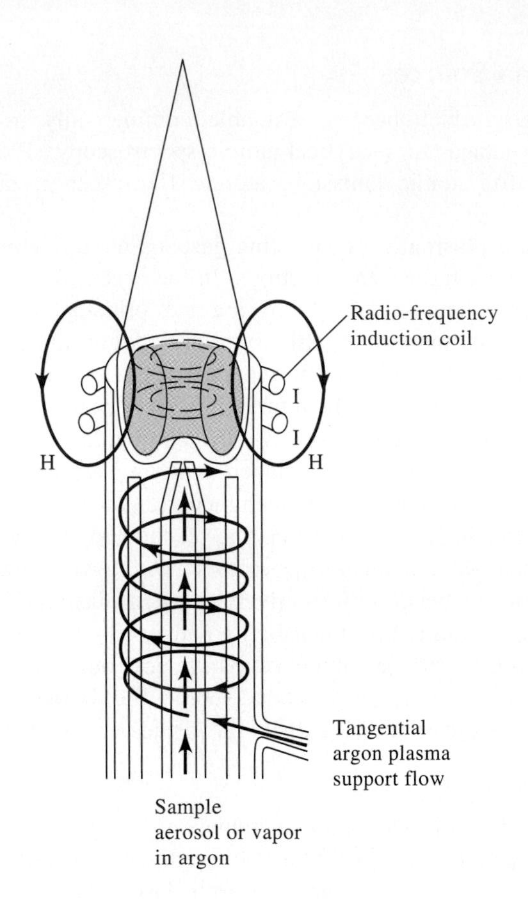

Figure 28-7 Inductively coupled plasma source. (From V. A. Fassel, *Science,* **1978,** *202,* 185. Reprinted with permission. Copyright 1978 by the American Association for the Advancement of Science.)

spectrometry because the ions could be easily extracted from the top of the torch into the high-vacuum region of the mass spectrometer. More recently, axial torches have become available for emission spectrometry. Several companies manufacture torches that can be switched from axial to radial viewing geometry in atomic emis-

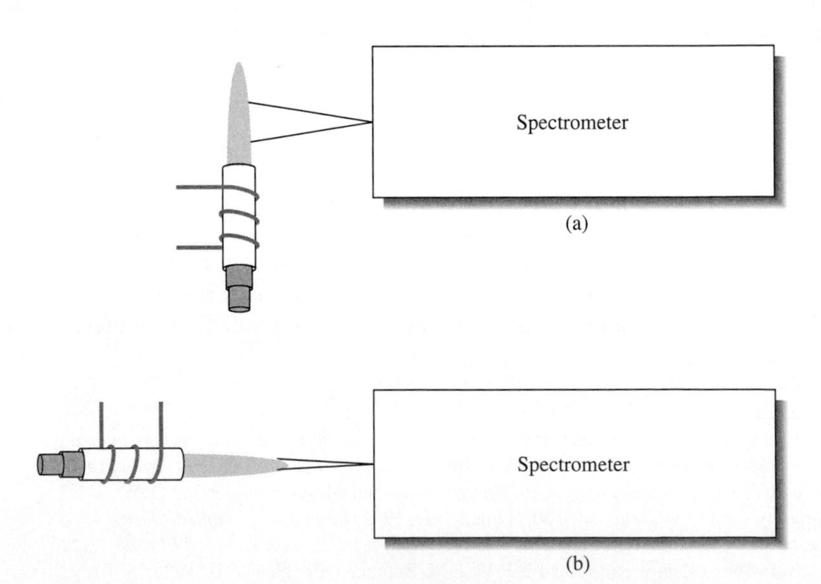

Figure 28-8 Viewing geometries for ICP sources. (a) Radial geometry used in ICP atomic emission spectrometers; (b) axial geometry used in ICP mass spectrometers and in several ICP atomic emission spectrometers.

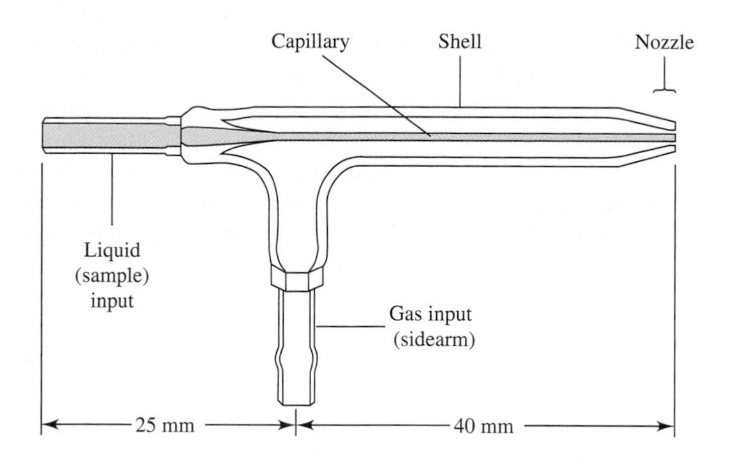

Figure 28-9 The Meinhard nebulizer. The nebulizing gas flows through an opening that surrounds the capillary concentrically. This causes a reduced pressure at the tip and aspiration of the sample. The high-velocity gas at the tip breaks up the solution into a mist or spray of various-sized droplets. (Courtesy of J. Meinhard Associates, Inc.)

sion spectrometry. Radial geometry provides better stability and precision, while axial geometry is used to achieve lower detection limits.

During the 1980s, low-flow, low-power torches appeared on the market. Typically, these torches require a total argon flow of less than 10 L/min and require less than 800 W of radio-frequency power.

Sample Introduction Samples can be introduced into the ICP by argon flowing at about 1 L/min through the central quartz tube. The sample can be an aerosol, a thermally generated vapor, or a fine powder. The most common means of sample introduction is the concentric glass nebulizer shown in Figure 28-9. The sample is transported to the tip by the **Bernoulli effect.** This transport process is called **aspiration.** The high-velocity gas breaks up the liquid into fine droplets of various sizes, which are then carried into the plasma.

Another popular type of nebulizer has a cross-flow design. Here a high-velocity gas flows across a capillary tip at right angles, causing the same Bernoulli effect. Often, in this type of nebulizer, the liquid is pumped through the capillary with a peristaltic pump. Many other types of nebulizers are available for higher efficiency nebulization, for nebulization of samples with high solids content, and for production of ultrafine mists.

Plasma Appearance and Spectra The typical plasma has a very intense, brilliant white opaque core topped by a flame-like tail. The core, which extends a few millimeters above the tube, produces a spectral continuum with the atomic spectrum of argon superimposed. The continuum is typical of ion-electron recombination reactions and **bremsstrahlung,** which is continuum radiation produced when charged particles are slowed or stopped.

In the region 10 to 30 mm above the core, the continuum fades and the plasma becomes slightly transparent. Spectral observations are generally made 15 to 20 mm above the induction coil, where the temperatures can be as high as 5000 to 6000 K. Here, the background radiation consists primarily of Ar lines, OH band emission, and some other molecular bands. Many of the most sensitive analyte lines in this region of the plasma are from ions such as Ca^+, Cd^+, Cr^+, and Mn^+. Above this second region, the "tail flame" can be observed when easily excited elements such as sodium or cesium are introduced. Temperatures in this region are similar to those in an ordinary flame ($\approx$3000 K). This lower temperature region can be used to determine easily excited elements such as alkali metals.

Analyte Atomization and Ionization By the time the analyte atoms and ions reach the observation point in the plasma, they have spent about 2 ms in the plasma at temperatures ranging from 6000 to 8000 K. These times and temperatures are two to three times greater than those attainable in the hottest combustion flames (acetylene/nitrous oxide). As a consequence, desolvation and vaporization are essentially complete, and the atomization efficiency is quite high. Therefore, there are fewer chemical interferences in ICPs than in combustion flames. Surprisingly, ionization interference effects are small or nonexistent because the large concentration of electrons from the ionization of argon maintains a more-or-less constant electron concentration in the plasma.

Several other advantages are associated with the ICP when compared with flames and other plasma sources. Atomization occurs in a chemically inert environment, in contrast to flames, where the environment is violent and highly reactive. In addition, the temperature cross-section of the plasma is relatively uniform. The plasma also has a rather thin optical path length, which minimizes self-absorption (see Section 28C-2). As a consequence, calibration curves are usually linear over several orders of magnitude of concentration. Ionization of analyte elements can be significant in typical ICPs. This has led to the use of the ICP as an ionization source for mass spectrometry, as discussed in Section 28F. One significant disadvantage of the ICP is that it is not very tolerant of organic solvents. Carbon deposits tend to build up on the quartz tube, which can lead to cross-contamination and clogging.

Direct Current and Other Plasma Sources

Direct-current plasma jets were first described in the 1920s and have been systematically investigated as sources for emission spectroscopy. In the early 1970s, the first commercial direct current plasma (DCP) was introduced. The source was quite popular, particularly among soil scientists and geochemists for multielement analysis.

Figure 28-10 is a diagram of a commercially available dc plasma source for the excitation of emission spectra. This plasma-jet source consists of three electrodes arranged in an inverted Y configuration. A graphite anode is located in each arm of the Y, and a tungsten cathode is located at the inverted base. Argon flows from the two anode blocks toward the cathode. The plasma jet is formed when the cathode is momentarily brought into contact with the anodes. Ionization of the argon occurs, and the current that develops ($\approx$14 A) generates additional ions to sustain itself indefinitely. The temperature is more than 8000 K in the arc core and about 5000 K in the viewing region. The sample is aspirated into the area between the two arms of the Y, where it is atomized and excited and its spectrum is viewed.

Spectra produced by the DCP tend to have fewer lines than those produced by the ICP, and the lines formed in the DCP are largely from atoms rather than ions. Sensitivities achieved with the DCP appear to range from an order of magnitude lower than to about the same as those obtainable with the ICP. The reproducibilities of the two systems are similar. Significantly less argon is required for the dc plasma, and the auxiliary power supply is simpler and less expensive. The DCP is able to handle organic solutions and aqueous solutions with a high solids content better than the ICP. Sample volatilization is often incomplete with the DCP, however, because of the short residence times in the high-temperature region. Also, the optimum viewing region with the DCP is quite small, so optics have to be carefully aligned to magnify the source image. In addition, the graphite electrodes must be replaced every few hours, whereas the ICP requires little maintenance.

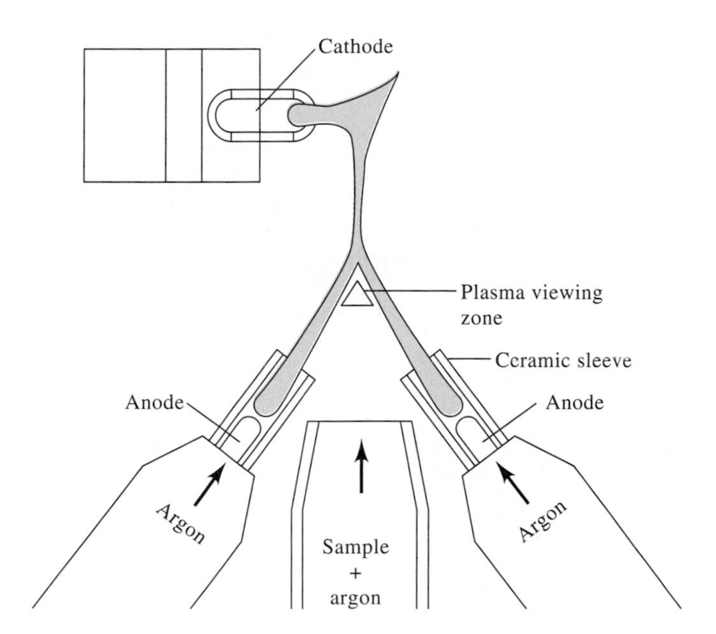

Figure 28-10 Diagram of a three-electrode dc plasma jet. Two separate dc plasmas have a single common cathode. The overall plasma burns in the form of an upside-down Y. Sample can be introduced as an aerosol from the area between the two graphite anodes. Observation of emission in the region beneath the strongly emitting plasma core avoids much of the plasma background emission.

28B-3 Flame Atomizers

A flame atomizer contains a pneumatic nebulizer, which converts the sample solution into a mist, or aerosol, that is then fed into a burner. The same type of nebulizers that are used with ICPs are used with flame atomizers. The concentric nebulizer is the most popular. In most atomizers, the high-pressure gas is the oxidant, and the aerosol containing oxidant is subsequently mixed with the fuel.

The burners used in flame spectroscopy are most often premixed, laminar flow burners. Figure 28-11 is a diagram of a typical commercial laminar-flow burner for atomic absorption spectroscopy that employs a concentric tube nebulizer. The aerosol flows into a **spray chamber,** where it encounters a series of baffles that remove all but the finest droplets. As a result, most of the sample collects in the bottom of the spray chamber, where it is drained to a waste container. Typical solution flow rates are 2 to 5 mL/min. The sample spray is also mixed with fuel and oxidant gas in the spray chamber. The aerosol, oxidant, and fuel are then burned in a slotted burner, which provides a flame that is usually 5 or 10 cm in length.

Laminar flow burners of the type shown in Figure 28-11 provide a relatively quiet flame and a long path length. These properties tend to enhance sensitivity for atomic absorption and reproducibility. The mixing chamber in this type of burner contains a potentially explosive mixture, which can be ignited by flashback if the flow rates are not sufficient. Note that the burner in Figure 28-11 is equipped with pressure relief vents for this reason.

◀ Modern flame atomic absorption instruments use laminar flow burners almost exclusively.

Properties of Flames

When a nebulized sample is carried into a flame, desolvation of the droplets occurs in the **primary combustion zone,** which is located just above the tip of the burner, as shown in Figure 28-12. The resulting finely divided solid particles are carried to a region in the center of the flame called the **inner cone.** Here, in this hottest part of the flame, the particles are vaporized and converted to gaseous atoms, elementary ions, and molecular species (see Figure 28-6). Excitation of atomic emission spectra also takes place in this region. Finally, the atoms, molecules, and ions

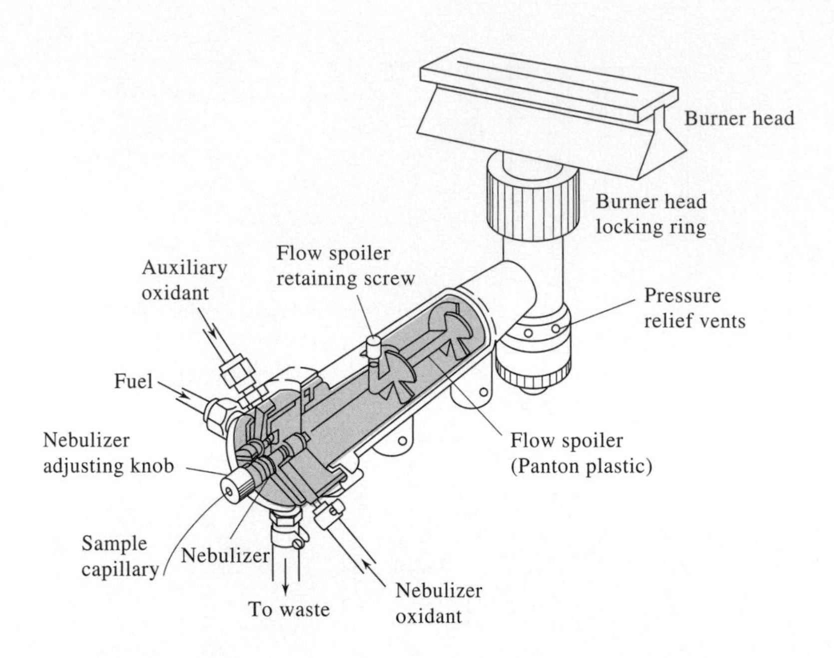

Figure 28-11 A laminar-flow burner used in flame atomic absorption spectroscopy. (Courtesy of Perkin-Elmer Corporation, Norwalk, CT.)

are carried to the outer edge, or **outer cone,** where oxidation may occur before the atomization products disperse into the atmosphere. Because the velocity of the fuel/oxidant mixture through the flame is high, only a fraction of the sample undergoes all these processes; indeed, a flame is not a very efficient atomizer.

Types of Flames Used in Atomic Spectroscopy

Table 28-2 lists the common fuels and oxidants employed in flame spectroscopy and the approximate range of temperatures realized with each of these mixtures. Note that temperatures of 1700°C to 2400°C are obtained with the various fuels when air serves as the oxidant. At these temperatures, only easily excitable species such as the alkali and alkaline earth metals produce usable emission spectra. For

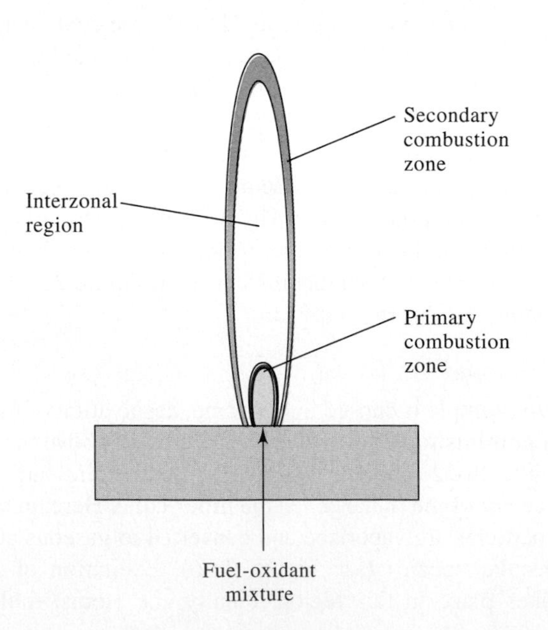

Figure 28-12 Regions of a flame.

heavy metal species, which are less readily excited, oxygen or nitrous oxide must be employed as the oxidant. These oxidants produce temperatures of 2500°C to 3100°C with common fuels.

TABLE 28-2

Flames Used in Atomic Spectroscopy

Fuel and Oxidant	Temperature, °C
*Gas/Air	1700–1900
*Gas/O_2	2700–2800
H_2/air	2000–2100
H_2/O_2	2500–2700
†C_2H_2/air	2100–2400
†C_2H_2/O_2	3050–3150
†C_2H_2/N_2O	2600–2800

*Propane or natural gas
†Acetylene

Effects of Flame Temperature

Both emission and absorption spectra are affected in a complex way by variations in flame temperature. In both cases, higher temperatures increase the total atom population of the flame and thus the sensitivity. With certain elements, such as the alkali metals, however, this increase in atom population is more than offset by the loss of atoms by ionization.

To a large extent, flame temperature determines the efficiency of atomization—that is, the fraction of the analyte that is desolvated, vaporized, and converted to free atoms or ions, or both. The flame temperature also determines the relative number of excited and unexcited atoms in a flame. In an air/acetylene flame, for example, the ratio of excited to unexcited magnesium atoms can be computed to be about 10^{-8}, whereas in an oxygen/acetylene flame, which is about 700°C hotter, this ratio is about 10^{-6}. Hence, from an excitation standpoint, control of temperature is very important in flame emission methods. For example, with a 2500°C flame, a temperature increase of 10°C causes the number of sodium atoms in the excited $3p$ state to increase by about 3%. In contrast, the corresponding *decrease* in the much larger number of ground-state atoms is only about 0.002%. Therefore, at first glance, emission methods, based as they are on the population of *excited atoms,* require much closer control of flame temperature than do absorption procedures, in which the analytical signal depends on the number of *unexcited atoms.* In practice, however, because of the temperature dependence of the atomization step, both methods show similar dependencies.

The number of unexcited atoms in a typical flame exceeds the number of excited atoms by a factor of 10^3 to 10^{10} or more. This suggests that absorption methods should show lower detection limits than emission methods. In fact, however, several other variables also influence detection limits, and the two methods tend to complement each other in this regard. Table 28-3 illustrates this point.

Absorption and Emission Spectra in Flames

Both atomic and molecular emission and absorption can be measured when a sample is atomized in a flame. A typical flame-emission spectrum was shown in Figure 24-19. Atomic emissions in this spectrum are made up of narrow lines, such as that for sodium at about 330 nm, potassium at approximately 404 nm, and calcium at 423 nm. Atomic spectra are thus called **line spectra.** Also present are emission bands that result from excitation of molecular species such as MgOH, MgO, CaOH, and OH. Here, vibrational transitions superimposed on electronic transitions produce

◀ The width of atomic emission lines in flames is on the order of 10^{-3} nm. The width can be measured with an interferometer.

TABLE 28-3

Comparison of Detection Limits for Various Elements by Flame Atomic Absorption and Flame Atomic Emission Methods*

Flame Emission Shows Lower DLs	DLs About the Same	AA Shows Lower DLs
Al, Ba, Ca, Eu, Ga, Ho, In, K, La, Li, Lu, Na, Nd, Pr, Rb, Re, Ru, Sm, Sr, Tb, Tl, Tm, W, Yb	Cr, Cu, Dy, Er, Gd, Ge, Mn, Mo, Nb, Pd, Rh, Sc, Ta, Ti, V, Y, Zr	Ag, As, Au, B, Be, Bi, Cd, Co, Fe, Hg, Ir, Mg, Ni, Pb, Pt, Sb, Se, Si, Sn, Te, Zn

*Adapted with permission from E. E. Pickett and S. R. Koirtyohann, *Anal. Chem.*, **1969**, *41,* 42A. Copyright American Chemical Society.

closely spaced lines that are not completely resolved by the spectrometer. Because of this, molecular spectra are often referred to as **band spectra.**

Atomic absorption spectra are seldom recorded because a high-resolution spectrometer or an interferometer would be required. Such spectra have much the same general appearance as Figure 24-19, with both atomic and molecular absorption components. The vertical axis in this case is absorbance rather than relative power.

Ionization in Flames

All elements ionize to some degree in a flame, which leads to a mixture of atoms, ions, and electrons in the hot medium. For example, when a sample containing barium is atomized, the equilibrium

► Ionization of an atomic species in a flame is an equilibrium process that can be treated by the law of mass action.

$$Ba \rightleftharpoons Ba^+ + e^-$$

is established in the inner cone of the flame. The position of this equilibrium depends on the temperature of the flame and the total concentration of barium as well as on the concentration of the electrons produced from the ionization of *all elements* present in the sample. At the temperatures of the hottest flames (>3000 K), nearly half of the barium is present in ionic form. The emission and absorption spectra of Ba and Ba^+ are, however, totally different from one another. Thus, in a high-temperature flame, two spectra for barium appear: one for the atom and one for its ion. Flame temperature again plays an important role in determining the fraction of the analyte ionized.

► The spectrum of an atom is entirely different from that of its ion.

28B-4 Electrothermal Atomizers

Electrothermal atomizers, which first appeared on the market in about 1970, generally provide enhanced sensitivity because the entire sample is atomized in a short period and the average residence time of the atoms in the optical path is a second or more.[5] Also, samples are introduced into a confined-volume furnace, which means that they are not diluted nearly as much as they would be in a plasma or flame. Electrothermal atomizers are used for atomic absorption and atomic fluorescence measurements but have not been applied generally to emission work. They are, however, used to vaporize samples in inductively coupled plasma emission spectroscopy.

With electrothermal atomizers, a few microliters of sample are first deposited in the furnace with a syringe or an autosampler. Next a programmed series of heating events occurs. The steps are **drying, ashing,** and **atomization.** During the drying step, the sample is evaporated at a relatively low temperature, usually 110°C. The temperature is then increased to 300°C to 1200°C, and the organic matter is ashed or converted to H_2O and CO_2. After ashing, the temperature is rapidly increased to perhaps 2000°C to 3000°C, which causes the sample to vaporize and atomize; atomization of the sample occurs in a period of a few milliseconds to seconds. The absorption or fluorescence of the atomized particles is then measured in the region immediately above the heated surface.

[5]For detailed discussions of electrothermal atomizers, see B. E. Erickson, *Anal. Chem.*, **2000,** *72,* 543A; *Electrothermal Atomization for Analytical Atomic Spectrometry,* K. W. Jackson, Ed. New York: Wiley, 1999; D. J. Buther and J. Sneddon, *A Practical Guide to Graphite Furnace Atomic Absorption Spectrometry.* New York: Wiley, 1998; C. W. Fuller, *Electrothermal Atomization for Atomic Absorption Spectroscopy.* London: The Chemical Society, 1978.

Atomizer Designs

Commercial electrothermal atomizers are small, electrically heated tubular furnaces. Figure 28-13a is a cross-sectional view of a commercial electrothermal atomizer. Atomization occurs in a cylindrical graphite tube that is open at both ends and has a central hole for introduction of sample. The tube is about 5 cm long and has an internal diameter of somewhat less than 1 cm. The interchangeable graphite tube fits snugly into a pair of cylindrical graphite electrical contacts located at the two ends of the tube. These contacts are held in a water-cooled metal housing. Two inert gas streams are provided. The external stream prevents the entrance of outside air and consequent incineration of the tube. The internal stream flows into the two ends of the tube and out the central sample port. This stream not only excludes air but also serves to carry away vapors generated from the sample matrix during the first two heating stages.

Figure 28-13b illustrates the L'vov platform, which is often used in graphite furnaces. The platform is also graphite and is located beneath the sample entrance port. The sample is evaporated and ashed on this platform in the usual way. When the tube temperature is raised rapidly, however, atomization is delayed, since the sample is no longer directly on the furnace wall. As a consequence, atomization occurs in an environment in which the temperature is not changing so rapidly. More reproducible signals are obtained as a result.

Several other designs of electrothermal atomizers are available commercially.

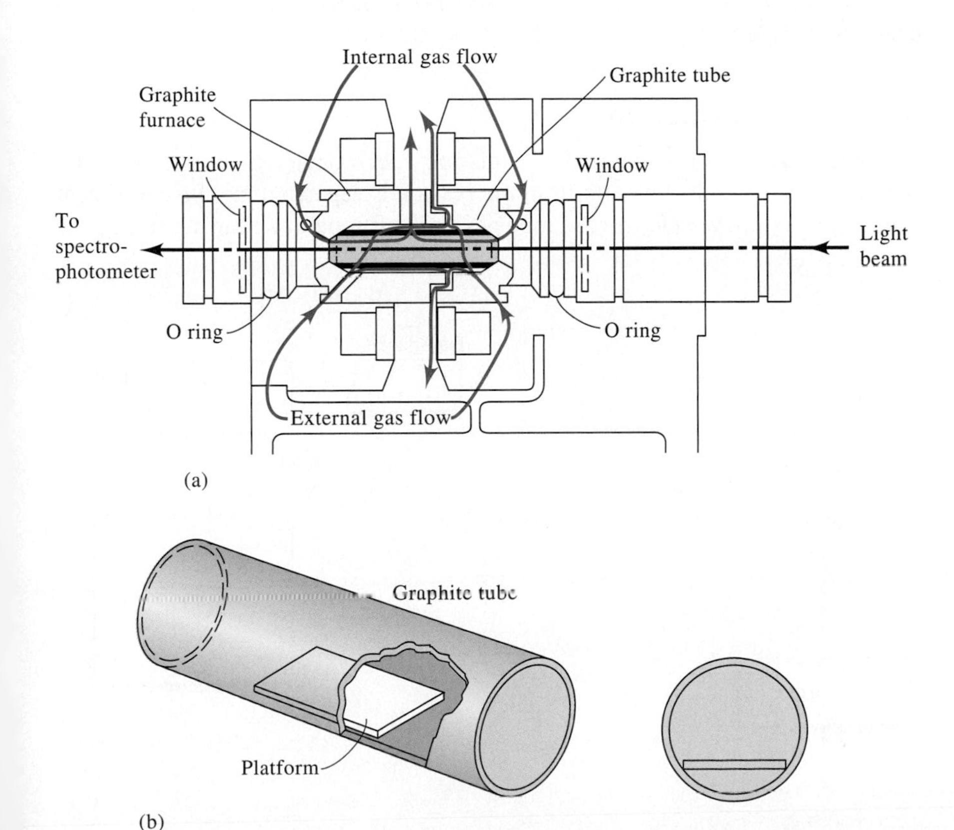

Figure 28-13 (a) Cross-sectional view of a graphite furnace atomizer. (b) The L'vov platform and its position in the graphite furnace. (Part a, courtesy of the Perkin-Elmer Corp., Norwalk, CT; part b, reprinted with permission from W. Slavin, *Anal. Chem.*, **1982,** *54,* 689A. Copyright 1982, American Chemical Society.)

Output Signals

The output signals in electrothermal atomic absorption are transient, not the steady-state signals seen with flame atomization. The atomization step produces a pulse of atomic vapor that lasts only a few seconds at most. The absorbance of the vapor is measured during this stage.

28B-5 Other Atomizers

Many other types of atomization devices have been used in atomic spectroscopy. Gas discharges operated at reduced pressure have been investigated as sources of atomic emission and as ion sources for mass spectrometry. The **glow discharge** is generated between two planar electrodes in a cylindrical glass tube filled with gas to a pressure of a few torr. High-powered lasers have been employed to ablate samples and to cause **laser-induced breakdown.** In the latter technique, dielectric breakdown of a gas occurs at the laser focal point.

A **dielectric** is a material that does not conduct electricity. By applying high voltages or radiation from a high-powered laser, a gas can be made to break down into ions and electrons, a phenomenon known as **dielectric breakdown.**

In the early days of atomic spectroscopy, dc and ac arcs and high-voltage sparks were popular for use in excitation of atomic emission. Such sources have almost entirely been replaced by the ICP.

28C ATOMIC EMISSION SPECTROMETRY

Atomic emission spectrometry is widely used in elemental analysis. The ICP is now the most popular source for emission spectrometry, although the DCP and flames are still used in some situations.

28C-1 Instrumentation

The block diagram of a typical ICP emission spectrometer is shown in Figure 28-14. Atomic or ionic emission from the plasma is separated into its constituent wavelengths by the wavelength isolation device. This separation can take place in a

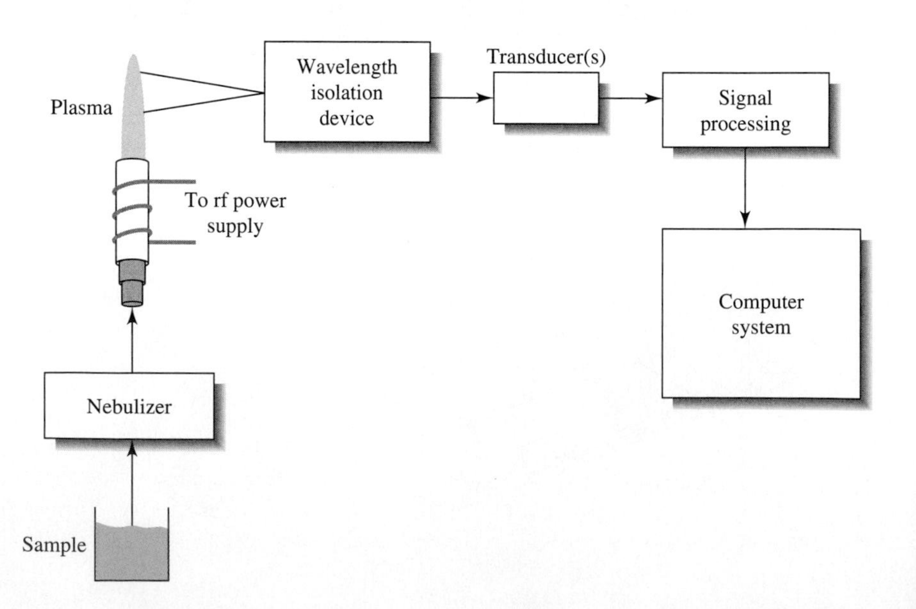

Figure 28-14 Block diagram of a typical ICP atomic emission spectrometer.

monochromator, a **polychromator,** or a **spectrograph.** The monochromator isolates one wavelength at a time at a single exit slit, while a polychromator isolates several wavelengths simultaneously at multiple exit slits. The spectrograph provides a large aperture at its output to allow a range of wavelengths to exit. The isolated radiation is converted into electrical signals by a single transducer, multiple transducers, or an array detector. The electrical signals are then processed and provided as input to the computer system.

Flame emission spectrometers and DCP emission spectrometers follow the same block diagram except that a flame or DCP is substituted for the ICP of Figure 28-14. Flame spectrometers most often isolate a single wavelength, while DCP spectrometers may isolate multiple wavelengths with a polychromator.

Wavelength Isolation

Emission spectrometry is often used for multielement determinations. There are two types of instruments generally available for this purpose. The **sequential spectrometer** uses a monochromator and scans to different emission lines in sequence. Usually, the wavelengths to be used are set by the user in a computer program, and the monochromator rapidly slews from one wavelength to the next. Alternatively, monochromators can scan a range of wavelengths. True **simultaneous spectrometers** use polychromators or spectrographs. The **direct reading spectrometer** uses a polychromator with as many as 64 detectors located at exit slits in the focal plane. Several modern spectrometers use spectrographs and one or more array detectors to monitor multiple wavelengths simultaneously. Some can even combine a scanning function with a spectrograph function to bring different wavelength regions to an array detector. The dispersive devices in these spectrometers can be gratings, grating/prism combinations, or echelle gratings. Simultaneous instruments are usually more expensive than sequential systems.

For routine flame-emission determinations of alkali metals and alkaline earth elements, simple filter photometers often suffice. A low-temperature flame is employed to prevent excitation of most other metals. As a consequence, the spectra are simple, and interference filters can be used to isolate the desired emission lines. Flame emission was once widely used in the clinical laboratory for the determination of sodium and potassium. These methods have largely been replaced by methods using ion-selective electrodes (see Section 21D).

Radiation Transducers

Single-wavelength instruments most often use photomultiplier transducers, as do direct reading spectrometers. The charge coupled device (CCD) has now become very popular as an array detector for simultaneous and some sequential spectrometers. Such devices are available with more than 1 million pixels to allow a fairly wide wavelength coverage. One commercial instrument uses a segmented array, charge coupled device detector to allow more than one wavelength region to be monitored simultaneously.

Computer Systems and Software

Commercial spectrometers now come with powerful computers and software. Most of the newer ICP emission systems provide software that can assist in wavelength selection, calibration, background correction, interelement correction, spectral deconvolution, standard additions calibration, quality control charts, and report generation.

28C-2 Sources of Nonlinearity in Atomic Emission Spectrometry

Quantitative results in atomic emission spectrometry are usually based on the method of external standards (see Section 8C-2). For many reasons, we desire calibration curves to be linear or at least to follow a predicted relationship. At high concentrations, the major cause of nonlinearity when resonance transitions are used is **self-absorption.** Even at high concentrations, the majority of the analyte atoms are in the ground state, with only a small fraction being excited. When the excited analyte atoms emit, the emitted photons can be absorbed by ground-state analyte atoms, since these have the appropriate energy levels to absorb. In media where the temperature is not homogeneous, resonance lines can be severely broadened and can even have a dip in the center due to a phenomenon known as **self-reversal.** In flame emission, self-absorption is usually seen at solution concentrations between 10 and 100 μg/mL. In plasmas, self-absorption is often not seen until concentrations are higher, owing to the lower optical path length for absorption in the plasma.

At low concentrations, ionization of the analyte can cause nonlinearity in calibration curves. With ICP and DCP sources, the high electron concentrations in the plasma tend to act as a buffer against changes in the extent of ionization of the analyte with concentration. Ionic emission lines are often used with the ICP, and these are not very susceptible to further ionization. Changes in atomizer characteristics (such as flow rate, temperature, and efficiency) with analyte concentration can also be a cause of nonlinearity.

Flame emission methods often show linearity over two or three decades in concentration. ICP and DCP sources can show very high linear ranges, often four to five decades in concentration.

28C-3 Interferences in Plasma and Flame Atomic Emission Spectroscopy

Many of the interference effects caused by concomitants are similar in plasma and flame atomic emission. Some techniques, however, may be prone to certain interferences and may exhibit freedom from others. The interference effects are conveniently divided into blank or additive interferences and analyte or multiplicative interferences.

Blank Interferences

A **blank** or **additive interference** produces an effect that is independent of the analyte concentration. These effects could be reduced or eliminated if a perfect blank could be prepared and analyzed under the same conditions. A **spectral interference** is an example. In emission spectroscopy, any element other than the analyte that emits radiation within the band-pass of the wavelength selection device or that causes stray light to appear within the band-pass causes a blank interference.

Spectral interferences are examples of blank interferences. They produce an effect independent of the analyte level.

An example of a blank interference is the effect of Na emission at 285.28 nm on the determination of Mg at 285.21 nm. With a moderate-resolution spectrometer, any sodium in the sample will cause high readings for magnesium unless a blank with the correct amount of sodium is subtracted. Such line interferences can, in principle, be reduced by improving the resolution of the spectrometer. The user rarely has the opportunity to change the spectrometer resolution, however. In multielement spectrometers, measurements at multiple wavelengths can be used at times to determine the correction factors to apply for an interfering species. Such interelement corrections are commonplace with modern computer-controlled ICP spectrometers.

Molecular band emission can also cause a blank interference. This is particularly troublesome in flame spectrometry, where the lower temperature and reactive atmosphere are more likely to produce molecular species. As an example, a high concentration of Ca in a sample can produce band emission from CaOH, which can cause a blank interference if it occurs at the analyte wavelength. Usually, improving the resolution of the spectrometer will not reduce band emission, since the narrow analyte lines are superimposed on a broad molecular emission band. Flame or plasma background radiation is generally well compensated by measurements on a blank solution.

Analyte Interferences

Analyte interferences change the magnitude of the analyte signal itself. Such interferences are usually not spectral in nature but rather physical or chemical effects.

Chemical, physical, and ionization interferences are examples of **analyte interferences.** These influence the magnitude of the analyte signal itself.

Physical interferences can alter the aspiration, nebulization, desolvation, and volatilization processes. Substances in the sample that change the solution viscosity, for example, can alter the flow rate and the efficiency of the nebulization process. Combustible constituents, such as organic solvents, can change the atomizer temperature and thus affect the atomization efficiency indirectly.

Chemical interferences are usually specific to particular analytes. They occur in the conversion of the solid or molten particle after desolvation into free atoms or elementary ions. Constituents that influence the volatilization of analyte particles cause this type of interference and are often called **solute volatilization interferences.** For example, in some flames the presence of phosphate in the sample can alter the atomic concentration of calcium in the flame owing to the formation of relatively nonvolatile complexes. Such effects can sometimes be eliminated or moderated by the use of higher temperatures. Alternatively, **releasing agents,** which are species that react preferentially with the interferent and prevent its interaction with the analyte, can be used. For example, the addition of excess Sr or La minimizes the phosphate interference on calcium because these cations form stronger phosphate compounds than Ca and release the analyte.

Releasing agents are cations that react selectively with anions and prevent them from interfering in the determination of a cationic analyte.

Protective agents prevent interference by preferentially forming stable but *volatile* species with the analyte. Three common reagents for this purpose are EDTA, 8-hydroxyquinoline, and APDC (the ammonium salt of 1-pyrrolidinecarbodithioc acid). For example, the presence of EDTA has been shown to minimize or eliminate interferences by silicate, phosphate, and sulfate in the determination of calcium.

Substances that alter the ionization of the analyte also cause **ionization interferences.** The presence of an easily ionized element, such as K, can alter the extent of ionization of a less easily ionized element, such as Ca. In flames, relatively large effects can occur unless an easily ionized element is purposely added to the sample in relatively large amounts. These **ionization suppressants** contain elements such as K, Na, Li, Cs, or Rb. When ionized in the flame, these elements produce electrons, which then shift the ionization equilibrium of the analyte to favor neutral atoms.

An **ionization suppressant** is a readily ionized species that produces a high concentration of electrons in a flame and represses ionization of the analyte.

28C-4 Applications

The ICP has become the most widely used source for emission spectroscopy. Its success stems from its high stability, low noise, low background, and freedom from many interferences. The ICP is, however, relatively expensive to purchase and to operate. Users require extensive training to operate and maintain these instruments. Modern computerized systems with their sophisticated software have eased the burden substantially, however.

The ICP is widely used in determining trace metals in environmental samples, such as drinking water, waste water, and groundwater supplies. The ICP is also used for determining trace metals in petroleum products, in foodstuffs, in geological samples, in biological materials, and in industrial quality control. The DCP has found a significant niche in trace metal determinations in soil and geological samples. Flame emission is still used in some clinical laboratories for determining Na and K.

Simultaneous multielement determinations using plasma sources have gained in popularity. Such determinations make it possible to form correlations and to reach conclusions that were impossible with single-element determinations. For example, trace metal determinations can aid in determining the origins of petroleum products found in oil spills or in identifying sources of pollution.

28D ATOMIC ABSORPTION SPECTROMETRY

Flame atomic absorption spectroscopy (AAS) is currently the most widely used of all the atomic methods listed in Table 28-1 because of its simplicity, effectiveness, and relatively low cost. The technique was introduced in 1955 by Walsh in Australia and by Alkemade and Milatz in Holland.[6] The first commercial atomic absorption (AA) spectrometer was introduced in 1959, and use of the technique grew explosively after that. Atomic absorption methods were not widely used until that time because of problems created by the very narrow widths of atomic absorption lines, as discussed in Section 28A-1.

28D-1 Line Width Effects in Atomic Absorption

► The widths of atomic absorption lines are much less than the effective bandwidths of most monochromators.

No ordinary monochromator is capable of yielding a band of radiation as narrow as the width of an atomic absorption line (0.002 to 0.005 nm). As a result, the use of radiation that has been isolated from a continuum source by a monochromator inevitably causes instrumental departures from Beer's law (see the discussion of instrument deviations from Beer's law in Section 24C-3). In addition, since the fraction of radiation absorbed from such a beam is small, the detector receives a signal that is less attenuated (that is, $P \rightarrow P_0$) and the sensitivity of the measurement is reduced. This effect is illustrated by the lower curve in Figure 24-17 (page 733).

The problem created by narrow absorption lines was surmounted by the use of radiation from a source that emits not only a *line of the same wavelength* as the one selected for absorption measurements but also one that is *narrower.* For example, a mercury vapor lamp is selected as the external radiation source for the determination of mercury. Gaseous mercury atoms electrically excited in such a lamp return to the ground state by *emitting* radiation with wavelengths that are identical to the wavelengths *absorbed* by the analyte mercury atoms in the flame. Since the lamp is operated at temperatures and pressures lower than those of the flame, the Doppler and pressure broadening of the mercury emission lines from the lamp is less than

[6] A. Walsh, *Spectrochim. Acta,* **1955,** *7,* 108; C. Th. J. Alkemade and J. M. W. Milatz, *J. Opt. Soc. Am.,* **1955,** *45,* 583.

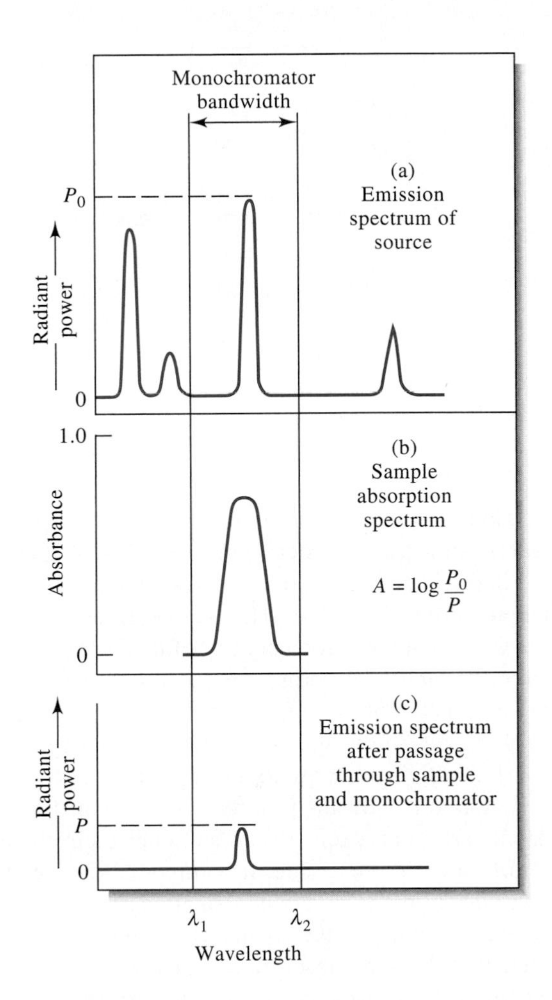

Figure 28-15 Atomic absorption of a narrow emission line from a source. The source lines in (a) are very narrow. One line is isolated by a monochromator. The line is absorbed by the broader absorption line of the analyte in the flame (b) resulting in attenuation (c) of the source radiation. Since most of the source radiation occurs at the peak of the absorption line, Beer's law is obeyed.

the corresponding broadening of the analyte absorption lines in the hot flame that holds the sample. The effective bandwidths of the lines emitted by the lamp are, therefore, significantly less than the corresponding bandwidths of the absorption lines for the analyte in the flame.

Figure 28-15 illustrates the strategy generally used to measure absorbances in atomic absorption methods. Figure 28-15a shows four narrow *emission* lines from a typical atomic absorption source. Also shown is how one of these lines is isolated by a filter or monochromator. Figure 28-15b shows the flame *absorption spectrum* for the analyte between the wavelengths λ_1 and λ_2; note that the width of the absorption line in the flame is significantly greater than the width of the emission line from the lamp. As shown in Figure 28-15c, the radiant power of the incident beam P_0 has been decreased to P by passage through the sample. Since the bandwidth of the emission line from the lamp is now significantly less than the bandwidth of the absorption line in the flame, log P_0/P is likely to be linearly related to concentration.

28D-2 Instrumentation

The instrumentation for AA can be fairly simple, as shown in Figure 28-16 for a single-beam AA spectrometer.

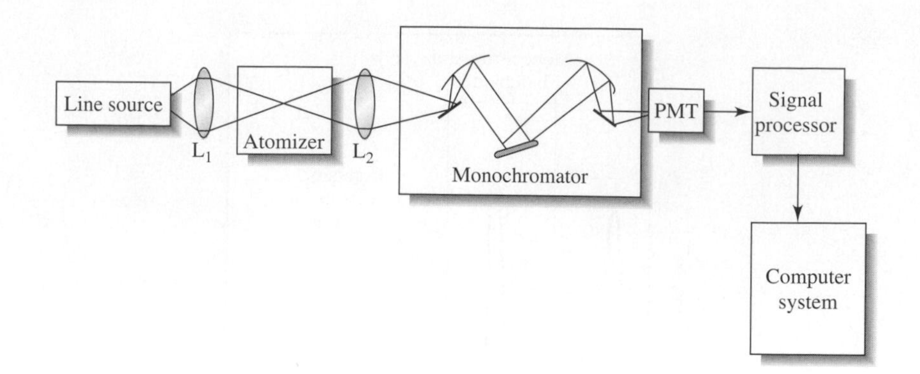

Figure 28-16 Block diagram of a single-beam atomic absorption spectrometer. Radiation from a line source is focused on the atomic vapor in a flame or an electrothermal atomizer. The attenuated source radiation then enters a monochromator, which isolates the line of interest. Next, the radiant power from the source, attenuated by absorption, is measured by the photomultiplier tube (PMT). The signal is then processed and directed to a computer system for output.

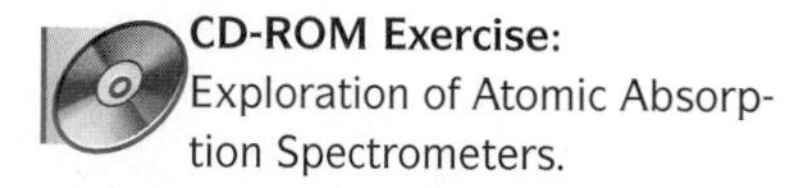
CD-ROM Exercise: Exploration of Atomic Absorption Spectrometers.

Line Sources

The most useful radiation source for atomic absorption spectroscopy is the **hollow-cathode lamp,** shown schematically in Figure 28-17. It consists of a tungsten anode and a cylindrical cathode sealed in a glass tube containing an inert gas, such as argon, at a pressure of 1 to 5 torr. The cathode either is fabricated from the analyte metal or serves as a support for a coating of that metal.

The application of about 300 V across the electrodes causes ionization of the argon and generation of a current of 5 to 10 mA as the argon cations and electrons migrate to the two electrodes. If the potential is sufficiently large, the argon cations strike the cathode with sufficient energy to dislodge some of the metal atoms and thereby produce an atomic cloud; this process is called **sputtering.** Some of the sputtered metal atoms are in an excited state and emit their characteristic wavelengths as they return to the ground state. It is important to recall that the atoms producing emission lines in the lamp are at a significantly lower temperature and pressure than the analyte atoms in the flame. Thus, the emission lines from the lamp are broadened less than the absorption peaks in the flame. The sputtered metal atoms eventually diffuse back to the cathode surface or to the walls of the lamp and are deposited.

Sputtering is a process in which atoms or ions are ejected from a surface by a beam of charged particles.

Hollow-cathode lamps for about 70 elements are available from commercial sources. For certain elements, high-intensity lamps are available. These provide an intensity that is about an order of magnitude higher than that of normal lamps. Some hollow-cathode lamps are fitted with a cathode containing more than one element; such lamps provide spectral lines for the determination of several species. The development of the hollow-cathode lamp is widely regarded as the single most important event in the evolution of atomic absorption spectroscopy.

► Hollow-cathode lamps made atomic absorption spectroscopy practical.

In addition to hollow-cathode lamps, **electrodeless-discharge lamps** are useful sources of atomic line spectra. These lamps are often one to two orders of

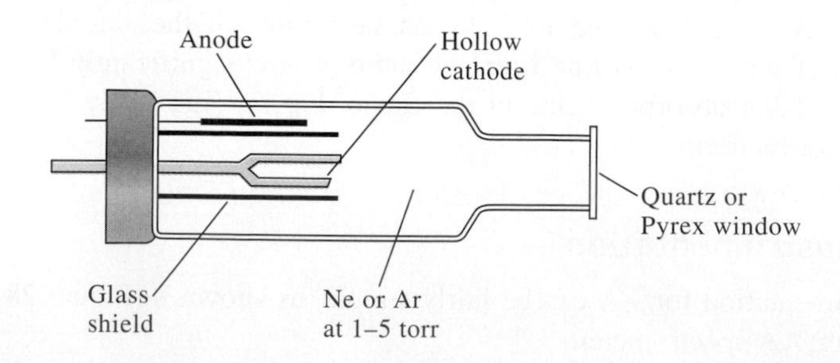

Figure 28-17 Diagram of a hollow-cathode lamp.

magnitude more intense than their hollow-cathode counterparts. A typical lamp is constructed from a sealed quartz tube containing an inert gas, such as argon, at a pressure of a few torr and a small quantity of the analyte metal (or its salt). The lamp contains no electrode but instead is energized by an intense field of radiofrequency or microwave radiation. The argon ionizes in this field, and the ions are accelerated by the high-frequency component of the field until they gain sufficient energy to excite (by collision) the atoms of the metal whose spectrum is sought.

Electrodeless-discharge lamps are available commercially for several elements. They are particularly useful for elements such as As, Se, and Te, for which hollow-cathode lamp intensities are low.

Source Modulation

In an atomic absorption measurement, it is necessary to discriminate between radiation from the hollow-cathode or electrodeless-discharge lamp and radiation from the atomizer. Much of the latter is eliminated by the monochromator, which is always located between the atomizer and the detector. The thermal excitation of a fraction of the analyte atoms in a flame, however, produces radiation of the wavelength at which the monochromator is set. Because such radiation is not removed, it acts as a potential source of interference.

The effect of analyte emission is overcome by **modulating** the output from the hollow-cathode lamp so that its intensity fluctuates at a constant frequency. The detector thus receives an alternating signal from the hollow-cathode lamp and a continuous signal from the flame and converts these signals into the corresponding types of electric current. A relatively simple electronic system then eliminates the unmodulated dc signal produced by the flame and passes the ac signal from the source to an amplifier and finally to the readout device.

Modulation is defined as the changing of some property of a carrier wave by the desired signal in such a way that the carrier wave can be used to convey information about the signal. Properties that are typically altered are frequency, amplitude, and wavelength. In AAS, the source radiation is amplitude modulated, but the background and analyte emission are not and are observed as dc signals.

Modulation can be accomplished by interposing a motor-driven circular chopper *between the source and the flame,* as shown in Figure 28-18. Segments of the metal chopper have been removed so that radiation passes through the device half the time and is reflected the other half. Rotation of the chopper at a constant speed causes the beam reaching the flame to vary periodically from zero intensity to some maximum intensity and then back to zero. As an alternative, the power supply for the source can be designed to pulse the hollow-cathode lamps in an alternating manner.

◀ Modulation of the source by using a beam chopper or by pulsing it electronically is widely used to convert the source radiation to an alternating form.

Complete Atomic Absorption Instruments

An atomic absorption instrument contains the same basic components as an instrument designed for molecular absorption measurements, as shown in Figure 28-16 for a single-beam system. Both single- and double-beam instruments are offered by numerous manufacturers. The range of sophistication and the cost (upward from a few thousand dollars) are both substantial.

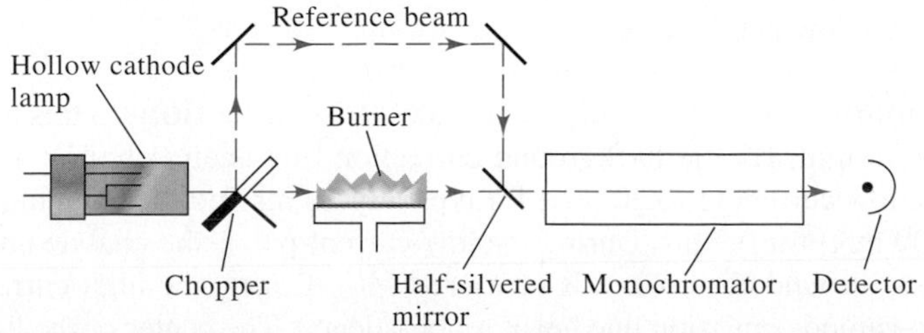

Hollow cathode lamp
Reference beam
Burner
Chopper
Half-silvered mirror
Monochromator
Detector

Figure 28-18 Optical paths in a double-beam atomic absorption spectrophotometer.

Photometers At a minimum, an instrument for atomic absorption spectroscopy must be capable of providing a sufficiently narrow bandwidth to isolate the line chosen for a measurement from other lines that may interfere with or diminish the sensitivity of the method. A photometer equipped with a hollow-cathode source and filters is satisfactory for measuring concentrations of the alkali metals, which have only a few widely spaced resonance lines in the visible region. A more versatile photometer is sold with readily interchangeable interference filters and lamps. A separate filter and lamp are used for each element. Satisfactory results for the determination of 22 metals are claimed.

Spectrophotometers Most measurements in AAS are made with instruments equipped with an ultraviolet/visible grating monochromator. Figure 28-18 is a schematic of a typical double-beam instrument. Radiation from the hollow-cathode lamp is chopped and mechanically split into two beams, one of which passes through the flame; the other passes around the flame. A half-silvered mirror returns both beams to a single path by which they pass alternately through the monochromator to the detector. The signal processor then separates the ac signal generated by the chopped light source from the dc signal produced by the flame. The logarithm of the ratio of the reference and sample components of the ac signal is then computed and sent to a computer or readout device for display as absorbance.

Background Correction

Absorption by the flame atomizer itself as well as by concomitants introduced into the flame or electrothermal atomizer can cause serious problems in atomic absorption. Rarely are there interferences from absorption of the analyte line by other atoms since the hollow-cathode lines are so narrow. Molecular species can absorb the radiation and cause errors in AA measurements, however.

The total measured absorbance A_T in AA is the sum of the analyte absorbance A_A plus the background absorbance A_B:

$$A_T = A_A + A_B \tag{28-1}$$

Background correction schemes attempt to measure A_B in addition to A_T and to obtain the true analyte absorbance by subtraction ($A_A = A_T - A_B$).

Continuum source background correction uses a deuterium lamp to obtain an estimate of the background absorbance. A hollow-cathode lamp obtains the total absorbance. The corrected absorbance is then obtained by calculating the difference between the two.

Continuum Source Background Correction A popular background correction scheme in commercial AA spectrometers is the continuum lamp technique. Here a deuterium lamp and the analyte hollow cathode are directed through the atomizer at different times. The hollow-cathode lamp measures the total absorbance A_T, while the deuterium lamp provides an estimate of the background absorbance A_B. The computer system or processing electronics calculates the difference and reports the background-corrected absorbance. This method has limitations for elements with lines in the visible range because the D_2 lamp intensity becomes quite low in this region.

Smith-Hieftje background correction uses a single hollow-cathode lamp pulsed with first a low current and then a high current. The low-current mode obtains the total absorbance, while the background is estimated during the high-current pulse. Read the interview at the beginning of Part V to learn more about Gary Hieftje and his work.

Pulsed Hollow-Cathode Lamp Background Correction In this technique, often called Smith-Hieftje background correction, the analyte hollow cathode is pulsed at a low current (5 to 20 mA) for typically 10 ms and then at a high current (100 to 500 mA) for 0.3 ms. During the low-current pulse, the analyte absorbance plus the background absorbance is measured (A_T). During the high-current pulse, the hollow-cathode emission line becomes broadened. The center of the line can be strongly self-absorbed so that much of the line at the analyte wavelength is missing. Hence, during the high-current pulse, a good estimate of the background absorbance A_B is obtained. The instrument computer then calculates the difference, which is an estimate of A_A, the true analyte absorption.

Zeeman Effect Background Correction Background correction with electrothermal atomizers can be done by means of the Zeeman effect. Here a magnetic field splits normally degenerate spectral lines into components with different polarization characteristics. Analyte and background absorption can be separated because of their different magnetic and polarization behaviors.

28D-3 Flame Atomic Absorption

Flame AA provides a sensitive means of determining some 60 to 70 elements. This method is well suited for routine measurements by relative unskilled operators. The major drawback of AA is its single-element-at-a-time nature, imposed by the need for a different lamp for each element.

Region of the Flame for Quantitative Measurements

Figure 28-19 shows the absorbance of three elements as a function of distance above the burner head. For magnesium and silver, the initial rise in absorbance is a consequence of the longer exposure to the heat, which leads to a greater concentration of atoms in the radiation path. The absorbance for magnesium, however, reaches a maximum near the center of the flame and then falls off as oxidation of the magnesium to magnesium oxide takes place. This effect is not seen with silver because this element is much more resistant to oxidation. For chromium, which forms very stable oxides, maximum absorbance lies immediately above the burner. For this element, oxide formation begins as soon as chromium atoms are formed.

It is clear from Figure 28-19 that the part of a flame to be used in an analysis must vary from element to element and that the position of the flame with respect to the source must be reproduced closely during calibration and analysis. Generally, the flame position is adjusted to yield a maximum absorbance reading.

Quantitative Analysis

Frequently, quantitative analyses are based on external-standard calibration (see Section 8C-2). In atomic absorption, departures from linearity are encountered more often than in molecular absorption. Thus, analyses should *never* be based on the

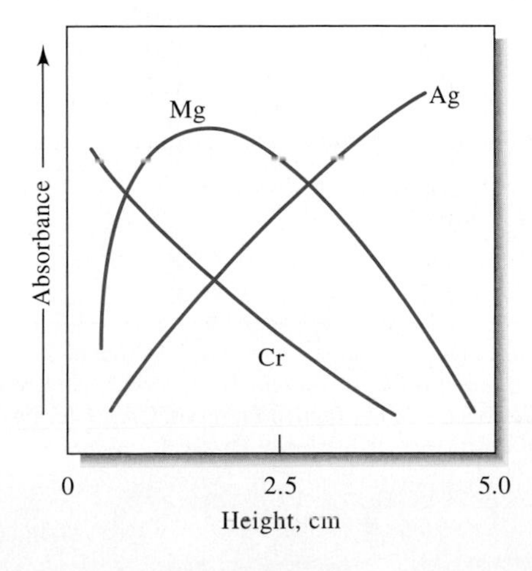

Figure 28-19 Absorbance versus height above the burner for three elements in flame AAS.

measurement of a single standard with the assumption that Beer's law is being followed. In addition, the production of an atomic vapor involves sufficient uncontrollable variables to warrant measuring the absorbance of at least one standard solution each time an analysis is performed. Often, two standards are employed whose absorbances bracket that of the unknown. Any deviation of the standard from its original calibration value can then be applied as a correction to the analytical results.

Standard addition methods, discussed in Section 8C-3, are also used extensively in atomic spectroscopy to compensate for differences between the composition of the standards and the unknowns.

Detection Limits and Accuracy

Column 2 of Table 28-4 shows detection limits for a number of common elements determined by flame atomic absorption and compares them with those obtained with other atomic spectroscopic methods. Under usual conditions, the relative error of flame absorption analysis is on the order of 1% to 2%. With special precautions, this figure can be lowered to a few tenths of 1%. Note that flame AA detection limits are generally better than flame AE detection limits except for the easily excited alkali metals.

28D-4 Atomic Absorption with Electrothermal Atomization

Electrothermal atomizers offer the advantage of unusually high sensitivity for small volumes of sample. Typically, sample volumes between 0.5 and 10 μL are employed; under these circumstances, absolute detection limits typically lie in the

TABLE 28-4

Detection Limits (ng/mL) for Some Elements by Atomic Spectroscopy*

Element	Flame AA	Electrothermal AA[†]	Flame Emission	ICP Emission	ICP-MS
Ag	3	0.02	20	0.2	0.003
Al	30	0.2	5	0.2	0.06
Ba	20	0.5	2	0.01	0.002
Ca	1	0.5	0.1	0.0001	2
Cd	1	0.02	2000	0.07	0.003
Cr	4	0.06	5	0.08	0.02
Cu	2	0.1	10	0.04	0.003
Fe	6	0.5	50	0.09	0.45
K	2	0.1	3	75	1
Mg	0.2	0.004	5	0.003	0.15
Mn	2	0.02	15	0.01	0.6
Mo	5	1	100	0.2	0.003
Na	0.2	0.04	0.1	0.1	0.05
Ni	3	1	600	0.2	0.005
Pb	5	0.2	200	1	0.007
Sn	15	10	300	1	0.02
V	25	2	200	8	0.005
Zn	1	0.01	200	0.1	0.008

*Values taken from V. A. Fassel and R. N. Knisely, *Anal. Chem.*, **1974,** *46,* 111A; J. D. Ingle, Jr., and S. R. Crouch, *Spectrochemical Analysis.* Upper Saddle River, NJ: Prentice-Hall, 1988; C. W. Fuller, *Electrothermal Atomization for Atomic Absorption Spectroscopy.* London: The Chemical Society, 1977; *Ultrapure Water Specifications. Quantitative ICP-MS Detection Limits.* Fremont, CA, Balazs Analytical Services, 1993. With permission.

[†]Based on a 10-μL sample.

picogram range. In general, electrothermal AA detection limits are best for the more volatile elements. Detection limits for electrothermal AA vary considerably from one manufacturer to the next because they depend on atomizer design and atomization conditions.

The relative precision of electrothermal methods is generally in the range of 5% to 10%, compared with the 1% or better that can be expected for flame or plasma atomization. Furthermore, furnace methods are slow and typically require several minutes per element. Still another disadvantage is that chemical interference effects are often more severe with electrothermal atomization than with flame atomization. A final disadvantage is that the analytical range is low, usually less than two orders of magnitude. Consequently, electrothermal atomization is ordinarily applied only when flame or plasma atomization provides inadequate detection limits or when sample sizes are extremely limited.

Another AA method applicable to volatile elements and compounds is the cold-vapor technique. Mercury is a volatile metal and can be determined by the method described in Feature 28-1. Other metals form volatile metal hydrides that can also be determined by the cold-vapor technique.

FEATURE 28-1

Determining Mercury by Cold-Vapor Atomic Absorption Spectroscopy

Our fascination with mercury began when prehistoric cave dwellers discovered the mineral cinnabar (HgS) and used it as a red pigment. Our first written record of the element came from Aristotle, who described it as "liquid silver" in the fourth century B.C. Today there are thousands of uses for mercury and its compounds in medicine, metallurgy, electronics, agriculture, and many other fields. Because it is a liquid metal at room temperature, mercury is used to make flexible and efficient electrical contacts in scientific, industrial, and household applications. Thermostats, silent light switches, and fluorescent light bulbs are but a few examples of electrical applications.

A useful property of metallic mercury is that it forms amalgams with other metals, which have a host of uses. For example, metallic sodium is produced as an amalgam by electrolysis of molten sodium chloride. Dentists use a 50% amalgam with an alloy of silver for fillings.

The toxicological effects of mercury have been known for many years. The bizarre behavior of the Mad Hatter in Lewis Carroll's *Alice in Wonderland* (Figure 28F-1) was a

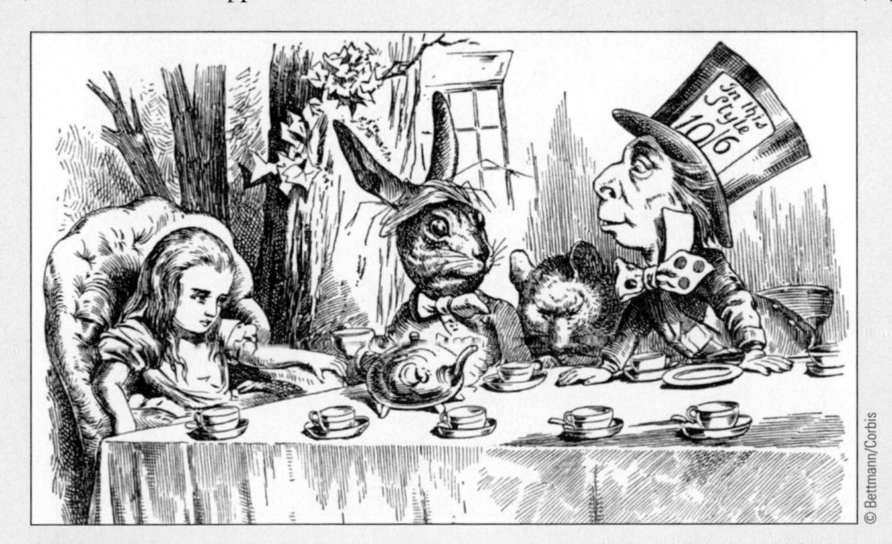

© Bettmann/Corbis

Figure 28F-1 Tea party from *Alice in Wonderland.*

(continued)

result of the effects of mercury and mercury compounds on the Hatter's brain. Mercury that has been absorbed through the skin and lungs destroys brain cells, which are not regenerated. Hatters of the 19th century used mercury compounds in processing fur to make felt hats. These workers and workers in other industries have suffered the debilitating symptoms of mercurialism, such as loosening of teeth, tremors, muscle spasms, personality changes, depression, irritability, and nervousness.

The toxicity of mercury is complicated by its tendency to form both inorganic and organic compounds. Because inorganic mercury is relatively insoluble in body tissues and fluids, it is expelled from the body about 10 times faster than organic mercury. Organic mercury, usually in the form of alkyl compounds such as methyl mercury, is somewhat soluble in fatty tissue such as the liver. Methyl mercury accumulates to toxic levels and is expelled from the body quite slowly. Even experienced scientists must take extreme precautions in handling organo-mercury compounds. In 1997, Dr. Karen Wetterhahn of Dartmouth College died as a result of mercury poisoning despite being one of the world's leading experts on the handling of methyl mercury.

Mercury concentrates in the environment, as illustrated in Figure 28F-2. Inorganic mercury is converted to organic mercury by anaerobic bacteria in sludge deposited at the bottom of lakes, streams, and other bodies of water. Small aquatic animals consume the organic mercury and are in turn eaten by larger life forms. As the element moves up the food chain from microbes to shrimp to fish and, ultimately, to larger animals such as swordfish, the mercury becomes ever more concentrated. Some sea creatures such as oysters may concentrate mercury by a factor of 100,000. At the top of the food chain, the concentration of mercury reaches levels as high as 20 ppm. The Food and Drug Administration has set a legal limit of 1 ppm in fish for human consumption. As a result, mercury levels in some areas threaten local fishing industries. The Environmental Protection Agency has set a limit of 1 ppb of mercury in drinking water, and the Occupational Safety and Health Administration has set a limit of 0.1 mg/m^3 in air.

Analytical methods for the determination of mercury play an important role in monitoring the safety of food and water supplies. One of the most useful methods is based on the atomic absorption by mercury of 253.7-nm radiation. Color plate 18 shows the striking absorption of

Figure 28F-2 Biological concentration of mercury in the environment.

ultraviolet light by mercury vapor that forms over the metal at room temperature. Figure 28F-3 shows an apparatus that is used to determine mercury by atomic absorption at room temperature.[7]

A sample suspected of containing mercury is decomposed in a hot mixture of nitric acid and sulfuric acid, which converts the mercury to the +2 state. The Hg(II) compounds are reduced to the metal with a mixture of hydroxylamine sulfate and tin(II) sulfate. Air is then pumped through the solution to carry the resulting mercury-containing vapor through the drying tube and into the observation cell. Water vapor is trapped by Drierite in the drying tube so that only mercury vapor and air pass through the cell. The monochromator of the atomic absorption spectrophotometer is tuned to a band around 254 nm. Radiation from the 253.7-nm line of the mercury hollow-cathode lamp passes through the quartz windows of the observation cell, which is placed in the light path of the instrument. The absorbance is directly proportional to the concentration of mercury in the cell, which is in turn proportional to the concentration of mercury in the sample. Solutions of known mercury concentration are treated in a similar way to calibrate the apparatus. This method depends on the low solubility of mercury in the reaction mixture and its appreciable vapor pressure, which is 2×10^{-3} torr at 25°C. The sensitivity of the method is about 1 ppb, and it is used to determine mercury in foods, metals, ores, and environmental samples. This method has the advantages of sensitivity, simplicity, and room temperature operation.

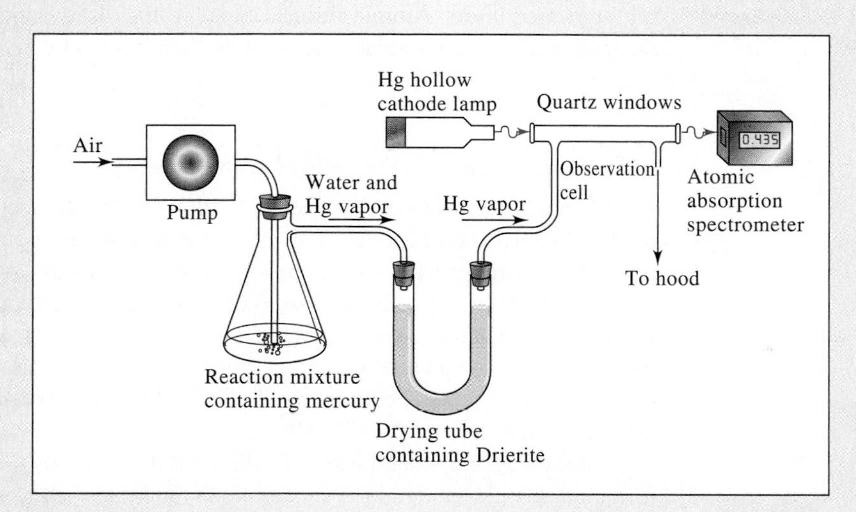

Figure 28F-3 Apparatus for cold-vapor atomic absorption determination of mercury.

28D-5 Interferences in Atomic Absorption

Flame atomic absorption is subject to many of the same chemical and physical interferences as flame atomic emission (see Section 28C-2). Spectral interferences by elements that absorb at the analyte wavelength are rare in AA. Molecular constituents and radiation scattering can cause interferences, however. These are often corrected by the background correction schemes discussed in Section 28D-2. In some cases, if the source of interference is known, an excess of the interferent can be added to both the sample and the standards. The added substance is sometimes called a **radiation buffer.**

A **radiation buffer** is a substance that is added in large excess to both samples and standards to swamp out the effect of matrix species and thus to minimize interference.

[7] W. R. Hatch and W. L. Ott, *Anal. Chem.*, **1968,** *40,* 2085.

28E ATOMIC FLUORESCENCE SPECTROMETRY

Atomic fluorescence spectrometry (AFS) is the newest of the optical atomic spectroscopic methods. As in atomic absorption, an external source is used to excite the element of interest. Instead of measuring the attenuation of the source, however, the radiation emitted as a result of absorption is measured, often at right angles to avoid measuring the source radiation.

► Despite its potential advantages of high sensitivity and selectivity, atomic fluorescence spectrometry has never been commercially successful. Difficulties can be attributed partly to the lack of reproducibility of the high-intensity sources required and to the single-element nature of AFS.

Atomic fluorescence with conventional hollow-cathode or electrodeless-discharge sources has not shown significant advantages over atomic absorption or atomic emission. As a consequence, the commercial development of atomic fluorescence instrumentation has been quite slow. Sensitivity advantages have been shown, however, for elements such as Hg, Sb, As, Se, and Te.

Laser-excited atomic fluorescence spectrometry is capable of extremely low detection limits, particularly when combined with electrothermal atomization. Detection limits in the femtogram (10^{-15} g) to attogram (10^{-18} g) range have been shown for many elements. Commercial instrumentation has not been developed for laser-based AFS, probably because of its expense and the nonroutine nature of high-powered lasers. Atomic fluorescence has the disadvantage of being a single-element method unless tunable lasers with their inherent complexities are used.

28F ATOMIC MASS SPECTROMETRY[8]

Atomic mass spectrometry has been around for many years, but the introduction of the inductively coupled plasma in the 1970s and its subsequent development for mass spectrometry[9] led to its successful commercialization by several instrument companies. Today inductively coupled plasma mass spectrometry (ICP-MS) is a widely used technique for the simultaneous determination of more than 70 elements in a few minutes. Some other sources, such as the glow discharge, are also used for atomic mass spectrometry. Because the ICP predominates, however, the discussion here focuses on ICP-MS.

A block diagram of a typical ICP-MS instrument is shown in Figure 28-20. Ions formed in the plasma are introduced into the mass analyzer, where they are sorted according to mass-to-charge ratio and detected. Solution samples are introduced into the plasma through a nebulizer, as in ICP atomic emission. Solids are either dissolved in solution or introduced directly by laser ablation methods. Gases can be introduced directly.

28F-1 Mass Spectrometer Interface

A major problem exists in extracting ions from the plasma. While an ICP operates at atmospheric pressure, a mass spectrometer operates at high vacuum, typically less than 10^{-6} torr. The interface region between the ICP and the mass spectrometer is thus critical to ensure that a substantial fraction of the ions produced are transported into the mass analyzer. The interface generally consists of two metal

[8]For additional information, see D. A. Skoog, F. J. Holler, and T. A. Nieman, *Principles of Instrumental Analysis,* 5th ed., Chapter 11. Belmont, CA: Brooks/Cole, 1998.

[9]R. S. Houk, V. A. Fassel, G. D. Flesch, H. J. Svec, A. L. Gray, and C. E. Taylor, *Anal. Chem.,* **1980,** *52,* 2283.

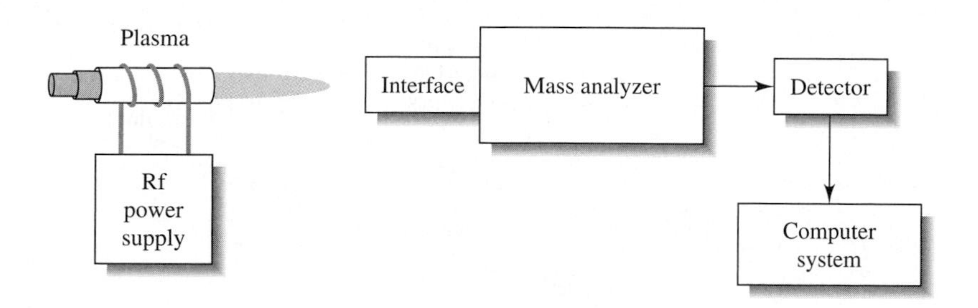

Figure 28-20 Block diagram of an ICP mass spectrometer system.

cones, called the **sampler** and the **skimmer.** Each cone has a small orifice (≈1 mm) to allow the ions to pass through to ion optics, which guide them into the mass analyzer.[10] The beam introduced into the mass spectrometer has about the same ionic composition as the plasma region from which the ions are extracted. Background ions include Ar^+, ArO^+, ArH^+, H_2O^+, O^+, O_2^+ and Ar_2^+, as well as argon adducts with metals. In addition, some polyatomic ions from constituents in the sample are also found in ICP mass spectra. Such background ions can interfere with the determination of analytes.

28F-2 Mass Analyzers

The most popular mass analyzers for ICP-MS have been quadrupole, magnetic-sector, and double-focusing analyzers, although time-of-flight analyzers are also used. These analyzers vary in resolution, throughput, and scanning time. The resolution of a mass analyzer is defined as:

$$R = m/\Delta m \tag{28-2}$$

where m is the nominal mass and Δm is the mass difference that can be just resolved. A resolution of 100 means that unit mass (1 Da) can be distinguished at a nominal mass of 100.

The quadrupole mass analyzer consists of four cylindrical rods, as illustrated in Figure 28-21. Quadrupole analyzers are basically mass filters that allow only ions of a certain mass-to-charge *(m/z)* ratio to pass. Ion motion in electric fields forms the basis of separation. Rods opposite each other are connected to dc and radio-frequency (rf) voltages. With proper adjustment of the voltages, a stable path is created for ions of a certain *m/z* ratio to pass through the analyzer to the detector. The mass spectrum is obtained by scanning the voltages applied to the rods. Quadrupole analyzers have relatively high throughput but lack resolution. Unit mass (1 Da) is the typical resolution of a quadrupole analyzer. This low resolution is often inadequate to separate monatomic species from polyatomic ions of similar *m/z* values.

Magnetic sector instruments are also used in ICP-MS. Here separation is based on the deflection of ions in a magnetic field. The trajectories that the ions take depend on their *m/z* values. Typically, the magnetic field is scanned to bring ions of different *m/z* value to a detector. Double-focusing instruments are also commercially available for ICP-MS. Here an electric sector precedes the magnetic sector. The electrostatic field serves to focus a beam of ions having only a narrow range of

[10]For more information, see R. S. Houk, *Acc. Chem. Res.*, **1994,** *27,* 333.

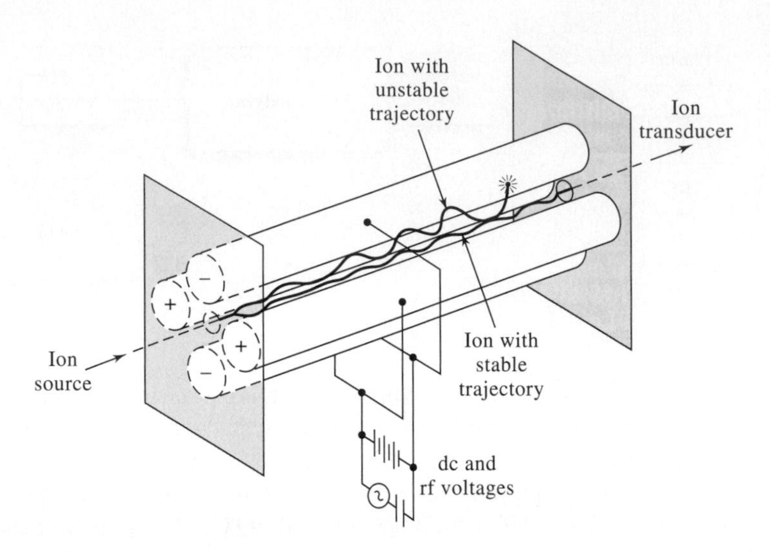

Figure 28-21 A quadrupole mass analyzer.

kinetic energies onto a slit that leads to the magnetic sector. Such instruments can have resolution values as high as 10,000. Some commercial instruments allow operation in a low-resolution mode ($R \approx 300$), a medium-resolution mode ($R \approx 4000$), and a high-resolution mode ($R \approx 10{,}000$). High-resolution instruments are significantly more expensive than quadrupole instruments. They usually allow much better separation of the ions of interest from background ions, however, and achieve superior detection limits.

Another approach is time-of-flight (TOF) mass spectrometry. Here a packet of ions is rapidly sampled; the ions enter a field-free region with nearly identical kinetic energies. The time required for the ions to reach a detector is inversely related to the ion mass. That is, ions with low m/z arrive at the detector more rapidly than those with high m/z. Each m/z value is then detected in sequence. Even so, analysis times are typically on the order of microseconds.

28F-3 Transducers

The most common transducers for ICP-MS are electron multipliers. The discrete dynode electron multiplier operates much like the photomultiplier transducer for ultraviolet/visible radiation, discussed in Section 25A-4. Electrons strike a cathode, where secondary electrons are emitted. These are attracted to dynodes that are each held at a successively higher positive voltage. Electron multipliers with up to 20 dynodes are available. These devices can multiply the signal strength by a factor of up to 10^7.

Continuous dynode electron multipliers are also popular. These are trumpet-shaped devices made of glass heavily doped with lead. A potential of 1.8 to 2 kV is imposed across the length of the device. Ions that strike the surface eject electrons that skip along the inner surface, ejecting more electrons with each impact.

28F-4 Interferences in Inductively Coupled Plasma Mass Spectrometry

Interferences in ICP-MS are of two classes: spectroscopic interferences and matrix effects. Spectroscopic interferences occur when an ionic species in the plasma has the same m/z value as an analyte ion. The majority of these interferences are from polyatomic ions, elements having isotopes with essentially the same mass, doubly charged ions, and refractory oxide ions.[11] High-resolution spectrometers can reduce or eliminate many of these interferences.

Matrix effects become noticeable at concomitant concentrations greater than about 500 to 1000 μg/mL. Usually, these effects cause a reduction in the analyte signal, although enhancements are sometimes observed. Generally, such effects can be minimized by diluting the sample, by altering the introduction procedure, or by separating the interfering species.

◀ High-resolution mass analyzers, such as double-focusing analyzers, can reduce or eliminate many spectral interferences in ICP-MS.

28F-5 Applications of Inductively Coupled Plasma Mass Spectrometry

ICP-MS is well suited for multielement analysis and for determinations such as isotope ratios. The technique has a wide dynamic range, typically four orders of magnitude, and produces spectra that are, in general, simpler and easier to interpret than optical-emission spectra. ICP-MS is finding widespread use in the semiconductor and electronics industries, in geochemistry, in environmental analyses, in biological and medical research, and in many other areas.

Detection limits for ICP-MS are listed in Table 28-4, where they are compared with those from several other atomic spectrometric methods. Most elements can be detected well below the part-per-billion level. Quadrupole instruments typically allow ppb detection for their entire mass range. High-resolution instruments can routinely achieve sub–part-per-trillion detection limits because the background levels in these instruments are extremely low.

Quantitative analysis is normally performed by preparing calibration curves using external standards. To compensate for instrument drifts, instabilities, and matrix effects, an internal standard is usually added to the standards and to the sample. Multiple internal standards are sometimes used to optimize matching of the characteristics of the standard to those of various analytes.

For simple solutions where the composition is known or the matrix can be matched well between samples and standards, accuracy can be better than 2% for analytes at concentrations 50 times the detection limit. For solutions of unknown composition, an accuracy of 5% is typical.

WEB WORKS

Do a Google search to find the Laboratory for Spectrochemistry at Indiana University. Find a list of research projects being done in this laboratory on plasma mass spectrometry. Click on one of the projects dealing with time-of-flight mass spectrometry. Describe in detail the purpose of the project, the instrumentation used, and any results obtained.

[11]For additional discussion of interferences in ICP-MS, see K. E. Jarvis, A. L. Gray, and R. S. Houk, *Handbook of Inductively Coupled Plasma Mass Spectrometry,* Ch. 5. New York: Blackie, 1992; G. Horlick and Y. Shao, in *Inductively Coupled Plasmas in Analytical Atomic Spectrometry,* 2nd ed., A Montaser and D. W. Golightly, Eds., pp. 571–596. New York: VCH-Wiley, 1992.

QUESTIONS AND PROBLEMS

*28-1. Describe the basic differences between atomic emission and atomic absorption spectroscopy.

28-2. Define
*(a) atomization.
(b) pressure broadening.
*(c) Doppler broadening.
(d) nebulizer.
*(e) plasma.
(f) hollow-cathode lamp.
*(g) sputtering.
(h) ionization suppressor.
*(i) spectral interference.
(j) chemical interference.
*(k) radiation buffer.
(l) releasing agent.
*(m) quadrupole mass filter.
(n) electron multiplier.

*28-3. Why is atomic emission more sensitive to flame instability than atomic absorption or fluorescence?

28-4. Why are ionization interferences usually not as severe in the ICP as they are in flames?

*28-5. Why are monochromators of a higher resolution found in ICP atomic emission spectrometers than in flame atomic absorption spectrometers?

28-6. Why is source modulation employed in atomic absorption spectroscopy?

*28-7. In flame AA with a hydrogen/oxygen flame, the absorbance for iron decreased in the presence of large concentrations of sulfate ion.
(a) Suggest an explanation for this observation.
(b) Suggest three possible methods of overcoming the potential interference of sulfate in a quantitative determination of iron.

28-8. Why are the lines from a hollow-cathode lamp generally narrower than the lines emitted by atoms in a flame?

*28-9. Name four characteristics of inductively coupled plasmas that make them suitable for atomic emission and atomic mass spectrometry.

28-10. Why is the ICP rarely used for atomic absorption measurements?

*28-11. Why are detection limits in ICP-MS generally lower with double-focusing spectrometers than with quadrupole mass spectrometers?

28-12. Discuss the differences that result in ICP atomic emission and ICP-MS when the plasma is viewed axially rather than radially.

*28-13. In the atomic absorption determination of uranium, a linear relationship is found between the absorbance at 351.5 nm and the concentration in the range of 500 to 2000 ppm of U. At lower concentrations, the relationship becomes nonlinear unless about 2000 ppm of an alkali metal salt are introduced. Explain.

28-14. What is the purpose of an internal standard in ICP-MS?

*28-15. A 5.00-mL sample of blood was treated with trichloroacetic acid to precipitate proteins. After centrifugation, the resulting solution was brought to pH 3 and extracted with two 5-mL portions of methyl isobutyl ketone containing the lead-complexing agent APCD. The extract was aspirated directly into an air/acetylene flame and yielded an absorbance of 0.502 at 283.3 nm. Five-milliliter aliquots of standard solutions containing 0.400 and 0.600 ppm of lead were treated in the same way and yielded absorbances of 0.396 and 0.599. Find the concentration of lead in the sample in ppm assuming that Beer's law is followed.

28-16. The chromium in a series of steel samples was determined by ICP emission spectroscopy. The spectrometer was calibrated with a series of standards containing 0, 2.0, 4.0, 6.0, and 8.0 μg $K_2Cr_2O_7$ per milliliter. The instrument readings for these solutions were 3.1, 21.5, 40.9, 57.1, and 77.3, respectively, in arbitrary units.
(a) Plot the data.
(b) Find the equation for the regression line.
(c) Calculate standard deviations for the slope and the intercept of the line in (b).
(d) The following data were obtained for replicate 1.00-g samples of cement dissolved in HCl and diluted to 100.0 mL after neutralization.

	Emission Readings			
	Blank	Sample A	Sample B	Sample C
Replicate 1	5.1	28.6	40.7	73.1
Replicate 2	4.8	28.2	41.2	72.1
Replicate 3	4.9	28.9	40.2	spilled

Calculate the percent Cr_2O_3 in each sample. What are the absolute and relative standard deviations for the average of each determination?

28-17. The copper in an aqueous sample was determined by atomic absorption flame spectrometry. First, 10.0 mL of the unknown were pipetted into each of five 50.0-mL volumetric flasks. Various volumes of a standard containing 12.2 ppm Cu were added to the flasks, and the solutions were then diluted to volume.

Unknown, mL	Standard, mL	Absorbance
10.0	0.0	0.201
10.0	10.0	0.292
10.0	20.0	0.378
10.0	30.0	0.467
10.0	40.0	0.554

(a) Plot absorbance as a function of volume of standard.

*(b) Derive an expression relating absorbance to the concentrations of the standards and unknowns (c_s and c_x) and to the volumes of the standards and unknowns (V_s and V_x), as well as to the volume to which the solutions were diluted (V_t).

*(c) Derive expressions for the slope and the intercept of the straight line obtained in (a) in terms of the variables listed in (b).

(d) Show that the concentration of the analyte is given by the relationship $c_x = bc_s/mV_x$, where m and b are the slope and the intercept of the straight line in (a).

*(e) Determine values for m and b by the method of least squares.

(f) Calculate the standard deviation for the slope and the intercept in (e).

*(g) Calculate the copper concentration in ppm Cu in the sample using the relationship given in (d).

28-18. Challenge Problem. Sea water samples were examined by ICP-MS in a multielement study. Vanadium was one of the elements determined. Standard solutions in a synthetic sea water matrix were prepared and determined by ICP-MS. The following results were obtained:

Concentration, pg/mL	Intensity, Arbitrary Units
0.0	2.1
2.0	5.0
4.0	9.2
6.0	12.5
8.0	17.4
10.0	20.9
12.0	24.7

(a) Determine the least-squares regression line.

(b) Determine the standard deviations of the slope and intercept.

(c) Test the hypothesis that the slope is equal to 2.00.

(d) Test the hypothesis that the intercept is equal to 2.00.

(e) Three sea water solutions gave readings for V of 3.5, 10.7, and 15.9. Determine their concentrations and the standard deviation of their concentrations.

(f) Determine the 95% confidence limits for the three unknowns in part (e).

(g) Estimate the limit of detection for determining V in sea water from the data (see Section 8D-1). Use a k value of 3 in your DL estimation.

(h) The second sea water sample with a reading of 10.7 units was a certified reference standard with a known concentration of 5.0 pg/mL. What was the absolute percent error in its determination?

(i) Test the hypothesis that the value determined in part (e) for the second sea water sample (reading of 10.7) is identical to the certified concentration of 5.0 pg/mL.

InfoTrac College Edition

For additional readings, go to InfoTrac College Edition, your online research library, at

http://infotrac.thomsonlearning.com

PART VI

Kinetics and Separations

Chapter 29
Kinetic Methods of Analysis

Chapter 30
Introduction to Analytical Separations

Chapter 31
Gas Chromatography

Chapter 32
High-Performance Liquid Chromatography

Chapter 33
Miscellaneous Separation Methods

A conversation with *Isiah M. Warner*

Jim Zeist, Louisiana State University Relations

On the surface, Isiah M. Warner's background is similar to that of any other academic chemist. He had an early interest in science and chose to major in chemistry in college. What is different about Warner is that all the schools he attended—elementary school, high school, and even college—were segregated. He received his B.S. in chemistry from Southern University, an historically black university, then worked for Battelle Labs during the Vietnam War. After 5 years, his desire to be the one doing the thinking led him to graduate school at the University of Washington, then to faculty positions at Texas A&M University and at Emory University in Atlanta. He currently holds an endowed chair in chemistry and is Vice Chancellor for Strategic Initiatives at Louisiana State University (LSU). Among his many awards and honors are the Presidential Award for Excellence in Science, Mathematics and Engineering Mentoring (1997) and the AAAS Lifetime Mentor Award.

Warner's research involves fundamental studies in analytical chemistry as well as the development and application of new methods—chemical, instrumental, and mathematical—for analytical measurements. His goal is to provide improved methodology for the analyses of complex systems. Although his interests span general analytical chemistry, many of his studies focus on environmental analyses.

Q: What was your earliest analytical experience?

A: I had an innate interest in science. We used kerosene lamps, and when I was two, I was curious to know what the chemical was that was causing the light to glow. I opened up the cabinet where the kerosene was stored, and I tasted it. That was my first analytical experience! I ended up in the hospital for several days while they tried to pump the kerosene out of me.

Q: Having grown up in the South, did you have any experiences with segregation?

A: Here in Louisiana schools were segregated, as they were through most of the South. Our textbooks were hand-me-downs from the white schools, and we had very poor equipment. This was a disadvantage in terms of content, but the advantage I had was that my teachers believed in me. They told me that I had a gift and that there were no limitations. That kind of mentoring encouraged me to go beyond the textbooks. I had a voracious appetite for material outside the classroom. I didn't let my circumstances hold me back.

Q: How did you decide to study chemistry?

A: I had a full scholarship to Southern lined up, and my high school English teacher told me about that school's summer program in chemistry. Based on her recommendation, I got in. I did very well, and at the end of the session, the chairperson of chemistry said that if I majored in chemistry, I wouldn't have to take the freshman chemistry course. I thought that was a good deal, so that's how I chose my major. As an undergraduate, I did research in organic chemistry. From then on, I was hooked on research.

Q: Do you also have any experience in industry?

A: Working in industry had a lot to do with the times. It was the height of the Vietnam War, and student deferments were no longer being given. A large proportion of African-Americans were being drafted, and my draft board in Louisiana told me that they were going to get me no matter what. Battelle Labs in Hanford, Washington, was an Atomic Energy Commission contractor and gave me a deferral. I'd never lived up north, and it was my first time in an integrated environment. It was quite an adjustment. I did technician work in analytical chemistry, but as a technician you're not often given the opportunity to think for yourself. After 5 years at Battelle, I had a need to get my Ph.D. I wanted to be the one at the top doing the thinking.

Q: Where did you do your graduate work?

A: The best school in the area was the University of Washington. My mother-in-law had moved to Washington state, and, since I was in graduate school and my wife and I had a son and a niece living with us, it was important to have family to help us. In graduate school, I was one of two African-American students in chemistry, but I had few problems. I had an advantage since I'd worked in industry and I was more mature than most of the other students.

I'm where I am because there were key mentors in place for me who, despite outdated texts and equipment, told me that I could achieve. I owe them a lot, and the way I repay them is to work with the next generations that follow me.

Q: Now you're back at LSU. Do you feel that it's changed over the years?

A: The faculty and administration at LSU have made me feel very welcome. This is not at all the LSU I remember from my childhood—this is a new LSU in a new era! We are now the nation's number one producer of African-American Ph.D.'s in chemistry. We produced 10 last year when the entire country produced only 60 to 70. All of this has occurred since I arrived in 1992. As we've gotten more African-Americans to recognize LSU as a place where they'll feel comfortable, the quality of our students has gone through the roof! Now, as Vice Chancellor for Strategic Initiatives, I'm working to increase the number of African-American graduate students and faculty throughout the university. If we can do it in chemistry, it should be simpler in other fields.

Q: What is the focus of your current lab work?

A: We're trying to develop new spectroscopic techniques for probing the guest/host interactions of chiral drugs with novel chiral polymers developed in my laboratory. Chirality is the left- or right-handedness of molecules, and it is very important in living systems; amino acids are chiral. The bodies of living organisms are selective to chirality. For example, our bodies use only the L form of amino acids and reject the D form. With sugars, we only use the D form. An example is the drug thalidomide, in which both the L and D forms are present. At one time, the drug was given to pregnant women to combat morning sickness. One form was medicinally beneficial, but the other caused babies to be born without arms and legs and with other severe problems. Since that time, the Food and Drug Administration has recognized that chiral drugs need to be monitored carefully because, while one form might be beneficial, the other form might be deleterious. In my lab, we want to quantify chirality using fluorescence anisotropy to measure differences in the interactions of two different forms of the drug with a chiral reagent. This work can be related directly to the chromatography that we do.

Q: Are you also studying the effect of chiral pesticides on the environment?

A: Like drugs, many pesticides and herbicides are chiral. When pesticides are synthesized, both forms are made, but only one form is typically useful. After application, both forms bleed off into the water. We're looking at the degradation products of these compounds in water systems. If both forms are in the pesticide that is used and all life forms tend to be selective in their interactions with chiral molecules, will bacteria just chew up one form? And if it's the good form they chew up, will that enhance the relative concentration of the bad form and create an environmental problem?

Q: You are also pursuing research on heart plaque formation, correct?

A: I'm working with a number of other chemists to understand the formation of heart plaque. We're each working on different aspects of the problem using different tools. One thing we're looking at is the chemistry of a bypass patient's native arteries and the bypass arteries. As it turns out, the native artery is reflective of the person's lifelong chemistry, and the bypass artery is reflective of the person's chemistry since the bypass. We're comparing these two chemistries. We hope to learn if changes in body chemistry that occur later in life cause the development of heart plaque. If we can figure out the cause of the plaque, we can figure out the mechanisms of its formation.

Q: Finally, what are your thoughts on teaching and mentoring?

A: I like trying to activate young minds in the classroom and the research lab. Research is not distinct from teaching; it's teaching students how to create new knowledge. If I can activate a young mind to go beyond the textbook, I find that exciting. My wife says that I'm the only person she knows who absolutely loves his job. I love working with students and seeing them transform from naive young people to highly skilled chemists with companies fighting to hire them. To help them go through that transition gives me a great deal of joy. I feel that if I or someone like me were not here, many of them wouldn't make it.

I'm where I am because there were key mentors in place for me who, despite outdated texts and equipment, told me that I could achieve. I owe them a lot, and the way I repay them is to work with the next generations that follow me. Because of the high concentration of African-American students studying chemistry at LSU, I am often recruited to mentor minorities. I am a mentor partly just by being here, partly for the three or four African-American students that I advise, but also for the other students throughout the department. When the African-American students have problems, they're in my office talking to me. Just being here and relating my own experiences will often help them through their problems. ■

CHAPTER 29

Kinetic Methods of Analysis

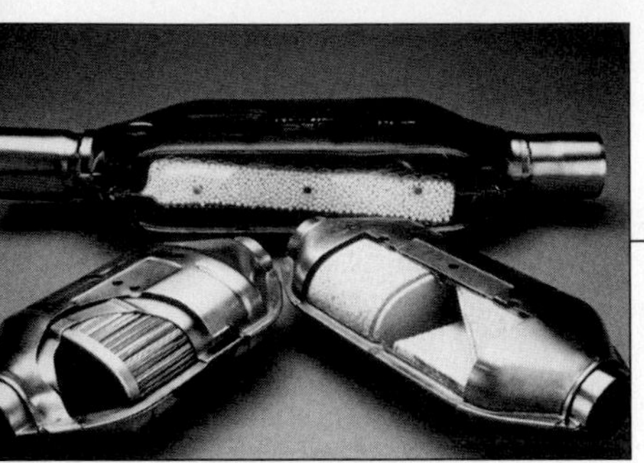

© Document General Motors/Reuter R/Corbis Sygma

Today's automobiles are equipped with a three-way catalytic converter to lower the emissions of nitrogen oxides, unburned hydrocarbons, and carbon monoxide to acceptable levels. The converter must oxidize CO and unburned hydrocarbons to CO_2 and H_2O, and it must reduce nitrogen oxides to N_2 gas. Hence, two different catalysts are used, an oxidation catalyst and a reduction catalyst. Three different converter styles are shown. Many cars use the honeycomb catalyst structure shown at the lower right to maximize the exposure of the catalysts to the exhaust stream. The catalysts are usually metals such as platinum, rhodium, or palladium.

The amount of catalyst can be determined by measuring how much the rate of a chemical reaction is affected. Catalytic methods, which are among the most sensitive of all analytical methods, are used for trace analysis of metals in the environment, organics in a variety of samples, and enzymes in biological systems.

In **kinetic methods of analysis,** measurements are made while net changes are occurring in the extent of the reaction. In **equilibrium methods of analysis,** measurements are made under conditions of equilibrium or steady state.

Kinetic methods of analysis differ in a fundamental way from the equilibrium, or thermodynamic, methods we have dealt with in previous chapters. In ***kinetic methods of analysis,*** *measurements are made under dynamic conditions in which the concentrations of reactants and products are changing as a function of time. In contrast, thermodynamic methods are performed on systems that have come to equilibrium or steady state, so that concentrations are static.*

The distinction between the two types of methods is illustrated in Figure 29-1, which shows the progress over time of the reaction

$$A + R \rightleftharpoons P \tag{29-1}$$

where A *represents the analyte,* R *the reagent, and* P *the product. Thermodynamic methods operate in the region beyond time* t_e, *when the bulk concentrations of reactants and product have become constant and the chemical system is at equilibrium. In contrast, kinetic methods are carried out during the time interval from 0 to* t_e, *when reactant and product concentrations are changing continuously.*

Selectivity in kinetic methods is achieved by choosing reagents and conditions that produce differences in the rates *at which the analyte and potential interferences react. Selectivity in thermodynamic methods is realized by choosing reagents and conditions that create differences in equilibrium constants.*

Kinetic methods greatly extend the number of chemical reactions that can be used for analytical purposes because they permit the use of reactions that are too slow or too incomplete for thermodynamic-based procedures. Kinetic methods can be based on complexation reactions, acid-base reactions, redox reactions, and others. Many kinetic methods are based on catalyzed reactions. In one type of catalytic method, the analyte is the catalyst and is determined from its effect on an

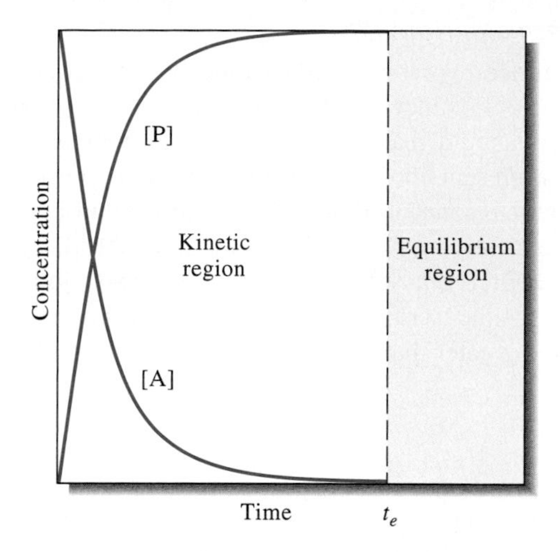

Figure 29-1 Change in concentration of analyte [A] and product [P] as a function of time. Until time t_e, the analyte and product concentrations are continuously changing. This is the kinetic regime. After t_e, the analyte and product concentrations are static.

indicator reaction involving reactants or products that are easily measured. Such methods are among the most sensitive in the chemist's repertoire. In another catalyzed reaction, the catalyst is introduced to hasten the reaction between analyte and reagent. This approach is often highly selective, or even specific, particularly when an enzyme serves as the catalyst.

Undoubtedly, the most widespread use of kinetic methods is in biochemical and clinical laboratories, where the number of analyses based on kinetics exceeds those based on thermodynamics.[1]

29A RATES OF CHEMICAL REACTIONS

This section provides a brief introduction to chemical kinetics, which is needed to understand the basis for kinetic methods of analysis.

29A-1 Reaction Mechanisms and Rate Laws

The **mechanism** of a chemical reaction consists of a series of chemical equations describing the individual elementary steps that lead to the formation of products from reactants. Much of what chemists know about mechanisms has been gained from studies in which the rate at which reactants are consumed or products are formed is measured as a function of such variables as reactant and product concentration, temperature, pressure, pH, and ionic strength. Such studies lead to an empirical **rate law** that relates the reaction rate to the concentrations of reactants, products, and intermediates at any instant. Mechanisms are derived by postulating a series of elementary steps that are chemically reasonable and consistent with the empirical rate law. Often, such mechanisms are further tested by doing studies designed to discover or monitor any transient intermediate species predicted by the mechanism.

The **rate law** for a reaction is an experimentally determined relationship between the rate of a reaction and the concentration of reactants, products, and other species such as catalysts, activators, and inhibitors.

[1] H. O. Mottola, *Kinetic Aspects of Analytical Chemistry.* New York: Wiley, 1988.

Concentration Terms in Rate Laws

Rate laws are algebraic expressions consisting of concentration terms and constants, which often look somewhat like an equilibrium-constant expression (see Equation 29-2). You should realize, however, that the square-bracketed terms in a rate expression represent molar concentrations *at a particular instant* rather than equilibrium molar concentrations (as in equilibrium-constant expressions). This meaning is frequently emphasized by adding a subscript to show the time to which the concentration refers. Thus, $[A]_t$, $[A]_0$, and $[A]_\infty$ indicate the concentration of A at time t, time zero, and infinite time, respectively. Infinite time is regarded as any time greater than required for equilibrium to be achieved. That is, $t_\infty > t_e$ in Figure 29-1.

▶ Molar concentrations are still symbolized with square brackets. In the context of kinetic methods, however, their numerical values change with time.

Reaction Order

Let us assume that the empirical rate law for the general reaction shown as Equation 29-1 is found by experiment to take the form

$$\text{rate} = -\frac{d[A]}{dt} = -\frac{d[R]}{dt} = \frac{d[P]}{dt} = k[A]^m[R]^n \qquad (29\text{-}2)$$

where the rate is the derivative of the concentration of A, R, or P with respect to time. Note that the first two rates carry a negative sign because the concentrations of A and R decrease as the reaction proceeds. In this rate expression, k is the **rate constant,** m is the **order of the reaction with respect to A,** and n is the **order of the reaction with respect to R.** The **overall order of the reaction** is $p = m + n$. Thus, if $m = 1$ and $n = 2$, the reaction is said to be first order in A, second order in R, and third order overall.

▶ Because A and R are being used up, the rates of change of [A] and [R] with respect to time are negative.

Units for Rate Constants

Since reaction rates are always expressed in terms of concentration per unit time, the units of the rate constant are determined by the overall order p of the reaction according to the relation

$$\frac{\text{concentration}}{\text{time}} = (\text{units of } k)(\text{concentration})^p$$

where $p = m + n$. Rearranging leads to

$$\text{units of } k = (\text{concentration})^{1-p} \times \text{time}^{-1}$$

Thus, the units for a first-order rate constant are s^{-1}, and the units for a second-order rate constant are $M^{-1}\,s^{-1}$.

▶ The units of the rate constant k for a first-order reaction are s^{-1}.

29A-2 The Rate Law for First-Order Reactions

The simplest case in the mathematical analysis of reaction kinetics is the spontaneous irreversible decomposition of a species A:

▶ Radioactive decay is an example of spontaneous decomposition.

$$A \xrightarrow{k} P \qquad (29\text{-}3)$$

The reaction is first order in A, and the rate is

$$\text{rate} = -\frac{d[\text{A}]}{dt} = k[\text{A}] \tag{29-4}$$

Pseudo–First-Order Reactions

A first-order decomposition reaction per se is generally of no use in analytical chemistry because an analysis is ordinarily based on reactions involving at least two species, an analyte and a reagent.[2] Usually, however, the rate law for a reaction involving two species is sufficiently complex that simplifications are needed for analytical purposes. In fact, the majority of useful kinetic methods are performed under conditions that permit the chemist to simplify complex rate laws to a form analogous to Equation 29-4. A higher order reaction that is carried out so that such a simplification is feasible is termed a **pseudo–first-order reaction.** Methods of converting higher order reactions to pseudo–first-order reactions are dealt with in later sections.

Mathematics Describing First-Order Behavior

Because the vast majority of kinetic determinations are performed under pseudo–first-order conditions, it is worthwhile to examine in detail some of the characteristics of reactions having rate laws that approximate Equation 29-4.

By rearranging Equation 29-4, we obtain

$$\frac{d[\text{A}]}{[\text{A}]} = -kdt \tag{29-5}$$

The integral of this equation from time zero, when $[\text{A}] = [\text{A}]_0$, to time t, when $[\text{A}] = [\text{A}]_t$, is

$$\int_{[A]_0}^{[A]_t} \frac{d[\text{A}]}{[\text{A}]} = -k\int_0^t dt$$

Evaluation of the integrals gives

$$\ln\frac{[\text{A}]_t}{[\text{A}]_0} = -kt \tag{29-6}$$

Finally, by taking the exponential of both sides of Equation 29-6, we obtain

$$\frac{[\text{A}]_t}{[\text{A}]_0} = e^{-kt} \qquad \text{or} \qquad [\text{A}]_t = [\text{A}]_0 e^{-kt} \tag{29-7}$$

This integrated form of the rate law gives the concentration of A as a function of the initial concentration $[\text{A}]_0$, the rate constant k, and the time t. A plot of this relationship is depicted in Figure 29-1. Example 29-1 illustrates the use of this equation in finding a reactant concentration at a particular time.

[2]Radioactive decay is an exception to this statement. The technique of neutron activation analysis is based on the measurement of the spontaneous decay of radionuclides created by irradiation of a sample in a nuclear reactor.

EXAMPLE 29-1

A reaction is first order with $k = 0.0370\ \text{s}^{-1}$. Calculate the concentration of reactant remaining 18.2 s after initiation of the reaction if its initial concentration is 0.0100 M.

Substituting into Equation 29-7 gives

$$[\text{A}]_{18.2} = (0.0100\ \text{M})e^{-(0.0370\ \text{s}^{-1}) \times (18.2\ \text{s})} = 0.00510\ \text{M}$$

When the rate of a reaction is being followed by the rate of appearance of a product P rather than the rate of disappearance of analyte A, it is useful to modify Equation 29-7 to relate the concentration of P at time t to the initial analyte concentration $[\text{A}]_0$. The concentration of A at any time is equal to its original concentration minus the concentration of product (when 1 mol of product forms for 1 mol of analyte). Thus

$$[\text{A}]_t = [\text{A}]_0 - [\text{P}]_t \tag{29-8}$$

Substituting this expression for $[\text{A}]_t$ into Equation 29-7 and rearranging gives

$$[\text{P}]_t = [\text{A}]_0(1 - e^{-kt}) \tag{29-9}$$

A plot of this relationship is also shown in Figure 29-1.

The form of Equations 29-7 and 29-9 is that of a pure exponential, which appears widely in science and engineering. A pure exponential in this case has the useful characteristic that equal elapsed times give equal fractional decreases in reactant concentration or increases in product concentration. As an example, consider a time interval $t = \tau = 1/k$, which on substitution into Equation 29-7 gives

► The fraction of reactant used (or product formed) in a first-order reaction is the same for any given period of time.

$$[\text{A}]_\tau = [\text{A}]_0 e^{-k\tau} = [\text{A}]_0 e^{-k/k} = (1/e)[\text{A}]_0$$

and likewise for a period $t = 2\tau = 2/k$,

$$[\text{A}]_{2\tau} = (1/e^2)[\text{A}]_0$$

and so on for successive periods, as shown in Figure 29-2.

► CHALLENGE: Derive an expression for $t_{1/2}$ in terms of τ.

The period $\tau = 1/k$ is sometimes referred to as the **natural lifetime** of species A. During time τ, the concentration of A decreases to $1/e$ of its original value. A second period, from $t = \tau$ to $t = 2\tau$, produces an equivalent fractional decrease in concentration to $1/e$ of the value at the beginning of the second interval, which is $(1/e)^2$ of $[\text{A}]_0$. A more familiar example of this property of exponentials is found in the half-life $t_{1/2}$ of radionuclides. During a period $t_{1/2}$, half of the atoms in a sample of a radioactive element decay to products; a second period of $t_{1/2}$ reduces the amount of the element to one quarter of its original number, and so on for succeeding periods. Regardless of the time interval chosen, equal elapsed times produce equal fractional decreases in reactant concentration for a first-order process.

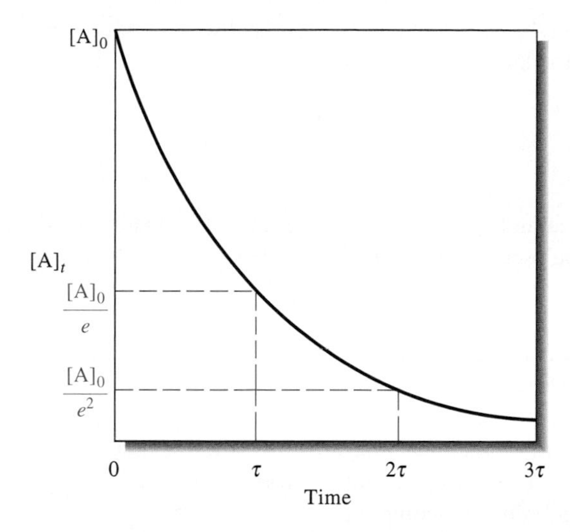

Figure 29-2 Progress curve for a first-order reaction showing that equal elapsed times produce equal fractional decreases in analyte concentration.

EXAMPLE 29-2

Calculate the time required for a first-order reaction with $k = 0.0500\ s^{-1}$ to proceed to 99.0% completion.

For 99.0% completion, $[A]_t/[A]_0 = (100 - 99)/100 = 0.010$; substitution into Equation 29-6 then gives

$$\ln 0.010 = -kt = -(0.0500\ s^{-1})t$$

$$t = \frac{\ln 0.010}{0.0500\ s^{-1}} = 92\ s$$

29A-3 Rate Laws for Second-Order and Pseudo–First-Order Reactions

Consider a typical analytical reaction in which 1 mol of analyte A reacts with 1 mol of reagent B to give a single product P. For now, we assume the reaction is irreversible and write

$$A + R \xrightarrow{k} P \tag{29-10}$$

If the reaction occurs in a single elementary step, the rate is proportional to the concentration of each of the reactants, and the rate law is

$$-\frac{d[A]}{dt} = k[A][R] \tag{29-11}$$

The reaction is first order in each of the reactants and second order overall. If the concentration of R is chosen such that $[R] \gg [A]$, the concentration of R changes very little during the course of the reaction, and we can write $k[R]$ = constant = k'. Equation 29-11 is then rewritten as

$$-\frac{d[A]}{dt} = k'[A] \tag{29-12}$$

► Second-order or other higher order reactions can usually be made pseudo–first-order through control of experimental conditions.

which is identical in form to the first-order case of Equation 29-4. Hence, the reaction is said to be pseudo–first-order in A (see Example 29-3).

EXAMPLE 29-3

For a pseudo–first-order reaction in which the reagent is present in 100-fold excess, find the relative error resulting from the assumption that $k[R]$ is constant when the reaction is 40% complete.

The initial concentration of the reagent can be expressed as

$$[R]_0 = 100[A]_0$$

At 40% reaction, 60% of A remains. Thus,

$$[A]_{40\%} = 0.60[A]_0$$

$$[R]_{40\%} = [R]_0 - 0.40[A]_0 = 100[A]_0 - 0.40[A]_0 = 99.6[A]_0$$

Assuming pseudo–first-order behavior, the rate at 40% reaction is

$$-\frac{d[A]_{40\%}}{dt} = k[R]_0[A]_{40\%}$$

The true rate at 40% reaction is $k(99.6[A]_0)(0.60[A]_0)$. Thus, the relative error is

$$\frac{k(100[A]_0)(0.60[A]_0) - k(99.6[A]_0)(0.60[A]_0)}{k(99.6[A]_0)(0.60[A]_0)} = 0.004 \qquad \text{(or 0.4 \%)}$$

As Example 29-3 shows, the error associated with the determination of the rate of a pseudo–first-order reaction with a 100-fold excess of reagent is quite small. A 50-fold reagent excess leads to a 1% error, which is usually deemed acceptable in kinetic methods. Moreover, the error is even less significant at times when the reaction is less than 40% complete.

Reactions are seldom completely irreversible, and a rigorous description of the kinetics of a second-order reaction that occurs in a single step must take into account the reverse reaction. The rate of the reaction is the difference between the forward rate and the reverse rate:

$$-\frac{d[A]}{dt} = k_1[A][R] - k_{-1}[P]$$

where k_1 is the second-order rate constant for the forward reaction and k_{-1} is the first-order rate constant for the reverse reaction. In deriving this equation, we have assumed for simplicity that a single product is formed, but more complex cases can be described as well.[3] As long as conditions are maintained such that k_{-1} or [P], or both, is relatively small, the rate of the reverse reaction is negligible, and little error is introduced by the assumption of pseudo–first-order behavior.

Spreadsheet Summary In Chapter 13 of *Applications of Microsoft® Excel in Analytical Chemistry,* the first exercise explores the properties of first- and second-order reactions. The time behavior of both types of reactions is considered, and linear plotting methods are studied. Conditions needed to obtain pseudo–first-order behavior are also investigated.

29A-4 Catalyzed Reactions

Catalyzed reactions, particularly those in which enzymes serve as catalysts, are widely used for the determination of a variety of biological and biochemical species as well as a number of inorganic cations and anions. We shall therefore use enzyme-catalyzed reactions to illustrate catalytic rate laws and to show how these rate laws can be reduced to relatively simple algebraic relationships, such as the pseudo–first-order Equation 29-12. These simplified relationships can then be used for analytical purposes.

◀ Enzymes are high-molecular-weight molecules that catalyze reactions in biological systems. They can serve as highly selective analytical reagents.

Enzyme-Catalyzed Reactions

Enzymes are high-molecular-weight protein molecules that catalyze reactions of importance in biology and biomedicine. Feature 29-1 discusses the basic features of enzymes. Enzymes are particularly useful as analytical reagents because of their selectivity. Consequently, they are widely used in the determination of molecules with which they combine when acting as catalysts. Such molecules are usually designated as **substrates.** In addition to the determination of substrates, enzyme-catalyzed reactions are employed for the determination of activators, inhibitors, and, of course, enzymes themselves.[4]

The species acted on by an enzyme is called a **substrate.** Species that enhance the rate of a reaction but do not take part in the stoichiometric reaction are called **activators**. Species that do not participate in the stoichiometric reaction but decrease the reaction rate are called **inhibitors.**

The behavior of a large number of enzymes is consistent with the general mechanism

$$E + S \underset{k_{-1}}{\overset{k_1}{\rightleftharpoons}} ES \xrightarrow{k_2} P + E \tag{29-13}$$

In this so-called **Michaelis-Menten mechanism,** the enzyme E reacts reversibly with the substrate S to form an enzyme-substrate complex ES. This complex then decomposes irreversibly to form the product(s) and the regenerated enzyme. The rate law for this mechanism assumes one of two forms, depending on the relative rates of the two steps. In the most general case, the rates of the two steps are fairly comparable in magnitude. Here, ES decomposes as rapidly as it is

[3]See J. H. Espenson, *Chemical Kinetics and Reaction Mechanisms,* 2nd ed., pp. 49–52. New York: McGraw Hill, 1995.

[4]For a recent review of catalyzed reactions for kinetic methods, see S. R. Crouch, A. Scheeline, and E.W. Kirkor, *Anal. Chem.,* **2000,** *72,* 53R.

FEATURE 29-1

Enzymes

Enzymes are proteins that catalyze reactions necessary to sustain life. Like other proteins, enzymes consist of chains of amino acids. The structural formulas of a few important amino acids are shown in Figure 29F-1. Molecules formed by linking two or more amino acids are called **peptides.** Each amino acid in a peptide is called a **residue.** Molecules with many amino acid linkages are **polypeptides,** and those with long polypeptide chains are **proteins.** Enzymes differ from other proteins in that a specific area of the structure, called the active site, assists in the catalysis. As a result, enzyme catalysis is often quite specific, favoring a particular substrate over other closely related compounds.

The protein structure is very important to its function. The **primary structure** is the sequence of amino acids in the protein. The **secondary structure** is the shape that the polypeptide chain assumes. Two types of secondary structures exist, the α-helix and the β-pleated sheet. The α-helix depicted in Figure 29F-2 is the most common shape adopted by animal proteins. In this structure, the helical shape is maintained by hydrogen bonds between neighboring residues. The β-pleated sheet structure is shown in Figure 29F-3. In this structure, the peptide chain is nearly fully extended, and the hydrogen bonding is between parallel sections of peptide chains rather than between close neighbors, as in the α-helix. The β-pleated sheet structure is found in fibers like silk.

The tertiary structure is the overall three-dimensional shape into which the α-helix or β-pleated sheet folds as a result of interactions between residues far apart in the primary structure. Proteins may also have a quaternary structure, which describes how polypeptide chains stack together in a multichain protein.

The effectiveness of an enzyme as a catalyst is called the **enzyme activity.** The activity is closely linked to the three-dimensional shape of the protein, particularly to its active site. In general, the active site is a part of the protein that binds the substrate. The specificity of the enzyme depends to a large extent on the structure in the active site region. One explanation of the role of the active site is the "lock and key" model. Here, the precise stereochemical fit of the substrate to the active site is deemed responsible for

glycine (gly): $H_2N-CH(H)-C(=O)-OH$

valine (val): $H_2N-CH(CH(CH_3)CH_3)-C(=O)-OH$

alanine (ala): $H_2N-CH(CH_3)-C(=O)-OH$

serine (ser): $H_2N-CH(CH_2OH)-C(=O)-OH$

tyrosine (tyr): $H_2N-CH(CH_2C_6H_4OH)-C(=O)-OH$

cysteine (cys): $H_2N-CH(CH_2SH)-C(=O)-OH$

Figure 29F-1 Some important amino acids. There are 20 different amino acids found in nature.

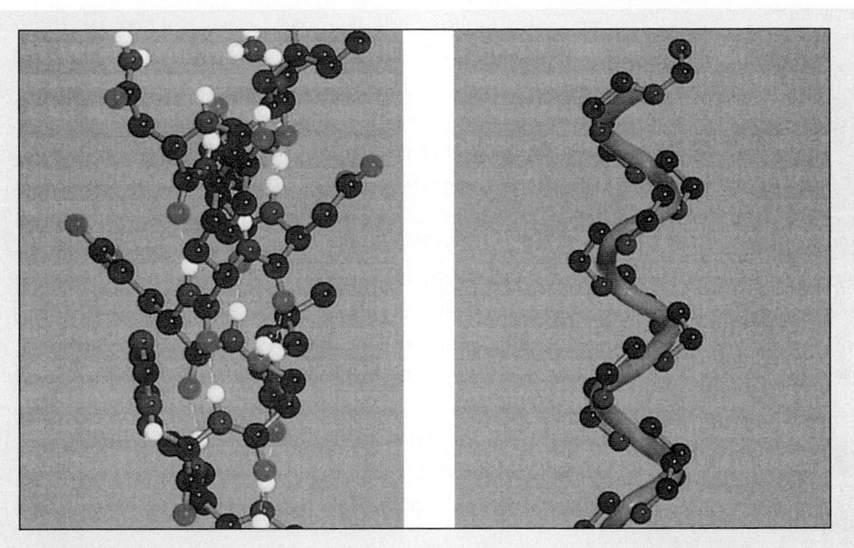

Figure 29F-2 The α-helix. In the left model, the hydrogen bonding between neighboring amino acid residues is shown, leading to the helical structure. In the right model, only the atoms in the polypeptide chain are shown to reveal more clearly the helical structure. (From D. L. Reger, S. R. Goode, and E. E. Mercer, *Chemistry: Principles and Practice.* Belmont, CA: Brooks/Cole, 1993.)

the specificity of the catalysis. Several more complex models, such as the induced-fit model, have been proposed.

A huge number of enzymes have been discovered, but only a fraction of these have been isolated and purified. The commercial availability of some of the most useful enzymes has spurred a good deal of interest in their analytical use. Enzymes have been covalently bonded to solid supports or have been encapsulated in gels and membranes to make them reusable and to decrease the analysis cost.

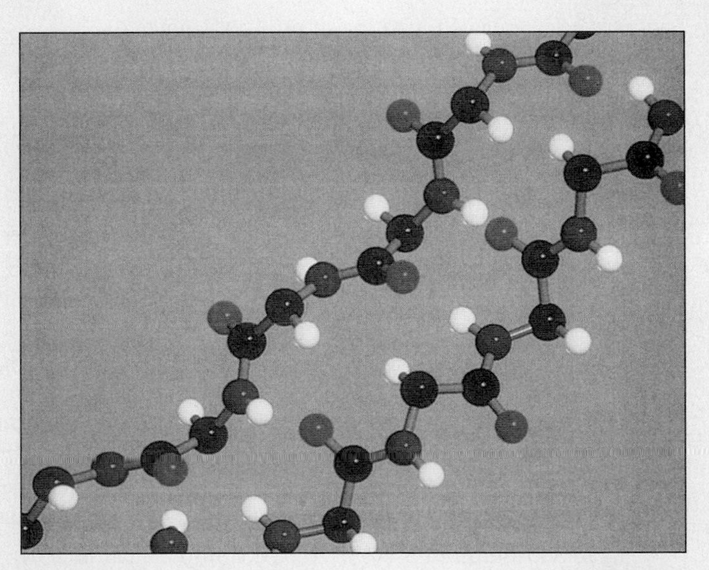

Figure 29F-3 The β-pleated sheet. Note that hydrogen bonding is between different sections of a polypeptide chain or between different chains, leading to a more extended structure. (From D. L. Reger, S. R. Goode, and E. E. Mercer, *Chemistry: Principles and Practice.* Belmont, CA: Brooks/Cole, 1993.)

formed, and its concentration can be assumed to be small and relatively constant throughout much of the reaction. If the second step is considerably slower than the first (case 1), the reactants and ES are always essentially at equilibrium. This so-called equilibrium case is readily derived from the general case. In the sections that follow, we show that in both cases, the reaction conditions can be arranged to yield simple relationships between rate and analyte concentration.

Steady-State Case

In the most general treatment, the rate law corresponding to the mechanism of Equation 29-13 is derived by using the **steady-state approximation.** In this approximation, the concentration of ES is presumed to be small and relatively constant throughout the reaction. The enzyme-substrate complex forms in the first step with rate constant k_1. It decomposes by two paths: the reverse of the first step (rate constant k_{-1}) and the second step to form product (rate constant k_2). Assuming that [ES] stays constant throughout is the same as assuming that the rate of change of [ES], $d[\mathrm{ES}]/dt$, is zero. Thus, mathematically, the steady-state assumption is written

$$\frac{d[\mathrm{ES}]}{dt} = k_1[\mathrm{E}][\mathrm{S}] - k_{-1}[\mathrm{ES}] - k_2[\mathrm{ES}] = 0 \tag{29-14}$$

In Equation 29-14, the concentrations of enzyme [E] and substrate refer to the free concentrations at any time t. Usually, we want to express the rate law in terms of the total concentration of enzyme, which is known or measurable. By mass balance, the total (initial) enzyme concentration $[\mathrm{E}]_0$ is given by

$$[\mathrm{E}]_0 = [\mathrm{E}] + [\mathrm{ES}] \tag{29-15}$$

The rate of formation of product is given by

$$\frac{d[\mathrm{P}]}{dt} = k_2[\mathrm{ES}] \tag{29-16}$$

If we solve Equation 29-14 for [ES], we obtain

$$[\mathrm{ES}] = \frac{k_1[\mathrm{E}][\mathrm{S}]}{k_{-1} + k_2} \tag{29-17}$$

If we now substitute for [E] the expression given in Equation 29-15 and re-solve for [ES], we obtain

$$[\mathrm{ES}] = \frac{k_1[\mathrm{E}]_0[\mathrm{S}]}{k_{-1} + k_2 + k_1[\mathrm{S}]} \tag{29-18}$$

Substituting this value for [ES] into Equation 29-16 and rearranging leads to the rate law

$$\frac{d[\mathrm{P}]}{dt} = \frac{k_2[\mathrm{E}]_0[\mathrm{S}]}{\dfrac{k_{-1} + k_2}{k_1} + [\mathrm{S}]} = \frac{k_2[\mathrm{E}]_0[\mathrm{S}]}{K_\mathrm{m} + [\mathrm{S}]} \tag{29-19}$$

where the term $K_m = (k_{-1} + k_2)/k_1$ is known as the **Michaelis constant.** Equation 29-19 is often called the **Michaelis-Menten equation.** From Equation 29-17, it can be seen that the Michaelis constant K_m is given by

$$K_m = \frac{k_{-1} + k_2}{k_1} = \frac{[E][S]}{[ES]} \tag{29-20}$$

The Michaelis constant is quite similar to the equilibrium constant for the dissociation of the enzyme-substrate complex. It is sometimes referred to as a **pseudo–equilibrium constant** since the k_2 in the numerator prevents it from being a "true" equilibrium constant. The Michaelis constant is usually expressed in units of millimoles/liter (mM) and ranges from 0.01 to 100 mM for many enzymes, as can be seen in Table 29-1.

The rate equation given in Equation 29-19 can be simplified so that the reaction rate is proportional to either enzyme or substrate concentration. For example, if the concentration of substrate is large enough to greatly exceed the Michaelis constant, $[S] \gg K_m$, Equation 29-19 reduces to

$$\frac{d[P]}{dt} = k_2[E]_0 \tag{29-21}$$

◀ To determine enzymes, the substrate concentration should be large compared with the Michaelis constant, $[S] \gg K_m$.

Under these conditions, when the rate is independent of substrate concentration, the reaction is said to be **pseudo–zero-order** in substrate, and the rate is directly proportional to the concentration of enzyme. The enzyme is said to be **saturated** with substrate.

When conditions are such that the concentration of S is small or K_m is relatively large, then $[S] \ll K_m$, and Equation 29-19 simplifies to

$$\frac{d[P]}{dt} = \frac{k_2}{K_m}[E]_0[S] = k'[S]$$

TABLE 29-1
Michaelis Constants for Some Enzymes

Enzyme	Substrate	K_m, mM
Alkaline phosphatase	*p*-Nitrophenylphosphate	0.1
Catalase	H_2O_2	25
Hexokinase	Glucose	0.15
	Fructose	1.5
Creatine phosphokinase	Creatine	19
Carbonic anhydrase	HCO_3^-	9.0
Chymotrypsin	*n*-Benzoyltyrosinamide	2.5
	n-Formyltyrosinamide	12.0
	n-Acetyltyrosinamide	32
	Glycyltyrosinamide	122
Glucose oxidase	Glucose saturated with O_2	0.013
Lactate dehydrogenase	Lactate	8.0
	Pyruvate	0.125
L-amino acid oxidase	L-leucine	1.0
Urease	Urea	2.0
Uricase	Uric acid saturated with O_2	0.0175

D-glucose D-fructose

Molecular models of glucose and fructose. Glucose and fructose are important monosaccharides. Glucose is a polyhydroxyaldehyde, while fructose is a polyhydroxyketone. Glucose is the primary fuel for biological cells. Fructose is the major sugar in fruits and vegetables. Both sugars are substrates for one or more enzymes.

where $k' = k_2[E]_0/K_m$. Hence, the kinetics are first order in substrate. To use this equation to determine analyte concentrations, it is necessary to measure $d[P]/dt$ at the beginning of the reaction, where $[S] \approx [S]_0$, so that

$$\frac{d[P]}{dt} \approx k'[S]_0 \tag{29-22}$$

► To determine substrates, conditions should be arranged such that the substrate concentration is low compared with the Michaelis constant: $[S] \ll K_m$.

The regions where Equations 29-21 and 29-22 are applicable are illustrated in Figure 29-3, in which the initial rate of an enzyme-catalyzed reaction is plotted as a function of substrate concentration. When substrate concentration is low, Equation 29-22, which is linear in substrate concentration, governs the shape of the curve. It is this region that is used to determine the amount of substrate present.

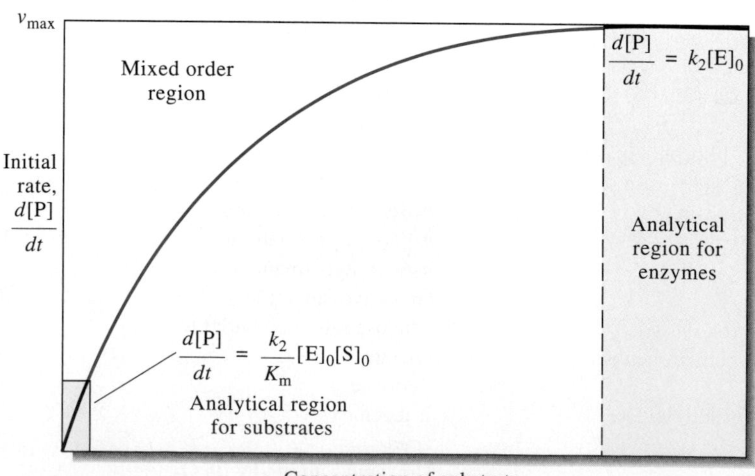

Figure 29-3 Plot of initial rate of product formation as a function of substrate concentration, showing the parts of the curve useful for the determination of substrate and enzyme.

If we wish to determine the amount of enzyme, the region of high substrate concentration is employed—where Equation 29-21 applies—and the rate is independent of substrate concentration. The limiting rate of the reaction at large values of [S] is the maximum rate that can be achieved at a given enzyme concentration, v_{max}, as indicated in the figure. It can be shown that the value of the substrate concentration at exactly $v_{max}/2$ is equal to the Michaelis constant K_m. Example 29-4 illustrates the use of the Michaelis-Menten equation.

EXAMPLE 29-4

The enzyme urease, which catalyzes the hydrolysis of urea, is widely used to determine urea in blood. Details of this application are given in Feature 29-3 on page 901. The Michaelis constant for urease at room temperature is 2.0 mM, and $k_2 = 2.5 \times 10^4\ s^{-1}$ at pH 7.5. (a) Calculate the initial rate of the reaction when the urea concentration is 0.030 mM and the urease concentration is 5.0 μM, and (b) find v_{max}.

(a) From Equation 29-19,

$$\frac{d[P]}{dt} = \frac{k_2[E]_0[S]}{k_m + [S]}$$

At the beginning of the reaction, $[S] = [S]_0$ and

$$\frac{d[P]}{dt} = \frac{(2.5 \times 10^4\,s^{-1})(5.0 \times 10^{-6}\,M)(0.030 \times 10^{-3}\,M)}{2.0 \times 10^{-3}\,M + 0.030 \times 10^{-3}\,M}$$

$$= 1.8 \times 10^{-3}\,M\,s^{-1}$$

(b) Figure 29-3 reveals that $d[P]/dt = v_{max}$ when the concentration of substrate is large and Equation 29-21 applies. Thus,

$$d[P]/dt = v_{max} = k_2[E]_0 = (2.5 \times 10^4\,s^{-1})(5.0 \times 10^{-6}\,M) = 0.125\,M\,s^{-1}$$

The Equilibrium Case

The equilibrium case is readily obtained from the general, steady-state case just discussed. When the conversion of ES to products is slow compared with the reversible first step of Equation 29-13, the first step is essentially at equilibrium throughout. Mathematically, this occurs when k_2 is much smaller than k_{-1}. Under these conditions, Equation 29-19 becomes

$$\frac{d[P]}{dt} = \frac{k_2[E]_0[S]}{\frac{k_{-1}}{k_1} + [S]} = \frac{k_2[E]_0[S]}{K + [S]} \qquad (29\text{-}23)$$

where the constant K is now a true equilibrium constant given by $K = k_{-1}/k_1$. Note that the form of Equation 29-23 is identical to the Michaelis-Menten equation (see Equation 29-19). There is only a subtle difference in the definitions of K_m and K.

Hence, substrate and enzyme concentrations can be determined in the same manner as for the steady-state case for enzyme reactions in which k_2 is small and the equilibrium assumption holds. Enzyme concentrations are determined under conditions in which the substrate concentration is large, while substrate concentrations are determined when $[S] \ll K$.

There are many more complex mechanisms for enzyme reactions involving reversible reactions, multiple substrates, activators, and inhibitors. Techniques for modeling and analyzing these systems are available.[5]

Although our discussion thus far has been concerned with enzymatic methods, an analogous treatment for ordinary catalysis gives rate laws that are similar in form to those for enzymes. These expressions often reduce to the first-order case for ease of data treatment, and many examples of kinetic-catalytic methods are found in the literature.[6]

Spreadsheet Summary The second exercise in Chapter 13 of *Applications of Microsoft® Excel in Analytical Chemistry* involves enzyme catalysis. A linear transformation is made so that the Michaelis constant, K_m, and the maximum velocity, v_{max}, can be determined from a least-squares procedure. The nonlinear regression method is used with Excel's Solver to find these parameters by fitting them into the nonlinear Michaelis-Menten equation.

29B DETERMINING REACTION RATES

Several methods are used for the determination of reaction rates. In this section, we describe some of these methods and when they are used in the course of a reaction.

29B-1 Experimental Methods

A **fast reaction** is 50% complete in 10 s or less.

The method by which reaction rates are measured depends on whether the reaction of interest is fast or slow. A reaction is generally regarded as fast if it proceeds to 50% of completion in 10 s or less. Analytical methods based on fast reactions generally require special equipment that permits rapid mixing of reagents and fast recording of data, as discussed in Feature 29-2.

FEATURE 29-2

Fast Reactions and Stopped-Flow Mixing

One of the most popular and reliable methods of carrying out rapid reactions is stopped-flow mixing. In this technique, streams of reagent and sample are mixed rapidly, and the flow of mixed solution is stopped suddenly. The progress of the reaction is then monitored at a position slightly downstream from the mixing point. The apparatus shown in Figure 29F-4 is designed to perform stopped-flow mixing.

[5]See, for example, I. H. Segel, *Enzyme Kinetics*. New York: Wiley, 1975; C. F. Lam, *Techniques for the Analysis and Modelling of Enzyme Kinetic Mechanisms*. Chichester: Research Studies Press-Wiley, 1981.

[6]See D. Perez-Bendito and M. Silva, *Kinetic Methods in Analytical Chemistry*. New York: Halsted Press-Wiley, 1988; H. A. Mottola, *Kinetic Aspects of Analytical Chemistry*. New York: Wiley, 1988.

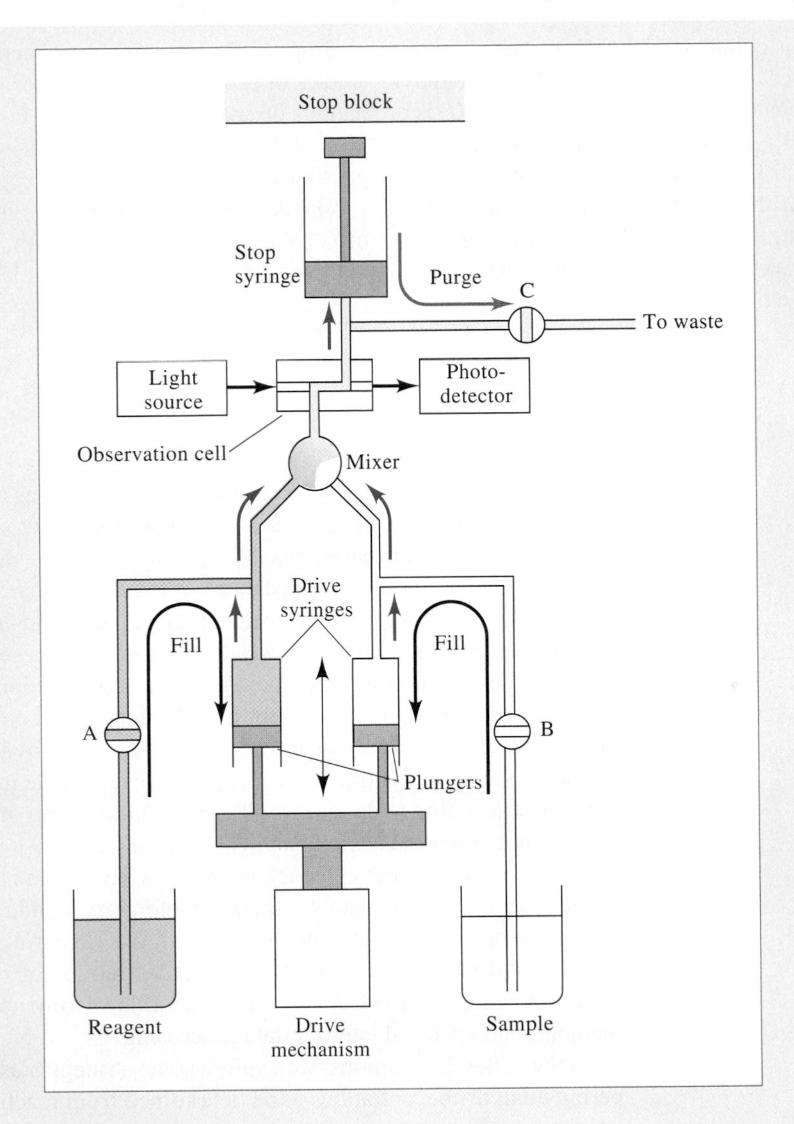

Figure 29F-4 Stopped-flow mixing apparatus.

To illustrate the operation of this apparatus, we begin with the drive syringes filled with reagent and sample and with valves A, B, and C closed. The stop syringe is empty. The drive mechanism is then activated to move the drive syringe plungers forward rapidly. The reagent and sample pass into the mixer, where they are mixed, and immediately into the observation cell, as indicated by the colored arrows. The reaction mixture then passes into the stop syringe. Eventually, the stop syringe fills and the stop syringe plunger strikes the stop block. This event causes the flow to cease almost instantly, with a recently mixed plug of solution in the observation cell. In this example, the observation cell is transparent so that a light beam can pass to make absorption measurements. In this way, the progress of the reaction can be monitored. All that is required is that the dead time, or the time between the mixing of reagents and the arrival of the sample in the observation cell, be short relative to the time required for the reaction to proceed to completion. For well-designed systems in which the turbulent flow of the mixer provides very rapid and efficient mixing, the dead time is on the order of 2 to 4 ms. Thus, first-order or pseudo–first-order reactions with $\tau \approx 25$ ms ($k \approx 40\ \text{s}^{-1}$) can be examined using the stopped-flow technique.

(continued)

When the reaction is complete, valve C is opened, and the stop syringe plunger is then pushed down to purge the stop syringe of its contents (gray arrow). Valve C is closed, valves A and B are opened, and the drive mechanism is moved down to fill the drive syringes with solution (black arrows). At this point, the apparatus is ready for another rapid mixing experiment. The entire apparatus can be placed under the control of a computer, which can also collect and analyze the reaction rate data.

Stopped-flow mixing has been used for fundamental studies of rapid reactions and for routine kinetic determinations of analytes involved in fast reactions. The principles of fluid dynamics that make stopped-flow mixing possible and the solution-handling capabilities of this and similar devices are used in many contexts to automatically mix solutions and measure analyte concentrations in numerous industrial and clinical laboratories.

If a reaction is sufficiently slow, conventional methods of analysis can be used to determine the concentration of a reactant or product as a function of time. Often, however, the reaction of interest is too rapid for many static measurement techniques—that is, concentrations change appreciably during the measurement process. Under these circumstances, either the reaction must be stopped (quenched) while the measurement is made or an instrumental technique that records concentrations continuously as the reaction proceeds must be employed. In the former case, an aliquot is removed from the reaction mixture and is rapidly quenched by mixing with a reagent that combines with one of the reactants to stop the reaction. Alternatively, quenching is accomplished by lowering the temperature rapidly to slow the reaction to an acceptable level for the measurement step. Unfortunately, quenching techniques tend to be laborious and often time consuming and are thus not widely used for analytical purposes.

The most convenient approach to obtain kinetic data is to monitor the progress of the reaction continuously by spectrophotometry, conductometry, potentiometry, or some other instrumental technique. With the advent of inexpensive computers, instrumental readings proportional to concentrations of reactants or products, or both, are often recorded directly as a function of time, stored in the computer's memory, and retrieved later for data processing.

In the following sections, we explore some strategies used in kinetic methods to permit analyte concentrations to be determined from reaction progress plots.

29B-2 Types of Kinetic Methods

Kinetic methods are classified according to the type of relationship that exists between the measured variable and the analyte concentration.

The Differential Method

In the **differential method,** concentrations are computed from reaction rates by means of a differential form of a rate expression. Rates are determined by measuring the slope of a curve relating analyte or product concentration to reaction time. To illustrate, let us substitute $[A]_t$ from Equation 29-7 for [A] in Equation 29-4:

$$\text{rate} = -\left(\frac{d[\text{A}]}{dt}\right) = k[\text{A}]_t = k[\text{A}]_0 e^{-kt} \qquad (29\text{-}24)$$

As an alternative, the rate can be expressed in terms of the product concentration. That is,

$$\text{rate} = \left(\frac{d[\mathrm{P}]}{dt}\right) = k[A]_0 e^{-kt} \qquad (29\text{-}25)$$

Equations 29-24 and 29-25 show the dependence of the rate on k, t, and, most important, $[\mathrm{A}]_0$, the initial concentration of the analyte. At any fixed time t, the factor $ke^{-kt}$is a constant, and the rate is directly proportional to the initial analyte concentration. Example 29-5 illustrates the use of the differential method to calculate the initial analyte concentration.

EXAMPLE 29-5

The rate constant for a pseudo–first-order reaction is $0.156\ \mathrm{s}^{-1}$. Find the initial concentration of the reactant if its rate of disappearance 10.00 s after the initiation of the reaction is $2.79 \times 10^{-4}\ \mathrm{M\ s}^{-1}$.

The proportionality constant ke^{-kt} is

$$ke^{-kt} = (0.156\ \mathrm{s}^{-1})e^{-(0.156\ \mathrm{s}^{-1})(10.00\ \mathrm{s})} = 3.28 \times 10^{-2}\ \mathrm{s}^{-1}$$

Rearranging Equation 29-24 and substituting numerical values, we have

$$\begin{aligned}[\mathrm{A}]_0 &= \text{rate}/ke^{-kt}\\ &= (2.79 \times 10^{-4}\ \mathrm{M\ s}^{-1})/(3.28 \times 10^{-2}\ \mathrm{s}^{-1})\\ &= 8.51 \times 10^{-3}\ \mathrm{M}\end{aligned}$$

The choice of the time at which a reaction rate is measured is often based on such factors as convenience, the existence of interfering side reactions, and the inherent precision of making the measurement at a particular time. It is often advantageous to make the measurement near $t = 0$ because this portion of the exponential curve is nearly linear (see, for example, the initial parts of the curves in Figure 29-1) and the slope is readily estimated from the tangent to the curve. Moreover, if the reaction is pseudo–first-order, such a small amount of excess reagent is consumed that no error arises from changes in k that result from changes in reagent concentration. Finally, the relative error in determining the slope is minimal at the beginning of the reaction because the slope is at a maximum in this region.

Figure 29-4 illustrates how the differential method is used to determine the concentration of an analyte $[\mathrm{A}]_0$ from experimental rate measurements for the reaction shown as Equation 29-1. The solid curves in Figure 29-4a are plots of the experimentally measured product concentration [P] as a function of reaction time for four standard solutions of A. These curves are used to prepare the differential calibration plot shown in Figure 29-4b. To obtain the rates, tangents are drawn to each of the curves in Figure 29-4a at a time near zero (dashed lines in part a). The slopes of the tangents are then plotted as a function of [A], giving the straight line shown in Figure 29-4b. Unknowns are treated in the same way, and analyte concentrations are determined from the calibration curve.

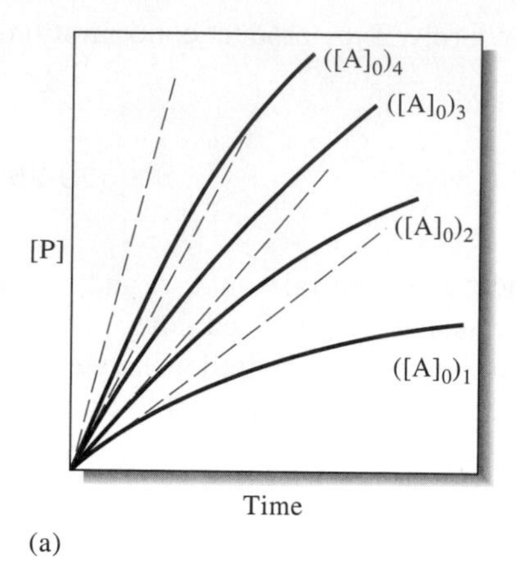

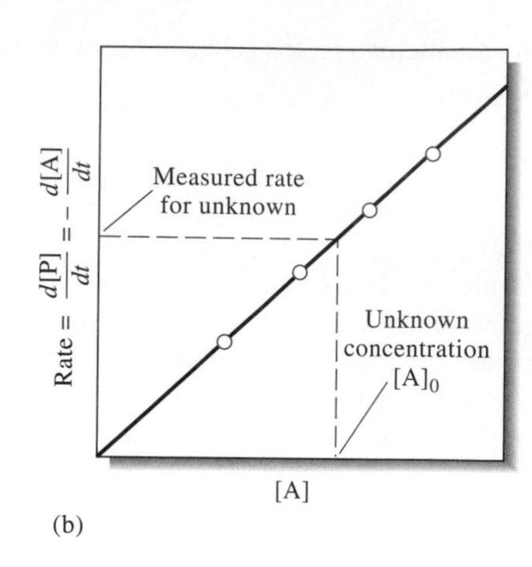

Figure 29-4 A plot of data for the determination of A by the differential method. (a) Solid lines are the experimental plots of product concentration as a function of time for four initial concentrations of A. Dashed lines are tangents to the curve at $t \rightarrow 0$. (b) A plot of the slopes obtained from the tangents in (a) as a function of the analyte concentration.

Of course, it is not necessary to record the entire rate curve, as has been done in Figure 29-4a, since only a small portion of the plot is used to measure the slope. As long as sufficient data points are collected to determine the initial slope precisely, time is saved and the entire procedure is simplified. More sophisticated data-handling procedures and numerical analysis of the data make possible high-precision rate measurements at later times as well; under certain circumstances such measurements are more accurate and more precise than those made near $t = 0$.

Integral Methods

In contrast to the differential method, **integral methods** take advantage of integrated forms of rate laws, such as those shown by Equations 29-6, 29-7, and 29-9.

Graphical Methods Equation 29-6 may be rearranged to give

$$\ln [A]_t = -kt + \ln [A]_0 \tag{29-26}$$

Thus, a plot of the natural logarithm of experimentally measured concentrations of A (or P) as a function of time should yield a straight line with a slope of $-k$ and a y-intercept of $\ln [A]_0$. Use of this procedure for the determination of nitromethane is illustrated in Example 29-6.

EXAMPLE 29-6

The data in the first two columns of Table 29-2 were recorded for the pseudo–first-order decomposition of nitromethane in the presence of excess base. Find the initial concentration of nitromethane and the pseudo–first-order rate constant for the reaction.

Computed values for the natural logarithms of nitromethane concentrations are shown in the third column of Table 29-2. The data are plotted in Figure 29-5. A least-squares analysis of the data (see Section 8C-2) leads to an intercept b of

$$b = \ln[CH_3NO_2]_0 = -5.129$$

which after exponentiation gives

$$[CH_3NO_2]_0 = 5.92 \times 10^{-3}\ M$$

The least-squares analysis also gives the slope of the line m, which in this case is

$$m = -1.62 = -k$$

and thus,

$$k = 1.62\ s^{-1}$$

TABLE 29-2

Data for the Decomposition of Nitromethane

Time, s	$[CH_3NO_2]$, M	$\ln[CH_3NO_2]$
0.25	3.86×10^{-3}	−5.557
0.50	2.59×10^{-3}	−5.956
0.75	1.84×10^{-3}	−6.298
1.00	1.21×10^{-3}	−6.717
1.25	0.742×10^{-3}	−7.206

Fixed-Time Methods Fixed-time methods are based on Equation 29-7 or 29-9. The former can be rearranged to

$$[A]_0 = \frac{[A]_t}{e^{-kt}} \tag{29-27}$$

The simplest way of employing this relationship is to perform a calibration experiment with a standard solution that has a known concentration $[A]_0$. After a carefully measured reaction time t, $[A]_t$ is determined and is used to evaluate the constant e^{-kt} by means of Equation 29-27. Unknowns are then analyzed by measuring $[A]_t$ after exactly the same reaction time and employing the calculated value for e^{-kt} to compute the analyte concentrations.

Equation 29-27 is easily modified for the situation in which [P] is measured experimentally rather than [A]. Equation 29-9 can be rearranged to solve for $[A]_0$. That is,

$$[A]_0 = \frac{[P]_t}{1 - e^{-kt}} \tag{29-28}$$

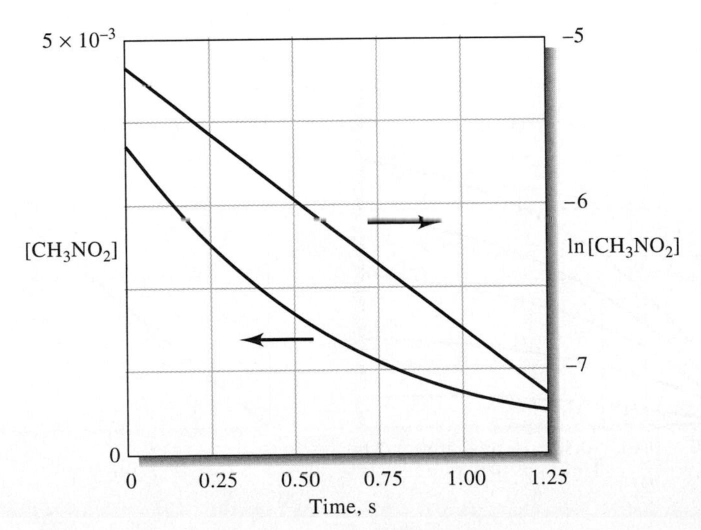

Figure 29-5 Plots of nitromethane concentration and the natural logarithm of nitromethane concentration as a function of time. The data are from Example 29-6.

A more desirable approach to the use of Equation 29-27 or 29-28 is to measure [A] or [P] at two times, t_1 and t_2. For example, if the product concentration is determined, we can write

$$[\mathrm{P}]_{t_1} = [\mathrm{A}]_0 (1 - e^{-kt_1})$$
$$[\mathrm{P}]_{t_2} = [\mathrm{A}]_0 (1 - e^{-kt_2})$$

Subtracting the first equation from the second and rearranging yields

$$[\mathrm{A}]_0 = \frac{[\mathrm{P}]_{t_2} - [\mathrm{P}]_{t_1}}{e^{-kt_1} - e^{-kt_2}} = C([\mathrm{P}]_{t_2} - [\mathrm{P}]_{t_1}) \qquad (29\text{-}29)$$

The reciprocal of the denominator, C, is constant for fixed t_1 and t_2.

► A major advantage of kinetic methods is their immunity to errors resulting from long-term drift of the measurement system.

The use of Equation 29-29 has a fundamental advantage common to most kinetic methods—that is, the absolute determination of concentration or of a variable proportional to concentration is unnecessary. The difference between two concentrations is proportional to the initial concentration of the analyte.

An important example of an uncatalyzed method is the fixed-time method for the determination of thiocyanate ion based on spectrophotometric measurements of the red iron(III) thiocyanate complex. The reaction in this application is

$$\mathrm{Fe^{3+}} + \mathrm{SCN^-} \underset{k_{-1}}{\overset{k_1}{\rightleftharpoons}} \underset{\text{red}}{\mathrm{Fe(SCN)^{2+}}}$$

Under conditions of excess Fe^{3+}, the reaction is pseudo–first-order in SCN^-. The curves in Figure 29-6a show the increase in absorbance due to the appearance of $Fe(SCN)^{2+}$ versus time following the rapid mixing of 0.100 M Fe^{3+} with various concentrations of SCN^- at pH 2. Since the concentration of $Fe(SCN)^{2+}$ is related to the absorbance by Beer's law, the experimental data can be used directly without conversion to concentration. Thus, the change in absorbance ΔA between times t_1 and t_2 is computed and plotted versus $[SCN^-]_0$, as in Figure 29-6b. Unknown concentrations are then determined by evaluating ΔA under the same experimental conditions and obtaining the concentration of thiocyanate ion from the calibration curve or the least-squares equation.

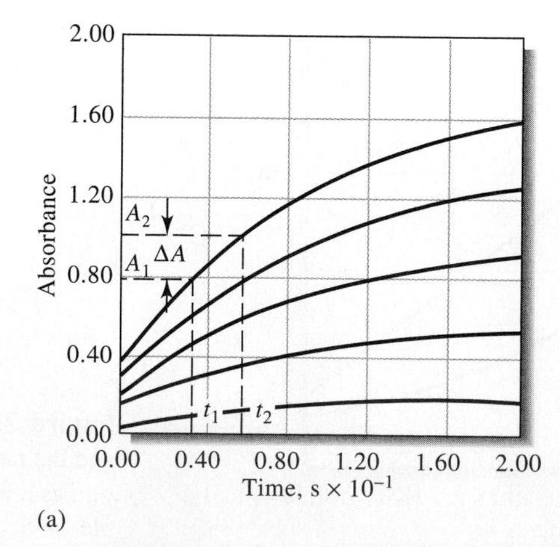

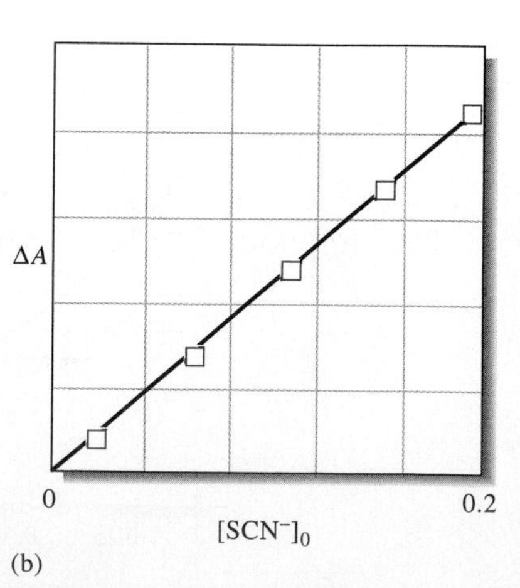

Figure 29-6 (a) Absorbance due to the formation of $Fe(SCN)^{2+}$ as a function of time for five concentrations of SCN^-. (b) A plot of the difference in absorbance ΔA at times t_2 and t_1 as a function of SCN^- concentration.

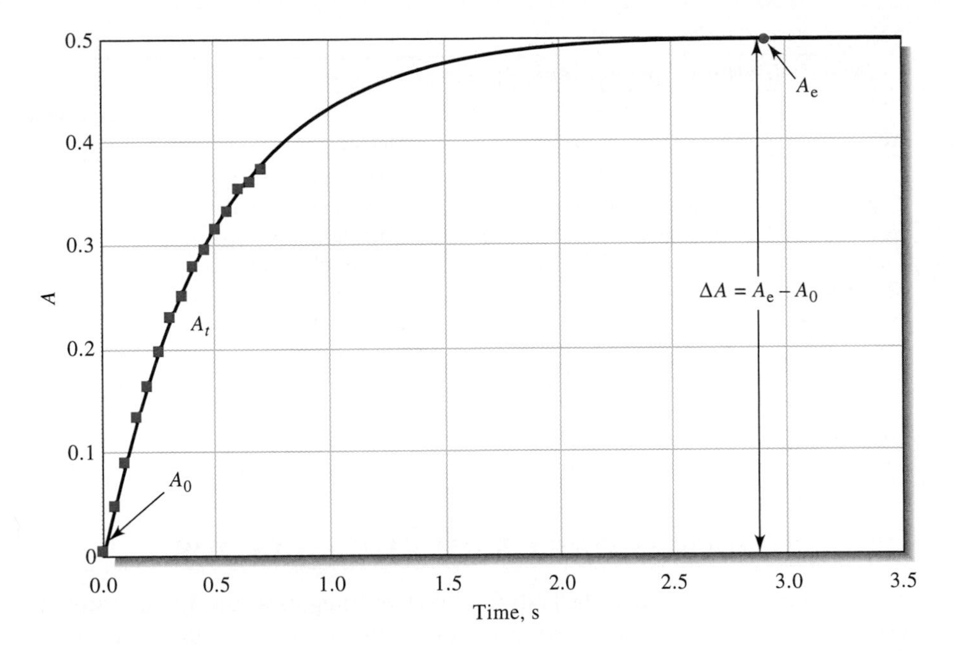

Figure 29-7 The predictive approach in kinetic methods. A mathematical model, shown as the blue squares, is used to fit the response, shown as the solid line, during the kinetic regime of a reaction. The model is then used to predict the equilibrium value of the signal, A_e, which is related to the analyte concentration. In the example shown, the absorbance is plotted vs. time and the early-time data is used to predict A_e, the equilibrium value, shown as the blue circle. (Reprinted with permission from G. L. Mieling and H. L. Pardue, *Anal. Chem.*, **1978,** *50,* 1611. Copyright 1978 American Chemical Society.)

Fixed-time methods are advantageous because the measured quantity is directly proportional to the analyte concentration and because measurements can be made *at any time* during the progress of first-order reactions. When instrumental methods are used to monitor reactions by means of fixed-time procedures, the precision of the analytical results approaches the precision of the instrument used.

Curve-Fitting Methods With computers attached to instruments, fitting a mathematical model to the concentration or signal versus time curve is straightforward. These techniques compute values of the model parameters, including the initial concentration of analyte, that "best fit" the data. The most sophisticated of these methods use the parameters of the model to estimate the value of the equilibrium or steady-state response. These methods can provide error compensation because the equilibrium position is less sensitive to such experimental variables as temperature, pH, and reagent concentrations. Figure 29-7 illustrates the use of this approach to predict the equilibrium absorbance from data obtained during the kinetic regime of the response curve. The equilibrium absorbance is then related to the analyte concentration in the usual way.

The computer enables many innovative techniques for kinetic methods. Some recent error compensation methods do not require prior knowledge of the reaction order for the system employed but instead use a generalized model. Still other methods calculate the model parameters as the data are collected instead of employing batch processing methods.

Spreadsheet Summary In the final exercise of Chapter 13 of *Applications of Microsoft® Excel in Analytical Chemistry,* the initial-rate method is explored for determining the concentration of an analyte. Initial rates are determined from a linear least-squares analysis and are used to establish a calibration curve and equation. An unknown concentration is determined.

TABLE 29-3

Catalytic Methods for Inorganic Species

Analyte	Indicator Reaction	Detection Method	Detection Limit, ng/mL
Cobalt	Catechol + H_2O_2	Spectrophotometry	3
Copper	Hydroquinone + H_2O_2	Spectrophotometry	0.2
Iron	$H_2O_2 + I^-$	Potentiometry	50
Mercury	$Fe(CN)_6^{4-} + C_6H_5NO$	Spectrophotometry	60
Molybdenum	$H_2O_2 + I^-$	Spectrophotometry	10
Bromide	Decomposition of BrO_3^-	Spectrophotometry	3
Chloride	$Fe^{2+} + ClO_3^-$	Spectrophotometry	100
Cyanide	Reduction of *o*-dinitrobenzene	Spectrophotometry	100
Iodide	Ce(IV) + As(III)	Potentiometry	0.2
Oxalate			

29C APPLICATIONS OF KINETIC METHODS

The reactions used in kinetic methods fall into two categories: **catalyzed** or **uncatalyzed.** As noted earlier, catalyzed reactions are the most widely used because of their superior sensitivity and selectivity. Uncatalyzed reactions are used to advantage when high-speed, automated measurements are required or when the sensitivity of the detection method is great.[7]

29C-1 Catalytic Methods

The Determination of Inorganic Species

Many inorganic cations and anions catalyze indicator reactions—that is, reactions whose rates are readily measured by instrumental methods, such as absorption spectrophotometry, fluorescence spectrometry, or electrochemistry. Conditions are then employed such that the rate is proportional to the concentration of catalyst, and, from the rate data, the concentration of catalyst is determined. Such catalytic methods often allow extremely sensitive detection of the catalyst concentration. Kinetic methods based on catalysis by inorganic analytes are widely applicable. For example, the literature in this area lists more than 40 cations and 15 anions that have been determined by a variety of indicator reactions.[8] Table 29-3 gives catalytic methods for several inorganic species along with the indicator reactions used, the method of detection, and the detection limit.

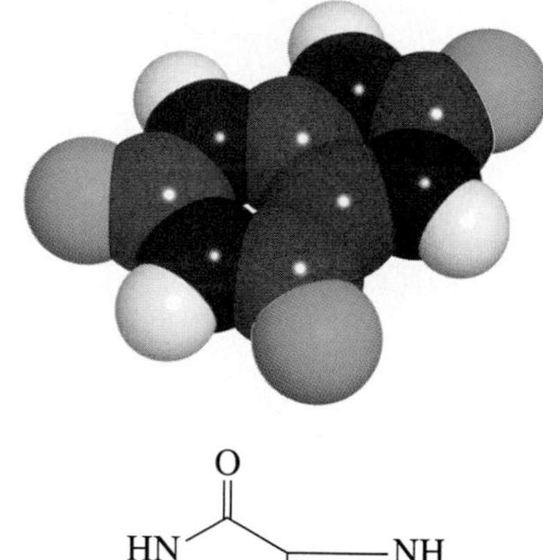

Molecular model of uric acid. Uric acid is essential to the digestive process. If the body produces too much uric acid or if not enough is excreted, however, high levels of uric acid in the blood can lead to crystals of sodium urate being concentrated in the joints and tendons. These cause the inflammation, pressure, and severe pain associated with gouty arthritis, or gout.

The Determination of Organic Species

Without question, the most important applications of catalyzed reactions to organic analyses involve the use of enzymes as catalysts. These methods have been used for the determination of both enzymes and substrates and serve as the basis for many of the routine and automated screening tests performed by thousands in clinical laboratories throughout the world. Many different enzyme substrates have been determined with enzyme-catalyzed reactions. Table 29-4 lists some of the sub-

[7]For reviews of applications of kinetic methods, see H. O. Mottola, *Kinetic Aspects of Analytical Chemistry,* pp. 88–121. New York: Wiley, 1988; D. Perez-Bendito and M. Silva, *Kinetic Methods in Analytical Chemistry,* pp. 31–189. New York: Halsted Press-Wiley, 1988.

[8]M. Kopanica and V. Stara, in *Comprehensive Analytical Chemistry,* G. Svehla, Ed., Vol. XVIII, pp. 11–227. New York: Elsevier, 1983.

TABLE 29-4
Important Substrates

Substrate	Enzyme	Application
Ethanol	Alcohol dehydrogenase	Law enforcement, alcoholism
Galactose	Galactose oxidase	Diagnosis of galactosemia
Glucose	Glucose oxidase	Diagnosis of diabetes
Lactose	Lactase	Food products
Maltose	α-Glucosidase	Food products
Penicillin	Penicillinase	Pharmaceutical preparations
Phenol	Tyrosinase	Water and wastewaters
Sucrose	Invertase	Food products
Urea	Urease	Diagnosis of liver and kidney disease
Uric acid	Uricase	Diagnosis of gout, leukemia, lymphoma

Molecular model of sucrose. Sucrose is a disaccharide, consisting of two monosaccharide units linked together. One of the units in sucrose is a glucose ring (6-member) and the other is a fructose ring (5-member). Sucrose is common table sugar.

strates that are determined in various applications.[9] One important application is the determination of the quantity of urea in blood; this is called the blood urea nitrogen (BUN) test. A description of this determination is given in Feature 29-3.

FEATURE 29-3

The Enzymatic Determination of Urea

The determination of urea in blood and urine is frequently carried out by measuring the rate of hydrolysis of urea $CO(NH_2)_2$ in the presence of the enzyme urease. The equation for the reaction is

$$CO(NH_2)_2 + 2H_2O \xrightarrow{\text{urease}} NH_4^+ + HCO_3^-$$

As suggested in Example 29-4, urea can be determined by measuring the initial rate of production of the products of this reaction. The high selectivity of the enzyme permits the use of nonselective detection methods, such as electrical conductance, for initial rate measurements. There are commercial instruments that operate on this principle. The sample is mixed with a small amount of an enzyme-buffer solution in a conductivity cell. The maximum rate of increase in conductance is measured within 10 s of mixing, and the concentration of urea is determined from a calibration curve consisting of a plot of maximum initial rate as a function of urea concentration. The precision of the instrument is on the order of 2% to 5% for concentrations in the physiological range of 2 to 10 mM.

Another method of following the rate of urea hydrolysis is based on a specific-ion electrode for ammonium ions (see Section 21D). Here, the production of NH_4^+ is monitored potentiometrically and is used to obtain the reaction rate. In yet another approach, the urease can be immobilized on the surface of a pH electrode and the rate of change of pH monitored. Many enzymes have now been immobilized onto supports such as gels, membranes, tubing walls, glass beads, polymers, and thin films. **Immobilized enzymes** often show enhanced stability over their soluble counterparts. In addition, they can be reused often for hundreds or thousands of analyses.

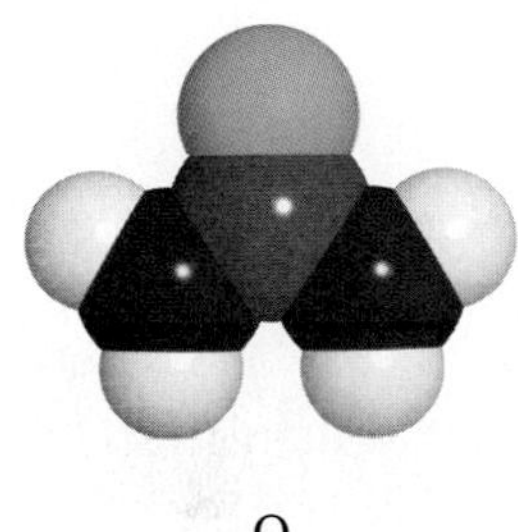

$$H_2N{-}\overset{\overset{\displaystyle O}{\|}}{C}{-}NH_2$$

Molecular model of urea. Urea is the diamide of carbonic acid. It is excreted by mammals as a waste product from the metabolism of protein.

◀ Enzymes can be immobilized by entrapment in a gel, by adsorption to a solid support, or by covalent bonding to a solid.

[9]For more information, see G. G. Guilbault, *Analytical Uses of Immobilized Enzymes.* New York: Dekker, 1984; P. W. Carr and L. D. Bowers, *Immobilized Enzymes in Analytical and Clinical Chemistry.* New York: Wiley, 1980.

A number of inorganic species can also be determined by enzyme-catalyzed reactions. These species include ammonia, hydrogen peroxide, carbon dioxide, and hydroxylamine as well as nitrate, phosphate, and pyrophosphate ions.

► Kinetic methods are necessary to determine enzyme activities since the enzyme is a catalyst and affects only the reaction rate.

Kinetic methods have been described for the quantitative determination of several hundred enzymes. Some of the enzymes that are important in diagnosing liver diseases are serum glutamic-oxaloacetic transaminase (SGOT), serum glutamate pyruvate transaminase (SGPT), and lactate dehydrogenase (LDH). Elevated levels of SGOT, SGPT, and LDH can also occur after heart attacks. These enzymes and creatine phosphokinase are often diagnostic for myocardial infarction. Other enzymes of interest in diagnosis include such hydrolases as amylase, lipase, and alkaline phosphatase; phosphohexose isomerase; and aldolase.

In addition, some two dozen inorganic cations and anions are known to decrease the rates of certain enzyme-catalyzed indicator reactions. These *inhibitors* can thus be determined from the decrease in rate brought about by their presence.

► Enzymes can be used for the determination of activators and inhibitors. Activators increase the reaction rate, while inhibitors decrease the rate.

Enzyme activators are substances, often inorganic ions, that are required for certain enzymes to become active as catalysts. Activators can be determined by their effect on the rates of enzyme-catalyzed reactions. For example, it has been reported that magnesium at concentrations as low as 10 ppb can be determined in blood plasma based on activation by this ion of the enzyme isocitric dehydrogenase.

29C-2 Uncatalyzed Reactions

As noted earlier, kinetic methods based on uncatalyzed reactions are not nearly as widely used as those in which a catalyst is involved. We have already described two of these methods (pages 896 and 898).

Generally, uncatalyzed reactions are useful when selective reagents are employed in conjunction with sensitive detection methods. For example, the selectivity of complexing agents can be controlled by adjusting the pH of the medium in the determination of metal ions, as discussed in Section 17D-8. Sensitivity can be achieved through the use of spectrophotometric detection to monitor reagents that form complexes with large molar absorptivities. The determination of Cu^{2+}, presented in Problem 29-13, is an example. A highly sensitive alternative is to select complexes that fluoresce so that the rate of change of fluorescence can be used as a measure of analyte concentration (see Problem 29-14).

The precision of both noncatalytic and catalytic kinetic methods depends on such experimental conditions as pH, ionic strength, and temperature. With careful control of these variables, relative standard deviations of 1% to 10% are typical. Automation of kinetic methods and computerized data analysis can often improve the relative precision to 1% or less.

29C-3 The Kinetic Determination of Components in Mixtures

An important application of kinetic methods is in the determination of closely related species in mixtures, such as alkaline earth cations or organic compounds with the same functional groups. For example, suppose two species A and B react with a common excess reagent to form products under pseudo–first-order conditions:

$$A + R \xrightarrow{k_A} P$$

$$B + R \xrightarrow{k_B} P'$$

Generally, k_A and k_B differ from each other. Thus, if $k_A > k_B$, A is depleted before B. It is possible to show that if the ratio k_A/k_B is greater than about 500, the consumption of A is approximately 99% complete before 1% of B is used up. Thus, a differential determination of A with no significant interference from B is possible provided that the rate is measured shortly after mixing.

When the ratio of the two rate constants is small, determination of both species is still possible by more complex methods of data treatment. Many of these methods use chemometric, multivariate calibration techniques similar to those described in Feature 8-3. The details of **multicomponent kinetic methods** are beyond the scope of this text.[10]

WEB WORKS

Go to **http://chemistry.brookscole.com/skoogfac/.** From the Chapter Resources menu, choose Web Works. Locate the Chapter 29 section, and you will find links to several instrument makers who produce glucose analyzers based on enzymatic reactions. Find one company that makes a spectrophotometric analyzer and one that makes an electrochemical analyzer. Compare and contrast the features of the two instruments.

QUESTIONS AND PROBLEMS

29-1. Define the following terms as they are used in kinetic methods of analysis.
- *(a) order of a reaction
- (b) pseudo–first-order
- *(c) enzyme
- (d) substrate
- *(e) Michaelis constant
- (f) differential method
- *(g) integral method
- (h) indicator reaction

29-2. The analysis of a multicomponent mixture by kinetic methods is sometimes referred to as a "kinetic separation." Explain the significance of this term.

***29-3.** Explain why pseudo–first-order conditions are used in most kinetic methods.

29-4. List three advantages of kinetic methods. Can you think of two possible limitations of kinetic methods when compared with equilibrium methods?

***29-5.** Develop an expression for the half-life of the reactant in a first-order process in terms of the rate constant k.

29-6. Find the natural lifetime in seconds for first-order reactions corresponding to
- *(a) $k = 0.351$.
- (b) $k = 6.62$.
- *(c) $[A]_0 = 1.06$ M and $[A]_t = 0.150$ M at $t = 4125$ s.
- (d) $[P]_\infty = 0.176$ M and $[P]_t = 0.0423$ M at $t = 9.62$ s. (Assume 1 mol of product is formed for each mole of analyte reacted.)
- *(e) half-life $t_{1/2} = 15.8$ years.
- (f) $t_{1/2} = 0.478$ s.

29-7. Find the first-order rate constant for a reaction that is 55.8% complete in
- *(a) 0.0100 s.
- (b) 0.100 s.
- *(c) 1.00 s.
- (d) 5280 s.
- *(e) 26.8 μs.
- (f) 8.86 ns.

29-8. Calculate the number of lifetimes τ required for a pseudo–first-order reaction to achieve the following levels of completion:

[10]For some applications of kinetic methods to multicomponent mixtures, see H. O. Mottola, *Kinetic Aspects of Analytical Chemistry,* pp. 122–148. New York: Wiley, 1988; D. Perez-Bendito and M. Silva, *Kinetic Methods in Analytical Chemistry,* pp. 172–189. New York: Halsted Press-Wiley, 1988.

(a) 10%.
(b) 50%.
(c) 90%.
(d) 99%.
(e) 99.9%.
(f) 99.99%.

29-9. Find the number of half-lives required to reach the levels of completion listed in Problem 29-8.

29-10. Find the relative error associated with the assumption that k' is invariant during the course of a pseudo–first-order reaction under the following conditions.

	Extent of Reaction, %	Excess of Reagent
*(a)	1	5×
(b)	1	10×
*(c)	1	50×
(d)	1	100×
*(e)	5	5×
(f)	5	10×
*(g)	5	100×
(h)	63.2	5×
*(i)	63.2	10×
(j)	63.2	50×
*(k)	63.2	100×

29-11. Show that for an enzyme reaction obeying Equation 29-19, the substrate concentration for which the rate equals $v_{max}/2$ is equal to K_m.

***29-12.** Equation 29-19 can be rearranged to produce the equation

$$\frac{1}{d[P]/dt} = \frac{K_m}{v_{max}[S]} + \frac{1}{v_{max}}$$

where $v_{max} = k_2[E]_0$, the maximum velocity when [S] is large.

(a) Suggest a way to employ this equation in the construction of a calibration (working) curve for the enzymatic determination of substrate.
(b) Describe how the resulting working curve can be used to find K_m and v_{max}.

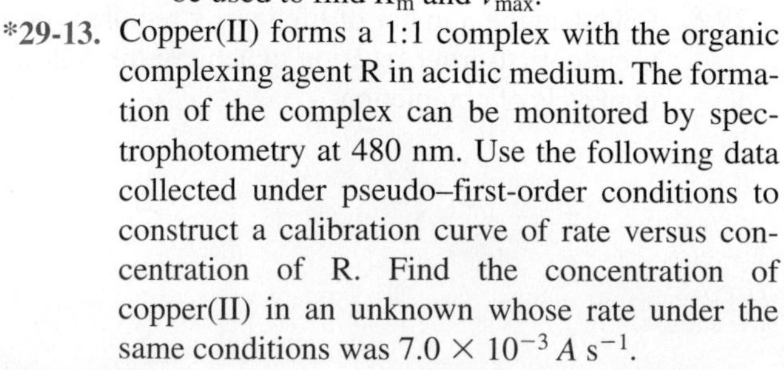

***29-13.** Copper(II) forms a 1:1 complex with the organic complexing agent R in acidic medium. The formation of the complex can be monitored by spectrophotometry at 480 nm. Use the following data collected under pseudo–first-order conditions to construct a calibration curve of rate versus concentration of R. Find the concentration of copper(II) in an unknown whose rate under the same conditions was 7.0×10^{-3} A s^{-1}.

$c_{Cu^{2+}}$, ppm	Rate, A s^{-1}
3.0	3.6×10^{-3}
5.0	5.4×10^{-3}
7.0	7.9×10^{-3}
9.0	1.03×10^{-2}

29-14. Aluminum forms a 1:1 complex with 2-hydroxy-1-naphthaldehyde-*p*-methoxybenzoylhydraxonal that exhibits fluorescence emission at 475 nm. Under pseudo–first-order conditions, a plot of the initial rate of the reaction (emission units per second) versus the concentration of aluminum (in μM) yields a straight line described by the equation

$$\text{rate} = 1.74c_{Al} - 0.225$$

Find the concentration of aluminum in a solution that exhibits a rate of 0.76 emission units per second under the same experimental conditions.

***29-15.** The enzyme monoamine oxidase catalyzes the oxidation of amines to aldehydes. For tryptamine, K_m for the enzyme is 4.0×10^{-4} M and $v_{max} = k_2[E]_0 = 1.6 \times 10^{-3}$ μM/min at pH 8. Find the concentration of a solution of tryptamine that reacts at a rate of 0.22 μM/min in the presence of monoamine oxidase under these conditions. Assume that [tryptamine] $\ll K_m$.

29-16. The following data represent the product concentrations versus time during the initial stages of pseudo–first-order reactions with different initial concentrations of analyte $[A]_0$.

t, s	[P] M				
0	0.00000	0.00000	0.00000	0.00000	0.00000
10	0.00004	0.00018	0.00027	0.00037	0.00014
20	0.00007	0.00037	0.00055	0.00073	0.00029
50	0.00018	0.00091	0.00137	0.00183	0.00072
100	0.00036	0.00181	0.00272	0.00362	0.00144
$[A]_0$, M	0.01000	0.05000	0.07500	0.10000	unknown

For each concentration of analyte, find the average initial reaction rate for the five time slots given. Plot the initial rate versus the concentration of analyte. Obtain the least-squares slope and intercept of the plot and determine the unknown concentration.

Hint: A good way to calculate the initial rate for a given analyte concentration is to find $\Delta[P]/\Delta t$ for the 0 to 10 s interval, the 10 to 20 s interval, the 20 to 50 s interval, and the 50 to 100 s interval; then average the four values obtained. Alternatively, the least-squares slope of a plot of [P] vs. t for the 0 to 100 s interval can be used.

***29-17.** Use Excel to calculate the product concentrations versus time for a pseudo–first-order reaction with $k' = 0.015\ s^{-1}$ and $[A]_0 = 0.005$ M. Use times of 0.000 s, 0.001 s, 0.01 s, 0.1 s, 0.2, s 0.5 s, 1.0 s, 2.0 s, 5.0 s, 10.0 s, 20.0 s, 50.0 s, 100.0 s, 200.0 s, 500.0 s, and 1000.0 s. From the two earliest time values, find the "true" initial rate of the reaction. Determine approximately what percentage of completion of the reaction occurs before the initial rate drops to (a) 99% and (b) 95% of the true value.

29-18. Challenge Problem. The hydrolysis of *N*-glutaryl-L-phenylalanine-*p*-nitroanilide (GPNA) by the enzyme α-chymotrypsin (CT) to form *p*-nitroaniline and *N*-glutaryl-L-phenylalanine follows the Michaelis-Menten mechanism in its early stages.

(a) Show that Equation 29-19 can be manipulated to give the following transformation:

$$\frac{1}{v_i} = \frac{K_m}{v_{max}[S]_0} + \frac{1}{v_{max}}$$

where v_i is the initial rate, $(d[P]/dt)_i$, v_{max} is $k_2[E]_0$; and $[S]_0$ is the initial GPNA concentration. This equation is often called the Lineweaver-Burke equation. A plot of $1/v_i$ vs. $1/[S]_0$ is called a Lineweaver-Burke plot.

(b) For [CT] = 4.0×10^{-6} M, use the following results and the Lineweaver-Burke plot to determine K_m, v_{max}, and k_2.

$[GPNA]_0$, mM	v_i, $\mu M\ s^{-1}$
0.250	0.037
0.500	0.063
10.0	0.098
15.0	0.118

(c) Show that the Michaelis-Menten equation for the initial rate can be transformed to give the Hanes-Woolf equation:

$$\frac{[S]}{v_i} = \frac{[S]_0}{v_{max}} + \frac{K_m}{v_{max}}$$

Use a Hanes-Woolf plot of the data in part (b) to determine K_m, v_{max}, and k_2.

(d) Show that the Michaelis-Menten equation for the initial rate can be transformed to give the Eadie-Hofster equation:

$$v_i = -\frac{K_m v_i}{[S]_0} + v_{max}$$

Use an Eadie-Hofster plot of the data in part (b) to determine K_m, v_{max}, and k_2.

(e) Comment on which of these plots should be most accurate for determining K_m and v_{max} under the circumstances given. Justify!

(f) The substrate GPNA is to be determined in a biological sample using the data in part (b) to construct a calibration curve. Three samples were analyzed under the same conditions as part (b) and gave initial rates of 0.069, 0.102, and 0.049 $\mu M\ s^{-1}$. What were the GNPA concentrations in these samples?

InfoTrac College Edition

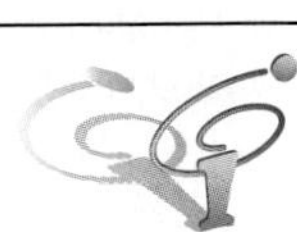

For additional readings, go to InfoTrac College Edition, your online research library, at

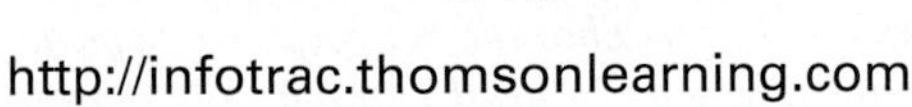

CHAPTER 30

Introduction to Analytical Separations

Charles E. Rotkin/Corbis

Separations are extremely important in synthesis, industrial chemistry, the biomedical sciences, and chemical analyses. Shown in the photo is a petroleum refinery. The first step in the refining process is to separate petroleum into fractions on the basis of boiling point in large distillation towers. The petroleum is fed into a large still, and the mixture is heated. The materials with the lowest boiling points vaporize first. The vapor moves up the tall distillation column or tower, where it recondenses into a much purer liquid. By regulating the temperatures of the still and the column, one can control the boiling-point range of the fraction condensed.

Analytical separations occur on a much smaller laboratory scale than the industrial-scale distillation shown. The separation methods introduced in this chapter include precipitation, distillation, extraction, ion exchange, and various chromatographic techniques.

Few, if any, measurement techniques used for chemical analysis are specific for a single chemical species; as a consequence, an important part of most analyses is dealing with foreign species that either attenuate the signal from the analyte or produce a signal that is indistinguishable from that of the analyte. A substance that affects an analytical signal or the background is called an ***interference*** *or an* ***interferent.***

An **interferent** is a chemical species that causes a systematic error in an analysis by enhancing or attenuating the analytical signal or the background.

Several methods can be used to deal with interferences in analytical procedures, as discussed in Section 8C-3. ***Separations*** *isolate the analyte from potentially interfering constituents. In addition, techniques such as matrix modification, masking, dilution, and saturation are often used to offset the effects of interferents. The internal standard and standard addition methods can sometimes be employed to compensate for or reduce interference effects. In this chapter, we focus on separation methods, which are the most powerful and widely used methods of treating interferences.*

The basic principles of a separation are depicted in Figure 30-1.[1] *As shown, separations can be complete or partial. The separation process involves transport of material and spatial redistribution of the components. We should note that a separation always requires energy, because the reverse process,* mixing *at constant volume, is spontaneous, being accompanied by an increase in entropy. Separations can be* preparative *or* analytical. *We focus here on analytical separations, although many of the same principles are involved in preparative separations.*

[1]See J. C. Giddings, *Unified Separation Science*, pp. 1–7. New York: Wiley, 1991.

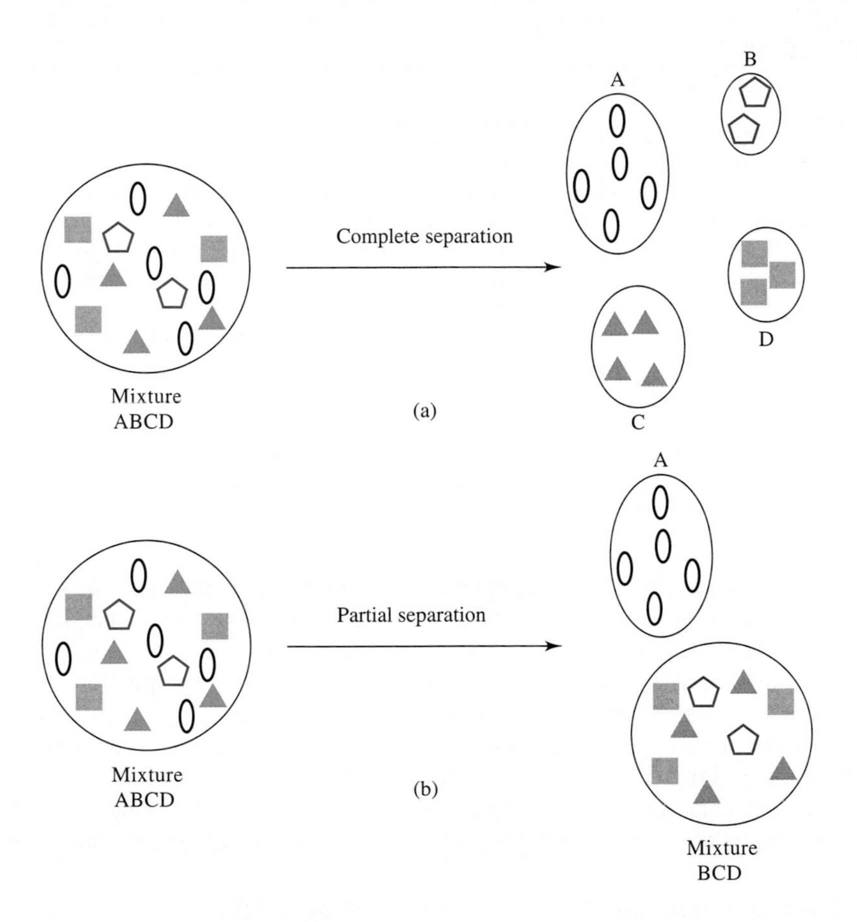

Figure 30-1 Separation principles. In (a), a mixture of four components is completely separated so that each component occupies a different spatial region. In (b), a partial separation is shown. Here species A is isolated from the remaining mixture of B, C, and D. The reverse of the separation process shown is mixing at constant volume.

The goals of an analytical separation are usually to eliminate or reduce interferences so that quantitative analytical information can be obtained about complex mixtures. Separations can also allow identification of the separated constituents if appropriate correlations are made or a structurally sensitive measurement technique, such as mass spectrometry, is used. With techniques such as chromatography, quantitative information is obtained nearly simultaneously with the separation. In other procedures, the separation step is distinct and quite independent of the measurement step that follows.

Table 30-1 lists a variety of separation methods that are in common use, including (1) chemical or electrolytic precipitation, (2) distillation, (3) solvent extraction, (4) ion exchange, (5) chromatography, (6) electrophoresis, and (7) field-flow fractionation. The first four are discussed in Sections 30A through 30E of this chapter. An introduction to chromatography is presented in Section 30F. Chapters 31 and 32 deal with gas and liquid chromatography, respectively, while Chapter 33 deals with electrophoresis, field-flow fractionation, and other separation methods.

30A SEPARATION BY PRECIPITATION

Separations by precipitation require large solubility differences between analyte and potential interferences. The theoretical feasibility of this type of separation can be determined by solubility calculations, such as those shown in Section 11C. Unfortunately, several other factors may preclude the use of precipitation to

TABLE 30-1
Separation Methods

Method	Basis of Method
Mechanical phase separation	
Precipitation and filtration	Difference in solubility of compounds formed
Distillation	Difference in volatility of compounds
Extraction	Difference in solubility in two immiscible liquids
Ion exchange	Difference in interaction of reactants with ion-exchange resin
Chromatography	Difference in rate of movement of a solute through a stationary phase
Electrophoresis	Difference in migration rate of charged species in an electric field
Field-flow fractionation	Difference in interaction with a field or gradient applied perpendicular to transport direction

achieve a separation. For example, the various coprecipitation phenomena described in Section 12A-5 may cause extensive contamination of a precipitate by an unwanted component, even though the solubility product of the contaminant has not been exceeded. Likewise, the rate of an otherwise feasible precipitation may be too slow to be useful for a separation. Finally, when precipitates form as colloidal suspensions, coagulation may be difficult and slow, particularly when the isolation of a small quantity of a solid phase is attempted.

Many precipitating agents have been employed for quantitative inorganic separations. Some of the most generally useful are described in the sections that follow.

30A-1 Separations Based on Control of Acidity

Enormous differences exist in the solubilities of the hydroxides, hydrous oxides, and acids of various elements. Moreover, the concentration of hydrogen or hydroxide ions in a solution can be varied by a factor of 10^{15} or more and can be readily controlled by the use of buffers. As a consequence, many separations based on pH control are, in theory, available to the chemist. In practice, these separations can be grouped into three categories: (1) those made in relatively concentrated solutions of strong acids, (2) those made in buffered solutions at intermediate pH values, and (3) those made in concentrated solutions of sodium or potassium hydroxide. Table 30-2 lists common separations that can be achieved by control of acidity.

TABLE 30-2
Separations Based on Control of Acidity

Reagent	Species Forming Precipitates	Species Not Precipitated
Hot concd HNO_3	Oxides of W(VI), Ta(V), Nb(V), Si(IV), Sn(IV), Sb(V)	Most other metal ions
NH_3/NH_4Cl buffer	Fe(III), Cr(III), Al(III)	Alkali and alkaline earths, Mn(II), Cu(II), Zn(II), Ni(II), Co(II)
$HOAc/NH_4OAc$ buffer	Fe(III), Cr(III), Al(III)	Cd(II), Co(II), Cu(II), Fe(II) Mg(II), Sn(II), Zn(II)
$NaOH/Na_2O_2$	Fe(III), most +2 ions, rare earths	Zn(II), Al(III), Cr(VI), V(V), U(VI)

30A-2 Sulfide Separations

With the exception of the alkali metals and the alkaline earth metals, most cations form sparingly soluble sulfides whose solubilities differ greatly from one another. Because it is relatively easy to control the sulfide ion concentration of an aqueous solution of H_2S by adjustment of pH (see Section 11C-2), separations based on the formation of sulfides have found extensive use. Sulfides can be conveniently precipitated from homogeneous solution, with the anion being generated by the hydrolysis of thioacetamide (see Table 12-1).

◀ Recall from Equation 11-42 that $[S^{2-}] = \dfrac{1.2 \times 10^{-22}}{[H_3O^+]^2}$

The ionic equilibria influencing the solubility of sulfide precipitates were considered in Section 11C-2. These treatments, however, may not always produce realistic conclusions about the feasibility of separations, because of coprecipitation and the slow rates at which some sulfides form. For these reasons, chemists often rely on previous results or empirical observations to indicate whether a given separation is likely to be successful.

Table 30-3 shows some common separations that can be accomplished with hydrogen sulfide through control of pH.

30A-3 Separations by Other Inorganic Precipitants

No other inorganic ions are as generally useful for separations as hydroxide and sulfide ions. Phosphate, carbonate, and oxalate ions are often employed as precipitants for cations, but their behavior is nonselective; therefore, separations must ordinarily precede their use.

Chloride and sulfate are useful because of their highly selective behavior. The former is used to separate silver from most other metals, and the latter is frequently employed to isolate a group of metals that includes lead, barium, and strontium.

30A-4 Separations by Organic Precipitants

Selected organic reagents for the isolation of various inorganic ions were discussed in Section 12D-3. Some of these organic precipitants, such as dimethylglyoxime, are useful because of their remarkable selectivity in forming precipitates with a few ions. Others, such as 8-hydroxyquinoline, yield slightly soluble compounds with a host of cations. The selectivity of this sort of reagent is due to the wide range of solubility among its reaction products and also to the fact that the precipitating reagent is ordinarily an anion that is the conjugate base of a weak acid. Thus, separations based on pH control can be realized, just as with hydrogen sulfide.

TABLE 30-3
Precipitation of Sulfides

Elements	Conditions of Precipitation*	Conditions for No Precipitation*
Hg(II), Cu(II), Ag(I)	1, 2, 3, 4	
As(V), As(III), Sb(V), Sb(III)	1, 2, 3	4
Bi(III), Cd(II), Pb(II), Sn(II)	2, 3, 4	1
Sn(IV)	2, 3	1, 4
Zn(II), Co(II), Ni(II)	3, 4	1, 2
Fe(II), Mn(II)	4	1, 2, 3

*1 = 3 M HCl; 2 = 0.3 M HCl; 3 = buffered to pH 6 with acetate; 4 = buffered to pH 9 with $NH_3/(NH_4)_2S$.

30A-5 Separation of Species Present in Trace Amounts by Precipitation

A problem often encountered in trace analysis is isolation of the species of interest, which may be present in microgram quantities, from the major components of the sample. Although such a separation is sometimes based on a precipitation, the techniques required differ from those used when the analyte is present in generous amounts.

Several problems attend the quantitative separation of a trace element by precipitation, even when solubility losses are not important. Supersaturation often delays formation of the precipitate, and coagulation of small amounts of a colloidally dispersed substance is often difficult. In addition, it is likely that an appreciable fraction of the solid will be lost during transfer and filtration. To minimize these difficulties, a quantity of some other ion that also forms a precipitate with the reagent is often added to the solution. The precipitate from the added ion is called a **collector** and carries the desired minor species out of solution. For example, in isolating manganese as the sparingly soluble manganese dioxide, a small amount of iron(III) is frequently added to the analyte solution before the introduction of ammonia as the precipitating reagent. The basic iron(III) oxide carries down even the smallest traces of the manganese dioxide. Other examples include the use of basic aluminum oxide as a collector of trace amounts of titanium and the use of copper sulfide for collection of traces of zinc and lead. Many other collectors are described by Sandell and Onishi.[2]

A **collector** is used to remove trace constituents from solution.

A collector may entrain a trace constituent as a result of similarities in their solubilities. Other collectors function by coprecipitation, in which the minor component is adsorbed on or incorporated into the collector precipitate as the result of mixed-crystal formation. Clearly, the collector must not interfere with the method selected to determine the trace component.

30A-6 Separation by Electrolytic Precipitation

Electrolytic precipitation is a highly useful method of accomplishing separations. In this process, the more easily reduced species, either the wanted or the unwanted component of the sample, is isolated as a separate phase. This method becomes particularly effective when the potential of the working electrode is controlled at a predetermined level (see Section 22B).

The mercury cathode (page 648) has found wide application in the removal of many metal ions prior to the analysis of the residual solution. In general, metals more easily reduced than zinc are conveniently deposited in the mercury, leaving such ions as aluminum, beryllium, the alkaline earth metals, and the alkali metals in solution. The potential required to decrease the concentration of a metal ion to any desired level is readily calculated from polarographic data.

30A-7 Salt-Induced Precipitation of Proteins

A common way to separate proteins is by adding a high concentration of salt. This procedure is termed **salting out** the protein. The solubility of protein molecules shows a complex dependence on pH, temperature, the nature of the protein, and the

---

[2]E. B. Sandell and H. Onishi, *Colorimetric Determination of Traces of Metals,* 4th ed., pp. 709–721. New York: Interscience, 1978.

concentration of the salt used. At low salt concentrations, solubility usually increases with increasing salt concentration. This **salting in effect** is explained by the Debye-Hückel theory. The counterions of the salt surround the protein, and the screening results in a decrease in the electrostatic attraction of protein molecules for each other. This in turn leads to increasing solubility with increasing ionic strength.

At high concentrations of salt, however, the repulsive effect of like charges is reduced, as are the forces leading to solvation of the protein. When these forces are reduced enough, the protein precipitates and salting out is observed. Ammonium sulfate is an inexpensive salt and is widely used because of its effectiveness and high inherent solubility.

At high concentrations, protein solubility, S, is given by the following empirical equation:

$$\log S = C - K\mu \qquad (30\text{-}1)$$

where C is a constant that is a function of pH, temperature, and the protein; K is the salting-out constant that is a function of the protein and the salt used; and μ is the ionic strength.

Proteins are commonly least soluble at their isoelectric points. Hence, a combination of high salt concentration and pH control is used to achieve salting out. Protein mixtures can be separated by a stepwise increase in the ionic strength. Care must be taken with some proteins because ammonium sulfate can denature the protein. Alcoholic solvents are sometimes used in place of salts. They reduce the dielectric constant and subsequently reduce solubility by lowering protein-solvent interactions.

30B SEPARATION OF SPECIES BY DISTILLATION

Distillation is widely used to separate volatile analytes from nonvolatile interferents. A common example is the separation of nitrogen analytes from many other species by conversion of the nitrogen to ammonia, which is then distilled from basic solution. Other examples include separating carbon as carbon dioxide and sulfur as sulfur dioxide.

30C SEPARATION BY EXTRACTION

The extent to which solutes, both inorganic and organic, distribute themselves between two immiscible liquids differs enormously, and these differences have been used for decades to accomplish separations of chemical species. This section considers applications of the distribution phenomenon to analytical separations.

30C-1 Principles

The partition of a solute between two immiscible phases is an equilibrium phenomenon that is governed by the **distribution law.** If the solute species A is allowed to distribute itself between water and an organic phase, the resulting equilibrium may be written as

$$A(aq) \rightleftharpoons A(org)$$

where the letters in parentheses refer to the aqueous and organic phases, respectively. Ideally, the ratio of activities for A in the two phases will be constant and independent of the total quantity of A; that is, at any given temperature,

$$K = \frac{(a_A)_{org}}{(a_A)_{aq}} \approx \frac{[A]_{org}}{[A]_{aq}} \tag{30-2}$$

where $(a_A)_{org}$ and $(a_A)_{aq}$ are the activities of A in each of the phases and the bracketed terms are molar concentrations of A. The equilibrium constant K is known as the **distribution constant.** As with many other equilibria, under many conditions molar concentrations can be substituted for activities without serious error. Generally, the numerical value for K approximates the ratio of the solubility of A in each solvent.

Distribution constants are useful because they permit us to calculate the concentration of an analyte remaining in a solution after a certain number of extractions. They also provide guidance as to the most efficient way to perform an extractive separation. Thus, we can show (see Feature 30-1) that for the simple system described by Equation 30-2, the concentration of A remaining in an aqueous solution after i extractions with an organic solvent ($[A]_i$) is given by the equation

$$[A]_i = \left(\frac{V_{aq}}{V_{org}K + V_{aq}}\right)^i [A]_0 \tag{30-3}$$

where $[A]_i$ is the concentration of A remaining in the aqueous solution after extracting V_{aq} mL of the solution with an original concentration of $[A]_0$ with i portions of the organic solvent, each with a volume of V_{org}. Example 30-1 illustrates how this equation can be used to decide on the most efficient way to perform an extraction.

EXAMPLE 30-1

The distribution constant for iodine between an organic solvent and H_2O is 85. Find the concentration of I_2 remaining in the aqueous layer after extraction of 50.0 mL of 1.00×10^{-3} M I_2 with the following quantities of the organic solvent: (a) 50.0 mL; (b) two 25.0-mL portions; (c) five 10.0-mL portions.

Substitution into Equation 30-3 gives

$$\text{(a)}\quad [I_2]_1 = \left(\frac{50.0}{(50.0 \times 85) + 50.0}\right)^1 \times 1.00 \times 10^{-3} = 1.16 \times 10^{-5}\text{ M}$$

$$\text{(b)}\quad [I_2]_2 = \left(\frac{50.0}{(25.0 \times 85) + 50.0}\right)^2 \times 1.00 \times 10^{-3} = 5.28 \times 10^{-7}\text{ M}$$

$$\text{(c)}\quad [I_2]_5 = \left(\frac{50.0}{(10.0 \times 85) + 50.0}\right)^5 \times 1.00 \times 10^{-3} = 5.29 \times 10^{-10}\text{ M}$$

► It is always better to use several small portions of solvent to extract a sample than to extract with one large portion.

Note the increased extraction efficiencies that result from dividing the original 50 mL of solvent into two 25-mL or five 10-mL portions.

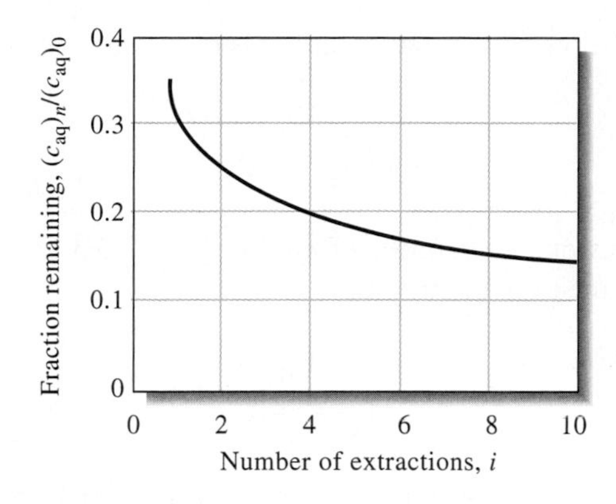

Figure 30-2 Plot of Equation 30-3 assuming that $K = 2$ and $V_{aq} = 100$ mL. The total volume of the organic solvent was assumed to be 100 mL, so that $V_{org} = 100/n_i$.

Figure 30-2 shows that the improved efficiency of multiple extractions falls off rapidly as a total fixed volume is subdivided into smaller and smaller portions. Clearly, little is to be gained by dividing the extracting solvent into more than five or six portions.

FEATURE 30-1

Derivation of Equation 30-3

Consider a simple system that is described by Equation 30-2. Suppose n_0 mmol of the solute A in V_{aq} mL of aqueous solution is extracted with V_{org} mL of an immiscible organic solvent. At equilibrium, n_1 mmol of A will remain in the aqueous layer, and $(n_0 - n_1)$ mmol will have been transferred to the organic layer. The concentrations of A in the two layers will then be

$$[A]_1 = \frac{n_1}{V_{aq}}$$

and

$$[A]_{org} = \frac{(n_0 - n_1)}{V_{org}}$$

Substitution of these quantities into Equation 30-2 and rearrangement gives

$$n_1 = \left(\frac{V_{aq}}{V_{org}K + V_{aq}}\right)n_0$$

Similarly, the number of millimoles, n_2, remaining after a second extraction with the same volume of solvent will be

$$n_2 = \left(\frac{V_{aq}}{V_{org}K + V_{aq}}\right)n_1$$

(continued)

Substitution of the previous equation into this expression gives

$$n_2 = \left(\frac{V_{aq}}{V_{org}K + V_{aq}}\right)^2 n_0$$

By the same argument, the number of millimoles, n_i, that remain after i extractions is given by the expression

$$n_i = \left(\frac{V_{aq}}{V_{org}K + V_{aq}}\right)^i n_0$$

Finally, this equation can be written in terms of the initial and final concentrations of A in the aqueous layer by substituting the relationships

$$n_i = [A]_i V_{aq} \quad \text{and} \quad n_0 = [A]_0 V_{aq}$$

Thus,

$$[A]_i = \left(\frac{V_{aq}}{V_{org}K + V_{aq}}\right)^i [A]_0$$

which is Equation 30-3.

30C-2 Extracting Inorganic Species

An extraction is frequently more attractive than a precipitation method for separating inorganic species. The processes of equilibration and separation of phases in a separatory funnel are less tedious and time consuming than conventional precipitation, filtration, and washing.

Separating Metal Ions as Chelates

Many organic chelating agents are weak acids that react with metal ions to give uncharged complexes that are highly soluble in organic solvents such as ethers, hydrocarbons, ketones, and chlorinated species (including chloroform and carbon tetrachloride).[3] Most uncharged metal chelates, however, are nearly insoluble in water. Similarly, the chelating agents themselves are often quite soluble in organic solvents but of limited solubility in water.

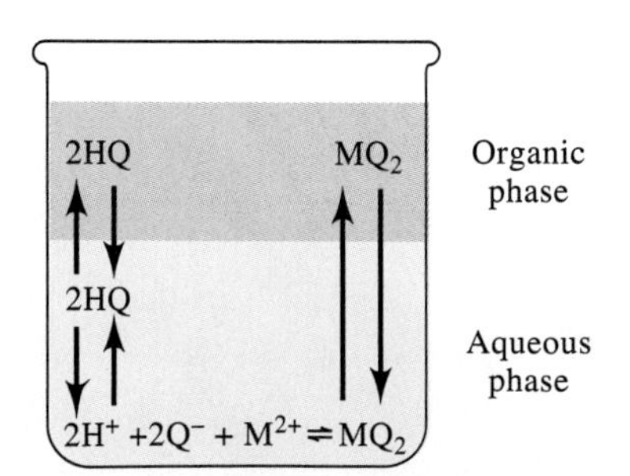

Figure 30-3 Equilibria in the extraction of an aqueous cation M^{2+} into an immiscible organic solvent containing 8-hydroxyquinoline.

Figure 30-3 shows the equilibria that develop when an aqueous solution of a divalent cation, such as zinc(II), is extracted with an organic solution containing a large excess of 8-hydroxyquinoline (see Section 12D-3 for the structure and reactions of this chelating agent). Four equilibria are shown. The first involves distribution of the 8-hydroxyquinoline, HQ, between the organic and aqueous layers. The second is acid dissociation of HQ to give H^+ and Q^- ions in the aqueous layer. The third equilibrium is the complex-formation reaction giving MQ_2. Fourth is distribution of the chelate between the two solvents. If it were not for the fourth equilib-

[3]The use of chlorinated solvents is decreasing because of concerns about their health effects and their possible role in ozone layer depletion.

rium, MQ_2 would precipitate out of the aqueous solution. The overall equilibrium is the sum of these four reactions, or

$$2HQ(org) + M^{2+}(aq) \rightleftharpoons MQ_2(org) + 2H^+(aq)$$

The equilibrium constant for this reaction is

$$K' = \frac{[MQ_2]_{org}[H^+]^2_{aq}}{[HQ]^2_{org}[M^{2+}]_{aq}}$$

Ordinarily, HQ is present in the organic layer in large excess with respect to M^{2+} in the aqueous phase so that $[HQ]_{org}$ remains essentially constant during the extraction. The equilibrium-constant expression can then be simplified to

$$K'[HQ]^2_{org} = K = \frac{[MQ_2]_{org}[H^+]^2_{aq}}{[M^{2+}]_{aq}}$$

or

$$\frac{[MQ_2]_{org}}{[M^{2+}]_{aq}} = \frac{K}{[H^+]^2_{aq}}$$

Thus, we see that the ratio of concentration of the metal species in the two layers is inversely proportional to the square of the hydrogen ion concentration of the aqueous layer. Equilibrium constants K vary widely from metal ion to metal ion; these differences often make it possible to selectively extract one cation from another by buffering the aqueous solution at a level where one is extracted nearly completely and the second remains largely in the aqueous phase.

Several useful extractive separations with 8-hydroxyquinoline have been developed. In addition, numerous chelating agents that behave in a similar way are described in the literature.[4] As a consequence, pH-controlled extractions provide a powerful method of separating metallic ions.

Extraction of Metal Chlorides and Nitrates

A number of inorganic species can be separated by extraction with suitable solvents. For example, a single ether extraction of a 6 M hydrochloric acid solution will cause better than 50% of several ions to be transferred to the organic phase; included among these are iron(III), antimony(V), titanium(III), gold(III), molybdenum(VI), and tin(IV). Other ions, such as aluminum(III) and the divalent cations of cobalt, lead, manganese, and nickel, are not extracted.

Uranium(VI) can be separated from such elements as lead and thorium by ether extraction of a solution that is 1.5 M in nitric acid and saturated with ammonium nitrate. Bismuth and iron(III) are also extracted to some extent from this medium.

30C-3 Solid-Phase Extraction

There are several limitations to liquid-liquid extractions. With extractions from aqueous solutions, the solvents that can be used must be immisicible with water

[4]For example, see J. A. Dean, in *Analytical Chemistry Handbook,* p. 2.24. New York: McGraw-Hill, 1995.

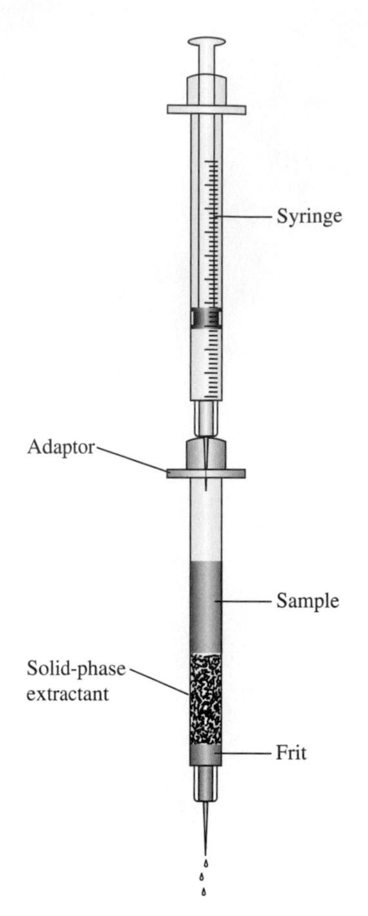

Figure 30-4 Solid-phase extraction performed in a small cartridge. The sample is placed in the cartridge and pressure is applied via a syringe plunger. Alternatively, a vacuum can be used to pull the sample through the extracting agent.

and must not form emulsions. Another difficulty is that liquid-liquid extractions use relatively large volumes of solvent, which can cause a problem with waste disposal. Also, most extractions are performed manually and as such are somewhat slow and tedious.

Solid-phase extraction, or liquid-solid extraction, can overcome several of these problems.[5] Solid-phase extraction techniques use membranes or small disposable syringe-barrel columns or cartridges. A hydrophobic organic compound is coated or chemically bonded to powdered silica to form the solid extracting phase. The compounds can be nonpolar, moderately polar, or polar. For example, an octadecyl (C_{18}) bonded silica (ODS) is a common packing. The functional groups bonded to the packing attract hydrophobic compounds in the sample by van der Waals interactions and extract them from the aqueous solution.

A typical cartridge system for solid-phase extractions is shown in Figure 30-4. The sample is placed in the cartridge, and pressure is applied by the syringe or from an air or nitrogen line. Alternatively, a vacuum can be used to pull the sample through the extractant. Organic molecules are then extracted from the sample and concentrated in the solid phase. They can later be displaced from the solid phase by a solvent such as methanol. By extracting the desired components from a large volume of water and then flushing them out with a small volume of solvent, the components can be concentrated. Preconcentration methods are often necessary for trace analytical methods. For example, solid-phase extractions are used in determining organic constituents in drinking water by methods approved by the Environmental Protection Agency. In some solid-phase extraction procedures, impurities are extracted into the solid phase, while compounds of interest pass through unretained.

In addition to packed cartridges, solid-phase extraction can be carried out by using small membranes or extraction disks. These have the advantages of reducing extraction time and reducing solvent use. Solid-phase extraction can be done in continuous flow systems, which can automate the preconcentration process.

A related technique, called **solid-phase microextraction,** uses a fused silica fiber coated with a nonvolatile polymer to extract organic analytes directly from aqueous samples or from the headspace above the samples.[6] The analyte partitions between the fiber and the liquid phase. The analytes are then desorbed thermally in the heated injector of a gas chromatograph (see Chapter 31). The extracting fiber is mounted in a holder that is much like an ordinary syringe. This technique combines sampling and sample preconcentration in a single step.

30D SEPARATING IONS BY ION EXCHANGE

► In the ion-exchange process, ions held on an ion-exchange resin are exchanged for ions in a solution brought into contact with the resin.

Ion exchange is a process by which ions held on a porous, essentially insoluble solid are exchanged for ions in a solution that is brought into contact with the solid. The ion-exchange properties of clays and zeolites have been recognized and

[5]For more information, see *Solid-Phase Extraction: Principles, Techniques and Applications,* N. J. K. Simpson, Ed. New York: Dekker, 2000; J. S. Fritz, *Analytical Solid-Phase Extraction.* New York: Wiley, 1999; E. M. Thurman and M. S. Mills, *Solid-Phase Extraction: Principles and Practice.* New York: Wiley, 1998.

[6]For more information, see *Solid-Phase Microextraction: A Practical Guide,* S. A. S. Wercinski, Ed. New York: Dekker, 1999; *Applications of Solid Phase Microextraction,* J. Pawliszyn, Ed. London: Royal Society of Chemistry, 1999.

studied for more than a century. Synthetic ion-exchange resins were first produced in 1935 and have since found widespread application in water softening, water deionization, solution purification, and ion separation.

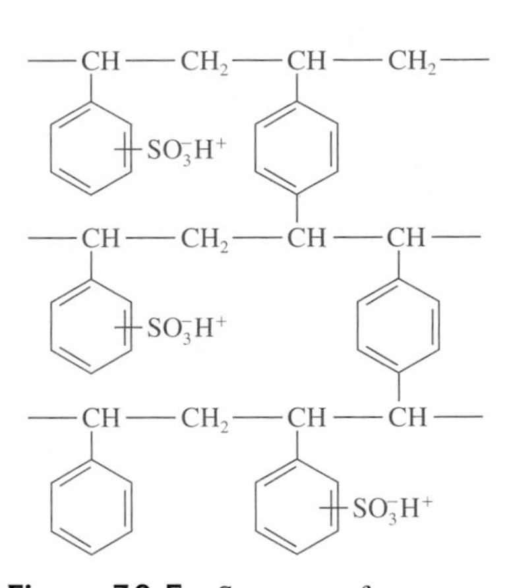

Figure 30-5 Structure of a cross-linked polystyrene ion-exchange resin. Similar resins are used in which the $-SO_3^-H^+$ group is replaced by $-COO^-H^+$, $-NH_3^+OH^-$, and $-N(CH_3)_3^+OH^-$ groups.

30D-1 Ion-Exchange Resins

Synthetic ion-exchange resins are high-molecular-weight polymers that contain large numbers of an ionic functional group per molecule. Cation-exchange resins contain acidic groups, while anion-exchange resins have basic groups. Strong-acid type exchangers have sulfonic acid groups ($-SO_3^-H^+$) attached to the polymeric matrix (Figure 30-5) and have wider application than weak-acid type exchangers, which owe their action to carboxylic acid ($-COOH$) groups. Similarly, strong-base anion exchangers contain quaternary amine [$-N(CH_3)_3^+OH^-$] groups, while weak-base types contain secondary or tertiary amines.

Cation exchange is illustrated by the equilibrium

$$\underset{\text{solid}}{xRSO_3^-H^+} + \underset{\text{soln}}{M^{x+}} \rightleftharpoons \underset{\text{solid}}{(RSO_3^-)_xM^{x+}} + \underset{\text{soln}}{xH^+}$$

where M^{x+} represents a cation and R represents *that part of a resin molecule that contains one sulfonic acid group.* The analogous equilibrium involving a strong-base anion exchanger and an anion A^{x-} is

$$\underset{\text{solid}}{xRN(CH_3)_3^+OH^-} + \underset{\text{soln}}{A^{x-}} \rightleftharpoons \underset{\text{solid}}{[RN(CH_3)_3^+]_xA^{x-}} + \underset{\text{soln}}{xOH^-}$$

30D-2 Ion-Exchange Equilibria

Ion-exchange equilibria can be treated by the law of mass action. For example, when a dilute solution containing calcium ions is passed through a column packed with a sulfonic acid resin, the following equilibrium is established:

$$Ca^{2+}(aq) + 2H^+(res) \rightleftharpoons Ca^{2+}(res) + 2H^+(aq)$$

for which an equilibrium constant K' is given by

$$K' = \frac{[Ca^{2+}]_{res}[H^+]_{aq}^2}{[Ca^{2+}]_{aq}[H^+]_{res}^2} \tag{30-4}$$

As usual, the bracketed terms are molar concentrations (strictly, activities) of the species in the two phases. Note that $[Ca^{2+}]_{res}$ and $[H^+]_{res}$ are molar concentrations of the two ions *in the solid phase.* In contrast to most solids, however, these concentrations can vary from zero to some maximum value when all the negative sites on the resin are occupied by only one species.

Ion-exchange separations are ordinarily performed under conditions in which one ion predominates in *both* phases. Thus, in the removal of calcium ions from a dilute and somewhat acidic solution, the calcium ion concentration will be much smaller than that of hydrogen ion in both the aqueous and resin phases; that is,

$$[Ca^{2+}]_{res} \ll [H^+]_{res}$$

and

$$[Ca^{2+}]_{aq} \ll [H^+]_{aq}$$

As a consequence, the hydrogen ion concentration is essentially constant in both phases, and Equation 30-4 can be rearranged to

$$\frac{[Ca^{2+}]_{res}}{[Ca^{2+}]_{aq}} = K' \frac{[H^+]^2_{res}}{[H^+]^2_{aq}} = K \quad (30\text{-}5)$$

where K is a distribution constant analogous to the constant that governs an extraction equilibrium (see Equation 30-2). Note that K in Equation 30-5 represents the affinity of the resin for calcium ion relative to another ion (here, H^+). In general, where K for an ion is large, a strong tendency for the resin phase to retain that ion exists; where K is small, the opposite is true. Selection of a common reference ion (such as H^+) permits a comparison of distribution constants for various ions on a given type of resin. Such experiments reveal that polyvalent ions are much more strongly retained than singly charged species. Within a given charge group, differences that exist among values for K appear to be related to the size of the hydrated ion as well as other properties. Thus, for a typical sulfonated cation-exchange resin, values of K for univalent ions decrease in the order $Ag^+ > Cs^+ > Rb^+ > K^+ > NH_4^+ > Na^+ > H^+ > Li^+$. For divalent cations, the order is $Ba^{2+} > Pb^{2+} > Sr^{2+} > Ca^{2+} > Ni^{2+} > Cd^{2+} > Cu^{2+} > Co^{2+} > Zn^{2+} > Mg^{2+} > UO_2^{2+}$.

30D-3 Applications of Ion-Exchange Methods

Ion-exchange resins are used to eliminate ions that would otherwise interfere with an analysis. For example, iron(III), aluminum(III), and many other cations tend to coprecipitate with barium sulfate during the determination of sulfate ion. Passage of a solution containing sulfate through a cation-exchange resin results in the retention of these cations and the release of an equivalent number of hydrogen ions. Sulfate ions pass freely through the column and can be precipitated as barium sulfate from the effluent.

Another valuable application of ion-exchange resins involves concentrating ions from a very dilute solution. Thus, traces of metallic elements in large volumes of natural waters can be collected on a cation-exchange column and subsequently liberated from the resin by treatment with a small volume of an acidic solution; the result is a considerably more concentrated solution for analysis by atomic absorption or ICP emission spectrometry (see Chapter 28).

The total salt content of a sample can be determined by titrating the hydrogen ion released as an aliquot of sample passes through a cation exchanger in the acidic form. Similarly, a standard hydrochloric acid solution can be prepared by diluting to known volume the effluent resulting from treatment of a cation-exchange resin with a known mass of sodium chloride. Substitution of an anion-exchange resin in its hydroxide form will permit the preparation of a standard base solution. Ion-exchange resins are also widely used in household water softeners, as discussed in Feature 30-2.

As shown in Section 32D, ion-exchange resins are particularly useful for the chromatographic separation of both inorganic and organic ionic species.

FEATURE 30-2
Home Water Softeners

Hard water is water that is rich in the salts of calcium, magnesium, and iron. The cations of hard water combine with fatty acid anions from soap to form insoluble salts known as **curd** or **soap curd.** In areas with particularly hard water, these precipitates can be seen as gray rings around bathtubs and sinks.

One method of solving the problem of hard water in homes is to exchange the calcium, magnesium, and iron cations for sodium ions, which form soluble fatty acid salts. A commercial water softener consists of a tank containing an ion-exchange resin, a storage reservoir for sodium chloride, and various valves and regulators for controlling the flow of water, as shown in Figure 30F-1. During the charging or regeneration cycle, concentrated salt water from the reservoir is directed through the ion-exchange resin, where the resin sites are occupied by Na^+ ions.

$$\underset{\text{solid}}{(RSO_3^-)_xM^{x+}} + \underset{\text{water}}{xNa^+} \rightleftharpoons \underset{\text{solid}}{xRSO_3^-Na^+} + \underset{\text{water}}{M^{x+}} \quad \text{(regeneration)}$$

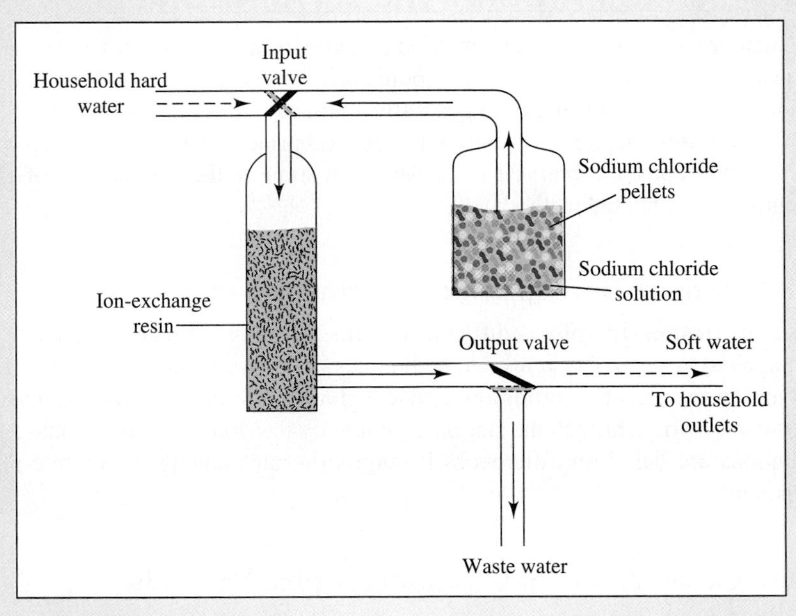

Figure 30F-1 Schematic of a home water softener. During the charging cycle, the valves are in the positions shown. Salt water from the storage reservoir passes through the ion-exchange resin to waste. Sodium ions from the salt water exchange with ions on the resin to leave the resin in the sodium form. During water use, the valves switch and hard water passes through the resin where the calcium, magnesium, and iron cations replace the sodium ions attached to the resin.

The M^{x+} cations (calcium, magnesium, or iron) released are sent to waste during this cycle.

After the regeneration cycle, the valves controlling the inlet to the ion-exchange resin and the outlet from the resin change so that water from the

(continued)

household supply passes through the resin and out to the household faucets. When the hard water passes through the resin, the M^{x+} cations are exchanged for Na^+ ions, and the water is softened.

$$\underset{\text{solid}}{x\text{RSO}_3^-\text{Na}^+} + \underset{\text{water}}{\text{M}^{x+}} \rightleftharpoons \underset{\text{solid}}{(\text{RSO}_3^-)_x\text{M}^{x+}} + \underset{\text{water}}{x\text{Na}^+} \quad \text{(household use)}$$

With use, the ion-exchange resin gradually accumulates the cations from the hard water. Hence, the softener must be periodically recharged by passing salt water through it and venting the hard-water ions to waste. After softening, soaps are much more effective because they remain dispersed in the water and do not form soap curds. Potassium chloride is also used instead of sodium chloride and is particularly advantageous for people on a sodium-restricted diet. Potassium chloride for water softeners is, however, more expensive than sodium chloride.

30E CHROMATOGRAPHIC SEPARATIONS

Chromatography is a technique in which the components of a mixture are separated based on differences in the rates at which they are carried through a fixed or stationary phase by a gaseous or liquid mobile phase.

Chromatography is a widely used method that allows the separation, identification, and determination of the chemical components in complex mixtures. No other separation method is as powerful and generally applicable as is chromatography.[7] The remainder of this chapter is devoted to the general principles that apply to all types of chromatography. Chapters 31 to 33 deal with some of the applications of chromatography and related methods.

30E-1 General Description of Chromatography

The **stationary phase** in chromatography is fixed in place either in a column or on a planar surface.

The **mobile phase** in chromatography moves over or through the stationary phase, carrying with it the analyte mixture. The mobile phase may be a gas, a liquid, or a supercritical fluid.

The term **chromatography** is difficult to define rigorously because the name has been applied to several systems and techniques. All these methods, however, have in common the use of a **stationary phase** and a **mobile phase.** Components of a mixture are carried through the stationary phase by the flow of a mobile phase, and separations are based on differences in migration rates among the mobile-phase components.

30E-2 Classification of Chromatographic Methods

► **Planar chromatography** and **column chromatography** are based on the same types of equilibria.

► Gas chromatography and supercritical fluid chromatography require the use of a column. Only liquid mobile phases can be used on planar surfaces.

Chromatographic methods are of two basic types. In **column chromatography,** the stationary phase is held in a narrow tube, and the mobile phase is forced through the tube under pressure or by gravity. In **planar chromatography,** the stationary phase is supported on a flat plate or in the pores of a paper. Here the mobile phase moves through the stationary phase by capillary action or under the influence of gravity. We deal here only with column chromatography.

As shown in the first column of Table 30-4, chromatographic methods fall into three categories based on the nature of the mobile phase: liquid, gas, and supercritical fluid. The second column of the table reveals that there are five types of liquid

[7]General references on chromatography include P. Sewell and B. Clarke, *Chromatographic Separations.* New York: Wiley, 1988; *Chromatographic Theory and Basic Principles,* J. A. Jonsson, Ed. New York: Marcel Dekker, 1987; A. Braithwaite and F. J. Smith, *Chromatographic Methods,* 5th ed. London: Blackie, 1996.

TABLE 30-4

Classification of Column Chromatographic Methods

General Classification	Specific Method	Stationary Phase	Type of Equilibrium
Gas chromatography (GC)	Gas-liquid (GLC)	Liquid adsorbed or bonded to a solid surface	Partition between gas and liquid
	Gas-solid	Solid	Adsorption
Liquid chromatography (LC)	Liquid-liquid, or partition	Liquid adsorbed or bonded to a solid surface	Partition between immiscible liquids
	Liquid-solid, or adsorption	Solid	Adsorption
	Ion exchange	Ion-exchange resin	Ion exchange
	Size exclusion	Liquid in interstices of a polymeric solid	Partition/sieving
	Affinity	Group-specific liquid bonded to a solid surface	Partition between surface liquid and mobile liquid
Supercritical fluid chromatography (SFC) (mobile phase: supercritical fluid)		Organic species bonded to a solid surface	Partition between supercritical fluid and bonded surface

chromatography and two types of gas chromatography that differ in the nature of the stationary phase and the types of equilibria between phases.

30E-3 Elution in Column Chromatography

Figure 30-6 shows how two components of a sample, A and B, are resolved on a packed column by **elution.** The column consists of a narrow-bore tubing packed with a finely divided inert solid that holds the stationary phase on its surface. The mobile phase occupies the open spaces between the particles of the packing. Initially, a solution of the sample containing a mixture of A and B in the mobile phase is introduced at the head of the column as a narrow plug, as shown in Figure 30-6 at time t_0. Here, the two components distribute themselves between the mobile phase and the stationary phase. Elution then occurs by forcing the sample components through the column by continuously adding fresh mobile phase.

Elution is a process in which solutes are washed through a stationary phase by the movement of a mobile phase. The mobile phase that exits the column is termed the **eluate.**

With the first introduction of fresh mobile phase, the **eluent**—the portion of the sample contained in the mobile phase—moves down the column, where further partitioning between the mobile phase and the stationary phase occurs (time t_1). Partitioning between the fresh mobile phase and the stationary phase takes place simultaneously at the site of the original sample.

An **eluent** is a solvent used to carry the components of a mixture through a stationary phase.

Further additions of solvent carry solute molecules down the column in a continuous series of transfers between the two phases. Because solute movement can occur only in the mobile phase, the average *rate* at which a solute migrates *depends on the fraction of time it spends in that phase.* This fraction is small for solutes that are strongly retained by the stationary phase (component B in Figure 30-6, for example) and large when retention in the mobile phase is more likely (component A). Ideally, the resulting differences in rates cause the components in a mixture to separate into **bands,** or **zones,** along the length of the column (see Figure 30-7). Isolation of the separated species is then accomplished by passing a sufficient quantity of mobile phase through the column to cause the individual bands to pass out the end (to be *eluted* from the column), where they can be collected or detected (times t_3 and t_4 in Figure 30-6).

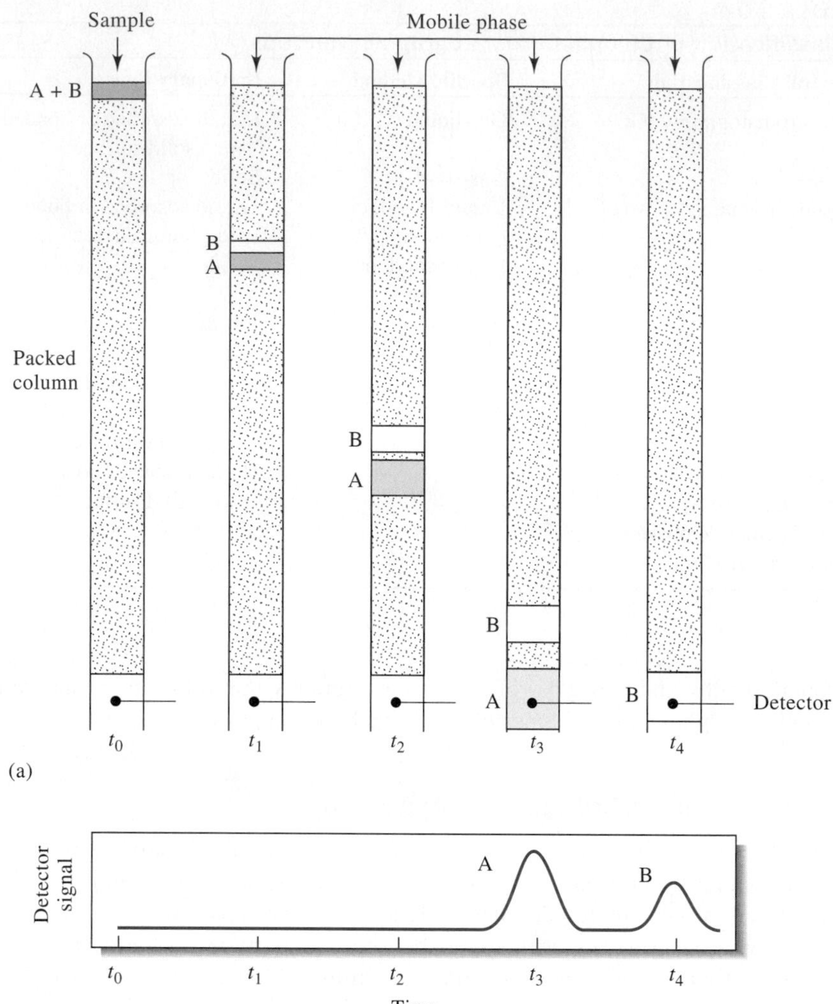

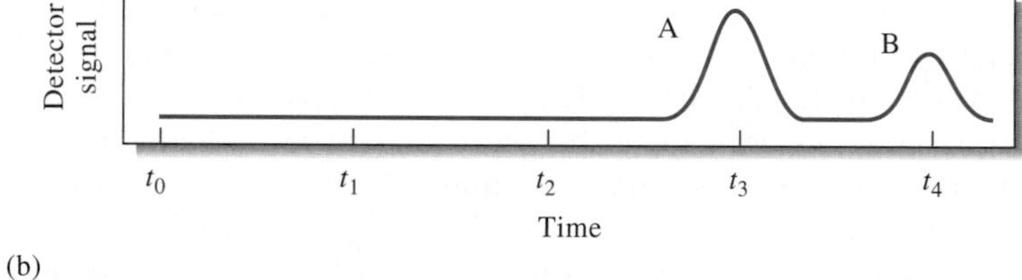

Figure 30-6 (a) Diagram showing the separation of a mixture of components A and B by column elution chromatography. (b) The detector signal at the various stages of elution shown in (a).

Chromatograms

If a detector that responds to solute concentration is placed at the end of the column during elution and its signal is plotted as a function of time (or of volume of added mobile phase), a series of peaks is obtained, as shown in the lower part of Figure 30-6. Such a plot, called a **chromatogram,** is useful for both qualitative and quan-

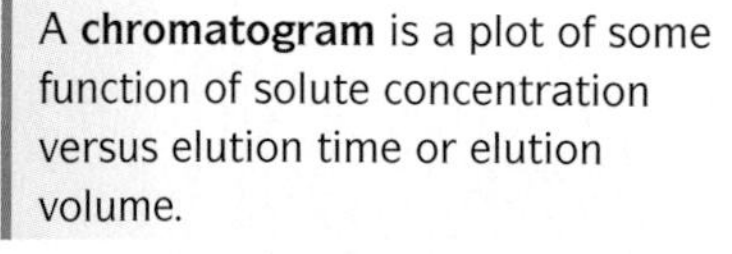

A **chromatogram** is a plot of some function of solute concentration versus elution time or elution volume.

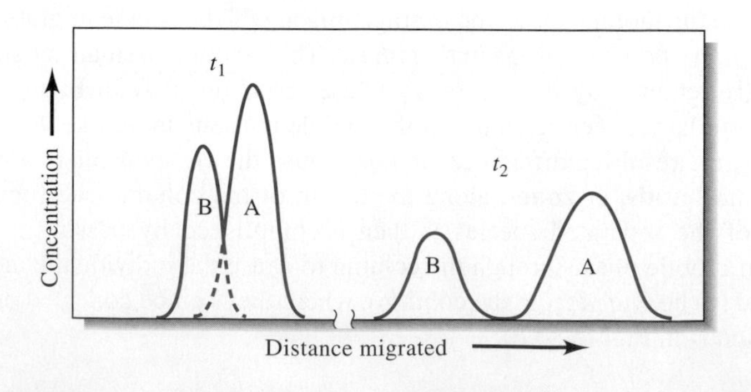

Figure 30-7 Concentration profiles of solute bands A and B at two different times in their migration down the column in Figure 30-6. The times t_1 and t_2 are indicated in Figure 30-6.

titative analysis. The positions of the peaks on the time axis can be used to identify the components of the sample; the areas under the peaks provide a quantitative measure of the amount of each species.

Chromatography was invented by the Russian botanist Mikhail Tswett shortly after the turn of the 20th century. He employed the technique to separate various plant pigments, such as chlorophylls and xanthophylls, by passing solutions of these species through glass columns packed with finely divided calcium carbonate. The separated species appeared as colored bands on the column, which accounts for the name he chose for the method (from the Greek *chroma* meaning "color" and *graphein* meaning "to write").

Methods of Improving Column Performance

Figure 30-7 shows concentration profiles for the bands containing solutes A and B on the column in Figure 30-6 at time t_1 and at a later time t_2.[8] Because B is more strongly retained by the stationary phase than is A, B lags during the migration. Clearly, the distance between the two increases as they move down the column. At the same time, however, broadening of both bands takes place, which lowers the efficiency of the column as a separating device. While band broadening is inevitable, conditions can often be found where it occurs more slowly than band separation. Thus, as shown in Figure 30-7, a clean separation of species is possible provided that the column is sufficiently long.

Several chemical and physical variables influence the rates of band separation and band broadening. As a consequence, improved separations can often be realized by the control of variables that either increase the rate of band separation or decrease the rate of band spreading. These alternatives are illustrated in Figure 30-8.

The variables that influence the relative rates at which solutes migrate through a stationary phase are described in the next section. Following this discussion, we turn to those factors that play a part in zone broadening.

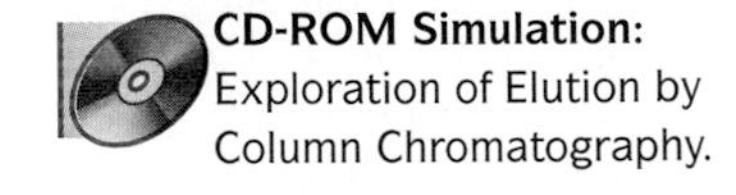

CD-ROM Simulation: Exploration of Elution by Column Chromatography.

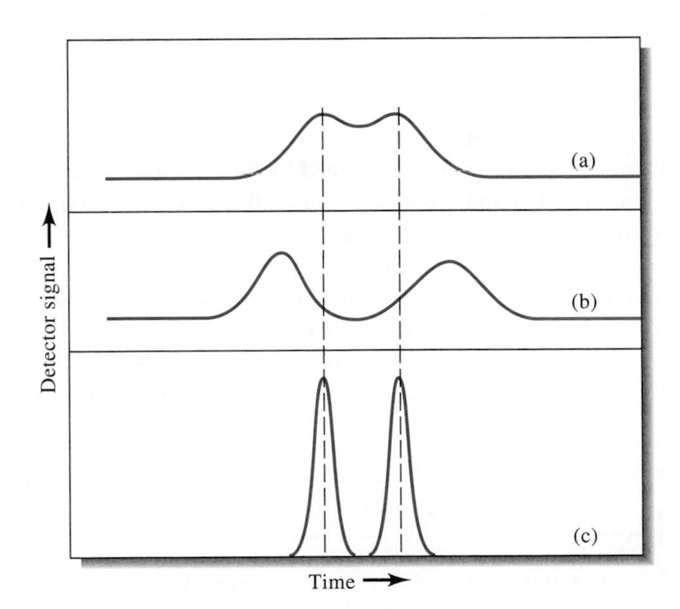

Figure 30-8 Two-component chromatogram illustrating two methods of improving separation: (a) original chromatogram with overlapping peaks; (b) improvement brought about by an increase in band separation; (c) improvement brought about by a decrease in the widths.

[8]Note that the relative positions of the bands for A and B in the concentration profile in Figure 30-7 appear to be reversed from their positions in the lower part of Figure 30-6. The difference is that the abscissa represents distance along the column in Figure 30-7, but time in Figure 30-6. Thus, in Figure 30-6, the *front* of a peak lies to the left and the *tail* to the right; in Figure 30-7, the reverse is true.

30E-4 Migration Rates of Solutes

The effectiveness of a chromatographic column in separating two solutes depends in part on the relative rates at which the two species are eluted. These rates in turn are determined by the ratios of the solute concentrations in each of the two phases.

Distribution Constants

All chromatographic separations are based on differences in the extent to which solutes are distributed between the mobile and stationary phases. For the solute species A, the equilibrium involved is described by the equation

$$A(mobile) \rightleftharpoons A(stationary) \tag{30-6}$$

The **distribution constant** for a solute in chromatography is equal to the ratio of its molar concentration in the stationary phase to its molar concentration in the mobile phase.

The equilibrium constant K_c for this reaction is called a **distribution constant,** which is defined as

$$K_c = \frac{(a_A)_S}{(a_A)_M} \tag{30-7}$$

where $(a_A)_S$ is the activity of solute A in the stationary phase and $(a_A)_M$ is the activity in the mobile phase. We often substitute c_S, the molar analytical concentration of the solute in the stationary phase, for $(a_A)_S$ and c_M for its molar analytical concentration in the mobile phase, $(a_A)_M$. Hence, we often write Equation 30-7 as

$$K_c = \frac{c_S}{c_M} \tag{30-8}$$

Ideally, the distribution constant is constant over a wide range of solute concentrations; that is, c_S is directly proportional to c_M.

Retention Times

The **dead time** (void time) t_M is the time it takes for an unretained species to pass through a chromatographic column. All components spend this amount of time in the mobile phase. Separations are based on the different times t_S that components spend in the stationary phase.

Figure 30-9 is a simple chromatogram made up of just two peaks. The small peak on the left is for a species that is *not* retained by the stationary phase. The time t_M between sample injection and the appearance of this peak is sometimes called the **dead** or **void time.** The dead time provides a measure of the average rate of migration of the mobile phase and is an important parameter in identifying analyte peaks. All components spend time t_M in the mobile phase. To aid in measuring t_M, an unretained species can be added if one is not already present in the sample or the mobile phase. The larger peak on the right in Figure 30-9 is that of an analyte species. The time required for this zone to reach the detector after sample injection is called the

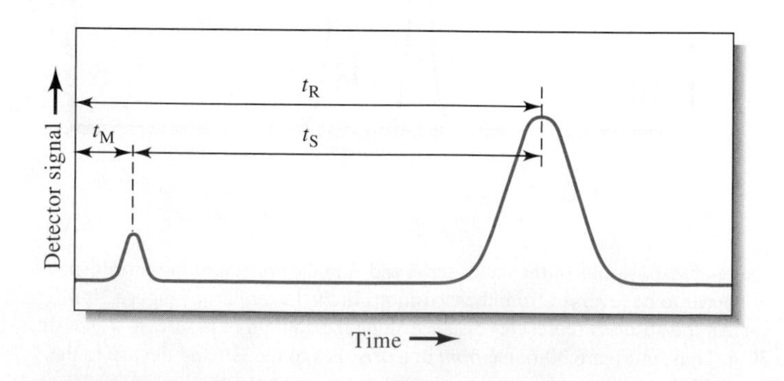

Figure 30-9 A typical chromatogram for a two-component mixture. The small peak on the left represents a solute that is not retained on the column and so reaches the detector almost immediately after elution is begun. Thus, its retention time t_M is approximately equal to the time required for a molecule of the mobile phase to pass through the column.

retention time and is represented by the symbol t_R. The analyte has been retained because it spends a time t_S in the stationary phase. The retention time is then

The **retention time** t_R is the time between injection of a sample and the appearance of a solute peak at the detector of a chromatographic column.

$$t_R = t_S + t_M \tag{30-9}$$

The average linear rate of solute migration, $\bar{v}$ (usually cm/s), is

$$\bar{v} = \frac{L}{t_R} \tag{30-10}$$

where L is the length of the column packing. Similarly, the average linear velocity, u, of the mobile-phase molecules is

$$u = \frac{L}{t_M} \tag{30-11}$$

The Relationship between Volumetric Flow Rate and Linear Flow Velocity

Experimentally, in chromatography the mobile phase flow is usually characterized by the volumetric flow rate, F (cm^3/min), at the column outlet. For an open tubular column, F is related to the linear velocity at the column outlet u_o

$$F = u_o A = u_o \times \pi r^2 \tag{30-12}$$

where A is the cross-sectional area of the tube (πr^2). For a packed column, the entire volume of the column is not available to the liquid, and so Equation 30-12 must be modified to

$$F = \pi r^2 u_o \varepsilon \tag{30-13}$$

where ε is the fraction of the total column volume available to the liquid (column porosity).

The Relationship between Migration Rate and Distribution Constant

To relate the rate of migration of a solute to its distribution constant, we express the rate as a fraction of the velocity of the mobile phase:

$$\bar{v} = u \times \text{fraction of time solute spends in mobile phase}$$

This fraction, however, equals the average number of moles of solute in the mobile phase at any instant divided by the total number of moles of solute in the column:

$$\bar{v} = u \times \frac{\text{moles of solute in mobile phase}}{\text{total moles of solute}}$$

The total number of moles of solute in the mobile phase is equal to the molar concentration, c_M, of the solute in that phase multiplied by its volume, V_M. Similarly, the number of moles of solute in the stationary phase is given by the product of the concentration c_S of the solute in the stationary phase and its volume, V_S. Therefore,

$$\bar{v} = u \times \frac{c_M V_M}{c_M V_M + c_S V_S} = u \times \frac{1}{1 + c_S V_S / c_M V_M}$$

Substitution of Equation 30-8 into this equation gives an expression for the rate of solute migration as a function of its distribution constant as well as a function of the volumes of the stationary and mobile phases:

$$\bar{v} = u \times \frac{1}{1 + K_c V_S / V_M} \tag{30-14}$$

The two volumes can be estimated from the method by which the column is prepared.

The Retention Factor, k

The retention factor is an important experimental parameter that is widely used to compare the migration rates of solutes on columns.[9] For solute A, the retention factor k_A is defined as

$$k_A = \frac{K_A V_S}{V_M} \tag{30-15}$$

where K_A is the distribution constant for solute A. Substitution of Equation 30-15 into Equation 30-14 yields

$$\bar{v} = u \times \frac{1}{1 + k_A} \tag{30-16}$$

To show how k_A can be derived from a chromatogram, we substitute Equations 30-10 and 30-11 into Equation 30-16:

$$\frac{L}{t_R} = \frac{L}{t_M} \times \frac{1}{1 + k_A} \tag{30-17}$$

This equation rearranges to

$$k_A = \frac{t_R - t_M}{t_M} = \frac{t_S}{t_M} \tag{30-18}$$

The **retention factor** k_A for solute A is related to the rate at which A migrates through a column. It is the amount of time a solute spends in the stationary phase relative to the time it spends in the mobile phase.

As shown in Figure 30-9, t_R and t_M are readily obtained from a chromatogram. A retention factor much less than unity means that the solute emerges from the column at a time near that of the void time. When the retention factor is larger than perhaps 20 to 30, elution times become inordinately long. Ideally, separations are performed under conditions in which the retention factors for the solutes in a mixture lie in the range of 1 to 5.

► Ideally, the retention factors for analytes in a sample are between 1 and 5.

[9] In the older literature, this constant was called the capacity factor and was symbolized by k'. In 1993, however, the IUPAC Committee on Analytical Nomenclature recommended that this constant be termed the *retention factor* and be symbolized by k.

Retention factors in gas chromatography can be varied by changing the temperature and the column packing, as discussed in Chapter 31. In liquid chromatography, retention factors can often be manipulated to give better separations by varying the composition of the mobile phase and the stationary phase, as illustrated in Chapter 32.

The Selectivity Factor

The **selectivity factor** α of a column for the two solutes A and B is defined as

The **selectivity factor** α for solutes A and B is defined as the ratio of the distribution constant of the more strongly retained solute (B) to the distribution constant for the less strongly held solute (A).

$$\alpha = \frac{K_B}{K_A} \quad (30\text{-}19)$$

where K_B is the distribution constant for the more strongly retained species B and K_A is the constant for the less strongly held or more rapidly eluted species A. According to this definition, *α is always greater than unity.*

Substitution of Equation 30-15 and the analogous equation for solute B into Equation 30-19 provides a relationship between the selectivity factor for two solutes and their retention factors:

◀ The selectivity factor for two analytes in a column provides a measure of how well the column will separate the two.

$$\alpha = \frac{k_B}{k_A} \quad (30\text{-}20)$$

where k_B and k_A are the retention factors for B and A, respectively. Substitution of Equation 30-18 for the two solutes into Equation 30-20 gives an expression that permits the determination of α from an experimental chromatogram:

$$\alpha = \frac{(t_R)_B - t_M}{(t_R)_A - t_M} \quad (30\text{-}21)$$

In Section 30E-7, we show how we use the retention factor to compute the resolving power of a column.

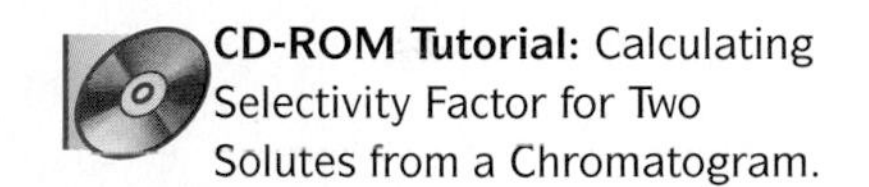
CD-ROM Tutorial: Calculating Selectivity Factor for Two Solutes from a Chromatogram.

30E-5 Band Broadening and Column Efficiency

The efficiency of a chromatographic column is affected by the amount of band broadening that occurs as a compound passes through the column. Before defining column efficiency in more quantitative terms, let us examine the reasons that bands become broader as they move down a column.

The Rate Theory of Chromatography

The **rate theory of chromatography** describes the shapes and breadths of elution bands in quantitative terms based on a random-walk mechanism for the migration of molecules through a column. A detailed discussion of the rate theory is beyond the scope of this text. We can, however, give a qualitative picture of why bands broaden and what variables improve column efficiency.[10]

If you examine the chromatograms shown in this and the next chapter, you will see that the elution peaks look very much like the Gaussian or normal error curves

[10]For more information, see J. C. Giddings, *Unified Separation Science,* pp. 94–96. New York: Wiley, 1991.

that you encountered in Chapters 6 and 7. As shown in Section 6A-2, normal error curves are rationalized by assuming that the uncertainty associated with any single measurement is the summation of a much larger number of small, individually undetectable and random uncertainties, each of which has an equal probability of being positive or negative. In a similar way, the typical Gaussian shape of a chromatographic band can be attributed to the additive combination of the random motions of the various molecules as they move down the column. We assume in the following discussion that a narrow zone has been introduced so that the injection width is not the limiting factor determining the overall width of the band that elutes. It is important to realize that the widths of eluting bands can never be *narrower* than the width of the injection zone.

It is instructive to consider a single solute molecule as it undergoes many thousands of transfers between the stationary and mobile phases during elution. Residence time in either phase is highly irregular. Transfer from one phase to the other requires energy, and the molecule must acquire this energy from its surroundings. Thus, the residence time in a given phase may be transitory after some transfers and relatively long after others. Recall that movement down the column can occur *only while the molecule is in the mobile phase.* As a consequence, certain particles travel rapidly by virtue of their accidental inclusion in the mobile phase for a majority of the time, whereas others lag because they happen to be incorporated into the stationary phase for a greater-than-average length of time. The result of these random individual processes is a symmetrical spread of velocities around the mean value, which represents the behavior of the average analyte molecule.

As shown in Figure 30-10, some chromatographic peaks are nonideal and exhibit **tailing** or **fronting.** In the former case, the tail of the peak, appearing to the right on the chromatogram, is drawn out while the front is steepened. With fronting, the reverse is the case. A common cause of tailing and fronting is a distribution con-

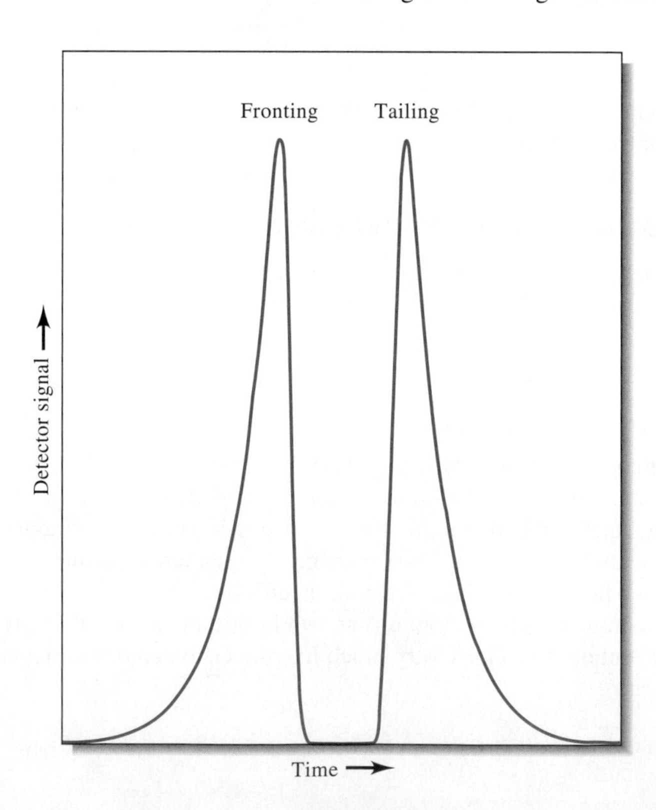

Figure 30-10 Illustration of fronting and tailing in chromatographic peaks.

stant that varies with concentration. Fronting also arises when the amount of sample introduced onto a column is too large. Distortions of this kind are undesirable because they lead to poorer separations and less reproducible elution times. In the discussion that follows, tailing and fronting are assumed to be minimal.

A Quantitative Description of Column Efficiency

Two related terms are widely used as quantitative measures of chromatographic column efficiency: (1) **plate height** H and (2) **plate count** or **number of theoretical plates** N. The two are related by the equation

$$N = \frac{L}{H} \tag{30-22}$$

where L is the length (usually in centimeters) of the column packing. The efficiency of chromatographic columns increases as the plate count N becomes greater and as the plate height H becomes smaller. Enormous differences in efficiencies are encountered in columns as a result of differences in column type and in mobile and stationary phases. Efficiencies in terms of plate numbers can vary from a few hundred to several hundred thousand; plate heights ranging from a few tenths to one thousandth of a centimeter or smaller are not uncommon.

In Section 6B-2, we pointed out that the breadth of a Gaussian curve is described by the standard deviation σ and the variance σ^2. Because chromatographic bands are usually Gaussian and because the efficiency of a column is reflected in the breadth of chromatographic peaks, the variance per unit length of column is used by chromatographers as a measure of column efficiency. That is, the column efficiency H is defined as

$$H = \frac{\sigma^2}{L} \tag{30-23}$$

This definition of column efficiency is illustrated in Figure 30-11a, which shows a column having a packing L cm in length. Above this schematic (Figure 30-11b) is a plot showing the distribution of molecules along the length of the column at the moment that the analyte peak reaches the end of the packing (that is, at the retention time). The curve is Gaussian, and the locations of $L + 1\sigma$ and $L - 1\sigma$ are indicated as broken vertical lines. Note that L carries units of centimeters and σ^2 units

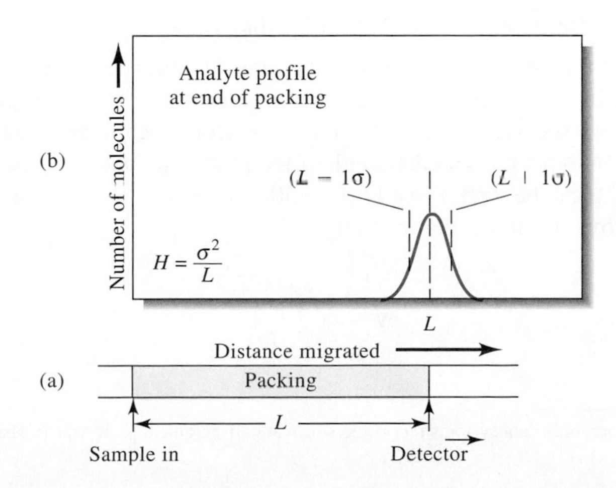

Figure 30-11 Definition of plate height, $H = \sigma^2/L$. In (a), the column length is shown as the distance from the sample entrance point to the detector. In (b), the Gaussian distribution of sample molecules is shown.

of centimeters squared; thus, H represents a linear distance in centimeters as well (see Equation 30-23). In fact, the plate height can be thought of as the length of column that contains a fraction of the analyte that lies between L and $L - \sigma$. Because the area under a normal error curve bounded by $\pm\sigma$ is about 68% of the total area (page 113), the plate height, as defined, contains 34% of the analyte.

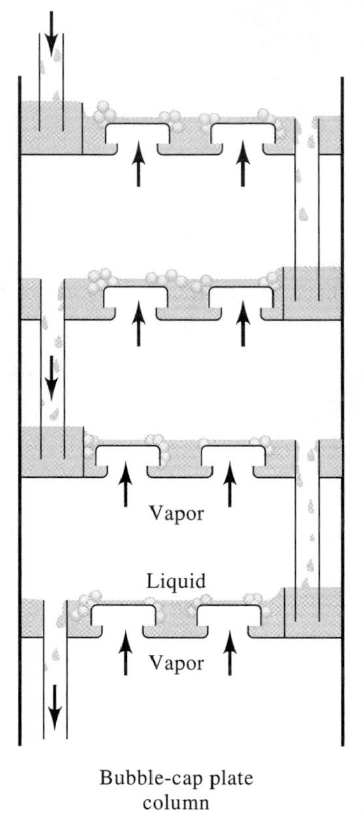

FEATURE 30-3

What Is the Source of the Terms **Plate** and **Plate Height?**

The 1952 Nobel Prize was awarded to two Englishmen, A. J. P. Martin and R. L. M. Synge, for their work in the development of modern chromatography. In their theoretical studies, they adapted a model that was first developed in the early 1920s to describe separations on fractional distillation columns. Fractionating columns, which were first used in the petroleum industry to separate closely related hydrocarbons, consisted of numerous interconnected bubble-cap plates (see Figure 30F-2) at which vapor-liquid equilibria were established when the column was operated under reflux conditions.

Martin and Synge treated a chromatographic column as if it were made up of a series of contiguous bubble-cap–like plates within which equilibrium conditions always prevail. This plate model successfully accounts for the Gaussian shape of chromatographic peaks as well as for factors that influence differences in solute migration rates. The plate model does not adequately account for zone broadening, however, because of its basic assumption that equilibrium conditions prevail throughout a column during elution. This assumption can never be valid in the dynamic state that exists in a chromatographic column, where phases are moving past one another at such a pace that sufficient time is not available for equilibration.

Because the plate model is not a very good representation of a chromatographic column, we strongly urge you to (1) avoid attaching any special significance to the terms *plate* and *plate height* and (2) view these terms as designators of column efficiency that are retained for historic reasons only and not because they have physical significance. Unfortunately, these terms are so well entrenched in the chromatographic literature that their replacement by more appropriate designations seems unlikely, as least in the near future.

Figure 30F-2 Plates in a fractionating column.

Experimental Determination of the Number of Plates in a Column

The number of theoretical plates, N, and the plate height, H, are widely used in the literature and by instrument manufacturers as measures of column performance. Figure 30-12 shows how N can be determined from a chromatogram. Here, the retention time of a peak t_R and the width of the peak at its base W (in units of time) are measured. It can be shown (see Feature 30-4) that the number of plates can then be computed by the simple relationship

$$N = 16\left(\frac{t_R}{W}\right)^2 \qquad (30\text{-}24)^{11}$$

[11]Many chromatographic data systems report the width at half-height, $W_{1/2}$, in which case $N = 5.54(t_R/W_{1/2})^2$.

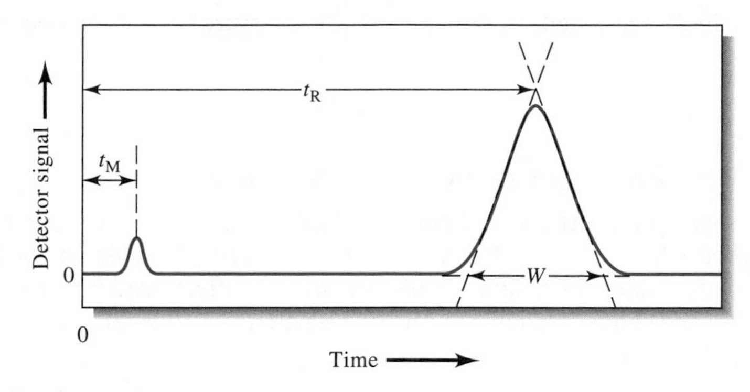

Figure 30-12 Determination of the number of plates, $N = 16\left(\frac{t_R}{W}\right)^2$.

FEATURE 30-4
Derivation of Equation 30-24

The variance of the peak shown in Figure 30-12 has units of seconds squared because the abscissa is time in seconds (or sometimes in minutes). This time-based variance is usually designated as τ^2 to distinguish it from σ^2, which has units of centimeters squared. The two standard deviations τ and σ are related by

$$\tau = \frac{\sigma}{L/t_R} \tag{30-25}$$

where L/t_R is the average linear velocity of the solute in centimeters per second.

Figure 30-12 illustrates a simple means of approximating τ from an experimental chromatogram. Tangents at the inflection points on the two sides of the chromatographic peak are extended to form a triangle with the baseline. The area of this triangle can be shown to be approximately 96% of the total area under the peak. In Section 6B-2, it was shown that about 96% of the area under a Gaussian peak is included within plus or minus two standard deviations ($\pm 2\sigma$) of its maximum. Thus, the intercepts shown in Figure 30-12 occur at approximately $\pm 2\tau$ from the maximum, and $W = 4\tau$, where W is the magnitude of the base of the triangle. Substituting this relationship into Equation 30-25 and rearranging yields

$$\sigma = \frac{LW}{4t_R}$$

Substitution of this equation for σ into Equation 30-23 gives

$$H = \frac{LW^2}{16t_R^2} \tag{30-26}$$

To obtain N, we substitute into Equation 30-22 and rearrange to get

$$N = 16\left(\frac{t_R}{W}\right)^2$$

Thus, N can be calculated from two time measurements, t_R and W; to obtain H, the length of the column packing L must also be known.

To obtain H, the length of the column L is measured and Equation 30-23 is applied.

30E-6 Variables That Affect Column Efficiency

Band broadening reflects a loss of column efficiency. The slower the rate of mass-transfer processes occurring while a solute migrates through a column, the broader the band at the column exit. Some of the variables that affect mass-transfer rates are controllable and can be exploited to improve separations. Table 30-5 lists the most important of these variables.

The Effect of Mobile-Phase Flow Rate

The extent of band broadening depends on the length of time that the mobile phase is in contact with the stationary phase, which in turn depends on the flow rate of the mobile phase. For this reason, efficiency studies have generally been carried out by determining H (by means of Equation 30-26) as a function of mobile-phase velocity. The plots for liquid chromatography and for gas chromatography shown in Figure 30-13 are typical of the data obtained from such studies. While both show a minimum in H (or a maximum in efficiency) at low linear flow rates, the minimum for liquid chromatography usually occurs at flow rates that are well below those for gas chromatography. Often, these flow rates are so low that the minimum H is not observed for liquid chromatography under normal operating conditions.

Linear flow rate and **volumetric flow rate** are two different but related quantities. The linear flow rate is related to the volumetric flow rate by the cross-sectional area and porosity (packed column) of the column (see Equations 30-12 and 30-13).

Generally, liquid chromatograms are obtained at lower linear flow rates than gas chromatograms. Furthermore, as shown in Figure 30-13, plate heights for liquid chromatographic columns are an order of magnitude or more smaller than those encountered with gas chromatographic columns. Offsetting this advantage is the fact that it is impractical to employ liquid chromatographic columns that are longer than about 25 to 50 cm because of high pressure drops. In contrast, gas chromatographic columns may be 50 m or more in length. Consequently, the total number of plates, and thus overall column efficiency, is usually superior with gas chromatographic columns.

Theory of Band Broadening

Over the past 40 years, an enormous amount of theoretical and experimental effort has been devoted to developing quantitative relationships describing the effects of the experimental variables listed in Table 30-5 on plate heights for various types of

TABLE 30-5
Variables That Influence Column Efficiency

Variable	Symbol	Usual Units
Linear velocity of mobile phase	u	cm s^{-1}
Diffusion coefficient in mobile phase*	D_M	cm^2 s^{-1}
Diffusion coefficient in stationary phase*	D_S	cm^2 s^{-1}
Retention factor (see Equation 30-18)	k	unitless
Diameter of packing particles	d_p	cm
Thickness of liquid coating on stationary phase	d_f	cm

*Increases as temperature increases and viscosity decreases.

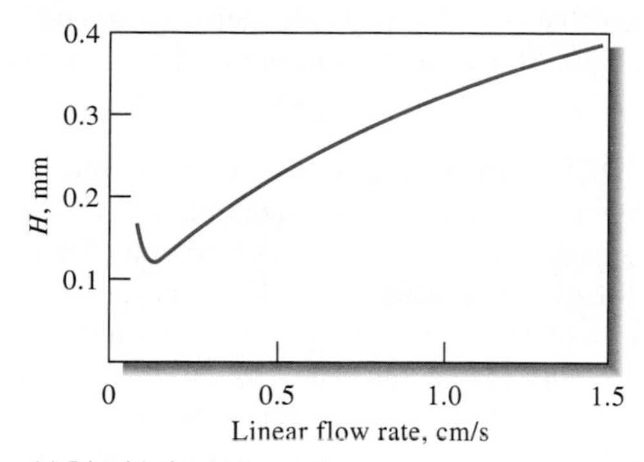

(a) Liquid chromatography

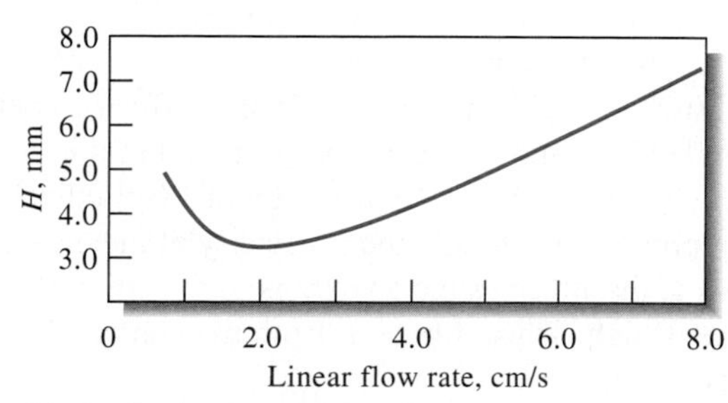

(b) Gas-liquid chromatography

Figure 30-13 Effect of mobile-phase flow rate on plate height for (a) liquid chromatography and (b) gas chromatography.

columns. Perhaps a dozen or more expressions for calculating plate height have been put forward and applied with various degrees of success. None of these is entirely adequate to explain the complex physical interactions and effects that lead to zone broadening and thus lower column efficiencies. Some of the equations, though imperfect, have been of considerable use, however, in pointing the way toward improved column performance. One of these is presented here.

The efficiency of capillary chromatographic columns and packed chromatographic columns at low flow velocities can be approximated by the expression

$$H = \frac{B}{u} + C_S u + C_M u \quad (30\text{-}27)$$

where H is the plate height in centimeters and u is the linear velocity of the mobile phase in centimeters per second.[12] The quantity B is the **longitudinal diffusion coefficient,** while C_S and C_M are **mass-transfer coefficients** for the stationary and mobile phases, respectively.

At high flow velocities in packed columns where flow effects dominate diffusion, the efficiency can be approximated by

$$H = A + \frac{B}{u} + C_S u \quad (30\text{-}28)$$

[12]S. J. Hawkes, *J. Chem. Educ.*, **1983,** *60,* 393.

Theoretical studies of zone broadening in the 1950s by Dutch chemical engineers led to the **van Deemter equation**, which can be written in the form

$$H = A + B/u + Cu$$

where the constants A, B, and C are coefficients of multiple-path effects, longitudinal diffusion, and mass transfer, respectively. Today, we consider the van Deemter equation to be appropriate only for packed columns at high flow velocities. For other cases, Equation 30-27 is usually a better description.

where A is a coefficient that describes multiple-path effects (eddy diffusion), as discussed later. Equation 30-28 is equivalent to the well-known **van Deemter equation,** which is often used to describe chromatographic efficiency.

The Longitudinal Diffusion Term B/u Diffusion is a process in which species migrate from a more concentrated part of a medium to a more dilute region. The rate of migration is proportional to the concentration difference between the regions and to the **diffusion coefficient** D_M of the species. The latter, which is a measure of the mobility of a substance in a given medium, is a constant for a given species equal to the velocity of migration under a unit concentration gradient.

In chromatography, longitudinal diffusion results in the migration of a solute from the concentrated center of a band to the more dilute regions on either side (that is, toward and opposed to the direction of flow). Longitudinal diffusion is a common source of band broadening in gas chromatography, where the rate at which molecules diffuse is high. The phenomenon is of little significance in liquid chromatography, where diffusion rates are much smaller. The magnitude of the B term in Equation 30-27 is largely determined by the diffusion coefficient D_M of the analyte in the mobile phase and is directly proportional to this constant.

As shown by Equation 30-27, the contribution of longitudinal diffusion to plate height is inversely proportional to the linear velocity of the eluent. Such a relationship is not surprising, inasmuch as the analyte is in the column for a briefer period when the flow rate is high. Thus, diffusion from the center of the band to the two edges has less time to occur.

The initial decreases in H shown in both curves in Figure 30-13 are a direct consequence of longitudinal diffusion. Note that the effect is much less pronounced in liquid chromatography because of the much lower diffusion rates in a liquid mobile phase. The striking difference in plate heights shown by the two curves in Figure 30-13 can also be explained by considering the relative rates of longitudinal diffusion in the two mobile phases. That is, diffusion coefficients in gaseous media are orders of magnitude larger than in liquids. Thus, band broadening occurs to a much greater extent in gas chromatography than in liquid chromatography.

► Diffusion coefficients in gases are usually about 1000 times larger than diffusion coefficients in liquids.

The Stationary Phase Mass-Transfer Term $C_S u$ When the stationary phase is an immobilized liquid, the mass-transfer coefficient is directly proportional to the square of the thickness of the film on the support particles, d_f^2, and inversely proportional to the diffusion coefficient, D_S, of the solute in the film. These effects can be understood by realizing that both reduce the average frequency at which analyte molecules reach the interface where transfer to the mobile phase can occur. That is, with thick films, molecules must on the average travel farther to reach the surface, and with smaller diffusion coefficients, they travel slower. The consequence is a slower rate of mass transfer and an increase in plate height.

When the stationary phase is a solid surface, the mass-transfer coefficient C_S is directly proportional to the time required for a species to be adsorbed or desorbed, which in turn is inversely proportional to the first-order rate constant for the processes.

The Mobile Phase Mass-Transfer Term $C_M u$ The mass-transfer processes that occur in the mobile phase are sufficiently complex that we do not yet have a complete quantitative description. We have a good qualitative understanding of the variables affecting zone broadening from this cause, however, and this understanding has led to vast improvements in all types of chromatographic columns.

The mobile phase mass-transfer coefficient C_M is known to be inversely proportional to the diffusion coefficient of the analyte in the mobile phase, D_M. For packed columns, C_M is proportional to the square of the particle diameter of the packing material, d_p^2. For capillary columns, C_M is proportional to the square of the column diameter d_c^2 and is a function of the flow rate.

The contribution of mobile phase mass transfer to plate height is the product of the mass-transfer coefficient C_M (which is a function of solvent velocity) as well as the velocity of the solvent itself. Thus, the net contribution of $C_M u$ to plate height is not linear in u (see the curve labeled $C_M u$ in Figure 30-15) but bears a complex dependency on solvent velocity.

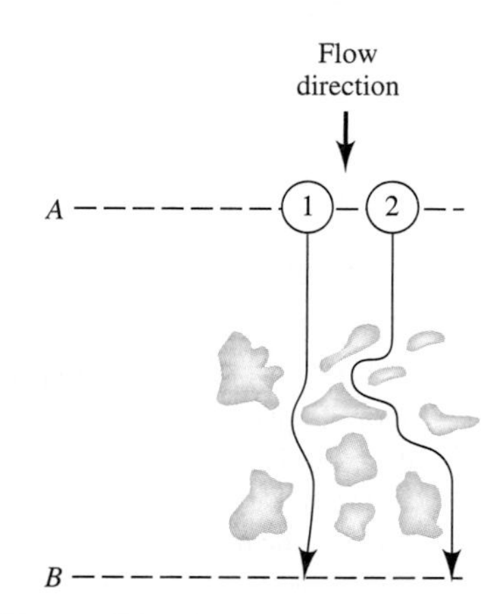

Figure 30-14 Typical pathways of two molecules during elution. Note that the distance traveled by molecule 2 is greater than that traveled by molecule 1. Thus, molecule 2 will arrive at *B* later than molecule 1.

Zone broadening in the mobile phase is due in part to the multitude of pathways by which a molecule (or an ion) can find its way through a packed column. As shown in Figure 30-14, the lengths of these pathways may differ significantly; thus, the residence times in the column for molecules of the same species are also variable. Solute molecules reach the end of the column over a certain time interval, which leads to a broadened band. This multiple-path effect, which is sometimes called **eddy diffusion,** would be independent of solvent velocity if it were not partially offset by ordinary diffusion, which results in molecules being transferred from a stream following one pathway to a stream following another. If the velocity of flow is very low, a large number of these transfers will occur, and each molecule in its movement down the column will sample numerous flow paths, spending a brief time in each. As a consequence, the rate at which each molecule moves down the column tends to approach that of the average. Thus, at low mobile-phase velocities, the molecules are not significantly dispersed by the multiple-path effect. At moderate or high velocities, however, sufficient time is not available for diffusion averaging to occur, and band broadening due to the different path lengths is observed. At sufficiently high velocities, the effect of eddy diffusion becomes independent of flow rate.

◀ Pathways for the mobile phase through the column are numerous and have different lengths.

Superimposed on the eddy diffusion effect is one that arises from stagnant pools of the mobile phase retained in the stationary phase. Thus, when a solid serves as the stationary phase, its pores are filled with *static* volumes of mobile phase. Solute molecules must then diffuse through these stagnant pools before transfer can occur between the *moving* mobile phase and the stationary phase. This situation applies not only to solid stationary phases but also to liquid stationary phases immobilized on porous solids because the immobilized liquid does not usually fully fill the pores.

The presence of stagnant pools of mobile phase slows the exchange process and results in a contribution to the plate height that is directly proportional to the mobile phase velocity and inversely proportional to the diffusion coefficient for the solute in the mobile phase. An increase in internal volume then accompanies increases in particle size.

◀ Static pools of solvent contribute to increases in *H*.

Effect of Mobile Phase Velocity on Terms in Equation 30-27 Figure 30-15 shows the variation of the three terms in Equation 30-27 as a function of mobile phase velocity. The top curve is the summation of these various effects. Note that an optimum flow rate exists at which the plate height is at a minimum and the separation efficiency is at a maximum.

Summary of Methods of Reducing Band Broadening For packed columns, one variable that affects column efficiency is the diameter of the particles making up the packing. For capillary columns, the diameter of the column itself is an important variable. The effect of particle diameter is demonstrated by the data shown in Figure 30-16 for gas chromatography. A similar plot for liquid chromatography is shown in Figure 32-1. To take advantage of the effect of column diameter, narrower and narrower columns have been used in recent years.

◀ For packed columns, band broadening is minimized by small particle diameters. For capillary columns, small column diameters reduce band broadening.

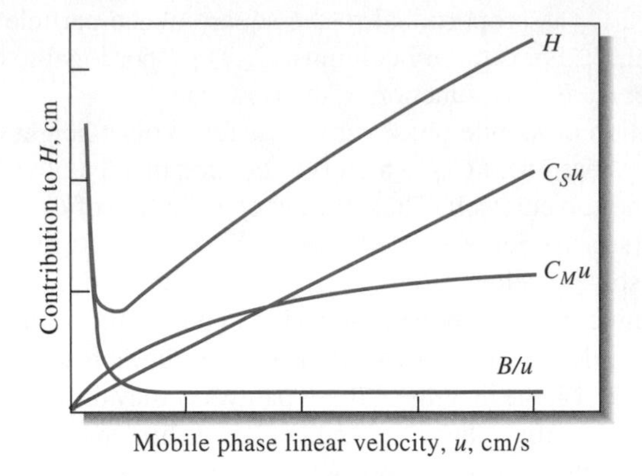

Figure 30-15 Contribution of various mass-transfer terms to plate height. $C_S u$ arises from the rate of mass transfer to and from the stationary phase; $C_M u$ comes from a limitation in the rate of mass transfer in the mobile phase; and B/u is associated with longitudinal diffusion.

With gaseous mobile phases, the rate of longitudinal diffusion can be reduced appreciably by lowering the temperature and thus the diffusion coefficient. The consequence is significantly smaller plate heights at lower temperatures. This effect is usually not noticeable in liquid chromatography because diffusion is slow enough that the longitudinal diffusion term has little effect on overall plate height.

► The diffusion coefficient D_M has a greater effect in gas chromatography than in liquid chromatography.

With liquid stationary phases, the thickness of the layer of adsorbed liquid should be minimized since C_S in Equation 30-27 is proportional to the square of this variable.

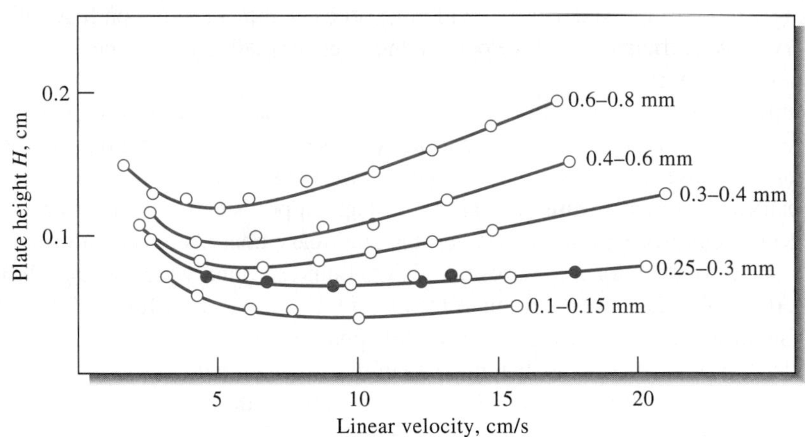

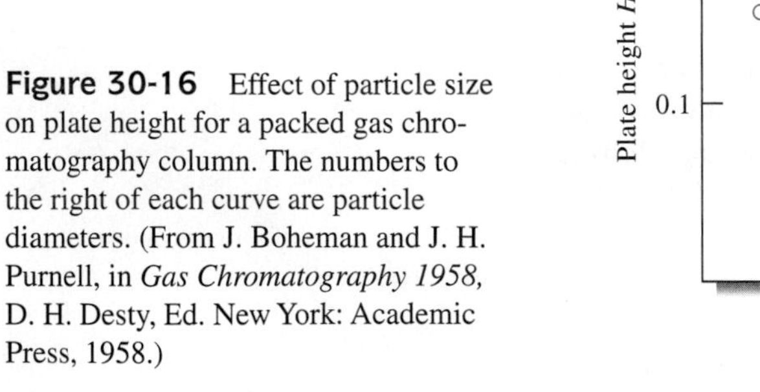

Figure 30-16 Effect of particle size on plate height for a packed gas chromatography column. The numbers to the right of each curve are particle diameters. (From J. Boheman and J. H. Purnell, in *Gas Chromatography 1958,* D. H. Desty, Ed. New York: Academic Press, 1958.)

30E-7 Column Resolution

The **resolution** of a chromatographic column is a quantitative measure of its ability to separate analytes A and B.

The **resolution** R_s of a column tells us how far apart two bands are relative to their widths. The resolution provides a quantitative measure of the ability of the column to separate two analytes. The significance of this term is illustrated in Figure 30-17, which consists of chromatograms for species A and B on three columns with different resolving powers. The resolution of each column is defined as

$$R_s = \frac{\Delta Z}{\frac{W_A}{2} + \frac{W_B}{2}} = \frac{2\Delta Z}{W_A + W_B} = \frac{2[(t_R)_B - (t_R)_A]}{W_A + W_B} \tag{30-29}$$

where all the terms on the right side are as defined in the figure.

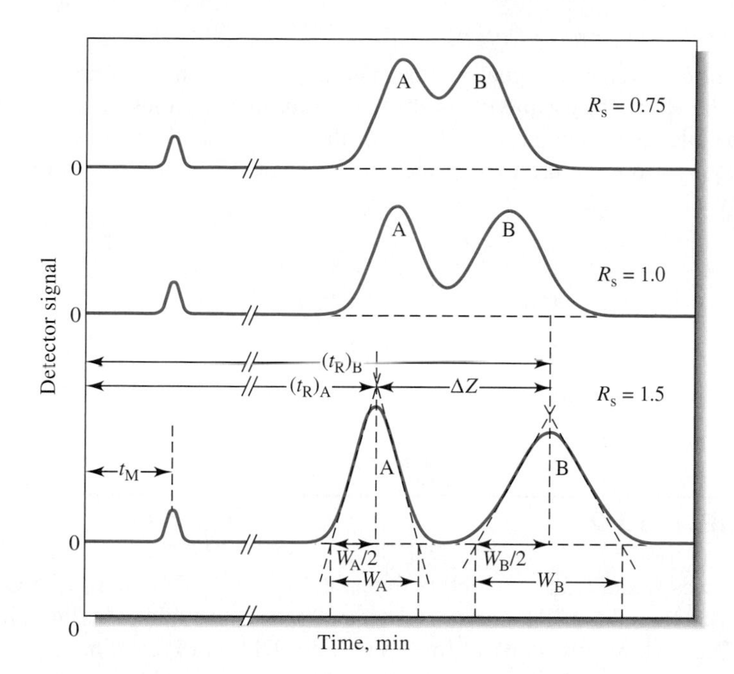

Figure 30-17 Separation at three resolution values: $R_s = 2\Delta Z/(W_A + W_B)$.

CD-ROM Tutorial: Calculating Column Resolution from a Chromatogram.

It is evident from Figure 30-17 that a resolution of 1.5 gives an essentially complete separation of A and B, whereas a resolution of 0.75 does not. At a resolution of 1.0, zone A contains about 4% B and zone B contains about 4% A. At a resolution of 1.5, the overlap is about 0.3%. The resolution for a given stationary phase can be improved by lengthening the column, thus increasing the number of plates. An adverse consequence of the added plates, however, is an increase in the time required for separating the components.

Effect of Retention Factor and Selectivity Factor on Resolution

A useful equation is readily derived that relates the resolution of a column to the number of plates it contains as well as to the retention and selectivity factors of a pair of solutes on the column. Thus, it can be shown[13] that for the two solutes A and B in Figure 30-17, the resolution is given by the equation

$$R_s = \frac{\sqrt{N}}{4}\left(\frac{\alpha - 1}{\alpha}\right)\left(\frac{k_B}{1 + k_B}\right) \tag{30-30}$$

where k_B is the retention factor of the slower moving species and α is the selectivity factor. This equation can be rearranged to give the number of plates needed to realize a given resolution:

$$N = 16R_s^2\left(\frac{\alpha}{\alpha - 1}\right)^2\left(\frac{1 + k_B}{k_B}\right)^2 \tag{30-31}$$

[13]See D. A. Skoog, F. J. Holler, and T. A. Nieman, *Principles of Instrumental Analysis,* 5th ed., p. 689. Belmont, CA: Brooks/Cole, 1998.

Effect of Resolution on Retention Time

As mentioned earlier, the goal in chromatography is the highest possible resolution in the shortest possible elapsed time. Unfortunately, these goals tend to be incompatible, and a compromise between the two is usually necessary. The time $(t_R)_B$ required to elute the two species in Figure 30-17 with a resolution of R_s is given by

$$(t_R)_B = \frac{16R_s^2H}{u}\left(\frac{\alpha}{\alpha - 1}\right)^2\frac{(1 + k_B)^3}{(k_B)^2} \qquad (30\text{-}32)$$

where u is the linear velocity of the mobile phase.

EXAMPLE 30-2

Substances A and B have retention times of 16.40 and 17.63 min, respectively, on a 30.0-cm column. An unretained species passes through the column in 1.30 min. The peak widths (at base) for A and B are 1.11 and 1.21 min, respectively. Calculate (a) the column resolution, (b) the average number of plates in the column, (c) the plate height, (d) the length of column required to achieve a resolution of 1.5, and (e) the time required to elute substance B on the column that gives an R_s value of 1.5.

(a) Employing Equation 30-29, we find

$$R_s = \frac{2(17.63 - 16.40)}{1.11 + 1.21} = 1.06$$

(b) Equation 30-24 permits computation of N:

$$N = 16\left(\frac{16.40}{1.11}\right)^2 = 3493 \quad \text{and} \quad N = 16\left(\frac{17.63}{1.21}\right)^2 = 3397$$

$$N_{avg} = \frac{3493 + 3397}{2} = 3445$$

(c) $H = \dfrac{L}{N} = \dfrac{30.0}{3445} = 8.7 \times 10^{-3}\,\text{cm}$

(d) k and α do not change greatly with increasing N and L. Thus, substituting N_1 and N_2 into Equation 30-30 and dividing one of the resulting equations by the other yield

$$\frac{(R_s)_1}{(R_s)_2} = \frac{\sqrt{N_1}}{\sqrt{N_2}}$$

where the subscripts 1 and 2 refer to the original and longer columns, respectively. Substituting the appropriate values for N_1, $(R_s)_1$, and $(R_s)_2$ gives

$$\frac{1.06}{1.5} = \frac{\sqrt{3445}}{\sqrt{N_2}}$$

$$N_2 = 3445\left(\frac{1.5}{1.06}\right)^2 = 6.9 \times 10^3$$

But

$$L = NH = 6.9 \times 10^3 \times 8.7 \times 10^{-3} = 60 \text{ cm}$$

(e) Substituting $(R_s)_1$ and $(R_s)_2$ into Equation 30-32 and dividing yield

$$\frac{(t_R)_1}{(t_R)_2} = \frac{(R_s)_1^2}{(R_s)_2^2} = \frac{17.63}{(t_R)_2} = \frac{(1.06)^2}{(1.5)^2}$$

$$(t_R)_2 = 35 \text{ min}$$

Thus, to obtain the improved resolution, the column length and, consequently, the separation time must be doubled.

Optimization Techniques

Equations 30-30 and 30-32 serve as guides in choosing conditions that lead to a desired degree of resolution with a minimum expenditure of time. An examination of these equations reveals that each is made up of three parts. The first describes the efficiency of the column in terms of $\sqrt{N}$ or H. The second, which is the quotient containing α, is a selectivity term that depends on the properties of the two solutes. The third component is the retention factor term, which is the quotient containing k_B; the term depends on the properties of both the solute and the column.

Variation in Plate Height As shown by Equation 30-30, the resolution of a column improves as the square root of the number of plates it contains increases. Example 30-2e reveals, however, that increasing the number of plates is expensive in terms of time unless the increase is achieved by reducing the plate height and not by increasing column length.

Methods of minimizing plate height, discussed in Section 30E-6, include reducing the particle size of the packing material, the diameter of the column, and the thickness of the liquid film. Optimizing the flow rate of the mobile phase is also helpful.

Variation in the Retention Factor Often, a separation can be improved significantly by manipulation of the retention factor k_B. Increases in k_B generally

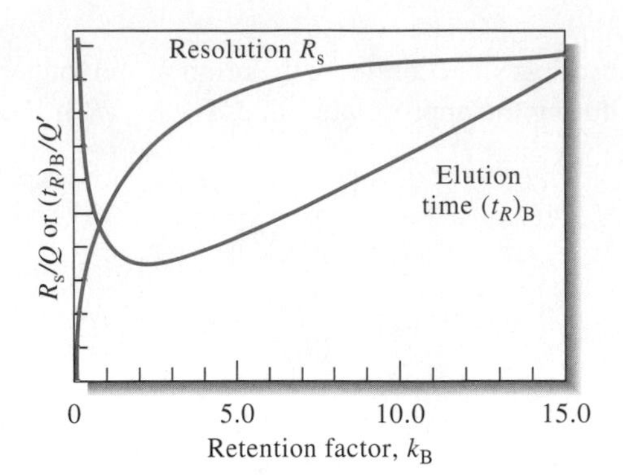

Figure 30-18 Effect of retention factor k_B on resolution R_s and elution time $(t_R)_B$. It is assumed that Q and Q' remain constant with variations in k_B.

enhance resolution (but at the expense of elution time). To determine the optimum range of values for k_B, it is convenient to write Equation 30-30 in the form

$$R_s = Q\left(\frac{k_B}{1 + k_B}\right)$$

and Equation 30-32 as

$$(t_R)_B = Q'\left(\frac{(1 + k_B)^3}{(k_B)^2}\right)$$

where Q and Q' contain the rest of the terms in the two equations. Figure 30-18 is a plot of R_s/Q and $(t_R)_B/Q'$ as a function of k_B, assuming Q and Q' remain approximately constant. It is clear that values of k_B greater than about 10 should be avoided because they provide little increase in resolution but markedly increase the time required for separations. The minimum in the elution-time curve occurs at $k_B \approx 2$. Often, then, the optimal value of k_B lies in the range from 1 to 5.

Usually, the easiest way to improve resolution is by optimizing k. For gaseous mobile phases, k can often be improved by temperature changes. For liquid mobile phases, changes in the solvent composition often permit manipulation of k to yield better separations. An example of the dramatic effect that relatively simple solvent changes can bring about is demonstrated in Figure 30-19. Here, modest variations in the methanol/water ratio convert unsatisfactory chromatograms (a and b) to ones with well-separated peaks for each component (c and d). For most purposes, the chromatogram shown in (c) is best since it shows adequate resolution in minimum time. The retention factor is also influenced by the stationary-phase film thickness.

Variation in the Selectivity Factor Optimizing k and increasing N are not sufficient to give a satisfactory separation of two solutes in a reasonable time when α approaches unity. A means must be sought to increase α while maintaining k in the range of 1 to 10. Several options are available; in decreasing order of desirability, as determined by promise and convenience, the options are (1) changing the composition of the mobile phase, (2) changing the column temperature, (3) changing the composition of the stationary phase, and (4) using special chemical effects.

An example of the use of option one has been reported for the separation of anisole ($C_6H_5OCH_3$) and benzene.[14] With a mobile phase that was a 50% mixture

[14] L. R. Snyder and J. J. Kirkland, *Introduction to Modern Liquid Chromatography,* 2nd ed., p. 75. New York: Wiley, 1979.

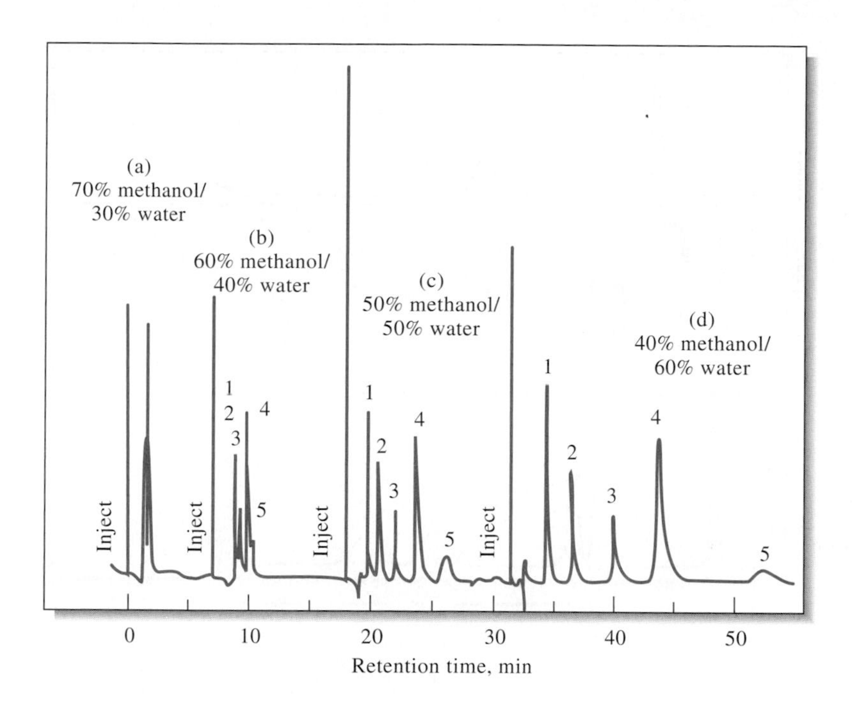

Figure 30-19 Effect of solvent variation on chromatograms. Analytes: (1) 9,10-anthraquinone; (2) 2-methyl-9,10-anthraquinone; (3) 2-ethyl-9,10-anthraquinone; (4) 1,4-dimethyl-9,10-anthraquinone; (5) 2-*t*-butyl-9,10-anthraquinone.

of water and methanol, k was 4.5 for anisole and 4.7 for benzene, while α was only 1.04. Substitution of an aqueous mobile phase containing 37% tetrahydrofuran gave k values of 3.9 and 4.7 and an α value of 1.20. Peak overlap was significant with the first solvent system and negligible with the second.

A less convenient but often highly effective method of improving α while maintaining values for k in their optimal range is to alter the chemical composition of the stationary phase. To take advantage of this option, most laboratories that carry out chromatographic separations frequently maintain several columns that can be interchanged with a minimum of effort.

Increases in temperature usually cause increases in k but have little effect on α values in liquid-liquid and liquid-solid chromatography. In contrast, with ion-exchange chromatography, temperature effects can be large enough to make exploration of this option worthwhile before resorting to a change in column packing material.

A final method of enhancing resolution is to incorporate into the stationary phase a species that complexes or otherwise interacts with one or more components of the sample. A well-known example of the use of this option occurs when an adsorbent impregnated with a silver salt is used to improve the separation of olefins. The improvement is a consequence of the formation of complexes between the silver ions and unsaturated organic compounds.

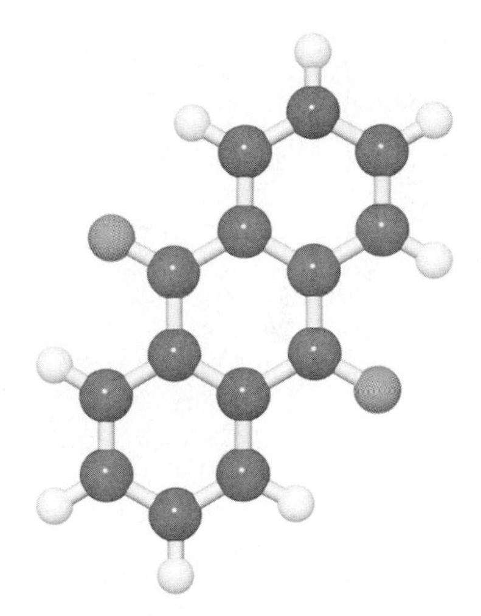

Molecular model of 9,10-anthraquinone.

The General Elution Problem

Figure 30-20 shows hypothetical chromatograms for a six-component mixture made up of three pairs of components with widely different distribution constants and thus widely different retention factors. In chromatogram a, conditions have been adjusted so that the retention factors for components 1 and 2 (k_1 and k_2) are in the optimal range of 1 to 5. The factors for the other components are far larger than the optimum, however. Thus, the bands corresponding to components 5 and 6 appear only after an inordinate length of time has passed; furthermore, the bands are so broad that they may be difficult to identify unambiguously.

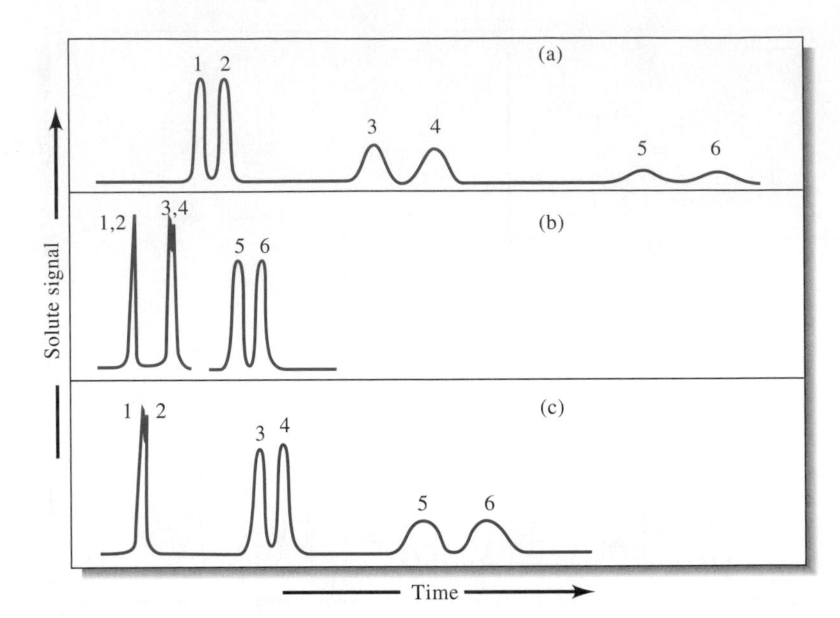

Figure 30-20 The general elution problem in chromatography.

As shown in chromatogram b, changing conditions to optimize the separation of components 5 and 6 bunches the peaks for the first four components to the point where their resolution is unsatisfactory. Here, however, the total elution time is ideal.

The phenomenon illustrated in Figure 30-20 is encountered often enough to be given a name: the **general elution problem.** A common solution to this problem is to change conditions that determine the values of k as the separation proceeds. These changes can be performed in a stepwise manner or continuously. Thus, for the mixture shown in Figure 30-20, conditions at the outset could be those producing chromatogram a. Immediately after the elution of components 1 and 2, conditions could be changed to those that are optimal for separating components 3 and 4 (as in chromatogram c). With the appearance of peaks for these components, the elution could be completed under the conditions used for producing chromatogram b. Often, such a procedure leads to satisfactory separation of all the components of a mixture in minimal time.

For liquid chromatography, variations in k are brought about by varying the composition of the mobile phase during elution. Such a procedure is called **gradient elution** or **solvent programming.** Elution under conditions of constant mobile-phase composition is called **isocratic elution.** For gas chromatography, the temperature can be changed in a known fashion to bring about changes in k. This **temperature programming** mode can help achieve optimal conditions for many separations.

30E-8 Applications of Chromatography

Chromatography is a powerful and versatile tool for separating closely related chemical species. In addition, it can be employed for the qualitative identification and quantitative determination of separated species. Examples of the applications of the various types of chromatography are given in Chapters 31 and 32.

Spreadsheet Summary In Chapter 14 of *Applications of Microsoft® Excel in Analytical Chemistry,* several exercises involving chromatography are suggested. In the first, a chromatogram of a three-component mixture is simulated. The resolution, number of theoretical plates, and retention times are varied, and their effect on the chromatograms is noted. The number of theoretical plates needed to achieve a given resolution is the subject of another exercise. A spreadsheet is constructed to find N for various retention factors of a two-component mixture. Nonideal peak shapes are also investigated. An exponentially modified Gaussian curve is investigated as a function of the time constant of the exponential. Finally, the optimization of chromatographic methods is illustrated by plotting the van Deemter equation for various flow velocities, longitudinal diffusion, and mass-transfer coefficient values.

WEB WORKS

Use the Google search engine to perform a search on peak tailing in reverse-phase liquid chromatography. Describe the phenomenon and discuss ways in which tailing can be minimized. Also perform a search on temperature effects in liquid chromatography. Describe how temperature influences chromatographic separations. Based on what you learn, would temperature programming appear valuable as an aid to separation in liquid chromatography? Why or why not?

QUESTIONS AND PROBLEMS

***30-1.** What is a masking agent and how does it function?

30-2. What two events accompany the separation process?

***30-3.** Name three methods based on mechanical phase separation.

30-4. How do strong and weak acid synthetic ion-exchange resins differ in structure?

30-5. Define
- *(a) elution.
- (b) mobile phase.
- *(c) stationary phase.
- (d) partition ratio.
- *(e) retention time.
- (f) retention factor.
- *(g) selectivity factor.
- (h) plate height.

30-6. List the variables that lead to zone broadening in chromatography.

***30-7.** What is the difference between gas-liquid and liquid-liquid chromatography?

30-8. What is the difference between liquid-liquid and liquid-solid chromatography?

***30-9.** Describe a method of determining the number of plates in a column.

30-10. Name two general methods of improving the resolution of two substances on a chromatographic column.

***30-11.** The distribution constant for X between *n*-hexane and water is 9.6. Calculate the concentration of X remaining in the aqueous phase after 50.0 mL of 0.150 M X are treated by extraction with the following quantities of *n*-hexane:
- (a) one 40.0-mL portion
- (b) two 20.0-mL portions
- (c) four 10.0-mL portions
- (d) eight 5.00-mL portions

30-12. The distribution coefficient for Z between *n*-hexane and water is 6.25. Calculate the percent of Z remaining in 25.0 mL of water that was originally 0.0600 M in Z after extraction with the following volumes of *n*-hexane:
- (a) one 25.0-mL portion
- (b) two 12.5-mL portions
- (c) five 5.00-mL portions
- (d) ten 2.50-mL portions

***30-13.** What volume of *n*-hexane is required to decrease the concentration of X in Problem 30-11 to 1.00 ×

10^{-4} M if 25.0 mL of 0.0500 M X are extracted with
(a) 25.0-mL portions?
(b) 10.0-mL portions?
(c) 2.0-mL portions?

30-14. What volume of *n*-hexane is required to decrease the concentration of Z in Problem 30-12 to 1.00×10^{-5} M if 40.0 mL of 0.0200 M Z are extracted with
(a) 50.0-mL portions of *n*-hexane?
(b) 25.0-mL portions?
(c) 10.0-mL portions?

***30-15.** What is the minimum distribution coefficient that permits removal of 99% of a solute from 50.0 mL of water with
(a) two 25.0-mL extractions with toluene?
(b) five 10.0-mL extractions with toluene?

30-16. If 30.0 mL of water that is 0.0500 M in Q are to be extracted with four 10.0-mL portions of an immiscible organic solvent, what is the minimum distribution coefficient that allows transfer of all but the following percentages of the solute to the organic layer:
*(a) 1.00×10^{-4}?
(b) 1.00×10^{-3}?
(c) 1.00×10^{-2}?

***30-17.** A 0.150 M aqueous solution of the weak organic acid HA was prepared from the pure compound, and three 50.0-mL aliquots were transferred to 100.0-mL volumetric flasks. Solution 1 was diluted to 100.0 mL with 1.0 M $HClO_4$; solution 2 was diluted to the mark with 1.0 M NaOH, and solution 3 was diluted to the mark with water. A 25.0-mL aliquot of each was extracted with 25.0 mL of *n*-hexane. The extract from solution 2 contained no detectable trace of A-containing species, indicating that A^- is not soluble in the organic solvent. The extract from solution 1 contained no ClO_4^- or $HClO_4$ but was found to be 0.0454 M in HA (by extraction with standard NaOH and back-titration with standard HCl). The extract from solution 3 was found to be 0.0225 M in HA. Assume that HA does not associate or dissociate in the organic solvent, and calculate
(a) the distribution ratio for HA between the two solvents.
(b) the concentration of the *species* HA and A^- in aqueous solution 3 after extraction.
(c) the dissociation constant of HA in water.

30-18. To determine the equilibrium constant for the reaction

$$I_2 + 2SCN^- \rightleftharpoons I(SCN)_2^- + I^-$$

25.0 mL of a 0.0100 M aqueous solution of I_2 were extracted with 10.0 mL of $CHCl_3$. After extraction, spectrophotometric measurements revealed that the I_2 concentration *of the aqueous layer* was 1.12×10^{-4} M. An aqueous solution that was 0.0100 M in I_2 and 0.100 M in KSCN was then prepared. After extraction of 25.0 mL of this solution with 10.0 mL of $CHCl_4$, the concentration of I_2 *in the $CHCl_3$ layer* was found from spectrophotometric measurement to be 1.02×10^{-3} M.
(a) What is the distribution constant for I_2 between $CHCl_3$ and H_2O?
(b) What is the formation constant for $I(SCN)_2^-$?

***30-19.** The total cation content of natural water is often determined by exchanging the cations for hydrogen ions on a strong-acid ion-exchange resin. A 25.0-mL sample of natural water was diluted to 100 mL with distilled water, and 2.0 g of a cation-exchange resin were added. After stirring, the mixture was filtered, and the solid remaining on the filter paper was washed with three 15.0-mL portions of water. The filtrate and washings required 15.3 mL of 0.0202 M NaOH to give a bromocresol green end point.
(a) Calculate the number of milliequivalents of cation present in exactly 1.00 L of sample. (Here, the equivalent weight of a cation is its formula weight divided by its charge.)
(b) Report the results in terms of milligrams of $CaCO_3$ per liter.

30-20. An organic acid was isolated and purified by recrystallization of its barium salt. To determine the equivalent weight of the acid, a 0.393-g sample of the salt was dissolved in about 100 mL of water. The solution was passed through a strong-acid ion-exchange resin, and the column was then washed with water; the eluate and washings were titrated with 18.1 mL of 0.1006 M NaOH to a phenolphthalein end point.
(a) Calculate the equivalent weight of the organic acid.
(b) A potentiometric titration curve of the solution resulting when a second sample was treated in the same way revealed two end points: one at pH 5 and the other at pH 9. What is the molecular weight of the acid?

***30-21.** Describe the preparation of exactly 2.00 L of 0.1500 M HCl from primary-standard grade NaCl using a cation-exchange resin.

30-22. An aqueous solution containing $MgCl_2$ and HCl was analyzed by first titrating a 25.00-mL aliquot to a bromocresol green end point with 18.96 mL of 0.02762 M NaOH. A 10.00-mL aliquot was then diluted to 50.00 mL with distilled water and passed through a strong-acid ion-exchange resin. The eluate and washings required 36.54 mL of the NaOH solution to reach the same end point. Report the molar concentrations of HCl and $MgCl_2$ in the sample.

***30-23.** An open tubular column used for gas chromatography had an inside diameter of 0.25 mm. A volumetric flow rate of 1.0 mL/min was used. Find the linear flow velocity in cm/s at the column outlet.

30-24. A packed column in gas chromatography had an inside diameter of 5.0 mm. The measured volumetric flow rate at the column outlet was 50 mL/min. If the column porosity was 0.45, what was the linear flow velocity in cm/s?

***30-25.** The following data are for a liquid chromatographic column:

Length of packing	24.7 cm
Flow rate	0.313 mL/min
V_M	1.37 mL
V_S	0.164 mL

A chromatogram of a mixture of species A, B, C, and D provided the following data:

	Retention Time, min	Width of Peak Base (*W*), min
Nonretained	3.1	—
A	5.4	0.41
B	13.3	1.07
C	14.1	1.16
D	21.6	1.72

Calculate
(a) the number of plates from each peak.
(b) the mean and the standard deviation for *N*.
(c) the plate height for the column.

30-26. From the data in Problem 30-25, calculate for A, B, C, and D
(a) the retention factor.
(b) the distribution constant.

***30-27.** From the data in Problem 30-25, calculate for species B and C
(a) the resolution.
(b) the selectivity factor.
(c) the length of column necessary to separate the two species with a resolution of 1.5.
(d) the time required to separate the two species on the column in part (c).

30-28. From the data in Problem 30-25, calculate for species C and D
(a) the resolution.
(b) the length of column necessary to separate the two species with a resolution of 1.5.

***30-29.** The following data were obtained by gas-liquid chromatography on a 40-cm packed column:

Compound	t_R, min	*W*, min
Air	1.9	—
Methylcyclohexane	10.0	0.76
Methylcyclohexene	10.9	0.82
Toluene	13.4	1.06

Calculate
(a) an average number of plates from the data.
(b) the standard deviation for the average in (a).
(c) an average plate height for the column.

30-30. Referring to Problem 30-29, calculate the resolution for
(a) methylcyclohexene and methylcyclohexane.
(b) methylcyclohexene and toluene.
(c) methylcyclohexane and toluene.

***30-31.** If a resolution of 1.5 is desired in separating methylcyclohexane and methylcyclohexene in Problem 30-29,
(a) how many plates are required?
(b) how long must the column be if the same packing is employed?
(c) what is the retention time for methylcyclohexene on the column in part (b)?

30-32. If V_S and V_M for the column in Problem 30-29 are 19.6 and 62.6 mL, respectively, and a nonretained air peak appears after 1.9 min, calculate
(a) the retention factor for each compound.
(b) the distribution constant for each compound.
(c) the selectivity factor for methylcyclohexane and methylcyclohexene.

***30-33.** From distribution studies, species M and N are known to have water/hexane distribution constants of 5.93 and 6.11 respectively ($K = [M]_{aq}/[M]_{hex}$). The two species are to be separated by elution with hexane in a column packed with silica gel containing adsorbed water. The ratio V_S/V_M for the packing is 0.398.
(a) Calculate the retention factor for each solute.
(b) Calculate the selectivity factor.
(c) How many plates are needed to provide a resolution of 1.5?
(d) How long a column is needed if the plate height of the packing is 1.9×10^{-3} cm?

(e) If a flow rate of 6.50 cm/min is employed, how long will it take to elute the two species?

30-34. Repeat the calculations in Problem 30-33 assuming that $K_M = 5.81$ and $K_N = 6.20$.

30-35. Challenge Problem. A chromatogram of a two-component mixture on a 25-cm packed liquid chromatography column is shown in the figure. The flow rate was 0.40 mL/min.

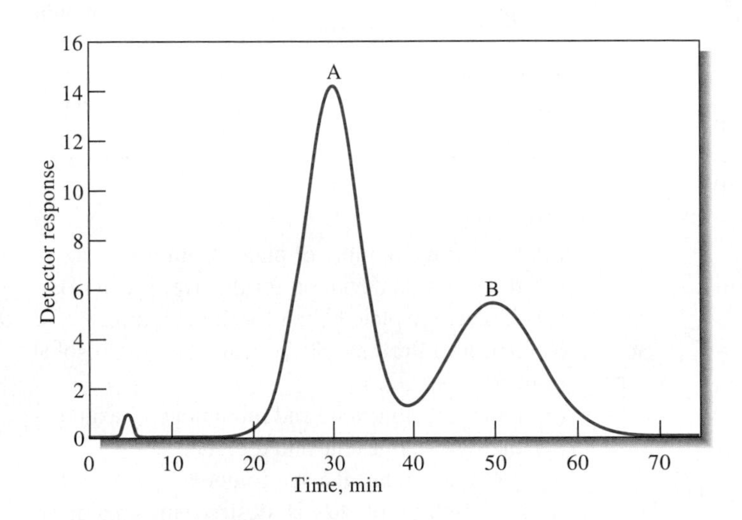

(a) Find the times that components A and B spend in the stationary phase.
(b) Find the retention times for A and B.
(c) Determine the retention factors for the two components.
(d) Find the full widths of each peak and the full width at half maximum values.
(e) Find the resolution of the two peaks.
(f) Find the average number of plates for the column.
(g) Find the average plate height.
(h) What column length would be needed to achieve a resolution of 1.75?
(i) What time would be required to achieve the resolution in part (h)?
(j) Assume that the column length is fixed at 25 cm and the packing material is fixed. What measures could you take to increase the resolution to achieve baseline separation?
(k) Are there any measures you could use to achieve a better separation in a shorter time with the same column as in part (j)?

InfoTrac College Edition

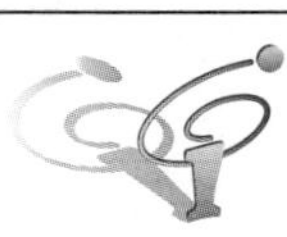

For additional readings, go to InfoTrac College Edition, your online research library, at

http://infotrac.thomsonlearning.com

CHAPTER 31

Gas Chromatography

Gas chromatography is one of the most widely used techniques for qualitative and quantitative analysis. The photo shows a capillary column useful in gas chromatographic determinations at temperatures exceeding 400°C. Such high-temperature applications require special stationary phases and tubing that will not decompose. The column shown has tubing made from stainless steel.

This chapter considers gas chromatography in detail, including the columns and stationary phases that are most widely used. Although this chapter is primarily concerned with gas-liquid chromatography, there is a brief discussion of gas-solid chromatography.

Restek Corp., Bellefonte, PA

In gas chromatography, the components of a vaporized sample are separated as a consequence of being partitioned between a mobile gaseous phase and a liquid or a solid stationary phase held in a column.[1] In performing a gas chromatographic separation, the sample is vaporized and injected onto the head of a chromatographic column. Elution is brought about by the flow of an inert gaseous mobile phase. In contrast to most other types of chromatography, the mobile phase does not interact with molecules of the analyte; its only function is to transport the analyte through the column.

Two types of gas chromatography are encountered: ***gas-liquid chromatography*** *(GLC) and* ***gas-solid chromatography*** *(GSC). Gas-liquid chromatography finds widespread use in all fields of science; its name is usually shortened to* ***gas chromatography*** *(GC). Gas-solid chromatography is based on a solid stationary phase in which retention of analytes occurs because of physical adsorption. Gas-solid chromatography has limited application owing to semipermanent retention of active or polar molecules and severe tailing of elution peaks (a consequence of the nonlinear character of adsorption process). Thus, this technique has not found widespread application except in the separation of certain low-molecular-weight gaseous species; we discuss this method briefly in Section 31D.*

In **gas-liquid chromatography,** the mobile phase is a gas, whereas the stationary phase is a liquid that is retained on the surface of an inert solid by adsorption or chemical bonding.

In **gas-solid chromatography,** the mobile phase is a gas, whereas the stationary phase is a solid that retains the analytes by physical adsorption. Gas-solid chromatography permits the separation and determination of low-molecular-mass gases, such as air components, hydrogen sulfide, carbon monoxide, and nitrogen oxides.

Gas-liquid chromatography is based on partitioning of the analyte between a gaseous mobile phase and a liquid phase immobilized on the surface of an inert solid packing or on the walls of capillary tubing. The concept of gas-liquid chromatography was first enunciated in 1941 by Martin and Synge, who were also responsible for the development of liquid-liquid partition chromatography. More

[1]For detailed treatment of GC, see J. Willet, *Gas Chromatography.* New York: Wiley, 1987; R. L. Grob, Ed., *Modern Practice of Gas Chromatography,* 3rd ed. New York: Wiley, 1995; R. P. W. Scott, *Introduction to Analytical Gas Chromatography,* 2nd ed. New York: Marcel Dekker, 1997; H. M. McNair and J. M. Miller, *Basic Gas Chromatography.* New York: Wiley, 1998.

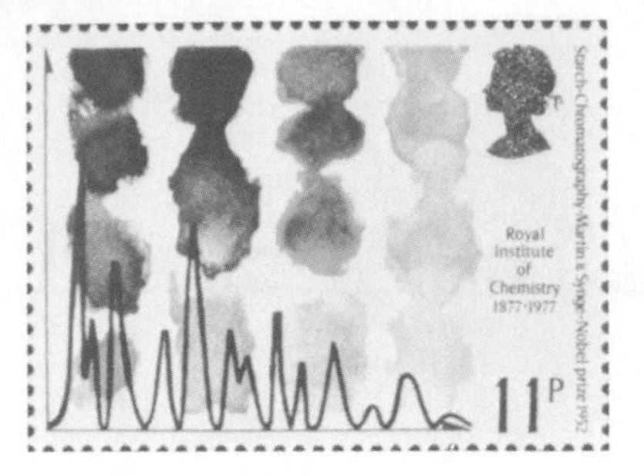

Stamp honoring biochemists Archer J. P. Martin (1910–2002) and Richard L. M. Synge (1914–present), who were awarded the 1952 Nobel Prize in chemistry for their contributions to the development of modern chromatography.

than a decade was to elapse, however, before the value of gas-liquid chromatography was demonstrated experimentally and this technique began to be used as a routine laboratory tool. In 1955, the first commercial apparatus for gas-liquid chromatography appeared on the market. Since that time, the growth in applications of this technique has been phenomenal. Currently, several hundred thousand gas chromatographs are in use throughout the world.

31A INSTRUMENTS FOR GAS-LIQUID CHROMATOGRAPHY

Many changes and improvements in gas chromatographic instruments have appeared in the marketplace since their commercial introduction. In the 1970s, electronic integrators and computer-based data processing equipment became common. The 1980s and 1990s saw the use of computers for automatic control of most instrument parameters, such as column temperature, flow rates, and sample injection; development of very high-performance instruments at moderate costs; and perhaps most important, the development of open tubular columns that are capable of separating components of complex mixtures in relatively short times. Today, over 50 instrument manufacturers offer about 150 different models of gas-chromatographic equipment at costs that vary from perhaps $1000 to more than $50,000. The basic components of a typical instrument for performing gas chromatography are shown in Figure 31-1 and are described briefly in this section.

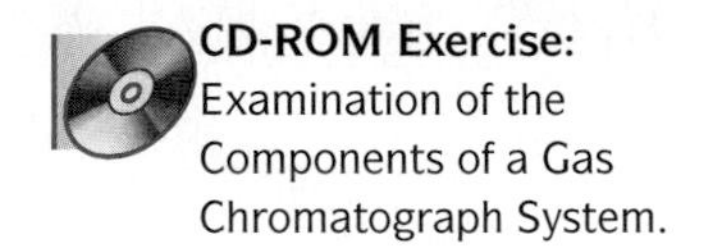

CD-ROM Exercise: Examination of the Components of a Gas Chromatograph System.

31A-1 Carrier Gas System

The mobile-phase gas in gas chromatography is called the **carrier gas** and must be chemically inert. Helium is the most common mobile-phase gas, although argon, nitrogen, and hydrogen are also used. These gases are available in pressurized tanks. Pressure regulators, gauges, and flow meters are required to control the flow rate of the gas.

Flow rates are normally controlled by a two-stage pressure regulator at the gas cylinder and some sort of pressure regulator or flow regulator mounted in the chromatograph. Inlet pressures usually range from 10 to 50 psi (lb/in^2) greater than room pressure, which lead to flow rates of 25 to 150 mL/min with packed columns and 1 to 25 mL/min for open-tubular capillary columns. Generally, it is assumed that flow rates will be constant if the inlet pressure remains constant. Flow rates can

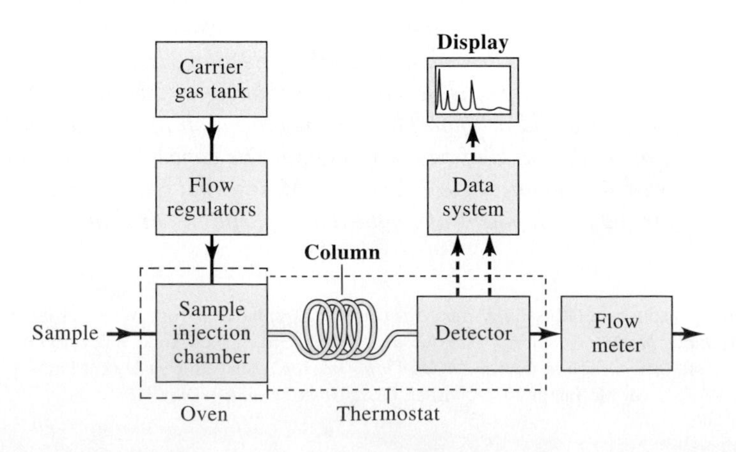

Figure 31-1 Block diagram of a typical gas chromatograph.

be established by a rotometer at the column head; this device, however, is not as accurate as the simple soap-bubble meter shown in Figure 31-2. Usually, the flow meter is located at the end of the column, as shown in Figure 31-1. A soap film is formed in the path of the gas when a rubber bulb containing an aqueous solution of soap or detergent is squeezed; the time required for this film to move between two graduations on the buret is measured and converted to volumetric flow rate (see Figure 31-2). Note that volumetric flow rates and linear flow velocities are related by Equation 30-12 or Equation 30-13.

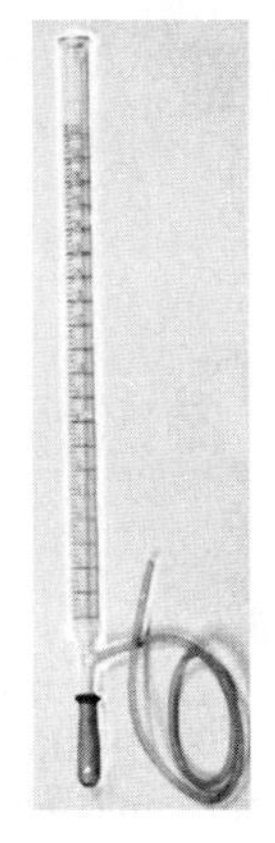

Figure 31-2 A soap-bubble flow meter. (Courtesy of Varian, Inc., Walnut Creek, CA.)

31A-2 Sample Injection System

Column efficiency requires that the sample be of a suitable size and be introduced as a "plug" of vapor; slow injection or oversized samples cause band spreading and poor resolution. Calibrated microsyringes, such as those shown in Figure 31-3, are used to inject liquid samples through a rubber or silicone diaphragm, or septum, into a heated sample port located at the head of the column. The sample port (Figure 31-4) is ordinarily kept at about 50°C greater than the boiling point of the least volatile component of the sample. For ordinary packed analytical columns, sample sizes range from a few tenths of a microliter to 20 μL. Capillary columns require samples that are smaller by a factor of 100 or more. A sample splitter is often needed with capillary columns to deliver a small known fraction (1:100 to 1:500) of the injected sample, with the remainder going to waste. Commercial gas chromatographs intended for use with capillary columns incorporate such splitters; they also allow for splitless injection when packed columns are used.

For quantitative work, more reproducible sample sizes for both liquids and gases are obtained by means of a sample valve, such as that shown in Figure 31-5. With such devices, sample sizes can be reproduced to better than 0.5% relative. Solid samples are introduced as solutions or, alternatively, are sealed into thin-walled vials that can be inserted at the head of the column and punctured or crushed from the outside.

Figure 31-3 A set of microsyringes for sample injection. (Courtesy of Varian, Inc., Walnut Creek, CA.)

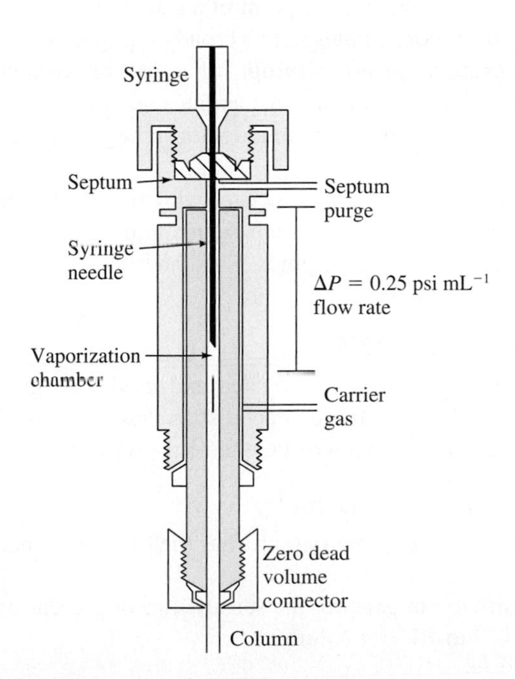

Figure 31-4 Cross-sectional view of a microflash vaporizer direct injector.

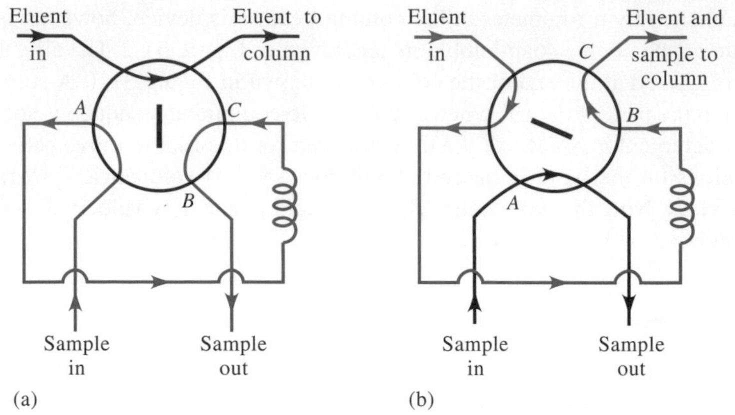

Figure 31-5 A rotary sample valve: Valve position (a) is for filling the sample loop *ACB;* position (b) is for introduction of sample into column.

31A-3 Column Configurations and Column Ovens

Two general types of columns are encountered in gas chromatography: **packed columns** and **open tubular,** or **capillary columns.** In the past, the vast majority of gas chromatographic analyses used packed columns. For most current applications, packed columns have been replaced by the more efficient and faster open tubular columns.

Chromatographic columns vary in length from less than 2 m to 50 m or more. They are constructed of stainless steel, glass, fused silica, or Teflon. To fit into an oven for thermostating, they are usually formed into coils having diameters of 10 to 30 cm (Figure 31-6). A detailed discussion of columns, column packings, and stationary phases is found in Section 31B.

Column temperature is an important variable that must be controlled to a few tenths of a degree for precise work. Thus, the column is ordinarily housed in a thermostated oven. The optimum column temperature depends on the boiling point of the sample and the degree of separation required. Roughly, a temperature equal to or slightly above the average boiling point of a sample results in a reasonable elution time (2 to 30 min). For samples with a broad boiling range, it is often desirable to employ **temperature programming,** whereby the column temperature is increased either continuously or in steps as the separation proceeds. Figure 31-7 shows the improvement in a chromatogram brought about by temperature programming.

Temperature programming in gas chromatography involves increasing the column temperature continuously or in steps during elution.

In general, optimum resolution is associated with minimal temperature; the cost of lowered temperature, however, is an increase in elution time and therefore the time required to complete an analysis. Figures 31-7a and 31-7b illustrate this principle.

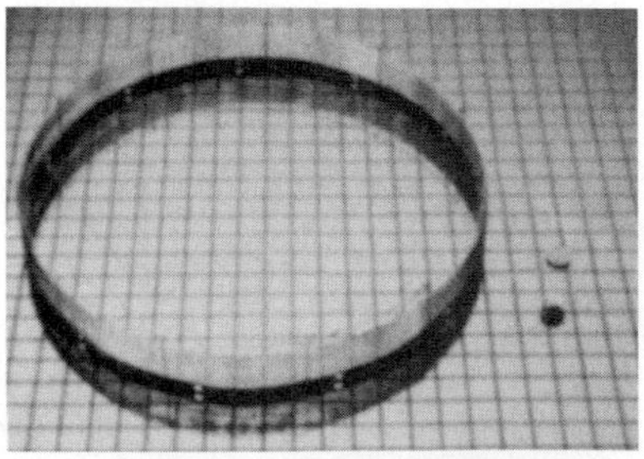

Figure 31-6 A 25-m fused-silica capillary column. (Courtesy of Varian, Inc., Walnut Creek, CA.)

31A-4 Detection Systems

Dozens of detectors have been investigated and used with gas chromatographic separations. We first describe the characteristics that are most desirable in a gas chromatographic detector and then discuss the most widely used detection systems.

Characteristics of the Ideal Detector

The ideal detector for gas chromatography has the following characteristics:

1. Adequate sensitivity. In general, the sensitivities of present-day detectors lie in the range of 10^{-8} to 10^{-15} g solute/s.

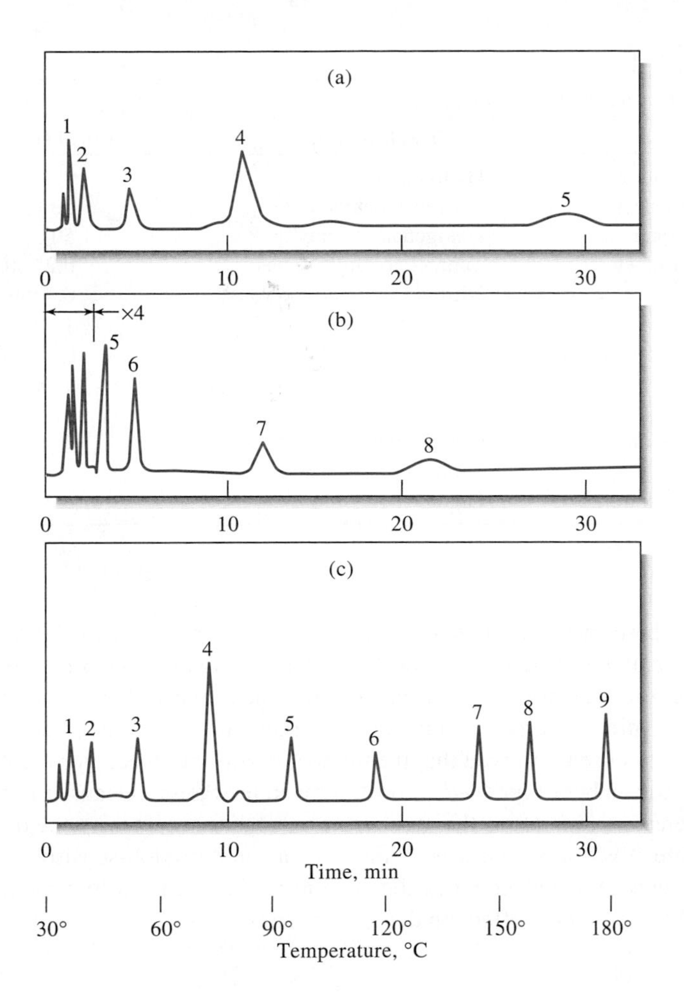

Figure 31-7 Effect of temperature on gas chromatograms. (a) Isothermal at 45°C; (b) isothermal at 145°C; (c) programmed at 30°C to 180°C. (From W. E. Harris and H. W. Habgood, *Programmed Temperature Gas Chromatography,* p. 10. New York: Wiley, 1966. Reprinted with permission of the author.)

2. Good stability and reproducibility.
3. A linear response to solutes that extends over several orders of magnitude.
4. A temperature range from room temperature to at least 400°C.
5. A short response time that is independent of flow rate.
6. High reliability and ease of use. The detector should, to the greatest extent possible, be foolproof in the hands of inexperienced operators.
7. Similarity in response toward all solutes or, alternatively, a highly predictable and selective response toward one or more classes of solutes.
8. Nondestructive of sample.

Needless to say, no current detector exhibits all these characteristics. Some of the more common detectors are listed in Table 31-1. Four of the most widely used detectors are described in the paragraphs that follow.

Flame Ionization Detectors

The flame ionization detector (FID) is the most widely used and generally applicable detector for gas chromatography. With a detector such as that shown in Figure 31-8, effluent from the column is directed into a small air/hydrogen flame. Most organic compounds produce ions and electrons when pyrolyzed at the temperature of an air/hydrogen flame. Detection involves monitoring the current produced by

TABLE 31-1
Gas Chromatographic Detectors

Type	Applicable Samples	Typical Detection Limit
Flame ionization	Hydrocarbons	0.2 pg/s
Thermal conductivity	Universal detector	500 pg/mL
Electron capture	Halogenated compounds	5 fg/s
Mass spectrometer	Tunable for any species	0.25–100 pg
Thermionic	Nitrogen and phosphorous compounds	0.1 pg/s (P) 1 pg/s (N)
Electrolytic conductivity (Hall)	Compounds containing halogens, sulfur, or nitrogen	0.5 pg Cl/s 2 pg S/s 4 pg N/s
Photoionization	Compounds ionized by UV radiation	2 pg C/s
Fourier transform IR	Organic compounds	0.2 to 40 ng

collecting these charge carriers. A few hundred volts applied between the burner tip and a collector electrode located above the flame serve to collect the ions and electrons. The resulting current ($\sim 10^{-12}$ A) is then measured with a picoammeter.

The ionization of carbon compounds in a flame is a poorly understood process, although it has been observed that the number of ions produced is roughly proportional to the number of *reduced* carbon atoms in the flame. Because the flame ionization detector responds to the number of carbon atoms entering the detector per unit of time, it is a *mass-sensitive,* rather than a concentration-sensitive, device. As a consequence, this detector has the advantage that changes in flow rate of the mobile phase have little effect on detector response.

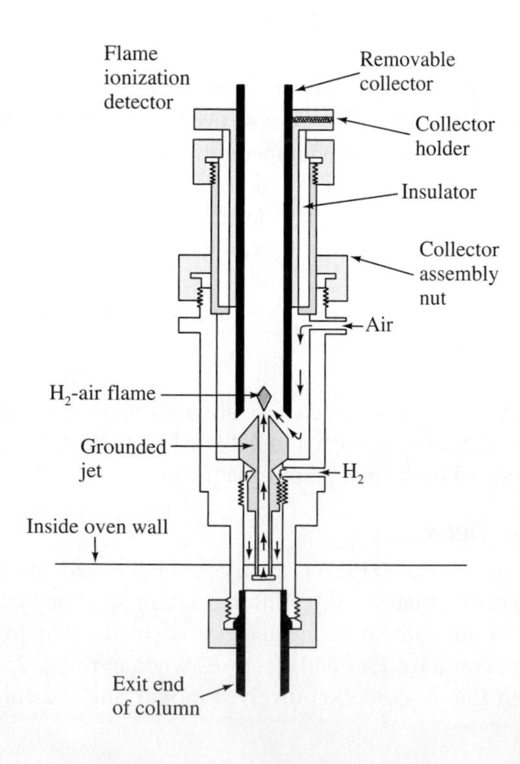

Figure 31-8 A typical flame ionization detector. (Courtesy of Agilent Technologies, Palo Alto, CA.)

Functional groups, such as carbonyl, alcohol, halogen, and amine, yield fewer ions or none at all in a flame. In addition, the detector is insensitive toward noncombustible gases such as H_2O, CO_2, SO_2, and NO_x. These properties make the flame ionization detector a most useful general detector for the analysis of most organic samples, including those that are contaminated with water and the oxides of nitrogen and sulfur.

The flame ionization detector exhibits a high sensitivity ($\sim10^{-13}$ g/s), large linear response range ($\sim10^7$), and low noise. It is generally rugged and easy to use. A disadvantage of the flame ionization detector is that it destroys the sample during the combustion step.

Thermal Conductivity Detectors

The **thermal conductivity detector** (TCD), which was one of the earliest detectors for gas chromatography, still finds wide application. This device consists of an electrically heated source whose temperature at constant electric power depends on the thermal conductivity of the surrounding gas. The heated element may be a fine platinum, gold, or tungsten wire (Figure 31-9a) or, alternatively, a small thermistor. The electrical resistance of this element depends on the thermal conductivity of the gas. Twin detectors are ordinarily used, one located ahead of the sample injection chamber and the other immediately beyond the column; alternatively, the gas stream can be split. The detectors are incorporated into two arms of a simple bridge circuit (see Figure 31-9) such that the thermal conductivity of the carrier gas is

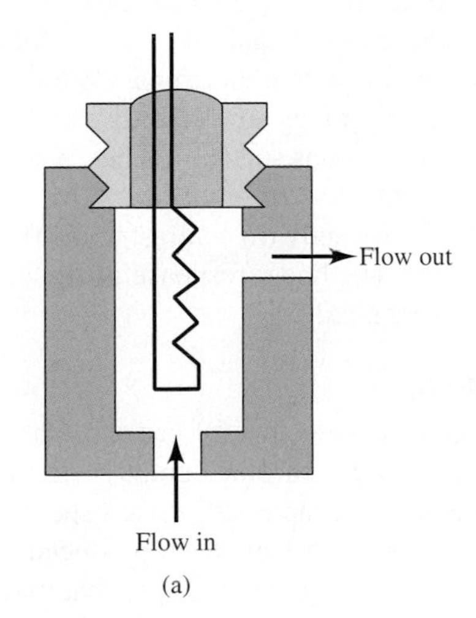

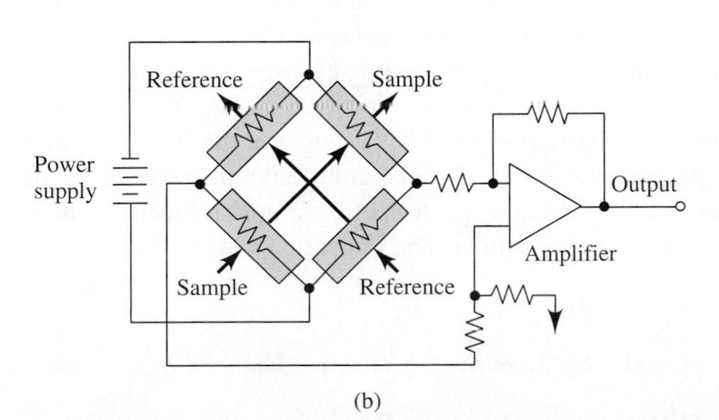

Figure 31-9 Schematic of (a) a thermal conductivity detector cell and (b) an arrangement of two sample detector cells and two reference detector cells (From J. Hinshaw, *LC-GC,* **1990,** *8,* 298. With permission.)

canceled. In addition, the effects of variations in temperature, pressure, and electric power are minimized. The thermal conductivities of helium and hydrogen are roughly 6 to 10 times greater than those of most organic compounds. Thus, even small amounts of organic species cause relatively large decreases in the thermal conductivity of the column effluent, which results in a marked rise in the temperature of the detector. Detection by thermal conductivity is less satisfactory with carrier gases whose conductivities closely resemble those of most sample components.

The advantages of the thermal conductivity detector are its simplicity, its large linear dynamic range (about five orders of magnitude), its general response to both organic and inorganic species, and its nondestructive character, which permits collection of solutes after detection. The chief limitation of the thermal conductivity detector is its relatively low sensitivity. Other detectors exceed this sensitivity by factors of 10^4 to 10^7.

Electron Capture Detectors

The electron capture detector (ECD) has become one of the most widely used detectors for environmental samples because this detector selectively responds to halogen-containing organic compounds, such as pesticides and polychlorinated biphenyls. In this detector, the sample eluate from a column is passed over a radioactive β-emitter, usually nickel-63. An electron from the emitter causes ionization of the carrier gas (often nitrogen) and the production of a burst of electrons. In the absence of organic species, a constant standing current between a pair of electrodes results from this ionization process. The current decreases markedly, however, in the presence of organic molecules containing electronegative functional groups that tend to capture electrons. Compounds such as halogens, peroxides, quinones, and nitro groups are detected with high sensitivity. The detector is insensitive to functional groups such as amines, alcohols, and hydrocarbons.

Electron capture detectors are highly sensitive and have the advantage of not altering the sample significantly (in contrast to the flame ionization detector, which consumes the sample). The linear response of the detector, however, is limited to about two orders of magnitude.

Mass Spectrometry

One of the most powerful detectors for gas chromatography is the mass spectrometer. The combination of gas chromatography and mass spectrometry is known as **GC/MS.**[2] As discussed in Chapter 28, a mass spectrometer measures the mass-to-charge ratio (m/z) of ions that have been produced from the sample. Most of the ions produced are singly charged ($z = 1$), so that mass spectrometrists often speak of measuring the mass of ions when mass-to-charge ratio is actually measured.

A block diagram of a typical molecular-mass spectrometer is shown in Figure 31-10. Sample molecules enter the mass spectrometer through an inlet system. In the case of GC, the sample is in the form of a vapor, and the inlet must interface between the atmospheric pressure GC system and the low-pressure (10^{-5} to 10^{-8} torr) mass spectrometer system. An elaborate vacuum system is needed to maintain the low pressure. In the mass spectrometer, sample molecules enter an ionization source, which ionizes the sample. The ionization sources for molecular mass spec-

---

[2]See M. McMaster and C. McMaster, *GC/MS: A Practical User's Guide.* New York: Wiley-VCH, 1998.

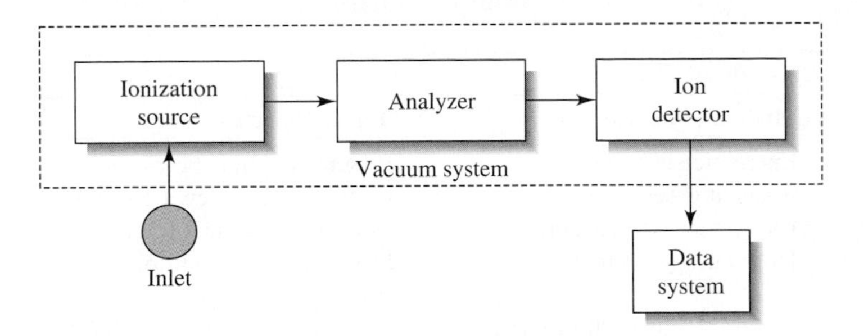

Figure 31-10 Block diagram of a mass spectrometer. The sample enters the ionization source through an inlet system. Sample molecules are converted to ions and are often fragmented in the ionization source. The ions then pass into an analyzer, where they are separated according to their mass-to-charge ratio. Next, the separated ions strike an ion detector, where they produce an electrical signal that is recorded and plotted by the data system.

trometry are energetic enough to break chemical bonds in the sample molecules but not so energetic as to decompose the sample molecules into their constituent atoms, as is done in atomic mass spectrometry (see Chapter 28). The ionization sources in GC/MS produce fragments, which can also be ionized. Hence, leaving the ion source are ions of the sample molecules, called **molecular ions,** fragment ions, and un-ionized molecules. The uncharged molecules and fragments are normally pumped out of the ion source by the vacuum pumps used to produce the low-pressure environment. The next section of the mass spectrometer is an analyzer stage. The analyzer serves to sort the ions according to their m/z values, just as in atomic mass spectrometry (see Section 28F-2). The separated ions are then detected, and a plot of the ion intensity versus m/z value is produced by the data system.

The mass spectrum of a simple molecule, CO_2, is shown in Figure 31-11. Note that several fragment ions are present. Breaking a C—O bond in the molecular ion leads to CO^+ (m/z = 28) and O^+ (m/z = 16). Loss of both oxygen atoms leads to C^+ (m/z = 12). Only positive ions are present in this example. Negative ions can also be produced and detected.

Inlet Systems In addition to GC inlets, samples can be introduced to a molecular mass spectrometer in many ways. Solids can be placed on the tip of a rod,

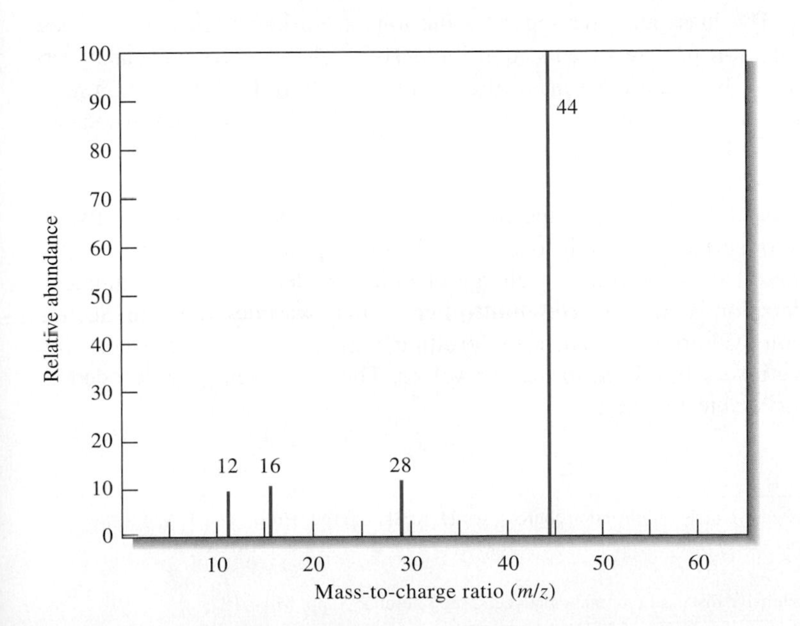

Figure 31-11 Mass spectrum of CO_2. Note that the molecular ion appears at m/z = 44 (C = 12, O = 16). Fragment ions appear at m/z values of 28, 16, and 12. These correspond to CO^+, O^+, and C^+, respectively.

TABLE 31-2

Ionization Sources for Molecular Mass Spectrometry

Basic Type	Name and Acronym	Method of Ionization	Type of Spectra
Gas phase	Electron impact (EI)	Energetic electrons	Fragmentation patterns
	Chemical ionization (CI)	Reagent gaseous ions	Proton adducts, few fragments
Desorption	Fast atom bombardment (FAB)	Energetic atomic beam	Molecular ions and fragments
	Matrix-assisted laser desorption/ ionization (MALDI)	High-energy photons	Molecular ions, multiply charged ions
	Electrospray ionization (ESI)	Electric field produces charged spray, which desolvates	Multiply charged molecular ions

inserted into the vacuum chamber, and evaporated or sublimed by heating. Liquids can be introduced through special controlled-flow inlets or they can be desorbed from a surface on which they are coated as a thin film. In general, samples for molecular mass spectrometry must be pure because the fragmentation that occurs causes the mass spectrum of mixtures to be very difficult to interpret. Gas chromatography is an ideal way to introduce mixtures because the components are separated from the mixture by the GC prior to introduction to the mass spectrometer.

Ionization Sources Several different types of ionization sources are available for molecular mass spectrometry. The most widely used sources are listed in Table 31-2.[3] One of the most common is the electron impact (EI) source. In this source, molecules are bombarded with a high-energy beam of electrons. This produces positive ions, negative ions, and neutral species. The positive ions are directed toward the analyzer by electrostatic repulsion.

In EI, the electron beam is so energetic that many fragments are produced. These fragments, however, are very useful in identifying the molecular species entering the mass spectrometer. Only electron impact and chemical ionization are used with GC/MS.

Analyzers The mass analyzer separates the ions according to their m/z values. The most common analyzers are listed in Table 31-3.[4] The most common analyzers for GC/MS are the quadrupole mass filter and the ion trap. High-resolution mass spectrometers use the double-focusing analyzer, the ion-cyclotron resonance analyzer, or the time-of-flight analyzer.

Ion Detectors In most mass spectrometers, ions are detected after collisions with a detector surface. The collisions cause electrons, photons, or other ions to be emitted. These can be detected by charge or radiation detectors. For example, a common detector is the **electron multiplier,** which was described in Section 28F-3. In ion-cyclotron resonance, the orbiting ions induce a signal whose frequencies are inversely related to the m/z values. The frequencies are decoded by Fourier transform techniques.

[3]For a more extensive table of ionization sources, see D. A. Skoog, F. J. Holler, and T. A. Nieman, *Principles of Instrumental Analysis,* 5th ed., p. 500. Belmont, CA: Brooks/Cole, 1998.

[4]For a more extensive discussion of mass analyzers, see reference 3, pp. 514–518.

TABLE 31-3

Common Mass Analyzers for Mass Spectrometry

Basic Type	Analysis Principle
Magnetic sector	Deflection of ions in a magnetic field. Ion trajectories depend on m/z value.
Double focusing	Electrostatic focusing followed by magnetic field deflection. Trajectories depend on m/z values.
Quadrupole	Ion motion in dc and radio-frequency fields. Only certain m/z values are passed.
Ion trap	Storage of ions in space defined by ring and end cap electrodes. Electric field sequentially ejects ions of increasing m/z values.
Ion-cyclotron resonance	Trapping of ions in cubic cell under influence of trapping voltage and magnetic field. Orbital frequency related inversely to m/z value.
Time-of-flight	Equal kinetic energy ions enter drift tube. Drift velocity and thus arrival time at detector depend on mass.

Complete GC/MS Instrument The schematic of a complete GC/MS system is shown in Figure 31-12. Here the sample is injected into the capillary GC (see Section 31B-1), and the effluent enters the inlet of a quadrupole mass spectrometer. The molecules are then fragmented and ionized by the source, are mass analyzed, and are detected by the electron multiplier.

In GC/MS, the mass spectrometer scans the masses repetitively during a chromatographic experiment. If the chromatographic run is 10 minutes, for example, and a scan is taken each second, 600 mass spectra are recorded. The data can be analyzed by the data system in several different ways. First, the ion abundances in each spectrum can be summed and plotted as a function of time to give a **total-ion chromatogram.** This plot is similar to a conventional chromatogram. One can also display the mass spectrum at a particular time during the chromatogram to identify the species that is eluting at that time. Finally, one can select a single m/z value and monitor it throughout the chromatographic experiment, a technique known as **selected-ion monitoring.**

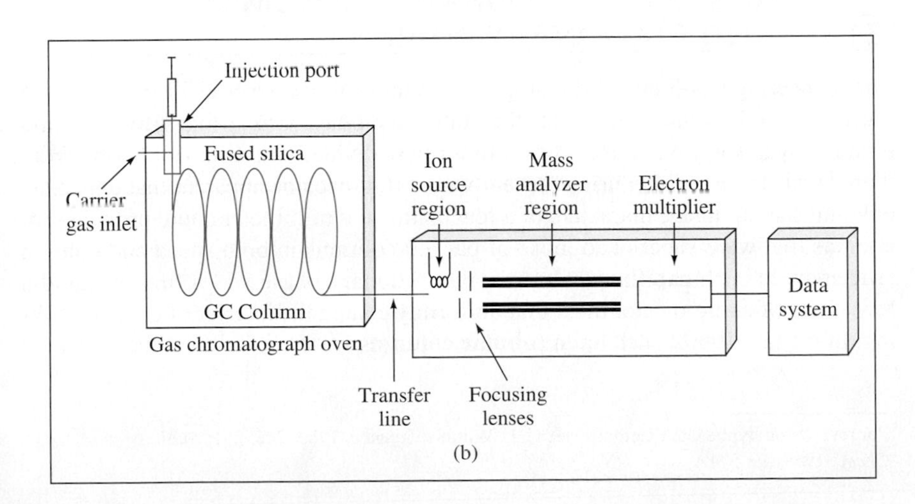

Figure 31-12 Schematic of a typical capillary GC/MS instrument. The effluent from the GC is passed into the inlet of a mass spectrometer, where the molecules in the gas are fragmented, ionized, analyzed, and detected.

Other Types of Detectors

Other important GC detectors include the thermionic detector, the electrolytic conductivity or Hall detector, and the photoionization detector. The thermionic detector is similar in construction to the FID. With the thermionic detector, nitrogen- and phosphorus-containing compounds produce increased currents in a flame in which an alkali metal salt is vaporized. The thermionic detector is widely used for organophosphorus pesticides and pharmaceutical compounds.

With the electrolytic conductivity detector, compounds containing halogens, sulfur, or nitrogen are mixed with a reaction gas in a small reactor tube. The products are next dissolved in a liquid, which produces a conductive solution. The change in conductivity as a result of the presence of the active compound is then measured. In the photoionization detector, molecules are photoionized by ultraviolet radiation. The ions and electrons produced are then collected with a pair of biased electrodes, and the resulting current is measured. The detector is often used for aromatic and other molecules that are easily photoionized.

Gas chromatography is often coupled with the selective techniques of spectroscopy and electrochemistry. We have discussed GC/MS, but gas chromatography can also be combined with several other techniques, such as infrared spectroscopy and nuclear magnetic resonance spectroscopy, to provide the chemist with powerful tools for identifying the components of complex mixtures. These combined techniques are sometimes called **hyphenated methods.**[5]

Hyphenated methods couple the separation capabilities of chromatography with the qualitative and quantitative detection capabilities of spectral methods.

In early hyphenated methods, the eluates from the chromatographic column were collected as separate fractions in a cold trap, and a nondestructive, nonselective detector was used to indicate their appearance. The composition of each fraction was then investigated by nuclear magnetic resonance, infrared, or mass spectrometry or by electroanalytical measurements. A serious limitation to this approach was the very small (usually micromolar) quantity of solute contained in a fraction.

Most modern hyphenated methods monitor the effluent from the chromatographic column continuously by spectroscopic methods. The combination of two techniques based on different principles can achieve tremendous selectivity. Today's computer-based GC instruments incorporate large databases for comparing spectra and identifying compounds.

31B GAS CHROMATOGRAPHY COLUMNS AND STATIONARY PHASES

The pioneering gas-liquid chromatographic studies in the early 1950s were carried out on packed columns in which the stationary phase was a thin film of liquid retained by adsorption on the surface of a finely divided, inert solid support. From theoretical studies made during this early period, it became apparent that unpacked columns having inside diameters of a few tenths of a millimeter could provide separations that were superior to those of packed columns in both speed and column efficiency. In such **capillary columns,** the stationary phase was a film of liquid a few tenths of a micrometer thick that uniformly coated the interior of capillary tubing. In the late 1950s, such **open tubular columns** were constructed; the predicted

[5]For reviews on hyphenated methods, see C. L. Wilkins, *Science,* **1983,** *222,* 251; *Anal. Chem.,* **1989,** *59,* 571A.

performance characteristics were confirmed experimentally in several laboratories, with open tubular columns having 300,000 plates or more being described.[6]

Despite such spectacular performance characteristics, capillary columns did not gain widespread use until more than two decades after their invention. The reasons for the delay were several, including small sample capacities, fragility of columns, mechanical problems associated with sample introduction and connection of the column to the detector, difficulties in coating the column reproducibly, short lifetimes of poorly prepared columns, tendencies of columns to clog, and patents, which limited commercial development to a single manufacturer. (The original patent expired in 1977.) By the late 1970s, these problems had become manageable, and several instrument companies began to offer open tubular columns at a reasonable cost. As a consequence, we have seen a major growth in the use of capillary columns since then.

31B-1 Capillary or Open Tubular Columns

Open tubular, or capillary, columns are of two basic types: **wall-coated open tubular** (WCOT) and **support-coated open tubular** (SCOT) columns.[7] Wall-coated columns are simply capillary tubes coated with a thin layer of the stationary phase. In support-coated open tubular columns, the inner surface of the capillary is lined with a thin film (~30 μm) of a support material, such as diatomaceous earth. This type of column holds several times as much stationary phase as does a wall-coated column and thus has a greater sample capacity. Generally, the efficiency of a SCOT column is less than that of a WCOT column but significantly greater than that of a packed column.

Early WCOT columns were constructed of stainless steel, aluminum, copper, or plastic. Subsequently, glass was used. Often, the glass was etched with gaseous hydrochloric acid, strong aqueous hydrochloric acid, or potassium hydrogen fluoride to give a rough surface, which bonded the stationary phase more tightly. The most widely used capillary columns are **fused-silica open tubular columns** (FSOT columns). Fused-silica capillaries are drawn from specially purified silica that contains minimal amounts of metal oxides. These capillaries have much thinner walls than their glass counterparts. The tubes are given added strength by an outside protective polyimide coating, which is applied as the capillary tubing is being drawn. The resulting columns are quite flexible and can be bent into coils with diameters of a few inches. Figure 31-6 shows a fused-silica open tubular column. Silica open tubular columns are available commercially and offer several important advantages such as physical strength, much lower reactivity toward sample components, and flexibility. For most applications, they have replaced the older-type WCOT glass columns.

◀ **Fused-silica open tubular columns (FSOT columns)** are currently the most widely used GC columns.

The most widely used silica open tubular columns have inside diameters of 0.32 and 0.25 mm. Higher resolution columns are also sold with diameters of 0.20 and

---

[6]In 1987, a world record for length of an open tubular column and number of theoretical plates was set, as attested in the *Guinness Book of World Records,* by Chrompack International Corporation of the Netherlands. The column was a fused-silica column drawn in one piece and having an internal diameter of 0.32 mm and a length of 2100 m, or 1.3 miles. The column was coated with a 0.1-m film of polydimethyl siloxane. A 1300-m section of this column contained more than 2 million plates.

[7]For a detailed description of open tubular columns, see M. L. Lee, F. J. Yang, and K. D. Bartle, *Open Tubular Column Gas Chromatography: Theory and Practice.* New York: Wiley, 1984.

TABLE 31-4
Properties and Characteristics of Typical GC Columns

	Type of Column			
	FSOT*	WCOT†	SCOT‡	Packed
Length, m	10–100	10–100	10–100	1–6
Inside diameter, mm	0.1–0.3	0.25–0.75	0.5	2–4
Efficiency, plates/m	2000–4000	1000–4000	600–1200	500–1000
Sample size, ng	10–75	10–1000	10–1000	10–10^6
Relative pressure	Low	Low	Low	High
Relative speed	Fast	Fast	Fast	Slow
Flexible?	Yes	No	No	No
Chemical inertness	Best	⟶	⟶	Poorest

*Fused-silica open tubular column.
†Wall-coated open tubular column.
‡Support-coated open tubular column (also called porous layer open tubular or PLOT).

0.15 mm. Such columns are more troublesome to use and are more demanding on the injection and detection systems. Thus, a sample splitter must be used to reduce the size of the sample injected onto the column, and a more sensitive detector system with a rapid response time is required.

Recently, 530-μm capillaries, sometimes called **megabore columns,** have appeared on the market. These columns will tolerate sample sizes that are similar to those for packed columns. The performance characteristics of megabore open tubular columns are not as good as those of smaller diameter columns but are significantly better than those of packed columns.

Table 31-4 compares the performance characteristics of fused-silica capillary columns with other types of wall-coated columns as well as with support-coated and packed columns.

31B-2 Packed Columns

Present-day packed columns are fabricated from glass or metal tubing; they are typically 2 to 3 m long and have inside diameters of 2 to 4 mm. These tubes are densely packed with a uniform, finely divided packing material, or solid support, that is coated with a thin layer (0.05 to 1 μm) of the stationary liquid phase. The tubes are ordinarily formed into coils with diameters of roughly 15 cm to permit convenient thermostating in an oven.

Solid Support Materials

The packing, or solid support in a packed column, serves to hold the liquid stationary phase in place so that as large a surface area as possible is exposed to the mobile phase. The ideal support consists of small, uniform, spherical particles with good mechanical strength and a specific surface area of at least 1 m^2/g. In addition, the material should be inert at elevated temperatures and be uniformly wetted by the liquid phase. No substance that meets all these criteria perfectly is yet available.

The earliest, and still the most widely used, packings for gas chromatography were prepared from naturally occurring diatomaceous earth, which consists of the skeletons of thousands of species of single-celled plants that inhabited ancient lakes and seas. (Figure 31-13 is an enlarged photo of a diatom obtained with a scanning electron microscope.) These support materials are often treated chemically with dimethylchlorosilane, which gives a surface layer of methyl groups. This treatment reduces the tendency of the packing to adsorb polar molecules.

Figure 31-13 A photomicrograph of a diatom. Magnification 5000×.

Particle Size of Supports

As shown in Figure 30-16 (page 936), the efficiency of a gas chromatographic column increases rapidly with decreasing particle diameter of the packing. The pressure difference required to maintain an acceptable flow rate of carrier gas, however, varies inversely as the square of the particle diameter; the latter relationship has placed lower limits on the size of particles employed in gas chromatography because it is not convenient to use pressure differences that are greater than about 50 psi. As a result, the usual support particles are 60 to 80 mesh (250 to 170 μm) or 80 to 100 mesh (170 to 149 μm).

31B-3 Liquid Stationary Phases

Desirable properties of the immobilized liquid phase in a gas-liquid chromatographic column include (1) *low volatility* (ideally, the boiling point of the liquid should be at least 100°C higher than the maximum operating temperature for the column); (2) *thermal stability;* (3) *chemical inertness;* and (4) *solvent characteristics* such that k and α (see Section 30E-4) values for the solutes to be resolved fall within a suitable range.

Many liquids have been proposed as stationary phases in the development of gas-liquid chromatography. Currently less than a dozen are in common use. The proper choice of stationary phase is often crucial to the success of a separation. Qualitative guidelines exist for making this choice, but in the end, the best stationary phase can only be determined in the laboratory.

The retention time for an analyte on a column depends on its distribution constant, which in turn is related to the chemical nature of the liquid stationary phase. To separate various sample components, their distribution constants must be sufficiently different to accomplish a clean separation. At the same time, these constants must not be extremely large or extremely small because the former leads to prohibitively long retention times and the latter results in such short retention times that separations are incomplete.

To have a reasonable residence time in the column, an analyte must show some degree of compatibility (solubility) with the stationary phase. Here, the principle of "like dissolves like" applies, where "like" refers to the polarities of the analyte and the immobilized liquid. Polarity is the electrical field effect in the immediate vicinity of a molecule and is measured by the dipole moment of the species. Polar stationary phases contain functional groups such as —CN, —CO, and —OH. Hydrocarbon-type stationary phases and dialkyl siloxanes are nonpolar, whereas polyester phases are highly polar. Polar analytes include alcohols, acids, and amines; solutes of medium polarity include ethers, ketones, and aldehydes. Saturated hydrocarbons are nonpolar. Generally, the polarity of the stationary phase should match that of the sample components. When the match is good, the order of elution is determined by the boiling point of the eluents.

◀ The polarities of common organic functional groups in increasing order are as follows: aliphatic hydrocarbons < olefins < aromatic hydrocarbons < halides < sulfides < ethers < nitro compounds < esters, aldehydes, ketones < alcohols, amines < sulfones < sulfoxides < amides < carboxylic acids < water.

Some Widely Used Stationary Phases

Table 31-5 lists the most widely used stationary phases for both packed and open tubular column gas chromatography in order of increasing polarity. These six liquids can probably provide satisfactory separations for 90% or more of the samples encountered by the scientist.

TABLE 31-5

Some Common Liquid Stationary Phases for Gas-Liquid Chromatography

Stationary Phase	Common Trade Name	Maximum Temperature, °C	Common Applications
Polydimethyl siloxane	OV-1, SE-30	350	General-purpose nonpolar phase, hydrocarbons, polynuclear aromatics, steroids, PCBs
5% Phenyl-polydimethyl siloxane	OV-3, SE-52	350	Fatty acid methyl esters, alkaloids, drugs, halogenated compounds
50% Phenyl-polydimethyl siloxane	OV-17	250	Drugs, steroids, pesticides, glycols
50% Trifluoropropyl-polydimethyl siloxane	OV-210	200	Chlorinated aromatics nitroaromatics, alkyl substituted benzenes
Polyethylene glycol	Carbowax 20M	250	Free acids, alcohols, ethers, essential oils, glycols
50% Cyanopropyl-polydimethyl siloxane	OV-275	240	Polyunsaturated fatty acids, rosin acids, free acids, alcohols

Five of the liquids listed in Table 31-5 are polydimethyl siloxanes that have the general structure

$$R-\underset{R}{\overset{R}{|}}{Si}-O-\left[\underset{R}{\overset{R}{|}}{Si}-O\right]_n-\underset{R}{\overset{R}{|}}{Si}-R$$

In the first of these, polydimethyl siloxane, the —R groups are all $—CH_3$, giving a liquid that is relatively nonpolar. In the other polysiloxanes shown in the table, a fraction of the methyl groups are replaced by functional groups such as phenyl ($—C_6H_5$), cyanopropyl ($—C_3H_6CN$), and trifluoropropyl ($—C_3H_6CF_3$). The percentage descriptions in each case give the amount of substitution of the named group for methyl groups on the polysiloxane backbone. Thus, for example, 5% phenyl polydimethyl siloxane has a phenyl ring bonded to 5% by number of the silicon atoms in the polymer. These substitutions increase the polarity of the liquids to various degrees.

The fifth entry in Table 31-5 is a polyethylene glycol with the structure

$$HO—CH_2—CH_2—(O—CH_2—CH_2)_n—OH$$

It finds widespread use in the separation of polar species.

Bonded and Cross-Linked Stationary Phases

Commercial columns are advertised as having bonded or cross-linked stationary phases, or both. The purpose of bonding and cross-linking is to provide a longer lasting stationary phase that can be rinsed with a solvent when the film becomes contaminated. With use, untreated columns slowly lose their stationary phase owing to "bleeding," in which a small amount of immobilized liquid is carried out of the column during the elution process. Bleeding is exacerbated when a column must be rinsed with a solvent to remove contaminants. Chemical bonding and cross-linking inhibit bleeding.

Bonding involves attaching a monomolecular layer of the stationary phase to the silica surface of the column by a chemical reaction. For commercial columns, the nature of the reaction is ordinarily proprietary.

Cross-linking is carried out in situ after a column is coated with one of the polymers listed in Table 31-5. One way of cross-linking is to incorporate a peroxide into the original liquid. When the film is heated, a reaction between the methyl groups in the polymer chains is initiated by a free-radical mechanism. The polymer molecules are then cross-linked through carbon-to-carbon bonds. The resulting films are less extractable and have considerably greater thermal stability than do untreated films. Cross-linking has also been initiated by exposing the coated columns to gamma radiation.

Film Thickness

Commercial columns are available with stationary phases that vary in thickness from 0.1 to 5 μm. Film thickness primarily affects the retentive character and the capacity of a column, as discussed in Section 30E-6. Thick films are used with highly volatile analytes because such films retain solutes for a longer time, thus providing a greater time for separation to take place. Thin films are useful for separating species of low volatility in a reasonable length of time. For most applications with 0.25- or 0.32-mm columns, a film thickness of 0.25 μm is recommended. With megabore columns, 1- to 1.5-μm films are often used. Today, columns with 8-μm films are marketed.

31C APPLICATIONS OF GAS-LIQUID CHROMATOGRAPHY

Gas-liquid chromatography is applicable to species that are appreciably volatile and thermally stable at temperatures up to a few hundred degrees Celsius. An enormous number of compounds of interest possess these qualities. Consequently, gas chromatography has been widely applied to the separation and determination of the components in a variety of sample types. Figure 31-14 shows chromatograms for a few such applications.

31C-1 Qualitative Analysis

Gas chromatograms are widely used to establish the purity of organic compounds. Contaminants, if present, are revealed by the appearance of additional peaks; the areas under these peaks provide rough estimates of the extent of contamination. The technique is also useful for evaluating the effectiveness of purification procedures.

In theory, GC retention times should be useful for identifying components in mixtures. In fact, however, the applicability of such data is limited by the number of variables that must be controlled to obtain reproducible results. Nevertheless, gas chromatography provides an excellent means of confirming the presence or absence of a suspected compound in a mixture, provided that an authentic sample of the substance is available. No new peaks in the chromatogram of the mixture should appear on addition of the known compound, and enhancement of an existing peak should be observed. The evidence is particularly convincing if the effect can be duplicated on different columns and at different temperatures. On the other hand, because a chromatogram provides but a single piece of information about each species in a mixture (the retention time), the application of the technique to the qualitative analysis of complex samples of unknown composition is limited.

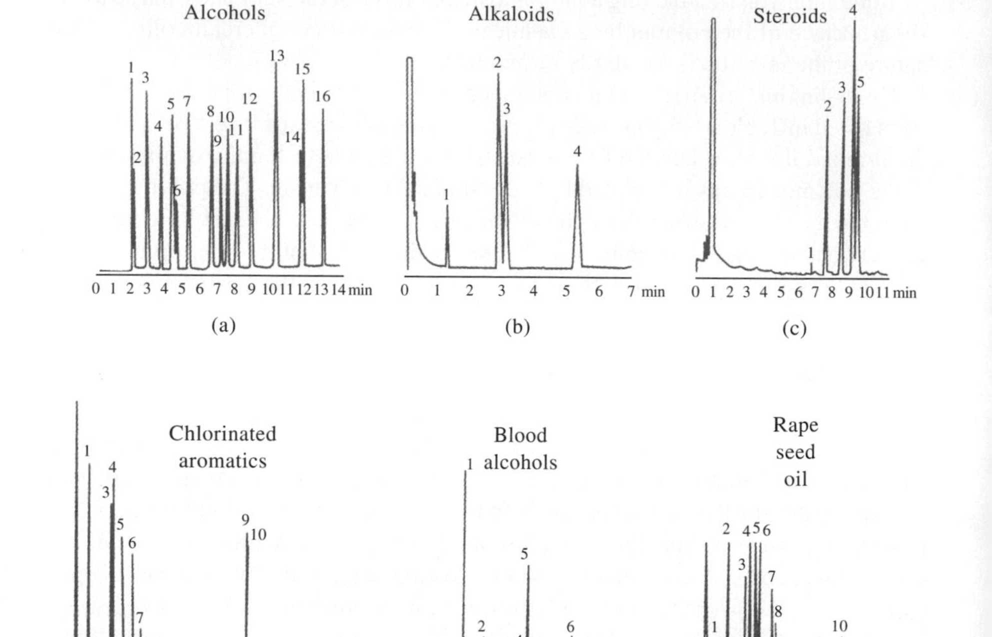

Figure 31-14 Typical chromatograms from open tubular columns coated with (a) polydimethyl siloxane, (b) 5% (phenylmethyldimethyl) siloxane, (c) 50% (phenylmethyldimethyl) siloxane, (d) 50% poly(trifluoropropyl-dimethyl) siloxane, (e) polyethylene glycol, and (f) 50% poly(cyanopropyl-dimethyl) siloxane. (Courtesy of J & W Scientific.)

This limitation has been largely overcome by linking chromatographic columns directly with ultraviolet, infrared, and mass spectrometers. The resulting hyphenated instruments are powerful tools for identifying the components of complex mixtures (see Section 31A-4). An example of the use of mass spectroscopy combined with gas chromatography for the identification of constituents in blood is given in Feature 31-1.

Although a chromatogram may not lead to positive identification of the species in a sample, it often provides sure evidence of the *absence* of a species. Thus, failure of a sample to produce a peak at the same retention time as a standard obtained under identical conditions is strong evidence that the compound in question is absent (or present at a concentration below the detection limit of the procedure).

31C-2 Quantitative Analysis

Gas chromatography owes its enormous growth in part to its speed, simplicity, relatively low cost, and wide applicability to separations. It is doubtful, however, that GC would have become so widely used were it not able to provide quantitative information about separated species as well.

Quantitative GC is based on comparison of either the height or the area of an analyte peak with that of one or more standards. If conditions are properly controlled, both these parameters vary linearly with concentration. Peak area is independent of the broadening effects discussed earlier. From this standpoint, therefore, area is a more satisfactory analytical parameter than peak height. Peak heights are more easily measured, however, and, for narrow peaks, more accurately determined. Most modern chromatographic instruments are equipped with computers

FEATURE 31-1

Use of GC/MS to Identify a Drug Metabolite in Blood[8]

A comatose patient was suspected of taking an overdose of a prescription drug, glutethimide (Doriden), because an empty prescription bottle had been found nearby. A gas chromatogram was obtained of a blood plasma extract, and two peaks were found, as shown in Figure 31F-1. The retention time for peak 1 corresponded to the retention time of glutethimide, but the compound responsible for peak 2 was not known. The possibility that the patient had taken another drug was considered. The retention time for peak 2 under the conditions used did not correspond to any other drug available to the patient, however, or to a known drug of abuse. Hence, gas chromatography/mass spectrometry was called on to establish the identity of peak 2 and to confirm the identity of peak 1 before treating the patient.

The plasma extract was subjected to GC/MS analysis, and the mass spectrum depicted in Figure 31F-2a confirmed that peak 1 was due to glutethimide. A peak in the mass spectrum at a mass-to-charge ratio of 217 is the correct ratio for the glutethimide molecular ion, and the mass spectrum was identical to that from a known sample of glutethimide. The mass spectrum of peak 2, however, showed a molecular ion peak at a mass-to-charge ratio of 233, as shown in Figure 31F-2b. This differs from the molecular ion of glutethimide by 16 mass units. Several other peaks in the mass spectrum from GC peak 2 differed from those of glutethimide by 16 mass units, indicating incorporation of oxygen into the glutethimide molecule. This led the scientists to believe that peak 2 was due to a 4-hydroxy metabolite of the parent drug.

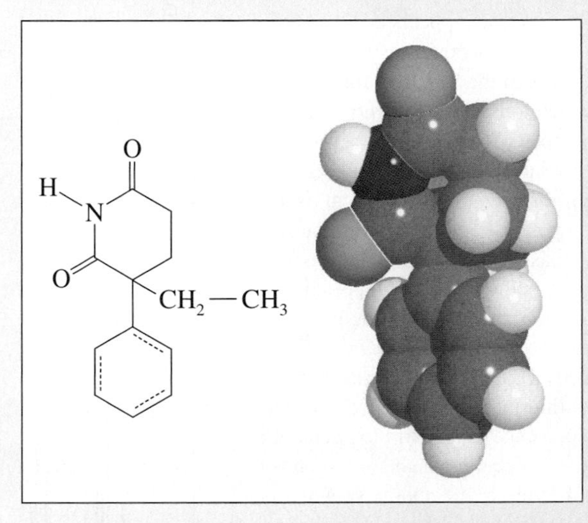

Structure and molecular model of glutethimide.

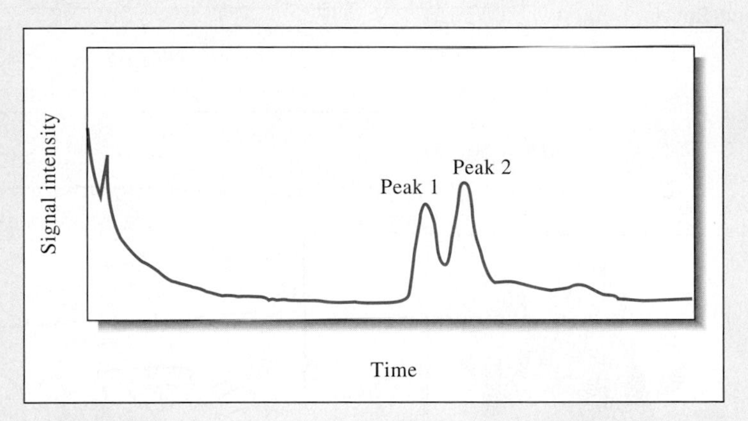

Figure 31F-1 Gas chromatogram of a blood plasma extract from a drug overdose victim. Peak 1 was at the appropriate retention time to be glutethimide, but the compound responsible for peak 2 was unknown until GC/MS was done.

(continued)

[8]From J. T. Watson, *Introduction to Mass Spectrometry,* 3rd ed., pp. 22–25. New York: Lippincott-Raven, 1997.

Figure 31F-2 (a) Mass spectrum obtained during peak 1 of the gas chromatogram in Figure 31F-1. This mass spectrum is identical with that of glutethimide. (b) Mass spectrum obtained during peak 2 of the gas chromatogram in Figure 31F-1. The fragmentation of the two compounds produces ions that are separated in the mass spectrometer. Each peak in the mass spectrum appears at a mass-to-charge ratio (m/z) corresponding to the mass of the fragment for single charged ions. Peak A at $m/z = 217$ in the upper spectrum (a) corresponds to the molar mass of glutethimide, and the mass spectrum is identical to that for a pure sample of the compound. The mass spectrum thus conclusively identifies the suspect compound as glutethimide. Peak B in the lower spectrum (b) appears at $m/z = 233$, exactly 16 units more massive than glutethimide. This evidence suggests the presence of an extra oxygen atom in the molecule, which corresponds to the 4-hydroxy metabolite shown below. (From J. T. Watson, *Introduction to Mass Spectrometry,* 3rd ed., p. 24. Philadelphia: Lippincott-Raven, 1997. With permission.)

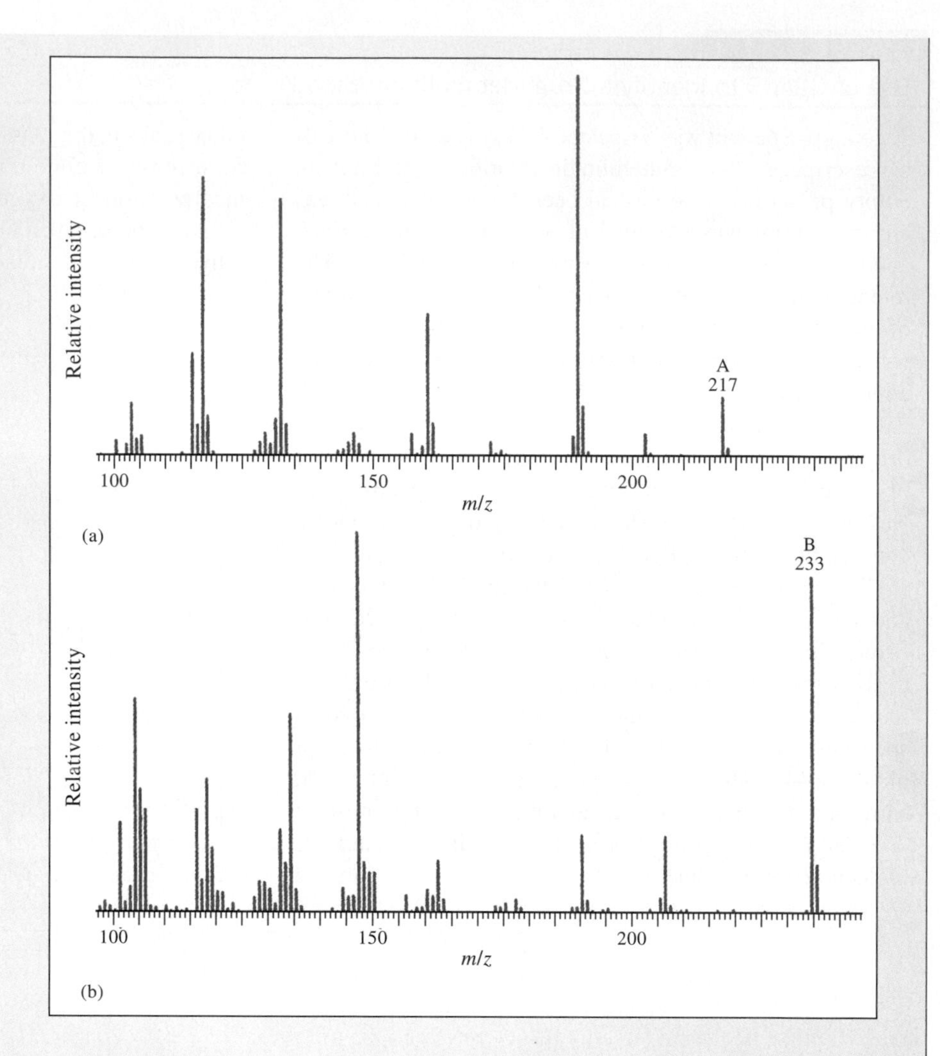

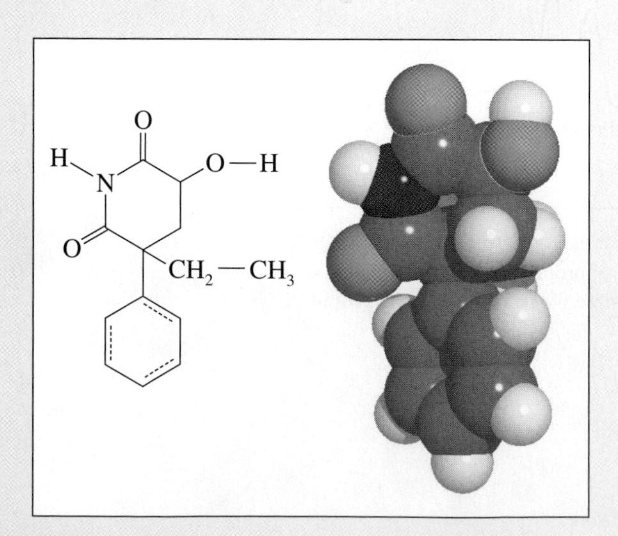

Structure and molecular model of 4-hydroxy metabolite.

FOR THE CHEMIST ON THE GO:
LAPTOP GAS CHROMATOGRAPH /MASS SPECTROMETER

An acetic anhydride derivative of the peak 2 material was then prepared and found to be identical to the acetate derivative of 4-hydroxy-2-ethyl-2-phenylglutarimide, the metabolite on page 966. This metabolite was known to exhibit toxicity in animals. The patient was then subjected to hemodialysis, which removed the polar metabolite more rapidly than the less polar parent drug. Soon thereafter, the patient regained consciousness.

that provide measurements of relative peak areas. If such equipment is not available, a manual estimate must be made. A simple method that works well for symmetrical peaks of reasonable widths is to multiply peak height by the width at one-half peak height.

Calibration with Standards

The most straightforward method of quantitative gas-chromatographic analysis involves the preparation of a series of standard solutions that approximate the composition of the unknown (external standard method). Chromatograms for the standards are then obtained, and peak heights or areas are plotted as a function of concentration to obtain a working curve. A plot of the data should yield a straight line passing through the origin; quantitative analyses are based on this plot. Frequent standardization is necessary for highest accuracy.

The Internal Standard Method

The highest precision for quantitative GC is obtained using internal standards because the uncertainties introduced by sample injection, flow rate, and variations in column conditions are minimized. In this procedure, a carefully measured quantity of an internal standard is introduced into each standard and sample (see Section 8C-3), and the ratio of analyte peak area (or height) to internal-standard peak area (or height) is used as the analytical parameter (see Example 31-1). For this method to be successful, it is necessary that the internal standard peak be well separated from the peaks of all other components in the sample. It must appear close to the analyte peak, however. Of course, the internal standard should be absent in the sample to be analyzed. With a suitable internal standard, precisions of 0.5% to 1% relative are reported.

EXAMPLE 31-1

Gas chromatographic peaks can be influenced by a variety of instrumental factors. We can often compensate for variations in these factors by using the internal standard method. Here, we add the same amount of internal standard to mixtures containing known amounts of the analyte and to the samples of unknown analyte concentration. We then calculate the ratio of peak height (or area) for the analyte to that of the internal standard.

The data shown in the table were obtained during the determination of a C_7 hydrocarbon with a closely related compound added to each standard and to the unknown as an internal standard.

(continued)

Percent Analyte	Peak Height Analyte	Peak Height, Internal Standard
0.05	18.8	50.0
0.10	48.1	64.1
0.15	63.4	55.1
0.20	63.2	42.7
0.25	93.6	53.8
Unknown	58.9	49.4

Construct a spreadsheet to determine the peak height ratio of the analyte to the internal standard and plot this versus the analyte concentration. Determine the concentration of the unknown and its standard deviation.

The spreadsheet is shown in Figure 31-15. The data are entered into columns A through C as shown. In cells D4 through D9, the peak height ratio is calculated by the formula shown in documentation cell A22. A plot of the calibration curve is also shown in the figure. The linear regression statistics are calculated in cells B11 through B20 using the same approach as described in Section 8C-2. The statistics are calculated by the formulas in documentation cells A23 through A31. The percentage of the analyte in the unknown is found to be 0.163 ± 0.008.

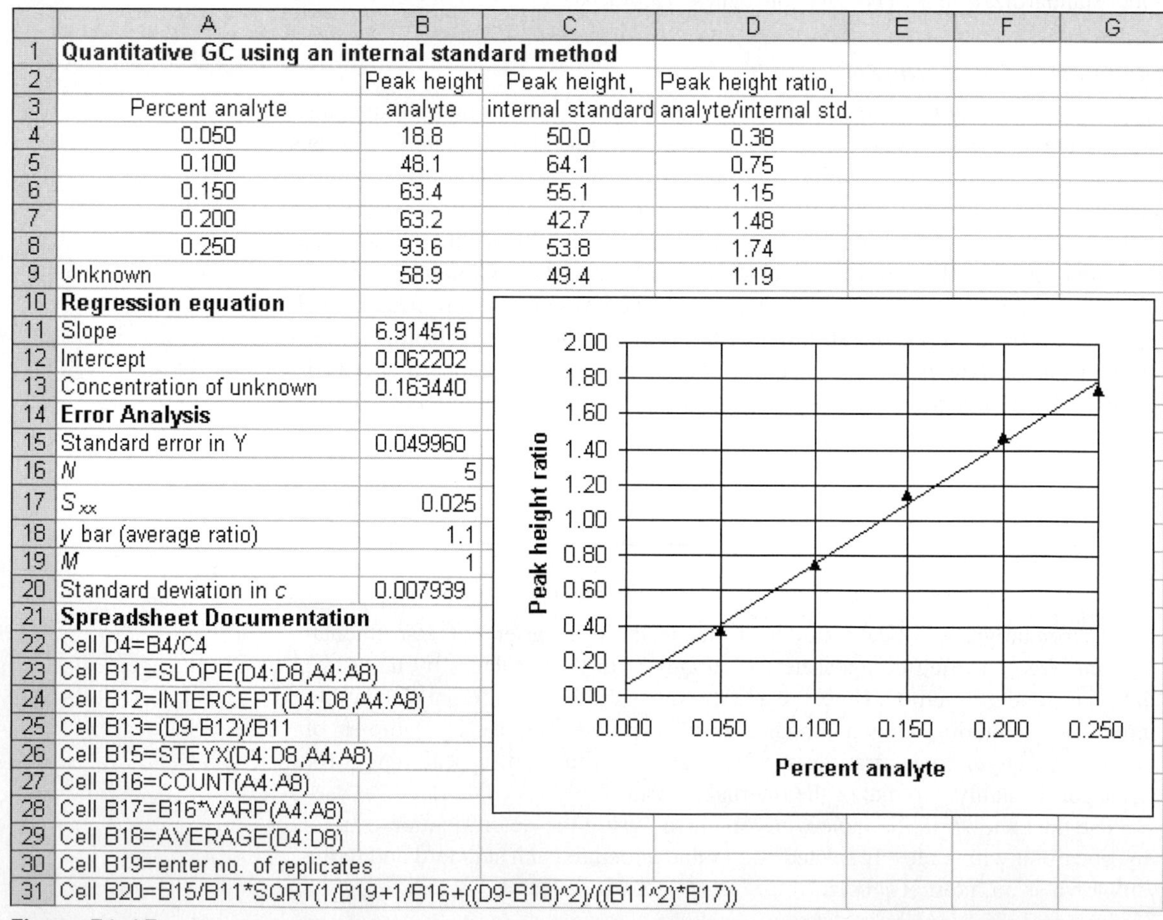

	A	B	C	D	E	F	G
1	**Quantitative GC using an internal standard method**						
2		Peak height	Peak height,	Peak height ratio,			
3	Percent analyte	analyte	internal standard	analyte/internal std.			
4	0.050	18.8	50.0	0.38			
5	0.100	48.1	64.1	0.75			
6	0.150	63.4	55.1	1.15			
7	0.200	63.2	42.7	1.48			
8	0.250	93.6	53.8	1.74			
9	Unknown	58.9	49.4	1.19			
10	**Regression equation**						
11	Slope	6.914515					
12	Intercept	0.062202					
13	Concentration of unknown	0.163440					
14	**Error Analysis**						
15	Standard error in Y	0.049960					
16	*N*	5					
17	S_{xx}	0.025					
18	*y* bar (average ratio)	1.1					
19	*M*	1					
20	Standard deviation in *c*	0.007939					
21	**Spreadsheet Documentation**						
22	Cell D4=B4/C4						
23	Cell B11=SLOPE(D4:D8,A4:A8)						
24	Cell B12=INTERCEPT(D4:D8,A4:A8)						
25	Cell B13=(D9-B12)/B11						
26	Cell B15=STEYX(D4:D8,A4:A8)						
27	Cell B16=COUNT(A4:A8)						
28	Cell B17=B16*VARP(A4:A8)						
29	Cell B18=AVERAGE(D4:D8)						
30	Cell B19=enter no. of replicates						
31	Cell B20=B15/B11*SQRT(1/B19+1/B16+((D9-B18)^2)/((B11^2)*B17))						

Figure 31-15 Spreadsheet to illustrate the internal standard method for the GC determination of a C_7 hydrocarbon.

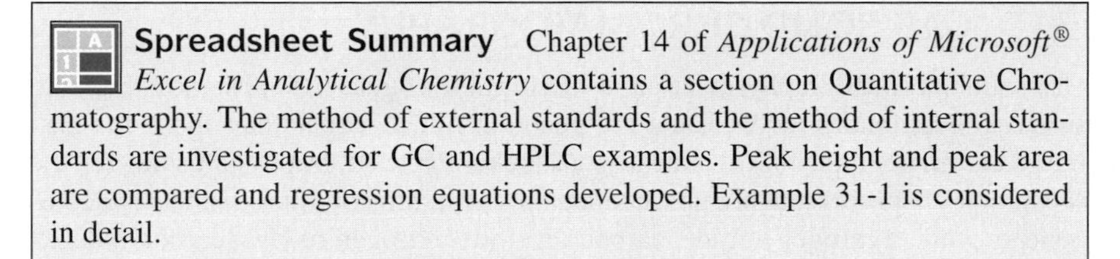

Spreadsheet Summary Chapter 14 of *Applications of Microsoft® Excel in Analytical Chemistry* contains a section on Quantitative Chromatography. The method of external standards and the method of internal standards are investigated for GC and HPLC examples. Peak height and peak area are compared and regression equations developed. Example 31-1 is considered in detail.

FEATURE 31-2

High-Speed Gas Chromatography[9]

Gas chromatography has often focused on achieving ever-higher resolution in order to separate more and more complex mixtures. In most separations, conditions are varied to separate the most difficult-to-separate pair of components, the so-called **critical pair.** Many of the components of interest, under these conditions, are highly overseparated. The basic idea of high-speed GC is that for many separations of interest, higher speed can be achieved, albeit at the expense of some selectivity and resolution.

To see how to arrange conditions for high-speed separations, we can write Equation 30-17 as

$$\frac{L}{t_R} = u \times \frac{1}{1 + k_n} \quad (31\text{-}1)$$

where k_n is the retention factor for the last component of interest in the chromatogram. If we rearrange Equation 31-1 and solve for the retention time of the last component of interest, we obtain

$$t_R = \frac{L}{u} \times (1 + k_n) \quad (31\text{-}2)$$

Equation 31-2 tells us that we can achieve faster separations by using short columns, higher-than-usual carrier gas velocities, and small retention factors. The price to be paid is reduced resolving power, caused by increased band broadening and reduced peak capacity (that is, the number of peaks that will fit in the chromatogram).

Sacks and coworkers at the University of Michigan[10] have been designing instrumentation and chromatographic conditions to optimize separation speed at the lowest cost in terms of resolution and peak capacity. They have designed systems to achieve tunable columns and high-speed temperature programming. A tunable column is a series combination of a polar and a nonpolar column. Figure 31F-3 shows the separation of 12 compounds prior to initiation of a programmed temperature ramp and 19 compounds after the temperature program was begun. The total time required was 140 s. These workers have also been using high-speed GC with mass spectrometry detection, including time-of-flight detection.[11]

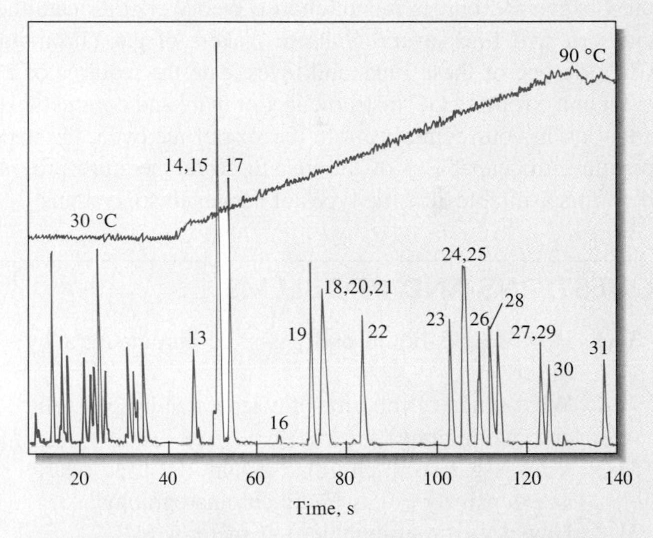

Figure 31F-3 High-speed chromatogram obtained with isothermal operation (30°C) for 37 s followed by a 35°C/min temperature ramp to 90°C. (Reprinted with permission from H. Smith and R. D. Sacks, *Anal. Chem.*, **1998,** *70,* 4960. Copyright 1998 American Chemical Society.)

[9]For a review, see R. Sacks, H. Smith, and M. Nowak, *Anal. Chem.*, **1998,** *70,* 29A.

[10]H. Smith and R. D. Sacks, *Anal. Chem.*, **1998,** *70,* 1960.

[11]C. Leonard and R. Sacks, *Anal. Chem.*, **1999,** *71,* 5177.

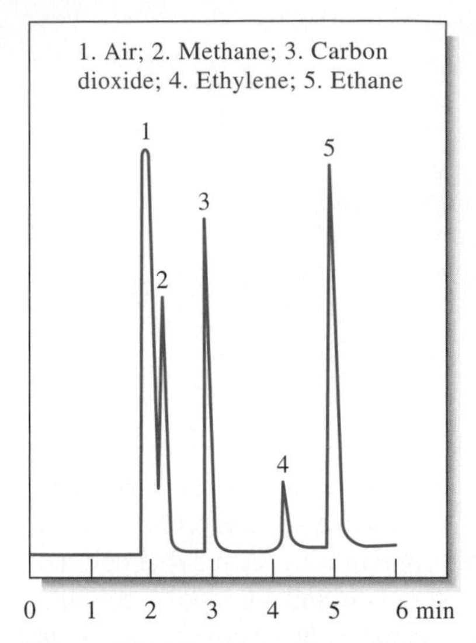

Figure 31-16 Typical gas-solid chromatogram on a PLOT column.

31D GAS-SOLID CHROMATOGRAPHY

Gas-solid chromatography is based on adsorption of gaseous substances on solid surfaces. Distribution coefficients are generally much larger than those for gas-liquid chromatography. Consequently, gas-solid chromatography is useful for the separation of species that are not retained by gas-liquid columns, such as the components of air, hydrogen sulfide, carbon disulfide, nitrogen oxides, carbon monoxide, carbon dioxide, and the rare gases.

Gas-solid chromatography is performed with both packed and open tubular columns. For the latter, a thin layer of the adsorbent is affixed to the inner walls of the capillary. Such columns are sometimes called **porous-layer open tubular columns,** or PLOT columns. Figure 31-16 shows a typical application of a PLOT column.

WEB WORKS

Direct your web browser to **http://chemistry.brookscole.com/skoogfac/.** From the Chapter Resources menu, choose Web Works. Locate the Chapter 31 section, and you will find several links to makers of gas chromatographic instruments. Click on one of these links and investigate the features of a premium GC instrument and a routine GC instrument. Compare and contrast these features. Pay close attention in your comparison to the size of the oven, the uncertainty in oven temperature, the capability of the unit to do temperature programming, the types of detectors available, and the types of data analysis systems.

QUESTIONS AND PROBLEMS

***31-1.** How do gas-liquid and gas-solid chromatography differ?

31-2. What kinds of mixtures are separated by gas-solid chromatography?

***31-3.** Why is gas-solid chromatography not used nearly as extensively as gas-liquid chromatography?

31-4. How does a soap-bubble flow meter work?

***31-5.** What is a chromatogram?

31-6. What is meant by temperature programming in gas chromatography?

***31-7.** Describe the physical differences between open tubular and packed columns. What are the advantages and disadvantages of each?

31-8. What variables must be controlled if satisfactory quantitative data are to be obtained from chromatograms?

***31-9.** What is the packing material used in most packed gas chromatographic columns?

31-10. Describe the principle on which each of the following gas chromatography detectors are based: (a) thermal conductivity, (b) flame ionization, (c) electron capture, (d) thermionic, and (e) photoionization.

***31-11.** What are the principal advantages and the principal limitations of each of the detectors listed in Problem 31-10?

31-12. What are *hyphenated* gas-chromatographic methods? Briefly describe three hyphenated methods.

***31-13.** What are megabore open-tubular columns? Why are they used?

31-14. How do the following open-tubular columns differ?
(a) PLOT columns
(b) WCOT columns
(c) SCOT columns

31-15. What properties should the stationary-phase liquid for gas chromatography possess?

31-16. What are the advantages of fused-silica capillary columns compared with glass or metal columns?

***31-17.** What is the effect of stationary-phase film thickness on gas chromatograms?

31-18. Why are gas-chromatographic stationary phases often bonded and cross-linked? What do these terms mean?

***31-19.** List the variables that lead to (a) band broadening and (b) band separation in gas-liquid chromatography.

31-20. One method for the quantitative determination of the concentration of constituents in a sample analyzed by gas chromatography is area normalization. Here, complete elution of all the sample constituents is necessary. The area of each peak is then measured and corrected for differences in detector response to the different eluates. This correction involves dividing the area by an empirically determined correction factor. The concentration of the analyte is found from the ratio of its corrected area to the total corrected area of all peaks. For a chromatogram containing three peaks, the relative areas were found to be 16.4, 45.2, and 30.2, in order of increasing retention time. Calculate the percentage of each compound if the relative detector responses were 0.60, 0.78, and 0.88, respectively.

***31-21.** Peak areas and relative detector responses are to be used to determine the concentration of the five species in a sample. The area-normalization method described in Problem 31-20 is to be used. The relative areas for the five gas-chromatographic peaks are given in the table. Also shown are the relative responses of the detector. Calculate the percentage of each component in the mixture.

Compound	Relative Peak Area	Relative Detector Response
A	32.5	0.70
B	20.7	0.72
C	60.1	0.75
D	30.2	0.73
E	18.3	0.78

31-22. For the data given in Example 31-1, compare the method of external standards to the internal standard method. Plot the analyte peak height versus percent analyte and determine the unknown without using the internal standard results. Are your results any more precise using the internal standard method? If so, give some possible reasons.

31-23. Challenge Problem. Cinnamaldehyde is the component responsible for cinnamon flavor. It is also a potent antimicrobial compound present in essential oils (see M. Friedman, N. Kozukue, and L. A. Harden, *J. Agric. Food Chem.*, **2000,** *48,* 5702). The GC response of an artificial mixture containing six essential oil components and methyl benzoate as an internal standard is shown in the figure.

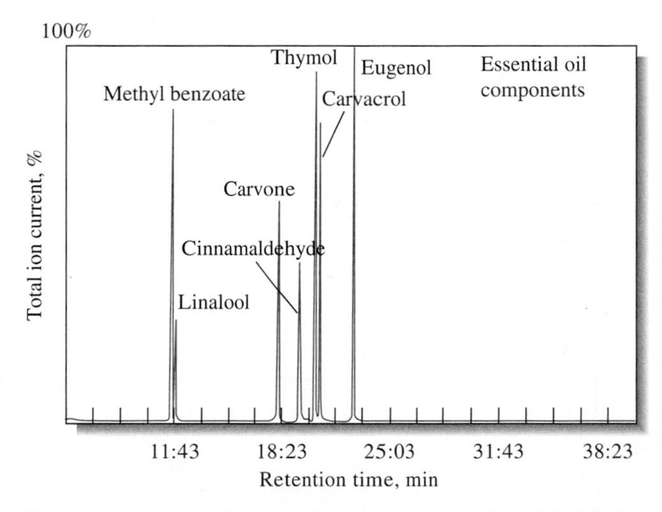

Gas chromatogram. (Reprinted with permission from M. Friedman, N. Kozukuc, and L.A. Harden; *J. Agric. Fed. Chem.*, **2000,** *48,* 570. Copyright 2000 American Chemical Society.)

(a) The following figure is an idealized enlargement of the region near the cinnamaldehyde peak.

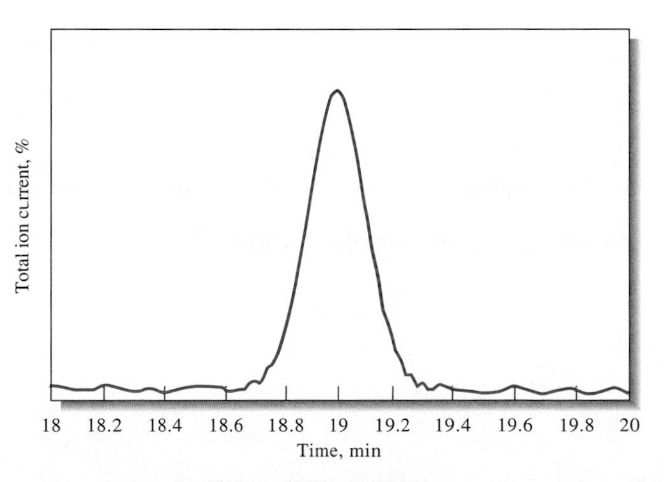

Enlarged chromatogram.

Determine the retention time for cinnamaldehyde.

(b) From the figure in part (a), determine the number of theoretical plates for the column.

(c) The fused-silica column was 0.25 mm × 30 cm with a 0.25-μm film. Determine the height equivalent to a theoretical plate from the data in parts (a) and (b).

(d) Quantitative data were obtained by using methyl benzoate as the internal standard. The following results were obtained for calibration curves of cinnamaldehyde, eugenol, and thymol. The values under each component represent the peak area of the component divided by the peak area of the internal standard.

Concentration, mg sample/ 200 μL	Cinnamaldehyde	Eugenol	Thymol
0.50		0.4	
0.65			1.8
0.75	1.0	0.8	
1.10		1.2	
1.25	2.0		
1.30			3.0
1.50		1.5	
1.90	3.1	2.0	4.6
2.50	4.0		5.8

Determine the calibration curve equations for each component. Include the R^2 values.

(e) From the data in part (d), determine which of the components has the highest calibration curve sensitivity. Which has the lowest?

(f) A sample containing the three essential oils in part (d) gave the peak areas relative to the internal standard area: cinnamaldeyde, 2.6; eugenol, 0.9; thymol, 3.8. Determine the concentrations of each of the oils in the sample and the standard deviations in concentration.

(g) A study was made of the decomposition of cinnamaldehyde in cinnamon oil. The oil was heated for various times at different temperature. These data were obtained.

Temp, °C	Time, min	% Cinnamaldehyde
25, initial		90.9
40	20	87.7
	40	88.2
	60	87.9
60	20	72.2
	40	63.1
	60	69.1
100	20	66.1
	40	57.6
	60	63.1
140	20	64.4
	40	53.7
	60	57.1
180	20	62.3
	40	63.1
	60	52.2
200	20	63.1
	40	64.5
	60	63.3
210	20	74.9
	40	73.4
	60	77.4

Use ANOVA to determine whether temperature has an effect on the decomposition of cinnamaldehyde. In the same way, determine if time of heating has an effect.

(h) With the data in part (g), assume that decomposition begins at 60°C. Test the hypothesis that there is no effect of temperature or time.

InfoTrac College Edition

For additional readings, go to InfoTrac College Edition, your online research library, at

http://infotrac.thomsonlearning.com

CHAPTER 32

High-Performance Liquid Chromatography

High-performance liquid chromatography has become an indispensable analytical tool. The crime labs in forensic and police dramas on television, such as *CSI*, *CSI Miami*, *Crossing Jordan*, and *Law and Order*, often use HPLC in the processing of evidence. The photo shows actress Marg Helgenberger (Catherine Willows) of *CSI* in the laboratory preparing samples for HPLC analysis.

This chapter considers the theory and practice of HPLC, including partition, adsorption, ion-exchange, size-exclusion, affinity, and chiral chromatography. HPLC has applications not only in forensics but also in biochemistry, environmental science, food science, pharmaceutical chemistry, and toxicology.

High-performance liquid chromatography (HPLC) is the most versatile and widely used type of elution chromatography. The technique is used by chemists to separate and determine species in a variety of organic, inorganic, and biological materials. In liquid chromatography, the mobile phase is a liquid solvent containing the sample as a mixture of solutes. The types of high-performance liquid chromatography are often classified by separation mechanism or by the type of stationary phase. These include (1) ***partition,*** *or* ***liquid-liquid, chromatography;*** *(2)* ***adsorption,*** *or* ***liquid-solid, chromatography;*** *(3)* ***ion-exchange,*** *or* ***ion chromatography;*** *(4)* ***size-exclusion chromatography;*** *(5)* ***affinity chromatography,*** *and (6)* ***chiral chromatography.***

Early liquid chromatography was performed in glass columns having inside diameters of perhaps 10 to 50 mm. The columns were packed with 50- to 500-cm lengths of solid particles coated with an adsorbed liquid that formed the stationary phase. To ensure reasonable flow rates through this type of stationary phase, the particle size of the solid was kept larger than 150 to 200 μm; even then, flow rates were a few tenths of a milliliter per minute, at best. Attempts to speed up this classic procedure by application of vacuum or pressure were not effective because increases in flow rates were accompanied by increases in plate heights and accompanying decreases in column efficiency.

Early in the development of the theory of liquid chromatography, it was recognized that large decreases in plate heights would be realized if the particle size of packings were reduced. This effect is shown by the data in Figure 32-1. Note that the minimum shown in Figure 30-13a (page 933) is not reached in any of these plots. The reason for this difference is that diffusion in liquids is much slower than in gases; consequently, its effect on plate heights is observed only at extremely low flow rates.

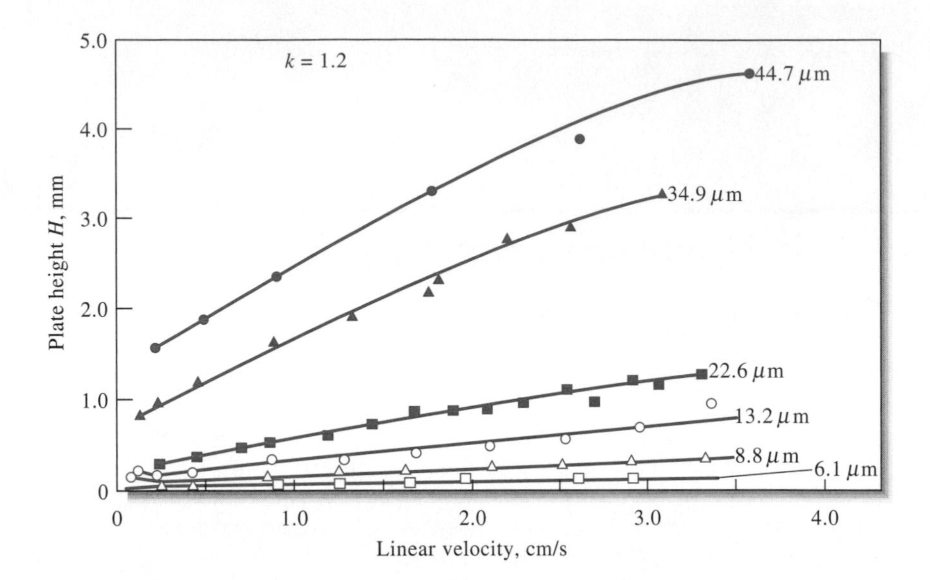

Figure 32-1 Effect of particle size of packing and flow rate on plate height in liquid chromatography. (From R. E. Majors, *J. Chromatogr. Sci.*, **1973,** *11,* 92. Reproduced from the *Journal of Chromatographic Science* by permission of Preston Publications, a Division of Preston Industries, Inc.)

Not until the late 1960s was the technology developed for producing and using packings with particle diameters as small as 3 to 10 μm. This technology required instruments capable of much higher pumping pressures than the simple devices that preceded them. Simultaneously, detectors were developed for continuous monitoring of column effluents. The term high-performance liquid chromatography *is often employed to distinguish this technology from the simple column chromatographic procedures that preceded them.*[1] *Simple column chromatography, however, still finds considerable use for preparative purposes.*

Applications of the most widely used types of HPLC for various analyte species are shown in Figure 32-2. Note that the various types of liquid chromatography tend to be complementary insofar as applications are concerned. For example, for analytes having molecular masses greater than 10,000, one of the two size-exclusion methods is often used: gel permeation for nonpolar species and gel filtration for polar or ionic compounds. For ionic species having lower molar masses, ion-exchange chromatography is generally the method of choice. Smaller polar but nonionic species are best handled by partition methods.

High-performance liquid chromatography, HPLC, is a type of chromatography that employs a liquid mobile phase and a very finely divided stationary phase. To obtain satisfactory flow rates, the liquid must be pressurized to several hundred or more pounds per square inch.

32A INSTRUMENTATION

Pumping pressures of several hundred atmospheres are required to achieve reasonable flow rates with packings in the 3- to 10-μm size range, which are common in modern liquid chromatography. As a consequence of these high pressures, the equipment for high-performance liquid chromatography tends to be considerably more elaborate and expensive than that encountered in other types of chromatography. Figure 32-3 is a diagram showing the important components of a typical HPLC instrument.

[1]For a detailed discussion of HPLC systems, see L. R. Snyder and J. J. Kirkland, *Introduction to Modern Liquid Chromatography,* 3rd ed. New York: Wiley, 1996 S. Lindsay, *High Performance Liquid Chromatography.* New York: Wiley, 1992; R. P. W. Scott, *Liquid Chromatography for the Analyst.* New York: Marcel Dekker, 1995.

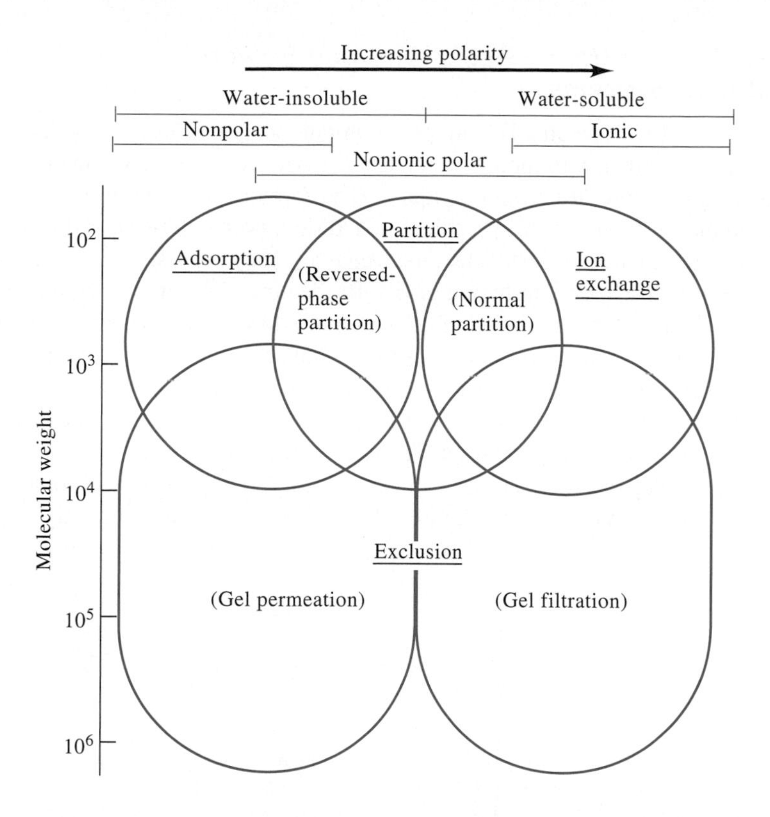

Figure 32-2 Applications of liquid chromatography. Note that the types of chromatography on the right side of the diagram are best suited for polar compounds. Techniques toward the bottom of the diagram are best suited for species of high molecular mass. (From D. L. Saunders, in *Chromatography,* 3rd ed., E. Heftmann, Ed., p. 81. New York: Van Nostrand Reinhold, 1975.)

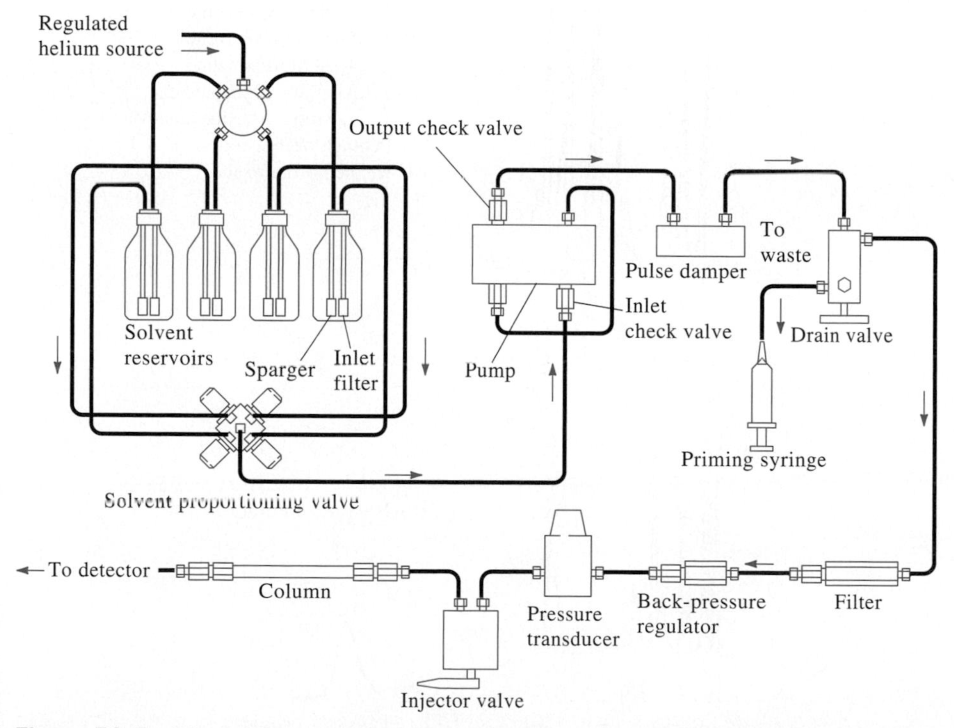

Figure 32-3 Block diagram showing components of a typical apparatus for HPLC. (Courtesy of Perkin-Elmer Corp., Norwalk, CT.)

32A-1 Mobile-Phase Reservoirs and Solvent Treatment Systems

A modern HPLC apparatus is equipped with one or more glass reservoirs, each of which contains 500 mL or more of a solvent. Provisions are often included to remove dissolved gases and dust from the liquids. The former produce bubbles in the column and thereby cause band spreading; in addition, both bubbles and dust interfere with the performance of most detectors. Degassers may consist of a vacuum pumping system, a distillation system, a device for heating and stirring, or, as shown in Figure 32-3, a system for *sparging,* in which the dissolved gases are swept out of solution by fine bubbles of an inert gas that is not soluble in the mobile phase.

Sparging is a process in which dissolved gases are swept out of a solvent by bubbles of an inert, insoluble gas.

An elution with a single solvent or a solvent mixture of constant composition is **isocratic.** In **gradient elution,** two (and sometimes more) solvent systems that differ significantly in polarity are used. The ratio of the two solvents is varied in a preprogrammed way during the separation, sometimes continuously and sometimes in a series of steps. As shown in Figure 32-4, gradient elution frequently improves

An **isocratic elution** in HPLC is one in which the solvent composition remains constant.

A **gradient elution** in HPLC is one in which the composition of the solvent is changed continuously or in a series of steps.

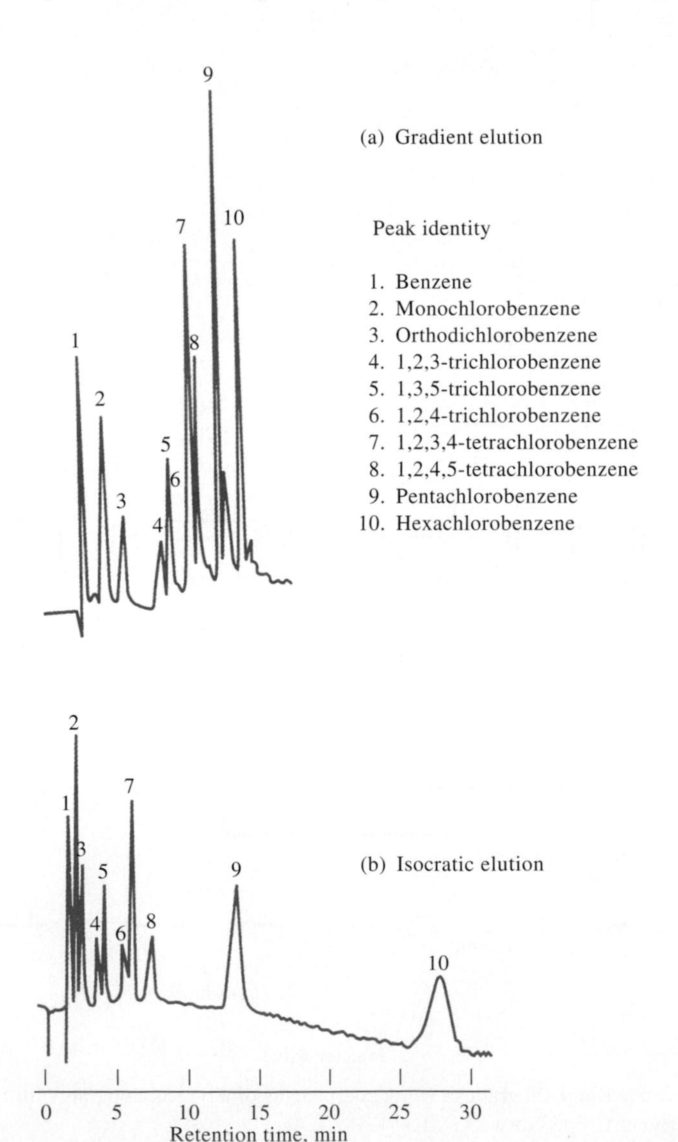

Figure 32-4 Improvement in separation efficiency by gradient elution. (From J. J. Kirkland, Ed., *Modern Practice of Liquid Chromatography,* p. 88. New York: Interscience, 1971.)

separation efficiency, just as temperature programming helps in gas chromatography. Modern HPLC instruments are often equipped with proportioning valves that introduce liquids from two or more reservoirs at ratios that can be varied continuously (see Figure 32-3).

32A-2 Pumping Systems

The requirements for liquid chromatographic pumps include (1) ability to generate pressures of up to 6000 psi (lb/in^2), (2) pulse-free output, (3) flow rates ranging from 0.1 to 10 mL/min, (4) flow reproducibilities of 0.5% relative or better, and (5) resistance to corrosion by a variety of solvents. The high pressures generated by liquid chromatographic pumps are not an explosion hazard because liquids are not very compressible. Thus, rupture of a component results only in solvent leakage. Such leakage may constitute a fire or environmental hazard with some solvents, however.

There are three major types of pumps: the screw-driven syringe type, the reciprocating pump, and the pneumatic or constant-pressure pump. Syringe-type pumps produce a pulse-free delivery whose flow rate is readily controlled; they suffer, however, from a lack of capacity (~250 mL) and are inconvenient when solvents must be changed. Figure 32-5 shows the most widely used type of pump, the reciprocating pump. This device consists of a small cylindrical chamber that is filled and then emptied by the back-and-forth motion of a piston. The pumping motion produces a pulsed flow that must be subsequently damped. Advantages of reciprocating pumps include small internal volume, high output pressure (up to 10,000 psi), ready adaptability to gradient elution, and constant flow rates, which are largely independent of column back-pressure and solvent viscosity. Most modern commercial chromatographs employ a reciprocating pump.

Some instruments use a pneumatic pump, which in its simplest form consists of a collapsible solvent container housed in a vessel that can be pressurized by a compressed gas. Pumps of this type are simple, inexpensive, and pulse free; they suffer from limited capacity and pressure output, however, and from pumping rates that depend on solvent viscosity. In addition, they are not adaptable to gradient elution.

32A-3 Sample Injection System

The most widely used method of sample introduction in liquid chromatography is based on a sampling loop such as that shown in Figure 32-6. These devices are an

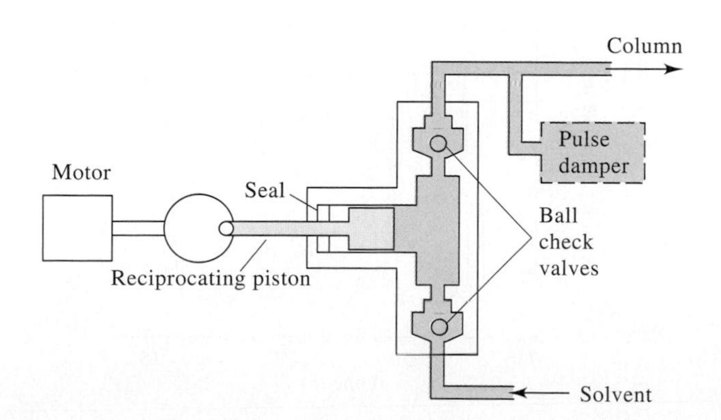

Figure 32-5 A reciprocating pump for HPLC.

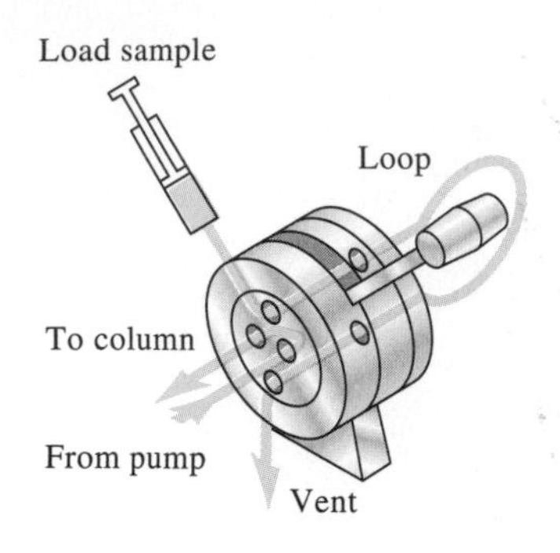

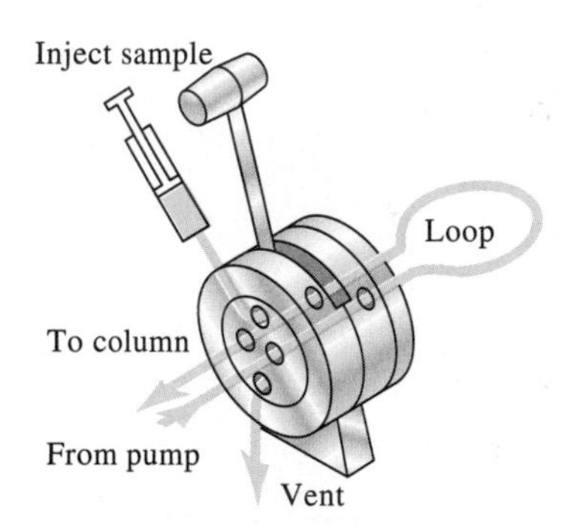

Figure 32-6 A sampling loop for liquid chromatography. (Courtesy of Beckman Coulter, Fullerton, CA.)

integral part of some liquid chromatography equipment. Often, interchangeable loops are available to provide a choice of sample sizes ranging from 5 to 500 μL. The reproducibility of injections with a typical sampling loop is a few tenths of a percent relative. Many HPLC instruments incorporate an autosampler with an automatic injector. These can inject continuously variable volumes.

32A-4 Columns for High-Performance Liquid Chromatography

Liquid chromatographic columns are usually constructed from stainless steel tubing, although glass or Tygon tubing is sometimes employed for lower pressure applications (<600 psi). Most columns range in length from 10 to 30 cm and have inside diameters of 2 to 5 mm. Column packings typically have particle sizes of 3 to 10 μm. Columns of this type provide 40,000 to 60,000 plates/m. Recently, microcolumns have become available with inside diameters of 1 to 4.6 mm and lengths of 3 to 7.5 cm. These columns, which are packed with 3- or 5-μm particles, contain as many as 100,000 plates/m and have the advantage of speed and minimal solvent consumption. This latter property is of considerable importance because the high-purity solvents required for liquid chromatography are expensive to purchase and to dispose of after use. Figure 32-7 illustrates the speed with which a separation can be performed on this type of column. Here, eight components of diverse type are separated in about 15 s. The column is 4 cm in length and has an inside diameter of 4 mm; it is packed with 3-μm particles.

The most common packing for liquid chromatography is prepared from silica particles, which are synthesized by agglomerating submicron silica particles under conditions that lead to larger particles with highly uniform diameters. The resulting particles are often coated with thin organic films, which are chemically or physically bonded to the surface. Other packing materials include alumina particles, porous polymer particles, and ion-exchange resins.

Guard Columns

Often, a short guard column is positioned ahead of the analytical column to increase the life of the analytical column by removing particulate matter and contaminants from the solvents. In addition, in liquid-liquid chromatography, the guard column serves to saturate the mobile phase with the stationary phase so that losses of the stationary phase from the analytical column are minimized. The com-

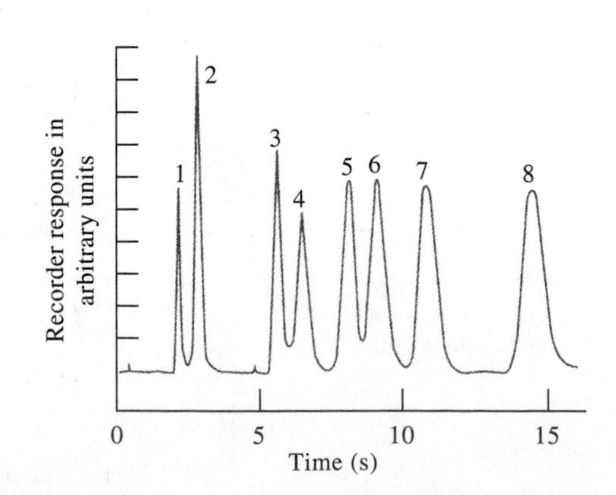

Figure 32-7 High-speed isocratic separation. Column dimensions: 4-cm length, 0.4 cm inside diameter; packing: 3-μm spherisorb; mobile phase: 4.1% ethyl acetate in *n*-hexane. Compounds: (1) *p*-xylene, (2) anisole, (3) benzyl acetate, (4) dioctyl phthalate, (5) dipentyl phthalate, (6) dibutyl phthalate, (7) dipropyl phthalate, (8) diethyl phthalate. (From R. P. W. Scott, *Small Bore Liquid Chromatography Columns: Their Properties and Uses,* p. 156. New York: Wiley, 1984. This material is used by permission of Wiley-Liss, Inc., a subsidiary of John Wiley & Sons, Inc.)

position of the guard-column packing should be similar to that of the analytical column; the particle size is usually larger, however, to minimize pressure drop.

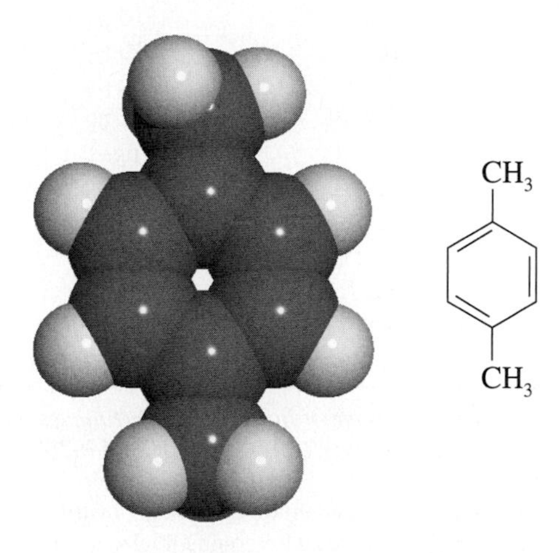

Molecular model of *p*-xylene. There are three xylene isomers: ortho, meta, and para. Para-xylene is used for the production of artificial fibers. Xylol is a mixture of the three isomers and is used as a solvent.

Column Thermostats

For many applications, close control of column temperature is not necessary, and columns are operated at room temperature. Often, however, better chromatograms are obtained by maintaining column temperatures constant to a few tenths of a degree Celsius. Most modern commercial instruments are now equipped with heaters that control column temperatures to a few tenths of a degree from near-ambient to 150°C. Columns may also be fitted with water jackets fed from a constant-temperature bath to give precise temperature control.

32A-5 Detectors

Detectors for HPLC must have low dead volume to minimize extra-column band broadening. The detector should be small and compatible with liquid flow. No highly sensitive, universal detector system, such as those for gas chromatography, is available for high-performance liquid chromatography. Thus, the detector used will depend on the nature of the sample. Table 32-1 lists some of the common detectors and their properties.

The most widely used detectors for liquid chromatography are based on absorption of ultraviolet or visible radiation (Figure 32-8). Both photometers and spectrophotometers specifically designed for use with chromatographic columns are available from commercial sources. The former often make use of the 254- and 280-nm lines from a mercury source because many organic functional groups absorb in the region. Deuterium sources or tungsten-filament sources with interference filters also provide a simple means of detecting absorbing species. Some modern instruments are equipped with filter wheels that contain several interference filters, which can be rapidly switched into place. Spectrophotometric detectors are considerably more versatile than photometers and are also widely used in high-performance instruments. Modern instruments use diode-array detectors that

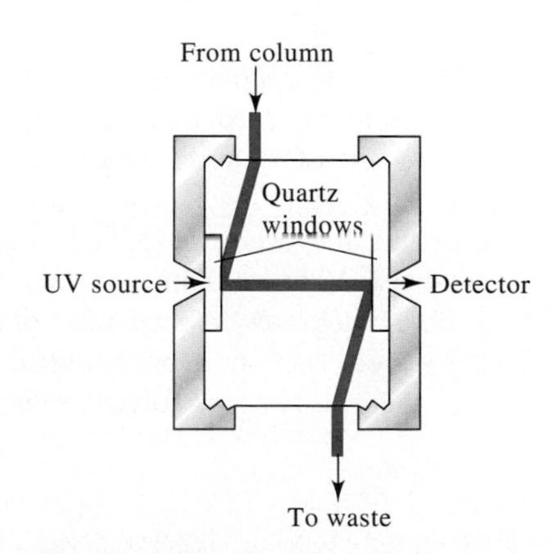

Figure 32-8 A UV-visible detector for HPLC.

TABLE 32-1

Performance of HPLC Detectors*

HPLC Detector	Commercially Available	Mass LOD† (typical)	Linear Range‡ (decades)
Absorbance	Yes	10 pg	3–4
Fluorescence	Yes	10 fg	5
Electrochemical	Yes	100 pg	4–5
Refractive index	Yes	1 ng	3
Conductivity	Yes	100 pg–1 ng	5
Mass spectrometry	Yes	<1 pg	5
FTIR	Yes	1 μg	3
Light scattering	Yes	1 μg	5
Optical activity	No	1 ng	4
Element selective	No	1 ng	4–5
Photoionization	No	<1 pg	4

*From manufacturer's literature; *Handbook of Instrumental Techniques for Analytical Chemistry,* F. Settle, Ed. Upper Saddle River, NJ: Prentice-Hall, 1997; E. S. Yeung and R. E. Synovec, *Anal. Chem.,* **1986,** *58,* 1237A.
†Mass LODs (limits of detection) are dependent on compound, instrument, and HPLC conditions, but those given are typical values with commercial systems when available.
‡Typical values from the sources above.

can display an entire spectrum as an analyte exits the column. Use of a combination of HPLC and a mass spectrometry detector is currently becoming quite popular. Such liquid chromatography/mass spectrometry systems can identify the analytes exiting from the HPLC column,[2] as discussed in Feature 32-1.

FEATURE 32-1

Liquid Chromatography (LC)/Mass Spectrometry (MS) and LC/MS/MS

The combination of liquid chromatography and mass spectrometry would seem to be an ideal merger of separation and detection. Just as in gas chromatography, a mass spectrometer could identify species as they elute from the chromatographic column. There are major problems, though, in the coupling of these two techniques. A gas-phase sample is needed for mass spectrometry, while the output of the LC column is a solute dissolved in a solvent. As a first step, the solvent must be vaporized. When vaporized, however, the LC solvent produces a gas volume that is 10 to 1000 times greater than the carrier gas in gas chromatography. Hence, most of the solvent must also be removed. There have been several devices developed to solve the problems of solvent removal and LC column interfacing. Today, the most popular approach is to use a low-flow-rate atmospheric pressure ionization technique. The block diagram of a typical LC/MS system is shown in Figure 32F-1. The HPLC system is typically a nanoscale capillary LC system with flow rates in the μL/min range. Alternatively, some interfaces allow flow rates as high as 1 to 2 mL/min, which is typical of conventional HPLC conditions. The most common ionization sources are electrospray ionization and atmospheric pressure chemical ionization (see Section 31A-4). The combination of HPLC and mass spectrometry gives high selectivity, since unresolved peaks can be isolated by monitoring only a selected mass. The LC/MS technique can provide fingerprinting of a particular eluate instead of relying on reten-

[2]See R. Willoughby, E. Sheehan, S. Mitrovich, *A Global View of LC/MS.* Pittsburgh: Global View Publishing, 1998; W. M. A. Niessen, *Liquid Chromatography-Mass Spectrometry,* 2nd Ed. New York: Dekker, 1999.

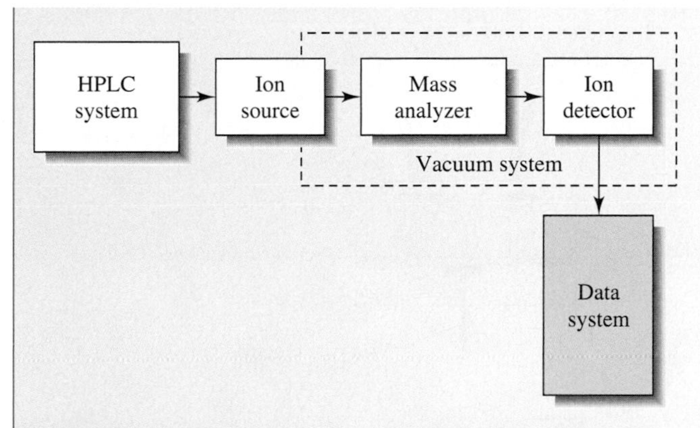

Figure 32F-1 Block diagram of an LC/MS system. The effluent from the LC column is introduced to an atmospheric pressure ionization source, such as an electrospray or chemical ionization. The ions produced are sorted by the mass analyzer and detected by the ion detector.

tion time, as in conventional HPLC. The combination also can give molecular mass and structural information and accurate quantitative analysis.[3]

For some complex mixtures, the combination of LC and MS does not provide enough resolution. In recent years, it has become feasible to couple two or more mass analyzers together in a technique known as tandem mass spectrometry.[4] When combined with LC, the tandem mass spectrometry system is called an LC/MS/MS instrument.[5]

Tandem mass spectrometers are either triple quadrupole systems (the collision cell is also a quadrupole) or quadrupole ion-trap spectrometers. A triple quadrupole mass spectrometry system is shown in Figure 32F-2. Here, the first quadrupole acts as a mass filter to select the ion of interest. This ion is then fragmented by collision with an inert gas in the collision cell. The final quadrupole mass analyzes the fragments produced. The triple quadrupole system can be operated in other modes. For example, if the first quadrupole is operated as a wide mass filter to transmit a wide range of ions and no collision gas is present in the collision cell, the instrument is operating as an LC/MS system. The instrument can be operated by scanning one or both quadrupoles to produce mass spectra of the fragments of ions selected by the first quadrupole as that quadrupole is scanned.

To attain higher resolution than can be achieved with a quadrupole, the final mass analyzer in a tandem MS system can be a time-of-flight mass spectrometer. Sector mass spectrometers can also be combined to give tandem systems. Ion cyclotron resonance and ion-trap mass spectrometers can be operated in such a way as to provide not just two stages of mass analysis, but *n* stages. Such MS^n systems provide the analysis steps sequentially within a single mass analyzer. These have been combined with LC systems in LC/MS^n instruments.

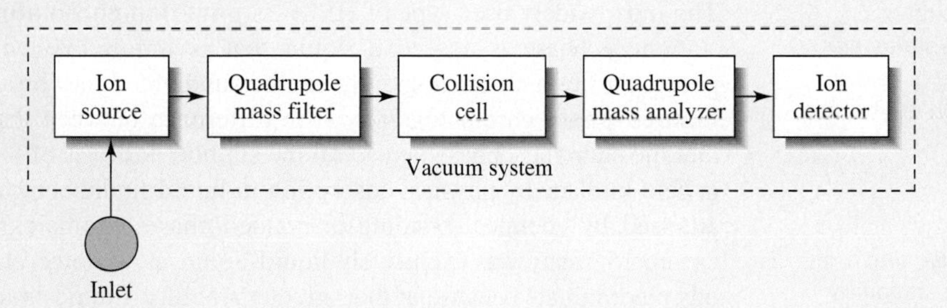

Figure 32F-2 A tandem mass spectrometry system. The ions produced in the source are filtered in the first quadrupole so that only a selected ion passes through to the collision cell. A collision gas in this cell causes fragmentation of the selected ion. The fragment masses are sorted by the quadrupole mass analyzer and detected. Usually, the collision cell is also a quadrupole operated in such a way that the fragment ions are directed into the mass analyzer.

[3]For a review of commercial LC/MS systems, see B. E. Erickson, *Anal. Chem.*, **2000,** *72,* 711A.
[4]For a description of commercial tandem mass spectrometers, see D. Noble, *Anal. Chem.*, **1995,** *67,* 265A.
[5]For recent developments in LC/MS/MS, see R. Thomas, *Spectroscopy,* **2001,** *16,* 28.

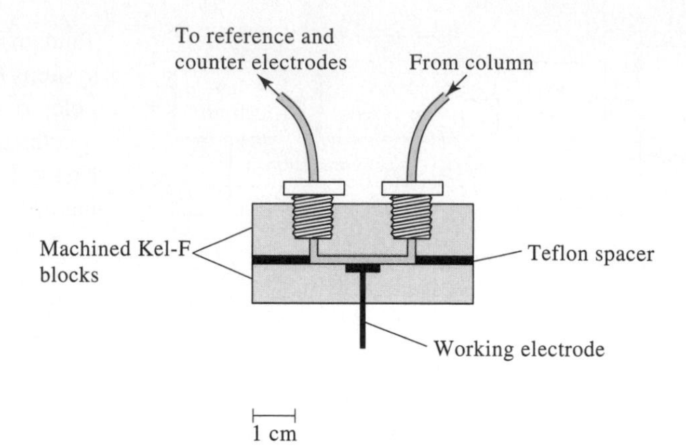

Figure 32-9 Amperometric thin-layer cell for HPLC.

Another detector, which has found considerable application, is based on the changes in the refractive index of the solvent that is caused by analyte molecules. In contrast to most of the other detectors listed in Table 32-1, the refractive index detector is general rather than selective and responds to the presence of all solutes. The disadvantage of this detector is its somewhat limited sensitivity. Several electrochemical detectors that are based on potentiometric, conductometric, and voltammetric measurements have also been introduced. An example of an amperometric detector is shown in Figure 32-9.

32B HIGH-PERFORMANCE PARTITION CHROMATOGRAPHY

In **liquid-liquid partition chromatography,** the stationary phase is a solvent that is held in place by adsorption on the surface of packing particles.

In **liquid-bonded-phase partition chromatography,** the stationary phase is an organic species that is attached to the surface of the packing particles by chemical bonds.

The most widely used type of HPLC is **partition chromatography,** in which the stationary phase is a second liquid that is immiscible with the liquid mobile phase. Partition chromatography can be subdivided into **liquid-liquid** and **liquid-bonded-phase** chromatography. The difference between the two lies in the way that the stationary phase is held on the support particles of the packing. The liquid is held in place by physical adsorption in liquid-liquid chromatography, while it is attached by chemical bonding in bonded-phase chromatography. Early partition chromatography was exclusively liquid-liquid; now, however, bonded-phase methods predominate because of their greater stability. Liquid-liquid packings are today relegated to certain special applications.

32B-1 Bonded-Phase Packings

Most bonded-phase packings are prepared by reaction of an organochlorosilane with the —OH groups formed on the surface of silica particles by hydrolysis in hot, dilute hydrochloric acid. The product is an organosiloxane. The reaction for one such SiOH site on the surface of a particle can be written as

$$-\overset{|}{\underset{|}{Si}}-OH + Cl-\overset{CH_3}{\underset{CH_3}{\overset{|}{\underset{|}{Si}}}}-R \longrightarrow -\overset{|}{\underset{|}{Si}}-O-\overset{CH_3}{\underset{CH_3}{\overset{|}{\underset{|}{Si}}}}-R$$

where R is often a straight-chain octyl or octyldecyl group. Other organic functional groups that have been bonded to silica surfaces include aliphatic amines, ethers, and nitriles, as well as aromatic hydrocarbons. Thus, many different polarities for the bonded stationary phase are available.

Bonded-phase packings have the advantage of markedly greater stability than physically held stationary phases. With the latter, periodic recoating of the solid surfaces is required because the stationary phase is gradually dissolved away in the mobile phase. Furthermore, gradient elution is not practical with liquid-liquid packings, again because of losses by solubility in the mobile phase. The main disadvantage of bonded-phase packings is their somewhat limited sample capacity.

32B-2 Normal- and Reversed-Phase Packings

Two types of partition chromatography are distinguishable based on the relative polarities of the mobile and stationary phases. Early work in liquid chromatography was based on highly polar stationary phases such as triethylene glycol or water; a relatively nonpolar solvent such as hexane or *i*-propyl ether then served as the mobile phase. For historic reasons, this type of chromatography is now called **normal-phase chromatography.** In **reversed-phase chromatography,** the stationary phase is nonpolar, often a hydrocarbon, and the mobile phase is a relatively polar solvent (such as water, methanol, acetonitrile, or tetrahydrofuran).[6]

In **normal-phase partition chromatography,** the stationary phase is polar and the mobile phase nonpolar. In **reversed-phase partition chromatography,** the polarity of these phases is reversed.

In normal-phase chromatography, the *least* polar component is eluted first; *increasing* the polarity of the mobile phase *decreases* the elution time. In contrast, with reversed-phase chromatography, the *most* polar component elutes first, and *increasing* the mobile-phase polarity *increases* the elution time.

◄ In normal-phase chromatography, the least polar analyte is eluted first. In reversed-phase chromatography, the least polar analyte is eluted last.

It has been estimated that more than three quarters of all HPLC separations are currently performed with reversed-phase, bonded, octyl- or octyldecyl-siloxane packings. With such preparations, the long-chain hydrocarbon groups are aligned parallel to one another and perpendicular to the surface of the particle, giving a brush-like, nonpolar, hydrocarbon surface. The mobile phase used with these packings is often an aqueous solution containing various concentrations of such solvents as methanol, acetonitrile, or tetrahydrofuran.

Ion-pair chromatography is a subset of reversed-phase chromatography in which easily ionizable species are separated on reversed-phase columns. In this type of chromatography, an organic salt containing a large organic counter-ion, such as a quaternary ammonium ion or alkyl sulfonate, is added to the mobile phase as an ion-pairing reagent. Two mechanisms of separation are postulated. In the first, the counter-ion forms an uncharged ion pair with a solute ion of opposite charge in the mobile phase. This ion pair then partitions into the nonpolar stationary phase, giving differential retention of solutes based on the affinity of the ion pair for the two phases. Alternatively, the counter-ion is retained strongly by the normally neutral stationary phase and imparts a charge to this phase. Separation of organic solute ions of the opposite charge then occurs by formation of reversible ion-pair complexes, with the more strongly retained solutes forming the strongest complexes with the stationary phase. Some unique separations of both ionic and nonionic compounds in the same sample can be accomplished by

Molecular model of octyldecyl-siloxane.

[6]For a detailed discussion of reversed-phase HPLC, see A. M. Krstulovic and P. R. Brown, *Reversed-Phase High-Performance Liquid Chromatography.* New York: Wiley, 1982.

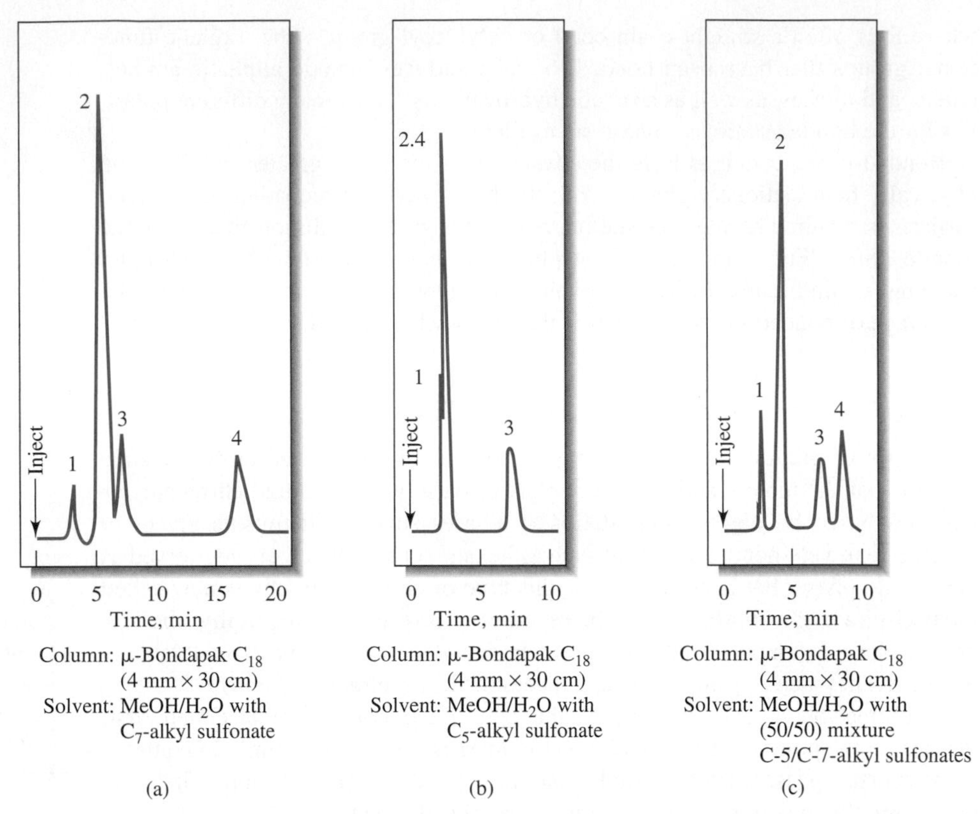

Figure 32-10 Chromatograms illustrating separations of mixtures of ionic and nonionic compounds by ion-pair chromatography. Compounds: (1) niacinamide, (2) pyridoxine, (3) riboflavin, (4) thiamine. At pH 3.5, niacinamide is strongly ionized, while riboflavin is nonionic. Pyridoxine and thiamine are weakly ionized. Column: μ-Bondapak, C_{18}, 4 mm × 30 cm. Mobile phase: (a) $MeOH/H_2O$ with C_7-alkyl sulfonate; (b) $MeOH/H_2O$ with C_5-alkyl sulfonate; (c) $MeOH/H_2O$ with 1:1 mixture of C_5- and C_7-alkyl sulfonate. (Courtesy of Waters Corp., Milford, MA.)

this form of partition chromatography. Figure 32-10 illustrates the separation of ionic and nonionic compounds using alkyl sulfonates of various chain lengths as ion-pairing agents. Note that a mixture of C_5- and C_7-alkyl sulfonates gives the best separation results.

► The order of polarities of common mobile phase solvents are water > acetonitrile > methanol > ethanol > tetrahydrofuran > propanol > cyclohexane > hexane.

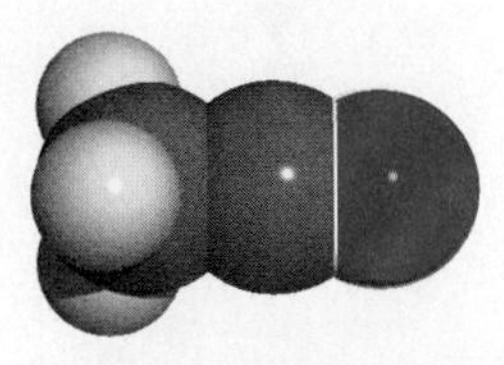

Molecular model of acetonitrile. Acetonitrile ($CH_3C{\equiv}N$) is a widely used organic solvent. Its use as an LC mobile phase stems from its being more polar than methanol but less polar than water.

32B-3 Choice of Mobile and Stationary Phases

Successful partition chromatography requires a proper balance of intermolecular forces among the three participants in the separation process—the analyte, the mobile phase, and the stationary phase. These intermolecular forces are described qualitatively in terms of the relative polarity possessed by each of the three components. In general, the polarities of common organic functional groups in increasing order are aliphatic hydrocarbons < olefins < aromatic hydrocarbons < halides < sulfides < ethers < nitro compounds < esters ≈ aldehydes ≈ ketones < alcohols ≈ amines < sulfones < sulfoxides < amides < carboxylic acids < water.

As a rule, most chromatographic separations are achieved by matching the polarity of the analyte to that of the stationary phase; a mobile phase of considerably different polarity is then used. This procedure is generally more successful

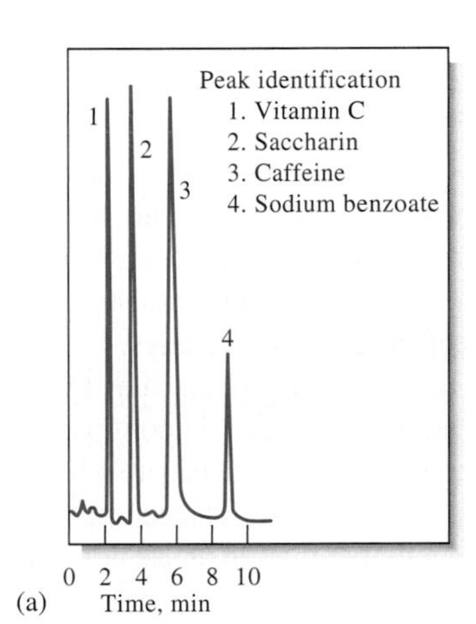

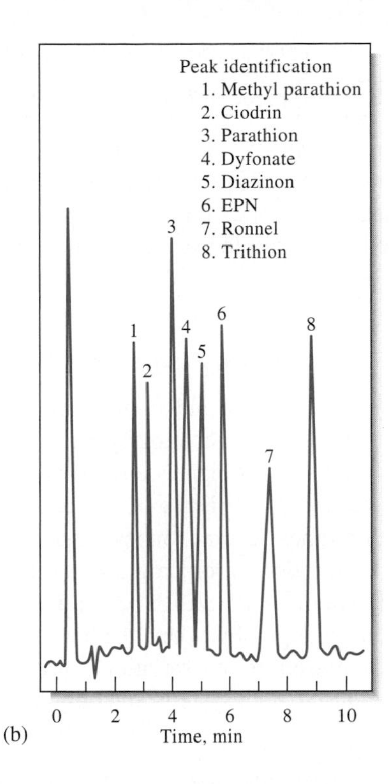

Figure 32-11 Typical applications of bonded-phase chromatography. (a) Soft-drink additives. Column: 4.6 × 250 mm packed with polar (nitrile) bonded-phase packing. Isocratic elution with 6% HOAc/94% H_2O. Flow rate: 1.0 mL/min. (Courtesy of BTR Separations, a DuPont ConAgra affiliate.) (b) Organophosphate insecticides. Column 4.5 × 250 mm packed with 5-μm C_8 bonded-phase particles. Gradient elution: 67% CH_3OH/33% H_2O to 80% CH_3OH/20% H_2O. Flow rate: 2 mL/min. Both used 254-nm UV detectors.

than one in which the polarities of the analyte and the mobile phase are matched but are different from that of the stationary phase. Here, the stationary phase often cannot compete successfully for the sample components; retention times then become too short for practical application. At the other extreme is the situation where the polarities of the analyte and stationary phase are too much alike; here, retention times become inordinately long.

32B-4 Applications

Figure 32-11 illustrates typical applications of bonded-phase partition chromatography for separating soft-drink addititives and organophosphate insecticides. Table 32-2 further illustrates the variety of samples to which the technique is applicable.

TABLE 32-2
Typical Applications of High-Performance Partition Chromatography

Field	Typical Mixtures Separated
Pharamaceuticals	Antibiotics, sedatives, steroids, analgesics
Biochemicals	Amino acids, proteins, carbohydrates, lipids
Food products	Artificial sweeteners, antioxidants, aflatoxins, additives
Industrial chemicals	Condensed aromatics, surfactants, propellants, dyes
Pollutants	Pesticides, herbicides, phenols, polychlorinated biphenyls (PCBs)
Forensic chemistry	Drugs, poisons, blood alcohol, narcotics
Clinical medicine	Bile acids, drug metabolites, urine extracts, estrogens

32C HIGH-PERFORMANCE ADSORPTION CHROMATOGRAPHY

In **adsorption chromatography**, analyte species are adsorbed onto the surface of a polar packing.

The pioneering work in chromatography was based on adsorption of analyte species on a solid surface. Here, the stationary phase is the surface of a finely divided polar solid. With such a packing, the analyte competes with the mobile phase for sites on the surface of the packing, and retention is the result of adsorption forces.

32C-1 Stationary and Mobile Phases

Finely divided silica and alumina are the only stationary phases that find extensive use in adsorption chromatography. Silica is preferred for most (but not all) applications because of its higher sample capacity and its wider range of useful forms. The adsorption characteristics of the two substances parallel one another. For both, retention times become longer as the polarity of the analyte increases.

► In adsorption chromatography, the mobile plase is usually an organic solvent or a mixture of organic solvents; the stationary phase is finely divided particles of silica or alumina.

In adsorption chromatography, the only variable that affects the distribution coefficient of analytes is the composition of the mobile phase (in contrast to partition chromatography, where the polarity of the stationary phase can also be varied). Fortunately, enormous variations in retention and thus resolution accompany variations in the solvent system, and only rarely is a suitable mobile phase not available.

32C-2 Applications of Adsorption Chromatography

Currently, liquid-solid HPLC is used extensively for the separations of relatively nonpolar, water-insoluble organic compounds with molecular masses that are less than about 5000. A particular strength of adsorption chromatography, which is not shared by other methods, is its ability to resolve isomeric mixtures such as meta and para substituted benzene derivatives.

32D ION-EXCHANGE CHROMATOGRAPHY

In Section 30D, we described some of the applications of ion-exchange resins to analytical separations. In addition, these materials are useful as stationary phases for liquid chromatography, where they are used to separate charged species.[7] In most cases, conductivity measurements are used to detect eluents.

Two types of ion chromatography are currently in use: **suppressor-based** and **single-column.** They differ in the method used to prevent the conductivity of the eluting electrolyte from interfering with the measurement of analyte conductivities.

32D-1 Ion Chromatography Based on Suppressors

► The conductivity detector is well suited for ion chromatography.

Conductivity detectors have many of the properties of the ideal detector. They can be highly sensitive, they are universal for charged species, and, as a general rule, they respond in a predictable way to concentration changes. Furthermore, such detectors are simple to operate, inexpensive to construct and maintain, easy to

[7] For brief reviews of ion chromatography, see J. S. Fritz, *Anal. Chem.*, **1987,** *59,* 335A; P. R. Haddad, *Anal. Chem.*, **2001,** *73,* 266A. For a detailed description of the method, see H. Small, *Ion Chromatography.* New York: Plenum Press, 1989; D. T. Gjerde and J. S. Fritz, *Ion Chromatography,* 3rd ed. New York: A. Heuthig, 2000.

miniaturize, and ordinarily give prolonged, trouble-free service. The only limitation to the use of conductivity detectors, which delayed their general application to ion chromatography until the mid-1970s, was the high electrolyte concentrations required to elute most analyte ions in a reasonable time. As a consequence, the conductivity from the mobile-phase components tends to swamp that from the analyte ions, thus greatly reducing the detector sensitivity.

In 1975, the problem created by the high conductance of eluents was solved by the introduction of an **eluent suppressor column** immediately following the ion-exchange column.[8] The suppressor column is packed with a second ion-exchange resin that effectively converts the ions of the eluting solvent to a molecular species of limited ionization without affecting the conductivity due to analyte ions. For example, when cations are being separated and determined, hydrochloric acid is chosen as the eluting reagent, and the suppressor column is an anion-exchange resin in the hydroxide form. The product of the reaction in the suppressor is water. That is,

$$H^+(aq) + Cl^-(aq) + resin^+OH^-(s) \rightarrow resin^+Cl^-(s) + H_2O$$

The analyte cations are not retained by this second column.

For anion separations, the suppressor packing is the acid form of a cation-exchange resin, and sodium bicarbonate or carbonate is the eluting agent. The reaction in the suppressor is

$$Na^+(aq) + HCO_3^-(aq) + resin^-H^+(s) \rightarrow resin^-Na^+(s) + H_2CO_3(aq)$$

The largely undissociated carbonic acid does not contribute significantly to the conductivity.

An inconvenience associated with the original suppressor columns was the need to regenerate them periodically (typically, every 8 to 10 hr) to convert the packing back to the original acid or base form. Recently, however, micromembrane suppressors that operate continuously have become available.[9] For example, when sodium carbonate or bicarbonate is to be removed, the eluent is passed over a series of ultra-thin cation-exchange membranes that separate it from a stream of acidic regenerating solution that flows continuously in the opposite direction. The sodium ions from the eluent exchange with hydrogen ions on the inner surface of the exchanger membrane and then migrate to the other surface for exchange with hydrogen ions from the regenerating reagent. Hydrogen ions from the regeneration solution migrate in the reverse direction, thus preserving electrical neutrality.

Figures 32-12 and 32-13 show applications of ion chromatography based on a suppressor column and conductometric detection. In each, the ions were present in the parts-per-million range; the sample size was 50 μL in one case and 20 μL in the other. The method is particularly important for anion analysis because there is no other rapid and convenient method of handling mixtures of this type.

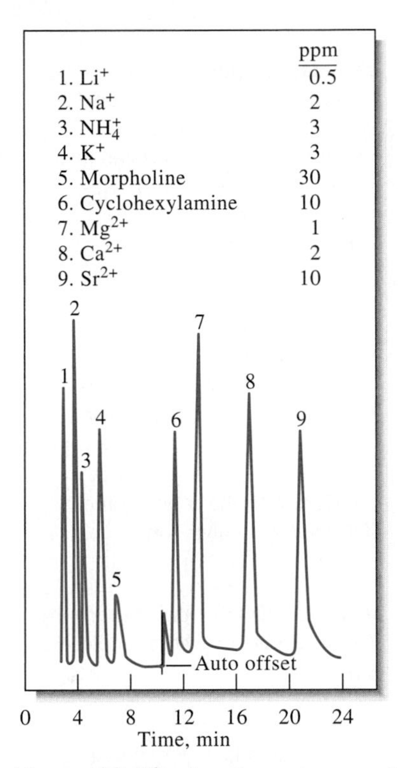

Figure 32-12 Ion chromatogram of a mixture of cations. (Courtesy of Dionex, Sunnyvale, CA.)

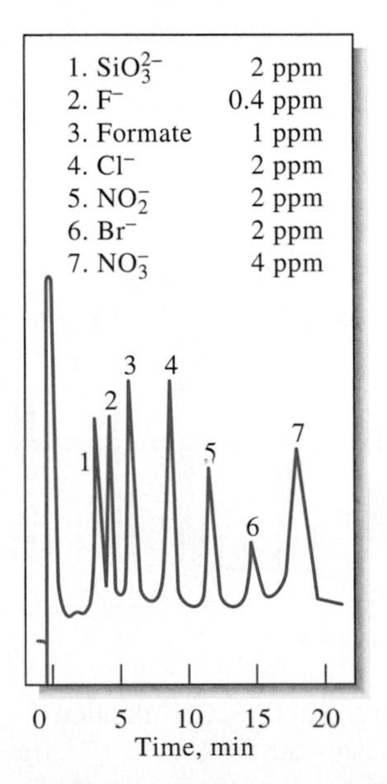

Figure 32-13 Ion chromatogram of a mixture of anions. (Courtesy of Dionex, Sunnyvale, CA.)

32D-2 Single-Column Ion Chromatography

Recently, commercial ion-chromatography instrumentation that requires no suppressor column has become available. This approach depends on the small differences in conductivity between sample ions and the prevailing eluent ions. To

[8] H. Small, T. S. Stevens, and W. C. Bauman, *Anal. Chem.*, **1975,** *47,* 1801.

[9] For a description of this device, see G. O. Franklin, *Amer. Lab.*, **1985** (3), 71.

In **suppressor-based ion chromatography,** the ion-exchange column is followed by a **suppressor column** or by a **suppressor membrane** that converts an ionic eluent into a non-ionic species that does not interfere with the conductometric detection of analyte ions.

In **single-column ion-exchange chromatography,** analyte ions are separated on a low-capacity ion exchanger by means of a low-ionic-strength eluent that does not interfere with the conductometric detection of analyte ions.

In **size-exclusion chromatography,** fractionation is based on molecular size.

Gel filtration is a type of size-exclusion chromatography in which the packing is hydrophilic. It is used to separate polar species.

Gel permeation is a type of size-exclusion chromatography in which the packing is hydrophilic. It is used to separate nonpolar species.

amplify these differences, low-capacity exchangers are used that permit elution with solutions with low electrolyte concentrations. Furthermore, eluents of low conductivity are chosen.[10]

Single-column ion chromatography offers the advantage of not requiring special equipment for suppression. It is a somewhat less sensitive method of determining anions than are suppressor column methods, however.

32E SIZE-EXCLUSION CHROMATOGRAPHY

Size-exclusion, or gel, chromatography is the newest of the liquid chromatographic procedures. It is a powerful technique that is particularly applicable to high-molecular-weight species.[11]

32E-1 Column Packings

Packings for size-exclusion chromatography consist of small (~10 μm) silica or polymer particles containing a network of uniform pores into which solute and solvent molecules can diffuse. While in the pores, molecules are effectively trapped and removed from the flow of the mobile phase. The average residence time of analyte molecules depends on their effective size. Molecules that are significantly larger than the average pore size of the packing are excluded and thus suffer no retention; that is, they travel through the column at the rate of the mobile phase. Molecules that are appreciably smaller than the pores can penetrate throughout the pore maze and are thus entrapped for the greatest time; they are last to elute. Between these two extremes are intermediate-size molecules whose average penetration into the pores of the packing depends on their diameters. The fractionation that occurs within this group is directly related to molecular size and, to some extent, molecular shape. Note that size-exclusion separations differ from the other chromatographic procedures in the respect that no chemical or physical interactions between analytes and the stationary phase are involved. Indeed, every effort is made to avoid such interactions because they lead to impaired column efficiencies.

Numerous size-exclusion packings are on the market. Some are hydrophilic for use with aqueous mobile phases; others are hydrophobic and are used with nonpolar organic solvents. Chromatography based on the hydrophilic packings is sometimes called **gel filtration,** while techniques based on hydrophobic packings are termed **gel permeation.** With both types of packings, many pore diameters are available. Ordinarily, a given packing will accommodate a 2- to 2.5-decade range of molecular weight. The average molecular weight suitable for a given packing may be as small as a few hundred or as large as several million.

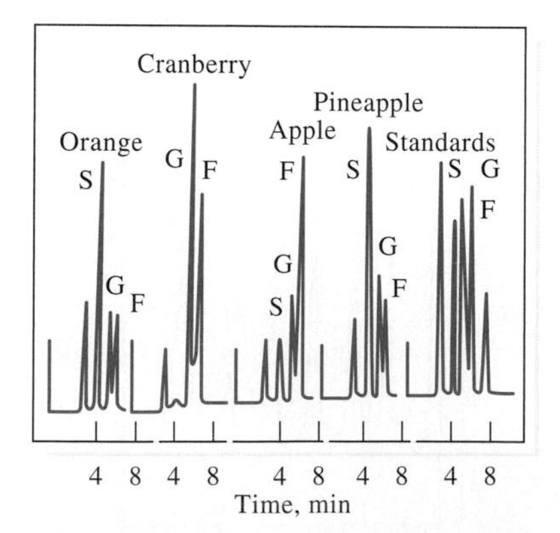

Figure 32-14 Gel-filtration chromatogram for glucose (G), fructose (F), and sucrose (S) in canned juices. (Courtesy of BTR Separations, a DuPont ConAgra affiliate.)

32E-2 Applications

Figures 32-14 and 32-15 illustrate typical applications of size-exclusion chromatography. In the chromatograms shown in Figure 32-14, a hydrophilic packing was used to exclude molecular weights greater than 1000. Several sugars in canned

[10]See R. M. Becker, *Anal Chem.,* **1980,** *52,* 1510; J. R. Benson, *Amer. Lab.,* **1985,** (6), 30; T. Jupille, *Amer. Lab.,* **1986,** (5), 114.

[11]For monographs on this subject, see *Size Exclusion Chromatography,* B. J. Hunt and S. R. Holding, Eds. New York: Chapman and Hall, 1988; *Handbook of Size Exclusion Chromatography,* C. S. Wu, Ed. New York: Dekker, 1995; *Column Handbook for Size Exclusion Chromatography,* C. S. Wu, Ed. San Diego: Academic Press, 1999.

juices were separated. The chromatogram in Figure 32-15 was obtained with a hydrophobic packing in which the eluent was tetrahydrofuran. The sample was a commercial epoxy resin in which each monomer unit had a molecular weight of 280 (n = number of monomer units).

Another important application of size-exclusion chromatography is the rapid determination of the molecular mass or the molecular mass distribution of large polymers or natural products. The key to such determinations is an accurate molecular mass calibration. Calibrations can be accomplished by means of standards of known molecular mass (peak position method) or by the "universal calibration method." The latter relies on the principle that the product of the intrinsic molecular viscosity η and molecular mass $\mathcal{M}$ is proportional to hydrodynamic volume (effective volume including solvation sheath). Ideally, molecules are separated in size-exclusion chromatography according to hydrodynamic volume. Hence, a universal calibration curve can be obtained by plotting log $[\eta\mathcal{M}]$ versus the retention volume, V_r, where $V_r = t_r \times F$. Alternatively, absolute calibration can be achieved by using a molar mass-sensitive detector such as a low-angle, light-scattering detector.

Feature 32-2 illustrates how size-exclusion chromatography can be used in the separation of fullerenes.

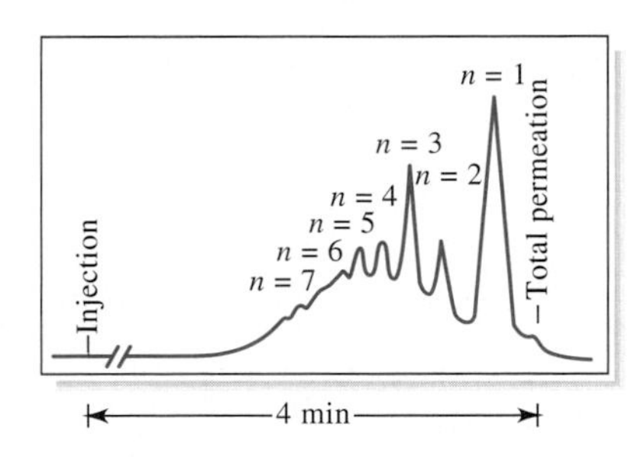

Figure 32-15 Gel-permeation separation of components in an epoxy resin. (Courtesy of BTR Separations, a DuPont ConAgra affiliate.)

FEATURE 32-2

Buckyballs: The Chromatographic Separation of Fullerenes

Our ideas about the nature of matter are often profoundly influenced by chance discoveries. No event in recent memory has captured the imagination of both the scientific community and the public as much as the serendipitous discovery in 1985 of the soccerball-shaped molecule C_{60}. This molecule, illustrated in Figure 32F-3, its cousin C_{70}, and other similar molecules discovered since 1985 are called *fullerenes,* or, more commonly, *buckyballs.*[12] The compounds are named in honor of the famous architect, R. Buckminster Fuller, who designed many geodesic dome buildings having the same hexagonal/pentagonal structure as buckyballs. Since their discovery, thousands of research groups throughout the world have studied various chemical and physical properties of these highly stable molecules. They represent a third allotropic form of carbon besides graphite and diamond.

The preparation of buckyballs is almost trivial. When an ac arc is established between two carbon electrodes in a flowing helium atmosphere, the soot that is collected is rich in C_{60} and C_{70}. Although the preparation was easy, the separation and purification of more than a few milligrams of C_{60} proved tedious and expensive. Relatively

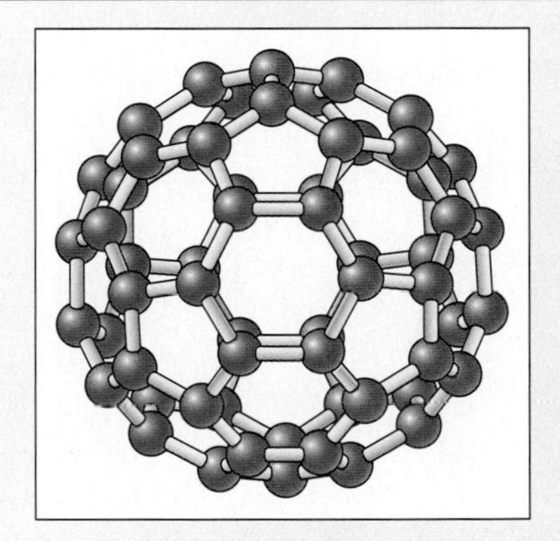

Figure 32F-3 Buckminster fullerene, C_{60}.

large quantities of buckyballs have been separated using size-exclusion chromatography.[13] Fullerenes are extracted from soot, prepared as mentioned above, and injected on a 199 mm × 30 cm, 500-Å Ultrastyragel column (Waters Corp., Milford, MA), using toluene as the mobile phase and ultraviolet/visible detection following separation. A typical chromatogram is shown in Figure 32F-4. The

(continued)

[12] R. F. Curl and R. E. Smalley, *Sci. Am.,* **1991,** *265* (4), 54.

[13] M. S. Meier and J. P. Selegue, *J. Org. Chem.,* **1992,** *57,* 1924; A. Gugel and K. Mullen, *J. Chromatogr.,* **1993,** *628,* 23.

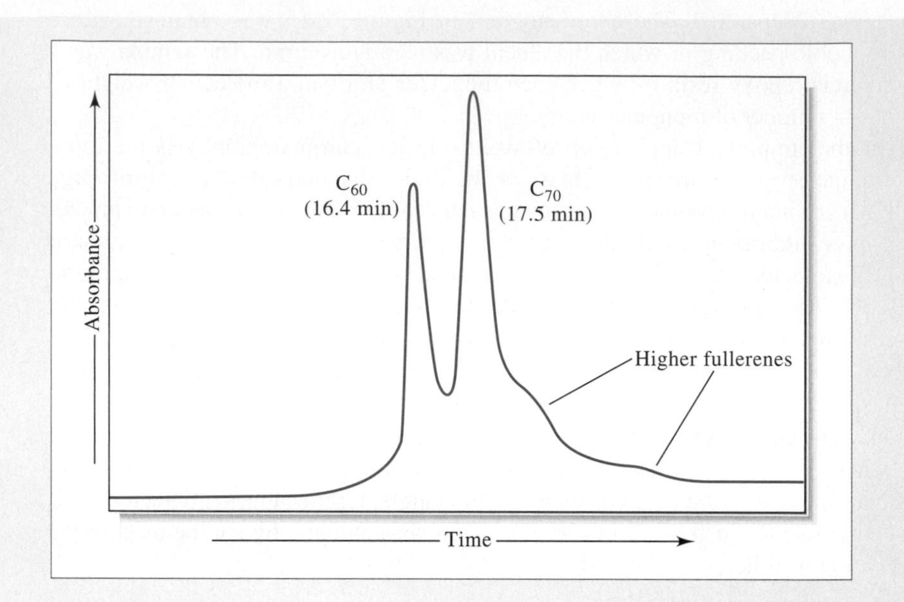

Figure 32F-4 Separation of fullerenes.

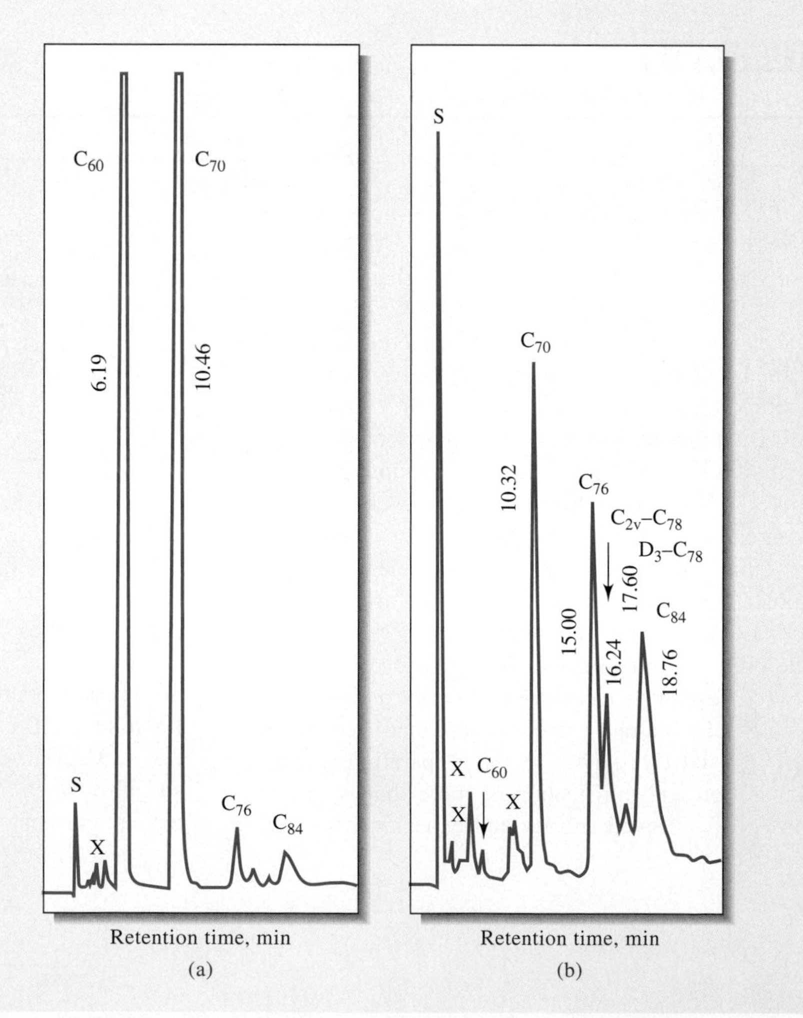

Figure 32F-5 Chromatograms of whole soot extract (a) and a higher fullerene fraction (b) obtained with a polymeric ODS column and an acetonitrile:toluene mobile phase. (Reprinted with permission from F. Diederich and R. L. Whetten, *Acc. Chem. Res.*, **1995,** *25,* 121. Copyright 1995 American Chemical Society.)

peaks in the chromatogram are labeled with their identities and retention times.

Note that C_{60} elutes before C_{70} and the higher fullerenes. This is contrary to what we expect; the smallest molecule, C_{60}, should be retained more strongly than C_{70} and the higher fullerenes. It has been suggested that the interaction between the solute molecules and the gel is on the surface of the gel rather than in its pores. Since C_{70} and the higher fullerenes have larger surface areas than C_{60}, the higher fullerenes are retained more strongly on the surface of the gel and are thus eluted after C_{60}. With an automated apparatus, this method of separation may be used to prepare several grams of 99.8% pure C_{60} from 5 to 10 g of a C_{60} to C_{70} mixture in a 24-hour period. These quantities of C_{60} can then be used to prepare and study the chemistry and physics of derivatives of this interesting and unusual form of carbon.

Recently, in HPLC separations of fullerenes, the octadecyl silica (ODS) bonded stationary phase has been extensively used.[14] Both polymeric and monomeric ODS phases have been used, and these provide a higher selectivity than other phases. Figure 32F-5 shows the preparative separation of whole soot extract and a higher fullerenes fraction on a polymeric ODS column. These were among the first separations of the individual higher fullerenes. Note the excellent resolution compared with the size-exclusion separation of Figure 32F-4.

32F AFFINITY CHROMATOGRAPHY

Affinity chromatography involves covalently bonding a reagent, called an **affinity ligand,** to a solid support.[15] Typical affinity ligands are antibodies, enzyme inhibitors, or other molecules that reversibly and selectively bind to analyte molecules in the sample. When the sample passes through the column, only the molecules that selectively bind to the affinity ligand are retained. Molecules that do not bind pass through the column with the mobile phase. After the undesired molecules are removed, the retained analytes can be eluted by changing the mobile-phase conditions.

The stationary phase for affinity chromatography is a solid such as agarose or a porous glass bead to which the affinity ligand is immobilized. The mobile phase in affinity chromatography has two distinct roles. First, it must support the strong binding of the analyte molecules to the ligand. Second, once the undesired species are removed, the mobile phase must weaken or eliminate the analyte-ligand interaction so that the analyte can be eluted. Often, changes in pH or ionic strength are used to change the elution conditions during the two stages of the process.

Affinity chromatography has the major advantage of extraordinary specificity. The primary use is the rapid isolation of biomolecules during preparative work.

32G CHIRAL CHROMATOGRAPHY

Tremendous advances have been made in recent years in separating compounds that are nonsuperimposable mirror images of each other, called **chiral compounds.**

[14]K. Jinno, H. Ohta, and Y. Sato, in *Separation of Fullerenes by Liquid Chromatography,* K. Jinno, Ed., Ch. 3. London: Royal Society of Chemistry, 1999.

[15]For details on affinity chromatography, see R. R. Walton, *Anal. Chem.,* **1985,** *57,* 1097A; *Handbook of Affinity Chromatography,* T. Kline, Ed. New York: Dekker, 1993; *Analytical Affinity Chromatography,* I. M. Chaiken, Ed. Boca Raton, FL: CRC Press, 1987.

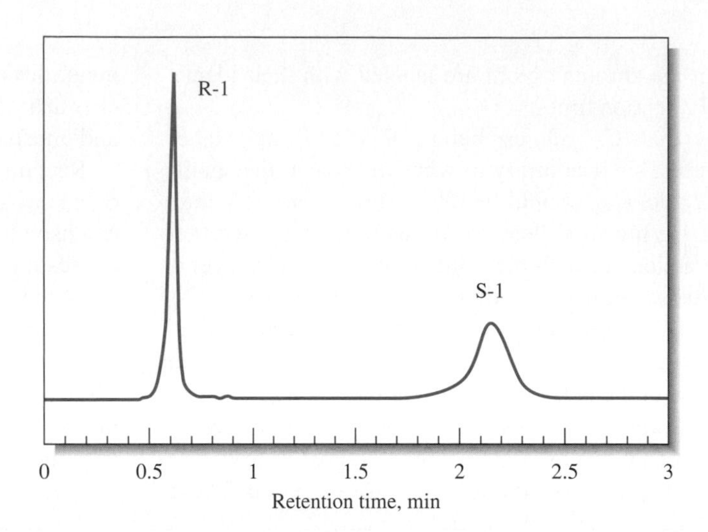

Figure 32-16 Chromatogram of a racemic mixture of *N*-(1-Naphthyl)leucine ester **1** on a dinitrobenzene-leucine chiral stationary phase. The *R* and *S* enantiomers are seen to be well separated. Column: 4.6 × 50 mm; mobile phase, 20% 2-propanol in hexane; flow rate: 1.2 mL/min; UV detector at 254 nm. (Reprinted with permission from L. H. Bluhm, Y. Wang, and T. Li, *Anal. Chem.*, **2000,** *72,* 5201. Copyright 2000 American Chemical Society.)

Such mirror images are called **enantiomers.** Either chiral mobile-phase additives or chiral stationary phases are required for these separations.[16] Preferential complexation between the chiral resolving agent (additive or stationary phase) and one of the isomers results in a separation of the enantiomers. The **chiral resolving agent** must have chiral character itself to recognize the chiral nature of the solute.

A **chiral resolving agent** is a chiral mobile-phase additive or a chiral stationary phase that preferentially complexes one of the enantiomers.

Chiral stationary phases have received the most attention.[17] Here, a chiral agent is immobilized on the surface of a solid support. Several different modes of interaction can occur between the chiral resolving agent and the solute.[18] In one type, the interactions are due to attractive forces such as those between π bonds, hydrogen bonds, or dipoles. In another type, the solute can fit into chiral cavities in the stationary phase to form inclusion complexes. No matter what the mode, the ability to separate these very closely related compounds is of extreme importance in many fields. Figure 32-16 shows the separation of a racemic mixture of an ester on a chiral stationary phase. Note the excellent resolution of the *R* and *S* enantiomers.

32H COMPARISON OF HIGH-PERFORMANCE LIQUID CHROMATOGRAPHY AND GAS CHROMATOGRAPHY

Table 32-3 provides a comparison between high-performance liquid chromatography and gas-liquid chromatography. When either is applicable, gas-liquid chromatography offers the advantage of speed and simplicity of equipment. On the other hand, high-performance liquid chromatography is applicable to nonvolatile substances (including inorganic ions) and thermally unstable materials, whereas gas-liquid chromatography is not. Often, the two methods are complementary.

[16] *Chiral Separations: Applications and Technology,* S. Ahuja, Ed. Washington: American Chemical Society, 1996; S. Ahuja, *Chiral Separations by Chromatography.* New York: Oxford University Press, 2000.

[17] For a recent review of chiral stationary phases, see D. W. Armstrong and B. Zhang, *Anal. Chem.,* **2001,** *73,* 557A.

[18] For a review of chiral interactions, see M. C. Ringo and C. E. Evans, *Anal. Chem.,* **1998,** *70,* 315A.

TABLE 32-3

Comparison of High-Performance Liquid Chromatography and Gas-Liquid Chromatography

Characteristics of both methods
- Efficient, highly selective, widely applicable
- Only small sample required
- May be nondestructive of sample
- Readily adapted to quantitative analysis

Advantages of HPLC
- Can accommodate nonvolatile and thermally unstable compounds
- Generally applicable to inorganic ions

Advantages of GC
- Simple and inexpensive equipment
- Rapid
- Unparalleled resolution (with capillary columns)
- Easily interfaced with mass spectrometry

Spreadsheet Summary Chapter 15 of *Applications of Microsoft® Excel in Analytical Chemistry* begins with an exercise treating the resolution of overlapped Gaussian peaks. The overlapped chromatogram, the response, is modeled as the sum of Gaussian curves. Initial estimates are made for the model parameters. Excel calculates the residuals, the difference between the response and the model, and the sum of the squares of the residuals. Excel's Solver is then used to minimize the sum of the squares of the residuals, while displaying the results of each iteration.

WEB WORKS

Connect to **http://chemistry.brookscole.com/skoogfac/**. From the Chapter Resources menu, choose Web Works and locate the Chapter 32 section. Find the link to LC-GC magazine. From the LC-GC home page, search for articles on LC/MS. Find an article, written in 2001, that compares mass analyzers for LC/MS applications. What are the most common ionization sources used for LC/MS? Describe any differences in mass range and mass resolution between quadrupole, time-of-flight, and ion-trap (Fourier transform) mass analyzers. Do these three mass analyzers show any differences in qualitative and quantitative analysis?

QUESTIONS AND PROBLEMS

32-1. List the types of substances to which each of the following chromatographic methods is most applicable:
- *(a) gas-liquid
- (b) liquid partition
- *(c) ion-exchange
- (d) liquid adsorption
- *(e) gel permeation
- (f) gel filtration
- *(g) gas-solid

32-2. Define
- *(a) isocratic elution.
- (b) gradient elution.
- *(c) stop-flow injection.

(d) reversed-phase packing.
*(e) normal-phase packing.
(f) ion pairing chromatography.
*(g) ion chromatography.
(h) eluent suppressor column.
*(i) gel filtration.
(j) gel permeation.

32-3. Indicate the order in which the following compounds would be eluted from an HPLC column containing a reversed-phase packing:
*(a) benzene, diethyl ether, *n*-hexane
(b) acetone, dichloroethane, acetamide

32-4. Indicate the order of elution of the following compounds from a normal-phase packed HPLC column:
*(a) ethyl acetate, acetic acid, dimethylamine
(b) propylene, hexane, benzene, dichlorobenzene

***32-5.** Describe the fundamental difference between adsorption and partition chromatography.

32-6. Describe the fundamental difference between ion-exchange and size-exclusion chromatography.

***32-7.** Describe the difference between gel-filtration and gel-permeation chromatography.

32-8. What types of species can be separated by HPLC but not by GC?

***32-9.** Describe the various kinds of pumps used in high-performance liquid chromatography. What are the advantages and disadvantages of each?

32-10. Describe the differences between single-column and suppressor-column ion chromatography.

***32-11.** Mass spectrometry is an extremely versatile detection system for gas chromatography. Interfacing an HPLC system to a mass spectrometer is a much more difficult task, however. Describe the major reasons why it is more difficult to combine HPLC with mass spectrometry than it is to combine GC with mass spectrometry.

32-12. Which of the GC detectors in Table 31-1 are suitable for HPLC? Why are some of these unsuitable for HPLC?

***32-13.** The ideal detector for GC is described in Section 31A-4. Which of the eight characteristics of an ideal GC detector are applicable to HPLC detectors? What additional characteristics would be added to describe the ideal HPLC detector?

32-14. Although temperature does not have nearly the effect on HPLC separations that it has on GC separations, it nonetheless can play an important role. Discuss how and why temperature might or might not influence the following separations:
(a) a reversed-phase chromatographic separation of a steroid mixture
(b) an adsorption chromatographic separation of a mixture of closely related isomers

***32-15.** Two components in an HPLC separation have retention times that differ by 15 s. The first peak elutes in 9.0 min, and the peak widths are approximately equal. The dead time t_M was 65 s. Use a spreadsheet to find the minimum number of theoretical plates needed to achieve the following resolution, R_s, values: 0.50, 0.75, 0.90, 1.0, 1.10, 1.25, 1.50, 1.75, 2.0, 2.5. How would the results change if peak 2 were twice as broad as peak 1?

32-16. An HPLC method was developed for the separation and determination of ibuprofen in rat plasma samples as part of a study of the time course of the drug in laboratory animals. Several standards were chromatographed, and these results were obtained:

Ibuprofen Concentration, μg/mL	Relative Peak Area
0.5	5.0
1.0	10.1
2.0	17.2
3.0	19.8
6.0	39.7
8.0	57.3
10.0	66.9
15.0	95.3

Next, a 10 mg/kg sample of ibuprofen was administered orally to a laboratory rat. Blood samples were drawn at various times after administration of the drug and subjected to HPLC analysis. The following results were obtained:

Time, hr	Peak Area
0	0
0.5	91.3
1.0	80.2
1.5	52.1
2.0	38.5
3.0	24.2
4.0	21.2
6.0	18.5
8.0	15.2

Find the concentration of ibuprofen in the blood plasma for each of the times given and plot the concentration versus time. On a percentage basis, during what half-hour period (1st, 2nd, 3rd, etc.) is most of the ibuprofen lost?

32-17. Challenge Problem. Assume for simplicity that the HPLC plate height, H, can be given by Equation 30-27 as

$$H = \frac{B}{u} + C_S u + C_M u = \frac{B}{u} + Cu$$

where $C = C_S + C_M$.

(a) By using calculus to find the minimum H, show that the optimum velocity u_{opt} can be expressed as

$$u_{opt} = \sqrt{\frac{B}{C}}$$

(b) Show that this leads to a minimum plate height H_{min}, given by

$$H_{min} = 2\sqrt{BC}$$

(c) Under some conditions for chromatography, C_S is negligible compared with C_M. For packed LC columns, C_M is given by

$$C_M = \frac{\omega d_p^2}{D_M}$$

where ω is a dimensionless constant, d_p is the particle size of the column packing, and D_M is the diffusion coefficient in the mobile phase. The B coefficient can be expressed as

$$B = 2\gamma D_M$$

where γ is also a dimensionless constant. Express u_{opt} and H_{min} in terms of D_M, d_p, and the dimensionless constants γ and ω.

(d) If the dimensionless constants are on the order of unity, show that u_{opt} and H_{min} can be expressed as

$$u_{opt} \approx \frac{D_M}{d_p} \quad \text{and} \quad H_{min} \approx d_p$$

(e) Under the conditions in part (d), how could the plate height be reduced by one third? What would happen to the optimum velocity under these conditions? What would happen to the number of theoretical plates N for the same-length column?

(f) For the conditions in part (e), how could you maintain the same number of theoretical plates while reducing the plate height by one third?

(g) The preceding discussion assumes that band broadening occurs within the column. Name two sources of extra-column band broadening that might also contribute to the overall width of LC peaks.

CHAPTER 33

Miscellaneous Separation Methods

© AFP/Corbis

Capillary electrophoresis (CE) has been assuming an increasingly important role in forensic DNA identification. In the World Trade Center disaster, the materials collected at the site were trucked and barged to the Fresh Kills Landfill in the center of Staten Island. Human remains were then segregated and used to obtain DNA evidence. Capillary electrophoresis was often the tool of choice in the identification process. CE is particularly useful when only small amounts of sample are available and when samples may have degraded with time. CE has been used to identify DNA in bone, blood, semen, saliva, and hair.

This chapter treats several separation methods that are not readily classified, including supercritical fluid chromatography, paper chromatography, capillary electrophoresis, capillary electrochromatography, and field-flow fractionation. The use of CE for DNA sequencing is the subject of a feature in the electrophoresis section of this chapter.

In this chapter, we discuss several additional methods of performing analytical separations: supercritical-fluid chromatography, thin-layer and paper chromatography, capillary electrophoresis, capillary electrochromatography, and field-flow fractionation.

33A SUPERCRITICAL-FLUID CHROMATOGRAPHY

Supercritical-fluid chromatography (SFC), in which the mobile phase is a supercritical fluid, is a hybrid of gas and liquid chromatography that combines some of the best features of each. For certain applications, it appears to be clearly superior to both gas-liquid and high-performance liquid chromatography.[1]

A **supercritical fluid** is the physical state of a substance held above its critical temperature.

The **critical temperature** is the temperature above which a substance cannot be liquified.

33A-1 Important Properties of Supercritical Fluids

A **supercritical fluid** is formed whenever a substance is heated above its **critical temperature.** Above the critical temperature, a substance can no longer be

[1]T. L. Chester and J. D. Pinkston, *Anal. Chem.*, **2002,** *74,* 2901; T. L. Chester and J. D. Pinkston, *Anal. Chem.*, **2000,** *72,* 129R; T. L. Chester, J. D. Pinkston, and D. B. Raynie, *Anal. Chem.*, **1998,** *70,* 301R; K. Anton and C. Berger, Eds., *Supercritical Fluid Chromatography with Packed Columns. Techniques and Applications.* New York: Dekker, 1998; M. Caude and D. Thiebaut, Eds., *Practical Supercritical Fluid Chromatography and Extraction.* Amsterdam: Harwood, 2000.

TABLE 33-1

Comparison of Properties of Supercritical Fluids, Liquids, and Gases*

Property	Gas (STP)	Supercritical Fluid	Liquid
Density, g/cm^3	$(0.6–2) \times 10^{-3}$	0.2–0.5	0.6–2
Diffusion coefficient, cm^2/s	$(1–4) \times 10^{-1}$	$10^{-3}–10^{-4}$	$(0.2–2) \times 10^{-5}$
Viscosity, $g\ cm^{-1}\ s^{-1}$	$(1–3) \times 10^{-4}$	$(1–3) \times 10^{-4}$	$(0.2–3) \times 10^{-2}$

*All data order of magnitude only.

condensed to a liquid by simply applying pressure. For example, carbon dioxide is a supercritical fluid at temperatures above 31°C. In this state, the molecules of carbon dioxide act independently of one another, just as they do in a gas.

As shown by the data in Table 33-1, the physical properties of a supercritical fluid can be remarkably different from the same properties in either the liquid or the gaseous state. For example, the density of a supercritical fluid is typically 200 to 400 times greater than that of the corresponding gas and approaches that of the substance in its liquid state. The properties compared in Table 33-1 are those that are of importance in gas, liquid, and supercritical-fluid chromatography.

◀ The density of a supercritical fluid is 200 to 400 times that of its gaseous state, and it is nearly as dense as its liquid state.

An important property of supercritical fluids and one that is related to their high densities (0.2 to 0.5 g/cm^3) is their ability to dissolve large nonvolatile molecules. For example, supercritical carbon dioxide readily dissolves *n*-alkanes containing 5 to 22 carbon atoms, di-*n*-alkylphthalates in which the alkyl groups contain 4 to 16 carbon atoms, and various polycyclic aromatic hydrocarbons consisting of several rings.[2]

◀ Supercritical fluids tend to dissolve large, nonvolatile molecules.

Critical temperatures for fluids used in chromatography vary widely, from about 30°C to more than 200°C. Lower critical temperatures are advantageous in chromatography from several standpoints. For this reason, much of the work to date has focused on the supercritical fluids shown in Table 33-2. Note that these temperatures and the pressures at these temperatures are well within the operating conditions of ordinary high-performance liquid chromatography (HPLC).

TABLE 33-2

Properties of Some Supercritical Fluids

Fluid	Critical Temperature, °C	Critical Pressure, atm	Critical Point Density, g/mL	Density at 400 atm, g/mL
CO_2	31.3	72.9	0.47	0.96
N_2O	36.5	71.7	0.45	0.94
NH_3	132.5	112.5	0.24	0.40
n-Butane	152.0	37.5	0.23	0.50

Reprinted with permission from M. L. Lee and K. E. Markides, *Science,* **1987,** *235,* 1345. Copyright 1987 American Association for the Advancement of Science. Data taken from *Matheson Gas Data Book and CRC Handbook of Chemistry and Physics.*

[2]Certain important industrial processes are based on the high solubility of organic species in supercritical carbon dioxide. For example, this medium has been employed to extract caffeine from coffee beans to produce decaffeinated coffee and to extract nicotine from cigarette tobacco.

33A-2 Instrumentation and Operating Variables

Instruments for supercritical-fluid chromatography are similar in design to high-performance liquid chromatographs except that in SFC there are provisions for controlling and measuring the column pressure. Several manufacturers began to offer apparatus for supercritical-fluid chromatography in the mid-1980s.[3]

The Effect of Pressure

► Gradient elution can be achieved in SFC by systematically changing the column pressure or the density of the supercritical fluid.

The density of a supercritical fluid increases rapidly and nonlinearly with pressure increases. Density increases also change retention factors (k) and thus elution times. For example, the elution time for hexadecane is reported to decrease from 25 to 5 min as the pressure of carbon dioxide is raised from 70 to 90 atm. An effect similar to that of temperature programming in gas chromatography and gradient elution in HPLC can be achieved by linearly increasing the column pressure or by regulating the pressure to obtain linear density increases. Figure 33-1 illustrates the improvement in chromatograms realized by pressure programming. The decompression of fluids as they travel through the column can give rise to temperature changes that can affect separations and thermodynamic measurements.

Columns

► Very long columns can be used in SFC because the viscosity of supercritical fluids is so low.

Both packed columns and open tubular columns are used in supercritical fluid chromatography. Packed columns can provide more theoretical plates and can handle larger sample volumes than open tubular columns. Because of the low viscosity of supercritical media, columns can be much longer than those used in liquid chromatography, and column lengths of 10 to 20 m with inside diameters of 50 or 100 μm are common. For difficult separations, columns 60 m in length or longer have

Sample: 1. cholesteryl octanoate
2. cholesteryl decylate
3. cholesteryl laurate
4. cholesteryl myristate
5. cholesteryl palmitate
6. cholesteryl stearate
Column: DB – 1
Mobile phase: CO_2
Temperature: 90°C
Detector: FID

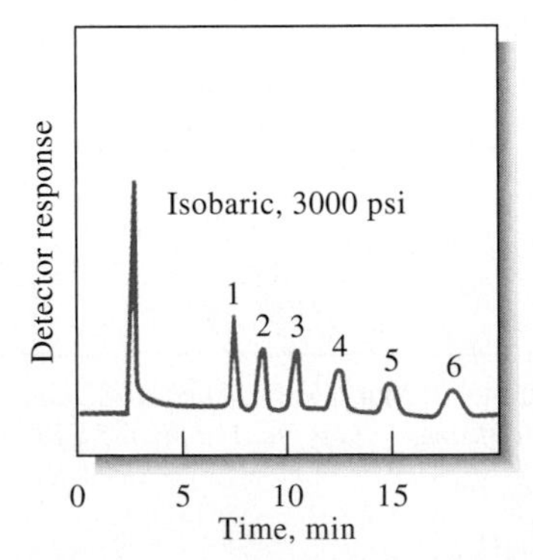

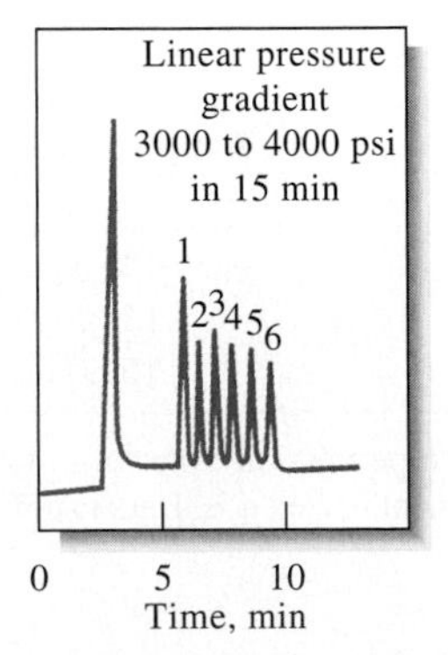

Figure 33-1 Effect of pressure programming in supercritical-fluid chromatography. Note the shorter time for the pressure-gradient chromatogram on the right compared with the isobaric chromatogram on the left. (Courtesy of Brownlee Labs, Santa Clara, CA.)

[3]For descriptions of several commercial instruments for SFC, see F. Wach, *Anal. Chem.*, **1994,** *66,* 369A; B. Erikson, *Anal. Chem.*, **1997,** *69,* 683A.

been used. Well over 100,000 plates can be achieved with packed columns. Open tubular columns are similar to the fused-silica open tubular (FSOT) columns described on page 959.

Many of the column coatings used in liquid chromatography have been applied to supercritical-fluid chromatography as well. Typically, these are polysiloxanes (see Section 31B-3) that are chemically bonded to the surface of silica particles or to the inner silica wall of capillary tubing. Film thicknesses are 0.05 to 0.4 μm.

Mobile Phases

The most widely used mobile phase for supercritical-fluid chromatography is carbon dioxide. It is an excellent solvent for a variety of nonpolar organic molecules. In addition, it transmits in the ultraviolet range and is odorless, nontoxic, readily available, and remarkably inexpensive relative to other chromatographic solvents. Its critical temperature of 31°C and its pressure of 73 atm at the critical temperature permit a wide selection of temperatures and pressures without exceeding the operating limits of modern high-performance liquid chromatography equipment. In some applications, polar organic modifiers, such as methanol, are introduced in small concentrations ($\approx$1%) to modify alpha values for analytes.

A number of other substances have served as mobile phases in supercritical chromatography, including ethane, pentane, dichlorodifluoromethane, diethyl ether, and tetrahydrofuran.

Detectors

A major advantage of supercritical-fluid chromatography is that the sensitive and universal detectors of gas-liquid chromatography are applicable to this technique as well. For example, the convenient flame ionization detector of gas-liquid chromatography can be applied by simply allowing the supercritical carrier to expand through a restrictor and into an air-hydrogen flame, where ions formed from the analytes are collected at biased electrodes, giving rise to an electrical current.

33A-3 Supercritical-Fluid Chromatography Versus Other Column Methods

The information in Table 33-1, and other data as well, reveal that several physical properties of supercritical fluids are intermediate between the properties of gases and liquids. As a consequence, this new type of chromatography combines some of the characteristics of both gas and liquid chromatography. Thus, like gas chromatography, supercritical-fluid chromatography is inherently faster than liquid chromatography because of the lower viscosity and higher diffusion rates in the mobile phase. High diffusivity, however, leads to longitudinal band spreading, which is a significant factor with gas but not with liquid chromatography. Thus, the intermediate diffusivities and viscosities of supercritical fluids result in faster separations than are achieved with liquid chromatography accompanied by less zone spreading than is encountered in gas chromatography.

Figure 33-2 shows plots of plate heights H as a function of average linear velocity $\bar{u}$ in cm/s for high-performance liquid chromatography and supercritical-fluid chromatography. In both cases, the solute was pyrene, and the stationary phase was a reversed-phase octadecyl silane maintained at 40°C. The mobile phase for HPLC was acetonitrile and water, while for SFC the mobile phase was carbon dioxide. These conditions yielded about the same retention factor (k) for both mobile phases. Note that the minimum in plate height occurred at a flow rate of 0.13 cm/s

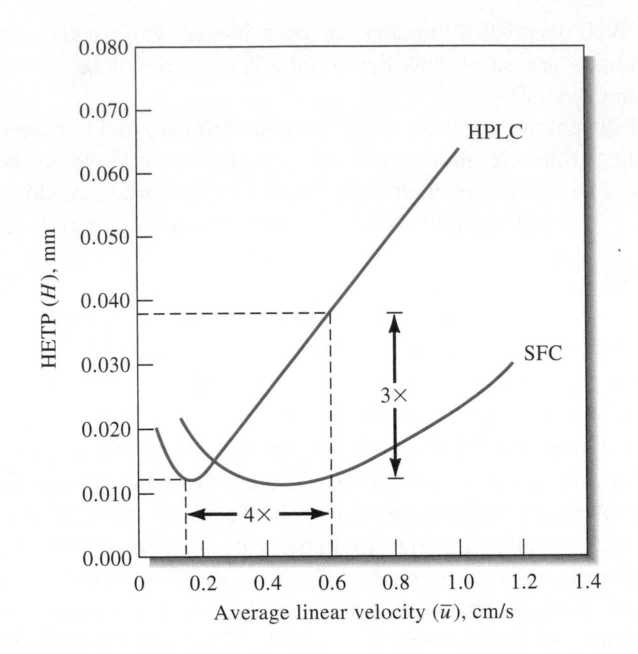

Figure 33-2 Performance characteristics of a 5-μm ODS column when elution is carried out with a conventional mobile phase (HPLC) and supercritical carbon dioxide SFC. (From D. R. Gere, *Application Note 800-3*. Hewlett-Packard Corp., Palo Alto, CA, 1983.)

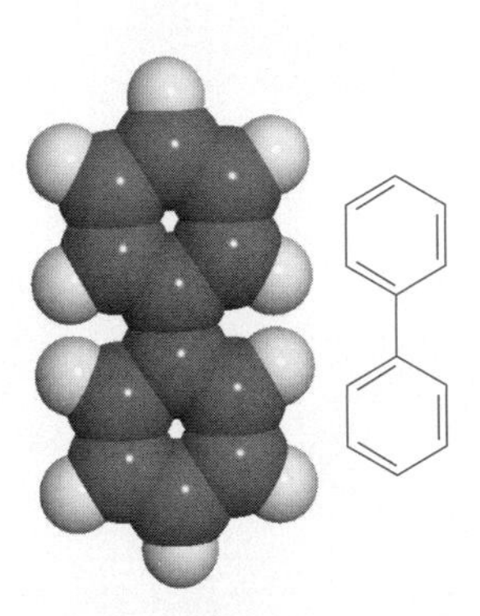

Molecular model and structure of biphenyl, a hazardous aromatic hydrocarbon. It is used as an intermediate in the production of emulsifiers, brighteners, plastics, and many other compounds. Biphenyl has been used as a heat transfer medium in heating fluids, as a dye carrier for textiles and copying paper, and as a solvent in pharmaceutical preparations. Paper impregnated with biphenyl is used in citrus fruit packaging to reduce fruit damage by fungus. Short-term exposure causes eye and skin irritation and toxic effects on the liver, kidneys, and nervous system. Long-term exposure causes kidney damage to laboratory animals and may affect the central nervous system in humans.

with the HPLC and 0.40 cm/s for the SFC. The consequence of this difference is shown in Figure 33-3, where these same conditions are used for the separation of pyrene from biphenyl. Note that the HPLC separation required more than twice the time of the SFC separation.

Despite its advantages, SFC has not gained widespread acceptance because of the complexity and cost of the instrumentation and the lack of applications for which it provides unique information. SFC still fills an important gap in the separations world, however, and provides a significant link between HPLC and gas chromatography.

33A-4 Applications

Supercritical-fluid chromatography appears to have a potential niche in the spectrum of column chromatographic methods because it is applicable to a class of compounds that is not readily amenable to either gas-liquid or liquid chromatography. These compounds include species that are nonvolatile or thermally unstable and, in addition, contain no chromophoric groups that can be used for photometric detection. Separation of these compounds is possible with supercritical-fluid chromatography at temperatures below 100°C; furthermore, detection is readily carried out by means of the highly sensitive flame ionization detector. It is also noteworthy that supercritical columns have the added advantage of being much easier to interface with mass spectrometers than liquid chromatographic columns.

► SFC with flame ionization detection works very well for nonvolatile or thermally unstable compounds that have no chromophores for photometric detection.

33B PLANAR CHROMATOGRAPHY

Planar chromatographic methods include **thin-layer chromatography** (TLC), **paper chromatography** (PC), and **electrochromatography.** Each makes use of a flat, relatively thin layer of material that is either self-supporting or is coated on a glass, plastic, or metal surface. The mobile phase moves through the stationary phase by capillary action, sometimes assisted by gravity or an electrical potential. Planar chromatography was once called two-dimensional chromatography, although the term has now come to signify the coupling of two chromatographic techniques with different separation mechanisms.

Currently, most planar chromatography is based on the thin-layer technique, which is faster, has better resolution, and is more sensitive than its paper counterpart. This section is devoted to thin-layer methods. Capillary electrochromatography is described in Section 33D.

33B-1 The Scope of Thin-Layer Chromatography

In terms of theory, the types of stationary and mobile phases, and applications, thin-layer and liquid chromatography are remarkably similar. In fact, thin-layer plates can be profitably used to develop optimal conditions for separations by column liquid chromatography. The advantages of following this procedure are the speed and low cost of the exploratory thin-layer experiments. Some chromatographers have taken the position that thin-layer experiments should always precede column experiments.

Thin-layer chromatography has become the workhorse of the drug industry for the all-important determination of product purity. It has also found widespread use in clinical laboratories and is the backbone of many biochemical and biological studies. Finally, it finds extensive use in industrial laboratories.[4] As a consequence of these many areas of application, TLC remains a very important technique.

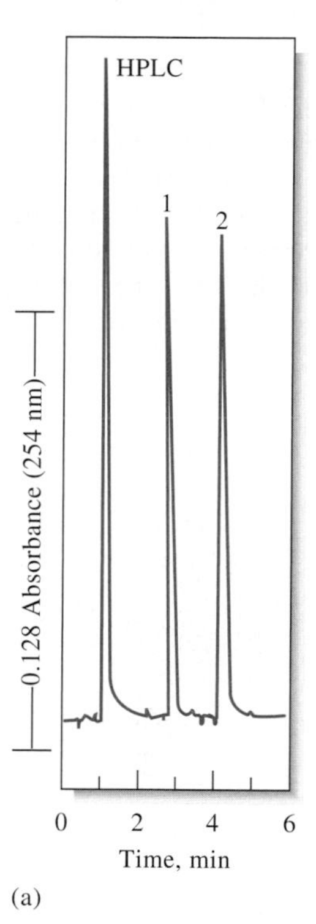

33B-2 Principles of Thin-Layer Chromatography

Typical thin-layer separations are performed on a glass plate that is coated with a thin and adherent layer of finely divided particles; this layer constitutes the stationary phase. The particles are similar to those described in the discussion of adsorption, normal- and reversed-phase partition, ion-exchange, and size-exclusion column chromatography. Mobile phases are also similar to those employed in high-performance liquid chromatography.

Preparation of Thin-Layer Plates

A thin-layer plate is prepared by spreading an aqueous slurry of the finely ground solid onto the clean surface of a glass or plastic plate or a microscope slide. Often, a binder is incorporated into the slurry to enhance adhesion of the solid particles to the glass and to one another. The plate is then allowed to stand until the layer has set and adheres tightly to the surface; for some purposes, it may be heated in an oven for several hours. Several chemical supply houses offer precoated plates of various kinds.

Plate Development

Plate development is the process in which a sample is carried through the stationary phase by a mobile phase. It is analogous to elution in liquid chromatography. The most common way of developing a plate is to place a drop of the sample near

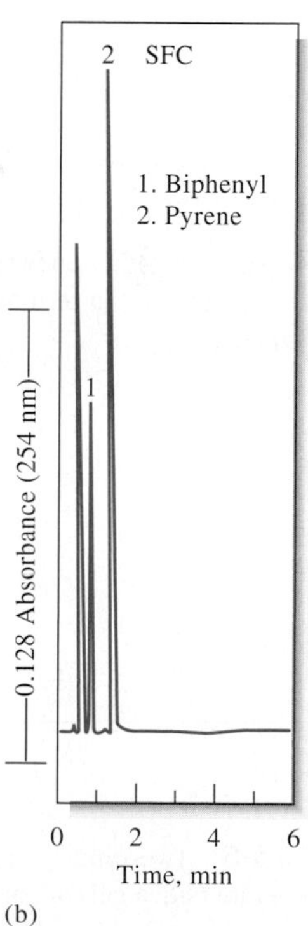

Figure 33-3 Separation of pyrene and biphenyl by (a) HPLC and (b) SFC. (Reprinted with permission from D. R. Gere, *Science*, **1983,** *222,* 255. Copyright 1983 American Association for the Advancement of Science.)

[4]Two monographs devoted to the principles and applications of thin-layer chromatography are B. Fried and J. Sherma, *Thin Layer Chromatography,* 4th ed. New York: Dekker, 1999; R. Hamilton and S. Hamilton, *Thin-Layer Chromatography.* New York: Wiley, 1987. For recent reviews, see J. Sherma, *Anal. Chem.,* **2002,** *74,* 2653; J. Sherma, **2000,** *72,* 9R; C. F. Poole and S. K. Poole, *Anal. Chem.,* **1994,** *66,* 27A.

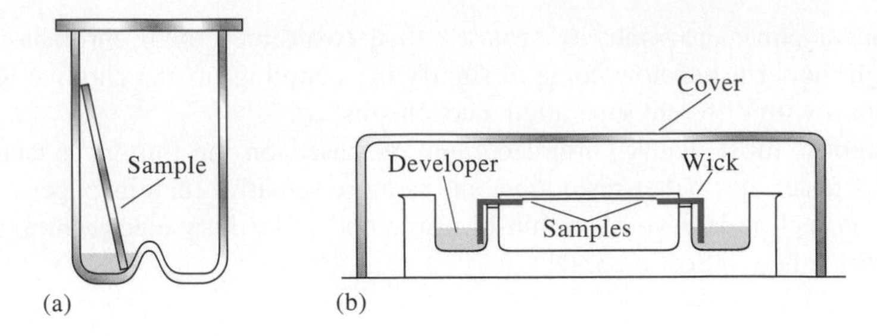

Figure 33-4 (a) Ascending-flow developing chamber. (b) Horizontal-flow developing chamber, in which samples are placed on both ends of the plate and developed toward the middle, thus doubling the number of samples that can be accommodated.

one edge of the plate (most plates have dimensions of 5 × 20 or 20 × 20 cm) and mark its position with a pencil. After the sample solvent has evaporated, the plate is placed in a closed container saturated with vapors of the developing solvent. One end of the plate is immersed in the developing solvent, with care being taken to avoid direct contact between the sample and the developer (Figure 33-4). After the developer has traversed one half or two thirds of the length of the plate, the plate is removed from the container and is dried. The positions of the components are then determined in any of several ways.

Figure 33-5 illustrates the separation of amino acids in a mixture by developing in two directions **(two-dimensional thin-layer chromatography).** The sample was placed in one corner of a square plate, and the plate was developed in the ascending direction with solvent A. This solvent was then removed by evaporation, and the plate was rotated 90 degrees, following which ascending development with solvent B was performed. After solvent removal, the positions of the amino acids were determined by spraying with ninhydrin, a reagent that forms a pink to purple product with amino acids. The spots were identified by comparison of their positions with those of standards.

Locating Analytes on the Plate

The process of locating analytes on a thin-layer plate is often termed **visualization.**

Several methods are employed to locate sample components after separation. Two common methods that can be applied to most organic mixtures involve spraying with a solution of iodine or sulfuric acid, both of which react with organic compounds to yield dark products. Several specific reagents (such as ninhydrin) are also useful for locating separated species.

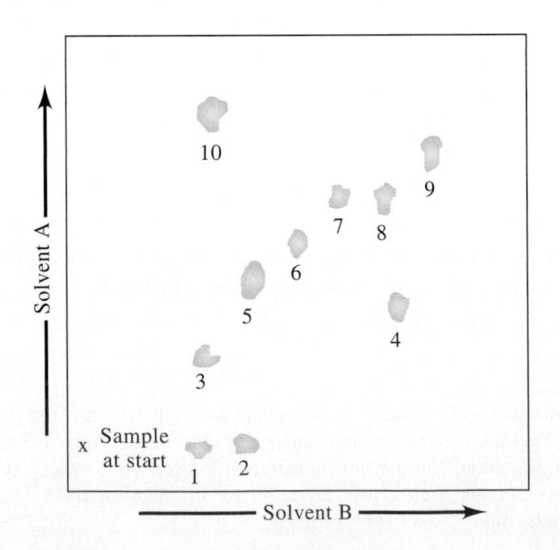

Figure 33-5 Two-dimensional thin-layer chromatogram (silica gel) of some amino acids. Solvent A: toluene/2-chloroethanol/pyridine. Solvent B: chloroform/benzyl alcohol/acetic acid. Amino acids: (1) aspartic acid, (2) glutamic acid, (3) serine, (4) β-alanine, (5) glycine, (6) alanine, (7) methionine, (8) valine, (9) isoleucine, (10) cysteine.

Another method of detection is based on incorporating a fluorescent material into the stationary phase. After development, the plate is examined under ultraviolet light. The sample components quench the fluorescence of the material so that all of the plate fluoresces except where the nonfluorescing sample components are located.

33B-3 Paper Chromatography

Separations by paper chromatography are performed in the same way as those on thin-layer plates. The papers are manufactured from highly purified cellulose, with close control over porosity and thickness. Such papers contain sufficient adsorbed water to make the stationary phase aqueous. Other liquids can be made to displace the water, however, thus providing a different type of stationary phase. For example, paper treated with silicone or paraffin oil permits reversed-phase paper chromatography, in which the mobile phase is a polar solvent. Also available commercially are special papers that contain an adsorbent or an ion-exchange resin, thus permitting adsorption and ion-exchange paper chromatography.

33C CAPILLARY ELECTROPHORESIS[5]

Electrophoresis is a separation method based on the differential rates of migration of charged species in an applied dc electric field. This separation technique for macrosize samples was first developed by the Swedish chemist Arne Tiselius in the 1930s for the study of serum proteins; he was awarded the 1948 Nobel Prize for his work.

Electrophoresis separations are based on the different rates at which charged species migrate in an electric field.

Electrophoresis on a macro scale has been applied to a variety of difficult analytical separation problems: inorganic anions and cations, amino acids, catecholamines, drugs, vitamins, carbohydrates, peptides, proteins, nucleic acids, nucleotides, polynucleotides, and numerous other species. A particular strength of electrophoresis is its unique ability to separate charged macromolecules of interest to biochemists, biologists, and clinical chemists. For many years, electrophoresis has been the powerhouse method of separating proteins (enzymes, hormones, antibodies) and nucleic acids (DNA, RNA), for which it offers unparalleled resolution.[6]

Until the appearance of capillary electrophoresis, electrophoretic separations were not carried out in columns but were performed in a flat stabilized medium such as paper or a porous semisolid gel. Remarkable separations were realized in such media, but the technique was slow, tedious, and required a good deal of operator skill. In the early 1980s, scientists began to explore the feasibility of performing these same separations on micro amounts of sample in fused silica capillary tubes. Their results proved promising in terms of resolution, speed, and potential for automation. As a consequence, capillary electrophoresis (CE) has become an important tool for a wide variety of analytical separation problems and is the only type of electrophoresis that we will consider.

[5]For additional discussion of the principles, instrumentation, and applications of capillary electrophoresis, see M. G. Khaledi, Ed., *High-Performance Capillary Electrophoresis: Theory, Techniques and Applications.* New York: Wiley, 1998: P. Camilleri, Ed., *Capillary Electrophoresis: Theory and Practice.* Boca Raton, FL: CRC Press, 1993; R. Weinberger, *Practical Capillary Electrophoresis.* New York: Academic Press, 2000.

[6]See S. Hu and N. J. Dovichi, *Anal. Chem.*, **2002,** *74,* 2833; S. N. Krylov and N. J. Dovichi, *Anal. Chem.*, **2000,** *72,* 111R.

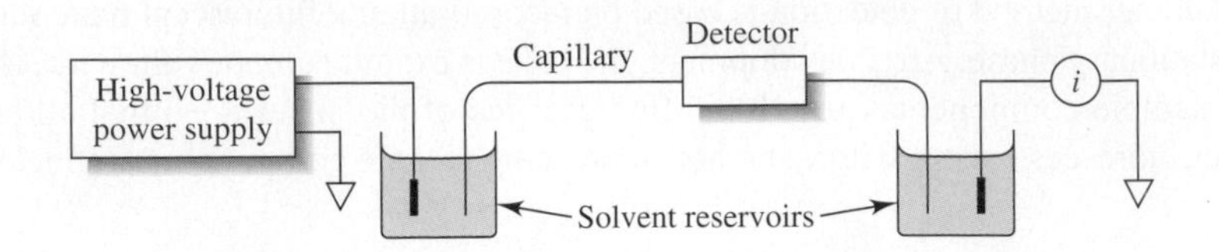

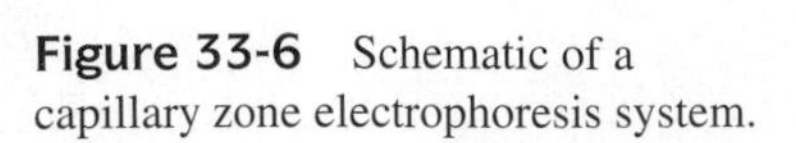
Figure 33-6 Schematic of a capillary zone electrophoresis system.

33C-1 Instrumentation for Capillary Electrophoresis

► Capillary electrophoresis instruments are relatively simple.

As shown in Figure 33-6, the instrumentation for capillary electrophoresis is simple.[7] A buffer-filled fused-silica capillary, typically 10 to 100 μm in internal diameter and 40 to 100 cm long, extends between two buffer reservoirs that also hold platinum electrodes. Sample introduction is performed at one end and detection at the other. A potential of 5 to 30 kV dc is applied across the two electrodes. The polarity of this high voltage can be as indicated in Figure 33-6 or can be reversed to allow rapid separation of anions.

Sample introduction is often accomplished by pressure injection, in which one end of the capillary is inserted into a vessel containing the sample. The vessel is then raised briefly above the level of the capillary to force sample into the tube. Alternatively, a vacuum is applied to the detector end of the tubing. Introduction may also be carried out by electroosmotic flow, which is described in the next section.

Because the separated analytes move past a common point in most types of capillary electrophoresis, detectors are similar in design and function to those described for HPLC. Table 33-3 lists several of the detection methods that have been reported for capillary electrophoresis. The second column of the table shows representative detection limits for these detectors.

TABLE 33-3
Detectors for Capillary Electrophoresis

Type of Detector	Representative Detection Limit* (attomoles detected)
Spectrometry	
Absorption†	1–1000
Fluorescence	1–0.01
Thermal lens†	10
Raman†	1000
Chemiluminescence†	1–0.0001
Mass spectrometry	1–0.01
Electrochemical	
Conductivity†	100
Potentiometry†	1
Amperometry	0.1

Sources: B. Huang, J. J. Li, L. Zhang, J. K. Cheng, *Anal. Chem.*, **1996,** *68,* 2366; S. C. Beale, *Anal. Chem.*, **1998,** *70,* 279R; S. N. Krylov and N. J. Dovichi, *Anal. Chem.*, **2000,** *72,* 111R; S. Hu and N. J. Dovichi, *Anal. Chem.*, **2002,** *74,* 2833.
*Detection limits quoted have been determined with injection volumes ranging from 18 pL to 10 nL.
†Mass detection limit converted from concentration detection limit using a 1-nL injection volume.

[7] For a review of current commercially available capillary electrophoresis instruments, see L. DeFrancesco, *Anal. Chem.*, **2001,** *73,* 497A.

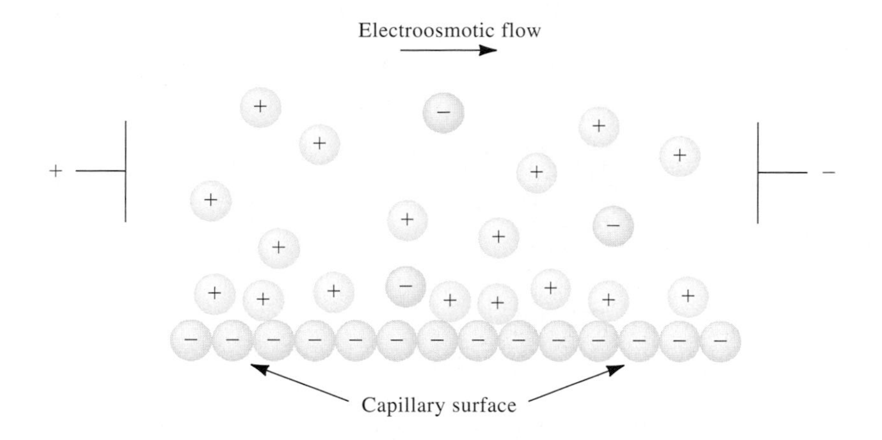

Figure 33-7 Charge distribution at a silica/capillary interface and resulting electroosmotic flow. (From A. G. Ewing, R. A. Wallingford, and T. M. Olefirowicz, *Anal. Chem.*, **1989,** *61,* 298A.)

33C-2 Electroosmotic Flow

A unique feature of capillary electrophoresis is **electroosmotic flow.** When a high voltage is applied across a fused-silica capillary tube containing a buffer solution, electroosmotic flow usually occurs in which the solvent migrates toward the cathode. The rate of migration can be substantial. For example, a 50-mM pH 8 buffer has been found to flow through a 50-cm capillary toward the cathode at approximately 5 cm/min with an applied potential of 25 kV.[8]

As shown in Figure 33-7, the cause of electroosmotic flow is the electrical double layer that develops at the silica/solution interface. At pH values higher than 3, the inside wall of a silica capillary is negatively charged owing to ionization of the surface silanol groups (Si—OH). Buffer cations congregate in an electrical double layer adjacent to the negative surface of the silica capillary. The cations in the diffuse outer layer to the double layer are attracted toward the cathode, or negative electrode, and since they are solvated, they drag the bulk solvent along with them. As shown in Figure 33-8, electroosmosis leads to bulk solution flow with a flat profile across the tube, because flow originates at the walls of the tubing. This profile is in contrast to the laminar (parabolic) profile that is observed with the pressure-driven flow encountered in HPLC. Because the profile is essentially flat, electroosmotic flow does not contribute significantly to band broadening the way pressure-driven flow does in liquid chromatography.

◀ The profile of electroosmotic flow is nearly flat, which minimizes band broadening.

The rate of electroosmotic flow is generally greater than the electrophoretic migration velocities of the individual ions and effectively becomes the mobile-phase pump of capillary zone electrophoresis. Even though analytes migrate according to their charges within the capillary, the electroosmotic flow rate is usually sufficient to sweep all positive, neutral, and even negative species toward the same end of the

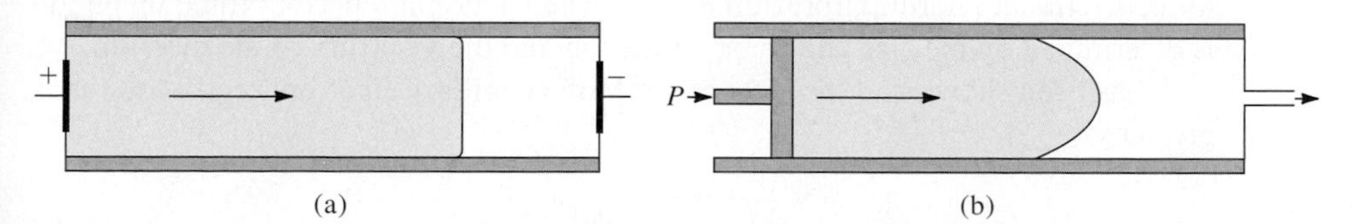

Figure 33-8 Flow profiles for liquids under (a) electroosmotic flow and (b) pressure-induced flow.

[8]J. D. Olechno, J. M. Y. Tso, J. Thayer, and A. Wainright, *Amer. Lab.*, **1990,** *22*(17), 51.

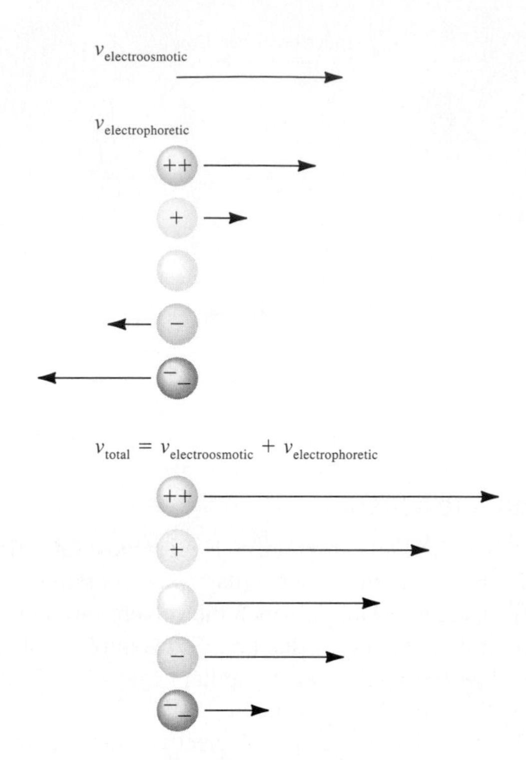

Figure 33-9 Velocities in the presence of electroosmotic flow. The length of the arrow next to an ion indicates the magnitude of its velocity; the direction of the arrow indicates the direction of motion. The negative electrode would be to the right and the positive electrode to the left of this section of solution.

capillary so that all can be detected as they pass by a common point (Figure 33-9). The resulting **electropherogram** looks like a chromatogram but with narrower peaks.

Electroosmosis is often desirable in certain types of capillary electrophoresis, but in other types it is not. Electroosmotic flow can be minimized by coating the inside capillary wall with a reagent like trimethylchlorosilane to eliminate the surface silanol groups.

33C-3 The Basis for Electrophoretic Separations

The migration rate v of an ion in an electric field is given by

$$v = \mu_e E = \mu_e \cdot \frac{V}{L} \tag{33-1}$$

The **electrophoretic mobility** is the ratio of the migration rate of an ion to the applied electric field.

where E is the electric field strength in volts per centimeter, V is the applied voltage, L is the length of the tube between electrodes, and μ_e is the **electrophoretic mobility,** which is proportional to the charge on the ion and inversely proportional to the frictional retarding force on the ion. The frictional retarding force on an ion is determined by the size and shape of the ion and the viscosity of the medium.

It has been shown that the plate count N of a capillary electrophoresis column is given by

$$N = \frac{\mu_e V}{2D} \tag{33-2}$$

where D is the diffusion coefficient of the solute (cm^2/s). Because resolution increases with plate count, it is desirable to use high applied voltages to achieve high-resolution separations. Note that for electrophoresis, contrary to the situation in chromatography, the plate count does not increase with the column length.

Typically, capillary electrophoresis plate counts are 100,000 to 200,000 at the usual applied voltages.

33C-4 Applications of Capillary Electrophoresis[9]

Capillary electrophoretic separations are performed in several ways called modes. These include **isoelectric focusing, isotachophoresis,** and **capillary zone electrophoresis** (CZE). We consider here only modes of capillary zone electrophoresis in which the buffer composition is constant throughout the region of the separation. The applied field causes each of the different ionic components of the mixture to migrate according to its own mobility and to separate into zones, which may be completely resolved or may be partially overlapped. Completely resolved zones have regions of buffer between them. The situation is analogous to elution column chromatography, in which regions of mobile phase are located between zones containing separated analytes.

Separation of Small Ions

For most electrophoretic separations of small ions, the smallest analysis time results when the analyte ions move in the same direction as the electroosmotic flow. Thus, for cation separations, the walls of the capillary are untreated, and the electroosmotic flow and the cation movement are toward the cathode. For the separation of anions, however, the electroosmotic flow is usually reversed by treating the walls of the capillary with an alkyl ammonium salt, such as cetyl trimethylammonium bromide. The positively charged ammonium ions become attached to the negatively charged silica surface and in turn create a negatively charged double layer of solution, which is attracted toward the anode, reversing the electroosmotic flow.

In the past, the most common method of analysis of small anions has been ion-exchange chromatography. For cations, the preferred techniques have been atomic absorption spectroscopy and inductively coupled plasma emission spectroscopy. Recently, however, capillary electrophoretic methods have begun to compete with these traditional methods for small ion analysis. Several major reasons for adoption of electrophoretic methods have been recognized: lower equipment costs, smaller sample size requirements, much greater speed, and better resolution.

The initial cost of equipment and the expense of maintenance for electrophoresis is generally significantly lower than for ion chromatography and atomic spectroscopy. Thus, commercial electrophoretic instruments are marketed in the price range of $10,000 to $65,000.[10]

Sample sizes for electrophoresis are in the nanoliter range, whereas microliter or larger samples are usually needed for other types of small ion analysis. Thus, electrophoretic methods are more sensitive than the other methods on a mass basis (but usually not on a concentration basis).

◀ Samples can be a few nanoliters in volume in capillary electrophoresis.

[9]For reviews of applications of capillary electrophoresis and electrochromatography, see D. R. Baker, *Capillary Electrophoresis: An Introduction.* New York: Wiley, 1995; *Handbook of Capillary Electrophoresis,* 2nd ed., J. P. Landers, Ed. Boca Raton, FL: CRC Press, 1997; R. Weinberger, *Practical Capillary Electrophoresis.* New York: Academic Press, 2000; S. N. Krylov and N. J. Dovichi, *Anal. Chem.,* **2000,** *72,* 111R; S. Hu and N. J. Dovichi, *Anal. Chem.,* **2002,** *74,* 2833.

[10]See L. DeFrancesco, *Anal. Chem.,* **2001,** *73,* 497A.

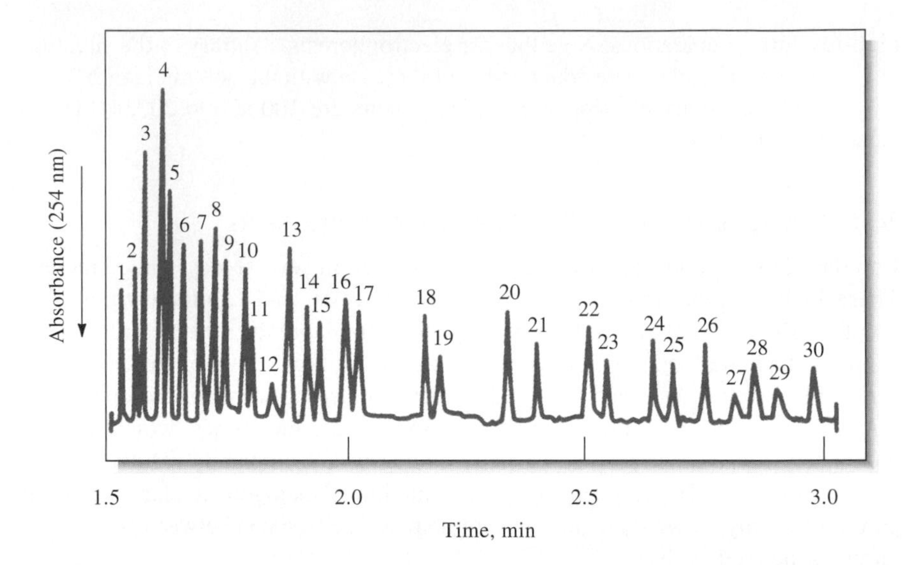

Figure 33-10 Electropherogram showing the separation of 30 anions. Capillary internal diameter: 50 μm (fused silica). Detection: indirect UV, 254 nm. Peaks: 1 = thiosulfate (4 ppm), 2 = bromide (4 ppm), 3 = chloride (2 ppm), 4 = sulfate (4 ppm), 5 = nitrite (4 ppm), 6 = nitrate (4 ppm), 7 = molybdate (10 ppm), 8 = azide (4 ppm), 9 = tungstate (10 ppm), 10 = monofluorophosphate (4 ppm), 11 = chlorate (4 ppm), 12 = citrate (2 ppm), 13 = fluoride (1 ppm), 14 = formate (2 ppm), 15 = phosphate (4 ppm), 16 = phosphite (4 ppm), 17 = chlorite (4 ppm), 18 = galactarate (5 ppm), 19 = carbonate (4 ppm), 20 = acetate (4 ppm), 21 = ethanesulfonate (4 ppm), 22 = propionate (5 ppm), 23 = propanesulfonate (4 ppm), 24 = butyrate (5 ppm), 25 = butanesulfonate (4 ppm), 26 = valerate (5 ppm), 27 = benzoate (4 ppm), 28 = *l*-glutamate (5 ppm), 29 = pentanesulfonate (4 ppm), 30 = *d*-gluconate (5 ppm). (Reprinted from W. A. Jones and P. Jandik, *J. Chromatogr.,* **1991,** *546,* 445, with permission of Elsevier Science.)

Figure 33-10 illustrates the unsurpassed quickness and resolution of electrophoretic separations of small anions. Here, 30 anions were separated cleanly in just over 3 minutes. Typically, an ion-exchange separation of only three or four anions could be accomplished in this brief time period. Figure 33-11 further illustrates the speed at which separations can be carried out. Here, 19 cations were separated in less than 2 minutes.

Separation of Molecular Species

A variety of small synthetic herbicides, pesticides, and pharmaceuticals that are ions or can be derivatized to yield ions have been separated and analyzed by CZE. Figure 33-12 is illustrative of this type of application; three anti-inflammatory drugs, which are carboxylic derivatives, are separated in less than 15 min.

Proteins, amino acids, and carbohydrates have all been separated in minimum times by CZE. In the case of neutral carbohydrates, the separations are preceded by formation of negatively charged borate complexes. The separation of protein mixtures is illustrated by Figure 33-13. Feature 33-1 discusses the use of capillary electrophoresis arrays for DNA sequencing.

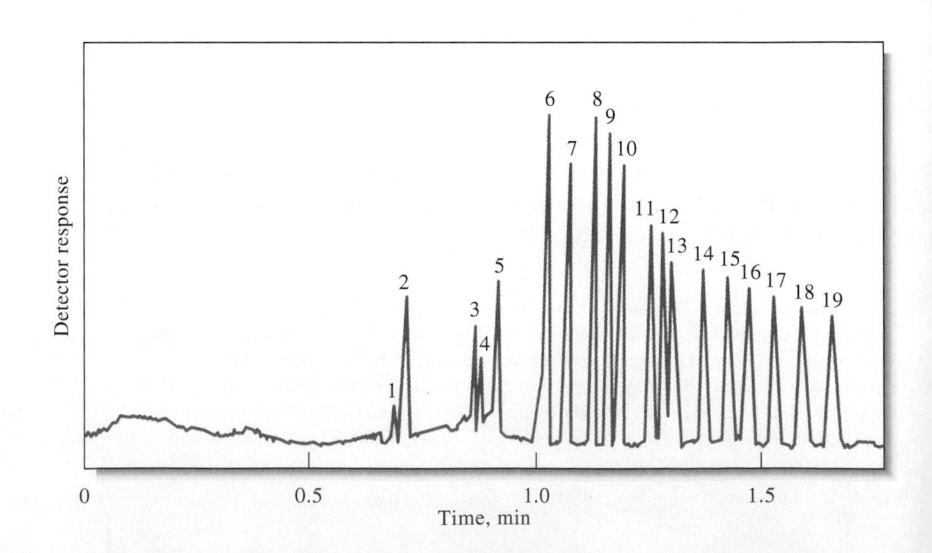

Figure 33-11 Separation of alkali, alkaline earths, and lanthanides. Capillary: 36.5 cm × 75 μm fused silica, + 30 kV. Injection: hydrostatic, 20 s at 10 cm. Detection: indirect UV, 214 nm. Peaks: 1 = rubidium (2 ppm), 2 = potassium (5 ppm), 3 = calcium (2 ppm), 4 = sodium (1 ppm), 5 = magnesium (1 ppm), 6 = lithium (1 ppm), 7 = lanthanum (5 ppm), 8 = cerium (5 ppm), 9 = praseodymium (5 ppm), 10 = neodymium (5 ppm), 11 = samarium (5 ppm), 12 = europium (5 ppm), 13 = gadolinium (5 ppm), 14 = terbium (5 ppm), 15 = dysprosium (5 ppm), 16 = holmium (5 ppm), 17 = erbium (5 ppm), 18 = thulium (5 ppm), 19 = ytterbium (5 ppm). (From P. Jandik, W. R. Jones, O. Weston, and P. R. Brown, *LC-GC,* **1991,** *9,* 634. With permission.)

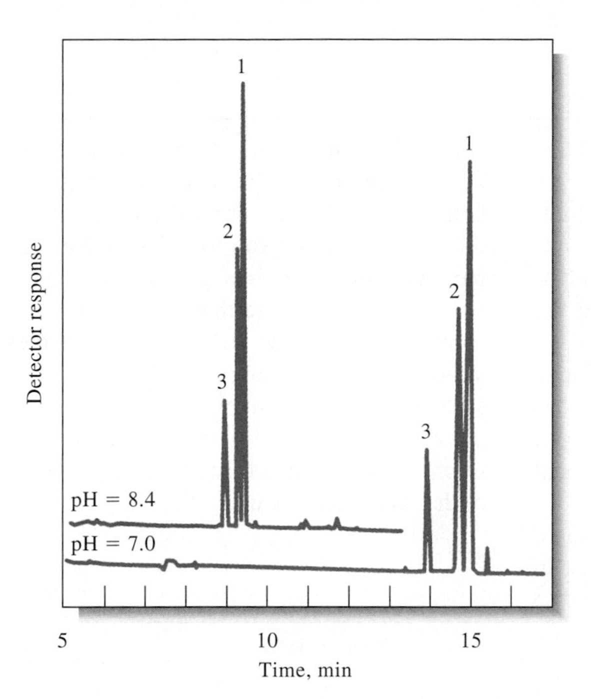

Figure 33-12 Separation of anti-inflammatory drugs by CZE. Detection: UV at 215 nm. Analytes: (1) naproxen, (2) ibuprofen, (3) tolmetin. (From A. Wainright, *J. Microcolumn. Sep.*, **1990,** *2,* 166. Reprinted by permission of John Wiley & Sons, Inc.)

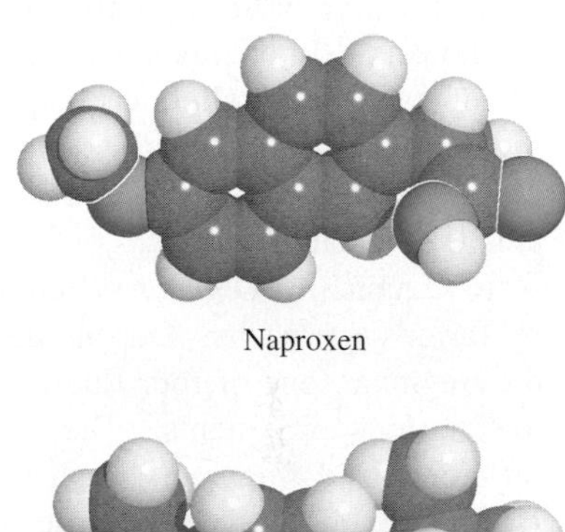

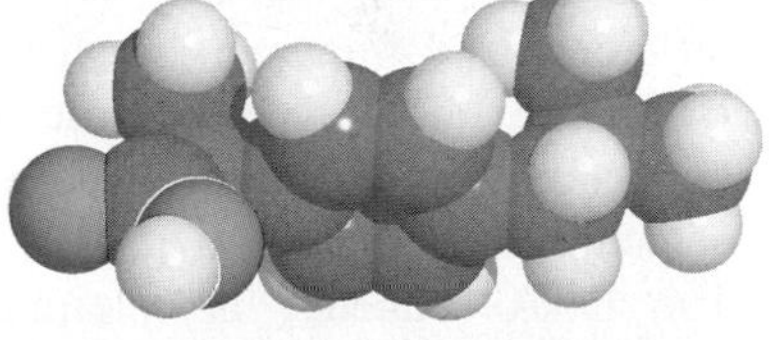

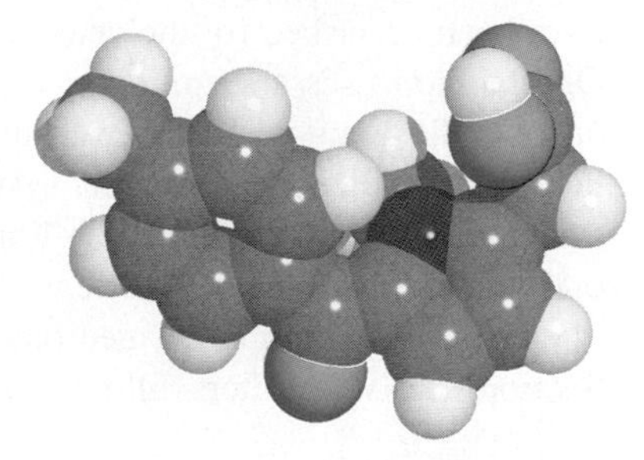

Molecular models of anti-inflammatory drugs: naproxen, ibuprofen and tolmetin. These nonsteroidal anti-inflammatory agents are thought to relieve pain by inhibiting the synthesis of prostaglandins, which are involved in the perception of pain and the production of fever and inflammation. Ibuprofen is also known as Motrin, Advil, and Nuprin. Naproxen sodium is Aleve, and tolmetin is Tolectin. Each has been used to treat symptoms of arthritis and to relieve the pain caused by gout; bursitis; tendinitis; sprains, strains, and other injuries; and menstrual cramps. Ibuprofen and naproxen are available over-the-counter in the United States.

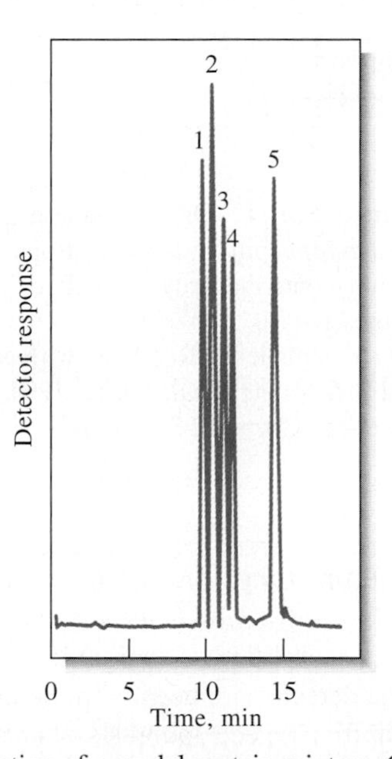

Figure 33-13 CZE separation of a model protein mixture. Conditions: pH 2.7 buffer; absorbance detection at 214 nm; 22 kV, 10 μA. Peaks identified in the following table.

Model Proteins Separated at pH 2.7

Peak No.	Proteins	Molecular Weight	Isoelectric Point, pH
1	Cytochrome c	12,400	10.7
2	Lysozyme	14,100	11.1
3	Trypsin	24,000	10.1
4	Trypsinogen	23,700	8.7
5	Trypsin inhibitor	20,100	4.5

FEATURE 33-1

Capillary Array Electrophoresis in DNA Sequencing

A major goal of the human genome project is to determine the order of occurrence of the four bases, adenine (A), cytosine (C), guanine (G), and thymine (T), in DNA molecules. The sequence defines an individual's genetic code. The need for sequencing DNA has spawned the development of several new analytical instruments. Among the most attractive of these approaches is capillary array electrophoresis.[11] In this technique, as many as 96 capillaries are operated in parallel. The capillaries are filled with a separation matrix, normally a linear polyacrylamide gel. The capillaries have inner diameters of 35 to 75 μm and are 30 to 60 cm in length.

In sequencing, DNA extracted from cells is fragmented by various approaches. Depending on the terminal base in the fragment, one of four fluorescent dyes is attached to the various fragments. The sample contains many different-sized fragments, each with a fluorescent label. Under the influence of the electrophoretic field, lower molecular weight fragments move faster and arrive at the detector sooner than higher molecular weight fragments. The DNA sequence is determined by the dye color sequence of the eluting fragments. Lasers are used to excite the dye fluorescence. Several different techniques have been described for the detection of the fluorescence. One method uses a scanning system such that the capillary bundle is moved relative to the excitation laser and the four-wavelength detection system. In the detection system illustrated in Figure 33F-1, a laser beam is focused onto the capillary array by a lens. The region that is illuminated by the laser is imaged onto a CCD detector (see Section 25A-4). Filters allow wavelength selection to detect the four colors. Simultaneous separation of 11 DNA fragments in 100 capillaries has been reported.[12] Other designs include sheath-flow detector systems and a detector that uses two diode lasers for excitation. Commercial instrumentation is available with prices ranging from $85,000 to more than $300,000.[13] Future developments should include miniaturization of these devices by lab-on-a-chip technology and improvements in detection systems. Such miniature systems will eventually become portable and able to be used in the field. Capillary electrophoresis has played a major role in identifying remains of the World Trade Center disaster.

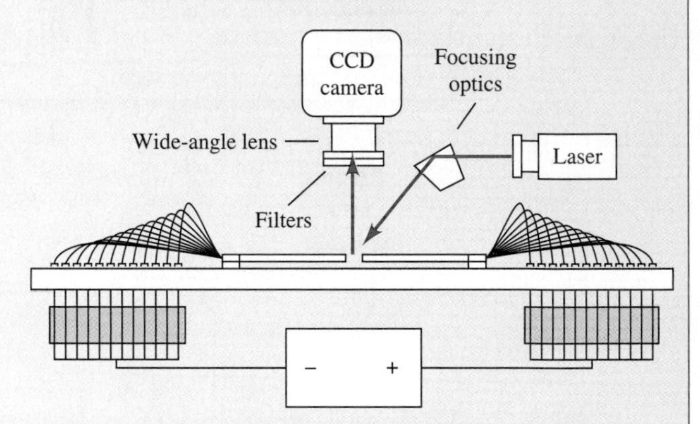

Figure 33F-1 On-column laser fluorescence detection system for capillary array electrophoresis. A laser is focused as a line onto the array of capillaries at a 45° angle. The fluorescence is filtered and detected by a CCD camera through a wide-angle lens. (Reprinted with permission from K. Ueno and E. S. Yeung, *Anal. Chem.*, **1994,** *66,* 1424. Copyright 1994 American Chemical Society.)

Spreadsheet Summary In Chapter 15 of *Applications of Microsoft® Excel in Analytical Chemistry,* capillary electrophoresis data are used to determine the mobilities of inorganic ions. Measurements of the arrival times of ions at the detector are used with the known mobility of Na^+ to determine mobilities. Capillary electrophoresis results are also used to determine pK_a values of several weak organic acids. Linear regression analysis is used to find the pK_a values from measurements of arrival times at different buffer pH values.

[11] For a review, see I. Kheterpal and R. A. Mathies, *Anal. Chem.*, **1999,** *71,* 31A.

[12] K. Ueno and E. S. Yeung, *Anal. Chem.*, **1994,** *66,* 1424.

[13] For a review of commercial sequences, see J. P. Smith and V. Hinson-Smith, *Anal. Chem.*, **2001,** *73,* 327A.

33D CAPILLARY ELECTROCHROMATOGRAPHY

Capillary electrochromatography (CEC) is a hybrid of HPLC and capillary electrophoresis that offers some of the best features of the two methods.[14] Like HPLC, it is applicable to the separation of neutral species. Like CE, however, it provides highly efficient separations on microvolumes of sample solution without the need for the high-pressure pumping system required for HPLC. In CEC, a mobile phase is transported across a stationary phase by electroosmotic flow. As shown in Figure 33-8, electroosmotic pumping leads to a flat plug profile rather than the parabolic profile that results from pressure-induced flow. The flat profile of osmotic pumping leads to narrow bands and thus high separation efficiencies.

33D-1 Packed Column Electrochromatography

Electrochromatography based on packed columns is the least mature of the various electroseparation techniques. In this method, a polar solvent is usually driven by electroosmotic flow through a capillary that is packed with a reversed-phase HPLC packing. Separations depend on the distribution of the analyte species between the mobile phase and the liquid stationary phase held on the packing. Figure 33-14 shows a typical electrochromatogram for the separation of 16 polyaromatic hydrocarbons in a 33-cm-long capillary having an inside diameter of 75 μm. The mobile phase consisted of acetonitrile in a 4-mM sodium borate solution. The stationary phase consisted of 3-μm octadecylsilica particles.

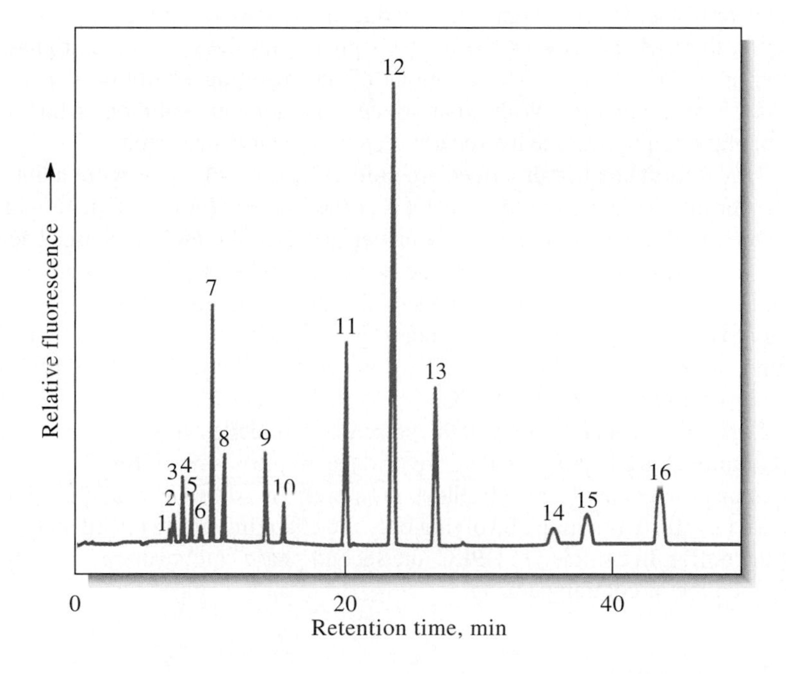

Figure 33-14 Electrochromatogram showing the electrochromatographic separation of 16 PAHs ($\sim 10^{-6}$ to 10^{-8} M of each compound). The peaks are identified as follows: (1) naphthalene, (2) acenaphthylene, (3) acenaphthene, (4) fluorene, (5) phenanthrene, (6) anthracene, (7) fluoranthene, (8) pyrene, (9) benz[*a*]anthracene, (10) chrysene, (11) benzo[*b*]fluoranthene, (12) benzo[*k*]fluoranthene, (13) benzo[*a*]pyrene, (14) dibenz[*a,h*]anthracene, (15) benzo[*ghi*]perylene, and (16) indeno[1,2,3-*cd*]pyrene. (Reprinted with permission from C. Yan, R. Dadoo, H. Zhao, D. J. Rakestroaw, and R. N. Zare, *Anal. Chem.*, **1995,** *67,* 2026. Copyright 1995 American Chemical Society.)

[14] For a discussion of this method, see L. A. Colon, Y. Guo, and A. Fermier, *Anal. Chem.*, **1997,** *69,* 461A.

33D-2 Micellar Electrokinetic Capillary Chromatography

The capillary electrophoretic methods we have described thus far are not applicable to the separation of uncharged solutes. In 1984, however, Terabe and collaborators[15] described a modification of the method that permitted the separation of low-molecular-weight aromatic phenols and nitro compounds with equipment such as shown in Figure 33-6. This technique involves introduction of a surfactant at a concentration level at which **micelles** form. Micelles form in aqueous solutions when the concentration of an ionic species having a long-chain hydrocarbon tail is increased above a certain level called the **critical micelle concentration** (CMC). At this point, the surfactant begins to form spherical aggregates made up to 40 to 100 ions with their hydrocarbon tails in the interior of the aggregate and their charged ends exposed to water on the outside. Micelles constitute a stable second phase that can incorporate nonpolar compounds in the hydrocarbon interior of the particles, thus *solubilizing* the nonpolar species. Solubilization is commonly encountered when a greasy material or surface is washed with a detergent solution.

Micelles are spherical aggregates with hydrocarbon tails in the interior and charged ends on the exterior exposed to water. See margin note on page 835.

Capillary electrophoresis carried out in the presence of micelles is termed **micellar electrokinetic capillary chromatography** and is given the acronym MECC or MEKC. In this technique, surfactants are added to the operating buffer in amounts that exceed the critical micelle concentration. For most applications to date, the surfactant has been sodium dodecyl sulfate (SDS). The surface of an ionic micelle of this type has a large negative charge, which gives it a large electrophoretic mobility. Most buffers, however, exhibit such a high electroosmotic flow rate toward the negative electrode that the anionic micelles are carried toward that electrode also, but at a much reduced rate. Thus, during an experiment, the buffer mixture consists of a faster moving aqueous phase and a slower moving micellar phase. When a sample is introduced into this system, the components distribute themselves between the aqueous phase and the hydrocarbon phase in the interior of the micelles. The positions of the resulting equilibria depend on the polarity of the solutes. With polar solutes, the aqueous solution is favored; with nonpolar compounds, the hydrocarbon environment is preferred.

The phenomena just described are quite similar to what occurs in a liquid partition chromatographic column except that the "stationary phase" is moving along the length of the column at a much slower rate than the mobile phase. The mechanism of separations is identical in the two cases and depends on differences in distribution constants for analytes between the mobile aqueous phase the hydrocarbon **pseudostationary phase.** The process is thus true chromatography; hence, the name micellar electrokinetic capillary *chromatography.* Figure 33-15 illustrates two typical separations by MECC.

Capillary chromatography in the presence of micelles appears to have a promising future. One advantage that this hybrid technique has over HPLC is much higher column efficiencies (100,000 plates or more). In addition, changing the second phase in MECC is simple, involving only the changing of the micellar composition of the buffer. In contrast, in HPLC, the second phase can be altered only by changing the type of column packing.

[15] S. Terabe, K. Otsuka, K. Ichikawa, A. Tsuchiya, and T. Ando, *Anal. Chem.*, **1984,** *56,* 111; S. Terabe, K. Otsuka, and T. Ando, *Anal. Chem.*, **1985,** *57,* 841. See also, K. R. Nielsen and J. P. Foley, in *Capillary Electrophoresis,* P. Camilleri, Ed. Ch. 4., Boca Raton, FL: CRC Press, 1993.

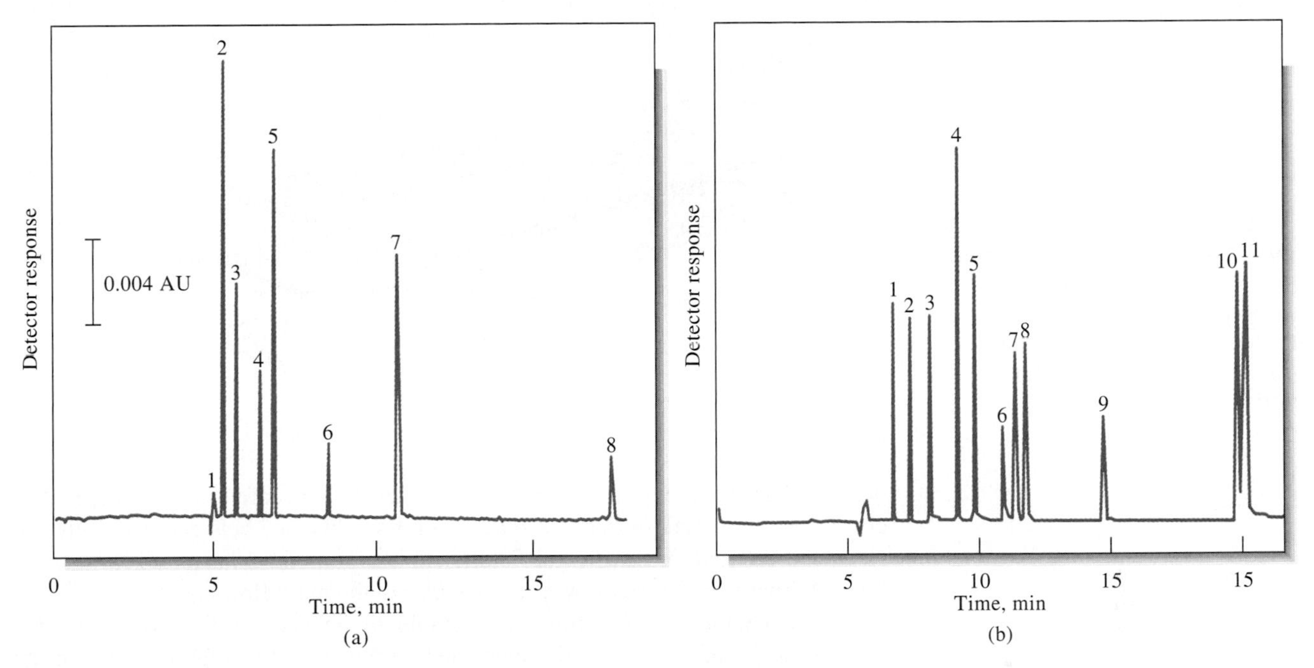

Figure 33-15 Typical separation by MECC. (a) Some test compounds: 1 = methanol, 2 = resorcinol, 3 = phenol, 4 = *p*-nitroaniline, 5 = nitrobenzene, 6 = toluene, 7 = 2-naphthol, 8 = Sudan III; capillary, 50-μm inside diameter, 500 mm to the detector; applied voltage, ca. 15 kV; detection UV absorption at 210 nm. (b) Analysis of a cold medicine: 1 = acetaminophen, 2 = caffeine, 3 = sulpyrinc, 4 = naproxen, 5 = guaiphenesin, 10 = noscapine, 11 = chloropheniramine and tipepidine; applied voltage, 20 kV; capillary, as in (a); detection UV absorption at 220 nm. (From S. Terabe, *Trends Anal. Chem.*, **1989,** *8,* 129.)

Spreadsheet Summary In the final exercise in Chapter 15 of *Applications of Microsoft® Excel in Analytical Chemistry,* micellar electrokinetic capillary chromatography is used to determine the critical micelle concentration (CMC) of a surfactant. An equation is developed to relate the retention factor to the CMC. Measured retention times are then used to determine the CMC from a regression analysis.

33E FIELD-FLOW FRACTIONATION

Field-flow fractionation (FFF) describes a group of analytical techniques that are becoming quite useful in the separation and characterization of dissolved or suspended materials such as polymers, large particles, and colloids. Although the FFF concept was first described by Giddings in 1966,[16] only recently have practical applications and advantages over other methods been shown.[17]

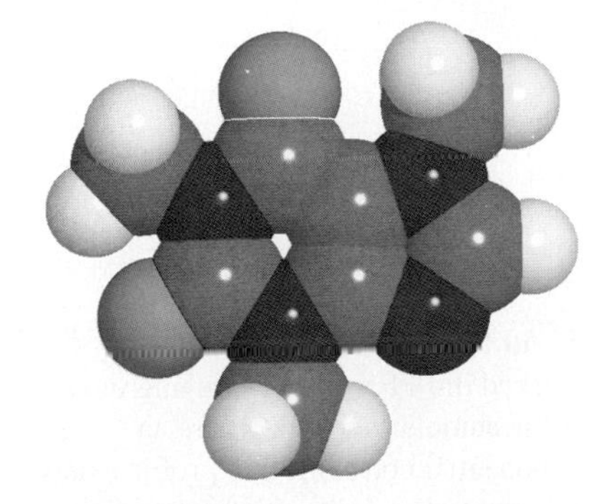

Molecular model of caffeine. Caffeine stimulates the cerebral cortex by inhibiting an enzyme that inactivates a certain form of adenosine triphosphate, the molecule that supplies energy. Caffeine occurs in coffee, tea, and cola drinks.

33E-1 Separation Mechanisms

Separations in FFF occur in a thin ribbon-like flow channel such as that shown in Figure 33-16. The channel is typically 25 to 100 cm in length and 1 to 3 cm in

[16]J. C. Gidding, *Sep. Sci.*, **1966,** *1,* 123.

[17]For a review of FFF methods, see J. C. Giddings, *Anal. Chem.*, **1995,** *67,* 592A.

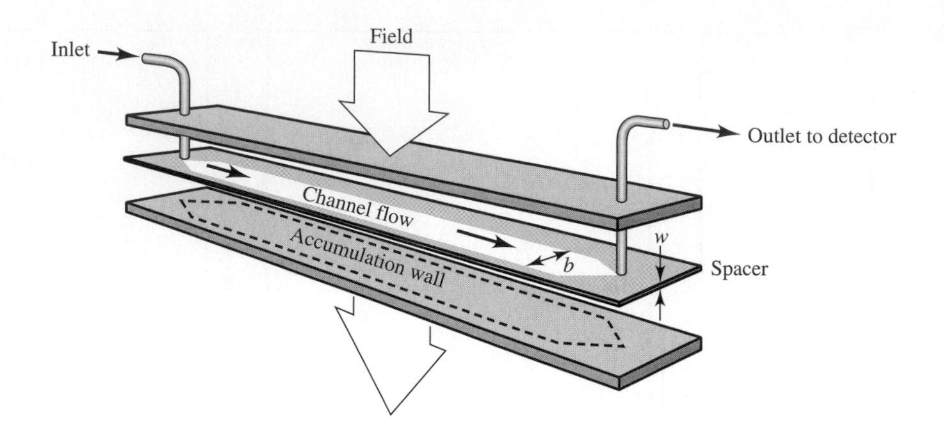

Figure 33-16 Schematic diagram of FFF flow channel sandwiched between two walls. An external field (electrical, thermal, centrifugal) is applied perpendicular to the flow direction.

breadth. The thickness of the ribbon-like structure is usually 50 to 500 μm. The channel is usually cut from a thin spacer and is sandwiched between two walls. An electrical, thermal, or centrifugal field is applied perpendicular to the flow direction. Alternatively, a cross-flow perpendicular to the main flow can be used.

In practice, the sample is injected at the inlet to the channel. The external field is next applied across the face of the channel, as illustrated in Figure 33-16. In the presence of the field, sample components migrate toward the **accumulation wall** at a velocity determined by the strength of the interaction of the component with the field. Sample components rapidly reach a steady-state concentration distribution near the accumulation wall, as shown in Figure 33-17. The mean thickness of the component layer l is related to the diffusion coefficient of the molecule D and to the field-induced velocity u toward the wall. The faster the component moves in the field, the thinner the layer near the wall. The larger the diffusion coefficient, the thicker the layer. Since the sample components have different values of D and u, the mean layer thickness will vary among components.

Once components have reached their steady-state profiles near the accumulation wall, the channel flow is begun. The flow is laminar, resulting in the parabolic profile shown on the left in Figure 33-17. The main carrier flow has its highest velocity in the center of the channel and its lowest velocity near the walls. Components that interact strongly with the field are compressed very near the wall, as shown by component A in Figure 33-18. Here they are eluted by slow-moving solvent. Components B and C protrude more into the channel and experience a higher solvent velocity. The elution order is thus C, then B, then A. Components that are separated by FFF flow through an ultraviolet-visible absorption, refractive index, or fluores-

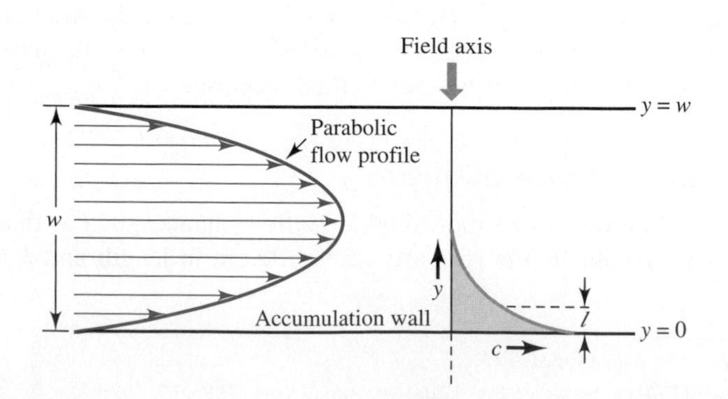

Figure 33-17 When the field is applied in FFF, components migrate to the accumulation wall, where an exponential concentration profile exists, as seen on the right. Components extend a distance y into the channel. The average thickness of the layer is l, which differs for each component. The main channel flow is then turned on, and the parabolic flow profile of the eluting solvent is shown on the left.

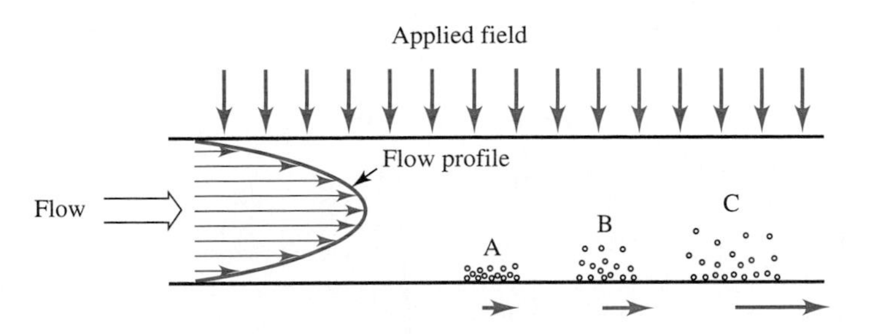

Figure 33-18 Three components A, B, and C are shown compressed against the accumulation wall in FFF to different amounts because of different interactions with the external field. When the flow is begun, component A experiences the lowest solvent velocity because it is the closest to the wall. Component B protrudes more into the channel, where it experiences a higher flow velocity. Component C, which interacts the least with the field, experiences the highest solvent flow velocity and thus is displaced the most rapidly by the flow.

cence detector located at the end of the flow channel. Detectors are similar to those used in HPLC separations. The separation results are revealed by a plot of detector response versus time, called a *fractogram,* which is similar to a chromatogram in chromatography.

33E-2 Field-Flow Fractionation Methods

Different FFF subtechniques result from the application of different types of fields or gradients.[18] To date, the methods that have been employed are **sedimentation, electrical, thermal,** and **flow FFF.**

In **field-flow fractionation,** components that interact strongly with the applied field are driven to the accumulation wall. Carrier flow elutes components protruding into the channel prior to those compressed near the accumulation wall.

Sedimentation Field-Flow Fractionation

Sedimentation FFF is by far the most widely used form. In this technique, the channel is coiled and made to fit inside a centrifuge basket, as illustrated in Figure 33-19. Components with the highest mass and density are driven to the wall by the sedimentation (centrifugation) force and elute last. Low-mass species are eluted first. There is relatively high selectivity between particles of different size in sedimentation FFF. A separation of polystyrene beads of various diameters by sedimentation FFF is shown in Figure 33-20.

Courtesy of Postnova Analytics Germany

Figure 33-19 Sedimentation FFF apparatus.

[18]For a discussion of the various FFF methods, see J. C. Giddings, *Unified Separation Science,* Ch. 9. New York: Wiley, 1991; M. E. Schimpf, K. Caldwell, and J. C. Giddings, Eds., *Field-Flow Fractionation Handbook.* New York: Wiley, 2000.

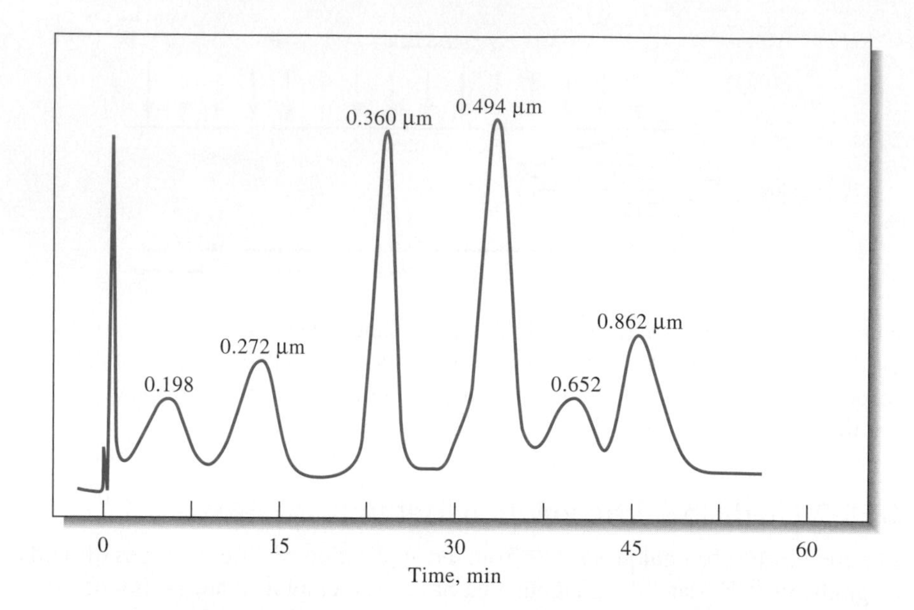

Figure 33-20 Fractogram illustrating separation of polystyrene beads of various diameters by sedimentation FFF. The channel flow rate was 2 mL/min. (Courtesy of FFFractionation, LLC, Salt Lake City, UT.)

Because the centrifugation forces are relatively weak for small molecules, sedimentation FFF is most applicable for molecules with molecular weights exceeding 10^6. Such systems as polymers, biological macromolecules, natural and industrial colloids, emulsions, and subcelluar particles appear to be amenable to separation by sedimentation FFF.

Electrical Field-Flow Fractionation

In electrical FFF, an electric field is applied perpendicular to the flow direction. Retention and separation occur based on electrical charge. Species with the highest charge are driven most effectively toward the accumulation wall. Species of lower charge are not as compacted and protrude more into the higher flow region. Hence, species of the lowest charge are eluted first, with highly charged species retained the most.

Because electrical fields are quite powerful, even small ions should be amenable to separation by electrical FFF. Electrolysis effects have limited the applications of this method to the separation of mixtures of proteins and other large molecules, however.

Thermal Field-Flow Fractionation

In thermal FFF, a thermal field is employed perpendicular to the flow direction by forming a temperature gradient across the FFF channel. The temperature difference induces thermal diffusion, in which the velocity of movement is related to the thermal diffusion coefficient of the species.

Thermal FFF is particularly well-suited for the separation of synthetic polymers with molecular weights in the range of 10^3 to 10^7. The technique has significant advantages over size-exclusion chromatography for high-molecular-weight polymers. Low-molecular-weight polymers, however, appear to be better separated by size exclusion methods. In addition to polymers, particles and colloids have been separated by thermal FFF.[19]

[19]P. M. Shiundu, G. Liu, and J. C. Giddings, *Anal. Chem.*, **1995,** *67,* 2705.

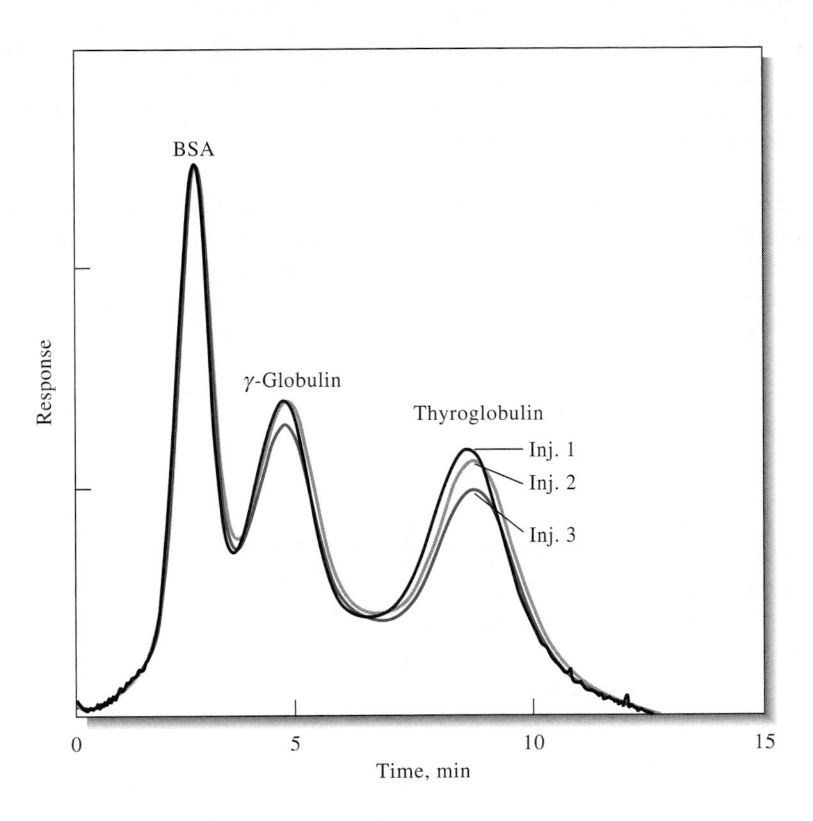

Figure 33-21 Separation of three proteins by flow FFF. Three separate injections are shown. In the experiment shown, the sample was concentrated at the head of the channel by means of an opposing flow. (Reprinted with permission from H. Lee, S. K. R. Williams, and J. C. Giddings, *Anal. Chem.*, **1998,** *70,* 2495. Copyright 1998 American Chemical Society.)

Flow Field-Flow Fractionation

Perhaps the most versatile of all the FFF subtechniques is flow FFF, in which the external field is replaced by a slow cross-flow of the carrier liquid. The perpendicular flow transports material to the accumulation wall in a nonselective manner. Steady-state layer thicknesses are different for various components, however, because they depend not only on the transport rate but also on molecular diffusion. Exponential distributions of differing thicknesses are formed, as in normal FFF.

Flow FFF has been applied to the separation of proteins, synthetic polymers, and a variety of colloidal particles. Figure 33-21 illustrates the separation of three proteins by flow FFF. The reproducibility is illustrated by the fractograms for the three injections.

33E-3 Advantages of Field-Flow Fractionation over Chromatographic Methods

Field-flow fractionation appears to have several advantages over ordinary chromatographic methods for some applications. First, no packing material or stationary phase is needed for separation to occur. In some chromatographic systems, there may be undesirable interactions between the packing material or stationary phase and the sample constituents. Some solvents or sample materials adsorb or react with the stationary phase or its support. Macromolecules and particles are particularly prone to such adverse interactions.

The geometry and flow profiles involved in FFF are well characterized. Likewise, the effects of most external fields can be readily modeled. As a consequence, fairly exact theoretical predictions of retention and plate height can be made in FFF. Chromatographic predictions are still rather inexact in comparison.

Finally, the external field governs FFF retention. With electrical, centrifugal, and flow FFF, the perpendicular forces can be varied rapidly and in a time-programmed fashion. This imparts to FFF a certain versatility in adapting to different types of samples. Likewise, methods can be readily optimized for resolution and separation speed.

Although field-flow fractionation is a fairly recent addition to analytical separation methods, it has been shown to be highly complementary to chromatography. The FFF methods are best suited at present for macromolecules and particles, which are for the most part beyond the molecular weight range of chromatographic methods. Chromatographic methods are clearly superior for low-molecular-weight substances, however.

WEB WORKS

Connect to our Web Site at **http://chemistry.brookscole.com/skoogfac/.** From the Chapter Resources menu, choose Web Works. Locate the Chapter 33 section and click on the link to the Agilent Technologies Web site. Find the application brief entitled *Analysis of human rhinovirus (common cold virus) in viral preparations by CZE.* What type of detector was used in this study? What type of injection was used, and how long was it applied? What was the applied voltage? What background electrolyte was used?

QUESTIONS AND PROBLEMS

33-1. List the types of substances to which each of the following separation methods is most applicable:
- *(a) supercritical-fluid chromatography
- (b) thin-layer chromatography
- *(c) capillary zone electrophoresis
- (d) thermal FFF
- *(e) flow FFF
- (f) micellar electrokinetic capillary chromatography

33-2. Define
- *(a) supercritical fluid.
- (b) critical point.
- *(c) two-dimensional thin-layer chromatography.
- (d) electrophoretic mobility.
- *(e) critical micelle concentration.
- (f) sedimentation FFF.

***33-3.** What properties of a supercritical fluid are important in chromatography?

33-4. How do instruments for supercritical-fluid chromatography differ from those for (a) HPLC and (b) GC?

***33-5.** Describe the effect of pressure on supercritical-fluid chromatograms.

33-6. List some of the advantages of supercritical CO_2 as a mobile phase for chromatographic separations.

***33-7.** What important property of supercritical fluids is related to their high densities?

33-8. Compare supercritical-fluid chromatography with other column chromatographic methods.

***33-9.** For supercritical carbon dioxide, predict the effect that the following changes will have on the elution time in an SFC experiment.
- (a) Increase the flow rate (at constant temperature and pressure).
- (b) Increase the pressure (at constant temperature and flow rate).
- (c) Increase the temperature (at constant pressure and flow rate).

33-10. What is electroosmotic flow? Why does it occur?

***33-11.** Suggest a way in which electroosmotic flow might be repressed.

33-12. Why does pH affect the separation of amino acids by electrophoresis?

***33-13.** What is the principle of separation by capillary zone electrophoresis?

33-14. A certain inorganic cation has an electrophoretic mobility of 4.31×10^{-4} cm^2 s^{-1} V^{-1}. This same ion has a diffusion coefficient of 9.8×10^{-6} cm^2 s^{-1}. If this ion is separated by capillary zone electrophoresis with a 50.0-cm capillary, what is the expected plate count N at applied voltages of
(a) 5.0 kV?
(b) 10.0 kV?
(c) 30.0 kV?

***33-15.** The cationic analyte of Problem 33-14 was separated by capillary zone electrophoresis in a 50.0-cm capillary at 10.0 kV. Under the separation conditions, the electroosmotic flow rate was 0.85 mm s^{-1} toward the cathode. If the detector were placed 40.0 cm from the injection end of the capillary, how long would it take in minutes for the analyte cation to reach the detector after the field is applied?

33-16. What is the principle of micellar electrokinetic capillary chromatography? How does it differ from capillary zone electrophoresis?

***33-17.** Describe a major advantage of micellar electrokinetic capillary chromatography over conventional liquid chromatography.

33-18. Three large proteins are ionized at the pH at which an electrical FFF separation is carried out. If the ions are designated A^{2+}, B^{+}, and C^{3+}, predict the order of elution.

***33-19.** What determines the elution order in sedimentation FFF?

33-20. List the major advantages and limitations of FFF compared with chromatographic methods.

33-21. Challenge Problem Doxorubicin (DOX) is a widely used anthracycline that has been effective in the treatment of leukemia and breast cancer in humans (A. B. Anderson, C. M. Ciriaks, K. M. Fuller, and E. A. Ariaga, *Anal. Chem.*, **2003,** *75,* 8). Unfortunately, side effects such as liver toxicity and drug resistance have been reported. In a recent study, Anderson et al. used laser-induced fluorescence (LIF) as a detection mode for capillary electrophoresis to investigate metabolites of DOX in single cells and subcellular fractions. The following are results similar to those obtained by Anderson et al. for quantifying doxorubicin by LIF. The CE peak areas were measured as a function of the DOX concentration to construct a calibration curve.

DOX Concentration, nM	Peak Area
0.10	0.10
1.00	0.80
5.00	4.52
10.00	8.32
20.00	15.7
30.00	26.2
50.00	41.5

(a) Find the equation for the calibration curve and the standard deviations of the slope and intercept. Find the R^2 value.
(b) Rearrange the equation found in part (a) to express concentration in terms of the measured area.
(c) The limit of detection for DOX was found to be 3×10^{-11} M. If the injection volume was 100 pL, what was the LOD in moles?
(d) Two samples of unknown DOX concentration were injected, and peak areas of 11.3 and 6.97 were obtained. What were the concentrations and their standard deviations?
(e) Under certain conditions, the DOX peak required 300 s to reach the LIF detector. What time would be required if the applied voltage were doubled? What time would be required if the capillary length were doubled at the same applied voltage?
(f) The capillary used in part (e) under normal conditions had a plate count of 100,000. What would N be if the capillary length were doubled at the same applied voltage? What would N be if the applied voltage were doubled at the original capillary length?
(g) For a 40.6-cm-long capillary of inside diameter 50 μm, what would the plate height be for a capillary with $N = 100{,}000$?
(h) For the same capillary as in part (g), what is the variance σ^2 of a typical peak?

InfoTrac College Edition

For additional readings, go to InfoTrac College Edition, your online research library, at

http://infotrac.thomsonlearning.com

PART VII

Practical Aspects of Chemical Analysis

This chapter is only available as an Adobe Acrobat® PDF file on the **Analytical Chemistry CD-ROM** enclosed in this book or on our Web site at **http://chemistry.brookscole.com/skoogfac/.**

A conversation with *Julie Leary*

Photo by The Picture People. Courtesy of Julie Leary.

Julie Leary grew up in a small, economically depressed mill town in the eastern United States. As the first member of her extended family to go to college, she received her B.A. in psychology from the University of Massachusetts. She was not happy in psychology, however, and soon discovered a love of chemistry. She went back to college for a B.S. in chemistry at the Lowell Technical Institute, and again for her Ph.D. in analytical chemistry at the Massachusetts Institute of Technology. After 1 year there as a postdoctoral researcher, she moved to Berkeley to join the faculty at the University of California, where she is currently Adjunct Full Professor and Director of the College of Chemistry Analytical Facilities—and a long way from the small mill town in which she grew up. Leary received the 2000 Biemann medal from the American Society for Mass Spectrometry, an award presented early in a scientist's career. It honors her for using metal ligand coordination to carbohydrates for stereochemical analysis.

Q: We understand that you originally studied psychology. How did you like that field?

A: I started out with a bachelor's degree in psychology and worked with Korsakoff syndrome patients. (Korsakoff syndrome is a neurological disorder characterized by severe amnesia.) I did not enjoy this position and ended up as a secretary for a biomedical research company. I became very interested in the topic and took an intensive summer course in organic chemistry. I did well and really enjoyed it, so I decided to pursue a career in chemistry.

Q: What was your subsequent training in chemistry?

A: After receiving my chemistry degree, I became a technical specialist at SmithKline Clinical Laboratories. I was involved with the development of a new New York State Department of Health laboratory for drug testing. The lab was designed for drug overdose analysis in the serum and urine of hospital patients as well as for determining drug levels in race horses. I gained a considerable amount of practical experience, but it was very clear to me that if I wanted an upper level position in industry or an academic appointment at a research university, I would need a doctoral degree. After being in the workforce and making a considerable salary, it was a difficult decision to go back to school. But it was well worth it.

Q: Do you now primarily teach?

A: I'm currently Adjunct Professor and Director of the College of Chemistry Analytical Facilities. Most of the time, I wear two hats. Half of my time is spent conducting research with graduate students and postdoctoral students, and the other half is overseeing the analytical facilities. I also teach one semester each academic year. In my administrative position, I set and watch the budgets for each of the five instrumentation facilities (NMR, X-ray diffraction, microanalysis, computer graphics, and mass spectrometry), oversee personnel, and institute instrument proposals when needed. For example, I determine what equipment we need to keep our facilities on the cutting edge. If a certain piece of new equipment is needed, I begin the process of organizing and writing a proposal to send to the National Science Foundation or the National Institutes of Health to obtain funding. As an adjunct professor, I mentor a research group, obtain extramural funding, and serve on various academic committees.

Q: One of your interests is mass spectrometry. Can you explain this topic?

A: Mass spectrometry provides the molecular weight of a compound, and high-resolution mass spectrometry allows you to determine the exact mass to within four significant figures. There are several different types of mass spectrometers—we have seven or eight at Berkeley—and several different ways in which you can ionize a sample, such as fast-atom bombardment or electrospray ionization. Basically, you introduce a sample into the instrument in such a way that it gives you information about its molecular weight. Then, using various sophisticated methods to perturbate the electrons, you take an ion that represents the molecular weight and force it to fall apart into its component pieces. After that, you can use the high-resolution data to work backward to determine the elemental composition. When you cause a compound to fall apart into its component pieces, you can get information about how the compound is put together.

Q: How have you been using mass spectrometry on cell surfaces?

A: We're characterizing carbohydrates on cell surfaces. This is important because many diseases are brought on by cell-to-cell interactions. Compounds on cell surfaces are used for communication to other cells to start or stop certain biochemical processes. In particular, we're looking at carbohydrates on the surfaces of bacteria. Signaling factors on bacteria can trigger responses in humans either to initiate an immune response or to let the bacteria infiltrate. With the goal of characterizing these carbohydrates, we've developed a method utilizing metals and metal ligands that are then synthesized onto the carbohydrates—they're used as tags on the carbohydrates. This allows us to obtain stereochemical information using mass spectrometry. The metal ligand helps lock carbohydrates into a confirmation in the gas phase so that solution memory is retained in the gas phase. This is the work for which I won the Biemann medal.

The use of analytical chemistry is turning up the heat in cell biology. Major advances are being made in understanding the genome and the proteome, and the techniques that underlie all these new discoveries derive from analytical chemistry.

Q: Are you also involved in synthetic chemistry?
A: During investigations of the metal-ligated oligomers, we discovered a compound that was unique; when coordinated to glucose, it gave a ligand bridging two glucoses. We isolated and purified that compound and sent it to the National Cancer Institute. They are interested in looking at pure compounds to screen against their 60-cancer-cell line to look for chemotherapeutic agents. Our compound showed activity against cell lines of breast and ovarian cancer. In fact, it showed four times more activity than tamoxifen against breast cancer in vitro. This set us on the path of looking for the target mechanism for the compound's chemotherapeutic effects. Our preliminary data indicate that it binds to the estrogen receptor. This is leading us into a whole synthetic area outside of analytical chemistry to make various analogues to test efficacy against the estrogen receptor.

Q: How do you use mass spectrometry to measure kinetic constants?
A: Our lab is currently very involved in measuring the kinetic constants of possible inhibitors to various enzymes using mass spectrometry. Using a mass spectrometry–based method, we can measure K_m, V_{max}, and K_i of enzymes, substrates, and inhibitors without the use of a calibration curve. We're now screening a variety of combinatorial libraries synthesized by the Bertozzi group at Berkeley. The compound libraries are intended to generate one or more inhibitors of some of the more important sulfotransferase enzymes. For example, one of the enzymes is estrogen sulfotransferase, which involves sulfation onto estrodial. This is affiliated with the onset of ovarian cancer. Once we identify the inhibitors, we will measure the kinetic constants.

Q: In your estimation, what is the value of analytical chemistry?
A: The use of analytical chemistry is turning up the heat in cell biology. Major advances are being made in understanding the genome and the proteome, and the techniques that underlie all these new discoveries derive from analytical chemistry. Mass spectrometry has been especially important in the area of proteomics, and now considerable attention is given to the subspecialty of analytical chemistry in most biochemistry textbooks. Analytical chemistry really permeates all areas of science.

Q: Have you seen any changes involving women in science?
A: When I was growing up, women were taught to be nurturers, and they had a tendency to migrate toward those careers that involved being caretakers. We were not encouraged to pursue careers in science. When I was a graduate student, there were few women around to talk to about my chemistry and my life as a female grad student. The "up" side to this was that I either talked to my male colleagues or I talked to no one. The number of women in science has changed considerably during the past 20 years. When I first attended our national mass spectrometry meeting in 1980, there were only a handful of women attending; now a third or maybe more of the 3000 members are women. I have tried to encourage women not to be timid or afraid about pursuing a career in analytical chemistry if that's what they want to do. There are so many avenues available.

Q: How do you juggle being a chemist and a mom?
A: In both being a mother and holding a demanding career position, two things are really important. In the kind of job I have, you need to set tasks and prioritize your time; thus, good organizational skills are imperative. It is very rewarding to know that at the week's end you have accomplished most of your objectives for that week. So setting reasonable goals and reaching them is important both psychologically and realistically. The second thing that is extremely important if you are married and have a family is having a spouse who is willing to support you and carry half the load. In our family, the chores and child care are evenly distributed, 50–50. Without that kind of support, it is extremely difficult—if not impossible—to be successful and effective at work and still make your family a priority.

CHAPTER 34

Analysis of Real Samples

© NASA/Corbis

The analysis of real samples, such as the soil and rock samples brought back to the earth from the moon by the Apollo astronauts, is usually quite complex compared with the analysis of materials studied in laboratory courses. As discussed in this chapter, the choice of analytical method for real materials is not simple, often requiring consultation of the literature, modification of existing methods, and extensive testing to determine method validity.

The photo shows one of the Apollo astronauts taking a core sample of the lunar soil. Such samples were valuable in determining the geological history of the moon and its relationship to the history of the earth.

Very early in this text (Section 1C), we pointed out that a quantitative analysis involves a sequence of steps: (1) selecting a method, (2) sampling, (3) preparing a laboratory sample, (4) defining replicate samples by mass or volume measurements, (5) preparing solutions of the samples, (6) eliminating interferences, (7) completing the analysis by performing measurements that are related in a known way to analyte concentration, and (8) computing the results and estimating their reliability.

Thus far we have focused largely on steps 6, 7, and 8 and to a lesser extent on steps 2 and 4. We have chosen this emphasis not *because the earlier steps are unimportant or easy. In fact, the preliminary steps may be more difficult and time-consuming than the two final steps of an analysis and may be greater sources of error.*

The reasons for postponing a discussion of the preliminary steps to this point are pedagogical. Experience has shown that it is easier to introduce students to analytical techniques by having them first perform measurements on simple materials for which no method selection is required and for which problems with sampling, sample preparation, and sample dissolution are either nonexistent or easily solved. Thus, we have been largely concerned so far with measuring the concentration of analytes in simple aqueous solutions that have few interfering species.

34A REAL SAMPLES

Determining an analyte in a simple sample is often easier than in a complex material because the number of variables that must be controlled is small and the tools available are numerous and easy to use. Also, with simple systems, our knowledge of the chemical and measurement principles allows us to anticipate problems and to correct for them.

In fact, academic and industrial chemists are usually interested in materials that are not, as a rule, simple. To the contrary, most analytical samples are complex mixtures of species; in some cases, hundreds of species. Such materials are frequently far from ideal in matters of solubility, volatility, stability, and homogeneity, and many steps must precede the final measurement step. Indeed, the final measurement may be easier and less time-consuming than any of the preceding steps.

For example, we showed in earlier chapters that the calcium ion concentration of an aqueous solution is readily determined by titration with a standard EDTA solution or by potential measurements with a specific-ion electrode. Alternatively, the calcium content of a solution can be determined either from atomic absorption or atomic emission measurements or by the precipitation of calcium oxalate followed by weighing or titrating with a standard solution of potassium permanganate.

All these methods can be used to determine the calcium content of a simple salt, such as the carbonate. Chemists are seldom interested in the calcium content of calcium carbonate, however. More likely, what is needed is the percentage of this element in a sample of animal tissue, a silicate rock, or a piece of glass. The analysis thus acquires a new level of complexity. For example, none of these materials is soluble in water or dilute aqueous reagents. Before calcium can be determined, therefore, the sample must be decomposed by high-temperature treatment with concentrated reagents. Unless care is taken, we could lose some calcium during this step, or, equally bad, we could introduce some calcium as a contaminant because of the relatively large quantities of reagent usually needed for sample decomposition.

Even after the sample has been decomposed to give a solution containing calcium ions, the procedures mentioned in the two previous paragraphs cannot ordinarily be applied immediately to complete the analysis because the reactions or properties used are not specific to calcium. Thus, a sample of animal tissue, silicate rock, or glass almost surely contains one or more components that also react with EDTA, act as a chemical interference in an atomic-absorption measurement, or form a precipitate with oxalate ion. In addition, the high ionic strength resulting from the reagents used for sample decomposition would complicate a direct potentiometric measurement. Because of these complications, several additional operations are required to eliminate interferences before the final measurement is made.

We have chosen the term **real samples** to describe materials such as those in the preceding illustration. In this context, most of the samples encountered in an elementary quantitative analysis laboratory course definitely are not real but rather are homogeneous, stable, readily soluble, and chemically simple. Also, there are well-established and thoroughly tested methods for their analysis. There is considerable value in introducing analytical techniques with such materials because they permit you to concentrate on the mechanical aspects of an analysis. Even experienced analysts use such samples when learning a new technique, calibrating an instrument, or standardizing solutions.

Real samples are far more complex than most of those that you encounter in the instructional laboratory.

In the real world, determining the compositions of real samples frequently demands more intellectual skill and chemical intuition than mechanical aptitude. Often, a compromise must be struck between the available time and the accuracy that is judged necessary. We are frequently happy to settle for an accuracy of one or two parts per hundred instead of one or two parts per thousand, knowing that ppt accuracy may require several hours or even days of additional effort. In fact, with complex materials, even parts-per-hundred accuracy may be unrealistic.

The difficulties encountered in the analysis of real samples stem from their complexity. As a result, the literature may not contain a well-tested analytical route for the kind of sample under consideration. For such cases, an existing procedure must be modified to take into account compositional differences between the current sample and the original samples. Alternatively, an entirely new analytical method might need to be developed. In either case, the number of variables that must be taken into account usually increases exponentially with the number of species contained in the sample.

As an example, contrast the problems associated with the inductively coupled plasma atomic emission analysis of calcium carbonate with those for a *real* calcium-containing sample. In the former, the number of components is small and the variables likely to affect the results are reasonably few. Principal among the variables are the physical losses of analyte due to the evolution of carbon dioxide when the sample is dissolved in acid; the effect of the anion of the acid and of the radio-frequency power on the intensity of the calcium emission line; the position of the plasma with respect to the entrance slit to the spectrometer; and the quality of the standard calcium solutions used for calibration.

Determining calcium in a real sample such as a bone or a silicate rock is far more complex, since the sample is insoluble in ordinary solvents and contains a dozen or more species. The silicate rock sample, for example, can only be dissolved by fusion at a high temperature with a large excess of a reagent such as sodium carbonate. Physical loss of the analyte is possible during this treatment unless suitable precautions are taken. Furthermore, the introduction of calcium from the excess sodium carbonate or the fusion vessel is of real concern. Following fusion, the sample and reagent are dissolved in acid. With this step, all the variables affecting the calcium carbonate sample are operating, but, in addition, a host of new variables are introduced because of the dozens of components in the sample matrix. Now, measures must be taken to minimize instrumental and chemical interference brought about by the presence of various anions and cations in the solution being introduced into the plasma.

The analysis of a real substance is often a challenging problem requiring knowledge, intuition, and experience. The development of a procedure for such materials is a demanding task even for an experienced chemist.

34B CHOICE OF ANALYTICAL METHOD

The choice of a method for the analysis of a complex substance requires good judgment based on sound knowledge of the advantages and limitations of the various analytical tools available. In addition, a familiarity with the literature of analytical chemistry is essential. We cannot be very explicit concerning how an analytical method is selected because there is no single best way that applies under all circumstances. We can, however, suggest a systematic approach to the problem and present some generalities that can aid in making intelligent decisions.

34B-1 Definition of the Problem

► The objectives of an analysis must be clearly defined before the work begins.

A first step, which must precede any choice of method, involves a clear definition of the analytical problem. The method of approach selected will be largely governed by the answers to the following questions:

What is the concentration range of the species to be determined?
What degree of accuracy is desired?
What other components are present in the sample?

What are the physical and chemical properties of the gross sample?
How many samples are to be analyzed?

The concentration range of the analyte may well limit the number of feasible methods. If, for example, we wish to determine an element present at the parts-per-billion or parts-per-million level, gravimetric or volumetric methods can generally be eliminated, and spectrometric, potentiometric, and other more sensitive methods become likely candidates. For components in the parts-per-billion and parts-per-million range, even small losses resulting from coprecipitation or volatility and contamination from reagents and apparatus become major concerns. In contrast, if the analyte is a major component of the sample, these considerations are less important, and a classical analytical method may well be preferable.

The answer to the question of required accuracy is vitally important in the choice of method and in the way it is performed because the time required to complete an analysis increases greatly with demands for higher accuracy. Thus, to improve the reliability of analytical results from 2% to 0.2% relative may require an increase in the analysis time by a factor of 100 or more. Consequently, we should always carefully consider the degree of accuracy really needed *before* undertaking an analysis.

◀ The time required to carry out an analysis increases, often in an exponential manner, with the desired level of accuracy.

The demands for accuracy frequently dictate the procedure chosen for an analysis. For example, if the allowable error in the determination of aluminum is only a few parts per thousand, a gravimetric procedure is probably required. If an error of 50 ppt can be tolerated, however, a spectroscopic or electroanalytical approach may be preferable.

The way in which an analysis is carried out is also affected by accuracy requirements. If precipitation with ammonia is chosen for the analysis of a sample containing 20% aluminum, the presence of 0.2% iron is of serious concern if accuracy in the parts-per-thousand range is demanded, and a preliminary separation of the two elements is necessary. If an error of 50 ppt is tolerable, however, the separation of iron is not necessary. This tolerance can also govern other aspects of the method. For example, 1-g samples can be weighed to perhaps 10 mg and certainly no closer than 1 mg. In addition, less care is needed in transferring and washing the precipitate and in other time-consuming operations of the gravimetric method. The intelligent use of shortcuts is not a sign of carelessness but a recognition of realities in matters of time and effort. The question of accuracy, then, must be clearly resolved before beginning an analysis.

◀ Often, you can save considerable time by the use of permissible shortcuts in an analytical procedure.

To choose a method for the determination of one or more species in a sample, it is necessary to know what other elements or compounds are present. If you are lacking such information, a qualitative analysis must be undertaken to identify components that are likely to interfere in the various methods under consideration. As we have noted repeatedly, most analytical methods are based on reactions and physical properties that are shared by several elements or compounds. Thus, measurement of the concentration of a given element by a method that is simple and straightforward in the presence of one group of elements or compounds may require many tedious and time-consuming separations in the presence of others. A solvent suitable for one combination of compounds may be totally unsatisfactory when applied to another. It is very important to know the approximate chemical composition of the sample before selecting a method for the quantitative determination of one or more of its components.

◀ It is frequently necessary to identify the components of a sample before undertaking a quantitative analysis.

We must also consider the physical state of the sample to determine whether it must be homogenized, whether volatility losses are likely, and whether its composition may change under laboratory conditions because of the absorption or loss of water. We must also determine how to decompose or dissolve the sample without loss of analyte. Preliminary tests of one sort or another may be needed to provide this type of information.

Finally, the number of samples to be analyzed is an important criterion in selecting a method. If there are many samples, considerable time can be spent in calibrating instruments, preparing reagents, assembling equipment, and investigating shortcuts, since the cost of these operations can be spread over the large number of samples. If, however, a few samples at most are to be analyzed, a longer and more tedious procedure involving a minimum of these preparatory operations may prove to be the wiser choice from the economic standpoint.

Once we have answered the preliminary questions, we can then consider possible approaches to the problem. Sometimes, based on past experience, the route to be followed is obvious. In other instances, we must speculate on problems that are likely to be encountered in the analysis and how they can be solved. By this time, some methods probably will have been eliminated from consideration and others put on the doubtful list. Ordinarily, however, we first turn to the analytical literature to profit from the experience of others.

► A little extra time spent in the library can save a tremendous amount of time and effort in the laboratory.

34B-2 Investigating the Literature

A list of reference books and journals concerned with various aspects of analytical chemistry appears in Appendix 1. This list is not exhaustive but rather one that is adequate for most work. It is divided into several categories. In many instances, the division is arbitrary since some works could be logically placed in more than one category.

We usually begin a search of the literature by referring to one or more of the treatises on analytical chemistry or to those devoted to the analysis of specific types of materials. In addition, it is often helpful to consult a general reference work relating to the compound or element of interest. From this survey, a clearer picture of the problem at hand may develop, including the steps that are likely to be difficult, the separations that must be made, and the pitfalls to be avoided. Occasionally, all the answers needed or even a set of specific instructions for the analysis may be found. Alternatively, journal references that lead directly to this information may be discovered. Sometimes, we find only a general notion of how to proceed. Several possible methods may appear suitable; others may be eliminated. At this point, it is often helpful to consider reference works concerned with specific substances or specific techniques. The various analytical journals may be consulted. Monographs on methods of completing the analysis are often valuable in deciding among several possible techniques.

► The technology for computer-based scientific information retrieval provides an efficient means of surveying the analytical literature. For example, complete archives of *all* American Chemical Society journals have recently become available on-line.

A major problem in using analytical journals is locating articles pertinent to the problem at hand. The various reference books are useful since most contain many references to the original journals. The key to a thorough search of the literature, however, is *Chemical Abstracts.* Manual searches involve the expenditure of a great deal of time and are often made unnecessary by consulting reliable reference works. Computer-aided literature searches have greatly minimized the time required for a careful literature search.

34B-3 Choosing or Devising a Method

After defining the problem and investigating the literature for possible approaches, we must next decide on the route to be followed in the laboratory. If the choice is simple and obvious, analysis can be undertaken directly. Frequently, however, the decision requires the exercise of considerable judgment and ingenuity; experience, an understanding of chemical principles, and perhaps intuition all come into play.

If the substance to be analyzed occurs widely, the literature survey usually yields several alternative methods for the analysis. Economic considerations may dictate a method that will yield the desired reliability with the least expenditure of time and effort. As mentioned earlier, the number of samples to be analyzed is often a determining factor in the choice.

Investigation of the literature does not invariably reveal a method designed specifically for the type of sample in question. Ordinarily, however, we will encounter procedures for materials that are at least similar in composition to the one in question. We must then decide whether the variables introduced by the differences in composition are likely to have any influence on the results. This judgment can be difficult, and we may still be uncertain as to the effects. Experiments in the laboratory may be the only way of making a wise decision.

◀ Preliminary laboratory testing may be needed to evaluate proposed changes to established methods.

If we conclude that existing procedures are not applicable, consideration must be given to modifications that may overcome the problems imposed by the variation in composition. Again, the complexity of the chemical system may dictate that we can propose only tentative alterations. Whether these modifications will accomplish their purpose without introducing new difficulties can be determined only in the laboratory.

After giving due consideration to existing methods and their modifications, we may decide that none fits the problem and an entirely new procedure must be developed. In doing so, all the facts about the chemical and physical properties of the analyte must be organized and given consideration. Several possible ways of performing the desired measurement may become evident from this information. Each possibility must then be examined critically, with consideration given to the influence of the other components in the sample as well as to the reagents that must be used for solution or decomposition. At this point, we must try to anticipate sources of error and possible interferences due to interactions among sample components and reagents; it may be necessary to devise strategies to circumvent such problems. The conclusion of such a preliminary survey generally produces one or more tentative methods worth testing. Usually, the feasibility of some of the steps in the procedure cannot be determined without preliminary laboratory testing. Certainly, critical evaluation of the entire procedure can come only from careful laboratory work.

34B-4 Testing the Procedure

Once a procedure for an analysis has been selected, we must decide whether it can be applied directly to the problem at hand, or it must be tested. The answer to this question is not simple and depends on a number of considerations. If the method chosen is the subject of a single literature reference, or at most a few, there may be real value in a preliminary laboratory evaluation. With experience, we become more and more cautious about accepting claims regarding the accuracy and applicability of a new method. All too often, statements found in the literature tend to be overly optimistic; a few hours spent in testing the procedure in the laboratory may be enlightening.

Whenever a major modification of a standard procedure is undertaken or an attempt is made to apply it to a type of sample different from that for which it was designed, a preliminary laboratory test is advisable. The effects of such changes simply cannot be predicted with certainty.

Finally, a newly devised procedure must be extensively tested before it is adapted for general use. We now consider the means by which a new method or a modification of an existing method can be tested for reliability.

The Analysis of Standard Samples

The best way to evaluate an analytical method is to analyze one or more standard samples whose analyte composition is reliably known. For this technique to be effective, however, it is essential that the standards closely resemble the samples to be analyzed with respect to both the analyte concentration range and the overall composition.

Occasionally, standards suitable for method testing can be synthesized by thoroughly homogenizing weighed quantities of pure compounds. Such a procedure is generally inapplicable, however, when the samples to be analyzed are complex, such as biological materials, soil samples, and many forensic samples.

► The National Institute of Standards and Technology is an important source for standard reference materials. For literature describing standard reference materials, see the references in Footnotes 4 and 5 in Chapter 5. Also see http://www.nist.gov.

Section 8D-3 discusses the general methods for validating analytical results. The National Institute of Standards and Technology sells a variety of standard reference materials that have been specifically prepared for validation purposes.[1] Most standard reference materials are substances commonly encountered in commerce or in environmental, pollution, clinical, biological, or forensic studies. The concentration of one or more components in these materials is certified by the Institute based on measurements using (1) a previously validated reference method, (2) two or more independent reliable measurement methods, or (3) results from a network of cooperating laboratories that are technically competent and thoroughly familiar with the material being tested. More than 1200 of these materials are available, including such substances as ferrous and nonferrous metals; ores, ceramics, and cements; environmental gases, liquids, and solids; primary and secondary chemicals; clinical, biological, and botanical samples; fertilizers; and glasses. Several industrial concerns also offer various kinds of standard materials designed for validating analytical procedures.

When standard reference materials are not available, the best we can do is to prepare a solution of known concentration whose composition approximates that of the sample after it has been decomposed and dissolved. Obviously, such a standard gives no information at all concerning the fate of the substance being determined during the important decomposition and solution steps.

Using Other Methods

The results of an analytical method can sometimes be evaluated by comparison with data obtained from an entirely different method, particularly if we have prior knowledge about the reliability of the reference method. The second method should be based on chemical or instrumental principles that differ as much as possible from the method being considered. Since it is unlikely that the same errors influence both methods, if we obtain results comparable to the reference method, we can usually conclude that our new method is satisfactory. Such a conclusion does not apply to those aspects of the two methods that are similar.

Standard Addition to the Sample

When standard reference materials and different analytical methods are not applicable, the standard-addition method may prove useful. Here, in addition to being

[1]See U.S. Department of Commerce, *NIST Standard Reference Materials Catalog,* 1998–99 Ed., NIST Special Publication 260-98-99. Washington, D.C.: U.S. Government Printing Office, 1998. More recent information can be found on the NIST Web site at http://www.nist.gov.

used to analyze the sample, the proposed procedure is tested against portions of the sample to which known amounts of the analyte have been added. The effectiveness of the method can then be established by evaluating the extent of recovery of the added quantity. The standard-addition method may reveal errors arising from the way the sample was treated or from the presence of the other elements or compounds in the matrix.

The **standard-addition method** is described in Section 8C-3. Applications of standard-addition methods are presented in Chapters 21, 26, and 28.

34C ACCURACY IN THE ANALYSIS OF COMPLEX MATERIALS

To provide a clear idea of the accuracy that can be expected in the analysis of a complex material, data on the determination of four elements in a variety of materials are presented in Tables 34-1 to 34-4. These data were taken from a much larger set of results collected by W. F. Hillebrand and G. E. F. Lundell of the National Bureau of Standards and published in the first edition of their classical book on inorganic analysis.[2]

TABLE 34-1

Determination of Iron in Various Materials*

Materials	Iron, %	Number of Analysts	Average Absolute Error	Average Relative Error, %
Soda-lime glass	0.064 (Fe_2O_3)	13	0.01	15.6
Cast bronze	0.12	14	0.02	16.7
Chromel	0.45	6	0.03	6.7
Refractory	0.90 (Fe_2O_3)	7	0.07	7.8
Manganese bronze	1.13	12	0.02	1.8
Refractory	2.38 (Fe_2O_3)	7	0.07	2.9
Bauxite	5.66	5	0.06	1.1
Chromel	22.8	5	0.17	0.75
Iron ore	68.57	19	0.05	0.07

*From W. F. Hillebrand and G. E. F. Lundell, *Applied Inorganic Analysis,* p. 878. New York: Wiley, 1929. Reprinted by permission of Mrs. Ernst D. Lundell.

TABLE 34-2

Determination of Manganese in Various Materials*

Material	Manganese, %	Number of Analysts	Average Absolute Error	Average Relative Error, %
Ferro-chromium	0.225	4	0.013	5.8
Cast iron	0.478	8	0.006	1.3
	0.897	10	0.005	0.56
Manganese bronze	1.59	12	0.02	1.3
Ferro-vanadium	3.57	12	0.06	1.7
Spiegeleisen	19.93	11	0.06	0.30
Manganese ore	58.35	3	0.06	0.10
Ferro-manganese	80.67	11	0.11	0.14

*From W. F. Hillebrand and G. E. F. Lundell, *Applied Inorganic Analysis,* p. 880. New York: Wiley, 1929. Reprinted by permission of Mrs. Ernst D. Lundell.

[2]W. F. Hillebrand and G. E. F. Lundell, *Applied Inorganic Analysis,* pp. 874–887. New York: Wiley, 1929.

TABLE 34-3
Determination of Phosphorus in Various Materials*

Material	Phosphorus, %	Number of Analysts	Average Absolute Error	Average Relative Error, %
Ferro-tungsten	0.015	9	0.003	20
Iron ore	0.014	31	0.001	2.5
Refractory	0.069 (P_2O_5)	5	0.011	16
Ferro-vanadium	0.243	11	0.013	5.4
Refractory	0.45	4	0.10	22
Cast iron	0.88	7	0.01	1.1
Phosphate rock	43.77 (P_2O_5)	11	0.5	1.1
Synthetic mixtures	52.18 (P_2O_5)	11	0.14	0.27
Phosphate rock	77.56 [$Ca_3(PO_4)_2$]	30	0.85	1.1

*From W. F. Hillebrand and G. E. F. Lundell, *Applied Inorganic Analysis,* p. 882. New York: Wiley, 1929. Reprinted by permission of Mrs. Ernst D. Lundell.

TABLE 34-4
Determination of Potassium in Various Materials*

Material	Potassium Oxide, %	Number of Analysts	Average Absolute Error	Average Relative Error, %
Soda-lime glass	0.04	8	0.02	50
Limestone	1.15	15	0.11	9.6
Refractory	1.37	6	0.09	6.6
	2.11	6	0.04	1.9
	2.83	6	0.10	3.5
Lead-barium glass	8.38	6	0.16	1.9

*From W. F. Hillebrand and G. E. F. Lundell, *Applied Inorganic Analysis,* p. 883. New York: Wiley, 1929. Reprinted by permission of Mrs. Ernst D. Lundell.

The materials analyzed were naturally occurring substances and items of commerce; they were specially prepared to give uniform and homogeneous samples and were distributed among chemists who were, for the most part, actively engaged in the analysis of similar materials. The analysts were allowed to use the methods they considered most reliable and best suited for the problem at hand. In most instances, special precautions were taken, and the results were consequently better than could be expected from the average routine analysis.

The numbers in the second column of Tables 34-1 to 34-4 are best values, obtained by the most painstaking analysis for the measured quantity. Each is considered to be the true value for calculation of the absolute and relative errors shown in the fourth and fifth columns. The fourth column was obtained by discarding extremely divergent results, determining the deviation of the remaining individual data from the best value (second column), and averaging these deviations. The fifth column was obtained by dividing the data in the fourth column by the best value (second column) and multiplying by 100%.

The results shown in these tables are typical of the data for 26 elements reported in the original publication. We conclude that (1) analyses reliable to a few tenths of a percent relative are the exception rather than the rule in the analysis of complex mixtures by ordinary methods and (2) unless we are willing to invest an inordinate amount of time in the analysis, errors on the order of 1% or 2% must be accepted. If the sample contains less than 1% of the analyte, we must expect even larger relative errors.

TABLE 34-5
Standard Deviation of Silica Results*

Year Reported	Sample Type	Number of Results	Standard Deviation (% Absolute)
1931	Glass	5	0.28†
1951	Granite	34	0.37
1963	Tonalite	14	0.26
1970	Feldspar	9	0.10
1972	Granite	30	0.18
1972	Syenite	36	1.06
1974	Granodiorite	35	0.46

*From S. Abbey, *Anal. Chem.*, **1981,** *53,* 529A.
†0.09 after eliminating one result.

The data in Tables 34-1 through 34-4 show that the accuracy obtainable in the determination of an element is greatly dependent on the nature and complexity of the substrate. Thus, the relative error in the determination of phosphorus in two phosphate rocks was 1.1%; in a synthetic mixture, it was only 0.27%. The relative error in an iron determination in a refractory was 7.8%; in a manganese bronze having about the same iron content, it was only 1.8%. In this example, the limiting factor in the accuracy was not in the completion step but rather in the dissolution of the samples and the elimination of interferences.

The data in the first four tables are more than 70 years old, and it is tempting to think that analyses carried out with more modern tools and additional experience are likely to be significantly better in terms of accuracy and precision. A study by S. Abbey suggests that this assumption is not valid, however.[3] For example, the data in Table 34-5, which were taken from his paper, reveal no significant improvement in silicate analyses of standard reference glass and rock samples in the 43-year period from 1931 to 1974. Indeed, the standard deviation among participating laboratories appears to be larger in later years.

◀ Fundamental sources of systematic and random error that were with us 70 years ago are still with us today.

The data in Tables 34-1 through 34-5 show that we are well advised to adopt a critical attitude regarding the accuracy of analytical results on real samples, even if we perform the analysis ourselves.

WEB WORKS

WWWWWWWWWW

Go to **http://chemistry.brookscole.com/skoogfac/.** From the Chapter Resources menu, choose Web Works. Locate the Chapter 34 section, click on the link for NIST, and find the pages dealing with Standard Reference Materials (SRMs). Look under Health Care and Nutrition. Find the Clinical Laboratory Materials available as SRMs. Find the information on glucose in frozen human serum and look up the Certificate of Analysis. Determine the relative uncertainties (as defined by NIST) of the glucose concentrations in mg/dL for the three different levels available.

InfoTrac College Edition

For additional readings, go to InfoTrac College Edition, your online research library, at

http://infotrac.thomsonlearning.com

[3]S. Abbey, *Anal. Chem.*, **1981,** *53,* 529A.

CHAPTER 35

Preparing Samples for Analysis

Carl Iwasaki/TimePix

The particle size of laboratory samples is often reduced prior to analysis by crushing and grinding operations. The techniques used in the laboratory are similar to those used in large-scale operations, such as the V-mixer/grinder used in a uranium plant shown in the photo. A V-mixer for laboratory use is described in Section 35A. In addition, this chapter considers several other methods of preparing samples for analysis, including various pulverizing and mixing methods. The chapter also considers the forms that moisture takes in solid samples and the methods of drying these samples.

In Section 8B, we considered the statistics involved in sampling and sample handling. In this chapter, we consider some of the details of preparing laboratory samples. In addition, the influence of moisture on samples and the determination of water in samples are discussed.

35A PREPARING LABORATORY SAMPLES

In Section 8B-4, we presented the statistical considerations involved in reducing the particle size of the gross sample so as to obtain a laboratory sample. Here, some of the specific techniques are described.

35A-1 Crushing and Grinding Samples

► Crushing and grinding the sample often change its composition.

A certain amount of crushing and grinding is usually required to decrease the particle size of solid samples. Because these operations tend to alter the composition of the sample, the particle size should be reduced no more than is required for homogeneity (see Section 8B-4) and ready attack by reagents.

Several factors can cause appreciable changes in sample composition as a result of grinding. The heat inevitably generated can cause losses of volatile components. In addition, grinding increases the surface area of the solid and thus increases its susceptibility to reaction with the atmosphere. For example, it has been observed that the iron(II) content of a rock may be decreased by as much as 40% during grinding—apparently a direct result of the iron being oxidized to the +3 state.

Frequently, the water content of a sample is altered substantially during grinding. Increases are observed as a consequence of the increased specific surface area that accompanies a decrease in particle size (page 322). The increased surface area leads to greater amounts of adsorbed water. For example, the water content of a piece of porcelain changed from 0 to 0.6% when it was ground to a fine powder.

In contrast, decreases in the water content of hydrates often take place during grinding as a result of localized frictional heating. For example, the water content

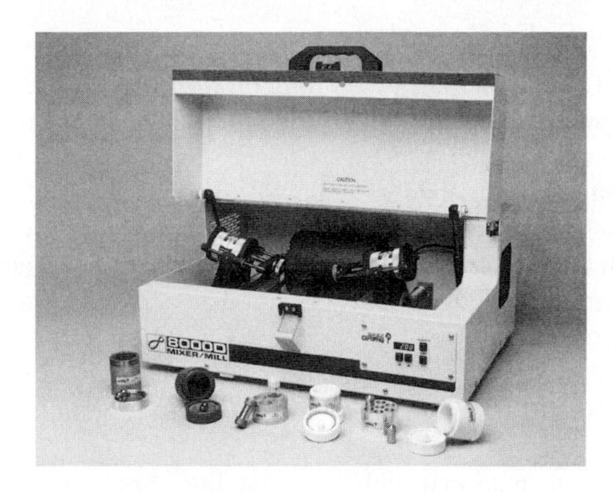

Figure 35-1 A commercial mixer/mill for pulverizing and blending samples. Stainless steel, ceramic, and tungsten carbide vials and mixing balls are available. Also shown are multiple sample adaptors. (Courtesy of Spex Industries, Metuchen, NJ.)

of gypsum ($CaSO_4 \cdot 2H_2O$) decreased from about 21% to 5% when the compound was ground to a fine powder.

Differences in hardness of the component can also introduce errors during crushing and grinding. Softer materials are ground to fine particles more rapidly than are hard ones and may be lost as dust as the grinding proceeds. In addition, flying fragments tend to contain a higher fraction of the harder components.

Intermittent screening often increases the efficiency of grinding. Screening involves shaking the ground sample on a wire or cloth sieve that will pass particles of a desired size. The residual particles are then returned for further grinding; the operation is repeated until the entire sample passes through the screen. The hardest materials, which often differ in composition from the bulk of the sample, are last to be reduced in particle size and are thus last through the screen. Therefore, grinding must be continued until every particle has been passed if the screened sample is to have the same composition as it had before grinding and screening.

◀ Crushing and grinding must be continued until the entire sample passes through a screen of the desired mesh size.

A serious contamination error can arise during grinding and crushing due to mechanical wear and abrasion of the grinding surfaces. Even though these surfaces are fabricated from hardened steel, agate, or boron carbide, contamination of the sample is nevertheless occasionally encountered. The problem is particularly acute in analyses for minor constituents.

◀ Mechanical abrasion of the surfaces of the grinding device can contaminate the sample.

A variety of tools are employed for reducing the particle size of solids, including jaw crushers and disk pulverizers for large samples containing large lumps, ball mills for medium-sized samples and particles, and various types of mortars for small amounts of material.

The **ball mill** is a useful device for grinding solids that are not too hard. It consists of a porcelain crock with a capacity of perhaps two liters that can be sealed and rotated mechanically. The container is charged with approximately equal volumes of the sample and flint or porcelain balls with a diameter of 20 to 50 mm. Grinding and crushing occur as the balls tumble in the rotating container. A finely ground and well-mixed powder can be produced in this way.

A commercial laboratory **mixer/mill** is shown in Figure 35-1 along with several mixing vials. The unit combines back-and-forth shaking with lateral motion for vigorous grinding of samples. The **Plattner diamond mortar,** shown in Figure 35-2, is used to crush hard, brittle materials. It is constructed of hardened tool steel and consists of a base plate, a removable collar, and a pestle. The sample is placed on the base plate inside the collar. The pestle is then fitted into place and is struck by several blows with a hammer, which reduces the solid to a fine powder that is collected on glazed paper after the apparatus has been disassembled.

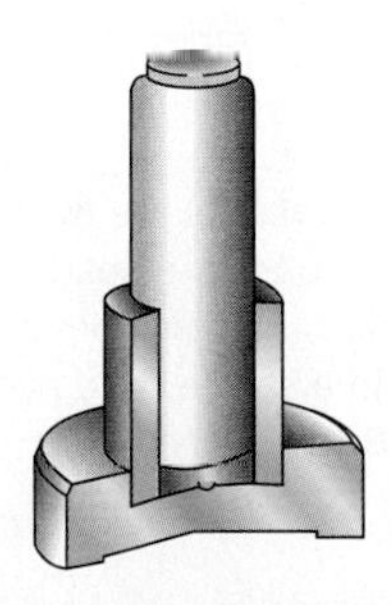

Figure 35-2 A Plattner diamond mortar.

35A-2 Mixing Solid Samples

It is essential that solid materials be thoroughly mixed to ensure random distribution of the components in the analytical samples. A common method of mixing powders involves rolling the sample on a sheet of glazed paper. A pile of the substance is placed in the center and is mixed by lifting one corner of the paper enough to roll the particles of the sample to the opposite corner. This operation is repeated many times, with the four corners of the sheet being lifted alternately.

Effective mixing of solids is also accomplished by rotating the sample for some time in a ball mill or a twin-shell V-blender. The latter consists of two connected cylinders that form a V-shaped container for the sample. As the blender is rotated, the sample is split and recombined with each rotation, leading to highly efficient mixing.

► Finely ground materials may segregate after standing for a long time.

It is worthwhile to note that with long-standing, finely ground homogeneous materials may segregate on the basis of particle size and density. For example, analyses of layers of a set of student unknowns that had not been used for several years revealed a regular variation in the analyte concentration from top to bottom of the container. Apparently, segregation occurred as a consequence of vibrations and of density differences in the sample components.

35B MOISTURE IN SAMPLES

Laboratory samples of solids often contain water that is in equilibrium with the atmosphere. As a consequence, unless special precautions are taken, the composition of the sample depends on the relative humidity and ambient temperature at the time it is analyzed. To cope with this variability in composition, it is common practice to remove moisture from solid samples prior to weighing or, if this is not possible, to bring the water content to some reproducible level that can be duplicated later if necessary. Traditionally, drying was accomplished by heating the sample in a conventional oven or a vacuum oven or by storing in a desiccator at a fixed humidity. These processes were carried out until the material became constant in mass. These treatments were time consuming, often requiring several hours or even several days. To speed up sample drying, microwave ovens or infrared lamps are currently used for sample preparation.[1] Several companies now offer equipment for this type of sample treatment (see Section 36C).

An alternative to drying samples before beginning an analysis is to determine the water content when the samples are weighed for analysis so that the results can be corrected to a dry basis. In any event, many analyses are preceded by some sort of preliminary treatment designed to take into account the presence of water.

35B-1 Forms of Water in Solids

Essential Water

Essential water is the water that is an integral part of a solid chemical compound in a stoichiometric amount in a stable solid hydrate such as $BaCl_2 \cdot 2H_2O$.

Essential water forms an integral part of the molecular or crystalline structure of a compound in its solid state. Thus, the water of crystallization in a stable solid hydrate (for example, $CaC_2O_4 \cdot 2H_2O$ and $BaCl_2 \cdot 2H_2O$) qualifies as a type of essential water. **Water of constitution** is a second type of essential water; it is found in compounds that yield stoichiometric amounts of water when heated or

[1]For a comparison of the reproducibility of these various methods of drying, see E. S. Berry, *Anal. Chem.*, **1988,** *60,* 742.

otherwise decomposed. Examples of this type of water are found in potassium hydrogen sulfate and calcium hydroxide, which when heated come to equilibrium with the moisture in the atmosphere, as shown by the reactions

$$2KHSO_4(s) \rightleftharpoons K_2S_2O_7(s) + H_2O(g)$$

$$Ca(OH)_2(s) \rightleftharpoons CaO(s) + H_2O(g)$$

Water of constitution is water that is formed when a pure solid is decomposed by heat or other chemical treatment.

Nonessential Water

Nonessential water is retained by the solid as a consequence of physical forces. It is not necessary for characterization of the chemical constitution of the sample and therefore does not occur in any sort of stoichiometric proportion.

Nonessential water is the water that is physically retained by a solid.

Adsorbed water is a type of nonessential water that is retained on the surface of solids. The amount adsorbed is dependent on humidity, temperature, and the specific surface area of the solid. Adsorption of water occurs to some degree on all solids.

A second type of nonessential water is called **sorbed water** and is encountered with many colloidal substances, such as starch, protein, charcoal, zeolite minerals, and silica gel. In contrast to adsorption, the quantity of sorbed water is often large, amounting to as much as 20% or more of the total mass of the solid. Solids containing even this amount of water may *appear* to be perfectly dry powders. Sorbed water is held as a condensed phase in the interstices or capillaries of the colloidal solid. The quantity contained in the solid is greatly dependent on temperature and humidity.

A third type of nonessential moisture is **occluded water,** liquid water entrapped in microscopic pockets spaced irregularly throughout solid crystals. Such cavities often occur in minerals and rocks (and in gravimetric precipitates).

35B-2 The Effect of Temperature and Humidity on the Water Content of Solids

In general, the concentration of water in a solid tends to decrease with increasing temperature and decreasing humidity. The magnitude of these effects and the rate at which they manifest differ considerably according to the manner in which the water is retained.

Relative humidity is the ratio of the vapor pressure of water in the atmosphere to its vapor pressure in air that is saturated with moisture. At 25°C, the partial pressure of water in saturated air is 23.76 torr. Thus, when air contains water at a partial pressure of 6 torr, the relative humidity is

$$\frac{6.00}{23.76} = 0.253 \text{ (or the percent relative humidity is 25.3\%)}$$

Compounds Containing Essential Water

The chemical composition of a compound containing essential water is dependent on temperature and relative humidity. For example, anhydrous barium chloride tends to take up atmospheric moisture to give one of two stable hydrates, depending on temperature and relative humidity. The equilibria involved are

$$BaCl_2(s) + H_2O(g) \rightleftharpoons BaCl_2 \cdot H_2O(s)$$

$$BaCl_2 \cdot H_2O(s) + H_2O(g) \rightleftharpoons BaCl_2 \cdot 2H_2O(s)$$

At room temperature and at a relative humidity between 25% and 90%, $BaCl_2 \cdot 2H_2O$ is the stable species. Since the relative humidity in most laboratories is well within these limits, the essential water content of the dihydrate is ordinarily independent of atmospheric conditions. Exposure of either $BaCl_2$ or $BaCl_2 \cdot H_2O$ to these conditions causes compositional changes that ultimately lead to formation of the dihydrate. On a very dry winter day (relative humidity <25%), however, the situation changes; the dihydrate becomes unstable with respect to the atmosphere, and

◀ The essential water content of a compound depends on the temperature and relative humidity of its surroundings.

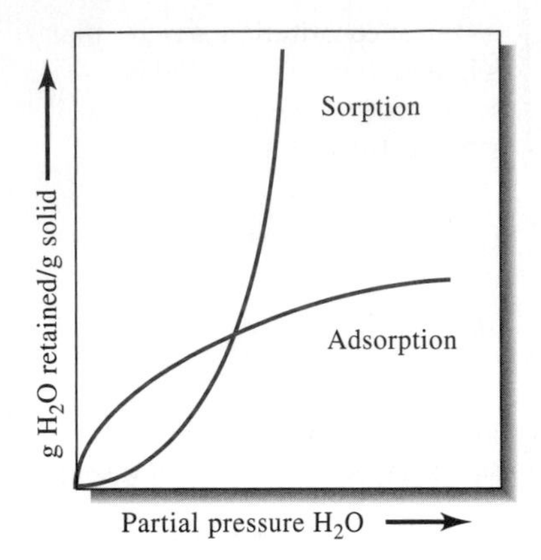

Figure 35-3 Typical adsorption and sorption isotherms.

a molecule of water is lost to form the new stable species $BaCl_2 \cdot H_2O$. At relative humidities less than about 8%, both hydrates lose water and the anhydrous compound is the stable species. Thus, we can see that the composition of a sample containing essential water depends greatly on the relative humidity of its environment.

Many hydrated compounds can be converted to the anhydrous condition by oven-drying at 100°C to 120°C for an hour or two. Such treatment often precedes an analysis of samples containing hydrated compounds.

Compounds Containing Adsorbed Water

Adsorbed water resides on the surface of the particles of a material

Figure 35-3 shows an **adsorption isotherm,** in which the mass of the water adsorbed on a typical solid is plotted against the partial pressure of water in the surrounding atmosphere. The diagram indicates that the extent of adsorption is particularly sensitive to changes in water vapor pressure at low partial pressures.

The amount of water adsorbed on a solid decreases as the temperature of the solid increases and generally approaches zero when the solid is heated above 100°C. Adsorption or desorption of moisture usually occurs rapidly, with equilibrium often being reached after 5 or 10 min. The speed of the process is often observable during the weighing of finely divided anhydrous solids, when a continuous increase in mass will occur unless the solid is contained in a tightly stoppered vessel.

Compounds Containing Sorbed Water

Sorbed water is contained within the interstices of the molecular structure of a colloidal compound.

The quantity of moisture sorbed by a colloidal solid varies tremendously with atmospheric conditions, as shown in Figure 35-3. In contrast to the behavior of adsorbed water, however, the sorption process may require days or even weeks to attain equilibrium, particularly at room temperature. Also, the amounts of water retained by the two processes are often quite different from each other. Typically, adsorbed moisture amounts to a few tenths of a percent of the mass of the solid, whereas sorbed water can amount to 10% to 20%.

The amount of water sorbed in a solid also decreases as the solid is heated. Complete removal of this type of moisture at 100°C is by no means a certainty, however, as indicated by the drying curves for an organic compound shown in Figure 35-4. After this material was dried for about 70 min at 105°C, its mass apparently became constant. Note, however, that additional moisture was removed by further increasing the temperature. Even at 230°C, dehydration was probably not complete. Commercial vapor sorption analyzers can automate the acquisition of moisture sorption and desorption isotherms.

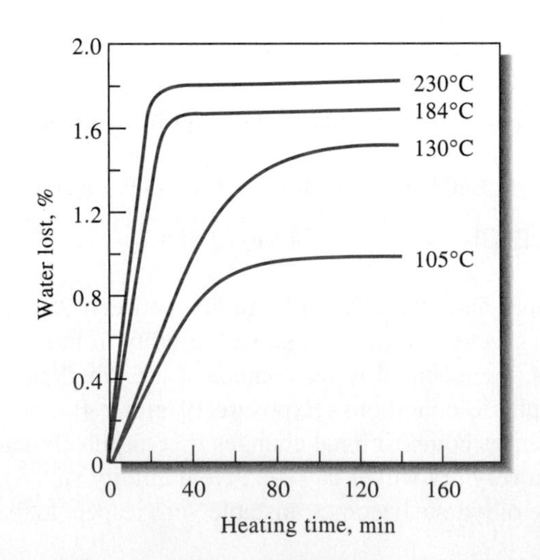

Figure 35-4 Removal of sorbed water from an organic compound at various temperatures. (Data with permission from C. O. Willits, *Anal. Chem.*, **1951,** *23,* 1058. Copyright 1958 American Chemical Society.)

Compounds Containing Occluded Water

Occluded water is not in equilibrium with the atmosphere and is therefore insensitive to changes in humidity. Heating a solid containing occluded water may cause a gradual diffusion of the moisture to the surface, where it evaporates. Frequently, heating is accompanied by **decrepitation,** in which the crystals of the solid are suddenly shattered by the steam pressure created from moisture contained in the internal cavities.

Occluded water is trapped in random microscopic pockets of solids, particularly minerals and rocks.

Decrepitation is a process in which a crystalline material containing occluded water suddenly explodes during heating because of a buildup in internal pressure resulting from steam formation.

35B-3 Drying the Analytical Sample

How we deal with moisture in solid samples depends on the information desired. When the composition of the material on an as-received basis is needed, the principal concern is that the moisture content not be altered as a consequence of grinding or other preliminary treatment and storage. If such changes are unavoidable or probable, it is often advantageous to determine the mass loss on drying by some reproducible procedure (say, heating to constant mass at 105°C) immediately after the sample is received. Then, when the time arrives for the analysis to be performed, the sample is again dried at this temperature so that the data can be corrected back to the original basis.

We have already noted that the moisture content of some substances is substantially changed by variations in humidity and temperature. Colloidal materials containing large amounts of sorbed moisture are particularly susceptible to the effects of these variables. For example, the moisture content of a potato starch has been found to vary from 10% to 21% as a consequence of an increase in relative humidity from 20% to 70%. With substances of this sort, comparable analytical data from one laboratory to another or even within the same laboratory can be achieved only by carefully specifying a procedure for taking the moisture content into consideration. For example, samples are frequently dried to constant mass at 105°C or at some other specified temperature. Analyses are then performed and results reported on this dry basis. While such a procedure may not render the solid completely free of water, it usually lowers the moisture content to a reproducible level.

35C DETERMINING WATER IN SAMPLES

Often, the only sure way to obtain a result on a dry basis is to determine the moisture in a set of samples taken concurrently with the samples that are to be analyzed. There are several methods of determining water in solid samples. The simplest involves determining the mass loss after the sample has been heated at 100°C to 110°C (or some other specified temperature) until the mass of the dried sample becomes constant. Unfortunately, this simple procedure is not at all specific for water, and large positive systematic errors occur in samples that yield volatile decomposition products (other than water) when they are heated. This method can also yield negative errors when applied to samples containing sorbed moisture (for example, see Figure 35-4). Modern thermal analysis methods, such as thermogravimetric analysis, differential thermal analysis, and differential scanning calorimetry, are also widely used in studying the loss of water and various decomposition reactions in solid samples.[2]

[2]See D. A. Skoog, F. J. Holler, and T. A. Nieman, *Principles of Instrumental Analysis,* Ch. 31. Belmont, CA: Brooks/Cole, 1998.

Several highly selective methods have been developed for the determination of water in solid and liquid samples. One of these, the Karl Fischer method, is described in Section 20C-5. Several others are described in the monographs by Mitchell and Smith.[3]

WEB WORKS

Set your browser to **http://chemistry.brookscole.com/skoogfac/.** From the Chapter Resources menu, choose Web Works. Locate the Chapter 35 section, and find the link for the Spex CertiPrep *Handbook of Sample Preparation and Handling.* Find the Pulverizing and Blending section. Look up information on pulverizing and blending. Describe how the shatterbox grinder pulverizes samples. How does the shatterbox differ from a mixer/mill? What kinds of samples are ground in freezer/mills?

QUESTIONS AND PROBLEMS

***35-1.** Describe some of the errors that might arise during the sample preparation stage.

35-2. Differentiate between
- *(a) sorbed water, adsorbed water, and occluded water.
- (b) water of crystallization and water of constitution.
- *(c) essential water and nonessential water.
- (d) the gross sample and the laboratory sample (recall Section 8B).

35-3. Why is it usually wise to decrease the particle size of the gross sample before producing the laboratory sample for analysis?

***35-4.** What types of contamination and changes in composition can occur during crushing and grinding?

35-5. Challenge Problem. Two different sample preparation methods are compared on the same sample to determine if the method results in differences.

Method 1, ppm Pb	Method 2, ppm Pb
10.5	9.7
11.7	10.8
11.1	9.9
10.6	11.8
11.4	10.2
10.2	9.8
10.4	9.6

(a) Determine whether the means of Pb determinations differ at the 95% confidence level for the two methods.

(b) The true mean for this sample is known from prior experience to be 11.3 ppm Pb. Is there a difference at the 95% confidence level between the mean of method 1 and the true mean? Method 2?

(c) If the means in Part (a) above do not differ at the 95% confidence level, can one of the methods still differ from the true value if the other does not? Why or why not?

(d) From the data presented, determine if the value of 11.8 by method 2 is an outlier at the 95% confidence level.

(e) How far away from the mean value would a method 1 result have to be to be considered an outlier?

InfoTrac College Edition

For additional readings, go to InfoTrac College Edition, your online research library, at

http://infotrac.thomsonlearning.com

[3]J. J. Mitchell, Jr., and D. M. Smith, *Aquametry,* 2nd ed., Vols. 1–3. New York: Wiley, 1977–1980.

CHAPTER 36

Decomposing and Dissolving the Sample

Microwave digestion systems have become very popular for decomposing samples. The photo shown is a closed-vessel microwave digestion system for high-pressure digestions. A microwave oven with a built-in fume exhaust system is shown along with sample trays that contain up to 12 samples. Teflon sample vessels can be operated at temperatures up to 2300°C and 625 psi.

This chapter considers methods of decomposing and dissolving real samples. Acid decomposition, microwave, combustion, and fusion methods are considered.

Courtesy of Aurora Instruments, Vancouver, BC, Canada

Most analytical measurements are performed on solutions (usually aqueous) of the analyte. While some samples dissolve readily in water or aqueous solutions of the common acids or bases, others require powerful reagents and rigorous treatment. For example, when sulfur or halogens are to be determined in an organic compound, the sample must be subjected to high temperatures and potent reagents to rupture the strong bonds between these elements and carbon. Similarly, drastic conditions are usually required to destroy the silicate structure of a siliceous mineral and to free the ions for analysis.

The proper choice among the various reagents and techniques for decomposing and dissolving analytical samples can be critical to the success of an analysis, particularly when refractory substances are involved or the analyte is present in trace amounts. In this chapter, we first consider the types of errors that can arise in decomposing and dissolving an analytical sample. We then describe four general methods of decomposing solid and liquid samples to obtain an aqueous solution of analytes. The four methods include (1) heating with aqueous strong acids (or occasionally bases) in open vessels; (2) microwave heating with acids; (3) high-temperature ignition in air or oxygen; and (4) fusion in molten salt media.[1] These methods differ in the temperature at which they are carried out and the strengths of the reagents used.

A **refractory substance** is a material that is resistant to heat and attack by strong chemical agents.

[1]For an extensive discussion of this topic, see R. Bock, *A Handbook of Decomposition Methods in Analytical Chemistry.* New York: Wiley, 1979; Z. Sulcek and P. Povondra, *Methods of Decomposition in Inorganic Analysis.* Boca Raton, FL: CRC Press, 1989; J. A. Dean, *Analytical Chemistry Handbook,* Section 1.7. New York: McGraw-Hill, 1995.

36A SOURCES OF ERROR IN DECOMPOSITION AND DISSOLUTION

We encounter several sources of error in the sample decomposition step. In fact, such errors often limit the accuracy that can be achieved in an analysis. The sources of these errors include the following:

► Ideally, the reagent selected should dissolve the entire sample, not just the analyte.

1. **Incomplete dissolution of the analytes.** Ideally, the sample treatment should dissolve the sample completely. Attempts to leach analytes quantitatively from an insoluble residue are usually not successful because portions of the analyte are retained within the residue.
2. **Losses of the analyte by volatilization.** An important concern in dissolving samples is the possibility that some portion of the analyte may volatilize. For example, carbon dioxide, sulfur dioxide, hydrogen sulfide, hydrogen selenide, and hydrogen telluride are generally volatilized when a sample is dissolved in strong acid, while ammonia is often lost when a basic reagent is used. Similarly, hydrofluoric acid reacts with silicates and boron-containing compounds to produce volatile fluorides. Strong oxidizing solvents often cause the evolution of chlorine, bromine, or iodine; reducing solvents may lead to the volatilization of such compounds as arsine, phosphine, and stibine.

   A number of elements form volatile chlorides that are partially or completely lost from hot hydrochloric acid solutions. Among these are the chlorides of tin(IV), germanium(IV), antimony(III), arsenic(III), and mercury(II). The oxychlorides of selenium and tellurium also volatilize to some extent from hot hydrochloric acid. The presence of chloride ion in hot concentrated sulfuric or perchloric acid solutions can cause volatilization losses of bismuth, manganese, molybdenum, thallium, vanadium, and chromium.

   Boric acid, nitric acid, and the halogen acids are lost from boiling aqueous solutions. Certain volatile oxides can also be lost from hot acidic solutions, including the tetroxides of osmium and ruthenium and the heptoxide of rhenium.
3. **Introduction of the analyte as a solvent contaminant.** Ordinarily, the mass of solvent required to dissolve a sample exceeds the mass of sample by one or two orders of magnitude. As a consequence, the presence of analyte species in the solvent even at small concentrations may lead to significant error, particularly when the analyte is present in trace amounts in the sample.
4. **Introduction of contaminants from reaction of the solvent with vessel walls.** This source of error is often encountered in decompositions that involve high-temperature fusions. Again, this source of error becomes of particular concern in trace analyses.

36B DECOMPOSING SAMPLES WITH INORGANIC ACIDS IN OPEN VESSELS

The most common reagents for open-vessel decomposition of inorganic analytical samples are the mineral acids. Much less frequently, ammonia and aqueous solutions of the alkali metal hydroxides are used. Ordinarily, a suspension of the sample in the acid is heated by flame or a hot plate until the dissolution is judged to be complete by the total disappearance of a solid phase. The temperature of the decomposition is the boiling (or decomposition) point of the acid reagent.

36B-1 Hydrochloric Acid

Concentrated hydrochloric acid is an excellent solvent for inorganic samples but finds limited application in the decomposition of organic materials. It is widely used to dissolve many metal oxides as well as metals more easily oxidized than hydrogen; often, it is a better solvent for oxides than the oxidizing acids. Concentrated hydrochloric acid is about 12 M. On heating, however, HCl gas is lost until a constant-boiling 6 M solution remains (boiling point about 110°C).

36B-2 Nitric Acid

Hot concentrated nitric acid is a strong oxidant that dissolves all common metals with the exception of aluminum and chromium, which become passive to this reagent owing to surface oxide formation. When alloys containing tin, tungsten, or antimony are treated with the hot reagent, slightly soluble hydrated oxides, such as $SnO_2 \cdot 4H_2O$, form. After coagulation, these colloidal materials can be separated from other metallic species by filtration.

Hot nitric acid alone or in combination with other acids and oxidizing agents, such as hydrogen peroxide and bromine, is widely used to decompose organic samples for determination of their trace metal content. This decomposition process, which is called **wet ashing,** converts the organic sample to carbon dioxide and water. Unless the process is carried out in a closed vessel, nonmetallic elements, such as the halogens, sulfur, and nitrogen, are completely or partially lost by volatilization.

Wet ashing is the process of oxidative decomposition of organic samples by liquid oxidizing reagents such as HNO_3, H_2SO_4, $HClO_4$, or mixtures of these acids.

36B-3 Sulfuric Acid

Many materials are decomposed and dissolved by hot concentrated sulfuric acid, which owes part of its effectiveness as a solvent to its high boiling point (about 340°C). Most organic compounds are dehydrated and oxidized at this temperature and are thus eliminated from samples as carbon dioxide and water by this wet ashing treatment. Most metals and many alloys are attacked by the hot acid.

36B-4 Perchloric Acid

Hot concentrated perchloric acid, a potent oxidizing agent, attacks a number of iron alloys and stainless steels that are not affected by other mineral acids. Care must be taken in using this reagent, however, because of its *potentially explosive nature.* The cold concentrated acid is not explosive, nor are heated dilute solutions. *Violent explosions occur, however, when hot concentrated perchloric acid comes into contact with organic materials or easily oxidized inorganic substances.* Because of this property, the concentrated reagent should be heated only in special hoods that are lined with glass or stainless steel, are seamless, and have a fog system for washing down the walls with water. A perchloric acid hood should always have its own fan system, one that is independent of all other systems.[2]

Perchloric acid is marketed as the 60% to 72% acid. A constant-boiling mixture (72.4% $HClO_4$) is obtained at 203°C.

[2]See A. A. Schilt, *Perchloric Acid and Perchlorates.* Columbus, OH: G. Frederick Smith Chemical Company, 1979.

36B-5 Oxidizing Mixtures

More rapid wet ashing can sometimes be obtained by the use of mixtures of acids or by the addition of oxidizing agents to a mineral acid. **Aqua regia,** a mixture containing three volumes of concentrated hydrochloric acid and one volume of nitric acid, is well known. The addition of bromine or hydrogen peroxide to mineral acids often increases their solvent action and hastens the oxidation of organic materials in the sample. Mixtures of nitric and perchloric acid are also useful for this purpose and less dangerous than perchloric acid alone. With this mixture, however, care must be taken to avoid evaporation of all the nitric acid before oxidation of the organic material is complete. *Severe explosions and injuries have resulted from failure to observe this precaution.*

36B-6 Hydrofluoric Acid

The primary use of hydrofluoric acid is for the decomposition of silicate rocks and minerals in the determination of species other than silica. In this treatment, silicon is evolved as the tetrafluoride. After decomposition is complete, the excess hydrofluoric acid is driven off by evaporation with sulfuric acid or perchloric acid. Complete removal is often essential to the success of an analysis because fluoride ion reacts with several cations to form extraordinarily stable complexes that interfere with the determination of the cations. For example, precipitation of aluminum (as $Al_2O_3 \cdot xH_2O$) with ammonia is incomplete if fluoride is present even in small amounts. Frequently, it is so difficult and time-consuming to remove the last traces of fluoride ion from a sample that the attractive features of hydrofluoric acid as a solvent are negated.

Hydrofluoric acid finds occasional use in conjunction with other acids in attacking steels that dissolve with difficulty in other solvents. Because hydrofluoric acid is extremely toxic, dissolution of samples and evaporation to remove excess reagent *should always be carried out in a well-ventilated hood.* Hydrofluoric acid *causes serious damage and painful injury* when brought into contact with the skin. Its effects may not become evident until hours after exposure. If the acid comes into contact with the skin, the affected area should be immediately washed with copious quantities of water. Treatment with a dilute solution of calcium ion, which precipitates fluoride ion, may also be of help.

36C MICROWAVE DECOMPOSITIONS

The use of microwave ovens for the decomposition of both inorganic and organic samples, first proposed in the mid-1970s, is now an important method of sample preparation.[3] Microwave digestions can be carried out in either closed or open vessels, but closed vessels are more popular because of the higher pressures and higher temperatures that can be achieved.

One of the main advantages of microwave decomposition compared with conventional methods using a flame or a hot plate (regardless of whether an open or a

[3]For more detailed discussions of microwave sample preparation and commercial instrumentation, see H. M. Kingston and S. J. Haswell, *Microwave-Enhanced Chemistry: Fundamentals, Sample Preparation and Applications.* Washington, DC: American Chemical Society, 1997; B. E. Erickson, *Anal. Chem.*, **1998,** *70,* 467A–471A; R. C. Richter, D. Link, and H. M. Kingston, *Anal. Chem.*, **2001,** *73,* 31A–37A.

closed container is used) is speed. Typically, microwave decompositions of even difficult samples can be accomplished in 5 to 10 minutes. In contrast, the same results require several hours when carried out by heating over a flame or a hot plate. The difference is due to the different mechanism by which energy is transferred to the molecules of the solution in the two methods. Heat transfer is by conduction in the conventional method. Because the vessels used in conductive heating are usually poor conductors, time is required to heat the vessel and then transfer the heat to the solution by conduction. Furthermore, because of convection within the solution, only a small fraction of the liquid is maintained at the temperature of the vessel and thus at its boiling point. In contrast, microwave energy is transferred directly to all the molecules of the solution nearly simultaneously without heating the vessel. Thus, boiling temperatures are reached throughout the entire solution very quickly.

As noted earlier, an advantage of using closed vessels for microwave decomposition is the higher temperature that develops as a consequence of the increased pressure. In addition, because evaporative losses are avoided, significantly smaller amounts of reagent are used, thus reducing interference by reagent contaminants. A further advantage of decompositions of this type is that loss of volatile components of samples is virtually eliminated. Finally, closed-vessel microwave decompositions are often easy to automate, thus reducing operator time required to prepare samples for analysis.

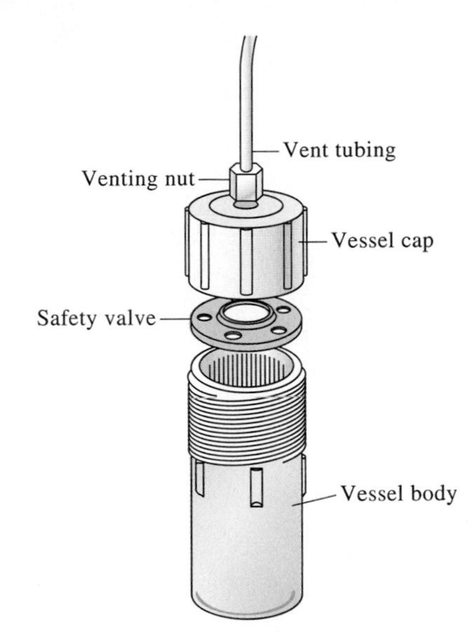

Figure 36-1 A moderate-pressure vessel for microwave decomposition. (Courtesy of CEM Corp., Matthews, NC.)

36C-1 Vessels for Moderate-Pressure Digestions

Microwave digestion vessels are constructed from low-loss materials that are transparent to microwaves. These materials must also be thermally stable and resistant to chemical attack by the various acids used for decomposition. Teflon is a nearly ideal material for many of the acids commonly used for dissolutions. It is transparent to microwaves, has a melting point of about 300°C, and is not attacked by any of the common acids. Sulfuric and phosphoric acids, however, have boiling points above the melting point of Teflon, which means that care must be exercised to control the temperature during decompositions. For these acids, quartz or borosilicate glass vessels are sometimes used in place of Teflon containers. Quartz or glass vessels have the disadvantage, however, of being attacked by hydrofluoric acid, a reagent that is often used to decompose silicates and refractory alloys.

Figure 36-1 is a schematic of a commercially available closed digestion vessel designed for use in a microwave oven. It consists of a Teflon body, a cap, and a safety relief valve designed to operate at 120 ± 10 psi. At this pressure, the safety valve opens and then reseals.

36C-2 High-Pressure Microwave Vessels

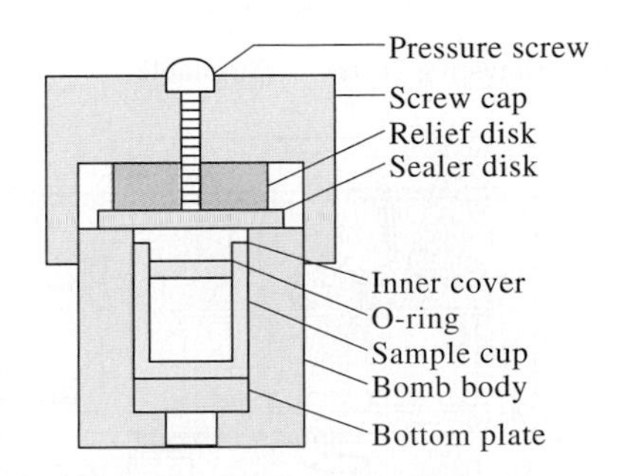

Figure 36-2 A bomb for high-pressure microwave decomposition. (Courtesy of Parr Instrument Co., Moline IL.)

Figure 36-2 is a schematic of a commercial microwave bomb designed to operate at 80 atm, or about 10 times the pressure that can be tolerated by the moderate-pressure vessels described in the previous section. The maximum recommended temperature with this device is 250°C. The heavy-wall bomb body is constructed of a polymeric material that is transparent to microwaves. The decomposition is carried out in a Teflon cup supported in the bomb body. The microwave bomb incorporates a Teflon O-ring in the liner cap that seats against a narrow rim on the exterior of the liner and its cap when the retaining jacket is screwed into place. When overpressurization occurs, the O-ring distorts, and the excess pressure then compresses the sealer disk, which allows the gases to escape into the surroundings. The sample is compromised when this occurs. The internal pressure in the bomb can be judged roughly by the distance that the pressure screw protrudes from the cap. This microwave bomb is particularly useful for dissolving highly refractory materials that are incompletely decomposed in the moderate-pressure vessel described earlier.

When alloys and metals are digested in high-pressure microwave vessels, there is a risk of explosion caused by the production of hydrogen gas. Common polymeric liner materials may not be capable of reaching the temperatures needed to fully decompose organic materials. Another limitation is that most high-pressure vessels are limited in sample size to less than 1 g of material. It is also necessary to allow time for cool down and depressurization.

36C-3 Atmospheric Pressure Digestions

The limitations of closed-vessel microwave digestion systems just noted have led to the development of atmospheric pressure units, often called open-vessel systems. These systems do not have an oven but instead use a focused microwave cavity. They can be purged with gases and equipped with tubing to allow for the insertion and removal of reagents. There is no longer a safety concern due to gas-forming reactions during the digestion process since the systems operate at atmospheric pressure. There are even flow-through systems available for on-line dissolution prior to introducing the samples into flames or ICPs for atomic spectroscopic determinations.

36C-4 Microwave Ovens

Figure 36-3 is a schematic of a microwave oven designed to heat simultaneously 12 of the moderate-pressure vessels described in Section 36C-1. The vessels are held on a turntable that rotates continuously through 360 degrees so that the average energy received by each of the vessels is approximately the same.

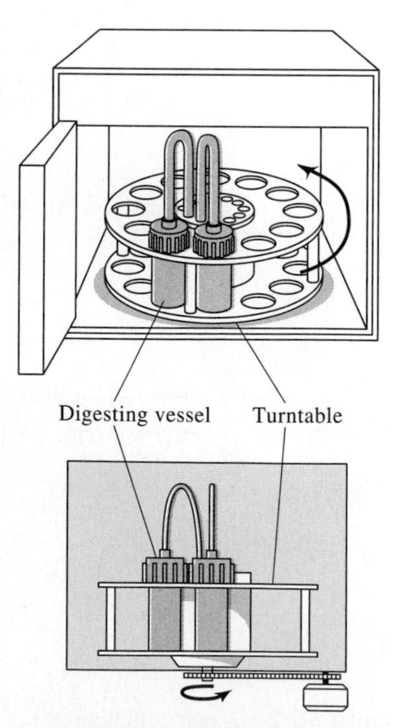

Figure 36-3 A microwave oven designed for use with 12 vessels of the type shown in Figure 36-1. (Courtesy of CEM Corp., Matthews, NC.)

36C-5 Microwave Furnaces

Recently, microwave furnaces have been developed for performing fusions and for dry ashing samples containing large amounts of organic materials before acid dissolution. These furnaces consist of a small chamber constructed of silicon carbide that is surrounded by quartz insulation. When microwaves are focused on this chamber, temperatures of 1000°C are reached in 2 minutes. The advantage of this type of furnace relative to a conventional muffle furnace is the speed at which high temperatures are reached. In contrast, conventional muffle furnaces are usually operated continuously because of the time required to get them up to temperature. Furthermore, with a microwave furnace, there are no burned-out heating coils such as are frequently encountered with conventional furnaces. Finally, the operator is not exposed to high temperatures when samples are introduced or removed from the furnace. A disadvantage of the microwave furnace is the small volume of the heating cavity, which accommodates only an ordinary-size crucible.

36C-6 Applications of Microwave Decompositions

During the past 25 years, hundreds of reports have appeared in the literature regarding the use of closed-vessel decompositions carried out in microwave ovens with the reagents described in Section 36B. These applications fall into two categories:

(1) oxidative decompositions of organic and biological samples (wet ashing) and (2) decomposition of refractory inorganic materials encountered in industry. In both cases, this new technique is replacing older, conventional methods because of the large economic gains that result from significant savings in time. Atmospheric-pressure digestions have also become popular in recent years, and their applications are on the increase.

36D COMBUSTION METHODS FOR DECOMPOSING ORGANIC SAMPLES[4]

36D-1 Combustion over an Open Flame (Dry Ashing)

Dry ashing is the process of oxidizing an organic sample with oxygen or air at high temperature, leaving the inorganic component for analysis.

The simplest method of decomposing an organic sample prior to determining the cations it contains is to heat the sample over a flame in an open dish or crucible until all carbonaceous material has been oxidized to carbon dioxide. Red heat is often required to complete the oxidation. Analysis of the nonvolatile components follows dissolution of the residual solid. Unfortunately, there is always substantial uncertainty about the completeness of recovery of supposedly nonvolatile elements from a dry-ashed sample. Some losses probably result from the entrainment of finely divided particulate matter in the convection currents around the crucible. In addition, volatile metallic compounds may be lost during the ignition. For example, copper, iron, and vanadium are appreciably volatilized when samples containing porphyrin compounds are ashed.

Although dry ashing is the simplest method of decomposing organic compounds, it is often the least reliable. It should not be used unless tests have demonstrated its applicability to a given type of sample.

36D-2 Combustion-Tube Methods

Several common and important elemental components of organic compounds are converted to gaseous products as a sample is pyrolyzed in the presence of oxygen. With suitable apparatus, it is possible to trap these volatile compounds quantitatively, thus making them available for the analysis of the element of interest. The heating is commonly performed in a glass or quartz combustion tube through which a stream of carrier gas is passed. The stream transports the volatile products to the parts of the apparatus where they are separated and retained for the measurement; the gas may also serve as the oxidizing agent. Elements susceptible to this type of treatment are carbon, hydrogen, nitrogen, the halogens, sulfur, and oxygen.

Automated combustion-tube analyzers are now available for the determination of either carbon, hydrogen, and nitrogen or carbon, hydrogen, and oxygen in a single sample.[5] The apparatus requires essentially no attention by the operator, and the analysis is complete in less than 15 minutes. In one such analyzer, the sample is ignited in a stream of helium and oxygen and passes over an oxidation catalyst consisting of a mixture of silver vanadate and silver tungstate. Halogens and sulfur are removed with a packing of silver salts. A packing of hot copper is located at the end

[4]For a thorough treatment of this topic, see T. S. Ma and R. C. Rittner, *Modern Organic Elemental Analysis*. New York: Marcel Dekker, 1979.

[5]For a description of these instruments, see Chapters 2, 3, and 4 of the reference in footnote 4.

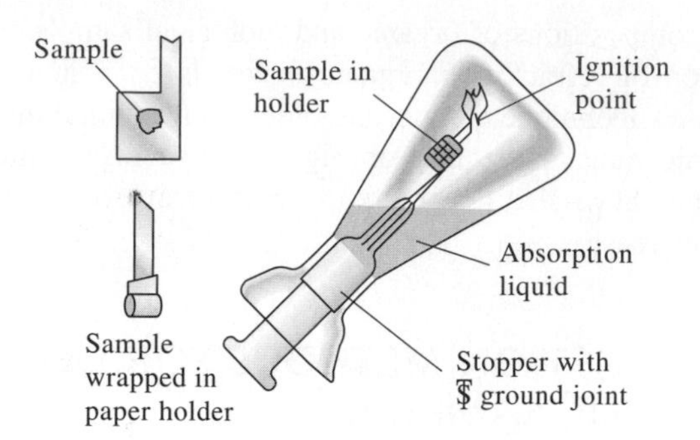

Figure 36-4 Schöniger combustion apparatus. (Courtesy Thomas Scientific, Swedesboro, NJ.)

of the combustion train to remove oxygen and convert nitrogen oxides to nitrogen. The exit gas, consisting of a mixture of water, carbon dioxide, nitrogen, and helium, is collected in a glass bulb. The analysis of this mixture is accomplished with three thermal-conductivity measurements (see Section 31A-4). The first is made on the intact mixture; the second is made on the mixture after water has been removed by passage of the gas through a dehydrating agent; and the third is made on the mixture after carbon dioxide has been removed by an absorbent. The relationship between thermal conductivity and concentration is linear, and the slope of the curve for each constituent is established by calibration with a pure compound such as acetanilide.

36D-3 Combustion with Oxygen in a Sealed Container

A relatively straightforward method for the decomposition of many organic substances involves combustion with oxygen in a sealed container. The reaction products are absorbed in a suitable solvent before the reaction vessel is opened; they are subsequently analyzed by ordinary methods.

A remarkably simple apparatus for performing such oxidations was suggested by Schöniger (Figure 36-4).[6] It consists of a heavy-walled flask of 300- to 1000-mL capacity fitted with a ground-glass stopper. Attached to the stopper is a platinum gauze basket that holds 2 to 200 mg of sample. If the substance to be analyzed is a solid, it is wrapped in a piece of low-ash filter paper cut in the shape shown in Figure 36-4. Liquid samples are weighed into gelatin capsules, which are then wrapped in a similar fashion. The paper tail serves as the ignition point.

A small volume of an absorbing solution (often sodium carbonate) is placed in the flask, and the air in the flask is displaced by oxygen. The tail of the paper is ignited, the stopper is quickly fitted into the flask, and the flask is inverted to prevent the escape of the volatile oxidation products. The reaction ordinarily proceeds rapidly, being catalyzed by the platinum gauze surrounding the sample. During combustion, the flask is shielded to minimize damage in case of explosion.

After cooling, the flask is shaken thoroughly and disassembled, and the inner surfaces are carefully rinsed. The analysis is then performed on the resulting solution. This procedure has been applied to the determination of halogens, sulfur, phosphorus, fluorine, arsenic, boron, carbon, and various metals in organic compounds.

[6] W. Schöniger, *Mikrochim. Acta,* **1955,** *123;* **1956,** *869.* See also the review articles by A. M. G. MacDonald, in *Advances in Analytical Chemistry and Instrumentation,* C. N. Reilley, Ed., Vol. 4, p. 75. New York: Interscience, 1965.

36E DECOMPOSITION OF INORGANIC MATERIALS BY FLUXES

Many common substances—notably silicates, some mineral oxides, and a few iron alloys—are attacked slowly, if at all, by the methods just considered. In such cases, recourse to use of a fused-salt medium is indicated. Here, the sample is mixed with an alkali metal salt, called the *flux,* and the combination is then fused to form a water-soluble product called the *melt.* Fluxes decompose most substances by virtue of the high temperatures required for their use (300°C to 1000°C) and the high concentrations of reagents brought into contact with the sample.

Where possible, we tend to avoid fluxes because of the possible danger and several disadvantages. Among these is the possible contamination of the sample by impurities in the flux. This possibility is exacerbated by the relatively large amount of flux (typically at least 10 times the sample mass) required for a successful fusion. Moreover, the aqueous solution that results when the melt from a fusion is dissolved has a high salt content, which may cause difficulties in the subsequent steps of the analysis. In addition, the high temperatures required for a fusion increase the danger of volatilization losses. Finally, the container in which the fusion is performed is almost inevitably attacked to some extent by the flux; again, contamination of the sample is the result.

◀ While they are very effective solvents, fluxes introduce high concentrations of ionic species to aqueous solutions of the melt.

For a sample containing only a small fraction of material that dissolves with difficulty, it is common practice to use a liquid reagent first; the undecomposed residue is then isolated by filtration and is fused with a relatively small quantity of a suitable flux. After cooling, the melt is dissolved and combined with the major portion of the sample.

36E-1 Carrying Out a Fusion

The sample, in the form of a very fine powder, is mixed intimately with perhaps a tenfold excess of the flux. Mixing is usually carried out in the crucible in which the fusion is to be performed. The time required for fusion can range from a few minutes to hours. The production of a clear melt signals completion of the decomposition, although often this condition is not always obvious.

When the fusion is complete, the mass is allowed to cool slowly; just before solidification, the crucible is rotated to distribute the solid around the walls to produce a thin layer of melt that is easy to dislodge.

36E-2 Types of Fluxes

With few exceptions, the common fluxes used in analysis are compounds of the alkali metals. Alkali metal carbonates, hydroxides, peroxides, and borates are basic fluxes used to attack acidic materials. The acidic fluxes are pyrosulfates, acid fluorides, and boric oxide. If an oxidizing flux is required, sodium peroxide can be used. As an alternative, small quantities of the alkali nitrates or chlorates can be mixed with sodium carbonate.

The properties of the common fluxes are summarized in Table 36-1.

Sodium Carbonate

Silicates and certain other refractory materials can be decomposed by heating to 1000°C to 1200°C with sodium carbonate. This treatment generally converts the cationic constituents of the sample to acid-soluble carbonates or oxides; the

TABLE 36-1
Common Fluxes

Flux	Melting Point, °C	Type of Crucible for Fusion	Type of Substance Decomposed
Na_2CO_3	851	Pt	Silicates and silica-containing samples, alumina-containing samples, sparingly soluble phosphate and sulfates
Na_2CO_3 + an oxidizing agent, such as KNO_3, $KClO_3$, or Na_2O_2	—	Pt (not with Na_2O_2), Ni	Samples requiring an oxidizing environment; that is, samples containing S, As, Sb, Cr, etc.
$LiBO_2$	849	Pt, Au, glassy carbon	Powerful basic flux for silicates most minerals, slags, ceramics
NaOH or KOH	318 380	Au, Ag, Ni	Powerful basic fluxes for silicates, silicon carbide, and certain minerals (main limitation is purity of reagents)
Na_2O_2	Decomposes	Fe, Ni	Powerful basic oxidizing flux for sulfides; acid-insoluble alloys of Fe, Ni, Cr, Mo, W, and Li; platinum alloys; Cr, Sn, Zr; minerals
$K_2S_2O_7$	300	Pt, porcelain	Acidic flux for slightly soluble oxides and oxide-containing samples
B_2O_3	577	Pt	Acidic flux for silicates and oxides where alkali metals are to be determined
$CaCO_3 + NH_4Cl$	—	Ni	On heating the flux, a mixture of CaO and $CaCl_2$ is produced; used to decompose silicates for determining alkali metals

nonmetallic constituents are converted to soluble sodium salts. Carbonate fusions are normally carried out in platinum crucibles.

Potassium Pyrosulfate

Potassium pyrosulfate is a potent acidic flux that is particularly useful for attacking the more intractable metal oxides. Fusions with this reagent are performed at about 400°C; at this temperature, the slow evolution of the highly acidic sulfur trioxide takes place:

$$K_2S_2O_7 \rightarrow K_2SO_4 + SO_3(g)$$

Potassium pyrosulfate can be prepared by heating potassium hydrogen sulfate:

$$2KHSO_4 \rightarrow K_2S_2O_7 + H_2O$$

Lithium Metaborate

Lithium metaborate, $LiBO_2$, by itself or mixed with lithium tetraborate, finds considerable use in attacking refractory silicate and alumina minerals, particularly for atomic absorption spectroscopy (AAS), ICP emission, and X-ray absorption or emission determinations. These fusions are generally carried out in graphite or platinum crucibles at about 900°C. The glass that results on cooling the melt can be used directly for X-ray fluorescence measurements. It is also readily soluble in mineral acids. After dissolving the melt, boric oxide is removed by evaporating the solution to dryness with methyl alcohol; methyl borate, $B(OCH_3)_3$, distills in the process.

WEB WORKS

Direct your Web browser to **http://chemistry.brookscole.com/skoogfac/.** From the Chapter Resources menu, choose Web Works. Locate the Chapter 36 section, and find the links for microwave digestion systems. Look up information on open-vessel microwave digestion systems versus closed-vessel systems. Summarize the advantages and disadvantages of these two approaches.

QUESTIONS AND PROBLEMS

***36-1.** Differentiate between wet ashing and dry ashing.

36-2. What is a flux? When is it used?

***36-3.** What fluxes are suitable for the determination of alkali metals in silicates?

36-4. What flux is commonly used for the decomposition of certain refractory oxides?

***36-5.** Under what conditions is the use of perchloric acid likely to be dangerous?

36-6. How are organic compounds decomposed for the determination of

*(a) halogens?

(b) sulfur?

*(c) nitrogen?

(d) heavy-metal species?

36-7. Give three major advantages of microwave decompositions.

36-8. What are the major limitations of closed-vessel, high-pressure microwave digestions?

InfoTrac College Edition

For additional readings, go to InfoTrac College Edition, your online research library, at

http://infotrac.thomsonlearning.com.

GLOSSARY

A

Absolute error An accuracy measurement equal to the numerical difference between an experimental measurement and its true (or accepted) value.

Absolute standard deviation A precision estimate based on the deviations between individual members in a set and the mean of that set (see Equation 6-4).

Absorbance, *A* The logarithm of the ratio between the initial power of a beam of radiation P_0 and its power after it has traversed an absorbing medium, *P*. $A = \log(P_0/P)$.

Absorption A process in which a substance is incorporated or assimilated within another; also, a process in which a beam of electromagnetic radiation is attenuated during passage through a medium.

Absorption of electromagnetic radiation Processes in which radiation causes transitions in atoms and molecules to excited states; the absorbed energy is lost, usually as heat, as the excited species return to their ground states.

Absorption filter A colored medium (usually glass) that transmits a relatively narrow band of the visible spectrum.

Absorption spectrum A plot of absorbance as a function of wavelength.

Absorptivity, *a* The proportionality constant in the Beer's law equation, $A = abc$, where *b* is the path length of radiation (usually in cm) and *c* is the concentration of the absorbing species (usually in mol/L). Thus, *a* has the units of length^{-1} concentration^{-1}.

Accuracy A measure of the agreement between an analytical result and the true or accepted value for the measured quantity; this agreement is measured in terms of error.

Acid dissociation constant, K_a The equilibrium constant for the dissociation reaction of a weak acid.

Acid error The tendency of a glass electrode to register anomalously high pH response in highly acidic media.

Acidic flux A salt that exhibits acidic properties in the molten state; used to convert refractory substances into water-soluble products.

Acid rain Rainwater that has been rendered acidic from absorption of airborne nitrogen and sulfur oxides produced mainly by mankind.

Acids Species that are capable of donating protons to other species that in turn are capable of accepting these protons.

Acid salt A conjugate base that possesses an acidic hydrogen.

Activity, *a* The effective concentration of a participant in a chemical equilibrium; the activity of a species is given by the product of the molar equilibrium concentration of the species and its activity coefficient.

Activity coefficient, γ_X A unitless quantity whose numerical value depends on the ionic strength of a solution; it is the proportionality constant between activity and concentration.

Adsorbed water Nonessential water that is held on the surface of solids.

Adsorption A process in which a substance becomes physically bound to the surface of a solid.

Adsorption chromatography A separation technique in which a solute equilibrates between the eluent and the surface of a finely divided adsorbed solid.

Agar A polysaccharide that forms a conducting gel with electrolyte solutions; used in salt bridges to provide electric contact between dissimilar solutions without mixing.

Air damper A device that hastens achievement of equilibrium by the beam of a mechanical analytical balance; also called a *dashpot.*

Aliquot A volume of liquid that is a known fraction of a larger volume.

Alkaline error The tendency of many glass electrodes to provide an anomalously low pH response in highly alkaline environments.

Alpha (α) value The ratio between the concentration of a particular species and the analytical concentration of the solute from which it is derived.

Alumina The common name for aluminum oxide. In a finely divided state, used as a stationary phase in adsorption chromatography; also finds application as a support for a liquid stationary phase in HPLC.

Amines Derivatives of ammonia with one or more organic groups replacing hydrogen.

Amino acids Weak organic acids that also contain basic amine groups; the amine group is α to the carboxylic acid group in amino acids derived from proteins.

Ammonium-1-pyrrolidinecarbodithiolate (APDC) A protective agent in atomic spectroscopy that forms volatile species with an analyte.

Amperometric titration A method based on applying a constant potential to a working electrode and recording the resulting current; a linear segment curve is obtained.

Amperostat An instrument that maintains a constant current; used for coulometric titrations.

Amphiprotic substances Species that can either donate protons or accept protons, depending on the environment.

Amylose A component of starch, the β-form of which is a specific indicator for iodine.

Analysis of Variance (ANOVA) A collection of statistical procedures for analysis of responses from experiments. Single-factor *ANOVA* allows comparison of more than two means of populations.

Analyte The species in the sample about which analytical information is sought.

Analytical balance An instrument for the accurate determination of mass.

Analytical molarity, c_X The moles of solute, X, that have been dissolved in sufficient solvent to give 1.000 liter of solution; also numerically equal to the number of millimoles of solute per milliliter of solution. Compare with *species molarity.*

Angstrom, Å A unit of length equal to 1×10^{-10} meter.

Angular dispersion, $dr/d\lambda$ A measure of the change in the angle of reflection or refraction of radiation by a prism or grating as a function of wavelength.

Anhydrone® Trade name for magnesium perchlorate, a drying agent.

Anion exchange resins High-molecular-weight polymers to which amine groups are bonded. They permit the exchange of anions in solution for hydroxide ions from the exchanger.

Anode The electrode of an electrochemical cell at which oxidation occurs.

Aqua regia A mixture containing three volumes of concentrated hydrochloric acid and one volume of nitric acid; a potent oxidizing solution.

Argentometric titration A titration in which the reagent is a solution of a silver salt (usually $AgNO_3$).

Arithmetic mean Synonymous with *mean* or *average.*

Asbestos A fibrous mineral, some varieties of which are carcinogenic; once used as a filtering medium in a Gooch crucible but currently subject to stringent regulation.

Ashing The process whereby an organic material is combusted in air. See also *dry ashing* and *wet ashing.*

Ashless filter paper Paper produced from cellulose fibers that have been treated to eliminate inorganic species, thus leaving no residue when ashed.

Aspiration The process by which a sample solution is drawn by suction in atomic spectroscopy.

Aspirator A device for sucking fluid through a medium.

Assay The process of determining how much of a given sample is the material indicated by its name.

Asymmetry potential A small potential that results from slight differences between the two surfaces of a glass membrane electrode.

Atomic absorption The process by which unexcited atoms in a flame, furnace, or plasma absorb characteristic radiation from a source and attenuate the radiant power of the source.

Atomic absorption spectroscopy (AAS) An analytical method that is based on atomic absorption.

Atomic emission The emission of radiation by atoms that have been excited in a plasma, a flame, or an electric arc or spark.

Atomic emission spectroscopy (AES) An analytical method based on atomic emission.

Atomic fluorescence Radiant emission from atoms that have been excited by absorption of electromagnetic radiation.

Atomic fluorescence spectroscopy (AFS) An analytical method based on atomic fluorescence.

Atomic mass unit A unit of mass based on 1/12 of the mass of the most abundant isotope of carbon, ^{12}C; equal to 1 Dalton.

Atomization The process of producing an atomic gas by applying energy to a sample.

Atomizer A device such as a plasma, a flame, or a furnace that produces an atomic vapor.

Attenuation In absorption spectroscopy, a decrease in the power of a beam of radiant energy.

Attenuator A device for diminishing the radiant power in the beam of an optical instrument.

Autocatalysis A condition in which the product of a reaction catalyzes the reaction itself.

Autoprotolysis A process in which a solvent undergoes self-dissociation.

Auxiliary balance A generic term for a balance that is less sensitive but more rugged than an analytical balance; synonymous with *laboratory balance.*

Average A number obtained by summing the values in a set and dividing the sum by the number of data points in the set. Synonymous with *mean* or *arithmetic mean.*

Average current Polarographic current determined by dividing the total charge accumulated by a mercury drop by its lifetime.

Average linear velocity, u The length, L, of a chromatographic column divided by the time, t_M, required for an unretained species to pass through the column.

Azo indicators A group of acid/base indicators that have in common the structure R—N=N—R′.

B

Back-titration The titration of an excess of a standard solution that has reacted completely with an analyte.

Ball mill A device for decreasing the particle size of the laboratory sample.

Band Ideally, a Gaussian-shaped distribution of (1) adjacent wavelengths encountered in spectroscopy or (2) the amount of a compound as it exits from a chromatographic or an electrophoretic column.

Band broadening The tendency of zones to spread as they pass through a chromatographic column; caused by various diffusion and mass transfer processes.

Band spectrum A molecular spectrum made up of one or more wavelength regions in which spectral lines are numerous and close together owing to rotational and vibrational transitions.

Bandwidth Usually, the range of wavelengths or frequencies of a spectral absorption or emission peak at half the height of the peak; the range passed by a wavelength isolation device.

Base dissociation constant, K_b The equilibrium constant for the reaction of a weak base with water.

Base region of a flame The region in which the solvent is evaporated, leaving the analyte as a finely divided solid.

Bases Species that are capable of accepting protons from donor (acid) species.

Basic flux A substance with basic characteristics in the molten state; used to solubilize refractory samples, principally silicates.

Beam The principal moving part of a mechanical analytical balance.

Beam arrest A mechanism that lifts the beam from its bearing surface when an analytical balance is not in use or when the load is being changed.

Beam splitter A device for dividing radiation from a monochromator such that one portion passes through the sample while the other passes through the blank.

Beer's law The fundamental relationship for the absorption of radiation by matter; that is, $A = abc$, where a is the absorptivity, b is the path length of the beam of radiation, and c is the concentration of the absorbing species.

Bernoulli effect In atomic spectroscopy, the mechanism by which sample droplets are aspirated into a plasma or flame.

β-amylose That component of starch that serves as a specific indicator for iodine.

Bias The tendency to skew estimates in the direction that favors the anticipated result. Also used to describe the effect of a *systematic error* on a set of measurements. Also a dc voltage that is used to polarize a circuit element.

Blackbody radiation Continuous radiation produced by a heated solid.

Blank determination The process of performing all steps of an analysis in the absence of sample; used to detect and compensate for systematic errors in an analysis.

Bolometer A detector for infrared radiation based on changes in resistance with changes in temperature.

Bonded-phase packings In HPLC, a support medium to which a liquid stationary phase is chemically bonded.

Bonded stationary phase A liquid stationary phase that is chemically bonded to the support medium.

Boundary potential, E_b The resultant of two potentials that develop at the surfaces of a glass membrane electrode.

Brønsted-Lowry acids and bases An acid of this type is defined as a proton donor and a base as a proton acceptor; the loss of a proton by an acid results in the formation of a species that is a potential proton acceptor, or *conjugate base* of the parent acid.

Buffer capacity The number of moles of strong acid (or strong base) needed to alter the pH of 1.00 L of a buffer solution by 1.00 unit.

Buffer solutions Solutions that tend to resist changes in pH as the result of dilution or the addition of small amounts of acids or bases.

Bumping The sudden and often violent boiling of a liquid that results from local overheating.

Buoyancy The displacement of the medium (ordinarily air) by an object, producing an apparent loss of mass; a significant source of error when the densities of the object and the comparison standards (weights) differ.

Buret A graduated tube from which accurately known volumes can be dispensed.

Burners Sources of heat for laboratory operations or for flame atomic spectroscopy.

C

Calibration The empirical determination of the relationship between a measured quantity and a known reference or standard value; used to establish analytical signal versus concentration relationships in a calibration or working curve.

Calomel The compound Hg_2Cl_2.

Calomel electrode A versatile reference electrode based on the half-reaction $Hg_2Cl_2(s) + 2e^- \rightleftharpoons 2Hg(l) + 2Cl^-$

Capillary column A small-diameter chromatographic column for GC or HPLC, fabricated of metal, glass, or fused silica. For GC, the stationary phase is a thin coating of liquid on the interior wall of the tube; for HPLC, capillary columns are often packed.

Capillary constant The product $m^{2/3}t^{1/6}$, where m is the mass of mercury (mg) delivered in time t(s) by a dropping mercury electrode.

Capillary electrophoresis High-speed, high-resolution electrophoresis performed in capillary tubes or in microchips.

Carbonate error A systematic error caused by absorption of carbon dioxide by standard solutions of base that will be used in the titration of weak acids.

Carrier gas The mobile phase for gas chromatography.

Catalytic method Analytical method for determining the concentration of a catalyst based on measuring the rate of a catalyzed reaction.

Catalytic reaction A reaction whose progress toward the equilibrium state is hastened by a substance that is not consumed in the overall process.

Cathode In an electrochemical cell, the electrode at which reduction takes place.

Cathode depolarizer A substance that is more easily reduced than hydrogen ion; used to prevent codeposition of hydrogen during an electrolysis.

Cathodic stripping analysis An electrochemical method in which the analyte is deposited by oxidation into a small-volume electrode and later stripped off by reduction.

Cation-exchange resins High-molecular-weight polymers to which acidic groups are bonded; these resins permit the substitution of cations in solution for hydrogen ions from the exchanger.

Cell A term with several meanings. (1) In statistics, the combination of adjacent data for display in a histogram. (2) In electrochemistry, an array consisting of a pair of electrodes immersed in solutions that are in electrical contact; the electrodes are connected externally by a metallic conductor. (3) In spectroscopy, the container that holds the sample in the light path of an optical instrument. (4) In an electronic balance, a system of constraints that assure alignment of the pan. (5) In a spreadsheet, a location at the intersection of a row and a column where data can be placed.

Cells without liquid junction Electrochemical cells in which both anode and cathode are immersed in a common electrolyte.

Charge-balance equation An expression relating the concentrations of anions and cations based on charge neutrality in a given solution.

Charge coupled device (CCD) A solid-state two-dimensional detector array used for spectroscopy and imaging.

Charge-injection device (CID) A solid-state photodetector array used in spectroscopy.

Charge transfer complexes Complexes that are made up of an electron donor group and an electron acceptor group; absorption of radiation by these complexes involves a transfer of electrons from the donor to the acceptor.

Charging current A positive or a negative nonfaradaic current resulting from a surplus or a deficiency of electrons in a mercury droplet at the instant of detachment.

Chelating agents Substances with multiple sites available for coordinate bonding with metal ions; such bonding typically results in the formation of five- or six-membered rings.

Chelation The reaction between a metal ion and a chelating reagent.

Chemical Abstracts A major source of chemical information worldwide; entered through an extensive system of indexes or a computer database.

Chemical deviations from Beer's law Deviations from Beer's law that result from association or dissociation of the absorbing species or reaction with the solvent, producing a product that absorbs differently from the analyte; in atomic spectroscopy, chemical interactions of the analyte with interferents that affect the absorption properties of the analyte.

Chemical equilibrium A dynamic state in which the rates of forward and reverse reactions are identical; a system in equilibrium will not spontaneously depart from this condition.

Chemiluminescence The emission of energy as electromagnetic radiation during a chemical reaction.

Chopper A mechanical device that alternately transmits and blocks radiation from a source.

Chromatogram A plot of analyte concentration signal as a function of elution time or elution volume.

Chromatograph An instrument for carrying out chromatographic separations.

Chromatographic bands The distribution (ideally Gaussian) of the concentration of eluted species about a central value; the result of variations in the time that analyte species reside in the mobile phase.

Chromatographic zones Synonymous with *chromatographic bands.*

Chromatography A term for methods of separation based on the interaction of species with a stationary phase while they are being transported by a mobile phase.

Clark oxygen sensor A voltammetric sensor for dissolved oxygen.

Coagulation The process whereby particles with colloidal dimensions are caused to form larger aggregates.

Coefficient of variation (CV) The relative standard deviation, expressed as a percentage.

Colloidal suspension A mixture (commonly of a solid in a liquid) in which the particles are so finely divided that they have no tendency to settle.

Colorimeter An optical instrument for the measurement of electromagnetic radiation in the visible region of the spectrum.

Column chromatography A chromatographic method in which the stationary phase is held within or on the surface of a narrow tube, and the mobile phase is forced through the tube, where compound separation occurs; compare with *planar chromatography.*

Column efficiency A measure of the degree of broadening of a chromatographic band; often expressed in terms of plate height, H, or the number of theoretical plates, N. Insofar as the distribution of analyte is Gaussian within the band, the plate height is given by the variance, σ^2, divided by the length, L, of the column.

Column resolution, R Measures the capability of a column to separate two analyte bands.

Common-ion effect The shift in the position of equilibrium caused by the addition of a participating ion.

Complex formation The process whereby a species with one or more unshared electron pairs forms coordinate bonds with metal ions.

Concentration-based equilibrium constant, K' The equilibrium constant based on molar equilibrium concentrations; the numerical value of K' depends on the ionic strength of the medium.

Concentration polarization The deviation of the electrode potential in an electrochemical cell from its equilibrium or Nernstian value on the passage of current as a result of slow transport of species to and from the electrode surface.

Concentration profile The distribution of analyte concentrations with time as they emerge from a chromatographic column; also, the time behavior of reactants or products during a chemical reaction.

Conduction of electricity The movement of charge by ions in solution, by electrochemical reaction at the surfaces of electrodes, or by movement of electrons in metals.

Conductometric detector A detector for charged species; finds use in ion chromatography.

Confidence interval Defines bounds about the experimental mean within which—with a given probability—the true mean should be located.

Confidence limits The values that define the confidence interval.

Conjugate acid/base pairs Species that differ from one another by one proton.

Constant-boiling HCl Solutions of hydrochloric acid with concentrations that depend on the atmospheric pressure.

Constant error A systematic error that is independent of the size of the sample taken for analysis; its effect on the results of an analysis increases as the sample size decreases.

Constant mass The condition in which the mass of an object is no longer altered by heating or cooling.

Constructive interference Increase in the amplitude of a wave in regions where two or more wave fronts are in phase with one another.

Continuous source A source that emits radiation continuously in time.

Continuum source A source that emits a spectral continuum of wavelengths; examples include tungsten filament lamps and deuterium lamps used in absorption spectroscopy.

Continuum spectrum Radiation consisting of a band of wavelengths as opposed to discrete lines. Incandescent solids provide continuum output *(blackbody radiation)* in the visible and infrared regions; deuterium and hydrogen lamps yield continuum spectra in the ultraviolet region.

Control chart A plot that demonstrates statistical control of a product or a service as a function of time.

Control circuit A three-electrode electrochemical apparatus that maintains a constant potential between the working electrode and the reference electrode; see *potentiostat.*

Controlled potential methods Electrochemical methods that use a potentiostat to maintain a constant potential between the working electrode and a reference electrode.

Convection The transport of a species in a liquid or gaseous medium by stirring, mechanical agitation, or temperature gradients.

Coordination compounds Species formed between metal ions and electron-pair donating groups; the product may be anionic, neutral, or cationic.

Coprecipitation The carrying down of otherwise soluble species either within a solid or on the surface of a solid as it precipitates.

Coulomb, C The quantity of charge provided by a constant current of one ampere in one second.

Coulometer A device that permits measurement of the quantity of charge. Electronic coulometers evaluate the integral of the current/time curve; chemical coulometers are based on the extent of reaction in an auxiliary cell.

Coulometric titration A type of coulometric analysis that involves measurement of the time needed for a constant current to produce enough reagent to react completely with an analyte.

Counter electrode The electrode that with the working electrode forms the electrolysis circuit in a three-electrode cell.

Counter-ion layer A region of solution surrounding a colloidal particle within which there exists a quantity of ions sufficient to balance the charge on the surface of the particle.

Creeping The tendency of some precipitates to spread over a wetted surface.

Critical temperature The temperature above which a substance can no longer exist in the liquid state, regardless of pressure.

Cross-linked stationary phase A polymer stationary phase in a chromatographic column in which covalent bonds link different strands of the polymer, thus creating a more stable phase.

Crystalline membrane electrode Electrode in which the sensing element is a crystalline solid that responds selectively to the activity of an ionic analyte.

Crystalline precipitates Solids that tend to form as large, easily filtered particles.

Crystalline suspensions Particles with greater-than-colloidal dimensions temporarily dispersed in a liquid.

Current, *i* The amount of electrical charge that passes through an electrical circuit per unit time; units are amperes, A.

Current density The current per unit area of an electrode in A/m^2.

Current efficiency Measure of the effectiveness of a quantity of electricity in bringing about an equivalent amount of chemical change in an analyte; coulometric methods require 100% current efficiency.

Current maxima Anomalous peaks in the current of a polarographic cell; can often be eliminated by the introduction of surface active agents.

Current-to-voltage converter A device for converting an electric current into a voltage that is proportional to the circuit.

Cuvette The container that holds the analyte in the light path in absorption spectroscopy.

D

Dalton Synonymous with *atomic mass unit.*

Dark currents Small currents that occur even when no radiation is reaching a photometric transducer.

Dashpot Synonymous with *air damper* in an analytical balance.

dc Plasma (DCP) spectroscopy A method that makes use of an electrically induced argon plasma to excite the emission spectra of analyte species.

Dead time In *column chromatography,* the time, t_M, required for an unretained species to traverse the column; in stopped-flow kinetics, the time between the mixing of reactants and the arrival of the mixture at the observation cell.

Debye-Hückel equation An expression that permits calculation of activity coefficients in media with ionic strengths less than 0.1.

Debye-Hückel limiting law A simplified form of the Debye-Hückel equation, applicable to solutions in which the ionic strength is less than 0.01.

Decantation The transfer of supernatant liquid and washings from a container to a filter without disturbing the precipitated solid in the container.

Decrepitation The shattering of a crystalline solid as it is heated; caused by vaporization of occluded water.

Degrees of freedom The number of members in a statistical sample that provide an independent measure of the precision of the set.

Dehydration The loss of water by a solid.

Dehydrite® Trade name for magnesium perchlorate, a drying agent.

Density The ratio of the mass of an object to its volume.

Depletion layer A nonconductive region in a reverse-biased semiconductor.

Depolarizer An additive that undergoes reaction at an electrode in preference to an otherwise undesirable process. See *cathode depolarizer.*

Derivative titration curve A plot of the change in the quantity measured per unit volume against the volume of titrant added; a derivative curve displays a maximum where there is a point of inflection in a conventional titration curve. See also *second derivative curve.*

Desiccants Drying agents.

Desiccator A container that provides a dry atmosphere for the storage of samples, crucibles, and precipitates.

Destructive interference A decrease in amplitude of waves resulting from the superposition of two or more wave fronts that are not in phase with one another.

Detection limit The minimum amount of analyte that a system or a method is capable of measuring.

Detector A device that responds to some characteristic of the system under observation and converts that response into a measurable signal.

Determinate error A class of errors that at least, in principle, has a known cause; synonymous with *systematic error.*

Deuterium lamp A source that provides a spectral continuum in the ultraviolet region of the spectrum; radiation results from application of about 40 V to a pair of electrodes housed in a deuterium atmosphere.

Devarda's alloy An alloy of copper, aluminum, and zinc; used to reduce nitrates and nitrites to ammonia in a basic medium.

Deviation The difference between an individual measurement and the mean (or median) value for a set of data.

Diatomaceous earth The siliceous skeletons of unicellular algae; used as a solid support, GC.

Differentiating solvents Solvents in which differences in the strengths of solute acids or bases are enhanced. Compare with *leveling solvents.*

Diffraction order, n Integer multiples of a wavelength at which constructive interference occurs.

Diffusion The migration of species from a region of high concentration in a solution to a more dilute region.

Diffusion coefficient (*polarographic, D, chromatographic,* D_m) A measure of the mobility of a species in units of cm^2/s.

Diffusion current, i_d**.** The limiting current in voltammetry when diffusion is the major form of mass transfer.

Digestion The practice of maintaining an unstirred mixture of freshly formed precipitate and solution from which it was formed at temperatures just below boiling; results in improved purity and particle size.

Dimethylglyoxime A precipitating reagent that is specific for nickel(II). Its formula is $CH_3(C{=}NOH)_2CH_3$.

Diode array detector A silicon chip that accommodates numerous photodiodes; provides the capability to collect data from entire spectral regions simultaneously. Usually contains 64 to 4096 photodiodes arranged linearly.

Diphenylthiocarbazide A chelating reagent, also known as *dithizone;* adducts with cations are sparingly soluble in water but are readily extracted with organic solvents.

Dissociation The splitting of molecules of a substance, commonly into two simpler entities.

Distribution constant The equilibrium constant for the distribution of an analyte in two immiscible solvents; approximately equal to the ratio of the equilibrium molar concentrations in the two solvents.

Dithizone Synonymous with *diphenylthiocarbazide.*

Doping The intentional introduction of traces of group III or group V elements to increase the semiconductor properties of a silicon or germanium crystal.

Doppler broadening Absorption or emission of radiation by a species in rapid motion, resulting in a broadening of spectral lines; a wavelength that is slightly shorter or longer than nominal is received by the detector, depending on the direction of motion of the species in the light path.

Double-beam instrument An optical instrument design that eliminates the need to alternate blank and analyte solutions manually in the light path. A *beam splitter* partitions the radiation in a double beam in space spectrometer; a *chopper* directs the beam alternately between blank and analyte in a double beam in time instrument.

Double precipitation Synonymous with *reprecipitation.*

Drierite® Trade name for calcium sulfate, a drying agent.

Dropping mercury electrode An electrode in which mercury is forced through a capillary, producing regular drops.

Dry ashing The elimination of organic matter from a sample by direct heating in air.

Dumas method A method of analysis based on the combustion of nitrogen-containing organic samples by CuO to convert the nitrogen to N_2, which is then measured volumetrically.

Dynamic methods Synonymous with *kinetic methods;* concerned with the changes that occur with time in chemical systems. Contrast with *static methods.*

Dynode An intermediate electrode in a photomultiplier tube.

E

Echelle grating A grating that is blazed with reflecting surfaces that are larger than the nonreflecting faces.

Eddy diffusion Diffusion of solutes that contributes to broadening of chromatographic bands, the result of differences in the pathways for solutes as they traverse a column.

EDTA An abbreviation of *ethylenediaminetetraacetic acid,* a chelating agent widely used for complex formation titrations. Its formula is $(HOOCCH_2)_2NCH_2CH_2N(CH_2COOH)_2$.

Effective bandwidth The bandwidth of a monochromator or an interference filter at which the transmittance is 50% that at the nominal wavelength.

Electric double layer Refers to the charge on the surface of a colloidal particle and the counter-ion layer that balances this charge; also, the charged layer on the surface of the working electrode in voltammetry.

Electroanalytical methods A large group of methods that have in common the measurement of an electrical property of the system that is proportional to the amount of analyte in the sample.

Electrochemical cell An array consisting of two electrodes, each of which is in contact with an electrolyte solution. Typically, the two electrolytes are in electrical contact through a *salt bridge;* an external metal conductor connects the two electrodes.

Electrochemical reversibility The ability of some cell processes to reverse themselves when the direction of the current is reversed; in an irreversible cell, reversal of current causes a different reaction at one or both electrodes.

Electrode A conductor at the surface of which electron transfer to or from the surrounding solution takes place.

Electrodeless discharge lamp A source of atomic line spectra that is powered by radio-frequency or microwave radiation.

Electrode of the first kind A metallic electrode whose potential is proportional to the logarithm of the concentration (strictly, activity) of a cation (or the ratio of cations) derived from the electrode metal.

Electrode of the second kind A metallic electrode whose response is proportional to the logarithm of the concentration (strictly, activity) of an anion that forms either a sparingly soluble species or stable complexes with a cation (or the ratio of cations) derived from the electrode metal.

Electrode potential The potential of an electrochemical cell in which the electrode of interest is the right-hand electrode and the standard hydrogen electrode is the left-hand electrode.

Electrogravimetric analysis A branch of gravimetric analysis that involves measuring the mass of species deposited on an electrode of an electrochemical cell.

Electrolysis circuit In a three-electrode arrangement, a dc source and a voltage divider to permit regulation of the potential between the working electrode and the counter electrode.

Electrolyte effect The dependence of numerical values for equilibrium constants on the ionic strength of the solution.

Electrolytes Solute species whose aqueous solutions conduct electricity.

Electrolytic cell An electrochemical cell that requires an external source of energy to drive the cell reaction. Compare with *galvanic cell.*

Electromagnetic radiation A form of energy with properties that can be described in terms of waves or, alternatively, as particulate photons, depending on the method of observation.

Electromagnetic spectrum The power or intensity of electromagnetic radiation plotted as a function of wavelength or frequency.

Electronic balance A balance in which an electromagnetic field supports the pan and its contents; the current needed to restore the loaded pan to its original position is proportional to the mass on the pan.

Electronic transition The promotion of an electron from one electronic state to a second electronic state, and conversely.

Electroosmotic flow The net flow of bulk liquid in an applied electric field.

Electrophoresis A separation method based on the differential rates of migration of charged species in an electric field.

Electrothermal analyzer Any of several devices that form an atomized gas containing an analyte in the light path of an instrument by electrical heating; used for atomic absorption and atomic fluorescence measurements.

Eluent A mobile phase in chromatography that is used to carry solutes through a stationary phase.

Eluent suppressor column In ion chromatography, a column downstream from the analytical column where ionic eluents are converted to nonconducting species, while analyte ions remain unaffected.

Elution chromatography Describes processes in which analytes are separated from one another on a column owing to differences in the time that they are retained in the column.

Emission spectrum The collection of spectral lines or bands that are observed when species in excited states relax by giving off their excess energy as electromagnetic radiation.

Empirical formula The simplest whole-number combination of atoms in a molecule.

End point An observable change during titration that signals that the amount of titrant added is chemically equivalent to the amount of analyte in the sample.

Enzymatic sensor A membrane electrode that has been coated with an immobilized enzyme; the electrode responds to the amount of analyte in the sample.

Enzyme substrate complex (ES) The intermediate formed in the process

$$\text{Enzyme (E)} + \text{substrate (S)} \rightleftharpoons \text{ES} \rightarrow \text{product (P)} + \text{E}$$

Eppendorf pipet A type of micropipet that delivers adjustable volumes of liquid.

Equal-arm balance An analytical balance equipped with a beam that supports two pans equidistant from the fulcrum—one for the load, the other to accommodate an equal mass of known weights.

Equilibrium-constant expression An algebraic statement that describes the equilibrium relationship among the participants in a chemical reaction.

Equilibrium molarity The concentration of a solute species (in mol/L or mmol/mL); synonymous with *species molarity.*

Equivalence point That point in a titration at which the amount of standard titrant added is chemically equivalent to the amount of analyte in the sample.

Equivalence-point potential The electrode potential of the system in an oxidation/reduction titration when the amount of titrant that has been added is chemically equivalent to the amount of analyte in the sample.

Equivalent For an oxidation/reduction reaction, that weight of a species that can donate or accept 1 mole of electrons; for an acid/base reaction, that weight of a species that can donate or accept 1 mole of protons.

Equivalent of chemical change The mass of a species that is directly or indirectly equivalent to one faraday (6.02×10^{23} electrons).

Equivalent weight A specialized basis for expressing mass in chemical terms similar to, but different from, *molar mass.* As a consequence of definition, one equivalent of an analyte reacts with one equivalent of a reagent, even if the stoichiometry of the reaction is not one to one.

Error The difference between an experimental measurement and its accepted value.

Essential water Water in a solid that exists in a fixed amount, either within the molecular structure *(water of constitution)* or within the crystalline structure *(water of crystallization).*

Ethylenediaminetetraacetic acid Probably the most versatile reagent for complex formation titrations; forms chelates with most cations. See *EDTA.*

Excitation The promotion of an atom, an ion, or a molecule to a state that is more energetic than a lower energy state.

Excitation spectrum In fluorescence spectroscopy, a plot of fluorescence intensity as a function of excitation wavelength.

Exhaustive extraction A cycle in which an organic solvent, after percolation through an aqueous phase containing the solute of interest, is distilled, condensed, and again passed through the aqueous phase.

F

Faradaic current An electric current produced by oxidation/reduction processes in an electrochemical cell.

Faraday, F The quantity of electricity associated with 6.022×10^{23} electrons.

Fast reaction Reaction that is half complete in 10 seconds or less.

Ferroin A common name for the 1,10-phenanthroline-iron(II) complex, which is a versatile redox indicator. Its formula is $(C_{12}H_8N_2)_3Fe^{2+}$.

Flame emission spectroscopy Method that uses a flame to cause an atomized analyte to emit its characteristic emission spectrum; also known as flame photometry.

Flame ionization detector (FID) A detector for gas chromatography based on the collection of ions produced during the pyrolysis of organic analytes in a flame.

Fluorescence Radiation produced by an atom or a molecule that has been excited by photons to a singlet excited state.

Fluorescence bands Groups of fluorescence lines that originate from the same excited electronic state.

Fluorescence spectrum A plot of fluorescence intensity versus wavelength in which the excitation wavelength is held constant.

Fluorometer A filter instrument for quantitative fluorescence measurements.

Fluxes Substances that in the molten state possess acidic or basic properties; used to solubilize the analyte in refractory samples.

Focal plane A surface on which dispersed radiation from a prism or a grating is focused.

Formality, F The number of formula masses of solute contained in each liter of solution; synonymous with *analytical molarity.*

Formal potential, $E^{0\prime}$ The electrode potential for a couple when the analytical concentrations of all participants are unity and the concentrations of other species in the solution are defined.

Formula weight The summation of atomic masses in the chemical formula of a substance; synonymous with gram formula weight and *molar mass.*

Fourier transform spectrometer A spectrometer in which an interferometer and Fourier transformation are used to obtain a spectrum.

Frequency, ν, of electromagnetic radiation The number of oscillations per second; has units of hertz (Hz), which is one oscillation per second.

Fritted-glass crucible A filtering crucible equipped with a porous glass bottom; also called a *sintered-glass crucible.*

Fronting Describes a nonideal chromatographic peak in which the early portions tend to be drawn out; compare with *tailing.*

***F*-test** A statistical method that permits comparison of the variances of two sets of measurements.

Fused-silica open tubular (FSOT) column A wall-coated gas chromatography column that has been fabricated from purified silica.

G

Galvanic cell An electrochemical cell that provides energy during its operation; synonymous with *voltaic cell.*

Galvanostat Synonymous with *amperostat.*

Gas chromatography (GC) Separation methods that make use of a gaseous mobile phase and a liquid or a solid stationary phase.

Gas electrode An electrode that involves the formation or consumption of a gas during its operation.

Gas-sensing probe An indicator/reference electrode system that is isolated from the analyte solution by a hydrophobic membrane. The membrane is permeable to a gas; the potential is proportional to the gas content of the analyte solution.

Gaussian distribution A theoretical bell-shaped distribution of results obtained for replicate measurements that are affected by random errors.

GC/MS A combined technique in which a mass spectrometer is used as a detector for gas chromatography.

Gel filtration chromatography A type of *size exclusion chromatography* that makes use of a hydrophilic packing; used to separate polar species.

Gel permeation chromatography A type of *size exclusion chromatography* that makes use of a hydrophobic packing; used to separate nonpolar species.

General elution problem The compromise between elution time and resolution; addressed through *gradient elution* (for liquid chromatography) or *temperature programming* (for gas chromatography).

General redox indicators Indicators that respond to changes in E_{system}.

Ghosts Double images in the output of a grating, the result of imperfections in the ruling engine used in its preparation.

Glass electrode An electrode in which a potential develops across a thin glass membrane, which provides a measure of the pH of the solution in which the electrode is immersed.

Gooch crucible A porcelain filtering crucible; filtration is accomplished by means of a glass fiber mat or a layer of asbestos fiber.

Gradient elution In liquid chromatography, the systematic alteration of mobile phase composition to optimize the chromatographic resolution of the components in a mixture.

Graphical kinetic methods Methods of determining reaction rates from plots of the concentration of a reactant or a product as a function of time.

Grating A device consisting of closely spaced grooves that is used to disperse polychromatic radiation by diffracting it into its component wavelengths.

Gravimetric analysis A group of analytical methods in which the amount of analyte is established through the measurement of the mass of a pure substance containing the analyte.

Gravimetric factor, GF The stoichiometric ratio between the analyte and the solid weighed in a gravimetric analysis.

Gravimetric titrimetry Titrations in which the mass of standard titrant is measured rather than volume; the concentration of titrant is expressed in mmol/g of solution (rather than the more familiar mmol/mL).

Gross error An occasional error, neither random nor systematic, that results in the occurrence of a questionable outlier result.

Gross sample A representative portion of a whole analytical sample; with further treatment, becomes the laboratory sample.

Ground state The lowest energy state of an atom or a molecule.

Guard column A precolumn located ahead of an HPLC column; the composition of the packing in the guard column is selected to extend the useful lifetime of the analytical column by removing particulate matter and contaminants and by saturating the eluent with the stationary phase.

H

Half-cell potential The potential of an electrochemical half-cell measured with respect to the standard hydrogen electrode.

Half-life, $t_{1/2}$ The time interval during which the amount of reactant has decreased by one half.

Half-reaction A method of portraying the oxidation or the reduction of a species; a balanced equation that shows the oxidized and reduced forms of a species, any H_2O or H^+ needed to balance the hydrogen and oxygen atoms in the system, and the number of electrons required to balance the charge.

Half-wave potential, $E_{1/2}$ The potential (ordinarily vs. SCE) at which the current of a voltammetric wave is one half the limiting current.

Hanging mercury drop electrode (HMDE) A microelectrode that can concentrate traces of metals by electrolysis into a small volume; the analysis is completed by voltammetric stripping of the metal from the mercury drop.

Heat detector A device that is sensitive to changes in the temperature of its surroundings; used to monitor infrared radiation.

Height equivalent of a theoretical plate, H (HETP) A measure of chromatographic column efficiency; equal to the length of a column divided by the number of theoretical plates in the column.

Henderson-Hasselbalch equation An expression used by biochemists to calculate the pH of a buffer solution; pH = $pK_a + \log (c_{NaA}/c_{HA})$, where pK_a is the negative logarithm of the dissociation constant for the acid and c_{NaA} and c_{HA} are the molar concentrations of the compounds making up the buffer.

High-performance adsorption chromatography Synonymous with *liquid-solid chromatography;* see also *adsorption chromatography.*

High-performance ion-exchange chromatography See *ion chromatography.*

High-performance liquid chromatography (HPLC) Column chromatography in which the mobile phase is a liquid, often forced through a stationary phase by pressure.

High-performance size exclusion chromatography See *size exclusion chromatography.*

Histogram A bar graph in which replicate results are grouped according to ranges of magnitude along the horizontal axis and by frequency of occurrence on the vertical axis.

Hollow-cathode lamp A source used in atomic absorption spectroscopy that emits sharp lines for a single element or sometimes for several elements.

Holographic grating A grating that has been produced by optical interference on a coated glass plate rather than by mechanical ruling.

Homogeneous precipitation A technique in which a precipitating agent is generated slowly throughout a solution of an analyte to yield a dense and easily filtered precipitate for gravimetric analysis.

Hundred percent T adjustment Adjustment of an optical absorption instrument to register 100% T with a suitable blank in the light path.

Hydrodynamic voltammetry Voltammetry performed with the analyte solution in constant motion relative to the electrode surface; produced by pumping the solution past a stationary electrode or by moving the electrode through the solution.

Hydrogen lamp A continuum source of radiation in the ultraviolet range that is similar in structure to a deuterium lamp.

Hydronium ion The hydrated proton whose symbol is H_3O^+.

8-Hydroxyquinoline A versatile chelating reagent; used in gravimetric analysis, in volumetric analysis, as a protective reagent in atomic spectroscopy, and as an extracting reagent; also known as *oxine.* Its formula is HOC_9H_6N.

Hygroscopic glass A glass that absorbs minute amounts of water on its surface; hygroscopicity is an essential property in the membrane of a glass electrode.

Hyphenated methods Methods involving the combination of two or more types of instrumentation; the product is an instrument with greater capabilities than any one instrument alone.

Hypothesis testing The process of testing a tentative assertion with various statistical tests. See *t-test, F-test, Q-test,* and *ANOVA.*

I

Ilkovic equation An equation that relates the diffusion current to the variables that affect it, i.e., the number of electrons involved (n) in the reaction with the analyte, the square root of the diffusion coefficient ($D^{1/2}$), the mass flow rate of mercury ($m^{2/3}$) and the drop time ($t^{1/6}$) of the dropping mercury electrode.

Immobilized enzyme reactor Tubular reactor or detector surface on which an enzyme has been attached by adsorption, covalent bonding, or entrapment.

Indeterminate error Synonymous with *random error.*

Indicator electrode An electrode whose potential is related to the logarithm of the activity of one or more species in contact with the electrode.

Indicator reaction, kinetics A fast reaction involving an indicator species that can be used to monitor the reaction of interest.

Inductively coupled plasma (ICP) spectroscopy A method that makes use of an inert gas (usually argon) plasma formed by the absorption of radio-frequency radiation to atomize and excite a sample for atomic emission spectroscopy.

Inert electrode An electrode that responds to the potential of the system, E_{system}, and is not otherwise involved in the cell reaction.

Infrared radiation Electromagnetic radiation in the 0.78 to 300 μm range.

Inhibitor, catalytic A species that decreases the rate of an enzyme-catalyzed reaction.

Initial rate methods Kinetic methods based on measurements made near the onset of a reaction.

Inner-filter effect Phenomenon causing nonlinear fluorescence calibration curves as a result of excessive absorption of the incident beam or the emitted beam.

Instrumental deviations from Beer's law Departures from linearity between absorbance and concentration that are attributable to the measuring device.

Integral methods Kinetic methods based on an integrated form of the rate law.

Intensity, I, of electromagnetic radiation The power per unit solid angle; often used synonymously with radiant power, P.

Intercept, b, of a regression line The y value in a regression line when the x value is zero; in an analytical calibration curve, the hypothetical value of the analytical signal when the concentration of analyte is zero.

Interference filter An optical filter that provides narrow bandwidths owing to constructive interference.

Interference order, n An integer that along with the thickness and the refractive index of the dielectric material determines the wavelength transmitted by an interference filter.

Interferences Species that affect the signal on which an analysis is based.

Interferometer A nondispersive device that obtains spectral information through constructive and destructive interference; used in Fourier transform infrared instruments.

Internal standard A known quantity of a species with properties similar to an analyte that is introduced into solutions of the standard and the unknown; the ratio of the signal from the internal standard to the signal from the analyte serves as the basis for the analysis.

International Union of Pure and Applied Chemistry (IUPAC) An international organization devoted to developing definitions and usages for the worldwide chemical community.

Ion chromatography An HPLC technique based on the partitioning of ionic species between a liquid mobile phase and a solid polymeric ionic exchanger; also called ion-exchange chromatography.

Ion-exchange resin A high-molecular-weight polymer to which a large number of acidic or basic functional groups have been bonded. Cationic resins permit the exchange of hydrogen ion for cations in solution; anionic resins substitute hydroxide ion for anions.

Ionic strength, μ A property of a solution that depends on the total concentration of ions in the solution as well as on the charge carried by each of these ions; that is, $\mu = \frac{1}{2}\Sigma c_i Z_i^2$, where c_i is the molar concentration of each ion and Z_i is its charge.

Ionization suppressor In atomic spectroscopy, an easily ionized species, such as potassium, that is introduced to suppress the ionization of the analyte.

***IR* drop** The potential drop across a cell due to resistance to the movement of charge; also known as the *ohmic potential drop.*

Irreversible cell An electrochemical cell in which the chemical reaction as a galvanic cell is different from that which occurs when the current is reversed.

Irreversible electrochemical reaction A reaction that yields a poorly defined voltammogram caused by the irreversibility of electron transfer at the electrode.

Isocratic elution Elution with a single solvent; compare with *gradient elution.*

Isoelectric point The pH at which an amino acid has no tendency to migrate under the influence of an electric field.

IUPAC convention A set of definitions relating to electrochemical cells and their potentials; also known as the *Stockholm convention.*

J

Jones reductor A column packed with amalgamated zinc; used for the prereduction of analytes.

Joule A unit of work equal to a newton-meter.

Junction potential The potential that develops at the interface between solutions with dissimilar composition; synonymous with *liquid junction* potential.

K

Karl Fischer reagent A reagent for the titrimetric determination of water.

Kilogram The base unit of mass in the SI system.

Kinetic methods Analytical methods based on relating the kinetics of a reaction to the analyte concentration.

Kinetic polarization Nonlinear behavior of an electrochemical cell caused by the slowness of the reaction at the surface of one or both electrodes.

Kjeldahl flask A long-necked flask used for the digestion of samples with hot, concentrated sulfuric acid.

Kjeldahl method A titration method for the determination of nitrogen in organic compounds in which the nitrogen is converted to ammonia, which is then distilled and determined by a neutralization titration.

Knife edge The nearly friction-free contact between the moving components of a mechanical analytical balance.

L

Laboratory balance Synonymous with *auxiliary balance.*

Laminar flow Streamline flow in a liquid near and parallel to a solid boundary. In a tube, this results in a parabolic flow profile; near an electrode surface, this results in parallel layers of liquid that slide by one another.

Least-squares method A statistical method of obtaining the parameters of a mathematical model (such as the equation for a straight line) by minimizing the sum of the square of the differences between the experimental points and the points predicted by the model.

Le Châtelier principle A statement that the application of a stress to a chemical system at equilibrium will result in a shift in the position of the equilibrium that tends to relieve the stress.

Leveling solvents Solvents in which the strength of solute acids or bases tend to be the same; compare with *differentiating solvents.*

Levitation As applied to electronic balances, the suspension of the pan of the balance in air by a magnetic field.

Ligand A molecule or an ion with at least one pair of unshared electrons available for coordinate bonding with cations.

Limiting current, i_l Current plateau reached in voltammetry when the electrode reaction rate is limited by the rate of mass transfer.

Linear scan voltammetry Electroanalytical methods that involve measurement of the current in a cell as the electrode potential is linearly increased or decreased with time; the basis for *hydrodynamic voltammetry* and *polarography.*

Linear segment curve A titration curve in which the end point is obtained by extrapolating linear regions well before and after the equivalence point; useful for reactions that do not strongly favor the formation of products.

Line source In atomic spectroscopy, a radiation source that emits sharp atomic lines characteristic of the analyte atoms; see *hollow-cathode lamp* and *electrodeless discharge lamp.*

Liquid bonded-phase chromatography Partition chromatography that makes use of a stationary phase that is chemically bonded to the column packing.

Liquid junction The interface between two liquids with different compositions.

Liquid-liquid chromatography Chromatography in which the mobile and stationary phases are liquids.

Liquid-solid chromatography Chromatography in which the mobile phase is a liquid and the stationary phase is a polar solid; synonymous with *adsorption chromatography.*

Liter One cubic decimeter or 1000 cubic centimeters.

Loading error An error in the measurement of a voltage due to current being drawn by the measuring device; occurs when the measuring device has resistance that is comparable to that of the voltage source being measured.

Longitudinal diffusion coefficient, *B* A measure of the tendency for analyte species to migrate from regions of high concentration to regions of lower concentration; contributes to band broadening in chromatography.

Longitudinal diffusion term, *B/u* A term in chromatographic band-broadening models that accounts for longitudinal diffusion.

Lower control limit, LCL The lower boundary that has been set for satisfactory performance of a process or measurement.

Luminescence Radiation resulting from photoexcitation (photoluminescence), chemical excitation (chemiluminescence), or thermal excitation (thermoluminescence).

L'vov platform Device for the electrothermal atomization of samples in atomic absorption spectroscopy.

M

Macrobalance An analytical balance with a capacity of 160 to 200 g and a precision of 0.1 mg.

Majority carrier The species principally responsible for the transport of electricity in a semiconductor.

Masking agent A reagent that combines with and inactivates matrix species that would otherwise interfere with the determination of an analyte.

Mass An invariant measure of the amount of matter in an object.

Mass-action effect The shift in the position of equilibrium through the addition or removal of a participant in the equilibrium; see also *Le Châtelier principle.*

Mass-balance equation An expression that relates the equilibrium concentrations of various species in a solution to one another and to the analytical concentration of the various solutes.

Mass-sensitive detector, chromatography A detector that responds to the mass of analyte, such as the *flame ionization detector.*

Mass spectrometry Methods based on forming ions in the gas phase and separating them on the basis of mass-to-charge ratio.

Mass-transfer coefficients, C_S, C_M Terms that account for mass transfer in the stationary and mobile phases in chromatography; mass transfer effects contribute to *band broadening.*

Mass transport The movement of species through a solution caused by diffusion, convection, and electrostatic forces.

Matrix The medium that contains an analyte.

Mean Synonymous with *arithmetic mean* and *average;* used to report what is considered the most representative value for a set of measurements.

Mean activity coefficient, $\gamma_{\pm}$ An experimentally measured activity coefficient for an ionic compound. It is not possible to resolve the mean activity coefficient into values for the individual participants.

Measuring pipet A pipet calibrated to deliver any desired volume up to its maximum capacity; compare with *volumetric pipet.*

Mechanical entrapment The incorporation of impurities within a growing crystal.

Mechanism of reaction The elementary steps involved in the formation of products from reactants.

Median The central value in a set of replicate measurements. For an odd number of data points, there are an equal number of points above and below the median; for an even number of data points, the median is the average of the central pair.

Megabore column An open tubular column that can accommodate samples that are larger than those in an ordinary packed column.

Melt The fused mass produced by the action of a flux; usually a fused salt.

Membrane electrode An indicator electrode whose response is due to ion-exchange processes on each side of a thin membrane.

Meniscus The curved surface displayed by a liquid held in a vessel.

Mercury electrode A static or dropping electrode used in voltammetry.

Mercury film electrode An electrode that has been coated with a thin layer of mercury; used in place of a *hanging mercury drop electrode* in anodic stripping analysis.

Metal oxide field effect transistor (MOSFET) A semiconductor device; when suitably coated can be used as an ion-selective electrode.

Method uncertainty, s_m The standard deviation associated with a measurement method; a factor, with the sampling standard deviation, in determining the overall standard deviation of an analysis.

Michaelis constant A collection of constants in the rate equation for enzyme kinetics; a measure of the dissociation of the enzyme-substrate complex.

Microanalytical balance An analytical balance with a capacity of 1 to 3 g and a precision of 0.0001 mg.

Microelectrode An electrode with dimensions on the micrometer scale; used in voltammetry.

Microgram, μg 1×10^{-6} g.

Microliter, μL 1×10^{-6} L.

Microporous membrane A hydrophobic membrane with a pore size that permits the passage of gases and is impermeable to other species; the sensing element of a *gas-sensing probe.*

Migration In electrochemistry, mass transport due to electrostatic attraction or repulsion; in chromatography, mass transport in the column.

Migration rate, $\bar{v}$ The rate at which an analyte traverses a chromatographic column.

Milligram, mg 1×10^{-3} g or 1×10^{-6} kg.

Milliliter, mL 1×10^{-3} L.

Millimole, mmol 1×10^{-3} mol.

Mixed-crystal formation A type of coprecipitation encountered in crystalline precipitates in which some of the ions in the analyte crystals are replaced by nonanalyte ions.

Mobile phase In chromatography, a liquid or a gas that carries analytes through a liquid or solid stationary phase.

Mobile phase mass-transfer coefficient, $C_M u$ A quantity that affects band broadening and thus plate height; nonlinear in solvent velocity u and influenced by the diffusion coefficient of the analyte, the particle size of the stationary phase, and the inside diameter of the column.

Modulation Process of superimposing the analytical signal on a carrier wave. In amplitude modulation, the carrier wave magnitude varies according to the variations in the analytical signal; in frequency modulation, the carrier wave frequency varies with the analytical signal.

Mohr's salt A common name for iron(II) ammonium sulfate hexahydrate.

Molar absorptivity, ε The proportionality constant in Beer's law; $\varepsilon = A/bc$, where A is the absorbance, b is the path length in centimeters, and c is the concentration in moles per liter; characteristic of the absorbing species.

Molarity, M The number of moles of a species contained in one liter of solution or the number of millimoles contained in one milliliter.

Molar mass, $\mathcal{M}$ The mass, in grams, of one mole of a chemical substance.

Mole The amount of substance contained in 6.022×10^{23} particles of that substance.

Molecular absorption The absorption of ultraviolet, visible, and infrared radiation brought about by quantized transitions in molecules.

Molecular fluorescence The process whereby singlet excited-state electrons in molecules return to a lower quantum state, with the resulting energy being given off as electromagnetic radiation.

Molecular formula A formula that includes structural information in addition to the number and identity of the atoms in a molecule.

Molecular weight Synonymous with molecular mass.

Monochromatic radiation Ideally, electromagnetic radiation that consists of a single wavelength; in practice, a very narrow band of wavelengths.

Monochromator A device for resolving polychromatic radiation into its component wavelengths.

Mother liquor The solution that remains following the precipitation of a solid.

Muffle furnace A heavy-duty oven capable of maintaining temperatures in excess of 1100°C.

N

Nanometer, nm 1×10^{-9} m.

National Institute of Standards and Technology (NIST) An agency of the U.S. Department of Commerce; formerly the *National Bureau of Standards* (NBS); a major source for primary standards and analyzed standard reference materials.

Natural lifetime, τ The radiative lifetime of an excited state; the time period during which the concentration of the reactant in a first-order process decreases to $1/e$ of its original value.

Nebulization The transformation of a liquid into a spray of small droplets.

Nernst diffusion layer, δ A thin layer of stagnant liquid at the surface of an electrode; caused by friction between the surface and the liquid that flows past the surface.

Nernst equation A mathematical expression that relates the potential of an electrode to the activities of those species in solution that are responsible for the potential.

Nernst glower A source of infrared radiation that consists of a cylinder of zirconium and yttrium oxides heated to a high temperature by passage of an electrical current.

Nichrome A nickel/chromium alloy; when incandescent, a source of infrared radiation.

Noise Random fluctuations of an analytical signal that result from a large number of uncontrolled variables affecting the signal; any signal that interferes with detection of the analyte signal.

Nominal wavelength The principal wavelength provided by a wavelength selection device.

Nonessential water Water that is retained in or on a solid by physical, rather than chemical, forces.

Normal error curve A plot of a Gaussian distribution of the frequency of results from random errors in a measurement.

Normal hydrogen electrode (NHE) Synonymous with *standard hydrogen electrode.*

Normality, c_N The number of equivalent weights of a species in one liter of solution.

Normal-phase chromatography A type of partition chromatography that involves a polar stationary phase and a nonpolar mobile phase; compare with *reversed-phase chromatography.*

Nucleation A process involving formation of very small aggregates of a solid during precipitation.

Null hypothesis A claim that a characteristic of a single population is equal to some specified value or that two or more population characteristics are identical; statistical tests are devised to validate or invalidate the null hypothesis with a specified level of probability.

Number of theoretical plates, N A characteristic of a chromatographic column used to describe its efficiency.

O

Occluded water Nonessential water that has been entrained in a growing crystal.

Occlusion The physical entrainment of soluble impurities in a growing crystal.

Occupational Safety and Health Administration (OSHA) A federal agency charged with assuring safety in the laboratory and the workplace.

Oesper's salt Common name for iron(II) ethylenediamine sulfate tetrahydrate.

Ohmic potential drop Synonymous with *IR drop.*

Open tubular column A capillary column of glass or fused silica used in gas chromatography; the walls of the tube are coated with a thin layer of the stationary phase.

Operational amplifier A versatile analog electronic amplifier for performing mathematical tasks and for conditioning output signals from instrument transducers.

Optical instruments A broad term for instruments that measure absorption, emission, or fluorescence by analyte species based on ultraviolet, visible, or infrared radiation.

Optical methods Synonymous with *spectrochemical methods.*

Optical wedge A device used in optical spectroscopy whose transmission decreases linearly along its length.

Order of reaction The exponent associated with the concentration of a species in the rate law for that reaction.

Outlier A result that appears at odds with the other members in a data set.

Overall reaction order A summation of the exponents for the species appearing in the rate law for a chemical reaction.

Overall standard deviation, s_o The square root of the sum of the variance of the measurement process and the variance of the sampling step.

Overpotential, overvoltage, Π Excess voltage necessary to produce current in a polarized electrochemical cell.

Oxidant Synonymous with *oxidizing agent.*

Oxidation The loss of electrons by a species in an oxidation/reduction reaction.

Oxidation potential The potential of an electrode process that is written as an oxidation.

Oxidizing agent A substance that acquires electrons in an oxidation/reduction reaction.

Oxine A common name for 8-hydroxyquinoline.

Oxygen wave At the dropping mercury electrode, oxygen produces two waves, the first due to formation of peroxide, the second due to further reduction to water; this can be an interference in the determination of other species but is used in the determination of dissolved oxygen.

P

Packed columns Chromatographic columns packed with porous materials to provide a large surface area for interaction with analytes in the mobile phase.

Pan arrest A device to support the pans of a balance when a load is being placed on them.

Parallax Apparent change in position of an object as a result of the movement of the observer; results in systematic errors in reading burets, pipets, and meters with pointers.

Particle growth A stage in the precipitation of solids.

Particle properties of electromagnetic radiation Behavior that is consistent with radiation acting as small particles or *quanta* of energy.

Partition chromatography A type of chromatography based on the distribution of solutes between a liquid mobile phase and a liquid stationary phase retained on the surface of a solid.

Partition coefficient An equilibrium constant for the distribution of a solute between two immiscible liquid phases; see *distribution constant.*

Parts per million, ppm A convenient method of expressing the concentration of a solute species that exists in trace amounts; for dilute aqueous solutions, ppm is synonymous with milligrams of solute per liter of solution.

Peak area, peak height Properties of peak-shaped signals that can be used for quantitative analysis; used in chromatography, electrothermal atomic absorption, and other techniques.

Peptization A process in which a coagulated colloid returns to its dispersed state.

Period of electromagnetic radiation The time required for successive peaks of an electromagnetic wave to pass a fixed point in space.

pH The negative logarithm of the hydrogen-ion activity of a solution.

Phosphorescence Emission of light from an excited triplet state; phosphorescence is slower than fluorescence and may occur over several minutes.

Phosphorus pentoxide, P_2O_5 A drying agent.

Photoconductive cell A detector of electromagnetic radiation whose electrical conductivity increases with the intensity of radiation impinging on it.

Photodecomposition The formation of new species from molecules excited by radiation; one of several ways by which excitation energy is dissipated.

Photodiode (1) A vacuum tube consisting of a wire anode and a photosensitive surface that produces an electron for each photon absorbed on the surface. (2) A reverse-biased silicon semiconductor that produces electrons and holes when irradiated by electromagnetic radiation. The resulting current provides a measure of the number of photons per second striking the device.

Photodiode array A linear array of photodiodes that can detect multiple wavelengths simultaneously; see *diode array detector.*

Photoelectric colorimeter A photometer that responds to visible radiation.

Photoelectron An electron released by the absorption of a photon striking a photoemissive surface.

Photoionization detector A chromatographic detector that uses intense ultraviolet radiation to ionize analyte species; the resulting currents, which are amplified and recorded, are proportional to analyte concentration.

Photometer An instrument for the measurement of absorbance that incorporates a filter for wavelength selection and a photon detector.

Photomultiplier tube A sensitive detector of electromagnetic radiation; amplification is accomplished by a series of dynodes that produce a cascade of electrons for each photon received by the tube.

Photon detector A generic term for transducers that convert an optical signal to an electrical signal.

Photons Energy packets of electromagnetic radiation; also known as *quanta.*

Phototube A transducer consisting of a photoemissive cathode, a wire anode, and a power supply to maintain a suitable potential between the electrodes.

Phthalein indicators Acid/base indicators derived from phthalic anhydride, the most common of which is phenolphthalein.

pIon meter An instrument that directly measures the concentration (strictly, activity) of an analyte; consists of a specific ion indicator electrode, a reference electrode, and a potential-measuring device.

Pipet A device to permit the transfer of known volumes of solution from one container to another.

Pixel A single detector element on a diode array detector or a charge transfer detector.

Planar chromatography The term used to describe chromatographic methods that make use of a flat stationary phase; the mobile phase migrates across the surface by gravity or capillary action.

Plasma A gaseous medium that owes its conductivity to appreciable amounts of ions and electrons.

Plate height, *H* A quantity describing the efficiency of a chromatographic column.

Platinum electrode Used extensively in electrochemical systems in which an inert metallic electrode is required.

Plattner diamond mortar A device for crushing small amounts of brittle materials.

Pneumatic detector A heat detector that is based on changes in the pressure that a gas exerts on a flexible diaphragm.

***p-n* junction diode** A semiconductor device containing a junction between electron-rich and electron-deficient regions; permits current in one direction only.

Polarization (1) In an electrochemical cell, a phenomenon in which the magnitude of the current is limited by the low rate of the electrode reactions (kinetic polarization) or the slowness of transport of reactants to the electrode surface (concentration polarization). (2) The process of causing electromagnetic radiation to vibrate in a definite pattern.

Polarogram The current/voltage plot obtained from polarographic measurements.

Polarography Voltammetry with a dropping mercury electrode.

Polychromatic radiation Electromagnetic radiation consisting of more than one wavelength; compare with *monochromatic radiation.*

Polyfunctional acids and bases Species that contain more than one acidic or basic functional group.

Population mean, μ The mean value for a population of data; the true value for a quantity that is free of systematic error.

Population of data The total number of values (sometimes infinite) that a measurement could take; also referred to as a *universe of data.*

Population standard deviation, σ A precision parameter based on a population of data.

Porous layer open tube (PLOT) column A capillary column for gas-solid chromatography in which a thin layer of the stationary phase is adsorbed on the walls of the column.

Potentiometric titration A titrimetric method involving measurement of the potential between a reference electrode and an indicator electrode as a function of titrant volume.

Potentiometry That branch of electrochemistry concerned with the relationship between the potential of an electrochemical cell and the concentration of the contents of the cell.

Potentiostat An electronic device that alters the applied potential so that the potential between a working electrode and a reference electrode is maintained at a fixed value.

Potentiostatic methods Electrochemical methods that use a controlled potential between the working electrode and a reference electrode.

Power, *P*, of electromagnetic radiation The energy that reaches a given area per second; often used synonymously with intensity, although the two are not precisely the same.

Precipitation from homogeneous solution Synonymous with *homogeneous precipitation.*

Precipitation methods of analysis Gravimetric and titrimetric methods involving the formation (or less frequently, the disappearance) of a precipitate.

Precision A measure of the internal agreement among a set of replicate observations.

Premix burner Burner in which gases are mixed prior to combustion.

Pressure broadening An effect that increases the width of an atomic spectral line; caused by collisions among atoms that result in slight variations in their energy states.

Primary absorption Absorption of the excitation beam in fluorescence or phosphorescence spectroscopy; compare with *secondary absorption.*

Primary adsorption layer Charged layer of ions on the surface of a solid, resulting from the attraction of lattice ions for ions of opposite charge in the solution.

Primary standard A highly pure chemical compound that is used to prepare or determine the concentrations of standard solutions for titrimetry.

Prism A transparent, prism-shaped solid that disperses polychromatic radiation into its component wavelengths by refraction.

Proportional error An error whose magnitude increases as the sample size increases.

Protective agent In atomic spectroscopy, species that form soluble complexes with the analyte and thereby prevent the formation of compounds that have low volatility.

Pseudo-order reactions Chemical systems in which the concentration of a reactant (or reactants) is large and essentially invariant with respect to that of the component (or components) of interest.

Pulse polarography Voltammetric methods that periodically impose a pulse on the linearly increasing excitation voltage; the difference in measured current, Δi, yields a peak whose height is proportional to the analyte concentration.

p-Value An expression of the concentration of a solute species as its negative logarithm; the use of p-values permits expression of enormous ranges of concentration in terms of relatively small numbers.

Pyroelectric detector A thermal detector based on the temperature-dependent potential that develops between electrodes separated by a pyroelectric material.

Q

Q test A statistical test that indicates—with a specified level of probability—whether an outlying measurement in a set of replicate data is a member of a given Gaussian distribution.

Quality assessment A protocol to assure that quality control methods are providing the information needed to evaluate satisfactory performance of a product or a service.

Quality assurance A protocol designed to demonstrate that a product or a service is meeting criteria that have been established for satisfactory performance.

Quanta Synonymous with *photons.*

Quantum yield of fluorescence The fraction of absorbed photons that are emitted as fluorescence photons.

Quenching (1) Process by which molecules in an excited state lose energy to other species without fluorescing. (2) An action that brings about the cessation of a chemical reaction.

R

Radiation buffers Potential interferents that are intentionally added in large amounts to samples and standards to swamp out their effects on atomic emission measurements.

Random errors Uncertainties resulting from the operation of small uncontrolled variables that are inevitable as measurement systems are extended to and beyond their limits.

Range, *w*, of data The difference between extreme values in a set of data; synonymous with *spread.*

Rate constant, *k* A proportionality constant in the rate expression.

Rate-determining step The slow step in the sequence of elementary reactions making up a mechanism.

Rate law The empirical relationship describing the rate of a reaction in terms of the concentrations of participating species.

Rate theory A theory that accounts for the shapes of chromatographic peaks.

Reaction order Synonymous with *order of reaction.*

Reagent-grade chemicals Highly pure chemicals that meet the standards of the Reagent Chemical Committee of the American Chemical Society.

Redox Synonymous with *oxidation/reduction.*

Redox electrode An inert electrode that responds to the electrode potential of the system.

Reducing agent The species that supplies electrons in an oxidation/reduction reaction.

Reductant Synonymous with *reducing agent.*

Reduction The process whereby a species acquires electrons.

Reduction potential The potential of an electrode process expressed as a reduction; synonymous with *electrode potential.*

Reductor A column packed with a granular metal through which a sample is passed to prereduce an analyte.

Reference electrode An electrode whose potential relative to the standard hydrogen electrode is known and against which potentials of unknown electrodes may be measured; the potential of a reference electrode is completely independent of the analyte concentration.

Reference standards Complex materials that have been extensively analyzed; a prime source for these standards is the National Institute of Standards and Technology (NIST).

Reflection The return of radiation from a surface.

Reflection grating An optical body that disperses polychromatic radiation into its component wavelengths. Consists of lines ruled on a reflecting surface; dispersion is the result of constructive and destructive interference.

Refractive index The ratio of the velocity of electromagnetic radiation in a vacuum to its velocity in some other medium.

Refractory materials Substances that resist attack by ordinary laboratory acids or bases; brought into solution by high-temperature fusion with a flux.

Regression analysis A statistical technique for determining the parameters of a model; see also *least-squares method.*

Relative electrode potential The potential of an electrode with respect to another (ordinarily the standard hydrogen electrode or saturated calomel electrode).

Relative error The error in a measurement divided by the true (or accepted) value for the measurement; often expressed as a percentage.

Relative humidity The ratio, often expressed as a percentage, between the ambient vapor pressure of water and its saturated vapor pressure at a given temperature.

Relative standard deviation (RSD) The standard deviation divided by the mean value for a set of data; when expressed as a percentage, the relative standard deviation is referred to as the *coefficient of variation.*

Relative supersaturation The difference between the instantaneous *(Q)* and the equilibrium *(S)* concentrations of a solute in a solution, divided by *S;* provides general guidance as to the particle size of a precipitate formed by addition of reagent to an analyte solution.

Relaxation The return of excited species to a lower energy level; the process is accompanied by the release of excitation energy as heat or luminescence.

Releasing agent In atomic absorption spectroscopy, species introduced to combine with sample components that would otherwise interfere by forming compounds of low volatility with the analyte.

Replica grating An impression of a master grating; used as the dispersing element in most grating instruments, owing to the high cost of a master grating.

Replicate samples Portions of a material, of approximately the same size, that are carried through an analysis at the same time and in the same way.

Reprecipitation A method of improving the purity of precipitates involving formation and filtration of the solid, followed by redissolution and re-formation of the precipitate.

Residual The difference between the value predicted by a model and the experimental value.

Residual current Nonfaradaic currents due to impurities and to charging of the electrical double layer.

Resolution, R_s Measures the ability of a chromatographic column to separate two analytes; defined as the difference between the retention times for the two peaks divided by their average widths.

Resonance fluorescence Fluorescence emission at a wavelength that is identical with the excitation wavelength.

Resonance line A spectral line resulting from a resonance transition.

Resonance transition A transition to and from the ground electronic state.

Retention factor, *k* A term used to describe the migration of a species through a chromatographic column. Its numerical value is given by $k = (t_R - t_M)/t_M$, where t_R is the retention time for a peak and t_M is the dead time; also called the *capacity factor.*

Retention time, t_R In chromatography, the time between sample injection on a chromatographic column and the arrival of an analyte peak at the detector.

Reversed-phase chromatography A type of liquid-liquid partition chromatography that makes use of a nonpolar stationary phase and a polar mobile phase; compare with *normal-phase chromatography.*

Reversible cell An electrochemical cell in which electron transfer is rapid in both directions.

Rheostat A type of voltage divider.

Rotational states Quantized states associated with the rotation of a molecule about its center of mass.

Rotational transition A change in quantized rotational energy states in a molecule.

Rubber policeman A small length of rubber tubing that has been crimped on one end; used to dislodge adherent particles of precipitate from beaker walls.

S

Salt The ionic species formed by the reaction of an acid and a base.

Salt bridge A device in an electrochemical cell that allows conduction of electricity between the two electrolyte solutions while minimizing mixing of the two.

Salt effect Influence of ions on the activities of reagents.

Salt-induced precipitation Technique used to precipitate proteins. At low salt concentration, adding salt increases solubility (salting-in effect), whereas high salt concentrations induce precipitation (salting-out effect).

Sample of data A finite group of replicate measurements.

Sample matrix The medium that contains an analyte.

Sample mean, $\bar{x}$ The arithmetic average of a finite set of measurements.

Sample splitter A device that permits the introduction of small and reproducible portions of sample to a chromatographic column. In capillary gas chromatography, a reproducible fraction of the injected sample is introduced onto the column, while the remaining portion goes to waste.

Sample standard deviation, s A precision estimate based on deviations of individual data from the mean, $\bar{x}$, of a data sample; also referred to as the *standard deviation.*

Sampling The process of collecting a small portion of a material whose composition is representative of the bulk of the material from which it was taken.

Sampling loop A small piece of tubing used in chromatography that has a sampling valve to inject small quantities of sample.

Sampling uncertainty, s_s The standard deviation associated with the taking of a sample; a factor—with the method uncertainty—in determining the overall standard deviation of an analysis.

Sampling valve A rotary valve used to inject small portions of a sample onto a chromatographic column; usually used in conjunction with a *sampling loop.*

Saponification The cleavage of an ester group to regenerate the alcohol and the acid from which the ester was derived.

Saturated calomel electrode (SCE) A reference electrode that can be formulated as $Hg|Hg_2Cl_2(sat), KCl(sat)||$. Its half-reaction is

$$Hg_2Cl_2(s) + 2e^- \rightleftharpoons 2Hg(l) + 2Cl^-$$

Schöniger apparatus A device for the combustion of samples in an oxygen-rich environment.

Second derivative curve A plot of $\Delta^2E/\Delta V^2$ for a potentiometric titration; the function undergoes a change in sign at the inflection point in a conventional titration curve.

Secondary absorption Absorption of the emitted radiation in fluorescence or phosphorescence spectrometry; compare with *primary absorption.*

Secondary standard A substance whose purity has been established and verified by chemical analysis.

Sector mirror A disk with portions that are partially mirrored and partially nonreflecting; when rotated, directs radiation from the monochromator of a double-beam spectrophotometer alternately through the sample and the reference cells.

Selectivity The tendency for a reagent or an instrumental method to react with or respond similarly to only a few species.

Selectivity coefficient, $k_{A,B}$ The selectivity coefficient for a specific ion electrode is a measure of the relative response of the electrode to ions A and B.

Selectivity factor, α In chromatography, $\alpha = K_B/K_A$, where K_B is the distribution constant for a less strongly retained species and K_A is the constant for a more strongly retained species.

Self-absorption A process in which analyte molecules absorb radiation emitted by other analyte molecules.

Semiconductor A material with electrical conductivity that is intermediate between a metal and an insulator.

Semimicroanalytical balance A balance with a capacity of about 30 g and a precision of 0.01 mg.

Servo system A device in which a small error signal is amplified and used to return the system to a null position.

Sigmoid curve An S-shaped curve; typical of the plot of the p-function of an analyte versus the volume of reagent in titrimetry.

Signal-to-noise ratio, S/N The ratio of the mean analyte output signal to the standard deviation of the signal.

Significant figure convention A system of imparting to the reader information concerning the reliability of numerical data; in general, all digits known with certainty, plus the first uncertain digit, are considered significant.

Silica Common name for silicon dioxide; used in the manufacture of crucibles and the cells for optical analysis and as a chromatographic support medium.

Silicon photodiode A photon detector based on a reverse-biased silicon diode; exposure to radiation creates new holes and electrons, thereby increasing photocurrent.

Silver-silver chloride electrode A widely encountered reference electrode, which can be formulated as $Ag|AgCl(s), KCl(xM)||$. The half-reaction for the electrode is

$$AgCl(s) + e^- \rightleftharpoons Ag(s) + Cl^-(xM)$$

Single-beam instruments Photometric instruments that use only one beam; they require the operator to position the sample and the blank alternately in a single light path.

Single-electrode potential Synonymous with *relative electrode potential.*

Single-pan balance An unequal-arm balance with the pan and weights on one side of the fulcrum and an air damper on the other; the weighing operation involves removal of standard weights in an amount equal to the mass of the object on the pan.

Sintered-glass crucible Synonymous with *fritted-glass crucible.*

SI units An international system of measurement that makes use of seven base units; all other units are derived from these seven units.

Size exclusion chromatography A type of chromatography in which the packing is a finely divided solid having a uniform pore size; separation is based on the size of analyte molecules.

Slope, *m*, of a calibration line A parameter of the linear model $y = mx + b$; determined by regression analysis.

Soap-bubble meter A device for measuring gas flow rates in gas chromatography.

Solubility-product constant, K_{sp} A numerical constant that describes equilibrium in a saturated solution of a sparingly soluble ionic salt.

Soluble starch β-amylose, an aqueous suspension of which is a specific indicator for iodine.

Solvent programming The systematic alteration of mobile-phase composition to optimize migration rates of solutes in a chromatographic column.

Sorbed water Nonessential water that is retained in the interstices of solid materials.

Sparging The removal of an unwanted dissolved gas by aeration with an inert gas.

Special-purpose chemicals Reagents that have been specially purified for a particular end use.

Species molarity The equilibrium concentration of a species expressed in moles per liter and symbolized with square brackets [ ]; synonymous with *equilibrium molarity.*

Specific gravity, sp gr The ratio of the density of a substance to that of water at a specified temperature (ordinarily 4°C).

Specific indicator A species that reacts with a particular species in an oxidation/reduction titration.

Specific surface area The ratio between the surface area of a solid and its mass.

Specificity Refers to methods or reagents that respond or react with one and only one analyte.

Spectra Plots of absorbance, transmittance, or emission intensity as a function of wavelength, frequency, or wavenumber.

Spectral interference Emission or absorption by species other than the analyte within the band-pass of the wavelength selection device; causes a blank interference.

Spectrochemical methods Synonymous with *spectrometric methods.*

Spectrofluorometer A fluorescence instrument that employs monochromators for selecting excitation and emission wavelengths; in some cases, hybrid instruments employ a filter and a monochromator.

Spectrograph An optical instrument equipped with a dispersing element, such as a grating or a prism, that allows a range of wavelengths to strike a spatially sensitive detector such as a diode array, charge coupled device, or photographic plate.

Spectrometer An instrument equipped with a monochromator or a polychromator, a photodetector, and an electronic readout that displays a number proportional to the intensity of an isolated spectral band.

Spectrometric methods Methods based on the absorption, emission, or fluorescence of electromagnetic radiation that is related to the amount of analyte in the sample.

Spectrophotometer A spectrometer designed for the measurement of the absorption of ultraviolet, visible, or infrared radiation. The instrument includes a source of radiation, a monochromator, and an electrical means of measuring the ratio of the intensities of the sample and reference beams.

Spectrophotometric titration A titration monitored by ultraviolet/visible spectrometry.

Spectroscope An optical instrument similar to a spectrometer except that spectral lines can be observed visually.

Spectroscopy A general term used to describe techniques based on the measurement of absorption, emission, or luminescence of electromagnetic radiation.

Spread, *w*, of data A precision estimate; synonymous with *range.*

Sputtering The process whereby an atomic vapor is produced by collisions with excited ions on a surface such as the cathode in a hollow-cathode lamp.

Square-wave polarography A variety of *pulse polarography.*

Standard-addition method A method of determining the concentration of an analyte in a solution. Small measured increments of the analyte are added to the sample solution, and instrument readings are recorded after one or more additions. The method compensates for some matrix interferences.

Standard deviation, σ or s A measure of how closely replicate data cluster around the mean; in a normal distribution, 67% of the data points can be expected to lie within one standard deviation of the mean.

Standard deviation about regression, s_r The standard deviation based on deviations from a least-square straight line.

Standard electrode potential, E^0 The potential (relative to the standard hydrogen electrode) of a half reaction written as a reduction when the activities of all reactants and products are unity.

Standard error of the mean, σ_m or s_m The standard deviation divided by the square root of the number of measurements in the set.

Standard hydrogen electrode (SHE) A gas electrode consisting of a platinized platinum electrode immersed in a solution that has a hydrogen ion activity of 1.00 and is kept saturated with hydrogen at a pressure of 1.00 atm. Its potential is assigned a value of 0.000 V at all temperatures.

Standardization Determination of the concentration of a solution by calibration, directly or indirectly, with a primary standard.

Standard reference materials (SRMs) Samples of various materials in which the concentration of one or more species is known with very high certainty.

Standard solution A solution in which the concentration of a solute is known with high reliability.

Static methods Methods based on observation of systems in equilibrium; compare with *kinetic methods.*

Stationary phase In chromatography, a solid or an immobilized liquid on which analyte species are partitioned during passage of a mobile phase.

Stationary phase mass-transfer term, $C_S u$ A measure of the rate at which an analyte molecule enters and is released from the stationary phase.

Statistical control The condition in which performance of a product or a service is deemed within bounds that have been set for quality assurance; defined by upper and lower control limits.

Statistical sample A finite set of measurements, drawn from a population of data, often from an infinite number of possible measurements.

Steady-state approximation The assumption that the concentration of an intermediate in a multistep reaction remains essentially constant with time.

Stirrup The link between the beam of a mechanical balance and its pan (or pans).

Stockholm convention A set of conventions relating to electrochemical cells and their potentials; also known as the *IUPAC convention.*

Stoichiometry Refers to the combining ratios among molar quantities of species in a chemical reaction.

Stokes shifts Differences in wavelengths of incident and emitted or scattered radiation.

Stop-flow injection In high-performance liquid chromatography, introduction of the sample at the head of the column while solvent flow is temporarily discontinued.

Stopped-flow mixing A technique in which the reactants are mixed rapidly and the course of the reaction is monitored downstream after the flow has stopped.

Stray radiation Radiation of a wavelength other than the wavelength selected for optical measurement.

Strong acids and strong bases Acids and bases that are completely dissociated in a particular solvent.

Strong electrolytes Solutes that are completely dissociated into ions in a particular solvent.

Student *t* test See ***t* test.**

Substrate (1) A substance acted on, usually by an enzyme. (2) A solid on which surface modifications are made.

Successive approximations A procedure for solving higher order equations through the use of intermediate estimates of the quantity sought.

Sulfide separations The use of sulfide precipitation to separate cations.

Sulfonic acid group —RSO_3H.

Supercritical fluid A substance that is maintained above its critical temperature; its properties are intermediate between those of a liquid and those of a gas.

Supercritical fluid chromatography Chromatography involving a supercritical fluid as the mobile phase.

Supersaturation A condition in which a solution temporarily contains an amount of solute that exceeds its equilibrium solubility.

Support-coated open tubular (SCOT) columns Capillary gas chromatography columns whose interior walls are lined with a solid support.

Supporting electrolyte A salt added to the solution in a voltammetric cell to eliminate migration of the analyte to the electrode surface.

Suppressor-based chromatography A chromatographic technique involving a column or a membrane located between the analytical column and a conductivity detector; its purpose is to convert ions of the eluting solvent into nonconducting species while passing ions of the sample.

Surface adsorption The retention of a normally soluble species on the surface of a solid.

Swamping The introduction of a potential interferent to both calibration standards and the solution of the analyte in order to minimize the effect of the interferent in the sample matrix.

Systematic error Errors that have a known source; they affect measurements in one and only one way, and can, in principle, be accounted for. Also called *determinate error* or *bias.*

T

0% *T* adjustment A calibration step that eliminates dark current and other background signals from the response of a spectrophotometer.

100% *T* adjustment Adjustment of a spectrophotometer to register 100% transmittance with a blank in the light path.

Tailing A nonideal condition in a chromatographic peak in which the latter portions are drawn out; compare with *fronting.*

Tare A counterweight used on an analytical balance to compensate for the mass of a container.

Temperature programming The systematic adjustment of column temperature in gas chromatography to optimize migration rates for solutes.

THAM *tris*-(hydroxymethyl) aminomethane, a primary standard for bases; its formula is $(HOCH_2)_3CNH_2$.

Thermal conductivity detector A detector used in gas chromatography that depends on measuring the thermal conductivity of the column eluent.

Thermal detector An infrared detector that produces heat as a result of absorption of radiation.

Thermionic detector (TID) A detector for gas chromatography similar to a flame ionization detector; particularly sensitive for analytes that contain nitrogen or phosphorus.

Thermistor A temperature-sensing semiconductor; used in some bolometers.

Thermodynamic equilibrium constant, *K* The equilibrium constant expressed in terms of the activities of all reactants and products.

TISAB (total ionic strength adjustment buffer) A solution used to swamp the effect of electrolytes on direct potentiometric analyses.

Titration The procedure whereby a standard solution reacts with known stoichiometry with an analyte to the point of chemical

equivalence, which is measured experimentally as the end point. The volume or the mass of the standard needed to reach the end point is used to calculate the amount of analyte present.

Titration error The difference between the titrant volume needed to reach an end point in a titration and the theoretical volume required to obtain an equivalence point.

Titrator An instrument that performs titrations automatically.

Titrimetry The process of systematically introducing an amount of titrant that is chemically equivalent to the quantity of analyte in a sample.

Transducer A device that converts a chemical or physical phenomenon into an electrical signal.

Transfer pipet Synonymous with *volumetric pipet.*

Transition pH range The span of acidities (frequently about 2 pH units) over which an acid/base indicator changes from its pure acid color to that of its conjugate base.

Transition potential The range in E_{system} over which an oxidation/reduction indicator changes from the color of its reduced form to that of its oxidized form.

Transmittance, T The ratio of the power, P, of a beam of radiation after it has traversed an absorbing medium to its original power, P_0; often expressed as a percentage:

$$\%T = (P/P_0) \times 100\%.$$

Transverse wave A wave motion in which the direction of displacement is perpendicular to the direction of propagation.

Triple-beam balance A rugged laboratory balance that is used to weigh approximate amounts.

TRIS Synonymous with *THAM.*

***t*-test** A statistical test used to decide whether an experimental value equals a known or theoretical value or whether two or more experimental values are identical with a given level of confidence; used with s and $\bar{x}$ when σ and μ are not available.

Tungsten filament lamp A convenient source of visible and near-infrared radiation.

Tungsten-halogen lamp A tungsten lamp that contains a small amount of I_2 within a quartz envelope that allows the lamp to operate at a higher temperature; brighter than a conventional tungsten filament lamp.

Turbulent flow Describes the random motion of liquid in the bulk of a flowing solution; compare with *laminar flow.*

Tyndall effect The scattering of radiation by particles in a solution or a gas that have colloidal dimensions.

U

Ultramicroelectrode Synonymous with *microelectrode.*

Ultraviolet/visible detector, HPLC Detector for high-performance liquid chromatography that uses ultraviolet/visible absorption to monitor eluted species as they exit a chromatographic column.

Ultraviolet/visible region The region of the electromagnetic spectrum between 180 and 780 nm; associated with electronic transitions in atoms and molecules.

Universe of data Synonymous with a *population of data.*

V

Valinomycin An antibiotic that has also found application in a membrane electrode for potassium.

van Deemter equation An equation that expresses plate height in terms of eddy diffusion, longitudinal diffusion, and mass transport.

Variance, σ^2 or s^2 A precision estimate consisting of the square of the standard deviation. Also a measure of column performance; given the symbol τ^2 where the abscissa of the chromatogram has units of time.

V-blender A device that is used to thoroughly mix dry samples.

Velocity of electromagnetic radiation, v *In vacuo,* 3×10^{10} cm/sec.

Vernier An aid for making estimates between graduation marks on a scale.

Vibrational relaxation A very efficient process in which excited molecules relax to the lowest vibrational level of an electronic state.

Vibrational transitions Transitions between vibrational states of an electronic state that are responsible for infrared absorption.

Visible radiation That portion of the electromagnetic spectrum (380 to 780 nm) to which the human eye is responsive.

Volatilization The process of converting a liquid (or a solid) to the vapor state.

Volatilization method of analysis A variant of the gravimetric method based on mass loss caused by heating or ignition.

Voltage divider A resistive network that provides a fraction of the input voltage at its output.

Voltaic cell Synonymous with *galvanic cell.*

Voltammetric wave Synonymous with *voltammogram.*

Voltammetry A group of electroanalytical methods that measure current as a function of the voltage applied to a working electrode.

Voltammogram A plot of current as a function of the potential applied to a working electrode.

Volume percent (v/v) The ratio of the volume of a liquid to the volume of its solution, multiplied by 100%.

Volumetric flask A container for preparing precise volumes of solution.

Volumetric methods Methods of analysis in which the final measurement is a volume of a standard titrant needed to react with the analyte in a known quantity of sample.

Volumetric pipet A device that will deliver a precise volume from one container to another; also called a *measuring pipet.*

W

Walden reductor A column packed with finely divided silver granules; used to prereduce analytes.

Wall-coated open tubular (WCOT) column A capillary column coated with a thin layer of stationary phase.

Water of constitution Essential water that is derived from the molecular composition of the species.

Water of crystallization Essential water that is an integral part of the crystal structure of a solid.

Wavelength, of electromagnetic radiation, λ The distance between successive maxima (or minima) of a wave.

Wavelength selector A device that limits the range of wavelengths used for an optical measurement.

Wavenumber, $\bar{\nu}$ The reciprocal of wavelength; has units of cm^{-1}.

Wave properties, electromagnetic radiation Behavior of radiation as an electromagnetic wave.

Weak acid/conjugate base pairs In the Brønsted-Lowry view, solute pairs that differ from one another by one proton.

Weak acids and weak bases Acids and bases that are only partially dissociated in a particular solvent.

Weak electrolytes Solutes that are incompletely dissociated into ions in a particular solvent.

Weighing bottle A lightweight container for the storage and weighing of analytical samples.

Weighing by difference The process of weighing a container plus the sample, followed by weighing the container after the sample has been removed.

Weighing form In gravimetric analysis, the species collected whose mass is proportional to the amount of analyte in the sample.

Weight The attraction between an object and its surroundings, terrestrially, the earth.

Weight molarity, M_w The concentration of titrant expressed as millimoles per gram.

Weight percent (w/w) The ratio of the mass of a solute to the mass of its solution, multiplied by 100%.

Weight titrimetry Synonymous with *gravimetric titrimetry.*

Weight/volume percent (w/v) The ratio of the mass of a solute to the volume of solution in which it is dissolved, multiplied by 100%.

Wet ashing The use of strong liquid oxidizing reagents to decompose the organic matter in a sample.

Windows, of cells Surfaces of cells through which radiation passes.

Z

Zero percent *T* adjustment A calibration step that compensates for dark current in the response of a spectrophotometer.

Zimmermann-Reinhardt reagent A solution of manganese(II) in concentrated H_2SO_4 and H_3PO_4 that prevents the induced oxidation of chloride ion by permanganate during the titration of iron(II).

Zones, chromatographic Synonymous with *chromatographic bands.*

Zwitterion The species that results from the transfer in solution of a proton from an acidic group to an acceptor site on the same molecule.

APPENDIX 1

The Literature of Analytical Chemistry

Treatises

As used here, the term *treatise* means a comprehensive presentation of one or more broad areas of analytical chemistry.

N. H. Furman and F. J. Welcher, Eds., *Standard Methods of Chemical Analysis,* 6th ed. New York: Van Nostrand, 1962–1966. In five parts; largely devoted to specific applications.

I. M. Kolthoff and P. J. Elving, Eds., *Treatise on Analytical Chemistry.* New York: Wiley, 1961–1986. Part I, 2nd ed. (14 volumes) is devoted to theory; Part II (17 volumes) deals with analytical methods for inorganic and organic compounds; Part III (4 volumes) treats industrial analytical chemistry.

Robert A. Meyers, Ed., *Encyclopedia of Analytical Chemistry: Applications, Theory and Instrumentation.* New York: Wiley, 2000. A 15-volume reference work for all areas of analytical chemistry.

B. W. Rossiter and R. C. Baetzold, Eds., *Physical Methods of Chemistry,* 2nd ed. New York: Wiley, 1986–1993. This series consists of 12 volumes devoted to various types of physical and chemical measurements performed by chemists.

C. L. Wilson and D. W. Wilson, Eds., *Comprehensive Analytical Chemistry.* New York: Elsevier, 1959–2003. To 2003, 39 volumes of this work have appeared.

Official Methods of Analysis

These publications are often single volumes that provide a useful source of analytical methods for the determination of specific substances in articles of commerce. The methods have been developed by various scientific societies and serve as standards in arbitration as well as in the courts.

Standard Methods for the Examination of Water and Wastewater, 20th ed., L. S. Clesceri, A. E. Greenberg, and A. D. Eaton, Eds. New York: American Public Health Association, 1998.

Annual Book of ASTM Standards. Philadelphia: American Society for Testing Materials. This 70-volume work is revised annually and contains methods for both physical testing and chemical analysis. Volumes 3.05 and 3.06, Analytical Chemistry for Metals, Ores and Related Materials, are particularly useful sources.

C. A. Watson, *Official and Standardized Methods of Analysis,* 3rd ed. London: Royal Society of Chemistry, 1994.

Official Methods of Analysis, 17th ed. Washington, DC: Association of Official Analytical Chemists, 2002. This is a very useful source of methods for the analysis of such materials as drugs, food, pesticides, agricultural materials, cosmetics, vitamins, and nutrients.

Review Serials

The reviews listed below are general reviews in the field. In addition, there are specific review serials devoted to advances in areas such as chromatography, electrochemistry, mass spectrometry, and many others.

Analytical Chemistry, Fundamental Reviews, American Chemical Society, Washington, DC. These reviews appear in even-numbered years in the June 15 issue of *Analytical Chemistry.* Many of the significant developments occurring in the past 2 years in several areas of analytical chemistry are covered.

Analytical Chemistry, Application Reviews, American Chemical Society, Washington, DC. These reviews appear in odd-numbered years in the June 15 issue of *Analytical Chemistry.* The articles are devoted to recent analytical work in specific areas, such as water analysis, clinical chemistry, petroleum products, and air pollution.

Critical Reviews in Analytical Chemistry, CRC Press, Boca Raton, FL. This publication appears quarterly and provides in-depth articles covering the newest developments in the analysis of biochemical substances.

Reviews in Analytical Chemistry, Freund Publishing, Tel Aviv. A journal devoted to reviews in the field.

Tabular Compilations

A. J. Bard, R. Parsons, and T. Jordan, Eds., *Standard Potentials in Aqueous Solution.* New York: Marcel Dekker, 1985.

J. A. Dean, *Analytical Chemistry Handbook.* New York: McGraw-Hill, 1995.

A. E. Martell and R. M. Smith, *Critical Stability Constants.* New York: Plenum Press, 1974–1989. Six volumes.

G. Milazzo, S. Caroli, and V. K. Sharma, *Tables of Standard Electrode Potential.* New York: Wiley, 1978.

Advanced Analytical and Instrumental Textbooks

J. N. Butler, *Ionic Equilibrium: A Mathematical Approach.* Reading, MA: Addison-Wesley, 1964.

J. N. Butler, Ionic *Equilibrium: Solubility and pH Calculations.* New York: Wiley, 1998.

G. D. Christian and J. E. O'Reilly, *Instrumental Analysis,* 2nd ed. Boston: Allyn and Bacon, 1986.

W. B. Guenther, *Unified Equilibrium Calculations.* New York: Wiley, 1991.

H. A. Laitinen and W. E. Harris, *Chemical Analysis,* 2nd ed. New York: McGraw-Hill, 1975.

F. A. Settle, Ed., *Handbook of Instrumental Techniques for Analytical Chemistry.* Upper Saddle River, NJ: Prentice-Hall, 1997.

D. A. Skoog, F. J. Holler, and T. A. Nieman, *Principles of Instrumental Analysis,* 5th ed. Philadelphia: Saunders College Publishing, 1998.

H. Strobel and W. R. Heineman, *Chemical Instrumentation: A Systematic Approach,* 3rd ed. Boston: Addison-Wesley, 1989.

Monographs

Hundreds of monographs devoted to specialized areas of analytical chemistry are available. In general, these are authored by experts and are excellent sources of information. Representative monographs in various areas are listed here.

Gravimetric and Titrimetric Methods

M. R. F. Ashworth, *Titrimetric Organic Analysis.* New York: Interscience, 1965. Two volumes.

R. deLevie, *Aqueous Acid-Base Equilibria and Titrations.* Oxford: Oxford University Press, 1999.

L. Erdey, *Gravimetric Analysis.* Oxford: Pergamon, 1965.

J. S. Fritz, *Acid-Base Titration in Nonaqueous Solvents.* Boston: Allyn and Bacon, 1973.

W. F. Hillebrand, G. E. F. Lundell, H. A. Bright, and J. I. Hoffman. *Applied Inorganic Analysis,* 2nd ed. New York: Wiley, 1953, reissued 1980.

I. M. Kolthoff, V. A. Stenger, and R. Belcher, *Volumetric Analysis.* New York: Interscience, 1942–1957. Three volumes.

T. S. Ma and R. C. Ritner, *Modern Organic Elemental Analysis.* New York: Marcel Dekker, 1979.

L. Safarik and Z. Stransky, *Titrimetric Analysis in Organic Solvents.* Amsterdam: Elsevier, 1986.

E. P. Serjeant, *Potentiometry and Potentiometric Titrations.* New York: Wiley, 1984.

W. Wagner and C. J. Hull, *Inorganic Titrimetric Analysis.* New York: Marcel Dekker, 1971.

Organic Analysis

S. Siggia and J. G. Hanna, *Quantitative Organic Analysis via Functional Groups,* 4th ed. New York: Wiley, 1979.

F. T. Weiss, *Determination of Organic Compounds: Methods and Procedures.* New York: Wiley-Interscience, 1970.

Spectrometric Methods

D. F. Boltz and J. A. Howell, *Colorimetric Determination of Nonmetals,* 2nd ed. New York: Wiley-Interscience, 1978.

Jose A. C. Broekaert, *Analytical Atomic Spectrometry with Flames and Plasmas.* Weinheim: Cambridge University Press: Wiley-VCH, 2002.

S. J. Hill, *Inductively Coupled Plasma Spectrometry and Its Applications.* Boca Raton, FL: CRC Press, 1999.

J. D. Ingle and S. R. Crouch, *Spectrochemical Analysis.* Upper Saddle River, NJ: Prentice-Hall, 1988.

L. H. J. Lajunen, *Spectrochemical Analysis by Atomic Absorption and Emission.* Cambridge: Royal Society of Chemistry, 1992.

J. R. Lakowiz, *Principles of Fluorescence Spectroscopy.* Plenum Press, 1999.

A. Montaser and D. W. Golightly, Eds., *Inductively Coupled Plasmas in Analytical Atomic Spectroscopy,* 2nd ed. New York: Wiley-VCH, 1992.

A. Montaser, Ed., *Inductively Coupled Plasma Mass Spectrometry.* New York: Wiley, 1998.

E. B. Sandell and H. Onishi, *Colorimetric Determination of Traces of Metals,* 4th ed. New York: Wiley, 1978–1989. Two volumes.

S. G. Schulman, Ed., *Molecular Luminescence Spectroscopy.* New York: Wiley, 1985. In two parts.

F. D. Snell, *Photometric and Fluorometric Methods of Analysis.* New York: Wiley, 1978–1981. Two volumes.

Electroanalytical Methods

A. J. Bard and L. R. Faulkner, *Electrochemical Methods,* 2nd ed. New York: Wiley, 2001.

P. T. Kissinger and W. R. Heinemann, Eds., *Laboratory Techniques in Electroanalytical Chemistry,* 2nd ed. New York: Marcel Dekker, 1996.

J. J. Lingane, *Electroanalytical Chemistry,* 2nd ed. New York: Interscience, 1954.

D. T. Sawyer, A. Sobkowiak, and J. L. Roberts, Jr. *Experimental Electrochemistry for Chemists,* 2nd ed. New York: Wiley, 1995.

J. Wang, *Analytical Electrochemistry.* New York: Wiley, 2000.

Analytical Separations

K. Anton and C. Berger, Eds., *Supercritical Fluid Chromatography with Packed Columns, Techniques and Applications.* New York: Dekker, 1998.

M. Caude and D. Thiebaut, Eds. *Practical Supercritical Fluid Chromatography and Extraction.* Amsterdam: Harwood, 2000.

P. Camilleri, Ed., *Capillary Electrophoresis: Theory and Practice.* Boca Raton, FL: CRC Press, 1993.

B. Fried and J. Sherma, *Thin Layer Chromatography,* 4th ed. New York: Marcel Dekker, 1999.

J. C. Giddings, *Unified Separation Science.* New York: Wiley, 1991.

E. Katz, *Quantitative Analysis Using Chromatographic Techniques.* New York: Wiley, 1987.

M. McMaster and C. McMaster, *GC/MS: A Practical User's Guide.* New York: Wiley–VCH, 1998.

H. M. McNair and J. M Miller, *Basic Gas Chromatography.* New York: Wiley, 1998.

W. M. A. Niessen, *Liquid Chromatography-Mass Spectrometry,* 2nd ed. New York: Marcel Dekker, 1999.

M. E. Schimpf, K. Caldwell, and J. C. Giddings, Eds., *Field-Flow Fractionation Handbook.* New York: Wiley, 2000.

R. P. W. Scott, *Introduction to Analytical Gas Chromatography,* 2nd ed. New York: Marcel Dekker, 1997.

R. P. W. Scott, *Liquid Chromatography for the Analyst.* New York: Marcel Dekker, 1995.

R. M. Smith, *Gas and Liquid Chromatography in Analytical Chemistry.* New York: Wiley, 1988.

L. R. Snyder and J. J. Kirkland, *Introduction to Modern Liquid Chromatography,* 3rd. ed. New York: Wiley, 1996.

R. Weinberger, *Practical Capillary Electrophoresis.* New York: Academic Press, 2000.

Miscellaneous

R. G. Bates, *Determination of pH: Theory and Practice,* 2nd ed. New York: Wiley, 1973.

R. Bock, *A Handbook of Decomposition Methods in Analytical Chemistry.* New York: Wiley, 1979.

G. D. Christian and J. B. Callis, *Trace Analysis.* New York: Wiley, 1986.

J. L. Devore and N. R. Farnum, *Applied Statistics for Engineers and Scientists.* Belmont, CA: Duxbury Press at Brooks Cole Publishing Co., 1999.

H. A. Mottola, *Kinetic Aspects of Analytical Chemistry.* New York: Wiley, 1988.

D. Perez-Bendito and M. Silva, *Kinetic Methods in Analytical Chemistry.* New York: Halsted Press–Wiley, 1988.

D. D. Perrin, *Masking and Demasking Chemical Reactions.* New York: Wiley, 1970.

W. Rieman and H. F. Walton, *Ion Exchange in Analytical Chemistry.* Oxford: Pergamon, 1970.

J. Ruzicka and E. H. Hansen, *Flow Injection Analysis,* 2nd ed. New York: Wiley, 1988.

J. T. Watson, *Introduction to Mass Spectrometry,* 3rd ed. New York: Lippincott-Raven, 1997.

Periodicals

Numerous journals are devoted to analytical chemistry; these are primary sources of information in the field. Some of the best known titles are listed here. The boldface portion of the title is the *Chemical Abstracts* abbreviation for the journal.

***Amer**ican **Lab**oratory*
***Analyst**, The*
***Anal**ytical and **Bioanal**ytical **Chem**istry*
***Anal**ytical **Biochem**istry*
***Anal**ytical **Chem**istry*
Anal**ytica **Chim**ica **Acta
***Anal**ytical **Lett**ers*
***Appl**ied **Spectrosc**opy*
***Clin**ical **Chem**istry*
***Int**ernational **J**ournal of **Mass Spectrom**etry*
***Instr**umentation **Sci**ence and **Tech**nology*
***J**ournal of the **Am**erican **Soc**iety for **Mass Spectrom**etry*
***J**ournal of the **Assoc**iation of **Off**icial **Anal**ytical **Chem**ists*
***J**ournal of **Chromatogr**aphic **Sci**ence*
***J**ournal of **Chromatogr**aphy*
***J**ournal of **Electroanal**ytical **Chem**istry*
***J**ournal of **Liq**uid **Chromatogr**aphy and Related Techniques*
***J**ournal of **Microcol**umn **Separ**ations*
***Microchem**ical **J**ournal*
Mikrochim**ica **Acta
***Separ**ation **Sci**ence*
Spectrochim**ica **Acta
Talanta

APPENDIX 2

Solubility Product Constants at 25°C

Compound	Formula	K_{sp}	Notes
Aluminum hydroxide	$Al(OH)_3$	3×10^{-34}	
Barium carbonate	$BaCO_3$	5.0×10^{-9}	
Barium chromate	$BaCrO_4$	2.1×10^{-10}	
Barium hydroxide	$Ba(OH)_2 \cdot 8H_2O$	3×10^{-4}	
Barium iodate	$Ba(IO_3)_2$	1.57×10^{-9}	
Barium oxalate	BaC_2O_4	1×10^{-6}	
Barium sulfate	$BaSO_4$	1.1×10^{-10}	
Cadmium carbonate	$CdCO_3$	1.8×10^{-14}	
Cadmium hydroxide	$Cd(OH)_2$	4.5×10^{-15}	
Cadmium oxalate	CdC_2O_4	9×10^{-8}	
Cadmium sulfide	CdS	1×10^{-27}	
Calcium carbonate	$CaCO_3$	4.5×10^{-9}	Calcite
	$CaCO_3$	6.0×10^{-9}	Aragonite
Calcium fluoride	CaF_2	3.9×10^{-11}	
Calcium hydroxide	$Ca(OH)_2$	6.5×10^{-6}	
Calcium oxalate	$CaC_2O_4 \cdot H_2O$	1.7×10^{-9}	
Calcium sulfate	$CaSO_4$	2.4×10^{-5}	
Cobalt(II) carbonate	$CoCO_3$	1.0×10^{-10}	
Cobalt(II) hydroxide	$Co(OH)_2$	1.3×10^{-15}	
Cobalt(II) sulfide	CoS	5×10^{-22}	α
	CoS	3×10^{-26}	β
Copper(I) bromide	$CuBr$	5×10^{-9}	
Copper(I) chloride	$CuCl$	1.9×10^{-7}	
Copper(I) hydroxide*	Cu_2O*	2×10^{-15}	
Copper(I) iodide	CuI	1×10^{-12}	
Copper(I) thiocyanate	$CuSCN$	4.0×10^{-14}	
Copper(II) hydroxide	$Cu(OH)_2$	4.8×10^{-20}	
Copper(II) sulfide	CuS	8×10^{-37}	
Iron(II) carbonate	$FeCO_3$	2.1×10^{-11}	
Iron(II) hydroxide	$Fe(OH)_2$	4.1×10^{-15}	
Iron(II) sulfide	FeS	8×10^{-19}	
Iron(III) hydroxide	$Fe(OH)_3$	2×10^{-39}	
Lanthanum iodate	$La(IO_3)_3$	1.0×10^{-11}	
Lead carbonate	$PbCO_3$	7.4×10^{-14}	
Lead chloride	$PbCl_2$	1.7×10^{-5}	
Lead chromate	$PbCrO_4$	3×10^{-13}	
Lead hydroxide	$PbO^{\dagger}$	8×10^{-16}	Yellow
	$PbO^{\dagger}$	5×10^{-16}	Red
Lead iodide	PbI_2	7.9×10^{-9}	
Lead oxalate	PbC_2O_4	8.5×10^{-9}	$\mu = 0.05$
Lead sulfate	$PbSO_4$	1.6×10^{-8}	
Lead sulfide	PbS	3×10^{-28}	
Magnesium ammonium phosphate	$MgNH_4PO_4$	3×10^{-13}	
Magnesium carbonate	$MgCO_3$	3.5×10^{-8}	

continues

Compound	Formula	K_{sp}	Notes
Magnesium hydroxide	$Mg(OH)_2$	7.1×10^{-12}	
Manganese carbonate	$MnCO_3$	5.0×10^{-10}	
Manganese hydroxide	$Mn(OH)_2$	2×10^{-13}	
Manganese sulfide	MnS	3×10^{-11}	Pink
	MnS	3×10^{-14}	Green
Mercury(I) bromide	Hg_2Br_2	5.6×10^{-23}	
Mercury(I) carbonate	Hg_2CO_3	8.9×10^{-17}	
Mercury(I) chloride	Hg_2Cl_2	1.2×10^{-18}	
Mercury(I) iodide	Hg_2I_2	4.7×10^{-29}	
Mercury(I) thiocyanate	$Hg_2(SCN)_2$	3.0×10^{-20}	
Mercury(II) hydroxide	HgO‡	3.6×10^{-26}	
Mercury(II) sulfide	HgS	2×10^{-53}	Black
	HgS	5×10^{-54}	Red
Nickel carbonate	$NiCO_3$	1.3×10^{-7}	
Nickel hydroxide	$Ni(OH)_2$	6×10^{-16}	
Nickel sulfide	NiS	4×10^{-20}	α
	NiS	1.3×10^{-25}	β
Silver arsenate	Ag_3AsO_4	6×10^{-23}	
Silver bromide	AgBr	5.0×10^{-13}	
Silver carbonate	Ag_2CO_3	8.1×10^{-12}	
Silver chloride	AgCl	1.82×10^{-10}	
Silver chromate	$AgCrO_4$	1.2×10^{-12}	
Silver cyanide	AgCN	2.2×10^{-16}	
Silver iodate	$AgIO_3$	3.1×10^{-8}	
Silver iodide	AgI	8.3×10^{-17}	
Silver oxalate	$Ag_2C_2O_4$	3.5×10^{-11}	
Silver sulfide	Ag_2S	8×10^{-51}	
Silver thiocyanate	AgSCN	1.1×10^{-12}	
Strontium carbonate	$SrCO_3$	9.3×10^{-10}	
Strontium oxalate	SrC_2O_4	5×10^{-8}	
Strontium sulfate	$SrSO_4$	3.2×10^{-7}	
Thallium(I) chloride	TlCl	1.8×10^{-4}	
Thallium(I) sulfide	Tl_2S	6×10^{-22}	
Zinc carbonate	$ZnCO_3$	1.0×10^{-10}	
Zinc hydroxide	$Zn(OH)_2$	3.0×10^{-16}	Amorphous
Zinc oxalate	ZnC_2O_4	8×10^{-9}	
Zinc sulfide	ZnS	2×10^{-25}	α
	ZnS	3×10^{-23}	β

Most of these data were taken from A. E. Martell and R. M Smith, *Critical Stability Constants,* Vol. 3–6. New York: Plenum, 1976–1989. In most cases, the ionic strength was 0.0 and the temperature 25°C.

*$Cu_2O(s) + H_2O \rightleftharpoons 2Cu^+ + 2OH^-$

†$PbO(s) + H_2O \rightleftharpoons Pb^{2+} + 2OH^-$

‡$HgO(s) + H_2O \rightleftharpoons Hg^{2+} + 2OH^-$

APPENDIX 3

Acid Dissociation Constants at 25°C

Acid	Formula	K_1	K_2	K_3
Acetic acid	CH_3COOH	1.75×10^{-5}		
Ammonium ion	NH_4^+	5.70×10^{-10}		
Anilinium ion	$C_6H_5NH_3^+$	2.51×10^{-5}		
Arsenic acid	H_3AsO_4	5.8×10^{-3}	1.1×10^{-7}	3.2×10^{-12}
Arsenous acid	H_3AsO_3	5.1×10^{-10}		
Benzoic acid	C_6H_5COOH	6.28×10^{-5}		
Boric acid	H_3BO_3	5.81×10^{-10}		
1-Butanoic acid	$CH_3CH_2CH_2COOH$	1.52×10^{-5}		
Carbonic acid	H_2CO_3	4.45×10^{-7}	4.69×10^{-11}	
Chloroacetic acid	$ClCH_2COOH$	1.36×10^{-3}		
Citric acid	$HOOC(OH)C(CH_2COOH)_2$	7.45×10^{-4}	1.73×10^{-5}	4.02×10^{-7}
Dimethyl ammonium ion	$(CH_3)_2NH_2^+$	1.68×10^{-11}		
Ethanol ammonium ion	$HOC_2H_4NH_3^+$	3.18×10^{-10}		
Ethyl ammonium ion	$C_2H_5NH_3^+$	2.31×10^{-11}		
Ethylene diammonium ion	$^+H_3NCH_2CH_2NH_3^+$	1.42×10^{-7}	1.18×10^{-10}	
Formic acid	$HCOOH$	1.80×10^{-4}		
Fumaric acid	*trans*-$HOOCCH{:}CHCOOH$	8.85×10^{-4}	3.21×10^{-5}	
Glycolic acid	$HOCH_2COOH$	1.47×10^{-4}		
Hydrazinium ion	$H_2NNH_3^+$	1.05×10^{-8}		
Hydrazoic acid	HN_3	2.2×10^{-5}		
Hydrogen cyanide	HCN	6.2×10^{-10}		
Hydrogen fluoride	HF	6.8×10^{-4}		
Hydrogen peroxide	H_2O_2	2.2×10^{-12}		
Hydrogen sulfide	H_2S	9.6×10^{-8}	1.3×10^{-14}	
Hydroxyl ammonium ion	$HONH_3^+$	1.10×10^{-6}		
Hypochlorous acid	$HOCl$	3.0×10^{-8}		
Iodic acid	HIO_3	1.7×10^{-1}		
Lactic acid	$CH_3CHOHCOOH$	1.38×10^{-4}		
Maleic acid	*cis*-$HOOCCH{:}CHCOOH$	1.3×10^{-2}	5.9×10^{-7}	
Malic acid	$HOOCCHOHCH_2COOH$	3.48×10^{-4}	8.00×10^{-6}	
Malonic acid	$HOOCCH_2COOH$	1.42×10^{-3}	2.01×10^{-6}	
Mandelic acid	$C_6H_5CHOHCOOH$	4.0×10^{-4}		
Methyl ammonium ion	$CH_3NH_3^+$	2.3×10^{-11}		
Nitrous acid	HNO_2	7.1×10^{-4}		
Oxalic acid	$HOOCCOOH$	5.60×10^{-2}	5.42×10^{-5}	
Periodic acid	H_5IO_6	2×10^{-2}	5×10^{-9}	
Phenol	C_6H_5OH	1.00×10^{-10}		
Phosphoric acid	H_3PO_4	7.11×10^{-3}	6.32×10^{-8}	4.5×10^{-13}
Phosphorous acid	H_3PO_3	3×10^{-2}	1.62×10^{-7}	
o-Phthalic acid	$C_6H_4(COOH)_2$	1.12×10^{-3}	3.91×10^{-6}	
Picric acid	$(NO_2)_3C_6H_2OH$	4.3×10^{-1}		
Piperidinium ion	$C_5H_{11}NH^+$	7.50×10^{-12}		
Propanoic acid	CH_3CH_2COOH	1.34×10^{-5}		

continues

Acid	Formula	K_1	K_2	K_3
Pyridinium ion	$C_5H_5NH^+$	5.90×10^{-6}		
Pyruvic acid	$CH_3COCOOH$	3.2×10^{-3}		
Salicylic acid	$C_6H_4(OH)COOH$	1.06×10^{-3}		
Sulfamic acid	H_2NSO_3H	1.03×10^{-1}		
Succinic acid	$HOOCCH_2CH_2COOH$	6.21×10^{-5}	2.31×10^{-6}	
Sulfuric acid	H_2SO_4	Strong	1.02×10^{-2}	
Sulfurous acid	H_2SO_3	1.23×10^{-2}	6.6×10^{-8}	
Tartaric acid	$HOOC(CHOH)_2COOH$	9.20×10^{-4}	4.31×10^{-5}	
Thiocyanic acid	HSCN	0.13		
Thiosulfuric acid	$H_2S_2O_3$	0.3	2.5×10^{-2}	
Trichloroacetic acid	Cl_3CCOOH	3		
Trimethyl ammonium ion	$(CH_3)_3NH^+$	1.58×10^{-10}		

Most data are for zero ionic strength. (From A. E. Martell and R. M. Smith, *Critical Stability Constants,* Vol. 1–6. New York Plenum Press, 1974–1989.)

APPENDIX 4

Formation Constants at 25 °C

Ligand	Cation	log K_1	log K_2	log K_3	log K_4	Ionic Strength
Acetate (CH_3COO^-)	Ag^+	0.73	−0.9			0.0
	Ca^{2+}	1.18				0.0
	Cd^{2+}	1.93	1.22			0.0
	Cu^{2+}	2.21	1.42			0.0
	Fe^{3+}	3.38*	3.1*	1.8*		0.1
	Hg^{2+}	$\log K_1K_2 = 8.45$				0.0
	Mg^{2+}	1.27				0.0
	Pb^{2+}	2.68	1.40			0.0
Ammonia (NH_3)	Ag^+	3.31	3.91			0.0
	Cd^{2+}	2.55	2.01	1.34	0.84	0.0
	Co^{2+}	1.99*	1.51	0.93	0.64	0.0
		$\log K_5 = 0.06$	$\log K_6 = -0.74$			0.0
	Cu^{2+}	4.04	3.43	2.80	1.48	0.0
	Hg^{2+}	8.8	8.6	1.0	0.7	0.5
	Ni^{2+}	2.72	2.17	1.66	1.12	0.0
		$\log K_5 = 0.67$	$\log K_6 = -0.03$			0.0
	Zn^{2+}	2.21	2.29	2.36	2.03	0.0
Bromide (Br^-)	Ag^+		$Ag^+ + 2Br^- \rightleftharpoons AgBr_2^-$	$\log K_1K_2 = 7.5$		0.0
	Hg^{2+}	9.00	8.1	2.3	1.6	0.5
	Pb^{2+}	1.77				0.0
Chloride (Cl^-)	Ag^+		$Ag^+ + 2Cl^- \rightleftharpoons AgCl_2^-$	$\log K_1K_2 = 5.25$		0.0
			$AgCl_2^- + Cl^- \rightleftharpoons AgCl_3^{2-}$	$\log K_3 = 0.37$		0.0
	Cu^+		$Cu^+ + 2Cl^- \rightleftharpoons CuCl_2^-$	log = 5.5*		0.0
	Fe^{3+}	1.48	0.65			0.0
	Hg^{2+}	7.30	6.70	1.0	0.6	0.0
	Pb^{2+}		$Pb^{2+} + 3Cl^- \rightleftharpoons PbCl_3^-$	$\log K_1K_2K_3 = 1.8$		0.0
	Sn^{2+}	1.51	0.74	−0.3	−0.5	0.0
Cyanide (CN^-)	Ag^+		$Ag^+ + 2CN^- \rightleftharpoons Ag(CN)_2^-$	$\log K_1K_2 = 20.48$		0.0
	Cd^{2+}	6.01	5.11	4.53	2.27	0.0
	Hg^{2+}	17.00	15.75	3.56	2.66	0.0
	Ni^{2+}		$Ni^{2+} + 4CN^- \rightleftharpoons Ni(CN)_4^-$	$\log K_1K_2K_3K_4 = 30.22$		0.0
	Zn^{2+}	$\log K_1K_2 = 11.07$		4.98	3.57	0.0
EDTA	See Table 17-3, page 464.					
Fluoride (F^-)	Al^{3+}	7.0	5.6	4.1	2.4	0.0
	Fe^{3+}	5.18	3.89	3.03		0.0
Hydroxide (OH^-)	Al^{3+}		$Al^{3+} + 4OH^- \rightleftharpoons Al(OH)_4^-$	$\log K_1K_2K_3K_4 = 33.4$		0.0
	Cd^{2+}	3.9	3.8			0.0
	Cu^{2+}	6.5				0.0
	Fe^{2+}	4.6				0.0
	Fe^{3+}	11.81	11.5			0.0
	Hg^{2+}	10.60	11.2			0.0
	Ni^{2+}	4.1	4.9	3		0.0
	Pb^{2+}	6.4	$Pb^{2+} + 3OH^- \rightleftharpoons Pb(OH)_3^-$		$\log K_1K_2K_3 = 13.9$	0.0
	Zn^{2+}	5.0	$Zn^{2+} + 4OH^- \rightleftharpoons Zn(OH)_4^{2-}$		$\log K_1K_2K_3K_4 = 15.5$	0.0

continues

Ligand	Cation	log K_1	log K_2	log K_3	log K_4	Ionic Strength
Iodide (I^-)	Cd^{2+}	2.28	1.64	1.0	1.0	0.0
	Cu^+		$Cu^+ + 2I^- \rightleftharpoons CuI_2^-$		log $K_1K_2 = 8.9$	0.0
	Hg^{2+}	12.87	10.95	3.8	2.2	0.5
	Pb^{2+}		$Pb^{2+} + 3I^- \rightleftharpoons PbI_3^-$		log $K_1K_2K_3 = 3.9$	0.0
			$Pb^{2+} + 4I^- \rightleftharpoons PbI_4^{2-}$		log $K_1K_2K_3K_4 = 4.5$	0.0
Oxalate ($C_2O_4^{2-}$)	Al^{3+}	5.97	4.96	5.04		0.1
	Ca^{2+}	3.19				0.0
	Cd^{2+}	2.73	1.4	1.0		1.0
	Fe^{3+}	7.58	6.23	4.8		1.0
	Mg^{2+}	3.42(18°C)				
	Pb^{2+}	4.20	2.11			1.0
Sulfate (SO_4^{2-})	Al^{3+}	3.89				0.0
	Ca^{2+}	2.13				0.0
	Cu^{2+}	2.34				0.0
	Fe^{3+}	4.04	1.34			0.0
	Mg^{2+}	2.23				0.0
Thiocyanate (SCN^-)	Cd^{2+}	1.89	0.89	0.1		0.0
	Cu^+	$Cu^+ + 3SCN^- \rightleftharpoons Cu(SCN)_3^{2-}$		log $K_1K_2K_3 = 11.60$		0.0
	Fe^{3+}	3.02	0.62*			0.0
	Hg^{2+}	log $K_1K_2 = 17.26$		2.7	1.8	0.0
	Ni^{2+}	1.76				0.0
Thiosulfate ($S_2O_3^{2-}$)	Ag^+	8.82*	4.7	0.7		0.0
	Cu^{2+}	log $K_1K_2 = 6.3$				0.0
	Hg^{2+}	log $K_1K_2 = 29.23$		1.4		0.0

Data from A. E. Martell and R. M. Smith, *Critical Stability Constants,* Vol. 3–6. New York: Plenum Press, 1974–1989.
*20°C.

APPENDIX 5

Standard and Formal Electrode Potentials

Half-Reaction	E^0, V*	Formal Potential, V†
Aluminum		
$Al^{3+} + 3e^- \rightleftharpoons Al(s)$	−1.662	
Antimony		
$Sb_2O_5(s) + 6H^+ + 4e^- \rightleftharpoons 2SbO^+ + 3H_2O$	+0.581	
Arsenic		
$H_3AsO_4 + 2H^+ + 2e^- \rightleftharpoons H_3AsO_3 + H_2O$	+0.559	0.577 in 1 M HCl, $HClO_4$
Barium		
$Ba^{2+} + 2e^- \rightleftharpoons Ba(s)$	−2.906	
Bismuth		
$BiO^+ + 2H^+ + 3e^- \rightleftharpoons Bi(s) + H_2O$	+0.320	
$BiCl_4^- + 3e^- \rightleftharpoons Bi(s) + 4Cl^-$	+0.16	
Bromine		
$Br_2(l) + 2e^- \rightleftharpoons 2Br^-$	+1.065	1.05 in 4 M HCl
$Br_2(aq) + 2e^- \rightleftharpoons 2Br^-$	+1.087‡	
$BrO_3^- + 6H^+ + 5e^- \rightleftharpoons \frac{1}{2}Br_2(l) + 3H_2O$	+1.52	
$BrO_3^- + 6H^+ + 6e^- \rightleftharpoons Br^- + 3H_2O$	+1.44	
Cadmium		
$Cd^{2+} + 2e^- \rightleftharpoons Cd(s)$	−0.403	
Calcium		
$Ca^{2+} + 2e^- \rightleftharpoons Ca(s)$	−2.866	
Carbon		
$C_6H_4O_2$ (quinone) $+ 2H^+ + 2e^- \rightleftharpoons C_6H_4(OH)_2$	+0.699	0.696 in 1 M HCl, $HClO_4$, H_2SO_4
$2CO_2(g) + 2H^+ + 2e^- \rightleftharpoons H_2C_2O_4$	−0.49	
Cerium		
$Ce^{4+} + e^- \rightleftharpoons Ce^{3+}$		+1.70 in 1 M $HClO_4$; +1.61 in 1 M HNO_3; 1.44 in 1 M H_2SO_4
Chlorine		
$Cl_2(g) + 2e^- \rightleftharpoons 2Cl^-$	+1.359	
$HClO + H^+ + e^- \rightleftharpoons \frac{1}{2}Cl_2(g) + H_2O$	+1.63	
$ClO_3^- + 6H^+ + 5e^- \rightleftharpoons \frac{1}{2}Cl_2(g) + 3H_2O$	+1.47	
Chromium		
$Cr^{3+} + e^- \rightleftharpoons Cr^{2+}$	−0.408	
$Cr^{3+} + 3e^- \rightleftharpoons Cr(s)$	−0.744	
$Cr_2O_7^{2-} + 14H^+ + 6e^- \rightleftharpoons 2Cr^{3+} + 7H_2O$	+1.33	
Cobalt		
$Co^{2+} + 2e^- \rightleftharpoons Co(s)$	−0.277	
$Co^{3+} + e^- \rightleftharpoons Co^{2+}$	+1.808	
Copper		
$Cu^{2+} + 2e^- \rightleftharpoons Cu(s)$	+0.337	
$Cu^{2+} + e^- \rightleftharpoons Cu^+$	+0.153	
$Cu^+ + e^- \rightleftharpoons Cu(s)$	+0.521	
$Cu^{2+} + I^- + e^- \rightleftharpoons CuI(s)$	+0.86	
$CuI(s) + e^- \rightleftharpoons Cu(s) + I^-$	−0.185	

continues

Half-Reaction	E^0, V*	Formal Potential, V†
Fluorine		
$F_2(g) + 2H^+ + 2e^- \rightleftharpoons 2HF(aq)$	+3.06	
Hydrogen		
$2H^+ + 2e^- \rightleftharpoons H_2(g)$	0.000	−0.005 in 1 M HCl, $HClO_4$
Iodine		
$I_2(s) + 2e^- \rightleftharpoons 2I^-$	+0.5355	
$I_2(aq) + 2e^- \rightleftharpoons 2I^-$	+0.615‡	
$I_3^- + 2e^- \rightleftharpoons 3I^-$	+0.536	
$ICl_2^- + e^- \rightleftharpoons \frac{1}{2}I_2(s) + 2Cl^-$	+1.056	
$IO_3^- + 6H^+ + 5e^- \rightleftharpoons \frac{1}{2}I_2(s) + 3H_2O$	+1.196	
$IO_3^- + 6H^+ + 5e^- \rightleftharpoons \frac{1}{2}I_2(aq) + 3H_2O$	+1.178‡	
$IO_3^- + 2Cl^- + 6H^+ + 4e^- \rightleftharpoons ICl_2^- + 3H_2O$	+1.24	
$H_5IO_6 + H^+ + 2e^- \rightleftharpoons IO_3^- + 3H_2O$	+1.601	
Iron		
$Fe^{2+} + 2e^- \rightleftharpoons Fe(s)$	−0.440	
$Fe^{3+} + e^- \rightleftharpoons Fe^{2+}$	+0.771	0.700 in 1 M HCl; 0.732 in 1 M $HClO_4$; 0.68 in 1 M H_2SO_4
$Fe(CN)_6^{3-} + e^- \rightleftharpoons Fe(CN)_6^{4-}$	+0.36	0.71 in 1 M HCl; 0.72 in 1 M $HClO_4$, H_2SO_4
Lead		
$Pb^{2+} + 2e^- \rightleftharpoons Ps(s)$	−0.126	−0.14 in 1 M $HClO_4$; −0.29 in 1 M H_2SO_4
$PbO_2(s) + 4H^+ + 2e^- \rightleftharpoons Pb^{2+} + 2H_2O$	+1.455	
$PbSO_4(s) + 2e^- \rightleftharpoons Pb(s) + SO_4^{2-}$	−0.350	
Lithium		
$Li^+ + e^- \rightleftharpoons Li(s)$	−3.045	
Magnesium		
$Mg^{2+} + 2e^- \rightleftharpoons Mg(s)$	−2.363	
Manganese		
$Mn^{2+} + 2e^- \rightleftharpoons Mn(s)$	−1.180	
$Mn^{3+} + e^- \rightleftharpoons Mn^{2+}$		1.51 in 7.5 M H_2SO_4
$MnO_2(s) + 4H^+ + 2e^- \rightleftharpoons Mn^{2+} + 2H_2O$	+1.23	
$MnO_4^- + 8H^+ + 5e^- \rightleftharpoons Mn^{2+} + 4H_2O$	+1.51	
$MnO_4^- + 4H^+ + 3e^- \rightleftharpoons MnO_2(s) + 2H_2O$	+1.695	
$MnO_4^- + e^- \rightleftharpoons MnO_4^{2-}$	+0.564	
Mercury		
$Hg_2^{2+} + 2e^- \rightleftharpoons 2Hg(l)$	+0.788	0.274 in 1 M HCl; 0.776 in 1 M $HClO_4$; 0.674 in 1 M H_2SO_4
$2Hg^{2+} + 2e^- \rightleftharpoons Hg_2^{2+}$	+0.920	0.907 in 1 M $HClO_4$
$Hg^{2+} + 2e^- \rightleftharpoons Hg(l)$	+0.854	
$Hg_2Cl_2(s) + 2e^- \rightleftharpoons 2Hg(l) + 2Cl^-$	+0.268	0.244 in sat'd KCl; 0.282 in 1 M KCl; 0.334 in 0.1 M KCl
$Hg_2SO_4(s) + 2e^- \rightleftharpoons 2Hg(l) + SO_4^{2-}$	+0.615	
Nickel		
$Ni^{2+} + 2e^- \rightleftharpoons Ni(s)$	−0.250	
Nitrogen		
$N_2(g) + 5H^+ + 4e^- \rightleftharpoons N_2H_5^+$	−0.23	
$HNO_2 + H^+ + e^- \rightleftharpoons NO(g) + H_2O$	+1.00	
$NO_3^- + 3H^+ + 2e^- \rightleftharpoons HNO_2 + H_2O$	+0.94	0.92 in 1 M HNO_3
Oxygen		
$H_2O_2 + 2H^+ + 2e^- \rightleftharpoons 2H_2O$	+1.776	
$HO_2^- + H_2O + 2e^- \rightleftharpoons 3OH^-$	+0.88	
$O_2(g) + 4H^+ + 4e^- \rightleftharpoons 2H_2O$	+1.229	
$O_2(g) + 2H^+ + 2e^- \rightleftharpoons H_2O_2$	+0.682	
$O_3(g) + 2H^+ + 2e^- \rightleftharpoons O_2(g) + H_2O$	+2.07	
Palladium		
$Pd^{2+} + 2e^- \rightleftharpoons Pd(s)$	+0.987	

continues

Half-Reaction	E^0, V*	Formal Potential, V†
Platinum		
$PtCl_4^{2-} + 2e^- \rightleftharpoons Pt(s) + 4Cl^-$	+0.73	
$PtCl_6^{2-} + 2e^- \rightleftharpoons PtCl_4^{2-} + 2Cl^-$	+0.68	
Potassium		
$K^+ + e^- \rightleftharpoons K(s)$	−2.925	
Selenium		
$H_2SeO_3 + 4H^+ + 4e^- \rightleftharpoons Se(s) + 3H_2O$	+0.740	
$SeO_4^{2-} + 4H^+ + 2e^- \rightleftharpoons H_2SeO_3 + H_2O$	+1.15	
Silver		
$Ag^+ + e^- \rightleftharpoons Ag(s)$	+0.799	0.228 in 1 M HCl; 0.792 in 1 M $HClO_4$; 0.77 in 1 M H_2SO_4
$AgBr(s) + e^- \rightleftharpoons Ag(s) + Br^-$	+0.073	
$AgCl(s) + e^- \rightleftharpoons Ag(s) + Cl^-$	+0.222	0.228 in 1 M KCl
$Ag(CN)_2^- + e^- \rightleftharpoons Ag(s) + 2CN^-$	−0.31	
$Ag_2CrO_4(s) + 2e^- \rightleftharpoons 2Ag(s) + CrO_4^{2-}$	+0.446	
$AgI(s) + e^- \rightleftharpoons Ag(s) + I^-$	−0.151	
$Ag(S_2O_3)_2^{3-} + e^- \rightleftharpoons Ag(s) + 2S_2O_3^{2-}$	+0.017	
Sodium		
$Na^+ + e^- \rightleftharpoons Na(s)$	−2.714	
Sulfur		
$S(s) + 2H^+ + 2e^- \rightleftharpoons H_2S(g)$	+0.141	
$H_2SO_3 + 4H^+ + 4e^- \rightleftharpoons S(s) + 3H_2O$	+0.450	
$SO_4^{2-} + 4H^+ + 2e^- \rightleftharpoons H_2SO_3 + H_2O$	+0.172	
$S_4O_6^{2-} + 2e^- \rightleftharpoons 2S_2O_3^{2-}$	+0.08	
$S_2O_8^{2-} + 2e^- \rightleftharpoons 2SO_4^{2-}$	+2.01	
Thallium		
$Tl^+ + e^- \rightleftharpoons Tl(s)$	−0.336	−0.551 in 1 M HCl; −0.33 in 1 M $HClO_4$, H_2SO_4
$Tl^{3+} + 2e^- \rightleftharpoons Tl^+$	+1.25	0.77 in 1 M HCl
Tin		
$Sn^{2+} + 2e^- \rightleftharpoons Sn(s)$	−0.136	−0.16 in 1 M $HClO_4$
$Sn^{4+} + 2e^- \rightleftharpoons Sn^{2+}$	+0.154	0.14 in 1 M HCl
Titanium		
$Ti^{3+} + e^- \rightleftharpoons Ti^{2+}$	−0.369	
$TiO^{2+} + 2H^+ + e^- \rightleftharpoons Ti^{3+} + H_2O$	+0.099	0.04 in 1 M H_2SO_4
Uranium		
$UO_2^{2+} + 4H^+ + 2e^- \rightleftharpoons U^{4+} + 2H_2O$	+0.334	
Vanadium		
$V^{3+} + e^- \rightleftharpoons V^{2+}$	−0.255	
$VO^{2+} + 2H^+ + e^- \rightleftharpoons V^{3+} + H_2O$	+0.359	
$V(OH)_4^+ + 2H^+ + e^- \rightleftharpoons VO^{2+} + 3H_2O$	+1.00	1.02 in 1 M HCl, $HClO_4$
Zinc		
$Zn^{2+} + 2e^- \rightleftharpoons Zn(s)$	−0.763	

*G. Milazzo, S. Caroli, and V. K. Sharma, *Tables of Standard Electrode Potentials.* London: Wiley, 1978.

†E. H. Swift and E. A. Butler, *Quantitative Measurements and Chemical Equilibria.* New York: Freeman, 1972.

‡These potentials are hypothetical because they correspond to solutions that are 1.00 M in Br_2 or I_2. The solubilities of these two compounds at 25°C are 0.18 M and 0.0020 M, respectively. In saturated solutions containing an excess of $Br_2(l)$ or $I_2(s)$, the standard potentials for the half-reaction $Br_2(l) + 2e^- \rightleftharpoons 2Br^-$ or $I_2(s) + 2e^- \rightleftharpoons 2I^-$ should be used. In contrast, at Br_2 and I_2 concentrations less than saturation, these hypothetical electrode potentials should be employed.

APPENDIX 6

Use of Exponential Numbers and Logarithms

Scientists frequently find it necessary (or convenient) to use exponential notation to express numerical data. A brief review of this notation follows.

Exponential Notation

An exponent is used to describe the process of repeated multiplication or division. For example, 3^5 means

$$3 \times 3 \times 3 \times 3 \times 3 = 3^5 = 243$$

The power 5 is the exponent of the number (or base) 3; thus, 3 raised to the fifth power is equal to 243.

A negative exponent represents repeated division. For example, 3^{-5} means

$$\frac{1}{3} \times \frac{1}{3} \times \frac{1}{3} \times \frac{1}{3} \times \frac{1}{3} = \frac{1}{3^5} = 3^{-5} = 0.00412$$

Note that changing the sign of the exponent yields the *reciprocal* of the number; that is,

$$3^{-5} = \frac{1}{3^5} = \frac{1}{243} = 0.00412$$

It is important to note that a number raised to the first power is the number itself, and any number raised to the zero power has a value of 1. For example,

$$4^1 = 4$$
$$4^0 = 1$$
$$67^0 = 1$$

Fractional Exponents

A fractional exponent symbolizes the process of extracting the root of a number. The fifth root of 243 is 3; this process is expressed exponentially as

$$(243)^{1/5} = 3$$

Other examples are

$$25^{1/2} = 5$$
$$25^{-1/2} = \frac{1}{25^{1/2}} = \frac{1}{5}$$

The Combination of Exponential Numbers in Multiplication and Division

Multiplication and division of exponential numbers having the same base are accomplished by adding and subtracting the exponents. For example,

$$3^3 \times 3^2 = (3 \times 3 \times 3)(3 \times 3) = 3^{(3+2)} = 3^5 = 243$$

$$3^4 \times 3^{-2} \times 3^0 = (3 \times 3 \times 3 \times 3)\left(\frac{1}{3} \times \frac{1}{3}\right) \times 1 = 3^{(4-2+0)} = 3^2 = 9$$

$$\frac{5^4}{5^2} = \frac{5 \times 5 \times 5 \times 5}{5 \times 5} = 5^{(4-2)} = 5^2 = 25$$

$$\frac{2^3}{2^{-1}} = \frac{(2 \times 2 \times 2)}{1/2} = 2^4 = 16$$

Note that in the last equation the exponent is given by the relationship

$$3 - (-1) = 3 + 1 = 4$$

Extraction of the Root of an Exponential Number

To obtain the root of an exponential number, the exponent is divided by the desired root. Thus,

$$(5^4)^{1/2} = (5 \times 5 \times 5 \times 5)^{1/2} = 5^{(4/2)} = 5^2 = 25$$

$$(10^{-8})^{1/4} = 10^{(-8/4)} = 10^{-2}$$

$$(10^9)^{1/2} = 10^{(9/2)} = 10^{4.5}$$

The Use of Exponents in Scientific Notation

Scientists and engineers are frequently called upon to use very large or very small numbers for which ordinary decimal notation is either awkward or impossible. For example, to express Avogadro's number in decimal notation would require 21 zeros following the number 602. In scientific notation the number is written as a multiple of two numbers, the one number in decimal notation and the other expressed as a power of 10. Thus, Avogadro's number is written as 6.02×10^{23}. Other examples are

$$4.32 \times 10^3 = 4.32 \times 10 \times 10 \times 10 = 4320$$

$$4.32 \times 10^{-3} = 4.32 \times \frac{1}{10} \times \frac{1}{10} \times \frac{1}{10} = 0.00432$$

$$0.002002 = 2.002 \times \frac{1}{10} \times \frac{1}{10} \times \frac{1}{10} = 2.002 \times 10^{-3}$$

$$375 = 3.75 \times 10 \times 10 = 3.75 \times 10^2$$

It should be noted that the scientific notation for a number can be expressed in any of several equivalent forms. Thus,

$$4.32 \times 10^3 = 43.2 \times 10^2 = 432 \times 10^1 = 0.432 \times 10^4 = 0.0432 \times 10^5$$

The number in the exponent is equal to the number of places the decimal must be shifted to convert a number from scientific to purely decimal notation. The shift is to the right if the exponent is positive and to the left if it is negative. The process is reversed when decimal numbers are converted to scientific notation.

Arithmetic Operations with Scientific Notation

The use of scientific notation is helpful in preventing decimal errors in arithmetic calculations. Some examples follow.

Multiplication

Here, the decimal parts of the numbers are multiplied and the exponents are added; thus,

$$\begin{aligned} 420{,}000 \times 0.0300 &= (4.20 \times 10^5)(3.00 \times 10^{-2}) \\ &= 12.60 \times 10^3 = 1.26 \times 10^4 \end{aligned}$$

$$\begin{aligned} 0.0060 \times 0.000020 &= 6.0 \times 10^{-3} \times 2.0 \times 10^{-5} \\ &= 12 \times 10^{-8} = 1.2 \times 10^{-7} \end{aligned}$$

Division

Here, the decimal parts of the numbers are divided; the exponent in the denominator is subtracted from that in the numerator. For example,

$$\frac{0.015}{5000} = \frac{15 \times 10^{-3}}{5.0 \times 10^3} 3.0 \times 10^{-6}$$

Addition and Subtraction

Addition or subtraction in scientific notation requires that all numbers be expressed to a common power of 10. The decimal parts are then added or subtracted, as appropriate. Thus,

$$\begin{aligned} &2.00 \times 10^{-11} + 4.00 \times 10^{-12} - 3.00 \times 10^{-10} \\ &\qquad = 2.00 \times 10^{-11} + 0.400 \times 10^{-11} - 30.0 \times 10^{-11} \\ &\qquad = -2.76 \times 10^{-10} = -27.6 \times 10^{-11} \end{aligned}$$

Raising to a Power a Number Written in Exponential Notation

Here, each part of the number is raised to the power separately. For example,

$$\begin{aligned} (2 \times 10^{-3})^4 &= (2.0)^4 \times (10^{-3})^4 = 16 \times 10^{-(3 \times 4)} \\ &= 16 \times 10^{-12} = 1.6 \times 10^{-11} \end{aligned}$$

Extraction of the Root of a Number Written in Exponential Notation

Here, the number is written in such a way that the exponent of 10 is evenly divisible by the root. Thus,

$$\begin{aligned} (4.0 \times 10^{-5})^{1/3} &= \sqrt[3]{40 \times 10^{-6}} = \sqrt[3]{40} \times \sqrt[3]{10^{-6}} \\ &= 3.4 \times 10^{-2} \end{aligned}$$

Logarithms

In this discussion, we will assume that the reader has available an electronic calculator for obtaining logarithms and antilogarithms of numbers. (The key for the antilogarithm function on most calculators is designated as 10^x.) It is desirable, however, to understand what a logarithm is as well as some of its properties. The discussion that follows provides this information.

A logarithm (or log) of a number is the power to which some base number (usually 10) must be raised in order to give the desired number. Thus, a logarithm is an exponent of the base 10. From the discussion in the previous paragraphs about exponential numbers, we can draw the following conclusions with respect to logs:

1. The logarithm of a product is the sum of the logarithms of the individual numbers in the product.

$$\log (100 \times 1000) = \log 10^2 + \log 10^3 = 2 + 3 = 5$$

2. The logarithm of a quotient is the difference between the logarithms of the individual numbers.

$$\log (100/1000) = \log 10^2 - \log 10^3 = 2 - 3 = -1$$

3. The logarithm of a number raised to some power is the logarithm of the number multiplied by that power.

$$\log (1000)^2 = 2 \times \log 10^3 = 2 \times 3 = 6$$

$$\log (0.01)^6 = 6 \times \log 10^{-2} = 6 \times (-2) = -12$$

4. The logarithm of a root of a number is the logarithm of that number divided by the root.

$$\log (1000)^{1/3} = \frac{1}{3} \times \log 10^3 = \frac{1}{3} \times 3 = 1$$

The following examples illustrate these statements:

$$\log 40 \times 10^{20} = \log 4.0 \times 10^{21} = \log 4.0 + \log 10^{21}$$
$$= 0.60 + 21 = 21.60$$

$$\log 2.0 \times 10^{-6} = \log 2.0 + \log 10^{-6} = 0.30 + (-6) = -5.70$$

For some purposes it is helpful to dispense with the subtraction step shown in the last example and report the log as a *negative* integer and a *positive* decimal number; that is,

$$\log 2.0 \times 10^{-6} = \log 2.0 + \log 10^{-6} = \bar{6}.30$$

The last two examples demonstrate that the logarithm of a number is the sum of two parts, a *characteristic* located to the left of the decimal point and a *mantissa* that lies to the right. The characteristic is the logarithm of 10 raised to a power and serves to indicate the location of the decimal point in the original number when that number is expressed in decimal notation. The mantissa is the logarithm of a number in the range between 0.00 and 9.99 ... Note that the mantissa is *always positive.* As a consequence, the characteristic in the last example is -6 and the mantissa is $+0.30$.

APPENDIX 7

Volumetric Calculations Using Normality and Equivalent Weight

The *normality* of a solution expresses the number of equivalents of solute contained in 1 L of solution or the number of milliequivalents in 1 mL. The equivalent and milliequivalent, like the mole and millimole, are units for describing the amount of a chemical species. The former two, however, are defined so that we may state that at the equivalence point in *any* titration,

$$\text{no. meq analyte present} = \text{no. meq standard reagent added} \quad \text{(A7-1)}$$

or

$$\text{no. eq analyte present} = \text{no. eq standard reagent added} \quad \text{(A7-2)}$$

As a consequence, stoichiometric ratios such as those described in Section 13C-3 (page 343) need not be derived every time a volumetric calculation is performed. Instead, the stoichiometry is taken into account by how the equivalent or milliequivalent weight is defined.

7A-1 The Definitions of Equivalent and Milliequivalent

In contrast to the mole, the amount of a substance contained in one equivalent can vary from reaction to reaction. Consequently, the weight of one equivalent of a compound can never be computed *without reference to a chemical reaction* in which that compound is, directly or indirectly, a participant. Similarly, the normality of a solution can never be specified *without knowledge about how the solution will be used.*

Equivalent Weights in Neutralization Reactions

One equivalent weight of a substance participating in a neutralization reaction is that amount of substance (molecule, ion, or paired ion such as NaOH) that either reacts with or supplies 1 mol of hydrogen ions *in that reaction.*[1] A milliequivalent is simply 1/1000 of an equivalent.

The relationship between equivalent weight (eqw) and the molar mass ($\mathcal{M}$) is straightforward for strong acids or bases and for other acids or bases that contain a single reaction hydrogen or hydroxide ion. For example, the equivalent weights of

◀ Once again we find ourselves using the term *weight* when we really mean *mass.* The term *equivalent weight* is so firmly engrained in the literature and vocabulary of chemistry that we retain it in this discussion.

[1]An alternative definition, proposed by the International Union of Pure and Applied Chemistry, is as follows: An equivalent is "that amount of substance, which, in a specified reaction, releases or replaces that amount of hydrogen that is combined with 3 g of carbon-12 in methane $^{12}CH_4$" (see *Information Bulletin* No. 36, International Union of Pure and Applied Chemistry, August 1974). This definition applies to acids. For other types of reactions and reagents, the amount of hydrogen referred to may be replaced by the equivalent amount of hydroxide ions, electrons, or cations. The reaction to which the definition is applied must be specified.

potassium hydroxide, hydrochloric acid, and acetic acid are equal to their molar masses because each has but a single reactive hydrogen ion or hydroxide ion. Barium hydroxide, which contains two identical hydroxide ions, reacts with two hydrogen ions in any acid/base reaction, and so its equivalent weight is one half its molar mass:

$$\text{eqw Ba(OH)}_2 = \frac{\mathcal{M}_{\text{Ba(OH)}_2}}{2}$$

The situation becomes more complex for acids or bases that contain two or more reactive hydrogen or hydroxide ions with different tendencies to dissociate. With certain indicators, for example, only the first of the three protons in phosphoric acid is titrated:

$$H_3PO_4 + OH^- \rightarrow H_2PO_4^- + H_2O$$

With certain other indicators, a color change occurs only after two hydrogen ions have reacted:

$$H_3PO_4 + 2OH^- \rightarrow HPO_4^{2-} + 2H_2O$$

For a titration involving the first reaction, the equivalent weight of phosphoric acid is equal to the molar mass; for the second, the equivalent weight is one half the molar mass. (Because it is not practical to titrate the third proton, an equivalent weight that is one third the molar mass is not generally encountered for H_3PO_4.) If it is not known which of these reactions is involved, an unambiguous definition of the equivalent weight for phosphoric acid *cannot be made.*

Equivalent Weights in Oxidation/Reduction Reactions

The equivalent weight of a participant in an oxidation/reduction reaction is that amount that directly or indirectly produces or consumes 1 mol of electrons. The numerical value for the equivalent weight is conveniently established by dividing the molar mass of the substance of interest by the change in oxidation number associated with its reaction. As an example, consider the oxidation of oxalate ion by permanganate ion:

$$5C_2O_4^{2-} + 2MnO_4^- + 16H^+ \rightarrow 10CO_2 + 2Mn^+ + 8H_2O \qquad \text{(A7-3)}$$

In this reaction, the change in oxidation number of manganese is 5 because the element passes from the +7 to the +2 state; the equivalent weights for MnO_4^- and Mn^{2+} are therefore one fifth their molar masses. Each carbon atom in the oxalate ion is oxidized from the +3 to the +4 state, leading to the production of two electrons by that species. Therefore, the equivalent weight of sodium oxalate is one half its molar mass. It is also possible to assign an equivalent weight to the carbon dioxide produced by the reaction. Since this molecule contains but a single carbon atom and since that carbon undergoes a change in oxidation number of 1, the molar mass and equivalent weight of the two are identical.

It is important to note that in evaluating the equivalent weight of a substance, *only* its change in oxidation number during the titration is considered. For example, suppose the manganese content of a sample containing Mn_2O_3 is to be determined by a titration based on the reaction given in Equation A7-3. The fact that each manganese in the Mn_2O_3 has an oxidation number of +3 plays no part in determining equivalent weight. That is, we must assume that by suitable treatment, all the manganese is oxidized to the +7 state before the titration is begun. Each manganese from the Mn_2O_3 is then reduced from the +7 to the +2 state in the titration step. The equivalent weight is thus the molar mass of Mn_2O_3 divided by $2 \times 5 = 10$.

As in neutralization reactions, the equivalent weight for a given oxidizing or reducing agent is not invariant. Potassium permanganate, for example, reacts under some conditions to give MnO_2:

$$MnO_4^- + 3e^- + 2H_2O \rightarrow MnO_2(s) + OH^-$$

The change in the oxidation state of manganese in this reaction is from +7 to +4, and the equivalent weight of potassium permanganate is now equal to its molar mass divided by 3 (instead of 5, as in the earlier example).

Equivalent Weights in Precipitation and Complex Formation Reactions

The equivalent weight of a participant in a precipitation or a complex formation reaction is that weight which reacts with or provides one mole of the *reacting* cation if it is univalent, one-half mole if it is divalent, one-third mole if it is trivalent, and so on. It is important to note that the cation referred to in this definition is always *the cation directly involved in the analytical reaction* and not necessarily the cation contained in the compound whose equivalent weight is being defined.

EXAMPLE A7-1

Define equivalent weights for $AlCl_3$ and BiOCl if the two compounds are determined by a precipitation titration with $AgNO_3$:

$$Ag^+ + Cl^- \rightarrow AgCl(s)$$

In this instance, the equivalent weight is based on the number of moles of *silver ions* involved in the titration of each compound. Since 1 mol of Ag^+ reacts with 1 mol of Cl^- provided by one-third mole of $AlCl_3$, we can write

$$\text{eqw } AlCl_3 = \frac{\mathcal{M}_{AlCl_3}}{3}$$

Because each mole of BiOCl reacts with only 1 Ag^+ ion,

$$\text{eqw BiOCl} = \frac{\mathcal{M}_{BiOCl}}{1}$$

Note that Bi^{3+} (or Al^{3+}) being trivalent has no bearing because the definition is based *on the cation involved in the titration:* Ag^+.

A7-2 The Definition of Normality

The normality c_N of a solution expresses the number of milliequivalents of solute contained in 1 mL of solution or the number of equivalents contained in 1 L. Thus, a 0.20 N hydrochloric acid solution contains 0.20 meq of HCl in each milliliter of solution or 0.20 eq in each liter.

The normal concentration of a solution is defined by equations analogous to Equation 4-2. Thus, for a solution of the species A, the normality $c_{N(A)}$ is given by the equations

$$c_{N(A)} = \frac{\text{no. meq A}}{\text{no. mL solution}} \tag{A7-4}$$

$$c_{N(A)} = \frac{\text{no. eq A}}{\text{no. L solution}} \tag{A7-5}$$

A7-3 Some Useful Algebraic Relationships

Two pairs of algebraic equations, analogous to Equations 13-1 and 13-2 as well as 13-3 and 13-4 in Chapter 13, apply when normal concentrations are used:

$$\text{amount A} = \text{no. meq A} = \frac{\text{mass A (g)}}{\text{meqw A (g/meq)}} \tag{A7-6}$$

$$\text{amount A} = \text{no. eq A} = \frac{\text{mass A (g)}}{\text{eqw A (g/eq)}} \tag{A7-7}$$

$$\text{amount A} = \text{no. meq A} = V\,(\text{mL}) \times c_{N(A)}(\text{meq/mL}) \tag{A7-8}$$

$$\text{amount A} = \text{no. eq A} = V\,(\text{L}) \times c_{N(A)}(\text{eq/L}) \tag{A7-9}$$

A7-4 Calculation of the Normality of Standard Solutions

Example A7-2 shows how the normality of a standard solution is computed from preparatory data.

EXAMPLE A7-2

Describe the preparation of 5.000 L of 0.1000 N Na_2CO_3 (105.99 g/mol) from the primary-standard solid, assuming the solution is to be used for titrations in which the reaction is

$$CO_3^{2-} + 2H^+ \rightarrow H_2O + CO_2$$

Applying Equation A7-9 gives

$$\text{amount } Na_2CO_3 = V\text{ soln (L)} \times c_{N(Na_2CO_3)}(\text{eq/L})$$
$$= 5.000\text{ L} \times 0.1000\text{ eq/L} = 0.5000\text{ eq } Na_2CO_3$$

Rearranging Equation A7-7 gives

$$\text{mass } Na_2CO_3 = \text{no. eq } Na_2CO_3 \times \text{eqw } Na_2CO_3$$

But 2 eq of Na_2CO_3 are contained in each mole of the compound; therefore,

$$\text{mass } Na_2CO_3 = 0.5000 \text{ eq } Na_2CO_3 \times \frac{105.99 \text{ g } Na_2CO_3}{2 \text{ eq } Na_2CO_3} = 26.50 \text{ g}$$

Therefore, dissolve 26.50 g in water and dilute to 5.000 L.

It is worth noting that when the carbonate ion reacts with two protons, the weight of sodium carbonate required to prepare a 0.10 N solution is just one half that required to prepare a 0.10 M solution.

A7-5 The Treatment of Titration Data With Normalities

Calculation of Normalities from Titration Data

Examples A7-3 and A7-4 illustrate how normality is computed from standardization data. Note that these examples are similar to Examples 13-4 and 13-5 in Chapter 13.

EXAMPLE A7-3

Exactly 50.00 mL of an HCl solution required 29.71 mL of 0.03926 N $Ba(OH)_2$ to give an end point with bromocresol green indicator. Calculate the normality of the HCl.

Note that the molarity of $Ba(OH)_2$ is one half its normality. That is,

$$c_{Ba(OH)_2} = 0.03926 \frac{\text{meq}}{\text{mL}} \times \frac{1 \text{ mmol}}{2 \text{ meq}} = 0.01963 \text{ M}$$

Because we are basing our calculations on the milliequivalent, we write

$$\text{no. meq HCl} = \text{no. meq } Ba(OH)_2$$

The number of milliequivalents of standard is obtained by substituting into Equation A7-8:

$$\text{amount } Ba(OH)_2 = 29.71 \text{ mL } Ba(OH)_2 \times 0.03926 \frac{\text{meq } Ba(OH)_2}{\text{mL } Ba(OH)_2}$$

To obtain the number of milliequivalents of HCl, we write

$$\text{amount HCl} = (29.71 \times 0.03926) \text{ meq } Ba(OH)_2 \times \frac{1 \text{ meq HCl}}{1 \text{ meq } Ba(OH)_2}$$

(continued)

Equating this result to Equation A7-8 yields

$$\text{amount HCl} = 50.00 \text{ mL} \times c_{N(HCl)}$$
$$= (29.71 \times 0.03926 \times 1) \text{ meq HCl}$$
$$c_{N(HCl)} = \frac{(29.71 \times 0.03926 \times 1) \text{ meq HCl}}{50.00 \text{ mL HCl}} = 0.02333 \text{ N}$$

EXAMPLE A7-4

A 0.2121-g sample of pure $Na_2C_2O_4$ (134.00 g/mol) was titrated with 43.31 mL of $KMnO_4$. What is the normality of the $KMnO_4$ solution? The chemical reaction is

$$2MnO_4^- + 5C_2O_4^{2-} + 16H^+ \rightarrow 2Mn^{2+} + 10CO_2 + 8H_2O$$

By definition, at the equivalence point in the titration,

$$\text{no. meq } Na_2C_2O_4 = \text{no. meq } KMnO_4$$

Substituting Equations A7-8 and A7-6 into this relationship gives

$$V_{KMnO_4} \times c_{N(KMnO_4)} = \frac{\text{mass } Na_2C_2O_4 \text{ (g)}}{\text{meqw } Na_2C_2O_4 \text{ (g/meq)}}$$

$$43.31 \text{ mL } KMnO_4 \times c_{N(KMnO_4)} = \frac{0.2121 \text{ g } Na_2C_2O_4}{0.13400 \text{ g } Na_2C_2O_4/2 \text{ meq}}$$

$$c_{N(KMnO_4)} = \frac{0.2121 \text{ g } Na_2C_2O_4}{43.31 \text{ mL } KMnO_4 \times 0.1340 \text{ g } Na_2C_2O_4/2 \text{ meq}}$$
$$= 0.073093 \text{ meq/mL } KMnO_4 = 0.07309 \text{ N}$$

Note that the normality found here is five times the molarity computed in Example 13-5.

Calculation of the Quantity of Analyte from Titration Data

The examples that follow illustrate how analyte concentrations are computed when normalities are involved. Note that Example A7-5 is similar to Example 13-6 in Chapter 13.

EXAMPLE A7-5

A 0.8040-g sample of an iron ore was dissolved in acid. The iron was then reduced to Fe^{2+} and titrated with 47.22 mL of 0.1121 N (0.02242 M) $KMnO_4$ solution. Calculate the results of this analysis in terms of (a) percent Fe (55.847 g/mol) and (b) percent Fe_3O_4 (231.54 g/mol). The reaction of the analyte with the reagent is described by the equation

$$MnO_4^- + 5Fe^{2+} + 8H^+ \rightarrow Mn^{2+} + 5Fe^{3+} + 4H_2O$$

(a) At the equivalence point, we know that

$$\text{no. meq KMnO}_4 = \text{no. meq Fe}^{2+} = \text{no. meq Fe}_3\text{O}_4$$

Substituting Equations A7-8 and A7-6 leads to

$$V_{\text{KMnO}_4}(\cancel{\text{mL}}) \times c_{\text{N(KMnO}_4)}(\text{meq}/\cancel{\text{mL}}) = \frac{\text{mass Fe}^{2+}(\cancel{g})}{\text{meqw Fe}^{2+}(\cancel{g}/\text{meq})}$$

Substituting numerical data into this equation gives, after rearranging,

$$\text{mass Fe}^{2+} = 47.22\ \cancel{\text{mL KMnO}_4} \times 0.1121 \frac{\cancel{\text{meq}}}{\cancel{\text{mL KMnO}_4}} \times \frac{0.055847\ \text{g}}{1\ \cancel{\text{meq}}}$$

Note that the milliequivalent weight of the Fe^{2+} is equal to its millimolar mass. The percentage of iron is

$$\text{percent Fe}^{2+} = \frac{(47.22 \times 0.1121 \times 0.055847)\ \text{g Fe}^{2+}}{0.8040\ \text{g sample}} \times 100\%$$
$$= 36.77\%$$

(b) Here,

$$\text{no. meq KMnO}_4 = \text{no. meq Fe}_3\text{O}_4$$

and

$$V_{\text{KMnO}_4}(\cancel{\text{mL}}) \times c_{\text{N(KMnO}_4)}(\text{meq}/\cancel{\text{mL}}) = \frac{\text{mass Fe}_3\text{O}_4(\cancel{g})}{\text{meqw Fe}_3\text{O}_4\ (\cancel{g}/\text{meq})}$$

Substituting numerical data and rearranging give

$$\text{mass Fe}_3\text{O}_4 = 47.22\ \cancel{\text{mL}} \times 0.1121 \frac{\cancel{\text{meq}}}{\cancel{\text{mL}}} \times 0.23154 \frac{\text{g Fe}_3\text{O}_4}{3\ \cancel{\text{meq}}}$$

Note that the milliequivalent weight of Fe_3O_4 is one third its millimolar mass because each Fe^{2+} undergoes a one-electron change and the compound is converted to $3Fe^{2+}$ before titration. The percentage of Fe_3O_4 is then

$$\text{percent Fe}_3\text{O}_4 = \frac{(47.22 \times 0.1121 \times 0.23154/3)\ \text{g Fe}_3\text{O}_4}{0.8040\ \text{g sample}} \times 100\%$$
$$= 58.81\%$$

Note that the answers to this example are identical to those in Example 13-6.

EXAMPLE A7-6

A 0.4755-g sample containing $(NH_4)_2C_2O_4$ and inert compounds was dissolved in water and made alkaline with KOH. The liberated NH_3 was distilled into 50.00 mL of 0.1007 N (0.05035 M) H_2SO_4. The excess H_2SO_4 was back-titrated with 11.13 mL of 0.1214 N NaOH. Calculate the percentage of N (14.007 g/mol) and of $(NH_4)_2C_2O_4$ (124.10 g/mol) in the sample.

At the equivalence point, the number of milliequivalents of acid and base are equal. In this titration, however, two bases are involved: NaOH and NH_3. Thus,

$$\text{no. meq } H_2SO_4 = \text{no. meq } NH_3 + \text{no. meq NaOH}$$

After rearranging,

$$\text{no. meq } NH_3 = \text{no. meq N} = \text{no. meq } H_2SO_4 - \text{no. meq NaOH}$$

Substituting Equations A7-6 and A7-8 for the number of milliequivalents of N and H_2SO_4, respectively, yields

$$\frac{\text{mass N (g)}}{\text{meqw N (}\cancel{g}\text{/meq)}} = 50.00\,\cancel{\text{mL } H_2SO_4} \times 0.1007\,\frac{\text{meq}}{\cancel{\text{mL } H_2SO_4}} - 11.13\,\cancel{\text{mL NaOH}} \times 0.1214\,\frac{\text{meq}}{\cancel{\text{mL NaOH}}}$$

$$\text{mass N} = (50.00 \times 0.1007 - 11.13 \times 0.1214)\,\cancel{\text{meq}} \times 0.014007\text{ g N/}\cancel{\text{meq}}$$

$$\text{percent N} = \frac{(50.00 \times 0.1007 - 11.13 \times 0.1214) \times 0.014007\text{ g N}}{0.4755\text{ g sample}} \times 100\% = 10.85\%$$

The number of milliequivalents of $(NH_4)_2C_2O_4$ is equal to the number of milliequivalents of NH_3 and N, but the milliequivalent weight of the $(NH_4)_2C_2O_4$ is equal to one half its molar mass. Thus,

$$\text{mass } (NH_4)_2C_2O_4 = (50.00 \times 0.1007 - 11.13 \times 0.1214)\,\cancel{\text{meq}} \times 0.12410\text{ g/2 }\cancel{\text{meq}}$$

$$\text{percent } (NH_4)_2C_2O_4 = \frac{(50.00 \times 0.1007 - 11.13 \times 0.1214) \times 0.06205\,\cancel{\text{g}(NH_4)_2C_2O_4}}{0.4755\,\cancel{\text{g sample}}} \times 100\% = 48.07\%$$

APPENDIX 8

Compounds Recommended for the Preparation of Standard Solutions of Some Common Elements*

Element	Compound	Molar Mass	Solvent†	Notes
Aluminum	Al metal	26.98	Hot dil HCl	a
Antimony	$KSbOC_4H_4O_6 \cdot \frac{1}{2}H_2O$	333.93	H_2O	c
Arsenic	As_2O_3	197.84	dil HCl	i,b,d
Barium	$BaCO_3$	197.35	dil HCl	
Bismuth	Bi_2O_3	465.96	HNO_3	
Boron	H_3BO_3	61.83	H_2O	d,e
Bromine	KBr	119.01	H_2O	a
Cadmium	CdO	128.40	HNO_3	
Calcium	$CaCO_3$	100.09	dil HCl	i
Cerium	$(NH_4)_2Ce(NO_3)_6$	548.23	H_2SO_4	
Chromium	$K_2Cr_2O_7$	294.19	H_2O	i,d
Cobalt	Co metal	58.93	HNO_3	a
Copper	Cu metal	63.55	dil HNO_3	a
Fluorine	NaF	41.99	H_2O	b
Iodine	KIO_3	214.00	H_2O	i
Iron	Fe metal	55.85	HCl, hot	a
Lanthanum	La_2O_3	325.82	HCl, hot	f
Lead	$Pb(NO_3)_2$	331.20	H_2O	a
Lithium	Li_2CO_3	73.89	HCl	a
Magnesium	MgO	40.31	HCl	
Manganese	$MnSO_4 \cdot H_2O$	169.01	H_2O	g
Mercury	$HgCl_2$	271.50	H_2O	b
Molybdenum	MoO_3	143.94	1 M NaOH	
Nickel	Ni metal	58.70	HNO_3, hot	a
Phosphorus	KH_2PO_4	136.09	H_2O	
Potassium	KCl	74.56	H_2O	a
	$KHC_8H_4O_4$	204.23	H_2O	i,d
	$K_2Cr_2O_7$	294.19	H_2O	i,d
Silicon	Si metal	28.09	NaOH, concd	
	SiO_2	60.08	HF	j
Silver	$AgNO_3$	169.87	H_2O	a
Sodium	NaCl	58.44	H_2O	i
	$Na_2C_2O_4$	134.00	H_2O	i,d
Strontium	$SrCO_3$	147.63	HCl	a
Sulfur	K_2SO_4	174.27	H_2O	
Tin	Sn metal	118.69	HCl	

continues

Element	Compound	FW	Solvent†	Notes
Titanium	Ti metal	47.90	H_2SO_4; 1 : 1	a
Tungsten	$Na_2WO_4 \cdot 2H_2O$	329.86	H_2O	h
Uranium	U_3O_8	842.09	HNO_3	d
Vanadium	V_2O_5	181.88	HCl, hot	
Zinc	ZnO	81.37	HCl	a

*The data in this table were taken from a more complete list assembled by B. W. Smith and M. L. Parsons, *J. Chem, Educ.*, **1973,** *50,* 679. Unless otherwise specified, compounds should be dried to constant weight at 110°C.

†Unless otherwise specified, acids are concentrated analytical grade.

[a]Conforms well to the criteria listed in Section 12A-2 and approaches primary-standard quality.

[b]Highly toxic.

[c]Loses $\frac{1}{2}H_2O$ at 110°C. After drying, molar mass = 324.92. The dried compound should be weighed quickly after removal from the desiccator.

[d]Available as a primary standard from the National Institute of Standards and Technology.

[e]H_3BO_3 should be weighed directly from the bottle. It loses 1 mole H_2O at 100°C and is difficult to dry to constant weight.

[f]Absorbs CO_2 and H_2O. Should be ignited just before use.

[g]May be dried at 110°C without loss of water.

[h]Loses both waters at 110°C. Molar mass = 293.82. Keep in desiccator after drying.

[i]Primary standard.

[j]HF is highly toxic and dissolves glass.

APPENDIX 9

Derivation of Error Propagation Equations

In this appendix we derive several equations that permit the calculation of the standard deviation for the results from various types of arithmetical computations.

A9-A Propagation of Measurement Uncertainties

The calculated result for a typical analysis ordinarily requires data from several independent experimental measurements, each of which is subject to a random uncertainty and each of which contributes to the net random error of the final result. For the purpose of showing how such random uncertainties affect the outcome of an analysis, let us assume that a result y is dependent on the experimental variables, $a, b, c, \ldots$, each of which fluctuates in a random and independent way. That is, y is a function of $a, b, c, \ldots$, so we may write

$$y = f(a, b, c, \ldots) \qquad \text{(A9-1)}$$

The uncertainty dy_i is generally given in terms of the deviation from the mean or $(y_i - \bar{y})$, which will depend on the size and sign of the corresponding uncertainties $da_i, db_i, dc_i, \ldots$. That is,

$$dy_i = (y_i - \bar{y}) = f(da_i, db_i, dc_i, \ldots)$$

The variable in dy as a function of the uncertainties in $a, b, c, \ldots$ can be derived by taking the total differential of Equation A9-1. That is,

$$dy = \left(\frac{\partial y}{\partial a}\right)_{b,c,\ldots} da + \left(\frac{\partial y}{\partial b}\right)_{a,c,\ldots} db + \left(\frac{\partial y}{\partial c}\right)_{a,b,\ldots} dc + \cdots \qquad \text{(A9-2)}$$

To develop a relationship between the standard deviation of y and the standard deviations of a, b, and c for N replicate measurements, we employ Equation 6-4 (p. 115), which requires that we square Equation A9-2, sum between $i = 0$ and $i = N$, divide by $N - 1$, and take the square root of the result. The square of Equation A9-2 takes the form

$$(dy)^2 = \left[\left(\frac{\partial y}{\partial a}\right)_{b,c,\ldots} da + \left(\frac{\partial y}{\partial b}\right)_{a,c,\ldots} db + \left(\frac{\partial y}{\partial c}\right)_{a,b,\ldots} dc + \cdots\right]^2 \qquad \text{(A9-3)}$$

This equation must then be summed between the limits of $i = 1$ to $i = N$.

In squaring Equation A9-2, two types of terms emerge from the right-hand side of the equation: (1) square terms and (2) cross terms. Square terms take the form

$$\left(\frac{\partial y}{\partial a}\right)^2 da^2, \left(\frac{\partial y}{\partial b}\right)^2 db^2, \left(\frac{\partial y}{\partial c}\right)^2 dc^2, \ldots$$

Square terms are always positive and can, therefore, *never* cancel when summed. In contrast, cross terms may be either positive or negative in sign. Examples are

$$\left(\frac{\partial y}{\partial a}\right)\left(\frac{\partial y}{\partial b}\right) da\,db, \quad \left(\frac{\partial y}{\partial a}\right)\left(\frac{\partial y}{\partial c}\right) da\,dc, \ldots$$

If da, db, and dc represent *independent* and *random uncertainties,* some of the cross terms will be negative and others positive. Thus, the *sum of all such terms should approach zero,* particularly when N is large.

As a consequence of the tendency of cross terms to cancel, the sum of Equation A9-3 from $i = 1$ to $i = N$ can be assumed to be made up exclusively of square terms. This sum then takes the form

$$\Sigma\,(dy_i)^2 = \left(\frac{\partial y}{\partial a}\right)^2 \Sigma\,(da_i)^2 + \left(\frac{\partial y}{\partial b}\right)^2 \Sigma\,(db_i)^2 + \left(\frac{\partial y}{\partial c}\right)^2 \Sigma\,(dc_i)^2 + \cdots \qquad \text{(A9-4)}$$

Dividing through by $N - 1$ gives

$$\frac{\Sigma\,(dy_i)^2}{N-1} = \left(\frac{\partial y}{\partial a}\right)^2 \frac{\Sigma\,(da_i)^2}{N-1} + \left(\frac{\partial y}{\partial b}\right)^2 \frac{\Sigma\,(db_i)^2}{N-1} + \left(\frac{\partial y}{\partial c}\right)^2 \frac{\Sigma\,(dc_i)^2}{N-1} + \cdots \qquad \text{(A9-5)}$$

From Equation 6-4, however, we see that

$$\frac{\Sigma\,(dy_i)^2}{N-1} = \Sigma\,\frac{(y_i - \bar{y})^2}{N-1} = s_y^2$$

where s_y^2 is the variance of y. Similarly,

$$\frac{\Sigma\,(dc_i)^2}{N-1} = \frac{\Sigma\,(a_i - \bar{a})^2}{N-1} = s_a^2$$

and so forth. Thus, Equation A9-5 can be written in terms of the variances of the variables; that is,

$$s_y^2 = \left(\frac{\partial y}{\partial a}\right)^2 s_a^2 + \left(\frac{\partial y}{\partial b}\right)^2 s_b^2 + \left(\frac{\partial y}{\partial c}\right)^2 s_c^2 + \ldots \qquad \text{(A9-6)}$$

A9-B The Standard Deviation of Computed Results

In this section, we employ Equation A9-6 to derive relationships that permit calculation of standard deviations for the results produced by five types of arithmetic operations.

A9-B.1 Addition and Subtraction

Consider the case where we wish to compute the quantity y from the three experimental quantities a, b, and c by means of the equation

$$y = a + b - c$$

We assume that the standard deviations for these quantities are s_y, s_a, s_b, and s_c. Applying Equation A9-6 leads to

$$s_y^2 = \left(\frac{\partial y}{\partial a}\right)_{b,c}^2 s_a^2 + \left(\frac{\partial y}{\partial b}\right)_{a,c}^2 s_b^2 + \left(\frac{\partial y}{\partial c}\right)_{a,b}^2 s_c^2$$

The partial derivatives of y with respect to the three experimental quantities are

$$\left(\frac{\partial y}{\partial a}\right)_{b,c} = 1; \quad \left(\frac{\partial y}{\partial b}\right)_{a,c} = 1; \quad \left(\frac{\partial y}{\partial c}\right)_{a,b} = -1$$

Therefore, the variance of y is given by

$$s_y^2 = (1)^2 s_a^2 + (1)^2 s_b^2 + (-1)^2 s_c^2 = s_a^2 + s_b^2 + s_c^2$$

or the standard deviation of the result is given by

$$s_y = \sqrt{s_a^2 + s_b^2 + s_c^2} \tag{A9-7}$$

Thus, the *absolute* standard deviation of a sum or difference is equal to the square root of the sum of the squares of the *absolute* standard deviation of the numbers making up the sum or difference.

A9-B.2 Multiplication and Division

Let us now consider the case where

$$y = \frac{ab}{c}$$

The partial derivatives of y with respect to a, b, and c are

$$\left(\frac{\partial y}{\partial a}\right)_{b,c} = \frac{b}{c}; \quad \left(\frac{\partial y}{\partial b}\right)_{a,c} = \frac{a}{c}; \quad \left(\frac{\partial y}{\partial c}\right) = -\frac{ab}{c^2}$$

Substituting into Equation A9-6 gives

$$s_y^2 = \left(\frac{b}{c}\right)^2 s_a^2 + \left(\frac{a}{c}\right)^2 s_b^2 + \left(-\frac{ab}{c^2}\right)^2 s_c^2$$

Dividing this equation by the square of the original equation ($y^2 = a^2b^2/c^2$) gives

$$\frac{s_y^2}{y^2} = \frac{s_a^2}{a^2} + \frac{s_b^2}{b^2} + \frac{s_c^2}{c^2}$$

or

$$\frac{s_y}{y} = \sqrt{\left(\frac{s_a}{a}\right)^2 + \left(\frac{s_b}{b}\right)^2 + \left(\frac{s_c}{c}\right)^2} \tag{A9-8}$$

Thus, for products and quotients, the *relative* standard deviation of the result is equal to the sum of the squares of the *relative* standard deviation of the number making up the product or quotient.

A9-B.3 Exponential Calculations

Consider the following computation

$$y = a^x$$

Here, Equation A9-6 takes the form

$$s_y^2 = \left(\frac{\partial a^x}{\partial y}\right)^2 s_a^2$$

or

$$s_y = \frac{\partial a^x}{\partial y} s_a$$

But

$$\frac{\partial a^x}{\partial y} = xa^{(x-1)}$$

Thus

$$s_y = xa^{(x-1)}s_a$$

and dividing by the original equation ($y = a^x$) gives

$$\frac{s_y}{y} = \frac{xa^{(x-1)}s_a}{a^x} = x\frac{s_a}{a} \qquad \text{(A9-9)}$$

Thus, the relative error of the result is equal to the relative error of numbers to be exponentiated, multiplied by the exponent.

It is important to note that the error propagated in taking a number to a power is different from the error propagated in multiplication. For example, consider the uncertainty in the square of 4.0(±0.2). Here the relative error in the result (16.0) is given by Equation A9-9

$$s_y/y = 2 \times (0.2/4) = 0.1 \qquad \text{or} \qquad 10\%$$

Consider now the case when y is the product of two *independently measured* numbers that by chance happen to have values of a = 4.0(±0.2) and b = 4.0(±0.2). In this case the relative error of the product ab = 16.0 is given by Equation A9-8:

$$s_y/y = \sqrt{(0.2/4)^2 + (0.2/4)^2} = 0.07 \quad \text{or} \quad 7\%$$

The reason for this apparent anomaly is that in the second case the sign associated with one error can be the same or different from that of the other. If they happen to be the same, the error is identical to that encountered in the first case, where the signs *must* be the same. In contrast, the possibility exists that one sign could be positive and the other negative, in which case the relative errors tend to cancel one another. Thus, the probable error lies between the maximum (10%) and zero.

A9-B.4 Calculation of Logarithms

Consider the computation

$$y = \log_{10} a$$

In this case, we can write Equation A9-6 as

$$s_y^2 = \left(\frac{\partial \log_{10} a}{\partial y}\right)^2 s_a^2$$

But

$$\frac{\partial \log_{10} a}{\partial y} = \frac{0.434}{a}$$

and

$$s_y = 0.434 \frac{s_a}{a} \qquad \text{(A9-10)}$$

Thus, the absolute standard deviation of a logarithm is determined by the *relative* standard deviation of the number.

A9-B.5 Calculation of Antilogarithms

Consider the relationship

$$y = \text{antilog}_{10}\, a = 10^a$$

$$\left(\frac{\partial y}{\partial a}\right) = 10^a \log_e 10 = 10^a \ln 10 = 2.303 \times 10^a$$

$$s_y^2 = \left(\frac{\partial y}{\partial a}\right)^2 s_a^2$$

or

$$s_y = \frac{\partial y}{\partial a} s_a = 2.303 \times 10^a s_a$$

Dividing by the original relationship gives

$$\frac{s_y}{y} = 2.303 s_a \qquad \text{(A9-11)}$$

Thus, the *relative* standard deviation of the antilog of a number is determined by the absolute standard deviation of the number.

ANSWERS TO SELECTED QUESTIONS AND PROBLEMS

Chapter 3

3-1. (a), SQRT returns a positive square root; (b), SUM adds numbers in a range of cells; (c), PI returns pi to 15 digits; (d), FACT returns the factorial of a number; (e), EXP returns *e,* the natural logarithm base, raised to a power; (f), LOG returns logarithm of a number to the base specified.

3-4. `=MID(D4,2,FIND("(",D4)-2)`

3-6. `=MID(D2,FIND("(",D2,1)+1,1)`

Chapter 4

4-1. **(a)** The *millimole* is the amount of an elementary species, such as an atom, an ion, a molecule, or an electron that contains 6.02×10^{23}

$$6.02 \times 10^{23} \frac{\text{particles}}{\cancel{\text{mole}}} \times 10^{-3} \frac{\cancel{\text{mole}}}{\text{millimole}} = 6.02 \times 10^{20} \frac{\text{particles}}{\text{millimole}}$$

(c) The millimolar mass of a species is the mass in grams of one millimole of the species.

4-3. $1\text{ L} = 10^{-3}\text{ m}^3$

$$1\text{ M} = 1\frac{\text{mol}}{\text{L}} = 1\frac{\text{mol}}{10^{-3}\text{ m}^3}$$

$1\text{ Å} = 10^{-10}\text{ m}$

4-4. **(a)** 320 kHz **(c)** 843 mmol **(e)** 89.6 μm

4-5. 5.98×10^{22} Na^+ ions

4-7. **(a)** 0.0712 mol **(b)** 8.73×10^{-4} mol **(c)** 0.0382 mol **(d)** 1.31×10^{-3} mol

4-9. **(a)** 6.5 mmol **(b)** 41.6 mmol **(c)** 8.47×10^{-3} mmol **(d)** 1165.6 mmol

4-11. **(a)** 4.90×10^4 mg **(b)** 2.015×10^4 mg **(c)** 1.80×10^6 mg **(d)** 2.37×10^6 mg

4-13. **(a)** 2.22×10^3 mg **(b)** 472.8 mg

4-14. **(a)** 2.51 g **(b)** 2.88×10^{-3} g

4-15. **(a)** pNa = 1.077 pCl = 1.475 pOH = 1.298 **(c)** pH = 0.222 pCl = 0.096 pZn = 0.996 **(e)** pK = 5.836 pOH = 6.385 $pFe(CN)_6$ = 6.582

4-16. **(a)** 1.7×10^{-5} M **(c)** 0.30 M **(e)** 4.8×10^{-8} M **(g)** 2.04 M

4-17. **(a)** pNa = pBr = 1.699 pH = pOH = 7.00 **(c)** pBa = 2.46 pOH = 2.15 pH = 11.85 **(e)** pCa = 2.17 pBa = 2.12 pCl = 1.54 pH = pOH = 7.00

4-18. **(a)** 2.14×10^{-10} M **(c)** 0.92 M **(e)** 1.66 M **(g)** 0.99 M

4-19. **(a)** $[Na^+] = 4.79 \times 10^{-2}$ M $[SO_4^{2-}] = 2.87 \times 10^{-3}$ M **(b)** pNa = 1.320 pSO_4 = 2.543

4-21. **(a)** 1.037×10^{-2} M **(b)** 1.037×10^{-2} M **(c)** 3.11×10^{-2} M **(d)** 0.288% (w/v) **(e)** 0.777 mmol Cl^- **(f)** 405 ppm **(g)** 1.984 **(h)** 1.507

4-23. **(a)** 0.281 M **(b)** 0.844 M **(c)** 68.0 g/L

4-25. **(a)** Dissolve 23.8 g EtOH in water and dilute to 500 mL. **(b)** Mix 23.8 g EtOH with 476.2 g water. **(c)** Dissolve 23.8 mL EtOH and dilute to 500 mL.

4-27. Dilute 300 mL reagent to 750 mL.

4-29. **(a)** Dissolve 6.37 g $AgNO_3$ in water and dilute to 500 mL. **(b)** Dilute 47.5 ml of 6.00 M HCl to 1.0 L. **(c)** Dissolve 2.98 g $K_4Fe(CN)_6$ in water and dilute to 400 mL. **(d)** Dilute 216 mL of $BaCl_2$ solution to 600 mL. **(e)** Dilute 20.3 mL of the concentrated reagent to 2.00 L. **(f)** Dissolve 1.67 g Na_2SO_4 in water and dilute to 9.00 L.

4-31. 5.01 g

4-33. **(a)** 0.09218 g CO_2 **(b)** 0.0312 M HCl

4-35. **(a)** 1.505 g SO_2 **(b)** 0.0595 M $HClO_4$

4-37. 2930 mL $AgNO_3$

Chapter 5

5-1. **(a)** *Constant errors* are the same magnitude regardless of sample size. *Proportional errors* are proportional in size to sample size. **(c)** The *mean* is the sum of the results in a set divided by the number of results. The *median* is the central value for a set of data.

5-2. (1) Random temperature fluctuations causing random changes in the length of the metal rule; (2) uncertainties from moving and positioning the rule twice; (3) personal judgment in reading the rule; (4) vibrations in the table and/or rule; (5) uncertainty in locating the rule perpendicular to the edge of the table.

5-3. *Instrumental error, method error,* and *personal error.*

5-5. (1) Incorrect calibration of the pipet; (2) temperature different from calibration temperature; (3) incorrect filling of the pipet (overshooting or undershooting the mark).

5-7. Constant and proportional errors.

5-8. **(a)** = −0.06% **(c)** −0.2%

5-9. **(a)** 17 g **(c)** 4 g

5-10. **(a)** 0.08% **(b)** 0.4% **(c)** 0.16%

5-11. **(a)** −1.0% **(c)** −0.10%

5-12.

	Mean	Median	Deviation from Mean	Mean Deviation
(a)	0.0106	0.0105	0.0004, 0.0002, 0.0001	0.0002
(c)	190	189	2, 0, 4, 3	2
(e)	39.59	39.64	0.24, 0.02, 0.34. 0.09	0.17

Chapter 6

6-1. **(a)** The numerical difference between the highest and lowest value

(c) All digits known with certainty plus the first uncertain digit.

6-2. **(a)** Sample standard deviation s is

$$s = \sqrt{\frac{\sum_{i=1}^{N}(x_i - \bar{x})^2}{N - 1}}$$

Sample variance is s^2.

(c) *Accuracy* represents the agreement between a measured value and the true or accepted value. *Precision* describes the agreement among measurements made in exactly the same way.

6-3. **(a)** In statistics, a sample is a small set of replicate measurements. In chemistry, a sample is a portion of a material used for analysis.

6-5. Probability between 0 and $1\sigma = 0.683/2 = 0.342$; probability between 0 and $2\sigma = 0.954/2 = 0.477$. Probability between 1σ and $2\sigma = 0.477 - 0.342 = 0.135$.

6-7.

	(a) Mean	(b) Median	(c) Spread	(d) Std Dev	(e) CV
A	3.1	3.1	1.0	0.4	12%
C	0.825	0.803	0.108	0.051	6.2%
E	70.53	70.64	0.44	0.22	0.31%

6-8.

	Absolute Error	Relative Error, ppt
A	0.10	33
C	−0.006	−7
E	0.48	6.9

6-9.

	s_y	CV	y
(a)	0.03	−2%	1.44(±0.03)
(c)	0.14×10^{-16}	1.8%	$7.5(\pm 0.1) \times 10^{-16}$
(e)	0.5×10^{-2}	6.9%	$7.6(\pm 0.5) \times 10^{-2}$

6-10.

	s_y	CV	y
(a)	0.3×10^{-9}	−4%	$6.7(\pm 0.3) \times 10^{-9}$
(c)	3	25%	12(±3)
(e)	25	50%	50(±25)

6-11.

	s_y	CV	y
(a)	0.0065	0.18%	−3.70(±0.01)
(c)	0.11	0.7%	13.8(±0.1)

6-12. **(a)** $s_y = 0.02 \times 10^{-10}$, CV = 1.9%, $y = 1.06(\pm 0.02) \times 10^{-10}$

6-13. $2.2(\pm 0.1) \times 10^5$ L

6-15. ±8 K

6-17. **(a)**

Sample	Mean	Standard Deviation
1	5.12	0.08
2	7.11	0.12
3	3.99	0.12
4	4.74	0.10
5	5.96	0.11

(b) $s_{pooled} = 0.11\%$

(c) s_{pooled} is a weighted average of the individual estimates of σ. It uses all the data from the five samples. The reliability of s improves with the number of results.

6-19. $s_{pooled} = 0.29\%$.

Chapter 7

7-1. The mean of 5 measurements $\bar{x}$ is a better estimate of the true value μ than any single measurement because the distribution of means is narrower than the distribution of individual results.

7-3. **(a)** As the sample size, N, increases, the confidence interval decreases in proportion to $\sqrt{N}$.

(b) As the desired confidence level increases, the confidence interval increases.

(c) As the standard deviation, s, increases, the confidence interval increases in direct proportion.

7-4.

	A	C	E
$\bar{x}$	3.1	0.82	70.53
s	0.37	0.05	0.22
95% CI	0.46	0.08	0.34

7-5. Set A 95% CI = 0.18; set C 95% CI = 0.009; set E 95% CI = 0.15

7-7. **(a)** 80% CI = 18.5 ± 3.1 μg/mL; 95% CI = 18.5 ± 4.7 μg/mL

(b) 80% CI = 18.5 ± 2.2 μg/mL; 95% CI = 18 ± 3.3 μg/mL

(c) 80% CI = 18.5 ± 1.5 μg/mL; 95% CI = 18.5 ± 2.4 μg/mL

7-9. 95%, 10 measurements; 99%, 17 measurements

7-11. **(a)** 3.22 ± 0.15 meq/L **(b)** 3.22 ± 0.06 meq/L

7-13. **(a)** 12 measurements

7-15. For C, no systematic error; for H, systematic error is indicated.

7-17. H_0: $\mu = 5.0$ ppm; H_a: $\mu < 5.0$ ppm. Accept H_0, reject H_a.

7-19. H_0: $\mu = 1.0$ ppb; H_a: $\mu < 1.0$ ppb. Type I error, we reject H_0 when it is actually true. Type II, we accept H_0 when it is false.

7-21. **(a)** H_0: $\mu = 7.03$ ppm; H_a: $\mu < 7.03$ ppm. Type I error, we reject H_0 and decide a systematic error is present when it is not. Type II, we accept H_0 and decide there is no systematic error when one is present. One-tailed test.

(c) H_0: $\sigma^2_{AA} = \sigma^2_{EC}$; H_a: $\sigma^2_{EC} < ^2_{AA.}$ One-tailed test. Type I, we decide that AA results are less precise than electrochemistry (EC) results, when the precision is the same. Type II, we decide that the precision is the same when the electrochemistry results are more precise.

7-23. **(a)** Paired t test to cancel the variation in samples and focus on the method differences.

(b) H_0: $\mu_d = 0$, H_a: $\mu_A \neq 0$, where μ_d is the mean difference between the methods. We reject H_0 at the 95% confidence level.
(c) No, H_0 would be rejected at 90%, 95%, and 99% confidence levels.

7-25. (a)

Variation Source	SS	df	MS	F
Between soils	0.2768	2	0.1384	17.09
Within soils	0.0972	12	0.0081	
Total	0.374	14		

(b) $H_0 = \mu_{samp1} = \mu_{samp2} = \mu_{samp3}$; H_a: at least two of the means differ.
(c) Reject H_0 and conclude the soils differ.

7-27. (a) $H_0 = \mu_{LabA} = \mu_{LabB} = \mu_{LabC} = \mu_{LabD} = \mu_{LabE}$; H_a: at least two of the means differ.
(b) $F = 6.9485$. At 95%, $F_{0.05,4,10} = 3.48$. Laboratories differ. At 99%, laboratories differ. At 99.9% confidence level, laboratories are the same.
(c) Laboratories A, C, and E differ from laboratory D, but laboratory B does not. Laboratories E and A differ from laboratory B, but laboratory C does not. No significant difference exists between laboratories E and A.

7-29. (a) $\mu_{Des1} = \mu_{Des2} = \mu_{Des3} = \mu_{Des4}$; H_a: at least two of the means differ.
(b) Accept H_0 and conclude there is no difference.
(c) No differences.

7-31. (a) Cannot reject 41.27 value.
(b) Reject 7.388 value.

7-33. Cannot reject 4.60 ppm value.

Chapter 8

8-1. (1) Identification of the population from which the sample is to be drawn, (2) collection of a gross sample, and (3) reduction of the gross sample to a small quantity of homogeneous material for analysis.

8-3. Factors depend on heterogeneity of the material, the particle size at which heterogeneity begins, and the uncertainty in composition that can be tolerated.

8-5. (a) 26% (b) 14 ± 6 (c) 1.69×10^3

8-7. (a) RSD = 0.12 or 12% (b) 220 bottles (c) 190 ± 40 (d) 1.5×10^3 bottles

8-9. Scheme A will have the lower variance.

8-11. No. There was insufficient sample.

8-13. (b) Slope = 0.23, intercept = 0.16
(d) $R^2 = 0.9834$, adjusted $R^2 = 0.9779$, $F = 177.6$. Regression is significant (large F). The R^2 value measures the fraction of the variation explained by the regression. The adjusted R^2 indicates the price to pay for adding an additional parameter.
(f) $s = 0.73$, CV = 6.3%

8-15. (a) Slope = 5.57, intercept = 0.90
(d) 1.69 mmol/L
(f) unknown = 3.93 mmol/L, $s_c = 0.08$, CV = 2.03%; for four measurements, $s_c = 0.05$, CV = 1.26%.

8-17. (a)

E, mV	−log c
106	0.69897
115	1.100179
121	1.199971
139	1.500313
153	1.69897
158	1.899629
174	2.100179
182	2.199971
187	2.400117
211	2.69897
220	2.899629
226	3

(c) 95% CL for $m = 55.37 \pm 3.91$; 95% CL for $b = 58.04 \pm 8.11$
(e) Standard error = 4.29, $R = 0.995$, multiple $R = 0.995$

8-19. (a) Slope = 0.0225, intercept = 0.044, $R^2 = 0.6312$, $F = 5.135$. Plot is not highly linear.
(b) Slope = 0.0486, intercept = 0.0106, $R^2 = 0.9936$, $F = 540.84$. Linearity is much better. Taking the ratio compensates for systematic errors that affect both the sample and the internal standard.
(c) 9.46% water

8-21. (a) 0.96 μg/mL (b) 0.81 μg/mL, 19% error

8-23. (a) For 5.0 ng/mL, $\bar{A} = 0.0489$, $s = 0.0028$; for 10.0 ng/mL, $\bar{A} = 0.0979$, $s = 0.0054$.
(b) $A = 0.0098c_{Zn} - 0.00002$; calibration sensitivity = 0.0098 $(ng/mL)^{-1}$, analytical sensitivity at 5.0 ng/mL = 3.47 $(ng/mL)^{-1}$.
(c) DL = 1.44 ng/mL
(d) $c_{Zn} = 6.63$ ng/ml, $s = 0.005$

8-25. Mean 50.3, $\sigma = 2.2$, UCL = 56.9, LCL = 43.7; process was always in control.

Chapter 9

9-1. (a) A *weak electrolyte* only partially ionizes in a solvent. $NaHCO_3$ is an example.
(c) The *conjugate acid of a Brønsted-Lowry base* is the species formed when the base accepts a proton. NH_4^+ is the conjugate acid of the base NH_3.
(e) An *amphiprotic solvent* can act either as an acid or as a base. Water is an example.
(g) *Autoprotolysis* is self-ionization of a solvent to produce both a conjugate acid and a conjugate base.
(i) The *Le Châtelier principle* states that the position of an equilibrium always shifts in such a direction to relieve a stress applied to the system.

9-2. (a) An *amphiprotic solute* is a chemical species that can act either as an acid or a base. The dihydrogen phosphate ion, $H_2PO_4^-$, is an example of an amphiprotic solute.
(c) A leveling solvent is one in which a series of acids (or bases) all dissociate completely. Water is an example, since strong acids like HCl and $HClO_4$ ionize completely.

9-3. For an aqueous equilibrium in which water is a participant, the concentration of water is normally so much larger than the concentrations of the other reactants or products that it can be assumed to be constant and independent of the equilibrium position. Thus, its concentration is included in the equilibrium constant. For a pure solid, the concentration of the chemical species in the solid phase is constant. As long as some solid exists as a second phase, its effect on the equilibrium is constant and is included within the equilibrium constant.

9-4.

	Acid	Conjugate Base
(a)	HOCl	OCl^-
(c)	NH_4^+	NH_3
(e)	$H_2PO_4^-$	HPO_4^{2-}

9-6. **(a)** $2H_2O \rightleftharpoons H_3O^+ + OH^-$
(c) $2CH_3NH_2 \rightleftharpoons CH_3NH_3^+ + CH_3NH^-$

9-7. **(a)** $K_b = \dfrac{K_w}{K_a} = \dfrac{1.00 \times 10^{-14}}{2.31 \times 10^{-11}} = 4.33 \times 10^{-4} = \dfrac{[C_2H_5NH_3^+][OH^-]}{[C_2H_5NH_2]}$

(c) $K_a = 5.90 \times 10^{-6} = \dfrac{[H_3O^+][C_5H_5N]}{[C_5H_5NH^+]}$

(e) $\beta_3 = K_1K_2K_3 = 2 \times 10^{-21} = \dfrac{[H_3O^+]^3[AsO_4^{3-}]}{[H_3AsO_4]}$

9-8. **(a)** $K_{sp} = [Cu^+][I^-]$
(b) $K_{sp} = [Pb^{2+}][Cl^-][F^-]$
(c) $K_{sp} = [Pb^{2+}][I^-]^2$

9-10. **(b)** $K_{sp} = 3.2 \times 10^{-13}$ **(d)** $K_{sp} = 1.0 \times 10^{-15}$

9-13. **(a)** 1.0×10^{-9} M **(b)** 0.3 M

9-15. **(a)** 0.0125 M **(b)** 7.0×10^{-3} M
(c) 4.8×10^{-7} M **(d)** 6.1×10^{-6} M

9-17. **(a)** $PbI_2 > BiI_3 > CuI > AgI$ in water.
(b) $PbI_2 > CuI > AgI > BiI_3$ in 0.10 M NaI.
(c) $PbI_2 > BiI_3 > CuI > AgI$ in a 0.01 M solution of the solute cation.

9-20. **(a)** $[H_3O^+] = 3.0 \times 10^{-5}$ M, $[OH^-] = 3.3 \times 10^{-10}$ M
(c) $[OH^-] = 6.3 \times 10^{-3}$ M, $[H_3O^+] = 1.6 \times 10^{-12}$ M
(e) $[OH^-] = 2.6 \times 10^{-4}$ M, $[H_3O^+] = 3.9 \times 10^{-11}$ M
(g) $[H_3O^+] = 5.24 \times 10^{-4}$ M, $[OH^-] = 1.91 \times 10^{-11}$ M

9-21. **(a)** $[H_3O^+] = 1.10 \times 10^{-2}$ M
(b) $[H_3O^+] = 1.17 \times 10^{-8}$ M
(e) 1.46×10^{-4} M

9-23. *Buffer capacity* of a solution is defined as the number of moles of a strong acid (or a strong base) that causes 1.00 L of a buffer to undergo a 1.00-unit change in pH.

9-25. Since the ratios of the amounts of weak acid to conjugate base are identical, the three solutions will have the same pH. They will differ in buffer capacity, however, with **(a)** having the greatest and **(c)** the least.

9-26. **(a)** Malic acid/sodium hydrogen malate.
(c) NH_4Cl/NH_3

9-27. 15.5 g sodium formate

9-29. 194 mL HCl

Chapter 10

10-1. **(a)** *Activity,* a_A, is the effective concentration of a chemical species A in solution. The *activity coefficient,* γ_A, is the numerical factor necessary to convert the molar concentration of the chemical species A to activity: $a_A = \gamma_A[A]$.
(b) The *thermodynamic equilibrium constant* refers to an ideal system within which each chemical species is unaffected by any others. A *concentration equilibrium constant* takes into account the influence exerted by solute species on one another. The thermodynamic equilibrium constant is based on activities of reactants and products and is independent of ionic strength; the concentration equilibrium constant is based on molar concentrations of reactants and products.

10-3. **(a)** Ionic strength should decrease.
(b) Ionic strength should be unchanged.
(c) Ionic strength should increase.

10-5. The initial slope is steeper because multiply charged ions deviate from ideality more than singly charged ions.

10-7. **(a)** 0.16 **(c)** 1.2

10-8. **(a)** 0.20 **(c)** 0.073

10-10. **(a)** 1.7×10^{-12} **(c)** 7.6×10^{-11}

10-11. **(a)** 5.2×10^{-6} M **(b)** 6.3×10^{-6} M
(c) 9.53×10^{-12} M **(d)** 1.5×10^{-7} M

10-12. **(a)** (1) 1.4×10^{-6} M (2) 1.0×10^{-6} M
(b) (1) 2.1×10^{-3} M (2) 1.3×10^{-3} M
(c) (1) 2.9×10^{-5} M (2) 1.0×10^{-5} M
(d) (1) 1.4×10^{-5} M (2) 2.0×10^{-6} M

10-13. **(a)** (1) 2.2×10^{-4} M (2) 1.8×10^{-4} M
(b) (1) 1.7×10^{-4} M (2) 1.2×10^{-4} M
(c) (1) 3.3×10^{-8} M (2) 6.6×10^{-9} M
(d) (1) 1.3×10^{-3} M (2) 7.8×10^{-4} M

10-14. **(a)** −19% **(c)** −40% **(e)** −46%

10-15. **(a)** 32%

Chapter 11

11-3. A charge-balance equation is derived by relating the concentration of cations and anions such that no. mol/L positive charge = no. mol/L negative charge. For a doubly charged ion, such as Ba^{2+}, the concentration of charge for each mole is twice the molar *concentration.* That is, no. mol/L positive charge = $2[Ba^{2+}]$. For Fe^{3+} it is three times the molar concentration. Thus, the molar concentration of all multiply charged species is always multiplied by the charge in a charge-balance equation.

11-4. **(a)** $0.20 = [H_3AsO_4] + [H_2AsO_4^-] + [HAsO_4^{2-}] + [AsO_4^{3-}]$
(c) $0.0500 + 0.100 = [ClO^-] + [HClO]$
(e) $0.100 = [Na^+] = [OH^-] + 2[Zn(OH)_4^{2-}]$
(g) $[Ca^{2+}] = ½([F^-] + [HF])$

11-6. **(a)** 2.1×10^{-4} M **(c)** 2.1×10^{-4} M

11-7. **(a)** 1.65×10^{-4} M **(c)** 2.75×10^{-5} M

11-8. **(a)** 5.1×10^{-9} M **(b)** 5.1×10^{-12} M

11-10. **(a)** 0.1 M

11-11. 1.4×10^{-5} M

11-13. **(a)** $Cu(OH)_2$ precipitates first
(b) 9.8×10^{-10} M
(c) 9.6×10^{-9} M

11-15. (a) 8.3×10^{-11} M (b) 1.6×10^{-11} M
(c) 1.3×10^{4} (d) 1.3×10^{4}
11-17. 1.877 g
11-19. (a) 0.0101 M; 49% (b) 7.14×10^{4}; 70%

Chapter 12

12-1. (a) A *colloidal precipitate* consists of solid particles with dimensions that are less than 10^{-4} cm. A *crystalline precipitate* consists of solid particles with dimensions that are at least 10^{-4} cm or greater. As a consequence, crystalline precipitates settle rapidly, whereas colloidal precipitates remain suspended in solution unless caused to agglomerate.
(c) *Precipitation* is the process by which a solid phase forms and is carried out of solution when the solubility product of a chemical species is exceeded. *Coprecipitation* is a process in which normally soluble compounds are carried out of solution during the formation of a precipitate.
(e) *Occlusion* is a type of coprecipitation in which a compound is trapped within a pocket formed during rapid crystal formation. *Mixed-crystal formation* is also a type of coprecipitation in which a contaminant ion replaces an ion in the crystal lattice.

12-2. (a) *Digestion* is a process in which a precipitate is heated in the presence of the solution from which it was formed (the *mother liquor*). Digestion improves the purity and filterability of the precipitate.
(c) In *reprecipitation,* the filtered solid precipitate is redissolved and then reformed from the new solution. Because the concentration of the impurity in the new solution is lower, the second precipitate contains less coprecipitated impurity.
(e) The *counter-ion layer* describes a layer of solution surrounding a charged particle that contains a sufficient excess of oppositely charged ions to balance the surface charge on the particle.
(g) *Supersaturation* describes an unstable state in which a solution contains higher solute concentration than a saturated solution. Supersaturation is relieved by precipitation of excess solute.

12-3. A *chelating agent* is an organic compound that contains two or more electron-donor groups located in such a configuration that five- or six-membered rings are formed when the donor groups complex a cation.

12-5. (a) positive charge (b) adsorbed Ag^{+} (c) NO_3^-

12-7. *Peptization* is the process by which a coagulated colloid returns to its original dispersed state as a consequence of a decrease in the electrolyte concentration of the solution in contact with the precipitate. Peptization can be avoided by washing the coagulated colloid with an electrolyte solution rather than pure water.

12-9. *Note:* $\mathcal{M}$ stands for molar or atomic mass in the equations below.

(a) $\text{mass } SO_2 = \text{mass } BaSO_4 \times \dfrac{\mathcal{M}_{SO_2}}{\mathcal{M}_{BaSO_4}}$

(c) $\text{mass In} = \text{mass } In_2O_3 \times \dfrac{2\mathcal{M}_{In}}{\mathcal{M}_{In_2O_3}}$

(e) $\text{mass CuO} = \text{mass } Cu_2(SCN)_2 \times \dfrac{2\mathcal{M}_{CuO}}{\mathcal{M}_{Cu_2(SCN)_2}}$

(i) $\text{mass } Na_2B_4O_7 \cdot 10H_2O = \text{mass } B_2O_3 \times \dfrac{\mathcal{M}_{Na_2B_4O_7 \cdot 10H_2O}}{2\mathcal{M}_{B_2O_3}}$

12-10. 60.59% KCl
12-12. 0.828 g $Cu(IO_3)_2$
12-14. 0.778 g AgI
12-18. 17.23% C
12-20. 41.46% Hg_2Cl_2
12-22. 38.74% NH_3
12-24. 0.550 g $BaSO_4$
12-26. (a) 0.239 g sample
(b) 0.494 g AgCl
(c) 0.406 g sample
12-28. 4.72% Cl^-, 27.05% I^-
12-30. 0.498
12-32. (a) 0.369 g $Ba(IO_3)_2$ (b) 0.0149 g $BaCl_2 \cdot 2H_2O$

Chapter 13

13-2. (a) The *millimole* is the amount of an elementary species, such as an atom, an ion, a molecule, or an electron. A millimole contains 10^{-3} moles or

$$6.02 \times 10^{23} \frac{\text{particles}}{\text{mole}} \times 10^{-3} \frac{\text{mole}}{\text{millimole}} = 6.02 \times 10^{20} \frac{\text{particles}}{\text{millimole}}$$

(c) The *stoichiometric ratio* is the molar ratio between two species that appear in a balanced chemical equation.

13-3. (a) The *equivalence point* in a titration is that point at which sufficient titrant has been added to be equivalent stoichiometrically to the amount of analyte initially present. The *end point* is the point at which an observable physical change signals the equivalence point.

13-4. The Fajans determination of chloride involves a direct titration, while the Volhard approach requires two standard solutions and a filtration step to eliminate AgCl.

13-5. (a) $\dfrac{1 \text{ mol } H_2NNH}{2 \text{ mol } I_2}$ (c) $\dfrac{1 \text{ mol } Na_2B_4O_7 \cdot 10H_2O}{2 \text{ mol } H^+}$

13-6. In contrast to Ag_2CO_3 and AgCN, the solubility of AgI is unaffected by the acidity. In addition, AgI is less soluble than AgSCN. The filtration step is thus unnecessary in the determination of iodide, whereas it is needed in the determination of carbonate or cyanide.

13-8. (a) Dissolve 6.37 g $AgNO_3$ in water and dilute to 500 mL.
(b) Dilute 108.3 mL of 6.00 M HCl to 2.00 L.
(c) Dissolve 6.22 g $K_4Fe(CN)_6$ in water and dilute to 750 mL.
(d) Dilute 115 mL of 0.500 M $BaCl_2$ to 600 mL with water.
(e) Dilute 25 mL of the commercial reagent to a volume of 2.0 L.
(f) Dissolve 1.67 g Na_2SO_4 in water and dilute to 9.00 L.

13-10. 8.190×10^{-2} M
13-12. 0.06581 M

13-13. 0.1799 M $HClO_4$; 0.1974 M NaOH
13-15. 0.09537 M
13-16. 116.7 mg analyte
13-18. 4.61% As_2O_3
13-19. The stoichiometry is 1:1, so only one of the seven chlorines is titrated.
13-21. **(a)** 1.19×10^{-2} M $Ba(OH)_2$ **(b)** 2.2×10^{-5} M
(c) Rel. error = −3 ppt; absolute error = -3.0×10^{-5} M
13-23. 15.60 mg saccharin/tablet
13-26. 21.5% CH_2O
13-27. 0.4348% warfarin
13-29. 10.6% Cl^-; 55.65% ClO_4^-
13-31.

Vol $AgNO_3$, mL	$[Ag^+]$	pAg
5.00	1.6×10^{-11} M	10.80
40.00	7.1×10^{-7} M	6.15
45.00	2.6×10^{-3} M	2.30

13-32. **(a)** 0.81%**(c)** 2.5%

Chapter 14

In the answers in this chapter, (Q) indicates that the answer was obtained by solving the quadratic equation.

14-1. **(a)** The initial pH of the NH_3 solution will be less than that for the solution containing NaOH. With the first addition of titrant, the pH of the NH_3 solution will decrease rapidly and then level off and become nearly constant throughout the middle part of the titration. In contrast, additions of standard acid to the NaOH solution will cause the pH to decrease gradually and nearly linearly until the equivalence point is approached. The equivalence-point pH for the NH_3 solution will be well below 7, whereas for the NaOH solution, it will be exactly 7.
(b) Beyond the equivalence point, the pH is determined by the excess titrant, and the curves are identical.
14-3. The limited sensitivity of the eye to small color differences requires that there be roughly a 10-fold excess of one form of the indicator for a color change to be seen. This corresponds to a pH range of ±1 pH unit about the p*K* of the indicator.
14-5. Because the reactions with strong acids or bases are more complete than those of their weaker counterparts. Sharper end points result.
14-6. **(a)** NaOCl **(c)** CH_3NH_2
14-7. **(a)** HIO_3 **(c)** pyruvic acid
14-9. 3.24
14-11. **(a)** 14.94
14-12. **(a)** 12.94
14-13. −0.607
14-15. 7.04(Q)
14-18. **(a)** 1.05 **(b)** 1.05 **(c)** 1.81
(d) 1.81 **(e)** 12.60
14-19. **(a)** 1.30 **(b)** 1.37
14-21. **(a)** 4.26 **(b)** 4.76 **(c)** 5.76
14-23. **(a)** 11.12 **(b)** 10.62 **(c)** 9.53 (Q)
14-25. **(a)** 12.04 (Q) **(b)** 11.48 (Q) **(c)** 9.97 (Q)
14-27. **(a)** 1.94 **(b)** 2.45 **(c)** 3.52
14-29. **(a)** 2.41 (Q) **(b)** 8.35 **(c)** 12.35
(d) 3.84
14-32. **(a)** 3.85 **(b)** 4.06 **(c)** 2.63 (Q)
(d) 2.10 (Q)

14-34. **(a)** 0.00 **(c)** −1.000
(e) −0.500 **(g)** 0.000
14-37. **(b)** −0.141
14-38.

V_{HCl}	pH	V_{HCl}	pH
0.00	13.00	49.00	11.00
10.00	12.82	50.00	7.00
25.00	12.52	51.00	3.00
40.00	12.05	55.00	2.32
45.00	11.72	60.00	2.04

14-39. The indicator should change color in the range of pH 6.5 to 9.8. Cresol purple (range 7.6 to 9.2 from Table 14-1) should be quite suitable.
14-41.

	(a)	(c)
Vol, mL	pH	pH
0.00	2.09 (Q)	3.12
5.00	2.38 (Q)	4.28
15.00	2.82 (Q)	4.86
25.00	3.17 (Q)	5.23
40.00	3.76 (Q)	5.83
45.00	4.11 (Q)	6.18
49.00	4.85 (Q)	6.92
50.00	7.92	8.96
51.00	11.00	11.00
55.00	11.68	11.68
60.00	11.96	11.96

14-42. **(a)**

Vol HCl, mL	pH	Vol HCl, mL	pH
0.00	11.12	49.00	7.55
10.00	10.20	50.00	5.27
15.00	9.61	51.00	3.00
25.00	9.24	55.00	2.32
40.00	8.64	60.00	2.04
45.00	8.29		

14-43.

	(a)	(c)
Vol, mL	pH	pH
0.00	2.80	4.26
5.00	3.65	6.57
15.00	4.23	7.15
25.00	4.60	7.52
40.00	5.20	8.12
49.00	6.29	9.21
50.00	8.65	10.11
51.00	11.00	11.00
55.00	11.68	11.68
60.00	11.96	11.96

14-44. **(a)** $\alpha_0 = 0.215$; $\alpha_1 = 0.785$
(c) $\alpha_0 = 0.769$; $\alpha_1 = 0.231$
(e) $\alpha_0 = 0.917$; $\alpha_1 = 0.083$
14-45. 6.61×10^{-2} M
14-47. Lactic, pH = 3.61, [HA] = 0.0768, $[A^-]$ = 0.0432, α_1 = 0.360
Sulfamic, [HA] = 0.095, $[A^-]$ = 0.155, α_0 = 0.380, α_1 = 0.620

Chapter 15

15-1. The HPO_4^{2-} ion is such a weak acid ($K_a = 4.5 \times 10^{-13}$) that the change in pH in the vicinity of the third equivalence point is too small to be observable.

15-2. (a) approximately neutral (c) neutral (e) basic (g) acidic

15-4. bromocresol green

15-6. (a) bromocresol green (c) cresol purple (e) bromocresol green (g) phenolphthalein

15-7. (a) 1.90 (Q) (c) 1.64 (Q) (e) 4.21 (Q)

15-8. (a) 4.63 (c) 4.28 (e) 9.80

15-9. (a) 11.99 (Q) (c) 9.70 (e) 12.58 (Q)

15-10. (a) 2.01 (Q) (b) 7.18 (c) 10.63 (d) 2.55 (Q) (e) 2.06 (Q)

15-12. (a) 1.54 (Q) (b) 1.99 (Q) (c) 12.07 (Q) (d) 12.01 (Q)

15-14. (a) $[SO_3^{2-}]/[HSO_3^-] = 15.2$
(b) $[HCit^{2-}]/[Cit^{3-}] = 2.5$
(c) $[HM^-]/[M^{2-}] = 0.498$
(d) $[HT^-]/[T^{2-}] = 0.232$

15-16. 50.2 g

15-18. (a) 2.11 (Q) (b) 7.38

15-20. Mix 442 mL of 0.300 M Na_2CO_3 with (1000−442) = 558 mL of 0.200 M HCl.

15-22. Mix 704 mL of 0.400 M HCl with 296 mL of 0.500 M Na_3AsO_4.

15-25. The volume to the first end point would have to be smaller than one-half the total volume to the second end point because in the titration from the first to second end point, both analytes are titrated, whereas to the first end point, only the H_3PO_4 is titrated.

15-27.

V_{acid}	pH	V_{acid}	pH
0.00	13.00	35.00	7.98
10.00	12.70	44.00	6.70
20.00	12.15	45.00	4.68
24.00	11.43	46.00	2.68
25.00	10.35	50.00	2.00
26.00	9.26		

15-29. (a) $\dfrac{[H_3AsO_4][HAsO_4^{2-}]}{[H_2AsO_4^-]^2} = 1.9 \times 10^{-5}$

15-30. $\dfrac{[NH_3][HOAc]}{[NH_4^+][OAc^-]} = 3.26 \times 10^{-5}$

15-31.

	pH	D	α_0	α_1	α_2	α_3
(a)	2.00	1.112×10^{-4}	0.899	0.101	3.94×10^{-5}	
	6.00	5.500×10^{-9}	1.82×10^{-4}	0.204	0.796	
	10.00	4.379×10^{-9}	2.28×10^{-12}	2.56×10^{-5}	1.000	
(c)	2.00	1.075×10^{-6}	0.931	6.93×10^{-2}	1.20×10^{-4}	4.82×10^{-9}
	6.00	1.882×10^{-14}	5.31×10^{-5}	3.96×10^{-2}	0.685	0.275
	10.00	5.182×10^{-15}	1.93×10^{-16}	1.44×10^{-9}	2.49×10^{-4}	1.000
(e)	2.00	4.000×10^{-4}	0.250	0.750	1.22×10^{-5}	
	6.00	3.486×10^{-9}	2.87×10^{-5}	0.861	0.139	
	10.00	4.863×10^{-9}	2.06×10^{-12}	6.17×10^{-4}	0.999	

Chapter 16

16-1. Carbon dioxide is not strongly bonded by water molecules, and thus is readily volatilized from aqueous media. When dissolved in water, gaseous HCl molecules are fully dissociated into H_3O^+ and Cl^-, which are nonvolatile.

16-3. Primary standard Na_2CO_3 can be obtained by heating primary standard grade $NaHCO_3$ for about an hour at 270° to 300°C. The reaction is
$2\,NaHCO_3(s) \rightarrow Na_2CO_3(s) + H_2O(g) + CO_2(g)$

16-5. For a 40-mL titration, 0.16 g $KH(IO_3)_2$ are required while 0.045 g of HBz are needed. The weighing error would be less for $KH(IO_3)_2$. A second reason is that the titration error would be less because $KH(IO_3)_2$ is a strong acid and HBz is not.

16-8. (a) Dissolve 17 g KOH and dilute to 2.0 L.
(b) Dissolve 9.5 g $Ba(OH)_2 \cdot 8H_2O$ and dilute to 2.0 L.
(c) Dilute about 120 mL of the reagent to 2.0 L.

16-10. (a) 0.1026 M (b) $s = 0.00039$, CV = 0.38% (c) spread = 0.00091

16-12. (a) 0.1388 M (b) 0.1500 M

16-14. (a) 0.08387 M (b) 0.1007 M (c) 0.1311 M

16-16. (a) 0.28 to 0.36 g Na_2CO_3
(c) 0.85 to 1.1 g HBz
(e) 0.17 to 0.22 g TRIS

16-17.

mL HCl	SD TRIS	SD Na_2CO_3	SD $Na_2B_4O_7 \cdot H_2O$
20.00	0.00004	0.00009	0.00003
30.00	0.00003	0.00006	0.00002
40.00	0.00002	0.00005	0.00001
50.00	0.00002	0.00004	0.00001

16-19. 0.1217 g H_2T/100 mL

16-21. (a) 46.25% $Na_2B_4O_7$
(b) 87.67% $Na_2B_4O_7 \cdot 10H_2O$
(c) 32.01% B_2O_3
(d) 9.94% B

16-23. 24.4% HCHO

16-25. 7.079% active ingredient

16-27. $MgCO_3$ with a molar mass of 84.31 seems a likely candidate.

16-29. 3.35×10^3 ppm

16-31. 6.333% P

16-32. 13.33% analyte

16-33. 25.98% neohetramine

16-35. 3.92% N

16-37. (a) 10.09% N (c) 47.61% $(NH_4)_2SO_4$

16-39. 15.23% $(NH_4)_2SO_4$ and 24.39% NH_4NO_3

16-40. 69.84% KOH; 21.04% K_2CO_3; 9.12% H_2O

16-42.

(a)	(b)	(c)	(d)
9.07 mL HCl	18.15 mL HCl	19.14 mL HCl	9.21 mL HCl
13.61 mL HCl	27.22 mL HCl	23.93 mL HCl	12.27 mL HCl
22.68 mL HCl	36.30 mL HCl	28.71 mL HCl	21.48 mL HCl
36.30 mL HCl	45.37 mL HCl	38.28 mL HCl	24.55 mL HCl

16-44. **(a)** 4.314 mg NaOH/mL
(b) 7.985 mg Na_2CO_3/mL and 4.358 mg $NaHCO_3$/mL
(c) 3.455 mg Na_2CO_3/mL and 4.396 mg NaOH/mL
(d) 8.215 mg Na_2CO_3/mL
(e) 13.46 mg $NaHCO_3$/mL

16-46. The equivalent weight of an acid is that weight of the pure material that contains one mole of titratable protons in a specified reaction. The equivalent weight of a base is that weight of a pure compound that consumes one mole of protons in a specified reaction.

16-48. **(a)** 0.4598 M HOAc **(b)** 2.75% HOAc

Chapter 17

17-1. **(a)** A *chelate* is a cyclic complex consisting of a metal ion and a reagent that contains two or more electron donor groups located in such a position that they can bond with the metal ion to form a heterocyclic ring structure.
(c) A *ligand* is a species that contains one or more electron pair donor groups that tend to form bonds with metal ions.
(e) A *conditional formation constant* is an equilibrium constant for the reaction between a metal ion and a complexing agent that applies only when the pH and/or the concentration of other complexing agents are carefully specified.
(g) *Water hardness* is the concentration of calcium carbonate that is equivalent to the total molar concentration of all the multivalent metal carbonates in the water.

17-3. Multidentate ligands usually form more stable complexes than unidentate ligands. They often form only a single complex with the cation, simplifying their titration curves and making end-point detection easier.

17-5. **(a)** $Zn(NH_3)_6^{2+}$ **(b)** $AgCl_2^-$ **(c)** $Cu(SO_4)_2^{2-}$
(d) $Fe(C_2O_4)_3^{3-}$ **(e)** $Fe(CN)_6^{4-}$

17-7. **(a)** $\alpha_1 = \dfrac{K_a}{[H^+] + K_a}$

(b) $\alpha_2 = \dfrac{K_{a1}K_{a2}}{[H^+]^2 + K_{a1}[H^+] + K_{a1}K_{a2}}$

(c) $\alpha_3 = \dfrac{K_{a1}K_{a2}K_{a3}}{[H^+]^3 + K_{a1}[H^+]^2 + K_{a1}K_{a2}[H^+] + K_{a1}K_{a2}K_{a3}}$

17-9. $\beta_3' = (\alpha_2)^3\,\beta_3 = \dfrac{[Fe(Ox)_3^{3-}]}{[Fe^{3+}](c_T)^3}$

17-11. $\beta_n = \dfrac{[ML_n]}{[M][L]^n}$

Taking logarithms of both sides gives $\beta_n = \log[ML_n] - \log[M] - n\log[L]$

Converting the right side to p-functions, $\log\beta_n = pM + npL - pML_n$

17-13. 0.00845 M

17-15. **(a)** 40.25 mL **(c)** 48.63 mL **(e)** 28.37 mL

17-17. 3.195%

17-19. 1.228% Tl_2SO_4

17-21. 184.0 ppm Fe^{3+} and 213.0 ppm Fe^{2+}

17-23. 55.16% Pb and 44.86% Cd

17-25. 99.7% ZnO and 0.256% Fe_2O_3

17-27. 64.68 ppm K^+

17-29. 8.518% Pb, 24.86% Zn, 64.08% Cu, and 2.54% Sn

17-31. **(a)** 2.8×10^4 **(b)** 3.0×10^6 **(c)** 4.9×10^7

17-34. **(b)** 350.0 ppm

Chapter 18

18-1. **(a)** *Oxidation* is a process in which a species loses one or more electrons.
(c) A *salt bridge* is a device that provides electrical contact but prevents mixing of dissimilar solutions in an electrochemical cell.
(e) The *Nernst equation* relates the potential to the concentrations (strictly, activities) of the participants in an electrochemical half-cell.

18-2. **(a)** The *electrode potential* is the potential of an electrochemical cell in which a standard hydrogen electrode acts as the reference electrode on the left and the half-cell of interest is on the right.
(c) The *standard electrode potential* for a half-reaction is the potential of a cell consisting of the half-reaction of interest on the right and a standard hydrogen electrode on the left. The activities of all the participants in the half-reaction are specified as having a value of unity. The standard electrode potential is always a *reduction potential.*

18-3. **(a)** *Reduction* is the process whereby a substance acquires electrons; a *reducing agent* is a supplier of electrons.
(c) The *anode* of a cell is the electrode at which oxidation occurs. The *cathode* is the electrode at which reduction occurs.
(e) The *standard electrode potential* is the potential of a cell in which the standard hydrogen electrode acts as the reference electrode on the left and all participants in the right-hand electrode process have unit activity. The *formal potential* differs in that the molar *concentrations* of all the reactants and products are unity and the concentration of other species in the solution are carefully specified.

18-4. The first standard potential is for a solution saturated with I_2, which has an $I_2(aq)$ activity significantly less than 1. The second potential is for a *hypothetical* half-cell in which the $I_2(aq)$ activity is unity. Although the second cell is hypothetical, it is nevertheless useful for calculating electrode potentials for solutions that are undersaturated in I_2.

18-5. To keep the solution saturated with $H_2(g)$. Only then is the hydrogen activity constant and the electrode potential constant and reproducible.

18-7. **(a)** $2Fe^{3+} + Sn^{2+} \rightarrow 2Fe^{2+} + Sn^{4+}$
(c) $2NO_3^- + Cu(s) + 4H^+ \rightarrow 2NO_2(g) + 2H_2O + Cu^{2+}$
(e) $Ti^{3+} + Fe(CN)_6^{3-} + H_2O \rightarrow TiO^{2+} + Fe(CN)_6^{4-} + 2H^+$

(g) $2Ag(s) + 2I^- + Sn^{4+} \rightarrow 2AgI(s) + Sn^{2+}$
(i) $5HNO_2 + 2MnO_4^- + H^+ \rightarrow 5NO_3^- + 2Mn^{2+} + 3H_2O$

18-8. (a) Oxidizing agent Fe^{3+}; $Fe^{3+} + e^- \rightleftharpoons Fe^{2+}$
Reducing agent Sn^{2+}; $Sn^{2+} \rightleftharpoons Sn^{4+} + 2e^-$
(c) Oxidizing agent NO_3^-; $NO_3^- + 2H^+ + e^- \rightleftharpoons NO_2(g) + H_2O$
Reducing agent Cu; $Cu(s) \rightleftharpoons Cu^{2+} + 2e^-$
(e) Oxidizing agent $Fe(CN)_6^{3-}$; $Fe(CN)_6^{3-} + e^- \rightleftharpoons Fe(CN)_6^{4-}$
Reducing agent Ti^{3+}; $Ti^{3+} + H_2O \rightleftharpoons TiO^{2+} + 2H^+ + e^-$
(g) Oxidizing agent Sn^{4+}; $Sn^{4+} + 2e^- \rightleftharpoons Sn^{2+}$
Reducing agent Ag; $Ag(s) + I^- \rightleftharpoons AgI(s) + e^-$
(i) Oxidizing agent MnO_4^-; $MnO_4^- + 8H^+ + 5e^- \rightleftharpoons Mn^{2+} + 4H_2O$
Reducing agent HNO_2; $HNO_2 + H_2O \rightleftharpoons NO_3^- + 3H^+ + 2e^-$

18-9. (a) $MnO_4^- + 5VO^{2+} + 11H_2O \rightarrow Mn^{2+} + 5V(OH)_4^+ + 2H^+$
(c) $Cr_2O_7^{2-} + 3U^{4+} + 2H^+ \rightarrow 2Cr^{3+} + 3UO_2^{2+} + H_2O$
(e) $IO_3^- + 5I^- + 6H^+ \rightarrow 3I_2 + H_2O$
(g) $HPO_3^{2-} + 2MnO_4^- + 3OH^- \rightarrow PO_4^{3-} + 2MnO_4^{2-} + 2H_2O$
(i) $V^{2+} + 2V(OH)_4^+ + 2H^+ \rightarrow 3VO^{2+} + 5H_2O$

18-11. (a) $AgBr(s) + e^- \rightleftharpoons Ag(s) + Br^-$ $\quad V^{2+} \rightleftharpoons V^{3+} + e^-$
$Ti^{3+} + 2e^- \rightleftharpoons Ti^+$ $\quad Fe(CN)_6^{4-} \rightleftharpoons Fe(CN)_6^{3-} + e^-$
$V^{3+} + e^- \rightleftharpoons V^{2+}$ $\quad Zn \rightleftharpoons Zn^{2+} + 2e^-$
$Fe(CN)_6^{3-} + e^- \rightleftharpoons Fe(CN)_6^{4-}$ $\quad Ag(s) + Br^- \rightleftharpoons AgBr(s) + e^-$
$S_2O_8^{2-} + 2e^- \rightleftharpoons 2SO_4^{2-}$ $\quad Ti^+ \rightleftharpoons Ti^{3+} + 2e^-$

(b), (c)	E^0
$S_2O_8^{2-} + 2e^- \rightleftharpoons 2SO_4^{2-}$	2.01
$Ti^{3+} + 2e^- \rightleftharpoons Ti^+$	1.25
$Fe(CN)_6^{3-} + e^- \rightleftharpoons Fe(CN)_6^{4-}$	0.36
$AgBr(s) + e^- \rightleftharpoons Ag(s) + Br^-$	0.073
$V^{3+} + e^- \rightleftharpoons V^{2+}$	−0.256
$Zn^{2+} + 2e^- \rightleftharpoons Zn(s)$	−0.763

18-13. (a) 0.297 V (b) 0.190 V (c) −0.152 V (d) 0.048 V (e) 0.007 V

18-16. (a) 0.78 V (b) 0.198 V (c) −0.355 V (d) 0.210 V (e) 0.177 V (f) 0.86 V

18-18. (a) −0.280 V, anode
(b) −0.090 V, anode
(c) 1.003 V, cathode
(d) 0.171 V, cathode
(e) −0.009 V, anode

18-20. 0.390 V

18-22. −0.96 V

18-24. −1.25 V

18-25. 0.13 V

Chapter 19

19-1. The electrode potential of a system is the electrode potential of all half-cell processes at equilibrium in the system.

19-4. For points before equivalence, potential data are computed from the analyte standard potential and the analytical concentrations of the analyte and its reaction product(s). Post–equivalence-point data are based on the standard potential for the titrant and its analytical concentrations. The equivalence-point potential is computed from the two standard potentials and the stoichiometric relation between the analyte and titrant.

19-6. An asymmetric titration curve will be encountered whenever the titrant and the analyte react in a ratio that is not 1:1.

19-8. (a) 0.452 V, oxidation on the left, reduction on the right.
(b) 0.031 V, oxidation on the left, reduction on the right.
(c) 0.414 V, oxidation on the left, reduction on the right.
(d) −0.401 V, reduction on the left, oxidation on the right.
(e) −0.208 V, reduction on the left, oxidation on the right.
(f) 0.724 V, oxidation on the left, reduction on the right.

19-9. (a) 0.631 V (c) 0.331 V

19-11. (a) 2.2×10^{17} (c) 3×10^{22} (e) 9×10^{37} (g) 2.4×10^{10}

19-14. (a) phenosafranine
(c) indigo tetrasulfonate or methylene blue
(e) erioglaucin A (g) none

19-15.

	E, V		
Vol, mL	**(a)**	**(c)**	**(e)**
10.00	−0.292	0.32	0.316
25.00	−0.256	0.36	0.334
49.00	−0.156	0.46	0.384
49.90	−0.097	0.52	0.414
50.00	0.017	0.95	1.17
50.10	0.074	1.17	1.48
51.00	0.104	1.20	1.49
60.00	0.133	1.23	1.50

Chapter 20

20-1. (a) $2Mn^{2+} + 5S_2O_8^{2-} + 8H_2O \rightarrow 10SO_4^{2-} + 2MnO_4^- + 16H^+$
(b) $NaBiO_3(s) + 2Ce^{3+} + 4H^+ \rightarrow BiO^+ + 2Ce^{4+} + 2H_2O + Na^+$
(c) $H_2O_2 + U^{4+} \rightarrow UO_2^{2+} + 2H^+$
(d) $V(OH)_4^+ + Ag(s) + Cl^- + 2H^+ \rightarrow VO^{2+} + AgCl(s) + 3H_2O$
(e) $2MnO_4^- + 5H_2O_2 + 6H^+ \rightarrow 5O_2 + 2Mn^{2+} + 8H_2O$
(f) $ClO_3^- + 6I^- + 6H^+ \rightarrow 3I_2 + Cl^- + 3H_2O$

20-3. Only in the presence of Cl^- is Ag a sufficiently good reducing agent to be useful for prereductions. With Cl^- present, the half-reaction in a Walden reductor is

$$Ag(s) + Cl^- \rightarrow AgCl(s) + e^-$$

The excess HCl increases the tendency of this reaction to occur by the common ion effect.

20-5. $UO_2^{2+} + 2Ag(s) + 4H^+ + 2Cl^- \rightleftharpoons U^{4+} + 2AgCl(s) + H_2O$

20-7. Standard solutions of reductants find somewhat limited use because of their susceptibility to air oxidation.

20-8. Standard $KMnO_4$ solutions are seldom used to titrate solutions containing HCl because of the tendency of MnO_4^- to oxidize Cl^- to Cl_2, thus causing overconsumption of Mn_4^-.

20-10. $2MnO_4^- + 3Mn^{2+} + 2H_2O \rightarrow 5MnO_2(s) + 4H^+$

20-13. $4MnO_4^- + 2H_2O \rightarrow \underset{\text{brown}}{4MnO_2(s)} + 3O_2 + 4OH^-$

20.15. Iodine is not sufficiently soluble in water to produce a useful standard reagent. It is quite soluble in solutions containing excess I^- because of formation of triiodide.

20-17. $S_2O_3^{2-} + H^+ \rightarrow HSO_3^- + S(s)$

20-19. $BrO_3^{2-} + \underset{\text{excess}}{6I^-} + 6H^+ \rightarrow Br^- + 3I_2 + 3H_2O$

$I_2 + 2S_2O_3^{2-} \rightarrow 2I^- + S_4O_6^{2-}$

20-21. $2I_2 + N_2H_4 \rightarrow N_2 + 4H^+ + 4I^-$

20-23. **(a)** 0.1142 M Ce^{4+}
(c) 0.02284 M MnO_4^-
(e) 0.02855 M IO_3^-

20-24. Dissolve 3.677 g $K_2Cr_2O_7$ in water and dilute to 500.0 mL.

20-26. Dissolve about 16 g of $KMnO_4$ in water and dilute to 2.0 L.

20-28. 0.01636 M $KMnO_4$

20-30. 0.0622 M $Na_2S_2O_3$

20-32. **(a)** 16.03% Sb
(b) 22.37% Sb_2S_3

20-34. 9.38% thiourea

20-35. **(a)** 32.08% Fe
(b) 45.86% Fe_2O_3

20-37. 0.03867 M H_2NOH

20-39. 50.78% $KClO_3$

20-41. 0.651% As_2O_3

20-43. 4.33% C_2H_5SH

20-45. 2.524% KI

20-46. 69.07% Fe and 21.07% Cr

20-48. 0.5622 g Tl

20-49. 10.4 ppm SO_2

20-51. 19.5 ppm H_2S

20-53. 0.0412 mg O_2/mL sample

Chapter 21

21-1. **(a)** An *indicator electrode* is an electrode used in potentiometry that responds to variations in the activity of an analyte ion or molecule.
(c) An *electrode of the first kind* is a metal electrode that responds to the activity of its cation in solution.

21-2. **(a)** A *liquid junction potential* is the potential that develops across the interface between two solutions having different electrolyte compositions.

21-3. **(a)** An *electrode of the first kind* for Hg(II) would take the form

$\| Hg^{2+}(x\ M) | Hg$

$$E_{Hg} = E^0_{Hg} - \frac{0.0592}{2}\log\frac{1}{[Hg^{2+}]}$$

$$= E^0_{Hg} + \frac{0.0592}{2}pHg$$

(b) An *electrode of the second kind* for EDTA would take the form

$\| HgY^{2-}(yM), Y^{4-}(xM) | Hg$

$$E_{Hg} = K - \frac{0.0592}{2}\log[Y^{4-}] = K + \frac{0.0592}{2}pY$$

$$\text{where } K = E^0 - \frac{0.0592}{2}\log\frac{1}{a_{HgY^{2-}}}$$

$$\approx 0.21 - \frac{0.0592}{2}\log\frac{1}{[HgY^{2-}]}$$

21-5. The potential arises from the difference in positions of dissociation equilibria on each of the two surfaces. These equilibria are described by

$$\underset{\text{membrane}}{H^+Gl^-} \rightleftharpoons \underset{\text{soln}}{H^+} + \underset{\text{membrane}}{Gl^-}$$

The surface exposed to the solution having the higher H^+ concentration becomes positive with respect to the other surface. This charge difference, or potential, serves as the analytical parameter when the pH of the solution on one side of the membrane is held constant.

21-7. Uncertainties include (1) the acid error in highly acidic solutions, (2) the alkaline error in strongly basic solutions, (3) the error that arises when the ionic strength of the calibration standards differs from that of the analyte solution, (4) uncertainties in the pH of the standard buffers, (5) nonreproducible junction potentials with solutions of low ionic strength, and (6) dehydration of the working surface.

21-9. The *alkaline error* arises when a glass electrode is employed to measure the pH of solutions having pH values in the 10 to 12 range or greater. In the presence of alkali ions, the glass surface becomes responsive to not only hydrogen ions but also alkali metal ions. Measured pH values are low as a result.

21-11. **(a)** The *asymmetry potential* in a membrane arises from differences in the composition or structure of the inner and outer surfaces. These differences may arise from contamination of one of the surfaces, wear and abrasion, and strains set up during manufacture.
(c) The *junction potential* in a glass/calomel electrode system develops at the interface between the saturated KCl solution in the salt bridge and the sample solution. It is caused by charge separation created by the differences in the rates at which ions migrate across the interface.

21-12. The direct potentiometric determination of pH provides a measure of the equilibrium concentration of hydronium ions in the sample. A potentiometric titration provides information on the amount of reactive protons, both ionized and nonionized in the sample.

21-15. **(a)** 0.354 V
(b) SCE $\| IO_3^-(xM), AgIO_3(\text{sat'd}) | Ag$
(c) $pIO_3 = \dfrac{E_{cell} - 0.110}{0.0592}$
(d) $pIO_3 = 3.11$

21-17. **(b)** SCE $\| I^-(xM), AgI(\text{sat'd}) | Ag$
(d) SCE $\| PO_4^{3-}(xM), Ag_3PO_4(\text{sat'd}) | Ag$

21-19. **(a)** 5.86
(c) 4.09

21-20. 4.53

21-21. **(a)** 12.46, 3.48×10^{-13}
(b) 4.15, 7.14×10^{-5}
(c) For part (a), pH range, 12.43 to 12.49; a_{H^+} range, 3.22 to 3.76×10^{-13}.
For part (b), pH range, 4.11 to 4.18; a_{H^+} range, 6.61 to 7.72×10^{-5}.

21-22. $\mathcal{M}_{HA} = 250$ g/mol

21-24.

Vol Ce(IV), mL	*E* vs. SCE, V	Vol Ce(IV), mL	*E* vs. SCE, V
5.00	0.58	50.00	0.80
10.00	0.59	50.01	0.98
15.00	0.60	50.05	1.02
25.00	0.61	50.10	1.04
40.00	0.63	50.20	1.05
49.00	0.66	50.30	1.06
49.50	0.67	50.40	1.07
49.60	0.67	50.50	1.08
49.70	0.67	51.00	1.10
49.80	0.68	60.00	1.15
49.90	0.69	75.00	1.18
49.95	0.70	90.00	1.19
49.99	0.72		

21-26. 3.2×10^{-4} M

Chapter 22

22-1. **(a)** *Concentration polarization* is a condition in which the current in an electrochemical cell is limited by the rate at which reactants are brought to or removed from the surface of one or both electrodes. *Kinetic polarization* is a condition in which the current is limited by the rate at which electrons are transferred between the electrode surfaces and the reactants in solution. For either type, the current is no longer linearly related to the cell potential.
(c) Both the *coulomb* and the *faraday* are units describing the quantity of charge. The coulomb is the quantity transported by one ampere of current in one second; the Faraday is equal to 96,485 coulombs, or one mole of electrons.

22-2. **(a)** *Current density* is the current at an electrode divided by the surface area of that electrode. It usually has the units of amperes per square centimeter.
(c) A *coulometric titration* is an electroanalytical method in which a constant current of known magnitude generates a reagent that reacts with the analyte. The time required to generate enough reagent to complete the reaction is measured.
(e) *Current efficiency* is a measure of agreement between the number of faradays of current and the number of moles of reactant oxidized or reduced at a working electrode.

22-3. *Diffusion,* which arises from concentration differences between the electrode surface and the bulk of solution. *Migration,* which results from electrostatic attract or repulsion. *Convection,* which results from stirring, vibration, or temperature differences.

22-5. Both kinetic and concentration polarization cause the potential of an electrode to be more negative than the thermodynamic value. Concentration polarization results from the slow rate at which reactants or products are transported to or away from the electrode surfaces. Kinetic polarization arises from the slow rate of the electrochemical reactions at the electrode surfaces.

22-8. Kinetic polarization is often encountered when the product of a reaction is a gas, particularly when the electrode is a soft metal such as mercury, zinc, or copper. It is likely to occur at low temperatures and high current densities.

22-10. Temperature, current density, complexation of the analyte, and codeposition of a gas influence the physical properties of an electrolytic deposit.

22-12. **(a)** A *galvanostat* is an instrument that provides a constant current to an electrolysis cell.
(b) A *potentiostat* controls the applied potential to maintain a constant potential between the working electrode and a reference electrode.

22-13. In controlled-potential coulometry, the working electrode potential is maintained at a constant value with respect to a reference electrode. In constant-current coulometry, the cell is operated so that the current is maintained at a constant value.

22-14. The species produced at the counter electrode are potential interferences by reacting with the products at the working electrode. Isolation of one from the other is ordinarily required.

22-16. **(b)** 6.2×10^{16} cations

22-17. **(a)** −0.738 V
(c) −0.337 V

22-18. −0.913 V

22-20. **(a)** −0.676 V
(b) $IR = -0.36$ V
(c) −0.154 V
(d) −1.67 V

22-22. **(a)** −0.94 V
(b) $IR = -0.35$ V
(c) −2.09 V
(d) −2.37 V

22-24. **(a)** $[BiO^+] = 5 \times 10^{-28}$ M
(b) 0.103 V

22-26. **(a)** Separation is impossible
(b) Separation is feasible
(c) Separation is feasible with a galvanic cell.

22-28. **(a)** 16.0 min
(b) 5.3 min

22-30. 196.0 g/eq

22-32. 68.3 ppm $CaCO_3$

22-34. 5.14% nitrobenzene

22-35. 23.0 ppm phenol

22-39. 50.9 μg aniline

22-40. 2.73×10^{-4} g quinine

Chapter 23

23-1. **(a)** *Voltammetry* is an analytical technique that is based on measuring the current that develops at a small electrode as the applied potential is varied. *Polarography* is a particular type of voltammetry in which a dropping mercury electrode is used.
(c) As shown in Figures 23-18 and 23-21, *differential pulse polarography* and *square wave polarography* differ in the type of pulse sequence used.
(e) A *residual current* in voltammetry is comprised of a nonfaradaic charging current and a current due to impurities. A *limiting current* is a constant faradaic current that is limited by the rate at which a reactant is brought to the electrode surface.

(g) *Turbulent flow* is a type of liquid flow that has no regular pattern. *Laminar flow* is a type of liquid flow in which layers of liquid slide by one another. It is characterized by a parabolic flow profile.

23-3. To minimize the *IR* drop that can distort voltammograms.

23-5. Most organic electrode processes involve hydrogen ions. Unless buffered solutions are used, marked pH changes can occur at the electrode surface as the reaction proceeds.

23-7. A plot of E_{appl} versus $\log \frac{i}{i_l - i}$ should yield a straight line having a slope of $-0.0592/n$. Thus n is readily obtained from the slope.

23-8. (a) -0.059 V

23-10. (a) 0.67%
(b) 0.13%
(c) 0.40%

23-11. (a) 0.369 mg/mL
(c) 0.144 mg/mL

23-13. At about + 0.1 V, the anodic reaction $2Hg + Br^- \rightarrow Hg_2Br_2(s) + 2e^-$ begins. The limiting current occurs at more positive potentials than about + 0.17 V. This wave is useful for the determination of Br^-, since the diffusion current should be directly proportional to $[Br^-]$.

23-15. Concentration of Pb =1.42 mM; standard deviation = 0.015 mM.

23-18. Because the analyte can be removed from a relatively large volume of solution and concentrated in a small volume. After concentration, the potential is reversed and all the analyte that has been deposited can be rapidly oxidized or reduced, producing a large current.

Chapter 24

24-1. A solution of $Cu(NH_3)_4^{2+}$ is blue because this ion absorbs yellow radiation and transmits blue radiation unchanged.

24-3. Failure to use monochromatic radiation, existence of stray radiation, experimental uncertainties in measurement of low absorbances, molecular interaction at high absorbances, concentration-dependent association or dissociation reactions.

24-6. (a) 1.01×10^{18} Hz
(c) 4.809×10^{14} Hz
(e) 8.00×10^{13} Hz

24-7. (a) 253.0 cm
(c) 222 cm

24-9. 5.41×10^4 cm^{-1} to 3.33×10^3 cm^{-1}; 1.62×10^{15} Hz to 1.00×10^{14} Hz

24-10. $\nu = 1.28 \times 10^{18}$ Hz; $E = 8.46 \times 10^{-16}$ J

24-12. (a) 464 nm

24-13. (a) cm^{-1} ppm^{-1}
(c) cm^{-1} %$^{-1}$

24-14. (a) 92.3%
(c) 49.0%
(e) 32.1%

24-15. (a) 0.644 (c) 0.502
(e) 0.987

24-18. (a) $\%T = 67.3$, $c = 4.07 \times 10^{-5}$ M, $c_{ppm} = 8.13$ ppm, $a = 2.11 \times 10^{-2}$ cm^{-1} ppm^{-1}
(c) $\%T = 30.2$, $c = 6.54 \times 10^{-5}$ M, $c_{ppm} = 13.1$ ppm, $a = 3.97 \times 10^{-2}$ cm^{-1} ppm^{-1}
(e) $A = 0.638$, $\%T = 23.0$, $c_{ppm} = 342$ ppm, $a = 1.87 \times 10^{-2}$ cm^{-1} ppm^{-1}
(g) $\%T = 15.9$, $c = 1.68 \times 10^{-4}$ M, $\varepsilon = 3.17 \times 10^3$ L mol^{-1} cm^{-1}, $a = 1.58 \times 10^{-2}$ cm^{-1} ppm^{-1}
(i) $c = 2.62 \times 10^{-5}$ M, $A = 1.281$, $b = 5.00$ cm, $a = 4.89 \times 10^{-2}$ cm^{-1} ppm^{-1}

24-19. 2.33×10^3 L mol^{-1} cm^{-1}

24-21. (a) 0.262
(b) 0.525
(c) 54.6% and 29.9%
(d) 0.564

24-22. 0.186

24-24. (a) 0.474
(b) 33.6%
(c) 1.35×10^{-5}
(d) 2.00 cm

24-25. A′ = 1.81; % error = −1.37%

24-27. (a) Slope = 0.206, intercept = 0.024, standard error in $y = 0.002366$, unknown = 2.50 ppm P, standard deviation = 0.013 ppm P
(b) 0.135 g
(c) 3.88 mM

Chapter 25

25-1. (a) *Filters* provide low-resolution wavelength selection suitable for quantitative work. *Monochromators* produce high resolution for qualitative and quantitative work. With monochromators, the wavelength can be varied continuously, whereas this is not possible with filters.
(c) *Phototubes* consist of a single photoemissive surface (cathode) and an anode in an evacuated envelope. They exhibit low dark current but have no inherent amplification. *Photomultipliers* have built-in gains and thus have very high sensitivities. They suffer from somewhat larger dark currents.

25-3. Photons in the infrared region of the spectrum do not have sufficient energy to cause photoemission from the cathode of a photomultiplier.

25-5. *Tungsten/halogen lamps* contain a small amount of iodine in the evacuated quartz envelope that contains the tungsten filament. The iodine prolongs the life of the lamp and permits it to operate at a higher temperature. The iodine combines with gaseous tungsten that sublimes from the filament and causes the metal to be redeposited, thus adding to the life of the lamp.

25-7. (a) 0.73 μm (730 nm)
(c) 1.45 μm (1450 nm).

25-9. (a) 1010 nm for 2870 K and 967 nm for 3000 K.
(b) 386 W/cm^2 for 2870 K and 461 W/cm^2 for 3000 K.

25-11. The 0% transmittance is measured with no light reaching the detector and compensates for any dark currents. The 100% transmittance adjustment is made with a blank in the light path and compensates for any absorption or reflection losses in the cell and optics.

25-13. Fourier transform IR spectrometers have the advantages over dispersive instruments of higher speed and sensitivity, better light-gathering power, more accurate and precise wavelength settings, simpler mechanical design, and elimination of stray light and IR emission.

25-14. **(a)** $\%T = 23.4$, $A = 0.632$
(c) $T = 0.055$

25-15. **(b)** $A = 0.266$
(d) 0.294

25-17. A *photon detector* produces a current or voltage as a result of the emission of electrons from a photosensitive surface when struck by photons. A *heat detector* consists of a darkened surface to absorb infrared radiation and produce a temperature increase. A thermal transducer produces an electrical signal whose magnitude is related to the temperature and thus the intensity of the infrared radiation.

25-19. An *absorption spectrometer* requires a separate radiation source and a sample compartment that holds containers for the sample and blank. With an *emission spectrometer,* the sample is introduced directly into a hot plasma or flame where excitation and emission occur.

25-21. **(a)** The *dark current* is the small current that develops in a radiation transducer in the absence of radiation.
(c) *Scattered radiation* in a monochromator is unwanted radiation that reaches the exit slit as a result of reflection and scattering. Its wavelength usually differs from that of the radiation reaching the slit from the dispersing element.
(e) The *majority carrier* in a semiconductor is the mobile charge carrier in either *n*-type or *p*-type materials. For *n*-type, the majority carrier is the electron, while in *p*-type, the majority carrier is a positively charged hole.

25-22. **(a)** $t = 1.69\ \mu m$
(b) $\lambda_2 = 2.27\ \mu m$, $\lambda_3 = 1.51\ \mu m$

Chapter 26

26-1. **(a)** *Spectrophotometers* use a grating or a prism to provide narrow bands of radiation, while *photometers* use filters for this purpose. The advantages of spectrophotometers are greater versatility and the ability to obtain entire spectra. The advantages of photometers are simplicity, ruggedness, higher light throughput and low cost.
(c) *Diode-array spectrophotometers* detect the entire spectral range essentially simultaneously and can produce a spectrum in less than a second. *Conventional spectrophotometers* require several minutes to scan the spectrum. Accordingly, diode-array instruments can be used to monitor processes that occur on fast time scales. Their resolution is usually lower than that of a conventional spectrophotometer.

26-3. Electrolyte concentration, pH, temperature.

26-5. $c \geq 1.1 \times 10^{-5}$ M and $\leq 9.7 \times 10^{-5}$ M

26-7. $c \geq 1.4 \times 10^{-4}$ M and $\leq 2.8 \times 10^{-3}$ M

26-9. **(a)** $T = 0.398$; $A = 0.400$
(b) $T = 0.158$

26-10. **(b)** $A = 0.471$
(d) $T = 0.114$

26-13. The absorbance should decrease approximately linearly with titrant volume until the end point. After the end point, the absorbance becomes independent of titrant volume.

26-16. 0.200 ppm Fe

26-18. 129 ppm Co and 132 ppm Ni

26-20. **(a)** $A = 0.492$
(c) $A = 0.190$

26-21. **(a)** $A = 0.301$
(b) $A = 0.413$
(c) $A = 0.491$

26-22. For solution A, pH = 5.60; for solution C, pH = 4.80

26-25. **(a)** $c_P = 2.08 \times 10^{-4}$ M, $c_Q = 4.91 \times 10^{-5}$ M
(c) $c_P = 8.37 \times 10^{-5}$ M, $c_Q = 6.10 \times 10^{-5}$ M
(e) $c_P = 2.11 \times 10^{-4}$ M, $c_Q = 9.65 \times 10^{-5}$ M

26-26. **(b)** $A = 0.03949\ c_{Fe} - 0.001008$
(c) $s_m = 1.1 \times 10^{-4}$ and $s_b = 2.7 \times 10^{-3}$

26-27.

	c_{Fe}, ppm	S_c, rel % 1 Result	S_c, rel % 3 Results
(a)	3.65	2.8	2.1
(c)	1.75	6.1	4.6
(e)	38.3	0.27	0.20

26-28. **(a)** 1:1 complex
(b) $\varepsilon = 1.35 \times 10^4$ L mol^{-1} cm^{-1}

26-30. **(a)** 1:1 complex

26-32. **(b)** slope = 0.05406, intercept = 0.57036
(d) $c_x = 4.22 \times 10^{-5}$ M

Chapter 27

27-1. **(a)** *Resonance fluorescence* is observed when excited atoms emit radiation of the same wavelength as that used to excite them.
(c) *Internal conversion* is the nonradiative relaxation of a molecule from a low-energy vibrational level of an excited electronic state to a high-energy vibrational level of a lower electronic state.

27-3. **(a)** Fluorescein because of its greater structural rigidity due to the bridging —O— groups.

27-5. Organic compounds containing aromatic rings often exhibit fluorescence. Rigid molecules or multiple ring systems tend to have large quantum yields of fluorescence, while flexible molecules generally have lower quantum yields.

27-8. Most fluorescence instruments are double beam to compensate for fluctuations in the analytical signal due to variations in source intensity.

27-10. **(b)** $y = 22.3\,x + 0.0004$ or $I_{rel} = 22.3c_{NADH} + 0.0004$
(d) 0.540 μM NADH
(e) 1.5%

27-12. 533 mg Q

Chapter 28

28-1. In *atomic emission spectroscopy,* the radiation source is the sample itself. The energy for excitation of analyte atoms is supplied by a plasma, a flame, an oven, or an electric arc or spark. The signal is the measured intensity of the source at the wavelength of interest. In *atomic absorption spectroscopy,* the radiation source is usually a line source such as a hollow cathode lamp, and the signal is the absorbance. The latter is calculated from the radiant power of the source and the resulting power after the radiation has passed through the atomized sample.

28-2 **(a)** *Atomization* is a process in which a sample, often in solution, is volatilized and decomposed to form an atomic vapor.

(c) *Doppler broadening* is an increase in the width of the atomic lines caused by the Doppler effect in which atoms moving toward a detector absorb or emit wavelengths that are slightly shorter than those absorbed or emitted by atoms moving at right angles to the detector. The reverse effect is observed for atoms moving away from the detector.

(e) A *plasma* is a conducting gas that contains a large concentration of ions and/or electrons.

(g) *Sputtering* is a process in which atoms of an element are dislodged from the surface of a cathode by bombardment by a stream of inert gas ions that have been accelerated toward the cathode by a high electric potential.

(i) A *spectral interference* in atomic spectroscopy occurs when a spectral line of an element in the sample matrix overlaps that of the analyte.

(k) A *radiation buffer* is a substance that is added in large excess to both standards and samples in atomic spectroscopy to prevent the presence of that substance in the sample matrix from having an appreciable effect on the results.

(m) A *quadrupole mass filter* consists of 4 cylindrical rods that allow only ions of a certain mass-to-charge (m/z) ratio to pass. With proper adjustment of the voltages applied to the rods, a stable path is created for ions of a certain m/z ratio to pass through the analyzer to the detector.

28-3. In atomic emission spectroscopy, the analytical signal is produced by the relatively small number of *excited* atoms or ions, whereas in atomic absorption the signal results from absorption by the much larger number of *unexcited species.* Any small change in flame conditions dramatically influences the number of *excited species,* whereas such changes have a much smaller effect on the number of *unexcited species.*

28-5. The resolution and selectivity in ICP emission comes primarily from the monochromator. As a result, a high-resolution monochromator can isolate the analyte spectral line from lines of concomitants and background emission. It can thus reduce spectral interferences. In atomic absorption spectrometry, the resolution comes primarily from the very narrow hollow cathode lamp emission. The monochromator must only isolate the emission line of the analyte element from lines of impurities and the fill gas, and from background emission from the atomizer. A much lower resolution is needed for this purpose.

28-7. **(a)** Sulfate ion forms complexes with Fe(III) that are not readily volatilized and converted to free atoms. Thus, the concentration of iron atoms is lower in the presence of sulfate.

(b) A releasing agent that forms more stable complexes with sulfate than iron forms could be added. A protective agent, such as EDTA, that forms a stable but volatile complex with Fe(III) could be introduced. A higher temperature flame could be used.

28-9. The temperatures are high, which favors the formation of atoms and ions. Sample residence times are long so that desolvation and vaporization are essentially complete. The atoms and ions are formed in a nearly chemically inert environment. The high and relatively constant electron concentration leads to fewer ionization interferences.

28-11. The higher resolution of the double focusing spectrometer allows the ions of interest to be better separated from background ions than with a relative low resolution quadrupole spectrometer. The higher signal-to-background ratio of the double focusing instrument leads to lower detection limits than with the quadrupole instrument.

28-13. Deviations from linearity at low concentrations are often the result of significant ionization of the analyte. When a high concentration of an easily ionized metal salt is added, the ionization of the analyte is suppressed because of the electrons produced by ionization of the metal.

28-15. 0.504 ppm Pb.

28-17. **(b)** $A = kc_s\frac{V_s}{V_t} + kc_x\frac{V_x}{V_t}$, where c_s and c_x are the concentrations of Cu in the standard and unknown, respectively, and V_s and V_x are the volumes of standard and unknown. The total volume $V_t = V_s + V_x$

(c) m = slope = kc_s/V_t; b = intercept = kc_xV_x/V_t

(e) $m = 8.81 \times 10^{-3}$; $b = 0.202$

(g) 28.0 ppm Cu

Chapter 29

29-1. **(a)** The *order of a reaction* is the numerical sum of the exponents of the concentration terms in the rate law for the reaction.

(c) *Enzymes* are high-molecular-weight organic molecules that catalyze reactions of biological importance.

(e) The *Michaelis constant* is an equilibrium-like constant for the dissociation of the enzyme-substrate complex. It is defined by the equation $K_m = (k_{-1} + k_2)/k_1$, where k_1 and k_{-1} are the rate constants for the forward and reverse reactions in the formation of the enzyme-substrate complex. The term k_2 is the rate constant for the dissociation of the complex to give products.

(g) *Integral methods* use integrated forms of the rate equations to calculate concentrations from kinetic data.

29-3. *Pseudo-first order* conditions are used in kinetic methods because under these conditions the reaction rate is directly proportional to the concentration of the analyte.

29-5. $t_{1/2} = \ln 2/k = 0.693/k$

29-6. **(a)** $t = 2.85$ s

(c) 2.112×10^3 s

(e) 7.19×10^8 s

29-7. **(a)** 58.3 s^{-1}

(c) 0.583 s^{-1}

(e) 2.18×10^4 s^{-1}

29-10. **(a)** 0.2% **(c)** 0.02% **(e)** 1.0%

(g) 0.05% **(i)** 6.7% **(k)** 0.64%

29-12. **(a)** Plot 1/Rate versus 1/[S] for known [S] to give a linear calibration curve. Measure rate for unknown [S], calculate 1/Rate and 1[S]$_{unknown}$ from the working curve and find [S]$_{unknown}$.

(b) The intercept of the calibration curve is $1/v_{max}$ and the slope is K_m/v_{max}. Use the intercept to calculate K_m = slope/intercept, and v_{max} = 1/intercept.

29-13. 6.2 ppm

29-15. 5.5×10^{-2} M

29-17. **(a)** Approximately 2% completion

(b) approximately 12%

Chapter 30

30-1. A *masking agent* is a complexing agent that reacts selectively with one or more components of a solution to prevent them from interfering in an analysis.

30-3 Precipitation, extraction, distillation, ion exchange.

30-5. **(a)** *Elution* is a process in which species are washed through a chromatographic column by additions of fresh mobile phase.

(c) The *stationary phase* in chromatography is a solid or liquid phase that is fixed in place. The mobile phase then passes over or through the stationary phase.

(e) The *retention time* for an analyte is the time interval between its injection onto a column and its appearance at the detector at the other end of the column.

(g) The *selectivity factor* α of a column toward two species is given by the equation $\alpha = K_B/K_A$, where K_B is the distribution constant for the more strongly retained species B and K_A is the constant for the less strongly held or more rapidly eluting species A.

30-7. In gas-liquid chromatography, the mobile phase is a gas, whereas in liquid-liquid chromatography, it is a liquid.

30-9. Determine the retention time t_R for a solute and the width of the solute peak at its base, W. The number of plates N is then $N = 16(t_R/W)^2$.

30-11. **(a)** 1.73×10^{-2} M **(b)** 6.40×10^{-3} M
(c) 2.06×10^{-3} M **(d)** 6.89×10^{-4} M

30-13 **(a)** 75 mL **(b)** 40 mL **(c)** 22 mL

30-15. **(a)** $K = 18.0$ **(b)** $K = 7.56$

30-16. **(a)** $K = 91.9$

30-17. **(a)** $K = 1.53$
(b) $[HA]_{aq} = 0.0147$ M; $[A^-]_{aq} = 0.0378$ M
(c) $K_a = 9.7 \times 10^{-2}$

30-19. **(a)** 12.4 meq cation/L sample
(b) 6.19×10^2 mg $CaCO_3$/L

30-21. Dissolve 17.53 g of NaCl in about 100 mL water and pass through the column packed with a cation exchange resin in its acid form. Wash with several hundred milliliters of water, collecting the liquid from the original solution and the washings in a 2.00-L volumetric flask. Dilute to the mark and mix well.

30-23. 2037 cm/s

30-25. **(a)** A 2775; B 2474; C 2363; D 2523
(b) $N = 2.5 \times 10^3$ and $s = 0.2 \times 10^3$
(c) $H = 0.0097$

30-27. **(a)** $R_s = 0.72$ **(b)** $\alpha_{C,B} = 1.1$
(c) $L = 108$ cm **(d)** $(t_R)_2 = 62$ min

30-29. **(a)** $\overline{N} = 2.7 \times 10^3$ plates **(b)** $s = 140$ plates
(c) $\overline{H} = 0.015$ cm/plate

30-31. **(a)** $N_2 = 4.7 \times 10^3$ plates **(b)** $L = 69$ cm
(c) $(t_R)_2 = 19$ min

30-33. **(a)** $k_M = 2.36$; $k_N = 2.43$ **(b)** $\alpha = 1.03$
(c) $N = 8.3 \times 10^4$ **(d)** $L = 157$ cm
(e) $(t_R)_N = 83$ min

Chapter 31

31-1. In *gas-liquid chromatography,* the stationary phase is a liquid that is immobilized on a solid. Retention of sample constituents involves equilibria between a gaseous and a liquid phase. In *gas-solid chromatography,* the stationary phase is a solid surface that retains analytes by physical adsorption. Here separation involves adsorption equilibria.

31-3. Gas-solid chromatography has limited application because active or polar compounds are retained more or less permanently on the packings. In addition, severe tailing is often observed owing to the nonlinear character of the physical adsorption process.

31-5. A *chromatogram* is a plot of detector response, which is proportional to analyte concentration or mass, as a function of time.

31-7. In *open tubular columns,* the stationary phase is held on the inner surface of a capillary, whereas in *packed columns,* the stationary phase is supported on particles that are contained in a glass or metal tube. Open tubular columns contain an enormous number of plates that permit rapid separations of closely related species. They suffer from small sample capacities.

31-9. The typical column packing is made of diatomaceous earth particles having diameters from 250 to 170 μm or 170 to 149 μm.

31-11. **(a)** Advantages of thermal conductivity: general applicability, large linear range, simplicity, nondestructive.
Disadvantage: low sensitivity.

(b) Advantages of flame ionization: high sensitivity, large linear range, low noise, ruggedness, ease of use, and response that is largely independent of flow rate.
Disadvantage: destructive.

(c) Advantages of electron capture: high sensitivity, selectivity toward halogen-containing compounds and several others, nondestructive.
Disadvantage: small linear range.

(d) Advantages of thermionic detector: high sensitivity for compounds containing nitrogen and phosphorus, good linear range.
Disadvantages: destructive, not applicable for many analytes.

(e) Advantages of photoionization: versatility, nondestructive, large linear range.
Disadvantage: not widely available, expensive.

31-13. Megabore columns are open tubular columns that have a greater inside diameter (530 μm) than typical open tubular columns (150 to 320 μm).

31-15. The stationary phase liquid should have low volatility, good thermal stability, chemical inertness and solvent characteristics that provide suitable retention factor and selectivity for the separation.

31-17. Film thickness influences the rate at which analytes are carried through the column, with the rate increasing as the thickness is decreased. Less band broadening is encountered with thin films.

31-19. **(a)** Band broadening arises from very high or very low flow rates, large particles making up packing, thick layers of stationary phase, low temperatures, and slow injection rates.

(b) Band separation is enhanced by maintaining conditions so that k lies in the range of 1 to 10, using small particles for packing, limiting the amount of stationary phase so that particle coatings are thin, and injecting the sample rapidly.

31-21. A = 21.1%, B = 13.1%, C = 36.4%, D = 18.8%, and E = 10.7%.

Chapter 32

32-1. (a) Substances that are somewhat volatile and are thermally stable.
(c) Substances that are ionic.
(e) High-molecular-weight compounds that are soluble in nonpolar solvents.
(g) Low-molecular-weight gases.

32-2. (a) In an *isocratic elution,* the solvent composition is held constant throughout the elution.
(c) In a *stop-flow injection,* the flow of solvent is stopped, a fitting at the head of the column is removed, and the sample is injected directly onto the head of the column. The fitting is then replaced and pumping is resumed.
(e) In a *normal-phase packing,* the stationary phase is quite polar and the mobile phase is relatively nonpolar.
(g) In *ion chromatography,* the stationary phase is an ion-exchange resin, and detection is ordinarily accomplished by a conductivity detector.
(i) *Gel filtration* is a type of size-exclusion chromatography in which the packings are hydrophilic, and eluents are aqueous. It is used for separating high-molecular-weight polar compounds.

32-3. (a) Diethyl ether, benzene, *n*-hexane

32-4. (a) Ethyl acetate, dimethylamine, acetic acid

32-5. In *adsorption chromatography,* separations are based on adsorption equilibria between the components of the sample and a solid surface. In *partition chromatography,* separations are based on distribution equilibria between two immiscible liquids.

32-7. *Gel filtration* is a type of size-exclusion chromatography in which the packings are hydrophilic and the eluents are aqueous. It is used for separating high-molecular-weight polar compounds. *Gel permeation chromatography* is a type of size-exclusion chromatography in which the packings are hydrophobic and the eluents are nonaqueous. It is used for separating high-molecular-weight nonpolar species.

32-9. *Pneumatic pumps* are simple, inexpensive, and pulse free. They consist of a collapsible solvent container housed in a vessel that can be pressurized by a compressed gas. This pump has limited capacity and pressure output and is not adaptable to gradient elution. The pumping rate depends on solvent viscosity.

Screw-driven syringe pumps consist of a large syringe in which the piston is moved by a motor-driven screw. They are pulse free, and the rate of delivery is easily varied. They suffer from lack of capacity and are inconvenient when solvents must be changed.

Reciprocating pumps are versatile and widely used. They consist of a small cylindrical chamber that is filled and then emptied by the back-and-forth motion of a piston. Advantages include small internal volume, high output pressures, adaptability to gradient elution, and constant flow rates that are independent of viscosity and back pressure. The pulsed output must be damped.

32-11. A gas-phase sample is needed for mass spectrometry. The output of the LC column is a solute dissolved in a solvent, whereas the output of the GC column is a gas and thus directly compatible. As a first step in LC/MS, the solvent must be vaporized. When vaporized, however, the LC solvent produces a gas volume that is 10 to 1000 times greater than the carrier gas in GC. Hence, most of the solvent must also be removed.

32-13. The ideal HPLC detector would have all the same characteristics as listed for the ideal GC detector. In addition, the HPLC detector should have low dead volume and be compatible with the liquid flows and pressures encountered in HPLC.

32-15.

R_s	N
0.50	5476
0.75	12321
0.90	17742
1.0	21904
1.1	26504
1.25	34225
1.50	49284
1.75	67081
2.0	87616
2.5	136900

Chapter 33

33-1. (a) Nonvolatile or thermally unstable species that contain no chromophoric groups.
(c) Inorganic anions and cations, amino acids, catecholamines, drugs, vitamins, carbohydrates, peptides, proteins, nucleic acids, nucleotides, and polynucleotides.
(e) Proteins, synthetic polymers, and colloidal particles.

33-2. (a) A *supercritical fluid* is a substance that is maintained above its critical temperature so that it cannot be condensed into a liquid no matter how great the pressure.
(c) In *two-dimensional thin-layer chromatography,* development is carried out with two solvents that are applied successively at right angles to one another.
(e) The *critical micelle concentration* is the level above which surfactant molecules begin to form spherical aggregates made up to 40 to 100 ions with their hydrocarbon tails in the interior of the aggregate and their charged ends exposed to water on the outside.

33-3. The properties of a supercritical fluid that are important in chromatography include its density, its viscosity, and the rates at which solutes diffuse in it. The magnitude of each of these lies intermediate between a typical gas and a typical liquid.

33-5. Pressure increases the density of a supercritical fluid, which causes the retention factor k for analytes to change. Generally, increases in pressure reduce the retention times of solutes.

33-7. Their ability to dissolve large nonvolatile molecules, such as large *n*-alkanes and polycyclic aromatic hydrocarbons.

33-9. (a) An increase in flow rate results in a decrease in retention time.
(b) An increase in pressure results in a decrease in retention time.
(c) An increase in temperature results in a decrease in density of supercritical fluids and thus an increase in retention time.

33-11. Electroosmotic flow can be repressed by reducing the charge on the interior of the capillary by chemical treatment of the surface.

33-13. Under the influence of an electric field, mobile ions in solution are attracted or repelled by the negative potential of one of the electrodes. The rate of movement toward or away from a negative electrode is dependent on the net charge on the analyte and the size and shape of analyte molecules. These properties vary from species to species. Hence, the rate at which molecules migrate under the influence of the electric field vary, and the time it takes them to traverse the capillary varies, making separations possible.

33-15. 3.9 min

33-17. Higher column efficiencies and the ease with which the pseudostationary phase can be altered.

33-19. Particle size and mass.

Chapter 35

35-1. Invalid sampling, loss of sample during weighing or dissolution, contamination by impurities in reagents, and changes in composition due to varying moisture content.

35-2. **(a)** *Sorbed water* is that held as a condensed liquid phase in the capillaries of a colloid. *Adsorbed water* is that retained by adsorption on the surface of a finely ground solid. *Occluded water* is that held in cavities distributed irregularly throughout a crystalline solid.

(c) *Essential water* is chemically bound water that occurs as an integral part of the molecular or crystalline structure of a compound in its solid state. *Nonessential water* is that retained by a solid as a consequence of physical forces.

35-4. Losses of volatile components as a result of heating, reaction with the atmosphere, changes in water content, losses as dust, and contamination due to mechanical wear and abrasion of the grinding surfaces.

Chapter 36

36-1. *Dry ashing* is carried out by igniting the sample in air or sometimes in oxygen. *Wet ashing* is done by heating the sample in an aqueous medium containing such oxidizing agents as H_2SO_4, $HClO_4$, HNO_3, H_2O_2, or some combination of these.

36-3. B_2O_3 or $CaCO_3/NH_4Cl$,

36-5. When hot concentrated $HClO_4$ comes in contact with organic materials or other oxidizable substances, explosions are highly probable.

36-6. **(a)** Samples for halogen determination may be decomposed in a Schöniger combustion flask, combusted in a tube furnace in a stream of oxygen, or fused in a peroxide bomb.

(c) Samples for nitrogen determination are decomposed in hot concentrated H_2SO_4 in a Kjeldahl flask or oxidized by CuO in a tube furnace in the Dumas method.

INDEX

Page numbers in boldface refer to specific laboratory directions that are included only on the companion CD-ROM and on the Web. Page numbers followed by t refer to tabular entries; page numbers followed by ss refer to spreadsheet exercises. Page numbers preceded by A refer to appendix references, and those preceded by CP refer to color plate numbers.

A

B

D

E

F

G

H

I

J

K

L

M

N

O

P

Q

R

S

T

X

Y

Z

INTERNATIONAL ATOMIC MASSES

Element	Symbol	Atomic Number	Atomic Mass
Actinium	Ac	89	(227)
Aluminum	Al	13	26.981538
Americium	Am	95	(243)
Antimony	Sb	51	121.76
Argon	Ar	18	39.948
Arsenic	As	33	74.9216
Astatine	At	85	(210)
Barium	Ba	56	137.327
Berkelium	Bk	97	(247)
Beryllium	Be	4	9.012182
Bismuth	Bi	83	208.98038
Bohrium	Bh	107	(264)
Boron	B	5	10.811
Bromine	Br	35	79.904
Cadmium	Cd	48	112.411
Calcium	Ca	20	40.078
Californium	Cf	98	(251)
Carbon	C	6	12.0107
Cerium	Ce	58	140.116
Cesium	Cs	55	132.90545
Chlorine	Cl	17	35.453
Chromium	Cr	24	51.9961
Cobalt	Co	27	58.93320
Copper	Cu	29	63.546
Curium	Cm	96	(247)
Dubnium	Db	105	(262)
Dysprosium	Dy	66	162.50
Einsteinium	Es	99	(252)
Erbium	Er	68	167.259
Europium	Eu	63	151.964
Fermium	Fm	100	(257)
Fluorine	F	9	18.9984032
Francium	Fr	87	(223)
Gadolinium	Gd	64	157.25
Gallium	Ga	31	69.723
Germanium	Ge	32	72.64
Gold	Au	79	196.96655
Hafnium	Hf	72	178.49
Hassium	Hs	108	(277)
Helium	He	2	4.002602
Holmium	Ho	67	164.93032
Hydrogen	H	1	1.00794
Indium	In	49	114.818
Iodine	I	53	126.90447
Iridium	Ir	77	192.217
Iron	Fe	26	55.845
Krypton	Kr	36	83.798
Lanthanum	La	57	138.9055
Lawrencium	Lr	103	(262)
Lead	Pb	82	207.2
Lithium	Li	3	6.941
Lutetium	Lu	71	174.967
Magnesium	Mg	12	24.3050
Manganese	Mn	25	54.938049
Meitnerium	Mt	109	(268)
Mendelevium	Md	101	(258)
Mercury	Hg	80	200.59
Molybdenum	Mo	42	95.94
Neodymium	Nd	60	144.24
Neon	Ne	10	20.1797
Neptunium	Np	93	(237)
Nickel	Ni	28	58.6934
Niobium	Nb	41	92.90638
Nitrogen	N	7	14.0067
Nobelium	No	102	(259)
Osmium	Os	76	190.23
Oxygen	O	8	15.9994
Palladium	Pd	46	106.42
Phosphorus	P	15	30.973761
Platinum	Pt	78	195.078
Plutonium	Pu	94	(244)
Polonium	Po	84	(210)
Potassium	K	19	39.0983
Praseodymium	Pr	59	140.90765
Promethium	Pm	61	(145)
Protactinium	Pa	91	231.03588
Radium	Pa	88	(226)
Radon	Rn	86	(222)
Rhenium	Re	75	186.207
Rhodium	Rh	45	102.90550
Rubidium	Rb	37	85.4678
Ruthenium	Ru	44	101.07
Rutherfordium	Rf	104	(261)
Samarium	Sm	62	150.36
Scandium	Sc	21	44.955910
Seaborgium	Sg	106	(266)
Selenium	Se	34	78.96
Silicon	Si	14	28.0855
Silver	Ag	47	107.8682
Sodium	Na	11	22.989770
Strontium	Sr	38	87.62
Sulfur	S	16	32.065
Tantalum	Ta	73	180.9479
Technetium	Tc	43	(99)
Tellurium	Te	52	127.60
Terbium	Tb	65	158.92534
Thallium	Tl	81	204.3833
Thorium	Th	90	232.0381
Thulium	Tm	69	168.93421
Tin	Sn	50	118.710
Titanium	Ti	22	47.867
Tungsten	W	74	183.84
Uranium	U	92	238.02891
Vanadium	V	23	50.9415
Xenon	Xe	54	131.293
Ytterbium	Yb	70	173.04
Yttrium	Y	39	88.90585
Zinc	Zn	30	65.409
Zirconium	Zr	40	91.224

The values given in parentheses are the atomic mass numbers of the isotopes of the longest known half-life.